QUANTUM FIELD THEORY

Lectures of Sidney Coleman

Foreword by David Kaiser

QUANTUM FIELD THEORY

Lectures of Sidney Coleman

Foreword by David Kaiser

Editors

Bryan Gin-ge Chen
Leiden University, the Netherlands

David Derbes
The Laboratory Schools
University of Chicago, USA

David Griffiths
Reed College, USA

Brian Hill
Saint Mary's College of California, USA

Richard Sohn
Kronos Inc., Lowell, MA

Yuan-Sen Ting
Harvard University, USA

World Scientific

NEW JERSEY • LONDON • SINGAPORE • BEIJING • SHANGHAI • HONG KONG • TAIPEI • CHENNAI • TOKYO

Published by

World Scientific Publishing Co. Pte. Ltd.
5 Toh Tuck Link, Singapore 596224
USA office: 27 Warren Street, Suite 401-402, Hackensack, NJ 07601
UK office: 57 Shelton Street, Covent Garden, London WC2H 9HE

Library of Congress Control Number: 2018041457

British Library Cataloguing-in-Publication Data
A catalogue record for this book is available from the British Library.

LECTURES OF SIDNEY COLEMAN ON QUANTUM FIELD THEORY

ISBN 978-981-4632-53-9
ISBN 978-981-4635-50-9 (pbk)

For any available supplementary material, please visit
https://www.worldscientific.com/worldscibooks/10.1142/9371#t=suppl

Printed in Singapore

to Diana Coleman

and for all of Sidney's students—past, present, and future

Contents

Foreword

Generations of theoretical physicists learned quantum field theory from Sidney Coleman. Hundreds attended his famous lecture course at Harvard University — the lecture hall was usually packed with listeners well beyond those registered for Physics 253 — while many more encountered photocopies of handwritten notes from the course or saw videos of his lectures long after they had been recorded. Coleman's special gift for exposition, and his evident delight for the material, simply could not be matched. A Coleman lecture on quantum field theory wasn't merely part of a course; it was an adventure.

Sidney Coleman was born in 1937 and grew up in Chicago. He showed keen interest in science at an early age, and won the Chicago Science Fair while in high school for his design of a rudimentary computer. He studied physics as an undergraduate at the Illinois Institute of Technology, graduating in 1957, and then pursued his doctorate in physics at the California Institute of Technology. At Caltech Coleman befriended Sheldon Glashow (then a postdoc), took courses from Richard Feynman, and wrote his dissertation under the supervision of Murray Gell-Mann. In 1961, as he was completing his dissertation, Coleman moved to Harvard as the Corning Lecturer and Fellow. He joined the Physics Department faculty at Harvard soon after that, and remained a member of the faculty until his retirement in 2006.[1]

For more than thirty years, Coleman led Harvard's group in theoretical high-energy physics. Colleagues and students alike came to consider him "the Oracle."[2] At one point Coleman's colleague, Nobel laureate Steven Weinberg, was giving a seminar in the department. Coleman had missed the talk and arrived during the question-and-answer session. Just as Coleman entered the room, Weinberg replied to someone else, "I'm sorry, but I don't know the answer to that question." "I do," Coleman called out from the back. "What was the question?" Coleman then listened to the question and answered without hesitation.[3]

Coleman had an off-scale personality, inspiring stories that colleagues and former students frequently still share. He kept unusual hours, working late into the night; at one point he

[1] Howard Georgi, "Sidney Coleman, March 7, 1937 – November 18, 2007," *Biographical Memoirs of the National Academy of Sciences* (2011), available at `http://www.nasonline.org/publications/biographical-memoirs/memoir-pdfs/coleman-sidney.pdf`.

[2] Quoted in Roberta Gordon, "Sidney Coleman dies at 70," *Harvard Gazette* (29 November 2007).

[3] David H. Freedman, "Maker of worlds," *Discover* (July 1990): 46–52, on p. 48.

complained to a colleague that he had been "dragged out of bed four hours before my usual time of rising (i.e., at 8 o'clock in the morning) to receive your telegram."[4] Indeed, he had refused to teach a course at 9 a.m., explaining, "I can't stay up that late."[5] Coleman's penchant for chain-smoking — even while lecturing — made at least one journalist marvel that Coleman never mistook his chalk for his cigarette.[6] In 1978, *Harvard Magazine* published a profile of Coleman, to which he took exception. As he wrote to the editors:

> Gentlemen:
>
> In your September-October issue, I am described as "a wild-looking guy, with scraggly black hair down to his shoulders and the worst slouch I've ever seen. He wears a purple polyester sports jacket."
>
> This allegation is both false to fact and damaging to my reputation; I must insist upon a retraction.
>
> The jacket in question is wool. All my purple jackets are wool.[7]

Little wonder that Coleman was often described as a superposition of Albert Einstein and comedian Woody Allen.[8]

Coleman had an extraordinary talent for wordplay as well as physics, and a lively, spontaneous wit. Once, while a journalist was preparing a feature article about him, Coleman received a telephone call in his office; the caller had misdialed. "No, I'm not Frank," the journalist captured Coleman replying, "but I'm not entirely disingenuous either."[9] Coleman frequently sprinkled literary and historical allusions throughout his writings, published articles and ephemeral correspondence alike. Writing to a colleague after a recent visit, for example, Coleman noted that the reimbursement he had received did not cover some of his travel expenses: "Samuel Gompers was once asked, 'What does Labor want?' He replied, 'More.'"[10] The lecture notes in this volume likewise include passing nods to Pliny the Elder, the plays of Molière, Sherlock Holmes stories, and more. He took the craft of writing quite seriously, at one point advising a friend, "Literary investigation by bombarding a manuscript with prepositions is as obsolete as electrical generation by beating a cat with an amber rod."[11]

Early in Coleman's career, colleagues began to admire his unusual skill in the lecture hall. Just a few years after joining Harvard's faculty, he was in such high demand for the

---

[4] Sidney Coleman to Antonino Zichichi, 5 November 1970. Coleman's correspondence is in the possession of his widow, Diana Coleman. Many of the letters quoted here will appear in the collection, *Theoretical Physics in Your Face: Selected Correspondence of Sidney Coleman*, ed. Aaron Wright, Diana Coleman, and David Kaiser (Singapore: World Scientific, forthcoming).

[5] Quoted in Gordon, "Sidney Coleman dies at 70."

[6] Freedman, "Maker of worlds," p. 48.

[7] Sidney Coleman to the editors of *Harvard Magazine*, 10 October 1978. The profile appeared in Timothy Noah, "Four good teachers," *Harvard Magazine* **80**, no. 7 (September-October 1978): 96–97. My thanks to Tiffany Nichols for retrieving a copy of the original article.

[8] Freedman, "Maker of worlds," p. 48.

[9] Quoted in Freedman, "Maker of worlds," p. 48.

[10] Sidney Coleman to Geoffrey West, 10 October 1978. Samuel Gompers, who founded the American Federation of Labor (AFL), had been a major figure in the American labor movement during the late nineteenth and early twentieth centuries.

[11] Sidney Coleman to Avram Davidson, 6 June 1977. Davidson (1923–1993), a longtime friend of Coleman's, was an award-winning science fiction author.

summer-school lecture circuit that he had to turn down more requests than he could accept.[12] He became a regular lecturer at the annual Ettore Majorana summer school in Erice, Italy, and developed a warm friendship with its organizer, Antonino Zichichi. Usually Coleman volunteered topics on which he planned to lecture at Erice — not infrequently using the summer course as an opportunity to teach himself material he felt he had not quite mastered yet — though sometimes Zichichi assigned topics to Coleman. Preparing for the 1969 summer school, for example, Zichichi pressed him, "Please stop refusing to lecture on the topic I have assigned to you. It is not my fault if you are among the few physicists who can lecture [on] anything."[13]

Coleman worked hard to keep his lectures fresh for his listeners, putting in significant effort ahead of time on organization and balance. He described his method to a colleague in 1975: "The notes I produce while preparing a lecture are skeletal in the extreme, nothing but equations without words." That way he could be sure to hit his main points while keeping most of his exposition fairly spontaneous.[14] The light touch on his first pass-through meant that Coleman needed to expend significant effort after the lectures were given, converting his sparse notes into polished prose that could be published in the summer-school lecture-note volumes. He often confided to colleagues that his "slothful" ways kept him from submitting manuscripts of his lecture notes on time.[15] Likely for that reason he shunned repeated invitations from publishers to write a textbook. "Not even the Symbionese Liberation Army would be able to convert me to writing an elementary physics text," he replied to one eager editor.[16]

Luckily for him — indeed, luckily for us — Coleman's lecture course on quantum field theory at Harvard was videotaped during the 1975–76 academic year, with no need for him to write up his notes. Filming a lecture course back then was quite novel, so much so that Coleman felt the need to explain why the large camera and associated equipment were perched in the back of the lecture hall. "The apparatus you see around here is part of a CIA surveillance project," he joked at the start of his first lecture, drawing immediate laughter from the students. He continued: "I fall within their domain because I read *JETP Letters*," inciting further laughter.[17] Hardly a spy caper, the videotapes were actually part of an experiment in educational technology.[18]

News of the tapes spread, and soon they were in high demand well beyond Cambridge. Colleagues wrote to Coleman, asking if they could acquire copies of the videotapes for their

---

[12] See, e.g., Sidney Coleman to Jack Steinberger, 6 October 1966.

[13] Antonino Zichichi to Sidney Coleman, 4 March 1969; cf. Coleman to Zichichi, 26 May 1967. Coleman republished several of his Erice lectures in his book, *Aspects of Symmetry: Selected Erice Lectures* (New York: Cambridge University Press, 1985).

[14] Sidney Coleman to Luis J. Boya, 18 April 1975.

[15] See, e.g., Sidney Coleman to Gian Carlo Wick, 8 October 1970; Coleman to Zichichi, 5 November 1970.

[16] Sidney Coleman to Gavin Borden, 12 March 1976. The "Symbionese Liberation Army" was a group of left-wing radicals that perpetrated several high-profile acts between 1973–75 in the United States, most famously kidnapping the wealthy publishing heiress Patty Hearst and allegedly "brainwashing" her into supporting their cause. See Jeffrey Toobin, *American Heiress: The Wild Saga of the Kidnapping, Crimes, and Trial of Patty Hearst* (New York: Doubleday, 2016).

[17] The American Institute of Physics began translating Soviet physics journals into English in 1955, including the *Journal of Experimental and Theoretical Physics* (*JETP*), as part of a Cold War effort to stay current on Soviet scientists' advances. See David Kaiser, "The physics of spin: Sputnik politics and American physicists in the 1950s," *Social Research* **73** (Winter 2006): 1225–1252.

[18] Coleman participated in other experiments involving videotaped lectures around the same time: Sheldon A. Buckler (Vice President of Polaroid Corporation) to Sidney Coleman, 16 April 1974; Peter Wensberg (Senior Vice President of Polaroid Corporation) to Sidney Coleman, 27 February 1975; and Sidney Coleman to Steven Abbott, 27 February 1976.

own use, from as far away as Edinburgh and Haifa.[19] Coleman's administrative assistant explained to one interested colleague in 1983 that the tapes had begun to "deteriorate badly" from overuse, yet they remained "in great demand even if those in use are in poor condition."[20] Years later, in 2007, Harvard's Physics Department arranged for the surviving videotapes to be digitized, and they are now available, for free, on the Department's website.[21] As David Derbes explains in his Preface, the editorial team made extensive use of the videos while preparing this volume.

* * *

Physicists knitted together what we now recognize as (nonrelativistic) quantum mechanics in a flurry of papers during the mid-1920s. The pace was extraordinary. Within less than a year — between July 1925 and June 1926 — Werner Heisenberg submitted his first paper on what would become known as "matrix mechanics," Erwin Schrödinger independently developed "wave mechanics," several physicists began to elucidate their mathematical equivalence, and Max Born postulated that Schrödinger's new wavefunction, ψ, could be interpreted as a probability amplitude. A few months after that, in March 1927, Heisenberg submitted his now-famous paper on the uncertainty principle.[22]

In hindsight, many physicists have tended to consider that brief burst of effort as a capstone, the end of a longer story that stretched from Max Planck's first intimations about blackbody radiation, through Albert Einstein's hypothesis about light quanta, to Niels Bohr's model of the atom and Louis de Broglie's suggestive insights about matter waves. To leading physicists at the time, however, the drumbeat of activity during the mid-1920s seemed to herald the start of a new endeavor, not the culmination of an old one. Already in 1926 and 1927, Werner Heisenberg, Pascual Jordan, Wolfgang Pauli, Paul Dirac and others were hard at work trying to quantize the electromagnetic field, and to reconcile quantum theory with special relativity. They had begun to craft quantum field theory.[23]

Those early efforts quickly foundered, as a series of divergences bedeviled physicists' calculations. By the early 1930s, theorists had identified several types of divergences — infinite self-energies, infinite vacuum polarization — which seemed to arise whenever they tried to incorporate the effects of "virtual particles" in a systematic way. Some leaders, like Heisenberg, called for yet another grand, conceptual revolution, as sweeping as the disruptions of 1925–27 had been, which would replace quantum field theory with some new, as-yet unknown framework. No clear candidate emerged, and before long physicists around the world found their attention

---

[19] David J. Wallace to Sidney Coleman, 12 June 1980; J. Avron to Sidney Coleman, 12 July 1983.

[20] Blanche F. Mabee to J. Avron, 19 August 1983; see also David J. Wallace to John B. Mather, 24 June 1980.

[21] `https://www.physics.harvard.edu/events/videos/Phys253`

[22] Many of the original articles are reprinted (in English translation) in B. L. van der Waerden, ed., *Sources of Quantum Mechanics* (Amsterdam: North-Holland, 1967). See also Max Jammer, *The Conceptual Development of Quantum Mechanics* (New York: McGraw-Hill, 1966); Olivier Darrigol, *From c-Numbers to q-Numbers: The Classical Analogy in the History of Quantum Theory* (Berkeley: University of California Press, 1992); and Mara Beller, *Quantum Dialogue: The Making of a Revolution* (Chicago: University of Chicago Press, 1999).

[23] Many important papers from this effort are reprinted (in English translation) in Arthur I. Miller, ed., *Early Quantum Electrodynamics: A Source Book* (New York: Cambridge University Press, 1994). See also Silvan S. Schweber, *QED and the Men Who Made It: Dyson, Feynman, Schwinger, and Tomonaga* (Princeton: Princeton University Press, 1994), chap. 1; Tian Yu Cao, *Conceptual Developments of 20th Century Field Theories* (New York: Cambridge University Press, 1997), chaps. 6–8; and the succinct historical introduction in Steven Weinberg, *The Quantum Theory of Fields*, vol. 1 (New York: Cambridge University Press, 1995), chap. 1.

absorbed by the rise of fascism and the outbreak of World War II.[24]

Soon after the war, a younger generation of physicists returned to the challenge of quantum field theory and its divergences. Many had spent the war years working on various applied projects, such as radar and the Manhattan Project, and had developed skills in wringing numerical predictions from seemingly intractable equations — what physicist and historian of science Silvan (Sam) Schweber dubbed, "getting the numbers out." Some had gained crash-course experience in engineers' effective-circuit approaches while working on radar; others had tinkered with techniques akin to Green's functions to estimate rates for processes like neutron diffusion within a volume of fissile material.[25] After the war, these younger physicists were further intrigued and inspired by new experimental results, likewise made possible by the wartime projects. In the late 1940s, experimental physicists like Willis Lamb and Isidor Rabi — using surplus equipment from the radar project and exploiting newfound skills in manipulating microwave-frequency electronics — measured tiny but unmistakeable effects, including a miniscule difference between the energy levels of an electron in the $2s$ versus $2p$ states of a hydrogen atom, and an "anomalous" magnetic moment of the electron, ever-so-slightly larger than the value predicted by Dirac's equation.[26]

Prodded by what seemed like tantalizing evidence of the effects of virtual particles, young theorists like Julian Schwinger and Richard Feynman worked out various "renormalization" techniques in 1947 and 1948, with which to tame the infinities within quantum electrodynamics (QED). They soon learned that Schwinger's approach was remarkably similar to ideas that Sin-itiro Tomonaga and colleagues had developed independently in Tokyo, during the war. Early in 1949, meanwhile, Freeman Dyson demonstrated a fundamental, underlying equivalence between the Tomonaga–Schwinger approach and Feynman's distinct-looking efforts, and further showed that renormalization should work at arbitrary perturbative order in QED — a remarkable synthesis at least as potent, and as surprising, as the earlier demonstrations had been, two decades earlier, that Heisenberg's and Schrödinger's approaches to quantum theory were mathematically equivalent.[27]

Dyson became adept at teaching the new approach to quantum field theory. Hectographed copies of his lecture notes from a 1951 course at Cornell University quickly began to circulate.[28] The unpublished notes provided a template for the first generation of textbooks on quantum field theory, written after the great breakthroughs in renormalization: books like Josef Jauch and Fritz Rohrlich's *The Theory of Photons and Electrons* (1955) and Silvan Schweber's massive *Introduction to Relativistic Quantum Field Theory* (1961), culminating in the pair of textbooks by James Bjorken and Sidney Drell, *Relativistic Quantum Mechanics* (1964) and *Relativistic Quantum Fields* (1965).[29]

---

[24] See especially Schweber, *QED and the Men Who Made It*, chap. 2.

[25] Schweber, *QED and the Men Who Made It*, pp. xii, 452 and chaps. 7–8; see also Julian Schwinger, "Two shakers of physics: Memorial lecture for Sin-itiro Tomonaga," in *The Birth of Particle Physics*, ed. L. M. Brown and L. Hoddeson (New York: Cambridge University Press, 1983), pp. 354–375; and Peter Galison, "Feynman's war: Modelling weapons, modelling nature," *Stud. Hist. Phil. Mod. Phys.* **29** (1998): 391–434.

[26] See especially Schweber, *QED and the Men Who Made It*, chap. 5.

[27] See the articles reprinted in Julian Schwinger, ed., *Selected Papers on Quantum Electrodynamics* (New York: Dover, 1958). See also Schweber, *QED and the Men Who Made It*, chaps. 6–9; and David Kaiser, *Drawing Theories Apart: The Dispersion of Feynman Diagrams in Postwar Physics* (Chicago: University of Chicago Press, 2005), chaps. 2–3.

[28] Kaiser, *Drawing Theories Apart*, pp. 81–83.

[29] J. M. Jauch and F. Rohrlich, *The Theory of Photons and Electrons* (Reading, MA: Addison-Wesley, 1955);

Yet the field did not stand still; new puzzles soon demanded attention. In 1957, for example, experimentalist Chien-Shiung Wu and her colleagues demonstrated that parity symmetry was violated in weak-force interactions, such as the β-decay of cobalt-60 nuclei: nature really did seem to distinguish between right-handed and left-handed orientations in space. Theorists T. D. Lee and C. N. Yang had hypothesized that the weak nuclear force might violate parity, and, soon after Wu's experiment, Murray Gell-Mann, Richard Feynman, and others published models of such parity-violating interactions within a field-theory framework.[30] Yet their models suffered from poor behavior at high energies, which led others, including Sidney Bludman, Julian Schwinger, and Sheldon Glashow to return to suggestive hints from Yang and Robert Mills: perhaps nuclear forces were mediated by sets of force-carrying particles — and perhaps those particles obeyed a nontrivial gauge symmetry, with more complicated structure than the simple U(1) gauge symmetry that seemed to govern electrodynamics.[31]

Mathematical physicists like Hermann Weyl had first explored gauge theories early in the 20th century, when thinking about the structure of spacetime in the context of Einstein's general theory of relativity. Decades later, in the mid-1950s, Yang and Mills, Robert Shaw, Ryoyu Utiyama, and Schwinger suggested that nontrivial gauge symmetries could help physicists parse the nuclear forces.[32] Yet applying such ideas to nuclear forces remained far from straightforward. For one thing, nuclear forces clearly had a finite range, which seemed to imply that the corresponding force-carrying particles should have a large mass. But inserting such mass terms by hand within the field-theoretic models violated the very gauge symmetries that those particles were meant to protect. These challenges drove several theorists to investigate spontaneous symmetry breaking in gauge field theories during the late 1950s through the mid-1960s, culminating in what has come to be known as the "Higgs mechanism."[33]

Another major challenge concerned how to treat strongly coupled particles, including the flood of nuclear particles — cousins of the familiar protons and neutrons — that physicists began to discover with their hulking particle accelerators. Dyson observed in 1953 that hardly a month went by without the announcement that physicists had discovered a new particle.[34] Whereas electrons and photons interacted with a relatively small coupling constant, $e^2 \sim 1/137$ (in appropriate units), many of the new particles seemed to interact strongly with each other, with coupling constants $g^2 \gg 1$. The small size of e^2 had been critical to the perturbative approaches of Tomonaga, Schwinger, Feynman, and Dyson; how could anyone perform a systematic calculation among strongly coupled particles? For just this reason, Feynman

---

S. S. Schweber, *An Introduction to Relativistic Quantum Field Theory* (Evanston, IL: Row, Peterson, 1961); J. D. Bjorken and S. Drell, *Relativistic Quantum Mechanics* (New York: McGraw-Hill, 1964); Bjorken and Drell, *Relativistic Quantum Fields* (New York: McGraw-Hill, 1965). Decades later, Dyson's 1951 lecture notes were beautifully typeset by David Derbes and published by World Scientific, so they are readily available today: Freeman Dyson, *Advanced Quantum Mechanics*, ed. David Derbes, 2nd ed. (Singapore: World Scientific, 2011).

[30] See, e.g., Allan Franklin, *The Neglect of Experiment* (New York: Cambridge University Press, 1986), chap. 1.

[31] Sidney Coleman described some of this work in a magnificent, brief essay: Coleman, "The 1979 Nobel Prize in Physics," *Science* **206** (14 December 1979): 1290–1292. See also the helpful discussion in Peter Renton, *Electroweak Interactions: An Introduction to the Physics of Quarks and Leptons* (New York: Cambridge University Press, 1990), chap. 5.

[32] Several original papers are available in Lochlainn O'Raifeartaigh, ed., *The Dawning of Gauge Theory* (Princeton: Princeton University Press, 1997).

[33] See, e.g., L. M. Brown, R. Brout, T. Y. Cao, P. Higgs, and Y. Nambu, "Panel discussion: Spontaneous breaking of symmetry," in *The Rise of the Standard Model: Particle Physics in the 1960s and 1970s*, ed. L. Hoddeson, L. Brown, M. Riordan, and M. Dresden (New York: Cambridge University Press, 1997), pp. 478–522; and Cao, *Conceptual Developments of 20th Century Field Theories*, chaps. 9–10.

[34] Freeman Dyson, "Field theory," *Scientific American* **188** (April 1953): 57–64, on p. 57.

himself cautioned Enrico Fermi in December, 1951, "Don't believe any calculation in meson theory which uses a Feynman diagram!"[35]

Some theorists, like Murray Gell-Mann and Yuval Ne'eman, sought to make headway by deploying symmetry arguments and tools from group theory. Gell-Mann introduced his famous "Eightfold Way" in 1961, for example, to try to understand certain regularities among nuclear particles by arranging them in various arrays, sorted by quantum numbers like isospin and hypercharge.[36] Others, such as Geoffrey Chew, embarked on an even more ambitious program to replace quantum field theory altogether. Chew announced at a conference in June 1961 that quantum field theory was "sterile with respect to the strong interactions" and therefore "destined not to die but just to fade away." He and his colleagues focused on an "autonomous S-matrix program," eschewing all talk of Lagrangians, virtual particles, and much of the apparatus that Dyson had so patiently assembled for making calculations in QED.[37]

Amid the turmoil and uncertainty, quantum field theory was never quite as dead as theorists like Chew liked to proclaim. Nonetheless, its status among high-energy physicists seemed far less settled in 1970 than it had been in 1950. The tide turned back toward field theory's advocates during the mid-1970s, driven by several important developments. First was the construction of a unified model of the electromagnetic and weak interactions, accomplished independently by Sheldon Glashow, Steven Weinberg, and Abdus Salam. Though they had published their work in the mid-1960s, it only attracted sustained attention from the community after Gerard 't Hooft and Martinus Veltman demonstrated in 1971–72 that such gauge field theories could be renormalized in a systematic way. (As Coleman observed, 't Hooft's work revealed "Weinberg and Salam's frog to be an enchanted prince.") Soon after that, in 1973–74, teams of experimentalists at CERN and Fermilab independently found evidence of weak neutral currents, as predicted by the Glashow–Weinberg–Salam theory.[38]

Meanwhile, other theorists, led by Yoichiro Nambu, Murray Gell-Mann, and Harald Fritzsch, developed quantum chromodynamics: a well-defined scheme for treating strong interactions among quarks and gluons, developed in analogy to QED but incorporating the kind of nontrivial gauge structure at the heart of the Glashow–Weinberg–Salam electroweak theory. The demonstration in 1973 by Coleman's student David Politzer and independently by David Gross and Frank Wilczek that the effective coupling strength between quarks and gluons in this model should *decrease* at short distances (or, correspondingly, high energies) — which came to be known as "asymptotic freedom" — breathed new life into field-theoretic approaches to the strong interactions.[39] Before long, the distinct threads of electroweak unification and

---

[35] Richard Feynman to Enrico Fermi, 19 December 1951, as quoted in Kaiser, *Drawing Theories Apart*, p. 201; see also *ibid.*, pp. 197–206; and L. Brown, M. Dresden, and L. Hoddeson, eds., *Pions to Quarks: Particle Physics in the 1950s* (New York: Cambridge University Press, 1989).

[36] M. Gell-Mann and Y. Ne'eman, eds., *The Eightfold Way* (New York: W. A. Benjamin, 1964).

[37] Quoted in Kaiser, *Drawing Theories Apart*, p. 306. See also G. F. Chew, *S-Matrix Theory of Strong Interactions* (New York: W. A. Benjamin, 1961); Chew, *The Analytic S Matrix: A Basis for Nuclear Democracy* (New York: W. A. Benjamin, 1966); and Kaiser, *Drawing Theories Apart*, chaps. 8–9.

[38] Coleman, "The 1979 Nobel Prize in Physics," 1291. See also Martinus Veltman, "The path to renormalizability," in Hoddeson *et al.*, *The Rise of the Standard Model*, pp. 145–178; and Gerard 't Hooft, "Renormalization of gauge theories," in *ibid.*, pp. 179–198. On the experimental detection of weak neutral currents, see Peter Galison, *How Experiments End* (Chicago: University of Chicago Press, 1983), chap. 4. Glashow, Weinberg, and Salam shared the Nobel Prize in Physics in 1979; Veltman and 't Hooft shared the Nobel Prize in Physics in 1999.

[39] David Gross, "Asymptotic freedom and the emergence of QCD," in Hoddeson *et al.*, *The Rise of the Standard Model*, pp. 199–232. Politzer, Gross, and Wilczek shared the Nobel Prize in Physics in 2004.

quantum chromodynamics were knitted into a single "Standard Model" of particle physics, a model built squarely within the framework of quantum field theory.[40]

* * *

Even as physicists' pursuit of field-theoretic techniques outstripped the template of perturbative QED during the 1960s, Dyson's crisp pedagogical model, which had been honed in the era of QED's great successes, continued to dominate in the classroom. Nobel laureate David Gross, for example, recalled his first course on quantum field theory at Berkeley in 1965, in which he and his fellow students were taught "that *field theory equals Feynman rules*": quantum field theory was still taught as if all that mattered were clever techniques for performing perturbative calculations.[41]

Coleman plotted a different course when he began teaching quantum field theory at Harvard a few years later. Early in the first semester, his students would practice drawing Feynman diagrams for perturbative calculations, to be sure. But in Coleman's classroom, quantum field theory would no longer be taught as a mere grab-bag of perturbative techniques. Coleman's course, in turn, helped to reinvigorate the study of quantum field theory more generally, following a protracted period when its fate seemed far from clear.

One obvious distinction between Coleman's pedagogical approach and Dyson's was an emphasis upon group theory and gauge symmetries. Coleman, after all, had written his dissertation at Caltech on "The Structure of Strong Interaction Symmetries," working closely with Gell-Mann just at the time that Gell-Mann introduced his "Eightfold Way." Coleman incorporated group-theoretic techniques into his teaching rather early, devoting his first set of summer-school lectures at Erice to the topic in 1966 (drawing extensively from his dissertation); Howard Georgi likewise recalls learning group theory from a course that Coleman taught at Harvard around the same time. Coleman continued to refine his presentation over the years. By the mid-1970s, he devoted several weeks of his course on quantum field theory to non-Abelian groups like SU(3) and their role in gauge field theories.[42]

Second was an emphasis on path-integral techniques. Although Feynman had developed path integrals in his Ph.D. dissertation and published on them in the 1940s, they had garnered virtually no space in the textbooks on quantum field theory published during the 1950s and 1960s. Nonetheless, several theorists began to recognize the power and elegance of path-integral techniques over the course of the 1960s, especially for tackling models with nontrivial gauge structure.[43] When Coleman began teaching his course on quantum field theory, he featured functional integration and path-integral methods prominently.

Third was an emphasis on spontaneous symmetry breaking. Coleman liked to joke with

---

[40] Laurie Brown, Michael Riordan, Max Dresden, and Lillian Hoddeson, "The Rise of the Standard Model, 1964–1979," in Hoddeson *et al.*, *The Rise of the Standard Model*, pp. 3–35.

[41] Gross, "Asymptotic freedom and the emergence of QCD," p. 202.

[42] Sidney Coleman, *The Structure of Strong Interaction Symmetries* (Ph.D. dissertation, Caltech, 1962); Coleman, "An introduction to unitary symmetry" (1966), reprinted in Coleman, *Aspects of Symmetry*, chap. 1; Georgi, "Sidney Coleman," p. 4.

[43] Richard Feynman, "Spacetime approach to non-relativistic quantum mechanics," *Rev. Mod. Phys.* **20** (1948): 367–387; L. D. Faddeev and V. N. Popov, "Feynman diagrams for the Yang–Mills field," *Phys. Lett. B* **25** (1967): 29–30. See also Schweber, *QED and the Men Who Made It*, pp. 389–397; Veltmann, "The path to renormalizability," pp. 158–159; Gross, "Asymptotic freedom and the emergence of QCD," pp. 201–202.

his students about his curious inability to predict what would become the most important developments in the field. Indeed, handwritten lecture notes from 1990 record him explaining:

> At crucial moments in the history of physics, I have often said about a new idea that I think it must be wrong. When the quark model was proposed, I thought it was wrong; likewise the Higgs mechanism. That's a good sign. If I say something isn't worth paying attention to, it probably isn't worth paying attention to. If I say it's *wrong*, then the idea merits careful examination — it may be important.[44]

Peter Higgs himself recalled that when he visited Harvard in the spring of 1966 to present his work, Coleman was ready to pounce, so certain was he that Higgs's work must be mistaken.[45]

For all the joking, however, Coleman rapidly became a leading expert on spontaneous symmetry breaking and one of its best-known expositors. He lectured on the subject at Erice and incorporated extensive material on symmetry breaking in his Harvard course on quantum field theory. Not only that: together with his graduate student Erick Weinberg, Coleman extended the idea to symmetries of an effective potential that could be broken by radiative corrections (known today as "Coleman–Weinberg" symmetry breaking), and later, with Curt Callan and Frank De Luccia, Coleman explored "the fate of the false vacuum," laying crucial groundwork for our modern understanding of early-universe cosmology.[46]

Coleman's lecture course on quantum field theory thus moved well beyond the earlier pedagogical tradition modeled on Dyson's notes and typified by Bjorken and Drell's *Relativistic Quantum Fields*. The differences lay not just in topics covered, but in underlying spirit. Coleman presented quantum field theory as a capacious framework, with significant nonperturbative structure. His style was neither overly rigorous nor narrowly phenomenological, offering an introduction to the Standard Model with an emphasis on general principles. In that way, his course remained more accessible and less axiomatic than many of the books that began to appear in the 1980s, such as Claude Itzykson and Jean-Bernard Zuber's compendious *Quantum Field Theory* (1980) or Ta-Pei Cheng and Ling-Fong Li's more specialized *Gauge Theory of Elementary Particle Physics* (1984).[47]

In at least one significant way, however, Coleman's pedagogical approach remained closer in spirit to Dyson's lectures than more recent developments. Renormalizability retained a special place for Coleman: models were adjudicated at least in part on whether divergences could be systematically removed for processes involving arbitrarily high energies. This, after all, was how (in Coleman's telling) 't Hooft's results had ennobled the Glashow–Weinberg–Salam model. Though Coleman was deeply impressed by Kenneth Wilson's work on the

---

[44] Transcribed from p. 210 of the handwritten lecture notes for Physics 253B, spring 1990.

[45] Peter Higgs in "Panel discussion: Spontaneous breaking of symmetry," p. 509. See also Higgs, "My life as a boson," available at `http://inspirehep.net/record/1288273/files/MyLifeasaBoson.pdf`.

[46] Coleman, "Secret symmetry: An introduction to spontaneous symmetry breakdown and gauge fields" (1973), reprinted in Coleman, *Aspects of Symmetry*, chap. 5. See also S. Coleman and E. J. Weinberg, "Radiative corrections as the origin of spontaneous symmetry breaking," *Phys. Rev. D* **7** (1973): 1888–1910; Coleman, "The fate of the false vacuum, I: Semiclassical theory," *Phys. Rev. D* **15** (1977): 2929–2936; C. G. Callan, Jr. and S. Coleman, "The fate of the false vacuum, II: First quantum corrections," *Phys. Rev. D* **16** (1977): 1762–1768; and S. Coleman and F. De Luccia, "Gravitational effects on and of vacuum decay," *Phys. Rev. D* **21** (1980): 3305–3315.

[47] Claude Itzykson and Jean-Bernard Zuber, *Quantum Field Theory* (New York: McGraw-Hill, 1980); Ta-Pei Cheng and Ling-Fong Li, *Gauge Theory of Elementary Particle Physics* (New York: Oxford University Press, 1984).

renormalization group — late in 1985, he remarked upon "Ken Wilson's double triumph" of uniting the study of field theory and critical phenomena — Coleman never fully adopted the viewpoint of effective field theory.[48] In effective field theories, physicists allow for an infinite tower of nonrenormalizable interaction terms — all terms consistent with some underlying symmetries — and calculate resulting processes for energy scales below some threshold, Λ. Though effective field theory techniques have become central to research in many areas of high-energy physics over the past three decades, today's popular textbooks on quantum field theory still rarely devote much space to the topic — so Coleman's lecture course continues to enjoy excellent company.[49]

* * *

I took the two-semester course Physics 253 with Sidney Coleman during the 1993–94 academic year, my first year in graduate school. For each semester, a large percentage of students' grades in the class derived from how well they did on a final exam: a 72-hour take-home exam that Coleman distributed on a Friday afternoon. My recollections of the weekend I spent working on the exam that fall semester are a bit hazy — much sweating, a bit of cursing, and very little sleep — and in the end, I ran out of time before I could complete the last problem. Sleep deprived, desperate, and more than a little inspired by Coleman's own sense of humor, I decided to appeal to a variant of the CPT theorem, on which we had focused so dutifully in class. In haste I scribbled down, "The final result follows from CBT: the Coleman Benevolence Theorem." A few days later I got back the marked exam. In Coleman's inimitable, blocky handwriting he had scrawled, "Be careful: This has not been experimentally verified."[50]

Pace Coleman's warning, there is plenty of evidence of his benevolence. I was struck recently, for example, by a questionnaire that he filled out in 1983, in preparation for his 30th high school reunion. By that time he had been elevated to an endowed professorship at Harvard and elected a member of the U.S. National Academy of Sciences. Yet in the space provided on the form for "occupation or profession," Coleman wrote, simply, "teacher."

And quite a teacher he was. Throughout his career, he supervised 40 Ph.D. students. He routinely shared his home telephone number with students (undergraduates and graduate students alike), encouraging them to call him at all hours. "Don't worry about disturbing me if you call me at home," he advised a group of undergraduates in the early 1990s. "I'm

---

[48] Sidney Coleman to Mirdza E. Berzins, 19 December 1985. Other leading field theorists shared Coleman's emphasis on renormalizability at the time. See, e.g., Steven Weinberg, "The search for unity: Notes for a history of quantum field theory," *Daedalus* **106** (Fall 1977): 17–35; cf. Weinberg, "Effective field theory, past and future," *Proceedings of Science* (CD09): 001, arXiv:0908.1964 [hep-th].

[49] For interesting historical perspectives on the shift to (nonrenormalizable) effective field theory approaches, see Tian Yu Cao, "New philosophy of renormalization: From the renormalization group equations to effective field theories," in *Renormalization: From Lorentz to Landau (and Beyond)*, ed. L. M. Brown (New York: Springer, 1993), pp. 87–133; and Silvan S. Schweber, "Changing conceptualization of renormalization theory" in *ibid.*, pp. 135–166. For brief introductions to effective field theory, see Anthony Zee, *Quantum Field Theory in a Nutshell*, 2nd ed. (Princeton: Princeton University Press, 2010), chap. VIII.3; and Steven Weinberg, *The Quantum Theory of Fields*, vol. 2 (New York: Cambridge University Press, 1996), chap. 19. See also Iain W. Stewart's course on "Effective Field Theory," available for free on the edX platform at `https://www.edx.org/course/effective-field-theory-mitx-8-eftx`.

[50] Despite my flub on the exam that first semester, Coleman kindly agreed to serve on my dissertation committee.

home most nights and usually stay up until 4 a.m. or so." That sort of dedication left an impression. "I would like to thank you for providing me with the best academic course I have ever encountered throughout my college career," wrote one undergraduate upon completing a course on quantum mechanics with Coleman. "I commend you on your excellent preparation for each and every lecture, your availability and helpfulness to each student, and particularly, your concern that the student develop his understanding and interest in the subject." Another undergraduate, who had taken Physics 253, wrote to Coleman a few years later that he was "one of the very best teachers I have had in my life."[51]

That spirit infuses this volume. Producing these lecture notes has been an enormous labor of love, initiated by David Derbes and brought to fruition thanks to the tireless efforts of a large editorial team, with special contributions from Bryan Gin-ge Chen, David Derbes, David Griffiths, Brian Hill, Richard Sohn, and Yuan-Sen Ting. This volume is proof that Sidney Coleman inspired benevolence enough to go around.[52]

David Kaiser

Germeshausen Professor of the History of Science
and Professor of Physics
Massachusetts Institute of Technology

---

[51] Sidney Coleman, memo to undergraduate advisees, 6 September 1991; Robert L. Veal to Sidney Coleman, 23 June 1974; Mark Carter to Sidney Coleman, 3 July 1981.

[52] It is a pleasure to thank David Derbes for inviting me to contribute this Foreword to the volume, and to Diana Coleman for sharing copies of Professor Coleman's correspondence. I am also grateful to Feraz Azhar, David Derbes, David Griffiths, Matthew Headrick, Richard Sohn, Jesse Thaler, and Aaron Wright for helpful comments on an earlier draft.

Preface

Sidney Coleman was not only a leader in theoretical particle physics, but also a hugely gifted and dedicated teacher. In his courses he found just the right balance between rigor and intuition, enlivened by wit, humor and a deep store of anecdotes about the history of physics. Very often these were first-hand accounts; if he wasn't a participant, he was an eyewitness. He made many important contributions to particle theory, but perhaps his most lasting contribution will prove to be his teaching. For many years he gave a series of celebrated summer school courses at the International Center for Scientific Culture "Ettore Majorana" in Erice, Sicily, under the directorship of Antonino Zichichi. A collection of these was published as a book, *Aspects of Symmetry*, by Cambridge University Press in 1985. This work, Prof. Coleman's only previous book, is now recognized as a classic.

Over three decades, Prof. Coleman taught Physics 253, the foundation course on quantum field theory to Harvard's graduate students in physics.[53] Many of the top American theoretical particle physicists learned quantum field theory in this course. Alas, he died much too young, at the age of 70, before he took the time away from his research to write the corresponding textbook. Brian Hill, one of Prof. Coleman's graduate students at Harvard (and the Teaching Fellow for the course for three years), had taken very careful notes of the course's first seven months from the fall of 1986, about one and a half semesters. He edited and rewrote these after every class. Xeroxes of Brian's handwritten notes were made available at Harvard for later classes, and served for nearly two decades as a *de facto* textbook for the first part of the course. In 2006, Bryan Gin-ge Chen, a Harvard undergraduate in physics, asked Brian if he could typeset his notes with the standard software LaTeX. Bryan got through Lecture 11. Yuan-Sen Ting, an undergraduate overseas, followed up in 2010, completing the typesetting of Brian's notes through Lecture 28. (Yuan-Sen completed a PhD at Harvard in astrophysics; Bryan moved to Penn for his in physics.) These notes were posted in 2011, with Brian Hill's

[53] Harvard offered Physics 253, on relativistic quantum theory, for decades before it became Prof. Coleman's signature course. For example, in 1965–66 Julian Schwinger taught the class, then called "Advanced Quantum Theory". Prof. Coleman also taught the course, in 1968 (and perhaps earlier). In 1974 it was renamed "Quantum Field Theory". He first taught this course in 1975–76, repeated it on and off until 1986, and thereafter taught it annually through the fall of 2002. He was to have taught the second semester in 2003, but his health did not permit it; those duties fell to Nima Arkani-Hamed (now at IAS, Princeton). I am very grateful to Marina Werbeloff for this information, which she gleaned at my request by searching through fifty years of course catalogs. I thank her for this and for much other assistance, without which the project likely would not have been completed.

introduction, at the arXiv, a free online repository for papers in physics, mathematics and many other subjects. I found them in the summer of 2013.

Like many, I wrote to Yuan-Sen and Bryan to express my thanks, and asked: Might we see the second semester some day? Yuan-Sen wrote back to say that unfortunately they did not have a copy of any second semester notes. I am a high school teacher, and I have been privileged to teach some remarkably talented young men and women, a few of whom later took Prof. Coleman's course. One of these, Matthew Headrick at Brandeis, had not only a set of second semester notes (from a second graduate student, who wishes to remain anonymous), but also a complete set of homework problems and solutions. He kindly sent these to me. I got in touch with Yuan-Sen and Bryan and suggested we now type up the second semester together with homework problems and solutions. Yuan-Sen was trying to finish his thesis, and Bryan had taken a position in Belgium. While they couldn't add to the work they'd already done, they offered their encouragement. Yuan-Sen also suggested that I get in touch with his colleague, Richard Sohn, who had been of significant help to them while typing up Brian's notes. Richard had gone through the notes carefully and corrected typos and a few minor glitches in note-taking or the lectures themselves. Even more enticing, Harvard's Physics Department, perhaps recognizing that something special was taking place, had videotaped Prof. Coleman's entire course in 1975–76. (The cameraman was Martin Roček, now at Stony Brook.) This was an experiment, as Prof. Coleman himself remarks at the very beginning of the first lecture. Other courses were videotaped, but it's noteworthy that, according to Marina Werbeloff, Harvard's Physics Librarian, only Prof. Coleman's tapes continued to circulate for thirty years. Aware that the frequently borrowed VHS tapes were starting to deteriorate, she had them digitized. In 2007, Maggie McFee, then head of the department's Computer Services, set up a small server to post these online in 2008. Perhaps the two semesters of notes and the videos could provide enough to put together something like the book that Prof. Coleman might have written himself. Some years earlier I had stumbled onto copies of Freeman Dyson's famous Cornell notes ("Advanced Quantum Mechanics", 1951) at MIT's website, and with Prof. Dyson's permission had typeset these with LaTeX for the arXiv. Soon thereafter World Scientific contacted Prof. Dyson and me to publish the notes as a book. Yuan-Sen wondered if World Scientific would be interested in publishing Coleman's lectures. I emailed Lakshmi Narayanan, my liaison for Prof. Dyson's notes. Indeed, World Scientific was very interested.

Now began a lengthy series of communications with all the interested parties. Neither Richard nor I sought royalties. Prof. Coleman's widow Diana Coleman is alive and the deserving party. She was happy to allow us to proceed. I got in touch both with Brian Hill and with the author of the second semester notes; each graciously agreed to our using their invaluable notes for this project. Through the kindness of Ms. Werbeloff, who responded after I asked Harvard about using the videotapes, I got in touch with Masahiro Morii, the chair of Harvard's Physics Department, who obtained approval from Harvard's Intellectual Property Department for us to use the videos. Ms. Werbeloff arranged to have the digitized video files transferred to a hard drive I sent to her. I cloned the returned drive and sent that to Richard. David Kaiser of MIT, a physicist and historian of physics, and also a former graduate student of Prof. Coleman's, generously agreed to write a foreword for these lectures. Additionally, Prof. Kaiser carefully read the manuscript and provided many corrections. He and Richard visited Ms. Coleman in Cambridge and got from her xeroxes of Prof. Coleman's own class notes, a priceless resource, particularly as these seem to be from the same year as the videotapes. Through Matt Headrick, I was able to contact the authors of the 1997–98 homework solutions, two of Prof. Coleman's graduate teaching assistants, David Lee and Nathan Salwen. They not only gave their permission for their solutions to be used, but generously provided the LaTeX source. Finally, we obtained a second set of lecture notes from Peter Woit at Columbia. Richard and I set to work from five separate records of the course: Brian Hill's and the anonymous graduate student's class notes; Prof. Coleman's own notes; class notes from Peter Woit; and our transcriptions of the videotaped lecture notes. Richard, far more conversant with modern field theory than I, would tackle the second semester, while I would start folding in my transcriptions of Prof. Coleman's videotaped lectures into the Hill–Ting–Chen notes, together with homework and solutions. To be sure, there are gaps in nearly all our accounts of the course (though Brian Hill's notes are complete, there are pages missing in Prof. Coleman's notes, quite a few electronic glitches in the forty-year old videotapes, and so on), but we seem to have a pretty complete record. The years are different (1975–76 for the videotapes and for Prof. Coleman's own lecture notes, 1978–79 for Peter Woit's notes, 1986–87 for Brian Hill's notes, and spring 1990 for the anonymous graduate student's), but the correspondence between these, particularly in the first semester, is remarkably close.[54] All of the contributions have been strictly voluntary; we have done this work out of respect and affection for Sidney Coleman.

Richard and I had been at work for about six months, when David Griffiths, who earned his PhD with Prof. Coleman, found the Hill–Chen–Ting notes at the arXiv, and wrote Yuan-Sen and Bryan to ask about the second semester. Yuan-Sen forwarded the email to me, and I wrote back. Prof. Griffiths, now emeritus at Reed College and the author of several widely admired physics textbooks, also wanted to see Prof. Coleman's course notes turned into a book. He has been an unbelievably careful and valuable critic, catching many of our mistakes, suggesting perhaps a hundred editorial changes per chapter, clarifications or alternative solutions in the homework, and generally improving the work enormously. Many of the last chapters were read by Prof. Jonathan L. Rosner, University of Chicago, who cleared up several misunderstandings. The responsibility for all errors, of course, rests with the last two editors, Richard and me.

The editors are profoundly grateful to all who have so generously offered their time, their expertise and their work to this project. We are particularly grateful to the talented staff at World Scientific for their hard work and their immense patience. We hope that were Prof. Coleman alive today, he would be pleased with our second-order efforts. They are only an approximation. This book can never be the equal of what he might have done, but we hope we have captured at least a little of his magic. May later generations of physics students learn, as so many before them have learned, from one of the best teachers our science has known.

David Derbes
The Laboratory Schools
The University of Chicago

[54] Very late in the project, we obtained a set of class notes, problems and exams from 2000–01, courtesy of another former student, Michael A. Levin of the University of Chicago. Through Michael we were able to get in touch with Prof. Coleman's last Teaching Fellow (1999–2002), Daniel Podolsky of the Israel Institute of Technology (Technion), Haifa. Daniel had *two* sets of typed notes for the course; his own, beautifully LaTeX'ed, which Michael had originally provided, and other notes (but missing many equations) from spring, 1999. A Harvard student hired by Prof. Coleman recorded the lectures as she took notes, and typed them up at home. The following summer Daniel worked with Prof. Coleman to edit the notes, but they did not get very far. Daniel's notes (both sets) were used primarily to check our completed work, but in a few places we have incorporated some very valuable insights from them.

Frequently cited references

Abers & Lee *GT* — Ernest S. Abers and Benjamin W. Lee, "Gauge Theories", *Phys. Lett.* **9C** (1973) 1–141.

Arfken & Weber *MMP* — George B. Arfken and Hans J. Weber, *Mathematical Methods for Physicists*, 6th ed., Elsevier, 2005.

Bjorken & Drell *RQM* — James D. Bjorken and Sidney D. Drell, *Relativistic Quantum Mechanics*, McGraw-Hill, 1961.

Bjorken & Drell *Fields* — James D. Bjorken and Sidney D. Drell, *Relativistic Quantum Fields*, McGraw-Hill, 1962.

Cheng & Li *GT* — Ta-Pei Cheng and Ling-Fong Li, *Gauge Theory of Elementary Particle Physics*, Oxford U. P., 1985.

Close *IP* — Frank Close, *The Infinity Puzzle*, Basic Books, 2011, 2013.

Coleman *Aspects* — Sidney Coleman, *Aspects of Symmetry: Selected Erice Lectures*, Cambridge U. P., 1985.

Crease & Mann *SC* — Robert P. Crease and Charles C. Mann, *The Second Creation: Makers of the Revolution in 20th-Century Physics*, Collier-Macmillan Publishing, 1986. Reprinted by Rutgers U. P., 1996.

Goldstein *et al. CM* — Herbert Goldstein, Charles P. Poole, and John L. Safko, *Classical Mechanics*, 3rd. ed., Addison-Wesley, 2001.

Gradshteyn & Ryzhik *TISP* — Izrael S. Gradshteyn and Iosif M. Ryzhik, *Table of Integrals, Series and Products*, 4th. ed., Academic Press, 1965.

Greiner & Müller *GTWI* — Walter Greiner and Berndt Müller, *Gauge Theory of Weak Interactions*, 4th ed., Springer, 2009.

Greiner & Müller *QMS* — Walter Greiner and Berndt Müller, *Quantum Mechanics: Symmetries*, 2nd ed., Springer, 1994.

Greiner & Reinhardt *FQ* — Walter Greiner and Joachim Reinhardt, *Field Quantization*, Springer, 1996.

Greiner & Reinhardt *QED* — Walter Greiner and Joachim Reinhardt, *Quantum Electrodynamics*, 4th ed., Springer, 2009.

Greiner *et. al QCD* — W. Greiner, Stefan Schramm, and Eckart Stein, *Quantum Chromodynamics*, 2nd rev. ed., Springer, 2002.

Griffiths *EP*	David Griffiths, *Introduction to Elementary Particles*, 2nd rev. ed., Wiley-VCH, 2016.
Itzykson & Zuber *QFT*	Claude Itzykson and Jean-Bernard Zuber, *Quantum Field Theory*, McGraw-Hill, 1980. Reprinted by Dover Publications, 2006.
Jackson *CE*	J. David Jackson, *Classical Electrodynamics*, 3rd ed., John Wiley, 1999.
Landau & Lifshitz *QM*	Lev D. Landau and Evgeniĭ M. Lifshitz, *Quantum Mechanics: Non-Relativistic Theory*, 2nd ed., Addison-Wesley, 1965.
Lurié *P&F*	David Lurié, *Particles and Fields*, Interscience Publishers, 1968.
PDG 2016	C. Patrignani *et al.* (Particle Data Group), *Chin. Phys. C* **40** (2016) 100001, `http://pdg.lbl.gov`.
Peskin & Schroeder *QFT*	Michael E. Peskin and Daniel V. Schroeder, *An Introduction to Quantum Field Theory*, Westview Press, 1995
Rosten *Joys*	Leo Rosten, *The New Joys of Yiddish*, rev. Lawrence Bush, Harmony, 2003.
Ryder *QFT*	Lewis H. Ryder, *Quantum Field Theory*, 2nd ed., Cambridge U. P., 1996.
Schweber *QED*	Silvan S. Schweber, *QED and the Men Who Made It*, Princeton U. P., 1994.
Schweber *RQFD*	Silvan S. Schweber, *An Introduction to Relativistic Quantum Field Theory*, Row, Peterson & Co., 1960. Reprinted by Dover Publications, 2005.
Schwinger *QED*	Julian Schwinger, ed., *Selected Papers on Quantum Electrodynamics*, Dover Publications, 1959.
Weinberg *QTF1*	Steven Weinberg, *The Quantum Theory of Fields I: Foundations*, Cambridge U. P., 1995, 1995.
Weinberg *QTF2*	Steven Weinberg, *The Quantum Theory of Fields II: Modern Applications*, Cambridge U. P., 1996.
Zee *GTN*	Anthony Zee, *Group Theory in a Nutshell for Physicists*, Princeton U. P., 2016.
Zee *QFTN*	Anthony Zee, *Quantum Field Theory in a Nutshell*, 2nd ed., Princeton U. P., 2010.

Index of useful formulae

A note on the problems

The classes in Physics 253a (fall) and Physics 253b (spring) ran as two ninety-minute lectures per week. Students were assigned problem sets (from one to four problems) nearly every week, and given solutions to them after the due date. As this book has fifty chapters, it seemed reasonable to include twenty-five problem sets. These include all the assigned problems from 1997–98 (with two exceptions),[1] some additional problems from other years that were not assigned in 1997–98, and a handful of final examination questions. In 1975–76, Coleman began the second semester material a little early, in the last part of the last lecture of 253a, Chapter 25. This material was moved forward into Chapter 26, which marks the beginning of the second semester, and an approximate dividing line for the provenance of the problems. Usually, those from 253a are placed before Chapter 26; those from 253b, after. Problems 14 are transitional: though assigned in the second semester, they involve first semester material.

The editors obtained complete sets of assigned problems and examinations (and their solutions) from the year 1978–79, the years 1980–82, and the year 1986–87 from Diana Coleman (via David Kaiser); 1990–91, from Matthew Headrick; 1997–98 from Matthew Headrick, Nathan Salwen and David Lee; 2000–01 from Michael Levin; and examination questions from 1988–2000 from Daniel Podolsky. John LoSecco provided a problem cited in the video of Lecture 50 (Problem 4 on the 1975a Final) and its solution, which appears here as Problem 15.4. In fact, only a few assigned problems from these other years do not appear in this book. Most of the problems were used over and over throughout the roughly thirty years that Coleman taught the course; sometimes a problem used in an examination was assigned for homework in later years, or *vice versa*. The solutions were written up by Teaching Fellows (notably by Brian Hill, but very probably some are due to Ian Affleck, John LoSecco, Bernard Grossman, Katherine Benson, Vineer Bhansali, Nathan Salwen, David Lee, and Daniel Podolsky, among many others unknown to us). Some solutions, particularly to the exam questions, are by Coleman himself. It's hard to know the authorship of many solutions—the same problems assigned ten years apart often have essentially identical solutions, though in different handwriting, and we may not have the original author's work. Now and

[1] [Eds.] Two questions were omitted, as the videotaped lectures of 1975–76, on which the text is based, include their solutions: (1997a 2.3), on the form of the energy-momentum tensor for a scalar field; and (1998b 10.1), on the mixing angle for the ρ and ω eigenstates of the mass-squared matrix for the $J^P = 1^-$ meson octet. The first is worked out in §5.5, (5.52)–(5.58); the second appears in §39.3, (39.19)–(39.35).

then the editors have added a little to a problem's solution, but most of the solutions are presented just as they were originally.

A century ago, it was customary in British mathematics textbooks to cite the provenance (if known) of problems; e.g., an exercise taken from the Cambridge Mathematical Tripos was indicated by the abbreviation "MT" and the year of the examination. Here, *(1998a 2.3)* indicates Problem 2.3 assigned in the fall of 1998. To aid the reader in finding a particular problem, succinct statements of them are given below. (Incidentally, the Paracelsus epigraph[2] in Problems 1 comes from the first 253a assignment in 1978.)

1.1 Show that the measure $\dfrac{d^3\mathbf{p}}{(2\pi)^3 2\omega_{\mathbf{p}}}$ is Lorentz invariant.

1.2 Show that $\langle 0|T(\phi(x)\phi(y))|0\rangle$ is a Green's function for the Klein–Gordon equation.

1.3 Show that the variance of $\phi(x)$ over large regions is tiny, and roughly classical; over small regions, it fluctuates as a typical quantum system.

2.1 Find the dimensions of various Lagrangians in d spacetime dimensions.

2.2 Rework Problem 1.3 from the point of view of dimensional analysis.

2.3 Obtain the Maxwell equations from the Maxwell Lagrangian, $\mathscr{L} = -\frac{1}{4}F^{\mu\nu}F_{\mu\nu}$.

2.4 Obtain both the canonical energy-momentum tensor, and an improved, symmetric version, for the Maxwell field.

3.1 Obtain Schrödinger's equation from a given Lagrangian.

3.2 Quantize the theory in 3.1.

3.3 Show that only two of the symmetries $\{P, C, T\}$ have corresponding unitary operators for this theory.

3.4 Examine dilation invariance for the massless Klein–Gordon theory.

4.1 Evaluate the real constant α in Model 1 in terms of its only Wick diagram, •——•.

4.2 Demonstrate various properties of the coherent states of a single harmonic oscillator.

4.3 Obtain expectation values of an operator in terms of a generalized delta function.

5.1 Evaluate $\langle \mathbf{p}|\mathrm{S}-1|\mathbf{p}'\rangle$ for the pair model, and show that its S-matrix is unitary, i.e. $\langle \mathbf{p}|\mathrm{S}^\dagger\mathrm{S}-1|\mathbf{p}'\rangle = 0$.

6.1 Let Model 3 describe kaon decay into pions ($K \sim \phi$, $\pi \sim \psi$), and determine the value of g/m_K to one significant digit.

6.2 Compute $d\sigma/d\Omega$ (c.o.m. frame) for elastic $N\overline{N}$ scattering in Model 3 to lowest order in g.

6.3 Compute $d\sigma/d\Omega$ (c.o.m. frame) for $N + \overline{N} \to 2\pi$ in Model 3 to lowest order in g.

6.4 Determine the behavior of the S-matrix in a free scalar field under an anti-unitary operator (as required for CPT symmetry).

7.1 Determine the two-particle density of states factor in an arbitrary frame of reference.

7.2 Calculate the decay $A \to B + C + D$ in a theory of four scalar fields $\{A, B, C, D\}$ if A is massive but the other three are massless, with $\mathscr{L}' = gABCD$.

7.3 Determine the density of states factor for particle decay if the universe is filled with a thermal distribution of mesons at a temperature T.

8.1 Replace a free Klein–Gordon field ϕ by $\phi = A + \frac{1}{2}gA^2$, and show that to $\mathcal{O}(g^2)$ the sum of all A-A scattering graphs vanishes.

9.1 Calculate the imaginary part of the renormalized meson self-energy $\widetilde{\Pi}'(p^2)$ in Model 3 to $\mathcal{O}(g^2)$.

9.2 Compute the Model 3 vertex $-i\widetilde{\Gamma}'(p^2, p'^2, q^2)$ to $\mathcal{O}(g^3)$ as an integral over two Feynman parameters, for $p^2 = p'^2 = m^2$.

---

[2] [Eds.] Theophrastus von Hohenheim (1493–1541), known as Paracelsus, a pioneering Swiss physician, alchemist, and astrologer.

10.1 Calculate the renormalized "nucleon" self-energy, $\widetilde{\Sigma}'(p^2)$, in Model 3 to $\mathcal{O}(g^2)$, expressing the answer as an integral over a single Feynman parameter.

10.2 Verify the Lie algebra of the Lorentz group's generators using the defining representation of the group.

11.1 Find the positive energy helicity eigenstates of the Dirac equation.

11.2 Work out trace identities for various products of Dirac gamma matrices.

12.1 Attempt the canonical quantization of a free Klein–Gordon field $\phi(x)$ with anticommutators, and show that the Hilbert space norm of $\langle\phi|\{\theta,\theta^\dagger\}|\phi\rangle$ cannot be positive.

12.2 Compute, to lowest nontrivial order, $d\sigma/d\Omega$ (c.o.m. frame) for the scattering $N+\phi\to N+\phi$ if $\mathscr{L}' = g\overline{\psi}\psi\phi$ (the "scalar" theory).

12.3 Compute, to lowest nontrivial order, $d\sigma/d\Omega$ (c.o.m. frame) for the scattering $N+\overline{N}\to N+\overline{N}$ if $\mathscr{L}' = g\overline{\psi}i\gamma_5\psi\phi$ (the "pseudoscalar" theory).

13.1 In the "pseudoscalar" theory, $\mathscr{L}' = g\overline{\psi}i\gamma_5\psi\phi$, calculate the renormalized "nucleon" self-energy, $\widetilde{\Sigma}'(k^2)$, to $\mathcal{O}(g^2)$. Leave your answer in terms of an integral over a single Feynman parameter.

13.2 In the same theory, compute the renormalized meson self-energy, $\widetilde{\Pi}'(k^2)$, to $\mathcal{O}(g^2)$. Again, leave your answer in terms of an integral over a single Feynman parameter. Check that the imaginary part of this quantity has the correct (negative) sign.

14.1 Given an interaction Hamiltonian of four-fermion interactions whose S-matrix is CPT-invariant, show that the Hamiltonian is itself invariant. Investigate under what circumstances it is invariant under the sub-symmetries of CPT, e.g., PT and P.

14.2 Derive the superficial degree of divergence D for a general Feynman graph in d spacetime dimensions.

14.3 In the generalized "pseudoscalar" theory, $\mathscr{L}' = g\overline{N}i\gamma_5\boldsymbol{\tau}\boldsymbol{\cdot}\boldsymbol{\pi}N$, calculate various $N+\pi\to N'+\pi'$ amplitudes, using isospin invariance.

14.4 Show that to $\mathcal{O}(g^2)$, the original "pseudoscalar" theory's scattering amplitudes coincide with those of a second Lagrangian, $\mathscr{L}'' = \mu^{-1}\left[ag\overline{\psi}\gamma^\mu\gamma_5\psi\partial_\mu\phi + bg^2\overline{\psi}\psi\phi^2\right]$, for appropriate choices of a and b.

15.1 Investigate the (four dimensionally) longitudinal solutions of the free Proca Lagrangian, and construct its Hamiltonian. Show that for an appropriate identification of A_0 and its conjugate momentum with ϕ and π of the Klein–Gordon equation, the Hamiltonians of the two theories are *identical.*

15.2 Construct the Hamiltonian of a free, massive vector in terms of its creation and annihilation operators.

15.3 Let A_μ be a vector of mass μ be coupled to two Dirac fields ψ_1 and ψ_2 of mass m_1 and m_2, respectively, according to the interaction Lagrangian $\mathscr{L}' = gA_\mu(\overline{\psi}_1\gamma^\mu\psi_2 + \overline{\psi}_2\gamma^\mu\psi_1)$. Compute the decay width Γ for $\psi_1\to\psi_2+\gamma$ if $m_1 > m_2+\mu$ to lowest nonvanishing order.

15.4 Compute elastic meson–meson scattering to $\mathcal{O}(g^2)$ in a scalar theory with the interaction $\mathscr{L}' = -(1/4!)g\phi'^4 - (1/4!)C\phi'^4$ in terms of the Mandelstam variables s, t, and u. Define the counterterm C by the requirement that $i\mathcal{A} = -ig$ when all four mesons are on the mass shell, at the symmetry point where the Mandelstam variables all equal $4\mu^2/3$.

16.1 A Dirac field is minimally coupled to a Proca field of mass μ. Compute, to lowest nontrivial order, the amplitude for elastic fermion–antifermion scattering, and show that the part proportional to $k^\mu k^\nu/\mu^2$ vanishes.

16.2 In this same theory, compute the amplitude for elastic vector–spinor scattering to lowest nontrivial order, and show that if the meson's spin vector ε^μ is aligned with its four-momentum k^μ (for either the outgoing or incoming vector), the amplitude vanishes. Repeat the calculation, substituting a scalar for the spinor.

16.3 Two Dirac fields A and B of masses m_A and m_B interact with a complex charged scalar field C of mass m_C according to the Lagrangian $\mathscr{L}' = g(\overline{A}i\gamma_5BC + \overline{B}i\gamma_5AC^*)$. Let the fields be minimally coupled to a Proca field, and let their charges (in units of e) be q_A, q_B, and q_C, such that $q_A = q_B + q_C$. Show that the amplitude for $\gamma + A\to B+C$ vanishes to lowest order (eg) if the Proca spin is aligned with its four-momentum.

17.1 A scalar field is quadratically coupled to a source J, i.e., with an interaction term $\frac{1}{2}J\phi^2$. From Chapter 27, it can be shown that $\langle 0|\mathrm{S}|0\rangle_J = (\det[A-i\epsilon]/\det[K-i\epsilon])^{-1/2}$, where $A = (\Box^2+\mu^2-J)$, and $K = (\Box^2+\mu^2)$. Show that you obtain the same result by summing Feynman graphs.

17.2 Using functional integrals, determine the photon propagator $D^C_{\mu\nu}$ in Coulomb gauge, $\boldsymbol{\nabla}\boldsymbol{\cdot}\mathbf{A} = 0$.

17.3 Compute, to $\mathcal{O}(e^2)$, the invariant Feynman amplitude for electron–electron scattering in both the Coulomb and Feynman gauge, and show that the final answers are the same.

18.1 Compute, to $\mathcal{O}(e^2)$, the renormalized photon self-energy $\widetilde{\Pi}'_{\mu\nu}(p^2)$ in the theory of a charged Dirac field minimally coupled to a massless photon. Write the answer as an integral over a single Feynman parameter, and handle the divergences with Pauli–Villars regulator fields.

18.2 Compute, to $\mathcal{O}(e^2)$, the renormalized photon self-energy $\widetilde{\Pi}'_{\mu\nu}(p^2)$ in the theory of a charged spinless meson minimally coupled to a photon. Write the answer as an integral over a single Feynman parameter, and handle the divergences with dimensional regularization.

18.3 Add to the standard Maxwell Lagrangian the interaction term $\mathscr{L}' = -\frac{1}{2}\lambda(\partial_\mu A^\mu + \sigma A_\mu A^\mu)^2$ and a ghost Lagrangian $\mathscr{L}_{\text{ghost}}$. Determine the latter, the ghost propagator, and the Feynman rules for the ghost vertices.

19.1 Carry out computations for a charged scalar particle minimally coupled to a massless photon parallel to those earlier calculations for a charged Dirac particle: Ward identity and its verification at tree level, identification of the normalized charge with the physical charge, determination of $F_1(q^2)$ and $F_2(q^2)$.

19.2 Compute the decay width Γ for the process $\psi_1 \to \psi_2 + \gamma$ if $m_1 > m_2$, for the Lagrangian $\mathscr{L}' = g\overline{\psi}_2\sigma_{\mu\nu}\psi_1 F^{\mu\nu}$ + h.c.

20.1 The errors on the anomalous magnetic moments, and hence on $1 + F_2(0)$, of the electron and the muon are 3×10^{-11} and 8×10^{-9}, respectively. What bounds do these place on a hypothetical massive photon whose mass M is much greater than the muon's mass?

20.2 Express the renormalized photon propagator (in Landau gauge) in terms of its spectral representation with spectral function $\rho(k^2)$, and show that the hadronic contribution $\rho_H(a^2)$ is proportional, to $\mathcal{O}(e^4)$ and $\mathcal{O}(e^2m^2/a^2)$, to the total cross-section σ_T for e^+-$e^- \to$ hadrons.

21.1 Show that the two SU(3)-invariant quartic self-couplings of the pseudoscalar octet, $\text{Tr}(\phi^4)$ and $(\text{Tr}(\phi^2))^2$, are proportional to each other.

21.2 Show that the magnetic moments within the SU(3) decuplet are proportional to the charge.

21.3 Assuming that the magnetic moments of quarks are proportional to their charges, $\boldsymbol{\mu} = \kappa q\boldsymbol{\sigma}$, where $\boldsymbol{\sigma}$ is the vector of Pauli matrices, determine the ratio of the proton and neutron magnetic moments, and compare with experiment.

22.1 Consider the scattering of two distinct, spinless particles below inelastic threshold. Find the relation between the s-wave scattering length a and the invariant Feynman amplitude, $\mathcal{A}$, evaluated at threshold.

22.2 Consider a massless neutrino and an electron coupled to a Proca field W of mass M. For the process $\nu+\overline{\nu} \to W+\overline{W}$, there are nine independent amplitudes. Find them. Some are well-behaved at high energy, but others grow without limit. Which are which? Show that all amplitudes become well-behaved by the addition of new terms to the Lagrangian, $\mathscr{L}'' = \overline{e}'(i\partial\!\!\!/ - M)e' + f(W^*_\mu\overline{e}'\gamma^\mu(1+\gamma_5)\nu + W_\mu\overline{\nu}(1-\gamma_5)\gamma^\mu e')$, if f is chosen proportional to g.

23.1 A charged scalar ψ of mass m is minimally coupled to the photon. A second massless neutral meson ϕ is coupled through the term $\mathscr{L}' = g\phi\epsilon^{\mu\nu\lambda\sigma}F_{\mu\nu}F_{\lambda\sigma}$. Determine $d\sigma/d\Omega$ to $\mathcal{O}(e^2g^2)$ for the process $\gamma + \psi \to \phi + \psi$.

23.2 Starting from the Goldstone model, find a solution $\phi(z)$ of the field equations, such that $\phi(\pm\infty) = \pm a$. These solutions could represent "domain walls" in the early universe. Find the energy of these domain walls in terms of the Goldstone parameters λ and a.

24.1 Verify an approximation (44.51) used in the derivation of the scalar field's effective potential.

24.2 Consider the full Yukawa theory of a triplet of pions and the nucleon doublet, with isospin-invariant interaction $\mathscr{L}' = -ig\overline{N}\gamma_5\boldsymbol{\tau}\bullet\boldsymbol{\Phi}N$ and a quartic pion self-coupling $\frac{1}{4}\lambda(\boldsymbol{\Phi}\bullet\boldsymbol{\Phi})^2 + \mathscr{L}_{CT}$. Let the fields now be minimally coupled to a massless photon, and determine the contributions to the proton and neutron form factors $F_2(0)$.

24.3 Consider the Goldstone model minimally coupled to a Proca field with mass μ_0. What is the mass of this "photon" after the symmetry breaks? Does the Goldstone boson survive, and if so, what is its mass?

25.1 A free Proca field of mass μ is coupled to a real scalar field ϕ of mass m by the interaction Lagrangian $\mathscr{L}' = gA^\mu A_\mu\phi$. There are nine independent amplitudes for the process $A+\phi \to A+\phi$, some well-behaved at high energy, and some not. Find them. Which are which? Show that all become well-behaved with the addition of a new term, $h\phi^2 A_\mu A^\mu$, for an appropriate choice of h.

25.2 From the infinitesimal form of the non-Abelian gauge transformation, determine the finite (integrated) form, and show that its corresponding unitary matrix $U(s)$ satisfies a particular differential equation.

25.3 Compute $k'_\mu M^\mu$ for the elastic scattering of non-Abelian gauge bosons off Dirac particles in the tree approximation (i.e., to $\mathcal{O}(g^2)$) where M^μ is the matrix element of a conserved current, by setting $\varepsilon'^*_\mu = k'_\mu$.

25.4 Compute, to $\mathcal{O}(g^2)$, elastic vector–scalar scattering in the Abelian Higgs model, for the case in which both the initial and final vector mesons have zero helicity, but at fixed scattering angle $\theta\,(\neq \pi, 0)$. Show that the amplitude approaches a limit at high energy, even though some individual graphs grow with energy.

1

Adding special relativity to quantum mechanics

1.1 Introductory remarks

This is Physics 253, a course in relativistic quantum mechanics. This subject has a notorious reputation for difficulty, and as this course progresses, you will see this reputation is well deserved. In non-relativistic quantum mechanics, rotational invariance simplifies scattering problems. Why does adding in special relativity, to include Lorentz invariance, complicate quantum mechanics?

The addition of relativity is necessary at energies $E \geq mc^2$. At these energies the reaction

$$p + p \to p + p + \pi^0$$

is possible. At slightly higher energies, the reaction

$$p + p \to p + p + p + \overline{p}$$

can occur. The exact solution of a high energy scattering problem necessarily involves many-particle processes.

You might think that for a given E, only a finite number, maybe only a small number, of processes actually contribute. But you already know from non-relativistic quantum mechanics that this isn't true. For example, if a perturbation δV is added to the Hamiltonian H, the ground state energy E_0 changes according to the rule

$$E_0 \to E_0 + \delta E_0 \quad \text{where} \quad \delta E_0 = \langle 0|\delta V|0\rangle + \sum_n \frac{|\langle 0|\delta V|n\rangle|^2}{E_0 - E_n} \tag{1.1}$$

Intermediate states of *all* energies contribute, suppressed by energy denominators.

For highly accurate calculations at low energy, it's reasonable to include relativistic effects of order $(v/c)^2$. Intermediate states with extra particles will contribute corrections of the same order:

$$\frac{\text{(typical energies in problem)}}{\text{(typical energy denominator)}} \sim \frac{E}{mc^2} \sim \frac{mv^2}{mc^2} = \left(\frac{v}{c}\right)^2 \tag{1.2}$$

As a general conclusion, the corrections of relativistic kinematics and the corrections from multi-particle intermediate states are comparable; relativity forces you to consider many-body problems.

There are however very special cases, due to the specific dynamics involved, where the kinematic effects of relativity are considerably larger than the effects of pair states. One of these is the hydrogen atom. That's why Dirac's theory[1] gives excellent results to order $(v/c)^2$ for the hydrogen atom, even without considering pair production and multi-particle intermediate states. This is a fluke.[2] Dirac's success was a good thing because it told people that the basic ideas were right, but it was a bad thing because it led people to spend a lot of time worrying about one-particle, two-particle, and three-particle theories, because they didn't realize the hydrogen atom was a very special system. We will see that you cannot have a consistent relativistic picture without pair production.

Units

Because we're doing relativistic (c) quantum mechanics ($\hbar$), we choose units such that

$$\hbar = c = 1 \tag{1.3}$$

This leaves us with one unit free. Typically we will choose it in a given problem to be the mass of an interesting particle, which we will then set equal to one. We'll never get into any problems with that. Just remember that an ordinary macroscopic motion like scratching your head has infinitesimal velocity and astronomical angular momentum! Consequently, in terms of dimensions,

$$[m] = [E] = [T]^{-1} = [L]^{-1} \tag{1.4}$$

Also, it's useful to know

$$(1\,\text{fermi})^{-1} \approx 197\,\text{MeV}; \qquad m_e \approx 0.5\,\text{MeV} = 7.8 \times 10^{20}\,\text{s}^{-1} = 2.6 \times 10^{10}\,\text{cm}^{-1} \tag{1.5}$$

We will say things like the inverse Compton wavelength of the proton is "1 GeV".

Lorentz invariance

The arena for all the physics we're going to do is Minkowski space, flat spacetime in which there are a bunch of points labeled by four coordinates. We write these coordinates as a **4-vector**:

$$x^\mu = (x^0, x^1, x^2, x^3) = (t, \mathbf{x}) \tag{1.6}$$

Sometimes I will suppress the index μ when there's no possibility of confusion and simply write x^μ as x. This is not the only four-component object we will deal with. In classical mechanics there is also the momentum of a particle, which we can call p^μ;

$$p^\mu = (p^0, \mathbf{p}) \tag{1.7}$$

The zeroth component of this 4-vector, the time component, has a special name: the energy. The space component $\mathbf{p}$ is of course called the momentum, and sometimes I will write p^μ as p. I can indiscriminately write p as k, because $\hbar = 1$. The time component k^0 of k^μ is

---

[1] [Eds.] P. A. M. Dirac, "The Quantum Theory of the Electron", *Proc. Roy. Soc. Lond. A* **117** (1928) 610–624; "The Quantum Theory of the Electron. Part II", *Proc. Roy. Soc. Lond. A* **118** (1928) 351–361.

[2] [Eds.] See H. Bethe and E. Salpeter, *Quantum Mechanics of One- and Two-Electron Atoms*, Plenum Publishing, 1977, p.77 and references therein; republished by Dover Publications, 2008; and M. E. Rose, *Relativistic Electron Theory*, Wiley, 1961, pp. 193–196. Rose explicitly shows the suppression of positron density near the hydrogen nucleus as $|\mathbf{p}| \to 0$, and ascribes this suppression to Coulomb repulsion acting on positrons.

the frequency ω. Any *contravariant* 4-vector a^μ can be written as $a^\mu = (a^0, a^i) = (a^0, \mathbf{a})$; similarly the *covariant* 4-vector $a_\mu = (a_0, a_i) = (a^0, -\mathbf{a})$. The four-dimensional inner product $a \cdot b$ between two 4-vectors a^μ, b^ν is

$$a \cdot b \equiv a^\mu b_\mu = a_\mu b^\mu = a^0 b_0 + a^1 b_1 + a^2 b_2 + a^3 b_3 = a^0 b^0 - \mathbf{a} \bullet \mathbf{b} \tag{1.8}$$

where $\mathbf{a} \bullet \mathbf{b}$ is the usual 3-vector inner product. We will as above adopt the so-called Einstein summation convention, I presume familiar to you, where sums over repeated indices are implied. This inner product is invariant under Lorentz transformations. Please note I have adopted the "west coast" metric signature,[3] $(+ - - -)$. The inner product of a 4-vector a^μ with itself usually will be written a^2;

$$a^2 \equiv a^\mu a_\mu = a^0 a_0 - \mathbf{a} \bullet \mathbf{a} \tag{1.9}$$

The inner product can also be written as

$$g_{\mu\nu} a^\mu b^\nu \tag{1.10}$$

where the **metric tensor** $g_{\mu\nu}$ is defined by

$$g_{00} = 1 = -g_{11} = -g_{22} = -g_{33} \tag{1.11}$$

This object is used to lower indices;

$$g_{\mu\nu} A^\nu = A_\mu \tag{1.12}$$

It is convenient to have an object to raise indices as well. We define the metric tensor with upper indices as the inverse matrix to the metric tensor with lower indices;

$$g_{\mu\lambda} g^{\lambda\nu} = \delta^\nu_\mu \tag{1.13}$$

where δ^ν_μ is the conventional Kronecker delta,

$$\delta^\nu_\mu = \begin{cases} 1, & \text{if } \mu = \nu \\ 0, & \text{if } \mu \neq \nu \end{cases} \tag{1.14}$$

This is an easy equation to solve; $g^{\mu\nu}$ is numerically equal to $g_{\mu\nu}$ if we have units such that $c = 1$.

Lorentz transformations on 4-vectors will be denoted by 4×4 matrices Λ^μ_ν. These act on 4-vectors as follows:

$$\Lambda\colon x^\mu \to x^{\mu\,\prime} = \Lambda^\mu_\nu x^\nu \equiv \Lambda x \tag{1.15}$$

Because of the invariance of the inner product,

$$\Lambda a \cdot \Lambda b = a \cdot b \tag{1.16}$$

The Lorentz transformations form a *group* in the mathematical sense: The product of any two Lorentz transformations is a Lorentz transformation, the inverse of a Lorentz transformation

[3] [Eds.] The official text for the course was the two-volume set *Relativistic Quantum Mechanics* and *Relativistic Quantum Fields* (hereafter, *RQM* and *Fields*, respectively) by James D. Bjorken and Sidney D. Drell, McGraw-Hill, 1964 and 1965, respectively. Coleman said this (in 1975) about the books: "I will try to keep my notational conventions close to those of Bjorken and Drell. It's the best available. People like it by an objective test: it is the book most frequently stolen from the Physics Research Library."

is a Lorentz transformation and so on. This group has a name: O(3, 1). The O stands for the orthogonal group. The (3, 1) means that it's not quite an orthogonal group because three of the terms in an inner product have one sign and the fourth has the other. This group is in fact a little too big for our purposes, because it includes transformations which are not invariances of nature: parity and time reversal which as you probably know are broken by the weak interactions. We will restrict ourselves in this course strictly to the *connected* Lorentz group, those Lorentz transformations which can be obtained from the identity by continuous changes. Thus we exclude things like parity and time reversal. Mathematicians call the connected[4] Lorentz group, SO(3, 1), with the S meaning "special", in the sense that the determinant of the matrix equals 1. If we were talking about rotations, we would be looking not at all orthogonal transformations, but rotations in the proper sense, excluding reflections. Every element of the full Lorentz group can be written as a product of an element of the connected Lorentz group with one of the following: $\{1, P, T, PT\}$. The parity operator P reflects all three-space components,

$$P\colon \mathbf{x} \to -\mathbf{x} \tag{1.17}$$

The time reversal operator T reflects the time t; $T\colon t \to -t$, and PT is the product of these. By Lorentz invariance we will mean invariance under SO(3, 1).

Under the action of the Lorentz group, 4-vectors fall into three classes: **timelike**, **spacelike** and **null** (or **lightlike**). These terms describe the invariant square of a 4-vector a^μ;

$$a^\mu \text{ is called } \begin{cases} \textit{timelike}, \text{ if } a^2 > 0 \\ \textit{spacelike}, \text{ if } a^2 < 0 \\ \textit{null}, \text{ if } a^2 = 0 \end{cases} \tag{1.18}$$

The same terms are applied to 4-vector inner products. Given two 4-vectors x and y, the invariant square of the difference $(x - y)$ between them, $(x^\mu - y^\mu)(x_\mu - y_\mu) = (x - y)^2$, will be called the **separation** or the **interval**.

Actually the world is supposed to be invariant under a larger, though no more glamorous, group, which contains the homogeneous Lorentz group as well as space-time translations; this is the Poincaré group. Nobody found that exciting because invariance under translations was known in Newton's time. Nevertheless we will have occasion to consider this larger group. Its elements are labeled by a Lorentz transformation Λ and a 4-vector a. They act on space-time points by Lorentz transformation and translation through a.

Conventions on integration, differentiation and special functions

The fundamental differential operator is denoted ∂_μ, defined to be

$$\partial_\mu = \frac{\partial}{\partial x^\mu} = \left(\frac{\partial}{\partial x^0}, \frac{\partial}{\partial x^i}\right) = \left(\frac{\partial}{\partial t}, \nabla\right) \tag{1.19}$$

It acts on functions of space and time. Note that I have written the operator with a lower index, while I have written x^μ with an upper index. This is correct. The operator ∂_μ does not transform like a contravariant vector a^μ, but instead like a covariant vector a_μ. The easy way

[4] [Eds.] Strictly speaking, the connected Lorentz group is the *orthochronous* Lorentz group, $SO^+(3, 1)$, the subgroup of SO(3, 1) preserving the sign of the zeroth component of a 4-vector.

to remember this is to observe that

$$\partial_\nu x^\mu = \frac{\partial x^\mu}{\partial x^\nu} = \delta^\mu_\nu \tag{1.20}$$

by definition. If we wrote both the operator and the coordinate with lower indices, we should have a g rather than a δ on the right-hand side. An object almost as important as the Laplace operator ∇^2 is the **d'Alembert operator** ∂^2, which we'll write as $\Box^2$,

$$\Box^2 = \partial^2 = \partial^\mu \partial_\mu = (\partial^0)^2 - \nabla^2 \tag{1.21}$$

This is a Lorentz invariant differential operator.[5]

Now for integration. When I don't put any upper or lower limits on an integral, I mean that the integral is to run from $-\infty$ to ∞. In particular, a four-dimensional integral over the components of a 4-vector a^μ;

$$\int d^4a \equiv \int_{-\infty}^{\infty} da^0 \int_{-\infty}^{\infty} da^1 \int_{-\infty}^{\infty} da^2 \int_{-\infty}^{\infty} da^3 \tag{1.22}$$

Delta functions over more than one variable will be written as $\delta^{(3)}(\mathbf{x})$ for three dimensions or $\delta^{(4)}(x)$ for four dimensions. If we define the Fourier transform $\widetilde{F}(k)$ of a function $F(x)$ as

$$\boxed{\widetilde{F}(k) \equiv \int d^4x\, F(x) e^{ik\cdot x}} \tag{1.23}$$

where k and x are both 4-vectors, then

$$\boxed{F(x) = \int \frac{d^4k}{(2\pi)^4} \widetilde{F}(k) e^{-ik\cdot x}} \tag{1.24}$$

I will try to adopt the convention that every dk (or dp) has a denominator of 2π. This will unfortunately lead me to writing down square roots of 2π at intermediate stages. But I will craftily arrange matters so that in the end all factors of dk will carry denominators of 2π, and there will be no other place a 2π comes from. That's important. Sometimes we get sloppy, and act like $1 = -1 = 2\pi$ and $1/(2\pi) = \text{1̶} =$ "one-bar" or something. Well, suppose you predict a result from a beautiful theory. Someone asks if it is measurable, and you say, yes it is. You're going to feel pretty silly if they spend a million and a half dollars to do the measurement and can't find it because you've put a $(2\pi)^2$ in a numerator when it should have been in the denominator...

There's one last function I will occasionally use, $\theta(x)$, the theta function.[6] The theta function is defined by

$$\theta(x) = \begin{cases} 1 & \text{if } x > 0 \\ 0 & \text{if } x < 0 \end{cases} \tag{1.25}$$

---

[5] [Eds.] Most authors write $\Box$ for the d'Alembertian, rather than $\Box^2$. Coleman used $\Box^2$, so that's what is used here.

[6] [Eds.] Also denoted $H(x)$, and frequently called the Heaviside step function, after Oliver Heaviside (1850–1925) who used it extensively. See H. Jeffreys, *Operational Methods in Mathematical Physics*, Cambridge Tracts in Mathematics and Mathematical Physics No. 23, Cambridge U. P., 1927, p. 10.

Its value at the jump, $x = 0$, will be irrelevant in every place we use the function. The derivative of the theta function is a delta function;

$$\frac{d\theta(x)}{dx} = \delta(x) \tag{1.26}$$

We are now ready to investigate our very first example of a relativistic quantum system.

1.2 Theory of a single free, spinless particle of mass μ

The state of a spinless particle is completely specified by its momentum, and the components of momentum form a complete set of commuting variables:[7]

$$\mathbf{P}\,|\mathbf{p}\rangle = \mathbf{p}\,|\mathbf{p}\rangle \tag{1.27}$$

The states are normalized by the condition

$$\langle \mathbf{p}|\mathbf{p}'\rangle = \delta^{(3)}(\mathbf{p} - \mathbf{p}') \tag{1.28}$$

The statement that these kets $|\mathbf{p}\rangle$ form a complete set of states, and that there are no others, is written

$$1 = \int d^3\mathbf{p}\ |\mathbf{p}\rangle\langle\mathbf{p}| \tag{1.29}$$

so that any state $|\psi\rangle$ can be expanded in terms of these;

$$|\psi\rangle = \int d^3\mathbf{p}\,\psi(\mathbf{p})\,|\mathbf{p}\rangle \qquad \text{where } \psi(\mathbf{p}) \equiv \langle\mathbf{p}|\psi\rangle \tag{1.30}$$

If we were doing non-relativistic quantum mechanics, we'd finish describing the theory by giving the Hamiltonian H, and thus the time evolution of the states; $H|\mathbf{p}\rangle = (|\mathbf{p}|^2/2\mu)\,|\mathbf{p}\rangle$.

For relativistic quantum mechanics, we take instead

$$H|\mathbf{p}\rangle = \sqrt{|\mathbf{p}|^2 + \mu^2}\,|\mathbf{p}\rangle \equiv \omega_{\mathbf{p}}\,|\mathbf{p}\rangle \tag{1.31}$$

That's it, the theory of a single free, spinless particle, made relativistic.

How do we know that this theory is Lorentz invariant? Just because it contains one relativistic formula does not necessarily mean it is relativistic. The theory is not manifestly Lorentz invariant. The theory is however manifestly rotationally and translationally invariant. Let's be more precise about this.

Translation invariance

To any active translation specified by a given 4-vector a^μ, there should be a linear operator $U(a)$ satisfying these conditions:

$$U(a)U(a)^\dagger = 1, \quad \text{to preserve probability amplitudes} \tag{1.32}$$

$$U(0) = 1 \tag{1.33}$$

$$U(a)U(b) = U(a+b) \tag{1.34}$$

[7] [Eds.] Because $\mathbf{p} = \hbar\mathbf{k}$, and in our units $\hbar = 1$, we could equally well use kets $|\mathbf{k}\rangle$; $\mathbf{p}$ and $\mathbf{k}$ both stand for momentum.

The operator U satisfying these conditions is $U(a) = e^{iP\cdot a}$ where $P^\mu = (H, \mathbf{P})$.

Aside. I've laid out this material in pedagogical, not logical order. The logical order would be to state:

1. We want to set up a translationally invariant theory of a spinless particle. The theory will contain unitary translation operators $U(\mathbf{a})$.

2. Define P^i as
$$P^i = i\left.\frac{\partial U(\mathbf{a})}{\partial a^i}\right|_{\mathbf{a}=0} \tag{1.35}$$
From (1.34), $[P_i, P_j] = 0$; and from (1.32), $\mathbf{P} = \mathbf{P}^\dagger$.

3. Declare P^i to be a complete set and classify the states by momentum.

4. Define $H = \sqrt{|\mathbf{P}|^2 + \mu^2}$, and thus give the time evolution.

Continuing with the pedagogical order:

States described by kets are transformed by $U(a) = e^{iP\cdot a}$ as follows:
$$U(a)\,|0\rangle = |a\rangle \tag{1.36}$$
where $|x\rangle$ means a state centered at x^μ; $|0\rangle$ means a state centered at the origin. Operators O transform as
$$O(x+a) = U(a)O(x)U(a)^\dagger \tag{1.37}$$
and expectation values transform as
$$\langle a|O(x+a)|a\rangle = \langle 0|O(x)|0\rangle \tag{1.38}$$
Reducing the transformations to space translations,
$$\begin{aligned} U(\mathbf{a}) &= e^{-i\mathbf{P\cdot a}} \\ e^{-i\mathbf{P\cdot a}}\,|\mathbf{q}\rangle &= |\mathbf{q}+\mathbf{a}\rangle \\ e^{-i\mathbf{P\cdot a}}\,O(\mathbf{x})\,e^{i\mathbf{P\cdot a}} &= O(\mathbf{x}+\mathbf{a}) \end{aligned} \tag{1.39}$$
Only operators localized in space transform according to this rule. The position operator $\widehat{\mathbf{q}}$ does not:
$$\begin{aligned} \widehat{\mathbf{q}}\,e^{-i\mathbf{P\cdot a}}\,|\mathbf{q}\rangle &= (\mathbf{q}+\mathbf{a})\,|\mathbf{q}+\mathbf{a}\rangle \\ e^{i\mathbf{P\cdot a}}\,\widehat{\mathbf{q}}\,e^{-i\mathbf{P\cdot a}}\,|\mathbf{q}\rangle &= (\mathbf{q}+\mathbf{a})\,|\mathbf{q}\rangle \\ \Rightarrow e^{i\mathbf{P\cdot a}}\,\widehat{\mathbf{q}}\,e^{-i\mathbf{P\cdot a}} &= \widehat{\mathbf{q}}+\mathbf{a} \end{aligned} \tag{1.40}$$
which looks like the opposite of the operator transformation rule (1.39) given above. The operator $\widehat{\mathbf{q}}$ is not an operator localized at $\mathbf{q}$, so there is no reason for these last two equations to look alike.

Rotational invariance

Given a rotation $R \in \mathrm{SO}(3)$, there should be a unitary operator $U(R)$ satisfying these conditions:

$$U(R)U(R)^\dagger = 1 \tag{1.41}$$
$$U(1) = 1 \tag{1.42}$$
$$U(R_1)U(R_2) = U(R_1R_2) \tag{1.43}$$

Denote a transformed ket by $|\psi'\rangle = U(R)\,|\psi\rangle$, and require for any $|\psi\rangle$ the rule

$$\langle\psi'|\mathbf{P}|\psi'\rangle = R\,\langle\psi|\mathbf{P}|\psi\rangle \tag{1.44}$$

so we get

$$U(R)^\dagger\mathbf{P}U(R) = R\mathbf{P} \tag{1.45}$$
$$U(R)^\dagger HU(R) = H \tag{1.46}$$

A $U(R)$ satisfying all these properties is given by

$$U(R)\,|\mathbf{p}\rangle = |R\mathbf{p}\rangle \tag{1.47}$$

That (1.42) and (1.43) are satisfied is trivial. To prove (1.41), insert a complete set between U and $U^\dagger$:

$$\begin{aligned} U(R)U(R)^\dagger &= U(R)\left[\int d^3\mathbf{p}\ |\mathbf{p}\rangle\langle\mathbf{p}|\right]U(R)^\dagger = \int d^3\mathbf{p}\,(U\,|\mathbf{p}\rangle)(\langle\mathbf{p}|\,U^\dagger) \\ &= \int d^3\mathbf{p}\ |R\mathbf{p}\rangle\langle R\mathbf{p}| \end{aligned} \tag{1.48}$$

Let $\mathbf{p}' = R\mathbf{p}$; the Jacobian is 1, so $d^3\mathbf{p}' = d^3\mathbf{p}$, and

$$U(R)U(R)^\dagger = \int d^3\mathbf{p}\,|R\mathbf{p}\rangle\langle R\mathbf{p}| = \int d^3\mathbf{p}'\ |\mathbf{p}'\rangle\langle\mathbf{p}'| = 1 \tag{1.49}$$

To prove (1.45), write

$$\begin{aligned} U(R)^\dagger\mathbf{P}U(R) &= U(R)^{-1}\mathbf{P}(U(R)^{-1})^\dagger && \text{by (1.41)} \\ &= U(R^{-1})\mathbf{P}U(R^{-1})^\dagger && \text{by (1.42) and (1.43)} \\ &= U(R^{-1})\mathbf{P}\int d^3\mathbf{p}\ |\mathbf{p}\rangle\langle\mathbf{p}|\,U(R^{-1})^\dagger \\ &= U(R^{-1})\int d^3\mathbf{p}\,\mathbf{p}\,|\mathbf{p}\rangle\langle\mathbf{p}|\,U(R^{-1})^\dagger \\ &= \int d^3\mathbf{p}\,\mathbf{p}\,|R^{-1}\mathbf{p}\rangle\langle R^{-1}\mathbf{p}| && (\text{Let } \mathbf{p} = R\mathbf{p}';\ d^3\mathbf{p} = d^3\mathbf{p}') \\ &= \int d^3\mathbf{p}'\,R\mathbf{p}'\,|\mathbf{p}'\rangle\langle\mathbf{p}'| \\ &= R\mathbf{P} \end{aligned} \tag{1.50}$$

The proof of (1.46) is left to you.

Constructing Lorentz invariant kets

Our study of rotations provides a template for studying Lorentz invariance. Suppose a silly physicist took for normalized three-momentum states the kets $|\mathbf{p}\rangle_S$ defined by

$$|\mathbf{p}\rangle_S = \sqrt{1+p_z^2}\,|\mathbf{p}\rangle \tag{1.51}$$

These kets are normalized by the condition

$$_S\langle\mathbf{p}|\mathbf{p}'\rangle_S = (1+p_z^2)\,\delta^{(3)}(\mathbf{p}-\mathbf{p}') \tag{1.52}$$

The completeness relation is

$$1 = \int d^3\mathbf{p}\,\frac{1}{1+p_z^2}\,|\mathbf{p}\rangle_S\,{}_S\langle\mathbf{p}| \tag{1.53}$$

If our silly physicist now took $U_S(R)\,|\mathbf{p}\rangle_S = |R\mathbf{p}\rangle_S$, his proofs of (1.41), (1.45) and (1.46) would break down, because

$$d^3\mathbf{p}\,\frac{1}{1+p_z^2} \neq d^3\mathbf{p}'\,\frac{1}{1+p_z'^2} \qquad \text{i.e., } d^3\mathbf{p}\,\frac{1}{1+p_z^2} \text{ is } \textit{not} \text{ a rotationally invariant measure.} \tag{1.54}$$

Let's apply this lesson. The usual 3-space normalization, $\langle\mathbf{p}|\mathbf{p}'\rangle = \delta^{(3)}(\mathbf{p}-\mathbf{p}')$, is a silly normalization for Lorentz invariance; $d^3\mathbf{p}$ is not a Lorentz invariant measure. We want a Lorentz invariant measure on the hyperboloid $p^2 = (p^0)^2 - |\mathbf{p}|^2 = \mu^2$, $p^0 > 0$. The measure

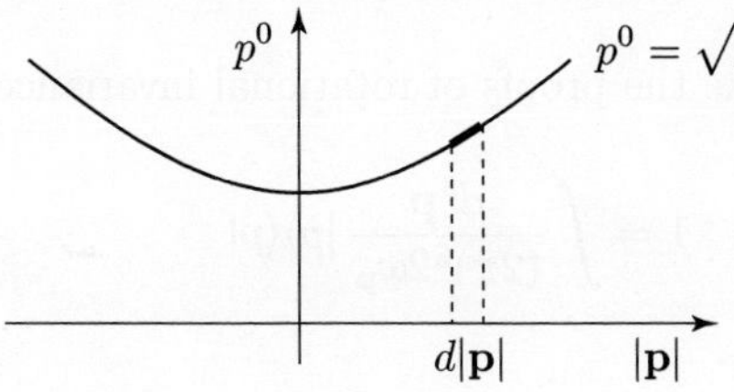

Figure 1.1: Restricting $|d\mathbf{p}|$ to the invariant hyperboloid $p^2 = \mu^2$

d^4p is Lorentz invariant. To restrict it to the hyperboloid, multiply it by the Lorentz invariant factor $\delta(p^2-\mu^2)\theta(p^0)$. That yields our relativistic measure on the hyperboloid[8]

$$\int_{p^0=-\infty}^{\infty} dp^0\left\{d^3\mathbf{p}\,\delta(p^2-\mu^2)\theta(p^0)\right\} = \frac{d^3\mathbf{p}}{2\omega_{\mathbf{p}}} \tag{1.55}$$

where

$$\omega_{\mathbf{p}} = \sqrt{|\mathbf{p}|^2+\mu^2}, \quad p^\mu = (\omega_{\mathbf{p}},\mathbf{p}) \tag{1.56}$$

[8] [Eds.] The equality follows from the identity $\delta(f(x)) = \sum_i \frac{\delta(x-a_i)}{|f'(a_i)|}$ where $\{a_i\}$ are the zeroes of $f(x)$. Then

$$\delta(p^2-\mu^2) = \delta((p^0)^2-\omega_{\mathbf{p}}^2) = \frac{\delta(p^0-\omega_{\mathbf{p}})}{2\omega_{\mathbf{p}}} + \frac{\delta(p^0+\omega_{\mathbf{p}})}{2\omega_{\mathbf{p}}}$$

The $\theta(p^0)$ factor kills the second delta function, and integrating over p^0 gives just the factor $(2\omega_{\mathbf{p}})^{-1}$, times the remaining $d^3\mathbf{p}$. Similarly, one can show $d^3\mathbf{x}\,\delta(|\mathbf{x}|^2-R^2) = \frac{1}{2}R\sin\theta\,d\theta\,d\phi$.

Later on, we'll want factors of 2π to come out right in Feynman diagrams, so we'll take for our relativistically normalized kets $|p\rangle$

$$|p\rangle = \sqrt{(2\pi)^3}\sqrt{2\omega_{\mathbf{p}}}\,|\mathbf{p}\rangle \tag{1.57}$$

so that

$$1 = \frac{1}{(2\pi)^3}\int d^4p\,\delta(p^2-\mu^2)\theta(p^0)\,|p\rangle\langle p| = \int d^3\mathbf{p}\,|\mathbf{p}\rangle\langle\mathbf{p}| \tag{1.58}$$

From the graph of the hyperbola, it looks like the factor multiplying $d^3\mathbf{p}$ ought to get larger as $|\mathbf{p}|$ gets large. This is an illusion, caused by graphing on Euclidean paper. It's the same illusion that occurs in the Twin Paradox: Though the moving twin's path appears longer, in fact that twin's proper time is shorter.

Now let's demonstrate Lorentz invariance. Given any Lorentz transformation Λ, define

$$U(\Lambda)\,|p\rangle = |\Lambda p\rangle \tag{1.59}$$

The unitary operator $U(\Lambda)$ satisfies these conditions:

$$U(\Lambda)U(\Lambda)^\dagger = 1 \tag{1.60}$$

$$U(1) = 1 \tag{1.61}$$

$$U(\Lambda_1)U(\Lambda_2) = U(\Lambda_1\Lambda_2) \tag{1.62}$$

$$U(\Lambda)^\dagger P U(\Lambda) = \Lambda P \tag{1.63}$$

The proofs of these are exactly like the proofs of rotational invariance, using the completeness relation

$$1 = \int \frac{d^3\mathbf{p}}{(2\pi)^3 2\omega_{\mathbf{p}}}\,|p\rangle\langle p| \tag{1.64}$$

and the invariance of the measure,

$$\frac{d^3\mathbf{p}}{(2\pi)^3 2\omega_{\mathbf{p}}} = \frac{d^3\mathbf{p}'}{(2\pi)^3 2\omega_{\mathbf{p}'}} \tag{1.65}$$

1.3 Determination of the position operator X

We have a fairly complete theory, except that we don't know where anything is; a particle could be at the origin or at the Andromeda galaxy. In non-relativistic quantum mechanics, if a particle is in an eigenstate of a position operator X, its position x is its eigenvalue. Can we construct a position operator, X, for our system? Fortunately we can write down some general conditions about such an operator, conditions we can all agree are perfectly reasonable, which will be enough to specify this operator uniquely.[9] And then there will be a surprise, because we'll find out that this uniquely specified operator is totally unsatisfactory! There will be a physical reason for that.

[9] [Eds.] See §22 "Schrödinger's Representation", in P. A. M. Dirac, *The Principles of Quantum Mechanics*, 4th ed. revised, Oxford U. P., 1967.

What conditions do we want our X operator to satisfy? These conditions will *not* involve Lorentz invariance, but only invariance under rotations and translations in space:

$$X = X^\dagger \tag{1.66}$$

$$U(\mathbf{a})^\dagger X U(\mathbf{a}) = e^{i\mathbf{P}\cdot\mathbf{a}}\, X\, e^{-i\mathbf{P}\cdot\mathbf{a}} = X + \mathbf{a} \tag{1.67}$$

$$U(R)^\dagger X U(R) = RX \tag{1.68}$$

We impose the first condition because x is an observable. The second condition is the rule (1.40). The third condition says that X transforms as a 3-vector, so we might as well write it as $\mathbf{X}$ or its components as X^i. Then, by taking $\partial/\partial a_i$ of the second condition and evaluating at $a_i = 0$, we get the usual commutator $i[P_i, X_j] = \delta_{ij}$. Now you see a new origin for this familiar equation.

From the commutator, we can deduce something about X^i;

$$X^i = i\frac{\partial}{\partial p_i} + R^i \tag{1.69}$$

where R^i is a remainder that must commute with P^j in order to give us the right result. We know that this expression for X^i has the right commutation relations. Now to find R^i.

We know something about our system. We know that the three components P^i are a complete set of commuting operators. From non-relativistic quantum mechanics, we know that anything that commutes with a complete set of commuting operators must be a function of those operators. Therefore R^i must be some function of the P^i's. According to the third condition (1.68), X^i must transform as a 3-vector, and so must R^i. That tells us R^i must be of the form

$$R^i = p^i F(|\mathbf{p}|^2) \tag{1.70}$$

where $F(|\mathbf{p}|^2)$ is an unknown function of $|\mathbf{p}|^2$. But any such function of this form is a gradient of some scalar function $G(|\mathbf{p}|^2)$; that is,

$$p^i F(|\mathbf{p}|^2) = \frac{\partial G(|\mathbf{p}|^2)}{\partial p_i} \tag{1.71}$$

This specifies the position operator to be

$$X^i = i\frac{\partial}{\partial p_i} + \frac{\partial G(|\mathbf{p}|^2)}{\partial p_i} \tag{1.72}$$

We can do more. We can eliminate the remainder term entirely by changing the phase of the P states:

$$|\mathbf{p}\rangle \to |\mathbf{p}\rangle_G = e^{iG(|\mathbf{p}|^2)}\,|\mathbf{p}\rangle \tag{1.73}$$

I'm perfectly free to make that reassignment. It does not affect the physics of theses states. These are still eigenstates of P_i with eigenvalues p_i, they are still eigenstates of H with eigenvalues $\sqrt{|\mathbf{p}|^2 + \mu^2}$, and they are still normalized in the same way. This is a unitary transformation; call it $U(G)$:

$$|\mathbf{p}\rangle \to U(G)\,|\mathbf{p}\rangle \tag{1.74}$$

and so the operators change accordingly:

$$\mathbf{X} \to U(G)^\dagger \mathbf{X} U(G) = \mathbf{X}_G \tag{1.75}$$

The only formula this transformation affects in all we have done so far is the expression for X^i on $|\mathbf{p}\rangle$. Now we have

$$\begin{aligned} X^i_G\,|\mathbf{p}\rangle_G &= e^{-iG(|\mathbf{p}|^2)}\left(i\frac{\partial}{\partial p_i} + \frac{\partial G(|\mathbf{p}|^2)}{\partial p_i}\right)e^{iG(|\mathbf{p}|^2)}\,|\mathbf{p}\rangle_G \\ &= e^{-iG(|\mathbf{p}|^2)}e^{iG(|\mathbf{p}|^2)}\left(i\frac{\partial}{\partial p_i} + \frac{\partial G(|\mathbf{p}|^2)}{\partial p_i} - \frac{\partial G(|\mathbf{p}|^2)}{\partial p_i}\right)|\mathbf{p}\rangle_G \\ &= i\frac{\partial}{\partial p_i}\,|\mathbf{p}\rangle_G \end{aligned} \tag{1.76}$$

Thus the *unique* candidate for the X^i operator—providing one chooses the phase for the eigenstates appropriately—is nothing more nor less than the good old-fashioned X^i operator in non-relativistic quantum mechanics, the operator which in P space is $i\partial/\partial p_i$. Let's make this choice, and drop the G subscripts from now on.

Now that we have found our X^i operator, we know where our particle is. Or do we? Let's do a thought experiment. If we really have, in a relativistic theory, a well-defined position operator, we should be able to say of our particle that it does not travel faster than light. That is, we can start out with a state where our particle is sharply localized, say at the origin,[10] allow that state to evolve in time (according to the Schrödinger equation, since we know the Hamiltonian), and see if at some later time there is a non-zero probability for the particle to have moved faster than the speed of light. We have all the equipment; we need only do the computation. Let's do it.

We start out with a state $|\psi\rangle$ localized at the origin at time $t = 0$, i.e.,

$$\langle\mathbf{x}|\psi\rangle = \delta^{(3)}(\mathbf{x}) \tag{1.77}$$

Because the X^i operator is its usual self, we can make use of the usual relation[11]

$$\langle\mathbf{x}|\mathbf{p}\rangle = \frac{1}{(2\pi)^{3/2}}e^{i\mathbf{p}\cdot\mathbf{x}} \tag{1.78}$$

and so at $t = 0$

$$\langle\mathbf{p}|\psi\rangle = \int \langle\mathbf{p}|\mathbf{x}\rangle\,d^3\mathbf{x}\,\langle\mathbf{x}|\psi\rangle = \int \frac{e^{-i\mathbf{p}\cdot\mathbf{x}}}{(2\pi)^{3/2}}\,\delta^{(3)}(\mathbf{x})\,d^3\mathbf{x} = \frac{1}{(2\pi)^{3/2}} \tag{1.79}$$

We wish to compute the probability amplitude for the particle to be found at position $\mathbf{x}$ at time t, which by the general rules of quantum mechanics is given by

$$\langle\mathbf{x}|e^{-iHt}|\psi\rangle \tag{1.80}$$

---

[10] By translational invariance and superposition, we could easily get the evolution of *any* configuration from this calculation.

[11] [Eds.] Consider $\langle\mathbf{x}|X^i|\mathbf{p}\rangle$. If we let the operator operate to the left, we get

$$\langle\mathbf{x}|X^i|\mathbf{p}\rangle = x^i\,\langle\mathbf{x}|\mathbf{p}\rangle$$

but operating to the right,

$$\langle\mathbf{x}|X^i|\mathbf{p}\rangle = i\frac{\partial}{\partial p_i}\,\langle\mathbf{x}|\mathbf{p}\rangle$$

so the quantity $\langle\mathbf{x}|\mathbf{p}\rangle$ satisfies the differential equation $i\partial/\partial p_i\,\langle\mathbf{x}|\mathbf{p}\rangle = x^i\,\langle\mathbf{x}|\mathbf{p}\rangle$. Then $\langle\mathbf{x}|\mathbf{p}\rangle = Ce^{i\mathbf{p}\cdot\mathbf{x}}$, where C may depend on x (but not p). By considering $\langle\mathbf{x}|P^i|\mathbf{p}\rangle$, you can show C is a constant, and we set $C = 1/(2\pi)^{3/2}$ for convenience. See Dirac, *op. cit.*, §23, "The momentum representation".

The operators X^i and P^j are the same; only the Hamiltonian is novel. Thus we can do the computation, we just put the pieces together:

$$\langle\mathbf{x}|e^{-iHt}|\psi\rangle = \int d^3\mathbf{p}\,\langle\mathbf{x}|e^{-iHt}|\mathbf{p}\rangle\,\langle\mathbf{p}|\psi\rangle \;\; = \int \frac{d^3\mathbf{p}}{(2\pi)^3}\, e^{i\mathbf{p}\cdot\mathbf{x}}\, e^{-i\omega_{\mathbf{p}}t} \tag{1.81}$$

because $H\,|\mathbf{p}\rangle = \omega_{\mathbf{p}}\,|\mathbf{p}\rangle$, and so $\langle\mathbf{x}|e^{-iHt}|\mathbf{p}\rangle = e^{-i\omega_{\mathbf{p}}t}\,\langle\mathbf{x}|\mathbf{p}\rangle = e^{-i\omega_{\mathbf{p}}t+i\mathbf{p}\cdot\mathbf{x}}/(2\pi)^{3/2}$. Compute the integral in the usual way by going over to polar coordinates,

$$\begin{aligned}\int \frac{d^3\mathbf{p}}{(2\pi)^3}\, e^{i\mathbf{p}\cdot\mathbf{x}}\, e^{-i\omega_{\mathbf{p}}t} &= \int_0^\infty \frac{p^2 dp}{(2\pi)^3} e^{-i\omega_p t} \int_0^\pi e^{ipr\cos\theta} \sin\theta\, d\theta \int_0^{2\pi} d\phi \\ &= \frac{1}{(2\pi)^2} \int_0^\infty dp\, p\, e^{-i\omega_p t} \frac{\left(e^{ipr} - e^{-ipr}\right)}{ir} = -i\frac{1}{(2\pi)^2 r} \int_{-\infty}^\infty dp\, p\, e^{ipr - i\omega_p t}\end{aligned} \tag{1.82}$$

(letting $r = |\mathbf{x}|$, $p = |\mathbf{p}|$ and $\omega_p = \sqrt{p^2 + \mu^2}$.) This is a messy integral, full of oscillations. It's difficult to tell if it vanishes outside the light cone, or not. Remember, in our units, the speed of light is 1. Since we started out with $r = 0$ at $t = 0$, if the particle is traveling faster than light, the probability amplitude for $r > t$ will be non-zero.

To calculate the integral, we extend p to complex values, and let $p \to z = x + iy$. We'll take the x axis as part of a contour C, and close the contour with a large semicircular arc above or below the x-axis. Our integrand is not however an analytic function of p, because the function $\omega_p = \sqrt{p^2 + \mu^2}$ has branch points at $p = \pm i\mu$, and thus also a branch line connecting these two points. I choose to write the branch cuts as extending from $+i\mu$ up along the positive imaginary axis, and from $-i\mu$ down along the negative imaginary axis. If we distort the semicircular contour C to avoid the branch cuts, the integrand is an analytic function within the region bounded by the distorted contour, as shown below.

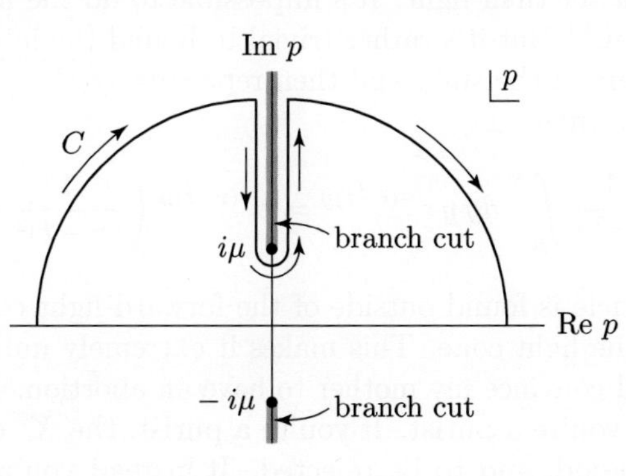

Figure 1.2: Contour for evaluating the integral $\int_{-\infty}^{\infty} dp\, p\, e^{ipr - i\omega_p t}$

Since the integrand is analytic within C, the integral along C, say counter-clockwise, gives zero. The original integral along the x axis then equals the rest of the integral in the opposite sense, going clockwise along the large arcs, going down on the left side of the branch cut, and up on the right side. Along the upper branch cut, p is parametrized by iy. The value of ω_p is discontinuous across this branch cut; its value on either side of the branch cut is[12]

$$\omega_p = \begin{cases} i\sqrt{y^2-\mu^2}, & x = 0+ \\ -i\sqrt{y^2-\mu^2}, & x = 0- \end{cases} \tag{1.83}$$

Along the large arcs, p is parametrized by $Re^{i\theta} = R\cos\theta + iR\sin\theta$, where $\pi \geq \theta \geq \pi/2$ on the left-hand arc, and $\pi/2 \geq \theta \geq 0$ on the right-hand arc. The integrand involves $e^{ipr-i\omega_p t}$ which is bounded, since $r > t$, by $e^{-Rr\sin\theta}$. Consequently in the limit $R \to \infty$, the large arcs contribute nothing to the integral. The small arc likewise contributes nothing in the limit as the small circle's radius goes to zero. Then

$$\langle \mathbf{x}|e^{-iHt}|\psi\rangle = \frac{i}{(2\pi)^2 r}\left[\int_\infty^\mu dy\, y\, e^{-ry-\sqrt{y^2-\mu^2}t} + \int_\mu^\infty dy\, y\, e^{-ry+\sqrt{y^2-\mu^2}t}\right] \tag{1.84}$$

Though the ω_p part of the exponential is damped on the left side of the cut, the exponential increases on the right side. However since $r > t$, the strictly damped part of the exponential, $-ry$, dominates over the increasing part $+\sqrt{y^2-\mu^2}t$. Changing the limits in the first term gives

$$\begin{aligned} \langle \mathbf{x}|e^{-iHt}|\psi\rangle &= \frac{i}{(2\pi)^2 r}\int_\mu^\infty dy\, y\, e^{-ry}\left[e^{\sqrt{y^2-\mu^2}t} - e^{-\sqrt{y^2-\mu^2}t}\right] \\ &= \frac{i}{2\pi^2 r}\int_\mu^\infty dy\, y\, e^{-ry}\sinh(\sqrt{y^2-\mu^2}t) \end{aligned} \tag{1.85}$$

This is bad news, boys and girls, because this integrand is a product of positive terms. Therefore the integral is *not* zero, and our particle always has uncertain nonzero probability amplitude for traveling faster than the speed of light. So the particle can move faster than light and thus backwards in time, with all the associated paradoxes. I hope you understand the titanic meaning of that statement.

Things are not so bad, however, as you would think. The particle doesn't have *much* of a probability of traveling faster than light. It's impossible to do the integral, which means the answer is a Bessel function.[13] But it's rather trivial to bound the integral by keeping only the increasing exponential part of the sinh, and then replacing $\sqrt{y^2-\mu^2}$ by y. This will give us an overestimate. We have then

$$\langle \mathbf{x}|e^{-iHt}|\psi\rangle < i\frac{1}{2\pi^2 r}\int_\mu^\infty dy\, y\, e^{-(r-t)y} = e^{-(r-t)\mu}\left(\frac{1}{(r-t)^2} + \frac{\mu}{(r-t)}\right) \tag{1.86}$$

The chance that the particle is found outside of the forward light cone falls off exponentially as you get farther from the light cone. This makes it extremely unlikely that, for example, I could go back in time and convince my mother to have an abortion. But if it is at all possible, it is still unacceptable if you're a purist. If you're a purist, the X^i operator we have defined is absolutely rotten, no good, and to be rejected. If instead you're a slob, it's not so bad,

---

[12] [Eds.] For the details, see the example on pp. 71–73 of *Mathematics for Physicists*, Phillipe Dennery and André Krzywicki, Harper and Row, 1967, republished by Dover Publications, 1996, or Example 2 in Chap. 7, pp. 202–205 of *Complex Variables and Applications*, Ruel V. Churchill and James Ward Brown, McGraw-Hill, 1974.

[13] [Eds.] Coleman is joking. *Mathematica* fails to find a closed form for this integral, so it isn't really a Bessel function.

because the amplitude of finding the particle outside of the forward light cone is rather small. It's exponentially damped, and if we go a few factors of $1/\mu$, a few of the particle's Compton wavelengths away from the light cone, the amplitude comes down quite a bit.

What we have discovered is that we cannot get a precise determination of where the particle is. But if we're only concerned with finding the particle to within a few of its own Compton wavelengths, in practice things are not so bad. In principle, the inability to localize a single particle is a disaster. How does nature get out of this disaster? Is there a physical basis for an escape? Yes, there is.

Suppose I attempt to localize a particle in the traditional *gedanken* experiment methods of Niels Bohr. (In fact, this argument is due to Niels Bohr.[14]) I build an impermeable box with moveable sides. I put the particle inside it. I turn the crank, like the Spanish Inquisition, and the sides of the box squeeze down. It appears that I can localize the particle as sharply as I want. What goes wrong? What could relativity possibly have to do with this?

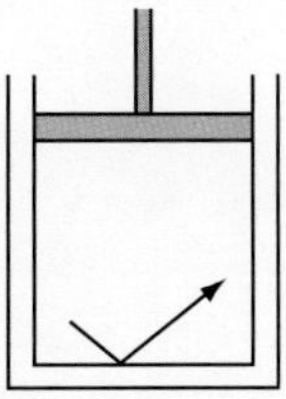

Figure 1.3: Particle in a box with a movable wall

The point is this. If I try to localize the particle within a space of dimensions L on the order of its own Compton wavelength, $L \sim \mathcal{O}(1/\mu)$, then not relativity, but our old reliable friend the Uncertainty Principle comes into play and tells us

$$\Delta p \gtrsim \mathcal{O}(\mu). \tag{1.87}$$

If the dispersion in p is on the order of μ, then so must p itself be at least the order of μ. Then we have enough energy in the box to produce pairs.

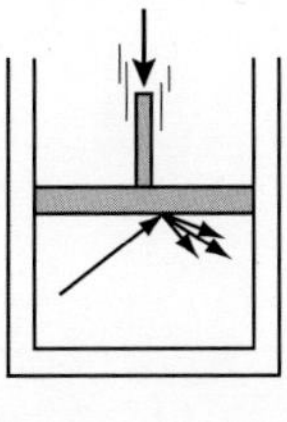

Figure 1.4: Particle squeezed in the box

Like the worm Ouroboros,[15] this section ends where it began, with pair production. If we squeeze the particle down more and more, we must have more and more uncertainty in

---

[14] [Eds.] N. Bohr and L. Rosenfeld, "Field and Charge Measurements in Quantum Electrodynamics", *Phys. Rev.* **78** (1950) 794–798.

[15] [Eds.] *The Worm Ouroboros* is a fantasy novel by E. R. Eddison, published in 1922; J. R. R. Tolkien was an admirer. The ouroboros (Greek οὐρά, "tail" + βόρος, "devouring") is the image of a snake or dragon eating its own tail. Originally ancient Egyptian, it entered western tradition via Greece, and came to be associated

momentum. If we have a large spread in momentum, there must be a probability for having a large energy inside the box. If we have a large energy inside the box, we know there's *something* inside the box, but we don't know it's a *single* particle. It could be three particles, or five, or seven. The moral of the story is that we cannot satisfactorily localize the particle in a single-particle theory.

So we can localize something, but what we're localizing is not a single particle. Because of the phenomena of pair production, not only is momentum complementary to position, but particle number is complementary to position. If we make a very precise measurement of position, we'll have a very big spread in momentum and therefore, because pair production takes place, we do not know how many particles we have. *Relativistic causality is inconsistent with a single-particle quantum theory.* The real world evades the conflict through pair production. That's the physical reason for the mathematics we've just gone through. This leads to our next topic, a discussion of *many* free particles.

with European alchemy during the Middle Ages. It is often used as a symbol of the eternal cycle of death and rebirth. See *A Dictionary of Symbols*, J. E. Cirlot, Dover Publications, 2002, pp. 15, 48, 87, 246–247. The German chemist August Kekulé reported that a dream of the ouroboros led him to propose the structure of the benzene ring: O. Theodor Benfey, trans., "August Kekulé and the Birth of the Structural Theory of Chemistry in 1858", *J. Chem. Ed.* **35** (1958) 21–23.

2

The simplest many-particle theory

In the last section we fiddled around with the theory of a single, relativistic spinless particle. We found some things that will be useful to us in the remainder of this course, like the Lorentz transformation properties of the particle, and some things that served only to delineate dead ends, e.g., we could not define a satisfactory X^i operator. When we tried to localize a particle, we found that the particle moved faster than the speed of light. At the end of the lecture I pointed out that the problem of localizing a particle could be approached from another viewpoint. Instead of staying within the theory of a single particle, we could imagine an idealized example of the real world in which pair creation occurs. We discovered that we couldn't localize a particle in a box. If the box was too small, it wasn't full of a single particle, it was full of pairs. This motivates us to investigate a slightly more complicated system, a system consisting of an arbitrary number of free, relativistic spinless particles. The investigation of this system will occupy this whole section. The problem of localization should be in the back of our minds but I won't say anything about it.

2.1 First steps in describing a many-particle state

The general subject is called **Fock space.**[1] That is the name for the Hilbert space, the space of states that describes the system we're going to talk about. We'll discover that when we first write down Fock space it will be extremely ugly and awkward. We will have to do a lot of work to find an efficient bookkeeping algorithm to enable us to manipulate Fock space without going crazy. The bookkeeping will be managed though the algebra of objects called *annihilation and creation operators*, which may be familiar to you from an earlier course in quantum mechanics.

The devices I'm going to introduce here—although we will use them exclusively for the purposes of relativistic quantum mechanics—are not exclusively applied to that. There are frequently systems in many-body theory and in statistical mechanics where the number of particles is not fixed. In statistical physics we wish frequently to consider the so-called grand canonical ensemble, where we average over states with different numbers of particles in them, the number fluctuating around a value determined by the chemical potential. In solid state

[1] [Eds.] V. Fock, "Konfigurationsraum und zweite Quantelung" (Configuration space and second quantization), *Zeits. f. Phys.* **75** (1932) 622–627.

physics, there's typically a lot of electrons in a solid but we are usually interested only in the electrons that have stuck their heads above the Fermi sea, the conduction electrons. The number of these electrons can change as electrons drop in and out of the Fermi sea. So the methods are of wider applicability. In order to keep that clear, I will use our non-relativistic normalization of states and non-relativistic notation and just switch to relativity at the end when we want to talk about Lorentz transformation properties.

Let me remind you of the Hilbert space of non-relativistic one-particle states we had before: the momentum kets $|\mathbf{p}\rangle$ labeled by a basis vector $\mathbf{p}$, and normalized with a delta function,

$$\langle\mathbf{p}|\mathbf{p}'\rangle = \delta^{(3)}(\mathbf{p}-\mathbf{p}') \tag{2.1}$$

which is standard for plane waves. These are simultaneous eigenstates of the Hamiltonian, H, with eigenvalues $\omega_{\mathbf{p}} = \sqrt{|\mathbf{p}|^2+\mu^2}$ (but that's not going to be relevant here), and of the momentum operator $\mathbf{P}$, with eigenvalues $\mathbf{p}$;

$$H|\mathbf{p}\rangle = \omega_{\mathbf{p}}|\mathbf{p}\rangle \qquad \mathbf{P}|\mathbf{p}\rangle = \mathbf{p}|\mathbf{p}\rangle \tag{2.2}$$

They also of course have well defined Lorentz transformation properties, which we talked about last time, but I'm not going to focus on that for the moment. In the last section this was a complete set of basis vectors for our Hilbert space; a general state was a linear combination of these states. But now we are after a bigger Hilbert space, so these will be just a subset of the basis vectors. I will call these "one-particle basis vectors". We are considering a situation where we look at the world and maybe we find one particle in it, but maybe we find two or three or four, and maybe we find some linear combination of these situations. Therefore we need more basis vectors. In particular we need two-particle basis vectors. I'll write down the construction for them, and then I'll just write down an "*et cetera*" for the remainder (the three-particle states, the four-particle states, ...)

A two-particle state describes two independent particles, and will be labeled by the momenta of the two particles, which I will call $\mathbf{p}_1$ and $\mathbf{p}_2$, which can be any two 3-vectors. (Don't confuse these subscripts with vector indices, the index labels the particle.) We will assume that our spinless particles are identical bosons, and therefore to incorporate Bose[2] statistics we label the state $|\mathbf{p}_1,\mathbf{p}_2\rangle$ which in fact is the same state designated by $|\mathbf{p}_2,\mathbf{p}_1\rangle$. It doesn't matter whether the first particle has one momentum and the second the other, or *vice versa.* We will normalize the states again with traditional delta function normalization:

$$\langle\mathbf{p}_1,\mathbf{p}_2|\mathbf{p}_1',\mathbf{p}_2'\rangle = \delta^{(3)}(\mathbf{p}_1-\mathbf{p}_1')\,\delta^{(3)}(\mathbf{p}_2-\mathbf{p}_2') + \delta^{(3)}(\mathbf{p}_1-\mathbf{p}_2')\,\delta^{(3)}(\mathbf{p}_2-\mathbf{p}_1') \tag{2.3}$$

The states are orthogonal, unless the two momenta involved are equal, either for one permutation or the other. We have to include both those terms or else we'll have a contradiction with the normalization equation for a single particle. These states are eigenstates of the Hamiltonian and their energies are of course the sum of the energies associated with the two individual particles, and they are eigenstates of the momentum operator, and their momentum is the sum of the two momenta of the two individual particles:

$$H|\mathbf{p}_1,\mathbf{p}_2\rangle = (\omega_{\mathbf{p}_1}+\omega_{\mathbf{p}_2})|\mathbf{p}_1,\mathbf{p}_2\rangle \qquad \mathbf{P}|\mathbf{p}_1,\mathbf{p}_2\rangle = (\mathbf{p}_1+\mathbf{p}_2)|\mathbf{p}_1,\mathbf{p}_2\rangle \tag{2.4}$$

[2] I was recently informed by a colleague from subcontinental India that this name should be pronounced "Bōsh", and I'll try to train myself to pronounce it correctly. But bosons are still "bōsäns".

The extension to three particles is just "*et cetera*". While "*et cetera*" is of course the end of the story, we have not really started with the beginning of the story. There is one thing we have left out. It is possible that we look upon the world and we find a probability for there being *no* particles. And therefore we need to add at least one basis vector to this infinite string, to wit, a no-particle basis vector, a single state, to account for a possibility that there are no particles around at all. We will denote this state $|0\rangle$. It is called the **vacuum state**. We will assume that the vacuum state is *unique*. This state $|0\rangle$ is of course an eigenstate of the energy with eigenvalue zero, and simultaneously an eigenstate of momentum with eigenvalue zero:

$$H\,|0\rangle = 0 \qquad \mathbf{P}\,|0\rangle = 0 \tag{2.5}$$

The vacuum state is Lorentz invariant, $U(\Lambda)\,|0\rangle = |0\rangle$. All observers agree that the state with no particles is the state with no particles. We will normalize it to 1,

$$\langle 0|0\rangle = 1. \tag{2.6}$$

This normalization is conventional for a discrete eigenstate of the Hamiltonian, one which is not part of the continuum. Please do not confuse the vacuum state with the zero vector in Hilbert space, which is not a state at all, having probability zero associated with it, nor with the state of a single particle with 3-momentum $\mathbf{p}$ equal to zero, $|\mathbf{0}\rangle$. That ket is denoted with the vector $\mathbf{0}$.

We now have a complete catalog of basis vectors. A general state $|\Psi\rangle$ in Fock space will be some linear combination of these basis vectors:

$$|\Psi\rangle = \psi_0\,|0\rangle + \int \psi_1(\mathbf{p})\,|\mathbf{p}\rangle\,d^3\mathbf{p} + \frac{1}{2!}\int \psi_2(\mathbf{p}_1,\mathbf{p}_2)\,|\mathbf{p}_1,\mathbf{p}_2\rangle\,d^3\mathbf{p}_1 d^3\mathbf{p}_2 + \cdots \tag{2.7}$$

This is some number, some probability amplitude times the no-particle state, the vacuum, plus the integral of some function $\psi_1(\mathbf{p})$ times the ket $|\mathbf{p}\rangle$ plus a one over 2! inserted by convention—I'll explain the reason for that convention—times the integral over two momenta of a function of both momenta times the two-particle ket, with dots indicating the three-particle, the four-particle *et cetera* states, going on forever.

I should explain the factor of $\frac{1}{2!}$. Since the state $|\mathbf{p}_1,\mathbf{p}_2\rangle$ is the same as the state $|\mathbf{p}_2,\mathbf{p}_1\rangle$, without any loss of generality we can choose

$$\psi_2(\mathbf{p}_1,\mathbf{p}_2) = \psi_2(\mathbf{p}_2,\mathbf{p}_1) \tag{2.8}$$

That is to say we can choose a Bose wave function for two bosons to be symmetric in the two arguments. I then insert the $\frac{1}{2!}$ to take account of the fact that I am counting the same state with the same coefficient twice when I integrate once over $\mathbf{p}_1$ and once over $\mathbf{p}_2$ in one order, and then in the other order. Likewise, successive terms for the three-particle or four-particle states will have corresponding factors of $\frac{1}{3!}$, $\frac{1}{4!}$ *et cetera.*

The squared norm $|\Psi|^2$ of the state $|\Psi\rangle$ is

$$|\Psi|^2 = \langle\Psi|\Psi\rangle = |\psi_0|^2 + \int d^3\mathbf{p}\,|\psi_1(\mathbf{p})|^2 + \frac{1}{2!}\int d^3\mathbf{p}_1\,d^3\mathbf{p}_2\,|\psi_2(\mathbf{p}_1,\mathbf{p}_2)|^2 + \cdots \tag{2.9}$$

The state $|\Psi\rangle$ exists and is normalizable as always only if $|\Psi|^2 < \infty$, so we can multiply it by a constant and make its norm 1, and speak about probabilities in a sensible way.

Well, in a sense we have solved our problem. We have described the space of states we want to talk about. But we have described it in a singularly awkward and ugly way. To describe a state, we need an infinite string of wave functions: the zero particle wave function, a function of a single variable, a function of two variables, a function of three variables, a function of four variables *et cetera, ad nauseam.* Handling a system of this kind by conventional Schrödinger equation techniques, describing the dynamics by an interaction operator made up out of q's and d/dp's and some sort of incredible integral-differential operator that mixes up functions of two or three or four or any number of variables with other such functions is a quick route to insanity. We need to find some simpler way of describing the system.

2.2 Occupation number representation

In order to minimize problems that arise when one is playing with delta functions and things like that I will brutally mutilate the physics by putting the system in a periodic box. So we will have only discrete values of the momentum to sum over instead of continuous values to integrate over. Of course this is a dreadful thing to do: It destroys Lorentz invariance; in fact it even destroys rotational invariance. But it's just a pedagogic device. In a little while I'll let the walls of the box go to infinity and we'll be back in the continuum case.

With the system in a periodic box, we imagine the momenta restricted to a discrete set of values which are labeled by

$$\mathbf{p} = \left(\frac{2\pi n_x}{L}, \frac{2\pi n_y}{L}, \frac{2\pi n_z}{L}\right) \tag{2.10}$$

The box is a cube of length L; the numbers n_x, n_y and n_z are integers. Instead of filling out 3-space, the momenta span a cubic lattice. Since we have discrete states we can use ordinary normalization rather than delta function normalization. For example, in the one-particle states

$$\langle \mathbf{p}|\mathbf{p}'\rangle = \delta_{\mathbf{p}\mathbf{p}'} \tag{2.11}$$

the Kronecker delta equaling 1 if $\mathbf{p} = \mathbf{p}'$, and zero otherwise. Integrations in the continuum case are replaced by sums over the whole lattice of the allowed momenta. These will be discrete, infinite sums.

In this box we can label our basis states in a somewhat different way than we have labeled them up to now. In our previous analysis we haven't exploited Bose statistics much; it's been rather *ad hoc.* We tell how many particles there are, we imagine the particles are distinguishable, we give the momentum of the first particle, the momentum of the second, the third, *et cetera*; and then we say as an afterthought that it's the same as giving the same set of momenta in a different permutation. Now as you all know from elementary quantum statistical mechanics where you count states in a box, there is a much simpler way of describing the basis states. We can describe our basis states by saying how many particles there are with this momentum, how many particles are there with that momentum, how many with some other momentum. We can describe our states by giving them *occupation numbers* $N(\mathbf{p})$, a function from the lattice of allowed one-particle momenta into the integers which is simply the number of particles with momentum $\mathbf{p}$. Obviously this is exactly equivalent; it describes not only the same Hilbert space but the same set of basis vectors as we've described before, providing of course we have the condition that

$$\sum_{\mathbf{p}} N(\mathbf{p}) < \infty \tag{2.12}$$

No fair writing down a state where there is one particle with each momentum! That's not in our counting, and anyway it would be a state of infinite energy.

This is not a change of basis in the normal sense, but just a relabeling of the same basis in a different way. We can label our states by an infinite string of integers $\{N(\mathbf{p})\}$, a sort of super-matrix[3] which you can imagine as this three-dimensional lattice whose axes are p_x, p_y, and p_z, with integer numbers $N(p_x, p_y, p_z)$ sitting on every lattice point. Of course, most of the numbers will be zero. I'll write such a labeling this way, with a curly bracket, just to remind you that this is not a state labeled by a single number $N(\mathbf{p})$ and a single vector $\mathbf{p}$, but by this matrix of integers.

The advantage of the occupation number labeling is of course that the Bose statistics is exploited, taken care of automatically. When I say there is one particle with this momentum and one particle with that momentum, I have described the state; I don't have to say which is the first particle and which is the second. In terms of this labeling the Hamiltonian has a very simple form:

$$H = \sum_{\mathbf{p}} \omega_{\mathbf{p}} N(\mathbf{p}) \tag{2.13}$$

The energy of the many-particle state is the sum of the energies of the individual particles. The momentum likewise has a very simple form:

$$\mathbf{P} = \sum_{\mathbf{p}} \mathbf{p}\, N(\mathbf{p}) \tag{2.14}$$

Staring at the expression (2.13) for the energy, we notice something that wasn't obvious in the other way of writing things: First, the energy is a sum of independent terms, one for each value of $\mathbf{p}$, and second, within each independent term we have a sequence of equally spaced energies separated by $\omega_{\mathbf{p}}$. We can have zero times $\omega_{\mathbf{p}}$, 1 times $\omega_{\mathbf{p}}$, 2 times $\omega_{\mathbf{p}}$, and so on. Such a structure of energy spacings is of course familiar to us: It's what occurs in the harmonic oscillator. In fact this is exactly like the summation we would get if we had an infinite assembly of uncoupled harmonic oscillators, each with frequency $\omega_{\mathbf{p}}$, except that the zero-point energy, the $\frac{1}{2}\omega_{\mathbf{p}}$, is missing. But other than that, this looks, both in the numbers of states and their energies, exactly like an infinite assembly of uncoupled harmonic oscillators. The two systems are completely different. In our many-particle theory, $N(\mathbf{p})$ tells us how many particles are present with momentum $\mathbf{p}$. In a system of harmonic oscillators, $N(\mathbf{p})$ gives the excitation level of the oscillator labeled by $\mathbf{p}$. Still, let us pursue this clue. And in order that we will all know the same things about the harmonic oscillator, I will now digress into a brief review on this topic. Most people will have seen this material in any previous quantum mechanics course. I apologize, but theoretical physics is defined as a sequence of courses, each of which discusses the harmonic oscillator.[4]

[3] [Eds.] In other words an infinite, rank 3 array, with integer matrix elements $N_{p_x\, p_y\, p_z}$, $p_i = 2\pi n_i/L, n_i = 1, 2, \ldots$

[4] [Eds.] A variation of this remark attributed to Coleman is: "The career of a young theoretical physicist consists of treating the harmonic oscillator at ever-increasing levels of abstraction."

2.3 Operator formalism and the harmonic oscillator

Consider a single harmonic oscillator. The momentum p and the position q are now not numbers, they are quantum operators obeying the much-beloved commutation relations

$$[q, p] = i \tag{2.15}$$

The Hamiltonian is[5]

$$H = \tfrac{1}{2}\omega(p^2 + q^2 - 1) \tag{2.16}$$

I subtract 1 here to adjust the zero of my good old-fashioned harmonic oscillator so that the ground state has energy zero. The Hamiltonian will then look exactly like one of the terms in the sum (2.13), not just qualitatively like it.

Now the famous way of solving this system is to introduce the operator a and its adjoint $a^\dagger$,

$$a = \frac{1}{\sqrt{2}}(q + ip) \qquad a^\dagger = \frac{1}{\sqrt{2}}(q - ip) \tag{2.17}$$

It is easy to compute the commutator $[a, a^\dagger]$:

$$[a, a^\dagger] = \tfrac{1}{2}[q + ip, q - ip] = [q, -ip] = 1. \tag{2.18}$$

We get a contribution only from the cross-terms, both of which give equal contributions and cancel out the $\frac{1}{2}$ that comes from squaring the $\frac{1}{\sqrt{2}}$. It is also easy to rewrite the Hamiltonian in terms of a and $a^\dagger$, since

$$q = \frac{1}{\sqrt{2}}(a + a^\dagger) \qquad p = \frac{1}{i\sqrt{2}}(a - a^\dagger) \tag{2.19}$$

Then

$$H = \tfrac{1}{2}\omega[p^2 + q^2 - 1] = \tfrac{1}{2}\omega[aa^\dagger + a^\dagger a - 1] = \omega a^\dagger a = \omega N \tag{2.20}$$

As promised, this expression for the harmonic oscillator Hamiltonian looks exactly like one of the terms in (2.13); we need only confirm that $N = a^\dagger a$ is a number operator. From these two equations, (2.18) and (2.20), plus one additional assumption, one can reconstruct the entire state structure of the harmonic oscillator. I will assume the ground state is unique. Without this assumption I would not know for example that I was dealing with a spinless harmonic oscillator; it might be a particle of spin 17, where the spin never enters the Hamiltonian. Then I would get twice 17 + 1 or 35 duplicates of a harmonic oscillator, corresponding to the various values of the z components of the spin. The assumption of a unique ground state will take

---

[5] [Eds.] This form of the Hamiltonian is the result of a canonical transformation of the usual harmonic oscillator Hamiltonian,

$$H = \frac{P^2}{2m} + \frac{1}{2}m\omega^2 Q^2$$

with $[Q, P] = i$, namely

$$p = P/\sqrt{m\omega}; \qquad q = \sqrt{m\omega}Q$$

This canonical transformation preserves the commutator, $[q, p] = i$ and leads to the form

$$H = \tfrac{1}{2}\omega(q^2 + p^2)$$

See Goldstein *et al.* *CM*, pp. 377–381.

care of the possibility of other dynamical variables being around that would give us a multiply degenerate ground state. Let's now determine the system. I presume you've all seen it before.

Compute the commutators of H with a and $a^\dagger$;

$$[H, a] = \omega[a^\dagger a, a] = \omega[a^\dagger, a]a = -\omega a \tag{2.21}$$

$$[H, a^\dagger] = \omega[a^\dagger a, a^\dagger] = \omega a^\dagger [a, a^\dagger] = \omega a^\dagger \tag{2.22}$$

Let us consider some energy eigenstate of this system. I will assume it's labeled by its energy, and denote it in the following way, $|E\rangle$, where of course

$$H\,|E\rangle = E\,|E\rangle \tag{2.23}$$

Now consider H acting on the state $a^\dagger\,|E\rangle$. By the equation above,

$$Ha^\dagger\,|E\rangle = [H, a^\dagger]\,|E\rangle + a^\dagger H\,|E\rangle = \omega a^\dagger\,|E\rangle + a^\dagger E\,|E\rangle = (E+\omega)a^\dagger\,|E\rangle \tag{2.24}$$

Thus, given a state $|E\rangle$ of energy E, I can obtain a state of energy $E+\omega$ by applying $a^\dagger$. I can draw a spectroscopic diagram, Figure 2.1.

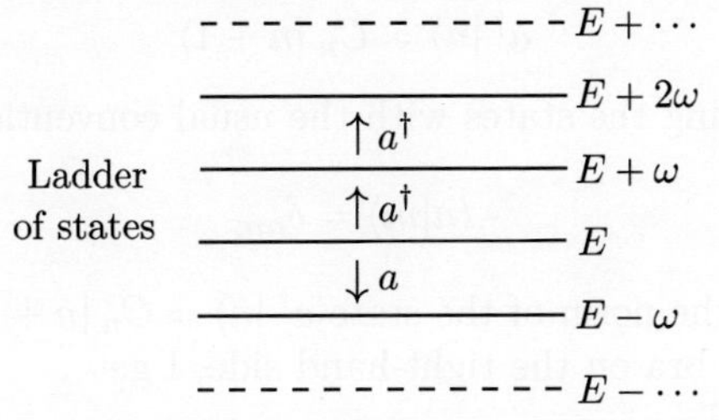

Figure 2.1: Traveling up and down the ladder of energy states

And of course I can build this ladder up forever by successive applications of $a^\dagger$. By the same reasoning applied to a, I obtain a similar equation:

$$Ha\,|E\rangle = [H, a]\,|E\rangle + aH\,|E\rangle = -\omega a\,|E\rangle + aE\,|E\rangle = (E-\omega)a\,|E\rangle \tag{2.25}$$

By applying a I can go down the ladder. For this reason $a^\dagger$ and a are called "raising" and "lowering" operators because they raise and lower the energy. Can we go up and down forever?

I don't know yet about going up, but about going down I can say something. The Hamiltonian is the product of an operator and its adjoint, and therefore it always has non-negative expectation values and non-negative eigenvalues. So the energy must be bounded below. There must be a place where I can no longer continue going down. Let me write the lowest energy eigenstate, the ground state, as $|E_0\rangle$, which by assumption is unique. Now there's no fighting this equation (2.25): Applying a to $|E_0\rangle$ gives me a state which is an eigenstate with energy $E_0 - \omega$;

$$Ha\,|E_0\rangle = (E_0 - \omega)a\,|E_0\rangle \tag{2.26}$$

On the other hand by assumption there is no eigenstate with energy lower than E_0. The only way these apparent contradictions can be reconciled is if

$$a\,|E_0\rangle = 0 \tag{2.27}$$

because $a\,|E_0\rangle = 0$ satisfies the equation (2.26) for any value of E_0. This of course determines the energy of the ground state, because $H = a^\dagger a$, and therefore

$$H\,|E_0\rangle = \omega a^\dagger a\,|E_0\rangle = 0 = E_0\,|E_0\rangle \;\Rightarrow\; E_0 = 0 \tag{2.28}$$

Therefore the ground state, by assumption unique, is a state of energy 0, and I will relabel the ground state:

$$|E_0\rangle \equiv |0\rangle \tag{2.29}$$

meaning, the ground state is the state of zero energy. All the other states of the system have energies which are integer multiples of ω because the ladder has integer spacing. We can label these $|n\rangle$, i.e.,

$$H\,|n\rangle = \omega N\,|n\rangle = n\omega\,|n\rangle \tag{2.30}$$

We obtain the states $|n\rangle$ from systematic application of $a^\dagger$ on the ground state:

$$|n\rangle \propto (a^\dagger)^n\,|0\rangle \tag{2.31}$$

Equation (2.30) follows from (2.31) and commuting H with $(a^\dagger)^n$, confirming $N = a^\dagger a$ as a number operator. Let's say

$$a^\dagger\,|n\rangle = C_n\,|n+1\rangle \tag{2.32}$$

and obtain C_n by normalizing the states with the usual convention,

$$\langle n|m\rangle = \delta_{nm} \tag{2.33}$$

If I compute the square of the norm of the state $a^\dagger\,|n\rangle = C_n\,|n+1\rangle$, the inner product of this ket with the corresponding bra on the right-hand side, I get

$$\begin{aligned}\langle n|aa^\dagger|n\rangle &= |C_n|^2\,\langle n+1|n+1\rangle = |C_n|^2\\ &= \langle n|a^\dagger a + 1|n\rangle = \langle n|(H/\omega)+1|n\rangle = (n+1)\,\langle n|n\rangle = (n+1)\end{aligned} \tag{2.34}$$

That determines C_n up to a phase. I have not yet made any statement that determines the relative phases of the various energy eigenstates, and I am free to choose the phase so that $\{C_n\}$ are real:

$$C_n = \sqrt{n+1} \tag{2.35}$$

We then have the fundamental expression for the action of $a^\dagger$ on an arbitrary state $|n\rangle$,

$$a^\dagger\,|n\rangle = \sqrt{n+1}\,|n+1\rangle \tag{2.36}$$

By similar reasoning or by direct application of the definition of the adjoint, we determine

$$a\,|n\rangle = \sqrt{n}\,|n-1\rangle \tag{2.37}$$

I have snuck something over on you. I have talked as if the ladder of states, built out of successive applications of $a^\dagger$ on the ground state, is the entire space of energy states. You know that is true for the harmonic oscillator, but I haven't proved it using just the algebra of the a's and $a^\dagger$'s. So let's demonstrate that.

If we have an operator A which commutes with both p and q, then A must be a multiple of the identity:

$$\text{If } [p,A] = 0 \text{ and } [q,A] = 0 \text{ then } A = \lambda I \tag{2.38}$$

where λ is some constant.[6] Say there is a state $|\psi\rangle$ which has a component not on the ladder. Presumably there is a projection operator, $\mathcal{P}$ which projects $|\psi\rangle$ onto the ladder. Since $a^\dagger$ and a keep a state on the ladder, it must be that

$$[\mathcal{P}, a] = [\mathcal{P}, a^\dagger] = 0 \tag{2.39}$$

But as q and p can be written as linear combinations of a and $a^\dagger$, we can say

$$[\mathcal{P}, p] = [\mathcal{P}, q] = 0 \Rightarrow \mathcal{P} = \lambda I \tag{2.40}$$

The projection operator is proportional to the identity, so there are no parts of any ket $|\psi\rangle$ not on the ladder; there are no other states except those already found.

Two or three or four decoupled harmonic oscillators can be handled in exactly the same way. We simply have two or three or four sets of raising and lowering operators. The Hamiltonian is the sum of expressions of this form over the various sets. Conversely, if we have a system with the structure of a harmonic oscillator, with equally spaced energy eigenstates, we can define operators a and $a^\dagger$ for each set, and then regain the algebraic structure and the expression for the Hamiltonian and complete the system in that way.

This completes the discussion of the harmonic oscillator. Its entire structure follows from these algebraic statements (2.18)–(2.25) and the mild assumption of minimality, that there is only one ground state.

2.4 The operator formalism applied to Fock space

Now let us turn to the particular system we have: An infinite assembly of harmonic oscillator-like objects, one for every point in our momentum space lattice. The analogs, the mathematical equivalents to the harmonic oscillator excitation numbers are the occupation numbers. Therefore we can define raising and lowering operators on the system, $a^\dagger_{\mathbf{p}}$ and $a_{\mathbf{p}}$, one for every lattice point, that is to say, for every value of $\mathbf{p}$.

The lowering operators associated with different oscillators have nothing to do with each other:

$$[a_{\mathbf{p}}, a_{\mathbf{p}'}] = 0 \tag{2.41}$$

The raising operators associated with different oscillators have nothing to do with each other:

$$[a^\dagger_{\mathbf{p}}, a^\dagger_{\mathbf{p}'}] = 0 \tag{2.42}$$

The raising and lowering operators for two oscillators have the conventional commutators

$$[a_{\mathbf{p}}, a^\dagger_{\mathbf{p}'}] = \delta_{\mathbf{p}\mathbf{p}'} \tag{2.43}$$

equalling 1 if they describe the same oscillator, and commuting otherwise.

The Hamiltonian is the sum of the Hamiltonians for each of the individual oscillators,

$$H = \sum_{\mathbf{p}} \omega_{\mathbf{p}}\, a^\dagger_{\mathbf{p}} a_{\mathbf{p}} \tag{2.44}$$

[6] [Eds.] If A commutes with q, then it is either a constant or a function of q. But if it is a function of q, it cannot commute with p. So it must be a constant, i.e., a multiple of the identity. This is an application of Schur's lemma. See Thomas F. Jordan, *Linear Operators for Quantum Mechanics*, Dover Publications, 2006, pp. 69–70.

The oscillators are labeled by the index $\mathbf{p}$, a 3-vector on the lattice, and each has energy $\omega_{\mathbf{p}}$. We haven't talked about the momentum operator in our discussion of a single oscillator, but of course it will be given by an expression precisely similar to the Hamiltonian. The factor multiplying $\omega_{\mathbf{p}}$, $a^\dagger_{\mathbf{p}}a_{\mathbf{p}}$, has eigenvalues $N(\mathbf{p})$, so the momentum operator is

$$\mathbf{P} = \sum_{\mathbf{p}} \mathbf{p}\, a^\dagger_{\mathbf{p}} a_{\mathbf{p}} \tag{2.45}$$

This set of equations defines Fock space in the same way as the corresponding set of equations defines the single oscillator. The only change is a change in nomenclature.

As you'll recall, the thing that corresponded to the excitation level of an oscillator was the number of particles bearing momentum $\mathbf{p}$. We will no longer call $a^\dagger_{\mathbf{p}}$ and $a_{\mathbf{p}}$ "raising" and "lowering" operators, respectively. We will call them **creation and annihilation operators** because applying $a^\dagger_{\mathbf{p}}$ raises an equivalent oscillator, that is to say, *adds one particle* of momentum $\mathbf{p}$; applying $a_{\mathbf{p}}$ lowers an equivalent oscillator, i.e., *removes one particle* of momentum $\mathbf{p}$. Another term will be changed from that of the oscillator problem. We normally do not call the simultaneous ground state of all the oscillators "the ground state"; we call it as I have told you, *the vacuum state.* The vacuum state is defined by the equation that any annihilation operator applied to it gives zero:

$$a_{\mathbf{p}}\,|0\rangle = 0 \tag{2.46}$$

The advantage of these algebraic equations over the original definition of Fock space is great. You see here they take only a few lines. The original definition filled a page or so. As shown by the argument with the oscillators, they give you the complete structure of the space: they tell you what the states are, they tell you what their normalizations are, they tell you the energy and momentum of any desired state. So we have made progress, by reducing many equations to a few.

I am now going to blow up the box, letting $L \to \infty$, and attempt to go to the continuum limit. I will not attempt to go to the continuum limit in the occupation number or equivalent oscillator formalism. That is certainly possible but it involves refined mathematical concepts. Instead of a direct product of individual oscillators spaces, we would get a sort of integral-direct product, a horrible mess. The point is that we can generalize these algebraic equations directly. These contain the entire content of the system. We can generalize them simply by taking a step backward from what we did to get to the box in the first place: we replaced all Dirac delta functions by Kronecker deltas, and all integrals by sums. If we undo this, replacing sums with integrals and Kronecker deltas with Dirac deltas, we will get a system that gives us continuum Fock space. I'll check that it works. I won't check every step because most of it is pretty obvious, but I'll check a few examples for you.

I'm going to define the system purely algebraically just as I defined the oscillator and Fock space for a box purely algebraically. There are a fundamental set of operators, $a^\dagger_{\mathbf{p}}$ and $a_{\mathbf{p}}$ for *any* value of $\mathbf{p}$ now, not just integer values defined on the lattice, and they obey these equations;

$$\boxed{\begin{aligned}[][a_{\mathbf{p}}, a_{\mathbf{p}'}] &= [a^\dagger_{\mathbf{p}}, a^\dagger_{\mathbf{p}'}] = 0 \\ [a_{\mathbf{p}}, a^\dagger_{\mathbf{p}'}] &= \delta^{(3)}(\mathbf{p}-\mathbf{p}')\end{aligned}} \tag{2.47}$$

The Hamiltonian is

$$H = \int d^3\mathbf{p}\, \omega_{\mathbf{p}}\, a^\dagger_{\mathbf{p}} a_{\mathbf{p}} = \int d^3\mathbf{p}\, \omega_{\mathbf{p}}\, N(\mathbf{p}) \tag{2.48}$$

and the total momentum operator is

$$\mathbf{P} = \int d^3\mathbf{p}\, \mathbf{p}\, a^\dagger_\mathbf{p} a_\mathbf{p} = \int d^3\mathbf{p}\, \mathbf{p}\, N(\mathbf{p}) \tag{2.49}$$

where

$$N(\mathbf{p}) = a^\dagger_\mathbf{p} a_\mathbf{p} \tag{2.50}$$

is the number operator. That's it. These statements (2.47)–(2.50), together with the technical assumption that the ground state of the system is unique, will define continuum Fock space in the same way the precisely parallel statements defined Fock space for particles in a box. Let's check that for a few simple states.

First, the ground state of the system, the vacuum, which is assumed to be unique, is defined by

$$a_\mathbf{p} |0\rangle = 0 \tag{2.51}$$

for all $\mathbf{p}$. Directly from the expressions for the energy and the momentum, this state is an eigenstate of the energy with eigenvalue zero, and of the momentum, with eigenvalue zero. Of course the algebraic structure doesn't tell us how we normalize the vacuum. That's a matter of convention, and we will choose that convention to be the same as before,

$$\langle 0|0\rangle = 1 \tag{2.52}$$

the vacuum state has norm 1. To make one-particle states we apply creation operators to the vacuum; that is a one-particle state of momentum $\mathbf{p}$. If all we were working from were the previous algebraic equations for the harmonic oscillators, according to (2.36) the one-particle state of momentum $\mathbf{p}$ would be obtained like this:

$$a^\dagger_\mathbf{p} |0\rangle = C_1 |\mathbf{p}\rangle = \sqrt{0+1}\, |\mathbf{p}\rangle = |\mathbf{p}\rangle \tag{2.53}$$

Let's assume this is right for the continuum Fock space, and compute the norm of this state:

$$\langle \mathbf{p}'|\mathbf{p}\rangle = \langle 0|a_{\mathbf{p}'} a^\dagger_\mathbf{p}|0\rangle \tag{2.54}$$

We have our fundamental commutation relations and so we will commute;

$$\begin{aligned} \langle \mathbf{p}'|\mathbf{p}\rangle &= \langle 0|a_{\mathbf{p}'} a^\dagger_\mathbf{p}|0\rangle \\ &= \langle 0|[a_{\mathbf{p}'}, a^\dagger_\mathbf{p}]|0\rangle + \langle 0|a^\dagger_\mathbf{p} a_{\mathbf{p}'}|0\rangle \\ &= \langle 0|\delta^{(3)}(\mathbf{p}-\mathbf{p}')|0\rangle + 0 = \delta^{(3)}(\mathbf{p}-\mathbf{p}') \end{aligned} \tag{2.55}$$

The first term is $\delta^{(3)}(\mathbf{p}-\mathbf{p}')$ times the norm of the vacuum which is one. The second term is zero because $a_{\mathbf{p}'}$ acting on the vacuum is zero; every annihilation operator acting on the vacuum is zero. Thus the state has the right norm. This looks good.

What about the energy of the single-particle states? Well, it's the same story:

$$H a^\dagger_\mathbf{p} |0\rangle = \int d^3\mathbf{p}'\omega_{\mathbf{p}'} a^\dagger_{\mathbf{p}'} a_{\mathbf{p}'} a^\dagger_\mathbf{p} |0\rangle = \int d^3\mathbf{p}' \left(\omega_{\mathbf{p}'} [a^\dagger_{\mathbf{p}'} a_{\mathbf{p}'}, a^\dagger_\mathbf{p}] |0\rangle + a^\dagger_\mathbf{p} H |0\rangle\right) \tag{2.56}$$

We know the commutations required to compute this; all $a^\dagger_\mathbf{p}$ commute with each other, all $a_\mathbf{p}$ commute with each other, the commutation of any $a^\dagger_\mathbf{p}$ and any $a_\mathbf{p}$ is a delta function:

$$[a^\dagger_{\mathbf{p}'} a_{\mathbf{p}'}, a^\dagger_\mathbf{p}] = a^\dagger_{\mathbf{p}'} [a_{\mathbf{p}'}, a^\dagger_\mathbf{p}] = a^\dagger_{\mathbf{p}'}\, \delta^{(3)}(\mathbf{p}-\mathbf{p}') \tag{2.57}$$

and H on the vacuum is zero. Then

$$Ha^\dagger_{\mathbf{p}}|0\rangle = \int d^3\mathbf{p}'\omega_{\mathbf{p}'}a^\dagger_{\mathbf{p}'}\delta^{(3)}(\mathbf{p}-\mathbf{p}')|0\rangle = \omega_{\mathbf{p}}a^\dagger_{\mathbf{p}}|0\rangle \tag{2.58}$$

which also looks good. *Et cetera* for the momentum, *et cetera* for the two-particle states, the three-particle states and so on. Here is an example of a two-particle state, just to write down its definition,

$$|\mathbf{p}_1,\mathbf{p}_2\rangle = a^\dagger_{\mathbf{p}_1}a^\dagger_{\mathbf{p}_2}|0\rangle \tag{2.59}$$

This state $|\mathbf{p}_1,\mathbf{p}_2\rangle$ is of course automatically equal to the state $|\mathbf{p}_2,\mathbf{p}_1\rangle$ because the two creation operators commute. It doesn't matter what order you put them in. The Bose statistics are taken account of automatically. I leave it to you to go through the necessary commutators to show that the state has the same normalization as the one we wrote down before, and that it has the right energy and the right momentum. The operations are exactly parallel to the operations I've done explicitly for the single particle state.

To summarize where we have gone: The algebraic equations plus the technical assumption that there exists a unique vacuum state, the ground state of the system, completely specify everything about Fock space we initially wrote down formally. This is obviously a great advantage; it's much simpler to manipulate these annihilation and creation operators than it would be to manipulate a number plus a function of one variable plus a function of two variables plus a function of three variables *et cetera.*

Now there are two further points I want to make before we leave the topic of Fock space and go on to our next topic. One is a point for mathematical purists. Those of you who are not mathematical purists may snooze while I make this point. In the technical sense of Hilbert space theories these $a^\dagger_{\mathbf{p}}$ and $a_{\mathbf{p}}$ we have introduced are *not operators* because when applied to an arbitrary state they can give you a non-normalizable result. For instance, $a^\dagger_{\mathbf{p}}|0\rangle$ is a plane wave $|\mathbf{p}\rangle$, and a plane wave is not normalizable ($\langle\mathbf{p}|\mathbf{p}\rangle = \delta^{(3)}(\mathbf{0}) = \infty$). Occasionally while browsing through *Physical Review*, or more likely through *Communications in Mathematical Physics*, you may come across people not talking about these things as operators—they are purists—but as "operator valued distributions." A "distribution", to a mathematician, means something like a delta function, which is not itself a function, but it becomes a function when it is smeared, integrated over in a product with some nice smoothing function. These $a^\dagger_{\mathbf{p}}$ and $a_{\mathbf{p}}$ are operator-valued distributions labeled by indices $\mathbf{p}$, and the things that are really sensible operators are smeared combinations like $\int d^3\mathbf{p}\, f(\mathbf{p})a^\dagger_{\mathbf{p}}$ where $f(\mathbf{p})$ is some nice smooth function. That creates a particle in a normalizable state, a wave packet state, with $f(\mathbf{p})$ the momentum space wave function describing its shape. And that's the thing that people who are careful about their mathematics like to talk about.[7] I am not a person who is careful about his mathematics; I won't use that language. But in case you run across it in some other course, you should know that there are people who use this language and this is the reason why they use it. They prefer to talk about the smeared combinations rather than the $a_{\mathbf{p}}$'s themselves.

Secondly—and this is not for purists, this is for real—since we're back in infinite space, we can sensibly talk about Lorentz transformations again. To complete this section, I should specify how Lorentz transformations are defined in terms of the creation and annihilation

[7] [Eds.] For an accessible, slim and inexpensive book about distributions, also known as "generalized functions", see M. J. Lighthill, *An Introduction to Fourier Analysis and Generalised Functions*, Cambridge U. P., 1958.

operators. The essential trick is to observe that just as we defined the relativistically normalized states, so we can define relativistically normalized creation operators that when applied to the vacuum will create states with the correct relativistic normalization.

I will call these operators $\alpha(p)$ and $\alpha^\dagger(p)$. The momentum index p is now a 4-vector, but the fourth component is constrained just as before: $p^0 = \omega_{\mathbf{p}}$. These creation operators are defined by

$$\alpha^\dagger(p) = (2\pi)^{3/2}\sqrt{2\omega_{\mathbf{p}}}\,a^\dagger_{\mathbf{p}} \tag{2.60}$$

Operating on the vacuum, this operator makes the same state as $a^\dagger_{\mathbf{p}}$ does. It creates the same particle, but with relativistic normalization and not just the $\delta^{(3)}(\mathbf{p}-\mathbf{p}')$ normalization (see (1.57)):

$$\alpha^\dagger(p)\,|0\rangle = (2\pi)^{3/2}\sqrt{2\omega_{\mathbf{p}}}\,a^\dagger_{\mathbf{p}}\,|0\rangle = (2\pi)^{3/2}\sqrt{2\omega_{\mathbf{p}}}\,|\mathbf{p}\rangle = |p\rangle \tag{2.61}$$

Of course there is also a relativistic annihilation operator which we can write down just by taking the adjoint,

$$\alpha(p) = (2\pi)^{3/2}\sqrt{2\omega_{\mathbf{p}}}\,a_{\mathbf{p}} \tag{2.62}$$

These operators, as you can convince yourself, transform simply under Lorentz transformations. We can determine the Lorentz and translation properties of these operators $\alpha^\dagger(p)$ and $\alpha(p)$ from the assumed transformations of the kets. First, consider the vacuum. It's obvious that the vacuum is Lorentz invariant, since it is the unique state in our whole Fock space of zero energy and zero momentum, and that's a Lorentz invariant statement. So $U(\Lambda)$ acting on the vacuum must give us the vacuum, since U is unitary and does not change the norm:

$$U(\Lambda)\,|0\rangle = |0\rangle \tag{2.63}$$

Then, for a single particle state $|p\rangle$, assume that

$$U(\Lambda)\,|p\rangle = |\Lambda p\rangle \tag{2.64}$$

and for a multi-particle state,

$$U(\Lambda)\,|p_1, p_2, \ldots, p_n\rangle = |\Lambda p_1, \Lambda p_2, \ldots, \Lambda p_n\rangle \tag{2.65}$$

From (2.64), we determine how $\alpha^\dagger(p)$ behaves under a Lorentz transformation:

$$\begin{aligned}|\Lambda p\rangle = U(\Lambda)\,|p\rangle = U(\Lambda)\alpha^\dagger(p)\,|0\rangle &= U(\Lambda)\alpha^\dagger(p)(U^\dagger(\Lambda)U(\Lambda))\,|0\rangle && \text{because } U^\dagger = U^{-1}\\ &= U(\Lambda)\alpha^\dagger(p)U^\dagger(\Lambda)\,|0\rangle && \text{because } U(\Lambda)\,|0\rangle = |0\rangle\\ &= \alpha^\dagger(\Lambda p)\,|0\rangle && \text{by definition; } |\Lambda p\rangle = \alpha^\dagger(\Lambda p)\,|0\rangle\end{aligned}$$

so

$$\alpha^\dagger(\Lambda p) = U(\Lambda)\alpha^\dagger(p)U^\dagger(\Lambda) \tag{2.66}$$

That is to say, $\alpha^\dagger(\Lambda p)$ is the creation operator of the transformed 4-momentum. And of course taking the adjoint equation

$$U(\Lambda)\alpha(p)U^\dagger(\Lambda) = \alpha(\Lambda p) \tag{2.67}$$

Just to check that these are right, let's compute the transformation acting on a multi-particle state, $|p, p_1, p_2, \ldots, p_n\rangle$. We have

$$\begin{aligned}U(\Lambda)\,|p, p_1, p_2, \ldots, p_n\rangle &= U(\Lambda)\alpha^\dagger(p)\,|p_1, p_2, \ldots, p_n\rangle = U(\Lambda)\alpha^\dagger(p)U^\dagger(\Lambda)U(\Lambda)\,|p_1, p_2, \ldots, p_n\rangle\\ &= U(\Lambda)\alpha^\dagger(p)U^\dagger(\Lambda)\,|\Lambda p_1, \Lambda p_2, \ldots, \Lambda p_n\rangle\\ &= \alpha^\dagger(\Lambda p)\,|\Lambda p_1, \Lambda p_2, \ldots, \Lambda p_n\rangle\\ &= |\Lambda p, \Lambda p_1, \Lambda p_2, \ldots, \Lambda p_n\rangle\end{aligned}$$

which is the desired result. The same argument would have worked for any p_i, or for that matter any set of the p_i's, since the kets are symmetric in the p_i's. Here's another way to think about the transformation of a multi-particle state:

$$\begin{aligned} U(\Lambda)\left|p_1, p_2, \ldots, p_n\right\rangle &= U(\Lambda)\alpha^\dagger(p_1)\alpha^\dagger(p_2)\cdots\alpha^\dagger(p_n)\left|0\right\rangle \\ &= U(\Lambda)\alpha^\dagger(p_1)U^\dagger(\Lambda)U(\Lambda)\alpha^\dagger(p_2)U^\dagger(\Lambda)\cdots U(\Lambda)\alpha^\dagger(p_n)U^\dagger(\Lambda)U(\Lambda)\left|0\right\rangle \\ &= \alpha^\dagger(\Lambda p_1)\alpha^\dagger(\Lambda p_2)\cdots\alpha^\dagger(\Lambda p_n)\left|0\right\rangle \\ &= \left|\Lambda p_1, \Lambda p_2, \ldots, \Lambda p_n\right\rangle \end{aligned}$$

So this system, defined by the operators $\alpha(p)$ and $\alpha(p)^\dagger$, admits unitary Lorentz transformations, as of course it should, because it's the same system we were talking about before. The action of these Lorentz transformations can be defined if we wish by these equations (2.63)–(2.67). That enables us to tell how every state Lorentz transforms.

Likewise the translation properties of the creation and annihilation operators are easily found from the transformations of the kets. The unitary operators of translations are $U(a) = e^{iP\cdot a}$. Because

$$P^\mu\left|0\right\rangle = 0 \tag{2.68}$$

and

$$P^\mu\left|p_1, p_2, \ldots, p_n\right\rangle = (p_1^\mu + p_2^\mu + \cdots + p_n^\mu)\left|p_1, p_2, \ldots, p_n\right\rangle = \left(\sum p_i^\mu\right)\left|p_1, p_2, \ldots, p_n\right\rangle \tag{2.69}$$

we get

$$U(a)\left|0\right\rangle = \left|0\right\rangle \tag{2.70}$$

and

$$U(a)\left|p_1, p_2, \ldots, p_n\right\rangle = e^{ia\cdot\sum p_i}\left|p_1, p_2, \ldots, p_n\right\rangle \tag{2.71}$$

A derivation analogous to the Lorentz transformation leads to the translational properties of $\alpha^\dagger(p)$ and $\alpha(p)$,

$$\begin{aligned} e^{iP\cdot x}\alpha^\dagger(p)e^{-iP\cdot x} &= e^{ip\cdot x}\alpha^\dagger(p) \\ e^{iP\cdot x}\alpha(p)e^{-iP\cdot x} &= e^{-ip\cdot x}\alpha(p) \end{aligned} \tag{2.72}$$

In the next section I return to the question which inspired us. (Actually, what inspired us is the fact that quantum electrodynamics predicts the right anomalous magnetic moment of the electron, but we won't get to that until the second half of this course![8]) What inspired us in this elegant but historically false line of reasoning, to consider an infinite, great big Hilbert space in the first place was the problem of localization. In the next section I will talk about localization from another tack, not about localizing particles, but instead about localizing *observations*. Incidentally, note that these operators $\alpha(p)$ and $\alpha^\dagger(p)$ depend on time as well as space. We are working in the Heisenberg representation, in which the states are constant but the operators depend on time, rather than the Schrödinger representation in which the operators are time-independent but the states evolve in time.

---

[8] [Eds.] §34.3, pp. 743–749.

3

Constructing a scalar quantum field

3.1 Ensuring relativistic causality

In ordinary, non-relativistic quantum mechanics, every Hermitian operator is an observable. Given anything measurable, any physicist, if she is only crafty enough, can manage to think up some apparatus that measures it. She measures the position x with a bubble chamber, she measures the momentum p with a bending magnet, she may have to be a real genius to measure a symmetrized product of p^4 times x^8, but in principle there's nothing to keep her from measuring that. It's only a matter of skill. *I* cannot measure the length of my own foot, but that's only due to my lack of skill. The words "Hermitian operator" and "observable" are synonymous, and we do not bother to introduce distinctions about what an idealized observer can measure and a Hermitian operator.

In particular, every observer can measure every operator and therefore every observer can measure non-commuting operators. One can measure σ_x and also measure σ_y. Now you know the measurement of non-commuting observables does not commute. If I have an electron and I measure its σ_x, and I turn my back for a moment, and Carlo Rubbia,[1] having just come in on a jet plane, sneaks into the room—he does a lot of experiments—and measures something that commutes with σ_x, like p_x, and then sneaks out again before I can turn around, I won't notice any difference when I measure σ_x a second time. If on the other hand when Carlo comes in he measures σ_y, then I will notice a *big* difference. My system will no longer be in the eigenstate of σ_x in which I had carefully prepared it; it will now be in an eigenstate of σ_y, and I will notice the change. I will know that someone has made a measurement even if I keep my eyes closed.

If we say every observer can measure every observable, even in the most idealized sense of "observer" and "observable", then we encounter problems in a relativistic theory. I after all have a finite spatial extent, and my travels, far and wide as they are, occupy only a finite spatial extent. And, alas, the human condition states that I also have only a finite temporal extent. There is some region of space and time within which all the experiments I can do are isolated. The earth goes around the sun, so my spatial extent is perhaps the diameter of the

[1] [Eds.] Rubbia shared the 1984 Physics Nobel Prize with Simon van der Meer, for experimental work leading to the discoveries of the $W^{\pm}$ and Z^0 (see §48.2). At the time of these lectures, Rubbia was commuting between CERN and Harvard.

solar system in width, and, if I give up smoking soon, maybe 75 years in length, but...that's it. Now let us imagine another observer similarly localized, say somewhere in the Andromeda galaxy, and he has a similar life expectancy and spatial extent. If both he and I can measure all observables, then he can measure an observable that doesn't commute with an observable I can measure. Therefore, instead of Carlo Rubbia sneaking into the room after I've done my experiment, he can just stay in Andromeda and do his experiment on a non-commuting observable. Just as if Carlo had come sneaking into the room, I would notice that my results had changed, and thus would deduce that he has made a measurement. That's *impossible* because the only way to get information from him there in Andromeda to me here on earth between the time of my two measurements is for information to travel faster than the speed of light. Therefore it *cannot* be that I can measure everything, and it cannot be that he can measure everything. There must be some things that I can measure and some things that he can measure, and it must be that everything that I can measure commutes with everything that he can measure. Otherwise he could send information faster than the speed of light, namely, the information that he has measured an observable that does not commute with an observable that I have just measured. The reasoning is abstract, but I hope simple and clear.

Even if we are going to be generous in our idealization, we have to realize that somehow in any sensible relativistic quantum mechanical theory, there must be some things that can be measured by people who are constrained to live in a certain spacetime region and some things they cannot measure. Within every region of space and time, out of the whole set of Hermitian operators, there must be only some of them that those people can measure. If this were not so we would run into contradictions between the most general principles of quantum mechanics—the interpretation rules that tell us how Hermitian operators are connected with observations—and the principle of Einstein causality, that information cannot travel faster than the speed of light. I have said a lot of words. Let me try to make them precise.

Say we have two regions of space and time, R_1 and R_2, open sets of points if you want to be mathematically precise. These regions are such that no information can get from R_1 to R_2 without traveling faster than the speed of light. Mathematically the condition can be expressed like this. If x_1 is any four-dimensional point in R_1, and x_2 is any four-dimensional point in R_2, then the square of the distance between these points is negative:

$$(x_2 - x_1)^2 < 0. \tag{3.1}$$

Recall (1.18) that two points satisfying this condition are said to be separated by a *spacelike* interval. Every point in our region R_1 is spacelike separated from every point in our region R_2.

Regions R_1 and R_2 are spacelike separated:

$x_1 \in R_1$ and $x_2 \in R_2 \Rightarrow (x_2 - x_1)^2 < 0$

Figure 3.1: Spacelike separated regions

Let $\mathcal{O}_1$ be any observable that can be measured in R_1. Likewise, let $\mathcal{O}_2$ be any observable that can be measured in R_2. Our theory must contain a rule for associating observables with spacetime regions:

$$\text{If } (x_2 - x_1)^2 < 0, \text{ then } [\mathcal{O}_1, \mathcal{O}_2] = 0. \tag{3.2}$$

Every Hermitian operator that is measurable in the region R_1, even according to our most abstract and generalized sense of measurement, must commute with every Hermitian operator that is measurable in region R_2. It's *got* to be so. This conclusion comes just from the general conditions of quantum mechanics and the statement that no information can travel faster than the speed of light. This is a very severe, very curious kind of restriction to place on a theory, in addition to all the usual dynamics, imposing a rule associating observables with spacetime regions. We've certainly never encountered it in quantum mechanics before; we've never encountered such a rule in relativistic quantum mechanics, either. Have we encountered it in relativistic classical mechanics? Well, in the relativistic theory of a point particle we haven't; but we have in classical electrodynamics.

Maxwell's equations have fields in them, say the electric field as a function of x^μ, a spacetime point. You know very well what you can measure if you're stuck in the spacetime region R_1 and all you have is an electrometer. You can measure $\mathbf{E}(\mathbf{x}, t)$ in the region R_1, and that's it. To phrase it perhaps more precisely, depending on the shape and size of the pith balls in your electrometer, you can measure some sort of average of the field smeared over some function $f(x)$ which is to vanish if x is not in R_1. Of course, if you have more than just pith balls, if you have dielectrics and other electromagnetic materials, you might be able to measure squares of the electric field or more. For example, you might be able to measure

$$\int d^4x_1 d^4x_2\, f(x_1, x_2) E_i(x_1) E_j(x_2) \tag{3.3}$$

where $f(x_1, x_2) = 0$ if x_1 and/or x_2 is outside of R_1. That is to say there *is* an entity in familiar classical physics that does enable us, in a natural way, to associate observations with definite spacetime regions: the electromagnetic field. What you can measure are the electric and magnetic fields, or perhaps combinations of them, *in* that region, but not *outside* that region. We can't design an apparatus right here to measure the electric field right now over there in Andromeda. This gives us a clue as to how to associate observables with spacetime regions in relativistic quantum mechanics. What we need is the quantum analog of something like the electromagnetic field. We have to find a *field*, $\phi(x)$, or maybe a bunch of them, $\phi^a(x)$, operator-valued—because we're now in quantum mechanics, and observables are operators—functions of space and time. Then the observables—I'm just pretending to guess now, but it's a natural guess—the observables we can measure in a region R_1 are things that are built up out of the field (or fields, if there are many fields involved), restricted to the spacetime region R_1. What is strongly suggested is that quantum mechanics and relativistic causality force us to introduce quantum fields. In fact, relativistic quantum mechanics is practically synonymous with quantum field theory.

So one way (perhaps not the only way) of implementing Einstein causality—that nothing goes faster than the speed of light—within the framework of a relativistic quantum theory is to construct a quantum field, the hero of this course—one of the two heroes, I should say; the other is the S matrix, but we won't get to that for a while. The center of all our interest, the hero *sans peur et sans reproche*[2] of this course, is the quantum field. That will give us a definition of what it means to localize observations. Once we have that definition, we won't have to worry about what it means to localize a particle. Forget that, that's irrelevant. If we know where the *observations* are, we don't have to know where the *particles* are. If we

---

[2] [Eds.] Originally this French phrase described the "perfect knight" Pierre Terrail, Chevalier de Bayard (1473–1524), "without fear and without flaw".

know where the Geiger counter is, and if we know what it means when we say the Geiger counter responds, the implications of doing a measurement with the Geiger counter change. Ultimately, what we want to describe are observations, not particles. The observables we build as functions of the fields will commute for spacelike separations if the fields commute for spacelike separations. If we can construct a quantum field, we will settle two problems at once. We will see how our theory can be made consistent with the principle of causality, and we will make irrelevant the question of where the particles are. If we can't do it with fields, we'll have to think again. But we will be able to do it with fields. I want to remind you of something I said at the end of the last section. These fields will depend not only on space but on time. That is, we are working in the Heisenberg picture.

3.2 Conditions to be satisfied by a scalar quantum field

We will try to build our observables from a complete set of N commuting quantum fields $\phi^a(x)$, $a = 1, \ldots, N$, each field an operator-valued function of points x^μ in spacetime. We will construct our fields out of creation and annihilation operators. What I am going to do is write down a set of conditions that we want our fields to satisfy, so that they give us a definition of locality. Some of these conditions will be inevitable; any field we can imagine must satisfy these conditions. Some of them will just be simplifying conditions. I'll look for simple examples first, and then if I fail in my search for simple examples—but in fact I won't fail—we can imagine systematically loosening those conditions and looking for more complicated examples.

These five conditions will determine the form of the fields:

1. $[\phi^a(x), \phi^b(y)] = 0$ if $(x-y)^2 < 0$, to guarantee that observables in spacelike separated regions commute.
2. $\phi^a(x) = \phi^a(x)^\dagger$. The fields are to be Hermitian (and so observable).
3. $e^{-iP\cdot y}\phi^a(x)e^{iP\cdot y} = \phi^a(x-y)$. The fields transform properly under translations.
4. $U(\Lambda)^\dagger \phi^a(x) U(\Lambda) = \phi^a(\Lambda^{-1}x)$. The fields transform properly *as scalars* under Lorentz transformations.
5. The fields are assumed to be *linear* combinations of the operators,

$$\phi^a(x) = \int d^3\mathbf{p}\,\left[F^a_{\mathbf{p}}(x)a_{\mathbf{p}} + G^a_{\mathbf{p}}(x)a^\dagger_{\mathbf{p}}\right] \tag{3.4}$$

I want to say more about these conditions before we apply them.

Let's start with the first and second conditions. We will frequently have occasion to deal—not so much in this lecture, but in subsequent lectures—with non-Hermitian fields. Of course, only a Hermitian operator can be an observable but sometimes it's convenient to sum together two Hermitian operators with an i in the summation to make a non-Hermitian operator. So I might as well tell you now that if I talk about a non-Hermitian field I mean its real and imaginary parts, or, this being quantum mechanics, its Hermitian and anti-Hermitian parts, are separately observables, to take account of the possibility of non-Hermitian fields. In other words,

$$[\phi^a(x)^\dagger, \phi^b(y)^\dagger] = 0 \text{ if } (x-y)^2 < 0 \tag{3.5}$$

That's just tantamount to saying the Hermitian and anti-Hermitian parts of the fields are all considered as a big set of fields, and all obey the first condition.

Now let's talk about the third and fourth conditions, on the transformation properties of the fields. We know in the specific case we're looking at, Fock space, how spacetime translations and Lorentz transformations act on the states. This should tell us something about how these transformations act on the fields. Let's try to figure out what that something is, by considering the more limited transformations of space translations and ordinary rotations. First, space translations.

We can conceive of an operator valued field even in non-relativistic quantum mechanics. Suppose for example we have an electron gas or the Thomas–Fermi model. We can think of the electron density at every space point as being a field, an observable, an operator. It's a rather trivial operator, of course, delta functions summed over the individual electrons, but it's an operator that's a function of position. Let's call this operator $\rho(\mathbf{x})$. The point $\mathbf{x}$ is not a quantum variable, it is just the point at which we are asking the question "What is the electron density?" If we have some arbitrary state $|\psi\rangle$, I'll write the function $f(\mathbf{x})$ for the expectation value of $\rho(\mathbf{x})$ in that state $|\psi\rangle$:

$$f(\mathbf{x}) = \langle\psi|\rho(\mathbf{x})|\psi\rangle \tag{3.6}$$

Now suppose we spatially translate the state. I define

$$|\psi'\rangle = e^{-i\mathbf{P}\cdot\mathbf{a}}|\psi\rangle \tag{3.7}$$

This is the state where I've picked up the whole boxful of electrons and moved it to the right by a distance $\mathbf{a}$. Now what of the expectation value? It becomes

$$\langle\psi'|\rho(\mathbf{x})|\psi'\rangle \tag{3.8}$$

If there's any sense in the world at all, this must be $f(\mathbf{x}-\mathbf{a})$. Perhaps that minus sign requires a little explanation.

Let's say I plot $f(\mathbf{x})$ peaked near the origin. Now if I translate things by a distance $\mathbf{a}$ to the right, I get the second plot:

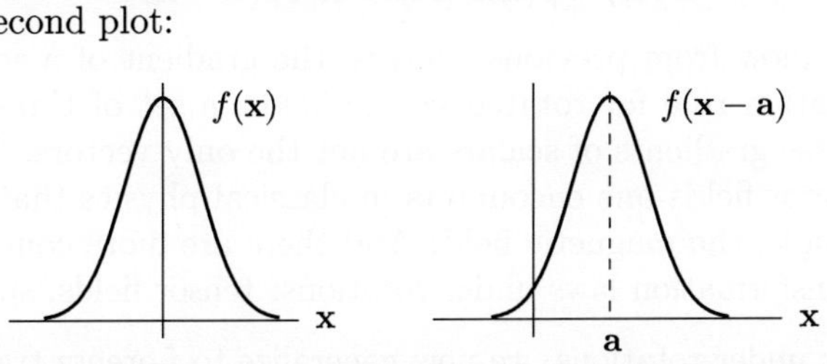

Figure 3.2: Expectation values for a state $|\psi\rangle$ and the state $|\psi'\rangle$ translated by $\mathbf{a}$

The value of $f(\mathbf{x}-\mathbf{a})$ is peaked at $\mathbf{x}=\mathbf{a}$ if $f(\mathbf{x})$ is peaked at the origin. That is the correct sign for moving the state over from being centered at the origin to being centered at $\mathbf{a}$. And that's why there is a minus sign in this equation and not a plus sign. Now of course

$$f(\mathbf{x}-\mathbf{a}) = \langle\psi'|\rho(\mathbf{x})|\psi'\rangle = \langle\psi|e^{i\mathbf{P}\cdot\mathbf{a}}\rho(\mathbf{x})e^{-i\mathbf{P}\cdot\mathbf{a}}|\psi\rangle \tag{3.9}$$

by the definition of $|\psi'\rangle$. On the other hand, rewriting (3.6) for $\mathbf{x}\to\mathbf{x}-\mathbf{a}$, we can say

$$f(\mathbf{x}-\mathbf{a}) = \langle\psi|\rho(\mathbf{x}-\mathbf{a})|\psi\rangle \tag{3.10}$$

Since $|\psi\rangle$ is an arbitrary state, and a Hermitian operator is completely determined by its expectation values in an arbitrary state, we can eliminate $|\psi\rangle$ and write

$$e^{i\mathbf{P}\cdot\mathbf{a}}\rho(\mathbf{x})e^{-i\mathbf{P}\cdot\mathbf{a}} = \rho(\mathbf{x}-\mathbf{a}) \tag{3.11}$$

in agreement with (1.39). This is simply the statement that if you translate the fields as you translate the states, this is what you find. We generalize this in the obvious way to translations in Minkowski space, where we have both purely spatial translations and time translations, and that is our third condition;

$$e^{-iP\cdot a}\phi(x)e^{iP\cdot a} = \phi(x-a) \tag{3.12}$$

This last equation is in fact four equations at once. Three of them correspond to the three space components of a and are just the previous equation rewritten. The fourth I obtained by generalization, but it should not be unfamiliar to you. The fourth equation, the one where a points purely in the time direction, is simply the integrated form of the Heisenberg equation of motion, since P^0 is the Hamiltonian.[3]

As far as condition 4 goes, let's first consider how a set of fields in general transforms under an ordinary rotation:

$$\phi^a(\mathbf{x}) \xrightarrow{R} \phi^{a\prime}(\mathbf{x}') = R^a_b\phi^b(R^{-1}\mathbf{x}) \tag{3.13}$$

However, if the fields transform as *scalars*, $R^a_b = \delta^a_b$. That is to say,

$$U(R)^\dagger\rho(\mathbf{x})U(R) = \rho(R^{-1}\mathbf{x}) \tag{3.14}$$

The R^{-1} appears here for the same reason that the $-a$ appeared in the previous argument. If the expectation value is peaked at a given point, the transformed expectation value will be peaked at the rotated point.

That's the transformation for a *scalar* field, like $\rho(\mathbf{x})$, but not every field is a scalar. If we were to consider, for example, $\nabla\rho(\mathbf{x})$, the *gradient* of ρ, we would discover as an elementary exercise that

$$U(R)^\dagger\nabla\rho(\mathbf{x})U(R) = R\,\nabla\rho(R^{-1}\mathbf{x}) \tag{3.15}$$

As you undoubtedly know from previous courses, the gradient of a scalar is a *vector*, and this is the transformation rule for rotated vector fields, a set of three operators for every spatial point. Of course gradients of scalars are not the only vectors. There are all sorts of three-dimensional vector fields one encounters in classical physics that are not gradients of scalar fields, for example, the magnetic field. And there are more complicated objects with more complicated transformation laws under rotations: tensor fields, spinor fields, etc.

From the behavior under rotations, we now generalize to Lorentz transformations, just as we generalized the space translation behavior to spacetime translations. In general, a set of fields $\phi^a(x)$ will transform as

$$\phi^a(x) \xrightarrow{\Lambda} \phi^{a\prime}(x') = S^a_b(\Lambda)\phi^b(\Lambda^{-1}x) \tag{3.16}$$

[3] [Eds]. Let $a^\mu = (dt,0,0,0)$ be an infinitesimal translation in time. Then with $P^0 = H$,

$$e^{-iP\cdot a}\phi(x)e^{iP\cdot a} = \phi(x-a) \approx (1-iHdt)\phi(\mathbf{x},t)(1+iHdt) = \phi(\mathbf{x},t) - i[H,\phi]dt \approx \phi(\mathbf{x},t) - dt\,\frac{d\phi(\mathbf{x},t)}{dt}$$

(expanding the right-hand side in a Taylor series) or $i[H,\phi(x)] = d\phi(x)/dt$, which is just the Heisenberg equation of motion.

Again, if the fields transform as *scalars*,

$$\phi^{a\prime}(x') = U(\Lambda)^\dagger \phi^a(x) U(\Lambda) = \phi^a(\Lambda^{-1}x) \tag{3.17}$$

The Λ^{-1} appears here for the same reason that the R^{-1} appeared before. If the expectation value is peaked at a given point, the transformed expectation value will be peaked at the Lorentz transformed point. One can consider the Lorentz transformation of much more complicated objects: tensor fields, spinor fields, etc. (In particular, the gradient $\partial^\mu \phi^a$ transforms as a Lorentz 4-vector if ϕ^a is a Lorentz scalar.) However the scalar field is certainly the simplest possibility, and therefore for my fourth condition I will assume my fields transform like scalars under Lorentz transformations. This is an assumption of pure simplicity. If this doesn't lead to a viable theory, we'll have to consider fields with more complicated transformation laws.

So conditions 1 and 2 are universal, absolutely necessary, while conditions 3 and 4 are just simplifying assumptions. We can think of unitary transformations acting in two separate ways: as transformations on the states, such that $|\psi\rangle \to U|\psi\rangle$, *or* as transformations on the operators, $A \to U^\dagger A U$; but *not* both.[4]

Condition 5 is a super-simplifying condition. We have these a's and $a^\dagger$'s floating around, so we'll make a *very* simplifying assumption that the $\phi^a(x)$ are linear combinations of the $a_{\mathbf{p}}$'s and $a^\dagger_{\mathbf{p}}$'s. If the linear combinations prove insufficient, we'll consider quadratic and higher powers of the operators. But this won't be necessary.

3.3 The explicit form of the scalar quantum field

In order to exploit these five conditions, I'll have to remind you of the properties of the $a_{\mathbf{p}}$ and $a^\dagger_{\mathbf{p}}$ operators. We worked all these out in the last lecture, so I'll just write them down. First, the algebra of the creation and annihilation operators:

$$[a_{\mathbf{p}}, a^\dagger_{\mathbf{p}'}] = \delta^{(3)}(\mathbf{p} - \mathbf{p}') \tag{3.18}$$

$$[a_{\mathbf{p}}, a_{\mathbf{p}'}] = [a^\dagger_{\mathbf{p}}, a^\dagger_{\mathbf{p}'}] = 0 \tag{3.19}$$

Then, for translations,

$$e^{iP\cdot x} a^\dagger_{\mathbf{p}} e^{-iP\cdot x} = e^{ip\cdot x} a^\dagger_{\mathbf{p}} \qquad e^{iP\cdot x} a_{\mathbf{p}} e^{-iP\cdot x} = e^{-ip\cdot x} a_{\mathbf{p}} \tag{3.20}$$

and finally for Lorentz transformations,

$$U(\Lambda) a_{\mathbf{p}} U^\dagger(\Lambda) = a_{\Lambda\mathbf{p}} \qquad U(\Lambda) a^\dagger_{\mathbf{p}} U^\dagger(\Lambda) = a^\dagger_{\Lambda\mathbf{p}} \tag{3.21}$$

To construct the fields, I'll now use these properties of the $a_{\mathbf{p}}$'s and $a^\dagger_{\mathbf{p}}$'s, and the five conditions in reverse order. Condition 1 is the hardest to check.

First we'll satisfy condition 5. I'll simply try to find the most general ϕ without an index a. If I find several such solutions, I'll call them ϕ^1, ϕ^2, and so on. We can start with $\phi(0)$ and use condition 3 to shift $\phi(0) \to \phi(x)$;

$$\phi(x) = e^{iP\cdot x} \phi(0) e^{-iP\cdot x} \tag{3.22}$$

[4] [Eds.] One can define a transformation U in terms of its action on the *states*, and then check that it acts correctly on the operators, or one can define it in terms of its action on the *operators*, and check that it has the proper effect on states. But one should not simply *assume* that it works both ways.

It will be convenient to write the fields in terms of $\alpha(p)$ and $\alpha^\dagger(p)$. There is no harm and much to be gained by putting in the Lorentz invariant measure. So the most general form looks like this:

$$\phi(0) = \int \frac{d^3\mathbf{p}}{(2\pi)^3(2\omega_\mathbf{p})} \left[f_p\, \alpha(p) + g_p\, \alpha^\dagger(p)\right] \tag{3.23}$$

where f_p and g_p are some unknown functions of p. As always these functions f_p and g_p depend not on four independent variables p^μ, but only three.

At this stage f_p and g_p are arbitrary functions, so there's an infinite number of solutions to condition 5, which is not surprising. They're not restricted to be Hermitian, or to be complex conjugates of each other. We get more information about f_p and g_p by examining Lorentz invariance.

A special case of condition 4 tells us

$$U(\Lambda)\phi(0)U^\dagger(\Lambda) = \phi(0) \text{ because } \Lambda 0 = 0 \tag{3.24}$$

Applying the transformation to $\phi(0)$ I find

$$\begin{aligned}
\phi(0) &= \int \frac{d^3\mathbf{p}}{(2\pi)^3(2\omega_\mathbf{p})} \left[f_p\, \alpha(p) + g_p\, \alpha^\dagger(p)\right] \\
&= U(\Lambda) \int \frac{d^3\mathbf{p}}{(2\pi)^3(2\omega_\mathbf{p})} \left[f_p\, \alpha(p) + g_p\, \alpha^\dagger(p)\right] U^\dagger(\Lambda) \\
&= \int \frac{d^3\mathbf{p}}{(2\pi)^3(2\omega_\mathbf{p})} [f_p \underbrace{U(\Lambda)\alpha(p)U^\dagger(\Lambda)}_{\alpha(\Lambda p)} + g_p \underbrace{U(\Lambda)\alpha^\dagger(p)U^\dagger(\Lambda)}_{\alpha^\dagger(\Lambda p)}] \\
&= \int \frac{d^3\mathbf{p}}{(2\pi)^3(2\omega_\mathbf{p})} \left[f_p \alpha(\Lambda p) + g_p \alpha^\dagger(\Lambda p)\right]
\end{aligned} \tag{3.25}$$

Note that $U(\Lambda)$ goes right through f_p and g_p like beet through a baby. Now I define $p' = \Lambda p$, and I can write the integration over p'. The Lorentz invariant measure is the same, so it doesn't change at all. The only thing that changes is

$$f_p \to f_{\Lambda^{-1}p} \quad g_p \to g_{\Lambda^{-1}p} \tag{3.26}$$

so that

$$\phi(0) = \int \frac{d^3\mathbf{p}}{(2\pi)^3(2\omega_\mathbf{p})} \left[f_{\Lambda^{-1}p}\, \alpha(p) + g_{\Lambda^{-1}p}\, \alpha^\dagger(p)\right] \tag{3.27}$$

Comparing this with the first expression above for $\phi(0)$, the two integrands must be equal. But the $\alpha(p)$'s and $\alpha^\dagger(p)$'s are linearly independent operators, therefore the coefficients must be equal, and I deduce

$$f_p = f_{\Lambda^{-1}p} \text{ and } g_p = g_{\Lambda^{-1}p} \text{ for any } p \text{ and any } \Lambda \tag{3.28}$$

The values of p are constrained to lie on the upper invariant mass hyperboloid I drew previously (Figure 1.1). It follows from special relativity that I can get from any point on this hyperboloid to any other on it by a Lorentz transformation. Because relativity can change the value of p without changing the values of f_p and g_p, they have the same value for *all* values of p. That implies

$$f_p = f \quad \text{and } g_p = g \tag{3.29}$$

where f and g are *constants* to be determined. So conditions 5 and 4 have taken us pretty far; we are down to two unknown constants. What about $\phi(x)$? Will that involve other constants? No, because if I use condition 3, replacing x with 0 and a with $-x$, I obtain

$$\phi(x) = e^{iP\cdot x}\phi(0)e^{-iP\cdot x} \tag{3.30}$$

which of course I can compute since I have an expression for $\phi(0)$, and I know how the operators $\alpha(p)$ and $\alpha^\dagger(p)$ transform (the same as the $a_{\mathbf{p}}$ and $a^\dagger_{\mathbf{p}}$, see (3.20)). Then

$$\begin{aligned}\phi(x) &= e^{iP\cdot x}\int\frac{d^3\mathbf{p}}{(2\pi)^3(2\omega_{\mathbf{p}})}\left[f\,\alpha(p)+g\,\alpha^\dagger(p)\right]e^{-iP\cdot x}\\ &= \int\frac{d^3\mathbf{p}}{(2\pi)^3(2\omega_{\mathbf{p}})}\left[f\,e^{iP\cdot x}\alpha(p)e^{-iP\cdot x}+g\,e^{iP\cdot x}\alpha^\dagger(p)e^{-iP\cdot x}\right]\\ &= \int\frac{d^3\mathbf{p}}{(2\pi)^3(2\omega_{\mathbf{p}})}\left[fe^{-ip\cdot x}\alpha(p)+ge^{ip\cdot x}\alpha^\dagger(p)\right]\end{aligned} \tag{3.31}$$

Here is $\phi(x)$, and I still only need the two arbitrary constants f and g. I haven't used all of the content of conditions 3, 4 and 5; for example, I've only used condition 4 at the origin. But I leave it as a trivial exercise for you to show that every expression of this form satisfies the conditions 3, 4 and 5 for *all* x and all a. You can almost read it off, I think.

So let's summarize the situation before we apply conditions 1 and 2. A general field satisfying 3, 4 and 5 can be written as the sum of two independent fields, which I will call $\phi^{(+)}(x)$ and $\phi^{(-)}(x)$. The fields $\phi^{(+)}(x)$ and $\phi^{(-)}(x)$ are the coefficients of f and g, but I'll write them not in terms of the α's but the a's, since it's easier to compute the commutators using the a's:

$$\phi(x) = f\phi^{(+)}(x) + g\phi^{(-)}(x) \tag{3.32}$$

where

$$\phi^{(+)}(x) = \int\frac{d^3\mathbf{p}}{(2\pi)^{3/2}\sqrt{2\omega_{\mathbf{p}}}}a_{\mathbf{p}}e^{-ip\cdot x} \qquad \phi^{(-)}(x) = \int\frac{d^3\mathbf{p}}{(2\pi)^{3/2}\sqrt{2\omega_{\mathbf{p}}}}a^\dagger_{\mathbf{p}}e^{ip\cdot x} \tag{3.33}$$

and as usual when the 4-vector p appears, its time component is $\omega_{\mathbf{p}}$. Note that $\phi^{(-)}(x)^\dagger = \phi^{(+)}(x)$. The assignment of $\phi^{(-)}(x)$ to the field involving $a^\dagger_{\mathbf{p}}$ seems completely bananas but it was established by Heisenberg and Pauli,[5] on the basis that $\phi^{(-)}(x)$ only involves p^0's with negative frequencies, i. e. with a sign of $+ip^0x_0$ in the exponential's argument, and similarly with $\phi^{(+)}(x)$.[6]

Now to apply condition 2, hermiticity. Two independent Hermitian combinations are

$$\phi^1(x) = \phi^{(+)}(x) + \phi^{(-)}(x) \qquad \phi^2(x) = i(\phi^{(+)}(x) - \phi^{(-)}(x)) \tag{3.34}$$

These are two independent cases of the most general choice satisfying condition 2:

$$\phi(x) = e^{i\theta}\phi^{(+)}(x) + e^{-i\theta}\phi^{(-)}(x) \tag{3.35}$$

---

[5] [Eds.] Werner Heisenberg and Wolfgang Pauli, "Zur Quantendynamik der Wellenfelder" (On the quantum dynamics of wave fields) *Zeits. f. Phys.* **56** (1929) 1–61, "Zur Quantendynamik der Wellenfelder II", *Zeits. f. Phys.* **59** (1930) 168–190.

[6] [Eds.] Schweber *RQFT*, p. 167.

where θ could in principle be any real number.

Now to satisfy condition 1. There are three possible outcomes:

Possibility A: Two independent solutions, $\phi^1(x)$ and $\phi^2(x)$, which commute with themselves and *with each other*. *Any* combination $a\phi^1(x) + b\phi^2(x)$ is observable, with a and b real constants.

Possibility B: Only the *single* combination $e^{i\theta}\phi^{(+)}(x) + e^{-i\theta}\phi^{(-)}(x)$ is observable. The most general Hermitian combination is, aside from an irrelevant multiplying factor, some complex number of magnitude 1 times $\phi^{(+)}(x)$ plus that complex number's conjugate times $\phi^{(-)}(x)$.

Possibility C: The program crashes. We'll need to weaken condition 5 or think harder.

So either we have two fields or we have one field, and if we have one field, it must be of this form (3.35) to be Hermitian. Actually we can shorten our work a bit by realizing that we can get rid of the phase factor by redefining $a_{\mathbf{p}}$ and $a^\dagger_{\mathbf{p}}$:

$$a_{\mathbf{p}} \to e^{i\theta} a_{\mathbf{p}} \qquad a^\dagger_{\mathbf{p}} \to e^{-i\theta} a^\dagger_{\mathbf{p}} \tag{3.36}$$

If we make such a redefinition, that changes no prior equation before the definition of $\phi(x)$. Then I might as well consider equivalently *Possibility B′*:

$$\textit{Possibility } B'\text{: } \phi = \phi^{(+)}(x) + \phi^{(-)}(x)$$

We really only have two independent possibilities to consider: Possibility A, in which we say both $\phi^{(+)}(x)$ and $\phi^{(-)}(x)$ are local fields (i.e., they commute for spacetime separations), and Possibility B' in which we say just the sum of $\phi^{(+)}(x)$ and $\phi^{(-)}(x)$ is observable, and the difference is not observable.

Now we will look at these two possibilities systematically. Everything we have to compute to check A, we also have to compute to check B'. So let's start with A. We want to see that everything commutes with itself for spacelike separation. If A is true, then $\phi^1(x)$ must commute with $\phi^2(y)$, and each must commute with itself. For example, we must have the commutator $[\phi^1(x), \phi^2(y)]$ equal to zero for spacelike separations. Is it?

$$\begin{aligned}
[\phi^1(x), \phi^2(y)] &= i[\phi^{(+)}(x) + \phi^{(-)}(x),\, \phi^{(+)}(y) - \phi^{(-)}(y)] \\
&= i[\phi^{(+)}(x), \phi^{(+)}(y)] - i[\phi^{(+)}(x), \phi^{(-)}(y)] \\
&\quad + i[\phi^{(-)}(x), \phi^{(+)}(y)] - i[\phi^{(-)}(x), \phi^{(-)}(y)] \\
&\stackrel{?}{=} 0
\end{aligned} \tag{3.37}$$

For $\phi^{(+)}(x)$ with $\phi^{(+)}(y)$, spacelike, shmacelike; they all involve nothing but annihilation operators, and all annihilation operators commute with each other no matter what we multiply them by, so that's zero. By similar reasoning, or by taking the adjoint, the same thing goes for $\phi^{(-)}(x)$ with $\phi^{(-)}(y)$.

Now we come to the crunch. Let's compute

$$\begin{aligned}
[\phi^{(+)}(x), \phi^{(-)}(y)] &= \left[\int \frac{d^3\mathbf{p}}{(2\pi)^{3/2}\sqrt{2\omega_\mathbf{p}}} a_\mathbf{p} e^{-ip\cdot x}, \int \frac{d^3\mathbf{p}'}{(2\pi)^{3/2}\sqrt{2\omega_{\mathbf{p}'}}} a^\dagger_{\mathbf{p}'} e^{ip'\cdot y}\right] \\
&= \int \frac{d^3\mathbf{p}}{(2\pi)^{3/2}\sqrt{2\omega_\mathbf{p}}} \int \frac{d^3\mathbf{p}'}{(2\pi)^{3/2}\sqrt{2\omega_{\mathbf{p}'}}} e^{-ip\cdot x} e^{ip'\cdot y} \left[a_\mathbf{p}, a^\dagger_{\mathbf{p}'}\right] \\
&= \int \frac{d^3\mathbf{p}}{(2\pi)^{3/2}\sqrt{2\omega_\mathbf{p}}} \int \frac{d^3\mathbf{p}'}{(2\pi)^{3/2}\sqrt{2\omega_{\mathbf{p}'}}} e^{-ip\cdot x} e^{ip'\cdot y} \delta^{(3)}(\mathbf{p}-\mathbf{p}') \\
&= \int \frac{d^3\mathbf{p}}{(2\pi)^3(2\omega_\mathbf{p})} e^{-ip\cdot(x-y)} \equiv \Delta_+(x-y;\mu^2)
\end{aligned} \tag{3.38}$$

This function is one of a series of similar functions that will turn up again and again in our investigations. This one is actually a Neumann function or something, but its name in quantum field theory is Δ_+. It's a function of the *difference* of the spacetime points, $(x-y)$, as is obvious from the expression, and the mass, from the definition of $\omega_\mathbf{p}$ and the value of p^0. To keep things short, since we're only worried about one mass, I'll suppress the μ^2 and just call this $\Delta_+(x-y)$. If we were worrying about several different types of particles with different masses we would have to distinguish between the different Δ_+'s.

You might expect that $\Delta_+(x)$ is a Lorentz scalar function:

$$\Delta_+(x) = \Delta_+(\Lambda x) \tag{3.39}$$

This is indeed true. The argument of the exponential is a Lorentz scalar, and the factors have come together to make the Lorentz invariant measure. The Lorentz invariance of $\Delta_+(x)$ will be a useful fact to us later. Another useful relation is

$$[\phi^{(-)}(x), \phi^{(+)}(y)] = -\Delta_+(y-x) \tag{3.40}$$

which follows easily from the definition of $\Delta_+(x-y)$.

The real question we want to ask now is: Does $\Delta_+(x) = 0$ if $x^2 < 0$? If so, we're home free. Otherwise we have to look at Possibility B'. Well, it doesn't. We know it doesn't from Chapter 1. If I take the time derivative of $\Delta_+(x)$, that cancels out the ω_p in the denominator,

$$\frac{\partial}{\partial x^0}\Delta_+(x) = -\frac{i}{2}\int \frac{d^3\mathbf{p}}{(2\pi)^3} e^{-ip\cdot x} \tag{3.41}$$

and we get precisely the integral (1.81) we had to consider in Chapter 1 when I wondered whether particles could travel faster than the speed of light. Now if a function vanishes for all spacelike x^2, its time derivative surely vanishes for all spacelike x^2. By the explicit computation of Chapter 1, its time derivative doesn't vanish, so *the function doesn't vanish*, either. The answer to the question "Is $\Delta_+(x) = 0$ for spacelike x^2?" is "No". (Never waste a calculation!) Possibility A is thrown into the garbage pail, and we turn to the only remaining hope, Possibility B'. If B' also gets thrown into the garbage pail not only this lecture but this entire course will end in disaster!

Here we only have one field so we only have one commutator to check. Now fortunately since this $\phi(x)$ is $\phi^{(+)}(x) + \phi^{(-)}(x)$, the commutator is the sum of four terms we have already computed:

$$\begin{aligned}[\phi(x),\phi(y)] &= [\phi^{(+)}(x)+\phi^{(-)}(x),\phi^{(+)}(y)+\phi^{(-)}(y)]\\ &= [\phi^{(+)}(x),\phi^{(-)}(y)]+[\phi^{(-)}(x),\phi^{(+)}(y)]\\ &= \Delta_+(x-y)-\Delta_+(y-x)\equiv i\Delta(x-y)\end{aligned} \tag{3.42}$$

This $i\Delta(x-y)$ is a new Lorentz invariant function (using the notation of Bjorken and Drell;[7] the conventions differ from text to text). Like $\Delta_+(x-y)$, $i\Delta(x-y)$ depends on the square of the mass μ^2, but if there's only a single type of particle around, we don't need to write it. Does this expression equal zero for spacelike separations, $(x-y)^2<0$? Yes, and we can see this without any calculation. A spacelike vector can be turned into its negative by a Lorentz transformation,[8] so

$$\Delta_+(x-y)=\Delta_+(y-x) \quad \text{if } (x-y)^2<0 \tag{3.43}$$

and so

$$[\phi(x),\phi(y)]=i\Delta(x-y)=0 \quad \text{if } (x-y)^2<0 \tag{3.44}$$

Possibility B' thus escapes the garbage pail, and we don't have to consider Possibility C. Our single free scalar quantum field of mass μ is then written

$$\boxed{\phi(x)=\int\frac{d^3\mathbf{p}}{(2\pi)^{3/2}\sqrt{2\omega_{\mathbf{p}}}}\left(a_{\mathbf{p}}e^{-ip\cdot x}+a^\dagger_{\mathbf{p}}e^{ip\cdot x}\right)} \tag{3.45}$$

or in terms of the $\alpha(p)$'s and $\alpha^\dagger(p)$'s,

$$\phi(x)=\int\frac{d^3\mathbf{p}}{(2\pi)^3(2\omega_{\mathbf{p}})}\left(\alpha(p)e^{-ip\cdot x}+\alpha^\dagger(p)e^{ip\cdot x}\right) \tag{3.46}$$

(Particularly important equations will be boxed.) We have constructed the scalar field. It is the object that observables are built from. Now we take off in a new direction.

3.4 Turning the argument around: the free scalar field as the fundamental object

Several times in the course of our development we have introduced auxiliary objects like the annihilation and creation operators and then showed that the whole theory could be defined in terms of their properties. I would now like to show that the whole theory can be reconstructed from certain properties of the free quantum field. I will have to derive those properties. The structure we have built is rigid and strong enough to be inverted. We can make the top story the foundation.

---

[7] [Eds.] Bjorken & Drell *Fields*, Appendix C.

[8] [Eds.] This would *not* be true for a *timelike* vector. Proper Lorentz transformations move a timelike vector x^μ satisfying $x^2=\kappa^2$, inside the light cone, (x_1^μ and x_2^μ in the diagram) around the upper and lower hyperboloids $t=\pm\sqrt{|\mathbf{r}|^2+\kappa^2}$, respectively, but cannot change the sign of t, and so cannot transform a forward pointing vector like x_1^μ into a backward pointing vector like x_2^μ. By contrast, proper Lorentz transformations move a spacelike vector x^μ satisfying $x^2=-\kappa^2$, outside the light cone, (x_3^μ and $-x_3^\mu$) around on the hyperbolic sheet $|\mathbf{r}|^2-t^2=\kappa^2$. Since both x_3^μ and $-x_3^\mu$ lie on the same sheet, a spacelike vector can always be Lorentz transformed into its negative.

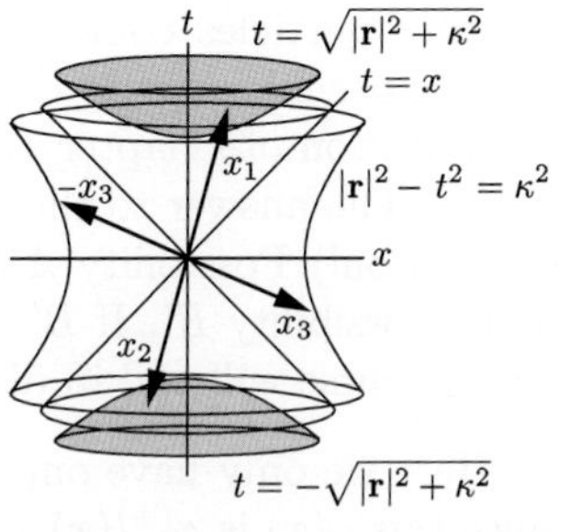

The first property is trivial to demonstrate, that $\phi(x)$ obeys this differential equation:[9]

$$\Box^2\phi(x) + \mu^2\phi(x) = 0 \qquad (\Box^2 \equiv \partial^\mu\partial_\mu) \tag{3.47}$$

This is just a statement that in momentum space p^2 equals μ^2. This is most easily shown by rewriting (3.46) in terms of the explicitly Lorentz invariant measure:

$$\phi(x) = \int \frac{d^4p}{(2\pi)^3}\,\delta(p^2-\mu^2)\,\theta(p^0)\left(\alpha(p)e^{-ip\cdot x} + \alpha^\dagger(p)e^{ip\cdot x}\right) \tag{3.48}$$

If I differentiate twice with respect to x^μ I obtain

$$\partial^\mu\partial_\mu\phi(x) = \int \frac{d^4p}{(2\pi)^3}\,\delta(p^2-\mu^2)\,\theta(p^0)(-p^2)\left(\alpha(p)e^{-ip\cdot x} + \alpha^\dagger(p)e^{ip\cdot x}\right) \tag{3.49}$$

so that

$$\Box^2\phi(x) + \mu^2\phi(x) = \int \frac{d^4p}{(2\pi)^3}\,\delta(p^2-\mu^2)\,\theta(p^0)(-p^2+\mu^2)\left(\alpha(p)e^{-ip\cdot x} + \alpha^\dagger(p)e^{ip\cdot x}\right) \tag{3.50}$$

The product $x\delta(x)$ is identically zero, so the product $(-p^2+\mu^2)\delta(p^2-\mu^2)$ guarantees the integrand vanishes, and $\phi(x)$ satisfies the differential equation.

The equation (3.47) has a famous name. It is called the **Klein–Gordon equation.**[10] As you might guess from the name, it was first written down by Schrödinger,[11] but he didn't know what to do with it. He wrote it down as a relativistic analog of the free Schrödinger equation. Recall that Schrödinger's original equation comes from the replacement $E \to i\hbar\partial/\partial t$, $\mathbf{p} \to -i\hbar\nabla$ into a Newtonian expression for the energy, now regarded as an operator equation acting on a wave function. Schrödinger, no dummy, knew the relativistic expression for energy and made the same substitutions into that. Then he said "Arrgh!" or the German equivalent, because he observed that the solutions had both positive and negative frequencies. And he said, "If this is a one-particle wave equation we are in the soup because we only want positive energies. We don't want negative energies!" We have encountered this equation not as a one-particle wave equation—that's the wrong context, that's garbage—but as an equation in quantum field theory where particles may be created and annihilated.

We already have the second property:

$$[\phi(x),\phi(y)] = i\Delta(x-y) = \int \frac{d^3\mathbf{p}}{(2\pi)^3(2\omega_\mathbf{p})}\left[e^{-ip\cdot(x-y)} - e^{ip\cdot(x-y)}\right] = 0 \text{ if } (x-y)^2 < 0 \tag{3.51}$$

These two equations (3.47) and (3.51), as I'll sketch out, completely define the Hilbert–Fock space and everything else. We *postulate* these two equations, together with the assumption

---

[9] [Eds.] To remind the reader: Though most authors let $\Box \equiv \partial^\mu\partial_\mu$, Coleman writes $\Box^2$.

[10] [Eds.] Walter Gordon, "Der Comptoneffekt nach der Schrödingerschen Theorie" (The Compton effect according to Schrödinger's theory), *Zeits. f. Phys.* **40** (1926) 117–133; Oskar Klein, "Elektrodynamik und Wellenmechanik vom Standpunkt des Korrespondenzprizips" (Electrodynamics and wave mechanics from the standpoint of the Correspondence Principles"), *Zeits. f. Phys.* **41** (1927) 407–422. According to Klein's obituary ("Oskar Klein", *Physics Today* **30** (1977) 67–88, written by his son-in-law, Stanley Deser), Klein symmetrically anticipated Schrödinger's more familiar equation, but was prevented from publishing it by a long illness.

[11] [Eds.] Erwin Schrödinger, "Quantisierung als Eigenwertproblem (Viete Mitteilung)" (Quantization as an eigenvalue problem, part 4.), *Ann. Physik* **81** (1926) 109–139. English translation in *Collected Papers on Wave Mechanics*, E. Schrödinger, AMS Chelsea Publishing, 2003. See equation (36).

of hermiticity ($\phi(x) = \phi(x)^\dagger$) and the scalar field's behavior under translations and Lorentz transformations (conditions 3 and 4 on p. 34.) In this way the scalar field which we've introduced as an auxiliary variable can just as well be thought of as the object that defines the theory.

We begin with the Klein–Gordon equation, (3.47). We can write the solution to it in its most general form (the factors of $(2\pi)^{-3/2}(2\omega_{\mathbf{p}})^{-1/2}$ are included for later convenience),

$$\phi(x) = \int \frac{d^3\mathbf{p}}{(2\pi)^{3/2}\sqrt{2\omega_{\mathbf{p}}}} \left(a_{\mathbf{p}} e^{-ip\cdot x} + b_{\mathbf{p}} e^{ip\cdot x} \right) \tag{3.52}$$

This is the most general expression for a solution of the Klein–Gordon equation with unknown Fourier components $a_{\mathbf{p}}$ and $b_{\mathbf{p}}$. The condition of hermiticity requires that $b_{\mathbf{p}} = a_{\mathbf{p}}^\dagger$, so that the most general solution is just what we had before, (3.45):

$$\phi(x) = \int \frac{d^3\mathbf{p}}{(2\pi)^{3/2}\sqrt{2\omega_{\mathbf{p}}}} \left(a_{\mathbf{p}} e^{-ip\cdot x} + a_{\mathbf{p}}^\dagger e^{ip\cdot x} \right) \tag{3.53}$$

We could now deduce the commutators of $a_{\mathbf{p}}$ and $a_{\mathbf{p}}^\dagger$ uniquely by substituting this expression into the second equation (3.51) and Fourier transforming the result. Once we have observed that the commutators are unique, we don't have to go through the whole calculation because we already know one commutator of $a_{\mathbf{k}}$ with $a_{\mathbf{k}}^\dagger$ that is consistent with everything else, the delta function $\delta^{(3)}(\mathbf{k} - \mathbf{k}')$, as in (3.18).

Finally from condition 3,

$$\phi(x - a) = e^{-iP\cdot a}\phi(x)e^{iP\cdot a} \tag{3.54}$$

we can deduce the commutators of the $a_{\mathbf{k}}$ and $a_{\mathbf{k}}^\dagger$ with the P^i's and the Hamiltonian simply by differentiation. For example, differentiating the previous equation gives the Heisenberg equation of motion

$$-\left.\frac{\partial\phi(x-a)}{\partial a^0}\right|_{a=0} = \frac{\partial\phi(x)}{\partial t} = i[H, \phi(x)] \tag{3.55}$$

Plugging the expression (3.45) in gives

$$\begin{aligned} \frac{\partial}{\partial t}\int \frac{d^3\mathbf{p}}{(2\pi)^{3/2}\sqrt{2\omega_{\mathbf{p}}}} \left(a_{\mathbf{p}} e^{-ip\cdot x} + a_{\mathbf{p}}^\dagger e^{ip\cdot x} \right) &= \int \frac{d^3\mathbf{p}}{(2\pi)^{3/2}\sqrt{2\omega_{\mathbf{p}}}} \left(-i\omega_{\mathbf{p}} a_{\mathbf{p}} e^{-ip\cdot x} + i\omega_{\mathbf{p}} a_{\mathbf{p}}^\dagger e^{ip\cdot x} \right) \\ &= \int \frac{d^3\mathbf{p}}{(2\pi)^{3/2}\sqrt{2\omega_{\mathbf{p}}}} \left(i[H, a_{\mathbf{p}}] e^{-ip\cdot x} + i[H, a_{\mathbf{p}}^\dagger] e^{ip\cdot x} \right) \end{aligned} \tag{3.56}$$

which gives, by Fourier transformation, the commutators of $a_{\mathbf{p}}$ and $a_{\mathbf{p}}^\dagger$ with the Hamiltonian, telling us that $a_{\mathbf{p}}^\dagger$ is an energy raising operator, identical to (2.22), and $a_{\mathbf{p}}$ is an energy lowering operator, the same as (2.21). And off we go! Just as in the middle of the last section, we can reconstruct all of Fock space on the basis of this operator algebra.

So that was a sketch, not a proof. I've leapt from mountain peak to mountain peak without going through the valleys but I hope the logic is clear. Of course this procedure does not give us a zero of energy, the energy of the ground state, but that's just a matter of convention. We can always define that by convention to be zero.

Now this is not all. We go on because (3.51) can be weakened. This commutator, our condition 1 (p. 34) can be replaced by two separate equations, two new commutators, say 1′(a) and 1′(b). The first specifies the commutator of $\phi(\mathbf{x}, t)$ with $\phi(\mathbf{y}, t)$. That is to say, the time components of the two points x and y are to be taken as equal; this is the so-called **equal time commutator**. The result is a definite function which we can compute. The second will be the equal time commutator of $\dot{\phi}(\mathbf{x}, t)$ with $\phi(\mathbf{y}, t)$, where the dot always means time derivative. This will equal something else, again a definite numerical function which we will shortly compute.[12]

Why do I say that condition 1 can be replaced by conditions 1′(a) and 1′(b)? Well, it's because the Klein–Gordon equation is a differential equation second-order in the time. I can operate on the commutator with $\Box_x^2 = \partial_x^2$, considering the variable y as fixed. Therefore I can just bring the operator through and use the Klein–Gordon equation. Consequently

$$(\Box_x^2 + \mu^2)[\phi(x), \phi(y)] = 0 \tag{3.57}$$

We know the solution of the second-order differential equation for arbitrary values of the argument if we know its value and its first time derivative at some fixed time: the initial value conditions. We need only compute the equation 1′(a)

$$[\phi(\mathbf{x}, t), \phi(\mathbf{y}, t)] = i\Delta(\mathbf{x} - \mathbf{y}) \tag{3.58}$$

and equation 1′(b)

$$[\dot{\phi}(\mathbf{x}, t), \phi(\mathbf{y}, t)] = i\frac{\partial}{\partial x^0}\Delta(x - y)\Big|_{x^0 = y^0} \tag{3.59}$$

for some fixed time t, integrate away as in all books on differential equations, and we will know the solution uniquely. That will be sufficient—because we know $i\Delta(x - y)$ obeys a differential equation, second-order in time—to compute these commutators and $i\Delta(x - y)$ for all times. So let's calculate. From (3.51),

$$[\phi(\mathbf{x}, t), \phi(\mathbf{y}, t)] = \int \frac{d^3\mathbf{p}}{(2\pi)^3(2\omega_{\mathbf{p}})} \left(e^{-i\mathbf{p}\cdot(\mathbf{x}-\mathbf{y})} - e^{i\mathbf{p}\cdot(\mathbf{x}-\mathbf{y})}\right) = 0 \tag{3.60}$$

because the integrand is an odd function. Equation 1′(b) is also easily computed:

$$\begin{aligned}
[\dot{\phi}(\mathbf{x}, t), \phi(\mathbf{y}, t)] &= \int \frac{d^3\mathbf{p}}{(2\pi)^3(2\omega_{\mathbf{p}})} \left(-i\omega_{\mathbf{p}} e^{-i\mathbf{p}\cdot(\mathbf{x}-\mathbf{y})} - i\omega_{\mathbf{p}} e^{i\mathbf{p}\cdot(\mathbf{x}-\mathbf{y})}\right) \\
&= -i\int \frac{d^3\mathbf{p}}{(2\pi)^3(2)} \left(e^{-i\mathbf{p}\cdot(\mathbf{x}-\mathbf{y})} + e^{i\mathbf{p}\cdot(\mathbf{x}-\mathbf{y})}\right) \\
&= -i\int \frac{d^3\mathbf{p}}{(2\pi)^3} e^{-i\mathbf{p}\cdot(\mathbf{x}-\mathbf{y})} = -i\delta^{(3)}(\mathbf{x} - \mathbf{y})
\end{aligned} \tag{3.61}$$

because the integrand is an even function.

As I've argued, conditions 1′(a), 1′(b) and 2 are sufficient to reconstruct the whole theory. The field which we introduced as an auxiliary entity not only gives us a definition of locality

12 [Eds.] The commutators can be restricted to equal times, because spacelike vectors can always be Lorentz transformed to purely spatial vectors, with zero time components. With the 4-vector $x - y$ transformed to a purely spatial vector, $x^0 - y^0 = 0$. So the restriction of spacelike separation can be replaced by the weaker condition $x^0 = y^0$. See note 8, p. 42.

consistent with the dynamics we had before, but in fact all the dynamics we had earlier can be expressed in terms of this field: it obeys the Klein–Gordon equation, it is Hermitian, it satisfies these two equal time commutation relations. Your homework problems ask you to play with these equations to develop certain identities that will be useful to us later on in the course.

3.5 A hint of things to come

I'm now going into mystic and visionary mode, to remind you that these equations look very similar to some equations you might very well have encountered before, in mechanics and in non-relativistic quantum mechanics: good old canonical commutation relations and canonical quantization. We have a set of equations for the Heisenberg picture operators in non-relativistic quantum mechanics. There's normally an $\hbar$ in these equations but I've set $\hbar$ equal to 1. In fact, there is a third that comes with the first two:

$$[q^a(t), q^b(t)] = 0 \tag{3.62}$$

$$[p^a(t), q^b(t)] = -i\delta^{ab} \tag{3.63}$$

$$[p^a(t), p^b(t)] = 0 \tag{3.64}$$

Now the first two of these equations bear a certain structural similarity to the equations (3.60) and (3.61) if I identify $\dot{\phi}(\mathbf{x}, t)$ with p^a and $\phi(\mathbf{x}, t)$ with q^b. Instead of the discrete indices a and b labeling the various coordinates I have continuous indices $\mathbf{x}$ and $\mathbf{y}$, and as a consequence of that, instead of a Kronecker delta I have a Dirac delta function, but otherwise they look very similar.

To test that vague similarity let me try to compute the analog of the third equation. If I identify p^a with $\dot{\phi}(\mathbf{x}, t)$ in the system I have to compute

$$[\dot{\phi}(\mathbf{x}, t), \dot{\phi}(\mathbf{y}, t)] \tag{3.65}$$

which, if the analogy holds, should equal zero. This is nearly the same computation as before;

$$\begin{aligned}
[\dot{\phi}(\mathbf{x}, t), \dot{\phi}(\mathbf{y}, t)] &= \int \frac{d^3\mathbf{p}}{(2\pi)^3(2\omega_\mathbf{p})} \left(-\omega_\mathbf{p}^2 e^{-i\mathbf{p}\cdot(\mathbf{x}-\mathbf{y})} + \omega_\mathbf{p}^2 e^{i\mathbf{p}\cdot(\mathbf{x}-\mathbf{y})}\right) \\
&= \int \frac{d^3\mathbf{p}}{(2\pi)^3(2)}(-\omega_\mathbf{p}) \left(e^{-i\mathbf{p}\cdot(\mathbf{x}-\mathbf{y})} - e^{i\mathbf{p}\cdot(\mathbf{x}-\mathbf{y})}\right) = 0
\end{aligned} \tag{3.66}$$

because the integrand is again an odd function. The commutator *does* equal zero, which looks awfully like the third equation. To summarize,

$$\boxed{[\phi(\mathbf{x}, t), \phi(\mathbf{y}, t)] = [\dot{\phi}(\mathbf{x}, t), \dot{\phi}(\mathbf{y}, t)] = 0} \tag{3.67a}$$

$$\boxed{[\phi(\mathbf{x}, t), \dot{\phi}(\mathbf{y}, t)] = i\delta^{(3)}(\mathbf{x} - \mathbf{y})} \tag{3.67b}$$

Therefore, there seems to be some vague connection with the system we have developed without ever talking about canonical equal time commutation relations and the canonical quantization method. Maybe. Or maybe I'm just dribbling on at the mouth. But there seems to be a certain suggestive structural similarity. In the next section, I will exploit that similarity in a much more systematic way. It's going to take two or three minutes for me to explain what that systematic way is.

With the new method I will develop this entire system by a completely different and independent line of approach. This method will be the method of canonical quantization, or as I will describe it somewhat colorfully, the "method of the missing box".

At the start of the next section I will review, in my characteristic lightning fashion, the introductory parts arising from material I assume you all know, the mechanics of Lagrange and Hamilton. You also may or may not know that you can generalize classical particle theory consistent with an infinite number of degrees of freedom or a continuous infinity of degrees of freedom and write down Hamiltonians and Lagrangians for classical field theory. There is also a standard procedure for getting from classical particle theory to non-relativistic quantum mechanics which I will review. What we will attempt to do in the second half of the next lecture is fill in the "missing box", to get to the thing you don't know anything about, or I pretend you don't know anything about, quantum field theory. We will again arrive at the same system, but by another path.

Problems 1

He who can do nothing,
understands nothing.

PARACELSUS

1.1 Some people were not happy with the method I used in class to show

$$\frac{d^3\mathbf{p}}{(2\pi)^3 2\omega} = \frac{d^3\mathbf{p}'}{(2\pi)^3 2\omega'} \tag{P1.1}$$

where $\mathbf{p}$ and $\mathbf{p}'$ are single-particle 3-momenta connected by a Lorentz transformation, while ω and ω' are the associated energies. Show the equation is true directly, just by using the elementary calculus formula for the change in a volume element under a change of coordinates. (HINT: The equation is obviously true for rotations, so you need only to check it for Lorentz boosts (to frames of reference moving at different speeds). Indeed, you need only check it for a boost in the z-direction.)

(1997a 1.1)

1.2 This problem and **1.3** deal with the **time-ordered product**, an object which will play a central role in our development of diagrammatic perturbation theory later in the course.

The time-ordered product of two fields, $A(x)$ and $B(y)$, is defined by

$$T(A(x)B(y)) = \begin{cases} A(x)B(y) & \text{if } x_0 > y_0 \\ B(y)A(x) & \text{if } y_0 > x_0 \end{cases} \tag{P1.2}$$

Using *only* the field equation and the equal time commutation relations, show that, for a free scalar field of mass μ,

$$(\Box_x^2 + \mu^2)\,\langle 0|T(\phi(x)\phi(y))|0\rangle = c\,\delta^{(4)}(x-y) \tag{P1.3}$$

and find c, the constant of proportionality.

(1997a 1.2)

1.3 Show that

$$\langle 0|T(\phi(x)\phi(y))|0\rangle = \lim_{\epsilon\to 0+} \int \frac{d^4p}{(2\pi)^4} e^{-ip\cdot(x-y)} \frac{-c}{p^2 - \mu^2 + i\epsilon} \tag{P1.4}$$

The limit symbol indicates that ϵ goes to zero from above, i.e., through positive values. (If ϵ were not present, the integral would be ill-defined, because it would have poles in the domain of integration.) (HINTS: Do the p_0 integration first, and compare your result with the expression for the left-hand side obtained by inserting the explicit form of the field (3.45). Treat the cases $x^0 > y^0$ and $x^0 < y^0$ separately.)

(1997a 1.3)

1.4 In a quantum theory, most observables do not have a definite value in the ground state of the theory. For a general observable A, a reasonable measure of this quantum spread in the ground state value of A is given by the ground state variance of A, defined by

$$\text{var}\,A \equiv \langle (A - \langle A\rangle)^2\rangle = \langle A^2\rangle - \langle A\rangle^2 \tag{P1.5}$$

where the brackets $\langle \cdots \rangle$ indicate the ground state expectation value.

In the theory of a free scalar field $\phi(x)$ of mass μ, define the observable

$$A(a) = \frac{1}{(a\sqrt{\pi})^3} \int d^3\mathbf{x}\, \phi(\mathbf{x}, 0)\, e^{-|\mathbf{x}|^2/a^2} \tag{P1.6}$$

where a is some length. Note that the Gaussian has been normalized so that its space integral is 1; thus this is a smoothed-out version of the field averaged over a region of size a. Express the ground state (vacuum) variance of $A(a)$ as an integral over a single variable. You are not required to evaluate this integral except in the limiting cases of very small a and very large a. In both these limits you should find

$$\operatorname{var} A(a) = \alpha a^{\beta} + \ldots \tag{P1.7}$$

where α and β are constants you are to find (with different values for the different limits), and the ... denote terms negligible in the limit compared to the term displayed. You should find that $\operatorname{var} A(a)$ goes to zero for large a while it blows up for small a. Speaking somewhat loosely, on large scales the average field is almost a classical variable, while on small scales quantum fluctuations are enormous.

(1997a 1.4)

Solutions 1

1.1 Consider Λ, a boost in the z-direction:

$$\Lambda\colon (\omega, p_x, p_y, p_z) \to (\omega', p'_x, p'_y, p'_z) = (\omega\cosh\chi + p_z\sinh\chi, p_x, p_y, p_z\cosh\chi + \omega\sinh\chi) \tag{S1.1}$$

where $\tanh\chi = v/c = v$ (in units where $c = 1$.) The change of volume element is given by the Jacobian determinant,

$$d^3\mathbf{p} \to d^3\mathbf{p}' = J\, d^3\mathbf{p} \tag{S1.2}$$

where

$$J = \begin{vmatrix} \dfrac{\partial p'_x}{\partial p_x} & \dfrac{\partial p'_x}{\partial p_y} & \dfrac{\partial p'_x}{\partial p_z} \\ \dfrac{\partial p'_y}{\partial p_x} & \dfrac{\partial p'_y}{\partial p_y} & \dfrac{\partial p'_y}{\partial p_z} \\ \dfrac{\partial p'_z}{\partial p_x} & \dfrac{\partial p'_z}{\partial p_y} & \dfrac{\partial p'_z}{\partial p_z} \end{vmatrix} = \begin{vmatrix} 1 & 0 & 0 \\ 0 & 1 & 0 \\ \dfrac{\partial\omega}{\partial p_x}\sinh\chi & \dfrac{\partial\omega}{\partial p_y}\sinh\chi & \cosh\chi + \dfrac{\partial\omega}{\partial p_z}\sinh\chi \end{vmatrix} = \cosh\chi + \frac{\partial\omega}{\partial p_z}\sinh\chi \tag{S1.3}$$

But

$$\frac{\partial\omega}{\partial p_z} = \frac{\partial}{\partial p_z}\sqrt{p_x^2 + p_y^2 + p_z^2 + \mu^2} = \frac{p_z}{\omega} \tag{S1.4}$$

so

$$J = \cosh\chi + \frac{p_z}{\omega}\sinh\chi \tag{S1.5}$$

Using (S1.1) for ω', and (S1.5) for the Jacobian gives

$$\frac{d^3\mathbf{p}'}{2\omega'} = \frac{(\cosh\chi + (p_z/\omega)\sinh\chi)\, d^3\mathbf{p}}{2(\omega\cosh\chi + p_z\sinh\chi)} = \frac{d^3\mathbf{p}}{2\omega}$$ ■

1.2 The Heaviside theta function, or step function, $\theta(x)$, is defined by

$$\theta(x) = \begin{cases} 1 & \text{if } x > 0 \\ 0 & \text{if } x < 0 \end{cases}$$

The extension to $\theta(x - a)$, where a is a constant, should be clear. Its derivative is a delta function:

$$\frac{d\theta(x-a)}{dx} = \delta(x - a) \tag{S1.6}$$

Using theta functions, we can write the time-ordered product (P1.2) of two operators $A(x)$, $B(y)$ like this:

$$T(A(x)B(y)) = \theta(x_0 - y_0)A(x)B(y) + \theta(y_0 - x_0)B(y)A(x)$$

The d'Alembertian $\Box^2$ (the 4-vector equivalent of ∇^2, the Laplacian) is

$$\Box^2 = \partial^\mu\partial_\mu = \frac{\partial^2}{\partial x_0^2} - \nabla^2$$

Look at the first partial derivative with respect to x_0:

$$\frac{\partial T(\phi(x)\phi(y))}{\partial x_0} = \delta(x_0 - y_0)\phi(x)\phi(y) + \theta(x_0 - y_0)\frac{\partial\phi(x)}{\partial x_0}\phi(y) - \delta(y_0 - x_0)\phi(y)\phi(x) + \theta(y_0 - x_0)\phi(y)\frac{\partial\phi(x)}{\partial x_0} \tag{S1.7}$$

Delta functions are even: $\delta(x-y) = \delta(y-z)$. Also, as $\delta(x-a) = 0$ unless $x = a$, $f(x)\delta(x-a) = f(a)\delta(x-a)$. The two terms involving delta functions can be written

$$\begin{aligned}\delta(x_0 - y_0)\phi(x)\phi(y) - \delta(y_0 - x_0)\phi(y)\phi(x) &= \delta(x_0 - y_0)[\phi(x), \phi(y)] \\ &= \delta(x_0 - y_0)\ [\phi(x), \phi(y)]\Big|_{x_0 = y_0} = 0\end{aligned} \tag{S1.8}$$

because the equal time commutator of the two fields equals zero. (But see the *Alternative solution* below!) Then

$$\frac{\partial T(\phi(x)\phi(y))}{\partial x_0} = \theta(x_0 - y_0)\frac{\partial\phi(x)}{\partial x_0}\phi(y) + \theta(y_0 - x_0)\phi(y)\frac{\partial\phi(x)}{\partial x_0} \tag{S1.9}$$

The second derivative goes much the same way,

$$\begin{aligned}\frac{\partial^2 T(\phi(x)\phi(y))}{\partial x_0^2} &= \delta(x_0 - y_0)\frac{\partial\phi(x)}{\partial x_0}\phi(y) + \theta(x_0 - y_0)\frac{\partial^2\phi(x)}{\partial x_0^2}\phi(y) \\ &\quad - \delta(y_0 - x_0)\phi(y)\frac{\partial\phi(x)}{\partial x_0} + \theta(y_0 - x_0)\phi(y)\frac{\partial^2\phi(x)}{\partial x_0^2} \\ &= \delta(x_0 - y_0)\left[\frac{\partial\phi(x)}{\partial x_0}, \phi(y)\right]\Bigg|_{x_0 = y_0} + \theta(x_0 - y_0)\frac{\partial^2\phi(x)}{\partial x_0^2}\phi(y) + \theta(y_0 - x_0)\phi(y)\frac{\partial^2\phi(x)}{\partial x_0^2} \\ &= -i\delta^{(4)}(x-y) + \theta(x_0 - y_0)\frac{\partial^2\phi(x)}{\partial x_0^2}\phi(y) + \theta(y_0 - x_0)\phi(y)\frac{\partial^2\phi(x)}{\partial x_0^2}\end{aligned}$$

because $[\dot\phi(\mathbf{x}), \phi(\mathbf{y})] = -i\delta^{(3)}(\mathbf{x} - \mathbf{y})$ at equal times. The Laplacian does not act on the θ functions, so

$$-\nabla^2(T(\phi(x)\phi(y)) = \theta(x_0 - y_0)(-\nabla^2\phi(x))\phi(y) + \theta(y_0 - x_0)\phi(y)(-\nabla^2\phi(x))$$

and consequently

$$\begin{aligned}(\Box^2 + \mu^2)T(\phi(x)\phi(y)) &= -i\delta^{(4)}(x-y) + \theta(x_0 - y_0)[(\Box^2 + \mu^2)\phi(x)]\phi(y) + \theta(y_0 - x_0)\phi(y)[(\Box^2 + \mu^2)\phi(x)] \\ &= -i\delta^{(4)}(x-y)\end{aligned}$$

because ϕ satisfies the Klein–Gordon equation, $\Box^2\phi(x) + \mu^2\phi(x) = 0$. Then

$$(\Box^2 + \mu^2)\,\langle 0|T(\phi(x)\phi(y))|0\rangle = \langle 0| - i\delta^{(4)}(x-y)|0\rangle = -i\delta^{(4)}(x-y)$$

in agreement with (P1.3), and the constant $c = -i$. ■

Alternative Solution. A purist might object to setting the quantity $\delta(x_0 - y_0)[\phi(x), \phi(y)]$ equal to zero. After all, the differential equation is second-order, and maybe we should carry the second time derivative all the way through; $\Box^2 = \partial_0^2 - \nabla^2$. Then

$$\begin{aligned}(\Box^2 + \mu^2)T(\phi(x)\phi(y)) &= \partial_0^2\left(\theta(x_0 - y_0)\phi(x)\phi(y) + \theta(y_0 - x_0)\phi(y)\phi(x)\right) \\ &\quad + \theta(x_0 - y_0)\left(-\nabla^2\phi(x) + \mu^2\phi(x)\right)\phi(y) + \theta(y_0 - x_0)\phi(y)\left(-\nabla^2\phi(x) + \mu^2\phi(x)\right)\end{aligned}$$

Let's look carefully at the first line of the previous equation. We can write

$$\begin{aligned}\partial_0^2\left(\theta(x_0 - y_0)\phi(x)\phi(y)\right) &= \partial_0\left(\delta(x_0 - y_0)\phi(x)\phi(y) + \theta(x_0 - y_0)\dot\phi(x)\phi(y)\right) \\ &= \partial_0\delta(x_0 - y_0)\,\phi(x)\phi(y) + 2\delta(x_0 - y_0)\dot\phi(x)\phi(y) + \theta(x_0 - y_0)\,\ddot\phi(x)\phi(y)\end{aligned}$$

Delta functions really only make sense in the context of being under an integral sign, multiplying some suitably smooth function. If we integrate $\partial_0\delta(x_0 - y_0)\,\phi(x)\phi(y)$ with respect to x_0 and use integration by parts, assuming that $\phi(x) \to 0$ as $x_0 \to \pm\infty$, then we can say

$$\partial_0\delta(x_0 - y_0)\,\phi(x)\phi(y) = -\delta(x_0 - y_0)\,\dot\phi(x)\phi(y)$$

Using this identity, we have

$$\partial_0^2\left(\theta(x_0-y_0)\,\phi(x)\phi(y)\right)=\delta(x_0-y_0)\,\dot\phi(x)\phi(y)+\theta(x_0-y_0)\,\ddot\phi(x)\phi(y)$$

Plug this (and a similar expression with $\phi(x)$ and $\phi(y)$ swapped, and an extra $-$ sign) into the original equation to obtain

$$\begin{aligned}(\Box^2+\mu^2)T(\phi(x)\phi(y))&=\theta(x_0-y_0)\left[(\Box^2+\mu^2)\phi(x)\right]\phi(y)+\theta(y_0-x_0)\phi(y)\left[(\Box^2+\mu^2)\phi(x)\right]\\&\quad+\delta(x_0-y_0)\,[\dot\phi(x),\phi(y)]\\&=-i\delta^{(4)}(x-y)\end{aligned}$$

which gives the same result. ■

1.3 We need to show that for $x_0>y_0$,

$$\langle 0|\phi(x)\phi(y)|0\rangle=\lim_{\epsilon\to 0+}\int\frac{d^4p}{(2\pi)^4}e^{-ip\cdot(x-y)}\frac{-c}{p^2-\mu^2+i\epsilon}\qquad\text{(S1.10)}$$

and for $y_0>x_0$,

$$\langle 0|\phi(y)\phi(x)|0\rangle=\lim_{\epsilon\to 0+}\int\frac{d^4p}{(2\pi)^4}e^{-ip\cdot(x-y)}\frac{-c}{p^2-\mu^2+i\epsilon}\qquad\text{(S1.11)}$$

The right-hand sides of (S1.10) and (S1.11) are the same. Swap x and y (so that now $x_0>y_0$) and obtain

$$\langle 0|\phi(x)\phi(y)|0\rangle=\lim_{\epsilon\to 0+}\int\frac{d^4p}{(2\pi)^4}e^{-ip\cdot(y-x)}\frac{-c}{p^2-\mu^2+i\epsilon}\qquad\text{(S1.12)}$$

The only difference between (S1.10) and (S1.12) is the sign of the exponential's argument. But if we take $p\to -p$, nothing changes except the sign of the argument: (S1.10) and (S1.12), and hence also (S1.11), are equivalent. (There's a second argument that's worth seeing; it will be given at the end.) Let's work on (S1.10) first. In the product of $\phi(x)\phi(y)$, there will be four separate products of creation and annihilation operators, aa, $aa^\dagger$, $a^\dagger a$ and $a^\dagger a^\dagger$. Sandwiched between vacuum states, only the second term survives, because $a\,|0\rangle=\langle 0|\,a^\dagger=0$. Because $\langle 0|a^\dagger_{\mathbf q}a_{\mathbf p}|0\rangle=0$, we can write

$$\langle 0|a_{\mathbf p}a^\dagger_{\mathbf q}|0\rangle=\langle 0|[a_{\mathbf p},a^\dagger_{\mathbf q}]|0\rangle=\delta^{(3)}(\mathbf p-\mathbf q)$$

Then

$$\begin{aligned}\langle 0|\phi(x)\phi(y)|0\rangle&=\langle 0|\int\frac{d^3\mathbf p}{\sqrt{(2\pi)^3 2\omega_{\mathbf p}}}a_{\mathbf p}e^{-ip\cdot x}\int\frac{d^3\mathbf q}{\sqrt{(2\pi)^3 2\omega_{\mathbf q}}}a^\dagger_{\mathbf q}e^{iq\cdot y}|0\rangle\\&=\langle 0|\int\frac{d^3\mathbf p}{\sqrt{(2\pi)^3 2\omega_{\mathbf p}}}e^{-ip\cdot x}\int\frac{d^3\mathbf q}{\sqrt{(2\pi)^3 2\omega_{\mathbf q}}}e^{iq\cdot y}\delta^{(3)}(\mathbf p-\mathbf q)|0\rangle\end{aligned}$$

The integrals sandwiched between the vacuum states are c-numbers, so the integrals merely multiply the inner product $\langle 0|0\rangle=1$. Either integral can be done quickly owing to the delta function. Performing the $\mathbf q$ integral gives

$$\langle 0|\phi(x)\phi(y)|0\rangle=\int\frac{d^3\mathbf p}{(2\pi)^3 2\omega_{\mathbf p}}e^{-ip\cdot(x-y)}\qquad\text{(S1.13)}$$

Now to work on the right-hand side of (S1.10), substituting in the value $c=-i$ found in Problem 1.2:

$$\int\frac{d^4p}{(2\pi)^4}e^{-ip\cdot(x-y)}\frac{i}{p^2-\mu^2+i\epsilon}=\int\frac{d^3\mathbf p}{(2\pi)^3}e^{i\mathbf p\cdot(\mathbf x-\mathbf x)}\int\frac{dp_0}{2\pi}\frac{ie^{-ip_0(x_0-y_0)}}{p_0^2-\omega_{\mathbf p}^2+i\epsilon}\qquad\text{(S1.14)}$$

Rewrite the p_0 integral:

$$\int\frac{dp_0}{2\pi}\frac{ie^{-ip_0(x_0-y_0)}}{p_0^2-\omega_{\mathbf p}^2+i\epsilon}=\int\frac{dp_0}{2\pi}\frac{ie^{-ip_0(x_0-y_0)}}{(p_0-\sqrt{\omega_{\mathbf p}{}^2-i\epsilon})(p_0+\sqrt{\omega_{\mathbf p}{}^2-i\epsilon})}$$

Because ϵ is small, we can write

$$\sqrt{\omega_{\mathbf p}{}^2-i\epsilon}\approx\omega_{\mathbf p}-\tfrac12 i\epsilon/\omega_{\mathbf p}\equiv\omega_{\mathbf p}-i\eta$$

where η is also a small quantity: $\eta\to 0$ as $\epsilon\to 0$. Rewrite the p_0 integral once again;

$$\int\frac{dp_0}{2\pi}\frac{ie^{-ip_0(x_0-y_0)}}{p_0^2-\omega_{\mathbf p}^2+i\epsilon}=\int\frac{dp_0}{2\pi}\frac{ie^{-ip_0(x_0-y_0)}}{(p_0-(\omega_{\mathbf p}-i\eta))(p_0+(\omega_{\mathbf p}-i\eta))}$$

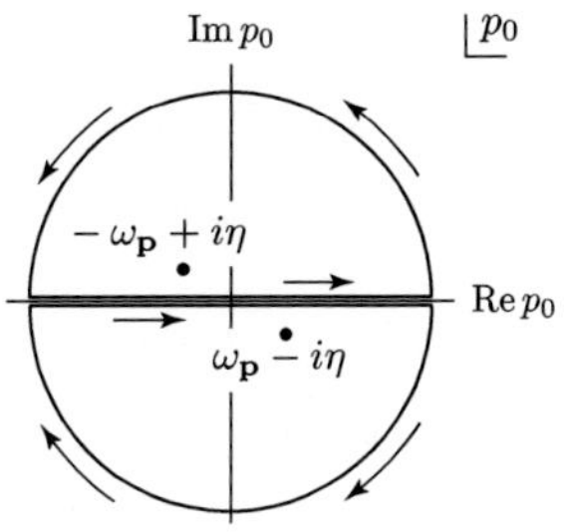

Figure S1.1: Contours for the p_0 integral in $\langle 0|T(\phi(x)\phi(y)|0\rangle$

Use Cauchy's integral formula to evaluate this, by extending p_0 to the complex plane. We'll use a contour which has a large semicircular arc of radius R and a diameter along the real axis; we need to choose the upper or the lower contour. There are two poles, at $\pm(\omega_{\mathbf{p}}-i\eta)$. For case (S1.10), $x_0>y_0$, the quantity in the exponential will be negative if Im p_0 is negative, so that the semicircular arc of radius R will contribute nothing as $R\to\infty$. That means we take the contour below the real axis, enclosing only the root $\omega_{\mathbf{p}}-i\eta$. Then by Cauchy's formula

$$\int\frac{dp_0}{2\pi}\frac{ie^{-ip_0(x_0-y_0)}}{(p_0-(\omega_{\mathbf{p}}-i\eta))(p_0+(\omega_{\mathbf{p}}-i\eta))}=(-1)\times 2\pi i\times\frac{1}{2\pi}\frac{ie^{-i(\omega_{\mathbf{p}}-i\eta)(x_0-y_0)}}{2(\omega_{\mathbf{p}}-i\eta)}$$

the extra factor of (-1) coming because the bottom contour is *clockwise*. We can now safely take the limit $\eta\to 0$, and the p^0 integral gives

$$\lim_{\eta\to 0+}\frac{e^{-i(\omega_{\mathbf{p}}-i\eta)(x_0-y_0)}}{2(\omega_{\mathbf{p}}-i\eta)}=\frac{e^{-i\omega_{\mathbf{p}}(x_0-y_0)}}{2\omega_{\mathbf{p}}}$$

Put this back into the original integral (S1.14) to obtain

$$\lim_{\epsilon\to 0+}\int\frac{d^4p}{(2\pi)^4}e^{-ip\cdot(x-y)}\frac{i}{p^2-\mu^2+i\epsilon}=\int\frac{d^3\mathbf{p}}{(2\pi)^3}e^{i\mathbf{p}\cdot(\mathbf{x}-\mathbf{y})}\frac{e^{-i\omega_{\mathbf{p}}(x_0-y_0)}}{2\omega_{\mathbf{p}}}=\int\frac{d^3\mathbf{p}}{(2\pi)^3 2\omega_{\mathbf{p}}}e^{-ip\cdot(x-y)}\tag{S1.15}$$

The right-hand side of (S1.15) is identical to the right-hand side of (S1.13), so the left-hand side of (S1.15) must equal the left-hand side of (S1.13). That establishes case (S1.10) (and (S1.11) also, since they're equivalent).

Case (S1.11) can also be done on its own. Now $y_0>x_0$. By symmetry, we can write down at once the equivalent of (S1.13):

$$\langle 0|\phi(y)\phi(x)|0\rangle=\int\frac{d^3\mathbf{p}}{(2\pi)^3 2\omega_{\mathbf{p}}}e^{-ip\cdot(y-x)}\tag{S1.16}$$

The right side of (S1.10) is the same as the right side of (S1.11). The p_0 integral is the same as before, but now $y_0>x_0$. That means the imaginary part of p_0 has to be *positive* in order to guarantee that the semicircular arc contributes nothing. Now we take the *upper* contour, *counter-clockwise*, which encloses only the root $-\omega_{\mathbf{p}}+i\eta$. Then

$$\int\frac{dp_0}{2\pi}\frac{ie^{-ip_0(x_0-y_0)}}{(p_0-(\omega_{\mathbf{p}}-i\eta))(p_0+(\omega_{\mathbf{p}}-i\eta))}=2\pi i\times\frac{1}{2\pi}\frac{ie^{i(\omega_{\mathbf{p}}-i\eta)(x_0-y_0)}}{-2(\omega_{\mathbf{p}}-i\eta)}\to\frac{e^{i\omega_{\mathbf{p}}(x_0-y_0)}}{2\omega_{\mathbf{p}}}$$

as $\eta\to 0$. Not surprisingly, the sign of the exponential's argument changes from the previous calculation. Substitute this back into (S1.14) to obtain

$$\lim_{\epsilon\to 0+}\int\frac{d^4p}{(2\pi)^4}e^{-ip\cdot(x-y)}\frac{i}{p^2-\mu^2+i\epsilon}=\int\frac{d^3\mathbf{p}}{(2\pi)^3}e^{i\mathbf{p}\cdot(\mathbf{x}-\mathbf{y})}\frac{e^{i\omega_{\mathbf{p}}(x_0-y_0)}}{2\omega_{\mathbf{p}}}\tag{S1.17}$$

Unfortunately, the sign of the exponentials do not now match up to give the inner product of two 4-vectors; we'd need the space parts to be negative. That's easy to arrange: Let $\mathbf{p}\to-\mathbf{p}$. Equation (S1.17) becomes

$$\lim_{\epsilon\to 0+}\int\frac{d^4p}{(2\pi)^4}e^{-ip\cdot(x-y)}\frac{i}{p^2-\mu^2+i\epsilon}=\int\frac{d^3\mathbf{p}}{(2\pi)^3}e^{-i\mathbf{p}\cdot(\mathbf{x}-\mathbf{y})}\frac{e^{i\omega_{\mathbf{p}}(x_0-y_0)}}{2\omega_{\mathbf{p}}}=\int\frac{d^3\mathbf{p}}{(2\pi)^3 2\omega_{\mathbf{p}}}e^{ip\cdot(x-y)}\tag{S1.18}$$

The right-hand side of (S1.16) is the same as the right-hand side of (S1.18), so the left-hand side of (S1.16) is the same as the left-hand side of (S1.18). That establishes (S1.11). ■

1.4 We first notice that

$$\langle A\rangle = \frac{1}{(a\sqrt{\pi})^3}\int d^3\mathbf{x}\; e^{-|\mathbf{x}|^2/a^2}\,\langle 0|\phi(\mathbf{x},0)|0\rangle = 0 \tag{S1.19}$$

since $\phi(\mathbf{x},0)$ is linear in $a_{\mathbf{p}}$ and $a_{\mathbf{p}}^\dagger$, and $\langle 0|a_{\mathbf{p}}|0\rangle = \langle 0|a_{\mathbf{p}}^\dagger|0\rangle = 0$. So $\mathrm{var}\,A = \langle A^2\rangle$. To calculate $\langle A^2\rangle$, notice that A^2 involves $\langle 0|\phi(\mathbf{x},0)\phi(\mathbf{y},0)|0\rangle$. In the product of the two ϕ's, the only non-zero term will be of the form $a_{\mathbf{p}}a_{\mathbf{q}}^\dagger$, where $\mathbf{q}$ is a dummy momentum variable. Then

$$\langle A^2\rangle = \frac{1}{\pi^3a^6}\iint d^3\mathbf{x}\,d^3\mathbf{y}\,e^{-(|\mathbf{x}|^2+|\mathbf{y}|^2)/a^2}\int\frac{d^3\mathbf{p}}{(2\pi)^{3/2}\sqrt{2\omega_{\mathbf{p}}}}\int\frac{d^3\mathbf{q}}{(2\pi)^{3/2}\sqrt{2\omega_{\mathbf{q}}}}e^{i\mathbf{p}\cdot\mathbf{x}}e^{-i\mathbf{q}\cdot\mathbf{y}}\,\langle 0|a_{\mathbf{p}}a_{\mathbf{q}}^\dagger|0\rangle \tag{S1.20}$$

The vacuum expectation value can be rewritten as

$$\langle 0|a_{\mathbf{p}}a_{\mathbf{q}}^\dagger|0\rangle = \langle 0|[a_{\mathbf{p}},a_{\mathbf{q}}^\dagger]|0\rangle = \delta^{(3)}(\mathbf{p}-\mathbf{q}) \tag{S1.21}$$

because $\langle 0|a_{\mathbf{q}}^\dagger a_{\mathbf{p}}|0\rangle = 0$. Then integrating over $\mathbf{q}$ with the help of the delta function,

$$\langle A^2\rangle = \frac{1}{\pi^3a^6}\int\frac{d^3\mathbf{p}}{(2\pi)^3 2\omega_{\mathbf{p}}}\left(\int d^3\mathbf{x}\,e^{-(|\mathbf{x}|^2/a^2)+i\mathbf{p}\cdot\mathbf{x}}\right)\left(\int d^3\mathbf{y}\,e^{-(|\mathbf{y}|^2/a^2)-i\mathbf{p}\cdot\mathbf{y}}\right) \tag{S1.22}$$

The integrals over $\mathbf{x}$ and $\mathbf{y}$ have the same form. Looking at the integral over $\mathbf{x}$,

$$\int d^3\mathbf{x}\,e^{-(\mathbf{x}^2/a^2)+i\mathbf{p}\cdot\mathbf{x}} = \prod_{i=1}^{3}\int dx_i\,e^{-(x_i^2/a^2)+ip_ix_i} = \prod_{i=1}^{3}a\sqrt{\pi}\,e^{a^2(ip_i)^2/4} = a^3(\pi)^{3/2}e^{-a^2|\mathbf{p}|^2/4}$$

using the identity $\int e^{-cx^2+bx}\,dx = \left(\sqrt{\pi/c}\right)e^{b^2/4c}$ with $c = 1/a^2$ and $b = ip_i$. Then

$$\langle A^2\rangle = \frac{1}{\pi^3a^6}\int\frac{d^3\mathbf{p}}{(2\pi)^3 2\omega_{\mathbf{p}}}(\pi a^2)^3e^{-a^2|\mathbf{p}|^2/2} = \int_0^\infty\frac{4\pi p^2\,dp}{(2\pi)^3 2\sqrt{p^2+\mu^2}}e^{-\frac{1}{2}a^2p^2} = \frac{1}{4\pi^2a^2}\int_0^\infty\frac{e^{-u^2/2}u^2\,du}{\sqrt{u^2+\mu^2a^2}}$$

Now to consider the limits. As $a\to 0$,

$$\langle A^2\rangle \to \frac{1}{4\pi^2a^2}\int_0^\infty e^{-u^2/2}u\,du = \frac{1}{4\pi^2a^2} = \alpha_0a^{\beta_0} \;\Rightarrow\; \alpha_0 = \frac{1}{4\pi^2},\ \beta_0 = -2 \tag{S1.23}$$

As $a\to\infty$,

$$\langle A^2\rangle \to \frac{1}{4\pi^2\mu a^3}\int_0^\infty e^{-u^2/2}u^2\,du = \frac{1}{4\pi^2\mu a^3}\sqrt{\frac{\pi}{2}} = \alpha_\infty a^{\beta_\infty} \;\Rightarrow\; \alpha_\infty = \frac{1}{4\pi^2\mu}\sqrt{\frac{\pi}{2}},\ \beta_\infty = -3 \tag{S1.24}$$

Just as claimed, as $a\to\infty$, the variance tends to zero; as $a\to 0$, when quantum fluctuations are expected to be enormous, the variance tends to infinity. ■

4

The method of the missing box

In the last lecture I told you we would find the same object, the quantum field, we had found a few minutes earlier, by a rather lengthy sequence of investigations, using a totally different method which I described in my characteristic colorful way as the method of the missing box. The method may be illustrated by this diagram:

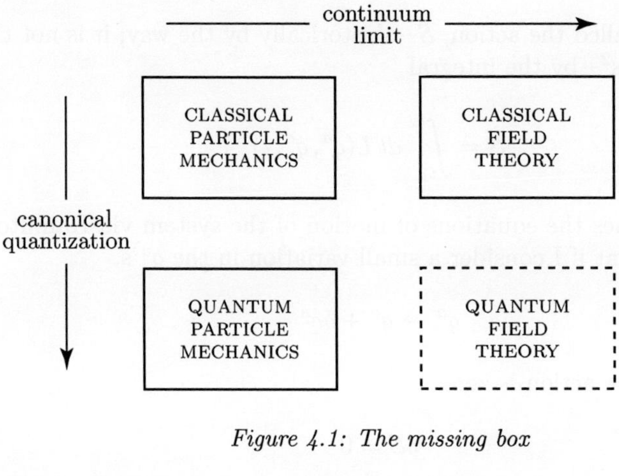

Figure 4.1: The missing box

I presume that three of these boxes are familiar to you. I will give brief summaries of them, complete but fast, in the first half of this lecture. We start out at the upper left corner with classical particle mechanics, summarize that, and, moving down, summarize how that extends to quantum particle mechanics. If you've had a good course in non-relativistic quantum mechanics, you know that there is a standard procedure for getting from classical particle theory to quantum theory, which I will review, called *canonical quantization*. Just to remind you of what that is I'll say it in great detail: You write the system in Hamiltonian form and you set commutation relations between the classical p's and q's. This leads to quantum particle theory. We also can move across, to the right, and summarize how classical particle mechanics is extended to systems with infinite numbers of degrees of freedom, indeed continuously infinite numbers of degrees of freedom, i. e., classical field theory: the classical theory of Maxwell's equations, of sound waves in a fluid, of elasticity in an elastic solid; classical continuum theory. What we will attempt to do in the second half of this lecture is to fill in the missing box, *quantum field theory.* As the arrows show it can either be viewed as the result of applying canonical quantization to classical field theory, following the arrow down; or alternatively, by following the arrow across from quantum particle theory, generalizing to systems with a continuous infinity of degrees of freedom. In the language of the algebraic topologists, this is a commutative diagram.

4.1 Classical particle mechanics

Classical particle mechanics deals with systems characterized by dynamical variables, ordinary real number functions of time called generalized coordinates. I will denote these as

$$q^a(t), \quad a = 1, 2, \ldots N.$$

In the simplest system these may be the coordinates x^i of an assembly of N particles moving in 3-space where i goes from 1 to $3N$. These could represent the three Cartesian coordinates or the three spherical coordinates of each of the particles. Lagrangian systems are those whose dynamics are determined by a function L called the Lagrangian. It depends on the q^a's and their time derivatives, which I indicate with a dot, $\dot{q}^a$, and possibly explicitly on the time:

$$L = L(q^1, q^2, \ldots, q^N, \dot{q}^1, \dot{q}^2, \ldots, \dot{q}^N, t)$$

We define a functional[1] called the action, $\mathcal{S}$—ahistorically by the way; it is not the action first introduced by Maupertuis[2]—by the integral

$$\mathcal{S} = \int_{t_1}^{t_2} dt\, L(q^a, \dot{q}^a, t) \tag{4.1}$$

The Lagrangian determines the equations of motion of the system via Hamilton's Principle, which is the statement that if I consider a small variation in the q^a's,

$$q^a \to q^a + \delta q^a \tag{4.2}$$

the resulting change in the action is zero:

$$\delta\mathcal{S} = 0 \tag{4.3}$$

---

[1] [Eds.] In this book, a *functional* $F[f]$ is a function F of a function f, mapping f to a number, real or complex, and will be realized by an integral.

[2] [Eds.] See Ch. IX, §100 in Edmund T. Whittaker, *A Treatise on the Analytical Dynamics of Particles and Rigid Bodies*, Cambridge U. P., 1959. Maupertuis' action, introduced in 1744, is the integral $\int \dot{q}(\partial L/\partial\dot{q})\, dt$. Whittaker says Maupertuis' Principle of Least Action was actually established by Euler. Lagrange's equations were introduced in his *Mécanique analytique* in 1788. For Hamilton's introduction of his equations see W. R. Hamilton, "On the application to dynamics of a general mathematical method previously applied to optics", *Report of the British Association for the Advancement of Science*, 4th meeting, 1834, pp. 513–518. Lanczos, citing Cayley, says that Lagrange and Cauchy anticipated Hamilton; see Cornelius Lanczos, *The Variational Principles of Mechanics*, 4th ed., University of Toronto Press, 1970, p. 168. See also Whittaker, *op. cit.*, Ch. X, §109 and Arthur Cayley, "Report on the Recent Progress of Theoretical Dynamics", in his *Collected Papers*, Cambridge U. P., 1890, v. III, pp. 156–204 for further references. What is now universally called "the action" was originally called "Hamilton's first principal function". See v. II, Lecture 19, p. 19-8 in Richard P. Feynman, Robert B. Leighton and Matthew Sands, *The Feynman Lectures on Physics (the New Millennium edition)*, Basic Books, 2010.

I use δ to indicate an infinitesimal variation; the Weierstrass revolution in calculus has not yet reached this lecture: we are Newtonians. The variations are subject to the restriction that they vanish at both endpoints of the integration;

$$\delta q^a(t_1) = \delta q^a(t_2) = 0 \tag{4.4}$$

From Hamilton's Principle one can derive equations of motion by the standard methods of the calculus of variations. One simply computes $\delta\mathcal{S}$ for a general change δq^a:

$$\delta\mathcal{S} = \int_{t_1}^{t_2} dt\left[\frac{\partial L}{\partial q^a}\delta q^a + \frac{\partial L}{\partial \dot{q}^a}\delta \dot{q}^a\right] \tag{4.5}$$

And here I have made a slight notational simplification by adopting the Einstein summation convention over the index a, so I don't have to write a couple of sigmas. As it will turn up again and again in our equations, I will define p_a, the *canonical momentum conjugate* to q_a,

$$p_a \equiv \frac{\partial L}{\partial \dot{q}^a} \tag{4.6}$$

(By the way I've arranged my upper and lower indices so that things look like they do in relativity: Differentiation with respect to an object with a lower index gives you an object with an upper index and vice versa. It's just a matter of definition.) From the definition of δq^a, of course $\delta\dot{q}^a = \dfrac{d}{dt}\delta q^a$. By substitution and integration of the last term by parts we obtain

$$\delta\mathcal{S} = \int_{t_1}^{t_2} dt\left(\left[\frac{\partial L}{\partial q^a} - \frac{dp_a}{dt}\right]\delta q^a\right) + p_a\delta q^a\Big|_{t_1}^{t_2} \tag{4.7}$$

Since δq^a are supposed to be arbitrary infinitesimal functions which vanish at the boundaries, the last term above equals zero and the quantity inside the square brackets must vanish everywhere. Thus we obtain the equations of motion

$$\frac{\partial L}{\partial q^a} - \frac{dp_a}{dt} = 0 \tag{4.8}$$

These are the Euler–Lagrange equations. I will not do specific examples. I presume you've all seen how this works out for particles and a system of particles with potentials and velocity dependent forces and all of those things. This gets us halfway through the first box. I will now discuss the Hamiltonian formulation.

We consider the expression defined by the *Legendre transformation*[3]

$$H \equiv p_a\dot{q}^a - L \tag{4.9}$$

H is called the *Hamiltonian*. It can be thought of as a function not of the q^a's and the $\dot{q}^a$'s, which is the natural way to write the right-hand side, but of the q^a's and the p_a's and possibly also of time. I will just tell you something about the Hamiltonian which you may remember from classical mechanics, though in fact we will prove it in the course of another investigation later on. If the Lagrangian is independent of time, then the Hamiltonian is identical with the

[3] [Eds.] Goldstein *et al. CM*, Section 8.1, pp. 334–338.

energy of the system, a conserved quantity whose conservation comes from invariance of the Lagrangian under time translation.

Let us consider the change in the Hamiltonian when we vary the q^a's and the $\dot{q}^a$'s (or equivalently, the p_a's and the q^a's) at a fixed time:

$$\begin{aligned}\delta H &= \frac{\partial H}{\partial p_a}\delta p_a + \frac{\partial H}{\partial q^a}\delta q^a \\ &= \dot{q}^a \delta p_a + p_a \delta \dot{q}^a - \frac{\partial L}{\partial q^a}\delta q^a - \frac{\partial L}{\partial \dot{q}^a}\delta \dot{q}^a\end{aligned} \tag{4.10}$$

the sum on a always implied. This is just the Chain Rule for differentiation. The second and fourth terms cancel, and $\partial L/\partial q^a = \dot{p}_a$. Because we can vary the p_a's and q^a's independently, we can now read off Hamilton's equations,

$$\frac{\partial H}{\partial p_a} = \dot{q}^a \qquad \frac{\partial H}{\partial q^a} = -\dot{p}_a \tag{4.11}$$

I presume they are also familiar to you, and I shall not bother to give specific examples.

This is a standard derivation, but I should like to make a point that is sometimes not made in elementary texts. We will have to confront it several times, not in this lecture but in subsequent lectures. In order to go from the Lagrangian formulation to the Hamilton formulation there are certain conditions which the p_a's and q^a's must obey as functions of the $\dot{p}_a$'s and the $\dot{q}^a$'s. The p_a's and q^a's must be **complete** and **independent**. Tacitly I'm assuming that these functions, the q^a's and the $\dot{q}^a$'s, have two properties. I assume, first, that it is possible to write the Hamiltonian as a function of just the q^a's and the p_a's. Maybe that's not so. In most simple cases it *is* so, but it's very hard to prove in general that it is *always* so, because I can write examples where it's *not* so. So this is the condition which we will call *completeness*. If the set of the q^a's and the $\dot{q}^a$'s is *complete*, it is possible to express the q^a's and the $\dot{q}^a$'s as functions of the q^a's and the p_a's at least to such an extent that it is possible to write the Hamiltonian as just functions of the q^a's and the p_a's. By *independent* I mean that I can make small variations of the q^a's and the p_a's at any time by appropriately choosing the variations of the q^a's and the $\dot{q}^a$'s independently. If I couldn't make such small variations, if there were some constraint coming from the definition of the p_a's that kept me from varying them all independently, then I couldn't get from (4.10) to (4.11), because I couldn't vary them one at a time.

To give a specific example where the q^a's and the p_a's are complete but *not* independent, consider a particle of mass m constrained to move on the surface of a sphere of unit radius. If you know any classical mechanics at all, you know there are two ways of doing this problem. You have three dynamical variables, three components of the position vector $\mathbf{x}$ of the particle. You can of course go to some coordinates in which you have only two variables, such as spherical coordinates. Then you don't have any equation of constraint and off you go by the standard methods. Alternatively you can keep all three coordinates and write things in terms of Lagrange multipliers. That is to say you can write a Lagrangian,

$$L = \tfrac{1}{2}m\dot{\mathbf{x}}^2 + \lambda(\mathbf{x}^2 - 1)$$

by the method of Lagrange multipliers, which I hope you all know—if not, take five minutes to read the appropriate section in Chapter 2 of Goldstein.[4] If I stick this last term in a

[4] [Eds.] Goldstein *et al. CM*, Section 2.4, pp. 45–50.

Lagrangian I get precisely equivalent equations to the Lagrangian in two coordinates without the constraint. By varying with respect to λ, I obtain the equation of constraint, and by varying with respect to the three other variables I obtain the equations of motion with the force of constraint on the right-hand side. From the viewpoint of mechanics this constrained Lagrangian is just as good as the other. However it does not allow passage to a Hamiltonian form by the usual procedure: p_λ, the canonical momentum associated with the variable λ, happens to be *zero*. There is no $\dot{\lambda} = d\lambda/dt$ in the Lagrangian, and λ is not an independent variable. I cannot get the Hamilton equations of motion involving the three components of $\mathbf{x}$ and their conjugate momenta and λ and its conjugate momentum because I cannot vary with respect to p_λ, which is zero by definition; zero is not an independently variable quantity. The equation (4.10) is true, but the equation (4.11), which appears to be such an evident consequence of it, is false. Things break down because the generalized coordinates aren't independent. There is *no* method of Lagrange multipliers in the Hamiltonian formulation of mechanics.

This is just something to keep in the back of your mind because all of the examples we will do in this lecture—in fact, everything—will be complete and independent. But then in later lectures we'll get to things where they're not. And if you have a Lagrangian system in such a form where you do *not* have a bunch of independent variables, then you have to beat on it, in the same way as we beat on this example, by eliminating Lagrange multipliers until you get it into shape where you can go to Hamiltonian form. This completes for the moment my discussion of the first box, classical particle mechanics.

4.2 Quantum particle mechanics

We go now to the second box, quantum particle mechanics and canonical quantization. I'm going to explain the arrow leading from classical mechanics to quantum mechanics, and something about what lies at the end of the arrow. Of course it will not be everything in quantum mechanics; that's a course by itself!

Canonical quantization is a uniform procedure for obtaining a quantum mechanical system from a given classical mechanical system in Hamiltonian form, by turning a crank. It is certainly not the only way of getting quantum mechanical systems. For example, when you took quantum mechanics, you didn't take care of the theory of electron spin by starting out with the classical theory of spinning electrons and canonically quantizing it. However it is *a* way and it has certain advantages. I will first explain the prescription and then the ambiguities that inevitably plague canonical quantization. Finally I will explain its advantages.

The quantum mechanical system has a complete set of dynamical variables that are the q's and the p's of the classical system. I will abuse notation by using the same letters for the quantum variables as the classical variables, instead of writing them with capitals or writing them with a subscript "op" or something. The classical dynamical variables obey the canonical *Poisson brackets*[5]

$$\{q^a, q^b\} = 0 = \{p_a, p_b\}; \quad \{q^a, p_b\} = \delta^a_b \tag{4.12}$$

We replace these dynamical variables by time-dependent (Heisenberg picture) operator-valued functions, which obey these universal commutators, independent of the system:

$$[q^a(t), q^b(t)] = 0 = [p_a(t), p_b(t)]; \quad [q^a(t), p_b(t)] = i\delta^a_b \tag{4.13}$$

5 [Eds.] Goldstein *et al.* *CM*, Section 9.5, pp. 388–396.

(Traditionally there is a factor of $\hbar$ on the right-hand side of the last equation, but we're keeping $\hbar = 1$.) The commutators are trivial except for the (q, p) commutators, and for that matter the (q, p) commutators are also pretty trivial. We assume that the set of q^a and p_a are Hermitian (and hence observable), and complete.

The Hamiltonian of the quantum system is the same as the classical Hamiltonian, but now it is a function of the operators q^a and p_a;

$$H = H(p_1, p_2, \ldots, p_N, q^1, q^2, \ldots, q^N, t) \tag{4.14}$$

Please notice that the prescription for constructing the Hamiltonian is inherently ambiguous. It doesn't tell you what order you are to put the q^a's and p_a's in, when you write out the expression for H. In the classical expression it doesn't matter if you write $p^2q^2 + q^2p^2$ or if you write $2pq^2p$, but in the quantum theory it does make a difference. I choose this particular example because the ambiguity cannot be resolved just by saying a quantum Hamiltonian should be Hermitian. This is just an ambiguity that we have to live with. The prescription of replacing the classical p's and q's by their quantum counterparts does not define a unique theory except in especially simple cases. In general there is no way to resolve ordering ambiguities. If we write the commutator with traditional units,

$$q^a p_b = p_b q^a + i\hbar\delta^a_b \rightarrow p_b q^a + (\text{a negligible quantity in the classical limit}) \tag{4.15}$$

so there are no ordering ambiguities in the classical limit.

For this reason we always try and write our quantum systems in terms of the coordinates of our classical system before canonical quantization, so that the ordering ambiguity causes the least damage for particles moving in a potential. (We usually quantize the system directly in Cartesian coordinates. If we are then to do a transformation to spherical coordinates, we do that after we have quantized the system, after we have written down the Schrödinger equation.) Why do we do this? It is an ambiguous rule. Why on earth would any sane person or even an inspired madman have written down this particular rule rather than some others? Well, historically the only motivation for connecting a classical mechanical system with a quantum system was the Correspondence Principle, the statement that the quantum system in some sense should reproduce the classical system if, for some set of experiments concerning that system, classical physics gives a good description.

The operator that generates infinitesimal time evolutions of the quantum system is the classical Hamiltonian function of the quantum q^a and p_a operators. For any operator $A(t)$,

$$\frac{dA}{dt} = i[H, A] + \frac{\partial A}{\partial t} \tag{4.16}$$

the last term appearing only if the operator has an explicit time dependence in addition to the implicit time-dependence arising from the q^a's and p_a's. In particular, we can rederive the Heisenberg equations (4.11),

$$\begin{aligned} \frac{dq^a}{dt} &= i[H, q^a] = i\left(-i\frac{\partial H}{\partial p_a}\right) = \frac{\partial H}{\partial p_a} \\ \frac{dp_a}{dt} &= i[H, p_a] = i\left(i\frac{\partial H}{\partial q^a}\right) = -\frac{\partial H}{\partial q^a} \end{aligned} \tag{4.17}$$

Let me explain the second step. Because of the canonical commutation relations, taking the commutator of q^a or p_a with any function of the p's and q's amounts to differentiation with respect to the conjugate variable, times a factor of $\pm i$. For example, taking the commutator of a monomial such as $q^b p_c p_d q^e$ with q^a, we get

$$\begin{aligned}[][q^b p_c p_d q^e, q^a] &= q^b p_c [p_d, q^a] q^e + q^b [p_c, q^a] p_d q^e \\ &= -i\delta^a_d q^b p_c q^e - i\delta^a_c q^b p_d q^e = -i\frac{\partial}{\partial p_a}\left(q^b p_c p_d q^e\right)\end{aligned} \tag{4.18}$$

If there is a single p_a in the expression, I get a 1 from the commutator, if there is a p_a^2 I get $2p_a$'s, if there's a p_a^3 I get $3p_a^2$, etc. Thus the quantum mechanical definition of the Hamiltonian tells us that $\dot{q}^a$ equals $i[H, q^a]$, which is $\partial H/\partial p_a$. Since p_a is just another operator, likewise $\dot{p}_a$ is just the commutator $i[H, p_a]$ which because of the minus sign when I switch around the canonical commutator gives us $-\partial H/\partial q^a$. Canonical quantization is a prescription that guarantees that the Heisenberg equations of motion for the quantum mechanical system are identical in form with the Hamilton equations for the corresponding classical system. This is an expression of the Correspondence Principle.

Consider a state in which classical mechanics offers a good description, a state where at least for the duration of our experiment, $\langle q^n\rangle$, the expectation value of q^n, equals $\langle q\rangle^n$, the nth power of the expectation value of q, within our experimental accuracy—we don't know that q is statistically distributed—and likewise the expectation value $\langle p^n\rangle$ of the nth power of p is $\langle p\rangle^n$, the nth power of the expectation value of p. Then by taking the expectation value of the quantum equations of motion, we observe that they equal, via the mean values of the particle position and momentum, the classical equations of motion. Of course, if the state does not obey that classical condition, if it is not (within our experimental accuracy) a sharp wave packet in both p and q, then quantum mechanics gives different results from the classical physics.

This concludes my rather brief discussion of the arrow descending from the first box, classical mechanics to quantum mechanics, the second box. We have taken care of classical mechanics and quantum mechanics in one half hour. Well, of course there is a lot more to be said about these systems and we'll return to them occasionally to get clues to say some of those things. But that's the only part of them I will need for this lecture.

4.3 Classical field theory

Now we come to something that might be novel to some of you: the extension from classical particle mechanics to classical field theory. In general the only difference between classical particle mechanics and classical field theory is that in one case the variables are finite in number, and in the other case one has an infinite number of variables. The infinite number of the dynamical variables, say in Maxwell's electromagnetic theory, are labeled sometimes by a continuum index. Instead of worrying about the position of the first particle, the position of the second particle, the position of the nth particle, one worries about the value of the electromagnetic field at every spatial point and the value of the magnetic field at every spatial point.

That is to say instead of having $q^a(t)$ one has a set of fields $\phi^a(\mathbf{x}, t)$. I make no assumptions about their Lorentz transformation properties or even about the Lorentz invariance of the theory at this moment; I will shortly. These fields may be components of vectors or scalars or tensors or spinors or whatever; I don't care about that right now. The important thing to remember in the analogy is that in going from the first box to the third, it is not that t

is analogous to the quadruplets x^i and t, but that a, the index that labels the variables, is analogous to a and $\mathbf{x}$.

It is sometimes a handy mnemonic to think of $x = (t, \mathbf{x})$ as a generalization of t, but that is not the right way to think about it. For example we are used to giving initial value data at a fixed time t in classical particle mechanics. In classical field theory, we need initial value data not at some fixed time t and some $\mathbf{x}$, but at a fixed time t and *all* $\mathbf{x}$. That $\mathbf{x}$ is continuous is in fact irrelevant because if I wanted to—although I shan't—I could just as well trade these variables for their Fourier coefficients in terms of some orthonormal basis. And then I would have a discrete set, say harmonic oscillator wave functions, and I would have a discrete variable replacing $\mathbf{x}$. The big difference is that the index is *infinite* in range, not finite. I will stay with $\mathbf{x}$ because I presume you all know that in manipulating functions of variables it doesn't matter whether you use a discrete basis or a continuum basis to describe them, whether you use harmonic oscillator wave functions or delta functions. With a discrete basis you have a Kronecker delta and with a continuum basis you have a Dirac delta. Otherwise the rules are exactly the same.

In classical particle mechanics you have a bunch of dynamical variables $q^a(t)$ which evolve in time. In classical field theory you have a bunch of dynamical variables $\phi^a(\mathbf{x}, t)$ that evolve in time interacting with each other. In classical particle mechanics the individual dynamical variables are labeled by the discrete index a; in classical field theory the individual dynamical variables are labeled by both the discrete index a and the continuous index $\mathbf{x}$. We can summarize the correspondence like this:

$$\begin{aligned} q^a(t) &\leftrightarrow \phi^a(\mathbf{x}, t) \\ t &\leftrightarrow t \\ a &\leftrightarrow a, \mathbf{x} \end{aligned} \tag{4.19}$$

In general I have some Lagrangian that is determined by some complicated functions of the ϕ^a's at every spatial point and their time derivatives and I just *go*, carrying on with the system. However I will instantly make a simplification.

In the final analysis we are interested only in Lorentz invariant theories. If we have an action S that is the integral of something that is local in time, it seems that it should also be the integral of something that is local in space, because space and time are on the same footing in Lorentz transformations. Likewise since the integrand involves time derivatives of only the first order, it should only involve first order space derivatives. Therefore we'll instantly limit the general framework (which I have not even written down) to the special case in which the Lagrangian—the ordinary Lagrangian L in the sense of the first box—will be the integral over 3-space of something called a **Lagrangian density**, $\mathscr{L}$:

$$L = \int \mathscr{L}\, d^3\mathbf{x} \tag{4.20}$$

This is in general some function of $\phi^1, \ldots \phi^N$, some function of $\partial_\mu \phi^1, \ldots \partial_\mu \phi^N$ and possibly some function of the spacetime position x. We will indeed consider Lagrangians that depend explicitly on the position x when we consider systems subject to external forces.

This is a specialization. There are of course many non-Lorentz invariant theories that follow these criteria: first order in space and time derivatives, integral over $d^3\mathbf{x}$, and so on. Most of the theories that describe hydrodynamics, a continuum system, do. Most of the theories that describe elasticity are of this form. But there are many that do not. For example, if we consider the vibrations of an electrically charged crystal, we have to insert the Coulomb interaction between the different parts of the crystal which is not expressible as an integral of a single spatial density of the crystal variables; it's a double integral involving the Coulomb Green's function. But we will restrict our attention to this form.

When we write down an expression of this form—whenever we have an infinite number of degrees of freedom—we have to worry a lot about questions of convergence. I will of course behave in typical physicist slob fashion and avoid such worry simply by ignoring these questions. But it should be stipulated that this object, L, is well defined. It is tacitly assumed that all the ϕ's go to zero as $\mathbf{x}$ goes to infinity. We will only consider configurations of that sort. Otherwise the Lagrangian would be a divergent quantity, and everything we do would be evidently nonsense. So without saying more about it, I will establish a rule that we assume whenever possible, whenever necessary, that not only the ϕ's are sufficiently differentiable so that we can do all the derivatives we want to do, but also that they go to zero sufficiently rapidly so we can do all the integration by parts we want to do. I leave it to mathematicians to worry about how rapid "sufficiently" is.

We define the Lagrangian L as

$$L = \int d^3\mathbf{x}\,\mathscr{L}(\phi^a(x), \partial_\mu\phi^a(x), x) \tag{4.21}$$

and the action S as

$$S = \int_{t_1}^{t_2} dt\,L = \int d^4x\,\mathscr{L}(\phi^a(x), \partial_\mu\phi^a(x), x) \tag{4.22}$$

(but the time integration is limited). We can derive the Euler–Lagrange equations from this expression for the action.

It's useful now to treat all four coordinates as analogous to t. If the Lagrangian density is a Lorentz invariant, the Euler–Lagrange equations will be Lorentz covariant; Lorentz invariance is now manifest. Treating the four coordinates equally is a bad thing to do for Hamiltonian dynamics but a good thing to do for this particular problem; it will allow us to do all four (if necessary) integrations by parts in one fell swoop. So let's do that and derive the Euler–Lagrange equations:

$$0 = \delta S = \int d^4x\left(\frac{\partial\mathscr{L}}{\partial\phi^a}\delta\phi^a + \frac{\partial\mathscr{L}}{\partial(\partial_\mu\phi^a)}\delta(\partial_\mu\phi^a)\right) \tag{4.23}$$

Observe that $\delta(\partial_\mu\phi^a)$ equals $\partial_\mu(\delta\phi^a)$. I can now perform integration by parts. In the space derivative, the space boundary term vanishes by my assumption that everything goes to zero at spatial infinity. In the time derivative, the time boundary term vanishes not from that assumption but from the universal condition attached to Hamilton's Principle, that I only consider variations that are zero at the initial and final times; $\delta\phi^a(\mathbf{x}, t_1) = \delta\phi^a(\mathbf{x}, t_2) = 0$. Then

$$0 = \delta S = \int d^4x\left[\frac{\partial\mathscr{L}}{\partial\phi^a} - \partial_\mu\left(\frac{\partial\mathscr{L}}{\partial(\partial_\mu\phi^a)}\right)\right]\delta\phi^a \tag{4.24}$$

Following closely upon my development in particle mechanics I will simply define an entity called π^μ_a;

$$\pi^\mu_a \equiv \frac{\partial\mathscr{L}}{\partial(\partial_\mu\phi^a)} \tag{4.25}$$

Since $\delta\phi^a$ is an arbitrary function (aside from going to zero at spatial infinity and at the time boundaries) I deduce, just as in the particle mechanics case, the Euler–Lagrange equations of motion,

$$\boxed{\frac{\partial \mathscr{L}}{\partial \phi^a} - \partial_\mu \pi^\mu_a = 0} \tag{4.26}$$

These are the same Euler–Lagrange equations of motion derived from the same Hamilton's Principle as in the other case. All that we have changed is to have had an infinite number of variables, and to have specified that the Lagrangian depended on these variables in a rather restricted way. The quantities π^μ_a should not be thought of as a 4-vector generalization of p_a. The correspondence is actually

$$\pi^0_a(\mathbf{x}, t) \leftrightarrow p_a(t) \tag{4.27}$$

Now you may not be as familiar with these equations as with their particle mechanics analogues. So let me here pause from my general discussion to do a specific example. Once I do that specific example maybe there won't be as many questions about the general discussion as there would be if I asked for questions now.

EXAMPLE. *A Lagrangian density $\mathscr{L}$ for a single real scalar field*

I want to construct a simple example. Well, first, the simplest thing I can imagine is one real scalar field, $\phi(x) = \phi^*(x)$, instead of a whole bunch of fields. Secondly, simple here really means that the equations of motion are linear. That requires a Lagrangian density $\mathscr{L}$ quadratic in ϕ and $\partial_\mu\phi$, because the equations of motion come from differentiating the Lagrangian. I'll assume a quadratic Lagrangian so I'll get linear equations of motion. And, thirdly, since I want the equations of motion eventually to be Lorentz invariant I want $\mathscr{L}$ to be a Lorentz scalar. That looks like a good set of criteria for constructing a simple example. Here is the most general of the simple Lagrangians we can construct:

$$\mathscr{L} = \pm\tfrac{1}{2}\left(a\partial_\mu\phi\,\partial^\mu\phi + b\phi^2\right) \tag{4.28}$$

Of course this determines the example completely. I've put a one half in front for later simplifications. There is some unknown real coefficient a times $\partial_\mu\phi\,\partial^\mu\phi$. That's the only Lorentz invariant term I can make that's quadratic in $\partial_\mu\phi$. I can't make anything Lorentz invariant out of ϕ and $\partial_\mu\phi$. If I multiply them together I just get a vector. And finally I can have some other coefficient b times ϕ squared, where a and b are arbitrary numbers. Now I hate to work with more arbitrary coefficients than I need, so I will instantly make a simplification that comes from redefining ϕ;

$$\phi \to \phi' = \phi\sqrt{|a|} \tag{4.29}$$

If we rewrite the Lagrangian in terms of ϕ', the Lagrangian becomes

$$\mathscr{L} = \pm\tfrac{1}{2}\left(\partial_\mu\phi'\,\partial^\mu\phi' + (b/a)\phi'^2\right) \tag{4.30}$$

From now on I will drop the primes and just call this field ϕ. So in fact we have in this Lagrangian just two elements of arbitrariness, an arbitrary real number (b/a), and the discrete choice about whether we choose the $+$ sign or the $-$ sign. We'll later see that this discrete choice is determined by the requirement that the energy must be positive. That's sort of obvious because the Hamiltonian is linearly related to the Lagrangian. So if I take minus the Lagrangian I'll get minus the Hamiltonian. If it's positive in one case, it's going to be negative in the other. And if it is positive in no cases, if the energy cannot be bounded in either case, I wouldn't have looked at this example!

Now let's use our general machine. Defining

$$\pi^\mu = \frac{\partial \mathscr{L}}{\partial(\partial_\mu \phi)} = \pm \partial^\mu \phi \tag{4.31}$$

the Euler–Lagrange equations become, since $\partial\mathscr{L}/\partial\phi = \pm(b/a)\phi$,

$$\partial_\mu \pi^\mu = \pm(b/a)\phi \tag{4.32}$$

The one half is canceled by the fact that we're differentiating squares. Or, plugging in the definition of π^μ,

$$\partial_\mu \partial^\mu \phi = (b/a)\phi \tag{4.33}$$

which is rather similar to the Klein–Gordon equation that materialized in the latter part of last lecture. This of course is another reason why I chose this particular example.

Let us now go to the question of the Hamiltonian form. I'll postpone the Hamilton equations of motion for a while and just try and derive the Hamiltonian in its guise as the energy. The question is, what is the analog of p? Well, it's pretty obvious what the p is. You recall that one way of defining p was by a partial derivative. You could say

$$dL = p_a d\dot{q}^a + \cdots \tag{4.34}$$

the dots indicating the other term which contains no time derivative. That's the definition of p_a; it's the thing that multiplies $d\dot{q}_a$. Now going over to functionals, there's an unfortunate change in notation that really makes no sense: we use a wiggly delta, δ, instead of a straight d, but of course it's the same concept, the infinitesimal change of the dependent variable under the infinitesimal change of the independent variable;

$$\delta \int d^3\mathbf{x}\, \mathscr{L} = \int d^3\mathbf{x}\, \frac{\partial \mathscr{L}}{\partial(\dot{\phi}^a)}\, \delta\dot{\phi}^a(\mathbf{x},t) + \cdots = \int d^3\mathbf{x}\, \pi_a(\mathbf{x},t)\, \delta\dot{\phi}^a(\mathbf{x},t) + \cdots \tag{4.35}$$

the dots representing terms with no time derivatives. What their explicit forms are, I don't care. Some have gradients and some have nothing differentiated, but they don't have any time derivatives. Hence the thing that is the analog of p_a, in fact the thing that *is* p_a for an infinite number of degrees of freedom, is

$$\pi_a = \frac{\partial \mathscr{L}}{\partial(\dot{\phi}^a)} \tag{4.36}$$

which is the canonical momentum. This expression is also equal to our previous π^μ_a, (4.25), with μ set equal to zero;

$$\pi^0_a = \frac{\partial \mathscr{L}}{\partial(\partial_0 \phi^a)} = \pi_a \tag{4.37}$$

So it's the time component of π^μ_a which is the generalized version of the canonical momentum, sometimes called the *canonical momentum density*. Parallel to this equation

$$H = p_a \dot{q}^a - L \tag{4.38}$$

is this equation

$$H = \int d^3\mathbf{x} \left(\pi_a \dot{\phi}^a - \mathscr{L}\right) \tag{4.39}$$

indeed, they are the same equation. In the former all the summations are absorbed into the summation convention; in the latter half the summations are absorbed into the summation convention and the other half are written as integrals. The expression

$$\mathscr{H} = \pi_a \dot{\phi}^a - \mathscr{L} \tag{4.40}$$

is called the **Hamiltonian density**; it's the thing you have to integrate to get the Hamiltonian.

The fact that we obtain the Hamiltonian, the total energy in the time-independent case, as an integral over $\mathbf{x}$ at fixed time is of course not surprising. To find out how much energy there is in the world, you add up the energy in every little infinitesimal volume. Let's apply these formulas to our simple example, (4.30);

$$\mathscr{L} = \pm\tfrac{1}{2}\left[(\dot{\phi})^2 - |\nabla\phi|^2 + (b/a)\phi^2\right] \tag{4.41}$$

the minus sign coming from our metric. The canonical momentum density π is the zero component of $\pi^\mu = \partial\mathscr{L}/\partial(\partial_\mu\phi)$, so

$$\pi = \pm\dot{\phi} \tag{4.42}$$

i.e., $\partial_0\phi(\mathbf{x}, t)$. Therefore the Hamilton density $\mathscr{H}$ is

$$\begin{aligned} \mathscr{H} &= \left[\pi\dot{\phi} \mp \tfrac{1}{2}\partial_\mu\phi\,\partial^\mu\phi \mp \tfrac{1}{2}(b/a)\phi^2\right] \\ &= \pm\left[\tfrac{1}{2}\pi^2 + \tfrac{1}{2}|\nabla\phi|^2 - \tfrac{1}{2}(b/a)\phi^2\right] \end{aligned} \tag{4.43}$$

We choose the + sign, to ensure that the π^2 cannot become arbitrarily large and negative; we want the energy to be bounded below. And if we don't want the ϕ^2 term to become arbitrarily large and negative, we had better choose (b/a) to be less than zero, a fact that I will express by writing (b/a) as minus the square of a real number, μ; $(b/a) = -\mu^2$. Thus our equations now have only one unknown quantity in them, the positive number μ^2, if we're to have positive energies. Here is what we have in that case:

$$\mathscr{L} = \tfrac{1}{2}\left(\partial_\mu\phi\,\partial^\mu\phi - \mu^2\phi^2\right) \tag{4.44}$$

$$\mathscr{H} = \tfrac{1}{2}\left(\pi^2 + |\nabla\phi|^2 + \mu^2\phi^2\right) \tag{4.45}$$

The equations of motion become

$$\Box^2\phi + \mu^2\phi = 0 \tag{4.46}$$

which is just the Klein–Gordon equation. Note that the Hamiltonian is the sum of three positive terms.

We could now go on and write down the classical Hamilton equations of motion in the general case and then proceed to canonical quantization. However time is running on and I will do things in one fell swoop. I will describe canonical quantization immediately. After all, this classical field is just the same as the classical particle system, except that a runs over an infinite range symbolized by the two variables a and $\mathbf{x}$. So that part about the Correspondence Principle in the whole song and dance I gave about going from classical mechanics to quantum mechanics should still be true. Therefore, I will now describe the "missing box": quantum field theory.

4.4 Quantum field theory

We simply write down the corresponding canonical commutators for the quantum field, just as we did to go from classical mechanics to quantum mechanics:

$$\boxed{\begin{aligned}&[\phi^a(\mathbf{x},t),\phi^b(\mathbf{y},t)] = [\pi_a(\mathbf{x},t),\pi_b(\mathbf{y},t)] = 0\\ &[\phi^a(\mathbf{x},t),\pi_b(\mathbf{y},t)] = i\delta^a_b\delta^{(3)}(\mathbf{x}-\mathbf{y})\end{aligned}} \tag{4.47}$$

We know that we should have $[q,p] = i\delta$. Which delta? Well, for discrete indices, a Kronecker delta; for continuous indices, a Dirac delta. The quantum Hamiltonian H is the integral

$$H = \int d^3\mathbf{x}\,\mathscr{H}$$

where $\mathscr{H}$ is a function of $\phi^1, \phi^2, \ldots$ (and their spatial derivatives); $\pi_1, \pi_2, \ldots$, and possibly also explicitly of $\mathbf{x}$ and t, though not in our simple example. But we might consider systems with external forces.

The set (4.47) is essentially the same set (4.13) we wrote down to find quantum particle mechanics. It's not even a generalization; the only generalization is to an infinite number of degrees of freedom. Since I never worried about whether my sums on a were infinite or finite in all my formal manipulations, I don't have to go through the computations again. They are the same computations. The only change is notational. For continuous indices we write a sum as an integral, but every operation is the same once you learn that transcription rule. The advantage of this procedure is that it reproduces the classical field theory in the limit where classical mechanics is supposed to be valid. There's just a lot more p's and q's. Otherwise there is no difference.

Let us check this with our specific example by explicitly deriving the Heisenberg equations of motion and seeing that they give us the Euler–Lagrange equations. I won't bother to write down the equal time commutators for our specific example because they are these equations (4.47) with the a's and b's erased, because there is only one ϕ and there is only one π. Okay? So let's do it with the example.

$$H = \tfrac{1}{2}\int d^3\mathbf{x}\left(\pi(\mathbf{x},t)^2 + \left|\nabla\phi(\mathbf{x},t)\right|^2 + \mu^2\phi(\mathbf{x},t)^2\right) \tag{4.48}$$

There is a universal rule (4.16) for computing the time derivative of any operator. We used that rule to compute the Heisenberg equations of motion in the particle case. I will now use this rule to compute them for π and ϕ, just as we computed them for p and q.

I'll start out with ϕ because that's easier. I will do this in tedious detail to pay my dues so that every subsequent such calculation I can do with lightning-like rapidity. The only thing in the Hamiltonian that ϕ does not commute with is π. The rule says

$$\dot{\phi}(\mathbf{x},t) = i[H,\phi(\mathbf{x},t)] = i\int d^3\mathbf{y}\,\tfrac{1}{2}[\pi(\mathbf{y},t)^2,\phi(\mathbf{x},t)] = i\int d^3\mathbf{y}\,\pi(\mathbf{y},t)(-i\delta^3(\mathbf{y}-\mathbf{x})) = \pi(\mathbf{x},t) \tag{4.49}$$

just using the rule $[a,b^2] = b[a,b] + [a,b]b$. This equation should be no surprise to you. It is one of the two Hamilton equations;

$$\dot{\phi}(\mathbf{x},t) = \pi(\mathbf{x},t) \tag{4.50}$$

Secondly I will compute $\dot{\pi}(\mathbf{x},t)$ by the same universal Heisenberg equation of motion, $\dot{\pi}(\mathbf{x},t) = i[H,\pi]$. Now there are two terms with which π does not commute: the gradient term and the ϕ^2 term. Let's write things out.

$$\dot{\pi}(\mathbf{x},t) = i[H,\pi(\mathbf{x},t)] = i\int d^3\mathbf{y}\,\tfrac{1}{2}\left\{\left[|\nabla\phi(\mathbf{y},t)|^2,\pi(\mathbf{x},t)\right] + \mu^2\left[\phi(\mathbf{y},t)^2,\pi(\mathbf{x},t)\right]\right\} \tag{4.51}$$

We have a factor of -1 different from the previous equation, since we are now reversing the order of the commutator of π with ϕ. The $\frac{1}{2}$ is again canceled because we're always commuting with squares. We get

$$\begin{aligned} i[H,\pi(\mathbf{x},t)] &= i\int d^3\mathbf{y}\,\left\{\nabla\phi(\mathbf{y},t)\boldsymbol{\cdot}\nabla(i\delta^3(\mathbf{y}-\mathbf{x})) + \mu^2\phi(\mathbf{y},t)(i\delta^3(\mathbf{y}-\mathbf{x}))\right\} \\ &= \nabla^2\phi(\mathbf{x},t) - \mu^2\phi(\mathbf{x},t) \end{aligned} \tag{4.52}$$

I have used the fact that the commutator of π with $\nabla\phi$ is proportional to the gradient of the delta function, which follows from differentiating the commutator with respect to $\mathbf{y}$. The integral is also trivial, though not quite so trivial as before, because we have to do an integration by parts. But it is one I think we can do by eye. This expression should be $\dot{\pi}(\mathbf{x},t)$. Plugging in from (4.50) to eliminate π and write a differential equation in terms of ϕ we obtain

$$\ddot{\phi}(\mathbf{x},t) = \nabla^2\phi(\mathbf{x},t) - \mu^2\phi(\mathbf{x},t) \tag{4.53}$$

which is of course the classical equation of motion, the Klein–Gordon equation.

Thus we have checked, in our specific example, the consistency of the procedure, and shown that the Heisenberg equations of motion yield the classical Euler–Lagrange equations of motion, at least up to ordering ambiguities which are rather trivial for linear equations of motion.[6]

Now we have obtained the Heisenberg equations of motion, the Klein–Gordon equation and the equal time commutators for our free scalar field in two different ways. These two methods define the same system. As I said, from here on in I could go through everything I did in the first three lectures running backwards and show that the system defines an assembly of free, spinless Bose particles, Fock space, the whole routine. One way occupied the first three lectures and the other took only one lecture. Actually if I had started out this way I would have had to run over a lot of the material in the first three lectures in the opposite order so it might have taken me two and a half lectures rather than one.

In any event we have two methods. One method is full of physical insight, I hope. I tried to put as much physical insight into it as I could. We built the many-particle space out of the one-particle space. We knew why we wanted to look for a field. It wasn't because Heisenberg told us we had to look for a field. We had some physical reasons for it. We constructed the field, we found it was unique under certain simplifying assumptions, we deduced its properties and then we showed everything was characterized in terms of the field. The other method is completely free of physical insight. We have this mechanical device like a pasta machine: the canonical quantization procedure. You feed in the dough at one end, you feed in the classical

---

[6] By the way much of the material in this lecture is covered in Chapters 11 and 12 of Bjorken and Drell, the first two chapters of volume II, in a somewhat different way so you might want to look at that. You don't need to look at it but you might want to.

theory, and the rigatoni, the quantum theory, comes out at the other. It's totally mechanical. When you're done you have a set of equations that you hope characterizes the system but you've got a lot of work to do to find their physical interpretation.

Well, since I've characterized these two methods praising the first so much and being so pejorative about the second, you should not be surprised when I tell you that in the remainder of the course we will use the second method almost exclusively. The reason is very simple. The first method we could go through because we already understood everything. It was just a system of free particles in a box or on an infinite space. We already had access to a complete solution to the physics; we already knew the whole spectrum of the theory. If we had tried to apply the first method to an interacting system we wouldn't be able to get off the ground, because we would have to know in advance the exact spectrum of the theory. Here if we want to introduce interactions in the canonical method, at least formally, we just write 'em down. For example, here's an interaction:

$$\lambda\phi^4(\mathbf{x},t) \tag{4.54}$$

We have a free theory, $\mathscr{L}(\phi,\partial^\mu\phi)$, equation (4.44), and I'll throw in this interaction. Better give it a minus sign so the classical energy at least will be positive:

$$\mathscr{L}\to\mathscr{L}-\lambda\phi^4(\mathbf{x},t) \tag{4.55}$$

There it is! There is an interaction between the system's fields, okay? We could do canonical quantization at least formally, if there are no problems with summing over infinite numbers of variables (and in fact we'll see there are, but that particular nightmare lies far in our future). We get a theory that looks like it has a nice energy bounded below, it looks Lorentz invariant, everything commutes for spacelike separations because they commute for equal times, and the whole thing is Lorentz invariant. So it's got all the general features we want it to have. And it looks like particles can scatter off of each other because if we do old-fashioned Born perturbation theory, the expansion of the interaction term will involve two annihilation operators and two creation operators. At the first order in perturbation theory, you can go from one two-particle state to another two-particle state, two into two scattering. At the second order, we'll get two-particle states into four-particle states and into six-particle states: pair production! So there it is! We may not know what it *means*, but at least it's a cheap way of constructing an interacting field theory that obeys all of our general assumptions. Of course this means there's a lot of work to be done. Why did I write down this interaction with a power 4 and not the power $\frac{5}{6}$? Well, you'll learn why I didn't write down $\frac{5}{6}$; there's a reason for it. But you won't learn that till later on.[7] But at least we wound up with some equations to play with that don't look as if they have any evident inconsistencies with the general principles of relativity and causality. So we can begin investigating the properties of such theories. It is just such an investigation that will occupy the next several lectures or indeed, essentially the remainder of the first term of the course.

4.5 Normal ordering

I have one more thing I want to say about the free field. Let's do another consistency check for our system. Since we have ϕ's and $\dot{\phi}$'s $=\pi$'s that obey the canonical commutators and obey the Klein–Gordon equation we can, as sketched out in the last lecture, express the field in

[7] [Eds.] See §16.4.

terms of annihilation and creation operators. Just as a consistency check, let us take such an expression, plug it into this expression (4.45) for the Hamiltonian density and see if we get the same thing, equation (2.48), for the energy as a function of annihilation and creation operators as we found before, for the Fock space of spinless particles. Here's the Hamiltonian again,

$$H = \tfrac{1}{2}\int d^3\mathbf{x}\,\left(\pi(\mathbf{x},t)^2 + \left|\nabla\phi(\mathbf{x},t)\right|^2 + \mu^2\phi(\mathbf{x},t)^2\right) \tag{4.56}$$

and let's write $\phi(\mathbf{x},t)$ once again in terms of its Fourier expansion, equation (3.45), separating out the space and time parts,

$$\phi(\mathbf{x},t) = \int \frac{d^3\mathbf{p}}{(2\pi)^{3/2}\sqrt{2\omega_\mathbf{p}}}\left(a_\mathbf{p}e^{-i\omega_\mathbf{p}t+i\mathbf{p}\cdot\mathbf{x}} + a_\mathbf{p}^\dagger e^{i\omega_\mathbf{p}t-i\mathbf{p}\cdot\mathbf{x}}\right)$$

This defines the operators $a_\mathbf{p}$ and $a_\mathbf{p}^\dagger$. Our game is to plug this expression into the Hamiltonian (recalling that $\pi = \dot{\phi}$), do the space integral, and see if we get a familiar result. This will lead to a triple integral, but we can do some of the integrations in very short order. Look, for example, at the first term only,

$$\begin{aligned}\tfrac{1}{2}\int d^3\mathbf{x}\,\pi(\mathbf{x},t)^2 = \tfrac{1}{2}\iiint \frac{d^3\mathbf{x}\,d^3\mathbf{p}\,d^3\mathbf{p}'}{2(2\pi)^3\sqrt{\omega_\mathbf{p}\omega_{\mathbf{p}'}}}&\left(-i\omega_\mathbf{p}a_\mathbf{p}e^{-i\omega_\mathbf{p}t+i\mathbf{p}\cdot\mathbf{x}} + i\omega_\mathbf{p}a_\mathbf{p}^\dagger e^{i\omega_\mathbf{p}t-i\mathbf{p}\cdot\mathbf{x}}\right)\times\\ &\left(-i\omega_{\mathbf{p}'}a_{\mathbf{p}'}e^{-i\omega_{\mathbf{p}'}t+i\mathbf{p}'\cdot\mathbf{x}} + i\omega_{\mathbf{p}'}a_{\mathbf{p}'}^\dagger e^{i\omega_{\mathbf{p}'}t-i\mathbf{p}'\cdot\mathbf{x}}\right)\end{aligned} \tag{4.57}$$

We'll get four terms in multiplying out the a's and $a^\dagger$'s, all involving exponentials like $e^{\pm i\mathbf{x}\cdot(\mathbf{p}\pm\mathbf{p}')}$. The space integral is done easily, producing a delta function in momentum,[8] which allows us to do the integral over $\mathbf{p}'$ quickly,

$$\begin{aligned}\tfrac{1}{2}\int d^3\mathbf{x}\,\pi(\mathbf{x},t)^2 = \tfrac{1}{2}\int \frac{d^3\mathbf{p}}{2\omega_\mathbf{p}}\Big[&-\omega_\mathbf{p}^2\left(a_\mathbf{p}a_{-\mathbf{p}}e^{-2i\omega_\mathbf{p}t} + a_\mathbf{p}^\dagger a_{-\mathbf{p}}^\dagger e^{2i\omega_\mathbf{p}t}\right)\\ &+\omega_\mathbf{p}^2\left(a_\mathbf{p}a_\mathbf{p}^\dagger + a_\mathbf{p}^\dagger a_\mathbf{p}\right)\Big]\end{aligned} \tag{4.58}$$

because $\omega_\mathbf{p} = \omega_{-\mathbf{p}}$. The other two terms in the Hamiltonian can now be done by eye,

$$\tfrac{1}{2}\int d^3\mathbf{x}\,\left|\nabla\phi(\mathbf{x},t)\right|^2 = \tfrac{1}{2}\int \frac{d^3\mathbf{p}}{2\omega_\mathbf{p}}|\mathbf{p}|^2\left(a_\mathbf{p}a_{-\mathbf{p}}e^{-2i\omega_\mathbf{p}t} + a_\mathbf{p}^\dagger a_{-\mathbf{p}}^\dagger e^{2i\omega_\mathbf{p}t} + a_\mathbf{p}a_\mathbf{p}^\dagger + a_\mathbf{p}^\dagger a_\mathbf{p}\right) \tag{4.59}$$

$$\tfrac{1}{2}\int d^3\mathbf{x}\,\mu^2\phi(\mathbf{x},t)^2 = \tfrac{1}{2}\int \frac{d^3\mathbf{p}}{2\omega_\mathbf{p}}\mu^2\left(a_\mathbf{p}a_{-\mathbf{p}}e^{-2i\omega_\mathbf{p}t} + a_\mathbf{p}^\dagger a_{-\mathbf{p}}^\dagger e^{2i\omega_\mathbf{p}t} + a_\mathbf{p}a_\mathbf{p}^\dagger + a_\mathbf{p}^\dagger a_\mathbf{p}\right) \tag{4.60}$$

What will I get for the Hamiltonian? I will now do this in one fell swoop having so well organized my computation:

$$\begin{aligned}H = \tfrac{1}{2}\int \frac{d^3\mathbf{p}}{2\omega_\mathbf{p}}\Big[&\left(a_\mathbf{p}a_{-\mathbf{p}}e^{-2i\omega_\mathbf{p}t} + a_\mathbf{p}^\dagger a_{-\mathbf{p}}^\dagger e^{2i\omega_\mathbf{p}t}\right)\\ &\times\underbrace{\left(-\omega_\mathbf{p}^2 + |\mathbf{p}|^2 + \mu^2\right)}_{=0} + \left(a_\mathbf{p}a_\mathbf{p}^\dagger + a_\mathbf{p}^\dagger a_\mathbf{p}\right)\underbrace{\left(\omega_\mathbf{p}^2 + |\mathbf{p}|^2 + \mu^2\right)}_{=2\omega_\mathbf{p}^2}\Big]\end{aligned} \tag{4.61}$$

---

[8] [Eds.] $\int d^3\mathbf{x}\,e^{\pm i\mathbf{p}\cdot\mathbf{x}} = (2\pi)^3\delta^{(3)}(\mathbf{p})$

We observe that there is a certain simplification here. For example this first term is zero, because the factor $(-\omega_{\mathbf{p}}^2+|\mathbf{p}|^2+\mu^2)$ is zero. Of course we could've checked that out on *a priori* grounds. We know the equations of motion should tell us the Hamiltonian is independent of the time. If it is independent of the time it is not going to have any factors like these time dependent exponentials. The second term has this other factor, $(\omega_{\mathbf{p}}^2+|\mathbf{p}|^2+\mu^2)$. It doesn't simplify so drastically but it still simplifies to $2\omega_{\mathbf{p}}^2$. Therefore, we have

$$H=\tfrac{1}{2}\int d^3\mathbf{p}\,\left(a_{\mathbf{p}}a_{\mathbf{p}}^{\dagger}+a_{\mathbf{p}}^{\dagger}a_{\mathbf{p}}\right)\omega_{\mathbf{p}} \tag{4.62}$$

This is almost but not quite what we expected, (2.48):

$$H=\int d^3\mathbf{p}\,\left(a_{\mathbf{p}}^{\dagger}a_{\mathbf{p}}\right)\omega_{\mathbf{p}} \tag{4.63}$$

The expression (4.62) differs from what we wanted by a constant ... and, surprise, that constant is *infinite*. Because (4.62) is of course

$$H=\int d^3\mathbf{p}\,\left(a_{\mathbf{p}}^{\dagger}a_{\mathbf{p}}+\tfrac{1}{2}[a_{\mathbf{p}},a_{\mathbf{p}}^{\dagger}]\right)\omega_{\mathbf{p}}=\int d^3\mathbf{p}\,\left(a_{\mathbf{p}}^{\dagger}a_{\mathbf{p}}+\tfrac{1}{2}\delta^{(3)}(0)\right)\omega_{\mathbf{p}} \tag{4.64}$$

The result of commuting $[a_{\mathbf{p}},a_{\mathbf{p}}^{\dagger}]$ gives $\delta^{(3)}(\mathbf{p}-\mathbf{p})=\delta^3(0)$. It's only the first term we want. We don't like that second term.

Now what can we say about this aside from making expressions of disgust? This infinity is no big deal for two reasons. First, you can't measure absolute energies, only differences, so it's stupid to worry about what the zero point energy is. This occurs even in elementary physics. We usually put interaction energies equal to zero when particles are infinitely far apart, but for some potentials you can't do that, and you have to choose the zero of energy somewhere else. There was some fast talking you let me get away with at the end of last lecture, probably because you were tired. I said: "We've got the equal time commutators of the Hamiltonian with the canonical variables, the equations of motion. Because these tell you the commutators of the annihilation and creation operators with the Hamiltonian, they determine everything except for the zero point of the energy, which we don't care about." Well, that's still true. They have determined everything except for the zero point of the energy. And if we still want to say we don't care about it we can say "infinite, schminfinite"; it's just a constant, so I can drop it. I can always put the zero of the energy wherever I want.

In general relativity, the absolute value of the energy density *does* matter. Einstein's equations,

$$R_{\mu\nu}-\tfrac{1}{2}g_{\mu\nu}R=-8\pi GT_{\mu\nu} \tag{4.65}$$

couple directly to the energy density T_{00}. Indeed, introducing a change in the vacuum energy density, in a covariant way like this

$$T_{\mu\nu}\to T_{\mu\nu}-\Lambda g_{\mu\nu} \tag{4.66}$$

is just a way of changing the cosmological constant Λ, a term introduced by Einstein and repudiated by him ten years later. No astronomer has ever observed a non-zero cosmological constant. We won't talk about why the cosmological constant is zero in this course. They don't explain it in any course given at Harvard because nobody knows why it is zero.[9]

Secondly, we can see physically why the second term comes in if we think of the analogy between this system and a harmonic oscillator. We have an infinite assembly of harmonic oscillators here but we wrote things just as if the individual Hamiltonians were $p^2 + q^2$; we haven't got the extra term of -1 as in (2.16). Therefore, we get the zero point energies in the expression for the individual oscillators. And since there is an infinite number of oscillators we get a summed infinite zero point energy. It's doubly infinite: infinite because of $\delta^{(3)}(0)$ and infinite because $\int d^3\mathbf{p}\,\omega_{\mathbf{p}}$ is infinite. Generally there are two types of infinities: *infrared infinities*, which disappear if we put the world in a box (the $\delta^{(3)}(0)$ would be replaced by the volume of the box); and *ultraviolet infinities*, due to arbitrarily high frequencies. The bad term here has both types of infinities.

An alternative way of saying the same thing is that canonical quantization gives you the right answers up to ordering ambiguities, and the only problem here is the order. I will use my freedom to get rid of ordering ambiguities by defining those terms ordered in another way. This idea, although it sounds silly and brings universal ridicule, is in fact a profitable way to proceed. I will therefore define an unconventional way of ordering expressions made only out of free fields which I will call **normal ordering.** I'll write down that definition and then I'll show you that normal ordering defines the right ordering. By the way the most significant feature of this calculation is I'm being very cavalier about the treatment of infinite quantities. And if you think it's bad in this lecture, just wait!

Let $\{\phi^{a_1}(x_1), \dots, \phi^{a_n}(x_n)\}$ be a set of free scalar fields. There may be a whole bunch of them with different masses and so on. The **normal-ordered product** of the fields, indicated by colons on either side,

$$:\phi^{a_1}(x_1)\phi^{a_2}(x_2)\cdots\phi^{a_i}(x_i): \tag{4.67}$$

means that this is not to be interpreted as the ordinary product, but instead is the expression reordered with all annihilation operators on the right and *a fortiori* all creation operators on the left.

That is the definition of normal ordering, of this normal ordered product of a string of free fields. I don't have to tell you the order of the annihilation operators because they all commute with each other. Just break every field up into its annihilation and creation parts, and you shove all the annihilation parts on the right. If the expression involves a sum of products, each of those terms is redefined by sticking all the annihilation operators on the right. This seems like a dumb definition. Nevertheless, take my word for it, this concept will be very useful to us in the sequel. This enables us to write down the proper formula for the

---

[9] [Eds.] Applied to the universe as a whole, Einstein's equations imply that its size is not static. Einstein found this conclusion unacceptable, and introduced Λ to keep the size fixed. Edwin Hubble's discovery in 1929, establishing the universe's expansion, apparently removed the need for Λ. In his posthumously published autobiography, George Gamow wrote "Much later, when I was discussing cosmological problems with Einstein, he remarked that the introduction of the cosmological constant was the biggest blunder of his life." (G. Gamow, *My World Line*, Viking, 1970, p. 44.) Gamow's account seems to be the only record of Einstein's repudiation of Λ. But things are not so simple. In 1998, two teams measuring supernova distances discovered that the expansion of the universe is accelerating, consistent with $\Lambda > 0$. For this discovery Saul Perlmutter, Adam Riess, and Brian P. Schmidt were awarded the Nobel Prize in 2011. The observational value on Λ is, in "natural" units where $G = h = c = 1$, on the order of $(10^{-3}\text{ eV})^4$: A. Zee, *Einstein's Gravity in a Nutshell*, Princeton U. P. 2008, p. 359; *PDG* 2016, p. 349 quotes a value for $\rho_\Lambda = (2.3 \times 10^{-3}\text{ eV})^4$. (In natural units, $1\text{ eV} = 1.76 \times 10^{-36}$ kg, $1\text{ eV}^{-1} = 1.97 \times 10^{-7}$ m, and in conventional units, $\Lambda = (8\pi G/c^4)\rho_\Lambda \sim 10^{-69}\text{ s}^2/\text{m}^4 \sim 10^{-52}\text{ m}^{-2}$.)

Hamiltonian in terms of local fields:

$$H = \tfrac{1}{2}\int d^3\mathbf{x}\,\left(:\pi^2 + |\nabla\phi|^2 + \mu^2\phi^2:\right) \tag{4.68}$$

That just tells us that whenever we run across the product of an a and an $a^\dagger$ we put the a on the right and therefore the adjoint $a^\dagger$ on the left. What could be simpler? To advance this elaborate definition just to take care of what I said in words five minutes ago may seem extremely silly to you, but we will use the normal ordered product again and again in this course. This is the first occasion we have had to use it and so I introduced it here. The name is a little bit bad because "normal order product" causes some students to get confused and weak in the head. They think you start out with the ordinary product and then you apply an operation to it called normal ordering. That is not so. This whole symbol, the string of operators and the colons, define something just as AB defines the product of the operator A and the operator B. In particular, "normal order" should not be interpreted as a verb, because it leads to contradictions. Suppose, for example, you attempted to normal order an equation, like this:

$$a^\dagger a = aa^\dagger - 1 \overset{?}{\rightarrow} :a^\dagger a: \,=\, :aa^\dagger: -1 \;\Rightarrow\; a^\dagger a = a^\dagger a - 1 \;\Rightarrow\; 0 = -1 \tag{4.69}$$

We don't "normal order" equations. Normal ordering is not derived from the ordinary product any more than the cross product is derived from the dot product.

The divergent zero-point energy is the first infinity encountered in this course. We'll encounter more ferocious infinities later on. We ran into this one because we asked a dumb question, a physically uninteresting question, about an unobservable quantity. Later on we'll have to think harder about what we've done wrong to get rid of troublesome infinities.

This concludes what I wish to say about canonical quantization of the free scalar field. If we wanted to get as quickly as possible to applications of quantum field theory, we'd develop scattering theory and perturbation theory next. But first we are going to get some more exact results from field theory.

Next lecture we'll go through the connection between symmetries and conservation laws. We'll talk about energy and momentum and angular momentum and the friends of angular momentum that come when you have Lorentz invariance. We'll talk about parity and time reversal, all for scalar fields. We'll talk about internal symmetries like isospin, found in systems of π mesons (which are scalar particles, and so within our domain). And we'll talk about the discrete internal symmetries like charge conjugation and so on, all on the level of formal classical field theory made quantum by canonical quantization.

5

Symmetries and conservation laws I. Spacetime symmetries

Last lecture we discussed canonical quantization and how it established correspondences between classical field theories and quantum field theories. We also talked about how those correspondences had to be taken *cum grano salis* because they included ordering ambiguities. At the last moment we had to check to make sure that we could order things in such a way that everything went through all right.

Today I would like to begin a sequence of lectures that will exploit that correspondence by studying the connection in classical physics between symmetries and conservation laws, and extending that to quantum physics by the canonical quantization procedure. We will thus obtain explicit expressions for objects like the momentum or the angular momentum, *et cetera*, in field theory, even including for example interactions like $\lambda\phi^4$. Of course, these expressions we find will also have to be taken with a grain of salt. We always have to check that we can make sense out of them by appropriately ordering things and we will do that check first. We will begin with typical cases for the free field theory.

Having cleared my conscience by telling you that nothing is to be trusted, I will now conduct the entire lecture as if everything can be trusted, without worrying about fine points.

5.1 Symmetries and conservation laws in classical particle mechanics

As always I will begin with classical particle mechanics and consider a general Lagrangian involving a set of dynamical variables and their time derivatives, and perhaps explicitly the time,

$$L(q^1, \cdots q^n, \dot{q}^1, \cdots, \dot{q}^n, t)$$

I would like to consider some one-parameter family of continuous transformations on these dynamical variables. I will assume for every real number λ I have defined some transformation

$$\lambda : q^a(t) \to q^a(t;\lambda) \tag{5.1}$$

that turns the old motion of the system into some new motion parameterized by the number λ. I will always assume we have chosen the zero of λ such that $q^a(t,0) = q^a(t)$.

As a specific example let's consider a transformation for a particular class of systems, an assembly of point particles, say. I'll give them different masses. The Lagrangian is

$$L = \tfrac{1}{2}\sum_r m_r \dot{\mathbf{x}}^r \cdot \dot{\mathbf{x}}^r + \sum_{r>s} V^{(r,s)}(|\mathbf{x}^r - \mathbf{x}^s|) \tag{5.2}$$

That's the conventional kinetic energy, plus some potential energy $V^{(r,s)}$ depending only on the differences between the positions $\mathbf{x}^r$ and $\mathbf{x}^s$ of the rth and s^{th} particles, respectively. The sort of transformation I want to consider for this system is a spatial translation along some particular direction, to wit, the transformation

$$\mathbf{x}^r \to \mathbf{x}^r + \mathbf{e}\lambda \tag{5.3}$$

where $\mathbf{e}$ is some fixed vector. I translate all the particles by an amount λ along the direction $\mathbf{e}$. Other examples of one-parameter families of transformations which we frequently find it profitable to consider in classical mechanics are time translations, rotations about a fixed axis and Lorentz transformations in a fixed direction. We will talk about all of these, and others, in the course of time.

Now we return to the general case. It will be convenient to study infinitesimal transformations,

$$q^a \to q^a + (Dq^a)\, d\lambda \tag{5.4}$$

q^a goes into q^a plus an object I will call Dq^a times $d\lambda$, the infinitesimal change in the parameter λ, where

$$Dq^a \equiv \left.\frac{\partial q^a}{\partial \lambda}\right|_{\lambda=0} \tag{5.5}$$

If I know how q^a transforms I know how $\dot{q}^a$ transforms, since it is just the time derivative of q^a. Thus $D\dot{q}^a$, the infinitesimal change of $\dot{q}^a$, defined in the same way, is d/dt of Dq^a, as we see just by differentiating (5.4) with respect to t; λ is a constant and t-independent. We also know how the Lagrangian transforms:

$$DL = \frac{\partial L}{\partial q^a} Dq^a + \frac{\partial L}{\partial \dot{q}^a} D\dot{q}^a = \frac{\partial L}{\partial q^a} Dq^a + p_a \frac{dDq^a}{dt} \tag{5.6}$$

We will always call the expression $\partial L/\partial \dot{q}^a$ the canonical momentum, p_a. Similarly we know how any function of q^a's and $\dot{q}^a$'s transforms under either the finite or the infinitesimal version of the transformation.

Definition. We will call a transformation a **symmetry** if and only if

$$DL = \frac{dF}{dt} \tag{5.7}$$

for some function $F(q^a, \dot{q}^a, t)$. This equality must hold for arbitrary functions $q^a(t)$, which need not satisfy the equations of motion.

Most transformations are *not* symmetries. Why do I adopt such a peculiar definition? Well, our intuitive idea of a symmetry is that a symmetry is a transformation that does not affect the dynamics. When we say a theory is invariant under, say, a space translation, we mean if we take a motion picture of the system, and if we then space translate the initial conditions, we get the space translated motion picture. Certainly this would be true if the

Lagrangian were unchanged by this transformation. But it could also be true if the change DL in the Lagrangian were of the form dF/dt, because a change of this form simply adds a boundary term to the action integral. And as we saw in our derivation of the Euler–Lagrange equations we can add boundary terms to the action integral at will without affecting the form of the Euler–Lagrange equations. Explicitly,[1]

$$\mathcal{S}' = \int_{t_1}^{t_2} L'(q^{a\prime}, \dot{q}^{a\prime}, t)\, dt = \int_{t_1}^{t_2} dt \left[L(q^a, \dot{q}^a, t) + \frac{dF}{dt} \right] = \mathcal{S} + F(q_2^a, \dot{q}_2^a, t_2) - F(q_1^a, \dot{q}_1^a, t_1) \tag{5.8}$$

Since $\mathcal{S}'$ and $\mathcal{S}$ differ only by a quantity which equals zero on variation, the conditions $\delta\mathcal{S}' = 0$ and $\delta\mathcal{S} = 0$ give equations of motion with the same form.

Whenever one has such an infinitesimal symmetry (in a Lagrangian) one has a conservation law. This amazing general theorem which I will now prove is called **Noether's Theorem.**[2] In fact the proof is practically already done.

I will prove it by explicitly constructing a formula for the conserved quantity, a function of the q^a's and $\dot{q}^a$'s which as a consequence of the Euler–Lagrange equations is independent of time. I will call this conserved quantity Q, in general; Q for "quantity":

$$Q = p_a Dq^a - F \tag{5.9}$$

This is a universal definition (notice we are using the summation convention). I will now show that this quantity is independent of time:

$$\frac{dQ}{dt} = \dot{p}_a Dq^a + p_a \frac{dDq^a}{dt} - \frac{dF}{dt} \tag{5.10}$$

Now I will use the Euler–Lagrange equation, which tells us that $\dot{p}_a = \partial L/\partial q_a$. We have two expressions for DL. The first one, (5.6), tells us that the sum of the first two terms in (5.10) is DL. The definition of a symmetry, (5.7), tells us that the last term in (5.10) is $-DL$. Therefore, the sum of the three terms is equal to zero, and $dQ/dt = 0$.

So this equation, (5.9), is the magic, universal formula. Given a one-parameter family of symmetries, (5.4), first you extract an infinitesimal symmetry, (5.5), and then from the infinitesimal symmetry you extract a conservation law. (There is no guarantee that $Q \neq 0$, or that for each independent symmetry we'll get another independent Q. In fact, the construction fails to produce a Q for gauge symmetries.[3] The rules are universal and of general applicability. I will give three examples.)

---

[1] [Eds.] See L. D. Landau and E. M. Lifshitz, *Mechanics*, §2, "The principle of least action", p. 4.

[2] [Eds.] The theorem was stated and proved by Emmy Noether in 1915 while helping David Hilbert with general relativity, and published by her in 1918. See E. Nöther, "Invariante Variationsprobleme", *Nachr. d. Königs. Gesellsch. d. Wiss. zu Göttingen, Math-phys. Klasse* (1918) 235–257. English translation by M. A. Tavel, "Invariant Variation Problem", *Transport Theory and Statistical Physics* **1** (1971) 183–207, and LaTeX'ed by Frank Y. Wang at arXiv:physics/0503066v1. See also Dwight E. Neuenschwander, *Emmy Noether's Wonderful Theorem*, rev. ed., Johns Hopkins Press, 2017.

[3] [Eds.] See note 2, p. 579. Coleman may have meant to say "gauge invariance" for "gauge symmetries". As note 2 makes clear, Coleman did *not* regard gauge invariance as a *symmetry*. On the other hand, (global) phase invariance *does* lead to conserved quantities.

EXAMPLE 1. For the Lagrangian (5.2), space translation of all the particles through a fixed vector, $\mathbf{e}$:

$$\begin{aligned} \mathbf{x}^r &\to \mathbf{x}^r + \lambda \mathbf{e} \\ D\mathbf{x}^r &= \left.\frac{\partial \mathbf{x}^r}{\partial \lambda}\right|_{\lambda=0} = \mathbf{e} \\ DL &= 0 \\ F &= 0 \end{aligned} \tag{5.11}$$

The Lagrangian is unchanged under these translations because V depends only on the differences between positions, and all are translated by the same amount, $\lambda\mathbf{e}$. F of course for this particular example is zero, because the Lagrangian is unchanged under these translations, and therefore $DL = dF/dt = 0$. From (5.9), the conserved quantity is

$$Q = \mathbf{e} \bullet \sum_r m_r \dot{\mathbf{x}}^r \tag{5.12}$$

This quantity Q is the sum of the canonical momenta $\mathbf{p}_r$ dotted with $\mathbf{e}$, the change in the corresponding coordinate. By this method we obtain an infinite number of conservation laws, for there are an infinite number of choices of $\mathbf{e}$. But in fact they can all be written as a linear combination of three linearly independent conservation laws which we obtain by taking $\mathbf{e}$ to be the unit vector along each coordinate axis, and therefore we actually obtain only three conservation laws,

$$\frac{d\mathbf{p}}{dt} = 0 \tag{5.13}$$

where

$$\mathbf{p} = \sum m_r \dot{\mathbf{x}}^r \tag{5.14}$$

This expression is not peculiar to the Lagrangian (5.2). Whenever we have a Lagrangian from which we get conserved quantities from spatial translation invariance, whether or not the system looks anything like a collection of point particles, we'll call the conserved quantity the momentum, $\mathbf{p}$. The expression (5.14) for the momentum would not be so simple if the Lagrangian contained velocity dependent forces, but the conservation laws would nevertheless exist.

EXAMPLE 2. A general Lagrangian $L(q^a, \dot{q}^a)$ where I only assume that it is independent of the time: $\partial L/\partial t = 0$. Look at time translation:

$$\begin{aligned} q^a &\to q^a(t+\lambda) \\ Dq^a &= \left.\frac{\partial q^a}{\partial \lambda}\right|_{\lambda=0} = \dot{q}^a \\ DL &= \frac{\partial L}{\partial q^a}\dot{q}^a + \frac{\partial L}{\partial \dot{q}^a}\ddot{q}^a = \frac{dL}{dt} - \frac{\partial L}{\partial t} = \frac{dL}{dt} \\ F &= L \end{aligned} \tag{5.15}$$

The only time dependence in the Lagrangian is that through the q^a's and their time derivatives. Therefore F equals L, because F is that which when differentiated with respect to time gives you the change in the Lagrangian. The conserved quantity is (summing over a)

$$Q = p_a Dq^a - F = p_a \dot{q}^a - L = E \tag{5.16}$$

Whenever we get a conserved quantity from time translation invariance, we'll call the conserved quantity the energy, E. It is related to time translation as the momenta are related to space translations. It is also sometimes called the Hamiltonian, H, when written as a function of the p's and the q's. I'm sure this is familiar material to those of you who have taken a standard undergraduate mechanics course.

EXAMPLE 3. Again using the Lagrangian (5.2), consider a rotation about an axis $\mathbf{e}$ through an angle λ:

$$\begin{aligned}\mathbf{x}^r &\to R(\lambda, \mathbf{e})\mathbf{x}^r \\ D\mathbf{x}^r &= \left.\frac{\partial \mathbf{x}^r}{\partial \lambda}\right|_{\lambda=0} = \mathbf{e} \times \mathbf{x}^r \\ DL &= 0 \\ F &= 0\end{aligned} \tag{5.17}$$

This Lagrangian is rotationally invariant, so $DL = 0$ and $F = 0$, as in Example 1. The conserved quantity is

$$Q = \sum_r \mathbf{p}_r \cdot (\mathbf{e} \times \mathbf{x}^r) = \mathbf{e} \cdot \sum_r \mathbf{x}^r \times \mathbf{p}_r = \mathbf{e} \cdot \mathbf{J} \tag{5.18}$$

Again, taking $\mathbf{e}$ to be a unit vector along a coordinate axis, we obtain three conservation laws, one for each component of angular momentum, $\mathbf{J}$. Whenever we get conserved quantities from rotational invariance, we'll call the conserved quantities the angular momentum.

There is nothing here that was not already in the Euler–Lagrange equations. What Noether's theorem provides us with is a "turn the crank" method for obtaining conservation laws from a variety of theories. Before this theorem, the existence of conserved quantities, like the energy, had to be noticed from the equations of motion in each new theory. Noether's theorem organizes conservation laws. It explains, for example, why a variety of theories, including ones with velocity-dependent potentials, all have a conserved Hamiltonian, or energy, as in Example 2.

5.2 Extension to quantum particle mechanics

Now when we quantize the theory, when we engage in canonical quantization, an amusing extra feature appears. I will state a theorem which I will not prove, or more properly, will prove only for a restricted class of theories. Most of the cases we will consider will belong to this class. When we come to one that does not fall under the restriction we will check the theorem by explicit computation.

In the quantum theory there is a peculiar closing of the circle. In classical mechanics and in quantum mechanics *modulo*[4] ordering ambiguities, whenever we have an infinitesimal symmetry we have a conservation law, a conserved quantity. In quantum theory the circle closes: We can use the conserved quantity to re-create the infinitesimal symmetry. Specifically,

$$[q^a, Q] = iDq^a \tag{5.19}$$

[4] [Eds.] Slang (American?) for the prepositional "except for", just as 5 *modulo* 3 = 2 (5 equals a multiple of 3, except for the remainder of 2). This usage occurs about a dozen times in the lectures.

That is to say, the conserved quantity Q is the *generator* of the infinitesimal transformation, something in fact we have already exploited in our general discussions for the components of the energy and momentum. This is obviously true if both Dq^a and F are independent of $\dot{q}^a$, because in that case the only term in Q (defined in (5.9)) that does not commute with q^a is p_a and the commutator manifestly gives the desired result:

$$[q^a, Q] = [q^a, p_b Dq^b - F] = [q^a, p_b]Dq^b = i\delta^a_b Dq^b = iDq^a$$

It is not so obvious that (5.19) holds if Dq^a or F involve the $\dot{q}^a$'s. It is nevertheless true but I don't want to go through the trouble of proving the general result. We have up to now seen one case where it is not obviously true. That one case is time translation, where Dq_a does involve the $\dot{q}^a$'s and so does F. But the equation is nevertheless true because in that case Q is the Hamiltonian, and (5.19) is the Heisenberg equation of motion:

$$[q^a, H] = i\dot{q}^a$$

I have gone fast because I presume this material is mainly familiar to you.[5]

5.3 Extension to field theory

So much for classical particle mechanics and quantum particle mechanics. We now turn to classical field theory. As with the special class of classical field theories I discussed last lecture, I have a Lagrangian density that depends on a set of fields ϕ^a, their derivatives $\partial_\mu\phi^a$, and perhaps explicitly on the spacetime location x^μ. I will construct my notation in such a way that when things become relativistic, the notation will be right for the relativistic case, but I will not assume Lorentz invariance until I tell you we now assume Lorentz invariance. So do not be misled by the appearance of upper and lower indices and things like that, into thinking I'm assuming Lorentz invariance at a stage when I'm not.

Now in one sense there is no work to be done because our only general formula, (5.9), goes through without alteration. It's just that instead of a sum on a discrete index we have a sum on a discrete index and an integral on a continuous index. In another sense however we get extra information because the dynamics are so very special, because the Lagrangian is obtained by integrating a local density point by point in space. And we will see not only a global conservation law that tells us the total quantity of Q is unchanged, we will also be able to localize the amount of Q and see Q flowing from one part of space to another part of space in such a way that the total quantity of Q is unchanged. That's a feature of the special structure of the class of theories we are looking at, that the Lagrangian is obtained by integrating a local function of the fields. We can see these extra features in electromagnetism.

---

[5] [Eds.] A question was asked: "Can you extend your remarks in the more general case, [when Dq^a and F involve the $\dot{q}^a$'s] that up to ordering..." Coleman responds: "No, if you don't worry about ordering it is true. It can be proven formally if you don't worry about ordering." The student follows up: "Are there cases where there simply doesn't exist any ordering?" Coleman replies: "Yeah, there are cases even where this breaks down, that dQ/dt is zero. We won't run into any such cases but they exist. Quantum field theorists call them by the pejorative name of **anomalies**. There is a whole lore about when they exist and when they don't, there's an elaborate theory, but it's on a much greater level of sophistication. We'll talk about that. I can't tell you the conditions under which this general formula (5.19) is true or false in ϕ^4 theory because we don't even know how to make sense out of ϕ^4 theory yet. We don't know how to order the ϕ^4 term. We'll play with it formally as if we did; and then later on when we learn more about it we'll see that most of the formal playing can be redeemed. But at the moment I can't say anything."

Electromagnetism possesses a conserved quantity Q, the charge, the integral of the charge density ρ:

$$Q = \int_V d^3\mathbf{x}\,\rho(\mathbf{x},t)$$

There is also a current density, $\mathbf{j}$, and a much stronger statement of charge conservation than $\dfrac{dQ}{dt} = 0$. Local charge conservation says

$$\frac{\partial \rho}{\partial t} + \nabla\boldsymbol{\cdot}\mathbf{j} = 0$$

Integrate this equation over any volume V with boundary S to get

$$\frac{dQ}{dt} = -\int_V d^3\mathbf{x}\,\nabla\boldsymbol{\cdot}\mathbf{j} = -\int_S d^2S\,\hat{\mathbf{n}}\boldsymbol{\cdot}\mathbf{j}$$

using Gauss's theorem. This equation says that you can see the charge change in any volume by watching the current flowing out of the volume. Imagine two stationary, opposite charges separated in space, suddenly winking out of existence at some time t' with nothing happening anywhere else, as in Figure 5.1. You can't have this. This picture satisfies global charge

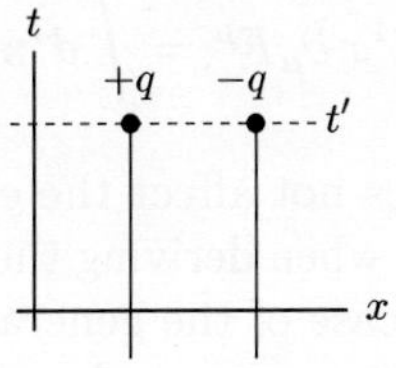

Figure 5.1: Two charges winking out of existence

conservation, but violates local charge conservation. You have to be able to account for the change in charge in any volume, and there would have to be a flow of current in between the two charges. Even if there were not a current and a local conservation law, we could invoke Lorentz invariance to show this scenario is impossible. In another frame the charges do not disappear simultaneously, and for at least a moment, global charge conservation is violated. Field theory, which embodies the idea of *local* measurements, should have *local* conservation laws.

Well, let's try and just go through the same arguments in this case as we went through before. Our dynamical variables are now a set of fields, $\phi^a(x)$, and we consider a one-parameter set of transformations of them,

$$\phi^a(x) \to \phi^a(x,\lambda) \qquad (5.20)$$

with $\phi^a(x,0) = \phi^a(x)$. We define as before

$$D\phi^a = \left.\frac{\partial \phi^a}{\partial \lambda}\right|_{\lambda=0} \qquad (5.21)$$

Definition. We consider an infinitesimal transformation a **symmetry** if and only if

$$D\mathscr{L} = \partial_\mu F^\mu \qquad (5.22)$$

That is to say, the change in the Lagrange density is the divergence of some four-component object $F^\mu(\phi^a, \partial_\mu\phi^a, x)$. This equality must hold for arbitrary $\phi^a(x)$, not necessarily satisfying the equations of motion.

This is an obvious generalization of our condition in the particle case, (5.7). The integral of the divergence also vanishes from the action principle; the time derivative disappearing for the reasons I have stated and the space derivative disappearing because we always assume everything goes to zero sufficiently rapidly in space so we can integrate by parts. Of course, the F of the previous discussion can be obtained from this more general expression. Consider the change in the Lagrangian, L,

$$DL = D\int d^3\mathbf{x}\,\mathscr{L} = \int d^3\mathbf{x}\,\partial_\mu F^\mu = \frac{d}{dt}\int d^3\mathbf{x}\,F^0 \tag{5.23}$$

The space derivatives disappear by integration by parts, and the time derivative can be pulled out of the integral. So the F of our previous discussion, (5.7), exists in this case and it is simply the space integral of F^0,

$$F = \int d^3\mathbf{x}\,F^0 \tag{5.24}$$

As in (5.8), the variation in the action results in boundary terms which can be discarded,

$$\delta S = \int d^4x\,D\mathscr{L} = \int d^4x\,\partial_\mu F^\mu = \int d^3\mathbf{x}\,\Big[F^0(\mathbf{x},t_2) - F^0(\mathbf{x},t_1)\Big] \tag{5.25}$$

Thus a symmetry transformation does not affect the equations of motion (we consider only variations that vanish at the endpoints when deriving the equations of motion). So the previous case, classical mechanics, is a special case of the general theory. However we can do more, as I announced earlier. Let me do that "more" now by following a path parallel to the earlier discussion leading up to (5.6).

I will compute $D\mathscr{L}$ for a field theory;[6]

$$D\mathscr{L} = \frac{\partial\mathscr{L}}{\partial\phi^a}D\phi^a + \pi^\mu_a\partial_\mu(D\phi^a) \tag{5.26}$$

(The quantities π^μ_a were defined in (4.25); the $\mu = 0$ components are the canonical momenta.) Parallel to the earlier discussion I will define a four-component object which I will call J^μ,

$$\boxed{J^\mu \equiv \pi^\mu_a D\phi^a - F^\mu} \tag{5.27}$$

(This is not necessarily a Lorentz 4-vector, because I'm not making any assumptions about Lorentz transformation properties.) There is an obvious parallelism between the definition (5.9) of a global object, Q, and this definition (5.27) of the four local objects, the four components of J^μ.

I will now show that the Euler–Lagrange equations of motion imply something interesting about the divergence $\partial_\mu J^\mu$ of this object, J^μ:

$$\partial_\mu J^\mu = (\partial_\mu\pi^\mu_a)D\phi^a + \pi^\mu_a\partial_\mu D\phi^a - \partial_\mu F^\mu$$

---

[6] $D(\partial^\mu\phi^a)$ is of course equivalent to $\partial^\mu(D\phi^a)$.

By the Euler–Lagrange equations of motion,

$$\partial_\mu \pi^\mu_a = \frac{\partial \mathscr{L}}{\partial \phi^a}$$

and everything else I will copy down unchanged,

$$\partial_\mu J^\mu = \frac{\partial \mathscr{L}}{\partial \phi^a} D\phi^a + \pi^\mu_a \partial_\mu D\phi^a - \partial_\mu F^\mu \tag{5.28}$$

Just as before we have two expressions for $D\mathscr{L}$. One of them, (5.26), is the sum of the first two terms in (5.28). The other one occurs in the definition of F^μ, (5.22). So we get

$$\partial_\mu J^\mu = 0 \tag{5.29}$$

Thus we arrive at Noether's Theorem applied to field theory: For every infinitesimal symmetry of this special type, (5.22)—this is a specialization of our previous formalism, just as this formula, (5.27) is a specialization of our previous formalism—we obtain something that we can call a **conserved current**. I will explain what that means in a moment.

Now for the physical interpretation of this. I will define J^0 as the density of stuff. What the stuff is depends on what symmetry we are considering. I will call $\mathbf{J}$, the space part of this, the current of stuff. I will now show that the words I have attached to these objects, density for J^0 and current for $\mathbf{J}$, have a simple and direct physical interpretation involving stuff flowing around through space in the course of time.

Let me take any ordinary volume V in space—not in spacetime—which has a surface S, as shown in Figure 5.2. The equation (5.29) we have derived tells us

$$\partial_0 J^0 + \nabla \cdot \mathbf{J} = 0$$

Integrating this equation over the volume V, I find

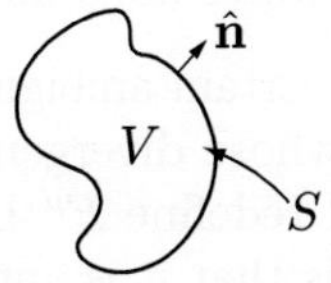

Figure 5.2: A volume V, its surface S, and a unit normal $\hat{\mathbf{n}}$

$$\partial_0 \int_V d^3\mathbf{x}\, J^0 = -\int_V d^3\mathbf{x}\, \nabla \cdot \mathbf{J} = -\int_S d^2S\, \hat{\mathbf{n}} \cdot \mathbf{J} \tag{5.30}$$

by Gauss's theorem. The last term is the integral over the surface S. The $(-)$ sign indicates the outward pointing normal vector $\hat{\mathbf{n}}$, the standard Gauss's theorem notation, dotted into $\mathbf{J}$. This equation verifies the interpretation I have given you, because it says if I take any volume that's got a certain amount of stuff in it, the net amount of stuff changes with time depending on how much stuff is flowing out of the boundaries. Notice that the signs are right: If $\mathbf{J}$ is pointing outwards that means stuff is flowing out, so this time derivative is negative and indeed there in (5.30) is the minus sign.

Of course this means, since stuff only leaves one volume in order to appear in an adjacent volume, that the total quantity of stuff is *conserved*, assuming of course that everything will go smoothly to zero at infinity so we don't have a current at infinity. Then

$$\partial_0 \int d^3\mathbf{x}\, J^0 = \partial_0 Q = 0 \tag{5.31}$$

So Q is independent of time. This is in fact just our general result again. Remember our definition of J^0. Then Q is the integral of J^0:

$$Q = \int d^3\mathbf{x}\, J^0 = \int d^3\mathbf{x}\, \pi^0_a D\phi^a - \int d^3\mathbf{x}\, F^0 = \int d^3\mathbf{x}\, \pi^0_a D\phi^a - F \tag{5.32}$$

(Notice that π^0_a is just the thing we previously called π_a, the conjugate momentum density, and the integral of F^0 is the previous F.)

This is just our previous formula, (5.9). The total conserved quantity is pDq summed over everything which in this case means both summed and integrated, minus the quantity F. So of course the general case contains all the consequences of the special case, which is what you would expect for special cases and general cases. But it contains more: Not only do we have a global conservation law that the total quantity of stuff is unchanged, we have a local conservation law that tells us we can watch stuff floating around, $\mathbf{J}$, and we have localized stuff, J^0. But there is a subtlety we need to address.

5.4 Conserved currents are not uniquely defined

Let's gather our basic equations,

$$\begin{aligned} J^\mu &= \pi^\mu_a D\phi^a - F^\mu \\ D\mathscr{L} &= \partial_\mu F^\mu \\ \partial_\mu J^\mu &= 0 \end{aligned}$$

Okay. There in summary is everything we've done until now.

There is, even in classical physics, a certain ambiguity present in the definition of the stuff Q, the current J^μ and the object F^μ whose divergence is the change $D\mathscr{L}$ in the Lagrange density. The reason is this. Suppose I redefine F^μ by adding to it the divergence of some object $A^{\mu\nu}$, where all I say about $A^{\mu\nu}$ is that it is antisymmetric:

$$F^\mu \to F^\mu + \partial_\nu A^{\mu\nu} \quad \text{where } A^{\mu\nu} = -A^{\nu\mu} \tag{5.33}$$

We defined F^μ through its divergence, $\partial_\mu F^\mu$; we have not defined F^μ itself. Under (5.33) the divergence itself goes as

$$\partial_\mu F^\mu \to \partial_\mu F^\mu + \partial_\mu \partial_\nu A^{\mu\nu} \tag{5.34}$$

Now ∂_μ and ∂_ν commute with each other, and $A^{\mu\nu}$ is antisymmetric, so

$$\partial_\mu F^\mu + \partial_\mu \partial_\nu A^{\mu\nu} = \partial_\mu F^\mu \tag{5.35}$$

So our new F^μ satisfies the defining equation just as well as our old F^μ. However this changes the definition (5.27) of the current, J^μ:

$$J^\mu \to J^\mu - \partial_\nu A^{\mu\nu} \tag{5.36}$$

because we've added something to F^μ and therefore we've subtracted something from the current. So we have another definition of the current that is just as good as our old definition, in terms of local density of stuff and the flow of stuff. On the other hand, I didn't call your attention to any such ambiguity in particle theory and indeed there was none. So we would expect that the definition of the total charge is unchanged. Let's verify that. Our charge transforms under (5.33) like this:

$$Q = \int d^3\mathbf{x}\, J^0 \to \int d^3\mathbf{x}\left(J^0 + \partial_i A^{0i}\right) \tag{5.37}$$

Why did I only write $\partial_i A^{0i}$, instead of $\partial_\nu A^{0\nu}$? Shouldn't I have $\partial_0 A^{00}$ in addition? Well, yes, but A^{00} is zero, because $A^{\mu\nu}$ is antisymmetric.

Now, the second term of (5.37) is a space integral of a space derivative and therefore it equals zero by integration by parts, assuming, as we always do, that everything goes to zero rapidly enough at infinity to enable us to integrate by parts as many times as we want to. Therefore, although we have an infinite family of possible definitions of the local current, this ambiguity gets washed out when we integrate J^0 to obtain the total quantity of stuff.

Some textbooks try to avoid this point, or nervously rub one foot across the other leg and natter about the best definition or the optimum definition, or what is it that unambiguously fixes the definition of a four-component current, J^μ. And the right answer is, of course, there's *nothing* to natter about, there's nothing to be disturbed about. It is something to be *pleased* about. If we have many objects that satisfy desirable general criteria, then that's *better* than having just one. And in a special case when we want to add some extra criteria, then we might be able to pick one out of this large set that satisfies, in addition to the general criteria, the special criteria we want for our immediate purposes. If we only had one object for the current, we would be stuck. We might not be able to make it work. The more freedom you have, the better. So, there are many of them? Good! We live with many of them. It doesn't affect the definition of the globally conserved quantities. It's like being passed a plate of cookies and someone starts arguing about which is the best cookie. They're all edible! And when we come to particular purposes, we may well want to redefine our currents by adding the derivative of an antisymmetric tensor to make things look especially nice for some special purpose we may have in mind.

5.5 Calculation of currents from spacetime translations

I'm now going to apply this general machinery to the particular cases of spacetime translations and Lorentz transformations. It will just be plug-in and crank, both long and tedious, because for the spatial translations I'll have a lot of indices floating around, and for Lorentz transformations I will have even more indices floating around. So I will cover the board with indices, and you will all feel nauseous, but... I gotta do the computation.

We want to apply the general formula, (5.27), first to the case where our theory is translation invariant, that is to say where the Lagrangian density $\mathscr{L}$ does not depend explicitly on x, and then to the case when our theory is Lorentz invariant, that is to say when the Lagrangian density is a Lorentz scalar.

First, we will study spacetime translations. We've discussed these transformations earlier for particle mechanics. We know the globally conserved quantities we will get out of this are the momentum and the energy. Since in field theory we always get densities as well, we will actually recover the density of energy, which we found last lecture (the Hamiltonian density), and the density of momentum, and also obtain a current of energy showing how energy flows and a current of momentum. The sort of transformation we wish to consider is

$$\phi^a(x) \to \phi^a(x + \lambda e) \tag{5.38}$$

where e_ρ is some constant four-component object. I put the index ρ in the lower position just to make some later equations look simple. The infinitesimal transformation—no assumptions about the Lorentz transformation properties of ϕ^a at this stage, they could be the components of a vector—is of course obtained by differentiating with respect to λ at $\lambda = 0$,

$$D\phi^a = \left.\frac{d\phi^a}{d\lambda}\right|_{\lambda=0} \tag{5.39}$$

which gives an expression which I will write

$$D\phi^a = e_\rho \partial^\rho \phi^a(x) \tag{5.40}$$

What we expect to get from here is a set of conserved currents that depend linearly on e_ρ. We have to compute the actual coefficients of e_ρ using the formula (5.27). Since this is an invariance of the Lagrangian, the currents will be e_ρ dotted into some object, $T^{\rho\mu}$,

$$J^\mu = e_\rho T^{\rho\mu} \tag{5.41}$$

I'm using Lorentz invariant notation but I'm not assuming anything. This is just the most general linear function of e_ρ. We will of course find that we get an infinite number of conservation laws this way because we have an infinite choice of e_ρ's, but we only have four linearly independent ones. Therefore we will obtain actually four conservation laws for the four values of the index ρ. They will be of the form

$$\partial_\mu T^{\rho\mu} = 0 \tag{5.42}$$

because we have four independent infinitesimal transformations. That's just $\partial_\mu J^\mu$ with the e_ρ factored out. The object we will obtain in this way has a name. It is called the **canonical energy-momentum tensor**. It is called "canonical" because it is what we get by plugging into our general formula. It's called a tensor because although we haven't talked about Lorentz invariance, it is sort of obvious by counting indices that in a Lorentz invariant theory it will be a tensor field.

The energy-momentum tensor is not unique. Different energy-momentum tensors may be obtained by adding the divergence of an antisymmetric object $A^{\rho\mu\lambda}$:

$$\theta^{\rho\mu} = T^{\rho\mu} + \partial_\lambda(A^{\rho\mu\lambda}), \quad A^{\rho\mu\lambda} = -A^{\rho\lambda\mu}, \tag{5.43}$$

so that $\theta^{\rho\mu}$, like $T^{\rho\mu}$, has zero divergence in its last index:

$$\partial_\mu \theta^{\rho\mu} = \partial_\mu T^{\rho\mu} + \partial_\mu \partial_\lambda(A^{\rho\mu\lambda}) = 0 \tag{5.44}$$

The second term vanishes because $\partial_\mu \partial_\lambda$ is symmetric in μ and λ, and $A^{\rho\mu\lambda}$ is antisymmetric in those indices. There are many different energy-momentum tensors in the literature. There's a tensor of Belinfante,[7] there is a tensor which I had a hand in inventing[8] that is very useful to consider if you were playing with conformal transformations, but we won't talk about any of that. We will just talk about this one, since this is not a lecture on the 42 different energy-momentum tensors that occur in the literature. And of course they all unambiguously define the same conserved quantities when you integrate. These conserved quantities are called P^ρ,

$$P^\rho = \int d^3\mathbf{x}\, T^{\rho 0} \tag{5.45}$$

They are called P^ρ because for space translations one gets the conservation of momentum and for time translations one gets the conservation of energy. Those are the objects, energy and momentum, which one normally sticks together in a single four-component object. So this is the general outline of what has to happen. The only thing we have to do is to compute $T^{\rho\mu}$ explicitly.

Now we have the general formulas. We have $D\phi^a$, (5.40). The only thing we need to compute is $D\mathscr{L}$, (5.26). Well, by assumption everything is translationally invariant. The only spacetime dependence of $\mathscr{L}$ is via the field, so

$$D\mathscr{L} = e_\rho \left[\frac{\partial \mathscr{L}}{\partial \phi^a} \partial^\rho \phi^a + \frac{\partial \mathscr{L}}{\partial(\partial^\nu \phi^a)} \partial^\rho (\partial^\nu \phi^a) \right] = e_\rho \partial^\rho \mathscr{L} \tag{5.46}$$

This is not as it stands the divergence of something, it's the gradient of something. But it's easy enough to make it a divergence. One simply writes this as

$$D\mathscr{L} = \partial_\mu (g^{\mu\rho} e_\rho \mathscr{L}) \tag{5.47}$$

That's the rule for raising indices. Note that ∂_μ commutes with e_ρ because e is a constant vector, and with $g^{\mu\nu}$ which is a constant tensor. Thus we have the object we have called F^μ, (5.22),

$$F^\mu = g^{\mu\rho} e_\rho \mathscr{L} \tag{5.48}$$

We can use our general formula, (5.27) to construct the conserved current,

$$J^\mu = \pi^\mu_a e_\rho \partial^\rho \phi^a - g^{\mu\rho} e_\rho \mathscr{L} \tag{5.49}$$

We obtain the tensor $T^{\rho\mu}$ by factoring out e_ρ, (5.41),

$$\boxed{T^{\rho\mu} = \pi^\mu_a \partial^\rho \phi^a - g^{\mu\rho} \mathscr{L}} \tag{5.50}$$

This is the general formula for the canonical energy-momentum tensor. Notice there is no reason for it to be a symmetric tensor. It turns out to be a symmetric tensor for simple theories, but in general we should distinguish between the indices ρ and μ. The first term doesn't have any obvious symmetry between ρ and μ. There *is* this symmetry for the free field theory we talked about, because π^μ_a was just $\partial^\mu \phi^a$. But in general $T^{\rho\mu}$ will not be symmetric.

---

[7] [Eds.] Often called the Belinfante-Rosenfeld tensor. See F. J. Belinfante, "On the current and density of the electric charge, the energy, the linear momentum and the angular momentum of arbitrary fields", *Physica* **viii** (1940) 449–474, and L. Rosenfeld, "Sur la tenseur d'impulsion-énergie", (On the momentum-energy tensor), *Mém. Acad. Roy. Belg. Soc.* **18** (1940) 1–30. This tensor is symmetric, as required by general relativity.

[8] [Eds.] C. G. Callan, S. Coleman and R. Jackiw, "A New, Improved Energy Momentum Tensor", *Ann. Phys.* **59** (1970) 42–73. This tensor is traceless, as required in conformally invariant theories.

The index μ plays the role of the general index in our discussion of currents. If it is a time index, 0, you get a density. If it is a space index, any of $\{1,2,3\}$, you get a current. The index ρ tells you what you get a density of and what you get a current of in each particular case. When ρ is zero, you get the density of energy or the current of energy, depending on the value of μ. When ρ is a space index, you get the density of the space component of momentum or the current of that space component of momentum.

Just to check that we haven't made any errors, let us look at T^{00} which should be the density of energy; density because the second index μ is zero, energy because the first index ρ is zero:

$$T^{00} = \pi^0_a \partial^0 \phi^a - g^{00}\mathscr{L} = \pi_a \dot{\phi}^a - \mathscr{L} \tag{5.51}$$

This is simply the Hamiltonian density, (4.40), which we arrived at last lecture. So indeed this is the quantity which when integrated over all space gives you the total energy.

To make another check, let's compute the total momentum, for a case where we know what is going on, by integrating the density of momentum over all space. The case where we know what is going on is that of a single free *quantum* field of mass μ. There is only one π^μ, which I remind you is $\partial^\mu\phi$, equations (4.25) and (4.44). The density of momentum is T^{i0}, $\rho = i$ because we're looking at momentum, $\mu = 0$ because we're looking at a density, and is therefore

$$T^{i0} = \pi^0 \partial^i \phi - g^{0i}\mathscr{L} = (\partial^0\phi)(\partial^i\phi) \tag{5.52}$$

Just to check that this is right, the total momentum $\mathbf{P}$ should be obtained by integrating this quantity,

$$\mathbf{P} = -\int d^3\mathbf{x}\,(\partial^0\phi)(\nabla\phi) \tag{5.53}$$

The minus sign is there because T^{i0} has ∂^i with an upper index, and ∇ is ∂_i with a lower index. When we raise the space index we get a minus sign from the metric.

Now let's actually evaluate this component (5.53) for the free quantum field, plugging in our famous expression (3.45) in terms of annihilation and creation operators,

$$\phi(x) = \int \frac{d^3\mathbf{p}}{(2\pi)^{3/2}\sqrt{2\omega_\mathbf{p}}} \left(a_\mathbf{p} e^{-ip\cdot x} + a^\dagger_\mathbf{p} e^{ip\cdot x}\right) \tag{3.45}$$

Let's see if we get our conventional momentum, up to possible ordering trouble such as we encountered with the Hamiltonian. This is a consistency check. Well, the calculation is almost like the calculation of the Hamiltonian at the end of the last lecture (p. 72), and therefore we can use the same shortcuts as there.

$$\begin{aligned}-\int d^3\mathbf{x}\,(\partial_0\phi)(\nabla\phi) = -\iiint \frac{d^3\mathbf{x}\,d^3\mathbf{p}\,d^3\mathbf{p}'}{2(2\pi)^3\sqrt{\omega_\mathbf{p}\omega_{\mathbf{p}'}}} &\left(-i\omega_\mathbf{p} a_\mathbf{p} e^{-i\omega_\mathbf{p}t + i\mathbf{p}\cdot\mathbf{x}} + i\omega_\mathbf{p} a^\dagger_\mathbf{p} e^{i\omega_\mathbf{p}t - i\mathbf{p}\cdot\mathbf{x}}\right) \times \\ &\left(i\mathbf{p}' a_{\mathbf{p}'} e^{-i\omega_{\mathbf{p}'}t + i\mathbf{p}'\cdot\mathbf{x}} - i\mathbf{p}' a^\dagger_{\mathbf{p}'} e^{i\omega_{\mathbf{p}'}t - i\mathbf{p}'\cdot\mathbf{x}}\right)\end{aligned} \tag{5.54}$$

The x integral and the $(2\pi)^3$ will be killed in making two delta functions, $\delta^{(3)}(\mathbf{p}-\mathbf{p}')$ and $\delta^{(3)}(\mathbf{p}+\mathbf{p}')$. That takes care of one $\mathbf{p}$ integral, say $\mathbf{p}'$, and we will end up with a single $\mathbf{p}$ integral. I gave you a general argument last time why the terms with two creation operators and two annihilation operators should vanish, so I won't even bother to compute them this time. They'll still have coefficients that oscillate in time and therefore must go out because of the conservation equation.[9] So I'll just compute the coefficients of $a_{\mathbf{p}}a_{\mathbf{p}}^{\dagger}$ and $a_{\mathbf{p}}^{\dagger}a_{\mathbf{p}}$, and I get

$$\mathbf{P} = -\int \frac{d^3\mathbf{p}}{2\omega_{\mathbf{p}}}\left[(-i\omega_{\mathbf{p}})(-i\mathbf{p})(a_{\mathbf{p}}a_{\mathbf{p}}^{\dagger}) + (i\omega_{\mathbf{p}})(i\mathbf{p})(a_{\mathbf{p}}^{\dagger}a_{\mathbf{p}})\right] = \tfrac{1}{2}\int d^3\mathbf{p}\left(a_{\mathbf{p}}a_{\mathbf{p}}^{\dagger} + a_{\mathbf{p}}^{\dagger}a_{\mathbf{p}}\right)\mathbf{p} \quad (5.55)$$

As before with the Hamiltonian (4.62), this is not the right expression; the first term is out of order for our convention of having the annihilation operators to the right and therefore we will commute the first term. We get

$$\mathbf{P} = \int d^3\mathbf{p}\left(a_{\mathbf{p}}^{\dagger}a_{\mathbf{p}} + \tfrac{1}{2}[a_{\mathbf{p}}, a_{\mathbf{p}}^{\dagger}]\right)\mathbf{p} \quad (5.56)$$

$$= \int d^3\mathbf{p}\left(a_{\mathbf{p}}^{\dagger}a_{\mathbf{p}}\right)\mathbf{p} + \tfrac{1}{2}\int d^3\mathbf{p}\,\delta^{(3)}(0)\,\mathbf{p} \quad (5.57)$$

Here if I'm willing to be especially cavalier with infinities I can simply say well, this second integral in (5.57) is the integral of an odd function of $\mathbf{p}$, albeit a divergent integral with a divergent coefficient, and therefore it gives me zero. If I'm willing to be more precise I mumble something about ordering ambiguities and say that in the quantum theory the proper result is not the expression (5.55), but this expression,

$$\mathbf{P} = \int d^3\mathbf{p}\left(a_{\mathbf{p}}^{\dagger}a_{\mathbf{p}}\right)\mathbf{p} \quad (5.58)$$

with normal ordering. In either case we certainly have no more troubles than we have with the Hamiltonian and we have less if you're willing to accept that dumb argument about the integral of an odd function being zero. And we got the right answer with the right sign. So that suggests that the formulas we have derived in the general case are not total nonsense.

5.6 Lorentz transformations, angular momentum and something else

We've gone through the machine for spacetime translations. Obviously the next step is the other universal conservation law, from Lorentz transformations (including both rotations and boosts). Here there is a technical obstacle we have to surmount. We don't have an explicit expression for a Lorentz transformation matrix as we do for spatial translation. It's some 4×4 matrix Λ that obeys some godawful constraint.[10] Therefore we can't directly find the infinitesimal transformation by differentiating with respect to the parameters of the Lorentz transformation because we don't have a parameterized form of a Lorentz matrix. I will avoid this problem by writing down the conditions that an infinitesimal transformation be an infinitesimal Lorentz transformation, and we'll find the infinitesimal Lorentz transformation directly from these conditions. In the first instance Lorentz transformations are defined as acting on spacetime points, so let us consider an infinitesimal transformation acting on a spacetime point, and see what conditions make it a Lorentz transformation.

So we consider the infinitesimal form of (1.15),

$$\Lambda\colon x^{\mu} \to x^{\mu} + \epsilon^{\mu\nu}x_{\nu}d\lambda \quad (5.59)$$

[9] [Eds.] The term involving two annihilation operators is $\frac{1}{2}\int d^3\mathbf{p}\,\mathbf{p}\,a_{\mathbf{p}}a_{-\mathbf{p}}e^{-2i\omega_{\mathbf{p}}t}$. The quantity multiplying $\mathbf{p}$ is manifestly even, while $\mathbf{p}$ is odd, and so the integral vanishes. The same argument applies to the term involving two creation operators.

[10] [Eds.] In matrix terms, $\Lambda^T g\Lambda = g$, or in components, $\Lambda^{\mu}{}_{\sigma}\, g_{\mu\nu}\, \Lambda^{\nu}{}_{\rho} = g_{\sigma\rho}$.

Now I've got to be very careful how I put my upper and lower indices. That is certainly the most general linear transformation on x^μ. I could have put the index ν on the ϵ downstairs and the second ν on the x upstairs and find the same thing, of course, but I choose to do it this way because otherwise if I had one index upstairs and one downstairs I would go batty trying to figure out which index on ϵ was the first index and which was the second. By keeping both of ϵ's indices upstairs I don't have that problem.

A second vector, y_μ, under the same transformation but lowering all the indices, goes into

$$y_\mu \to y_\mu + \epsilon_{\mu\nu} y^\nu d\lambda \tag{5.60}$$

This infinitesimal transformation is a Lorentz transformation if $x^\mu y_\mu$ is unchanged (1.16) for general x and y. Substituting,

$$x^\mu y_\mu \to x^\mu y_\mu + \epsilon^{\mu\nu} x_\nu y_\mu d\lambda + \epsilon_{\mu\nu} y^\nu x^\mu d\lambda \tag{5.61}$$

and because the transformation is infinitesimal we only retain terms to first order in $d\lambda$.

In order to compare the second term to the third, I will lower the indices on $\epsilon^{\mu\nu}$ and raise them on x and y. But of course when I raise the coordinate indices I get the ν on the x and the μ on the y. That's not good for comparison, so I'll exchange μ with ν. They're just summation indices and it doesn't matter what we call them. Then we get

$$x^\mu y_\mu \to x^\mu y_\mu + \left(\epsilon_{\mu\nu} + \epsilon_{\nu\mu}\right) x^\mu y^\nu d\lambda \tag{5.62}$$

Now for this to be a Lorentz transformation, the sum must equal $x^\mu y_\mu$. That's the definition of a Lorentz transformation, it doesn't affect the inner product. Therefore, since x and y are perfectly general and the coefficient of the term bilinear in y and x is $\epsilon_{\mu\nu} + \epsilon_{\nu\mu}$ I find

$$\epsilon_{\mu\nu} + \epsilon_{\nu\mu} = 0 \tag{5.63}$$

That is to say, $\epsilon_{\mu\nu}$ is an antisymmetric matrix. You could write ϵ with both indices upper or with both lower; either way ϵ is an antisymmetric matrix although a different antisymmetric matrix because of the intervention of the metric tensor. If you write it with one upper and one lower index, it's something horribly ugly; it's not antisymmetric at all.

So an infinitesimal Lorentz transformation is characterized by a 4×4 antisymmetric matrix. Let's just check if this makes sense when counting parameters. A 4×4 antisymmetric matrix has $\frac{1}{2}4 \times (4-1) = 6$ independent entries. That's just right, because there are six parameters in Lorentz transformations: three parameters to describe the three axes about which one can rotate, and three to describe each direction in which one can perform pure Lorentz boosts.

Let's consider the case where

$$\epsilon^{12} = 1 = -\epsilon^{21} \tag{5.64}$$

all other matrix entries zero. In that case I find from the formula (5.59)

$$\begin{aligned} x^1 &\to x^1 + \epsilon^{12} x_2 d\lambda = x^1 - x^2 d\lambda, \\ x^2 &\to x^2 + \epsilon^{21} x_1 d\lambda = x^2 + x^1 d\lambda \end{aligned} \tag{5.65}$$

(Raising a space index gives a minus sign.) Only x^1 and x^2 get changed. Equation (5.65) is the infinitesimal form of the rotation

$$\begin{aligned} x^1 &\to x^1 \cos\lambda - x^2 \sin\lambda, \\ x^2 &\to x^2 \cos\lambda + x^1 \sin\lambda \end{aligned} \tag{5.66}$$

Notice that (5.65) is what you get by differentiating (5.66) with respect to λ and setting λ to zero, in accordance with equations (5.4) and (5.5),

$$\begin{aligned} Dx^1 &= -x^2 \\ Dx^2 &= x^1 \end{aligned} \tag{5.67}$$

So $\epsilon^{\mu\nu}$ with non-zero components only for the indices 1 and 2 corresponds to what one usually calls a rotation about the z-axis; x^3 and x^0 are of course are unchanged.

To take another example, consider

$$\epsilon^{10} = 1 = -\epsilon^{01} \tag{5.68}$$

all other entries zero. Here only x^0 and x^1 get changed;

$$\begin{aligned} x^0 &\to x^0 + \epsilon^{01} x_1 d\lambda = x^0 + x^1 d\lambda, \\ x^1 &\to x^1 + \epsilon^{10} x_0 d\lambda = x^1 + x^0 d\lambda \end{aligned} \tag{5.69}$$

In the first expression, we raise the x index 1, gaining a minus sign, but there's also a minus sign in ϵ^{01}. In the second, there's no minus sign in ϵ^{10}, and there's no minus sign from raising the index, so

$$\begin{aligned} Dx^1 &= x^0 \\ Dx^0 &= x^1 \end{aligned} \tag{5.70}$$

This is the infinitesimal form of

$$\begin{aligned} x^0 &\to x^0 \cosh\lambda + x^1 \sinh\lambda, \\ x^1 &\to x^1 \cosh\lambda + x^0 \sinh\lambda \end{aligned} \tag{5.71}$$

which is a Lorentz boost along the x^1 direction. Please notice how the signs of the metric tensor take care of the sign differences between finite rotations and finite Lorentz transformations, one using trigonometric functions and the other using hyperbolic functions, just by introducing minus signs at the appropriate moment. So it all works out; it all takes care of itself.

Now we come to the dirty work of figuring out the implications of all this for a field theory, with scalar fields only. I have not yet written down the Lorentz transformation properties of fields other than scalars. That's the only thing I know how to Lorentz transform. However, just as for the case of translations we can write down some things in general. We know we will obtain a conserved current, J^μ. We know it must be linear in ϵ and therefore I will write things as

$$J^\mu = \tfrac{1}{2}\epsilon_{\lambda\rho} M^{\lambda\rho\mu} \tag{5.72}$$

The $\frac{1}{2}$ is there to prevent double counting. Since ϵ is antisymmetric, with no loss of generality I can define $M^{\lambda\rho\mu}$ to be antisymmetric in the indices ρ and λ. If it had a symmetric part, that would vanish in the summation on the indices ρ and λ. And therefore I put a $\frac{1}{2}$ in here because really I'm counting twice; I'm counting M^{01} once when I sum it with ϵ_{01}, counting it again when I sum it with ϵ_{10}. Since $\epsilon_{\lambda\rho}$ is constant and perfectly general aside from the antisymmetry condition, I know (from (5.29)) that

$$\partial_\mu M^{\lambda\rho\mu} = 0 \tag{5.73}$$

Therefore I will obtain six global conservation laws,

$$J^{\lambda\rho} = \int d^3\mathbf{x}\, M^{\lambda\rho 0} \tag{5.74}$$

Remember it's μ that plays our general role here, λ and ρ are just along for the ride to multiply the ϵ which I have factored out.[11] The 4×4 antisymmetric tensor $M^{\lambda\rho 0}$ will give us six conservation laws. Three of these should be old friends of ours, the conservation of angular momentum. We know for example that if we look at ϵ_{12} we get z rotations which lead to the conservation of the z component of angular momentum. So J^{12}, aside from a sign or normalization factor, should be identical with the third component of angular momentum, J^{23} with the first, J^{31} with the second, because those are the conservation laws you get from those rotations. On the other hand the (01), (02), (03) components of J will be new objects that will give us new conservation laws to associate with Lorentz invariance, laws we have not previously studied. We will see what those conservation laws are at the end of this lecture. The computation will be hairy, because I've got three indices to keep track of. I hope I have organized it in such a way that it will not be too bad. But now let's compute.

We're only considering scalar fields, so I will study[12]

$$\Lambda\colon\ \phi^a(x) \rightarrow \phi^a(\Lambda^{-1}x) \tag{5.75}$$

$(\Lambda^{-1}x)^\rho$ is to be an infinitesimal Lorentz transformation, to wit

$$(\Lambda^{-1}x)^\rho = x^\rho - \epsilon^{\rho\sigma}x_\sigma d\lambda = x^\rho + \epsilon^{\sigma\rho}x_\sigma d\lambda \tag{5.76}$$

(See (5.59).) Therefore $D\phi^a$ is obtained by expanding out to first order in $d\lambda$ and dividing by $d\lambda$,

$$D\phi^a = \epsilon^{\sigma\rho}x_\sigma\partial_\rho\phi^a \tag{5.77}$$

Since I chose to write (5.72) in terms of lower indices on $\epsilon_{\lambda\rho}$ I will drop my indices and raise them again:

$$D\phi^a = \epsilon_{\sigma\rho}x^\sigma\partial^\rho\phi^a \tag{5.78}$$

I know this drives some people crazy. When I was in graduate school a friend of mine, Gerry Pollack, now a distinguished worker on noble gas crystals, once said to me, "I'm so bad at tensor analysis that whenever I raise an index I get a hernia." Nevertheless, you will have to acquire facility with these things, although this is about as many indices as you will ever have to manipulate in this course.

Now by the same token, since we are assuming Lorentz invariance, that is to say, we assume $\mathscr{L}$ is a Lorentz scalar,

$$D\mathscr{L} = \epsilon_{\sigma\rho}x^\sigma\partial^\rho\mathscr{L} \tag{5.79}$$

I will choose to write this as

$$D\mathscr{L} = \partial^\rho(\epsilon_{\sigma\rho}x^\sigma\mathscr{L}) \tag{5.80}$$

---

[11] [Eds.] When Coleman says that a tensor $T^{\lambda\rho\cdots\mu}$ is conserved, he means $\partial_\mu T^{\lambda\rho\cdots\mu} = 0$. He always puts a conserved 4-tensor's conserved index *farthest to the right*, in the last position, and always denotes this index as μ. These conventions, particularly the first, are unusual.

[12] [Eds.] This looks strange, but it's correct. Under $\Lambda\colon x^\mu \rightarrow x'^\mu = \Lambda^{\mu\nu}x_\nu$, the transformation induced in a field ϕ is $\phi(x) \rightarrow \phi'(x') = S(\Lambda)\phi(\Lambda^{-1}x)$, where $S(\Lambda)$ is a matrix depending on the tensorial character of ϕ. For a scalar, $S(\Lambda)$ equals 1. See (3.16).

because bringing the ∂^ρ through the constant $\epsilon_{\sigma\rho}$ does no harm, nor does bringing the ∂^ρ through x^σ. Since $\partial^\rho x^\sigma$ is $g^{\rho\sigma}$, symmetric in ρ and σ, the term vanishes upon summation with the antisymmetric $\epsilon_{\rho\sigma}$. By the same trick as before, used to go from (5.46) to (5.47), I can write this as

$$D\mathscr{L} = \partial_\mu(g^{\mu\rho}\epsilon_{\sigma\rho}x^\sigma\mathscr{L}) = \partial_\mu F^\mu \tag{5.81}$$

Now we have all we need to get the whole thing, the conserved current J^μ, (5.27). We have the change $D\phi^a$ in the field and we have the change in the Lagrangian written as the divergence of something, F^μ. We can put the whole thing together and get J^μ,

$$J^\mu = \pi^\mu_a D\phi^a - F^\mu = \epsilon_{\sigma\rho}x^\sigma\left[\pi^\mu_a\partial^\rho\phi^a - g^{\mu\rho}\mathscr{L}\right] \tag{5.82}$$

That is straight substitution. Now we may notice that for the special case of scalar fields, this particular combination is one we have seen before, aside from the x. It's simply the definition (5.50) of the canonical energy-momentum tensor, $T^{\rho\mu}$:

$$T^{\rho\mu} = \pi^\mu_a\partial^\rho\phi^a - g^{\mu\rho}\mathscr{L}$$

So in terms of the energy-momentum tensor, the conserved current is

$$J^\mu = \epsilon_{\sigma\rho}x^\sigma T^{\rho\mu} \tag{5.83}$$

This is not the end of the story; $x^\sigma T^{\rho\mu}$ is not antisymmetric in σ and ρ, and the symmetric part of it is irrelevant, since $\epsilon_{\rho\sigma}$ is antisymmetric. To construct $M^{\rho\sigma\mu}$, (5.72), I should antisymmetrize the product $x^\sigma T^{\rho\mu}$ in σ and ρ, and write

$$J^\mu = \tfrac{1}{2}\epsilon_{\sigma\rho}\left(x^\sigma T^{\rho\mu} - x^\rho T^{\sigma\mu}\right) \tag{5.84}$$

and therefore

$$M^{\sigma\rho\mu} = x^\sigma T^{\rho\mu} - x^\rho T^{\sigma\mu} \tag{5.85}$$

I want to talk about the meaning of this. The derivation may have put you to sleep but if it didn't, it should have been totally straightforward, step-by-step, plug-in and crank. A lot of indices to take care of but we took care of them.

This tensor $M^{\sigma\rho\mu}$ is a collection of six objects labeled by the antisymmetric pair of indices σ and ρ, each of which has four components labeled by the index μ. Each of them is respectively, depending upon the value of μ, a current of stuff or a density of stuff. Let us compute a typical component of this thing for various values of σ and ρ to see if the expressions for these conserved quantities are physically reasonable or physically preposterous.

Let us compute J^{12}, (5.74). This is

$$J^{12} = \int d^3\mathbf{x}\left[x^1T^{20} - x^2T^{10}\right] \tag{5.86}$$

Now this is a very reasonable expression for the z component of the angular momentum, which was what this object should be. I am simply saying I have a density of the two-component of momentum P^2 distributed throughout space given by T^{20}, and also the density of the one-component of momentum P^1 given by T^{10}. To find the total angular momentum I just take $\mathbf{x}$ in a cross product with the density of momentum and integrate it: x^1 times the density of the two-component of momentum minus x^2 times the density of the one-component of momentum. That's the normal thing you would write down for the total angular momentum of a continuum system, a fluid or a rigid body or something like that, where you have a momentum density. More properly, I should say it's the *orbital* angular momentum, if we think quantum mechanically for a moment. And the reason for that is because we're considering a set of spinless particles. If we had vector or tensor or spinor fields, we would have extra terms in $D\phi^a$, (5.77), that would generate extra terms in $M^{\rho\sigma\mu}$. These could be identified as the *spin* contribution to the angular momentum, that which does not come from $\mathbf{x}\times\mathbf{p}$. However I won't bother to do that out in detail, I just wanted to show you a particular case.

Now what about the funny components—the ones we haven't talked about before or perhaps haven't seen before in a non-relativistic theory—the conserved quantities like J^{10}?

$$J^{10} = \int d^3\mathbf{x}\,\left[x^1T^{00} - x^0T^{10}\right] \tag{5.87}$$

Well, that also has a definite meaning and it is *not* a surprise conservation law. You might think it's some new law, the conservation of zilch,[13] never seen before! Not true. Notice that this is a very peculiar conservation law in comparison to the others. It explicitly involves x^0, the time. We've never seen a conservation law explicitly involving the time before. That however has an advantage. It means we can bring the x^0 out through the integral sign and write J^{10} as

$$J^{10} = \int d^3\mathbf{x}\,\left[x^1T^{00}\right] - t\int d^3\mathbf{x}\;T^{10} = \int d^3\mathbf{x}\,\left[x^1T^{00}\right] - tP^1 \tag{5.88}$$

Now, what does d/dt of this thing say?

$$\frac{dJ^{10}}{dt} = 0 = \frac{d}{dt}\int d^3\mathbf{x}\,\left[x^1T^{00}\right] - P^1 \tag{5.89}$$

You have seen the non-relativistic analog of this formula. This is simply the law of steady motion of the center-of-mass.

For a system of point particles or a continuum system, if you recall, you define the center-of-mass as the integral of the mass density $\rho(\mathbf{x},t)$ times the position $\mathbf{x}$, divided by the total mass M,

$$\mathbf{x}_{\text{cm}} = \frac{1}{M}\int d^3\mathbf{x}\,\left[\mathbf{x}\rho(\mathbf{x},t)\right] \tag{5.90}$$

The time derivative of the center-of-mass, the velocity of the center-of-mass, is a constant, equal to the total momentum $\mathbf{P}$ divided by the total mass M,

$$\mathbf{v}_{\text{cm}} = \frac{d\mathbf{x}_{\text{cm}}}{dt} = \frac{d}{dt}\left(\frac{1}{M}\int d^3\mathbf{x}\,\left[\mathbf{x}\rho(\mathbf{x},t)\right]\right) = \frac{\mathbf{P}}{M} \tag{5.91}$$

Equation (5.89) is the relativistic analog of the x^1 component of that law, (5.91), multiplied by the total mass. The only change is precisely the change you would expect if you have seen Einstein's headstone,[14] $E = mc^2$, and remember we're working in units where $c = 1$. Instead

[13] [Eds.] Although it looks Yiddish, "zilch" ("nothing, zero") apparently derives from a fictional insignificant person (in Yiddish, a *nebbish*; see Rosten *Joys*, p. 387), "Mr. Zilch", who appears in a 1920s-era comic magazine, *Ballyhoo*. Coleman seems to be using it as a generic synonym for some unimportant quantity. This usage appears a few times in this book.

[14] [Eds.] Coleman is joking. Einstein has neither a grave nor a headstone. His body was cremated and the ashes scattered in an unknown location, as he wished. On the other hand, Boltzmann's headstone (in Vienna) has $S = k\log W$ on it. See also note 33, p. 749.

of the mass density and the law of steady motion of the center-of-mass, we have the energy density, T^{00}, and therefore we have the law of steady motion of the *center of energy*. The center of energy of a relativistic continuum system moves in a straight line with velocity $\mathbf{P}/E$, where E is the total energy. The x component of that law is the same as (5.89) divided by E. Therefore the three conservation laws which we get from Lorentz transformations are not new conservation laws at all, but simply the relativistic generalization of the old non-relativistic law of steady motion of the center-of-mass, trivially generalized to become the law of steady motion of the center of energy. The conserved quantities J^{i0} corresponding to Lorentz boosts can be written

$$J^{i0} = ER^i - tP^i \quad \text{where } R^i = \frac{1}{E}\int d^3\mathbf{x}\, x^i T^{00} \quad \text{and } E = \int d^3\mathbf{x}\, T^{00} \tag{5.92}$$

The quantities R^i are the components of the center of energy. The J^{i0} are the Lorentz partners of the components of angular momentum, and the law of steady motion of the center of energy is the Lorentz partner of the law of the conservation of angular momentum.

You don't normally think of the law of steady motion of the center-of-mass (or energy) as a conservation law because you don't normally think of conserved quantities as explicitly involving t, but these do, and this *is* a conservation law. And that's the end of this lecture.

Next lecture we will go on and talk about less familiar symmetries and less familiar conservation laws.

Problems 2

2.1 Even though we have set $\hbar = c = 1$, we can still do dimensional analysis, because we still have one unit left, mass (or 1/length). In d space-time dimensions (1 time and $d-1$ space), what is the dimension (in mass units) of a canonical free scalar field, ϕ? (Work it out from the equal-time commutation relations.) Still in d dimensions, the Lagrangian density for a scalar field with self-interactions might be of the form

$$\mathscr{L} = \tfrac{1}{2}(\partial_\mu\phi\,\partial^\mu\phi) - \sum_{n\geq 2} a_n\phi^n \tag{P2.1}$$

What is the dimension (again in mass units) of the Lagrangian density? The action? The coefficients a_n? (As a check, whatever the value of d, a_2 had better have the dimensions of $(\text{mass})^2$.

(1997a 2.1)

2.2 Dimensional analysis can sometimes give us very quickly results that would otherwise require tedious computations. In Problem 1.4, I defined the observable

$$A(a) = (a\sqrt{\pi})^{-3}\int d^3\mathbf{x}\,\phi(\mathbf{x},0)\,e^{-|\mathbf{x}|^2/a^2}$$

where $\phi(x)$ was a free scalar field of mass μ, and a was some length. I defined the variance of A as $\operatorname{var} A = \langle A^2\rangle - \langle A\rangle^2$, and I asked you to show that for small a,

$$\operatorname{var} A(a) = \alpha a^\beta + \dots$$

and to find the coefficients α and β. (I also asked you to study things for large a, but that's not relevant to this problem.) If we're working at very small distances, it's reasonable to assume that the Compton wavelength $h/\mu c$ might as well be infinite, that is to say, we might as well replace μ by zero. In this case, the coefficient β is completely determined by dimensional analysis.

(a) For a general dimension d (with a $(d-1)$ dimensional Gaussian replacing the three-dimensional one in the definition of $A(a)$), find β. Check your result by showing that it reproduces the answer to Problem 1.4 for $d = 4$.

(b) What if instead of ϕ we had the energy density, T^{00}? (Again, take $\mu = 0$.)

(1997a 2.2)

2.3 In class thus far all my examples have involved scalar fields. Here's a vector field theory for you to explore: Consider the classical theory of a real vector field, A_μ, with dynamics defined by the Lagrangian density

$$\mathscr{L} = -\tfrac{1}{4}(\partial_\mu A_\nu - \partial_\nu A_\mu)(\partial^\mu A^\nu - \partial^\nu A^\mu) \tag{P2.2}$$

Derive the Euler–Lagrange equations. Show that if we define[1]

$$F_{\mu\nu} = \partial_\mu A_\nu - \partial_\nu A_\mu \tag{P2.3}$$

[1] [Eds.] This definition differs by a sign from that given in (14.1), p. 68 in Bjorken & Drell *Fields*.

and further define two 3-vectors **E** and **B** by

$$\mathbf{E} \equiv (F^{10}, F^{20}, F^{30}) \qquad \mathbf{B} \equiv (F^{32}, F^{13}, F^{21}) \tag{P2.4}$$

then **E** and **B** obey the free (empty space) Maxwell's equations in rationalized units (with neither 4π's nor ϵ_0's.)

(1997a 2.4)

2.4 Use the procedure explained in Chapter 5 to construct $T^{\mu\nu}$, the energy-momentum tensor, for the theory of the proceeding problem. This turns out to be a rather ugly object; $T^{\mu\nu}$ is not equal to $T^{\nu\mu}$ and T^{00} is not the usual electromagnetic energy density, $\frac{1}{2}(|\mathbf{E}|^2 + |\mathbf{B}|^2)$. However, as I explained in class, we can always construct a new energy-momentum tensor that gives the same energy and momentum as the old one by adding the divergence of an antisymmetric object.

Show that if we define

$$\theta^{\nu\mu} = T^{\nu\mu} + a\partial_\lambda(A^\nu F^{\mu\lambda}) \tag{P2.5}$$

then, for an appropriate choice of the constant a, $\theta^{\nu\mu} = \theta^{\mu\nu}$, and θ^{00} is the usual energy density, $\frac{1}{2}\left(|\mathbf{E}|^2 + |\mathbf{B}|^2\right)$. Find this value of a.

(1997a 2.5)

Solutions 2

2.1 As in Lecture 1, define

$$[M] \equiv \text{units of mass}$$
$$[L] \equiv \text{units of length}$$
$$[T] \equiv \text{units of time}$$
$$[E] \equiv \text{units of energy}$$

and let $[A]$ denote the units of the quantity A. Then

$$[c] = \frac{[L]}{[T]} = 1 \Rightarrow [L] = [T] \tag{S2.1}$$

$$[\hbar] = [E][T] = [M][c]^2[T] = [M][T] = 1 \Rightarrow [M] = 1/[L] \tag{S2.2}$$

We also have

$$[\partial_\mu] = [\partial^\mu] = [L]^{-1} = [M]$$

Since

$$\int d^n\mathbf{x}\, \delta^{(n)}(\mathbf{x}) = 1$$

for any (integer) power n, and $[d^n\mathbf{x}] = [L]^n$, it follows

$$[\delta^{(n)}(\mathbf{x})] = [L]^{-n} = [M]^n \tag{S2.3}$$

Following the hint, consider the equal-time commutator (3.61),

$$[\phi(\mathbf{x},t), \dot{\phi}(\mathbf{y},t)] = i\delta^{(d-1)}(\mathbf{x}-\mathbf{y})$$

It follows

$$[\phi][\dot{\phi}] = [L]^{-1}[\phi][\phi] = [M][\phi]^2 = [\delta^{(d-1)}] = [M]^{d-1} \;\Rightarrow\; [\phi] = [M]^{(d/2)-1} \tag{S2.4}$$

The units of the Lagrangian density $\mathscr{L}$ can be deduced from the kinetic term, $\frac{1}{2}(\partial_\mu\phi\,\partial^\mu\phi)$;

$$[\mathscr{L}] = [\partial_\mu\phi]^2 = ([M][\phi])^2 = [M]^d \tag{S2.5}$$

The action S is the integral over all space-time of the Lagrangian density, so

$$[S] = [\int d^d\mathbf{x}\,\mathscr{L}] = [L]^d[M]^d = 1 \tag{S2.6}$$

To find the units of a_n, note that all the terms of the Lagrangian density must have the same units, so

$$[a_n\phi^n] = [M]^d = [a_n][\phi]^n = [a_n]([M]^{(d/2)-1})^n \;\Rightarrow\; [a_n] = [M]^{d+n-\frac{1}{2}nd} \tag{S2.7}$$

We were asked to check that the units of a_2 should be equal to (mass)^2, whatever the value of d. According to (S2.7), $[a_2] = [M]^2$, independent of d. The interpretation of μ as a mass in the Klein–Gordon Lagrangian density (4.44) is consistent with its units. ■

2.2 (a) The $d-1$ dimensional Gaussian is just the $d-1$ product of individual Gaussians, so ($[a]$ has the units $[L]$)

$$\int d^{d-1}\mathbf{x}\, e^{-|\mathbf{x}|^2/a^2} = \left(\int dx\, e^{-x^2/a}\right)^{d-1} = (a\sqrt{\pi})^{d-1}$$

To normalize the observable in $d-1$ dimensions, we have to redefine $A(a)$ as

$$A(a) = (a\sqrt{\pi})^{1-d}\int d^{d-1}\mathbf{x}\,\phi(\mathbf{x},0)e^{-\mathbf{x}^2/a^2}$$

By definition,

$$\operatorname{var} A = \langle A^2\rangle - \langle A\rangle^2 \quad \text{so} \quad [\operatorname{var} A] = [A]^2$$

and

$$[A(a)]^2 = \left([a]^{1-d}[L]^{d-1}[\phi]\right)^2 = [\phi]^2 = ([M]^{(d/2)-1})^2 = [M]^{d-2} \tag{S2.8}$$

If as before we take for small a

$$\operatorname{var} A(a) = \alpha a^{\beta} + \dots$$

then

$$[\operatorname{var} A] = [\alpha a^{\beta}] = [M]^{d-2} \tag{S2.9}$$

We know α is a constant and therefore independent of a, the only variable with dimensions. Consequently α has to have no units, and so, because a has the units of $[L] = [M]^{-1}$

$$[\operatorname{var} A(a)] = [a^{\beta}] = [M]^{d-2} = [M]^{-\beta} \Rightarrow \beta = 2-d \tag{S2.10}$$

In the solution to Problem 1.4, we found $\beta_0 = -2$ for small a, which agrees with this result. ■

(b) The canonical energy-momentum tensor is defined by (5.50), and its component T^{00} is

$$T^{00} = \pi_a\dot{\phi}^a - \mathscr{L}$$

(summation on a); this is also the Hamiltonian density $\mathscr{H}$ (see (4.40)). Then $[T^{00}] = [\mathscr{L}] = [M]^d = [\mathscr{H}]$. If we define

$$A_{\mathscr{H}}(a) = (a\sqrt{\pi})^{1-d}\int d^{d-1}\mathbf{x}\,\mathscr{H}(\mathbf{x},0)e^{-\mathbf{x}^2/a^2} \tag{S2.11}$$

we get

$$[\operatorname{var} A_{\mathscr{H}}(a)] = [A_{\mathscr{H}}(a)]^2 = [\mathscr{H}]^2 = [M]^{2d} \tag{S2.12}$$

If we set

$$\operatorname{var} A_{\mathscr{H}}(a) = \alpha_{\mathscr{H}} a^{\beta_{\mathscr{H}}}$$

then by the previous reasoning, since a has the units of $[L] = [M]^{-1}$, we find $\beta_{\mathscr{H}} = -2d$. We note that the fluctuations of the energy density grow more rapidly at small distances than those of the field itself. ■

2.3 We start with the Lagrangian density (P2.2)

$$\mathscr{L} = -\tfrac{1}{4}(\partial_\mu A_\nu - \partial_\nu A_\mu)(\partial^\mu A^\nu - \partial^\nu A^\mu)$$

The Euler–Lagrange equations are

$$\frac{\partial\mathscr{L}}{\partial A^\sigma} - \frac{\partial}{\partial x_\lambda}\left(\frac{\partial\,\mathscr{L}}{\partial(\partial^\lambda A^\sigma)}\right) = 0$$

The first term is identically zero. Using the identity

$$\frac{\partial(\partial^\mu A^\nu)}{\partial(\partial^\lambda A^\sigma)} = \delta^\mu_\lambda\delta^\nu_\sigma$$

the Euler–Lagrange equations are

$$\partial^\lambda\left[\tfrac{1}{2}\left(\delta^\mu_\lambda\delta^\nu_\sigma - \delta^\nu_\lambda\delta^\mu_\sigma\right)(\partial_\mu A_\nu - \partial_\nu A_\mu)\right] = 0$$

The quantity in the square brackets becomes, multiplying it all out, the antisymmetric $F_{\lambda\sigma}$:

$$F_{\lambda\sigma} = \partial_\lambda A_\sigma - \partial_\sigma A_\lambda$$

and so the Euler–Lagrange equations become

$$\partial^\lambda F_{\lambda\sigma} = 0 \quad \text{or equivalently,} \quad \boxed{\partial_\lambda F^{\lambda\sigma} = 0} \tag{S2.13}$$

These represent four different equations. First, let's look at $\sigma = 0$:

$$\begin{aligned}\partial_\lambda F^{\lambda 0} &= \partial_0 F^{00} + \partial_1 F^{10} + \partial_2 F^{20} + \partial_3 F^{30} \\ &= \partial_1 F^{10} + \partial_2 F^{20} + \partial_3 F^{30} \quad (\text{because } F^{00} = 0) \\ &= \nabla \cdot \mathbf{E} = 0\end{aligned} \tag{S2.14}$$

if, as the problem suggests, we call $F^{i0} = E^i$. (Recall $\nabla^i = \partial_i = \partial/\partial x^i$.) That is Gauss's Law in empty space. Now consider $\sigma = i$, in particular, let's say $\sigma = 1$. Then

$$\partial_\lambda F^{\lambda 1} = \partial_0 F^{01} + \partial_2 F^{21} + \partial_3 F^{31} \tag{S2.15}$$

(the term $\partial_1 F^{11}$ is identically zero, since $F^{11} = 0$). Following the identification in the original problem, $B^1 = F^{32}$, $B^2 = F^{13}$, and $B^3 = F^{21}$, and using the antisymmetry of $F^{\mu\nu}$, we have $F^{01} = -E^1 = -E_x$. Then this equation (S2.15) becomes

$$-\frac{\partial E_x}{\partial t} + \frac{\partial B_z}{\partial y} - \frac{\partial B_y}{\partial z} = 0 \Rightarrow (\nabla \times \mathbf{B})_x = \left(\frac{\partial \mathbf{E}}{\partial t}\right)_x \tag{S2.16}$$

which is the x component of Ampère's Law. Similarly, $i = 2$ and $i = 3$ are the y and z components of Ampère's Law.

The identification of the components of $F^{\mu\nu}$ with the electric and magnetic fields is an easy consequence of identifying the 4-vector A^μ with a four-component object $(\phi, \mathbf{A})$, the electric potential ϕ and the magnetic vector potential $\mathbf{A}$. Then

$$\begin{aligned}F^{i0} &= \partial^i A^0 - \partial^0 A^i = -\left(\nabla\phi + \dot{\mathbf{A}}\right)^i = E^i \\ F^{ij} &= \partial^i A^j - \partial^j A^i = \epsilon^{ijk} B_k \quad \text{where } B^k = (\nabla \times \mathbf{A})^k\end{aligned} \tag{S2.17}$$

The Euler–Lagrange equations give half of Maxwell's equations, Ampère's Law and Gauss's Law, but not the other half. Those can be obtained from the **Bianchi identities**,

$$\partial^\lambda F^{\mu\nu} + \partial^\mu F^{\nu\lambda} + \partial^\nu F^{\lambda\mu} = 0 \tag{S2.18}$$

which follow easily from the definition of $F^{\mu\nu}$ as a sort of four-dimensional "curl" of the 4-vector A^μ. The Bianchi identities are non-zero only when $\{\lambda, \mu, \nu\}$ are all different, so there are only four non-vanishing components. Let one of the indices be zero, and the other two be $\{1, 2\}$. Then (recall $\partial^i = -\nabla^i$)

$$\partial^0 F^{12} + \partial^1 F^{20} + \partial^2 F^{01} = 0 = -\frac{\partial B_z}{\partial t} - \frac{\partial E_y}{\partial x} + \frac{\partial E_x}{\partial y} \tag{S2.19}$$

which is the z component of Faraday's Law, $\nabla \times \mathbf{E} = -\partial \mathbf{B}/\partial t$. The set $\{0, i, j\}$ give all three components of Faraday's Law. If none of the indices are zero, there is only one non-vanishing component,

$$\partial^1 F^{23} + \partial^2 F^{31} + \partial^3 F^{12} = 0 = \frac{\partial B_x}{\partial x} + \frac{\partial B_y}{\partial y} + \frac{\partial B_z}{\partial z} = \nabla \cdot \mathbf{B} \tag{S2.20}$$

the last of Maxwell's equations. ■

2.4 Using the results of Problem 2.3, we have from the definition of the canonical energy-momentum tensor (5.50)

$$\begin{aligned}T^{\nu\mu} &= \frac{\partial \mathscr{L}}{\partial(\partial_\mu A_\lambda)} \partial^\nu A_\lambda - g^{\mu\nu} \mathscr{L} \\ &= -F^{\mu\lambda} \partial^\nu A_\lambda + \tfrac{1}{4} g^{\mu\nu} F^{\lambda\sigma} F_{\lambda\sigma}\end{aligned}$$

The first term is not symmetric in $\{\mu, \nu\}$. Following the suggested prescription, we add the divergence of an antisymmetric tensor,

$$\theta^{\nu\mu} = T^{\nu\mu} + a\partial_\lambda(F^{\mu\lambda} A^\nu)$$

We already know from (5.44) that $\partial_\mu \theta^{\nu\mu} = 0$. We need to determine the value for a so that $\theta^{\mu\nu}$ is symmetric.

Because of the boxed Euler–Lagrange equations (S2.13) above, $\partial_\lambda F^{\mu\lambda} = 0$, so

$$\partial_\lambda (F^{\mu\lambda} A^\nu) = F^{\mu\lambda} \partial_\lambda A^\nu \tag{S2.21}$$

and the new tensor becomes

$$\begin{aligned}\theta^{\nu\mu} &= -F^{\mu\lambda}\partial^{\nu}A_{\lambda} + \tfrac{1}{4}g^{\mu\nu}F^{\lambda\sigma}F_{\lambda\sigma} + aF^{\mu\lambda}\partial_{\lambda}A^{\nu} \\ &= -F^{\mu\lambda}\left(\partial^{\nu}A_{\lambda} - a\partial_{\lambda}A^{\nu}\right) + \tfrac{1}{4}g^{\mu\nu}F^{\lambda\sigma}F_{\lambda\sigma} \\ &= -F^{\mu\lambda}g_{\lambda\sigma}\left(\partial^{\nu}A^{\sigma} - a\partial^{\sigma}A^{\nu}\right) + \tfrac{1}{4}g^{\mu\nu}F^{\lambda\sigma}F_{\lambda\sigma}\end{aligned} \tag{S2.22}$$

If we choose $a = 1$, the term in the parentheses is just $F^{\nu\sigma}$, and the resulting tensor is symmetric:

$$\theta^{\nu\mu} = -g_{\lambda\sigma}F^{\mu\lambda}F^{\nu\sigma} + \tfrac{1}{4}g^{\mu\nu}F^{\lambda\sigma}F_{\lambda\sigma} \tag{S2.23}$$

The other problem with T^{00} is that it fails to give the correct energy density for Maxwell's theory. What about θ^{00}? Let's see:

$$\begin{aligned}\theta^{00} &= -g_{\lambda\sigma}F^{0\lambda}F^{0\sigma} + \tfrac{1}{4}g^{00}F^{\lambda\sigma}F_{\lambda\sigma} = -F^{0i}F_{0i} + \tfrac{1}{4}\left(F^{0i}F_{0i} + F^{i0}F_{i0} + F^{ij}F_{ij}\right) \\ &= -\tfrac{1}{2}F^{0i}F_{0i} + \tfrac{1}{2}\left(F^{23}F_{23} + F^{31}F_{31} + F^{12}F_{12}\right) = \tfrac{1}{2}\left(|\mathbf{E}|^2 + |\mathbf{B}|^2\right)\end{aligned} \tag{S2.24}$$

as desired. ■

6

Symmetries and conservation laws II. Internal symmetries

I would like to continue the discussion of symmetries and conservation laws that we began last lecture by considering a new class of continuous transformations. From these we will extract the associated conserved currents and the associated global conservation laws, like conservation of electric charge, conservation of baryon number and conservation of lepton number which we have not yet considered in detail. This new class of symmetries is not universal; they occur only in specific theories whose Lagrangians[1] have special properties. We believe on good experimental grounds that if we attempt to explain the world with a field theory, that theory had better be translationally invariant and Lorentz invariant. Those symmetries led to the conservation of P^μ and $J^{\mu\nu}$. However, some field theories which people have invented to understand the world turn out to have larger groups of symmetries than just those associated with the Poincaré group. These symmetries commute with spacetime translations and with Lorentz transformations, and so we expect that the conserved quantities Q associated with them will be Lorentz scalars. These new symmetries are given the somewhat deceptive name of **internal symmetries**. "Internal" historically meant that somehow you were doing something to the interior structure of the particle; you were not moving it about in space or rotating it. The word is deceptive because, as you will see, it applies to theories of structureless particles, in particular, to free field theories. Nevertheless the nomenclature is standard, and I will use it. For us, "internal" will mean "non-geometrical". These internal symmetries will *not* relate fields at different spacetime points, but only transform fields at the *same* spacetime point into one another. Conservation laws are the best guide for looking for theories that actually describe the world, because the existence of a conservation law is a qualitative fact that greatly constrains the form of the Lagrangian.

6.1 Continuous symmetries

EXAMPLE 1. SO(2)

As a simple example of a theory that possesses an internal symmetry, let me take a theory involving a set of scalar fields, all of them free and all of them with the same mass and—this is the simplest nontrivial case—I will let the index a range over only two values, $a = \{1, 2\}$. The

[1] We have left particle mechanics behind, and I'll often use Lagrangian to mean the Lagrangian density, $\mathscr{L}$.

Lagrangian is

$$\begin{aligned}\mathscr{L} &= \tfrac{1}{2}\left(\partial^\mu\phi^a\partial_\mu\phi^a - \mu^2\phi^a\phi^a\right)\\ &= \tfrac{1}{2}(\partial^\mu\phi^1\partial_\mu\phi^1 - \mu^2\phi^1\phi^1) + \tfrac{1}{2}(\partial^\mu\phi^2\partial_\mu\phi^2 - \mu^2\phi^2\phi^2)\end{aligned} \tag{6.1}$$

So this is simply a sum of two free Lagrangians, each of them for a free scalar field of mass μ.

Now this Lagrangian possesses a rather obvious symmetry. Since everything involves the quadratic form $\phi^a\phi^a$, it is invariant under a group that is isomorphic to the two-dimensional rotation group of Euclidean geometry, SO(2). This will describe not two-dimensional rotations in the x-y plane, or in the y-z plane but in the 1-2 plane between the fields ϕ^1 and ϕ^2. To be specific, for any λ, if I make the transformation

$$\begin{aligned}\phi^1 &\to \phi^1\cos\lambda + \phi^2\sin\lambda\\ \phi^2 &\to \phi^2\cos\lambda - \phi^1\sin\lambda\end{aligned} \tag{6.2}$$

the Lagrangian is obviously unchanged.

This is a symmetry of this particular sample Lagrangian. It is not connected in any way with geometry; it's not a spatial translation and it's not a Lorentz transformation. I could write more complicated Lagrangians which possess the same symmetry. For example, I could add to this any power of $\phi^a\phi^a$ times some negative constant, (negative, so it will come out with a positive sign in the energy), like the quadratic

$$\mathscr{L} \to \mathscr{L}' = \mathscr{L} - g(\phi^a\phi^a)^2 \tag{6.3}$$

or a term in $\phi^a\phi^a$ cubed or to the fifth power. The new Lagrangian would still be invariant under this transformation because $\phi^a\phi^a$ is invariant under this transformation: the sum of the squares is preserved by rotations.

Now let us extract the consequences of this symmetry. Let's feed it into our general machinery, turn the crank and see what happens. In terms of the general formula (5.21),

$$\begin{aligned}D\phi^1 &= \phi^2\\ D\phi^2 &= -\phi^1\end{aligned} \tag{6.4}$$

We need the derivatives (4.31),

$$\pi_1^\mu = \frac{\partial\mathscr{L}}{\partial(\partial_\mu\phi^1)} = \partial^\mu\phi^1; \quad \pi_2^\mu = \frac{\partial\mathscr{L}}{\partial(\partial_\mu\phi^2)} = \partial^\mu\phi^2 \tag{6.5}$$

We also need the four-component object I called F^μ last time, defined by (5.22),

$$D\mathscr{L} = \partial_\mu F^\mu \tag{6.6}$$

This Lagrangian is unchanged, so $F^\mu = 0$. We construct the current by our general formula, (5.27),

$$J^\mu = \pi_a^\mu D\phi^a - F^\mu = (\partial^\mu\phi^1)\phi^2 - (\partial^\mu\phi^2)\phi^1 \tag{6.7}$$

This is the formal classical expression. As we see, this current is conserved, $\partial_\mu J^\mu = 0$, because both ϕ^1 and ϕ^2 satisfy the Klein–Gordon equation with the same mass μ. We will later investigate whether or not the formal expression has to be normal ordered, in the case when $g = 0$. What we have to do to make sense of the theory when g is *not* equal to zero is a subject we will investigate much later in the course. The associated conserved quantity Q is the integral of the zero component of this current.

Let's compute Q in the case where $g = 0$. And I remind you once again of our expression (3.45) for the free fields,

$$\phi^a(x) = \int \frac{d^3\mathbf{p}}{(2\pi)^{3/2}\sqrt{2\omega_\mathbf{p}}} \left(a_\mathbf{p}^{(a)} e^{-ip\cdot x} + a_\mathbf{p}^{(a)\dagger} e^{ip\cdot x} \right) \tag{6.8}$$

I should really have a draftsman write this formula on a piece of cardboard which I could nail up above the blackboard. The creation and annihilation operators satisfy the relations

$$\begin{aligned} [a_\mathbf{p}^{(a)}, a_{\mathbf{p}'}^{(b)}] &= [a_\mathbf{p}^{(a)\dagger}, a_{\mathbf{p}'}^{(b)\dagger}] = 0 \\ [a_\mathbf{p}^{(a)}, a_{\mathbf{p}'}^{(b)\dagger}] &= \delta^{ab}\delta^{(3)}(\mathbf{p}-\mathbf{p}') \end{aligned} \tag{6.9}$$

We compute Q by our usual tricks. It's exactly the same calculation as the others we have done (e.g., the calculation of $\mathbf{P}$, (5.53) through (5.58)),

$$Q = \int d^3\mathbf{x}\, \left[(\partial_0\phi^1)\phi^2 - (\partial_0\phi^2)\phi^1 \right] \tag{6.10}$$

$$= i\int d^3\mathbf{p}\, \left[a_\mathbf{p}^{(1)\dagger} a_\mathbf{p}^{(2)} - a_\mathbf{p}^{(2)\dagger} a_\mathbf{p}^{(1)} \right] \tag{6.11}$$

Once again there's no need to keep track of the product of two annihilation operators or two creation operators. On *a priori* grounds these products must vanish because their coefficients involve oscillating factors that have no hope of canceling, and Q is supposed to be time independent. I have written it already in normal ordered form, not that it matters here. There's no need to worry about the order of the operators because a type 1 operator and a type 2 operator always commute.

The expression (6.11) for the charge is very nice. It has all the properties you would expect for an internal symmetry. It commutes with the energy, (2.48); it commutes with the momentum, (2.49); and it annihilates the vacuum:

$$Q\,|0\rangle = 0 \tag{6.12}$$

And as we'll see shortly (§6.2), it is also Lorentz invariant, because it is the space integral of the time component of a conserved current (6.7). The expression is nice, however it is hardly transparent. On the other hand, the charge Q is not diagonal with respect to the operators $\{a_\mathbf{p}^{(a)}\}$ and $\{a_\mathbf{p}^{(b)\dagger}\}$:

$$[Q, a_\mathbf{p}^{(a)}] = -i\epsilon^{ab} a_\mathbf{p}^{(b)} \qquad [Q, a_\mathbf{p}^{(a)\dagger}] = -i\epsilon^{ab} a_\mathbf{p}^{(b)\dagger} \tag{6.13}$$

(where $\epsilon^{12} = -\epsilon^{21} = 1, \epsilon^{11} = \epsilon^{22} = 0$). The first term in the integrand (6.11) replaces a type 2 particle with a type 1 particle; the second term acts *vice versa* with 2 replacing 1.

One can make things much simpler by defining new annihilation and creation operators which are linear combinations of our original $a_\mathbf{p}^{(a)}$ and $a_\mathbf{p}^{(b)\dagger}$. We will define $b_\mathbf{p}$ and $b_\mathbf{p}^\dagger$ as

$$\begin{aligned} b_\mathbf{p} &\equiv \frac{1}{\sqrt{2}}\left(a_\mathbf{p}^{(1)} + i a_\mathbf{p}^{(2)} \right) \\ b_\mathbf{p}^\dagger &\equiv \frac{1}{\sqrt{2}}\left(a_\mathbf{p}^{(1)\dagger} - i a_\mathbf{p}^{(2)\dagger} \right) \end{aligned} \tag{6.14}$$

(the $\sqrt{2}$ is there for a reason that will become clear shortly). Likewise I will define $c_{\mathbf{p}}$ and $c^\dagger_{\mathbf{p}}$ as the other obvious combinations,

$$\begin{aligned} c_{\mathbf{p}} &= \frac{1}{\sqrt{2}}\left(a^{(1)}_{\mathbf{p}} - ia^{(2)}_{\mathbf{p}}\right) \\ c^\dagger_{\mathbf{p}} &= \frac{1}{\sqrt{2}}\left(a^{(1)\dagger}_{\mathbf{p}} + ia^{(2)\dagger}_{\mathbf{p}}\right) \end{aligned} \tag{6.15}$$

These are also annihilation and creation operators. They create particles in states that are linear combinations of state 1 and state 2. It is easy to check that they, too, obey the commutators for annihilation and creation operators. All the commutators vanish except for

$$[b_{\mathbf{p}}, b^\dagger_{\mathbf{p}'}] = [c_{\mathbf{p}}, c^\dagger_{\mathbf{p}'}] = \delta^{(3)}(\mathbf{p} - \mathbf{p}') \tag{6.16}$$

I inserted the $\sqrt{2}$ in the denominators so that this would come out equal to $\delta^{(3)}(\mathbf{p} - \mathbf{p}')$ rather than twice that. If it is not obvious to you that all the other commutators are zero, let me show you. Any annihilation operator, $b_{\mathbf{p}}$ or $c_{\mathbf{p}}$, commutes with any other annihilation operator, since both of these are linear combinations of commuting operators. For the same reason, any creation operator, $b^\dagger_{\mathbf{p}}$ or $c^\dagger_{\mathbf{p}}$, commutes with any other creation operator. So let's check the annihilation operator $b_{\mathbf{p}}$ with the creation operator $c^\dagger_{\mathbf{p}'}$,

$$\begin{aligned} [b_{\mathbf{p}}, c^\dagger_{\mathbf{p}'}] &= \tfrac{1}{2}[a^{(1)}_{\mathbf{p}} + ia^{(2)}_{\mathbf{p}}, a^{(1)\dagger}_{\mathbf{p}'} + ia^{(2)\dagger}_{\mathbf{p}'}] = \tfrac{1}{2}[a^{(1)}_{\mathbf{p}}, a^{(1)\dagger}_{\mathbf{p}'}] - \tfrac{1}{2}[a^{(2)}_{\mathbf{p}}, a^{(2)\dagger}_{\mathbf{p}'}] \\ &= \tfrac{1}{2}\delta^{(3)}(\mathbf{p} - \mathbf{p}') - \tfrac{1}{2}\delta^{(3)}(\mathbf{p} - \mathbf{p}') = 0 \end{aligned} \tag{6.17}$$

The other combination also commutes:

$$[b^\dagger_{\mathbf{p}}, c_{\mathbf{p}'}] = 0. \tag{6.18}$$

The b's and c's obey the same algebra as the $a^{(1)}$'s and $a^{(2)}$'s because, for any given value of $\mathbf{p}$, the b's and c's are annihilation and creation operators for orthogonal single particle states. There are two states which we called, arbitrarily, the type 1 meson and the type 2 meson. Whenever we have a degenerate subspace of states, we are perfectly free to choose a different orthogonal linear combination to be our basis vectors. Here we have chosen a linear combination of a type 1 meson with a type 2 meson to be a b-type meson, and the orthogonal linear combination to be a c-type meson. If we have a Fock space with two degenerate kinds of particles—with the same mass, μ—it doesn't matter which two independent vectors we choose to be our fundamental mesons.

Why do I choose these combinations, (6.14) and (6.15)? I could just as well have chosen the coefficients of $a_{\mathbf{p}}$ and $a^\dagger_{\mathbf{p}}$ to be $\sin\theta$ and $\cos\theta$ for the b's, and $\cos\theta$ and $-\sin\theta$ for the $b^\dagger$'s and so on; the algebra would have worked out the same. Well, I choose these combinations because both the expression of the charge Q and the algebra of Q with the b's and c's work out particularly simply.

By substitution, you can see pretty easily that

$$Q = \int d^3\mathbf{p}\, \left[b^\dagger_{\mathbf{p}} b_{\mathbf{p}} - c^\dagger_{\mathbf{p}} c_{\mathbf{p}}\right] = N_b - N_c \tag{6.19}$$

where N_b and N_c are the number of b-type and c-type mesons, respectively, in a given state (see (2.50)). Then you will be able to check by eyeball that unlike the type 1 and type 2 mesons, the b and c type mesons *are* eigenstates of Q; Q *is* diagonal with respect to these mesons:

$$[Q, b_{\mathbf{p}}] = -b_{\mathbf{p}} \qquad [Q, b_{\mathbf{p}}^\dagger] = b_{\mathbf{p}}^\dagger \tag{6.20}$$

$$[Q, c_{\mathbf{p}}] = c_{\mathbf{p}} \qquad [Q, c_{\mathbf{p}}^\dagger] = -c_{\mathbf{p}}^\dagger \tag{6.21}$$

Thus we have a much simpler interpretation of Q. We have diagonalized Q by writing this expression (6.19), and made it easy to see what basis vectors diagonalize Q. As a result, we have two kinds of particles with different values of Q. The value of Q does not depend on the momentum of the particle. A b type, whatever its momentum, carries a value $Q = +1$, and the other, the c type, carries a value of $Q = -1$. The two kinds are like particles and antiparticles, with the same mass but opposite charge. We see in (6.19) that Q is simply N_b minus N_c. This is very similar to electric charge. These particles for example could be π^+ and π^- mesons, and Q could be the electric charge. The total charge of the system is obtained by counting the number of particles of one kind and subtracting the number of particles of the other kind. For this reason I called this Q "charge", but we haven't deduced the conservation of electric charge or anything like that. I have simply cooked up an arbitrary example with a symmetry leading to a conservation law that has some structural resemblance to the conservation of electric charge. I said "π^+ and π^- mesons", but I could just as well have said "electrons and positrons", aside from the fact that electrons and positrons have spin. Q needn't be electric charge. If we were considering electrons and positrons, I could have let Q be **lepton number** instead of electric charge. Lepton number also has this kind of structure.

In terms of the new operators, we can write the Hamiltonian as (see (4.63))

$$H = \int d^3\mathbf{p}\, \omega_{\mathbf{p}} \left[b_{\mathbf{p}}^\dagger b_{\mathbf{p}} + c_{\mathbf{p}}^\dagger c_{\mathbf{p}} \right] \tag{6.22}$$

This expression is easily obtained from the sum of the two free field Hamiltonians for ϕ^1 and ϕ^2 by substitution. I've introduced these combinations of the original $a^{(1)}$'s and $a^{(2)}$'s to simplify the representation of the charge Q in terms of annihilation and creation operators.

Aside: Complex fields

I would like to digress now, in a direction that really has nothing to do with symmetries. I would like to talk about putting together two real fields to make a complex field. The simple, diagonal expression of the charge suggests that maybe we should make this complex combination not just on the level of the annihilation and creation operators, but on the level of the fields themselves. That might make things look even simpler. Therefore let me define a new field ψ, complex and non-Hermitian, and its adjoint, ψ^*,

$$\begin{aligned} \psi &= \frac{1}{\sqrt{2}}\left(\phi^1 + i\phi^2\right) \\ \psi^* &= \frac{1}{\sqrt{2}}\left(\phi^1 - i\phi^2\right) \end{aligned} \tag{6.23}$$

Properly I should write $\psi^\dagger$ for the adjoint, but the star $(*)$ is traditionally used for this purpose in the literature. In terms of creation and annihilation operators, ψ and ψ^* are written

$$\begin{aligned}\psi(x) &= \int \frac{d^3\mathbf{p}}{(2\pi)^{3/2}\sqrt{2\omega_\mathbf{p}}} \left[b_\mathbf{p} e^{-ip\cdot x} + c_\mathbf{p}^\dagger e^{ip\cdot x}\right] \\ \psi^*(x) &= \int \frac{d^3\mathbf{p}}{(2\pi)^{3/2}\sqrt{2\omega_\mathbf{p}}} \left[b_\mathbf{p}^\dagger e^{ip\cdot x} + c_\mathbf{p} e^{-ip\cdot x}\right]\end{aligned} \tag{6.24}$$

Our old fields ϕ^1 and ϕ^2 have rather messy commutators with Q. If you commute either with Q you get the other with some coefficient:

$$[Q, \phi^a(x)] = -i\epsilon^{ab}\phi^b(x) \tag{6.25}$$

Note that this equation follows the general rule for charges and symmetries, (5.19),

$$[Q, \phi^a(x)] = -iD\phi^a(x) \tag{6.26}$$

that the conserved charge generates the transformation. The new fields ψ and ψ^* have neat commutators with Q:

$$\begin{aligned}[Q, \psi] &= -\psi \\ [Q, \psi^*] &= \psi^*\end{aligned} \tag{6.27}$$

Like every free field, this ψ is an operator that can both annihilate and create. It has a definite charge changing property. It always lowers the charge by 1, either by annihilating a b particle with charge $+1$ or by creating a c particle with charge -1. Likewise ψ^* always raises the charge, either annihilating a c particle of charge -1 or creating a b particle with charge $+1$.

The new fields ψ and ψ^* have very interesting equal-time commutators. The fields ϕ^1 and ϕ^2 commute with themselves and with each other at equal-times. Because they are linear combinations of ϕ^1 and ϕ^2, ψ and ψ^* also commute with themselves and with each other at equal-times:

$$[\psi(\mathbf{x}, t), \psi(\mathbf{y}, t)] = [\psi^*(\mathbf{x}, t), \psi^*(\mathbf{y}, t)] = [\psi(\mathbf{x}, t), \psi^*(\mathbf{y}, t)] = 0 \tag{6.28}$$

More interesting is $\psi(\mathbf{x}, t)$ with $\partial_0\psi(\mathbf{y}, t)$. That also happens to be zero, because it will involve the commutator of $b_\mathbf{p}$ with $c_{\mathbf{p}'}^\dagger$, and from (6.17) these commute:

$$[\psi(\mathbf{x}, t), \dot\psi(\mathbf{y}, t)] = 0 \tag{6.29}$$

The adjoint of this commutator, $[\psi^*(\mathbf{x}, t), \dot\psi^*(\mathbf{y}, t)]$, involves the commutator of $b_\mathbf{p}^\dagger$ with $c_{\mathbf{p}'}$. But from (6.18), it also equals zero,

$$[\psi^*(\mathbf{x}, t), \dot\psi^*(\mathbf{y}, t)] = 0 \tag{6.30}$$

Indeed the only non-zero equal-time commutators are $\psi(\mathbf{x}, t)$ with $\partial_0\psi^*(\mathbf{y}, t)$ and $\psi^*(\mathbf{x}, t)$ with $\partial_0\psi(\mathbf{y}, t)$,

$$[\psi(\mathbf{x}, t), \dot\psi^*(\mathbf{y}, t)] = [\psi^*(\mathbf{x}, t), \dot\psi(\mathbf{y}, t)] = i\delta^{(3)}(\mathbf{x} - \mathbf{y}) \tag{6.31}$$

Of course since they are linear combinations of ϕ^1 and ϕ^2, ψ and ψ^* also obey the Klein–Gordon equation,

$$\Box^2\psi + \mu^2\psi = 0 \tag{6.32}$$

$$\Box^2\psi^* + \mu^2\psi^* = 0 \tag{6.33}$$

Now why did I bother to do all this, to rewrite the theory of two scalar fields in terms of a complex field and its conjugate? Well, to recast the Lagrangian (6.1) in terms of ψ and ψ^*. We can just as well write

$$\mathscr{L} = (\partial_\mu \psi^* \partial^\mu \psi) - \mu^2 \psi^* \psi \tag{6.34}$$

If we look at this theory's structure, equations (6.28)–(6.33), and read it backwards, it looks very much as if these are equations we *could* have found by doing something that, at first glance, seems extremely silly. If we had started out with this Lagrangian, (6.34), and treated ψ and ψ^* as if they were *independent variables*, and not in fact each other's complex conjugate, it would have seemed the ultimate in dumb procedure. But let's proceed anyway.

By varying the Lagrangian with respect to ψ^*, we obtain the Klein–Gordon equation for ψ,

$$\frac{\partial \mathscr{L}}{\partial \psi^*} - \partial_\mu \left(\frac{\partial \mathscr{L}}{\partial (\partial_\mu \psi^*)} \right) = -\mu^2 \psi - \Box^2 \psi = 0 \tag{6.35}$$

and by varying with respect to ψ we would obtain the Klein–Gordon equation for ψ^*. Treating ψ and ψ^* as independent variables, we find that the canonical momentum to ψ is $\partial_0 \psi^*$,

$$\pi_\psi^\mu = \frac{\partial \mathscr{L}}{\partial (\partial_\mu \psi)} = \partial^\mu \psi^* \tag{6.36}$$

Likewise, the canonical momentum conjugate to ψ^* is $\partial_0 \psi$, expressed in the adjoint equation which I won't bother to write down. Canonical quantization then leads to (6.31). For the other commutators, we would find that ψ and ψ^* commute at equal times because they are q type variables, and that ψ and $\partial_0 \psi$ commute at equal times because they are the q for one variable and the p for another variable. So had we been foolish enough to write the Lagrangian in terms of complex fields to begin with, and to treat ψ and ψ^* as if they were independent, we would have obtained, in this particular instance at least, exactly the same results as we obtained by doing things correctly, treating ϕ^1 and ϕ^2 as real independent variables. My motivation may have been baffling, but I went through this sequence of computations to make this point.

So it turns out it is *not* dumb to treat ψ and ψ^* as independent. I will begin—I will not complete it, because once you've seen how the first part of it goes, the rest of it will be a trivial exercise—to show that you will *always* get the right results if you have a Lagrangian expressed in terms of complex fields, and simply treat ψ and ψ^* *as if* they were independent. I sketch out why it is legitimate as far as the derivation of the Euler–Lagrange equations goes. Once you've seen my method you will see that the same method can be carried through to obtain the Hamiltonian form, the equal-time commutators, and so on.

Suppose I have a Lagrangian that depends on a set of fields ψ and ψ^*, complex conjugates of each other, and also on the gradients of ψ and ψ^*,

$$\mathscr{L} = \mathscr{L}(\psi, \psi^*, \partial^\mu \psi, \partial^\mu \psi^*) \tag{6.37}$$

For most practical purposes this Lagrangian is set up so that the action integral is real, guaranteeing that the Hamiltonian will be Hermitian when we're all done with quantization. That's not going to be necessary to any of the proofs I'm going to give, but I might as well point it out. (This restriction is not practical in all cases; a real Lagrangian is not a completely general function of these variables.) If I were to go through the variational procedure that leads to the Euler–Lagrange equations, varying both ψ and ψ^*, I would obtain

$$\delta S = \int d^4x \, (A \, \delta\psi + A^* \delta\psi^*) = 0 \tag{6.38}$$

This is the integral of some god-awful mess obtained by doing all my integration by parts, some coefficient I'll just call A to indicate I'm not concerned about its structure, times $\delta\psi$, plus the conjugate god-awful mess, A^* times $\delta\psi^*$. Nobody can fault me on that.

Now if I were foolishly to treat $\delta\psi$ and $\delta\psi^*$ as independent, that is, if I were to consider the variation

$$\delta\psi = (\text{an arbitrary function}),\ \delta\psi^* = 0 \tag{6.39}$$

that would be obvious nonsense, because ψ^* is the conjugate of ψ; I can't vary them independently. I would obtain an equation of motion which says

$$A = 0. \tag{6.40}$$

but saying nothing about A^*. Likewise by making $\delta\psi^*$ arbitrary and $\delta\psi = 0$, I would obtain $A^* = 0$ and those would be my two Euler–Lagrange equations of motion. This is obviously illegitimate. I cannot vary ψ without simultaneously varying ψ^* because they're conjugates. On the other hand, what I certainly can do is choose matters such that

$$\delta\psi = \delta\psi^* \tag{6.41}$$

with $\delta\psi$ real. From this I deduce

$$A + A^* = 0 \tag{6.42}$$

That's legitimate. Alternatively, I could just as well arrange things such that $\delta\psi$ is pure imaginary,

$$\delta\psi = -\delta\psi^* \tag{6.43}$$

from which I deduce

$$A - A^* = 0 \tag{6.44}$$

The net result of the equations (6.42) and (6.44) is

$$A = A^* = 0 \tag{6.45}$$

The consequences of this correct procedure are exactly the same as the consequences of the manifestly silly procedure. I leave it as an exercise to carry out all the other steps of the canonical program, the introduction of canonical momenta and the Hamiltonian, and the working out of canonical commutation relations, to show that in general it all comes out the same as if one had treated the ψ and ψ^* as independent variables.

Just to show how this goes, I'll work out the form of the current in the ψ-ψ^* formalism. Let's remember how our transformations work in our original basis, (6.2). For every λ,

$$\begin{aligned} \phi^1 &\to \phi^1 \cos\lambda + \phi^2 \sin\lambda \\ \phi^2 &\to \phi^2 \cos\lambda - \phi^1 \sin\lambda \end{aligned} \tag{6.46}$$

so that

$$\begin{aligned} \psi &= \frac{1}{\sqrt{2}}\left(\phi^1 + i\phi^2\right) \to \frac{1}{\sqrt{2}}\left(\phi^1 + i\phi^2\right)(\cos\lambda - i\sin\lambda) = e^{-i\lambda}\psi \\ \psi^* &= \frac{1}{\sqrt{2}}\left(\phi^1 - i\phi^2\right) \to \frac{1}{\sqrt{2}}\left(\phi^1 - i\phi^2\right)(\cos\lambda + i\sin\lambda) = e^{i\lambda}\psi^* \end{aligned} \tag{6.47}$$

The group defined by the symmetry (6.47) is called U(1), the **unitary group** in one dimension. It has the same algebraic structure as SO(2); mathematicians call these two groups "isomorphic". We have a very simple expression for $D\psi$, (see (5.21))

$$D\psi = \left.\frac{\partial\psi}{\partial\lambda}\right|_{\lambda=0} = -i\psi \tag{6.48}$$

Likewise by considering the transformation properties of ψ^*, or by taking the conjugate of this equation,

$$D\psi^* = i\psi^* \tag{6.49}$$

These are two of the ingredients we need to construct the canonical current. The others are the respective conjugate momenta:

$$\begin{aligned}\pi^\mu_\psi &= \frac{\partial\mathscr{L}}{\partial(\partial_\mu\psi)} = \partial^\mu\psi^* \\ \pi^\mu_{\psi^*} &= \frac{\partial\mathscr{L}}{\partial(\partial_\mu\psi^*)} = \partial^\mu\psi\end{aligned} \tag{6.50}$$

The Lagrangian (6.34) is obviously invariant under the symmetry (6.47) and F^μ of course is still equal to zero whether we express the Lagrangian in terms of ϕ^1 and ϕ^2 or in terms of ψ and ψ^*:

$$J^\mu = \pi^\mu_\psi D\psi + \pi^\mu_{\psi^*} D\psi^* = -i(\partial^\mu\psi^*)\psi + i(\partial^\mu\psi)\psi^* \tag{6.51}$$

By inspection, $\partial_\mu J^\mu = 0$; the current is conserved.

On the classical level, it is an elementary substitution to show that (6.51) is the same current (6.7) as before; work it out if you don't believe me. On the quantum level for free fields it is, in this case, necessary to normal order so that we get the same results as before. When we write things as (6.51), ψ and $\partial^\mu\psi^*$ do not commute, and we have to normal order to make sure that all of our annihilation and creation operators are in the right place. This concludes the discussion of complex fields.

EXAMPLE 2. SO(n)

We've discussed a very simple example involving internal symmetries in which there were only two fields. There was a digression on the method of complex fields, which enabled us to simplify somewhat the representations of things in that case. We can get much more complicated internal symmetry structures simply by returning to our original expression (6.1) with real fields,

$$\mathscr{L} = \tfrac{1}{2}\Big(\partial^\mu\phi^a\partial_\mu\phi^a - \mu^2\phi^a\phi^a\Big) \tag{6.52}$$

and possibly some interaction, say

$$-g(\phi^a\phi^a)^2 \tag{6.53}$$

but now a runs not from 1 to 2, but from 1 to some number n, your choice.

In the same way that the previous theory was invariant under SO(2), this Lagrangian (6.52) is invariant under SO(n), the connected group of all orthogonal transformations on n real variables. We can imagine these fields as being labeled by vectors in some n-dimensional space, an abstract, internal space. For every rotation in that n-dimensional space, there is a transformation of the fields among themselves that would leave the Lagrangian invariant.

We can go through the elaborate procedure of constructing the currents, but I hope you have learned enough about the n-dimensional rotation group to know that a complete and independent set of infinitesimal transformations are rotations in the 1-2 plane, rotations in the 2-3 plane, rotations in the 1-3 plane, etc.,[2] each of which is something we have already done, with a slight relabeling. Therefore we will obtain $\frac{1}{2}n(n-1)$ conserved currents, because we have n choices for the first number that labels the plane, $n-1$ choices for the second, and as it doesn't matter what we call first or second when we're labeling, we divide by 2. The form of these currents, say the current corresponding to rotation in the a-b plane, will be exactly the same as before,

$$J_\mu^{[ab]} = (\partial_\mu \phi^a)\phi^b - (\partial_\mu \phi^b)\phi^a = -J_\mu^{[ba]} \tag{6.54}$$

That's the same expression as (6.7), with 1 and 2 replaced by a and b. As you see, only $\frac{1}{2}n(n-1)$ of these currents are independent, because when $a = b$ the current is zero, and if I interchange a and b, I just get minus the same current.

There is no analog in this more complicated case for the trick of complex fields. The reason is very simple. That trick was based on diagonalizing the charge. Here I can hardly expect to diagonalize all the charges simultaneously, because the corresponding transformations do not commute; a 1-2 rotation does not commute with a 2-3 rotation. Therefore the corresponding charges should not commute. Still, in some cases it is convenient to pick one of the $\frac{1}{2}n(n-1)$ charges as a "nice" charge, and arrange the fields to diagonalize this one charge, the others remaining ugly and awkward. For example, for $n = 3$, when we have three degenerate particles, it is frequently convenient to diagonalize arbitrarily the 1-2 charge. We introduce the fields

$$\begin{aligned} \psi &= (\phi^1 + i\phi^2)/\sqrt{2} \\ \psi^* &= (\phi^1 - i\phi^2)\sqrt{2} \\ \phi_0 &= \phi^3 \end{aligned} \tag{6.55}$$

The field ψ lowers the 1-2 charge Q^{12} by one unit, ψ^* raises it by one unit and ϕ_0 doesn't change it at all. The fields change the charge or not because ψ either creates negatively charged particles or annihilates positively charged particles, *vice versa* for ψ^*, and ϕ_0 creates and annihilates neutral particles, whence the subscript "nought". This notation occurs most frequently when we consider the system of the three π mesons. In the absence of electromagnetism and the weak interactions, as you probably know from other courses, all three pions would be degenerate in mass. Indeed this is due to a group with precisely the structure of SO(3) called the **isospin group** that acts in exactly the prescribed way on the three pions.[3] We pluck out the 1-2 subgroup because when we introduce electromagnetism the full SO(3) invariance is broken, but a subgroup SO(2) remains, and this 1-2 invariance corresponds to the conservation of electric charge. The charged pions have the same mass, but the neutral π^0 has a different mass. Of course the π mesons interact with a lot of other particles, so there's plenty of "+ ..." in the Lagrangian.

---

[2] [Eds.] See Goldstein *et al. CM*, Section 4.8, "Infinitesimal Rotations", pp. 163–171, or Greiner & Müller *QMS*, Section 1.8, "Rotations and their Group Theoretical Properties", pp. 35–37.

[3] [Eds.] Isospin is equally well described by the group SU(2), which is locally isomorphic to SO(3). See note 37, p. 791.

6.2 Lorentz transformation properties of the charges

Before I leave the general discussion on continuous symmetries there is one gap in our arguments. I have not discussed the Lorentz transformation properties of the conserved quantities associated with these currents. You expect from the examples I have given you that the Lorentz transformation properties are the same as those of the currents except that one index is absent. That is to say, if the current transforms like an nth-rank tensor, we want to show that the conserved quantity transforms like an $(n-1)^{\text{st}}$-rank tensor. Where we have a current that transforms like a 4-vector, the associated conserved quantity is charge, which is a scalar. When we have a two-index object, for example the energy-momentum tensor, the associated conserved quantity is a one-index object, the total four-momentum. When we have a three-index object, such as $M^{\nu\lambda\mu}$ that I talked about before, the angular momentum currents, the associated object $J^{\nu\lambda}$ is a two-index object, which appears to be a tensor, but we haven't proved that it is a tensor. We've proved it in particular cases by writing explicit expressions for these objects and showing how they transform, whereupon it is manifest that they transform in the desired way. But we haven't shown it in general. So let me now attack the general problem: If we know how the current transforms, how does the associated charge transform? I will do in detail in the case of a 4-vector current, J^μ. Once I do it, you will visualize with your mind's eye, by adding extra indices to the equations, how the whole thing works out for the energy-momentum tensor and the angular momentum current.

I have a conserved current, $\partial_\mu J^\mu = 0$. I will assume it transforms, as in the case of an internal symmetry, like a vector field. That is to say under a Lorentz transformation Λ,

$$J^\mu(x) \xrightarrow{\Lambda} J^{\mu\prime}(x') = \Lambda^\mu_\nu J^\nu(\Lambda^{-1}x) \tag{6.56}$$

Remember when we were discussing field transformation laws, we said that there's always an inverse operator in the argument, but the un-inverted thing outside. This will be my only input. This equation could be in classical physics, in which we take some field configuration and Lorentz transform it, and the current transforms in this way. Or it could be in quantum mechanics where this transformation is effected by a unitary transformation. Since all the equations I manipulate will be linear in J^μ, it will be irrelevant whether J^μ is a c-number field or an operator field.

Now we define Q as

$$Q \equiv \int d^3\mathbf{x}\, J^0(\mathbf{x}, 0) \tag{6.57}$$

We can define Q at any time, since the charge is independent of time by the conservation equation. So just for notational convenience I will choose $t = 0$. Because we know how J^μ transforms, we know how Q transforms. It will go into some object which we'll compute in a moment and which I will denote by Q'. We wish to ask, "Is $Q' = Q$?" That is to say, in this case, is the space integral of the time component of the current a scalar? To make this demonstration work, I will have to rewrite the expression for the charge in a way that makes its Lorentz transformation properties more evident:

$$Q = \int d^4x\, \delta(n\cdot x)\, n\cdot J(x) \tag{6.58}$$

turning the space integral (6.57) into a four dimensional integral with $x^0 = 0$. The 4-vector $n_\mu = (1, 0, 0, 0)$ is the unit vector pointing in the time direction, so $n \cdot x = x^0$. The expression

$n \cdot J(x)$ is simply a fancy way of writing J^0. An equivalent way of writing the same thing is

$$Q = \int d^4x \, \partial_\mu \theta(n \cdot x) \, J^\mu(x) \tag{6.59}$$

because the space derivative of this theta function is zero, and the time derivative gives us $\delta(x_0)$;

$$\partial_\mu \theta(n \cdot x) = n_\mu \delta(n \cdot x) \tag{6.60}$$

This form (6.59) may make you feel a little nervous because it looks like we can make Q equal zero by integrating by parts. But we do not have control over the time boundary conditions, and $\theta(x) = 1$ for all positive x, so we can't get rid of the boundary term in the time integration by parts.

I will now write a corresponding expression for Q';

$$Q' = \int d^4x \, \delta(n \cdot x) \, n \cdot \Lambda J(\Lambda^{-1}x) \tag{6.61}$$

We transform the fields, and then do the same experiment on the transformed field configuration. We're taking an active view of transformations. We do *not* change the integration surface at the same time. The experimenter is not transformed; that would be a no-no. If we changed both the current and the integration surface, we would obviously get the same answer. So we are measuring Q', the same Q defined in exactly the same way for the transformed current. We have $n \cdot J$ in (6.58), but in (6.61) we have $n \cdot (\Lambda J)$, that's $\Lambda^0_\nu J^\nu(\Lambda^{-1}x)$, written out in compressed notation. That is the same integral of the same component, the time component, for the current corresponding to the transformed field configuration.

In this form it is easy to see how to make Q' look more comparable to Q. We define

$$x = \Lambda x', \quad n = \Lambda n', \tag{6.62}$$

and so, by Lorentz invariance of the inner product,

$$n \cdot x = \Lambda x' \cdot \Lambda n' = n' \cdot x' \quad \text{and} \quad n \cdot \Lambda J = \Lambda n' \cdot \Lambda J = n' \cdot J \tag{6.63}$$

We plug these into our integral (6.61), and we find

$$\begin{aligned} Q' &= \int d^4x' \, \delta(n' \cdot x') \, n' \cdot J(x') \\ &= \int d^4x \, \delta(n' \cdot x) \, n' \cdot J(x) \\ &= \int d^4x \, \partial_\mu \theta(n' \cdot x) \, J^\mu(x) \end{aligned} \tag{6.64}$$

In the first step, we use the invariance of d^4x under a Lorentz transformation, and in the second step, we simply change the variable of integration, as is our privilege, we can call it what we please. The third step is just the same reasoning as gets us from (6.58) to (6.59). Now the only difference between the expressions for Q and Q' is that n has been redefined. For Q', the surface of integration is $t' = 0$, and we take $n' \cdot J$ in the t' direction. Our active transformation has had the exact same effect as if we had made a passive transformation, changing coordinates to $x' = \Lambda^{-1}x$. It's the same old story, the difference between an *alias*, another name, and an *alibi*, another place. The former corresponds to a passive transformation and the latter to an active transformation.[4]

To show $Q = Q'$, we will compute $Q - Q'$, and see that it equals zero:

$$Q - Q' = \int d^4x\,(\partial_\mu[\theta(n\cdot x) - \theta(n'\cdot x)])J^\mu(x) \tag{6.65}$$

Now integration by parts is legitimate, because we can drop the surface terms, the integral over dS_μ:

$$\begin{aligned} Q - Q' &= -\int d^4x\,\left[\theta(n\cdot x) - \theta(n'\cdot x)\right]\partial_\mu J^\mu(x) + \int d^4x\,\partial_\mu\Big(\left[\theta(n\cdot x) - \theta(n'\cdot x)\right]J^\mu(x)\Big) \\ &= -\int d^4x\,\left[\theta(n\cdot x) - \theta(n'\cdot x)\right]\partial_\mu J^\mu(x) + \int dS_\mu\,\left[\theta(n\cdot x) - \theta(n'\cdot x)\right]J^\mu(x) \end{aligned} \tag{6.66}$$

In the surface integral, the quantity in brackets, although not zero, certainly goes to zero at any fixed x as $t \to \infty$, because eventually $n\cdot x$ becomes positive and $n'\cdot x$ also becomes positive, each θ function equals 1, and the difference vanishes. Likewise as $t \to -\infty$, eventually both arguments become negative and each becomes zero.

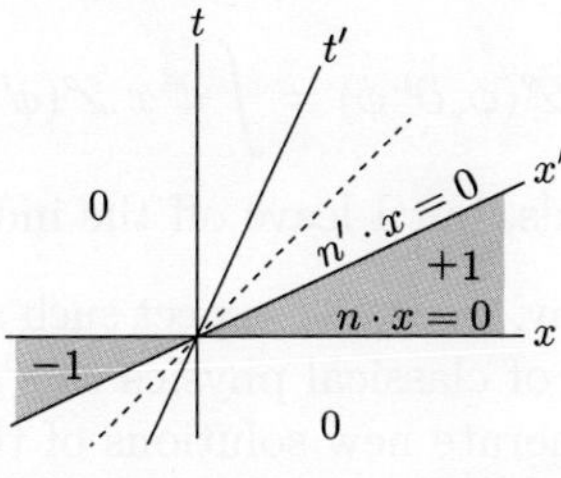

Figure 6.1: The spacetime surfaces $n\cdot x = 0$ and $n'\cdot x = 0$

Here's spacetime, showing the surface $n\cdot x = 0$ and the surface $n'\cdot x = 0$, some Lorentz transformed plane. Okay? The difference of the two θ functions is $+1$ in this shaded region on the right, where you're above the $n\cdot x$ surface but below the $n'\cdot x$ surface; the difference is -1 in the shaded region on the left, where you're above and below in the opposite order; zero when you're above both surfaces, so both θ's equal $+1$, and zero when you're below both surfaces, so both θ's equal zero. Therefore, I can integrate by parts in time without worrying about boundary terms, as the surface integral goes to zero. So

$$Q - Q' = -\int d^4x\,\left[\theta(n\cdot x) - \theta(n'\cdot x)\right]\partial_\mu J^\mu(x) \tag{6.67}$$

which equals zero, as $\partial_\mu J^\mu = 0$. Thus $Q' = Q$. **QED**

I've constructed this argument[5] so that you can readily see that hardly anything is changed

---

[4] [Eds.] See pp. 18–19 and p. 36 in Greiner & Müller *QMS*.

[5] [Eds.] For an extended version of this argument, see Eugene J. Saletan and Alan H. Cromer, *Theoretical Mechanics*, J. Wiley & Sons, (1971), pp. 282–283. In the literature this argument is sometimes called "Laue's theorem", after Max von Laue (Physics Nobel Prize 1914, x-ray diffraction). See M. Laue, "Zur Dynamik der

if I had had a tensor current, say $T^{\lambda\mu}$, instead of a vector current.[6] If we'd had a tensor current, the only difference would have been an extra index with an extra Lorentz matrix on it which I would never ever have had to play with. This matrix would simply have been carried through all of these equations, playing no role in any of my manipulations, except to emerge at the end, to tell me that P^μ was a 4-vector.

6.3 Discrete symmetries

Of course, there are all sorts of symmetries in nature that are *not* continuous, not part of some connected group that contains the identity transformation. Among them are such old friends from non-relativistic quantum mechanics as parity and time reversal. So we will now study discrete symmetries.

A **discrete symmetry** is a transformation where

$$\phi(x) \to \phi'(x) \tag{6.68}$$

but there's no parameter in the transformation; it simply doesn't appear. There's no such thing as a parity transformation by 7°; there is only parity: either there is space reflection or there is no space reflection. It's not like a rotation. We will assume these things are symmetries in the usual sense. That is to say that, at least for appropriately chosen boundaries, the action is invariant:

$$\int d^4x\, \mathscr{L}(\phi, \partial^\mu\phi) = \int d^4x\, \mathscr{L}(\phi', \partial^\mu\phi') \tag{6.69}$$

Of course, there may be many fields, but I leave off the indices out of sheer laziness.

Now in a rough and heuristic way, we would expect such a transformation to be a symmetry of classical physics. And in terms of classical physics this symmetry does what a symmetry always does: it enables you to generate new solutions of the equations of motion out of old solutions. But in general it is not connected with a conservation law, as continuous symmetries are. In quantum mechanics there will be no Hermitian operator associated with these things, to generate the infinitesimal transformation, for the excellent reason that there *is* no infinitesimal transformation. We would nevertheless expect that there would be a unitary operator that effects the *finite* transformation. Indeed though the argument is rough and ready, everything is determined from the action by appropriate variations in canonical quantization and so on. The action is the same for ϕ as it is for ϕ'. We should find a one-to-one correspondence between the Hilbert space we get by doing things in terms of ϕ and the Hilbert space we get by doing things in terms of ϕ', since step by step, every step's the same. If the transformation doesn't change the action it can't change the quantum mechanics; and that means there's a unitary transformation that turns ϕ into ϕ'. You know this argument is rough because it's a lie for time reversal, where there is no unitary transformation, but we won't get to that until the

Relativitätstheorie" (On the dynamics of relativity theory), *Ann. Phys.* **35** (1911) 524–542. Von Laue (he gained the "von" through his father, in 1913) was courageously public in his fierce opposition to the Nazis. Lanczos writes, "Years after the Second World War an eminent physicist from Germany visited [Einstein] in Princeton. As he was about to leave, he asked Einstein whether he wanted to send greetings to his old friends in Germany. 'Grussen Sie Laue', was Einstein's answer: 'Greetings to Laue'. 'Yes', said the visitor, 'I shall be happy to convey these greetings. But you know very well, Professor Einstein, that you have many other friends in Germany'. Einstein pondered for a moment, then he repeated: 'Grussen Sie Laue'." (C. Lanczos, *The Einstein Decade 1905–1915*, Paul Elek Scientific Books, (1971), p. 23.)

[6] [Eds.] See note 11, p. 94.

end of this lecture. This is just a rough argument for the sake of orientation. Let's do some particular cases where we can see simply what is going on and tell whether or not there is a unitary transformation.

Charge conjugation

The first case I will turn to is our good old example of two free fields of the same mass, (6.1),

$$\mathscr{L} = \tfrac{1}{2}\left(\partial^\mu \phi^a \partial_\mu \phi^a - \mu^2 \phi^a \phi^a\right)$$

On a formal level everything I say will also be true if there's an interaction, say of this form, for example:

$$\mathscr{L}_I = -g(\phi^a \phi^a)^2$$

I said that this system was SO(2) invariant but in fact it has a larger invariance group of internal symmetries, including a discrete internal symmetry. It has full O(2) invariance. That is to say it is invariant not just under proper rotations but under improper rotations; not just under rotations but also under reflections. We've already studied all the consequences of the rotations. And since every reflection is the product of some standard reflection and a rotation, we might as well just consider one standard reflection which I will choose to be the transformation

$$\begin{aligned} \phi^1 &\to \phi^1 \\ \phi^2 &\to -\phi^2 \end{aligned} \tag{6.70}$$

At least in the free field case, where we can explore the Hilbert space completely, and even in the general case, if we are willing to extract from non-relativistic quantum mechanics a statement that any operation that doesn't change the canonical commutators is unitarily implementable, we can see that there is a unitary transformation that effects (6.70). In the free case, $g = 0$, we just read off from (6.70) that if there is a unitary transformation U such that

$$\begin{aligned} \phi^1 &\to U^\dagger \phi^1 U = \phi^1 \\ \phi^2 &\to U^\dagger \phi^2 U = -\phi^2 \end{aligned} \tag{6.71}$$

then U operates on the annihilation operators like this:

$$\begin{aligned} a_{\mathbf{p}}^{(1)} &\to U^\dagger a_{\mathbf{p}}^{(1)} U = a_{\mathbf{p}}^{(1)} \\ a_{\mathbf{p}}^{(2)} &\to U^\dagger a_{\mathbf{p}}^{(2)} U = -a_{\mathbf{p}}^{(2)} \end{aligned} \tag{6.72}$$

and the same thing for the creation operators just by taking the adjoint.

A unitary transformation that does the job in the free case acts on states with a definite number of particles of type 1 and a definite number particles of type 2 by multiplying the state by (-1), or equivalently $e^{i\pi}$, raised to the number operator N_2 of 2-type particles, where

$$N_2 = \int d^3\mathbf{p}\, a_{\mathbf{p}}^{(2)\dagger} a_{\mathbf{p}}^{(2)} \tag{6.73}$$

Then

$$U\left|p_1, p_2, \ldots, p_n\right\rangle = (-1)^{N_2}\left|p_1, p_2, \ldots, p_n\right\rangle \tag{6.74}$$

That obviously has the desired property, and works just as well on the fields:

$$U^\dagger \phi^1 U = (-1)^{N_2} \phi^1 (-1)^{N_2} = \phi^1; \quad U^\dagger \phi^2 U = (-1)^{N_2} \phi^2 (-1)^{N_2} = -\phi^2 \tag{6.75}$$

The first equation follows because N_2 commutes with ϕ^1. The second equation is true because ϕ^2 will either create or annihilate a type 2 meson, and hence change their number by 1.

This unitary transformation is perhaps more simply expressed in terms of the b's and the c's. First, recall the definition (6.23) of the complex field ψ and its conjugate ψ^*. Then

$$\psi = \frac{1}{\sqrt{2}}(\phi^1 + i\phi^2) \overset{U}{\rightarrow} \frac{1}{\sqrt{2}}(\phi^1 - i\phi^2) = \psi^*, \quad \psi^* \overset{U}{\rightarrow} \psi \tag{6.76}$$

Equally well you could say that this U acting on any state turns all the b-type particles into c-type particles and all the c-type particles into b-type particles. From equations (6.14) and (6.15),

$$\begin{aligned} b_{\mathbf{p}} &= \frac{1}{\sqrt{2}}(a_{\mathbf{p}}^{(1)} + ia_{\mathbf{p}}^{(2)}) \overset{U}{\rightarrow} \frac{1}{\sqrt{2}}(a_{\mathbf{p}}^{(1)} - ia_{\mathbf{p}}^{(2)}) = c_{\mathbf{p}} \\ c_{\mathbf{p}} &\overset{U}{\rightarrow} b_{\mathbf{p}} \end{aligned} \tag{6.77}$$

Such a transformation is called **charge conjugation**. "Conjugation" is a bad word; it sounds like it shouldn't be unitary. After all, complex conjugation is not a unitary operation. Perhaps it would better be called "particle–antiparticle exchange", because the transformation exchanges particles and antiparticles, π^+'s and π^-'s for example. We normally call this symmetry C, and put a little subscript C on the unitary operator, U_C, to tell you that that's the transformation it's associated with. We can rewrite the transformations on $b_{\mathbf{p}}$ and $c_{\mathbf{p}}$ in a compact form,

$$C\colon \begin{Bmatrix} b_{\mathbf{p}} \\ c_{\mathbf{p}} \end{Bmatrix} \rightarrow U_C^\dagger \begin{Bmatrix} b_{\mathbf{p}} \\ c_{\mathbf{p}} \end{Bmatrix} U_C = \begin{Bmatrix} c_{\mathbf{p}} \\ b_{\mathbf{p}} \end{Bmatrix} \tag{6.78}$$

As I said before, in general a unitary operator is not an observable, and therefore we normally don't get a conserved quantity even though the unitary operator may commute with the Hamiltonian. However there is one special case in which a unitary operator *does* give us a conserved quantity, and that is when the unitary operator is itself Hermitian. This happens in the case of charge conjugation because operating twice with U_C is just the identity. Applying C once, you turn every b-type particle into a c-type particle, and then applying it a second time you turn it back again into a b-type particle;

$$U_C^2 = 1 \tag{6.79}$$

Because U_C is also unitary, $U_C^\dagger = U_C^{-1}$, and so

$$U_C U_C^\dagger = 1 \Rightarrow U_C^\dagger = U_C \tag{6.80}$$

That is to say U_C is both unitary and Hermitian. That is rather obvious in terms of the C operator's action on type 1 and type 2 particles, where the eigenvalues were $+1$ and -1, numbers that are both of modulus one and real. Note that from (6.19),

$$C\colon Q \rightarrow U_C^\dagger Q U_C = -Q \tag{6.81}$$

(this is the Q associated with the continuous group SO(2)). And so in this particular case, even though this transformation is not associated with a continuous symmetry, we can divide states up into C-eigenstates, because C is also a Hermitian operator. This is usually not done in practice except when you have equal numbers of particles and antiparticles, considering states of a π^+-π^- system for example. The terminology we now use for particles connected by this kind of transformation, to have equal numbers of particles and antiparticles, is **even** and **odd** under charge conjugation, depending upon whether the wave function is symmetric or antisymmetric under exchange of the π^+ and π^- variables. Since charge conjugation commutes with the Hamiltonian, the notion of even or odd under charge conjugation can be used to deduce consequences for transition amplitudes. Actually we won't do that for π^+'s and π^-'s because you gain no information there that you don't gain from parity, but we will use it for electrons and positrons, where you do gain additional information.

I haven't *deduced* particle–antiparticle symmetry. I have simply given an example of a theory which I cooked up to possess a symmetry that is structurally similar to a symmetry I know exists in nature by experiment, just to show you how such a symmetry could arise within the context of Lagrangian field theory.[7]

Parity

As my next example, I would like to discuss **parity**. Parity changes the signs of the spatial coordinates, leaving the time coordinate untouched:

$$P\colon \begin{cases} \mathbf{x} \to -\mathbf{x} \\ t \to t \end{cases} \tag{6.82}$$

Parity is closely related to reflection (say, reflection in the x-y plane, which would take $z \to -z$ and leave x, y and t unchanged). A parity transformation is the same thing as a reflection (in any plane) followed by a rotation about the normal to that plane by 180°. So a theory with rotational symmetry is parity-invariant if and only if it is reflection-invariant. But parity is an improper rotation (its determinant equals -1), and parity invariance is *not* implied by rotational invariance alone. Nevertheless, until the discovery by Wu and her group[8] that parity was violated in beta decay, it was universally assumed that any realistic physical theory would be parity-symmetric.

An ordinary scalar (mass m, for example) is invariant under parity, while an ordinary 3-vector, like velocity, $\mathbf{v}$, changes sign:

$$P\colon m \to m, \quad P\colon \mathbf{v} \to -\mathbf{v} \tag{6.83}$$

On the other hand, a cross-product of two vectors (the angular momentum $\mathbf{L} = \mathbf{r} \times \mathbf{p}$, say) picks up *two* minus signs, and the scalar triple product $w = \mathbf{a} \cdot (\mathbf{b} \times \mathbf{c})$ is a scalar that changes sign:

$$P\colon \mathbf{L} \to \mathbf{L}, \quad P\colon w \to -w \tag{6.84}$$

---

[7] [Eds.] A student asks about the *CPT* Theorem. Coleman responds: "*CPT* is very different. That's something we won't get to until very late in this course if we bother to do it all. Just from general assumptions of field theory—Lorentz invariance, the positivity of the energy, and locality (the fact that fields commute at spacelike separations)—without making any assumption about the form of the Lagrangian, or even whether things are derived from a Lagrangian, you can show that there is *CPT* invariance. This is the famous *CPT* Theorem. Although one after another—parity, time reversal, and charge conjugation—have fallen to experimenters, the combined symmetry *CPT* remains unbroken, and we believe the reason is the *CPT* Theorem. Indeed one of the most revolutionary experimental results conceivable—well, violation of conservation of energy would also be pretty revolutionary—would be that *CPT* had been found to be violated. If that were so, we would not only have to sacrifice all the particular theories with which we hope to explain the world; we would have to sacrifice the general ideas, including the idea of a Lagrangian field theory and indeed the general idea of local fields. It would be back to Lecture 1; this whole course would be canceled out!"

[8] [Eds.] C. S. Wu, E. Ambler, R. W. Hayward, D. D. Hoppes, and R. P. Hudson, "Experimental Test of Parity Conservation in Beta Decay", *Phys. Rev.* **105** (1957) 1413–1415.

We call these *axial* vectors and *pseudo*scalars, respectively, because of their anomalous behavior under parity.

In a field theory we can have scalar fields, pseudoscalar fields, vector fields, axial vector fields, and so on. Moreover, if there are several fields, they can be mixed by the parity transformation. In this sense the parity transformation is intrinsically ambiguous: it takes $\mathbf{x}$ into $-\mathbf{x}$ (and t into t), but what *else* it does is a matter of convention and convenience, though we will assume that its action is always linear:

$$P\colon \phi^a(\mathbf{x},t) \to M^a_b\phi^b(-\mathbf{x},t) \tag{6.85}$$

(summing on repeated indices). Parity turns the fields at a point $(\mathbf{x},t)$ into some linear combination of the fields at the point $(-\mathbf{x},t)$. A theory is parity invariant if the action is unchanged by *some* transformation of the form (6.85), but it is not always obvious how we should choose the coefficients M^a_b. Parity can be very strange and I hope to amuse you by cooking up a bunch of theories, some of which have no actual resemblance to nature, in which P takes peculiar forms.

EXAMPLE 3. *Scalar field with a quartic interaction*

Let's look at a scalar field with a quartic interaction:

$$\mathscr{L}^{(1)} = \tfrac{1}{2}(\partial_\mu\phi)^2 - \tfrac{1}{2}\mu^2\phi^2 - g\phi^4 \tag{6.86}$$

(I am tired of writing $\partial^\mu\phi\partial_\mu\phi$; you know what the first term means.) This obviously possesses a parity invariance,

$$P\colon \phi(\mathbf{x},t) \to \phi(-\mathbf{x},t) \tag{6.87}$$

This transformation changes the Lagrangian

$$P\colon \mathscr{L}^{(1)}(\mathbf{x},t) \to \mathscr{L}^{(1)}(-\mathbf{x},t) \tag{6.88}$$

but it doesn't change the action. In the case of $g = 0$, it is implemented by the unitary transformation

$$P\colon \begin{Bmatrix} a_{\mathbf{p}} \\ a^\dagger_{\mathbf{p}} \end{Bmatrix} \to U^\dagger_P \begin{Bmatrix} a_{\mathbf{p}} \\ a^\dagger_{\mathbf{p}} \end{Bmatrix} U_P = \begin{Bmatrix} a_{-\mathbf{p}} \\ a^\dagger_{-\mathbf{p}} \end{Bmatrix} \tag{6.89}$$

The parity transformation turns either a creation or an annihilation operator with momentum $\mathbf{p}$ into a creation or annihilation operator with momentum $-\mathbf{p}$. The proof is simple: Apply (6.87) to the definition (6.8) of the free fields, and then change the integration variable $\mathbf{p}$ into the integration variable $-\mathbf{p}$. This turns $\mathbf{x}$ into $-\mathbf{x}$ and doesn't change t. Thus parity takes a particle going, say, this way, $\rightarrow$, and turns it into particle going *this* way, $\leftarrow$, the usual thing that parity does in non-relativistic particle physics. Acting on the basis states,

$$U_P\,|\mathbf{p}_1, \mathbf{p}_2, \ldots, \mathbf{p}_n\rangle = |-\mathbf{p}_1, -\mathbf{p}_2, \ldots, -\mathbf{p}_n\rangle \tag{6.90}$$

There is an alternative parity transformation, which I will call P',

$$P'\colon \phi(\mathbf{x},t) \to -\phi(-\mathbf{x},t) \tag{6.91}$$

This transformation is also an invariance of our Lagrangian (6.86) if (6.87) is, because our Lagrangian is invariant under $\phi \to -\phi$ (a trivial internal symmetry closely corresponding to what we did to ϕ^2, (6.70), in the discussion of charge conjugation), and the product of two symmetries is a symmetry. The transformation law (6.87) is called the **scalar transformation law**, and (6.91) is called the **pseudoscalar transformation law**. The unitary transformation $U_{P'}$ is given by

$$U_{P'} = (-1)^N U_P \tag{6.92}$$

where N is the number of pseudoscalar fields being acted on. Likewise, on a basis state describing n pseudoscalar particles,

$$U_{P'} \left|\mathbf{p}_1, \mathbf{p}_2, \ldots, \mathbf{p}_n\right\rangle = (-1)^n \left|-\mathbf{p}_1, -\mathbf{p}_2, \ldots, -\mathbf{p}_n\right\rangle \tag{6.93}$$

The first important point of this example is that it is merely a matter of convention for a particular theory whether you say ϕ is a scalar field or ϕ is a pseudoscalar field. Whenever there is an internal symmetry in a theory, I can multiply one definition of parity by an element of the internal symmetry group, discrete or continuous, and get another definition of parity. This theory has two symmetries, among others; one which is C-like and one which is P-like. The product CP is a symmetry; and which you call parity and which you call charge conjugation or $\phi \to -\phi$ times parity is a matter of taste; nobody can fault you. What is important is the total group of symmetries admitted by Lagrangians, from which one draws all sorts of physical consequences, not what names one attaches to individual members. As long as you have one possible definition of parity, and you have internal symmetries around, you can always adopt a new convention and new nomenclature. You can take the product of one of those internal symmetries and parity and call that parity, and call your original parity the product of your new parity and the inverse internal symmetry. Nobody can stop you and nobody should, as long as when you are writing your papers or giving your lectures, you are clear about what convention you are using.

Of course if the Lagrangian does not have the internal symmetry then you might end up with a unique definition of parity because there will be no internal symmetries from which you can multiply parity.

EXAMPLE 4. *Cubic and quartic scalar interactions together*

Consider the Lagrangian

$$\mathscr{L}^{(2)} = \tfrac{1}{2}(\partial_\mu \phi)^2 - \tfrac{1}{2}\mu^2\phi^2 - g\phi^4 - h\phi^3 \tag{6.94}$$

If I take $\mathscr{L}^{(1)}$, the same Lagrangian as before, and add to it a term $h\phi^3$, then $\phi \to -\phi$ (in the sense of (6.91)) is no longer a good definition of parity nor is it a symmetry. In this case the only sensible definition of parity is the scalar law, without the minus sign; the pseudoscalar won't work. You can call the pseudoscalar transformation "parity" if you want, but then you have got yourself into the position of saying this theory is not parity conserving, which is a silly thing to say. In nature, in the real world, sometimes there is no good definition of parity. There is *no* way of defining parity so that the weak interactions preserve parity.

If you throw away the weak interactions you have a lot of internal symmetries: the commuting one parameter groups corresponding to electron number, muon number, nucleon number, electric charge, and strangeness. The relative parity of the electron and the muon is a matter of convention. You can always multiply muon number into your definition of parity to change the parity of the muon and the muon neutrino and nothing else. The relative parity of the electron and the proton is a matter of convention, as is that of the proton and the Λ

hyperon; you can multiply strangeness into that definition of parity. Usually these conventions are established to make all those relative parities +1, but that's just convention.

I have shown you an example where the scalar transformation is an okay definition of parity but the pseudoscalar is not. I will now construct examples where it goes the other way. These examples are rather unnatural, involving scalar fields. When finally we talk about fermions, we will find we can write very simple interactions that have this property, but it can also be shown with scalar fields. To do so, I have to write down a grotesque sort of interaction, using the four-dimensional Levi–Civita tensor, $\epsilon_{\mu\nu\rho\sigma}$, which is completely antisymmetric (like its three-dimensional cousin ϵ_{ijk}), and $\epsilon_{0123} = 1$. With this and with four 4-vectors, one can form a Lorentz scalar but it will have funny parity properties. I will now give an example of how to make something where the pseudoscalar law is forced on us if we hope to have the Lagrangian invariant.

EXAMPLE 5. *Coupling via* $\epsilon_{\mu\nu\rho\sigma}$

$$\mathscr{L}^{(3)} = \tfrac{1}{2}\sum_{a=1}^{4}\Big[(\partial_\mu\phi^a)^2 - \mu_a^2(\phi^a)^2\Big] - \lambda\epsilon_{\mu\nu\rho\sigma}\partial^\mu\phi^1\partial^\nu\phi^2\partial^\rho\phi^3\partial^\sigma\phi^4 \tag{6.95}$$

If we were to declare all four fields to transform as scalars, then the Levi–Civita term breaks parity because, as you will notice, every term involves one time index and three space indices. The space derivatives change sign under parity, and the time derivatives do not. We pile up three minus signs when we parity transform this object, which is a disaster since three minus signs change the sign of this term. We have to declare that one of the fields is pseudoscalar and three of the fields are scalar, or vice-versa. Since we have total freedom to make the whole large group of internal transformations, it's a matter of taste which one (or three) of the four we call pseudoscalar. That is just a matter of how we multiply an internal symmetry by a parity.

EXAMPLE 6. *The last example, plus a sum of cubic terms*

$$\mathscr{L}^{(4)} = \mathscr{L}^{(3)} - h\sum_{a=1}^{4}(\phi^a)^3 \tag{6.96}$$

There is *no* good definition of parity for $\mathscr{L}^{(4)}$. I have to have a minus sign in one (or three) of the fields to make $\mathscr{L}^{(3)}$ work out all right, but then the new term, in h, is disastrous, with a sign of -1. On the other hand if I choose all the fields to be scalar, to get the new term to work out, it breaks the invariance of $\mathscr{L}^{(3)}$. Whether I choose scalar or pseudoscalar fields, it doesn't matter; there is *no* symmetry that can be interpreted as parity for this Lagrangian.

Now this demonstration might lead you to think the only possible effect of a parity transformation is a plus sign or a minus sign, where the particles have intrinsic positive parity or intrinsic negative parity. I will now give an example where the only possible definition of parity has an i in it. This will be super-grotesque and will involve a complex scalar field, ψ.

EXAMPLE 7. *Modifying the last example by adding new fields*

$$\begin{aligned}\mathscr{L}^{(5)} = \sum_{a=1}^{4}\left[\tfrac{1}{2}(\partial_\mu\phi^a)^2 - \tfrac{1}{2}\mu_a^2(\phi^a)^2 - h(\phi^a)^3\right] + \partial_\mu\psi^*\partial^\mu\psi - \mu_5^2\psi^*\psi \\ -\lambda\epsilon^{\mu\nu\rho\sigma}\partial_\mu\phi^1\partial_\nu\phi^2\partial_\rho\phi^3\partial_\sigma\phi^4\left[(\psi)^2 + (\psi^*)^2\right]\end{aligned} \tag{6.97}$$

My free Lagrangian now has five fields in it, four real scalar fields ϕ^a with some four masses μ_a, and a complex scalar field ψ with some fifth mass μ_5. We still have the h term that keeps us from letting the scalar fields be pseudoscalar. Now, however, I've multiplied this last term by the sum of the squares of ψ and ψ^*. The sesquilinear form in the fields with an epsilon tensor is not one we will encounter in any of the theories we will take seriously, but it's still an amusing example. Though grotesque, it's got all the properties we want: it's Hermitian, and if we are creative, it will have a legitimate parity. The four real fields can be taken as scalars, so all the terms except for the last are all right. We need the last to go into +1 times itself. That will happen for the last term even with scalars, provided

$$U_P^\dagger\left\{\begin{matrix}\psi(\mathbf{x},t)\\ \psi^*(\mathbf{x},t)\end{matrix}\right\}U_P = \left\{\begin{matrix}i\psi(-\mathbf{x},t)\\ -i\psi^*(-\mathbf{x},t)\end{matrix}\right\} \tag{6.98}$$

Since $\psi(\mathbf{x},t)$ goes into $i\psi(-\mathbf{x},t)$, the square of ψ supplies the missing minus sign for the epsilon term from i squared, and the same is true for ψ^*. The other terms in ψ and ψ^* are unchanged by (6.98).

This is just for fun, but it is an example where parity is so strange that, as you can readily convince yourself, for this grotesque theory this is the only possible parity that will work. And in this case, things are so strange that the square of parity is not even 1 on the complex field. If you ever read a paper in which someone says on general *a priori* grounds the square of parity must be +1, send him a postcard telling him about this example. He may say, "Oh, I wouldn't call that parity," but then you would say he was being pretty foolish, because if the world really were like this, there would be this very useful symmetry that turns observables at the point $\mathbf{x}$ into observables at the point $-\mathbf{x}$, putting all sorts of restrictions on scattering cross-sections and energy levels and all the things a symmetry usually does, and its square happens not to be one. If he doesn't want to call that parity, what is he going to call it?[9]

Time reversal

Now of the famous discrete symmetries known and loved by physicists, I have left one undiscussed: time reversal. Time reversal is rather peculiar in that unlike all the other symmetries we have discussed until now, it is not represented by a unitary operator; it is represented by an *anti*-unitary operator.

---

[9] [Eds.] A student asks: Why are we concentrating on *linear* transformations? Coleman replies: "The linear functions come from the fact that in all our examples, the kinetic energy is a standard quadratic form. That means nonlinear transformations will turn the kinetic energy from a quadratic function of the fields to a messy function of the fields. We could rewrite things in an ordinary, perfectly symmetric theory with two fields in it, ϕ^1 and ϕ^2. Out of sheer perversity I could introduce fields $\phi^{1\prime} = \phi^1$ and, say, $\phi^{2\prime} = (\phi^2 + a\phi^1)^2$. And then my kinetic energy would look rather disgusting and my ordinary isospin transformations discussed earlier that turned ϕ^1 into ϕ^2 would look like horrible nonlinear transformations. That's a silly thing to do but it is not absolutely forbidden. So there is nothing sacred about linear transformation laws of fields. It's the *bilinear* structure of the kinetic energy that makes linear transformation laws of such interest to us."

Consider a particle in one dimension moving in a potential. The classical theory is invariant under the time reversal transformation

$$T\colon \begin{cases} q(t) \to q(-t) \\ p(t) \to -p(-t) \end{cases} \tag{6.99}$$

While $q(t)$ goes into $q(-t)$, $p(t)$ goes into $-p(-t)$ because p is proportional to $\dot{q}$. That is to say if you take a motion picture of this classical system and run the reel for the projector backwards, you will obtain a motion perfectly consistent with the classical equations of motion. One's first guess is that there should be a unitary operator, which I'll call U_T, that effects this transformation in the quantum theory:

$$U_T^\dagger \begin{Bmatrix} q(t) \\ p(t) \end{Bmatrix} U_T \stackrel{?}{=} \begin{Bmatrix} q(-t) \\ -p(-t) \end{Bmatrix} \tag{6.100}$$

This, however, leads one into a grinding contradiction almost immediately. We know from the canonical commutators that, at equal times,

$$[q(t), p(t)] = i \tag{6.101}$$

Apply U_T to the right-hand side of the commutator and $U_T^\dagger$ to the left side, for the time $t = 0$:

$$U_T^\dagger q(0) U_T U_T^\dagger p(0) U_T - U_T^\dagger p(0) U_T U_T^\dagger q(0) U_T = -q(0)p(0) + p(0)q(0) = -[q(0), p(0)] \tag{6.102}$$

which is unfortunately not i but $-i$. It looks like we would have to give up our canonical commutation relations to implement time reversal. Thus we have obtained an immediate contradiction with our hypothesis, so the answer to this is not "What is the operator?", but instead, "There *is* no (unitary) operator."

There is a second contradiction. We expect that U_T, if it exists, should reverse time evolution, i.e.,

$$U_T^\dagger e^{-iHt} U_T = e^{iHt} \tag{6.103}$$

Take d/dt of both sides of this equation at $t = 0$ to obtain

$$U_T^\dagger(-iH)U_T = iH \tag{6.104}$$

Canceling the i's, we see that H and $-H$ are related by a unitary transformation. Operators so related must have the same spectrum, and yet they cannot both be bounded from below! A unitary time reversal operator makes no sense whatsoever. The resolution of these difficulties is well known: Time reversal is not a unitary operator but an anti-unitary operator. As I will prove, anti-unitary operators are also anti-linear.

Before getting into anti-unitary operators, let's review the properties of unitary operators. Unfortunately, one reason the Dirac notation is so wonderful is that a lot of facts about linear operators are embedded in it subliminally. Anti-linear operators are therefore difficult to describe in Dirac notation. So instead of using bras and kets I will use an alternative notation. I will label states by lowercase Latin letters: a, b, ... These are vectors in Hilbert space. And instead of talking about the inner product $\langle a|b\rangle$ I will write that as (a, b). Complex numbers will be denoted by Greek letters, α, β, ... and operators will be denoted by capital Latin letters, A, B, ...

An operator U is *unitary* if two conditions are met: it is invertible, and for any two vectors a and b in Hilbert space, the Hilbert space inner product (a, b) is preserved:

$$(Ua, Ub) = (a, b) \tag{6.105}$$

Thus U preserves the norm. (The simplest unitary operator is 1.) An operator U is *linear* if for any two complex numbers α and β and any two vectors a and b in Hilbert space,

$$U(\alpha a + \beta b) = \alpha U a + \beta U b \tag{6.106}$$

The condition (6.105) is sufficient to show that U is linear, by a variation on a theorem to be shown below. The *adjoint* $A^\dagger$ of a linear operator A is defined by

$$(a, A^\dagger b) = (Aa, b) \tag{6.107}$$

It's easy to show that if U is unitary, then $U^\dagger = U^{-1}$:

$$(a, U^{-1}b) = (Ua, UU^{-1}b) = (Ua, b) = (a, U^\dagger b) \tag{6.108}$$

the first step following from U being unitary. A transformation of the states $a \to Ua$ can be thought of as a transformation on the operators:

$$(a, Ab) \to (Ua, AUb) = (a, U^\dagger AUb) \;\Rightarrow\; A \to U^\dagger AU \tag{6.109}$$

An anti-unitary operator is an invertible operator, traditionally represented by an omega, Ω (one of the few instances of felicitous notation in theoretical physics, as an omega is a U upside down), defined by

$$(\Omega a, \Omega b) = (a, b)^* = (b, a) \tag{6.110}$$

The product of two anti-unitary operators is a unitary operator, the product of an anti-unitary object and a unitary object is anti-unitary, and so on. The multiplication table is shown in Figure 6.2.

	U	Ω
U	U	Ω
Ω	Ω	U

Figure 6.2: Multiplication table for Ω and U

Such operators Ω certainly exist. A simple example which obeys all of these conditions in one-dimensional quantum mechanics is complex conjugation K of the Schrödinger wave function. The complex conjugate of a linear superposition of two wave functions is the superposition of the complex conjugate with complex conjugate coefficients:

$$K(\alpha\psi_1 + \beta\psi_2) = \alpha^*\psi_1^* + \beta^*\psi_2^* \tag{6.111}$$

Likewise if I complex conjugate both factors in the inner product I complex conjugate the inner product:

$$(K\psi_1, K\psi_2) = (\psi_1^*, \psi_2^*) = (\psi_2, \psi_1) = (\psi_1, \psi_2)^* \tag{6.112}$$

A useful fact (especially conceptually) is that any anti-unitary operator Ω can be written as the product of a unitary operator U and the complex conjugation operator K: $\Omega = UK$. It's easy to prove this by construction: take $U = \Omega K$.

An operator A (not necessarily invertible) is called *anti-linear* if

$$A(\alpha a + \beta b) = \alpha^* Aa + \beta^* Ab \tag{6.113}$$

To show that an anti-unitary operator must be also anti-linear, consider the inner product of

$$\Omega(\alpha a + \beta b) - (\alpha^* \Omega a + \beta^* \Omega b) \tag{6.114}$$

with itself. If this is equal to zero, the positive-definite inner product implies that the original state is zero, i.e.,

$$\Omega(\alpha a + \beta b) = \alpha^* \Omega a + \beta^* \Omega b \tag{6.115}$$

It suffices to multiply out the nine terms of the inner product (6.114) and apply the relation (6.110) to remove all instances of Ω, e.g.,

$$(\alpha^* \Omega a, \Omega(\alpha a + \beta b)) = \alpha(\Omega a, \Omega(\alpha a + \beta b)) = \alpha(\alpha a + \beta b, a) \tag{6.116}$$

and then expanding the five terms containing $(\alpha a + \beta b)$. Sure enough, you obtain zero, thus establishing (6.115). (The analogous proof that unitary operators are necessarily linear only uses properties of the inner product, and is even easier.)

The transformation of the states under an anti-unitary operator Ω, $a \to \Omega a$, can also be thought of as a transformation of the Hermitian operators in the theory, though in a more limited sense. Consider the expectation value of a Hermitian operator A acting on the state a. It transforms under Ω as

$$(a, Aa) \to (\Omega a, A\Omega a) = (A\Omega a, \Omega a) = (\Omega\Omega^{-1} A\Omega a, \Omega a) = (a, \Omega^{-1} A\Omega a) \tag{6.117}$$

So the transformation may alternatively be thought of as $A \to \Omega^{-1} A \Omega$. We do not write $\Omega^\dagger A \Omega$ because $\Omega^\dagger$ is not even defined for anti-unitary operators.

The resolution of the contradictions, (6.102) and (6.104), is that time reversal is effected by an anti-unitary operator. For the first contradiction, (6.102),

$$\Omega_T^{-1}\, [q(0), p(0)]\, \Omega_T = -[q(0), p(0)] = \Omega_T^{-1}\, i\, \Omega_T = -i \tag{6.118}$$

because whenever we drag a complex number through an anti-unitary operator we complex conjugate it. Thus the right- and left-hand sides of the equation match and the contradiction disappears. Indeed, for this particular problem, it is easy to find the anti-unitary operator that effects time reversal: it is complex conjugation in the x representation. That turns x into x and it turns p which is $-i\partial/\partial x$ into $-p$ because of the i.

As for the second contradiction, (6.104),

$$\Omega_T^{-1}(-iH)\Omega_T = iH = \Omega_T^{-1}(-i)\Omega_T\Omega_T^{-1} H\Omega_T = i\Omega_T^{-1} H\Omega_T \;\Rightarrow\; \Omega_T^{-1} H\Omega_T = H \tag{6.119}$$

which resolves the second contradiction, provided H is invariant under time-reversal. So much for a lightning summary of the situation in non-relativistic quantum mechanics.

You may have heard of Wigner's beautiful theorem,[10] which tells you that, up to phases, an operator that preserves the norm of the inner product must be either unitary or anti-unitary. (It is *not* necessary to preserve inner products; they aren't measurable. Only the *probabilities* are measurable.) In the study of symmetries, as Wigner pointed out, all one really has to consider on *a priori* grounds are unitary and anti-unitary operators; there is no need worrying that someday we will find a symmetry that is implemented by a quasi-unitary operator or some other entity not yet thought of by mathematicians. Simply put, Wigner's theorem says that if F is a continuous transformation mapping some Hilbert space $\mathcal{H}$ into itself, and if F preserves probabilities, then F must be the product of a phase and a unitary or an anti-unitary operator. That is, if $a, b \in \mathcal{H}$, then

$$|(F(a), F(b)|^2 = |(a, b)|^2 \;\Rightarrow\; F(a) = e^{i\phi(a)} \times \begin{cases} U, & \text{a unitary operator, if } F \text{ is unitary} \\ \Omega, & \text{an anti-unitary operator, otherwise} \end{cases}$$

where $\phi : \mathcal{H} \rightarrow \mathbb{R}$.

We now wish to take our standard field theoretic system, the free scalar field of mass μ, and find a time reversal operator. So we are interested in the system defined by the Lagrange density

$$\mathscr{L} = \tfrac{1}{2}\left(\partial^\mu \phi\right)^2 - \tfrac{1}{2}\mu^2\phi^2 \tag{6.120}$$

I pick this one because we can explicitly write the operators on the state space. What I said about parity also applies to time reversal; I can multiply the time reversal operator by any internal symmetry and obtain an equally good time reversal operator. Let's try to figure out what Ω_T must be, working directly with the states, the opposite direction from which we worked before, and then show what Ω_T does to the fields. In a relativistic theory, it is more convenient to study Ω_{PT} than Ω_T, that is to say the product of parity and time reversal. The reason is very simple. Acting on x^μ, PT multiplies all four components by -1. This operation commutes with the Lorentz group. Time reversal multiplies only t by -1, singling out one component of the 4-vector x^μ, and does not mesh well with Lorentz transformations.

Now, what do we expect the combined symmetry PT to do to a single-particle state? Well, if I have a particle whose momentum vector is represented by an arrow, $\rightarrow$, parity will reverse the sign, and make it $\leftarrow$; but time reversal will reverse it again from $\leftarrow$ to $\rightarrow$. So I expect PT to do *nothing* to the momentum of the particle. Therefore I define the anti-unitary operator Ω_{PT} acting on a complete set of basis states (assuming that $\Omega_{PT}\,|0\rangle = |0\rangle$)

$$\Omega_{PT}\,|\mathbf{p}_1, \mathbf{p}_2, \cdots, \mathbf{p}_n\rangle = |\mathbf{p}_1, \mathbf{p}_2, \cdots, \mathbf{p}_n\rangle \tag{6.121}$$

For either kind of operator, unitary or anti-unitary, if you specify its action on a complete orthonormal basis, you have specified it everywhere. Notice that this does not imply (as it would for a unitary operator) that $\Omega_{PT} = 1$, because it's an anti-unitary operator, and therefore, although it turns these states into themselves, it doesn't turn i times these states

[10] [Eds.] Eugene Wigner, *Group Theory and Its Application to the Quantum Mechanics of Atomic Spectra*, Academic Press, 1959, Appendix to Chap. 20, "Electron Spin", pp. 233–236, and Chap. 26, "Time Inversion", pp. 325–348; Weinberg *QTF1*, Chap. 2, Appendix A, pp. 91–96. Wigner, a childhood friend of John von Neumann and trained as a chemical engineer, was instrumental in the construction of the Chicago nuclear pile (2 December 1942), and, with Alvin M. Weinberg, wrote the book on the design of subsequent reactors. Perhaps the leading proponent of group theoretical methods in quantum mechanics, he shared the 1963 Physics Nobel with Maria Goeppert-Mayer (until 2018, the only other woman Physics Laureate besides Marie Curie) and J. Hans D. Jensen. Wigner was Dirac's brother-in-law; Dirac married Wigner's sister Margit in 1937.

into themselves; it turns them into $-i$ times these states. Okay, that's our guess. I've defined an anti-unitary operator which is a symmetry if there ever was one; it commutes with Lorentz transformations, the Hamiltonian, and the momentum; that's surely good enough to be a symmetry..[11] Let's figure out what it does to the fields, ϕ.

Well, let's begin with the annihilation and creation operators. The formulas that define the annihilation and creation operators only involve real numbers, and Ω_{PT} does nothing to $\mathbf{p}$, so one easily deduces that

$$\Omega_{PT} a_{\mathbf{p}} = a_{\mathbf{p}} \Omega_{PT} \tag{6.122}$$

Equivalently, multiplying from the left by Ω_{PT}^{-1} we get

$$a_{\mathbf{p}} = \Omega_{PT}^{-1} a_{\mathbf{p}} \Omega_{PT} \tag{6.123}$$

It sure looks like Ω_{PT} acts like 1 so far. By the same reasoning we have a similar equation with $a_{\mathbf{p}}^{\dagger}$ replacing $a_{\mathbf{p}}$. Now what about the field? Here comes the cute trick. The field, as you recall from (6.8), is

$$\phi(x) = \int \frac{d^3\mathbf{p}}{(2\pi)^{3/2}\sqrt{2\omega_{\mathbf{p}}}} \left(a_{\mathbf{p}} e^{-ip\cdot x} + a_{\mathbf{p}}^{\dagger} e^{ip\cdot x} \right) \tag{6.124}$$

Now when I apply Ω_{PT}^{-1} and Ω_{PT} to this, what happens? Well, nothing happens to the $d^3\mathbf{p}$, nothing happens to the $(2\pi)^{3/2}$, nothing happens to the $\sqrt{2\omega_{\mathbf{p}}}$, nothing happens to the $a_{\mathbf{p}}$. But ahh, the $e^{-ip\cdot x}$ gets complex conjugated, and likewise the $e^{ip\cdot x}$, so I get $\phi(-x)$, which is exactly what I would want for a PT operation—it turns the field at the spacetime point x^{μ} into the field at the spacetime point $-x^{\mu}$:

$$PT\colon \phi(x) \to \Omega_{PT}^{-1} \phi(x) \Omega_{PT} = \phi(-x) \tag{6.125}$$

The operator Ω_{PT} is *not* acting like 1, the identity, because the operator is *anti*-unitary. Any equation defining an operator in terms of the states where it only has real matrix elements will commute with Ω_{PT}, but not if the elements are complex or imaginary.

This concludes the discussion of time reversal. Because we were dealing with scalar particles, the discussion was rather simple. Much later in this course when we deal with particles with spin, time reversal will be somewhat more complicated, just as it is with spin in non-relativistic quantum mechanics. This also concludes the general discussion of symmetry. Our next topic is the beginning of perturbation theory.

---

[11] [Eds.] Coleman will state later (§22.4) that the Klein–Gordon equation is invariant under PT. He doesn't prove this, but it's obvious. The KG operator is second order in both $\mathbf{x}$ and t, so it is invariant. We've seen that $\phi(x) \to \phi(-x)$. By the Chain Rule, the Klein–Gordon equation is thus invariant under PT.

7

Introduction to perturbation theory and scattering

We are now going to turn to a topic that in one guise or other will occupy us for the rest of the semester, the topic of perturbation theory and scattering. This will lead us to Feynman diagrams, complicated homework problems, worries about renormalization, and everything else. But we begin at the beginning.

I want to divide the problem into two pieces: perturbation theory, and scattering, at least on our first go-through. First I will discuss perturbation theory: How one solves quantum dynamics in perturbation theory, how one finds the transition matrix or whatever you wish to discuss, between states at finite times in perturbation theory. Next, I will discuss the asymptotic problem: Given such a solution, how does one extract from it scattering matrix elements. So first I'll discuss perturbative dynamics. After that I will discuss scattering.

7.1 The Schrödinger and Heisenberg pictures

I begin by reminding you of the two pictures that play such a large role in ordinary quantum mechanics, the Schrödinger and Heisenberg pictures. I will put little subscripts on things, S or H, to indicate whether we are in the **Schrödinger picture** or the **Heisenberg picture**, respectively. First, the Schrödinger picture. In the Schrödinger picture, the fundamental dynamical variables, the p's and the q's, are time-independent:

$$q_S(t) = q_S(0) = q_S \qquad p_S(t) = p_S(0) = p_S \tag{7.1}$$

I'll speak as if there's only one p and one q, just to simplify equations, but everything I say will be true if there are a million p's and a million q's. The states, on the other hand are time-dependent, and obey the Schrödinger equation

$$i\frac{d}{dt}\,|\psi(t)\rangle_S = H(p_S, q_S, t)\,|\psi(t)\rangle_S \tag{7.2}$$

The Hamiltonian H depends on p_S, q_S and perhaps also t.

The fundamental dynamical problem in the Schrödinger picture is this: Given the state $|\psi(t')\rangle$ at any time t', determine the state at a later time t. We define an operator $U(t,t')$, called **the time evolution operator**, by this equation,

$$|\psi(t)\rangle_S = U(t,t')\,|\psi(t')\rangle_S \tag{7.3}$$

That is to say, the U operator takes the state at time t' and produces a state at time t. $U(t,t')$ is a linear operator since the Schrödinger equation is a linear equation, and is a unitary operator,

$$U^{-1}(t,t') = U^{\dagger}(t,t') \tag{7.4}$$

because the Schrödinger equation conserves probability. The operator $U(t,t')$ obeys what we might call a sort of group property, a composition law

$$U(t,t')U(t',t'') = U(t,t'') \tag{7.5}$$

That is to say if I go first from time t'' to time t', and then from time t' to time t, that's the same as going from t'' to t in one fell swoop. The U matrix also obeys a differential equation, the Schrödinger equation,

$$i\frac{\partial}{\partial t}U(t,t') = H(p_S, q_S, t)U(t,t') \tag{7.6}$$

with the initial condition

$$U(t',t') = 1 \tag{7.7}$$

This differential equation is a direct consequence of the Schrödinger equation (7.2). Notice that the initial condition and the composition law imply

$$U(t,t') = U^{-1}(t',t) \tag{7.8}$$

Solving dynamics in the Schrödinger picture is equivalent to finding this U operator. If H is simply a function of p_S and q_S, that is to say, if H does not depend explicitly on t, then we can at least write a formal expression for the U matrix,

$$U(t,t') = e^{-iH(p_S,q_S)(t-t')} \tag{7.9}$$

Things get more complicated if H is time-dependent. For the time being, we'll assume H is time-independent.

The Heisenberg picture is the same as with a time-dependent unitary transformation. In the Heisenberg picture, the states are defined to be time-independent. Just so we can compare the two pictures, we identify the Heisenberg states with the Schrödinger states at the arbitrarily chosen time $t = 0$:

$$|\psi(t)\rangle_H = |\psi(0)\rangle_H = |\psi(0)\rangle_S \tag{7.10}$$

so that

$$|\psi(0)\rangle_H = e^{iH(p_S,q_S)t}\,|\psi(t)\rangle_S \tag{7.11}$$

In the Heisenberg picture, on the other hand, the fundamental p and q operators are defined to be time-dependent. In particular,

$$q_H(t) = U(t,0)^{\dagger}q_H(0)U(t,0) = U(t,0)^{\dagger}q_S(0)U(t,0) \tag{7.12}$$

because we identify $q_H(0)$ with $q_S(0)$,

$$q_H(0) = q_S(0) = q_S \tag{7.13}$$

and likewise for p. I won't bother to write down the equation for p. The reason we define things this way is that a Heisenberg picture operator $A_H(t)$ evaluated between Heisenberg picture states at any particular time is equivalent to the corresponding Schrödinger picture operator $A_S(t)$ evaluated between Schrödinger picture states at the same time:

$$_S\langle\psi(t)|A_S(t)|\psi(t)\rangle_S = {}_H\langle\psi(t)|A_H(t)|\psi(t)\rangle_H \tag{7.14}$$

It's just in one case you've got the U operator on the states and in the other case you've got the U operator on the operators, but the combined expression is the same. The correspondence between $A_S(t)$ and $A_H(t)$ follows:

$$\begin{aligned} {}_H\langle\psi(t)|A_H(t)|\psi(t)\rangle_H &= {}_S\langle\psi(0)|A_H(t)|\psi(0)\rangle_S \\ &= {}_S\langle\psi(t)|U^\dagger(0,t)A_H(t)U(0,t)|\psi(t)\rangle_S \\ &= {}_S\langle\psi(t)|A_S(t)|\psi(t)\rangle_S \\ \therefore A_H(t) &= U^\dagger(t,0)A_S(t)U(t,0) = U(0,t)A_S(t)U^\dagger(0,t) \end{aligned} \tag{7.15}$$

From this, we find the time evolution of the fundamental operators $p_H(t)$ and $q_H(t)$ in the Heisenberg picture:

$$\begin{aligned} \frac{d}{dt}p_H(t) &= U^\dagger(t,0)p_S(-iH(p_S,q_S))U(t,0) + U^\dagger(t,0)(iH(p_S,q_S)p_S)U(t,0) \\ &= iU^\dagger(t,0)[H(p_S,q_S),p_S]U(t,0) \\ &= i[H(p_H,q_H,t),p_H(t)] \end{aligned} \tag{7.16}$$

This is general quantum dynamics, independent of perturbation theory.

7.2 The interaction picture

I would now like to turn to perturbation theory computations of the U operator. Notice please that solving the dynamics in the Heisenberg picture is tantamount to solving it in the Schrödinger picture: they are both equivalent to finding the U operator.

We will consider a class of problems where the Hamiltonian $H(p,q,t)$ is the sum of a free Hamiltonian, H_0, let's say in the Schrödinger picture, and a Hamiltonian H' that may or may not depend on the time,

$$H = H_0(p,q) + H'(p,q,t) \tag{7.17}$$

Ultimately we are interested in real-world dynamics, where the total Hamiltonian is time-independent. But it's frequently useful, when we're doing some approximations to the real world, to consider time-dependent Hamiltonians. For example, if we have an electron in a synchrotron, we don't normally want to have to solve the quantum mechanics of the synchrotron. We could do it that way, but it's inconvenient, and we normally consider the synchrotron as a time-dependent pattern of classical external electric and magnetic fields acting on the electron. And therefore I will consider time-dependent interaction Hamiltonians. We assume that we could solve the problem exactly if it were not for H'. We wish to get a power series expansion for the dynamics in terms of H'. That's our problem. We can go first-order in H', second-order in H', etc. If you want, you can put a hypothetical small coupling constant in front of H', and say we are finding a power series expansion in that coupling constant, but I won't bother to do that.

This is most easily done by going to a special picture called the **interaction picture** (also known as the **Dirac picture**), which is sort of halfway between the Schrödinger picture and the Heisenberg picture.[1] We move from the Schrödinger to the interaction picture with the same kind of transformation that takes us from the Schrödinger picture to the Heisenberg picture, but now using *only* the free part of the Hamiltonian:

$$q_I(t) = e^{iH_0(p_S,q_S)t} q_S(t) e^{-iH_0(p_S,q_S)t} \tag{7.18}$$

where $q_S(t) = q_S(0)$, and $p_I(t)$ similarly. Of course, we must also change the states,

$$|\psi(t)\rangle_I = e^{iH_0(p_S,q_S)t} |\psi(t)\rangle_S \tag{7.19}$$

and the operators,

$$A_I(t) = e^{iH_0(p_S,q_S)t} A_S(t) e^{-iH_0(p_S,q_S)t} = U_0(t,0)^\dagger A_S(t) U_0(t,0) \tag{7.20}$$

This ensures

$$_I\langle\psi(t)|A_I(t)|\psi(t)\rangle_I = {}_S\langle\psi(t)|A_S(t)|\psi(t)\rangle_S \tag{7.21}$$

The advantage of the interaction picture is this: If H' were zero, H would equal H_0, and $|\psi(t)\rangle_I$ in the interaction picture would be independent of time, because it would be the Heisenberg picture; (7.19) would reduce to (7.11). Thus all the time dependence of $|\psi(t)\rangle_I$ comes from the presence of the interaction.

We can derive a differential equation for $|\psi(t)\rangle_I$ which we will then attempt to solve iteratively in perturbation theory:

$$\begin{aligned}
\frac{d}{dt}|\psi(t)\rangle_I &= e^{iH_0(p_S,q_S)t}\left(iH_0(p_S,q_S) - iH(p_S,q_S,t)\right)|\psi(t)\rangle_S \\
&= e^{iH_0(p_S,q_S)t}\left(-iH'_S(p_S,q_S,t)\right)e^{-iH_0(p_S,q_S)t}|\psi(t)\rangle_I \\
&= -iH'(p_I,q_I,t)\,|\psi(t)\rangle_I
\end{aligned} \tag{7.22}$$

where

$$H'(p_I,q_I,t) \equiv H_I(t) = e^{iH_0(p_S,q_S)t} H'_S(p_S,q_S,t) e^{-iH_0(p_S,q_S)t} \tag{7.23}$$

$H'_S(p_S,q_S,t)$ can be expanded as a power series in p_S and q_S, and factors of $e^{-iH_0t}e^{iH_0t}$ can be inserted everywhere to turn p_S and q_S into p_I and q_I. $H_I(t)$ is the same function of the interaction picture p's and q's as the Schrödinger interaction Hamiltonian H'_S is of the Schrödinger picture p's and q's. As promised, the time evolution of $|\psi(t)\rangle_I$ goes to zero when H' goes to zero.

This equation, (7.22), is the *key* equation. By solving it iteratively, we will obtain the solution to the time evolution problem as a power series in H_I. We will always use perturbation theory for the case where H_0 is time-independent. On the other hand, please notice that even if H'_S is time-independent, H_I might well be time-dependent because of the time dependence of p_I and q_I. In all the cases we will treat, H_I will be a polynomial, e.g., $\lambda\phi^4$. This equation (7.23) is true *modulo* ordering ambiguities if H_I is any function of the p's and q's.

We solve (7.22) by introducing the interaction picture operator $U_I(t,t')$, defined by the equation

$$|\psi(t)\rangle_I = U_I(t,t')\,|\psi(t')\rangle_I \tag{7.24}$$

[1] [Eds.] See Schweber *RQFT*, Section 11.c, "The Dirac Picture", pp. 316–325; J. J. Sakurai, S. F. Tuan, ed., *Modern Quantum Mechanics*, rev. ed., Addison-Wesley, 1994, p. 319.

This is of course just like the ordinary $U(t,t')$ operator. You give me the state of the system at a time, oh, 100 BCE, and, by operating on it with U_I, I will tell you the state of the system now. It obeys equations similar to the ordinary U. It's unitary:

$$U_I^{-1}(t,t') = U_I^\dagger(t,t') \tag{7.25}$$

and, just as with the earlier U, one can get from t'' to t by going through an intermediate time t',

$$U_I(t,t')U_I(t',t'') = U_I(t,t'') \tag{7.26}$$

From these two equations one can derive a third as in the earlier case:

$$U_I(t,t') = U_I^{-1}(t',t) \tag{7.27}$$

The earlier equation (7.8) wasn't useful to us, but this one will be.

U_I is not an independent entity; it is given in terms of the ordinary U. Let's look at $t = 0$ when all of our pictures coincide:

$$|\psi(0)\rangle_H = |\psi(0)\rangle_I = |\psi(0)\rangle_S\,; \qquad A_H(0) = A_I(0) = A_S(0) \tag{7.28}$$

just as in passing from the Schrödinger to the Heisenberg picture. From (7.3)

$$U(t,0)\,|\psi(0)\rangle_S = |\psi(t)\rangle_S \tag{7.29}$$

Moreover, from (7.19) and (7.24),

$$|\psi(t)\rangle_I = e^{iH_0 t}\,|\psi(t)\rangle_S = U_I(t,0)\,|\psi(0)\rangle_I \tag{7.30}$$

Then, from the identity of the kets at $t = 0$,

$$U_I(t,0) = e^{iH_0 t}U(t,0) = e^{iH_0 t}e^{-iHt} \tag{7.31}$$

For other times, things can be reconstructed using the known properties of the U's. For example,

$$U_I(t,t') = U_I(t,0)U_I^\dagger(t',0) = e^{iH_0 t}U(t,0)U^\dagger(t',0)e^{-iH_0 t'} = e^{iH_0 t}U(t,t')e^{-iH_0 t'} \tag{7.32}$$

Finally, (from (7.22) and (7.24)) U_I obeys a differential equation

$$\boxed{i\frac{\partial}{\partial t}U_I(t,t') = H_I(t)U_I(t,t')} \qquad \text{with the boundary condition } U_I(t',t') = 1 \tag{7.33}$$

just as in the development in the Schrödinger picture.

7.3 Dyson's formula

Our task now will be to solve this differential equation, (7.33). That is, we want to find a formal power series solution for it which is equivalent to solving dynamics in the interaction picture, and, by formula (7.31), to solving the dynamics in *any* picture. If we were doing the very simplest kind of quantum mechanical system, with a one-dimensional Hilbert space, then the solution would be simply

$$U_I(t,t') = \exp\left(-i\int_{t'}^{t} dt''\,H_I(t'')\right) \tag{7.34}$$

Unfortunately, H_I is not a one-by-one matrix; it isn't even an infinity-by-infinity matrix in most cases, and H_I's at different times do not commute with each other. So this formula is false. If we attempt the differentiation to make things work out, we'll find, after we differentiate, that we get all sorts of factors inside other factors, which we can't drag out to the left. I will take care of this difficulty by introducing a new ordering, called **time ordering**, rather parallel to the normal ordering we saw earlier.

Given a sequence of operators $A_1(t_1), A_2(t_2), \ldots, A_n(t_n)$ labeled by the times, I define the **time-ordered product** $T(A_1(t_1)A_2(t_2)\ldots A_n(t_n))$ of the string of operators as

$$T(A_1(t_1)A_2(t_2)\ldots A_n(t_n)) = A_{j_1}(t_{j_1})A_{j_2}(t_{j_2})\ldots A_{j_n}(t_{j_n}) \qquad \text{where } t_{j_1} > t_{j_2} > t_{j_3} \cdots > t_{j_n} \tag{7.35}$$

the same string of operators rearranged, such that the operator with the latest time is on the far left, then the next latest time, then the next, and so on. The convention, thank God, has a simple mnemonic, "later on the left", easy to remember. If two or more times are equal, then the time ordered product is in fact ambiguous. There are cases where we have to worry about that ambiguity, if the two operators do not commute at equal times. In the exponential for U_I, however, we will apply the time ordering to factors of H_I, and since H_I commutes with itself at equal times, there is no problem. You have seen this time ordering before, for two operators. I defined it in the first homework assignment (Problems 1.2 and 1.3, p. 49). That earlier definition agrees with this one for the case when there are only two operators.

The time-ordering symbol shares many features with the normal-ordering symbol. For example, the order in which you write the operators down inside the brackets is completely irrelevant, since the actual order in which we are to multiply them is determined by their times, not by the order in which they are written. As with the normal-order product, I must warn you the time-ordering prescription is not, "Compute the ordinary product and then do some mysterious operation to it, called time ordering". It is a new way of interpreting those symbols as they are written. I say this to keep you from getting into contradictions. Suppose you have two free fields, $\phi(t_1)$ and $\phi(t_2)$. The time-ordered product of the commutator of these two is zero, but the commutator is a number, and how can the time-ordered product of a number be zero? That's false reasoning. Time ordering a product means: "Rearrange the terms and *then* evaluate the product."

I will now demonstrate that the correct solution to our problem is the following beautiful formula, due to Dyson[2]

$$\boxed{U_I(t,t') = T\exp\left(-i\int_{t'}^{t} dt''\, H_I(t'')\right)} \tag{7.36}$$

almost the same formally as (7.34), but (7.36) defines a completely different operator because everything is to be time ordered. This is called **Dyson's formula**. I will say a little about the meaning of the formula and then show you that it solves the equation. This formula is valid

---

[2] [Eds.] Freeman J. Dyson, "The Radiation Theories of Tomonaga, Schwinger, and Feynman", *Phys. Rev.* **75** (1949) 486–502; see equation (32). (Dyson denotes time ordering in this article by the symbol P; see equation (29).) Coleman adds, "Without the use of the time ordering notation, this formula for $U_I(t,t')$ was written down by Dirac 15 years before Dyson wrote it this way." He is probably referring to P. A. M. Dirac, "The Lagrangian in Quantum Mechanics", *Phys. Zeits. Sowjetunion* **3** (1933) 64–72. Both Dyson's and Dirac's articles are reprinted in Schwinger *QED*. For the historical background, see Schwinger's preface to this collection, and Schweber *QED*. A careful proof of (7.36) is given in Greiner & Reinhardt *FQ*, Section 8.3, pp. 215–219.

only if t is greater than t'. It is not true otherwise. Fortunately that presents no difficulties because if we know how to compute U_I for one ordering, we know from (7.25) and (7.27) how to compute it for the other ordering, by taking the adjoint.

This formula is *only* interpretable as a formal power series. It's not saying, "Compute the integral, find out what operator is exponentiated, and then do something." I will write out the first three terms in the power series just to emphasize that:

$$U_I(t,t') = 1 - i\int_{t'}^{t} dt_1\, H_I(t_1) + \frac{(-i)^2}{2!}T\left(\int_{t'}^{t}\int_{t'}^{t} dt_1 dt_2\, H_I(t_1)\, H_I(t_2)\right) + \ldots \tag{7.37}$$

The first term is 1, and the time-ordering symbol does nothing to that. The second term involves only a single operator, so again the time-ordering symbol carries no force. The third term involves two integrals from t' to t, and here I can't drop the time-ordering symbol because I have two operators and two times. Over half the range of integration where t_1 is greater than t_2, this symbol is to be written first $H_I(t_1)$ then $H_I(t_2)$. Over the other half of the range of integration where t_1 is less than t_2 the two operators are to be flipped.

Now why is this time-ordered power series the solution to the differential equation (7.33)? It certainly obeys the boundary conditions: it's equal to one when $t = t'$ because the integrals are all zero, and the series reduces to the first term only. Let's evaluate its time derivative:

$$\frac{\partial U_I(t,t')}{\partial t} = T\left(-iH_I(t)\exp\left(-i\int_{t'}^{t} dt''\, H_I(t'')\right)\right) \tag{7.38}$$

Inside the time-ordering symbol everything commutes, so in doing our differentiation we don't have to worry about the orders of the operators. We will get just what we would get by differentiating naïvely, to wit, everything inside the time-ordering symbol: $H_I(t)$ times the time-ordered exponential; the time-ordering symbol takes care of all the ordering for us. Now comes the beauty part: t is the absolute latest time that occurs anywhere in the expression because the integral runs from t' to t and t is greater than t'. Therefore the Hamiltonian $H_I(t)$ has the latest time of any of the operators that occur within the time ordering, and latest is left-est! The Hamiltonian is always on the left in every term in the power series expansion, so we can write

$$\frac{\partial U_I(t,t')}{\partial t} = -iH_I(t)\, T\left(\exp\left(-i\int_{t'}^{t} dt''\, H_I(t'')\right)\right) = -iH_I(t)U_I(t,t') \tag{7.39}$$

That is precisely the differential equation for which we sought a solution, so the argument is complete. If the question is how do we do time-dependent perturbation theory, to find the dynamics as a formal power series in an interaction, the answer is Dyson's formula. Although perfectly valid, Dyson's formula is rather schematic, and we will beat on it quite a bit using all sorts of combinatorial tricks to find efficient computational rules. The entire contents of time-dependent perturbation theory is in this formula, (7.36).

7.4 Scattering and the S-matrix

The next problem we will discuss is scattering theory. I presume you have taken a course in non-relativistic quantum mechanics, and so you have a general idea of the shape of non-relativistic scattering theory. I would like to review some features of that theory, just to emphasize certain points to see what the *beau idéal* of a scattering theory should be, what criteria should we choose, and then we will try to construct a description of scattering in relativistic quantum mechanics. We will emphasize features important for our purposes that might not have been considered important in non-relativistic quantum mechanics, and so may be new to you.

What I mean by an ideal scattering theory is a description of scattering, of what information you have to drag out of the dynamics to compute cross-sections or something, that makes *no reference whatsoever* to perturbation theory. Then, if you could solve the problem exactly, if you could find the U matrix, you'd have a machine. You'd feed a Lagrangian in, you'd turn the crank, and you would fill out the cross-sections. You cannot solve for the U matrix exactly in typical cases. You might for example only be able to solve for it in perturbation theory. But that's all right. If you have an approximate U matrix, you put it into exactly the same machine, you turn the crank and out comes an approximation for the cross-section.

So let's consider a non-relativistic Hamiltonian,

$$H = \frac{p^2}{2m} + V(x) \tag{7.40}$$

This is really the simplest case, and I will assume that $V(x)$ goes to zero, say, faster than $1/x^2$, so we don't have to worry about long-range forces. (I think it suffices to say it goes to zero faster than $1/(x \log x)$, but forget that.)

Characteristic of a scattering problem is that a quite complicated motion at finite time interpolates between simple motion, according to the free Schrödinger equation, in the far past and the far future. Say I have this potential, $V(x)$, localized say in the vicinity of my overflowing ashtray. In the very, very far past, far from this potential, I prepare a wave packet. I allow this wave packet to move through space towards the potential. It goes along as if it were a free wave packet, until it (or its fringes, since it's spreading out) intersects the potential. Then it goes bananas, it wiggles and bounces around in quite complicated ways. And then, after a while, fragments of the wave packet fly out in various directions. If I then look in the very far future I have just a bunch of free wave packets now all moving away from the potential. The problem with scattering is a problem of connecting the simple motion in the far past with the simple motion in the far future. Let us frame these words in equations.

Since I talked about wave packets, I better look at the Schrödinger picture. Let $|\psi(t)\rangle$ be a solution of the free Schrödinger equation

$$|\psi(t)\rangle = e^{-iH_0(t-t')}\,|\psi(t')\rangle = U_0(t,t')\,|\psi(t')\rangle \tag{7.41}$$

where

$$H_0 = \frac{p^2}{2m} \tag{7.42}$$

This ket $|\psi(t)\rangle$ represents the wave packet I have prepared in the far past. If there were no potential, it would just evolve according to the free Schrödinger equation. In the very far past, because the wave packet is very far from the potential, it evolves according to the free Schrödinger equation. The ket $|\psi(t)\rangle$, and the other solutions to the free Hamiltonian, H_0, belong to a Hilbert space, $\mathcal{H}_0$. The solutions to H, the actual Hamiltonian of the world, belong to a Hilbert space $\mathcal{H}$. Somewhere in $\mathcal{H}$ there is a corresponding state that, in the distant past, looks very much like $|\psi(t)\rangle$. We will call this state $|\psi(t)\rangle^{\text{in}}$. It is a solution of the exact

Schrödinger equation that represents what the wave packet really does:

$$|\psi(t)\rangle^{\text{in}} = e^{-iH(t-t')}\,|\psi(t')\rangle^{\text{in}} \tag{7.43}$$

The two states $|\psi(t)\rangle$ and $|\psi(t)\rangle^{\text{in}}$ are connected by the requirement that if I look in the very far past, I can't tell the difference between them. That is to say,

$$\lim_{t\to-\infty}\left\| e^{-iH_0 t}\,|\psi\rangle - e^{-iHt}\,|\psi\rangle^{\text{in}}\right\| = 0 \tag{7.44}$$

(where $|\psi\rangle = |\psi(0)\rangle$ and $|\psi\rangle^{\text{in}} = |\psi(0)\rangle^{\text{in}}$). The norm of the difference of the states goes to zero as $t \to -\infty$. The operation of associating the appropriate state $|\psi\rangle^{\text{in}} \in \mathcal{H}$ to a given state $|\psi\rangle \in \mathcal{H}_0$ in the limit $t \to -\infty$ can be called "in-ing".

I emphasize that $|\psi(t)\rangle^{\text{in}}$ is a genuinely normalizable wave packet state. I can't make $|\psi(t)\rangle^{\text{in}}$ a plane wave, because a plane wave state has no norm. And physically, it doesn't make any sense to talk about a plane wave. It doesn't make any difference whether you go to the far past or the far future. Because a plane wave has infinite spatial extent, it never gets away from the scattering center.

The distinction between past and future makes a great deal of difference to human beings, but not so much to quantum dynamics, so we need to consider the far future as well. Given another state $|\phi(t)\rangle \in \mathcal{H}_0$, there is another state in $\mathcal{H}$ that looks a great deal like $|\phi(t)\rangle$ in the far future, which we'll call $|\phi(t)\rangle^{\text{out}}$. This ket is also a solution of the exact Schrödinger equation:

$$|\phi(t)\rangle^{\text{out}} = e^{-iH(t-t'')}\,|\phi(t'')\rangle^{\text{out}} \tag{7.45}$$

In the far future, these two corresponding states cannot be distinguished:

$$\lim_{t\to\infty}\left\| e^{-iH_0 t}\,|\phi\rangle - e^{-iHt}\,|\phi\rangle^{\text{out}}\right\| = 0 \tag{7.46}$$

The operation of associating the appropriate state $|\phi\rangle^{\text{out}} \in \mathcal{H}$ to a given state $|\phi\rangle \in \mathcal{H}_0$ in the limit $t \to \infty$ can be called "out-ing".

For every free motion there is a physical motion that looks like it in the far past and another physical motion that looks like it in the far future.[3] In-ing and out-ing connect the free solution to the exact solution in the far past and the far future, respectively, turning free motions into physical motions. We use free particle states as descriptors, to describe actual interacting particle states. We know how to associate a state with these descriptors by these correspondences.

Think of classical scattering in a potential. The analog of a free motion would be a straight-line motion in classical mechanics. Figure 7.1 shows some motion of the particle, when there is no potential. That's the analog of $|\psi(t)\rangle$. If the potential is restricted to some finite space-time region, the real motion of the particle looks like Figure 7.2. The particle enters the potential and it deviates from that, and then it comes out, again moving freely. At the lower right is $|\psi(t)\rangle^{\text{in}}$, the exact motion that looks like $|\psi(t)\rangle$ in the far past. At the upper left is $|\phi(t)\rangle^{\text{out}}$, the exact motion that looks like $|\phi(t)\rangle$ in the far future. The in and out states are exact solutions of the real Hamiltonian at all times. In scattering theory, we are trying to

[3] [Eds.] See John R. Taylor, *Scattering Theory: The Quantum Theory of Non-relativistic Collisions*, Dover Publications (2006), Section 2-c, "The Asymptotic Condition", pp. 28–31.

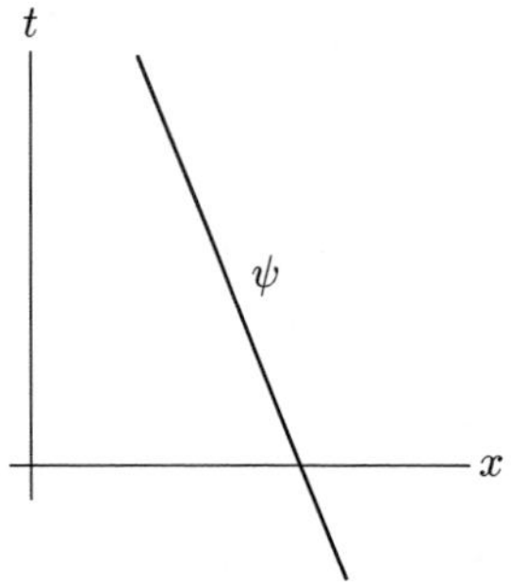

Figure 7.1: The free ψ $(V = 0)$

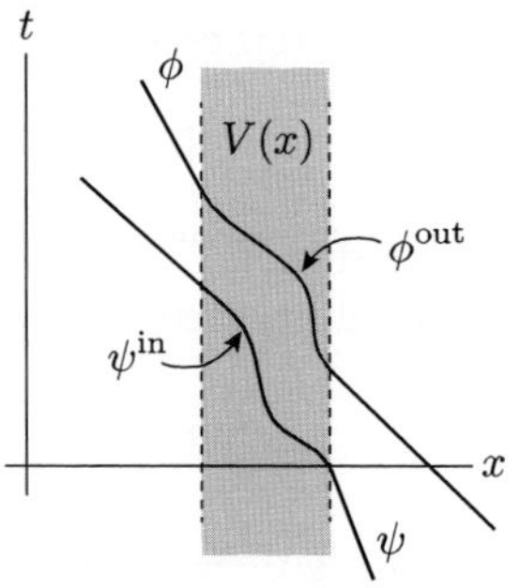

Figure 7.2: ψ^{in} and ϕ^{out}, asymptotic to the free ψ and ϕ

find the probability, and hence the amplitude, that a given state looking like $|\psi\rangle$ in the far past will look like $|\phi\rangle$ in the far future, namely ${}^{\text{out}}\langle\phi|\psi\rangle^{\text{in}}$. (Notice that we don't have to put a t in this expression because both $|\psi\rangle^{\text{in}}$ and $|\phi\rangle^{\text{out}}$ evolve according to the exact Schrödinger equation, and their inner product is independent of time.) The correspondences between $|\psi\rangle^{\text{in}}$ and $|\psi\rangle$ and between $|\phi\rangle^{\text{out}}$ and $|\phi\rangle$ allow us to define a very important operator in $\mathcal{H}_0$, the **scattering matrix** S, which acts between the descriptor states.[4] We define the S-matrix by the equation[5]

$$\boxed{\langle\phi|\mathrm{S}|\psi\rangle \equiv {}^{\text{out}}\langle\phi|\psi\rangle^{\text{in}}} \tag{7.47}$$

The S-matrix obeys certain conditions. For example,

$$\mathrm{SS}^\dagger = \mathrm{S}^\dagger\mathrm{S} = 1 \tag{7.48}$$

That is, the scattering matrix conserves probability. It also conserves energy, if this is a time-independent problem:

$$[\mathrm{S}, H_0] = 0 \tag{7.49}$$

Notice the H is H_0, because the energy operator acts on the descriptors, the states that move according to the free equation. The operator S turns free states of a given energy into other free states of a given energy. You prove this by computing the expectation value of the energy (or any power of the energy) in the far past, when you can't tell the in state from the free state, and computing it again in the far future, when you can't tell the out state from the free state, and requiring that these values be the same.

So much for the scattering theory of a single particle and a potential. I've gone through it in a dogmatic way without proving any of these equations because I presume you have seen them before. We're not going to use all this formalism, by the way, in relativistic theory, at least not for the time being.

---

[4] [Eds.] The S-matrix was introduced by John A. Wheeler: "On the Mathematical Description of Light Nuclei by the Method of Resonating Group Structure", *Phys. Rev.* **32** (1937) 1107–1122; see his equation (31). It was extended and refined by Heisenberg: W. Heisenberg, "Die 'beobachtbaren Größen' in der Theorie der Elementarteilchen" (The "observable sizes" in the theory of elementary particles), *Zeits. f. Phys.* **120** (1943) 513–538; part II, *Zeits. f. Phys.* **120** (1943) 673–702.

[5] [Eds.] To distinguish between the action and the S-matrix, different fonts are used for these quantities: the action by $\mathcal{S}$, and the S-matrix by S.

Now it should be emphasized that in this way of linking states, it looks like there's a connection with perturbation theory, with breaking up the Hamiltonian into an H_0 and a V. This is not so. And the easiest way to demonstrate that is to consider another simple system.

Let's consider three particles, all with the same mass, with central potentials between them:

$$H = \sum_{i=1}^{3} \frac{p_i^2}{2m} + V_{12}(|\mathbf{x}_1 - \mathbf{x}_2|) + V_{23}(|\mathbf{x}_2 - \mathbf{x}_3|) + V_{13}(|\mathbf{x}_1 - \mathbf{x}_3|) \tag{7.50}$$

where the arguments of the potentials are the usual differences between the centers of the particles. Let me assume that V_{12} all by itself could make a bound state. The center of mass Schrödinger equation is

$$-\frac{1}{2\mu}\nabla^2\psi_0(r) + V_{12}(r)\psi_0(r) = \epsilon\psi_0(r) \tag{7.51}$$

where μ is the reduced mass, and $r = |\mathbf{x}_1 - \mathbf{x}_2|$. It has one bound state, and none of the other potentials make bound states, they're all repulsive, and this one has only one. This could be, aside from the long-range nature of the forces (and the hypothetical equality of the masses), a proton, an electron and a neutral π^0 meson. There is no binding between the proton and pion, nor between the electron and the pion, but there *is* binding between the proton and the electron, to make a hydrogen atom. Of course, the hydrogen atom has an infinite number of bound states.

Now if we seek for descriptors here, we find things fall into two channels, one in which, in the far past, the states look like three free particles, and the other looking like one free particle and one bound state of the 1 and 2 particles. Both of those can happen since 1 and 2 can bind. Therefore we have two kinds of states, type I with corresponding in and out states like this:

$$|\mathbf{p}_1, \mathbf{p}_2, \mathbf{p}_3\rangle_{\mathrm{I}}^{\text{in, out}} \tag{7.52}$$

labeling these as type I states. These are solutions to the exact Schrödinger equation that in the far past and the far future look like three widely separated particles. For them, H_0 is just

$$H_0 = \sum_{i=1}^{3} \frac{p_i^2}{2m} \tag{7.53}$$

We also have states of type II: orthogonal, exact solutions of the Schrödinger equation for which a complete basis could be specified by giving the momentum of the third particle which doesn't bind, and the combined momentum $\mathbf{p}$ of the 1-2 pair (with respect to the center of mass), which is in a bound state

$$|\mathbf{p}, \mathbf{p}_3\rangle_{\mathrm{II}}^{\text{in, out}} \tag{7.54}$$

For these, the Hamiltonian is

$$H_0 = \frac{p^2}{2\mu} + \frac{p_3^2}{2m} + V_{12}(r) + \frac{p_{\text{cm}}^2}{4m} \tag{7.55}$$

If $V_{12}(r)$ is not in the Hamiltonian, these type II free states will not time-evolve in the appropriate way. It's V_{12} that keeps them held together; without this potential, the 1 and 2 particles will fly away from each other. In this case, there are two alternatives for the free Hamiltonian for the definition of the in and out states, depending on what kind of states we look at. All the in states of type I are orthogonal to all the in states of type II, and the same is true of the out states. If a state looks like three widely separated particles in the far past it is also not going to look like one free particle and one bound state in the far past. Its probability for doing that is zero. On the other hand the in states of type II are not orthogonal to the out states of type I or *vice versa*: ionization can occur. You can scatter a free particle off of a bound state in the far past and get three free particles in the far future. So this shows the situation to be more complicated, and the complication has nothing to do with perturbation theory.

Now, what are we looking for? What is the *beau idéal*, the grail of a quantum field theory, to describe relativistic quantum scattering? What sort of in states and out states do we expect to have? Well, fortunately we have locality and all that. We imagine that if we have a particle of type zilch,[6] we can have two widely separated particles of type zilch, or three widely separated particles of type zilch, so we would expect that our descriptor states would belong to a Fock space for a bunch of free particles. That would correspond to 1, 2, 3 ... particles of various kinds, moving in toward each other or moving away from each other in the far past, all in appropriate wave packets. What kind of particles should be there? Well, all the stable particles in the world, whatever they are! That's a *big* list of particles. There's electrons, and neutrinos, and there are hydrogen atoms in their ground states, and there are photons, and there are alpha particles and there are ashtrays. (That's a stable system; I don't think I've ever seen an ashtray decay—it has a lot of excited states, you can put dimples in it and everything, but it's a stable system. Fortunately, we have to go to quite a high center-of-mass energy before we begin to worry about ashtray–anti-ashtray production.) They should *all* be there, and there should be a great big Fock space that describes states of, say, one electron, 17 photons, 14 protons, 4 alpha particles, and 6 ashtrays. And then there would be some S-matrix that connects one to the other.[7]

To describe a scattering theory that is capable of handling the situation is a tall order. After setting up these high hopes, I will make you all groan by describing the simple way we are going to do scattering theory for our first run through. This first description of scattering theory will obviously be inadequate. We will eventually develop a description that in principle will enable us to handle a situation of this complexity. In practice, of course, it's a different story, just as in practice it's a very difficult thing to compute ionization in any sensible approximation. But we will develop a description where, if we did know the time evolution exactly, we would be able to compute all scattering matrix elements exactly. This description will however take quite a long time to develop.

There are many features of the general description that are rather obscure, if you are working with no specific examples to think back on. And so I will begin with the crudest and simplest description of scattering, the most ham-handed possible. Then, as we go along doing examples, we will find places where this description clearly has to be fixed up. To make our Model A work,[8] we will add a tail fin here, and change the carburetor there. After we've gained a lot of experience with this jerry-built jalopy, I will go through a sequence of one or two lectures on a very high level of abstraction, where I explain what the *real* scattering

---

[6] [Eds.] See note 13, p. 96.

[7] I didn't list the neutron. There's a reason for that. Neutrons aren't stable; they last 15 minutes on the average. We never find, in the very far future, a neutron coming out; we find an electron, a proton and an anti-neutrino coming out, but not a neutron.

[8] [Eds.] The Ford Model A, sold from 1927–1931, was the successor to Ford's Model T automobile.

theory is like. I do things this way so that you can get a lot of particular examples under your belt before we fly off into a Never Never Land, or really, an Ever Ever Land, of abstraction.

I will now explain the incredibly crude approximation we will use. I will take $H_I(t)$ and bluntly multiply it by a function of time $f(t, T, \Delta)$ depending on the time, t, and two numbers T and Δ:

$$H = H_0 + H_I(t) \to H_0 + f(t, T, \Delta) H_I(t) \tag{7.56}$$

This function $f(t, T, \Delta)$ will provide a so-called adiabatic turning on and off of the interaction. It will be equal to 1 for a long time which I will call T, and then it will slowly make its way to zero over a time I will call Δ. This function is illustrated by Figure 7.3. Why have I stuck

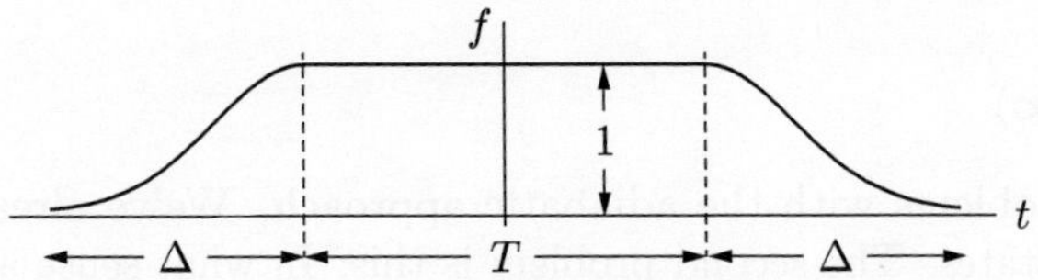

Figure 7.3: The adiabatic function $f(t, T, \Delta)$

this artificial function in my theory? Well, if we think of particle scattering in a potential, this approximation makes the computation of the S-matrix rather simple. In the far past when $f(t, T, \Delta)$ is zero, the theory is not in some sense asymptotically equal to a free theory, it is *exactly* equal to a free theory. So we have a free wave packet going along on its way to the potential. While it's on its way to the potential, we turn on the interaction, but it doesn't know that until it reaches the potential. And then it reaches the potential and scatters, and goes off in fragments. After the fragments have all flown away, we carefully turn off the potential again. Again, the wave packet fragments don't notice that, because they're away from the potential.

For a scattering of particles in a potential, we have a very simple formula for the S-matrix. We don't have to worry about in states and out states, because the in states *are* the states in the far past, and the out states are the states in the far future:

$$\begin{aligned} |\psi(-\infty)\rangle^{\text{in}} &= \lim_{t' \to -\infty} e^{iH_0 t'} e^{-iHt'} |\psi\rangle = \lim_{t' \to -\infty} U_I(0, t') |\psi\rangle \\ |\phi(\infty)\rangle^{\text{out}} &= \lim_{t'' \to \infty} e^{iH_0 t''} e^{-iHt''} |\phi\rangle = \lim_{t'' \to \infty} U_I(0, t'') |\phi\rangle \end{aligned} \tag{7.57}$$

In the far past and the far future, $f(t) = 0$ and $\mathcal{H} = \mathcal{H}_0$, and the Hamiltonian that gives the evolution of the asymptotically simple states, H_0, is the full Hamiltonian, H. So the S-matrix can be written

$$\mathrm{S} = \lim_{\substack{T \to \infty \\ \Delta \to \infty \\ (\Delta/T) \to 0}} U_I(\infty, -\infty) \tag{7.58}$$

We want the limits $T \to \infty$ and $\Delta \to \infty$. We keep the interaction on for a longer and longer time, and turn it on and off more and more adiabatically. Δ/T goes to zero in the limit, so at the fringes, the transient terms we would expect to get from the boundaries are trivial compared to the terms we get from $U_I(\infty, -\infty)$, while keeping the potential on. The interaction picture is highly suitable to our purposes, because it takes out the factors of $e^{iH_0 t}$ that are in the free evolution of the initial and final states. There is no harm to the physics in computing the S-matrix this way for particle scattering in a potential. This approach may lack something in elegance. Instead of solving the real problem, you solve a substitute problem with an adiabatic turning on and off function, and then you let the turning on and off go away. But it certainly corresponds to all the physics we would think would be there.

Here's why (7.58) is true.[9] By the definition of the S-matrix, (7.47), in the Schrödinger picture,

$$\begin{aligned}\langle\phi|\mathrm{S}|\psi\rangle &= {}^{\mathrm{out}}\langle\phi|\psi\rangle^{\mathrm{in}} = \lim_{\substack{t\to-\infty\\ t'\to\infty}} \langle\phi|U_I^\dagger(0,t')U_I(0,t)|\psi\rangle = \lim_{\substack{t\to-\infty\\ t'\to\infty}} \langle\phi|U_I(t',0)U_I(0,t)|\psi\rangle \\ &= \lim_{\substack{t\to-\infty\\ t'\to\infty}} \langle\phi|U_I(t',t)|\psi\rangle = \langle\phi|U_I(\infty,-\infty)|\psi\rangle \qquad (7.59)\\ \therefore\ \mathrm{S} &= U_I(\infty,-\infty)\end{aligned}$$

There are two problems with the adiabatic approach. We've already talked about the problem with bound states. The second problem is this: In what sense are the particles really non-interacting when they're far from each other? Haven't we all heard about those virtual photons that surround charged particles, and stuff like that? Well, we'll eventually worry about that question in detail, but for now let me say this. In slightly racy language, the electron without its cloud of photons is called a "bare" electron, and with its cloud of photons, a "dressed" electron. The scattering process goes like this: In the far past, a bare electron moves freely along. A billion years before it is to interact, it leisurely dresses itself. Then it moves along for a long time as a dressed electron, briefly interacts with another (dressed) electron, and moves away for a long time, still dressed. Then it leisurely undresses. For the time being, though, we will adopt this supremely simple-minded definition, (7.58), of the S-matrix, because it enables us to make immediate contact with time-dependent perturbation theory, and start computing things.

As we compute things, we will find that indeed this method is too simple-minded. We will have to fix it up systematically, but we will discover how to do that. Meanwhile, we will be doing lots of calculations and gaining lots of experience, developing our intuition. And then finally we will junk the Model A altogether, and replace it with the supreme model of scattering theory. So that is the outline of what we will be doing. Next time, we will begin exploring our simple-minded model by developing a sequence of algorithms, starting from Dyson's formula, to evaluate the U_I matrix in terms of diagrams.

---

9 [Eds.] See Schweber *RQFT*, Section 11.c, "The Dirac Picture", pp. 316–318.

Problems 3

In the first three problems, you are asked to apply the methods of the last few weeks to a non-relativistic field theory, defined by the Lagrangian

$$\mathscr{L} = i\psi^*\partial_0\psi + b\nabla\psi^*\cdot\nabla\psi \tag{P3.1}$$

where b is some real number. As your investigation proceeds, you should discover an old friend hiding inside a new formalism. (This Lagrange density is not real, but that's all right: The action integral *is* real; the effect of complex conjugation is undone by integration by parts.)

3.1 Consider $\mathscr{L}$ as defining a classical field theory.

(a) Find the Euler–Lagrange equations.

(b) Find the plane-wave solutions, those for which $\psi = e^{i(\mathbf{p}\cdot\mathbf{x}-\omega t)}$, and find ω as a function of $\mathbf{p}$.

(c) Although this theory is not Lorentz-invariant, it is invariant under spacetime translations and the internal symmetry transformation

$$\psi \to e^{-i\lambda}\psi, \qquad \psi^* \to e^{i\lambda}\psi^* \tag{P3.2}$$

Thus it possesses a conserved energy, a conserved linear momentum, and a conserved charge associated with the internal symmetry. Find these quantities as integrals of the fields and their derivatives. Fix the sign of b by demanding the energy be bounded below.

(As explained in class, in dealing with complex fields, you just turn the crank, ignoring the fact that ψ and ψ^* are complex conjugates. Everything should turn out all right in the end: The equation of motion for ψ will be the complex conjugate of that for ψ^*, and the conserved quantities will be real. WARNING: Even though this is a non-relativistic problem, our formalism is set up with relativistic conventions. Don't miss minus signs associated with raising and lowering spatial indices.)

(1997a 3.1)

3.2 (a) Canonically quantize the theory. (HINT: You may be bothered by the fact that the momentum conjugate to ψ^* vanishes. Don't be. Because the equations of motion are first-order in time, a complete and independent set of initial-value data consists of ψ and its conjugate momentum, $i\psi^*$, alone. It is only on these that you need to impose the canonical quantization conditions.)

(b) Identify appropriately normalized coefficients in the expansion of the fields in terms of plane wave solutions with annihilation and/or creation operators.

(c) Write the energy, linear momentum and internal-symmetry charge in terms of these operators. (Normal-order freely.)

(1997a 3.2)

3.3 For a relativistic complex scalar field, I constructed in class a unitary charge-conjugation operator, U_C, a unitary parity operator, U_P, and an anti-unitary time-reversal operator, Ω_T, such that

$$\begin{aligned} U_C^\dagger \psi(\mathbf{x},t)U_C &= \psi^*(\mathbf{x},t), \\ U_P^\dagger \psi(\mathbf{x},t)U_P &= \psi(-\mathbf{x},t), \quad \text{and} \\ \Omega_T^{-1}\psi(\mathbf{x},t)\Omega_T &= \psi(\mathbf{x},-t) \end{aligned} \tag{P3.3}$$

For the theory at hand, only two of these three operators exist. Which two? Construct them (that is to say, define them in terms of their action on the creation and annihilation operators).

(1997a 3.3)

3.4 The Lagrangian of a free, massless scalar field

$$\mathscr{L}(x) = \tfrac{1}{2}\partial^\mu\phi(x)\partial_\mu\phi(x)$$

possesses a one-parameter family of symmetry transformations, called scale transformations, or **dilations**, defined by

$$\mathcal{D}\colon \phi(x) \to e^\lambda \phi(e^\lambda x) \tag{P3.4}$$

(a) Show that the action $\mathcal{S} = \int d^4x\,\mathscr{L}(x)$ is invariant under this transformation.

(b) Compute the associated conserved current and the conserved quantity, Q.

(c) Compute the commutator of Q with ϕ, and show that this obeys the assertion, following (5.19), that the conserved quantity Q is the generator of the infinitesimal transformation:

$$[\phi, Q] = iD\phi \tag{P3.5}$$

(d) Compute the commutator of Q with the components of the four-momentum, P^μ, and show that

$$[Q, P^\mu] = iP^\mu \tag{P3.6}$$

(You are not required to write things in terms of annihilation and creation operators, nor need you worry about whether the formal expression for Q should be normal ordered.)[1]

(1991a 6)

[1] [Eds.] In the context of quantum mechanics, the dilation operator Q is represented by $x^\mu P_\mu$ and is often denoted D. It is easy to see that $[D, P^\mu] = iP^\mu$. Because D does not commute with P^μ, it does not commute with P^2, and so only massless theories can have dilation invariance. The dilation operator, together with the Poincaré group operators $\{P^\mu, M^{\mu\nu}\}$ and four "special conformal operators" $\{K^\mu\}$, form the 15 parameter **conformal group**, the group that leaves invariant the square of a lightlike 4-vector. In addition to the usual Poincaré commutators

$$\begin{aligned} [P_\lambda, M_{\mu\nu}] &= i(g_{\lambda\mu}P_\nu - g_{\lambda\nu}P_\mu) \\ [M_{\alpha\beta}, M_{\mu\nu}] &= i(g_{\alpha\mu}M_{\nu\beta} - g_{\alpha\nu}M_{\mu\beta} + g_{\beta\mu}M_{\alpha\nu} - g_{\beta\nu}M_{\alpha\mu}) \end{aligned}$$

we have

$$\begin{aligned} [P_\mu, P_\nu] = [K_\mu, K_\nu] = [D, M_{\mu\nu}] &= 0 \\ [D, P_\mu] &= iP_\mu \\ [D, K_\mu] &= -iK_\mu \\ [K_\mu, P_\nu] &= 2i(g_{\mu\nu}D - M_{\mu\nu}) \\ [K_\lambda, M_{\mu\nu}] &= i(g_{\lambda\mu}K_\nu - g_{\lambda\nu}K_\mu) \end{aligned}$$

The conformal group was discovered in 1909 by Harry Bateman ("The conformal transformations of a space of four dimensions and their applications to geometrical optics", *Proc. Lond. Math. Soc.* **7**, s.2, (1909), 70–89), and later that year was shown by Ebenezer Cunningham to be the largest group of transformations leaving Maxwell's equations invariant ("The principle of relativity in electrodynamics and an extension thereof", *Proc. Lond. Math. Soc.* **8**, s.2, (1909), 77–98).

Solutions 3

3.1 The Lagrange density $\mathscr{L}$ has the form

$$\mathscr{L} = i\psi^* \partial_0 \psi + b\nabla\psi^* \cdot \nabla\psi \tag{S3.1}$$

Treating ψ and ψ^* as independent fields, the Euler–Lagrange equations are, for ψ,

$$\frac{\partial \mathscr{L}}{\partial \psi^*} - \partial_\mu \frac{\partial \mathscr{L}}{\partial(\partial_\mu \psi^*)} = i\partial_0\psi - 0 - \partial_i(b\partial_i\psi) = i\dot{\psi} - b\nabla^2\psi = 0$$

and for ψ^*,

$$\frac{\partial \mathscr{L}}{\partial \psi} - \partial_\mu \frac{\partial \mathscr{L}}{\partial(\partial_\mu \psi)} = 0 - \partial_0(i\psi^*) - \partial_i(b\partial_i\psi^*) = -i\dot{\psi}^* - b\nabla^2\psi^* = 0$$

That answers (a). As expected, these equations are complex conjugates of each other. For $b < 0$ (and we will see shortly that this condition is necessary) the first equation is nothing but the time-dependent Schrödinger equation for a free particle of mass $m = -1/(2b)$. To find plane wave solutions, set

$$\psi = Ae^{i(\mathbf{p}\cdot\mathbf{x} - \omega t)}$$

and plug into the equations of motion. We find

$$i\dot{\psi} - b\nabla^2\psi = (\omega + b|\mathbf{p}|^2)\psi = 0 \Rightarrow \omega = -b|\mathbf{p}|^2 \tag{S3.2}$$

That answers (b). To answer (c), recall the definition (5.50),

$$T^{\nu\mu} = \pi^\mu_a \partial^\nu \phi^a - g^{\mu\nu}\mathscr{L} = \frac{\partial \mathscr{L}}{\partial(\partial_\mu\psi)}\partial^\nu\psi + \frac{\partial \mathscr{L}}{\partial(\partial_\mu\psi^*)}\partial^\nu\psi^* - g^{\mu\nu}\mathscr{L}$$

Note that $\pi^0_\psi = i\psi^*$, and $\pi^0_{\psi^*} = 0$. For T^{00} we obtain

$$\begin{aligned} T^{00} &= \pi^0_\psi \partial^0\psi + \pi^0_{\psi^*}\partial^0\psi^* - \mathscr{L} \\ &= i\psi^*\dot{\psi} + 0 - i\psi^*\dot{\psi} - b\nabla\psi^*\cdot\nabla\psi = -b\nabla\psi^*\cdot\nabla\psi \end{aligned}$$

The space integral of T^{00} gives the Hamiltonian:

$$H = -b\int d^3\mathbf{x}\,\nabla\psi^*\cdot\nabla\psi$$

The integrand is positive definite. If the energy is to be bounded from below, we have to take $b < 0$. To make the analogy with the Schrödinger equation explicit, set $b = -1/2m$. Then using integration by parts,

$$H = -\frac{1}{2m}\int d^3\,\mathbf{x}\,\psi^*\nabla^2\psi \tag{S3.3}$$

which should be familiar as the expectation value of the Hamiltonian for a free particle in the Schrödinger theory. For $\nu = i$, we find for the momentum density (recall $\nabla = \partial_i = -\partial^i$)

$$T^{i0} = \frac{\partial \mathscr{L}}{\partial(\partial_0\psi)}\partial^i\psi + \frac{\partial \mathscr{L}}{\partial(\partial_0\psi^*)}\partial^i\psi^* = -i\psi^*\nabla\psi$$

and so the momentum is

$$\mathbf{P} = -i\int d^3\mathbf{x}\,\psi^*\nabla\psi \tag{S3.4}$$

which is the expectation value of the momentum for a free particle in the Schrödinger theory. For the internal symmetry,

$$\begin{aligned}
\psi &\to e^{-i\lambda}\psi \qquad & D\psi &= \left.\frac{d\psi}{d\lambda}\right|_{\lambda=0} = -i\psi \\
\psi^* &\to e^{i\lambda}\psi^* \qquad & D\psi^* &= \left.\frac{d\psi^*}{d\lambda}\right|_{\lambda=0} = i\psi^*
\end{aligned} \tag{S3.5}$$

To construct the conserved current, use (5.27),

$$J^\mu = \pi^\mu_a D\phi^a - F^\mu$$

Here, $F^\mu = 0$ because the Lagrange density is invariant under the symmetry. Then

$$J^0 = \pi^0_\psi D\psi + \pi^0_{\psi^*} D\psi^* = i\psi^*(-i\psi) + 0 = \psi^*\psi$$

so

$$Q = \int d^3\mathbf{x}\,\psi^*\psi \tag{S3.6}$$

In the usual single-particle Schrödinger equation, the integral of the square of the wave function is used to determine its normalization. If the norm is constant, probability is conserved. In the language of quantum field theory, as we will see in the next problem, Q is associated with the number of particles. ■

3.2 (a) The classical field theory of the $\mathscr{L}$ (S3.1) resembles the Schrödinger theory for a single free particle. What happens in the context of quantum field theory? Since the Euler–Lagrange equations are first-order in time, ψ and its conjugate momentum $\pi = i\psi^*$ form a complete set of initial-value data. Impose the canonical commutation relations:

$$[\psi(\mathbf{x},t),\psi(\mathbf{y},t)] = [\psi^*(\mathbf{x},t),\psi^*(\mathbf{y},t)] = 0 \tag{S3.7}$$

$$[\psi(\mathbf{x},t), i\psi^*(\mathbf{y},t)] = i\delta^{(3)}(\mathbf{x}-\mathbf{y}) \tag{S3.8}$$

(b) Try a Fourier expansion, following (3.45),

$$\psi(\mathbf{x},t) = \int d^3\mathbf{p}\left(f(\mathbf{p})\,a_\mathbf{p}e^{i(\mathbf{p}\cdot\mathbf{x}-\omega_\mathbf{p}t)} + g(\mathbf{p})\,a^\dagger_\mathbf{p}e^{-i(\mathbf{p}\cdot\mathbf{x}-\omega_\mathbf{p}t)}\right)$$

$$\psi^*(\mathbf{x},t) = \int d^3\mathbf{p}\left(f^*(\mathbf{p})\,a^\dagger_\mathbf{p}e^{-i(\mathbf{p}\cdot\mathbf{x}-\omega_\mathbf{p}t)} + g^*(\mathbf{p})\,a_\mathbf{p}e^{i(\mathbf{p}\cdot\mathbf{x}-\omega_\mathbf{p}t)}\right)$$

where $f(\mathbf{p})$ and $g(\mathbf{p})$ are functions to be determined, and (from (S3.2)) $\omega_\mathbf{p} = |\mathbf{p}|^2/2m$. If we assume the relations (3.18) and (3.19),

$$[a_\mathbf{p}, a^\dagger_{\mathbf{p}'}] = \delta^{(3)}(\mathbf{p}-\mathbf{p}') \tag{3.18}$$

$$[a_\mathbf{p}, a_{\mathbf{p}'}] = [a^\dagger_\mathbf{p}, a^\dagger_{\mathbf{p}'}] = 0 \tag{3.19}$$

then we find

$$[\psi(\mathbf{x},t),\psi^*(\mathbf{y},t)] = \int d^3\mathbf{p}\left(|f(\mathbf{p})|^2 e^{i\mathbf{p}\cdot(\mathbf{x}-\mathbf{y})} - |g(\mathbf{p})|^2 e^{-i\mathbf{p}\cdot(\mathbf{x}-\mathbf{y})}\right)$$

This must equal $\delta^3(\mathbf{x}-\mathbf{y})$ to satisfy the canonical commutation relation (S3.8). In the original expression (3.45), $f(\mathbf{p}) = g(\mathbf{p}) = 1/(2\pi)^{(3/2)}\sqrt{2\omega_\mathbf{p}}$, and the equal-time commutator of two fields vanishes (4.47), as required. That won't do here. There is a clue, however, in the original wording of the problem: "Identify appropriately normalized coefficients in the expansion of the fields in terms of plane wave solutions with annihilation *and/or* creation operators." We can satisfy the canonical commutation relation by the choice of coefficients

$$f(\mathbf{p}) = \frac{1}{(2\pi)^{3/2}}, \qquad g(\mathbf{p}) = 0$$

This choice also ensures that (S3.7) holds, because ψ contains only annihilation operators, and ψ^* contains only creation operators.

(c) Obtain the expressions for H, P and Q by plugging in the expressions for ψ and ψ^* into (S3.3), (S3.4) and (S3.6), respectively:

$$\begin{aligned} H &= -\int d^3\mathbf{x}\,\frac{1}{2m}\psi^*(\mathbf{x},t)\nabla^2\psi(\mathbf{x},t) \\ &= \int \frac{d^3\mathbf{x}}{(2\pi)^3}\,d^3\mathbf{p}\,d^3\mathbf{p}'\,\frac{|\mathbf{p}'|^2}{2m}\,a^\dagger_{\mathbf{p}}e^{-i(\mathbf{p}\cdot\mathbf{x}-\omega_{\mathbf{p}}t)}\,a_{\mathbf{p}'}e^{i(\mathbf{p}'\cdot\mathbf{x}-\omega_{\mathbf{p}'}t)} \\ &= \int d^3\mathbf{p}\,d^3\mathbf{p}'\,\frac{|\mathbf{p}'|^2}{2m}\,a^\dagger_{\mathbf{p}}e^{i(\omega_{\mathbf{p}}-\omega_{\mathbf{p}'})t}\,a_{\mathbf{p}'}\delta^{(3)}(\mathbf{p}-\mathbf{p}') \\ &= \int d^3\mathbf{p}\,\frac{|\mathbf{p}|^2}{2m}\,a^\dagger_{\mathbf{p}}a_{\mathbf{p}} \end{aligned} \tag{S3.9}$$

That's the Hamiltonian. The momentum $\mathbf{P}$ goes the same way:

$$\begin{aligned} \mathbf{P} &= -i\int d^3\mathbf{x}\,\psi^*(\mathbf{x},t)\nabla\psi(\mathbf{x},t) \\ &= -i\int \frac{d^3\mathbf{x}}{(2\pi)^3}\,d^3\mathbf{p}\,d^3\mathbf{p}'\,i\mathbf{p}'\,a^\dagger_{\mathbf{p}}e^{-i\mathbf{p}\cdot\mathbf{x}-\omega_{\mathbf{p}}t)}\,a_{\mathbf{p}'}e^{i(\mathbf{p}'\cdot\mathbf{x}-\omega_{\mathbf{p}'}t)} \\ &= \int d^3\mathbf{p}\,d^3\mathbf{p}'\,\mathbf{p}'\,a^\dagger_{\mathbf{p}}e^{i(\omega_{\mathbf{p}}-\omega_{\mathbf{p}'})t}\,a_{\mathbf{p}'}\delta^{(3)}(\mathbf{p}-\mathbf{p}') \\ &= \int d^3\mathbf{p}\,\mathbf{p}\,a^\dagger_{\mathbf{p}}a_{\mathbf{p}} \end{aligned} \tag{S3.10}$$

Finally, the charge Q:

$$Q = \int d^3\mathbf{x}\,\psi^*(\mathbf{x},t)\psi(\mathbf{x},t) = \int \frac{d^3\mathbf{x}}{(2\pi)^3}\,d^3\mathbf{p}\,d^3\mathbf{p}'\,a^\dagger_{\mathbf{p}}e^{-i(\mathbf{p}\cdot\mathbf{x}-\omega_{\mathbf{p}}t)}\,a_{\mathbf{p}'}e^{i(\mathbf{p}'\cdot\mathbf{x}-\omega_{\mathbf{p}'}t)} = \int d^3\mathbf{p}\,a^\dagger_{\mathbf{p}}a_{\mathbf{p}} \tag{S3.11}$$

This is the theory of a set of free, non-relativistic, identical bosons, all with mass m. Each boson has momentum $\mathbf{p}$ and energy $E = \omega = |\mathbf{p}|^2/2m$. The conserved charge Q is the number of bosons. Note that all of the operators H, $\mathbf{P}$ and Q are time-independent. ■

3.3 First, define a parity transformation as in (6.89):

$$P\colon \begin{Bmatrix} a_{\mathbf{p}} \\ a^\dagger_{\mathbf{p}} \end{Bmatrix} \to U^\dagger_P \begin{Bmatrix} a_{\mathbf{p}} \\ a^\dagger_{\mathbf{p}} \end{Bmatrix} U_P = \begin{Bmatrix} a_{-\mathbf{p}} \\ a^\dagger_{-\mathbf{p}} \end{Bmatrix}$$

Then

$$U^\dagger_P\psi(\mathbf{x},t)U_P = \int \frac{d^3\mathbf{p}}{\sqrt{(2\pi)^3}}\,a_{-\mathbf{p}}e^{i(\mathbf{p}\cdot\mathbf{x}-\omega_{\mathbf{p}}t)} = \int \frac{d^3\mathbf{p}}{\sqrt{(2\pi)^3}}\,a_{\mathbf{p}}e^{i(-\mathbf{p}\cdot\mathbf{x}-\omega_{\mathbf{p}}t)} = \psi(-\mathbf{x},t) \tag{S3.12}$$

The measure $d^3\mathbf{p}$ over all momenta is invariant under the reflection $\mathbf{p}\to-\mathbf{p}$, the energy is a quadratic function of $\mathbf{p}$ and so invariant under reflection, and the last integral is by definition $\psi(-\mathbf{x},t)$. In exactly the same way

$$U^\dagger_P\psi^*(\mathbf{x},t)U_P = \int \frac{d^3\mathbf{p}}{\sqrt{(2\pi)^3}}\,a^\dagger_{-\mathbf{p}}e^{-i(\mathbf{p}\cdot\mathbf{x}-\omega_{\mathbf{p}}t)} = \int \frac{d^3\mathbf{p}}{\sqrt{(2\pi)^3}}\,a^\dagger_{\mathbf{p}}e^{-i(-\mathbf{p}\cdot\mathbf{x}-\omega_{\mathbf{p}}t)} = \psi^*(-\mathbf{x},t) \tag{S3.13}$$

There's no problem with a unitary parity operator, U_P. Now for time reversal. Because $T\colon \mathbf{p}\to-\mathbf{p}$, it follows that we should have (recalling that Ω_T is *anti*-unitary)

$$\Omega_T^{-1}\begin{Bmatrix} a_{\mathbf{p}} \\ a^\dagger_{\mathbf{p}} \end{Bmatrix}\Omega_T = \begin{Bmatrix} a_{-\mathbf{p}} \\ a^\dagger_{-\mathbf{p}} \end{Bmatrix} \tag{S3.14}$$

Then

$$\begin{aligned} \Omega_T^{-1}\psi(\mathbf{x},t)\Omega_T &= \int \frac{d^3\mathbf{p}}{\sqrt{(2\pi)^3}}\,a_{-\mathbf{p}}\left(e^{i(\mathbf{p}\cdot\mathbf{x}-\omega_{\mathbf{p}}t)}\right)^* = \int \frac{d^3\mathbf{p}}{\sqrt{(2\pi)^3}}\,a_{-\mathbf{p}}e^{-i(\mathbf{p}\cdot\mathbf{x}-\omega_{\mathbf{p}}t)} \\ &= \int \frac{d^3\mathbf{p}}{\sqrt{(2\pi)^3}}\,a_{\mathbf{p}}e^{i(\mathbf{p}\cdot\mathbf{x}+\omega_{\mathbf{p}}t)} = \psi(\mathbf{x},-t) \end{aligned} \tag{S3.15}$$

as desired. The field $\psi^*(\mathbf{x},t)$ transforms in exactly the same way;

$$\Omega_T^{-1}\psi^*(\mathbf{x},t)\Omega_T = \psi^*(\mathbf{x},-t) \tag{S3.16}$$

So there's no problem with an anti-unitary operator Ω_T. The problem is with charge conjugation. A unitary charge conjugation operator U_C, if it exists, would transform ψ into ψ^*, and vice-versa:

$$U_C^\dagger \psi(\mathbf{x},t) U_C = \psi^*(\mathbf{x},t); \quad U_C^\dagger \psi^*(\mathbf{x},t) U_C = \psi(\mathbf{x},t) \tag{S3.17}$$

The canonical commutation relation (S3.7) says, dividing out the i,

$$[\psi(\mathbf{x},t), \psi^*(\mathbf{y},t)] = \delta^{(3)}(\mathbf{x}-\mathbf{y}) \tag{S3.18}$$

Then

$$U_C^\dagger[\psi(\mathbf{x},t), \psi^*(\mathbf{y},t)]U_C = U_C^\dagger \delta^{(3)}(\mathbf{x}-\mathbf{y}) U_C = \delta^{(3)}(\mathbf{x}-\mathbf{y}) \tag{S3.19}$$

But

$$\begin{aligned} U_C^\dagger[\psi(\mathbf{x},t), \psi^*(\mathbf{y},t)]U_C &= U_C^\dagger \psi(\mathbf{x},t)\psi^*(\mathbf{y},t) U_C - U_C^\dagger \psi^*(\mathbf{y},t)\psi(\mathbf{x},t) U_C \\ &= [U_C^\dagger \psi(\mathbf{x},t) U_C, U_C^\dagger \psi^*(\mathbf{y},t) U_C] \\ &= [\psi^*(\mathbf{y},t), \psi(\mathbf{x},t)] = -\delta^{(3)}(\mathbf{y}-\mathbf{x}) = -\delta^{(3)}(\mathbf{x}-\mathbf{y}) \end{aligned} \tag{S3.20}$$

This is a contradiction. There is no such operator U_C for this theory.

This model is very much like the complex Klein–Gordon theory, with three exceptions: the energy is non-relativistic, it lacks a charge conjugation operator, and there are no antiparticles. The charge Q counts simply the number of particles, rather than the difference between particles and antiparticles. ■

3.4 (a) Let $y^\mu \equiv e^\lambda x^\mu$. Then the transformation on $\partial^\mu \phi(x)$ becomes

$$\mathcal{D}\colon \frac{\partial \phi(x)}{\partial x^\mu} \to \frac{\partial(e^\lambda \phi(e^\lambda x))}{\partial x^\mu} = e^\lambda \frac{\partial \phi(y)}{\partial y^\nu}\frac{\partial y^\nu}{\partial x^\mu} = e^{2\lambda}\frac{\partial \phi(y)}{\partial y^\mu}$$

Thus $\mathcal{D}\colon \mathscr{L}(x) \to e^{4\lambda}\mathscr{L}(y)$. Then the action becomes

$$S = \int d^4x\, \mathscr{L}(x) \to \int e^{-4\lambda} d^4y\, e^{4\lambda}\mathscr{L}(y) = \int d^4y\, \mathscr{L}(y)$$

Relabeling the dummy variable y to x, the action is manifestly invariant under dilations.

(b) From (5.21),

$$D\phi(x) = \left.\frac{d\phi}{d\lambda}\right|_{\lambda=0} = \left.\left[e^\lambda \phi(e^\lambda x) + e^\lambda \frac{\partial \phi(e^\lambda x)}{\partial x^\alpha} e^\lambda x^\alpha\right]\right|_{\lambda=0} = (1 + x\cdot\partial)\phi(x) \tag{S3.21}$$

Then using (5.26) and (4.25),

$$\begin{aligned} D\mathscr{L} &= \frac{\partial \mathscr{L}}{\partial \phi} D\phi + \pi^\mu \partial_\mu (D\phi) = \pi^\mu \partial_\mu(D\phi) = \partial^\mu\phi\, \partial_\mu\left[(1 + x^\alpha \partial_\alpha)\phi(x)\right] \\ &= \partial^\mu \phi \left[\partial_\mu \phi + \delta^\alpha_\mu \partial_\alpha \phi + x^\alpha \partial_\alpha \partial_\mu \phi\right] \\ &= 2\partial^\mu\phi\,\partial_\mu\phi + \partial^\mu\phi\, x\cdot\partial(\partial_\mu\phi) \\ &= 2\partial^\mu\phi\,\partial_\mu\phi + \tfrac{1}{2} x\cdot\partial\left(\partial^\mu\phi\,\partial_\mu\phi\right) \end{aligned} \tag{S3.22}$$

But $\partial_\mu x^\alpha = \delta^\alpha_\mu$, so $\partial_\mu x^\mu = 4$. Then

$$D\mathscr{L} = \tfrac{1}{2}\left(4\,\partial^\mu\phi\,\partial_\mu\phi + x\cdot\partial\left(\partial^\mu\phi\,\partial_\mu\phi\right)\right) = \partial_\mu\left(\tfrac{1}{2}x^\mu \partial^\alpha\phi\,\partial_\alpha\phi\right) = \partial_\mu\left(x^\mu\mathscr{L}\right) \equiv \partial_\mu F^\mu \tag{S3.23}$$

where $F^\mu = x^\mu \mathscr{L}$. The Noetherian current is defined by (5.27),

$$J^\mu \equiv \pi^\mu_a D\phi^a - F^\mu = \partial^\mu\phi\,(1 + x\cdot\partial)\phi - x^\mu\mathscr{L} = (\partial^\mu\phi)\phi + x_\alpha(\partial^\mu\phi\,\partial^\alpha\phi - g^{\alpha\mu}\mathscr{L}) \tag{S3.24}$$

and the conserved charge Q is the space integral of the zeroth component,

$$Q = \int d^3\mathbf{x}\left[(\partial^0\phi)\phi + x_\alpha(\partial^0\phi\,\partial^\alpha\phi - g^{\alpha 0}\mathscr{L})\right] = \int d^3\mathbf{x}\left[\pi\phi + \pi(x\cdot\partial)\phi - \tfrac{1}{2}x^0(\partial^\mu\phi\,\partial_\mu\phi)\right] \tag{S3.25}$$

where $\pi \equiv \pi^0$ (see (4.27)).

(c) We need to show $[\phi(y), Q] = iD\phi(y)$. The charge Q is time-independent, so we can take its time to be the same as y^0, the time of $\phi(y)$. Then, using the equal-time commutators (3.60) and (3.61),

$$\begin{aligned}
[\phi(y), Q] &= \int d^3\mathbf{x} \left\{ [\phi(y), \pi(x)\phi(x)] + [\phi(y), \pi(x)(x\cdot\partial)\phi(x)] - \tfrac{1}{2}[\phi(y), x^0 \partial^\mu\phi(x)\,\partial_\mu\phi(x)] \right\} \\
&= \int d^3\mathbf{x} \left\{ [\phi(y), \pi(x)]\phi(x) + [\phi(y), \pi(x)](x\cdot\partial)\phi(x) + \pi(x)x_\alpha[\phi(y), \partial^\alpha\phi(x)] - x^0 g^{\mu\nu}[\phi(y), \partial_\mu\phi]\partial_\nu\phi \right\} \\
&= \int d^3\mathbf{x}\, i\delta^{(3)}(\mathbf{y}-\mathbf{x}) \left\{ \phi(x) + (x\cdot\partial)\phi(x) + \pi(x)x_\alpha g^{\alpha 0} - x^0 g^{\mu\nu} g_{\mu 0}\partial_\nu\phi(x) \right\} \\
&= \int d^3\mathbf{x}\, i\delta^{(3)}(\mathbf{y}-\mathbf{x}) \left\{ \phi(x) + (x\cdot\partial)\phi(x) + \pi(x)x^0 - x^0\partial_0\phi(x) \right\} \\
&= i(1 + y\cdot\partial)\phi(y) = iD\phi(y)
\end{aligned} \tag{S3.26}$$

as required.

(d) We need to calculate $[P^\mu, Q]$. The expression for P^μ, (5.45), is the component $T^{\mu 0}$ of the canonical energy-momentum tensor density $T^{\mu\rho}$ given by (5.50), $T^{\mu\rho} = \pi^\rho\partial^\mu\phi - g^{\rho\mu}\mathscr{L}$, so that

$$P^\mu = \int d^3\mathbf{x}\, T^{\mu 0} = \int d^3\mathbf{x} \left[\pi(x)\partial^\mu\phi(x) - g^{0\mu}\mathscr{L} \right] \tag{S3.27}$$

We will need the commutators of Q and the derivatives $\partial^\mu\phi$. The easiest approach is to differentiate the commutator $[\phi, Q]$ because Q is a constant. (Alternatively, we could proceed as in (b), but that is a lot more work.) Then

$$[\partial^\mu\phi(x), Q] = \partial^\mu[\phi(x), Q] = \partial^\mu(iD\phi(x)) = i\partial^\mu(1 + x\cdot\partial)\phi(x) = i(2 + x\cdot\partial)\,\partial^\mu\phi(x) \tag{S3.28}$$

and in particular,

$$[\pi(x), Q] = i(2 + x\cdot\partial)\pi(x) \tag{S3.29}$$

Using these relations,

$$\begin{aligned}
[P^\mu, Q] &= \int d^3\mathbf{x} \left\{ [\pi(x)\partial^\mu\phi(x), Q] - g^{0\mu}[\mathscr{L}, Q] \right\} \\
&= \int d^3\mathbf{x} \left\{ \pi(x)[\partial^\mu\phi(x), Q] + [\pi(x), Q]\partial^\mu\phi(x) - \tfrac{1}{2}g^{0\mu}\left\{ \partial_\alpha\phi(x)[\partial^\alpha\phi(x), Q] + [\partial_\alpha\phi(x), Q]\partial^\alpha\phi(x) \right\} \right\} \\
&= i\int d^3\mathbf{x} \left\{ 4\pi(x)\partial^\mu\phi(x) + (x\cdot\partial)\left(\pi(x)\partial^\mu\phi(x)\right) - \tfrac{1}{2}g^{0\mu}\left[4\partial_\alpha\phi(x)\partial^\alpha\phi(x) + (x\cdot\partial)\left(\partial_\alpha\phi(x)\partial^\alpha\phi(x)\right) \right] \right\} \\
&= 4i\int d^3\mathbf{x} \left[\pi(x)\,\partial^\mu\phi(x) - g^{0\mu}\mathscr{L} \right] + i\int d^3\mathbf{x}\,(x\cdot\partial)\left(\pi(x)\,\partial^\mu\phi(x) - g^{0\mu}\mathscr{L} \right)
\end{aligned} \tag{S3.30}$$

The second term can be written

$$\begin{aligned}
i\int d^3\mathbf{x}\,(x\cdot\partial)\left(\pi(x)\,\partial^\mu\phi(x) - g^{0\mu}\mathscr{L} \right) &= ix^0\partial_0 \int d^3\mathbf{x} \left[\pi(x)\,\partial^\mu\phi(x) - g^{0\mu}\mathscr{L} \right] \\
&\quad + i\int d^3\mathbf{x}\,(x^j\partial_j) \left[\pi(x)\,\partial^\mu\phi(x) - g^{0\mu}\mathscr{L} \right] \\
&= ix^0\partial_0 P^\mu + i\int d^3\mathbf{x}\,\partial_j \left(x^j \left[\pi(x)\,\partial^\mu\phi(x) - g^{0\mu}\mathscr{L} \right] \right) \\
&\quad - i\int d^3\mathbf{x}\,(\partial_j x^j) \left[\pi(x)\,\partial^\mu\phi(x) - g^{0\mu}\mathscr{L} \right] \\
&= -3i\int d^3\mathbf{y} \left[\pi(y)\,\partial^\mu\phi - g^{0\mu}\mathscr{L} \right]
\end{aligned} \tag{S3.31}$$

because P^μ is time-independent, the second integral is a divergence to be transformed into a surface integral at infinity, and $\partial_i y^i = 3$. Then

$$[P^\mu, Q] = (4i - 3i)\int d^3\mathbf{x} \left[\pi(x)\,\partial^\mu\phi(x) - g^{0\mu}\mathscr{L} \right] = iP^\mu \tag{S3.32}$$

which was to be shown. ■

8

Perturbation theory I. Wick diagrams

We begin with the expression for the S-matrix introduced last time, (7.58), written in terms of Dyson's formula, (7.36),

$$\mathrm{S} = U_I(\infty, -\infty) = T \exp\left(-i \int_{-\infty}^{\infty} dt\, H_I(t) \right) \tag{8.1}$$

We will use good old quantum field theory to evaluate this object, applied to three specific examples, model theories, which we will discuss at various times throughout these lectures.

8.1 Three model field theories

Here are our three models.

MODEL 1:

$$\mathscr{L} = \tfrac{1}{2}(\partial^\mu \phi)(\partial_\mu \phi) - \tfrac{1}{2}\mu^2\phi^2 - g\rho(x)\phi(x) \tag{8.2}$$

Model 1 is a scalar field, $\phi(x)$, interacting with some spacetime-dependent c-number function, $\rho(x)$, which we may vary experimentally as we wish. I will assume, to make everything simple, that $\rho(x)$ goes to zero as x goes to infinity in either a spacelike or timelike direction. The variable g is a free parameter called the **coupling constant**. I could of course absorb g in $\rho(x)$. But later on, I would like to study what happens if I increase g while keeping $\rho(x)$ fixed.

I choose this Lagrangian because the field obeys the equation of motion

$$(\Box^2 + \mu^2)\phi(x) = -g\rho(x) \tag{8.3}$$

This equation is very similar to the fundamental equation of electrodynamics in the Lorenz[1] gauge,

$$\Box^2 A^\mu = -eJ^\mu \tag{8.4}$$

where J^μ is the electromagnetic current. In the real world, the electromagnetic current is some complicated function of the fields of charged particles. You've seen how to construct them,

[1] [Eds.] Often rendered "Lorentz", after Hendrik A. Lorentz (1853–1928), but in fact due to Ludvig V. Lorenz (1829–1891). See J. D. Jackson and L. B. Okun, "Historical Roots of Gauge Invariance", *Rev. Mod. Phys.* (2001) **73**, 663–680.

(5.27). It's frequently convenient, however, to consider a simpler problem, where J^μ is just some c-number function under our experimental control. We could move large charged bodies around on tracks in some classical way, changing the current. This makes light, photons in the quantum theory. Model 1 describes a theory analogous to the electromagnetic field (which we don't yet know how to quantize) in an external current, a scalar field for a meson in an external current.

We know from electromagnetic theory that this current J^μ makes light. We also know that light is photons, so this current makes photons. We would expect, in the analogous case, that when we wiggle the source $\rho(x)$—turn it on and off and shake it around—we should shake off mesons. We will try to compute exactly how many mesons are shaken off and in what states. This will be our simplest model, because there is no need here to invoke an adiabatic turning on and off function. The real honest-to-goodness physics of the problem with $\rho(x)$ automatically turns itself on and off by assumption.

MODEL 2:

$$\mathscr{L} = \tfrac{1}{2}(\partial^\mu\phi)(\partial_\mu\phi) - \tfrac{1}{2}\mu^2\phi^2 - g\rho(\mathbf{x})\phi(x) \tag{8.5}$$

Our second model is exactly the same as Model 1, except that we restrict ρ to be a function of $\mathbf{x}$ only, independent of time. Analytically, Model 2 is somewhat simpler, but physically it requires a bit more thought. Again I'll assume $\rho(\mathbf{x})$ goes to zero as rapidly as necessary to make any of our integrals converge as $\mathbf{x} \to \infty$.

Model 2 is analogous to good old electrostatics:: Given a static charge distribution or a constant current distribution, compute the electromagnetic field it makes. In Model 2 we have a static source. We don't know at this stage what's going to happen. Maybe mesons will scatter off this static source, and it will act like a potential in which they move, we'll see. This problem requires slightly more sophisticated thought, as we will see, because here we will indeed have to put in a turning on and off function; the physics doesn't turn itself off.

MODEL 3:

$$\mathscr{L} = \tfrac{1}{2}(\partial^\mu\phi)(\partial_\mu\phi) - \tfrac{1}{2}\mu^2\phi^2 + \partial^\mu\psi^*\partial_\mu\psi - m^2\psi^*\psi - g\phi\psi^*\psi \tag{8.6}$$

The third model involves two fields, one neutral, ϕ, and one charged, ψ, which is a linear combination of two other scalar fields, ϕ^1 and ϕ^2, as in (6.23). As the coupling constant g goes to zero, we have three free particles: a particle and its antiparticle from the terms in $\psi^*\psi$, and a single neutral particle from the terms in ϕ^2. In the last term, we have a coupling between them.

The equation of motion for the ϕ field is

$$(\Box^2 + \mu^2)\phi = -g\psi^*\psi \tag{8.7}$$

This is beginning to look like the real thing. Aside from the fact that nothing has spin, and I haven't put in any derivatives or tensor indices, this is very similar in its algebraic structure to what we would expect for real electrodynamics. In real electrodynamics, the current J^μ is not prescribed, but is due to the presence of charged particles. Here the electromagnetic field mimicked by the ϕ field is coupled to a quadratic function in the fields of the charged particles. If Model 2 can be described as quantum meso-statics, Model 3 is quantum meso-dynamics.

The equation of motion for the ψ field is

$$(\Box^2 + m^2)\psi = -g\psi\phi \tag{8.8}$$

This model is also very similar to Yukawa's theory of the interaction between mesons and nucleons.[2] These fields play an important role in the theory of nuclear forces. And so I will sometimes refer to the ψ and ψ^* particles as nucleons and antinucleons respectively, and the quanta created by the ϕ as mesons. They are of course scalar nucleons and scalar mesons. Actually we had better not push this theory too far (we'll only do low orders in perturbation theory with it). The Hamiltonian contains the term $g\phi\psi^*\psi$, which is not bounded below for either sign of g.

We will attempt to evaluate the U_I matrix (and thus the scattering matrix) in all these cases by Dyson's formula, written as

$$U_I(\infty, -\infty) = T\exp\left(-i\int d^4x\,\mathscr{H}_I(x)\right) = \text{S} \tag{8.9}$$

The integral in the exponential is equivalent to $\int dt\, H_I(t)$. The interaction Hamiltonian density for Model 1 is

$$\mathscr{H}_I^{(1)} = g\rho(x)\phi(x) \tag{8.10}$$

Note that since we are always working in the interaction representation, $\phi(x) = \phi_I(x)$. For Models 2 and 3, we must put in the adiabatic function $f(t)$,

$$\mathscr{H}_I^{(2)} = gf(t)\rho(\mathbf{x})\phi(x) \tag{8.11}$$

$$\mathscr{H}_I^{(3)} = gf(t)\psi^*\psi\phi \tag{8.12}$$

For Models 1 and 2, we have to take ρ real in order that H_I be Hermitian. In all three cases, we will attempt to analyze the problem by interaction picture perturbation theory.

So these are the three models we're going to play with. I should tell you in advance that it will turn out that for Models 1 and 2, we will be able to sum our perturbation theory and solve the models *exactly*. That should not surprise you because the Heisenberg equations of motion are linear, and anything that involves linear equations of motion is an exactly soluble system.

8.2 Wick's theorem

Our general trick will be an algorithm for turning time ordered products into normal ordered products and some extra terms. Time ordered products are not defined for *every* string of field operators; they are only defined for strings of field operators that have time labels on them. Normal ordered products are not defined for any string of operators; they are only defined for strings of free fields. Fortunately in Dyson's formula, we have both things: operators with time labels, and operators which are free fields. So it makes sense to talk about writing those things alternatively, in terms of time ordered products and normal ordered products. This is a useful thing to do, because it's very easy to compute the matrix elements of normal ordered products once you have them.

For example, consider Model 1, with $\mathscr{H}_I = g\rho(x)\phi(x)$. At the nth order of perturbation theory, we will have a string of n ϕ's. If we sandwich the normal ordering of this string between two-particle states,

$$\langle p_1', p_2'|\, {:}\phi(x_1)\phi(x_2)\cdots\phi(x_n){:}\, |p_1, p_2\rangle \tag{8.13}$$

[2] [Eds.] See note 11, p. 193.

the expression must equal zero for $n > 4$. In that case, each term will contain the product of five operators at least. Each will have either too many annihilation operators, three or more, or too many creation operators. If the former, then let the operators act on the state on the right. Two of these get you to the vacuum at best, but the third annihilates the state. If the latter, the product has too many creation operators, whereupon acting on the state on the left, the same arguments apply and again the state is annihilated. All the normal ordered products that involve more than four field operators are of no interest to us.

What happens with (8.13) if $n = 4$? All that can happen is that the annihilation part of two of the field operators must annihilate the two initial particles, taking you down to the vacuum, and then two others spit out the two final particles bringing you back to the final two-particle state. If we can find an algorithm for turning a time ordered product of operators into a normal ordered product of those operators, plus perhaps some c-number terms, we will have gone a long way in making the successive terms of this perturbation expansion easier to compute, and minimizing the amount of operator algebra we have to play with.

Fortunately there *is* such an algorithm, due to Wick.[3] To explain it, I will have to give some definitions. Let $A(x)$ and $B(y)$ be free fields. (We're always dealing with free fields, since we're always in the interaction picture.) Define an object called the **contraction** of $A(x)$ and $B(y)$, as the difference between the time ordered product and the normal ordered product of the two fields:

$$\overbracket{A(x)B}(y) = T(A(x)B(y)) - :A(x)B(y): \tag{8.14}$$

For free fields, I can prove that the contraction is a c-number. We will evaluate it for the cases we need, that is to say for two ϕ's, a ϕ and a ψ, a ψ and a ψ^*, etc.

To prove the contraction is a c-number I will assume for the moment that $x_0 > y_0$. The corresponding formula when $x_0 < y_0$ will follow by the same reasoning. In this case,

$$T(A(x)B(y)) = A(x)B(y) \quad (\text{because } x_0 > y_0) \tag{8.15}$$

Break each field up into its creation and annihilation parts,

$$A(x) = A^{(+)}(x) + A^{(-)}(x); \qquad B(y) = B^{(+)}(y) + B^{(-)}(y) \tag{8.16}$$

where $A^{(-)}$ and $B^{(-)}$ contain each field's respective creation operators, while $A^{(+)}$ and $B^{(+)}$ contain the annihilation operators (see the discussion following (3.33) on p. 39). Then

$$T(A(x)B(y)) = A^{(+)}(x)B^{(+)}(y)+A^{(-)}(x)B^{(+)}(y)+A^{(-)}(x)B^{(-)}(y)+A^{(+)}(x)B^{(-)}(y) \tag{8.17}$$

There are four terms in the product $A(x)B(x)$, and three of them are already normal ordered. The only one that is not normal ordered is the last. Therefore the right-hand side is the normal ordered product, plus a commutator:

$$T(A(x)B(y)) = \, :A(x)B(y): \, + \, [A^{(+)}(x), B^{(-)}(y)] \tag{8.18}$$

The commutator is a c-number (see (3.38)):

$$[A^{(+)}(x), B^{(-)}(y)] = \begin{cases} \Delta_+(x-y), & \text{if } A = B \\ 0, & \text{if } A \neq B \end{cases} \tag{8.19}$$

---

[3] [Eds.] Gian-Carlo Wick, "The Evaluation of the Collision Matrix", *Phys. Rev.* **80** (1950) 268–272.

A similar argument goes for $x_0 < y_0$, so that we can write

$$\begin{aligned}\overbrace{A(x)A(y)} &= \theta(x_0 - y_0)[A^{(+)}(x), A^{(-)}(y)] + \theta(y_0 - x_0)[A^{(+)}(y), A^{(-)}(x)] \\ &= \theta(x_0 - y_0)\Delta_+(x - y) + \theta(y_0 - x_0)\Delta_+(y - x)\end{aligned} \tag{8.20}$$

That tells us that the contraction is a c-number.

We can write another expression for the contraction simply by taking the ground state expectation value of (8.14) above:

$$\begin{aligned}\overbrace{A(x)B(y)} = \langle 0|\overbrace{A(x)B(y)}|0\rangle &= \langle 0|\, T(A(x)B(y))\,|0\rangle \;-\; \langle 0|\!:\!A(x)B(y)\!:\!|0\rangle \\ &= \langle 0|\, T(A(x)B(y))\,|0\rangle\end{aligned} \tag{8.21}$$

because, by design, a normal ordered product always has zero vacuum expectation value. Consequently,

$$\overbrace{\phi(x)\phi(y)} = \langle 0|\, T(\phi(x)\phi(y))\,|0\rangle \tag{8.22}$$

By an amazing "coincidence", the right-hand side of this equation is something which you computed in your first homework (see Problem 1.3). I will save all of us the time to work it out again, and just remind you

$$\langle 0|\, T(\phi(x)\phi(y))\,|0\rangle = \lim_{\epsilon\to 0^+} \int \frac{d^4p}{(2\pi)^4} e^{-ip\cdot(x-y)} \frac{i}{p^2 - \mu^2 + i\epsilon} \tag{8.23}$$

Whenever I write an ϵ in the denominator in the future, you will need to remember that we are to take the expression in the limit $\epsilon \to 0^+$. Although you didn't do this for ψ, it's essentially the same calculation, and it's very easy to see that

$$\overbrace{\psi^*(x)\psi(y)} = \overbrace{\psi(x)\psi^*(y)} = \lim_{\epsilon\to 0^+} \int \frac{d^4p}{(2\pi)^4} e^{-ip\cdot(x-y)} \frac{i}{p^2 - m^2 + i\epsilon} \tag{8.24}$$

You get two equal terms from the ϕ^1 and ϕ^2, but that 2 is canceled by the $\sqrt{2}$ in the definitions of ψ and ψ^*. All other contractions equal zero:

$$\overbrace{\psi(x)\phi(y)} = \overbrace{\psi^*(x)\phi(y)} = \overbrace{\psi(x)\psi(y)} = \overbrace{\psi^*(x)\psi^*(y)} = 0 \tag{8.25}$$

That's how it goes for two fields. Wick proved the same procedure works for a string of fields. We'll want two pieces of notation before diving into the proof. First, suppose we have a normal ordered string of fields, and want to contract two which are not immediately adjacent. Then

$$:\!A(x)\overbrace{B(y)C(z)D(w)}\!: \;\equiv\; :\!A(x)C(z)\!:\, \overbrace{B(y)D(w)} \tag{8.26}$$

(the contraction on the left connects $B(y)$ and $D(w)$)

And, just for short, write

$$\phi^{a_1}(x_1)\phi^{a_2}(x_2)\cdots\phi^{a_n}(x_n) \equiv \phi_1\phi_2\cdots\phi_n \tag{8.27}$$

With those two conventions established, let's state **Wick's Theorem**:

$$\boxed{\begin{aligned} T(\phi_1\phi_2\cdots\phi_n) \;=\;& :\!\phi_1\phi_2\cdots\phi_n\!: \\ &+ \;:\!\overbrace{\phi_1\phi_2}\cdots\phi_n\!: \;+\; \text{(all other terms with one contraction)} \\ &+ \;:\!\overbrace{\phi_1\phi_2}\overbrace{\phi_3\phi_4}\cdots\phi_n\!: \;+\; \text{(all other terms with two contractions)} \\ &+\cdots+\; \text{(all terms with } \tfrac{1}{2}n \text{ or } \tfrac{1}{2}(n-1) \text{ contractions)}\end{aligned}} \tag{8.28}$$

If n is even, the last terms contain $\frac{1}{2}n$ contractions, otherwise they contain $\frac{1}{2}(n-1)$ contractions and a single field. It's perhaps not surprising that you obtain all the terms on the right-hand side of (8.28). A remarkable and graceful feature of the theorem is that each term appears exactly once, with coefficient +1.

Proof. The proof proceeds by induction on n. Let $W(\phi_1\phi_2\cdots\phi_n)$ denote the right-hand side of (8.28). It is trivially true that $T(\phi_1) = W(\phi_1)$, because both sides simply equal ϕ_1. We've already established

$$T(\phi_1\phi_2) = :\phi_1\phi_2: + :\overbracket{\phi_1\phi_2}: = W(\phi_1\phi_2) \tag{8.29}$$

By the induction hypothesis, assume $T(\phi_1\phi_2\cdots\phi_{n-1}) = W(\phi_1\phi_2\cdots\phi_{n-1})$. If we can show $T(\phi_1\phi_2\cdots\phi_n) = W(\phi_1\phi_2\cdots\phi_n)$, we're done. Without loss of generality, we can relabel the fields, such that $x_1^0 \geq x_2^0 \geq x_3^0 \geq \cdots \geq x_n^0$, and suppose we have then, by the induction hypothesis,

$$T(\phi_2\phi_3\cdots\phi_n) = W(\phi_2\phi_3\cdots\phi_n) \tag{8.30}$$

The job now is to show $T(\phi_1\phi_2\cdots\phi_n) = W(\phi_1\phi_2\cdots\phi_n)$. Multiply both sides of (8.30) by ϕ_1:

$$\phi_1 T(\phi_2\phi_3\cdots\phi_n) = \phi_1 W(\phi_2\phi_3\cdots\phi_n) \tag{8.31}$$

The left-hand side of this equation is $T(\phi_1\phi_2\cdots\phi_n)$, because x_1^0 is larger than all the other times. The right-hand side is

$$\begin{aligned}\phi_1 W(\phi_2\phi_3\cdots\phi_n) &= \left(\phi_1^{(+)} + \phi_1^{(-)}\right) W(\phi_2\phi_3\cdots\phi_n)\\ &= \phi_1^{(-)} W(\phi_2\phi_3\cdots\phi_n) + W(\phi_2\phi_3\cdots\phi_n)\phi_1^{(+)} + [\phi_1^{(+)}, W(\phi_2\phi_3\cdots\phi_n)]\end{aligned} \tag{8.32}$$

W contains two types of elements, normal ordered strings and contractions. All of the terms in (8.32) are normal ordered. The first two terms on the right-hand side contain all contractions that do not involve ϕ_1, as well as the remainder (if any) of uncontracted fields in normal order. Within the commutator, all the purely c-number terms, if any, will commute with $\phi_1^{(+)}$. The other terms will produce all the contractions that do involve ϕ_1. Either a contraction involves ϕ_1, or it does not. Therefore, the right-hand side of (8.32) is a normal ordered series containing all possible contractions of the n fields: it is equal to $W(\phi_1\phi_2\cdots\phi_n)$. **QED**

I leave it as an exercise to show that Wick's theorem can also be written in the form

$$T(\phi_1\phi_2\cdots\phi_n) = :\exp\left(\tfrac{1}{2}\sum_{i,j=1}^{n} \overbracket{\phi_i\phi_j}\frac{\partial}{\partial\phi_i}\frac{\partial}{\partial\phi_j}\right)\phi_1\phi_2\cdots\phi_n: \tag{8.33}$$

8.3 Dyson's formula expressed in Wick diagrams

Wick's theorem is very nice, but we are going to find something even better: We're going to find a diagrammatic rule for representing every term in the **Wick expansion**, i.e., the application of Wick's theorem to the Taylor expansion of Dyson's formula (8.1). Instead of having to write complicated contractions, we can just write simple looking diagrams. These are not yet the famous Feynman diagrams.[4] I am introducing these objects *ad hoc*, to make

[4] [Eds.] In an interview with Charles Weiner of the American Institute of Physics, Richard Feynman said, "I was working on the self-energy of the electron, and I was making a lot of these pictures to visualize the

the eventual passage to Feynman diagrams (in Chapter 10) as painless as possible. I will call these objects Wick diagrams. They differ from Feynman diagrams because Wick diagrams represent operators and Feynman diagrams represent matrix elements. Most textbooks go directly to the Feynman diagrams. I find the combinatorics gets too complicated that way. I will explain this diagrammatic rule using our third example, Model 3, (8.6), which has the most complicated interaction Hamiltonian, (8.12), of the three models we are considering:

$$\mathscr{H}_I = gf(t)\psi^*\psi\phi \tag{8.12}$$

We'll keep the compressed notation, writing $\phi(x_1)$ as ϕ_1, etc., to simplify things. Dyson's formula (8.9), the thing we have to study, is the time-ordered exponential of the expression

$$-i\int d^4x\, \mathscr{H}_I \tag{8.34}$$

A typical term in Dyson's formula arising in n^{th} order of perturbation theory will involve a product of n copies of this expression (8.12), integrated over points $x_1, x_2, \ldots, x_n$. Let's look at the second-order term:

$$\frac{(-ig)^2}{2!}\int d^4x_1\, d^4x_2\, f(t_1)f(t_2)T(\psi_1^*\psi_1\phi_1\psi_2^*\psi_2\phi_2) \tag{8.35}$$

I will draw a diagram, starting with dots labeled 1, 2, and so on, indicating x_1, x_2 etc. The number of dots is the order in perturbation theory to which you are going. Associated with

Figure 8.1: Two points, for the second-order term in Dyson's formula

each dot, we will draw an arrow going into the dot, and an arrow going out from the dot, and a line without an arrow on it at all, one end attached to the dot. An arrow going in corresponds to the factor ψ; an arrow going out is a ψ^*, and the plain line corresponds to ϕ. We draw these at point 1 to associate the fields at x_1, and similarly for the second dot. In this

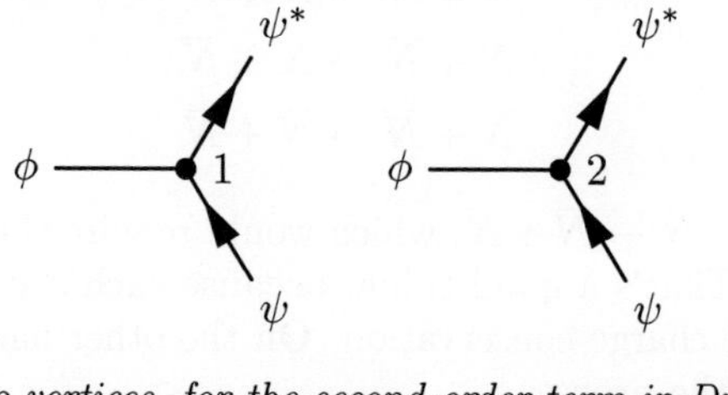

Figure 8.2: Two vertices, for the second-order term in Dyson's formula

way I can associate various terms that occur in the expansion with a pattern of dots, with three lines coming out from each dot.

---

various terms and thinking about the various terms, that a moment occurred—I remember distinctly—when I looked at these, and they looked very funny to me. They were funny-looking pictures. And I did think consciously: Wouldn't it be funny if this turns out to be useful, and the *Physical Review* would be all full of these funny-looking pictures? It would be very amusing." Quoted by Schweber *QED*, p. 434. For a history of the introduction and dispersion of Feynman diagrams, see David Kaiser, *Drawing Theories Apart*, U. of Chicago Press, 2005.

What is the prescription for contractions? Whenever two fields are contracted, for x_1 and x_2, say, we will join the appropriate lines from those two dots. We can either join a straight line with a straight line if there's a $\phi - \phi$ contraction, or we can join the head of an arrow with the tail of an arrow if there's a $\psi^* - \psi$ contraction. For example, the Wick expansion of the second-order contribution (8.35) includes the term

$$\frac{(-ig)^2}{2!}\int d^4x_1\, d^4x_2\, f(t_1)f(t_2) :\psi_1^*\psi_1\overbracket{\phi_1\psi_2^*\psi_2\phi_2}: \tag{8.36}$$

Associated with this term is the diagram in Figure 8.3. The term (8.36) can contribute to a

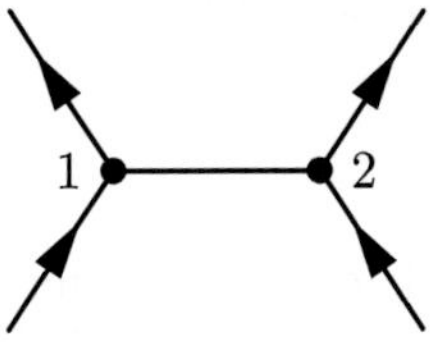

Figure 8.3: Second-order diagram for Model 3, with $\phi - \phi$ contraction

variety of physical processes. The operator

$$:\psi_1^*\psi_1\psi_2^*\psi_2: \tag{8.37}$$

contains, within the ψ field, operators that can annihilate a "nucleon", N, as well as operators that can create an "antinucleon", $\overline{N}$, while the operators within the ψ^* field can create N and annihilate $\overline{N}$ (see (6.24).) Consequently the amplitude

$$\langle\,\text{final two nucleon state}\,|:\psi_1^*\psi_1\psi_2^*\psi_2:|\,\text{initial two nucleon state}\,\rangle \tag{8.38}$$

will not be zero, because there are two annihilation operators in the two ψ fields to destroy the two nucleons in the initial state, and two creation operators in the two ψ^* fields to create two nucleons in the final state. The term (8.36) thus contributes to these reactions:

$$\begin{aligned} N + N &\to N + N \\ N + \overline{N} &\to N + \overline{N} \\ \overline{N} + \overline{N} &\to \overline{N} + \overline{N} \end{aligned} \tag{8.39}$$

It cannot contribute to $N + N \to \overline{N} + \overline{N}$, which would require the ψ field to create N and the ψ^* field to annihilate $\overline{N}$. That's a good thing, because such a process would break the U(1) symmetry and thus violate charge conservation. On the other hand, it looks like the operator (8.37) *could* contribute to the process

$$\text{vacuum} \to N + N + \overline{N} + \overline{N}$$

which does not violate charge conservation, but it *does* violate energy-momentum conservation. That would be a disaster. The coefficient of the term after integrating over x_1 and x_2 had better turn out to be zero.

As a second example, another term in the Wick expansion of the second-order contribution is

$$\frac{(-ig)^2}{2!}\int d^4x_1\, d^4x_2\, f(t_1)f(t_2) :\overbracket{\psi_1^*\psi_1\phi_1\psi_2^*\psi_2}\phi_2: \tag{8.40}$$

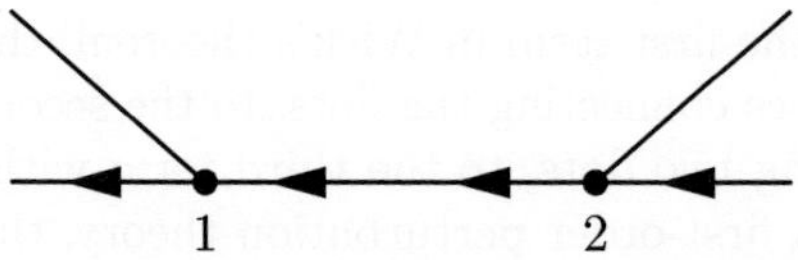

Figure 8.4: Second-order diagram for Model 3, with $\psi^* - \psi$ contraction

For the diagram corresponding to this term, see Figure 8.4. Here we have an operator $:\psi_1\phi_1\psi_2^*\phi_2:$ containing an uncontracted ψ, an uncontracted ψ^* and two uncontracted ϕ's. This particular operator could contribute, for example, to the processes

$$\begin{gathered} N+\phi \to N+\phi \\ \overline{N}+\phi \to \overline{N}+\phi \\ \overline{N}+N \to \phi+\phi \end{gathered} \tag{8.41}$$

That is, "nucleon" plus meson (remember, "nucleons" are what our ψ fields annihilate) go to "nucleon" plus meson, because the operator $\psi_1\phi_1\psi_2^*\phi_2$ contains a nucleon annihilation operator, a meson annihilation operator, a nucleon creation operator (in ψ^*) and a meson creation operator. It could also make a contribution to the matrix elements of the process "antinucleon" plus meson goes into "antinucleon" plus meson, because every term that contains a nucleon creation operator also contains an antinucleon annihilation operator. Or, for example, it could contribute to the process where $\phi+\phi$ go into N plus $\overline{N}$, or N plus $\overline{N}$ go into $\phi+\phi$, picking annihilation and creation operators in the right way. Notice that we can't have $N \to \phi+\phi+N$, because of energy-momentum conservation.

Just as we can draw a diagram from the corresponding expression in the Wick expansion, so we can write down the Wick expansion term from a given diagram. For example, consider the diagram in Figure 8.3. Reading this diagram and remembering what the theory is, with the rules given earlier about drawing the diagrams, we can write down what is going on. This is a second-order perturbation diagram because there are two vertices. Each vertex contributes a term $(-ig)$, and a term of $1/2!$ comes from the expansion of the exponential. We have $d^4x_1\,d^4x_2$ because we've got two d^4x's; two vertices. The internal line corresponds to a contraction of the two ϕ operators, and this is the only contraction. The external lines show two ψ fields (the inward going arrows) and two ψ^* fields (the outward going arrows.) So the remainder of the operator must correspond to the normal ordered product $:\psi_1^*\psi_1\psi_2^*\psi_2:$ of the "nucleon" field operators. Therefore we recover the associated operator (8.36),

$$\frac{(-ig)^2}{2!}\int d^4x_1\,d^4x_2\,f(t_1)f(t_2)\,:\psi_1^*\psi_1\psi_2^*\psi_2:\,\overbrace{\phi_1\phi_2} \tag{8.36}$$

Given any term in the Wick expansion, we can find the corresponding Wick diagram, and vice-versa:

The Wick diagrams are in 1: 1 correspondence with the terms in the Wick expansion.

The *entire* Wick expansion may be represented by a series of diagrams, *every* possible diagram, though some may evaluate to zero. For example, for Model 3, the terms in Wick's theorem of 17th order consist of all diagrams with 17 dots, with all lines connecting them drawn in all possible ways, ranging from the first term in Wick's theorem, the normal ordered product of $17 \times 3 = 51$ fields, with no lines connecting the dots, to the second term with one contraction, diagrams with one line joining two dots, to the third term with two contractions, with two lines joining the dots, etc. In first-order perturbation theory, the Wick expansion involves a product of three operators, and has two terms,

$$W(\psi, \psi^*, \phi) = :\psi^*\psi\phi: + \overbrace{\psi^*\psi}\phi \tag{8.42}$$

and so two diagrams, but both turn out to vanish by energy-momentum conservation. This is a product of three field operators. The first term has nothing contracted, and vanishes unless we've stupidly chosen our meson mass to be so large it can decay into nucleon and antinucleon. The second term vanishes again by energy-momentum conservation because you can't build a one meson state that has the same energy and momentum as the vacuum state.

Some terms in the Wick expansion contribute nothing. For example, this term

$$\frac{(-ig)^2}{2!} \int d^4x_1\, d^4x_2\, f(t_1)f(t_2) :\overbrace{\psi_1^*\psi_1\phi_1\psi_2^*}\psi_2\phi_2: \tag{8.43}$$

is zero, because the contraction $\overbrace{\psi^*\psi^*}$ is zero. So we will never write down a diagram like Figure 8.5: the arrows must line up with the same orientation. That means we can shorten the

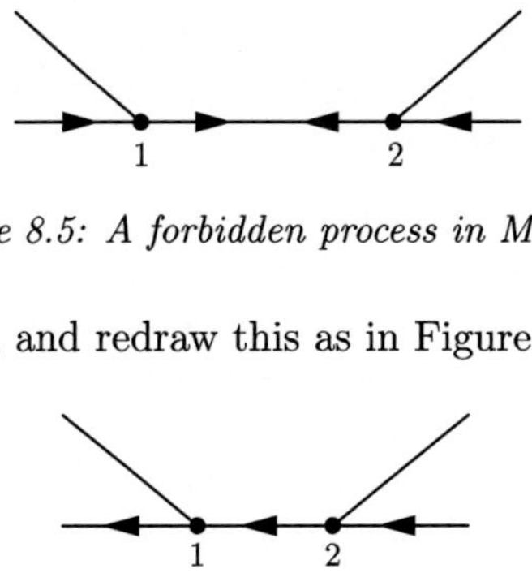

Figure 8.5: A forbidden process in Model 3

middle two arrows in Figure 8.4, and redraw this as in Figure 8.6. Notice only the topological

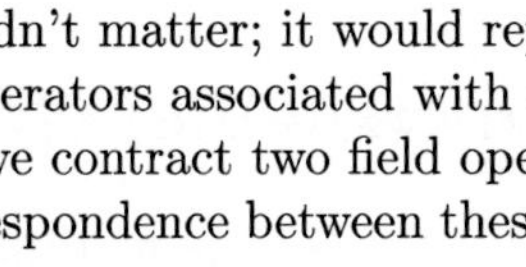

Figure 8.6: Second-order diagram (Figure 8.4) for Model 3, with $\psi^* - \psi$ contraction, redrawn

character of these diagrams is important. If I could have written a term twisted upside down or bent around upon itself, it wouldn't matter; it would represent the same term. It's enough that we represent the three field operators associated with each integration point by an object as shown in Figure 8.2, and when we contract two field operators, we join their corresponding lines. So we have a one-to-one correspondence between these diagrams and the terms in Wick's theorem. Because the terms

$$:\overbrace{\psi_1^*\psi_1\phi_1\psi_2^*\psi_2}\phi_2: \quad \text{and} \quad :\psi_1^*\overbrace{\psi_1\phi_1\psi_2^*}\psi_2\phi_2: \tag{8.44}$$

are distinct, so are their diagrams, as shown in Figure 8.7. After integration over d^4x_1 and d^4x_2, however, the operators corresponding to these diagrams are the same. Just to remind you, these Wick diagrams are *not* Feynman diagrams, but they are most of the way to them. Feynman diagrams do not have labeled points, but they will have labeled momenta on the external lines.

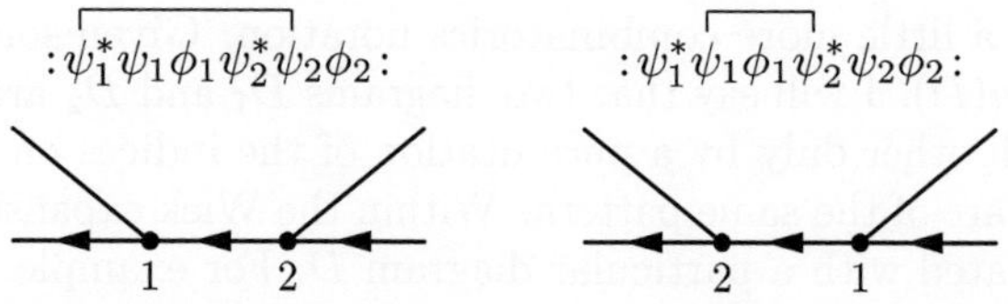

Figure 8.7: Each term in the Wick expansion gets its own diagram

Some Wick diagrams do not have *any* external lines. Those are the terms where everything is contracted. We will discover what they mean in the course of time. For example, this term also occurs in second-order perturbation theory for Model 3:

$$\frac{(-ig)^2}{2!} \int d^4x_1\, d^4x_2\, f(t_1) f(t_2) :\psi_1^* \psi_1 \phi_1 \psi_2^* \psi_2 \phi_2: \tag{8.45}$$

The appropriate diagram is given in Figure 8.8:

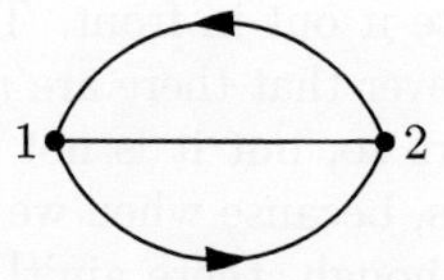

Figure 8.8: The diagram from the operator $:\psi_1^* \psi_1 \phi_1 \psi_2^* \psi_2 \phi_2:$

Here I have contracted the ϕ at 1 with the ϕ at 2. I can join an undirected line to an undirected line because there is a non-zero ϕ–ϕ contraction. I can join the head of an arrow to a tail of the arrow because there's a non-zero ψ–ψ^* contraction. It would be incorrect to draw a diagram in which I connected the head of an arrow to the head of an arrow because that would be a ψ^*–ψ^* contraction, which vanishes. You might think that there is a second diagram, with the labels 1 and 2 switched. But that is exactly the same as Figure 8.8 rotated through 180° in the plane of the page. There is only one way to contract all the fields. That's what Wick's theorem says: Make all possible contractions. This means simply that we draw diagrams with all possible connections. Diagrams with no external lines are perhaps a little unexpected, but they're there because Wick's theorem tells you they're there.

8.4 Connected and disconnected Wick diagrams

Having given you a headache over Wick's theorem and then over the diagrammatic representation of Wick's theorem, I will now give you even more of a headache by manipulating these diagrams in certain ways. It's obvious that if we attempted to compute all these diagrams individually and then sum them up, we would do the same computation several times. For example, in Figure 8.7, as I emphasized, the diagram on the right, with 1 and 2 interchanged, is not the same as the original diagram on the left: it represents a different term in the integrand. However the integrals are identical because we end up integrating over x_1 and x_2, and that will give us exactly the same answer for both diagrams once we're done integrating. Remember, we apply Wick's theorem *before* we integrate. Indeed, any other diagram we obtain from a given diagram by merely permuting the indices will give us the same result, because all that the indices on the vertices tell us is what we call x_1 and what we call x_2, and we're integrating over all of them in the end.

So I will introduce a little more combinatorics notation. Given some diagram D, let the number of vertices be $n(D)$. I will say that two diagrams D_1 and D_2 are "of the same pattern" if they differ from each other only by a permutation of the indices on the vertices. The two diagrams in Figure 8.7 are of the same pattern. Within the Wick expansion of (8.1) are various operators $O(D)$ associated with a particular diagram D. For example, let D be the diagram in Figure 8.3. The operator $O(D)$ associated with it is (8.36), but multiplied by the factorial 2! for reasons that will become clear:

$$O(D) = (-ig)^2 \int d^4x_1\, d^4x_2\, f(t_1)f(t_2)\, \psi_1^*\psi_1\overline{\phi_1\psi_2^*\psi_2\phi_2} \tag{8.46}$$

For any diagram D of a given pattern and its associated operator $O(D)$, introduce the operator

$$\frac{:O(D):}{n(D)!} \tag{8.47}$$

I'm going to pay special attention to the factor of $n(D)!$. Factorials are always important in combinatoric discussions so I write it out in front. There are $n(D)!$ ways of rearranging the indices. This does not mean however that there are $n(D)!$ different diagrams of the same pattern. It would be lovely if it were so, but it is not so. In the case of Figure 8.3, there are $n(D)! = 2! = 2$ different diagrams, because when we exchange 2 and 1 we get a different diagram. In the case of Figure 8.8 though, there ain't! I need to introduce the symmetry number, $S(D)$, equal to the number of permutations of indices that do not change anything. For example, in the case of Figure 8.8, exchanging the indices 1 and 2 doesn't change a thing; $S(D) = 2$. For a second example, consider the diagrams in Figure 8.9. Diagram (a) is not distinct from diagram (b), or from two other cyclic permutations. But these *are* distinct from similar diagrams with non-cyclic permutations. For this diagram, $S(D) = 4$.

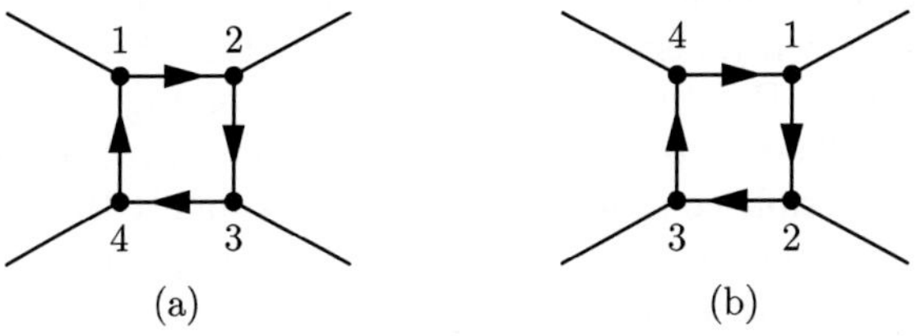

Figure 8.9: A fourth-order contribution in Model 3

A more complicated example is shown in Figure 8.10. This diagram contributes to nucleon–nucleon scattering in the sixth order of perturbation theory. This diagram has $S(D)$ equal to 2. There are only two permutations of the indices that don't change anything, corresponding to switching all of the bottom indices with all the top indices, or rotating the diagram about the horizontal dashed line. You see that vertex 1 plays exactly the same role as vertex 2, contract meson at 1 with meson on 2, 5 and 6 play exactly the same role as 4 and 3.

Once I have taken account of this, say by declaring 4 to be the top vertex of the nucleon loop, then all the others are completely determined. Once I decide which of 4 and 5 is 4 and which is 5, then I have everything labeled uniquely, and all other permutations of the indices will reproduce different terms in the Wick expansion. You can play around, if you enjoy these sorts of combinatoric games, trying to invent diagrams with $S(D) = 3$, or 6, and so on, for all sorts of things.

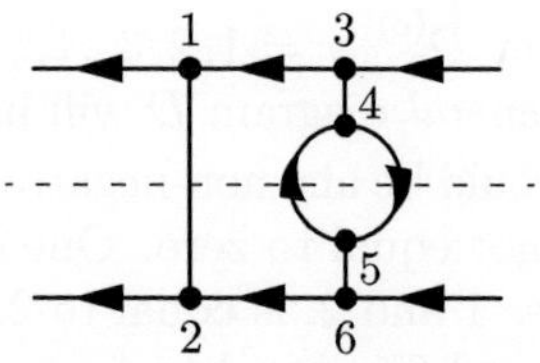

Figure 8.10: A sixth-order contribution in Model 3

How many distinct terms do we get with each pattern? There are $n(D)!/S(D)$ terms. If we permute the indices in all possible ways we get $n(D)!$ different things, but we're over-counting by $S(D)$. Summing over a whole pattern—everything of the same pattern as a given diagram—yields

$$\begin{pmatrix}\text{sum of all diagrams}\\ \text{in a given pattern}\end{pmatrix} = \frac{n(D)!}{S(D)}\frac{:O(D):}{n(D)!} = \frac{:O(D):}{S(D)} \tag{8.48}$$

Therefore the $n(D)!$ gets knocked down into simply $S(D)$. Well, it looks a bit complicated but we've saved ourselves labor. If we were really going to compute this diagram, there are 6! different permutations and it would really be rather stupid to compute all 720 different diagrams.

All the diagrams I've written down up to now are **connected**. "Connected" means (in any theory) that the diagram is in one piece; all the parts of the diagram are contiguous to at least one other part at a vertex. People sometimes confuse "connected" with *contracted.* You can have a connected diagram without a contraction, as shown in Figure 8.11.

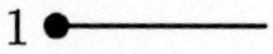

Figure 8.11: A first-order contribution in Model 1

But you can imagine a disconnected diagram. Here is one that arises in fourth order.

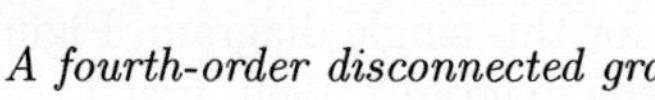

Figure 8.12: A fourth-order disconnected graph in Model 1

This is a perfectly reasonable Wick diagram. Anything I can draw, as long as I don't connect the head of an arrow with the head of an arrow (or tail to tail), is acceptable. Here is a more complicated diagram with three disconnected components.

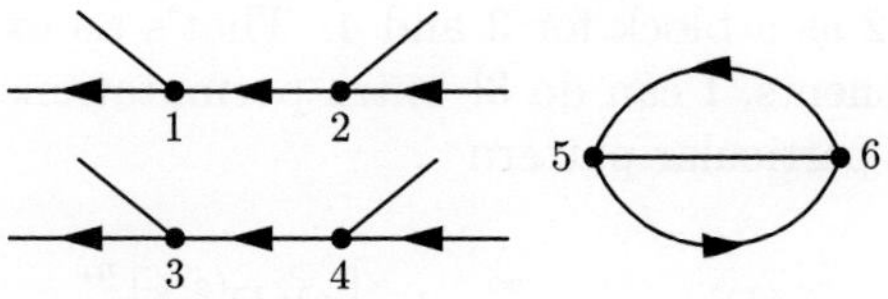

Figure 8.13: A sixth-order disconnected graph in Model 3

Now we come to a marvelous theorem involving Wick diagrams. I will state it first:

$$\boxed{\sum \text{all Wick diagrams} = :\exp\Big(\sum \text{connected Wick diagrams}\Big):} \tag{8.49}$$

I have to define some variables. Let $D_r^{(c)}$, $r = 1, 2, 3, \ldots, \infty$ be a complete set of *connected* diagrams, one of each pattern. A *general* diagram D will have some integer n_r components of the pattern of $D_r^{(c)}$, where the n_r's could be any non-negative integer. For example, the diagram in Figure 8.13 has two of the n_r's not equal to zero. One of them, the one corresponding to the connected diagram with vertices 1 and 2, is equal to 2, and the one corresponding to the connected diagram with vertices 5 and 6 is equal to 1.

I'm going to try to write what a general diagram gives us, from its individual connected parts, in terms of the operators associated with all the diagrams of each pattern. After all, it is pretty easy. (This will be the last of our combinatoric exercises.) Consider the graph in Figure 8.13. The operator in the piece containing vertices 1 and 2 has an integral over $d^4x_1\, d^4x_2$, and we've only got functions of x_1 and x_2 in the integrand. The next piece goes the same way, with functions only of x_3 and x_4, and the final piece likewise with x_5 and x_6. The entire expression for the diagram splits into three *factors*: the diagram yields an operator which is, apart from the combinatoric factor, a product of other operators. From the first piece we get some operator from doing the x_1-x_2 integral, some operator from doing the x_3-x_4 integral from the second, some operator from doing the x_5-x_6 integral from the third. So for this one diagram, we get a single operator squared, and another operator once. That's characteristic of disconnected diagrams: the operators associated with them are simply the normal ordered products of the operators associated with the individual connected components. The contribution for a disconnected diagram $D^{(d)}$ with connected components $D_r^{(c)}$ may be written as

$$:O(D^{(d)}): \;=\; :\prod_{r=1}^{\infty}\left[O(D_r^{(c)})\right]^{n_r}: \tag{8.50}$$

In fact this holds not only for a disconnected diagram $D^{(d)}$, but for a general diagram D. If D is connected, the product involves only a single term: $n_r = 1$ for that single diagram, and $n_r = 0$ for all other diagrams.

Now, what about the symmetry number $S(D)$ for all the diagrams of a particular pattern? Consider the combinatoric factor for this single diagram, Figure 8.13. How many permutations can I make that will not change the diagram? Well, first I could permute the indices. Within each component I can certainly permute the indices just as if that component were there all by itself. Therefore I get the product on r of $1/S(D_r^{(c)})^{n_r}$. I can do it in the first component, I can do it in the second, I can do it in the third, but now I can do one thing more. If I have two identical components, I can bodily exchange the indices in the first component and those in the second, with 1 and 2 as a block for 3 and 4. That's an extra permutation. And if I have three identical components, I can do 3! extra permutations. Therefore I have for the sum of all diagrams D of a particular pattern

$$\frac{:O(D):}{S(D)} \;=\; :\prod_{r=1}^{\infty}\frac{1}{n_r!}\left[\frac{O(D_r^{(c)})}{S(D_r^{(c)})}\right]^{n_r}: \tag{8.51}$$

We are now in a position to get a very simple expression for the matrix U_I which is the sum of all diagrams. Here in (8.51) I've got an expression for a diagram in terms of the operators attached to connected diagrams. The final stroke, and the end of the combinatorics calisthenics for the moment, is to recognize that $U_I(\infty, -\infty)$ is the sum over *all possible* patterns. That is to say,

$$U_I(\infty,-\infty) = \sum \text{all Wick diagrams} = \sum_{n_1=0}^{\infty}\sum_{n_2=0}^{\infty}\cdots :\prod_{r=1}^{\infty}\frac{1}{n_r!}\left[\frac{O(D_r^{(c)})}{S(D_r^{(c)})}\right]^{n_r}: \tag{8.52}$$

Now we can commute the sum and the product, to obtain

$$\sum \text{all Wick diagrams} = :\prod_{r=1}^{\infty}\sum_{n_r=0}^{\infty}\frac{1}{n_r!}\left[\frac{O(D_r^{(c)})}{S(D_r^{(c)})}\right]^{n_r}: \tag{8.53}$$

The sum on each of the n_r's simply gives us the famous formula for the exponential:

$$:\prod_{r=1}^{\infty}\sum_{n_r=0}^{\infty}\frac{1}{n_r!}\left[\frac{O(D_r^{(c)})}{S(D_r^{(c)})}\right]^{n_r}: = :\prod_{r=1}^{\infty}\exp\left(\frac{O(D_r^{(c)})}{S(D_r^{(c)})}\right): \tag{8.54}$$

Everything is inside the normal ordering symbols so I don't have to worry about how the operators go. By another easy manipulation we can write

$$:\prod_{r=1}^{\infty}\exp\left(\frac{O(D_r^{(c)})}{S(D_r^{(c)})}\right): = :\exp\sum_{r=1}^{\infty}\left(\frac{O(D_r^{(c)})}{S(D_r^{(c)})}\right): \tag{8.55}$$

and thus

$$U_I(\infty,-\infty) = \sum \text{all Wick diagrams} = :\exp\sum_{r=1}^{\infty}\left(\frac{O(D_r^{(c)})}{S(D_r^{(c)})}\right): \quad \textbf{QED} \tag{8.56}$$

Now we can forget about all of our combinatorics. We have this one wonderful master theorem which is obviously not special in any way to some particular theory, that the sum of *all* Wick diagrams, the matrix $U_I(\infty,-\infty)$, is in fact simply the normal ordered exponential of the sum of the connected diagrams. It was a long journey, but it was worth it. This is a very nice theorem to have, and it is important. Actually it is more important in statistical mechanics and condensed matter physics than it is in our present study of quantum field theory. In statistical mechanics, you study the operator $e^{-\beta H}$, and in particular its trace, Tr $e^{-\beta H}$, the partition function. The operator $e^{-\beta H}$ is, after all, not that different in its algebraic structure from the operator e^{-iHt}. Typically you compute the partition function in perturbation theory, and then you take its logarithm to get the free energy, the quantity you really want. This identity, (8.56), is the key to getting a direct perturbative expansion for the free energy, rather than having to first compute the partition function in perturbation theory, and then compute its logarithm by a horrible operation. The free energy is just the sum of the connected diagrams.

8.5 The exact solution of Model 1

I will now use the formula (8.56) to solve Model 1, whose interaction Hamiltonian density is

$$\mathscr{H}_I = g\rho(x)\phi(x) \tag{8.57}$$

where ρ is some spacetime function that goes to zero in all directions as rapidly as we please. In Model 1 there are also diagrams. The vertices look much simpler. The primitive vertex out

Figure 8.14: Diagram D_1 in Model 1

of which all diagrams are built is just a single line with a single vertex because there is only one ϕ field with each H_I. I'll call this diagram D_1.

This still means we can make a lot of diagrams. For example, I could make a diagram of 42nd order by joining forty-two of those vertices, one on top of another. The set of Wick diagrams is infinite, but there are only two connected Wick diagrams. D_1 is the first.

A second diagram, D_2, looks like this:

Figure 8.15: Diagram D_2 in Model 1

If you have a pattern of vertices such that only one line can come out of any one of them, you can only draw two connected diagrams, D_1 and D_2. Each of them is the only diagram of their pattern. D_1 has only the single figure, so its symmetry number $S(D_1)$ equals 1. All the diagrams D_1 correspond to the operator

$$O_1 = -ig \int d^4x_1\, \rho(x_1)\phi(x_1) \tag{8.58}$$

The $-ig$ comes from (8.9). D_2 has its symmetry number equal to 2. That is to say, if you exchange 2 and 1, you get the same barbell, just flipped around. The operator corresponding to all of the diagrams D_2 is

$$O_2 = (-ig)^2 \int d^4x_1 d^4x_2\, \overbracket{\phi(x_1)\phi(x_2)}\rho(x_1)\rho(x_2) \tag{8.59}$$

There are no operators left in O_2; it is equal to some complex number $-\alpha + i\beta$ which you'll compute in a homework problem:

$$O_2 = -\alpha + i\beta \quad (\text{where } \alpha > 0) \tag{8.60}$$

By our general theorem, (8.56), we have a closed form expression for $U_I(\infty, -\infty)$:

$$U_I(\infty, -\infty) \;=\; :\!\exp\left(\frac{O_1}{1!} + \frac{O_2}{2!}\right)\!: \;=\; :\!\exp\left(\tfrac{1}{2}O_2\right)\exp(O_1)\!: \;=\; e^{\frac{1}{2}(-\alpha+i\beta)}\; :\!\exp(O_1)\!: \tag{8.61}$$

This is the complete expression for the S-matrix as a sum of normal ordered terms. The first factor is a complex number we'll call A, whose magnitude $|A|$ is an overall normalization constant which I will determine later by a consistency argument. (We won't care about its phase.)

$$A = e^{\frac{1}{2}(-\alpha+i\beta)} \tag{8.62}$$

As I told you, Model 1 is exactly soluble. There may be fifty ways to solve it exactly. It has linear equations of motion, and anything with linear equations of motion is essentially an assembly of harmonic oscillators. An assembly of harmonic oscillators can always be solved by any method you wish. Few are the methods so powerless that they cannot successfully treat an assembly of harmonic oscillators.

Now let's evaluate the expression for $:\exp(O_1):$. After all, ϕ is a free field, so we know what ϕ is in terms of annihilation and creation operators, namely our old formula (3.45),

$$\phi(x) = \int \frac{d^3\mathbf{p}}{(2\pi)^{3/2}\sqrt{2\omega_\mathbf{p}}} \left(a_\mathbf{p} e^{-ip\cdot x} + a_\mathbf{p}^\dagger e^{ip\cdot x} \right) \tag{3.45}$$

Define the Fourier transform $\tilde{\rho}(p)$ of $\rho(x)$ as (this is the same definition as (1.23))

$$\boxed{\tilde{\rho}(p) = \int d^4x\, e^{ip\cdot x} \rho(x)} \tag{8.63}$$

That is, for a function $f(t)$ of time and a function $g(\mathbf{x})$ of space,

$$\boxed{\tilde{f}(\omega) = \int dt\, e^{i\omega t} f(t)}\,; \qquad \boxed{\tilde{g}(\mathbf{p}) = \int d^3\mathbf{x}\, e^{-i\mathbf{p}\cdot\mathbf{x}} g(\mathbf{x})} \tag{8.64}$$

Then (note $\tilde{\rho}(-p) = \tilde{\rho}(p)^*$)

$$O_1 = -ig \int d^4x\, \rho(x)\phi(x) = -ig \int \frac{d^3\mathbf{p}}{(2\pi)^{3/2}\sqrt{2\omega_\mathbf{p}}} [a_\mathbf{p}\, \tilde{\rho}(\mathbf{p}, \omega_\mathbf{p})^* + a_\mathbf{p}^\dagger\, \tilde{\rho}(\mathbf{p}, \omega_\mathbf{p})] \tag{8.65}$$

(Remember, the four components of p^μ are not free; p^0 equals $\omega_\mathbf{p}$.) To keep from writing the complicated expression (8.65) for O_1 over and over again, I will write[5]

$$O_1 = \int d^3\mathbf{p} \left[-h(\mathbf{p})^*\, a_\mathbf{p} + h(\mathbf{p})\, a_\mathbf{p}^\dagger \right] \tag{8.66}$$

where $h(\mathbf{p})$ is defined by

$$h(\mathbf{p}) \equiv \frac{-ig\tilde{\rho}(\mathbf{p}, \omega_\mathbf{p})}{(2\pi)^{3/2}\sqrt{2\omega_\mathbf{p}}} \tag{8.67}$$

It's important to observe that if $\rho(x)$ is non-zero but its Fourier transform $\tilde{\rho}(p)$ vanishes on the mass shell,[6] when $p^0 = \omega_\mathbf{p}$, then *nothin'* happens. This is simply the law of conservation of energy-momentum, and the diagrammatic observation that the operator O_1 makes mesons one at a time.[7] The amount of energy and momentum drawn off from the source must be consistent with the meson energy-momentum relation. If $\tilde{\rho}(\mathbf{p}, \omega_\mathbf{p})$ is zero, even if O_1 has a lot of other Fourier components that aren't zero, off the mass shell, it's not going to be able make a meson. If $h(\mathbf{p})$ is non-zero, O_1 can make mesons.

Let's examine the simplest case. We start out with the vacuum state, turn on our source, wiggle it around, oscillate it, and mesons come flying out. How many mesons? To answer this,

---

[5] [Eds.] Coleman's $f(\mathbf{p})$ has been changed to $h(\mathbf{p})$ to avoid confusion with the adiabatic function $f(t)$.

[6] [Eds.] The mass shell is the four-dimensional hyperboloid $p^2 = \mu^2$.

[7] [Eds.] At the beginning of the next lecture, a student asks about this remark. Coleman replies, "I said we could understand that [the four-momentum restricted to the mass shell value] physically by looking at the structure of the diagrams, which we could interpret as saying that mesons were made one at a time. If we had had an interaction like $\rho\phi^2$, then we would have diagrams like this:

with two ϕ's coming to a single vertex. Then we would not have found the same mass shell constraint, because you could add the two momenta of these produced mesons together to make practically anything in Fourier space. It would not be only the value on the mass shell that would be relevant."

we've got to compute $U_I(\infty, -\infty)$ on the ground state of the free field, because, by assumption, that's the system's starting condition. That's the experiment we wish to do. This gives us

$$
\begin{aligned}
|\psi\rangle = S\,|0\rangle = U_I(\infty,-\infty)\,|0\rangle &= A :\!\exp\left[\int d^3\mathbf{p}\,h(\mathbf{p})\,a^\dagger_{\mathbf{p}}\right]\exp\left[-\int d^3\mathbf{p}\,h(\mathbf{p})^*\,a_{\mathbf{p}}\right]:|0\rangle \\
&= A :\!\exp\left[\int d^3\mathbf{p}\,h(\mathbf{p})\,a^\dagger_{\mathbf{p}}\right]\sum_{n=0}^{\infty}\frac{1}{n!}\left(-\int d^3\mathbf{p}\,h(\mathbf{p})^*\,a_{\mathbf{p}}\right)^n : |0\rangle \\
&= A\exp\left[\int d^3\mathbf{p}\,h(\mathbf{p})\,a^\dagger_{\mathbf{p}}\right]|0\rangle
\end{aligned}
\tag{8.68}
$$

The $a_{\mathbf{p}}$'s and the normal ordering symbol take care of each other. Because of normal ordering, the $a_{\mathbf{p}}$'s are on the right where they meet the vacuum and get turned into zero. Only the first term in the sum, equal to 1, survives. We're left with just those terms that have nothing but $a^\dagger_{\mathbf{p}}$'s in them. The $a^\dagger_{\mathbf{p}}$'s all commute with each other, so I no longer have to write the colon.

In Chapter 2, I defined a two-particle wave function $|\mathbf{p}_1, \mathbf{p}_2\rangle$ (see (2.59)). The extension to an n-particle state is straightforward:

$$
|\mathbf{p}_1, \mathbf{p}_2, \dots, \mathbf{p}_n\rangle = a^\dagger_{\mathbf{p}_1} a^\dagger_{\mathbf{p}_2} \cdots a^\dagger_{\mathbf{p}_n}\,|0\rangle \tag{8.69}
$$

We write the general state $|\psi\rangle$ as

$$
|\psi\rangle = \sum_{n=0}^{\infty}\frac{1}{n!}\int d^3\mathbf{p}_1\,d^3\mathbf{p}_2\cdots d^3\mathbf{p}_n\,\psi^{(n)}(\mathbf{p}_1,\mathbf{p}_2,\dots,\mathbf{p}_n)\,|\mathbf{p}_1,\mathbf{p}_2,\dots,\mathbf{p}_n\rangle \tag{8.70}
$$

where

$$
\psi^{(n)}(\mathbf{p}_1,\mathbf{p}_2,\dots,\mathbf{p}_n) \equiv \langle\mathbf{p}_1,\mathbf{p}_2,\dots,\mathbf{p}_n|\psi\rangle = \langle\mathbf{p}_1,\mathbf{p}_2,\dots,\mathbf{p}_n|S|0\rangle \tag{8.71}
$$

Comparing (8.68) with (8.70) we have

$$
\begin{aligned}
\psi^{(0)} &= A \\
\psi^{(1)} &= Ah(\mathbf{p}) \\
\psi^{(2)} &= Ah(\mathbf{p}_1)h(\mathbf{p}_2) \\
&\ \ \vdots \\
\psi^{(n)} &= Ah(\mathbf{p}_1)h(\mathbf{p}_2)\cdots h(\mathbf{p}_n)
\end{aligned}
\tag{8.72}
$$

But what happened to the factor of 2! in the second term? That disappeared because there are two possibilities for the two-particle state. Either the first creation operator in the integral creates $|\mathbf{p}_1\rangle$ and the second creates $|\mathbf{p}_2, \mathbf{p}_1\rangle$, or vice-versa: the states $|\mathbf{p}_2, \mathbf{p}_1\rangle$ and $|\mathbf{p}_1, \mathbf{p}_2\rangle$ are the same. This symmetry cancels the 2! from the exponential. In fact, the symmetry cancels the $n!$ factor in the nth term.

The probability $P(n)$ of finding n mesons in the final state is given by

$$
P(n) = \frac{1}{n!}\int d^3\mathbf{p}_1\,d^3\mathbf{p}_2\cdots d^3\mathbf{p}_n\,|\psi^{(n)}(\mathbf{p}_1,\mathbf{p}_2,\dots,\mathbf{p}_n)|^2 \tag{8.73}
$$

(The divisor $n!$ prevents over-counting.) Substituting in from (8.72),

$$
P(n) = |A|^2\frac{1}{n!}\left[\int d^3\mathbf{p}\,|h(\mathbf{p})|^2\right]^n \tag{8.74}
$$

It is now easy to sum up $P(n)$. Of course, the sum of $P(n)$ over all n must be one; that is the conservation of probability. Put another way, we demand the unitarity of the S-matrix. Then $|\psi\rangle$ will have norm 1, as it is equal to $S\,|0\rangle$, the result of a unitary matrix acting on a ket of norm 1. Therefore

$$1=\sum_{n=0}^{\infty}P(n)=|A|^2\sum_{n=0}^{\infty}\frac{1}{n!}\left(\int d^3\mathbf{p}\,|h(\mathbf{p})|^2\right)^n=|A|^2\exp\left(\int d^3\mathbf{p}\,|h(\mathbf{p})|^2\right) \tag{8.75}$$

so that

$$|A|^2=\exp\left(-\int d^3\mathbf{p}\,|h(\mathbf{p})|^2\right) \tag{8.76}$$

That's the consistency argument. In (8.62), we defined $A=e^{\frac{1}{2}(-\alpha+i\beta)}$, and thus

$$\alpha=\int d^3\mathbf{p}\,|h(\mathbf{p})|^2 \tag{8.77}$$

Substituting into (8.74), $P(n)$, the probability of finding n particles in the final state, is then given by

$$P(n)=e^{-\alpha}\frac{\alpha^n}{n!} \tag{8.78}$$

the famous Poisson distribution.

Thus we find, in this radiation process, the probability of finding n mesons—what a high-energy physicist would call the "multiplicity distribution"—is a Poisson distribution. What is the average number of mesons produced? That's also an interesting question. Or as we say, what is the mean multiplicity? If you do the experiment a billion times, what is the average number $\langle N\rangle$ of mesons made each time?

$$\langle N\rangle=\sum_{n=0}^{\infty}nP(n)=\sum_{n=1}^{\infty}nP(n)=e^{-\alpha}\sum_{n=1}^{\infty}\frac{\alpha^n}{(n-1)!}=\alpha \tag{8.79}$$

That's just standard fun and games with the Poisson distribution. So this quantity $\alpha=\int d^3\mathbf{p}\,|h(\mathbf{p})|^2$ is in fact the mean multiplicity. Because α is proportional to g^2, the square of the coupling constant, the probability $P(n')$ of any particular number n' of mesons decreases as g increases, but $\langle N\rangle$ increases.

The n-particle states we make are very simple. Well, it is a very simple theory. The n-particle states are all determined in terms of the one-particle state, and the wave function for the n mesons is just a product of the n single meson wave functions. It's as close to an uncorrelated state as you can get, *modulo* the conditions imposed by Bose statistics. This kind of state occurs in quantum optics. In the corresponding optical problem, you have some big piece of charged matter moving up and down. The photon state turns out to be this kind of state, and so a peculiar optical terminology is used to describe such states: they are called "coherent states". These are characteristic not just of classical sources, but of all conditions where the source that is making the mesons or the photons can be effectively treated as classical. For example, if we have a charged particle passing through matter, it's slowed down by the fact that it is ionizing atoms, and hence it gives off a lot of photons. In extreme cases, these photons produce the so-called Cherenkov radiation. The very energetic photons know that the charged particle is not just a classical source, because they give it a gigantic recoil whenever it emits one of those very energetic photons. But from the viewpoint of not so energetic photons, what we call "soft" photons, the piece of matter is enormously heavy, essentially a classical object that does not recoil. So the soft part of the photon spectrum emitted in the passage of a charged particle through matter is a coherent state pattern. The bending of a charged particle in a magnetic field also qualifies as a coherent state pattern.

Coherent states of the harmonic oscillator are

$$|\lambda\rangle \equiv e^{\lambda a^\dagger}\,|0\rangle \tag{8.80}$$

where $a^\dagger$ and a are respectively the usual harmonic oscillator raising and lowering operators, (2.17). These states diagonalize a:

$$a\,|\lambda\rangle = ae^{\lambda a^\dagger}\,|0\rangle = [a, e^{\lambda a^\dagger}]\,|0\rangle = \lambda e^{\lambda a^\dagger}\,|0\rangle = \lambda\,|\lambda\rangle \tag{8.81}$$

The coherent states $|\lambda\rangle$ in Model 1 are

$$|\lambda\rangle \equiv :e^{\lambda O_1}:\,|0\rangle = \exp\left(\lambda\int d^3\mathbf{p}\,h(\mathbf{p})a^\dagger_{\mathbf{p}}\right)|0\rangle \tag{8.82}$$

These states are also eigenvectors of $\phi^+(x)$ with eigenvalue λ_ϕ:

$$\lambda_\phi = \lambda\int\frac{d^3\mathbf{p}}{(2\pi)^{3/2}\sqrt{2\omega_\mathbf{p}}}\,e^{-ip\cdot x}h(\mathbf{p}) \tag{8.83}$$

Except for a factor of $1/n!$, the state $|\lambda\rangle$ has an n particle part which is just the product of n one-particle states. The expectation values $\langle x\rangle = \langle\lambda|x|\lambda\rangle$ and $\langle p\rangle = \langle\lambda|p|\lambda\rangle$ oscillate sinusoidally like the classical variables.[8]

Let's now compute the average energy, produced in the process where we start off with the vacuum state, wiggle the scalar source around, turn it off, and then see how many mesons are left. The average energy, the expectation value of the Hamiltonian in the final state, is

$$\begin{aligned}\langle E\rangle = \langle\psi|H|\psi\rangle &= \sum_{n=0}^{\infty}\frac{1}{n!}\int d^3\mathbf{p}_1\cdots d^3\mathbf{p}_n\,\langle\psi|H|\mathbf{p}_1,\mathbf{p}_2,\ldots,\mathbf{p}_n\rangle\,\langle\mathbf{p}_1,\mathbf{p}_2,\ldots,\mathbf{p}_n|\psi\rangle\\ &= \sum_{n=0}^{\infty}\frac{1}{n!}\int d^3\mathbf{p}_1\cdots d^3\mathbf{p}_n\,|\psi^{(n)}|^2(\omega_{\mathbf{p}_1}+\omega_{\mathbf{p}_2}+\cdots+\omega_{\mathbf{p}_n})\end{aligned} \tag{8.84}$$

the $n!$ because we don't want to over-count states. Otherwise we would be counting the state $\psi^{(2)}(\mathbf{p}_1,\mathbf{p}_2)$ and the state $\psi^{(2)}(\mathbf{p}_2,\mathbf{p}_1)$ separately. That is a bad thing to do, because they are the same state. The expression (8.84) can be simplified because everything is symmetric. We can just as well write

$$\langle E\rangle = \sum_{n=0}^{\infty}\frac{1}{n!}\int d^3\mathbf{p}_1\cdots d^3\mathbf{p}_n\,e^{-\alpha}|h(\mathbf{p}_1)|^2\cdots|h(\mathbf{p}_n)|^2\,n\,\omega_{\mathbf{p}_1} \tag{8.85}$$

in terms of one of the $\omega_\mathbf{p}$'s, say $\omega_{\mathbf{p}_1}$, as the others give $n-1$ equal contributions, Since the first term is zero, when n equals zero, I can write the summation from $n-1=0$ to ∞; the

[8] [Eds.] For more about coherent states, see Problem 4.2, p. 175, and the references at the end of its solution.

term with $n = 0$ does not contribute. The integral is simple to do, because $(n-1)$ of the integrals give us α, (8.77). So we obtain

$$\langle E \rangle = \sum_{n=1}^{\infty} \frac{\alpha^{n-1}}{(n-1)!} e^{-\alpha} \int d^3\mathbf{p}\, |h(\mathbf{p})|^2\, \omega_{\mathbf{p}} \tag{8.86}$$

Of course, the summation is nothing but a fancy way of writing 1. So we have a simple expression for the mean energy emitted in our process. It is simply

$$\langle E \rangle = \int d^3\mathbf{p}\, |h(\mathbf{p})|^2\, \omega_{\mathbf{p}} \tag{8.87}$$

The mean momentum can be obtained by an identical computation with $\mathbf{p}$'s replacing $\omega_{\mathbf{p}}$'s, and that is equal to

$$\langle \mathbf{p} \rangle = \int d^3\mathbf{p}\, |h(\mathbf{p})|^2\, \mathbf{p} \tag{8.88}$$

This completes for the moment our analysis of Model 1. We'll return to it later and find out some other things about it.

In the next lecture I will go on to Model 2, which is also exactly soluble.

Problems 4

4.1 In class we studied

$$\mathscr{L} = \tfrac{1}{2}(\partial_\mu \phi)^2 - \tfrac{1}{2}\mu^2\phi^2 - g\rho(x)\phi(x)$$

and found an operator for $U_I(\infty, -\infty)$ as the product of a constant, A (which I wrote as $e^{\frac{1}{2}(-\alpha+i\beta)}$), and a known operator. In class we found α by a self-consistency argument. Find α by evaluating the real part of the relevant diagram, Figure (8.15) on p. 168:

and show that this agrees with what we found in class. You may find the following formula useful:

$$\lim_{\epsilon\to 0}\left[\frac{1}{x+i\epsilon} - \frac{1}{x-i\epsilon}\right] = -2\pi i\delta(x) \tag{P4.1}$$

(1997a 4.1)

4.2 In solving Model 1 in class, I mentioned the idea of a coherent state.[1] Although we won't use coherent states much in this course, they do have applications in all sorts of odd corners of physics, and working out their properties is an instructive exercise in manipulating annihilation and creation operators.

It suffices to study a single harmonic oscillator; the generalization to a free field (= many oscillators) is trivial. Let

$$H = \tfrac{1}{2}(p^2 + q^2)$$

and, as usual, let us define

$$a = \tfrac{1}{\sqrt{2}}(q + ip) \qquad a^\dagger = \tfrac{1}{\sqrt{2}}(q - ip)$$

Define the coherent state $|z\rangle$ by

$$|z\rangle = N e^{za^\dagger}|0\rangle \tag{P4.2}$$

where z is a complex number and N is a real, positive normalization factor (dependent on z), chosen such that $\langle z|z\rangle = 1$.

(a) Find N.

(b) Compute $\langle z|z'\rangle$.

(c) Show that $|z\rangle$ is an eigenstate of the annihilation operator a, and find its eigenvalue. (Do not be disturbed by finding non-orthogonal eigenvectors with complex eigenvalues: a is not a Hermitian operator.)

(d) The set of all coherent states for all values of z is obviously complete. Indeed, it is overcomplete: The energy eigenstates can all be constructed by taking successive derivatives at $z = 0$, so the coherent states

[1] [Eds.] Roy J. Glauber, "Photon correlations", *Phys. Rev. Lett.* **10** (1963) 83–86. Glauber won the 2005 Nobel Prize in Physics for research in optical coherence.

with z in some small, real interval around the origin are already enough. Show that, despite this, there is an equation that looks something like a completeness relation, namely

$$1 = \alpha \int d(\mathrm{Re}\, z)\, d(\mathrm{Im}\, z)\, e^{-\beta z^* z}\, |z\rangle\, \langle z| \tag{P4.3}$$

and find the real constants α and β.

(e) Show that if $F(p, q)$ is any polynomial in the two canonical variables,

$$\langle z| :F(p, q): |z\rangle = F(\overline{p}, \overline{q}) \tag{P4.4}$$

where $\overline{p}$ and $\overline{q}$ are real numbers. Find $\overline{p}$ and $\overline{q}$ in terms of z and z^*.

(f) The statement that $|z\rangle$ is an eigenstate of a with known eigenvalue (part (c), above) is, in the q-representation, a first-order differential equation for $\langle q|z\rangle$, the position-space wave function of $|z\rangle$. Solve this equation and find this wave function. (Don't bother with normalization factors here.)

(1997a 4.2)

4.3 Let K be a Hermitian operator, and $|\psi\rangle$ a state of norm 1. Given a function $f(K)$ of K, its expectation value in the state $|\psi\rangle$ is defined by

$$\langle f(K)\rangle \equiv \langle\psi|f(K)|\psi\rangle \tag{P4.5}$$

Suppose we introduce the function $\eta(k)$ of a real variable k:

$$\eta(k) \equiv \langle\delta(K-k)\rangle = \langle\psi|\delta(K-k)|\psi\rangle \tag{P4.6}$$

Then (as you can easily show)

$$\langle f(K)\rangle = \int dk\, f(k)\, \eta(k) \tag{P4.7}$$

This works in ordinary quantum mechanics as well as in quantum field theory. Find $\eta(k)$ for the vacuum state of a free scalar field of mass m, if

$$K = \int d^3\mathbf{x}\, g(\mathbf{x})\phi(\mathbf{x}, 0) \tag{P4.8}$$

and $g(\mathbf{x})$ is some infinitely differentiable c-number function that goes to zero rapidly at infinity. You should find that $\eta(k)$ a Gaussian whose width is proportional to the integral of the square of the Fourier transform of $g(\mathbf{x})$.

HINTS:

(a) Express the delta function as a Fourier transform,

$$\delta(z) = \int_{-\infty}^{\infty} \frac{dq}{2\pi} e^{-iqz}$$

(b) The results of Problem 4.1, and the discussion from (8.62) to (8.77) may be helpful. You may assume $\beta = 0$ in (8.62).

Comment: That the answer is a Gaussian should be no surprise. After all, the theory is really just that of an assembly of uncoupled harmonic oscillators.

(1986a 11)

Solutions 4

4.1 Recall how α was defined (see (8.59) and (8.60)):

$$O_2 = -\alpha + i\beta = (-ig)^2 \int d^4x\, d^4y\, \overline{\phi(x)\phi(y)}\rho(x)\rho(y)$$

Using the expression (8.23) for the contraction,

$$\begin{aligned}
\alpha &= -\lim_{\epsilon\to 0+} \text{Re}\left[(-ig)^2 \int d^4x\, d^4y\, \frac{d^4p}{(2\pi)^4}\frac{ie^{-ip\cdot(x-y)}}{p^2-\mu^2+i\epsilon}\rho(x)\rho(y)\right] \\
&= \lim_{\epsilon\to 0+} \text{Re}\left[g^2 \int \frac{d^4p}{(2\pi)^4}\frac{i}{p^2-\mu^2+i\epsilon}\int d^4x\, e^{-ip\cdot x}\rho(x)\int d^4y\, e^{ip\cdot y}\rho(y)\right] \\
&= \lim_{\epsilon\to 0+} \text{Re}\left[g^2 \int \frac{d^4p}{(2\pi)^4}\frac{i}{p^2-\mu^2+i\epsilon}\tilde{\rho}(p)^*\tilde{\rho}(p)\right] \\
&= \lim_{\epsilon\to 0+} -\text{Im}\left[g^2 \int \frac{d^4p}{(2\pi)^4}\frac{1}{p^2-\mu^2+i\epsilon}\tilde{\rho}(p)^*\tilde{\rho}(p)\right]
\end{aligned} \tag{S4.1}$$

because for any complex number $z = a + ib$, $\text{Re}(iz) = -\text{Im}(z)$. To make use of the hint (P4.1), note that

$$\text{Im}\left[\frac{1}{a+i\epsilon}\right] = -\frac{\epsilon}{a^2+\epsilon^2} = \frac{i}{2}\left[\frac{1}{a-i\epsilon} - \frac{1}{a+i\epsilon}\right]$$

Substituting,

$$\begin{aligned}
\alpha &= \frac{i}{2}\lim_{\epsilon\to 0+}\left[g^2\int\frac{d^4p}{(2\pi)^4}\left(\frac{1}{p^2-\mu^2+i\epsilon} - \frac{1}{p^2-\mu^2-i\epsilon}\right)\tilde{\rho}(p)^*\tilde{\rho}(p)\right] \\
&= \frac{ig^2}{2}\int\frac{d^4p}{(2\pi)^4}\tilde{\rho}(p)^*\tilde{\rho}(p)\left(-2\pi i\delta(p^2-\mu^2)\right) \quad \text{(using the hint)} \\
&= \frac{g^2}{2}\int\frac{d^4p}{(2\pi)^3}\tilde{\rho}(p)^*\tilde{\rho}(p)\delta(p_0^2-\omega_{\mathbf{p}}^2) \\
&= \frac{g^2}{2}\int\frac{d^4p}{(2\pi)^3}\tilde{\rho}(p)^*\tilde{\rho}(p)\left[\frac{\delta(p_0-\omega_{\mathbf{p}})}{|p_0+\omega_{\mathbf{p}}|} + \frac{\delta(p_0+\omega_{\mathbf{p}})}{|p_0-\omega_{\mathbf{p}}|}\right] \quad \text{(note 8, p. 9)} \\
&= \frac{g^2}{2}\int\frac{d^3\mathbf{p}}{(2\pi)^3}\left[\frac{\tilde{\rho}(\mathbf{p},\omega_{\mathbf{p}})^*\tilde{\rho}(\mathbf{p},\omega_{\mathbf{p}})}{2\omega_{\mathbf{p}}} + \frac{\tilde{\rho}(\mathbf{p},-\omega_{\mathbf{p}})^*\tilde{\rho}(\mathbf{p},-\omega_{\mathbf{p}})}{2\omega_{\mathbf{p}}}\right]
\end{aligned}$$

By definition, $\tilde{\rho}(\mathbf{p},\omega_{\mathbf{p}}) = \int d^4x\, e^{-ip\cdot x}\rho(x)$, so

$$\tilde{\rho}(\mathbf{p},-\omega_{\mathbf{p}}) = \int d^4x\, e^{i\mathbf{p}\cdot\mathbf{x} - i\omega_{\mathbf{p}}x_0}\rho(x) = \tilde{\rho}(-\mathbf{p},\omega_{\mathbf{p}})^* = \tilde{\rho}(-\mathbf{p},\omega_{-\mathbf{p}})^*$$

because $\rho(x) = \rho(x)^*$ and $\omega_{\mathbf{p}} = \omega_{-\mathbf{p}}$. Substituting,

$$\alpha = \frac{g^2}{2}\left[\left(\int \frac{d^3\mathbf{p}}{(2\pi)^3}\frac{\tilde{\rho}(\mathbf{p},\omega_{\mathbf{p}})^*\tilde{\rho}(\mathbf{p},\omega_{\mathbf{p}})}{2\omega_{\mathbf{p}}}\right) + \left(\int \frac{d^3\mathbf{p}}{(2\pi)^3}\frac{\tilde{\rho}(-\mathbf{p},\omega_{-\mathbf{p}})\tilde{\rho}(-\mathbf{p},-\omega_{-\mathbf{p}})^*}{2\omega_{-\mathbf{p}}}\right)\right]$$

$$= g^2 \int \frac{d^3\mathbf{p}}{(2\pi)^3}\frac{\tilde{\rho}(\mathbf{p},\omega_{\mathbf{p}})^*\tilde{\rho}(\mathbf{p},\omega_{\mathbf{p}})}{2\omega_{\mathbf{p}}} \quad (\mathbf{p} \to -\mathbf{p} \text{ in the second integral}) \tag{S4.2}$$

$$= \int d^3\mathbf{p}\, h(\mathbf{p})^* h(\mathbf{p}) = \int d^3\mathbf{p}\, |h(\mathbf{p})|^2$$

using (8.67), in agreement with (8.77). ■

4.2 We have to do (a) and (b) at the same time. Let a properly normalized oscillator energy eigenfunction be denoted $|n\rangle$, n an integer. Recall (2.36):

$$a^\dagger |n\rangle = \sqrt{(n+1)}\, |n+1\rangle \tag{2.36}$$

so $(a^\dagger)^n |0\rangle = \sqrt{n!}\, |n\rangle$. Then (we are told to take N to be real)

$$|z\rangle = Ne^{za^\dagger} |0\rangle = N \sum_n \frac{z^n}{n!}(a^\dagger)^n |0\rangle = N \sum_n \frac{z^n}{\sqrt{n!}} |n\rangle \tag{S4.3}$$

The inner product of two such states will be

$$\langle z|z'\rangle = N(z)N(z') \sum_m \sum_n \frac{(z^*)^m z^n}{\sqrt{m!}\sqrt{n!}} \langle m|n\rangle = N(z)N(z') \sum_m \sum_n \frac{(z^*)^m (z')^n}{\sqrt{m!}\sqrt{n!}}\delta_{mn}$$

$$= N(z)N(z') \sum_n \frac{(z^* z')^n}{n!} = N(z)N(z')e^{z^* z'}$$

Set the norm of the coherent state vectors $\langle z|z\rangle$ equal to 1 to obtain

$$N(z) = e^{-\frac{1}{2}|z|^2}$$

That answers (a). Then the inner product of two vectors gives

$$\langle z|z'\rangle = e^{-\frac{1}{2}(|z|^2 + |z'|^2)} e^{z^* z'} \tag{S4.4}$$

which answers (b).

(c) To show $|z\rangle$ is an eigenvector of a, recall (2.37),

$$a\, |n\rangle = \sqrt{n}\, |n-1\rangle \tag{2.37}$$

Then operating with a,

$$a\, |z\rangle = aN(z) \sum_{n=0} \frac{z^n}{\sqrt{n!}} |n\rangle = N(z) \sum_{n=1} \frac{z^n}{\sqrt{n!}} \sqrt{n}\, |n-1\rangle = zN(z) \sum_{n=1} \frac{z^{n-1}}{\sqrt{(n-1)!}} |n-1\rangle = z\, |z\rangle$$

The kets $|z\rangle$ are eigenvectors of a with eigenvalue z.

A more elegant approach is to recall that for canonically conjugate variables u and v, when $[u, v] = 1$, then

$$[u, f(v)] = \frac{\partial f(v)}{\partial v}$$

Since $[a, a^\dagger] = 1$, it follows

$$a\, |z\rangle = ae^{za^\dagger} |0\rangle = [a, e^{za^\dagger}]\, |0\rangle + e^{za^\dagger} a\, |0\rangle = \frac{\partial e^{za^\dagger}}{\partial a^\dagger} |0\rangle + 0 = ze^{za^\dagger} |0\rangle = z\, |z\rangle$$

(d) The problem states that derivatives of the coherent states $|z\rangle$ in the neighborhood of $z = 0$ generate the energy eigenstates $|n\rangle$. Then the $|z\rangle$'s form a complete set, because the energy eigenstates are a complete set. In fact, the $|z\rangle$'s are "overcomplete", because $\langle z|z'\rangle \neq 0$ even when $z \neq z'$. It isn't clear that the problem asks us to demonstrate this first statement, and indeed it's not straightforward to do so.

The difficulty arises because the normalization constant $N = e^{\frac{1}{2}|z|^2}$ depends on the function $|z|^2$ which does not have a derivative everywhere. However, its derivative *does* exist at the origin, and *only* at the origin, where it equals zero.[1] If it is permissible to regard *all* the derivatives of N as equal to zero at the origin, the demonstration proceeds like this:

$$\frac{\partial^m}{\partial z^m}|z\rangle\bigg|_{z=0} = N\frac{\partial^m}{\partial z^m}\sum_{n=0}\frac{z^n}{\sqrt{n!}}|n\rangle\bigg|_{z=0} = N\sum_{n=0}\frac{n(n-1)(n-2)\cdots(n-m+1)z^{n-m}}{\sqrt{n!}}|n\rangle\bigg|_{z=0}$$

$$= N\sum_{n=m}\frac{n!\,0^{n-m}}{(n-m)!\sqrt{n!}}|n\rangle = N\sqrt{m!}\,|m\rangle \text{ as claimed.}$$

We are now asked to find α and β such that

$$1 = \alpha\int d\mathrm{Re}(z)\,d\mathrm{Im}(z)\,e^{-\beta|z|^2}\,|z\rangle\langle z|$$

Write $z = x + iy$, and use the form (S4.3) for the kets (and the appropriate bras):

$$1 = \alpha\sum_{n,m}\frac{1}{\sqrt{n!}\sqrt{m!}}\int dx\,dy\,e^{-\beta(x^2+y^2)}e^{-(x^2+y^2)}(x+iy)^n(x-iy)^m\,|n\rangle\langle m|$$

Go to polar coordinates: $x + iy = re^{i\theta}$, and $dx\,dy = r\,dr\,d\theta$. Then

$$1 = \alpha\sum_{n,m}\frac{1}{\sqrt{n!}\sqrt{m!}}\int r\,dr\,d\theta\,e^{-(\beta+1)r^2}(r)^{n+m}e^{i(n-m)\theta}\,|n\rangle\langle m|$$

The θ integral is

$$\int_0^{2\pi} d\theta\,e^{i(n-m)\theta} = 2\pi\,\delta_{nm}$$

so

$$1 = 2\pi\alpha\sum_{n=0}^{\infty}\frac{1}{n!}\left[\int_0^{\infty} dr\,e^{-(\beta+1)r^2}r^{2n+1}\right]|n\rangle\langle n|$$

Let $(\beta+1)r^2 = u$. Then the r integral becomes

$$\int_0^{\infty} dr\,e^{-(\beta+1)r^2}r^{2n+1} = \tfrac{1}{2}(\beta+1)^{-(n+1)}\int_0^{\infty} du\,u^n e^{-u} = \tfrac{1}{2}n!(\beta+1)^{-(n+1)}$$

Plugging this result in, and using the standard equation $1 = \sum|n\rangle\langle n|$, we find

$$1 = \pi\alpha\sum_{n=0}^{\infty}(\beta+1)^{-(n+1)}\,|n\rangle\langle n| = \sum_{n=0}^{\infty}|n\rangle\langle n|$$

That is, $\beta = 0$ and $\alpha = 1/\pi$. That answers (d).

(e) Start with a general monomial, $p^m q^n$. From (2.19),

$$q = \tfrac{1}{\sqrt{2}}\left(a + a^\dagger\right) \qquad p = -i\tfrac{1}{\sqrt{2}}\left(a - a^\dagger\right)$$

we have

$$:p^m q^n: \;=\; :\left(-i\tfrac{1}{\sqrt{2}}\left(a - a^\dagger\right)\right)^m\left(\tfrac{1}{\sqrt{2}}\left(a + a^\dagger\right)\right)^n: \;=\; \sum_{i,j}C_{ij}(a^\dagger)^i a^j \tag{S4.5}$$

for some undetermined coefficient matrix C_{ij}; the exact values do not matter for this argument. Equation (S4.5) is an identity for any c-number variables x and y;

$$\left(-i\tfrac{1}{\sqrt{2}}(x-y)\right)^m\left(\tfrac{1}{\sqrt{2}}(x+y)\right)^n = \sum_{i,j}C_{ij}y^i x^j$$

Then

$$\langle z|:p^m q^n:|z\rangle = \langle z|\sum_{i,j}C_{ij}(a^\dagger)^i a^j|z\rangle = \sum_{i,j}C_{ij}(z^*)^i z^j$$

$$= \left(-i\tfrac{1}{\sqrt{2}}(z-z^*)\right)^m\left(\tfrac{1}{\sqrt{2}}(z+z^*)\right)^n = (\sqrt{2}\,\mathrm{Im}\,z)^m(\sqrt{2}\,\mathrm{Re}\,z)^n$$

[1] [Eds.] R. V. Churchill and J. W. Brown, *Complex Variables and Applications*, 4th ed., Mc-Graw Hill, 1984, p. 40.

Any normal-ordered polynomial $:F(p,q):$ is simply a linear combination $\sum c_{mn} :p^m q^n:$ of such monomials, and

$$\langle z| :F(p,q): |z\rangle = \langle z| \sum c_{mn} :p^m q^n: |z\rangle = \sum c_{mn} \langle z| :p^m q^n: |z\rangle$$
$$= \sum c_{mn}(\sqrt{2}\,\mathrm{Im}\, z)^m(\sqrt{2}\,\mathrm{Re}\, z)^n = F(\sqrt{2}\,\mathrm{Im}\, z, \sqrt{2}\,\mathrm{Re}\, z)$$

That is, $\overline{p} = \sqrt{2}\,\mathrm{Im}\, z$, and $\overline{q} = \sqrt{2}\,\mathrm{Re}\, z$.

(f) The kets $|z\rangle$ are eigenvectors of a: $a\,|z\rangle = z\,|z\rangle$, so

$$\langle q|a|z\rangle = z\,\langle q|z\rangle = \langle q|\frac{1}{\sqrt{2}}(q+ip)|z\rangle = \frac{1}{\sqrt{2}}q\,\langle q|z\rangle + \frac{1}{\sqrt{2}}\frac{\partial}{\partial q}\,\langle q|z\rangle$$

Try the solution $\langle q|z\rangle = e^{f(q)}$. Then

$$qe^{f(q)} + \frac{df}{dq}e^{f(q)} = \sqrt{2}\,ze^{f(q)}$$

Divide out $e^{f(q)}$ to obtain

$$\frac{df}{dq} = -q + \sqrt{2}z \;\Rightarrow\; f(q) = -\tfrac{1}{2}q^2 + \sqrt{2}zq + C$$

so

$$\langle q|z\rangle = e^C \exp\left(-\tfrac{1}{2}q^2 + \sqrt{2}zq\right)$$

At $q = 0$, we have $\langle 0|z\rangle = e^C$, so

$$\langle q|z\rangle = \langle 0|z\rangle \exp\left(-\tfrac{1}{2}q^2 + \sqrt{2}zq\right)$$ ■

(For more about coherent states, see J. J. Sakurai, *Modern Quantum Mechanics*, Addison-Wesley, 1985, p. 97; Problem 2.18, p. 147, and references therein; D. J. Griffiths, *Introduction to Quantum Mechanics*, 2nd ed., Cambridge U. P., Problem 3.35, p. 127; and W. Greiner, *Quantum Mechanics: Special Chapters*, Springer, 1998, Section 1.5, pp. 16–20.)

4.3 Following Hint (a),

$$\eta(k) = \langle 0|\delta(K-k)|0\rangle = \int \frac{dq}{2\pi}\,\langle 0|\exp\left[-iq\left(\int d^3\mathbf{x}\, g(\mathbf{x})\phi(\mathbf{x},0) - k\right)\right]|0\rangle \tag{S4.6}$$

which can be written suggestively as

$$\eta(k) = \int \frac{dq}{2\pi}e^{iqk}\,\langle 0|\exp\left[-iq\left(\int d^4x\, G(x)\phi(x)\right)\right]|0\rangle \quad \text{where} \quad G(x) = g(\mathbf{x})\delta(t)$$

Because there is actually no time-dependent operator in the expression, we can just as well write

$$\eta(k) = \int \frac{dq}{2\pi}e^{iqk}\,\langle 0|T\exp\left[-iq\left(\int d^4x\, G(x)\phi(x)\right)\right]|0\rangle \tag{S4.7}$$

Now it so happens that we have already worked out this matrix element, in Model 1 (see (7.59), (8.9), and (8.10)):

$$\langle 0|S|0\rangle = \langle 0|T\exp\left[-ig\left(\int d^4x\, \rho(x)\phi(x)\right)\right]|0\rangle = A\,\langle 0|\exp\left(\int d^3\mathbf{p}\, h(\mathbf{p})^* a^\dagger_{\mathbf{p}}\right)|0\rangle = A \tag{S4.8}$$

(only the zeroth term in the power series for the exponential survives). The form of A comes from (8.62):

$$A = e^{\bullet\!-\!\!-\!\bullet} = e^{\frac{1}{2}O_2} = e^{\frac{1}{2}(-\alpha+i\beta)} \tag{S4.9}$$

and (S4.2) with (8.67),

$$\alpha = \int d^3\mathbf{p}\,|h(\mathbf{p})|^2 = g^2\int \frac{d^3\mathbf{p}}{(2\pi)^3 2\omega_{\mathbf{p}}}|\tilde{\rho}(\mathbf{p},\omega_{\mathbf{p}})|^2 \to q^2\int \frac{d^3\mathbf{p}}{(2\pi)^3 2\omega_{\mathbf{p}}}|\widetilde{G}(\mathbf{p},\omega_{\mathbf{p}})|^2 \tag{S4.10}$$

substituting $g \to q$ and $\rho(x) \to G(x)$. Using (8.63) and (8.64),

$$\widetilde{G}(\mathbf{p},\omega_{\mathbf{p}}) = \int d^4x\, e^{-ip\cdot x}G(x) = \int dt\, d^3\mathbf{x}\, e^{-i\omega_{\mathbf{p}}\mathbf{p}\cdot\mathbf{x}}g(\mathbf{x})\delta(t) = \int d^3\mathbf{x} e^{i\mathbf{p}\cdot\mathbf{x}}g(\mathbf{x}) = \tilde{g}(\mathbf{p}) \tag{S4.11}$$

so

$$\alpha \to q^2 \int \frac{d^3\mathbf{p}}{(2\pi)^3 2\omega_{\mathbf{p}}} |\tilde{g}(\mathbf{p})|^2 = q^2\sigma \quad \text{where} \quad \sigma \equiv \int \frac{d^3\mathbf{p}}{(2\pi)^3 2\omega_{\mathbf{p}}} |\tilde{g}(\mathbf{p})|^2 \tag{S4.12}$$

Plugging this into (S4.7), and assuming, from Hint (b), that $\beta = 0$, we have

$$\eta(k) = \int \frac{dq}{2\pi} e^{iqk} e^{-\frac{1}{2}q^2\sigma} = e^{-k^2/2\sigma} \int \frac{dq}{2\pi} \exp\left[-\tfrac{1}{2}\sigma\left(q - \frac{ik}{\sigma}\right)^2\right] = \frac{1}{\sqrt{2\pi\sigma}} e^{-k^2/2\sigma} \tag{S4.13}$$

which is indeed a Gaussian, whose width σ is proportional to the integral of the square of the Fourier transform of $g(x)$. ■

Alternative solution. Let M and N be two operators. The Baker–Campbell–Hausdorff formula[2] says

$$e^M e^N = e^Z \quad \text{where } Z = M + N + \tfrac{1}{2}[M,N] + \tfrac{1}{12}\left([M,[M,N]] + [N,[N,M]]\right) + \cdots \tag{S4.14}$$

If $[M,N]$ is a c-number, or otherwise commutes with M and N, the formula for Z truncates after three terms, and

$$e^{M+N} = e^M e^N e^{-\frac{1}{2}[M,N]} \tag{S4.15}$$

The field ϕ (see (3.45)), written more conveniently in terms of $\phi^{\pm}$ (see (3.33)), is $\phi(x) = \phi^+(x) + \phi^-(x)$, so the exponent in (S4.6) can be expressed as the sum of two operators:

$$-iq\int d^3\mathbf{x}\, g(\mathbf{x})\phi(\mathbf{x},0) = \underbrace{-iq\int d^3\mathbf{x}\, g(\mathbf{x})\phi^-(\mathbf{x},0)}_{M} \underbrace{-iq\int d^3\mathbf{x}\, g(\mathbf{x})\phi^+(\mathbf{x},0)}_{N} \tag{S4.16}$$

From (3.38), the commutator is

$$\begin{aligned}
[M,N] &= -q^2 \int d^3\mathbf{x}\, d^3\mathbf{y}\, g(\mathbf{x})g(\mathbf{y})\, [\phi^{(-)}(\mathbf{x},0), \phi^{(+)}(\mathbf{y},0)] \\
&= q^2 \int d^3\mathbf{x}\, d^3\mathbf{y}\, g(\mathbf{x})g(\mathbf{y}) \int \frac{d^3\mathbf{p}}{(2\pi)^3 2\omega_{\mathbf{p}}} e^{i\mathbf{p}\cdot(\mathbf{x}-\mathbf{y})} \\
&= q^2 \int \frac{d^3\mathbf{p}}{(2\pi)^3 \omega_{\mathbf{p}}} \int d^3\mathbf{x}\, e^{-i\mathbf{p}\cdot\mathbf{x} g(\mathbf{x}) \int d^3\mathbf{y}}\, e^{i\mathbf{p}\cdot\mathbf{y}} g(\mathbf{y}) \\
&= q^2 \int \frac{d^3\mathbf{p}}{(2\pi)^3 \omega_{\mathbf{p}}} |\tilde{g}(\mathbf{p})|^2 = q^2\sigma, \quad \text{a c-number}
\end{aligned} \tag{S4.17}$$

From (S4.15)

$$\langle 0|e^{M+N}|0\rangle = \langle 0|e^M e^N|0\rangle\, e^{-\frac{1}{2}q^2\sigma} = e^{-\frac{1}{2}q^2\sigma} \tag{S4.18}$$

because M includes only $a^\dagger_{\mathbf{p}}$, so $\langle 0|\, e^M = \langle 0|$, and similarly $e^N\,|0\rangle = |0\rangle$, and the rest of the problem goes as before. ■

---

[2] [Eds.] Often invoked, rarely cited. See Example 1.2, pp. 20–27 and Exercise 1.3, pp. 27–29 in Greiner & Reinhardt *FQ*. The formula predates quantum mechanics by a quarter century. John E. Campbell, "On a law of combination of operators (second paper)", *Proc. Lond. Math. Soc.* **29**(1) (1897) 14–32; Henry F. Baker, "Alternants and continuous groups", *Proc. Lond. Math. Soc.* (Ser. 2) **3** (1905) 24–47; Felix Hausdorff, "Die symbolische Exponentialformel in der Gruppentheorie" (The symbolic exponential formula in group theory), *Ber. Verh. Sächs. Akad. Wiss. Leipzig* **58** (1906) 19–48. Reprinted in Hausdorff's *Gesammelte Werke*, Band IV, *Analysis, Algebra und Zahlentheorie*, Springer, 2002, pp. 431–460. Baker calls commutators "alternants".

9

Perturbation theory II. Divergences and counterterms

Now I turn to our Model 2, whose Hamiltonian density

$$\mathscr{H}_I = g\phi(x)\rho(\mathbf{x}) \tag{9.1}$$

is exactly the same as in Model 1, *except* that $\rho(\mathbf{x})$ is now time-independent. This interaction doesn't actually turn off in the far past and the far future. To fit it into our somewhat clumsy formulation of scattering theory, we have to insert an adiabatic switching function $f(t)$ that turns the interaction on and off by hand:[1]

$$H_I(t) = g\int d^3\mathbf{x}\,\phi(x)\rho(\mathbf{x}) \to f(t,T,\Delta)H_I(t) \tag{9.2}$$

The field $\phi(x)$ is the interaction picture $\phi_I(x)$, but I won't write the subscript I on the field. A plot of the adiabatic function $f(t)$ was given earlier, in Figure 7.3, but for convenience I'll draw it again. The left dashed line occurs at $t = -T/2$, and the right dashed line at $t = T/2$.

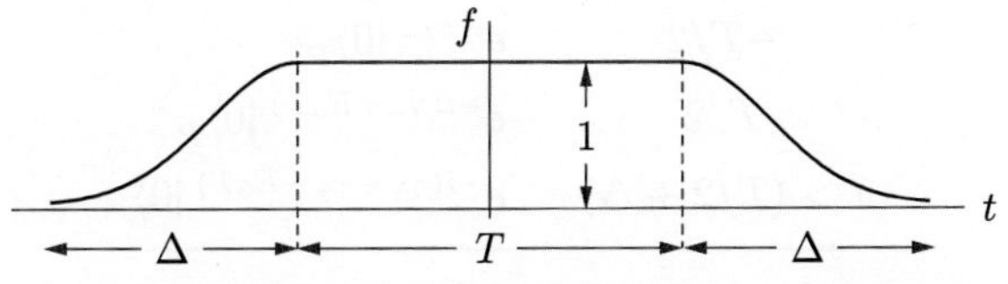

Figure 9.1: The adiabatic function $f(t, T, \Delta)$

The function slowly rises during a time interval Δ from 0 to the value 1 at $t = -T/2$, stays at 1 until $t = T/2$, then it goes down to zero in a way that is supposed to be symmetric with its rise.

---

[1] [Eds.] Localized particles are described by wave packets, but because scattering in terms of wave packets is mathematically awkward, initial and final states are usually represented by plane waves. The use of plane waves leads to mathematical ambiguities if the interaction does not go to zero sufficiently rapidly as $t \to \pm\infty$. These ambiguities are removed by introducing an adiabatic switching function $f(t)$, often of the form $e^{-\epsilon|t|}$. See Lurié *P&F*, pp. 213–214.

9.1 The need for a counterterm in Model 2

Something peculiar occurs in this model, and it shows us that we have been a bit too sanguine about the harmlessness of an adiabatic function's turning the interaction on and off. If we compute the S-matrix using our formula, we find, doing the naive calculation, that there are terms that depend on the time T in a nontrivial way, terms which do not go to zero in the limit $T \to \infty$. We should have

$$\lim_{T\to\infty} \langle 0|U_I(T,-T)|0\rangle = \langle 0|\mathrm{S}|0\rangle = 1 \tag{9.3}$$

but that's not what happens when we have the adiabatic function in our interaction Hamiltonian. Let me explain the physics of why that happens. I will show you how to cure it, and then we will solve the model by summing up the diagrams.

Let me first introduce some notation. We use $|0\rangle$ to represent the ground state of the non-interacting theory. Therefore H_0 on $|0\rangle$ equals zero:

$$H_0\,|0\rangle = 0 \tag{9.4}$$

Of course the real physical theory also has a ground state, $|0\rangle_P$, whose energy E_0 is *not* likely to be zero:

$$H\,|0\rangle_P = (H_0 + H_I)\,|0\rangle_P = E_0\,|0\rangle_P \tag{9.5}$$

This energy arises in the theory *not* from the adiabatic function $f(t)$, but just from the extra term H_I added to its Hamiltonian. Here, $|0\rangle_P$ is the actual ground state of the interacting system *without* the adiabatic $f(t)$, or with $f(t) = 1$, if you prefer. Generally when we add an interaction term to a Hamiltonian, not only does the ground state wave function change, but the ground state energy also changes. So the new ground state $|0\rangle_P$ will have some energy which I will call E_0.

Now let's make a chart of how Model 2's ground state evolves in the Schrödinger picture.

time, t	*Schrödinger state*
$t < -(T/2+\Delta)$	$\|0\rangle$
$-T/2$	$e^{-i\gamma_-}\,\|0\rangle_P$
$T/2$	$e^{-i(\gamma_- + E_0 T)}\,\|0\rangle_P$
$t > (T/2+\Delta)$	$e^{-i(\gamma_+ + \gamma_- + E_0 T)}\,\|0\rangle$

We start out with the ground state of the non-interacting theory, at the beginning of time. Up to the time $-T/2 - \Delta$, nothing has happened because the Hamiltonian H_0 is a non-interacting Hamiltonian, and

$$|0\rangle_{t<-(T/2+\Delta)} = U(t, -(T/2+\Delta))\,|0\rangle = e^{-iH_0\left(t-(T/2+\Delta)\right)}\,|0\rangle = |0\rangle \tag{9.6}$$

The ground state doesn't even acquire a phase, because its energy is zero. We then slowly turn on the interaction over a time Δ, to reach its full strength at the time $-T/2$. By the adiabatic theorem[2] we expect the ground state $|0\rangle$ of the non-interacting system to move

[2] [Eds.] Leonard I. Schiff, *Quantum Mechanics*, 3rd ed., McGraw-Hill, 1968. See Section 35, "Methods for time-dependent problems", pp. 279–292.

smoothly from $t = -(T/2) + \Delta$ to $t = -(T/2)$ into the ground state $|0\rangle_P$ of the interacting system with probability 1. I haven't established any phase conventions for the state, so we might get the physical vacuum $|0\rangle_P$ with some phase, which I will write simply as $e^{-i\gamma_-}$ where γ_- is some real number:

$$|0\rangle_{t=-(T/2)} = \exp\left(-i\int_{-(T/2+\Delta)}^{-(T/2)} dt\,\left[H_0 + f(t)H_I(t)\right]\right)|0\rangle = e^{-i\gamma_-}\,|0\rangle_P \tag{9.7}$$

Between $-T/2$ and $T/2$, the system evolves in time according to the full interacting Hamiltonian. The state $|0\rangle_P$ is an eigenstate of the full interacting Hamiltonian, so it gains a new phase, winding up as $e^{-i\gamma_-}e^{-iE_0T}\,|0\rangle_P$:

$$|0\rangle_{t=(T/2)} = U_I(T/2, -T/2)e^{-i\gamma_-}\,|0\rangle_P = e^{-i\gamma_-}e^{-iHT}\,|0\rangle_P = e^{-i\gamma_-}e^{-iE_0T}\,|0\rangle_P \tag{9.8}$$

Finally we reach the time $T/2$, and again the adiabatic hypothesis takes over from $t = T/2$ to $t = T/2 + \Delta$. The physical state $|0\rangle_P$ turns back into the state $|0\rangle$ associated with the free Hamiltonian, H_0, but with a new phase factor which I'll call γ_+. The state becomes $e^{-i(\gamma_+ + \gamma_- + E_0T)}\,|0\rangle$, an exponential factor times the non-interacting vacuum state, the "bare" vacuum as we sometimes say. This is a straightforward computation in the Schrödinger picture, using the adiabatic theorem of non-relativistic quantum mechanics, which, if we're lucky, should be true in this instance. Incidentally, according to time-reversal invariance, the phases γ_- and γ_+ should be equal.

The Schrödinger state at time $t = -\infty$ is $|0\rangle$. At time $t = \infty$, it is $e^{-i(\gamma_+ + \gamma_- + E_0T}|0\rangle$. Writing the state at $t = \infty$ in terms of the U matrix, the time-evolution matrix, we find

$$\langle 0|U(\infty, -\infty)|0\rangle = e^{-i(\gamma_+ + \gamma_- + E_0T)} \tag{9.9}$$

We have an equation, (7.31), that tells us that $U_I(t, 0)$ is $e^{-iH_0t}U(t, 0)$. By taking the adjoint, $U_I(0, t)$ equals $U(0, t)e^{iH_0t}$. We see, writing $U(\infty, -\infty)$ as $U(\infty, 0)U(0, -\infty)$ that

$$\langle 0|U(\infty, -\infty)|0\rangle = \langle 0|U(\infty, 0)U(0, -\infty)|0\rangle = \langle 0|U_I(\infty, 0)U_I(0, -\infty)|0\rangle = \langle 0|\mathrm{S}|0\rangle \tag{9.10}$$

since $|0\rangle$ is an eigenstate of H_0 with eigenvalue zero. Consequently,

$$\langle 0|\mathrm{S}|0\rangle = e^{-i(\gamma_+ + \gamma_- + E_0T)} \tag{9.11}$$

Now this is just dumb. In the theory without the artificially introduced $f(t)$, this can't possibly be the S-matrix element between the initial ground state and the final ground state. In the real theory, without the $f(t)$, T does not appear, so you can hardly get an answer that depends on T. The sensible way to define this S-matrix element is to say that its vacuum expectation value is 1. You start out with a static source with no mesons going in, it just lies there like a lump. At the end of time, there are no mesons coming out. In this analysis, we have obtained a spurious phase factor. The origin of that spurious phase factor is my hand-waving argument that when you turn on the interaction, the system is going adiabatically from the free particle states to the corresponding in states. I forgot about phases! The states can develop phases. And if we have a mismatch between the vacuum state energy of the free theory, and the corresponding vacuum state energy of the interacting theory, then we will get a spurious phase factor, as we have seen. If we can rid ourselves of the mismatch, we'll eliminate the problem.

Now there's a very simple way of getting rid of the unwanted phase factor and obtaining a correct theory, by adding an extra term to our interaction Hamiltonian, called a **counterterm.** I will eliminate the phase factor for the ground state and then worry about whether there are corresponding spurious phase factors for the other states of the theory.[3]

I write

$$H_I \to \left[g \int d^3\mathbf{x}\, \rho(\mathbf{x})\phi(\mathbf{x},t) - a\right] f(t) \tag{9.12}$$

I have added to the Hamiltonian a new extra term, little a. It's just a number. It is called a counterterm, because it is designed to counteract our error. I will choose the value of a so that the phase factor we found in (9.11) is completely canceled:

$$a \int dt f(t) = a\left(T + \mathcal{O}(\Delta)\right) = \gamma_+ + \gamma_- + E_0 T \tag{9.13}$$

In other words, I choose a such as to force

$$\langle 0|U_I(\infty, -\infty)|0\rangle = 1 \tag{9.14}$$

This equation determines the counterterm. Thus a is *not* a free constant, and I do not have to go beyond the scattering perturbation theory I have previously developed to compute it. I can just compute it self-consistently, order by order, in perturbation theory for the U_I matrix simply by imposing, in whatever order in computation or whatever approximation I am doing, this condition (9.14), which fixes a.

Of course, we can also compute a as a by-product of our computation of the S-matrix. That's interesting, because in the limit as T goes to infinity,

$$\lim_{T\to\infty} aT(1 + \mathcal{O}(\tfrac{\Delta}{T})) = \lim_{T\to\infty} (\gamma_+ + \gamma_- + E_0 T) \tag{9.15}$$

and therefore

$$a = E_0 \tag{9.16}$$

My counterterm a is identified with E_0 in the limit of large T. If I happen to be interested in the numerical value of the ground state energy, I can compute it, because in the limit of large T, a is equal to the ground state energy, E_0.

So we've done two things with this counterterm. We have eliminated our error in mismatching the phases, i.e., mismatching the energies for the ground state, and we have found a way to use the U_I matrix to compute the ground state energy, if we want to do that. Adding the counterterm is a good thing to do. It cures our disease, and also gives us a bonus, the ground state energy.

---

[3] [Eds.] There are in principle *two* reasons to add a counterterm in Models 2 and 3: to deal with the factor T arising from the adiabatic function, and to ensure that the vacuum energy is the same with and without the interaction. Neither of these motivations applies to Model 1. By assumption $\rho(\mathbf{x},t) \to 0$ as $t \to \pm\infty$ all by itself, so there is no need to add the adiabatic function, and T does not appear. Then, as is evident from (8.85), $\langle 0|H_0|0\rangle = \langle 0|H|0\rangle = 0$; the sum reduces to a single term proportional to $n = 0$. That is, the Model 1 interaction does not change the vacuum energy, so a is not needed here, either.

9.2 Evaluating the S matrix in Model 2

In the case of Model 2, once we have matched the ground state energies for the interacting and non-interacting systems, there should be no problems for the other states of the theory, because all the other states presumably consist of 1, 2, 3, 4, etc., meson wave packets impinging on ρ. And if we go to the very far past, those states are away from ρ, and therefore they should add exactly as much to the energy as they would in a free field theory. On the other hand, we don't expect this to happen in Model 3.

In Model 3, the particles are interacting even when they are far away from ρ—there is no ρ in fact, but instead $\psi^*\psi$—and even when they are far away from each other. In that case we should expect an energy mismatch for the states with real mesons in them as well as for just the ground state. However in Model 2, knock on wood, we anticipate that this counterterm will take care of all the phase factors caused by any energy mismatch. If it doesn't, we will discover that soon enough, as we explicitly compute the S-matrix. If it indeed involves terms that don't go to constants as T approaches infinity, I will know that my confident statement was wrong.

I know that many are uncomfortable when I give general arguments. You've been trained for years that if the argument involves an equation, you just accept it, but if it involves words, you don't understand it. But now we're going to do the computation.[4] We have our Hamiltonian, (9.12). We have the condition (9.14) that fixes a. We remember that as $T \to \infty$, $a = E_0$.

We now have three connected Wick diagrams. Two are exactly the same as in the previous model: D_1, which we talked about before,

Figure 9.2: Diagram D_1 in Model 2

and D_2, which is just a number. We will calculate it.

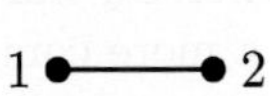

Figure 9.3: Diagram D_2 in Model 2

Now, because we have a new term in the Hamiltonian, a, we have a third diagram, D_3, which I'll represent by a cross:

Figure 9.4: Diagram D_3 in Model 2 for the counterterm $+ia$

It doesn't have any lines on it. Its contribution as a connected diagram is simply $+ia$, and its symmetry number is 1.

---

[4] [Eds.] A student asks: Is the reason why you made a general argument because we're going to do renormalization? Coleman replies: "Yeah, we're going to get there. We're going to talk about renormalization, in a little while, or at least part of it. We won't get to wave function and charge renormalization for a few weeks. But we'll talk about mass renormalization."

As before, define the operators corresponding to these diagrams. For the first,

$$O_1 = -ig \int d^3\mathbf{x}\, dt\, \rho(\mathbf{x})\phi(x)f(t) \tag{9.17}$$

For the second diagram,

$$O_2 = (-ig)^2 \int d^4x_1 d^4x_2\, \overbrace{\phi(x_1)\phi(x_2)}\rho(\mathbf{x}_1)\rho(\mathbf{x}_2)f(t_1)f(t_2) \tag{9.18}$$

and finally, for the third,

$$O_3 = ia \int dt\, f(t) \tag{9.19}$$

As before, the S-matrix can be written

$$\mathrm{S} = U_I(\infty, -\infty) = \,:\!\exp\left(\frac{O_1}{1!} + \frac{O_2}{2!} + \frac{O_3}{1!}\right)\!: \tag{9.20}$$

or, somewhat symbolically,

$$\mathrm{S} = \,:\!e^{(1)+(2)+(3)}\!: \;=\; e^{(2)+(3)} :\!e^{(1)}\!: \tag{9.21}$$

The contributions of D_2 and D_3 are pure numbers, so normal ordering is unnecessary for them. Only these two diagrams contribute to the vacuum-to-vacuum U_I matrix element $\langle 0|\mathrm{S}|0\rangle$, given by the exponential of their contributions. This is, by the definition of a, equal to one, so the contributions of D_2 and D_3 sum to zero:

$$(2) + (3) = 0 \tag{9.22}$$

Therefore, if we are interested in calculating the ground state energy, we just have to calculate D_2. That will fix a, and a is the ground state energy.

However, if we are not interested in computing the ground state energy, but only the S-matrix element, we need compute neither D_2 nor D_3, since their sum is zero. This is in general what will happen even if we have a more complicated theory with such a counterterm. The effect of the counterterm will be to cancel all Wick diagrams with no external lines, because the sum of all those diagrams makes precisely a phase factor which by assertion is to be canceled by a.

So to get the S-matrix we need only calculate D_1. Let's go.

$$U_I(\infty, -\infty) \;=\; :\!\exp\left(-ig \int d^3\mathbf{x}\, dt\, f(t)\rho(\mathbf{x})\phi(x)\right)\!: \tag{9.23}$$

The argument of the exponential, O_1, is exactly the same as in Model 1, (8.58), except for the time independence of $\rho(\mathbf{x})$ and the adiabatic function $f(t)$. Putting in the explicit form (3.45) of $\phi(x)$, the previous four-dimensional Fourier transform (8.63) for $\rho(\mathbf{x}, t)$ now factors into a three-dimensional Fourier transform and a one-dimensional one:

$$\begin{aligned} O_1 &= -ig \int d^3\mathbf{x}\, dt\, \rho(\mathbf{x})f(t) \int \frac{d^3\mathbf{p}}{(2\pi)^{3/2}\sqrt{2\omega_{\mathbf{p}}}} \left(e^{-ip\cdot x}a_{\mathbf{p}} + e^{ip\cdot x}a^\dagger_{\mathbf{p}}\right) \\ &= -ig \int \frac{d^3\mathbf{p}}{(2\pi)^{3/2}\sqrt{2\omega_{\mathbf{p}}}} \left(\tilde{f}(\omega_{\mathbf{p}})\tilde{\rho}(\mathbf{p})a_{\mathbf{p}} + \tilde{f}(\omega_{\mathbf{p}})^*\,\tilde{\rho}(\mathbf{p})^*\,a^\dagger_{\mathbf{p}}\right) \end{aligned} \tag{9.24}$$

There's our old $\tilde{\rho}(\mathbf{p}, \omega_{\mathbf{p}})$, now a product of two terms, times $a_{\mathbf{p}}$, just as before, (see (8.65)), plus the Hermitian conjugate.[5]

Well, what does this tell us? Look again at the graph of $f(t)$, Figure 9.1: It approaches a constant function equal to 1 for large T. So for a large T, its Fourier transform, $\tilde{f}(\omega_{\mathbf{p}})$,

$$\tilde{f}(\omega_{\mathbf{p}}) = \int dt\, e^{-i\omega_{\mathbf{p}}t} f(t) \tag{9.25}$$

approaches the Fourier transform of 1, or 2π times a delta function. If we plot $\tilde{f}(\omega_{\mathbf{p}})$ against

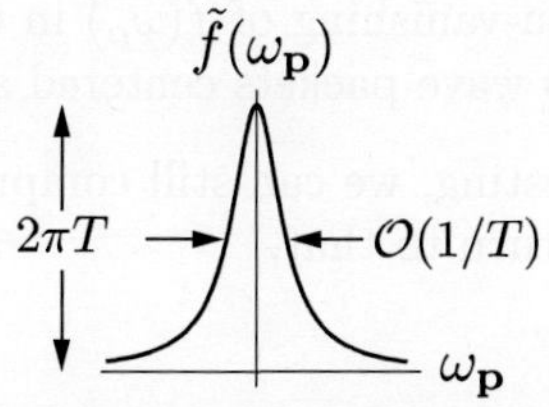

Figure 9.5: The Fourier transform $\tilde{f}(\omega_{\mathbf{p}})$ of $f(t)$

$\omega_{\mathbf{p}}$, we'll get some very highly peaked function with its spread on the order of $1/T$, and a height on the order of $2\pi T$, to make the total area equal to 2π. As T grows larger and larger, $\tilde{f}(\omega_{\mathbf{p}})$ gets narrower and higher, and eventually becomes 2π times a delta function concentrated at $\omega_{\mathbf{p}} = 0$.

Now this has interesting implications. Since $\omega_{\mathbf{p}}$ is always greater than μ, $\tilde{f}(\omega_{\mathbf{p}})$ goes to zero for any $\omega_{\mathbf{p}}$ of interest, because it is concentrated at the origin, and has a spread only $\mathcal{O}(1/T)$. Eventually $1/T$ gets much less than μ, so

$$\lim_{T\to\infty} O_1 = 0 \tag{9.26}$$

Therefore

$$\lim_{T\to\infty} S = \lim_{T\to\infty} \; : e^{(1)} : \; = \; e^0 = 1 \tag{9.27}$$

As T goes to infinity, the S-matrix goes to the exponential of zero, which is 1. This S-matrix is indeed a unitary matrix and completely free of dependence on T, as required. Of course it's physically rather uninteresting. It's as if we have this lump $\rho(\mathbf{x})$ sitting there, and we send a meson to scatter off of it, the meson doesn't scatter! It just goes right on by ...

That the Model 2 S-matrix turns out to be equal to 1 can be explained with much the same physical argument we used to describe the production of mesons in Model 1. Following (8.67), I argued that the Model 1 operator O_1 vanishes unless $\tilde{\rho}(\mathbf{p}, \omega_{\mathbf{p}})$ is non-zero on the mass shell ($\omega_{\mathbf{p}} = \sqrt{|\mathbf{p}|^2 + \mu^2}$). Additionally, in Model 1, mesons were absorbed or emitted by the source one at a time, because of the corresponding Diagram 1. In Model 2 we have the same Diagram 1, and we have an example of a non-zero function $f(t)$ whose Fourier transform $\tilde{f}(\omega_{\mathbf{p}})$ vanishes on the mass shell. In fact $\tilde{f}(\omega_{\mathbf{p}})$ vanishes everywhere except for a tiny neighborhood of $\omega_{\mathbf{p}} = 0$, which does not include any part of the mass hyperboloid. Again, the mesons are either absorbed or emitted by the source one at a time. A time independent source like $\rho(\mathbf{x})$

[5] [Eds.] Note that $\tilde{f}(-\omega_{\mathbf{p}}) = \tilde{f}(\omega_{\mathbf{p}})^*$, and $\tilde{\rho}(-\mathbf{p}) = \tilde{\rho}(\mathbf{p})^*$.

cannot transfer energy; it can only transfer momentum. That means it can't absorb or emit a meson, because those processes require energy transfer. A meson always has non-zero energy. So the S-matrix is identically equal to 1, and there is no scattering in Model 2.

This theory is a complete washout as far as scattering is concerned. While this was easy to see in the formalism we have built up, it was obscure when people were evaluating this same model theory in the Born approximation. Not until the discovery of miraculous cancellations of all the fourth-order terms in the Born series did people realize that they should try to prove the S-matrix for this model was identically equal to 1, to all orders.[6]

This result holds in the massless case as well. Since there is no scattering for all $\mathbf{p} \neq 0$, you have only to prove that the non-vanishing of $\tilde{f}(\omega_{\mathbf{p}})$ in the neighborhood of $\omega_{\mathbf{p}} = 0$, a set of measure zero, does not screw up wave packets centered about $\mathbf{p} = 0$.

Even if the S-matrix is uninteresting, we can still compute the ground state energy. That may be interesting. So let us now turn to that.

9.3 Computing the Model 2 ground state energy

Let's write down the condition that these two diagrams, D_2 and D_3, cancel:

$$\lim_{T\to\infty} \left[\frac{O_2}{2!} + O_3\right] = 0 \tag{9.28}$$

where O_2 and O_3 are given by (9.18) and (9.19), respectively. Using the identity (see (8.22) and (8.23))

$$\overline{\phi(x_1)\phi(x_2)} = \lim_{\epsilon\to 0+} \int \frac{d^4p}{(2\pi)^4} e^{-ip\cdot(x_1-x_2)} \frac{i}{p^2 - \mu^2 + i\epsilon} \tag{9.29}$$

the contribution of D_2 can be written

$$\begin{aligned}\frac{O_2}{2!} &= \lim_{\epsilon\to 0+} \frac{-ig^2}{2!} \int \frac{d^4p}{(2\pi)^4} \frac{1}{\omega_{\mathbf{p}}^2 - |\mathbf{p}|^2 - \mu^2 + i\epsilon} |\tilde{\rho}(\mathbf{p})|^2 |\tilde{f}(\omega_{\mathbf{p}})|^2 \\ &= \lim_{\epsilon\to 0+} \frac{-ig^2}{2!} \int \frac{d^3\mathbf{p}}{(2\pi)^3} |\tilde{\rho}(\mathbf{p})|^2 \int \frac{d\omega_{\mathbf{p}}}{2\pi} \frac{|\tilde{f}(\omega_{\mathbf{p}})|^2}{\omega_{\mathbf{p}}^2 - |\mathbf{p}|^2 - \mu^2 + i\epsilon}\end{aligned} \tag{9.30}$$

Let us now go to the limit of large T, because that's what we have to do to compute the energy. In this limit, $\tilde{f}(\omega_{\mathbf{p}})$ is sharply peaked about $\omega_{\mathbf{p}} = 0$. That means in the second integral, we can simply replace $\omega_{\mathbf{p}}$ with the value 0. With this replacement, the denominator will never equal zero, so we no longer need the $i\epsilon$ nor the limit, and we can write

$$\frac{O_2}{2!} = \frac{ig^2}{2!} \int \frac{d^3\mathbf{p}}{(2\pi)^3} \frac{|\tilde{\rho}(\mathbf{p})|^2}{|\mathbf{p}|^2 + \mu^2} \int \frac{d\omega_{\mathbf{p}}}{2\pi} |\tilde{f}(\omega_{\mathbf{p}})|^2 \tag{9.31}$$

---

[6] [Eds.] For an older approach to Models 1 and 2, see Gregor Wentzel, *Quantum Theory of Fields*, trans. C. Houtermans and J. M. Jauch, Interscience, 1947, Chap. II, §7, "Real fields with sources", pp. 37–48; republished by Dover Publications, 2003.

Now we invoke a famous relation, **Parseval's theorem**[7]:

$$\int \frac{d\omega_{\mathbf{p}}}{2\pi} |\tilde{f}(\omega_{\mathbf{p}})|^2 = \int dt |f(t)|^2 \tag{9.32}$$

As $f(t)$ has the value 1 for the interval $(-T/2, T/2)$ we can say that its square is also equal to 1 in that region, and

$$\int dt |f(t)|^2 = T + \mathcal{O}(\Delta) = T(1 + \mathcal{O}(\Delta/T)) \tag{9.33}$$

From (9.20) and (9.22), we require $O_2/2! = -O_3$, which is, in the limit as $T \to \infty$,

$$O_3 = ia \int dt\, f(t) = iaT(1 + \mathcal{O}(\Delta/T)) \tag{9.34}$$

Setting $O_2/2! = -O_3$ we have

$$T(1 + \mathcal{O}(\Delta/T)) \frac{ig^2}{2!} \int \frac{d^3\mathbf{p}}{(2\pi)^3} \frac{|\tilde{\rho}(\mathbf{p})|^2}{|\mathbf{p}|^2 + \mu^2} = -iaT(1 + \mathcal{O}(\Delta/T)) \tag{9.35}$$

The T's cancel, the i's cancel, so I get a real energy, which is a relief. The ground state energy is given by the formula

$$a = E_0 = -\frac{g^2}{2} \int \frac{d^3\mathbf{p}}{(2\pi)^3} \frac{|\tilde{\rho}(\mathbf{p})|^2}{|\mathbf{p}|^2 + \mu^2} \tag{9.36}$$

This is in a sense the final and complete answer to our problem. It tells us what the ground state energy is. Note that the sign is negative, as we should expect. There's a general theorem that if you add a term to the Hamiltonian with zero expectation value in the unperturbed ground state, then that always lowers the energy. That's a trivial consequence of the variational principle. The term we have added is linear in ϕ, and therefore has zero expectation value in the unperturbed ground state. If the sign had not come out negative I would have been very disturbed.

It's worth a little work to transform this formula (9.36) from momentum space into position space. It can be written as[8]

$$E_0 = -\tfrac{1}{2} g^2 \int d^3\mathbf{x}\, d^3\mathbf{y}\, \rho(\mathbf{x}) V(\mathbf{x} - \mathbf{y}) \rho(\mathbf{y}) \tag{9.37}$$

where

$$V(\mathbf{x}) = \int \frac{d^3\mathbf{p}}{(2\pi)^3} \frac{e^{i\mathbf{p}\cdot\mathbf{x}}}{|\mathbf{p}|^2 + \mu^2} \tag{9.38}$$

---

[7] [Eds.] Some reserve the name "Parseval's theorem"' for the Fourier series version of this theorem, and call the Fourier integral version "Plancherel's theorem". See Gilbert Strang, *Introduction to Applied Mathematics*, Wellesley-Cambridge Press, 1986, p. 313; or Philippe Dennery and André Krzywicki, *Mathematics for Physicists*, Harper & Row, 1967, p. 224, Theorem 2. Others make no distinction between the discrete and continuous cases, and call both versions "Parseval's theorem", e.g., Philip M. Morse and Herman Feshbach, *Methods of Theoretical Physics*, Part I, McGraw-Hill, 1953, p. 456, or Richard Courant and David Hilbert, *Methods of Mathematical Physics*, vol. II, Interscience, 1962, p. 794.

[8] [Eds.] See note 5, p. 189.

I called this model "quantum meso-statics", because $\rho(\mathbf{x})$ is a sort of classical version of "nucleon density", just like the classical charge distributions that enter in electrostatics. So I've written the energy of the system in a form that looks very much like the energy of an electrostatic system:[9]

$$E = \tfrac{1}{2}\int d^3\mathbf{x}\, d^3\mathbf{y}\, \frac{\rho(\mathbf{x})\rho(\mathbf{y})}{|\mathbf{x}-\mathbf{y}|} \tag{9.39}$$

The $\frac{1}{2}$ is also there in electrostatics. There is a minus sign in (9.37), whereas in electrostatics there's a plus sign. The potential (9.38) represents an attractive potential between our infinitesimal elements of "nucleonic charge", rather than the repulsive one as in electrostatics. Also, the integrand $1/(|\mathbf{p}|^2+\mu^2)$ of (9.38) is not the Fourier transform of the Coulomb potential $1/|\mathbf{x}-\mathbf{y}|$ of electrostatics, but something different, representing the interaction between two infinitesimal elements of "nucleonic charge", as opposed to electric charge.

The integral (9.38) for $V(\mathbf{x})$ can be performed in the usual way.[10] Let $|\mathbf{p}| = p$, and $|\mathbf{x}| = r$. Then

$$\begin{aligned} V(r) &= \int_0^\infty \frac{dp\,p^2}{(2\pi)^3(p^2+\mu^2)}\int_0^\pi d\theta\, e^{ipr\cos\theta}\sin\theta\int_0^{2\pi} d\phi \\ &= \frac{1}{(2\pi)^2}\int_0^\infty \frac{dp\,p^2}{(p^2+\mu^2)}\left(\frac{e^{ipr}-e^{-ipr}}{ipr}\right) \\ &= -\frac{i}{(2\pi)^2 r}\int_{-\infty}^\infty \frac{dp\,p e^{ipr}}{(p^2+\mu^2)} \end{aligned} \tag{9.40}$$

The last integral can be done by Cauchy's theorem. The integrand has two poles, at $p = \pm i\mu$. Because r is always positive, I can safely complete the contour of integration in the upper half p plane where the exponential decreases unbearably rapidly; see Figure 9.6.

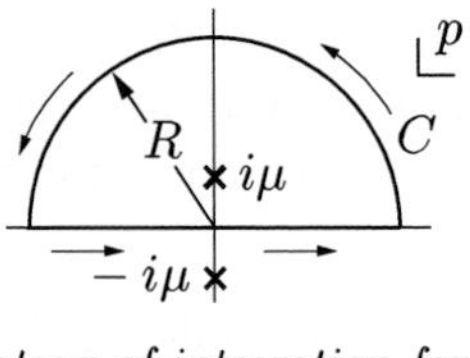

Figure 9.6: Contour of integration for V(r) in Model 2

Now all I have within the contour of integration is a single pole, $p = i\mu$. I can evaluate the integral by Cauchy's residue formula:

$$\begin{aligned} V(r) &= -\frac{i}{(2\pi)^2 r}\int_C \frac{dp\,p e^{ipr}}{(p+i\mu)(p-i\mu)} = -\frac{i}{(2\pi)^2 r}\left[2\pi i\left(\frac{i\mu e^{-\mu r}}{2i\mu}\right)\right] \\ &= \frac{1}{4\pi r}e^{-\mu r} \end{aligned} \tag{9.41}$$

[9] [Eds.] Jackson *CE*, p. 41, equation (1.52).

[10] [Eds.] In the video for Lecture 9, Coleman remarks that his students from Physics 251 (the Harvard graduate course in non-relativistic quantum mechanics) could "probably wake up screaming while doing this integral. But for the benefit of those of you who have missed that golden experience", he goes through the calculation in detail, adding, "this kind of integral is very useful in doing the hydrogen atom and all sorts of such things."

which is known as the **Yukawa potential**.[11]

Thus the infinitesimal elements of this quantity $\rho(\mathbf{x})$, which we have called "nucleonic charge density", have an interaction energy proportional to a Yukawa potential. Notice that the singularity of the Yukawa potential at $r = 0$ is the same is as the singularity of the Coulomb potential at $r = 0$. Of course, the large r behavior is very different. The Yukawa potential falls off rapidly with distance, being essentially negligible when r is several times greater than $1/\mu$, that is to say, when r is several times greater than the Compton wavelength ($h/\mu c$, in conventional units) of the meson, a meson which doesn't scatter, but is still responsible for the force between the elements of nuclear matter.

We could model a two-nucleon system like this:

$$\rho(\mathbf{x}) = \text{“}\delta^{(3)}(\mathbf{x} - \mathbf{x}_1)\text{”} + \text{“}\delta^{(3)}(\mathbf{x} - \mathbf{x}_2)\text{”} \tag{9.42}$$

where the "$\delta^{(3)}(\mathbf{x})$"s are similar to delta functions. They are highly-peaked functions which vanish outside of a small interval around $\mathbf{x}$. The nucleons are localized in neighborhoods of $\mathbf{x}_1$ and $\mathbf{x}_2$. Substituting (9.42) and (9.41) into (9.37),

$$E_0 = -g^2 \frac{e^{-\mu|\mathbf{x}_1 - \mathbf{x}_2|}}{4\pi|\mathbf{x}_1 - \mathbf{x}_2|} + (\text{term independent of } \mathbf{x}_1 \text{ and } \mathbf{x}_2) \tag{9.43}$$

This force is *attractive* between like charges, and short-range, and so has some of the essential features of the real nuclear force. That the force here is attractive turns out to be an example of a general rule: For forces mediated by the exchange of even-spin particles, like particles attract; for forces mediated by the exchange of odd-spin particles, like particles repel. This force is mediated by the exchange of zero-spin bosons, so it is attractive.

Notice also that if we had specified $\rho(\mathbf{x})$ as a point charge (or a collection of point charges),

$$\rho(\mathbf{x}) = \delta^{(3)}(\mathbf{x}) \tag{9.44}$$

then just as in electrostatics the energy would be infinite. That's an important observation:

$$\text{As } \rho(\mathbf{x}) \to \delta^{(3)}(\mathbf{x}),\ E_0 \to \infty \tag{9.45}$$

This divergence is called an **ultraviolet divergence**, because in $\mathbf{p}$-space it corresponds to the integral blowing up at high $|\mathbf{p}|$. If $\rho(\mathbf{x})$ is a delta function, then $\tilde{\rho}(\mathbf{p})$ is a constant, and the integral (9.36) blows up like $d^3\mathbf{p}/|\mathbf{p}|^2$.

This divergence, appearing in the term in (9.43) and *not* depending on the positions of the nucleons, is nothing to worry about. If nuclear matter need not be an assembly of point particles, then $\rho(\mathbf{x})$ need not be a delta function. Even if there were some fixed number of point particles, say seven of them moving about on little tracks, the terms coming from the self-energy of the particles are totally irrelevant. You cannot measure that energy. It exerts no force. It doesn't change as you move the particles apart. The only term that you actually measure is the part that depends on the separation between the particles—the only thing you can adjust—and that part is of course perfectly finite, if they're at finite distances from each other.

---

[11] [Eds.] Hideki Yukawa (1907–1981), Nobel Prize in Physics 1949. See "On the Interaction of Elementary Particles. I.", *Proc. Phys.-Math. Soc. Japan* **17** (1935) 48–57. Reprinted in D. M. Brink, *Nuclear Forces*, Pergamon Press, 1965.

I wanted to emphasize in this model that first, we get a Yukawa force, and second, we get an ultraviolet divergence if we go towards the point-particle limit. This may cause us some troubles when we finally get to Model 3, where the interaction is $\phi\psi^*\psi$ without any integrating functions to smear things out. Our nucleons there are not like the nucleons here. They're real particles that can recoil and be produced, but they still definitely interact with the ϕ field at a single point, and therefore we might get an infinite energy shift which we would have to worry about.

9.4 The ground state wave function in Model 2

We can compute not only the ground state energy, but also the ground state wave function, an expansion of the physical vacuum $|0\rangle_P$ into the basis states $|\mathbf{p}_1, \ldots, \mathbf{p}_n\rangle$, eigenstates of the non-interacting Hamiltonian H_0. Thus we want to calculate the quantities

$$\langle \mathbf{p}_1, \ldots, \mathbf{p}_n | 0\rangle_P \tag{9.46}$$

This is just an exercise to show that restricting ourselves to time-dependent perturbation theory is not as restrictive as you might think. We can do all the things we usually do in non-relativistic quantum mechanics with time-independent perturbation theory. In particular we can construct the ground state wave function.

We use the interaction Hamiltonian

$$H_I(t) = g \int d^3\mathbf{x}\, \phi(\mathbf{x}, t)\rho(\mathbf{x}) f(t) \tag{9.47}$$

When we studied the interaction turning on and off adiabatically, we said that in the large T, large Δ limit,

$$U(0, -\infty)\, |0\rangle = |0\rangle_P \tag{9.48}$$

That's just the statement that the U operator up to $t = 0$, halfway along the way after the interaction has been turned on, times the bare vacuum $|0\rangle$, equals the physical vacuum $|0\rangle_P$, times a phase factor. This phase factor is of no physical interest, and I won't bother writing it down. Because $e^{-iH_0 t}$ makes no difference to the ground state, (9.48) is equivalent to (see (7.31))

$$U_I(0, -\infty)\, |0\rangle = |0\rangle_P \tag{9.49}$$

Now let us consider, for anything that's adiabatically turned on,

$$f(t) = e^{\epsilon t} \text{ for } t < 0. \tag{9.50}$$

As usual we'll consider the limit $\epsilon \to 0^+$. If we extend $f(t)$ for positive t in the following rather discontinuous way,

$$f(t) = \begin{cases} e^{\epsilon t} & \text{if } t < 0 \\ 0 & \text{if } t > 0 \end{cases} \tag{9.51}$$

(this function is graphed in Figure 9.7) then we can write (9.49) as

$$U_I(\infty, -\infty)\, |0\rangle = |0\rangle_P \tag{9.52}$$

and therefore

$$\langle \mathbf{p}_1, \ldots, \mathbf{p}_n | 0\rangle_P = \langle \mathbf{p}_1, \ldots, \mathbf{p}_n | U_I(\infty, -\infty) | 0\rangle \tag{9.53}$$

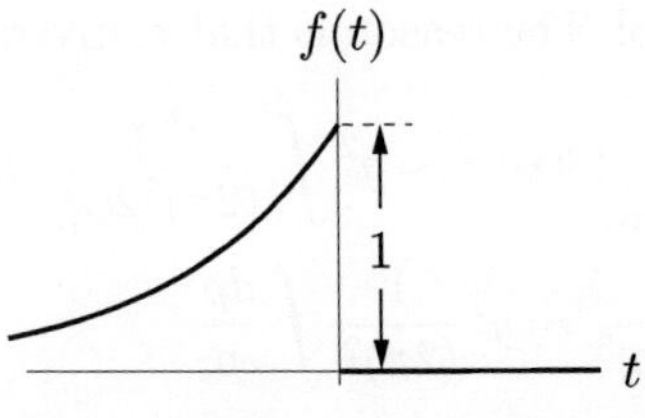

Figure 9.7: The extended adiabatic function $f(t)$

Now we know how to compute *that*. Indeed, we learned how to compute it last lecture, when we were looking at Model 1. It's just that now the space and time dependence of the source, $\rho(\mathbf{x})f(t)$, are somewhat peculiar.

This expression (9.53) gives the expansion of the physical ground state in terms of appropriate wave functions of the non-interacting Hamiltonian, or as we say in our somewhat colorful way, the amplitude for finding n bare mesons in the physical ground state. That's the confusing language people use to describe the expansion of the interacting system's ground state in energy eigenstates of the non-interacting system. The ground state just lies there. There are *no* particles moving around in it.

We can apply the results of Model 1—(8.62), (8.71), and (8.72)—to write

$$\langle \mathbf{p}_1, \ldots, \mathbf{p}_n | U_I(\infty, -\infty)|0\rangle = e^{\frac{1}{2}(-\alpha+i\beta)} h(\mathbf{p}_1)^* h(\mathbf{p}_2)^* \cdots h(\mathbf{p}_n)^* \tag{9.54}$$

where, analogous to (8.67),

$$h(\mathbf{p}) = \frac{-ig\tilde{\rho}(\mathbf{p})\tilde{f}(\omega_{\mathbf{p}})}{(2\pi)^{3/2}\sqrt{2\omega_{\mathbf{p}}}} \tag{9.55}$$

The expression (9.54) is always a product, whatever the form of $\rho(x)$. The Fourier transform of the adiabatic function (9.51) is

$$\tilde{f}(\omega_{\mathbf{p}}) = \int_{-\infty}^{\infty} dt\, e^{-i\omega_{\mathbf{p}}t} f(t) = \int_{-\infty}^{0} dt\, e^{-i\omega_{\mathbf{p}}t} e^{\epsilon t} = \frac{1}{\epsilon - i\omega_{\mathbf{p}}} \to \frac{i}{\omega_{\mathbf{p}}} \text{ as } \epsilon \to 0 \tag{9.56}$$

Then the expression for α, analogous to (8.77), becomes

$$\alpha = \int d^3\mathbf{p}\, |h(\mathbf{p})|^2 = g^2 \int \frac{d^3\mathbf{p}}{(2\pi)^3 2\omega_{\mathbf{p}}} |\tilde{\rho}(\mathbf{p})|^2 |\tilde{f}(\omega_{\mathbf{p}})|^2 \underset{\epsilon\to 0}{\to} g^2 \int \frac{d^3\mathbf{p}}{(2\pi)^3 2\omega_{\mathbf{p}}^3} |\tilde{\rho}(\mathbf{p})|^2 \tag{9.57}$$

The probability of finding n bare mesons is the probability amplitude squared for the physical ground state having a component in the n meson subspace of the non-interacting Hamiltonian,

$$P(n) = \frac{e^{-\alpha}\alpha^n}{n!} \tag{9.58}$$

the Poisson distribution we had before.

Something very interesting happens to the expansion (8.70) of the ground state wave function, if we consider a point particle: $\rho(\mathbf{x})$ goes to a delta function, and $\tilde{\rho}(\mathbf{p})$ becomes a constant. The expansion blows up! The reason is that α diverges logarithmically (at large $|\mathbf{p}|$):

$$\begin{aligned}\alpha &= g^2\int\frac{d^3\mathbf{p}}{(2\pi)^3 2\omega_\mathbf{p}^3}|\tilde{\rho}(\mathbf{p})|^2 \sim g^2\int\frac{d^3\mathbf{p}}{(2\pi)^3 2\omega_\mathbf{p}^3}\quad\text{as } \rho(\mathbf{x})\to\delta^{(3)}(\mathbf{x})\\ &\sim g^2\int\frac{d^3\mathbf{p}}{(2\pi)^3 2|\mathbf{p}|^3}\sim g^2\frac{1}{(2\pi)^2}\int\frac{dp}{p}\end{aligned}\tag{9.59}$$

This isn't as bad a divergence as the energy, which, as you'll recall, went at high $|\mathbf{p}|$ like $d^3\mathbf{p}/|\mathbf{p}|^2\sim dp$. Still, $\alpha\to\infty$ as $\rho(\mathbf{x})$ approaches a delta function. So what do we make of that?

Recall from last time that we found

$$\langle N\rangle = \sum_{n=0}^{\infty} nP(n) = \alpha \tag{8.79}$$

The average number $\langle N\rangle$ of bare mesons in the theory gets very, very large as the source gets more and more concentrated. On the other hand, the probability $P(n)$ of finding any given number n of bare mesons goes to zero as the source becomes a point. As $\rho(\mathbf{x})$ goes to a point in position space, or $\tilde{\rho}(\mathbf{p})$ goes to a constant in Fourier space, α and the peak of the Poisson distribution zoom out towards infinity. That's disgusting behavior. It's a good thing that in the future we won't worry about computing things like the difference between the ground state energies of the interacting and non-interacting Hamiltonians for a single particle, or the amplitude for finding the non-interacting ground state in the interacting ground state. Nobody really should worry about those questions, because in real models with realistic theories, you don't have the freedom to turn off the interaction. You don't have the freedom to find out what the energy of the one-electron state would be, if there were no electromagnetic interaction, because, although we give ourselves considerable airs at times, we do not have the power to change the electromagnetic interaction, say the fine-structure constant, by one jot or tittle.

Fortunately those things which are physically measurable in this theory—for example, the interaction energy between two separated point charges—do not display such pathological ultraviolet divergences. So, (knock on wood), maybe we'll be lucky. Maybe we can get by with our theory of point particles even if it turns out to include all sorts of disgusting infinities. Perhaps those infinities won't enter into any physically observable quantities; perhaps they will. Probably it depends on what the theory is. We'll have to wait and see.

9.5 An infrared divergence

There is another divergence implicit in the integral for α which has nothing to do with what I have been discussing. It is perhaps best expressed not by thinking of this integral as an example of Model 2, expanding the ground state, but as an example of Model 1, where we have just chosen a particular form of $f(t)$, one where we turn things on very slowly and then turn them off abruptly. In this case the formula (9.57) has another kind of divergence, not dependent on how $\rho(x)$ is distributed. It has to do with the mass of the meson. The formula for α blows up as μ goes to zero, unless $\tilde{\rho}(\mathbf{p})$ vanishes at $|\mathbf{p}|=0$:

$$\lim_{\mu\to 0}\alpha = \infty \quad (\text{if } \tilde{\rho}(0)\neq 0) \tag{9.60}$$

If $\mu=0$, then $\omega_\mathbf{p}=|\mathbf{p}|$, and at the low-energy end the integral blows up like $d^3\mathbf{p}/|\mathbf{p}|^3$. For obvious reasons, this is called an **infrared divergence**. Since we will eventually have to confront theories of massless particles that are indeed radiated in interaction processes—in particular we will have to confront the theory of photons—it is perhaps worth saying a few words about this divergence.

This divergence is also unphysical. Let's call our massless mesons "photons" for a moment, abusing language. If we have a source which we build up slowly and turn off abruptly, on the average we will radiate an infinite number of "photons". That's rather silly. This example is very far from being a real photon experiment, but in a real photon experiment there is a detector, say a photomultiplier tube. You will never read a report that says, "We observed an infinite number of counts..." Although an infinite number of *photons* are radiated in this process, only a finite amount of *energy* is radiated, because the formula for the expectation value $\langle E\rangle$ of the energy

$$\langle E\rangle = \int d^3\mathbf{p}\,|h(\mathbf{p})|^2\,\omega_{\mathbf{p}} \tag{8.87}$$

has an extra factor of $\omega_{\mathbf{p}}$, as we showed at the end of last lecture. Putting in the factors we have

$$\langle E\rangle = g^2\int \frac{d^3\mathbf{p}}{(2\pi)^3 2\omega^3_{\mathbf{p}}}|\tilde{\rho}(\mathbf{p})|^2\omega_{\mathbf{p}} \tag{9.61}$$

This integral does not diverge as μ goes to zero. At small $|\mathbf{p}|$ it behaves as $d^3\mathbf{p}/|\mathbf{p}|^2$, which is perfectly convergent.

What has happened recalls Zeno's paradox of Achilles and the tortoise. You have a finite amount of energy to distribute, but photons are massless. You can give smaller and smaller amounts of energy to each photon. You could give half the energy to one photon, and a quarter of the energy to another photon, an eighth of the energy to a third photon, a 16th of the energy to a fourth, and so distribute a finite amount of energy among an infinite number of photons.

Most of the photons from this infinite number, indeed the overwhelming majority, had arbitrarily low energy. That means they had very, very long wavelengths. The actual experimental apparatus, a photomultiplier tube or a radar antenna or anything else at all, however you are detecting your photons, has a low-frequency cut-off. If the photon is sufficiently soft that the electromagnetic radiation has a sufficiently large wavelength, then you cannot detect it with any finite experimental apparatus. To detect a photon that has a wavelength of a thousand light years, you need a radar antenna that is a thousand light years on a side. Those are not found in your average high-energy physics laboratory! The reason we got an infinite answer again, in the extreme limit $\mu \to 0$, is because we were asking a unphysical question, just as unphysical as asking about the energy of a point source if we turned off the interaction. These are unphysical questions. How many photons would we detect if we had an experimental apparatus that could detect any photon, no matter how long its wavelength? That is an impossible question. If we asked a different question, what is the average number of photons we can detect if our experimental apparatus can only detect photons of momentum greater than a certain threshold $|\mathbf{p}|_{min}$, then it is easy to see that in the integral for α we would not integrate all the way down to zero, but just down to our low-energy experimental cut-off. And then, even as μ went to zero, α would not go to infinity, but to a finite value.[12] Once

[12] [Eds.] Coleman is referring to *the infra-red divergence.* This famous problem is not discussed in this book. The classic treatment is due to Bloch and Nordsieck: F. Bloch and A. Nordsieck, "Note on the Radiation Field of the Electron", *Phys. Rev.* **50** (1937) 54–59. A fuller explanation is given in J. M. Jauch and F. Rohrlich,

again, we're saved! It's a real *Perils of Pauline* story.[13] If we're sloppy, and ask questions that are empirically unanswerable, we get, in extreme—but physically reasonable—limits, nonsense answers. If we're careful and restrict ourselves only to asking questions corresponding to experiments we can really do, then we get finite answers, even in those extreme limits.

So far, in our simple theories, the divergences have restricted themselves to unobservable quantities, and thus kept in quarantine. Such theories are called **renormalizable**. Whether that situation will prevail when we go to more complicated theories than the ones we have at hand here, is a question that will be resolved only by future investigation, which I will begin next lecture, when we start to tackle Model 3.

The Theory of Photons and Electrons, 2nd expanded ed., Springer-Verlag, 1976, Section 16-1, pp. 390–405, or Bjorken & Drell *RQM*, pp. 162–176, and Bjorken & Drell *Fields*, pp. 202–207. It should perhaps be mentioned that the first edition of Jauch and Rohrlich (1955) was among the very first American textbooks to teach the use of Feynman diagrams; see David Kaiser, *Drawing Theories Apart: The dispersion of Feynman diagrams in postwar physics*, U. Chicago Press, 2005, Chapter 7, pp. 253–263.

[13] [Eds.] A series of short, silent World War I-era movies shown before a full-length feature, with the title heroine in a succession of grave dangers from week to week, only to be rescued in the nick of time.

Problem 5

5.1 The pair model, invented by G. Wentzel[1], is a variant on Model 2 in which there is a bilinear interaction of the meson field with a time-independent c-number source, instead of a linear one. This is more complicated than Model 2, but the theory is still exactly soluble, because it is still just a quadratic Hamiltonian. Unlike Model 2, in this model scattering (but only elastic scattering) can occur.

The Hamiltonian for the theory is of the form $H = H_0 + H_I$, where H_0 is the standard Hamiltonian for a free scalar field of mass μ. The interaction Hamiltonian H_I is

$$H_I = \tfrac{1}{2} g \left(\int d^3\mathbf{x}\, \rho(\mathbf{x}) \phi(\mathbf{x}, t) \right)^2$$

where g is a positive constant, and $\rho(\mathbf{x})$ is some smooth, real function of space only that goes to zero rapidly at infinity. (Note that the interaction here is *not* the integral of a local density, but the *square* of such an integral.)

(a) Compute $\langle \mathbf{p} | (\mathrm{S} - 1) | \mathbf{p}' \rangle$, the scattering matrix element between (non-relativistically normalized) one-meson states, by summing up all the connected Wick diagrams (shown below). Start with Dyson's formula (8.1), and use Wick's theorem (8.28) to evaluate the relevant terms. Don't worry about $f(t)$ or any counterterms.

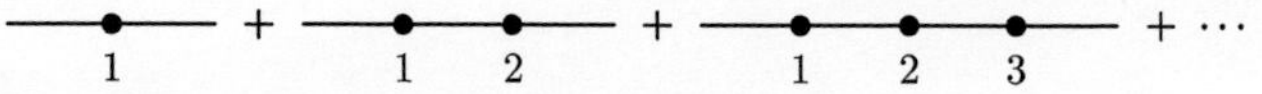

Show that

$$\langle \mathbf{p} | \mathrm{S} - 1 | \mathbf{p}' \rangle = \tilde{\rho}(\mathbf{p})^* \tilde{\rho}(\mathbf{p}') F(\omega_{\mathbf{p}}) \delta(\omega_{\mathbf{p}} - \omega_{\mathbf{p}'}) \tag{P5.1}$$

where $F(\omega_{\mathbf{p}})$ is a function you are to compute in terms of an integral over $|\tilde{\rho}(\mathbf{p})|^2$.

(b) The pair model has no non-vanishing Wick diagrams for one particle going into more than one particle; thus the S matrix restricted to one-particle initial and final states should be unitary. Explicitly verify this. That is, show explicitly that $\mathrm{S}^\dagger \mathrm{S} = 1$ for two one-particle states:

$$\langle \mathbf{p} | \mathrm{S}^\dagger \mathrm{S} - 1 | \mathbf{p}' \rangle = 0 \tag{P5.2}$$

Many comments:

(1) In addition to the diagrams shown, there are also diagrams with no uncontracted fields, (i.e., no external legs), but you don't have to worry about them; they're cancelled in the computation of the S matrix by the ground state energy counterterm, just as in Model 2.

(2) Note that every vertex in the diagrams represents a seven-dimensional integral: two three-dimensional spatial integrals and one time integral.

(3) I've only drawn one diagram of each pattern. There are others, obtained by permuting the labels 1, 2, $\dots$, n.

---

[1] [Eds.] "Zur Paartheorie der Kernkräfte" (Towards a pair theory of nuclear forces), *Helv. Phys. Acta* **15** (1942) 111–126.

(4) Even after you assign the labels, there are still 2^n identical terms, because there are two choices at each vertex, of which field gets contracted which way. This cancels the $1/2^n$ from the nth power of H_I in Dyson's formula.

(5) Don't get involved with mathematical niceties. Assume that $\rho(\mathbf{x})$ is sufficiently smooth and falls off sufficiently rapidly as $|\mathbf{x}| \to \pm\infty$ to justify any manipulations you wish to make, that all power series converge, etc.

(6) The answer involves an integral over $\mathbf{p}$ defined in terms of $\tilde{\rho}(\mathbf{p})$, the Fourier transform of $\rho(\mathbf{x})$. It's not possible to simplify this integral for general $\rho(\mathbf{x})$; don't waste your time by trying to do so. On the other hand, if you have more complicated things than this (double integrals, unsummed infinite series, etc.), you have more to do.

(7) Don't assume $\rho(\mathbf{x})$ is spherically symmetric.

(1997a 5.1)

Solution 5

5.1 (a) The interaction Hamiltonian is

$$H_I = \tfrac{1}{2}g\left(\int d^3\mathbf{x}\,\rho(\mathbf{x})\phi(\mathbf{x},t)\right)\left(\int d^3\mathbf{y}\,\rho(\mathbf{y})\phi(\mathbf{y},t)\right) \tag{S5.1}$$

The matrix element of interest is

$$\langle\mathbf{p}|S-1|\mathbf{p}'\rangle = \langle\mathbf{p}|\,T\left[\exp\left(-i\int dt\,H_I\right)\right]-1\,|\mathbf{p}'\rangle = \sum_{n=1}^{\infty}\frac{(-i)^n}{n!}\int dt_1\cdots dt_n\,\langle\mathbf{p}|\,T(H_I(t_1)\cdots H_I(t_n))\,|\mathbf{p}'\rangle \tag{S5.2}$$

From Wick's Theorem (8.28), the relevant terms are

$$\langle\mathbf{p}|S-1|\mathbf{p}'\rangle = \sum_{n=1}^{\infty}\frac{(-ig)^n}{2^n n!}\int dt_1\,d^3\mathbf{x}_1\,d^3\mathbf{y}_1\cdots dt_n\,d^3\mathbf{x}_n\,d^3\mathbf{y}_n\,\rho(\mathbf{x}_1)\rho(\mathbf{y}_1)\cdots\rho(\mathbf{x}_n)\rho(\mathbf{y}_n)\,\times$$

$$\langle\mathbf{p}|:\phi(\mathbf{x}_1,t_1)\overbrace{\phi(\mathbf{y}_1,t_1)\phi(\mathbf{x}_2,t_2)}\overbrace{\phi(\mathbf{y}_2,t_2)\phi(\mathbf{x}_3,t_3)}\cdots\overbrace{\phi(\mathbf{y}_{n-1},t_{n-1})\phi(\mathbf{x}_n,t_n)}\phi(\mathbf{y}_n,t_n):|\mathbf{p}'\rangle$$

$$+\ \text{permutations}$$

We deal with the permutations in two steps. First, from Coleman's comment (4), we can cancel the factor of $1/2^n$ (because we can swap x_i and y_i.) Next, there are n pairings ($n-1$ contractions plus one uncontracted pair). These can be arranged in any order, so there are $n!$ ways. This cancels the factor of $1/n!$. Then

$$\langle\mathbf{p}|S-1|\mathbf{p}'\rangle = \sum_{n=1}^{\infty}(-ig)^n\int dt_1\,d^3\mathbf{x}_1\,d^3\mathbf{y}_1\cdots dt_n\,d^3\mathbf{x}_n\,d^3\mathbf{y}_n\,\rho(\mathbf{x}_1)\rho(\mathbf{y}_1)\cdots\rho(\mathbf{x}_n)\rho(\mathbf{y}_n)\,\times$$

$$\langle\mathbf{p}|:\phi(\mathbf{x}_1,t_1)\overbrace{\phi(\mathbf{y}_1,t_1)\phi(\mathbf{x}_2,t_2)}\overbrace{\phi(\mathbf{y}_2,t_2)\phi(\mathbf{x}_3,t_3)}\cdots\overbrace{\phi(\mathbf{y}_{n-1},t_{n-1})\phi(\mathbf{x}_n,t_n)}\phi(\mathbf{y}_n,t_n):|\mathbf{p}'\rangle$$

The contractions are c-numbers, and they can be moved outside the inner product, leaving

$$\langle\mathbf{p}|:\phi(\mathbf{x}_1,t_1)\phi(\mathbf{y}_n,t_n):|\mathbf{p}'\rangle$$

In the notation of (3.33),

$$:\phi(x)\phi(y): \;=\; :(\phi^+(x)+\phi^-(x))(\phi^+(y)+\phi^-(y)): \;=\; \phi^+(x)\phi^+(y)+\phi^-(y)\phi^+(x)+\phi^-(x)\phi^+(y)+\phi^-(x)\phi^-(y)$$

(recall that the annihilation operators are in ϕ^+, and the creation operators in ϕ^-). Sandwiched between $\langle\mathbf{p}|$ and $|\mathbf{p}'\rangle$, the first and last terms give zero. We have already accounted for the two different orderings of x and y, so the normal ordering simply replaces $\phi(\mathbf{x}_1,t_1)$ by $\phi^-(\mathbf{x}_1,t_1)$, and $\phi(\mathbf{y}_n,t_n)$ by $\phi^+(\mathbf{y}_n,t_n)$:

$$\langle\mathbf{p}|S-1|\mathbf{p}'\rangle = \sum_{n=1}^{\infty}(-ig)^n\int dt_1\,d^3\mathbf{x}_1\,d^3\mathbf{y}_1\cdots dt_n\,d^3\mathbf{x}_n\,d^3\mathbf{y}_n\,\rho(\mathbf{x}_1)\rho(\mathbf{y}_1)\cdots\rho(\mathbf{x}_n)\rho(\mathbf{y}_n)\,\times$$

$$\langle\mathbf{p}|\phi^-(\mathbf{x}_1,t_1)\phi^+(\mathbf{y}_n,t_n)|\mathbf{p}'\rangle\,\overbrace{\phi(\mathbf{y}_1,t_1)\phi(\mathbf{x}_2,t_2)}\overbrace{\phi(\mathbf{y}_2,t_2)\phi(\mathbf{x}_3,t_3)}\cdots\overbrace{\phi(\mathbf{y}_{n-1},t_{n-1})\phi(\mathbf{x}_n,t_n)}$$

From (2.53) and (3.33),

$$\langle \mathbf{p}|\phi^-(\mathbf{x}_1,t_1)\phi^+(\mathbf{y}_n,t_n)|\mathbf{p}'\rangle = \langle 0|a_{\mathbf{p}}\left(\int \frac{d^3\mathbf{k}}{(2\pi)^{3/2}\sqrt{2\omega_{\mathbf{k}}}}a^\dagger_{\mathbf{k}}e^{ik\cdot x_1}\right)\left(\int \frac{d^3\mathbf{k}'}{(2\pi)^{3/2}\sqrt{2\omega_{\mathbf{k}'}}}a_{\mathbf{k}'}e^{-ik'\cdot y_n}\right)a^\dagger_{\mathbf{p}'}|0\rangle$$
$$= \int \frac{d^3\mathbf{k}}{(2\pi)^{3/2}\sqrt{2\omega_{\mathbf{k}}}}e^{ik\cdot x_1}\int \frac{d^3\mathbf{k}'}{(2\pi)^{3/2}\sqrt{2\omega_{\mathbf{k}'}}}e^{-ik'\cdot y_n}\langle 0|a_{\mathbf{p}}a^\dagger_{\mathbf{k}}a_{\mathbf{k}'}a^\dagger_{\mathbf{p}'}|0\rangle \tag{S5.3}$$

Since $\langle 0|\, a^\dagger_{\mathbf{k}} = a_{\mathbf{k}'}\,|0\rangle = 0$,

$$\langle 0|a_{\mathbf{p}}a^\dagger_{\mathbf{k}}a_{\mathbf{k}'}a^\dagger_{\mathbf{p}'}|0\rangle = \langle 0|[a_{\mathbf{p}},a^\dagger_{\mathbf{k}}][a_{\mathbf{k}'},a^\dagger_{\mathbf{p}'}]|0\rangle = \delta^{(3)}(\mathbf{p}-\mathbf{k})\,\delta^{(3)}(\mathbf{k}'-\mathbf{p}') \tag{S5.4}$$

so

$$\langle \mathbf{p}|\phi^-(\mathbf{x}_1,t_1)\phi^+(\mathbf{y}_n,t_n)|\mathbf{p}'\rangle = \frac{e^{ip\cdot x_1}}{(2\pi)^{3/2}\sqrt{2\omega_{\mathbf{k}}}}\frac{e^{-ip'\cdot y_n}}{(2\pi)^{3/2}\sqrt{2\omega_{\mathbf{k}'}}} \tag{S5.5}$$

Using (9.29) for the expression of the contractions,

$$\langle \mathbf{p}|\mathrm{S}-1|\mathbf{p}'\rangle = \sum_{n=1}^{\infty}(-ig)^n\int dt_1\, d^3\mathbf{x}_1\, d^3\mathbf{y}_1\cdots dt_n\, d^3\mathbf{x}_n\, d^3\mathbf{y}_n\, \rho(\mathbf{x}_1)\rho(\mathbf{y}_1)\cdots\rho(\mathbf{x}_n)\rho(\mathbf{y}_n)\times$$
$$\frac{e^{ip\cdot x_1}}{\sqrt{(2\pi)^3 2\omega_{\mathbf{p}}}}\frac{e^{-ip'\cdot y_n}}{\sqrt{(2\pi)^3 2\omega_{\mathbf{p}'}}}\prod_{i=1}^{n-1}\int\frac{d^4k_i}{(2\pi)^4}\frac{ie^{-ik_i\cdot(y_i-x_{i+1})}}{k_i^2-\mu^2+i\epsilon}$$

Next, we do the space integrals, using (8.64):

$$\int d^3\mathbf{x}_1\,\rho(\mathbf{x}_1)e^{ip\cdot x_1} = e^{i\omega_{\mathbf{p}}t_1}\tilde\rho(\mathbf{p})^*, \qquad \int d^3\mathbf{x}_{i+1}\,\rho(\mathbf{x}_{i+1})e^{ik_i\cdot x_{i+1}} = e^{ik_i^0 t_{i+1}}\tilde\rho(\mathbf{k}_i)^*$$
$$\int d^3\mathbf{y}_i\,\rho(\mathbf{y}_i)e^{-ik_i\cdot y_i} = e^{-ik_i^0 t_i}\tilde\rho(\mathbf{k}_i), \qquad \int d^3\mathbf{y}_n\,\rho(\mathbf{y}_n)e^{-ip'\cdot y_n} = e^{-i\omega_{\mathbf{p}'}t_n}\tilde\rho(\mathbf{p}')$$

(for $i = 1,\dots,n-1$). Then

$$\langle \mathbf{p}|S-1|\mathbf{p}'\rangle = \sum_{n=1}^{\infty}\frac{(-ig)^n\tilde\rho(\mathbf{p})^*\tilde\rho(\mathbf{p}')}{(2\pi)^3\sqrt{2\omega_{\mathbf{p}}2\omega_{\mathbf{p}'}}}\int dt_1\cdots dt_n\, e^{i\omega_{\mathbf{p}}t_1}e^{-i\omega_{\mathbf{p}'}t_n}\prod_{i=1}^{n-1}\frac{d^4k_i}{(2\pi)^4}\frac{|\tilde\rho(\mathbf{k}_i)|^2\, ie^{-ik_i^0(t_i-t_{i+1})}}{k_i^2-\mu^2+i\epsilon}$$

Now do all the time integrals:

$$\int dt_1\, e^{i(\omega_{\mathbf{p}}-k_1^0)t_1} = 2\pi\delta(\omega_{\mathbf{p}}-k_1^0)$$
$$\int dt_n\, e^{i(k_{n-1}^0-\omega_{\mathbf{p}'})t_n} = 2\pi\delta(k_{n-1}^0-\omega_{\mathbf{p}'})$$
$$\int dt_i\, e^{i(k_{i-1}^0-k_i^0)t_i} = 2\pi\delta(k_{i-1}^0-k_i^0)\quad (\text{for } i=2,\dots,i-1)$$

These time integrals yield a product of delta functions which simplifies enormously:

$$\delta(\omega_{\mathbf{p}}-k_1^0)\,\delta(k_1^0-k_2^0)\,\delta(k_2^0-k_3^0)\cdots\delta(k_{n-2}^0-k_{n-1}^0)\,\delta(k_{n-1}^0-\omega_{\mathbf{p}'}) = \delta(\omega_{\mathbf{p}}-\omega_{\mathbf{p}'})\prod_{i=1}^{n-1}\delta(\omega_{\mathbf{p}}-k_i^0)$$

because $\delta(x-a)\,\delta(x-b) = \delta(x-a)\,\delta(a-b)$. This leaves

$$\langle \mathbf{p}|\mathrm{S}-1|\mathbf{p}'\rangle = 2\pi\delta(\omega_{\mathbf{p}}-\omega_{\mathbf{p}'})\frac{\tilde\rho(\mathbf{p})^*\tilde\rho(\mathbf{p}')}{(2\pi)^3 2\omega_{\mathbf{p}}}\sum_{n=1}^{\infty}(-ig)^n i^{n-1}\prod_{i=1}^{n-1}\int\frac{d^4k_i}{(2\pi)^4}\frac{|\tilde\rho(\mathbf{k}_i)|^2}{k_i^2-\mu^2+i\epsilon}2\pi\delta(\omega_{\mathbf{p}}-k_i^0)$$

Now do all the k_i^0 integrals. This sends every k_i^0 to $\omega_{\mathbf{p}}$:

$$\langle \mathbf{p}|\mathrm{S}-1|\mathbf{p}'\rangle = -ig\,\delta(\omega_{\mathbf{p}}-\omega_{\mathbf{p}'})\frac{\tilde\rho(\mathbf{p})^*\tilde\rho(\mathbf{p}')}{(2\pi)^2 2\omega_{\mathbf{p}}}\sum_{n=1}^{\infty}\prod_{i=1}^{n-1}g\int\frac{d^3\mathbf{k}_i}{(2\pi)^3}\frac{|\tilde\rho(\mathbf{k}_i)|^2}{\omega_{\mathbf{p}}^2-|\mathbf{k}_i|^2-\mu^2+i\epsilon}$$

The remaining $\mathbf{k}_i$ integrals are all identical. Define

$$G(\omega_{\mathbf{p}}) = g\int\frac{d^3\mathbf{k}}{(2\pi)^3}\frac{|\tilde\rho(\mathbf{k})|^2}{\omega_{\mathbf{p}}^2-|\mathbf{k}|^2-\mu^2+i\epsilon} = g\int\frac{d^3\mathbf{k}}{(2\pi)^3}\frac{|\tilde\rho(\mathbf{k})|^2}{\omega_{\mathbf{p}}^2-\omega_{\mathbf{k}}^2+i\epsilon} \tag{S5.6}$$

Then we have

$$\langle \mathbf{p}|S-1|\mathbf{p}'\rangle = -\frac{ig\tilde{\rho}(\mathbf{p})^*\tilde{\rho}(\mathbf{p}')}{8\pi^2\omega_{\mathbf{p}}}\delta(\omega_{\mathbf{p}}-\omega_{\mathbf{p}'})\sum_{n=1}^{\infty}[G(\omega_{\mathbf{p}})]^{n-1} \tag{S5.7}$$

Summing the (assumed convergent!) geometric series gives

$$\langle \mathbf{p}|S-1|\mathbf{p}'\rangle = -\frac{ig\tilde{\rho}(\mathbf{p})^*\tilde{\rho}(\mathbf{p}')}{8\pi^2\omega_{\mathbf{p}}}\delta(\omega_{\mathbf{p}}-\omega_{\mathbf{p}'})\frac{1}{1-G(\omega_{\mathbf{p}})} \tag{S5.8}$$

Comparing this with (P5.1), we conclude

$$F(\omega_{\mathbf{p}}) = -\frac{ig}{8\pi^2\omega_{\mathbf{p}}}\frac{1}{1-G(\omega_{\mathbf{p}})} \tag{S5.9}$$

with $G(\omega_{\mathbf{p}})$ given by (S5.6). ∎

Alternate solution (S. Coleman)

The problem can be solved graphically much more quickly. The fundamental vertex is

$$\overset{\leftarrow p' \quad \leftarrow p}{\bullet\rule{1cm}{0.4pt}} = -ig(2\pi)\delta(\omega_{\mathbf{p}}-\omega_{\mathbf{p}'})\tilde{\rho}(\mathbf{p})\tilde{\rho}(\mathbf{p}')^* \tag{S5.10}$$

where

$$\tilde{\rho}(\mathbf{p}) = \int d^3x\, e^{i\mathbf{k}\cdot\mathbf{x}}\rho(\mathbf{x}) = \tilde{\rho}(-\mathbf{k})^*$$

The propagator is, as before,

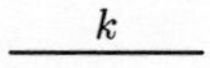

$$\rule{2cm}{0.4pt}\overset{k}{} = \frac{i}{k^2-\mu^2+i\epsilon}$$

Then

$$\langle \mathbf{p}'|S-1|\mathbf{p}\rangle = \rule{1cm}{0.4pt}\bullet\rule{0.5cm}{0.4pt} + \rule{0.5cm}{0.4pt}\bullet\rule{0.5cm}{0.4pt}\bullet\rule{0.5cm}{0.4pt} + \rule{0.5cm}{0.4pt}\bullet\rule{0.4cm}{0.4pt}\bullet\rule{0.4cm}{0.4pt}\bullet\rule{0.5cm}{0.4pt} + \cdots$$ 

$$= -ig(2\pi)\delta(\omega_{\mathbf{p}}-\omega_{\mathbf{p}'})\tilde{\rho}(\mathbf{p})\tilde{\rho}(\mathbf{p}')^*\underbrace{\frac{1}{2\omega_{\mathbf{p}}(2\pi)^3}}_{\substack{\text{non-rel.}\\\text{normalized}}}\sum_{n=0}^{\infty}\underbrace{\left[-ig\int\frac{d^3k}{(2\pi)^3}|\tilde{\rho}(\mathbf{k})|^2\frac{i}{\omega_{\mathbf{p}}^2-|\mathbf{k}|^2-\mu^2+i\epsilon}\right]^n}_{G(\omega_{\mathbf{p}})}$$

$$= \frac{-ig\tilde{\rho}(\mathbf{p})\tilde{\rho}(\mathbf{p}')^*\delta(\omega_{\mathbf{p}}-\omega_{\mathbf{p}'})}{8\pi^2\omega_{\mathbf{p}}(1-G(\omega_{\mathbf{p}}))} \tag{S5.11}$$

in agreement with the earlier answer, (S5.8). ∎

(b) We need to show

$$\langle \mathbf{p}|SS^\dagger-1|\mathbf{p}'\rangle = 0 \tag{P5.2}$$

Some preliminary identities. First,

$$SS^\dagger - 1 = (S-1) + (S^\dagger-1) + (S-1)(S^\dagger-1)$$

and

$$\langle \mathbf{p}|SS^\dagger-1|\mathbf{p}'\rangle = \langle \mathbf{p}|S-1|\mathbf{p}'\rangle + \langle \mathbf{p}|S^\dagger-1|\mathbf{p}'\rangle + \langle \mathbf{p}|(S-1)(S^\dagger-1)|\mathbf{p}'\rangle \tag{S5.12}$$

From (S5.8),

$$\langle \mathbf{p}|S^\dagger-1|\mathbf{p}'\rangle = \langle \mathbf{p}'|S-1|\mathbf{p}\rangle^* = \frac{ig\tilde{\rho}(\mathbf{p}')\tilde{\rho}(\mathbf{p})^*}{8\pi^2\omega_{\mathbf{p}}}\delta(\omega_{\mathbf{p}}-\omega_{\mathbf{p}'})\frac{1}{1-G(\omega_{\mathbf{p}})^*}$$

so

$$\langle \mathbf{p}|S^\dagger-1|\mathbf{p}'\rangle + \langle \mathbf{p}|S-1|\mathbf{p}'\rangle = \frac{ig\tilde{\rho}(\mathbf{p})^*\tilde{\rho}(\mathbf{p}')}{8\pi^2\omega_{\mathbf{p}}}\delta(\omega_{\mathbf{p}}-\omega_{\mathbf{p}'})\left[\frac{1}{1-G(\omega_{\mathbf{p}})^*} - \frac{1}{1-G(\omega_{\mathbf{p}})}\right]$$

$$= \frac{ig\tilde{\rho}(\mathbf{p})^*\tilde{\rho}(\mathbf{p}')}{8\pi^2\omega_{\mathbf{p}}|1-G(\omega_{\mathbf{p}})|^2}\delta(\omega_{\mathbf{p}}-\omega_{\mathbf{p}'})\,g\int\frac{d^3k}{(2\pi)^3}|\tilde{\rho}(\mathbf{k})|^2\left[\frac{1}{\omega_{\mathbf{p}}^2-\omega_{\mathbf{k}}^2-i\epsilon} - \frac{1}{\omega_{\mathbf{p}}^2-\omega_{\mathbf{k}}^2+i\epsilon}\right]$$

Using (P4.1) and footnote 8, p. 9 we can write

$$\lim_{\epsilon\to 0}\frac{1}{\omega_{\mathbf{p}}^2-\omega_{\mathbf{k}}^2-i\epsilon} - \frac{1}{\omega_{\mathbf{p}}^2-\omega_{\mathbf{k}}^2+i\epsilon} = 2\pi i\delta(\omega_{\mathbf{p}}^2-\omega_{\mathbf{k}}^2) = 2\pi i\left(\frac{\delta(\omega_{\mathbf{p}}-\omega_{\mathbf{k}})}{2\omega_{\mathbf{p}}} + \frac{\delta(\omega_{\mathbf{p}}+\omega_{\mathbf{k}})}{2\omega_{\mathbf{p}}}\right)$$

Because $\omega_{\mathbf{p}}$ and $\omega_{\mathbf{k}}$ are both positive, $\delta(\omega_{\mathbf{p}} + \omega_{\mathbf{k}}) = 0$. Then

$$\langle \mathbf{p}|\mathrm{S}^\dagger - 1|\mathbf{p}'\rangle + \langle \mathbf{p}|\mathrm{S} - 1|\mathbf{p}'\rangle = \frac{ig\tilde{\rho}(\mathbf{p})^*\tilde{\rho}(\mathbf{p}')}{8\pi^2\omega_{\mathbf{p}}|1 - G(\omega_{\mathbf{p}})|^2}\delta(\omega_{\mathbf{p}} - \omega_{\mathbf{p}'})\left[\frac{ig}{8\pi^2\omega_{\mathbf{p}}}\int d^3\mathbf{k}\,|\tilde{\rho}(\mathbf{k})|^2\delta(\omega_{\mathbf{p}} - \omega_{\mathbf{k}})\right] \tag{S5.13}$$

The remaining term of (S5.12) can be expressed as

$$\langle \mathbf{p}|(\mathrm{S} - 1)(\mathrm{S}^\dagger - 1)|\mathbf{p}'\rangle = \int d^3\mathbf{k}\,\langle \mathbf{p}|(\mathrm{S} - 1)|\mathbf{k}\rangle\,\langle \mathbf{k}|(\mathrm{S}^\dagger - 1)|\mathbf{p}'\rangle = \int d^3\mathbf{k}\,\langle \mathbf{k}|\mathrm{S} - 1|\mathbf{p}\rangle^*\,\langle \mathbf{k}|\mathrm{S} - 1|\mathbf{p}'\rangle$$

$$= \frac{ig\tilde{\rho}(\mathbf{p})^*\tilde{\rho}(\mathbf{p}')}{8\pi^2\omega_{\mathbf{p}}(1 - G(\omega_{\mathbf{p}})^*)(1 - G(\omega_{\mathbf{p}'}))}\left[\frac{-ig}{8\pi^2}\int \frac{d^3\mathbf{k}}{\omega_{\mathbf{p}'}}\,\tilde{\rho}(\mathbf{k})\delta(\omega_{\mathbf{p}} - \omega_{\mathbf{k}})\tilde{\rho}(\mathbf{k})^*\delta(\omega_{\mathbf{p}'} - \omega_{\mathbf{k}})\right]$$

$$= \frac{ig\tilde{\rho}(\mathbf{p})^*\tilde{\rho}(\mathbf{p}')}{8\pi^2\omega_{\mathbf{p}}|1 - G(\omega_{\mathbf{p}})|^2}\delta(\omega_{\mathbf{p}} - \omega_{\mathbf{p}'})\left[\frac{-ig}{8\pi^2\omega_{\mathbf{p}}}\int d^3\mathbf{k}\,|\tilde{\rho}(\mathbf{k})|^2\delta(\omega_{\mathbf{p}} - \omega_{\mathbf{k}})\right] \tag{S5.14}$$

The right-hand side of (S5.12) is equal to the sum of (S5.13) and (S5.14), but these cancel, so the left-hand side of (S5.12) is zero. That establishes (P5.2). The pair model S matrix is unitary. ■

(For more about the pair model, see Schweber *RQFT*, Section 12c, "Other simple models", p. 371, and references therein.)

10

Mass renormalization and Feynman diagrams

I now want to consider our third model,

$$\mathscr{H}_I = g\psi^*\psi\phi f(t) \tag{10.1}$$

There are two new features that come up here. First, there are new problems arising in the same way that the energy-shift problem arose last lecture. These problems are subsumed under the term **mass renormalization**. Mass renormalization is unfortunately a term used in two senses, both for the phenomenon that occurs, and for the prescription we follow to deal with the phenomenon, a prescription for adding counterterms to the Lagrangian. It's a peculiar linguistic situation in which the disease and the cure have the same name. The second new feature is that our Wick graphs will not have the extraordinarily simple structure they had in the previous case. In Model 3 we no longer have a finite family of connected Wick graphs, but an infinite family. We therefore have no hope of computing the S-matrix in closed form as we did for the previous two examples, at least not by these methods, nor by any methods known to man or woman. I will not speak about alien life-forms; they may be cleverer.[1] All we can do is settle down with a specific matrix element for a particular scattering process and compute it, order by order in perturbation theory, until we reach the limits of our computational abilities. It will prove convenient to use not Wick diagrams, but another kind of diagram, a *Feynman diagram* (also called a *Feynman graph*), that represents the contribution of a Wick operator to a particular matrix element. These two topics will occupy this lecture. I'd first like to begin by discussing mass renormalization.

10.1 Mass renormalization in Model 3

This subject has an interesting history. Let me begin with a rigid sphere immersed in a fluid, a problem considered by George Green of Nottingham, he for whom Green's functions are named. He published the results of his investigation on the motion of a pendulum in an ideal fluid, in 1834, in the *Transactions of the Royal Society of Edinburgh*.[2] Green's problem can be posed in the following way.

---

[1] [Eds.] Coleman was a passionate fan of science fiction.

[2] [Eds.] George Green, "Researches on the vibrations of pendulums in fluid media", *Trans. Roy. Soc. Edin.* **13** (1834) 54–63. Reprinted in *Mathematical Papers of George Green*, ed. N. M. Ferrers, AMS/Chelsea Publishing Company, 1970. Green (1793–1841), by profession a miller and almost entirely self-taught in mathematics and

Suppose I have a rigid sphere of volume V, say a small, spherical zeppelin filled with hydrogen or some other very light gas, immersed in a perfect fluid of density ρ, with zero viscosity. Let's say that

$$m_0 = \tfrac{1}{10}\rho V \tag{10.2}$$

so that the sphere has a mass of one tenth of the volume it displaces. Now if we do an elementary statics calculation on this object, there is a gravitational force $m_0 g$ pulling downwards, and an Archimedian buoyancy force $10\,m_0 g$ pushing upwards. If we let go of the sphere, we should observe an upward acceleration of the object, the net force over its mass, equal to $9g$.

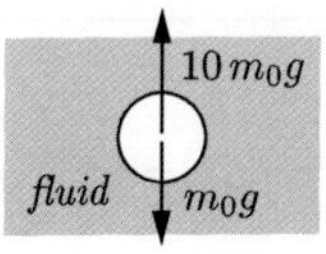

Figure 10.1: Sphere in fluid

Now if you've ever let go of a ping-pong ball which you have held at the bottom of a swimming pool, or in a sink, you will know that this is grossly in error. The ping-pong ball does not go up with an acceleration of $9g$. You might at first ascribe this effect to fluid friction. But that can't be so during the early stage of the motion, because all such frictional forces are proportional to the velocity. Until the system builds up some substantial velocity, friction cannot be important. It's important in the late stages of motion as the ping-pong ball approaches terminal velocity, but not in the early stages.

Green discovered while he was doing the small vibration problem (which should be good enough for the early stages of the motion) a remarkable result, which I will quote: "Hence in this last case [of a spherical mass] we shall have the true time of the pendulum's vibration, if we suppose it to move *in vacuo*, and then simply conceive its mass augmented by half that of an equal volume of the fluid, whilst the moving force with which it is actuated is diminished by the whole weight of the same volume of fluid." Green's result says that there was actually an *effective* mass—what we might call the *physical* mass, the only mass we could measure if we couldn't take the ping-pong ball out of the water; if, say, the universe were filled with water. That effective mass m is

$$m = m_0 + \tfrac{1}{2}\rho V \tag{10.3}$$

and equal to $6\,m_0$, if the fluid's density is ten times the sphere's. Consequently the ping-pong ball's acceleration in the perfect fluid should be

$$a = \frac{F_{net}}{m} = \frac{9m_0 g}{6m_0} = \tfrac{3}{2}g \tag{10.4}$$

a much more reasonable result than the $9g$ we obtained naïvely.

The physical explanation for this phenomenon was given a decade later in a review article by Stokes,[3] well known as the inventor of Stokes' theorem. Stokes pointed out that if you

physics, was completely unknown when he self-published *An essay on the application of mathematical analysis to the theories of electricity and magnetism* in 1828. Einstein declared the *Essay* twenty years ahead of its time. After the success of his *Essay*, Green was urged to attend Cambridge University, and did so, entering as an undergraduate at the age of 39. No portrait or other likeness of Green is known. See D. Mary Cannell, *George Green, Mathematician and Physicist 1793–1841*, The Athlone Press, 1993; Julian Schwinger, "The Greening of Quantum Field Theory: George and I", `https://arxiv.org/pdf/hep-ph/9310283.pdf`.

[3] [Eds.] G. G. Stokes, "Memoir in some cases of fluid motion", *Trans. Camb. Phil. Soc.* **VIII** (1849) 105–137. The paper was presented on May 29, 1843. Reprinted in v. I of Stokes' *Mathematical and Physical Papers*, Cambridge U. P., 1880.

imagine a rigid sphere moving through a fluid with some velocity $\mathbf{v}$, the fluid can't just stand there, because it's got to move to get around the sphere. As you know, there is a pattern of flow set up in the fluid, which you might have looked at in earlier courses. If we were to calculate the total momentum $\mathbf{p}_{\text{total}}$ of this equilibrium configuration, we would find that the momentum is $m_0\mathbf{v}$ plus the fluid momentum, which is expressed in terms of the zeroth and first Legendre polynomials only, as I recall, to get the velocity potential for the fluid. You integrate this velocity potential to get the momentum in the fluid. If you do this integral, you obtain an answer $m\mathbf{v}$, where m is defined in (10.3):

$$\mathbf{p}_{\text{total}} = m_0\mathbf{v} + \mathbf{p}_{\text{fluid}} = m_0\mathbf{v} + \tfrac{1}{2}\rho V\mathbf{v} + \mathcal{O}(|\mathbf{v}|^2) = m\mathbf{v} + (\text{higher orders of } |\mathbf{v}|) \tag{10.5}$$

The response of the system to a small external force is the derivative of the total momentum with respect to the velocity, and you obtain m, not m_0. Thus what we have here is a system something like our ping-pong ball, interacting with a continuum system, in this case an ideal fluid. And we find the mass of the system is changed, as a result of its interactions with the continuum.[4]

The next time this idea enters physics is in the electron theory of Lorentz,[5] much later in the 19th century, and Abraham's work[6] on the electron theory of Lorentz. Lorentz thought of the electron as a rigid body of some characteristic radius r carrying a charge, e. He observed that if you computed the momentum of such a body in steady motion—nowadays we know about relativity, and we do it more easily just by computing the mass—you would obtain not only the energy of the body at rest, but also the energy of the attached Coulomb field integrated over all space. This contribution will be equal to some constant k, depending upon whether it's a spherical shell of charge or a uniformly distributed sphere of charge, times e^2/r:

$$E = m_0c^2 + \frac{ke^2}{r} \tag{10.6}$$

If you put an electron on a scale, you will not weigh the electron by itself, but the electron with its associated electromagnetic field. Your scale tells you the combined mass of the two things:

$$m = m_0 + \frac{ke^2}{rc^2} \tag{10.7}$$

Likewise, if you attempt to accelerate an electron, you are not only putting the electron into steady motion, you are moving the associated Coulomb field. You don't leave the Coulomb field behind when you give the electron a little push; the field moves with it. Therefore you get not just the momentum of the moving m_0, a rigid body, but also the momentum of the electromagnetic field that moves with it.

---

[4] [Eds.] See also Lev D. Landau and Evgeniĭ M. Lifshitz, *Fluid Mechanics*, Pergamon Press, 1966, §11 and the following Problem 1, pp. 31–36; Kerson Huang, *Statistical Mechanics*, 2nd ed., John Wiley & Sons, 1987, Section 5.9, "Examples in Hydrodynamics", pp. 117–119.

[5] [Eds.] H. A. Lorentz, *The Theory of Electrons and its Applications to the Phenomena of Light and Radiant Heat (a course of lectures delivered at Columbia University, New York, March and April, 1906)*, 2nd ed., B. G. Teubner, 1916. Reprinted by Dover Publications, 2011.

[6] [Eds.] For some background on the electron theory of Max Abraham and H. A. Lorentz, see *Intellectual Mastery of Nature*, v. 2, C. Jungnickel and R. McCormmach, U. of Chicago Press, 1990, pp. 231–241. Abraham's revision of A. Föppl's influential text *Theorie der Elektrizität* is regarded as a classic, and was itself revised by R. Becker. Though an expert on relativity, Abraham believed in the luminiferous aether.

Thus in general whenever we have a particle interacting with a continuum system, its mass is changed from what it would be if the interaction with the continuum system were not present, whether the continuum system is the classical hydrodynamic field or Maxwell's electrodynamics. (We didn't really need this historical introduction; I just can't resist talking about Green and Stokes.)

Now let's consider the theory we have to worry about. I'll just focus for the moment on the meson mass. We have a Lagrangian

$$\mathscr{L} = \tfrac{1}{2}(\partial^\mu\phi)^2 - \tfrac{1}{2}\mu_0^2\phi^2 - g\psi^*\psi\phi + \cdots \tag{10.8}$$

plus nucleon terms, denoted by the dots, which I won't bother to write down at the moment. In honor of the previous discussion, I've written the quantity that multiplies ϕ^2 as μ_0^2, rather than μ^2, because after all μ_0 is the mass in the absence of any interactions.

Now there is absolutely no reason to believe that in the presence of the interaction, the square of the meson mass will be μ_0^2. That is unquestionably what the meson mass would be, if the interaction were turned off. We solved that theory,[7] and we found out that the coefficient of ϕ^2 is the meson mass. However, just as interactions with the hydrodynamic field and the electromagnetic field change the effective mass, respectively, of ping-pong balls immersed in water, and charged shells within an electromagnetic field, so we would expect the interactions with the nucleons to change the mass of the meson. If we were able to solve this theory exactly, and we looked at the one-meson state with momentum zero, we've no reason to expect its mass to be μ_0. It will be some dynamically determined number, and it's a complicated computation to figure out what it is. So the actual, *physical* mass of the meson, μ^2, is in general *not* equal to μ_0^2:

$$\mu^2 \neq \mu_0^2 \quad \text{in the presence of the interaction.} \tag{10.9}$$

This is not only an interesting phenomenon, it is also a problem for a scattering theory in the same way the energy mismatch for the vacuum state was a problem. If we arrange to turn on the interaction adiabatically, the mass, and therefore the energy, of a single-meson state coming in will change. Even an isolated single-meson state, even if it's far from anything, will develop a phase just as the vacuum state developed a phase, in the course of turning the interaction on and off. When we compute the one-meson-to-one-meson S-matrix element, we should find it equal to 1 for the same reason the vacuum-to-vacuum S-matrix element is 1. If the universe is empty except for a single meson, the meson is not going to scatter off of anything, it's just going to go on. In fact we will not get 1, but instead some preposterous phase factor involving the length of time T during which we've kept the interaction on.

We will avoid that difficulty by introducing counterterms. Consider the following Lagrangian:

$$\mathscr{L} = \tfrac{1}{2}(\partial_\mu\phi)^2 - \tfrac{1}{2}\mu^2\phi^2 + \partial^\mu\psi^*\partial_\mu\psi - m^2\psi^*\psi + f(t)\left[-g\psi^*\psi\phi + a + \tfrac{1}{2}b\phi^2 + c\psi^*\psi\right] \tag{10.10}$$

so that the interaction Hamiltonian density is

$$\mathscr{H}_I = -f(t)\left[-g\psi^*\psi\phi + a + \tfrac{1}{2}b\phi^2 + c\psi^*\psi\right] \tag{10.11}$$

[7] [Eds.] See §4.4.

Note that I've written μ^2, not μ_0^2, and m^2 instead of m_0^2. I still need my old-fashioned vacuum counterterm a (I'll come back and make further remarks about that). I added a counterterm $\frac{1}{2}b\phi^2$, which will have to do with the mass of the meson, and a counterterm $c\psi^*\psi$, to do with the mass of the "nucleon". It doesn't matter whether these are taken to be positive or negative; there is no standard convention about their signs.

The functions of these b and c counterterms are to adjust matters so that the masses of the meson and the "nucleon" stay the same as I turn on the interaction, just as the function of the a counterterm is to adjust matters so the energy of the vacuum stays the same when I turn on the interaction. The mass of the meson begins to change, because there's an interaction. That won't bother me. I'll just turn on b at the same time with just the bare mass, μ_0^2, keeping in step, so that the physical mass always stays equal to μ^2. It's μ^2 when the interaction is off, and μ^2 when the interaction is on. I should say it's μ^2 in the average sense. It ranges so that the phase mismatch integrates to zero, just as for the vacuum state I arrange matters so the phase mismatch integrates to zero. The same procedure holds for the mass of the "nucleon".[8]

I should make one technical remark. Please notice here we have added a to the Hamiltonian density, not to the Hamiltonian as before. The reason is that the vacuum is a homogeneous system in space, of infinite spatial extent, so we would expect not to find a finite total energy shift, but instead an energy shift per unit volume, just as if we had an infinite crystal and changed the strength of the electromagnetic interactions a tiny bit. The energy of the whole crystal would change by an infinite amount because it's infinite! It's the energy per cubic centimeter that we hope to change by a finite amount, and since this is a spatially infinite system, I have added my counterterm to the Hamiltonian density rather than to the Hamiltonian.

Now these three additional terms I have added, a, b and c, are of course not free parameters. They are completely determined. The counterterm a is determined by the requirement that $\langle 0|\mathrm{S}|0\rangle$ equals one:

$$\langle 0|\mathrm{S}|0\rangle = 1 \;\Rightarrow\; a \tag{10.12}$$

I can obtain a to any order in perturbation theory by computing the vacuum-to-vacuum matrix element to that order in perturbation theory. The b counterterm is determined by the condition that there be no phase mismatch between one-meson states, $|\mathbf{q}\rangle$ and $|\mathbf{q}'\rangle$,

$$\langle \mathbf{q}|\mathrm{S}|\mathbf{q}'\rangle = \delta^{(3)}(\mathbf{q}-\mathbf{q}') \;\Rightarrow\; b \tag{10.13}$$

This condition determines b completely. I compute the phase I would get for the one-meson-to-one-meson amplitude, and I force it to be one. If I have computed that phase to some order in perturbation theory, I've fixed b to that order in perturbation theory. Likewise, this fixes c, again to any order in my expansion:

$$\langle \mathbf{p}|\mathrm{S}|\mathbf{p}'\rangle = \delta^{(3)}(\mathbf{p}-\mathbf{p}') \;\Rightarrow\; c \tag{10.14}$$

where $|\mathbf{p}\rangle$ and $|\mathbf{p}'\rangle$ are one-"nucleon" states. Unfortunately my notation is not well enough developed so that you can see at a glance whether a given ket is a one-nucleon state or a one-meson state. (I'll use $\mathbf{p}$'s for nucleon momenta and $\mathbf{q}$'s for meson momenta, as a visual

[8] [Eds.] The reader may be wondering if counterterms are to be added at every difficulty, and if the addition of these terms is going to have unwanted side-effects. In renormalizable theories, the number of counterterms is finite, and their addition will not alter the physics. Much more will be said about renormalization later in the course, in Chapters 15, 16, 25, and 33.

aid.) Furthermore, not only do these conditions fix these counterterms, as they are in principle computable quantities, but they allow me to answer questions. For example, assuming this were a realistic theory of the world, I can ask: what is the bare mass of the meson if I know its physical mass? How much of its mass is due to its interactions and how much of its mass was given to us by God before the interactions are turned on? I can compute that bare mass. I see from the terms in the Lagrangian that

$$\mu_0^2 = \mu^2 - b \quad \text{and} \quad m_0^2 = m^2 - c \tag{10.15}$$

So if I want to compute the masses, I have a systematic way of computing them order by order in perturbation theory.

You may wonder: Is this all? Have I gotten rid of all mismatches in phase, energy and mass? Well, of course we can't really tell until I do computations, or else put our scattering theory on a firmer foundation than we have now, with this dumb $f(t)$ function. But it looks plausible. I've arranged things so that there's no energy mismatch between the interacting vacuum and the bare vacuum; nor between a physical one-meson state and a bare one-meson state, or between a physical one-nucleon state and a bare one-nucleon state. If I have a scattering state that's 32 mesons and 47 nucleons all coming in from the far past, all thousands of light years away from each other, then the energy of the multiparticle state is simply the sum of the energies of the single particles. That's an empirical fact. With these counterterms I have arranged that the energies of the single particles are all coming out right, so the energy of the multiparticle state should also come out right. It looks like these three counterterms are sufficient to take care of all of our problems of mismatch. Later on we will discover if this is right or not, after we have put our scattering theory on a firmer foundation. Then we will see just how many we need.

But for the moment things look good, so keeping my fingers crossed I will run with this Lagrangian. That takes care of the first topic.

10.2 Feynman rules in Model 3

Now for the second topic. We know what our Lagrangian looks like, and now I'm going to talk about the diagrammatic representation of that Lagrangian. I will now explain what Feynman diagrams are.

I wish to compute matrix elements of the S-matrix between particular states. Remember, every term in the Wick expansion will in general contribute to many independent scattering processes depending upon whether we use the loose external lines to create a particle or annihilate a particle. I would now like to write down a different sort of diagram for matrix elements. For example, let's consider a process in which a nucleon with momentum $\mathbf{p}_1$ plus a nucleon with momentum $\mathbf{p}_2$ goes into a nucleon with momentum $\mathbf{p}_1'$ plus a nucleon with momentum $\mathbf{p}_2'$, to order $\mathcal{O}(g^2)$, just for simplicity:

$$N(\mathbf{p}_1) + N(\mathbf{p}_2) \to N(\mathbf{p}_1') + N(\mathbf{p}_2') \tag{10.16}$$

Let's consider the $\mathcal{O}(g^2)$ contribution (the lowest non-zero order) to the S-matrix elements

$$\langle p_1', p_2' | (\mathrm{S} - 1) | p_1, p_2 \rangle \tag{10.17}$$

There's always a 1 term so I'll just subtract that out. These kets are relativistically normalized states. There will be a variety of Wick diagrams that may contribute to (10.17). I'll write down the ones to order g^2, neglecting the effects of the counterterms for the moment. (I'll talk about the counterterms later.)

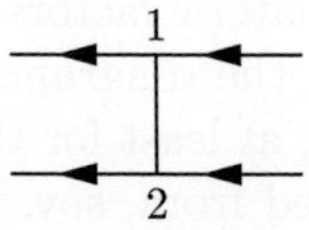

Figure 10.2: $\mathcal{O}(g^2)$ Wick diagram for Model 3 nucleon–nucleon scattering

Just to remind you, the arrow going into a vertex corresponds to a field ψ, the arrow coming out of a vertex corresponds to a field ψ^*, and the line at a vertex without an arrow corresponds to a field ϕ.

The term of $\mathcal{O}(g^2)$ in S − 1 is

$$\tfrac{1}{2}(-ig)^2 \int d^4x_1\, d^4x_2\, T\left(\psi(x_1)^*\psi(x_1)\phi(x_1)\psi(x_2)^*\psi(x_2)\phi(x_2)\right) \qquad (10.18)$$

Notice that there is no sign of the adiabatic function, $f(t)$. After all the hoopla about $f(t)$, we will (knock heavily on wood!) simply go to the limit $f(t) \to 1$. I think I've taken account of all the residual effects that come from $f(t)$ with my renormalization counterterms. Later on we will worry a great deal about whether this is legitimate.[9]

In the original Wick diagrams, it didn't matter how I had the external lines sticking out from the diagram. The external lines of the new diagrams will be oriented, following particular conventions. All the fields that are going to annihilate a particle in the initial state I will put on the right, where the initial state is.[10] All the fields that are going to create a particle in the final state, I'll put on the left, where the final state is. Then I will label the external lines with the momentum of the particles they are annihilating and creating. For example, I'll write down two typical diagrams, (a) and (b), for this process (there are actually four; the other two, (c) and (d), are obtained by permuting the vertices in (a) and (b), respectively):

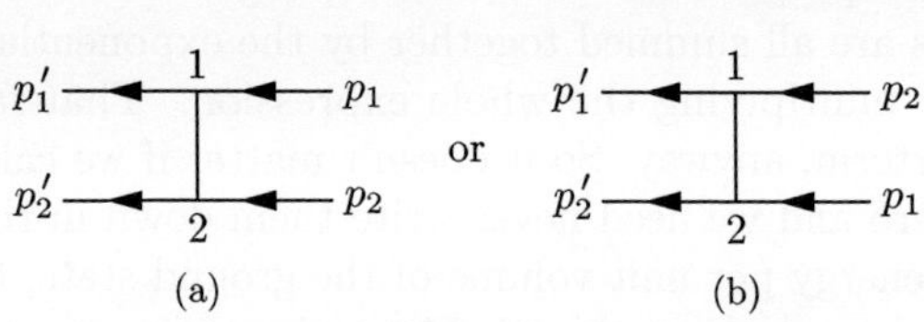

Figure 10.3: $\mathcal{O}(g^2)$ Momentum diagrams for Model 3 nucleon–nucleon scattering

In Diagram (a), I can use the free nucleon field at 1 to annihilate a nucleon of momentum p_1 and the free nucleon field at 2 to annihilate a nucleon of momentum p_2. I can use the free antinucleon field at 1 to create a nucleon of momentum p_1' and the free antinucleon field at 2 to create a nucleon of momentum p_2'. Thus the initial state $|p_1, p_2\rangle$ goes into the final state $|p_1', p_2'\rangle$. There are of course 3 other ways of doing this, even with only this single Wick diagram. I could for example produce an alternative Diagram (b) with p_1 and p_2 swapped,

---

[9] [Eds]. See note 3, p. 186. Even if $f(t)$ has been set equal to 1, a is still needed in Model 3 to ensure that the physical vacuum's energy is equal to zero.

[10] [Eds.] Coleman puts *initial* states on the *right* and *final* states on the *left*, in effect choosing a time axis running right to left. So his diagrams should be read right to left. Though unconventional, this choice aligns with matrix elements $\langle f|\mathrm{S}|i\rangle$.

where I use the field at 1 to annihilate the nucleon state with momentum p_2 and the field at 2 to annihilate the nucleon state with momentum p_1, the reverse of the previous situation.

I would like to discuss first the combinatoric factors associated with these kinds of diagrams, and second, how you actually evaluate the diagrams. The combinatoric factors are much simpler than they are for Wick diagrams, at least for this theory. The reason is very simple. If we look at diagrams of this sort obtained from, say, Diagram (a) by permuting the indices, we notice that all the vertices are uniquely labeled, assuming for the moment that p_1 is not equal to p_2, and p'_1 is not equal to p'_2. (I'm not excluding forward scattering; p_1 could equal p'_1. All I'm excluding is scattering at threshold which is after all only a single point, and we can always get to it by continuity. The two four-momenta are equal near a threshold at the center of mass, where the particles are mutually at rest with respect to one another.)

Vertex 1 is uniquely labeled as the vertex where the nucleon of momentum p_1 is absorbed. With a few exceptions, once I have labeled one vertex uniquely in a diagram, all the other vertices are uniquely labeled. In the present example, 2 is a vertex you get to by following the meson line from the vertex where p_1 is absorbed. If it were a much more complicated diagram you could just trace through it: follow the nucleon line along the arrow, follow the nucleon line against the arrow, follow the meson line. As soon as you label one vertex uniquely every other vertex is labeled uniquely by such a set of instructions. The corresponding diagram we'd get by permuting 1 and 2 would be a different term—still the same term in the Wick expansion, but a completely different contribution, though numerically equal—and would precisely cancel out the 2! in Dyson's formula, and in general cancel the $n!$ in a complicated diagram. Henceforth we *erase the labeling on the vertices* and just drop the factor of $1/n!$. Diagrams of this sort, without labeled vertices but with labeled ends, are called **Feynman diagrams**.

For this theory, Model 3, the one over $n!$ in Dyson's formula is canceled for Feynman diagrams, except if you are considering a diagram that contains a disconnected component with no external lines; these contribute to vacuum-to-vacuum amplitudes. Within that disconnected component there's nothing that's absorbing anything or emitting anything, and you may have trouble labeling vertices uniquely. However this isn't going to trouble us, mostly because those disconnected components are all summed together by the exponentiation theorem, (8.49), to make a numerical factor multiplying the whole expression. That factor is supposed to be canceled by the a counterterm, anyway. So it doesn't matter if we calculate them correctly or not, as they sum up to zero and we need never write them down in the first place. If however we want to calculate the energy per unit volume of the ground state, following our calculation in the last lecture, we do have to keep the combinatoric structure straight. But if we're only interested in computing S-matrix elements, all of those things cancel among themselves, and we don't have to worry about them.

You can have residual combinatoric factors left over in other theories. For example, a ϕ^4 interaction spells trouble because we would have identical lines emerging from each vertex, and we would have to think a little bit more carefully. In ϕ^4 theory there would be four meson lines coming in or out of each vertex, and they all look the same. You follow one in, and I say, follow the meson line out, and you say, *which* meson line? There are three going out. You can get into trouble. But I chose a nice model, without that complication. Fortunately meson–nucleon theory and quantum electrodynamics, which will occupy us for some time, are very similar in their combinatoric structure, and in QED also the $1/n!$ factors disappear. You may, however, have leftover symmetry numbers even in Feynman diagrams. It depends on the theory. That takes care of the combinatorics.

Now we come to the actual evaluation of these diagrams. If I do one of them, you will see how all of them go. So let me do Diagram (a). The only term in the Wick expansion of (10.18) that can contribute to two nucleons scattering into two nucleons (hereafter, $NN \to NN$) is

$$\tfrac{1}{2}(-ig)^2 \int d^4x_1\, d^4x_2\ :\psi(x_1)^*\psi(x_1)\psi(x_2)^*\psi(x_2):\ \overline{\phi(x_1)\phi(x_2)} \tag{10.19}$$

First we have to compute the amplitude for uncontracted fields absorbing and emitting a meson and a nucleon. For example (see (2.53) and (6.24)),

$$\langle 0|\psi(x)|p\rangle = e^{-ip\cdot x} \tag{10.20}$$

That amplitude is very simple, because we have relativistically normalized states, with a factor of $(2\pi)^{3/2}\sqrt{2\omega_{\mathbf{p}}}$ in their normalization (see (1.57)). Now you see why I originally put in that factor. I said then that we'd want factors of 2π to come out right in Feynman diagrams. This normalization guarantees that the free field matrix element to annihilate a single nucleon is simply $e^{-ip\cdot x}$. The same holds for absorption of a meson, emission of a nucleon, emission of an antinucleon, etc. Then

$$\begin{aligned}\langle p_1'p_2'|\ :\psi(x_1)^*\psi(x_1)\psi(x_2)^*\psi(x_2):\ |p_1p_2\rangle &= \langle p_1'p_2'|\psi(x_1)^*\psi(x_2)^*\psi(x_1)\psi(x_2)|p_1p_2\rangle \\ &= \langle p_1'p_2'|\psi(x_1)^*\psi(x_2)^*|0\rangle\,\langle 0|\psi(x_1)\psi(x_2)|p_1p_2\rangle\end{aligned} \tag{10.21}$$

The nucleon annihilation operators in $\psi(x_1)$ and $\psi(x_2)$ have to be used to annihilate the two incoming nucleons with momenta p_1 and p_2, and the corresponding creation operators in $\psi(x_1)^*$ and $\psi(x_2)^*$ have to be used to create outgoing nucleons with momenta p_1' and p_2', so as not to get a zero inner product. We have

$$\langle 0|\psi(x_1)\psi(x_2)|p_1p_2\rangle = \underbrace{e^{-ip_1\cdot x_1-ip_2\cdot x_2}}_{\substack{p_1 \text{ absorbed at } x_1\\ p_2 \text{ absorbed at } x_2}} + \underbrace{e^{-ip_1\cdot x_2-ip_2\cdot x_1}}_{\substack{p_1 \text{ absorbed at } x_2\\ p_2 \text{ absorbed at } x_1}} \tag{10.22}$$

$$\langle p_1'p_2'|\psi(x_1)^*\psi(x_2)^*|0\rangle = \underbrace{e^{ip_1'\cdot x_1+ip_2'\cdot x_2}}_{\substack{p_1' \text{ emitted at } x_1\\ p_2' \text{ emitted at } x_2}} + \underbrace{e^{ip_1'\cdot x_2+ip_2'\cdot x_1}}_{\substack{p_1' \text{ emitted at } x_2\\ p_2' \text{ emitted at } x_1}} \tag{10.23}$$

and so

$$\begin{aligned}&\langle p_1'p_2'|\psi^*(x_1)\psi^*(x_2)|0\rangle\,\langle 0|\psi(x_1)\psi(x_2)|p_1p_2\rangle \\ &\quad= \left(e^{ip_1'\cdot x_1+ip_2'\cdot x_2}+e^{ip_1'\cdot x_2+ip_2'\cdot x_1}\right)\left(e^{-ip_1\cdot x_1-ip_2\cdot x_2}+e^{-ip_1\cdot x_2-ip_2\cdot x_1}\right) \\ &\quad= \underbrace{\left[e^{ix_1\cdot(p_1'-p_1)+ix_2\cdot(p_2'-p_2)}\right]}_{\text{Diagram (a)}} + \underbrace{\left[e^{ix_1\cdot(p_1'-p_2)+ix_2\cdot(p_2'-p_1)}\right]}_{\text{Diagram (b)}} + (x_1 \leftrightarrow x_2)\end{aligned} \tag{10.24}$$

The term $(x_1 \leftrightarrow x_2)$ means that two other terms appear in (10.24) that are exactly the same as the two shown, but with these variables swapped. These terms correspond to Diagram (c) and Diagram (d) (not drawn), identical to Diagram (a) and Diagram (b), respectively, but with permuted vertices $(1 \leftrightarrow 2)$.

The contribution to S − 1 from Diagram (a) is an integral over x_1 and x_2. Because of this integration, Diagram (c) is equivalent to Diagram (a) and makes the same contribution. We can thus regard these two diagrams, with the vertex labels erased, as a *single* Feynman diagram, and simply drop the 1/2! factor in Dyson's formula, as we talked about earlier. That takes care of all the loose, uncontracted fields. I still have the contraction of $\phi(x_1)$ and $\phi(x_2)$.

Earlier we found a rather simple expression for this contraction, (9.29):

$$\overbrace{\phi(x_1)\phi(x_2)} = \int \frac{d^4q}{(2\pi)^4} e^{-iq\cdot(x_1-x_2)} \frac{i}{q^2-\mu^2+i\epsilon} \tag{9.29}$$

If we insert this expression in (10.19), we notice that a great simplification occurs, because all of the x integrals are trivial. They simply give us delta functions:

$$\begin{aligned}\langle p_1', p_2'|(\mathrm{S}-1)|p_1, p_2\rangle &= (-ig)^2 \int \frac{d^4q}{(2\pi)^4} \frac{i}{q^2-\mu^2+i\epsilon} \int d^4x_1\, d^4x_2\, e^{ix_1\cdot(p_1'-p_1-q)+ix_2\cdot(p_2'-p_2+q)} \\ &\quad + (p_1 \leftrightarrow p_2) \\ &= (-ig)^2(2\pi)^4 \int d^4q \frac{i}{q^2-\mu^2+i\epsilon} \delta^{(4)}(p_1'-p_1-q)\,\delta^{(4)}(p_2'-p_2+q) \\ &\quad + (p_1 \leftrightarrow p_2)\end{aligned} \tag{10.25}$$

Because $\delta(x-a)\delta(x-b) = \delta(b-a)\delta(x-b)$, we can write (10.25) as

$$\begin{aligned}\langle p_1', p_2'|(\mathrm{S}-1)|p_1, p_2\rangle = (-ig)^2(2\pi)^4\delta^{(4)}(p_1'+p_2'-p_1-p_2) \times \\ \int d^4q \frac{i}{q^2-\mu^2+i\epsilon} \left[\delta^{(4)}(p_1'-p_1-q) + \delta^{(4)}(p_2'-p_1+q)\right]\end{aligned} \tag{10.26}$$

All we're left with is an easy integral over the momentum q of the internal meson line. If we define the **invariant Feynman amplitude** $\mathcal{A}_{fi}$ by

$$\boxed{\langle f|(\mathrm{S}-1)|i\rangle = (2\pi)^4\delta^{(4)}(p_f-p_i)\, i\mathcal{A}_{fi}} \tag{10.27}$$

then the amplitude $\mathcal{A}_{fi}$ for the $\mathcal{O}(g^2)$ contribution to NN → NN scattering can be written

$$\mathcal{A}_{fi} = \underbrace{(-ig)^2 \frac{1}{(p_1-p_1')^2-\mu^2+i\epsilon}}_{\text{Diagram (a)}} + \underbrace{(-ig)^2 \frac{1}{(p_1-p_2')^2-\mu^2+i\epsilon}}_{\text{Diagram (b)}} \tag{10.28}$$

With these two terms go two pictures (see Figure 10.4) and two stories. The story that goes with Diagram (a) is this: A nucleon with momentum p_1 comes in and interacts at a point. Out comes a nucleon with momentum p_1' and a "virtual" meson with momentum q. This "virtual" meson then interacts with a nucleon p_2, and out comes a nucleon with momentum p_2'. The interaction points x_1 and x_2 can occur anywhere, and so they are integrated over all possible values. Furthermore, the **virtual** meson, unlike a *real* meson, can have *any* 4-momentum q, so q is to be integrated over all possible values, though as you can see from the factor $(q^2-\mu^2+i\epsilon)^{-1}$, q likes to be on the mass shell, with $q^0 = \pm\sqrt{|\mathbf{q}|^2+\mu^2}$. The story belonging to Diagram (b) is much the same, except that the roles of the nucleons with momenta p_1 and p_2 are reversed.

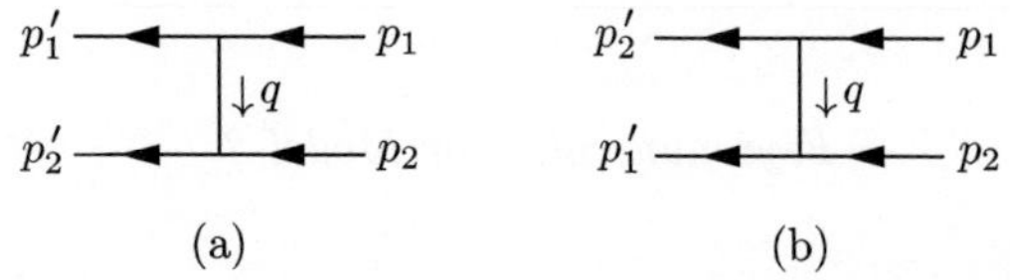

Figure 10.4: $\mathcal{O}(g^2)$ Feynman diagrams for Model 3 nucleon–nucleon scattering

Fairy tales like this helped Feynman think about quantum electrodynamics. In our formalism, they are little more than stories, but in the *path integral* formulation of quantum mechanics, these fairy tales gain some justification, as we will see in the second half of the course. The words not only match the pictures, they parallel the mathematics.[11] What we have done for Diagram (a) is completely general. We could have been working out a much more complicated diagram, of arbitrary complexity.

Given a Lagrangian, there is a set of rules for drawing diagrams and associating factors with elements of the diagrams, to produce amplitudes for physical processes. These are the famous **Feynman rules**. So let me now write down (in the box on p. 216) the Feynman rules for this theory, with initial states on the *right* and final states on the *left*.[12] Notice that the vertex and the counterterms have energy-momentum conserving delta functions. The vertex contains a factor of $(-ig)$ coming from the expansion, and the meson and nucleon counterterms contain factors of ib and ic, respectively. You follow the flow of momentum around the diagram like current flowing around an electrical circuit. The sum of momenta flowing into a vertex is the sum of momenta that flows out, much like current into and out of a junction in a circuit. The difference between what flows in and what flows out—which should be zero—is the argument of the delta function.

The a counterterm is just a number. I don't need a special diagrammatic rule for that. The a counterterm has no momentum associated with it, and its delta function has an argument of zero. If the system were in a box, the term $(2\pi)^4\delta^{(4)}(0)$ would turn into VT, the volume of spacetime in the box. This counterterm is designed to cancel all the vacuum bubble diagrams, those without external lines, which you will see also have a factor of $\delta^{(4)}(0)$. Like the counterterm a, the counterterms b and c will be expressed as infinite power series in the coupling constant g. I will explain to you shortly how we determine them order by order.

A minor technical point here. You might think there should be a factor of $\frac{1}{2}$ in the meson counterterm because the term in the interaction Hamiltonian (10.11) is $-\frac{1}{2}b\phi^2$. But you have two possible terms, depending on which ϕ you are going to contract in the forward direction as you move along the internal line, and which ϕ you are going to contract in the backward direction. There are always two choices, and those two choices cancel out the $\frac{1}{2}$.

That's it! That's every diagram for Model 3. To calculate things, you just take what you need from all of this stuff, stick it together for the appropriate process, and you get a big integral. Everything is fixed by the momentum on the external lines, which affects the momentum on the internal lines via the delta functions.

---

[11] [Eds.] When Feynman presented his work at the 1948 Pocono conference, Bohr responded that if Feynman would ascribe classical trajectories to electrons and photons, he had completely misunderstood quantum mechanics: Schweber *QED*, pp. 344–345.

[12] [Eds.] A reminder: in Feynman diagrams, time conventionally flows from left to right. Coleman's time runs from *right to left*.

Feynman rules for Model 3

1. For external lines $\left\{\begin{matrix}\text{incoming}\\ \text{outgoing}\end{matrix}\right\}$, momenta are directed $\left\{\begin{matrix}\text{in}\\ \text{out}\end{matrix}\right\}$.
2. Assign a directed momentum to every internal line.
3. For every ... Write ...

(a) internal meson line

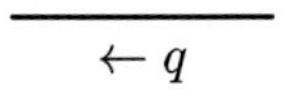

$$\int \frac{d^4q}{(2\pi)^4}\frac{i}{q^2-\mu^2+i\epsilon}$$

(b) internal nucleon line

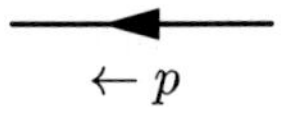

$$\int \frac{d^4p}{(2\pi)^4}\frac{i}{p^2-m^2+i\epsilon}$$

(c) vertex

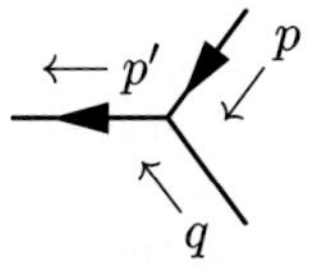

$$-ig\,(2\pi)^4\,\delta^{(4)}(p'-q-p)$$

(d) meson mass counterterm

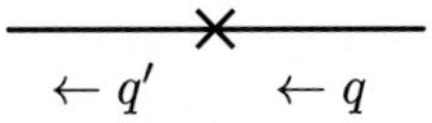

$$ib\,(2\pi)^4\,\delta^{(4)}(q'-q)$$

(e) nucleon mass counterterm

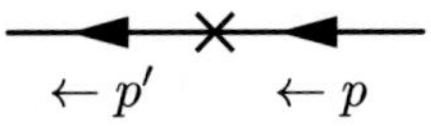

$$ic\,(2\pi)^4\,\delta^{(4)}(p'-p)$$

(f) vacuum counterterm ×

$$ia\,(2\pi)^4\,\delta^{(4)}(0)$$

4. Integrate over all internal momenta (lines belonging to "virtual particles").

These rules are very simple. They are also cleverly arranged. They enable you directly to compute the S-matrix element, to any arbitrary order in perturbation theory, between any set of relativistically normalized states, with any number of incoming mesons and any number of nucleons, and any number of outgoing mesons and nucleons. Just draw all possible diagrams with the appropriate number of vertices, write down for each of these diagrams an expression given by these rules, and do the integrals, to the best of your ability. In many cases the integrals are trivial, and in other cases they are complicated.

So for the time being we can forget Wick's theorem, we can forget Dyson's formula, we can forget fields. All we have is a sequence of rules, like the rules of arithmetic, for computing any contribution to any order for any scattering process in this field theory. Please notice these rules have been arranged to take care of one of the important practical problems of theoretical physics: keeping track of the 2π's. The only factors of 2π which appear anywhere in these rules are due to a $1/(2\pi)$ for every momentum integral, and a 2π for every delta function. There is no problem keeping track of the 2π's (there may be some left over).

In most of the diagrams that we have written down so far, all of the internal momenta are fixed by the four-momentum conserving delta functions. It is often trivial, as it is for Diagram (a), to get rid of all the internal integrals, and just be left with one delta function expressing overall four-momentum conservation. We should expect to see such a factor in an S-matrix element for a theory with spacetime translation invariance. On the other hand, one can write down diagrams, say one of this structure,

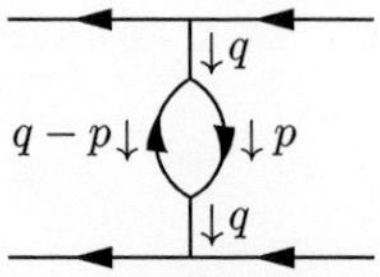

Figure 10.5: A $\mathcal{O}(g^4)$ contribution to $N + N \to N + N$ involving a virtual meson → virtual $N + \overline{N}$

where the internal momenta are not fixed completely by energy-momentum conservation. The virtual meson splits into a virtual nucleon and antinucleon, which recombine to make a virtual meson, which then hits the second original nucleon. I can always add an extra momentum, $+p$ to the right nucleon line and $-p$ to the left antinucleon line, and everything is still conserved. So diagrams that have internal closed loops will still have residual integrals. Note that positive momentum flow will often be indicated by the lighter arrows *off* a line: down ↓ or up ↑, to the left ← or right →. The Feynman arrows ——◀—— are *on* the lines, and point *inwards*, toward a vertex for a fermion ψ, and *outwards*, away from a vertex, for an anti-fermion ψ^*.

For any such diagram we can imagine a metaphorical interpretation, and we sometimes attach words to it. We say these *virtual processes* conserve energy and momentum because of the four-dimensional delta function that appears at every vertex. The funny thing about *virtual particles* is that they need not be on the mass shell. They can have *any* four-momentum, and you have to integrate over all possible four-momenta, given the factor[13]

$$\boxed{\widetilde{\Delta}_F(q^2) = \frac{i}{q^2 - \mu^2 + i\epsilon}} \tag{10.29}$$

This interpretation is due to Feynman who, by fiddling around, by a miracle, got these rules before Dyson and Wick. The miracle was *genetic*: Feynman was a genius. Factors like (10.29) give the probability amplitude, in this metaphorical language, for the virtual particle going between two vertices, as in Diagram (a). They describe how a virtual particle propagates from one vertex to another. For this reason, they are called **Feynman propagators**. The language, I stress, is purely metaphorical. If you don't want to use it, don't use it. But then you'll find 90% of the physicists in the world will be unintelligible to you when they give seminars. It's very convenient, but it should not be taken too seriously.

We have derived these rules without any talk about virtual particles, or summing up probability amplitudes for the propagation of virtual particles.[14] We've derived them just from the standard operations of non-relativistic quantum mechanics and a lot of combinatorics.[15]

---

[13] [Eds.] Some define the Feynman propagator without the i in the numerator, cf. Bjorken & Drell *Fields*, p. 42, equation (12.71). *Caveat lector!* Note: Coleman wrote $\widetilde{D}(q^2)$ for $\widetilde{\Delta}_F(q^2)$; this notation is used in §15.3. See also problem 1.3, p. 49.

[14] [Eds.] Again Coleman foreshadows Feynman's *sum over histories*, the path integral formulation of quantum mechanics; see the aside on p. 656.

[15] [Eds.] A student asks: "How do we know that all this is right?" Coleman replies: "Sure: *experiment*. You

10.3 Feynman diagrams in Model 3 to order g^2

I will begin going systematically through all the Feynman diagrams that arise in our model theory to order g^2, or at least those allowed by energy-momentum conservation, one at a time. They come together in families, so it's not so tedious. I won't finish the survey in this lecture. With each there will be a little point of physics I would like to discuss.

$\mathcal{O}(g)$ diagrams. There are only two:

1.1 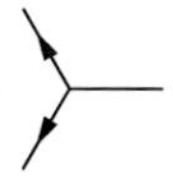 $= 0$ if $\mu < 2m$

1.2 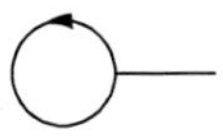 $= 0$

I won't bother to write the actual values of the external momenta until it is time to compute things. Diagram 1.1 above represents the decay of a meson into a nucleon–antinucleon pair. It is zero, unless we choose the physical mass μ of the meson to be larger than twice the mass m of the nucleon. I will talk later about what happens when we do make that choice, but for the moment I will assume $\mu < 2m$. Then our meson is stable, and there is a genuine asymptotic meson state, and that diagram vanishes by energy-momentum conservation. Nor can any of the other processes of a real nucleon decaying into a real nucleon and a real meson, all on the mass shell, occur. These processes are well known to be impossible, no matter how we choose the masses. Diagram 1.2 is even less likely. It represents the vacuum spontaneously absorbing a meson. That is also equal to zero by energy-momentum conservation, because of the energy-momentum conserving delta function that comes out in front. Those are the two diagrams of order g.

$\mathcal{O}(g^2)$ diagrams. There are twenty-three (or seventeen, accounting for symmetries):

2.1 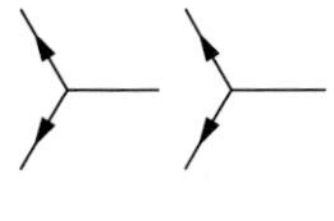 $= 0$ if $\mu < 2m$

Diagram 2.1 is Diagram 1.1 above, doubled. It's two mesons decaying and is again equal to zero if $\mu < 2m$. Of all $\mathcal{O}(g^2)$ diagrams, this one has the most external lines: six. I'll now start counting at the other end, and go up from zero external lines to four.

Diagram 2.2 consists of three separate diagrams, two contributions to the vacuum energy, and the counterterm:

2.2 (a)  (b) (c) $\times$ (computed to $\mathcal{O}(g^2)$)

---

could have asked the same question in classical mechanics: How do you know, Mr. Newton, that gravity is proportional to $1/r^2$, rather than proportional to r, as Mr. Hooke suggests? It's unambiguous to check a theory if the couplings are weak, and you can do perturbation theory. If the couplings are strong, and perturbation theory is useless, then it's not at all unambiguous. It's a very hard job that's still in progress to try and figure out the theory that explains the strong interactions. In electrodynamics, at least, we can make predictions, and we can check, by experiment, that the Lagrangian and the rules we write down describe reality."

There are vacuum self-energy corrections from 2.2 (a) and 2.2 (b), but there's also an a term. These are the only contributions of order g^2 to $\langle 0|S|0\rangle$. The counterterm is fixed by the requirement that these three contributions sum to zero, so that there are no $\mathcal{O}(g^2)$ terms in the vacuum-to-vacuum amplitude:

$$\langle 0|S|0\rangle = 1 \text{ to } \mathcal{O}(g^2) \tag{10.30}$$

That is to say, the vacuum-to-vacuum amplitude has no corrections. Of course, the a term is an *infinite power series*; this equation is just the $\mathcal{O}(g^2)$ expression of (10.12), and determines the a term only to second order in g. If I wanted to know the vacuum energy per unit volume, a, to $\mathcal{O}(g^2)$ I could calculate it from these diagrams.

Onwards! Diagrams with two external lines (mesons first, then nucleons):

2.3 (a) 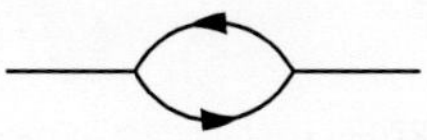 (b) 

The first diagram is interesting if I want to compute the bare mass of the meson to order g^2. If I don't, the condition that fixes the meson mass renormalization counterterm b is precisely the condition that these two diagrams sum to zero, to $\mathcal{O}(g^2)$. If they didn't, there would be a nonzero correction of order g^2 to the one-meson-to-one-meson matrix element, and there shouldn't be:

$$\langle \mathbf{q}|S-1|\mathbf{q}'\rangle = 0 \quad \text{to } \mathcal{O}(g^2) \tag{10.31}$$

Again, this is just the $\mathcal{O}(g^2)$ statement of (10.13).

2.4 (a)  (b)  (c) 

It's nearly the same story for the nucleon, and the same answer, except that now there are three diagrams. The first two diagrams taken together will give the bare mass of the nucleon to $\mathcal{O}(g^2)$. The condition that fixes the nucleon mass counterterm c is that the contributions of these three terms sum to zero, to $\mathcal{O}(g^2)$:

$$\langle \mathbf{p}|S-1|\mathbf{p}'\rangle = 0 \quad \text{to } \mathcal{O}(g^2) \tag{10.32}$$

(the $\mathcal{O}(g^2)$ statement of (10.14)).

We've gone through a large number of Feynman diagrams with hardly any labor. Now we are left only with the diagrams of $\mathcal{O}(g^2)$ with four external lines. Diagrams with four external lines describe seven separate processes, but we'll only look at four of these, because the other three can be obtained by a particular symmetry from the first four. Two of the lines must be incoming and two outgoing, otherwise energy-momentum conservation will make them vanish trivially. We cannot have a single particle go into three, or the vacuum go into four. First I'll write down these processes and then I will write down the diagrams. If you don't yet have these Feynman rules in your head, you soon will.

We could have nucleon–nucleon scattering, Figure 10.7:

2.5 $\quad N + N \to N + N$

We could also have antinucleon–antinucleon scattering, but that's connected to nucleon–nucleon scattering by C, the charge-conjugation operator, since our theory does have charge conjugation invariance:

$$N + N \to N + N \quad \overset{C}{\Longleftrightarrow} \quad \overline{N} + \overline{N} \to \overline{N} + \overline{N} \tag{10.33}$$

So I'm not going to bother to discuss antinucleon–antinucleon scattering, since it is diagram for diagram identical with nucleon–nucleon scattering. We could have nucleon–antinucleon scattering,

2.6 $N + \overline{N} \to N + \overline{N}$

C doesn't help me much here. We could have nucleon–meson scattering,

2.7 $N + \phi \to N + \phi$

which is connected by C to antinucleon–antimeson scattering:

$$N + \phi \to N + \phi \quad \overset{C}{\Longleftrightarrow} \quad \overline{N} + \phi \to \overline{N} + \phi \tag{10.34}$$

I'm only writing down the processes that conserve charge. Remember, the nucleons have charge one, the mesons have charge zero. And finally, we could have nucleon–antinucleon annihilation into meson plus meson,

2.8 $\overline{N} + N \to \phi + \phi$

That process is connected by time reversal, T, to meson plus meson makes a nucleon–antinucleon pair:

$$\overline{N} + N \to \phi + \phi \quad \overset{T}{\Longleftrightarrow} \quad \phi + \phi \to N + \overline{N} \tag{10.35}$$

So I won't bother to write that one down.

You may be wondering about the process $\phi + \phi \to \phi + \phi$, meson-meson scattering. This process occurs in Model 3, but it is $\mathcal{O}(g^4)$:

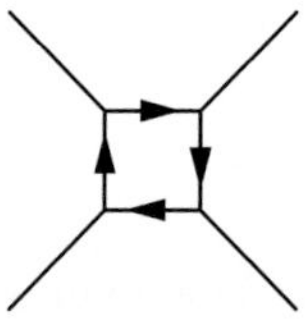

Figure 10.6: Lowest order contribution in Model 3 to $\phi + \phi \to \phi + \phi$ scattering, $\mathcal{O}(g^4)$

so we won't discuss it now.

Thus we have but four processes of $\mathcal{O}(g^2)$ with four external legs to consider. For each of these, we will find two Feynman diagrams which we will have to sum up. I would like to write down those two Feynman diagrams for all four processes, eight Feynman diagrams in all. I will discuss the physical meaning of each term in the perturbation expansion, because each is interesting.

In order to simplify matters, I will use the notation of (10.27), which I'll repeat here:

$$\langle p_1' p_2' | S - 1 | p_1 p_2 \rangle = (2\pi)^4 \delta^{(4)}(p_f - p_i) i\mathcal{A}_{fi} \tag{10.27}$$

The i is there by convention,[16] so our relativistic scattering amplitude $\mathcal{A}_{fi}$ will have the same phase of the amplitude $f(\theta)$ as defined in non-relativistic quantum mechanics.[17]

10.4 $\mathcal{O}(g^2)$ nucleon–nucleon scattering in Model 3

Now let's look at the diagrams corresponding to process 2.5, nucleon–nucleon scattering. We've already looked at these in Fig. 10.4 and even found the invariant amplitude (10.28). I'll draw them again, with a variation:

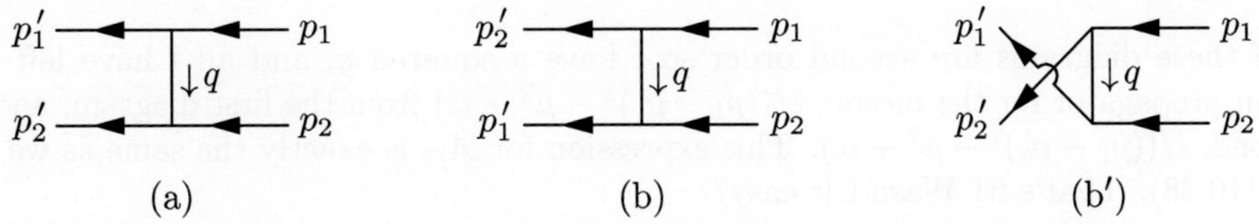

Figure 10.7: $\mathcal{O}(g^2)$ Feynman diagrams for Model 3 $NN \to NN$

There are the two diagrams I've written down before, (a) and (b), and a new one, (b′), you sometimes see. People tired of writing p's whenever they draw a diagram sometimes leave the p's off of these diagrams. They start with (a), and to let you know two of the p's are exchanged in the other one, they sometimes write (b′) instead of (b), stealing a notational device from electrical engineering. The drawing indicates that you're to put the momenta in by yourself at the same places on the two diagrams. Then the terms will take care of themselves. The diagrams (b) and (b′) are the same diagram, just written with the lines twisted around.

Though we've already found the invariant amplitude for this process, it's worth doing again quickly with the Feynman rules. They give for Diagram (a) the contribution

$$\begin{aligned}\int \frac{d^4q}{(2\pi)^4}(-ig)(2\pi)^4\delta^{(4)}(p_1'+q-p_1)\frac{i}{q^2-\mu^2+i\epsilon}(-ig)(2\pi)^4\delta^{(4)}(p_2'-q-p_2)\\ = -g^2(2\pi)^4\delta^{(4)}(p_2'+p_1'-p_2-p_1)\frac{i}{(p_1-p_1')^2-\mu^2+i\epsilon}\end{aligned} \tag{10.36}$$

Momenta are *positive* leaving a vertex and *negative* entering. All of our internal momenta are fixed, so there are no leftover integrations, no leftover delta functions except the one delta function for overall energy-momentum conservation. The internal momentum q in Diagram (a) is fixed by the delta function to be $p_1 - p_1'$ or equivalently $p_2' - p_2$ (they're the same) and the internal momentum in Diagram (b) is fixed to be $p_1 - p_2'$. The term from Diagram (b) is

[16] I would be just as happy if the convention were otherwise but I'm not going to change all the literature. I once gave a course in which I adamantly refused to put in that dumb i by convention, and proved the Optical Theorem for the *real* parts of scattering amplitudes. It got pretty silly at the end. I wound up putting the i back in.

[17] [Eds.] In non-relativistic quantum mechanics, one describes an incoming particle by a plane wave $\psi = e^{ikz}$. Far from the scattering center, the scattered particle is described by a spherical wave, $f(\theta)e^{ikr}/r$. See David J. Griffiths, *Introduction to Quantum Mechanics*, 2nd ed., Prentice-Hall, 2004, p. 401, or Landau & Lifshitz, *QM*, p. 469.

added to that of Diagram (a), to give for the amplitude $\mathcal{A}_{fi}$

$$(2\pi)^4\delta^{(4)}(p_2'+p_1'-p_2-p_1)i\mathcal{A}_{fi}$$
$$= -g^2(2\pi)^4\delta^{(4)}(p_2'+p_1'-p_2-p_1)\left[\frac{i}{(p_1-p_1')^2-\mu^2+i\epsilon}+\frac{i}{(p_1-p_2')^2-\mu^2+i\epsilon}\right] \tag{10.37}$$

Dividing out the common factors (except for the i), all that is left in this case is

$$i\mathcal{A}_{fi} = -g^2\left[\frac{i}{(p_1-p_1')^2-\mu^2+i\epsilon}+\frac{i}{(p_1-p_2')^2-\mu^2+i\epsilon}\right] \tag{10.38}$$

Both of these diagrams are second order so I have a squared g, and all I have left is the Feynman propagator for the meson, $i/((p_1-p_1')^2-\mu^2+i\epsilon)$ from the first diagram, and from the second, $i/((p_1-p_2')^2-\mu^2+i\epsilon)$. This expression for $\mathcal{A}_{fi}$ is exactly the same as we found before, (10.28). That's it! Wasn't it easy?

I would now like to discuss the meaning of these two terms. After all, relativistic quantum mechanics is, among other things, supposed to approach non-relativistic quantum mechanics in the low-energy regime. We have all done, I hope, many Born approximation computations in non-relativistic quantum mechanics. Have we ever seen an expression like (10.38) before?

Well, it's easiest to see the connection between our amplitude and the Born approximation in the center-of-momentum frame. We have Lorentz invariance, so why not use the center-of-momentum frame?

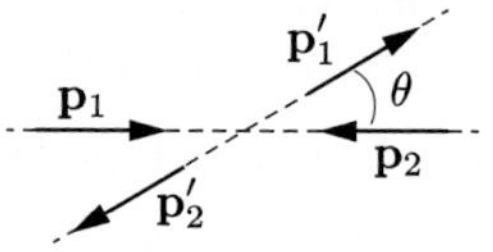

Figure 10.8: $NN \to NN$ scattering in the center of momentum

In the center-of-momentum frame, the three-momenta $\mathbf{p}_1$ and $\mathbf{p}_2$ of the incoming particles are equal and opposite, so we can write

$$\mathbf{p}_1 = -\mathbf{p}_2 = \mathbf{e}p \quad \text{where } p = |\mathbf{p}_1| = |\mathbf{p}_2| \quad \text{and } \mathbf{e} \text{ is a unit vector.} \tag{10.39}$$

The four-momenta are thus

$$\begin{aligned} p_1 &= (\sqrt{p^2+m^2}, \mathbf{e}p) \\ p_2 &= (\sqrt{p^2+m^2}, -\mathbf{e}p) \end{aligned} \tag{10.40}$$

The energies of the outgoing particles are the same, and the magnitudes of the outgoing momenta are the same magnitude as p; they just have a different direction, with a different unit vector $\mathbf{e}'$. That is, the new four-momenta are

$$\begin{aligned} p_1' &= (\sqrt{p^2+m^2}, \mathbf{e}'p) \\ p_2' &= (\sqrt{p^2+m^2}, -\mathbf{e}'p) \end{aligned} \tag{10.41}$$

The denominator $(p_1-p_1')^2$ works out like this:

$$(p_1-p_1')^2 = (\sqrt{p^2+m^2}-\sqrt{p^2+m^2})^2 - p^2(\mathbf{e}-\mathbf{e}')^2 = -2p^2(1-\cos\theta) = -\Delta^2 \tag{10.42}$$

The angle θ is the scattering angle, and $\Delta = |\boldsymbol{\Delta}|$, where $\boldsymbol{\Delta}$ is the non-relativistic momentum transfer,

$$\boldsymbol{\Delta} \equiv \mathbf{p}_1 - \mathbf{p}_1' \tag{10.43}$$

The other denominator is

$$(p_1 - p_2')^2 = -p^2(\mathbf{e} + \mathbf{e}')^2 = -2p^2(1 + \cos\theta) = -\Delta_c^2 \tag{10.44}$$

where the non-relativistic cross momentum transfer $\boldsymbol{\Delta}_c$ is defined by

$$\boldsymbol{\Delta}_c \equiv \mathbf{p}_1 - \mathbf{p}_2' \tag{10.45}$$

The cross momentum transfer is the momentum transfer that would arise if you considered the particle we have arbitrarily labeled as 2 as the descendent of the particle we have labeled as 1, rather than 1 being the descendent of 1.

Substituting this in, we find the invariant amplitude $\mathcal{A}_{fi}$ in the center-of-momentum frame,

$$\mathcal{A}_{fi} = g^2 \left[\frac{1}{\Delta^2 + \mu^2} + \frac{1}{\Delta_c^2 + \mu^2} \right] \tag{10.46}$$

We can now drop the $i\epsilon$ because the denominators are positive. All the i's and minus signs cancel. This is the same expression as we obtained earlier, (10.28), just written in a special coordinate frame, the center-of-momentum frame. It should now look much more familiar to you.

People were scattering nucleons off nucleons long before quantum field theory was around, and at low energies, they could describe scattering processes adequately with non-relativistic quantum mechanics. The non-relativistic amplitude $\mathcal{A}_{fi\,\mathrm{NR}}$ for scattering is proportional to an integral of the potential,

$$\mathcal{A}_{fi\,\mathrm{NR}} = \langle f|V|i\rangle \propto \int d^3\mathbf{r}\, V(r)\, e^{-i\boldsymbol{\Delta}\cdot\mathbf{r}} = \widetilde{V}(\Delta) \tag{10.47}$$

in the lowest order of perturbation theory. This is the famous **Born approximation**, the lifeblood of non-relativistic quantum scattering. As we found earlier (see the discussion leading to (9.41)),

$$\widetilde{V}(\Delta) \propto g^2 \frac{1}{\Delta^2 + \mu^2} \quad \text{if } V(r) = V_{\text{Yukawa}}(r) = g^2 \frac{e^{-\mu r}}{r} \tag{10.48}$$

What we have obtained as the first term in relativistic perturbation theory, the term of lowest nontrivial order, is *precisely* what we would have obtained if we had used non-relativistic perturbation theory to compute the scattering amplitude for a Yukawa potential, to lowest nontrivial order. This is in perfect agreement with what we discovered last lecture. We found, to use Feynman's language, the exchange of a virtual meson between two *static* sources $\rho(\mathbf{x})$ produced a Yukawa potential. Here the exchange of a virtual meson between two *moving* sources, two actual particles, produces a scattering amplitude that would be produced in this order of perturbation theory by a Yukawa potential.

What about the second term, where Δ is replaced by Δ_c? It, too, has an analog in non-relativistic quantum theory. In non-relativistic scattering theory involving two identical particles, where you have to take account of the symmetry, it's very convenient to introduce something called the *exchange operator*, $\mathscr{E}$, which when acting on a two-particle wave function exchanges the two particles:

$$\mathscr{E}\psi(r_1, r_2) = \psi(r_2, r_1) \tag{10.49}$$

If we consider a non-relativistic scattering problem in which

$$V(r) \propto V_{\text{Yukawa}}(r)\,\mathscr{E} \tag{10.50}$$

then we will find

$$\langle f|V|i\rangle \propto \int d^3\mathbf{r}\, V_{\text{Yukawa}}(r)\, e^{-i\boldsymbol{\Delta}_c\boldsymbol{\cdot}\mathbf{r}} = \widetilde{V}(\Delta_c) \tag{10.51}$$

So the term with Δ would come from an ordinary Yukawa potential, and the term with Δ_c would come from an *exchange* Yukawa potential. That we get both a Yukawa potential and an exchange Yukawa potential is not surprising, because these *are* identical particles. The scattering amplitude must be invariant under the interchange of p_1' and p_2'. That is to say, if the first term in (10.46) is present, the second term also must be present, because there is no way of telling apart the configurations in which you have exchanged the particles from the configurations in which you have not. And since we are working in a formulation of many-particle theory, quantum field theory of scalar particles, where Bose statistics is automatic, it must automatically come out having the right symmetry properties. The presence of the first term demands the presence of the second; the interaction must have the form

$$V(r) = V_{\text{Yukawa}}(r)(1 + \mathscr{E}) \tag{10.52}$$

That takes care of process 2.5, nucleon–nucleon scattering.

Next lecture I will begin to discuss nucleon–antinucleon scattering with similar arguments. We will find some things in common and some things different, and continue with the other processes. Then I will discuss some mysterious connections that exist between these processes; in particular I will discuss crossing and CPT invariance on the level of second-order perturbation theory. I will then go on to a dull but unfortunately necessary kinematic exercise of how to connect S-matrix elements to cross-sections, which are what experimentalists publish, after all...

11

Scattering I. Mandelstam variables, CPT and phase space

The four problems in the next assignment[1] are all on material that you either already know or will know at the end of this lecture, or perhaps at the very beginning of next lecture; I'm not quite sure how far we'll get. The fourth one has a little interest to it, and the other three are just dull, dumb computation. I encourage you to do them because the only way you will learn to manipulate Feynman diagrams is by doing one Feynman calculation through from beginning to end and keeping track of all the π's and all the other factors we're going to talk about. As Feynman said in a lecture when I was a graduate student: "If you say ya understand the subject, but you don't know where to put the π's and minus signs, then you don't understand *nuttin'*."[2]

11.1 Nucleon–antinucleon scattering

As you recall, at the end of last lecture, we began our study of four scattering processes. The first one we considered was nucleon–nucleon scattering where "nucleon" should be imagined with invisible quotation marks around it in our model. There we discovered two graphs, each of which had a clear non-relativistic analog. One corresponded to the Born term for scattering in the direct Yukawa potential, and the other corresponded to the Born term for scattering in the exchange Yukawa potential. Of course for nucleon–nucleon scattering, if one is there, the other has to be there, just because of Bose statistics.

---

[1] [Eds.] Problems 6, p. 261.

[2] [Eds.] Coleman delivers this in a New York accent, adding, "I try to get his tone of voice, but it's difficult. . . ." (Coleman did his doctorate on unitary symmetry under Murray Gell-Mann at Caltech, and took graduate courses there from Richard Feynman.) A moment later, a student asks: "Did Feynman refer to them as Feynman diagrams?" Coleman replies: "No. *Drawings* I think he called them," waving his hand as if to say, "Don't get me started...", to much laughter. Then he adds: "When Feynman and Schwinger did much of this work in parallel, it was said that Feynman's papers were written in such a way as to make you believe *anyone* could have done the computation, and Schwinger's were written in such a way as to make you believe that *only Schwinger* could have done it. Schwinger did *not* use diagrammatic methods." Julian Schwinger (1918–1994) shared the 1965 Nobel Prize with Feynman and Shin'ichirō Tomonaga (1906–1979) for advances in quantum electrodynamics. In 1980, Schwinger described his reaction in 1947 to the introduction of Feynman diagrams "Like the silicon chip of more recent years, Feynman was bringing computation to the masses." J. Schwinger, "Renormalization theory of quantum electrodynamics: an individual view", in Laurie M. Brown and Lillian Hoddeson, eds., *The Birth of Particle Physics*, Cambridge U. P. 1983, p. 343.

Now let's go on to our next process,

2.6 $N + \overline{N} \to N + \overline{N}$

the class of diagrams that contribute to $N\overline{N}$ scattering. As before, there are two, denoted (a) and (b); see Figure 11.1.

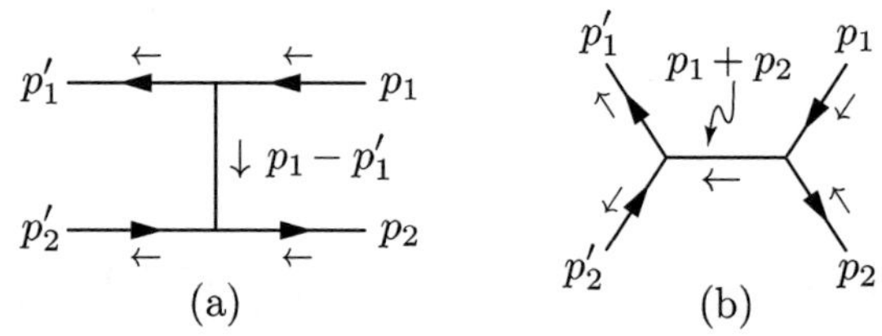

Figure 11.1: $\mathcal{O}(g^2)$ Feynman diagrams for Model 3 $N\overline{N} \to N\overline{N}$

In both diagrams, the incoming nucleon line, drawn at upper right and pointing in, toward the vertex, has momentum p_1; the incoming antinucleon line, at lower right, pointing out, away from the vertex, has momentum p_2. (Remember, I draw a line pointing outward to indicate an incoming ψ^* field, which can either create a nucleon or annihilate an antinucleon. Similarly, I draw a line pointing inward, towards the vertex, to indicate an outgoing ψ^* field.) The outgoing nucleon has momentum p_1', and the outgoing antinucleon has momentum p_2'. Once again writing down the graph is mechanical. The internal momentum q is completely determined by the conservation of energy-momentum at each vertex. In Diagram (a) it is equal to $p_1 - p_1'$ or equivalently $p_2' - p_2$. In Diagram (b), it is equal to $p_1 + p_2$ since both p_1 and p_2 are coming in to that vertex. Thus the amplitude, by the same reasoning as before, is

$$i\mathcal{A}_{fi} = (-ig)^2 \left[\frac{i}{(p_1 - p_1')^2 - \mu^2 + i\epsilon} + \frac{i}{(p_1 + p_2)^2 - \mu^2 + i\epsilon} \right] \tag{11.1}$$

These are the Feynman graphs obtained just by following the Feynman rules. The delta functions all take care of themselves except for one overall $(2\pi)^4\delta^{(4)}(p_2' + p_1' - p_2 - p_1)$ to conserve energy-momentum, which as you recall from the end of last lecture I factored out when I defined $\mathcal{A}_{fi}$ in (10.27). You should be able to write formulas down by eye like this yourself. So much for the expression. Now for the interpretation.

The first term has the same interpretation as last time. It corresponds to the non-relativistic Born approximation for a Yukawa potential of range μ^{-1}. It's exactly the same as the first term we had for NN scattering, process 2.5 (see §10.4). Unlike the amplitude for that process, here the second term is not an exchange potential; it's not $p_1 - p_2'$ or anything like that, it is $p_1 + p_2$. Of course since a nucleon and an antinucleon are not identical particles, there is no reason why a Yukawa potential should be accompanied in this process by an exchange potential. We can understand its physical meaning if we observe that in the center-of-momentum frame, in which the total three-momentum is zero,

$$p_1 + p_2 = (p_1^0 + p_2^0, \mathbf{p}_1 + \mathbf{p}_2) = (E_T, 0) \tag{11.2}$$

E_T is the total energy of the original two-particle system. So

$$(p_1 + p_2)^2 - \mu^2 = E_T^2 - \mu^2 \tag{11.3}$$

This contribution to the amplitude has a pole—in fact, *two* poles, $E_T = \pm\mu$—which are presumably below the threshold for creating a nucleon–antinucleon pair. As $\mu < 2m$, the denominator can never equal zero, and we need not worry about the $i\epsilon$.

We have not talked about partial-wave analysis. If I did a partial-wave decomposition of nucleon–antinucleon scattering, these poles would occur in the s-wave amplitude. You don't have to know much about Legendre polynomials to understand why that is so: $1/E_T^2$ is rotationally invariant, so the only factor of $P_\ell(\cos\theta)$ which occurs is for ℓ equals zero, the constant Legendre polynomial. The amplitude has no angular dependence at all, and therefore it is pure s-wave.

Now, do we encounter such poles in non-relativistic perturbation theory? The answer is yes, we do; typically not in lowest order but in second order. In the non-relativistic formula (10.47) I wrote $\mathcal{A}_{fi}$ as proportional to the first Born approximation, ("proportional", because we hadn't worked out the kinematic factors, yet). To the second approximation,[3]

$$\mathcal{A}_{fi} \propto \langle f|V|i\rangle + \sum_n \frac{\langle f|V|n\rangle\,\langle n|V|i\rangle}{E_n - E_i} + \cdots \tag{11.4}$$

The first term is the good old first Born approximation, and then there is the almost as good old second Born approximation summed over a complete set of energy eigenstates. Now, if there is in our unperturbed problem an isolated energy eigenstate $|n\rangle$ lying below the threshold at $E_{\mathbf{p}} = E_i = E_n$ for the continuum states in the scattering, such that $\langle n|V|i\rangle \neq 0$, then we will get a pole from the second-order formula. (This is an unusual situation in potential scattering, but not in field theory.) Of course we don't see this pole in physical scattering; we have to analytically continue below the physical region. Furthermore, as is obvious from the structure of this expression, the pole occurs in the partial wave with $\ell = \ell_n$, which has the same angular momentum as the state $|n\rangle$. If I expand the second approximation out in terms of angular momentum eigenstates and if V is rotationally symmetric, I will only get a nonzero matrix element if the angular momentum of this state is equal to that of the partial wave I am studying. The second Born approximation thus reveals that the pole, or at least one of these two poles, at $E_T = \mu$, corresponds to an energy eigenstate.

This is the non-relativistic analog of the pole in the relativistic scattering at $E_T = \mu$. It is exactly what we would get in a non-relativistic problem in which there was an isolated energy eigenstate, in addition to continuum states, before we turned on our potential V. The pole at $E_T = -\mu$ that comes along with the pole $E_T = \mu$ is without a non-relativistic analog, but that's not surprising. After all, if I make μ very close to $2m$, the pole at $E_T = \mu$ might well be within the expected domain of validity of non-relativistic physics, but the other pole at $E_T = -\mu$ is at least 2μ below threshold, or in non-relativistic units, $2\mu c^2$ below threshold, quite a long distance out to trust non-relativistic physics. Once again this second term is, aside from various kinematic factors, not a novel phenomenon of relativistic theory, but simply a conventional energy-eigenstate pole, which has precisely the location, and precisely the angular dependence, that one would expect from the non-relativistic formula.

Up to this stage, we have found nothing new. We've learned things, such as the right ways to generalize some non-relativistic phenomena to a particular relativistic problem, but we have found no relativistic phenomena that are without non-relativistic analogs. Now we go on to the next process. Again we will find nothing fundamentally new—just Yukawa potential-like terms, exchange Yukawa potential-like terms, and energy eigenstate pole-like terms. Let's work it out.

---

[3] [Eds.] See Marvin L. Goldberger and Kenneth M. Watson, *Collision Theory*, Dover Publications, 2004, p. 306, equation (376.b) and Philip M. Morse and Herman Feshbach, *Methods of Theoretical Physics*, Part 2, Mc-Graw Hill, 1953, p. 1077, equation (9.3.49).

11.2 Nucleon–meson scattering and meson pair creation

There are two processes left, and we're going to give the last, $N + \overline{N} \to \phi + \phi$, very little attention. Next to last is nucleon–meson scattering:

2.7 $N + \phi \to N + \phi$

Here are the diagrams that contribute to nucleon–meson scattering. Antinucleon–meson scattering is trivially related to this by charge conjugation, and the amplitudes are the same. There are two diagrams, though the second can be written as either (b) or (b'):

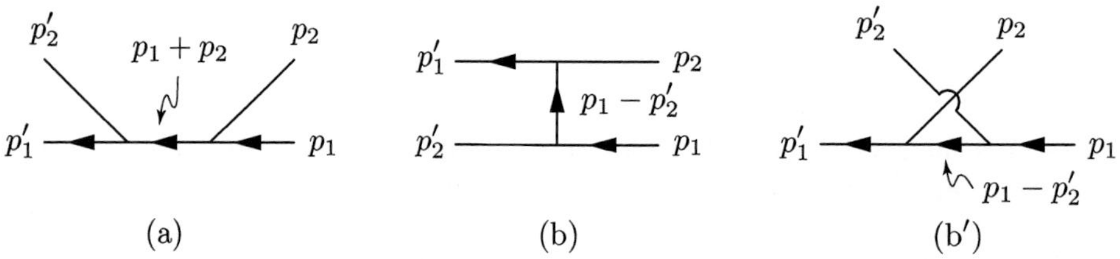

Figure 11.2: $\mathcal{O}(g^2)$ Feynman diagrams for Model 3 $N\phi \to N\phi$

Once again the internal momentum is fixed, $p_1 + p_2$ for Diagram (a) and $p_1 - p_2'$ for Diagram (b). The invariant amplitude is

$$i\mathcal{A}_{fi} = (-ig)^2 \left[\frac{i}{(p_1 + p_2)^2 - m^2 + i\epsilon} + \frac{i}{(p_1 - p_2')^2 - m^2 + i\epsilon} \right] \tag{11.5}$$

Please notice that the propagator mass is m^2, not μ^2 this time, because it is an internal nucleon line, not an internal meson line.

Now I need not belabor the first graph, Diagram (a), which describes an energy-eigenstate pole, just like the earlier Diagram (b) in nucleon–antinucleon scattering (Figure 11.1). The only difference is that now the energy eigenstate is a nucleon, appearing in the meson–nucleon channel, rather than the meson appearing in the nucleon–antinucleon channel. The arguments are, except for replacing μ by m, word for word the same as those I have just given.

The second graph, Diagram (b) in Figure 11.2, looks like an exchange Yukawa potential. It's rather odd in terms of non-relativistic scattering theory to see an exchange potential *without* a direct potential, but after all, mesons and nucleons are not identical particles. If nucleon and antinucleon can have a direct potential without an exchange potential, apparently meson and nucleon can have an exchange potential without a direct one. It is slightly different kinematically from the exchange potentials we discussed in the cases of nucleon–nucleon and nucleon–antinucleon scattering, because its range in the center-of-momentum frame is energy dependent. As I will now demonstrate, this arises because the meson and nucleon have different masses.

In the center-of-momentum frame,

$$\mathbf{p}_1 = -\mathbf{p}_2 = \mathbf{e}p \quad \text{where} \quad \text{where } p = |\mathbf{p}_1| = |\mathbf{p}_2| \quad \text{and } \mathbf{e} \text{ is a unit vector.} \tag{11.6}$$

The four-momenta are

$$\begin{aligned} p_1 &= (\sqrt{p^2+m^2}, \mathbf{e}p) \quad \text{(for the nucleon)} \\ p_2 &= (\sqrt{p^2+\mu^2}, -\mathbf{e}p) \quad \text{(for the meson)} \\ p_1' &= (\sqrt{p^2+m^2}, \mathbf{e}'p) \\ p_2' &= (\sqrt{p^2+\mu^2}, -\mathbf{e}'p) \end{aligned} \tag{11.7}$$

The denominator of the exchange term then becomes

$$\begin{aligned} (p_1 - p_2')^2 - m^2 &= (\sqrt{p^2+m^2} - \sqrt{p^2+\mu^2})^2 - 2p^2(1+\cos\theta) - m^2 \\ &= (\sqrt{p^2+m^2} - \sqrt{p^2+\mu^2})^2 - \Delta_c^2 - m^2 \end{aligned} \tag{11.8}$$

where, as in (10.45), $\mathbf{\Delta}_c = \mathbf{p}_1 - \mathbf{p}_2'$ is the cross momentum transfer. Unlike the case of nucleon–nucleon scattering, the energy terms in meson–nucleon scattering do not cancel, because $\mu \neq m$. This affects the range of the potential, because it is dependent on the Fourier transform of the amplitude.

In nucleon–nucleon scattering, we had

$$\text{(exchange amplitude)} \propto \frac{1}{\Delta_c^2 + \mu^2} \Rightarrow V(r) \propto \frac{e^{-\mu r}}{r} \tag{11.9}$$

That is,

$$\text{(exchange amplitude)} \propto \frac{1}{\Delta_c^2 + (1/R)^2} \Rightarrow V(r) \propto \frac{e^{-r/R}}{r} \tag{11.10}$$

so that the reciprocal of the mass μ serves as a range parameter, R. In nucleon–meson scattering, however,

$$\text{(exchange amplitude)} \propto \frac{1}{\Delta_c^2 + m^2 - \left(\sqrt{p^2+m^2} - \sqrt{p^2+\mu^2}\right)^2} \tag{11.11}$$

The reciprocal of R^2, formerly μ^2, is now $m^2 - \left(\sqrt{p^2+m^2} - \sqrt{p^2+\mu^2}\right)^2$. Consider the limits:

$$\lim_{p\to\infty} R^{-2} = \lim_{p\to\infty} \left[m^2 - (\sqrt{p^2+m^2} - \sqrt{p^2+\mu^2})^2\right] = m^2 \tag{11.12}$$

The inverse of the range parameter, R^{-1}, goes to m as $p \to \infty$, as if two equal-mass particles were exchanging an object of mass m, just as in the usual Yukawa potential. On the other hand,

$$\lim_{p\to 0} R^{-2} = \lim_{p\to 0} \left[m^2 - (\sqrt{p^2+m^2} - \sqrt{p^2+\mu^2})^2\right] = 2m\mu - \mu^2 \tag{11.13}$$

R^{-1} becomes smaller than m as $p \to 0$. (The right hand side is positive, because we have chosen $2m > \mu$. If this were not the case, the meson would not be stable.) Thus we have an exchange potential with an energy-dependent range, a novelty from the point of non-relativistic physics.

If we attempt to give some reality to this system, neglecting the fact that real nucleons have spin, by imagining the nucleon–meson mass ratio is that of a real nucleon and a real π

meson, that is to say something like 7:1, R^{-2} of the potential at low energies would be on the order of

$$2m\mu - \mu^2 = m^2\left[\frac{2\mu}{m} - \left(\frac{\mu}{m}\right)^2\right] \approx \tfrac{13}{49}m^2 \sim \tfrac{1}{4}m^2 \tag{11.14}$$

The Yukawa potential for nucleon–π meson scattering at high energies, ignoring spin, would go roughly as

$$V(r) \propto \begin{cases} e^{-mr}/r & \text{as } p \to \infty \\ e^{-\frac{1}{2}mr}/r & \text{as } p \to 0 \end{cases} \tag{11.15}$$

so the potential has much longer range at low energy than at high energy. At any given energy it's like the Born approximation for a Yukawa potential, but the value of this R parameter changes with the energy. It's a purely kinematic effect; there's no mystery to it. In the spinless case, this might have a significant effect on low energy meson–nucleon dynamics. In the spinless case, an exchange potential has the same effect as a direct potential in even partial waves, but the opposite effect in odd partial waves. Therefore we have a potential of rather long range in this problem which is attractive in sign for the even partial waves, and repulsive in sign for the odd partial waves. If we wish to be a little ambitious, and imagine we could turn up the potential just a slight amount while still using these ideas from perturbation theory—a dangerous step, but let's take it—we would expect in this case, because the potential is an exchange potential, to perhaps begin seeing bound states in the even partial waves, but never in the odd. If it were a direct potential, we would of course see bound states in all partial waves.

This is in fact very close to the dynamics of actual meson–nucleon low energy scattering for complicated reasons that we won't get to until quite late in the course. There is a potential between meson and nucleon caused by the exchange of a nucleon, rather like a Yukawa potential of quite long range at low energy, because μ is a very small number compared to m. When we take account of all the spin and isospin factors, it turns out to be attractive in odd partial waves and repulsive in even, and it isn't quite strong enough to make a bound state, but it is strong enough to make a nearly-bound state, or resonance, which is the famous Δ or N* resonance: an unstable p-wave state in the pion–nucleon system with a mass of 1232 GeV. We've now got about half the physics required to establish that. The parts we don't have are the kinematics involving spin, which we will get to in the course of time, and a good reason why we should trust lowest-order perturbation theory. We'll see not too much later why we should trust it, at least for the long-range part of the potential, because it doesn't get corrected. I have gotten ahead of my systematic analysis of lowest-order Feynman graphs, but I thought I would give you a taste of future physics.

The eighth process, and the last, of this class of $\mathcal{O}(g^2)$ Feynman graphs, is nucleon plus antinucleon goes into two mesons:

2.8 $\overline{N} + N \to \phi + \phi$

I won't bother to treat this in detail, since this process involves absolutely no novel features, and anyway I've given this to you as a homework problem (Problem 6.3, p. 261).

There are two graphs, shown in Figure 11.3. The second is the same as the first, with p_2' and p_1' interchanged. We can now just stare at these from what we have learned already without even writing down the expression and say, "Aha!" Diagram (a) is a direct Yukawa potential with energy-dependent range because the mass of the nucleon is not the mass of the

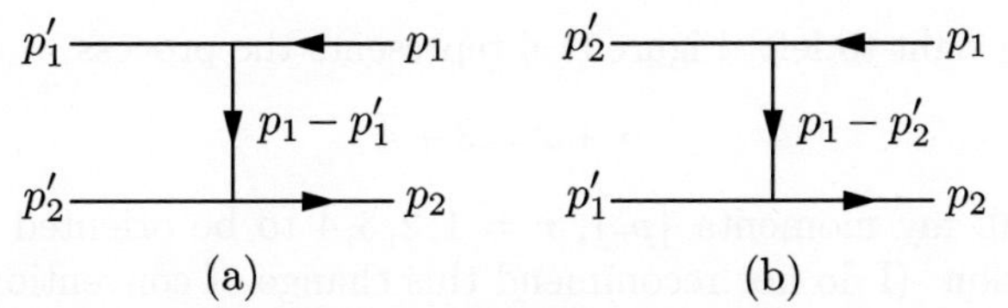

Figure 11.3: $\mathcal{O}(g^2)$ Feynman diagrams for Model 3 $N\overline{N} \to \phi\,\phi$

meson, and Diagram (b) is an exchange Yukawa potential with energy-dependent range. And of course the direct and the exchange potentials must come in together because, even though nucleon and antinucleon are not identical particles, meson and meson are. So if you have one graph, you must have the other graph.

That concludes the discussion, our first runthrough of the twenty-odd lowest non-vanishing Feynman diagrams to $\mathcal{O}(g^2)$ that arise in this theory.

11.3 Crossing symmetry and *CPT* invariance

We have identified three kinds of phenomena that arise in lowest-order perturbation theory, and have labeled them by names corresponding to the entities they become in the non-relativistic limit, to wit: direct Yukawa potential, exchange Yukawa potential, and energy-eigenstate pole. Although no one of these things is in any sense a relativistic novelty, there *is* a relativistic novelty, which I would now like to discuss. These three things are in fact aspects of *one* thing. I would like to explain how they are connected. It goes under the name of **crossing**. It is sometimes called **crossing symmetry**. You should put the word "symmetry" in quotes, because it has nothing to do with symmetries and conservation laws in the sense we've discussed them.

In order to discuss crossing, I have to introduce a slightly different notation than the one I have been using until now, and a slightly more general field theory. Just to keep things straight, consider a field theory in which there are four different kinds of particles, call them 1, 2, 3 and 4, none of which are equal to their antiparticles. That's the most general case. These particles have various trilinear interactions, and can exchange various charged and neutral mesons making the sort of Born approximation graphs we have been talking about. I would like to consider a general graph involving one of each of these particles. I don't know what's going on inside, and I don't want to specify the process at the moment, so I draw a blob. By convention I will choose to arrange all my lines so they all go inward.

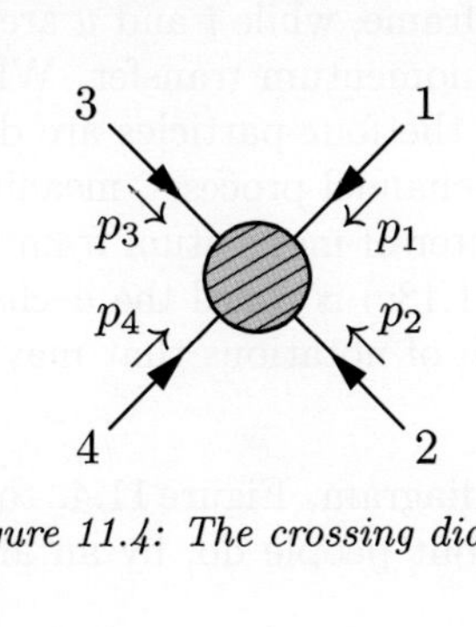

Figure 11.4: The crossing diagram

If we read it from right to left, Figure 11.4 represents the process

$$1 + 2 = \overline{3} + \overline{4} \tag{11.16}$$

I will also arrange all my momenta $\{p_r\}$, $r = 1, 2, 3, 4$ to be oriented inward, contrary to my previous convention. (I do not recommend this change of convention in general, but it's suitable for this particular discussion.) With this convention,

$$p_1 + p_2 + p_3 + p_4 = 0 \tag{11.17}$$

We will tell which particle is incoming and which particle is outgoing not by saying whether it's on the right or left but simply by checking whether the zeroth component p_r^0 is positive or negative. If it is positive it is an incoming particle; if negative it's outgoing. If it's really an outgoing particle, then the inward-oriented momentum is on the bottom mass hyperboloid, not the top.

Thus this blob in my new compressed notation could describe a variety of processes:

$$1 + 2 \to \overline{3} + \overline{4} \quad \text{(reading right to left)} \tag{11.18a}$$

$$1 + 3 \to \overline{2} + \overline{4} \quad \text{(by rotating the diagram 90° clockwise)} \tag{11.18b}$$

$$1 + 4 \to \overline{2} + \overline{3} \quad \text{(by reflecting the bottom half of the diagram horizontally)} \tag{11.18c}$$

I'm not telling you what the process is. You can deduce that only when you know the values of the p's—which ones have positive zeroth components and which ones have negative zeroth components. Of course, there are three other processes this could describe, the charge conjugates of these processes, with $\overline{1}$ replacing 1, and so on. I won't write those down for reasons that will become clear shortly. We'll get to them. I'm not assuming in any way that the interactions between these particles conserve charge or parity or are time-reversal invariant or anything like that. We're going to be very general.

No matter which process I am describing, it is very convenient to introduce an over-complete set of three kinematic variables to describe the system. Of course, for any given process we only need two, the energy and the scattering angle (in the center of momentum reference frame). Nevertheless, for reasons that will become clear, I want to introduce *three*:

$$s = (p_1 + p_2)^2 = (p_3 + p_4)^2 \tag{11.19a}$$

$$t = (p_1 + p_3)^2 = (p_2 + p_4)^2 \tag{11.19b}$$

$$u = (p_1 + p_4)^2 = (p_2 + p_3)^2 \tag{11.19c}$$

Any *two* of the three constitute a *complete set* of relativistic invariants, and any invariant can be expressed in terms of these. For the process (11.18a), drawn in Figure 11.4, s is the energy in the center-of-momentum frame, while t and u are minus the squares of momentum transfer, one direct and one cross momentum transfer. Which one I call direct and which I call crossed is a matter of taste, if the four particles are different. For this reason, process (11.18a) is sometimes called "the s–channel process", meaning it is the channel for which s is interpreted as the energy in the center-of-momentum frame. For the same reason (11.18b) is called the t–channel process, and (11.18c) is called the u–channel process. There is no physics in any of this. This is just a bunch of notations that may seem to you to be over-complex until we get to the pay-off.

Suppose we read the crossing diagram, Figure 11.4, top to bottom. In this channel—I shouldn't really call it a channel, but people do, by an abuse of language—in this process

(11.18b), the variable t is the energy in the center-of-momentum frame, while s and u are momentum transfers, and *vice versa* with (11.18c) and u. The variables s, t and u are called **Mandelstam variables** after Stanley Mandelstam.[4]

Because only two relativistic invariants are needed, and here we have three, there must be some formula relating s, t and u. Let's derive this relationship. We have

$$\begin{aligned} 2(s+t+u) &= (p_1+p_2)^2 + (p_3+p_4)^2 + (p_1+p_3)^2 + (p_2+p_4)^2 + (p_1+p_4)^2 + (p_2+p_3)^2 \\ &= 3\sum_i p_i \cdot p_i + 2\sum_{i>j} p_i \cdot p_j \\ &= 3\sum_i m_i^2 + 2\sum_{i>j} p_i \cdot p_j \end{aligned} \tag{11.20}$$

However I have an additional piece of information. Because all the p_i are oriented inward, the total momentum flowing into the diagram is zero (11.17), and so is its square:

$$0 = \Big(\sum_i p_i\Big)^2 = \sum_i p_i \cdot p_i + 2\sum_{i>j} p_i \cdot p_j = \sum_i m_i^2 + 2\sum_{i>j} p_i \cdot p_j \tag{11.21}$$

Subtracting (11.21) from (11.20) and dividing by two we find the rather pleasant and symmetric constraint

$$s+t+u = \sum_i m_i^2 \tag{11.22}$$

This expresses in a rather neat and simple way the dependence of the three variables.

We can represent the symmetric dependence graphically in a simple way. Let s, t and u be perpendicular axes as shown. The relationship (11.22) describes a plane:

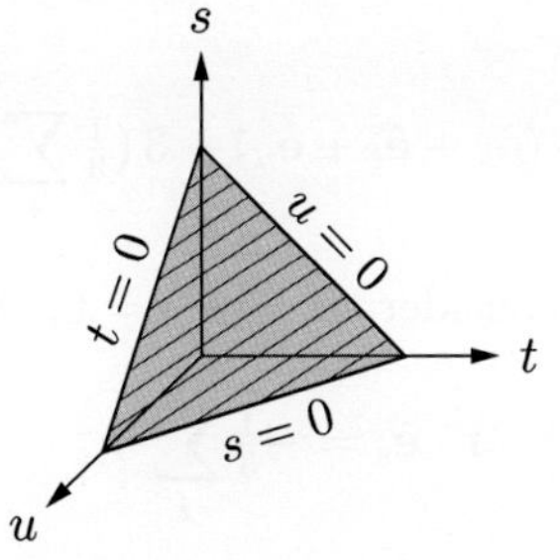

Figure 11.5: The Mandelstam plane $s+t+u = \sum_i m_i^2$

We can indicate the values of s, t and u by a point in this plane, since there are really only two independent variables. Take the origin of the plane to be the center of the equilateral triangle bounded by the lines $s = 0$, $t = 0$ and $u = 0$. I don't want to destroy the symmetry between s, t and u by, say, picking s and t, and declaring u the independent one. Instead, I introduce three unit vectors in the plane, $\hat{\mathbf{e}}_s$, $\hat{\mathbf{e}}_t$, and $\hat{\mathbf{e}}_u$, as shown in Figure 11.6. The unit vector $\hat{\mathbf{e}}_s$ is perpendicular to the line $s = 0$, $\hat{\mathbf{e}}_t$ is perpendicular to the line $t = 0$, and $\hat{\mathbf{e}}_u$ is perpendicular to the line $u = 0$.

---

[4] [Eds.] S. Mandelstam, "Determination of the Pion–Nucleon Scattering Amplitude from Dispersion Relations and Unitarity", *Phys. Rev.* **112** (1958) 1344–1360.

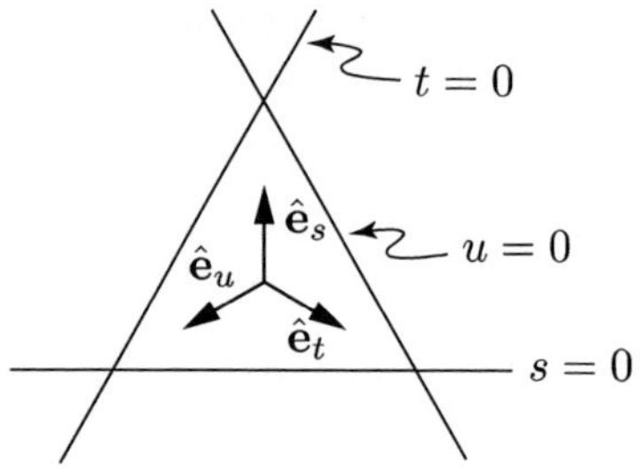

Figure 11.6: The Mandelstam plane and its unit vectors

The angle between any two of these unit vectors is $2\pi/3$, and they have the property that

$$\hat{\mathbf{e}}_s + \hat{\mathbf{e}}_t + \hat{\mathbf{e}}_u = 0 \tag{11.23}$$

Each unit vector has a square of 1, and an inner product with the other two unit vectors equal to $-\frac{1}{2}$.

As you can show for yourself, the vector $\mathbf{r}$ (Figure 11.7) from the origin to the point (s, t, u) can be written

$$\mathbf{r} = \tfrac{2}{3}(s\hat{\mathbf{e}}_s + t\hat{\mathbf{e}}_t + u\hat{\mathbf{e}}_u) \tag{11.24}$$

If we dot $\mathbf{r}$ with any Mandelstam unit vector, say $\hat{\mathbf{e}}_s$, we obtain the constraint

$$s = \mathbf{r} \cdot \hat{\mathbf{e}}_s + \tfrac{1}{3}\sum_i m_i^2 \tag{11.25}$$

and likewise for t and u. So every point in the plane is associated with a triplet of numbers s, t and u which obey (11.22),

$$s + t + u = \mathbf{r} \cdot (\hat{\mathbf{e}}_s + \hat{\mathbf{e}}_t + \hat{\mathbf{e}}_u) + 3\left(\tfrac{1}{3}\sum_i m_i^2\right) = \sum_i m_i^2 \tag{11.26}$$

Just to show you how this works, consider the line $s = 0$, the set of all points whose vectors $\mathbf{r}$ satisfy the relation

$$\mathbf{r} \cdot \hat{\mathbf{e}}_s = -\tfrac{1}{3}\sum_i m_i^2 \tag{11.27}$$

The lines $t = 0$ and $u = 0$ are similar. For a given point on the Mandelstam plane, s is the perpendicular distance from the point to the line $s = 0$, t is the perpendicular distance to the line $t = 0$ and u is the perpendicular distance to the line $u = 0$. Given a point $\mathbf{r}$, if you want to know what s, t and u are, you just have to draw the three perpendiculars to these three lines and measure the distances; see Figure 11.7.[5]

This is a very useful plot, not only in this problem but in problems like three-particle decays where you'd like very much to express things in terms of a symmetric set of variables, especially if the decay involves three identical particles, say three neutral pions. The energies of the three pions would be a useful set of variables, but they're constrained: the sum of those

[5] [Eds.] Adapted from John M. Ziman, *Elements of Advanced Quantum Theory*, Cambridge U. P., 1969, p. 205.

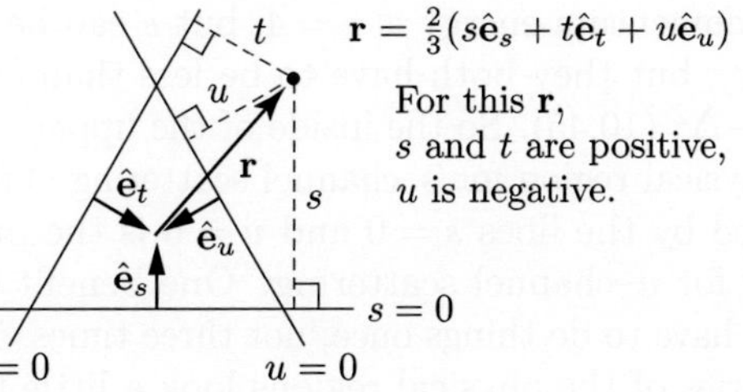

Figure 11.7: The variables s, t and u for a point $\mathbf{r}$

energies has to be the energy of the decaying particle. In the case of three-particle decays, this is called a **Dalitz plot**,[6] and we'll say something about it next lecture. The case we're considering is called the **Mandelstam–Kibble** plot. It was no doubt invented by Euclid; it's nothing but classical geometry. So far, this plot is just a way of representing three constrained variables. Let's get back to our three scattering processes.

Not every point in this plane corresponds to a physical scattering process. For example, if I pick a random point where all three values s, t and u are positive, (say, near the origin), that would correspond to *no* physical scattering process, because two of them have to be the squares of a momentum transfer, which is either zero or negative. Let's sketch out the regions of the Mandelstam–Kibble plot that correspond to our three physical scattering processes, as in Figure 11.8. In general the boundaries are rather complicated and involve solving quartic equations, so just to give you an idea of what they look like, I will restrict myself to the case where all of the masses are equal. For convenience I will choose $m_r^2 = 1$, for each r. The four particles may still be distinct.

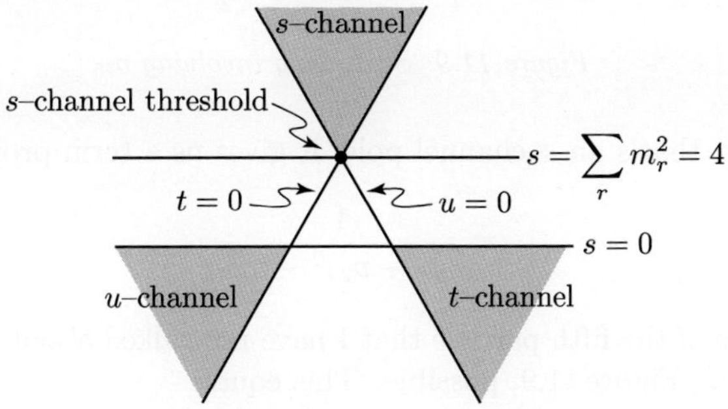

Figure 11.8: The Mandelstam–Kibble plot

In the center-of-momentum frame, $p_1 + p_2 = (E_1 + E_2, \mathbf{0})$. The physical region for the s channel is

$$s = (p_1 + p_2)^2 = (E_1 + E_2)^2 \geq (m_1 + m_2)^2 = 4 \tag{11.28}$$

[6] [Eds.] Richard H. Dalitz (1925–2006) was a particle physicist from Australia, and a student of Rudolf Peierls at Birmingham. Soon after Peierls went to Oxford, he invited Dalitz to join him. There Dalitz taught Christopher Llewellyn-Smith, a future director general of CERN, and many others. Dalitz was one of the early proponents of quarks as physical entities, and not merely mathematical abstractions. Dalitz plots are introduced in R. H. Dalitz, "On the analysis of τ-meson data and the nature of the τ meson", *Phil. Mag.* **44** (1953) 1068–1080, and "Decay of τ Mesons of Known Charge", *Phys. Rev.* **94** (1954) 1046–1051.

The threshold center-of-momentum energy is $s = 4$, but s can be above threshold. The other variables t and u can vary, but they both have to be less than or equal to zero: one is $-\Delta^2$ (10.43), and the other is $-\Delta_c^2$ (10.45). So the inside of the upper triangle bounded by the lines $t = 0$ and $u = 0$ is the physical region for s–channel scattering. Likewise inside the lower right triangular region bounded by the lines $s = 0$ and $u = 0$ is the physical region for t–channel scattering, and similarly for u–channel scattering. One benefit of this way of studying the scattering is that we only have to do things once, not three times. If our particles have unequal masses, then the boundaries of the physical regions look a little bit more complicated. They curve around, wiggle and bend. Of course, they asymptotically approach this plot as the energy gets large compared to the masses. The shaded regions never overlap, because there's no possible way that you can have a process being in two channels at the same time, say the physical s–channel and the physical t–channel.

Up until now we've been living in one or the other of these shaded regions. In our theory of mesons and nucleons, the masses are different, and so our kinematics are not those of equal masses, and the regions aren't quite so simple in shape. But we've been living in one or the other of those shaded regions and have systematically gone dancing, exploring things in each of these allowed regions. The actual amplitudes we have obtained are however defined for *all* s, t and u. They are meromorphic functions of the invariants. In particular, consider a process, say as shown in Figure 11.9, involving a fifth type of particle, different from the other four, being exchanged between two others.

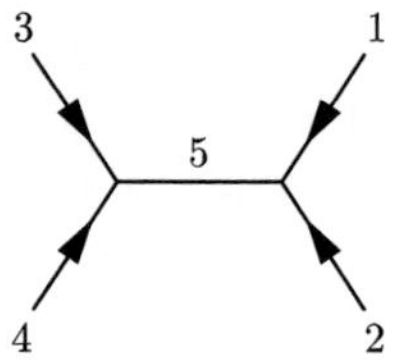

Figure 11.9: Scattering involving m_5

Reading right to left, that's an s–channel pole; it gives us a term proportional to

$$\frac{1}{(p_1 + p_2)^2 - m_5^2} \tag{11.29}$$

where m_5 is the mass of the fifth particle that I have not talked about yet. I introduced it just to make that diagram, Figure 11.9, possible. This equals

$$\frac{1}{s - m_5^2} \tag{11.30}$$

That's a meromorphic function. The pole is located at $s = m_5^2$ which had better be below the 1–2 threshold.

The line $s = m_5^2$ is where that function has a pole. Unfortunately I cannot draw it as a point in the complex plane because I would need two complex variables, which are hard to draw; it's a four-dimensional graph. But fortunately the location of the pole is on the real part of that plane and is everywhere along the line $s = m_5^2$.

That amplitude (11.30) is analytically defined for all s, t and u, aside from right on top of the pole, of course. What does that amplitude look like? In the s–channel, it looks like an

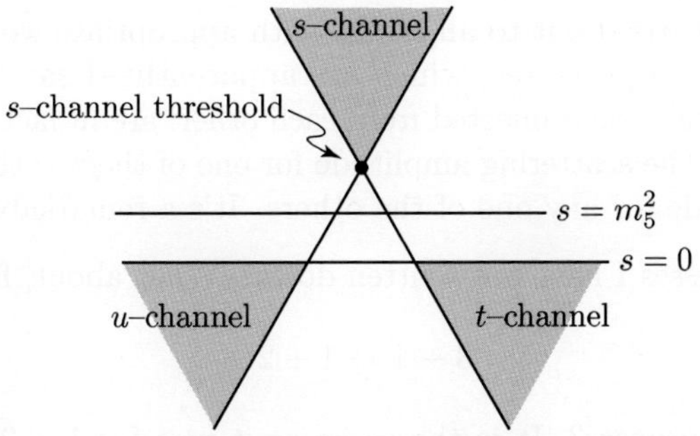

Figure 11.10: The Mandelstam–Kibble plot, showing $s = m_5^2$

energy-eigenstate pole. What does it look like if I'm in the t–channel? Well, I just analytically continue the same amplitude. In the t–channel, reading the crossing diagram, Figure 11.4, from top to bottom, the incoming particles are 1 and 3. Read this way, s is a momentum transfer (squared) in the t–channel, and t is the center-of-momentum energy (squared). So exactly the same meromorphic function down in the lower right shaded region, the t–channel, looks like a momentum-transfer pole, i.e., a direct Yukawa potential. And over in the lower left shaded region, the u–channel, it looks like an exchange Yukawa potential. That is, the three classes of phenomena we have been discussing—the Yukawa potential, the exchange potential and the energy-eigenstate pole—are in fact simply three aspects of the *same* meromorphic function restricted to three disconnected regions of the complex s–t–u plane. A direct Yukawa potential in this sense is simply the analytic continuation of an energy-eigenstate pole, and so is an exchange Yukawa potential. These processes are no more independent entities than are the functions $\sin z$ and $i \sinh z$, objects that look very different, but are the same meromorphic functions restricted to two different real environments in the complex plane.

Therefore we have this property, unfortunately called *crossing symmetry*, but all the same a remarkable feature of scattering theory. By reading the crossing diagram from top to bottom we're crossing the line from the past into the future; by reading it from bottom to top, we're crossing from the future into the past. These three processes appear to be completely different. The same process in our model could be nucleon–antinucleon annihilation into meson–meson pair production, reading top to bottom; reading right to left, it could be meson–nucleon scattering. Nevertheless, they're connected by analytic continuation. In fact the three different phenomena we have discussed are manifestations of a single meromorphic function in this s–t–u plane. This is something we do *not* see in non-relativistic physics: the Yukawa potential in non-relativistic physics has no connection with an energy-eigenstate pole for some other process. The regions are physically separated by energies $\sim m^2$. When we go to the non-relativistic limit, (including a c^2 in the mass terms), and the three regions become very far apart: mc^2 is large, and we can't analytically continue from one region to another. They become disconnected as c goes to infinity.

To what extent do we expect this crossing symmetry to be an artifact of our lowest-order theory? Well, certainly we expect things to be much more complicated when we go to higher orders, because we know even in non-relativistic physics the scattering amplitudes in general do not only have poles, they also have branch cuts. So presumably there will be all sorts of cuts floating around this complex two-variable plane, and we'll have to worry when we analytically continue: Do we go above the cut or below the cut, and which way do we go around? I assure you, the analysis can be carried out to all levels, with appropriate worrying, as we will do in the second semester. These processes, which are apparently—I say it again because it's so important—*apparently* totally disconnected from each other, are in fact connected by a process of analytic continuation. The scattering amplitude for one of them is the analytic continuation of the scattering amplitude for any one of the others. It's a remarkable fact.

What about the processes I have not written down? What about, for example, the process

$$3 + 4 \to \overline{1} + \overline{2} \tag{11.31}$$

Well, what is s for this process? It is the same as it was for $1 + 2 \to \overline{3} + \overline{4}$. That's just changing the signs of all the momenta; it doesn't affect these quadratic invariants. These processes correspond to the same point in the Mandelstam plane. If I change the sign of all four momenta, changing all of my incoming particles into outgoing antiparticles, and all of my outgoing particles into incoming antiparticles, I haven't changed a single thing. In fact, this equality has nothing to do with analytic continuation, and nothing to do with lowest-order perturbation theory. Because my Feynman diagrams only involve quadratic functions of the p's, all my Feynman rules to all orders in perturbation theory for the theory in question are manifestly unchanged if I make the transformation

$$p_r \to -p_r \tag{11.32}$$

for all r. We could have all sorts of complex coupling constants that would break parity invariance, and we could have terms that involve $\psi + i\psi^*$ floating around somewhere that would break charge conjugation invariance. We could even have parity-violating terms involving derivative interactions. I haven't told you yet how to derive the Feynman rules for derivative interactions, but in momentum space the derivatives are replaced by momenta, and we might expect the interactions to change sign. But if there's any grace in the world, the rules should involve at the vertices an epsilon tensor $\epsilon_{\mu\nu\rho\sigma}$ with four momenta. And since the epsilon tensor has four indices in four dimensions, and therefore involves four momenta, when I change the sign of all momenta, interior as well as exterior, that term in epsilon, multiplied by $(-1)^4$, is not going to change sign, either. It's special to four dimensions, but it's still true that the derivative coupling won't change when $p \to -p$. It looks like any Lorentz invariant interaction I can write down will be invariant under changing the sign of all the momenta.

In general, then, for any Lorentz invariant interaction, to all orders of perturbation theory, amplitudes are unchanged if I take all the momenta and change their signs, that is to say, if I take every incoming particle and turn it into an outgoing antiparticle with exactly the same three-momentum. With our convention, this means multiplying the four momenta by -1. This invariance is called ***CPT* symmetry** for reasons I will shortly make clear. It says the scattering amplitude for a given process is exactly the same as the scattering amplitude for the reverse process where all the incoming particles are turned into outgoing antiparticles and *vice versa*, to all orders of perturbation theory. This is the ***CPT* theorem**.[7] It is just a consequence of Lorentz invariance, but it is a remarkable result.

---

[7] [Eds.] Usually, the *CPT* theorem states that if a local quantum field theory is Lorentz invariant and the usual connection between spin and statistics holds, then the theory is invariant under the combination of operations *CPT*. J. Schwinger, "Theory of Quantized Fields, I.", *Phys. Rev.* **82** (1951) 914–92 (reprinted in Schwinger *QED*); W. Pauli, "Exclusion Principle, Lorentz group and reflection of space-time and charge", in *Niels Bohr and the Development of Physics*, pp. 30–51, McGraw-Hill, 1955; Gerhart Lüders, "Proof of the *CPT* theorem", *Ann. Phys. (NY)* **2** (1957) 1–15.

The *CPT* theorem says that the world may violate parity—we've written down examples. It may violate charge-conjugation invariance. It may violate time reversal. It's trivial to write down examples to do that. But, if the world is Lorentz invariant, it *cannot* violate *CPT*. Notice this is not like parity, for which there are phase ambiguities, or charge conjugation or time reversal individually. There is *no* phase ambiguity, there are no minus signs. This theorem not only tells you that there *is* a *CPT* symmetry, it tells you *what it does*, at least for theories only involving scalar particles, which is all we can handle now. *CPT* symmetry turns an incoming nucleon into an outgoing antinucleon with a plus sign. It turns an incoming K^+ meson into an outgoing K^- meson. As we will see, the theorem can be generalized to spinor particles; *CPT* does something to their spin. It's called *CPT*, because it combines the operations of time reversal, charge conjugation and parity taken together. It changes incoming particles to outgoing particles, which is what time reversal T does; it changes particles to antiparticles, which is what charge conjugation C does; and it changes the sign of space variables, which is what parity P does. Notice it does not change the sign of three-momentum, because of the combined action of TP.

To give a specific example, consider once again the s–channel process

$$1 + 2 \rightarrow \bar{3} + \bar{4} \tag{11.18a}$$

The amplitude for this process can be written as

$$\mathcal{A}_{fi} = \mathcal{A}(p_1, p_2, -p_3, -p_4) \tag{11.33}$$

where $\mathcal{A}$ is a function of the particles' four momenta. If we charge conjugate this process with the operator C, we get a related process with an amplitude $\mathcal{A}^C_{fi}$,

$$C\colon 1 + 2 \rightarrow \bar{3} + \bar{4} \longrightarrow \bar{1} + \bar{2} \rightarrow 3 + 4 \tag{11.34}$$

If a theory is invariant under charge conjugation, then these amplitudes are the same, but in general they won't be the same. Now let's consider the charge-conjugated s–channel process under T, the time-reversal operator. If you run a movie backwards, the products of a reaction become the reactants, and *vice versa.* What once went north now goes south, and what once went up now goes down: velocities are reversed. So time reversal does two things: it switches the role of incoming and outgoing particles, and it reverses the direction of velocities. Finally, we apply the parity operator P to the time-reversed, charge-conjugated s–channel process. What this does is undo the reversal of velocities without swapping the roles of the incoming and outgoing particles:

$$TP\colon \bar{1} + \bar{2} \rightarrow 3 + 4 \longrightarrow 3 + 4 \rightarrow \bar{1} + \bar{2} \tag{11.35}$$

The amplitude for this process can be written as

$$\mathcal{A}^{TPC}_{fi} = \mathcal{A}(-p_1, -p_2, p_3, p_4) \tag{11.36}$$

The original s–channel process and its *CPT*-transform occupy the *same* point on the Mandelstam–Kibble plot, and so the change of $p_r \rightarrow -p_r$ cannot change anything. That is,

$$\mathcal{A}_{fi} = \mathcal{A}^{TPC}_{fi} \tag{11.37}$$

I want to emphasize the importance of *CPT* invariance. If an experiment were found to violate *CPT*, it would not be like the downfall of parity in the original Wu-Ambler experiments,[8] nor like the violation of CP invariance (and hence T individually) in the Fitch-Cronin experiments[9] on K^0 decays. All that happened then was that someone said, "Well, that just means it's not a CP-conserving theory." So we write down our possible Lagrangians, and all those terms we crossed out before, because they were CP-violating, we now leave in. That's not a revolution. We just go back and fix things up with a CP-violating interaction, and that's it. But if CPT violation is observed in the laboratory, that means Lagrangian field theory in general is cooked! Out the window! We'd have to start afresh. *That* would be a revolution.

11.4 Phase space and the S matrix

So much for grand abstract themes and powerful, beautiful general theorems. We now have to begin a bit of dirty work. We now have a deep understanding of everything about S-matrix elements, except how to connect them to cross-sections, which is what experimentalists measure. And we have to get that kinematics right if we want to understand what we're doing. So from a high plane of abstraction, we descend to a valley of practicality and go through the manipulations required to find the formulas that turn S-matrix elements into differential cross-sections, $d\sigma/d\Omega$.

This is purely a kinematic problem, and there are two ways of approaching it. One is to be extra-careful and consider a realistic scattering experiment with wave packets. That's the right way to do it, but it takes a long time. So I will do it fast and dirty by putting everything in a box, computing my scattering amplitude in the box, and then letting the size of the box go to infinity. I'm also going to turn things on and off in time. If you use wave packets, the box and turning things on and off in time are unnecessary and awkward. The box is there to make the kinematics simple by replacing integrals by discrete sums. So I put the world in a cubical box of side L, and volume $V = L^3$, with momenta given by

$$\mathbf{p} = \frac{2\pi}{L}(n_x, n_y, n_z) \qquad (n_x, n_y, n_z \text{ are integers}) \tag{11.38}$$

and I put in an adiabatic function $f(t)$, turned on for a time T. I will then choose the one-particle states in the theory's Hilbert space to be box normalized as we discussed earlier, in §2.2, treating Fock space in a box,

$$\langle \mathbf{p}|\mathbf{p}'\rangle = \delta_{\mathbf{p}\mathbf{p}'} \tag{11.39}$$

The momenta $\mathbf{p}$ and $\mathbf{p}'$ run over the discrete set allowed by the box. This is also the commutator of the creation and annihilation operators,

$$[a_{\mathbf{p}}, a^\dagger_{\mathbf{p}'}] = \delta_{\mathbf{p}\mathbf{p}'} \tag{11.40}$$

I have not described the expansion of the free field in a box before, but it's easy to see what it is:

$$\phi(x) = \sum_{\mathbf{p}} \frac{1}{\sqrt{V}}\frac{1}{\sqrt{2E_{\mathbf{p}}}}\left(a_{\mathbf{p}}e^{-ip\cdot x} + a^\dagger_{\mathbf{p}}e^{ip\cdot x}\right) \tag{11.41}$$

[8] [Eds.] See footnote 8, p. 121.

[9] [Eds.] J. H. Christenson, J. W. Cronin, V. L. Fitch and R. Turlay, "Evidence for the 2π decay of the K^0_2 meson", *Phys. Rev. Lett.* **13** (1964) 138–140. In 1980 James Cronin and Val Fitch won the Physics Nobel Prize for this work.

Instead of an integral over $\mathbf{p}$, we have a sum on $\mathbf{p}$, and in the denominator $\sqrt{2E_{\mathbf{p}}}$, instead of the square root of $2\omega_{\mathbf{p}}$ we had before, in (3.45). Instead of the $1/(2\pi)^{3/2}$ which is appropriate for a delta-function normalization, we have a $1/\sqrt{V}$ appropriate for box normalization. It is easy to see that this is right by checking that it gives the right equal time commutators (3.61) between $\phi(\mathbf{x},t)$ and $\dot{\phi}(\mathbf{y},t)$.[10]

Now since everything is in a box and the interaction is only going on for a finite time, I can directly calculate the transition probability between a given initial state and a given final state:

$$(\text{transition probability}) = \left|\langle f|S-1|i\rangle\right|^2 \tag{11.42}$$

We subtract the one so if there is no interaction then there's no transition. I will restrict myself in these lectures to two kinds of initial states. I could consider a one-particle initial state:

$$|i\rangle = |\mathbf{p}\rangle \tag{11.43}$$

That's one of the nice advantages of turning the interaction on and off, because then I can get a crude but serviceable theory of decay processes, to wit, I put the particle in the box, turn on the interaction, and watch it decay into various final states. I might have an unstable particle in my model, and I will develop rules for calculating the lifetime of such an unstable particle. I may also want to consider a scattering processing in which I have two particles. Your first thought might be to write

$$|i\rangle \stackrel{?}{=} |\mathbf{p}_1,\mathbf{p}_2\rangle \tag{11.44}$$

But that's not correct. Each particle has probability one of being somewhere in the box. If (11.44) were correct, as I let the box get bigger and bigger the probability of finding the second particle in the neighborhood of the first particle goes to zero, and therefore I should expect the transition amplitude to go to zero as the box expands, which is not correct. The right way to normalize this initial state is to set

$$|i\rangle = \sqrt{V}\,|\mathbf{p}_1,\mathbf{p}_2\rangle \tag{11.45}$$

Then we can say one of the particles has probability 1 of being in the box and the second particle has probability 1 of being in any unit volume. As we let the box get bigger and bigger, the probability of the second particle being near the first particle stays constant. We don't want them not to scatter simply because there is no chance of getting within any appreciable distance of each other, and we put in the $\sqrt{V}$ to take care of that.

Now we are ready to go. I will write down the expression for the matrix element of $S-1$ for any given final state (some collection of particles with specified momenta). I'm not going to restrict the final states; they could be two-particle states or they could be 32-particle states. What will I have? Well, I want to write this so it looks as much as possible like (10.27), the formula we found in the relativistic case. I'll write it like this:

$$\langle f|S-1|i\rangle = i\mathcal{A}^{VT}_{fi}(2\pi)^4\delta^{(4)}_{VT}(p_f-p_i)\prod_{\text{final}}\left[\frac{1}{\sqrt{2E_{\mathbf{p}_f}}}\frac{1}{\sqrt{V}}\right]\prod_{\text{initial}}\left[\frac{1}{\sqrt{2E_{\mathbf{p}_i}}}\right]\frac{1}{\sqrt{V}} \tag{11.46}$$

[10] [Eds.] $(1/L)\sum_{n=-\infty}^{\infty}e^{-i(2\pi n/L)x} = (1/L)\sum_p e^{-ipx} = \delta(x)$ is the Fourier series expansion of the delta function.

I'll have an invariant amplitude, not quite the same as before, because it involves a sum on **p**'s, rather than a continuous integral. I'll indicate that by putting in a superscript VT, indicating it represents factors from the box volume and the time during which the interaction is turned on. The invariant amplitude $\mathcal{A}_{fi}^{VT}$ is so constructed that

$$\lim_{V,T\to\infty} \mathcal{A}_{fi}^{VT} = \mathcal{A}_{fi} \tag{11.47}$$

There is also a factor that looks like a delta function but isn't, quite. I'll write down what that is in a moment.

The first three factors on the right of (11.46) look like the relativistic form. That would be all there was if the states $|\mathbf{p}\rangle$ in the box were normalized the same way that relativistic states $|p\rangle$ are normalized, but they ain't! There are extra factors you have never seen before, coming from the states' normalization. Instead of (10.20), we have

$$\langle 0|\psi(x)|\mathbf{p}\rangle = \frac{1}{\sqrt{2E_{\mathbf{p}}}}\frac{1}{\sqrt{V}}e^{-ip\cdot x} \tag{11.48}$$

I've got to put in this energy denominator and $\sqrt{V}$ factor for each of the annihilation and creation operators, and take the product over all the final particles. The product over all initial particles is a little different. We get $1/\sqrt{2E_{\mathbf{p}_i}}$ for each particle's energy, but we only have one factor of $1/\sqrt{V}$ whether we have one or two particles in the initial state. If there is only one particle, we get a $1/\sqrt{V}$ factor. If there are two, we have $(1/\sqrt{V})^2$, but one factor will be canceled by the unconventional normalization (11.45) we used to define the initial two-particle state. That is, there will be a single factor of $1/\sqrt{V}$ whether the initial state is one particle or two particles.

Finally, let's address the new delta function, $\delta_{VT}^{(4)}(p_i - p_f)$. We've got to be a little bit more careful about this, because when we calculate the probability (11.42), we're going to get its square. If we say $\delta_{VT}^{(4)}(p_i - p_f)$ approaches a delta function, that doesn't mean its square goes to the square of a delta function, because the square of a delta function is garbage. Let me write down an explicit expression for this thing:

$$(2\pi)^4\delta_{VT}^{(4)}(p) \equiv \int_V d^3\mathbf{x}\int_{-T/2}^{T/2} dt\, f(t)\, e^{ip\cdot x} \tag{11.49}$$

That's how our energy-momentum-conserving delta function came out before, by doing an x integral. Here we're also doing an x integral, but we're doing it only over a restricted volume of space and a restricted duration of time. Sure enough, this is a highly peaked function that goes to an ordinary delta function as V and T go to infinity:

$$\lim_{V,T\to\infty}\int_V d^3\mathbf{x}\int_{-T/2}^{T/2} dt\, f(t)\, e^{ip\cdot x} = \int d^4x\, e^{ip\cdot x} = (2\pi)^4\delta^{(4)}(p) \tag{11.50}$$

There's no question about that. But that's not what we're interested in. As I said, we're interested in its *square.*

Well, its square is also going to approximate a delta function, because if something is highly peaked, its square is also highly peaked. We'll get a delta function again, but with what coefficient? A very short computation turning a Fourier transform into an integral over position space using Parseval's theorem (9.32), shows us that this is $(2\pi)^4 VT$:

$$\begin{aligned}
\int \frac{d^4p}{(2\pi)^4}\left|(2\pi)^4\delta^{(4)}_{VT}(p)\right|^2 &= \int_V d^3\mathbf{x}\left[\int_V d^3\mathbf{x}'\int\frac{d^3\mathbf{p}}{(2\pi)^3}e^{-i\mathbf{p}\cdot(\mathbf{x}-\mathbf{x}')}\right]\int\frac{d\omega_\mathbf{p}}{2\pi}\left|\int_{-T/2}^{T/2}dt\,e^{i\omega_\mathbf{p}t}f(t)\right|^2\\
&= \int_V d^3\mathbf{x}\int_V d^3\mathbf{x}'\,\delta^{(3)}(\mathbf{x}-\mathbf{x}')\int\frac{d\omega_\mathbf{p}}{2\pi}|\tilde{f}(\omega_\mathbf{p})|^2\\
&= \int_V d^3\mathbf{x}\int dt\,|f(t)|^2 = VT
\end{aligned}\tag{11.51}$$

Therefore we should write

$$\lim_{V,T\to\infty}\left[(2\pi)^4\delta^{(4)}_{VT}(p)\right]^2 = (2\pi)^4\,VT\,\delta^{(4)}(p) \tag{11.52}$$

Now we are ready to do the limit except for one thing. When we compute the square, we get the transition probability to a fixed final state. This is of course a dumb thing to look at as the volume of the box goes to infinity, because the allowed values of $\mathbf{p}$ are little dots lying on a cubic lattice.

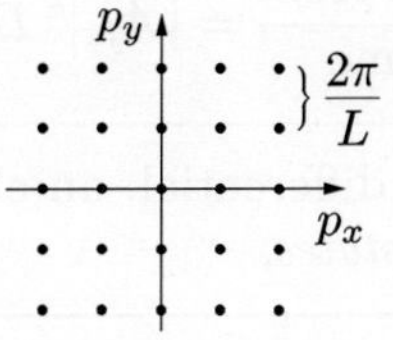

Figure 11.11: Density of allowed momentum values

If we focus on some small volume of $\mathbf{p}$ space, we get more and more dots as the size L of the box increases: $p_i \propto 1/L$, so the separation between dots inside the volume of $\mathbf{p}$ space decreases. The number of states in a small volume $d^3\mathbf{p}$ goes like

$$\left(\frac{n_x n_y n_z}{p_x p_y p_z}\right) d^3\mathbf{p} = \left(\frac{L}{2\pi}\right)^3 d^3\mathbf{p} = \frac{V d^3\mathbf{p}}{(2\pi)^3} \tag{11.53}$$

Therefore if we want something that has a smooth limit (as V and $T\to\infty$), what we should look at is a differential transition probability,

$$\begin{pmatrix}\text{differential transition}\\ \text{probability}\end{pmatrix} = \begin{pmatrix}\text{transition}\\ \text{probability}\end{pmatrix}\times\prod_{\text{final}}\frac{V d^3\mathbf{p}_f}{(2\pi)^3} \tag{11.54}$$

which is the probability for going to some fixed volume of final state space. We don't want to compute the *total* transition probability, we want to compute the transition probability integrated over some small region of final state space.

We are now in a position to stick all of this together and allow V and T to go to infinity.

Now comes the crunch: Will V disappear? Putting all the factors together we have

$$\begin{aligned}\begin{pmatrix}\text{diff. transition}\\ \text{probability}\end{pmatrix} &= |\langle f|S-1|i\rangle|^2 \times \prod_{\text{final}} \frac{Vd^3\mathbf{p}_f}{(2\pi)^3} \\ &= \left|\mathcal{A}_{fi}^{VT}\right|^2 (2\pi)^4\, VT\,\delta^{(4)}(p_f - p_i)\left[\prod_{\text{final}} \frac{1}{2E_f}\frac{1}{V}\frac{Vd^3\mathbf{p}_f}{(2\pi)^3}\right]\left[\prod_{\text{initial}}\left(\frac{1}{2E_i}\right)\right]\frac{1}{V}\end{aligned} \tag{11.55}$$

Ta-daa! All the factors of V cancel, so there's no problem in going to the limit $V \to \infty$. It looks like we didn't make any errors. Well that's rather nice, isn't it? There's our old Lorentz-invariant measure coming back again. We've still got the factor of T, but of course we expect that if we keep the interaction on forever, and have particles described by plane-wave states, they will go on interacting forever. So we divide by T, and taking the limits, we can write the differential transition probability per time as

$$\begin{pmatrix}\text{differential transition}\\ \text{probability}\\ \text{per unit time}\end{pmatrix} = (2\pi)^4\delta^{(4)}(p_f - p_i)\left|\mathcal{A}_{fi}\right|^2 \prod_{\text{final}} \frac{d^3\mathbf{p}_f}{(2\pi)^3 2E_f} \prod_{\text{initial}} \frac{1}{2E_i} \tag{11.56}$$

This is the master formula, sometimes written as

$$\boxed{\frac{\text{diff. trans. prob.}}{\text{unit time}} = \left|\mathcal{A}_{fi}\right|^2 D \prod_{\text{initial}} \frac{1}{2E_i}} \tag{11.57}$$

where D is an invariant phase space differential, an element of volume in final state space, called *the relativistic density of final states,*

$$\boxed{D = (2\pi)^4\delta^{(4)}(p_i - p_f)\prod_{\text{final}} \frac{d^3\mathbf{p}_f}{(2\pi)^3 2E_f}} \tag{11.58}$$

D should really be called *the final state measure.* It's like the density of final states that you always have to play with when you do time-dependent perturbation theory in non-relativistic physics. Of course it's energy-momentum conserving, so there's $(2\pi)^4\delta^{(4)}(p_i - p_f)$ for the total incoming momentum which is determined by what the initial state is, minus the total outgoing momentum which is the sum of the four-momenta for all the outgoing particles. That tells you there's *no* probability for making a final state which doesn't conserve energy and momentum.

The master formula is easy to remember. The density of final state factors is the one thing that's unnatural. It's there to make things have the right Lorentz transformation properties; for example, so a moving particle will decay more slowly than a stationary particle. Please notice I have gone to great care to arrange these conventions so there is no problem remembering where the 2π's go. You may think this is a silly thing to be proud of, if you've never tried to do a Feynman calculation in another convention. In the Feynman rules and in the density-of-states factor, there is one and only one origin of a 2π in a denominator, and there is one and only one origin of a 2π in the numerator: Every 2π in the denominator is associated with a differential dp in the numerator; every 2π in the numerator is associated with a delta function of p.

At the beginning of the next lecture I will apply these rules to obtain specific formulas for scattering into two particles, scattering into three particles, decay processes, etc. Once I'm done with that, you will be prepared to do the homework.

12

Scattering II. Applications

I will devote this lecture to the systematic exploitation of the formulas (11.57) and (11.58). This will be a nice change, because it will not involve any new ideas and therefore you don't have to think too hard. I will apply these formulas to five straightforward (or even pedestrian) topics. First I will discuss decay processes. After decay processes, I'll talk about cross-sections and I'll explicitly evaluate D for two-particle final states in the center-of-momentum frame. I'll discuss the famous Optical Theorem that connects the imaginary part of the forward scattering amplitude to the total cross-section. Finally I will discuss D for three particle final states and say a little bit more about those Dalitz plots that are so useful.

12.1 Decay processes

Let us begin with decay processes. We start out with some one-particle state that would be stable were it not for its interactions, turn on the interactions and watch it decay. We know the rate at which it decays from our master magic formula, (11.57). There's only one particle in the initial state. The differential transition probability for decaying into some n-particle final state is conventionally called $d\Gamma$. It is given by our master formula,

$$d\Gamma = \frac{\text{diff. decay prob.}}{\text{unit time}} = \frac{1}{2E_{\mathbf{p}}}|\mathcal{A}_{fi}|^2 D \tag{12.1}$$

$E_{\mathbf{p}}$ is the energy of the initial particle, with momentum $\mathbf{p}$. I don't bother to put an index on it because there is only one particle. The amplitude for the decay is $\mathcal{A}_{fi}$; this is to be multiplied by the invariant phase space differential, the density of final states D. The differential transition property is clearly a differential of something, and that is contained in the factor D. If the particle decays into a final state containing three particles, D will include a factor like $d^3\mathbf{p}_1\, d^3\mathbf{p}_2\, d^3\mathbf{p}_3$.

The total transition probability per unit time, Γ, is obtained by integrating $d\Gamma$ over all final momenta and summing over all possible final states, if there are many final particle states into which this thing can decay—three mesons, two mesons, a nucleon and antinucleon etc.[1] Γ is typically evaluated for the incoming particle at rest (its momentum equal to zero). When

[1] [Eds.] In conventional units, $\Gamma = \hbar/\tau$, where τ is the particle's mean lifetime. Γ has the units of energy.

you see a table of Γ's, it doesn't say, "This is Γ for the particle moving at one third the speed of light". Then the energy is that of the particle at rest, $E_{\mathbf{p}} = m$ or μ, whatever the mass of the incoming particle is, so

$$\frac{\text{rest decay prob.}}{\text{unit time}} = \Gamma = \frac{1}{2m} \sum_{\substack{\text{final}\\ \text{states}}} \int |\mathcal{A}_{fi}|^2 D \tag{12.2}$$

This way of writing the formula makes it very clear what the decay amplitude is for a moving particle. The quantity $\int |\mathcal{A}_{fi}|^2 D$ is Lorentz invariant, because D is a Lorentz invariant measure, and $\mathcal{A}_{fi}$ is a Lorentz invariant object. If we evaluate this expression for incoming momentum equal to something else, $\mathbf{p}$, then the only difference is the factor in front, $2m$ being replaced by $2E_{\mathbf{p}}$. That is,

$$\frac{\text{decay prob.}}{\text{unit time}} = \frac{m}{E_{\mathbf{p}}}\Gamma = \frac{d\tau}{dt}\Gamma \tag{12.3}$$

This is of course just what we would expect from time dilation. This equation expresses the fact that a moving π meson decays more slowly than a stationary π meson. The faster it moves, the more slowly it decays. That helps to explain the physical meaning, at least in this case, of that mysterious factor of $1/2E_{\mathbf{p}}$ for the initial particle, the only thing that is not Lorentz invariant in our expression. It damn well had better be there, otherwise we would have predicted that a moving particle decayed at the same rate as a stationary particle, which would be bad news both from the viewpoint of relativity theory and from the viewpoint of experiment.

12.2 Differential cross-section for a two-particle initial state

If we have a beam of particles impinging on a stationary target, or if we have a target moving into a beam of particles, the differential cross-section is defined as the transition probability per unit time per unit flux. By definition, the differential element $d\sigma$ of the cross-section is

$$d\sigma = \frac{\text{diff. trans. prob}}{(\text{unit time}) \times (\text{unit flux})} = \frac{1}{4E_1E_2}|\mathcal{A}_{fi}|^2 D \frac{1}{|\mathbf{v}_1 - \mathbf{v}_2|} \tag{12.4}$$

That is to say, we divide the differential transition probability per unit time by the flux of particles impinging on the target. The energies E_1 and E_2, and the velocities $\mathbf{v}_1$ and $\mathbf{v}_2$, are those of the initial particles. The unit flux, the number of beam particles hitting the target particle per unit time per unit area, is the difference in the three-velocities of the incoming particles:

$$(\text{unit flux}) = |\mathbf{v}_1 - \mathbf{v}_2| \tag{12.5}$$

Let's understand this factor. Our normalization convention is such that we can think of one of the particles as having probability one for being somewhere in the box, and the other particle having probability one for being in a given unit volume,

$$|i\rangle = \sqrt{V}\,|\mathbf{p}_1, \mathbf{p}_2\rangle \tag{12.6}$$

Suppose the particle with momentum $\mathbf{p}_1$ presents some area, A, to the particle beam with momentum $\mathbf{p}_2$. The normal to the area A is parallel to the direction $\mathbf{v}_1 - \mathbf{v}_2$. See Figure 12.1.

In a time t, the area normal to $\mathbf{v}_1 - \mathbf{v}_2$ sweeps out a volume $|\mathbf{v}_1 - \mathbf{v}_2|At$. The particle flux is defined to be

$$(\text{flux}) = \frac{(\text{particle density}) \times V}{(\text{area}) \times (\text{time})} = \frac{|\mathbf{v}_1 - \mathbf{v}_2|At}{At} = |\mathbf{v}_1 - \mathbf{v}_2| \tag{12.7}$$

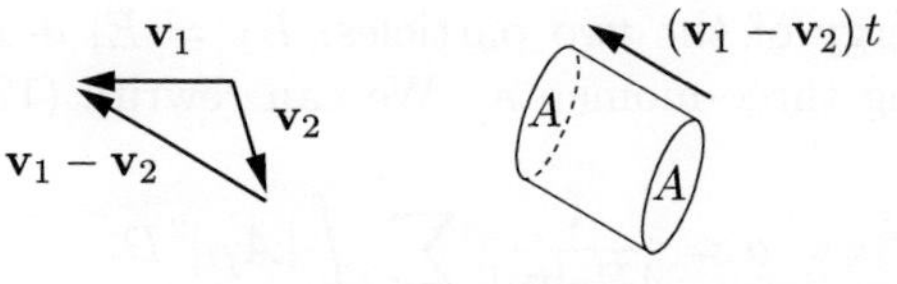

Figure 12.1: Flux for a two-particle initial state

So the differential cross-section (12.4) is our general formula (11.57) divided by the flux, (12.5).

I can think of one of these particles as being the target. Which of these two $\mathbf{p}$'s I associate the square root of V with is a matter of taste, but let me consider the first one as being the target, somewhere in the box. The second one is the beam. It has probability one for being someplace in the box. I have the target moving through the box with velocity $\mathbf{v}_1$, while the beam moves through the box with velocity $\mathbf{v}_2$. The probability flux hitting the target is $\mathbf{v}_1 - \mathbf{v}_2$, the ordinary *non-relativistic* difference of velocities. I emphasize that. A friend of mine who once did a thesis on neutrino–neutrino scattering, a rather abstract subject but of some cosmological interest, had a factor of 4 error in his thesis because he said, "Oh, they're relativistic particles; their relative velocity here must be c." It is not. For neutrino–neutrino scattering, it is $2c$, if they head into each other. We'll see that that's consistent with relativity also. If I turn on my stopwatch for one second and ask how much beam has passed the target in that one second, the answer is one over $\mathbf{v}_1 - \mathbf{v}_2$ worth of beam. That's unambiguous.

The total cross-section σ is obtained by summing and integrating over the final states:

$$\sigma = \underbrace{\frac{1}{|\mathbf{v}_1 - \mathbf{v}_2|}\frac{1}{4E_1E_2}}_{\text{not Lorentz invariant}} \underbrace{\sum_{\substack{\text{final}\\ \text{states}}} \int |\mathcal{A}_{fi}|^2 D}_{\text{Lorentz invariant}} \tag{12.8}$$

We have an evidently non-Lorentz invariant factor in front, $(4E_1E_2|\mathbf{v}_1 - \mathbf{v}_2|)^{-1}$, times a Lorentz invariant. I will now discuss the Lorentz transformation properties of the factor in front to demonstrate that they are what one would think they should be.

Consider a special form of (12.8) in a Lorentz frame in which the two particles are moving head on towards each other, or one is catching up to the other, where the two three-momenta are aligned. I will consider

$$\begin{aligned} p_1 &= (E_1, p_{1x}, 0, 0) \\ p_2 &= (E_2, p_{2x}, 0, 0) \end{aligned} \tag{12.9}$$

where I've chosen the coordinate system such that the x axis is aligned with $\mathbf{v}_1$, and so also with either $\mathbf{v}_2$ or $-\mathbf{v}_2$. Because $\mathbf{v} = \mathbf{p}/E$,

$$E_1E_2|\mathbf{v}_1 - \mathbf{v}_2| = E_1E_2\left|\frac{p_{1x}}{E_1} - \frac{p_{2x}}{E_2}\right| = |p_{1x}E_2 - p_{2x}E_1| \tag{12.10}$$

For later computations it will be useful to have the expression for (12.10) in the center-of-momentum frame. Then

$$p_{1x} = |\mathbf{p}_i| \quad p_{2x} = -|\mathbf{p}_i| \tag{12.11}$$

and (12.10) simply becomes

$$|p_{1x}E_2 - p_{2x}E_1| = E_T|\mathbf{p}_i| \tag{12.12}$$

where E_T is the total energy of the two particles, $E_T = E_1 + E_2$, and p_i is the common magnitude of the incoming three-momenta. We can rewrite (12.8) for σ in the center-of-momentum frame,

$$\sigma = \frac{1}{4E_T|\mathbf{p}_i|} \sum_{\substack{\text{final}\\ \text{states}}} \int |\mathcal{A}_{fi}|^2 D \tag{12.13}$$

We all know from non-relativistic physics the geometrical picture of the total cross-section, and why it is called a cross-section. We have a beam of particles heading in one direction, which we will call the negative x axis, and some object, off of which the beam is scattering, heading in the direction of the positive x-axis:

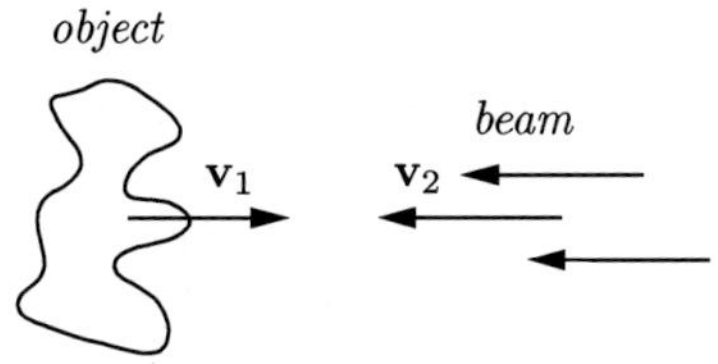

Figure 12.2: Object in beam

The total cross-section gives the total probability of scattering, because it is simply the geometrical cross-section presented by the object. In the classical picture, if the beam hits the object, it scatters, and if it misses the object, it doesn't. This picture, Figure 12.2, would indicate that the total cross-section should be the same in any Lorentz frame that preserves the expression (12.10). If I make a Lorentz transformation which is a boost along the x direction, that preserves the expression. That is to say, a Lorentz transformation restricted to t and x, a rotation in the $(0, 1)$ plane, will change the appearance of the target in the x direction by Lorentz contraction, but it won't change its perpendicular dimensions, so it won't change its cross-section. Of course, if I make another kind of Lorentz transformation, then things are different; then I am distorting the particle in the direction that the beam sees, and then I shouldn't expect the total cross-section to be invariant.

Now let's check that this is so. Is (12.10) invariant under Lorentz transformations restricted to the $(0, 1)$ plane, or is it not? Well, it's obvious that it is, once I've got things in this form, because

$$|p_{1x}E_2 - p_{2x}E_1| = |\epsilon^{23\mu\nu}p_{1\mu}p_{2\nu}| \tag{12.14}$$

where $2, 3$ here are Lorentz indices, equal to y, z, respectively; $\epsilon^{\rho\lambda\mu\nu}$ is the completely antisymmetric object we talked about before. The right-hand side of (12.14) has only two non-vanishing terms if we fix $\rho = 2$ and $\lambda = 3$: a term where $\mu = 0$ and $\nu = 1$, and another where $\mu = 1$ and $\nu = 0$. These terms have a minus sign between them, and they give you the two terms on the left-hand side. Or maybe they give you the terms with the sign reversed; it doesn't matter, it's the absolute value. The right-hand side of (12.14) is obviously invariant under Lorentz transformations restricted to the $(0, 1)$ plane, because the 2 and 3 indexed variables don't change and the rest is a Lorentz invariant sum. So this expression (12.10) is okay.

The total cross-section does what it should. If in any Lorentz frame where the 3-momenta are along the x axis, you compute the total cross-section, you get the same result as in any other Lorentz frame in which the 3-momenta are along the x axis. Thus we have established that cross-sections Lorentz contract as they should Lorentz contract: not at all, if you make your Lorentz transformation along the direction of the beam. Please notice once again that the mysterious factors of $1/E$ that come into our formula for the transition probability per unit time are essential for the result to come out right.

12.3 The density of final states for two particles

We now turn to the third topic. It would be very nice to have a more compact formula for the density of final states than the awful expression (11.58). So let me now compute D for a two-particle final state, where the initial particles are in the center-of-momentum frame:

$$\mathbf{p}_T = \mathbf{p}_1 + \mathbf{p}_2 = \mathbf{0}; \qquad E_T = E_1 + E_2 \tag{12.15}$$

I'll do it in this frame because it's the simplest case. (It's also pretty easy to do it in some other frame; see Problem 7.1.) In the case of two final particles with momenta $\mathbf{p}_3$ and $\mathbf{p}_4$,

$$D = (2\pi)^4 \delta^{(4)}(p_i - p_f) \frac{d^3\mathbf{p}_3}{(2\pi)^3 2E_3} \frac{d^3\mathbf{p}_4}{(2\pi)^3 2E_4} \tag{12.16}$$

I'll split the four-dimensional energy-momentum-conserving delta function into two factors,

$$\delta^{(4)}(p_i - p_f) = \delta^{(3)}(\mathbf{p}_3 + \mathbf{p}_4 - \mathbf{0})\, \delta(E_3 + E_4 - E_T) \tag{12.17}$$

where E_T is the total incoming energy. We now want to cancel out some of the delta functions against some of the differentials. The easy one to do first is the integration over $\mathbf{p}_4$, and we use the $\delta^{(3)}(\mathbf{p}_3 + \mathbf{p}_4)$ to cancel it out. That is to say, if we integrate D with any function, doing the integral is the same as replacing $\mathbf{p}_4$ by $-\mathbf{p}_3$, canceling the $d^3\mathbf{p}_4$ and canceling the delta function. So (12.16) becomes

$$D = (2\pi)\, \delta(E_3 + E_4 - E_T) \frac{d^3\mathbf{p}_3}{(2\pi)^3 4E_3 E_4} \tag{12.18}$$

where $\mathbf{p}_4$ is now constrained to be $-\mathbf{p}_3$. And of course that means E_4 is also a function of $\mathbf{p}_3$.

I now go to angular variables, and write $d^3\mathbf{p}_3 = |\mathbf{p}_3|^2 d|\mathbf{p}_3| d\Omega_3$. Then

$$D = (2\pi)\, \delta(E_3 + E_4 - E_T) \frac{|\mathbf{p}_3|^2 d|\mathbf{p}_3| d\Omega_3}{(2\pi)^3 4E_3 E_4} \tag{12.19}$$

Now I'll cancel the delta function of the incoming energy by integrating over $d|\mathbf{p}_3|$, which fixes $|\mathbf{p}_3| = |\mathbf{p}_4|$. We'll need the important rule (note 8, p. 9), which I presume you recall,

$$\delta(f(x)) = \frac{1}{|f'(x_0)|}\, \delta(x - x_0) \tag{12.20}$$

where x_0 is the root of $f(x) = 0$. (If there are several zeros, you get a sum of such terms, one from each zero, but there won't be in this case.) Performing the integration over $|\mathbf{p}_3|$, I get

$$D = \frac{|\mathbf{p}_3|^2 d\Omega_3}{16\pi^2 E_3 E_4} \left| \frac{\partial(E_3 + E_4)}{\partial |\mathbf{p}_3|} \right|^{-1} \tag{12.21}$$

I don't need the absolute value here since these are positive quantities. We have

$$E_3^2 = |\mathbf{p}_3|^2 + m_3^2; \qquad E_4^2 = |\mathbf{p}_4|^2 + m_4^2 = |\mathbf{p}_3|^2 + m_4^2 \tag{12.22}$$

By differentiation, $E_3\,dE_3 = |\mathbf{p}_3|\,d|\mathbf{p}_3|$ and $E_4\,dE_4 = |\mathbf{p}_3|\,d|\mathbf{p}_3|$, so

$$\left(\frac{\partial(E_3+E_4)}{\partial|\mathbf{p}_3|}\right)^{-1} = \left(|\mathbf{p}_3|\left(\frac{1}{E_3}+\frac{1}{E_4}\right)\right)^{-1} = \frac{E_3E_4}{|\mathbf{p}_3|(E_3+E_4)} \tag{12.23}$$

If we substitute (12.23) into (12.21), one factor of $|\mathbf{p}_3|$ cancels and the product E_3E_4 cancels. Let's identify $E_3 + E_4 = E_T$, the total final energy, and $|\mathbf{p}_3|$ as the magnitude $|\mathbf{p}_f|$ of the momentum of either final particle in the center-of-momentum frame. We obtain a formula important enough for me to put in a box:

$$\boxed{D = \frac{1}{16\pi^2}\frac{|\mathbf{p}_f|d\Omega_f}{E_T}} \tag{12.24}$$

Please notice that the magnitudes of the initial particles' momenta $|\mathbf{p}_i|$ and the final particles' momenta $|\mathbf{p}_f|$ in the center-of-momentum frame are different if the final and the initial particles have different masses. The factor $d\Omega_f$ describes the solid angle associated with $d^3\mathbf{p}_f$.

EXAMPLE 1. *Calculating $d\sigma/d\Omega$ in the center-of-momentum frame.*

To compute $d\sigma/d\Omega$ in the center-of-momentum frame, we return to (12.4), and substitute in (12.12) and (12.24):

$$d\sigma = \frac{1}{4E_1E_2}\frac{1}{|\mathbf{v}_1-\mathbf{v}_2|}|\mathcal{A}_{fi}|^2D = \frac{1}{4E_T|\mathbf{p}_i|}|\mathcal{A}_{fi}|^2\frac{1}{16\pi^2}\frac{|\mathbf{p}_f|d\Omega_f}{E_T} \tag{12.25}$$

so that

$$\boxed{\frac{d\sigma}{d\Omega} = \frac{1}{64\pi^2E_T^2}\frac{|\mathbf{p}_f|}{|\mathbf{p}_i|}|\mathcal{A}_{fi}|^2} \tag{12.26}$$

I should make two remarks. Please notice the factor of $|\mathbf{p}_f|$ over $|\mathbf{p}_i|$. In an inelastic process, the masses of the final particles are different from those of the initial particles, and so $|\mathbf{p}_f|$ does not equal $|\mathbf{p}_i|$. Even though time reversal may tell us that the amplitude for an inelastic process is the same as the amplitude for the time-reversed process, this does *not* mean the cross-section for the process is the same as the cross-section for the time-reversed process. We have $|\mathbf{p}_f|$ over $|\mathbf{p}_i|$ in one case and $|\mathbf{p}_i|$ over $|\mathbf{p}_f|$ in another, so that even if the amplitudes are the same, the differential cross-sections will *not* be the same. This will be a familiar result from non-relativistic physics, if you've ever studied things like the scattering of electrons off atoms. These collisions can excite the atom, and occur as both exothermic and endothermic reactions. Thus for example in our model the total cross-section for nucleon–antinucleon annihilation into meson plus meson is not the same as a total cross-section for meson–meson production of a nucleon–antinucleon pair, even though the amplitudes are identical.

Note that for an exothermic reaction, we can have $\mathbf{p}_i = 0$ but $\mathbf{p}_f \neq 0$. That means $d\sigma/d\Omega$ and hence σ can be infinite even when the amplitude $\mathcal{A}_{fi}$ is finite. It is simple to understand the meaning of the ratio with $\mathbf{p}_i$ in the denominator. As $\mathbf{p}_i \to 0$, i.e., approaching threshold, the two particles spend more time near each other, increasing the likelihood of interaction. Engineers slow down neutrons in atomic piles to minimize the denominator, and maximize the chance of neutron capture in the pile.

Second remark: We can now compare this expression (12.26) for the relativistic differential cross-section to what we find in non-relativistic physics, to make the correspondence between our non-relativistic and relativistic notational conventions. That's a useful thing to do if we ever want to check that things have the right non-relativistic limit. In non-relativistic physics, we also define a scattering amplitude, using the normalization conventions convenient for non-relativistic physics. In non-relativistic quantum mechanics, we always have elastic scattering by a potential, with $p_f = p_i = p$. The scattering amplitude, usually called $f(p, \cos\theta)$, is given by a very simple formula,

$$\frac{d\sigma}{d\Omega} = |f(p,\cos\theta)|^2 \tag{12.27}$$

In the elastic case, comparing (12.27) with (12.26), we see the connection between the relativistic and the non-relativistic scattering amplitude,

$$f(p,\cos\theta) = \pm\frac{1}{8\pi E_T}\mathcal{A}_{fi} \tag{12.28}$$

(up to a phase). When I derive the Optical Theorem, we'll see that the sign should be positive (and that there is no phase factor). This is the scattering amplitude as conventionally defined in non-relativistic potential scattering, and we now see how it is related to our relativistic scattering amplitude.

EXAMPLE 2. *Calculating Γ for $\phi \to N\overline{N}$ in Model 3.*

We have assumed that $\mu < 2m$, so this process should not occur. For a moment, let's relax this constraint. We have, ignoring the adiabatic function,

$$\mathscr{H}_I = g\psi^*\psi\phi \tag{12.29}$$

The relevant Feynman diagram for the decay of a meson at rest into a nucleon–antinucleon pair is Diagram 3(c) on p. 216:

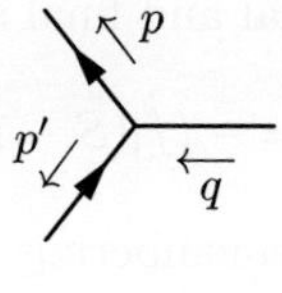

Figure 12.3: Vertex for Model 3

The contribution of this graph is

$$-ig(2\pi)^4\delta^{(4)}(p+p'-q) \tag{12.30}$$

so that, to first order in g, the invariant amplitude for meson decay into a nucleon–antinucleon pair is given by

$$(2\pi)^4\delta^{(4)}(p+p'-q)i\mathcal{A}_{fi} = (2\pi)^4\delta^{(4)}(p+p'-q)(-ig) \tag{12.31}$$

That is,

$$i\mathcal{A}_{fi} = -ig \tag{12.32}$$

It couldn't be simpler. From (12.2) and (12.24), to $\mathcal{O}(g^2)$, the decay rate Γ of the muon is

$$\Gamma = \frac{1}{2\mu}\int |\mathcal{A}_{fi}|^2\frac{1}{16\pi^2}\frac{|\mathbf{p}_f|\,d\Omega_f}{E_T} = \frac{g^2}{8\pi\mu E_T}|\mathbf{p}_f| \tag{12.33}$$

In the center-of-momentum frame, $|\mathbf{p}_1| = |\mathbf{p}_2| = |\mathbf{p}_f|$, $E_T = 2\sqrt{|\mathbf{p}_f|^2 + m^2}$, and $E_T = \mu$. Then

$$|\mathbf{p}_f| = \sqrt{(\tfrac{1}{2}E_T)^2 - m^2} = \tfrac{1}{2}\sqrt{\mu^2 - 4m^2} \tag{12.34}$$

and

$$\Gamma = \frac{g^2}{16\pi\mu^2}\sqrt{\mu^2 - 4m^2} \tag{12.35}$$

Clearly, this is imaginary unless $\mu > 2m$. That is, the meson is stable if $\mu < 2m$, as we have assumed.

12.4 The Optical Theorem

For simplicity I will assume I am working in a theory in which there is only one kind of particle, say a meson. Then, when I sum over all final states, I don't have to complicate my notations unnecessarily to indicate summing over mesons *and* nucleons *and* antinucleons. The generalization of my arguments to more complicated theories will be trivial.

I start out with this famous equation, expressing the unitarity of the S-matrix:

$$S^\dagger S = 1 \tag{12.36}$$

I will deduce the Optical Theorem as a consequence of this equation. Our invariant amplitudes are defined in terms of $S - 1$, but it is pretty easy to find an equation corresponding to (12.36) in terms of $S - 1$:

$$\begin{aligned}(S^\dagger - 1)(S - 1) &= (S-1)^\dagger(S-1) = S^\dagger S - S^\dagger - S + 1 = 2 - S - S^\dagger \\ &= -[(S-1) + (S-1)^\dagger]\end{aligned} \tag{12.37}$$

I now evaluate this identity between initial and final states, $|i\rangle$ and $|f\rangle$, respectively:

$$\langle f|(S-1)^\dagger(S-1)|i\rangle = -\langle f|(S-1)|i\rangle - \langle f|(S-1)^\dagger|i\rangle \tag{12.38}$$

I'll begin with the right-hand side. Remembering (10.27),

$$\begin{aligned}\langle f|(S-1)|i\rangle &= i\mathcal{A}_{fi}(2\pi)^4\delta^{(4)}(p_f - p_i) \\ \langle f|(S-1)^\dagger|i\rangle &= -i\mathcal{A}^*_{if}(2\pi)^4\delta^{(4)}(p_f - p_i)\end{aligned} \tag{12.39}$$

(because of the adjoint, I complex conjugate the amplitude and swap its indices). So we have

$$-\langle f|(S-1)|i\rangle - \langle f|(S-1)^\dagger|i\rangle = (2\pi)^4\delta^{(4)}(p_f - p_i)(i\mathcal{A}^*_{if} - i\mathcal{A}_{fi}) \tag{12.40}$$

The left-hand side of (12.38) evaluated between $|f\rangle$ and $|i\rangle$ I will write in terms of a complete set $|m\rangle$ of intermediate states. I'm assuming there is only one kind of particle, so the intermediate states $|m\rangle$ are r-particle states: $|m\rangle = |q_1, \ldots, q_r\rangle$. Then, using (1.64),

$$\begin{aligned}\langle f|(S-1)^\dagger(S-1)\,|i\rangle &= \sum_m \langle f|(S-1)^\dagger|m\rangle\,\langle m|(S-1)|i\rangle \\ &= \sum_{r=1}^{\infty}\frac{1}{r!}\int \frac{d^3\mathbf{q}_1}{(2\pi)^3 2E_1}\cdots\frac{d^3\mathbf{q}_r}{(2\pi)^3 2E_r}\,\langle f|(S-1)^\dagger|q_1,\ldots,q_r\rangle\,\langle q_1,\ldots,q_r|(S-1)|i\rangle\end{aligned} \tag{12.41}$$

I divide by $r!$ to keep from over-counting the states. That's simply the left-hand side of (12.38) written in terms of the sum of a complete set of intermediate states.

We simplify this expression as before:

$$\begin{aligned}\langle q_1,\dots,q_r|(S-1)|i\rangle &= i\mathcal{A}_{mi}(2\pi)^4\delta^{(4)}(p_m-p_i)\\ \langle f|(S-1)^\dagger|q_1,\dots,q_r\rangle &= -i\mathcal{A}^*_{mf}(2\pi)^4\delta^{(4)}(p_f-p_m)\end{aligned}\tag{12.42}$$

The left-hand side of (12.38) can then be written

$$\begin{aligned}\langle f|\,(S-1)^\dagger(S-1)\,|i\rangle &= \sum_{r=1}^{\infty}\frac{1}{r!}\prod_{n=1}^{r}\int\frac{d^3\mathbf{q}_n}{(2\pi)^3 2E_n}\,\mathcal{A}^*_{mf}\mathcal{A}_{mi}(2\pi)^8\delta^{(4)}(p_m-p_i)\delta^{(4)}(p_f-p_m)\\ &= (2\pi)^4\delta^{(4)}(p_f-p_i)\sum_{r=1}^{\infty}\frac{1}{r!}\prod_{n=1}^{r}\int\frac{d^3\mathbf{q}_n}{(2\pi)^3 2E_n}\,\mathcal{A}^*_{mf}\mathcal{A}_{mi}(2\pi)^4\delta^{(4)}(p_m-p_i)\end{aligned}\tag{12.43}$$

where we replace $\delta^{(4)}(p_m-p_i)\,\delta^{(4)}(p_f-p_m)$ by $\delta^{(4)}(p_f-p_i)\,\delta^{(4)}(p_f-p_m)$, and take the common delta function factor outside the sum. Comparing the right-hand side (12.40) of (12.38) with the left-hand side (12.4), I have $(2\pi)^4\delta^{(4)}(p_f-p_i)$ on both sides of the equation, so I can divide it out:

$$i\mathcal{A}^*_{if}-i\mathcal{A}_{fi}=\sum_{r=1}^{\infty}\frac{1}{r!}\prod_{n=1}^{r}\int\frac{d^3\mathbf{q}_n}{(2\pi)^3 2E_n}\,\mathcal{A}^*_{mf}\mathcal{A}_{mi}(2\pi)^4\delta^{(4)}(p_m-p_i)\tag{12.44}$$

Once I have divided out the delta function, I can safely set $|i\rangle=|f\rangle$, because I will no longer encounter an infinity. Let's do that. Then

$$i\mathcal{A}^*_{ii}-i\mathcal{A}_{ii}=\sum_{r=1}^{\infty}\frac{1}{r!}\prod_{n=1}^{r}\int\frac{d^3\mathbf{q}_n}{(2\pi)^3 2E_n}\,|\mathcal{A}_{mi}|^2(2\pi)^4\delta^{(4)}(p_m-p_i)\tag{12.45}$$

What do we get? The left-hand side of this equation is twice the imaginary part of $\mathcal{A}_{ii}$. On the right-hand side, I get something very interesting. Comparison with (11.58) shows

$$(2\pi)^4\delta^{(4)}(p_m-p_i)\prod_{n=1}^{r}\int\frac{d^3\mathbf{q}_n}{(2\pi)^3 2E_n}=D\tag{12.46}$$

In particular, say that the initial state has two particles:

$$|i\rangle=|\mathbf{p}_i,-\mathbf{p}_i\rangle\tag{12.47}$$

Then the right-hand side of (12.45) becomes a statement about the cross-section:

$$\sum_{r=1}^{\infty}\frac{1}{r!}(2\pi)^4\delta^{(4)}(p_m-p_i)\prod_{n=1}^{r}\int\frac{d^3\mathbf{q}_n}{(2\pi)^3 2E_n}\,|\mathcal{A}_{mi}|^2=\sum_{\substack{\text{final}\\\text{states}}}\int D\,|\mathcal{A}_{fi}|^2=4E_T|\mathbf{p}_i|\sigma\tag{12.48}$$

using (12.13). That is to say, it *is* the total cross-section, except for that funny factor $4|\mathbf{p}_i|E_T$. Setting the two sides equal, we can divide out a common factor of 2.

Thus after excruciatingly dull labor we have arrived, as a consequence of the unitarity of the S-matrix, at the famous Optical Theorem,

$$\boxed{\operatorname{Im}\mathcal{A}_{ii}=2E_T|\mathbf{p}_i|\sigma}\tag{12.49}$$

This asserts that the imaginary part of the relativistic forward scattering amplitude equals twice the total energy times the momentum of the particles in the center-of-momentum frame times the total cross-section. It doesn't matter if the particles are incoming or outgoing, because the elastic scattering amplitude in the forward direction is the same for the initial states as the final states. It is just a consequence of the unitarity of the S-matrix, and therefore it is a very general result.

Incidentally, since the right-hand side of (12.49) is zero for $NN \to NN$ scattering up till $\mathcal{O}(g^4)$, it follows that $\text{Im}\,\mathcal{A}_{ii} = 0$ up to $\mathcal{O}(g^4)$ in Model 3. This proves that the forward scattering for $NN \to NN$ is real at $\mathcal{O}(g^2)$ in Model 3. In fact, Feynman amplitudes are real to $\mathcal{O}(g^2)$ for scalar fields in all directions.

Just to check, let's compare this with the equally famous Optical Theorem of non-relativistic scattering theory.[2] Recalling that $|\mathbf{p}_f| = |\mathbf{p}_i| = |\mathbf{p}|$ for elastic scattering, we have

$$\text{Im}\, f(|\mathbf{p}|, \cos\theta)\Big|_{\theta=0} = \frac{|\mathbf{p}|}{4\pi}\sigma \tag{12.50}$$

That is, the imaginary part of $f(|\mathbf{p}|, \cos\theta)$ in the forward direction, is equal, by some elementary algebra, to $|\mathbf{p}|/4\pi$ times σ. Comparing (12.49) and (12.50), we see that

$$\text{Im}\, f(|\mathbf{p}|, \cos\theta)\Big|_{\theta=0} = \frac{1}{8\pi E_T}\text{Im}\,\mathcal{A}_{ii} \tag{12.51}$$

Looking back to (12.28), and taking the imaginary part of both sides, we conclude that the + sign is appropriate, and that there is no phase factor:

$$f(|\mathbf{p}|, \cos\theta) = \frac{1}{8\pi E_T}\mathcal{A}_{fi} \tag{12.52}$$

Indeed, we didn't have to go through this argument for relativistic scattering, because the Optical Theorem of non-relativistic scattering theory is true whether or not the theory is relativistically invariant. We've shown that it is just a consequence of unitarity. It's an amusing exercise, in case you had not seen it in non-relativistic quantum mechanics, to see it come out here.

12.5 The density of final states for three particles

I will now consider phase space for three-body final states, just to show you that these integrals are not particularly difficult. I will again work in the center-of-momentum frame:

$$\mathbf{p}_i = \mathbf{0} \;\Rightarrow\; \mathbf{p}_3 = -(\mathbf{p}_1 + \mathbf{p}_2) \tag{12.53}$$

Since you've already seen me do one of these calculations in gory detail, I will skip immediately to something you can see by eyeball. I'll write down D from our general formula, (11.58):

$$D = (2\pi)^4\delta^{(4)}(p_i - p_f)\,\frac{1}{(2\pi)^9}\frac{d^3\mathbf{p}_1\, d^3\mathbf{p}_2\, d^3\mathbf{p}_3}{8E_1E_2E_3} \tag{12.54}$$

[2] [Eds.] L. I. Schiff, *Quantum Mechanics*, 3rd ed., McGraw-Hill, 1968, p. 137; or Landau & Lifshitz, *QM*, p. 476.

First, we will take care of the 2π's. There are three factors of $(2\pi)^3$ in the denominator, one for each of the final particles, and there is a $(2\pi)^4$ in the overall energy-momentum-conserving delta function, so the net is $(2\pi)^{-5}$. There is an energy denominator, $2E$ for each final particle, that gives us $(8E_1E_2E_3)^{-1}$. There are $d^3\mathbf{p}_1$ and $d^3\mathbf{p}_2$ which I will immediately write in polar form. There's a $d^3\mathbf{p}_3$ which I will cancel off against the space part of the delta function, and need not write down. That gives

$$D = \delta(E_1 + E_2 + E_3 - E_T)\frac{1}{(2\pi)^5}\frac{p_1^2\,dp_1\,d\Omega_1\,p_2^2\,dp_2\,d\Omega_2}{8E_1E_2E_3} \tag{12.55}$$

And there is a remaining delta function of $E_1 + E_2 + E_3 - E_T$. The hard part will be doing the integral to get rid of E_3 which will cancel out one of our other variables.

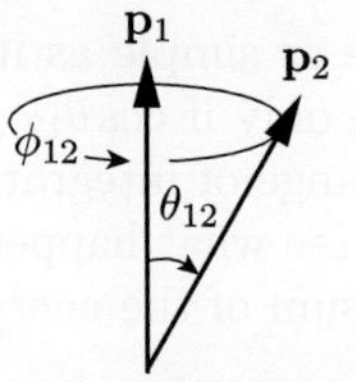

Figure 12.4: The angles θ_{12} (latitude) and ϕ_{12} (azimuth)

Let's look at the angular integrals (Figure 12.4). Here's $\mathbf{p}_1$, which I'll take to be my z direction. Off someplace is $\mathbf{p}_2$, and between them is the angle θ_{12}. I'll hold $\mathbf{p}_1$ fixed, and integrate over Ω_2 first. The angular differentials can be written as

$$d\Omega_1\,d\Omega_2 = d\Omega_1\,d\phi_{12}\,d(\cos\theta_{12}) \tag{12.56}$$

where θ_{12} is the relative angle between $\mathbf{p}_1$ and $\mathbf{p}_2$, and ϕ_{12} is the azimuthal angle of $\mathbf{p}_2$ relative to $\mathbf{p}_1$.

We write the angular integrals in this way because the *only* variable that depends on θ_{12} when we keep the other variables fixed is E_3. We can thus cancel the energy delta function against the integral in θ_{12}:

$$\begin{aligned} E_3(\cos\theta_{12}) &= \sqrt{\mathbf{p}_3^2 + m_3^2} = \sqrt{(\mathbf{p}_1 + \mathbf{p}_2)^2 + m_3^2} \\ &= \sqrt{|\mathbf{p}_1|^2 + |\mathbf{p}_2|^2 + 2|\mathbf{p}_1||\mathbf{p}_2|\cos\theta_{12} + m_3^2} \end{aligned} \tag{12.57}$$

Using the rule about delta functions of functions,

$$\delta(E_1 + E_2 + E_3(\cos\theta_{12}) - E_T) = \delta(\cos\theta_{12} - \xi)\frac{1}{E_3'(\xi)} \tag{12.58}$$

where ξ is the value of $\cos\theta_{12}$ that ensures $E_1 + E_2 + E_3 = E_T$. The derivative is easy:

$$E_3'(\xi) = \left.\frac{\partial E_3}{\partial\cos\theta_{12}}\right|_{\cos\theta_{12}=\xi} = \frac{|\mathbf{p}_1||\mathbf{p}_2|}{E_3} \tag{12.59}$$

The integral over θ_{12} is trivial. Carrying that out, we obtain

$$\begin{aligned} D &= \int \delta(\cos\theta_{12} - \xi)\frac{E_3}{|\mathbf{p}_1||\mathbf{p}_2|}\frac{|\mathbf{p}_1|^2\,d|\mathbf{p}_1|\,d\Omega_1\,|\mathbf{p}_2|^2\,d|\mathbf{p}_2|\,d(\cos\theta_{12})\,d\phi_{12}}{(2\pi)^5\,8E_1E_2E_3} \\ &= \frac{|\mathbf{p}_1|\,d|\mathbf{p}_1|\,|\mathbf{p}_2|\,d|\mathbf{p}_2|\,d\Omega_1\,d\phi_{12}}{8E_1E_2(2\pi)^5} \end{aligned} \tag{12.60}$$

This expression becomes especially simple once you recall the famous result[3] that $|\mathbf{p}|\, d|\mathbf{p}| = E\, dE$, so

$$|\mathbf{p}_1|\, d|\mathbf{p}_1| = E_1\, dE_1; \qquad |\mathbf{p}_2|\, d|\mathbf{p}_2| = E_2\, dE_2 \tag{12.61}$$

Even more of the denominators cancel out when I make those substitutions, and I find a remarkably simple expression for the density factor in three-body phase space:

$$D = \frac{1}{8(2\pi)^5}\, dE_1\, dE_2\, d\Omega_1\, d\phi_{12} \tag{12.62}$$

We have now run out of delta functions, and can't carry out any more integrations. That's as far as one can go in the general case. To calculate lifetimes or cross-sections in a specific case, we need to know how the amplitude for that process depends on the various 4-momenta.

In truth, the situation is not quite as simple as it may appear: E_1 and E_2 are severely restricted because step (12.58) is true only if $\cos\theta_{12}$ goes from -1 to 1, so that the zero in the delta function occurs within the range of integration. That is to say, E_1 and E_2 are not allowed to range freely. Indeed we can see what happens at one extreme where the vectors $\mathbf{p}_1$ and $\mathbf{p}_2$ are aligned, and $\theta_{12} = 0$. The sum of the energies is

$$E_T = \sqrt{|\mathbf{p}_1|^2 + m_1^2} + \sqrt{|\mathbf{p}_2|^2 + m_2^2} + \sqrt{|\mathbf{p}_1 + \mathbf{p}_2|^2 + m_3^2} \tag{12.63}$$

The quantity $|\mathbf{p}_1 + \mathbf{p}_2|$ is the biggest $|\mathbf{p}_3|$ can get. And that upper limit in the integration had better be greater than or equal to E_T, which in turn must be greater than or equal to the lower limit of the integration,

$$E_T = \sqrt{|\mathbf{p}_1|^2 + m_1^2} + \sqrt{|\mathbf{p}_2|^2 + m_2^2} + \sqrt{|\mathbf{p}_1 - \mathbf{p}_2|^2 + m_3^2} \tag{12.64}$$

These boundaries define an awful, hideous-looking region in (p_1, p_2) space, or equivalently (E_1, E_2) space. One can work at it and beat on it and you will end up with a cubic equation involving E_1 and E_2 to determine the boundaries of this region, but that is still pretty terrible.

So there is some ugly-looking region in (E_1, E_2) space, or in a Dalitz plot, something that looks, for example, like Figure 12.5. The dots denote events. It's not quite as monstrous as that; I believe it's convex. But in general there is some monstrous blob where kinematics allow the final particles to come out.

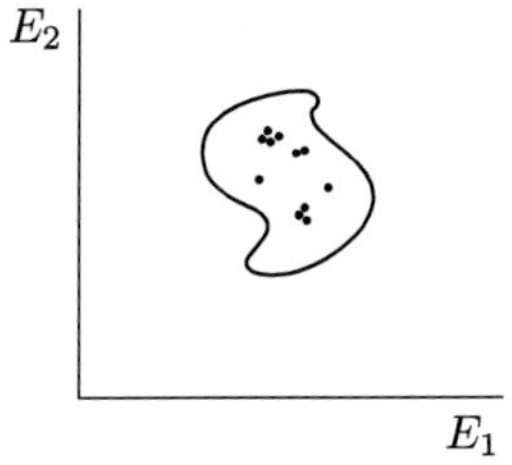

The dots represent events.

Figure 12.5: Dalitz plot for E_1 and E_2 for three-particle final states

Although you do have a simple thing to integrate, you have to integrate it over terrible boundaries of integration. This prospect causes strong men to weep, brave women to quail and sensible people to go to their nearest digital computer.

[3] Just differentiate $E^2 = |\mathbf{p}|^2 + m^2$, to get $2E\, dE = 2|\mathbf{p}|\, d|\mathbf{p}|$.

An especially interesting application of the formula (12.62) occurs in the decay of a spinless particle. If a spinless particle decays into three spinless particles, or indeed if a particle with spin decays into three particles with spin, but you average over all the spins, so there's no preferred direction, then obviously the differential decay amplitude $d\Gamma$ doesn't depend on the angular variables ϕ_{12} and Ω_1. Of course, that's not true if you have a particle with a definite spin, or if you have two particles coming in, because then there's a specified direction in the problem (the direction of the spin, or the direction along which the two particles approach one another). For the decay of a spinless particle, you might as well go ahead and do the integral over Ω_1 and ϕ_{12} of D:

$$D = \frac{1}{32\pi^3} dE_1 dE_2 \tag{12.65}$$

In such a case, you will frequently find people making plots in E_1 and E_2, or the symmetric $E_1E_2E_3$ diagram, analogous to the Mandelstam diagram. They will put little dots whenever they've observed a decay event, as shown in Figure 12.5. This is very groovy,[4] because you can directly read off the squares of the invariant matrix elements without going through any kinematic computations. They're proportional to the density of dots, with a proportionality factor we know to be $1/(32\pi^3)$. It's very nice, when you're trying to see if experiment and theory fit, not to have to do any complicated phase space computations.

12.6 A question and a preview

As you now know, this course is divided into two kinds of lectures: those that are totally understandable and inexpressibly boring, and those that have exciting ideas in them—well, perhaps I'm giving myself airs—but are absolutely incomprehensible. Next time we will turn to the second kind of lecture.

We're going to try and redeem our scattering theory and all our Feynman graphs by re-establishing things in such a way that the adiabatic function $f(t)$ does not appear, and doing things straight, in the real world. This will involve a long sequence of arguments. Just the beginning of this topic will occupy a couple of lectures. Working out all the details that follow from it, which will involve us in all sorts of strange things with strange names like wave function renormalization, will take another two lectures. So it's going to take us a lot of time, and it's going to begin in a rather abstract way. We will start by investigating, in our old framework of scattering theory, what seems to be a silly question. I'll tell you what the silly question is now, although the investigation won't proceed until the next lecture.

The silly question is: *What is the meaning of a Feynman diagram when the external lines are off the mass shell?* External lines represent real particles, whose 4-momenta satisfy $p^\mu p_\mu = m^2$; they are said to be "on the mass shell". Internal lines, by contrast, represent *virtual* particles, which do *not* lie on the mass shell. A Feynman diagram gives us the scattering amplitude when the external lines are *on* the mass shell. However, if I take a Feynman diagram, let's say a particularly complicated and grotesque-looking diagram for meson–meson scattering, as in Figure 12.6, the Feynman rules as I wrote them don't say the external lines have to be on the mass shell. If I were some sort of maniac, I could compute this diagram with the external lines *not* obeying $p^\mu p_\mu = m^2$. Does a Feynman diagram with the external lines off the mass shell have any meaning? Well, it has a sort of primitive meaning that one can see. It could

[4] [Eds.] "Groovy" is antiquated American slang from the 1960's, meaning "excellent"; here, "welcome", or "a good thing".

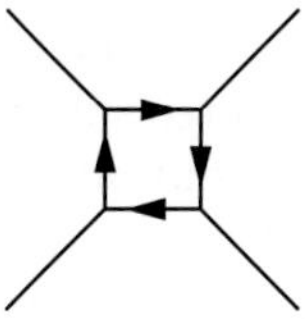

Figure 12.6: $\mathcal{O}(g^4)$ $\phi + \phi \to \phi + \phi$ *scattering*

be the internal part of some more complicated Feynman diagram, as in Figure 12.7: Yeah,

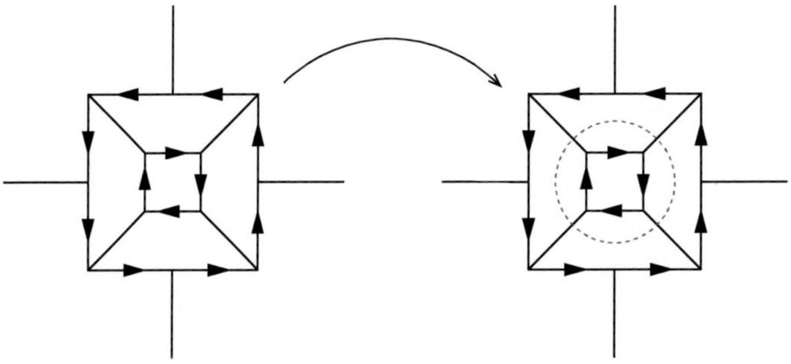

Figure 12.7: $\mathcal{O}(g^{12})$ $\phi + \phi \to \phi + \phi$ *scattering*

that's a homework problem. (Ha! I'm just kidding.) If I'm to have any hope of evaluating this diagram, I might try to put a dotted line around this inner part, evaluate it, get some function of the four 4-momenta coming in at the four vertices of the inner square, plug that dotted inner part as a black box into the bigger diagram, and then do the big integrals. So at least here is some sense of talking about Feynman diagrams with the external lines off the mass shell: It might be the internal part of a more complicated Feynman diagram. In my Feynman rules, although the outer lines of the larger diagram have to be on the mass shell, these lines on the smaller diagram don't, because they're internal lines.

Next lecture I will show that within the framework of our old scattering theory, these Feynman diagrams with lines off the mass shell can be given two other meanings, aside from the rather trivial meaning I have assigned to them originally by drawing this dotted circle. The second meaning is that these Feynman graphs, or rather, the sum of all Feynman graphs with a given number of external lines, can be related to objects called Green's functions that determine the response of the system to a particular kind of external source. In particular, if I take the Hamiltonian density for a system, and combine together, say, Model 1 and Model 3, there's a lot of interactions in the Model 3 $\mathscr{H}_I$, but there's also an external source,

$$\mathscr{H}_I \to \mathscr{H}_I + \rho(x)\phi(x) \tag{12.66}$$

Were we to compute the vacuum-to-vacuum matrix elements in the presence of ρ, we could make particles with the source, as we did in Model 1. But now it won't be so simple, because the particles are interacting. We have a new interaction in the theory caused by ρ. I'll write down its analytic form next lecture. And that new interaction in turn makes new Feynman diagrams:[5]

[5] [Eds.] Figure 12.8 does not appear in the video of Lecture 12, but it does in Coleman's handwritten notes

Figure 12.8: Feynman graph for an interaction term $\rho(x)\phi(x)$

If I am to compute what happens in this combined theory, say to fourth order in ρ and some order in the coupling constant, as shown in Figure 12.9, one of the diagrams I will encounter will be the Feynman diagram shown in Figure 12.6, the thing with the circle around it in Figure 12.7. But now, because of the interaction with ρ, denoted by the dots, the formerly external lines are internal lines, and therefore have to be integrated off the mass shall. I will work out the details of that next lecture and develop a second meaning of these graphs, with external lines off the mass shell, as Green's functions, in the primitive sense of George Green, a function that tells you the response of a system when you kick it, the system's response to an external source.

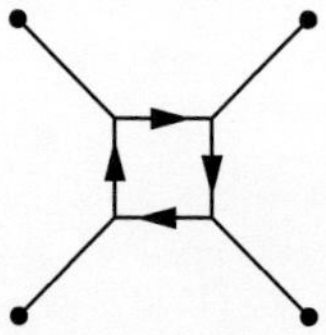

Figure 12.9: $\phi + \phi \to \phi + \phi$ scattering to $\mathcal{O}(\rho^4)$

I will then give a third meaning. I will show that in fact these things express a certain property of the Heisenberg fields, the exact solutions to the Heisenberg equations of motion. I will then assemble these three things and write down a formula that is really just a statement that you get a scattering amplitude by taking a Feynman graph with lines off the mass shell and putting the lines on the mass shell. I will connect that to a certain expression constructed of the Heisenberg fields. That expression will turn out to have no reference to our original adiabatic turning-on-and-off function $f(t)$, and that will be the starting point of our new investigation of scattering theory. I will then attempt, by going through considerable contortions and waving my hands at a ferocious rate, to justify that expression without talking about the turning-on-and-off function, and thus getting a formulation of scattering theory that has nothing to do with turning on and off. That is the outline of the next couple of lectures.

---

for October 28, 1986. In the video of Lecture 13, Coleman says that this Feynman graph was calculated when talking about Model 1. In fact, the *Wick* diagram was calculated in Chap. 8 (see (8.66)), but not the Feynman graph. The $\mathcal{O}(g^0)$ matrix element is

$$\langle 0|S|p\rangle = \langle 0| - i\int d^4x\, \rho(x)\phi(x)|p\rangle = -i\,\langle 0|\int d^4x\, e^{-ip\cdot x}\rho(x)|0\rangle = -i\tilde{\rho}(p)$$

Problems 6

6.1 The most common decay mode of the short-lived neutral kaon (mass 498 MeV) is into two charged pions (mass 140 MeV). For this process, $\Gamma = 0.773 \times 10^{10}\ \mathrm{s}^{-1}$. Make the silly assumption that the only interactions between pions and kaons are those of Model 3 of the lectures, with the kaon playing the role of the meson and the pion of the "nucleon", and compute, from these experimental data, the magnitude of the dimensionless quantity g/m_K, to one significant digit. Can you see why this is called a "weak interaction"?

Comments:

(1) Actually, the silly assumption is irrelevant: by Lorentz invariance, the matrix element for this process, a, is just a number; the center-of-momentum energy is the kaon mass, and, by rotational invariance in the c.o.m. frame, a cannot depend on the angle of the outgoing pions. What we are really doing is computing a, without any dynamical assumptions at all.

(2) Take $\hbar = 6.58 \times 10^{-22}$ MeV-s.

(1997a 6.1)

6.2 In Model 3, compute, to lowest non-vanishing order in g, the center-of-momentum differential cross-section and the total cross section for "nucleon"–"antinucleon" elastic scattering.

(1997a 6.2)

6.3 Do the same for "nucleon–antinucleon" annihilation into two mesons. WARNING: Don't double-count the final states.

(1997a 6.3)

6.4 In class, I showed that to every order of perturbation theory, the invariant Feynman amplitude was unchanged under multiplication of all 4-momenta by -1, and I claimed that this was equivalent to invariance of the S matrix under an anti-unitary operator, Ω_{CPT}. In this problem, you're asked to work out explicitly what an anti-unitary symmetry implies about the S matrix, to verify (or perhaps refute) my claim. For notational simplicity, we'll work with time reversal, Ω_T; the extension to TCP is trivial.

Let us denote the action of time reversal on a state $|a\rangle$ by $|a^T\rangle$:

$$|a^T\rangle = \Omega_T\,|a\rangle \tag{P6.1}$$

Thus, in the theory of a free scalar field,

$$\text{if } |a\rangle = |\mathbf{p}_1, \mathbf{p}_2, \ldots, \mathbf{p}_n\rangle\,, \text{ then } |a^T\rangle = |-\mathbf{p}_1, -\mathbf{p}_2, \ldots, -\mathbf{p}_n\rangle \tag{P6.2}$$

Assume that in the interacting theory,

$$\Omega_T\,|a\rangle^{\text{in}} = |a^T\rangle^{\text{out}} \tag{P6.3}$$

and also assume a like equation with "in" and "out" interchanged.

(a) Show from the definition of the S matrix in terms of in and out states that this rule implies

$$\langle a|S|b\rangle = \langle b^T|S|a^T\rangle \tag{P6.4}$$

(Note that this is a sensible equation, in that both the left- and right-hand sides are linear functions of $|b\rangle$ and anti-linear functions of $|a\rangle$.)

(b) Get the same result as (P6.4), starting from the fundamental formula (7.59) of our adiabatic scattering theory,

$$\langle a|S|b\rangle = \langle a|U_I(\infty, -\infty)|b\rangle \tag{7.59}$$

assuming, of course, that the interaction is invariant under time reversal:

$$\Omega_T H_I(t)\Omega_T^{-1} = H_I(-t) \tag{P6.5}$$

(1997a 6.4)

Solutions 6

6.1 The relevant Feynman graph is Figure 12.3 on p. 251. The relevant equation is (12.35). The coupling constant g has units of $[L]^{-1}$ or MeV, and so does Γ. The experimental value of Γ is given in s^{-1}. To get the units right, we have to put $\hbar$ in: $\Gamma \to \hbar\Gamma$. Then

$$\hbar\Gamma = \frac{g^2}{16\pi m_K^2}\sqrt{m_K^2 - 4m_\pi^2} \tag{S6.1}$$

Solving for g/m_K gives

$$\frac{g}{m_K} = \sqrt{\frac{16\pi\hbar\Gamma}{\sqrt{m_K^2 - 4m_\pi^2}}} = 8\times 10^{-7} \tag{S6.2}$$

The estimate helps to explain why this is a "weak" interaction, with g/m_K on the order of 10^{-5} smaller than $\alpha = e^2/\hbar c$. ■

6.2 The formula for the differential cross-section is given by (12.26),

$$\frac{d\sigma}{d\Omega} = \frac{1}{64\pi^2 E_T^2}\frac{|\mathbf{p}_f|}{|\mathbf{p}_i|}|\mathcal{A}_{fi}|^2 \tag{12.26}$$

Because the collision is elastic (between particles of identical mass), we have $|\mathbf{p}_i| = |\mathbf{p}_f|$, and the differential cross-section becomes

$$\frac{d\sigma}{d\Omega} = \frac{1}{64\pi^2 E_T^2}|\mathcal{A}_{fi}|^2 \tag{S6.3}$$

The amplitude comes from the two Feynman graphs in Figure 11.1 and is given by (11.1), with $p_1' \to p_3$ and $p_2' \to p_4$, respectively,

$$i\mathcal{A}_{fi} = (-ig)^2\left[\frac{i}{(p_1-p_3)^2 - \mu^2 + i\epsilon} + \frac{i}{(p_1+p_2)^2 - \mu^2 + i\epsilon}\right] \tag{11.1}$$

In the center-of-momentum frame we have

$$\begin{aligned} p_1 &= (\sqrt{|\mathbf{p}_i|^2 + m^2}, \mathbf{p}_i) \qquad p_2 = (\sqrt{|\mathbf{p}_i|^2 + m^2}, -\mathbf{p}_i) \\ p_3 &= (\sqrt{|\mathbf{p}_i|^2 + m^2}, \mathbf{p}_f) \qquad p_4 = (\sqrt{|\mathbf{p}_i|^2 + m^2}, -\mathbf{p}_f) \end{aligned} \tag{S6.4}$$

The total energy $E_T = 2\sqrt{|\mathbf{p}_i|^2 + m^2}$, so

$$E_T^2 = 4(|\mathbf{p}_i|^2 + m^2) \tag{S6.5}$$

Let θ be the angle between $\mathbf{p}_i$ and $\mathbf{p}_f$, and ϕ be the azimuthal angle about the $\mathbf{p}_i$ axis. Then

$$(p_1 - p_3) = (0, \mathbf{p}_i - \mathbf{p}_f) \Rightarrow (p_1-p_3)^2 = -(\mathbf{p}_i - \mathbf{p}_f)^2 = -2|\mathbf{p}_i|^2(1-\cos\theta) \tag{S6.6}$$

and similarly

$$(p_1 + p_2) = (E_T, \mathbf{0}) \Rightarrow (p_1+p_2)^2 = E_T^2 \tag{S6.7}$$

Plugging these in to the amplitude,

$$\mathcal{A}_{fi} = -g^2 \left[\frac{-1}{2|\mathbf{p}_i|^2(1-\cos\theta) + \mu^2 - i\epsilon} + \frac{1}{E_T^2 - \mu^2 + i\epsilon} \right] \tag{S6.8}$$

We can safely drop the $i\epsilon$'s, because neither denominator will become zero. Then plugging into (S6.3),

$$\frac{d\sigma}{d\Omega} = \frac{g^4}{64\pi^2 E_T^2} \left[\frac{2(|\mathbf{p}_i|^2(1-\cos\theta)+\mu^2) - E_T^2}{(E_T^2-\mu^2)(2|\mathbf{p}_i|^2(1-\cos\theta)+\mu^2)} \right]^2 \tag{S6.9}$$

To obtain the cross-section, we need to integrate this over the solid angle, $d\Omega = -d\phi\, d\cos\theta$. There is no ϕ dependence, so we can do that by inspection, to obtain 2π. Pulling out the constant terms, and writing $\cos\theta = z$, the cross-section is

$$\sigma = \frac{g^4}{32\pi E_T^2(E_T^2-\mu^2)^2} \int_{-1}^{1} dz \left[\frac{2(|\mathbf{p}_i|^2(1-z)+\mu^2) - E_T^2}{(2|\mathbf{p}_i|^2(1-z)+\mu^2)} \right]^2 \tag{S6.10}$$

The integral is easily done with the substitution $u = 2|\mathbf{p}_i|^2(1-z) + \mu^2$. Then

$$\begin{aligned} \sigma &= \frac{g^4}{64\pi p_i^2 E_T^2(E_T^2-\mu^2)^2} \int_{\mu^2}^{4|\mathbf{p}_i|^2+\mu^2} du \left[1 - \frac{(E_T^2-\mu^2)}{u} \right]^2 \\ &= \frac{g^4}{16\pi E_T^2(E_T^2-\mu^2)^2} \left[1 - \frac{(E_T^2-\mu^2)}{2|\mathbf{p}_i|^2} \ln\left(\frac{4|\mathbf{p}_i|^2+\mu^2}{\mu^2} \right) + \frac{(E_T^2-\mu^2)^2}{\mu^2(4|\mathbf{p}_i|^2+\mu^2)} \right] \end{aligned} \tag{S6.11}$$

The cross-section has a finite limit as $|\mathbf{p}_i| \to 0$, $\sigma \to \dfrac{g^4}{16\pi m^2 \mu^4} \left(\dfrac{\mu^2 - 2m^2}{4m^2 - \mu^2} \right)^2$. ■

6.3 The relevant diagrams are the two Feynman graphs in Figure 11.3. Let's redraw these, to let q, q' stand for the meson momenta, and p, p' for the "nucleon" and "antinucleon" momenta, respectively:

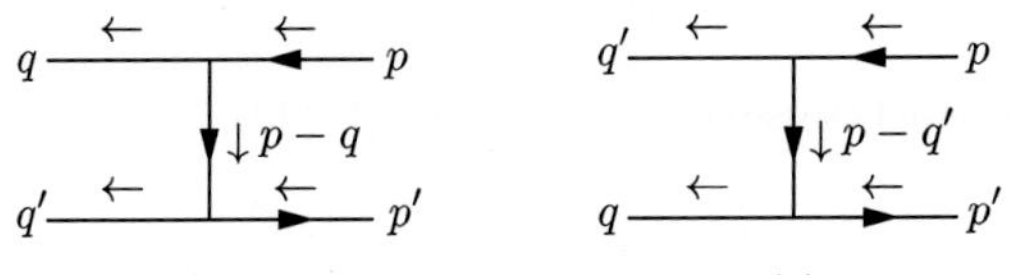

(a) (b)

The amplitude is

$$i\mathcal{A}_{fi} = (-ig)^2 \left[\frac{i}{(p-q)^2 - m^2 + i\epsilon} + \frac{i}{(p-q')^2 - m^2 + i\epsilon} \right] \tag{S6.12}$$

In the center-of-momentum frame we have

$$\begin{aligned} p &= (\sqrt{|\mathbf{p}|^2+m^2}, \mathbf{p}) \qquad & p' &= (\sqrt{|\mathbf{p}|^2+m^2}, -\mathbf{p}) \\ q &= (\sqrt{|\mathbf{q}|^2+\mu^2}, \mathbf{q})) \qquad & q' &= (\sqrt{|\mathbf{q}|^2+\mu^2}, -\mathbf{q})) \end{aligned} \tag{S6.13}$$

The total energy can be written in two equivalent forms, by energy conservation:

$$E_T^2 = 4(|\mathbf{p}|^2 + m^2) = 4(|\mathbf{q}|^2 + \mu^2) \tag{S6.14}$$

This means that $|\mathbf{q}|$ can be written as $\sqrt{|\mathbf{p}|^2 + m^2 - \mu^2}$, but we'll express our answers using both $|\mathbf{p}|$ and $|\mathbf{q}|$. We can also write

$$\begin{aligned} p &= (\tfrac{1}{2}E_T, \mathbf{p}) \qquad & p' &= (\tfrac{1}{2}E_T, -\mathbf{p}) \\ q &= (\tfrac{1}{2}E_T, \mathbf{q})) \qquad & q' &= (\tfrac{1}{2}E_T, -\mathbf{q}) \end{aligned} \tag{S6.15}$$

Let θ equal the angle between $\mathbf{p}$ and $\mathbf{q}$. Then

$$(p-q) = (0, \mathbf{p}-\mathbf{q}) \Rightarrow (p-q)^2 = 2|\mathbf{p}||\mathbf{q}|\cos\theta - (|\mathbf{p}|^2 + |\mathbf{q}|^2) \tag{S6.16}$$

Similarly,

$$(p-q') = (0, \mathbf{p}+\mathbf{q}) \Rightarrow (p-q')^2 = -2|\mathbf{p}||\mathbf{q}|\cos\theta - (|\mathbf{p}|^2 + |\mathbf{q}|^2) \tag{S6.17}$$

Both these quantities are negative definite, so neither contribution to the amplitude gives a pole, and once again the $i\epsilon$ terms may be dropped. Then

$$\mathcal{A}_{fi} = -g^2\left[\frac{1}{2|\mathbf{p}||\mathbf{q}|\cos\theta - (|\mathbf{p}|^2+|\mathbf{q}|^2) - m^2} + \frac{1}{-2|\mathbf{p}||\mathbf{q}|\cos\theta - (|\mathbf{p}|^2+|\mathbf{q}|^2) - m^2}\right] \tag{S6.18}$$

Mindful of the warning, recognize that the two final states $|q,q'\rangle$ and $|q',q\rangle$ are the same, and divide by 2 to prevent overcounting. Then

$$\frac{d\sigma}{d\Omega} = \frac{g^4}{32\pi^2 E_T^2}\frac{|\mathbf{q}|}{|\mathbf{p}|}\left[\frac{(|\mathbf{p}|^2+|\mathbf{q}|^2+m^2)}{(2|\mathbf{p}||\mathbf{q}|\cos\theta)^2 - (|\mathbf{p}|^2+|\mathbf{q}|^2+m^2)^2}\right]^2 \tag{S6.19}$$

As before, to obtain σ, we integrate this over the solid angle. Once again there is no ϕ dependence, so the $d\phi$ integration gives 2π. Writing $\cos\theta = z$, the cross-section is (note that the limits have been halved, as the integrand is even)

$$\sigma = \frac{g^4(|\mathbf{p}|^2+|\mathbf{q}|^2+m^2)^2}{16\pi E_T^2}\frac{|\mathbf{q}|}{|\mathbf{p}|}\int_0^1 dz\frac{1}{\left[(2|\mathbf{p}||\mathbf{q}|z)^2 - (|\mathbf{p}|^2+|\mathbf{q}|^2+m^2)^2\right]^2} \tag{S6.20}$$

The integral is of the form

$$\int\frac{dx}{\left[(ax)^2-b^2\right]^2} = \frac{1}{2b^2}\left[\frac{1}{ab}\tanh^{-1}\left(\frac{ax}{b}\right) - \frac{x}{(ax)^2-b^2}\right] + C \tag{S6.21}$$

This identity may be obtained by differentiation of the standard integral

$$\int\frac{dx}{(ax)^2-b^2} = -\frac{1}{ab}\tanh^{-1}\left(\frac{ax}{b}\right) + C$$

with respect to b. Using the expression (S6.21), we obtain, with $a = 2|\mathbf{p}||\mathbf{q}|$ and $b = (|\mathbf{p}|^2+|\mathbf{q}|^2+m^2)$,

$$\sigma = \frac{g^4}{16\pi E_T^2}\frac{|\mathbf{q}|}{|\mathbf{p}|}\left[\frac{1}{2|\mathbf{p}||\mathbf{q}|(|\mathbf{p}|^2+|\mathbf{q}|^2+m^2)}\tanh^{-1}\left(\frac{2|\mathbf{p}||\mathbf{q}|}{(|\mathbf{p}|^2+|\mathbf{q}|^2+m^2)}\right) - \frac{1}{(2|\mathbf{p}||\mathbf{q}|)^2 - (|\mathbf{p}|^2+|\mathbf{q}|^2+m^2)^2}\right] \tag{S6.22}$$

You can substitute $|\mathbf{q}| = \sqrt{|\mathbf{p}|^2+m^2-\mu^2}$, if you prefer. Note that $\sigma\to\infty$ as $|\mathbf{p}|\to 0$. ■

6.4 (a) The S-matrix is defined by (7.47):

$$\langle\phi|S|\psi\rangle \equiv {}^{\text{out}}\langle\phi|\psi\rangle^{\text{in}} \tag{7.47}$$

For clarity, introduce the inner product

$$(a,b) = \langle a|b\rangle \tag{S6.23}$$

Then

$$\langle a|S|b\rangle = (a^{\text{out}}, b^{\text{in}})$$

Using the anti-unitarity of Ω_T, (see (6.110))

$$(\Omega_T a, \Omega_T b) = (a^T, b^T) = (b,a) \tag{S6.24}$$

so in particular

$$\langle a|S|b\rangle = (a^{\text{out}}, b^{\text{in}}) = (\Omega_T b^{\text{in}}, \Omega_T a^{\text{out}}) \tag{S6.25}$$

The statement of the problem says that we are to assume

$$\Omega_T|b\rangle^{\text{in}} = |b^T\rangle^{\text{out}}; \qquad \Omega_T|a\rangle^{\text{out}} = |a^T\rangle^{\text{in}}$$

Then

$$\langle a|S|b\rangle = (\Omega_T b^{\text{in}}, \Omega_T a^{\text{out}}) = (b^{T\,\text{out}}, a^{T\,\text{in}}) = {}^{\text{out}}\langle b^T|a^T\rangle^{\text{in}} = \langle b^T|S|a^T\rangle \tag{S6.26}$$

which was to be shown. ■

(b) Using Dyson's expansion (7.36) for the S matrix,

$$\langle a|S|b\rangle = \langle a|U_I(\infty,-\infty)|b\rangle = \sum_n \frac{(-i)^n}{n!}(a, T\left[\int dt_1 \dots dt_n\, H_I(t_1 \cdots H_{t_n})\right] b)$$

$$= \sum_n (-i)^n (a, \int_{t_1>t_2>\cdots>t_n} dt_1 \dots dt_n \left[H_I(t_1)\cdots H_I(t_n)\right] b)$$

$$= \sum_n (-i)^n (\Omega_T \left[\int_{t_1>t_2>\cdots>t_n} dt_1 \dots dt_n\, H_I(t_1)\cdots H_I(t_n)\right] b, \Omega_T a) \tag{S6.27}$$

$$= \sum_n (-i)^n (\int_{t_1>t_2>\cdots>t_n} dt_1 \dots dt_n \left[\Omega_T H_I(t_1)\Omega_T^{-1}\Omega_T \cdots \Omega_T H_I(t_n)\Omega_T^{-1}\right] \Omega_T b, \Omega_T a) \tag{S6.28}$$

$$= \sum_n (-i)^n (\int_{t_1>t_2>\cdots>t_n} dt_1 \dots dt_n \left[H_I(-t_1)\cdots H_I(-t_n)\right] b^T, a^T) \tag{S6.29}$$

$$= \sum_n (-i)^n (b^T, \int_{t_1>t_2>\cdots>t_n} dt_1 \dots dt_n \left[H_I(-t_1)\cdots H_I(-t_n)\right]^\dagger a^T) \tag{S6.30}$$

$$= \sum_n (-i)^n (b^T, (\int_{t_1>t_2>\cdots>t_n} dt_1 \dots dt_n \left[H_I(-t_n)\cdots H_I(-t_1)\right] a^T) \tag{S6.31}$$

Change variables: let $-t_i = \tau_i$. We have to adjust the inequalities: if $t_i > t_j$, then $-t_i < -t_j$. There will be an extra -1 from each change of the variables of integration, and a second -1 from changing the implicit limits of integration, e.g., $\int_{t_{i+1}}^{t_i} \to -\int_{t_i}^{t_{i+1}}$, or an overall change of $(-1)^2$ for each integral:

$$\langle a|S|b\rangle = (b^T, \sum_n (-i)^n \int_{\tau_1<\tau_2<\cdots<\tau_n} d\tau_1 \dots d\tau_n \left[H_I(\tau_n)\cdots H_I(\tau_1)\right] a^T) \tag{S6.32}$$

$$= (b^T, \sum_n (-i)^n \int_{\tau_n>\cdots>\tau_2>\tau_1} d\tau_n \dots d\tau_1 \left[H_I(\tau_n)\cdots H_I(\tau_1)\right] a^T) \tag{S6.33}$$

$$= (b^T, U_I(\infty,-\infty)a^T) = \langle b^T|S|a^T\rangle \tag{S6.34}$$

which was to be shown. ■

13

Green's functions and Heisenberg fields

We will now consider diagrams with external lines off the mass shell. Although much of what we say will not be restricted to our model theory, Model 3, I'll use that continually as an example. Here is a typical Feynman graph (the same one we looked at last time, Figure 12.6), which I choose to evaluate not just with lines on the mass shell but with lines *off* the mass shell.

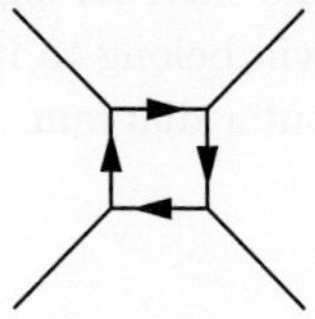

Figure 13.1: $\mathcal{O}(g^4)$ $\phi + \phi \rightarrow \phi + \phi$ *scattering*

That is an interesting object because it could be an internal part of a more complicated Feynman graph, as I explained at the end of the last lecture. For simplicity I will deal only with graphs with external meson lines. The extension to graphs with both external mesons and nucleon lines, or more complicated kinds of theories when you have 17 kinds of particles in 17 different kinds of external lines, is trivial. I will not assume that the only particles in the theory are the mesons, just that the only graphs we're going to look at are those with external meson lines.

13.1 The graphical definition of $\widetilde{G}^{(n)}(k_i)$

I define the *four-point function* $\widetilde{G}^{(4)}(k_1, k_2, k_3, k_4)$ to be the sum of *all* graphs with four external meson lines to *all* orders of perturbation theory. I will indicate this sum graphically by a shaded blob, as in Figure 13.2.

All of the external momenta are labeled. As in our discussion of crossing, all k's are oriented inward and, by energy-momentum conservation, their sum must be zero. Because we're off the mass shell, and dealing with spacelike as well as timelike momenta, there's no point in adopting any other orientation convention.

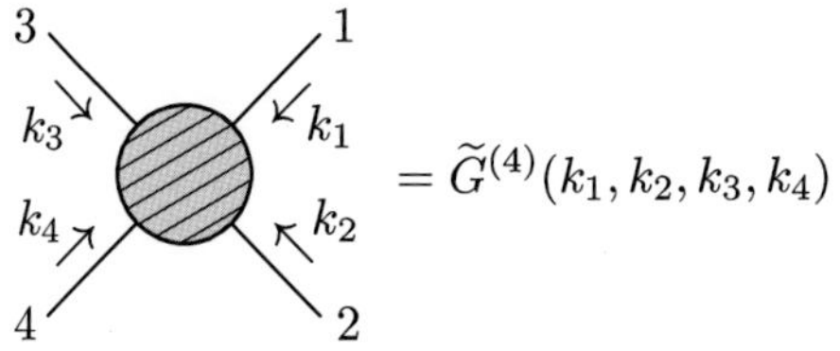

Figure 13.2: Graphical representation of $\widetilde{G}^{(4)}(k_1, \ldots, k_4)$

I have some freedom about how to define these graphs. I define them to include: all *connected* graphs, all *disconnected* graphs, all delta functions, including the overall energy-momentum conserving delta function which we previously have been factoring out of our Feynman graphs, and all propagators (including those on the external lines). The disconnected graphs are rather trivial for the four-point function $\widetilde{G}^{(4)}(k_i)$ (see Figure 13.4), but of course we're going to consider things with more than four lines shortly. Putting the propagators on the external lines is just a convenience if the blob is a internal part of some more complicated graph, like this one:

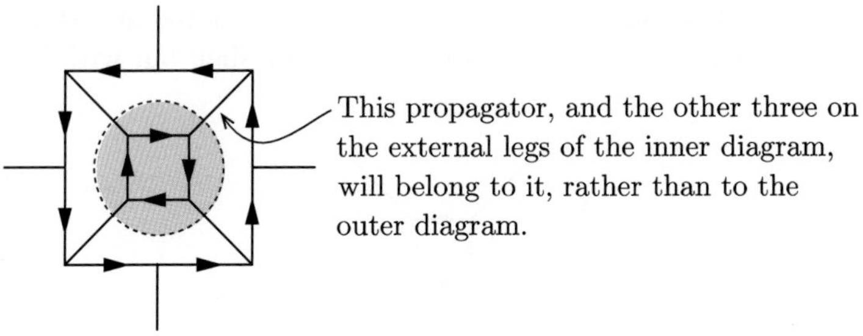
This propagator, and the other three on the external legs of the inner diagram, will belong to it, rather than to the outer diagram.

Figure 13.3: The blob as an internal graph

I draw a dotted line about the internal part I'm studying. It's a matter of convenience whether I put the propagators on these lines inside the blob, within the dotted line, or outside the blob. I'll put them inside the blob. That will turn out later to be convenient.

To give a definite example, let me write down the first few graphs that contribute in our theory to $\widetilde{G}^{(4)}(k_1, \ldots, k_4)$, the first few that are inside the blob. You could have zeroth-order contributions in which all that happens is that the four lines go right through the blob and don't interact at all, plus two permutations depending on whether I match up k_1 with k_2, k_3 or k_4. And there would be fourth-order corrections, including Figure 13.1 and its friends, and higher-order corrections:

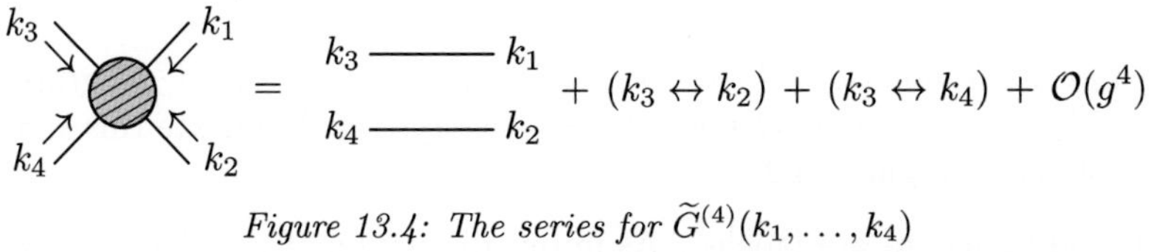

Figure 13.4: The series for $\widetilde{G}^{(4)}(k_1, \ldots, k_4)$

(Meson–meson scattering in our theory begins with $\mathcal{O}(g^4)$, although nucleon scattering processes begin in $\mathcal{O}(g^2)$.)

Analytically this equation is

$$\widetilde{G}^{(4)}(k_1,k_2,k_3,k_4) = (2\pi)^4\delta^{(4)}(k_1+k_3)\frac{i}{k_1^2-\mu^2+i\epsilon}(2\pi)^4\delta^{(4)}(k_2+k_4)\frac{i}{k_4^2-\mu^2+i\epsilon}+\cdots \tag{13.1}$$

the dots indicating two permutations corresponding to the two other ways I can pair up momenta with k_3, plus terms of order g^4. The momenta in the delta functions are plus because all the momenta are oriented inward. The k_1^2 could just as well be k_3^2, equally the k_4^2 could be k_2^2; it doesn't matter because of the delta functions.

If you have an expression for $\widetilde{G}^{(4)}$ off the mass shell, you have it on the mass shell as well, simply by putting the lines on the mass shell. We can, if we know $\widetilde{G}^{(4)}$, compute the corresponding S-matrix element. In the particular case we're studying at the moment, we have

$$\langle k_3,k_4|S-1|k_1,k_2\rangle = \prod_{r=1}^{4}\left[(-i)(k_r^2-\mu^2)\right]\widetilde{G}^{(4)}(-k_3,-k_4,k_1,k_2) \tag{13.2}$$

with the momenta k_1, k_2, k_3 and k_4 now on the mass shell. The product of the factors $k_r^2-\mu^2$ is to cancel out the four propagators we've put on the outer lines by convention; we now take them off. The argument of $\widetilde{G}^{(4)}$ is symmetric; how I arrange the momenta doesn't matter. I'll say $\widetilde{G}^{(4)}(-k_3,-k_4,k_1,k_2)$. What results is just our old formula for the S-matrix element again. Please notice that the three disconnected graphs I wrote down that arise in zeroth order make no contribution to the S-matrix, as indeed they should not, because they each have only two propagators, as in (13.1), two pole factors, but we have four factors of zero in front of them, and therefore they get completely canceled out.

So this is our rule. If you have $\widetilde{G}^{(4)}(k_i)$, the Feynman diagrams on the mass shell are obtained by taking the Feynman diagrams off the mass shell, canceling out the propagators we put in by convention, and putting the lines on the mass shell. We define $\widetilde{G}^{(n)}(k_i)$ in exactly the same way for n external lines (restricted here to mesons), directed inwards:

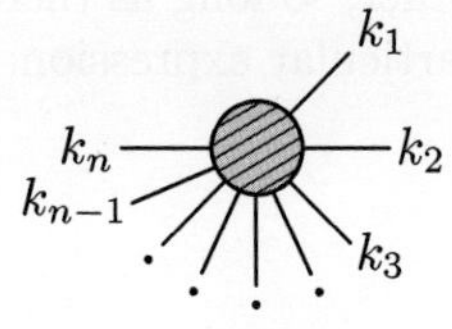

$= \widetilde{G}^{(n)}(k_1,k_2,\ldots,k_n) =$ sum of all graphs with n external lines (13.3)

As with $\widetilde{G}^{(4)}$, the functions $\widetilde{G}^{(n)}$ follow these conventions:

1. The momenta k_i are oriented inward.
2. The external lines include propagators $(k_i^2-\mu^2+i\epsilon)^{-1}$.
3. All 4-momentum conserving delta functions are included.
4. All connected graphs are included.
5. All disconnected graphs are included.

As you might guess from the twiddle, we can define $\widetilde{G}^{(n)}(k_i)$ as the Fourier transform of some object, $G^{(n)}(x_1,\ldots,x_n)$:

$$G^{(n)}(x_1,\ldots,x_n) = \int\frac{d^4k_1}{(2\pi)^4}\cdots\frac{d^4k_n}{(2\pi)^4}e^{-ik_1\cdot x_1-\cdots-ik_n\cdot x_n}\widetilde{G}^{(n)}(k_1,\ldots,k_n) \tag{13.4}$$

Since all of the $\widetilde{G}^{(n)}$'s are even functions of the momenta, it hardly matters what signs I use for the exponents in the Fourier transform, but I want to be consistent in my notation, defined in (8.63).

13.2 The generating functional $Z[\rho]$ for $G^{(n)}(x_i)$

We can attach a second meaning to $\widetilde{G}^{(n)}(k_i)$ by changing the Hamiltonian of our theory to consider a combined version of, for instance, Model 3, or some general theory involving a scalar field, and Model 1. That is to say, we can take $\mathscr{H}$ and imagine changing it, by adding to it:

$$\mathscr{H} \to \mathscr{H} + \rho(x)\phi(x) \tag{13.5}$$

where $\rho(x)$ as usual is some smooth function that vanishes as x goes to infinity. Then if we are to compute $\langle 0|\mathrm{S}|0\rangle$ (or any S-matrix element) in the presence of ρ, we have a new diagram in our theory which I could indicate by a dot, with a single line coming out of it: If I orient

Figure 13.5: Feynman graph for an interaction term $\rho(x)\phi(x)$

the momentum k to move outwards, so it will fit onto other things where k is going inwards, it is easily seen to be $-i\tilde{\rho}(-k)$: Or, since ρ is a real function, this could just as well be written

Figure 13.6: Feynman graph for an interaction term $\rho(x)^*\phi(x)$

$-i\tilde{\rho}(k)^*$. That is the value of that vertex we obtained in Model 1.[1]

If we now consider the matrix element $\langle 0|\mathrm{S}|0\rangle$ in the presence of this source, ρ, we can expand things in a power series of our new vertex, imagining we have already summed up all powers and all of our old vertices. For example, to fourth order in ρ, what we get is shown in Figure 13.7. This blob is precisely the same, Figure 13.4, as we defined before. You have the four lines coming out, and they can do whatever they want with each other, so long as there's no ρ involved, because we're only going to fourth order in ρ for this particular expression.

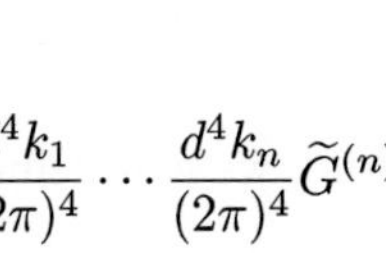

Figure 13.7: $\widetilde{G}^{(4)}(k_1, \dots, k_4)$ with $\tilde{\rho}(k)$

We define $\langle 0|\mathrm{S}|0\rangle$ in the presence of the source ρ to be a functional of ρ, a numerical function of ρ, which we call $Z[\rho]$:

$$\langle 0|\mathrm{S}|0\rangle_\rho = Z[\rho] \equiv 1 + \sum_{n=1}^{\infty} \frac{(-i)^n}{n!} \int \frac{d^4k_1}{(2\pi)^4} \cdots \frac{d^4k_n}{(2\pi)^4} \widetilde{G}^{(n)}(k_1, \dots, k_n)\tilde{\rho}(-k_1)\cdots\tilde{\rho}(-k_n) \tag{13.6}$$

We say "function*al*" rather than "function", because of a dumb convention that a numerical function of a function is called a functional. Often the convention is to use square brackets for the argument of a functional: $Z[\rho]$.

---

[1] [Eds.] See note 5, p. 258.

There is a residual combinatoric factor of $1/n!$ because this is a vacuum-to-vacuum diagram, so our usual arguments that all the $n!$'s cancel do not apply. Why this factor is $1/n!$ is easy to see. If we imagine restricting ourselves to the case where the first ρ gives up momentum k_1, the second gives a momentum k_2, etc., then all of our lines are well-defined, and we have no factor of $1/n!$. On the other hand, when we integrate over all k's in this expression, we overcount each those terms $n!$ times, corresponding to the $n!$ permutations of a given set of k's, and therefore we need a $1/n!$ to cancel it out. I know combinatoric arguments are often not clear the first time you hear them, but after a little thought, they become clear.

This formula (13.6) is so set up that it can also be written as a formula in position space simply by invoking a generalization of Parseval's Theorem, (9.32),

$$\int dx\, f(x)g(x) = \int \frac{dk}{2\pi}\tilde{f}(k)\tilde{g}(k)^* = \int \frac{dk}{2\pi}\tilde{f}(k)\tilde{g}(-k) \tag{13.7}$$

the last equality following if $g(x)$ is a real function. Then, since $\rho(x)$ is a real function,

$$Z[\rho] = 1 + \sum_{n=1}^{\infty}\frac{(-i)^n}{n!}\int d^4x_1\cdots d^4x_n\, G^{(n)}(x_1,\ldots,x_n)\rho(x_1)\cdots\rho(x_n) \tag{13.8}$$

The $G^{(n)}(x_1\ldots x_n)$'s now reveal their *second* meaning, as **Green's functions**, objects that give the response of a system (in this case, the vacuum) to an external perturbation (here, $\rho(x)\phi(x)$). George Green of Nottingham introduced the concept in the early 19th century for a linear system, so he only had a single Green's function. Now we have a system that has a possible nonlinear response, and therefore we have an infinite power series in powers of ρ. That's why we denote these functions with G's, in honor of Green.[2]

An amusing feature of the formula (13.8) is that all physical information about the system, at least concerning experiments involving mesons, is embedded in the single functional $Z[\rho]$. If you know $Z[\rho]$ for an arbitrary ρ, then you know the $G^{(n)}$'s. And if you know the $G^{(n)}$'s, then you know the scattering amplitudes. It's a fat chance you'll know $Z[\rho]$ for an arbitrary ρ. Nevertheless, it's sometimes formally very useful. Instead of manipulating the infinite string of objects on the right-hand side of (13.8), it can be easier to work with the single object $Z[\rho]$. We'll give some examples of that.

$Z[\rho]$ is sometimes called a *generating functional.* This terminology comes from the theory of special functions. In working with, say, Legendre polynomials, it's convenient to have a generating function, a single function of two variables. When you do a power series expansion in one of the variables, you obtain the Legendre polynomials as the coefficients of the powers of the variables. That's useful in proving things about special functions. $Z[\rho]$ is the same sort of thing: If we expand $Z[\rho]$ out in a power series of the ρ's, the coefficients are the Green's

[2] [Eds.] Feynman's propagators were the first systematic use of Green's functions in quantum field theory. R. P. Feynman, "The Theory of Positrons", *Phys. Rev.* **76** (1949) 749–759. The introduction of sources to obtain them was pioneered by Schwinger in a series of papers: "On gauge invariance and vacuum polarization", *Phys. Rev.* **82** (1951) 664–679; "The theory of quantized fields I.", *Phys. Rev.* **82** (1951) 914–927; "The theory of quantized fields II.", *Phys. Rev.* **91** (1953) 713–728. All of these papers may be found in Schwinger *QED*. For accessible introductions to Green's functions, see J. W. Dettman, *Mathematical Methods in Physics and Engineering*, 2nd ed., McGraw-Hill, 1969, Chap. 5; and F. W. Byron and R. W. Fuller, *Mathematics of Classical and Quantum Physics*, Addison-Wesley, 1970, Chap. 7. Both of these texts have been reprinted by Dover Publications.

functions:

$$G^{(n)}(x_1,\ldots,x_n) = i^n \left.\frac{\partial^n Z[\rho]}{\partial\rho(x_1)\cdots\partial\rho(x_n)}\right|_{\rho(x)=0} \tag{13.9}$$

You can play cunning tricks with these generating functionals. Although that's not really the point of this lecture, I cannot resist a digression. It is easy to write down a generating functional that gives you not the full set of Green's functions but only the *connected* Green's functions:

$$Z_c[\rho] = 1 + \sum_{n=1}^{\infty} \frac{(-i)^n}{n!} \int d^4x_1 \cdots d^4x_n \, G_c^{(n)}(x_1,\ldots,x_n)\,\rho(x_1)\cdots\rho(x_n) = \ln Z[\rho] \tag{13.10}$$

where the c means that the expression includes connected graphs only. That's our old exponentiation theorem, (8.49). Remember, the sum of all Feynman graphs for $\langle 0|\mathrm{S}|0\rangle$ is the exponential of the sum of the *connected* Feynman graphs. This relation is often written as

$$\boxed{Z[\rho] = \exp\{iW[\rho]\}} \tag{13.11}$$

where $iW[\rho] = Z_c[\rho]$ is the sum of the *connected* Feynman graphs. So if you want the generating functional for the **connected Green's functions**, the sums of the connected graphs, you just take the logarithm of $Z[\rho]$. We won't use this formula immediately, but it is so cute and its demonstration so easy, I could not resist putting it down.

13.3 Scattering without an adiabatic function

Thus far our discussion of Green's functions and the generating functional has been in the framework of our old theory, where the interaction Hamiltonian is adiabatically turned on and off with the function $f(t)$. The reason I've gone through these manipulations is to get a formulation that I can extend to the case where $f(t)$ is abolished, i.e., set equal to one. We forget about all of our old theories, and start afresh on the problem of computing the S-matrix.

We begin with

$$f(t) = 1 \tag{13.12}$$

That is, we now set $f(t)$ always and forever equal to one. No more will we talk about an adiabatic turning-on-and-off function. I can however still take my Hamiltonian and add to it a source term involving $\rho(x)$, a c-number space-time function which I control:

$$\mathscr{H} \to \mathscr{H} + \rho(x)\phi(x) \tag{13.13}$$

I now redefine $Z[\rho]$ as the amplitude for getting from the physical vacuum $|0\rangle^P$ to the physical vacuum, in the presence of the source ρ:

$$Z[\rho] = {}^P\langle 0|U_I(\infty,-\infty)|0\rangle^P \tag{13.14}$$

where $U_I(\infty,-\infty)$ is the *Schrödinger* picture U_I operator. This is different from (13.6) because, for the moment, I don't want to talk about bare vacua. The physical vacuum is the real vacuum, the ground state of the Hamiltonian. I will assume I have normalized my theory so that the physical vacuum has energy zero:

$$H\,|0\rangle^P = 0 \tag{13.15}$$

(The Hamiltonian H in (13.15) does not include $\rho(x)$. If it did, it would be a time-dependent Hamiltonian, and there would be no well-defined ground state.) We will introduce a normalizing constant to give the physical vacuum norm 1:

$$ {}^{P}\langle 0|0\rangle^{P} = 1 \tag{13.16} $$

Equation (13.14) is our new definition of $Z[\rho]$. There is no bare vacuum $|0\rangle$ in the picture. I have this real, honest to goodness theory, without the artifice of the adiabatic function. I make the theory even more complicated by adding the term $\rho(x)\phi(x)$. I start with the vacuum state. I then wiggle my source $\rho(x)$ and I ask, what is the amplitude that I'm still in the vacuum state? I don't write (13.14) in terms of the S-matrix because I don't know yet what the S-matrix is (remember, in Section 7.4, we introduced the function $f(t)$ to facilitate the definition of the S-matrix). As before, I define $\widetilde{G}^{(n)}(k_i)$ and $G^{(n)}(x_i)$ as successive terms in a power series expansion of $Z[\rho]$ in powers of ρ.

I now want to ask two questions.

Question 1. Are the $\widetilde{G}^{(n)}(k_i)$'s given by the formal sum of the Feynman graphs, as with our first scattering theory?

$Z[\rho]$ does not have the same definition as before, but of course it's not the same theory: we no longer have an adiabatic turning-on-and-off function. This is a question linked to perturbation theory. (Whether the sum converges is not a question I strive to answer in this lecture or indeed in this course.) We will answer this question shortly, and the answer is yes.

Question 2. Is (13.2) still true in the new theory, without the adiabatic function $f(t)$?

It is clear that the object on the right-hand side of (13.2) is well-defined without reference to perturbation theory, without expansion in any coupling constant lurking inside $\mathscr{H}$. Maybe we found this object by being a genius in summing up perturbation theory; maybe an angel flying through the window gave it to us on a golden tablet. To put the second question another way: Do we get the S-matrix element from (13.2)? This question has nothing to do with perturbation theory. The full answer will have to wait till next time, but I'll tell you now: it is *almost* true. There is a correction.

This program will give us what I described in an earlier lecture as a real scattering theory: one where you have a formula, (13.2), that tells you how to extract the S-matrix elements if you can solve the dynamics exactly: if you can obtain the $\widetilde{G}^{(n)}$'s. You could find them from perturbation theory (the answer to the first question) and thus develop perturbation theory for S-matrix elements. However, if you have some other approximate method for solving the dynamics—a variational method, Regge poles, dispersion relations, maybe some brand new method from the latest issue of *Physical Review Letters*—it doesn't matter; it just means you have a different approximation for the right-hand side. This formula (13.2) is exact (apart from the correction), and you can feed in the $\widetilde{G}^{(n)}$'s from your preferred method to get the approximation for the S-matrix element.

We'll actually construct in and out states, with a certain amount of hand-waving, to find the S-matrix element as the inner product of an in state and an out state, as in (7.47), when I was sketching out non-relativistic scattering theory. I will then show that the S-matrix element is, aside from a correction factor, given by the right-hand side of (13.2). The correction factor is called **wave function renormalization.** We will have defined the S-matrix without an adiabatic turning-on-and-off function.

13.4 Green's functions in the Heisenberg picture

I now turn to Question 1. I will first develop a formula, independent of perturbation theory, for these Green's functions that will be extremely useful for comparison with the corresponding series from perturbation theory. I have a Hamiltonian, $\mathscr{H} + \rho\phi$. I will investigate this Hamiltonian not by Wick's theorem, but by Dyson's formula. I'll split it up in a rather peculiar way:

$$\mathscr{H} \to \mathscr{H} + \rho(x)\phi(x) \equiv \text{“}\mathscr{H}_0\text{''} + \mathscr{H}_I \tag{13.17}$$

treating the source term $\rho(x)\phi(x)$ as the interaction Hamiltonian $\mathscr{H}_I$, and the original Model 3 Hamiltonian $\mathscr{H}$ as if it were $\mathscr{H}_0$. I've put quotes around it temporarily, because later I'm going to break $\mathscr{H}$ up into the original free Hamiltonian $\mathscr{H}_0$ plus the Model 3 interaction, which I'll call $\mathscr{H}'$;

$$\mathscr{H}' = g\psi^*(x)\psi(x)\phi(x) \tag{13.18}$$

We have the freedom to do this, because Dyson's formula says we can divide things up into a free part and an interacting part any way we please. In this way of doing things, when $\rho = 0$, the interaction picture field is the real, honest to goodness Heisenberg field, because the interaction picture field $\phi_I(x)$ is always the Heisenberg field $\phi_H(x)$ when you throw away the interaction Hamiltonian:

$$\text{“}\phi_I(x)\text{''} = \phi_H(x) \quad \text{when } \rho(x) = 0 \tag{13.19}$$

Thus we can apply Dyson's formula to compute $Z[\rho]$ in exactly the same way as we used it to obtain the S-matrix in Model 3, (8.1) (though we will put $Z[\rho]$ in terms of the U_I matrix, as we haven't talked about the S-matrix, yet):

$$Z[\rho] = {}^P\langle 0|U_I(\infty,-\infty)|0\rangle^P = {}^P\langle 0|\, T\exp\left[-i\int d^4x\, \rho(x)\phi_H(x)\right]|0\rangle^P \tag{13.20}$$

This is the time-ordered exponential of $(-i)$ times the integral of the interaction Hamiltonian (with quotes understood) of fields in the interaction picture (again, with quotes understood). It's the same old Dyson formula; I've just broken things up into a free part and an interacting part in a different way. $Z[\rho]$ can be expanded as a sequence of powers in ρ. I can't use Wick's theorem because the Heisenberg field doesn't have c-number commutators for arbitrary separations. But I can still expand the power series:

$$Z[\rho] = \sum_{n=0}^{\infty} \frac{(-i)^n}{n!} \int d^4x_1 \cdots d^4x_n\, \rho(x_1)\cdots\rho(x_n)\, {}^P\langle 0|T\left(\phi_H(x_1)\cdots\phi_H(x_n)\right)|0\rangle^P \tag{13.21}$$

Comparing this formula (13.21) for $Z[\rho]$ with the previous expression (13.8), we see that

$$\boxed{G^{(n)}(x_1,\ldots,x_n) = {}^P\langle 0|T\left(\phi_H(x_1)\cdots\phi_H(x_n)\right)|0\rangle^P} \tag{13.22}$$

So we have a *third* meaning for the blobs, the Green's functions $G^{(n)}(x_1,\ldots,x_n)$: They are, in position space, simply the physical vacuum expectation values of the time-ordered product of a string of Heisenberg fields $\phi_H(x_1),\ldots,\phi_H(x_n)$. In (13.21), you've got $G^{(n)}$ defined as in (13.8), except that $Z[\rho]$ is now given in terms of the physical vacuum $|0\rangle^P$ and the U_I operator, instead of the bare vacuum $|0\rangle$ and the S-matrix. The expressions are term by term equal, if we make the identification (13.22). All the other factors, the minus i's and the n!'s, come out right. Of course, that is a consequence of choosing the right notational conventions originally.

This is one side of Question 1. We've defined $G^{(n)}(x_i)$ in (13.22). The other side of that question is: What corresponding quantities $G^{(n)\,\text{Feyn}}(x_1,\dots,x_n)$ do we get by summing up Feynman graphs? Are they the same?

Let $Z[\rho]^{\text{Feyn}}$ denote $Z[\rho]$ as we would compute it by summing the Feynman graphs, and the quantities $G^{(n)\,\text{Feyn}}(x_1\dots x_n)$ will be defined to be the coefficients of powers of ρ, as before. We will show that $Z[\rho]^{\text{Feyn}}$ is equal to the original $Z[\rho]$. The expression for $Z[\rho]^{\text{Feyn}}$ is

$$Z[\rho]^{\text{Feyn}} = \lim_{\substack{t_+\to+\infty \\ t_-\to-\infty}} \frac{\langle 0|\,T\exp[-i\int_{t_-}^{t_+} d^4x\,(\mathscr{H}_I + \rho\phi_I)]|0\rangle}{\langle 0|\,T\exp[-i\int_{t_-}^{t_+} d^4x\,\mathscr{H}_I]|0\rangle} \tag{13.23}$$

where $\mathscr{H}_I$ is the old Model 3 interaction Hamiltonian,

$$\mathscr{H}_I = g\psi^*(x)\psi(x)\phi(x) \tag{13.24}$$

We only restrict the time limits of the integral; the space limits go from $-\infty$ to $+\infty$. The numerator approaches the vacuum expectation value of $U_I(\infty,-\infty)$, the sum of all the Feynman graphs for the vacuum-to-vacuum transition in the presence of ρ. The denominator is the same thing without the ρ term, the sum of all vacuum-to-vacuum graphs in the absence of ρ. It cancels out the disconnected vacuum bubbles that may be in our graphs. You may say, "Oh, there's no need to do that because we've got our counterterm to normalize the energy properly, and the disconnected vacuum bubbles are removed." That's what I said earlier. But that applies to a theory with an adiabatic function. As we will see in a moment, this denominator indeed cancels out the disconnected vacuum bubbles in the real theory, *without* an adiabatic function.[3]

Please notice it is the bare vacuum appearing in (13.23), and not the physical vacuum. In our derivation of the Feynman rules, we used the interaction picture fields, free fields. We shoved all the free particle $a_{\mathbf{p}}$'s to the right, and all the free particle $a_{\mathbf{p}}^\dagger$'s to the left where they vanished, because they encountered the bare vacuum. To show that $G^{(n)\,\text{Feyn}}(x_1,\dots,x_n)$ is the same as the real $G^{(n)}(x_1,\dots,x_n)$, we will need to figure out what turns the bare vacuum into the physical vacuum.

We expand $Z[\rho]^{\text{Feyn}}$ in powers of ρ, and obtain

$$Z[\rho]^{\text{Feyn}} = \sum_{n=0}^{\infty} \frac{(-i)^n}{n!} \int d^4x_1\cdots d^4x_n\,\rho(x_1)\cdots\rho(x_n)\,G^{(n)\,\text{Feyn}}(x_1,\dots,x_n) \tag{13.25}$$

where

$$G^{(n)\,\text{Feyn}}(x_1,\dots,x_n) = \lim_{\substack{t_+\to+\infty \\ t_-\to-\infty}} \frac{\langle 0|\,T\exp[-i\int_{t_-}^{t_+} d^4x\,\mathscr{H}_I]\,\phi_I(x_1)\cdots\phi_I(x_n)|0\rangle}{\langle 0|\,T\exp[-i\int_{t_-}^{t_+} d^4x\,\mathscr{H}_I]|0\rangle} \tag{13.26}$$

Both sides of (13.25) are symmetric under interchange of the arguments x_1 to x_n. With no loss of generality I will take these things to be time-ordered; to wit, t_1, the time part of x_1

---

[3] [Eds.] Also, to agree with (13.6), we should have $Z[\rho] = 1$ when $\rho = 0$. The denominator ensures this.

to be greater than or equal to t_2, the time part of x_2, all the way down to t_n. Since t_+ and t_- are going to plus and minus infinity, I might as well begin evaluating my limit when t_+ is greater than all of the t_i's and t_- is less than all of the t_i's. In this case the time ordering of the objects within the numerator is rather trivial. We can write the numerator as (using the definition (7.36) of $U_I(t,t')$)

$$\lim_{\substack{t_+\to+\infty \\ t_-\to-\infty}} \langle 0|\, T\exp\left[-i\int_{t_1}^{t_+} d^4x\,\mathscr{H}_I\right]\phi_I(x_1)\exp\left[-i\int_{t_2}^{t_1} d^4x\,\mathscr{H}_I\right]\times$$
$$\times\,\phi_I(x_2)\cdots\phi_I(x_n)\exp\left[-i\int_{t_-}^{t_n} d^4x\,\mathscr{H}_I\right]|0\rangle \tag{13.27}$$
$$= \lim_{\substack{t_+\to+\infty \\ t_-\to-\infty}} \langle 0|U_I(t_+,t_1)\phi_I(x_1)U_I(t_1,t_2)\phi_I(x_2)\cdots U_I(t_{n-1},t_n)\phi_I(x_n)U_I(t_n,t_-)|0\rangle$$

The denominator is simply $\langle 0|U_I(t_+,t_-)|0\rangle$, so the Feynman Green's functions can be written

$$G^{(n)\,\text{Feyn}}(x_1\ldots x_n) = \lim_{\substack{t_+\to+\infty \\ t_-\to-\infty}} \frac{\langle 0|U_I(t_+,t_1)\phi_I(x_1)\cdots U_I(t_{n-1},t_n)\phi_I(x_n)U_I(t_n,t_-)|0\rangle}{\langle 0|U_I(t_+,t_-)|0\rangle} \tag{13.28}$$

The group property (7.26) of the U_I's tells me that

$$U_I(t_{i-1},t_i) = U_I(t_{i-1},0)U_I(0,t_i) \tag{13.29}$$

We also know that we can use the U_I to find the Heisenberg fields in terms of the interaction picture fields. The correspondence (7.15) between Heisenberg and Schrödinger picture operators says

$$\phi_H(t,\mathbf{x}) = U(0,t)\phi_S(t,\mathbf{x})U(0,t)^\dagger = U(t,0)^\dagger\phi_S(t,\mathbf{x})U(t,0) \tag{13.30}$$

where, from (7.9), $U(t,t') = e^{-iH(t-t')}$. We also have a correspondence (7.20) between the Schrödinger and interaction pictures,

$$\phi_I(t,\mathbf{x}) = U_0^\dagger(t,0)\phi_S(t,\mathbf{x})U_0(t,0) = e^{iH_0t}\phi_S(t,\mathbf{x})e^{-iH_0t} = U_0(0,t)\phi_S(t,\mathbf{x})U_0^\dagger(0,t) \tag{13.31}$$

Combining (13.30) and (13.31) we obtain

$$\phi_H(t,\mathbf{x}) = e^{iHt}e^{-iH_0t}\phi_I(t,\mathbf{x})e^{iH_0t}e^{-iHt} = U_I(0,t)\phi_I(x)U_I^\dagger(0,t) = U_I(0,t)\phi_I(x)U_I(t,0) \tag{13.32}$$

from (7.31).

We see now that we can get at least part of (13.22) in the Feynman expression (13.26), if we break up each of the U_I's into going from one time t_i to zero, and then from zero to the next time t_{i+1}. We find associated with each ϕ_I exactly those operators required to turn it into a ϕ_H, and we will obtain a string of Heisenberg fields:

$$\begin{aligned} U_I(t_{i-1},t_i)\phi_I(x_i)U_I(t_i,t_{i+1}) &= U_I(t_{i-1},0)U_I(0,t_i)\phi_I(x_i)U_I(t_i,0)U_I(0,t_{i+1}) \\ &= U_I(t_{i-1},0)\phi_H(x_i)U_I(0,t_{i+1}) \end{aligned} \tag{13.33}$$

We can thus write

$$G^{(n)\,\text{Feyn}}(x_1\ldots x_n) = \lim_{\substack{t_+\to+\infty \\ t_-\to-\infty}} \frac{\langle 0|U_I(t_+,0)\phi_H(x_1)\cdots\phi_H(x_n)U_I(0,t_-)|0\rangle}{\langle 0|U_I(t_+,0)U_I(0,t_-)|0\rangle} \tag{13.34}$$

There are no U_I's in between, it's just a string of ϕ_H's, time-ordered because of our convention. I've broken up the denominator in the same way.

We are halfway there. We have almost the same expression here in (13.34) as we have in our definition (13.22). Things are automatically time-ordered by our convention on how we've arranged the x's. We've regained the Heisenberg fields. The only thing is, instead of the physical vacuum, we have this funny quantity, the bare vacuum, and a leftover U_I matrix. The algebra may be dull, but I hope it is not obscure. There are, it's true, technical difficulties when one has derivative interactions, when π's as well as ϕ's enter the interaction Hamiltonian H_I. I will ignore those technical difficulties for the moment. Much later on, when we encounter realistic theories with derivative interactions, like the electrodynamics of spinless mesons, I will devote a lecture to straightening everything out for derivative interactions.

We now have to worry about what happens as t_+ and t_- go to $\pm\infty$. Much as we hate to do it, there will be times in this course when we have to think seriously about limits, and this is one of them.

We have two limits. We'll take them one at a time. It will later turn out that it doesn't matter what order we take them in. We'll hold t_+ fixed, and consider the limit as $t_- \to -\infty$. First, the numerator. Regard the bra $\langle 0|\, U_I(t_+,0)\phi_H(x_1)\cdots\phi_H(x_n)$ as a fixed state $\langle\psi|$ for the moment:

$$\langle\psi| \equiv \langle 0|\, U_I(t_+,0)\phi_H(x_1)\cdots\phi_H(x_n) \tag{13.35}$$

We can do the same thing for the denominator, letting $\langle 0|\, U_I(t_+,0)$ be the fixed state $\langle\chi|$. We have

$$\langle\psi|U_I(0,t_-)|0\rangle = \langle\psi|e^{iHt_-}e^{-iH_0t_-}|0\rangle = \langle\psi|e^{iHt_-}|0\rangle \tag{13.36}$$

because the bare vacuum is an eigenstate of the free Hamiltonian with eigenvalue 0. There is a complete set of states, $|n\rangle$, of the Hamiltonian H;

$$H\,|n\rangle = E_n\,|n\rangle \tag{13.37}$$

In particular, the physical vacuum $|0\rangle^P$ is one of these states, with

$$H\,|0\rangle^P = 0 \tag{13.38}$$

Every state of this set *except* the physical vacuum $|0\rangle^P$ is a continuum state. I'll separate that out. If we now insert this complete set into (13.36), we obtain

$$\begin{aligned}\langle\psi|e^{iHt_-}|0\rangle &= \sum_n \langle\psi|n\rangle\,\langle n|e^{iHt_-}|0\rangle = \langle\psi|0\rangle^{P\,P}\langle 0|\,e^{iHt_-}\,|0\rangle + \sum_{n\neq|0\rangle^P}\langle\psi|n\rangle\,\langle n|e^{iHt_-}|0\rangle \\ &= \langle\psi|0\rangle^{P\,P}\langle 0|0\rangle + \sum_{n\neq|0\rangle^P} e^{-iE_nt_-}\,\langle\psi|n\rangle\,\langle n|0\rangle\end{aligned} \tag{13.39}$$

The sum on n is really an integral, but I use standard quantum mechanics conventions and write it as a sum. Thus our limit becomes

$$\lim_{t_-\to-\infty}\langle\psi|U_I(0,t_-)|0\rangle = \langle\psi|0\rangle^{P\,P}\langle 0|0\rangle + \lim_{t_-\to-\infty}\sum_{n\neq|0\rangle^P} e^{-iE_nt_-}\,\langle\psi|n\rangle\,\langle n|0\rangle \tag{13.40}$$

What do we have here, in the sum? We have a continuum integral of oscillating terms. There are one-particle, two-particle, three-particle energy eigenstates, but they're in the middle of a continuum. Now there's a well known theorem from Fourier analysis that says a continuum integral of oscillating terms goes to zero, as t goes to infinity: all the oscillations cancel out. This is known as the Riemann–Lebesgue lemma.[4] Physically, the Riemann–Lebesgue lemma says that if you take the inner product of a state with a fixed state in some fixed region and wait long enough, the only trace of that state that will remain is its true vacuum component. All the one-particle states and multiparticle states will have run away.

Consequently

$$\lim_{t_- \to -\infty} \langle\psi|U_I(0,t_-)|0\rangle = \langle\psi|0\rangle^{P\,P}\langle 0|0\rangle \tag{13.41}$$

This result states that the time limit makes the bare vacuum into the physical vacuum. The denominator goes the same way. By exactly the same reasoning, the Riemann–Lebesgue lemma applies to the other limit, as t_+ goes to infinity, with the result

$$\begin{aligned} G^{(n)\,\text{Feyn}}(x_1\ldots x_n) &= \lim_{\substack{t_+\to+\infty\\ t_-\to-\infty}} \frac{\langle 0|U_I(t_+,0)\phi_H(x_1)\cdots\phi_H(x_n)U_I(0,t_-)|0\rangle}{\langle 0|U_I(t_+,0)U_I(0,t_-)|0\rangle} \\ &= \lim_{\substack{t_+\to+\infty\\ t_-\to-\infty}} \frac{\langle 0|0\rangle^{P\,P}\langle 0|U_I(t_+,0)\phi_H(x_1)\cdots\phi_H(x_n)U_I(0,t_-)|0\rangle^{P\,P}\langle 0|0\rangle}{\langle 0|0\rangle^{P\,P}\langle 0|0\rangle^{P\,P}\langle 0|0\rangle} \\ &= \lim_{\substack{t_+\to+\infty\\ t_-\to-\infty}} {}^{P}\langle 0|U_I(t_+,0)\phi_H(x_1)\cdots\phi_H(x_n)U_I(0,t_-)|0\rangle^{P} \qquad (13.42) \\ &= G^{(n)}(x_1\ldots x_n) \end{aligned}$$

because the factors $\langle 0|0\rangle^{P}$ and ${}^{P}\langle 0|0\rangle$ cancel, and the norm of the physical vacuum is 1. The time ordering symbol is of course irrelevant because we have arranged things so that x_i is later than x_{i+1} for all values of i.

We have answered Question 1 in the positive. This thing we get by summing up all the Feynman graphs is indeed the actual Green's function as we have defined it.This result is tricky but it's pretty. The tricky part is this: by taking the time limit, we wash out everything except the real physical vacuum state.

13.5 Constructing in and out states

We turn now to Question 2. How do we construct the S-matrix without the adiabatic function? Given these $G^{(n)}$'s, how do we compute the S-matrix in terms of them? This question has nothing to do with perturbation theory, and nothing to do with breaking the Hamiltonian up into two parts. We won't be able to answer it until next time. We first have to figure out how to construct in and out states.

Since I will always be working in the Heisenberg picture, for the remainder of this lecture and the first part of next lecture, I will denote $\phi_H(x)$ just by $\phi(x)$, the Heisenberg picture field:

$$\phi(x) \equiv \phi_H(x) \tag{13.43}$$

---

[4] [Eds.] See M. Spivak, *Calculus*, 3rd ed., Publish or Perish, 1994, Problem 15.26, p. 317, or W. Rudin, *Real and Complex Analysis*, McGraw-Hill, 1970, p, 103. The coefficients of $e^{-iE_n t}$ do not need to be continuous, but only integrable.

Also, as the physical vacuum $|0\rangle^P$ is the only vacuum we'll be talking about, we will set

$$|0\rangle \equiv |0\rangle^P \tag{13.44}$$

The physical vacuum satisfies these conditions:

$$\begin{aligned} P^\mu |0\rangle &= 0 \\ \langle 0|0\rangle &= 1 \end{aligned} \tag{13.45}$$

The vacuum is an eigenstate of the energy and momentum operators, with eigenvalues zero, and it is normalized to one. I assume we have physical one-meson states $|p\rangle$ in our theory. (If the meson is unstable, there's no point in trying to compute meson–meson scattering matrix elements.) I will relativistically normalize them:

$$\langle p|p'\rangle = (2\pi)^3 2\omega_{\mathbf{p}}\, \delta^{(3)}(\mathbf{p}-\mathbf{p}') \tag{13.46}$$

These states are eigenstates of the momentum operator:

$$P^\mu |p\rangle = p^\mu |p\rangle \quad \text{where} \quad p^0 = \omega_{\mathbf{p}} = \sqrt{|\mathbf{p}|^2 + \mu^2} \tag{13.47}$$

where μ is the real, physical mass of a real meson. Those are just notational conventions. I won't write down the normalization for a two-meson state now, because a two-meson state could be an in state or an out state, and they aren't the same things; a state that looks like two mesons in the far past may look like a nucleon and an antinucleon in the far future. One of the problems we're going to confront is how to construct those states. We'll have troubles enough with just the vacuum and the one-particle states.

Because the computations we're going to go through are long, I should give you an overview of what we're going to do. We're going to be inspired by our previous limiting process. There we saw how we could pluck out the vacuum state by taking some object involving finite times, and going to a limit. I'm going to do this same sort of thing again. Since our field operators are interacting, they're not going to make *only* one-particle states when they hit the vacuum. They'll make one-particle states, two-particle states, three-particle states, and 72-particle states; they're capable of doing a lot. We're going to make several definitions to construct a limit in time that will enable me to get, from the field operator hitting the vacuum, *only* the one-particle part. If I can do that, I will have crafted something like a creation operator for a single particle. And then I will be able to use these "creation operators" next time to create states that look like two-particle states, either in the far past or the far future, by making a time limit go to $-\infty$ or ∞, respectively. All that will be shown in detail. Our first job is to find a time limit that makes exclusively a one-particle state.

We will need some conventions about the scale of our field. I'm going to work with these Heisenberg fields, without using any details of the equations of motion, just the fact that there *are* equations of motion; $\phi(x)$ is a local scalar field. I'm not even going to say this field obeys the canonical commutators. I will require two normalization conditions.

The first condition concerns the (physical) vacuum expectation value of the Heisenberg field. By translational invariance, (condition 3, just before (3.4)) this will be independent of x:

$$\langle 0|\phi(x)|0\rangle = \langle 0|e^{iP\cdot x}\phi(0)e^{-iP\cdot x}|0\rangle = \langle 0|\phi(0)|0\rangle \tag{13.48}$$

I require my field to have a vacuum expectation value of zero. If it is not zero, I will redefine the field, subtracting from it the constant $\langle 0|\phi(0)|0\rangle$:

$$\phi(x) \to \phi'(x) = \phi(x) - \langle 0|\phi(0)|0\rangle \;\Rightarrow\; \langle 0|\phi'(x)|0\rangle = 0 \tag{13.49}$$

Second, I need to specify the normalization of the one-particle matrix element. Because these one-particle states are momentum eigenstates,

$$\langle k|\phi'(x)|0\rangle = \langle k|e^{iP\cdot x}\phi'(0)e^{-iP\cdot x}|0\rangle = e^{ik\cdot x}\,\langle k|\phi'(0)|0\rangle \tag{13.50}$$

Since Lorentz transformations don't change $\phi'(0)$, or change any one-meson state to any other one-meson state, the coefficient $\langle k|\phi'(0)|0\rangle$ of $e^{ik\cdot x}$ must be Lorentz invariant, and so can depend only on k^2. Presumably, the one-particle state is on its mass shell. Then $k^2 = \mu^2$, and $\langle k|\phi'(0)|0\rangle$ is a constant. By convention this constant is denoted by $\sqrt{Z_3}$;

$$\langle k|\phi'(0)|0\rangle = \sqrt{Z_3} \tag{13.51}$$

(The notation comes from one of Dyson's classic papers[5] on quantum electrodynamics, in which he defined three quantities $\{Z_1, Z_2, Z_3\}$. If we were to treat a one-nucleon state the same as a one-meson state, the equivalent constant for a one-nucleon state would be called Z_2. We won't get to Z_1 for weeks, so don't worry about it.) Now redefine $\phi'(x)$ by

$$\phi'(x) = Z_3^{-1/2}(\phi(x) - \langle 0|\phi(0)|0\rangle) \equiv Z_3^{-1/2}\phi_s(x) \tag{13.52}$$

where $\phi_s(x)$ is the subtracted field. I will assume Z_3 is not zero, so this definition makes sense. Then $\phi'(x)$ has the property that it has the same matrix element between the physical vacuum and the renormalized one-particle state as a free field has between the bare vacuum and the bare one-particle state:

$$\langle k|\phi'(x)|0\rangle = \frac{1}{\sqrt{Z_3}}\,\langle k|\phi(x)|0\rangle = \frac{1}{\sqrt{Z_3}}\,\langle k|\phi(0)|0\rangle\, e^{ik\cdot x} = e^{ik\cdot x} \tag{13.53}$$

These two conditions, (13.49) and (13.52), are just matters of definition. $\phi'(x)$ is called the **renormalized field**, if $\phi(x)$ is the canonical field, obeying canonical commutators. It's called "renormalized" for an obvious reason: we have changed the normalization. Z_3 is called, for reasons so obscure and so embedded in the early history of quantum electrodynamics I don't want to describe them, "the wave function renormalization constant". It *should* be called "the field renormalization constant".

I can now tell you what the "almost" in the answer to Question 2 means. Even without the adiabatic function, the naive formula (13.2) is *almost* right. The only correction is that the Green's functions are those of the *renormalized* fields, not those of the ordinary fields. These Green's functions differ from the earlier versions by powers of $Z_3^{-1/2}$. In due course, we will establish the right formula for the renormalized fields.

The renormalized fields have been scaled in such a way that if all they did was create and annihilate single-particle states when hitting the vacuum, they would do so in exactly the same way as a free field. They do more than that, however, and therefore we've got to define a limiting procedure. It's actually not so bad. Most of our work will consist of writing down a bunch of definitions, and then investigating their implications.

Unfortunately, we would get into a lot of trouble if we were to try and do limits involving plane wave states, so I would like to develop some notation for normalizable wave packet states:

$$|f\rangle \equiv \int \frac{d^3\mathbf{k}}{(2\pi)^3 2\omega_{\mathbf{k}}} F(\mathbf{k})\,|k\rangle \tag{13.54}$$

[5] [Eds.] Freeman J. Dyson, "The S matrix in quantum electrodynamics", *Phys. Rev.* **75** (1949) 1736–1755, and reprinted in Schwinger *QED*. The constants Z_1, Z_2 and Z_3 are introduced on p. 1750.

Associated with each of these wave packets is a function $f(x)$,

$$f(x) \equiv \int \frac{d^3\mathbf{k}}{(2\pi)^3 2\omega_{\mathbf{k}}} F(\mathbf{k}) e^{-ik\cdot x} \tag{13.55}$$

which is obtained with exactly the same integral as the ket $|f\rangle$, but whose integrand has $e^{-ik\cdot x}$ instead of the ket $|k\rangle$. For reasons that will become clear in a moment, I don't want to denote $F(\mathbf{k})$ by $\tilde{f}(\mathbf{k})$. This function $f(x)$ is a positive frequency solution to the free Klein-Gordon equation:

$$(\Box^2 + \mu^2) f(x) = 0 \tag{13.56}$$

We also have

$$\langle k'|f\rangle = \int \frac{d^3\mathbf{k}}{(2\pi)^3 2\omega_{\mathbf{k}}} F(\mathbf{k}) \langle k'|k\rangle = \int \frac{d^3\mathbf{k}}{(2\pi)^3 2\omega_{\mathbf{k}}} F(\mathbf{k}) (2\pi)^3 2\omega_{\mathbf{k}'}\, \delta^{(3)}(\mathbf{k} - \mathbf{k}') = F(\mathbf{k}') \tag{13.57}$$

Furthermore, if the one-particle state $|f\rangle$ goes to a plane wave state $|k\rangle$, $F(\mathbf{k}')$ goes to $(2\pi)^3 2\omega_{\mathbf{k}}\, \delta^{(3)}(\mathbf{k} - \mathbf{k}')$, and $f(x)$ goes to the plane wave solution $e^{-ik\cdot x}$. I've arranged a one-to-one mapping such that our relativistically normalized states correspond to plane waves with no factors in front of them.

I'm now going to define an operator that at first glance looks absolutely disgusting:

$$\phi'^f(t) \equiv i \int d^3\mathbf{x} \left[\phi'(x)\, \partial_0 f(x) - f(x)\, \partial_0 \phi'(x)\right] \tag{13.58}$$

Remember, $\phi'(x)$ is a Heisenberg field, a function of $\mathbf{x}$ and t; this produces a function of t only. We can say some things about this object. In particular, we know its vacuum-to-vacuum matrix element:

$$\langle 0|\phi'^f(t)|0\rangle = 0 \tag{13.59}$$

We can also work out its one-particle matrix elements:

$$\begin{aligned} \langle k|\phi'^f(t)|0\rangle &= i \int d^3\mathbf{x} \int \frac{d^3\mathbf{k}'}{(2\pi)^3 2\omega_{\mathbf{k}'}} F(\mathbf{k}') \left[-i\omega_{\mathbf{k}'} e^{-ik'\cdot x} - e^{-ik'\cdot x}\partial_0\right] \langle k|\phi'(x)|0\rangle \\ &= i \int d^3\mathbf{x}\, e^{ik\cdot x} \int \frac{d^3\mathbf{k}'}{(2\pi)^3 2\omega_{\mathbf{k}'}} F(\mathbf{k}') e^{-ik'\cdot x} \left[-i\omega_{\mathbf{k}'} - i\omega_{\mathbf{k}}\right] \\ &= i \left[\frac{-2i\omega_{\mathbf{k}}}{2\omega_{\mathbf{k}}}\right] F(\mathbf{k}) = F(\mathbf{k}) = \langle k|f\rangle \end{aligned} \tag{13.60}$$

(using (13.53) in the second step), so that, as part of an inner product with a one-particle bra, we can say

$$\phi'^f(t)\, |0\rangle = |f\rangle \tag{13.61}$$

A calculation analogous to (13.60), but differing in one crucial minus sign, gives

$$\langle 0|\phi'^f(t)|k\rangle = 0 \tag{13.62}$$

Thus this operator $\phi'^f(t)$ has *time-independent* matrix elements from vacuum to vacuum, and from vacuum to any one-particle state; the time-dependent phases cancel in (13.60). In fact, if we just restrict ourselves to the one-particle subspace at any given time, $\phi'^f(t)$ is like a creation operator for the normalized state $|f\rangle$.

What about a multiparticle state? Suppose I take a state $|n\rangle$ with two or more particles, and total momentum p_n^μ;

$$P^\mu |n\rangle = p_n^\mu |n\rangle \tag{13.63}$$

The matrix element of the state $|n\rangle$ with our new creation operator $\phi'^f(t)$ can be worked out in exactly the same way. There is a small complication in that we don't know the normalization of $\langle n|\phi'(x)|0\rangle$:

$$\langle n|\phi'(x)|0\rangle = \langle n|e^{iP\cdot x}\phi'(0)e^{-iP\cdot x}|0\rangle = e^{ip_n\cdot x}\langle n|\phi'(0)|0\rangle \tag{13.64}$$

and we don't know what $\langle n|\phi'(0)|0\rangle$ is, yet. In terms of this quantity,

$$\begin{aligned}\langle n|\phi'^f(t)|0\rangle &= i\int d^3\mathbf{x}\,[\partial_0 f - f\partial_0]\,e^{ip_n\cdot x}\langle n|\phi'(0)|0\rangle \\ &= i\int\frac{d^3\mathbf{k}}{(2\pi)^3 2\omega_\mathbf{k}}F(\mathbf{k})(-i\omega_\mathbf{k} - iE_n)\langle n|\phi'(0)|0\rangle\int d^3\mathbf{x}\,e^{-i(k-p_n)\cdot x} \\ &= \left[\frac{\omega_{\mathbf{p}_n}+E_n}{2\omega_{\mathbf{p}_n}}\right]F(\mathbf{p}_n)\langle n|\phi'(0)|0\rangle\,e^{-i(\omega_{\mathbf{p}_n}-E_n)t}\end{aligned} \tag{13.65}$$

The real killer is in the exponential, $e^{-i(\omega_{\mathbf{p}_n}-E_n)t}$. A multiparticle state always has energy $E_n > \omega_{\mathbf{p}_n}$, more energy than a single particle state with momentum $\mathbf{p}_n$. For example, a two-meson state with $\mathbf{p}_n = 0$ can have any energy from $E = 2\mu$ to infinity. The one-meson state with $\mathbf{p} = 0$ has energy $E = \mu$. So the exponential will provide the same sort of oscillatory factor as we saw in (13.40). Thus we can use the same argument with the operator $\phi'^f(t)$ as we did with the U_I matrix, (13.41):

$$\lim_{t\to\pm\infty}\langle n|\phi'^f(t)|0\rangle = 0 \tag{13.66}$$

by the Riemann–Lebesgue lemma, provided $|n\rangle$ is a multiparticle state.

Let $\langle\psi|$ be a fixed, normalizable state, and consider the limit as $t\to\pm\infty$ of the matrix element $\langle\psi|\phi'^f(t)|0\rangle$:

$$\begin{aligned}\lim_{t\to\pm\infty}\langle\psi|\phi'^f(t)|0\rangle &= \lim_{t\to\pm\infty}\sum_n\langle\psi|n\rangle\langle n|\phi'^f(t)|0\rangle \qquad (13.67)\\ &= \lim_{t\to\pm\infty}\Big[\langle\psi|0\rangle\langle 0|\phi'^f(t)|0\rangle + \sum_{\substack{n,\text{ single}\\ \text{particle}}}\langle\psi|n\rangle\langle n|\phi'^f(t)|0\rangle + \sum_{\substack{n,\text{ multi-}\\ \text{particle}}}\langle\psi|n\rangle\langle n|\phi'^f(t)|0\rangle\Big]\end{aligned}$$

For any fixed state $|\psi\rangle$ sitting on the left of the operator, the matrix element with the vacuum state will give us nothing, by (13.59); the matrix elements with the one-particle states will give us $F(\mathbf{k})$, independent of time, by (13.60); and everything else in the whole wide world will give us oscillations which vanish, by (13.66). Thus

$$\lim_{t\to\pm\infty}\langle\psi|\phi'^f(t)|0\rangle = \int\frac{d^3\mathbf{k}}{(2\pi)^3 2\omega_\mathbf{k}}\langle\psi|k\rangle\langle k|\phi'^f(x)|0\rangle = \int\frac{d^3\mathbf{k}}{(2\pi)^3 2\omega_\mathbf{k}}\langle\psi|k\rangle F(\mathbf{k}) = \langle\psi|f\rangle \tag{13.68}$$

So this is exactly analogous to the formula (13.60) we found with the one-particle state $|k\rangle$ sitting on the left. The operator just projects out the part $F(\mathbf{k})$ and gives you $\langle\psi|f\rangle$. That is, we have something that can act either in the far past or the far future as a creation operator for a normalizable state $|f\rangle$.

An analogous calculation gives

$$\lim_{t \to \pm\infty} \langle 0|\phi'^f(t)|\psi\rangle = 0 \tag{13.69}$$

because the arguments of the exponentials add and never cancel, for every single particle or multiparticle momentum eigenstate.

Now this procedure looks very tempting as a prescription for constructing two particle in states and two particle out states, and to find S-matrix elements. I will yield to that temptation at the beginning of the next lecture.

Problems 7

7.1 In class we derived (12.24), the two-particle density of states factor, D, in the center-of-momentum frame, $\mathbf{P}_T = \mathbf{0}$,

$$D = \frac{1}{16\pi^2}\frac{|\mathbf{p}_f|\, d\Omega_f}{E_T} \tag{P7.1}$$

where the notation is as explained in §12.2. Find the formula that replaces this one if $\mathbf{P}_T \neq \mathbf{0}$. *Comment:* Although the center-of-momentum frame is certainly the simplest one in which to work, sometimes we want to do calculations in other frames, for example, the "lab frame", in which one of the two initial particles is at rest.

(1997a 7.1)

7.2 Let A, B, C, and D be four real scalar fields, with dynamics determined by the Lagrangian density

$$\mathscr{L} = \tfrac{1}{2}\left[(\partial_\mu A)^2 - m^2A^2 + (\partial_\mu B)^2 + (\partial_\mu C)^2 + (\partial_\mu D)^2\right] + gABCD \tag{P7.2}$$

where m and g are positive real numbers. Note that A is massive while B, C, and D are massless. Thus the decay of the A into the other three is kinematically allowed. Compute, to the lowest non-vanishing order of perturbation theory, the total decay width of the A. What would the answer be if the interaction were instead gAB^3? Hint: The trick here is to find the kinematically allowed region in the $E_B - E_C$ plane. Some of the constraints are obvious: E_B and E_C must be positive, as must $E_D = m - E_B - E_C$. One is a little less obvious: $\cos\theta_{BC}$ (called θ_{12} in class) must be between -1 and 1.

(1997a 7.2)

7.3 In class I discussed how to compute the decay of a particle into a number of spinless mesons, assuming the universe was empty of mesons before the decay. Sometimes (for example, in cosmology), we wish to compute the decay of a particle (at rest), not into an empty universe, but into one that is filled with a thermal distribution of mesons at a temperature T. This is not hard to do, if we treat the mesons in the final state as non-interacting particles (frequently a very reasonable approximation), and assume there are no other particles of the same type as the initial particle in the initial distribution of particles. (This frequently happens in cosmology. For example, the initial particles could be very massive and were produced [in thermal equilibrium] at an early epoch when the temperature is very high. The expansion of the universe rapidly brings these particles out of equilibrium and reduces their density to a negligible value. They then decay in an environment consisting of a hot gas of much less massive particles.) Show that in this case, the only change in the formalism presented in class is that the density of states factor, D, has an additional multiplicative factor, $f(E/kT)$, for each final meson, where E is the meson energy, k is Boltzmann's constant and f is a function you are to find.

Possibly useful information: (1) For any system in thermal equilibrium, the probability of finding the system in its nth energy eigenstate is proportional to $\exp(-E_n/kT)$. (2) For a single harmonic oscillator, $\langle n|a^\dagger|n-1\rangle = \sqrt{n}$ (see (2.36)).

Cultural note: There are problems in which one has to use the results of Problem 7.1 together with those of this problem (extended to the case in which there is an initial-state thermal distribution, as well as a final-state one). One famous example is the scattering of high energy cosmic ray protons off the 3 K cosmic microwave background radiation.

(1997a 7.3)

Solutions 7

7.1 Let the total energy be E_T and the total momentum $\mathbf{P}_T$, and let the 3-momenta of the final particles be $\mathbf{k}$ and $\mathbf{q}$. Let the corresponding masses and energies be m_k, E_k and m_q, E_q. The two-particle density of states is, from (12.16),

$$D = (2\pi)^4 \frac{d^3\mathbf{k}}{(2\pi)^3 2E_k} \frac{d^3\mathbf{q}}{(2\pi)^3 2E_q} \delta(E_k + E_q - E_T)\, \delta^{(3)}(\mathbf{k} + \mathbf{q} - \mathbf{P}_T) \tag{S7.1}$$

Integrate over $\mathbf{q}$, using the final delta function:

$$D = (2\pi)^4 \frac{d^3\mathbf{k}}{(2\pi)^6 4E_k E_q} \delta(E_k + E_q - E_T)\bigg|_{\mathbf{q}=\mathbf{P}_T-\mathbf{k}} \tag{S7.2}$$

Let θ be the angle between $\mathbf{k}$ and $\mathbf{P}_T$, and ϕ the azimuthal angle of $\mathbf{k}$ about the $\mathbf{P}_T$ axis. Then

$$D = \frac{|\mathbf{k}|^2 d|\mathbf{k}|\, d(\cos\theta)\, d\phi}{(2\pi)^2 4E_k E_q} \delta(E_k + E_q - E_T)\bigg|_{\mathbf{q}=\mathbf{P}_T-\mathbf{k}} \tag{S7.3}$$

In the argument of the delta function only E_q depends on θ. Using the identity in Footnote 8 on p. 9,

$$\delta(E_k + E_q - E_T) = \left|\frac{\partial E_q}{\partial \cos\theta}\right|^{-1} \delta(\cos\theta - \cos\theta_k) \tag{S7.4}$$

where θ_k is the value of θ at which $(E_k + E_q - E_T) = 0$. Now $\mathbf{q} = \mathbf{P}_T - \mathbf{k}$ means

$$E_q^2 = m_q^2 + (\mathbf{P}_T - \mathbf{k})^2 = m_q^2 + |\mathbf{P}_T|^2 + |\mathbf{k}|^2 - 2|\mathbf{P}_T||\mathbf{k}|\cos\theta \tag{S7.5}$$

so

$$\frac{\partial E_q}{\partial \cos\theta} = -\frac{|\mathbf{P}_T||\mathbf{k}|}{E_q} \tag{S7.6}$$

and hence

$$\delta(E_k + E_q - E_T) = \frac{E_q}{|\mathbf{P}_T||\mathbf{k}|} \delta(\cos\theta - \cos\theta_k) \tag{S7.7}$$

We can now do the $\cos\theta$ integral:

$$D = \frac{|\mathbf{k}|^2 d|\mathbf{k}|\, d\phi}{16\pi^2 E_k E_q} \frac{E_q}{|\mathbf{P}_T||\mathbf{k}|} = \frac{dE_k\, d\phi}{16\pi^2 |\mathbf{P}_T|} \tag{S7.8}$$

(using $|\mathbf{k}|\, d|\mathbf{k}| = E_k\, dE_k$ in the final step). Everything (including whatever may multiply D) is to be evaluated at $\mathbf{q} = \mathbf{P}_T - \mathbf{k}$, and $\theta = \theta_k$. Determine θ_k by putting $E_q = E_T - E_k$ into (S7.5),

$$(E_T - E_k)^2 = m_q^2 + |\mathbf{P}_T|^2 + |\mathbf{k}|^2 - 2|\mathbf{P}_T||\mathbf{k}|\cos\theta_k \tag{S7.9}$$

and solving for $\cos\theta_k$:

$$\cos\theta_k = \frac{(m_q^2 + |\mathbf{P}_T|^2 + |\mathbf{k}|^2) - (E_T - E_k)^2}{2|\mathbf{P}_T||\mathbf{k}|} = \frac{m_q^2 + 2E_T E_k - (E_T^2 - |\mathbf{P}_T|^2) - m_k^2}{2|\mathbf{P}_T|\sqrt{E_k^2 - m_k^2}} \tag{S7.10}$$

For $\mathbf{P}_T \neq 0$, the density of states is given by (S7.8), with the restrictions noted above, and θ_k given by (S7.10). (For the decay of a particle of mass M, $E_T^2 - |\mathbf{P}_T|^2 = M^2$.) ■

7.2 The decay width Γ is given by (12.2),

$$\Gamma = \frac{1}{2m}\int |\mathcal{A}_{fi}|^2 D \tag{S7.11}$$

The amplitude $\mathcal{A}_{fi}$ is given graphically and analytically by

$i\mathcal{A}_{fi}$ = 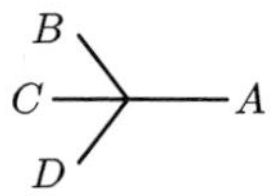 $= -ig$ (S7.12)

The density of states factor is given, for a final state of three spinless particles, by (12.65),

$$D = \frac{1}{32\pi^3} dE_B\, dE_C \tag{S7.13}$$

so

$$\Gamma = \frac{g^2}{64m\pi^3}\int dE_B\, dE_C = \frac{g^2}{64m\pi^3} \times (\text{kinematically allowed area in } E_B - E_C \text{ plane}) \tag{S7.14}$$

The task now is to determine the kinematically allowed region. By conservation of momentum and energy,

$$\mathbf{p}_A = 0 = \mathbf{p}_B + \mathbf{p}_C + \mathbf{p}_D \tag{S7.15}$$

$$E_A = m = E_B + E_C + E_D \tag{S7.16}$$

Also, because B, C, and D are massless, we have

$$E_B = |\mathbf{p}_B|, \quad E_C = |\mathbf{p}_C|, \quad E_D = |\mathbf{p}_D| \tag{S7.17}$$

By the Triangle Inequality,

$$\big||\mathbf{p}_B| - |\mathbf{p}_C|\big| \leq |\mathbf{p}_D| \leq |\mathbf{p}_B| + |\mathbf{p}_C| \tag{S7.18}$$

so that, substituting,

$$|E_B - E_C| \leq m - E_B - E_C \leq E_B + E_C \tag{S7.19}$$

Add $E_B + E_C$ to each, and divide by 2, to obtain

$$\max(E_B, E_C) \leq \tfrac{1}{2}m \leq E_B + E_C \tag{S7.20}$$

The allowed region in the $E_B - E_C$ plane is triangular, with an area of $\frac{1}{8}m^2$:

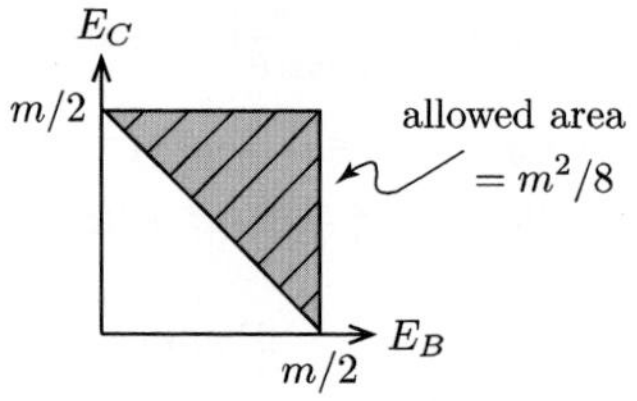

Plugging this area into the decay width gives

$$\Gamma = \frac{g^2}{64m\pi^3} \times \tfrac{1}{8}m^2 = \frac{g^2 m}{512\pi^3} \tag{S7.21}$$

Observe the large difference from the naive "dimensional analysis" guess of g^2m; $512\pi^3 \approx 16{,}000$.

If the interaction had been gAB^3 instead of $gABCD$, there would have been no distinction between the three fields B, C, and D. We would have had

$$i\mathcal{A}_{fi} = 3!(-ig) \tag{S7.22}$$

and we might have naively expected the amplitude to increase by $(3!)^2$. But integrating over all final states now over-counts by a factor of 3!, since the outgoing particles are indistinguishable. We must divide by 3!. The new answer is

$$\Gamma = \frac{(3!)^2 g^2}{3!64m\pi^3} \times \tfrac{1}{8}m^2 = \frac{3!g^2 m}{512\pi^3} = \frac{3g^2 m}{256\pi^3} \tag{S7.23}$$

This decay width is 3! or six times the earlier value. ■

7.3 We will use the index i as a generic label to distinguish mesons according to their particle type as well as their momentum. For simplicity, we can imagine that we are working in a box with discrete momenta. We will consider the decay of a particle of type 0 into a set of particles of types $1, 2, \ldots, j$. By assumption, the original state has only a single particle of type 0, but may have n_i of type i. The relevant matrix elements of the Wick expression for this process are of the form

$$\langle f|a_1^\dagger a_2^\dagger \cdots a_j^\dagger a_0|i\rangle \tag{S7.24}$$

Let us assume that the 'background' state has n_i particles of type i. Recalling that

$$\langle n_i+1|a_i^\dagger|n_i\rangle = \sqrt{n_i+1} \tag{S7.25}$$

we see that the decay amplitude $\mathcal{A}_{fi}$ is enhanced by a factor

$$\prod_{i=1}^{j} \sqrt{n_i+1} \tag{S7.26}$$

when compared with the analogous process with no background states. So the probability of transition, proportional to $|\mathcal{A}_{fi}|^2$, will be enhanced over the vacuum probability of decay by the square of the factor (S7.26).

The probability that there are n_i quanta of type i is, by a standard thermodynamic argument,

$$P(n_i) = \frac{e^{-\beta n_i E_i}}{\sum_{n_i} e^{-\beta n_i E_i}} = e^{-\beta n_i E_i}\left(1 - e^{-\beta E_i}\right) \tag{S7.27}$$

where as usual $\beta = 1/kT$. The overall probability of decay is

$$\begin{aligned}
P_{\text{therm}} &= \sum_{\text{states}} (\text{probability of state})(\text{transition probability in a state}) \\
&= \sum_{\text{states}} \prod_{i=1}^{j} e^{-\beta n_i E_i}\left(1 - e^{-\beta E_i}\right)(n_i+1)P_{\text{vac}} \\
&= \prod_{i=1}^{j} \sum_{n_i=0}^{\infty} \left(1 - e^{-\beta E_i}\right) e^{\beta E_i} e^{-(n_i+1)\beta E_i}(n_i+1)P_{\text{vac}} \\
&= \prod_{i=1}^{j} \left(-1 + e^{-\beta E_i}\right) e^{\beta E_i} \frac{\partial}{\partial(\beta E_i)} \sum_{n_i=0}^{\infty} e^{-(n_i+1)\beta E_i} P_{\text{vac}} \\
&= \prod_{i=1}^{j} \left(-1 + e^{-\beta E_i}\right) e^{\beta E_i} \frac{\partial}{\partial(\beta E_i)} \frac{e^{-\beta E_i}}{1 - e^{-\beta E_i}} P_{\text{vac}} \\
&= \prod_{i=1}^{j} \frac{1}{1 - e^{-\beta E_i}} P_{\text{vac}}
\end{aligned} \tag{S7.28}$$

This shows that the decay width has an extra factor of $(1 - e^{-\beta E_i})^{-1}$ for each mode created by the decay process. We would get the same result if we were to change the density of states, (11.58), by the substitution

$$\frac{d^3\mathbf{p}}{(2\pi)^3 2\omega_{\mathbf{p}}} \to \frac{d^3\mathbf{p}}{(2\pi)^3 2\omega_{\mathbf{p}}} f(\omega_{\mathbf{p}}/kT) \tag{S7.29}$$

where $f(\omega_{\mathbf{p}}/kT) = \dfrac{1}{1 - e^{-\omega_{\mathbf{p}}/kT}}$. ■

14

The LSZ formalism

Let me summarize some of the things we said last time, and the question we are trying to answer. With every normalizable one-particle state $|f\rangle$ we have associated a function $F(\mathbf{k})$

$$|f\rangle = \int \frac{d^3\mathbf{k}}{(2\pi)^3(2\omega_{\mathbf{k}})} F(\mathbf{k})\,|k\rangle \tag{14.1}$$

Likewise I associated with the same state a function $f(x)$

$$f(x) = \int \frac{d^3\mathbf{k}}{(2\pi)^3(2\omega_{\mathbf{k}})} F(\mathbf{k})\,e^{-ik\cdot x} \tag{14.2}$$

which is a positive frequency solution of the Klein-Gordon equation,

$$(\Box^2 + \mu^2)f = 0 \tag{14.3}$$

So I have a one to one mapping between normalizable states and solutions of the Klein-Gordon equation. For my renormalized field operator $\phi'(x)$, I defined a function of time $\phi'^f(t)$ as

$$\phi'^f(t) = i\int d^3\mathbf{x}\,\left[\phi'(x)\,\partial_0 f(x) - f(x)\,\partial_0\phi'(x)\right] \tag{14.4}$$

and I showed that, for any fixed, normalizable state $|\psi\rangle$

$$\lim_{t\to\pm\infty} \langle\psi|\phi'^f(t)|0\rangle = \langle\psi|f\rangle \tag{14.5}$$

It will also be important that

$$|f\rangle \to |k\rangle \quad \text{as} \quad f(x) \to e^{-ik\cdot x} \tag{14.6}$$

That is: as $f(x)$ goes to a plane wave, the state $|f\rangle$ goes to a relativistically normalized momentum eigenstate $|k\rangle$. That was the conclusion of everything we investigated last lecture. We showed that this operator $\phi'^f(t)$ in the limit as the time t goes to either positive or negative infinity was, so to speak, a one-particle creation operator. Of course, at intermediate times it is by no means a one-particle creation operator. It makes, as would any smeared version of the field operators at fixed time, not just a single-particle state, but two-particle states, three-particle states ..., *ad infinitum*, at least if we investigate in higher and higher orders of perturbation theory. Only in the limit $t \to \pm\infty$ do we cancel out all the multiparticle terms that would in principle contribute to this matrix element at any finite time, because of the *non*-cancellation of phases. So that's where we wound up.

14.1 Two-particle states

We can of course get some related formulas from the result (14.5). For example, if we put the vacuum on the other side,

$$\lim_{t\to\pm\infty} \langle 0|\phi'^f(t)|\psi\rangle = 0 \tag{14.7}$$

then as we found in (13.69), even for the one-particle states we have a phase mismatch: in this matrix element, all of the phases have a positive frequency and never cancel. So this limit is zero. All the phases mismatch, which is again what you would expect if this asymptotic limit is producing something like a creation operator. A creation operator does indeed annihilate the vacuum on the left. Of course, we have certain trivial equations that follow from (14.5) just by taking the adjoint;

$$\lim_{t\to\pm\infty} \langle 0|\phi'^{f\dagger}(t)|\psi\rangle = \langle f|\psi\rangle \tag{14.8}$$

The operator $\phi'^{f\dagger}(t)$ is not Hermitian, because there's an explicit i in the definition (14.4); moreover, $f(x)$ is not a real function. The adjoint equation has a limit of zero:

$$\lim_{t\to\pm\infty} \langle \psi|\phi'^{f\dagger}(t)|0\rangle = 0 \tag{14.9}$$

Again, this is what you would expect if ϕ'^f is a creation operator, because $\phi'^{f\dagger}$ should then be an annihilation operator. This is just what an annihilation operator does: it makes a one-particle state from the vacuum on the left, and kills the vacuum on the right.

Now we come to the great leap of faith. I assume I have two functions $F_1(\mathbf{k})$ and $F_2(\mathbf{k})$, which are associated with nice, normalized, non-interacting wave packet states $|f_1\rangle$ and $|f_2\rangle$, respectively, in the sense of (14.1). We require that the functions $F_1(\mathbf{k})$ and $F_2(\mathbf{k})$ have no common support in momentum space. That is,

$$F_1(\mathbf{k})F_2(\mathbf{k}) = 0 \text{ for each } \mathbf{k} \tag{14.10}$$

By making this statement, we are leaving out only a negligible region of phase space. When we eventually let the kets $|f_1\rangle$ and $|f_2\rangle$ go to plane wave states, this restriction will exclude just the configurations with two collinear momenta, which correspond to scattering at threshold in the center-of-mass frame, a case we excluded in our other analysis also. Thus one of these kets is associated with a one-particle state which is going off in some direction, and the other is associated with a one-particle state going off in another direction. I'll call the functions and states associated with $F_1(\mathbf{k})$ and $F_2(\mathbf{k})$, $f_1(x)$, $f_2(x)$ and $|f_1\rangle$, $|f_2\rangle$, respectively.

I now want to consider what happens if I take the limit

$$\lim_{t\to+\infty} \langle \psi|\phi'^{f_2}(t)|f_1\rangle \tag{14.11}$$

the operator $\phi'^{f_2}(t)$ acting on not the vacuum now, but on the state $|f_1\rangle$. Well, (14.11) is a matrix element, which we can think about in either the Schrödinger or the Heisenberg picture: matrix elements are matrix elements, even though these are all Heisenberg fields. Let's think about this operation in the Schrödinger picture. I have a state $|f_1\rangle$, described by some wave packet, say with the center of the wave packet traveling in some direction. I wait for some very large future time, say, a billion years. If I wait long enough, that wave packet has gotten very very far away, maybe several galaxies over in the original direction. Now I come into this room. I have an operator which if I applied it to the vacuum would make a state $|f_2\rangle$. If I were now to go a billion light years in the opposite direction, carrying this operator, and hit the vacuum with it there, it would make a single-particle state with distribution f_2. That's the physics of what is going on.

So let me ask a question. What happens if I apply it *not* to the vacuum, but to the state that has that other particle over there, way beyond the Andromeda galaxy, two million light years away? Well, if there's any sense in the world whatsoever, the fact that that other particle is on the other side of the Andromeda galaxy should be completely *irrelevant.* I'd have to travel to the other side of Andromeda to see it's there. As far as I'm concerned, I don't know in the whole region of spacetime in which I'm working that I haven't got the vacuum state. The particle that is really there, that is secretly there, I can hardly expect to see in any experiment I can do, because it is all the way over on the other side of Andromeda. It *can't* affect what I'm doing in this room, or what I'm doing two million light years away in the other direction. I am making a state by this operation that is effectively a two-particle state, with the two particles in the far future moving away from each other, one going in one direction and the other going in another direction. Therefore, I assert this limit should exist and should give the definition of a two-particle out state, a state that in the far future looks like two particles moving away from each other:

$$\lim_{t\to+\infty} \langle\psi|\phi'^{f_2}(t)|f_1\rangle = \langle\psi|f_1, f_2\rangle^{\text{out}} \tag{14.12}$$

That's an argument, not a proof. If you want a mathematical proof you have to read a long paper by Klaus Hepp;[1] but it is physically very reasonable. The only thing I am incorporating is that there is some rough idea of localization in this theory, some sort of approximation to position. And if there's a particle on the other side of the Andromeda galaxy traveling away from me, I'll never know it.

In fact the analysis can be extended to collinear momenta, but it requires much more complicated reasoning, and the result is not even on the rigorous level of Hepp's argument. The physics is clear, even if the momenta are collinear, because wave packets tend to spread out. If I wait long enough, I'll have a negligible probability for the first particle to be anywhere near the second particle even though the centers of the wave packets are moving in the same direction. So it turns out it's also true for collinear momenta. The limit will be a little slower, because in spreading out, they're not moving away from each other as fast as they would if their motions were pointing in different directions.

Of course, if our limit were for the time to approach minus infinity, all the arguments would be exactly the same, but time reversed. Instead of an "out", I would have an "in":

$$\lim_{t\to-\infty} \langle\psi|\phi'^{f_2}(t)|f_1\rangle = \langle\psi|f_1, f_2\rangle^{\text{in}} \tag{14.13}$$

Thus we have the prescription for constructing in states and out states, states that look like two-particle states in the far past, and states that look like two-particle states in the far future. I use two-particle states only for simplicity. After I go through all the agonies I will go through for two particles scattering into two particles, if you wish you can extend the arguments to two into three or two into four or seven into eighteen. We can construct states that are indeed asymptotic states.

---

[1] [Eds.] Klaus Hepp, "On the connection between the LSZ and Wightman quantum field theory", *Comm. Math. Phys.* **1** (1965) 95–111.

14.2 The proof of the LSZ formula

We're now in a position to answer Question 2 (p. 273). As I told you, the answer to Question 2 is *almost* "Yes": the relation (13.2) needs to be modified by replacing the ϕ fields with the *renormalized* fields, the ϕ' fields. Analogous to $G^{(n)}(x_1,\dots,x_n)$ and $\widetilde{G}^{(n)}(k_1,\dots,k_n)$ defined in (13.4) and (13.22), we now define

$$G'^{(n)}(x_1,\dots,x_n) = \int \frac{d^4k_1}{(2\pi)^4}\cdots\frac{d^4k_n}{(2\pi)^4} e^{ik_1\cdot x_1+\cdots+ik_n\cdot x_n}\,\widetilde{G}'^{(n)}(k_1,\dots,k_n) \tag{14.14}$$

with

$$\boxed{G'^{(n)}(x_1,\dots,x_n) = \langle 0|T\left(\phi'(x_1)\cdots\phi'(x_n)\right)|0\rangle} \tag{14.15}$$

Let's look at a specific example, the four-point function $\widetilde{G}'^{(4)}(k_1,\dots,k_4)$:

$$\widetilde{G}'^{(4)}(k_1,\dots,k_4) = \int d^4x_1\cdots d^4x_4\, e^{-(ik_1\cdot x_1+\cdots+ik_4\cdot x_4)}G'^{(4)}(x_1,\dots,x_4) \tag{14.16}$$

with

$$G'^{(4)}(x_1,\dots,x_4) = \langle 0|T(\phi'(x_1)\cdots\phi'(x_4)|0\rangle = Z_3^{-2}G^{(4)}(x_1,\dots,x_4) \tag{14.17}$$

That's the renormalized Green's function, $G'^{(4)}(x_1,\dots,x_4)$, the physical vacuum expectation value of a string of renormalized Heisenberg fields, and $G^{(4)}(x_1,\dots,x_4)$ is the old Green's function; there's a factor of $Z_3^{-1/2}$ for each renormalized field. The question we want to test is whether the *renormalized* version of (13.2) is true:

$$\langle k_3,k_4|S-1|k_1,k_2\rangle \stackrel{?}{=} (-i)^4\prod_{r=1}^{4}(k_r^2-\mu^2)\widetilde{G}'^{(4)}(-k_3,-k_4,k_1,k_2) \tag{14.18}$$

We want to prove this relation, due originally to Lehmann, Symanzik and Zimmerman,[2] and known as the **LSZ reduction formula**.

What I will actually prove is the analog of (14.18) for wave packet states of the form (14.12) and (14.13). Scattering is physically defined only for wave packet states; a plane wave state never gets far away from the interaction because it has uniform probability density over all space. Let the final states be characterized by two non-overlapping wave packets $|g_1\rangle$ and $|g_2\rangle$ analogous to the non-overlapping initial wave packets $|f_1\rangle$ and $|f_2\rangle$. Then what I will prove is that

$$\begin{aligned}\langle g_1,g_2|S-1|f_1,f_2\rangle \stackrel{?}{=} (i)^4\int d^4x_1\cdots d^4x_4\, g_1^*(x_1)g_2^*(x_2)f_1(x_3)f_2(x_4)\\ \times\prod_{r=1}^{4}(\Box_r^2+\mu^2)\,\langle 0|T\left(\phi'(x_1)\cdots\phi'(x_4)\right)|0\rangle\end{aligned} \tag{14.19}$$

I've put in a question mark for the time being, to indicate we haven't yet proved it at this stage.

---

[2] [Eds.] Harry E. Lehmann, Kurt Symanzik and Wolfhart Zimmerman, "Zur Formulierung quantisierter Feldtheorien", (Toward the formulation of quantized field theories) *Nuovo Cim.* ser. 10, **1** (1955) 205–225.

Now this does reduce to the statement (14.18) when we allow the f's and g's to go to plane waves as stated in (14.6). When I make the f's and g's plane waves, I simply get the definition of the Fourier transform in (14.19), with the momenta associated with g_1 and g_2 replaced by minus their natural value because I'm complex conjugating. Operating on a function of position space with $(\Box^2 + \mu^2)$ is the same thing as multiplying that function in momentum space by $(-k^2 + \mu^2)$, and that produces the propagator factors in (14.18) except for a minus sign, which is taken care of by replacing the $(-i)$ in (14.18) by the i in (14.19). The sign of the i's doesn't matter for the four-point function, because i^4 is $(-i)^4$; but I want to construct the arguments so you can see easily how trivial the generalization is to n particles in and m particles out. So if I prove (14.19), I will be home. We'll start with the left-hand side, and transform it into the right-hand side.

Now, in order to study the limit of (14.19) as the wave packets turn into plane waves, I will establish a useful lemma. Say we have a function $f(x)$ which is a solution of the Klein-Gordon equation, and which goes to zero rapidly as $|x| \to \infty$. That is, the wave packet $|f\rangle$ to which $f(x)$ corresponds is a nice, normalizable wave function that dies away at infinity sufficiently rapidly that integration by parts on spatial derivatives on $f(x)$ is legitimate. We can't say the same for time derivatives, because this thing is evolving in time. Let $A(x)$ be another quantity which could be a single field, or maybe a string of operators, with the dependence on the other variables besides x suppressed. If I define, in analogy with $\phi'^f(t)$ in (14.4),

$$A^f(t) \equiv i \int d^3\mathbf{x}\, [A(\partial_0 f) - f(\partial_0 A)] \tag{14.20}$$

then the lemma says

$$i \int d^4x\, f(x)(\Box^2 + \mu^2)A(x) = \left(\lim_{t\to-\infty} - \lim_{t\to\infty} \right) A^f(t) \tag{14.21}$$

The proof is straightforward, starting with the left-hand side of the lemma:

$$\begin{aligned}
i \int d^4x\, f(x)(\partial^2 + \mu^2)A(x) &= i \int d^4x\, f(\partial_0^2 A - \nabla^2 A + \mu^2 A) \\
&= i \int d^4x\, \Big[f(\partial_0^2 A) + A(-\nabla^2 f + \mu^2 f) \Big] \\
&= i \int d^4x\, \Big[f(\partial_0^2 A) - A(\partial_0^2 f) \Big] = i \int d^4x\, \partial_0 \left[f(\partial_0 A) - A(\partial_0 f) \right] \\
&= i \int dt\, \partial_0 \int d^3\mathbf{x}\, [f(\partial_0 A) - A(\partial_0 f)] = -\int dt\, \partial_0 A^f(t)
\end{aligned} \tag{14.22}$$

Now few things are easier to do than the time integral of a time derivative, and so we obtain

$$i \int d^4x\, f(x)(\Box^2 + \mu^2)A(x) = -\int_{-\infty}^{\infty} dt\, \partial_0 A^f(t) = \left(\lim_{t\to-\infty} - \lim_{t\to\infty} \right) A^f(t) \tag{14.23}$$

by the Fundamental Theorem of Calculus, **QED**.

We can establish a similar equation for the conjugate function f^*. We'll now assume A is some Hermitian operator, $A = A^\dagger$, and note that

$$A^{f\dagger}(t) = -i \int d^3\mathbf{x}\, [A(\partial_0 f^*) - f^*(\partial_0 A)] \tag{14.24}$$

Then as you can show easily

$$i\int d^4x\, f^*(x)(\Box^2+\mu^2)A(x) = \left(\lim_{t\to\infty} - \lim_{t\to-\infty}\right) A^{f\dagger}(t) \tag{14.25}$$

There is a sign flip for the adjoint.

Armed with this lemma we can now turn the formidable expression (14.19) into a grotesque sequence of limits. Let's do the x_4 integration. Using the lemma, we have

$$\begin{aligned}\langle g_1, g_2|S-1|f_1, f_2\rangle \stackrel{?}{=} \left(\lim_{t_4\to-\infty} - \lim_{t_4\to\infty}\right)(i)^3\int d^4x_1\cdots d^4x_3\, g_1^*(x_1)g_2^*(x_2)f_1(x_3)\\ \times\prod_{r=1}^{3}(\Box_r^2+\mu^2)\langle 0|T\left(\phi'^{f_2}(t_4)\phi'(x_1)\cdots\phi'(x_3)\right)|0\rangle\end{aligned} \tag{14.26}$$

You might think I have swindled you because I've slipped the ϕ'^{f_2} in here, and in doing so, pushed a time derivative $\partial/\partial t_4$ past the time ordering symbol. As you know from Problem 1.2, that may give me a term involving an equal time delta function, but that's irrelevant in this limit, because if I keep $\{t_1, t_2, t_3\}$ fixed and send t_4 to either plus or minus infinity, t_4 is not the same time as any of the other three times, and therefore I can push the time derivative through the time ordering symbol without losing (or gaining) anything.

Continuing in this way, we can turn (14.26) into

$$\begin{aligned}&\langle g_1, g_2|S-1\,|f_1, f_2\rangle\\ &\stackrel{?}{=}\prod_{r=1}^{2}\left[\lim_{t_r\to\infty} - \lim_{t_r\to-\infty}\right]\prod_{s=3}^{4}\left[\lim_{t_s\to-\infty} - \lim_{t_s\to\infty}\right]\langle 0|T\left(\phi'^{g_1\dagger}(t_1)\phi'^{g_2\dagger}(t_2)\phi'^{f_1}(t_3)\phi'^{f_2}(t_4)\right)|0\rangle\end{aligned} \tag{14.27}$$

It doesn't matter which one I integrate first. That's easy to show, if you assume that the time ordered product has any sort of reasonable large distance fall-off, if it's a tempered distribution or something like that. If we had reduced the integrals in some other order, we would have had a different order of limits. In fact, all 4! orderings lead to the same result. As an exercise you can do the limits in any other order, and see that you get exactly the same answer.

The successive limits are actually duck soup.[3] Let's do one of them and see what happens. Let's do the t_4 limit. We've got two terms, one for t_4 goes to $-\infty$ and the other for t_4 goes to $+\infty$, with the arguments $\{x_1, x_2, x_3\}$ held fixed. I have $\phi'^{g_1\dagger}(t_1)$, $\phi'^{g_2\dagger}(t_2)$, $\phi'^{f_1}(t_3)$, and $\phi'^{f_2}(t_4)$. Now, what happens in the limit as t_4 goes to $-\infty$? Looking only at this part, we have

$$\lim_{t_4\to-\infty}\langle 0|T\left(\phi'^{g_1\dagger}(t_1)\phi'^{g_2\dagger}(t_2)\phi'^{f_1}(t_3)\phi'^{f_2}(t_4)\right)|0\rangle \tag{14.28}$$

Well, $t_4 = -\infty$ is certainly earlier than any finite times. The time ordering symbol says that as t_4 goes to $-\infty$, $\phi'^{f_2}(t_4)$ goes all the way over on the right, where it encounters the vacuum

[3] [Eds.] "Duck soup" is idiomatic American English for "a task easily accomplished" (and also the title of a Marx Brothers movie (1933)), synonymous with "a piece of cake" or "a snap".

and, from (14.5), makes the state $|f_2\rangle$:

$$\begin{aligned}\lim_{t_4\to-\infty}\langle 0|\,T\left(\phi'^{g_1\dagger}(t_1)\phi'^{g_2\dagger}(t_2)\phi'^{f_1}(t_3)\phi'^{f_2}(t_4)\right)|0\rangle \\ = \langle 0|T\left(\phi'^{g_1\dagger}(t_1)\phi'^{g_2\dagger}(t_2)\phi'^{f_1}(t_3)\right)\phi'^{f_2}(-\infty)|0\rangle \\ = \langle 0|T\left(\phi'^{g_1\dagger}(t_1)\phi'^{g_2\dagger}(t_2)\phi'^{f_1}(t_3)\right)|f_2\rangle\end{aligned} \tag{14.29}$$

That takes care of the $t_4 \to -\infty$ limit. What about $t_4 \to +\infty$? Well, plus infinity is later than any finite time and therefore $\phi'^{f_2}(t_4)$ is situated all the way over on the left, where it hits the vacuum and, from (14.7), gives us zero:

$$\begin{aligned}\lim_{t_4\to\infty}\langle 0|\,T\Big(\phi'^{g_1\dagger}(t_1)\phi'^{g_2\dagger}(t_2)\phi'^{f_1}(t_3)\phi'^{f_2}(t_4)\Big)|0\rangle \\ = \langle 0|\phi'^{f_2}(\infty)\,T\Big(\phi'^{g_1\dagger}(t_1)\phi'^{g_2\dagger}(t_2)\phi'^{f_1}(t_3)\Big)|0\rangle = 0\end{aligned} \tag{14.30}$$

I don't bother to specify whether $|f_2\rangle$ is an in state or an out state, because for a one-particle state, they're the same: one particle just sits there, or travels along; it doesn't have anything to scatter off of. So far, so good. We're getting there.

Let's look at the t_3 limit. That's much the same story. When $t_3 \to \infty$, it is the latest time, and thus the time ordering puts ϕ'^{f_1} on the extreme left, where it hits the vacuum and produces zero. When $t_3 \to -\infty$, ϕ'^{f_1} goes against the state $|f_2\rangle$ on the right, and according to (14.13) produces the state $|f_1, f_2\rangle^{\text{in}}$. (It's definitely an in state, because both creation operators were at a time of minus infinity.) We wind up with

$$\langle g_1, g_2|\,S-1\,|f_1, f_2\rangle \overset{?}{=} \prod_{r=1}^{2}\left[\lim_{t_r\to\infty} - \lim_{t_r\to-\infty}\right]\langle 0|\,T\left(\phi'^{g_1\dagger}(t_1)\phi'^{g_2\dagger}(t_2)\right)|f_1, f_2\rangle^{\text{in}} \tag{14.31}$$

Now let's look at the t_2 limits. There's a limit as t_2 goes to $+\infty$. Because of the time ordering symbol, the operator $\phi'^{g_2\dagger}(t_2)$ ends up on the extreme left where, from (14.8), it makes a one-particle state, $\langle g_2|$. Ignoring the t_1 limits for a moment, the right-hand side of (14.31) becomes

$$\langle g_2|\,\phi'^{g_1\dagger}(t_1)|f_1, f_2\rangle^{\text{in}} - \lim_{t_2\to-\infty}\langle 0|\,T\left(\phi'^{g_1\dagger}(t_1)\phi'^{g_2\dagger}(t_2)\right)|f_1, f_2\rangle^{\text{in}} \tag{14.32}$$

Unfortunately, there's a term left over. In the limit as $t_2 \to -\infty$, the operator $\phi'^{g_2\dagger}(t_2)$ winds up against the state $|f_1, f_2\rangle$. We don't know what that is, but I'll denote it by $|\psi\rangle$:

$$|\psi\rangle = \lim_{t_2\to-\infty}\phi'^{g_2\dagger}(t_2)\,|f_1, f_2\rangle^{\text{in}} \tag{14.33}$$

With only one operator left, there is no more need for the time ordering symbol. Putting in the last limits, the right-hand side of (14.31) becomes

$$\begin{aligned}\left[\lim_{t_1\to\infty} - \lim_{t_1\to-\infty}\right]\left(\langle g_2|\,\phi'^{g_1\dagger}(t_1)|f_1, f_2\rangle^{\text{in}} - \langle 0|\phi'^{g_1\dagger}(t_1)|\psi\rangle\right) \\ = {}^{\text{out}}\langle g_2, g_1|f_1, f_2\rangle^{\text{in}} - \langle g_1|\psi\rangle - {}^{\text{in}}\langle g_1, g_2|f_1, f_2\rangle^{\text{in}} + \langle g_1|\psi\rangle \\ = \langle g_2, g_1|S|f_1, f_2\rangle - \langle g_1, g_2|f_1, f_2\rangle \\ \overset{\checkmark}{=} \langle g_1, g_2|S-1|f_1, f_2\rangle \quad \textbf{QED}\end{aligned} \tag{14.34}$$

In the second step, recall (14.8): $\lim_{t\to+\infty} \langle 0|\phi'^{f\dagger}|\psi\rangle = \lim_{t\to-\infty} \langle 0|\phi'^{f\dagger}|\psi\rangle = \langle f|\psi\rangle$.

Now, what have we got? Aside from an ordering of g_1 and g_2, irrelevant because these are Bose particles, we have proved, at the cost of rather lengthy calculation, exactly what we set out to prove. Let's summarize where we are. We've addressed the two questions raised in the last chapter (p. 273). We have answered Question 1: we correctly compute the Green's functions for the unnormalized fields by summing up the Feynman diagrams. And we have answered Question 2: we correctly get S-matrix elements by putting the Green's function lines on the mass shell and multiplying by factors of $(k^2 - \mu^2)$ to get rid of the extra propagators for the renormalized Green's functions. That's why I said the answer to Question 2 was not "Yes" but "Almost". We have to use the Fourier transforms of the *renormalized* Green's functions, the vacuum expectation values of the time-ordered product of the *renormalized* fields, to get the right S-matrix elements.

Now if we just consider the answer to Question 2 in isolation, without worrying about how we compute things, we also see that we have what I described in an earlier lecture as the *beau idéal* of a scattering theory: a way of finding the S-matrix elements from the finite time dynamics, without resorting to any approximation procedure. That is given by the LSZ formula (14.18), from which I can now erase the question mark: it is correct. I don't know why it's called a reduction formula, maybe because you get some reduced information from a Green's function by only looking at its mass-shell value in Fourier space.

The mathematical expression (14.19) makes sense even with the f and g wave packets replaced by plane waves. We'll make that expression the definition of an $(\mathrm{S}-1)$ matrix element for plane waves. Of course we only get something physically measurable when we smear out the plane waves into wave packets. This situation is analogous to the expression (9.39) for electrostatic energy. No one can build a point charge, and so no one can make a charge distribution that directly measures the Coulomb potential. All you can do is measure E_0 for various charge distributions. Then you can abstract the notion of an interaction between two point charges. The formula analogous to (9.39) is

$$\langle g_1, g_2|S-1|f_1, f_2\rangle = \int \prod_{i=1}^{4} \frac{d^3\mathbf{k}_i}{(2\pi)^3 2\omega_{\mathbf{k}_i}} G_1^*(\mathbf{k}_3) G_2^*(\mathbf{k}_4) F_1(\mathbf{k}_1) F_2(\mathbf{k}_2) \langle k_3, k_4|S-1|k_1, k_2\rangle \tag{14.35}$$

The only thing that was required in deriving the LSZ reduction is that somehow we could get our hands on a local field with a non-zero vacuum to one-particle matrix element, that makes *some* kind of particle out of the vacuum. It can make any other kind of junk it wants, as long as it has a non-zero matrix element. We don't demand that the field satisfy the canonical commutation relations. It could be something like ϕ^2. That might be a good one to look at for a two-particle state. Or maybe if that doesn't work, $\partial^\mu\phi\,\partial_\mu\phi$, or if that one doesn't work, maybe it's a 72-particle state, maybe we want to look at $\phi^{70}\,\partial^\mu\,\phi\partial_\mu\phi$. As long as we can find such a local field for making the desired particle out of the vacuum, we know in principle how to calculate the S-matrix element. In practice it's just as much a mess as before. It's very hard to find out that there is a 72-particle bound state in a theory, let alone to compute its mass. That's what we need, since we've got to get the exact mass in the LSZ formula; the field has to obey the real Klein-Gordon equation with the real physical mass. Using different fields would change the Green's functions off mass shell, but would have no effect on S-matrix elements.

The assumptions I mentioned explicitly in deriving the LSZ reduction formula have no reference whatsoever to whether the particle we're talking about, the meson, is a fundamental particle or a composite particle. It does have an explicit reference to the fact that it is a spinless particle. That's because we've only set up the formalism for scalar fields, but it is fairly obvious that if I have a particle of spin one I can play the same sort of game with the vector field, etc. Therefore this formula (14.18) contains, in addition to the correct version of perturbation theory, the answer to how we compute S-matrix elements for composite particles like hydrogen atoms or nuclei, or blackboard erasers. To compute this Green's function is no easy job: it's a complicated mess. But in principle we have solved the problem. We shifted the field to get rid of its vacuum-to-vacuum expectation value, scaled it to put its vacuum to one-particle matrix element in standard form and off we went. There are no problems of principle. There are, as usual, the enormously difficult problems of practice, which you know about if you've ever looked at the scattering of molecules off molecules, or problems of that kind. There's no problem in *defining* the S-matrix, although there may be severe problems in *computing* it. So we have found a formulation of scattering theory that in principle is capable of describing any conceivable situation.

Other formulas can be derived using methods of the same type as those used to derive the LSZ formula. For example, one can stop "half way" in the reduction formula and obtain

$$\begin{aligned}\langle k_3, k_4|S-1\,|k_1,k_2\rangle \\ = \int d^4x_3\,d^4x_4\,e^{i(k_3\cdot x_3+k_4\cdot x_4)}(i)^2(\Box_3^2+\mu^2)(\Box_4^2+\mu^2)\,\langle 0|T(\phi'(x_3)\phi'(x_4))|k_1,k_2\rangle^{\text{in}}\end{aligned} \tag{14.36}$$

This method is used to derive theorems about the production of "soft" (low energy) pions and photons. We can also use LSZ methods to derive expressions for the matrix elements of fields between in and out states. For example,

$$\begin{aligned}{}^{\text{out}}\langle k_1,\ldots,k_n|A(x)|0\rangle = i^n\int d^4x_1\cdots d^4x_n\,\Big[e^{ik_1\cdot x_1+\cdots+ik_n\cdot x_n} \\ \times\prod_{r=1}^{n}(\Box_r^2+\mu^2)\,\langle 0|T(\phi'(x_1)\cdots\phi'(x_n)A(x))|0\rangle\Big]\end{aligned} \tag{14.37}$$

Of course, this is really just an abstraction of the relation

$$\begin{aligned}{}^{\text{out}}\langle f_1,\ldots,f_n|A(x)|0\rangle = i^n\int d^4x_1\cdots d^4x_n\,f^*(x_1)\cdots f^*(x_n) \\ \times\prod_{r=1}^{n}(\Box_r^2+\mu^2)\langle 0|T(\phi'(x_1)\cdots\phi'(x_n)A(x))\,|0\rangle\end{aligned} \tag{14.38}$$

In the same way that we showed that the right-hand side of (14.19) is equal to the right-hand side of (14.27), we can show that the right-hand side of (14.38) is equal to

$$\prod_{r=1}^{n}\left[\lim_{t_r\to\infty}-\lim_{t_r\to-\infty}\right]\langle 0|T\left(\phi'^{f_1\dagger}(t_1)\cdots\phi'^{f_n\dagger}(t_n)A(x)\right)|0\rangle \tag{14.39}$$

and these limits evaluate to the left-hand side of (14.38).

I should say that at the moment there is in principle no need for counterterms, except for the trivial vacuum energy counterterm. We do want to define our theory so the energy of the vacuum is zero. But aside from that, there is no need to introduce any counterterm. If we could solve the theory exactly, we could write down the Lagrangian in its original form in terms of bare masses and unrenormalized fields, compute all the Green's functions exactly, compute the physical mass of the particles so we know what mass to use in the reduction formula, compute the vacuum to one-particle matrix elements as we want to know how to rescale the fields, crank our answer into the reduction formula, and off we go! In practice the counterterms will come back again and I will talk about that shortly. But they come back again as a matter of convenience, and not as a question of necessity.

14.3 Model 3 revisited

Let me return to our highly unrealistic example, Model 3. By the way, none of what we've done so far has anything specifically to do with our particular example; it's completely general. In its full glory, our example looks like this:

$$\mathscr{L} = \tfrac{1}{2}(\partial_\mu\phi)^2 - \tfrac{1}{2}\mu_0^2\phi^2 + \partial_\mu\psi^*\partial^\mu\psi - m_0^2\psi^*\psi - g_0\psi^*\psi\phi + \text{const.} \tag{14.40}$$

The quantity μ_0 is the bare mass of the meson, which may have absolutely no connection with the physical mass μ of the meson. Similarly, m_0 is the bare mass of the nucleon, and m is its physical mass. The constant g_0 is something I will call a bare coupling constant, some parameter that characterizes the theory. I'll just stick a nought on it, you'll learn why in a moment. We'll compute the conventionally defined coupling constant g from g_0, and then invert the equation to eliminate g_0, which is not directly measured, from all other quantities of interest. The Lagrangian may include a trivial constant, to adjust the zero of the energy to come out right. I won't even bother to give it a special name, at this stage.

In principle, we could compute everything in this theory. After we had managed to solve the theory by some analytic *tour de force*, we could then determine the actual physical masses and the renormalized fields. In this case, since we have two kinds of particles around, mesons and nucleons, we have two kinds of renormalized fields. We have the renormalized meson field defined as before:

$$\phi' = Z_3^{-1/2}\left[\phi - \langle 0|\phi|0\rangle\right] \tag{13.52}$$

Here $|0\rangle$ is the real physical vacuum, the only vacuum we're talking about. By the way, I have tacitly assumed in all of this that Z_3 was chosen to be a positive real number. We are always free to do this: it's just a statement about how we choose the phase of the one-meson states. Thinking back, I realize that I assumed ϕ' was Hermitian if ϕ was Hermitian, and that's not true if Z_3 is not real. The renormalized meson field is determined by the statements (see (13.53))

$$\langle 0|\phi'|0\rangle = 0 \tag{14.41}$$

$$\langle k|\phi'(0)|0\rangle = 1 \tag{14.42}$$

where $\langle k|$ is a one-meson state, the lightest state, other than the vacuum, of charge zero in the theory.

We have to renormalize the nucleon fields likewise with an independent renormalization constant, Z_2:

$$\psi' = Z_2^{-1/2}\psi \tag{14.43}$$

This constant is determined by similar equations. First,

$$\langle 0|\psi'|0\rangle = 0 \tag{14.44}$$

There's no need to add a constant to the nucleon field, because the vacuum expectation value of ψ is automatically zero as a consequence of electric charge conservation. The nucleon field carries electric charge one and therefore can hardly connect the vacuum with the vacuum. So we don't have to bother shifting to impose this condition: the symmetries of the theory impose it for us. And of course a condition similar to the meson's scale (14.42) holds for the nucleon,

$$\langle p|\psi'(0)|0\rangle = 1 \tag{14.45}$$

where $\langle p|$ is a one-antinucleon state.

There's no need to impose a similar condition for $\psi'^\dagger$ making one nucleon because this matrix element is guaranteed to be identical to (14.45) by charge conjugation invariance, indeed by TCP invariance. So it's the same story as we've talked about before, except that we have two wave function renormalization constants because we have two kinds of fields in the theory.

We still have renormalization of various quantities, the masses and the coupling constant. The physical meson mass μ is in general not equal to the bare meson mass μ_0, the physical nucleon mass m is in general not equal to the bare nucleon mass m_0. We also have to renormalize the coupling constant. If this were a realistic theory—which it ain't—the physical coupling constant g would in general not be equal to the bare coupling constant g_0. If this were some real interaction like electrodynamics or the weak interaction, the coupling constant in those little tables circulated by the Particle Data Group would be *the* coupling constant as defined by some standard experiment set up by an IUPAP committee,[4] say the Coulomb interaction between two distantly separated charges, or perhaps pion–nucleon scattering at a certain point for the strong interactions, or beta decay for the weak interactions. There's no reason why the answer to that standard experiment should be g_0. The answer to the standard experiment might be something else. So whatever it is that appears in the tables as a result of experimental measurement as the physical coupling constant is certainly not equal to g_0, unless the experiment has been incredibly cunningly chosen. Of course, these quantities are not entirely unrelated.

For example, if the theory is free, when $g_0 = 0$, Z_2 and Z_3 are equal to one. In the interacting theory, they might have had corrections of order g_0 (in fact, as we will see, the corrections are of order g_0^2). But they certainly reduce to one as g_0 goes to zero. Likewise m^2 is m_0^2 plus corrections of order g_0. We surely want any sensible definition of the coupling constant as physically defined to reduce to the coefficient in the Lagrangian for very weak coupling, so we will normally accept from that IUPAP committee as a sensible definition only one such that $g = g_0$ with perhaps higher order corrections.

Now in principle we could solve the theory in the following way. We could do perturbation theory, which would give us the Green's functions for unrenormalized ψ's and ϕ's, up to some finite order, as a power series expansion in g_0 with m_0 and μ_0 held fixed. We could then determine the physical masses as functions of m_0 and μ_0 and g_0, all the wave function renormalization constants as functions of m_0 and μ_0 and g_0, and the result of that standard experiment defined by that IUPAP committee as functions of these parameters. We could then adjust the values of these bare parameters to give the right answer as multiplied by the Z's and compute their scattering matrix elements. That's possible, but it's also an enormous pain in the neck. You are computing the wrong Green's functions in terms of the power series in the wrong coupling constant with the wrong masses held fixed. For practical purposes we'd

[4] [Eds.] The International Union of Pure and Applied Physics.

like to do an expansion in a realistic theory like quantum electrodynamics, not in the bare charge, but in the actual charge that is measured, with the physical masses held fixed and in terms of the renormalized Green's functions, which are the things we're after at the end. This is purely a practical question of convenience. It has nothing to do with one of principle.

We can avoid the wrong expansion by rewriting the Lagrangian (14.40) in terms of these renormalized quantities with things left over.

$$\mathscr{L} = \tfrac{1}{2}(\partial_\mu\phi')^2 - \tfrac{1}{2}\mu^2\phi'^2 + \partial^\mu\psi'^*\partial_\mu\psi' - m^2\psi'^*\psi' - g\psi'^*\psi'\phi' + (\text{leftover stuff}) \qquad (14.46)$$

There is a lot of leftover stuff, because the Lagrangian (14.40) written in terms of the *bare* quantities isn't equal to this Lagrangian (14.46) written in terms of the *physical* quantities, *without* the leftover stuff. We take all the leftover parts and sum them up into counterterms:

$$(\text{leftover stuff}) = \mathscr{L}_{CT}$$

The expression $\mathscr{L}_{CT}$ looks pretty horrible:

$$\mathscr{L}_{CT} = A\phi' + \tfrac{1}{2}B(\partial_\mu\phi')^2 - \tfrac{1}{2}C\phi'^2 + D\partial_\mu\psi'^*\partial^\mu\psi' - E\psi'^*\psi' - F\psi'^*\psi'\phi' + \text{const}' \qquad (14.47)$$

There'll be some coefficient linear in ϕ' that will come from shifting the quadratic term because (13.52) ϕ is proportional to ϕ' plus a constant. A, B, C, D, E and F and the new value of the constant are given simply by requiring the two Lagrangians to be equal, although these formulas will be of absolutely no interest to us. If you work things out, A is $-Z_3^{1/2}\mu_0^2\,\langle 0|\phi|0\rangle$, B is $Z_3 - 1$, C is $-\mu^2 + Z_3\,\mu_0^2$, and so on.

Now the general strategy is this. Please notice that all of these coefficients are going to be things at least of order g, and the coefficient F is going to be at least of order g^2, because of how we've defined things. And therefore our strategy will be to treat these as we treated the counterterms before, that is to say, to compute everything treating the set $\{A, \ldots, F\}$ as free parameters and then to fix them by imposing our renormalization conditions, the conditions that define the renormalized mass and renormalized scale of the fields. Notice we have just enough conditions to do this. If we ignore the (constant) and (constant)$'$ we have six counterterms $\{A, \ldots, F\}$ and we have six renormalization conditions:

Renormalization conditions for Model 3

1. $\langle 0|\phi'|0\rangle = 0$ fixes A
2. $\langle q|\phi'(0)|0\rangle = 1$ fixes B
3. The physical meson mass, μ, fixes C
4. $\langle p|\psi'(0)|0\rangle = 1$ fixes D
5. The physical nucleon mass, m, fixes E
6. The definition of g fixes F

(The condition $\langle 0|\psi'|0\rangle = 0$ is automatic, so we don't have to impose it.)

So we systematically go out order by order in perturbation theory treating the set of constants $\{A, \ldots, F\}$ as free parameters. To any fixed order in perturbation theory we impose the values of the set of constants by asserting our six renormalization conditions. We then determine $\{A, \ldots, F\}$ self-consistently as a power series in the physical coupling constant, and we have achieved our desired end. We have turned a stupid, although in principle valid, perturbation series for the wrong Green's functions in terms of the wrong coupling constant with the wrong masses held fixed, into a systematic perturbation expansion for the right Green's functions in terms of the right coupling constant with the right masses held fixed.

Now we're almost ready to begin doing computations to see how this formalism works out. ("Almost ready" means we'll get to it next time.) But before we do that, there are two questions which we have to consider. The first problem is this. Considering $\mathscr{L}_{CT}$ (14.47) as part of the interaction, we have introduced derivative interactions (the terms proportional to B and D), and our whole formalism is set up for non-derivative interactions. So there is an awkward but necessary technical point we have to investigate: What is the effect of a derivative interaction? That is, what does it do to the Feynman rules? We've got vertices in the theory corresponding to terms in the Lagrangian that have derivatives in them, and that will give us all sorts of problems. It changes the definition of the π's (they're no longer $\partial_0\phi$'s), and everything gets horribly messed up. So we've got to worry about those derivative interactions. That's a trivial problem which I'll take care of this lecture.

The second problem confronting us is that our renormalization conditions are not well set up to be systematically applied in perturbation theory: they're not phrased in terms of Green's functions. The second condition doesn't have to do with Green's functions. The fifth condition, that m is the physical mass of the particle, is not phrased in terms of Green's functions. Our whole Feynman apparatus is set up for computing Green's functions. If we really want to build a smoothly running, well oiled machine where we can just grind out calculations without any thought, or better yet, write a computer program that will do it for us, we would like to phrase the renormalization conditions in terms of properties of certain Green's functions. We'd like to say that the sum of all Feynman graphs of a certain kind vanish, or equals one, or something. That's equivalent to the equation that gives us the scale of the field. We haven't got that yet. So these are the two tasks before us that we have to complete before we can automate this scheme, and be able to compute without thought. We begin with the first task, derivative interactions.

14.4 Guessing the Feynman rules for a derivative interaction

The general formalism for derivative interactions is an incredible combinatoric mess. Things really get awful. The coupling constant enters into the definition of the canonical momentum:

$$\pi_\mu = \frac{\partial \mathscr{L}}{\partial(\partial^\mu\phi)} \neq \partial_\mu\phi \quad \text{if } \mathscr{H}_I = \mathscr{H}_I(\partial_\mu\phi) \tag{14.48}$$

The interaction Hamiltonian has the canonical momentum in it, and therefore you have all sorts of problems about what the time ordered product of a string of interaction Hamiltonians and a string of fields means, because they no longer commute at equal times:

$$T(\partial_\mu\phi(x)\cdots) \neq \partial_\mu T(\phi(x)\cdots) \tag{14.49}$$

In particular, it is no longer true that $\mathscr{H}_I = -\mathscr{L}_I$:

$$\mathscr{H}_I \neq -\mathscr{L}_I \tag{14.50}$$

Things are just too horrible to contemplate, at least at this stage of our development. After we've hardened our heads, karate fashion, by banging them on some difficult problems, we will return to the question of derivative interactions and straighten everything out for them. But I really don't want to get into that whole complicated subject to handle such a simple derivative interaction as this sort, like the term proportional to D. After all, it's not really much of an interaction, it's just a term quadratic in the fields. Therefore what I will do is *guess* the Feynman rules appropriate to this derivative interaction, and then try and show you that my guess is okay by doing some simple examples. When I'm done, it will be obvious how to generalize the results to this theory, and we will see that the generalization gives the desired results.

The first theory I will look at will be the simplest of all theories, a free scalar field:

$$\mathscr{L} = \tfrac{1}{2}(\partial_\mu \phi)^2 - \tfrac{1}{2}\mu^2\phi^2 \tag{14.51}$$

This doesn't have a derivative interaction; indeed, it doesn't have any interaction at all. But we can fake matters so it looks like it has a derivative interaction by introducing a new variable ϕ':

$$\phi = \sqrt{Z_3}\phi' \tag{14.52}$$

Of course ϕ' is not a renormalized field in the standard sense here; perhaps I should not call this constant Z_3. The field ϕ is already perfectly adequately normalized. Nevertheless to remind you of the problem it is connected with, I will call this quantity Z_3. It's some arbitrary constant, maybe one plus a squared coupling constant. If we rewrite this theory in terms of ϕ', we obtain

$$\mathscr{L} = \tfrac{1}{2}(\partial_\mu \phi')^2 - \tfrac{1}{2}\mu^2\phi'^2 + (Z_3 - 1)\left[\tfrac{1}{2}(\partial_\mu \phi')^2 - \tfrac{1}{2}\mu^2\phi'^2\right] \tag{14.53}$$

For a simple example, I'll take $(Z_3 - 1)$ to be equal to g^2, a coupling constant, and I will call the term in $(Z_3 - 1)$ an interaction. I'd like to get a Feynman rule for that vertex. We had a similar term in the free Lagrangian of Model 3, a counterterm which I wrote as $\frac{1}{2}b\phi^2$. But this term was not in the interaction Lagrangian, and in any case it did not have a derivative. When we were looking at Model 3, we were still in a state of primal innocence, unaware of wave function renormalization; we only considered mass renormalization. In Model 3, this term had the Feynman graph and Feynman rule (see p. 216, 3.(d))

$\frac{1}{2}b\phi^2$ in $\mathscr{L}$: 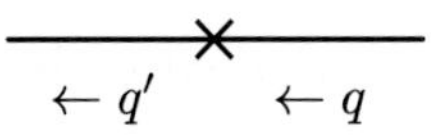

$ib\,(2\pi)^4\,\delta^{(4)}(q' - q)$

Here I have an interaction

$$\mathscr{L}_I = \tfrac{1}{2}(Z_3 - 1)\left[(\partial_\mu \phi')^2 - \mu^2\phi'^2\right] \tag{14.54}$$

which I'll indicate by this:

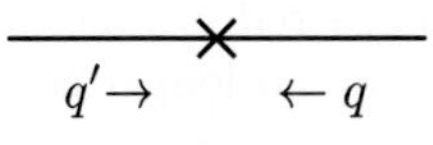

Figure 14.1: Meson derivative interaction

It's the only interaction in the theory. I give the two lines momenta q and q', both oriented inward. The interaction has two parts, one coming from the μ^2 term, and one coming from the $(\partial_\mu\phi)^2$ term. We worked out the result of the μ^2 before, as written above. By analogy we get a Feynman rule that looks like this:

$$\tfrac{1}{2}(Z_3-1)\left[(\partial_\mu\phi')^2-\mu^2\phi'^2\right]: \quad i(Z_3-1)(2\pi)^4\,\delta^{(4)}(q'+q)\left[(\cdots)-\mu^2\right] \tag{14.55}$$

That's how we treated our old mass counterterm, and it's unquestionably what descends from the μ^2 part of the interaction.

But what do we get from the second term, $(\partial_\mu\phi')^2$? God only knows. Though we are not divine, we are allowed to guess. Since we see μ^2 here, I will guess q^2 belongs where we now have $(\cdots)$. This is just a sheer, blind guess:

$$\tfrac{1}{2}(Z_3-1)\left[(\partial_\mu\phi')^2-\mu^2\phi'^2\right] \quad \stackrel{?}{\Longrightarrow} \quad i(Z_3-1)(2\pi)^4\,\delta^{(4)}(q'+q)\left[q^2-\mu^2\right] \tag{14.56}$$

No derivatives, μ^2; two derivatives, q^2: total guesswork. We are *guessing* that a derivative ∂^μ in the interaction leads to a power of momentum q^μ in the Feynman rules, to within factors of $\pm i$:

$$\boxed{\partial^\mu(\text{field}) \text{ in } \mathscr{L}_I \stackrel{?}{\Longrightarrow} \pm i p^\mu_{\text{field}} \text{ in Feynman rules}} \tag{14.57}$$

Now I'll check my guess by computing $\widetilde{G}'^{(2)}$, the two-particle Green's function, by summing up the perturbation expansion in this object. Since I already know the exact form of $\widetilde{G}'^{(2)}$, there should be no problem in seeing whether the guess is right in this simple case.

First, the exact answer, from (14.14) and (14.15):

$$\widetilde{G}'^{(2)}(q,q') = Z_3^{-1}\widetilde{G}^{(2)}(q,q') = Z_3^{-1}(2\pi)^4\delta^{(4)}(q+q')\left(\frac{i}{q^2-\mu^2}\right) \tag{14.58}$$

That's the Green's function in the original free theory multiplied by the inverse of Z_3. (I omit the $i\epsilon$ out of sheer laziness.) This is exact. It doesn't depend on my guess.

Do we get the same thing by summing up diagrams? The series of diagrams is very simple; see Figure 14.2. There's a zeroth order diagram, the first correction, second-order correction, three of 'em on a line, etc. Those are the total collection of all Feynman diagrams with two

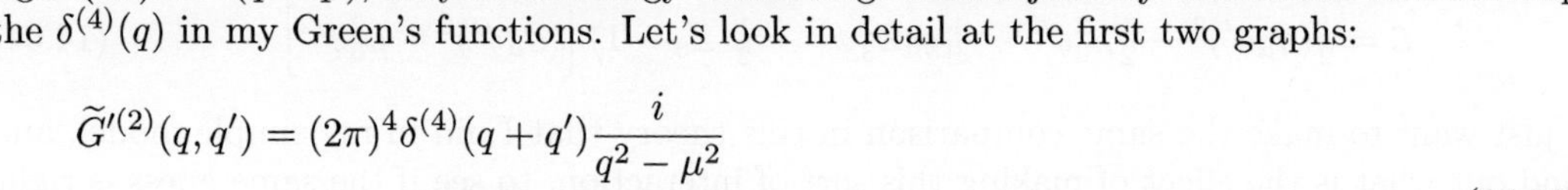

Figure 14.2: Perturbation series for the free scalar field $\widetilde{G}'^{(2)}$

external lines, with the interaction indicated by an $\times$ on the line. Now what do we have? First I get $(2\pi)^4\delta^{(4)}(q+q')$, they're all energy conserving. That's just my convention that I'll keep the $\delta^{(4)}(q)$ in my Green's functions. Let's look in detail at the first two graphs:

$$\widetilde{G}'^{(2)}(q,q') = (2\pi)^4\delta^{(4)}(q+q')\frac{i}{q^2-\mu^2} \times\left[1+\int\frac{d^4q_1}{(2\pi)^4}i(2\pi)^4\delta^4(q_1-q)(Z_3-1)(q_1^2-\mu^2)\frac{i}{q_1^2-\mu^2}+\cdots\right] \tag{14.59}$$

The first diagram in the series is just the free propagator times $(2\pi)^4$ times the delta function. No question there; that's the same diagram that emerged in the free theory. That will be a common factor. It's in all of these things if we count from the left because all have a propagator on the left: our Green's functions have propagators on the external legs.

What happens when I consider one vertex and one propagator? Well, we notice something rather peculiar. The factor of $(2\pi)^4$ from the interaction cancels the $(2\pi)^4$ in the denominator of the integration. The delta function from the interaction term is canceled by the integration; the internal momenta are the same. The i from the interaction term and the i in the propagator multiply to make a factor of (-1). The interaction term has a factor of $(q_1^2 - \mu^2)$ that exactly cancels the same factor in the propagator. The only factors not canceled are $(-1)(Z_3 - 1) = (1 - Z_3)$. The result of adding both one more interaction and one more propagator is simply to add a factor of $1 - Z_3$. The net result for the first two graphs is

$$\widetilde{G}'^{(2)}(q,q') = (2\pi)^4\delta^{(4)}(q+q')\frac{i}{q^2-\mu^2}\left[1 + (1 - Z_3) + \cdots\right] \tag{14.60}$$

Likewise the result of adding two more interactions and two more propagators is to add a factor of $(1 - Z_3)^2$. We know how to sum a geometric series:

$$\begin{aligned}\widetilde{G}'^{(2)}(q,q') &= (2\pi)^4\delta^{(4)}(q+q')\frac{i}{q^2-\mu^2}\left[1 + (1 - Z_3) + (1 - Z_3)^2 + \cdots\right] \\ &= (2\pi)^4\delta^{(4)}(q+q')\frac{i}{q^2-\mu^2}(1 - (1 - Z_3))^{-1} \\ &= Z_3^{-1}\widetilde{G}^{(2)}(q,q')\end{aligned} \tag{14.61}$$

in agreement with the exact result. As usual we take physicist license and don't worry about questions of convergence. The agreement leads us to believe our guess (14.57) about the Feynman rule for the derivative coupling is correct.

The guess also works for an interacting theory as I will now demonstrate. To avoid inessential complications of notation, I will use an interacting theory which has only a single field,

$$\mathcal{L} = \tfrac{1}{2}(\partial_\mu\phi)^2 - \tfrac{1}{2}\mu_0^2\phi^2 - \tfrac{1}{4!}g_0\phi^4 \tag{14.62}$$

I've included a factor of 4! to be in line with conventions. It's a nice interacting theory that's only got one field in it. Once again I define

$$\phi = \sqrt{Z_3}\phi' \tag{14.63}$$

I also choose that ϕ has a vacuum expectation value of zero, because the theory is symmetric under $\phi \to -\phi$. I won't bother to make the redefinitions that are responsible for mass and coupling constant renormalization, which in principle we'd have to do in this theory, just as in our simple model. I'll write exactly the same Lagrangian as

$$\mathcal{L} = \tfrac{1}{2}(\partial_\mu\phi')^2 - \tfrac{1}{2}\mu_0^2\phi'^2 - \tfrac{1}{4!}g_0Z_3^2\phi'^4 + \tfrac{1}{2}(Z_3 - 1)\left[(\partial_\mu\phi')^2 - \mu_0^2\phi'^2\right] \tag{14.64}$$

I just want to make the same comparison in this theory that I did in my simple model, and find out what is the effect of making this sort of interaction, to see if the same guess is right.

Suppose I have a graph for $\widetilde{G}^{(n)}$ in this model. For example, this is a graph for the unrenormalized $\widetilde{G}^{(4)}$, that is to say, computed using the first form of this Lagrangian:

I don't even care what the Feynman rules are for this graph. This is the only kind of graph with just two vertices that can possibly emerge if I use the first form of the Lagrangian.

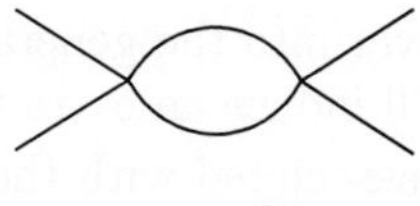

Figure 14.3: Graph for $\widetilde{G}^{(4)}$ in $\frac{1}{4!}g\phi^4$ theory

I will introduce a little topological notation. Such a graph has n external lines, I internal lines and V vertices; n, I and V are numbers associated with a particular graph. What are the connections between these quantities? They are not three free parameters. They are connected by the law of conservation of ends of lines. Every external line has one end that ends on a vertex. Every internal line has two ends that end on a vertex. Every vertex has four lines ending on it (because the theory has a term in ϕ^4). Then

$$n + 2I = 4V \tag{14.65}$$

A slight variant of this formula will turn out to be useful:

$$2V - I = \tfrac{1}{2}n \tag{14.66}$$

Now let's turn to the second form of the Lagrangian. From any one of these graphs can be generated an infinite family of graphs which differ simply by any arbitrary number of my new interaction, which I represent just as before, by a $\times$, on any one of these six lines. In Figure 14.4, I've added seven. I have to sum up this infinite family in order to find out the graph that corresponds to this, which, if my guess was right, should be the graph for $\widetilde{G}'^{(4)}$.

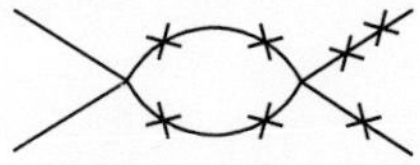

Figure 14.4: Graph for $\widetilde{G}'^{(4)}$ in $\frac{1}{4!}g\phi^4$ theory

The family of graphs for $\widetilde{G}'^{(n)}$ differs from the corresponding graphs for $\widetilde{G}^{(n)}$ in two ways. First, I have a different coefficient of the four-particle vertex. For $\widetilde{G}^{(n)}$, each four-particle vertex is multiplied by g_0; for $\widetilde{G}'^{(n)}$, it's multiplied by $Z_3^2 g_0$. So the net factor from the vertices is Z_3^{2V}. The comparison is really graph by graph, but we'll shortly see the factors that depend on anything except n disappear. Second, every line, internal or external, has all of these crosses sitting on it like crows on a telephone wire, any number, and we have to sum that up. I can put any number on each propagator. In Figure 14.4, I have six geometric series, which are all independent and which all sum up, because there are six lines internal or external in the diagram. I sum up all of those things, all the possibilities, 17 insertions on the first line, none on the fourth, 42 on the second, etc. From each I get an independent geometric series. Fortunately we did that summation earlier, and we discovered the summation just multiplied each propagator by Z_3^{-1}. Therefore we have Z_3^{-1} from every line, since we do it on both internal and external lines, because the Green's function has propagators on the external lines. That gives another factor of $Z_3^{-(n+I)}$. All in all, we have

$$\widetilde{G}^{(n)} = \widetilde{G}'^{(n)} Z_3^{2V} Z_3^{-(n+I)} = \widetilde{G}'^{(n)} Z_3^{2V-I-n} = \widetilde{G}'^{(n)} Z_3^{-n/2} \tag{14.67}$$

using our topological statement, $2V - I = \frac{1}{2}n$. This is of course exactly the result we would've obtained by substitution.

This argument obviously carries over into the general case. If I make this operation on a much more complicated Lagrangian, I'll induce an extra interaction proportional to $Z_3 - 1$. In each vertex I'll stick a power of $Z_3^{1/2}$ associated with the number of ϕ fields coming into that vertex. When I have an internal line, the result of summing all the crows on the telephone wire will give me a factor of Z_3^{-1}, which will precisely cancel the factors at the two vertices, between which the internal line goes. If I have an external line I still get a factor of Z_3^{-1} but it only goes into one vertex. Therefore it's only half canceled, and I'm left with an overall factor of $Z_3^{-1/2}$, so for the n vertices in $\widetilde{G}'^{(n)}$, the factor is $Z_3^{-n/2}$. The argument is trivially generalized to any theory. Once you see how this one works, you should be able to see how it works in any case. So we have indeed, without having to go through all the work it would take us to develop the general theory of derivative interactions, successfully guessed the right form (14.57) for this particular derivative interaction, at least, to within a sign. Next time we will deal with the second problem, expressing things in terms of Green's functions.

Problem 8

8.1 One consequence of our new formulation of scattering theory is that it doesn't matter much what local field you assign to a particle: any field that has a properly normalized vacuum to one-particle matrix element will give the right S matrix element. (See the discussion following (14.35).)

Consider the theory of a free scalar field,

$$\mathscr{L} = \tfrac{1}{2}(\partial_\mu \phi)^2 - \tfrac{1}{2}\mu^2\phi^2 \tag{P8.1}$$

Let us define a new field, A, by

$$\phi = A + \tfrac{1}{2}gA^2 \tag{P8.2}$$

In terms of A, the Lagrangian becomes

$$\mathscr{L} = \tfrac{1}{2}(\partial_\mu A)^2(1+gA)^2 - \tfrac{1}{2}\mu^2(A + \tfrac{1}{2}gA^2)^2 \tag{P8.3}$$

If you had been presented with this Lagrangian, and didn't know its origin, you would probably think it described a highly nontrivial theory, with complicated non-zero scattering amplitudes. Of course, you do know its origin, and thus you know that it must predict vanishing scattering. Verify this by actually summing up all the graphs that contribute to meson–meson elastic scattering in the A-field formulation, to lowest nontrivial order in g, i.e., g^2, and showing that the sum vanishes.

Comments:

(1) Our general theory does not tell us that the A field Green's functions are the same as the ϕ field Green's functions, so the amplitudes may not vanish if the external momenta are not on the mass shell.

(2) To the order in which we are working, we can completely ignore renormalization counterterms.

(3) This is a theory with derivative interactions. As discussed in class, this leads to potential problems: the interaction Lagrangian is not the same as minus the interaction Hamiltonian, and we can't pull time derivatives through the time-ordering symbol in Dyson's formula. Much later in this course, we shall study such theories using the methods of functional integration, and discover that (to this order in perturbation theory) these problems cancel. Take this on trust here; use the naive Feynman rules as explained in class. (That is to say, treat the theory as if the interaction Hamiltonian were minus the interaction Lagrangian, and as if every derivative ∂_μ became a factor of $-ip_\mu$ for an incoming momentum, and ip_μ for an outgoing one.)

(4) If you have an interaction proportional to A^4, there are 4! different ways of choosing which fields annihilate and which create which mesons. If you don't keep proper track of these (and similar) factors, you'll never get the right answer.

(5) A graph of order g^2 may contain either one vertex proportional to g^2 or two vertices each proportional to g.

(6) To get you started, here are the graphs you'll have to study (with various momenta on the external lines):

(a) 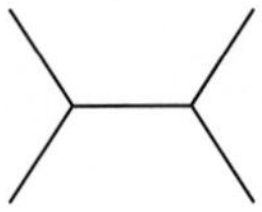

Each vertex here derived from

$$\mathscr{L}_I = \cdots + gA(\partial_\mu A)^2 - \tfrac{1}{2}g\mu^2 A^3 + \cdots$$

(b) 

Vertex here derived from

$$\mathscr{L}_I = \cdots + \tfrac{1}{2}g^2 A^2(\partial_\mu A)^2 - \tfrac{1}{8}\mu^2 g^2 A^4 + \cdots$$

(1997a 8.1)

Solution 8

8.1 In terms of the field A, the Lagrangian may be written

$$\mathscr{L} = \tfrac{1}{2}(\partial_\mu A)^2 - \tfrac{1}{2}\mu^2 A^2 + \mathscr{L}_I \tag{S8.1}$$

where

$$\mathscr{L}_I = g(A\partial_\mu A\,\partial^\mu A - \tfrac{1}{2}\mu^2 A^3) + g^2(\tfrac{1}{2}A^2\partial_\mu\,A\partial^\mu A - \tfrac{1}{8}\mu^2 A^4) \tag{S8.2}$$

Following the advice, we take $\mathscr{H}_I = -\mathscr{L}_I$. There is also a $-i$ in Dyson's formula. Corresponding to the vertex

$$g(A\partial_\mu A\,\partial^\mu A - \tfrac{1}{2}\mu^2 A^3) \tag{S8.3}$$

we have this graph, and, using the naive rule about derivatives—$\partial_\mu A \to -ik_\mu A$ for an incoming particle—this Feynman rule:

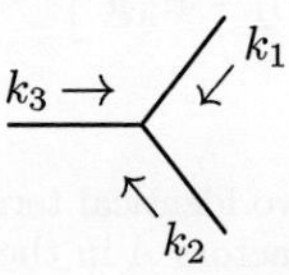

$$ig\left[2!\left((-ik_1)\cdot(-ik_2) + (-ik_1)\cdot(-ik_3) + (-ik_2)\cdot(-ik_3)\right) - 3!\tfrac{1}{2}\mu^2\right] \tag{S8.4}$$

The 2! arises from the symmetry of the two identical factors $\partial_\mu A$, and the 3! from the symmetry of the three identical factors A in the second term. The rule can be simplified by using the identity

$$(k_1 + k_2 + k_3)^2 = k_1^2 + k_2^2 + k_3^2 + 2k_1\cdot k_2 + 2k_1\cdot k_2 + 2k_2\cdot k_3 \tag{S8.5}$$

Moreover, in this case, $k_1 + k_2 + k_3 = 0$. Consequently the Feynman rule becomes

$$ig(k_1^2 + k_2^2 + k_3^2 - 3\mu^2) \tag{S8.6}$$

Consider a diagram like (6)(a) in the statement of the problem, for example,

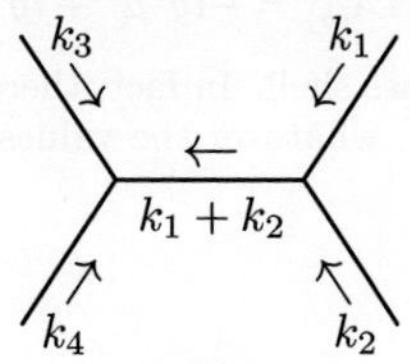

This diagram makes a contribution $i\mathcal{A}_{12,34}$ to the total $2 \to 2$ amplitude equal to

$$i\mathcal{A}_{12,34} = ig\left[k_1^2 + k_2^2 + (k_1+k_2)^2 - 3\mu^2\right]\frac{i}{(k_1+k_2)^2 - \mu^2}ig\left[k_3^2 + k_4^2 + (k_1+k_2)^2 - 3\mu^2\right] \tag{S8.7}$$

We are only looking for contributions on the mass shell, when $k_i^2 = \mu^2$, in which case

$$i\mathcal{A}_{12,34} = ig\left[(k_1+k_2)^2 - \mu^2\right]\frac{i}{(k_1+k_2)^2-\mu^2}ig\left[(k_1+k_2)^2-\mu^2\right] = -ig^2((k_1+k_2)^2-\mu^2) \quad \text{(S8.8)}$$

There are two other diagrams of the same form, obtained by permutations, making contributions $i\mathcal{A}_{13,24}$ and $i\mathcal{A}_{14,23}$. Adding these all together gives all contributions from diagrams of the form (6)(a),

$$i\mathcal{A}_{12,34} + i\mathcal{A}_{13,24} + i\mathcal{A}_{14,23} = -ig^2((k_1+k_2)^2 + (k_1+k_3)^2 + (k_2+k_3)^2 - 3\mu^2) \quad \text{(S8.9)}$$

This may not look very symmetric, but it is, because $k_1+k_2+k_3+k_4 = 0$, by virtue of the conventions about inward pointing momenta. That means

$$(k_1+k_2)^2 = \tfrac{1}{2}\left[(k_1+k_2)^2 + (k_3+k_4)^2\right] \quad \text{(S8.10)}$$

Making the equivalent substitution for the other two contributions gives, after a little algebra,

$$\begin{aligned}(k_1+k_2)^2 + (k_1+k_3)^2 + (k_2+k_3)^2 &= k_1^2+k_2^2+k_3^2+k_4^2 - (k_1+k_2+k_3+k_4)^2 \\ &= k_1^2+k_2^2+k_3^2+k_4^2\end{aligned} \quad \text{(S8.11)}$$

so that the total contribution from all of the diagrams of the form (6)(a) is

$$i\mathcal{A}_{12,34} + i\mathcal{A}_{13,24} + i\mathcal{A}_{14,23} = -ig^2\left[k_1^2+k_2^2+k_3^2+k_4^2-3\mu^2\right] \quad \text{(S8.12)}$$

If we evaluate this on the mass shell, then $k_i^2 = \mu^2$, and so

$$i\mathcal{A}_{(a)} = i\mathcal{A}_{12,34} + i\mathcal{A}_{13,24} + i\mathcal{A}_{14,23} = -ig^2\mu^2 \quad \text{(S8.13)}$$

Now for the graphs of the form (6)(b). Corresponding to the vertex

$$g^2(\tfrac{1}{2}A^2\partial_\mu A\partial^\mu A - \tfrac{1}{8}\mu^2A^4) \quad \text{(S8.14)}$$

we have this graph, and this Feynman rule:

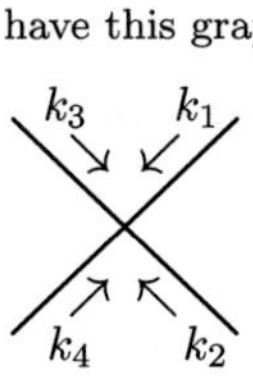

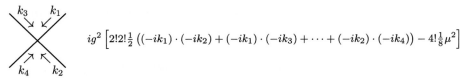

$$ig^2\left[2!2!\tfrac{1}{2}\left((-ik_1)\cdot(-ik_2) + (-ik_1)\cdot(-ik_3) + \cdots + (-ik_2)\cdot(-ik_4)\right) - 4!\tfrac{1}{8}\mu^2\right] \quad \text{(S8.15)}$$

Once again, the factorials arise from symmetry. There are two identical terms A and two identical terms $\partial_\mu A$, so we have a factor of 2! from each. The 4! from the symmetry of the four identical factors A in the second term. The sum of the $\binom{4}{2}$ products of the two different momenta is easily seen to satisfy the identity

$$\begin{aligned}-2\,(k_1\cdot k_2 + k_1\cdot k_3 + \cdots + k_2\cdot k_4) &= -(k_1+k_2+k_3+k_4)^2 + k_1^2+k_2^2+k_3^2+k_4^2 \\ &= k_1^2+k_2^2+k_3^2+k_4^2\end{aligned} \quad \text{(S8.16)}$$

because the sum of the 4-momenta is zero. The contribution $i\mathcal{A}_{(b)}$ from (6)(b) is then

$$i\mathcal{A}_{(b)} = ig^2(k_1^2+k_2^2+k_3^2+k_4^2-3\mu^2) \quad \text{(S8.17)}$$

On the mass shell, this is $ig^2\mu^2$. The total amplitude is thus

$$i\mathcal{A}_{(a)} + i\mathcal{A}_{(b)} = -ig^2\mu^2 + ig^2\mu^2 = 0 \quad \text{(S8.18)}$$

as required. There is no scattering on the mass shell. In fact, there is no scattering *off* the mass shell, because the contributions (S8.13) and (S8.17) cancel, whatever the values of the four 4-momenta. ■

15

Renormalization I. Determination of counterterms

At the end of the last lecture, we found ourselves with six counterterms in our Lagrangian, (14.47), and six renormalization conditions (box, p. 302) for fixing them: that the expectation value of the renormalized meson field in the vacuum state be zero, a wave function renormalization condition for the meson field, a wave function renormalization condition for the nucleon field, two conditions that the mass parameters appearing in our Lagrangian be the physical masses, and one condition to fix the coupling constant. These conditions determine the six counterterms order by order in perturbation theory.

15.1 The perturbative determination of A

Our perturbation theory is set up for the computation of Green's functions, and so we would like to phrase our six renormalization conditions in terms of Green's functions. At the moment only one of them, involving A, is immediately phrased in terms of Green's functions:

$$\langle 0|\phi'(0)|0\rangle = 0 \tag{15.1}$$

The physical vacuum expectation value of the renormalized meson field should be zero. This *is* a Green's function: it can be thought of as the vacuum expectation value of the time-ordered product of *one* field. Graphically, this condition is simply

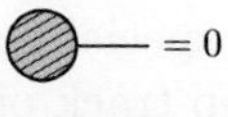

Figure 15.1: The renormalization condition for A

This makes it very easy to determine iteratively the renormalization counterterm A order by order in perturbation theory. In the Lagrangian there's a bunch of stuff, then there is $A\phi'$ plus a bunch of other stuff:

$$\mathscr{L} = \cdots + A\phi' + \cdots \tag{15.2}$$

Let's imagine I have A as a power series expansion in g. I will write this as

$$A = \sum_n A_n \qquad \text{where } A_n \propto g^n \tag{15.3}$$

Graphically:

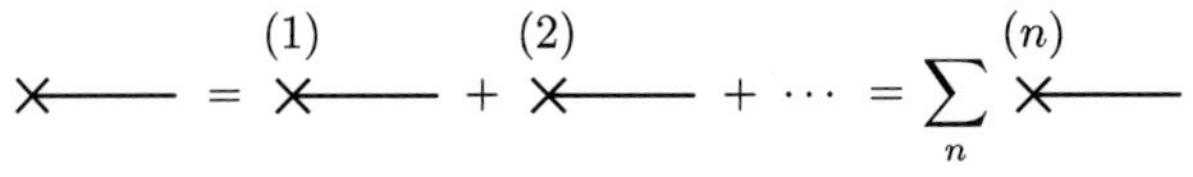

(15.4)

where the (n) over the vertex means you are taking the term proportional to g^n.

Now let us suppose that we have computed everything up to order $n-1$, all Feynman graphs for all Green's functions and, by some method which I have not yet explained, all counterterms; not only A, but $\{B, C, \ldots, F\}$ up to order $n-1$. I will show how that enables us by a computation to determine A_n.

The argument is very simple. We have our renormalization condition, Figure 15.1, which states that for all values of g, the blob equals zero. We will now compute this blob to $\mathcal{O}(g^n)$. There will be two terms:

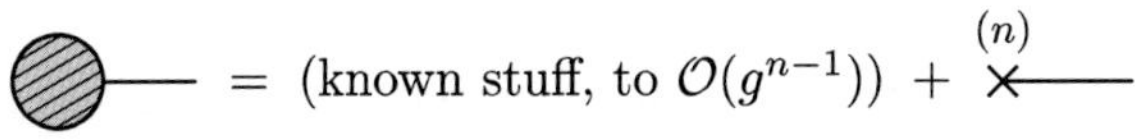

(15.5)

The first term will be all sorts of complicated Feynman diagrams that may well involve, as internal parts, all of the other counterterms $\{B, \cdots, F\}$ in lower order. Those terms are known in principle, because they only involve at their vertices counterterms of lower order. By assumption we know these counterterms: we are analytically muscular, and we can compute any Feynman diagram. Then there is one unknown object that contributes, and it contributes in only one way: the nth order of A. The nth order of A never appears as an internal part of some diagram of more complicated structure, because if it did, that diagram would be of one order higher than $n-1$. The whole thing sums to zero. Therefore this relation (15.5) fixes the nth order of A. So this is how we could iteratively determine the counterterm A and put up a table of its values: the first, the second, the third orders and so on, if we can do the same sort of trick for the other counterterms.

We'll later see that exactly the same thing will happen for all the other counterterms. We will phrase our renormalization conditions in such a way that a certain sum of graphs, a Green's function or an object defined in terms of Green's functions, is equal to zero. We will carefully choose a sum so that if we compute it to n^{th} order, the nth order counterterm will come in only in a simple form like this, plus known stuff, and then we'll have a systematic iterative procedure for computing the counterterms.

In fact, we hardly need A. There is a special feature for this rather simple counterterm, A, that means we don't even have to keep track of this table. This isn't true for the other counterterms. Suppose I consider any graph of the following structure:

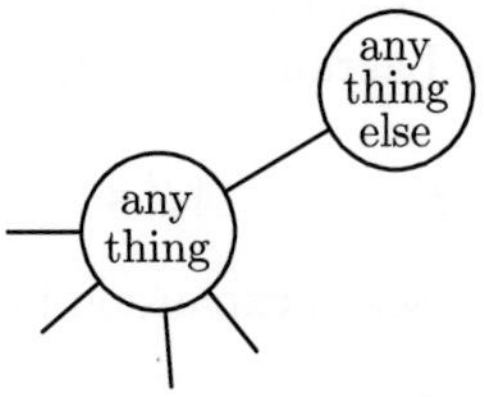

Figure 15.2: A tadpole diagram

These types of graphs are sometimes called **tadpole diagrams** for obvious reasons.[1] Here I've got absolutely anything inside this left blob, and I have a large number of lines or a small number of lines, it matters not, coming out. Then I have a line connecting the first blob to a second blob on the right, containing anything else—but with *no* external lines. That is to say, the graph has a topological structure of two parts connected by a single internal line such that, if I cut that line, the graph separates into two discrete pieces. Now if I sum over all the possible things I can put in for anything else, to a given order, without changing this part on the left, then I obtain the relation

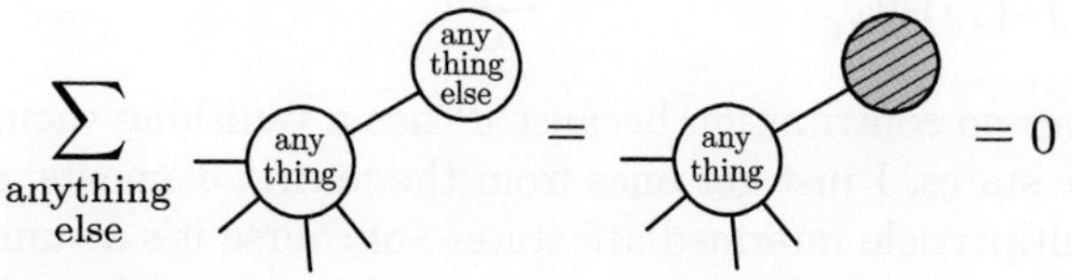

(15.6)

Summing up "anything else" gives the shaded blob, which, by the renormalization condition, is zero. So the net contribution of these tadpole diagrams is zero. Since it is only in graphs of the structure shown in Figure 15.2 that this counterterm appears, in fact we need not worry about the counterterm or about the renormalization condition. They're going to cancel out. All the tadpole graphs sum up to zero and so you can ignore them, just as you can ignore the graphs with disconnected vacuum components.

This demonstration was pretty trivial. That was a good thing because I was able to show the iterative establishment of counterterms in a simple context. I now turn to something much more complicated, the phrasing of the wave function renormalization and mass renormalization conditions in terms of Green's functions. Despite the added complications, we will be able to reach the end in a fairly short time.

15.2 The Källén-Lehmann spectral representation

I will begin by making a general study, with hardly any assumptions, of the two-point function, $\langle 0|\phi'(x)\phi'(y)|0\rangle$. I will derive some properties of this object. From this object of course I can reconstruct the Green's function $\tilde{G}'^{(2)}$ just by multiplying by theta functions. We use systematically the identity, true for any state $|n\rangle$,

$$\langle n|\phi'(y)|0\rangle = e^{ip_n\cdot y}\,\langle n|\phi'(0)|0\rangle \tag{15.7}$$

and in the case of a one-particle state $|p\rangle$,

$$\langle p|\phi'(0)|0\rangle = 1 \tag{15.8}$$

by our normalization condition, (14.42).

---

[1] [Eds.] S. Coleman and Sheldon L. Glashow, "Departures from the Eightfold Way", *Phys. Rev.* **134** (1964) B671–B681. The term was coined by Coleman. The *Physical Review* editors originally objected to this name, so Coleman offered "lollipop diagram" or "sperm diagram" as alternatives. The editors accepted "tadpole diagram". See Peter Woit, *Not Even Wrong*, Perseus Books, New York, 2006, p. 54. A tadpole diagram is a blob with only one line coming out: .

Now I will analyze this object by putting in a complete set of intermediate states and eliminating both the x and the y dependence by using (15.7):

$$\begin{aligned}\langle 0|\phi'(x)\phi'(y)|0\rangle &= \sum_n \langle 0|\phi'(x)|n\rangle\,\langle n|\phi'(y)|0\rangle = \sum_n e^{-ip_n\cdot(x-y)}\big|\langle n|\phi'(0)|0\rangle\big|^2 \\ &= \cancelto{0}{\big|\langle 0|\phi'(0)|0\rangle\big|^2} + \int \frac{d^3\mathbf{p}}{(2\pi)^3 2\omega_{\mathbf{p}}} e^{-ip\cdot(x-y)} \cancelto{1}{\big|\langle p|\phi'(0)|0\rangle\big|^2} + \sum_n{}' \big|\langle n|\phi'(0)|0\rangle\big|^2 e^{-ip_n\cdot(x-y)} \\ &= \int \frac{d^3\mathbf{p}}{(2\pi)^3 2\omega_{\mathbf{p}}} e^{-ip\cdot(x-y)} + \sum_n{}' \big|\langle n|\phi'(0)|0\rangle\big|^2 e^{-ip_n\cdot(x-y)}\end{aligned} \tag{15.9}$$

The vacuum state gives no contribution because ϕ' has a vanishing vacuum expectation value. From the one-particle states, I just get ones from the matrix elements, and obtain $e^{-ip\cdot(x-y)}$. The sum over the multiparticle intermediate states—of course it's a sum and an integral—has a prime to indicate we are excluding the vacuum and one-particle states. The first term is an object we have discussed before, (3.38), in connection with quantizing a free field, with $p^0 = \omega_{\mathbf{p}}$. There we called it $\Delta_+(x-y;\mu^2)$, where μ^2 is the physical mass of the meson:

$$\langle 0|\phi'(x)\phi'(y)|0\rangle = \Delta_+(x-y;\mu^2) + \sum_n{}' \big|\langle n|\phi'(0)|0\rangle\big|^2 e^{-ip_n\cdot(x-y)} \tag{15.10}$$

This big sum is going to give us some Lorentz invariant function of p_n which vanishes unless p^0 is on the upper hyperboloid; by assumption, we only have positive energy states in our theory. Therefore I will write it in the following way:

$$\sum_n{}' \big|\langle n|\phi'(0)|0\rangle\big|^2 e^{-ip_n\cdot(x-y)} = \int \frac{d^4q}{(2\pi)^3}\sigma(q^2)\,\theta(q^0)\,e^{-iq\cdot(x-y)} \tag{15.11}$$

The $(2\pi)^3$ is unfortunate, but I'll run into a convention clash with standard notation if I put a $(2\pi)^4$ there. (Our convention, violated here, is that every momentum integral is accompanied by a 2π in the denominator.) The theta function $\theta(q^0)$ ensures that things vanish, except on the upper hyperboloid. The function $\sigma(q^2)$ is defined by this:

$$\sigma(q^2)\,\theta(q^0) = \sum_n{}' (2\pi)^3\delta^{(4)}(q-p_n)\big|\langle n|\phi'(0)|0\rangle\big|^2 \tag{15.12}$$

If I stick this expression into the previous equation (15.11) I obviously get equation (15.10), just by doing the integral over q. We know that σ is a function of q^2, rather than q, because of Lorentz invariance: the sum over intermediate states is Lorentz invariant. Alternatively, the left-hand side of (15.11) must be a function only of $(x-y)^2$. So its Fourier transform should be a function only of q^2.

We know other general features about $\sigma(q^2)$. In perturbation theory we would expect that the lightest multiparticle states that can be made by a ϕ' field hitting the vacuum are either two mesons or a nucleon–antinucleon pair. So we would expect in perturbation theory that $\sigma(q^2)$ equals zero, for q^2 less than the minimum of $4m^2$ if the nucleon–antinucleon pair is lighter, or $4\mu^2$ if the two meson state is lighter:

$$\sigma(q^2) = 0 \text{ if } q^2 < \min(4\mu^2, 4m^2) \tag{15.13}$$

Of course this is just a perturbation statement. In the real theory, there might be bound states appearing which lie below either the meson–meson or nucleon–antinucleon threshold.

Say that the lightest particle in the theory is the meson. Then in the real theory, in any event

$$\sigma(q^2) = 0 \text{ if } q^2 < \mu^2 + \eta \quad \text{for some } \eta > 0 \tag{15.14}$$

The value of η depends on the energy of the bound state. If the bound state sinks below the one-meson state, then we call the bound state *the* one-meson state, because by definition the one-meson state is the lightest state with the quantum numbers of the meson. If they're right on top of each other, then we were making the wrong assumption about the spectrum: there are *two* one-meson states, and we have to rethink the whole thing. Additionally, because σ is defined in (15.12) as an integral of squares times positive terms, we also know that $\sigma(q^2)$ is always greater than or equal to zero:

$$\sigma(q^2) \geq 0 \tag{15.15}$$

These two facts, (15.14) and (15.15), will be very important to us in our subsequent development.

We can rewrite the expression (15.10) as follows:

$$\langle 0|\phi'(x)\phi'(y)|0\rangle = \Delta_+(x-y;\mu^2) + \int \frac{d^4q}{(2\pi)^3}\int_0^\infty da^2\,\delta(a^2-q^2)\sigma(a^2)\theta(q^0)e^{-iq\cdot(x-y)} \tag{15.16}$$

Here a^2 is a new dummy variable. Because of the delta function, this is just another way of writing $\sigma(q^2)\theta(q^0)$. The advantage is that I can now do the q integral, because what I have here is, for each fixed value of a, the expression that gives me $\Delta_+(x-y;a^2)$ for a free field of mass a^2. We have the definition (3.38),

$$\Delta_+(x-y;\mu^2) = \int \frac{d^3\mathbf{p}}{(2\pi)^3 2\omega_{\mathbf{p}}} e^{-ip\cdot(x-y)} \tag{15.17}$$

But we also have the relativistic measure (1.55):

$$\int_{p^0} d^4p\,\delta(p^2-\mu^2)\theta(p^0) = \frac{d^3\mathbf{p}}{2\omega_{\mathbf{p}}} \tag{15.18}$$

so we can write an alternative definition,

$$\Delta_+(x-y;\mu^2) = \int \frac{d^4p}{(2\pi)^3}\,\delta(p^2-\mu^2)\theta(p^0)e^{-ip\cdot(x-y)} \tag{15.19}$$

Thus (15.16) can be written

$$\langle 0|\phi'(x)\phi'(y)|0\rangle = \Delta_+(x-y;\mu^2) + \int_0^\infty da^2\,\sigma(a^2)\,\Delta_+(x-y;a^2) \tag{15.20}$$

That is to say, we've written the exact vacuum expectation value of the product of two fields as a superposition of free field vacuum expectation values, integrated over the mass spectrum of the theory. This is sometimes called **the spectral representation** for that reason. It is also called **the Källén-Lehmann spectral representation.**[2] You may see it in the literature

[2] [Eds.] Gunnar Källén, "On the Definition of the Renormalization Constants in Quantum Electrodynamics," *Helv. Phys. Acta* **25** (1952) 417–434; Harry Lehmann, "Über Eigenschaften von Ausbreitungsfunktionen und Renormierungskonstanten quantisierter Felder", (On the characteristics of fields quantized by propagation functions and renormalization constants), *Nuovo Cim.* **11**(4) (1954) 342–357. Källén (pronounced "chal-LANE") was a prominent Swedish quantum field theorist, the author of highly regarded textbooks on QED and elementary particle physics, and one of the first to join CERN's staff. He died in 1968 when the plane he was piloting crashed in Hannover, en route to Geneva from Malmö. Källén was 42.

written in the form

$$\langle 0|\phi'(x)\phi'(y)|0\rangle = \int_0^\infty da^2\, \rho(a^2)\, \Delta_+(x-y;a^2) \tag{15.21}$$

where $\rho(a^2)$ is of course equal to $\delta(\mu^2 - a^2) + \sigma(a^2)$. We won't use this form much.

I will now use the spectral representation, first, to get a representation of the commutator that will give us an interesting inequality, and second, to get a representation of the Green's function, the time-ordered product. Since we have everything represented as a linear superposition of free field quantities, we can simply go through all of our old free field manipulations appropriately superimposing them. Thus for example (see (3.42))

$$\langle 0|[\phi'(x),\phi'(y)]|0\rangle = i\Delta(x-y;\mu^2) + \int_0^\infty da^2\, \sigma(a^2)\, i\Delta(x-y;a^2) \tag{15.22}$$

We can now compute the vacuum expectation value of the equal time commutator. This is amusing because we know what the equal time commutator is, in terms of Z_3, since we know ϕ' in terms of canonical fields and Z_3: ϕ' is $Z_3^{-1/2}\phi_s$ (13.52), where the shifted field $\phi_s = \phi - \langle 0|\phi|0\rangle$. Since ϕ_s differs from ϕ only by a subtracted c-number (13.49), they have the same commutators, and so, from (3.61)

$$\langle 0|[\phi'(\mathbf{x},t),\dot{\phi}'(\mathbf{y},t)]|0\rangle = Z_3^{-1}\langle 0|[\phi_s(\mathbf{x},t),\dot{\phi}_s(\mathbf{y},t)]|0\rangle = Z_3^{-1} i\delta^{(3)}(\mathbf{x}-\mathbf{y}) \tag{15.23}$$

On the other hand we can get exactly the same thing from the spectral representation. We know from (3.59) and (3.61) that

$$i\frac{\partial}{\partial x^0}\,\Delta(x-y)\Big|_{x^0=y^0} = -i\delta^{(3)}(\mathbf{x}-\mathbf{y}) \tag{15.24}$$

Differentiating (15.22) with respect to y^0 and evaluating at equal times will give us

$$\langle 0|[\phi'(\mathbf{x},t),\dot{\phi}'(\mathbf{y},t)]|0\rangle = i\delta^{(3)}(\mathbf{x}-\mathbf{y})\left[1+\int_0^\infty da^2\,\sigma(a^2)\right] \tag{15.25}$$

Comparing these two expressions, (15.25) and (15.23), we find **Lehmann's sum rule**:

$$Z_3^{-1} = 1 + \int_0^\infty da^2\sigma(a^2) \tag{15.26}$$

Or, since $\sigma(a^2)$ is guaranteed to be non-negative,

$$Z_3^{-1} \geq 1 \tag{15.27}$$

Most likely, we should expect Z_3^{-1} to be greater than 1. It will only be equal to 1 if $\sigma(a^2)$ vanishes, which would be a pretty trivial field theory.[3] Equivalently, we can say

$$Z_3 \leq 1 \tag{15.28}$$

It is sometimes alleged that this statement (15.28) has a trivial explanation. After all, $Z_3^{1/2}$ is defined to equal $\langle k|\phi(0)|0\rangle$ where ϕ is the unrenormalized field (see (13.51)). I will

[3] [Eds.] If $\sigma(a^2) = 0$, the theory admits no states with $p^2 > \mu^2$, so no particle creation.

now tell you an argument that is a lie, but at least it will help you remember which way the sign goes between Z_3 and 1. People say, "Look, we know $\phi(0)$ hitting the vacuum makes a single bare particle and therefore $\langle k|\phi(0)|0\rangle$ is the amplitude for making a physical particle. So it's the inner product between a physical particle and a bare particle, which is less than or equal to one, like all inner products between appropriately normalized states. Therefore Z_3 is less than or equal to 1." This argument is a lie, of course, because $\phi(0)$ is scaled so that it has amplitude 1 for making a bare particle when applied to the *bare* vacuum, and here we are applying it to the *physical* vacuum. So the argument is completely useless. Nevertheless it'll help you remember the sign. By the way, this stuff is treated at enormous length in all the standard texts, including Bjorken and Drell.

15.3 The renormalized meson propagator $\widetilde{D}'$

However amusing we may have found our work thus far, we haven't gotten very close to expressing our renormalization conditions in terms of perturbation theory objects, i.e., in terms of Green's functions. So we will go on and compute the *renormalized* Green's functions. Of course once we know the vacuum expectation value of the unordered product of a pair of fields, we can obtain the two-particle Green's function $\widetilde{G}'^{(2)}(p,p')$ by a linear sequence of operations: permuting the arguments, and multiplying by theta functions, Fourier transforming etc. So we can just write down the answer. It's convenient to express things in terms of objects with the delta function factored out. So I'll write the expression for $\widetilde{G}'^{(2)}(p,p')$ as

$$\widetilde{G}'^{(2)}(p,p') \equiv (2\pi)^4\delta^{(4)}(p+p')\widetilde{D}'(p^2) \tag{15.29}$$

where the entity $\widetilde{D}'(p^2)$ is sometimes called **the renormalized propagator**. (Sometimes the prime indicates renormalized fields.) So I take the ordinary Feynman propagator, and now we'll put all sorts of corrections on it, to get what really happens when one meson goes into a blob and one meson comes out. $\widetilde{D}'(p^2)$ is, by the spectral representation, a linear superposition of free propagators with the same weighting function as with Δ, (15.22). That is to say,

$$\widetilde{D}'(p^2) = \frac{i}{p^2-\mu^2+i\epsilon} + \int_0^\infty da^2\,\sigma(a^2)\frac{i}{p^2-a^2+i\epsilon} \tag{15.30}$$

This spectral representation of $\widetilde{D}'(p^2)$ tells us something very interesting about the analytic properties of $\widetilde{D}'(p^2)$ considered as a function of complex p. As you see, for example, for all p^2's not on the positive real axis, this integral defines an analytic function of p^2. If p^2 is not on the real axis, the denominator never vanishes, so the function is well-defined and its derivative is also well-defined. Thus if I were to draw the complex p^2 plane,

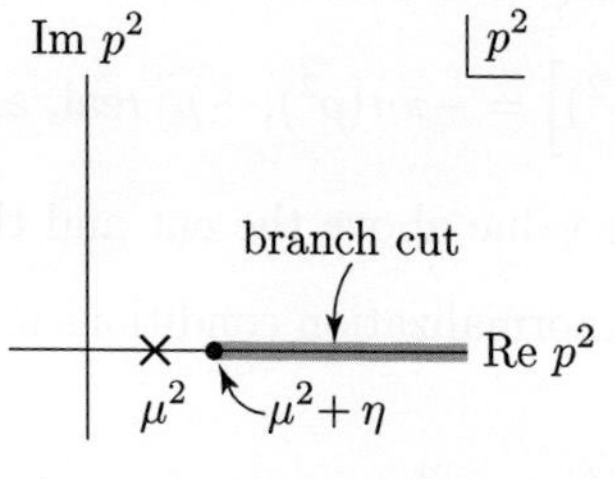

Figure 15.3: The analytic properties of $\widetilde{D}'(p^2)$ in the complex p^2 plane

$\widetilde{D}'(p^2)$ would be an analytic function in that plane, except for a pole at μ^2 and the branch cut, a line of singularities beginning from $\mu^2+\eta$, where $\sigma(p^2)$ begins to be non-zero, extending out presumably to infinity. The actual physical value of $\widetilde{D}'$ for real p is of course totally unambiguous. Along the branch cut, though, we have to say which side of the cut we're on. Feynman's $i\epsilon$ prescription tells us $\mu^2 = p^2 + i\epsilon$, which means we are *above* the cut in this analytic continuation of $\widetilde{D}'$. The original $\widetilde{D}'$, defined only for real p^2, is obtained by taking the analytic function onto the cut from above. Those of you who have studied the analytic properties of partial wave amplitudes in non-relativistic scattering theory will not find this analytic structure surprising.

We could get into troubles if the $\sigma(a^2)$ integral doesn't converge. That means the sum over intermediate states doesn't converge. The formula (15.12) is great if $\sigma(a^2)$ has any reasonable behavior. The swindle I've put on you is that if $\sigma(a^2)$ grows too rapidly at infinity, say like a power of a^2, then if you look at this function in position space, it's a well-defined distribution, but it's not true that a theta function times a distribution is necessarily a distribution. I will assume that in our case the time-ordered product is defined. That is the sort of thing purists have to worry about. Maybe we'll become purists when we get a deeper understanding of field theory, and we may go back and worry about that.

In principle, if we ever reach any trouble we will be quite willing to act like slobs. If we cannot justify our intermediate stages with what we've got, we will brutally truncate our theory by throwing away the high momentum modes, therefore making $\sigma(a^2)$ vanish beyond a certain point, a cutoff, and guaranteeing the convergence of everything. We'll just cut them out of the theory, bam! We will then have a sick, nonsensical theory. That's not real physics, but we'll just go ahead. When we finally get the S-matrix elements, if they have nice smooth limits as the cutoff goes away, we're happy. We will have reached a satisfactory result even if our intermediate stages are garbage. If they don't, there's no point worrying about mathematical rigor, because nothing we can do will make sense out of it. That's our general attitude whenever we run into trouble because of the high energy behavior of integrals. People untrained as carpenters who nevertheless build houses are called "wood butchers". The attitude I'm describing is that of a "physics butcher".

We actually know a little bit more about $-i\widetilde{D}'(p^2)$. It has what is called a Schwarz reflection principle in the theory of functions of a complex variable. It's easy to see from the spectral formula (15.30)

$$\left[-i\widetilde{D}'(p^2)\right]^* = -i\widetilde{D}'((p^2)^*)$$

Once we've multiplied by $-i$ to get rid of the i in the numerator, conjugating p^2 is the same as conjugating the function, in the domain of analyticity. If you're above the cut, this is the value below the cut. The discontinuity over the cut is therefore connected to the imaginary part of $\widetilde{D}'$. By a formula we used previously in a homework problem,[4] the imaginary part is given by

$$\text{Im}\left[-i\widetilde{D}'(p^2)\right] = -\pi\sigma(p^2), \quad p^2 \text{ real, and } p^2 > \mu^2 \tag{15.31}$$

That's the difference between the value above the cut and the value below the cut.

The mass and wave function renormalization conditions are embedded in a statement about the Green's functions:

$$\widetilde{D}'(p^2) = \frac{i}{p^2 - \mu^2 + i\epsilon} + (\text{a function analytic at } p^2 = \mu^2) \tag{15.32}$$

[4] [Eds.] See (P4.1), p. 175. The relevant formula is $\lim\limits_{\epsilon\to 0} \dfrac{1}{x + i\epsilon} = -i\pi\delta(x) + \dfrac{1}{x}$.

This equation contains our two renormalization conditions: for mass, that there *is* a pole at μ^2, and for the wave function, that the residue of that pole is i. If the field had been normalized differently, the residue of the pole would be $17i$ or $\frac{1}{3}i$ or something. This gives us, in principal, a way of determining the mass of the particle and the normalization of the field in terms of the properties of $\widetilde{D}'$, which is defined in terms of a Green's function.

This condition will have to be massaged a bit to put it into the best form for doing the computation we want to do. To that end I will define a special kind of Green's function, called a **one-particle irreducible Green's function**, denoted by **1PI**, for "one-particle irreducible", and indicated by a blob like this, with however many external lines coming out of it:

Figure 15.4: One-particle irreducible diagram

This is the sum of all connected graphs that cannot be disconnected by cutting a single (one particle) internal line. By convention, when we evaluate 1PI diagrams, we do not include the energy-momentum conserving delta function, nor external line propagators. These conventions will turn out to simplify our algebra with these things.

To give an example of what is and what is not a 1PI graph, take two diagrams from nucleon–antinucleon into nucleon–antinucleon scattering in Model 3, as shown in Figure 15.5. Diagram (a) is not 1PI because cutting (or removing) the internal line divides it into two separate parts. On the other hand, Diagram (b) is 1PI, because there is no way I can split it into two parts by cutting any *one* internal line: it still remains connected. I have to break at least *two* internal lines in Diagram (b) to make it fall apart.

Diagram (a): Not 1PI Diagram (b): 1PI

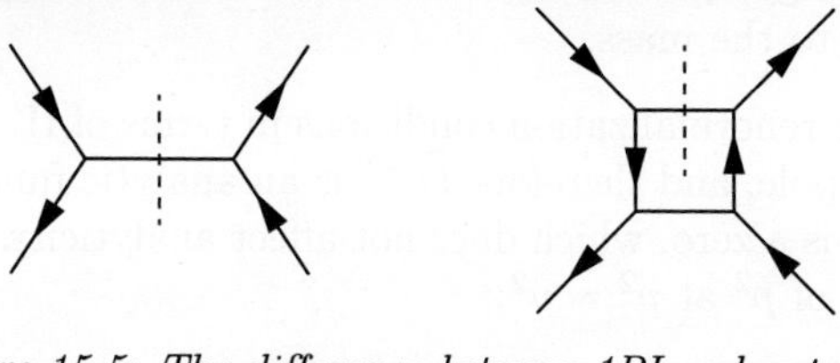

Figure 15.5: The difference between 1PI and not 1PI.

We can now define an object to express in simple terms our mass and wave function renormalization conditions. Looking at the 1PI 2 meson function, we define $i\widetilde{\Pi}'(p^2)$, a function of p^2 only, the sum over *all* 1PI diagrams:

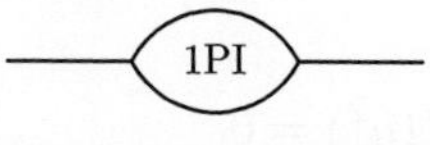

$$\equiv -i\widetilde{\Pi}'(p^2) \tag{15.33}$$

For reasons that will soon become clear, $\widetilde{\Pi}'(p^2)$ is called the **meson self-energy operator**.

Now let's look at the renormalized two-particle Green's function, $\widetilde{D}'$, and write it in terms of $\widetilde{\Pi}'$. This is a lovely process. We make drawings which are easy to manipulate, and then they turn into equations which are also easy to manipulate. What is the perturbation series for this object? Well, first we could have a single unadorned line, in zeroth order. Then we could have a one-particle irreducible diagram just sitting there, with the two external lines to give the propagators that we've left off by convention. After that, we could have a diagram that's actually one-particle reducible; that is to say, which I can cut someplace and make fall into two parts. If there's only one place where I can cut it, then on the left of the cut there must be something one-particle irreducible, and likewise on the right. Here I explicitly display the one line I can cut to make it fall into two parts; it's got to be cut somewhere between there as we go along. And then everything else by definition must be one-particle irreducible because there's only one place where I can cut it. If there are two places where I can cut it, ..., well, you see where we're going:

$$\widetilde{D}'(p^2) \equiv$$

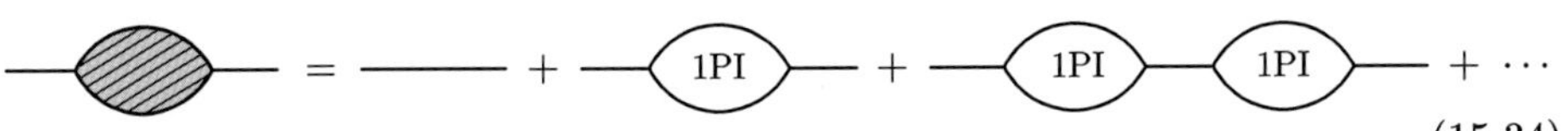

(15.34)

Now what does this say in equations? Factoring out the overall delta functions that occur everywhere on the left-hand side, and writing[5]

$$\widetilde{D}(p^2) = \frac{i}{p^2 - \mu^2 + i\epsilon} \tag{15.35}$$

as in (10.29), we have

$$\begin{aligned}\widetilde{D}'(p^2) &= \widetilde{D} + \widetilde{D}\left[-i\widetilde{\Pi}'(p^2)\right]\widetilde{D} + \widetilde{D}\left[-i\widetilde{\Pi}'(p^2)\right]\widetilde{D}\left[-i\widetilde{\Pi}'(p^2)\right]\widetilde{D} + \cdots \\ &= \frac{i}{p^2 - \mu^2 - \widetilde{\Pi}'(p^2) + i\epsilon}\end{aligned} \tag{15.36}$$

The geometric series sums up to a propagator with a mass term of $\mu^2 + \widetilde{\Pi}'(p^2)$. This is why $\widetilde{\Pi}'(p^2)$ is called the self-energy operator, or sometimes the self-energy function, or the self-mass function, because it adds to the mass.

We can now phrase our renormalization conditions in terms of $\widetilde{\Pi}'$. $\widetilde{D}'$ is an analytic function near $p^2 = \mu^2$ except for a pole, and therefore $\widetilde{D}'^{-1}$ is an analytic function near $p^2 = \mu^2$, period, since the inverse of a pole is a zero, which does not affect analyticity. Therefore $\widetilde{\Pi}'$ has a power series expansion in terms of p^2 at $p^2 = \mu^2$:

$$\widetilde{\Pi}'(p^2) = \widetilde{\Pi}'(\mu^2) + (p^2 - \mu^2)\left.\frac{d\widetilde{\Pi}'}{dp^2}\right|_{\mu^2} + \cdots \tag{15.37}$$

The value of $\widetilde{\Pi}'(\mu^2)$ must be zero. If it were not, from (15.36), we would not have a pole in $\widetilde{D}'(p^2)$ for $p^2 = \mu^2$. Thus we have

$$\widetilde{\Pi}'(\mu^2) = 0 \tag{15.38}$$

from the statement that the physical mass of the meson is μ^2. At the pole $p^2 = \mu^2$, the residue

[5] [Eds.] Reminder: Coleman used $\widetilde{D}(p^2)$ for what is usually denoted $\widetilde{\Delta}_F(p^2)$.

of $\widetilde{D}'$ must be i. Expanding the denominator,

$$p^2 - \mu^2 - \widetilde{\Pi}'(p^2) = p^2 - \mu^2 - \widetilde{\Pi}'(\mu^2) - (p^2 - \mu^2)\left.\frac{d\widetilde{\Pi}'}{dp^2}\right|_{\mu^2} + \cdots$$

$$= (p^2 - \mu^2)\left[1 - \left.\frac{d\widetilde{\Pi}'}{dp^2}\right|_{\mu^2}\right] + \cdots \tag{15.39}$$

because $\widetilde{\Pi}'(\mu^2) = 0$. Therefore the first derivative of $\widetilde{\Pi}'$ must be zero. If it were not zero, the residue would not be i, but instead i times the reciprocal of (1 plus the first derivative of $\widetilde{\Pi}'$). Consequently we must have

$$\left.\frac{d\widetilde{\Pi}'}{dp^2}\right|_{\mu^2} = 0 \tag{15.40}$$

These two statements about $\widetilde{\Pi}'$, that it and its derivative vanish at $p^2 = \mu^2$, are precisely equivalent to our two renormalization conditions, that $\widetilde{D}'$ has a pole at μ^2, and that its residue at this pole equals i.

Now the nice thing about these conditions is that they enable us to determine iteratively the mass and wave function renormalization constants in exactly the same way as we outlined for the A counterterm. Let's focus on the B and C counterterms. From (14.47),

$$\mathscr{L}_{CT} = \cdots + \tfrac{1}{2}B(\partial_\mu\phi')^2 - \tfrac{1}{2}C\phi'^2 + \cdots \tag{15.41}$$

These two terms lead together to an interaction which I'll indicate diagrammatically by a single cross:

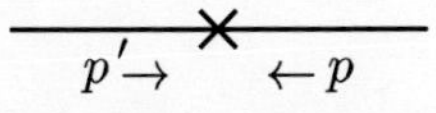

$$i(2\pi)^4\delta^{(4)}(p + p')\left[Bp^2 - C\right] \tag{15.42}$$

The interaction is determined in terms of both coefficients. There's a B part which, because of the derivative coupling, gives us iBp^2, and the C part which gives us $-iC$, as demonstrated at the end of last lecture. As before, we break this up into a power series in the coupling constant:

$$B = \sum_n B_n, \text{ and } B_n \propto g^n; \quad C = \sum_n C_n, \text{ and } C_n \propto g^n \tag{15.43}$$

In diagrams, we indicate each term of order g^n by (n):

 $= \sum_n$ 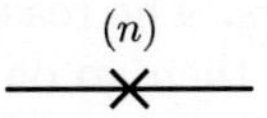 (15.44)

We assume that we know everything to order $n - 1$, and are about to compute things to order n. We have

$-i\widetilde{\Pi}'(p^2) =$  $=$ (known stuff, to $\mathcal{O}(g^{n-1})$) $+$ 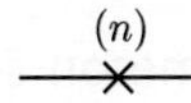 (15.45)

$$-i\widetilde{\Pi}'(p^2) = (\text{known stuff}) + i(B_n p^2 - C_n) \tag{15.46}$$

I assume we can compute the lower orders in perturbation theory (the "known stuff"); to determine B_n and C_n we impose the two constraints (15.38) and (15.40):

$$i(B_n\mu^2 - C_n) = -\left.(\text{known stuff})\right|_{\mu^2} \tag{15.47}$$

and

$$iB_n = -\frac{d}{dp^2}(\text{known stuff})\bigg|_{\mu^2} \tag{15.48}$$

All this goes through, *mutatis mutandis*, for the nucleon field, since our nucleon is not really that different from the meson, despite the name we've given it. It's just another scalar field. I won't bother to write down the whole spectral representation for the nucleon, but for the self-energy term we can write, analogous to the renormalized meson propagator,

$$\widetilde{S}'(\not{p}) \equiv \tag{15.49}$$

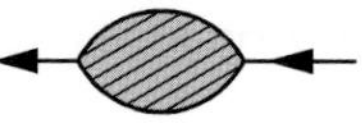

and the appropriate 1PI graph,

$$-i\widetilde{\Sigma}'(p^2) \equiv \tag{15.50}$$

The corresponding counterterms are, from (14.47),

$$\mathscr{L}_{CT} = \cdots + D\partial_\mu\psi'^*\partial^\mu\psi' - E\psi'^*\psi' + \cdots \tag{15.51}$$

The values of D and E, the counterterms associated with nucleon mass and wave function renormalization, respectively, are fixed by the two conditions

$$\widetilde{\Sigma}'(m^2) = 0 \tag{15.52a}$$

$$\frac{d\widetilde{\Sigma}'}{dp^2}\bigg|_{m^2} = 0 \tag{15.52b}$$

It's just the same thing written over again.

We found a very nice result when we were considering the A counterterm: we could simply ignore it. Indeed, we could ignore *all* graphs that contain tadpoles. Nothing nearly as nice happens here, unfortunately. We cannot ignore graphs that have these sorts of insertions on them if they occur in *internal* lines. But we *can* ignore these kinds of insertions if we are dealing with external lines *on the mass shell*—in particular if we are computing S-matrix elements. For an S-matrix element, with all external lines on the mass shell, we can ignore all corrections to external lines. The reason is that in getting an on-shell S-matrix element, we multiply by $(p^2 - \mu^2)$ and then go on to the mass shell, thus turning the external bare propagator into i. The result of all possible corrections to the external lines is just to turn the propagator $\widetilde{D}$ into $\widetilde{D}'$, which has a pole at the same place and a residue at the same place as the original propagator. So there's no need to bother with these corrections.

15.4 The meson self-energy to $\mathcal{O}(g^2)$

In principle, if I were to go on, I should now investigate the coupling constant renormalization, and therefore complete our program of writing down all the equations to determine all the renormalization constants iteratively. But just for variety, I'd like to do a simple computation, of the meson self-energy operator function $\widetilde{\Pi}'(p^2)$ to order g^2. This calculation doesn't require the coupling constant renormalization. Here you can see how the renormalizations work out. We'll also learn some little tricks about how to do the integrals which occur in renormalization.

Two graphs contribute to order g^2 to $\tilde{\Pi}'$:

$-i\tilde{\Pi}'(p^2) =$ 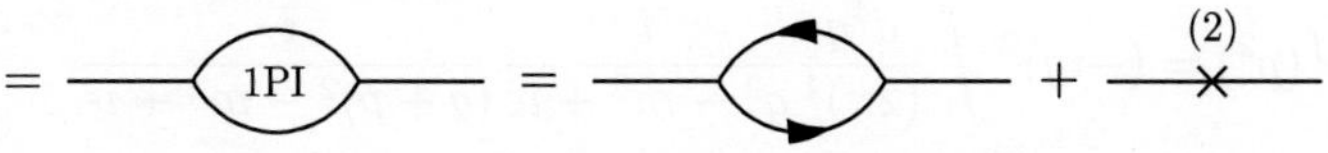

$+\ \mathcal{O}(g^3)$ (15.53)

One is the first Feynman graph containing a closed loop we are going to look at seriously, and the other is the g^2 contribution to the counterterms B and C, which we determine iteratively in terms of the other entities. We'll write these contributions as[6]

$$-i\tilde{\Pi}'(p^2) = -i\tilde{\Pi}^f(p^2) + i(B_2p^2 - C_2) \tag{15.54}$$

Now B_2 and C_2 are determined iteratively by the two conditions (15.38) and (15.40):

$$\tilde{\Pi}'(\mu^2) = 0 \ \Rightarrow\ C_2 = B_2\mu^2 - \tilde{\Pi}^f(\mu^2) \tag{15.55a}$$

$$\left.\frac{d\tilde{\Pi}'}{dp^2}\right|_{\mu^2} = 0 \ \Rightarrow\ B_2 = \left.\frac{d\tilde{\Pi}^f}{dp^2}\right|_{\mu^2} \tag{15.55b}$$

If I'm not interested in doing higher-order computations, I can eliminate the counterterms from (15.54) at once, and write

$$-i\tilde{\Pi}'(p^2) = -i\left[\tilde{\Pi}^f(p^2) - \tilde{\Pi}^f(\mu^2) - (p^2-\mu^2)\left.\frac{d\tilde{\Pi}^f}{dp^2}\right|_{\mu^2}\right] \tag{15.56}$$

That's obviously right. I've added a term proportional to p^2 and a constant term such that the total expression and its first derivative vanish at $p^2 = \mu^2$. If you want to compute the counterterms B_2 and C_2 to $\mathcal{O}(g^2)$, of course, you can do so just by comparing (15.54) to (15.56). But if I'm only interested in computing $\tilde{\Pi}'(p^2)$ to $\mathcal{O}(g^2)$, this expression (15.56) suffices.

Now let's do the computation. The important thing is to compute is $\tilde{\Pi}^f(p^2)$, the contribution from the closed nucleon loop. Then we'll plug it into (15.56) and get the real $\tilde{\Pi}'(p^2)$. Well, to do that, let's label our momenta:

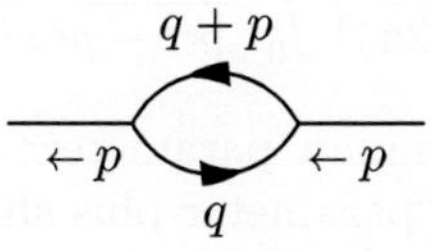

Figure 15.6: Nucleon loop $\mathcal{O}(g^2)$ contribution to $\tilde{\Pi}'(p^2)$

There is momentum p coming in. There's an unknown internal loop momentum which I'll call q. The momentum at the top is $q+p$, and the momentum going out is p. The internal momenta are oriented in the direction of the arrows (not that it matters, since all the propagators are even functions). The loop momentum q is not determined by energy-momentum conservation. It runs counter-clockwise around the loop, and we have to integrate over it.

---

[6] [Eds.] In the video of Chapter 23, Coleman says that the superscript f stands for for "Feynman", not "finite".

So we have

$$-i\widetilde{\Pi}^f(p^2) = (-ig)^2 \int \frac{d^4q}{(2\pi)^4} \frac{i}{q^2 - m^2 + i\epsilon} \frac{i}{(q+p)^2 - m^2 + i\epsilon} \tag{15.57}$$

There's a $(-ig)$ from each vertex, and two Feynman propagators, and we have to integrate over the unknown momentum. The propagator of the antinucleon line carries momentum q and the propagator of the nucleon line carries the momentum $q + p$. This is simply a straightforward application of the Feynman rules. As stated, this will be a function of p only. You may be getting a little antsy. Although many of this integral's properties are not obvious, one of them leaps out: it's divergent! (The denominator goes like q^4, at large q, while $d^4q \sim q^3 dq$, so the integral is logarithmically divergent.) For the time being we'll put on blinders. But we'll worry about that very soon.

The next stage is to manipulate this integral (15.57) using a famous formula due to Feynman:[7]

$$\int_0^1 dx \frac{1}{\left[ax + b(1-x)\right]^2} = \frac{1}{ab} \tag{15.58}$$

The variable x is called a **Feynman parameter**. This assumes that a and b are such that there is no pole in the domain of integration. The integral is simple to check, since the integrand is a rational function. We will apply this formula to our integral for $\widetilde{\Pi}^f(p^2)$ using the two Feynman denominators as b and a. Because of the $i\epsilon$'s they indeed satisfy the condition that the denominator never vanishes inside the domain of integration. Therefore I have

$$-i\widetilde{\Pi}^f(p^2) = g^2 \int \frac{d^4q}{(2\pi)^4} \int_0^1 dx \frac{1}{[q^2 + 2p\cdot qx + p^2x - m^2 + i\epsilon]^2} \tag{15.59}$$

To complete the square, let

$$k \equiv q + px \qquad \text{whence} \qquad k^2 = q^2 + 2p\cdot qx + p^2x^2 \tag{15.60}$$

Thus I can write the integral as

$$-i\widetilde{\Pi}^f(p^2) = g^2 \int \frac{d^4k}{(2\pi)^4} \int_0^1 \frac{dx}{[k^2 - p^2x^2 + p^2x - m^2 + i\epsilon]^2} \tag{15.61}$$

Rewriting integrals in terms of Feynman parameters and completing the square is often convenient. I call this rewriting the "parameter plus shift" trick. (In the next chapter I will show you how to do this when there are three or four or five lines running around the loop; then we'll need more than one Feynman parameter.) We will shortly transform this integral from Minkowski space into Euclidean space. That will make the integral awfully easy to do, because the Lorentz invariant integral will become rotationally invariant (in four Euclidean dimensions). You'll notice we've changed gears. From doing highbrow theory we're now grubbing around with integrals, but it's important you learn how to do both.

Now we have to face the fact that this integral is divergent. Actually we don't have to do that, because what we are really interested in is not $\widetilde{\Pi}^f$, but the whole thing in the square

[7] [Eds.] In a letter to Hans Bethe, Feynman called this identity "a swanky new scheme". Quoted in Schweber *QED*, p. 453.

brackets (15.56). If we just look at the first two terms, the difference $\widetilde{\Pi}^f(p^2) - \widetilde{\Pi}^f(\mu^2)$ goes like $1/q^6$ (inside the integrand) at high q, and therefore the integral converges. The last term, the derivative of $\widetilde{\Pi}^f$ with respect to p^2 is obviously convergent because the derivative drags down another power of q^2. Therefore $\widetilde{\Pi}'(p^2)$ is *finite.* What a surprise! And I mean it really *is* a surprise. We embarked upon this renormalization just to turn the wrong perturbation theory for the wrong quantities in terms of the wrong expansion parameter with the wrong masses held fixed, into the right perturbation theory for the right quantities with the right expansion parameter and the right masses held fixed, without ever bothering our little heads about the question of infinities. Renormalization turns out to reveal itself not as Clark Kent, but Superman, come to rescue us when we are confronted with this otherwise insuperable problem of divergences. We would have come to a screaming halt at this point if we had not renormalized our perturbation theory. As it turns out, however, this means that C_2 is in fact given by a divergent integral. Whether that's bad news or good is something we still have to worry about. But the quantity $\widetilde{\Pi}'(p^2)$, the only thing that's physically observable, is represented by a perfectly convergent integral.[8] Note that a second subtraction is *not* needed to render $\widetilde{\Pi}'(p^2)$ convergent in Model 3. Put another way, *only* the mass counterterm C is needed for the self-energy to be finite; in this theory, a finite $\widetilde{\Pi}'(p^2)$ does not require the wave-function counterterm B. We'll come back to this.

Will this continue to all orders in perturbation theory? Does this happen only in this theory, or in all theories? Well, those are interesting questions, but for the moment let us be thankful for what we have and continue with this computation. We will turn to those questions later.

15.5 A table of integrals for one loop

I will explain how one finishes doing the integral (15.61). This is one of a family of similar integrals that arise in one-loop integration. It is useful to have a table of such integrals. We'll derive this integral table now. That requires a little side story, and then we can assemble the whole thing and get the answer to the integral (15.61) for $\widetilde{\Pi}^f(p^2)$.

Let me suppose I have an integral of this form:

$$I_n = \int \frac{d^4q}{(2\pi)^4}\frac{1}{(q^2+a+i\epsilon)^n} \tag{15.62}$$

where a is some real number. We will normally consider the case $n \geq 3$, and n an integer, in which case the integral is convergent. However we frequently run across expressions with lesser values of n as parts of sums of terms such as here, such that the total thing is convergent, even though the individual terms are not. And therefore we should also provide in the integral tables values of this integral for $n = 1$ or 2, but those are to be taken *cum grano salis*, to be used only in convergent combinations.

To do this integral I am going to rotate the contour of the q_0 integration. First I'll write it out explicitly,

$$I_n = \int \frac{dq_0 d^3\mathbf{q}}{(2\pi)^4}\frac{1}{[q_0^2 - (|\mathbf{q}|^2 - a - i\epsilon)]^n} \tag{15.63}$$

[8] [Eds.] This calculation is duplicated, though with a somewhat different focus, in Lurié *P&F*, Section 6-4, pp. 266–274.

Let's consider where the singularities arise in the complex q_0 plane. We have two possibilities: $|\mathbf{q}|^2 - a$ can be greater than zero, or $|\mathbf{q}|^2 - a$ can be less than zero. It could also be equal to zero but that's trivial, as the two other cases go continuously into each other. In either case, the contour can be rotated as shown, so that it runs up the imaginary q_0 axis, because the rotation does not cross any poles. This is called a **Wick rotation.**[9] This rotation translates our integral from Minkowski space into Euclidean space.

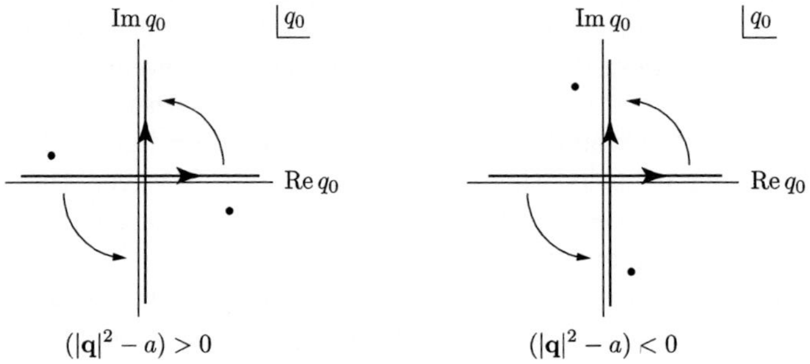

Figure 15.7: The Wick rotation for q_0

We define

$$q_0 = iq_4 \tag{15.64}$$

and therefore

$$q^2 = -q_E^2 = -(|\mathbf{q}|^2 + q_4^2) \tag{15.65}$$

$$d^4q = id^4q_E = idq_4d^3\mathbf{q} \tag{15.66}$$

Thus our integral becomes

$$I_n = i\int \frac{d^4q_E}{(2\pi)^4}\frac{1}{[-q_E^2 + a + i\epsilon]^n} \tag{15.67}$$

(I may still have to hold on to the $i\epsilon$ for $n = 2$.)

We now have a four-dimensional, spherically symmetric integral to do in Euclidean space. So we need another little piece of lore. Everyone knows how to do spherically symmetric integrals in ordinary three-dimensional space. How will we do spherically symmetric integrals in four-dimensional Euclidean space? Consider

$$\int d^4q_E f(q_E^2) \tag{15.68}$$

where $f(q_E^2)$ is any function of q_E^2. If I introduce a variable $z = q_E^2$, then I expect

$$\int d^4q_E f(q_E^2) = \alpha\int_0^\infty zdz\, f(z) \tag{15.69}$$

In other words, this integral should equal some constant α, arising from the angular integration, times the integral of zdz—that's from the r^3dr—times $f(z)$, integrated from zero to infinity.

[9] [Eds.] Gian-Carlo Wick, "Properties of Bethe-Salpeter wave functions", *Phys. Rev.* **95** (1954) 1124–1134.

But what is α? Since α is a universal constant, we could find its value by integrating any constant over spherical coordinates in four space, but that's a pain in the neck. We only have to evaluate this integral for a single function to find out what α is. I will look at the function $f = e^{-q_E^2}$ which is just the product of four Gaussians, one for each component:

$$\int d^4q_E f(q_E^2) = \int d^4q_E\, e^{-q_E^2} = \prod_{i=1}^{4} \int_{-\infty}^{\infty} dq_i\, e^{-q_i^2} = (\sqrt{\pi})^4 = \pi^2 \tag{15.70}$$

On the right-hand side I have

$$\alpha \int_0^\infty z dz\, e^{-z} = \alpha\Gamma(2) = \alpha \cdot 1! = \alpha \tag{15.71}$$

Therefore we have determined α without having to go to spherical coordinates in four-dimensional space, thank heavens: α is π^2. And we have the general rule:

$$\int d^4q_E\, f(q_E^2) = \pi^2 \int_0^\infty z dz f(z) \tag{15.72}$$

That's how you determine the volume of a sphere in 4-space without doing any work.[10]

Now we're in a position to derive the integral table, given below. I'll reserve for next time plugging the appropriate formula into the expression for $\widetilde{\Pi}^f(p^2)$ and then doing the integral. Actually we only need to do one integral from this table, the case $n = 1$. From (15.67),

$$I_1(a) = \int \frac{d^4q}{(2\pi)^4} \frac{1}{q^2 + a} = i \int \frac{d^4q_E}{(2\pi)^4} \frac{1}{-q_E^2 + a} \tag{15.73}$$

From that we can get all the others, by differentiating with respect to a. It will appear in a convergent combination, so we don't lose anything by truncating the integration at some high q^2, which I call Λ. I'll assume Λ is much greater than a. Then, using (15.72),

$$I_1(a) = -\frac{i}{16\pi^2} \int_0^\Lambda \frac{z\, dz}{z - a} \tag{15.74}$$

The integral is pretty simple because the numerator can be written as $z - a + a$. So I get

$$\int_0^\Lambda \frac{z\, dz}{z - a} = \Lambda + a \ln \frac{\Lambda - a}{(-a)} \tag{15.75}$$

which can be approximated as

$$\Lambda + a \ln \Lambda - a \ln(-a) + \mathcal{O}(a^2/\Lambda) \tag{15.76}$$

---

[10] [Eds.] The volume of an n-dimensional sphere of radius R is

$$\int_{\|\mathbf{q}\| \le R} d^n q = \frac{\pi^{n/2}}{\Gamma(\frac{n}{2} + 1)} R^n = \begin{cases} \dfrac{\pi^m}{m!} R^{2m}, & \text{for } n = 2m \\[2ex] \dfrac{\pi^m m!}{(2m+1)!} (2R)^{2m+1}, & \text{for } n = 2m + 1 \end{cases}$$

For $n = 4$, the volume is $\frac{1}{2}\pi^2 R^4$. The surface areas are obtained by differentiation with respect to R, e.g., for $n = 4$, the surface area is $2\pi^2 R^3$. See D. M. Y. Sommerville, *An Introduction to the Geometry of N Dimensions*, Dover Publications, 1958, pp. 135–6.

I can neglect the last term because Λ is supposed to be much larger than a.

Integral table for Feynman parametrized integrals

The Minkowski space integral,

$$I_n(a) = \int \frac{d^4q}{(2\pi)^4} \frac{1}{(q^2+a)^n} \tag{I.1}$$

with n integer and $\mathrm{Im}\, a > 0$, is given by

$$I_n(a) = i\left[16\pi^2(n-1)(n-2)a^{n-2}\right]^{-1} \tag{I.2}$$

for $n \geq 3$. For $n = 1, 2$,

$$I_1 = \frac{i}{16\pi^2}\, a \ln(-a) + \cdots \tag{I.3}$$

and

$$I_2 = -\frac{i}{16\pi^2} \ln(-a) + \cdots \tag{I.4}$$

where the dots indicate divergent terms that cancel when two such terms are subtracted, provided the total integrand vanishes for high q faster than q^{-4}.

Now if this is part of a convergent combination of terms that in fact do not depend on Λ, so that the integral doesn't depend on Λ, that means all terms and the individual integrands that depend on Λ must vanish in such a combination. That's what "convergent combination" means. So the two terms with explicit factors of Λ vanish in convergent combinations. The same is true however many such terms there are. If you now look at the entry in the integral table for I_1, you will see, with the appropriate insertions of i's and Euclidean rotations, π^2 from α and $(2\pi)^4$ from the denominator of I_1, what we have derived is just the I_1 entry. You can get I_2, I_3, ... by differentiating with respect to a. I leave that to you as an exercise. You are now in a position to derive the integral table for yourself in exactly the same way I did.[11]

Next time we will apply the integral table to complete our computation of $\widetilde{\Pi}'$ to second order. We will discuss coupling constant renormalization, talk about the marvelous properties of realistic pion–nucleon scattering and nucleon–nucleon scattering, and have a little more to say about renormalization in general.

[11] [Eds.] See also Appendix A.4, pp. 806–808 in Peskin & Schroeder *QFT*. Copies of this integral table were handed out in class over the years; handwritten at first but later typed.

16

Renormalization II. Generalization and extension

This lecture will be something of a smorgasbord, with a lot of little topics. First, I would like to complete the computation of the meson self-energy $\widetilde{\Pi}'(p^2)$ to $\mathcal{O}(g^2)$. We will check the calculation by looking at the analyticity properties of our result, and comparing them with what we would expect on general grounds. Next I will explain how you tackle graphs either with more lines on a single loop, or with more than one loop. I will show how we can perform systematically the Feynman trick for the associated integrals, and reduce everything to an integral over parameters. Then I will return to our renormalization program to consider coupling constant renormalization, the one renormalization we have not yet discussed in detail. Finally I will make a few not very deep remarks about whether renormalization gets rid of infinities for every theory, or only for certain special theories.

16.1 The meson self-energy to $\mathcal{O}(g^2)$, completed

The first topic I will just begin *in media res.* Using our integral table for I_2 on the parametric integral we had, (15.61),

$$-i\widetilde{\Pi}^f(p^2) = g^2 \int \frac{d^4k'}{(2\pi)^4} \int_0^1 \frac{dx}{[k'^2 - p^2x^2 + p^2x - m^2 + i\epsilon]^2} \tag{16.1}$$

we obtain

$$\widetilde{\Pi}^f(p^2) = \frac{g^2}{16\pi^2} \int_0^1 dx \ln(m^2 - p^2x(1-x) - i\epsilon) + \cdots \tag{16.2}$$

The dots represent irrelevant terms from the integral table, terms that vanish in a convergent combination, which we have. Recalling (15.56) we have, ignoring the irrelevant terms,

$$\widetilde{\Pi}'(p^2) = \widetilde{\Pi}^f(p^2) - \widetilde{\Pi}^f(\mu^2) - (p^2 - \mu^2)\left.\frac{d\widetilde{\Pi}^f}{dp^2}\right|_{\mu^2} \tag{16.3}$$

We plug (16.2) into this and get

$$\widetilde{\Pi}'(p^2) = \frac{g^2}{16\pi^2} \int_0^1 dx \left\{ \ln\Big(\frac{m^2 - p^2x(1-x) - i\epsilon}{m^2 - \mu^2x(1-x) - i\epsilon}\Big) + (p^2 - \mu^2)\frac{x(1-x)}{m^2 - \mu^2x(1-x) - i\epsilon} \right\} \tag{16.4}$$

We need not retain the $i\epsilon$ in $\widetilde{\Pi}^f(\mu^2)$ or its derivative, i.e., in the denominators of (16.4), and we shouldn't: $\widetilde{\Pi}^f(\mu^2)$ had better be a real number, otherwise something's gone drastically wrong with our computation. The terms B and C have to be real, because the Lagrangian is Hermitian. Indeed, since the maximum of $x(1-x)$ is $\frac{1}{4}$, and we assume as always that $(\mu < 2m)$ (otherwise the muon would be unstable, decaying into a nucleon–antinucleon pair), $m^2 - \mu^2 x(1-x)$ is positive definite, and the $-i\epsilon$ is never needed to avoid the singularity. So

$$\widetilde{\Pi}'(p^2) = \frac{g^2}{16\pi^2}\int_0^1 dx\left\{\ln\left(\frac{m^2 - p^2x(1-x) - i\epsilon}{m^2 - \mu^2 x(1-x)}\right) + (p^2-\mu^2)\frac{x(1-x)}{m^2-\mu^2x(1-x)}\right\} \tag{16.5}$$

This ugly expression is our final result for $\widetilde{\Pi}'(p^2)$ to $\mathcal{O}(g^2)$. Things don't get any prettier if you integrate. The x-integral is in fact elementary and can be found in a table of integrals.[1] I believe it gives you an inverse tangent, but don't take my word for it. I leave it for interested parties to carry out the integration.

As a consistency check, I would like to investigate the analytic properties of this integral. We want to be sure that the remaining $i\epsilon$ is unnecessary for $p^2 < 4m^2$, and that there is a cut at $4m^2$. After all, $\widetilde{\Pi}'(p^2)$ is linearly related to the inverse of $\widetilde{D}'(p^2)$ (see (15.36)), and therefore it should have the same analytic properties as $\widetilde{D}'(p^2)$, except of course $\widetilde{\Pi}'(p^2)$ doesn't have a pole at μ^2, where $\widetilde{D}'(p^2)$ has a pole: it has a zero. Therefore $\widetilde{\Pi}'(p^2)$ should be analytic, except for a cut along the positive real axis. I claim that in this order of perturbation theory, the cut begins at $4m^2$ (corresponding to a virtual nucleon–antinucleon pair), and not (as you might suppose) at $4\mu^2$ (corresponding to a pair of virtual mesons).

The argument that the cut begins at $4m^2$ just requires looking at a Feynman graph. From (15.31) the cut is associated with the function $\sigma(p^2)$ (see Figure 15.3), the amplitude for the meson field to make a state when applied to the vacuum. If we consider

$$\langle n|\phi'(0)|0\rangle \tag{16.6}$$

the graph for that will consist of, to first order in g, simply Figure 16.1 (a):

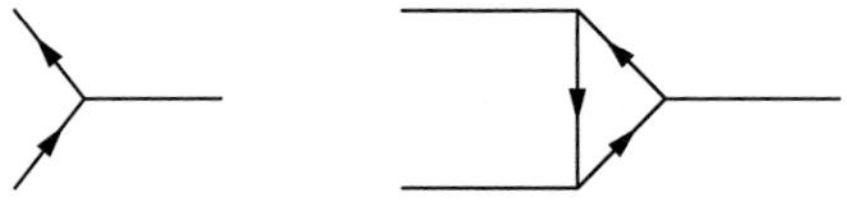

(a) (b)

Figure 16.1: $\mathcal{O}(g)$ and $\mathcal{O}(g^3)$ Feynman graphs for $\langle n|\phi'(0)|0\rangle$

The field ϕ' applied to the vacuum can make a nucleon–antinucleon state, and that's the only order g^2 contribution to the spectral representation, because the contribution from this graph gets squared to make the nucleon loop (Figure 15.6). The field ϕ' doesn't make a meson pair until order g^3, as in Figure 16.1 (b), so we won't get contributions from two-meson

---

[1] [Eds.] From *Mathematica*,

$$\int_0^1 dx\,\ln(m^2 - p^2x(1-x) - i\epsilon) = -2 + \ln(m^2 - i\epsilon) + \frac{2\sqrt{4m^2-p^2-4i\epsilon}}{p}\tan^{-1}\Big(\frac{p}{\sqrt{4m^2-p^2-4i\epsilon}}\Big)$$

if Re $\sqrt{4m^2-p^2-4i\epsilon} \neq 0$ and $\sqrt{p^2-4m^2-4i\epsilon} \notin$ Reals.

intermediate states in the spectral representation until we reach order g^6. We won't see them in $\mathcal{O}(g^2)$. Thus we expect $\Pi'(p^2)$ to be an analytic function of p^2 aside from a cut beginning at $4m^2$, as asserted.

Now let's work out the analytic properties of $\widetilde{\Pi}'(p^2)$. It's an analytic function except for the branch cut introduced by the cut in the logarithm; see Figure 16.2. This branch cut survives when we do the x integral.

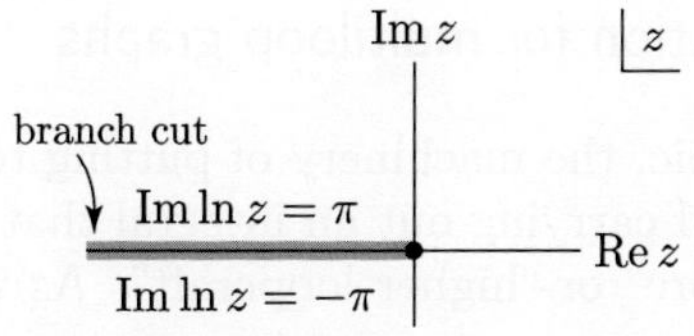

Figure 16.2: Branch cut for $\ln z$

The only part of the integral that we've got to study is the numerator of the logarithm, ignoring the $i\epsilon$. Troubles arise if the function is evaluated at the cut,

$$m^2 - p^2x(1-x) \le 0 \tag{16.7}$$

If this numerator is negative or zero for x in the range of integration, $0 \le x \le 1$, then we're on the branch line of the logarithm. When will the numerator be non-positive?

For $0 < x < 1$, $x(1-x)$ is always positive, so if the imaginary part of p^2 is not equal to zero, then the numerator has a non-zero imaginary part. At the boundary, when $x = 0$ or $x = 1$, the numerator is equal to m^2, and that's not a negative number. So there's no singularity if Im $p^2 \neq 0$. If p^2 lies along the negative real axis ($p^2 \le 0$), again there's no problem, because the numerator is positive. So the only case we have to worry about is p^2 real, and greater than 0.

Let's graph the numerator. It is of course an upward pointing parabola; the coefficient of x^2 is positive.

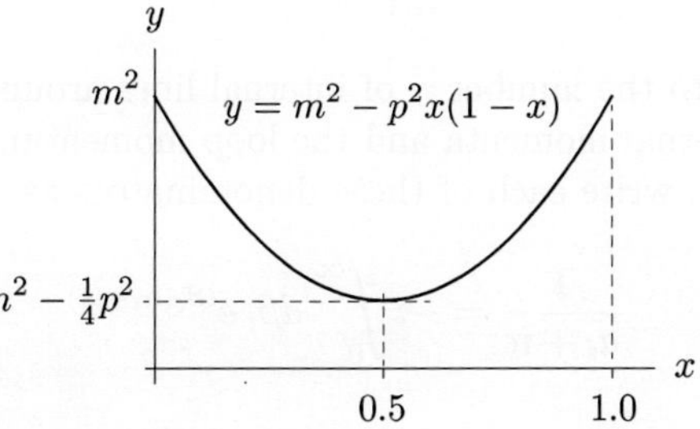

Figure 16.3: Graph of $m^2 - p^2x(1-x)$ *for* $0 \le x \le 1$

At $x = 0$ and $x = 1$, the numerator is equal to m^2. The numerator reaches its minimum value at $x = \frac{1}{2}$. To check that it is positive throughout the domain of integration, we need only check its value at $x = \frac{1}{2}$:

$$m^2 - p^2x(1-x)\Big|_{x=\frac{1}{2}} = m^2 - \tfrac{1}{4}p^2 \overset{?}{>} 0 \tag{16.8}$$

If $p^2 < 4m^2$, we are away from the cut, because then the argument of the logarithm is always positive, and we can drop the $i\epsilon$ in the numerator. On the other hand, if $p^2 \geq 4m^2$, there's a cut, because then the argument becomes non-positive, and we have to keep the $i\epsilon$ in our prescription. In this case it matters whether we've approached the real p^2 axis from above or below. We should shout in triumph, because these are exactly the analyticity properties we had anticipated on general grounds.

16.2 Feynman parametrization for multiloop graphs

We now turn to the second topic, the machinery of putting together many denominators to generalize Feynman's trick, and carrying out an integral that may have more than one loop in it. We can call this "loop lore" or "higher loopcraft". As we will see, there is essentially nothing new.

The first thing is to put together many denominators, all the denominators that run around a loop. We know how to do the parametrization when there are only two denominators. But of course there may be many of them, more than two. Even with only a single loop, we may have more than two propagators. For example this graph, Figure 16.4, which we've discussed before (see p. 267), has four propagators:

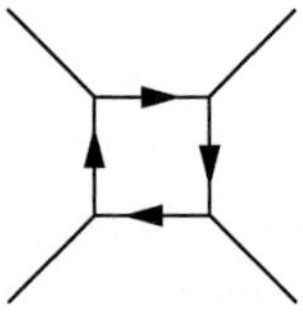

Figure 16.4: A single loop graph with four propagators

At the moment, we do not know how to do the integral associated with this graph.

Consider a product of n Feynman denominators:

$$\prod_{i=1}^{n} \frac{1}{a_i + i\epsilon} \tag{16.9}$$

The number i goes from 1 to the number n of internal lines around our loop. Each a_i is some function of the various external momenta and the loop momentum. I will derive a parametric expression for (16.9). First, write each of these denominators as

$$\frac{1}{a_i + i\epsilon} = -i \int_0^\infty d\beta_i \, e^{i\beta_i(a_i + i\epsilon)} \tag{16.10}$$

so that

$$\prod_{i=1}^{n} \frac{1}{a_i + i\epsilon} = (-i)^n \prod_{i=1}^{n} \int_0^\infty d\beta_i \, e^{i\beta_i(a_i + i\epsilon)} \tag{16.11}$$

I will multiply (16.11) by an integral B that is equal to 1, by the rules for integrating a delta function:

$$B = \int_0^\infty \frac{d\lambda}{\lambda} \delta\left(1 - \frac{\beta}{\lambda}\right) = \int_0^\infty \frac{d\lambda}{\lambda} \left|\frac{\beta}{\lambda^2}\right|^{-1} \delta(\lambda - \beta) = \int_0^\infty d\lambda \, \delta(\lambda - \beta) = 1 \tag{16.12}$$

where β is a positive constant. We choose $\beta = \sum_{i=1}^{n} \beta_i$, and rewrite (16.11) as follows:

$$\prod_{i=1}^{n} \frac{1}{a_i + i\epsilon} = (-i)^n \prod_{i=1}^{n} \int_0^\infty d\beta_i\, e^{i\beta_i(a_i+i\epsilon)} \int_0^\infty \frac{d\lambda}{\lambda}\, \delta\left(1 - \frac{\sum \beta_i}{\lambda}\right) \tag{16.13}$$

Changing the integration variables from β_i to $\alpha_i \equiv \beta_i/\lambda$, the right side becomes

$$(-i)^n \int_0^\infty d\lambda\, \lambda^{n-1} \int_0^1 d\alpha_1 \cdots d\alpha_n\, \delta(1 - \textstyle\sum \alpha_i)\, e^{i\lambda \sum \alpha_i(a_i+i\epsilon)} \tag{16.14}$$

(Because of the delta function, there is no contribution if any of the α_i's are greater than 1, so we might as well lower the upper limits of integration.) The λ integral is elementary:

$$\int_0^\infty d\lambda\, \lambda^{n-1} e^{i\lambda q} = i^n \frac{\Gamma(n)}{q^n} \tag{16.15}$$

and we conclude that

$$\boxed{\prod_{i=1}^{n} \frac{1}{a_i + i\epsilon} = \int_0^1 d\alpha_1 \cdots d\alpha_n \frac{(n-1)!}{\left[\sum \alpha_i(a_i + i\epsilon)\right]^n}\, \delta(1 - \textstyle\sum \alpha_i)} \tag{16.16}$$

This formula tells us how to write a product of Feynman denominators as one big super-Feynman denominator with parameters, raised to a power. The α's are *Feynman parameters*. They are the generalizations of the variable x in our previous formula, (15.58). The right-hand side of (16.16) looks like an integral over n parameters if there are n denominators, but of course the delta function makes one of the integrals trivial. In the case $n = 2$, you can let $\alpha_2 = y$ and $\alpha_1 = x$. Once you use the delta function to perform the y-integration, y becomes $1 - x$, and you obtain the earlier formula.

So (16.16) is the generalization to more lines. Please notice it is not clear *a priori* that the parametrization is always a good thing to do. It means that any one graph which starts out as an integral d^4k can be reduced to an integral essentially over $n - 1$ parameters, where n is the number of lines in the loop. This is obviously a good thing to do if there are four or fewer lines in the loop, as $4 - 1 < 4$. It is not obvious that parametrization is a good thing to do if there are five lines or more.

With the aid of this formula, and the integral table on p. 330, you can reduce any graph with only one loop to an integral over Feynman parameters. If we could do the remaining α_i integrals, we would be very happy people. But unfortunately it doesn't turn out that way. These are usually messy integrals that cannot be done in terms of elementary functions, except in simple cases. And that is why people who calculate things like the sixth order correction to the anomalous magnetic moment of the electron spend a lot of time programming computers.

This parameter technique can be generalized to graphs with more than one loop. We've seen how more lines are incorporated; I will now discuss more loops. That will complete the lore of doing Feynman integrals, at least for theories that involve only non-derivative interactions of scalar particles. For more complicated theories, it's pretty much the same, except that those integrals have factors in the numerator as well as in the denominator when all the dust settles.

As an example, suppose I take ϕ^4 theory:

$$\mathscr{L} = \mathscr{L}_0 - \frac{\lambda}{4!}\phi^4 \tag{16.17}$$

In this case the lowest order nontrivial contribution to the meson self-energy would involve a graph that looks like this:

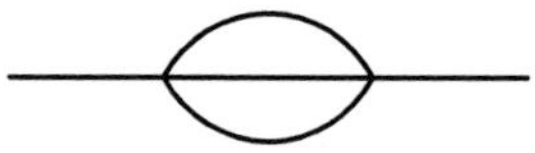

Figure 16.5: The lowest order graph for the meson self-energy in ϕ^4 theory

If we call the external momentum p, we have two possible momenta, k_1 say, running around the top loop and k_2 running around the bottom loop. The momentum on the lowest arc is then $p - k_1 - k_2$, all oriented from right to left:

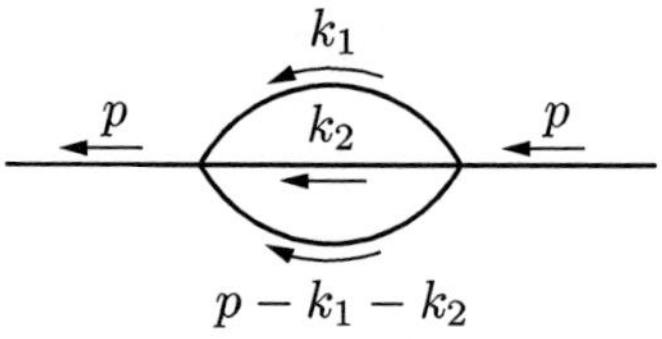

Figure 16.6: Momentum flow for a meson self-energy graph in ϕ^4 theory

Aside from the combinatorial factors (which are a great pain in the neck for ϕ^4 theories, since you have four identical fields at each vertex), and the constant numerical factors—the g's, the $(2\pi)^4$'s, and the i's—this graph is associated with an integral of the general form

$$I = \int d^4k_1\, d^4k_2\, \frac{1}{(k_1^2 - \mu^2)(k_2^2 - \mu^2)((p - k_1 - k_2)^2 - \mu^2)} \tag{16.18}$$

I've suppressed the $i\epsilon$'s.

Now let's consider the general case where we have ℓ loops, and a k_i for each loop: i in this case goes from 1 to ℓ. We also have a bunch of external momenta, p_j, how many there are depends on how many external lines there are; and we have in general, n internal lines. I will sketch how to do such an integral, using nothing but our integral table and the Feynman formula (16.16) to reduce it to an integral over Feynman parameters.

The first part of the trick is to use the Feynman formula to reduce all the internal lines simultaneously to one big denominator, as I've just sketched out. Thus we arrive at an integral of the following form (again I will suppress numerical factors, including the $(n-1)!$):

$$I = \int d^4k_1 \cdots d^4k_\ell \int_0^1 d\alpha_1\, d\alpha_2 \cdots d\alpha_n\, \delta(1 - \textstyle\sum \alpha)\, \frac{1}{D^n} \tag{16.19}$$

D is going to be some quadratic function of the k's, obtained by combining all the denominators. Every internal momentum is of course a linear function of the loop momenta, the k's, and the external momenta, the p's. In our example, the internal momenta are k_1, k_2, and $p - k_1 - k_2$. So D will be of the following form:

$$D = \sum_{i,j=1}^{\ell} A_{ij}\, k_i \cdot k_j + \sum_{i=1}^{\ell} B_i \cdot k_i + C \tag{16.20}$$

A_{ij} is a symmetric $\ell \times \ell$ matrix, linearly dependent on the Feynman parameters α_i, and independent of the external momenta p_j. If all $\alpha_i > 0$ (as is the case, within the region of integration), it can be shown that A_{ij} is invertible.[2] The B_i are a set of ℓ 4-vectors, linear in the α's and the external momenta p_i. In our example, one of the $B_i \cdot k_i$ might be $\alpha_3 p \cdot k_1$. C is a scalar depending linearly on the α's as well as on the squares of the p_i's and the squares of the masses appearing in the propagators. This is inevitably the general form that D will take.[3]

Further simplification can be made, because A_{ij} is invertible. We can perform a shift on the loop momenta, to remove the linear terms (involving the vectors B). That is to say, we can define

$$k'_i = k_i + \tfrac{1}{2}\sum_j A^{-1}_{ij}\,B_j \tag{16.21}$$

Substituting in for k_i turns the denominator into

$$D = \sum_{i,j}^{\ell} A_{ij}k'_i \cdot k'_j + C' \tag{16.22}$$

where the new scalar C' is given by

$$C' = C - \tfrac{1}{4}\sum_{i,j}^{\ell} B_i A^{-1}_{ij} B_j \tag{16.23}$$

We're just doing in general what we did for the one-loop integral. We've eliminated all the terms linear in the loop momentum in the denominator by defining new integration variables which are shifted versions of our old ones.

We can make the integral simpler yet by using the fact that A_{ij} is a symmetric matrix, and therefore we can diagonalize it. We can introduce new integration variables k''_i, linear combinations of the k''s corresponding to the eigenvalues a_i of A_{ij}, and thereby make A_{ij} diagonal. Since A_{ij} is a symmetric matrix, the transformation $k'_i \to k''_i$ is an orthogonal

---

[2] [Eds.] See Noboru Nakanishi, *Graph Theory and Feynman Integrals*, Gordon and Breach, 1971, theorem 7-2, p. 58.

[3] [Eds.] In response to a question, Coleman adds that the energy-momentum conserving delta functions at the vertices have been left out. In the example, the original graph had momenta as shown below.

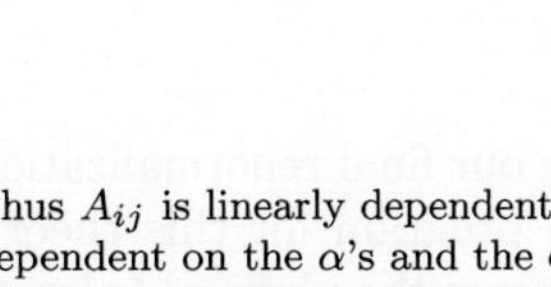

The two delta functions, $\delta^{(4)}(p - k_1 - k_2 - k_3)$ and $\delta^{(4)}(k_1 + k_2 + k_3 - p')$, at the vertices produce an overall energy-momentum conserving delta function $\delta^{(4)}(p' - p)$, and allow a trivial integration over one of the three loop momenta, k_3. Then D is, from multiplying out (16.18),

$$D = (\alpha_1 + \alpha_3)k_1^2 + (\alpha_2 + \alpha_3)k_2^2 + 2\alpha_3 k_1 \cdot k_2 - 2\alpha_3 p \cdot (k_1 + k_2) + \alpha_3 p^2 - (\alpha_1 + \alpha_2 + \alpha_3)\mu^2$$

Comparing with (16.20), one identifies

$$A_{11} = \alpha_1 + \alpha_3; \qquad A_{22} = \alpha_2 + \alpha_3; \qquad A_{12} = A_{21} = \alpha_3;$$
$$B_1 = B_2 = -2\alpha_3 p;$$
$$C = \alpha_3 p^2 - (\alpha_1 + \alpha_2 + \alpha_3)\mu^2$$

Thus A_{ij} is linearly dependent on the α's and independent of the external momenta, B_i are 4-vectors linearly dependent on the α's and the external momenta, and C is a scalar linearly dependent on the α's, the squares of the external momenta and the squares of the masses in the propagators, exactly as described. Note also that, as expected, $\det A \neq 0$.

transformation with determinant 1, and hence Jacobian equal to 1. The integral becomes

$$I = \int d^4k_1'' \cdots d^4k_\ell'' \int_0^1 d\alpha_1 \cdots d\alpha_n \delta(1 - \textstyle\sum_{i=1}^n \alpha_i) \frac{1}{(\sum_{i=1}^\ell a_i (k_i'')^2 + C')^n} \tag{16.24}$$

C', independent of the internal momenta k_i', is not changed by this transformation.

We can make one last transformation for one last simplification:

$$k_i''' = \sqrt{a_i} k_i'' \tag{16.25}$$

The a_i's are of course positive within the domain of integration, though there may be some places where they vanish at the boundaries of the integration. Thus we find our integral becomes

$$I = \int d^4k_1''' \cdots d^4k_\ell''' \int_0^\infty d\alpha_1 \cdots d\alpha_n \, \delta(1 - \textstyle\sum_i \alpha_i) \frac{1}{(\sum_{i=1}^\ell (k_i''')^2 + C')^n} \prod_{i=1}^\ell \frac{1}{(\sqrt{a_i})^4} \tag{16.26}$$

We see now that we didn't have to worry about analyzing the matrix A, because the product of the eigenvalues is just the determinant of the matrix. That is, the integral becomes, finally,

$$I = \int d^4k_1''' \cdots d^4k_\ell''' \int_0^1 d\alpha_1 \cdots d\alpha_n \frac{1}{(\det A)^2} \delta(1 - \textstyle\sum_{i=1}^n \alpha_i) \frac{1}{(\sum_{i=1}^\ell (k_i''')^2 + C')^n} \tag{16.27}$$

So you don't actually have to go through the diagonalization, you just have to be able to compute the determinant. You knew the value of C' before you ever diagonalized the matrix.

We now have the situation in the shape where we can systematically do all the k''' integrals, one right after another, just using our integral table. Whenever we do one of them, we'll knock n down by two (because $k^2 \to z$ after Wick rotation; see the table), and pick up a horrendous numerical factor, and we'll just keep on going until we do them all. By this algorithm, we can systematically reduce any integral arising from any Feynman graph, providing always, of course, it is a convergent graph, or arises in a convergent combination of graphs, in an integration over Feynman parameters equal in number to the number of internal lines. Thus for example for the graph I sketched out, the integral would be eight-dimensional in the first instance, over $d^4k_1\, d^4k_2$. Our prescription reduces it to a three-dimensional integral over three Feynman parameters, and one of those is trivial because of the delta function. It's not the world's most exciting subject, but if you are ever confronted with the problem of computing a multiloop graph like Figure 16.5, you will be happy that I have shown you this algorithm. I've arranged it so there are no numerical factors to memorize, just a procedure to understand, which you can work out afresh for every particular instance. In principle you can reduce any Feynman graph to an integral over Feynman parameters. At that point, typically, you are stuck, but you can always work it out numerically with a computer.

16.3 Coupling constant renormalization

I would now like to discuss briefly the condition that will determine our final renormalization constant. Remember we were going through the renormalization program for this theory, Model 3, and we had left one thing to fix: the condition that determines the physical value of g, a matter to be decided (on the basis of appropriate experiments) by a IUPAP committee (see p. 301), and which would eventually set the value of our last renormalization constant, F (p. 302):

$$\mathscr{L} = \cdots + F\psi^*\psi\phi \tag{16.28}$$

I will first state the definition, then show you how it works in fixing F iteratively, and finally explain how it can be connected, through a physically realizable experiment, to what looks at first glance like a totally unphysical object. To determine A, we studied the one-point Green's function. To determine B, C, D and E we studied a two-point Green's function. To study F, we have to study a three-point Green's function, with one ψ, one ψ^*, and one ϕ.

Define the object $-i\widetilde{\Gamma}'(p^2, p'^2, q^2)$ as this one-particle irreducible (1PI) graph,

$$-i\widetilde{\Gamma}'(p^2, p'^2, q^2) = \tag{16.29}$$

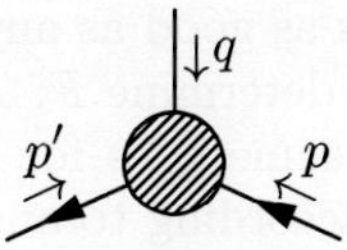

It is of course a Lorentz invariant function. (The $-i$ is included so that $\widetilde{\Gamma}' = g$ to lowest order.) Since the three momenta are arranged so that $p + p' + q = 0$, $-i\widetilde{\Gamma}'(p^2, p'^2, q^2)$ is a function really of only two independent vectors which we can take to be p and p', and therefore a function of three inner products: p^2, p'^2 and $p \cdot p'$. Actually, it will be more convenient for us to write $\widetilde{\Gamma}'$ as a function of p^2, p'^2 and q^2 (q^2 is linearly related to $p \cdot p'$ and the other two).

Up to third order in perturbation theory it's easy to see that there are only a very few graphs that contribute to this thing:

$$-i\widetilde{\Gamma}'(p^2, p'^2, q^2) = \quad + \quad + \quad (\text{to } \mathcal{O}(g^3)) \tag{16.30}$$

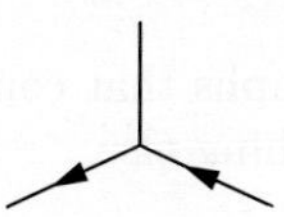

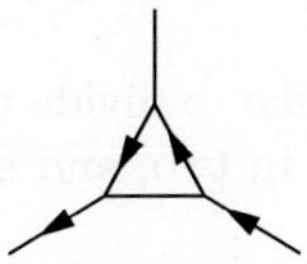

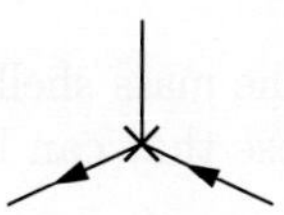

There is the first-order graph, with a contribution $-ig(2\pi)^4\delta^{(4)}(p + p' + q)$. Then there is a genuine monster of a third-order graph. And finally there may be a counterterm,

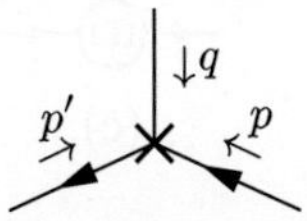

$$= -iF(2\pi)^4\delta^{(4)}(p + p' + q) \tag{16.31}$$

evaluated only to third order in perturbation theory. I've assigned the monster middle graph as a homework problem (Problem 9.2), to check that you understand the algorithms for doing integrals for loop graphs like these.

To define the renormalized coupling constant, I impose this condition:

$$\widetilde{\Gamma}'(m^2, m^2, \mu^2) \equiv g \tag{16.32}$$

This is the definition of the physical g: We set $\widetilde{\Gamma}' = g$ at the one point where all three lines are on the mass shell:

$$p^2 = p'^2 = m^2; \qquad q^2 = \mu^2 \tag{16.33}$$

To find a trio of 4-vectors satisfying these conditions, as well as the conservation of momentum $p + p' + q = 0$, some of the components have to be complex. This point cannot be attained by any physical scattering processes, as the meson is stable. It can be shown, however, that the domain of analyticity of $\widetilde{\Gamma}'$, considered as a function of three complex variables, is sufficiently large to define the analytic continuation of $\widetilde{\Gamma}'$ from any of its physically accessible regions to this point, (16.33), and $\widetilde{\Gamma}'$ is real there. (The homework problem asks you to check the reality of $\widetilde{\Gamma}'$ to third order.) The choice (16.32) is totally arbitrary, but it is the one we make for reasons I will explain shortly. This condition determines F iteratively, order by order in perturbation theory, in exactly the same way as the other counterterms. For example, because of (16.32), the sum of the last two graphs in (16.30) must cancel at $p^2 = p'^2 = m^2$, which determines F_3, the coupling constant counterterm F to third order. This completes our specification of renormalization conditions.

In principle, since the definition of the coupling constant is completely arbitrary, anything that gives g to lowest order is as good as anything else. That's the one condition I want to maintain, so I can iteratively determine F. Aside from that, any value of p^2 would do—m^2, $(m^2/\mu)^2$, whatever—and the same goes for p'^2 and q^2. At this level it's just a matter of reparametrizing the theory, according to another IUPAP committee, defining the coupling constant differently. Still, it is worth devoting a few minutes to explain why this particular definition (16.32) is useful, and is therefore used by many workers in the field, not for *this* theory, which is only a model, but for the corresponding real one, and other theories. The point is this. I'll show that the square of $\widetilde{\Gamma}'$ is a physically observable quantity if you do the right experiment. That's all we can hope for, because we can arbitrarily change the sign of g just by changing the sign of ϕ.

Consider the process of meson–nucleon scattering:

$$\phi + N \to \phi + N \tag{16.34}$$

with everything on the mass shell. I'd like to divide the graphs that contribute to this process into two classes: those that can be cut in two, and everything else.

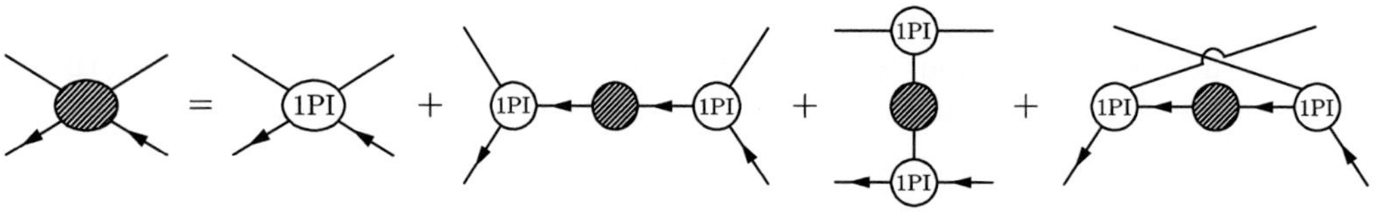

$\phi + N \to \phi + N$ (a) (b) (c) (d)

(16.35)

The unshaded blobs are one-particle irreducible graphs. The parts that can be cut in two look broadly like s, t and u–channel graphs, denoted (b), (c) and (d), respectively. Recalling that s is the center-of-mass energy for the meson–nucleon system, all of the graphs that can be cut in two by dividing a nucleon propagator are in (b). This graph on mass shell has the form

$$\text{(b)} \; = -i\widetilde{\Gamma}'(s, m^2, \mu^2)\widetilde{D}'(s)(-i\widetilde{\Gamma}'(m^2, s, \mu^2)) \tag{16.36}$$

and has a pole in it, at $s = m^2$. The full nucleon propagator is staring at us from the middle of the graph. As $s \to m^2$, all the graphs in (b) will certainly have a pole. The graphs in (a) we don't know anything about. However it seems plausible that they will *not* have a pole, because they don't have a propagator joining two parts of a graph. If you've got two or three particles running across the graph, we'll be integrating over all those propagators, and we'll not get poles, but cuts. I ask you to take on trust that *only* the graphs in (b) have poles at $s = m^2$, while the others, the graphs in (a), (c) and (d), are analytic at $s = m^2$, although they may have terrible singularities someplace else. We know that the graphs in (b) have poles at $s = m^2$. That *only* these have poles at $s = m^2$, is just a flat assertion I'm asking you to swallow. The graphs in (c) and in (d) presumably have poles at $t = \mu^2$ and $u = m^2$, respectively, but we don't expect these to have poles at $s = m^2$, and it's reasonable that they are analytic in s.

Every graph that can be cut in two by cutting a nucleon propagator is drawn as shown in (b), with incoming meson and nucleon lines meeting in a one-particle irreducible blob, the full nucleon propagator, another one-particle irreducible blob, and outgoing meson and nucleon lines. Why is this so? The incoming external lines are S-matrix elements, so they do not get any decorations. So it is obviously one-particle irreducible when you cut either external line. The line in the middle can be decorated as much as we please, so we decorate it in every possible way, and get the full propagator. We then go on to the next vertex.

These graphs have a pole at $s = m^2$. What is the residue of that pole? We happen to know it, because the blob in the middle is the renormalized propagator $D'(s) = i/(s - m^2)$. The blob on the right is $-i\widetilde{\Gamma}'(s, m^2, \mu^2)$, and on the left the blob is $-i\Gamma(m^2, s, \mu^2)$. To find the residue at the pole we have to evaluate the coefficient in (16.36) of $(s - m^2)$ at $s = m^2$. The vertices are both simply $-i\widetilde{\Gamma}'(m^2, m^2, \mu^2) = -ig$. The contribution from the propagator is just i. Everything else by assumption is analytic near $s = m^2$. That is,

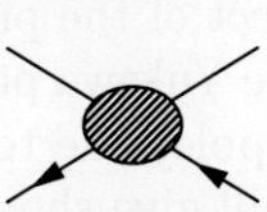

$$= \frac{-ig^2}{s - m^2} + \text{terms analytic at } s = m^2 \tag{16.37}$$

Thus we know how to determine g, or more properly g^2, physically. We look at meson–nucleon scattering. It is some function of s, and of course also the momentum transfer. We extrapolate in s below threshold numerically, or in principal by an analytic continuation, to the point $s = m^2$. We find a pole there at $s = m^2$, and we determine the residue of the pole. That is g^2, aside from the factor of $-i$. So that's how we physically define g.

Why did I say meson–nucleon scattering and not, for example, nucleon–nucleon scattering? No reason in the world. I can run through exactly the same reasoning for nucleon–nucleon scattering and I will now do it.

$$N + N \to N + N \tag{16.38}$$

I do exactly the same thing for the meson pole that we know from lowest order occurs in nucleon–nucleon scattering, the t–channel pole.

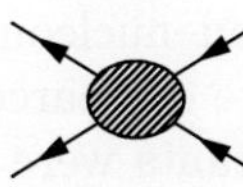

$=$

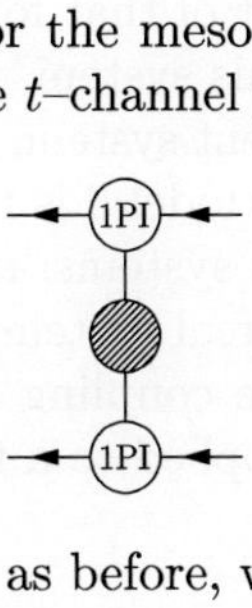

$$+ \text{ graphs without poles at } t = \mu^2 \tag{16.39}$$

By exactly the same reasoning as before, with t replacing s, I get

$$= \frac{-ig^2}{t - \mu^2} + \text{terms analytic at } t = \mu^2 \tag{16.40}$$

So I could just as well do nucleon–nucleon scattering, and look at the extrapolation to the pole at $t = \mu^2$, which of course is outside the physical region. In the physically accessible scattering region, t runs from zero to some number depending on the energy. But extrapolate to the pole of $t = \mu^2$ and then again you'll compute g^2. Notice these are two completely different experiments. It's not that they're related by crossing or anything; there's no way you can cross meson–nucleon scattering into nucleon–nucleon scattering. It's two different extrapolations for two completely different experiments. I claim that the two of them, when you massage them in two different ways, will end up giving you the same number. Now no one has done this in nature, because there are no scalar particles with these kinds of interactions in nature. But they have studied the pion–nucleon system. This is a real system, which is similar in its combinatoric structure to Model 3, although there are lots of Dirac matrices floating around at the vertices. I will tell you what happens. I should also emphasize that this is not a perturbation theory result. Although we have obtained it in the context of perturbation theory, this is true in all orders, the whole summed up theory.

Chew and Low[4] analyzed pion–nucleon scattering in the forward direction, where the best experiment was, analytically continued to the nucleon pole that exists in pion–nucleon scattering, and computed g^2. They got it to within two or three percent, because of the experimental inaccuracies. As I recall, for this system g^2 is 13.7, so you don't want to use perturbation theory.

Several years later Mike Moravcsik said, "Gee, there's a lot of data on nucleon–nucleon scattering. Wouldn't it be nice if we could extract out the effect of the pion pole?" In fact he knew this was possible. The longest range part of the force, the Yukawa potential with a range of the inverse of the pion mass, should come just from this pole, due to the pion exchange. And there will be cuts in t beginning someplace else which will give shorter range potentials. Moravcsik knew that there were tremendous phase shift tables on nucleon–nucleon scattering. He and his colleagues[5] made the first few phase shifts completely free parameters, to take care of the short range part of the potential, whatever it was, at low energies. The remaining scattering data were fit with the Born approximation. Why the Born approximation? Because when you go out to large phase shifts, you're very far from the center of the potential, so even if the coefficient in front of the Yukawa potential is large, it's still a weak force insofar as it affects the large phase shifts. With g^2 and the pion mass as free parameters, low energy nucleon–nucleon scattering with arbitrary phase shifts were fit for the first four or five partial waves, and the higher partial waves were fit with the Born approximation. And lo and behold, the actual pion mass came out, somewhere between 135 and 140 MeV, and the best fit coupling constant was, as I recall, within five or ten percent of that found in pion–nucleon scattering. The experimental errors were a little worse for this system, compared with those found by Chew and Low from looking at a completely different system, but the agreement between the values of g^2 was not bad. What we have done in Model 3 is to equate the coupling constant to residues arising from the poles in two different systems, nucleon–nucleon scattering and nucleon–meson scattering. And when two similar real systems were compared, pion–nucleon and nucleon–nucleon scattering, the values of the coupling constants were found to agree within a few percent. So our procedure, though applied to a toy model, seems to be checked by a real experiment.

---

[4] [Eds.] G. F. Chew and F. E. Low, "Effective-Range Approach to the Low-Energy p-Wave Pion-Nucleon Interaction", *Phys. Rev.* **101** (1956) 1570–1579. The coupling constant $f^2 = (4m^2/\mu^2)g^2$ was found to be 0.08, giving $g^2 = 14.5$ (with $m_N = 939$ MeV, and $\mu_\pi = 139.6$ MeV).

[5] [Eds.] Peter Cziffra, Malcolm H. MacGregor, Michael J. Moravcsik and Henry M. Stapp, "Modified Analysis of Nucleon-Nucleon Scattering. I. Theory and p–p Scattering at 310 MeV", *Phys. Rev.* **114** (1959) 880–886.

That takes care of topic three, coupling constant renormalization. You have now seen how in principle to compute an arbitrary graph in our model theory, including the effects of all renormalization counterterms, or at least reduce it to an integral over Feynman parameters.

16.4 Are all quantum field theories renormalizable?

The last topic I want to discuss is the relationship between renormalization and infinities. We have seen that in the low order graphs of Model 3, the renormalization constants eat the infinities. In fact we have more renormalization constants than we need to eat the infinities that occur in this theory. For example, we have a graph[6] associated with coupling constant renormalization, in $\widetilde{\Gamma}'(p^2, p'^2, q^2)$:

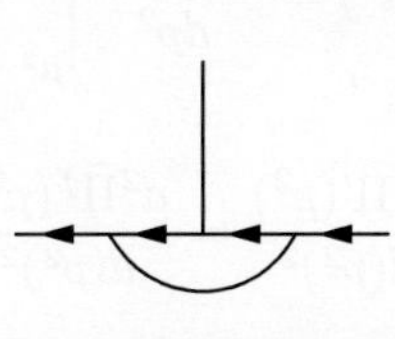

Figure 16.7: $\mathcal{O}(g^3)$ graph in Model 3

Its integral is finite; at high k it goes as d^4k/k^6 because there are three denominators around the loop. At least at $\mathcal{O}(g^3)$, the coupling constant counterterm, (16.31), is not needed to eat the infinities. It *is* required, of course, to give the beautiful result of g^2 measurable in two different experiments. (This extreme convergence is peculiar to Model 3 and other models where all the coupling constants have positive mass dimension, as we will see later.) Let's look at a few low order graphs for a somewhat more complicated theory, using only crude estimates, to see if the renormalization constants we have will eat the infinities or not.

As a first example let me take a single free scalar field, interacting with itself due to a quartic interaction:

$$\mathscr{L} = \mathscr{L}_o - \tfrac{1}{4!} g_0 \phi^4 \tag{16.41}$$

I've already written down several graphs from that theory. This theory is much more divergent in low orders than Model 3. For example, in order g_0^2, we get this graph which I wrote down earlier, the so-called "lip graph":[7]

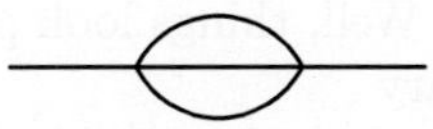

Figure 16.8: $\mathcal{O}(g_0^2)$ graph in ϕ^4 theory

At high k, this graph is quadratically divergent, not just logarithmically divergent, because you have $d^4k_1\, d^4k_2$ in the numerator, eight powers of k, and only six powers of k in the denominator, from the three propagators. Fortunately, in this theory we have more counterterms than we need. Remember when we were studying Model 3's cubic interaction $g\psi^*\psi\phi$, all we needed was a mass renormalization counterterm, C, ——×——, to make the self-energy finite; the additional subtraction caused by the wave function renormalization counterterm B was

---

[6] [Eds.] Though drawn differently, this is the same as the "monster" graph in (16.30), whose evaluation is Problem 9.2, p. 349.

[7] [Eds.] Often drawn as 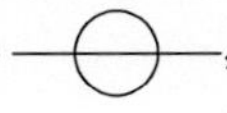, and called the "sunset" graph.

not needed.[8] It is an easy check in this theory, which you can do on the back of an envelope, that the ϕ^4 theory needs *both* renormalization counterterms to render things finite. But these two suffice. Before, in Model 3, the first subtraction turned a logarithmically divergent integral into a convergent integral. With a quartic interaction, the first subtraction turns a quadratic divergence into a logarithmic divergence, and the second subtraction turns the logarithmic divergence into a convergent integral. To put it another way, all we really need to know is the second derivative of this graph with respect to p^2, since its value and its first derivative at the renormalization point are fixed. Every time we differentiate with respect to p^2, we put an extra power of p^2 in the denominator. Do it twice, and you've made the integral convergent. Recall (16.3), and note that

$$\widetilde{\Pi}'(\mu^2) = \left.\frac{d\widetilde{\Pi}'(p^2)}{dp^2}\right|_{\mu^2} = 0 \tag{16.42}$$

so that

$$\frac{d^2\widetilde{\Pi}'(p^2)}{d(p^2)^2} = \frac{d^2\widetilde{\Pi}^f(p^2)}{d(p^2)^2} \tag{16.43}$$

We also have to the same order, g_0^2, a correction to the four-point function, as shown in Figure 16.9, plus crossed versions of this graph. We've seen this before (Figure 14.3). It's only logarithmically divergent. That will be canceled by the coupling constant renormalization counterterm shown in Figure 16.10, which one introduces in this theory. The counterterm just makes a single subtraction and treats the correction to the four-point function just like the treatment of $\widetilde{\Pi}'(p^2)$, and makes everything finite.

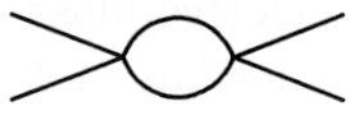

Figure 16.9: $\mathcal{O}(g_0^2)$ correction to $\widetilde{G}^4$ in ϕ^4 theory

Figure 16.10: $\mathcal{O}(g_0^2)$ four-point counterterm in ϕ^4 theory

At $\mathcal{O}(g_0^3)$, we have a graph like Figure 16.11. The integral associated with this graph is finite, because at high k it goes as d^4k/k^6. Well, things look pretty good. But it also looks like we are approaching some sort of boundary.

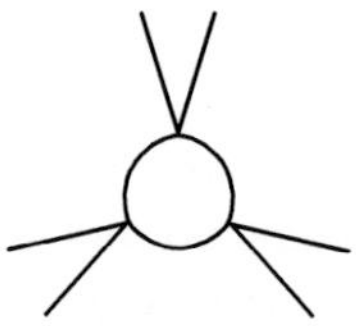

Figure 16.11: $\mathcal{O}(g_0^3)$ graph in ϕ^4 theory

Consider the fifth degree interaction

$$\mathscr{L} = \mathscr{L}_0 - \tfrac{1}{5!}g\phi^5 \tag{16.44}$$

[8] [Eds.] See the discussion on p. 327.

Here things are going to blow up in our faces. A Lagrangian is renormalizable only if all the counterterms required to remove infinities from Green's functions are terms of the same type as those present in the original Lagrangian. But that isn't going to happen here. Let's look at the simple one-loop graph in Figure 16.12.

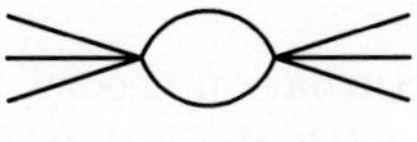

Figure 16.12: $\mathcal{O}(g^2)$ correction to $\widetilde{G}^{(5)}$ in ϕ^5 theory

This graph is logarithmically divergent, our good old d^4k/k^4 integral. To cancel this graph's divergence, we'd need a term that would give rise to a graph like Figure 16.13. This is a ϕ^6 counterterm. But there was *no* ϕ^6 term in our original Lagrangian, and so we'd have to add a term of *higher degree* to the Lagrangian than was originally present. We're stuck! This theory, even on the lowest level of renormalization, does not eliminate the infinities. Someone says, "Okay, wiseguy, I guessed the wrong theory. I agree the ϕ^5 theory is no good. But I'll put in a ϕ^6 term from the start!" Well, let's see what happens with this theory.

Figure 16.13: $\mathcal{O}(g^2)$ six-point counterterm in ϕ^5 theory

$$\mathscr{L} = \mathscr{L}_0 - \tfrac{1}{5!}g\phi^5 - \tfrac{1}{6!}h\phi^6 \tag{16.45}$$

Now we've got a ϕ^6 term that will admit a counterterm to cancel the logarithmic divergence from the graph in Figure 16.12. But with the ϕ^6 term, you also get these graphs, Figure 16.14 (a), arising at order h^2, and a cross-term graph, Figure 16.14 (b), of order gh. These would

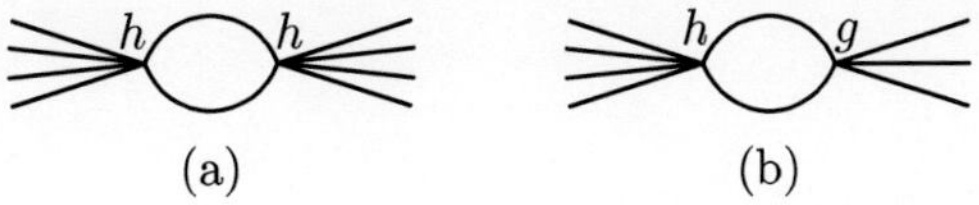

(a) (b)

Figure 16.14: Graphs arising from $g\phi^5$ and $h\phi^6$ terms

require new counterterms of ϕ^7 and ϕ^8 to cancel them. Well then, I guess I need ϕ^7 and ϕ^8 interactions in the Lagrangian, too:

$$\tfrac{1}{7!}j\phi^7 + \tfrac{1}{8!}k\phi^8 \tag{16.46}$$

But then you need counterterms to up to 12th order, which require new terms in the Lagrangian. It just keeps going up and up, an unending escalation of ambiguities! In order to cancel all of the divergences that arise generated by the ϕ^5 term, you need to add a ϕ^6 term. To cancel the divergences of the ϕ^6 term you have to introduce a ϕ^7 and a ϕ^8 term. To cancel the divergences of the ϕ^7 and ϕ^8 terms, you need up to ϕ^{12} terms, and it doesn't stop there. It never stops! As soon as we introduce the ϕ^5 term, it's like the single bite of the apple in the Garden of Eden, the whole thing collapses. The only way of making everything finite and eliminating all the divergences is to have an infinite string of coupling constants. Theories that cannot be made finite without introducing an infinite number of counterterms are called **non-renormalizable**. Theories where you need only a limited number of interactions to get a finite theory are called **renormalizable**. We have not shown that theories with interactions involving only three and four fields are renormalizable; we've just shown that nothing goes wrong in low order. There's a complicated theorem which I will talk about later to show that they are in fact renormalizable. But I'd like to postpone discussing that until we can handle fermions, and do everything at once.

At least from the viewpoint of perturbation theory, as soon as we introduce a little bit of ϕ^5, everything goes crazy. Notice that it's the infinities that make the situation drastically more constrained than in non-relativistic quantum mechanics. In non-relativistic quantum mechanics, you describe your dynamical degrees of freedom, and then you can write down the interaction between them pretty much as you please: two body forces, three body forces, four body forces, nothing goes wrong with any of those things as long as they aren't too pathological. Here, it's not so. If I have a single ϕ field, I can have a cubic term, and I can have a quartic term. And that's it. Anything else, the whole thing goes bananas. Whether that is because of our ignorance, or because such theories are really and truly nonsensical, at this time no one knows.

Next time we'll discuss what happens if the meson becomes heavier than twice the nucleon, and the meson becomes unstable.

Problems 9

9.1 (a) Compute, using (16.5), the imaginary part of the meson self-energy $\widetilde{\Pi}'(p^2)$ to $\mathcal{O}(g^2)$.

(b) Use the spectral representation (15.30) for $\widetilde{D}'(p^2)$ to show that in (15.36), to all orders,

$$\text{Im}\,\widetilde{\Pi}'(p^2) \propto |\widetilde{D}'(p^2)|^{-2}\sigma(p^2) \tag{P9.1}$$

and find the constant of proportionality. HINT: You may wish to use the hint in Problem 4.1, (P4.1).

(c) Compute $\sigma(p^2)$ to $\mathcal{O}(g^2)$, and verify that your answer to (a) is consistent with your answer to (b).

HINT: The spectral density $\sigma(p^2)$ is defined (15.12) in terms of the matrix element of a renormalized field between the vacuum and a multiparticle state. Use (14.37) and (14.15) to express this matrix element in terms of an integral over an appropriate renormalized Green's function.

(1986a 18)

9.2 (a) Compute the Model 3 vertex (16.29), $-i\widetilde{\Gamma}'(p^2, p'^2, q^2)$, to order g^3, as an integral over (two) Feynman parameters, for $p^2 = p'^2 = m^2$.

(b) Show that this is an analytic function of q^2 in the entire complex q^2 plane except for a cut along the positive real axis beginning at $q^2 = 4m^2$. (This function also has interesting analytic properties when all three arguments are complex, but untangling the analytic structure of a function of three complex variables is a bit too much work for a homework problem.)

(1986a 19)

Solutions 9

9.1 (a) From (16.5), discarding the $i\epsilon$'s in the denominators (dismissing the possibility that $\mu^2 > 4m^2$—see p. 331),

$$\widetilde{\Pi}'(p^2) = \frac{g^2}{16\pi^2}\int_0^1 dx \left\{ \ln\left(\frac{m^2 - p^2 x(1-x) - i\epsilon}{m^2 - \mu^2 x(1-x)}\right) + (p^2 - \mu^2)\frac{x(1-x)}{m^2 - \mu^2 x(1-x)} \right\} \tag{S9.1}$$

For real p^2, this expression has non-zero imaginary part only when the real part of the argument of the logarithm becomes negative. For $x \in [0,1]$, this can happen if $p^2 > 4m^2$. The $i\epsilon$ prescription tells us how to deal with the branch cut when $p^2 x(1-x) > m^2$. Let $\xi = p^2 x(1-x) - m^2$. We know ξ is real, because p^2 is real. Then

$$\text{Im } \ln(-\xi - i\epsilon) = \left\{ \begin{matrix} -\pi, & \xi > 0 \\ 0, & \xi < 0 \end{matrix} \right\} = -\pi\theta(\xi) \tag{S9.2}$$

So

$$\text{Im } \widetilde{\Pi}'(p^2) = -\frac{g^2}{16\pi}\int_0^1 dx\, \theta(p^2 x(1-x) - m^2) \tag{S9.3}$$

The function $-\xi(x)$ is an upside-down parabola with roots $x = \{x_1, x_2\} = \frac{1}{2}(1 \mp \sqrt{1 - (2m/p)^2})$, and positive between those roots. If $p^2 > 4m^2$, to $\mathcal{O}(g^2)$

$$\text{Im } \widetilde{\Pi}'(p^2) = -\frac{g^2}{16\pi}\int_{x_1}^{x_2} dx = -\frac{g^2}{16\pi}(x_2 - x_1) = -\frac{g^2}{16\pi}\sqrt{1 - \frac{4m^2}{p^2}} \tag{S9.4}$$

If $p^2 < 4m^2$, $\text{Im } \widetilde{\Pi}'(p^2) = 0$.

Though the problem specified beginning with (16.5), the imaginary part of $\widetilde{\Pi}'(p^2)$ can also be calculated from the integral in note 1, p. 332. Taking the limit $\epsilon \to 0$, we have

$$\widetilde{\Pi}'(p^2) = \frac{g^2}{16\pi^2}\left[-2 + \ln m^2 + 2\frac{\sqrt{4m^2 - p^2}}{p}\tan^{-1}\left(\frac{p}{\sqrt{4m^2 - p^2}}\right)\right] + \cdots \tag{S9.5}$$

where the dots indicate other real terms. Again, if $4m^2 > p^2$, this expression is real, and $\text{Im } \widetilde{\Pi}'(p^2) = 0$. So look at $p^2 > 4m^2$:

$$\begin{aligned} \text{Im } \widetilde{\Pi}'(p^2) &= \frac{g^2}{16\pi^2}\text{Im}\left[2i\frac{\sqrt{p^2 - 4m^2}}{p}\tan^{-1}\left(-i\frac{p}{\sqrt{p^2 - 4m^2}}\right)\right] \\ &= \frac{g^2}{8\pi^2}\frac{\sqrt{p^2 - 4m^2}}{p}\text{Re}\left[\tan^{-1}\left(-i\frac{p}{\sqrt{p^2 - 4m^2}}\right)\right] \\ &= \frac{g^2}{8\pi^2}\frac{\sqrt{p^2 - 4m^2}}{p}\cdot\left[-\frac{\pi}{2}\right] = -\frac{g^2}{16\pi}\sqrt{1 - \frac{4m^2}{p^2}} \end{aligned} \tag{S9.6}$$

exactly as before.[1]

(b) The spectral representation (15.30) for $\widetilde{D}'(p^2)$ says

$$-i\widetilde{D}'(p^2) = \frac{1}{p^2-\mu^2+i\epsilon} + \int_0^\infty da^2\,\sigma(a^2)\frac{1}{p^2-a^2+i\epsilon} \tag{S9.7}$$

We also have, from (15.36),

$$-i\widetilde{D}'(p^2) = \frac{1}{p^2-\mu^2-\widetilde{\Pi}'(p^2)+i\epsilon} \tag{S9.8}$$

Setting these two expressions for $-i\widetilde{D}'(p^2)$ equal to each other,

$$\frac{1}{p^2-\mu^2-\widetilde{\Pi}'(p^2)+i\epsilon} = \frac{1}{p^2-\mu^2+i\epsilon} + \int_0^\infty da^2\,\sigma(a^2)\frac{1}{p^2-a^2+i\epsilon} \tag{S9.9}$$

Using the hint (P4.1), we can write (in the limit as $\epsilon \to 0$)

$$\text{Im}\,\frac{1}{p^2-\mu^2+i\epsilon} = \frac{1}{2i}\left(\frac{1}{p^2-\mu^2+i\epsilon} - \frac{1}{p^2-\mu^2-i\epsilon}\right) = -\pi\delta(p^2-\mu^2) \tag{S9.10}$$

Take the imaginary part of both sides of (S9.9),

$$\text{Im}\,\frac{1}{p^2-\mu^2-\widetilde{\Pi}'(p^2)+i\epsilon} = -\pi\delta(p^2-\mu^2) - \pi\int_0^\infty da^2\,\delta(p^2-a^2)\sigma(a^2) \tag{S9.11}$$

There are two cases: $p^2 \neq \mu^2$, and $p^2 = \mu^2$. In the first case, we can drop the $i\epsilon$ on the left-hand side, because there will be no pole to avoid: μ is the *physical* mass of the meson, so by definition the pole is at $p^2 = \mu^2$; as a renormalization condition, $\widetilde{\Pi}'(\mu^2) = 0$ (see (15.38)). We can also drop the delta function on the right. Then

$$-\pi\sigma(p^2) = \frac{1}{2i}\left(\frac{1}{p^2-\mu^2-\widetilde{\Pi}'(p^2)} - \frac{1}{p^2-\mu^2-\widetilde{\Pi}'(p^2)^*}\right) = \frac{1}{2i}\left(\frac{\widetilde{\Pi}'(p^2)-\widetilde{\Pi}'(p^2)^*}{|p^2-\mu^2-\widetilde{\Pi}'(p^2)|^2}\right) = \text{Im}\,\widetilde{\Pi}'(p^2)|\widetilde{D}'(p^2)|^2 \tag{S9.12}$$

That is, for $p^2 \neq \mu^2$,

$$\text{Im}\,\widetilde{\Pi}'(p^2) = -\pi\frac{\sigma(p^2)}{|\widetilde{D}'(p^2)|^2} \tag{S9.13}$$

The constant of proportionality is $-\pi$. For the case $p^2 \to \mu^2$, the limit of the left-hand side (S9.11) is $-\pi\delta(p^2-\mu^2)$. All will be well if

$$\lim_{p^2\to\mu^2}\sigma(p^2) = 0$$

In perturbation theory, this is fine because we know (15.13) that $\sigma(p^2) = 0$ if $p^2 < \min(4\mu^2, 4m^2)$. If $\mu < m$, then $p^2 = \mu^2 < 4\mu^2$, and $\sigma(\mu^2) = 0$. If $m < \mu$, we also know that $\mu^2 < 4m^2$ (because the meson is stable), so again $\sigma(\mu^2) = 0$. Consequently, within perturbation theory, whatever the value of p^2, (S9.13) holds. (Another derivation will be given in Chapter 17; see (17.4).)

(c) We need to evaluate, to $\mathcal{O}(g^2)$, the spectral density $\sigma(q^2)$, (15.12):

$$\sigma(p^2)\,\theta(p^0) = \sum_n{}'(2\pi)^3\delta^{(4)}(p-q_n)\left|\langle n|\phi'(0)|0\rangle\right|^2 \tag{S9.14}$$

(The prime means that $|n\rangle$ is neither a single meson state nor the vacuum.) The kets $|n\rangle$ can be taken as out states; they are a complete set. Using (14.37) and (13.3), we can say

$$\begin{aligned}{}^{\text{out}}\langle k_1\cdots k_n|\phi'(y)|0\rangle &= i^n\int d^4x_1\cdots d^4x_n\,e^{ik_1\cdot x_1}\cdots e^{ik_n\cdot x_n}(\Box_1^2+\mu^2)\cdots(\Box_n^2+\mu^2)\times \\ &\quad\times\langle 0|T(\phi'(x_1)\cdots\phi'(x_n)\phi'(y))|0\rangle \qquad \text{(S9.15)} \\ &= i^n\int d^4x_1\cdots d^4x_n\,e^{ik_1\cdot x_1}\cdots e^{ik_n\cdot x_n}(\Box_1^2+\mu^2)\cdots(\Box_n^2+\mu^2)\,G'^{(n+1)}(x_1,x_2,\ldots,x_n,y)\end{aligned}$$

---

[1] [Eds.] For real x, $\text{Re}\tan^{-1}(-ix) = \begin{cases} -\dfrac{\pi}{2}, & \text{if } x > 1, \\ 0, & \text{if } 1 > x \geq 0 \end{cases}$

The equality between these two expressions comes from (14.15). We Fourier transform the Green's functions with the convention (13.4), and obtain

$$G'^{(n+1)}(x_1, x_2, \ldots, x_n, y) = \int \frac{d^4\ell_1}{(2\pi)^4} \cdots \frac{d^4\ell_n}{(2\pi)^4} \frac{d^4q}{(2\pi)^4} e^{i\ell_1\cdot x_1 + \cdots + i\ell_n \cdot x_n + iq\cdot y} \widetilde{G}'^{(n+1)}(\ell_1, \ldots, \ell_n, q) \quad \text{(S9.16)}$$

Substitute this into (S9.15) and obtain, after differentiation and integration,

$$^{\text{out}}\langle k_1 \cdots k_n|\phi'(y)|0\rangle = \int \frac{d^4q}{(2\pi)^4} e^{iq\cdot y} \frac{(k_1^2 - \mu^2)}{i} \cdots \frac{(k_n^2 - \mu^2)}{i} \widetilde{G}'^{(n+1)}(-k_1, -k_2, \ldots, -k_n, q) \quad \text{(S9.17)}$$

(We actually only need this for $y = 0$.) The Green's functions (13.3) contain the counterterms, an overall energy-momentum conserving $\delta^{(4)}(k_1 + k_2 + \cdots - q)$ (so we can do the integration easily), and propagators for all $n + 1$ external lines. The n factors of $(k_i^2 - \mu^2)/i$ cancel out all the outgoing particle propagators. Because the terms $^{\text{out}}\langle k_1 \cdots k_n|\phi'(y)|0\rangle$ appear squared in $\sigma(p^2)$, and we are only asked to calculate $\sigma(p^2)$ to $\mathcal{O}(g^2)$, we need only calculate the Green's functions to $\mathcal{O}(g)$. We are freed from computing counterterms, which are all zero to $\mathcal{O}(g)$ in Model 3. Nor need we worry about more than three legs on the Green's functions. Consider $\widetilde{G}'^{(7)}$. Say the meson goes straight through the blob. One of the possible contributions to the other parts of $\widetilde{G}'^{(7)}$ is the disconnected graph that looks like this:

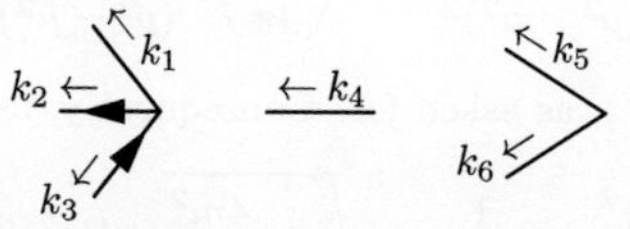

Each disconnected part has its own delta function, e.g., the graph at left has $\delta^{(4)}(k_1 + k_2 + k_3)$. The arguments of these delta functions will never equal zero, because all the k_i are outgoing, so these graphs do not contribute. The only exception is if all the other parts look like a single line, but those are excluded by the prime on the summation: we are not including single meson final states. So we needn't consider the meson simply passing through a blob. The only surviving possibilities include a meson with momentum q branching into a nucleon–antinucleon pair, plus perhaps extra disconnected parts. If there are extra disconnected parts, then as before, each contains a delta function whose argument will never equal zero. These parts, if they even appear, will contribute nothing. So we are left with one contribution to exactly one Green's function, $\widetilde{G}'^{(3)}(-k, -k', q)$ at order g, with this graph:

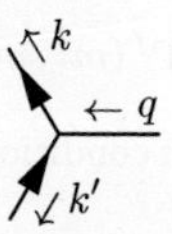

$$= (-ig)(2\pi)^4\delta^{(4)}(k + k' - q)\frac{i}{q^2 - \mu^2}\frac{i}{k^2 - m^2}\frac{i}{k'^2 - m^2} \quad \text{(S9.18)}$$

The outgoing nucleon has momentum k, the outgoing antinucleon has k' and the incoming meson has q. After we substitute this into (S9.17), the nucleon and antinucleon propagators cancel the momentum factors, leaving only a q integration to be performed (with $y = 0$ in the exponential's argument):

$$^{\text{out}}\langle k, k'|\phi'(0)|0\rangle = (-ig)\int d^4q\, \delta^{(4)}(k + k' - q)\frac{i}{q^2 - \mu^2} = -ig\frac{i}{(k + k')^2 - \mu^2} \quad \text{(S9.19)}$$

So to $\mathcal{O}(g^2)$,

$$\sigma(p^2)\theta(p^0) = \sum_n{}' \left|\frac{g}{(k + k')^2 - \mu^2}\right|^2 (2\pi)^3\delta^{(4)}(p - k - k') \quad \text{(S9.20)}$$

The summation in this case is a pair of integrations over k and k' over all states $|k, k'\rangle$ with one nucleon and one antinucleon:

$$\sigma(p^2)\theta(p^0) = \int \frac{d^3\mathbf{k}}{(2\pi)^3 2E_{\mathbf{k}}} \frac{d^3\mathbf{k}'}{(2\pi)^3 2E_{\mathbf{k}'}} \left|\frac{g}{(k + k')^2 - \mu^2}\right|^2 (2\pi)^3\delta^{(4)}(p - k - k') \quad \text{(S9.21)}$$

Because of the delta function, we can take the meson propagator out, with $k + k' = p$. Moreover, as k and k' are timelike vectors, their time components are positive, and so must be p^0. We can then set $\theta(p^0) = 1$, and write

$$\sigma(p^2) = \left|\frac{g}{p^2 - \mu^2}\right|^2 \frac{1}{2\pi} \int \frac{d^3\mathbf{k}}{(2\pi)^3 2E_{\mathbf{k}}} \frac{d^3\mathbf{k}'}{(2\pi)^3 2E_{\mathbf{k}'}} (2\pi)^4\delta^{(4)}(p - k - k') \quad \text{(S9.22)}$$

The extra factor of $1/2\pi$ allows us to write the coefficient of the delta function as $(2\pi)^4$, and the two integrals are now seen to be nothing but the density D, of final states for two particles, (12.16). Integrating over the angle we obtain (for the center-of-momentum frame) from (12.24)

$$D = \int \frac{d^3\mathbf{k}}{(2\pi)^3 2E_{\mathbf{k}}} \frac{d^3\mathbf{k}'}{(2\pi)^3 2E_{\mathbf{k}'}} (2\pi)^4 \delta^{(4)}(p - k - k') = \frac{1}{4\pi} \frac{|\mathbf{p}_f|}{E_T} \tag{S9.23}$$

The delta function says that $p = k + k'$. The total energy E_T is just $\sqrt{(k+k')^2} = \sqrt{p^2}$. In addition, we have $k = (k^0, \mathbf{k})$ and $k' = (k'^0, \mathbf{k}')$. Since both k and k' are on the mass shell, and we're in the center-of-momentum frame, we have $k' = (k^0, -\mathbf{k})$. Then

$$(p^0, \mathbf{p}) = (k^0, \mathbf{k}) + (k^0, -\mathbf{k}) = (2k^0, 0) \;\Rightarrow\; p^2 = 4(k^0)^2 = 4(|\mathbf{k}|^2 + m^2) \tag{S9.24}$$

So

$$D = \frac{1}{4\pi} \frac{|\mathbf{k}|}{E_T} = \frac{1}{4\pi} \frac{\sqrt{\frac{1}{4}p^2 - m^2}}{\sqrt{p^2}} = \frac{1}{8\pi} \sqrt{1 - \frac{4m^2}{p^2}} \tag{S9.25}$$

Putting all the factors together,

$$\sigma(p^2) = \frac{g^2}{2\pi} \frac{1}{(p^2 - \mu^2)^2} D = \left(\frac{g}{4\pi}\right)^2 \frac{1}{(p^2 - \mu^2)^2} \sqrt{1 - \frac{4m^2}{p^2}} \tag{S9.26}$$

which gives the value of $\sigma(p^2)$ to $\mathcal{O}(g^2)$, as asked for. Consequently, from (S9.13),

$$\text{Im}\,\widetilde{\Pi}'(p^2) = -\pi \left(\frac{g}{4\pi}\right)^2 \frac{1}{(p^2 - \mu^2)^2} \sqrt{1 - \frac{4m^2}{p^2}} |D'(p^2)|^{-2} = -\frac{g^2}{16\pi} \sqrt{1 - \frac{4m^2}{p^2}} \tag{S9.27}$$

because, to $\mathcal{O}(g)$, $D'(p^2) = (p^2 - \mu^2 + i\epsilon)^{-1}$. This is identical to (S9.4), as required. If $p^2 < 4m^2$, ${}^{\text{out}}\langle k, k'|\phi'(0)|0\rangle = 0$, so in that case, $\text{Im}\,\widetilde{\Pi}'(p^2) = 0$, as before. ∎

9.2 (a) The Model 3 vertex to $\mathcal{O}(g^3)$ is, for $p^2 = p'^2 = m^2$,

$-i\widetilde{\Gamma}'(m^2, m^2, q^2) =$ 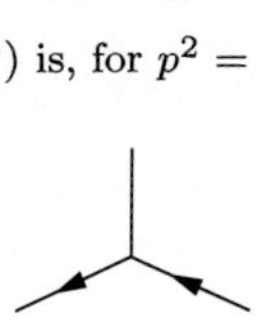 $+$ 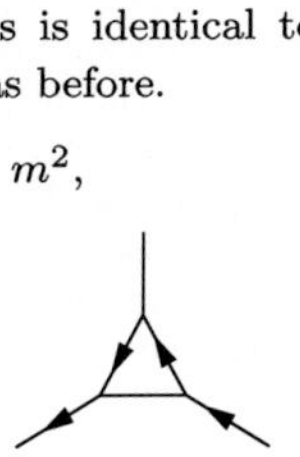 $+$ 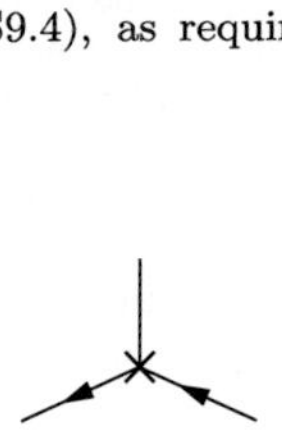 (S9.28)

$$= \quad -ig \quad - \quad i\Gamma^f(m^2, m^2, q^2) \quad + \quad i\Gamma^f(m^2, m^2, \mu^2)$$

The value of F_3, the counterterm F to $\mathcal{O}(g^3)$, is determined by the renormalization condition $-i\widetilde{\Gamma}'(m^2, m^2, \mu^2) = -ig$ to be equal to $i\Gamma^f(m^2, m^2, \mu^2)$. Redrawing the middle term,

$-i\Gamma^f(p^2, p'^2, q^2) =$ 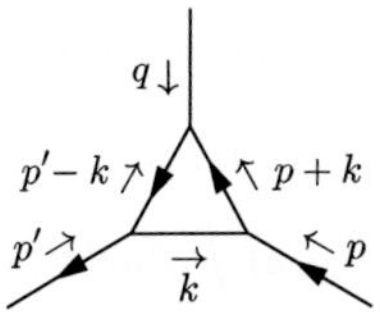 (S9.29)

Applying the Feynman rules (box, p. 216) to this graph,

$$-i\Gamma^f(p^2, p'^2, q^2) = (-ig)^3 \int \frac{d^4k}{(2\pi)^4} \frac{i}{k^2 - \mu^2 + i\epsilon} \frac{i}{(p+k)^2 - m^2 + i\epsilon} \frac{i}{(p'-k)^2 - m^2 + i\epsilon} \tag{S9.30}$$

First we follow the formula (16.16), and rewrite:

$$\begin{gathered} -i\Gamma^f(p^2, p'^2, q^2) = g^3 \int \frac{d^4k}{(2\pi)^4} \int_0^1 d\alpha_1\, d\alpha_2\, d\alpha_3\, \delta(1 - \alpha_1 - \alpha_2 - \alpha_3) \frac{(3-1)!}{D^3} \\ \text{where } D = ((p-k')^2 - m^2 + i\epsilon)\alpha_1 + ((p+k)^2 - m^2 + i\epsilon)\alpha_2 + (k^2 - \mu^2 + i\epsilon)\alpha_3 \end{gathered} \tag{S9.31}$$

We can integrate over α_3 easily because of the delta function. Rewriting $\alpha_1 = x$, $\alpha_2 = y$ and $\alpha_3 = 1 - x - y$, the integral becomes

$$-i\Gamma^f = 2g^3 \int_0^1 dx \int_0^{1-x} dy \int \frac{d^4k}{(2\pi)^4} \frac{1}{D^3} \tag{S9.32}$$

where

$$\begin{aligned} D &= (1-x-y)(k^2-\mu^2+i\epsilon) + x[(p'-k)^2 - m^2 + i\epsilon] + y[(p+k)^2 - m^2 + i\epsilon] \\ &= (k - p'x + py)^2 - p'^2x^2 - p^2y^2 + 2p\cdot p'xy + xp'^2 + yp^2 - (1-x-y)\mu^2 - (x+y)m^2 + i\epsilon \\ &= (k-p'x+py)^2 + a \end{aligned} \tag{S9.33}$$

Now we shift the variable of integration:

$$k \to k' = k - p'x + py \tag{S9.34}$$

The momentum integral becomes

$$\int \frac{d^4k'}{(2\pi)^4}\frac{1}{(k'^2+a)^3} \tag{S9.35}$$

Using the integral table on p. 330, the equation (I.2) for the case $n = 3$ gives

$$\int \frac{d^4k'}{(2\pi)^4}\frac{1}{(k'^2+a)^3} = \frac{i}{32\pi^2 a} \tag{S9.36}$$

We can simplify a a little bit. By momentum conservation $q = p + p'$. Squaring both sides we find $2p\cdot p' = q^2 - (p^2 + p'^2)$. Next, we are told to restrict our attention to $p^2 = p'^2 = m^2$. Then

$$\begin{aligned} a &= p'^2x(1-x) + p^2y(1-y) + 2p\cdot p'xy - (1-x-y)\mu^2 - (x+y)m^2 + i\epsilon \\ &= m^2x(1-x) + m^2y(1-y) + (q^2-2m^2)xy - (1-x-y)\mu^2 - (x+y)m^2 + i\epsilon \\ &= xyq^2 - m^2(x+y)^2 - (1-x-y)\mu^2 + i\epsilon \end{aligned} \tag{S9.37}$$

Finally, then,

$$\Gamma^f(m^2,m^2,q^2) = -\frac{g^3}{16\pi^2}\int_0^1 dx\int_0^{1-x} dy\,\frac{1}{xyq^2 - m^2(x+y)^2 - (1-x-y)\mu^2 + i\epsilon} \tag{S9.38}$$

and

$$\widetilde{\Gamma}'(m^2,m^2,q^2) = g + \Gamma^f(m^2,m^2,q^2) - \Gamma^f(m^2,m^2,\mu^2) \tag{S9.39}$$

(b) Where is $\widetilde{\Gamma}'(m^2,m^2,q^2)$ analytic in q^2? The only part of the vertex function that depends on q^2 is $\Gamma^f(m^2,m^2,q^2)$. This is by inspection an analytic function except where the denominator equals zero (hence the $i\epsilon$ in the denominator). The x, y integration runs over a triangular region as shown:

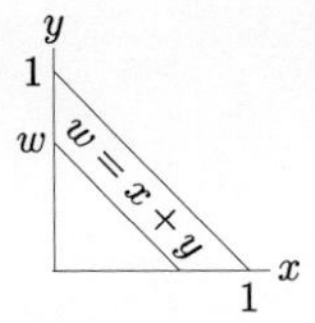

Because of the form of the denominator, it makes sense to reparametrize the integral as

$$\begin{aligned} w &= x + y, \quad w\in[0,1] \\ z &= x - y, \quad z \in [-w, w] \\ w^2 - z^2 &= 4xy \end{aligned}$$

Then we can rewrite the denominator as $(w^2 - z^2)q^2 - 4(w^2m^2 + (1-w)\mu^2)$. The function will cease to be analytic if

$$q^2 = \frac{4(w^2m^2 + (1-w)\mu^2)}{(w^2 - z^2)} \tag{S9.40}$$

Every pair (w, z) gives an excluded value of q^2. It's easy to see that for any value of w, z can be chosen arbitrarily close to w, so that q^2 can be made arbitrarily large. The minimum value of q occurs when the denominator takes its maximum value, namely for $z = 0$. Then

$$q^2_{\text{min},z} = 4m^2 + \frac{(1-w)}{w^2}\mu^2 \tag{S9.41}$$

and the least value of this in both w and z is clearly $w = 1$, so $q^2_{\text{min},z,w} = 4m^2$. Consequently the function is analytic except for a branch cut from $q^2 = 4m^2 \to \infty$, along the real axis. ■

17

Unstable particles

I'd like to review briefly what we've done in the past few chapters. We've gone through a whole lot of complicated technical reasoning, and you might not be able to see the forest for the trees or even see the trees for the leaves. I'd just like to devote a few moments to recapitulating the arguments of the last four chapters.

First, we gave a description of scattering theory on a basis that had nothing to do with any approximation scheme, nor with any artifactual features like adiabatic turning on and off. We proved, in my usual handwaving sense, the Lehmann, Symanzik and Zimmerman reduction formula, (14.19). That formula tells us if you know the Green's functions exactly, then you know the S-matrix elements exactly, and *a fortiori* if you have an approximation for the Green's functions, you have an approximation for the S-matrix elements. Second, we gave a new interpretation to our only approximation technique, Feynman perturbation theory. We showed that Feynman perturbation theory is in fact a perturbation theory for the Green's functions when the lines are off the mass shell, and we took account of all the trivial kinematic factors in the way I described. However, Feynman perturbation theory is actually *many* possible perturbation theories, because there are many possible ways of breaking a given Hamiltonian up into a free part and an interacting part. With each such break-up you get a different expansion. The most naive way of breaking up the Hamiltonian, gathering all the quadratic terms and calling them the free Hamiltonian, and simply taking the cubic and higher terms and calling them the interaction, leads to an expansion which, although perfectly valid (at least, as valid *a priori* as any other expansion) is not particularly useful. It is an expansion for the wrong Green's functions, those of the unrenormalized fields, in terms of the wrong coupling constant, the impossible-to-observe bare coupling constant, with the wrong parameters held fixed, and the unknowable bare masses. We corrected this difficulty by breaking up the Hamiltonian in a different way, where the free Hamiltonian was given in terms of the renormalized fields, the physical masses and the physical coupling constant. The difference between the free Hamiltonian in terms of the renormalized quantities and everything else is now the interaction. Such an expansion necessarily generates counterterms. Of course, there is no way of telling what the counterterms are until you insert into your computational scheme the definition of the parameters you call μ and the quantity you call ϕ', etc. Therefore we went through a long song and dance in which we found out how to define those things consistently within our perturbative scheme, that is to say, as properties of Green's functions. That gave us the definition of the physical masses and physical coupling constants for the renormalized fields, and allowed us to insert explicitly into our scheme the definitions of μ, m, ϕ', g, etc.

In the course of these developments we went on many excursions, and found many interesting things that will be useful outside of this computational scheme. Three of them in particular will be very important to us later. First was the spectral representation (15.16) for the propagator, which comes up in other contexts of physics. Second we learned the lore of loops, how to reduce any Feynman integral from a momentum integral to an integral over Feynman parameters. Third, we encountered and conquered, at least in low orders, the infinities that occur in Feynman perturbation theory. From this way of looking at things we found a surprise, that at least in low orders, for certain simple theories, the infinities could be absorbed by the renormalization counterterms. On the other hand, we found that it was easy to construct theories for which the infinities could *not* be absorbed by the renormalization counterterms, as I discussed at the end of the last chapter. Of course we did not prove that our theories with cubic and quartic interactions are renormalizable in the sense that *all* infinities that occur in *all* orders are eaten by the counterterms. We showed only that this holds in low order, and we postpone until a future date the question of whether it happens to all orders. That's the summary of what we went through.

17.1 Calculating the propagator $\widetilde{D}'$ for $\mu > 2m$

I would now like to turn to the remaining loose ends in Model 3. The one computation we have not redeemed in the new formulation of scattering theory is the one[1] where you computed the ratio g/m_K from the decay rate Γ of the K meson, represented by the field ϕ, when the mass $\mu = m_K$ of the ϕ was greater than twice the mass, $m = m_\pi$, of the "nucleon" field ψ:

$$\Gamma = \frac{g^2}{16\pi m_K^2}\sqrt{m_K^2 - 4m_\pi^2} \tag{17.1}$$

In our formulation of scattering theory, the concept of an unstable particle never occurs. There's an unstable particle? *Nu*, there's an unstable particle.[2] You just forget about the field associated with that particle. It will never appear in asymptotic states, because it decays long before the time gets to be plus or minus infinity, and you compute scattering matrix elements between stable particle states. Nevertheless, it is amusing to ask the question: What if someone with no physical intuition whatsoever decided to follow through our computational scheme, and chose μ, the renormalized mass of the meson, to be greater than $2m$?

$$\text{If } \mu > 2m, \text{ then } \ldots?$$

Surely our hypothetical person must run into trouble: We can't make a stable meson with mass greater than or equal to $2m$. But what is the *specific* trouble he will encounter? Well, before I answer that question, I'd like to derive an identity (S9.13) for the imaginary part of $\widetilde{\Pi}'$, which appears in the solution to Problem 9.1. This identity has nothing to do with whether or not the meson is stable, but it will be useful to us in our investigation.

---

[1] [Eds.] Problem 6.1, p. 261.

[2] [Eds.] The Yiddish word *nu* has a bewilderingly large number of meanings. In this context, it means "So what?" Rosten says that it is the word most frequently used in Yiddish, besides *oy* and the articles: Rosten *Joys*, pp. 271–272.

From (15.36),

$$p^2 - \mu^2 - \widetilde{\Pi}'(p^2) = \left[-i\widetilde{D}'(p^2)\right]^{-1} \tag{17.2}$$

We can deduce a formula for the imaginary part of $\widetilde{\Pi}'$ using the fact that for any complex number z, the imaginary part of z^{-1} is minus the imaginary part of z divided by the absolute value of z squared:

$$\text{Im } (z^{-1}) = -\frac{\text{Im } z}{|z|^2} \tag{17.3}$$

We have a formula (15.31) for the imaginary part of $-i\widetilde{D}'$,

$$\text{Im } (-i\widetilde{D}'(p^2)) = -\pi\sigma(p^2) \tag{15.31}$$

So we find

$$-\text{Im } \left[-i\widetilde{D}'(p^2)\right]^{-1} = \frac{\text{Im } (-i\widetilde{D}'(p^2))}{|\widetilde{D}'(p^2)|^2} = -\frac{\pi\sigma(p^2)}{|\widetilde{D}'(p^2)|^2} = \text{Im } \widetilde{\Pi}'(p^2) \tag{17.4}$$

in agreement with (S9.13). (We're always talking about real p^2 here, because (15.31) assumes real p^2.) Recall the definition, (15.12), of $\sigma(p^2)$:

$$\sigma(p^2)\,\theta(p^0) = \sum_n{}' (2\pi)^3\delta^{(4)}(p - q_n)\left|\langle n|\phi'(0)|0\rangle\right|^2 \tag{15.12}$$

(the prime indicating that neither a single meson state nor the vacuum state is included in the sum). So we can write (17.4) as

$$\text{Im } \widetilde{\Pi}'(p^2) = -\tfrac{1}{2}\left|\widetilde{D}'(p^2)\right|^{-2} \sum_n{}' \left|\langle n|\phi'(0)|0\rangle\right|^2 (2\pi)^4\delta^{(4)}(p - q_n) \tag{17.5}$$

Here of course not only is p^2 real but p^0 is greater than zero, otherwise Im $\widetilde{\Pi}'(p^2)$ would be zero. So we can drop the $\theta(p^0)$. We'll use this formula very shortly.

A significant feature of this result, as we saw both in the discussion following Figure 16.2 and in Problem 9.1, is that $\widetilde{\Pi}'(p^2)$ has an imaginary part for $p^2 > 4m^2$ in lowest nontrivial order in perturbation theory. Thus, if our hypothetical dumb cluck, attempting to carry through the renormalization program for $\mu^2 > 4m^2$, attempts to impose the condition

$$\widetilde{\Pi}'(\mu^2) = 0 \text{ with } \mu^2 > 4m^2 \tag{17.6}$$

he will arrive at a complex counterterm B_2, because he has to subtract not only a real number but also an imaginary number. He would have to be really dumb not to realize that something is going wrong in this case, because he should realize that he is adding to his Hamiltonian a non-Hermitian operator by allowing B_2 to be complex. Even if he doesn't know anything else, he should recognize that that's bad news. So imposing the condition (17.6) is not possible unless you want to play with theories with non-Hermitian Hamiltonians.

Of course this means that if $\mu^2 > 4m^2$, our standard renormalization conditions are not applicable. (Everything I say for $\widetilde{\Pi}'$ itself holds as well for its derivatives.) From the viewpoint of pure scattering theory, in fact, we don't need *any* renormalization conditions for this particle, because this particle is unstable, and therefore will never appear on the outside of the scattering process. So how we renormalize it is totally irrelevant. However, renormalization does also serve the purpose of eliminating infinities, and it would be nice to have some set of conventions that accomplish this. The conventions I'll choose to adopt in the remainder of this discussion are these:

$$\text{Re } \widetilde{\Pi}'(\mu^2) = 0 \tag{17.7}$$

$$\text{Re } \left. \frac{d\widetilde{\Pi}'}{dp^2} \right|_{p^2=\mu^2} = 0 \tag{17.8}$$

These conditions are arbitrary, but they enable us to fix our subtraction constant in a way that's continuous as μ^2 goes from below $4m^2$ to above $4m^2$. They're certainly not forced on us by any general principle. I adopt them because they will make the things we're going to discuss somewhat simpler. The conditions have no physics in them, but they are certainly allowed; I can fix those counterterms any way I want.

Let's use these conditions to compute what the propagator looks like to order g^2. We did it for the case when things are stable (see (15.36) and (16.4)), when $\mu^2 < 4m^2$. Let's now do it for the case where $\mu^2 > 4m^2$. I will restrict myself to the neighborhood of the point μ^2 on the real p^2 axis because I don't want to write down horrible equations. In fact I will treat $(p^2 - \mu^2)$ itself as a quantity of $\mathcal{O}(g^2)$. The obvious tool for looking at the propagator in this region is a power series expansion of (17.2):

$$\left[-i\widetilde{D}'(p^2)\right]^{-1} = (p^2-\mu^2) - \widetilde{\Pi}'(\mu^2) - (p^2-\mu^2)\left.\frac{d\widetilde{\Pi}'}{dp^2}\right|_{p^2=\mu^2} + \mathcal{O}(g^4) \tag{17.9}$$

The first term by hypothesis is $\mathcal{O}(g^2)$, so we keep that. The third term is in fact $\mathcal{O}(g^4)$. The factor $(p^2-\mu^2)$ is $\mathcal{O}(g^2)$. The derivative $d\widetilde{\Pi}'/dp^2$ evaluated at $p^2 = \mu^2$, like $\widetilde{\Pi}'(p^2)$, is also $\mathcal{O}(g^2)$. We know from our renormalization condition (17.7) that $\widetilde{\Pi}'(\mu^2)$ has vanishing real part:

$$\widetilde{\Pi}'(\mu^2) = \text{Re } \widetilde{\Pi}'(\mu^2) + i \,\text{Im } \widetilde{\Pi}'(\mu^2) = i\,\text{Im } \widetilde{\Pi}'(\mu^2) \tag{17.10}$$

So the inverse propagator becomes

$$\left[-i\widetilde{D}'(p^2)\right]^{-1} = (p^2-\mu^2) - i\,\text{Im } \widetilde{\Pi}'(\mu^2) + \mathcal{O}(g^4) \tag{17.11}$$

Now we happen to have a formula, (17.5), that will give us the imaginary part of $\widetilde{\Pi}'(\mu^2)$. To order g^2, the only diagram that contributes to this formula looks like this:

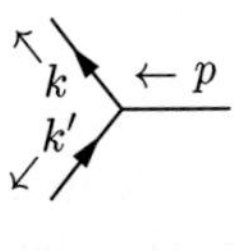

$$= -ig\frac{i}{p^2-\mu^2+i\epsilon} \tag{17.12}$$

The only thing the meson field can make to order g^2 is a nucleon–antinucleon pair. Since the meson field is off the mass shell, the value of this diagram is $-ig$ from the coupling constant, times $i/(p^2-\mu^2)$, the lowest-order propagator. That makes the contribution to order g^2. So (17.5) becomes

$$\begin{aligned}\text{Im } \widetilde{\Pi}'(p^2) &= -\tfrac{1}{2}|\widetilde{D}'(p^2)|^{-2}\left|-ig\frac{i}{p^2-\mu^2+i\epsilon}\right|^2 (2\pi)^4\delta^{(4)}(k+k'-p) \\ &= -\tfrac{1}{2}g^2(2\pi)^4\delta^{(4)}(k+k'-p)\end{aligned} \tag{17.13}$$

The $(p^2 - \mu^2)^2$ is canceled by the $|\widetilde{D}'(p^2)|^{-2}$, which must also be taken to lowest order in g^2. We wind up with the same diagram and much the same matrix element we computed[3] for the kaon decay. The contribution from (17.12) differs from that of the matrix element we looked at in our old dumb theory of unstable particle decay mainly by the factor of $(p^2 - \mu^2)^{-2}$, but we have, nicely enough, $|\widetilde{D}'|^{-2}$ in front to cancel that factor out. The limit $p^2 \to \mu^2$ of Im $\widetilde{\Pi}'(p^2)$ gives us, apart from the factor of $-\frac{1}{2}$, the rest frame amplitude for the decay. From (12.2),

$$\Gamma = \frac{1}{2\mu} \sum_{\substack{\text{final}\\ \text{states}}} \int |\mathcal{A}_{fi}|^2 D \tag{17.14}$$

so that

$$\lim_{p^2 \to \mu^2} |\widetilde{D}'(p^2)|^{-2} {\sum_n}' |\langle n|\phi'(0)|0\rangle|^2 (2\pi)^4 \delta^{(4)}(p - q_n) = \sum_{\substack{\text{final}\\ \text{states}}} \int |\mathcal{A}_{fi}|^2 D = 2\mu\Gamma \tag{17.15}$$

Then

$$\text{Im}\, \widetilde{\Pi}'(\mu^2) = -\tfrac{1}{2}(2\mu\Gamma) = -\mu\Gamma \tag{17.16}$$

and

$$\begin{aligned} \left[-i\widetilde{D}'(p^2)\right]^{-1} &= p^2 - \mu^2 - i\,\text{Im}\, \widetilde{\Pi}'(\mu^2) = p^2 - \mu^2 + i\mu\Gamma + \mathcal{O}(g^4) \\ &= p^2 - (\mu - \tfrac{1}{2}i\Gamma)^2 + \mathcal{O}(g^4) \end{aligned} \tag{17.17}$$

These last two expressions are equivalent, because Γ here is also $\mathcal{O}(g^2)$.

Thus we see that the effect of the interaction that enables this particle to decay is to displace the pole in the propagator from p^2 equals μ^2 to p^2 equals $(\mu - i\Gamma/2)^2$. Here's what it looks like on the complex plane.

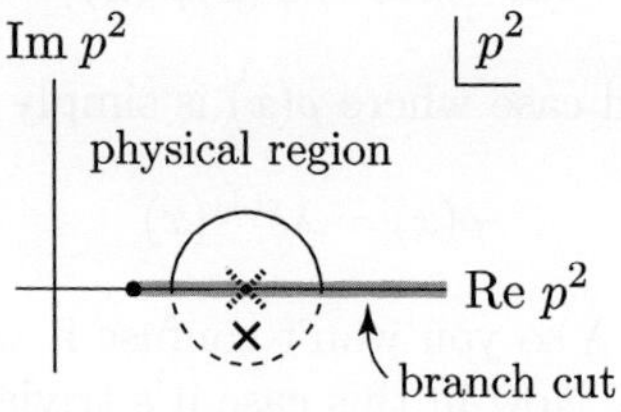

Figure 17.1: The pole μ^2 shifted to $(\mu - \frac{1}{2}i\Gamma)^2$

When you look at this drawing, you may think I have gone bananas. The dashed × on the cut is the pole μ^2. Now I've got a pole at $(\mu - i\Gamma/2)^2$ in the complex plane, even though I have proved to you that the propagator was analytic in the complex plane. The point is, however, that we're starting out above the cut, and doing a power series expansion there. So we get a power series expansion that's valid in some circle when we go above the cut, and when we go down, we go *through* the cut into the second sheet of this function of a complex variable. I indicate that by a dotted line. You should think of this as a circular disk of paper that is sitting on top of the cut, and then goes through the cut and extends onto the second sheet. So this pole is there, all right, but it's not in the function on the cut plane, but in what we would get if we were to analytically continue the function from above the cut, to the second sheet.

[3] [Eds.] Problem 6.1, p. 261. See also Figure 12.3, p. 251 and (12.30).

17.2 The Breit–Wigner formula

I would now like to investigate the physical consequences of what we have discovered in lowest order perturbation theory. In particular, I would like to discuss two features traditionally associated with unstable particles, the Breit–Wigner formula[4] and the exponential decay law, and see how they arise in this formalism. I will do this by means of two thought experiments. I will completely forget the perturbation theory origin of these formulas. My only assumption will be that I have some field with a propagator, of the form

$$\widetilde{D}'(p^2) = \frac{i}{p^2 - \mu^2 + i\mu\Gamma} \tag{17.18}$$

for some real parameters μ and Γ, and for some range of the real axis. Of course there's no need for the $i\epsilon$ now that I've put in the $i\mu\Gamma$. Certainly this expression for the propagator isn't true everywhere; it doesn't have the right analytic properties. But I assume there is some tiny stretch of the real axis where this is an excellent approximation for the propagator. That certainly happens in second-order perturbation theory, as we've seen, but it may well happen in other circumstances where there is no trace of a fundamental ϕ' particle. You could, for example, arrive at an unstable state because of some mechanism analogous to those which arise in non-relativistic quantum mechanics, where you have a bound state that's not quite bound. Only if you turned up the value of the coupling constant would this state become a bound state, but now it appears as an unstable state, or resonance. So the form (17.18) of the propagator is the only assumption I will make here.

The first thought experiment involves the momentum analysis of an idealized production process. I add to the Lagrangian, $\mathscr{L}$, that describes this theory, an external perturbation of the sort we've discussed so frequently before,

$$\mathscr{L} \rightarrow \mathscr{L} + \rho(x)\phi'(x) \tag{17.19}$$

I'll look at an extremely idealized case where $\rho(x)$ is simply a delta function:

$$\rho(x) = \lambda\delta^{(4)}(x) \tag{17.20}$$

I've called the coupling constant λ so you won't confuse it with any coupling constant that's already in the Lagrangian, like g. Now in this case it's trivial to determine the amplitude for going into any given final state. At space point 0, at time point 0, I bash the vacuum of the system with this field. I impulsively turn it on and turn it off, and see what comes out of the vacuum. To lowest order in λ, the amplitude for going to any given state $|n\rangle$ from the vacuum is simply

$$\mathcal{A}_{\text{vac}\rightarrow n} \propto \lambda \langle n|\phi'(0)|0\rangle + \mathcal{O}(\lambda^2) \tag{17.21}$$

ignoring any phase factor. I'll write this as a proportion, since I'm not particularly interested in 2π's or anything like that. Let k be any four-momentum that's allowed in the final states of the theory. Then the probability of producing a final state with momentum k, which I'll call $P(k)$, can be written as

$$P(k) \propto \lambda^2 \sum_n{}' \,|\langle n|\phi'(0)|0\rangle|^2 (2\pi)^4\delta^{(4)}(k - p_n) + \mathcal{O}(\lambda^3) \tag{17.22}$$

[4] [Eds.] Gregory Breit and Eugene Wigner, "Capture of Slow Neutrons", *Phys. Rev.* **49** (1936) 519–531.

The delta function is to extract just the part that has momentum k. I can keep λ as small as I want, so I can suppress those corrections (of order λ^3) as much as I like. There may be kinematic factors in here, but that's all washed into the proportionality sign.

Near $k^2 = \mu^2$, the sum in (17.22) has a very simple formula. It is precisely what appeared on the right-hand side of our expression for the imaginary part of the propagator (see (15.12) and (15.31)). So $P(k)$ is proportional to λ^2 times the imaginary part of $\widetilde{D}'(k^2)$:

$$P(k) \propto \lambda^2 \sigma(k^2) \propto -\lambda^2 \,\mathrm{Im}\,(-i\widetilde{D}'(k^2)) + \mathcal{O}(\lambda^3) \tag{17.23}$$

Now we have an *ansatz* (17.18) for $\widetilde{D}'(k^2)$, and we can easily find its imaginary part. We just plug that in:

$$P(k) \propto \lambda^2 \frac{\mu\Gamma}{(k^2-\mu^2)^2 + \mu^2\Gamma^2} + \mathcal{O}(\lambda^3) \tag{17.24}$$

If we graph $P(k)$ as a function of k^2, we get the characteristic Lorentzian line shape, or Breit–Wigner shape, very sharply peaked near the mass of the unstable particle with a width depending on the parameter Γ, as shown in Figure 17.2. This is the same result you find for scattering amplitudes in non-relativistic theories near an unstable energy level.

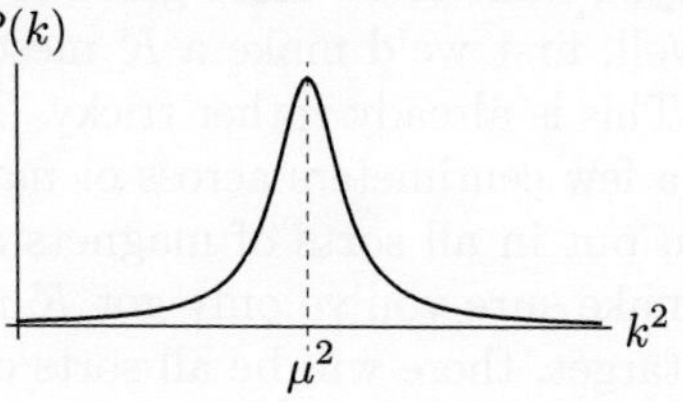

Figure 17.2: The Breit–Wigner distribution

You may be a bit disturbed if you remember that the full width at half maximum is supposed to be Γ, in the ordinary non-relativistic analysis, and here it looks like it's $2\mu\Gamma$, but that's simply because we've written things in terms of the squared invariant mass, k^2, of the state that's produced. If we wrote things in terms of the center-of-momentum energy of that state, that is to say, if we chose the four-vector k to be $(E, \mathbf{0})$, then

$$k^2 - \mu^2 = E^2 - \mu^2 = (E+\mu)(E-\mu) \approx 2\mu(E-\mu) \tag{17.25}$$

since E is supposed to be close to μ. Dropping the terms of $\mathcal{O}(\lambda^3)$,

$$P(E) \propto \lambda^2 \frac{\mu\Gamma}{4\mu^2[(E-\mu)^2 + \frac{1}{4}\Gamma^2]} \tag{17.26}$$

Aside from the factor of $4\mu^2$ in front, this is now the conventional Breit–Wigner denominator, with the $\frac{1}{4}$ where it is in the familiar formula. In terms of energy we have exactly the same kind of peak, and Γ is indeed the full width at half maximum, just as it should be in the conventional Breit–Wigner formula.

You have to be careful when you compute Γ's to higher orders, because as soon as you get diagrams that involve internal meson lines, you get funny things going on. Those meson lines can be on the mass shell in the range of integration, so pretty soon you will start computing amplitudes that have funny singularities in them. The computation cannot be carried out to all orders. The best way, if you want to calculate Γ to all orders for some exceptionally refined analysis, is to compute the propagator to all orders. That leads you to *no* ambiguities along the real axis, and you can extrapolate the propagator into the complex plane and see where the pole is.

That takes care of the Breit–Wigner formula, one of the two hallmarks in non-relativistic scattering theory of the occurrence of an unstable state. We have, I grant you, not done a general scattering experiment but a very idealized production experiment: we get a ϕ' hammer and hit the vacuum with it and see what comes out. But as expected, what comes out, the probability $P(E)$ as a function of the center-of-momentum energy, has the characteristic Breit–Wigner shape.

17.3 A first look at the exponential decay law

The second hallmark of unstable particles is the exponential decay law. This was the first principle used in the studies of radioactive nuclei once it was discovered in the late 19th century. A second thought experiment will enable us to see if the exponential decay law is in our model.

Suppose we conducted an experiment to measure the lifetime of some unstable particle, say a radioactive nucleus, or if you want to be more glamorous, a K meson or something like that. How would we do it? Well, first we'd make a K meson in some specific region, with some well-defined momentum. This is already rather tricky. Typically you send a high-energy accelerator beam into a target a few centimeters across or maybe a bit larger, and all sorts of junk comes out. And then you put in all sorts of magnets and various devices designed to filter the momentum, and to make sure you've only got K mesons coming out at the other end. When the beam hits the target, there will be all sorts of things coming out that you're not interested in, like π mesons and fragments of the target, several atoms, molecules boiling off the sides. You don't want anything but the K mesons. Then you move your K meson detector a certain distance down the beam, and see how the population of K mesons falls off as you move the detector further and further down the beam. And if you are lucky, the curve you will plot will have a nice exponential in it. By looking at the parameters of that exponential you will see the decay law.

Now that's a rather complicated experiment. Let's idealize it a bit. First we want something that makes things localized in a certain region of space and time and also localized in a certain region of momentum space. We'll use the same sort of production process as in the last discussion to do that, writing

$$\mathscr{L} \to \mathscr{L} + f(x)\phi'(x) \tag{17.27}$$

That's not a very realistic production process; people do not have things coupled to scalar fields in general, but it is one which we can manipulate very simply analytically. I want the function $f(x)$ to have two properties. I want $f(x)$ to be reasonably well localized in space and time, at the origin, so I know that I've made the particle at the target and not someplace else. And I also want its Fourier transform $\tilde{f}(k)$ to be reasonably well localized in k space, at some mean value of k I'll call $\bar{k}$. These properties are represented symbolically in Figure 17.3.

We can imagine $f(x)$ as a Gaussian in all four variables, although that won't be relevant. I want to make sure I'm looking at K mesons, and not two pion states or a fragment of the target or something like that coming out. So I'll arrange matters so that $\bar{k}^2$ is near μ^2. In particular I'll assume that $\tilde{f}(k)$ is sufficiently sharp in momentum space that throughout the relevant region of momentum for this production process, I can use the approximation (17.18)

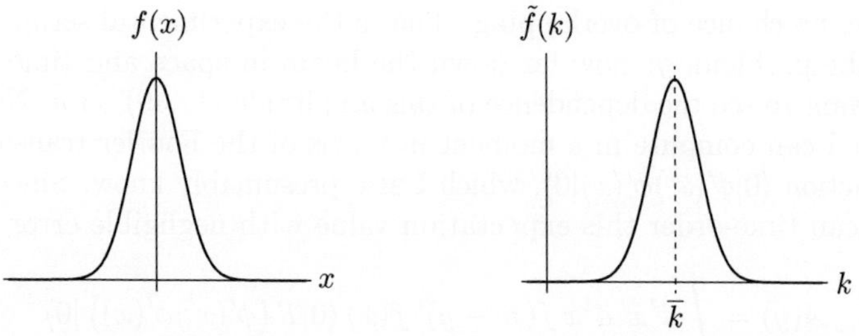

Figure 17.3: The function $f(x)$ and its Fourier transform $\tilde{f}(k)$

for the propagator. That's my single assumption. That may mean I have to have a fast target in position space, but we'll see that's no problem; I can certainly always do that. One last significant point: $f(x)$ is real, which of course implies that

$$\tilde{f}(k) = \tilde{f}(-k)^* \tag{17.28}$$

I'll make my state by taking the vacuum and hitting it with some source that makes a bunch of stuff:

$$\int d^4x\, f(x)\phi'(x)\,|0\rangle \tag{17.29}$$

That's the production apparatus. I have produced this state, maybe with some tiny coefficient, mainly vacuum, but vacuum I won't detect. Now I want to detect the state. What about the detection apparatus? As a theorist, I'm very economical. Instead of inventing new detection apparatus, I'll use the same thing as for production. After all, this formula (17.28) tells me that if $\tilde{f}(k)$ produces a given amplitude for making K mesons of momentum k, it produces an amplitude of the same magnitude for absorbing K mesons of the same momentum, k. So I'll just move myself down the beam to a spacetime point y, and set up a detection apparatus. The amplitude $\mathcal{A}(y)$ I wish to study is therefore

$$\mathcal{A}(y) = \langle 0| \int d^4x'\, d^4x\, f(x'-y)^*\phi'(x')f(x)\phi'(x)|0\rangle \tag{17.30}$$

Perhaps a spacetime diagram would be useful. Figure 17.4 shows a light cone and two spacetime regions (the shaded circles). In the neighborhood of the origin, there is a region in which $f(x)$ is concentrated and where the kaons will be produced.

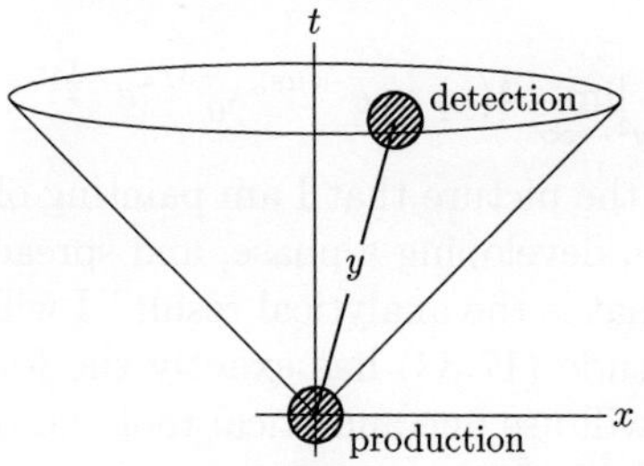

Figure 17.4: Spacetime diagram for production and detection of K mesons

Some huge, timelike distance away at a point y is a second region where $f(y)$ is sufficiently different from zero, and where we'll detect the kaons. We want y to be so far away that these regions have no chance of overlapping. That's the experimental setup. I have only one free variable in the problem, y: how far down the beam in space and time can I locate my detector. And I want to see the dependence of this amplitude (17.30) on y. Now of course this is something that I can compute in a moment in terms of the Fourier transform of $f(x)$ and the two-point function $\langle 0|\phi'(x')\phi'(x)|0\rangle$, which I also presumably know. Since y is far later in time than x, we can time-order this expectation value with negligible error. The amplitude becomes

$$\mathcal{A}(y) = \int d^4x'\, d^4x\, f(x'-y)^* f(x)\, \langle 0|T\left(\phi'(x')\phi'(x)\right)|0\rangle \tag{17.31}$$

Writing

$$\langle 0|T\left(\phi'(x')\phi'(x)\right)|0\rangle = \int \frac{d^4q}{(2\pi)^4} e^{-iq\cdot(x'-x)} \frac{i}{q^2-\mu^2+i\mu\Gamma} \tag{17.32}$$

(I can insert this approximation for the propagator legitimately because I assume $\tilde{f}(k)$ is concentrated near $\overline{k}$), the evaluation of $\mathcal{A}(y)$ is just a Fourier transform operation. We obtain, as you might have guessed,

$$\mathcal{A}(y) = \int \frac{d^4k}{(2\pi)^4} |\tilde{f}(k)|^2 e^{-ik\cdot y} \frac{i}{k^2-\mu^2+i\mu\Gamma} \tag{17.33}$$

Suppose we knew quantum mechanics, but didn't know anything about quantum field theory. If we were told about this experiment—we had a production apparatus that produced particles in a restricted range of momentum and a detection apparatus that only detected particles in the restricted range of momentum—what would be our naive guess for the asymptotic properties of this expression as y^2 goes to ∞? Let's approximate the kaon's proper time s_0 by

$$s_0 = \sqrt{y^2} \tag{17.34}$$

What would we expect the amplitude to look like as a function of s_0? Well, as an experimenter I would say I'd put in a big fat proportionality sign because I got this creation and detection apparatus from a theorist's workshop, so I don't know its properties or its resolution or anything like that. Then, the particle is traveling at proper time s_0 and it has mass μ, so I'd expect there to be a phase factor from the good old Schrödinger equation. But it's an unstable particle and it decays. I expect an exponential decay from the square of the amplitude, so half that magnitude in the amplitude itself. And if I'm a very sophisticated experimentalist I know from my studies of the Schrödinger equation that wave packets tend to spread out for large time t such that the amplitude goes down like $t^{-3/2}$. When y^2 is very far down beam, portions of the decay products miss the detector. So I would insert the factor of $s_0^{-3/2}$ in the amplitude to account for the spreading out of that wave packet. Then our naive guess looks something like this:

$$\lim_{y^2\to\infty} \mathcal{A}(y) \propto e^{-i\mu s_o} s_0^{-3/2} e^{-\frac{1}{2}\Gamma s_0} \tag{17.35}$$

This is a dumb guess, based on the picture that I am painting of an unstable particle, traveling around like an ordinary particle, developing a phase, and spreading out. But it's got this little extra feature: it decays. But what is the analytical result? I will show you that the asymptotic form for large y^2 of the amplitude (17.33) has exactly the form of our naive guess, (17.35). That requires some analysis. I will use one analytical tool and one trick. The analytical tool is the **method of stationary phase.**[5]

---

[5] [Eds.] See Section 8.2, pp. 229–234 of G. N. Watson, *A Treatise on Bessel Functions*, 2nd ed., Cambridge U. Press, 1966, and Sections 17.03–17.05, pp. 471–474, of Harold & Bertha S. Jeffreys, *Methods of Mathematical Physics*, Cambridge U. Press, 1946.

17.4 Obtaining the decay law by stationary phase approximation

If we have an integral

$$I = \int dt\, e^{i\theta(t)} g(t) \tag{17.36}$$

with a real function $\theta(t)$ that varies rapidly compared to the rate at which $g(t)$ varies, then in general the value of this integral will be zilch—nothing. Because $\theta(t)$ is oscillating rapidly, the exponential averages out to zero. The main contribution to the integral comes at points of stationary phase, where the derivative of $\theta(t)$ is zero. At such points $\theta(t)$ is not rapidly varying; it doesn't vary at all. Therefore the integral is dominated by stationary phase points. I will assume there is only one such point, t_0;

$$\frac{d\theta}{dt} = 0 \text{ for } t = t_0 \tag{17.37}$$

If there are several such points, you get a sum. People normally like to phrase this principle by putting a parameter λ in front of $\theta(t)$ and say "we're studying large λ." That's neither here nor there. There may or may not be an adjustable parameter in the theory. It's just that if $\theta(t)$ is varying very rapidly, and $g(t)$ is not, this is a good approximation. We therefore approximate the integral by its value near the stationary phase point. By the stationary phase approximation

$$I = e^{i\theta(t_0)} g(t_0) \int dt\, e^{\frac{i}{2}\theta''(t_0)(t-t_0)^2} \tag{17.38}$$

The integral is trivial: it is a complex version of a Gaussian, and it gives us

$$I = e^{i\theta(t_0)} g(t_0) \sqrt{\frac{2\pi}{|\theta''(t_0)|}}\, e^{i(\pi/4)\ \mathrm{sgn}(\theta''(t_0))} \tag{17.39}$$

If $\theta''(t_0) = 0$, we have to think again. We have to expand out to the quartic term, or cubic, whichever is the first non-vanishing one. We will apply this method to our amplitude, (17.33).

Our integral is of stationary phase form because we have a complex exponential with argument $k \cdot y$, and all four components of y are getting huge. So we have four integrals we can do by stationary phase. But before we can do that, we need to use a trick. We have a problem with the propagator in (17.33). We can use the approximation (17.18) only because we're also near $k^2 = \mu^2$, and therefore the phase of the denominator is also changing rapidly over the region of integration as we pass by the pole. It changes by 180°, very rapidly if Γ is very small. We certainly don't want to find an approximation that's good only for $y^2 = s_0$ very large compared to Γ^{-1}, because then we'll properly get zero for the value of the integral. We'd like to put the phase variation of the propagator into a form where we can treat it also by stationary phase. That's the trick. We write the propagator as an exponential integral:

$$\frac{i}{k^2 - \mu^2 + i\mu\Gamma} = \int_0^\infty \frac{ds}{2\mu}\, e^{i(s/2\mu)(k^2-\mu^2+i\mu\Gamma)} \tag{17.40}$$

The reason for the scaling of the integration variable s by 2μ will become clear later. This turns my four-dimensional integral into a five-dimensional integral, but what I gained from that is putting the propagator up in the exponential, where I can treat its phase variation by the stationary phase formula.

Using this trick, the amplitude (17.33) becomes

$$\mathcal{A}(y) = \frac{1}{(2\pi)^4 2\mu} \int_0^\infty ds \int d^4k\, |\tilde{f}(k)|^2 e^{-ik\cdot y + i(s/2\mu)(k^2-\mu^2+i\mu\Gamma)} \tag{17.41}$$

Now the first step is to do the four k integrals by stationary phase. There are just two phase factors that involve k, the product $k \cdot y$ and the quadratic term from the propagator, so $\theta(k) = -k \cdot y + (sk^2/2\mu)$. One finds easily $k_0^\alpha = (\mu/s)y^\alpha$, so $\theta(k_0) = -(\mu y^2/2s)$. Each of the four k integrals gives a factor

$$e^{i(\pi/4)\,\mathrm{sgn}(\theta''(k_0))}\sqrt{2\pi/|\theta''(k_0)|} = e^{i\pi/4}\sqrt{2\pi\mu/s}$$

and turns one component of k into the corresponding component of $k_0 = (\mu/s)y$, in $|\tilde{f}(k)|^2$ and in the exponent. Carrying out all four integrals, we have

$$\mathcal{A}(y) = e^{i\pi}\frac{\mu}{2(2\pi)^2}\int_0^\infty ds\, |\tilde{f}(k_0)|^2\, e^{-\frac{1}{2}\Gamma s}\frac{1}{s^2}e^{-i\theta(s)} \quad \text{where } \theta(s) = \frac{\mu y^2}{2s} + \frac{\mu s}{2} \tag{17.42}$$

That does the first stationary phase integral. Note the interpretation of $k = (\mu/s)y$. If you classically propagate a stable particle with 4-velocity $v^\alpha = k^\alpha/\mu$, in a proper time s, it will arrive at a point $y^\alpha = v^\alpha s$. Since $v_\alpha v^\alpha = 1$, it follows $s = \sqrt{y^2}$. This is just classical kinematics, but you see we have recovered it in the limit of large y from quantum field theory. Here, the conditions of stationary phase give an equation from classical mechanics.

Now we're ready to do the s integral, also by stationary phase. The phase is rapidly varying because it has this gigantic factor y^2 in it. We find easily $s_0 = \sqrt{y^2}$, $\theta(s_0) = \mu\sqrt{y^2}$, and $\theta''(s_0) = (\mu/s_0)$. Note that there is *no* stationary phase point if y^2 is spacelike, because as $y \to \infty$, $y^2 \to -\infty$, and there is no probability that a particle will be detected. I plug $\sqrt{2\pi/|\theta''(s_0)|} = \sqrt{2\pi s_0/\mu}$ into (17.42), and evaluate everything at $s_0 = \sqrt{y^2}$. Note that now $k_0 = \mu(y/s_0)$, and

$$\mathcal{A}(y) = -\sqrt{\frac{\mu}{32\pi^3}}\, e^{i\pi/4}\, |\tilde{f}(k_0)|^2\, e^{-i\mu s_0} e^{-\frac{1}{2}\Gamma s_0} s_0^{-3/2} \tag{17.43}$$

It looks like our dumb guess (17.35) was not so dumb after all. In the amplitude we see a number of factors common to the dumb guess. The square of the Fourier transform represents the factors that depend on the details of the experimental apparatus producing and detecting our unstable particles. There is a common factor, $e^{-i\mu s_0}$, giving the evolution of phase as this particle of mass μ marches along in time. There is the common exponential decay factor, and there is the common factor of one over $s_0^{3/2}$, the spreading of the wave packet. I hope you have understood the physical import of what we have obtained. We have *derived* the exponential decay law. The statement that the propagator in a certain region of k space can be approximated by the expression (17.18) is completely equivalent to the statement that under the physical circumstances in which we would expect to observe an exponential decay law, we *do* observe an exponential decay.

This is the cleanest derivation of the exponential decay law: the stationary phase approximation to the pole on an imaginary sheet. There are other derivations in the literature that are wrong. If you just Fourier transform the original expression for the amplitude $\mathcal{A}(y)$, you don't take into account the momentum cuts in the detection apparatus. There's a famous false statement in the literature that the decay law is not strictly exponential. It is true that there are satellite terms in the amplitude that are non-exponential. The interpretation of those satellite terms is that they are experimental background. Experimentalists know about them, and they take account of them. The exponential decay law is 100% valid.

Some people set up a thought experiment where they haven't been as careful as I have been to put a good momentum filter in at the beginning and at the end. If you do that kind of thought experiment then you get in fact a very large contribution for making two π mesons, with the mass say half that of the K mesons. Then the experimental apparatus you built up has a nice probability for detecting pairs of π mesons, as well as detecting K mesons, because you haven't got a sharp enough momentum filter, and then you get a mess. And doing the data analysis, you may be led to say the exponential decay is just an approximation. To avoid that, you've got to do the experiment so that you only get momentum near the Breit–Wigner peak. Then you suppress those unwanted things enormously: uncorrelated π mesons are randomly distributed in phase space, more or less. If you don't do that, then you get something that looks very different, and there are papers in the literature by very bright people many years ago, when this phenomenon was not so well understood, that said, "Hey, the exponential decay law should not be true. There should be terms, for example, that go as inverse powers of the time." That's what happens if you just Fourier transform the propagator without putting in these $\tilde{f}(k)$'s. When you Fourier transform the propagator, if you don't put in this momentum spread, if you just have a sharp position experiment, then you get an enormous contribution from the two π meson states because then you can make, in particular, two π mesons on threshold. Two π mesons on threshold have small relative velocity and therefore do not spread very much from each other. And if you work things out, you get a one over s to the sixth term coming in with a tiny coefficient. And that's just wrong. Even a physicist of the stature of Abdus Salam once thought, back in the 1950s, that the decay was not purely exponential.[6] That's the threshold singularity he was seeing, not the pole on the second sheet. So there's an error in the literature.

This concludes everything we are going to say in a world in which there are only scalar particles. Next time we will begin studying particles with spin.

[6] [Eds.] P. T. Matthews and Abdus Salam, "Relativistic Theory of Unstable Particles. II.", *Phys. Rev.* **115** (1959) 1079–1084. See Section 5, and equation (5.4).

18

Representations of the Lorentz Group

We will now put aside for a while those questions of Green's functions and factors of i and $k^2 - \mu^2$ that drove us crazy. We'll come back to them eventually, and generalize them to the case of spin one-half particles. Now we are going to look at a topic that has nothing to do with quantum field theory, but a lot to do with Lorentz transformations. We are going to construct the quantum theory of spin one-half particles and the Dirac equation. I do not wish to start out by saying, "Well, you all know the Dirac equation", and start covering the boards with gamma matrices. Instead, I want to derive the Dirac equation as a classical field theory, and then canonically quantize it by our standard machine to find out what's going on. Part of this discussion can be done in some generality. The general discussion will be useful for subsequent purposes and will also give us additional insight into the structure of the Dirac equation.

18.1 Defining the problem: Lorentz transformations in general

I will begin by asking what are the most general possible transformation properties of a finite number of fields under the Lorentz group, assuming they transform linearly; they just shuffle among themselves. Let Λ be an element of the Lorentz group, which is called by its friends SO(3, 1).[1] Say I have a set of fields $\phi_a(x)$, which transform under the Lorentz group according to the rule

$$U(\Lambda)^\dagger \phi_a(x) U(\Lambda) = D_{ab}(\Lambda)\phi_b(\Lambda^{-1}x) \tag{18.1}$$

($a = 1, 2, \ldots, n$; the sum on b is implied). What are the possible choices for the matrices D_{ab}? We know there are many choices. The fields ϕ_a could be Lorentz scalars, for which D_{ab} is the identity matrix. There are vector fields typified for us by the derivatives of scalar fields. For these fields, the matrices D_{ab} are the 4×4 Lorentz matrices $\Lambda^\mu{}_\nu$ themselves. There are tensor fields where D_{ab} are products of a bunch of those Lorentz matrices, one for each index, which are here all summed up in the super-index a, but what else is there? What are the possibilities? I will explore the constraints placed on the matrix D_{ab} by what we know

[1] [Eds.] SO(3, 1) is the group of orthogonal transformations with determinant 1 which preserve the square of the Minkowski norm,

$$x_0^2 - x_1^2 - x_2^2 - x_3^2.$$

about the matrix $U(\Lambda)$. In order to keep from writing indices when I don't really need to, I'll assemble ϕ into a big (column) vector and simply write (18.1) as

$$U(\Lambda)^\dagger\phi(x)U(\Lambda) = D(\Lambda)\phi(\Lambda^{-1}x) \tag{18.2}$$

where ϕ is some n component vector, and D is some $n \times n$ matrix.

The transformations U are constrained. If I have two Lorentz transformations, $U(\Lambda_1)$ and $U(\Lambda_2)$, then as I said much earlier (see (1.62)), U of the product should be the product of the U's for the individual ones:

$$U(\Lambda_1)U(\Lambda_2) = U(\Lambda_1\Lambda_2) \tag{18.3}$$

Actually this isn't quite right. It's impossible to rule out in quantum mechanics that (18.3) might need to be generalized to

$$U(\Lambda_1)U(\Lambda_2) = U(\Lambda_1\Lambda_2)e^{i\phi(\Lambda_1,\Lambda_2)} \tag{18.4}$$

We know it's not quite right even if only rotations are considered, let alone the full Lorentz group. It turns out that for the rotation group in three dimensions, SO(3), and for the Lorentz group, the phases can be removed except in **spinor representations**, where a rotation by π about any axis $\hat{\mathbf{n}}$ followed by a second such rotation results in a net multiplication by -1. I won't bother to write down the general definition of a ray representation, as it is called.[2] Spinor representations are used to describe spin-½ particles, and so we expect minus signs if there are spin-½ particles in the theory. Since God put spin-½ particles into the world, we must allow the occasional minus sign if we want to describe reality. We're going to be a little bit sloppier with spin-½ than we have been with Bose particles.

From (18.2) and (18.3) we obtain a constraint on D. It goes like this:

$$U(\Lambda_1\Lambda_2)^\dagger\phi(x)U(\Lambda_1\Lambda_2) = D(\Lambda_1\Lambda_2)\phi(\Lambda_2^{-1}\Lambda_1^{-1}x) \tag{18.5}$$

because the inverse of the product is the product of the inverses in the reverse order. Now let's write out the same thing using the product equation (18.3):

$$\begin{aligned} U(\Lambda_1\Lambda_2)^\dagger\phi(x)U(\Lambda_1\Lambda_2) &= U(\Lambda_2)^\dagger U(\Lambda_1)^\dagger\phi(x)U(\Lambda_1)U(\Lambda_2) \\ &= U(\Lambda_2)^\dagger D(\Lambda_1)\phi(\Lambda_1^{-1}x)U(\Lambda_2) \\ &= D(\Lambda_1)U(\Lambda_2)^\dagger\phi(\Lambda_1^{-1}x)U(\Lambda_2) \\ &= D(\Lambda_1)D(\Lambda_2)\phi(\Lambda_2^{-1}\Lambda_1^{-1}x) \end{aligned} \tag{18.6}$$

D is just a numerical matrix, U is an operator in Hilbert space; they have nothing to do with each other, and they commute. By inspection we obtain

$$D(\Lambda_1\Lambda_2) = D(\Lambda_1)D(\Lambda_2) \tag{18.7}$$

And again I tell you that strictly speaking, we are working with a looser condition than this, and occasional minus signs are also okay in this equation. It follows, if we let both Λ_1 and Λ_2 be the identity matrix, that

$$D(1) = 1 \tag{18.8}$$

[2] The only good reference I know is Valentine Bargmann's "On unitary ray representations of continuous groups", *Ann. Math.* **59** (1954) 1–46. I am not recommending that you study this article. We will get all the right results with much less effort, by being cavalier and lucky.

A **representation** of a group is a set of matrices, one associated with each group element, that obeys the same algebra as the group elements they represent:

$$\text{If } \Lambda_1\Lambda_2 = \Lambda_3 \text{ then } D(\Lambda_1)D(\Lambda_2) = D(\Lambda_3) \tag{18.9}$$

(If we were considering the ordinary rotation group or the 17-dimensional rotation group or the discrete group that describes the symmetries of a crystal, we would have the same equation (18.7) with Λ replaced by the appropriate symbol labeling the transformation in question.) It is also easy to demonstrate that

$$D(\Lambda^{-1}) = D(\Lambda)^{-1} \tag{18.10}$$

Thus the matrices D form a finite dimensional representation of the Lorentz group. The D matrices obey all the properties of their corresponding group elements, and you might reasonably think that from any set of D's you could reconstruct the group. But that's not necessarily so. Many of the group elements can be mapped into a single D, so that $D(\Lambda) = D(\Lambda')$ even if $\Lambda \neq \Lambda'$. The trivial prototypical example is to assign $D(\Lambda) = 1$ for all elements Λ. On the other hand, if $D(\Lambda) = D(\Lambda')$ *only* when $\Lambda = \Lambda'$, the representation is said to be **faithful**.

Our problem of finding all possible linear field transformation laws, involving only a finite set of fields and consistent with Lorentz invariance, is equivalent to finding all finite dimensional matrix representations of SO(3,1) satisfying the equations (18.7) and (18.8). That makes it sound like a very difficult problem. But as we'll see, it's a very easy problem. Once we have found these D representations, we can use them to construct field transformation laws, which, from now on, we'll think of not as being laws of the quantum theory, but laws for the transformation properties of a classical field before quantization. From these possible laws we'll then select out some particularly tasty looking fields with not too many components, capable of describing spin-½ particles. We'll attempt to construct quadratic Lagrangians out of them, so we'll get free field theories, and then try to develop a theory of free spin-½ particles. Please notice that I want the D matrices to be finite dimensional, but I'm not going to impose the constraint[3] that the D's be *unitary* ($D^\dagger = D^{-1}$). The U's, of course, have to be unitary, but that doesn't necessarily mean the D's are.

For example, the 3-vector representation of the (3-dimensional) rotation group, $D(R) = R$, is unitary. Consider a rotation about the z axis through an angle θ:

$$R = \begin{pmatrix} \cos\theta & -\sin\theta & 0 \\ \sin\theta & \cos\theta & 0 \\ 0 & 0 & 1 \end{pmatrix} \qquad R^\dagger = \begin{pmatrix} \cos\theta & \sin\theta & 0 \\ -\sin\theta & \cos\theta & 0 \\ 0 & 0 & 1 \end{pmatrix} \quad \Rightarrow \quad R^\dagger R = \mathbb{1} \tag{18.11}$$

On the other hand, the 4-vector representation of the Lorentz group, $D(\Lambda) = \Lambda$, is *not* unitary. Consider a boost along the x-axis by a speed v (as usual, $\gamma = (1-\beta^2)^{-1/2}$, and $\beta = v$):

$$\Lambda = \begin{pmatrix} \gamma & -\beta\gamma & 0 & 0 \\ -\beta\gamma & \gamma & 0 & 0 \\ 0 & 0 & 1 & 0 \\ 0 & 0 & 0 & 1 \end{pmatrix} \qquad \Lambda^\dagger = \begin{pmatrix} \gamma & -\beta\gamma & 0 & 0 \\ -\beta\gamma & \gamma & 0 & 0 \\ 0 & 0 & 1 & 0 \\ 0 & 0 & 0 & 1 \end{pmatrix} \quad \Rightarrow \quad \Lambda^\dagger\Lambda \neq \mathbb{1} \tag{18.12}$$

[3] [Eds.] Coleman could not impose this constraint even if he wanted to: the Lorentz group is non-compact, and there are no faithful, finite-dimensional, unitary irreducible representations of non-compact groups. See Ashok Das and Susumu Okubo, *Lie Groups and Lie Algebras for Physicists*, World Scientific, 2014, p. 47.

so the representation $D(\Lambda) = \Lambda$ is *not* unitary, even though U is a unitary operator down there in Hilbert space. So do not assume that we are looking only for unitary representations.

To find all matrix representations D obeying these equations is to answer a big question. It can be replaced by a smaller question, because there are two trivial ways of obtaining new representations from old. One is this. If $D(\Lambda)$ is a representation, so is

$$D(\Lambda)' = TD(\Lambda)T^{-1} \tag{18.13}$$

for any fixed matrix T, because it doesn't affect anything in the multiplication. If $D(\Lambda_1)(D\Lambda_2) = D(\Lambda_1\Lambda_2)$, the same is true for the transformed representations $D(\Lambda)'$:

$$\begin{aligned} D(\Lambda_1)'D(\Lambda_2)' &= TD(\Lambda_1)T^{-1}TD(\Lambda_2)T^{-1} \\ &= TD(\Lambda_1)D(\Lambda_2)T^{-1} = TD(\Lambda_1\Lambda_2)T^{-1} = D(\Lambda_1\Lambda_2)' \end{aligned} \tag{18.14}$$

If we have two representations related in this way, we write

$$D(\Lambda)' = TD(\Lambda)T^{-1} \Longleftrightarrow D(\Lambda) \sim D(\Lambda)' \tag{18.15}$$

and say that $D(\Lambda)$ is **equivalent** to $D(\Lambda)'$. This just corresponds to choosing different linear combinations of the ϕ_a's as the fundamental fields. We can generate an infinite number of new, equivalent representations from old ones. But it's trivial. We will restrict our problem to finding all finite dimensional, inequivalent representations of SO(3,1).

There is a second, trivial way of making new representations from old. Suppose I have two representations, $D^{(1)}(\Lambda)$ and $D^{(2)}(\Lambda)$, of dimensions n_1 and n_2, respectively. The dimension n describes both the number of fields involved and the size of the matrices, $n \times n$. I can build a new representation in the following way. I make a great big matrix

$$D(\Lambda) = \begin{bmatrix} D^{(1)}(\Lambda) & 0 \\ 0 & D^{(2)}(\Lambda) \end{bmatrix} \equiv D^{(1)}(\Lambda) \oplus D^{(2)}(\Lambda) \tag{18.16}$$

This matrix is called the **direct sum** of $D^{(1)}$ and $D^{(2)}$, and denoted $D^{(1)}(\Lambda) \oplus D^{(2)}(\Lambda)$. This, too, is a representation. When I multiply these things together, $D^{(1)}$ and $D^{(2)}$ never talk to each other; they just multiply independently. The dimension of this representation is the sum of the dimensions of the component representations:

$$\dim D(\Lambda) = \dim D^{(1)}(\Lambda) + \dim D^{(2)}(\Lambda) = n_1 + n_2 \tag{18.17}$$

I'm not interested in representations that can be written as direct sums. If I tell you I have a field theory that's Lorentz invariant with a scalar field, and I have another Lorentz invariant field theory with a vector field, it would not surprise you that I can build a Lorentz invariant field theory that has five fields in it, one scalar and the four components of the vector. If a representation D can be written as a direct sum of two representations of smaller dimensions, or is equivalent to a direct sum, we say $D(\Lambda)$ is **reducible**. If it is not reducible, then we say, to no one's surprise, that it is **irreducible**. Our task of finding all possible Lorentz transformation laws of fields has thus been reduced to the task of making a catalog of all inequivalent, irreducible finite dimensional representations of SO(3,1).

Now this is a problem that was solved for the rotation group SO(3) many years ago. It is part of the standard lore of quantum mechanics, though perhaps not in this language. Every quantum mechanics course has a section in it devoted to the subject of angular momentum, and there you saw this problem solved, although, like the man in Molière's play who didn't know he was speaking prose all his life,[4] you may not have known that you were in fact finding the irreducible inequivalent representations of the rotation group. By a wonderful fluke peculiar to living in $(3+1)$ dimensions, the representations of SO(3, 1) can be obtained rapidly from the representations of SO(3).

18.2 Irreducible representations of the rotation group

Let's now consider the related problem, finding all inequivalent irreducible representations for the rotation group.[5] SO(3) is the group of rotations in space (or, as the mathematicians would write, $\mathbb{R}^3$) about some axis by some angle. Every rotation matrix R can be labeled by an axis,[6] $\hat{\mathbf{n}}$, and an angle, θ:

$$R \in \mathrm{SO}(3)\colon R = R(\hat{\mathbf{n}}\theta) \quad 0 \leq \theta \leq \pi \tag{18.18}$$

Notice that the angle and the axis appear as a product. By convention, the angle is always chosen to be less than or equal to π; if you rotate by more than π, that's equivalent to rotating by the supplementary angle about the opposite axis. We will use the multiplication rules of the rotation group to gain information about the representations, D.

First, observe that if you have two rotations about the same axis with two different angles, the angles simply add:

$$R(\hat{\mathbf{n}}\theta')R(\hat{\mathbf{n}}\theta) = R(\hat{\mathbf{n}}(\theta' + \theta)) \tag{18.19}$$

So any representation, not necessarily irreducible, must satisfy

$$D(R(\hat{\mathbf{n}}\theta'))D(R(\hat{\mathbf{n}}\theta)) = D(R(\hat{\mathbf{n}}(\theta' + \theta))) \tag{18.20}$$

Let's differentiate this equation. (As usual I'm being a mathematical slob, and will assume everything is differentiable.) I will define

$$i\frac{\partial}{\partial\theta}D(R(\hat{\mathbf{n}}\theta))\bigg|_{\theta=0} \equiv \hat{\mathbf{n}}\cdot\mathbf{L} \tag{18.21}$$

The derivative of the representation evaluated at $\theta = 0$ must be some linear function of $\hat{\mathbf{n}}$. This defines a vector of "angular momentum" matrices $\mathbf{L}$, sometimes called the **generators** of infinitesimal rotations. If I differentiate (18.20) with respect to θ' and set θ' equal to zero I obtain[7]

$$\frac{\partial}{\partial\theta}D(R(\hat{\mathbf{n}}\theta)) = -i\hat{\mathbf{n}}\cdot\mathbf{L}D(R(\hat{\mathbf{n}}\theta)) \tag{18.22}$$

---

[4] [Eds.] Coleman is referring to Monsieur Jourdain, the title character of Molière's *Le Bourgeois Gentilhomme*, 1670. Jean-Baptiste Poquelin (1622–1673), known by his stage name Molière, is widely regarded as one of the greatest French writers.

[5] These are carefully constructed in a few pages, in a way that generalizes to other groups, beginning on p. 16 of the first edition of Howard Georgi's *Lie Algebras in Particle Physics*, Benjamin-Cummings (1982). Actually what are constructed there are the representations of the Lie algebra of SO(3) rather than the Lie group, but you'll see that is what we want. ([Eds.] See also Chapter 3 of the second edition, Perseus Press, 1999.)

[6] [Eds.] Coleman uses **e** for the axis. The notation was changed to avoid confusion with e, the base of natural logs.

[7] [Eds.] $\dfrac{\partial D(R(\hat{\mathbf{n}}(\theta'+\theta)))}{\partial\theta'} = \dfrac{\partial D(R(\hat{\mathbf{n}}(\theta'+\theta)))}{\partial\theta}$.

This differential equation is trivial to solve,[8] using the "initial condition" $D(R(0)) = 1$:

$$D(R(\hat{\mathbf{n}}\theta)) = e^{-i\hat{\mathbf{n}}\bullet\mathbf{L}\theta} \tag{18.23}$$

We've simplified our problem enormously. We don't have to work out $D(R(\hat{\mathbf{n}}\theta))$ for all axes $\hat{\mathbf{n}}$ and all angles θ. We just have to tabulate the three "angular momentum" matrices $\{L^i\}$, $i = \{1, 2, 3\}$. (There are 3 generators because SO(3) is a three parameter group. In general, the group SO(n) is described with $\frac{1}{2}n(n-1)$ parameters.)

Of course our concept of equivalence and reducibility also apply here. Two representations are equivalent if and only if the two $\mathbf{L}$'s are equivalent:

$$D \sim D' \iff \mathbf{L} \sim \mathbf{L}' \tag{18.24}$$

(in the sense of (18.13)). If the representation is a direct sum, so too are the $\mathbf{L}$'s:

$$D(R) = D^{(1)}(R) \oplus D^{(2)}(R) \iff \mathbf{L} = \mathbf{L}^{(1)} \oplus \mathbf{L}^{(2)} \tag{18.25}$$

So as far as checking for reducibility and equivalence, we might as well work with the $\mathbf{L}$'s as with the D's.

Let's work out the algebra of the matrices $\{L^i\}$. The transformation of a vector $\mathbf{v}$ under an infinitesimal rotation by θ about an axis $\hat{\mathbf{n}}$ is given by[9]

$$\mathbf{v} \to \mathbf{v} + \theta\hat{\mathbf{n}} \times \mathbf{v} + \mathcal{O}(\theta^2) \tag{18.26}$$

Moreover, the operators $\mathbf{L}$ transform as a vector:

$$D(R^{-1})\,\mathbf{L}D(R) = R\mathbf{L} \tag{18.27}$$

so for an infinitesimal transformation

$$(1 + i\hat{\mathbf{n}}\bullet\mathbf{L}\theta)\mathbf{L}(1 - i\hat{\mathbf{n}}\bullet\mathbf{L}\theta) = \mathbf{L} + \theta\hat{\mathbf{n}} \times \mathbf{L} + \mathcal{O}(\theta^2) \tag{18.28}$$

Equating terms of $\mathcal{O}(\theta)$ gives

$$i[\hat{\mathbf{n}}\bullet\mathbf{L}, \mathbf{L}] = \hat{\mathbf{n}} \times \mathbf{L} \tag{18.29}$$

Letting $\hat{\mathbf{n}}$ be $\hat{\mathbf{x}}$, $\hat{\mathbf{y}}$ or $\hat{\mathbf{z}}$, we obtain the famous angular momentum commutation relations

$$[L_i, L_j] = i\epsilon_{ijk}L_k \tag{18.30}$$

---

[8] [Eds.] Different components of $\mathbf{L}$ do not commute, but $i\hat{\mathbf{n}}\bullet\mathbf{L}$ *does* commute with $e^{-i\hat{\mathbf{n}}\bullet\mathbf{L}\theta}$ because $[i\hat{\mathbf{n}}\bullet\mathbf{L}, i\hat{\mathbf{n}}\bullet\mathbf{L}] = 0$.

[9] [Eds.] Equation (18.26) is the limiting case for infinitesimal θ of *Rodrigues' formula*,

$$\mathbf{v} \to \mathbf{v}' = \mathbf{v}\cos\theta + (\hat{\mathbf{n}} \times \mathbf{v})\sin\theta + \hat{\mathbf{n}}(\hat{\mathbf{n}}\bullet\mathbf{v})(1 - \cos\theta)$$

Alternatively, consider a rotation of $\mathbf{x} = (x, y, z)$ about the z axis though an infinitesimal angle θ:

$$x \to x\cos\theta - y\sin\theta = x - y\theta + \mathcal{O}(\theta^2)$$
$$y \to y\cos\theta + x\sin\theta = y + x\theta + \mathcal{O}(\theta^2)$$

which is the same as $\mathbf{x} \to \mathbf{x} + \theta\hat{\mathbf{z}} \times \mathbf{x} + \mathcal{O}(\theta^2)$.

sum on k implied. The generators $\mathbf{L}$ are said to form a representation of the **Lie algebra** of the rotation group; the D's form a representation of the **Lie group.**[10] Any finite dimensional set of matrices that form a representation of the rotation group necessarily lead to a triplet of finite dimensional matrices that obey the angular momentum commutation rules. Thus if we can find, up to equivalence and direct sum, all matrices that obey these commutation relations, we will have all representations of the rotation group. (We might find some things that *aren't* representations. I won't take the time to show you that the process is reversible.)

This problem was solved by Pauli.[11] Irreducible representations $D^{(s)}(R)$ are labeled by an index s called the **spin**:

$$D^{(s)}(R) = e^{-i\mathbf{L}^{(s)}\bullet\hat{\mathbf{n}}\theta} \tag{18.31}$$

where $\mathbf{L}^{(s)}$ is a triplet of matrices appropriate to the spin s. Let me recall a number of well-known facts about these matrices. The index s equals $\{0, \frac{1}{2}, 1, \dots\}$, etc. The dimension of the representation $D^{(s)}(R)$ is $2s+1$. The square of $\mathbf{L}^{(s)}$ is a multiple of the corresponding identity matrix, I:

$$\mathbf{L}^{(s)}\bullet\mathbf{L}^{(s)} = s(s+1)I \tag{18.32}$$

It is convenient to label eigenstates by eigenvalues m of $L_z = L^3$. I'll now switch to Dirac notation even though I've only got a finite dimensional space:

$$L_z\,|m\rangle = m\,|m\rangle\,; \qquad m = -s, -s+1, -s+2, \dots, s-1, s \tag{18.33}$$

The eigenvalue m takes as many values as the representation's dimension, $2s+1$. The first few matrices $\mathbf{L}^{(s)}$ are:

$$\begin{aligned} s = 0&: \ \mathbf{L}^{(0)} = 0 \\ s = \tfrac{1}{2}&: \ \mathbf{L}^{(\frac{1}{2})} = \tfrac{1}{2}\boldsymbol{\sigma} = \tfrac{1}{2}\left(\boldsymbol{\sigma}_x, \boldsymbol{\sigma}_y, \boldsymbol{\sigma}_z\right) \quad \text{(the Pauli } \sigma \text{ matrices)} \\ s = 1&: \ (L_i^{(1)})_{jk} = -i\epsilon_{ijk} \end{aligned} \tag{18.34}$$

(Note that the bold type for $\boldsymbol{\sigma}$ is doing double duty: for the vector nature of the triplet of sigmas, and also to remind you that each sigma is a 2×2 matrix.) For larger values of m, you can find worked out, in nearly every quantum mechanics textbook, the explicit matrix elements of L_x, L_y and L_z in this m basis. We can always choose our basis such that these matrices are Hermitian:

$$\mathbf{L}^{(s)} = \mathbf{L}^{(s)\dagger} \tag{18.35}$$

This is not surprising, as the eigenvalues are observables. So the representation matrices $D^{(s)}$ are *unitary*. This is a special feature of the rotation group, or indeed of any *compact*

---

[10] [Eds.] The reader has likely encountered the concepts of Lie groups and Lie algebras in earlier courses. Briefly, *Lie groups* are groups whose elements are analytic functions of one or more continuous parameters; every Lie group thus contains an infinite number of elements. The most familiar is probably SO(2), the group of rotations in a plane; each element is characterized by a single parameter, the angle through which the rotation is carried out. A Lie group can be constructed by the exponentiation of a set of parameters multiplying a finite set of generators, which among themselves satisfy the group's *Lie algebra*. See §36.2, note 8, p. 782 and note 16, p. 784.

[11] [Eds.] Coleman is probably referring to Pauli's paper, "Zur quantenmechanik des magnetischen elektrons", (On the quantum mechanics of magnetic electrons) *Zeits. f. Phys.* **43** (1927) 601–623, which introduces the Pauli matrices. Reprinted in L. C. Biedenharn and H. van Dam, *Quantum Theory of Angular Momentum*, Academic Press, 1965. English translation by David H. Delphenich online at `http://neo-classical-physics.info/electromagnetism.html`.

group, any group of finite "volume". It is *not* true of the Lorentz group, as we'll see. Finally, the analog of (18.7) is true for the integer values of s, but true only to within a phase for half-integer values of s: they are double-valued. For any rotation,

$$R(\theta_1\hat{\mathbf{n}})R(\theta_2\hat{\mathbf{n}}) = R((\theta_1+\theta_2)\hat{\mathbf{n}}) = I, \text{ if } \theta_1+\theta_2 = 2\pi \tag{18.36}$$

We should expect then that $D^{(s)}(R(2\pi\hat{\mathbf{n}})) = D^{(s)}(R(0)) = 1$. However, it turns out

$$D^{(s)}(R(2\pi\hat{\mathbf{n}})) = (-1)^{2s} \tag{18.37}$$

The representation is only good to within a factor of -1 for half-integer values of s. The double-valued character of the half-integer representations will not prevent our using them for physical purposes.

18.3 Irreducible representations of the Lorentz group

I will now go through this whole routine for the Lorentz group. You might expect this will take a substantial investment of time and effort. It is not so, by a fluke which we will soon discover. One subgroup of the Lorentz group is of course the rotation group. By an abuse of notation, I will indicate these rotations by the symbol R even though R is no longer a 3×3 matrix, but now a 4×4 matrix acting trivially on the time components of 4-vectors. Another subgroup of the Lorentz group concerns pure accelerations, or **boosts**. A boost $A(\hat{\mathbf{a}}\phi)$ along a given axis $\hat{\mathbf{a}}$ and velocity parameter ϕ (called the **rapidity**) is a pure Lorentz transformation that takes a particle at rest and changes its velocity to some new value along that axis.[12] For example, a boost $A(\hat{\mathbf{z}}\phi)$ along the $z = x^3$ direction by an amount ϕ is defined as

$$A(\hat{\mathbf{z}}\phi)\colon \begin{cases} x^0 \to x'^0 = x^0\cosh\phi - x^3\sinh\phi \\ x^3 \to x'^3 = x^3\cosh\phi - x^0\sinh\phi \end{cases} \tag{18.38}$$

while x^1 and x^2 are unchanged. The hyperbolic tangent of the rapidity ϕ is the new speed;[13]

$$\tanh\phi = v, \quad 0 \le \phi < \infty \tag{18.39}$$

This is easy to see by considering x'^3 to be the z component of the primed frame's origin. Then

$$x'^3 = 0 = x^3\cosh\phi - x^0\sinh\phi \;\Rightarrow\; \tanh\phi = \frac{x^3}{x^0} = \frac{z}{t} = v \tag{18.40}$$

It's standard special relativity lore that every Lorentz transformation can be written as a product of a rotation and an acceleration. If we know the representation matrices for the rotations and the accelerations, we know them for everything. As with the rotations, we have defined things with this angle ϕ so that two successive boosts by different hyperbolic angles ϕ and ϕ' along the same axis give a combined boost along the same axis:

$$A(\hat{\mathbf{a}}\phi)A(\hat{\mathbf{a}}\phi') = A(\hat{\mathbf{a}}(\phi+\phi')) \tag{18.41}$$

---

[12] [Eds.] In his lectures, Coleman used the same symbol, $\mathbf{e}$, for both the axis of rotations and the axis of boosts. To avoid possible confusion, the axis for a boost will be denoted by the unit vector $\hat{\mathbf{a}}$.

[13] [Eds.] The Lorentz group is non-compact because the parameter ϕ in (18.39) is unbounded, and the matrix elements sinh and cosh increase monotonically with ϕ.

Thus we can treat the rotations as we treated them before, and the accelerations in exactly the same way as the rotations, simply replacing R's by A's at appropriate points.

As before (18.21), define

$$\hat{\mathbf{n}}\bullet\mathbf{L} = i\frac{\partial}{\partial\theta}D(R(\hat{\mathbf{n}}\theta))\bigg|_{\theta=0} \tag{18.42}$$

and analogously

$$\hat{\mathbf{a}}\bullet\mathbf{M} = i\frac{\partial}{\partial\phi}D(A(\hat{\mathbf{a}}\phi))\bigg|_{\phi=0} \tag{18.43}$$

The $\{M_i\}$ will generate the boosts just as the $\{L_i\}$ generate the rotations. We find, with the initial conditions that $D(R(0)) = D(A(0)) = 1$, that

$$D(R(\hat{\mathbf{n}}\theta)) = e^{-i\hat{\mathbf{n}}\bullet\mathbf{L}\theta} \tag{18.44}$$

$$D(A(\hat{\mathbf{a}}\phi)) = e^{-i\hat{\mathbf{a}}\bullet\mathbf{M}\phi} \tag{18.45}$$

The next step is to figure out all the commutators of $\mathbf{L}$ and $\mathbf{M}$. If we know $\mathbf{L}$ and $\mathbf{M}$ we know the representation matrix for an arbitrary rotation and an arbitrary boost, and by multiplication, we can find the representation matrix for any general Lorentz transformation. I won't compute the commutators for you, but I'll write them down and try to make them plausible. For the rotation generators,

$$[L_i, L_j] = i\epsilon_{ijk}L_k \tag{18.46}$$

That of course is no news; these commutators are the same as (18.30) because the rotations are a subgroup.

$$[L_i, M_j] = i\epsilon_{ijk}M_k \tag{18.47}$$

This is not a big surprise; it's just the statement that $\mathbf{M}$ transforms like a vector under infinitesimal rotations, just like $\mathbf{L}$. Both (18.46) and (18.47) can be shown with the same method as (18.30). We also have

$$[M_i, L_j] = i\epsilon_{ijk}M_k \tag{18.48}$$

the minus sign from swapping i and j is compensated for by the minus sign from exchanging the two terms in the commutators. The only one you have to work at is

$$[M_i, M_j] = -i\epsilon_{ijk}L_k \tag{18.49}$$

Because $\{M_i\}$ transform as a 3-vector, the method used to derive (18.30) fails to produce (18.49), and I leave this as an exercise.[14] The minus sign in (18.49) is important. If we were

---

[14] [Eds.] The boost equivalent to Rodrigues' formula is

$$t \to t' = t\cosh\phi - (\hat{\mathbf{a}}\bullet\mathbf{x})\sinh\phi$$

$$\mathbf{x} \to \mathbf{x}' = \mathbf{x} - \hat{\mathbf{a}}t\sinh\phi + (\hat{\mathbf{a}}\bullet\mathbf{x})\hat{\mathbf{a}}(\cosh\phi - 1)$$

Let M_i generate a boost along the x^i axis. Under an infinitesimal boost along $\hat{\mathbf{x}}$, $x'^{\mu} = x^{\mu} + i\phi\,(M_1)^{\mu}{}_{\nu}\,x^{\nu} + \mathcal{O}(\phi^2)$:

$$\begin{pmatrix} t' \\ x' \\ y' \\ z' \end{pmatrix} = \begin{pmatrix} t \\ x \\ y \\ z \end{pmatrix} + \phi\begin{pmatrix} 0 & -1 & 0 & 0 \\ -1 & 0 & 0 & 0 \\ 0 & 0 & 0 & 0 \\ 0 & 0 & 0 & 0 \end{pmatrix}\begin{pmatrix} t \\ x \\ y \\ z \end{pmatrix} + \mathcal{O}(\phi^2)$$

The matrices M_2 and M_3 are found in the same way. It follows easily that, e.g., $[M_1, M_2] = -iL_3$. See Problem 10.2, p. 387.

doing the four-dimensional rotation group SO(4), rather than SO(3,1), we could've made almost the same definitions with sinh's and cosh's replaced by sines and cosines, and then we would have gotten a plus sign in this last commutator.

To show you that the commutators are at least self-consistent, let me remark that if the theory we are studying has not only Lorentz invariance but also parity invariance—it need not, of course—then we can figure out how **L** and **M** transform under parity. Parity commutes with rotations. And therefore

$$P\colon \mathbf{L} \to \mathbf{L} \tag{18.50}$$

On the other hand, parity switches the sign of a boost, because it transforms a velocity to its opposite. So **M** goes to minus **M**:

$$P\colon \mathbf{M} \to -\mathbf{M} \tag{18.51}$$

Please notice that these commutators are consistent with that, because they are unchanged by the replacements **L** into **L** and **M** into −**M**: (18.46) is totally unchanged; (18.47) gets a − sign on both the right- and left-hand side; and (18.49) gets two minus signs on the left-hand side and no change on the right-hand side.

I will now find, in a very few lines, all the irreducible representations of the Lorentz algebra. It's based on a special trick. If we were unfortunate enough to live in five-dimensional space, the trick wouldn't work. But fortunately we live in four-dimensional spacetime and the trick works. I define operators analogous to the raising and lowering operators you'll recall from quantum mechanics,

$$\mathbf{J}^{(\pm)} = \tfrac{1}{2}\left(\mathbf{L} \pm i\mathbf{M}\right) \qquad \text{so} \tag{18.52}$$

$$\mathbf{L} = \mathbf{J}^{(+)} + \mathbf{J}^{(-)} \tag{18.53}$$

$$\mathbf{M} = -i(\mathbf{J}^{(+)} - \mathbf{J}^{(-)}) \tag{18.54}$$

Let us compute the commutation rules for $J^{(+)}$ and $J^{(-)}$:

$$[J_i^{(+)}, J_j^{(+)}] = \tfrac{1}{4} i\epsilon_{ijk}\left(L_k + iM_k + iM_k + L_k\right) = i\epsilon_{ijk}J_k^{(+)} \tag{18.55}$$

The same result is obtained with (−) instead of (+) in both places:

$$[J_i^{(-)}, J_j^{(-)}] = i\epsilon_{ijk}J_k^{(-)} \tag{18.56}$$

What about $J_i^{(+)}$ with $J_j^{(-)}$? We find

$$[J_i^{(+)}, J_j^{(-)}] = \tfrac{1}{4} i\epsilon_{ijk}\left(L_k - iM_k + iM_k - L_k\right) = 0 \tag{18.57}$$

Thus $\{J_i^{(+)}\}$ and $\{J_i^{(-)}\}$ *commute.* We have reduced this apparently formidable algebra into two commuting angular momentum algebras. Exactly this problem arises in ordinary non-relativistic quantum mechanics, where we have both orbital and spin angular momentum, each of which obey the commutation rules of the rotation group, but which commute with each other.

It is now a snap to write down a complete set of irreducible, inequivalent representations of the Lorentz group. They are characterized by two independent spin quantum numbers, s_+ and s_-, one each for $\mathbf{J}^{(+)}$ and $\mathbf{J}^{(-)}$, respectively, and are written as

$$D^{(s_+,\,s_-)}(\Lambda), \quad s_\pm = 0, \tfrac{1}{2}, 1, \ldots \tag{18.58}$$

The squares of these operators $\mathbf{J}^{(+)}$ and $\mathbf{J}^{(-)}$ are multiples of the identity:

$$\mathbf{J}^{(+)} \cdot \mathbf{J}^{(+)} = s_+(s_+ + 1) \tag{18.59}$$

$$\mathbf{J}^{(-)} \cdot \mathbf{J}^{(-)} = s_-(s_- + 1) \tag{18.60}$$

The complete set of basis states is described by two numbers, m_+ and m_-, eigenvalues of $J_z^{(+)}$ and $J_z^{(-)}$, respectively:

$$J_z^{(\pm)} |m_+, m_-\rangle = m_\pm |m_+, m_-\rangle \tag{18.61}$$

The states $|m_+, m_-\rangle$ are simultaneous eigenstates of the commuting operators $J_z^{(+)}$ and $J_z^{(-)}$. The eigenvalues m_+ run from $-s_+, -s_+ + 1, \dots, s_+$, and the eigenvalues m_- run from $-s_-, -s_- + 1, \dots, s_-$. Hence

$$\dim D^{(s_+,\, s_-)}(\Lambda) = (2s_+ + 1)(2s_- + 1) \tag{18.62}$$

The dimension of $D^{(s_+,\, s_-)}(\Lambda)$ is also the number of basic vectors. To make things more explicit, consider the matrix element

$$\langle m'_+, m'_- |\mathbf{J}^{(+)}| m_+, m_-\rangle = \delta_{m_-, m'_-} \langle m'_+ |\mathbf{J}^{(+)}| m_+\rangle \tag{18.63}$$

$\mathbf{J}^{(+)}$ has nothing to do with m_-, so I simply get δ_{m_-, m'_-} times the matrix element $\langle m'_+ |\mathbf{J}^{(+)}| m_+\rangle$, full of square roots, which you will find in any elementary quantum mechanics book. The same equation holds if the plus and minus signs are swapped. We have two commuting "angular momenta", so there's no problem in finding all the irreducible, inequivalent (finite dimensional) representations of SO(3,1).

We can always choose things so that $\mathbf{J}^{(+)}$ and $\mathbf{J}^{(-)}$ are Hermitian matrices. $\mathbf{L}$, the sum of these, is indeed Hermitian, so the representations $D(R(\hat{\mathbf{n}}\theta))$ are unitary:

$$\mathbf{L} \text{ is Hermitian} \Rightarrow D(R(\hat{\mathbf{n}}\theta)) = e^{-i\hat{\mathbf{n}} \cdot \mathbf{L}\theta} \text{ is unitary} \tag{18.64}$$

The same is *not* true of $\mathbf{M}$ which is $-i$ times the difference of $\mathbf{J}^{(+)}$ and $\mathbf{J}^{(+)}$. So $\mathbf{M}$ is an anti-Hermitian matrix, and consequently the representations $D(A(\hat{\mathbf{a}}\phi))$ are *not* unitary:[15]

$$\mathbf{M} \text{ is anti-Hermitian} \Rightarrow D(A(\hat{\mathbf{a}}\phi)) = e^{-i\hat{\mathbf{a}} \cdot \mathbf{M}\phi} \text{ is } \textit{not} \text{ unitary} \tag{18.65}$$

18.4 Properties of the SO(3) representations $D^{(s)}$

Now that we have all of the representations of SO(3,1), we would like to know their properties. We can deduce a list of properties just by knowing some elementary facts about the rotation group, SO(3). From these I will derive properties of the representations of SO(3,1).

Complex conjugation

If I complex conjugate (this is not to be confused with taking the Hermitian adjoint) a representation of SO(3), or in fact of any group, I again obtain a representation

$$\text{If } D^{(s)}(R) \text{ is a representation of SO(3), then so is } D^{(s)}(R)^* \tag{18.66}$$

---

[15] [Eds.] Because the Lorentz group is non-compact, the faithful, finite dimensional representations $D(A(\hat{\mathbf{a}}\phi)) = e^{-i\hat{\mathbf{a}} \cdot \mathbf{M}\phi}$ cannot be unitary. See note 3 on p. 371.

because the product of two complex conjugated matrices is the complex conjugate of the product. Since there's only one irreducible representation of a given dimension, the complex conjugate must be equivalent to $D^{(s)}(R)$:

$$D^{(s)}(R) \sim D^{(s)}(R)^* \tag{18.67}$$

That is, for some matrix T

$$\begin{aligned} \left[e^{-i\mathbf{J}^{(s)} \bullet \hat{\mathbf{n}}\theta}\right]^* &= T\left[e^{-i\mathbf{J}^{(s)} \bullet \hat{\mathbf{n}}\theta}\right]T^{-1} \\ e^{i\mathbf{J}^{(s)*} \bullet \hat{\mathbf{n}}\theta} &= e^{-iT\mathbf{J}^{(s)}T^{-1} \bullet \hat{\mathbf{n}}\theta} \end{aligned} \tag{18.68}$$

and therefore we must have

$$\mathbf{J}^{(s)} \sim -\mathbf{J}^{(s)*} \tag{18.69}$$

This doesn't necessarily mean we can write the **J**'s as imaginary matrices. It just means that there is some transformation T such that

$$T\mathbf{J}^{(s)}T^{-1} = -\mathbf{J}^{(s)*} \tag{18.70}$$

(the *same* T, of course, for all three J_i's for a given s).

Direct product

If we have a set of fields that transform under a rotation as an irreducible representation $D^{(s_1)}(R)$, a vector, a spinor or something, and if we have a second set of fields that transform as some other irreducible representation $D^{(s_2)}(R)$, we can consider all products of components of the two fields. This defines a brand new representation of the group called the **direct product**, denoted by

$$D^{(s_1)}(R) \otimes D^{(s_2)}(R) \tag{18.71}$$

The dimension of the direct product is of course the product of the dimensions of the two representations. Because you're multiplying two things together, you have two indices to play with:

$$\dim\left[D^{(s_1)}(R) \otimes D^{(s_2)}(R)\right] = (2s_1+1)(2s_2+1) \tag{18.72}$$

This product is certainly a representation. But it's usually *not* an irreducible representation. There is a rule for finding how it breaks up into irreducible representations. It's equivalent to a direct sum which I will indicate this way,[16]

$$D^{(s_1)}(R) \otimes D^{(s_2)}(R) \sim \oplus \sum_{s=|s_1-s_2|}^{s_1+s_2} D^{(s)}(R) \tag{18.73}$$

The quantity on the right is a *direct* sum over s, as in (18.16), not a numerical sum. s goes from $|s_1 - s_2|$ to $s_1 + s_2$ by unit integer steps. This is the so-called rule for addition of angular momentum written in slightly more sophisticated language, and you should be familiar with it. Thus for example if I multiply together $D^{(\frac{1}{2})}$ times $D^{(\frac{1}{2})}$, the product of two spinors gives four objects, and I obtain a $D^{(0)} \oplus D^{(1)}$, a scalar and a vector, a one-dimensional object and a three-dimensional object.

[16] [Eds.] Sometimes called "the Clebsch–Gordan series" in the literature.

Exchange symmetry

There's also a sub-item we can add for the direct product. If $s_1 = s_2$, then it's a sensible question to ask what happens when you exchange them, since they transform in the same way. If $s_1 = s_2$, then $D^{(2s_1)}$, the irreducible representation of highest spin, is symmetric under exchange. That is probably familiar to you, but if not, it can be found in many quantum mechanics texts. Then $D^{(2s_1-1)}$ is antisymmetric under exchange, etc. These are three facts about the rotation group. I presume you've seen them before, though perhaps in different language. If they seem new, you may be suffering merely from linguistic impedance matching.

18.5 Properties of the SO(3,1) representations $D^{(s_+,\, s_-)}$

I will now take what we know about the representations $D^{(s)}(R)$ to discuss seven questions about the properties of the SO(3,1) representations $D^{(s_+,\, s_-)}(\Lambda)$. The discussions of these questions will be very brief, because we know the answers, we've just got to put things together and keep track of factors like i's.[17]

Complex conjugation

The equivalence (18.67) between a representation of SO(3) and its complex conjugate doesn't quite work for SO(3,1). $\mathbf{J}^{(+)}$ and $\mathbf{J}^{(-)}$ are ordinary rotation matrices, and for any particular value of s, they are equivalent to minus their conjugates by (18.70). Therefore $\mathbf{L}$, which is their sum, is equivalent to $-\mathbf{L}^*$:

$$\mathbf{L} \sim -\mathbf{L}^* \tag{18.74}$$

But $\mathbf{M}$, $-i$ times the difference of $\mathbf{J}^{(+)}$ and $\mathbf{J}^{(-)}$, is equivalent to $+\mathbf{M}^*$, because of the intervening i:

$$\mathbf{M}^* = \left[-i\left(\mathbf{J}^{(+)} - \mathbf{J}^{(-)}\right)\right]^* = i\left(\mathbf{J}^{(+)*} - \mathbf{J}^{(-)*}\right) \sim i\left(-\mathbf{J}^{(+)} + \mathbf{J}^{(-)}\right) \sim \mathbf{M} \tag{18.75}$$

$D^{(s)}(R)$ is equivalent to its complex conjugate because of the sign change of the generators $\mathbf{J}$. Here, the disgusting lack of sign change in $\mathbf{M}$ prevents $D^{(s_+,\, s_-)}$ from being equivalent to its conjugate. We can introduce a sign change in the right place if we exchange $\mathbf{J}^{(+)}$ and $\mathbf{J}^{(-)}$. This will not change the sign of $\mathbf{L}$, but it does change the sign of $\mathbf{M}$. Therefore we deduce

$$\left[D^{(s_+,\, s_-)}(\Lambda)\right]^* \sim D^{(s_-,\, s_+)}(\Lambda) \tag{18.76}$$

$D^{(s_+,\, s_-)}(\Lambda)$ is equivalent under complex conjugation to $D^{(s_-,\, s_+)}(\Lambda)$. The effects of complex conjugation can be canceled out up to an equivalence transformation by exchanging $\mathbf{J}^{(+)}$ and $\mathbf{J}^{(-)}$. Notice that there is some funny business going on. If I have a set of fields, and they transform in a certain way, their complex conjugates do not transform in the same way unless s_+ is equal to s_-.

---

[17] [Eds.] Near this point in the videotape of Lecture 18, a student yawns loudly. Coleman responds: "Come on, you can't say it's *boring*. It's not boring. As Dr. Johnson said, in another context, 'A man who is tired of group theory is tired of life, sir.' I made a killing. I get a good salary. It's all done with group theory! We were even thinking of advertising on matchbooks: 'Learn how to make $20,000 a year through group theory!' But then the job market collapsed, so the whole scheme fell apart..." (Samuel Johnson (1709–1784), to his friend and biographer James Boswell: "Sir, when a man is tired of London, he is tired of life." Entry for September 20, 1777 in J. Boswell, *The Life of Samuel Johnson, LL.D.*, 1791.)

Parity

Recall that parity turns $\mathbf{L}$ into $\mathbf{L}$ and $\mathbf{M}$ into $-\mathbf{M}$. The operation that turns $\mathbf{M}$ into $-\mathbf{M}$ again can be thought of as exchanging $\mathbf{J}^{(+)}$ and $\mathbf{J}^{(-)}$. Equivalently we could say, "Parity exchanges $\mathbf{J}^{(+)}$ and $\mathbf{J}^{(-)}$." Thus if we wish to have a parity-conserving theory involving only fields that transform according to the representation $(\frac{3}{2}, \frac{1}{2})$ we have the chance of the proverbial snowball in hell: Parity acting on a field that transforms like $D^{(s_+,\, s_-)}(\Lambda)$ must turn it into field that transforms like $D^{(s_-,\, s_+)}(\Lambda)$:

$$P\colon D^{(s_+,\, s_-)}(\Lambda) \to D^{(s_-,\, s_+)}(\Lambda) \tag{18.77}$$

On the other hand, parity plus complex conjugation turns a field into one that transforms in the same way. We will see later on that this property will make it easy for us to construct theories that are *CP* invariant, but neither *C* invariant nor *P* invariant, a nice thing for weak interaction theory. Onward!

Direct product

We've got two independent angular momenta. We add them together independently. There's no problem:

$$D^{(s_+^1,\, s_-^1)}(\Lambda) \otimes D^{(s_+^2,\, s_-^2)}(\Lambda) \sim \oplus \sum_{s_+} \sum_{s_-} D^{(s_+,\, s_-)}(\Lambda) \tag{18.78}$$

where s_+ goes by unit steps from $|s_+^1 - s_+^2|$ to $s_+^1 + s_+^2$ and s_- independently does the same, between $|s_-^1 - s_-^2|$ and $s_-^1 + s_-^2$. Here are two angular momenta that don't talk with each other. Add them together, and they still don't talk with each other.

Exchange symmetry

Exchange symmetry is a reasonable topic only if two representations are of the same spin, just as before: $s_\pm^1 = s_\pm^2$. Well if you exchange 'em, you exchange both the s_+ and the s_- parts, so it's symmetric if the two parts are individually symmetric or if the two parts are individually antisymmetric, and antisymmetric otherwise. Thus

$$\begin{aligned} s_+ &= 2s_+^1,\ s_- = 2s_-^1 \quad \text{is symmetric} \\ s_+ &= 2s_+^1 - 1,\ s_- = 2s_-^1 \quad \text{is antisymmetric} \end{aligned} \tag{18.79}$$

because it's antisymmetric in the first variable and symmetric in the second; likewise $s_+ = 2s_+^1$, $s_- = 2s_-^1 - 1$ is antisymmetric, etc; and

$$s_+ = 2s_+^1 - 1,\ s_- = 2s_-^1 - 1 \quad \text{is symmetric} \tag{18.80}$$

because it's antisymmetric in both variables.

The rotation subgroup of the Lorentz group

What happens when I look at just the rotations, at the SO(3) subgroup of the Lorentz group SO(3,1)? Any representation of a big group is a representation of any subgroup, but if it's an irreducible representation of the big group, it might not be an irreducible representation of the subgroup. Well,

$$D(R(\hat{\mathbf{n}}\theta)) = e^{-i(\mathbf{L}\,\bullet\,\hat{\mathbf{n}}\theta)} = e^{-i(\mathbf{J}^{(+)}+\mathbf{J}^{(-)})\,\bullet\,\hat{\mathbf{n}}\theta} \tag{18.81}$$

Thus if we just restrict ourselves to rotations, we can think of $\mathbf{J}^{(+)}$ and $\mathbf{J}^{(-)}$ as being like orbital angular momentum and spin angular momentum—it's as if we have coupled orbital angular momentum and spin angular momentum, and only consider the combined rotation group: simultaneous spin and orbital rotations by the same angle. This is just our direct product formula again, so I have

$$D^{(s_+,\, s_-)}(R) \sim \oplus \sum_{s=|s_+-s_-|}^{s_++s_-} D^{(s)}(R) \tag{18.82}$$

We'll see some examples in a moment.

How are vectors represented?

Where in our representations will we find a vector field like V_μ? A vector field transforms according to *some* representation of the Lorentz group. That representation is pretty obviously irreducible, so it must be somewhere in our catalog. What do we know about a vector? First, we know it's got four components, so the representation is four-dimensional:

$$(2s_+ + 1)(2s_- + 1) = 4 \tag{18.83}$$

Since both of these factors are integers, there are not many solutions. To be precise we have three possible solutions. First, we could have $s_+ = \frac{3}{2}$, $s_- = 0$. That gives a product of 4×1. But it's obviously no good because it is not equivalent to its complex conjugate, whereas a vector representation is certainly equivalent to its complex conjugate; we'd need to have $s_+ = s_-$. This representation also does not admit a parity (again, we'd need $s_+ = s_-$) and a vector certainly does. So this representation fails on two counts. And the representation $s_+ = 0$, $s_- = \frac{3}{2}$ is also ruled out.

Finally, we have $s_+ = s_- = \frac{1}{2}$. This is the only possibility, and as Sherlock Holmes used to say, therefore it is the right answer.[18] So a vector field transforms according to the four-dimensional irreducible representation $D^{(\frac{1}{2},\frac{1}{2})}$. Let's check that, by using our previous result, (18.82):

$$D^{(s_+,\, s_-)}(R) \sim D^{(1)}(R) \oplus D^{(0)}(R) \tag{18.84}$$

The direct sum goes from $|\frac{1}{2} - \frac{1}{2}|$ to $(\frac{1}{2} + \frac{1}{2})$ by integer steps, so there are only these two. The first, $D^{(1)}(R)$, is a spatial 3-vector, and the second is a scalar, a single number that doesn't transform under rotations. Is this indeed what happens to a Lorentz 4-vector when we restrict ourselves to rotations? It certainly is: the time component is unaffected by rotations, and the three space components transform like a 3-vector. So it all holds together; it checks.

How are tensors represented?

Once we have vectors, we can construct tensors, because tensors are direct products of vectors. (I'll only talk about rank 2 tensors.) Where are the tensors in our classification of representations of the Lorentz group? We can find them if we think about their properties. With our formula (18.78), we can figure out how rank two tensors like $T_{\mu\nu}$ transform. It doesn't matter whether I write upper or lower indices, of course. That's just an equivalence

[18] [Eds.] "How often have I said to you that when you have eliminated the impossible, whatever remains, *however improbable*, must be the truth?" Sherlock Holmes to Dr. John Watson. Sir Arthur Conan Doyle, *The Sign of Four*, Smith, Elder & Co, 1908, Chapter 6, "Sherlock Holmes gives a demonstration", p. 94.

transformation, with the metric $g_{\mu\nu}$ as the matrix that effects the equivalence transformation. There is a basis of all two index tensors, $T_{\mu\nu}$, for a 16-dimensional representation of the Lorentz group. If I take such a tensor and Lorentz transform it in the standard way I get 16 linearly independent objects that shuffle among themselves according to the Lorentz transformation I have made. The transformation of $T_{\mu\nu}$ defines some 16×16 matrix representation $D(\Lambda)$:

$$\dim D(\Lambda) = 16 \qquad \text{(for a rank 2 tensor representation)} \tag{18.85}$$

Its form depends on how I choose the basis for the 16-dimensional space of tensors. I want to find out what it is in terms of irreducible representations. A tensor is an object that transforms like the product of two vectors. So

$$\begin{aligned} D(\Lambda) &\sim D^{(\frac{1}{2},\,\frac{1}{2})}(\Lambda) \otimes D^{(\frac{1}{2},\,\frac{1}{2})}(\Lambda) \\ &\sim D^{(1,\,1)}(\Lambda) \oplus D^{(1,\,0)}(\Lambda) \oplus D^{(0,\,1)}(\Lambda) \oplus D^{(0,\,0)}(\Lambda) \end{aligned} \tag{18.86}$$

Let's check our dimensions. The dimension of $D^{(\frac{1}{2},\,\frac{1}{2})}(\Lambda)$ is 4, and $4 \times 4 = 16$, as required. The direct product is given by our product algorithm (18.78). For the rotation group, one half and one half gives you zero and one. Here we're doing two such sums independently and getting all possible combinations.

Now let's check that this is right by adding up the dimensions. Using (18.62) we have

$$\begin{aligned} \dim D^{(1,\,1)}(\Lambda) &= 9 \\ \dim D^{(1,\,0)}(\Lambda) &= 3 \\ \dim D^{(0,\,1)}(\Lambda) &= 3 \\ \dim D^{(0,\,0)}(\Lambda) &= 1 \end{aligned} \tag{18.87}$$

And indeed, $9+3+3+1$ is 16. We also know how these things transform under permutation of the indices $s^i_\pm$, $i = \{1, 2\}$. If we think of this $D^{(s_+,\,s_-)}$ as in (18.78), then from (18.79), $D^{(1,\,1)}(\Lambda)$ is symmetric under the exchange ($1 \leftrightarrows 2$), and the representations $D^{(1,\,0)}(\Lambda)$ and $D^{(1,\,0)}(\Lambda)$ are antisymmetric. Likewise, in agreement with (18.80), the representation $D^{(0,\,0)}(\Lambda)$ is symmetric because it's antisymmetric in both the indices. Thus the general theory of representations of the Lorentz group says that we should be able to break the 16-dimensional space up into a nine-dimensional subspace, two three-dimensional subspaces and a one-dimensional subspace. When we apply the Lorentz transformation, a tensor constructed out of basis tensors in any one of these subspaces goes into a tensor in the same subspace. Parts of the tensor in different subspaces don't talk to each other under Lorentz transformations; they each transform independently. That's what the direct sum means.

Let's try to figure out what this break-up is in traditional tensor language. Every rank 2 tensor $T_{\mu\nu}$ can be written unambiguously as the sum of a symmetric tensor, $S_{\mu\nu}$, and an antisymmetric tensor, $A_{\mu\nu}$:

$$T_{\mu\nu} = S_{\mu\nu} + A_{\mu\nu} \tag{18.88}$$

with

$$\begin{aligned} S_{\mu\nu} &= \tfrac{1}{2}(T_{\mu\nu} + T_{\nu\mu}) \\ A_{\mu\nu} &= \tfrac{1}{2}(T_{\mu\nu} - T_{\nu\mu}) \end{aligned} \tag{18.89}$$

Since the two indices $\{\mu, \nu\}$ transform identically, symmetric tensors transform into symmetric tensors, and antisymmetric tensors go into antisymmetric tensors under Lorentz transformations. So (18.89) is a Lorentz invariant break-up. Thus I have written my representation as a direct sum, and the Lorentz transformation can be written as a block diagonal matrix, with a part that acts on the space of symmetric tensors and a part that acts on the space of antisymmetric tensors. How many linearly independent components does a symmetric tensor have? For $n \times n$ matrices there are $\frac{1}{2}n(n+1)$ symmetric elements. For $n = 4$, we have a ten-dimensional subspace. The number of antisymmetric matrices fills a $16 - 10 = 6$-dimensional subspace. Let's check that with our algorithm. We have two symmetric subspaces, the nine-dimensional representation $D^{(1,\,1)}(\Lambda)$ and the one-dimensional representation $D^{(0,\,0)}(\Lambda)$. Then

$$D(\Lambda) \sim \underbrace{D^{(1,\,1)}(\Lambda) \oplus D^{(0,\,0)}(\Lambda)}_{\text{symmetric subspace of dim } 9+1} \quad \oplus \quad \underbrace{D^{(1,\,0)}(\Lambda) \oplus D^{(0,\,1)}(\Lambda)}_{\text{anti-sym. subspace of dim } 3+3} \tag{18.90}$$

The symmetric ten-dimensional subspace is written as a direct sum of a nine-dimensional subspace and a one-dimensional subspace; the antisymmetric six-dimensional subspace is written as the direct sum of two three-dimensional subspaces. So far, things are checking out.

Let's now consider a symmetric tensor, $S_{\mu\nu}$. If we think of $S_{\mu\nu}$ as a matrix, we can break it up into a traceless part and a part proportional to the metric tensor $g_{\mu\nu}$:

$$S_{\mu\nu} - \tfrac{1}{4}g_{\mu\nu}S^{\lambda}{}_{\lambda} + \tfrac{1}{4}g_{\mu\nu}S^{\lambda}{}_{\lambda} = \widehat{S}_{\mu\nu} + \tfrac{1}{4}g_{\mu\nu}S^{\lambda}{}_{\lambda} \tag{18.91}$$

$\widehat{S}_{\mu\nu} \equiv S_{\mu\nu} - \frac{1}{4}g_{\mu\nu}S^{\lambda}{}_{\lambda}$ is traceless, as you can quickly verify:

$$\widehat{S}^{\mu}{}_{\mu} = 0 \tag{18.92}$$

Thus we have broken up the ten-dimensional subspace of symmetric tensors into a nine-dimensional subspace of traceless, symmetric tensors, and a one-dimensional subspace of symmetric tensors proportional to $g_{\mu\nu}$. A tensor proportional to $g_{\mu\nu}$ stays proportional to $g_{\mu\nu}$ under a Lorentz transformation, and if it's traceless, it remains traceless after the transformation, because these are Lorentz invariant equations. So we have block diagonalized the representation.

The break-up of the antisymmetric tensor $A_{\mu\nu}$ is a little trickier, because we normally don't think of an antisymmetric tensor as being the sum of two three-component objects. You've played with antisymmetric tensors in electromagnetic theory, where the field vectors **E** and **B** combine[19] to form an antisymmetric tensor $F^{\mu\nu}$.You don't think of $F^{\mu\nu}$ as being broken up into the sum of two 3-component objects, each of which transforms only into itself under the action of the Lorentz group, because that's not true of **E** and **B**: they transform into *each other*. The mathematical reason you don't think of this division of $F^{\mu\nu}$ is that the representations $D^{(1,\,0)}(\Lambda)$ and $D^{(0,\,1)}(\Lambda)$ are not *real*; they're complex conjugates of each other, as in (18.76). The breakup of the six-dimensional subspace into two three-dimensional subspaces will in fact involve complex combinations of the components of the antisymmetric tensor. I'll demonstrate how that goes.

For any antisymmetric tensor $A_{\mu\nu}$, define its **dual**, $\star A_{\mu\nu}$:

$$\star A^{\mu\nu} = \tfrac{1}{2}\epsilon^{\mu\nu\lambda\sigma}A_{\lambda\sigma} \qquad (\epsilon^{0123} = +1) \tag{18.93}$$

I've put in a factor $\frac{1}{2}$ because in such a sum over two antisymmetric tensors, there is always double counting. This is a Lorentz invariant way of associating one antisymmetric tensor in a

[19] [Eds.] See Problem 2.3, p. 99.

linear way with another. Just to see what the dual looks like, consider a particular element of $\star A^{\mu\nu}$, say $\star A^{01}$:

$$\star A^{01} = \tfrac{1}{2}\epsilon^{01\lambda\sigma}A_{\lambda\sigma} = \epsilon^{0123}A_{23} = A_{23} \tag{18.94}$$

$\lambda = 2$ and $\sigma = 3$ and *vice versa* give the only non-zero combination; these are equal and you get A_{23}. Lowering the indices, $\star A^{01} = -\star A_{01}$, so

$$\star A_{01} = -A_{23} \tag{18.95}$$

Let's do it again. What is the double dual? Find the double dual of A^{23}:

$$\star\star A^{23} = \epsilon^{2301}\star A_{01} = \star A_{01} = -A_{23} = -A^{23} \tag{18.96}$$

because raising a pair of spatial indices does not change the sign of the tensor, and 2301 is an even permutation of 0123, so $\epsilon^{2301} = +1$. There is nothing special about the set of indices $(0,1)$ and $(2,3)$, so we find

$$\star\star A^{\mu\nu} = -A^{\mu\nu} \tag{18.97}$$

Now the operation of forming a dual of a tensor obviously commutes with all Lorentz transformations, since $\epsilon^{\mu\nu\alpha\beta}$ does,[20] and certainly lowering indices does. Therefore I have a linear operation, $\star$, defined on the six-dimensional space, with the property that its square is -1. I can form eigentensors of this operation, and the eigenvalues λ must have the values $\pm i$, since $\lambda^2 = -1$. That is to say, I can write any $A^{\mu\nu}$ as a linear combination of $A^{(+)}_{\mu\nu}$ and $A^{(-)}_{\mu\nu}$, where

$$A^{(\pm)}_{\mu\nu} = \tfrac{1}{2}\left(A_{\mu\nu} \pm i\star A_{\mu\nu}\right) \tag{18.98}$$

The tensors $A^{(\pm)}_{\mu\nu}$ are eigentensors of the dual operation:

$$\star A^{(\pm)}_{\mu\nu} = \mp i\, A^{(\pm)}_{\mu\nu} \tag{18.99}$$

Therefore we have these two kinds of objects, $A^{(+)}_{\mu\nu}$ and $A^{(-)}_{\mu\nu}$, each of which form a three-dimensional subspace of the six-dimensional space of antisymmetric rank 2 tensors. They are course the representations $D^{(1,\,0)}(\Lambda)$ and $D^{(0,\,1)}(\Lambda)$. I will not bother to work out which is which.

To summarize, a vector transforms according to representation $(\frac{1}{2}, \frac{1}{2})$; a scalar according to representation $(0,0)$; a traceless, symmetric tensor according to the representation $(1,1)$, an antisymmetric tensor according to the *reducible* representation $(1,0) \oplus (0,1)$, which we can reduce if we are willing to form complex combinations.

Next time we will start building field theories from some of the simple representations that we have found here, in particular $D^{(\frac{1}{2},\,0)}$ and $D^{(0,\,\frac{1}{2})}$, which we need for the Dirac equation.

---

[20] [Eds.] Strictly speaking, the Levi–Civita symbol $\epsilon^{\lambda\mu\alpha\beta}$ is a tensor *density*, and under Lorentz transformations

$$\epsilon^{\mu\nu\alpha\beta} \xrightarrow{\Lambda} (\det\Lambda)\,\epsilon^{\mu\nu\alpha\beta}$$

Under *proper* Lorentz transformations (SO(3,1)), the determinant equals 1, so there's no problem.

Problems 10

10.1 In Chapter 16, we computed $\widetilde{\Pi}'(p^2)$, the renormalized meson self-energy operator, to $\mathcal{O}(g^2)$, in Model 3. We expressed in (16.5) the answer as an integral over a single Feynman parameter, x, and we saw that $\widetilde{\Pi}'(p^2)$ was an analytic function of p^2, except for a cut running from $4m^2$ to ∞. In the same theory, compute the renormalized "nucleon" self-energy, $\widetilde{\Sigma}'(p^2)$, again to $\mathcal{O}(g^2)$. Express the answer as an integral over a single Feynman parameter, and show that this too is an analytic function of p^2, except for a cut running from a location you are to find, to ∞.

(1997a 9.1)

10.2 Verify the commutation relations (18.46)–(18.49), using the defining representation of the group, $D(\Lambda) = \Lambda$. For example,

$$D(R(\hat{\mathbf{x}}\theta)) = \begin{pmatrix} 1 & 0 & 0 & 0 \\ 0 & 1 & 0 & 0 \\ 0 & 0 & \cos\theta & -\sin\theta \\ 0 & 0 & \sin\theta & \cos\theta \end{pmatrix} \tag{P10.1}$$

and

$$D(A(\hat{\mathbf{x}}\phi)) = \begin{pmatrix} \cosh\phi & \sinh\phi & 0 & 0 \\ \sinh\phi & \cosh\phi & 0 & 0 \\ 0 & 0 & 1 & 0 \\ 0 & 0 & 0 & 1 \end{pmatrix} \tag{P10.2}$$

Expressions for rotations and boosts along the $\hat{\mathbf{y}}$ and $\hat{\mathbf{z}}$ directions can be found from these by cyclic permutation of x, y and z. Check by explicit computation that

$$[L_x, L_y] = iL_z \qquad [L_x, M_y] = iM_z \qquad [M_x, M_y] = -iL_x \tag{P10.3}$$

(The other relations follow from these by cyclic permutation.)

(1997a 9.2)

Solutions 10

10.1 The renormalized "nucleon" self-energy is, analogous to (15.56),

$$-i\widetilde{\Sigma}'(p^2) = -i\left[\Sigma^f(p^2) - \Sigma^f(m^2) - (p^2-m^2)\left.\frac{d\Sigma^f}{dp^2}\right|_{m^2}\right] \tag{S10.1}$$

where $-i\widetilde{\Sigma}'(p^2)$ is the sum of all two-point 1PI diagrams. At $\mathcal{O}(g^2)$, the only two-point 1PI diagram is shown below:

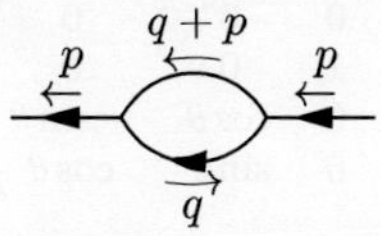

(This is diagram 2.4 (a), following (10.31).) The Model 3 Feynman rules (p. 216) give for this diagram

$$-i\Sigma^f(p^2) = (-ig)^2\int\frac{d^4q}{(2\pi)^4}\frac{i}{(q+p)^2-\mu^2+i\epsilon}\frac{i}{q^2-m^2+i\epsilon} \tag{S10.2}$$

Combining the denominators with a Feynman parameter x, we have

$$\begin{aligned}-i\Sigma^f(p^2) &= \frac{g^2}{(2\pi)^4}\int_0^1 dx\int\frac{d^4q}{[x((q+p)^2-\mu^2+i\epsilon)+(1-x)(q^2-m^2+i\epsilon)]^2}\\ &= \frac{g^2}{(2\pi)^4}\int_0^1 dx\int\frac{d^4q}{[q^2+2xq\cdot p+x(p^2+m^2-\mu^2)-m^2+i\epsilon]^2}\end{aligned} \tag{S10.3}$$

Shift the integration by setting $k = q + xp$:

$$-i\Sigma^f(p^2) = \frac{g^2}{(2\pi)^4}\int_0^1 dx\int\frac{d^4k}{[k^2-x^2p^2+x(p^2+m^2-\mu^2)-m^2+i\epsilon]^2} \tag{S10.4}$$

Using the integral table on p. 330, (I.4) gives us

$$-i\Sigma^f(p^2) = -i\frac{g^2}{16\pi^2}\int_0^1 dx\,\ln[x^2p^2-x(p^2+m^2-\mu^2)+m^2-i\epsilon] \tag{S10.5}$$

From (S10.1)

$$\widetilde{\Sigma}'(p^2) = \frac{g^2}{16\pi^2}\int_0^1 dx\left\{\ln\left[\frac{x^2p^2-x(p^2+m^2-\mu^2)+m^2-i\epsilon}{x^2m^2-x(2m^2-\mu^2)+m^2-i\epsilon}\right]-\frac{(x^2-x)(p^2-m^2)}{x^2m^2-x(2m^2-\mu^2)+m^2-i\epsilon}\right\} \tag{S10.6}$$

The question to be answered now concerns the branch cut. The shared denominator of the expressions between the curly brackets can be rewritten:

$$x^2m^2-x(2m^2-\mu^2)+m^2 = m^2(1-x)^2+x\mu^2 > 0 \quad \text{for } x\in[0,1] \tag{S10.7}$$

so we need not worry about the denominator. Then $\widetilde{\Sigma}'(p^2)$ has a branch cut discontinuity only should the numerator $f(x)$ of the fraction in the logarithm equal zero for some $x \in [0,1]$:

$$f(x) = x^2p^2 - x(p^2 + m^2 - \mu^2) + m^2 \stackrel{?}{=} 0 \tag{S10.8}$$

This function is a quadratic in x, so will either be concave up or down. It's easy to see that

$$f(0) = m^2 \qquad \text{and} \qquad f(1) = \mu^2 \tag{S10.9}$$

If $f(x)$ is concave down, there will never be a value $x \in [0,1]$ where $f(x) = 0$. So we need worry only about concave up, i.e., $p^2 > 0$. And in fact we know $p^2 \geq m^2$. We will have $f(x) \leq 0$ only if the minimum value of $f(x)$ is less than or equal to zero. So we need to find this minimum value:

$$\frac{df}{dx} = 0 \Rightarrow x = \tfrac{1}{2} + \frac{m^2 - \mu^2}{2p^2} \tag{S10.10}$$

so that

$$f(x)_{\min} = -\tfrac{1}{4}p^2 - \tfrac{1}{4}\frac{(m^2 - \mu^2)^2}{p^2} + \tfrac{1}{2}(m^2 + \mu^2) \tag{S10.11}$$

This minimum value will be less than or equal to zero only if

$$p^4 - 2p^2(m^2 + \mu^2) + (m^2 - \mu^2)^2 \geq 0 \tag{S10.12}$$

This is a quadratic in p^2, and the roots are

$$p^2 = (m \pm \mu)^2 \tag{S10.13}$$

The root $p^2 = (m - \mu)^2 < m^2$ is impossible (if $\mu > 2m$, the meson would be unstable), so $p^2 \geq (m + \mu)^2$ is the start of the branch cut. This is what we expect from the spectral representation. To $\mathcal{O}(g^2)$, the only particle state a nucleon field can make when applied to the vacuum is a state containing one meson and one nucleon. ■

10.2 For the defining representation of SO(3,1), we have the generators of rotations,

$$L_x = i\frac{d}{d\theta}D(R(\hat{\mathbf{x}}\theta))\bigg|_{\theta=0} = i\frac{d}{d\theta}\left.\begin{pmatrix} 1 & 0 & 0 & 0 \\ 0 & 1 & 0 & 0 \\ 0 & 0 & \cos\theta & -\sin\theta \\ 0 & 0 & \sin\theta & \cos\theta \end{pmatrix}\right|_{\theta=0} = \begin{pmatrix} 0 & 0 & 0 & 0 \\ 0 & 0 & 0 & 0 \\ 0 & 0 & 0 & -i \\ 0 & 0 & i & 0 \end{pmatrix} \tag{S10.14}$$

and similarly (it's just the cyclic permutations; $(L_i)_{jk} = -i\epsilon_{ijk}$, with $\epsilon_{123} = 1$, and $(L_i)_{0k} = (L_i)_{k0} = 0$)

$$L_y = \begin{pmatrix} 0 & 0 & 0 & 0 \\ 0 & 0 & 0 & i \\ 0 & 0 & 0 & 0 \\ 0 & -i & 0 & 0 \end{pmatrix} \qquad L_z = \begin{pmatrix} 0 & 0 & 0 & 0 \\ 0 & 0 & -i & 0 \\ 0 & i & 0 & 0 \\ 0 & 0 & 0 & 0 \end{pmatrix} \tag{S10.15}$$

The generators of the boosts are

$$M_x = i\frac{d}{d\phi}D(A(\hat{\mathbf{x}}\phi))\bigg|_{\phi=0} = i\frac{d}{d\phi}\left.\begin{pmatrix} \cosh\phi & \sinh\phi & 0 & 0 \\ \sinh\phi & \cosh\phi & 0 & 0 \\ 0 & 0 & 1 & 0 \\ 0 & 0 & 0 & 1 \end{pmatrix}\right|_{\phi=0} = \begin{pmatrix} 0 & i & 0 & 0 \\ i & 0 & 0 & 0 \\ 0 & 0 & 0 & 0 \\ 0 & 0 & 0 & 0 \end{pmatrix} \tag{S10.16}$$

and M_y, M_z similarly; $(M_i)^{\mu}{}_{\nu} = i(\delta^{\mu}_{i}g_{0\nu} - \delta^{\mu}_{0}g_{i\nu})$:

$$M_y = \begin{pmatrix} 0 & 0 & i & 0 \\ 0 & 0 & 0 & 0 \\ i & 0 & 0 & 0 \\ 0 & 0 & 0 & 0 \end{pmatrix} \qquad M_z = \begin{pmatrix} 0 & 0 & 0 & i \\ 0 & 0 & 0 & 0 \\ 0 & 0 & 0 & 0 \\ i & 0 & 0 & 0 \end{pmatrix} \tag{S10.17}$$

Then

$$[L_x, L_y] = \begin{pmatrix} 0 & 0 & 0 & 0 \\ 0 & 0 & 0 & 0 \\ 0 & -1 & 0 & 0 \\ 0 & 0 & 0 & 0 \end{pmatrix} - \begin{pmatrix} 0 & 0 & 0 & 0 \\ 0 & 0 & -1 & 0 \\ 0 & 0 & 0 & 0 \\ 0 & 0 & 0 & 0 \end{pmatrix} = \begin{pmatrix} 0 & 0 & 0 & 0 \\ 0 & 0 & 1 & 0 \\ 0 & -1 & 0 & 0 \\ 0 & 0 & 0 & 0 \end{pmatrix} = iL_z \tag{S10.18}$$

$$[L_x, M_y] = \begin{pmatrix} 0 & 0 & 0 & 0 \\ 0 & 0 & 0 & 0 \\ 0 & 0 & 0 & 0 \\ -1 & 0 & 0 & 0 \end{pmatrix} - \begin{pmatrix} 0 & 0 & 0 & 1 \\ 0 & 0 & 0 & 0 \\ 0 & 0 & 0 & 0 \\ 0 & 0 & 0 & 0 \end{pmatrix} = \begin{pmatrix} 0 & 0 & 0 & -1 \\ 0 & 0 & 0 & 0 \\ 0 & 0 & 0 & 0 \\ -1 & 0 & 0 & 0 \end{pmatrix} = iM_z \tag{S10.19}$$

$$[M_x, M_y] = \begin{pmatrix} 0 & 0 & 0 & 0 \\ 0 & 0 & -1 & 0 \\ 0 & 0 & 0 & 0 \\ 0 & 0 & 0 & 0 \end{pmatrix} - \begin{pmatrix} 0 & 0 & 0 & 1 \\ 0 & 0 & 0 & 0 \\ 0 & -1 & 0 & 0 \\ 0 & 0 & 0 & 0 \end{pmatrix} = \begin{pmatrix} 0 & 0 & 0 & 0 \\ 0 & 0 & -1 & 0 \\ 0 & 1 & 0 & 0 \\ 0 & 0 & 0 & 0 \end{pmatrix} = -iL_z \qquad \text{(S10.20)}$$

As the problem states, the other commutators can be found in the same way, or by cyclic permutation, in agreement with (18.46)–(18.49). ■

19

The Dirac Equation I. Constructing a Lagrangian

We are now in a position to take the simplest of the Lorentz group representations that have a chance of representing particles of spin-½ and making field theories with them. In the first instance we will consider field theories with linear equations of motion, so we'll have theories of free particles. After we quantize them, we'll start adding interaction terms, following the path of the first part of this course, and develop theories of interacting particles.

19.1 Building vectors out of spinors

We will want to construct a Lagrangian, and of course we want this Lagrangian to be a Lorentz scalar. The Lagrangian will have derivatives in it, which transform as vectors. So it would be good to see how we might construct a vector out of whatever we use to represent a spin-½ particle, in order to build Lorentz scalars as inner products between derivatives and these new vectors.

We know how to represent spin-½ as representations of the rotation group, with Pauli spinors. The simplest Lorentz group representations that could describe particles of spin-½ are the representations $D^{(\frac{1}{2},0)}(\Lambda)$ and $D^{(0,\frac{1}{2})}(\Lambda)$, which reduce to Pauli spinors when we restrict ourselves to the rotation group. The generators $\mathbf{L}$ of rotations for both these representations are

$$\mathbf{L} = \tfrac{1}{2}\boldsymbol{\sigma} = (\mathbf{J}^{(+)} + \mathbf{J}^{(-)}) \tag{19.1}$$

The generators $\mathbf{M}$ of boosts differ for the two representations:

$$\mathbf{M} = \pm\tfrac{i}{2}\boldsymbol{\sigma} = -i(\mathbf{J}^{(+)} - \mathbf{J}^{(-)}) \tag{19.2}$$

the plus sign applying to $D^{(\frac{1}{2},0)}(\Lambda)$ and the minus sign to $D^{(0,\frac{1}{2})}(\Lambda)$.

Thus for example, consider the two component objects u_+ or u_-, belonging to $D^{(\frac{1}{2},0)}(\Lambda)$ or $D^{(0,\frac{1}{2})}(\Lambda)$, respectively, and transforming accordingly under the Lorentz group. (For the moment we'll ignore the space and time dependence of the u's.) Under rotation about an axis $\hat{\mathbf{n}}$ through an angle θ, these transform just like a Pauli spinor:

$$R(\hat{\mathbf{n}}\theta)\colon u_\pm \to e^{-\frac{1}{2}i\boldsymbol{\sigma}\cdot\hat{\mathbf{n}}\theta}u_\pm \tag{19.3}$$

It doesn't matter which case we're looking at, u_+ or u_-, the generator $\mathbf{L}$ is always $\frac{1}{2}\boldsymbol{\sigma}$. On the other hand, under a boost along an axis $\hat{\mathbf{a}}$ with a speed $v = \tanh\phi$,

$$A(\hat{\mathbf{a}}\phi)\colon u_\pm \rightarrow e^{\pm\frac{1}{2}\boldsymbol{\sigma}\cdot\hat{\mathbf{a}}\phi}u_\pm \tag{19.4}$$

Please notice that the two objects $u_\pm$ transform differently under boosts. These are two component objects, each of which transforms according to some irreducible representation of the Lorentz group. They are called **Weyl spinors.**[1] And because parity exchanges fields belonging to $D^{(\frac{1}{2},0)}(\Lambda)$ and $D^{(0,\frac{1}{2})}(\Lambda)$,

$$P\colon u_\pm \rightarrow u_\mp \tag{19.5}$$

Let's see what we can build out of u_+ and $u_+^\dagger$ by putting together bilinear forms in u_+ and $u_+^\dagger$. Everything I say will go for the minus case, within trivial sign changes. Four linearly independent bilinear forms can be built out of u_+ and $u_+^\dagger$. Because u_+ transforms like $D^{(\frac{1}{2},0)}$, its conjugate $u_+^\dagger$ transforms like $D^{(0,\frac{1}{2})}$. Then the bilinear forms transform like

$$D^{(\frac{1}{2},0)}(\Lambda)\otimes D^{(0,\frac{1}{2})}(\Lambda) \sim D^{(\frac{1}{2},\frac{1}{2})}(\Lambda) \tag{19.6}$$

Whether we use u_+ or u_-, it doesn't matter: one is the conjugate and the other is not. And the product is of course simply $D^{(\frac{1}{2},\frac{1}{2})}$, which is the representation for a vector, as we've seen earlier. Therefore if I put together bilinear forms in u_+ and $u_+^\dagger$, the four independent bilinear forms should transform like the four components of a vector. Let's work out precisely what that vector is. I'll write it as the contravariant vector V^μ. There's only one possible choice for the time component:

$$V^0 = u_+^\dagger u_+ \tag{19.7}$$

That's certainly the only bilinear form which is a scalar under rotations, from the ordinary theory of spin-½ particles. Likewise, up to a multiplicative factor which I'll call η, there is only one possible choice for the three space components:[2]

$$V^i = \eta u_+^\dagger \boldsymbol{\sigma}^i u_+ \tag{19.8}$$

Our general formalism hasn't led us astray, at least so far. The four bilinear forms we can make can indeed be arranged into an SO(3) scalar and an SO(3) vector, which is what we want for a 4-vector.

Let's try to figure out how these bilinear forms transform under boosts by applying an acceleration, say about the z axis by a hyperbolic angle ϕ. Of course they must behave as a 4-vector for the appropriate choice of η, but it's amusing to work it out. First, we need the transformations of u_+ and $u_+^\dagger$:

$$A(\hat{\mathbf{z}}\phi)\colon \begin{cases} u_+ \rightarrow e^{\frac{1}{2}\boldsymbol{\sigma}_z\phi}u_+ \\ u_+^\dagger \rightarrow u_+^\dagger e^{\frac{1}{2}\boldsymbol{\sigma}_z\phi} \end{cases} \tag{19.9}$$

---

[1] [Eds.] Pronounced "vile". Noticing the curious reaction of his students, Coleman adds: "Not that they are disgusting, but that they were first explored by Hermann Weyl." (Hermann Weyl (1885–1955), among the great mathematicians of the twentieth century, also contributed to relativity and quantum theory. He was a colleague and friend of Schrödinger, Einstein and Emmy Noether, whose funeral oration he gave at Bryn Mawr College, 17 April 1935.)

[2] [Eds.] In the video of Lecture 19, Coleman uses α for the factor written here as η, to avoid confusion with the Dirac matrices, α_i, and the fine-structure constant, α.

The argument $iM_z = \pm\frac{1}{2}\boldsymbol{\sigma}_z$ in the exponential is now a Hermitian matrix. That's the difference between Lorentz transformations and rotations, which have an i in the exponential:

$$\begin{aligned} R\colon\ \left(e^{-\frac{1}{2}i\boldsymbol{\sigma}\bullet\hat{\mathbf{n}}\theta}\right)^\dagger &= e^{\frac{1}{2}i\boldsymbol{\sigma}\bullet\hat{\mathbf{n}}\theta} \\ A\colon\ \left(e^{\pm\frac{1}{2}\boldsymbol{\sigma}\bullet\hat{\mathbf{a}}\phi}\right)^\dagger &= e^{\pm\frac{1}{2}\boldsymbol{\sigma}\bullet\hat{\mathbf{a}}\phi} \end{aligned} \tag{19.10}$$

Now let's work out what happens to the four components of our putative vector and see if they indeed transform as components of a vector should transform. Well, we know how u_+ and $u_+^\dagger$ transform, so we just stick the transformed u's into (19.7):

$$V^0 = u_+^\dagger u_+ \to u_+^\dagger e^{\boldsymbol{\sigma}_z\phi}u_+ \tag{19.11}$$

The $\frac{1}{2}$ in the exponent disappears. This is a very easy matrix to compute, since the even powers in the power series expansion are proportional to one, $\boldsymbol{\sigma}^2$ being one, and the odd powers are proportional to $\boldsymbol{\sigma}_z$:

$$\begin{aligned} e^{\boldsymbol{\sigma}_z\phi} &= \mathbf{1} + \boldsymbol{\sigma}_z\phi + \tfrac{1}{2!}(\boldsymbol{\sigma}_z\phi)^2 + \tfrac{1}{3!}(\boldsymbol{\sigma}_z\phi)^3 + \cdots \\ &= \mathbf{1}(1 + \tfrac{1}{2!}\phi^2 + \tfrac{1}{4!}\phi^4 + \cdots) + \boldsymbol{\sigma}_z(\phi + \tfrac{1}{3!}\phi^3 + \tfrac{1}{5!}\phi^5 + \cdots) \end{aligned} \tag{19.12}$$

The even powers give us $\cosh\phi$, the odd powers give us $\sinh\phi$. This is

$$u_+^\dagger u_+ \to u_+^\dagger u_+ \cosh\phi + u_+^\dagger\boldsymbol{\sigma}_z u_+ \sinh\phi \tag{19.13}$$

which is the statement

$$V^0 \to V^0\cosh\phi + (1/\eta)V^3\sinh\phi \tag{19.14}$$

which is just what we want, if we choose $\eta = 1$. That is, we can identify a set of bilinear terms with a Lorentz 4-vector:

$$V^\mu = (V^0, \mathbf{V}) = (u_+^\dagger u_+, u_+^\dagger\boldsymbol{\sigma}u_+) \tag{19.15}$$

Let's check the other components, starting with V^3:

$$V^3 = u_+^\dagger\boldsymbol{\sigma}_z u_+ \to u_+^\dagger e^{\frac{1}{2}\boldsymbol{\sigma}_z\phi}\boldsymbol{\sigma}_z e^{\frac{1}{2}\boldsymbol{\sigma}_z\phi}u_+ \tag{19.16}$$

Here $\boldsymbol{\sigma}_z$ commutes with $\boldsymbol{\sigma}_z$, so I can use the same expansion again:

$$e^{\frac{1}{2}\boldsymbol{\sigma}_z\phi}\boldsymbol{\sigma}_z e^{\frac{1}{2}\boldsymbol{\sigma}_z\phi} = \boldsymbol{\sigma}_z e^{\boldsymbol{\sigma}_z\phi} = \boldsymbol{\sigma}_z(\cosh\phi + \boldsymbol{\sigma}_z\sinh\phi) = \boldsymbol{\sigma}_z\cosh\phi + \sinh\phi \tag{19.17}$$

and

$$u_+^\dagger\boldsymbol{\sigma}_z u_+ \to u_+^\dagger\boldsymbol{\sigma}_z u_+\cosh\phi + u_+^\dagger u_+\sinh\phi \tag{19.18}$$

so that

$$V^3 \to V^3\cosh\phi + V^0\sinh\phi \tag{19.19}$$

which is again the right answer.

What about V^1 or V^2? Those are supposed to be unchanged under an acceleration in the z direction. Well, V^1 or V^2 goes into

$$V^{1,2} = u_+^\dagger\boldsymbol{\sigma}_{x,y}u_+ \to u_+^\dagger e^{\frac{1}{2}\boldsymbol{\sigma}_z\phi}\boldsymbol{\sigma}_{x,y}e^{\frac{1}{2}\boldsymbol{\sigma}_z\phi}u_+ \tag{19.20}$$

Now $\boldsymbol{\sigma}_z$ anticommutes with either $\boldsymbol{\sigma}_x$ or $\boldsymbol{\sigma}_y$;

$$\{\boldsymbol{\sigma}_i, \boldsymbol{\sigma}_j\} \equiv \boldsymbol{\sigma}_i\boldsymbol{\sigma}_j + \boldsymbol{\sigma}_j\boldsymbol{\sigma}_i = 2\delta_{ij}\mathbf{1} \tag{19.21}$$

and therefore, when I bring a $\boldsymbol{\sigma}_z$ through a $\boldsymbol{\sigma}_x$ or a $\boldsymbol{\sigma}_y$, it gets turned into $-\boldsymbol{\sigma}_z$. So

$$e^{\frac{1}{2}\boldsymbol{\sigma}_z\phi}\boldsymbol{\sigma}_{x,y}e^{\frac{1}{2}\boldsymbol{\sigma}_z\phi} = \boldsymbol{\sigma}_{x,y}e^{-\frac{1}{2}\boldsymbol{\sigma}_z\phi}e^{\frac{1}{2}\boldsymbol{\sigma}_z\phi} = \boldsymbol{\sigma}_{x,y} \tag{19.22}$$

because the combination $e^{-\frac{1}{2}\boldsymbol{\sigma}_z\phi}e^{\frac{1}{2}\boldsymbol{\sigma}_z\phi}$ is known to its friends as 1. The result is

$$u_+^\dagger e^{\frac{1}{2}\boldsymbol{\sigma}_z\phi}\boldsymbol{\sigma}_{x,y}e^{\frac{1}{2}\boldsymbol{\sigma}_z\phi}u_+ = u_+^\dagger\boldsymbol{\sigma}_{x,y}u_+ \tag{19.23}$$

in other words,

$$V^{1,2} \to V^{1,2} \tag{19.24}$$

Thus everything works out just the way it should. Still, it is reassuring to see the marvelous algebra of the Pauli matrices doing our job for us, enabling us to construct, out of these two component objects u_+ and $u_+^\dagger$, a vector which has a sensible transformation law not only under rotations but under Lorentz transformations as well. The Weyl spinors u_+ and $u_+^\dagger$ don't transform like vectors, but more like square roots of vectors, as it were, because it is bilinear combinations of Weyl spinors that act like Lorentz vectors.

Exactly the same reasoning applies for u_-, except there is a minus sign in the $\boldsymbol{\sigma}$ matrix. If we were working with u_-, the corresponding vector object W^μ, a completely different vector from V^μ, would be

$$W^\mu = (W^0, \mathbf{W}) = (u_-^\dagger u_-, -u_-^\dagger\boldsymbol{\sigma}u_-) \tag{19.25}$$

The vectors V^μ and W^μ are products of a Weyl spinor and its adjoint. Which of the two different kinds of Weyl spinors you are working with affects only the sign of the space component of the vector.

Incidentally, the complex conjugate u_+^* is equivalent to u_-. Our starting point is (19.3). Complex conjugate this equation:

$$R(\hat{\mathbf{n}}\theta)\colon u_+^* \to e^{+\frac{1}{2}i\boldsymbol{\sigma}^*\bullet\hat{\mathbf{n}}\theta}u_+^* \tag{19.26}$$

The $\boldsymbol{\sigma}_i$ are not all real, and the $-i$ goes into i. Now use an identity:

$$\boldsymbol{\sigma}_y\boldsymbol{\sigma}^*\boldsymbol{\sigma}_y^{-1} = \boldsymbol{\sigma}_y\boldsymbol{\sigma}^*\boldsymbol{\sigma}_y = -\boldsymbol{\sigma} \tag{19.27}$$

The identity is easy to prove, because $\boldsymbol{\sigma}_y$ is the only imaginary $\boldsymbol{\sigma}_i$, and also the only $\boldsymbol{\sigma}$ matrix that commutes rather than anticommutes with $\boldsymbol{\sigma}_y$. We can make a similarity transformation using $T = \boldsymbol{\sigma}_y$,

$$\boldsymbol{\sigma}_y e^{+\frac{1}{2}i\boldsymbol{\sigma}^*\bullet\hat{\mathbf{n}}\theta}\boldsymbol{\sigma}_y^{-1} = e^{+\frac{1}{2}i\boldsymbol{\sigma}_y\boldsymbol{\sigma}^*\boldsymbol{\sigma}_y^{-1}\bullet\hat{\mathbf{n}}\theta} = e^{-\frac{1}{2}i\boldsymbol{\sigma}\bullet\hat{\mathbf{n}}\theta} \tag{19.28}$$

because $\boldsymbol{\sigma}_y^2 = 1$, and I can insert it in between every factor in the power series expansion of the exponential. This is a formal proof of the assertion made in (18.67). But we still have to look at the boosts. It's the same manipulation, starting with (19.4):

$$A(\hat{\mathbf{a}}\phi)\colon u_+^* \to e^{\frac{1}{2}\boldsymbol{\sigma}^*\bullet\hat{\mathbf{a}}\phi}u_+^* \tag{19.29}$$

Then making the same similarity transformation,

$$\boldsymbol{\sigma}_y e^{\frac{1}{2}\boldsymbol{\sigma}^*\bullet\hat{\mathbf{a}}\phi}\boldsymbol{\sigma}_y^{-1} = e^{-\frac{1}{2}\boldsymbol{\sigma}\bullet\hat{\mathbf{a}}\phi} \tag{19.30}$$

which is the appropriate matrix (19.4) for u_-. So u_+^* transforms in a way equivalent to the way u_- transforms, after a change of basis, with $T = \boldsymbol{\sigma}_y$.

19.2 A Lagrangian for Weyl spinors

Now let's try to build a free field theory using a u_+ object only. Let's promote these things from two component objects to two component *fields*, functions in space and time, and attempt to build a free classical field theory. I'll do the u_+ case in detail, and then I'll just tell you how the answers change if you have the u_- field instead. This will be our first stab at making a Lagrangian $\mathscr{L}$ for a spin-½ particle. Guided by our experience with scalar fields, we begin with some criteria for the theory:[3]

Criteria for a Weyl Lagrangian

1. $\mathscr{L}$ must be a Lorentz scalar, to guarantee Lorentz invariance.
2. $\mathscr{L}$ must have an internal symmetry, to give a conserved charge.
3. $\mathscr{L}$ must be bilinear in the fields, to give linear equations of motion.
4. $\mathscr{L}$ should have no more than two derivatives, for simplicity.
5. The action $\mathcal{S} = \int d^4x\, \mathscr{L}$ should be real; $\mathcal{S} = \mathcal{S}^*$.

The first requirement needs no discussion. What about the second? Every known spin-$\frac{1}{2}$ particle is associated with some conservation law; the conservation of baryon number or the conservation of lepton number, hence I might as well only consider free Lagrangians that obey that conservation law. So I will demand invariance under a phase transformation

$$u_+ \to e^{i\alpha} u_+ \qquad u_+^\dagger \to e^{-i\alpha} u_+^\dagger \tag{19.31}$$

with arbitrary α, since we know from our previous experience that it's phase transformations like these that give us conservation laws in scalar field theories. Third, I want to obtain linear equations of motion, so I want my Lagrangian to be bilinear in the fields. Since I also want it to be invariant under phase transformations, I want each term in the Lagrangian to contain one factor of u_+ (or its derivative) and one factor of its adjoint (or its derivative). That'll simultaneously give me linear equations of motion and guarantee invariance under the phase transformation. We can say more about the derivatives. In the scalar case I was able to get by with no more than two derivatives in the equations of motion, so to keep things simple, and following our general formalism, as a fourth condition I'll demand no more than two powers of derivatives in any term in the Lagrangian. Thus we can in principle have three kinds of terms in the Lagrangian:

$$\text{(a) } u_+^\dagger u_+ \qquad \text{(b) } u_+^\dagger \partial^\mu u_+ \qquad \text{(c) } u_+^\dagger \partial^\mu \partial^\nu u_+ \tag{19.32}$$

(With integration by parts, terms of type (c) could be replaced by terms with one derivative on $u_+^\dagger$ and one on u_+.) These are just generic. We don't know however whether these will obey the first condition, that $\mathscr{L}$ be a Lorentz scalar. What is consistent with constructing a scalar?

We've already shown there are four linear combinations of type (a). None of them transforms like a scalar; they transform like the four components of a 4-vector. And there isn't any way of putting together the four components of the vector to make a scalar that would be only

---

[3] [Eds.] We remind the reader that the action is denoted $\mathcal{S}$, and the S-matrix is denoted S.

bilinear in the u's. We *can* make a scalar, but it would have the form

$$V^\mu V_\mu = (u_+^\dagger u_+)^2 - (u_+^\dagger \boldsymbol{\sigma} u_+)\boldsymbol{\cdot}(u_+^\dagger \boldsymbol{\sigma} u_+) \tag{19.33}$$

which is quartic in the u's. So: no bilinear terms of type (a). That's pretty grim. We would expect from our previous experience that the mass term would show up as a quadratic term of type (a). It looks like we will only be able to construct a theory of massless particles. We also know that this theory will not conserve parity, because to get parity we need both a u_+ and a u_-. So we'll get a theory of massless particles that is incapable of expressing parity invariance. Well, after all, neutrinos exist.[4] Let's see where we can go with this, and later, we may try more complicated theories that have a chance of working for spin-½ particles other than neutrinos, like electrons or protons.

By the same token we can't include a term of type (c). The derivative operator is a vector, the bilinear forms all transform like vectors, and out of three vectors there is no way of building a scalar. You can build a vector, or some crazy kind of three index tensor, but you can't build a scalar.

Fortunately there *are* possible terms of type (b), because we can put together the vector index of a derivative with the index of the vector V^μ, (19.15), that we found before. An invariant Lorentz product of these two vectors can be written as

$$u_+^\dagger \partial_0 u_+ + u_+^\dagger \boldsymbol{\sigma}\boldsymbol{\cdot}\nabla u_+ \tag{19.34}$$

Here I've put together the index of ∂_μ with the index of V^μ and had the derivative act only on u_+. We could of course also put the derivative on $u_+^\dagger$, but if we're constructing an action out of this expression (19.34),

$$\mathcal{S} \propto \int d^4x \left[u_+^\dagger \partial_0 u_+ + u_+^\dagger \boldsymbol{\sigma}\boldsymbol{\cdot}\nabla u_+\right] \tag{19.35}$$

I can move the derivative with integration by parts:

$$\mathcal{S} \propto -\int d^4x \left[(\partial_0 u_+^\dagger) u_+ + (\nabla u_+^\dagger)\boldsymbol{\cdot}\boldsymbol{\sigma} u_+\right] \tag{19.36}$$

That's the same thing, aside from the minus sign. So in fact I have only one invariant, this object (19.34), which I can use to make a Lagrangian. Everything else is either not Lorentz invariant, or equivalent to (19.34) under integration by parts.

At the end of all this messing around, we find we have essentially a unique Lagrangian, aside from a scale factor in front:

$$\mathscr{L} \propto u_+^\dagger \partial_0 u_+ + u_+^\dagger \boldsymbol{\sigma}\boldsymbol{\cdot}\nabla u_+ \tag{19.37}$$

The magnitude of the proportionality constant can be absorbed by rescaling the u's, just as in the scalar case we analyzed so long ago. The adjoint of the integrand of (19.35) is the (positive) integrand of (19.36), but integration by parts of the Lagrangian (19.35) turns it into $-\mathscr{L}$. To satisfy the fifth criterion, that the action be real, the coefficient in front has to be

[4] [Eds.] In 1975, neutrinos were thought to be strictly massless.

purely imaginary. So we are left with just two choices, as in our earlier analysis of the scalar case:

$$\mathscr{L} = \pm i\left(u_+^\dagger \partial_0 u_+ + u_+^\dagger \boldsymbol{\sigma} \cdot \nabla u_+\right) \tag{19.38}$$

We won't be able to fix the plus or minus sign until we finally quantize this theory, put it in canonical form, compute the energy and see whether it is positive or negative. For the u_- case, the only difference would be a different sign for the gradient term:

$$\mathscr{L} = \pm i\left(u_-^\dagger \partial_0 u_- - u_-^\dagger \boldsymbol{\sigma} \cdot \nabla u_-\right) \tag{19.39}$$

As we'll see, this has a profound effect on the particles we finally get out of this theory. I now propose to explore this Lagrangian (19.38) first on the classical level, and next time on the quantum level.

19.3 The Weyl equation

Our first step is to derive the equations of motion which we get by varying the fields. The easiest variation to do is that with respect to $u_+^\dagger$ since we don't even have to do any integration by parts. We obtain the **Weyl equation**

$$(\partial_0 + \boldsymbol{\sigma} \cdot \nabla)u_+ = 0 \tag{19.40}$$

By varying with respect to u_+ and integrating by parts we just get the adjoint of this equation, as is usual for complex fields. That gives us no new information. Equation (19.40) is our equation of motion. It may not look Lorentz covariant, but it is. Of course for u_- we would get

$$(\partial_0 - \boldsymbol{\sigma} \cdot \nabla)u_- = 0 \tag{19.41}$$

We can gain some insight into the meaning of (19.40) by multiplying it on the left by the operator $\partial_0 - \boldsymbol{\sigma} \cdot \nabla$:

$$(\partial_0 - \boldsymbol{\sigma} \cdot \nabla)(\partial^0 + \boldsymbol{\sigma} \cdot \nabla)u_+ = 0 \tag{19.42}$$

The product is simple to work out. The cross terms cancel because ∂_0 commutes with $\boldsymbol{\sigma} \cdot \nabla$, and

$$(\boldsymbol{\sigma} \cdot \nabla)^2 = \boldsymbol{\sigma}_i \boldsymbol{\sigma}_j \partial_i \partial_j = \tfrac{1}{2}\{\boldsymbol{\sigma}_i, \boldsymbol{\sigma}_j\}\partial_i \partial_j = \nabla^2 \tag{19.43}$$

so we obtain

$$(\partial_0^2 - \nabla^2)u_+ = \Box^2 u_+ = 0 \tag{19.44}$$

which is of course just what we we expect for a massless particle. All plane wave solutions of this equation are of the form

$$u_+(x) = u_{\mathbf{p}} e^{\pm ip\cdot x} \tag{19.45}$$

The spinor $u_{\mathbf{p}}$ is constant (independent of x), and since $p^2 = 0$,

$$p^0 = |\mathbf{p}| \tag{19.46}$$

These plane waves are like those for massless scalar particles. I should make a remark, although this is anticipating things a bit. We should expect in the quantum theory that when we expand out the general solutions to the field equation in terms of these linearly independent plane wave solutions, the coefficients of the $e^{-ip\cdot x}$ terms will be annihilation operators, and the coefficients of $e^{+ip\cdot x}$ terms will be creation operators, both for mass zero particles. That's just a guess based on what we discovered for scalar theories.

Let's determine $u_{\mathbf{p}}$. For simplicity I will consider the case

$$\mathbf{p} = p^0\hat{\mathbf{z}} \quad \text{or} \quad p^\mu = p^0(1,0,0,1) \tag{19.47}$$

So we have a particle moving in the $+z$ direction. The sign of p^0 is irrelevant; it factors out of the equation. And indeed the magnitude of p^0 factors out of the equation. Plugging the plane wave solution (19.45) into the Weyl equation (19.40) for u_+, we obtain, dividing out $\pm ip^0$,

$$(1 - \boldsymbol{\sigma}_z)u_{\mathbf{p}} = 0 \tag{19.48}$$

That's a pretty easy equation to solve. If we use the standard representation of the Pauli matrices,

$$\boldsymbol{\sigma}_x = \begin{pmatrix} 0 & 1 \\ 1 & 0 \end{pmatrix} \qquad \boldsymbol{\sigma}_y = \begin{pmatrix} 0 & -i \\ i & 0 \end{pmatrix} \qquad \boldsymbol{\sigma}_z = \begin{pmatrix} 1 & 0 \\ 0 & -1 \end{pmatrix} \tag{19.49}$$

we find

$$u_{\mathbf{p}} = \begin{pmatrix} 1 \\ 0 \end{pmatrix} \tag{19.50}$$

This means that the Weyl equation has one linearly independent solution for each value of the 4-momentum on the light cone. Thus we would expect, when we quantize this theory, that we have one kind of particle for each momentum, described by one annihilation operator and one creation operator.

Let's make a guess about the quantum theory of this particle. In particular I'm interested in its spin. Well, I shouldn't really say "spin", because spin is a concept that applies only to particles with mass, because only for a particle of non-zero mass can we Lorentz transform to its rest frame and there compute its angular momentum, which is its spin. For a massless particle, there *is* no rest frame, so we can't talk about the spin. We can however talk about its **helicity**, the component of angular momentum along the direction of motion. That's perfectly reasonable and doesn't involve the rest frame. So let's try to compute J_z, for a one-particle state, $|p\rangle$, with momentum p, associated with this equation of motion, (19.40).

By comparison with what we found in the scalar theory, we'd expect to write the quantum field as a superposition of these solutions, some with annihilation operators, and some with creation operators. Therefore we should expect, aside from inessential normalization factors, that if we put the quantum field between the vacuum and this one-particle state, we would obtain something proportional to $u_{\mathbf{p}}e^{-ip\cdot x}$:

$$\langle 0|u_+(x)|p\rangle \propto u_{\mathbf{p}}e^{-ip\cdot x} \tag{19.51}$$

That's a straightforward transposition of what we discovered in the scalar theory. I'll be interested in this equation only at the point $x = 0$. That will suffice to allow us to determine the helicity:

$$\langle 0|u_+(0)|p\rangle \propto u_{\mathbf{p}} \tag{19.52}$$

Now we would expect that a particle moving in the z direction can always be chosen to be an eigenstate of helicity J_z. There's only one particle, so it must automatically turn out to be an eigenstate of J_z:

$$J_z\,|p\rangle = \lambda\,|p\rangle \tag{19.53}$$

where λ, the eigenvalue of J_z, is the helicity of the particle. So the unitary transformation

$$e^{-iJ_z\theta} = U(R(\hat{\mathbf{z}}\theta)) \tag{19.54}$$

that effects a rotation about the z axis by the angle θ in the Hilbert space of the theory, applied to the state $|k\rangle$, results in the eigenvalue equation

$$e^{-iJ_z\theta}\,|p\rangle = e^{-i\lambda\theta}\,|p\rangle \tag{19.55}$$

Then

$$e^{-i\lambda\theta}\,\langle 0|u_+(0)|p\rangle = \langle 0|u_+(0)e^{-iJ_z\theta}|p\rangle = \langle 0|e^{iJ_z\theta}u_+(0)e^{-iJ_z\theta}|p\rangle \tag{19.56}$$

(I assume the vacuum is rotationally invariant, so applying $e^{iJ_z\theta}$ to $|0\rangle$ shouldn't have any effect.) But we know from (18.1) and (19.3) that

$$\begin{aligned} e^{iJ_z\theta}u_+(0)e^{-iJ_z\theta} &= U(R(\hat{\mathbf{z}}\theta))^\dagger u_+(0)U(R(\hat{\mathbf{z}}\theta)) = D^{(\frac{1}{2},0)}(R(\hat{\mathbf{z}}\theta))u_+(0) \\ &= e^{-\frac{1}{2}i\boldsymbol{\sigma}_z\theta}u_+(0) \end{aligned} \tag{19.57}$$

and so, from (19.52),

$$e^{-i\lambda\theta}\,\langle 0|u_+(0)|p\rangle \propto e^{-i\lambda\theta}u_{\mathbf{p}} \tag{19.58}$$

$$\langle 0|e^{iJ_z\theta}u_+(0)e^{-iJ_z\theta}|p\rangle \propto e^{-\frac{1}{2}i\boldsymbol{\sigma}_z\theta}u_{\mathbf{p}} = e^{-\frac{1}{2}i\theta}u_{\mathbf{p}} \tag{19.59}$$

because $u_{\mathbf{p}}$ is an eigenstate of $\boldsymbol{\sigma}_z$ with eigenvalue $+1$. Comparing the two sides of (19.56), we see $\lambda = +\frac{1}{2}$. Thus the particles in this theory—if there *are* particles, if we can successfully quantize it—annihilated by u_+ have helicity $+\frac{1}{2}$, but this theory does *not* have particles of helicity $-\frac{1}{2}$. Such a theory is only possible when the particles are massless, and when parity is not conserved. If the particles had a mass, you could always transform to a reference frame traveling faster than the particle. In that frame, the particle would be going in the opposite direction, and hence with reversed helicity.

Of course there are two kinds of particles in this theory, because this is a charged field.[5] And therefore the field should not only annihilate particles of charge $+1$, but create antiparticles of charge -1, just as a charged scalar field does. That is to say, there will also be terms proportional to $e^{ip\cdot x}$, and they will be creation operators for *different* particles. These won't be the same as the original particles, because the field isn't real. To investigate the antiparticles, I have to put the antiparticle state on the left,

$$\langle p|\,u_+(0) \tag{19.60}$$

so it has a chance of being made. This will of course be proportional to the same $u_{\mathbf{p}}$ since it doesn't matter which sign of $e^{\pm ip\cdot x}$ I look at. For the antiparticle, however, the equation corresponding to (19.55) is

$$\langle p|\,e^{iJ_z\theta} = e^{+i\lambda'\theta}\,\langle p| \tag{19.61}$$

The antiparticle helicity is λ'. From this point on, the whole argument goes through in exactly the same way as before, except that $e^{-i\lambda\theta}$ is replaced by $e^{+i\lambda'\theta}$. Therefore we find

$$\lambda' = -\tfrac{1}{2} \tag{19.62}$$

The antiparticle has helicity $-\frac{1}{2}$. Thus our guess is that this theory, if we can successfully quantize it, will describe massless particles and their antiparticles. The massless particles

[5] See the second criterion in the box on p. 397. The current associated with this symmetry is just V^μ, given by (19.15), as you can check by our general formula (5.27) applied to the symmetry (19.31). I will not work out the current for the Weyl particles, but will do it for the massive particles described by spinor fields.

will carry one charge, and the antiparticles will carry the opposite charge. The *particles*, by definition those objects annihilated by u_+, will have helicity $\frac{1}{2}$, and the antiparticles, created by u_+, will have helicity $-\frac{1}{2}$. Similarly, particles created by $u_+^\dagger$ will have helicity $+\frac{1}{2}$, and particles destroyed by $u_+^\dagger$ will have helicity $-\frac{1}{2}$. Conventionally, a particle with helicity $+\frac{1}{2}$ is called "right-handed". If you could see a right-handed particle spinning as it came toward you, it would appear to spin in a counter-clockwise fashion.

For u_-, of course, because of the minus sign in the equations of motion everything gets switched around; $\boldsymbol{\sigma}_z$ gets replaced by $-\boldsymbol{\sigma}_z$, but otherwise nothing is changed. If we were to consider the theory of a u_- field, we would find the particles' helicity to be $-\frac{1}{2}$, and the antiparticles have helicity $+\frac{1}{2}$. Of course that's what you would expect, because the complex conjugate of a u_+ field is a u_- field. When we complex conjugate the fields we simply change the definition of what we call "particle" and what we call "antiparticle". This structure is no longer alien to physics, although it was when Hermann Weyl first proposed it.[6] Physicists dismissed this theory as the work of a dumb mathematician: there was no parity invariance, the particles were massless, this was nothing like our world. But in fact Weyl's theory describes precisely the massless neutrino with but one helicity (left-handed); the antineutrino has the opposite helicity. We haven't quantized this u_+ theory yet, but what I've described is what we would expect if quantization were to go like the quantization of scalar theories. It's possible for a massless particle to have only one helicity, if the theory is not invariant under parity. That's perfectly Lorentz invariant. If you introduce parity invariance, then the helicities have to come in pairs; if a particle can be in a given helicity state, it must be able to occupy a state with the opposite helicity. (A massive particle has to have every helicity between its maximum helicity and minus its maximum helicity by integer steps. But massless particles, if they are parity invariant, have only two helicity states.)

For example, photons have helicity $+1$ and -1. That's because electromagnetism is parity invariant, and if a photon has helicity $+1$, helicity -1 also has to exist. If electromagnetism were not parity invariant, it would be possible to conceive of a photon that has only one helicity, say $+1$. Because they're massless, they're allowed not to have helicity zero. You could always add to the electromagnetic field a massless scalar field, and call the three states you get this way "the photon". Then there would be helicity $+1$, -1 and 0. You might think it perverse to add a such a scalar field, and I would agree with you, but it is certainly possible. As Pauli said in a very similar context about only using irreducible representations in building a theory, "What God has put asunder, let no man join together."

19.4 The Dirac equation

This Weyl theory does a nice job with the free neutrino, but of course there are a lot of spin-$\frac{1}{2}$ particles in the world that are not massless. To get beyond massless particles and get something that has a chance of being a reasonable theory of the free electron or the free proton, we have to complicate our theory somewhat. We've explored everything we can reasonably do with a just a u_+ field or a u_- field. We also know that the interactions of the proton and the electron are parity-conserving up to an excellent approximation. So we will now try and make a parity-conserving theory. To make a parity-conserving theory, we need both a u_+ and a u_-

[6] [Eds.] H. Weyl, "Elektron und Gravitation", *Zeits. f. Phys.* **56** (1929) 330-352; English translation in L. O'Raifeartaigh, *The Dawning of Gauge Theory*, Princeton U. P., 1997, pp. 121–144; the Weyl equation appears on p. 142 (setting $f_p = 0$ in the absence of gravitational coupling).

field because parity turns u_+ into u_-, and *vice versa.* We can list some criteria for a theory of massive spinors:

> *Criteria for a Lagrangian with massive spinors*
>
> 1. $\mathscr{L}$ must be a Lorentz scalar, to guarantee Lorentz invariance.
> 2. $\mathscr{L}$ must be bilinear in the fields, to give linear equations of motion.
> 3. $\mathscr{L}$ must have an internal symmetry, to give a conserved charge.
> 4. $\mathscr{L}$ must be invariant under parity.
> 5. $\mathscr{L}$ should have no more than one derivative.
> 6. The action $\mathcal{S} = \int d^4x\, \mathscr{L}$ should be real; $\mathcal{S} = \mathcal{S}^*$.

I want $\mathscr{L}$ bilinear in the fields so I have linear equations of motion for the free field theory. I still want to preserve an invariance that corresponds to charge conjugation,

$$u_\pm \to e^{i\alpha} u_\pm \qquad u_\pm^\dagger \to e^{-i\alpha} u_\pm^\dagger \tag{19.63}$$

I don't want two conserved charges, so I don't want to say there's an independent phase transformation for u_+ and u_-. Then I want the Lagrangian to be invariant under a parity transformation and I'll assume in the first instance the most general form

$$u_+(\mathbf{x},t) \to e^{i\phi_1} u_-(-\mathbf{x},t), \qquad u_-(\mathbf{x},t) \to e^{i\phi_2} u_+(-\mathbf{x},t) \quad \text{for some } \phi_1 \text{ and } \phi_2 \tag{19.64}$$

I know parity interchanges u_+ and u_- but it might multiply them by a phase. I'm going to be as general as I can be. As we'll soon see, this generality is spurious, and we can pin ϕ_1 and ϕ_2 to be fixed numbers. I don't care if the square of the operation is not one. Any sort of transformation of this form I'll call "parity". In the fifth criterion, I will be a little more restrictive than I was in the preceding case, and assume no more than one derivative. I could assume two derivatives, but after all, in the previous case I got along just fine with one derivative, so let's try one derivative here.

Now let's write down the most general Lagrangian. Because of condition one, we have several kinds of terms. We could have $u_+^\dagger u_+$, either with or without a derivative in there somewhere, and likewise $u_-^\dagger u_-$:

$$u_+^\dagger u_+, \quad u_+^\dagger \partial_\mu u_+, \quad u_-^\dagger u_-, \quad u_-^\dagger \partial_\mu u_- \tag{19.65}$$

And we could have these terms,

$$u_+^\dagger u_-, \quad u_+^\dagger \partial_\mu u_-, \quad u_-^\dagger u_+, \quad u_-^\dagger \partial_\mu u_+ \tag{19.66}$$

Now we've already classified all the $u_+^\dagger u_+$ and $u_-^\dagger u_-$ terms. The only Lorentz invariants involving these terms, (19.38) and (19.39), have derivatives. How do these change under parity? Because

$$P\colon \nabla \to -\nabla \tag{19.67}$$

we have

$$P\colon \begin{cases} u_+^\dagger \partial_0 u_+ + u_+^\dagger \boldsymbol{\sigma}\cdot\nabla u_+ \;\to\; u_-^\dagger \partial_0 u_- - u_-^\dagger \boldsymbol{\sigma}\cdot\nabla u_- \\ u_-^\dagger \partial_0 u_- - u_-^\dagger \boldsymbol{\sigma}\cdot\nabla u_- \;\to\; u_+^\dagger \partial_0 u_+ + u_+^\dagger \boldsymbol{\sigma}\cdot\nabla u_+ \end{cases} \tag{19.68}$$

The phase factors (19.64) are irrelevant, because they cancel out in this combination. Parity transforms the two Lagrangians, (19.38) and (19.39), into each other. Consequently their sum is invariant. The real benefit for the relative minus sign between Lorentz transformation laws for u_+ and u_-, arising from (19.4) and leading to the different forms of 4-vectors in (19.15) and (19.25), is that we can build a parity invariant theory.

What about $u_+^\dagger u_-$? Whichever of u_+ or u_- transforms like $D^{(\frac{1}{2},0)}(\Lambda)$, the other transforms in the other way. But the adjoint takes care of that, and switches it back again. So we have to deal with things like this:

$$D^{(\frac{1}{2},0)}(\Lambda) \otimes D^{(\frac{1}{2},0)}(\Lambda) \sim D^{(0,0)}(\Lambda) \oplus D^{(1,0)}(\Lambda) \tag{19.69}$$

for which we get $D^{(0,0)}$, that's a scalar. That means we can build the scalar without any worries. But the second part, $D^{(1,0)}(\Lambda)$, is half of an antisymmetric tensor. This means that with this form we can't build anything with derivatives in it, because the derivative operator is a vector, and there's no vector here to dot things into. However we *can* build a non-derivative term, like this:

$$-mu_+^\dagger u_- \tag{19.70}$$

where m is an arbitrary complex number. This is the only combination that's a scalar under rotations, so it must be the combination that's a scalar under the full Lorentz group. Because we want the Lagrangian to be Hermitian, we add the other possibility, with the conjugate coefficient:

$$-m^* u_-^\dagger u_+ \tag{19.71}$$

Thus the most general Lagrangian satisfying our five conditions takes the form

$$\mathscr{L} = \pm\left[i\left(u_+^\dagger \partial_0 u_+ + u_+^\dagger \boldsymbol{\sigma}\cdot\nabla u_+ + u_-^\dagger \partial_0 u_- - u_-^\dagger \boldsymbol{\sigma}\cdot\nabla u_-\right) - mu_+^\dagger u_- - m^* u_-^\dagger u_+\right] \tag{19.72}$$

It involves a single arbitrary complex number m which I will shortly trade for a positive real number.

Now let's get rid of the phase in m. I can always redefine

$$u_+ = u'_+ e^{i\varphi} \tag{19.73}$$

I have the freedom to change variables when writing down my Lagrangian. If I substitute that in, and drop the primes, the terms in the derivatives aren't affected, but the terms in m and m^* are affected; their phases are changed. I can always absorb the phase of m in such a transformation into my definition of u_+, to make m real, and greater than or equal to zero. This changes the definition of parity, of course; ϕ_1 and ϕ_2 in (19.64) are changed. With m chosen to be real, the new definition of parity is

$$u_\pm(\mathbf{x}, t) \to u_\mp(-\mathbf{x}, t)e^{i\phi_1} \tag{19.74}$$

Once I've chosen the phase of u_+ so that m is real, I no longer have the freedom to assign phases differently to u_+ and u_-. Of course I still have an infinite set of possible choices for the phase ϕ_1, because I can always multiply parity by an internal symmetry and declare *that* to be parity. That is my privilege. It's only the *total* group of symmetries of the Lagrangian that counts, not what names we give to any individual member. I will define a standard parity which is simply the natural choice:

$$P\colon u_\pm(\mathbf{x}, t) \to u_\mp(-\mathbf{x}, t) \tag{19.75}$$

This is purely a convention of nomenclature. If I were perverse, I could've chosen any one of (19.74) to call parity. That, too, would be a symmetry of the Lagrangian.

We now have an ugly-looking Lagrangian, characterized by a single real number, m, and an overall sign choice, exactly as many free parameters as we had in the corresponding case of the free scalar field:

$$\mathscr{L} = \pm \left[i \left(u_+^\dagger \partial_0 u_+ + u_-^\dagger \partial_0 u_- + u_+^\dagger \boldsymbol{\sigma} \bullet \nabla u_+ - u_-^\dagger \boldsymbol{\sigma} \bullet \nabla u_- \right) - m(u_+^\dagger u_- + u_-^\dagger u_+) \right] \tag{19.76}$$

The equations of motion are scarcely more complicated than the Weyl equations:

$$i \left(\partial_0 + \boldsymbol{\sigma} \bullet \nabla \right) u_+ = m u_- \tag{19.77}$$

$$i \left(\partial_0 - \boldsymbol{\sigma} \bullet \nabla \right) u_- = m u_+ \tag{19.78}$$

Now it's pretty easy to see what to do with these equations. I multiply (19.77) on the left by $i(\partial_0 - \boldsymbol{\sigma} \bullet \nabla)$, and find

$$- \left(\partial_0^2 - \nabla^2 \right) u_+ = i \left(\partial_0 - \boldsymbol{\sigma} \bullet \nabla \right) m u_- = m \left[i \left(\partial_0 - \boldsymbol{\sigma} \bullet \nabla \right) u_- \right] = m^2 u_+ \tag{19.79}$$

That is, u_+ obeys the Klein-Gordon equation appropriate for a particle of mass m. Likewise we can start with (19.78), multiply by $i(\partial_0 + \boldsymbol{\sigma} \bullet \nabla)$ and in exactly the same way show that

$$- \left(\partial_0^2 - \nabla^2 \right) u_- = m^2 u_- \tag{19.80}$$

So I have not lied to you in my choice of the symbol "m" for this free field theory: m is indeed the mass of the quantum of the field. Further implications of the theory, what the spins of the particles are and so on, are topics for next time.

Our Lagrangian is the sum of a bunch of grotesque-looking terms. Let us simplify our notation somewhat by incorporating u_+ and u_- together into a single four-component object, ψ:

$$\psi = \begin{pmatrix} u_+ \\ u_- \end{pmatrix} \tag{19.81}$$

The top two components of ψ are the two components of u_+, and the bottom two are the components of u_-. I will define three 4×4 matrices, $\boldsymbol{\alpha}$, which are block diagonal with the Pauli sigma matrices $\boldsymbol{\sigma}$, $-\boldsymbol{\sigma}$ and zeros elsewhere:

$$\boldsymbol{\alpha} = \begin{pmatrix} \boldsymbol{\sigma} & \mathbf{0} \\ \mathbf{0} & -\boldsymbol{\sigma} \end{pmatrix} \tag{19.82}$$

I will define a fourth 4×4 matrix, β , which is block off diagonal:

$$\beta = \begin{pmatrix} \mathbf{0} & \mathbf{1} \\ \mathbf{1} & \mathbf{0} \end{pmatrix} \tag{19.83}$$

These matrices $\boldsymbol{\alpha}$ and β are chosen so that the ugly Lagrangian (19.76) has a rather simple form in terms of $\boldsymbol{\alpha}$, β and ψ:

$$\mathscr{L} = \pm \left[i\psi^\dagger (\partial_0 + \boldsymbol{\alpha} \bullet \nabla) \psi - m\psi^\dagger \beta \psi \right] \tag{19.84}$$

The equations of motion can be obtained by writing (19.77) and (19.78) in terms of $\boldsymbol{\alpha}$, β and ψ, but we can get them directly by varying the Lagrangian with respect to $\psi^\dagger$:

$$i\partial_0\psi + i\boldsymbol{\alpha}\cdot\nabla\psi = m\beta\psi \tag{19.85}$$

This equation is called the **Dirac equation**, though expressed here in a slightly different basis than that written down by Dirac in 1929. The forms of ψ, $\boldsymbol{\alpha}$ and β used here are called the **Weyl basis**, and the matrices $\boldsymbol{\alpha}$ and β are called **Dirac matrices**. Next time we will begin exploring the properties of the Dirac equation—finding out what the solutions are, making guesses about the properties of the particles represented by those solutions, and so on. Since we'll be spending a lot of time with the Dirac equation, we will want to develop a sequence of algorithms for handling its solutions as effectively as possible. Finally, we will quantize the Dirac theory, looking at the energy and establishing the correct sign of the Lagrangian.

20

The Dirac Equation II. Solutions

Having derived the Dirac equation, we're now going to manipulate it, study its solutions, and write it in different bases with new notation, all with the aim of making the free Dirac equation easy to work with. Later on, when we have to do complicated things with it and make it part of an interesting quantum field theory, we'll be able to do it efficiently.

20.1 The Dirac basis

The Dirac equation in the form we found it last time had a Lagrangian, (19.84):

$$\mathscr{L} = \pm\left[i\psi^\dagger\partial_0\psi + i\psi^\dagger\boldsymbol{\alpha}\cdot\nabla\psi - m\psi^\dagger\beta\psi\right] \tag{20.1}$$

where m was a positive real number, and

$$\boldsymbol{\alpha} = \begin{pmatrix} \boldsymbol{\sigma} & \mathbf{0} \\ \mathbf{0} & -\boldsymbol{\sigma} \end{pmatrix} \qquad \beta = \begin{pmatrix} \mathbf{0} & \mathbf{1} \\ \mathbf{1} & \mathbf{0} \end{pmatrix} \qquad \psi = \begin{pmatrix} u_+ \\ u_- \end{pmatrix} \tag{20.2}$$

Notice that these matrices are Hermitian,

$$\boldsymbol{\alpha}^\dagger = \boldsymbol{\alpha} \qquad \beta^\dagger = \beta \tag{20.3}$$

They also obey the **Dirac algebra** (also called the "Clifford algebra"[1])

$$\{\alpha_i, \alpha_j\} = 2\delta_{ij} \qquad \{\alpha_i, \beta\} = 0 \tag{20.4}$$

[1] [Eds.] W. K. Clifford, "Applications of Grassmann's Extensive Algebra", *Amer. Jour. Math.* v. 1 (1878) 350–358; reprinted (along with all of Clifford's papers) in *Mathematical Papers by William Kingdon Clifford*, ed. Robert Tucker, Macmillan, 1882. See also Appendix E, pp. 675–677 of Bernard de Wit and Jack Smith, *Field Theory in Particle Physics* v. 1, North-Holland, 1986. William Kingdon Clifford (1845–1879) was a British geometer who translated Bernhard Riemann's inaugural lecture (June 10, 1854) "Über die Hypothesen, welche der Geometrie zu Grunde liegen" (On the hypotheses which lie at the base of geometry), *Abhand. König. Gesell. Wiss. Gött.* **13** (1868) 133–150, into English: *Nature* **VIII** (1873), No. 183, pp. 14–17; No. 184, pp. 36–37. This work led to Clifford's brief speculative note anticipating Einstein's general relativity, "On the space-theory of matter", *Camb. Phil. Soc. Proc.* v.2, 1866–1876, Feb. 21, 1870, pp. 157–158, suggesting that matter curved space, which action might be the basis of gravity.

where the curly brackets indicate the anticommutator, the sum of the products in the two different orders; by definition

$$\{A, B\} \equiv AB + BA \tag{20.5}$$

Finally, the square of each α_i, and of β, is 1.

We can write the generators of Lorentz boosts and rotations in terms of $\boldsymbol{\alpha}$ and β. The Lorentz boost generator $\mathbf{M}$ is (19.2)

$$\mathbf{M} = \tfrac{i}{2}\boldsymbol{\alpha} \tag{20.6}$$

because the generator is $i\boldsymbol{\sigma}/2$ for u_+ and $-i\boldsymbol{\sigma}/2$ for u_-. The rotation generators are

$$\mathbf{L} = \tfrac{1}{2}\begin{pmatrix} \boldsymbol{\sigma} & 0 \\ 0 & \boldsymbol{\sigma} \end{pmatrix} \equiv \tfrac{1}{2}\boldsymbol{\Sigma} \tag{20.7}$$

because, from (19.3), u_+ and u_- transform the same way under rotations. Finally, from (19.5), parity exchanges u_+ and u_-, so we can write

$$P\colon \psi(\mathbf{x}, t) \to \beta\psi(-\mathbf{x}, t) \tag{20.8}$$

This formulation is specific to a particular basis: arranging the two components of u_+ and the two components of u_- into a four-component object ψ in a certain way, (19.81). The equations (20.6), (20.7) and (20.8) depend on the explicit form of $\boldsymbol{\alpha}$, β and ψ. The Dirac algebra (20.4), the boost commutators and the other parts of the Lorentz algebra, do not. If we had chosen our basis differently, if we had put u_+ and u_- together to make ψ in a different way, the explicit forms of $\boldsymbol{\alpha}$, β, $\mathbf{L}$ and $\mathbf{M}$ would be changed, but the equations expressing the Dirac and Lorentz algebras would not be. They would simply be expressed in terms of the $\boldsymbol{\alpha}$, β and $\mathbf{L}$ in the new basis. This particular basis is called the **Weyl representation** of the Dirac equation. (The word "representation" is a bit strange, since its usage here has little to do with group theory.) It is not the representation in which Dirac first wrote down the equation. He chose to write

$$\psi = \frac{1}{\sqrt{2}}\begin{pmatrix} u_+ + u_- \\ u_+ - u_- \end{pmatrix} \tag{20.9}$$

In such a basis, the $1/\sqrt{2}$ is inserted, so that the Lagrangian's term

$$i\psi^\dagger\partial_0\psi = iu_+^\dagger\partial_0 u_+ + iu_-^\dagger\partial_0 u_- \tag{20.10}$$

will be unchanged. In the **standard representation**, with this form of ψ,

$$\boldsymbol{\alpha} = \begin{pmatrix} \mathbf{0} & \boldsymbol{\sigma} \\ \boldsymbol{\sigma} & \mathbf{0} \end{pmatrix} \qquad \beta = \begin{pmatrix} \mathbf{1} & \mathbf{0} \\ \mathbf{0} & -\mathbf{1} \end{pmatrix} \tag{20.11}$$

The matrix β is block diagonal, $\boldsymbol{\alpha}$ is block off-diagonal, and $\mathbf{L}$ still has the form (20.7), because the sum and difference of u_+ and u_- transform in the same way under rotations as ordinary Pauli spinors. The Dirac representation is called the "standard representation" for historical reasons; it was the one first written down by Dirac. Aside from the explicit forms of ψ, $\boldsymbol{\alpha}$ and β, equations (20.2) versus (20.9) and (20.11), everything is the same in both representations. I don't expect you to see offhand what $\boldsymbol{\alpha}$ and β look like in the standard representation, but you can check it in a moment just by plugging them into the Lagrangian (20.1) and seeing that you get the same quadratic functions[2] of u_+ and u_-.

20.2 Plane wave solutions

The standard representation is especially convenient for finding explicit solutions to the Dirac equation in the limit of small $\mathbf{p}$. In that limit, the term in β dominates, and we have diagonalized β in this representation. Let me work out the plane wave solutions to the Dirac equation, (19.85):

$$i\partial_0\psi + i\boldsymbol{\alpha}\cdot\nabla\psi = \beta m\psi \tag{19.85}$$

I will look first at positive frequency solutions, so-called because they have a negative term in the exponential:[3]

$$\psi(\mathbf{x},t) = u_{\mathbf{p}}e^{-ip\cdot x} \tag{20.12}$$

(The four-component coefficient $u_{\mathbf{p}}$ is not to be confused with the two-component objects u_+ or u_-.) These are the solutions you would expect in the quantum field theory to multiply particle annihilation operators. (I'll talk about the solutions that would be associated with antiparticle creation operators later.) Since we know all solutions of the Dirac equation obey the Klein–Gordon equation with mass m, p^0 can be chosen to be $E_{\mathbf{p}}$;

$$E_{\mathbf{p}} = \sqrt{\mathbf{p}^2+m^2} \tag{20.13}$$

Plugging (20.12) into the Dirac equation, the i and the $-i$ cancel, and I obtain

$$\left[E_{\mathbf{p}} - \boldsymbol{\alpha}\cdot\mathbf{p}\right]u_{\mathbf{p}} = \beta m u_{\mathbf{p}} \tag{20.14}$$

When $\mathbf{p}=0$, this equation is particularly easy to solve in the standard representation. Then E becomes m, and we obtain

$$u_{\mathbf{0}} = \beta u_{\mathbf{0}} \tag{20.15}$$

This equation has two linearly independent solutions, $u_{\mathbf{0}}^{(r)}$, $r=\{1,2\}$. Explicitly, they are

$$u_{\mathbf{0}}^{(1)} = \sqrt{2m}\begin{pmatrix}1\\0\\0\\0\end{pmatrix} \qquad u_{\mathbf{0}}^{(2)} = \sqrt{2m}\begin{pmatrix}0\\1\\0\\0\end{pmatrix} \tag{20.16}$$

These solutions are normalized so that

$$u_{\mathbf{0}}^{(r)\,\dagger}u_{\mathbf{0}}^{(s)} = 2m\,\delta^{rs} \tag{20.17}$$

---

[2] [Eds.] The Dirac and Weyl representations are related by a similarity transformation,

$$\begin{aligned}\psi_D &= T\psi_W\\ \boldsymbol{\alpha}_D &= T\boldsymbol{\alpha}_W T^{-1} \qquad \text{where} \quad T = \frac{1}{\sqrt{2}}\begin{pmatrix}1 & 1\\ 1 & -1\end{pmatrix} = T^{-1}\\ \beta_D &= T\beta_W T^{-1}\end{aligned}$$

[3] [Eds.] From the Schrödinger equation, $H\psi = i\hbar(\partial\psi/\partial t) = \hbar\omega\psi$ has a positive eigenvalue $\hbar\omega$ if ψ has a time dependence of the form $\exp(-i\omega t)$.

The reason for this peculiar normalization convention will become clear shortly. Also,

$$u_{\mathbf{0}}^{(r)\,\dagger}\boldsymbol{\alpha}u_{\mathbf{0}}^{(s)} = \mathbf{0} \tag{20.18}$$

This results from $\boldsymbol{\alpha}$ being a block off-diagonal matrix. These two results can be put together into a suggestive form,

$$(u_{\mathbf{0}}^{(r)\,\dagger}u_{\mathbf{0}}^{(s)}, u_{\mathbf{0}}^{(r)\,\dagger}\boldsymbol{\alpha}u_{\mathbf{0}}^{(s)}) = (2m, \mathbf{0})\,\delta^{rs} \tag{20.19}$$

Maybe this normalization, which looks like it might be Lorentz invariant, will hold even for solutions with non-zero $\mathbf{p}$.

By guesswork identical to that used for the Weyl equation (§19.3), $u^{(1)}$ should be associated with the annihilation operator for a particle at rest with zero momentum and $J_z = +\frac{1}{2}$. The particle in this equation resembles an electron, so I'll call it an electron. The solution $u^{(2)}$ is the same thing with $J_z = -\frac{1}{2}$. Of course, that's the real reason we have two solutions: We have a theory of massive particles with spin one half, and we cannot get by with fewer than two solutions for each value of the momentum. I will not always use these two solutions called $u^{(1)}$ and $u^{(2)}$, because I may not be interested in the z axis; maybe I want to look at the x-axis. However I will always normalize my solutions so that this normalization convention (20.19) holds. Any linear combination of these $u^{(r)}$ will be just as good. So much for solutions at rest.

What about moving solutions, solutions associated with nonzero $\mathbf{p}$? It might be thought that we have to solve a more complicated equation to construct them. Actually, we don't. The theory has been constructed to be Lorentz invariant, so let's exploit this Lorentz invariance and obtain a solution associated with a nonzero $\mathbf{p}$ by applying a Lorentz boost to a solution associated with the zero $\mathbf{p}$. Thus I define (see (18.45))

$$u_{\mathbf{p}}^{(r)} = e^{-i\hat{\mathbf{a}}\cdot\mathbf{M}\phi}u_{\mathbf{0}}^{(r)} = e^{\frac{1}{2}\boldsymbol{\alpha}\cdot\hat{\mathbf{a}}\phi}u_{\mathbf{0}}^{(r)} \tag{20.20}$$

The operator in front of $u_{\mathbf{0}}^{(r)}$ is a Lorentz boost along an axis $\hat{\mathbf{a}}$ by a hyperbolic angle ϕ. The axis is chosen to boost the particle at rest in the direction of the desired momentum, $\mathbf{p}$,

$$\hat{\mathbf{a}} = \frac{\mathbf{p}}{|\mathbf{p}|} \tag{20.21}$$

and ϕ is chosen to boost it by the right amount,

$$\cosh\phi = \frac{E_{\mathbf{p}}}{m} \tag{20.22}$$

so that $\phi = 0$ when $E_{\mathbf{p}} = m$. The normalization conditions obeyed by these solutions are simple, since (20.19) form the space and time components of a 4-vector:

$$(u_{\mathbf{p}}^{(r)\,\dagger}u_{\mathbf{p}}^{(s)}, u_{\mathbf{p}}^{(r)\,\dagger}\boldsymbol{\alpha}u_{\mathbf{p}}^{(s)}) = 2(E_{\mathbf{p}}, \mathbf{p})\,\delta^{rs} \tag{20.23}$$

Just to be clear about this, let's work out the explicit case for the particle moving in the positive z direction. Then $\hat{\mathbf{a}} = \hat{\mathbf{z}}$ and the relevant matrix is α_z,

$$\alpha_z = \begin{pmatrix} 0 & \boldsymbol{\sigma}_z \\ \boldsymbol{\sigma}_z & 0 \end{pmatrix} \tag{20.24}$$

Because α_z^2 is 1, it's easy to compute

$$e^{\frac{1}{2}\alpha_z\phi} = \cosh\tfrac{1}{2}\phi + \alpha_z \sinh\tfrac{1}{2}\phi = \begin{pmatrix} \cosh\frac{1}{2}\phi & 0 & \sinh\frac{1}{2}\phi & 0 \\ 0 & \cosh\frac{1}{2}\phi & 0 & -\sinh\frac{1}{2}\phi \\ \sinh\frac{1}{2}\phi & 0 & \cosh\frac{1}{2}\phi & 0 \\ 0 & -\sinh\frac{1}{2}\phi & 0 & \cosh\frac{1}{2}\phi \end{pmatrix} \tag{20.25}$$

where

$$\cosh\tfrac{1}{2}\phi = \sqrt{\frac{\cosh\phi+1}{2}} = \frac{\sqrt{E_{\mathbf{p}}+m}}{\sqrt{2m}} \qquad \sinh\tfrac{1}{2}\phi = \sqrt{\frac{\cosh\phi-1}{2}} = \frac{\sqrt{E_{\mathbf{p}}-m}}{\sqrt{2m}} \tag{20.26}$$

Now you see the reason for the $\sqrt{2m}$ in the normalization (20.16): it's to cancel out those ugly denominators. Thus we find

$$u_{\mathbf{p}}^{(1)} = \begin{pmatrix} \sqrt{E_{\mathbf{p}}+m} \\ 0 \\ \sqrt{E_{\mathbf{p}}-m} \\ 0 \end{pmatrix} \tag{20.27}$$

A nice thing about (20.20) and the normalization (20.16) is that the solution (20.27) has a smooth limit as m goes to zero (though the method we have used to define these functions does not):

$$\lim_{\mathbf{p}\to\mathbf{0}} u_{\mathbf{p}} = u_{\mathbf{0}} \tag{20.28}$$

Thus we can smoothly take the limit as the particle mass becomes negligible, either because we're studying the physics of a massless fermion, or because we're doing a process where an electron (or something similar) gets produced at such a high energy that its mass is negligible.[4]

Using (20.27), we have

$$u^{(1)\dagger}u^{(1)} = (E_{\mathbf{p}}+m) + (E_{\mathbf{p}}-m) = 2E_{\mathbf{p}} \tag{20.29}$$

$$u^{(1)\dagger}\alpha_x u^{(1)} = u^{(1)\dagger}\alpha_y u^{(1)} = 0 \tag{20.30}$$

$$u^{(1)\dagger}\alpha_z u^{(1)} = 2\sqrt{E_{\mathbf{p}}^2 - m^2} = 2|\mathbf{p}| = 2p_z \tag{20.31}$$

so that, in agreement with (20.23),

$$(u^{(1)\dagger}u^{(1)}, u^{(1)\dagger}\boldsymbol{\alpha}u^{(1)}) = 2(E_{\mathbf{p}}, \mathbf{p}) \tag{20.32}$$

Similarly we can construct

$$u_{\mathbf{p}}^{(2)} = \begin{pmatrix} 0 \\ \sqrt{E_{\mathbf{p}}+m} \\ 0 \\ -\sqrt{E_{\mathbf{p}}-m} \end{pmatrix} \tag{20.33}$$

---

[4] This normalization convention is not that of Bjorken and Drell. ([Eds.] See Bjorken & Drell *RQM*, p. 31, equation (3.11).)

and we can work out that

$$(u^{(2)\dagger}u^{(2)}, u^{(2)\dagger}\boldsymbol{\alpha}u^{(2)}) = 2(E_{\mathbf{p}}, \mathbf{p}) \tag{20.34}$$

$$(u^{(1)\dagger}u^{(2)}, u^{(1)\dagger}\boldsymbol{\alpha}u^{(2)}) = (u^{(2)\dagger}u^{(1)}, u^{(2)\dagger}\boldsymbol{\alpha}u^{(1)}) = 0 \tag{20.35}$$

in agreement with (20.23).

Of course, everything I have said about the positive frequency solutions goes through, *mutatis mutandis*, for the negative frequency solutions. Writing $v_{\mathbf{p}}$ for these spinors,

$$\psi(\mathbf{x}, t) = v_{\mathbf{p}}e^{+ip\cdot x} \tag{20.36}$$

Once again

$$p^0 = E_{\mathbf{p}} = \sqrt{\mathbf{p}^2 + m^2} \tag{20.37}$$

When we finally quantize this theory, we expect these solutions to multiply creation operators for positrons, the antiparticles of electrons. We plug (20.36) into the Dirac equation and get an equation almost identical to (20.14):

$$\left[E_{\mathbf{p}} - \boldsymbol{\alpha}\cdot\mathbf{p}\right] v_{\mathbf{p}} = -\beta m v_{\mathbf{p}} \tag{20.38}$$

The minus sign on the β term is a reflection of the different sign of the exponential's argument. Once again the solution is most easily done in the case when $\mathbf{p} = 0$:

$$v_{\mathbf{0}} = -\beta v_{\mathbf{0}} \tag{20.39}$$

The negative frequency solutions are, like those of positive frequency, two in number. In the standard representation,

$$v_0^{(1)} = \sqrt{2m}\begin{pmatrix}0\\0\\0\\1\end{pmatrix} \qquad v_0^{(2)} = \sqrt{2m}\begin{pmatrix}0\\0\\1\\0\end{pmatrix} \tag{20.40}$$

Because we expect these to be the coefficients of creation operators rather than annihilation operators, the same sign switch in the eigenvalues that occurred when we were discussing helicity[5] occurs here, and we expect $v^{(1)}$ to multiply the creation operator for a positron with $J_z = -\frac{1}{2}$, while $v^{(2)}$ creates a positron with $J_z = +\frac{1}{2}$. These states are supposed to multiply a creation operator, and therefore the phase of the helicity gets switched, because the state is on the left, rather than the right, of the creation operator.

We define moving solutions in exactly the same way as before (see (20.20)):

$$v_{\mathbf{p}}^{(r)} = e^{\frac{1}{2}\boldsymbol{\alpha}\cdot\hat{\mathbf{a}}\phi}v_0^{(r)} \tag{20.41}$$

This gives

$$v_{\mathbf{p}}^{(1)} = \begin{pmatrix}0\\-\sqrt{E_{\mathbf{p}} - m}\\0\\\sqrt{E_{\mathbf{p}} + m}\end{pmatrix} \qquad v_{\mathbf{p}}^{(2)} = \begin{pmatrix}\sqrt{E_{\mathbf{p}} - m}\\0\\\sqrt{E_{\mathbf{p}} + m}\\0\end{pmatrix} \tag{20.42}$$

We have the same normalization condition among the v's as for the u's:

$$(v_{\mathbf{p}}^{\dagger\,(r)}v_{\mathbf{p}}^{(s)}, v_{\mathbf{p}}^{\dagger\,(r)}\boldsymbol{\alpha}v_{\mathbf{p}}^{(s)}) = 2(E_{\mathbf{p}}, \mathbf{p})\delta^{rs} \tag{20.43}$$

So much for the Dirac equation and its plane wave solutions.

---

[5] [Eds.] See pp. 400–401.

20.3 Pauli's theorem

So far we have discussed two different representations of the Dirac matrices, due to Weyl and Dirac. Of course there are an infinite number; any invertible 4×4 matrix can be used to transform one representation of the Dirac matrices into another (by a similarity transformation). But there are some properties of the Dirac matrices that are independent of one's choice of basis. In particular, the set of four matrices $\{\alpha_i, \beta\}$ obey the Dirac algebra

$$\{\alpha_i, \alpha_j\} = 2\delta_{ij} \qquad \{\alpha_i, \beta\} = 0 \tag{20.4}$$

It follows that the square of any α_i equals 1, and we require the same of β. I will prove a theorem[6] due to Pauli:

Theorem 20.1. *Any set of* 4×4 *matrices with unit squares obeying the Dirac algebra is equivalent to the Weyl representation.*

Actually the entire structure of the theory is embedded in these algebraic equations: *Any* equivalent set of 4×4 matrices defines physically the same Dirac equation as any other equivalent set. The set of Dirac matrices used is merely a matter of choice of basis for the four components of the Dirac field. So, implicit in this proof is a more significant result: *All irreducible representations of the Lorentz group symmetric under parity and containing only spin-1/2 particles are equivalent.* In effect, there is only *one* such representation, and all the others are related by similarity transformations. The proof I'll use is not the one given by Pauli, but one based on our analysis of the representations of the Lorentz group, since we already have a lot of useful theory from our earlier work.

The proof goes as follows. Define M_i and L_i by

$$M_i = \tfrac{1}{2} i\alpha_i \tag{20.44}$$

$$[M_i, M_j] = -i\epsilon_{ijk} L_k \tag{20.45}$$

I will now prove that the M's and the L's so defined obey the commutation relations for the set of generators of a representation of the Lorentz group, *as a consequence of the Dirac algebra* (20.4). The $[M_i, M_j]$ commutator is of course true by definition. Let's use it to determine one of the components of $\mathbf{L}$, say the x component:

$$-iL_x = [M_y, M_z] = -\tfrac{1}{4}\left(\alpha_y\alpha_z - \alpha_z\alpha_y\right) = -\tfrac{1}{2}\alpha_y\alpha_z \tag{20.46}$$

because α_y and α_z anticommute. So we have

$$L_x = -\tfrac{i}{2}\alpha_y\alpha_z \tag{20.47}$$

L_y and L_z are found by cyclic permutation.

Now let's check a typical $[L_i, M_j]$ commutator, say L_x with M_y. If we get this one right, we get the others by cyclic permutation.

$$[L_x, M_y] = \tfrac{1}{4}\alpha_y\alpha_z\alpha_y - \tfrac{1}{4}\alpha_y\alpha_y\alpha_z = -\tfrac{1}{2}\alpha_y\alpha_y\alpha_z = -\tfrac{1}{2}\alpha_z = iM_z \tag{20.48}$$

[6] [Eds.] W. Pauli, "Contributions mathématiques à la théorie des matrices de Dirac" (Mathematical contributions to the theory of Dirac's matrices), *Annales de l'Institut Henri Poincaré* **6**, n. 2 (1936) 109–136.

This is of course the right result for L_x with M_y (18.47). Let's check a typical $[L_i, L_j]$ commutator, say L_x with L_y:

$$[L_x, L_y] = -\tfrac{1}{4}\left(\alpha_y\alpha_z\alpha_z\alpha_x - \alpha_z\alpha_x\alpha_y\alpha_z\right) = -\tfrac{1}{4}\left(\alpha_y\alpha_x - \alpha_x\alpha_y\right) = \tfrac{1}{2}\alpha_x\alpha_y = iL_z \tag{20.49}$$

because α_z^2 is 1, and in the second term, the α_z at the left can be moved to the right through both α_y and α_x, which gives two sign changes. Thus all the commutators check. If I start out with four 4×4 matrices with unit squares obeying the Dirac algebra, I generate from them a four-dimensional representation of the Lorentz group. But which one?

There are many four-dimensional representations of the Lorentz group. For example, there are $D^{(\frac{3}{2},0)}$ and $D^{(\frac{1}{2},\frac{1}{2})}$, but neither of these can be the right one, because

$$L_z^2 = -\tfrac{1}{4}\alpha_x\alpha_y\alpha_x\alpha_y = +\tfrac{1}{4} \tag{20.50}$$

So the representation we've generated with the α's contains only the eigenvalues $\pm\frac{1}{2}$ of L_z. The representation $D^{(\frac{3}{2},0)}$ contains $L_z = \pm\frac{3}{2}$, and $D^{(\frac{1}{2},\frac{1}{2})}$ contains only integer values of L_z. There are thus only three possibilities:

$$D^{(\frac{1}{2},0)}(\Lambda) \oplus D^{(\frac{1}{2},0)}(\Lambda) \qquad D^{(0,\frac{1}{2})}(\Lambda) \oplus D^{(0,\frac{1}{2})}(\Lambda) \qquad D^{(0,\frac{1}{2})}(\Lambda) \oplus D^{(\frac{1}{2},0)}(\Lambda) \tag{20.51}$$

These are the only four-dimensional representations of the Lorentz group with the right eigenvalues of L_z.

We now use β to select from these three the unique representation generated by the α's. From the identity

$$\beta\boldsymbol{\alpha}\beta^{-1} = \beta\boldsymbol{\alpha}\beta = -\boldsymbol{\alpha} \tag{20.52}$$

we have (see (20.44)),

$$\beta\mathbf{M}\beta^{-1} = -\mathbf{M} \tag{20.53}$$

and

$$\beta\mathbf{L}\beta^{-1} = +\mathbf{L} \tag{20.54}$$

since $\mathbf{L}$ is bilinear in the α's (see (20.47)). The similarity transform of both $\mathbf{L}$ and $\mathbf{M}$ with β as the matrix T are exactly how these generators transform under parity, as in (18.50) and (18.51), respectively. Therefore β can be used to define a parity operation. The representation we've generated thus must be invariant under parity, and so equivalent to its parity transform. But under parity, as we have seen in (18.77), the two indices of $D^{(s_1,s_2)}$ are swapped. Of the three candidates in (20.51), only the last, $D^{(0,\frac{1}{2})}(\Lambda) \oplus D^{(\frac{1}{2},0)}(\Lambda)$, is equivalent to its parity transform.

So suppose I have some matrices $\boldsymbol{\alpha}$ and β, not necessarily the Dirac representation, which satisfy the Dirac algebra, and I've found a nonsingular matrix T which takes the three α_i to the Weyl basis for this representation:

$$T\boldsymbol{\alpha}T^{-1} = \boldsymbol{\alpha}_W = \begin{pmatrix} \boldsymbol{\sigma} & \mathbf{0} \\ \mathbf{0} & -\boldsymbol{\sigma} \end{pmatrix} \tag{20.55}$$

Since β must anticommute with $\boldsymbol{\alpha}$, and its square is one, whatever form it had before, its form β' after the similarity transform T must be[7]

$$T\beta T^{-1} = \beta' = \begin{pmatrix} \mathbf{0} & \lambda\cdot\mathbf{1} \\ \lambda^{-1}\cdot\mathbf{1} & \mathbf{0} \end{pmatrix} \tag{20.56}$$

The unknown $\lambda \neq 0$ multiplies a 2×2 identity matrix $\mathbf{1}$. I'm not assuming $\boldsymbol{\alpha}_W$ and β' are Hermitian. Now I make a second similarity transformation:

$$\boldsymbol{\alpha}_W \to S\boldsymbol{\alpha}_W S^{-1} \qquad \beta' \to S\beta' S^{-1} \tag{20.57}$$

where

$$S = \begin{pmatrix} \lambda\cdot\mathbf{1} & \mathbf{0} \\ \mathbf{0} & \mathbf{1} \end{pmatrix} \tag{20.58}$$

By elementary multiplication, this transformation doesn't do anything to $\boldsymbol{\alpha}_W$;

$$S\boldsymbol{\alpha}_W S^{-1} = \boldsymbol{\alpha}_W \tag{20.59}$$

but

$$S\beta' S^{-1} = \begin{pmatrix} \mathbf{0} & \mathbf{1} \\ \mathbf{1} & \mathbf{0} \end{pmatrix} = \beta_W \tag{20.60}$$

This similarity transform turns β' into its Weyl form, as desired. Therefore, one and the same transformation ST turns a given set of unit square 4×4 matrices satisfying the Dirac algebra into the Weyl representation, **QED**.

So what is the point of the theorem? If we want to write down Dirac matrices or the Dirac equation in some crazy basis, we don't have to construct this matrix $S' = ST$. We are guaranteed that any four unit square matrices satisfying the Dirac algebra will be connected to the standard matrices by *some* S'. Secondly, and more importantly, it distinguishes what is important from what is not. A lot of talented people, some of them Nobel laureates, did complicated computations in the early 1930s involving spin-½ particles. Typically this work was done by writing down explicit solutions of the Dirac equation. But they were doing things the hard way. There is no need to write down these solutions, because any desired calculation can be performed using only the algebra of the Dirac matrices. Messy as the anticommutation relations are, they are a lot less trouble than working with explicit 4×4 matrices. The whole structure of the theory lies in these anticommutation relations.

---

[7] [Eds.] Let

$$\beta' = \begin{pmatrix} \mathbf{a} & \mathbf{b} \\ \mathbf{c} & \mathbf{d} \end{pmatrix}$$

where $\{\mathbf{a},\dots,\mathbf{d}\}$ are all 2×2 matrices. Because $\{\beta', \boldsymbol{\alpha}_W\} = 0$, it follows

$$\{\mathbf{a}, \boldsymbol{\sigma}\} = \{\mathbf{d}, \boldsymbol{\sigma}\} = [\mathbf{b}, \boldsymbol{\sigma}] = [\mathbf{c}, \boldsymbol{\sigma}] = 0$$

The set $\{\mathbf{1}, \boldsymbol{\sigma}\} \equiv \sigma^\mu$ is a complete basis for 2×2 matrices, so we can write $\mathbf{a} = a_\mu \sigma^\mu$, and similarly for $\{\mathbf{b}, \mathbf{c}, \mathbf{d}\}$. Then it is easy to show that $a_\mu = d_\mu \equiv 0$, and $b_i = c_i = 0$. That is, only b_0 and c_0 are non-zero, so $\mathbf{b} = b_0\mathbf{1}$ and $\mathbf{c} = c_0\mathbf{1}$. Finally, in order that $\beta'^2 = \mathbb{1}$, we have to have $c_0 = b_0^{-1}$, so β' has to have the form (20.56).

20.4 The γ matrices

The manipulation of Dirac matrices is facilitated by a formalism introduced by Pauli (and automated further by Feynman). Pauli's scheme assumes the four Dirac matrices are Hermitian:

$$\boldsymbol{\alpha}^\dagger = \boldsymbol{\alpha} \qquad \beta^\dagger = \beta \tag{20.61}$$

Both the Weyl representation and the standard representation satisfy this criterion. We will assume our Dirac matrices are all Hermitian in the sequel. We now define a somewhat peculiar "adjoint" of ψ:

$$\overline{\psi} \equiv \psi^\dagger \beta \tag{20.62}$$

The quantity $\overline{\psi}$ is called the **Dirac adjoint**, though in fact it was introduced by Pauli. The motivation for this definition comes from the mass term, $m\psi^\dagger\beta\psi$, in the Dirac Lagrangian, which transforms as a scalar. In terms of the Dirac adjoint,

$$m\psi^\dagger\beta\psi = m\overline{\psi}\psi \tag{20.63}$$

The term $\psi^\dagger\beta\psi$ may have appeared a little awkward for a scalar; $\overline{\psi}\psi$ looks much more natural. With

$$(u_+ + u_-) = \begin{pmatrix}\zeta\\ \eta\end{pmatrix} \text{ and } (u_+ - u_-) = \begin{pmatrix}\xi\\ \chi\end{pmatrix} \tag{20.64}$$

in the Dirac basis (20.11),

$$\begin{aligned}\overline{\psi}\psi &= \frac{1}{2}\begin{pmatrix}u_+ + u_-\\ u_+ - u_-\end{pmatrix}^\dagger \begin{pmatrix}\mathbf{1} & \mathbf{0}\\ \mathbf{0} & -\mathbf{1}\end{pmatrix}\begin{pmatrix}u_+ + u_-\\ u_+ - u_-\end{pmatrix}\\ &= \tfrac{1}{2}\Big[(u_+ + u_-)^\dagger(u_+ + u_-) - (u_+ - u_-)^\dagger(u_+ - u_-)\Big]\\ &= u_+^\dagger u_- + u_-^\dagger u_+ = \tfrac{1}{2}[(|\zeta|^2 + |\eta|^2) - (|\xi|^2 + |\chi|^2)]\end{aligned} \tag{20.65}$$

(This follows even more simply in the Weyl basis.) This expression, the sum of two quantities minus the sum of two others, is a Lorentz scalar.

We would like to define a new adjoint $\overline{A}$ for any 4×4 matrix A, such that $\overline{(A\psi)}$ equals $\overline{\psi}\,\overline{A}$, just as for the ordinary adjoint. The obvious answer is

$$\overline{A} \equiv \beta A^\dagger \beta \tag{20.66}$$

The Lorentz matrices $D(\Lambda)$ that effect Lorentz transformations

$$\Lambda\colon \psi \to D(\Lambda)\psi \tag{20.67}$$

play well with the Dirac adjoint operation. Here,

$$D(\Lambda) \sim D^{(\frac{1}{2},0)}(\Lambda) \oplus D^{(0,\frac{1}{2})}(\Lambda) \tag{20.68}$$

Taking the Dirac adjoint of $D(\Lambda)\psi$, we have

$$\overline{(D(\Lambda)\psi)} = \overline{\psi}\,\overline{D}(\Lambda) \tag{20.69}$$

Because $\overline{\psi}\psi$ is a Lorentz scalar,

$$\Lambda\colon \overline{\psi}\psi \to \overline{\psi}\,\overline{D}(\Lambda)D(\Lambda)\psi = \overline{\psi}\psi \tag{20.70}$$

Since ψ is arbitrary, we deduce that

$$\boxed{\overline{D}(\Lambda)D(\Lambda) = 1} \tag{20.71}$$

Although the $D(\Lambda)$ are *not* unitary, they are "Dirac unitary". While they do not preserve the *conventional* quadratic form, the sum of the squares, as a unitary matrix does, they do preserve this unconventional quadratic form (20.65), the sum of two squares and the difference of two others.[8]

We saw earlier that two 4-vectors, V^μ and W^μ, could be constructed from bilinear products:

$$V^\mu = (u_+^\dagger u_+, u_+^\dagger \boldsymbol{\sigma} u_+) \tag{19.15}$$

$$W^\mu = (u_-^\dagger u_-, -u_-^\dagger \boldsymbol{\sigma} u_-) \tag{19.25}$$

The sum of these is also a 4-vector, and can be written (in either basis) as

$$V^\mu + W^\mu = U^\mu = (\psi^\dagger\psi, \psi^\dagger \boldsymbol{\alpha}\psi) \tag{20.72}$$

If we insert $\beta^2 = \mathbb{1}$ after $\psi^\dagger$, we can make the 4-vector nature of the bilinear product explicit:

$$(\psi^\dagger\beta\beta\psi, \psi^\dagger\beta\beta\boldsymbol{\alpha}\psi) = (\overline{\psi}\beta\psi, \overline{\psi}\beta\boldsymbol{\alpha}\psi) \equiv (\overline{\psi}\gamma^0\psi, \overline{\psi}\boldsymbol{\gamma}\psi) = \overline{\psi}\gamma^\mu\psi \tag{20.73}$$

where we define the four **Dirac gamma matrices**:

$$\boxed{\gamma^\mu = (\gamma^0, \gamma^i) \equiv (\beta, \beta\alpha^i)} \tag{20.74}$$

In the Dirac basis (20.11),

$$\gamma^0 = \begin{pmatrix} \mathbf{1} & \mathbf{0} \\ \mathbf{0} & -\mathbf{1} \end{pmatrix} \qquad \gamma^i = \begin{pmatrix} \mathbf{0} & \boldsymbol{\sigma}^i \\ -\boldsymbol{\sigma}^i & \mathbf{0} \end{pmatrix} \tag{20.75}$$

Under a Lorentz transformation,

$$\Lambda\colon \overline{\psi}\gamma^\mu\psi \to \overline{\psi}\overline{D}(\Lambda)\gamma^\mu D(\Lambda)\psi \tag{20.76}$$

but for a vector,

$$\Lambda\colon U^\mu \to \Lambda^\mu{}_\nu U^\nu \tag{20.77}$$

so we have to have

$$\boxed{\overline{D}(\Lambda)\gamma^\mu D(\Lambda) = \Lambda^\mu{}_\nu\gamma^\nu} \tag{20.78}$$

With a slight abuse of language, we say that the gamma matrices "transform as a vector". Actually, the matrices themselves don't transform *at all*; but sandwiched between two Dirac spinors, the quantity $\overline{\psi}\gamma^\mu\psi$ transforms as a vector.

From their definition,

$$(\gamma^0)^2 = \beta^2 = 1 \tag{20.79}$$

$$(\gamma^i)^2 = \beta\alpha^i\beta\alpha^i = -(\alpha^i)^2 = -1 \tag{20.80}$$

$$\{\gamma^0, \gamma^i\} = \gamma^0\gamma^i + \gamma^i\gamma^0 = \beta\beta\alpha^i + \beta\alpha^i\beta = 0 \tag{20.81}$$

$$\{\gamma^i, \gamma^j\} = \beta\alpha^i\beta\alpha^j + \beta\alpha^j\beta\alpha^i = -\{\alpha^i, \alpha^j\} = 0 \quad (i \neq j) \tag{20.82}$$

[8] [Eds.] Video 20 ends here, at 1:02:18. Typically classes ran for 90 minutes, and occasionally for as long as 115 minutes.

All of these relations can be summarized in one line:

$$\boxed{\{\gamma^\mu, \gamma^\nu\} = 2g^{\mu\nu}\mathbb{1}} \tag{20.83}$$

where $\mathbb{1}$ is a 4×4 identity matrix (which we usually don't bother to write explicitly). Other properties of the gamma matrices are:

$$(\gamma^0)^\dagger = \gamma^0; \qquad (\gamma^i)^\dagger = -\gamma^i; \qquad \overline{\gamma}^\mu = \gamma^\mu; \qquad (\gamma^\mu)^\dagger = \gamma_\mu \tag{20.84}$$

(The statements about $(\gamma^\mu)^\dagger$ hold only for Hermitian β and α^i.)

We can rewrite the Dirac Lagrangian (20.1) in terms of them:

$$\begin{aligned}\mathscr{L} &= \pm\left[i\psi^\dagger\partial_0\psi + i\psi^\dagger\boldsymbol{\alpha}\cdot\nabla\psi - m\psi^\dagger\beta\psi\right]\\ &= \pm\left[i\psi^\dagger\beta\beta\partial_0\psi + i\psi^\dagger\beta\beta\boldsymbol{\alpha}\cdot\nabla\psi - m\psi^\dagger\beta\psi\right]\\ &= \pm\left[i\overline{\psi}\beta\partial_0\psi + i\overline{\psi}\beta\boldsymbol{\alpha}\cdot\nabla\psi - m\overline{\psi}\psi\right]\\ &= \boxed{\pm\overline{\psi}(i\gamma^\mu\partial_\mu - m)\psi}\end{aligned} \tag{20.85}$$

Products of 4-vectors and gamma matrices occur frequently. Feynman introduced a useful shorthand for these products:

$$\gamma^\mu a_\mu \equiv \not{a} \tag{20.86}$$

(pronounced "a slash"). Then

$$(\not{a})^2 = a_\mu a_\nu \gamma^\mu\gamma^\nu = \tfrac{1}{2}a_\mu a_\nu\{\gamma^\mu, \gamma^\nu\} = a^\mu a_\mu = a^2 \tag{20.87}$$

and similarly

$$\{\not{a}, \not{b}\} = 2a^\mu b_\mu = 2a \cdot b \tag{20.88}$$

The Dirac Lagrangian can be rewritten in the slash notation,

$$\mathscr{L} = \pm\overline{\psi}(i\not{\partial} - m)\psi \tag{20.89}$$

and the equation of motion (from varying $\overline{\psi}$)

$$(i\not{\partial} - m)\psi = 0 \tag{20.90}$$

If we multiply on the left with $(i\not{\partial} + m)$, we obtain the Klein–Gordon equation:

$$(i\not{\partial} + m)(i\not{\partial} - m)\psi = 0 = -[(\not{\partial})^2 + m^2]\psi = -(\Box^2 + m^2)\psi \tag{20.91}$$

That is, each of the four components of ψ satisfies the Klein–Gordon equation.

20.5 Bilinear spinor products

We've seen that $\overline{\psi}\psi$ is a Lorentz scalar, and $\overline{\psi}\gamma^\mu\psi$ is a Lorentz vector. It is worthwhile to investigate the transformation character of other bilinear spinor products, with two or more gamma matrices sandwiched between them. As we'll see, there are sixteen[9] linearly

[9] [Eds.] By "bilinear product" we mean a linear combination of terms of the form $(\psi^{(i)})^*\psi^{(j)}$, $i, j = 1, \ldots, 4$ over the components of the spinor. There are (obviously) 16 such terms in all, and what we are doing here is collecting together the linear combinations with specific tensorial behavior.

independent bilinear forms: scalar (1 component), vector (4), antisymmetric tensor (6), axial vector (4), and pseudoscalar (1). First, though, a brief detour to consider the behavior of $\overline{\psi}\psi$ under parity.

Earlier we argued, in the paragraph following (20.54), that for Dirac spinors, $\beta = \gamma^0$ effects a parity transformation:

$$P\colon \psi(\mathbf{x}, t) \to \gamma^0\psi(-\mathbf{x}, t) \tag{20.92}$$

Taking the Dirac adjoint of this equation gives

$$\overline{\gamma^0\psi}(-\mathbf{x}, t) = \overline{\psi}(-\mathbf{x}, t)\overline{\gamma^0} = \overline{\psi}(-\mathbf{x}, t)\gamma^0 \tag{20.93}$$

so we have

$$P\colon \overline{\psi}(\mathbf{x}, t) \to \overline{\psi}(-\mathbf{x}, t)\gamma^0 \tag{20.94}$$

and consequently

$$P\colon \overline{\psi}\psi(\mathbf{x}, t) \to \overline{\psi}\psi(-\mathbf{x}, t) \tag{20.95}$$

That is to say, $\overline{\psi}\psi(\mathbf{x}, t)$ transforms as a scalar under parity. By the same token,

$$P\colon \overline{\psi}\gamma^\mu\psi(\mathbf{x}, t) \to \begin{cases} \overline{\psi}\gamma^0\psi(-\mathbf{x}, t), & \text{if } \mu = 0 \\ -\overline{\psi}\gamma^i\psi(-\mathbf{x}, t), & \text{if } \mu \neq 0 \end{cases} \tag{20.96}$$

which is exactly how you'd expect a vector to transform.

The next most complicated expression is

$$\overline{\psi}\gamma^\mu\gamma^\nu\psi = \tfrac{1}{2}\overline{\psi}\{\gamma^\mu, \gamma^\nu\}\psi + \tfrac{1}{2}\overline{\psi}[\gamma^\mu, \gamma^\nu]\psi = g^{\mu\nu}\overline{\psi}\psi + \tfrac{1}{2}\overline{\psi}[\gamma^\mu, \gamma^\nu]\psi \tag{20.97}$$

The symmetric part of the bilinear expression is nothing new (it's just the old scalar $\overline{\psi}\psi$); but the antisymmetric part is. It's conventional to define

$$\sigma^{\mu\nu} = \frac{i}{2}[\gamma^\mu, \gamma^\nu] \tag{20.98}$$

(The factor of i is included so that $\overline{\sigma}^{\mu\nu} = \sigma^{\mu\nu}$.) It's easy to verify that the bilinear expression $\overline{\psi}\sigma^{\mu\nu}\psi$ transforms as a **tensor**. Consider the transformation of $\overline{\psi}\gamma^\mu\gamma^\nu\psi$:

$$\Lambda\colon \overline{\psi}\gamma^\mu\gamma^\nu\psi \to \overline{\psi}\,\overline{D}(\Lambda)\gamma^\mu D(\Lambda)\overline{D}(\Lambda)\gamma^\nu D(\Lambda)\psi = \Lambda^\mu{}_\alpha\Lambda^\nu{}_\beta\overline{\psi}\gamma^\alpha\gamma^\beta\psi \tag{20.99}$$

The bilinear $\overline{\psi}\gamma^\mu\gamma^\nu\psi$ transforms as a tensor, so the commutator $\overline{\psi}[\gamma^\mu, \gamma^\nu]\psi = 2i\overline{\psi}\sigma^{\mu\nu}\psi$ must as well—it's an antisymmetric second-rank tensor.

What about the three gammas? Only if all three are different will the product differ from a single gamma matrix (to within a sign). Disallowing the products equivalent to a single gamma, the products of three gammas produce only four independent matrices. These Lorentz transform as the components of a 4-vector. For example, consider $\gamma^1\gamma^2\gamma^3$. We have

$$\gamma^1\gamma^2\gamma^3 = \gamma^0\gamma^0\gamma^1\gamma^2\gamma^3 \tag{20.100}$$

Each of the four independent matrices $\gamma^\lambda\gamma^\mu\gamma^\nu$ can be multiplied by the square of the "missing" gamma. That is, the four independent matrices can be written collectively as

$$\gamma^0\gamma^1\gamma^2\gamma^3\gamma^\mu \equiv -i\gamma_5\gamma^\mu \tag{20.101}$$

where the "fifth γ matrix", γ_5, the unique product of four gammas that does not reduce to the product of two gammas or the identity matrix, is defined[10] to be (with the convention $\epsilon_{0123} = +1$)

$$\gamma_5 \equiv i\gamma^0\gamma^1\gamma^2\gamma^3 = \frac{1}{4!}i\epsilon_{\mu\nu\alpha\beta}\gamma^\mu\gamma^\nu\gamma^\alpha\gamma^\beta = \begin{pmatrix} \mathbf{0} & \mathbf{1} \\ \mathbf{1} & \mathbf{0} \end{pmatrix} \text{ (Dirac basis)} \tag{20.102}$$

with properties

$$(\gamma_5)^2 = 1; \qquad (\gamma_5)^\dagger = \gamma_5 = -\overline{\gamma}_5; \qquad \{\gamma_5, \gamma^\mu\} = 0 \tag{20.103}$$

We'll sometimes work with $i\gamma_5$ in preference to γ_5, because

$$\overline{i\gamma_5} = i\gamma_5 \tag{20.104}$$

The quantity $i\overline{\psi}\gamma_5\psi$ transforms as a scalar under proper Lorentz transformations (with $\det\Lambda = 1$), but under parity,

$$P\colon i\overline{\psi}\gamma_5\psi(\mathbf{x}, t) \to i\overline{\psi}\gamma^0\gamma_5\gamma^0\psi(-\mathbf{x}, t) = -i\overline{\psi}\gamma_5\psi(-\mathbf{x}, t) \tag{20.105}$$

That is, $i\overline{\psi}\gamma_5\psi$ is a **pseudoscalar**. (It is also Hermitian, and can appear in a Lagrangian with a real coefficient.)

Finally, we have the bilinear product

$$\overline{\psi}\gamma^\mu\gamma_5\psi \tag{20.106}$$

Under proper Lorentz transformations, it behaves as a vector. But under parity,

$$P\colon \overline{\psi}\gamma^\mu\gamma_5\psi(\mathbf{x}, t) \to \begin{cases} -\overline{\psi}\gamma^0\gamma_5\psi(-\mathbf{x}, t), & \text{if } \mu = 0 \\ \overline{\psi}\gamma^i\gamma_5\psi(-\mathbf{x}, t), & \text{if } \mu \neq 0 \end{cases} \tag{20.107}$$

The quantity $\overline{\psi}\gamma^\mu\gamma_5\psi$ is thus an **axial vector**.

We have now found five bilinear spinor products transforming in distinct ways under parity and Lorentz transformations:

Label	*Product*		*Under P*	*Components*	*Lorentz*
S	$\overline{\psi}\psi$	$\to$	$\overline{\psi}\psi$	1	scalar
V	$(\overline{\psi}\gamma^0\psi, \overline{\psi}\gamma^i\psi)$	$\to$	$(\overline{\psi}\gamma^0\psi, -\overline{\psi}\gamma^i\psi)$	4	vector
T	$\overline{\psi}\sigma^{\mu\nu}\psi$	$\to$	$\overline{\psi}\sigma^{\mu\nu}\psi$	6	tensor
A	$(\overline{\psi}\gamma^0\gamma_5\psi, \overline{\psi}\gamma^i\gamma_5\psi)$	$\to$	$(-\overline{\psi}\gamma^0\gamma_5\psi, \overline{\psi}\gamma^i\gamma_5\psi)$	4	axial vector
P	$\overline{\psi}i\gamma_5\psi$	$\to$	$-\overline{\psi}i\gamma_5\psi$	1	pseudoscalar

Any 4×4 matrix can be expressed as a linear combination of 16 basis elements. These 16 bilinear products thus form a basis for any bilinear product. Ultimately we will build interactions out of these bilinear products.

---

[10] [Eds.] Coleman uses γ_5 and γ^5 interchangeably. Here only the lower index γ_5 will be used.

20.6 Orthogonality and completeness

We can express the normalization conditions (20.23) and (20.43) in terms of the gamma matrices:

$$\overline{u}_{\mathbf{p}}^{(r)}\gamma^{\mu}u_{\mathbf{p}}^{(s)} = 2p^{\mu}\delta^{rs} \tag{20.108a}$$

$$\overline{v}_{\mathbf{p}}^{(r)}\gamma^{\mu}v_{\mathbf{p}}^{(s)} = 2p^{\mu}\delta^{rs} \tag{20.108b}$$

By taking the product of these with p^{μ} we get

$$\overline{u}_{\mathbf{p}}^{(r)}\not{p}u_{\mathbf{p}}^{(s)} = 2m^2\delta^{rs} \tag{20.109a}$$

$$\overline{v}_{\mathbf{p}}^{(r)}\not{p}v_{\mathbf{p}}^{(s)} = 2m^2\delta^{rs} \tag{20.109b}$$

If we substitute the plane wave solutions (20.12) and (20.36) into the equations of motion (20.90), we obtain

$$(\not{p}-m)u_{\mathbf{p}}^{(r)} = 0 \Rightarrow \not{p}u_{\mathbf{p}}^{(r)} = mu_{\mathbf{p}}^{(r)} \tag{20.110a}$$

$$(\not{p}+m)v_{\mathbf{p}}^{(r)} = 0 \Rightarrow \not{p}v_{\mathbf{p}}^{(r)} = -mv_{\mathbf{p}}^{(r)} \tag{20.110b}$$

so that

$$\overline{u}_{\mathbf{p}}^{(r)}\not{p}u_{\mathbf{p}}^{(s)} = m\overline{u}_{\mathbf{p}}^{(r)}u_{\mathbf{p}}^{(s)} \tag{20.111a}$$

$$\overline{v}_{\mathbf{p}}^{(r)}\not{p}v_{\mathbf{p}}^{(s)} = -m\overline{v}_{\mathbf{p}}^{(r)}v_{\mathbf{p}}^{(s)} \tag{20.111b}$$

Comparing (20.109a) with (20.111a), and (20.109b) with (20.111b), we find

$$\overline{u}_{\mathbf{p}}^{(r)}u_{\mathbf{p}}^{(s)} = 2m\,\delta^{rs} \tag{20.112a}$$

$$\overline{v}_{\mathbf{p}}^{(r)}v_{\mathbf{p}}^{(s)} = -2m\,\delta^{rs} \tag{20.112b}$$

So the solutions $u_{\mathbf{p}}^{(1)}$ and $u_{\mathbf{p}}^{(2)}$ are Dirac orthogonal to each other, as are $v_{\mathbf{p}}^{(1)}$ and $v_{\mathbf{p}}^{(2)}$.

What about the mixed expressions $\overline{v}_{\mathbf{p}}^{(r)}u_{\mathbf{p}}^{(s)}$ and $\overline{u}_{\mathbf{p}}^{(r)}v_{\mathbf{p}}^{(s)}$? Taking the Dirac adjoint of (20.110a) gives

$$\overline{u}_{\mathbf{p}}^{(r)}\not{p} = m\overline{u}_{\mathbf{p}}^{(r)} \tag{20.113}$$

and multiplying on the right with $v_{\mathbf{p}}^{(s)}$, we have

$$\overline{u}_{\mathbf{p}}^{(r)}\not{p}v_{\mathbf{p}}^{(s)} = m\overline{u}_{\mathbf{p}}^{(r)}v_{\mathbf{p}}^{(s)} \tag{20.114}$$

However, (20.110b) says

$$\not{p}v_{\mathbf{p}}^{(s)} = -mv_{\mathbf{p}}^{(s)} \tag{20.115}$$

Multiplying this equation by $\overline{u}_{\mathbf{p}}^{(r)}$ on the left gives

$$\overline{u}_{\mathbf{p}}^{(r)}\not{p}v_{\mathbf{p}}^{(s)} = -m\overline{u}_{\mathbf{p}}^{(r)}v_{\mathbf{p}}^{(s)} \tag{20.116}$$

Comparing (20.114) with (20.116), it follows, as $m \neq 0$, that

$$\overline{u}_{\mathbf{p}}^{(r)}v_{\mathbf{p}}^{(s)} = 0 \tag{20.117}$$

and returning to (20.116), we have also

$$\overline{u}_{\mathbf{p}}^{(r)}\gamma^{\mu}v_{\mathbf{p}}^{(s)} = 0 \tag{20.118}$$

The positive and negative frequency spinors are Dirac orthogonal, and a vector made of positive and negative frequency spinors vanishes. So much for orthogonality.

In calculations to come, we will frequently need to evaluate expressions involving sums of spinors. Suppose we apply the operator $(\not{p} - m)$ to $u_{\mathbf{p}}^{(r)}$, and likewise $(\not{p} + m)$ to $v_{\mathbf{p}}^{(r)}$. Then

$$(\not{p} - m)u_{\mathbf{p}}^{(r)} = (m - m)u_{\mathbf{p}}^{(r)} = 0 \tag{20.119a}$$
$$(\not{p} + m)v_{\mathbf{p}}^{(r)} = (-m + m)v_{\mathbf{p}}^{(r)} = 0 \tag{20.119b}$$

Consider the sum (note the "backwards" order, with the column vector first and the row vector second)

$$A = \frac{1}{2m}\sum_{r=1}^{2} u_{\mathbf{p}}^{(r)}\overline{u}_{\mathbf{p}}^{(r)} \tag{20.120}$$

This 4×4 matrix A acts like a *projection operator.* If we apply A to a linear combination of $u_{\mathbf{p}}^{(r)}$ and $v_{\mathbf{p}}^{(r)}$, only the $u_{\mathbf{p}}^{(r)}$ component survives:

$$Au_{\mathbf{p}}^{(s)} = \sum_{r=1}^{2} u_{\mathbf{p}}^{(r)}\overline{u}_{\mathbf{p}}^{(r)}u_{\mathbf{p}}^{(s)} = 2mu_{\mathbf{p}}^{(s)} \tag{20.121a}$$
$$Av_{\mathbf{p}}^{(s)} = \sum_{r=1}^{2} u_{\mathbf{p}}^{(r)}\overline{u}_{\mathbf{p}}^{(r)}v_{\mathbf{p}}^{(s)} = 0 \tag{20.121b}$$

Well, we already have a 4×4 matrix with these properties:

$$(\not{p} + m)u_{\mathbf{p}}^{(r)} = 2mu_{\mathbf{p}}^{(r)} \qquad (\not{p} + m)v_{\mathbf{p}}^{(r)} = 0 \tag{20.122}$$

so we have two *completeness relations,*[11]

$$\sum_{r=1}^{2} u_{\mathbf{p}}^{(r)}\overline{u}_{\mathbf{p}}^{(r)} = \not{p} + m \tag{20.123a}$$
$$\sum_{r=1}^{2} v_{\mathbf{p}}^{(r)}\overline{v}_{\mathbf{p}}^{(r)} = \not{p} - m \tag{20.123b}$$

[11] [Eds.] The relations (20.123a) and (20.123b) can be checked explicitly, in the special case of $\mathbf{p} = (0, 0, p_z)$, from (20.27), (20.33) and (20.42), but it's tedious. For example, letting the indices on Dirac spinors and matrices run from 1 to 4,

$$(\not{p} + m)_{13} = \begin{pmatrix} (p_0 + m)\mathbf{1} & -\mathbf{p}\bullet\boldsymbol{\sigma} \\ \mathbf{p}\bullet\boldsymbol{\sigma} & (-p_0 + m)\mathbf{1} \end{pmatrix}_{13} = -p^1(\boldsymbol{\sigma}_1)_{11} - p^2(\boldsymbol{\sigma}_2)_{11} - p^3(\boldsymbol{\sigma}_3)_{11} = -p^3 = -|\mathbf{p}|$$

and

$$(u_{\mathbf{p}}^{(1)})_1(\overline{u}_{\mathbf{p}}^{(1)})_3 + (u_{\mathbf{p}}^{(2)})_1(\overline{u}_{\mathbf{p}}^{(2)})_3 = (\sqrt{E_{\mathbf{p}} + m})(-\sqrt{E_{\mathbf{p}} - m}) + (0)(0) = -\sqrt{E_{\mathbf{p}}^2 - m^2} = -|\mathbf{p}|$$

(the second following by the same reasoning as for the first) and two complementary **spinor projection operators** P_u and P_v:

$$P_u = \frac{1}{2m}\sum_{r=1}^{2} u_{\mathbf{p}}^{(r)}\overline{u}_{\mathbf{p}}^{(r)} = \frac{1}{2m}(\not{p} + m) \tag{20.124a}$$

$$P_v = -\frac{1}{2m}\sum_{r=1}^{2} v_{\mathbf{p}}^{(r)}\overline{v}_{\mathbf{p}}^{(r)} = -\frac{1}{2m}(\not{p} - m) \tag{20.124b}$$

These operators have the expected properties:

$$P_u\left(au_{\mathbf{p}}^{(r)} + bv_{\mathbf{p}}^{(r)}\right) = au_{\mathbf{p}}^{(r)} \tag{20.125a}$$

$$P_v\left(au_{\mathbf{p}}^{(r)} + bv_{\mathbf{p}}^{(r)}\right) = bv_{\mathbf{p}}^{(r)} \tag{20.125b}$$

Their sum is the identity:

$$P_u + P_v = \frac{1}{2m}(\not{p} + m + m - \not{p}) = \mathbb{1} \tag{20.126}$$

where $\mathbb{1}$ is the 4×4 identity matrix. Acting on either $u_{\mathbf{p}}^{(s)}$ or $v_{\mathbf{p}}^{(s)}$, they are orthogonal

$$P_u P_v u_{\mathbf{p}}^{(s)} = P_v P_u u_{\mathbf{p}}^{(s)} = P_u P_v v_{\mathbf{p}}^{(s)} = P_v P_u v_{\mathbf{p}}^{(s)} = 0 \tag{20.127}$$

and idempotent:

$$P_u P_u u_{\mathbf{p}}^{(s)} = \frac{1}{4m^2}\left(\not{p}\not{p} + m^2 + 2m\not{p}\right)u_{\mathbf{p}}^{(s)} = \frac{1}{2m}(\not{p} + m)u_{\mathbf{p}}^{(s)} = P_u u_{\mathbf{p}}^{(s)} \tag{20.128a}$$

$$P_v P_v v_{\mathbf{p}}^{(s)} = \frac{1}{4m^2}\left(\not{p}\not{p} + m^2 - 2m\not{p}\right)v_{\mathbf{p}}^{(s)} = -\frac{1}{2m}(\not{p} - m)v_{\mathbf{p}}^{(s)} = P_v v_{\mathbf{p}}^{(s)} \tag{20.128b}$$

These relations will be very helpful in computing processes with spin-½ particles.

Next time we will tackle the canonical quantization of the Dirac field, the calculation of the appropriate propagators and the Feynman rules.

Problems 11

11.1 For any $\mathbf{p}$, find two independent positive frequency solutions (i.e., u's, not v's) of the Dirac equation that are eigenstates of helicity, angular momentum along the direction of motion. (The solutions displayed in class are not helicity eigenstates unless $\mathbf{p}$ points along the z-axis.) Express the four components of u as explicit functions of θ and ϕ, the polar angles of the direction of motion. HINT: The helicity operator commutes with rotations.

(1997a 10.1)

11.2 The following identities are easy to see. For the last two, use the cyclic property of the trace, $\mathrm{Tr}\,(ABC) = \mathrm{Tr}\,(CAB)$:

$$\mathrm{Tr}\,(\mathbb{1}) = 4 \tag{P11.1}$$

$$\mathrm{Tr}\,(\not{a}) = \mathrm{Tr}\,(\gamma_5\gamma_5\not{a}) = \mathrm{Tr}\,(\gamma_5\not{a}\gamma_5) = -\mathrm{Tr}\,(\not{a}) = 0 \tag{P11.2}$$

$$\mathrm{Tr}\,(\not{a}\not{b}) = \tfrac{1}{2}\mathrm{Tr}\,(\not{a}\not{b} + \not{b}\not{a}) = 4a \cdot b \tag{P11.3}$$

Carry on. Compute $\mathrm{Tr}\,(\not{a}\not{b}\not{c})$, $\mathrm{Tr}\,(\not{a}\not{b}\not{c}\not{d})$, and the trace of up to four slashed vectors and one factor of γ_5. The last computation, of $\mathrm{Tr}\,(\not{a}\not{b}\not{c}\not{d}\gamma_5)$, will involve $\epsilon_{\alpha\beta\mu\nu}$. Just to make sure we are all working with the same sign conventions, choose $\epsilon_{0123} = +1$. (You can find the answers to these in any relativistic quantum theory text, but it's more fun, as well as more instructive, to work them out yourself.)

(1997a 10.2)

Solutions 11

11.1 Start with the helicity eigenstates, (20.27) and (20.33), for momenta in the z direction,

$$u^{(1)}_{p_z} = \begin{pmatrix} \sqrt{E_{\mathbf{p}}+m} \\ 0 \\ \sqrt{E_{\mathbf{p}}-m} \\ 0 \end{pmatrix} \qquad u^{(2)}_{\mathbf{p}} = \begin{pmatrix} 0 \\ \sqrt{E_{\mathbf{p}}+m} \\ 0 \\ -\sqrt{E_{\mathbf{p}}-m} \end{pmatrix} \tag{S11.1}$$

For momentum in an arbitrary direction (θ, ϕ), the unit vector $\hat{\mathbf{p}}$ is given by

$$\hat{\mathbf{p}} = (\sin\theta\cos\phi, \sin\theta\sin\phi, \cos\theta) \tag{S11.2}$$

To get from $\hat{\mathbf{p}} = \hat{\mathbf{z}}$ to $\hat{\mathbf{p}} = (\sin\theta\cos\phi, \sin\theta\sin\phi, \cos\theta)$, first we rotate about the y axis by θ, and then we rotate about the z axis by ϕ. From (20.7):

$$\mathbf{L} = \tfrac{1}{2}\begin{pmatrix} \boldsymbol{\sigma} & \mathbf{0} \\ \mathbf{0} & \boldsymbol{\sigma} \end{pmatrix} \tag{S11.3}$$

So the rotation operators (18.23) we need are

$$D(R(\hat{\mathbf{y}}\theta)) = e^{-iL_y\theta} = \cos\tfrac{1}{2}\theta - i\sin\tfrac{1}{2}\theta\begin{pmatrix} \boldsymbol{\sigma}_y & \mathbf{0} \\ \mathbf{0} & \boldsymbol{\sigma}_y \end{pmatrix} = \begin{pmatrix} \cos\frac{1}{2}\theta & -\sin\frac{1}{2}\theta & 0 & 0 \\ \sin\frac{1}{2}\theta & \cos\frac{1}{2}\theta & 0 & 0 \\ 0 & 0 & \cos\frac{1}{2}\theta & -\sin\frac{1}{2}\theta \\ 0 & 0 & \sin\frac{1}{2}\theta & \cos\frac{1}{2}\theta \end{pmatrix} \tag{S11.4}$$

$$D(R(\hat{\mathbf{z}}\phi)) = e^{-iL_z\phi} = \cos\tfrac{1}{2}\phi - i\sin\tfrac{1}{2}\phi\begin{pmatrix} \boldsymbol{\sigma}_z & \mathbf{0} \\ \mathbf{0} & \boldsymbol{\sigma}_z \end{pmatrix} = \begin{pmatrix} e^{-\frac{i}{2}\phi} & 0 & 0 & 0 \\ 0 & e^{\frac{i}{2}\phi} & 0 & 0 \\ 0 & 0 & e^{-\frac{i}{2}\phi} & 0 \\ 0 & 0 & 0 & e^{\frac{i}{2}\phi} \end{pmatrix} \tag{S11.5}$$

Then

$$u^{(1)}_{\mathbf{p}} = D(R(\hat{\mathbf{z}}\phi))D(R(\hat{\mathbf{y}}\theta))u^{(1)}_{p_z} = \begin{pmatrix} e^{-\frac{i}{2}\phi}\cos\frac{1}{2}\theta\sqrt{E_{\mathbf{p}}+m} \\ e^{\frac{i}{2}\phi}\sin\frac{1}{2}\theta\sqrt{E_{\mathbf{p}}+m} \\ e^{-\frac{i}{2}\phi}\cos\frac{1}{2}\theta\sqrt{E_{\mathbf{p}}-m} \\ e^{\frac{i}{2}\phi}\sin\frac{1}{2}\theta\sqrt{E_{\mathbf{p}}-m} \end{pmatrix} \tag{S11.6}$$

and

$$u^{(2)}_{\mathbf{p}} = D(R(\hat{\mathbf{z}}\phi))D(R(\hat{\mathbf{y}}\theta))u^{(2)}_{p_z} = \begin{pmatrix} -e^{-\frac{i}{2}\phi}\sin\frac{1}{2}\theta\sqrt{E_{\mathbf{p}}+m} \\ e^{\frac{i}{2}\phi}\cos\frac{1}{2}\theta\sqrt{E_{\mathbf{p}}+m} \\ e^{-\frac{i}{2}\phi}\sin\frac{1}{2}\theta\sqrt{E_{\mathbf{p}}-m} \\ -e^{\frac{i}{2}\phi}\cos\frac{1}{2}\theta\sqrt{E_{\mathbf{p}}-m} \end{pmatrix} \tag{S11.7}$$

The spinors $u_{\mathbf{p}}^{(1)}$ and $u_{\mathbf{p}}^{(2)}$ have positive and negative helicities, respectively, because the original $u_{p_z}^{(1)}$ and $u_{p_z}^{(2)}$ had those helicities, and the helicity operator commutes with the rotation operators. ■

11.2 First, it is easy to see that the trace of the product of three gamma matrices is zero:[1]

$$\mathrm{Tr}\,(\gamma^\alpha\gamma^\mu\gamma^\nu) = \mathrm{Tr}\,(\gamma_5\gamma_5\gamma^\alpha\gamma^\mu\gamma^\nu) = \mathrm{Tr}\,(\gamma_5\gamma^\alpha\gamma^\mu\gamma^\nu\gamma_5) = (-1)^3\mathrm{Tr}\,(\gamma_5\gamma_5\gamma^\alpha\gamma^\mu\gamma^\nu) = -\mathrm{Tr}\,(\gamma^\alpha\gamma^\mu\gamma^\nu) \tag{S11.8}$$

the third equality following from the cyclic property of the trace. By the same argument, the trace of the product of an odd number of gamma matrices is zero. Thus

$$\mathrm{Tr}\,(\not{a}\not{b}\not{c}) = 0 \tag{S11.9}$$

For an even number of gamma matrices, we use repeatedly the identity

$$\gamma^\mu\gamma^\nu = 2g^{\mu\nu} - \gamma^\nu\gamma^\mu \tag{S11.10}$$

The algorithm is simple: Using this identity, start with the rightmost gamma, work it through to the leftmost position, then use the cyclic property to return it to its original position. For example,

$$\mathrm{Tr}\,(\gamma^\mu\gamma^\nu) = 2g^{\mu\nu}\,\mathrm{Tr}\,(\mathbb{1}) - \mathrm{Tr}\,(\gamma^\nu\gamma^\mu) = 2g^{\mu\nu}\,\mathrm{Tr}\,(\mathbb{1}) - \mathrm{Tr}\,(\gamma^\mu\gamma^\nu) \tag{S11.11}$$

so that

$$\mathrm{Tr}\,(\gamma^\mu\gamma^\nu) = g^{\mu\nu}\,\mathrm{Tr}\,(\mathbb{1}) = 4g^{\mu\nu} \tag{S11.12}$$

and hence

$$\mathrm{Tr}\,(\not{a}\not{b}) = 4a\cdot b \tag{S11.13}$$

Let's do this for four gammas:

$$\begin{aligned}\mathrm{Tr}\,(\gamma^\alpha\gamma^\beta\gamma^\mu\gamma^\nu) &= 2g^{\mu\nu}\,\mathrm{Tr}\,(\gamma^\alpha\gamma^\beta) - \mathrm{Tr}\,(\gamma^\alpha\gamma^\beta\gamma^\nu\gamma^\mu)\\ &= 2g^{\mu\nu}\,\mathrm{Tr}\,(\gamma^\alpha\gamma^\beta) - 2g^{\beta\nu}\,\mathrm{Tr}\,(\gamma^\alpha\gamma^\mu) + 2g^{\alpha\nu}\,\mathrm{Tr}\,(\gamma^\beta\gamma^\mu) - \mathrm{Tr}\,(\gamma^\nu\gamma^\alpha\gamma^\beta\gamma^\mu)\end{aligned} \tag{S11.14}$$

The last term is the same as the original. Move it to the left-hand side of the equation, and divide by 2, to obtain

$$\begin{aligned}\mathrm{Tr}\,(\gamma^\alpha\gamma^\beta\gamma^\mu\gamma^\nu) &= g^{\mu\nu}\,\mathrm{Tr}\,(\gamma^\alpha\gamma^\beta) - g^{\beta\nu}\,\mathrm{Tr}\,(\gamma^\alpha\gamma^\mu) + g^{\alpha\nu}\,\mathrm{Tr}\,(\gamma^\beta\gamma^\mu)\\ &= 4g^{\mu\nu}g^{\alpha\beta} - 4g^{\beta\nu}g^{\alpha\mu} + 4g^{\alpha\nu}g^{\beta\mu}\end{aligned} \tag{S11.15}$$

Then

$$\mathrm{Tr}\,(\not{a}\not{b}\not{c}\not{d}) = 4(a\cdot b)(c\cdot d) - 4(a\cdot c)(b\cdot d) + 4(a\cdot d)(b\cdot c) \tag{S11.16}$$

Now we come to γ_5:

$$\gamma_5 = i\gamma^0\gamma^1\gamma^2\gamma^3 \tag{S11.17}$$

We have just shown, though, that the trace of four gammas must have two indices in common to have a non-vanishing trace. But there are *no* repeated indices in γ_5. So its trace is zero:

$$\mathrm{Tr}\,(\gamma_5) = 0 \tag{S11.18}$$

(This is also evident from its explicit form in the Dirac basis (20.102).) Notice that γ_5 is itself the product of an *even* number of gammas, so the trace of γ_5 times an odd number of gammas vanishes.

What about the trace of γ_5 with two gammas? If in $\gamma_5\gamma^\mu\gamma^\nu$, $\mu = \nu$, then the product reduces to $\pm\gamma_5$, and that trace vanishes. Say that $\mu \neq \nu$. Pick a value of α different from both μ and ν. Then (no sum on α)

$$\mathrm{Tr}\,(\gamma_5\gamma^\mu\gamma^\nu) = \pm\mathrm{Tr}\,(\gamma^\alpha\gamma^\alpha\gamma_5\gamma^\mu\gamma^\nu) = \mp\mathrm{Tr}\,(\gamma^\alpha\gamma_5\gamma^\mu\gamma^\nu\gamma^\alpha) = \mp\mathrm{Tr}\,(\gamma^\alpha\gamma^\alpha\gamma_5\gamma^\mu\gamma^\nu) \tag{S11.19}$$

So the trace of $\gamma_5\gamma^\mu\gamma^\nu$ vanishes for all choices of μ and ν:

$$\mathrm{Tr}\,(\not{a}\not{b}\gamma_5) = 0 \tag{S11.20}$$

Finally, what about four gammas with γ_5? Clearly, we need to have all four gammas be different. If any two are the same, then we have again two gammas with γ_5, whose trace vanishes. And if all are different, then their product is nothing but $\pm i\epsilon_{\mu\nu\rho\sigma}\gamma_5$. That is,

$$\mathrm{Tr}\,(\gamma_5\gamma_\alpha\gamma_\beta\gamma_\mu\gamma_\nu) = \pm i\epsilon_{\alpha\beta\mu\nu}\mathrm{Tr}\,(\gamma_5\gamma_5) = \pm 4i\epsilon_{\alpha\beta\mu\nu} \tag{S11.21}$$

We can determine the sign by looking at $\gamma_5\gamma_0\gamma_1\gamma_2\gamma_3 = i\gamma_5\gamma_5 = i$, so

$$\mathrm{Tr}\,(\gamma_5\gamma_\alpha\gamma_\beta\gamma_\mu\gamma_\nu) = 4i\epsilon_{\alpha\beta\mu\nu} \qquad \text{and} \qquad \mathrm{Tr}\,(\gamma_5\not{a}\not{b}\not{c}\not{d}) = 4i\epsilon_{\alpha\beta\mu\nu}a^\alpha b^\beta c^\mu d^\nu \tag{S11.22}$$

To sum up, the trace of γ_5 with fewer than 4 gammas is zero; and with four, it's an expression proportional to $\epsilon_{\alpha\beta\mu\nu}$. ■

---

[1] [Eds.] Call this the "gamma-5 trick".

21

The Dirac Equation III. Quantization and Feynman Rules

We will now canonically quantize the free Dirac Lagrangian. After that, we'll consider simple interacting models, and construct the Feynman rules for a theory involving fermions. We will refer to these fermions for the time being as nucleons, though the formalism will hold good for any $m \neq 0$, spin-$\frac{1}{2}$ particle, including electrons. Our interacting theories will involve ψ, $\overline{\psi}$, and a scalar field ϕ. We considered a similar theory in Model 3, but with spinless nucleons. In that theory, we had only three fields to worry about: ϕ, ψ and ψ^*. We now have nine: 4 components for ψ, 4 for $\overline{\psi}$, and ϕ. So we'll have more combinatorics to juggle, and we'll worry about some minus signs due to Fermi statistics.[1]

21.1 Canonical quantization of the Dirac field

From the Dirac Lagrangian,

$$\mathscr{L} = \pm\left[i\psi^\dagger\partial_0\psi + i\psi^\dagger\boldsymbol{\alpha}\boldsymbol{\cdot}\nabla\psi - m\psi^\dagger\beta\psi\right] \tag{20.1}$$

we obtain the canonical momentum

$$\pi_\psi \equiv \frac{\partial\mathscr{L}}{\partial(\partial_0\psi)} = \pm i\psi^\dagger \tag{21.1}$$

The Dirac spinor ψ has four components $\{\psi_a\}$, $a = 1,\ldots,4$, and each has its canonical momentum $\{\pi_\psi^a\}$. These four pairs form a complete and independent set of initial data, since the Dirac equation is linear. Following our usual program, we should impose the equal time commutation relations (4.47)

$$[\psi_a(\mathbf{x},t), \psi_b(\mathbf{y},t)] = 0 \tag{21.2}$$

$$[\psi_a^\dagger(\mathbf{x},t), \psi_b^\dagger(\mathbf{y},t)] = 0 \tag{21.3}$$

$$[\psi_a(\mathbf{x},t), \psi_b^\dagger(\mathbf{y},t)] = \pm\delta^{(3)}(\mathbf{x}-\mathbf{y})\delta_{ab} \tag{21.4}$$

[1] [Eds.] The videotape of Lecture 21 does not begin until §21.3. To make matters worse, Coleman's notes covering the first two sections of this chapter are also missing. Thus these first two sections are based entirely on Brian Hill's and Peter Woit's reliable notes, with some guessed-at interpolation by the editors. *Caveat lector!*

We will see, however, that Fermi statistics require that these conditions be modified. The Hamiltonian is (4.40),

$$\mathscr{H} = \sum_a \pi^a_\psi \partial_0 \psi_a - \mathscr{L} = \pm i\psi^\dagger \partial_0 \psi - \mathscr{L} = \pm \psi^\dagger \left(-i\boldsymbol{\alpha}\bullet\nabla + m\beta\right)\psi = \pm i\psi^\dagger \partial_0 \psi \tag{21.5}$$

the last equality following from the Euler–Lagrange equations (4.26). We express the Fermi fields in a manner analogous to the expression (6.24) for complex scalar fields,

$$\boxed{\psi(x) = \int \frac{d^3\mathbf{p}}{(2\pi)^{3/2}\sqrt{2E_\mathbf{p}}} \sum_{r=1}^{2} \left[b^{(r)}_\mathbf{p} u^{(r)}_\mathbf{p} e^{-ip\cdot x} + c^{(r)\dagger}_\mathbf{p} v^{(r)}_\mathbf{p} e^{ip\cdot x} \right]} \tag{21.6}$$

Each Fermi field ψ contains two spinor components $u^{(r)}_\mathbf{p}$ for positive frequency solutions, and two for negative frequency solutions, $v^{(r)}_\mathbf{p}$. The operators $b^{(r)}_\mathbf{p}$ multiplying the positive frequency solutions, in analogy with the charged scalar field expression, annihilate nucleons. (See §6.1.) Then the operators $c^{(r)\dagger}_\mathbf{p}$, multiplying negative frequency solutions, must create antinucleons. Of course $\psi^\dagger(y)$ is nearly the same, except that the operators and spinors appear as their adjoints, the signs of the exponentials are reversed, and x is replaced by y:

$$\psi^\dagger(y) = \int \frac{d^3\mathbf{p}}{(2\pi)^{3/2}\sqrt{2E_\mathbf{p}}} \sum_{r=1}^{2} \left[b^{(r)\dagger}_\mathbf{p} u^{(r)\dagger}_\mathbf{p} e^{ip\cdot y} + c^{(r)}_\mathbf{p} v^{(r)\dagger}_\mathbf{p} e^{-ip\cdot y} \right] \tag{21.7}$$

The operators $b^{(r)\dagger}_\mathbf{p}$ create nucleons, and the operators $c^{(r)}_\mathbf{p}$ annihilate antinucleons. It's hard to keep these operators straight, so here's a chart to summarize them:

Operator	*What it does*
$b^{(r)}_\mathbf{p}$	annihilates nucleons with momentum **p** and spin r
$c^{(r)}_\mathbf{p}$	annihilates antinucleons with momentum **p** and spin r
$b^{\dagger(r)}_\mathbf{p}$	creates nucleons with momentum **p** and spin r
$c^{\dagger(r)}_\mathbf{p}$	creates antinucleons with momentum **p** and spin r

What are the commutation relations for the creation and annihilation operators? To avoid having to play around with Fourier analysis, use the *ansatz*

$$[b^{(r)}_\mathbf{p}, b^{(s)\dagger}_{\mathbf{p}'}] = \delta^{rs}\delta^{(3)}(\mathbf{p}-\mathbf{p}')\,B(\mathbf{p}) \tag{21.8}$$

$$[c^{(r)\dagger}_\mathbf{p}, c^{(s)}_{\mathbf{p}'}] = \delta^{rs}\delta^{(3)}(\mathbf{p}-\mathbf{p}')\,C(\mathbf{p}) \tag{21.9}$$

The functions $B(\mathbf{p})$ and $C(\mathbf{p})$ are to be determined. We'll assume all other commutators between the b's and c's are zero. Then automatically (21.2) and (21.3) are satisfied, and

$$[\psi_a(\mathbf{x},t), \psi^\dagger_b(\mathbf{y},t)] = \int \frac{d^3\mathbf{p}}{(2\pi)^3 2E_\mathbf{p}} \left[\left(e^{i\mathbf{p}\bullet(\mathbf{x}-\mathbf{y})} B(\mathbf{p}) \sum_{r=1}^{2} u^{(r)}_{\mathbf{p}a} u^{(r)\dagger}_{\mathbf{p}b} \right) + \left(e^{-i\mathbf{p}\bullet(\mathbf{x}-\mathbf{y})} C(\mathbf{p}) \sum_{r=1}^{2} v^{(r)}_{\mathbf{p}a} v^{(r)\dagger}_{\mathbf{p}b} \right) \right] \tag{21.10}$$

From the completeness relations (20.123a) and (20.123b) (multiplied by γ^0 to convert $\overline{u}$ and $\overline{v}$ to $u^\dagger$ and $v^\dagger$, respectively),

$$\sum_{r=1}^{2} u^{(r)}_{\mathbf{p}\,a} u^{(r)\dagger}_{\mathbf{p}\,b} = E_{\mathbf{p}}\delta_{ab} - \mathbf{p}\cdot(\boldsymbol{\gamma}\gamma^0)_{ab} + m\gamma^0_{ab} \tag{21.11}$$

$$\sum_{r=1}^{2} v^{(r)}_{\mathbf{p}\,a} v^{(r)\dagger}_{\mathbf{p}\,b} = E_{\mathbf{p}}\delta_{ab} - \mathbf{p}\cdot(\boldsymbol{\gamma}\gamma^0)_{ab} - m\gamma^0_{ab} \tag{21.12}$$

Changing $\mathbf{p} \to -\mathbf{p}$ in the second integrand of (21.10), we obtain the canonical commutation relations (21.4) if we set $B(\mathbf{p}) = C(-\mathbf{p}) = \pm 1$; the terms proportional to $\mathbf{p}$ and m cancel, and

$$[\psi_a(\mathbf{x},t), \psi^\dagger_b(\mathbf{y},t)] = \pm\delta_{ab}\int \frac{d^3\mathbf{p}}{(2\pi)^3} e^{i\mathbf{p}\cdot(\mathbf{x}-\mathbf{y})} = \pm\delta_{ab}\delta^{(3)}(\mathbf{x}-\mathbf{y}) \tag{21.13}$$

But not all is well. Let's compute the energy:

$$\begin{aligned} H &= \int d^3\mathbf{x}\,\mathscr{H} = \pm\int d^3\mathbf{x}\; i\psi^\dagger\partial_0\psi \\ &= \pm\int \frac{d^3\mathbf{p}}{2E_{\mathbf{p}}} E_{\mathbf{p}} \Big[\sum_{r,s} 2E_{\mathbf{p}}\delta_{rs} b^{(r)\dagger}_{\mathbf{p}} b^{(s)}_{\mathbf{p}} - \sum_{r,s} 2E_{\mathbf{p}}\delta_{rs} c^{(r)}_{\mathbf{p}} c^{(s)\dagger}_{\mathbf{p}}\Big] \end{aligned} \tag{21.14}$$

where we've used the spinor relations (20.108a), (20.108b) and (20.118). Then

$$H = \pm\int d^3\mathbf{p}\, E_{\mathbf{p}} \sum_r \Big[b^{(r)\dagger}_{\mathbf{p}} b^{(r)}_{\mathbf{p}} - c^{(r)}_{\mathbf{p}} c^{(r)\dagger}_{\mathbf{p}}\Big] \tag{21.15}$$

No matter which sign we choose, this expression is not positive-definite, and hence not bounded below. If we choose the plus sign, the antinucleons carry negative energy. This is a mess. We didn't run into this problem in either the charged (6.22) or uncharged (4.63) scalar cases, where the bilinear combinations of creation and annihilation operators appear with positive signs, because the Hamiltonian for scalar fields is *quadratic* in the derivatives (4.56), while that for Dirac fields is *linear* in the derivatives (21.14).

The way out of these troubles was found by Jordan and Wigner.[2] We know that in ordinary quantum mechanics, multi-particle wave functions describing fermions have to change sign upon interchange of two particles, to enforce Pauli's exclusion principle. This antisymmetry suggests that we adopt the following scheme: We divide all quantities into two classes, Bose and Fermi. Bose fields are to be quantized as usual, with commutators. Fermi fields, on the other hand, are now to be quantized with *anticommutators* (20.5). We will assume that a Bose operator and a Fermi operator always commute with each other. That is, let Bose-type variables and their conjugate variables be denoted $\{q_a\}$ and $\{p_a\}$, and let Fermi-type variables and their conjugate variables be denoted $\{\theta_a\}$ and $\{\pi_a\}$. The new commutator rules are

$$[q_a, q_b] = [p_a, p_b] = 0; \qquad [q_a, p_b] = i\delta_{ab} \tag{21.16}$$

$$\{\theta_a, \theta_b\} = \{\pi_a, \pi_b\} = 0; \qquad \{\theta_a, \pi_b\} = i\delta_{ab} \tag{21.17}$$

$$[q_a, \theta_b] = [p_a, \theta_b] = [q_a, \pi_b] = [p_a, \pi_b] = 0 \tag{21.18}$$

---

[2] [Eds.] P. Jordan and E. Wigner, "Über das Paulische Äquivalenzverbot" (On the Pauli exclusion principle), *Zeits. f. Phys.* **47** (1928) 631–651; reprinted in Schwinger *QED*. The anticommutator is given in their equation (36), p. 639; see also the chart on p. 640.

Now you may be concerned that changing commutators to anticommutator is going to produce unwanted side effects somewhere. In fact, I have arranged things so that all our previous manipulations go through just as well for anticommutators as for commutators. Though we've determined $B(\mathbf{p}) = C(\mathbf{p}) = \pm 1$, we don't yet know what sign to choose. Consider

$$\theta(t) \equiv \int d^3\mathbf{x} \sum_a f_a(\mathbf{x})\psi_a(\mathbf{x},t) \tag{21.19}$$

for some test functions $f_a(\mathbf{x})$. Then for a state $|\phi\rangle$ we have

$$\langle\phi|\{\theta,\theta^\dagger\}|\phi\rangle = \langle\phi|\theta\theta^\dagger|\phi\rangle + \langle\phi|\theta^\dagger\theta|\phi\rangle = \|\,\theta^\dagger\,|\phi\rangle\|^2 + \|\,\theta\,|\phi\rangle\|^2 \geq 0 \tag{21.20}$$

but we also have (using (21.4), but with the anticommutator)

$$\{\theta,\theta^\dagger\} = \pm\int d^3\mathbf{x} \sum_a |f_a(\mathbf{x})|^2 \tag{21.21}$$

Only the + sign is consistent. So we must take $B = C = +1$. The revised equal time commutation relations for ψ and $\psi^\dagger$ are

$$\boxed{\begin{aligned}\{\psi_a(\mathbf{x},t),\psi_b(\mathbf{y},t)\} = \{\psi_a^\dagger(\mathbf{x},t),\psi_b^\dagger(\mathbf{y},t)\} &= 0 \qquad (21.22\text{a})\\ \{\psi_a(\mathbf{x},t),\psi_b^\dagger(\mathbf{y},t)\} &= \delta_{ab}\,\delta^{(3)}(\mathbf{x}-\mathbf{y}) \qquad (21.22\text{b})\end{aligned}}$$

From (21.22b) we find unambiguously

$$\boxed{\begin{aligned}\{b_{\mathbf{p}}^{(r)},b_{\mathbf{p}'}^{(s)\dagger}\} &= \delta^{rs}\delta^{(3)}(\mathbf{p}-\mathbf{p}') \qquad (21.23\text{a})\\ \{c_{\mathbf{p}}^{(r)},c_{\mathbf{p}'}^{(s)\dagger}\} &= \delta^{rs}\delta^{(3)}(\mathbf{p}-\mathbf{p}') \qquad (21.23\text{b})\end{aligned}}$$

and from (21.22a), we find that all other anticommutators vanish; in particular,

$$\boxed{\begin{aligned}\{b_{\mathbf{p}}^{(r)},b_{\mathbf{p}'}^{(s)}\} = \{b_{\mathbf{p}}^{(r)\dagger},b_{\mathbf{p}'}^{(s)\dagger}\} &= 0 \qquad (21.24\text{a})\\ \{c_{\mathbf{p}}^{(r)},c_{\mathbf{p}'}^{(s)}\} = \{c_{\mathbf{p}}^{(r)\dagger},c_{\mathbf{p}'}^{(s)\dagger}\} &= 0 \qquad (21.24\text{b})\end{aligned}}$$

Returning to the energy calculation, we now have, choosing the plus sign,

$$\begin{aligned}H &= \int d^3\mathbf{p}\,E_{\mathbf{p}}\sum_r \left[b_{\mathbf{p}}^{(r)\dagger}b_{\mathbf{p}}^{(r)} - c_{\mathbf{p}}^{(r)}c_{\mathbf{p}}^{(r)\dagger}\right]\\ &= \int d^3\mathbf{p}\,E_{\mathbf{p}}\sum_r \left[b_{\mathbf{p}}^{(r)\dagger}b_{\mathbf{p}}^{(r)} + c_{\mathbf{p}}^{(r)\dagger}c_{\mathbf{p}}^{(r)} - \delta^{(3)}(0)\right]\end{aligned} \tag{21.25}$$

As we've done before (see the discussion following (4.64)), we can redefine the zero of the energy, and discard the infinite constant. The energy of the Dirac field becomes

$$H = \int d^3\mathbf{p}\,E_{\mathbf{p}}\sum_r \left[b_{\mathbf{p}}^{(r)\dagger}b_{\mathbf{p}}^{(r)} + c_{\mathbf{p}}^{(r)\dagger}c_{\mathbf{p}}^{(r)}\right] \tag{21.26}$$

This is positive definite, and the problem of energy unbounded from below is solved.

Now, what about the states of the Dirac theory? Are they like those in the Fock space of the scalar theory (see Chapter 2), with a 0-particle subspace, a 1-particle subspace, and so on? For pedagogical simplicity, suppose that the theory only has particles, created by the Fermi operators $b^\dagger_{\mathbf{p}}$ and annihilated by $b_{\mathbf{p}}$, without antiparticles (and their associated operators, $c^\dagger_{\mathbf{p}}$ and $c_{\mathbf{p}}$). We'll also forget about spin, so our b operators will not carry the spinor indices. The only non-zero anticommutator is

$$\{b_{\mathbf{p}}, b^\dagger_{\mathbf{p}'}\} = \delta^{(3)}(\mathbf{p} - \mathbf{p}') \tag{21.27}$$

and the Hamiltonian is

$$H = \int d^3\mathbf{p}\, E_{\mathbf{p}} b^\dagger_{\mathbf{p}} b_{\mathbf{p}} \tag{21.28}$$

Using the identity

$$[AB, C] = A\{B, C\} - \{A, C\}B \tag{21.29}$$

we find $b_{\mathbf{q}}$ is an energy-lowering operator:

$$[H, b_{\mathbf{q}}] = -E_{\mathbf{q}} b_{\mathbf{q}} \tag{21.30}$$

and by the same argument $b^\dagger_{\mathbf{q}}$ is an energy-raising operator:

$$[H, b^\dagger_{\mathbf{q}}] = E_{\mathbf{q}} b^\dagger_{\mathbf{q}} \tag{21.31}$$

The equations for a single Dirac particle state are the same as for the scalar field (§2.4):

$$b_{\mathbf{q}}\,|0\rangle = 0; \qquad H\,|0\rangle = 0; \qquad b^\dagger_{\mathbf{q}}\,|0\rangle = |\mathbf{q}\rangle\,; \qquad H\,|\mathbf{q}\rangle = E_{\mathbf{q}}\,|\mathbf{q}\rangle \tag{21.32}$$

$$\langle \mathbf{q}'|\mathbf{q}\rangle = \langle 0|b_{\mathbf{q}'} b^\dagger_{\mathbf{q}}|0\rangle = \delta^{(3)}(\mathbf{q}' - \mathbf{q}) - \langle 0|b^\dagger_{\mathbf{q}} b_{\mathbf{q}'}|0\rangle = \delta^{(3)}(\mathbf{q}' - \mathbf{q}) \tag{21.33}$$

The multi-particle Dirac states look the same formally as multi-particle scalar field states:

$$|\mathbf{q}_1, \mathbf{q}_2, \ldots, \mathbf{q}_n\rangle = b^\dagger_{\mathbf{q}_1} b^\dagger_{\mathbf{q}_2} \cdots b^\dagger_{\mathbf{q}_n}\,|0\rangle \tag{21.34}$$

but they differ in an important respect. A two-particle boson state (2.59) is symmetric under interchange

$$|\mathbf{p}_1, \mathbf{p}_2\rangle = |\mathbf{p}_2, \mathbf{p}_1\rangle \tag{21.35}$$

because the boson creation operators commute. But a two-particle fermion state is antisymmetric, because the creation operators *anticommute*:

$$|\mathbf{q}_1, \mathbf{q}_2\rangle = b^\dagger_{\mathbf{q}_1} b^\dagger_{\mathbf{q}_2}\,|0\rangle = -b^\dagger_{\mathbf{q}_2} b^\dagger_{\mathbf{q}_1}\,|0\rangle = -\,|\mathbf{q}_2, \mathbf{q}_1\rangle \tag{21.36}$$

This antisymmetry enforces the exclusion principle: if $\mathbf{q}_1 = \mathbf{q}_2$, then

$$|\mathbf{q}_1, \mathbf{q}_1\rangle = b^\dagger_{\mathbf{q}_1} b^\dagger_{\mathbf{q}_1}\,|0\rangle = -b^\dagger_{\mathbf{q}_1} b^\dagger_{\mathbf{q}_1}\,|0\rangle = 0 \tag{21.37}$$

The square of any Fermi creation or annihilation operator is zero. The energies of the multi-particle states work out as we expect:

$$\begin{aligned} H\,|\mathbf{q}_1, \mathbf{q}_2\rangle &= H b^\dagger_{\mathbf{q}_1} b^\dagger_{\mathbf{q}_2}\,|0\rangle = [H, b^\dagger_{\mathbf{q}_1} b^\dagger_{\mathbf{q}_2}]\,|0\rangle \\ &= b^\dagger_{\mathbf{q}_1}[H, b^\dagger_{\mathbf{q}_2}]\,|0\rangle + [H, b^\dagger_{\mathbf{q}_1}] b^\dagger_{\mathbf{q}_2}\,|0\rangle \\ &= (E_{\mathbf{q}_1} + E_{\mathbf{q}_2})\,|\mathbf{q}_1, \mathbf{q}_2\rangle \end{aligned} \tag{21.38}$$

The properties of Fermi fields affect the extent to which they can be observed. Observables made from Bose fields commute at equal times, and so by Lorentz invariance commute for all spacelike separations. On the other hand, for Fermi fields ψ_a at equal times,

$$[\psi_a(\mathbf{x},t),\psi_b(\mathbf{y},t)] = 2\psi_a(\mathbf{x},t)\psi_b(\mathbf{y},t) \neq 0 \tag{21.39}$$

Observables that do not commute at spacelike separations are unphysical, so $\psi(x)$ is *not* an observable. Observables can only be made from products of an *even number* of Fermi fields, which *do* commute at spacelike intervals. Moreover, consider the behavior of Fermi fields under rotation. Under a rotation by 2π, a Fermi field changes sign. This is not the behavior of an observable. No meter on any experimental apparatus has ever given a different reading when the experiment was rotated by 2π. In some sense, a Fermi field ψ is the "square root" of an observable.

Finally, let's look at the classical limit of a Fermi field, as $\hbar \to 0$. There are in fact several limits. First, consider N *bosons*, all having the same energy $\hbar\omega$, in a box. The system has an energy $E = N\hbar\omega$. If we keep N, E and $|\mathbf{p}| = \hbar k$ fixed, and let $\hbar \to 0$, then $\omega \to \infty$ and $k \to \infty$. The wavelength $\lambda = 2\pi/k$ of the quanta goes to zero, and there will be no diffraction. This limit corresponds to the quanta acting like classical particles. Alternatively, we could keep all the variables (E, ω, k, and λ) *except* N fixed. The limit $\hbar \to 0$ corresponds to $N \to \infty$. In this limit, quantum granularity is lost, but the quanta still exhibit wave behavior. We could repeat the first limit with fermions, but not the second. In classical electromagnetic theory, we can have many photons in each mode. We can't do this with fermions, because of the Pauli exclusion principle. There is no classical wave behavior for them.

Formally, the limit $\hbar \to 0$ in the commutator algebra leads to a classical theory of commuting boson fields. But for fermion fields, in order to have agreement even at $\mathcal{O}(\hbar^0)$, we need *c-numbers that anticommute.* In the literature, anticommuting quantities (even if they are not c-numbers) are called *Grassmann variables*, or more formally, "elements of a Grassmann algebra".[3] Without such numbers, we can't even preserve the Heisenberg equations of motion in the limit $\hbar \to 0$.

21.2 Wick's theorem for Fermi fields

We're going to modify Model 3, our scalar meson–nucleon theory (8.6), treating the nucleons as Dirac fields. The Lagrangian takes the form

$$\mathscr{L} = \overline{\psi}(i\partial\!\!\!/ - m)\psi + \tfrac{1}{2}\partial_\mu\phi\,\partial^\mu\phi - \tfrac{1}{2}\mu^2\phi^2 - g\overline{\psi}\Gamma\psi\phi + \dots \tag{21.40}$$

where the dots indicate counterterms, which we'll neglect for now. The matrix Γ will be 1 if the field ϕ is a scalar, and $i\gamma_5$ if ϕ is a pseudoscalar. We'll use perturbation theory to study the interactions, so we'll need Dyson's formula, Wick's theorem and all that.

---

[3] [Eds.] Hermann Grassmann (1809–1877), German schoolteacher and polymath, invented exterior algebra in his *Die Lineale Ausdehnungslehre* ("The linear theory of extended magnitudes") (1844; second edition 1862). The term "Grassmann variable" for an anticommuting quantity may derive from F. A. Berezin, *The Method of Second Quantization*, Academic Press, 1966. Berezin describes (p. 5) anticommuting variables as "elements of a Grassmann algebra". For a brief description of Grassmann's work, see J. L. Coolidge, *A History of Geometrical Methods*, Dover Publications, 1963 and 2003, Ch. VI, §4, pp. 252–257, and D. Fearnley-Sander, "Hermann Grassmann and the Creation of Linear Algebra", *Amer. Math. Monthly*, **86** (1979) 809–817.

The first hurdle is time ordering. Suppose a point P at coordinates x is outside the origin's light cone, so that $x^2 < 0$. In the solid coordinates as shown in Figure 21.1, $x^0 > 0$, and for a scalar field ϕ,

$$T(\phi(0)\phi(x)) = \phi(x)\phi(0) \tag{21.41}$$

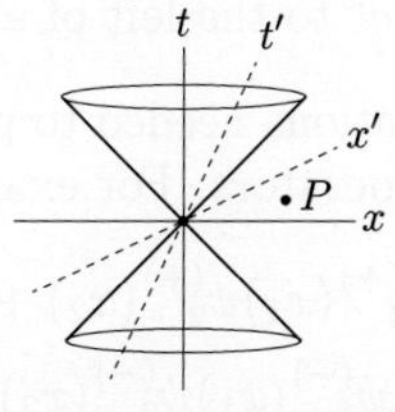

Figure 21.1: A point P in two coordinate systems

In a different frame of reference, indicated by the dashed coordinates, we have $x^0 < 0$, and so

$$T(\phi(0)\phi(x)) = \phi(0)\phi(x) \tag{21.42}$$

There is no problem when $x^2 < 0$, because for spacelike separations, $[\phi(x), \phi(0)] = 0$, and there's no problem for $x^2 > 0$, because then there's no ambiguity about which is earlier. On the other hand, suppose we were considering Fermi fields, $\psi_a(x)$. Then the time ordering is not Lorentz invariant. In the solid frame,

$$T(\psi_a(0)\psi_b(x)) = \psi_b(x)\psi_a(0) \tag{21.43}$$

but in the dashed frame,

$$T(\psi_a(0)\psi_b(x)) = \psi_a(0)\psi_b(x) \quad = -\psi_b(x)\psi_a(0), \text{ if } x^2 < 0 \tag{21.44}$$

because $\{\psi_a(0), \psi_b(x)\} = 0$ for spacelike separations.[4]

The way to patch this up is to put an extra minus sign into the definition of the time ordered product whenever the number p of permutations required to turn a product of Fermi fields into a time ordered product is odd. Define time ordering (7.35) on Fermi fields as

$$T(\psi_1(t_1)\psi_2(t_2)\cdots\psi_n(t_n)) = (-1)^p \psi_{j_1}(t_{j_1})\psi_{j_2}(t_{j_2})\cdots\psi_{j_n}(t_{j_n}) \; (t_{j_1} > t_{j_2} > \cdots > t_{j_n}) \tag{21.45}$$

For example,

$$T(\psi_a(x)\psi_b(y)) = -T(\psi_b(y)\psi_a(x)) = \begin{cases} \psi_a(x)\psi_b(y), & \text{if } x_0 > y_0 \\ -\psi_b(y)\psi_a(x), & \text{if } y_0 > x_0 \end{cases} \tag{21.46}$$

[4] [Eds.] Because the anticommutators (21.23a) and (21.23b) are the only ones involving the b's and the c's that do not vanish, the anticommutators $\{\psi_a(x), \psi_b(y)\}$ and $\{\psi_a^\dagger(x), \psi_b^\dagger(y)\}$ vanish for *all* values of x and y, not only for $(x-y)^2 < 0$. The case $\{\psi_a(x), \psi_b^\dagger(y)\}$ can be computed with the help of (21.11) and (21.12):

$$\{\psi_a(x), \psi_b^\dagger(y)\} = \gamma^0_{ab}(i\not\partial_x - m)\int \frac{d^3p}{(2\pi)^3 2E_{\mathbf{p}}}\left[e^{-ip\cdot(x-y)} - e^{ip\cdot(x-y)}\right] = \gamma^0_{ab}(i\not\partial_x - m)i\Delta(x-y)$$

But we know (3.51) that $i\Delta(x-y) = 0$ for $(x-y)^2 < 0$, so the same is true for $\{\psi_a(x), \psi_b^\dagger(y)\}$.

Dyson's formula (7.36) will not be a problem, because H_I will be quadratic in Fermi fields, and so all permutations will involve even powers. Normal ordering needs the same prescription. For products involving Fermi fields,

$$\begin{gathered} :\psi_1(x_1)\psi_2(x_2)\cdots\psi_n(t_n): = (-1)^p \psi_{j_1}(x_{j_1})\psi_{j_2}(x_{j_2})\cdots\psi_{j_n}(x_{j_n}) \\ (\text{all } b^\dagger,\, c^\dagger \text{ to the left of all } b,\, c) \end{gathered} \tag{21.47}$$

where again p is the number of permutations needed to put all the Fermi creation operators to the left of all the Fermi annihilation operators. For example, if ψ_1 and ψ_2 are Fermi fields,

$$\begin{aligned} :\psi_1(x_1)\psi_2(x_2): \;=\; & \psi_1^{(+)}(x_1)\psi_2^{(+)}(x_2) + \psi_1^{(-)}(x_1)\psi_2^{(+)}(x_2) \\ & + \psi_1^{(-)}(x_1)\psi_2^{(-)}(x_2) - \psi_2^{(-)}(x_2)\psi_1^{(+)}(x_1) \end{aligned} \tag{21.48}$$

where $\psi^{(+)}$ and $\psi^{(-)}$ are defined analogously to $\phi^{(+)}$ and $\phi^{(-)}$ in (3.33):

$$\begin{aligned} \psi^{(+)} &= \int \frac{d^3\mathbf{p}}{(2\pi)^{3/2}\sqrt{2E_\mathbf{p}}} \sum_{r=1}^{2} b_\mathbf{p}^{(r)} u_\mathbf{p}^{(r)} e^{-ip\cdot x} \\ \psi^{(-)} &= \int \frac{d^3\mathbf{p}}{(2\pi)^{3/2}\sqrt{2E_\mathbf{p}}} \sum_{r=1}^{2} c_\mathbf{p}^{(r)\dagger} v_\mathbf{p}^{(r)} e^{ip\cdot x} \end{aligned} \tag{21.49}$$

As with time-ordered products, Fermi fields anticommute within normal-ordered products:

$$:\psi_1(x_1)\psi_2(x_2): = -\,:\psi_2(x_2)\psi_1(x_1): \tag{21.50}$$

With the generalized time-ordered and normal-ordered products, Wick's theorem (8.28) can be proved for Fermi fields in exactly the same way as it was for Bose fields. We won't do that here, but we will show that the theorem holds for two Fermi fields ψ and χ. If, following (8.20), we define

$$\overbrace{\psi_a(x)\overline{\chi}_b}(y) = \theta(x_0 - y_0)\{\psi_a^{(+)}, \overline{\chi}_b^{(-)}\} - \theta(y_0 - x_0)\{\overline{\chi}_b^{(+)}, \psi_a^{(-)}\} \tag{21.51}$$

making due allowance for anticommutation, then Wick's theorem says

$$\overbrace{\psi_a(x)\overline{\chi}_b}(y) = T(\psi_a(x)\overline{\chi}_b(y)) - :\psi_a(x)\overline{\chi}_b(y): \tag{21.52}$$

We'll postpone the calculation of the contraction itself. We'll prove the theorem by cases. Suppose $x_0 > y_0$. Then, from (21.51),

$$\overbrace{\psi_a(x)\overline{\chi}_b}(y) = \{\psi_a^{(+)}(x), \overline{\chi}_b^{(-)}(y)\} = \psi_a^{(+)}(x)\overline{\chi}_b^{(-)}(y) + \overline{\chi}_b^{(-)}(y)\psi_a^{(+)}(x) \tag{21.53}$$

For $x_0 > y_0$, $T(\psi_a(x)\overline{\chi}_b(y)) = \psi_a(x)\overline{\chi}_b(y)$, so

$$\begin{aligned} T(\psi_a(x)\overline{\chi}_b(y)) = \; & \psi_a^{(+)}(x)\overline{\chi}_b^{(+)}(y) + \psi_a^{(+)}(x)\overline{\chi}_b^{(-)}(y) \\ & + \psi_a^{(-)}(x)\overline{\chi}_b^{(+)}(y) + \psi_a^{(-)}(x)\overline{\chi}_b^{(-)}(y) \end{aligned} \tag{21.54}$$

and, from (21.48),

$$\begin{aligned} :\psi_a(x)\overline{\chi}_b(y): \;=\; & \psi_a^{(+)}(x)\overline{\chi}_b^{(+)}(y) + \psi_a^{(-)}(x)\overline{\chi}_b^{(+)}(y) \\ & + \psi_a^{(-)}(x)\overline{\chi}_b^{(-)}(y) - \overline{\chi}_b^{(-)}(y)\psi_a^{(+)}(x) \end{aligned} \tag{21.55}$$

Subtracting $:\psi_a(x)\overline{\chi}_b(y):$ from $T(\psi_a(x)\overline{\chi}_b(y))$ gives

$$T(\psi_a(x)\overline{\chi}_b(y)) - :\psi_a(x)\overline{\chi}_b(y): = \psi_a^{(+)}(x)\overline{\chi}_b^{(-)}(y) + \overline{\chi}_b^{(-)}(y)\psi_a^{(+)}(x) = \overbrace{\psi_a(x)\overline{\chi}_b}(y) \tag{21.56}$$

So (21.52) is true if $x_0 > y_0$. If $y_0 > x_0$, all three expressions pick up an overall minus sign. From (21.51),

$$\overbrace{\psi_a(x)\overline{\chi}_b}(y) = -\{\overline{\chi}_b^{(+)}, \psi_a^{(-)}\} \tag{21.57}$$

but from (21.46) and (21.50), respectively,

$$T(\psi_a(x)\overline{\chi}_b(y)) = -\,\overline{\chi}_b(y)\psi_a(x) \tag{21.58}$$

$$:\psi_a(x)\overline{\chi}_b(y): = -\,:\overline{\chi}_b(y)\psi_a(x): \tag{21.59}$$

and the demonstration goes through exactly as before. That establishes the Fermi field version of Wick's theorem, at least for two fields.

21.3 Calculating the Dirac propagator

In the sort of theories we're looking at, such as (21.40), the free meson propagator is unaltered by the presence of Fermi fields. It's the contraction of two Fermi fields we need to compute:

$$\overbrace{\psi(x)\overline{\psi}}(y) = T(\psi(x)\overline{\psi}(y)) - :\psi(x)\overline{\psi}(y): \tag{21.60}$$

Three remarks need to be made. First, by reasoning parallel to that in the scalar case (pp. 156–157), the contraction of two Fermi fields is a c-number. The difference between the two orderings is just an anticommutator, which is always a c-number, because of the minus signs we've inserted into our definition of time ordering and normal ordering. Second, in order to keep from cluttering equations with indices, I will adopt this convention: Whatever the order of the operators, the order of the indices in this expression will be such that the $\overline{\psi}$ index is always *on the right.* That way we won't get a silly expression that is a 4×4 matrix for one ordering and a 1×1 matrix for another ordering. Third, because we've defined both the time-ordered product and the normal-ordered product to be antisymmetric in interchange of the two operators,

$$\overbrace{\psi(x)\overline{\psi}}(y) = -\overbrace{\overline{\psi}(y)\psi}(x) \tag{21.61}$$

exchanging the two operators in the contraction gives us a minus sign.

We have to write down once again the expression (21.6) for the free field, $\psi(x)$:

$$\psi(x) = \int \frac{d^3\mathbf{p}}{(2\pi)^{3/2}\sqrt{2E_\mathbf{p}}} \sum_{r=1}^{2} \left[b_\mathbf{p}^{(r)} u_\mathbf{p}^{(r)} e^{-ip\cdot x} + c_\mathbf{p}^{(r)\dagger} v_\mathbf{p}^{(r)} e^{ip\cdot x} \right] \tag{21.62}$$

And of course $\overline{\psi}(y)$ is almost the same thing:

$$\overline{\psi}(y) = \int \frac{d^3\mathbf{p}'}{(2\pi)^{3/2}\sqrt{2E_{\mathbf{p}'}}} \sum_{r'=1}^{2} \left[b_{\mathbf{p}'}^{(r')\dagger} \overline{u}_{\mathbf{p}'}^{(r')} e^{ip'\cdot y} + c_{\mathbf{p}'}^{(r')} \overline{v}_{\mathbf{p}'}^{(r')} e^{-ip'\cdot y} \right] \tag{21.63}$$

In such expressions, p^0 is always equal to $E_\mathbf{p}$.

Taking the vacuum expectation value of (21.60), the contraction is equal to the vacuum expectation value of the time ordered product alone, because the vacuum expectation value of the normal-ordered product vanishes:

$$\overbracket{\psi(x)\overline{\psi}}(y) = \langle 0|\overbracket{\psi(x)\overline{\psi}}(y)|0\rangle = \langle 0|T(\psi(x)\overline{\psi}(y))|0\rangle \tag{21.64}$$

We will compute this, for $x_0 > y_0$ and for $x_0 < y_0$, and then join the two results together. For $x_0 > y_0$ the $\overline{\psi}$ is on the right, where only the creation part, proportional to $b_{\mathbf{p}'}^{(r')\dagger}$, is relevant; the ψ is on the left where only the annihilation part, proportional to $b_{\mathbf{p}}^{(r)}$, contributes. So I obtain

$$\langle 0|T(\psi(x)\overline{\psi}(y))|0\rangle = \int \frac{d^3\mathbf{p}\, d^3\mathbf{p}'}{(2\pi)^3\sqrt{2E_{\mathbf{p}}}\sqrt{2E_{\mathbf{p}'}}} e^{-ip\cdot x + ip'\cdot y} \sum_{r,r'} \langle 0|b_{\mathbf{p}}^{(r)} b_{\mathbf{p}'}^{(r')\dagger}|0\rangle\, u_{\mathbf{p}}^{(r)}\overline{u}_{\mathbf{p}'}^{(r')} \tag{21.65}$$

But from (21.23a),

$$\langle 0|b_{\mathbf{p}}^{(r)} b_{\mathbf{p}'}^{(r')\dagger}|0\rangle = \langle 0|\{b_{\mathbf{p}}^{(r)}, b_{\mathbf{p}'}^{(r')\dagger}\}|0\rangle = \delta^{rr'}\delta^{(3)}(\mathbf{p}-\mathbf{p}') \tag{21.66}$$

so

$$\langle 0|T(\psi(x)\overline{\psi}(y))|0\rangle = \int \frac{d^3\mathbf{p}}{(2\pi)^3 2E_{\mathbf{p}}} e^{-ip\cdot(x-y)} \sum_r u_{\mathbf{p}}^{(r)}\overline{u}_{\mathbf{p}}^{(r)} \tag{21.67}$$

We have a wonderful identity, (20.123a), for the spinor sum: it is $\not{p} + m$. And therefore we can write this whole expression (for $x_0 > y_0$) as

$$\begin{aligned}\langle 0|T(\psi(x)\overline{\psi}(y))|0\rangle &= \int \frac{d^3\mathbf{p}}{(2\pi)^3 2E_{\mathbf{p}}} e^{-ip\cdot(x-y)}(\not{p} + m) \\ &= (i\not{\partial}_x + m)\int \frac{d^3\mathbf{p}}{(2\pi)^3 2E_{\mathbf{p}}} e^{-ip\cdot(x-y)}\end{aligned} \tag{21.68}$$

(I put an x on the derivative to show differentiation with respect to x and not y.) This integral is the same one we encountered while evaluating the scalar propagator (S1.13). If we imagine a hypothetical scalar field φ of mass m (which is *not* the scalar field coupled with the Fermi field in this theory), we can write, if $x_0 > y_0$,

$$\langle 0|T(\psi(x)\overline{\psi}(y))|0\rangle = (i\not{\partial}_x + m)\,\langle 0|T(\varphi(x)\varphi(y))|0\rangle \tag{21.69}$$

because the second integral in (21.68) is just equal to the expectation value of the time-ordered product of some scalar field, $\varphi(x)$. Equivalently

$$\overbracket{\psi(x)\overline{\psi}}(y) = (i\not{\partial}_x + m)\overbracket{\varphi(x)\varphi}(y) \tag{21.70}$$

I now turn to the case $y_0 > x_0$. The order of the operators is reversed:

$$\overbracket{\psi(x)\overline{\psi}}(y) = \langle 0|T(\psi(x)\overline{\psi}(y))|0\rangle = -\,\langle 0|\overline{\psi}(y)\psi(x)|0\rangle \tag{21.71}$$

Of course the order of the matrix indices is not reversed; otherwise the integral is exactly the same, though the integrand looks different:

$$\begin{aligned}\langle 0|T(\psi(x)\overline{\psi}(y))|0\rangle &= -\int \frac{d^3\mathbf{p}\,d^3\mathbf{p}'}{(2\pi)^3\sqrt{2E_{\mathbf{p}}}\sqrt{2E_{\mathbf{p}'}}}e^{-ip\cdot y+ip'\cdot x}\sum_{r,r'}\langle 0|c_{\mathbf{p}'}^{(r')}c_{\mathbf{p}}^{(r)\dagger}|0\rangle\, v_{\mathbf{p}}^{(r)}\overline{v}_{\mathbf{p}'}^{(r')}\\ &= -\int \frac{d^3\mathbf{p}\,d^3\mathbf{p}'}{(2\pi)^3\sqrt{2E_{\mathbf{p}}}\sqrt{2E_{\mathbf{p}'}}}e^{-ip\cdot y+ip'\cdot x}\sum_{r,r'}\delta^{rr'}\delta^{(3)}(\mathbf{p}-\mathbf{p}')v_{\mathbf{p}}^{(r)}\overline{v}_{\mathbf{p}'}^{(r')}\\ &= -\int \frac{d^3\mathbf{p}}{(2\pi)^3 2E_{\mathbf{p}}}e^{ip\cdot(x-y)}\sum_{r}v_{\mathbf{p}}^{(r)}\overline{v}_{\mathbf{p}}^{(r)} = -\int \frac{d^3\mathbf{p}}{(2\pi)^3 2E_{\mathbf{p}}}e^{ip\cdot(x-y)}(\not{p}-m)\end{aligned} \tag{21.72}$$

Swapping x and y makes the creation and annihilation parts change places, so I get the $c_{\mathbf{p}}^{(r)}$'s exchanged for the $b_{\mathbf{p}}^{(r)}$'s, and I get the sum on our $v_{\mathbf{p}}^{(r)}$'s instead of the $u_{\mathbf{p}}^{(r)}$'s, and the sign of the exponent changes. The spinor sum is, by (20.123b), $(\not{p}-m)$, and thus

$$\begin{aligned}\langle 0|T(\psi(x)\overline{\psi}(y))|0\rangle &= (i\not{\partial}_x+m)\int \frac{d^3\mathbf{p}}{(2\pi)^3 2E_{\mathbf{p}}}e^{ip\cdot(x-y)}\\ &= (i\not{\partial}_x+m)\int \frac{d^3\mathbf{p}}{(2\pi)^3 2E_{\mathbf{p}}}e^{-ip\cdot(x-y)}\end{aligned} \tag{21.73}$$

because the integral is unchanged under $p\to -p$. That is, the expression $\langle 0|T(\psi(x)\overline{\psi}(y))|0\rangle$ is exactly the same for $y_0 > x_0$ (21.73) as it is for $x_0 > y_0$ (21.68); for *all* times,

$$\overbracket{\psi(x)\overline{\psi}(y)} = (i\not{\partial}_x+m)\overbracket{\varphi(x)\varphi(y)} \tag{21.74}$$

From (8.23), we can write down immediately the Fourier transform of the scalar field contraction:

$$\begin{aligned}\overbracket{\psi(x)\overline{\psi}(y)} &= (i\not{\partial}_x+m)\int \frac{d^4p}{(2\pi)^4}e^{-ip\cdot(x-y)}\frac{i}{p^2-m^2+i\epsilon}\\ &= \int \frac{d^4p}{(2\pi)^4}e^{-ip\cdot(x-y)}\frac{i(\not{p}+m)}{p^2-m^2+i\epsilon}\end{aligned} \tag{21.75}$$

The effect of the $(i\not{\partial}_x+m)$ is to put $(\not{p}+m)$ in the numerator.

Thus the analog of the scalar field propagator $i/(p^2-m^2+i\epsilon)$ is, for a fermion field,

$$\boxed{\widetilde{S}_F = \frac{i(\not{p}+m)}{p^2-m^2+i\epsilon}} \tag{21.76}$$

Though we're dealing with a four-component field, it has only twice as many physical degrees of freedom as a charged boson field. The field has four components, but there are only two spin states for the particle, and two for the antiparticle. There are actually only two kinds of particles we can exchange, so we should have some kind of projection operator for the exchange of those particles, at least as p^2 approaches m^2, and we pick up the one-particle states. And as can be seen from (20.124a) and (20.123a), we've got the projection operator on the positive frequency states in the numerator.

Another way of understanding the propagator is to write it in an alternative form which you will frequently find in the literature. Since the only function of the $i\epsilon$ is to tell us how to control the pole, we can put a minus $i\epsilon$ in the numerator with no loss of generality:

$$\frac{i(\not{p}+m)}{p^2-m^2+i\epsilon}=\frac{i(\not{p}+m-i\epsilon)}{p^2-m^2+i\epsilon} \tag{21.77}$$

Because m is a positive number, $(m-i\epsilon)^2$ puts the pole in the same place $m^2-i\epsilon$ does. We can thus rewrite the denominator as

$$p^2-(m-i\epsilon)^2=(\not{p}-(m-i\epsilon))(\not{p}+(m-i\epsilon)) \tag{21.78}$$

Then[5]

$$\frac{i(\not{p}+m)}{p^2-m^2+i\epsilon}=\frac{i(\not{p}+m-i\epsilon)}{(\not{p}-(m-i\epsilon))(\not{p}+(m-i\epsilon))}=\frac{i}{\not{p}-m+i\epsilon} \tag{21.79}$$

(We can be a little cavalier here about matrix manipulations, because $\not{p}-m+i\epsilon$ commutes with $\not{p}+m-i\epsilon$.) In this form, the Feynman propagator for the Dirac theory very closely parallels the Feynman propagator for a scalar theory. In a scalar theory, the free Klein–Gordon equation in momentum space involves the operator p^2-m^2. The scalar Feynman propagator is i over this operator with the pole difficulty resolved by giving the mass a small (negative) imaginary part. The Dirac equation in momentum space involves the operator $(\not{p}-m)$. The fermion Feynman propagator is i over this operator with the pole ambiguity resolved by giving the mass a small (negative) imaginary part. In short,

$$\begin{aligned}\text{Klein–Gordon}: \Box^2+m^2 &\longrightarrow \text{propagator } \widetilde{\Delta}_F(p^2): \frac{i}{p^2-m^2+i\epsilon}\\ \text{Dirac}: i\not{\partial}-m &\longrightarrow \text{propagator } \widetilde{S}_F(\not{p}): \frac{i}{\not{p}-m+i\epsilon}\end{aligned} \tag{21.80}$$

We shall see later on, when I talk about quantization through functional integration, that the propagator is always, in a sense, the inverse of the operator $\mathscr{D}$ that appears in the free Lagrangian $\phi\mathscr{D}\phi$ (with $i\partial\to p$).

21.4 An example: Nucleon–meson scattering

Before writing down the Feynman rules in such a theory it's probably best to see how things work out by evaluating a particular diagram and watching how all the various factors fit together. We'll consider the Lagrangian (21.40), which describes a free Fermi field, a free meson field and an interaction between them:

$$\mathscr{L}_I=-g\bar{\psi}\Gamma\psi\phi \tag{21.81}$$

where Γ is either 1 or $i\gamma_5$. In the former case the theory is parity invariant if $\phi(x)$ is a scalar; in the latter case it is parity invariant if $\phi(x)$ is a pseudoscalar. For what I am going to do now, the choice of Γ is irrelevant. Let's consider a typical scattering process, for example nucleon plus meson goes into nucleon plus meson:

$$N+\phi\to N+\phi \tag{21.82}$$

A Feynman diagram (drawn in two ways) that contributes to this process to lowest order is shown in Figure 21.2. We adopt the same diagrammatic conventions as in the scalar

[5] [Eds.] In the literature, the Feynman propagator $i/(\not{p}-m+i\epsilon)$ for the Dirac field is often written as $\widetilde{S}_F(\not{p})$ or $S_F(\not{p})$. Bjorken and Drell define this propagator as $iS_F(\not{p})$; see Bjorken & Drell *RQM*, p. 93, equation (6.42).

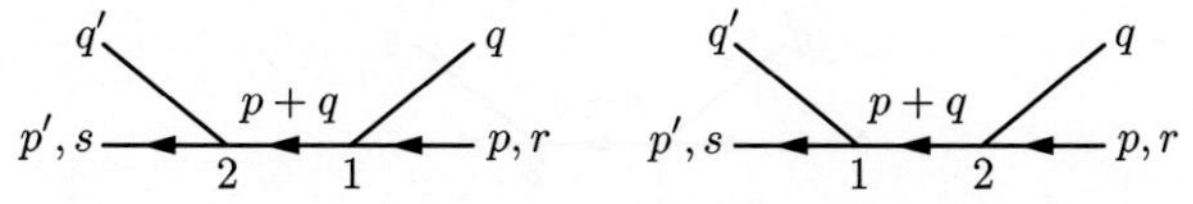

Figure 21.2: A diagram for lowest order nucleon–meson scattering

model, with the spinor charged nucleon field replacing the scalar charged nucleon field we had before (§8.3). The incoming nucleon and meson are characterized by the momenta p and q, respectively; the momenta p' and q' denote the respective outgoing momenta. And of course the nucleon is in some spin state. We're constructing S-matrix elements between states of definite spin, so I give an index r for p, and an index s for p' where $\{r, s\}$ equals 1 or 2, telling you whether the nucleon is spin up or spin down. Let's look at the order g^2 term in Dyson's formula, $S = Te^{-i\int d^4x\, \mathscr{H}_I}$:

$$\tfrac{1}{2!}(-ig)^2 \int d^4x_1\, d^4x_2\, T(\overline{\psi}_2\Gamma\psi_2\phi_2\overline{\psi}_1\Gamma\psi_1\phi_1) \tag{21.83}$$

The relevant terms in the Wick expansion corresponding to this diagram are

$$\tfrac{1}{2!}(-ig)^2 \int d^4x_1\, d^4x_2\, [:\overline{\psi}_2\Gamma\psi_2\phi_2\overline{\psi}_1\Gamma\psi_1\phi_1: + :\overline{\psi}_2\Gamma\psi_2\phi_2\overline{\psi}_1\Gamma\psi_1\phi_1:] \tag{21.84}$$

where the subscripts 1 and 2 indicate that the functions depend on x_1 and x_2, respectively. The picture for the second term looks identical to the first picture, except for an interchange $1 \leftrightarrow 2$ of the dummy variables. The second operator is the same as the first, and serves only to cancel the 2!; the two pictures are the same diagram written twice.

Let's write down the S-matrix element between the final state and the initial state coming from this term in the Wick expansion:

$$\langle f|\mathrm{S}-1|i\rangle = \langle p', s; q'|(-ig)^2 \int d^4x_1\, d^4x_2\; :\overline{\psi}_2\Gamma\psi_2\phi_2\overline{\psi}_1\Gamma\psi_1\phi_1: |p, r; q\rangle \tag{21.85}$$

Just as in the scalar case, we use relativistically normalized states (1.57), so we don't have to keep track of the factors of $(2\pi)^{3/2}$ and $1/\sqrt{2E_{\mathbf{p}}}$. Those are automatically taken care of as part of the density of states (11.58) in our rules for turning S-matrix elements into cross-sections. We then get the integral over $d^4x_1\, d^4x_2$ and a bunch of exponential factors. First, ϕ_1 is annihilating a meson in the initial state, so I obtain $e^{-iq\cdot x_1}$, and ψ_1 is annihilating the initial nucleon, so I have $e^{-ip\cdot x_1}$. Likewise everything is being created at x_2, so I have positive exponential factors for x_2: $e^{iq'\cdot x_2}e^{ip'\cdot x_2}$. We can drag ϕ_2 as we please inside the normal-ordered product, since it commutes with the ψ's. Then I will have an integral over some momentum k that occurs in the Fourier expansion of the propagator, the contraction of $\overline{\psi}_2$ and ψ_1.

Now we have to deal with the matrices and the spinors. Let's go in the order in which the integral is set up, from right to left. The annihilation of a meson carries nothing besides the exponential. Annihilation of the nucleon, however, carries the factor here of $u_{\mathbf{p}}^{(r)}$. That takes care of the first factors and the initial state. Then as we go along, there's a Γ. Then we've got the contraction, $i/(\not{k} - m + i\epsilon)$, followed by another Γ, and $\overline{u}_{\mathbf{p}'}^{(s)}$ from the final state:

$$\begin{aligned}\langle f|\mathrm{S}-1|i\rangle = (-ig)^2 \int \frac{d^4k}{(2\pi)^4}\, d^4x_1\, d^4x_2\, e^{i(q'+p')\cdot x_2} \Big[& e^{-i(q+p)\cdot x_1} e^{-ik\cdot(x_2-x_1)} \times \\ & \times\; \overline{u}_{\mathbf{p}'}^{(s)}\Gamma \frac{i}{\not{k} - m + i\epsilon}\Gamma u_{\mathbf{p}}^{(r)} \Big]\end{aligned} \tag{21.86}$$

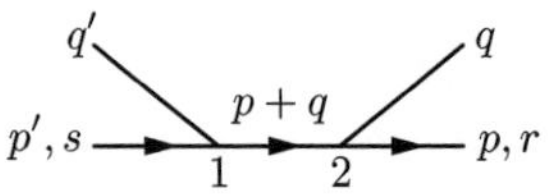

Figure 21.3: Antinucleon–meson scattering to lowest order

The x integrals are trivial, and as usual give us a $(2\pi)^4$ times an overall energy momentum conserving delta function, just as in the scalar case. The k integral is also trivial here, because of the delta functions:

$$\langle f|\mathrm{S}-1|i\rangle = (-ig)^2(2\pi)^4\delta^{(4)}(p+q-p'-q')\,\overline{u}^{(s)}_{\mathbf{p}'}\Gamma\frac{i}{\not p+\not q-m+i\epsilon}\Gamma u^{(r)}_{\mathbf{p}} \tag{21.87}$$

or, using the notation of (10.27),

$$i\mathcal{A}_{fi} = (-ig)^2\,\overline{u}^{(s)}_{\mathbf{p}'}\Gamma\frac{i}{\not p+\not q-m+i\epsilon}\Gamma u^{(r)}_{\mathbf{p}} \tag{21.88}$$

We have to be careful about the sign of the intermediate momentum, k. Should $k = p+q$ or $k = -(p+q)$? The fermion propagator is not invariant under change of sign of p, as the scalar propagator is. But it is clear from (21.86) that the plus sign is correct here.

This amplitude (21.88) is the sort of generalization you would expect even if you hadn't gone through the derivation. When you have a set of four fields, you also have a bunch of 4×4 matrices as you pass through a vertex and propagate something. Of course a matrix element is not a matrix, it is a number. So to make it a number, you need a column vector like $u^{(r)}_{\mathbf{p}}$ for the initial nucleon, and a row vector like $\overline{u}^{(s)}_{\mathbf{p}'}$ for the final nucleon.

Before I write down what happens in general, let me consider a second process nearly the same as nucleon–meson scattering, antinucleon–meson scattering:

$$\overline{N}+\phi\to\overline{N}+\phi$$

The diagram is shown in Figure 21.3. Many things are the same. The principal change is this. In the left diagram in Figure 21.2, I think of the vertex on the right, where the nucleon and the meson are annihilated, as point 1, and the vertex on the left as point 2. The diagram in Figure 21.3 comes from exactly the same term (21.84) in the Wick expansion. Now however the operator needed to annihilate the initial antinucleon is found in $\overline{\psi}_2$, and the field ψ_1 creates the final antinucleon. So I think of these vertices, read right to left, as 2, 1. Matrix multiplication still goes from 1 to 2 along the line. We obtain

$$\langle f|S-1|i\rangle = (-1)\cdot(-ig)^2(2\pi)^4\delta^{(4)}(p+q-p'-q')\,\overline{v}^{(r)}_{\mathbf{p}}\Gamma\frac{i}{-\not p-\not q-m+i\epsilon}\Gamma v^{(s)}_{\mathbf{p}'} \tag{21.89}$$

At the left-hand side of the matrix we have $\overline{v}^{r}_{\mathbf{p}}$, the factor associated with the annihilation of the initial antinucleon, and then we have a Γ. The propagator's denominator is $-(\not p+\not q)-m+i\epsilon$, because we switched x_1 and x_2, and so we've changed k into $-k$ in the energy–momentum conserving delta function. There's another Γ, and then there is the final antinucleon being created, which gives us $v^{(s)}_{\mathbf{p}'}$. But there is a new feature, a factor of -1 coming from Fermi statistics. In the Wick term (21.84), the field that annihilates the initial antinucleon, $\overline{\psi}_2$, is over on the left, where it shouldn't be. To put things in the right order, I have to switch the ψ_1 and $\overline{\psi}_2$. Within a normal-ordered product, the switching throws in a minus sign.

These two examples contain practically all the novel features we encounter in the Fermi theory. So I can now write down the Feynman rules for theories involving fermions.

21.5 The Feynman rules for theories involving fermions

I'll list the rules in three sections. First, I'll tell you what the factors are. Then I'll give the rules for handling the matrices. Finally I'll tell you what to do about the terrible Fermi minus signs.

Feynman rules for theories with fermions I. Factors

1. For every ... Write ...

(a) internal meson line 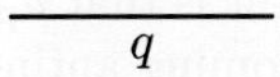

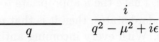

$\dfrac{i}{q^2 - \mu^2 + i\epsilon}$

(b) internal nucleon line 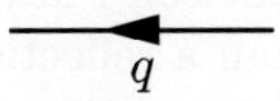

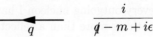

$\dfrac{i}{\not{q} - m + i\epsilon}$

(c) vertex 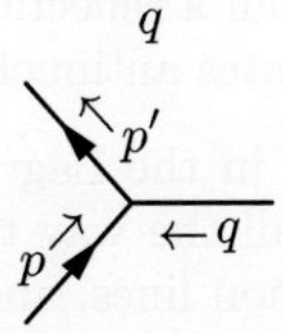

$-ig\Gamma$

2. Ensure momentum conservation at each vertex: $(2\pi)^4\,\delta^{(4)}(\sum p_{\text{out}} - \sum p_{\text{in}})$

3. Multiply by $\displaystyle\int \frac{d^4q}{(2\pi)^4}$ and integrate over each internal momentum.

4. Spinor factors:

- For every incoming nucleon, write a u.
- For every outgoing nucleon, write a $\overline{u}$.
- For every incoming antinucleon, write a $\overline{v}$.
- For every outgoing antinucleon, write a v.

We still have to take care of all the matrix and spinor factors. First I'll state the rules, and then I'll explain them with examples.

Feynman rules for theories with fermions II. Assembling the pieces

1. Along a fermion line:

   - Starting with the arrowhead, follow each fermion line backwards through the diagram, assembling factors as you go.

2. For a closed fermion loop:

   - Include a factor of (-1).
   - Take the trace of the product of Dirac factors.

Leaving aside the counterterms for the moment, the factors are pretty much the same as in Model 3 (box, p. 216). We assume that the initial state is on the right. The momentum orientation does not affect the meson propagator, because q^2 is $(-q)^2$, but it does affect the fermion propagator. We'll orient the momentum in the same direction as the arrow on the line for nucleons, and in the opposite direction for antinucleons. If you happen to find it convenient in a particular graph to orient a nucleon's momentum q the other way, that's fine by me, but then you must write the propagator as i over $(-\not{q} - m + i\epsilon)$. Every vertex, for example with q and p coming in, p' going out, gives us a factor $-ig\Gamma(2\pi)^4\delta^{(4)}(p+q-p')$, exactly the same as in the scalar theory, except that it's now a matrix because of the presence of Γ.

Some row vectors and column vectors are associated with initial and final fermions. For every incoming nucleon, I have the u appropriate to the nucleon's state. If the nucleon happens to be in one of our standard states, then it is that $u_{\mathbf{p}}^{(r)}$. If it's not, the state will be some linear combination of the u's. With every incoming antinucleon I have associated a $\overline{v}$, as shown in the last example. With every outgoing nucleon I have a $\overline{u}$ and with every outgoing antinucleon I associate a v. This is nothing more than a reflection of the fact that ψ annihilates nucleons and creates antinucleons, but $\overline{\psi}$ annihilates antinucleons and creates nucleons.

Because fermions appear bilinearly in the Lagrangian (we'll soon see that quartics are ruled out), a fermion line either goes all the way through a graph, or it appears in a loop. We have matrices associated with fermion lines, and row or column vectors associated with incoming or outgoing fermions. It doesn't matter which way the fermion line is going through the diagram. As we habitually write from left to write, we'll start at the head of the arrow and work against the arrows. At an incoming antinucleon, we write down a $\overline{v}$; at an outgoing nucleon, we write a $\overline{u}$. The next thing you encounter is a vertex. Write down the matrix Γ for the vertex. Then you get a propagator associated with the internal line, followed by another vertex with another Γ. This may repeat a number of times. When you get to the tail of the line, you arrive at either an incoming nucleon, and write a u, or an outgoing antinucleon, and a v. That's the order in which things come out in Wick's theorem.

Now, to assemble the pieces.

EXAMPLE. *A linear fermion graph*

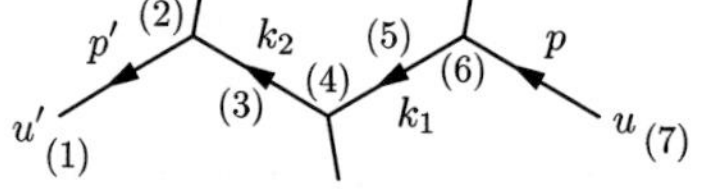

Figure 21.4: A linear fermion graph

Consider the graph shown in Figure 21.4. Numbering the various factors, the amplitude for this graph is

$$i\mathcal{A}_{fi} = \underset{(1)}{\overline{u}_{\mathbf{p}'}} \underset{(2)}{(-ig\Gamma)} \underset{(3)}{\frac{i}{\not{k}_2 - m + i\epsilon}} \underset{(4)}{(-ig\Gamma)} \underset{(5)}{\frac{i}{\not{k}_1 - m + i\epsilon}} \underset{(6)}{(-ig\Gamma)} \underset{(7)}{u_{\mathbf{p}}} \tag{21.90}$$

EXAMPLE. *A closed fermion loop*

Consider a completely closed fermion loop, as in Figure 21.5. The term in Wick's theorem that gives us this closed loop is, ignoring the factors of $(-ig)$,

$$:\overline{\psi}_1\Gamma\psi_1\overline{\psi}_2\Gamma\psi_2\overline{\psi}_3\Gamma\psi_3\overline{\psi}_4\Gamma\psi_4: \tag{21.91}$$

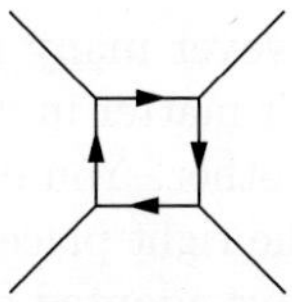

Figure 21.5: A closed fermion loop

This is an $\mathcal{O}(g^4)$ diagram describing two mesons → two mesons scattering. The factor of interest in this contribution to the process is

$$:\overline{\psi}_1\Gamma\psi_1\overline{\psi}_2\Gamma\psi_2\overline{\psi}_3\Gamma\psi_3\overline{\psi}_4\Gamma\psi_4: \,=\, -\,:\psi_4\overline{\psi}_1\Gamma\psi_1\overline{\psi}_2\Gamma\psi_2\overline{\psi}_3\Gamma\psi_3\overline{\psi}_4\Gamma: \tag{21.92}$$

To put the contraction between ψ_4 and $\overline{\psi}_1$ in Fermi fields (21.75), I have to move ψ_4 past seven other Fermi fields, so there's an extra minus sign. But this isn't all. Before taking the contractions, look at the Dirac indices (summation over repeated indices implied):

$$:\psi_{4h}\overline{\psi}_{1a}\Gamma_{ab}\psi_{1b}\overline{\psi}_{2c}\Gamma_{cd}\psi_{2d}\overline{\psi}_{3e}\Gamma_{ef}\psi_{3f}\overline{\psi}_{4g}\Gamma_{gh}: \tag{21.93}$$

The term between the colons is $M_{hh} = \text{Tr}(M)$ for the matrix $M = (\psi_4\overline{\psi}_1\Gamma\psi_1\overline{\psi}_2\Gamma\psi_2\overline{\psi}_3\Gamma\psi_3\overline{\psi}_4\Gamma)$. Thus the contribution is

$$-\text{Tr}\,(:\psi_4\overline{\psi}_1\Gamma\psi_1\overline{\psi}_2\Gamma\psi_2\overline{\psi}_3\Gamma\psi_3\overline{\psi}_4\Gamma:) \tag{21.94}$$

With a closed loop it doesn't matter where you start multiplying the matrices: The trace is invariant under cyclic permutations of the matrix factors. Start anywhere, and working against the arrows, write down the vertex and propagator matrices until you get back to where you started. In the product, you will have the makings of a contraction, but in the wrong order: the $\overline{\psi}$ on the left and the ψ on the right. And as I remarked (21.61), that is *minus* the contraction in the standard order. So our first minus sign rule is: For every closed fermion loop, include a factor of (-1). We will check this rule for consistency when we compute the meson self energy in this theory. There the factor of (-1) from Fermi statistics will be very important for the closed nucleon loop that occurs. As you will see, this factor is needed to guarantee that the imaginary part of the self energy has the proper sign, consistent with the spectral representation. If it weren't there, we would obtain an insane answer for the meson self energy.

In general, though, it can be tricky to get the signs right. We found an extra minus sign in antinucleon–meson scattering as compared with nucleon–meson scattering, because we had to switch around the operators. It is possible to give a sequence of rules for the result of switching around operators in the general case, but it's awkward. Say we have an initial state of 32 nucleons and 47 antinucleons, and a final state of six nucleons and seven antinucleons. You've got to establish rules about what you mean by a six-nucleon, seven-antinucleon state; you have to specify *in what order* they are created (you'll see why this is relevant when we work through an example). I will just make the simple statement that, as we see already from the string (21.92), *the normal-ordered operators* (for each particle) *always come in the order* $:\overline{\psi}\psi:$, where the ψ is associated with the tail of a line and the $\overline{\psi}$ with the head of a line. That's the only fact you have to remember. Whether you're going to use it to create or annihilate, the operator ψ is always associated with the beginning of the line; the operator $\overline{\psi}$ following is associated with the end of the line. However many lines traverse the diagrams in different directions, making hairpin turns, it doesn't matter in what order I put strings of $\overline{\psi}\psi$'s, because pairs of $\overline{\psi}\psi$'s always commute with each other. You do have to look to see if the annihilation operators and creation operators are in the right places, or if you have to switch them around, depending upon whatever ordering you have adopted for the creation of the initial state. Once you get the knack of it, these rules are not difficult to work with. Instead of saying r and

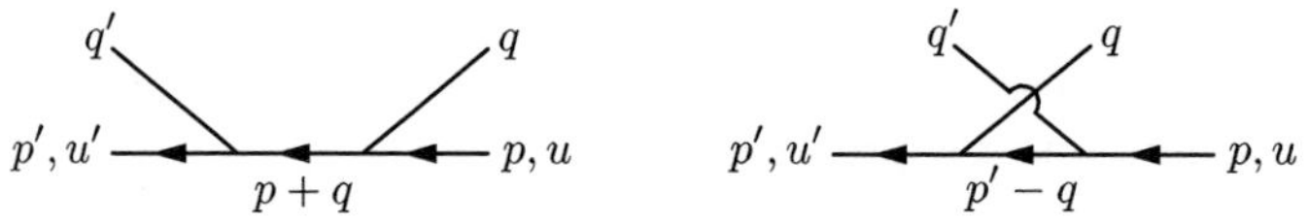

Figure 21.6: Meson–nucleon scattering to lowest order

s, I'll specify the spinors as u and u', which are linear combinations of $u^{(1)}$ and $u^{(2)}$. The internal momentum is fixed by energy momentum conservation. On the left it is $p+q$ running along the arrow, and on the right it is $p'-q$. To make things definite, I will choose $\Gamma = i\gamma_5$.

Let's write down the invariant amplitude for these two diagrams.

$$i\mathcal{A}_{fi} = (-ig)^2\overline{u}'\left[i\gamma_5\frac{i}{\not{p}+\not{q}-m+i\epsilon}i\gamma_5\right]u + (-ig)^2\overline{u}'\left[i\gamma_5\frac{i}{\not{p}'-\not{q}-m+i\epsilon}i\gamma_5\right]u \tag{21.95}$$

The order of matrix multiplication is with the head of the arrow on the left, the tail of the arrow on the right, a $\overline{u}$ for every outgoing particle, a u for every incoming particle. The second diagram, on the right, contributes nearly the same as the first. I have no Fermi minus signs; the expression (21.84) I get from Wick's theorem is $\overline{\psi}\cdots\psi$, with ψ and $\overline{\psi}$ in the right positions to annihilate the initial nucleon and to create the final nucleon, respectively.

We can simplify this. In this kinematic region we don't need to keep track of the ϵ's, and we can rationalize the denominators:

$$\mathcal{A}_{fi} = g^2\overline{u}'\gamma_5\left[\frac{\not{p}+\not{q}+m}{(p+q)^2-m^2}+\frac{\not{p}'-\not{q}+m}{(p'-q)^2-m^2}\right]\gamma_5 u \tag{21.96}$$

I can get rid of the γ_5's in a flash because γ_5 anticommutes with $\not{p}$ and $\not{q}$. So I just drag it through, and use $\gamma_5^2 = 1$:

$$\mathcal{A}_{fi} = g^2\overline{u}'\left[\frac{-\not{p}-\not{q}+m}{(p+q)^2-m^2}+\frac{-\not{p}'+\not{q}+m}{(p'-q)^2-m^2}\right]u \tag{21.97}$$

This expression in fact simplifies enormously. This is typically what happens in Feynman calculations. The calculations with spinors are horrible, but not so horrible as one would think naively, because of the spinors' properties. Here in the first term we have $\not{p}$ acting on a free particle spinor on the right, which I remind you carries momentum p. And therefore $\not{p}u = mu$, (20.110a). Likewise in the second term $\overline{u}'\not{p}'$ equals $\overline{u}'m$, (20.113). Then the $\not{p}$'s cancel the m's and we're left with

$$\mathcal{A}_{fi} = g^2\overline{u}'\not{q}u\left[\frac{1}{(p'-q)^2-m^2}-\frac{1}{(p+q)^2-m^2}\right] \tag{21.98}$$

It's rather pleasant once you get the knack of it, like doing a crossword puzzle. You just move things around to eliminate some factors when they're hitting solutions of the free Dirac equation.

Here's a second example, nucleon–nucleon scattering:

$$N + N \to N + N \tag{21.99}$$

The Fermi minus signs are a bit more complicated, though the Γ algebra is considerably

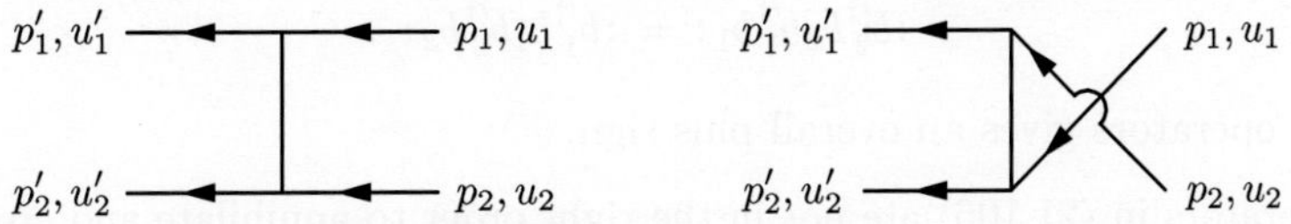

Figure 21.7: Nucleon–nucleon scattering to lowest order

simpler. I will write down the expression for the amplitude without determining the Fermi minus sign factors, which I'll just write as "(sign 1)" and "(sign 2)", which are going to be equal to +1 or −1. Indeed, I don't know what the factors are before I specify the initial state.

$$\begin{aligned} i\mathcal{A}_{fi} = (-ig)^2(\text{sign } 1)(\overline{u}_1' i\gamma_5 u_1)(\overline{u}_2' i\gamma_5 u_2)\frac{i}{(p_1 - p_1')^2 - \mu^2} \\ + (-ig)^2(\text{sign } 2)(\overline{u}_1' i\gamma_5 u_2)(\overline{u}_2' i\gamma_5 u_1)\frac{i}{(p_1 - p_2')^2 - \mu^2} \end{aligned} \tag{21.100}$$

Again there's no need to include the $i\epsilon$ factors.

First, let's talk through the graph on the left. From the top line, $(-ig)\overline{u}_1'(i\gamma_5)u_1$. That's all there is to it; there's just a vertex, there are no internal propagators. From the bottom line, $(-ig)\overline{u}_2' i\gamma_5 u_2$. The vertical line represents the meson propagator, $i/((p_1 - p_1')^2 - \mu^2)$. The second graph gives a similar expression. To determine the Fermi signs (1) and (2), we have to use the magic $\overline{\psi}\psi$ rule. I will label creation operators for the initial state simply as $b_1^\dagger$ and $b_2^\dagger$ to avoid writing a lot of **p** and (r) indices. Let's take the initial state to be

$$|i\rangle = b_1^\dagger b_2^\dagger\,|0\rangle \tag{21.101}$$

That is, nucleon number 2 is created first, and then nucleon number 1. The final state should be

$$|f\rangle = b_1'^\dagger b_2'^\dagger\,|0\rangle \tag{21.102}$$

If 1 equals 1′ and 2 equals 2′, the final state is the same as the initial state, not minus the initial state. Then the adjoint gives us

$$\langle f| = \langle 0|\, b_2' b_1' \tag{21.103}$$

We can now work out what happens using the $\overline{\psi}\psi$ rule. We always have the operator associated with the head of the line to the *left* of the operator associated with the tail of the line. Let's do the left graph in Figure 21.7 first. The tail of the top line is annihilating particle 1. The head of the line is creating particle 1′ and, from left to right, head goes before tail. From the Wick expansion, we have

$$:\overline{\psi'}_1\psi_1: \quad\to\quad :b_1'^\dagger b_1: \tag{21.104}$$

Likewise on the bottom line, head goes before tail, and we have

$$:\overline{\psi'}_2\psi_2: \quad\to\quad :b_2'^\dagger b_2: \tag{21.105}$$

That's the $\overline{\psi}\psi$ rule: heads before tails. The net result for the S-matrix element involves the factor

$$\langle f| :b_2'^{\dagger} b_2 b_1'^{\dagger} b_1 : |i\rangle \tag{21.106}$$

It doesn't matter in which order I write the two pairs of operators;

$$:b_2'^{\dagger} b_2 b_1'^{\dagger} b_1 : \;=\; :b_1'^{\dagger} b_1 b_2'^{\dagger} b_2 : \tag{21.107}$$

Permuting the operators gives an overall plus sign.

But the operators in (21.106) are not in the right order to annihilate and create the initial state (21.101) and the final state (21.103). The operator b_1 is in a great position to kill the incoming nucleon 1, but $b_1'^{\dagger}$ is not all the way over on the left to create the outgoing nucleon $1'$. Therefore I rearrange it by bringing $b_1'^{\dagger}$ over to the left. That requires two permutations, so it's an overall plus sign:

$$:b_2'^{\dagger} b_2 b_1'^{\dagger} b_1 : \;=\; :b_1'^{\dagger} b_2'^{\dagger} b_2 b_1 : \tag{21.108}$$

Now everything is in great position: b_1 can knock off $b_1^{\dagger}$, b_2 can then eliminate $b_2^{\dagger}$; $b_1'^{\dagger}$ can take care of b_1', and similarly $b_2'^{\dagger}$ can cancel off b_2'. In this case, the Fermi sign, (sign 1), equals $+1$. (It is not really as tedious as this. After you've gone through it two or three times, you can do it by eyeball.)

What about the second case, the rightmost diagram in (21.106)? Here we have on one line nucleon 1 being annihilated and nucleon $2'$ being created. On the other line we have 2 being annihilated and $1'$ being created. Using the $\overline{\psi}\psi$, head-before-tail rule, the corresponding Wick term is

$$:\overline{\psi}'_1 \psi_2 \overline{\psi}'_2 \psi_1 : \quad\rightarrow\quad :b_1'^{\dagger} b_2 b_2'^{\dagger} b_1 : \tag{21.109}$$

Again I want to put the operators in the correct position,

$$:b_1'^{\dagger} b_2'^{\dagger} b_2 b_1 : \tag{21.110}$$

The b_1 is still in the right place, and this time, so is the $b_1'^{\dagger}$. But the b_2 and $b_2'^{\dagger}$ need to switch places. When we permute the operators, we get a minus sign: (sign 2) $= (-1)$.

Unless one were extraordinarily clever, one could not have guessed the absolute sign of either of these two terms. One can, however, easily guess that the relative sign had to be -1. The statement that the relative sign is -1 is simply the statement that if one interchanges all the "1" and "2" labels, the total amplitude changes by a sign, just as we would expect for a scattering process involving Fermi particles. This is frequently a useful rule. Sometimes you don't have to work out the absolute sign if all you're going to do is square the amplitude at the end of the computation. Frequently Fermi statistics are good enough to tell you the relative signs between the various graphs. This is not always true. It doesn't work for example in meson–meson into nucleon–antinucleon scattering. But sometimes it's enough.

21.6 Summing and averaging over spin states

I have now told you all there is to say about the actual computation of S-matrix elements between particles in definite spin states. If you are interested in particular spin states, say states of definite helicity for the initial and final fermions, all you have to do is plug in the appropriate u's and $\overline{u}$'s for the initial and final particles and evaluate the matrix element. However there is a large class of experiments in which one is either uninterested in, or unable to measure, the spin of the initial or final states. One frequently does experiments with unpolarized beams of particles, and in which we choose not to measure the spin of the final nucleon. In such cases, one is frequently interested in cross-sections which are summed over final spins (since your apparatus responds whatever the final spin is), and averaged over initial spins (because you have a statistical distribution of initial spins in the incoming beam).

As a specific example, let's return to nucleon–meson scattering,

$$N + \phi \to N + \phi$$

We showed in (21.98) that the scattering amplitude was some function $F(E, \theta)$ times a bilinear spinor expression, $\overline{u}' \not{q} u$:

$$\mathcal{A}_{fi} = F(E, \theta)\, \overline{u}' \not{q} u \tag{21.111}$$

where E is the center-of-mass energy and θ is the scattering angle. We want to compute $|\mathcal{A}_{fi}|^2$, say between a definite polarization state r and a final one s. The initial spinor is characterized by momentum $\mathbf{p}$, and the final spinor by $\mathbf{p}'$. For these particular states,

$$\mathcal{A}_{sr} = F(E, \theta)\, \overline{u}^{(s)}_{\mathbf{p}'} \not{q} u^{(r)}_{\mathbf{p}} \tag{21.112}$$

What we want to do is square the amplitude, sum on r and divide by 2 (because we're averaging over the initial spins, two in number), and sum over s (the final spin). We use these facts: the function $F(E, \theta)$ is independent of the spins r and s; $\not{q}$ is "self-bar", because q^μ is real and the gamma matrices are self-bar (20.84); and the Hermitian adjoint of the bilinear spinor product is the same as its Dirac adjoint (20.66):

$$\left(\overline{u}^{(s)}_{\mathbf{p}'} \not{q} u^{(r)}_{\mathbf{p}}\right)^\dagger = u^{(r)\dagger}_{\mathbf{p}} \gamma^{\mu\dagger} \gamma^{0\dagger} u^{(s)}_{\mathbf{p}'} q_\mu = \overline{u}^{(r)}_{\mathbf{p}} \overline{\not{q}} u^{(s)}_{\mathbf{p}'} = \overline{u}^{(r)}_{\mathbf{p}} \not{q} u^{(s)}_{\mathbf{p}'} \tag{21.113}$$

Then

$$\begin{aligned} |\mathcal{A}|^2 \equiv \tfrac{1}{2} \sum_{r,s} |\mathcal{A}_{sr}|^2 &= \tfrac{1}{2} \sum_{r,s} |F(E, \theta)|^2 \left(\overline{u}^{(s)}_{\mathbf{p}'} \not{q} u^{(r)}_{\mathbf{p}}\right)^\dagger \overline{u}^{(s)}_{\mathbf{p}'} \not{q} u^{(r)}_{\mathbf{p}} \\ &= \tfrac{1}{2} |F(E, \theta)|^2 \sum_{r,s} \overline{u}^{(r)}_{\mathbf{p}} \not{q} u^{(s)}_{\mathbf{p}'}\, \overline{u}^{(s)}_{\mathbf{p}'} \not{q} u^{(r)}_{\mathbf{p}} \end{aligned} \tag{21.114}$$

Now we borrow a cunning idea due to Feynman,[6] which has saved generations of physicists from having to compute sixteen 4×4 matrix elements with explicit spinors and sum them all up. That's what they used to do, when they were doing this sort of computation back in the 1930's. He observed that a number can be thought of as a 1×1 matrix, and that a 1×1 matrix is equal to its trace. Thus

$$\sum_{r,s} \overline{u}^{(r)}_{\mathbf{p}} \not{q} u^{(s)}_{\mathbf{p}'}\, \overline{u}^{(s)}_{\mathbf{p}'} \not{q} u^{(r)}_{\mathbf{p}} = \mathrm{Tr}\left(\sum_{r,s} \overline{u}^{(r)}_{\mathbf{p}} \not{q} u^{(s)}_{\mathbf{p}'}\, \overline{u}^{(s)}_{\mathbf{p}'} \not{q} u^{(r)}_{\mathbf{p}}\right) \tag{21.115}$$

---

[6] [Eds.] R. P. Feynman, "The theory of positrons", *Phys. Rev.* **76** (1949) 749–759. See equation (36); "Sp" = *spur*, German for *trace*. See also R. P. Feynman, *Quantum Electrodynamics*, W. A. Benjamin, 1962, Lecture 23, "A method of summing matrix elements over spin states", pp. 112–114. The technique seems to have been first used by the Dutch theorist Hendrik B. G. Casimir (1909–2000), and is sometimes called "Casimir's trick": H. Casimir, "Über die Intensität der Streustrahlung gebundener Electronen" (On the intensity of radiation scattered by bound electrons), *Helv. Phys. Acta* **6** (1933) 287–305. See §4, p. 293; Griffiths *EP*, p. 251. Casimir's autobiography (*Haphazard Reality: Half a Century of Science*, Harper & Row, 1984) draws its title from a Bohr quote: "When telling a true story, one should not be overly influenced by the haphazard occurrences of reality."

The trace is invariant under cyclic permutation of factors, so I can write (21.115) as

$$\text{Tr}\left(\sum_{r,s} \overline{u}^{(r)}_{\mathbf{p}} \not{q} u^{(s)}_{\mathbf{p}'} \overline{u}^{(s)}_{\mathbf{p}'} \not{q} u^{(r)}_{\mathbf{p}}\right) = \text{Tr}\left(\sum_{r,s} \not{q} u^{(r)}_{\mathbf{p}} \overline{u}^{(r)}_{\mathbf{p}} \not{q} u^{(s)}_{\mathbf{p}'} \overline{u}^{(s)}_{\mathbf{p}'}\right) \tag{21.116}$$

I moved the factor $\not{q} u^{(r)}_{\mathbf{p}}$ from the rightmost position to the leftmost to make use of the wonderful completeness relation (20.123a):

$$\text{Tr}\left(\sum_{r,s} \overline{u}^{(r)}_{\mathbf{p}} \not{q} u^{(s)}_{\mathbf{p}'} \overline{u}^{(s)}_{\mathbf{p}'} \not{q} u^{(r)}_{\mathbf{p}}\right) = \text{Tr}\left(\not{q}(\not{p}+m)\not{q}(\not{p}'+m)\right) \tag{21.117}$$

Here is the redemption of that homework problem[7] on traces of Dirac matrices. You might have wondered why you were working out all those dumb trace identities. Recall that the trace of an odd number of γ matrices always vanishes. The traces of a product of two and of four slashed quantities are given by the identities (S11.13) and (S11.16), respectively:

$$\text{Tr}\left(\not{a}\not{b}\right) = 4(a\cdot b) \tag{S11.13}$$

$$\text{Tr}\left(\not{a}\not{b}\not{c}\not{d}\right) = 4[(a\cdot b)(c\cdot d) - (a\cdot c)(b\cdot d) + (a\cdot d)(b\cdot c)] \tag{S11.16}$$

So we're all set up for completing the computation:

$$\begin{aligned} |\mathcal{A}|^2 &= \tfrac{1}{2}\,|F|^2\,\text{Tr}\left(\not{q}(\not{p}+m)\not{q}(\not{p}'+m)\right) \\ &= 2|F|^2\left[m^2q^2 + (q\cdot p)(q\cdot p') - (q^2 p\cdot p') + (q\cdot p)(q\cdot p')\right] \end{aligned} \tag{21.118}$$

This can be simplified somewhat. The meson is on its mass shell, and therefore $q^2 = \mu^2$. So

$$|\mathcal{A}|^2 = 2|F|^2\left[\mu^2(m^2 - p\cdot p') + 2(q\cdot p)(q\cdot p')\right] \tag{21.119}$$

This expression, by trivial kinematic exercises that I won't bother to go through, can be reduced to functions of the only two invariants, the center-of-mass energy, E and the center-of-mass scattering angle, θ.

Now that we have our general formalism, we can discuss charge conjugation, time reversal, and TCP invariance. We'll do that next time, and then begin renormalization for theories involving fermion fields.

[7] [Eds.] Problem 11.2, p. 425.

Problems 12

12.1 When we attempted to quantize the free Dirac theory

$$\mathscr{L} = \pm\overline{\psi}(i\partial\!\!\!/ - m)\psi \tag{P12.1}$$

with canonical commutation relations, we encountered a disastrous contradiction (21.15) with the positivity of energy. We succeeded when we used canonical anticommutators (if we chose $(\pm)$ to be $+$). Much earlier we were able to quantize the free charged scalar field,

$$\mathscr{L} = \pm(\partial_\mu\phi^*\partial^\mu\phi - \mu^2\phi^*\phi) \tag{P12.2}$$

with canonical commutators (if we chose $(\pm)$ to be $+$). Attempt to quantize the free charged scalar field with (nearly) canonical anticommutators:

$$\begin{gathered}\{\phi(\mathbf{x},t),\phi(\mathbf{y},t)\} = \{\dot\phi(\mathbf{x},t),\dot\phi(\mathbf{y},t)\} = 0\\ \{\phi(\mathbf{x},t),\phi^*(\mathbf{y},t)\} = \{\dot\phi(\mathbf{x},t),\dot\phi^*(\mathbf{y},t)\} = 0\\ \{\phi(\mathbf{x},t),\dot\phi^*(\mathbf{y},t)\} = \lambda\,\delta^{(3)}(\mathbf{x}-\mathbf{y})\end{gathered} \tag{P12.3}$$

where λ is a (possibly complex) constant.

Show that one reaches a disastrous contradiction with the positivity of the norm in Hilbert space; that is to say, with (21.20):

$$\langle\phi|\{\theta,\theta^\dagger\}|\phi\rangle \geq 0 \tag{21.20}$$

for any operator θ and any state $|\phi\rangle$.

HINTS: (1) Canonical anticommutation implies that, even on the classical level, ϕ and ϕ^* are Grassmann variables. If you don't take proper account of this (especially in ordering terms when deriving the canonical momenta), you'll get hopelessly confused. (2) Dirac theory is successfully quantized using anticommutators; the sign of the Lagrangian is fixed by appealing to the positivity of the inner product in Hilbert space. If we attempt to quantize the theory using commutators, we get into trouble with the positivity of the energy. The Klein–Gordon theory is successfully quantized using commutators; the sign of the Lagrangian is fixed by appealing to the positivity of energy. So it's to be expected that we'd get into trouble, if we attempted to quantize the Klein–Gordon theory with anticommutators, with the positivity of the inner product.

(1997a 11.1)

12.2. Compute the differential cross-section $d\sigma/d\Omega$ in the center-of-mass frame, to lowest non-trivial order in perturbation theory, averaged over initial spins and summed over final spins, for meson–nucleon scattering in the "scalar" theory discussed in §21.4,

$$\mathscr{L}' = g\overline{\psi}\psi\phi \tag{P12.4}$$

NOTE: You are required only to compute $d\sigma/d\Omega$, not the total cross-section, for this problem and the next.

(1997a 11.2)

12.3 The same for nucleon–antinucleon scattering in the "pseudoscalar" theory,

$$\mathscr{L}' = g\overline{\psi}i\gamma_5\psi\phi \tag{P12.5}$$

Note: Since you are only interested in cross-sections, all you need to know is the relative sign between the two graphs; the absolute sign is irrelevant.

(1997a 11.3)

Solutions 12

12.1 From the Lagrangian

$$\mathscr{L} = \pm(\partial_\mu \phi^* \partial^\mu \phi - \mu^2 \phi^* \phi) \tag{S12.1}$$

we derive the canonical momentum to $\phi(x)$,

$$\pi = \frac{\partial \mathscr{L}}{\partial(\partial_0 \phi)} = \mp \partial^0 \phi^* = \mp \dot{\phi}^* \tag{S12.2}$$

Note that as $\phi(x)$ and $\phi^*(x)$ are regarded as Grassmann variables, when we move the derivative $\partial/\partial(\partial_0\phi)$ past $\partial_\mu \phi^*$, we pick up an extra minus sign. The question asks that we impose the (nearly) canonical anticommutation relations

$$\{\phi(\mathbf{x}, t), \dot{\phi}^*(\mathbf{y}, t)\} = \lambda\, \delta^{(3)}(\mathbf{x} - \mathbf{y}) \tag{S12.3}$$

As usual, expand $\phi(x)$ in terms of annihilation and creation operators,

$$\phi(x) = \int \frac{d^3\mathbf{p}}{(2\pi)^{3/2}\sqrt{2\omega_{\mathbf{p}}}} \left[b_{\mathbf{p}} e^{-ip\cdot x} + c^\dagger_{\mathbf{p}} e^{ip\cdot x} \right] \tag{S12.4}$$

Then

$$\begin{aligned} \phi(\mathbf{x}, 0) &= \int \frac{d^3\mathbf{p}}{(2\pi)^{3/2}\sqrt{2\omega_{\mathbf{p}}}} \left[b_{\mathbf{p}} e^{i\mathbf{p}\cdot\mathbf{x}} + c^\dagger_{\mathbf{p}} e^{-i\mathbf{p}\cdot\mathbf{x}} \right] \\ \dot{\phi}(\mathbf{y}, 0) &= i \int \frac{d^3\mathbf{p}'}{(2\pi)^{3/2}} \sqrt{\frac{\omega_{\mathbf{p}'}}{2}} \left[-b_{\mathbf{p}'} e^{i\mathbf{p}'\cdot\mathbf{y}} + c^\dagger_{\mathbf{p}'} e^{-i\mathbf{p}'\cdot\mathbf{y}} \right] \end{aligned} \tag{S12.5}$$

We can invert these relations to solve for $b_{\mathbf{p}}$ and $b^\dagger_{\mathbf{p}'}$:

$$\begin{aligned} b_{\mathbf{p}} &= \int \frac{d^3\mathbf{x}}{(2\pi)^{3/2}} e^{-i\mathbf{p}\cdot\mathbf{x}} \left[\sqrt{\frac{\omega_{\mathbf{p}}}{2}} \phi(\mathbf{x}, 0) + \frac{i}{\sqrt{2\omega_{\mathbf{p}}}} \dot{\phi}(\mathbf{x}, 0) \right] \\ b^\dagger_{\mathbf{p}'} &= \int \frac{d^3\mathbf{y}}{(2\pi)^{3/2}} e^{i\mathbf{p}'\cdot\mathbf{y}} \left[\sqrt{\frac{\omega_{\mathbf{p}'}}{2}} \phi^*(\mathbf{y}, 0) - \frac{i}{\sqrt{2\omega_{\mathbf{p}'}}} \dot{\phi}^*(\mathbf{y}, 0) \right] \end{aligned} \tag{S12.6}$$

Then

$$\begin{aligned} \{b_{\mathbf{p}}, b^\dagger_{\mathbf{p}'}\} &= \int \frac{d^3\mathbf{x}\, d^3\mathbf{y}}{(2\pi)^3} e^{-i\mathbf{p}\cdot\mathbf{x} + i\mathbf{p}'\cdot\mathbf{y}} \left[-\frac{i}{2}\lambda \sqrt{\frac{\omega_{\mathbf{p}}}{\omega_{\mathbf{p}'}}} + \frac{i}{2}\lambda^* \sqrt{\frac{\omega_{\mathbf{p}'}}{\omega_{\mathbf{p}}}} \right] \delta^{(3)}(\mathbf{x} - \mathbf{y}) \\ &= \frac{i}{2} \left[-\lambda \sqrt{\frac{\omega_{\mathbf{p}}}{\omega_{\mathbf{p}'}}} + \lambda^* \sqrt{\frac{\omega_{\mathbf{p}'}}{\omega_{\mathbf{p}}}} \right] \int \frac{d^3\mathbf{x}}{(2\pi)^3} e^{-i\mathbf{x}\cdot(\mathbf{p} - \mathbf{p}')} \\ &= \frac{i}{2}(\lambda^* - \lambda)\, \delta^{(3)}(\mathbf{p} - \mathbf{p}') = \mathrm{Im}(\lambda)\, \delta^{(3)}(\mathbf{p} - \mathbf{p}') \end{aligned} \tag{S12.7}$$

Similarly,

$$c_{\mathbf{p}} = \int \frac{d^3\mathbf{y}}{(2\pi)^{3/2}} e^{-i\mathbf{p}\cdot\mathbf{y}} \left[\sqrt{\frac{\omega_{\mathbf{p}}}{2}} \phi^*(\mathbf{y},0) + \frac{i}{\sqrt{2\omega_{\mathbf{p}}}} \dot{\phi}^*(\mathbf{y},0) \right]$$
$$c^\dagger_{\mathbf{p}} = \int \frac{d^3\mathbf{x}}{(2\pi)^{3/2}} e^{i\mathbf{p}\cdot\mathbf{x}} \left[\sqrt{\frac{\omega_{\mathbf{p}}}{2}} \phi(\mathbf{x},0) - \frac{i}{\sqrt{2\omega_{\mathbf{p}}}} \dot{\phi}(\mathbf{x},0) \right] \tag{S12.8}$$

and

$$\{c^\dagger_{\mathbf{p}}, c_{\mathbf{p}'}\} = \frac{i}{2}(\lambda - \lambda^*)\,\delta^{(3)}(\mathbf{p}-\mathbf{p}') = -\mathrm{Im}(\lambda)\,\delta^{(3)}(\mathbf{p}-\mathbf{p}') \tag{S12.9}$$

Consequently, $\langle 0|\{b_{\mathbf{p}}, b^\dagger_{\mathbf{p}'}\}|0\rangle$ and $\langle 0|\{c^\dagger_{\mathbf{p}}, c_{\mathbf{p}'}\}|0\rangle$ cannot *both* be positive, so the positive definite norm does not hold if we attempt to canonically quantize a scalar field with anticommutators. ■

12.2 Using the Feynman rules for fermions (box, p. 443), we see that the vertex involves two fermion lines and a meson line. The relevant Feynman diagrams are shown in Figure S12.1.

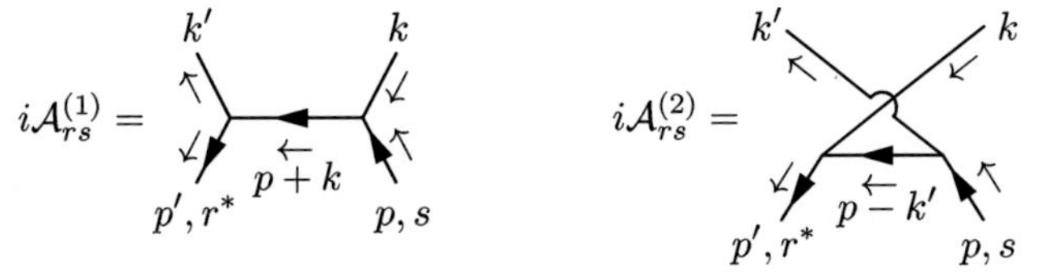

Figure S12.1: Graphs for lowest order meson–nucleon scattering

These two contributions add, because only bosons have been swapped;

$$i\mathcal{A}_{rs} = i\mathcal{A}^{(1)}_{rs} + i\mathcal{A}^{(2)}_{rs} \tag{S12.10}$$

Using (12.26) (averaging over the initial spins, and summing over the final spins),

$$\frac{d\sigma}{d\Omega} = \frac{1}{64\pi^2 E_T^2}\frac{|\mathbf{p}_f|}{|\mathbf{p}_i|}\,\tfrac{1}{2}\sum_{r,s}|\mathcal{A}_{rs}|^2 = \frac{1}{64\pi^2 E_T^2}\,\tfrac{1}{2}\sum_{r,s}|\mathcal{A}_{rs}|^2 \tag{S12.11}$$

because the scattering is elastic, so $|\mathbf{p}_f| = |\mathbf{p}_i|$. Using the Feynman rules, we find

$$i\mathcal{A}^{(1)}_{rs} = (ig)^2 \overline{u}^{(r)}_{\mathbf{p}'} \frac{i(\not{p} + \not{k} + m)}{(p+k)^2 - m^2} u^{(s)}_{\mathbf{p}} = -ig^2 \overline{u}^{(r)}_{\mathbf{p}'} \frac{(2m + \not{k})}{(p+k)^2 - m^2} u^{(s)}_{\mathbf{p}}$$
$$i\mathcal{A}^{(2)}_{rs} = (ig)^2 \overline{u}^{(r)}_{\mathbf{p}'} \frac{i(\not{p}' - \not{k} + m)}{(p'-k)^2 - m^2} u^{(s)}_{\mathbf{p}} = -ig^2 \overline{u}^{(r)}_{\mathbf{p}'} \frac{(2m - \not{k})}{(p'-k)^2 - m^2} u^{(s)}_{\mathbf{p}} \tag{S12.12}$$

For convenience, let's define

$$A = \frac{1}{(p+k)^2 - m^2} + \frac{1}{(p'-k)^2 - m^2}$$
$$B = \frac{1}{(p+k)^2 - m^2} - \frac{1}{(p'-k)^2 - m^2} \tag{S12.13}$$

Then

$$i\mathcal{A}_{rs} = -ig^2 \overline{u}^{(r)}_{\mathbf{p}'} (2mA + \not{k}B) u^{(s)}_{\mathbf{p}} \tag{S12.14}$$

The differential cross-section becomes

$$\frac{d\sigma}{d\Omega} = \frac{1}{64\pi^2 E_T^2}\,\tfrac{1}{2}\sum_{r,s}|\mathcal{A}_{rs}|^2 = \frac{1}{128\pi^2 E_T^2}\sum_{r,s}\left[i\mathcal{A}_{rs}\overline{i\mathcal{A}_{rs}} \right] \tag{S12.15}$$

Calculating the sum,

$$\begin{aligned}
\sum_{r,s}\left[i\mathcal{A}_{rs}\overline{i\mathcal{A}_{rs}}\right] &= g^4\sum_{r,s}\left[\overline{u}_{\mathbf{p}'}^{(r)}(2mA+\not{k}B)u_{\mathbf{p}}^{(s)}\overline{u}_{\mathbf{p}}^{(s)}(2mA+\not{k}B)u_{\mathbf{p}'}^{(r)}\right] \\
&= g^4\sum_r\sum_{i,j=1}^4\left[\left(\overline{u}_{\mathbf{p}'}^{(r)}\right)_i\left((2mA+\not{k}B)(\not{p}+m)(2mA+\not{k}B)\right)_{ij}\left(u_{\mathbf{p}'}^{(r)}\right)_j\right] \quad \text{(completeness)} \\
&= g^4\sum_{i,j=1}^4\left((2mA+\not{k}B)(\not{p}+m)(2mA+\not{k}B)\right)_{ij}\left(\not{p}'+m\right)_{ji} \quad \text{(completeness)} \qquad \text{(S12.16)} \\
&= g^4\,\mathrm{Tr}\left[4m^2A^2(\not{p}\not{p}'+m^2)+4m^2AB\not{k}(\not{p}+\not{p}')+B^2(m^2\not{k}\not{k}+\not{k}\not{p}\not{k}\not{p}')\right] \\
&= 4g^4\left[4m^2A^2(p\cdot p'+m^2)+4m^2ABk\cdot(p+p')+B^2(m^2\mu^2+2(k\cdot p)(k\cdot p')-\mu^2(p\cdot p'))\right]
\end{aligned}$$

(In the fourth step, we use the result that the trace of an odd number of γ's is zero.) We therefore have

$$\begin{aligned}
\frac{d\sigma}{d\Omega} = \frac{g^4}{32\pi^2E_T^2}&\left[4m^2A^2(p\cdot p'+m^2)+4m^2ABk\cdot(p+p')\right. \\
&\left.+B^2(m^2\mu^2+2(k\cdot p)(k\cdot p')-\mu^2(p\cdot p'))\right]
\end{aligned} \qquad \text{(S12.17)}$$

In the center-of-momentum frame (with the incident nucleon in the x direction and the outgoing nucleon in the xy plane),

$$\begin{aligned}
p^\mu &= (\sqrt{|\mathbf{p}|^2+m^2}, |\mathbf{p}|, 0, 0) \\
k^\mu &= (\sqrt{|\mathbf{p}|^2+\mu^2}, -|\mathbf{p}|, 0, 0) \\
p'^\mu &= (\sqrt{|\mathbf{p}|^2+m^2}, |\mathbf{p}|\cos\theta, |\mathbf{p}|\sin\theta, 0) \\
k'^\mu &= (\sqrt{|\mathbf{p}|^2+\mu^2}, -|\mathbf{p}|\cos\theta, -|\mathbf{p}|\sin\theta, 0)
\end{aligned} \qquad \text{(S12.18)}$$

and so

$$\begin{aligned}
p\cdot k &= \sqrt{|\mathbf{p}|^2+m^2}\sqrt{|\mathbf{p}|^2+\mu^2}+|\mathbf{p}|^2 \\
p'\cdot k &= \sqrt{|\mathbf{p}|^2+m^2}\sqrt{|\mathbf{p}|^2+\mu^2}+|\mathbf{p}|^2\cos\theta \\
p\cdot p' &= |\mathbf{p}|^2(1-\cos\theta)+m^2 \\
(p+k)^2 = E_T^2 &= \left[\sqrt{|\mathbf{p}|^2+m^2}+\sqrt{|\mathbf{p}|^2+\mu^2}\right]^2 \\
(p'-k)^2 &= m^2+\mu^2-2\sqrt{|\mathbf{p}|^2+m^2}\sqrt{|\mathbf{p}|^2+\mu^2}-2|\mathbf{p}|^2\cos\theta
\end{aligned} \qquad \text{(S12.19)}$$

If desired, we could put these factors into (S12.17), to get $d\sigma/d\Omega$ in terms of p and the scattering angle θ. ∎

12.3 The relevant Feynman diagrams are shown in Figure S12.2.

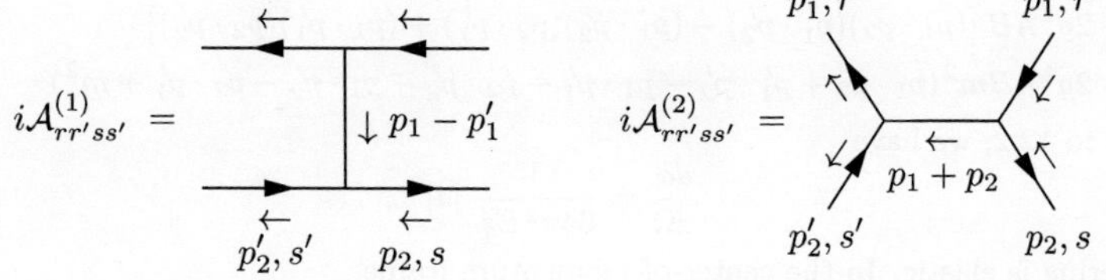

Figure S12.2: Graphs for lowest order nucleon–antinucleon scattering

The total amplitude for this scattering process can be written as

$$i\mathcal{A}_{rr'ss'} = \pm\left[i\mathcal{A}^{(1)}_{rr'ss'} - i\mathcal{A}^{(2)}_{rr'ss'}\right] \qquad \text{(S12.20)}$$

the relative minus sign coming from the exchange of the fermion lines for p_1' and p_2 between the diagrams. We can write down the amplitudes as (letting $u^{(r)}_{\mathbf{p}_1} \equiv u_1$, etc.)

$$\begin{aligned} i\mathcal{A}^{(1)}_{rr'ss'} &= (ig)^2 \left[\overline{u}_{1'} i\gamma_5 u_1\right] \frac{i}{(p_1 - p_1')^2 - \mu^2} \left[\overline{v}_2 i\gamma_5 v_{2'}\right] \\ i\mathcal{A}^{(2)}_{rr'ss'} &= (ig)^2 \left[\overline{u}_{1'} i\gamma_5 v_{2'}\right] \frac{i}{(p_1 + p_2)^2 - \mu^2} \left[\overline{v}_2 i\gamma_5 u_1\right] \end{aligned} \tag{S12.21}$$

The overall sign of $i\mathcal{A}_{rr'ss'}$ doesn't matter here, because it's going to be squared. So we can take

$$i\mathcal{A}_{rr'ss'} = ig^2 \left(-A \left[\overline{u}_{1'}\gamma_5 u_1\right]\left[\overline{v}_2\gamma_5 v_{2'}\right] + B \left[\overline{u}_{1'}\gamma_5 v_{2'}\right]\left[\overline{v}_2\gamma_5 u_1\right]\right) \tag{S12.22}$$

where

$$A = \frac{1}{(p_1 - p_1')^2 - \mu^2} \qquad B = \frac{1}{(p_1 + p_2)^2 - \mu^2} \tag{S12.23}$$

so

$$|i\mathcal{A}_{rr'ss'}|^2 = g^4 \begin{bmatrix} A^2 \left[\overline{u}_{1'}\gamma_5 u_1\right]\left[\overline{v}_2\gamma_5 v_{2'}\right]\left[\overline{v}_{2'}\gamma_5 v_2\right]\left[\overline{u}_1\gamma_5 u_{1'}\right] \\ -AB \left[\overline{u}_{1'}\gamma_5 u_1\right]\left[\overline{v}_2\gamma_5 v_{2'}\right]\left[\overline{u}_1\gamma_5 v_2\right]\left[\overline{v}_{2'}\gamma_5 u_{1'}\right] \\ -AB \left[\overline{u}_{1'}\gamma_5 v_{2'}\right]\left[\overline{v}_2\gamma_5 u_1\right]\left[\overline{v}_{2'}\gamma_5 v_2\right]\left[\overline{u}_1\gamma_5 u_{1'}\right] \\ +B^2 \left[\overline{u}_{1'}\gamma_5 v_{2'}\right]\left[\overline{v}_2\gamma_5 u_1\right]\left[\overline{u}_1\gamma_5 v_2\right]\left[\overline{v}_{2'}\gamma_5 u_{1'}\right] \end{bmatrix} \tag{S12.24}$$

We need to average over both pairs of initial spins and sum over the final spins, which we do with the trace theorems:

$$\frac{1}{4}\sum_{rr'ss'} |i\mathcal{A}_{rr'ss'}|^2 = \tfrac{1}{4}g^4 \begin{bmatrix} A^2\,\mathrm{Tr}[(\not{p}_1' + m)\gamma_5(\not{p}_1 + m)\gamma_5]\cdot\mathrm{Tr}[(\not{p}_2 - m)\gamma_5(\not{p}_2' - m)\gamma_5] \\ -AB\,\mathrm{Tr}[(\not{p}_1' + m)\gamma_5(\not{p}_1 + m)\gamma_5(\not{p}_2 - m)\gamma_5(\not{p}_2' - m)\gamma_5] \\ -AB\,\mathrm{Tr}[(\not{p}_1' + m)\gamma_5(\not{p}_2' - m)\gamma_5(\not{p}_2 - m)\gamma_5(\not{p}_1 + m)\gamma_5] \\ +B^2\,\mathrm{Tr}[(\not{p}_1' + m)\gamma_5(\not{p}_2' - m)\gamma_5]\cdot\mathrm{Tr}[(\not{p}_1 + m)\gamma_5(\not{p}_2 - m)\gamma_5] \end{bmatrix} \tag{S12.25}$$

Let's call this quantity $|\mathcal{A}|^2$. We can move the γ_5's past the gammas in the slashed momenta, and we find

$$|\mathcal{A}|^2 = \tfrac{1}{4}g^4 \begin{bmatrix} A^2\,\mathrm{Tr}[(\not{p}_1' + m)(\not{p}_1 - m)]\cdot\mathrm{Tr}[(\not{p}_2 - m)(\not{p}_2' + m)] \\ -AB\,\mathrm{Tr}[(\not{p}_1' + m)(\not{p}_1 - m)(\not{p}_2 - m)(\not{p}_2' + m)] \\ -AB\,\mathrm{Tr}[(\not{p}_1' + m)(\not{p}_2' + m)(\not{p}_2 - m)(\not{p}_1 - m)] \\ +B^2\,\mathrm{Tr}[(\not{p}_1' + m)(\not{p}_2' + m)]\cdot\mathrm{Tr}[(\not{p}_1 + m)(\not{p}_2 + m)] \end{bmatrix} \tag{S12.26}$$

The two middle terms are equal. (This is not obvious, but it's so.) Using the trace identities,

$$\begin{aligned} |\mathcal{A}|^2 &= 4g^4A^2(p_1\cdot p_1' - m^2)(p_2\cdot p_2' - m^2) + 4g^4B^2(p_1\cdot p_2 + m^2)(p_1'\cdot p_2' + m^2) \\ &\quad - 2g^4AB\left[(p_1\cdot p_2)(p_1'\cdot p_2') - (p_1\cdot p_2')(p_2\cdot p_1') + (p_1\cdot p_1')(p_2\cdot p_2')\right] \\ &\quad - 2g^4ABm^2(p_1\cdot p_2 + p_1'\cdot p_2' - p_1\cdot p_1' - p_2\cdot p_2' - p_1\cdot p_2' - p_2\cdot p_1' + m^2) \end{aligned} \tag{S12.27}$$

As in the solution to 12.2, we have

$$\frac{d\sigma}{d\Omega} = \frac{1}{64\pi^2 E_T^2}\,|a|^2 \tag{S12.28}$$

because the scattering is elastic. In the center-of-momentum frame,

$$\begin{aligned} p_1^\mu &= (\sqrt{|\mathbf{p}|^2 + m^2}, |\mathbf{p}|, 0, 0) \\ p_2^\mu &= (\sqrt{|\mathbf{p}|^2 + m^2}, -|\mathbf{p}|, 0, 0) \\ p_1'^\mu &= (\sqrt{|\mathbf{p}|^2 + m^2}, |\mathbf{p}|\cos\theta, |\mathbf{p}|\sin\theta, 0) \\ p_2'^\mu &= (\sqrt{|\mathbf{p}|^2 + m^2}, -|\mathbf{p}|\cos\theta, -|\mathbf{p}|\sin\theta, 0) \end{aligned} \tag{S12.29}$$

and so

$$
\begin{aligned}
(p_1 + p_2)^2 &= E_T^2 = 4(|\mathbf{p}|^2 + m^2) \\
(p_1 - p_1')^2 &= -4|\mathbf{p}|^2 \sin^2 \tfrac{1}{2}\theta \\
p_1 \cdot p_1' = p_2 \cdot p_2' &= 2|\mathbf{p}|^2 \sin^2 \tfrac{1}{2}\theta + m^2 \\
p_1 \cdot p_2' = p_2 \cdot p_1' &= 2|\mathbf{p}|^2 \cos^2 \tfrac{1}{2}\theta + m^2 \\
p_1 \cdot p_2 = p_1' \cdot p_2' &= 2|\mathbf{p}|^2 + m^2
\end{aligned}
\tag{S12.30}
$$

These factors go into (S12.28), to get $d\sigma/d\Omega$ in terms of p and the scattering angle θ. ■

22

CPT and Fermi fields

We are going to discuss for a Dirac theory the famous discrete symmetries of nature: parity P, charge conjugation C, and time reversal T. We've already talked a lot about parity (§6.3, §18.5, and §20.1; (18.50), (18.51), and (19.5)), but I will say more. As always in a relativistic theory, it's more convenient to discuss the product of parity and time reversal, PT. I will also prove, in the same diagrammatic way I proved it for purely scalar theories (§11.3), the CPT Theorem.

22.1 Parity and Fermi fields

In the theory of a single scalar field, we defined parity simply as $\mathbf{x}$ going into $-\mathbf{x}$ in the argument of the field:

$$P\colon \phi(\mathbf{x},t) \rightarrow \phi(-\mathbf{x},t) \tag{6.87}$$

However we realized that when we had a theory of more fields $\{\phi^a(\mathbf{x},t)\}$, we could have a more complicated definition. It is possible that the fields mix up among themselves, in addition to the space point changing:

$$P\colon \phi^a(\mathbf{x},t) \rightarrow M^a_b\phi^b(-\mathbf{x},t) \tag{6.85}$$

I gave several examples in Chapter 6 (pp. 122–125) where the matrix M_{ab} was diagonal: some of the fields were multiplied by $+1$, others by -1. The same is true for spinor fields. If we have a set $\{\psi^a(\mathbf{x},t)\}$ of spinor fields, we may not have the freedom to make individual phase transformations on each of them, but perhaps only on all of them collectively; there may be only one internal symmetry, not one for each field. In that case, we may have to mix the fields up among themselves to define parity, and multiply the different fields by different phase factors:

$$P\colon \psi_a(\mathbf{x},t) \rightarrow M_{ab}\psi_b(-\mathbf{x},t) \tag{22.1}$$

As an example, let's consider the theory of two spinor fields, ψ_A and ψ_B, interacting with a spinless field ϕ. The interaction

$$\mathscr{L}_1 = g_1\overline{\psi}_A i\gamma_5\psi_A\phi \tag{22.2}$$

is parity invariant if ψ_A transforms in the standard way:

$$P\colon \psi_A(\mathbf{x},t) \rightarrow \beta\psi_A(-\mathbf{x},t) \tag{20.8}$$

and ϕ is a pseudoscalar:

$$P\colon \phi(\mathbf{x},t) \to -\phi(-\mathbf{x},t) \tag{6.91}$$

Likewise a second interaction is invariant under parity if ψ_B transforms the same as ψ_A:

$$\mathscr{L}_2 = g_2\overline{\psi}_B i\gamma_5\psi_B\phi \tag{22.3}$$

Now I throw in a third interaction without a γ_5,

$$\mathscr{L}_3 = g_3(\overline{\psi}_B\psi_A + \overline{\psi}_A\psi_B)\phi \tag{22.4}$$

The theory described by

$$\mathscr{L}' = \mathscr{L}_1 + \mathscr{L}_2 + \mathscr{L}_3 \tag{22.5}$$

is *not* parity invariant, if ψ_B transforms the same as ψ_A. But there are other possible definitions of parity. Among others, this one works:

$$P\colon \begin{cases} \psi_A(\mathbf{x},t) \to \beta\psi_A(-\mathbf{x},t) \\ \psi_B(\mathbf{x},t) \to -\beta\psi_B(-\mathbf{x},t) \\ \phi(\mathbf{x},t) \to -\phi(-\mathbf{x},t) \end{cases} \tag{22.6}$$

While ψ_A keeps its standard transformation, and ϕ remains a pseudoscalar (6.91), we've changed ψ_B's transformation to include a minus sign. The Lagrangian (22.5) is invariant under this definition of parity.

Of course a definition of parity could also include a phase factor in the transformations of A and B:

$$\begin{aligned} \psi_A(\mathbf{x},t) &\to \beta e^{i\alpha}\psi_A(-\mathbf{x},t) \\ \psi_B(\mathbf{x},t) &\to -\beta e^{i\alpha}\psi_B(-\mathbf{x},t) \end{aligned} \tag{22.7}$$

That would still be all right, but here comes the usual ambiguity.[1] Whenever we have both an internal symmetry and one good definition of parity, we can just as well define parity anew, by multiplying the original parity by an internal symmetry. Which of these we choose to call parity is merely a matter of convention. We could describe the situation by saying perhaps that the B particle has opposite intrinsic parity to the A particle. A state of two A particles in an s-wave state would be even in parity, an eigenstate of parity with eigenvalue $(+1)$; the same would hold true for a state of two B particles in an s-wave state. But a state of an A particle and a B particle would be an eigenstate of parity with eigenvalue (-1).[2]

What about the parity of antiparticles? For a charged Bose field, the particle and the antiparticle have the same parity: whatever happens to the particle, whether it gets a $(+)$ sign or a $(-)$ sign, the antiparticle gets the same; ϕ and ϕ^* have the same parity transformation laws. What happens to the antiparticles in the theory of a Fermi field, $\psi(\mathbf{x},t)$?

The particles are associated with the fields via their annihilation and creation operators. We assume that there is a unitary operator U_P in Hilbert space (with no spinor indices) effecting this change of the field:

$$\boxed{P\colon \psi(\mathbf{x},t) \to U_P^\dagger\psi(\mathbf{x},t)U_P = \beta\psi(-\mathbf{x},t)} \tag{22.8}$$

---

[1] [Eds.] See the discussion following (6.93), page 123 and Example 2, page 123.

[2] [Eds.] The parity P of a state is the product of the intrinsic parities P_i of the i constituent particles times (-1) raised to the power of the angular momentum ℓ of the state: $P = \left(\prod_i P_i\right) \times (-1)^\ell$. See Bjorken & Drell *Fields*, Section 15.11, pp. 108–113.

That tells us how the unitary transformation associated with parity acts on the creation and annihilation operators, and therefore on the states. We have

$$\psi(\mathbf{x},t) = \int \frac{d^3\mathbf{p}}{(2\pi)^{3/2}\sqrt{2E_\mathbf{p}}} \sum_{r=1}^{2} \left[b_\mathbf{p}^{(r)} u_\mathbf{p}^{(r)} e^{-ip\cdot x} + c_\mathbf{p}^{(r)\dagger} v_\mathbf{p}^{(r)} e^{ip\cdot x} \right] \tag{21.6}$$

and

$$\begin{aligned} \psi(-\mathbf{x},t) &= \int \frac{d^3\mathbf{p}}{(2\pi)^{3/2}\sqrt{2E_\mathbf{p}}} \sum_{r=1}^{2} \left[b_\mathbf{p}^{(r)} u_\mathbf{p}^{(r)} e^{-iE_\mathbf{p}t - i\mathbf{p}\cdot\mathbf{x}} + c_\mathbf{p}^{(r)\dagger} v_\mathbf{p}^{(r)} e^{iE_\mathbf{p}t + i\mathbf{p}\cdot\mathbf{x}} \right] \\ &= \int \frac{d^3\mathbf{p}}{(2\pi)^{3/2}\sqrt{2E_\mathbf{p}}} \sum_{r=1}^{2} \left[b_{-\mathbf{p}}^{(r)} u_{-\mathbf{p}}^{(r)} e^{-ip\cdot x} + c_{-\mathbf{p}}^{(r)\dagger} v_{-\mathbf{p}}^{(r)} e^{ip\cdot x} \right] \end{aligned} \tag{22.9}$$

To work out the effects of U_P on the creation and annihilation operators, we need to know how the other factors of ψ, the spinors $u_\mathbf{p}^{(r)}$ and $v_\mathbf{p}^{(r)}$, transform. For spinors describing a particle or antiparticle at rest,

$$\beta u_\mathbf{0}^{(r)} = u_\mathbf{0}^{(r)} \tag{20.15}$$

$$\beta v_\mathbf{0}^{(r)} = -v_\mathbf{0}^{(r)} \tag{20.39}$$

The spinors $u_\mathbf{p}^{(r)}$ and $v_\mathbf{p}^{(r)}$ are related to their rest frame versions by the same Lorentz boost:

$$u_\mathbf{p}^{(r)} = e^{\frac{1}{2}\boldsymbol{\alpha}\cdot\hat{\mathbf{a}}\phi} u_\mathbf{0}^{(r)} \tag{20.20}$$

$$v_\mathbf{p}^{(r)} = e^{\frac{1}{2}\boldsymbol{\alpha}\cdot\hat{\mathbf{a}}\phi} v_\mathbf{0}^{(r)} \tag{20.41}$$

with $\hat{\mathbf{a}} = \mathbf{p}/|\mathbf{p}|$, and $\phi = \sinh^{-1}(|\mathbf{p}|/m)$. By the known anticommutation properties (20.4) of β and α_i, we can write

$$\beta u_\mathbf{p}^{(r)} = \beta e^{\frac{1}{2}\boldsymbol{\alpha}\cdot\hat{\mathbf{a}}\phi} u_0^{(r)} = e^{-\frac{1}{2}\boldsymbol{\alpha}\cdot\hat{\mathbf{a}}\phi} \beta u_0^{(r)} = e^{-\frac{1}{2}\boldsymbol{\alpha}\cdot\hat{\mathbf{a}}\phi} u_0^{(r)} = u_{-\mathbf{p}}^{(r)} \tag{22.10}$$

and by the same argument

$$\beta v_\mathbf{p}^{(r)} = -v_{-\mathbf{p}}^{(r)} \tag{22.11}$$

Rewriting (22.9),

$$\psi(-\mathbf{x},t) = \int \frac{d^3\mathbf{p}}{(2\pi)^{3/2}\sqrt{2E_\mathbf{p}}} \sum_{r=1}^{2} \left[b_{-\mathbf{p}}^{(r)} \beta u_\mathbf{p}^{(r)} e^{-ip\cdot x} - c_{-\mathbf{p}}^{(r)\dagger} \beta v_\mathbf{p}^{(r)} e^{ip\cdot x} \right] \tag{22.12}$$

or, multiplying both sides by β,

$$\beta\psi(-\mathbf{x},t) = \int \frac{d^3\mathbf{p}}{(2\pi)^{3/2}\sqrt{2E_\mathbf{p}}} \sum_{r=1}^{2} \left[b_{-\mathbf{p}}^{(r)} u_\mathbf{p}^{(r)} e^{-ip\cdot x} - c_{-\mathbf{p}}^{(r)\dagger} v_\mathbf{p}^{(r)} e^{ip\cdot x} \right] = U_P^\dagger \psi(\mathbf{x},t) U_P \tag{22.13}$$

Therefore

$$U_P^\dagger \begin{Bmatrix} b_\mathbf{p}^{(r)} \\ c_\mathbf{p}^{(r)\dagger} \end{Bmatrix} U_P = \begin{Bmatrix} b_{-\mathbf{p}}^{(r)} \\ -c_{-\mathbf{p}}^{(r)\dagger} \end{Bmatrix} \tag{22.14}$$

and taking the Hermitian conjugate of both sides,

$$U_P^\dagger \begin{Bmatrix} b_{\mathbf{p}}^{(r)\dagger} \\ c_{\mathbf{p}}^{(r)} \end{Bmatrix} U_P = \begin{Bmatrix} b_{-\mathbf{p}}^{(r)\dagger} \\ -c_{-\mathbf{p}}^{(r)} \end{Bmatrix} \tag{22.15}$$

Let $|\mathbf{p}, r; N\rangle$ be a nucleon state. Then

$$U_P^\dagger |\mathbf{p}, r; N\rangle = U_P^\dagger b_{\mathbf{p}}^{(r)\dagger} |0\rangle = U_P^\dagger b_{\mathbf{p}}^{(r)\dagger} U_P |0\rangle = b_{-\mathbf{p}}^{(r)\dagger} |0\rangle = |{-\mathbf{p}}, r; N\rangle \tag{22.16}$$

On the other hand, for an antinucleon state $|\mathbf{p}, r; \overline{N}\rangle$

$$U_P^\dagger |\mathbf{p}, r; \overline{N}\rangle = U_P^\dagger c_{\mathbf{p}}^{(r)\dagger} |0\rangle = U_P^\dagger c_{\mathbf{p}}^{(r)\dagger} U_P |0\rangle = -c_{-\mathbf{p}}^{(r)\dagger} |0\rangle = -\,|{-\mathbf{p}}, r; \overline{N}\rangle \tag{22.17}$$

Thus the unitary operator that effects parity acting on free particle states (or on in and out states, if we're talking about an interacting theory) will transform a one-nucleon state $|\mathbf{p}, r; N\rangle$ into $|{-\mathbf{p}}, r; N\rangle$ with eigenvalue $(+1)$; and a one-antinucleon state $|\mathbf{p}, r; \overline{N}\rangle$ into $|{-\mathbf{p}}, r; \overline{N}\rangle$ with eigenvalue (-1). That is to say, parity has the same properties in quantum field theory that it has in non-relativistic quantum mechanics: it changes the sign of the momentum, but does not affect the spin. However it gives opposite signs to a nucleon and an antinucleon; a nucleon and an antinucleon have *opposite* parity. Thus, for example, a nucleon and an antinucleon in an s-wave state has parity -1, while a nucleon and nucleon, or an antinucleon and an antinucleon in an s-wave has parity $+1$.

This has important experimental implications if you are dealing with a parity-conserving theory. As an example, let me consider the processes

$$p + \overline{p} \to \begin{cases} \pi^+ + \pi^- \\ \pi^0 + \pi^0 \end{cases} \tag{22.18}$$

at rest. (That is to say, the proton and antiproton are "at rest"—for example, slow antiprotons which we're sending into a block of ordinary matter.) We know from non-relativistic quantum mechanics that such exothermic reactions at small velocities of the incoming particle are dominated by the s-waves. This is also true in relativistic quantum mechanics. If there is no spatial momentum, then there is no spatial angular momentum: if $\mathbf{p}$ vanishes, $\mathbf{r} \times \mathbf{p}$ vanishes. That argument has nothing to do with relativity. At rest the process is dominated by s-waves, and therefore there are two relevant $p\overline{p}$ states:

$$p\overline{p}\colon \begin{cases} \ell = 0 \quad s = 0 \quad J = 0 \quad P = -1 \\ \ell = 0 \quad s = 1 \quad J = 1 \quad P = -1 \end{cases} \tag{22.19}$$

The total angular momentum J is conserved. Both of these $p\overline{p}$ states are parity eigenstates with eigenvalue -1. On the other hand, in the final state $\pi^+ + \pi^-$, the particle and antiparticle are bosons, and therefore they have the same intrinsic parity, whatever that may be. If the final state is $\pi^0 + \pi^0$, they obviously have the same parity. It turns out the pion is a pseudoscalar, with parity (-1). That's irrelevant, because the square in any case will be $+1$. So the parity of the final two pion states is determined by the value of ℓ:

$$2\pi\colon \begin{cases} \ell = 0 \quad J = 0 \quad P = +1 \\ \ell = 1 \quad J = 1 \quad P = -1 \end{cases} \tag{22.20}$$

(Note that the orbital angular momentum ℓ contributes a factor of $(-1)^\ell$ to the parity.) If one were to do this experiment with a polarized target and a polarized beam of antinucleons with $J = 0$, two pions would not be produced, because both angular momentum and parity must be conserved. The $J = 0$ state for the two pions is forbidden by conservation of parity:

$$J = 0\colon N\overline{N} \not\to 2\pi \tag{22.21}$$

The $J = 0$ $p\overline{p}$ state is forbidden from creating two pions, or indeed, any two particles of the same intrinsic parity; typically it goes into *three* pions. Only the $J = 1$ state for $p\overline{p}$ is allowed to make $\pi^+ + \pi^-$, or $\pi^0 + \pi^0$. This example demonstrates that what we have derived is not merely some formal convention, but something that actually carries experimental consequences.

22.2 The Majorana representation

The choice of the right coordinate system often simplifies a particular problem. So too with representations of the gamma matrices. Our discussion of charge conjugation will be facilitated by choosing to work in a representation in which all the gamma matrices are imaginary. Let me review what we found in the theory of a charged (complex) scalar field. There our starting point was the Klein–Gordon equation. The Klein–Gordon operator is *real.* Therefore if ϕ is a solution, so too is ϕ^*:

$$(\Box^2 + m^2)\phi = 0 \Longleftrightarrow (\Box^2 + m^2)\phi^* = 0 \tag{22.22}$$

We saw in (6.27) that there is a close connection between *charge* conjugation and *complex* conjugation: the complex conjugate of a complex field has the opposite charge. For a complex field ϕ, we can identify these two operations:

$$C\colon \begin{Bmatrix} \phi(x) \\ \phi^*(x) \end{Bmatrix} = \begin{Bmatrix} \phi^*(x) \\ \phi(x) \end{Bmatrix} \tag{22.23}$$

or in terms of creation and annihilation operators, from (6.78),

$$C\colon \begin{Bmatrix} b_{\mathbf{p}} \\ c_{\mathbf{p}} \end{Bmatrix} = \begin{Bmatrix} c_{\mathbf{p}} \\ b_{\mathbf{p}} \end{Bmatrix} \tag{22.24}$$

(and similarly for $b^\dagger_{\mathbf{p}}$ and $c^\dagger_{\mathbf{p}}$).

Now we have to deal with the Dirac equation. Is there a similar connection between complex conjugation and charge conjugation here? Let's look at complex conjugation.

$$(i\partial\!\!\!/ - m)\psi = 0 \overset{?}{\Longleftrightarrow} (i\partial\!\!\!/ - m)\psi^* = 0 \tag{22.25}$$

I write ψ^* rather than $\psi^\dagger$, meaning I will take the complex conjugate of each of the four components of the Dirac field, but I will not turn a column vector into a row vector. Likewise when I discuss the quantum theory I will use an asterisk (*) to mean the operator adjoint of each of the four operators. If you like, you can think of ψ^* as $(\psi^\dagger)^T$. I'm sorry for that notation, but I have no other symbol to use for just obtaining the adjoint of operators without turning column vectors into row vectors.

Is the charge conjugated Dirac equation true? It depends on the representation of the gamma matrices. If we can find a representation of the gamma matrices in which they are all imaginary,

$$\gamma_\mu = -\gamma^*_\mu \tag{22.26}$$

then the answer is "yes", because then the Dirac equation would be real, just like the Klein–Gordon equation. Given the symmetry in this representation—ψ^* is a solution if ψ is—we would be able to find a similar symmetry in any other representation just by making the right transformation. Do such representations (22.26) exist? They do. Their utility was first pointed out by Ettore Majorana[3] and they are called **Majorana representations**.

I will demonstrate the existence of a Majorana representation by constructing a set of four purely imaginary 4×4 matrices that obey the Dirac algebra. The trick is to write down our original standard representation (20.11) of β and α^i and shuffle them around (perhaps putting i's in certain places) so that the gamma matrices

$$\gamma^\mu = (\gamma^0, \gamma^i) = (\beta, \beta\alpha^i) \tag{20.74}$$

are all imaginary and everything obeys the right algebra. By Pauli's theorem (§20.3), we can find a similarity transformation T to swap α_2, the only imaginary matrix in the Dirac representation of $\{\beta, \alpha^i\}$, with β. This exchange makes all the gammas imaginary and preserves the algebra. Of course, this set of imaginary gammas is just one of an infinite number.[4] Such a similarity transformation is given by the unitary matrix

$$T = \frac{1}{\sqrt{2}}\begin{pmatrix} \boldsymbol{\sigma}_x & i\boldsymbol{\sigma}_z \\ -i\boldsymbol{\sigma}_z & \boldsymbol{\sigma}_x \end{pmatrix} = T^\dagger \tag{22.27}$$

With this transformation, we have

$$\begin{aligned} \alpha_2^M &= T^\dagger\alpha_2 T = \begin{pmatrix} \mathbf{1} & \mathbf{0} \\ \mathbf{0} & -\mathbf{1} \end{pmatrix} = \beta & \beta^M &= T^\dagger\beta T = \begin{pmatrix} \mathbf{0} & \boldsymbol{\sigma}_y \\ \boldsymbol{\sigma}_y & \mathbf{0} \end{pmatrix} = \alpha_2 \\ \alpha_1^M &= T^\dagger\alpha_1 T = \begin{pmatrix} \mathbf{0} & \boldsymbol{\sigma}_x \\ \boldsymbol{\sigma}_x & \mathbf{0} \end{pmatrix} = \alpha_1 & \alpha_3^M &= T^\dagger\alpha_3 T = \begin{pmatrix} \mathbf{0} & -\boldsymbol{\sigma}_z \\ -\boldsymbol{\sigma}_z & \mathbf{0} \end{pmatrix} = -\alpha_3 \end{aligned} \tag{22.28}$$

With this set $\{\beta^M, \alpha_i^M\}$ of matrices, the definition (20.74) leads to this Majorana representation:

$$\gamma^0 = \begin{pmatrix} \mathbf{0} & \boldsymbol{\sigma}_y \\ \boldsymbol{\sigma}_y & \mathbf{0} \end{pmatrix} \quad \gamma^1 = -i\begin{pmatrix} \boldsymbol{\sigma}_z & \mathbf{0} \\ \mathbf{0} & \boldsymbol{\sigma}_z \end{pmatrix} \quad \gamma^2 = \begin{pmatrix} \mathbf{0} & -\boldsymbol{\sigma}_y \\ \boldsymbol{\sigma}_y & \mathbf{0} \end{pmatrix} \quad \gamma^3 = -i\begin{pmatrix} \boldsymbol{\sigma}_x & \mathbf{0} \\ \mathbf{0} & \boldsymbol{\sigma}_x \end{pmatrix} \tag{22.29}$$

and for completeness,

$$\gamma_5 = i\gamma^0\gamma^1\gamma^2\gamma^3 = \begin{pmatrix} -\boldsymbol{\sigma}_y & \mathbf{0} \\ \mathbf{0} & \boldsymbol{\sigma}_y \end{pmatrix} \tag{22.30}$$

As you can check, the Majorana gammas satisfy the Dirac algebra:

$$\{\gamma^\mu, \gamma^\nu\} = 2g^{\mu\nu}\mathbb{1} \tag{20.83}$$

---

[3] [Eds.] Ettore Majorana (pronounced "Mah-yore-AHN-a") (1906–?), a brilliant Sicilian student of Fermi's, first postulated the existence of the neutron, but Fermi could not convince him to write the paper. In 1938 he boarded a ship from Naples to Palermo and was never seen again. The Erice summer school in Sicily where Coleman gave so many celebrated courses is named in Majorana's honor. João Magueijo has written a biography of Majorana, *A Brilliant Darkness*, Basic Books, 2009.

[4] [Eds.] In the Hill–Ting–Chen notes, Coleman chose a different Majorana representation,

$$\gamma^0 = \begin{pmatrix} \mathbf{0} & \boldsymbol{\sigma}_y \\ \boldsymbol{\sigma}_y & \mathbf{0} \end{pmatrix} \quad \gamma^1 = i\begin{pmatrix} \mathbf{1} & \mathbf{0} \\ \mathbf{0} & -\mathbf{1} \end{pmatrix} \quad \gamma^2 = i\begin{pmatrix} \mathbf{0} & \boldsymbol{\sigma}_x \\ \boldsymbol{\sigma}_x & \mathbf{0} \end{pmatrix} \quad \gamma^3 = i\begin{pmatrix} \mathbf{0} & \boldsymbol{\sigma}_z \\ \boldsymbol{\sigma}_z & \mathbf{0} \end{pmatrix}$$

They have the right squares, they anticommute with each other, and they are manifestly imaginary. Therefore there *is* a charge conjugation invariance, at least on a classical level (providing we treat the components of ψ as Grassmann variables).

Before I turn to charge conjugation, let me write down some general conclusions that follow from the choice of a Majorana representation. These will be useful not only for charge conjugation, but also in our discussion of time reversal, which looks simpler in a Majorana representation than in any other. I should emphasize that results derived in this particular representation will hold for all representation-invariant objects such as $\overline{\psi}\psi$. But the Majorana representation is advantageous even when we look at the properties of ψ and $\overline{\psi}$ individually. In this representation, the Lorentz transformations have the nice property that the matrices $D(\Lambda)$ (18.1) are *real*:

$$D(\Lambda) = D(\Lambda)^* \tag{22.31}$$

Just to convince you of this, I'll write down explicit expressions for the L's and the M's. Let's start with M_z, which, I remind you, is

$$M_z = \tfrac{1}{2}i\alpha_z = \tfrac{1}{2}i\gamma^0\gamma^3 \tag{20.44}$$

Since γ^0 and γ^3 are imaginary matrices, their product is real, and the i makes things imaginary:

$$M_z = -M_z^* \tag{22.32}$$

which holds, *mutatis mutandis*, for the other components of M. The representation of a boost along the z axis by a hyperbolic angle ϕ is, from (18.45),

$$D(A(\hat{\mathbf{z}}\phi)) = e^{-iM_z\phi} = \exp(\tfrac{1}{2}\gamma^0\gamma^3\phi) \tag{22.33}$$

This is a *real* matrix. As another example, let me take a rotation about the z axis. From (20.46),

$$L_z = i[M_x, M_y] = \tfrac{1}{2}i\gamma^1\gamma^2 \tag{22.34}$$

so again

$$L_z = -L_z^* \tag{22.35}$$

and likewise for the other components. A representation matrix for rotation about the z axis by an angle θ is, from (18.44),

$$D(R(\hat{\mathbf{z}}\theta)) = e^{-iL_z\theta} = \exp(\tfrac{1}{2}\gamma^1\gamma^2\theta) \tag{22.36}$$

which is again a real matrix. Therefore, the Lorentz matrices in the Majorana representation are *real*, as advertised.

22.3 Charge conjugation and Fermi fields

Now let's work out what charge conjugation does to the plane wave solutions of the free Dirac equation, the u's and the v's. The positive frequency solutions satisfy

$$(\not{p} - m)u_{\mathbf{p}}^{(r)} = 0 \tag{20.119a}$$

We are working now in a basis where the gamma matrices are imaginary. If I take the complex conjugate of this equation, the first term changes sign:

$$(-\not{p} - m)u_{\mathbf{p}}^{(r)*} = 0 = (\not{p} + m)u_{\mathbf{p}}^{(r)*} \tag{22.37}$$

Therefore the complex conjugate solution $u_{\mathbf{p}}^{(r)*}$ is a v-type solution (20.119b), and (22.37) invites the tentative identification

$$u_{\mathbf{p}}^{(r)*} \stackrel{?}{=} v_{\mathbf{p}}^{(r)} \tag{22.38}$$

Because the Lorentz transformations are real in this basis, *complex conjugation commutes with Lorentz transformations.* If we can show

$$u_{\mathbf{0}}^{(r)*} = v_{\mathbf{0}}^{(r)} \tag{22.39}$$

we can with a clear conscience remove the question mark in (22.38).

Let's take a particle at rest, and look at $u_{\mathbf{0}}^{(1)}$, which is supposed to be an eigenstate of L_z with eigenvalue $+\frac{1}{2}$. We can't just quote (20.16), because we're using a different set of gammas. We return to (20.15), $u_0 = \beta u_0$, using β^M instead of the standard β, and find two solutions:

$$u_{\mathbf{0}}^{(1)M} = \sqrt{m}\begin{pmatrix} 0 \\ 1 \\ -i \\ 0 \end{pmatrix} \qquad u_{\mathbf{0}}^{(2)M} = \sqrt{m}\begin{pmatrix} 1 \\ 0 \\ 0 \\ i \end{pmatrix} \tag{22.40}$$

(these can also be obtained from $u_{\mathbf{0}}^{(r)M} = Tu_{\mathbf{0}}^{(r)}$). Using the explicit form of (22.34) in the Majorana basis,

$$L_z = \tfrac{1}{2}i\gamma^1\gamma^2 = \tfrac{1}{2}i\begin{pmatrix} \mathbf{0} & \boldsymbol{\sigma}_x \\ -\boldsymbol{\sigma}_x & \mathbf{0} \end{pmatrix} \tag{22.41}$$

you can quickly check the eigenvalues of $u_{\mathbf{0}}^{(1)M}$ and $u_{\mathbf{0}}^{(2)M}$:

$$L_z u_{\mathbf{0}}^{(1)M} = \tfrac{1}{2}u_{\mathbf{0}}^{(1)M} \qquad L_z u_{\mathbf{0}}^{(2)M} = -\tfrac{1}{2}u_{\mathbf{0}}^{(2)M} \tag{22.42}$$

In exactly the same way we obtain the Majorana versions of the v solutions by working them out from (20.39), $v_0 = -\beta v_0$ and β^M:

$$v_{\mathbf{0}}^{(1)M} = \sqrt{m}\begin{pmatrix} 0 \\ 1 \\ i \\ 0 \end{pmatrix} \qquad v_{\mathbf{0}}^{(2)M} = \sqrt{m}\begin{pmatrix} 1 \\ 0 \\ 0 \\ -i \end{pmatrix} \tag{22.43}$$

(You can also obtain $-iv_{\mathbf{0}}^{(1)M} = Tv_{\mathbf{0}}^{(1)}$, $iv_{\mathbf{0}}^{(2)M} = Tv_{\mathbf{0}}^{(2)}$.) By inspection, we see that indeed

$$u_{\mathbf{0}}^{(r)M*} = v_{\mathbf{0}}^{(r)M} \tag{22.44}$$

which establishes (22.39), and hence also (22.38). The v's are also eigenstates of L_z:

$$L_z v_{\mathbf{0}}^{(1)M} = -\tfrac{1}{2}v_{\mathbf{0}}^{(1)M} \qquad L_z v_{\mathbf{0}}^{(2)M} = \tfrac{1}{2}v_{\mathbf{0}}^{(2)M} \tag{22.45}$$

That is, the eigenvalues of u's and v's take opposite signs for their complex conjugates. There's a simpler way to see this. If we take the complex conjugates of the first of (22.42), we obtain

$$\left(L_z u_{\mathbf{0}}^{(1)M}\right)^* = \tfrac{1}{2}u_{\mathbf{0}}^{(1)M*} = -L_z u_{\mathbf{0}}^{(1)M*} \Rightarrow L_z u_{\mathbf{0}}^{(1)M*} = -\tfrac{1}{2}u_{\mathbf{0}}^{(1)M*} \tag{22.46}$$

because L_z is imaginary. This makes sense physically. The spinor associated with the annihilation of a particle with z component of angular momentum $+\frac{1}{2}$ is $u^{(1)}$, with eigenvalue of $L_z = +\frac{1}{2}$. But it's the $v^{(1)}$ with eigenvalue of $L_z = -\frac{1}{2}$ that's associated with the creation of a particle with $L_z = +\frac{1}{2}$, the sign flip coming because one's got a creation operator and the other's got an annihilation operator.

In the Majorana representation we can rewrite the usual expression (21.6) as

$$\psi(\mathbf{x},t) = \int \frac{d^3\mathbf{p}}{(2\pi)^{3/2}\sqrt{2E_{\mathbf{p}}}} \sum_{r=1}^{2} \left[b_{\mathbf{p}}^{(r)} u_{\mathbf{p}}^{(r)} e^{-ip\cdot x} + c_{\mathbf{p}}^{(r)\dagger} u_{\mathbf{p}}^{(r)*} e^{ip\cdot x} \right] \tag{22.47}$$

replacing $v_{\mathbf{p}}^{(r)}$ with $u_{\mathbf{p}}^{(r)*}$. Likewise we can rewrite ψ^* ((21.7), but without the transpose):

$$\psi^*(\mathbf{x},t) = \int \frac{d^3\mathbf{p}}{(2\pi)^{3/2}\sqrt{2E_{\mathbf{p}}}} \sum_{r=1}^{2} \left[b_{\mathbf{p}}^{(r)\dagger} u_{\mathbf{p}}^{(r)*} e^{ip\cdot x} + c_{\mathbf{p}}^{(r)} u_{\mathbf{p}}^{(r)} e^{-ip\cdot x} \right] \tag{22.48}$$

Notice the similarity between the two expressions. If I define a unitary charge conjugation operator U_C such that

$$\boxed{C\colon \psi \to \psi_C = U_C^\dagger \psi(\mathbf{x},t) U_C = \psi^*(\mathbf{x},t)} \tag{22.49}$$

then

$$U_C^\dagger \psi(\mathbf{x},t) U_C = \int \frac{d^3\mathbf{p}}{(2\pi)^{3/2}\sqrt{2E_{\mathbf{p}}}} \sum_{r=1}^{2} \left[\left(U_C^\dagger b_{\mathbf{p}}^{(r)} U_C \right) u_{\mathbf{p}}^{(r)} e^{-ip\cdot x} + \left(U_C^\dagger c_{\mathbf{p}}^{(r)\dagger} U_C \right) u_{\mathbf{p}}^{(r)*} e^{ip\cdot x} \right] \tag{22.50}$$

Requiring that (22.48) be the same as (22.50), and comparing terms, I instantly deduce that

$$U_C^\dagger \left\{ \begin{matrix} b_{\mathbf{p}}^{(r)} \\ c_{\mathbf{p}}^{(r)} \end{matrix} \right\} U_C = \left\{ \begin{matrix} c_{\mathbf{p}}^{(r)} \\ b_{\mathbf{p}}^{(r)} \end{matrix} \right\} \tag{22.51}$$

By taking the adjoint of both sides, we see that the rules (22.51) apply equally to the creation operators,

$$U_C^\dagger \left\{ \begin{matrix} b_{\mathbf{p}}^{(r)\dagger} \\ c_{\mathbf{p}}^{(r)\dagger} \end{matrix} \right\} U_C = \left\{ \begin{matrix} c_{\mathbf{p}}^{(r)\dagger} \\ b_{\mathbf{p}}^{(r)\dagger} \end{matrix} \right\} \tag{22.52}$$

These rules applied to a many-particle state define a unitary operator, if we also require the reasonable condition that the vacuum is invariant under its action:

$$U_C\,|0\rangle = |0\rangle \tag{22.53}$$

You might have been worried about complex conjugation turning a positive frequency solution into a negative frequency solution. You may have thought "Uh oh, we're going to get something that exchanges annihilation and creation operators." It doesn't happen that way: annihilation operators stay annihilation operators, and creation operators stay creation operators. The operator C, as you can easily demonstrate, commutes with the free field Hamiltonian H, which is written as an integral of the sum of normal ordered products of annihilation and creation operators, $b_{\mathbf{p}}^{(r)\dagger} b_{\mathbf{p}}^{(r)}$ and $c_{\mathbf{p}}^{(r)\dagger} c_{\mathbf{p}}^{(r)}$; C merely exchanges the b's and c's, so a particle state and an antiparticle state have the same energy. Thanks to the way we've set up the correspondence, complex conjugation does not mix spin up and spin down states. It's exactly as if the spin up electron were a boson whose antiparticle was a spin up positron.

What does charge conjugation look like if we're *not* using a Majorana representation? So as not to jettison completely the approach that most books take to charge conjugation, I'll show you what this looks like in a general basis. Let ψ_S be a Dirac spinor in some other basis. Then there is a transformation S such that

$$\psi_S = S\psi_M \;\Leftrightarrow\; \psi_M = S^{-1}\psi_S \tag{22.54}$$

Then

$$C\colon \psi_M \to \psi_M^* = (S^{-1}\psi_S)^* = (S^{-1})^*\psi_S^* \tag{22.55}$$

Writing $\psi_M = S^{-1}\psi_S$, and multiplying both sides by S gives

$$C\colon \psi_S \to S(S^{-1})^*\psi_S^* \equiv \mathscr{C}\psi_S^* \tag{22.56}$$

The matrix $\mathscr{C} = S(S^{-1})^*$ appears explicitly if we're not working in a Majorana representation. This is all we'll have to say about $\mathscr{C}$, because the Majorana representation calculations we'll do are vastly simpler.

So much for the free field. To discuss an interacting theory we have to consider the charge conjugation properties of the various combinations of fields that describe the interactions, to see whether or not they commute with charge conjugation. All of the interactions we will deal with can be written in terms of the sixteen fundamental quadratic forms (chart, p. 420) built out of pairs of some spinor field and some Dirac adjoint field: $\overline{\psi}_A M\psi_B$, where ψ_A and ψ_B are two (perhaps different) Dirac fields and M is some 4×4 matrix, either 1 or γ_5 or one of the four γ^μ, etc. I will assume that ψ_A and ψ_B have the charge conjugation properties (22.49)

$$C\colon \begin{Bmatrix}\psi_A(x)\\ \psi_B(x)\end{Bmatrix} \to \begin{Bmatrix}\psi_A^*(x)\\ \psi_B^*(x)\end{Bmatrix} \tag{22.57}$$

It follows that

$$U_C\psi_A^*(x)U_C^\dagger = \psi_A(x), \qquad U_C^\dagger\psi_A^\dagger(x)U_C = \psi_A^T(x) \tag{22.58}$$

(where the superscript T denotes the transpose). Of course when I'm dealing with more than one field, everything I said about parity in this context (see the discussion following (22.7)) also applies to charge conjugation. There are cases in which theories do not look charge conjugation invariant if you give every field the same phase factor under charge conjugation, but by putting in an extra minus sign in front of one field or another, you can save things. I won't bother to show that here. I'll work out the charged conjugation properties of the quadratic forms assuming everything has this charge conjugation property. It's not hard to figure out what happens if I put a minus sign in one of these transformations.

I want to study charge conjugation in a quantum field theory, so I will assume the object $\overline{\psi}_A M\psi_B$ occurs in the interaction picture Hamiltonian, in the normal-ordered form $:\overline{\psi}_A M\psi_B:$ so I don't have to worry about delta functions appearing. Equivalently I can say that this is an object built out of those funny anticommuting objects that appear in the classical theory of Fermi fields. Products of Grassmann variables have the same combinatorial structure as normal-ordered Dirac fields: when I switch the two fields I get a minus sign. That's going to be important later on. Then

$$U_C^\dagger :\overline{\psi}_A M\psi_B: U_C = U_C^\dagger :\psi_A^\dagger\gamma^0 M\psi_B: U_C = :\psi_A^T\gamma^0 M\psi_B^*: \tag{22.59}$$

This object is not in a particularly nice form to express as $\overline{\psi}_A(\text{stuff})\psi_B$, but it *is* in a nice form to write as $\overline{\psi}_B(\text{stuff})\psi_A$, if we realize that every one-by-one matrix is equal to its transpose.

Therefore I can write

$$:\psi_A^T\gamma^0 M\psi_B^*: = (:\psi_A^T\gamma^0 M\psi_B^*:)^T = -:(\psi_B^*)^T M^T(\gamma^0)^T\psi_A: \tag{22.60}$$

When rearranging these things, because of the definition of the transpose of a product, I have to move ψ_B to the left of ψ_A. That gives me a minus sign. At the moment M is any 4×4 matrix, and its Lorentz transformation properties are irrelevant. On the other hand, γ^0 is a Hermitian imaginary matrix in the Majorana basis, so

$$(\gamma^0)^T = -\gamma^0 \tag{22.61}$$

Then

$$-:(\psi_B^*)^T M^T(\gamma^0)^T\psi_A: = :\psi_B^\dagger M^T\gamma^0\psi_A: = :\psi_B^\dagger\gamma^0\gamma^0 M^T\gamma^0\psi_A: = :\overline{\psi}_B\,\overline{M}^*\psi_A: \tag{22.62}$$

The M term is starred because (20.66) $\gamma^0 M^\dagger\gamma^0 = \overline{M}$; complex conjugation changes $M^\dagger$ to M^T. I "star" $M^\dagger$ to undo the complex conjugate of the adjoint. Thus I have the general "bar–star rule":

$$\boxed{C\colon :\overline{\psi}_A M\psi_B: \to :\overline{\psi}_B\,\overline{M}^*\psi_A:} \tag{22.63}$$

That has the structure you would expect for charge conjugation: it takes an operator that annihilates a B and creates an A, and turns it into an operator that annihilates an anti-A and creates an anti-B. Thus we can work out the 16 quadratic forms, just by using this rule.

Let's make a table of the bilinear forms and their behavior under charge conjugation:

Product		*Under C*	*Under P*
$\overline{\psi}\psi$	$\to$	$\overline{\psi}\psi$	scalar
$\overline{\psi}i\gamma_5\psi$	$\to$	$\overline{\psi}i\gamma_5\psi$	pseudoscalar
$\overline{\psi}\gamma^\mu\psi$	$\to$	$-\overline{\psi}\gamma^\mu\psi$	vector
$\overline{\psi}\gamma^\mu\gamma_5\psi$	$\to$	$\overline{\psi}\gamma^\mu\gamma_5\psi$	axial vector
$\overline{\psi}\sigma^{\mu\nu}\psi$	$\to$	$-\overline{\psi}\sigma^{\mu\nu}\psi$	tensor

First consider the scalar and pseudoscalar bilinears. In $\overline{\psi}\psi$, the matrix M is 1 which is both self-bar and real. Therefore $\overline{\psi}\psi$ goes into $\overline{\psi}\psi$. In other words, $\overline{\psi}\psi$ is even under charge conjugation, as you would expect, because the free Hamiltonian's mass term is $m\overline{\psi}\psi$, and that's certainly invariant under C. For the pseudoscalar, $M = i\gamma_5$. The matrix γ_5 is the product of four gamma matrices times i (20.102), so it is imaginary. It is also anti-self-bar, because $\overline{\gamma}_5 = \gamma^0\gamma_5^\dagger\gamma^0 = \gamma^0\gamma_5\gamma^0 = -\gamma_5$. Thus $i\gamma_5$ is real and self-bar, so $\overline{\psi}i\gamma_5\psi$ is also even under C.

For the vector bilinear, $M = \gamma^\mu$ is self-bar, but it is imaginary. So $\overline{\psi}\gamma^\mu\psi$ goes into $-\overline{\psi}\gamma^\mu\psi$, and the vector bilinear is *odd* under C. This should be no surprise. When ψ describes an electron, the bilinear $\overline{\psi}\gamma^\mu\psi$ is the electromagnetic current. The charge changes sign under charge conjugation if anything does, and so must the current. For the axial vector, things are different. The matrix $\gamma^\mu\gamma_5$ is self-bar, but γ^μ is imaginary, and so is γ_5: $\gamma^\mu\gamma_5$ is real. Then neither starring nor barring do anything to these matrices, and $\overline{\psi}\gamma^\mu\gamma_5\psi$ is the axial vector current, which is even under charge conjugation. Finally, the tensor bilinear product $\sigma^{\mu\nu}$ is $\frac{1}{2}i$ times the commutator of two gamma matrices. Therefore it is both self-bar and imaginary, and so it is odd under charge conjugation. The *derivations*, I remind you, are specific to the Majorana basis, because of course the properties of $\overline{M}^*$ depend on what basis you are in. But the *results* are basis-independent. Should you forget what's even and what's odd, and you want to rederive it, I recommend that you work in the Majorana basis.

Thus the two model theories we're looking at, with scalar and pseudoscalar interactions, $\overline{\psi}\psi\phi$ and $\overline{\psi}i\gamma_5\psi\phi$, respectively, are charge conjugation invariant, providing we define the scalar field to transform appropriately under C. On the other hand, we have a different sort of interaction that arises in classical electromagnetism: the interaction $J^\mu A_\mu$, where J^μ is the electric current and A_μ is the 4-vector potential. In quantum electrodynamics (which we will later discuss in detail) the coupling takes the form

$$\overline{\psi}\gamma^\mu\psi A_\mu \tag{22.64}$$

where A_μ is a vector field. This is charge conjugation invariant *only* if A_μ goes into $-A_\mu$:

$$C\colon A_\mu(x) \to -A_\mu(x) \tag{22.65}$$

The electromagnetic field is a real quantum field, and so its quanta (photons) are neutral particles. If you wish to define charge conjugation in such a way that electromagnetism is invariant under C, then the photon has to be odd under C; a one-photon state is multiplied by -1.[5]

This has interesting consequences for the properties of states, particularly those built up out of one particle and one antiparticle. (States built up out of two particles are of course turned into states built out of two antiparticles by charge conjugation, and who cares what the relative phase is—it's a completely different process. In any case, neither a two-particle state nor a two-antiparticle state can be a charge conjugation eigenstate.) Suppose I have such a particle/antiparticle state, let me call it $|\psi\rangle$:

$$|\psi\rangle = \int d^3\mathbf{p}\, d^3\mathbf{p}' \sum_{r,s} f_{rs}(\mathbf{p},\mathbf{p}')\, b_{\mathbf{p}}^{(r)\dagger} c_{\mathbf{p}'}^{(s)\dagger}\,|0\rangle \tag{22.66}$$

The function $f_{rs}(\mathbf{p},\mathbf{p}')$ is some smearing-out function to make ψ a nice, normalizable two-particle state. When I apply charge conjugation to ψ, I don't change anything in this expression except that $b \leftrightarrow c$:

$$U_C\,|\psi\rangle = \int d^3\mathbf{p}\, d^3\mathbf{p}' \sum_{r,s} f_{rs}(\mathbf{p},\mathbf{p}')\, c_{\mathbf{p}}^{(r)\dagger} b_{\mathbf{p}'}^{(s)\dagger}\,|0\rangle \tag{22.67}$$

These operators are in the wrong order for comparing the final state with the initial state, and I have to switch them around, which gives me the famous Fermi minus sign:

$$U_C\,|\psi\rangle = -\int d^3\mathbf{p}\, d^3\mathbf{p}' \sum_{r,s} f_{rs}(\mathbf{p},\mathbf{p}')\, b_{\mathbf{p}'}^{(s)\dagger} c_{\mathbf{p}}^{(r)\dagger}\,|0\rangle \tag{22.68}$$

Now r and s are just summation variables, $\mathbf{p}$ and $\mathbf{p}'$ are integration variables, so I can exchange r with s and $\mathbf{p}$ with $\mathbf{p}'$. Thus

$$U_C\,|\psi\rangle = -\int d^3\mathbf{p}\, d^3\mathbf{p}' \sum_{r,s} f_{sr}(\mathbf{p}',\mathbf{p})\, b_{\mathbf{p}}^{(r)\dagger} c_{\mathbf{p}'}^{(s)\dagger}\,|0\rangle = \pm\,|\psi\rangle \tag{22.69}$$

---

[5] [Eds.] This is perfectly reasonable: classically, the vector potential $\mathbf{A}$ is proportional to the source charge, so of course it changes sign under C.

as one would have guessed, the sign depending upon whether the smearing function $f_{rs}(\mathbf{p}', \mathbf{p})$ is symmetric or antisymmetric in its arguments. However, somewhat surprisingly,

$$\begin{aligned} f_{sr}(\mathbf{p}', \mathbf{p}) = -f_{rs}(\mathbf{p}, \mathbf{p}') &\iff U_C\,|\psi\rangle = +\,|\psi\rangle \\ f_{sr}(\mathbf{p}', \mathbf{p}) = f_{rs}(\mathbf{p}, \mathbf{p}') &\iff U_C\,|\psi\rangle = -\,|\psi\rangle \end{aligned} \tag{22.70}$$

it is the *anti*-symmetric smearing function that goes with the *even* state under charge conjugation, the eigenstate with eigenvalue +1, and the *symmetric* smearing function that creates the odd state. This is just a result of having to reorder the two creation operators.

EXAMPLE. *The decay of positronium*

Positronium is a bound state of e^+ and e^- in an s-wave, in the ground state; it's like the hydrogen atom, with a positron in place of the proton. As in our earlier discussion on nucleon–antinucleon annihilation (p. 462), there are two s-wave states available depending on the two spin states: $J = 1$, called *ortho-positronium*, and $J = 0$, called *para-positronium*; strangely, "para" means they're *anti*-parallel. If you have an electron captured by an proton, it quickly cascades down to a normal hydrogen atom in an s-wave state. Unlike the ground state of the hydrogen atom, the s-wave state of positronium is not stable, because it can decay into photons; the electron and positron can annihilate each other.

$$e^+ + e^- \to \gamma + \gamma \tag{22.71}$$

The $J = 1$ s-state is a totally symmetric wave function, and it is symmetric in spin and symmetric in space, and as we have just seen, *odd* under charge conjugation: $C = -1$.

$$e^+ e^-,\ J = 1\colon\ C = -1 \tag{22.72}$$

The other s-state, with $J = 0$, is symmetric in space but antisymmetric in spin and therefore it is *even* under charge conjugation:

$$e^+ e^-,\ J = 0\colon\ C = +1 \tag{22.73}$$

Photons have $C = -1$, and so the two photons have a net $C = +1$. Therefore the decay into two photons is *allowed* for the $J = 0$ state, but *forbidden* for the $J = 1$ state. The electromagnetic coupling is not strong. As you probably know, you get an extra factor in the amplitude of around 1/137 (typically more like $1/2\pi$ times 1/137) whenever you emit another photon; the probability is the square of the amplitude. Even without considerations of charge conjugation, the decay into two photons is much more probable than the decay into more than two photons. But because of C invariance, the $J = 1$ state must go into *three* photons, a final state of $C = -1$, but much more slowly than the $J = 0$ state going into two photons. So although both ground states of positronium are unstable, the $J = 0$ state is considerably less stable than the $J = 1$ state.[6]

The commutation relations of C and P are peculiar, even for a free Dirac theory. If, for example, I take a one-particle state $|\psi\rangle$, and apply first P and then C, I get minus the result

[6] [Eds.] The rates are:

$$\Gamma(\text{p-Ps} \to \gamma\gamma) = 7989.50(2)\,(\mu s)^{-1}$$
$$\Gamma(\text{o-Ps} \to \gamma\gamma\gamma) = 7.0382\,(\mu s)^{-1}$$

See A. Czarnecki and S. Karshenboim, "Decays of Positronium", *14th International Workshop on High Energy Physics and Quantum Field Theory*, 1999, Moscow; `https://arxiv.org/pdf/hep-ph/9911410.pdf`.

of applying first C and then P:

$$\begin{aligned} U_C U_P \|\psi\rangle &= U_C U_P \, b_{\mathbf{p}}^{(r)\dagger} U_P^\dagger U_C^\dagger \|0\rangle = U_C \, b_{-\mathbf{p}}^{(r)\dagger} U_C^\dagger \|0\rangle = c_{-\mathbf{p}}^{(r)\dagger} \|0\rangle \\ U_P U_C \|\psi\rangle &= U_P U_C \, b_{\mathbf{p}}^{(r)\dagger} U_C^\dagger U_P^\dagger \|0\rangle = U_P \, c_{\mathbf{p}}^{(r)\dagger} U_P^\dagger \|0\rangle = -c_{-\mathbf{p}}^{(r)\dagger} \|0\rangle \end{aligned} \tag{22.74}$$

(we assume the vacuum is parity and charge conjugation invariant). In the top line parity acts first on the nucleon creation operator, and reverses the direction of $\mathbf{p}$ (22.15); then the charge conjugation turns it into an antinucleon (22.51). In the second line charge conjugation first turns the nucleon into an antinucleon, and then parity reverses $\mathbf{p}$ *and introduces a minus sign* (22.15). The same thing happens with one-antinucleon states. In general, if I act on a state with an odd number of fermions in it, PC is $-CP$; if I act on a state with an even number (including zero), PC is $+CP$. This can be summed up by saying

$$U_C U_P = U_P U_C \times \begin{pmatrix} \text{a unitary operator associated with} \\ \text{rotation by } 360^\circ \text{ about any axis} \end{pmatrix} \tag{22.75}$$

since the rotation by 360° about any axis multiplies every individual fermion wave function by -1, and therefore is $+1$ acting on states with an even number of fermions, and -1 acting on a state with an odd number of fermions. That's a perfectly legitimate symmetry operator. Any rotation, including one by 360°, is a legitimate symmetry of the theory, so it's not surprising that it should turn up in the product of P and C. In other words,

$$U_C U_P = U_P U_C \, U(R(2\pi\hat{\mathbf{n}})) = U_P U_C \, e^{-2\pi i \hat{\mathbf{n}} \cdot \mathbf{L}} = U_P U_C \, (-1)^{N_f} \tag{22.76}$$

where N_f is the number of fermions in the state that CP is acting on.

22.4 *PT* invariance and Fermi fields

I will next discuss PT, which is always easier in a relativistic theory than T by itself, since the product of parity and time reversal commutes with Lorentz transformations. I will continue using a Majorana basis. This is convenient because there is a connection between C and PT via the CPT theorem, and what's sauce for C is sauce for PT.[7]

Again I will begin by looking at the scalar case. I remind you that the Klein–Gordon equation is invariant[8] under the combined actions of parity and time reversal:

$$(\Box^2 + \mu^2)\phi(x) = 0 \iff (\Box^2 + \mu^2)\phi(-x) = 0 \tag{22.77}$$

We want to ask if there's a similar symmetry in free Dirac theory,

$$(i\partial\!\!\!/ - m)\psi(x) = 0 \overset{?}{\iff} (i\partial\!\!\!/ - m)M\psi(-x) = 0 \tag{22.78}$$

I've put in a matrix M here which I'll try to figure out later. The answer is "yes", if we can cancel out the sign reversal caused by changing x to $-x$ with the matrix M, such that $\gamma^\mu M$ is $-M\gamma^\mu$. Of course there *is* such a matrix: it's γ_5 or some scalar multiple of γ_5. And to make

---

[7] [Eds.] For those readers whose first language is not English, this refers to an old English proverb, "What's sauce for the goose is sauce for the gander" (a gander is a male goose): What applies to one case applies to the other.

[8] [Eds.] See note 11, p. 130.

the CPT theorem come out right in the end (and not become the $iCPT$ theorem) I will define the matrix most people choose conventionally to effect PT:

$$PT\colon \psi(x) \to i\gamma_5\psi(-x) \tag{22.79}$$

For later purposes, I note that we're going to realize this symmetry with an anti-unitary operator. We suspect that there is an anti-unitary operator, Ω_{PT}, such that

$$\boxed{PT\colon \psi(x) \to \Omega_{PT}^{-1}\psi(x)\Omega_{PT} = i\gamma_5\psi(-x)} \tag{22.80}$$

Let me work out some of the properties of this hypothetical anti-unitary operator before I actually show you that it exists. Perhaps its most interesting property is its square:

$$\Omega_{PT}^{-2}\psi(x)\Omega_{PT}^{2} = \Omega_{PT}^{-1}\left(i\gamma_5\psi(-x)\right)\Omega_{PT} = i\gamma_5\Omega_{PT}^{-1}\psi(-x)\Omega_{PT} = (i\gamma_5)^2\psi(x) = -\psi(x) \tag{22.81}$$

Remember, in the Majorana representation $i\gamma^5$ is *real*; it slips neatly through the Ω_{PT}. Thus $(\Omega_{PT})^2$ is *not* equal to $\mathbb{1}$ though it must be unitary, since it is the square of an anti-unitary operator. It turns a Fermi field into minus itself. Of course, we know such an operator:

$$\Omega_{PT}^2 = U(R(2\pi\hat{\mathbf{n}})) \tag{22.82}$$

it's the old friend introduced last section, a rotation about any axis by 2π. You might think I have pulled a swindle, because that minus sign is only there because I chose M to be $i\gamma_5$ rather than γ_5. Might we have obtained $(\Omega_{PT})^2 = \mathbb{1}$ with no minus sign, by choosing γ_5? No, and I will demonstrate that.

Suppose I consider an alternative definition of PT,

$$\Omega'_{PT} = e^{i\theta}\Omega_{PT} \tag{22.83}$$

where $e^{i\theta}$ is an arbitrary phase factor. I can certainly do this for the free theory; that's an internal symmetry. Apply it twice:

$$\begin{aligned}\Omega'^{-2}_{PT}\psi(x)\Omega'^{2}_{PT} &= \Omega'^{-1}_{PT}\left(e^{i\theta}i\gamma_5\psi(-x)\right)\Omega'_{PT} = e^{-i\theta}i\gamma_5\Omega'^{-1}_{PT}\psi(-x)\Omega'_{PT}\\ &= e^{-i\theta}e^{i\theta}(i\gamma_5)^2\psi(x) = -\psi(x)\end{aligned} \tag{22.84}$$

Now when I apply Ω_{PT} a second time, I can slip the $i\gamma_5$ out through the external Ω_{PT} with no problem, because it's real. But when I bring the phase factor through an anti-unitary operator, it gets complex conjugated, and I get exactly the same thing as before. Thus Ω_{PT} has a square of -1 and there's no fighting it by putting in a phase factor or anything like that. You will still have a square of an operator that produces this U operator, the rotation by 2π.

At this point in our discussions of C and P, I worked through the transformations of all the bilinear forms. I won't do that with PT, because a homework problem[9] asks you to do some of that. But I will work out what happens to $\overline{\psi}\psi$, so the homework problem won't be too difficult.

It follows from the definition of an anti-unitary operator (6.110), that if $\Omega^{-1}A\Omega$ is A', where A is some ordinary linear operator, then $\Omega^{-1}A^\dagger\Omega$ is $A'^\dagger$. Let's apply this general rule to PT (22.80):

$$\Omega_{PT}^{-1}\psi^\dagger(x)\Omega_{PT} = \psi^\dagger(-x)(i\gamma_5)^\dagger = -i\psi^\dagger(-x)\gamma_5 \tag{22.85}$$

---

[9] [Eds.] Problem 14.1, p. 545.

Now let's look at $\overline{\psi}$:

$$\Omega_{PT}^{-1}\overline{\psi}(x)\Omega_{PT} = \Omega_{PT}^{-1}\psi^\dagger(x)\gamma^0\Omega_{PT} = (\Omega_{PT}^{-1}\psi^\dagger(x)\Omega_{PT})(-\gamma^0) \tag{22.86}$$

The γ^0 is just sitting there, so we can drag it through the Ω_{PT}. It's a numerical matrix, but it's *imaginary*, because we're working in a Majorana representation, so we get an additional minus sign. Therefore

$$\Omega_{PT}^{-1}\overline{\psi}(x)\Omega_{PT} = (i\psi^\dagger(-x)\gamma_5)\gamma^0 = -i\psi^\dagger(-x)\gamma^0\gamma_5 = -i\overline{\psi}(-x)\gamma_5 \tag{22.87}$$

This means (combining (22.80) and (22.87))

$$PT\colon \overline{\psi}\psi(x) \to \Omega_{PT}^{-1}\overline{\psi}(x)\psi(x)\Omega_{PT} = \overline{\psi}(-x)(-i\gamma_5)(i\gamma_5)\psi(-x) = \overline{\psi}\psi(-x) \tag{22.88}$$

This seems reasonable. After all, the term $\overline{\psi}\psi$ occurs in the free Lagrangian multiplying the mass, and you would expect it to have nice PT transformation properties. How about the kinetic term in the free Lagrangian?

$$\begin{aligned} PT\colon i\overline{\psi}(x)\gamma^\mu\partial_\mu\psi(x) &\to \Omega_{PT}^{-1}\left(i\overline{\psi}(x)\gamma^\mu\partial_\mu\psi(x)\right)\Omega_{PT} \\ &= (-i)(\overline{\psi}(-x))(-i\gamma_5)(-\gamma^\mu)(-\partial_\mu)(i\gamma_5)\psi(-x) \\ &= -i\overline{\psi}(-x)\gamma_5\gamma^\mu\partial_\mu\gamma_5\psi(-x) \\ &= i\overline{\psi}(-x)\gamma^\mu\partial_\mu\psi(-x) \end{aligned} \tag{22.89}$$

In comparison with $\overline{\psi}\psi$, I get three extra minus signs: one from the i, one from the imaginary γ^μ, and one from changing the sign of ∂_μ. There is a further factor of $-i\gamma_5$ from the transformation of $\overline{\psi}$, and a factor of $i\gamma_5$ from transforming ψ. Finally, moving the leftmost γ_5 through the γ^μ gives a fourth minus sign. So the end result is simply

$$PT\colon i\overline{\psi}\gamma^\mu\partial_\mu\psi(x) \to i\overline{\psi}(-x)\gamma^\mu\partial_\mu\psi(-x) \tag{22.90}$$

As expected, both terms in the free Lagrangian have the same transformation properties; otherwise we could hardly expect the free Dirac theory to be PT invariant. Please notice the critical role of the i. Because Ω_{PT} is an anti-unitary operator, the transformation properties of i times an operator are opposite to those of the operator without the i. Bringing the i through the Ω_{PT} has the nontrivial effect of introducing an extra minus sign. Thus if one is writing down interaction Hamiltonians and wishes to check that they conserve PT, or P or C, typically the restrictions implied by P or C are that certain coupling constants vanish, or are equal to others. The restriction implied by time reversal or PT invariance is usually that certain coupling constants are complex conjugates of other coupling constants.[10] Let me convince you briefly that I can construct an anti-unitary operator that does this job. I will rewrite (22.80) and put the $i\gamma_5$ over onto the other side, and for convenience I will replace x by $-x$:

$$-i\gamma_5\Omega_{PT}^{-1}\psi(-x)\Omega_{PT} = \psi(x) \tag{22.91}$$

I'll check this equation by writing down the expression for a free field and seeing if Ω_{PT} does sensible things to creation and annihilation operators. From (21.6)

$$\psi(x) = \int \frac{d^3\mathbf{p}}{(2\pi)^{3/2}\sqrt{2E_{\mathbf{p}}}}\sum_{r=1}^{2}\left[b_{\mathbf{p}}^{(r)}u_{\mathbf{p}}^{(r)}e^{-ip\cdot x} + \cdots\right] \tag{22.92}$$

[10] [Eds.] Again, see Problem 14.1, p. 545.

The c terms will turn out to follow the b terms with no alteration. Now let's transform the right-hand side of the expression using (22.91). I replace x by $-x$, which sends $-ip\cdot x$ to $ip\cdot x$. Then I apply the anti-unitary transformation, which turns $ip\cdot x$ back again to $-ip\cdot x$. And finally I multiply by $-i\gamma_5$. We obtain

$$\psi(x)=\int\frac{d^3\mathbf{p}}{(2\pi)^{3/2}\sqrt{2E_\mathbf{p}}}\sum_{r=1}^{2}\left[-i\gamma_5 u_\mathbf{p}^{(r)*}\left(\Omega_{PT}^{-1}b_\mathbf{p}^{(r)}\Omega_{PT}\right)e^{-ip\cdot x}+\cdots\right] \tag{22.93}$$

Note that $u_\mathbf{p}^{(r)}$ is complex conjugated, because Ω_{PT} is an anti-unitary operator. We now have the quantity, $\Omega_{PT}^{-1}b_\mathbf{p}^{(r)}\Omega_{PT}$, we wish to compute.

Let's check that this is sensible. It's only going to be sensible if the objects $-i\gamma_5 u^{(r)*}$ are also u's. Well, they *are*, because

$$(\not{p}-m)(-i\gamma_5 u_\mathbf{p}^{(r)*})=0 \tag{22.94}$$

We found, in our discussion of charge conjugation, that complex conjugating turns a u into a v (22.38)—that is, it changes the sign of the $\not{p}$ term. But dragging γ_5 through the Dirac operator changes its sign *again*, and the two minus signs cancel. So $-i\gamma_5 u_\mathbf{p}^{(r)*}$ is, to within some phase factor, some other u. *What* other u it is depends on how you've chosen your bases and the phases of the u's, which I don't want to go through in detail.[11] It implies some reshuffling of the b's, which defines the anti-unitary operator on the one-particle states, and thus, *a fortiori* on the many-particle states. I won't bother working it out in detail, except to make one remark. I will show by example that PT reverses the direction of spin. We've already seen ((22.42) and (22.46)) that the rest state $u_\mathbf{0}^{(1)}$ and its complex conjugates have opposite spins:

$$L_z u_\mathbf{0}^{(1)}=\tfrac{1}{2}u_\mathbf{0}^{(1)}\qquad L_z u_\mathbf{0}^{(1)*}=-\tfrac{1}{2}u_\mathbf{0}^{(1)*} \tag{22.95}$$

This follows because L_z as you'll recall is purely imaginary in the Majorana basis (22.35). L_z, a product of two gamma matrices, also commutes with γ_5, and therefore

$$L_z(-i\gamma_5 u_\mathbf{0}^{(1)*})=-\tfrac{1}{2}(-i\gamma_5 u_\mathbf{0}^{(1)*}) \tag{22.96}$$

If I start out with a state with spin up at rest, PT transforms it into a state with spin down at rest. There is a sign change. This is just what we would expect. It's reasonable that the combined result of parity and time reversal not change the momentum of a particle. Consider

---

[11] [Eds.] For completeness, here are the details for the Majorana representation (22.29):

$$-i\gamma_5=\begin{pmatrix} i\boldsymbol{\sigma}_y & \mathbf{0}\\ \mathbf{0} & -i\boldsymbol{\sigma}_y\end{pmatrix}$$

An easy calculation with the spinors (22.40) shows $-i\gamma_5 u_\mathbf{0}^{(1)*}=u_\mathbf{0}^{(2)}$, $-i\gamma_5 u_\mathbf{0}^{(2)*}=-u_\mathbf{0}^{(1)}$, and the same holds for the v's. A boost along the z direction involves α_3 (20.44) which commutes with $i\gamma_5$, so we can also say $-i\gamma_5 u_\mathbf{p}^{(1)*}=u_\mathbf{p}^{(2)}$, etc. If (22.93) is to equal (22.92), we require

$$\Omega_{PT}^{-1}\begin{Bmatrix} b_\mathbf{p}^{(1)}\\ b_\mathbf{p}^{(2)}\end{Bmatrix}\Omega_{PT}=\begin{Bmatrix} b_\mathbf{p}^{(2)}\\ -b_\mathbf{p}^{(1)}\end{Bmatrix}$$

The same relations hold for the $c_\mathbf{p}^{(r)}$'s and the respective adjoints, the $b_\mathbf{p}^{(r)\dagger}$'s and the $c_\mathbf{p}^{(r)\dagger}$'s. Once again, PT does nothing to the momentum but it does reverse the spin.

a particle moving in some direction. I change the direction of time, and the velocity is reversed. I make a parity transformation which changes $\mathbf{x}$ to $-\mathbf{x}$, and the particle is back to its original velocity. What about the spin? In non-relativistic theory we know that parity does not affect the angular momentum: $\boldsymbol{\sigma}$ and $\mathbf{L}$ both commute with parity. On the other hand, time reversal of course changes the sign of the spin. If it's rotating one way, and I run the motion picture backwards, it's whirling around the other way. And therefore the combined operation PT should change the sign of the spin, and we found that it *does*.

22.5 The *CPT* theorem and Fermi fields

I would now like to discuss the proof of the CPT theorem when spin-½ particles are involved. It's not going to be too difficult because we already did the hard part when we discussed this theorem for scalar particles (§11.3). Earlier we showed the CPT theorem is trivial, at least order by order in perturbation theory. (It's not at all trivial to prove it rigorously.) In perturbation theory, the CPT theorem is just the statement that if we reverse the sign of all the momenta in any Feynman graph, so that every incoming particle becomes an outgoing antiparticle, then the Feynman graph is unchanged. And that result was the CPT theorem:

$$\mathcal{A}(p_1, p_2, \cdots, p_n) = \mathcal{A}(-p_1, -p_2, \cdots, -p_n) \tag{22.97}$$

I would like to do the same thing here, but more cleverly than the last time, in a way that will also indicate how this proof can be generalized to particles of arbitrary spin. For simplicity, I will consider a graph involving a theory of nucleons and mesons, perhaps in our scalar or pseudoscalar theory, or maybe in some grotesque theory that doesn't obey parity, charge conjugation, time reversal, PC, CT, or TP. It could be some messy, horrible theory full of $\epsilon_{\mu\nu\rho\sigma}$'s and derivative couplings and God knows what. For simplicity I will restrict myself to a graph which has only one fermion line going through it:

$$N + \phi_1 + \phi_2 + \cdots \to N' + \phi_1' + \phi_2' + \ldots \tag{22.98}$$

(We'll see at the end what happens if there are 17 fermion lines.) I have this one fermion line,

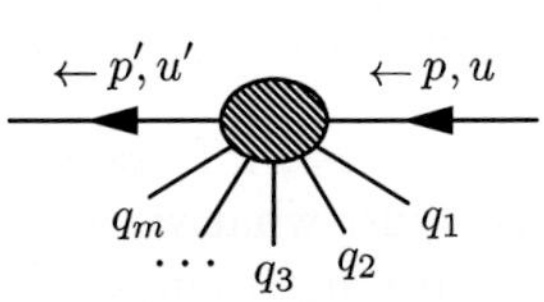

Figure 22.1: $N + \phi_1 + \phi_2 + \cdots \to N' + \phi_1' + \phi_2' + \ldots$

from which a bunch of meson lines come out. These mesons may have grotesquely complicated self interactions, but I don't care. There is some incoming spinor u, and some outgoing spinor u'. The amplitude for this graph is going to be a mess, but I can certainly write it in the following form:

$$\mathcal{A} = \overline{u}'\, M(p, p', q_1, q_2, q_3, \cdots)\, u \tag{22.99}$$

where

$$M = \int d^4k_1\, d^4k_2 \cdots d^4k_n \frac{N(p, p', q_1, q_2, q_3, \cdots, k_1, k_2, \cdots)}{D(p, p', q_1, q_2, \cdots, k_1, k_2, \cdots)} \tag{22.100}$$

The numerator N will be a function of p, q_1, $q_2, \ldots, p'$, full of gamma matrices. And then there will be a nasty Feynman denominator D full of $i\epsilon$'s and inner products of momenta—a function of practically everything in the whole graph, including what's in the shaded blob. Then this whole thing is going to be integrated over all the internal momenta. That's the general form of a Feynman amplitude.

No matter how grotesque the theory is, whatever the Feynman rules may be, we know that a Feynman amplitude (10.27) has got to be Lorentz invariant. Therefore this Feynman amplitude has to be equal to the same thing with everything Lorentz transformed (all the internal momenta, all the external momenta and all the spins):

$$\mathcal{A}_\Lambda = \overline{u}' D^{-1}(\Lambda)\, M(\Lambda p, \Lambda p', \Lambda q_1, \Lambda q_2, \Lambda q_3, \cdots)\, D(\Lambda) u \tag{22.101}$$

I should really write $\overline{D}(\Lambda)$ in place of $D^{-1}(\Lambda)$ here, but for reasons that will become clear I'll write the inverse, which is equivalent to $\overline{D}(\Lambda) = \gamma^0 D(\Lambda)^\dagger \gamma^0$. I should mention that in M, both the numerator and the denominator are Lorentz transformed, but only a purist bothers with the denominator, since it is expressed in terms of invariant inner products.

Now comes the cunning part. This expression can be analytically continued to a Lorentz transformation with *complex* parameters θ and ϕ. "Analytic continuation?" you say, "Ugh! That's always a big job; you have to show that you don't encounter cuts, poles, essential singularities..." None of that matters here! The denominator is manifestly analytically continuable, since it's a function of inner products that need not be Lorentz transformed (though you may put the Λ's in if you wish). The numerator has two separate types of factors, the $D(\Lambda)$'s and polynomials in the p's and q's. There's never any problem in analytically continuing a polynomial. $D(\Lambda)$ is an exponential function of gamma matrices times the parameters ϕ and θ, and exponential functions are analytic. That's why I wrote $D^{-1}(\Lambda)$ rather than $\overline{D}(\Lambda)$. The complex conjugate of an analytic function is not an analytic function, but its inverse *is*. As the $D(\Lambda)$ and $D^{-1}(\Lambda)$ are analytic, there is no problem with poles. Therefore $D(\Lambda)$ can be analytically continued to

$$D(\Lambda) = D(R(\hat{\mathbf{n}}\theta)) \cdot D(A(\hat{\mathbf{a}}\phi)) \quad \{\theta, \phi\} \in \mathbb{C} \tag{22.102}$$

This equation is true for real ϕ's and θ's, and it's evidently true by the principle of analytic continuation for complex ϕ's and θ's.

What do we get for our analytic continuation? Complex ϕ's and θ's give us something totally unphysical: complex momentum, disgusting. That may be useful in some other context, but it's not obvious that it's of much help here. There is however a *particular* complex Lorentz transformation that is extremely useful for proving the CPT theorem. And incidentally, by doing the proof this way, you will see how to generalize it to arbitrary spins.

Let's consider the complex Lorentz transformation

$$\Lambda = R(\hat{\mathbf{z}}\pi) \cdot A(i\hat{\mathbf{z}}\pi) \tag{22.103}$$

That's a product of a boost and a rotation (see the discussion following (18.40)). The rotation through the real angle 180° about the z axis is certainly reasonable. That changes the sign of x and y, but does nothing to z and t (note 9 on p. 374):

$$R(\hat{\mathbf{z}}\pi)\colon \begin{cases} p^0 \to p^0 \\ p^1 \to p^1 \cos\pi - p^2 \sin\pi = -p^1 \\ p^2 \to p^2 \cos\pi + p^1 \sin\pi = -p^2 \\ p^3 \to p^3 \end{cases} \tag{22.104}$$

In order to make the signs come out right, I'll set the parameter ϕ equal to $i\pi$, which makes a 180° "rotation" in the zt plane. Just to show you what this particular boost does, let's work out what the usual z axis boost does with a real rapidity ϕ (18.38):

$$A(\hat{\mathbf{z}}\phi)\colon \begin{cases} p^0 \to p^0\cosh\phi - p^3\sinh\phi \\ p^1 \to p^1 \\ p^2 \to p^2 \\ p^3 \to p^3\cosh\phi - p^0\sinh\phi \end{cases} \tag{22.105}$$

Now if I take ϕ to be $i\pi$,

$$\cosh(i\pi) = \cos\pi = -1 \qquad \sinh(i\pi) = i\sin\pi = 0 \tag{22.106}$$

and I get

$$A(i\hat{\mathbf{z}}\pi)\colon \begin{cases} p^0 \to -p^0 \\ p^1 \to p^1 \\ p^2 \to p^2 \\ p^3 \to -p^3 \end{cases} \tag{22.107}$$

Thus

$$\Lambda = R(\hat{\mathbf{z}}\pi)\cdot A(i\hat{\mathbf{z}}\pi) = -\mathbb{1} \tag{22.108}$$

So this Λ is in fact $-\mathbb{1}$. That's not a physical Lorentz transformation, but we can obtain it by analytic continuation from a real Lorentz transformation.

I have now done exactly what I did in the scalar theory: I have changed the sign of each and every momentum with this particular Lorentz transformation. At the same time, I'm also doing something to the spinors, because they are being transformed by the $D(\Lambda)$'s. Of course, I have to change the spinors, the u for an incoming nucleon must become a v for an outgoing antinucleon. How does $D(\Lambda)$ do that?

Earlier I wrote down the form of $D(\Lambda)$ for a boost along the z axis (22.33) and for a rotation about the z axis (22.36), in terms of gamma matrices. For the rotation I have

$$D(R(\hat{\mathbf{z}}\pi)) = \exp(\tfrac{1}{2}\gamma^1\gamma^2\pi) \tag{22.109}$$

For the acceleration, I have

$$D(A(i\hat{\mathbf{z}}\pi)) = \exp(i\tfrac{1}{2}\gamma^0\gamma^3\pi) \tag{22.110}$$

and for the Lorentz matrix,

$$D(\Lambda) = D(R(\hat{\mathbf{z}}\pi))\cdot D(A(i\hat{\mathbf{z}}\pi)) \tag{22.111}$$

Now let's work out what $D(\Lambda)$ is. Note that $\gamma^1\gamma^2$ is a matrix whose square is -1. Therefore

$$\begin{aligned} D(R(\hat{\mathbf{z}}\pi)) = \exp\left(\tfrac{1}{2}\gamma^1\gamma^2\pi\right) &= \mathbb{1}\left[1 - \tfrac{1}{2!}(\tfrac{1}{2}\pi)^2\ldots\right] + \gamma^1\gamma^2\left[(\tfrac{1}{2}\pi) - \tfrac{1}{3!}(\tfrac{1}{2}\pi)^3\ldots\right] \\ &= \gamma^1\gamma^2 \end{aligned} \tag{22.112}$$

because the bracketed even and odd terms equal $\cos\frac{1}{2}\pi$ and $\sin\frac{1}{2}\pi$, numbers known to their friends as 0 and 1, respectively. Likewise $i\gamma^0\gamma^3$ is a matrix whose square is -1 so

$$D(A(i\hat{\mathbf{z}}\pi)) = \exp\left(i\tfrac{1}{2}\gamma^0\gamma^3\pi\right) = i\gamma^0\gamma^3 \tag{22.113}$$

With a little rearrangement

$$D(\Lambda) = D(R(\hat{\mathbf{z}}\pi)) \cdot D(A(i\hat{\mathbf{z}}\pi)) = i\gamma^0\gamma^1\gamma^2\gamma^3 = \gamma_5 \tag{22.114}$$

The spinor transformation $D(\Lambda)$ for this particular Lorentz matrix $\Lambda = -\mathbb{1}$ is nothing but our old friend γ_5, whose inverse, by the way, is also γ_5. Therefore what we have shown is that our original amplitude (22.99) (which by Lorentz invariance is exactly the same as (22.101)) can be written as

$$\mathcal{A}_\Lambda = \overline{u}'\gamma_5\, M(-p, -p', -q_1, -q_2, -q_3, \cdots)\, \gamma_5 u \tag{22.115}$$

Notice that because γ_5 anticommutes with $\not{p}$, $\gamma_5 u$ can be interpreted as the Dirac wave function v associated with an outgoing antinucleon: if u is a positive frequency solution of the Dirac equation,

$$(\not{p} - m)u = 0 \tag{22.116}$$

then $\gamma_5 u$ is a negative frequency solution:

$$\gamma_5(\not{p} - m)u = 0 = -(\not{p} + m)\gamma_5 u \tag{22.117}$$

and the negative frequency solutions are v's. Likewise $\overline{u}$ was formerly associated with an outgoing nucleon, but $\overline{u}\gamma_5$ is just right to be associated with the incoming antinucleon; it acts like $\overline{v}$:

$$\overline{u}(\not{p} - m) = 0 \Rightarrow \overline{u}\gamma_5(\not{p} + m) = 0 \tag{22.118}$$

However, things are not quite right. The amplitude (22.115) is the Lorentz transform of the original amplitude, and therefore equal to it, but it is not yet the amplitude for the *CPT* reversed process, because of the Fermi minus sign. Remember, we've done nucleon scattering (example, p. 447) and antinucleon scattering (Problem 12.3). Antinucleon scattering is the *CPT* transformed version of nucleon scattering. And we found nucleon scattering and antinucleon scattering differed by a minus sign. That is, to make the minus sign come out right, we have to have

$$\mathcal{A}_{CPT} = (-1) \times \mathcal{A}_\Lambda \tag{22.119}$$

But there is an easy fix for this. To patch up the missing minus sign, we simply require

$$CPT\colon \begin{cases} u_{\mathbf{p}}^{(r)} \to i\gamma_5 u_{\mathbf{p}}^{(r)} \\ \overline{u}_{\mathbf{p}}^{(r)} \to \overline{u}_{\mathbf{p}}^{(r)} i\gamma_5 \end{cases} \tag{22.120}$$

Then, finally,

$$\mathcal{A}_{CPT} = -\mathcal{A}_\Lambda = \overline{u}' i\gamma_5\, M(-p, -p', -q_1, -q_2, -q_3, \cdots)\, i\gamma_5 u \tag{22.121}$$

The u for an incoming nucleon is replaced by $i\gamma_5 u$ which is a v, the appropriate object for an outgoing antinucleon, and, since $i\gamma_5$ is self-bar, likewise $\overline{u}$ for an outgoing nucleon is replaced by $\overline{u}i\gamma_5$, which is the same as $\overline{v}$, the appropriate object for an incoming antinucleon (see box, p. 443).

This whole argument generalizes if 72 nucleon lines go through the blob, or 35 antinucleon lines, or a line with a hairpin turn. Relative to the original process, you get a minus sign in the amplitude for the *CPT* reversed process for each external fermion line. In *any* Lorentz invariant theory of Dirac fields and scalar fields, you get exactly the same amplitude for a given process and the *CPT* reversed process if you follow this prescription:

How to obtain $\mathcal{A}_{CPT}$ from $\mathcal{A}$

- Every incoming particle becomes an antiparticle of the same momentum
- Every u for an incoming nucleon is replaced by $i\gamma_5 u$
- Every v for an outgoing antinucleon is replaced by $i\gamma_5 v$
- Every $\overline{u}$ for an outgoing nucleon is replaced by $\overline{u}i\gamma_5$
- Every $\overline{v}$ for an incoming antinucleon is replaced by $\overline{v}i\gamma_5$

The reason it's $i\gamma_5$ everywhere is because there is only *one* complex Lorentz transformation that effects all of these changes, so it's always the same matrix. The secret of the CPT theorem is analytic continuation to complex Lorentz transformations.

Now, if you happen to have a field theory of spin-3/2 particles on hand, you can see how to derive the CPT theorem for them. Just compute $D(\Lambda)$ for this complex Lorentz transformation, and insert an i because they're fermions. To prove the theorem for a spin 2 particle, on the other hand, we don't insert an i into $D(\Lambda)$, because they're bosons. That's all there is to it. You tell me how a particle transforms under the Lorentz group, and I will tell you the right matrix that goes into *CPT*. So, it's much simpler than either C or *PT*, and it's universal. If someone comes to you and says, "I have written a Lagrangian that is invariant under parity with the conventional phase, charge conjugation with the conventional phase, but in time reversal I have to insert an unconventional phase", you know that either this person has made an error, or else the theory as written is not Lorentz invariant.

Next time, I will sketch out how to do renormalization for a spinor theory.

23

Renormalization of spin-$\frac{1}{2}$ theories

This lecture is devoted to the subject for which you have no doubt all been waiting, the renormalization program for a theory involving spinor fields.

23.1 Lessons from Model 3

In Model 3, we discussed scalar nucleons. Here I will also work with a specific example, our meson-nucleon theory with γ_5 interactions, referred to in the antique literature as the "ps-ps theory", pseudoscalar mesons with pseudoscalar couplings:

$$\mathscr{L}' = \overline{\psi}(i\partial\!\!\!/ - m_0)\psi + \tfrac{1}{2}(\partial_\mu\phi)^2 - \tfrac{1}{2}\mu_0\phi^2 + g_0\overline{\psi}i\gamma_5\psi\phi \tag{23.1}$$

Since we're about to do renormalization, I warn you that this m_0 is not equal to the observed mass of the one nucleon state, μ_0 is not the observed meson mass and the coupling constant g_0 may not be the coupling constant as defined by some hypothetical experiment, say by looking for the t-channel pole in nucleon–nucleon scattering. Not only are these not the physically defined masses and coupling constants, but ψ and ϕ are not the most convenient fields for scattering theory. In the case of the meson field, we saw in Model 3 (13.52) that we had to introduce a wave function renormalization counterterm to get the right S-matrix elements. That goes through in exactly the same way as before (§14.2), because none of our general formulas depended on the presence of the nucleon field in the theory. All those expressions we found for the meson propagator and so on were true whether or not there were other fields in the theory. So we look at the vacuum expectation value of the meson field, which is of course independent of position by translational invariance.

$$\langle 0|\phi(x)|0\rangle = \langle 0|e^{ip\cdot x}\phi(0)e^{-ip\cdot x}|0\rangle = \langle 0|\phi(0)|0\rangle = 0 \quad \text{by parity invariance} \tag{23.2}$$

(Since the vacuum is parity even, the pseudoscalar meson field has vanishing vacuum expectation value, so we don't have to introduce a shift, as we did in Model 3 (13.52). With ϕ a scalar field rather than a pseudoscalar, with an interaction $g\overline{\psi}\psi\phi$, the field ϕ's vacuum expectation value would *not* necessarily vanish. If it did not, we would have to shift the field. This complication cannot arise in ps-ps theory, and so we avoid the complication.)

The next thing we did (13.51) was to look at the matrix element $\langle 0|\phi(0)|p\rangle$ (where $|p\rangle$ describes a one meson state). We defined

$$\langle 0|\phi(0)|p\rangle = \sqrt{Z_3} \tag{23.3}$$

By Lorentz invariance $\sqrt{Z_3}$ cannot depend on p, when the states are relativistically normalized (1.57) (as we did before and will do here). We introduced a field $\phi'(x)$:

$$\phi' = \frac{1}{\sqrt{Z_3}}\phi \tag{23.4}$$

that had "good" matrix elements—the same amplitude as a free field for making a one particle state out of the vacuum or annihilating a one particle state. By convention we defined the phase of the one particle states such that that Z_3 is a positive real number. This is the right field to use for the mesons in the reduction formula, and the one that gives us the correct S-matrix elements.

We want to do exactly the same thing for the nucleon field. The first step is to study the amplitude for annihilating a nucleon state or creating an antinucleon state. As before we need only study one matrix element, say the annihilation amplitude

$$\langle 0|\psi(x)|r,p\rangle \tag{23.5}$$

and again I will assume the states are relativistically normalized. The associated matrix element with the nucleon state on the left is connected to this one by complex conjugation. The corresponding expressions involving $\overline{\psi}$ are connected to these by charge conjugation invariance, which is preserved by this theory. If you're working with a theory that does not respect change conjugation invariance, the operators are connected by CPT. So in general this matrix element is the only one I have to study; the others can be found by symmetries. Because of Lorentz invariance, I can construct the state $|r,p\rangle$ by boosting a state at rest,

$$|r,p\rangle = U(A)\,|r,p_{(0)}\rangle \tag{23.6}$$

Here $p_{(0)}$ is *not* the time component of p^μ; it is a four-vector that corresponds to a particle with momentum 0:

$$p_{(0)} = (m,\mathbf{0}) \tag{23.7}$$

where m is the actual physical mass of the nucleon.

Let me first write down the *assumed* properties of the states. I assume that the spectrum of states will resemble that in the free theory. There may be bound states or something like that, but there is a stable physical meson which has the properties of the meson state in the free theory as far as parity and Lorentz transformation properties are concerned. It's a pseudoscalar: odd under parity. I assume there is a physical nucleon. Its mass is not equal to m_0, but it is a spin-½ particle carrying charge one and transforming under parity as the free nucleon does. So the states of the nucleon can all be obtained by applying appropriate Lorentz boosts to a nucleon at rest. The states at rest are two in number. By convention I'll choose $|1,p_{(0)}\rangle$ to represent a particle with spin up,

$$J_z\,|1,p_{(0)}\rangle = +\tfrac{1}{2}\,|1,p_{(0)}\rangle \tag{23.8}$$

$\mathbf{J}$ is the operator in Hilbert space that corresponds to total angular momentum. This state has the parity transformation property

$$U_P\,|1,p_{(0)}\rangle = +1\,|1,p_{(0)}\rangle \tag{23.9}$$

Parity does nothing to spin, and since the particle's at rest, it does nothing to the momentum. I will denote by $|2,p_{(0)}\rangle$ the state obtained by applying the lowering operator in the usual way for a spin-½ particle:

$$|2,p_{(0)}\rangle = (J_x - iJ_y)\,|1,p_{(0)}\rangle \tag{23.10}$$

That defines its phase and everything else. Thus by applying symmetry operators, I can generate everything from a single particle state at rest spinning up. That is, I need only study the matrix elements of the nucleon field at position zero in the spin up state and at rest:

$$\langle 0|\psi(0)|1,p_{(0)}\rangle \tag{23.11}$$

Once I've studied that, by Lorentz transformations and rotations I know the general matrix elements. The state $|1,p_{(0)}\rangle$ is the "descendent", in some sense, of the free nucleon. For the free field theory and the nucleon at rest we have (21.6)

$$\langle 0|\psi(0)_{\text{free}}|1,p_{(0)}\rangle = u_{\mathbf{0}}^{(1)} \tag{23.12}$$

Let's compute these matrix elements (23.11). Will they look like (23.12)? The field ψ is a four component object, so in principle there could be four matrix elements here, one for each of the four components of ψ. However we have two conditions which we can use to define the state: (23.8) tells us it's spin-½ and (23.9) says it's parity plus. Consider a rotation acting on $\psi(0)$:

$$\langle 0|e^{i\theta J_z}\psi(0)e^{-i\theta J_z}|1,p_{(0)}\rangle \tag{23.13}$$

where θ is an arbitrary angle. We can evaluate this expression in two ways. We can first apply the operator $e^{i\theta J_z}$ to the vacuum which is of course rotationally invariant, then apply $e^{-i\theta J_z}$ to the one nucleon state $|1,p_{(0)}\rangle$ and obtain by assumption $e^{-\frac{1}{2}i\theta}$ times the original matrix element:

$$\langle 0|e^{i\theta J_z}\psi(0)e^{-i\theta J_z}|1,p_{(0)}\rangle = e^{-\frac{1}{2}i\theta}\,\langle 0|\psi(0)|1,p_{(0)}\rangle \tag{23.14}$$

On the other hand, from the known transformation properties (19.3) of the field, this object is also equal to

$$e^{-iL_z\theta}\,\langle 0|\psi(0)|1,p_{(0)}\rangle \tag{23.15}$$

where L_z is the 4×4 matrix (20.7) that effects rotations while acting on the four components of ψ. (It's not the z component of the orbital angular momentum $\mathbf{L}$ in non-relativistic quantum mechanics. The distinction between spin and orbital angular momentum is not Lorentz invariant.) In order to match (23.14) with (23.15), the matrix element must be composed only of eigenstates of L_z with eigenvalue $+\frac{1}{2}$. Thus if we work in the standard representation,

$$L_z = \tfrac{1}{2}\begin{pmatrix}\boldsymbol{\sigma}_z & \mathbf{0}\\ \mathbf{0} & \boldsymbol{\sigma}_z\end{pmatrix} \tag{23.16}$$

then the matrix element must have the form

$$\langle 0|\psi(0)|1,p_{(0)}\rangle = \sqrt{2m}\begin{pmatrix}a\\0\\b\\0\end{pmatrix} \tag{23.17}$$

(I use the standard representation of the Dirac matrices because it is most convenient for discussing states at rest.)

At first glance it looks like we have two independent numbers in contrast to the scalar case, where we had only one number to characterize this matrix element (23.3). But of course we have not yet used parity. Let's apply that. Both the state on the right and the state on the left are by assumption eigenstates of parity with eigenvalue +1; the vacuum is certainly parity invariant, and we have said (23.9) the nucleon state has positive parity. So

$$\langle 0|U_P^\dagger \psi(0) U_P|1, p_{(0)}\rangle = \langle 0|\psi(0)|1, p_{(0)}\rangle \tag{23.18}$$

On the other hand, by the known parity transformation properties (20.8) of the nucleon field,

$$\langle 0|U_P^\dagger \psi(0) U_P|1, p_{(0)}\rangle = \langle 0|\beta\psi(0)|1, p_{(0)}\rangle = \beta\,\langle 0|\psi(0)|1, p_{(0)}\rangle \tag{23.19}$$

(Because we're at the origin, the parity change $\mathbf{x} \to -\mathbf{x}$ is irrelevant.) Using the explicit standard representation (20.75) of $\beta = \gamma^0$,

$$\beta\,\langle 0|\psi(0)|1, p_{(0)}\rangle = \beta\sqrt{2m}\begin{pmatrix} a \\ 0 \\ b \\ 0 \end{pmatrix} = \sqrt{2m}\begin{pmatrix} a \\ 0 \\ -b \\ 0 \end{pmatrix} \tag{23.20}$$

Consistency ((23.17), (23.18), (23.19), and (23.20)) requires that

$$b = 0 \tag{23.21}$$

and there is in fact only one unknown number, a, which (following (13.51)) I will define to be

$$a = \sqrt{Z_2} \tag{23.22}$$

By choosing the phase appropriately for the state $|1, p_{(0)}\rangle$, I can arrange that Z_2 is real and positive.

Thus we can arrange that the matrix element of the fully interacting field between the real physical one particle eigenstate and the vacuum is the same as the matrix element of the free field between a one particle state and the vacuum, if we rescale the field:

$$\psi' = \frac{1}{\sqrt{Z_2}}\psi \tag{23.23}$$

Then

$$\langle 0|\psi'(0)|1, p_{(0)}\rangle = u_{\mathbf{0}}^{(1)} \implies \langle 0|\psi'(x)|r, p\rangle = e^{-ip\cdot x}u_{\mathbf{p}}^{(r)} \tag{23.24}$$

simply by applying rotations and Lorentz transformations. That is to say, ψ' has the same matrix elements as the free field.

I will not go through the reduction formula afresh for the Fermi case because it's exactly the same as the derivation in the pure scalar case (§14.2); there are just more indices floating around and Fermi minus signs taking care of the conventions in our time ordered product. Once we had gone through this, we could then write things in terms of the rescaled fields, compute Feynman diagrams, put the external lines on the mass shell and obtain the correct S-matrix elements just as when we only had Bose fields to play with. None of the arguments we went through in our derivation of the LSZ formula is sensitive to spin.

Thus our task is the same as in Model 3 (§14.3): to rewrite the Lagrangian in terms of the physical masses, the physical coupling constants and the renormalized fields, thus generating a bunch of counterterms which we need to determine. In Model 3 (14.47), there were six counterterms (including one for a shift); here there are five:

$$\begin{aligned}\mathscr{L}' &= \tfrac{1}{2}(\partial_\mu\phi')^2 - \tfrac{1}{2}\mu^2\phi'^2 + \overline{\psi}'(i\not\partial - m)\psi' - g\overline{\psi}'i\gamma_5\psi'\phi' \\ &\quad + \tfrac{1}{2}A(\partial_\mu\phi')^2 - \tfrac{1}{2}B\phi'^2 + C\overline{\psi}'i\not\partial\psi' - D\overline{\psi}'\psi' - E\overline{\psi}'i\gamma_5\psi'\phi'\end{aligned} \tag{23.25}$$

We have to determine the counterterms $\{A, B, C, D, E\}$ iteratively in perturbation theory and then we can do calculations. After that, all we've got to do is multiplicatively renormalize ψ. That will produce matrix elements between the vacuum and one particle states identical to those of a free field, as they should be. The rest of the development is largely a repeat of what we did for the scalar nucleons in Model 3.

A digression on theories that do not conserve parity

Up till now I have used universal principles to construct our model theories: Lorentz invariance, translation invariance, rotational invariance (well, that's just a subgroup of Lorentz invariance), CPT to connect the matrix elements of ψ to those of $\overline{\psi}$. We learned last time that CPT was universal. But the one thing I have used that doesn't have the flavor of universality is parity. We might want to study theories in which parity is *not* conserved.

When I first taught this course back in the early Neolithic era, I didn't bother to discuss renormalization in a non-parity conserving theory, because the only such theories concerned weak interactions; it's only the weak interactions that violate parity. But at the time, there were no renormalizable theories of the weak interaction. So why should we bother to write down the prescription? The answers would be infinite and thus uninteresting. Well, times have changed, due to work done in this very building.[1] There are now renormalizable weak interaction theories rather more complicated than this,[2] but still I would like to make a digression about how this part of the analysis changes. It's really very simple and it will just take a few lines.

What if there is no parity invariance, and you can't use parity conservation to eliminate b? How then do we construct a field that has the same matrix elements as the free field? We are rescued by γ_5. In the standard representation

$$\gamma_5 = \begin{pmatrix} \mathbf{0} & \mathbf{1} \\ \mathbf{1} & \mathbf{0} \end{pmatrix} \tag{20.102}$$

As you'll recall (from (22.116) and (22.117)), γ_5 turns positive frequency solutions at rest into negative frequency solutions at rest. It anticommutes with β which determines the difference between these solutions, and it commutes with all the Lorentz generators. Therefore if I apply γ_5 to (23.17), I obtain an equation that has exactly same Lorentz transformation properties as

[1] [Eds.] Sheldon L. Glashow, "The Renormalizability of Vector Meson Interactions", *Nuc. Phys.* **10** (1958) 107–117; and "Partial-Symmetries of Weak Interactions", *Nuc. Phys.* **22** (1960) 579–588; Steven Weinberg, "A Model of Leptons", *Phys. Rev. Lett.* **19** (1967) 1264–66; J. Schwinger, "A Theory of the Fundamental Interactions", *Ann. Phys.* **2** (1957) 407–434. Glashow earned his doctorate under Schwinger. Harvard's Physics Department is housed in the Lyman Laboratory building, 17 Oxford St., Cambridge, MA.

[2] [Eds.] Abdus Salam, in *Elementary Particle Physics*, ed. N. Svartholm, Almqvist and Wiksell, Stockholm 1968; Weinberg, *ibid.*; Glashow, *ibid.* Glashow, Salam and Weinberg shared the 1979 Physics Nobel for their electroweak theory. A little later this theory was indeed shown to be renormalizable by Gerard 't Hooft and Martinus Veltman, whose work was recognized by another Physics Nobel, in 1999.

the original equation. Doing that gives

$$\langle 0|\gamma_5\psi(0)|1, p_{(0)}\rangle = \sqrt{2m}\begin{pmatrix} b \\ 0 \\ a \\ 0 \end{pmatrix} \tag{23.26}$$

We can construct ψ' as a Lorentz covariant linear combination of ψ and $\gamma_5\psi$ to produces the desired matrix element:

$$\psi' = \frac{a\psi - b\gamma_5\psi}{a^2 - b^2} \tag{23.27}$$

The combination $a\psi - b\gamma_5\psi$ knocks out the lower non-zero entry, and to make the other entry equal to 1, I have to divide by $a^2 - b^2$. Then the matrix element with ψ' will have just the $\sqrt{2m}$ in the first entry and zeros everywhere else:

$$\langle 0|\psi'(0)|1, p_{(0)}\rangle = \sqrt{2m}\begin{pmatrix} 1 \\ 0 \\ 0 \\ 0 \end{pmatrix} = u_{\mathbf{0}}^{(1)} \tag{23.28}$$

as we'd like. Of course, when I make that substitution in the Lagrangian, I get all sorts of parity non-conserving terms involving $\partial\!\!\!/\gamma_5$ and things like that, but I should, because I started out with a parity non-conserving Lagrangian. I'll have a lot more counterterms. I won't carry through the parity *non*-conserving case any further than this, but that's the general story. It just means life gets a little bit uglier; the *fewer* symmetries you have, the *more kinds* of counterterms you have to worry about.

Let's return to the main topic. We need to construct five renormalization conditions that will fix the five counterterms. These conditions will express the fact that the physical charge is whatever we decide to define it as, presumably by some nice experiment that's directly connected to physically observable quantities. We also require that the meson field is properly normalized, that the nucleon field is properly normalized, that the physical mass of the meson is μ, that the physical mass of the nucleon is m. Of course we've already gone through the analysis for the meson. Nothing we said about the meson field alone, and in particular about the corrected meson propagator (15.36) in Model 3, is altered by the fact that some of our intermediate states might have spin-½ particles in them. So that gives us two of the five terms we are looking at, from our previous analysis.[3] Let me remind you of what we found for the meson in Model 3.

As you recall (15.29), we began by studying the full Green's function for one meson in and one meson out (the Fourier transform of the time ordered product of two meson fields), which I defined in terms of the renormalized propagator $\widetilde{D}'$:

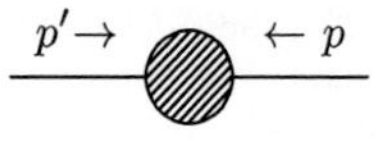

$$= \int d^4x\, d^4y\, e^{ip'\cdot x}e^{-ip\cdot y}\,\langle 0|T(\phi'(x)\phi'(y))|0\rangle$$
$$\equiv (2\pi)^4\delta^{(4)}(p+p')\widetilde{D}'(p^2) \tag{23.29}$$

[3] [Eds]. See (15.46)–(15.48), and (16.5). Note that the terms B and C in (14.47) correspond to the terms A and B in (23.25), respectively.

There is always an energy-momentum conserving delta function, with both p and p' considered positive if they point inward. By Lorentz invariance, what's left has to be a function of p^2 only. I then (15.30) derived a spectral representation for $\widetilde{D}'(p^2)$ by putting in a bunch of intermediate states:

$$\widetilde{D}'(p^2) = \frac{i}{p^2 - \mu^2 + i\epsilon} + \int_{\mu^2+\eta}^{\infty} da^2\, \sigma(a^2) \frac{i}{p^2 - a^2 + i\epsilon} \tag{23.30}$$

and summing over states. I got the first term from the contribution from the one-particle intermediate states. Then there was an unknown continuum which could be thought of as a continuous superposition of free particle propagators. The integral goes from wherever the lowest threshold is, $\mu^2 + \eta$, to infinity. As far as this field knows, the only difference between creating a discrete one particle state and a many particle state is that the masses are smeared out, rather than taking their values at a fixed point. We also found that $\sigma(a^2)$ was greater than or equal to zero. This meant that if we drew the complex p^2 plane, extending p^2 to complex values, $\widetilde{D}'(p^2)$ was an analytic function except for a cut beginning at the lowest threshold, and the pole at μ^2. The first term of (23.30) gives us a pole at $p^2 = \mu^2$, and the

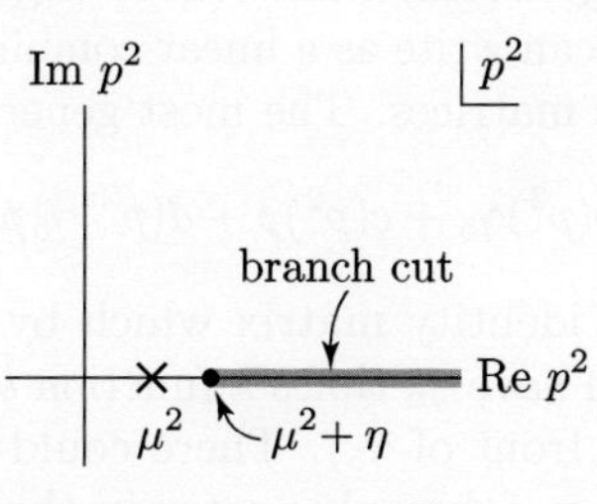

Figure 23.1: The analytic properties of $\widetilde{D}'(p^2)$ in the complex p^2 plane

second a function analytic apart from the cut.

We then folded this in with a purely diagrammatic analysis to look at the one-particle irreducible (1PI) part of $\widetilde{D}'(p^2)$, defined (15.33) to be $-i\widetilde{\Pi}'(p^2)$:

$$\equiv -i\widetilde{\Pi}'(p^2) \tag{23.31}$$

I then derived the equation (15.34)

$$\widetilde{D}'(p^2) \equiv$$

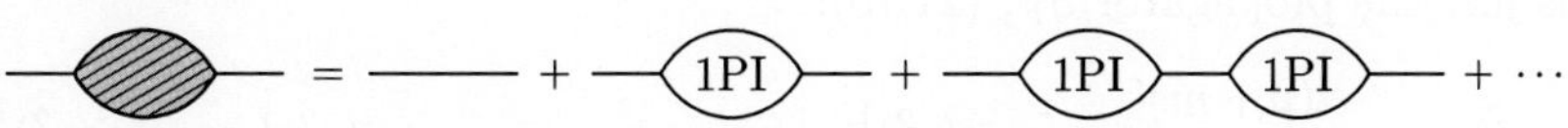

(23.32)

which is expressed analytically as (see (15.38) and (15.40))

$$\begin{aligned}\widetilde{D}'(p^2) &= D + D\left[-i\widetilde{\Pi}'(p^2)\right]D + D\left[-i\widetilde{\Pi}'(p^2)\right]D\left[-i\widetilde{\Pi}'(p^2)\right]D + \cdots \\ &= \frac{i}{p^2 - \mu^2 - \widetilde{\Pi}'(p^2) + i\epsilon}\end{aligned} \tag{15.36}$$

And from this, I deduced two renormalization conditions that fix what I am now calling the A and B counterterms,

$$\widetilde{D}'(p^2) \text{ has a pole at } \mu^2 \Rightarrow \widetilde{\Pi}'(\mu^2) = 0 \tag{23.33}$$

$$\text{the residue at this pole is } i \Rightarrow \left.\frac{d\widetilde{\Pi}'(p^2)}{dp^2}\right|_{\mu^2} = 0 \tag{23.34}$$

This is just a sketch of what we went through before, because I want to parallel each of these steps for the case of the Fermi field.

23.2 The renormalized Dirac propagator $\widetilde{S}'$

There are really no particularly grave complications except those caused by the fact that the Fermi field has four components. The first step is to study the one nucleon Green's function, the sum of all Feynman graphs with a one nucleon field and one antinucleon field. This is the spin-½ analog of the boson propagator $\widetilde{D}'(p^2)$, (23.29).

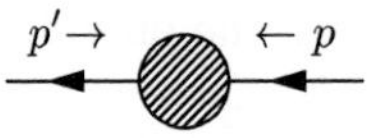

$$= \int d^4x\, d^4y\, e^{ip'\cdot x}e^{-ip\cdot y}\,\langle 0|T(\psi'(x)\overline{\psi'}(y))|0\rangle \\ \equiv (2\pi)^4\delta^{(4)}(p+p')\widetilde{S}'(p) \tag{23.35}$$

(For the time being, we'll write the argument of $\widetilde{S}'$ as p, rather than p^2.) Like every Green's function, $\widetilde{S}'(p)$ will have an energy–momentum conserving delta function in front of it, but now it is a 4×4 matrix, which we can write as a linear combination of a basis of 4×4 matrices, the sixteen combinations of Dirac matrices. The most general expression looks like this:

$$\widetilde{S}'(p) = a(p^2) + b(p^2)\gamma_5 + c(p^2)\not{p} + d(p^2)\gamma_5\not{p} + e(p^2)\sigma^{\mu\nu}p_\mu p_\nu \tag{23.36}$$

We could have a multiple of the identity matrix which by Lorentz invariance could be an arbitrary function $a(p^2)$. We could have γ_5 times a function $b(p^2)$ (I don't know if $b(p^2)$ is real, so I won't bother to put the i in front of γ_5). There could be a term γ^μ which for Lorentz invariance must be multiplied by p_μ, the only vector in the problem, times a function $c(p^2)$. Maybe there's a function $d(p^2)$ times $\not{p}\gamma_5$. Finally, I could have some function to $e(p^2)$ times $\sigma^{\mu\nu}$, but Lorentz invariance requires p_μ and p_ν dotted into it, and that drops out immediately, because $\sigma^{\mu\nu}$ is antisymmetric: $\sigma^{\mu\nu}p_\mu p_\nu = 0$.

Now, if the theory did not conserve parity, I could have all four of the surviving terms. But parity gives us an enormous simplification: the terms with the γ_5's have the wrong parity transformation properties. I don't have to work it out. It's obvious that however they transform under parity, they'll transform *opposite* to the two without the γ_5, $a(p^2)$ and $c(p^2)\not{p}$. These I know are right because they *have* to be there in zeroth order perturbation theory, when $\widetilde{S}'$ is just the propagator $\widetilde{S}_F$, (21.76):

$$\widetilde{S}'\Big|_0 = \widetilde{S}_F = \frac{i(\not{p}+m)}{p^2-m^2+i\epsilon} \Rightarrow c(p^2)\Big|_0 = \frac{i}{p^2-m^2+i\epsilon}; \quad a(p^2)\Big|_0 = mc(p^2)\Big|_0 \tag{23.37}$$

The terms in γ_5 must be wrong! So I'll just set them to zero because of parity. For the parity conserving case, I have just two unknown scalar functions,

$$\widetilde{S}'(p) = a(p^2) + c(p^2)\not{p} \tag{23.38}$$

Given these two functions $a(p^2)$ and $c(p^2)$, let me consider them for the moment as functions of some scalar variable z, a complex number:

$$\widetilde{S}'(z) = a(z^2) + c(z^2)z \tag{23.39}$$

Written in this way, $\widetilde{S}'(z)$ has an even part and an odd part. But if we recall the identity

$$\not{p}^2 = p^2 \tag{23.40}$$

we can write

$$\widetilde{S}'(\not{p}) = a(\not{p}^2) + c(\not{p}^2)\not{p} \tag{23.41}$$

Instead of considering this as two scalar coefficient functions, both of which are unchanged as p goes into $-p$, we can consider it as a single function, with no particular oddness or evenness properties, of the variable $\not{p}$. There's no ambiguity because there's only one matrix in the problem, and one matrix always commutes with itself. You don't have to worry about orderings. This is the place where things become much harder in the parity non-conserving case, because then we would have matrices around that wouldn't commute with each other. In particular, we would have γ_5 in the mix, and it commutes with almost nothing.

23.3 The spectral representation of $\widetilde{S}'$

We've now expressed the fermion Green's function in terms of $\widetilde{S}'(\not{p})$, the matrix function analogous to the scalar function $\widetilde{D}'(p^2)$ (15.30). The wrinkles are that $\widetilde{S}'(\not{p})$ is a function of the matrix variable $\not{p}$ rather than the scalar variable p^2, and is itself a matrix. That takes care of part one. Now part two: deriving the spectral representation of $\widetilde{S}'(\not{p})$.

First I will simply write down the spectral representation. It will look totally mysterious to you. And then I will explain where it came from. You'll be able to go through it in your head—if not now in front of me, when I'm looking at you with my beady eyes, later on. Here is the full expression:

$$\begin{aligned}\widetilde{S}'(\not{p}) = \frac{i}{\not{p} - m + i\epsilon} &+ i\int_{(m+\mu)^2}^{\infty} da^2\, \sigma_+(a^2)\, \frac{\not{p} + a}{p^2 - a^2 + i\epsilon} \\ &+ i\int_{(m+\mu)^2}^{\infty} da^2\, \sigma_-(a^2)\, \frac{\not{p} - a}{p^2 - a^2 + i\epsilon}\end{aligned} \tag{23.42}$$

The integrals go from a lower bound in perturbation theory of $(m + \mu)^2$ to infinity. In comparison with (23.30), the third term may be a bit of a surprise. I'll explain where it comes from.

Remember the derivation (§15.2) of the old spectral representation (15.30). We began with the unordered product, no time ordering symbol, and put in the complete set of intermediate states. From the one particle intermediate states, we got the same result as in the free theory, because as far as one particle states go, you can't tell the renormalized field from the free field. From the higher states we had all sorts of states that could be made by hitting the vacuum with $\phi(0)$. All those states had the same quantum numbers as a single particle. In the frame in which their spatial momentum $\mathbf{p}$ was zero, they were states of zero angular momentum. The only thing was that there was a continuous distribution of them rather than the single isolated point, and therefore we got a continuous smeared-out integral of one particle things. That's the fortune cookie size description of what we did there.

What happens here? We'll have a contribution from the one particle intermediate states when we expand the product of two field operators, a ψ and a $\overline{\psi}$, in terms of one particle intermediate states. That's the first term in (23.42). It'll give us the same result, $\widetilde{S}(\not{p})$, as in the free theory. Now we consider all the continuum states (whose mass begins, in perturbation theory at least, at $(m + \mu)^2$, the meson–nucleon threshold) that we can make by hitting the vacuum with $\overline{\psi}(0)$. We can make two kinds in the rest frame of the states. The two upper components of ψ which are even under parity can make states of spin-½ in their rest frame, and parity plus, or, using the notation J^P for angular momentum J and parity P, states of $\frac{1}{2}^+$. Those states are just like one nucleon states except they're smeared out in mass, and therefore we get a smeared out distribution, with an appropriate smearing function $\sigma_+(a^2)$, of one nucleon propagators, where, in analogy with (15.12),

$$\sigma_+(p^2)\theta(p^0)(\not{p}+m) \equiv (2\pi)^3 \sum_{n,\, J^P = 1/2^+}{}' \delta^{(4)}(p-p_n)\,\langle 0|\psi'(0)|n\rangle\,\langle n|\overline{\psi'}(0)|0\rangle \tag{23.43}$$

(The prime on the sum means that we are not including single particle states.)

On the other hand in the continuum there are states besides $\frac{1}{2}^+$. Even in perturbation theory we not only have a nucleon and a meson in a p wave state with $J^P = \frac{1}{2}^+$, but also nucleon–meson s wave states which are $\frac{1}{2}^-$. These $\frac{1}{2}^-$ states are only connected to the vacuum in their rest frame by the two *lower* components of ψ, which are *odd* under parity. They're eigenstates of β with eigenvalue -1, and they contribute the third term, with $\sigma_-(a^2)$, to $\widetilde{S}'(\not{p})$:

$$\sigma_-(p^2)\theta(p^0)(\not{p}-m) \equiv (2\pi)^3 \sum_{n,\, J^P = 1/2^-}{}' \delta^{(4)}(p-p_n)\,\langle 0|\psi'(0)|n\rangle\,\langle n|\overline{\psi'}(0)|0\rangle \tag{23.44}$$

The distribution functions σ_+ and σ_- are both non-negative by the positivity of the norm on Hilbert space. Note the presence of the projection operators in the definitions. You might worry that the projection operator $\not{p}-m$ (acting on the two lower components of ψ) is negative in a frame where $\mathbf{p} = 0$. But that's as things should be, because $\overline{\psi}$ has a minus sign in its definition for the two lower components in comparison with $\psi^\dagger$. So both projection operators produce a positive definite result. On the other hand, the functions can be zero:

$$\sigma_+ = \sigma_- = 0 \quad \text{if } p^2 < (m+\mu)^2 \tag{23.45}$$

That is, of course, if the momentum is below threshold to make a meson.

Equation (23.42) is sometimes written in an even more suggestive form obtained by *un*-rationalizing the denominators:

$$\widetilde{S}'(\not{p}) = \frac{i}{\not{p}-m+i\epsilon} + \int_{(m+\mu)^2}^{\infty} da^2\,\sigma_+(a^2)\,\frac{i}{\not{p}-a+i\epsilon} + \int_{(m+\mu)^2}^{\infty} da^2\,\sigma_-(a^2)\,\frac{i}{\not{p}+a-i\epsilon} \tag{23.46}$$

(Note that we always add a *negative* imaginary part to the mass.) In this form, we can discuss the analytic property of $\widetilde{S}'$ as a function of $\not{p}$—or rather, the analytic properties of $\widetilde{S}'(z)$, the function of a single variable obtained by replacing the 4×4 matrix $\not{p}$ by a complex number z. If we draw the complex z plane, as in Figure 23.2, we see that $\widetilde{S}'(z)$, no longer an even function, is an analytic function of z, except for a pole at $z = m$, the physical mass of the nucleon, and *two* branch cuts: from the σ_+ term, a cut that gives you singularities when $z \geq (m+\mu)$, and from the σ_- term a corresponding cut going off on the left-hand axis. This term becomes singular when $z \leq -(m+\mu)$. This looks a little bit different than Figure 23.1, the corresponding drawing for the meson. There's a left hand cut as well as a right hand cut, but the general features remain the same. The statement that the renormalized mass of the particle is m gives us the location of the pole, and the residue of the pole is given to us by the scale of the fields. Of course, Figure 23.2 is just the structure in perturbation theory. If bound states develop, there may be further poles around here someplace due to the bound states; on

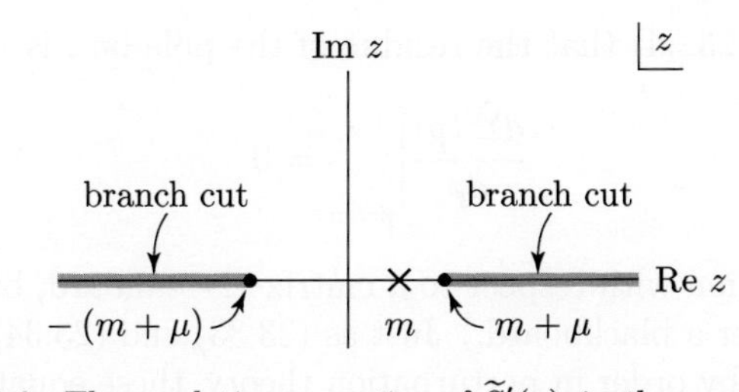

Figure 23.2: The analytic properties of $\widetilde{S}'(z)$ in the complex z plane

the right side if they're positive parity, and on the left side if they're negative parity. And if the bound states are sufficiently low you might move the location of the cuts, you might have a bound state that's lighter than the nucleon or the meson, in which case the cut will move down from $(m+\mu)$, etc. But the general features are as sketched.

23.4 The nucleon self-energy $\widetilde{\Sigma}'$

We are still following the road map provided by Model 3. We have renormalized the nucleon wave function. Then we wrote the Lagrangian in terms of the renormalized wave functions and the new counterterms. The goal is to obtain equations allowing us to compute the counterterms in perturbation theory. As in Model 3, we have obtained expressions for the renormalized nucleon propagator, both as a definition and in terms of a spectral representation. Finally, we have looked at the analyticity of the renormalized nucleon propagator from its spectral representation.

To keep the parallelism going, the next step is to define the one-particle irreducible (1PI) diagram occurring in $\widetilde{S}'(\not{p})$ and sum up the sequence of graphs, analogous to the role of $\widetilde{\Pi}'(p^2)$ in $\widetilde{D}'(p^2)$. I define a function of $\not{p}$ which is traditionally denoted $\widetilde{\Sigma}'(\not{p})$:

$$\xleftarrow{p}\text{(1PI)}\xleftarrow{p} \equiv -i\widetilde{\Sigma}'(\not{p}) \tag{23.47}$$

The 1PI graphs have the same transformation properties—Lorentz, parity, and so on—as the full sets of graphs. This definition for Fermi fields is parallel to the definition (23.31) of $\widetilde{\Pi}'(p^2)$, for Bose fields, except that $\widetilde{\Sigma}'(\not{p})$, like $\widetilde{S}'(\not{p})$, is a function of p^2 plus a function of p^2 times $\not{p}$, and will be written as a single function of $\not{p}$. Here, $\widetilde{\Sigma}'(\not{p})$ is the **nucleon self-energy**. We know that $\widetilde{S}'(\not{p})$ can be written as a geometric series in $-i\widetilde{\Sigma}'(\not{p})$. I won't even bother to write it down. It's exactly the same as (23.32), you just replace $\widetilde{D}'$ with $\widetilde{S}'$ and $\widetilde{\Pi}'$ with $\widetilde{\Sigma}'$. Everything is a function of matrices, and there's matrix multiplication but they're all functions of the single matrix $\not{p}$ so they all commute. So I obtain (15.36)

$$\widetilde{S}'(\not{p}) = \frac{i}{\not{p} - m + i\epsilon - \widetilde{\Sigma}'(\not{p})} \tag{23.48}$$

As in (23.33), the requirement that the renormalized propagator $\widetilde{S}'(\not{p})$ must have a pole at $\not{p} = m$ sets the condition

$$\widetilde{\Sigma}'(m) = 0 \tag{23.49}$$

The condition parallel to (23.34) that the residue of the pole be i is

$$\left.\frac{d\widetilde{\Sigma}'(\not{p})}{d\not{p}}\right|_{\not{p}=m} = 0 \tag{23.50}$$

(This notation—differentiation with respect to a matrix—is standard, but it affects some people like fingernails scraping over a blackboard.) Just as (23.33) and (23.34) enable us to determine A and B iteratively, order by order in perturbation theory, these equations (23.49) and (23.50) do the same for C and D in (23.25).

EXAMPLE. $\widetilde{\Sigma}'(\not{p})$ *to order* g^2

In our discussion of Model 3 (§15.4; §16.1), we computed the meson self-energy $\widetilde{\Pi}'(p^2)$ to order g^2. Following what I did then, I will now compute $\widetilde{\Sigma}'(\not{p})$ to order g^2, the first nontrivial order. This calculation will involve the manipulation of Dirac matrices as well as internal loops. I will not carry it out all the way. When I get things down to integrals that can be found in our integral table I will stop. But I will want to get to its matrix structure.

To order g^2, we have two kinds of graphs:

$-i\widetilde{\Sigma}'(\not{p}) =$ 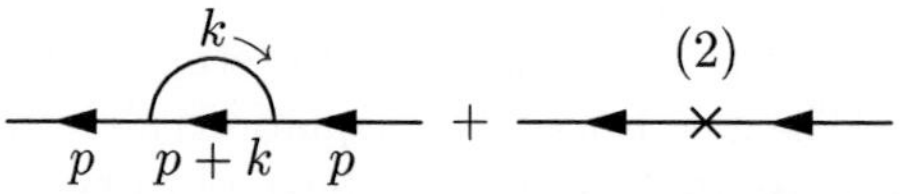

(23.51)

We have a closed loop, and we have the counterterm, evaluated to second order in g, using the same notation (15.54) as in Model 3. Thus we have two contributions:

$$-i\widetilde{\Sigma}'(\not{p}) = -i\Sigma^f(\not{p}) + iC_2\not{p} - iD_2 \tag{23.52}$$

Invoking the same guess as before,[4] the result of the derivative in the counterterm $C\overline{\psi}'i\not{\partial}\psi'$ is just a power of momentum.

The term $-i\Sigma^f(\not{p})$ is the contribution from the Feynman graph with the loop (the superscript f stands for "Feynman", not "finite"), and C_2 and D_2 are the counterterms to $\mathcal{O}(g^2)$. These can be eliminated from (23.52) using the two renormalization conditions (23.49) and (23.50):

$$\widetilde{\Sigma}'(\not{p}) = \Sigma^f(\not{p}) - \Sigma^f(m) - (\not{p}-m)\left.\frac{d\Sigma^f}{d\not{p}}\right|_{\not{p}=m} \tag{23.53}$$

This is the same reasoning that took us from (15.54) to (15.56), except that instead of p^2 and $p^2-\mu^2$, I have $\not{p}$ and $(\not{p}-m)$, respectively. Of course the computation is a bit different because it's a dynamically different theory and we have to manipulate matrices.

Just to remind you of the relevant part of the interaction Lagrangian (23.25), here it is:

$$\mathscr{L}'_{\widetilde{\Sigma}'} = -g\overline{\psi}'i\gamma_5\psi'\phi' + C\overline{\psi}'i\not{\partial}\psi' - D\overline{\psi}'\psi' \tag{23.54}$$

The graph for $-i\widetilde{\Sigma}'^f(\not{p})$ gives us

$$-i\Sigma^f(\not{p}) = (-ig)^2\int\frac{d^4k}{(2\pi)^4}\frac{i}{k^2-\mu^2+i\epsilon}i\gamma_5\frac{i(\not{p}+\not{k}+m)}{(p+k)^2-m^2+i\epsilon}i\gamma_5 \tag{23.55}$$

[4] [Eds.] See §14.4, (14.57) and Problem 8.1, Comment (3), p. 309.

We just have the boson propagator, and then, head before tail, $i\gamma_5$, the fermion propagator, $i\gamma_5$. No u and no $\overline{u}$ because this is not a scattering matrix element, nor a number; it is a part of a *propagator*, which is a 4×4 matrix.

Well, of course this is a divergent integral. We've got counterterms that are going to take care of that, although it's not going to be quite as easy as it was before. I can gather together all of the i's. I'll bring the γ_5 through, changing the sign of $\not{p}$ and $\not{k}$, multiplying the other γ_5, and becoming 1. This doesn't change the sign of m. (Of course, γ_5 commutes with 1.) Then

$$\Sigma^f(\not{p}) = -\frac{ig^2}{(2\pi)^4} \int d^4k \, \frac{1}{k^2 - \mu^2 + i\epsilon} \, \frac{(-\not{p} - \not{k} + m)}{(p+k)^2 - m^2 + i\epsilon} \tag{23.56}$$

The next step is always the same. I use the Feynman parametrization (15.58) with

$$a = (p+k)^2 - m^2 + i\epsilon \tag{23.57}$$

$$b = k^2 - \mu^2 + i\epsilon \tag{23.58}$$

to write the integral in a parametric form:

$$\Sigma^f(\not{p}) = -\frac{ig^2}{(2\pi)^4} \int d^4k \int_0^1 dx \, \frac{(-\not{p} - \not{k} + m)}{\left[k^2 + 2k \cdot px + p^2x - m^2x - \mu^2(1-x) + i\epsilon\right]^2} \tag{23.59}$$

Now I shift the momentum, exactly as before (15.60):

$$k = k' - px \tag{23.60}$$

We are hoping that the counterterm subtractions will be enough to make the integral convergent. Writing $\Sigma^f(\not{p})$ in terms of k',

$$\Sigma^f(\not{p}) = -\frac{ig^2}{(2\pi)^4} \int d^4k' \int_0^1 dx \, \frac{-\not{p}(1-x) + m - \not{k}'}{\left[k'^2 + p^2x(1-x) - m^2x - \mu^2(1-x) + i\epsilon\right]^2} \tag{23.61}$$

I should say that it's very difficult to avoid making mistakes. You write down a lot of mysterious equations, you erase a lot and curse a lot, that's how you do it. (Every course in Feynman diagrams is also a course in foul language.) Now we notice something useful: the denominator is an even function of k'—that's what the shift buys us—and we are integrating over all values of each component of k'^μ, from $-\infty$ to ∞. The $\not{k}'$ in the numerator is an odd function and thus irrelevant; it vanishes upon integration:

$$\Sigma^f(\not{p}) = -\frac{ig^2}{(2\pi)^4} \int d^4k' \int_0^1 dx \, \frac{-\not{p}(1-x) + m}{\left[k'^2 + p^2x(1-x) - m^2x - \mu^2(1-x) + i\epsilon\right]^2} \tag{23.62}$$

Thus we have explicitly displayed the result as some function of p^2 times $\not{p}$ plus some function of p^2 times the identity matrix, which is what we anticipated; it's the only form that can result if the theory is Lorentz invariant.

The complete $\widetilde{\Sigma}'(\not{p})$ is given by (23.53). It's only out of a misguided puritanism that I write down the whole thing, but I want to show you how it works out in full detail at least once:

$$\widetilde{\Sigma}'(\not{p}) = -\frac{ig^2}{(2\pi)^4} \int d^4k' \int_0^1 dx \left\{ \cdots \right\} \tag{23.63}$$

where the terms in the curly brackets are (remember: the last two terms are evaluated at $\not p = m$, so *a fortiori* $p^2 = m^2$)

$$\left\{\cdots\right\} = \left\{ \frac{-\not p(1-x)+m}{\left[k'^2 + p^2x(1-x) - m^2x - \mu^2(1-x) + i\epsilon\right]^2} - \frac{mx}{\left[k'^2 - m^2x^2 - \mu^2(1-x)+i\epsilon\right]^2} \right.$$
$$\left. -(\not p - m)\left[\frac{(x-1)}{\left[k'^2 - m^2x^2 - \mu^2(1-x)+i\epsilon\right]^2} + \frac{4m^2x^2(x-1)}{\left[k'^2 - m^2x^2 - \mu^2(1-x)+i\epsilon\right]^3}\right]\right\} \tag{23.64}$$

In our calculation of the Model 3 $\widetilde{\Pi}'(p^2)$, we discovered (p. 327) that the single subtraction of $\widetilde{\Pi}^f(p^2) - \widetilde{\Pi}^f(\mu^2)$ was enough to render $\widetilde{\Pi}'(p^2)$ finite; the derivative term proportional to $(p^2 - \mu^2)$ was separately finite. Model 3 is an example of a "super-renormalizable" theory, where we don't need all the subtractions. In the present case, the ps-ps theory, one subtraction isn't sufficient. If we look at each of the first two terms in (23.64), we see that the coefficient of $\not p$ is logarithmically divergent, as is the coefficient of m. And even after the subtraction, at large k', the sum of the first two terms goes as

$$\text{first two terms of } \widetilde{\Sigma}'(\not p) \sim \int d^4k' \frac{k'^4}{k'^8} \sim \int \frac{dk'}{k'} \tag{23.65}$$

which is still logarithmically divergent. There's no cancellation in the numerator, as there was in Model 3.

I can hardly praise this expression (23.64) for its beauty, but I can at least ask: Is it finite, or have I been leading you down the garden path and doing my computations in a non-renormalizable theory? I would not be so nasty. The last term, thank God, has no problem with divergences:

$$\text{last term of } \widetilde{\Sigma}'(\not p) \sim \int d^4k' \frac{1}{(k'^2)^3} \sim \int \frac{dk'}{k'^3} \tag{23.66}$$

So this is finite. How do all the other terms go? They all have denominators that go at high k' as k'^4, so they're all proportional to $\int d^4k'/k'^4$. That's the only divergent part at high k'; everything else converges at high energy. Combining all the numerators gives

$$\text{first three terms of } \widetilde{\Sigma}'(\not p) \sim \int d^4k' \frac{1}{k'^4} \times \left\{-\not p(1-x) + m - mx - (\not p - m)(x-1)\right\} \tag{23.67}$$

Now you will please notice that this is in fact *zero*. That is to say, $\widetilde{\Sigma}'(\not p)$ at high k' is *convergent*. I went through all the details to show you that it really works out.

Actually there is a quicker way to show that $\widetilde{\Sigma}'(\not p)$ is convergent. From (23.53), we see that $\widetilde{\Sigma}'(\not p)$ differs from $\Sigma^f(\not p)$ by a constant term and a term linear in $\not p$. Consequently

$$\frac{d^2\widetilde{\Sigma}'(\not p)}{d\not p^2} = \frac{d^2\Sigma^f(\not p)}{d\not p^2} \tag{23.68}$$

The second derivative of $\Sigma^f(\not p)$ completely determines $\widetilde{\Sigma}'$, just by integration. So if we know the second derivative of $\Sigma^f(\not p)$, we can get $\widetilde{\Sigma}'(\not p)$:

$$\frac{d^2\Sigma^f}{d\not p^2} \Rightarrow \widetilde{\Sigma}'(\not p) \tag{23.69}$$

We can quickly compute the second derivative. If it is finite, which we will show in a moment, $\widetilde{\Sigma}'(\not{p})$ is certainly going to be finite. Written in the crudest possible way,

$$\Sigma^f(\not{p}) \propto \int d^4k'\, i\gamma_5 \frac{1}{\not{p}+\not{k}'-m} i\gamma_5 \frac{1}{k'^2-\mu^2} \sim \int d^4k' \frac{1}{k'^3} \tag{23.70}$$

I've suppressed the $i\epsilon$'s. Whenever I differentiate $\Sigma^f(\not{p})$ with respect to $\not{p}$, I drag an extra $\not{k}'$ into the denominator. I differentiate it twice and I get d^4k over something that goes like one over the fifth power of k, which is obviously a convergent integral:

$$\frac{d^2\Sigma^f}{d\not{p}^2} \sim \int d^4k' \frac{1}{k'^5} \sim \int \frac{dk'}{k'^2} \tag{23.71}$$

That is the crude, slovenly argument that the integral is convergent. However, differentiating (23.70) is not the simplest way to compute $\widetilde{\Sigma}'(\not{p})$; if you really want to determine $\widetilde{\Sigma}'(\not{p})$, it's best to use (23.64) and the integral table in §15.5. That the second derivative (but not the first) is finite also tells us that we need *two* counterterms, and thus two subtractions, to make $\widetilde{\Sigma}'(\not{p})$ finite.

EXAMPLE. *A second look at $\widetilde{\Pi}'(p^2)$ to order g^2*

Let's look at the meson self-energy in the ps-ps theory. A similar remark about the divergence of the nucleon self-energy operator $\widetilde{\Sigma}'(\not{p})$ applies to the meson self-energy operator, $\widetilde{\Pi}'(p^2)$: the second derivative of $\widetilde{\Pi}^f(p^2)$ determines $\widetilde{\Pi}'(p^2)$. The $\mathcal{O}(g^2)$ diagram for $\widetilde{\Pi}'(p^2)$ is just Figure 15.6 again, plus the counterterm graph:

$-i\widetilde{\Pi}'(p^2) =$ 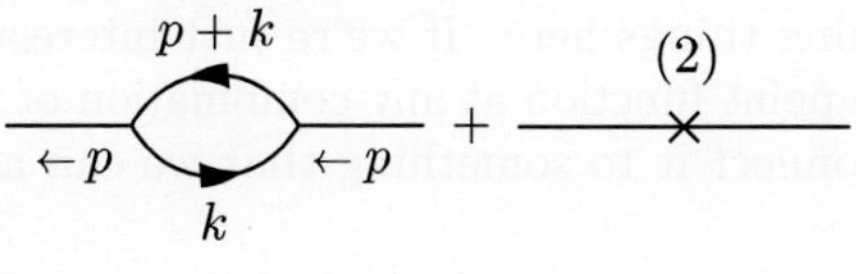 (23.72)

In Model 3, the "nucleons" were scalar particles, and the contribution (15.57) to $\widetilde{\Pi}^f(p^2)$ involved only scalar propagators. Now, of course, we get a different contribution to the fermion loop, following the Feynman rules:

$$\widetilde{\Pi}^f(p^2) \sim \int d^4k\, \mathrm{Tr}\left[i\gamma_5 \frac{1}{\not{p}+\not{k}-m} i\gamma_5 \frac{1}{\not{k}-m}\right] \tag{23.73}$$

Rationalizing the denominators, moving the $i\gamma_5$ through the numerators, and taking the trace, we get

$$\widetilde{\Pi}^f(p^2) \sim \int d^4k \frac{k^2}{k^4} \sim \int dk\, k \tag{23.74}$$

This integral looks like it is quadratically divergent. But whenever I differentiate with respect to p^2, I drag an extra k^2 into the denominator.[5] It is the second derivative of $\widetilde{\Pi}^f(p^2)$ with

---

[5] [Eds.] After canceling the $i\gamma_5$ and taking the trace,

$$\widetilde{\Pi}^f(p^2) \sim \int d^4k \frac{4(k^2+p\cdot k-m^2)}{[(p+k)^2-m^2][k^2-m^2]}$$

respect to p^2 that is relevant (compare (15.56) with (23.53)):

$$\frac{d^2\widetilde{\Pi}^f}{(dp^2)^2} \Rightarrow \widetilde{\Pi}' \tag{23.75}$$

because this is a boson expression. It's the value at $p^2 = \mu^2$ and the first derivative with respect to p^2 at $p^2 = \mu^2$ that enter into the renormalization equations. Once I differentiate with respect to p^2, I change the integral from quadratically divergent into logarithmically divergent because I put a k^2 into the denominator. A second differentiation with respect to p^2 turns it from logarithmically divergent to convergent. These two derivatives turn it from d^4k over k^2 to d^4k over k^6. The meson self-energy $\widetilde{\Pi}'(p^2)$ is right on the borderline of being renormalizable. *Two* subtractions are needed and two subtractions are what the renormalization prescription gives us. One derivative would turn this from being quadratically divergent to logarithmically divergent; not enough. Notice the marvelous way in which the renormalization program just scrapes by and saves us! Here where we're differentiating with respect to p^2, the first graph is quadratically divergent. We need those *four* powers of k in the denominator to make it convergent. The fermion self-energy is just linearly divergent, but we're only differentiating with respect to $\not{p}$. There we need those *two* powers of k in the denominator, not two powers of k^2, to save us.

23.5 The renormalized coupling constant

Finally, I will discuss the renormalization condition that fixes the last counterterm, E: the definition of the renormalized coupling constant g, order by order in perturbation theory. There are some cunning things here. If we're just interested in getting rid of infinities, we could define the three point function at any combination of momenta. But we want an elegant definition that will connect it to something that we can actually measure. That requires a little care.

You'll recall in Model 3 we defined the renormalized coupling constant $-i\widetilde{\Gamma}'(p^2, p'^2, q^2)$ as a one-particle-irreducible (1PI) graph:

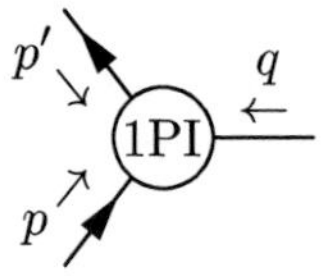

$$\equiv -i\widetilde{\Gamma}'(p^2, p'^2, q^2) \tag{16.29}$$

Using the Feynman parameter trick and shifting $k = k' - px$ exactly as before,

$$\widetilde{\Pi}^f(p^2) \sim 4\int_0^1 dx \int d^4k' \,\frac{k'^2 + k'\cdot p(1-2x) - p^2x(1-x) - m^2}{[k'^2 + p^2x(1-x) - m^2]^2}$$

As before, the term linear in k' in the numerator is odd, and can be dropped. The net result is

$$\widetilde{\Pi}^f(p^2) \sim 4\int_0^1 dx \int d^4k' \,\frac{k'^2 - p^2x(1-x) - m^2}{[k'^2 + p^2x(1-x) - m^2]^2}$$

This is a function of p^2, as required. Every differentiation with respect to p^2 reduces the integrand's power of k' by 2.

And we found (Problem 9.3, (S9.28), p. 352), to lowest order in perturbation theory

$$\widetilde{\Gamma}'(p^2, p'^2, q^2) = g + \mathcal{O}(g^3) \tag{23.76}$$

We fixed E by choosing to evaluate $\widetilde{\Gamma}'$ on the mass shell:

$$\widetilde{\Gamma}'(m^2, m^2, \mu^2) = g \tag{16.32}$$

where

$$p^2 = p'^2 = m^2; \quad q^2 = \mu^2 \tag{23.77}$$
$$p + p' + q = 0 \tag{23.78}$$

but these two equations cannot be satisfied unless some momentum components are complex.[6] We can't reach this point by a physical process. The only way to get there is by analytic continuation, and you had to accept on trust that we could do this. The great advantage of this definition is that when we discussed processes like meson–nucleon scattering with scalar mesons and scalar nucleons and all sorts of corrections to the vertices and the internal nucleon line, for example a process like the one shown in (23.79), the coefficient of the pole in the s channel below threshold (or the t channel unphysical pole in nucleon–nucleon scattering) was given directly in terms of g:

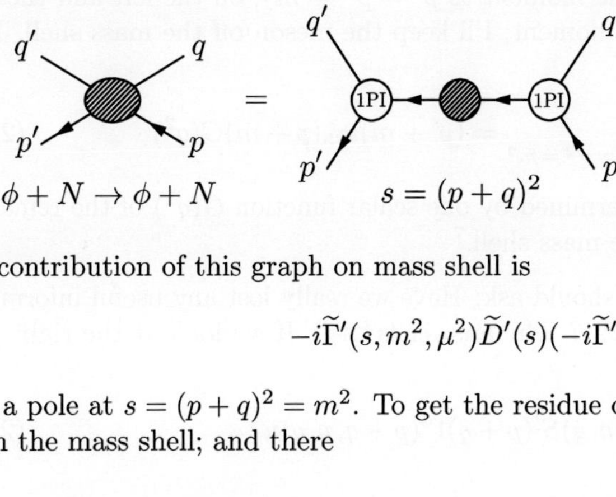

+ terms analytic at $s = m$ (23.79)

The contribution of this graph on mass shell is

$$-i\widetilde{\Gamma}'(s, m^2, \mu^2)\widetilde{D}'(s)(-i\widetilde{\Gamma}'(m^2, s, \mu^2)) \tag{23.80}$$

with a pole at $s = (p+q)^2 = m^2$. To get the residue of the pole, all the external lines had to be on the mass shell; and there

$$-i\widetilde{\Gamma}'(s, m^2, \mu^2)\widetilde{D}'(s)(-i\widetilde{\Gamma}'(m^2, s, \mu^2)) = (-ig)\frac{i}{(s - m^2)}(-ig) = -\frac{ig^2}{s - m^2} \tag{23.81}$$

(with the scalar "nucleon" propagator for $\widetilde{D}'(s)$). That made for a very nice definition, giving us something that was physically measurable in this hypothetical theory (see 16.37)).

Now I will try to do the same thing in our current theory, in which the nucleons are fermions described by Dirac spinors with complicated matrix structure. (I can't call it the "true" theory; the *true* theory is quarks and non-Abelian gauge mesons.) It's exactly the same

[6] [Eds.] In its rest frame, a real (on mass shell) nucleon cannot emit a real meson. Where would the energy come from? It *could* emit a *virtual* meson, but that meson's momentum would be complex. This is easy to show algebraically: $p'^2 = p^2 + q^2 + 2p \cdot q$, or $q^2 = -2p \cdot q = -2mE_{\mathbf{q}}$. This is impossible if $q^2 = \mu^2$.

diagram as (16.29), but now the nucleon lines are fermion lines and therefore $-i\widetilde{\Gamma}'(p',p,q)$ is a 4×4 matrix:

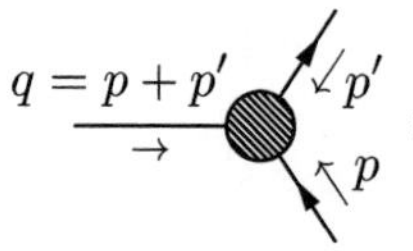

$$\equiv -i\widetilde{\Gamma}'(p',p,q) \tag{23.82}$$

In perturbation theory $\widetilde{\Gamma}'$ starts out very simple:

$$\widetilde{\Gamma}' = i\gamma_5 g + \mathcal{O}(g^3) \tag{23.83}$$

However a technical obstacle arises which must be surmounted. In general $\widetilde{\Gamma}'$ will contain all sorts of god-awful matrices because you have two vectors to play with: any two of p^μ, p'^μ and q^μ to dot into any of the sixteen Dirac matrices. We could have $\gamma_5 \not{p}$ or $\sigma_{\mu\nu}p^\mu p'^\nu$, and so on. Some of them may be thrown out by parity and other considerations, but most of them survive. $\widetilde{\Gamma}'$ could be a horrible object, requiring up to 16 different conditions for its determination. So we can't say that for some particular set of momenta $\widetilde{\Gamma}'$ is equal to $i\gamma_5$ times g, because the equations would be overdetermined. We can certainly adjust matters so the coefficient of γ_5 is what we want, by picking our counterterm as we please. But we've got all those other coefficients. Therefore I will look at a simpler object:

$$(\not{p}' + m)\widetilde{\Gamma}'(p',p,q)(\not{p}+m)\Big|_{p^2=p'^2=m^2} \tag{23.84}$$

I'll multiply $\widetilde{\Gamma}'(p',p,q)$, restricted for the moment to $p^2 = p'^2 = m^2$, on the left and the right by these projection operators. For the moment, I'll keep the meson off the mass shell. I will demonstrate presently that

$$(\not{p}' + m)\widetilde{\Gamma}'(p',p,q)(\not{p}+m)\Big|_{p^2=p'^2=m^2} = (\not{p}'+m)i\gamma_5(\not{p}+m)G(q^2) \tag{23.85}$$

That is to say, the object (23.84) is determined by one scalar function $G(q^2)$ of the remaining variable q, once I've set p and p' on the mass shell.[7]

Before I show that (23.85) is true, I should ask: Have we really lost any useful information by looking at (23.84) rather than (16.29)? The answer is "no". If we look at the right hand graph of (23.79), its contribution is

$$-\overline{u}'\widetilde{\Gamma}'(p',p+q,q)\widetilde{S}'(p+q)\widetilde{\Gamma}'(p+q,p,q)u \tag{23.86}$$

At the pole,

$$\widetilde{S}'(p+q) = \frac{i(\not{p}+\not{q}+m)}{s-m^2} \tag{23.87}$$

So the combination $(\not{p}+\not{q}+m)\widetilde{\Gamma}'(p+q,p,q)$ comes in automatically from the product $\widetilde{S}'(p+q)\widetilde{\Gamma}'(p+q,p,q)$. No such projection operator appears automatically on the right of $\widetilde{\Gamma}'(p+q,p,q)$, but we can slip it in without loss of generality, because

$$\frac{\not{p}+m}{2m}u = u \tag{23.88}$$

[7] [Eds.] In modern language, $G(q^2)$ is called a *running coupling constant*, dependent on momentum (or in position space, separation); see §50.4. It is impossible to put all three particles on their mass shell in the (nonphysical) process $N \to N + \pi$, so Coleman wants to look at the value of $G(q^2)$ in experimentally accessible processes like $N + N \to N + N$.

That is,

$$\widetilde{S}'(p+q)\widetilde{\Gamma}'(p+q,p,q)u \sim \frac{1}{s-m^2}\frac{1}{2m}\left[(\not{p}+\not{q}+m)\widetilde{\Gamma}'(p+q,p,q)(\not{p}+m)\right]u \tag{23.89}$$

The bracketed quantity has just the form of (23.84). The same argument goes through for the remaining factors in (23.79), $\overline{u}'\Gamma(p',p+q)$. So I may have restricted myself in what I can look at, destroying this marvelous, rich structure of 42 different combinations of Dirac matrices that could be in $\widetilde{\Gamma}'(p,p',q)$, by putting projection operators on either side of it. But I saved all the parts that are important when I'm computing the residue of the pole in the on mass shell scattering process.

Now let me give a quick demonstration that (23.85) is true. The right hand graph of (23.79) with the nucleons on the mass shell, but the meson off, can be thought of as a variety of processes, depending upon whether the nucleon lines are on the upper or lower mass hyperboloid. It could be the sort of matrix element we used when we were discussing decay processes, with $\phi(0)$ between the vacuum and a nucleon antinucleon out state:

$$^{\text{out}}\langle N,\overline{N}|\phi(0)|0\rangle \tag{23.90}$$

When some of the fields are on the mass shell, they become in states or out states, depending on where they are, and the others stay as fields. Or it could be that both are on the upper hyperboloid:

$$\langle N|\phi(0)|N\rangle \tag{23.91}$$

There I don't have to say in or out, depending on how I put them. Or, you know, the nucleon and the antinucleon on the right, in an in state. It doesn't matter.

Now all these processes are of course connected by analytic continuation. I say "of course", because it takes a month to prove it. But they're all obtained from the same function and it's just a matter of whether q^2 is timelike or spacelike, positive or negative, to say which process you're describing. So as far as counting invariants, I might as well count them with the process (23.90), to see how many numbers I need to describe this process, or any other related to it:

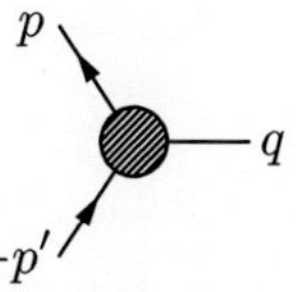

(23.92)

By Lorentz invariance I might as well look at the frame in which q carries a timelike momentum to make a real pair: the total momentum of the pair equals the momentum carried by the meson. So I'll make

$$q^\mu = (q^0, \mathbf{0}) \tag{23.93}$$

You hit the vacuum with the meson field; the field makes a pair. I choose to look in the Lorentz frame in which the pair is made at rest.

Now what do I know about the field? $\phi(0)$ is Lorentz invariant, in particular rotationally invariant, so the final state must have total angular momentum zero, and $\phi(0)$ is a pseudoscalar field. The state you make by hitting the vacuum with $\phi(0)$ is odd under parity: its parity equals -1, and its J^P equals 0^-. The states of the nucleon–antinucleon pairs come out with the given value of q^2. The two spin ½ particles in their final state can make $S = 1$ or $S = 0$. To conserve angular momentum, their J value must equal 0. Their possible values of J^P are:

$$J = 0 \quad L = 0 \quad S = 0 \quad P = -1 \tag{23.94}$$

$$J = 0 \quad L = 1 \quad S = 1 \quad P = 1 \tag{23.95}$$

But the second violates parity. Thus there's just one invariant amplitude, the amplitude for making this s-wave state, which is completely determined once I've given its center of mass energy q^2. If we didn't have parity invariance we'd have two. Thus this process is indeed described by a single function of q^2. All we're doing is counting states here. If there were 72 partial wave states it could go into, it would be described by 72 functions of q^2. We have only one. Therefore we are free to impose our one renormalization condition, to wit: at $q^2 = \mu^2$, $G(\mu^2) = g$:

$$G(q^2)\Big|_{q^2=\mu^2} = g \tag{23.96}$$

That is *the* definition of the physical coupling constant that corresponds to what we did in the boson case. The task is done.

Next time I will talk about isospin and how it fits in with field theory.

Problems 13

13.1 In the discussion (§23.4) of renormalization of the "pseudoscalar" theory,

$$\mathscr{L}' = g\overline{\psi}i\gamma_5\psi\phi \tag{P13.1}$$

we sketched a computation of the renormalized nucleon self-energy $\widetilde{\Sigma}'(p^2)$ to $\mathcal{O}(g^2)$. Complete the calculation. Again, leave the integral over the Feynman parameter undone.

(1991a 11.2)

13.2 In the same theory,

(a) Compute the renormalized meson self-energy, $\widetilde{\Pi}'(k^2)$, to lowest nonvanishing order in perturbation theory, $\mathcal{O}(g^2)$. Leave the integral over the Feynman parameter undone, just as it was left undone in our discussion of the same object (16.5) in Model 3. HINT: All you need for this problem are the conditions that fix B and C, the $\partial_\mu\phi\,\partial^\mu\phi$ and ϕ^2 counterterms in (14.47). These conditions are the same as in Model 3, (15.38) and (15.40):

$$\widetilde{\Pi}'(\mu^2) = \left.\frac{d\widetilde{\Pi}'(k^2)}{dk^2}\right|_{k^2=\mu^2} = 0 \tag{P13.2}$$

(b) We derived a formula for the imaginary part of $[-i\widetilde{D}'(k^2)]^{-1}$, for real k^2, in terms of the spectral function $\sigma(k^2)$,

$$-\text{Im}[-i\widetilde{D}'(k^2)]^{-1} = \text{Im}\,\widetilde{\Pi}'(k^2) = -\frac{\pi\sigma(k^2)}{|\widetilde{D}'(k^2)|^2} \tag{P13.3}$$

(For this equation in Model 3, see (S9.13) and (17.4).) Because we want $\sigma(k^2) > 0$ above the two-particle threshold, $k > 2m$, it follows directly that the imaginary part of the self-energy $\widetilde{\Pi}'(k^2)$ is negative for $k > 2m$. Check that in your calculation the imaginary part of the self-energy has the right (negative) sign, confirming the correctness of the rule, "a minus sign for every closed fermion loop" (or possibly only confirming that you've made an even number of sign errors.) NOTE: The minus sign rule (item 2 in the table on page 443) for closed fermion loops is essential in getting this sign right.

(1997a 11.4; 1991a 11.3)

Solutions 13

13.1 The renormalization conditions give us the self-energy (23.53)

$$\widetilde{\Sigma}'(\not{p}) = \Sigma^f(\not{p}) - \Sigma^f(m) - (\not{p} - m)\left.\frac{d\Sigma^f(\not{p}}{d\not{p}}\right|_{\not{p}=m} \tag{S13.1}$$

To $\mathcal{O}(g^2)$, $-i\Sigma^f(\not{p})$ is the amplitude for this diagram: 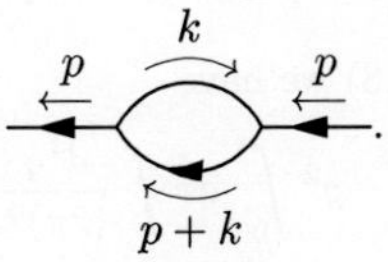

. Then (23.55)

$$\begin{aligned}
-i\Sigma^f(\not{p}) &= (-ig)^2 \int \frac{d^4k}{(2\pi)^4} \frac{i}{k^2 - \mu^2 + i\epsilon} i\gamma_5 \frac{i(\not{p} + \not{k} + m)}{(p+k)^2 - m^2 + i\epsilon} i\gamma_5 \\
&= g^2 \int \frac{d^4k}{(2\pi)^4} \frac{1}{k^2 - \mu^2 + i\epsilon} \frac{(\not{p} + \not{k} - m)}{(p+k)^2 - m^2 + i\epsilon} \\
&= g^2 \int_0^1 dx \int \frac{d^4k}{(2\pi)^4} \frac{(\not{p} + \not{k} - m)}{\left((1-x)(k^2 - \mu^2 + i\epsilon) + x[(p+k)^2 - m^2 + i\epsilon]\right)^2}
\end{aligned} \tag{S13.2}$$

The denominator can be simplified:

$$\begin{aligned}
(1-x)(k^2 - \mu^2 + i\epsilon) + x[(p+k)^2 - m^2 + i\epsilon] &= k^2 - (1-x)\mu^2 + xp^2 + 2xp\cdot k - xm^2 + i\epsilon \\
&= (k + xp)^2 + p^2x(1-x) - (1-x)\mu^2 - xm^2 + i\epsilon
\end{aligned} \tag{S13.3}$$

Shifting the momentum $k \to q = k + xp$ gives

$$-i\Sigma^f(\not{p}) = g^2 \int_0^1 dx \int \frac{d^4q}{(2\pi)^4} \frac{\not{p}(1-x) + \not{q} - m}{(q^2 + p^2x(1-x) - (1-x)\mu^2 - xm^2 + i\epsilon)^2} \tag{S13.4}$$

The linear term is odd, so it contributes nothing and may be discarded. Then using the integral table ((I.4), p. 330), we get

$$-i\Sigma^f(\not{p}) = g^2 \int_0^1 dx\, (\not{p}(1-x) - m)\left[\frac{-i}{16\pi^2}\right] \ln\Big((1-x)\mu^2 + xm^2 - x(1-x)p^2 - i\epsilon\Big) + \cdots \tag{S13.5}$$

Finally, from (S13.1),

$$\begin{aligned}
\widetilde{\Sigma}'(\not{p}) = \frac{g^2}{16\pi^2} \int_0^1 dx \Bigg\{ &(\not{p}(1-x) - m) \ln\Big(xm^2 + (1-x)\mu^2 - x(1-x)p^2 - i\epsilon\Big) \\
&+ mx \ln\Big((1-x)\mu^2 + x^2m^2 - i\epsilon\Big) \\
&-(\not{p} - m)\left[(1-x)\ln\Big((1-x)\mu^2 + x^2m^2 - i\epsilon\Big) + \frac{2m^2x^2(1-x)}{(1-x)\mu^2 + x^2m^2 - i\epsilon}\right]\Bigg\}
\end{aligned} \tag{S13.6}$$

We can simplify no further without integrating over the Feynman parameter, x. ■

13.2 (a) The renormalization conditions give us the renormalized self-energy (16.3)

$$-i\widetilde{\Pi}'(k^2) = -i\widetilde{\Pi}^f(k^2) + i\widetilde{\Pi}^f(\mu^2) + i(k^2-\mu^2)\left.\frac{d\widetilde{\Pi}^f(k^2)}{dk^2}\right|_{k^2=\mu^2} \tag{S13.7}$$

To $\mathcal{O}(g^2)$, $-i\widetilde{\Pi}^f(k^2)$ is the amplitude for this diagram: 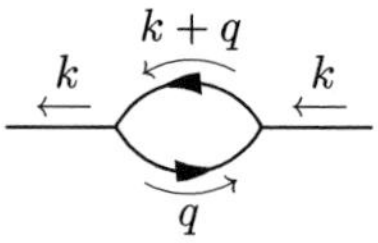

Using the Feynman rules in §21.5, being careful to put in a minus sign for the closed Fermi loop, we have

$$\begin{aligned}-i\widetilde{\Pi}^f(k^2) &= (-ig)^2\int\frac{d^4q}{(2\pi)^4}(-1)\cdot\mathrm{Tr}\left[\frac{i(\not{k}+\not{q}+m)}{((k+q)^2-m^2+i\epsilon)}\cdot i\gamma_5\cdot\frac{i(\not{q}+m)}{(q^2-m^2+i\epsilon)}\cdot i\gamma_5\right]\\ &= -g^2\int\frac{d^4q}{(2\pi)^4}\frac{4\left[(k+q)\cdot q-m^2\right]}{((k+q)^2-m^2+i\epsilon)(q^2-m^2+i\epsilon)}\end{aligned} \tag{S13.8}$$

Now we use Feynman's trick (16.16) to combine the denominators:

$$\begin{aligned}\frac{1}{((k+q)^2-m^2+i\epsilon)(q^2-m^2+i\epsilon)} &= \int_0^1 dx\,dy\,\delta(1-x-y)\frac{(2-1)!}{[x((k+q)^2-m^2+i\epsilon)+y(q^2-m^2+i\epsilon)]^2}\\ &= \int_0^1\frac{dx}{(q^2+2q\cdot kx+k^2x-m^2+i\epsilon)^2}\end{aligned} \tag{S13.9}$$

Plugging this expression into (S13.8) we have

$$-i\widetilde{\Pi}^f(k^2) = -g^2\int_0^1 dx\int\frac{d^4q}{(2\pi)^4}\frac{4\left[(k+q)\cdot q-m^2\right]}{(q^2+2q\cdot kx+k^2x-m^2+i\epsilon)^2} \tag{S13.10}$$

Now shift the momentum q;

$$q' = q+kx \tag{S13.11}$$

and after algebra,

$$-i\widetilde{\Pi}^f(k^2) = -g^2\int_0^1 dx\int\frac{d^4q'}{(2\pi)^4}\frac{4\left[q'^2+k\cdot q'(1-2x)-k^2x(1-x)-m^2\right]}{(q'^2+k^2x(1-x)-m^2+i\epsilon)^2} \tag{S13.12}$$

The denominator is even, and so the terms odd in q' in the numerator can be discarded:

$$-i\widetilde{\Pi}^f(k^2) = -g^2\int_0^1 dx\int\frac{d^4q'}{(2\pi)^4}\frac{4\left[q'^2-k^2x(1-x)-m^2\right]}{(q'^2+k^2x(1-x)-m^2+i\epsilon)^2} \tag{S13.13}$$

The integrand can be rewritten:

$$\frac{q'^2-k^2x(1-x)-m^2}{(q'^2+k^2x(1-x)-m^2+i\epsilon)^2} = \frac{1}{(q'^2+k^2x(1-x)-m^2+i\epsilon)} - \frac{2k^2x(1-x)}{(q'^2+k^2x(1-x)-m^2+i\epsilon)^2} \tag{S13.14}$$

and the integral becomes

$$-i\widetilde{\Pi}^f(k^2) = 4g^2\int_0^1 dx\int\frac{d^4q'}{(2\pi)^4}\left[\frac{2k^2x(1-x)}{(q'^2+k^2x(1-x)-m^2+i\epsilon)^2} - \frac{1}{(q'^2+k^2x(1-x)-m^2+i\epsilon)}\right] \tag{S13.15}$$

Now we consult the integral table (box, p. 330). From (I.4),

$$\int\frac{d^4q'}{(2\pi)^4}\frac{2k^2x(1-x)}{(q'^2+k^2x(1-x)-m^2+i\epsilon)^2} = -\frac{i}{16\pi^2}(2k^2x(1-x))\ln(-k^2x(1-x)+m^2-i\epsilon)+\cdots \tag{S13.16}$$

and from (I.3),

$$\int\frac{d^4q'}{(2\pi)^4}\frac{1}{(q'^2+k^2x(1-x)-m^2+i\epsilon)} = \frac{i}{16\pi^2}(k^2x(1-x)-m^2+i\epsilon)\ln(-k^2x(1-x)+m^2-i\epsilon)+\cdots \tag{S13.17}$$

Then

$$\widetilde{\Pi}^f(k^2) = \frac{g^2}{4\pi^2}\int_0^1 dx\,(3k^2x(1-x)-m^2+i\epsilon)\ln(-k^2x(1-x)+m^2-i\epsilon)+(\text{convergent terms}) \tag{S13.18}$$

(We can drop this $i\epsilon$; it doesn't affect the analytic properties of the expression.) Then from (S13.7)

$$\widetilde{\Pi}'(k^2) = \frac{g^2}{4\pi^2}\int_0^1 dx\left[(3k^2x(1-x)-m^2)\ln\left(\frac{m^2-k^2x(1-x)-i\epsilon}{m^2-\mu^2x(1-x)}\right)\right.$$
$$\left.+\frac{[3\mu^2x(1-x)-m^2]x(1-x)}{m^2-\mu^2x(1-x)}(k^2-\mu^2)\right] \tag{S13.19}$$

We dropped the $i\epsilon$ from the denominators, because below the two-particle threshold, $k^2=\mu^2<4m^2$, and the denominators will never equal zero:

$$x(1-x)\mu^2 \le \tfrac{1}{4}\mu^2 < m^2 \tag{S13.20}$$

That's as far as we can take this expression without integrating over the Feynman parameter, x. ■

(b) From (S13.18)

$$\widetilde{\Pi}^f(k^2) = \frac{g^2}{4\pi^2}\int_0^1 dx\,(3k^2x(1-x)-m^2)\ln(-k^2x(1-x)+m^2-i\epsilon) \tag{S13.21}$$

We need to investigate the sign of the imaginary part of this expression for $k^2>4m^2$. From (S9.2) we have

$$\text{Im}\,\ln(-a-i\epsilon) = -\pi\theta(a) \tag{S13.22}$$

The imaginary part of $\widetilde{\Pi}^f(k^2)$ will be zero unless

$$-k^2x(1-x)+m^2<0 \;\Rightarrow\; k^2 > \frac{m^2}{x(1-x)} \tag{S13.23}$$

In the region of integration, $x(1-x)$ takes its greatest value at $x=\frac{1}{2}$, and so we must have

$$k^2_{\text{min}} = \mu^2 > 4m^2 \tag{S13.24}$$

which is exactly the region of interest. Then $3k^2x(1-x)>m^2$, so the coefficient of the logarithm is positive, and consequently

$$\text{sgn}[\text{Im}\,\widetilde{\Pi}^f(k^2)] = \text{sgn}[\text{Im}\,\ln(-k^2x(1-x)+m^2-i\epsilon)] = \text{sgn}(-\pi) = -1 \tag{S13.25}$$

which was to be shown. ■

24

Isospin

Much of what I've talked about so far has a rather indirect connection with experiment. I'd like to discuss a set of experimental phenomena, though not in our main line of development, in which we can use the field theoretic ideas we have been developing, not in a precise numerical way but generally, as clues to construct a theory that describes experiment. This will not involve rigorous logical development, or even what passes for rigorous logical development by the standards of modern theoretical physics, but rather guesswork.[1]

24.1 Field theoretic constraints on coupling constants

The subject I would like to discuss is one that you have probably encountered previously, isotopic spin. The standard development of isotopic spin begins with the study of nuclear energy levels.[2] Consider a sequence of light nuclei containing the same total number of nucleons, but differing from one another by their charge, for example boron-12 (five protons and seven neutrons), carbon-12 (six of each), and nitrogen-12 (seven protons, five neutrons). (Nuclei with the same number of nucleons but different charges are called **mirror nuclei**, or **isobars**.) Inspecting Figure 24.1 we notice that[3] these states differ from each other merely by the exchange of protons and neutrons. The near-equality of energy levels suggests that the force between two protons is approximately equal to that between two neutrons; the nuclear force is independent of the identity of the nucleons.

Of course there are small differences in the energy levels, but these can be accounted for at least qualitatively by Coulomb corrections: the nitrogen nucleus has a charge of seven while the boron has a charge of five. If one looks at carbon-12, with six protons and six neutrons,

[1] [Eds.] This lecture does not appear either in the Hill–Ting–Chen notes, or in Coleman's notes. This chapter is based on the videotape of Lecture 24 and Peter Woit's notes.

[2] [Eds.] Werner Heisenberg, "Über den Bau der Atomkerne", (On the structure of atomic nuclei), *Zeits. f. Phys.* **77** (1932) 1–11, English translation in D. M. Brink, ed. *Nuclear Forces*, Pergamon Press, 1965.

[3] [Eds.] Figure 24.1 is based on Figure 9, p. 104 of Fay Ajzenberg-Selove, "Energy Levels of Light Nuclei, A = 11–12", *Nuc. Phys.* **A506** (1990) 1–158. The ground states of ^{12}B and ^{12}N correspond to an excited state of ^{12}C, about 15.1 MeV higher than its ground state. This has been set as the zero for the whole diagram, as in the original figure. See Steven Weinberg, *Lectures on Quantum Mechanics*, Cambridge U. P., 2013, p. 133, and Brink, *op. cit.*, Figure 7a, p. 59 for a similar (and quantitative) comparison between the energy levels of ^{11}C and ^{11}B.

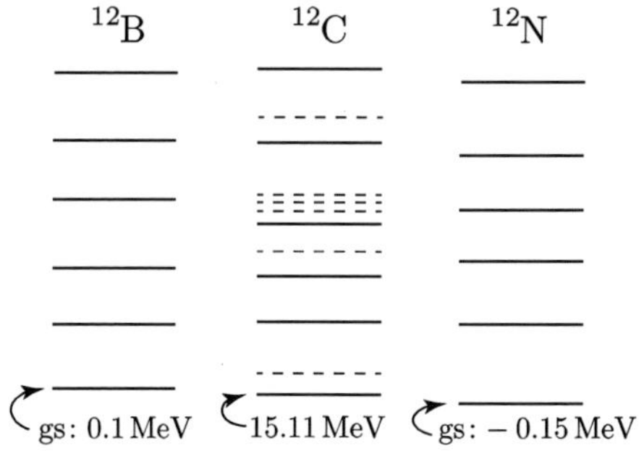

Figure 24.1: Nuclear energy levels for the isobars ^{12}B, ^{12}C and ^{12}N

one finds a similar energy level spectrum, except that there are states (indicated by dashed lines) that are not present in the other cases. This can be viewed as an effect of the exclusion principle, which is less restrictive than with seven protons and five neutrons or *vice versa.* If we imagine these as particles interacting in some sort of collective potential, there are energy levels that can be occupied in the case of carbon, but are forbidden by the exclusion principle in the other cases. This implies that the proton–neutron nuclear force is the same as the proton–proton or neutron–neutron force, except that there are states in which a proton and neutron can occupy that a proton and proton or neutron and neutron cannot. So I should compare energy levels in antisymmetric neutron–proton states only; there are no symmetric neutron–neutron or proton–proton states. This is all in neglect of electromagnetism and of course the weak interactions. (To determine the effects of the weak interactions on nuclear energy level spectra would require more sophisticated experiments.)

I said that the differences in energy levels between these mirror nuclei were qualitatively what would be expected from Coulomb corrections. Actually there are two effects that make the energy levels slightly different. Besides the Coulomb corrections there is the fact that the mass of the proton is not quite equal to the mass of the neutron. It's very tempting (and we will yield to that temptation here) to assume that the proton and neutron mass difference is *itself* an electromagnetic effect, and that the mass of the proton would be equal to that of the neutron, if we had the magical power to turn off electromagnetism (and the weak interactions). There is no real evidence for this, but the order of magnitude of the energy difference is roughly what you would expect on the basis of dimensional analysis, for a sphere of charge q and the approximate radius r of a nuclear particle,

$$\Delta E \sim \frac{q^2}{r} \tag{24.1}$$

It is a plausible idea.[4] No one has been able to calculate the proton and neutron mass difference.[5] We will see later what consequences can be drawn from this hypothesis, and how well it is supported by experiment.

[4] [Eds.] But it suggests a mass difference with the wrong sign: the proton is *less* massive than the neutron, and its electromagnetic energy should make the proton *more* massive.

[5] [Eds.] No one had managed to do it at the time of Coleman's lecture (1976), but a recent calculation of the *n-p* mass difference using QCD lattice gauge theory together with perturbative QED gives pretty good results: $m_n - m_p = +1.512$ MeV, about 17% larger than the empirical value of +1.293 MeV. See Sz. Borsanyi *et al.*, "Ab initio calculation of the neutron–proton mass difference", *Science* **347** (2015) 1452–1455. For a discussion of this result, see Frank Wilczek, "A weighty mass difference", *Nature* **520** (2015) 303–304.

So far I've made no use of field theory. Indeed this argument could be constructed by someone who only knew non-relativistic quantum mechanics. From field theory we know that at least part of the force between two nucleons, certainly the long-range part of the force, is due to the exchange of a π meson. Depending on what the scattering process is, this may be a π^0 meson or a charged π meson. I will assume, in the same spirit as before, that the pions all weigh the same in the absence of electromagnetism. (In reality there is of course a 4 or 5 MeV difference between the pion masses.) I will attempt to construct a field theory consistent with what is known about the pions and nucleons that would give a force due to the exchange of pions, and then investigate the constraints placed on the coupling constants. The field theory will involve a proton field and a neutron field, which I will denote simply by "p" and "n". These are four-component Dirac spinor fields. The p field is, by my usual convention, the field that annihilates the proton and creates the antiproton, and likewise for the n field. This theory will also involve a charged field for the pions, which I will call ϕ_+. This field annihilates π^+. It has a conjugate $\phi_+^\dagger$, which I'll call ϕ_-. There will also be ϕ_0, the field of the neutral pion, which is equal to its adjoint, $\phi_0 = \phi_0^\dagger$. I will assume that the pions interact with the nucleons through interactions of the sort we have been discussing. (Later I will indicate that this assumption can be relaxed.) The pions are empirically known to be pseudoscalar particles: they require γ_5 interactions:

$$\mathscr{L}' = g_P\overline{p}i\gamma_5 p\phi_0 + g_N\overline{n}i\gamma_5 n\phi_0 + g_C\overline{p}i\gamma_5 n\phi_+ + g_C^*\overline{n}i\gamma_5 p\phi_- + \cdots \tag{24.2}$$

That's the most general Lagrangian consistent with Lorentz invariance, parity and electric charge conservation that does not involve any derivative couplings and allows for trilinear pion–nucleon interaction. It involves three unknown parameters, the two real parameters g_P and g_N and the complex parameter g_C. (The free Lagrangian is assumed to be of standard form.) We can simplify matters somewhat by using our freedom to redefine the phases of the complex pion fields: we can absorb the phase of g_C into ϕ_+, and stipulate that g_C is greater than or equal to zero, and we can use our freedom to reverse the sign of ϕ_0 to arrange that g_P is greater than zero. The coupling constants g_P and g_N have to be real if the Lagrangian is to be Hermitian. The dots indicate renormalization counterterms, but we won't bother with them here.

One thing we know about these forces is that they are strong. Otherwise nuclei would fall apart because of the electromagnetic repulsion. In fact if we do scattering experiments, and define the renormalized "charge" in the manner I explained last time, then the coupling constants turn out to be absolutely gigantic. We'll see that they're all the same order of magnitude. A typical coupling constant g, from the analysis of pion–nucleon scattering gives

$$\frac{g^2}{4\pi} \sim 14.7 \tag{24.3}$$

This is a very large number, and therefore perturbation theory, the only analytic tool we have now, is completely useless for analyzing this problem. (In QED, the coupling constant $e^2/4\pi \sim 1/137$.) On the other hand we have one result that is independent of perturbation theory: the pole in the pion propagator lies at the physical mass of the pion. That's one of our renormalization conditions, and it's true to all orders in perturbation theory. If the total force is to be symmetric, i.e., the same between any two nucleons, we should at least have the part of it caused by the pion pole be symmetric. That is the force we get by doing ordinary perturbation theory to second order in any of these coupling constants. One condition that is certainly implied by this is that pp scattering equals nn scattering equals pn scattering in an antisymmetric state to lowest order in perturbation theory:

$$\mathcal{A}(pp \to pp) = \mathcal{A}(nn \to nn) = \mathcal{A}(pn \to pn) \quad \text{in an antisymmetric state to } \mathcal{O}(g^2) \qquad (24.4)$$

This is not an assumption that the strong interactions are weak: it is using what we know, that the lowest order graph in the neighborhood of the pole suffers no corrections because of our renormalization condition, and therefore gives us the exact scattering amplitude at that point. And if the exact scattering amplitude obeys these equalities, then in particular the residue at the poles should obey these equalities.

The next task will be to actually compute the lowest order scattering graphs. Of course we've done them in a simpler theory, one that has only one nucleon and one meson. We need to compute them as functions of the unknown parameters g_P, g_N and g_C, and see what restriction the assumption of nucleon symmetry places on those coupling constants.

We'll begin with pp scattering. We have an initial state $|i\rangle$ which is characterized by some spinor u_1 and some 4-momentum p_1, some spinor u_2 and some 4-momentum p_2 which for shorthand I will simply write as $1, 2$:

$$|i\rangle = |u_1, p_1; u_2, p_2\rangle \equiv |1, 2\rangle \qquad (24.5)$$

and similarly for the final state $|f\rangle$

$$|f\rangle = |u_1', p_1'; u_2', p_2'\rangle \equiv |1', 2'\rangle \qquad (24.6)$$

We wish to compute the scattering amplitude. There are two graphs: the direct graph, and the exchange graph:

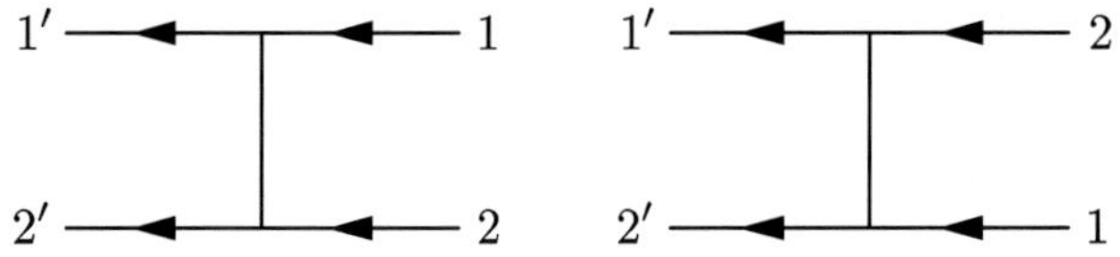

Figure 24.2: $\mathcal{O}(g^2)$ Feynman diagrams for pp scattering

The existence of the neutron and the charged pion are irrelevant here. Because these are two protons, the only particle that can enter is a π_0. These graphs will both be proportional to g_P^2 times some function f, a thing we've computed before (21.100). I will write it out explicitly in a moment (though we won't need its actual form) for $1, 2$ going into $1', 2'$, and then we have the exchange graph, which is the same function with 1 and 2 interchanged:

$$\mathcal{A}(pp \to pp) = g_P^2 \left[f(1', 2'; 1, 2) - f(1', 2'; 2, 1) \right] \qquad (24.7)$$

(see the example on p. 447 for an explanation of the minus sign) where

$$f(1', 2'; 1, 2) = -(\overline{u}_1' i\gamma_5 u_1) \frac{i}{(p_1 - p_1')^2 - \mu^2 + i\epsilon} (\overline{u}_2' i\gamma_5 u_2) \qquad (24.8)$$

Next we do neutron–neutron scattering. There again we have exactly the same two graphs with the π^0 exchanged and the amplitude is g_N^2 times the same thing. Thus from the statement that the pp force equals the nn force we derive the equation

$$g_P^2 = g_N^2 \qquad (24.9)$$

This admits of two possibilities:

$$\begin{aligned}\text{Possibility A}: \; g_P &= g_N \\ \text{Possibility B}: \; g_P &= -g_N\end{aligned} \tag{24.10}$$

(we chose $g_P \geq 0$ but g_N could still be negative).

The case of proton–neutron scattering is a bit more complicated because we have to construct an antisymmetric state. Let me first do it for a non-antisymmetrized state. So again I'll take $|i\rangle = |1, 2\rangle$. It doesn't matter whether the proton or the neutron creation operator comes first, as long as I adopt the same convention for the final state. Later I will construct the amplitude for the antisymmetric state. There are two possible graphs. In the graph on the left, the proton labeled by 1 and $1'$ and the neutron, labeled by 2 and $2'$, fly past each other with the exchange of a π^0. In the one on the right, the proton labeled by 1 comes in, emits a π^+, turning into a neutron, $2'$ and the neutron coming in, 2, absorbs the π^+ and turns into the proton, $1'$. (Of course I could just as well say that the original neutron emits a π^- and turns into a proton—it's the same graph.)

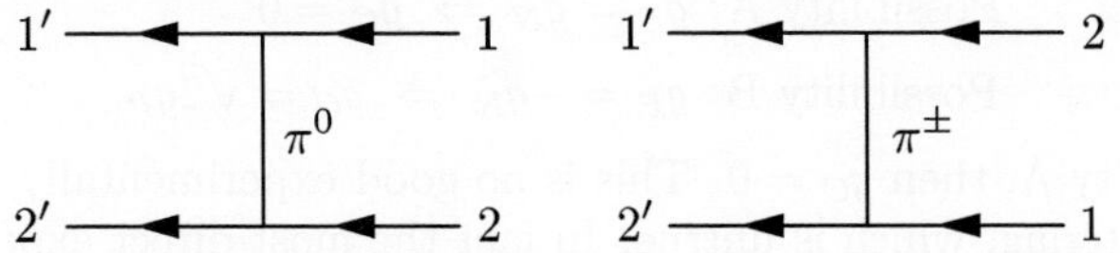

Figure 24.3: $\mathcal{O}(g^2)$ Feynman diagrams for pn scattering

These graphs are the same as those in Figure 24.2, except that the coupling constants enter differently:

$$\mathcal{A}(pn \to pn) = g_P g_N f(1', 2'; 1, 2) - g_C^2 f(1', 2'; 2, 1) \tag{24.11}$$

The minus sign arises because, just as before, the annihilation and creation operators have to be rearranged.

Now we have to construct the scattering amplitude for the antisymmetric state, properly normalized; the initial state is

$$|i\rangle = \frac{1}{\sqrt{2}} \left(|1, 2\rangle - |2, 1\rangle \right) \tag{24.12}$$

The first particle is still a proton, the second particle is still a neutron, we've just changed the momentum and spin labels. Similarly, the final state is

$$|f\rangle = \frac{1}{\sqrt{2}} \left(|1', 2'\rangle - |2', 1'\rangle \right) \tag{24.13}$$

There are four terms in the scattering amplitude. There will be direct terms coming from $|1, 2\rangle$ with $|1', 2'\rangle$, and from $|2, 1\rangle$ with $|2', 1'\rangle$. These will give me identical expressions that will cancel out the $\sqrt{2}$ in the denominator. So from these terms I will get just the same result (24.11) as before:

$$\mathcal{A}_D = g_P g_N f(1', 2'; 1, 2) - g_C^2 f(1', 2'; 2, 1) \tag{24.14}$$

Then there will be the exchange terms, from $|2, 1\rangle$ with $|1', 2'\rangle$ and from $|1, 2\rangle$ with $|2', 1'\rangle$. This contribution will have an overall minus sign because there's an explicit minus sign there.

Once again they will give identical contributions canceling out the $\sqrt{2}$, and the contribution a_E is obtained by making the appropriate exchange, and adding the minus sign:

$$\mathcal{A}_E = -\left[g_P g_N f(1',2';2,1) - g_C^2 f(1',2';1,2)\right] \tag{24.15}$$

To no one's surprise, we have an expression of similar form to (24.14). Adding the two gives

$$\mathcal{A}(pn \to pn)_{\text{antisym}} = (g_P g_N + g_C^2)\left[f(1',2';1,2) - f(1',2';2,1)\right] \tag{24.16}$$

Comparing (24.16) to (24.7) and (24.9), we find

$$g_P^2 = g_N^2 = g_P g_N + g_C^2 \tag{24.17}$$

This equation asserts that (at least in the neighborhood of the pole) proton–neutron scattering in the antisymmetric state is the same as proton–proton scattering and neutron–neutron scattering.

Depending upon whether we adopt Possibility A or Possibility B we obtain two solutions to (24.17)

$$\begin{aligned} &\text{Possibility A:}\ g_P = g_N \ \Rightarrow\ g_C = 0 \\ &\text{Possibility B:}\ g_P = -g_N \ \Rightarrow\ g_C = \sqrt{2}g_P \end{aligned} \tag{24.18}$$

If we adopt Possibility A, then $g_C = 0$. This is no good experimentally. It would mean that there is no $\pi^+ p$ scattering, which is untrue. In fact the most direct experimental evidence is that you have an obvious nucleon pole in pion photoproduction off nucleons, but I didn't want to rest the argument on that because we haven't yet discussed electrodynamics. In any event, $g_C = 0$ is in flat contradiction with experiment and must be rejected.

With Possibility B, $g_P = -g_N$, we deduce that $g_C = \sqrt{2}g_P$. Remember, we have chosen our phases so that both g_C and g_P are positive, so there is no sign ambiguity in taking the square root. Therefore we have found essentially the unique possibility. Our three unknown coupling constants have been reduced to one overall unknown coupling constant. It is therefore useful to change the notation slightly, and introduce a new coupling constant:

$$g = g_P = -g_N = \frac{g_C}{\sqrt{2}} \tag{24.19}$$

The form of our interaction Lagrangian then becomes

$$\mathscr{L}' = g\left[\phi_0(\overline{p}i\gamma_5 p - \overline{n}i\gamma_5 n) + \sqrt{2}\phi_+\overline{p}i\gamma_5 n + \sqrt{2}\phi_-\overline{n}i\gamma_5 p\right] \tag{24.20}$$

24.2 The nucleon and pion as isospin multiplets

This Lagrangian (24.20) has more symmetries than you might think. The square roots of 2 are a little ugly, but remember that we had a similar $\sqrt{2}$ when we defined a complex field (6.23) as a sum of two real fields. That suggests we should define fields ϕ_1 and ϕ_2:

$$\phi_\pm = \frac{1}{\sqrt{2}}(\phi_1 \mp i\phi_2) \tag{24.21}$$

In the same spirit I will relabel ϕ_0 and call it ϕ_3. The form of the interaction then becomes

$$\mathscr{L}' = g\left[\phi_3(\overline{p}i\gamma_5 p - \overline{n}i\gamma_5 n) + \phi_1(\overline{p}i\gamma_5 n + \overline{n}i\gamma_5 p) - i\phi_2(\overline{p}i\gamma_5 n - \overline{n}i\gamma_5 p)\right] \tag{24.22}$$

Define a vector $\mathbf{\Phi}$ by

$$\mathbf{\Phi} = \begin{pmatrix} \phi_1 \\ \phi_2 \\ \phi_3 \end{pmatrix} \tag{24.23}$$

I will also define an eight-component nucleon spinor N:

$$N = \begin{pmatrix} p \\ n \end{pmatrix} \tag{24.24}$$

N consists of a four-component Dirac spinor p sitting on top of the four-component Dirac spinor n. Likewise

$$\overline{N} = (\overline{p},\ \overline{n}) \tag{24.25}$$

I can write out my Lagrangian in terms of these objects. I'll first do the free Lagrangian. Note that

$$\partial^\mu \mathbf{\Phi} \cdot \partial_\mu \mathbf{\Phi} = (\partial_\mu \phi_1)^2 + (\partial_\mu \phi_2)^2 + (\partial_\mu \phi_3)^2 \tag{24.26}$$

so the free Lagrangian for the mesons can be written as

$$\mathscr{L}_\Phi = \tfrac{1}{2}\partial^\mu \mathbf{\Phi} \cdot \partial_\mu \mathbf{\Phi} - \tfrac{1}{2}\mu^2 \mathbf{\Phi} \cdot \mathbf{\Phi} \tag{24.27}$$

Similarly, the free Lagrangian for the nucleons can be written as the sum of the proton and neutron terms:

$$\mathscr{L}_N = \overline{p}(i\partial\!\!\!/ - m)p + \overline{n}(i\partial\!\!\!/ - m)n$$

which I will write, by an abuse of notation, as

$$\mathscr{L}_N = \overline{N}(i\partial\!\!\!/ - m)N \tag{24.28}$$

I will write the interaction similarly as

$$\mathscr{L}' = g\mathbf{\Phi} \cdot \overline{N} i\gamma_5 \boldsymbol{\tau} N \tag{24.29}$$

where $\boldsymbol{\tau}$ is a set of three 8×8 matrices, block diagonal with respect to the four Dirac components, chosen to reproduce the couplings in (24.22):

$$\tau_1 = \begin{pmatrix} \mathbb{0} & \mathbb{1} \\ \mathbb{1} & \mathbb{0} \end{pmatrix} \qquad \tau_2 = i\begin{pmatrix} \mathbb{0} & -\mathbb{1} \\ \mathbb{1} & \mathbb{0} \end{pmatrix} \qquad \tau_3 = \begin{pmatrix} \mathbb{1} & \mathbb{0} \\ \mathbb{0} & -\mathbb{1} \end{pmatrix} \tag{24.30}$$

where $\mathbb{0}$ is the 4×4 matrix whose elements are all zero, and $\mathbb{1}$ is the 4×4 identity matrix.

These three matrices are not strangers to us. They are precisely the three Pauli matrices, but 8-dimensional. Indeed this whole Lagrangian is revealed to be symmetric under a group that is isomorphic to the three-dimensional rotation group, SO(3), or equivalently SU(2), the group of unitary 2×2 matrices with determinant equal to 1. This group has nothing to do with spacetime geometry; it's a purely internal symmetry, acting on ϕ_1, ϕ_2, and ϕ_3 as well as on the neutron and proton field. The internal space in which the transformations are carried out will be called **isospace**. The transformation of the triplet of fields $\mathbf{\Phi}$ is

$$\mathbf{\Phi}(x) \to R(\hat{\mathbf{e}}\theta)\mathbf{\Phi}(x) \tag{24.31}$$

That is, $R(\hat{\mathbf{e}}\theta)$ is a 3×3 rotation matrix characterized by the axis $\hat{\mathbf{e}}$ and the angle θ which acts on $\mathbf{\Phi}$. The trio of fields $\mathbf{\Phi}$ transforms like a vector under this internal group, so we'll call it an **isovector**. The eight-component object N transforms as an **isospinor**:

$$N(x) \to e^{-\frac{1}{2}i\hat{\mathbf{e}} \cdot \boldsymbol{\tau}\theta} N(x) \tag{24.32}$$

under the same group. (Note that the generators of isospin rotations are $\frac{1}{2}\tau_i$, not τ_i, just as the generators of rotations for spin-1/2 are $\frac{1}{2}\sigma_i$.) Just to remind you that these transformations have nothing to do with space rotations, I've written the unchanged variable x as the argument of the fields. This group is called the **isospin group**, or sometimes the group of isotopic rotations; "spin", to emphasize that it is group-theoretically identical ("isomorphic", in mathematical language) to ordinary three-dimensional spin, and "iso-" to indicate that it connects together nuclear isobars.[6] We haven't done anything that might be called rigorous, but we've actually got a lot of symmetry out of a simple assumption, that the proton–proton force and the neutron–neutron force are the same as the neutron–proton force, in antisymmetric states. Using the field-theoretic idea that the long-range part of these forces is caused by the exchange of a pion, we have found that our theory is symmetric under a three-parameter continuous group of internal symmetries.

This is our third encounter with the three-dimensional rotation group. We came across it in its proper guise as SO(3) in §5.6. We met it again analyzing the representations of the Lorentz group (§18.3), when we were able to reduce the Lorentz group into two SO(3) factors. And now we see it a third time, as a purely internal symmetry group. This is very convenient for us, as we do not have to develop a new group theory for these different problems; we just have to continually apply the theory of the three-dimensional rotation group. Unfortunately this is the end of our luck: the next group we will encounter if we continue this line of development is SU(3), the group associated with the Eightfold Way of Gell-Mann and Ne'eman. And for that you have to learn some additional group theory and representation theory. It is *not* expressible in terms of SO(3).

Since this is an internal symmetry, we can apply our machinery (§5.3). I'll just sketch out the results. Since we have a three-parameter continuous group, we can deduce three conserved isospin currents, $\{J_1^\mu, J_2^\mu, J_3^\mu\}$. I'll label them with the index $i = 1, 2, 3$. Do not confuse this with the spatial vector index. It is easy to figure out that the pion field contributes a term

$$(\mathbf{\Phi} \times \partial^\mu \mathbf{\Phi})_i \tag{24.33}$$

That's just the isospin analogy of our old friend $\mathbf{r} \times \mathbf{p}$ in a new guise. The nucleon field contributes

$$\tfrac{1}{2}\overline{N}\gamma^\mu \tau_i N \tag{24.34}$$

That's the same form as the electromagnetic current of a charged fermion field, except there's a $\boldsymbol{\tau}$ to account for three isospin components. So the isospin current is

$$\mathbf{J}^\mu = (\mathbf{\Phi} \times \partial^\mu \mathbf{\Phi}) + \tfrac{1}{2}\overline{N}\gamma^\mu \boldsymbol{\tau} N \tag{24.35}$$

By integrating the time components J_i^0 we obtain three generators:

$$I_i = \int d^3\mathbf{x}\, J_i^0(\mathbf{x}, t) \tag{24.36}$$

These are all conserved quantities, neglecting as always electromagnetism and the weak interactions:

$$\partial_0 I_i = 0 \tag{24.37}$$

[6] It should really be called "isobaric spin", rather than "isotopic spin", but it isn't; it's called isotopic spin by a historically well-embedded slip of the tongue. ([Eds.] Isobars are nuclei with the same number of *nucleons*; isotopes are those with the same number of *protons*; isospin transformations conserve the number of nucleons, but not necessarily the number of protons.)

The three generators of isotopic spin, by the usual arguments, obey the algebra of the rotation group:

$$[I_i, I_j] = i\epsilon_{ijk} I_k \tag{24.38}$$

(sum on repeated indices implied).

For later purposes it will be convenient to introduce the raising and lowering operators I_+ and I_-:

$$I_\pm = I_1 \pm i I_2 \tag{24.39}$$

and write the algebra associated with the three-dimensional rotation group:

$$[I_3, I_\pm] = \pm I_\pm \tag{24.40}$$

This tells us that I_+ and I_- are I_3 raising operators and lowering operators, and

$$[I_+, I_-] = 2I_3 \tag{24.41}$$

These are just formulas which I copied out of the section of the non-relativistic quantum mechanics book that I happened to have on hand, changing J's to I's. I presume you are familiar with all of them.

The $\mathbf{\Phi}$ field transforms like a vector under isospin rotations. A field that transforms under rotations as a vector has $J = 1$; here we say $\mathbf{\Phi}$ carries $I = 1$. The nucleon field, transforming as an isospinor, has $I = \frac{1}{2}$. Once we have given the transformation properties of the fields, we know the transformation properties of the particles they create and annihilate. I will simply write down the table giving the total isospin I and the value of I_3:

Multiplet	*Particle*	I_3
$I = \frac{1}{2}$	p	$½$
	n	$-½$
$I = 1$	π^+	1
	π^0	0
	π^-	-1

The vacuum is of course an isoscalar with $I = 0$. The proton and neutron are both $I = \frac{1}{2}$ since they're created by hitting the vacuum with the $I = \frac{1}{2}$ nucleon field. Their values of I_3 are $+½$ and $-½$ as you can see simply by reading off the spinor:

$$I_3 N = \tfrac{1}{2}\tau_3 N = \tfrac{1}{2}\tau_3 \begin{pmatrix} p \\ n \end{pmatrix} = \tfrac{1}{2}\begin{pmatrix} p \\ -n \end{pmatrix} \tag{24.42}$$

The π^+, π^- and π^0 form an isotriplet. They dance among themselves under the action of the isospin group. The π^0 obviously has $I_3 = 0$, slightly less obviously the π^+ has $I_3 = 1$ and the π^- has $I_3 = -1$. To make sure things are right, we simply observe that we can turn a proton into a neutron plus a π^+ virtually. That's one of our couplings. For the isospin to add up, I_3 has to be conserved. The proton has $I_3 = +½$, the neutron has $I_3 = -½$, the π^+ had damn well better have $I_3 = +1$. I've gone through this rather briefly because it's essentially a direct copy of what I presume you have done many times for ordinary spin. The only novelty is replacing J, angular momentum or S, spin by I at appropriate places.

24.3 Experimental consequences of isospin conservation

Isospin is very restrictive and very easy to test experimentally beyond the simple NN scattering we started with. For example, let's consider pion–nucleon scattering.

There are three possible pion states and two possible nucleon states, so there are six possible initial states and six possible final states. In principle this would give us 36 scattering amplitudes. Of course many of them are zero because of the conservation of electric charge. So just to see how much additional information isospin gives us, let's first count how many amplitudes there are if we only insist on charge conservation. There is a unique initial state with $q = 2$, $\pi^+ p$, and the only thing it can scatter into is $\pi^+ p$. So there is only one amplitude, one function of space and spin variables, for the $q = 2$ channel. There are two states of charge 1, $\pi^0 p$ and $\pi^+ n$, and there are four scattering amplitudes if we consider scattering in the $q = 1$ channel: the elastic scattering of $\pi^0 p \to \pi^0 p$ and $\pi^+ n \to \pi^+ n$, and what is called charge exchange scattering, $\pi^0 p \to \pi^+ n$, and *vice versa*. Likewise for charge 0 we have two states, $\pi^- p$ and $\pi^0 n$, so again there are four amplitudes. And finally for charge -1 we have the unique possibility $\pi^- n \to \pi^- n$. This gives us 10 scattering amplitudes. Two pairs of these are connected if we also insist on time reversal invariance, because $\pi^0 p \to \pi^+ n$ is connected by time reversal to $\pi^+ n \to \pi^0 p$, and likewise $\pi^- p \to \pi^0 n$ is the reverse of $\pi^0 n \to \pi^- p$. This reduces the 10 amplitudes to 8. (Time reversal is a good symmetry for everything except the weak interactions.[7]) This classification can be summarized with a chart:

Charge	*States*	*Number of amplitudes*
2	$\pi^+ p$	1
1	$\pi^0 p$, $\pi^+ n$	4 (3 with T)
0	$\pi^- p$, $\pi^0 n$	4 (3 with T)
-1	$\pi^- n$	1

On the other hand, if we do an isospin analysis, a pion is isospin 1, and the nucleon is isospin ½. Combining these gives only two possible total isospins, $I = \frac{3}{2}$ and $I = \frac{1}{2}$. We have in fact only *two* possible kinds of final states, ignoring space and spin degrees of freedom, and only *two* amplitudes. All these 10 independent amplitudes, or 8 independent amplitudes, are some *linear combinations* of these two amplitudes, one for the isospin-½ channel and one for the isospin-3/2 channel, $\mathcal{A}_{1/2}$ and $\mathcal{A}_{3/2}$, respectively. Therefore isospin, even for the simple problem of pion–nucleon scattering (a very well measured process experimentally) produces enormous restrictions. It enables us to predict these eight independent scattering amplitudes in terms of two unknown functions of momentum and spin. It is a very restrictive assumption—*modulo* electromagnetic corrections, of course. In actual fact, especially for low momentum transfer scattering, electromagnetic corrections can be quite important. Although electromagnetism is weak, it acts over long ranges and therefore dominates the small momentum transfer part of the scattering amplitude. To experimentally check that all the amplitudes are linear combinations of $\mathcal{A}_{1/2}$ and $\mathcal{A}_{3/2}$, you must either restrict yourself to large momentum transfers or make explicit numerical corrections for electromagnetic effects. This just makes life harder for someone who wants to design an experiment to check isospin invariance; it does not affect the conclusion.

[7] [Eds.] The combined symmetry CPT is always good. Since CP is violated by the weak interactions (note 9, p. 240), T must be as well.

Let's explore a very simple application. There is a famous resonance in one of the easiest of scattering experiments to do, π^+p scattering, that occurs in the total cross-section. If you look at a plot of the total cross-section for π^+p as a function of the center-of-momentum energy, there's an enormous bump centered around 1232 MeV, with a width of around 100 MeV, an obvious resonance. It's the famous Delta resonance, Δ^{++}, with charge $+2e$. Aside from kinematic factors which are slowly varying over the width of this resonance, $\sigma(\pi^+p)$ is proportional to the imaginary part of the forward scattering amplitude for $\pi^+p \to \pi^+p$, by the Optical Theorem (12.49). But this amplitude in turn must be written completely in terms of the isospin-3/2 scattering amplitude $\mathcal{A}_{3/2}$, since π^+p is an isospin-3/2 state:

$$\sigma(\pi^+p) \propto \operatorname{Im}\mathcal{A}(\pi^+p \to \pi^+p) \propto \operatorname{Im}\mathcal{A}_{3/2} \tag{24.43}$$

We can compare this with pion–nucleon scattering in which the initial state is not purely $I = \frac{3}{2}$. Looking at our last chart, there are four possibilities: π^0p, π^+n, π^-n, or π^-p. It's always easier to do experiments with proton targets than with neutron targets, since hydrogen is easily available. There is no corresponding substance made out of neutrons. And it's always easier to do experiments with charged pions, because charged beams can be guided with magnets. Neutral pion beams are much harder to manipulate. So I will consider the π^-p amplitude. The initial state π^-p as we see from the multiplet table has $I_3 = -1/2$, and therefore we know from standard Clebsch–Gordan rules (just replacing J's by I's) that a π^-p state will be some linear combination of the state with $I = 3/2$ and $I_3 = -1/2$ and the state with $I = 1/2$ and $I_3 = -1/2$. I looked things up in a table, and I found the coefficients are $1/\sqrt{3}$ and $-\sqrt{2/3}$:

$$|\pi^-p\rangle = \tfrac{1}{\sqrt{3}}\,|I = \tfrac{3}{2}, I_3 = -\tfrac{1}{2}\rangle - \sqrt{\tfrac{2}{3}}\,|I = \tfrac{1}{2}, I_3 = -\tfrac{1}{2}\rangle \tag{24.44}$$

Thus if I compute the forward scattering amplitude for $\pi^-p \to \pi^-p$, I obtain

$$\mathcal{A}(\pi^-p \to \pi^-p) = \tfrac{1}{3}\mathcal{A}_{3/2} + \tfrac{2}{3}\mathcal{A}_{1/2} \tag{24.45}$$

Now I know there is a big resonance, the Δ^{++}, with isospin-3/2. I don't know of course whether there might be, by some incredible fluke, a *second* resonance with isospin-1/2 sitting at exactly the same point. Certainly the simplest assumption is that there isn't any such resonance, and therefore that the imaginary part of the isospin-1/2 amplitude will be relatively small. Then

$$\sigma(\pi^-p) \propto \operatorname{Im}\mathcal{A}(\pi^-p \to \pi^-p) \propto \tfrac{1}{3}\operatorname{Im}\mathcal{A}_{3/2} + (\text{small}) \approx \tfrac{1}{3}\sigma(\pi^+p) \tag{24.46}$$

That is, $\sigma(\pi^-p)$ should look the same as $\sigma(\pi^+p)$ but diminished in height. If I let h be the height of $\sigma(\pi^+p)$, then the height of $\sigma(\pi^-p)$ should be $\frac{1}{3}h$. And indeed if you actually look at the experimental data, which are available in all sorts of tables,[8] there *is* a corresponding peak in the π^-p amplitude, and this height *is* one third of the peak of $\sigma(\pi^+p)$. See Figure 24.4. Isospin is vindicated! The peak in the π^+p amplitude is a little more than 200 millibarns, and that of $\sigma(\pi^-p)$ is about 70 millibarns. It's a beautiful check.

[8] [Eds.] Figure 24.4 is based largely on the graph on p. 229 in Murray Gell-Mann and Kenneth M. Watson, "The Interactions Between π-Mesons and Nucleons", *Ann. Rev. Nuc. Sci.* **4** (1954) 219–270. Coleman earned his PhD under Gell-Mann at Caltech in 1962. The horizontal axis is the kinetic energy K_π of the pion in the laboratory frame (proton at rest); it is related to the total energy (proton plus pion) in the CM frame by $E_{\text{cm}} = \sqrt{(m+\mu)^2 + 2mK_\pi}$. During the lecture, Coleman explained that he had drawn the figures from memory. John LoSecco, now a professor at Notre Dame and the Teaching Fellow for the course in 1975–76, pulled out his copy of the Particle Data Group booklet, looked up the relevant graph, and held it aloft so that Coleman could consult it. Coleman joked that he'd lost his own copy, but they refused to send him a duplicate until there was a new printing.

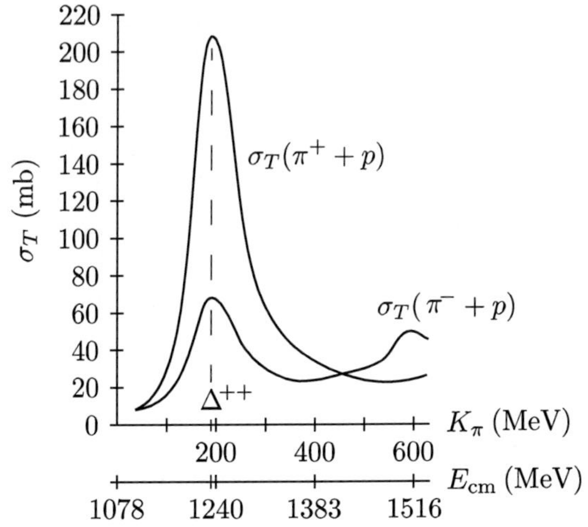

Figure 24.4: σ_T for π^+p and π^-p scattering, with Δ^{++} resonance

A second application of isospin follows if we fold together isospin with some earlier concepts. We know that composite states of spinless identical particles must be symmetric in the spatial variables. Once we introduce spin, the state must be either symmetric or antisymmetric in the product of space and spin variables, depending on whether the particles are bosons or fermions. It's interesting to ask what happens if we introduce isospin. The reasoning is very simple. If I have a two-particle state made out of bosons, say, I make that state by hitting the vacuum with two creation operators

$$a^\dagger_{(i')\,\mathbf{p}'}a^\dagger_{(i)\,\mathbf{p}}\,|0\rangle \tag{24.47}$$

The subscript (i) tells me what the isospin is, or more precisely what the value of I_3 is. Now since these are bosons, these creation operators commute and therefore *a fortiori* just as a consequence of our general formalism for handling multi-particle states, the state is symmetric under the interchange of all variables. So for a multi-particle state built of identical bosons, the state must be symmetric in space variables times spin variables times isospin variables. By the same reasoning, a multi-particle state built of identical fermions must be antisymmetric in space variables times spin variables times isospin variables. In the Fermi case this is sometimes called the **generalized Pauli principle**. It is simply a consequence of the algebra of creation and annihilation operators for independent particles, that they commute for bosons and anticommute for fermions. It is in fact totally free of dynamical assumptions. It is simply a consequence of our bookkeeping rules, which have no physical content. But it does have consequences; it makes some implications of isospin easy to see.

For example, suppose we have a particle called X, some unstable particle that decays into two pions in some charged combinations, I won't specify what:

$$X \to 2\pi \tag{24.48}$$

If X decays to some state with two pions, it must have some definite isospin, if we assume the interactions are isospin symmetric. If X has even J, the two pions, since they have no spin, must be in a state of even L, which is symmetric in space. And therefore since the overall state must be symmetric, it must also be even in isospin. That is to say, $I^{\rm tot}$ must be an even number. Thus even J must be associated with $I = \{0, 2\}$ and odd J must be associated with the antisymmetric isospin wave function, and the antisymmetric combination of two spin-1's is $I = 1$. For instance, a particle with angular momentum 1 and isospin 2 or 0 is forbidden from decaying into two pions, if its decay goes through the strong interactions. Conversely a particle with angular momentum 1 and isospin 1 (such as the ρ meson) is allowed to decay into two pions. As an example of the other case, there is the ω meson which has angular momentum 1 and isospin 0. Its principal decay mode is into three pions. There is a very tiny admixture of two pions, but that's believed to be due to the intervention of electromagnetism, which of course does not respect isospin invariance.

You may have noticed that I have started talking about particles other than pions and nucleons. We know that there's a host of strongly interacting particles. Indeed most particles participate in the strong interactions. Only a very few do not: electrons and muons, their neutrinos, the photon and the graviton, the intermediate vector bosons of the weak interactions.[9] Strongly interacting particles are called **hadrons**. Although we have done our analysis just with pions and nucleons, we know that any system of particles that participates in the strong interactions must observe isospin invariance. We know that from field theory and from the fact that the strong interactions are strong. After all, even if the particles are not pions and nucleons, even if we're only discussing pion–nucleon scattering, these other particles can occur as internal lines. For example, we could scatter a π meson off a proton and exchange a ρ meson, as in Figure 24.5, a particle we had barely mentioned until now. Or we could

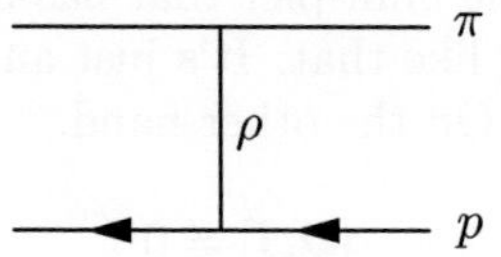

Figure 24.5: Scattering a π meson off a proton with ρ meson exchange

build complicated internal loops with Λ's and Σ's and what have you running around inside the loops. Now in general we can't compute these effects perturbatively, because these are strong interactions (though we *can*, for example, compute the residue of the ρ meson pole). Remember, the characteristic strength of the coupling constant is on the order of magnitude of 10. On the other hand, unless something miraculous is going on, there is no reason for believing these effects are small. Thus if there *were* other strongly interacting particles which did not respect isospin invariance, we would expect them to corrupt the isospin invariance of the pion–nucleon system, and we have seen nothing that suggests that. That's not a proof. Maybe some crazy dynamics comes along and makes all these effects, individually large in perturbation theory, sum up to be something small. If that were true, it would be very exciting. It would tell us something about very strong interactions, but it does not seem likely. And putting aside that possibility, we know that everything that interacts strongly with the pion–nucleon system must interact in a way that conserves isospin, so the isospin invariance of pion–nucleon interactions will not be corrupted. Remember, it's the *total* pion–nucleon force that is known to be isospin invariant, and therefore if the ρ interacted in an isospin-violating way, this would produce a force between pions and nucleons that did not obey the assumption

[9] [Eds.] The gauge bosons of the weak force, formerly "intermediate vector bosons" or IVB's, are now universally known by W^+, W^- and Z^0. All were found in a series of experiments at CERN in the period 1981–83. This work was recognized by the 1984 Physics Nobel Prize, awarded to Simon van der Meer and Carlo Rubbia, the experimental team's leaders.

we originally put down, an assumption that seems well supported by the data. That's a pretty vague argument, but it's powerful. Of course, this doesn't apply to particles that interact electromagnetically; the electromagnetic force is not isospin invariant. The strong interactions are isospin invariant because pion–nucleon scattering is isospin invariant. We could say the same using nucleon–nucleon scattering instead.

24.4 Hypercharge and G-parity

Are there any other quantities that we know are conserved exactly for the strong interactions? I don't mean angular momentum or linear momentum or things like that; I'm talking about internal symmetries. There are. We know of two conservation laws, internal symmetries, that hold good for all interactions, not just the strong interactions. These are **baryon number**, B, sometimes called "nucleon number", and Q, electric charge. There's also a discrete internal symmetry, charge conjugation, and we'll talk about that later. For the moment I just want to talk about these things that are associated with an infinitesimal phase transformation.

Baryon number is pretty trivial; it just goes along for the ride. As far as anyone knows, baryon number commutes with isospin:

$$[B, \mathbf{I}] = 0 \tag{24.49}$$

No one has ever observed an isotopic multiplet that has baryons and mesons mixed together, alternating in isospin or something like that. It's just an extra conservation law which we'll add in a little footnote at the end. On the other hand,

$$[Q, \mathbf{I}] \neq 0 \tag{24.50}$$

electric charge certainly doesn't commute with isospin, because electric charge is *not* constant within an isotopic multiplet, as we can see just from the proton–neutron case; they have different electric charges. Indeed if we restrict ourselves to a system of nucleons and pions, we can write down the commutators. The states we wrote, which were I_3 eigenstates, were also Q eigenstates:

$$[Q, I_3] = 0 \tag{24.51}$$

and raising I_3, going from a π^0 to a π^+ or from a π^- to a π^0 or from a neutron to a proton, also raises the electric charge. So the commutator of Q with $I_\pm$ equals $\pm I_\pm$:

$$[Q, I_\pm] = \pm I_\pm \tag{24.52}$$

(This is just for the system of pions and nucleons. We don't know that it's true in general.) That suggests we define an object equal to Q minus I_3, which is denoted $\frac{1}{2}Y$, the factor of $\frac{1}{2}$ being added to make Y an integer:

$$Y = 2\,(Q - I_3) \tag{24.53}$$

Y is called the **hypercharge**.[10]

---

[10] [Eds.] Coleman has not yet discussed the quantum number S, **strangeness**, associated with strange quarks. Hypercharge was introduced independently by Murray Gell-Mann, and by Kazuhiko Nishijima and Tadao Nakano as baryon number plus strangeness. The relation $Q = I_3 + \frac{1}{2}Y$ is often called the **Gell-Mann–**

Conservation of isospin and conservation of electric charge trivially imply conservation of hypercharge. As we have defined it, Y commutes with isospin, at least for the system of nucleons and pions:

$$[Y, I_i] = 0 \qquad \text{(for nucleons and pions)} \tag{24.54}$$

Of course that doesn't prove that Y commutes with isospin in general. It could be that this commutator of two conserved quantities, which must itself be a conserved quantity, is some conserved quantity X which *vanishes* for the particles we have discussed until now, but

$$[Y, I_i] = X_i \neq 0 \qquad \text{(for other hadrons)} \tag{24.55}$$

If that were so, we would have not five conservation laws—three components of isospin, baryon number and electric charge, or equivalently hypercharge—but six! There would also be the conservation of X which is something altogether different. That would be very nice; it would introduce an extra simplification into the theory of the strong interactions. In fact we would get *three* new conserved quantities. Because of the three components of isospin, X had better carry an isovector index. But it ain't so. We will accept the simplest possibility that this commutator is zero for *all* particles, not just evaluated between states made up out of pions and nucleons. We therefore deduce that hypercharge, like baryon number, commutes with isospin and is just something along for the ride.

The computation of hypercharge for a given particle is made easy by observing that the average value of I_3 over an isotopic multiplet is zero: we always have equal numbers of plus and minus factors. Therefore

$$\langle Y \rangle = 2\,\langle Q \rangle \qquad \text{(averages within an isotopic multiplet)} \tag{24.56}$$

The average hypercharge $\langle Y \rangle$ equals twice the average charge $\langle Q \rangle$ over the isotopic multiplet. But this average hypercharge is in fact the value for *each* member of the isotopic multiplet, because hypercharge is constant over the multiplet. For example, the nucleon multiplet, the proton and neutron, have charge 1 and charge 0 respectively, and thus both proton and neutron have a hypercharge of 1. The pion multiplet has average charge zero, so each pion has hypercharge of zero. This means that for the system of nucleons and pions, the conservation of hypercharge is really trivial, since this is precisely the value of the baryon number assigned to these particles. But this is *not* true if you look at other strongly interacting particles. For example, the Λ hyperon[11] is a particle with baryon number 1 and it is an isotopic singlet and is electrically neutral, so it has $Y = 0$. Likewise the K meson is an isodoublet with charges $+1$ and 0, so it has hypercharge 1 even though it has a baryon number zero. So in general if we consider strongly interacting particles beyond the proton and neutron, the conservation of hypercharge is independent from the conservation of baryon number and gives us useful constraints. Once again, let's summarize these results with a table:

Nishijima relation, and will be introduced in (35.52). See M. Gell-Mann, "The Interpretation of the New Particles as Displaced Charged Multiplets", *Nuov. Cim.* **4** (Supplement) (1956) 848–866; T. Nakano and K. Nishijima, "Charge Independence for V-Particles", *Prog. Theo. Phys.* **10** (1953) 581–582; K. Nishijima, "Some Remarks on the Even-Odd Rule", *Prog. Theo. Phys.* **12** (1954) 107–108; K. Nishijima, "Charge Independence Theory of V Particles," *Prog. Theo. Phys.* **13** (3) (1955) 285–304. See also Ryder *QFT*, pp. 14–15. For the historical background, see Crease & Mann *SC* pp. 177–179.

[11] [Eds.] "Hyperon" is an old-fashioned term for a baryon containing one or more strange quarks, i.e., with strangeness $S \neq 0$, but not any charm, top or bottom quarks. The term was in use before those other quarks were postulated. See note 10, p. 520.

Particle	I_z	Q	Y	B
p	$+1/2$	1	1	1
n	$-1/2$	0	1	1
π^+	1	1	0	0
π^0	0	0	0	0
π^-	-1	-1	0	0
K^+	$+1/2$	1	1	0
K^0	$-1/2$	0	1	0
Λ	0	0	0	1

So far then, the continuous part of the internal symmetry group of the strong interactions is a five-parameter group generated by exponentiating the three components of isospin, baryon number and hypercharge. Isospin obeys angular momentum commutation rules. Baryon number and hypercharge commute with each other and with isospin:

$$\begin{aligned}[I_i, I_j] &= i\epsilon_{ijk} I_k \\ [B, Y] = [B, I_i] &= [Y, I_i] = 0\end{aligned} \tag{24.57}$$

We also have a well-known *discrete* internal symmetry which we have discussed in some detail (§6.3, §11.3, and §22.3), charge conjugation. We have to work out the commutators of charge conjugation with isospin. It is perhaps easiest to start with the nucleon system. The J_3 isospin current is, from (24.34),

$$J_3^\mu = \tfrac{1}{2}(\overline{p}\gamma^\mu p - \overline{n}\gamma^\mu n) \tag{24.58}$$

This is the difference of two currents of the sort we have discussed before. We can work out how this changes under C from the chart on p. 469. The key equation is

$$C\colon \overline{\psi}_A\gamma^\mu\psi_B \to U_C^\dagger\overline{\psi}_A\gamma^\mu\psi_B U_C = -\overline{\psi}_B\gamma^\mu\psi_A \tag{24.59}$$

Then

$$C\colon J_3^\mu \to U_C^\dagger J_3^\mu U_C = U_C^\dagger \tfrac{1}{2}(\overline{p}\gamma^\mu p - \overline{n}\gamma^\mu n)U_C = -\tfrac{1}{2}(\overline{p}\gamma^\mu p - \overline{n}\gamma^\mu n) = -J_3^\mu \tag{24.60}$$

and therefore the integral of its zeroth component, the generator I_3 also changes sign. The J_1 current is much the same,

$$J_1^\mu = \tfrac{1}{2}(\overline{p}\gamma^\mu n + \overline{n}\gamma^\mu p) \tag{24.61}$$

It also turns into minus itself under charge conjugation:

$$C\colon J_1^\mu \to U_C^\dagger \tfrac{1}{2}\left(\overline{p}\gamma^\mu n + \overline{n}\gamma^\mu p\right) U_C = -\tfrac{1}{2}(\overline{n}\gamma^\mu p + \overline{p}\gamma^\mu n) = -J_1^\mu \tag{24.62}$$

and so does the generator I_1. But J_2^μ is different:

$$J_2^\mu = \tfrac{1}{2}i(\overline{p}\gamma^\mu n - \overline{n}\gamma^\mu p) \tag{24.63}$$

The difference does *not* come from the overall factor of i. If you work in the Majorana basis, you will find, surprisingly,

$$C\colon \overline{\psi}_A i\gamma^\mu\psi_B \to U_C^\dagger\overline{\psi}_A i\gamma^\mu\psi_B U_C = -\overline{\psi}_B i\gamma^\mu\psi_A \tag{24.64}$$

What's going on is that *only* J_2^μ is antisymmetric under the interchange of the spinors ψ_A and ψ_B on either side of the Dirac matrix; the others are symmetric under this interchange. So J_2^μ is *unchanged* under C:

$$C\colon J_2^\mu \to U_C^\dagger \tfrac{1}{2}(\overline{p}i\gamma^\mu n - \overline{n}i\gamma^\mu p)U_C = \tfrac{1}{2}(-\overline{n}i\gamma^\mu p + \overline{p}i\gamma^\mu n) = J_2^\mu \tag{24.65}$$

We should check that the same rules are obeyed by the pion part of the current, but in the interest of time I will assume that they are obeyed in general:

$$U_C^\dagger \begin{Bmatrix} I_1 \\ I_2 \\ I_3 \end{Bmatrix} U_C = \begin{Bmatrix} -I_1 \\ I_2 \\ -I_3 \end{Bmatrix} \tag{24.66}$$

(Again, there might be some additional term on the right-hand side that happens to vanish for pions and nucleons. That would be groovy: it would give us an additional conserved quantity. But unfortunately it is not so.)

This complicated set of rules, different for the three components I_i, makes charge conjugation a somewhat awkward object to work with if we are considering the strong interactions. It's convenient to define a new operation G, sometimes misleadingly called "G-parity" (it's got nothing to do with space reflection). It is defined as the product of charge conjugation times a rotation through 180° about the 2-axis in isospin space:

$$U_G = U_C e^{-i\pi I_2} \tag{24.67}$$

Note that the order of the factors is irrelevant, since U_C commutes with I_2. The motivation for this definition is that the 180° rotation about the isospin 2-axis changes the sign of both I_3 and I_1, but does nothing to I_2, and cancels out the effects of charge conjugation. So G, beautifully enough, commutes with all three components of isospin:

$$[G, I_i] = 0 \tag{24.68}$$

Thus we can assign to isotopic multiplets, provided they contain both particles and antiparticles (like the pion triplet), definite values of G.[12] Of course, G turns nucleons into antinucleons because it has charge conjugation in it. For the pions it's easy to see what happens. The π^0 is even under charge conjugation since among other things, it couples to $\overline{p}i\gamma_5 p$, a bilinear well known to be even under charge conjugation. The 180° rotation turns π^0 into $-\pi^0$. Since G commutes with isospin, what I say for the π^0 must be true for both the π^+ and the π^-. Therefore the pion field transforms as

$$G\colon \boldsymbol{\Phi} \to U_G^\dagger\, \boldsymbol{\Phi}\, U_G = -\boldsymbol{\Phi} \tag{24.69}$$

Thus the pion, the particle created by the pion field acting on the vacuum, is G odd, and therefore we obtain a useful selection rule from G-parity conservation: it tells us, for example, that the process $2\pi \to 3\pi$, pion production in π-π scattering, must be forbidden because the initial state has $G = +1$ and the final state has $G = -1$. This process is not forbidden by

[12] [Eds.] Time may have prevented Coleman from making the usefulness of G-parity as clear as he might have wished. The point is that charge conjugation invariance is of limited direct utility, because for a particle to be an eigenstate of C it must be electrically neutral. But if you combine C with an isospin rotation that turns, e.g., a π^+ into a π^-, then charged particles *can* be eigenstates of this joint operation. As it is conserved, you can immediately read off implications for physical processes. G-parity will be revisited in §35.4.

anything else. It is not forbidden by isospin conservation; you can indeed put together five vectors to make an isoscalar: stick three of them together with ϵ_{ijk}, and dot the last two to make a scalar. It is not forbidden by parity, even though the pion is pseudoscalar: this is a five-particle vertex, and therefore involves four independent 4-vectors, which we can stick together with $\epsilon_{\mu\nu\lambda\rho}$. It is not forbidden, to my knowledge, by anything *except* the combination of charge conjugation and isospin, that is to say by G-parity.

25

Coping with infinities: regularization and renormalization

Thus far we've been slovenly with renormalization, but we've gotten away with it. For example, returning to the ps–ps theory (23.1), we had for the vertex function

$$\widetilde{\Gamma}' = i\gamma_5 g + \mathcal{O}(g^3) \tag{23.83}$$

The two graphs contributing to $\widetilde{\Gamma}'$ at $\mathcal{O}(g^3)$ are shown in Figure 25.1. At high k, the integral

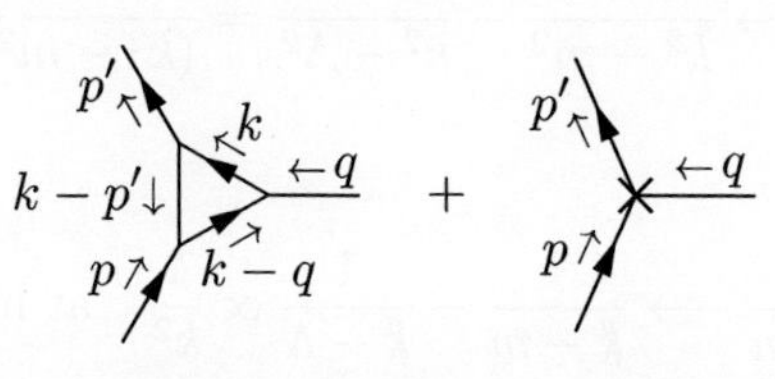

Figure 25.1: $\mathcal{O}(g^3)$ graphs for $\widetilde{\Gamma}'$ in ps-ps theory

for the left-hand diagram goes as

$$\int d^4k \, \frac{1}{(k-p')^2 - \mu^2 + i\epsilon}\gamma_5 \frac{1}{\not{k} - \not{q} - m + i\epsilon}\gamma_5 \frac{1}{\not{k} - m + i\epsilon}\gamma_5 \sim \gamma_5 \int \frac{d^4k}{k^4} \tag{25.1}$$

This is divergent, but it's only logarithmically divergent. The right-hand graph contributes the $\mathcal{O}(g^3)$ counterterm $-E_3\gamma_5$, (see (23.25)) which cancels the divergence.

So far, this is about as far as we've taken renormalization. Throwing around ill-defined quantities and finding that they always end up in convergent combinations isn't really enough. It's time to answer the fundamental question. Is renormalization necessary and sufficient to get rid of infinities?[1]

[1] [Eds]. Unfortunately the videotape of Lecture 25 starts about 70 minutes into the lecture. The first part of this chapter is interpolated from the Hill–Ting–Chen and Woit notes, and Coleman's own notes. See also Coleman's 1971 Erice lecture "Renormalization and symmetry: a review for non-specialists", Chapter 4 in Coleman *Aspects*. In 1976, copies of this chapter were handed out in class.

25.1 Regularization

No one knows of a quantum field theory that is nontrivial and finite. It was realized long ago that we should, as an intermediate step, render finite those quantities that are formally infinite *before* carrying out calculations, to avoid making *ad hoc* cancellations of infinite quantities. This process is called **regularization**. Typically it involves introducing a parameter, often denoted Λ, with $\Lambda < \infty$. At the end of the calculation, we restore the quantities to their original values (usually by taking the limit $\Lambda \to \infty$). If all goes well, there will be no trace of the intermediate regularization, so we can be reasonably confident of the results. Several regularization schemes have been introduced.

Method 1: Brute force

We simply throw away $\widetilde{\pi}(\mathbf{k})$ and $\widetilde{\phi}(\mathbf{k})$ for $|\mathbf{k}| > \Lambda$. Alternatively, we can put the entire theory in a box of finite size, reducing the degrees of freedom to a finite number. This procedure is admired by the mathematically inclined. The disadvantages are serious: we lose both Lorentz invariance and gauge invariance.

Method 2: Propagator modification

This procedure was pioneered by Feynman, and by Stueckelberg and Rivier[2] and developed extensively by Pauli and Villars,[3] by whose names it is often known. The idea is to replace propagators in the Feynman integrals by expressions that fall off fast enough at high momentum so that loop integrals will be finite. In the simplest version, we make the replacement

$$\frac{1}{k^2 - m^2} \longrightarrow \frac{1}{k^2 - m^2} - \frac{1}{k^2 - \Lambda^2} = \frac{m^2 - \Lambda^2}{(k^2 - m^2)(k^2 - \Lambda^2)} \tag{25.2}$$

which is $\mathcal{O}(1/k^4)$. Similarly,

$$\frac{1}{\not{k} - m} \longrightarrow \frac{1}{\not{k} - m} - \frac{1}{\not{k} - \Lambda} \propto \frac{1}{k^2} \quad \text{at high } k \tag{25.3}$$

Changing a propagator's momentum dependence from inverse square to inverse fourth power may not be enough to make some diagrams finite. More generally,

$$\frac{1}{k^2 - m^2} \longrightarrow \frac{1}{k^2 - m^2} + \sum_{i=1}^{N} \frac{c_i}{k^2 - M_i^2} \qquad M_i > \Lambda \tag{25.4}$$

We can look at the behavior of this for high k^2 by expanding each term; for instance:

$$\frac{1}{k^2 - m^2} = \frac{1}{k^2}\left(\frac{1}{1 - (m/k)^2}\right) = \frac{1}{k^2}\left[1 + \left(\frac{m}{k}\right)^2 + \left(\frac{m}{k}\right)^4 + \cdots\right] \tag{25.5}$$

The choice of the coefficients $\{c_i\}$ determines the high k behavior of the regularized propagator (25.4):

---

[2] [Eds.] Richard P. Feynman, "Relativistic Cut-Off for Quantum Electrodynamics", *Phys. Rev.* **74** (1948) 1430–1438; Dominique Rivier and Ernst C. G. Stueckelberg, "A Convergent Expression for the Magnetic Moment of the Neutron", *Phys. Rev.* **74** (1948) 218; Erratum, 986.

[3] [Eds.] Wolfgang Pauli and Felix Villars, "On Invariant Regularization in Relativistic Quantum Theory", *Rev. Mod. Phys.* **21** (1949) 434–444.

Condition	High k Behavior
$1+\sum_{i=1}^{N} c_i = 0$	$\sim \dfrac{1}{k^4}$
and $m^2+\sum_{i=1}^{N} M_i^2 c_i = 0$	$\sim \dfrac{1}{k^6}$
and $m^4+\sum_{i=1}^{N} M_i^4 c_i = 0$	$\sim \dfrac{1}{k^8}$
and $\cdots$	$\cdots$

and the pattern should be clear. The case $N = 1$, $c_1 = -1$ and $M_1 = \Lambda$ reproduces (25.2). With this procedure we can make the propagator fall off like any inverse polynomial in k^2. Notice that the M_i's have to have different values to solve for the c_i's.

There is also an operator form of the Pauli–Villars procedure. Suppose the original theory has the form

$$\mathscr{L} = \tfrac{1}{2}(\partial^\mu \phi')^2 - \tfrac{1}{2}\mu^2\phi'^2 + \mathscr{L}_I(\phi') \tag{25.6}$$

Introduce a Pauli–Villars **regulator field**, ϕ_1. The new Lagrangian is

$$\mathscr{L}' = \tfrac{1}{2}(\partial^\mu \phi')^2 + \tfrac{1}{2}(\partial^\mu \phi_1)^2 - \tfrac{1}{2}\mu^2\phi'^2 - \tfrac{1}{2}M^2\phi_1^2 + \mathscr{L}_I(\Phi) \qquad \text{where } \Phi = \phi' + i\phi_1 \tag{25.7}$$

so that the contraction $\overline{\Phi(x)\Phi}(0)$ is given by (see (9.29))

$$\overline{\Phi(x)\Phi}(0) = \int \frac{d^4k}{(2\pi)^4} e^{-ik\cdot x}\left(\frac{i}{k^2 - m^2 + i\epsilon} - \frac{i}{k^2 - M^2 + i\epsilon}\right) \tag{25.8}$$

The field ϕ_1 is a little strange, appearing as the imaginary part of Φ. Because Φ is not Hermitian, neither is it Lagrangian. Define N_1 as the number operator for ϕ_1 particles. Then

$$(-1)^{N_1}\phi_1 = \phi_1(-1)^{N_1+1} = -\phi_1(-1)^{N_1} \tag{25.9}$$

That is, ϕ_1 anticommutes with $(-1)^{N_1}$. We can gain some insight into what is going on by defining a new inner product,

$$\langle a|b\rangle_{\text{new}} = \langle a|(-1)^{N_1}|b\rangle \tag{25.10}$$

This metric is not positive definite. If the state $|a\rangle$ has an odd number of ϕ_1 particles in it,

$$\langle a|a\rangle_{\text{new}} = -\langle a|a\rangle < 0 \tag{25.11}$$

For states without ϕ_1 particles, the inner product is its old self. The purpose of the new inner product is to make Φ Hermitian. With it,

$$\langle a|\phi_1|b\rangle_{\text{new}} = -\langle b|\phi_1|a\rangle^*_{\text{new}} \tag{25.12}$$

so

$$(\phi_1)^\dagger_{\text{new}} = -\phi_1 \;\Rightarrow\; (\Phi)^\dagger_{\text{new}} = \Phi \tag{25.13}$$

In the old metric, which was positive definite, the Hamiltonian wasn't Hermitian, and thus didn't conserve probability. In the new metric, not positive definite, the Hamiltonian is Hermitian, probability *is* conserved, and the S-matrix is unitary. At the end of our calculations, we won't be interested in amplitudes that contain the phony ϕ_1 particles in initial or final states. When we take $M \to \infty$, we remove them, because energetically they cannot be produced. So we have every expectation that in this limit the resulting theory will be sensible.

Regulator fields have many desirable properties. They preserve Lorentz invariance and internal symmetries in theories with massive particles (though if the original theory is massless they may spoil some symmetries). They conserve probability in processes low in energy compared with the cut-off mass, and with some modification, they even preserve gauge invariance in QED. Finally, they are easy to introduce.

Method 3: Dimensional regularization

The idea here is to modify the number of spacetime dimensions from 4 to a *continuous variable*, d, chosen to make integrals (or sums) convergent. At the end, one takes the limit $d \to 4$. This procedure was proposed independently by several physicists[4] in 1972, but is usually associated with 't Hooft and Veltman.

Consider the integral

$$I = \int \frac{d^d k}{(k^2 + a^2)^n} \tag{25.14}$$

Here, k is taken to be a vector in d-dimensional Euclidean space:

$$k^2 \equiv k_1^2 + k_2^2 + \cdots + k_d^2$$

whereas for a vector in d-dimensional Minkowski space

$$k^2 \equiv k_0^2 - k_1^2 - k_2^2 - \cdots - k_{d-1}^2$$

The integral (25.14) is convergent if $n > d/2$. (We computed a similar integral (I.1) in $d = 4$ Minkowski space for the integral table on p. 330.) We go from Minkowski space to Euclidean space via a Wick rotation (15.7), to simplify the calculation and remove the poles. Later on, in §28.2, we'll use this trick to turn an oscillating exponential into a damped exponential, and thus guarantee convergence of certain integrals. To evaluate this integral, we use a trick to convert a denominator into an exponential. Recall the gamma function,

$$\Gamma(n) = \int_0^\infty dt\, t^{n-1} e^{-t} \tag{25.15}$$

If we change variables, letting $t = \alpha\lambda$, with α real and positive, then

$$\frac{1}{\alpha^n} = \frac{1}{\Gamma(n)} \int_0^\infty d\lambda\, \lambda^{n-1} e^{-\alpha\lambda} \tag{25.16}$$

In particular, letting $\alpha = (k^2 + a^2)$,

$$\frac{1}{(k^2 + a^2)^n} = \frac{1}{\Gamma(n)} \int_0^\infty d\lambda\, \lambda^{n-1}\, e^{-\lambda(k^2+a^2)}$$

[4] [Eds.] J. F. Ashmore, "A Method of Gauge-Invariant Regularization", *Lett. Nuovo Cim.* **4** (1972) 289–90; C. G. Bollini & J. J. Giambiagi, "Dimensional Renormalization: The Number of Dimensions as a Regularizing Parameter", *Nuovo Cim.* **12B** (1972) 20–26; G. M. Cicuta & E. Montaldi, "Analytic Renormalization Via Continuous Space Dimension", *Lett. Nuovo Cim.* **4** (1972) 329–32; Gerard 't Hooft & Martinus Veltman, "Regularization and Renormalization of Gauge Fields", *Nuc. Phys.* **B44** (1972) 189–213. The name 't Hooft is pronounced (approximately) as "ət HOAFT", to rhyme with "(u)t loaf(ed)".

so (25.14):

$$I = \frac{1}{\Gamma(n)} \int_0^\infty d\lambda\, \lambda^{n-1} \int d^d k\, e^{-\lambda(k^2+a^2)}$$

But

$$\int d^d k\, e^{-\lambda(k^2+a^2)} = e^{-\lambda a^2} \int d^d k\, e^{-\lambda k^2} = e^{-\lambda a^2} \prod_{i=1}^{d} \int_{-\infty}^{\infty} dk_i e^{-\lambda k_i^2}$$
$$= e^{-\lambda a^2} \left(\sqrt{\pi/\lambda}\right)^d$$

so

$$I = \frac{\pi^{d/2}}{\Gamma(n)} \int_0^\infty d\lambda\, \lambda^{n-(d/2)-1} e^{-\lambda a^2} = \frac{\pi^{d/2}}{\Gamma(n)} \frac{\Gamma(n - \frac{1}{2}d)}{a^{2n-d}} \tag{25.17}$$

(using (25.16) again, this time with $\alpha = a^2$ and $n \to n - (d/2)$). Or, returning to the original expression,

$$\int \frac{d^d k}{(k^2 + a^2)^n} = \frac{\pi^{d/2}}{a^{2n-d}} \frac{\Gamma(n - \frac{1}{2}d)}{\Gamma(n)} \tag{25.18}$$

(This expression is identical with (I.1) on p. 330 when $d = 4$, *modulo* factors of $(2\pi)^4$ and an i, from (15.66).) The idea is now to adopt this formula for *complex and continuous* values of d. If you stay away from even integers $d \geq 2n$, the expression is well-defined. You do renormalization with regularized quantities in terms of an arbitrary d, and only after obtaining expressions for the graphs plus counterterms in convergent combinations (with poles in $(d - 4)$ cancelling) do you take the limit $d \to 4$.

Technical issues arise in changing the number of spacetime dimensions, of course. For example, you can't maintain

$$\alpha = \frac{e^2}{4\pi} \approx \frac{1}{137} \tag{25.19}$$

as a *dimensionless* constant, because that is true only for $d = 4$. (Sometimes, "dimension" will be used as shorthand for the powers of mass, $[M]$, or inverse powers of $[L]$, length, of an object.) And how should we define a set of Dirac γ matrices in a different number of spacetime dimensions? This is a particular problem for γ_5. But even with a simpler theory like

$$\mathscr{L} = \tfrac{1}{2}(\partial^\mu \phi')^2 - \tfrac{1}{2}\mu^2 \phi'^2 - \tfrac{1}{4!}\lambda \phi'^4 + \mathscr{L}_{CT} \tag{25.20}$$

there are complications. The quantity $\int d^d x\, \mathscr{L}$ must be dimensionless, so $[\mathscr{L}] = [M]^d$. Because

$$[\phi'] = [M]^{\frac{1}{2}d-1} \tag{25.21}$$

(see Problem 2.1, p. 99 and (S2.4), p. 101), we know μ has dimension 1, just as mass does in four dimensions. However, we must have $[\lambda \phi'^4] = [M]^d$, so

$$[\lambda] = [M]^d ([M]^{\frac{1}{2}d-1})^{-4} = [M]^{4-d} \tag{25.22}$$

Only in four dimensions is λ dimensionless. To keep λ dimensionless as we change d, we introduce a parameter ν with the dimension of mass, $[M]^1$, and rewrite the interaction as

$$\mathscr{L}_I = \tfrac{1}{4!}\lambda \nu^{4-d} \phi'^4 \tag{25.23}$$

You might think that after we take the limit $d \to 4$, all ν dependence would go away. But that is not so, as the next example shows.

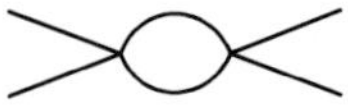

Figure 25.2: $\mathcal{O}(\lambda^2)$ correction to $\widetilde{G}^{(4)}$ in ϕ^4 theory

EXAMPLE. *An $\mathcal{O}(\lambda^2)$ contribution to the four-point function $\widetilde{G}^{(4)}$*

Consider the diagram below:

The contribution $\mathcal{A}$ from this diagram is proportional to

$$\mathcal{A} \propto (\lambda\nu^{4-d})^2 \int \frac{d^dk}{(2\pi)^d} \frac{1}{(k^2+a^2)^2} \tag{25.24}$$

where a contains masses, external momenta and perhaps Feynman parameters. I have suppressed the Feynman parameter integral (15.58). According to (25.18),

$$\mathcal{A} \propto \frac{1}{(2\pi)^d}(\lambda\nu^{4-d})^2 \frac{\pi^{d/2}}{\Gamma(2)}\Gamma(2-(d/2))a^{d-4} \tag{25.25}$$

Let $\epsilon \equiv 2-(d/2)$ be very small; then[5]

$$\Gamma(\epsilon) = \frac{1}{\epsilon} - \gamma + \mathcal{O}(\epsilon) \tag{25.26}$$

Perhaps substituting (25.26) into (25.25) and evaluating all but the pole term at $\epsilon = 0$ would give the finite part of this expression. In fact, more care is required. We pull out a factor of ν^{4-d}, the dimension of this Green's function; the remainder is dimensionless.[6]

Substituting (25.26) into (25.25),

$$\mathcal{A} \propto \frac{1}{(2\pi)^d}\lambda^2\nu^{2\epsilon}\frac{\pi^{d/2}}{\Gamma(2)}\Gamma(\epsilon)a^{-2\epsilon} = \frac{\lambda^2\pi^2}{(2\pi)^d}\left[\frac{1}{\epsilon} - \gamma + \mathcal{O}(\epsilon)\right]\left(\frac{\nu^2}{\pi a^2}\right)^{\epsilon} \tag{25.27}$$

Rewriting,

$$\left(\frac{\nu^2}{\pi a^2}\right)^{\epsilon} = \exp\left[\ln\left(\frac{\nu^2}{\pi a^2}\right)^{\epsilon}\right] = \exp\left[\epsilon\ln\left(\frac{\nu^2}{\pi a^2}\right)\right] \approx 1 + \epsilon\ln\left(\frac{\nu^2}{\pi a^2}\right) \tag{25.28}$$

---

[5] [Eds.] The singularities of $\Gamma(z)$ occur at $z = 0$ and negative integers. Writing $z = -s + \epsilon$ with s an integer and ϵ small,

$$\Gamma(-s+\epsilon) = \frac{(-1)^s}{s!}\left[\frac{1}{\epsilon} + \psi(s+1) + \mathcal{O}(\epsilon)\right]$$

See equation (3.17), p. 152 in Pierre Ramond, *Field Theory: A Modern Primer*, Benjamin, 1981. Derivations are given in Appendix 8D of Hagen Kleinert and Verena Schulte-Frohlinde, *Critical Properties of ϕ^4 Theories*, World Scientific, 2001, pp. 126–129; the formula is equation (8D.24); and in Ryder *QFT*, Appendix 9B, pp. 385–387. For our purposes, $s = 0$. The **digamma function** $\psi(s)$ is the derivative of the logarithm of $\Gamma(s)$:

$$\psi(s) = \frac{d}{ds}\ln\Gamma(s) = -\gamma + H_{s-1}$$

where γ is the Euler–Mascheroni constant), equal to 0.57721... and $H_s = 1 + \frac{1}{2} + \cdots + \frac{1}{s}$ is the harmonic series of order s; $H_0 = 0$, and so $\psi(1) = -\gamma$. See Julian Havil, *Gamma*, Princeton U. Press, 2003, p. 58.

[6] [Eds.] To lowest non-trivial order, $\widetilde{G}^{(4)} =$ 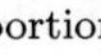, proportional to λ in four dimensions, and in d dimensions proportional to $\lambda\nu^{4-d}$. That sets the dimensions of all the terms in $\widetilde{G}^{(4)}$.

Putting it all together,

$$\mathcal{A} \propto \frac{\lambda^2\pi^2}{(2\pi)^d}\left[\frac{1}{\epsilon} - \gamma + \ln\left(\frac{\nu^2}{\pi a^2}\right) + \mathcal{O}(\epsilon)\right] \tag{25.29}$$

Had we set $\epsilon = 0$ (or equivalently, $d = 4$) prematurely, we would have lost the $\ln(\nu^2/\pi a^2)$ term.

A companion to dimensional regularization is called **minimal subtraction**, a method of determining counterterms. It makes no reference to the physical mass and physical coupling constants, so it is not good for comparison with experiment. Theorists like it *because* it makes no comparison with experiment, and because it is easy. It amounts to just throwing away the pole terms in the dimensionally regularized integrals.

Continuing with our example, we found (25.27) that (suppressing the Feynman parameter integral) the four point function gave a contribution proportional to

$$\nu^{4-d}\lambda^2\pi^2\left[\frac{1}{2-(d/2)} + (\text{finite as } d \to 4)\right] \tag{25.30}$$

The coefficient of this pole is unambiguous. Minimal subtraction says we introduce a counterterm in $\mathscr{L}_{CT}$ to cancel it:

$$\nu^{4-d}\lambda^2\pi^2\frac{1}{2-(d/2)}\,\tfrac{1}{4!}\phi'^4 \tag{25.31}$$

There's a systematic way to add counterterms, to which we now turn.

25.2 The BPHZ algorithm

To make things simple, we'll restrict our attention to theories describing only spin-0 and spin-½. To explain this iterative algorithm we need to introduce some useful terminology. We'll look at Lagrangians of the form

$$\mathscr{L} = \mathscr{L}_0 + \sum_i \mathscr{L}_i \qquad \text{where } \mathscr{L}_i \text{ are monomials containing } \begin{cases} f_i & \text{Fermi fields} \\ b_i & \text{Bose fields} \\ d_i & \text{derivatives} \end{cases} \tag{25.32}$$

and $\mathscr{L}_0$ is a sum of free Lagrangians. Here is a table of some typical Lagrangians with their f_i, b_i, and d_i values:

$\mathscr{L}_i$	f_i	b_i	d_i
ϕ^4	0	4	0
$\overline{\psi}\psi\phi$	2	1	0
$\overline{\psi}\gamma_\mu\gamma_5\psi\partial^\mu\phi$	2	1	1

We'll need these numbers shortly.

The superficial degree of divergence D

In the integral associated with any Feynman diagram, let P_N be the power of the momenta in the numerator and P_D the power in the denominator. For instance, every loop integral puts 4 powers of momenta into the numerator (in the form of d^4p); every boson propagator puts 2 powers into the denominator, every fermion propagator puts 1 power into the denominator, and every derivative puts a factor of p into the numerator. Then define D, the **superficial degree of divergence**, as

$$D \equiv P_N - P_D: \text{ If } \begin{cases} D < 0 & \text{the diagram is superficially convergent} \\ D = 0 & \text{the diagram is superficially logarithmically divergent} \\ D = 1 & \text{the diagram is superficially linearly divergent} \end{cases} \tag{25.33}$$

and so on. For example, consider the following diagrams. In ϕ^4 theory,

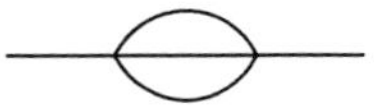

$$D = (\text{two loops}) - (\text{three propagators}) = 8 - 6 = 2 \tag{25.34}$$

which is superficially quadratically divergent. In pion–nucleon theory, we have

$$D = (\text{one loop}) - (\text{two propagators}) = 4 - 3 = 1 \tag{25.35}$$

superficially linearly divergent;

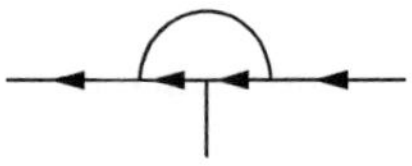

$$D = (\text{one loop}) - (\text{three propagators}) = 4 - 4 = 0 \tag{25.36}$$

superficially logarithmically divergent; and

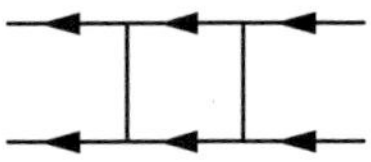

$$D = (\text{one loop}) - (\text{four propagators}) = 4 - 6 = -2 \tag{25.37}$$

which is superficially convergent. Well, why do I say "superficially"? Consider this diagram:

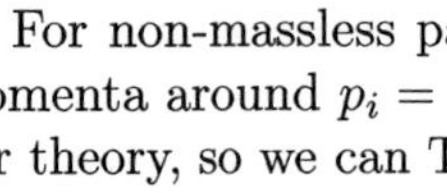

$$D = (\text{two loops}) - (\text{seven propagators}) = 8 - 10 = -2 \tag{25.38}$$

The rule says it's superficially convergent, but in fact it's divergent![7] Despite its inability to predict a diagram's divergence accurately, D will be very useful, as we'll see.

Taylor expansion about the point $p = 0$

For non-massless particle theory, Feynman diagrams are analytic functions of external momenta around $p_i = 0$. We will henceforth assume that there are no massless particles in our theory, so we can Taylor expand the expressions associated with our diagrams:

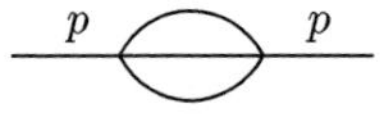

$$= a + bp^2 + \cdots \tag{25.39}$$

$$= A\gamma_5 + B\gamma_5 \not{p} + B'\gamma_5 \not{p}' + \cdots \tag{25.40}$$

[7] [Eds.] This is an application of *Weinberg's theorem*; the graph contains the divergent (25.35) as a subgraph, and so it too is divergent. See Bjorken & Drell *Fields*, p. 324.

The first diagram lacks a linear term (all the p terms come from propagators, p^2, or loop integrals, p^4); the zeroth and second-order terms are the first terms in the Taylor expansion about $p = 0$. In the second diagram, there are zeroth and first-order terms in both p and p'.

With these preliminaries out of the way, I can now describe the algorithm, originally applied to theories renormalized with cut-off parameters:

The BPHZ algorithm

1. Compute in perturbation theory to all orders until you reach a 1PI diagram with $D \geq 0$.
2. Add to $\mathscr{L}$ counterterms to cancel the terms in the graph's Taylor expansion (about zero) of order $\leq D$.
3. Return to 1, continuing to compute with the new, corrected $\mathscr{L}' = \mathscr{L} + \mathscr{L}_{CT}$.

The algorithm also explains exactly what form these terms take, as we'll see shortly. The algorithm appears in an article by Bogoliubov and Parasiuk.[8] The power of the algorithm derives from a theorem by Hepp.[9] Zimmerman[10] showed that all ultraviolet divergences are removed by the algorithm, so the procedure is known as **BPH** or **BPHZ renormalization**.

Hepp's theorem: Bogoliubov's algorithm removes all divergences (if the theory does not involve massless fields). The Green's functions resulting from the algorithm are independent of the cut-off Λ as $\Lambda \to \infty$, to all orders in perturbation theory, no matter what the regularization procedure.

The algorithm solves the problem of renormalization, since the counterterms are built up correctly. At each order of perturbation theory, the only new problems arise from new divergences connected to superficially divergent diagrams. Other divergences are taken care of automatically by earlier counterterms. We'll see how this works with specific examples.

25.3 Applying the algorithm

Instead of just stating theorems in a loud voice, I will now compute in a simple way the superficial degree of divergence of a particular Feynman graph in a theory of this kind. This will enable us to see the difference between a renormalizable and a non-renormalizable theory.[11] In addition, I will state a rule for constructing the counterterms based on the superficial degree of divergence.

I will need to define some terms. F_E is the number of external Fermi line in a graph. F_I is the number of internal Fermi lines. Likewise B_E is the number of external Bose lines, and B_I is the number of internal Bose lines. Let n_i be the number of vertices of type i; that is to say,

---

[8] [Eds.] Nikolai N. Bogoliubov and Ostap S. Parasiuk, "Über die Multiplikation der Kauselfunktionen in der Quantentheorie der Felder" (On the multiplication of causal functions in the quantum theory of fields), *Acta Math.* **97** (1957) 227–266.

[9] [Eds.] Klaus Hepp, "Proof of the Bogoliubov–Parasiuk Theorem on Renormalization", *Comm. Math. Phys.* **2** (1966) 301–326.

[10] [Eds.] Wolfhart Zimmerman, "Local Operator Products and Renormalization in Quantum Field Theory", pp. 399–589 in *Lectures on Elementary Particles and Quantum Field Theory (1970 Brandeis University Summer Institute in Theoretical Physics)*, v. 1, eds. Stanley Deser, Marc Grisaru, and Hugh Pendleton, MIT Press, 1970; "Convergence of Bogoliubov's method of renormalization in momentum space", *Comm. Math. Phys.* **15** (1969) 208–234.

[11] [Eds.] The videotape of Lecture 25 begins here.

coming from an interaction of the ith type in our effective Lagrangian (25.32), and as before d_i is the number of derivatives in that interaction.

I will write a formula for the superficial degree of divergence D of such a graph. This will simplify things enormously. First I'll just count powers. Every internal Bose line gives a factor of one over p^2 from the propagator and a factor of d^4p from our integration, or two powers of p in the numerator. Some of those will be reduced by delta functions at the vertices, but I'll take care of that later. Every internal Fermi line gives us one d^4p in the numerator and one power of p in the denominator, a total of three powers of p in the numerator. Every derivative interaction will give us one power of p in the numerator.

I've overcounted the internal momenta because not all of their integration variables are independent: every vertex has a delta function, and that knocks out four integration variables. These are all 1PI graphs and therefore *a fortiori* connected. However, there is one overall delta function left over for energy momentum conservation. So I've overcounted the number of delta function restraints by 4. This is simply a general formula for what I did before when I was counting numerators and denominators. Four powers of the internal momenta for each internal line of any kind, cut down by two for a Bose propagator, reduced by one for a Fermi propagator, knocked down by four for each delta function at a vertex, except for one delta function left over for overall energy momentum conservation. That one doesn't restrain the loop momentum. Then

$$D = 2B_I + 3F_I + \sum_i n_i d_i - 4\sum_i n_i + 4 \tag{25.41}$$

In this form the expression for D is a mess. Fortunately we can simplify it by using the laws of conservation of boson and fermion ends. Every external boson line has one end that winds up on a vertex and one end that is left hanging. Every internal boson line has two ends, each tied to a vertex. Every vertex of ith type has b_i boson ends tied to it. Then

$$B_E + 2B_I = \sum n_i b_i \tag{25.42}$$

This is the law of conservation of boson ends. Likewise there is a law of conservation of fermion line ends:

$$F_E + 2F_I = \sum n_i f_i \tag{25.43}$$

By elementary algebra we may eliminate the factors involving internal lines and only be left with factors involving the vertices and the external lines:

$$\begin{aligned} 2B_I &= \sum n_i b_i - B_E \\ 3F_I &= \tfrac{3}{2}\left(\sum n_i f_i - F_E\right) \end{aligned} \tag{25.44}$$

Substituting these into (25.41) gives

$$D = -B_E - \tfrac{3}{2}F_E + \sum n_i\left(b_i + \tfrac{3}{2}f_i + d_i - 4\right) + 4 \tag{25.45}$$

This formula is extremely nice because it tells us how much more divergent a graph becomes when we add an interaction of a given type. We can simplify it further if we define the **index of divergence**, δ_i, of an interaction Lagrangian $\mathscr{L}_i$:

$$\delta_i \equiv b_i + \tfrac{3}{2}f_i + d_i - 4 \tag{25.46}$$

Then

$$\boxed{D = -B_E - \tfrac{3}{2}F_E + \sum n_i\delta_i + 4} \tag{25.47}$$

I won't prove it, but this formula contains the explicit prescription for constructing counterterms:

$$\begin{cases} D & \text{is the number of derivatives in the counterterm} \\ B_E & \text{is the number of boson lines in the counterterm} \\ F_E & \text{is the number of fermion lines in the counterterm} \end{cases} \tag{25.48}$$

I am obviously going to have to introduce a lot of counter terms if I have an interaction in my theory with δ_i positive. Whenever I add an extra internal vertex of that type, I increase the superficial degree of divergence by one, I have to make more subtractions in my Taylor expansion, and I have to add therefore a counterterm with more derivatives. I want to give two specific examples to show you what's going on.

EXAMPLE. *A ϕ^4 interaction*

$$\mathscr{L} = \tfrac{1}{2}\partial_\mu\phi\,\partial^\mu\phi - \tfrac{1}{2}\mu^2\phi^2 - \tfrac{1}{4!}\lambda\phi^4 + \mathscr{L}_{CT} \tag{25.49}$$

I will figure out the counterterms iteratively by using the formula (25.47). There is only one interaction, the term in ϕ^4. For this interaction, the number b_1 of boson lines is 4, the number f_1 of fermion fields is zero, and the number d_1 of derivatives is zero. The index of divergence δ of this interaction is zero:

$$\delta_1 = b_1 + \tfrac{3}{2}f_1 + d_1 - 4 = 4 + \tfrac{3}{2}\cdot 0 + 0 - 4 = 0 \tag{25.50}$$

Now let us compute the superficial degree of divergence, D, from (25.47). No matter how many internal ϕ^4 vertices we have, even if I drew a complicated diagram that would cover a couple of blackboards, elementary algebraic counting shows this term $\delta_1 = 0$, so it contributes nothing. There are no fermions in the theory, so $F_E = 0$. The superficial divergence D is determined just by the number of external boson lines:

$$D = -B_E + 4 \tag{25.51}$$

Graphs with more than four external boson lines will always be superficially convergent, and by the Bogoliubov prescription will require no counterterm. I need consider only graphs with $B_E \leq 4$. Because the theory is invariant under $\phi \to -\phi$, we don't have to consider graphs with odd numbers of boson lines: they vanish. Thus we have to look only at three cases: $B_E = 0$, $B_E = 2$, and $B_E = 4$. The possibility $B_E = 0$, $D = 4$ is irrelevant: graphs with no external lines are vacuum to vacuum graphs, and we throw those away. The next case is $B_E = 2$, $D = 2$. This is of the form

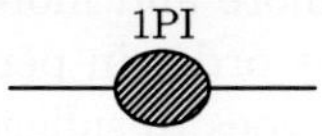

(25.52)

According to (25.48) we need a counterterm with two ϕ's and two derivatives. We will have to subtract out the first two terms in the Taylor series expansion. Therefore to any order in perturbation theory this graph will introduce counterterms proportional to ϕ^2, which will cancel the zeroth order term in the Taylor expansion with some coefficient depending on how far we've brought in perturbation theory and what the cutoff is, and some terms proportional to $(\partial^\mu\phi)^2$:

$$\mathscr{L}_{CT,1} = A\phi^2 + B(\partial^\mu\phi)^2 \tag{25.53}$$

Then I have $B_E = 4$, $D = 0$, corresponding to this graph:

(25.54)

The only counter term introduced here is

$$\mathscr{L}_{CT,\,2} = C\phi^4 \tag{25.55}$$

There are no derivatives because $D = 0$, and we only go to zeroth order in the Taylor expansion, so we have no powers of momentum. The counterterms are new interactions, and for consistency I should check that they don't change my divergence counting. The C counterterm is proportional to ϕ^4, so it, too has $\delta = 0$. The B term has two ϕ's and two derivatives so again has $\delta = 0$. The A term has only two ϕ's and no derivatives so has $\delta = -2$, which is groovy. They don't change the divergence counting of the original Lagrangian.

Thus in this theory, the counterterm Lagrangian is a sum, with some coefficients I have to compute, of a ϕ^2, a $(\partial^\mu\phi)^2$ and a ϕ^4:

$$\mathscr{L}_{CT} = A\phi^2 + B(\partial^\mu\phi)^2 + C\phi^4 \tag{25.56}$$

These counterterms can be interpreted in our usual way by rescaling the field, to make the $(\partial^\mu\phi)^2$ coefficient its usual self, $\frac{1}{2}$. We assemble the other two terms to define the bare mass and the bare coupling constants. The result therefore of Hepp's theorem applied to this example is to make all observable quantities in this theory, to any finite order in perturbation theory, independent of the cut-off, in the large cut-off limit. This can be done if the field is appropriately rescaled and if the bare parameters are chosen in an appropriate cut-off independent way, because the terms we have added are of the *same form* as the terms that were there in the first place. That is what we mean when we say a theory is renormalizable. The ϕ^4 interaction is a renormalizable theory. You choose the bare coupling constant in the appropriate cut-off independent way, the bare mass in an appropriate cut-off independent way, rescale the field in an appropriate cut-off independent way, and all the divergences will cancel to any finite order in perturbation theory. I want that point firmly in your head.

EXAMPLE. *A ϕ^5 interaction*

$$\mathscr{L}' \propto \phi^5 \qquad \delta = 1 \tag{25.57}$$

This interaction has *five* boson lines, and $\delta = 1$. When we consider a graph containing more and more of these ϕ^5 interactions, the superficial degree of divergence will get larger and larger in the graph's Taylor expansion about zero (see the example on p. 344). We'll have to make more subtractions and we need more and more different kinds of counter terms. Not only do the coefficients change order by order in perturbation theory, but their qualitative character changes as well. A graph that goes to sufficiently high order in the ϕ^5 interaction, a graph with two external boson lines, will have D equal to 1 million—that happens to millionth order in the ϕ^5 interaction—and therefore we would have to subtract something with two ϕ's and a million derivatives. This theory is non-renormalizable. We are off on the unending escalation of ambiguities that characterizes such theories. As we go to higher and higher orders in perturbation theory, we need more and more different kinds of counter terms that cannot be interpreted as simply a rescaling of ϕ and a redefinition of the parameters that occur in our original Lagrangian.

People sometimes say, "Well, so what? You've got a prescription that fits most things uniquely, you know the Bogoliubov prescription is unambiguous and tells you what those counter terms are." But the Bogoliubov prescription is arbitrary; Bogoliubov invented it to make the theorem easy to prove. You don't have to subtract at zero, you could subtract at some randomly chosen point of momentum space if you want to avoid all those thresholds. You could subtract different Green's functions at different points, you could subtract the second-order term in the Taylor expansion about the point zero, the third-order term in the Taylor expansion about some other point. The whole thing is just an *ad hoc* prescription to make the algorithm run simply. If you get a ϕ^{17} interaction, or $(\partial^\mu\phi)^{42}$ term that comes out as part of the counterterm prescription, there's no reason why it shouldn't have been there in the original Lagrangian. A non-renormalizable theory involves an *unlimited* number of terms and free parameters.

So this is the dividing line between renormalizable and non-renormalizable theories: either all the interactions have δ less than or equal to zero, or some of the interactions have δ greater than zero. If you have positive δ's you are cooked; it is a non-renormalizable theory. It is bad news. I don't know how to make sense out of them, and nobody else does, either. Every few years someone has an idea about how to deal with them and every few years he's shot down.[12]

I should say that renormalization makes a lot of people nervous, dealing with a theory that involves infinite quantities, the bare charge and the bare mass, in its Heisenberg equations of motion. Suppose at some future date the constructive field theorists conquer quantum electrodynamics in the sense of establishing a rigorous proof that shows if you put in a cut-off, the equations of motion have unique well-defined solutions, and those solutions have a definite limit as the cut-off goes to infinity, presuming you adjust the bare coupling constants and the bare masses appropriately as functions of the cut-off: they prove non-perturbatively what has been proved in perturbation theory. In that sense they construct a mathematically well-defined theory, albeit through a limiting procedure. Now there it is, a mathematically well-defined theory that obeys all the general assumptions you'd want a quantum field theory to obey: it's got local fields that commute for spacelike separations, it's Lorentz invariant, it has a particle spectrum, *et cetera*. Are you going to reject it out of hand just because you don't like the fact that it's defined through a limiting procedure? That was Bishop Berkeley's objection to the calculus. He said infinitesimals didn't exist. But calculus was later reformulated in terms of a limiting procedure, and you can formulate renormalization in terms of a limiting procedure, through regularization. Maybe God did things that way, with limiting procedures. And perhaps if there is a physical cut-off, it may be at some distance so small that it might as well not be there for all practical purposes. It might be that gravity in its mysterious way does something strange, although nobody knows how it could. But we can do dimensional analysis and see that the characteristic ("Planck") length of gravity is 10^{-33} centimeters, which is at least 10 orders of magnitude shorter than the current experimentally accessible range of distances. And if there is a cut-off at that distance, who cares?

---

[12] [Eds.] The video of Lecture 25 ends here. The remainder of this chapter comes from the first 36 minutes of the video of Lecture 26.

25.4 Survey of renormalizable theories for spin 0 and spin ½

For the type of theories we are considering, scalar fields and Dirac spinor fields, the degree of divergence is connected with the dimensionality of the interaction (or equivalently, with the dimensionality of the coupling constant that multiplies the interaction) in a relatively simple way. We can see that by elementary dimensional analysis. The derivative operator has dimensions of length to the inverse first power, or, in the units we are using, where mass and length have inverse dimensions, dimensions of mass to the first power:

$$[\partial^\mu] = [L]^{-1} = [M] \tag{25.58}$$

The action has the dimensions of Planck's constant; that is to say, it is dimensionless:

$$[\textstyle\int d^4x\,\mathscr{L}] = [L]^0 = [M]^0 \tag{25.59}$$

Since d^4x has the dimensions of L^4, the Lagrangian must have dimensions of length to the inverse fourth power, or equivalently, mass to the fourth power:

$$[\mathscr{L}] = [L]^{-4} = [M]^4 \tag{25.60}$$

The Lagrangian for a scalar field contains a kinetic term $(\partial_\mu\phi)^2$ with two derivatives and two ϕ's. This term must have dimensions of M^4, so

$$[\phi] = [L]^{-1} = [M] \tag{25.61}$$

By the same argument the spinor field has dimensions of mass to the 3⁄2 because its Lagrangian is $i\overline{\psi}\not{\partial}\psi$:

$$[\psi] = [L]^{-3/2} = [M]^{3/2} \tag{25.62}$$

Counting only the dimensions of the fields and the derivatives (ignoring whatever dimension any coupling constant has), the dimension (the power of M) of an interaction Lagrangian is

$$(b_i + \tfrac{3}{2}f_i + d_i) = \delta_i + 4 \tag{25.63}$$

That is, not including the coupling constant,

$$[\mathscr{L}_i] = [M]^{\delta_i+4} \tag{25.64}$$

(Let the *raw dimension* of an interaction be its dimension *without* including the coupling constant.) A check: if δ_i is zero, the dimension is 4. If you remember the rules for dimensions you also remember the rules for computing the index of divergence δ_i in powers of mass. Equivalently if you include the coupling constant and arrange matters so the whole Lagrange density has dimensions $[M]^4$, the dimension of the coupling constant is δ_i in units of inverse mass.

An interaction is said to be of *renormalizable type* if the index of divergence δ_i is less than or equal to zero. As we include more and more of these interactions going to higher order in perturbation theory, we do not increase the superficial degree of divergence D; we will not need to add more and more counterterms. It is possible to make a complete list of these renormalizable interactions in four dimensions. The minimum case of δ is -3. The case $\delta = -4$ is in principle possible with no derivatives, no fermions, and no bosons, but that's not much of an interaction; that's just adding a constant to the Lagrangian. Therefore we'll start with -3. Here the only possibility is a term linear in a scalar field, ϕ:

$$\delta = -3: \ \left\{\phi\right. \tag{25.65}$$

(We'll use ϕ and ψ generically. When I write ϕ in a theory with 21 different scalar fields in the Lagrangian, it could be any linear combination of 21 such terms.)

$$\delta = -2: \ \begin{cases} \phi^2 \\ \partial^\mu \phi \end{cases} \tag{25.66}$$

We can get by with one scalar field and one derivative, which is not particularly interesting since that's not Lorentz invariant, and it also vanishes by integration by parts; or with two scalar fields, ϕ^2. Again it could be $\phi_1\phi_2$, for instance, if there are two scalar fields.

$$\delta = -1: \ \begin{cases} \phi^3 \\ \phi\,\partial^\mu \phi \\ \overline{\psi}\psi, \ \overline{\psi}i\gamma_5\psi \end{cases} \tag{25.67}$$

At $\delta = -1$, things are a bit richer. We could have ϕ^3, or $\phi\,\partial^\mu\phi$ (which is not Lorentz invariant), we could have $\overline{\psi}\psi$ (or if our theory is not parity conserving, $\overline{\psi}i\gamma_5\psi$). These three kinds of interactions with δ strictly less than zero are sometimes called *super-renormalizable*. Although they require counterterms, they only require a finite number of them in perturbation theory. When you put in enough of these interactions D becomes negative and no new counterterms are required. Super-renormalizable theories are of course much nicer than merely renormalizable theories, because the divergent part of the perturbation series terminates. Unfortunately, at least in four dimensions the only super-renormalizable theories we can get are either trivial, in which the spinor products are the only interactions, or the energy is unbounded below, if we allow the ϕ^3 interaction without a ϕ^4 term to compensate for it. In fewer dimensions than four, of course, the counting is rather different, and you can find theories that are super-renormalizable with sensible energy spectra. These are nice models to look at if you want to do some rigorous mathematics and prove that a quantum field theory exists. There are still divergences to handle, but it's much easier than in four dimensions.

Finally we have $\delta = 0$:

$$\delta = 0: \ \begin{cases} \phi^4 \\ (\partial^\mu\phi)^2 \\ \overline{\psi}\not{\partial}\psi, \ \overline{\psi}\not{\partial}i\gamma_5\psi \\ \overline{\psi}\psi\phi, \ \overline{\psi}i\gamma_5\psi\phi \end{cases} \tag{25.68}$$

the genuinely renormalizable types of interactions. Here we can have ϕ^4; $(\partial^\mu\phi)^2$, two derivatives and two scalar fields; $\overline{\psi}\not{\partial}\psi$, the normal term that arises in the free Lagrangian; $\overline{\psi}\not{\partial}i\gamma_5\psi$ which you might encounter as a counterterm in a theory with parity-violating interactions; and finally the two kinds of Yukawa coupling, to a scalar or pseudoscalar field: $\overline{\psi}\psi\phi$ and $\overline{\psi}i\gamma_5\psi\phi$. That's it. That completes the list as far as the fields we have talked about. In a little while we will talk about what happens when you allow for vector fields and how this formalism is extended.

As you see, renormalizability is a very severe restriction. It's one of the striking differences between relativistic local quantum mechanics (i.e., quantum field theory), and non-relativistic quantum mechanics. In non-relativistic quantum mechanics, there is no *a priori* constraint of any sort on the interactions. There may be two body forces with arbitrary potentials, there may be three body forces, four body forces, etc. As far as anyone knows there is no general criterion that restricts in any significant way the interactions between the particles. You don't want them to be so singular that the energy is unbounded below and so on, but aside from that anything goes. In quantum field theory, if you accept renormalizability as a criterion that distinguishes sensible theories from nonsensical theories, or at least those theories about which we can say something significant beyond lowest order in perturbation theory from those we cannot, things are much more restricted. Once you have told me the number of spinless fields and the number of Dirac bispinor fields in the theory, I have only a finite number of free parameters which I can adjust, the coefficients of the renormalizable couplings.

Interactions of renormalizable type do not generate an infinite sequence of counterterms. However in normal parlance we may use the word "renormalizable" in a slightly stronger sense: we not only want the number of counterterms generated to be finite, but we want them all to be interpretable as *redefinitions* of parameters multiplying terms that *already occur* in our initial Lagrangian; all counterterms are of the same form as terms in the original Lagrangian. In the literature such a theory is called **strictly renormalizable**. For example, if we take a theory of scalar fields, as well as theories that can be generated from it by rescaling the fields (such as by a wave function renormalization counterterm), all the counterterms that arise in every order of the Bogoliubov iterative procedure can be reinterpreted as corrections to the "bare" parameters of the theory.

Thus for example in the *strict* sense of renormalizability, our good old friend, the Yukawa interaction with a pseudoscalar meson,

$$\mathscr{L}' = g\overline{\psi}i\gamma_5\psi\phi \tag{25.69}$$

with $\delta = 0$ (25.46), is *not* a renormalizable theory, because as we can see from our formula or just by counting, this graph for meson–meson scattering

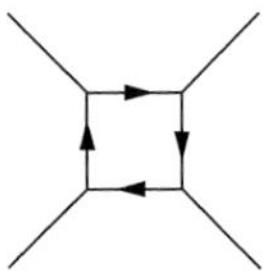

$$D = 0 \tag{25.70}$$

is logarithmically divergent: d^4k over over k^4. Equivalently, $B_E = 4$, $F_E = 0$, so (25.47) $D = 0$. On the other hand if I add a ϕ^4 interaction to the Yukawa interaction,

$$\mathscr{L}' = g\overline{\psi}i\gamma_5\psi\phi - \tfrac{1}{4!}\lambda\phi^4 \tag{25.71}$$

I have a ϕ^4 counterterm that can be used to cancel out this divergence, and it is easy to check that the theory is strictly renormalizable. The only divergent graphs are those that *can* be interpreted as redefining the parameters in the Lagrangian. In the sense of strict renormalization, there is no point in talking about Yukawa theory as a *one* parameter theory; it is a *two* parameter theory. You have to specify independently the Yukawa coupling g and the ϕ^4 coupling λ.

It's possible to give some general theorems that characterize large classes of strictly renormalizable theories, involving only a set of spin zero and spin ½ fields (you'll have to specify how many of each there are). I'll give three such theorems.

Theorem 1. The most general Lagrangian involving all interactions of *raw* dimension less than or equal to four, or equivalently, with δ less than or equal to zero, is strictly renormalizable.

How do I prove that? I start with (25.47), moving some of the terms over to the other side:

$$D + \tfrac{3}{2}F_E + B_E - 4 = \sum_i n_i \delta_i \tag{25.72}$$

Let a given diagram contain a divergence with $D \geq 0$ and index δ. According to step 2 of the Bogoliubov algorithm, I add counterterms to cancel the Taylor expansion of the divergence about $p = 0$ up to order D. F_E tells me the number of Fermi fields that I have to put into my counterterm, B_E tells me the number of boson fields, and D tells me the maximum number of derivatives I have to include to subtract the appropriate terms in the Taylor expansion. (We might not need to go as high as D, because it's possible that we already have counterterms to cancel that order from earlier in the algorithm.) That is, for any diagram

$$F_E = f \qquad B_E = b \qquad D \geq d \tag{25.73}$$

But then the left-hand side of (25.72) is just the formula that enters into the definition (25.46) of δ_i; it's the same combination. So the δ of this diagram is

$$\delta = b + \tfrac{3}{2}f + d - 4 \leq B_E + \tfrac{3}{2}F_E + D - 4 \tag{25.74}$$

Thus the δ of the counterterms for a diagram is always less than or equal to the sum of the δ's of the interactions in the diagram:

$$\delta \leq \sum_i n_i \delta_i \tag{25.75}$$

It's elementary algebra. I say less than or equal because I have to subtract all the terms in the Taylor series up to order D. Thus if my original Lagrangian contains all monomials with δ less than zero, every counterterm I introduce will be a monomial with δ less than zero, and therefore it can be reinterpreted as a renormalization of the coefficient of one of those monomials.

Theorem 2. The most general Lagrangian involving all interactions consistent with some internal symmetry or parity, of (raw) dimension less than or equal to four, or equivalently, with δ less than or equal to zero, is strictly renormalizable.

Unless I am so perverse as to choose a cut-off procedure that all by itself violates the internal symmetry or parity, parity-violating graphs or internal symmetry-violating graphs will not occur. Even though they may have a superficial degree of divergence (D) greater than or equal to zero, I will not have to make any subtractions for them because they are zero, and therefore all terms in their Taylor expansion are zero. Thus for example Yukawa theory with a ϕ^4 interaction

$$\mathscr{L}' = g\overline{\psi}i\gamma_5\psi\phi - \tfrac{1}{4!}\lambda\phi^4 \tag{25.76}$$

is, by this criterion, strictly renormalizable, because it represents the most general interaction between these kinds of fields consistent with parity. In principle, if it weren't for parity, I could have a $\overline{\psi}\psi\phi$ counterterm and a ϕ^3 counterterm or a term linear in ϕ as a counterterm, but those would all violate parity. Likewise the corresponding isospin and parity invariant Yukawa theory:

$$g\overline{N}\boldsymbol{\tau}i\gamma_5 N\cdot\boldsymbol{\Phi} - \tfrac{1}{4!}\lambda(\boldsymbol{\Phi}\cdot\boldsymbol{\Phi})^2 \tag{25.77}$$

The first term is our old Yukawa interaction (24.29) for a triplet of pions. Now just as before we have to add the possibility of a ϕ^4 interaction, but the only one that is consistent with isospin invariance for ϕ—the only way I can make an isoscalar without introducing derivatives—is $(\mathbf{\Phi} \bullet \mathbf{\Phi})^2$. This is the most general interaction of this form only involving terms with dimension less than or equal to 4, δ less than or equal to zero, which is invariant under both parity and under isospin rotations. It is therefore strictly renormalizable. The only kinds of counterterms we will encounter are terms of the same sort we had to begin with in the Lagrangian. It's really very simple. Of course, the reason it's very simple is because I cheated on you: I told you Hepp's theorem without telling you the proof. If I had gone through the proof of Hepp's theorem you wouldn't think it was so simple. But once you have that big theorem, everything else falls out.

Theorem 3. The conclusions of Theorem 2 remain true if any symmetry-breaking interaction is added to the Lagrangian, provided the symmetry-breaking interaction's (raw) dimension equals 1, 2 or 3. This result was discovered in 1970 by Symanzik. We will call this **Symanzik's rule.**[13]

The point here is that if you have an asymmetric interaction as well as a symmetric one but the asymmetric interaction is of low dimension, with a negative δ, by (25.75) it will only introduce asymmetric counterterms that also have a negative δ. For example if the only interaction in your theory that breaks the symmetry has $\delta \leq -2$, you will only get counterterms that violate your symmetry considerations of $\delta \leq -2$. And therefore you will never generate a higher value of δ than that of the original interaction. Such symmetry breaking is sometimes called "super-renormalizable symmetry breaking", or "soft symmetry breaking". For example we could break the symmetry in (25.77) by adding an unequal mass term for the π^0,

$$\mathscr{L} \to \mathscr{L} + \tfrac{1}{2}\epsilon(\phi^0)^2 \tag{25.78}$$

to give the π^0 a different mass than the π^+ and the π^-. That is a symmetry breaking term with $\delta = -2$, of dimension 2, and indeed it is the only possible symmetry breaking term of dimension 2 or less consistent with parity and charge conjugation, etc. This symmetry-breaking term will never generate any counterterms except those of the same form, also consistent with parity and charge conjugation. Thus for example it is perfectly consistent with renormalization within the framework of meson–nucleon theory to say that the theory is completely isospin symmetric except for a difference between the bare mass of the charged pions and the neutral pion.

The bare masses of the nucleons are the same because *that* counterterm is never forced on you. The counterterm has dimension 3, $\delta = -1$. All the bare couplings, of dimension 4, remain symmetric. Although this is a cute result, it unfortunately does not help us explain mass differences in nature (for example, between the neutron and the proton) on the basis of electromagnetism. That interaction, $\overline{\psi}\gamma^\mu\psi A_\mu$, is of dimension 4. This theorem is just what we don't want; we want something that goes the other way, in which the bare masses are the same, and it's the coupling that becomes asymmetric. That requires much more straining, and is not an easy result like Symanzik's theorem. It requires setting up a theory much more complicated than electromagnetism, called a *spontaneously broken gauge field theory.*

[13] [Eds.] Kurt Symanzik, "Renormalization of models with broken symmetry", pp. 263–278, in *Fundamental Interactions at High Energies (Coral Gables Conference on High Energy Physics II)*, eds. A. Perlmutter, G. J. Iverson & R. M. Williams, Gordon and Breach, 1970.

The Symanzik rule is useful in other cases. There are many established models of *chiral symmetry*, notably the sigma model of pion–nucleon interactions,[14] in which the symmetry is broken by a term linear in one of the scalar fields only. That is of course consistent with the Symanzik rule; that term has dimension one, and as we've seen the only possible term of dimension 1, $\delta = -3$ is a term linear in a scalar field.

This concludes my discussion for the moment. Of course we will have to return to the topic when we discuss electrodynamics.

[14] [Eds.] Benjamin W. Lee, *Chiral Dynamics*, Gordon and Breach, 1972. See also B. W. Lee, "Renormalization of the σ-Model", *Nuc. Phys.* **B9** (1969) 649–672.

Problems 14

14.1 Let ψ_A, ψ_B, ψ_C and ψ_D be four Dirac spinor fields. These fields interact with each other (and possibly with unspecified scalar and pseudoscalar fields) in some way that is invariant under P, C, and T, where these operations are defined in the "standard way" as discussed in Chapter 22:

$$U_P^\dagger \psi(\mathbf{x}, t) U_P = \beta\psi(-\mathbf{x}, t) \tag{22.8}$$

Likewise,

$$U_C^\dagger \psi(x) U_C = \psi^*(x) \tag{22.49}$$

in a Majorana basis (one in which $\gamma^\mu = -\gamma^{\mu *}$). Finally,

$$\Omega_{PT}^{-1}\psi(x)\Omega_{PT} = i\gamma_5\psi(-x) \tag{22.80}$$

again in a Majorana basis. Now let us consider adding a term to the Hamiltonian density,

$$\begin{aligned}\mathscr{H}' = g_1(\overline{\psi}_A\gamma^\mu\psi_B)(\overline{\psi}_C\gamma_\mu\psi_D) + g_2(\overline{\psi}_A\gamma^\mu\psi_B)(\overline{\psi}_C\gamma_\mu\gamma_5\psi_D) + g_3(\overline{\psi}_A\gamma^\mu\gamma_5\psi_B)(\overline{\psi}_C\gamma_\mu\psi_D)\\ + g_4(\overline{\psi}_A\gamma^\mu\gamma_5\psi_B)(\overline{\psi}_C\gamma_\mu\gamma_5\psi_D) + \text{Hermitian conjugate}\end{aligned} \tag{P14.1}$$

where the g_i's are (possibly complex) numbers.

(a) In class, we proved the CPT theorem for S-matrix elements. It would be really weird if the S-matrix were CPT invariant but the Hamiltonian density were not. Show that $\mathscr{H}'(0)$ is CPT-invariant regardless of what the g's are.

(b) Under what conditions on the g's is $\mathscr{H}'(0)$ invariant under C? Under P? Under T? PC? CT? TP? REMINDER: Ω_{PT} is anti-unitary.

(1998b 1.1)

14.2 In class I computed, in four dimensions, the superficial degree of divergence, D, for a general Feynman graph with F_E external Fermi line and B_E external Bose lines, in a theory where the Lagrangian was the sum of monomials in scalar fields, Dirac fields and their derivatives,

$$\mathscr{L} = \sum_i \mathscr{L}_i$$

The result (25.47) was

$$D = 4 - B_E - \tfrac{3}{2}F_E + \sum_i n_i\delta_i$$

where n_i is the number of vertices of ith type and δ_i, the index of divergence, is

$$\delta_i = \mathit{dim}\,\mathscr{L}_i - 4$$

$\mathit{dim}\,\mathscr{L}_i$ is the dimension of $\mathscr{L}_i$ in units of mass, not counting any dimensions attached to the coupling constants.

Derive the corresponding formulae in d dimensions for arbitrary positive integer d. For arbitrary d, what is the largest value of n for which ϕ^n is of renormalizable type? For what values of d is $(\overline{\psi}\psi)^2$ of renormalizable type?

Comments: In any number of dimensions the action $\mathcal{S}$ must be dimensionless, or the Lagrangian $\mathscr{L}$ must have dimension d (in mass units). Thus the mass dimension of both scalar and Dirac fields depend on d. Also, Dirac fields in d dimensions are just like Dirac fields in 4 dimensions, except for the number of components, which is irrelevant to our interest here.

(1998b 1.2)

14.3 In §21.4 we spent some time computing things for the theory of a Dirac bispinor Yukawa-coupled to a neutral pseudoscalar meson, described by the interaction Lagrangian

$$\mathscr{L}' = g\overline{\psi}i\gamma_5\psi\phi \tag{P14.2}$$

(This interaction was also the subject of Problems 12.3, 13.1, and 13.2.) To order g^2 the Feynman amplitude for the process $\phi + \psi \to \phi + \psi$ is given by the sum of two graphs:

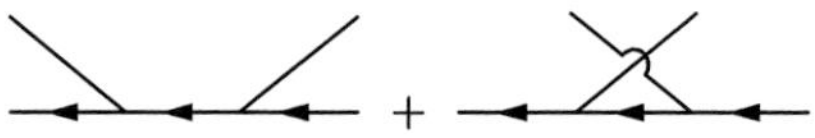

In equations

$$a = g^2M_1 + g^2M_2 \tag{P14.3}$$

where g^2M_1 and g^2M_2 are the contributions of the first and second graphs, respectively. (These are functions of momentum and spins, but we won't need their explicit forms for this problem.)

Now let us consider the isospin-invariant theory of pions and nucleons discussed in class (24.29),

$$\mathscr{L}' = g\overline{N}i\gamma_5\boldsymbol{\tau}\cdot\boldsymbol{\pi}N \tag{P14.4}$$

Compute to order g^2, in terms of g, M_1 and M_2, the amplitudes for the following processes:

1. $p + \pi^+ \to p + \pi^+$
2. $n + \pi^+ \to n + \pi^+$
3. $n + \pi^+ \to p + \pi^0$

Also compute $a_{1/2}$ and $a_{3/2}$, the scattering amplitudes for the pure $I = \frac{1}{2}$ and $I = \frac{3}{2}$ initial (and therefore final) states.

(1998b 5.3)

14.4 In this problem, you are to compare two theories of the interactions of mesons and nucleons. In both theories the free Lagrangian is the same:

$$\mathscr{L}_0 = \tfrac{1}{2}\partial_\mu\phi\partial^\mu\phi - \tfrac{1}{2}\mu^2\phi^2 + \overline{\psi}(i\partial\!\!\!/ - m)\psi \tag{P14.5}$$

The first theory was discussed in class, with a pseudoscalar Yukawa coupling:

$$\mathscr{L}'_{\rm I} = ig\overline{\psi}\gamma_5\psi\phi \tag{P14.6}$$

The second theory is defined by "gradient-coupling" and a quadratic coupling to the meson,

$$\mathscr{L}'_{\rm II} = \mu^{-1}\left[ag\overline{\psi}\gamma^\mu\gamma_5\psi\partial_\mu\phi + bg^2\overline{\psi}\psi\phi^2\right] \tag{P14.7}$$

Here a and b are real dimensionless constants; they are assumed to be independent of g, but may depend on the dimensionless ratio μ/m. Show that to lowest nontrivial order in perturbation theory—order g^2—the two theories predict the same scattering amplitudes for both meson–nucleon scattering and nucleon–nucleon scattering, if a and b are properly chosen. Find the proper choices. (Note that since we are free to redefine the sign of the meson field in the two theories independently, we can always by convention take both g and a to be positive.)

Remark. I have not yet derived the Feynman rules for derivative couplings in class, and I do not expect you to derive them from first principles for this problem. (But see Problem 8.1, comment (3), p. 309, and §14.4, (14.57).) Take the following on trust: An interaction of the form

$$(ag/\mu)\,\overline{\psi}\gamma^\mu\gamma_5\psi\partial_\mu\phi \tag{P14.8}$$

generates a vertex of the form

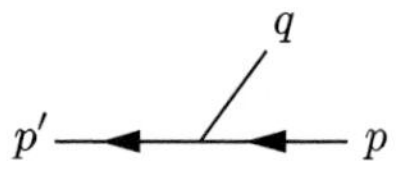

with which there is associated a factor

$$(ag/\mu)\, \not{q}\gamma_5\, (2\pi)^4 \delta^{(4)}(p + p' + q) \tag{P14.9}$$

where all momenta are directed *inward.*

(1980 253a Final, Problem 3; 2000 253a Final, Problem 1)

Solutions 14

14.1 (a) The x dependence is not at issue here, so we suppress the arguments of the fields. To begin with, let's study the transformation properties of the individual bilinear forms. From Chapter 20 (box, p. 420)

$$U_P^\dagger \begin{Bmatrix} \overline{\psi}_A\gamma^0\psi_B \\ \overline{\psi}_A\gamma^i\psi_B \\ \overline{\psi}_A\gamma^0\gamma_5\psi_B \\ \overline{\psi}_A\gamma^i\gamma_5\psi_B \end{Bmatrix} U_P = \begin{Bmatrix} +\overline{\psi}_A\gamma^0\psi_B \\ -\overline{\psi}_A\gamma^i\psi_B \\ -\overline{\psi}_A\gamma^0\gamma_5\psi_B \\ +\overline{\psi}_A\gamma^i\gamma_5\psi_B \end{Bmatrix} \tag{S14.1}$$

and from Chapter 22, (box, p. 469 and (22.63))

$$U_C^\dagger \begin{Bmatrix} \overline{\psi}_A\gamma^\mu\psi_B \\ \overline{\psi}_A\gamma^\mu\gamma_5\psi_B \end{Bmatrix} U_C = \begin{Bmatrix} -\overline{\psi}_B\gamma^\mu\psi_A \\ +\overline{\psi}_B\gamma^\mu\gamma_5\psi_A \end{Bmatrix} \tag{S14.2}$$

Under PT (22.87),

$$\Omega_{PT}^{-1}\overline{\psi}(x)\Omega_{PT} = -i\overline{\psi}(-x)(\gamma_5) \tag{S14.3}$$

so (in the Majorana basis, where γ^μ and γ_5 are imaginary)

$$\Omega_{PT}^{-1}\overline{\psi}_A\gamma^\mu\psi_B\Omega_{PT} = -i\overline{\psi}_A\gamma_5\gamma^{\mu *}i\gamma_5\psi_B = -\overline{\psi}_A\gamma_5\gamma^\mu\gamma_5\psi_B = \overline{\psi}_A\gamma^\mu\psi_B \tag{S14.4}$$

and

$$\Omega_{PT}^{-1}\overline{\psi}_A\gamma^\mu\gamma_5\psi_B\Omega_{PT} = -i\overline{\psi}_A\gamma_5\gamma^{\mu *}\gamma_5^* i\gamma_5\psi_B = -\overline{\psi}_A\gamma^\mu\gamma_5\psi_B \tag{S14.5}$$

From these results we can determine the effect of CPT. Define

$$\Omega_{CPT} \equiv U_C\Omega_{PT} \tag{S14.6}$$

Then

$$\begin{aligned} \Omega_{CPT}^{-1} \begin{Bmatrix} \overline{\psi}_A\gamma^\mu\psi_B \\ \overline{\psi}_A\gamma^\mu\gamma_5\psi_B \end{Bmatrix} \Omega_{CPT} &= \Omega_{PT}^{-1} \begin{Bmatrix} -\overline{\psi}_B\gamma^\mu\psi_A \\ \overline{\psi}_B\gamma^\mu\gamma_5\psi_A \end{Bmatrix} \Omega_{PT} = - \begin{Bmatrix} \overline{\psi}_B\gamma^\mu\psi_A \\ \overline{\psi}_B\gamma^\mu\gamma_5\psi_A \end{Bmatrix} \\ &= - \begin{Bmatrix} (\overline{\psi}_A\gamma^\mu\psi_B)^\dagger \\ (\overline{\psi}_A\gamma^\mu\gamma_5\psi_B)^\dagger \end{Bmatrix} \end{aligned} \tag{S14.7}$$

The last equality follows because

$$(\gamma^\mu)^\dagger = \gamma^0\gamma^\mu\gamma^0 \tag{S14.8}$$

By construction, $\mathscr{H}'^\dagger = \mathscr{H}'$. On the other hand, under CPT all the bilinears are turned into -1 times their adjoints. The Hamiltonian is built of *pairs* of bilinears, so the signs cancel, and

$$\Omega_{CPT}^{-1}\mathscr{H}'\Omega_{CPT}^{-1} = \mathscr{H}'^\dagger = \mathscr{H}' \tag{S14.9}$$

Thus the Hamiltonian is invariant under CPT, without conditions on the g_i. (If any of the g_i's are complex, the operator Ω_{CPT} has to be anti-unitary, to turn g_i into g_i^*.) ■

(b) Because of CPT invariance, the Hamiltonian is invariant under C if and only if it is invariant under PT, and similarly for the others. So there are really only three other cases to check:

$$\begin{aligned} P\,\text{invariance} &\Leftrightarrow CT\,\text{invariance} \\ C\,\text{invariance} &\Leftrightarrow PT\,\text{invariance} \\ T\,\text{invariance} &\Leftrightarrow PC\,\text{invariance} \end{aligned} \tag{S14.10}$$

Under P, or equivalently, under CT, the first term of the Hamiltonian transforms as follows:

$$\begin{aligned} U_P^\dagger \left[g_1(\overline{\psi}_A\gamma^\mu\psi_B)(\overline{\psi}_C\gamma_\mu\psi_D)\right]U_P &= U_P^\dagger \left[g_1(\overline{\psi}_A\gamma^0\psi_B)(\overline{\psi}_C\gamma_0\psi_D)\right]U_P + U_P^\dagger \left[g_1(\overline{\psi}_A\gamma^i\psi_B)(\overline{\psi}_C\gamma_i\psi_D)\right]U_P \\ &= g_1(\overline{\psi}_A\gamma^0\psi_B)(\overline{\psi}_C\gamma_0\psi_D) + g_1(-\overline{\psi}_A\gamma^i\psi_B)(-\overline{\psi}_C\gamma_i\psi_D) \\ &= g_1(\overline{\psi}_A\gamma^\mu\psi_B)(\overline{\psi}_C\gamma_\mu\psi_D) \end{aligned} \tag{S14.11}$$

So the first term is unchanged. The last term is likewise unchanged. However, the second and third terms pick up an overall minus sign. Thus

$$\begin{aligned} U_P^\dagger\mathscr{H}'U_P &= g_1(\overline{\psi}_A\gamma^\mu\psi_B)(\overline{\psi}_C\gamma_\mu\psi_D) - g_2(\overline{\psi}_A\gamma^\mu\psi_B)(\overline{\psi}_C\gamma_\mu\gamma_5\psi_D) \\ &\quad - g_3(\overline{\psi}_A\gamma^\mu\gamma_5\psi_B)(\overline{\psi}_C\gamma_\mu\psi_D) + g_4(\overline{\psi}_A\gamma^\mu\gamma_5\psi_B)(\overline{\psi}_C\gamma_\mu\gamma_5\psi_D) + \text{Herm. conj.} \end{aligned} \tag{S14.12}$$

The only way this can equal the original Hamiltonian is if $g_2 = g_3 = 0$.

Under C, or equivalently, under PT, the first term of the Hamiltonian transforms as follows:

$$\begin{aligned} U_C^\dagger \left[g_1(\overline{\psi}_A\gamma^\mu\psi_B)(\overline{\psi}_C\gamma_\mu\psi_D)\right]U_C &= g_1(-\overline{\psi}_B\gamma^\mu\psi_A)(-\overline{\psi}_D\gamma_\mu\psi_C) \\ &= g_1(\overline{\psi}_B\gamma^\mu\psi_A)(\overline{\psi}_D\gamma_\mu\psi_C) \\ &= g_1\left((\overline{\psi}_A\gamma^\mu\psi_B)(\overline{\psi}_C\gamma_\mu\psi_D)\right)^\dagger \end{aligned} \tag{S14.13}$$

The full Hamiltonian includes, as part of its Hermitian conjugate, the term $g_1^*\left((\overline{\psi}_A\gamma^\mu\psi_B)(\overline{\psi}_C\gamma_\mu\psi_D)\right)^\dagger$. The only way this can equal the transform of the first term is if $g_1^* = g_1$, i.e., g_1 is real. The same argument holds for g_4. The second and third terms pick up an extra minus sign under C. Consequently the Hamiltonian will be invariant under C, or PT, if g_1 and g_4 are real, and if g_2 and g_3 are imaginary.

Finally, let's consider T, or equivalently PC. We need to work out what happens to the bilinears under PC:

$$U_{PC}^\dagger \begin{Bmatrix} \overline{\psi}_A\gamma^0\psi_B \\ \overline{\psi}_A\gamma^i\psi_B \\ \overline{\psi}_A\gamma^0\gamma_5\psi_B \\ \overline{\psi}_A\gamma^i\gamma_5\psi_B \end{Bmatrix} U_{PC} = U_C^\dagger \begin{Bmatrix} +\overline{\psi}_A\gamma^0\psi_B \\ -\overline{\psi}_A\gamma^i\psi_B \\ -\overline{\psi}_A\gamma^0\gamma_5\psi_B \\ +\overline{\psi}_A\gamma^i\gamma_5\psi_B \end{Bmatrix} U_C = \begin{Bmatrix} -\overline{\psi}_B\gamma^0\psi_A \\ +\overline{\psi}_B\gamma^i\psi_A \\ -\overline{\psi}_B\gamma^0\gamma_5\psi_A \\ +\overline{\psi}_B\gamma^i\gamma_5\psi_A \end{Bmatrix} \tag{S14.14}$$

We see that both the axial vector and the vector terms transform as vectors (not axial vectors) under PC— except for the switch in ordering, which (as noted in (S14.7)) amounts to Hermitian conjugation. Therefore the Hamiltonian will be invariant under T, or under PC, if all the g_i's are real, because the Hamiltonian is built up of products of vectors and axial vectors.

To summarize,

$$\begin{aligned} P\,\text{inv.} \Leftrightarrow CT\,\text{inv.} &\Leftrightarrow g_2 = g_3 = 0 \\ C\,\text{inv.} \Leftrightarrow PT\,\text{inv.} &\Leftrightarrow g_1, g_4 \text{ real and } g_2, g_3 \text{ imaginary} \\ T\,\text{inv.} \Leftrightarrow PC\,\text{inv.} &\Leftrightarrow g_i \text{ all real} \end{aligned} \tag{S14.15}$$

(Most of the homework solutions in this book were generated by graduate students. Problems assigned as homework one year often became exam problems another year, and *vice versa*. In addition to being used as the first homework problem in Physics 253b in 1998, 14.1 appeared in the Physics 253a final in 1981. This solution is Coleman's, with a few extra steps.) ■

14.2. In d dimensions, the superficial degree of divergence of a Feynman diagram is

$$D = dL - 2B_I - F_I + \sum_i n_i d_i \tag{S14.16}$$

where

L is the number of loops, each putting d^dp into the integrand of the Feynman diagram

B_I is the number of internal scalar lines, each bringing $1/p^2$ at high p

F_I is the number of internal fermion lines, each bringing $1/p$ at high p

n_i is the number of interaction vertices of type i

d_i is the number of derivatives at vertices of type i, each bringing a factor of p

The only difference between this formula and the four-dimensional formula is that the coefficient of L is now d. The number of loops is

$$L = B_I + F_I - \Big[\sum_i n_i\Big] + 1 \tag{S14.17}$$

The $\sum_i n_i$ is for the d-momentum conserving δ functions at each vertex, and the 1 is for the overall d-momentum conserving δ function. Inserting (S14.17) into (S14.16) we get

$$D = (d-2)B_I + (d-1)F_I + \Big[\sum_i n_i(d_i - d)\Big] + d \tag{S14.18}$$

As discussed in lecture, we can count the number of scalar and fermion line-ends in two different ways and find the constraints

$$\begin{aligned} 2B_I + B_E &= \sum_i n_i b_i \\ 2F_I + F_E &= \sum_i n_i f_i \end{aligned} \tag{S14.19}$$

where B_E is the number of external scalar lines, F_E is the number of external fermion lines, b_i is the number of scalar fields in interaction i, and f_i is the number of fermion fields in interaction i. Combining (S14.18) and (S14.19), we find

$$D = \tfrac{1}{2}(2-d)B_E + \tfrac{1}{2}(1-d)F_E + \sum_i n_i\delta_i + d$$

where the index of divergence δ_i is

$$\delta_i = \tfrac{1}{2}(d-2)b_i + \tfrac{1}{2}(d-1)f_i + d_i - d$$

From the equal-time commutators or anticommutators, it follows that in d dimensions, a scalar field has mass dimensions $\frac{1}{2}(d-2)$, and a spinor field has mass dimensions $\frac{1}{2}(d-1)$. Therefore $\delta_i + d$ equals the mass dimensions of the interaction i (not including the dimension of the coupling parameter). The interaction is of renormalizable type when the index of divergence $\delta_i \le 0$, i.e., when the dimension of the interaction is less than or equal to d.

The index of divergence δ_i is more fundamental for this analysis than the superficial degree of divergence D because it focuses on individual interaction vertices and is not concerned with the number of loops, integration momenta, etc. We just look at what's going on at a vertex, and that tells us if the interaction is renormalizable or not.

The interaction ϕ^n has dimensions $\frac{1}{2}n(d-2)$ and is of renormalizable type if

$$\tfrac{1}{2}(d-2)n \le d.$$

For $d \le 2$, ϕ^n is of renormalizable type for all n. For $d \ge 3$, we must have

$$n \le \frac{2d}{d-2}$$

Thus there are no nontrivial interactions of this kind for $d \ge 6$. As a check, for $d = 4$, we get $n \le 4$, which we know is true from §25.3.

Finally, the interaction $(\overline{\psi}\psi)^2$ has dimensions $2(d-1)$ and is of renormalizable type when

$$2(d-1) \le d \;\Rightarrow\; d \le 2$$

■

14.3 Let the isovector of pion fields be denoted

$$\boldsymbol{\pi} = \begin{pmatrix} \pi^1 \\ \pi^2 \\ \pi^3 \end{pmatrix} = \begin{pmatrix} \frac{1}{\sqrt{2}}(\pi^+ + \pi^-) \\ \frac{1}{\sqrt{2}}i(\pi^+ - \pi^-) \\ \pi^0 \end{pmatrix} \tag{S14.20}$$

(see (24.21)). Writing $N = \begin{pmatrix} p \\ n \end{pmatrix}$, and with $\boldsymbol{\tau}$ the Pauli matrices, the interaction Lagrangian (P14.4) becomes (see 24.20))

$$\begin{aligned}\mathscr{L}' &= g\overline{N}i\gamma_5\boldsymbol{\tau}\cdot\boldsymbol{\pi}N = g(\overline{p},\overline{n})\begin{pmatrix} \pi^0 & \sqrt{2}\pi^+ \\ \sqrt{2}\pi^- & -\pi^0 \end{pmatrix} i\gamma_5 \begin{pmatrix} p \\ n \end{pmatrix} \\ &= g(\overline{p}i\gamma_5 p\pi^0 + \sqrt{2}\,\overline{p}i\gamma_5 n\pi^+ + \sqrt{2}\,\overline{n}i\gamma_5 p\pi^- - \overline{n}i\gamma_5 n\pi^0)\end{aligned} \tag{S14.21}$$

For reaction 1, $p + \pi^+ \to p + \pi^+$, the first graph cannot contribute, because there is no intermediate state with a charge of $+2$, but the second can:

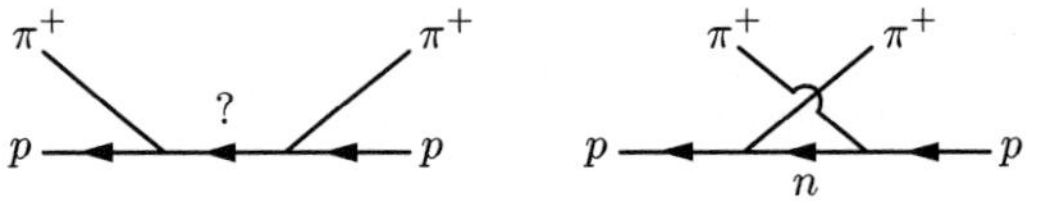

M_1 cannot contribute... but M_2 can

That is, the amplitude $a(p\pi^+ \to p\pi^+)$ is given by

$$a(p\pi^+ \to p\pi^+) = (\sqrt{2}g)^2 M_2 = 2g^2 M_2 \tag{S14.22}$$

Similarly, the second graph cannot contribute to reaction 2, $n + \pi^+ \to n + \pi^+$, and so

$$a(n\pi^+ \to n\pi^+) = (\sqrt{2}g)^2 M_1 = 2g^2 M_1 \tag{S14.23}$$

On the other hand, *both* graphs contribute to reaction 3, $n + \pi^+ \to p + \pi^0$:

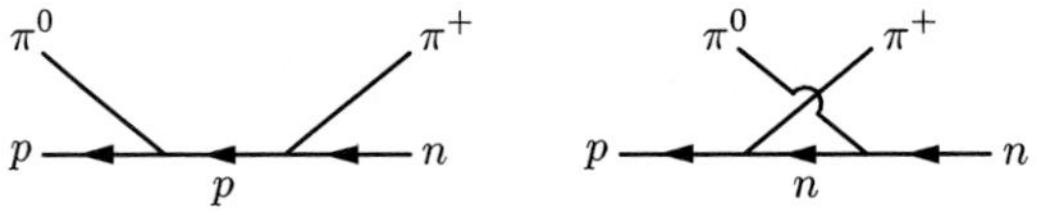

The amplitude now comes from two cross-terms,

$$a(n\pi^+ \to p\pi^0) = \sqrt{2}g^2 M_2 - \sqrt{2}g^2 M_1 = \sqrt{2}g^2(M_2 - M_1) \tag{S14.24}$$

Using the Clebsch–Gordan tables in the Particle Data Group's *Review of Particle Properties*,[1] we have (writing $|I, I_z\rangle$ for the isospin eigenstates)

$$\begin{aligned}|p\pi^+\rangle &= |\tfrac{3}{2},\tfrac{3}{2}\rangle \\ |n\pi^+\rangle &= \sqrt{\tfrac{1}{3}}\,|\tfrac{3}{2},\tfrac{1}{2}\rangle + \sqrt{\tfrac{2}{3}}\,|\tfrac{1}{2},\tfrac{1}{2}\rangle \\ |p\pi^0\rangle &= \sqrt{\tfrac{2}{3}}\,|\tfrac{3}{2},\tfrac{1}{2}\rangle - \sqrt{\tfrac{1}{3}}\,|\tfrac{1}{2},\tfrac{1}{2}\rangle\end{aligned} \tag{S14.25}$$

So

$$a(p\pi^+ \to p\pi^+) = a_{3/2} = 2g^2 M_2 \tag{S14.26}$$

$$a(n\pi^+ \to n\pi^+) = \tfrac{1}{3}a_{3/2} + \tfrac{2}{3}a_{1/2} = 2g^2 M_1 \tag{S14.27}$$

$$a(n\pi^+ \to p\pi^0) = \tfrac{1}{3}\sqrt{2}(a_{3/2} - a_{1/2}) = \sqrt{2}g^2(M_2 - M_1) \tag{S14.28}$$

That answers the original question. Note that we can solve for the amplitudes:

$$a_{3/2} = 2g^2 M_2 \qquad a_{1/2} = g^2(3M_1 - M_2) \tag{S14.29}$$

In §24.3 we compared $p\pi^+ \to p\pi^+$ with $p\pi^- \to p\pi^-$. In this second process, only the first graph contributes:

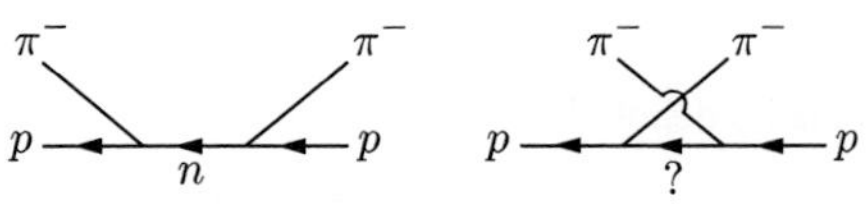

[1] [Eds.] *PDG* 2016, `http://pdg.lbl.gov/2016/reviews/rpp2016-rev-clebsch-gordan-coefs.pdf`

That is,

$$a(p\pi^- \to p\pi^-) = 2g^2 M_1 \tag{S14.30}$$

From (24.44) we have

$$|p\pi^-\rangle = \sqrt{\tfrac{1}{3}}\,|\tfrac{3}{2},-\tfrac{1}{2}\rangle - \sqrt{\tfrac{2}{3}}\,|\tfrac{1}{2},-\tfrac{1}{2}\rangle$$

Then, as we found earlier,

$$a(p\pi^- \to p\pi^-) = \tfrac{1}{3}a_{3/2} + \tfrac{2}{3}a_{1/2}$$

just as in (S14.27). Thus the results (S14.26) and (S14.27) are consistent with the arguments and results in §24.3. ■

14.4 (a) $N + N \to N + N$

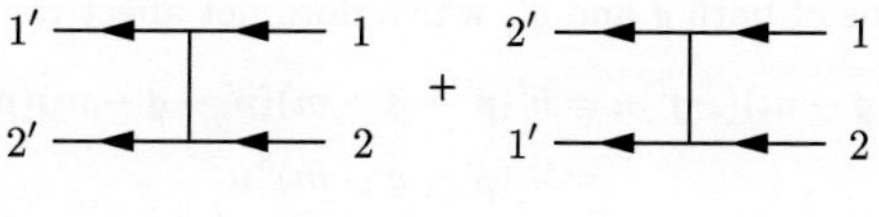

Theory I:

$$ia = -g^2\left[\overline{u}_{1'}i\gamma_5 u_1 \overline{u}_{2'}i\gamma_5 u_2 \frac{i}{(p_1 - p_{1'})^2 - \mu^2} - (1' \leftrightarrow 2')\right] \tag{S14.31}$$

Theory II:

$$ia = +\frac{g^2a^2}{\mu^2}\left[\overline{u}_{1'}(\not{p}_{1'} - \not{p}_1)\gamma_5 u_1 \overline{u}_{2'}(\not{p}_{2'} - \not{p}_1)\gamma_5 u_2 \frac{i}{(p_1 - p_{1'})^2 - \mu^2} - (1' \leftrightarrow 2')\right] \tag{S14.32}$$

But the fermions are on the mass shell, so

$$\overline{u}_{1'}(\not{p}_{1'} - \not{p}_1)\gamma_5 u_1 = \overline{u}_{1'}(\not{p}_{1'}\gamma_5 + \gamma_5\not{p}_1)u_1 = 2m\overline{u}_{1'}\gamma_5 u_1 \tag{S14.33}$$

and likewise $\overline{u}_{2'}(\not{p}_{2'} - \not{p}_1)\gamma_5 u_2 = 2m\overline{u}_{2'}\gamma_5 u_2$. Thus we obtain equality if

$$a = \frac{\mu}{2m} \tag{S14.34}$$

(b) $N + \phi \to N + \phi$

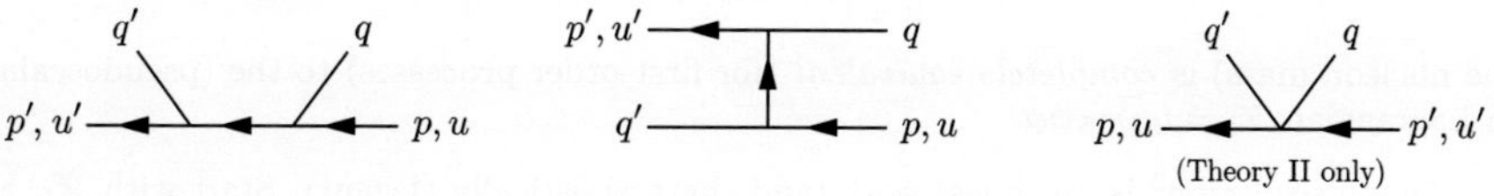

Theory I—as in the lecture (§21.4):

$$\begin{aligned} ia &= -ig^2\overline{u}'\left[i\gamma_5\frac{\not{p}+\not{q}+m}{s-m^2}i\gamma_5 + i\gamma_5\frac{\not{p}'-\not{q}+m}{u-m^2}i\gamma_5\right]u = -ig^2\overline{u}'\left[\frac{\not{p}-m+\not{q}}{s-m^2} + \frac{\not{p}'-m-\not{q}}{u-m^2}\right]u \\ &= -ig^2\overline{u}'\not{q}u\left[\frac{1}{s-m^2} - \frac{1}{u-m^2}\right] \quad \text{where } s = (p+q)^2,\ u = (p'-q)^2;\ \textit{cf.}\ (21.98) \end{aligned} \tag{S14.35}$$

Theory II:

$$ia = \frac{ia^2g^2}{\mu^2}\overline{u}'\left[(-\not{q}')\gamma_5\frac{\not{p}+\not{q}+m}{s-m^2}\not{q}\gamma_5 + \not{q}\gamma_5\frac{\not{p}'-\not{q}+m}{u-m^2}(-\not{q}')\gamma_5\right]u + \frac{2ig^2b}{\mu}\overline{u}'u \tag{S14.36}$$

where both q and $-q'$ are inwards; the last term has a factor of 2 for symmetry (there's a choice which ϕ emits, and which ϕ absorbs, mesons). Once again moving the γ_5's, and substituting $a = \mu/2m$,

$$ia = -\frac{ig^2}{4m^2}\overline{u}'\left[\not{q}'\frac{\not{p}+\not{q}-m}{s-m^2}\not{q} + \not{q}\frac{\not{p}'-\not{q}-m}{u-m^2}\not{q}'\right]u + \frac{2ig^2b}{\mu}\overline{u}'u \tag{S14.37}$$

Now $\overline{u}'(\not{p}'-m) = 0$, and so we can substitute $\overline{u}'(\not{p}'-m+\not{q}')$ for $\overline{u}'\not{q}'$, and similarly $\not{q}u = (\not{q}+\not{p}-m)u$. Then

$$\begin{aligned} \overline{u}'\not{q}'(\not{p}+\not{q}-m)\not{q}u &= \overline{u'}(\not{p}'+\not{q}'-m)(\not{p}+\not{q}-m)(\not{p}+\not{q}-m)u \\ &= \overline{u}'(\not{p}+\not{q}-m)^3u \end{aligned} \tag{S14.38}$$

because the overall delta function ensures that $p + q = p' + q'$. Expanding the cubic,

$$\begin{aligned}\overline{u}'(\not p + \not q - m)^3 u &= u\left[(\not p + \not q)^3 - 3(\not p + \not q)^2 m + 3(\not p + \not q)m^2 - m^3\right]u \\ &= u\left[(\not p + \not q)^2(\not p + \not q) - 3(\not p + \not q)^2 m + 3(\not p + \not q)m^2 - m^3\right]u \\ &= u\left[s(\not p + \not q) - 3sm + 3(\not p + \not q)m^2 - m^3\right]u\end{aligned} \tag{S14.39}$$

because $\not A^2 = A^2$. Then, because $\not p u = mu$, we have finally

$$\overline{u}'(\not p + \not q - m)^3 u = u\left(-2ms + 2m^3 + (s + 3m^2)\not q\right)u \tag{S14.40}$$

Similarly (first changing the signs of both $\not q$ and $\not q'$, which does not affect the product)

$$\begin{aligned}\overline{u}'(-\not q)(\not p' - \not q - m)(-\not q')u &= \overline{u}'(\not p' - \not q - m)(\not p' - \not q - m)(\not p - \not q' - m)u \\ &= \overline{u}'(\not p' - \not q - m)^3 u \\ &= \overline{u}'\left(-2um + 2m^3 - (u + 3m^2)\not q\right)u\end{aligned} \tag{S14.41}$$

Substituting (S14.39) and (S14.41) into (S14.37), we have

$$\begin{aligned}ia &= -\frac{ig^2}{4m^2}\overline{u}'\left[\left(-2m + \not q + 4m^2\frac{\not q}{s - m^2}\right) + \left(-2m - \not q - 4m^2\frac{\not q}{u - m^2}\right)\right]u + \frac{2ig^2 b}{\mu}\overline{u}'u \\ &= -ig^2\overline{u}'\not q u\left[\frac{1}{s - m^2} - \frac{1}{u - m^2}\right] + ig^2\overline{u}'u\left\{\frac{1}{m} + \frac{2b}{\mu}\right\}\end{aligned} \tag{S14.42}$$

Comparing the Theory I amplitude (S14.35) with the Theory II amplitude (S14.42), we see that the terms in square brackets are identical. If the two amplitudes are to agree, the terms in the curly brackets must vanish:

$$\frac{1}{m} + \frac{2b}{\mu} = 0 \Rightarrow b = -\frac{\mu}{2m} \tag{S14.43}$$

In short, the interaction Lagrangian

$$\mathscr{L}'_{\text{II}} = \frac{g}{2m}\left[\overline{\psi}\gamma^\mu\gamma_5\psi\partial_\mu\phi - g\overline{\psi}\psi\phi^2\right] \tag{S14.44}$$

(with m the nucleon mass) is *completely equivalent* (for first-order processes) to the (pseudoscalar Yukawa) interaction Lagrangian $\mathscr{L}_{\text{I}} = ig\overline{\psi}\gamma_5\psi\phi$. ■

Remark. Dyson's solution[2] is much less work (and characteristically elegant). Start with $\mathscr{L}_I$, and change the nucleon field:

$$\psi \to \exp\{i\alpha\gamma_5\phi\}\psi; \quad \overline{\psi} \to \overline{\psi}\exp\{i\alpha\gamma_5\phi\} \tag{S14.45}$$

Then $\overline{\psi}\psi \to \overline{\psi}\exp\{2i\alpha\gamma_5\phi\}\psi$, and $\overline{\psi}\gamma^\mu\psi$ is unchanged. Split $\mathscr{L} = \mathscr{L}_0 + \mathscr{L}_{\text{II}}$ into kinetic and potential terms, to find

$$\begin{gathered}\overline{\psi}\left(i\gamma^\mu\partial_\mu\right)\psi \to \overline{\psi}\left(i\gamma^\mu\partial_\mu\right)\psi - \alpha\overline{\psi}\gamma^\mu\gamma_5\psi\partial_\mu\phi \\ \overline{\psi}(-m + ig\gamma_5\phi)\psi \to \overline{\psi}(-m + ig\gamma_5\phi)\exp\{2i\alpha\gamma_5\phi\}\psi = \overline{\psi}(-m + ig\gamma_5\phi)(1 + 2i\alpha\gamma_5\phi - 2\alpha^2\phi^2)\psi + \mathcal{O}(\alpha^3)\end{gathered} \tag{S14.46}$$

Multiplying everything out and gathering terms gives

$$\mathscr{L} = \mathscr{L}_0 - \alpha\overline{\psi}\gamma^\mu\gamma_5\psi\partial_\mu\phi + i\overline{\psi}\gamma_5\psi\phi(g - 2\alpha m) + 2\alpha\overline{\psi}\psi\phi^2(\alpha m - g) - 2ig\alpha^2\overline{\psi}\psi\phi^2 + \mathcal{O}(\alpha^3) \tag{S14.47}$$

If we choose $\alpha = (g/2m)$, the linear term in ϕ goes away. A second change of variables $\phi \to -\phi$ gives

$$\mathscr{L} = \mathscr{L}_0 + \frac{g}{2m}\left[\overline{\psi}\gamma^\mu\gamma_5\psi\partial_\mu\phi - g\overline{\psi}\psi\phi^2\right] + \mathcal{O}(g^3) \tag{S14.48}$$

exactly as before.

---

[2] [Eds.] F. J. Dyson, "The Interactions of Nucleons with Meson Fields", *Phys. Rev.* **73** (1948) 929–930.

26

Vector fields

After scalar and spinor fields, the next case is the vector field. As the spin gets higher and higher, and the degrees of freedom increase, we have more and more indices to keep track of, but apart from that it will be pretty much a rerun of what we did for the scalar and spinor fields.

26.1 The free real vector field

I'll call the vector field $A^\mu(x)$ in honor of the most famous example: electrodynamics. I'll do the real case, $A_\mu = A_\mu^*$, because, as in the scalar case, the extension to complex fields is trivial. We begin as always by writing down the possible terms in the Lagrangian that are Lorentz scalars. Fortunately we can short circuit a lot of the stuff we did for spinors—I presume you know the Lorentz transformation properties of a vector.[1]

The first step is to write down the most general Lagrangian, quadratic in A_μ, with no more than two derivatives, to define the classical field theory. I will then restrict the parameters by requiring that the energy be positive, and quantize canonically. Here are the possible terms.

1. *No derivatives:* There is only one Lorentz invariant form:

$$A_\mu A^\mu \tag{26.1}$$

2. *One derivative:* I can build nothing, because with three vector indices, one from the derivative and two from the field, there's no possible way to make a scalar. We'll always have an uncontracted index somewhere.

3. *Two derivatives:* Things get more complicated. At first glance (and certainly with integration by parts) I can always arrange matters so that one field is differentiated once and the other field is differentiated once. There are apparently three possibilities:

$$(\partial_\mu A_\nu)(\partial^\mu A^\nu) \tag{26.2a}$$
$$(\partial^\mu A_\mu)(\partial^\nu A_\nu) \tag{26.2b}$$
$$(\partial^\nu A_\mu)(\partial^\mu A_\nu) \tag{26.2c}$$

[1] [Eds.] See §18.3.

As far as the Lagrangian goes, the third is the same as the second: using integration by parts, I can turn them into each other by switching the derivatives around (ignoring surface terms):

$$(\partial^\mu A_\mu)(\partial^\nu A_\nu) \Longleftrightarrow (\partial^\nu A_\mu)(\partial^\mu A_\nu) \quad \text{(these are equivalent via integration by parts)}$$

So in fact there are only three possible terms: (26.1), (26.2a) and (26.2b). That's slightly more complicated than the scalar field, where we had only two possible terms, but not much. I can rescale the fields to turn the coefficient of one of these terms into whatever I please, up to a sign. So I will write the most general form of the Lagrangian as

$$\mathscr{L} = \pm\tfrac{1}{2}\left[(\partial_\mu A_\nu)(\partial^\mu A^\nu) + a(\partial_\mu A^\mu)(\partial_\nu A^\nu) + bA_\nu A^\nu\right] \tag{26.3}$$

The factor $\frac{1}{2}$ will turn out later to be a convenient choice for the first term; there's some unknown real coefficient a in the second term, and some other real coefficient b in the third term, and that defines our Lagrangian. Higher order terms could be added, but then the Lagrangian would not describe a free field: free field Lagrangians mean *linear* equations of motion.

The next step is to vary the Lagrangian and derive the equations of motion:

$$-\partial^\mu\partial_\mu A_\nu - a\partial_\nu\partial_\mu A^\mu + bA_\nu = 0 \tag{26.4}$$

This is a messy equation and it's rather hard to see what particles of what mass are being described here. Let's just blithely go ahead and look for plane wave solutions:

$$A_\nu = \varepsilon_\nu e^{-ik\cdot x} \tag{26.5}$$

The four-vector ε_ν is called the **polarization vector**. Plugging this into the equations of motion gives

$$+k^2\varepsilon_\nu + ak_\nu\,\varepsilon\cdot k + b\varepsilon_\nu = 0 \tag{26.6}$$

We could write this as a 4×4 matrix in k acting on the vectors ε_μ, and find the eigenvectors in the usual way. Instead, we'll just read them off. This equation has two kinds of solutions: *longitudinal* solutions, where ε_μ is aligned along k_μ; and *transverse* solutions, with ε_μ perpendicular to k_μ.

(a) *Longitudinal:* $\varepsilon_\nu \propto k_\nu$.

$$\left[k^2 + ak^2 + b\right]k_\nu = 0; \qquad k^2 \equiv \mu_L^2 = -\frac{b}{1+a} \tag{26.7}$$

where μ_L is the mass of the longitudinal mode;

(b) *Transverse:* $\varepsilon\cdot k = 0$.

$$k^2\varepsilon_\nu + b\varepsilon_\nu = 0; \qquad k^2 \equiv \mu_T^2 = -b \tag{26.8}$$

where μ_T is the mass of the transverse mode.

So this theory is capable of describing two types of oscillations—longitudinal, with ε_μ parallel to k_μ, and transverse, with ε_μ perpendicular to k_μ. It's rather like the three dimensional theory of an elastic solid.[2] The longitudinal oscillations have one mass, and the transverse

[2] [Eds.] Phonon vibrations likewise have transverse and longitudinal modes. See, e.g., Charles Kittel, *Introduction to Solid State Physics*, 7th ed., J. Wiley, 1996, Chapter 4.

have another. Under a Lorentz transformation a longitudinal oscillation remains longitudinal and a transverse oscillation remains transverse. One would expect upon quantization that the longitudinal oscillations would correspond to scalar particles (one degree of freedom), and the transverse oscillations would correspond to spin one, (vector particles with three degrees of freedom). There are three independent vectors perpendicular (in a four dimensional sense) to a given four-vector like k_μ.[3] This should really be no surprise. We already know we could describe an ordinary scalar meson in terms of a vector field; to wit, its gradient $\partial_\mu\phi$. We *can* do it, but we don't particularly *want* to: we'd like to get a theory that when we quantize it describes vector particles *only*, without longitudinal oscillations.

Is it possible to arrange the parameters in our Lagrangian to suppress these longitudinal oscillations? Well, the answer is obvious: if we choose $a = -1$, so long as $b \neq 0$, our free wave equation has no longitudinal solutions; there are only transverse modes, with a mass $b \equiv -\mu^2$. The transverse solutions are the only things around; there is only one mass in the theory. Notice that this trick for suppressing the longitudinal solutions does not work when $-b = \mu^2 = 0$. In that case, if I set $a = -1$, then in fact I can have longitudinal oscillations of *any* mass. Instead of getting no solutions I have simply no restricting equation, because (26.7) becomes

$$k^2 - k^2 + 0 = 0 \tag{26.9}$$

which is unquestionably true, but it doesn't limit the longitudinal motions very much! (It's also possible to construct a theory with only longitudinal waves and no transverse waves. In this case there are both scalar and vector mesons with independent masses. This is not the easy way to do that. Why should you join together what God hath put asunder?)[4]

With these choices, $a = -1$ and $b = -\mu^2 \neq 0$, the Lagrangian becomes

$$\mathscr{L} = \pm\tfrac{1}{2}\left[(\partial_\mu A_\nu)(\partial^\mu A^\nu) - (\partial_\mu A^\mu)(\partial_\nu A^\nu) - \mu^2 A_\nu A^\nu\right] \tag{26.10}$$

Remember: the middle term is equivalent to $-(\partial_\mu A_\nu)(\partial^\nu A^\mu)$, plus surface terms. This Lagrangian is so simple I can't resist introducing notation to make it look obscure. Define the **field strength tensor**

$$\boxed{F_{\mu\nu} \equiv \partial_\mu A_\nu - \partial_\nu A_\mu} \tag{26.11}$$

This is the convention in modern high energy physics literature.[5,6] Then

$$\mathscr{L} = \pm\left[\tfrac{1}{4}F_{\mu\nu}F^{\mu\nu} - \tfrac{1}{2}\mu^2 A_\nu A^\nu\right] \tag{26.12}$$

The $\mu^2 \to 0$ limit of (26.12) describes free electromagnetism, written in relativistic form.[7] If I interpret A^0 as the scalar potential and A^i as the vector potential, (26.11) tells me that F^{ij} is

---

[3] [Eds.] For general results on the orthogonality of four vectors, see A. O. Barut, *Electrodynamics and the Classical Theory of Fields and Particles*, Macmillan, 1964, Chapter 1. Reprinted by Dover Publications, 1980.

[4] [Eds.] Oral tradition ascribes this quip ("What God hath put asunder, let no man join together.") to Pauli, in response to attempts by Weyl and Einstein (see footnote 3, p. 583) to unite electromagnetism with gravity.

[5] [Eds.] *Warning!* Bjorken and Drell define $F_{\mu\nu}$ as $F_{\mu\nu} \equiv \partial_\nu A_\mu - \partial_\mu A_\nu$, which differs by a minus sign; see Bjorken & Drell *Fields*, p. 68, equation (14.1).

[6] The tensor $F_{\mu\nu}$ is a *differential form*, so you can write it using the exterior derivative d as $F = dA$; but we won't. ([Eds.] See Ryder *QFT*, Section 2.9.)

[7] [Eds] See Problem 2.3, p. 99 and Jackson *CE*, Chap. 12; Ryder *QFT*, Sect. 3.3; Lev D. Landau and Evgeniĭ M. Lifshitz, *The Classical Theory of Fields*, 3rd rev. ed., Pergamon, 1971, §23 and §27.

the magnetic field, **B**, the curl of the vector potential. Likewise F^{0i} is the time derivative of the vector potential plus the gradient of the scalar potential, (-1) times the familiar formula for the electric field, **E**. As μ^2 goes to zero, the Lagrangian (26.12) becomes $\mathbf{E}^2 - \mathbf{B}^2$ times a factor, the familiar Lagrangian for free electromagnetic theory. We now see why we have no restraint on the longitudinal oscillations when $\mu^2 = 0$, because what we've been calling a longitudinal oscillation is equivalent in conventional electromagnetic theory to a **gauge transformation**. It is well known that you can add the four-gradient of a function $\lambda(x)$ to the four-potential A_μ, with no change to the physics:

$$A_\mu \to A'_\mu = A_\mu + \partial_\mu \lambda \quad \text{and} \quad F_{\mu\nu} \to F'_{\mu\nu} = \partial_\mu A'_\nu - \partial_\nu A'_\mu = F_{\mu\nu} \tag{26.13}$$

In the limit $\mu^2 \to 0$, the gauge invariance of electromagnetism pops into the theory described by (26.12), leading to some funny problems in quantizing the theory, as we will see. That doesn't happen in this theory for any other value of μ^2.

26.2 The Proca equation and its solutions

Just as an exercise, let's rederive the equations of motion (26.4) for the choices $a = -1$, $b = -\mu^2$

$$-\partial^\mu \partial_\mu A_\nu + \partial_\nu \partial_\mu A^\mu - \mu^2 A_\nu = 0 \tag{26.14}$$

from the Lagrangian (26.12). The field A_μ enters the Lagrangian four times: twice in each field tensor $F^{\mu\nu}$, and thus four times in its square. A variation of the Langrangian with respect to A_μ will produce a factor of 4 to cancel out the overall factor of $\frac{1}{4}$ multiplying $F_{\mu\nu}F^{\mu\nu}$. Written in terms of $F_{\mu\nu}$, the Euler–Lagrange equations are

$$\partial^\mu F_{\nu\mu} - \mu^2 A_\nu = 0 \tag{26.15}$$

This equation of motion, equivalent to (26.14), is called the **Proca equation.**[8] As the Klein–Gordon equation is to spin 0 and the Dirac equation is to spin $\frac{1}{2}$, so the Proca equation is to spin 1. Note: If $\mu = 0$, you get

$$\partial^\mu F_{\nu\mu} = 0 \tag{26.16}$$

which are two of the (empty space) Maxwell equations.[9]

Taking the divergence of the Proca equation, we get

$$\partial^\nu \partial^\mu F_{\nu\mu} - \mu^2 \partial^\nu A_\nu = 0 \tag{26.17}$$

The first term is zero because of the antisymmetry of $F_{\mu\nu}$. Assuming $\mu^2 \neq 0$, we see that the divergence of A_ν vanishes:

$$\partial^\nu A_\nu = 0 \qquad (\mu^2 \neq 0) \tag{26.18}$$

This equation is called the **Lorenz condition**. It ensures the suppression of the longitudinal waves. We've done this computation before in momentum space, (26.8), $k^\mu A_\mu = 0$. Now we see it again in position space.[10]

---

[8] [Eds.] Alexandru Proca, "Sur la théorie ondulaitoire des électrons positifs et négatifs" (On the wave theory of positive and negative electrons), *J. Phys. Radium* **7** (1936) 347–353. Proca, a Romanian-French physicist, was a student of de Broglie. See Y. Takahashi, *An Introduction to Field Quantization*, Pergamon, 1969 for a discussion of the Proca field as well as other less familiar fields, e.g., the Duffin–Kemmer–Petiau field and the Rarita–Schwinger spin-3/2 field, which comes up in supergravity theories. See also Ryder *QFT* Sections 2.8 and 4.5.

[9] [Eds.] The $\nu = 0$ component corresponds to Gauss's Law, $\nabla \bullet \mathbf{E} = 0$, and the $\nu = i$ components to the three components of Ampère's Law, $(\nabla \times \mathbf{B})_i = (\partial \mathbf{E}/\partial t)_i$.

[10] [Eds.] In the videotaped lectures, Coleman frequently calls it "Fourier space" instead of "momentum space".

Returning to the Proca equation in terms of the A's,

$$\partial^\mu \partial_\nu A_\mu - \partial^\mu \partial_\mu A_\nu - \mu^2 A_\nu = 0 \tag{26.19}$$

The first term is equal to $\partial_\nu \partial^\mu A_\mu$ which is zero. What remains is the Klein–Gordon equation:

$$(\Box^2 + \mu^2) A_\nu = 0 \tag{26.20}$$

The waves are transverse, with mass μ^2. There are four solutions, but only three are linearly independent because of the constraint (26.8). We now know in position space what we found earlier in momentum space.

As we did for the solutions of the Dirac equation (20.112a) and (20.112b), I'd like to establish some normalization conventions for the three independent solutions (labeled by r) to the Klein–Gordon equation,

$$A_\mu = \varepsilon_\mu^{(r)}(k)\, e^{-ik\cdot x} \qquad (r = 1, 2, 3) \tag{26.21}$$

with the three polarization vectors $\varepsilon_\mu^{(r)}$ orthogonal to k^μ, (26.18):

$$k^\mu \varepsilon_\mu^{(r)} = k \cdot \varepsilon^{(r)} = 0 \tag{26.22}$$

In the rest frame,

$$k = (\mu, \mathbf{0}) \tag{26.23}$$

and we choose the solutions to be orthonormal to each other, in this frame:

$$\varepsilon^{(r)*} \cdot \varepsilon^{(s)} = -\delta^{rs} \tag{26.24}$$

By Lorentz invariance, (26.24) is true in *any* inertial frame. These are three spacelike vectors, for instance, the three unit vectors in (x, y, z) perpendicular to timelike k. We get the $(-)$ sign in (26.24) from the metric. We have

$$k^\mu k_\mu = k \cdot k = \mu^2 \tag{26.25}$$

and we take $k^0 > 0$. In the rest frame, for example, we can choose the usual orthonormal space basis (for linear polarization):

$$\begin{aligned} \varepsilon^{(1)} &= (0, 1, 0, 0) \\ \varepsilon^{(2)} &= (0, 0, 1, 0) \\ \varepsilon^{(3)} &= (0, 0, 0, 1) \end{aligned} \tag{26.26}$$

or, for spin along the z axis (circular polarization):

$$\begin{aligned} \varepsilon^{(1)} &= \tfrac{1}{\sqrt{2}}(0, 1, i, 0) \\ \varepsilon^{(2)} &= \tfrac{1}{\sqrt{2}}(0, 1, -i, 0) \\ \varepsilon^{(3)} &= (0, 0, 0, 1) \end{aligned} \tag{26.27}$$

Here, the vectors $\varepsilon^{(1)}$ and $\varepsilon^{(2)}$ pick up the phase $e^{\pm i\theta}$ if rotated through θ about the z axis; they are eigenstates of J_z with $m = \pm 1$.

For the Dirac equation, normalization conditions (20.112a) and (20.112b) led to the completeness relations:

$$\sum_{r=1}^{2} u^r \overline{u}^r = \not{p} + m \tag{20.123a}$$

$$\sum_{r=1}^{2} v^r \overline{v}^r = \not{p} - m \tag{20.123b}$$

We also have a completeness relation for the vector field, easily derived. The analogous expression for the vector field is

$$P_{\mu\nu}(k) \equiv \sum_{r=1}^{3} \varepsilon_\mu^{(r)}(k)\, \varepsilon_\nu^{(r)*}(k) \tag{26.28}$$

What does this sum equal? By Lorentz invariance, it must be that

$$P_{\mu\nu}(k) = Ag_{\mu\nu} + Bk_\mu k_\nu \tag{26.29}$$

(where A and B are constants) because there are no other quantities that transform as rank 2 Lorentz tensors. We know from (26.22) that

$$P_{\mu\nu}k^\mu = \sum_{r=1}^{3} k^\mu \varepsilon_\mu^{(r)}(k)\varepsilon_\nu^{(r)*}(k) = 0 \;\Rightarrow\; A = -\mu^2 B \tag{26.30}$$

If we multiply $P_{\mu\nu}$ by $\varepsilon^{\nu(s)}$ we obtain from (26.24)

$$P_{\mu\nu}\varepsilon^{\nu(s)} = -\varepsilon_\mu^{(s)} \;\Rightarrow\; \mu^2 B = 1 \tag{26.31}$$

Consequently we have a completeness relation for the massive vector theory:

$$P_{\mu\nu} = \sum_{r=1}^{3} \varepsilon_\mu^{(r)} \varepsilon_\nu^{(r)*} = -g_{\mu\nu} + \frac{k_\mu k_\nu}{\mu^2} \tag{26.32}$$

To check this expression, consider the rest frame. If μ and ν are both space indices, i and j respectively, we have

$$P_{ij} = \sum_{r=1}^{3} \varepsilon_i^{(r)} \varepsilon_j^{(r)*} = -g_{ij} + 0 = \delta_{ij} \tag{26.33}$$

which can be confirmed by inspection from either (26.26), where $\varepsilon_i^{(r)} = \delta_i^r$, or (26.27). If either μ or ν is 0, the definition (26.28) gives $P_{\mu\nu} = 0$, since the time components of all the ε vectors from (26.26) or (26.27) are zero. And that's just what we get:

$$P_{00} = -g_{00} + \frac{k_0 k_0}{\mu^2} = -1 + \frac{\mu^2}{\mu^2} = 0, \quad P_{0i} = -g_{0i} + \frac{k_0 k_i}{\mu^2} = 0 \tag{26.34}$$

The projection operator $P_{\mu\nu}$ serves the same purpose as the projection operators (20.123a) and (20.123b), and will be just as useful when we need to sum over spins of vector mesons.

26.3 Canonical quantization of the Proca field

So much for the classical Proca equation and its plane wave solutions. We now turn to canonical quantization. There will be some complications quite apart from juggling all those indices. As usual it's convenient to break things up into space and time components: the Latin indices like i and j will run over the space indices only. This split will separate the p's and q's; the q's are the space components A_i of the vector field A_μ.

$$\mathscr{L} = \pm\left[\tfrac{1}{2}F_{0i}F^{0i} + \tfrac{1}{4}F_{ij}F^{ij} - \tfrac{1}{2}\mu^2 A_0A^0 - \tfrac{1}{2}\mu^2 A_iA^i\right] \tag{26.35}$$

The first $\frac{1}{4}$ turns into a $\frac{1}{2}$ because $F_{0i} = F_{i0}$. The only part with a time derivative is $F_{0i} = \partial_0 A_i - \partial_i A_0$; F^{ij} involves only space derivatives. The p's, the canonical momentum densities conjugate to A_i, are

$$\pi^i = \frac{\partial\mathscr{L}}{\partial(\partial_0 A_i)} = \pm F^{0i} \tag{26.36}$$

but—surprise!?—the momentum conjugate to A_0 is *zero*:

$$\pi^0 = \frac{\partial\mathscr{L}}{\partial(\partial_0 A_0)} = 0 \tag{26.37}$$

Is this a disaster or not? Well, that depends on whether or not the three quantities A_i and the three conjugate momenta F^{0i} are a complete and independent set of *initial value data* (IVD). I will demonstrate that in fact the entire set of initial value data is given in terms of the set $\{A_i, F^{0i}\}$ at a fixed time: we won't need π^0. That there is no momentum conjugate to A_0 is totally irrelevant. Even if there were a non-zero π^0, we'd have to throw it away; we already have a complete set.

To prove this, note that each component of A_μ obeys the Klein–Gordon equation (26.20). The field $\phi(\mathbf{x}, t_0)$ and its time derivative $\partial_0\phi(\mathbf{x}, t_0)$ at a fixed time t_0 provide a complete set of IVD for a field $\phi(\mathbf{x}, t)$ satisfying the Klein–Gordon equation, and we have four decoupled Klein–Gordon equations here. Our task is to show that at any given time, the quantities $\{A_i, \partial_0 A_i, A_0, \partial_0 A_0\}$ can be given in terms of $\{A_i\}$ and $\{F^{0i}\}$. If we can do that, we will have demonstrated that the six fields $\{A_i, F^{0i}\}$ suffice; we don't need eight fields, as one might have thought for the four decoupled Klein–Gordon equations. In particular, we don't need π_0.

We have $\{A_i\}$ and $\{F^{0i}\}$. That certainly includes A_i. From the constraint equation (26.18) we obtain $\partial^0 A_0$:

$$\partial^\mu A_\mu = 0 \Rightarrow \partial^0 A_0 = -\partial^i A_i \tag{26.38}$$

If I know A_i at a fixed time, I can certainly find its divergence at that time, and therefore I know $\partial^0 A_0$. Next, let's look at the full form of the equations of motion:

$$\partial_\mu F^{\nu\mu} - \mu^2 A^\nu = 0 \tag{26.15}$$

From the $\nu = 0$ component

$$\partial_i F^{0i} - \mu^2 A^0 = 0 \Rightarrow A^0 = (1/\mu^2)\,\partial_i F^{0i} \tag{26.39}$$

which determines A^0 in terms of space derivatives of the known quantities F^{0i}. Finally, from F_{0i} and A_0, we obtain $\partial_0 A_i$:

$$F_{0i} = \partial_0 A_i - \partial_i A_0 \Rightarrow \partial_0 A_i = \partial_i A_0 - F_{0i} \tag{26.40}$$

If I already know A^0, I can find its space derivatives, and I had F^{0i} from the start, so I can determine $\partial^0 A^i$. That completes the argument.

In short, once we have $A_i(\mathbf{x}, t_0)$ and $F_{0i}(\mathbf{x}, t_0)$ at some initial time t_0, that's all we need to compute $A_\mu(\mathbf{x}, t)$. The eight quantities $\{A_0, A_i, \partial_0 A_0, \partial_0 A_i\}$ can be found from the six quantities $\{A_i, F^{0i}\}$, and the equations of motion. This is really just transversality, expressed by the funny condition, $a = -1$. Had we not imposed transversality, we would have had four degrees of freedom. *With* that condition, we have only three independent Klein–Gordon equations, not four. We *should* need only six functions, two for each Klein–Gordon equation, not eight. And that's just what we found. This isn't a surprising result. Consider other field theories:

Theory	*IVD*	*Particle states*	*Description*
real Klein–Gordon	$\phi,\ \partial_0\phi$	1	particle
complex Klein–Gordon	$\left\{\begin{matrix}\phi,\ \partial_0\phi \\ \phi^*,\ \partial_0\phi^*\end{matrix}\right\}$	2	{particle, antiparticle}
Dirac	$\psi,\ \overline{\psi}$	4	{electron (up & down), positron (up & down)}
Proca	$A_i,\ F_{0i}$	3	particles

In every case,

$$\begin{pmatrix}\text{Number of} \\ \text{particle states}\end{pmatrix} = \begin{pmatrix}\text{Number of independent} \\ \text{plane wave solutions}\end{pmatrix} = \tfrac{1}{2}\begin{pmatrix}\text{Number of} \\ \text{pieces of IVD}\end{pmatrix} \tag{26.41}$$

Now let's compute the Hamiltonian density. As always,

$$\mathscr{H} = \pi^i \partial_0 A_i - \mathscr{L} = \pm F^{0i}\partial_0 A_i - \mathscr{L} \tag{4.40}$$

This is an awkward expression for our purposes, because we want to write everything in terms of p's and q's, i.e., in terms of F^{0i} and A_i. Using the identity

$$\partial_0 A_i = F_{0i} + \partial_i A_0 \tag{26.42}$$

the Hamiltonian density can be written

$$\mathscr{H} = \pm F^{0i}F_{0i} \pm F^{0i}\partial_i A_0 - \mathscr{L} \tag{26.43}$$

All we're really interested in is the Hamiltonian, the space integral of the Hamiltonian density $\mathscr{H}$. We can integrate (26.43) by parts, to rewrite the second term on the right:

$$F^{0i}\partial_i A_0 \overset{\text{space} \int}{\Longleftrightarrow} -(\partial_i F^{0i})A_0 = -\mu^2 A^0 A_0 \tag{26.44}$$

the last equality following from (26.39). Then

$$\begin{aligned}\mathscr{H} &= \pm\left[F_{0i}F^{0i} - \mu^2 A_0 A^0 - \tfrac{1}{2}F_{0i}F^{0i} - \tfrac{1}{4}F_{ij}F^{ij} + \tfrac{1}{2}\mu^2 A_0 A^0 + \tfrac{1}{2}\mu^2 A_i A^i\right] \\ &= \pm[\underbrace{\tfrac{1}{2}F_{0i}F^{0i}}_{\le 0} - \underbrace{\tfrac{1}{2}\mu^2 A_0 A^0}_{\le 0} - \underbrace{\tfrac{1}{4}F_{ij}F^{ij}}_{\le 0} + \underbrace{\tfrac{1}{2}\mu^2 A_i A^i}_{\le 0}]\end{aligned} \tag{26.45}$$

Of course if I really wanted to express the Hamiltonian as a function of p's (F^{i0}) and q's (A_i), I should write for A_0 the expression $(1/\mu^2)\partial_i F^{0i}$.

Now we come to the question of positivity of the energy. *Each* of these terms is individually negative. The first and fourth terms inside the brackets are each negative sums of squares, because $F_{0i} = -F^{0i}$ and $A_i = -A^i$. The second term inside the brackets is positive, because $A_0 = A^0$. The third term inside the brackets is a positive sum of squares, because $F_{ij} = F^{ij}$. Consequently the quantity in the square brackets is a sum of four negative terms. The Hamiltonian must be bounded from below, so the overall sign must be (−). With our sign ambiguity resolved, we can write for the Hamiltonian

$$\mathscr{H} = -\left[\tfrac{1}{2}F_{0i}F^{0i} - \mu^2 A_0 A^0 - \tfrac{1}{4}F_{ij}F^{ij} + \tfrac{1}{2}\mu^2 A_\mu A^\mu\right] \tag{26.46}$$

and for the Lagrangian,

$$\mathscr{L} = -\tfrac{1}{4}F_{\mu\nu}F^{\mu\nu} + \tfrac{1}{2}\mu^2 A_\nu A^\nu \tag{26.47}$$

This is pretty much like what we got for the scalar field, (4.44), except that the sign of the mass term looks wrong. Remember that the true dynamical variables, the q's and the p's, are $\{A_i\}$ and $\{F^{0i}\}$, respectively. Rewriting the Lagrangian we have

$$\begin{aligned}\mathscr{L} &= -\tfrac{1}{2}F_{0i}F^{0i} - \tfrac{1}{4}F_{ij}F^{ij} + \tfrac{1}{2}\mu^2 A_0 A^0 + \tfrac{1}{2}\mu^2 A_i A^i \\ &= +\tfrac{1}{2}(F_{0i})^2 - \tfrac{1}{4}F_{ij}F^{ij} + \tfrac{1}{2}\mu^2 (A_0)^2 - \tfrac{1}{2}\mu^2 (A_i)^2\end{aligned} \tag{26.48}$$

So it's just like the conventional expression in Lagrangian mechanics,

$$L = T - V \tag{26.49}$$

The term with time derivatives in T has a positive coefficient, as does the corresponding term in (26.48), $+\frac{1}{2}(F_{0i})^2$. Likewise one of the terms in V, the mass term $-\frac{1}{2}\mu^2(A_i)^2$, has a negative coefficient just as it should, introduced by the metric rather than by an explicit minus sign in front. The canonical momentum is given by (26.36), with the minus sign:

$$\pi^i = -F^{0i} = +F_{0i} = +F^{i0} \tag{26.50}$$

Let's canonically quantize this theory (4.47):

$$\left[A_i(\mathbf{x},t), A_j(\mathbf{y},t)\right] = 0 \tag{26.51}$$

$$\left[F^{i0}(\mathbf{x},t), A_j(\mathbf{y},t)\right] = -i\delta^i_{\ j}\,\delta^{(3)}(\mathbf{x}-\mathbf{y}) \tag{26.52}$$

$$\left[F^{i0}(\mathbf{x},t), F^{j0}(\mathbf{y},t)\right] = 0 \tag{26.53}$$

To shorten a lengthy calculation, I will write the field at any spacetime point in terms of a sequence of Fourier coefficients, with the usual measure:

$$A_\mu(x) = \sum_{r=1}^{3}\int \frac{d^3\mathbf{k}}{(2\pi)^{3/2}\sqrt{2\omega_{\mathbf{k}}}}\left[a_{\mathbf{k}}^{(r)}\varepsilon_\mu^{(r)}(k)e^{-ik\cdot x} + a_{\mathbf{k}}^{(r)\dagger}\varepsilon_\mu^{(r)*}(k)e^{+ik\cdot x}\right] \tag{26.54}$$

This is much like what we did for the Dirac field (21.6), except that this field A_μ is Hermitian. In place of the Dirac spinors we have the analogous polarization vectors $\varepsilon_\mu^{(r)}$ multiplying the operators $a_{\mathbf{k}}^{(r)\dagger}$ and $a_{\mathbf{k}}^{(r)}$. It's a long calculation to determine the commutation relations of the operators $a_{\mathbf{k}}^{(r)}$ and $a_{\mathbf{k}}^{(s)\dagger}$ which ensure the canonical commutation relations for A_i and F^{0j}. Rather than plug and chug, let's *guess* the commutation relations are the same as (2.47), for the scalar case:

$$\left[a_{\mathbf{k}}^{(r)}, a_{\mathbf{k}'}^{(s)\dagger}\right] = \delta^{(3)}(\mathbf{k}-\mathbf{k}')\,\delta^{rs} \tag{26.55}$$

and all others zero. In other words, the operators $a_{\mathbf{k}}^{(s)\dagger}$ and $a_{\mathbf{k}}^{(r)}$ are creation and annihilation operators. From these we will check the commutation relations, (26.51).

To confirm that these commutators (26.55) indeed give the canonical commutation relations (26.51) is a horrendous computation, because of all the indices. And therefore I'll be a little sneaky, and recall that when we did a free scalar field, we found the commutators for arbitrary times. Check that these commutation relations work by recalling the scalar theory:

$$\phi(x) = \int \frac{d^3\mathbf{k}}{(2\pi)^{3/2}\sqrt{2\omega_{\mathbf{k}}}} \left[a_{\mathbf{k}} e^{-ik\cdot x} + a_{\mathbf{k}}^{\dagger} e^{ik\cdot x}\right] \tag{3.45}$$

The commutation relation $[\phi(x), \phi(y)]$ for arbitrary time is

$$[\phi(x), \phi(y)] = i\Delta(x-y) \tag{3.51}$$

where $i\Delta(x-y)$ can be written, from (3.38) and (3.42), as

$$i\Delta(x-y) \equiv \int \frac{d^3\mathbf{k}}{(2\pi)^3 2\omega_{\mathbf{k}}} \left(e^{-ik\cdot(x-y)} - e^{ik\cdot(x-y)}\right) = \int d^3\mathbf{k}\,(\text{mess}) \tag{26.56}$$

Let $\partial_\mu^x = \partial/\partial x^\mu$. Then $\Delta(x-y)$ satisfies the following conditions for $x^0 = y^0$ (see (3.60), (3.61) and (3.66), respectively):

$$\Delta(x-y) = 0; \quad \partial_0^x \Delta(x-y) = -i\delta^{(3)}(\mathbf{x}-\mathbf{y}); \quad \partial_0^x \partial_0^y \Delta(x-y) = 0 \tag{26.57}$$

(These correspond to $[q_i, q_j] = 0$, $[p_i, q_j] = -i\delta_{ij}$, and $[p_i, p_j] = 0$.) The computation we have to do here is basically the same, except that after we've commuted everything, we also have a polarization sum on r. For each value of r, it's the same computation we did for the scalar field. By analogy

$$[A_\mu(x), A_\nu(y)] = \int d^3\mathbf{k}\,(\text{mess}) \sum_r \varepsilon_\mu^{(r)}(k)\,\varepsilon_\nu^{(r)*}(k) = \int d^3\mathbf{k}\,(\text{mess}) \left[-g_{\mu\nu} + \frac{k_\mu k_\nu}{\mu^2}\right] \tag{26.58}$$

using (26.32). So we'll get exactly the same Fourier transform as for the scalar field. The only thing is that in momentum space, the completeness relation will be stuck inside the integral. But it's easy to see what the Fourier transform produces, since multiplication by k_μ is Fourier-equivalent to differentiation: $k_\mu \to -i\partial_\mu$:

$$\begin{aligned}[A_\mu(x), A_\nu(y)] &= \int d^3\mathbf{k}\,(\text{mess}) \left[-g_{\mu\nu} + \frac{k_\mu k_\nu}{\mu^2}\right] = \left[-g_{\mu\nu} - \frac{\partial_\mu^x \partial_\nu^x}{\mu^2}\right] \int d^3\mathbf{k}\,(\text{mess}) \\ &= \left[-g_{\mu\nu} - \frac{\partial_\mu^x \partial_\nu^x}{\mu^2}\right] i\Delta(x-y)\end{aligned} \tag{26.59}$$

Without doing any computation at all, but just by being sneaky, we obtain the expression above. That saves a lot of labor. There's no reason to do a calculation twice when you've already done it once.

But does it give the correct commutation relations? Start with the easiest:

$$[A_i(\mathbf{x},t), A_j(\mathbf{y},t)] = \left[-g_{ij} - \frac{\partial_i^x \partial_j^x}{\mu^2}\right] i\Delta(\mathbf{x}-\mathbf{y},0) = 0 \tag{26.60}$$

because both $\Delta(x-y)$ and its gradient $\partial_i^x \Delta(x-y)$ vanish at $x_0 = y_0$. So that one checks: $[q_i, q_j] = 0$. (Note that A_0 does *not* commute with A_i at equal times. Because of (26.39), A_0 is some awful divergence of the canonical momentum.) Next, we have

$$\begin{aligned}
[F_{\lambda\mu}(x), A_\nu(y)] &= \partial_\lambda^x [A_\mu(x), A_\nu(y)] - \partial_\mu^x [A_\lambda(x), A_\nu(y)] \\
&= \partial_\lambda^x \left(-g_{\mu\nu} - \frac{\partial_\mu^x \partial_\nu^x}{\mu^2}\right) i\Delta(x-y) - \partial_\mu^x \left(-g_{\lambda\nu} - \frac{\partial_\lambda^x \partial_\nu^x}{\mu^2}\right) i\Delta(x-y) \\
&= \left(-g_{\mu\nu}\partial_\lambda^x + g_{\lambda\nu}\partial_\mu^x\right) i\Delta(x-y)
\end{aligned} \tag{26.61}$$

Then for $x_0 = y_0$,

$$\left[F^{i0}(x), A_j(y)\right]\Big|_{x_0=y_0} = \left(-\delta_j^0 \partial_x^i + \delta_j^i \partial_x^0\right) i\Delta(x-y)\big|_{x_0=y_0} = -i\delta_j^i \delta^{(3)}(\mathbf{x}-\mathbf{y}) \tag{26.62}$$

as required for the equal-time canonical commutation relations for F^{j0} and A_i; the $[p_i, q_j]$ commutator checks. Finally, for the last set (with $x_0 = y_0$)

$$\begin{aligned}
[F_{0i}(x), F_{0j}(y)] &= \partial_0^x \left[F_{j0}(y), A_i(x)\right] - \partial_i^x \left[F_{j0}(y), A_0(x)\right] \\
&= \left[\partial_0^x \left(-g_{0i}\partial_j^y + g_{ji}\partial_0^y\right) - \partial_i^x \left(-g_{00}\partial_j^y + g_{j0}\partial_0^y\right)\right] i\Delta(x-y)\big|_{x_0=y_0} \\
&= \left(g_{ji}\partial_0^x\partial_0^y + g_{00}\partial_i^x\partial_j^y\right) i\Delta(x-y)\big|_{x_o=y_0} \\
&= 0
\end{aligned} \tag{26.63}$$

The first term is zero because two time derivatives become $(\partial_0^x \partial_0^y)$ acting on $\Delta(x-y)$, which equals zero at equal times. The second term vanishes because $\Delta(x-y) = 0$ for $x_0 = y_0$, and so *a fortiori* does its gradient. That makes $[p_i, p_j] = 0$. We have verified the equal time canonical commutation relations. As an exercise, you should be able to show that, analogous to (2.48)

$$H = \int d^3\mathbf{x}\, \mathscr{H} = \int d^3\mathbf{k}\, \omega_{\mathbf{k}} \sum_r a_{\mathbf{k}}^{(r)\dagger} a_{\mathbf{k}}^{(r)} \tag{26.64}$$

(plus a divergent constant, usually dropped, disposed of by normal ordering).[11]

If however we try to canonically quantize the vector meson in the limit as $\mu^2 \to 0$, we come to a screeching halt. In the limit as $\mu^2 \to 0$, the Proca equation reduces to the Maxwell equations. We take A^0 to be the scalar potential, A^i to be the vector potential, and

$$F^{i0} = \partial^i A^0 - \partial^0 A^i = E^i \tag{26.65}$$

[11] [Eds.] See §4.5 for the scalar case. The calculation of the massive vector's Hamiltonian is the subject of Problem 15.2, p. 591.

just the usual definition of the electric field. Then $\partial_i F^{i0} = \nabla\cdot\mathbf{E} = 0$, according to Maxwell's equations (in empty space), so the commutator of $\partial_i F^{i0}$ with anything should be zero. But

$$[\partial_i F^{i0}(\mathbf{x},t), A_j(\mathbf{y},t)] = \partial_i^x [F^{i0}(\mathbf{x},t), A_j(\mathbf{y},t)] = -i\partial_j^x \delta^{(3)}(\mathbf{x}-\mathbf{y}) \neq 0 \tag{26.66}$$

What should be zero is not zero. We will solve this problem soon enough, but you should be aware of it. The problem does not arise if $\mu^2 \neq 0$, because the Proca equation component equivalent to Gauss's Law in empty space is

$$\partial_i F^{i0} = -\mu^2 A^0 \tag{26.67}$$

The commutator (26.66) that gave us trouble with $\mu^2 = 0$ is no longer a problem. Using (26.59),

$$\begin{aligned}\partial_i^x [F^{i0}(\mathbf{x},t), A_j(\mathbf{y},t)] &= -\mu^2 [A^0(\mathbf{x},t), A_j(\mathbf{y},t)] = -\mu^2 \left[-\delta_j^0 - \frac{\partial_x^0 \partial_j^x}{\mu^2}\right] i\Delta(\mathbf{x}-\mathbf{y},0) \\ &= -i\partial_j^x \delta(\mathbf{x}-\mathbf{y})\end{aligned} \tag{26.68}$$

Unlike (26.66), this result is perfectly consistent with the canonical commutation relations.

26.4 The limit $\mu \to 0$: a simple physical consequence

I want to discuss a topic more interesting than this dumb commutation computation (which, once you've done, you never have to think about again). Let's consider something with a little more physics in it: how an actual physical process is affected as the mass of this real vector meson goes to zero. I can't talk about a very complicated theory, because we have yet to compute the vector meson's propagator. However there is one theory I can discuss, the analog of our earlier scalar theory, Model 1.[12] That interaction (8.57) was between the field ϕ and a c-number source, $\rho(x)$:

$$\mathscr{H}_I = g\rho(x)\phi(x) \tag{26.69}$$

Here we'll have a c-number source J^μ, some arbitrary vector function of x vanishing at infinity, coupled to A_μ. That is, we write the Lagrangian as

$$\mathscr{L} = -\tfrac{1}{4}F_{\mu\nu}F^{\mu\nu} + \tfrac{1}{2}\mu^2 A_\nu A^\nu - J^\mu A_\mu \tag{26.70}$$

I can discuss the process where that c-number source emits one meson. This is the analog of

Figure 26.1: Emission of one vector meson by J^μ

the Feynman graph, Figure 8.14, that we talked about in Model 1. The wavy line in Figure 26.1 indicates the vector meson. Let me write down the equation of motion,

$$\partial^\nu F_{\mu\nu} - \mu^2 A_\mu + J_\mu = 0 \tag{26.71}$$

Does this have a sensible zero mass limit? Let's take the divergence of this equation, and obtain

$$\partial^\mu \partial^\nu F_{\mu\nu} - \mu^2 \partial^\mu A_\mu + \partial^\mu J_\mu = 0 \tag{26.72}$$

[12] [Eds.] See §8.5.

Now the first term is always zero, by the antisymmetry of $F_{\mu\nu}$. If this is to make any sense at all when I set $\mu^2 = 0$, I had better have $\partial^\mu J_\mu = 0$, otherwise everything will go bananas. So let's consider not a general external current but a *conserved* current, for which $\partial^\mu J_\mu = 0$. That's my one condition.

What is the amplitude for the emission of a meson of helicity type r (p. 400) and momentum k? We sort of know what happens from our old scalar model. There, the amplitude for one meson emission was proportional to $\tilde{\rho}(k)$, (8.67). Here, the amplitude $\mathcal{A}_{fi}$ must look something like this:

$$\mathcal{A}_{fi} \propto \varepsilon_\mu^{(r)\,*}\widetilde{J}^\mu(k) \tag{26.73}$$

Within kinematic factors, it *has* to look like this. To lowest nontrivial order, the amplitude must be linear in $\widetilde{J}^\mu(k)$, and it has to be a Lorentz scalar. There are only two other four-vectors available with which to build an invariant amplitude, $\varepsilon_\mu^{(r)\,*}$ and k_μ. But by our conservation condition we must also have

$$k_\mu \widetilde{J}^\mu = 0 \tag{26.74}$$

so the expression (26.73) is the only Lorentz scalar available. Let's consider a meson of specified momentum, and a nice, smooth function $\widetilde{J}^\mu(k)$. Take that momentum to point in the z direction with some magnitude $|\mathbf{k}|$, and of course it has energy $\sqrt{|\mathbf{k}|^2 + \mu^2}$.

$$k^\mu = (\sqrt{|\mathbf{k}|^2 + \mu^2}, 0, 0, |\mathbf{k}|) \tag{26.75}$$

If the amplitude for this process goes as the corresponding scalar field's (8.67), it has a factor $1/\sqrt{\omega_\mathbf{k}}$. Let's assume that is so. For small values of μ,

$$\omega_\mathbf{k} = \sqrt{|\mathbf{k}|^2 + \mu^2} = |\mathbf{k}|\left(1 + \mathcal{O}(\mu^2/|\mathbf{k}|^2)\right) \tag{26.76}$$

In the rest frame of the particle, this amplitude does not have a smooth limit as $\mu \to 0$. Instead let's consider the limit $(\mu/|\mathbf{k}|) \to 0$, for each of three independent kinds of mesons that can be emitted. Remember that the ε_μ's are restricted to be orthonormal spacelike vectors orthogonal to k^μ. Two are

$$\begin{aligned} \varepsilon^{(1)\,\mu} &= (0, 1, 0, 0) \\ \varepsilon^{(2)\,\mu} &= (0, 0, 1, 0) \end{aligned} \tag{26.77}$$

Linear combinations of these, $(1/\sqrt{2})(\varepsilon^{(1)\,\mu} \pm i\varepsilon^{(2)\,\mu})$, have helicity ± 1. Then there's one unchanged by rotations about the z axis, the third vector $\varepsilon^{(3)\,\mu}$ with helicity zero, which must be orthogonal to k^μ and to the other two ε_μ's, so it cannot have x or y components. It must look like this:

$$\varepsilon^{(3)\,\mu} = \frac{1}{\mu}(|\mathbf{k}|, 0, 0, \sqrt{|\mathbf{k}|^2 + \mu^2}) \tag{26.78}$$

(To satisfy (26.24) I have to divide by μ.) We're now ready to go. I'm going to show you that something very interesting happens to the amplitude for emitting a meson of this kind in the limit as $\mu \to 0$.

The amplitude for emitting a meson of type 3 goes as

$$\mathcal{A}_3 \sim \frac{1}{\mu}\left(|\mathbf{k}|\widetilde{J}^0 - \sqrt{|\mathbf{k}|^2 + \mu^2}\widetilde{J}^3\right) \tag{26.79}$$

From the conservation rule (26.74), with $\mathbf{k}$ parallel to $\hat{\mathbf{z}}$, we have

$$\sqrt{|\mathbf{k}|^2 + \mu^2}\widetilde{J}^0 - |\mathbf{k}|\widetilde{J}^3 = 0 \tag{26.80}$$

Putting $\widetilde{J}^3$ in terms of $\widetilde{J}^0$, we notice an amazing cancellation:

$$\mathcal{A}_3 \sim \frac{1}{\mu}\widetilde{J}^0\left(|\mathbf{k}| - \frac{|\mathbf{k}|^2+\mu^2}{|\mathbf{k}|}\right) \sim -\frac{\mu}{|\mathbf{k}|}\widetilde{J}^0 \tag{26.81}$$

No matter how small μ is, this is a system with three degrees of freedom. Everyone says the photon is massless. But suppose the photon had a mass of 10^{-23} of the electron's. This would be a very hard thing to determine experimentally. Some people say, "No, absolutely not! It would be trivial to detect experimentally because we know the real massless photon has only two degrees of freedom; polarized light and so on. If we took a hot oven and let things come to thermal equilibrium, because the walls are emitting and absorbing photons, we wouldn't get the Planck Law, but instead $\frac{3}{2}$ times the Planck Law." This is *garbage*. The amplitude for the oven walls to radiate a helicity zero photon, according to this current, goes to *zero* in the limit as $\mu/|\mathbf{k}| \to 0$. At every stage in the limiting process there are indeed three degrees of freedom just as you'd expect from a theory of massive vector mesons. But as $\mu/|\mathbf{k}| \to 0$, the amplitude for emitting the third photon goes to zero. If the photon mass is small enough, it will require twenty trillion years for that oven to reach thermal equilibrium![13]

Thus we see something very interesting. Our whole formalism collapses completely as $\mu \to 0$. But if we go to the end and compute the amplitude for a physically reasonable, very simple process—the emission of a single meson by an external source—we find it goes to what you would expect if you knew anything about electrodynamics, namely, the electrodynamic answer: A vector meson of mass 10^{-23} times the electron mass looks just like a photon. Shake your source, and despite the three degrees of freedom, except for a negligible factor, only helicity ± 1 mesons are emitted; there is *no* amplitude to emit helicity zero.

The black body law turns out not to be the best way to determine an upper limit for the photon mass. The best way is the Coulomb law which would be modified to a Yukawa law, or the analog for a dipole. Although there are magnetic fields over cosmic distances, these are not good because the Compton wavelength is messed up by interstellar plasma. The best measurement[14] was made from the magnetic dipole field of the earth, by the Explorer 12 satellite, *at night*, because in the daytime Explorer 12 was in the solar wind. In the night time the earth shields Explorer 12 from the solar wind, and the plasma effects are much smaller. The satellite was 10,000 km from the earth's center, and measured the dipole field with something like 10 or 15 percent accuracy. Therefore a number on the order of 10^7 m is the current best lower bound on the Compton wavelength of the photon. The Compton wavelength of the electron is 10^{-12} m, so this provides a upper bound to the photon's mass of about 10^{-19} of the electron's mass, or about 10^{-47} g.

26.5 Feynman rules for a real massive vector field

Let's briefly consider interactions of the massive vector field with either an electron (equivalent to quantum electrodynamics with a massive photon, "massive QED") or a charged scalar field

[13] [Eds.] See also L. de Broglie, *Mécanique Ondulaire du Photon et Théorie Quantique des Champs* (Wave Mechanics of the Photon and the Quantum Theory of Fields), 2nd ed., Gauthier-Villars 1957, Chapter V, §5; and L. Bass and E. Schrödinger, "Must the Photon Mass Be Zero?", *Proc. Roy. Soc. Lond. A* **232** (1955) 1–6.

[14] [Eds.] As of 1975! See V. L. Patel, "Structure of the equations of cosmic electrodynamics and the photon rest mass", *Phys. Lett.* **14** (1965) 105-106. Current bounds are about 10^5 times more stringent; see A. S. Goldhaber and M. M. Nieto, "Photon and graviton mass limits", *Rev. Mod. Phys.* **82** (2010) 939–979.

("massive charged scalar electrodynamics").[15] This is a good background for actual QED, which we will begin to tackle next time. We'll add an interaction between the real massive vector field with a Dirac field,

$$\mathscr{L}' = -e\overline{\psi}\gamma^{\mu}\Gamma\psi A_{\mu} \tag{26.82}$$

where $\Gamma = 1$, with A_{μ} a vector; or $\Gamma = i\gamma_5$, with A_{μ} an axial vector; and e is a coupling constant. (Interactions of this type occur in discussions of the Z_0 meson in electroweak theory, which we'll discuss near the end of this course.) The first thing we have to do is to work out the propagator for the massive vector meson.

As usual, we define the Wick contraction as the time-ordered vacuum expectation value:

$$\overline{A_{\mu}(x)A_{\nu}(y)} = \langle 0|\, T(A_{\mu}(x)A_{\nu}(y))\, |0\rangle \tag{26.83}$$

Consider first a *scalar* field of the same mass. Recall (3.38) the vacuum expectation value,

$$\langle 0|\phi(x)\phi(y)|0\rangle = \langle 0|[\phi^{(+)}(x), \phi^{(-)}(y)]|0\rangle = \int \frac{d^3\mathbf{k}}{(2\pi)^3(2\omega_{\mathbf{k}})}\, e^{-ik\cdot(x-y)} = \Delta_+(x-y) \tag{26.84}$$

By analogy,

$$\begin{aligned}\langle 0|A_{\mu}(x)A_{\nu}(y)|0\rangle &= \int \frac{d^3\mathbf{k}}{(2\pi)^3(2\omega_{\mathbf{k}})}\, e^{-ik\cdot(x-y)} \left[-g_{\mu\nu} + \frac{k_{\mu}k_{\nu}}{\mu^2}\right] \\ &= \left[-g_{\mu\nu} - \frac{\partial^x_{\mu}\partial^x_{\nu}}{\mu^2}\right]\Delta_+(x-y)\end{aligned} \tag{26.85}$$

The time-ordered vacuum expectation value of the scalar field is (P1.4)

$$\langle 0|\, T(\phi(x)\phi(y))\, |0\rangle = \theta(x^0 - y^0)\,\Delta_+(x-y) \;+\; (x \leftrightarrow y) = \int \frac{d^4k}{(2\pi)^4}\frac{i}{k^2-\mu^2+i\varepsilon}\, e^{-ik\cdot(x-y)} \tag{26.86}$$

and for the vector fields,

$$\langle 0|T(A_{\mu}(x)A_{\nu}(y))|0\rangle = \theta(x^0-y^0)\left[-g_{\mu\nu} - \frac{\partial^x_{\mu}\partial^x_{\nu}}{\mu^2}\right]\Delta_+(x-y) \;+\; (x\leftrightarrow y) \tag{26.87}$$

We'd now be ecstatically happy (well, let's just say mildly cheerful) if the time-ordered product were equal to this obvious guess:

$$\begin{aligned}\langle 0|T(A_{\mu}(x)A_{\nu}(y))|0\rangle &\stackrel{?}{=} \left[-g_{\mu\nu} - \frac{\partial^x_{\mu}\partial^x_{\nu}}{\mu^2}\right]\theta(x^0-y^0)\,\Delta_+(x-y) \;+\; (x\leftrightarrow y) \\ &= \int \frac{d^4k}{(2\pi)^4}\frac{i}{k^2-\mu^2+i\varepsilon}\left(-g_{\mu\nu} + \frac{k_{\mu}k_{\nu}}{\mu^2}\right) e^{-ik\cdot(x-y)}\end{aligned} \tag{26.88}$$

The $\stackrel{?}{=}$ indicates that it's not obvious we can bring the derivatives through the θ functions, which we'd need to do to express the time-ordered product as an integral. This is all right so long as μ and ν are not *both* equal to zero. If μ and ν are both space indices, there's no

---

[15] [Eds.] Parts of this section are based on class notes from 1999, provided by Daniel Podolsky.

problem because the θ functions depend only on time. When one of the indices is a space index and the other a 0, then pulling the $\partial_\mu\partial_\nu$ in front of the θ in (26.88) will give, apart from the right-hand side of (26.87), the term

$$\partial_i\left[\delta(x_0-y_0)\Delta_+(x-y)-(x\leftrightarrow y)\right] \tag{26.89}$$

Because of the delta function in front, $y-x$ has a vanishing time component, so it is a spacelike vector. Therefore by making a rotation by π, we can turn $x-y$ into $y-x$, while the function Δ_+ is itself rotation invariant. The two parts of the extra term cancel, so there's no problem here, either.

When both indices equal zero, it is *not* all right. If we could drag the time derivatives through the θ functions, we could write, for example,

$$(\Box^2+\mu^2)\left[\theta(x_0-y_0)\,\Delta_+(x-y)+(x\leftrightarrow y)\right]=\theta(x_0-y_0)\,(\Box^2+\mu^2)\Delta_+(x-y)+(x\leftrightarrow y) \tag{26.90}$$

On the left-hand side of this equation, we have the Klein–Gordon operator acting on the scalar Feynman propagator $\Delta_F(x-y)$, the Fourier transform of (10.29); that gives us $-i\delta^{(4)}(x-y)$. On the right-hand side we have the Klein–Gordon operator acting on $\Delta_+(x-y)$, that is, on the vacuum expectation value of the product of two fields—one at x and one at y—each of which obeys the Klein–Gordon equation. Therefore the right-hand side is zero, and (26.90) is equivalent to the statement that $-i\delta^{(4)}(x-y)=0$. Thus, in the case of $\nu=\mu=0$, we have a problem. It looks like the propagator is not Lorentz invariant, because its 00 component has an extra piece coming from the time derivatives on the θ functions. Well, this is certainly a Lorentz invariant theory; is there somewhere else where there could possibly be another extra piece to cancel this one?

You may remember (§26.3) that when we were doing the initial value problem for a free vector field, we discovered that A_0 was *not* an independent dynamical variable. If we break the interaction into two pieces,

$$\mathscr{L}'=-e\overline{\psi}\gamma^0\Gamma\psi A_0+e\overline{\psi}\gamma^i\Gamma\psi A_i \tag{26.91}$$

then the space term is like a "q" in the language of p's and q's, since the A_i's are independent dynamical variables, each with its canonical momentum. But A_0 is proportional to ∂_iF^{i0} by the Proca equation, and F_{i0} is a p-type variable; thus a term involving A_0, like $\overline{\psi}\gamma^0\Gamma\psi A_0$, is in fact a derivative interaction involving the p's. When we write the theory in Hamiltonian form, in terms of p's and q's, then this term must be involved in the Hamiltonian, but the other terms in ψ are not; $\mathscr{H}_I\neq-\mathscr{L}_I$. Then the Hamiltonian, the thing that appears in Dyson's formula, is not going to look Lorentz invariant, either. Maybe God is on our side, and these two difficulties cure each other. In fact during the heroic period of the late Forties it was shown that this desperate prayer is indeed answered: if you treat $\mathscr{H}_I$ naïvely as if it were equal to $-\mathscr{L}_I$, and the propagator as if it were equal to the right-hand side of (26.88), the troubles cancel.[16] Later on, we'll see how the troubles cancel when we develop more efficient methods for handing these kinds of theories (§29.4). For the moment accept on trust that everything works if you treat $\mathscr{H}_I=-\mathscr{L}_I$ and the propagator as

$$D_F^{\mu\nu}(x-y)=\int\frac{d^4k}{(2\pi)^4}\,\frac{i}{k^2-\mu^2+i\varepsilon}\left(-g_{\mu\nu}+\frac{k_\mu k_\nu}{\mu^2}\right)e^{-ik\cdot(x-y)} \tag{26.92}$$

[16] [Eds.] See also the example beginning with (27.75), p. 587, and note 7, p. 589.

The Feynman rules for the theory with the interaction Lagrangian (26.82) are set out in the box below.

Feynman rules for Massive QED

1. For every ... Write ...

   (a) internal vector line 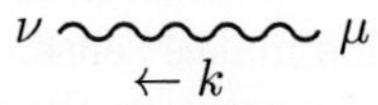

   $\left(-g^{\mu\nu}+\frac{k^\mu k^\nu}{\mu^2}\right)\frac{i}{k^2-\mu^2+i\epsilon}$

   (b) internal electron line 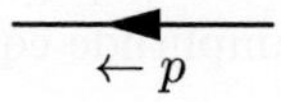

   $\frac{i}{\not p - m + i\epsilon}$

   (c) vertex 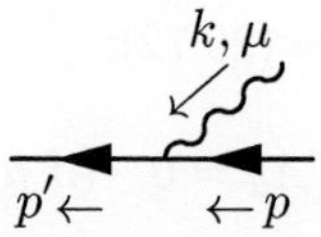

   $-ie\gamma^\mu\Gamma$

2. Ensure momentum conservation at each vertex: $(2\pi)^4\,\delta^{(4)}(\sum p_{\text{out}} - \sum p_{\text{in}})$

3. Multiply by $\displaystyle\int\frac{d^4q}{(2\pi)^4}$ and integrate over all internal momenta q.

4. Spinor factors:

   For every $\begin{Bmatrix}\text{incoming}\\ \text{outgoing}\end{Bmatrix}$ electron, a factor $\begin{Bmatrix}u\\ \overline{u}\end{Bmatrix}$; 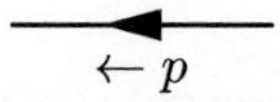

   for every $\begin{Bmatrix}\text{incoming}\\ \text{outgoing}\end{Bmatrix}$ positron, a factor $\begin{Bmatrix}\overline{v}\\ v\end{Bmatrix}$; 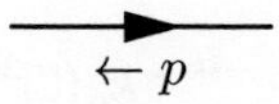

5. Polarization factors:

   For every $\begin{Bmatrix}\text{incoming}\\ \text{outgoing}\end{Bmatrix}$ vector meson, a factor $\begin{Bmatrix}\varepsilon_\mu\\ \varepsilon^{*\prime}_\mu\end{Bmatrix}$, with $\varepsilon\cdot k = 0$, $\varepsilon'\cdot k' = 0$.

In doing spin sums the completeness relation

$$\sum_{r=1}^{3}\varepsilon^{(r)}_\mu\varepsilon^{(r)*}_\nu = -g_{\mu\nu}+\frac{k_\mu k_\nu}{\mu^2} \tag{26.93}$$

is as useful as the Dirac sum rules (20.123a) and (20.123b).

As an example of these rules, consider the Compton effect with massive photons, as shown in Figure 26.2. (Here, $\Gamma = 1$.)

There are two graphs, with amplitude given by

$$i\mathcal{A}_{fi} = (-ie)^2\overline{u}'\left[\not\varepsilon'^*\frac{i}{\not p+\not k-m}\not\varepsilon+\not\varepsilon\frac{i}{\not p'-\not k-m}\not\varepsilon'^*\right]u \tag{26.94}$$

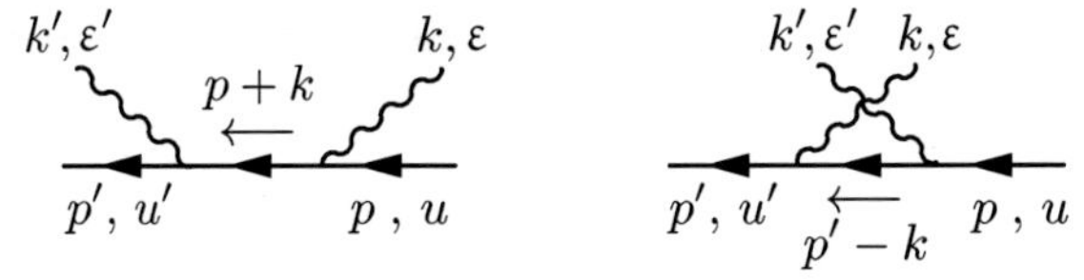

Figure 26.2: Compton scattering with massive photons

We don't need the pole-avoiding $i\epsilon$'s in the denominators. (Beware the notational ambiguity: $\not{\varepsilon}'^{*} \equiv \varepsilon^*_\mu \gamma^\mu$, *not* $(\varepsilon_\mu \gamma^\mu)^*$.) To see that this makes sense, we should expect that the amplitude for longitudinal massive photons would be zero: in the Compton effect, the initial and final states involve physical photons, and there are no physical longitudinal photons.[17] That is, if ε_μ is parallel to k_μ, we need to have the amplitude equal zero. It *is* zero, because of current conservation. Let's see how that goes.

From the equations of motion we have (26.72):

$$\partial^\mu J_\mu = \mu^2 \partial^\mu A_\mu \;\Rightarrow\; \partial^\mu A_\mu = 0 \quad (\text{if } \mu \neq 0 \text{ and } \partial^\mu J_\mu = 0) \tag{26.95}$$

First, a consistency check. Using the LSZ formalism (14.37),

$$\langle k', p'|S - 1|k, p\rangle = i\mathcal{A}_{fi} = i\int d^4x\, e^{-ik\cdot x}\, \varepsilon_\mu (\Box^2 + \mu^2)\, \langle k', p'|A^\mu(x)|p\rangle \tag{26.96}$$

With

$$\varepsilon_\mu = (1/\mu) k_\mu \tag{26.97}$$

(dividing by μ as in (26.78), for type 3 massive photons),

$$\begin{aligned}
\mathcal{A}_{fi} &\propto \int d^4x\, e^{-ik\cdot x}\, k_\mu (\Box^2 + \mu^2)\, \langle k', p'|A^\mu(x)|p\rangle \\
&\propto -\int d^4x\, \partial_\mu (e^{-ik\cdot x})\, (\Box^2 + \mu^2)\, \langle k', p'|A^\mu(x)|p\rangle \\
&\propto \int d^4x\, e^{-ik\cdot x}\, (\Box^2 + \mu^2)\, \langle k', p'|\partial_\mu A^\mu(x)|p\rangle \quad (\text{integration by parts}) \\
&\propto \int d^4x\, e^{-ik\cdot x}\, (\Box^2 + \mu^2)\, \langle k', p'|\partial_\mu J^\mu(x)|p\rangle \quad (\text{equations of motion}) \\
&= 0 \quad (\text{if the current is conserved})
\end{aligned} \tag{26.98}$$

(Note: this assumes the states $|k, p\rangle$ and $|k', p'\rangle$ describe particles that are on the mass shell, not virtual particles.) So that checks. What about the amplitude (26.94) itself? Substituting (26.97),

$$i\mathcal{A}_{fi} = \frac{(-ie)^2}{\mu} \overline{u}' \left[\not{\varepsilon}'^{*} \frac{i}{\not{p} + \not{k} - m} \not{k} + \not{k} \frac{i}{\not{p}' - \not{k} - m} \not{\varepsilon}'^{*} \right] u \tag{26.99}$$

---

[17] [Eds.] Coleman is using the term "longitudinal photon" to mean that its polarization 4-vector ε_μ is parallel to its four-momentum: $\varepsilon^\mu \propto k^\mu$. Usually this term describes a photon whose polarization *3*-vector is parallel to its direction of motion; $\boldsymbol{\varepsilon} \propto \mathbf{k}$.

The spinor u satisfies (20.110a) $(\not{p} - m)u = 0$; likewise (20.113) $\overline{u}'(\not{p}' - m) = 0$. Adding these zero terms, we can write

$$\mathcal{A}_{fi} = \frac{(-ie)^2}{\mu}\overline{u}'\left[\not{\epsilon}'^*\frac{\not{k}+\not{p}-m}{\not{p}+\not{k}-m} - \frac{-\not{k}+\not{p}'-m}{\not{p}'-\not{k}-m}\not{\epsilon}'^*\right]u = \frac{(-ie)^2}{\mu}\overline{u}'\left[\not{\epsilon}'^* - \not{\epsilon}'^*\right]u = 0 \quad (26.100)$$

Thus we have the physically reasonable result that the Compton amplitude for longitudinal photons is zero. This result also verifies that the current is conserved; otherwise, the LSZ computation would be inconsistent.

Next time we will talk about the interactions of vectors, massive and massless, as an introduction to quantum electrodynamics.

27

Electromagnetic interactions and minimal coupling

Last time I got started on one of the main theories of this course, quantum electrodynamics, by talking about the theory of a free massive vector meson. I will now address three topics, all dealing with the interactions of a vector field. I won't do more than introduce the last topic, which will be the subject of the next lecture.

First, I will talk about the classical Lagrangian theory of a vector meson field interacting with other fields. If the vector meson is either massless or has a very small mass, we can think of it as the photon, whose interactions with matter fields constitute electrodynamics.[1] To describe the interactions of vector fields, we need to discuss three things that turn out to be intimately related: Gauge invariance for the massless case, a conserved current for the massive case, and the minimal coupling prescription. At the end of this discussion we'll be in a position to write down the interaction of photons with an arbitrary system: a free meson field, a free fermion field, or with interacting meson and fermion fields. We won't yet be able to write down the Feynman rules for such a theory, because we will encounter a large number of purely technical problems. These problems make up the second topic of this lecture. They are certainly soluble by methods we already have. But if I attempted to solve them in that way, we would be led into a sequence of extremely narrow and complicated arguments which would involve us in a large number of combinatorial calisthenics that I would just as soon avoid. Therefore I will stop the discussion at that stage and move on to the third topic, to introduce a new technique of great generality, the method of *functional integrals*.

27.1 Gauge invariance and conserved currents

To remind you of the system we studied last time, we had a Lagrangian density of the Proca form for a massive vector meson

$$\mathscr{L}_P \equiv -\tfrac{1}{4}F_{\mu\nu}F^{\mu\nu} + \tfrac{1}{2}\mu^2 A_\nu A^\nu \tag{26.47}$$

where

$$F^{\mu\nu} = \partial^\mu A^\nu - \partial^\nu A^\mu \tag{27.1}$$

[1] [Eds.] Readers may wish to know in advance Coleman's strategy for discussing features of the massless vector field: to consider those features one after another in relation to a massive vector field, and then take the limit as its mass goes to zero.

plus possibly an interaction with an external current J^μ, of the form $-J^\mu A_\mu$, where J^μ was a c-number function of space and time. The equation of motion you get from varying this Lagrangian is the Proca equation with a current source:

$$\partial^\mu F_{\mu\nu} + \mu^2 A_\nu = J_\nu \tag{27.2}$$

(this is the same as (26.71), with the indices on $F_{\mu\nu}$ reversed).

To describe electromagnetism, we'll need to talk about massless vectors. In the limit $\mu \to 0$, the Proca equation reduces to the Maxwell equations. If we take the four-divergence of both sides, the first term vanishes because of the antisymmetry of $F_{\mu\nu}$:

$$\cancelto{0}{\partial^\nu\partial^\mu F_{\mu\nu}} + \mu^2\partial^\nu A_\nu = \partial^\nu J_\nu \tag{27.3}$$

so that

$$\partial^\nu A_\nu = (1/\mu^2)\partial^\nu J_\nu \tag{27.4}$$

Trouble ensues in the limit $\mu \to 0$. We discovered in the massive case that we could get a smooth limit of this theory as $\mu \to 0$ *only* if $\partial_\mu J^\mu = 0$, i.e., only if the vector meson is coupled to a *conserved* current. So we learned it was a good thing to have a conserved current. And considering the emission of a single meson by this conserved current acting on the vacuum, we discovered a very interesting fact: of the three helicity states of a massive vector meson, one of them completely decoupled as the mass of the vector meson μ goes to zero; the amplitude for the helicity = 0 state goes to zero linearly with the mass.

We'll take the working hypothesis that we can have a sensible limit $\mu \to 0$ only when $\partial^\nu J_\nu = 0$. If this is so, the limit $\mu \to 0$ of (27.2) *is* Maxwell's equations, with the field components

$$F^{i0} = E^i \quad \text{and} \quad F^{ij} = \epsilon^{ijk}B_k \tag{27.5}$$

Maxwell's equations emerge in rationalized Heaviside-Lorentz units with $c = 1$. These are the units God uses, so we'll use them, too.

I would like to consider a more general kind of theory in which the Lagrangian $\mathscr{L}$ has the same free form (26.47) as before, plus a contribution from the matter fields, with Lagrangian $\mathscr{L}'$, which may include an interaction of the vector meson with something else; scalar mesons, nucleons, what have you. I will represent the fields here generically by a big column vector ϕ,

$$\phi = \begin{pmatrix} \phi^1 \\ \phi^2 \\ \vdots \\ \phi^N \end{pmatrix} \tag{27.6}$$

The components ϕ^i will be scalars, components of spinors ψ and ψ^*, the pion fields and the proton fields and whatever else, perhaps even A^μ itself, and conceivably derivatives of A^μ. The Lagrangian is

$$\mathscr{L} = -\tfrac{1}{4}F_{\mu\nu}F^{\mu\nu} + \tfrac{1}{2}\mu^2 A_\nu A^\nu + \mathscr{L}'(\phi, \partial_\mu\phi, A_\mu, \partial_\nu A_\mu) = \mathscr{L}_P + \mathscr{L}' \tag{27.7}$$

When I vary this Lagrangian, I know what we get from the first two terms, the Proca Lagrangian $\mathscr{L}_P$, but I don't know what we get from the third, because I don't know what

$\mathscr{L}'$ is. Whatever we get, though, I can write like this:

$$\delta\mathscr{L} = \delta\mathscr{L}_P + \delta\mathscr{L}' = \left(\partial_\nu F^{\nu\mu} + \mu^2 A^\mu + \frac{\partial\mathscr{L}'}{\partial A_\mu}\right)\delta A_\mu \tag{27.8}$$

The stuff I get from varying $\mathscr{L}'$ with respect to A_μ I'll call $-J^\mu$:

$$\frac{\partial\mathscr{L}'}{\partial A_\mu} \equiv -J^\mu \tag{27.9}$$

so that the Euler–Lagrange equation for A^μ becomes

$$\partial_\nu F^{\nu\mu} + \mu^2 A^\mu - J^\mu = 0 \tag{27.10}$$

This is the same equation (27.2) that we had before, but now J^μ is some complicated function of ϕ and $\partial^\mu\phi$ and maybe even A^μ.

We discovered that it was a good thing in the massive vector theory that the external source current was conserved. I would like to arrange matters so that our new current, (27.9), is also conserved:

$$\partial_\mu J^\mu = 0 \tag{27.11}$$

It's a four-vector, after all, and it is the electromagnetic current in the massless case. *A priori* I could couple the vector meson to the other fields in many possible ways, but only some of them will yield current conservation as a consequence of the equations of motion. I would like to find ways of coupling A_μ to the fields ϕ so that (27.11) is true. So that's one problem: *find the right coupling.*

We have a second and apparently unrelated problem: *gauge invariance.* On the classical level, we can discuss the theory of a vector field with $\mu^2 = 0$, but in the massless case, we had trouble with canonical quantization. Now, in electromagnetism, one of the standard dogmas is that the electric and magnetic fields are all there is; the potentials do not make any difference. You can take the scalar and vector potentials, assembled into the four-vector A^μ, and you can add to them the gradient of any function $\chi(x)$ of space and time, but this transformation does not affect $F^{\mu\nu}$, which is the only real thing (see (26.11)):

$$\text{Gauge transformation:} \quad \begin{cases} A_\mu \xrightarrow{\chi} A_\mu + \partial_\mu\chi \\ F_{\mu\nu} \xrightarrow{\chi} F_{\mu\nu} \end{cases} \tag{27.12}$$

Phrased in terms of an infinitesimal transformation $\delta\chi$, if I make an infinitesimal variation in A^μ such as

$$\delta A_\mu = \partial_\mu \delta\chi \tag{27.13}$$

then the free Lagrangian, $-\frac{1}{4}F^{\mu\nu}F_{\mu\nu}$ is unchanged:

$$\delta\left[-\tfrac{1}{4}F^{\mu\nu}F_{\mu\nu}\right] = 0 \tag{27.14}$$

This is the so-called **gauge invariance** of the theory. In Maxwell's theory, you can choose this function $\delta\chi$ as you wish. You may choose the Coulomb gauge, or the Lorenz gauge, or some other gauge; *it don't matter.* No one ever reads a paper describing an experiment involving photons with a footnote that says, "This experiment was done in Coulomb gauge."

Because of this, one would like to arrange matters so that $\mathscr{L}'$ has the property that all the matter fields transform in such a way that $\delta\mathscr{L}' = 0$. That is, the A_μ's and the $F_{\mu\nu}$'s transform according to (27.12), and the matter fields ϕ transform in some other way dependent on χ:

$$\phi(x) \xrightarrow{\chi} \phi'(x) \tag{27.15}$$

We're trying to discover the transformation of $\phi(x)$ that will preserve, in the interacting theory, the desirable property of gauge invariance.

These two questions are ostensibly unrelated: the problem of getting a conserved current in the case of a massive photon, and the problem of preserving gauge invariance in the case of a massless photon. Although they look like they are unrelated, in fact they have the same solution. In particular I will show that *if $\mathscr{L}'$ preserves gauge invariance, it will also generate a conserved current.*

Suppose I have assigned some transformation properties (27.15) to ϕ such that $\mathscr{L}'$ is gauge invariant. I then transform the fields: $\delta A^\mu = \partial^\mu \delta\chi$, and $\delta\phi$ equals something—I'll have to figure out what that is—and compute $\delta\mathscr{L}_P$. I don't have to know how ϕ transforms explicitly. All I need to know is that $\delta\mathscr{L}' = 0$. The only term in $\mathscr{L}_P$ not gauge invariant is the vector meson's mass term, which transforms as

$$\delta(\tfrac{1}{2}\mu^2 A^\mu A_\mu) = \mu^2 A^\mu \partial_\mu \delta\chi \tag{27.16}$$

Hamilton's Principle tells me that the change in the action is zero for arbitrary variations of the fields which vanish on the boundaries of the region of integration. In particular, I can choose $\delta\chi$ to vanish on these boundaries. Integrating by parts we have

$$\delta S = \int d^4x\, \mu^2 A^\mu \partial_\mu \delta\chi = -\int d^4x\, \mu^2 \partial_\mu A^\mu \delta\chi \tag{27.17}$$

If the action is to be invariant for any choice of $\delta\chi$, for example a four-dimensional delta function, then we must have

$$\mu^2 \left(\partial_\mu A^\mu\right) = 0 \tag{27.18}$$

Then as a consequence of the equations of motion,

$$\mu^2 \neq 0 \Rightarrow \partial_\mu A^\mu = 0 \tag{27.19}$$

Now we return to the definition of J^μ. I know the equations of motion imply $\partial_\mu A^\mu = 0$. I also know (27.9)

$$J^\nu = \partial_\mu F^{\mu\nu} + \mu^2 A^\nu \tag{27.20}$$

I take the divergence of both sides:

$$\partial_\nu J^\nu = \partial_\nu \partial_\mu F^{\mu\nu} + \mu^2 \partial_\nu A^\nu \tag{27.21}$$

The first term on the right-hand side is zero because $F^{\mu\nu}$ is antisymmetric. The second term on the right-hand side is also zero, because I just proved it for $\mu^2 \neq 0$. Therefore I know the current is conserved, $\partial_\nu J^\nu = 0$, as a consequence of the equations of motion. Thus my two problems have a single solution. If I can arrange my matter Lagrangian $\mathscr{L}'$ such that it is gauge invariant, and then break gauge invariance *only* by giving the photon a mass, I will obtain a conserved current. To solve the gauge invariance problem for the massless photon is to solve the conserved current problem for the massive photon. It's straightforward algebra.

I should emphasize that gauge invariance ($\delta\mathscr{L}' = 0$ under (27.12) and (27.15)) is *not* a symmetry. The conserved current (5.27) associated with it (not to be confused with J^ν in (27.20)!) is *zero*, simply because the Lagrangian doesn't change at all.[2]

There is a big difference between gauge invariance and an internal symmetry, such as an isospin rotation in the theory of pions and nucleons. There *is* a difference between a proton and a neutron, and they are turned into each other by an isospin rotation. But there is *no* physical difference between the state of the electromagnetic field in Lorenz gauge or in Coulomb gauge or in any other gauge. Gauge invariance is like general coordinate invariance in general relativity, or like the statement that the contents of a physics paper are unchanged if it is translated from English into French. These are different descriptions of the *same* system, not different systems with symmetric dynamics. That's not clear from the analytic structure of the theory, but it's the meaning we attach to the physics in both cases. We could *test* isospin invariance in the real world (in situations where the electromagnetic force can be neglected), by doing proton–proton scattering and then doing neutron–neutron scattering. You would get two papers, one describing proton–proton scattering at 300 GeV, and the other neutron–neutron scattering at 300 GeV, and they would produce the same cross-section. That would be a test. But you will never read a paper saying photon–electron scattering in the Coulomb gauge and photon–electron scattering in the Lorenz gauge give you the same cross-section, and claiming to have verified gauge invariance.

So far I've shown that if $\mathscr{L}'$ is chosen to be gauge invariant, that has the desirable effect of preserving the gauge invariance of massless electrodynamics, and if we break the gauge invariance just by adding a photon mass term, we obtain a massive vector meson coupled to a conserved current. Now I will tackle the problem of constructing a Lagrangian invariant under the gauge transformation (27.12). The key to the construction is a prescription, called **minimal coupling**, that will generate extra terms, such that the resulting Lagrangian will automatically be gauge invariant. These terms will not involve the derivatives of A_μ, but only A_μ itself. I'll explain at the end where the word "minimal" comes from. The prescription is a machine, but it is not universally applicable. You give me a Lagrangian for matter (or whatever you want to call it) *without* electromagnetism, and, provided one condition is met, I'll generate the Lagrangian *including* the interactions with electromagnetism.

The necessary condition is that the matter Lagrangian $\mathscr{L}_m(\phi, \partial^\mu\phi)$ describing a set of fields—scalar, spinor, vector, whatever, but *not* including A_μ itself—has a one-parameter group of internal symmetries of the sort we talked about earlier (in Chapter 6). Under this group, an infinitesimal transformation of ϕ is given by

$$\delta\phi = -iQ\phi\,\delta\lambda \quad \text{where } Q \text{ is some matrix, and } \delta\lambda \text{ some parameter} \tag{27.22}$$

---

[2] [Eds.] In response to a student's question, Coleman reiterates, "No, it is *not* a symmetry. The conserved current associated with it is *zero*." For a somewhat different viewpoint but much the same conclusion, see S. Weinberg, "Dynamic and Algebraic Symmetries", pp. 290–393, in *Lectures on Elementary Particles and Quantum Field Theory (1970 Brandeis University Summer Institute in Theoretical Physics)*, v. 1, eds. Stanley Deser, Marc Grisaru, and Hugh Pendleton, MIT Press, 1970. Weinberg finds (equation (2.B.10)) that there *is* a non-zero conserved current,

$$J^\mu_\chi = \partial_\nu\left(\frac{\partial\mathscr{L}}{\partial(\partial_\nu A_\mu)}\chi\right)$$

but that it does not lead to a new charge independent of that found from global phase invariance with χ a constant. It is perhaps worth noting that Coleman's prescription requires setting $\chi = 0$. If that is done here, the current vanishes.

or, for a finite transformation,

$$\phi(x) \to \phi(x,\lambda) = e^{-iQ\lambda}\phi(x) \tag{27.23}$$

If this transformation leaves $\mathscr{L}_m$ unchanged, under normal circumstances it would enable us to deduce the existence of a conserved current, J^μ, (5.27):

$$J^\mu = -i\pi^\mu Q\phi \tag{27.24}$$

This seems a reasonable and necessary condition to get an interaction Lagrangian coupled to the photon. If the photon is coupled to the matter fields with a constant, say the electromagnetic coupling constant e, and if you have a conserved current when the photon is around, you can imagine it should stay conserved as $e \to 0$. So you want to start out with a theory that has a conserved current even before you include the photon. I'll give two examples.

EXAMPLE 1. The Lagrangian for a free Dirac field is

$$\mathscr{L}_m = \overline{\psi}(i\partial\!\!\!/ - m)\psi \tag{27.25}$$

If we assemble ψ and $\overline{\psi}$ into a two-component vector ϕ,

$$\phi = \begin{pmatrix} \psi \\ \overline{\psi} \end{pmatrix} \tag{27.26}$$

and take Q to be a matrix with eigenvalues ± 1,

$$Q = \begin{pmatrix} 1 & 0 \\ 0 & -1 \end{pmatrix} \tag{27.27}$$

then the spinor ψ has eigenvalue $+1$, and its Dirac adjoint has eigenvalue -1:

$$Q\psi = \psi; \quad Q\overline{\psi} = -\overline{\psi} \tag{27.28}$$

Following the formula (5.27) for the construction of the conserved current,

$$J^\mu = \frac{\partial \mathscr{L}_m}{\partial(\partial_\mu \phi)} D\phi = \overline{\psi}\gamma^\mu\psi \tag{27.29}$$

We already know that this current is conserved:

$$\partial_\mu(\overline{\psi}\gamma^\mu\psi) = (\partial_\mu\overline{\psi})\gamma^\mu\psi + \overline{\psi}\gamma^\mu(\partial_\mu\psi) = im\overline{\psi}\psi - im\overline{\psi}\psi = 0 \tag{27.30}$$

EXAMPLE 2. The Lagrangian for a free charged scalar field is

$$\mathscr{L}_m = \partial_\mu\phi^*\partial^\mu\phi - \mu^2\phi^*\phi \tag{27.31}$$

Again we'll say

$$Q\phi = -i\phi; \quad Q\phi^* = i\phi^* \tag{27.32}$$

The current associated with this symmetry is

$$J^\mu = (\partial^\mu\phi^*)(-i\phi) + (\partial^\mu\phi)(i\phi^*) = i[\phi^*\partial^\mu\phi - \phi\,\partial^\mu\phi^*] \tag{27.33}$$

This current is also conserved;

$$\partial_\mu(i[\phi^*\partial^\mu\phi - \phi\,\partial^\mu\phi^*] = i(\phi^*\Box^2\phi - \phi\,\Box^2\phi^*) = i(\phi^*\mu^2\phi - \phi\,\mu^2\phi^*) = 0 \tag{27.34}$$

Now to couple the vector meson. Our first try is to add the term $-eJ^\mu A_\mu$, where e is a coupling constant:

$$\mathscr{L} \stackrel{?}{=} \mathscr{L}_P + \mathscr{L}_m - eJ_\mu A^\mu = \mathscr{L}_P + \mathscr{L}' \tag{27.35}$$

where $\mathscr{L}_P$ is the Proca Lagrangian (26.47) and $\mathscr{L}_m$ (either (27.25) or (27.31)) is the free matter Lagrangian. Note that this requires a slight redefinition of the current (compare (27.9)):

$$\frac{\partial \mathscr{L}'}{\partial A_\mu} \equiv -eJ^\mu \tag{27.36}$$

Is the current still conserved? It is in Example 1, but not in Example 2. In Example 1, the equations of motion become

$$i\partial\!\!\!/\psi - m\psi - e\gamma^\mu\psi A_\mu = 0 \tag{27.37}$$

$$-i\partial_\mu\overline{\psi}\gamma^\mu - m\overline{\psi} - e\overline{\psi}\gamma^\mu A_\mu = 0 \tag{27.38}$$

Multiply the top equation by $\overline{\psi}$ on the left, the bottom by ψ on the right, and subtract these from each other to find

$$i\partial_\mu(\overline{\psi}\gamma^\mu\psi) = 0 = i\partial_\mu J^\mu \tag{27.39}$$

so the current remains conserved. In Example 2, however, the equations of motion become

$$\Box^2\phi + \mu^2\phi + 2ieA^\mu\partial_\mu\phi + ie\phi\,\partial_\mu A^\mu = 0 \tag{27.40}$$

$$\Box^2\phi^* + \mu^2\phi^* - 2ieA^\mu\partial_\mu\phi^* - ie\phi^*\,\partial_\mu A^\mu = 0 \tag{27.41}$$

Multiply the first by ϕ^* and the second by ϕ, and subtract the second from the first:

$$\begin{aligned} -i\partial_\mu J^\mu = \phi^*\Box^2\phi - \phi\,\Box^2\phi^* &= -2ieA_\mu(\phi^*\partial^\mu\phi + \phi\,\partial^\mu\phi^*) - 2ie\phi^*\phi\,\partial_\mu A^\mu \\ &= -2ie\partial^\mu(A_\mu\phi^*\phi) \end{aligned} \tag{27.42}$$

which is not necessarily zero. If we take the divergence of the vector meson equation of motion, we get, discarding the zero term $\partial_\mu\partial_\nu F^{\mu\nu}$,

$$\mu^2\partial_\mu A^\mu = e\partial_\mu J^\mu \tag{27.43}$$

We can substitute for the divergence of A^μ into the previous equation to obtain

$$-i\partial_\mu J^\mu = \phi^*\Box^2\phi - \phi\,\Box^2\phi^* = -2ieA_\mu(\phi^*\partial^\mu\phi + \phi\,\partial^\mu\phi^*) - 2i\frac{e^2}{\mu^2}\phi^*\phi\,\partial_\mu J^\mu \tag{27.44}$$

or rewriting,

$$-i\left(1 - 2\frac{e^2}{\mu^2}\phi^*\phi\right)\partial_\mu J^\mu = -2ieA_\mu(\phi^*\partial^\mu\phi + \phi\,\partial^\mu\phi^*) \tag{27.45}$$

We could try to iterate the Lagrangian to see how to add higher powers of e, but there's a better way.

27.2 The minimal coupling prescription

To motivate the minimal coupling prescription, consider this set of transformations on the matter fields, of exactly the same form (27.22) as before, but now with $\delta\lambda$ an arbitrary function of space and time:

$$\delta\phi = -iQ\phi\,\delta\lambda(x) \tag{27.46}$$

That is *not*, of course, an invariance of our Lagrangian, because if I compute $\delta(\partial^\mu\phi)$, I'll get *two* terms;

$$\delta(\partial^\mu\phi) = \partial^\mu(\delta\phi) = \partial^\mu(-iQ\phi\,\delta\lambda(x)) = -iQ(\partial^\mu\phi)\delta\lambda - iQ\phi(\partial^\mu\delta\lambda) \tag{27.47}$$

The first term is hunky dory, no problem there. But the second term is a disaster. However we've also got the electromagnetic field involved, which under an infinitesimal gauge transformation obeys (27.13). Consider a combination of the ordinary derivative ∂_μ acting on ϕ with a product of the vector field A_μ times ϕ:

$$\boxed{D_\mu\phi = \partial_\mu\phi + ieA_\mu Q\phi} \tag{27.48}$$

This expression is called the **covariant derivative**. Its transformation $\delta(D_\mu\phi)$ looks like this:

$$\begin{aligned}\delta(D_\mu\phi) &= \delta(\partial_\mu\phi + ieA_\mu Q\phi) = \delta(\partial_\mu\phi) + ie(\delta A_\mu)Q\phi + ieA_\mu Q(\delta\phi)\\ &= -iQ(\partial^\mu\phi)\delta\lambda - iQ\phi(\partial^\mu\delta\lambda) + ie(\partial^\mu\delta\chi)Q\phi + ieA_\mu Q(-iQ\phi\,\delta\lambda)\end{aligned} \tag{27.49}$$

We can make the second and third terms cancel if we choose

$$\delta\lambda = e\delta\chi \tag{27.50}$$

in which case

$$\delta(D_\mu\phi) = -iQ(\partial^\mu\phi + ieA_\mu Q\phi)\delta\lambda = -iQ(D_\mu\phi)\lambda \tag{27.51}$$

The covariant derivative of ϕ transforms in exactly the same way as ϕ, which is why it's called "covariant". The combined transformations are

$$\begin{cases}\delta\phi = -iQ\phi\,\delta\lambda\\ \delta A_\mu = \dfrac{1}{e}\,\partial_\mu\delta\lambda\end{cases} \tag{27.52}$$

or, for finite transformations,

$$\begin{cases}\phi \to e^{-iQ\lambda}\phi\\ A_\mu \to A_\mu + \dfrac{1}{e}\partial_\mu\lambda\end{cases} \tag{27.53}$$

This set of transformations is called collectively a "gauge transformation", though historically that term was applied, as in (27.12), only to the potentials. The infinitesimal parameter $\delta\lambda(x)$ is "local"; it is a function of space and time. These transformations emerged from Hermann Weyl's generalization of general relativity in 1918. He suggested that not only could you rotate local coordinate systems, but that there was no absolute standard of length. Weyl's theory used a *real* function in the exponent. He was wrong; but this idea has resurfaced as conformal invariance on a string's world sheet. Weyl called his theory *eichinvarianz*, or "scale invariance", but it was translated into English as "gauge invariance". It was reintroduced with an *imaginary* exponential, by F. London in 1927, who named it after Weyl's theory.[3] Gauge transformations have no "active" interpretation, only a "passive" one; it's just a change of coordinates.

The **minimal coupling prescription** is simple:

Replace all derivatives ∂_μ by covariant derivatives D_μ.

That is,

$$\mathscr{L} = \mathscr{L}_P + \mathscr{L}_m(\phi, \partial_\mu\phi) \to \mathscr{L}_P + \mathscr{L}_m(\phi, D_\mu\phi) \tag{27.54}$$

This Lagrangian is invariant under the transformations

$$\begin{aligned} \phi &\longrightarrow e^{-i\lambda(x)Q}\phi \\ A_\mu &\longrightarrow A_\mu + (1/e)\partial_\mu\lambda(x) \\ D_\mu\phi &\longrightarrow e^{-i\lambda(x)Q}D_\mu\phi \\ F_{\mu\nu} &\longrightarrow F_{\mu\nu} \end{aligned} \tag{27.55}$$

so the Lagrangian has a conserved current,

$$J^\mu = \pi^\mu(-iQ)\phi \tag{27.56}$$

The equations of motion become

$$\partial^\mu F_{\mu\nu} + \mu^2 A_\nu = -\frac{\partial\mathscr{L}_m}{\partial A^\nu} = -ie\frac{\partial\mathscr{L}}{\partial(\partial^\nu\phi)}Q\phi = eJ_\nu \tag{27.57}$$

No matter how complicated $\mathscr{L}$ is, the right-hand side of the Proca equation is a conserved current.

EXAMPLE 1, *revisited*

$$\begin{aligned} D_\mu\psi &= \partial_\mu\psi + ieA_\mu\psi \\ i\bar{\psi}\not{\partial}\psi &\longrightarrow i\bar{\psi}\not{\partial}\psi - eA_\mu\bar{\psi}\gamma^\mu\psi = i\bar{\psi}\not{\partial}\psi - eJ^\mu A_\mu \end{aligned} \tag{27.58}$$

which gives us the same conserved current as before (27.29).

---

[3] [Eds.] H. Weyl, "Gravitation und Elektrizität", *Sitzungs. Pruess. Akad. Wiss. Berlin* (1918) 465–480. English translation in L. O'Raifeartaigh, *The Dawning of Gauge Theory*, Princeton U. P., 1997. O'Raifeartaigh also includes an English translation of London's paper, F. London, "Quantenmechanische Deutung der Theorie von Weyl" (Quantum mechanical interpretation of Weyl's theory), *Zeits. f. Phys.* **42** (1927) 375, and likewise of Weyl's independent article, "Elektron und Gravitation", *Zeits. f. Phys.* **56** (1929) 330–352. See also J. D. Jackson and L. B. Okun, "Historical roots of gauge invariance", *Rev. Mod. Phys.* **73** (2001) 663–680, and H. Weyl, *Space-Time-Matter*, translated by Henry L. Brose, Dover Publications, 1952; in §16, *eichinvarianz* is translated as "calibration invariance". In typed notes for Feb. 11, 1999, Coleman suggested consulting Pauli's famous review article, "Relativitätstheorie" (1921), for a discussion of Weyl's theory: W. Pauli, *Theory of Relativity*, Pergamon, 1958, §65, pp. 192–202; note 20, p. 223; republished by Dover Publications, 1981. Pauli was all of 21 when he wrote the article.

EXAMPLE 2, *revisited*

Something more interesting happens in the scalar case. Minimal coupling

$$\partial^\mu \phi^* \partial_\mu \phi \longrightarrow (\partial^\mu \phi^* - ieA^\mu \phi^*)(\partial_\mu \phi + ieA_\mu \phi) \tag{27.59}$$

leads to the interaction

$$-ie[\phi^* \partial^\mu \phi - \phi\, \partial^\mu \phi^*] A_\mu + e^2 A^\mu A_\mu \phi^* \phi \tag{27.60}$$

including both a linear coupling of A_μ to the derivatives of ϕ and ϕ^*, and a quadratic coupling of A_μ to ϕ, and to a current (27.36)

$$J^\mu = -\frac{1}{e}\frac{\partial \mathscr{L}_m}{\partial A_\mu} = i[\phi^* \partial^\mu \phi - \phi\, \partial^\mu \phi^*] - 2eA^\mu \phi^* \phi \tag{27.61}$$

Notice that the terms in the electromagnetic current linear in the vector field A^μ do *not* cancel, as they did with the Dirac Lagrangian. Of course, you could say that the current must have an A^μ in it to be gauge invariant. The current (27.61) could also be written

$$J^\mu = i(\phi^* D^\mu \phi - \phi\, D^\mu \phi^*) \tag{27.62}$$

Aside from the corrections required to make the electromagnetic current gauge invariant, (27.61) has the same form as the conserved current (27.33) in the non-interacting case.

Is this current conserved? The scalar equations of motion become

$$\Box^2 \phi + m^2 \phi + 2ieA^\mu \partial_\mu \phi + ie\phi\, \partial_\mu A^\mu - e^2 A_\mu A^\mu \phi = 0 \tag{27.63}$$

$$\Box^2 \phi^* + m^2 \phi^* - 2ieA^\mu \partial_\mu \phi^* - ie\phi^*\, \partial_\mu A^\mu - e^2 A_\mu A^\mu \phi^* = 0 \tag{27.64}$$

Using the same trick as before, we find

$$\phi^* \Box^2 \phi - \phi \Box^2 \phi^* + 2ieA_\mu (\phi^* \partial^\mu \phi + \phi\, \partial^\mu \phi^*) + 2ie\phi^* \phi\, \partial_\mu A^\mu = 0 \tag{27.65}$$

If we now take the divergence of the minimal coupling current (27.61), we get

$$\partial_\mu J^\mu = i(\phi^* \Box^2 \phi - \phi \Box^2 \phi^*) - 2e(\partial_\mu A^\mu)\phi^* \phi - 2eA^\mu (\phi^* \partial_\mu \phi + \phi\, \partial_\mu \phi^*) \tag{27.66}$$

The right-hand side is just i times the left-hand side of the previous equation. Thus

$$\partial_\mu J^\mu = 0 \tag{27.67}$$

Although the original current (27.33), without the A^μ term, was *not* conserved, the current from minimal coupling *is* conserved. The minimal coupling prescription gives the right result, with minimal effort! The new current reproduces the derivative coupling term obtained before, but it also includes a term directly proportional to the vector meson A^μ. It's got to have an A^μ in it to make the derivative in the current covariant, as in (27.62). This latter term produces in the Lagrangian a quadratic non-derivative interaction,

$$e^2 A_\mu A^\mu \phi^* \phi \tag{27.68}$$

In its detailed structure the electrodynamics of charged scalar particles is very different from that of charged spinor particles, and the Feynman rules we'll get eventually in these two cases

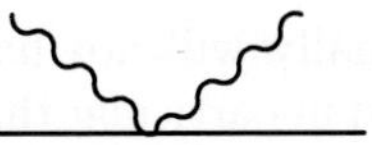

Figure 27.1: Seagull diagram in scalar electrodynamics

will also be quite different. In particular, the quadratic term will give rise to "seagull" diagrams, as shown in Figure 27.1. In both cases, the currents obtained with minimal coupling also have the desirable property that they remain conserved in the limit as the electromagnetic coupling constant goes to zero. Nevertheless, it's the same prescription for both. You give me a Lagrangian that has a conserved current in the absence of electromagnetism, and I will generate the gauge invariant Lagrangian that describes the coupling with electromagnetism, with the minimal coupling prescription. The electromagnetic current J^μ is given by (27.36).

It should not be thought that minimal coupling between a vector meson and a "matter" field is the *only* way to obtain a conserved current. For example, consider this Lagrangian, coupling a Dirac field to a vector meson:

$$\mathscr{L} = \mathscr{L}_P + \overline{\psi}(i\not{D} - m)\psi + a\overline{\psi}\sigma_{\mu\nu}\psi F^{\mu\nu} \tag{27.69}$$

where (20.98) $\sigma_{\mu\nu} = \frac{1}{2}i[\gamma_\mu, \gamma_\nu]$. The extra term is sometimes called a "Pauli term".[4] That's perfectly gauge invariant: $F^{\mu\nu}$ is gauge invariant, and $\overline{\psi}\sigma_{\mu\nu}\psi$ is gauge invariant without any funny business with A^μ; it doesn't involve any derivatives. That's why minimal coupling (the vector field-matter field coupling arising *only* as part of the covariant derivative) is called "minimal" coupling.[5] You could always complicate matters by including additional gauge invariant terms to the Lagrangian, which would therefore still yield conserved currents. This Lagrangian has a conserved current,

$$J_\mu = e\overline{\psi}\gamma_\mu\psi + 2a\,\partial^\nu(\overline{\psi}\sigma_{\mu\nu}\psi) \tag{27.70}$$

The second term comes from shifting the derivative in the Pauli term from A^μ to the product $\overline{\psi}\sigma_{\mu\nu}\psi$ with an integration by parts;

$$a\overline{\psi}\sigma_{\mu\nu}\psi F^{\mu\nu} = -2a\overline{\psi}\sigma_{\mu\nu}\psi\,\partial^\nu A^\mu \xrightarrow{\text{parts}\int} 2a\,\partial^\nu(\overline{\psi}\sigma_{\mu\nu}\psi)A^\mu \tag{27.71}$$

How might such a term arise? Consider a different Lagrangian:

$$\mathscr{L}' = (\partial^\mu\overline{\psi})\sigma_{\mu\nu}\partial^\nu\psi \xrightarrow{\text{parts}\int} -\overline{\psi}\sigma_{\mu\nu}\partial^\mu\partial^\nu\psi \tag{27.72}$$

Because $\sigma_{\mu\nu}$ is antisymmetric and $\partial^\mu\partial^\nu$ is symmetric, this vanishes trivially. Suppose instead that we first apply minimal coupling:

$$\mathscr{L}' = (D^\mu\overline{\psi})\sigma_{\mu\nu}D^\nu\psi \xrightarrow{\text{parts}\int} -\overline{\psi}\sigma_{\mu\nu}D^\mu D^\nu\psi \tag{27.73}$$

But $D^\mu D^\nu\psi$ is *not* symmetric, and this term does *not* vanish. In fact,

$$[D^\mu D^\nu - D^\nu D^\mu]\psi = ieF^{\mu\nu}\psi \tag{27.74}$$

---

[4] [Eds.] W. Pauli, "Relativistic Field Theories of Elementary Particles", *Rev. Mod. Phys.* **13** (1941) 203–232. The Pauli term appears in equation (91) and is defined in the previous (unnumbered) equation.

[5] [Eds.] The name is due to M. Gell-Mann, "The interpretation of new particles as displaced charge multiplets", *Nuovo Cimento* **4**, Supplement 2, (1956) 848–866.

In four-dimensional theories, we usually will not have to worry about these non-minimal interactions because typically they, and in particular this one, turn out to be non-renormalizable. Since we have not discussed the renormalization problem for electrodynamics—we don't even have the propagator for the free photon—this is a premature remark. I make it anyway.

27.3 Technical problems

I would now like to turn to the second topic: technical problems in quantizing electrodynamics. In principle everything is all set up. We know what our interactions are. We have this gigantic machine, canonical quantization with canonical commutators for bosons and anticommutators for fermions. All we have to do is grind through: develop a Hamiltonian, write down Dyson's formula, apply Wick's theorem, get the Feynman rules, use the renormalization prescription, pull out the S-matrix and happily start computing, say, the anomalous magnetic moment of the electron to order e^8.

Now why am I *not* going to launch into that progression? Well, things get pretty complicated. There are three reasons why they get complicated. They're all technical complications; none of them is insuperable and none of them really requires the panacea that will emerge a little later, but they all lead us into horrendous complications. One is the problem of gauge invariance, which arises when $\mu^2 = 0$. You remember that for $\mu^2 \neq 0$, we had no problem canonically quantizing the massive photon, nor any problem quantizing the free field theory. However for $\mu^2 = 0$, the canonical quantization program came (26.66) to a screeching halt; we couldn't do it. I can now reveal that the reason is gauge invariance.

The canonical quantization program, indeed the Hamiltonian formulation of classical mechanics, depends on having a complete and independent set of initial value data. Take, for example, massive vector meson theory (p. 562). Say I give you the fields, three $A_i(\mathbf{x}, t)$'s and the three $F^{0i}(\mathbf{x}, t)$'s, at time $t = 0$, and their first thirty-two derivatives at time $t = 0$, and you propose to tell me, by solving the equations of motion, what they are at any future time. In the Proca theory there's no problem. But in a gauge invariant theory this is impossible. Suppose you've determined $\{A_i(\mathbf{x}, t)\}$ for all subsequent times. I say these solutions cannot be unique. For I have the freedom to gauge transform this set of fields $\{A_i(\mathbf{x}, t)\}$ with a function $\chi(\mathbf{x}, t)$ whose derivatives $\partial_i\chi(\mathbf{x}, t)$ vanish at $t = 0$ and within a little slice of width ϵ around $t = 0$, but are nonzero for $t > \epsilon$. Then the transformed fields $\{A'_i(\mathbf{x}, t)\}$ (and their time derivatives) are the same as the original set at $t = 0$, and different at later times; in particular, say, at $t = 1$. Therefore I have two sets of fields, $\{A_i(\mathbf{x}, t)\}$ and $\{A'_i(\mathbf{x}, t)\}$, with *exactly the same* initial value data, but they are *different* at $t = 1$. You can't possibly get a unique solution for the initial value problem, and you can't possibly write electromagnetism in Hamiltonian form. It's just a consequence of gauge invariance. Were we to attempt canonical quantization of electromagnetism, we'd first have to impose a **gauge condition**, for example the *Lorenz gauge* $\partial^\mu A_\mu = 0$, or the *Coulomb gauge* $\nabla \cdot \mathbf{A} = 0$, or some other. But the Coulomb gauge destroys manifest Lorentz invariance, while the Lorenz gauge does not specify the fields completely.[6] Is the resulting theory Lorentz invariant, and are the results independent of the

[6] [Eds.] For canonical quantization of the Maxwell field in the Coulomb gauge, see Bjorken & Drell *Fields*, Chap. 14, pp. 68–80. For canonical quantization in the Lorenz gauge, see, e.g., Franz Mandl and Graham Shaw, *Quantum Field Theory*, John Wiley and Sons, 1984, Chap. 5, pp. 86–90. The original canonical quantization of the electromagnetic field is due to Enrico Fermi, "Sopra l'Elettrodinamica Quantistica" (On quantum electrodynamics), *Rend. Lincei* **5** (1929) 881–887, reprinted as paper 50 in v. I of Fermi's *Collected Papers*,

choice of gauge, as they must be? These questions are much easier to address if we quantize using a different method, that of *functional integrals*, to which we will soon turn.

The second problem I have alluded to many times: the problem of *derivative couplings*. I have always been sort of antsy whenever questions about derivative couplings came up in these lectures, and I mumbled "Hrmph, hrmph, we'll talk about that later." The standard tricks can be applied to theories with derivative couplings, and we'll have to confront such theories (such as scalar electrodynamics), but they lead to a terrible mess. I'll work out some details so you can stand on the precipice and look into this canyon full of garbage, to see exactly what would happen and what sort of problems we would run into if we tried to take seriously a theory with derivative couplings. (Before, in §14.4, we just guessed.) But we'll pull back, and not plunge into that pit.

EXAMPLE. *Pseudoscalar-spinor derivative coupling*

Let's consider a simple example, a spinor field ψ interacting with a pseudoscalar meson field ϕ via this interaction,

$$\mathscr{L}_I = f\overline{\psi}\gamma^\mu\gamma^5\psi\partial_\mu\phi \equiv fK^\mu\partial_\mu\phi \tag{27.75}$$

with some coupling constant f. I'll just let K^μ stand for $\overline{\psi}\gamma^\mu\gamma^5\psi$. Let's try to write this in Hamiltonian form. The Lagrangian is

$$\mathscr{L} = \tfrac{1}{2}(\partial^\mu\phi\partial_\mu\phi - m^2\phi^2) + \overline{\psi}(i\gamma^\mu\partial_\mu - m)\psi + fK^\mu\partial_\mu\phi \tag{27.76}$$

It has none of the special problems of electromagnetism. There's no momentum conjugate to $\overline{\psi}$, but we won't worry about that. The canonical momentum conjugate to ψ is $i\overline{\psi}\gamma^0$; no problem there. The canonical momentum conjugate to ϕ, however, looks a little funny:

$$\pi_f = \frac{\partial\mathscr{L}}{\partial(\partial_0\phi)} = \partial^0\phi + fK^0 \tag{27.77}$$

(Note the subscript f, to distinguish this quantity from the free canonical momentum $\pi = \partial^0\phi$.) This means the interaction Hamiltonian $\mathscr{H}_I$ will *not* equal minus the interaction Lagrangian, $\mathscr{L}_I$,

$$\mathscr{H}_I \neq -\mathscr{L}_I \tag{27.78}$$

because of the presence of this extra term, fK^0.

To be more explicit, if we focus on the terms in the Hamiltonian that involve time derivatives of ϕ, we'll find first

$$\pi_f(\partial^0\phi) = \pi_f(\pi_f - fK^0) \tag{27.79}$$

which is not simply $(\pi_f)^2$, as it would have been if we'd had a non-derivative coupling. Second, we can write the term in the Lagrangian involving time derivatives of ϕ like this:

$$\tfrac{1}{2}(\partial^0\phi)^2 + fK^0\partial_0\phi = \tfrac{1}{2}(\partial^0\phi + fK^0)^2 - \tfrac{1}{2}f^2(K^0)^2 = \tfrac{1}{2}\pi_f^2 - \tfrac{1}{2}f^2(K^0)^2 \tag{27.80}$$

---

ed. E. Segrè *et al.*, U of Chicago Press, 1962. A much expanded version of this work, in English, is E. Fermi, "Quantum Theory of Radiation", *Rev. Mod. Phys.* **4** (1932) 87–132, reprinted as paper 67 in Fermi's *Collected Papers*, v. I and in Schwinger, *QED*.

When we assemble the Hamiltonian using the standard formula, all the terms that do not involve the derivatives of ϕ will come together to give us the free Hamiltonian for the scalar and spinor fields, the original interaction Lagrangian, and one extra piece:

$$\begin{aligned}\mathscr{H} &= \tfrac{1}{2}(\pi_f^2 + (\nabla\phi)^2 + m^2\phi^2) + \overline{\psi}(\gamma\bullet\nabla\psi + m\psi) - fK^\mu\partial_\mu\phi - \tfrac{1}{2}f^2(K^0)^2 \\ &= \mathscr{H}_\phi + \mathscr{H}_\psi - \mathscr{L}_I - \tfrac{1}{2}f^2(\overline{\psi}\gamma^0\gamma^5\psi)^2 \\ &= \mathscr{H}_\phi + \mathscr{H}_\psi + \mathscr{H}_I\end{aligned} \tag{27.81}$$

In Dyson's formula you're interested in $\mathscr{H}_I$, the interaction Hamiltonian density in the interaction picture, written as a function of interaction picture fields. Here,

$$\mathscr{H}_I = -fK^\mu\partial_\mu\phi - \tfrac{1}{2}f^2(\overline{\psi}\gamma^0\gamma^5\psi)^2 \tag{27.82}$$

The term $-fK^\mu\partial_\mu\phi$ is a nice Lorentz invariant object to go up there in the exponential. But the last term, the little bastard, doesn't cancel out! It's as disgustingly non-Lorentz invariant as an object can be:

$$(\overline{\psi}\gamma^0\gamma^5\psi)^2 = (\psi^\dagger\gamma^5\psi)^2 \tag{27.83}$$

and it's sitting there in the interaction Hamiltonian, in the exponential in Dyson's formula, giving us what looks like a terrible, non-Lorentz invariant four point vertex involving four Fermi fields in our Feynman rules.

This is not the only difficulty that would come up in this theory. In addition to this four Fermi, non-covariant term in our interaction Hamiltonian in the interaction picture, we have also a non-covariant contraction term, as I will demonstrate. I remind you of the definition of a time-ordered product of two Bose operators, $A(x)$ and $B(y)$,

$$T(A(x)B(y)) = \theta(x_o - y_o)A(x)B(y) + \theta(y_o - x_o)B(y)A(x) \tag{27.84}$$

Suppose I wanted to compute the time derivative (with respect to x_0) of this thing. From differentiating the field operators I obtain simply the time-ordered product of the derivative $\partial_x^0 A(x)B(y)$—couldn't be nicer. But when I differentiate the theta functions, I get a delta function. So I get three terms,

$$\partial_x^0 T(A(x)B(y)) = T(\partial_x^0 A(x)\,B(y)) + \delta(x_o - y_0)A(x)B(y) - \delta(y_o - x_0)B(y)A(x) \tag{27.85}$$

or equivalently

$$\partial_x^0 T(A(x)B(y)) = T(\partial_x^0 A(x)\,B(y)) + \delta(x_o - y_0)[A(x), B(y)] \tag{27.86}$$

In the theory we wish to consider here, our interaction Hamiltonian not only involves ϕ's, but time derivatives of ϕ's. So we'll have to compute the contraction function for two time derivatives of ϕ's which we'll do using this identity. The contraction function is the vacuum expectation value of the time-ordered product, so I might as well work with the time-ordered product.

Let's consider two derivatives on the time-ordered product of two ϕ's (in the interaction picture),

$$\partial_x^0\partial_y^0 T(\phi_I(x)\,\phi_I(y)) \tag{27.87}$$

First bring the y time derivative through. We can do that with no problems because the equal time commutator that develops is the equal time commutator of $\phi(x)$ with $\phi(y)$, which is zero;

$$\partial_y^0 T(\phi_I(x)\,\phi_I(y)) = T(\phi_I(x)\partial_y^0\phi_I(y)) + \delta(x_o - y_0)[\phi_I(x), \phi_I(y)] = T(\phi_I(x)\partial_y^0\phi_I(y)) \tag{27.88}$$

I've still got the second derivative to worry about, and when I bring the x time derivative through, the extra term (the equal time commutator) is *not* zero, and I get

$$\begin{aligned}\partial_x^0 T(\phi_I(x)\partial_y^0\phi_I(y)) &= T(\partial_x^0\phi_I(x)\partial_y^0\phi_I(y)) + \delta(x_o - y_0)[\phi_I(x), \partial_y^0\phi_I(y)] \\ &= T(\pi_I(x)\pi_I(y)) + i\delta(x_o - y_0)\delta^{(3)}(\mathbf{x}-\mathbf{y})\end{aligned} \tag{27.89}$$

or,

$$\partial_x^0\partial_y^0 T(\phi_I(x)\,\phi_I(y)) = T(\pi_I(x)\pi_I(y)) + i\delta^{(4)}(x-y) \tag{27.90}$$

Thus we have a rather peculiar equation that we'll have to also feed into our Feynman rules in computing the contraction functions of $\partial^\mu\phi_I(x)$ with $\partial^\nu\phi_I(y)$. There's no problem with the space terms. Pitching terms onto the other side,

$$\langle 0|T(\partial_x^\mu\phi_I(x)\partial_y^\nu\phi_I(y))|0\rangle = \partial_x^\mu\partial_y^\nu i\Delta(x-y) - ig^{0\nu}g^{0\mu}\delta^{(4)}(x-y) \tag{27.91}$$

We have a non-covariant interaction Hamiltonian, and a non-covariant contraction function. The first term on the right is a nice covariant object, with the same Fourier transform as the Feynman propagator, with an extra couple of k's in the numerator, because the derivatives are outside, but then we've got this extra term which is disgusting.

I have led you to the edge of the precipice, but we're not going to plunge into that pit of garbage. Of course these two diseases turn out to be each other's cure. The theory is after all Lorentz invariant, and you must get Lorentz invariant answers finally. It turns out that the disgusting term in the interaction Hamiltonian cancels the disgusting term in the contraction function, after a horrendous amount of combinatorics that I'm not going to do. You can now see the sort of problems that we would have to deal with if we attempted to treat this theory in a straightforward way. I promise that you get all the correct answers this way, but to redeem that promise would require a lot of work.[7]

The third technical difficulty is the same problem we encountered attempting to compute the propagator or the Hamiltonian for even the massive photon field. Here it comes again. If you recall, doing the free theory, (see (26.45) and the sentence following) we had to eliminate A^0 from the Lagrangian before we could write down the Hamiltonian. The equation that eliminated it was the field equation evaluated for $\mu = 0$. In an interacting theory we'll have the modified field equation where this will now be a function of all those charged particle fields in the theory and therefore the equation we will have is

$$\mu^2 A^0 = \partial_i F^{i0} - J^0$$

In the course of eliminating A^0 we'll introduce terms in the interaction Hamiltonian of the form $(J^0)^2$, from squaring A^0, just like those ugly $(K^0)^2$ terms we have here. Likewise when computing the A^0 propagator, because A^0 is related to a canonical momentum, we'll have exactly the same sort of problems we had when we were computing $[\partial^0\phi(x), \partial^0\phi(y)]$. A^0 is related to the time derivative of A_i, and if we attempt to compute the A^0 propagator we'll get

---

[7] [Eds.] A similar example was given in §26.5. See also Section 4.4 of Chapter 4, "Secret Symmetry", pp. 154–156 in Coleman *Aspects*, and the paper cited there, Kuo-Shung Cheng, "Quantization of a General Dynamical System by Feynman's Path Integration Formulation", *J. Math. Phys.* **13** (1972) 1723–1726. (But see also note 7, p. 623.) For an explicit example showing the combinatorics and cancellation of two non-covariant pieces, see the discussion of scalar electrodynamics in Itzykson & Zuber *QFT*, pp. 282–285.

non-covariant terms there for the same reason we got them in the previous example. So even in spinor electrodynamics with a massive photon, with none of the other problems of derivative interactions, in the course of eliminating A^0 to set up the theory in Hamiltonian form, we will find almost as many troubles as in the previous theory. This is because A^0 is related to a momentum density just like $\partial^0\phi$, and squaring this term will give us a non-covariant interaction.

So it looks like we have a lot of problems. None are insuperable, in principle. If we just kept the faith, plugged along, did all our combinatorics right and brushed our teeth every day, we would no doubt arrive at the right answer. It just gets messier and messier. This is a good point for us to break off the discussion of electrodynamics and begin to discuss a method that allows us to organize some of the mess. Things will still be hairy once we have learned this method, but they will be considerably less hairy than if we'd attempted to solve the same problem by straightforward means. Thus I will begin next time the topic of a method which is very useful in doing complicated problems of this kind, the method of functional integrals.

Problems 15

15.1 In class (§26.1) I constructed a vector field theory for which the solutions were four-dimensionally transverse waves, and I quantized it to construct a theory of free vector mesons. In this problem you are asked to carry out the same program for the complementary theory, one for which the only solutions are four-dimensionally longitudinal.

Consider

$$\mathscr{L} = \pm\tfrac{1}{2}[(\partial_\nu A^\nu)^2 - \mu^2 A_\nu A^\nu] \tag{P15.1}$$

where μ^2 is a positive number. Derive the field equations. Show that the solutions are longitudinal waves of mass μ. Show that A_0 and its conjugate momentum are a complete set of initial value data. Construct the Hamiltonian in terms of these, and determine the overall sign of the Lagrangian by demanding that the Hamiltonian be bounded below. Show that if you make an appropriate identification of A_0 and its conjugate momentum with ϕ and π of Klein-Gordon theory, the Hamiltonians of the two theories are *identical.* (Thus there is no need to go any farther in the quantization program.)

(1998b 2.1)

15.2 In a theory of a free vector meson, compute the Hamiltonian as a function of annihilation and creation operators. Normal order freely.

Comment: I doubt if there is a single person who will be surprised by the answer to this problem. Nevertheless, it's fun to see how it comes out of that mess of F's and A's.

(1998b 2.2)

15.3 Let ψ_1 and ψ_2 be two Dirac fields, of mass m_1 and m_2 respectively, and let A_ν be a real vector meson field of mass μ. Let the interaction of these fields be

$$\mathscr{L}' = gA_\mu(\overline{\psi}_1\gamma^\mu\psi_2 + \overline{\psi}_2\gamma^\mu\psi_1) \tag{P15.2}$$

with g a real number. If $m_1 > m_2 + \mu$, this interaction will cause ψ_1 to decay into ψ_2 and A_ν. Compute the decay width Γ (12.33) for this process to lowest non-vanishing order in perturbation theory.

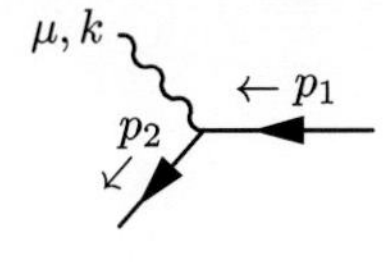

Comment: The vector meson in this problem is coupled to a non-conserved current, because $m_1 \neq m_2$. This leads to disaster when μ goes to zero. You should see this disaster in your answer.

(1998b 2.3)

15.4 Consider the theory of a Hermitian scalar field ϕ defined by

$$\mathscr{L} = \tfrac{1}{2}\partial^\mu\phi'\partial_\mu\phi' - \tfrac{1}{2}\mu^2\phi'^2 - \tfrac{1}{4!}g\phi'^4 - \tfrac{1}{4!}C\phi'^4 + \cdots \tag{P15.3}$$

Here μ is the renormalized mass, g is the positive renormalized coupling constant, C is the $\mathcal{O}(g^2)$ coupling constant renormalization counterterm, and the dots indicate the mass and wave-function counterterms. (I have not bothered to write these out explicitly because they are not needed for the problem at hand.)

To lowest order, the amplitude for two-particle elastic meson–meson scattering

$$p_1 + p_2 \to p_1' + p_2'$$

is associated with the single graph

The amplitude is given by

$$i\mathcal{A} = -ig \tag{P15.4}$$

(the 1/4! is cancelled by the 4! ways in which the four meson fields can annihilate and create the incoming and outgoing mesons). We define the renormalized coupling constant g (and determine, order by order, the counterterm C in perturbation theory) by insisting that the above equation (P15.4) be *exact* (and so, *no* contributions from higher-order graphs) when all four mesons are on the mass shell, at the symmetry point $s = t = u = 4\mu^2/3$. I remind you that for the scattering process

$$p + q \to p' + q'$$

the Mandelstam variables (11.19) are

$$s = (p+q)^2, \qquad t = (p-p')^2, \qquad u = (p-q')^2$$

Compute the meson–meson elastic scattering amplitude to order g^2. Express the answer as a function of s, t, and u. Although it is possible to do all integrals in the problem explicitly, it suffices to express your answer as a sum of terms, each of which is written as an integral over a single Feynman parameter. Check your answer by verifying that, to $\mathcal{O}(g^2)$, the forward scattering amplitude obeys the Optical Theorem (12.49). Note that to get the total cross-section to $\mathcal{O}(g^2)$, you only need the scattering amplitude to $\mathcal{O}(g)$, which I have given you. (HINT. Be careful not to double-count the final states when computing the total cross-section.)

(1975 253a Final, Problem 4; 1986 253a Final, Problem 2)

Solutions 15

15.1 We consider the Lagrangian density

$$\mathscr{L} = \tfrac{1}{2}s\left[\partial_\mu A^\mu \partial_\nu A^\nu - \mu^2 A_\nu A^\nu\right] \tag{P15.1}$$

where $s = \pm 1$, to be determined. Independent of s, the Euler-Lagrange equations give

$$\partial_\nu \partial^\mu A_\mu = -\mu^2 A_\nu \tag{S15.1}$$

As in (26.5), we look for plane wave solutions $A_\mu = \varepsilon_\mu e^{-ik\cdot x}$, and find

$$-k_\nu(k\cdot\varepsilon) = -\mu^2\varepsilon_\nu \Rightarrow \varepsilon_\nu = \frac{k\cdot\varepsilon}{\mu^2}k_\nu \tag{S15.2}$$

If $k\cdot\varepsilon = 0$, the only solutions are trivial. Since the polarization is parallel to the momentum, the plane-wave solutions we have found are longitudinal waves. Take the inner product of (S15.2) with k^ν to find

$$k\cdot\varepsilon = \frac{k\cdot\varepsilon}{\mu^2}k^2 \Rightarrow k^2 = \mu^2$$

The longitudinal waves have mass μ.

We now show that initial value data for A^0 and its conjugate momentum,

$$\pi_0 = \frac{\partial\mathscr{L}}{\partial(\partial_0 A^0)} = s\partial_\mu A^\mu \tag{S15.3}$$

are enough to determine A^ν at any time. (The momenta π_i conjugate to the A^i are all zero.) Taking the divergence of (S15.1), we obtain the Klein–Gordon equation for π_0:

$$\Box^2\pi_0 = -\mu^2\pi_0$$

It follows that π_0 is completely determined by initial value data for π_0 and $\partial_0\pi_0$. Setting $\nu = 0$ in (S15.1), we have

$$\partial_0\pi_0 = -s\mu^2 A_0 \tag{S15.4}$$

and thus initial value data for A^0 and π_0 completely determine π_0 at any time. Again from (S15.1),

$$A_\nu = -\frac{1}{\mu^2}\,\partial_\nu\partial_\mu A^\mu = -\frac{s}{\mu^2}\partial_\nu\pi_0, \tag{S15.5}$$

and therefore A_ν is completely determined as well.

Taking $\partial_0 A^0$ from (S15.3), the Hamiltonian density is given by

$$\mathscr{H} = \pi_0\partial_0 A^0 - \mathscr{L} = \pi_0\left[s\pi_0 - \partial_i A^i\right] - \tfrac{1}{2}s\left[\pi_0^2 - \mu^2 A^0 A_0 - \mu^2 A^i A_i\right]$$

Substituting the right-hand side of (S15.5) for A_i, we find

$$\begin{aligned}\mathscr{H} &= \tfrac{1}{2}s\pi_0^2 + \tfrac{1}{2}s\mu^2 A_0^2 + \frac{s}{\mu^2}\pi_0\partial_i\partial^i\pi_0 + \frac{s^3}{2\mu^2}(\partial_i\pi_0\partial^i\pi_0) \\ &= \tfrac{1}{2}s\pi_0^2 + \tfrac{1}{2}s\mu^2 A_0^2 - \frac{s}{\mu^2}(\partial_i\pi_0)(\partial^i\pi_0) + \cdots + \frac{s^3}{2\mu^2}(\partial_i\pi_0\partial^i\pi_0) \\ &= s\left[\tfrac{1}{2}\pi_0^2 + \tfrac{1}{2}\mu^2 A_0^2 + \frac{1}{2\mu^2}(\partial_i\pi_0)^2\right]\end{aligned} \tag{S15.6}$$

where we've used $s^2 = 1$, and the dots indicate a three-divergence $(s/\mu^2)\partial^i(\pi_0\partial_i\pi_0)$ which we can convert to a surface integral at infinity. Each term between the square brackets in (S15.6) is non-negative (note the lowered superscript i), and so is the Hamiltonian if we choose $s = +1$. Making that choice,

$$\mathscr{L} = \tfrac{1}{2}\left[\partial_\mu A^\mu \partial_\nu A^\nu - \mu^2 A_\nu A^\nu\right]$$

If we define

$$\phi = -\frac{1}{\mu}\pi_0 \qquad \pi = \mu A^0 \tag{S15.7}$$

then, in terms of these variables, the Hamiltonian becomes

$$\mathscr{H} = \tfrac{1}{2}\pi^2 + \tfrac{1}{2}(\partial_i\phi)^2 + \tfrac{1}{2}\mu^2\phi^2 \tag{S15.8}$$

which is indeed identical to the free Klein–Gordon theory, as was to be shown. The extra minus sign in (S15.7) attached to ϕ (or alternatively, to π) is needed to get the correct equations of motion,, e.g., $\partial_0\phi = \pi$, as follows from (S15.4). ■

15.2 We start with the Hamiltonian density (26.46) for a free massive vector meson, slightly rewritten:

$$H = \int d^3\mathbf{x}\,\mathscr{H} = \int d^3\mathbf{x} :\left[-\tfrac{1}{2}F^{0i}F_{0i} + \tfrac{1}{4}F^{ij}F_{ij} + \tfrac{1}{2}\mu^2 A^0 A_0 - \tfrac{1}{2}\mu^2 A^i A_i\right]:$$

Writing A^μ in terms of creation and annihilation operators, we have

$$A^\mu(x) = \sum_r \int \frac{d^3\mathbf{k}}{(2\pi)^{3/2}\sqrt{2\omega_\mathbf{k}}}\left[a_{\mathbf{k},r}\varepsilon_\mathbf{k}^{(r)\mu}e^{-ik\cdot x} + a^\dagger_{\mathbf{k},r}\varepsilon_\mathbf{k}^{(r)\mu *}e^{ik\cdot x}\right] \tag{S15.9}$$

$$F^{\mu\nu}(x) = i\sum_r \int \frac{d^3\mathbf{k}}{(2\pi)^{3/2}\sqrt{2\omega_\mathbf{k}}}\left[a_{\mathbf{k},r}(k^\nu\varepsilon_\mathbf{k}^{(r)\mu} - k^\mu\varepsilon_\mathbf{k}^{(r)\nu})e^{-ik\cdot x} - a^\dagger_{\mathbf{k},r}(k^\nu\varepsilon_\mathbf{k}^{(r)\mu *} - k^\mu\varepsilon_\mathbf{k}^{(r)\nu *})e^{ik\cdot x}\right] \tag{S15.10}$$

We recall that the polarization vectors are orthonormal:

$$\varepsilon^{(r)*}\cdot\varepsilon^{(s)} = -\delta^{rs} \tag{26.24}$$

If we define

$$\begin{aligned} a_\mathbf{k}^\mu &= \sum_r a_{\mathbf{k},r}\varepsilon_\mathbf{k}^{(r)\mu} \\ a_\mathbf{k}^{\dagger\mu} &= \sum_r a^\dagger_{\mathbf{k},r}\varepsilon_\mathbf{k}^{(r)\mu *}.\end{aligned} \tag{S15.11}$$

then we can write

$$\begin{aligned} A^\mu(x) &= \int \frac{d^3\mathbf{k}}{(2\pi)^{3/2}\sqrt{2\omega_\mathbf{k}}}\left[a_\mathbf{k}^\mu e^{-ik\cdot x} + a_\mathbf{k}^{\dagger\mu}e^{ik\cdot x}\right] \\ F^{\mu\nu}(x) &= i\int \frac{d^3\mathbf{k}}{(2\pi)^{3/2}\sqrt{2\omega_\mathbf{k}}}\left[(a_\mathbf{k}^\mu k^\nu - k^\mu a_\mathbf{k}^\nu)e^{-ik\cdot x} - (a_\mathbf{k}^{\dagger\mu}k^\nu - k^\mu a_\mathbf{k}^{\dagger\nu})e^{ik\cdot x}\right]\end{aligned} \tag{S15.12}$$

Because

$$a_\mathbf{k}^\mu k_\mu = \sum_r a_{\mathbf{k},r}\varepsilon_\mathbf{k}^{(r)\,\mu}k_\mu = 0,$$

it follows that

$$a_\mathbf{k}^i k_i = -a_\mathbf{k}^0 k_0. \tag{S15.13}$$

To save space and writing, let

$$d\mu_\mathbf{k} = \frac{d^3\mathbf{k}}{(2\pi)^{3/2}\sqrt{2\omega_\mathbf{k}}}; \qquad d\mu_{\mathbf{k}'} = \frac{d^3\mathbf{k}'}{(2\pi)^{3/2}\sqrt{2\omega_{\mathbf{k}'}}}$$

From (S15.12) and the judicious use of (S15.13) we find

$$\int d^3\mathbf{x}\, F^{0i}F_{0i} = -\int d^3\mathbf{x}\, d\mu_{\mathbf{k}}\, d\mu_{\mathbf{k}'} : \begin{bmatrix} (a^0_{\mathbf{k}}k^i - k^0 a^i_{\mathbf{k}})(a_{\mathbf{k}',0}k'_i - k'_0 a_{\mathbf{k}',i})e^{-ik\cdot x - ik'\cdot x} \\ -(a^0_{\mathbf{k}}k^i - k^0 a^i_{\mathbf{k}})(a^\dagger_{\mathbf{k}',0}k'_i - k'_0 a^\dagger_{\mathbf{k}',i})e^{-ik\cdot x + ik'\cdot x} \\ -(a^{\dagger 0}_{\mathbf{k}}k^i - k^0 a^{\dagger i}_{\mathbf{k}})(a_{\mathbf{k}',0}k'_i - k'_0 a_{\mathbf{k}',i})e^{+ik\cdot x - ik'\cdot x} \\ +(a^{\dagger 0}_{\mathbf{k}}k^i - k^0 a^{\dagger i}_{\mathbf{k}})(a^\dagger_{\mathbf{k}',0}k'_i - k'_0 a^\dagger_{\mathbf{k}',i})e^{+ik\cdot x + ik'\cdot x} \end{bmatrix} :$$

$$= -\int \frac{d^3\mathbf{k}}{2k_0} : \begin{bmatrix} ((|\mathbf{k}|^2 - 2k_0^2)a^0_{\mathbf{k}}a_{-\mathbf{k},0} + k_0^2 a^i_{\mathbf{k}}a_{-\mathbf{k},i})e^{-2ik^0t} + (2|\mathbf{k}|^2 - 4k_0^2)a^{\dagger 0}_{\mathbf{k}}a_{\mathbf{k},0} \\ -2k_0^2 a^\dagger_{\mathbf{k},i}a^i_{\mathbf{k}} + ((|\mathbf{k}|^2 - 2k_0^2)a^{\dagger 0}_{\mathbf{k}}a^\dagger_{-\mathbf{k},0} + k_0^2 a^{\dagger i}_{\mathbf{k}}a^\dagger_{-\mathbf{k},i})e^{+2ik^0t} \end{bmatrix} :$$

$$\int d^3\mathbf{x}\, F^{ij}F_{ij} = -\int d^3\mathbf{x} \int d\mu_{\mathbf{k}}\, d\mu_{\mathbf{k}'} : \begin{bmatrix} (a^i_{\mathbf{k}}k^j - k^i a^j_{\mathbf{k}})(a_{\mathbf{k}',i}k'_j - k'_i a_{\mathbf{k}',j})e^{-ik\cdot x - ik'\cdot x} \\ -(a^i_{\mathbf{k}}k^j - k^i a^j_{\mathbf{k}})(a^\dagger_{\mathbf{k}',i}k'_j - k'_i a^\dagger_{\mathbf{k}',j})e^{-ik\cdot x + ik'\cdot x} \\ -(a^{\dagger i}_{\mathbf{k}}k^j - k^i a^{\dagger j}_{\mathbf{k}})(a_{\mathbf{k}',i}k'_j - k'_i a_{\mathbf{k}',j})e^{+ik\cdot x - ik'\cdot x} \\ +(a^{\dagger i}_{\mathbf{k}}k^j - k^i a^{\dagger j}_{\mathbf{k}})(a^\dagger_{\mathbf{k}',i}k'_j - k'_i a^\dagger_{\mathbf{k}',j})e^{+ik\cdot x + ik'\cdot x} \end{bmatrix} :$$

$$= -\int \frac{d^3\mathbf{k}}{2k_0} : \begin{bmatrix} (-2k_0^2 a^0_{\mathbf{k}}a_{-\mathbf{k},0} + 2|\mathbf{k}|^2 a^i_{\mathbf{k}}a_{-\mathbf{k},i})e^{-2ik^0t} + 4k_0^2 a^{\dagger 0}_{\mathbf{k}}a_{\mathbf{k},0} \\ +4|\mathbf{k}|^2 a^\dagger_{\mathbf{k},i}a^i_{\mathbf{k}} + (-2k_0^2 a^{\dagger 0}_{\mathbf{k}}a^\dagger_{-\mathbf{k},0} + 2|\mathbf{k}|^2 a^{\dagger i}_{\mathbf{k}}a^\dagger_{-\mathbf{k},i})e^{+2ik^0t} \end{bmatrix} :$$

$$\int d^3\mathbf{x}\, A^0A_0 = \int d^3\mathbf{x} \int d\mu_{\mathbf{k}}\, d\mu_{\mathbf{k}'} : \begin{bmatrix} a^0_{\mathbf{k}}a_{\mathbf{k}',0}e^{-ik\cdot x - ik'\cdot x} + a^0_{\mathbf{k}}a^\dagger_{\mathbf{k}',0}e^{-ik\cdot x + ik'\cdot x} \\ +a^{\dagger 0}_{\mathbf{k}}a_{\mathbf{k}',0}e^{+ik\cdot x - ik'\cdot x} + a^{\dagger 0}_{\mathbf{k}}a^\dagger_{\mathbf{k}',0}e^{+ik\cdot x + ik'\cdot x} \end{bmatrix} :$$

$$= \int \frac{d^3\mathbf{k}}{2k_0} : \left[a^0_{\mathbf{k}}a_{-\mathbf{k},0}e^{-2ik^0t} + 2a^{\dagger 0}_{\mathbf{k}}a_{\mathbf{k},0} + a^{\dagger 0}_{\mathbf{k}}a^\dagger_{-\mathbf{k},0}e^{+2ik^0t}\right] :$$

$$\int d^3\mathbf{x}\, A^iA_i = \int d^3\mathbf{x} \int d\mu_{\mathbf{k}}\, d\mu_{\mathbf{k}'} : \begin{bmatrix} a^i_{\mathbf{k}}a_{\mathbf{k}',i}e^{-ik\cdot x - ik'\cdot x} + a^i_{\mathbf{k}}a^\dagger_{\mathbf{k}',i}e^{-ik\cdot x + ik'\cdot x} \\ +a^{\dagger i}_{\mathbf{k}}a_{\mathbf{k}',i}e^{+ik\cdot x - ik'\cdot x} + a^{\dagger i}_{\mathbf{k}}a^\dagger_{\mathbf{k}',i}e^{+ik\cdot x + ik'\cdot x} \end{bmatrix} :$$

$$= \int \frac{d^3\mathbf{k}}{2k_0} : \left[a^i_{\mathbf{k}}a_{-\mathbf{k},i}e^{-2ik^0t} + 2a^{\dagger i}_{\mathbf{k}}a_{\mathbf{k},i} + a^{\dagger i}_{\mathbf{k}}a^\dagger_{-\mathbf{k},i}e^{+2ik^0t}\right] :$$

There are eighteen terms in the Hamiltonian density. We know that H must be time-independent, and we know that the a and $a^\dagger$ operators are independent. So just to see how this goes, let's look only at the four integrands which are proportional to $a^0_{\mathbf{k}}a_{-\mathbf{k},0}e^{-2ik^0t}$. The coefficients of this factor in the Hamiltonian density are

$$\mathscr{H}|_{\text{coef}} = \underbrace{-\tfrac{1}{2}(-|\mathbf{k}|^2 + 2k_0^2)}_{-\frac{1}{2}F^{0i}F_{0i}} + \underbrace{\tfrac{1}{4}(2k_0^2)}_{\frac{1}{4}F^{ij}F_{ij}} + \underbrace{\tfrac{1}{2}\mu^2(1)}_{\frac{1}{2}\mu^2A^0A_0} + \underbrace{-\tfrac{1}{2}\mu^2(0)}_{-\frac{1}{2}\mu^2A^iA_i} = -\tfrac{1}{2}(k_0^2 - |\mathbf{k}|^2 - \mu^2) = 0 \qquad \text{(S15.14)}$$

The other time-dependent terms cancel similarly. The time-independent terms are

$$\begin{aligned} \mathscr{H}|_{\text{time-ind}} &= a^{\dagger 0}_{\mathbf{k}}a_{\mathbf{k},0}\left[-\tfrac{1}{2}(-2|\mathbf{k}|^2 + 4k_0^2) + \tfrac{1}{4}(-4k_0^2) + \tfrac{1}{2}\mu^2(2) - \tfrac{1}{2}\mu^2(0)\right] \\ &\quad + a^{\dagger i}_{\mathbf{k}}a_{\mathbf{k},i}\left[-\tfrac{1}{2}(2k_0^2) + \tfrac{1}{4}(-4|\mathbf{k}|^2) + \tfrac{1}{2}\mu^2(0) - \tfrac{1}{2}\mu^2(2)\right] \\ &= a^{\dagger 0}_{\mathbf{k}}a_{\mathbf{k},0}[-2k_0^2 - (k_0^2 - |\mathbf{k}|^2 - \mu^2)] + a^{\dagger i}_{\mathbf{k}}a_{\mathbf{k},i}[-k_0^2 - |\mathbf{k}|^2 - \mu^2] \\ &= (a^{\dagger 0}_{\mathbf{k}}a_{\mathbf{k},0} + a^{\dagger i}_{\mathbf{k}}a_{\mathbf{k},i})(-2k_0^2) = (a^{\dagger\mu}_{\mathbf{k}}a_{\mathbf{k},\mu})(-2k_0^2) \end{aligned} \qquad \text{(S15.15)}$$

From the definitions (S15.11) of a_μ and $a^\dagger_\mu$ and the normalization (26.24), it follows

$$a^{\dagger\mu}_{\mathbf{k}}a_{\mathbf{k},\mu} = -\sum_r a^{(r)\dagger}_{\mathbf{k}}a^{(r)}_{\mathbf{k}} \qquad \text{(S15.16)}$$

and so finally

$$H = \int \frac{d^3\mathbf{k}}{2k_0}(a^{\dagger\mu}_{\mathbf{k}}a_{\mathbf{k},\mu})(-2k_0^2) = \int d^3\mathbf{k}\,\omega_{\mathbf{k}} \sum_r a^{(r)\dagger}_{\mathbf{k}}a^{(r)}_{\mathbf{k}} \qquad \text{(S15.17)}$$

This result is not a surprise; it's surely what we all expected, in agreement with (26.64). ■

15.3 Applying the Feynman rules in the box on p. 571 to the diagram in the problem, we can write the amplitude for the decay process as

$$i\mathcal{A}_{fi} = ig\overline{u}_2^{s'}(p_2)\gamma^\nu u_1^s(p_1)\varepsilon_\nu^{(r)*}(k) \tag{S15.18}$$

where u_1^s and $u_2^{s'}$ are the polarization spinors corresponding to ψ_1, ψ_2 with momenta p_1 and p_2, and $\varepsilon_\nu^{(r)}(k)$ is the polarization vector of A_ν, with $p_1 = p_2 + k$. Summing over initial spins and averaging over final spins, we then have (21.115)

$$\begin{aligned}
\tfrac{1}{2}\sum_{s,s',r}|\mathcal{A}_{fi}|^2 &= \tfrac{1}{2}g^2\sum_{s,s',r}\overline{u}_1^s\gamma^\mu u_2^{s'}\overline{u}_2^{s'}\gamma^\nu u_1^s\varepsilon_\mu^{(r)}\varepsilon_\nu^{(r)*} = \tfrac{1}{2}g^2\left[\sum_{s,s}\text{Tr}\left(\gamma^\mu u_2^{s'}\overline{u}_2^{s'}\gamma^\nu u_1^s\overline{u}_1^s\right)\right]\sum_r\varepsilon_\mu^{(r)}\varepsilon_\nu^{(r)*} \\
&= \tfrac{1}{2}g^2\text{Tr}\left[\gamma^\mu(\not{p}_2+m_2)\gamma^\nu(\not{p}_1+m_1)\right]\left(-g_{\mu\nu}+\frac{(p_1-p_2)_\mu(p_1-p_2)_\nu}{\mu^2}\right) \\
&= \frac{g^2}{2\mu^2}\text{Tr}\left[(\not{p}_1-\not{p}_2)(\not{p}_2+m_2)(\not{p}_1-\not{p}_2)(\not{p}_1+m_1)\right] - \tfrac{1}{2}g^2\text{Tr}\left[\gamma^\mu(\not{p}_2+m_2)\gamma_\mu(\not{p}_1+m_1)\right] \\
&= \frac{g^2}{2\mu^2}A - \tfrac{1}{2}g^2 B
\end{aligned} \tag{S15.19}$$

letting A and B stand for the trace terms. Now we use the trace identities worked out in Problem 11.2, dropping terms with an odd number of γ's:

$$\begin{aligned}
A &= \text{Tr}\left[(\not{p}_1-\not{p}_2)\not{p}_2(\not{p}_1-\not{p}_2)\not{p}_1\right] + m_1m_2\text{Tr}\left[(\not{p}_1-\not{p}_2)(\not{p}_1-\not{p}_2)\right] \\
&= 8[(p_1-p_2)\cdot p_1][(p_1-p_2)\cdot p_2] - 4(p_1-p_2)^2(p_1\cdot p_2) + 4m_1m_2(p_1-p_2)^2 \\
&= 4(p_1\cdot p_2 + m_1m_2)(m_1-m_2)^2
\end{aligned} \tag{S15.20}$$

B can be simplified a little, because $\gamma^\mu(\not{p}_2+m_2)\gamma_\mu = -2\not{p}_2 + 4m_2$, so

$$\begin{aligned}
B &= \text{Tr}\left[(-2\not{p}_2+4m_2)(\not{p}_1+m_1)\right] = \text{Tr}\left[-2\not{p}_2\not{p}_1\right] + \text{Tr}\,(4m_2m_1) \\
&= -8p_1\cdot p_2 + 16m_1m_2
\end{aligned} \tag{S15.21}$$

That gives

$$\tfrac{1}{2}\sum_{s,s',r}|\mathcal{A}_{fi}|^2 = \frac{g^2}{2\mu^2}\left[4(p_1\cdot p_2+m_1m_2)(m_1-m_2)^2\right] - \tfrac{1}{2}g^2\left[16m_1m_2 - 8p_1\cdot p_2\right] \tag{S15.22}$$

In the rest frame of the decaying ψ_1, $p_1 = (m_1, \mathbf{0})$, and $p_1 = p_2 + k$, so eliminating k and E_γ,

$$|\mathbf{p}_2| = \frac{\sqrt{(m_1^2-(m_2+\mu)^2)(m_1^2-(m_2-\mu)^2)}}{2m_1}\ ; \qquad E_2 = \sqrt{|\mathbf{p}_2|^2+m_2^2} \tag{S15.23}$$

(Note that $|\mathbf{p}_2|$ is imaginary for $m_1 < m_2 + \mu$, as it should be.) The decay amplitude in ψ_1's rest frame becomes

$$\tfrac{1}{2}\sum_{s,s',r}|\mathcal{A}_{fi}|^2 = 2g^2m_1\left[\frac{(E_2+m_2)}{\mu^2}(m_1-m_2)^2 - 2(2m_1-E_2)\right] \tag{S15.24}$$

For the decay probability per unit time, Γ, we have

$$\frac{\text{rest decay prob.}}{\text{unit time}} = \Gamma = \frac{1}{2m}\sum_{\substack{\text{final}\\\text{states}}}\int|\mathcal{A}_{fi}|^2 D \tag{12.2}$$

where D is the density of final states, given by

$$D = \frac{1}{16\pi^2}\frac{|\mathbf{p}_f|d\Omega_f}{E_T} \tag{12.24}$$

Finally, from (12.33) (see also Example 2 on p. 251)

$$\Gamma = \frac{1}{8\pi}\frac{|\mathbf{p}_2|}{m_1^2}\tfrac{1}{2}\sum_{r,r',s}|\mathcal{A}_{fi}|^2 = \frac{g^2}{4\pi m_1}|\mathbf{p}_2|\left[\frac{(E_2+m_2)}{\mu^2}(m_1-m_2)^2 + 2(E_2-2m_1)\right] \tag{S15.25}$$

(The factor $\frac{1}{2}$ comes from averaging over final spins.) We see in this answer the foretold disaster as $\mu \to 0$, that Γ diverges. This is because the current $j^\mu = g(\overline{\psi}_1\gamma^\mu\psi_2 + \overline{\psi}_2\gamma^\mu\psi_1)$, to which the vector is coupled, is not conserved:

$$\begin{aligned}\partial^\mu j_\mu &= g(\overline{\psi}_1\not{p}_1\psi_2 + \overline{\psi}_1\not{p}_2\psi_2 + \overline{\psi}_2\not{p}_2\psi_1 + \overline{\psi}_2\not{p}_1\psi_1)\\ &= g(-m_1\overline{\psi}_1\psi_2 + m_2\overline{\psi}_1\psi_2 - m_2\overline{\psi}_2\psi_1 + m_1\overline{\psi}_2\psi_1)\\ &= g(m_2 - m_1)\left(\overline{\psi}_1\psi_2 - \overline{\psi}_2\psi_1\right)\end{aligned} \tag{S15.26}$$

The proximate cause of the non-conservation of the current is that $m_1 \neq m_2$, just as the problem stated. Indeed, if $m_1 = m_2$, not only is the current conserved, but the troublesome term in (S15.25), divergent as $\mu \to 0$, vanishes. The moral of this story is that coupling a current to a massless vector does not work unless the current is conserved. See the discussion on p. 579 about the necessity of a *conserved* current for the principle of minimal coupling to work. ■

15.4 To order g, the only scattering graph is

As stated, this leads to

$$i\mathcal{A} = -ig + \mathcal{O}(g^2) \tag{S15.27}$$

To $\mathcal{O}(g^2)$, there are three scattering graphs, and the counterterm:

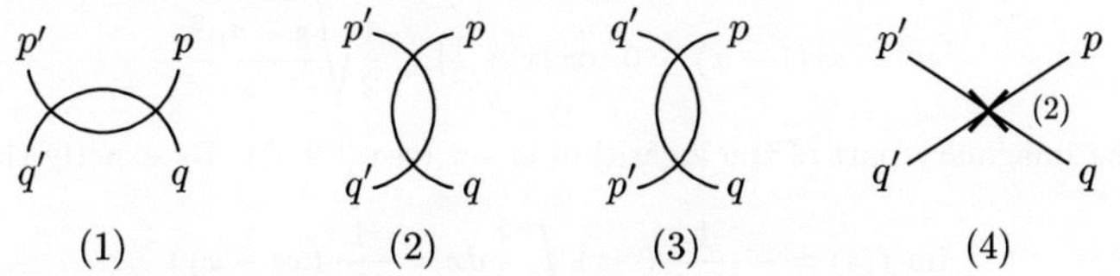

Graph (4) is the charge constant renormalization diagram to $\mathcal{O}(g^2)$; its value is

$$\mathcal{A}_4 = -C_2 \tag{S15.28}$$

The value C_2 is fixed by the renormalization condition

$$\mathcal{A}_1 + \mathcal{A}_2 + \mathcal{A}_3 - C_2 = 0 \quad \text{at } s = t = u = \tfrac{4}{3}\mu^2 \tag{S15.29}$$

Thus, to the desired order,

$$\mathcal{A} = -g + \sum_{i=1}^{3}\left[\mathcal{A}_i - \mathcal{A}_i\Big|_{s=t=u=(4\mu^2/3)}\right] + \mathcal{O}(g^3) \tag{S15.30}$$

All we need to do is to evaluate graphs (1), (2), and (3). Note that the subtraction eliminates the (logarithmic) divergence, so all the integrals are finite.

Let's look at graph (1):

$$i\mathcal{A}_1 = (-ig)^2\left(\tfrac{1}{2}\right)\int \frac{d^4k}{(2\pi)^4}\frac{i}{(k^2 - \mu^2 + i\epsilon)}\frac{i}{[(k+p+q)^2 - \mu^2 + i\epsilon]} \tag{S15.31}$$

The factor of (1/2) arises because the (1/4!) at the vertices are incompletely cancelled; there is no way of distinguishing the two internal lines.

Note that $\mathcal{A}_1$ is a function of $(p+q)^2 = s$ *only*. Likewise, $\mathcal{A}_2$ is the *same* function of $(p-q)^2 = t$, and $\mathcal{A}_3$ of $(p-q')^2 = u$. Furthermore, the integral is *identical* to one we did in class, the meson self-energy $-i\widetilde{\Pi}^f(p^2)$ in Model 3 (15.57), with p replaced by $p+q$ (remember, the "nucleons" in Model 3 were scalars). Using the result (16.2), we have

$$A_1(s) - A_1(s = 4\mu^2/3) = g^2 f(s) = -\frac{g^2}{32\pi^2}\int_0^1 dx\,\ln\left[\frac{\mu^2 - sx(1-x) - i\epsilon}{\mu^2 - 4\mu^2 x(1-x)/3}\right] \tag{S15.32}$$

Then

$$\mathcal{A} = -g + g^2(f(s) + f(t) + f(u)) + \mathcal{O}(g^3) \tag{S15.33}$$

Now to check the Optical Theorem,

$$\mathrm{Im}\,\mathcal{A}_{ii} = 2E_T|\mathbf{p}_i|\sigma \tag{12.49}$$

Using the standard formula (12.26) in the center of momentum frame,

$$\frac{d\sigma}{d\Omega} = \frac{g^2}{64\pi^2 s} + \mathcal{O}(g^3) \tag{S15.34}$$

and thus

$$\sigma = \tfrac{1}{2}\int d\Omega\,\frac{d\sigma}{d\Omega} = \frac{g^2}{32\pi s} + \mathcal{O}(g^3) \tag{S15.35}$$

the extra factor of $\frac{1}{2}$ coming from the scattering of identical particles. Since $s = E_T^2 = (2\sqrt{|\mathbf{p}_i|^2+\mu^2})^2 = 4(|\mathbf{p}_i|^2+\mu^2)$, we have

$$2|\mathbf{p}_i|E_T\sigma = \frac{g^2}{32\pi}\sqrt{\frac{s-4\mu^2}{s}} + \mathcal{O}(g^3) \tag{S15.36}$$

For forward scattering, $s \geq 4\mu^2$, $t = 0$, and $u = 4\mu^2 - s \leq 0$. That means $f(t)$ and $f(u)$ are *real*, and the imaginary part of $\mathcal{A}$ can come only from $f(s)$:

$$\mathrm{Im}\,\mathcal{A} = g^2\mathrm{Im}\,f(s) + \mathcal{O}(g^3) \tag{S15.37}$$

To investigate the imaginary part of $f(s)$, recall the nearly identical integral (S9.1) in Problem 9.1, p. 349. The integrand has an imaginary part only when

$$\mu^2 - sx(1-x) \leq 0, \text{ or } |x - \tfrac{1}{2}| \leq \frac{1}{2}\sqrt{\frac{s-4\mu^2}{s}}$$

In this region of x, the imaginary part of the logarithm is $-\pi$ (see (S9.2)). By exactly the same analysis used in Problem 9.1,

$$\mathrm{Im}\,f(s) = -\frac{1}{32\pi^2}(-\pi)\int_{x_1}^{x_2} dx = \frac{1}{32\pi}(x_2 - x_1) \tag{S15.38}$$

where

$$\{x_1, x_2\} = \frac{1}{2} \mp \frac{1}{2}\sqrt{\frac{s-4\mu^2}{s}} \tag{S15.39}$$

and so

$$\mathrm{Im}\,\mathcal{A} = \frac{g^2}{32\pi}\sqrt{\frac{s-4\mu^2}{s}} + \mathcal{O}(g^3) \tag{S15.40}$$

in agreement with (S15.36). The Optical Theorem is valid, to $\mathcal{O}(g^2)$. Note that

$$\lim_{s\to\infty} \mathrm{Im}\,\mathcal{A} = \frac{g^2}{32\pi} + \mathcal{O}(g^3) \tag{S15.41}$$

That is, to second order in g, $\mathrm{Im}\,\mathcal{A}$ approaches a positive, finite constant. ■

28

Functional integration and Feynman rules

The first version of functional integrals (integrals over an infinite number of dimensions), called "path integrals", was introduced into physics in the late 1940s by Feynman,[1] but these methods were not fully appreciated until the early 1960s. This method will give us enormous advantages, and enable us to settle with derivative coupling, superfluous variables, gauge invariance and so on. Functional integrals are sometimes called "integration over function spaces" or "integration over infinite dimensional spaces".[2] We all know how to do integrals over a one-dimensional space, or over n-dimensional space; I am now going to take n to infinity.

28.1 First steps with functional integrals

For the moment, we'll put aside the vector fields and their associated problems, and talk about what will in the first instance be purely a topic in mathematics (butcher grade—the way physicists do it), and then we'll eventually come back and develop a bunch of techniques using this mathematical method that will help us unravel things.

I begin with a simple one-dimensional integral, the Gaussian,

$$\int_{-\infty}^{\infty} dx\, e^{-\frac{1}{2}ax^2} = \sqrt{2\pi}\, a^{-\frac{1}{2}} \tag{28.1}$$

where a is a positive, real number to ensure damping of the integral at infinity. By analytic continuation, the identity is true whenever the integral converges. That is to say, a can be a

[1] [Eds.] R. P. Feynman, "Space-Time Approach to Non-Relativistic Quantum Mechanics", *Rev. Mod. Phys.* **20** (1948) 367–387, also reprinted in *Feynman's Thesis: A New Approach to Quantum Theory*, ed. Laurie M. Brown, World Scientific Press, 2005. See also Richard P. Feynman and Albert R. Hibbs, *Quantum Mechanics and Path Integrals*, McGraw–Hill, 1965; edited and corrected by Daniel F. Styer, Dover Publications, 2010. Feynman devised the technique to reformulate quantum mechanics. Much of this chapter restates, in different words, Section 4 of Coleman's 1973 Erice lecture, "Secret Symmetry", reprinted as Chapter 5 in Coleman *Aspects*. Copies of this lecture were handed out during this class.

[2] [Eds.] The American mathematician Norbert Wiener (1894–1964), investigating Brownian motion, had developed methods similar to Feynman's about a decade earlier. For an illuminating article about his work and its connections to Feynman's, see Mark Kac, "Wiener and integration in function spaces", *Bull. Amer. Math. Soc.* **72** (1966) 52–68. A brief introduction to function space, Lebesgue measure and generalized functions is given in Chap. III, pp. 179–255 of *Mathematics for Physicists*, Phillipe Dennery and André Krzywicki, Harper and Row, 1967, republished by Dover Publications, 1996.

complex number, so long as its real part is greater than zero. We can also do the n-dimensional version of this integral. Let $\mathbf{x}$ be a vector in this n-dimensional space. For some symmetric matrix A, I define

$$(\mathbf{x}, A\mathbf{x}) = \sum_{a,b=1}^{n} x_a A_{ab} x_b. \tag{28.2}$$

By making an orthogonal transformation to diagonalize A, I instantly find that

$$(\mathbf{x}, A\mathbf{x}) = \sum_{a=1}^{n} \lambda_a x_a^2 \tag{28.3}$$

so that

$$\int d^n\mathbf{x}\, e^{-\frac{1}{2}(\mathbf{x},A\mathbf{x})} = (2\pi)^{n/2} \prod_{i=1}^{n} (\lambda_i)^{-\frac{1}{2}} = (2\pi)^{n/2} (\det A)^{-\frac{1}{2}} \tag{28.4}$$

provided all of the eigenvalues λ_i are positive—or, by analytical continuation, if $\mathrm{Re}(\mathbf{x}, A\mathbf{x}) > 0$. That is enough to make the integral converge. These factors of 2π are irritating, so I will introduce a notational convention. I will write

$$(d\mathbf{x}) = \frac{d^n\mathbf{x}}{\sqrt{(2\pi)^n}} \tag{28.5}$$

and write the previous integral as

$$\int (d\mathbf{x}) e^{-\frac{1}{2}(\mathbf{x},A\mathbf{x})} = (\det A)^{-\frac{1}{2}} \tag{28.6}$$

Of course, if we can do a Gaussian integral, we can do a general quadratic form by completing the square. Consider the quadratic form

$$Q(\mathbf{x}) = \tfrac{1}{2}(\mathbf{x}, A\mathbf{x}) + (\mathbf{b}, \mathbf{x}) + c \tag{28.7}$$

where $\mathbf{b}$ is some n-vector. $Q(\mathbf{x})$ is minimized at $\mathbf{x} = \overline{\mathbf{x}}$;

$$\overline{\mathbf{x}} = -A^{-1}\mathbf{b} \tag{28.8}$$

Then

$$Q(\overline{\mathbf{x}}) = \tfrac{1}{2}(\overline{\mathbf{x}}, A\overline{\mathbf{x}}) + (\mathbf{b}, \overline{\mathbf{x}}) + c = -\tfrac{1}{2}(\mathbf{b}, A^{-1}\mathbf{b}) + c \tag{28.9}$$

and so

$$Q(\mathbf{x}) = Q(\overline{\mathbf{x}}) + \tfrac{1}{2}(\mathbf{x} - \overline{\mathbf{x}}, A[\mathbf{x} - \overline{\mathbf{x}}]) \tag{28.10}$$

Thus I find, with $\mathbf{y} = \mathbf{x} - \overline{\mathbf{x}}$,

$$\int (d\mathbf{x})\, e^{-Q(\mathbf{x})} = e^{-Q(\overline{\mathbf{x}})} \int (d\mathbf{y})\, e^{-\frac{1}{2}(\mathbf{y},A\mathbf{y})} = e^{-Q(\overline{\mathbf{x}})} (\det A)^{-\frac{1}{2}} \tag{28.11}$$

where $e^{-Q(\overline{\mathbf{x}})} = \exp\left[\tfrac{1}{2}(\mathbf{b}, A^{-1}\mathbf{b}) - c\right]$ is a constant.

Once we can do a general quadratic form, we can do a polynomial times a generalized Gaussian. If I have any polynomial $P(\mathbf{x})$ in $\mathbf{x}$, an expression of the form

$$\int (d\mathbf{x})\, P(\mathbf{x})\, e^{-Q(\mathbf{x})} \tag{28.12}$$

can be computed by taking derivatives;

$$\int (d\mathbf{x})\, P(\mathbf{x})\, e^{-Q(\mathbf{x})} = \int (d\mathbf{x})\, P\left(-\frac{\partial}{\partial \mathbf{b}}\right) e^{-Q(\mathbf{x})} \tag{28.13}$$

That is, whenever I see a component of $\mathbf{x}$ (x_1 or x_2 or x_{17}), I differentiate with respect to the same component of $\mathbf{b}$—$\partial/\partial b_1$ or whatever—this drags that component of $\mathbf{x}$ down from $e^{-Q(\mathbf{x})}$. Now I take the derivative outside the integral

$$\int (d\mathbf{x})\, P(\mathbf{x})\, e^{-Q(\mathbf{x})} = P\left(-\frac{\partial}{\partial \mathbf{b}}\right) \int (d\mathbf{x})\, e^{-Q(\mathbf{x})} = (\det A)^{-\frac{1}{2}}\, P\left(-\frac{\partial}{\partial \mathbf{b}}\right) e^{-Q(\overline{\mathbf{x}})} \tag{28.14}$$

For example,

$$\int (d\mathbf{x})\, x_1 x_5\, e^{-Q(\mathbf{x})} = (\det A)^{-\frac{1}{2}} \left(-\frac{\partial}{\partial b_1}\right)\left(-\frac{\partial}{\partial b_5}\right) e^{-Q(\overline{\mathbf{x}})} \tag{28.15}$$

So I have told you something you no doubt already know, although perhaps in a somewhat more compressed notation than you are used to: how to integrate Gaussians, generalized Gaussians, and polynomials times generalized Gaussians.

It will turn out for later purposes to be convenient to integrate over functions not only of real n-vectors but complex n-vectors. I don't have something fancy in mind involving contour integrals or anything like that, I just mean integrating over the real part and then integrating over the imaginary part. In particular, I'll take a complex vector and break it up into real and imaginary parts like this:

$$\mathbf{z} = \frac{1}{\sqrt{2}}(\mathbf{x} + i\mathbf{y}) \tag{28.16}$$

and similarly for $\mathbf{z}^*$. I've included the $\sqrt{2}$, for reasons that will become clear shortly.[3] I define

$$(d\mathbf{z}^*)(d\mathbf{z}) \equiv (d\mathbf{x})(d\mathbf{y}) \tag{28.17}$$

whence it follows for example that

$$\int (d\mathbf{z}^*)(d\mathbf{z})\, e^{-(\mathbf{z}^*,\, A\mathbf{z})} = (\det A)^{-1} \tag{28.18}$$

Well, it's pretty trivial. You diagonalize A, you write $\mathbf{z}$ in terms of $\mathbf{x}$ and $\mathbf{y}$, and the $\frac{1}{2}$ comes in automatically as I've arranged matters. I simply have one integral for the $\mathbf{x}$ and one integral for the $\mathbf{y}$. So each eigenvalue occurs twice, and I get the exponent now equal to -1 rather than $-\frac{1}{2}$. And similar formulas follow for generalized quadratics and polynomials times general quadratics.

Now comes the big leap of faith. I have arranged all the formulas so that the dimension n of the vector space over which I am integrating never appears explicitly. Therefore I am

---

[3] [Eds.] Square roots of 2 have already appeared in field theory, for similar reasons, when we decomposed a complex field. See the digression in §6.1, p. 109. Note that for ordinary two dimensional integrals, the Jacobian determinant $J = \partial(z, z^*)/\partial(x, y)$, with z and z^* defined as in (28.16), equals $-i$:

$$dz\, dz^* = J\, dx\, dy = -i dx\, dy$$

so that, to within a phase constant of norm 1, the identification $(d\mathbf{z}^*)(d\mathbf{z}) \equiv (d\mathbf{x})(d\mathbf{y})$ is unobjectionable.

going to simply extend these formulas to an *infinite*-dimensional space! This is a **functional integral**. We're just going to say these formulas *define* integrals of Gaussians, and polynomials times Gaussians, which will turn out to be practically everything we will need to do, over an infinite-dimensional space. Everything is exactly the same, except that the sums in (28.2) and (28.3) run not from 1 to n, but from 1 to ∞. Obviously this involves deep and subtle mathematical questions, about which I will say nothing.

More generally, you start out with an infinite-dimensional space, a Hilbert space of some sort. You have a quadratic form on it, defined by some infinite matrix, some positive definite operator. That's completely legitimate. Then you take that infinite-dimensional space, and you look at a finite-dimensional subspace. You compute the integral in that finite-dimensional subspace, and just restricting yourself to that subspace, you can compute the determinant. Absolutely no problem there. Then you let the finite-dimensional space get larger and larger, until it fills out the whole space, adding basis vector after basis vector one at a time. If there's a limit of the integral, it will be the limit of the determinant. If there isn't, that's our bad luck.

It's a deep question, which we leave for the mathematicians, to determine for which quadratic forms the limits of the integral and the determinant exist. Another deep question which we leave to the mathematicians is, if the limits exist for one way of filling up the Hilbert space, do they exist for another choice of basis vectors, where you fill the space out in another order? We'll just leave these questions alone, and blithely manipulate equations, assuming that everything will be okay unless something goes wrong. If we can compute the integral, no doubt it can be rigorously shown to exist. If we get zero or infinity or something like that, then we're going to be in trouble. We'll try to avoid that sort of thing. Since we're going to apply these things to field theory, of course we will get zero or infinity an awful lot of the time, but those will just be our old friends the ultraviolet divergences coming up again, and we can get rid of them by cutting off the theory in any one of the standard ways.

You can also do this for continuous spaces: the set of all functions in 4-space, for example, or all functions on a line. These can be turned into a discrete space by expanding the functions in terms of, say, harmonic oscillator wave functions. Likewise for the set of all functions in n dimensions. So there's no difference between a discretely infinite Hilbert space and a continuously infinite Hilbert space; that's just the difference between a discrete basis and a continuous basis.

Now there are two points I want to make. First, the sort of space over which these integrals are defined is a *very big* space. In fact, precisely *how* big it is doesn't matter. It could be a Hilbert space, it could be a bigger space, the space of all continuous functions. It hardly matters when we do the integral because of this exponential damping. If you throw in some finite number of basis vectors that are badly behaved in one way or another, the exponential damping will cut them out; they'll make a zero contribution to the integral.

Just to emphasize how big it is, I will consider an infinite-dimensional space, and the simplest possible Gaussian integral, for $A = 1$:

$$\int (d\mathbf{x})\, e^{-\frac{1}{2}(\mathbf{x},\mathbf{x})} = \prod_{r=1}^{\infty} \int \frac{dx_r}{\sqrt{2\pi}}\, e^{-\frac{1}{2}x_r^2} = 1 \tag{28.19}$$

Let's now consider a function which is not a polynomial, the step function θ_L localized on a gigantic box. Define

$$\theta_L(\mathbf{x}) = \begin{cases} 1, & \text{if } |x_r| \le L, \text{ where } L \text{ is some positive number, for all } r \\ 0, & \text{otherwise} \end{cases} \tag{28.20}$$

That is to say, this is a step function equal to 1 inside an infinite-dimensional hypercube of edge length $2L$, and equal to zero elsewhere. Then[4]

$$\int (d\mathbf{x})\, \theta_L(\mathbf{x})\, e^{-\frac{1}{2}(\mathbf{x},\mathbf{x})} = \prod_{r=1}^{\infty} \int_{-L}^{L} \frac{dx_r}{\sqrt{2\pi}}\, e^{-\frac{1}{2}x_r^2} \tag{28.21}$$

Now what is this quantity? Well, each of the infinite number of terms is identical, and each of them is a little bit less than 1, no matter how big L is, so long as $L < \infty$. Take an infinite product of terms, each of them $\frac{3}{4}$ or $\frac{7}{8}$ or $\frac{15}{16}$, you always get the same answer: zero! So this integral is completely well defined, and is equal to zero. Function space is *so big* that if you use $\theta_L(\mathbf{x})$ to define a measure on function space—it's a positive functional, and so defines a measure, by giving a volume to every set—then the measure of an infinite-dimensional box of side $2L$ is zero! There's a *lot* more outside than inside. That's a set of measure zero, like the rational numbers inside the real numbers with ordinary Lebesgue measure. Function space is a *very big* space. There's more outside the box than inside. If I take a slice on a straight line I've got a lot outside and little inside; if I take a box in a plane I have a lot more outside compared to inside; if I take a cube I've got even more outside than inside. When I go to infinite dimensional space I've got hardly any inside at all compared to outside. You can get lost in it if you try to be careful, so we won't be. You should be warned about that.

The second point has to do with the choice of a basis. Here I have used an infinite-dimensional space described in terms of a discrete basis. But I could equally well define an infinite-dimensional space in terms of a continuum basis. For example, the space could be the space of all real, integrable functions $\phi(x)$ defined on 4-space, as a Hilbert space. The inner product is

$$(\phi_1, \phi_2) = \int d^4x\, \phi_1(x)\phi_2(x) \tag{28.22}$$

I could define a quadratic form $Q[\phi]$ in terms of a c-number valued function $\phi(x)$ as

$$Q[\phi] = \tfrac{1}{2}\int d^4x\, d^4y\, \phi(x)A(x,y)\phi(y) + \int d^4x\, b(x)\phi(x) + c. \tag{28.23}$$

$A(x,y)$ is called an *integral kernel.* That's the same sort of thing as (28.7), only now defined in function space. This is a *functional,* a number-valued function that depends on $\phi(x)$.[5] If I take these functions $\phi(x)$ and expand them in a discrete basis, I will end up with an infinite-dimensional matrix, an infinite-dimensional vector, and an ordinary number. I could go through all my integration formulas, and at least formally, they would make sense. Whether they actually make sense would depend upon how astute I am in choosing the object $A(x,y)$.

---

[4] [Eds.] The function $(1/\sqrt{2\pi})\int_{-x}^{x} dt\, e^{-\frac{1}{2}t^2}$ is the error function, $\text{erf}(x)$, with $\lim_{x\to\infty} \text{erf}(x) = 1$. See H. Margenau and G. M. Murphy, *The Mathematics of Physics and Chemistry*, Van Nostrand Co., 1952, pp. 487–489.

[5] [Eds.] In the literature the dependence of a functional on its arguments is often in square brackets; see the sentences following (13.6).

28.2 Functional integrals in field theory

I will now explain a fundamental formula, which I will give in a form only partially defined, and which I will prove later on. Take a theory of a single classical scalar field ϕ with non-derivative interactions $\mathscr{L}'(\phi)$, and a linear coupling $J\phi$ with an external current J. Define the Lagrangian

$$\mathscr{L}(\phi, J) = \tfrac{1}{2}(\partial^\mu \phi)^2 - \tfrac{1}{2}\mu^2\phi^2 + \mathscr{L}'(\phi) + J\phi \tag{28.24}$$

and the classical action, $\mathcal{S}_c$

$$\mathcal{S}_c[\phi, J] = \int d^4x\, \mathscr{L}(\phi, J) \tag{28.25}$$

This functional depends on two c-number valued functions, $\phi(x)$ and $J(x)$. You give me a $\phi(x)$, you give me a $J(x)$, I can compute, with perhaps considerable labor, depending on how strange their forms are, the number $\mathcal{S}_c[\phi, J]$. In the quantum theory, obtained from this classical theory by canonical quantization, there is an object $Z[J]$ we have seen before, (13.6),

$$Z[J] = \langle 0|\mathrm{S}|0\rangle_J \tag{28.26}$$

where S is the S-matrix. It's the **generating functional** for the Green's functions. I will demonstrate, in a certain sense which I will make precise,

$$\boxed{Z[J] = N\int (d\phi)\, e^{i\mathcal{S}_c[\phi,\, J]}} \tag{28.27}$$

N is a normalization constant independent of J, adjusted such that

$$Z[0] = 1 \tag{28.28}$$

N is closely related to the disconnected vacuum-to-vacuum graphs which we divide out in Dyson's formula (see (13.23) and the discussion following). The precise sense will be, for our purposes, to every order in perturbation theory. When I expand out (28.27) in powers of $\mathscr{L}'(\phi)$, I will have a quadratic form in $\mathcal{S}_c$ (a Gaussian integral) times polynomials in ϕ. I know how to do Gaussians times polynomials, and I will prove, order by order in perturbation theory, that the right-hand side is equal to the left-hand side. I will have to make some subsidiary definitions to prove it, but I will prove it.

The advantage of doing things this way is that on the left-hand side we have an object, $Z[J]$, with all those commutators that were giving us so much trouble. But on the right-hand side, the object has *no* quantum objects, just *classical* fields which all commute with each other. They're just ordinary c-numbers, and I'm integrating over 'em. This will turn out to be an enormous advantage, and enable us to settle with a single stroke all the problems associated with derivative interactions, superfluous variables, gauge invariance and so on.

As it stands, it doesn't look like the action $\mathcal{S}_c[\phi, J]$, with the Lagrangian (28.24) is the sort of functional integral we can do safely, even without worrying about the infinite dimensions of function space, and even for a free field theory, $\mathscr{L}' = 0$. Instead of a nice positive definite quadratic form in the exponential, or at least a form with a positive definite real part, we have an overall factor of i multiplying $\mathcal{S}_c$: the exponential *oscillates*. Put simply, to make sense of (28.27), we have to continue both sides into *Euclidean space*: to obtain four-vectors with real space components, *imaginary* time components, and a quadratic form of definite sign. We discussed this earlier (§15.5) in the context of analytic continuation of Feynman integrals. I first have to demonstrate that the Green's functions of a quantum field theory can be continued into Euclidean space. Then I will show that the functional integral is well-defined in Euclidean space, (or as well-defined as our other functional integrals have been), and then I'll be able to show that the two sides are equal.

When we were doing loop integrations, we studied the propagator in the q_0 plane and found that it had poles in a typical case as shown in Figure 15.7; either $(|\mathbf{q}|^2 - a) > 0$ or $(|\mathbf{q}|^2 - a) < 0$. In either case, we did not cross any poles if we rotated our contour of integration onto the imaginary axis:

$$q_0 \to e^{i\alpha} q_0 \quad 0 \le \alpha \le \tfrac{\pi}{2} \tag{28.29}$$

leading to the Wick-rotated values

$$\begin{aligned} q_0 &= iq_4 \quad \text{(for loop momenta)} \\ p_0 &= ip_4 \quad \text{(for other momenta)} \end{aligned} \tag{28.30}$$

With these rotated time components, the Lorentz square k^2 of a momentum four-vector k^μ turned into a *negative* Euclidean square k_E^2:

$$k^2 = k_0^2 - \mathbf{k}^2 = -k_4^2 - \mathbf{k}^2 = -k_E^2 = -\sum_{i=1}^{4} k_i^2 \tag{28.31}$$

(We made this rotation after we had performed all the momentum shifts.) If we do this simultaneously to all the external momenta in the problem, and at the same time that we rotate the external momenta in the complex plane we rotate the internal momenta in the complex plane, to preserve energy-momentum conservation at every vertex, this obviously goes through. We end up with a Feynman integral that has no zeros in its denominators anywhere. Everything is the square of a Euclidean vector plus a positive mass squared. The function is not only well-defined, it is an analytic function of the external momenta. When we rotate our external energies in the complex plane this defines an analytic continuation in k-space.

On the left-hand side of (28.27), we have an expression in x-space. It's pretty easy to see what we have to do in x-space to keep things going right in k-space; we have to rotate x in the *opposite* direction:

$$x_0 \to e^{-i\pi/2} x_0 = -ix_4 \tag{28.32}$$

The phase factor in x_0 cancels the phase factor in k_0; otherwise the Fourier transform would develop an exponential blowup. The minus sign is going to be important in making our formulas come out right. So, to all orders in perturbation theory, we can define our Green's functions for Euclidean spacetime separations, where the formal connections between the complex variable x_0 and the real variable x_4 are given above.

It's also possible to give a direct position-space argument to demonstrate that everything can be continued to imaginary time, *without* recourse to perturbation theory. That's sufficiently amusing that I will give it. (See (13.25).)

We want to study a position space Green's function, (13.22):

$$G^{(n)}(x_1, \ldots, x_n) = \langle 0|T\left[\phi_H(x_1)\cdots\phi_H(x_n)\right]|0\rangle \tag{28.33}$$

For convenience, I will assume

$$x_1^0 \ge x_2^0 \ge \cdots \ge x_n^0 \tag{28.34}$$

so that I can drop the time-ordering symbol; things are already time-ordered. Now explicitly pull out the time dependence using the Heisenberg equations of motion:

$$\begin{aligned}\langle 0|[\phi_H(x_1)\cdots\phi_H(x_n)]|0\rangle = \langle 0|\big[\phi_H(\mathbf{x}_1,0)\,e^{-iH(x_1^0-x_2^0)}\phi_H(\mathbf{x}_2,0)\cdots\\ \times\, e^{-iH(x_{(n-1)}^0-x_n^0)}\phi_H(\mathbf{x}_n,0)\big]|0\rangle\end{aligned} \tag{28.35}$$

I can now investigate what happens when I attempt to analytically continue these to imaginary times. It's convenient to introduce a complete set of energy eigenstates

$$H\,|n\rangle = E_n\,|n\rangle\,,\qquad E_n \geq 0 \tag{28.36}$$

and insert a complete set

$$\sum_n |n\rangle\,\langle n| = 1 \tag{28.37}$$

between every pair of field operators; we get, for instance

$$e^{-iH(x_1^0-x_2^0)}\sum_n |n\rangle\,\langle n| = \sum_n e^{-iE_n(x_1^0-x_2^0)}\,|n\rangle\,\langle n| = \sum_n e^{-E_n(x_1^4-x_2^4)}\,|n\rangle\,\langle n| \tag{28.38}$$

Then

$$\begin{aligned}&\langle 0|[\phi_H(x_1)\cdots\phi_H(x_n)]|0\rangle\\ &= \sum_{n_1\cdots n_{r-1}} \langle 0|\phi_H(\mathbf{x}_1,0)|n_1\rangle\,\langle n_1|\phi_H(\mathbf{x}_2,0)|n_2\rangle\cdots\langle n_{r-1}|\phi_H(\mathbf{x}_n,0)|0\rangle\;e^{-E_{n_1}(x_1^4-x_2^4)}\ldots\end{aligned} \tag{28.39}$$

By assumption

$$x_j^0 - x_{j+1}^0 > 0 \quad \text{for all } j \tag{28.40}$$

so that, as we rotate the x_0's downward onto the lower imaginary axis, we get $(-i)\times(-i)$ in the exponent, a *damped* exponential. If the sum converged when the exponential oscillated, with the factor of i, it will converge even better as a damped exponential, with a factor of -1. In fact, this is not just a well-defined function of x but, because of the marvelous exponential, an *analytic function* of x. No matter how many times we differentiate with respect to some x_4, although we get more powers of E, that terrific damped exponential keeps things from blowing up. Notice that if we had tried to rotate in the other way, up to the positive imaginary x-axis, we would have gotten an increasing exponential and would have become extremely nervous at this point.

So, the left-hand side of (28.27) is a Euclidean generating functional, a completely well-defined object. To get an idea of what it looks like, let's compute it first for a free field theory.

28.3 The Euclidean $Z_0[J]$ for a free theory

We know how to compute the free theory generating functional $Z_0[J]$ for (28.24) when $\mathscr{L}' = 0$ in Minkowski space. This theory is nothing but (8.57), our old Model 1 of the three models we considered in the early part of the course (with the replacement $\rho(x)\to J(x)$ and setting $g = 1$). We found in §8.5 that

$$U_I(\infty,-\infty) = \mathrm{S} = e^{\frac{1}{2}(-\alpha+i\beta)}\,{:}\exp(O_1){:} \tag{8.61}$$

and so to within a phase

$$\langle 0|S|0\rangle = Z_0[J] = e^{-\frac{1}{2}\alpha} \tag{28.41}$$

with α given by (S4.1) (see the solution to Problem 4.1, p. 177):

$$Z_0[J] = \exp\left\{-\tfrac{1}{2}\int\frac{d^4k}{(2\pi)^4}\widetilde{J}(-k)\frac{i}{k^2-\mu^2+i\epsilon}\widetilde{J}(k)\right\} \tag{28.42}$$

(recall $\widetilde{J}^*(k) = \widetilde{J}(-k)$). $Z_0[J]$ is the exponential of the sum of connected graphs of the form

$\alpha =$  (28.43)

In the argument of the exponential, one i is from the propagator, two i's come from the $\widetilde{J}$'s and $\frac{1}{2}$ from the combinatoric factor. It will be convenient to write this in position space

$$\begin{aligned} Z_0[J] &= \exp\left\{-\tfrac{1}{2}\int d^4x\, d^4y\, J(x)\Delta_F(x-y)J(y)\right\} \\ &= \exp\left\{\tfrac{1}{2}\int d^4x\, J(x)\frac{i}{\Box^2+\mu^2-i\epsilon}J(x)\right\} \end{aligned} \tag{28.44}$$

where $\Delta_F(x-y)$ is the Feynman propagator (10.29) in position space,[6]

$$\Delta_F(x-y) = \frac{-i}{\Box_x^2+\mu^2-i\epsilon}\delta^{(4)}(x-y) \tag{28.45}$$

We go to Euclidean space by rotating $x_0 \to -ix_4$. The d'Alembertian operator $\Box^2$ and the four-dimensional measure d^4x are transformed to their Euclidean forms (with a subscript E):

$$\Box^2 = \partial_0^2 - \nabla^2 \to -\partial_4^2 - \nabla^2 = -\Box_E^2 \tag{28.46}$$

$$d^4x = dx^0 d^3\mathbf{x} \to -idx^4 d^3\mathbf{x} = -id^4x_E \tag{28.47}$$

Continuing $Z_0[J]$ into Euclidean space, we get

$$Z_E[J] = \exp\left\{\tfrac{1}{2}\int d^4x_E\, J(x)\frac{1}{-\Box_E^2+\mu^2}J(x)\right\} = \exp\left\{\tfrac{1}{2}(J,[-\Box_E^2+\mu^2]^{-1}J)\right\} \tag{28.48}$$

in our compact notation, treating $J(x)$ as a vector in an infinite-dimensional Hilbert space. (Note that there's no need for the $i\epsilon$ in the Euclidean propagator; $-\Box_E^2$ is like $+k_E^2$, so that $(-\Box_E^2+\mu^2)$ is a positive-definite operator.) This looks like what we get from a Gaussian integral; let's check that it is.

---

[6] [Eds.] $\Delta_F(x-y)$ can be thought of as a matrix element of an operator which is formally the inverse of the Klein–Gordon operator,

$$\Delta_F(x-y) = \langle x|\frac{-i}{-p^\mu p_\mu+\mu^2-i\epsilon}|y\rangle$$

Making the usual replacement $p^\mu \to i\partial^\mu$, and inserting a complete set of momentum eigenstates $|k\rangle$, we find

$$\Delta_F(x-y) = \int\frac{d^4k}{(2\pi)^4}\langle x|\frac{-i}{-p^\mu p_\mu+\mu^2-i\epsilon}|k\rangle\langle k|y\rangle = \int\frac{d^4k}{(2\pi)^4}\frac{i}{k^2-\mu^2+i\epsilon}e^{-ik\cdot(x-y)}$$

the usual expression. $\Delta_F(x)$ is sometimes written symbolically as $-i(\Box_x^2+\mu^2-i\epsilon)^{-1}$, as in the right-hand side of (28.48). This notation really means the middle or right-hand side of (28.44).

Start by writing the argument of the exponential in the right-hand side of (28.27) in Minkowski space. We rotate to Euclidean space and perform an integration by parts:

$$\begin{aligned} iS_c[\phi, J] &= i\int d^4x\, \left[\tfrac{1}{2}(\partial_0\phi)^2 - \tfrac{1}{2}(\nabla\phi)^2 - \tfrac{1}{2}\mu^2\phi^2 + J\phi\right] \\ &= \int d^4x_E\, \left[-\tfrac{1}{2}(\partial_4\phi)^2 - \tfrac{1}{2}(\nabla\phi)^2 - \tfrac{1}{2}\mu^2\phi^2 + J\phi\right] \\ &= -\int d^4x_E\, \left[\tfrac{1}{2}\phi(-\Box_E^2 + \mu^2)\phi - J\phi\right] \end{aligned} \tag{28.49}$$

Because $(-\Box_E^2 + \mu^2)$ is a positive-definite operator in Euclidean space, everything damps out nicely. We have a formula for doing Gaussian integrals with a positive-definite matrix or operator, the extension of (28.11):

$$\int (d\phi)\, \exp\left[-\tfrac{1}{2}(\phi, A\phi) + (b, \phi)\right] = \exp\left[\tfrac{1}{2}(b, A^{-1}b)\right]\, (\det A)^{-\frac{1}{2}} \tag{28.50}$$

The argument (28.49) in the exponential of the functional integral (28.27) of is of exactly this form with

$$A = -\Box_E^2 + \mu^2, \qquad b = J \tag{28.51}$$

Then

$$\begin{aligned} Z_E[J] &= N\int (d\phi)\, e^{iS_c[\phi,J]} \\ &= N\det(-\Box_E^2 + \mu^2)^{-\frac{1}{2}} \exp\left\{\tfrac{1}{2}\int d^4x_E\, J(x)\frac{1}{-\Box_E^2 + \mu^2}J(x)\right\} \end{aligned} \tag{28.52}$$

The normalization constant N is chosen so that $Z[0] = 1$:

$$N\det(-\Box_E^2 + \mu^2)^{-\frac{1}{2}} = 1 \tag{28.53}$$

The constant $N = \det(-\Box_E^2 + \mu^2)^{\frac{1}{2}}$ is divergent,[7] but so what? The determinant's divergence is a reflection of the infinite zero-point energy of the free theory if we don't normal order things. Then

$$Z_E[J] = \exp\left\{\tfrac{1}{2}\int d^4x_E J(x)\frac{1}{-\Box_E^2 + \mu^2}J(x)\right\} = \exp\left\{\tfrac{1}{2}(J, [-\Box_E^2 + \mu^2]^{-1}J)\right\} \tag{28.54}$$

the same as (28.48).

We have learned how to continue things into Euclidean space in a simple example, how to do a functional integral of an interesting sort, and we have verified the assertion that (28.27) is valid in the simple case of a free field. Although these integrals are really defined in Euclidean space, we will adopt a construction that treats them as if they are defined in Minkowski space.

---

[7] [Eds.] The calculation of $\det A = \det(-\Box_E^2 + \mu^2)$ is given in Greiner & Reinhardt *FQ*, pp. 377–378, equation (12.57). Using the identity $\det A = \exp(\mathrm{Tr}[\ln A])$, (see Arfken & Weber *MMP*, Chapter 3, "Determinants and Matrices", p. 224, equation (3.171) for a proof of this identity)

$$\det A = \mathrm{Tr}[\ln A] = \int d^4x_E \int \frac{d^4p_E}{(2\pi)^4}\ln(p_E^2 + \mu^2)$$

See also Problem 17.1, p. 679.

28.4 The Euclidean $Z[J]$ for an interacting field theory

Having verified (28.27) for the case of a free field, it's trivial to verify the formula for the case of an interacting field. We will show that our formula is precisely equivalent to Dyson's formula (8.9):

$$Z[J] = N'\,{}^{B}\langle 0|\,T\,[e^{-i\int d^4x\,\mathscr{H}_I}]|0\rangle^{B} = N'\,{}^{B}\langle 0|\,T\,[e^{i\int d^4x\,(\mathscr{L}'(\phi_I)+J\phi_I)}]|0\rangle^{B} \tag{28.55}$$

where $|0\rangle^B$ denotes the bare vacuum and ϕ_I is the field in the interaction picture. This is ordinary perturbation theory before the application of Wick's theorem. Using the trick (28.13), recast for functional integrals, we can write this as

$$\begin{aligned} Z[J] &= N' \exp\left[i\int d^4z\,\mathscr{L}'(-i\frac{\delta}{\delta J(z)})\right]\,{}^{B}\langle 0|\,T\,[e^{i\int d^4x\,J\phi_I}]|0\rangle^{B} \\ &= N' \exp\left[i\int d^4z\,\mathscr{L}'(-i\frac{\delta}{\delta J(z)})\right]\,Z_0[J] \end{aligned} \tag{28.56}$$

where

$$Z_0[J] = \exp\{-\tfrac{1}{2}(J, \frac{i}{-\Box^2-\mu^2+i\epsilon}J)\} \tag{28.57}$$

is the generating functional for noninteracting Green's functions. (We needn't worry about ordering of operators within the differentiation; the time ordering takes care of that.) This is the left-hand side of the functional integral, (28.27). Now for the right-hand side. Writing $\mathcal{S}_0$ for the free part plus the source term in the action,

$$\mathcal{S}_0[\phi, J] = \int d^4x\,[\mathscr{L}_0 + J\phi] \tag{28.58}$$

the functional integral on the right-hand side can be written as[8]

$$\begin{aligned} Z[J] &= N\int (d\phi)\,e^{i\mathcal{S}[\phi,J]} = N\int (d\phi)\,e^{i\mathcal{S}_0[\phi,J]}e^{i\int d^4y\,\mathscr{L}'(\phi(y))} \\ &= N\exp\left[i\int d^4y\,\mathscr{L}'(-i\frac{\delta}{\delta J(y)})\right]\underbrace{\int (d\phi)\,e^{i\mathcal{S}_0[\phi,J]}}_{N_0^{-1}Z_0[J]} \end{aligned} \tag{28.59}$$

It's the same trick as we used on the left-hand side. In one case, the exponential $e^{i\int d^4x\,J\phi_I}$ is interpreted as an operator inside a time ordering symbol; in the other case, the exponential $e^{i\mathcal{S}_0[\phi,J]}$ is a function of c-number fields inside an integral. But the trick works the same in either case. Equations (28.56) and (28.59) are the same if we choose

$$N' = \frac{N}{N_0} \tag{28.60}$$

Thus I have proved the startling assertion that I made earlier: that you can, for this particular kind of interaction, represent the generating functional, the thing that tells you everything you want to know about the theory, in terms of a functional integral.

---

[8] [Eds.] Remember that everything here is a classical quantity so we don't need to worry about commutators.

We have been doing integrals blithely in Minkowski space. It is easy to rotate from Minkowski to Euclidean space and back again. If there's ever an ambiguity in the Minkowski space integral, we have to be more careful. Such ambiguities can occur. If we had tried to do the free-field case directly in Minkowski space, we would have encountered $(\Box^2+\mu^2)^{-1}$, an ill-defined object. We would have to stick in an $i\epsilon$ to make it a well-defined object. There is nothing in our functional integral formulas to tell us if it's $+i\epsilon$ or $-i\epsilon$. The $i\epsilon$ is introduced automatically by continuing back from Euclidean to Minkowski space. We know which way we have to continue back on general principles, and that puts in the $+i\epsilon$. The functional integral is properly done in Euclidean space where there is no ambiguity in finding the inverse of the operator.

These formulas generalize to any theory of non-derivative interactions. Here are some simple extensions:

1. Set of scalar fields

$$\mathscr{L}=\tfrac{1}{2}(\partial_\mu\phi^a)(\partial^\mu\phi^a)-U(\phi^1\cdots\phi^n) \tag{28.61}$$

(summation implied over repeated indices). U contains both mass terms and a non-derivative interaction. The action is

$$\mathcal{S}_c[\phi^1,\cdots\phi^n,J_1,\cdots J_n]=\int d^4x\,(\mathscr{L}+J_a\phi^a) \tag{28.62}$$

and the generating functional is

$$Z[J_1,\cdots J_n]=\langle 0|\mathrm{S}|0\rangle_J=\int\prod_a(d\phi^a)e^{i\mathcal{S}_c[\phi^a,\,J_a]} \tag{28.63}$$

2. Complex fields

As usual, we assemble real fields pairwise into complex fields:

$$\left.\begin{matrix}\phi\\ \phi^*\end{matrix}\right\}=\frac{\phi_1\pm i\phi_2}{\sqrt{2}},\ \text{etc} \tag{28.64}$$

$$(d\phi_1)(d\phi_2)=(d\phi)(d\phi^*),\ \text{etc.} \tag{28.65}$$

This is not a big generalization, but it is convenient if $\mathscr{L}$ can be written in terms of complex fields.

3. Beyond Minkowski space

The formalism of functional integrals is not restricted only to four dimensions; we can have any integer number of dimensions. In particular, it works for (quantum) particle mechanics in one dimension, say, an assembly of harmonic oscillators with perhaps anharmonic interactions:

$$L=\tfrac{1}{2}\dot{q}^a\dot{q}^a-V(q^1\cdots q^n) \tag{28.66}$$

$$\mathcal{S}_c=\int dt\{L+J_a(t)q^a(t)\} \tag{28.67}$$

We can define the ground state to ground state Green's function in the usual way. The S-matrix is

$$\langle 0|\mathrm{S}|0\rangle_J=\int\prod_a(dq^a)\,e^{i\mathcal{S}_c[q^a,\,J_a]} \tag{28.68}$$

That's the restriction from integrating over the four dimensions of a field theory to integrating over no space dimensions and one time dimension for particles. The form (28.68) of the functional integral is in fact more general than (28.27), because if a runs over an infinite set, say the Fourier components of the field, we can always think of a field theory as a special case of a particle theory. The particle form only requires a restriction on how the time derivatives of the fields enter L; how the space derivatives of the fields enter L is irrelevant. V may be a complicated interaction between the Fourier components but it doesn't involve time derivatives. We will use the particle language when discussing derivative interactions next time, because it is more general.

28.5 Feynman rules from functional integrals

We have found a functional integral representation for the generating functional, $Z[J]$, which we have shown is parallel to the original development from Dyson's formula. Originally we went through a long journey from Dyson's formula to derive the Feynman rules: Wick's theorem, diagrammatic representation of the terms in the Wick expansion, we danced, we stood on our heads, finally we found the Feynman rules. I would like to demonstrate that we can also get the Feynman rules by directly manipulating the functional integral, just to show you the utility of functional integrals.

Recall our functional integral formula for $Z[J]$, (28.56), in the case of a single scalar field for the generating functional $Z[J]$:

$$Z[J] = N' \exp\left[i\int d^4z\, \mathscr{L}'(-i\frac{\delta}{\delta J(z)})\right] Z_0[J]$$

I'd like to write out $Z_0[J]$ a little more explicitly than before:

$$Z_0[J] = \exp\left\{-\tfrac{1}{2}\int d^4x\, d^4y\, J(x)\Delta_F(x-y)J(y)\right\} \tag{28.69}$$

The Feynman propagator $\Delta_F(x-y)$ is just shorthand for what we called the contraction in our earlier discussion. It's the Fourier transform of the propagator in momentum space:

$$\Delta_F(x-y) = \int \frac{d^4k}{(2\pi)^4} e^{-ik\cdot(x-y)} \frac{i}{k^2-\mu^2+i\epsilon} \tag{28.70}$$

Let's expand $Z_0[J]$. We get zero propagators, one propagator with two J's, two propagators with four J's, and so on:

$$\begin{aligned} Z_0[J] &= \exp\left\{\tfrac{1}{2}\int d^4x\, d^4y\, \big(iJ(x)\big)\,\Delta_F(x-y)\,\big(iJ(y)\big)\right\} \\ &= \sum_n \left(\tfrac{1}{2}\right)^n \frac{1}{n!}\int d^4x_1\cdots d^4x_n\, d^4y_1\cdots d^4y_n \Big[\big(iJ(x_1)\big)\,\Delta_F(x_1-y_1)\,\big(iJ(y_1)\big)\cdots \\ &\qquad \times \big(iJ(x_n)\big)\,\Delta_F(x_n-y_n)\,\big(iJ(y_n)\big)\Big] \end{aligned} \tag{28.71}$$

As an example, graphically the fourth-order term of (28.71) is represented as

$Z_0[J]^{(4)} =$ (28.72)

where there is a J attached to each end point and a Feynman propagator between end points.

Now for the other part of (28.56). The first exponential involves the expression

$$\mathscr{L}'\left(-i\frac{\delta}{\delta J(y)}\right) \tag{28.73}$$

Every $-i\delta/\delta J(y)$ will knock off an $iJ(x_k)$, using

$$\frac{\delta J(x_k)}{\delta J(y)} = \delta^{(4)}(x_k - y), \tag{28.74}$$

the generalization of $\partial x^a/\partial x^b = \delta^a{}_b$. If for example the interaction is a simple cubic interaction,

$$\mathscr{L}'(\phi) = g\phi^3 \tag{28.75}$$

then the first nontrivial term in (28.59) will involve

$$ig \int d^4y \left(-i\frac{\delta}{\delta J(y)}\right)^3 \tag{28.76}$$

acting on the diagrams in (28.72). When this term hits the set of graphs above, three of the end points will have the J's removed. That results in two different sets of graphs. The first looks like this:

$$\left(\frac{\delta}{\delta J} \text{ on each of three pieces}\right) = $$  (28.77)

and those three free end points will join together into a three-particle vertex:

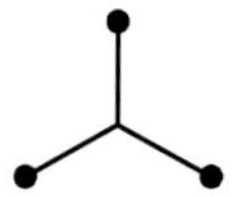

(28.78)

The whole thing will be multiplied by ig, and there will also be a combinatorial factor arising from the freedom of deciding which end points are differentiated and joined together. The second set is

$$\left(\frac{\delta^2}{\delta J^2} \text{ on one piece, } \frac{\delta}{\delta J} \text{ on another}\right) = $$  (28.79)

The three free ends are joined together to make a tadpole diagram:

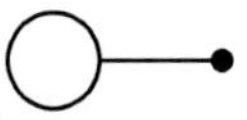

(28.80)

At higher orders in the expansion of (28.56), the functional integral recreates the Feynman rules without the need for any explicit normal ordering. For instance, in second order of $\mathscr{L}'$ we obtain diagrams including these:

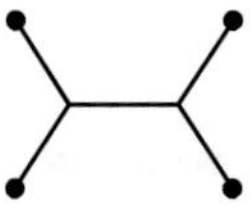

(28.81)

and

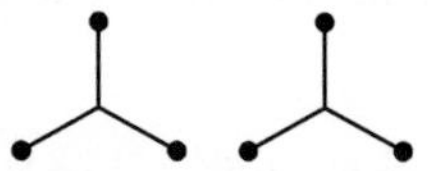

(28.82)

So we get a string of terms that we can represent as propagators with J's, and a second string that we can represent as vertices. Differentiating with respect to the J's will connect the free propagators in all possible ways to make the diagrams. In higher orders in the expansion of (28.59), the functional integral recreates the Feynman rules; this expression *is* Wick's theorem, in a very compact notation. The operation of contraction becomes here the operation of differentiation. Whatever the Gaussian is, it defines a propagator; the polynomial $\mathscr{L}'(\phi)$ defines the vertices. Then we just stick things together according to the Feynman rules.

Suppose that I have, by dint of hard work, an expression of the following form (it generalizes trivially from one field to many):

$$Z[J] \equiv \int [d\phi]\, e^{i[\frac{1}{2}(\phi, A\phi)+(J,\phi)+\mathcal{S}'(\phi)]} \tag{28.83}$$

where A is any differential operator, $\mathcal{S}'$ is any function of ϕ, with 47 derivatives in it and a non-local integral kernel of 17 variables, just some horrible mess. I write down naive interactions corresponding to $\mathcal{S}'$; there's no time ordering, because these are just classical c-number fields, and everything commutes with everything else. The propagator D_F is the inverse of A with the appropriate factor of i. It is the solution of this equation

$$AD_F(x-y) = i\delta^{(4)}(x-y) \tag{28.84}$$

the inverse operator written as an integral kernel of two variables, x and y. Any ambiguities that arise in inverting the differential equation are to be resolved by continuing into Euclidean space where the functional integral is really supposed to be defined; that will tell us where to put in the $i\epsilon$.

This equation (28.83) doesn't mean much yet, because the only classical field theory for which we know how to write the functional integral is the one for which we already know the Feynman rules. We will soon encounter more complicated field theories where we do not know the Feynman rules, but for which we will nevertheless be able to write the generating functional as a functional integral. Once we do that we can use (28.84) to find D_F as the inverse of A. We can forget the explicit formulas about integrating a Gaussian, or a polynomial times a Gaussian, etc. We will naively read off the interactions from the functional integral: a derivative gives a factor of momentum, and so on. This is one of the utilities of functional integrals. They will allow us to handle all the terrible problems with derivative interactions and anything else. If we can only get the theory in the form of (28.83), then we can just read off the Feynman rules. Never again will we have to worry about Dyson's formula, Wick's theorem or anything else. This will be the magic method.

28.6 The functional integral for massive vector mesons

As an example of how this method works its magic, I will use it in a case where we have not yet justified it. Let me suppose it is true for the theory of massive vector mesons, a theory for which we can write the generating functional as a functional integral. (I will prove that later on.) To give an example of this algorithm at work let me assume I have proved it and I will attempt to construct the propagator. To avoid confusion between the operator A and the vector field A_μ, I will write the vector field as B_μ. The propagator is given by the free action (26.47), with $F_{\mu\nu}$ given by (26.11):

$$\begin{aligned} S_0(B_\mu) &= \int d^4x \left[-\tfrac{1}{4}F_{\mu\nu}F^{\mu\nu} + \tfrac{1}{2}\mu^2 B_\mu B^\mu\right] \\ &= \int d^4x \left[-\tfrac{1}{2}\partial_\mu B_\nu \left(\partial^\mu B^\nu - \partial^\nu B^\mu\right) + \tfrac{1}{2}\mu^2 B_\mu B^\mu\right] \\ &= \int d^4x \left[\tfrac{1}{2}B_\mu \left(g^{\mu\nu}\Box^2 - \partial^\mu\partial^\nu + \mu^2 g^{\mu\nu}\right) B_\nu\right] \\ &\equiv \int d^4x \, \tfrac{1}{2}B_\mu A^{\mu\nu} B_\nu \end{aligned} \tag{28.85}$$

I have used the antisymmetry of $F_{\mu\nu}$ in going to the second line. The operator $A^{\mu\nu}$, which we have to invert as a matrix differential operator, is

$$A^{\mu\nu} = g^{\mu\nu}\Box^2 - \partial^\mu\partial^\nu + \mu^2 g^{\mu\nu} \tag{28.86}$$

The matrix Green's function for the differential operator $A^{\mu\nu}$ is defined by

$$A^{\mu\nu} D^P_{\nu\rho}(x-y) = i\delta^\mu{}_\rho \, \delta^{(4)}(x-y) \tag{28.87}$$

(The superscript P is for Proca.) If the Green's function is ambiguous, if the problem does not have a unique solution, we settle the ambiguity by adding an $i\epsilon$ so we can rotate into Euclidean space. This is the prescription generalized to a field with many components.

The solution to (28.87), like all differential equations with constant coefficients, is most easily found in momentum space:

$$D^P_{\mu\nu}(x) = \int \frac{d^4k}{(2\pi)^4} e^{-ik\cdot x} \widetilde{D}^P_{\mu\nu}(k) \tag{28.88}$$

Recalling that

$$\partial_\mu \to -ik_\mu \qquad \Box^2 \to -k^2 \tag{28.89}$$

we get, applying $\widetilde{A}^{\mu\nu}$ to (28.88),

$$\left(-k^2 g^{\mu\nu} + k^\mu k^\nu + \mu^2 g^{\mu\nu}\right) \widetilde{D}^P_{\nu\rho}(k) = i\delta^\mu{}_\rho \tag{28.90}$$

We have a 4×4 matrix equation that we have to invert. We break it up into the sum of two projection operators, the transverse and longitudinal projection operators, $P^T_{\mu\nu}$ and $P^L_{\mu\nu}$, respectively:

$$P^T_{\mu\nu} = g_{\mu\nu} - \frac{k_\mu k_\nu}{k^2} \qquad \text{(projection operator onto vectors orthogonal to } k^\mu\text{)} \tag{28.91}$$

$$P^L_{\mu\nu} = \frac{k_\mu k_\nu}{k^2} \qquad \text{(projection operator onto vectors aligned along } k^\mu\text{)} \tag{28.92}$$

These have the usual properties of projection operators—their sum equals the identity, they're idempotent, and they're orthogonal to each other:

$$P^T_{\mu\nu} + P^L_{\mu\nu} = g_{\mu\nu} \tag{28.93a}$$

$$P^T_{\mu\nu} P^{T\nu}{}_\rho = P^T_{\mu\rho} \tag{28.93b}$$

$$P^L_{\mu\nu} P^{L\nu}{}_\rho = P^L_{\mu\rho} \tag{28.93c}$$

$$P^T_{\mu\nu} P^{L\nu}{}_\rho = P^L_{\mu\nu} P^{T\nu}{}_\rho = 0 \tag{28.93d}$$

Since $\widetilde{A}_{\mu\nu}$ is a linear combination of $g^{\mu\nu}$ and $k^\mu k^\nu$, it can be written as a linear combination of the transverse and longitudinal projection operators:

$$\widetilde{A}_{\mu\nu} = \left(-k^2 g_{\mu\nu} + k_\mu k_\nu + \mu^2 g_{\mu\nu}\right) = (-k^2+\mu^2)P^T_{\mu\nu} + \mu^2 P^L_{\mu\nu} \tag{28.94}$$

Then it's very easy to solve (28.90): we just invert the coefficients of the two projection operators, and multiply by i:

$$\widetilde{D}^P_{\mu\nu}(k) = i\frac{P^T_{\mu\nu}}{-k^2+\mu^2} + i\frac{P^L_{\mu\nu}}{\mu^2} \tag{28.95}$$

There is an ambiguity in the first term which I will resolve shortly by going into Euclidean space. We can easily check that $\widetilde{D}^P_{\mu\nu}(k)$ is the inverse by multiplying by $\widetilde{A}_{\mu\nu}$:

$$\left((-k^2+\mu^2)P^{T\mu\nu} + \mu^2 P^{L\mu\nu}\right) i\left(\frac{P^T_{\nu\rho}}{-k^2+\mu^2} + \frac{P^L_{\nu\rho}}{\mu^2}\right) = i\left[P^{T\mu}{}_\rho + P^{L\mu}{}_\rho\right] = i\delta^\mu{}_\rho \tag{28.96}$$

The propagator is

$$\begin{aligned}\widetilde{D}^P_{\mu\nu}(k) &= i\left[\frac{g^{\mu\nu} - (k^\mu k^\nu)/k^2}{-k^2+\mu^2} + \frac{1}{\mu^2}\frac{k^\mu k^\nu}{k^2}\right] = \frac{-i}{k^2-\mu^2}\left[g^{\mu\nu} - \frac{k_\mu k_\nu}{\mu^2}\right] \\ &\to \frac{i}{k^2-\mu^2+i\epsilon}\left[-g^{\mu\nu} + \frac{k_\mu k_\nu}{\mu^2}\right]\end{aligned} \tag{28.97}$$

This *is* the correct expression for a massive vector field's propagator. (I had to resolve the ambiguity when $k^2 = \mu^2$ with an $i\epsilon$. This $i\epsilon$ is also necessary if I want to rotate the propagator into Euclidean space: I have to avoid any poles.)

Note that (28.97) has the same general form as we found in the case of a free fermion (21.76), or a free spinless particle (10.29), a fraction formed from a projection operator divided by the particle's momentum squared minus its mass squared. For free spinless particles, the numerator was 1, which we can think of as the projection operator onto the one physically allowed state. For the fermion we had $\not{p} + m$, the projection operator onto the physically allowed states. Here we have

$$-g_{\mu\nu} + \frac{k_\mu k_\nu}{\mu^2} \tag{28.98}$$

which is nothing but the projection operator $P^T_{\mu\nu}$ onto the three allowed transverse polarization vectors. All these expressions have the same form.

Even after we derive the Feynman rules using this magic method, we will still have troubles. For one thing, $(k_\mu k_\nu)/\mu^2$ does not appear to have a smooth limit as $\mu \to 0$. For another, $(k_\mu k_\nu)/\mu^2$ looks badly behaved as $k \to \infty$; maybe it spoils the renormalizability of the theory. It doesn't, but we will have to check that.

Next time I will redeem this computation by showing that I can indeed write the generating functional for electrodynamics as a functional integral, even when we have to eliminate degrees of freedom from $\mathscr{L}$ before we can write the integral in Hamiltonian form. We will also consider scalar electrodynamics, so we're going to have to worry about theories with derivative couplings. And I haven't said anything at all about using the functional integral formalism for fermions. That will lead us into rather peculiar waters. For bosons, we write the functional integral over

Bose fields, that is, classical c-number fields, the limit of quantum Bose fields as $\hbar \to 0$. This would lead you to suspect, if you were bold at guessing, that for a theory with fermions, you write things as a functional integral over classical Fermi fields. But what are classical Fermi fields? We will model these with Grassmann variables, anti-commuting c-numbers,[9] and we will learn how to do calculus with them.

[9] [Eds.] See note 3, p. 434.

29

Extending the methods of functional integrals

We're going to discuss four cute topics in functional integration that extend the method from Bose fields and non-derivative interactions to Fermi fields, derivative interactions, and constrained variables.

29.1 Functional integration for Fermi fields

So far, we've only dealt with functional integrals for bosonic systems, where all the dynamical variables, whether quantum mechanical or classical, have commutation properties, and are represented by c-number fields, the classical limits of Bose fields. To define functional integrals for fermionic systems, whose dynamical variables have anticommutation properties, we need to introduce anticommuting c-numbers, also known as **Grassmann variables.**[1]

Introduce the anticommuting quantities

$$\eta, \xi, \overline{\eta}, \overline{\xi} \tag{29.1}$$

with the properties that any two of them anticommute:[2]

$$\{\eta, \xi\} = \eta\xi + \xi\eta = 0, \quad \{\overline{\eta}, \xi\} = \overline{\eta}\xi + \xi\overline{\eta} = 0, \quad \text{etc.} \tag{29.2}$$

These imply that the square of any Grassmann variable vanishes:

$$\eta^2 = \overline{\eta}^2 = 0, \quad \text{etc.} \tag{29.3}$$

[1] [Eds.] See footnote 3 on p. 434. Here, "c-number" means "classical number", as opposed to "q-number", or "quantum number", typically an operator of some sort. The nomenclature is due to Dirac: P. A. M. Dirac, "Quantum Mechanics and a Preliminary Investigation of the Hydrogen Atom", *Proc. Roy. Soc.* **110** (1926) 561–579; the terms appear on p. 562. Initially "c-number" was meant to imply a commuting number, so the concept of an anticommuting c-number seems a contradiction in terms; as Coleman remarks (*Aspects*, p. 156), "Anticommuting c-numbers are notoriously objects that make strong men quail."

[2] [Eds.] The bar, suggesting a Dirac adjoint (20.62), is a little misleading; these objects are eventually going to be *scalar* fields that obey *Fermi* statistics! It would be less confusing to write them as η and η^*, but that is not Coleman's notation, which we follow here. (The bar may help remind the reader that η and $\overline{\eta}$, like Fermi fields, anti-commute.)

As a result, the Taylor expansion of any function of Grassmann variables will have only a finite number of terms. In particular,

$$e^{\eta} = 1 + \eta + \tfrac{1}{2!}\eta^2 + \tfrac{1}{3!}\eta^3 + \cdots = 1 + \eta \tag{29.4}$$

because the square of any Grassmann variable is zero.

I will impose some conditions on the integrals of these quantities.[3]

1. *Linearity:*

$$\begin{gathered}\int d\eta\,[\alpha F_1(\eta) + \beta F_2(\eta)] = \pm\alpha \int d\eta\, F_1(\eta) \pm \beta \int d\eta\, F_2(\eta) \\ \int d\eta\, d\overline{\eta}\,\big[\alpha F_1(\eta,\overline{\eta}) + \beta F_2(\eta,\overline{\eta})\big] = \alpha \int d\eta\, d\overline{\eta}\, F_1(\eta,\overline{\eta}) + \beta \int d\eta\, d\overline{\eta}\, F_2(\eta,\overline{\eta})\end{gathered} \tag{29.5}$$

The quantities α and β can be any constants (independent of η and $\overline{\eta}$), even Grassmann numbers; the minus signs are appropriate in the first integral if α and β are Grassmann variables. When we take α and β out of the double integral, there is no change of sign in either case, since they are going past both $d\eta$ and $d\overline{\eta}$.[4]

2. *Translation invariance:*

$$\int d\eta\, F(\eta) = \int d\eta\, F(\eta + \xi) \tag{29.6}$$

where ξ is another Grassmann variable. This is analogous to the translation invariance of the integral $\int_{-\infty}^{\infty} dx\, f(x)$; all Grassmann integrals are definite, with limits unchanged by translation.

3. *Normalization:* To make things simple later, I impose the normalization condition

$$\int d\eta\, d\overline{\eta}\, e^{\overline{\eta}\eta} = 1 \tag{29.7}$$

You may be surprised that I've chosen a normalization condition which, with ordinary numbers, would be an increasing exponential rather than a decreasing exponential. But of course that's irrelevant. Remember

$$\overline{\eta}\eta = -\eta\overline{\eta} \tag{29.8}$$

so who knows what's increasing and what's decreasing?

---

[3] [Eds.] For an accessible overview of the algebra and calculus of Grassmann variables, see L. D. Faddeev and A. A. Slavnov, *Gauge Fields: Introduction to Quantum Theory*, Benjamin/Cummings, 1982, Section 2.4, pp. 49–55. For more about integration of Grassmann variables, see Peskin & Schroeder *QFT*, pp. 299–300. The definitions $\int d\eta = 0$, $\int d\eta\,\eta = 1$ are due to Berezin: F. A. Berezin, *The Method of Second Quantization*, Academic Press, 1966, p. 52, equation (3.10).

[4] [Eds.] Coleman defines the integration measure in the opposite order, $d\overline{\eta}\, d\eta$, which differs by a minus sign from this choice. This will affect how some functions of the Grassmann variables are represented. One reason for reversing the order is that, for Grassmann variables, *integration is the same operation as differentiation.* With the order $d\eta\, d\overline{\eta}$, we have

$$\int d\eta\, d\overline{\eta}\, e^{\overline{\eta}\eta} = \frac{\partial}{\partial\eta}\frac{\partial}{\partial\overline{\eta}}\,[1 + \overline{\eta}\eta] = 1$$

Using Coleman's order, the derivatives give -1. In a moment he will use this integral (29.7) as a normalization condition.

I will now show that these three requirements taken together uniquely determine an integral of the form

$$\int d\eta\, d\overline{\eta}\, f(\eta,\overline{\eta})$$

The key point is that I can get a complete integral table for any function of η and $\overline{\eta}$ defined by a power series. The Taylor expansion of $g(\eta)$ can only involve two terms:

$$g(\eta) = A + B\eta \tag{29.9}$$

because all higher powers of η vanish. Likewise, the Taylor expansion of $f(\eta,\overline{\eta})$ can have only four terms, each proportional to one of these four expressions:

$$\{1, \eta, \overline{\eta}, \overline{\eta}\eta\} \tag{29.10}$$

Any other string of η's and $\overline{\eta}$'s will be zero if it has more than one η or more than one $\overline{\eta}$ in it. So the Grassmann integral table will have only four entries. Once I know how to integrate the set (29.10), I can integrate any function $f(\eta,\overline{\eta})$ defined by its Taylor series. So let's determine the table.

By condition 2,

$$\int d\eta\, g(\eta) = \int d\eta\, g(\eta+\xi) \tag{29.11}$$

Expanding g as a Taylor series,

$$\int d\eta\,(A+B\eta) = \int d\eta\,(A+B\eta+B\xi) \tag{29.12}$$

By condition 1,

$$B\xi\int d\eta = 0\ ,\ \text{so}\ \int d\eta = 0 \tag{29.13}$$

because ξ and B are arbitrary. By the same reasoning,

$$\int d\overline{\eta} = 0 \tag{29.14}$$

It follows

$$\begin{aligned}\int d\eta\, d\overline{\eta}\, 1 &= \int d\eta \left[\int d\overline{\eta}\, 1\right] = 0\\ \int d\eta\, d\overline{\eta}\, \eta &= -\int d\overline{\eta} \left[\int d\eta\, \eta\right] = 0\\ \int d\eta\, d\overline{\eta}\, \overline{\eta} &= \int d\eta \left[\int d\overline{\eta}\, \overline{\eta}\right] = 0\end{aligned} \tag{29.15}$$

because the quantities in the brackets are independent of the integration variables. Finally, by condition 3,

$$1 = \int d\eta\, d\overline{\eta}\, e^{\overline{\eta}\eta} = \int d\eta\, d\overline{\eta}\,(1+\overline{\eta}\eta) = \int d\eta\, d\overline{\eta}\, \overline{\eta}\eta = \left[\int d\eta\, \eta\right]\left[\int d\overline{\eta}\, \overline{\eta}\right] \tag{29.16}$$

Assuming $\int d\eta\, \eta > 0$, we get in fact two integral tables. For a single Grassmann variable,

$$\int d\eta \begin{Bmatrix}1\\ \eta\end{Bmatrix} = \begin{Bmatrix}0\\ 1\end{Bmatrix} \tag{29.17}$$

and for a Grassmann variable and its conjugate,

$$\int d\eta\, d\overline{\eta} \begin{Bmatrix} 1 \\ \eta \\ \overline{\eta} \\ \overline{\eta}\eta \end{Bmatrix} = \begin{Bmatrix} 0 \\ 0 \\ 0 \\ 1 \end{Bmatrix} \tag{29.18}$$

The rule for integration of Grassmann variables is simple: it's the *same operation as differentiation.*

We can now do more complicated integrals, such as a Gaussian. In this, a is an ordinary commuting number:

$$\int d\eta\, d\overline{\eta}\, e^{a\overline{\eta}\eta} = \int d\eta\, d\overline{\eta}\,(1 + a\overline{\eta}\eta) = a \tag{29.19}$$

Contrast this with the result we would have with ordinary complex numbers:

$$\int dz\, dz^{\star}\, e^{-az^{\star}z} = \frac{2\pi}{a} \tag{29.20}$$

In (29.19) we get an a; in (29.20) we get an a^{-1}. This can be viewed as a consequence of the Taylor series terminating after a finite number of terms. We will shortly give another explanation for this difference. The 2π is irrelevant; that's just because we've normalized the two integrals differently.

We are now ready to do a general Gaussian in many anticommuting variables, $\eta_1, \cdots, \eta_n$. Consider the quadratic form

$$(\overline{\eta}, A\eta) = \sum_{i,j=1}^{n} \overline{\eta}_i A_{ij} \eta_j = \sum_i a_i \overline{\eta}_i \eta_i \tag{29.21}$$

The matrix elements of A are commuting quantities, and the set $\{a_i\}$ are A's eigenvalues. (A may involve bilinear forms of anticommuting quantities that are not variables of integration.) Now define the n-dimensional Grassmann measure

$$(d\eta)(d\overline{\eta}) = d\eta_1 d\overline{\eta}_1 \cdots d\eta_n d\overline{\eta}_n = \prod_r d\eta_r d\overline{\eta}_r \tag{29.22}$$

so that

$$\int (d\eta)(d\overline{\eta})\, e^{(\overline{\eta}, A\eta)} = \int (d\eta)(d\overline{\eta}) \prod_r e^{a_r \overline{\eta}_r \eta_r} = \prod_r a_r = \det A \tag{29.23}$$

Here we get the determinant, $\det A$. Recall (28.18),

$$\int (d\mathbf{z})(d\mathbf{z}^{*})\, e^{-(\mathbf{z}^{*},\, A\mathbf{z})} = (\det A)^{-1}$$

The only difference between integrating Gaussians over anticommuting numbers and integrating Gaussians over commuting numbers is that in the first case the *determinant* appears and in the second case the *inverse* of the determinant appears. From this we can derive rules for integrating generalized quadratic forms, polynomials of anticommuting c-numbers times exponentials, etc.

Now we take the same bold, slovenly step we made in the commuting case to go from a finite-dimensional vector space over anticommuting numbers, where we have proved everything, and extend it blithely to an infinite-dimensional space, i.e., a function space. With the integral defined, does the same formula (28.59) give us the sum of all the Feynman diagrams for theories with Fermi fields? I will restrict myself to theories that are no higher than second order in Fermi fields, just for simplicity; this will be sufficient for our purposes.

Consider a Fermi field; it could be a single field ψ or a multi-component field

$$\psi = \begin{pmatrix} \psi_1 \\ \vdots \\ \psi_N \end{pmatrix}, \qquad \overline{\psi} = (\overline{\psi}_1, \cdots \overline{\psi}_N) \tag{29.24}$$

For notational simplicity I will just call it $\psi(x)$. Let $\phi(x)$ be a bosonic field. Assume

$$\mathcal{S} = \int d^4x \left[\overline{\psi}(x) A(\phi)\psi(x)\right] + \mathcal{S}_B(\phi) = (\overline{\psi}, A(\phi)\psi) + \mathcal{S}_B[\phi(x)] \tag{29.25}$$

$\mathcal{S}_B$ is the purely bosonic part. $A(\phi)$ is a combination of terms. It may involve the kinetic energy, $\overline{\psi}(i\not{\partial} - m)\psi$, and functions of scalar or pseudoscalar or vector fields in the theory, ϕ or A_μ or ϕ^{15} or whatever. As an example, we could have

$$\overline{\psi}(x)A(\phi)\psi(x) = \overline{\psi}(x)\left(i\not{\partial} - m + i\gamma_5\phi(x)\right)\psi(x) \tag{29.26}$$

I want to consider the functional integral

$$N\int (d\psi)(d\overline{\psi})e^{i\mathcal{S}} \propto \det A \tag{29.27}$$

to within some normalization constant. For the moment, we are only interested in the Fermi part of the integral, and we can treat the Bose fields as fixed. (We can do the Bose integrals after we've integrated over the Fermi fields.) This is the most general case. Does (29.27) equal what we would obtain, aside from a normalization constant, from conventional Feynman perturbation theory? With that approach, we know how to get the vacuum-to-vacuum amplitude: we would compute the connected Feynman diagrams, including the effects of whatever Bose fields are in A, treated as external fields. The big dot represents whatever vertex is generated by A:

$\langle 0|\mathrm{S}|0\rangle = \exp$ 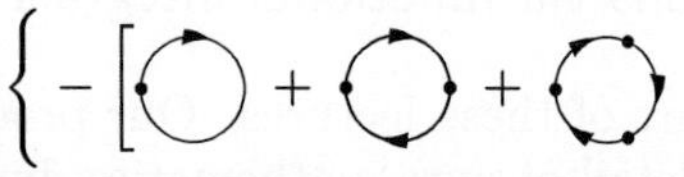 (29.28)

The overall minus sign arises because every one of these diagrams involves one and only one Fermi loop and therefore we get exactly one minus sign from each loop. That is the sum of the connected Feynman diagrams. The exponential of the sum of the connected Feynman diagrams is certainly the right answer for the vacuum-to-vacuum amplitude, $\langle 0|\mathrm{S}|0\rangle$. Is this the same as the fermionic functional integral?

$$N\int (d\psi)(d\overline{\psi})e^{i\mathcal{S}} \propto \det A \stackrel{?}{=} \langle 0|\mathrm{S}|0\rangle \tag{29.29}$$

It would be very nice if the question mark could be removed. It would make life simple in summing up all those loop diagrams, provided you can compute the determinant of an operator in an infinite-dimensional space. Can we do it without actually summing up the diagrams? Well, yes, we can.

Consider the case where ψ is a *Bose* field; a field which has exactly the same dynamics, exactly as many components, except we quantize it according to Bose statistics rather than Fermi statistics. That gives us a sick field theory, by the spin-statistics theorem.[5] Nevertheless, the Feynman rules for such a theory would be well-defined, and the sum of the Feynman diagrams would be well-defined. In the Bose case, we don't have the Fermi minus sign so we would get ((28.50) and (28.18))

$$N\int (d\psi)(d\psi^*)\, e^{iS} \propto (\det A)^{-1} \tag{29.30}$$

That's on the functional integral side. On the diagrammatic side we would get exactly the same exponential of exactly the same sum of diagrams, except we wouldn't have the Fermi minus sign. Otherwise everything is the same:

$$N\int (d\psi)(d\psi^*)\, e^{iS} \propto \exp\left\{\left[\;\cdots\; \right]\right\} \tag{29.31}$$

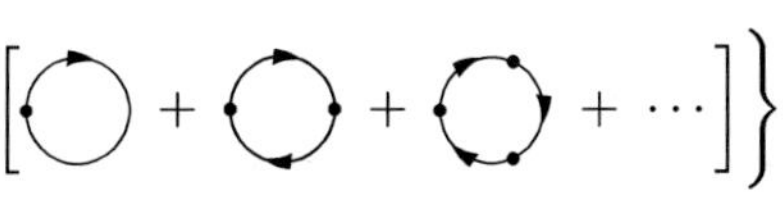

There are arrows on the Bose fields because they are still charged. The functional integral gives the right answer for bosons, assuming there are no derivative interactions, which we are assuming for the moment; we'll take care of that case later. Therefore $(\det A)^{-1}$ must equal the sum of the diagrams. But if $(\det A)^{-1}$ is equal to the sum of the diagrams, then if I stick a minus sign in the sum we get

$$\langle 0|\mathrm{S}|0\rangle = \det A \tag{29.32}$$

in the Fermi case, **QED**. The fact that an A appears in the Fermi case and an A^{-1} appears in the Bose case is merely a reflection of the Fermi minus sign for single fermion loops, which we obtained earlier from the complicated combinatorics of anticommuting Fermi fields.[6]

So far we've only shown that one horrendous expression is equal to another equally horrendous expression, but not what it's good for. When we start manipulating things we will see how useful it is.

29.2 Derivative interactions via functional integrals

We now enter the darkest part of these lectures. Our proofs will be both complicated and inadequate; all the fiddling detail of pure mathematics, but with none of its generality and rigor. I will restrict myself to the one case where the results I'm about to state have been carefully proven. I'll give a proof shortly but the proof should be considered to be between two very large quotation marks. Its combinatoric complexity will be matched only by its lack of rigor. I will consider a classical Lagrangian in particle, not field, language (sum on repeated indices):

$$L = \tfrac{1}{2}\dot{q}^a A_{ab}(q)\dot{q}^b + B_a(q)\dot{q}^a - V(q) + J_a q^a \tag{29.33}$$

[5] [Eds.] W. Pauli, "The Connection Between Spin and Statistics", *Phys. Rev.* **58** (1940) 716–722; reprinted in Schwinger *QED*.

[6] [Eds.] See §21.5.

It is no more than quadratic in derivatives so we can get the Hamiltonian without having to solve a quadratic or higher order equation to determine the p's:

$$p_a = \frac{\partial L}{\partial \dot{q}^a} = A_{ab}\dot{q}^b + B_a \Rightarrow \dot{q}^a = (A^{-1})^{ab}(p_b - B_b) \tag{29.34}$$

Therefore the Hamiltonian is

$$H = \tfrac{1}{2}p_a(A^{-1})^{ab}p_b + \text{(terms at most linear in the } p\text{'s)} \tag{29.35}$$

I will now state a result that I will first exploit and then come back and prove: The **Lagrangian form** of the generating functional is given by the expression[7]

$$\boxed{Z[J] = N\int\prod_a (dq^a)\,(\det A)^{\frac{1}{2}}\,e^{i\mathcal{S}}} \tag{29.36}$$

That peculiar factor, $(\det A)^{1/2}$, is a surprise. It is not going to give us any problems if the only derivative interaction is linear in the derivative, but it will give us problems if there are terms quadratic in the derivatives. This factor is less of a surprise if we write the action integral in Hamiltonian form:

$$\mathcal{S} = \int dt\,L = \int dt(p_a\dot{q}^a - H) = \mathcal{S}_H \tag{29.37}$$

These two are equal for solutions to the equations of motion. I would like to consider the right-hand term as a function of p and q, regarded as independent quantities, defined over a larger space of functions than the left-hand side. The term on the left is defined for arbitrary motions in q-space; the term on the right is defined for arbitrary motions in phase space, with twice as many dimensions. With this form of the action,

$$Z[J] = N\int\prod_a (dq^a)(dp_a)e^{i\mathcal{S}_H} \tag{29.38}$$

This is the **Hamiltonian form** of the generating functional.[8]

I claim that the Lagrangian and Hamiltonian forms of the generating functional are equal. H is at most a quadratic function of the p's, so with q held fixed the p integral is simply a Gaussian. We know how to evaluate a Gaussian: find the minimum of the expression and plug the answer back in;

$$\dot{q}^a = \frac{\partial H}{\partial p_a} \tag{29.39}$$

Putting this back in recreates the Lagrangian form of the action, by putting in the Hamilton equation that gives us p_a as a function of the $\dot{q}^a$'s. However, because the coefficient of the

[7] [Eds.] On p. 184 of *Aspects*, in Note 21 Coleman cites K. S. Cheng, "Quantization of a General Dynamical System by Feynman's Path Integration Formulation", *J. Math. Phys.* **13** (1972) 1723–26 for the derivation of (29.36). Starting from the Lagrangian $\frac{1}{2}g_{ij}\dot{q}^i\dot{q}^j$, the measure $\sqrt{\det(g_{ij})}\prod dq^i$, reminiscent of the general relativistic measure $\sqrt{-g}\,d^4x$, is simply written down, to keep the Lagrangian invariant under general coordinate transformations on the q^i; from it, a *different* result is obtained.

[8] [Eds.] See Appendix B of Richard P. Feynman, "An Operator Calculus Having Applications in Quantum Electrodynamics", *Phys. Rev.* **84** (1951) 108–128; and L. D. Faddeev, "The Feynman Integral for Singular Lagrangians", *Theo. Math. Phys.* **1** (1969) 1–13. Feynman's equation (16-a) is Faddeev's equation (1).

quadratic term in (29.35) is *not* a constant, we have to insert the determinant, to the ½ power, of the coefficients of the quadratic form, and that introduces the $(\det A)^{1/2}$ coming from the A^{-1} in H. The equation (29.36) is just the Gaussian integral in p-space coupled with

$$\tfrac{1}{2}p_a(A^{-1})^{ab}p_b + \cdots \tag{29.40}$$

So the Lagrangian and Hamiltonian forms of $Z[J]$ are equal. I still haven't "proved" that the Hamiltonian form actually gives us the generating functional, and so for the moment I ask you to take (29.36) on faith. We will manipulate (29.36) and show some of its consequences, make some comments about it, and after we've done a few examples we'll go back and prove it.

Two points should be made. First, the standard canonical methods of constructing a quantum theory from a classical theory suffer from ordering ambiguities: there are many ways we can order the p's and q's if H contains forms like (29.40). But the expressions above have no ordering ambiguities whatsoever. That is, the functional integral procedure *defines* one among many possible ways of ordering any given Hamiltonian of the stated type. We *could* figure out what way that is, but we won't in this course. But there *is* a way of ordering.[9] Second, the Hamiltonian form of the functional integral, although very useful (as you'll see), must be taken *cum grano salis*, because, unlike the Lagrangian form, it does not become well-defined when we rotate into Euclidean space. The $p_a\dot{q}^a$ term integrated dt picks up an i when $dt \to idt$, but we get a compensating i from $\dot{q}^a$, so it continues to oscillate even in Euclidean space, instead of becoming a damping factor. Indeed, it is possible to show that for a sufficiently complicated H, the answer we get depends on whether we integrate over the p's first or over the q's first. It is not a well-defined, uniformly convergent integral. It converges, even in Euclidean space, only because of cancellations of phases, and therefore the order in which we integrate over phase space may be important. We will ignore this point, which concerns only purists. If you are a purist, when you write it in Hamiltonian form (29.38), you must add a little footnote that says "integrate over the p's first." Once we integrate over the p's, we obtain the Lagrangian form of the path integral, which *does* become well-defined after we rotate into Euclidean space.

Integrating over the q's first instead of the p's first gives a different ordering for the Hamiltonian; the two orderings are not consistent. Eventually, when we go to a high enough order in perturbation theory we'll find different expressions for Green's functions, depending on the order of integration. As a simple example, consider the Hamiltonian

$$H = p^2 + q^2 + \lambda p^2 q^2 \tag{29.41}$$

This is a quadratic form in the q's with fixed p's and a quadratic form in the p's with fixed q's. So we can do either the p integral first or the q integral first explicitly and then be left with a mess, which we can evaluate perturbatively. The two prescriptions begin to differ at higher orders in λ.

In fact, most of our work will be done with the Lagrangian form of the functional integral, (29.36), with the explicit determinant sitting out there in front. That's a very nice form. But it has one deficiency: we can't evaluate it using the Feynman rules, because it's not the integral of an exponential. This problem can be eliminated by introducing extra fields that exponentiate the determinant. I will now explain how this is done.

[9] [Eds.] For an extensive review of the problem and some solutions, see S. Twareque Ali and Miroslav Engliš, "Quantization methods: a guide for physicists and analysts", *Rev. Math. Phys.* **17** (2005) 391–490.

29.3 Ghost fields

We have the expression

$$Z[J] = N \int \prod_a (dq^a)(\det A)^{1/2} e^{i\mathcal{S}} \tag{29.36}$$

We need to get $(\det A)^{1/2}$ up into the exponential. To do this I will introduce new fields, so-called **ghost fields**,[10] Fermi fields, η^a and $\overline{\eta}^a$, that put the determinant into the exponent, as in (29.27). They do not correspond to any dynamical degrees of freedom that are actually in the system; hence we give them the pejorative name *ghosts*. They have no physical interpretation. Given the action $\mathcal{S}$, we define the effective action

$$\mathcal{S}_{\text{eff}} = \mathcal{S} + \int dt\, \overline{\eta}^a (A^{1/2})_{ab}\, \eta^b \tag{29.42}$$

Then the functional integral becomes

$$\begin{aligned} Z[J] &= N \int \prod_a (dq^a)\, e^{i\mathcal{S}} (\det A)^{1/2} \\ &= N \int \prod_a (dq^a)(d\eta^a)(d\overline{\eta}^a)\, e^{i\mathcal{S}_{\text{eff}}} \end{aligned} \tag{29.43}$$

The ghost variables (they are not fields here, but we'll soon look at ghost fields) are just things we have stuck in to move the determinant from out in front, where we can't do anything with it, up into the exponential where we can evaluate it by the ordinary Feynman rules.

EXAMPLE. *Change of variables for a scalar field*

Let's do an example, cooked up so that we know the right answer in advance. Apart from the source term, it's a free field theory. That way we can check to see if all these prescriptions are correct.[11]

$$\mathscr{L} = \tfrac{1}{2}(\partial_\mu \phi)^2 - \tfrac{1}{2}\mu^2\phi^2 + J\phi \tag{29.44}$$

The source term is there because we'll be talking about Green's functions, not just S-matrix elements. Change the variables from ϕ to A:

$$\phi = A + \tfrac{1}{2} g A^2 \tag{29.45}$$

$$\partial_\mu \phi = (1 + gA)\partial_\mu A \tag{29.46}$$

[10] [Eds.] Ghost particles, or *fictitious particles*, the quanta of ghost fields, were introduced by Feynman to solve some problems encountered in quantizing gravitational and Yang–Mills fields (see footnote 5, p. 646) at the one loop level: R. P. Feynman, "The Quantum Theory of Gravitation", *Acta Phys. Polon.* **24** (1963) 697–722. Reprinted in *Selected Papers of Richard Feynman*, Laurie M. Brown, ed., World Scientific, 2000. DeWitt, who extended Feynman's idea to multiple loops, describes ghost particles' role this way: "The fictitious particles play a compensating role, canceling the effects around the closed loops of the non-transverse [gravitational and Yang–Mills] field modes ... Their presence is central to the preservation of the unitarity of the S-matrix and to the complete invariance of the theory under group transformations, as well as changes in supplementary conditions. In principle, they are needed even in electrodynamics. However, in that special case, the vertex [between the ghost particles and the photon] vanishes, owing to the Abelian character of the gauge group." (See the box on p. 1042, item (h); for an Abelian theory, $c^{abc} = 0$.) B. S. DeWitt, *Dynamical Theory of Groups and Fields*, Gordon and Breach, 1965, p. 227. (Also published as part of *Relativity, Groups and Topology* (Les Houches 1963), Gordon and Breach, 1964; p. 812.) See also B. S. DeWitt, "Quantum Theory of Gravity II. The Manifestly Covariant Theory", *Phys. Rev.* **162** (1967) 1195–1239 and "Quantum Theory of Gravity III. Applications of the Covariant Theory", *Phys. Rev.* **162** (1967) 1239–1256. Ghost *fields* are usually associated with the names of Faddeev and Popov; see Chapter 31, and note 8, p. 1036.

[11] [Eds.] See Problem 8.1, p. 309. The field A is *not* to be confused with the matrix A in (29.42).

The quantity g is a free parameter, which I will treat as a coupling constant. The physics is completely unaffected by this substitution. It was chosen for its algebraic simplicity. This transformation is not invertible, but that's irrelevant. We will be doing perturbation theory and the perturbation series is formally invertible. The Lagrangian becomes

$$\mathscr{L} = \tfrac{1}{2}(\partial_\mu A)^2(1+gA)^2 - \tfrac{1}{2}\mu^2(A+\tfrac{1}{2}gA^2)^2 + J(A+\tfrac{1}{2}gA^2) \tag{29.47}$$

This is the same theory we started with, though it looks like a horrible interacting theory. All the Green's functions obtained by computing $Z[J]$ and functionally differentiating with respect to J should be the same for the two different forms of $\mathscr{L}$. It is not obvious how this will work out; it looks as though there are cubic interactions, quartic interactions, and derivative interactions, all governed by g. All that must cancel. But we won't see it cancel if we naïvely read the Feynman rules off the second form of $\mathscr{L}$.

To get the Feynman rules we must look at an effective Lagrangian which involves a single ghost field, $\eta(x)$:

$$\mathscr{L}_{\text{eff}} = \mathscr{L} + \overline{\eta}\eta(1+gA) \tag{29.48}$$

Following (29.42), $(1+gA)$ is the square root of the coefficient of the derivative term in the Lagrangian (29.47). You see the unphysical ghostly nature of η and $\overline{\eta}$ if you look at their propagators, using our standard rules. The propagator is always i divided by the coefficient of the quadratic term. So we have, in addition to our normal fields, a propagator for the ghost field, denoted by a dotted line:

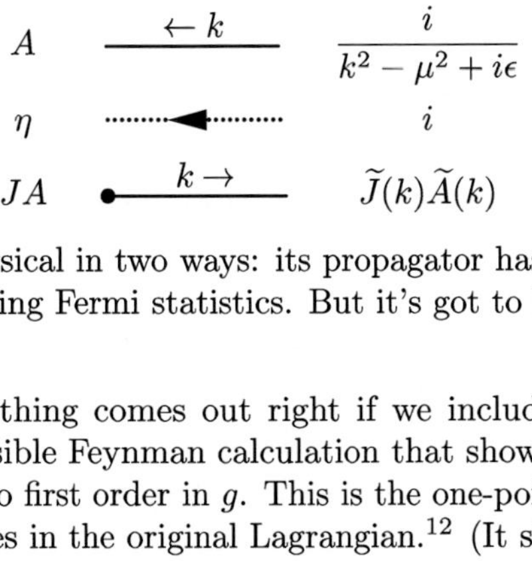

(29.49)

The ghost field $\eta(x)$ is unphysical in two ways: its propagator has no momentum dependence, and it is a spinless field obeying Fermi statistics. But it's got to be there to make everything come out right.

We can show that everything comes out right if we include the ghost fields, but not otherwise. The simplest possible Feynman calculation that shows this is the computation of $Z[J]$ to first order in J and to first order in g. This is the one-point (tadpole) function, which should vanish since it vanishes in the original Lagrangian.[12] (It should vanish to *all* orders in g, but that's beyond my computational abilities, and I suspect beyond your patience.) There are five terms of $\mathcal{O}(g)$ and $\mathcal{O}(J)$ in the effective Lagrangian:

$$JA; \qquad gA(\partial_\mu A)^2; \qquad -\tfrac{1}{2}g\mu^2A^3; \qquad \tfrac{1}{2}gJA^2; \qquad gA\overline{\eta}\eta \tag{29.50}$$

These give rise to four graphs of first order in g and J:

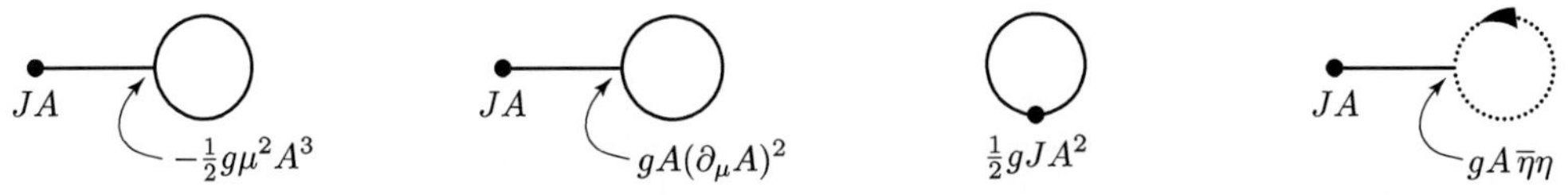

Graph 1 *Graph 2* *Graph 3* *Graph 4*

[12] [Eds.] See also Coleman *Aspects*, Ch. 4, "Secret Symmetry", Section 4.6, pp. 158–159.

All four are quadratically divergent—they go as d^4k/k^2—but we will sum them up nonetheless. The sum had better come out to be zero.

There are some factors common to all four diagrams. There is an overall i from the interaction Lagrangian. There is a factor of g for the interaction. There is a $\widetilde{J}(0)$, evaluated at $k=0$ since the propagator lines carry zero momentum: the momentum at the loop vertices is always zero, and momentum is conserved at the vertices. And there is an integration $\int d^4k/(2\pi)^4$ over loop momentum for each of the loops. After factoring out the common elements, we have:

1. In the first graph, one of the three A's at the vertex contracts with the A coupled to the source, to produce a zero-momentum propagator, $i/(-\mu^2)$; the other two contract to give the loop, and a meson propagator of $i/(k^2-\mu^2+i\epsilon)$. There is also a vertex which gives a factor of ig, but we've already factored out the g. Because the three A's at the vertex are indistinguishable, there are three choices of which of them is contracted with the source term A. This gives a net contribution of

$$-\tfrac{1}{2}(3)\mu^2\,\frac{i}{-\mu^2}\,i\,\frac{i}{k^2-\mu^2+i\epsilon}$$

2. In the second graph, the derivatives on A give terms of $\pm ik^\mu$ times a creation or annihilation operator. Since the meson line has momentum 0, the undifferentiated A at the vertex must be the one contracted with the A from JA, giving once again $i/(-\mu^2)$. If the differentiated A were contracted with it, we would get 0. The two $(\partial_\mu A)$'s are contracted together, giving $(ik_\mu)(-ik^\mu)=k^2$. And there's a factor of i from the vertex. The second diagram contributes

$$\frac{i}{-\mu^2}\,i\,\frac{ik^2}{k^2-\mu^2+i\epsilon}$$

3. The third graph is pretty simple. It has only one interaction, which we've already taken out. So all we have is the propagator for the internal loop times an explicit factor of $\frac{1}{2}$. We just have to contract the two fields with each other, and we get

$$\frac{1}{2}\,\frac{i}{k^2-\mu^2+i\epsilon}$$

4. The contribution from the fourth graph looks peculiar. There is a zero-q A propagator $i/(-\mu^2)$; there is an interaction, i, and the combinatorics for the vertex gives 1, there's no choice there. There is a ghost propagator around the loop, i. And finally there is a minus one for the loop because the ghosts are fermions. That gives

$$(-1)(1)i\,\frac{i}{-\mu^2}\,i$$

The sum of the four graphs is $ig\widetilde{J}(0)\displaystyle\int\frac{d^4k}{(2\pi)^4}\,\{\cdots\}$, where

$$\{\cdots\} = -\tfrac{3}{2}\mu^2\frac{i}{-\mu^2}\,i\,\frac{i}{k^2-\mu^2+i\epsilon}+\frac{i}{-\mu^2}\,i\,\frac{ik^2}{k^2-\mu^2+i\epsilon}+\frac{1}{2}\,\frac{i}{k^2-\mu^2+i\epsilon}-i\,\frac{i}{-\mu^2}\,i \tag{29.51}$$

The expression in the curly brackets can be simplified:

$$\{\cdots\} = \frac{i}{\mu^2(k^2-\mu^2+i\epsilon)}\left[-\tfrac{3}{2}\mu^2+k^2+\tfrac{1}{2}\mu^2-(k^2-\mu^2+i\epsilon)\right] \tag{29.52}$$

combining the last two terms with common denominators. In the limit $\epsilon \to 0$, this is zero, the right answer! Please notice the *absolute necessity* of the ghosts, without which we would not have gotten this to work out. It's the fourth term, with its fermion loop minus sign, that makes everything cancel. We are saved by the friendly ghosts.

29.4 The Hamiltonian form of the generating functional

Now let's "prove" the Hamiltonian form of the functional integral, (29.38). We'll do the Hamiltonian form rather than the Lagrangian, so we don't have to worry about constructing the Hamiltonian out of the Lagrangian. We want to show that it's equal to Dyson's formula (7.36) and Dyson's formula is given in terms of the Hamiltonian.

For simplicity consider a single scalar field.

$$\mathscr{H} = \tfrac{1}{2}\pi^2 + \tfrac{1}{2}(\nabla\phi)^2 + \tfrac{1}{2}\mu^2\phi^2 + \mathscr{H}'(\pi,\phi) - J\phi - K\pi \tag{29.53}$$

(The argument goes through without alteration, aside from a proliferation of indices, if there are many fields of various spins.) The field K, a source coupled to the canonical momentum π, will eventually be set to zero, but the term $K\pi$ will be useful at intermediate stages. Dyson's formula, universally valid, tells us that

$$Z[J,K] = \langle 0|S|0\rangle_{J,K} \propto \langle 0|T\exp\left\{-i\int d^4x\,\left[\mathscr{H}'(\pi_I,\phi_I) - J\phi_I - K\pi_I\right]\right\}|0\rangle \tag{29.54}$$

where the subscript I indicates the interaction picture. We use the proportionality symbol to avoid complications from N, N', etc. Now we see the advantage of introducing the source K: swapping the fields for the derivatives

$$\phi_I \leftrightarrow -i\frac{\delta}{\delta J} \quad\text{and}\quad \pi_I \leftrightarrow -i\frac{\delta}{\delta K} \tag{29.55}$$

we can take the $\mathscr{H}'$ outside and write this as

$$Z[J,K] \propto \exp\left\{-i\int d^4y\,\mathscr{H}'(-i\frac{\delta}{\delta J(y)}, -i\frac{\delta}{\delta K(y)})\right\}\langle 0|T\exp\left\{i\int d^4x\,\left[J\phi_I + K\pi_I\right]\right\}|0\rangle \tag{29.56}$$

(My conscience tells me I should tell you when I'm cheating. This is a swindle: $\delta/\delta J$ and $\delta/\delta K$ are commuting operators, but ϕ_I and π_I are not. Thus we have chosen some (unspecified) ordering. We will ignore this problem.) This takes care of Dyson's formula.

Now let's look at the functional integral, which I want to show is equal to Dyson's formula.

$$Z[J,K] \propto \int (d\phi)(d\pi)\exp\left\{i\int d^4x\,\left[\pi\dot{\phi} - \mathscr{H}\right]\right\} \tag{29.57}$$

with $\mathscr{H}$ given by (29.53). Again we pull out $\mathscr{H}'$, and obtain

$$\begin{aligned} Z[J,K] \propto{}& \exp\left\{-i\int d^4y\,\mathscr{H}'(-i\frac{\delta}{\delta J(y)}, -i\frac{\delta}{\delta K(y)})\right\} \\ &\times \int (d\phi)(d\pi)\exp\left\{i\int d^4x\,\left[\pi\dot{\phi} - \underbrace{(\tfrac{1}{2}\pi^2 + \tfrac{1}{2}(\nabla\phi)^2 + \tfrac{1}{2}\mu^2\phi^2)}_{\mathscr{H}_0} + J\phi + K\pi\right]\right\} \end{aligned} \tag{29.58}$$

Comparing (29.56) and (29.58), we see the same operator $\mathscr{H}'(-i\delta/\delta J(y), -i\delta/\delta K(y))$ acting on two different expressions. We will prove the theorem in general if we can prove it for a free field theory, augmented with J and K sources. So, are these two expressions equal?

$$\begin{aligned}&\langle 0|T\exp\left\{i\int d^4x\ [J\phi_I + K\pi_I]\right\}|0\rangle\\ &\quad\overset{?}{=}\int (d\phi)(d\pi)\exp\left\{i\int d^4x\left[\pi\dot{\phi} - \tfrac{1}{2}\pi^2 - \tfrac{1}{2}(\nabla\phi)^2 - \tfrac{1}{2}\mu^2\phi^2 + J\phi + K\pi\right]\right\}\end{aligned} \tag{29.59}$$

The π integral on the right-hand side is a pure Gaussian (28.50), with determinant 1. The quadratic form is

$$Q(\pi) = \tfrac{1}{2}\pi^2 - \pi\dot{\phi} - K\pi \tag{29.60}$$

with minimum $\overline{\pi}$ given by

$$\overline{\pi} = \dot{\phi} + K \tag{29.61}$$

The quadratic form at the minimum is

$$Q(\overline{\pi}) = -\tfrac{1}{2}\dot{\phi}^2 - \tfrac{1}{2}K^2 - K\dot{\phi} \tag{29.62}$$

so the Gaussian π integral gives

$$\begin{aligned}&\int (d\phi)\exp\left\{i\int d^4x\left[\underbrace{\tfrac{1}{2}\dot{\phi}^2 - \tfrac{1}{2}(\nabla\phi)^2 - \tfrac{1}{2}\mu^2\phi^2}_{\mathscr{L}_0} + J\phi + \underbrace{K\dot{\phi}}_{\substack{\text{int. by}\\ \text{parts}}} + \tfrac{1}{2}K^2\right]\right\}\\ &\quad= \int (d\phi)\exp\left\{i\int d^4x\left[\mathscr{L}_0 + (J - \dot{K})\phi + \tfrac{1}{2}K^2\right]\right\}\\ &\quad= \exp\left\{-\tfrac{1}{2}\int d^4x\, d^4y\left[\left(J(x) - \dot{K}(x)\right)\Delta_F(x-y)\left(J(y) - \dot{K}(y)\right)\right]\right\} e^{\frac{i}{2}\int d^4x\,(K(x))^2}\end{aligned} \tag{29.63}$$

Note that the last term in the exponential has no dependence on ϕ, so it can be taken out of the functional integral.

Now let's evaluate Dyson's formula:

$$Z_0[J,K] = \langle 0|T\exp\left\{i\int d^4x\ [J\phi_I + K\pi_I]\right\}|0\rangle \tag{29.64}$$

Both ϕ_I and π_I are free fields, each with its own propagator. The only kind of diagrams we will get will be ϕ_I–ϕ_I contractions, π_I–π_I contractions as well as the joint propagators, ϕ_I–π_I contractions. Thus we get

$$\begin{aligned}\langle 0|T\exp\left\{i\int d^4x\ [J\phi_I + K\pi_I]\right\}|0\rangle = \exp\Bigg\{-\int d^4x\, d^4y\Big[&\tfrac{1}{2}J(x)J(y)\overline{\phi_I(x)\phi_I(y)}\\ &+ J(x)K(y)\overline{\phi_I(x)\pi_I(y)} + \tfrac{1}{2}K(x)K(y)\overline{\pi_I(x)\pi_I(y)}\Big]\Bigg\}\end{aligned} \tag{29.65}$$

The $\frac{1}{2}$'s are because we can't tell the vertices $\phi_I(x)\phi_I(y)$ apart from $\phi_I(y)\phi_I(x)$, and the same holds true for the quadratic term in π_I. No such symmetry factor is needed for the $\phi_I\pi_I$ term.

Both ϕ_I and π_I are free fields, linear in a and $a^\dagger$, so by Wick's theorem, their contractions are all c-numbers. The ϕ contraction is straightforward: it is the Feynman propagator.

$$\overline{\phi_I(x)\phi_I(y)} = \langle 0|T(\phi_I(x)\phi_I(y))|0\rangle = \Delta_F(x-y) \tag{29.66}$$

In the interaction picture,

$$\pi_I(x) = \partial_0\phi_I(x) \tag{29.67}$$

To evaluate the contraction involving ϕ_I and π_I, use the identity (27.86)

$$\partial_x^0 T(A(x)B(y)) = T(\partial_x^0 A(x)\,B(y)) + \delta(x_0-y_0)[A(x),B(y)]$$

Because (3.60) $[\phi_I(\mathbf{x},t),\phi_I(\mathbf{y},t)]=0$,

$$\frac{\partial}{\partial y^0}T[\phi_I(x)\phi_I(y)] = T[\phi_I(x)\pi_I(y)] \tag{29.68}$$

so

$$\overline{\phi_I(x)\pi_I(y)} = \partial_0^y\Delta_F(x-y) \tag{29.69}$$

To evaluate the contraction involving two π_I's, use the identity (27.86) again,

$$\begin{aligned}\frac{\partial}{\partial x^0}T[\phi_I(x)\pi_I(y)] &= T[\pi_I(x)\pi_I(y)] + \delta(x_0-y_0)[\phi_I(\mathbf{x},t),\pi_I(\mathbf{y},t)]\\ &= T[\pi_I(x)\pi_I(y)] + i\delta^{(4)}(x-y)\end{aligned} \tag{29.70}$$

from (3.61). Then

$$\overline{\pi_I(x)\pi_I(y)} = \partial_0^x\partial_0^y\Delta_F(x-y) - i\delta^{(4)}(x-y) \tag{29.71}$$

We encountered a similar expression in (27.91). There, the first term was nicely covariant, but the second was disgustingly non-covariant, and something we didn't want. Here, though, it will be welcome.

Now the moment of truth:

$$\begin{aligned}&\langle 0|T\exp\left\{i\int d^4x\ [J\phi_I + K\pi_I]\right\}|0\rangle\\
&\quad= \exp\left\{-\int d^4x\,d^4y\left[\tfrac12 J(x)J(y)\overline{\phi_I(x)\phi_I(y)} + J(x)K(y)\overline{\phi_I(x)\pi_I(y)}\right.\right.\\
&\qquad\qquad \left.\left.+\ \tfrac12 K(x)K(y)\overline{\pi_I(x)\pi_I(y)}\right]\right\}\\
&\quad= \exp\left\{-\int d^4x\,d^4y\left[\tfrac12 J(x)J(y)\ \Delta_F(x-y) + J(x)K(y)\partial_0^y\Delta_F(x-y)\right.\right.\\
&\qquad\qquad \left.\left.+\ \tfrac12 K(x)K(y)(\partial_0^x\partial_0^y\Delta_F(x-y) - i\delta^{(4)}(x-y))\right]\right\}\\
&\quad= \exp\left\{-\int d^4x\,d^4y\left[\left(\tfrac12 J(x)J(y) - J(x)\dot K(y) + \tfrac12\dot K(x)\dot K(y)\right)\Delta_F(x-y)\right.\right.\\
&\qquad\qquad \left.\left.-\ \tfrac{i}{2}K(x)K(y)\delta^{(4)}(x-y)\right]\right\}\\
&\quad= \exp\left\{-\tfrac12\int d^4x\,d^4y\left[\left(J(x)-\dot K(x)\right)\Delta_F(x-y)\left(J(y)-\dot K(y)\right)\right]\right\}e^{\frac{i}{2}\int d^4x(K(x))^2}\end{aligned} \tag{29.72}$$

which agrees with (29.63). It works! This completes the demonstration that the functional integral really does give us everything, including the messy $\exp\{\frac{i}{2}\int d^4x(K(x))^2\}$ term that came out of the Hamiltonian form of the functional integral. It gives us the right result for a free field theory and therefore, if we accept this sloppy proof, it gives us the right results for an interacting theory. **QED**

There is one other useful rule for handling functional integrals. It's quite simple, and does not involve anything like these hairy complications. It will enable us to describe electrodynamics for a massive photon, including scalar electrodynamics. (We need new physics to take care of gauge invariance for real photons.)

29.5 How to eliminate constrained variables

Sometimes we encounter in Lagrangian systems dynamical variables for which the Euler–Lagrangian equations are not equations of motion but are simply equations of constraint. They tell us something about the initial value data but not how things develop in time. An example is the time component $A_0(x)$ in vector field theory. There are no terms involving $\partial_0 A^0$ in $\mathscr{L}$, so the A_0 equation of motion is simply an equation of constraint:

$$\frac{\partial\mathscr{L}}{\partial\partial_0 A^0} = 0 \Rightarrow \mu^2 A^0 = \partial_i F^{0i} \tag{29.73}$$

We frequently encounter Lagrangians of the form (written in terms of a single particle, for simplicity)

$$L = L_1(q,\dot{q}) + \tfrac{1}{2}ay^2 + b(q)y \tag{29.74}$$

The variable y appears quadratically with a coefficient a independent of q. Its time derivative $\dot{y}$ does not appear in L. Thus y is a constrained variable, not a dynamical variable. We must solve for y to eliminate it from the theory before we can obtain the Hamiltonian. The Euler–Lagrange equation for y is

$$ay = -b \quad \text{or} \quad y = -\frac{b}{a} \tag{29.75}$$

Thus we obtain for the Lagrangian

$$\overline{L} = L_1 - \frac{b^2}{2a} \tag{29.76}$$

This is the Lagrangian we have to use when working with the Hamiltonian form. The (almost trivial) point I want to make is this: Aside from a normalization factor,

$$\int (dq)(dy)\exp\left\{i\textstyle\int dt\, L\right\} \propto \int (dq)\exp\left\{i\textstyle\int dt\, \overline{L}\right\} \tag{29.77}$$

This is just our old rule (28.11) for doing a Gaussian. We evaluate the Gaussian at its minimum, which is precisely this prescription. (We pick up an irrelevant constant because a is independent of q.)

So if we have a constrained variable that enters the Lagrangian L at most quadratically and the coefficient of the quadratic term is independent of the other dynamical variables, we might as well integrate over it, which is equivalent to eliminating it. The two prescriptions are the same. If a *does* involve the other dynamical variables, then integrating over it *isn't* the same as eliminating it: a determinant will appear and we can handle that determinant in the usual way, with ghost fields. That won't occur in the cases we have at hand.

This procedure is easily extended to many variables, or a field. For example, suppose the Lagrangian for a system of particles is

$$L(q_1, \cdots, q_n, \dot{q}_1, \cdots, \dot{q}_n, y_1, \cdots, y_m) \tag{29.78}$$

The conditions

$$\frac{\partial L}{\partial y^a} = 0 \tag{29.79}$$

fix the y_i as functions of the q's and the $\dot{q}$'s, allowing us to eliminate the y's. If the Lagrangian is of the form

$$L = \tfrac{1}{2}\sum_{a,b} y^a A_{ab} y^b + \sum_a b_a y^a + L_1 \tag{29.80}$$

with A_{ab} independent of the q's and the $\dot{q}$'s, then functionally integrating over the y^a is equivalent to eliminating them. If the A_{ab} depend on the q's and the $\dot{q}$'s then we introduce a set of ghost fields $\{\eta^a, \overline{\eta}^a\}$:

$$\int (dy)(d\eta)(d\overline{\eta}) \exp\left\{i\int dt \left[L + \overline{\eta}^a A_{ab}^{1/2} \eta^b\right]\right\} \tag{29.81}$$

29.6 Functional integrals for QED with massive photons

Now we can apply these two rules (the simple rule for elimination of constrained variables and the rule for the Hamiltonian form of the functional integral) to extract and derive the complete Feynman rules for, first, spinor electrodynamics with charged spinor particles and, second, scalar electrodynamics with charged scalar particles, in both cases with a massive photon. The trick is just writing the same damned action in 14 different ways (actually, only in two different ways).

Let's begin with spinor electrodynamics. We already know a form for the action

$$\mathcal{S}_{\text{2nd}} = \int d^4x\{-\tfrac{1}{4}(\partial_\mu A_\nu - \partial_\nu A_\mu)^2 + \tfrac{1}{2}\mu^2 A^\nu A_\nu + \overline{\psi}(i\partial\!\!\!/ - eA\!\!\!/ - m)\psi\} \tag{29.82}$$

This is called the **second-order action** because it is quadratic in the A_μ derivatives. There is an equivalent form with many more constrained variables. This is called the **first-order action** because it only involves derivatives to the first power.

$$\begin{aligned}\mathcal{S}_{\text{1st}} &= \int d^4x \left\{\tfrac{1}{4}F_{\mu\nu}F^{\mu\nu} - \tfrac{1}{2}F_{\mu\nu}(\partial^\mu A^\nu - \partial^\nu A^\mu) + \tfrac{1}{2}\mu^2 A^\nu A_\nu + \overline{\psi}(i\partial\!\!\!/ - m - eA\!\!\!/)\psi\right\} \\ &= \mathcal{S}_{\text{2nd}} + \int d^4x\, \tfrac{1}{4}\left[F_{\mu\nu} - \partial_\mu A_\nu + \partial_\nu A_\mu\right]^2\end{aligned} \tag{29.83}$$

$F_{\mu\nu}$ and A_μ in this form are to be considered completely independent dynamical variables in the Lagrangian sense. The equation of motion for $F_{\mu\nu}$ is

$$F_{\mu\nu} = \partial_\mu A_\nu - \partial_\nu A_\mu \tag{29.84}$$

Plugging this into $\mathscr{L}$, a necessary step to get $\mathscr{H}$, we arrive back at the original form, (29.82).

Now the game goes like this. By our rule for eliminating constrained variables

$$\int \prod_{\mu\nu}(dF_{\mu\nu}) \prod_{\lambda}(dA_\lambda)\, e^{i\mathcal{S}_{\rm 1st}} = \int \prod_{\lambda}(dA_\lambda)\, e^{i\mathcal{S}_{\rm 2nd}} \tag{29.85}$$

just by eliminating the six components of $F_{\mu\nu}$, since the quadratic term has constant coefficients. On the other hand, we could equally well choose to eliminate some other variables from the theory. In particular, we could choose to eliminate F_{ij} and A_0, leaving F_{0i} and A_i. The F_{ij}^2 and A_0^2 terms have constant coefficients, and the $F_{ij}\,\partial^i A^0$ cross-terms involve no time derivatives. So these *four* variables $\{F_{ij}, A_0\}$ follow the rule for constrained variables just as the other six, $\{F_{\mu\nu}\}$, and we could write the same integral, just by choosing to integrate over a different bunch of variables first, as

$$\int \prod_{\mu\nu}(dF_{\mu\nu}) \prod_{\lambda}(dA_\lambda)\, e^{i\mathcal{S}_{\rm 1st}} = \int \prod_{j}(dF_{0j}) \prod_{i}(dA_i)\, e^{i\mathcal{S}_{\rm other}} \tag{29.86}$$

But what *is* $\mathcal{S}_{\rm other}$? Well, we've done this step before, in §26.3, and we found (26.48). If we eliminate F_{ij} and A_0 to write the action in terms of A_i and F_{0i} only, we get the Hamiltonian form $\mathcal{S}_{\mathscr{H}}$ of the action, because $\{A_i, F_{0i}\}$ are the canonical (q, p) pairs of the massive vector field:

$$\mathcal{S}_{\mathscr{H}} = \int d^4x\, \mathscr{L}(A_i, F_{0i}, \overline{\psi}, \psi) \tag{29.87}$$

and so we can write

$$\int \prod_{j}(dF_{0j}) \prod_{i}(dA_i)\, e^{i\mathcal{S}_{\mathscr{H}}} = \int \prod_{\mu\nu}(dF_{\mu\nu}) \prod_{\lambda}(dA_\lambda)\, e^{i\mathcal{S}_{\rm 1st}} = \int \prod_{\lambda}(dA_\lambda)\, e^{i\mathcal{S}_{\rm 2nd}} \tag{29.88}$$

Now the argument is complete: two things equal to the same thing are equal to each other. What $\mathcal{S}_{\rm 1st}$ is good for is to allow us to go from $\mathcal{S}_{\mathscr{H}}$ (the Hamiltonian form we get from canonical quantization but which is awful for Feynman rules) to $\mathcal{S}_{\rm 2nd}$ (which is not what we get from canonical quantization but which produces covariant-looking Feynman rules quite easily). The functional integral makes the change of variables easy. Each of these is the generating functional. From the second-order form

$$\int \prod_{\lambda}(dA_\lambda)\, e^{i\mathcal{S}_{\rm 2nd}}$$

with appropriate source terms, we can derive the Feynman rules by just reading off the propagators and the interactions from $\mathcal{S}$. The task is done; I've derived the Feynman rules for massive electrodynamics with a massive photon.

We begin to see the utility of the functional integral formalism. The point is not that it's particularly easy to evaluate a functional integral; the rules for evaluating a functional integral are just the Feynman rules. The point is that *the functional integral is particularly easy to manipulate.* Here we've gone from a formalism with 10 independent variables (six F's, four A)'s to a formalism with four independent variables (four A's) to a formalism with six independent variables (three A's and their conjugate momenta), and we do it just by writing down the equations. No fancy unitary operators, canonical transformations, etc. The functional integral with the Hamiltonian form ($\mathcal{S}_{\mathscr{H}}$) of the action is always right; that's our general theorem. The functional integral with the second-order action $\mathcal{S}_{\text{2nd}}$ gives us the naive Feynman rules. $\mathcal{S}_{\mathscr{H}}$ is trustworthy; $\mathcal{S}_{\text{2nd}}$ is useful; *and they're equivalent.* Therefore *the naïve Feynman rules are right.*

I will complete the analysis next time by using the exact same trick to handle the case where there are charged scalar fields. We could do it by elementary methods, but it would be a terrible mess. Not only do we have there the problem of the A_0–A_0 contraction, which we've taken care of without ever worrying about it, we've got derivative interactions, momentum-momentum contractions on top of the A–A contractions, A's with π's, and it would just look awful. It can be done; it was originally done that way, without functional integrals. It would take us a week to do that one problem right. Next time I will do it in two minutes. And then I will do some sample computations dealing with massive photon electrodynamics.

Problems 16

16.1 A massive vector meson (with the standard Proca free Lagrangian) is minimally coupled to a Dirac particle, with coupling constant e. Compute, to lowest nontrivial order in perturbation theory, the amplitude for elastic fermion-antifermion scattering and explicitly verify that the contribution of the term in the vector meson propagator proportional to $k_\mu k_\nu/\mu^2$ vanishes. (You are not asked to do spin sums or compute cross sections, or even to simplify the amplitude any more than is needed to demonstrate the desired result.)

(1998b 3.1)

16.2 (a) In the theory of the previous problem, compute the amplitude for elastic vector-spinor scattering, again to lowest nontrivial order. Verify that if the vector meson spin vector, ε_μ, is aligned with its four momentum k_μ, for either the incoming or the outgoing meson, the amplitude vanishes, even when the meson in question is off mass shell (but the other particles are on mass shell). (Parenthetical remark as above.) Of course, what you are verifying is that A_μ has vanishing divergence between initial and final states defined by the on-shell particles.

(b) The same problem, but this time with a scalar particle rather than a Dirac particle. Use the naïve Feynman rules (see the table on p. 644). But be sure to include the "seagull" diagram; otherwise you won't get the right answer.

(1998b 3.2)

16.3 Consider a theory of two charged Dirac fields A and B with masses m_A and m_B, and a complex charged scalar field C with mass m_C. These interact with a Yukawa-like coupling

$$\begin{aligned}\mathscr{L}' &= g'\overline{A}i\gamma_5 BC + (\text{Hermitian conjugate})\\ &= g(\overline{A}i\gamma_5 BC + \overline{B}i\gamma_5 AC^*)\end{aligned} \tag{P16.1}$$

where $g' = ge^{i\phi}$, and g is a positive (real) number. (The phase of g has been absorbed into the definition of C.) Now consider this theory minimally coupled to a massive "photon", with the three fields having charges (in units of e) q_A, q_B and q_C with $q_A = q_B + q_C$.

Scalar meson photoproduction, $\gamma + A \to B + C$, first occurs in order eg. As in the preceding problem, show that the amplitude for this process in this order vanishes if the "photon" spin is aligned with its 4-momentum. (Note that you have to sum three graphs.)

(1998b 3.3)

Solutions 16

16.1 The lowest-order diagrams for fermion–anti-fermion scattering are shown below. The amplitude is

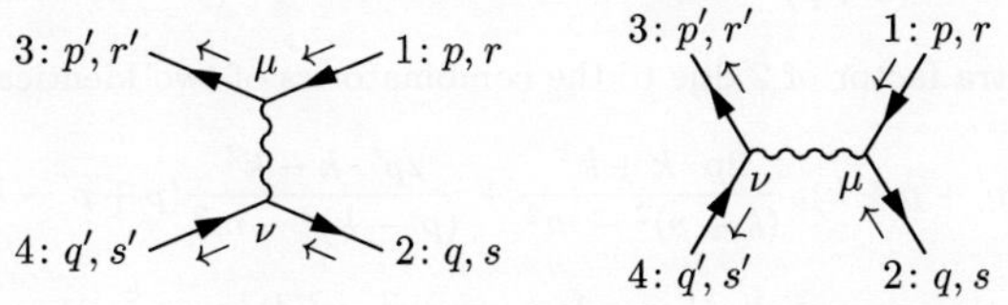

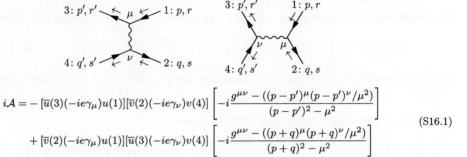

$$\begin{aligned} i\mathcal{A} = &- [\overline{u}(3)(-ie\gamma_\mu)u(1)][\overline{v}(2)(-ie\gamma_\nu)v(4)]\left[-i\frac{g^{\mu\nu} - ((p-p')^\mu(p-p')^\nu/\mu^2)}{(p-p')^2 - \mu^2}\right] \\ &+ [\overline{v}(2)(-ie\gamma_\mu)u(1)][\overline{u}(3)(-ie\gamma_\nu)v(4)]\left[-i\frac{g^{\mu\nu} - ((p+q)^\mu(p+q)^\nu/\mu^2)}{(p+q)^2 - \mu^2}\right] \end{aligned} \tag{S16.1}$$

The relative minus sign between the two terms is due to the exchange of external fermion lines. We note that

$$\overline{u}(3)(\not{p} - \not{p}')u(1) = \overline{u}(3)(m-m)u(1) = 0 \tag{S16.2}$$

$$\overline{v}(2)(\not{p} + \not{q})u(1) = \overline{v}(2)(m-m)u(1) = 0 \tag{S16.3}$$

Thus any term proportional to $(p-p')^\mu(p-p')^\nu$ or $(p+q)^\mu(p+q)^\nu$ vanishes. The amplitude simplifies to

$$i\mathcal{A} = -ie^2[\overline{u}(3)\gamma_\mu u(1)][\overline{v}(2)\gamma^\mu v(4)]\frac{1}{(p-p')^2 - \mu^2} + ie^2[\overline{v}(2)\gamma_\mu u(1)][\overline{u}(3)\gamma^\mu v(4)]\frac{1}{(p+q)^2 - \mu^2} \tag{S16.4}$$

This problem is very similar to the EXAMPLE on Coulomb scattering in §30.3, p. 646. ■

16.2 (a) The lowest-order diagrams for vector–spinor scattering are shown below. The amplitude is

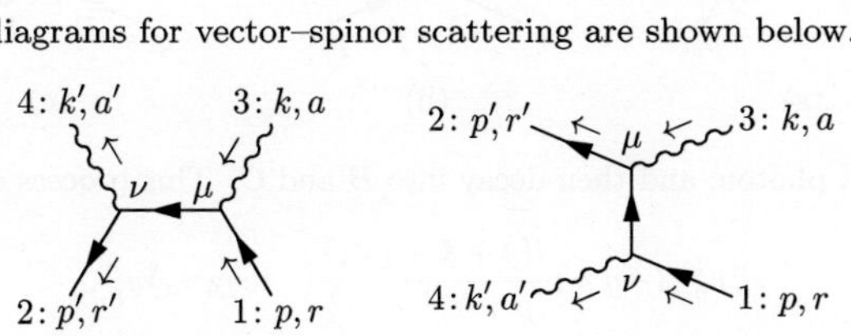

$$i\mathcal{A} = \varepsilon^{\nu *}_{a'}(4)\varepsilon^{\mu}_{a}(3)\overline{u}(2)(-ie\gamma_\nu)\frac{i(\not{k} + \not{p} + m)}{(k+p)^2 - m^2}(-ie\gamma_\mu)u(1) + \varepsilon^{\nu *}_{a'}(4)\varepsilon^{\mu}_{a}(3)\overline{u}(2)(-ie\gamma_\mu)\frac{i(\not{p}' - \not{k} + m)}{(p'-k)^2 - m^2}(-ie\gamma_\nu)u(1) \tag{S16.5}$$

Let the incoming meson's polarization vector be aligned with its momentum: $\varepsilon^\mu(3) = \lambda k^\mu$, for some constant λ. Then

$$i\mathcal{A} = -i\lambda e^2 \varepsilon^{\nu *}_{a'}(4)\overline{u}(2)\left[\gamma_\nu \frac{(\not{k} + \not{p} + m)}{(k+p)^2 - m^2}\not{k} + \not{k}\frac{(\not{p}' - \not{k} + m)}{(p'-k)^2 - m^2}\gamma_\nu\right]u(1) \tag{S16.6}$$

We use the same trick as in (30.40). Since $(\not{p}-m)u(1)=\overline{u}(2)(\not{p}'-m)=0$, we can rewrite (S16.6) as

$$\begin{aligned} i\mathcal{A} &= -i\lambda e^2\varepsilon^{\nu *}_{a'}(4)\overline{u}(2)\left[\gamma_\nu\frac{(\not{k}+\not{p}+m)}{(k+p)^2-m^2}(\not{k}+\not{p}-m)-(\not{p}'-m-\not{k})\frac{(\not{p}'-\not{k}+m)}{(p'-k)^2-m^2}\gamma_\nu\right]u(1) \\ &= -i\lambda e^2\varepsilon^{\nu *}_{a'}(4)\overline{u}(2)\left[\gamma_\nu\frac{(k+p)^2-m^2}{(k+p)^2-m^2}-\frac{(p'-k)^2-m^2)}{(p'-k)^2-m^2}\gamma_\nu\right]u(1)=0 \end{aligned} \tag{S16.7}$$

By similar arguments, we can also show that $i\mathcal{A}=0$ when $\varepsilon^{\nu *}(4)=\lambda k'^\nu$. ■

(b) The Feynman rules for vector-charged scalar interactions are given in the box on p. 644. The three relevant diagrams are shown below:

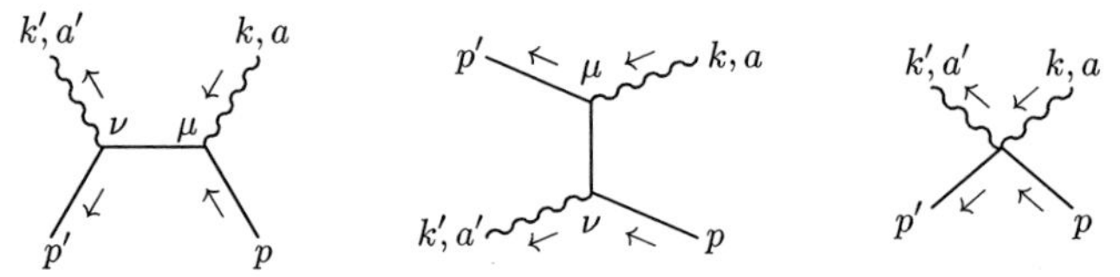

The amplitude is

$$i\mathcal{A}=-e^2\varepsilon^{\nu *}_{a'}\varepsilon^\mu_a\left[(p'+p+k)_\nu\frac{i}{(k+p)^2-m^2}(2p+k)_\mu+(2p'-k)_\mu\frac{i}{(p'-k)^2-m^2}(p+p'-k)_\nu+2ig_{\mu\nu}\right]$$

(the seagull term has an extra factor of 2 due to the combinatorics of two identical A's). When $\varepsilon^\mu_a=\lambda k^\mu$,

$$i\mathcal{A}=-i\lambda e^2\varepsilon^{\nu *}_{a'}\left[(p'+p+k)_\nu\frac{2p\cdot k+k^2}{(k+p)^2-m^2}+\frac{2p'\cdot k-k^2}{(p'-k)^2-m^2}(p+p'-k)_\nu-2k_\nu\right] \tag{S16.8}$$

Though the vector isn't on the mass shell, the scalars are: $p^2=(p')^2=m^2$. Using these constraints,

$$(k+p)^2-m^2=k^2+2p\cdot k \qquad (p'-k)^2-m^2=-2p'\cdot k+k^2 \tag{S16.9}$$

Plugging these expressions into the appropriate denominators,

$$\begin{aligned} i\mathcal{A} &= -i\lambda e^2\varepsilon^{\nu *}_{a'}\left[(p'+p+k)_\nu\frac{2p\cdot k+k^2}{k^2+2p\cdot k}+\frac{2p'\cdot k-k^2}{-2p'\cdot k+k^2}(p+p'-k)_\nu-2k_\nu\right] \\ &= -i\lambda e^2\varepsilon^{\nu *}_{a'}\left[(p'+p+k)_\nu-(p+p'-k)_\nu-2k_\nu\right]=0 \end{aligned} \tag{S16.10}$$

By similar arguments, we can also show that $i\mathcal{A}=0$ when $\varepsilon^{\nu *}_{a'}=\lambda k'^\nu$. ■

16.3 There are three ways the reaction $A+\gamma\to B+C$ can occur, as shown below.

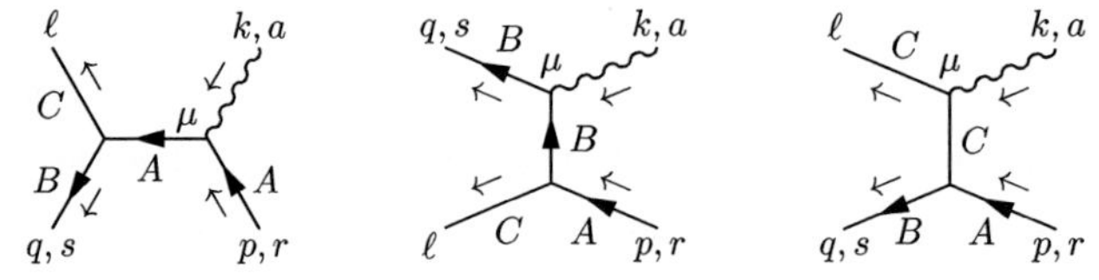

(a) (b) (c)

In graph (a), A can absorb a photon, and then decay into B and C. This process contributes to the amplitude $i\mathcal{A}$ a term

$$\varepsilon^\mu_a\overline{u}^B_{q,s}(-g\gamma_5)\frac{i(\not{p}+\not{k}+m_A)}{(p+k)^2-m_A^2}(-ieq_A\gamma_\mu)u^A_{p,r} \tag{S16.11}$$

Alternatively, as shown in graph (b), A can decay into B and C, and B can absorb the photon, contributing a term

$$\varepsilon^\mu_a\overline{u}^B_{q,s}(-ieq_B\gamma_\mu)\frac{i(\not{q}-\not{k}+m_B)}{(q-k)^2-m_B^2}(-g\gamma_5)u^A_{p,r} \tag{S16.12}$$

Finally, as shown in graph (c), A can decay into B and C, and C can absorb the photon. This gives

$$\varepsilon^\mu_a\overline{u}^B_{q,s}(-g\gamma_5)u^A_{p,r}\frac{i}{(p-q)^2-m_C^2}(-ieq_C)(p-q+\ell)_\mu \tag{S16.13}$$

The amplitude is the sum of these three terms. When $\varepsilon_a^\mu = \lambda k^\mu$, we obtain

$$i\mathcal{A} = \lambda e g \overline{u}_{q,s}^B \left[(\gamma_5) \frac{(\not{p} + \not{k} + m_A)}{(p+k)^2 - m_A^2} (-q_A \not{k}) - (q_B \not{k}) \frac{(\not{q} - \not{k} + m_B)}{(q-k)^2 - m_B^2} (\gamma_5) - (\gamma_5 q_C) \frac{(p-q+\ell) \cdot k}{(p-q)^2 - m_C^2} \right] u_{p,r}^A$$

We use the same trick as before: since $(\not{p} - m_A) u_{p,r}^A = \overline{u}_{q,s}^B (\not{q} - m_B) = 0$, we can add zero terms to the factors of $\not{k}$ in the amplitude to rewrite the factor in the brackets as

$$\begin{aligned}[\cdots] = \Bigg[&(\gamma_5) \frac{(\not{p} + \not{k} + m_A)(\not{p} + \not{k} - m_A)}{(p+k)^2 - m_A^2} (-q_A) \\ &- (q_B) \frac{(-(\not{q} - \not{k}) + m_B)(\not{q} - \not{k} + m_B)}{(q-k)^2 - m_B^2} (\gamma_5) \\ &- (\gamma_5 q_C) \frac{(p-q+\ell) \cdot k}{(p-q)^2 - m_C^2} \Bigg]\end{aligned}$$

Because $\ell + q = k + p$, $k = \ell + q - p$, and

$$k \cdot (p - q + \ell) = -((p-q) - \ell)((p-q) + \ell) = -(p-q)^2 + m_C^2$$

so that

$$[\cdots] = \gamma_5 \left[-q_A + q_B + q_C \right] = 0 \tag{S16.14}$$

We have to have $q_A = q_B + q_C$, to conserve charge at the ABC vertex. ■

30

Electrodynamics with a massive photon

Last time we used the technique of eliminating constrained variables to prove an important theorem in electrodynamics: that all three forms of the functional integral, the first-order form, the second-order form and the Hamiltonian form, are equal. From this we showed that the naive Feynman rules are valid for spinor electrodynamics with a massive photon. The theorem is also true for the electrodynamics of charged scalar particles interacting with a massive photon; let's demonstrate that.[1]

30.1 Obtaining the Feynman rules for scalar electrodynamics

If we attempted to treat scalar electrodynamics canonically we'd be in a terrible mess. We'd have all the problems associated with the derivative interactions of the scalar field[2] and the problems associated with the elimination of A_0 from the vector equations of motion[3] They are problems that can be handled, but they are messy. Here the functional methods pay off: we have only to write down three horrible equations, instead of many more.

We'll use the same trick as we used for spinors. We begin with the second-order form of the Proca Lagrangian (26.10), and the minimally coupled charged Klein–Gordon Lagrangian[4] that follows from (27.59):

$$\mathcal{S}_{\text{2nd}} = \int d^4x \left[(D_\mu\phi)^*(D^\mu\phi) - m^2\phi^*\phi - \tfrac{1}{2}\partial_\mu A_\nu(\partial^\mu A^\nu - \partial^\nu A^\mu) + \tfrac{1}{2}\mu^2 A^\mu A_\mu\right] \tag{30.1}$$

where, from (27.48),

$$\begin{aligned} D_\mu\phi &= (\partial_\mu + ieA_\mu)\phi \\ (D_\mu\phi)^* &= (\partial_\mu - ieA_\mu)\phi^* \end{aligned} \tag{30.2}$$

[1] [Eds.] The Feynman rules for scalar electrodynamics are treated in Greiner & Reinhardt *QED*, Section 8.4, pp. 434–435; H. Kleinert, *Particles and Quantum Fields*, World Scientific, 2015, Chapter 17.

[2] [Eds.] See the example on p. 587.

[3] [Eds.] See §29.5.

[4] [Eds.] We use m for the scalar meson mass because we are writing μ for the vector meson mass.

We can also write the theory in first-order form (with an action that involves no more than first derivatives):

$$\begin{aligned}\mathcal{S}_{\rm 1st} = \int d^4x\, \Big[&\tfrac{1}{4}F_{\mu\nu}F^{\mu\nu} - \tfrac{1}{2}F_{\mu\nu}(\partial^\mu A^\nu - \partial^\nu A^\mu) + \tfrac{1}{2}\mu^2 A_\mu A^\mu - \pi_\mu\pi^{\mu *} - m^2\phi^*\phi \\ &+ \pi^*_\mu(\partial^\mu + ieA^\mu)\phi + \pi_\mu(\partial^\mu - ieA^\mu)\phi^*\Big]\end{aligned} \tag{30.3}$$

This is a mess but it's a simple generalization of what we did with spinors, going from (29.82) to (29.83). The Euler–Lagrange equations of motion for $F_{\mu\nu}$, π_μ, and π^*_μ are trivial:

$$\begin{aligned}F_{\mu\nu} &= \partial_\mu A_\nu - \partial_\nu A_\mu \\ \pi_\mu &= (\partial_\mu + ieA_\mu)\phi \\ \pi^*_\mu &= (\partial_\mu - ieA_\mu)\phi^*\end{aligned} \tag{30.4}$$

$\mathcal{S}_{\rm 1st}$ has been cooked up to generate these equations. When we substitute them back into (30.3), we get the conventional form (30.1) of the Lagrangian. The second-order form has second derivatives, with no trace of $F_{\mu\nu}$, and π_μ as independent dynamical variables. We have discussed this process before (§29.5). It involves searching for the minimum of a quadratic form. Indeed, we see that the quadratic terms in the variables to be eliminated—$F_{\mu\nu}F^{\mu\nu}$ and $\pi_\mu\pi^{\mu *}$—have only constant coefficients. Thus, just as before,

$$\int \prod_\mu (d\pi_\mu) \prod_\nu (d\pi^*_\nu)(d\phi)(d\phi^*) \prod_\lambda (dA_\lambda) \prod_{\sigma\tau}(dF_{\sigma\tau})\, e^{i\mathcal{S}_{\rm 1st}} = \int \prod_\lambda (d\phi)(d\phi^*)(dA_\lambda)\, e^{i\mathcal{S}_{\rm 2nd}} \tag{30.5}$$

where $\mathcal{S}_{\rm 2nd}$ is the action in Lagrangian form, a function only of the variables displayed in the integration measure.

We could however choose to eliminate a different set of constrained variables: A_0, F_{ij}, π_i and π^*_i. We can do it by our trick if these terms enter the Lagrangian no more than quadratically, and if the coefficients of the quadratic terms are constants. None of the terms gives us problems since these *are* quadratic and *a fortiori* quadratic or linear or constant in the variables we wish to eliminate. The only terms that could give us problems are $\pi^*_\mu(\partial^\mu + ieA^\mu)\phi$ or $\pi_\mu(\partial^\mu - ieA^\mu)\phi^*$. There are no cross terms of any kind between any of the constrained variables. These terms *do* involve A_0 but only multiplied by π^*_0 or π_0 which we are not intending to eliminate, and likewise π_i or π^*_i, but only multiplied by A_i, which we are keeping. So $ie\pi^*_\mu A^\mu\phi$ and $-ie\pi^*_\mu A^\mu\phi^*$ are both linear in the terms we are going to lose upon substitution. We are left with F_{0i}, π_0, π^*_0, ϕ, ϕ^* and A_i. These are precisely the q's and p's, the fields and their conjugate momenta, of the Hamiltonian formulation. Writing the action in terms of *these*

q	p
A_i	F_{0i}
ϕ	π_0
ϕ^*	π^*_0

Table 30.1: Hamiltonian variables for scalar electrodynamics

variables is what it means to write the action in Hamiltonian form, as $\mathcal{S}_{\mathscr{H}}$, in terms of the independent variables of Hamiltonian dynamics:

$$\int \prod_\mu (d\pi_\mu) \prod_\nu (d\pi^*_\nu)(d\phi)(d\phi^*) \prod_\lambda (dA_\lambda) \prod_{\sigma\tau} (dF_{\sigma\tau})\, e^{iS_{1\text{st}}} = \int (d\phi)(d\phi^*) \prod_\lambda (dA_\lambda)\, e^{iS_{2\text{nd}}}$$
$$= \int (d\pi_0)(d\pi^*_0)(d\phi)(d\phi^*) \prod_i (dA_i) \prod_j (dF_{0j})\, e^{iS_{\mathscr{H}}} \tag{30.6}$$

The last form is the functional integral in the Hamiltonian form, so it is guaranteed to give us results equivalent to canonical quantization and Dyson's formula. All three forms are equal. The middle form gives us the naïve Feynman rules. Therefore, just as in charged spinor-massive photon theory, *the naive Feynman rules are true.* Every derivative interaction is simply a factor of p_μ, etc. We have redeemed the guess (14.57), as promised at the beginning of Chapter 28. We just read the Feynman rules off the Lagrangian, and the problem is solved.

Once we get the basic trick, the demonstration goes very fast. There are no worries about pulling time derivatives through time-ordered products. We solved that problem once for a free field (see (29.66) through (29.71)), but that was the only time we had to solve it, by dint of proving our general theorems when we talked about J's and K's. From now on, it's going to happen automatically; this formula is going to take care of everything. It is the same formula that works in quantum electrodynamics.

We will go through this line of reasoning again for gauge fields. We'll write the functional integral in two forms, obtained from each other by the most trivial of manipulations in functional integration language, although rather difficult manipulations if we attempt to do it in operator language. One form manifestly gives the right generating functional because it is the Hamiltonian form, but it is difficult to try to derive Feynman rules in this language—it doesn't even look covariant. We have π_0 but not π_i, F_{0i} but not F_{ij}. The other form looks nice and covariant and has nice simple Feynman rules. The two functional integrals are equal. In the Hamiltonian form it's easy to show everything is OK; in the Lagrangian form it's easy to derive Feynman rules. This has been a rather abstract stretch, so perhaps it's time to do some specific computations.

30.2 The Feynman rules for massive photon electrodynamics

I'll give the Feynman rules for massive electrodynamics with both scalar and spinor interactions, and then some low order computations with spinor electrodynamics, a somewhat simpler theory than scalar electrodynamics. There will be problems on scalar electrodynamics in the next homework (Problems 16).

Scalar electrodynamics with a massive photon

We've established that we can read off the Feynman rules from the second-order form of the Lagrangian, so let's start with that. Rewriting (30.1) and integrating twice by parts,

$$\begin{aligned}\mathscr{L} &= -\tfrac{1}{4}(\partial_\mu A_\nu - \partial_\nu A_\mu)(\partial^\mu A^\nu - \partial^\nu A^\mu) + \tfrac{1}{2}\mu^2 A_\mu A^\mu + (\partial_\mu - ieA_\mu)\phi^*(\partial^\mu + ieA^\mu)\phi - m^2\phi^*\phi \\ &= \underbrace{\tfrac{1}{2}A_\mu[g^{\mu\nu}(\Box^2 + \mu^2) - \partial^\mu\partial^\nu]A_\nu - \phi^*(\Box^2 + m^2)\phi}_{\mathscr{L}_0} + \underbrace{e^2 A_\mu A^\mu \phi^*\phi - ieA_\mu(\phi^*\partial^\mu\phi - (\partial^\mu\phi^*)\phi)}_{\mathscr{L}'}\end{aligned} \tag{30.7}$$

The massive vector propagator $D_F^{\mu\nu}(k)$ is given by (28.97), and the scalar propagator by the Feynman propagator $\widetilde{\Delta}_F(q)$, (10.29). I'll write down the complete set of Feynman rules for scalar-massive photon interactions (see also the box on p. 571).

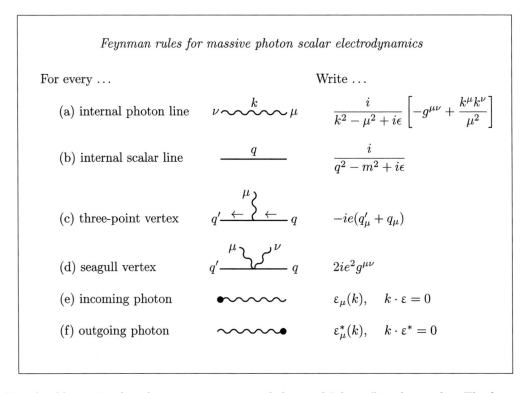

Feynman rules for massive photon scalar electrodynamics

For every ...		Write ...
(a) internal photon line	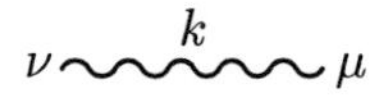	$\dfrac{i}{k^2-\mu^2+i\epsilon}\left[-g^{\mu\nu}+\dfrac{k^\mu k^\nu}{\mu^2}\right]$
(b) internal scalar line	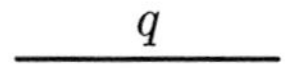	$\dfrac{i}{q^2-m^2+i\epsilon}$
(c) three-point vertex	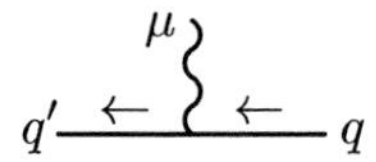	$-ie(q'_\mu+q_\mu)$
(d) seagull vertex	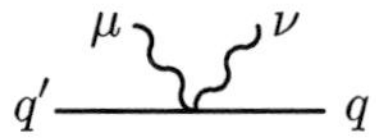	$2ie^2 g^{\mu\nu}$
(e) incoming photon		$\varepsilon_\mu(k), \quad k\cdot\varepsilon=0$
(f) outgoing photon		$\varepsilon^*_\mu(k), \quad k\cdot\varepsilon^*=0$

You should imagine that there are quotes around the word "photon" in these rules. The factors are easily read off the Lagrangian. For example, the three-point vertex arises from the term $-ieA_\mu(\phi^*\partial^\mu\phi-(\partial^\mu\phi^*)\phi)$. In momentum space, we pick up an i from the vertex, a factor of $-ie$ from the coefficient of the fields, and a factor of $(-iq^\mu - iq'^\mu)$ from the derivatives acting on ϕ and $-\phi^*$, respectively. That gives an overall factor of $-ie(q^\mu+q'^\mu)$. Similarly the four-point vertex has a factor of i from the vertex, e^2 from the coefficient, $g^{\mu\nu}$ from the two fields, and a factor of 2 from the combinatorics ($A^\nu A^\mu$ gives the same contribution as $A^\mu A^\nu$).

Spinor electrodynamics with a massive photon

In the same way, we can read off the Feynman rules from the Lagrangian for a Fermi field coupled to a massive photon. This is the sum of the free Proca Lagrangian (26.47), written in terms of the A^μ's, and the minimally coupled (27.58) Dirac Lagrangian:

$$\begin{aligned}\mathscr{L} &= -\tfrac{1}{4}(\partial_\mu A_\nu-\partial_\nu A_\mu)(\partial^\mu A^\nu-\partial^\nu A^\mu)+\tfrac{1}{2}\mu^2 A_\mu A^\mu+\overline{\psi}(i\partial\!\!\!/-m-eA\!\!\!/)\psi \\ &= \underbrace{\tfrac{1}{2}A_\mu\{g^{\mu\nu}[\Box^2+\mu^2]-\partial^\mu\partial^\nu\}A_\nu+\overline{\psi}(i\partial\!\!\!/-m)\psi}_{\mathscr{L}_0}+\underbrace{[-eA_\mu\overline{\psi}\gamma^\mu\psi]}_{\mathscr{L}'}\end{aligned} \tag{30.8}$$

This is a somewhat simpler theory than scalar electrodynamics, where we have to worry about the balance between the $e^2A_\mu^2$ interaction and the derivative interaction. The Feynman propagator $S_F(p\!\!\!/)$ for a fermion is given by (21.79). Here are the Feynman rules for spinor electrodynamics with a massive photon:

Feynman rules for fermions and massive photons

For every ...		Write ...
(a) internal photon line	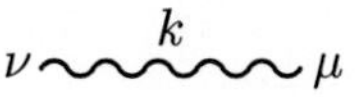	$\dfrac{i}{k^2-\mu^2+i\epsilon}\left[-g^{\mu\nu}+\dfrac{k^\mu k^\nu}{\mu^2}\right]$
(b) internal fermion line	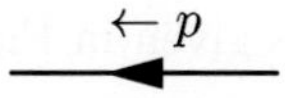	$\dfrac{i}{\not{p}-m+i\epsilon}$
(c) three-point vertex	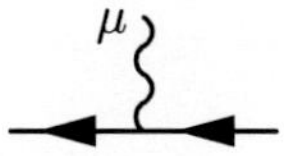	$-ie\gamma_\mu$
(d) incoming electron	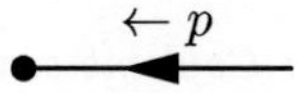	u_p
(e) outgoing electron	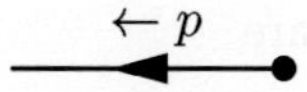	$\overline{u}_p$
(f) incoming positron	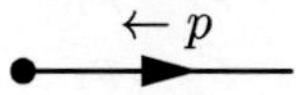	$\overline{v}_p$
(g) outgoing positron	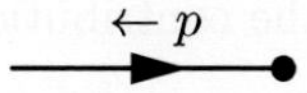	v_p
(h) incoming photon		$\varepsilon_\mu(k), \quad k\cdot\varepsilon=0$
(i) outgoing photon		$\varepsilon^*_\mu(k), \quad k\cdot\varepsilon^*=0$

There is a factor in both these sets of Feynman rules that at first glance should make you a bit nervous about a smooth passage to a zero-mass limit, to actual photons. In both the scalar and the spinor cases, the term

$$\frac{k_\mu k_\nu}{\mu^2}$$

in the vector propagator looks like bad news, in two ways. This term will give us trouble not only in going to the zero-mass limit, but also in keeping the theory renormalizable. At high energies, the propagator goes not like $1/k^2$, as with scalar propagators, but like k^2/k^2, which is simply 1. That certainly is going to make Feynman integrals much more divergent than they would be in a theory with scalar 'photons'. We're going to have to worry about that. In the low order computations, I will demonstrate that this term could have been crossed out without changing anything. In a future lecture, I will demonstrate that you can get rid of this part of the propagator altogether for a massive Abelian theory, with only one vector particle, like massive QED. For massive *Yang–Mills* theories, with more than one vector particle, you can't get rid of it. For non-Abelian gauge theories the situation is different. There the massless theory is *not* the limit of the massive theory.[5]

30.3 Some low order computations in spinor electrodynamics

We'll look at two processes, Coulomb scattering and Compton scattering.

EXAMPLE. *Coulomb scattering*

Let's consider the elastic scattering of two electrons, $e(p_1)+e(p_2) \to e(p_1')+e(p_2')$ to $\mathcal{O}(e^2)$. The topological structure of this process is given in Figure 30.1 (*cf.* Figure 21.7).

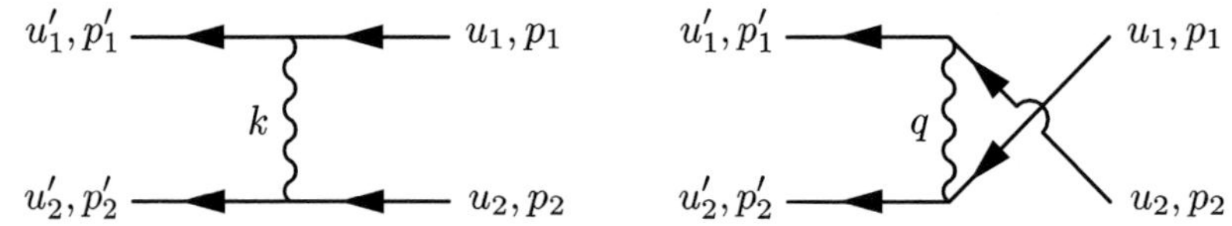

Figure 30.1: Coulomb scattering

The internal momenta in the two graphs are

$$\begin{aligned} k &= p_1 - p_1' \\ q &= p_2' - p_1 \end{aligned} \tag{30.9}$$

The invariant amplitude $\mathcal{A}_{fi}$ is a sum of the contributions $\mathcal{A}_1$ and $\mathcal{A}_2$ of the first and second graphs, respectively:

$$\begin{aligned} \mathcal{A}_{fi} &= \mathcal{A}_1 + \mathcal{A}_2 \\ i\mathcal{A}_1 &= -i(ie)^2(\overline{u}_1'\gamma^\mu u_1)(\overline{u}_2'\gamma^\nu u_2)\frac{[g_{\mu\nu} - (k_\mu k_\nu/\mu^2)]}{k^2-\mu^2} \\ i\mathcal{A}_2 &= +i(ie)^2(\overline{u}_1'\gamma^\mu u_2)(\overline{u}_2'\gamma^\nu u_1)\frac{[g_{\mu\nu} - (q_\mu q_\nu/\mu^2)]}{q^2-\mu^2} \end{aligned} \tag{30.10}$$

where the relative minus sign is due to the exchange of the two incoming identical fermions (see the discussion of nucleon-nucleon scattering in §21.5, pp. 447–448).

Let's focus on the apparently disastrous terms, with $k_\mu k_\nu/\mu^2$ or $q_\mu q_\nu/\mu^2$. These are free fermions, so

$$k_\mu(\overline{u}_1'\gamma^\mu u_1) = \overline{u}_1'(\not{p}_1 - \not{p}_1')u_1 = \overline{u}_1'(m-m)u_1 = 0 \tag{30.11}$$

Therefore the $k_\mu k_\nu/\mu^2$ term in $\mathcal{A}_1$ actually drops out, and the same thing happens to the $q_\mu q_\nu/\mu^2$ term in $\mathcal{A}_2$. (We will see later the general reason why these terms are *always* absent.)

[5] [Eds.] The reader is probably familiar with the terms "Yang–Mills" and "non-Abelian gauge theories", just as Coleman's students were in 1976. In 1954, Yang and Mills wrote a landmark paper generalizing Maxwell's theory of a single vector field to a theory of three vector fields transforming among themselves under the Lie group SU(2). The gauge invariance of electrodynamics is based upon the Lie group U(1). This group has only one generator, and so it is trivially Abelian: its generator commutes with itself. SU(2), with three non-commuting generators, is non-Abelian. The terms "Yang–Mills theory" and "non-Abelian gauge theory" are effectively synonymous, even though electrodynamics is a Yang–Mills theory, and general relativity, though a gauge theory, is not usually regarded as a Yang–Mills theory: C. N. Yang and R. Mills, "Conservation of Isotopic Spin and Isotopic Gauge Invariance", *Phys. Rev.* **96** (1954) 191–195; see also §46.2 and §47.3. For the different zero-mass limits of Abelian *vs.* non-Abelian gauge theories, see note 22, p. 1044.

Then

$$\mathcal{A}_1 = e^2(\overline{u}_1'\gamma_\mu u_1)(\overline{u}_2'\gamma_\nu u_2)\frac{g^{\mu\nu}}{k^2-\mu^2} \tag{30.12}$$

$$\mathcal{A}_2 = -e^2(\overline{u}_1'\gamma_\mu u_2)(\overline{u}_2'\gamma_\nu u_1)\frac{g^{\mu\nu}}{q^2-\mu^2} \tag{30.13}$$

Aside from the spin factors this is much like the exchange of a scalar meson, (21.100): the only difference is that we have a γ_μ instead of a γ_5. This is exactly the same as if we had exchanged four scalar mesons, one coupled to each of γ_0, γ_1, γ_2 and γ_3. It looks like there are four kinds of photons and that one of them, is very peculiar: it has a *negative* propagator, proportional to g^{00}. That's an illusion. From the orthogonality condition, $k\cdot\varepsilon = 0$, we see that there are only three transverse photons being exchanged; the projection operator doesn't make any difference. To obtain the differential cross-section, I would need to do the spin sum. I won't do that here. If you want to see that, it's in Bjorken and Drell.[6] I would instead like to talk about the zero-mass limit, which we can now take in a smooth way since we've gotten rid of the $k^\mu k^\nu/\mu^2$ terms in the vector propagator.

The zero-mass limit: $\mu^2 \to 0$.

There is first the fact that the forward peak, which typically occurs in lowest order scattering, has now moved onto the verge of the physical region.[7] The forward peak is infinite: $k^2 = 0$ in the forward direction and the denominator blows up. So the differential cross-section $d\sigma/d\Omega$ has an infinite peak in the forward direction. This is no surprise: exactly the same thing occurs in non-relativistic Coulomb scattering, under the Rutherford formula:[8]

$$\frac{d\sigma}{d\Omega} = \left(\frac{Z_1 Z_2 e^2}{2mv^2}\right)^2 \frac{1}{\sin^4(\theta/2)} \tag{30.14}$$

It is due to the long range nature of the Coulomb force. There is nothing particularly field theoretic about it; it's just what happens when we scatter charged particles.

Something else stands out in the $\mu \to 0$ limit. If you've had some previous exposure to QED, the limit of (30.12) or (30.13) may look a little strange. It looks like we are exchanging *four* photons, one of which has a negative sign in its propagator. This isn't the quantum electrodynamics you may have seen before, quantized in the Coulomb gauge, in which there are only *two* kinds of photons, the two three-dimensional transverse photons, with polarization vectors $\boldsymbol{\varepsilon}$ perpendicular to $\mathbf{k}$. In addition, something arises in the Coulomb gauge quantization

---

[6] [Eds.] See Bjorken & Drell *RQM*, pp. 102–106 for electron scattering in a Coulomb potential; they obtain the *Mott cross-section* for Rutherford scattering in equation (7.22):

$$\frac{d\sigma}{d\Omega} = \left(\frac{Z_1 Z_2 e^2}{2mv^2}\right)^2 \frac{1}{\sin^4(\theta/2)}\left(1 - v^2\sin^2(\theta/2)\right)$$

The Mott formula is the relativistic generalization of the Rutherford formula. For electron–electron scattering, called *Møller scattering*, see section 7.9, pp. 135–140. That cross-section, in the high-energy limit, is given in equation (7.84), p. 138.

[7] [Eds.] See §11.3. In terms of the Mandelstam variables, $k^2 = t$ and $q^2 = u$. In the limit that both t and u approach zero, the process becomes unphysical.

[8] [Eds.] See Problem 11.1, p. 397, in D. J. Griffiths, *Introduction to Quantum Mechanics*, 2nd ed., Cambridge U. P., 2016; Landau & Lifshitz, *QM*, §133, pp. 516–519.

that doesn't seem to appear in this formulation at all, an instantaneous action-at-a-distance Coulomb interaction.[9] Where does that come from?

Well, that Coulomb interaction is actually in (30.10); we'll see this by rewriting and reinterpreting the amplitude by purely algebraic means. To make life simple, so that we don't need to continually write $\overline{u}$ and u, we define the "currents" j^i_μ as

$$j_\mu^{(1,2)} = \overline{u}'_{(1,2)}\gamma_\mu u_{(1,2)} \tag{30.15}$$

From (30.11), each current is conserved:

$$k^\mu j_\mu^{(1,2)} = 0 \tag{30.16}$$

To see the non-relativistic Coulomb contribution, separate out the space and time parts:

$$k_0 j_0^{(1,2)} - \mathbf{k}\cdot\mathbf{j}^{(1,2)} = 0 \tag{30.17}$$

Write (30.12) as

$$\mathcal{A}_1 = \frac{e^2(j_0^{(1)}j_0^{(2)} - \mathbf{j}^{(1)}\cdot\mathbf{j}^{(2)})}{k_0^2 - |\mathbf{k}|^2} \tag{30.18}$$

(the same reasoning applies to $\mathcal{A}_2$, with a different set of j's). Separate the $\mathbf{j}^{(r)}$'s ($r = 1, 2$) into their spatially transverse and longitudinal parts:

$$\mathbf{j}^{(r)} = \mathbf{j}^{(r)\,T} + \mathbf{k}\,\frac{\mathbf{k}\cdot\mathbf{j}^{(r)}}{|\mathbf{k}|^2} \tag{30.19}$$

By construction,

$$\mathbf{k}\cdot\mathbf{j}^{(r)\,T} = 0 \tag{30.20}$$

and by current conservation,

$$\mathbf{j}^{(r)} = \mathbf{j}^{(r)\,T} + \mathbf{k}\,\frac{k_0 j_0^{(r)}}{|\mathbf{k}|^2} \tag{30.21}$$

Rewriting $\mathcal{A}_1$ in terms of $\mathbf{j}^{(r)}$,

$$\mathcal{A}_1 = e^2\left\{-\frac{\mathbf{j}^{(1)\,T}\cdot\mathbf{j}^{(2)\,T}}{k_0^2 - |\mathbf{k}|^2} + \frac{j_0^{(1)}j_0^{(2)}}{k_0^2 - |\mathbf{k}|^2}\left[1 - \frac{k_0^2}{|\mathbf{k}|^2}\right]\right\} = -e^2\,\frac{\mathbf{j}^{(1)\,T}\cdot\mathbf{j}^{(2)\,T}}{k_0^2 - |\mathbf{k}|^2} - e^2\,\frac{j_0^{(1)}j_0^{(2)}}{|\mathbf{k}|^2} \tag{30.22}$$

Calling the four photons "apples", we see that we have exchanged four apples for two apples and an orange.[10] The first term on the right side of (30.22) may be interpreted as the exchange of two transverse photons. There is the typical massless photon propagator in the denominator, and in the numerator the interaction between the traverse parts of the current. So there is only an interaction between two types of photons, because there are only two

[9] [Eds.] See Bjorken & Drell *Fields*, Chap. 14, pp. 68–81; the instantaneous Coulomb interaction part of the propagator is given in equation (14.55), p. 80; or Appendix A, p. 301 in J. J. Sakurai, *Advanced Quantum Mechanics*, Addison–Wesley, 1967. This instantaneous Coulomb interaction is also found in the *classical* solution of Maxwell's equations in Coulomb gauge. See the paragraph following equation (10.10), p. 441 in David J. Griffiths, *Introduction to Electrodynamics*, 4th ed., Pearson, 2013.

[10] [Eds.] What follows from here through (30.32) is based on class notes from 1999, supplied by Daniel Podolsky.

independent components to a transverse 3-vector field. In the second term (the orange), the coefficient of $j_0^{(1)} j_0^{(2)}$ is simply $-e^2/|\mathbf{k}|^2$, with *no* k_0 in the denominator. That is to say, it does not correspond to a time-dependent interaction, which would have k_0 in its Fourier transform, but to an *instantaneous* interaction. Notice also the appropriate sign change has taken place; previously the propagator had the wrong sign. And indeed, the Fourier transform of $1/|\mathbf{k}|^2$ is the Coulomb interaction $\propto 1/|\mathbf{r}|$ between the currents. Thus this amplitude, which appears to correspond to the exchange of four kinds of photons, one with the wrong sign, is indeed equivalent to the exchange of two transverse photons plus the instantaneous Coulomb interaction between the charge densities, just like standard QED in the Coulomb gauge.

It is worthwhile to compare this result with what we found in Model 2 (§9.3). If we let the interaction Lagrangian be

$$\mathscr{L}_I = -g\phi(x)\rho(x) \tag{30.23}$$

(without the counterterm a) then as we found earlier (9.36) for a time-independent $\rho(x) = \rho(\mathbf{x})$,

$$\begin{aligned} E_0 = -(i/T)\ln\langle 0|\mathrm{S}|0\rangle &= \tfrac{1}{2}(-ig)^2 \int \frac{d^3\mathbf{k}}{(2\pi)^3} \frac{|\tilde{\rho}(\mathbf{k})|^2}{|\mathbf{k}|^2+\mu^2} \\ &= \tfrac{1}{2}\int d^3\mathbf{x}\, d^3\mathbf{y}\, \rho(\mathbf{x}) \left[-g^2 \frac{e^{-\mu|\mathbf{x}-\mathbf{y}|}}{4\pi|\mathbf{x}-\mathbf{y}|} \right] \rho(\mathbf{y}) \end{aligned} \tag{30.24}$$

The quantity in the square brackets is the Yukawa potential $V(|\mathbf{x}-\mathbf{y}|)$; scalar exchange is *attractive* between identical particles. In the massive vector case,

$$\mathscr{L}_I = -eA_\mu J^\mu \tag{30.25}$$

we have

$$\ln\langle 0|\mathrm{S}|0\rangle = \tfrac{1}{2}(-ie)^2 \int \frac{d^4k}{(2\pi)^4} \widetilde{J}_\mu(k)\widetilde{J}_\nu(k)^* \frac{i}{k^2-\mu^2+i\epsilon}\left[-g^{\mu\nu} + \frac{k^\mu k^\nu}{\mu^2} \right] \tag{30.26}$$

Because the current is conserved, $\partial^\mu J_\mu = 0$, it follows that

$$k^\mu \widetilde{J}_\mu(k) = 0 \tag{30.27}$$

so that (in agreement with the general result) the second term in the propagator can be discarded, and we can safely set $\mu = 0$ to study the photon interaction:

$$\ln\langle 0|\mathrm{S}|0\rangle = \tfrac{1}{2}(-ie)^2 \int \frac{d^4k}{(2\pi)^4} \widetilde{J}_\mu(k)\widetilde{J}_\nu(k)^* \frac{i}{k^2+i\epsilon}\left[-g^{\mu\nu} \right] \tag{30.28}$$

If we're looking at electrostatics,

$$J_\mu(x) = (J_0(x), \mathbf{J}(x)) = (\rho(\mathbf{x}), \mathbf{0}) \tag{30.29}$$

Analogous to (30.24), (30.28) leads to

$$\begin{aligned} E_0 = -(i/T)\ln\langle 0|\mathrm{S}|0\rangle &= -\tfrac{1}{2}(-ie)^2 \int \frac{d^3\mathbf{k}}{(2\pi)^3} \frac{|\tilde{\rho}(\mathbf{k})|^2}{|\mathbf{k}|^2} \\ &= \tfrac{1}{2}\int d^3\mathbf{x}\, d^3\mathbf{y}\, \rho(\mathbf{x}) \left[\frac{e^2}{4\pi|\mathbf{x}-\mathbf{y}|} \right] \rho(\mathbf{y}) \end{aligned} \tag{30.30}$$

That is, the Coulomb potential

$$V(|\mathbf{x}-\mathbf{y}|) = \frac{e^2}{4\pi|\mathbf{x}-\mathbf{y}|} \tag{30.31}$$

is *repulsive* between identical charges, the extra minus sign coming from the g^{00} in the propagator.[11] Returning to (30.22), it's easy to see that the Coulomb interaction is *repulsive* if $j_0^{(1)}$ and $j_0^{(2)}$ have the same sign and attractive if $j_0^{(1)}$ and $j_0^{(2)}$ have opposite signs:

$$\begin{aligned}
\tfrac{1}{2}e^2\int\frac{d^4k}{(2\pi)^4}\frac{\tilde{j}_0^{(1)}(k)\tilde{j}_0^{(2)*}(k)}{|\mathbf{k}|^2} &= \tfrac{1}{2}e^2\int d^4x\,d^4y\int\frac{d^4k}{(2\pi)^4}\frac{1}{|\mathbf{k}|^2}j_0^{(1)}(x)e^{ik\cdot x}j_0^{(2)}(y)e^{-ik\cdot y}\\
&= \tfrac{1}{2}e^2\int d^4x\,d^4y\int\frac{d^3\mathbf{k}}{(2\pi)^3}e^{-i\mathbf{k}\bullet(\mathbf{x}-\mathbf{y})}\delta(x^0-y^0)\frac{1}{|\mathbf{k}|^2}j_0^{(1)}(x)j_0^{(2)}(y)\\
&= \tfrac{1}{2}e^2\int dt\,d^3\mathbf{x}\,d^3\mathbf{y}\,j_0^{(1)}(\mathbf{x},t)j_0^{(2)}(\mathbf{y},t)\int\frac{d^3\mathbf{k}}{(2\pi)^3}e^{-i\mathbf{k}\bullet(\mathbf{x}-\mathbf{y})}\frac{1}{|\mathbf{k}|^2}\\
&= \tfrac{1}{2}\int dt\,d^3\mathbf{x}\,d^3\mathbf{y}\,j_0^{(1)}(\mathbf{x},t)\frac{e^2}{4\pi|\mathbf{x}-\mathbf{y}|}j_0^{(2)}(\mathbf{y},t)
\end{aligned} \tag{30.32}$$

As the currents have the same time t in their arguments, the interaction is instantaneous.

Of course we've only been playing with low order diagrams. If we want to show that similar things are true in general we either have to crank up an enormous amount of combinatoric machinery, or establish some general formalism which enables us to short-circuit the combinatoric machinery, i.e., functional integration. I will do that later on, but I thought you should see how these things work out in particular diagrams before I show you the general argument.

EXAMPLE. *Compton scattering*

The next process I would like to discuss, although not in nearly as much detail as Coulomb scattering, is Compton scattering, $e(p)+\gamma(k,\varepsilon)\to e(p')+\gamma(k',\varepsilon')$. (See Bjorken and Drell for a fuller discussion.[12]) Aside from the extra indices this is just the same sort of thing as meson-nucleon scattering (see Figure 11.2, p. 228). There are two diagrams with the same

---

[11] [Eds.] The identical argument is given by Zee *QFTN*, pp. 32–33, with explicit citation of Coleman's QFT course. Incidentally, Zee earned his PhD under Coleman. The first demonstration that photon exchange leads to the Coulomb potential seems to have been given by V. A. Fock and Boris Podolsky, "On the quantization of electro-magnetic waves and the interaction of charges in Dirac's theory", *Phys. Zeits. Sowjetunion* **1** (1932) 801–817, following Dirac's demonstration of the one-dimensional (attractive!) result: P. A. M. Dirac, "Relativistic Quantum Mechanics", *Proc. Roy. Soc. Ser. A* **136** (1932) 453–464. Schweber writes, "In 1932, Fermi and Bethe set out to derive the interaction potential between two charged particles, including magnetic and retardation effects (H. Bethe und E. Fermi, "Über die Wechselwirkung von zwei Elektronen" (On the interaction of two electrons), *Zeits. Phys.* **77** (1932) 296-306; reprinted in *Enrico Fermi: Collected Papers*, v.1, ed. E. Segrè *et al.*, U Chicago Press, 1962)... Bethe and Fermi's aim was to reveal the relation between Møller's and Breit's approaches, and more important, to demonstrate how perturbation theory could be used to generate transparent results. It is clear from their derivation that Bethe and Fermi considered the force between the charged particles as arising form the exchange of the photons between them." Silvan S. Schweber, "Enrico Fermi and Quantum Electrodynamics, 1929–32", *Phys. Today* **55** (2002) 31–36. An expression equivalent to (30.30) is given in Gregor Wentzel, *Quantum Theory of Fields*, trans. C. Houtermans and J. M. Jauch, Interscience, 1949, p. 132, equation (17.38); the original German text was published in 1943.

[12] [Eds.] Bjorken & Drell *RQM*, pp. 127–132. See also the example in §26.5, p. 571.

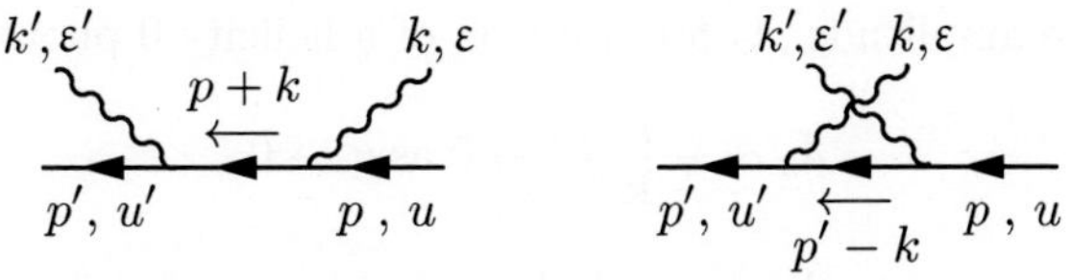

Figure 30.2: Compton scattering

topological structure. The invariant Feynman amplitude is

$$\begin{aligned} i\mathcal{A}_{fi} &= (-ie)^2 i[\overline{u}'\gamma_\mu \varepsilon'^{\mu *}\frac{1}{\not p + \not k - m}\gamma_\lambda \varepsilon^\lambda u + \overline{u}'\gamma_\lambda \varepsilon^\lambda \frac{1}{\not p - \not k' - m}\gamma_\mu \varepsilon'^{\mu *} u] \\ &= -ie^2[\overline{u}'\not\varepsilon'^{*}\frac{1}{\not p + \not k - m}\not\varepsilon u + \overline{u}'\not\varepsilon \frac{1}{\not p - \not k' - m}\not\varepsilon'^{*} u] \end{aligned} \tag{30.33}$$

(Note that $\not p - \not k' = \not p' - \not k$.) We don't need the $i\epsilon$ in the denominator because the pole is not in the physical region. We could rationalize the propagator denominators, commute γ matrices around, and use the fact that u and u' are free solutions to the Dirac equation ($\not p u = mu$) to simplify things. I shall not bore you with that; it's a standard computation that you can look up in Bjorken and Drell. Instead, I would again like to focus on the zero mass limit. We will find some interesting properties of this as $\mu^2 \to 0$.

The zero-mass limit: $\mu^2 \to 0$

Let's recall something from an earlier lecture (§26.4): the emission or absorption of a photon by an external current distribution j^μ. We found (26.73) the amplitude $\mathcal{A}_{fi}$ for the emission of a single photon was proportional to $\varepsilon^* \cdot \widetilde{j}$:

$$\mathcal{A}_{fi} \propto \varepsilon^*_\mu(k)\widetilde{j}^\mu(k) \tag{30.34}$$

We knew, because the external current was conserved, that

$$k_\mu \widetilde{j}^\mu(k) = 0 \tag{30.35}$$

We also showed[13] that for helicity 0,

$$\varepsilon_\mu = \frac{k_\mu}{\mu} + \mathcal{O}(\frac{\mu}{|\mathbf{k}|}) \tag{30.36}$$

[13] [Eds.] This argument is a little incomplete. For massive vector fields, the polarization vectors ε^μ are *orthogonal* to the 4-momentum: $k^\mu \varepsilon_\mu = 0$. With $k^\mu = (\omega_{\mathbf{k}}, 0, 0, |\mathbf{k}|)$, the polarization vector $\varepsilon^{(3)\mu}$ for a helicity 0-vector was given as

$$\varepsilon^{(3)\mu} = \frac{1}{\mu}(|\mathbf{k}|, 0, 0, \omega_{\mathbf{k}}) \tag{26.78}$$

However, as usual,

$$\omega_{\mathbf{k}} = \sqrt{|\mathbf{k}|^2 + \mu^2} = |\mathbf{k}|(1 + \mathcal{O}(\mu^2/|\mathbf{k}|^2))$$

so that we can write

$$\begin{aligned} \varepsilon^{(3)\mu} &= \frac{1}{\mu}(|\mathbf{k}|, 0, 0, |\mathbf{k}| + \mathcal{O}(\mu^2/|\mathbf{k}|)) \\ k^\mu &= (|\mathbf{k}| + \mathcal{O}(\mu^2/|\mathbf{k}|), 0, 0, |\mathbf{k}|) \\ \therefore k^\mu - \mu\varepsilon^{(3)\mu} &= \mathcal{O}(\mu^2/|\mathbf{k}|)(1, 0, 0, -1) \end{aligned}$$

Though $\varepsilon^{(3)\mu}$ and k^μ must be orthogonal for a massive vector, $\varepsilon^{(3)} \cdot k = 0$, $\varepsilon^{(3)\mu}$ and k^μ/μ are also *parallel* to within $\mathcal{O}(\mu/|\mathbf{k}|)$.

From this we found the amplitude $\mathcal{A}_3$ for emission of a helicity 0 photon,

$$\mathcal{A}_3 \sim -\frac{\mu}{|\mathbf{k}|}\widetilde{j}^0 \to 0 \text{ as } \mu \to 0 \tag{26.81}$$

The point of this exercise was to demonstrate how the theory of a massive photon with three helicity states goes over to the theory of a massless photon with two helicity states as the mass goes to zero. Is a corresponding thing true in our more complicated theory, full of interactions? The key equation is the analog of current conservation, (30.16), $k_\mu \widetilde{j}^\mu(k) = 0$.

How do we compute the amplitude for emission or absorption of a photon $a \to b + \gamma$ in a fully interacting field theory? We know how to compute that amplitude in the general theory: we do it via the LSZ reduction formula, (14.18). We reduce and reduce and reduce. Imagine that we have reduced everything until only the last photon is unreduced. Then the reduction formula gives the amplitude $\mathcal{A}_{fi}$ as

$$\mathcal{A}_{fi} \propto \varepsilon^{*\mu}(\Box^2 + \mu^2)\,\langle b|A_\mu(x)|a\rangle \tag{30.37}$$

The photon contributes its polarization vector $\varepsilon^{*\mu}$, and the Klein–Gordon operator takes care of the pole in the propagator. The unreduced field $A_\mu(x)$ is the exact Heisenberg field, the states are the exact physical states. The time-ordering symbol T is unnecessary here; we have only one field left. If we're taking account of renormalization, we should use the renormalized A'_μ here, with amplitude 1 for making a photon, but we can absorb the constant into the proportionality. The Heisenberg equation of motion,

$$\partial_\mu A^\mu = 0 \tag{30.38}$$

follows from the Euler–Lagrange equations and current conservation. The derivative ∂_μ commutes with the Klein–Gordon operator, and therefore we find

$$\partial_\mu(\Box^2 + \mu^2)\,\langle b|A^\mu(x)|a\rangle = 0 \tag{30.39}$$

This is precisely the same statement, in position space rather than in momentum space, that we used earlier ($k_\mu \widetilde{j}^\mu(k) = 0$) to prove the suppression of helicity zero photons. Therefore the argument for the suppression of helicity zero photons as $\mu \to 0$ should be as true in the full field theory as it was in the theory with a c-number source.

General arguments are always nice, but one sleeps better at night if one has made particular checks in simple cases. So just to make sure nothing is going wrong, let me attempt to check this formula, that $\mathcal{A}_{fi} = 0$ when $\varepsilon^\mu \propto k^\mu$, which is equivalent to checking the conservation equation. The conservation equation is exact whether or nor $\mu = 0$. The suppression of helicity zero states is a kinematic consequence of it as $\mu \to 0$. So let's take the expression for the amplitude, (30.33), plug in $\varepsilon^\mu = k^\mu/\mu$ and see if the amplitude vanishes or not. Looking at the numerators, we can say

$$\begin{aligned} \not{\varepsilon} u &= \not{k} u/\mu = (\not{p} + \not{k} - m)u/\mu \\ \overline{u}'\not{\varepsilon} &= \overline{u}'\not{k}/\mu = \overline{u}'(-\not{p}' + \not{k} + m)/\mu \end{aligned} \tag{30.40}$$

since $(\not{p} - m)u = \overline{u}'(\not{p}' - m) = 0$. We can substitute these expressions in the numerators, and the extra factors cancel the Feynman denominators. Thus, just as in (26.100), we find that the full term is

$$\mathcal{A}_{fi} = -(e^2/\mu)[\overline{u'}\not{\varepsilon}'^* u - \overline{u'}\not{\varepsilon}'^* u] = 0 \tag{30.41}$$

Thus the amplitude vanishes for any non-zero mass when $\varepsilon^\mu = k^\mu/\mu$ and therefore vanishes for helicity 0 states when $\mu \to 0$.

Just because we've proven that something is true in general does not mean it is true at the level of individual Feynman diagrams. When we know something is true in general for all values of e, it is true for all derivatives with respect to e at $e = 0$; that is, for all orders of perturbation theory. However, each order of perturbation theory is a *sum* of Feynman diagrams. It means that it must be true for that whole sum; it does *not* mean that it must be true for individual Feynman diagrams. In this case we must take account of *both* diagrams to get the proper cancellation. We can use the general formula (30.37); we don't have to go through the complicated combinatorics for diagrams of order e^4, e^6, etc., to prove the theorem. The exact statement is that if $\varepsilon^\mu = k^\mu/\mu$ then $\mathcal{A}_{fi} = 0$. The second part of the argument is that, for a helicity zero state, we have (30.36):

$$\varepsilon_\mu = \frac{k_\mu}{\mu} + \mathcal{O}(\frac{\mu}{|\mathbf{k}|}) = \frac{k_\mu}{\mu} + \mathcal{O}(\frac{\mu}{k_0}) \tag{30.42}$$

That's just kinematics. Write down the properly normalized helicity zero polarization. Then dotting k^μ/μ into the amplitude gives zero, plus a term proportional to μ/k_0, the mass over the energy. The only part of the argument that needs to be checked in the full field theory is the statement that with $\varepsilon^\mu = k^\mu/\mu$ that $\mathcal{A}_{fi} = 0$. We've given a general proof and a specific example. We can write the amplitude for emitting a photon in a particular spin state r as

$$\mathcal{A}_{fi} = \varepsilon_\mu^{(r)*} M^\mu \qquad (M^\mu \text{ a conserved 4-vector; } k^\mu M_\mu = 0) \tag{30.43}$$

It doesn't matter where this amplitude M^μ came from, as long as it obeys the condition that $k_\mu M^\mu = 0$. That shows the suppression of helicity zero photons, because those photons have $\varepsilon_\mu \propto k_\mu$.

Summing and averaging over photon spins

Before I depart from specific Feynman calculations I would like to make a few comments about summing over photon spins.[14] We typically have to do a spin sum over final states,

$$\sum_{r=1}^{3} |M|^2 = \sum_{r=1}^{3} M^{*\mu} \varepsilon_\mu^{(r)} \varepsilon_\nu^{(r)*} M^\nu = M^{*\mu}\left[-g^{\mu\nu} + \frac{k^\mu k^\nu}{\mu^2}\right] M^\nu = -M_\mu^* M^\mu \tag{30.44}$$

the second equality following from (26.93); $-g^{\mu\nu} + (k^\mu k^\nu/\mu^2)$ is the projection operator onto the three four-dimensional transverse vectors. It's just like the spinor sum $\not p + m$, the projection operator onto the positive energy spinors. The vector spin sums are considerably easier than the spinor sums; for one thing $k^\mu a_\mu = 0$, so the second term in the projection operator gives zero. Also we don't have the analog of all those ugly extra γ matrices to commute around. So we simply obtain $-M_\mu^* M^\mu$.

Likewise for averaging over initial photon spins, if one has an unpolarized beam. This polarization sum (26.93) is true whether the mass is large or small. If the mass is small we may think that we want to sum over only two spin states. But we might as well sum over the third, because the amplitude for emitting the third is negligible. If we have an unpolarized beam and we wish to average over initial spins, we get the above result (30.44) multiplied by

---

[14] [Eds.] Bjorken & Drell *RQM*, p. 125.

1/3, for the three photon states if $\mu \approx k_0$, or by 1/2 for the two photon states if $\mu \ll k_0$.[15] If $\mu \ll k_0$ (the current experimental bound has $\lambda_{\text{Compton}} \gg 10^4$ km for the photon),[16] even the most imperfect light bulb will not emit all three helicity states with indifference. It will in fact emit *no* helicity zero photons. We may think it's an unpolarized beam, but in fact it's just a random mixture of two polarizations, not three; we would make an error inserting a $\frac{1}{3}$. In practice, we don't have to worry about intermediate ranges of mass. Either we are talking about something like ρ mesons, which, if unpolarized, really have three polarization states, or photons, which have, for all practical purposes, two, even if the photon mass is not strictly zero but only, say, $10^{-30}\,m_{\text{electron}}$.

30.4 Quantizing massless electrodynamics with functional integrals

We are now going back to the wonderland of functional integrals, where I will adopt the lecture style of the Delphic Oracle.[17] We're going to begin a discussion of massless electrodynamics. Not, as we have treated it until now, as the limit of the massive theory as $\mu \to 0$, but *sui generis*, as a theory by itself, without embedding it in another family of Lagrangians, and try to quantize it. The Lagrangian is nearly the same as (30.8), but now there is no mass term for the photon:

$$\mathscr{L} = -\tfrac{1}{4}(\partial_\mu A_\nu - \partial_\nu A_\mu)^2 + \overline{\psi}(i\partial\!\!\!/ - m - eA\!\!\!/)\psi + (\text{source terms in } J(x)) \tag{30.45}$$

This theory cannot be directly quantized by naive canonical methods. We cannot eliminate A_0 when $\mu = 0$ and the whole canonical quantization program falls apart. That's because this theory has gauge invariance. Let's review.

Gauge transformations are not like ordinary internal symmetries. They do not turn one physical situation into a distinguishable physical situation with an identical scattering amplitude, like isospin transformations turn a proton into a neutron. Rather, they simply represent changes in the description, not actual changes of the state. More conventional transformations can be sensibly interpreted both actively, as changing the state, or passively, as changing the description. But a gauge transformation is *only* sensibly interpreted in the passive sense; it is a change in the description, like translating a physics paper from English into French.[18]

In order to canonically quantize the theory we must pick a gauge. If we don't have a condition telling us what gauge we are in then we don't have a well-defined initial value problem. No matter how many derivatives of fields we specify on the initial value surface, we can always make a gauge transformation; that is, the identity on the initial surface and not the identity at some future time. Therefore, we must adopt a condition that firmly and forever fixes the gauge, such as the Coulomb gauge condition $\nabla \cdot \mathbf{A} = 0$. Then we have eliminated, by convention and by fiat, the gauge degrees of freedom, and we have a well-defined initial

---

[15] [Eds.] Bjorken & Drell *RQM*, p. 131; Jackson *CE*, pp. 694–5; V. B. Berestetskiĭ, E. M. Lifshitz, and L. P. Pitaevskiĭ, *Quantum Electrodynamics*, 2nd ed., Pergamon, 1982, pp. 354–364.

[16] [Eds.] See §26.4.

[17] [Eds.] Coleman jokes: "Like her, I speak while breathing in a gas—not a natural gas, as she did, but tobacco smoke." In the videotaped 1975–6 lectures, Coleman typically smoked eight cigarettes during a ninety-minute lecture. The Oracle of Delphi, (700 BCE–400 CE), known as the Pythia, was the high priestess of Apollo, thought to be able to foretell the future.

[18] [Eds.] Coleman is reiterating a point he made originally on p. 579.

value problem. If we are lucky in our choice of gauge, we can crank out the entire canonical machinery, eliminate the constrained variables, impose canonical commutation rules and be able to compute everything. We hope that the choice of gauge doesn't matter as far as actually observable quantities go, i.e., we predict gauge invariant results. The computations may be simpler in one gauge than in another, but the final answers should be the same in *all* gauges. Right now this statement is an act of faith. It is hoped that gauge invariance of the classical theory which we are about to quantize will carry over at least that much into the canonically quantized theory, but this remains to be shown; we will show it next time.[19] That's the canonical viewpoint of quantization of theories with gauge invariance as expressed by Fermi and Dirac, circa 1929–1930.[20]

We could also take a functional integral viewpoint. We have not been thinking of the functional integral as a primary object but as something we derived from canonical quantization. But some young revolutionaries could just as well take functional integration as fundamental, and forget about canonical quantization: we have these magic formulas, let's just apply them. However, if they were to try that in this case, they would run into trouble. Let's press our luck and see how far we get.

We would try to find the Feynman propagator by inverting the quadratic part of $\mathscr{L}$. Following the development from (28.86) to (28.94), we break things up into transverse and longitudinal projection operators

$$P^T_{\mu\nu} = g_{\mu\nu} - \frac{k_\mu k_\nu}{k^2}, \quad P^L_{\mu\nu} = \frac{k_\mu k_\nu}{k^2} \tag{30.46}$$

Only the transverse part of the field enters $\mathscr{L}$ because only the antisymmetric derivative of the field appears in the Lagrangian,[21] and the longitudinal part doesn't enter into the quadratic part of the Lagrangian at all. That is, setting $\mu^2 = 0$ in (28.95),

$$A_{\mu\nu} = -k^2 P^T_{\mu\nu} = (-k^2)P^T_{\mu\nu} + (0)P^L_{\mu\nu} \tag{30.47}$$

Inverting the coefficients as in (28.94) to get the Feynman propagator, we obtain

$$\widetilde{D}^F_{\mu\nu}(k) = -i\frac{[g_{\mu\nu} - (k_\mu k_\nu/k^2)]}{k^2} + \frac{k_\mu k_\nu}{k^2}\frac{i}{0} \tag{30.48}$$

This is garbage; I can't do any computations with this disastrous propagator.

Many years ago, two young Russians, Faddeev and Popov, looked at this problem from the functional integral point of view and made a guess about what to do.[22] With that guess they

---

[19] [Eds.] For a full discussion of gauge transformations at the operator level, including indefinite-metric state space and the appearance of the Coulomb interaction, see K. Haller, "Gauge Problems in Spinor Quantum Electrodynamics", *Acta Phys. Austr.* **42** (1975) 163–215.

[20] [Eds.] E. Fermi, "Quantum Theory of Radiation", *Rev. Mod. Phys.*, **4** (1932) 87–132; P. A. M. Dirac, "The Quantum Theory of the Emission and Absorption of Radiation", *Proc. Roy. Soc. London*, **A114** (1927) 243–265. Both papers are reprinted in Schwinger *QED*.

[21] [Eds.] Since $F_{\mu\nu}$ is gauge invariant, it can only be transverse.

[22] [Eds.] Ludvig D. Faddeev and Victor N. Popov, "Feynman diagrams for the Yang–Mills Field", *Phys. Lett.* **25B** (1967) 29–30; V. N. Popov and L. D. Faddeev, "Perturbation Theory for Gauge-Invariant Fields", Fermilab report NAL-THY-57 (1972); reprinted in G. 't Hooft, ed., *50 Years of Yang–Mills Theory*, World Scientific, 2005; and L. D. Faddeev, "Introduction to Functional Methods", pp. 1–40, in *Methods in Field Theory* (Les Houches 1975), R. Balian and J. Zinn–Justin, eds., North-Holland, 1976. The Faddeev–Popov methods are discussed in Chapter 31.

were able to verify canonical quantization. However, in order to explain their guess I will have to tell you a little bit more about Feynman's original formulation of functional integrals.

Aside. A brief historical digression on Feynman's sum over histories

Feynman called his functional integrals **path integrals**, and the specific operation of calculating them **the sum over histories.**[23] He didn't do it in a source formalism with a generating functional. He wanted to compute actual transition matrix elements. We'll just write down his formula without proof, for the simplest example of a particle in a potential

$$H = \frac{p^2}{2m} + V(q) \tag{30.49}$$

Feynman wanted to compute the transition amplitude $\langle q_2|e^{-iH(t_2-t_1)}|q_1\rangle$ for the state where the particle was at position q_1 at time t_1, to the state where it was at position q_2 at time t_2. He showed that it could be written as[24]

$$\langle q_2|e^{-iH(t_2-t_1)}|q_1\rangle = \int (dq)\, e^{i\int_{t_1}^{t_2} L dt} \tag{30.50}$$

where the integration doesn't go over arbitrary functions in the range t_1 to t_2 but is restricted to run over functions that are held fixed at the end points, $q(t_1) = q_1$, $q(t_2) = q_2$, just as in Hamilton's formulation of Lagrangian mechanics. Feynman described this as a "sum over histories". He said this was a neat formulation of quantum mechanics, and indeed it was. We imagine the particle goes over all possible classical paths from the desired initial state q_1 at t_1 to the desired final state q_2 at t_2. We sum $e^{i\int L dt}$ over all possible paths to get the transition matrix element. The functional integral gives a precise meaning to the concept of summation. Briefly, we divide the time between t_1 and t_2 into N equal intervals of width Δt, with $t_2 = t_1 + N\Delta t$, and approximate a classical path as a connected set of linear segments. Two such are shown[25] in Figure 30.3. We can add in a source term and let $t_1 \to -\infty$, $t_2 \to \infty$, and then we see how to get our formulation. If we kept the end points q_1 and q_2 fixed, we would get the transition matrix element from some initial state $|q_1\rangle$ to some final state $|q_2\rangle$ over an infinite stretch of time.

We aren't keeping the end points fixed, but that doesn't matter. Our discussion (§7.3) of how Dyson's formula gave us the S-matrix elements included an argument, for field theories, that as we go to the far past and the far future, no matter what states we have on the right and the left, all that survives is the vacuum-to-vacuum transition; all the other parts are canceled by contributions of oscillating phases.[26] So aside from some normalization factor, we could do it with a particular q_1 and q_2, and get the vacuum-to-vacuum transition. Or, we

---

[23] [Eds.] See Richard P. Feynman and Albert R. Hibbs, *Quantum Mechanics and Path Integrals*, McGraw-Hill, 1965; edited and corrected by Daniel F. Styer, Dover Publications, 2010. Feynman had been looking for a way to base quantum mechanics not on the Hamiltonian, but the Lagrangian. A colleague, Herbert Jehle, told Feynman about Dirac's paper: P. A. M. Dirac, "The Lagrangian in Quantum Mechanics", *Phys. Zeits. Sowjetunion* **3** (1933) 64–72; reprinted in Schwinger *QED*. See also *Feynman's Thesis: A New Approach to Quantum Mechanics*, ed. Laurie M. Brown, World Scientific, 2006. For the development of the path integral method, including many historical references, see D. Derbes, "Feynman's Derivation of the Schrödinger Equation", *Am. J. Phys.* **64** (1996) 881–884.

[24] [Eds.] The development of (30.50) is given in pp. 60–62 of Ernest S. Abers and Benjamin W. Lee,"Gauge Theories", *Phys. Lett.* **C9** (1973), 1–145. (*Physics Letters C* subsequently became *Physics Reports.*)

[25] [Eds.] Figure 30.3 is based on Figure 5.3, p. 157 in Ryder *QFT*.

[26] [Eds.] See (13.40) and the discussion following.

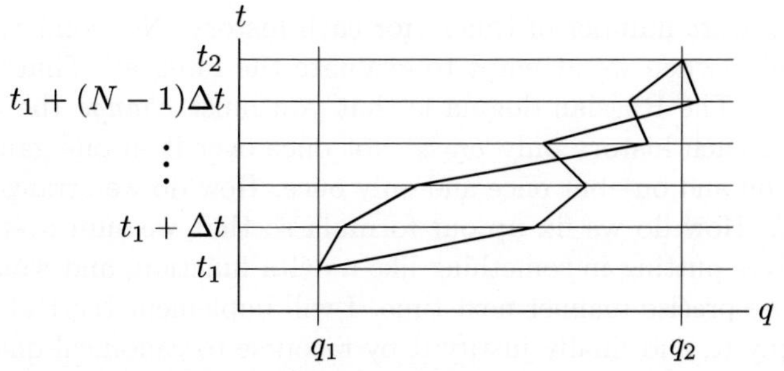

Figure 30.3: Two approximate paths from (q_1, t_1) to (q_2, t_2)

could let q_1 and q_2 be free, integrate over all possible q_1's and all possible q_2's, and then just change the normalization factor. So our form comes from Feynman's form. In fact, it's even better. We really work in Euclidean space where the non-vacuum states aren't canceled by some Riemann–Lebesgue argument, but by a decreasing exponential; it knocks them out even more forcefully.[27]

There's a second thing that we can see from this formulation that's not obvious from our formulation. We see why classical mechanics is important in the small $\hbar$ limit. From Feynman's formulation, we can see where Hamilton's Principle comes from by restoring the $\hbar$. The dimensions of $\hbar$, J-s, are those of an action, so Feynman's formulation should really read

$$\langle q_2 | e^{\{-(i/\hbar)H(t_2-t_1)\}} | q_1 \rangle = \int (dq)\, e^{\{(i/\hbar)\int_{t_1}^{t_2} L dt\}} \tag{30.51}$$

As $\hbar \to 0$, the phase factor on the right-hand side oscillates more and more rapidly. Rapidly oscillating integrals are dominated by points of stationary phase,[28] points where the phase is stationary when we vary the integration variables. In our case, the phase is the action $\mathcal{S}$ and the integration variable is the coordinate q, so we must vary it so that

$$\frac{\delta \mathcal{S}}{\delta q} = 0 \tag{30.52}$$

sticking to Feynman's boundary conditions $q(t_1) = q_1$, $q(t_2) = q_2$. This is nothing but Hamilton's Principle that picks out the classical motions. The reason that classical motions are important in the small $\hbar$ limit, according to Feynman, is because of the principle of stationary phase. They are the points where the phase, the action, is stationary.

We can now explain Faddeev and Popov's central idea. They said that putting the Lagrangian (30.45) into the functional integral was a very dumb thing to do, because Feynman says "sum over histories". If you have a gauge theory, then the same history, exactly the same motion for all observable quantities, may be represented by an infinite number of different fields, all of them connected to each other by a gauge transformation. So we're not summing over the histories in the right way. If we just stuck the Lagrangian, (30.45) into the functional integral and tried to sum over histories, we'd be summing over the same histories many, many

[27] [Eds.] See (28.39) and the discussion following.
[28] [Eds.] See §17.4.

times—in fact, an *infinite* number of times—for each history. No wonder, said Faddeev and Popov, we get infinity when we attempt to evaluate the integral! That's where the $\frac{1}{0}$, the infinity, comes from. The Russian dogma is that you must *change* the functional integral formula to sum over each history only *once*. Not once over it in one gauge, once over it in another gauge, and on and on—but once and only once. How do we arrange for the functional integral to do that? How do we fix up our formula so that we sum over each history only once? Well, it involves putting in something like a delta function, and what I mean by that I will explain in a more precise manner next time. I will implement the Faddeev–Popov idea in equations, then apply it, and finally justify it by recourse to canonical quantization.

31

The Faddeev–Popov prescription

At the end of the last lecture I was on the verge of describing the bright idea of Faddeev and Popov.[1] Their essential insight was this. Feynman tells us to sum over histories. However, if we just blindly perform the functional integral in a gauge theory, we are not summing over all histories once and only once; we are summing over each history many times, in all of its various gauge-transformed versions. But we *should* count each history *exactly once*. That was their idea. It was just a guess. We're going to formulate this guess in a precise form, explore its consequences and then prove it is true by showing that it is equivalent to canonical quantization for the theories of interest.

31.1 The prescription in a finite number of dimensions

In order to write down the guess of Faddeev and Popov, it's easier to start with a finite-dimensional analog where we integrate over only a finite-dimensional space, then generalize, in my usual brutal way, to a function space, by simply copying down some of the equations and changing some of the symbols. And then I will have arrived at their guess. The finite-dimensional model of what we're doing in a gauge invariant field theory is this: we have a function that depends on $n + m$ real variables

$$F(z_1, z_2, \cdots z_{n+m}) \tag{31.1}$$

That's the analog, in some sense, of the gauge invariant e^{iS}. (I've called them z_i, but they are in fact real variables.) We divide the z's up as follows:

$$\begin{aligned} x_r &= \{z_1, \cdots z_n\}, & r &= 1, \cdots n \\ y_s &= \{z_{n+1}, \cdots z_{n+m}\}, & s &= 1, \cdots m \end{aligned} \tag{31.2}$$

The idea is that F depends only the x_r's:

$$F(z_1, \cdots z_{n+m}) = F(x_r) \quad \text{or} \quad \frac{\partial F(z)}{\partial y_a} = 0 \tag{31.3}$$

---

[1] [Eds.] See note 22, p. 655, and note 10, p. 1037. The Faddeev–Popov technique is discussed in every modern QFT textbook. See Peskin & Schroeder *QFT*, Section 9.4, pp. 294–298.

The y's are like the gauge degrees of freedom: we can change the y's without changing the value of F, just as a gauge transformation does not change the value of $\exp\{i\mathcal{S}\}$.

Perhaps a picture will be helpful; see Figure 31.1. As I am restricted to a two-dimensional blackboard, there will be only one x and one y.

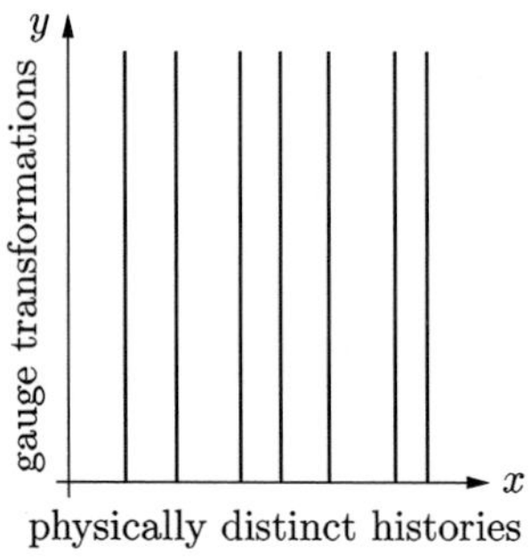

Figure 31.1: Gauge freedom described as motion along a line

Along any of these lines parallel to the y axis, the function F is a constant. They are the finite-dimensional analogs in the gauge system of the various motions that are connected together by gauge transformations. Gauge transformations are like translations in the y direction.[2]

Now we want to define an integral. We'll obviously get a divergent integral if we try to integrate this thing over all the z's. So we define the integral by integrating *only* over the x's:

$$I = \int \prod_{a=1}^{n} dx_a F(x) \tag{31.4}$$

That's certainly an integral that cuts each set of these equivalent points, each vertical line, only once, because I'm just integrating along the surface $y = 0$. Equivalently, we could write this as an integral over *all* the z's:

$$I = \int \prod_{a=1}^{n} dx_a\, F(x) = \int \prod_{a=1}^{n+m} dz_a\, F(z) \prod_{b=1}^{m} \delta(y_b) \tag{31.5}$$

where the delta function restricts us to $y_b = 0$. Of course we don't have to restrict ourselves to the flat surface $y = 0$. We could restrict the integral to some curved surface defined by

$$y_b = f_b(x_1, \cdots x_n) \tag{31.6}$$

which cuts the lines as shown in Figure 31.2, and integrate restricting the y_b's to that surface:

$$I = \int \prod_{a=1}^{n+m} dz_a\, F(z) \prod_{b=1}^{m} \delta(y_b - f_b) \tag{31.7}$$

[2] [Eds.] Coleman is using some concepts from differential geometry: *fiber bundles*. Connections over vector bundles (analogous to Christoffel symbols in general relativity) are another way of viewing Yang–Mills fields. See Sections 18.1c–18.2c, pp. 479–488 in Theodore Frankel, *The Geometry of Physics: An Introduction*, 3rd ed., Cambridge U.P., 2012.

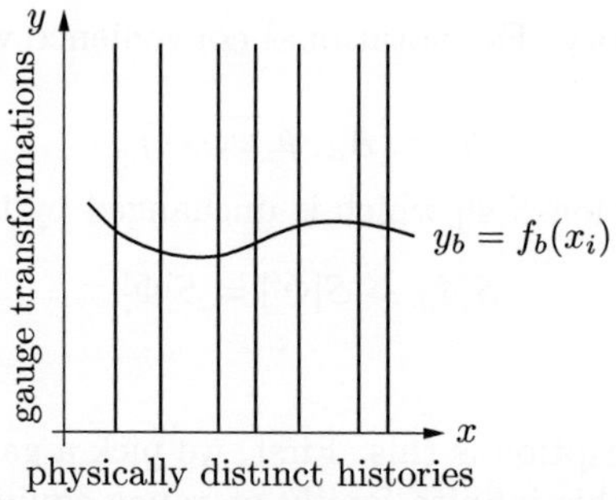

Figure 31.2: Fixing the values of the y's

That's the *same* integral as (31.5). It may not be convenient to parameterize the surface by $y_b = f_b(x_1, x_2 ...)$. It might be better to give the y's implicitly as functions of the x's; i.e., by a set of equations

$$G_b(z_1, \cdots z_{n+m}) = 0 \tag{31.8}$$

such that when we solve for the y's we get $y_b = f_b(x_1, \cdots x_n)$. A completely equivalent way to write I is

$$I = \int \prod_{a=1}^{n+m} dz_a F(z) \prod_{b=1}^{m} \delta(G_b)\Delta \tag{31.9}$$

The factor Δ is the Jacobian determinant to take account that we are integrating with respect to different variables;

$$\Delta = \det\left(\frac{\partial G_b}{\partial y_c}\right) \tag{31.10}$$

($\partial G_b / \partial y_c$ is an $m \times m$ matrix.) That just reproduces the same m-dimensional delta function as in (31.7). We've rewritten a very simple integral, where we just integrate over the x's, in a much more complicated form. But it is the same integral.[3]

31.2 Extending the prescription to a gauge field theory

Armed with this finite-dimensional knowledge, we can now describe the Faddeev–Popov prescription for a gauge field theory; in particular, for quantum electrodynamics.[4] In QED, we have a set of fields transforming in various ways under a gauge transformation parameterized by χ:

$$\begin{aligned} A_\mu \to A'_\mu &= A_\mu + \partial_\mu \chi \\ \psi \to \psi' &= e^{-ie\chi}\psi \\ \overline{\psi} \to \overline{\psi}' &= \overline{\psi} e^{ie\chi} \end{aligned} \tag{31.11}$$

These transformations describe physically equivalent situations: given any history (A_μ, ψ, etc) as a function of space and time, if I apply a gauge transformation I get a new set of functions

[3] [Eds.] The logic and the associated procedure in going from (31.9) to (31.10) is much the same as that used to connect both sides of (1.55). See footnote 8, p. 9.

[4] [Eds.] For a complete justification of the Faddeev–Popov *ansatz*, including the important demonstration that the determinant Δ is itself gauge invariant, see Ryder *QFT*, Section 7.2, pp. 245–255.

which describe the same history. For notational convenience we will assemble all the fields into a single big field Φ.

$$\Phi = (A_\mu, \psi, \overline{\psi}, \cdots) \tag{31.12}$$

We have a gauge invariant action $\mathcal{S}[\Phi]$ which is unchanged by this transformation

$$\mathcal{S}[\Phi] \to \mathcal{S}[\Phi'] = \mathcal{S}[\Phi] \tag{31.13}$$

corresponding to (31.3).

The Faddeev–Popov prescription is this. First, we pick a gauge. That is to say, we adopt some condition that, out of this infinite family of gauge-equivalent motions, picks out *one and only one.* This is equivalent to picking out an integration surface that passes through each of the lines in Figure 31.2 only once. Recall that a **gauge** is some condition $G(\Phi) = 0$, analogous to the earlier $G_b(z_i) = 0$, that you choose *to eliminate the freedom* to make gauge transformations.[5]

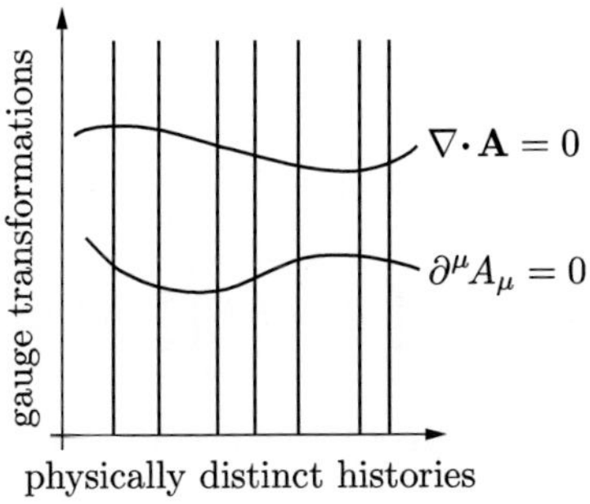

Figure 31.3: Fixing the gauge

Some standard gauges:

Coulomb gauge:

$$\nabla \cdot \mathbf{A} = 0 \tag{31.14}$$

(The Coulomb gauge is also known as the *radiation gauge.*) As you all know from your experience of classical electrodynamics, once you have adopted this condition, you have no further freedom to make gauge transformations, assuming that you impose the usual boundary condition that $\mathbf{A}$ falls off at infinity. (We'll review this argument in a moment.)

Axial gauge:

$$A_3 = 0 \tag{31.15}$$

This is a gauge we will find convenient for proving certain theorems, although it's terrible for computations; it destroys the manifest rotational invariance of the theory. It's called "axial gauge" because it picks out a certain coordinate, i.e., a certain axis.[6]

Lorenz gauge:[7]

$$\partial_\mu A^\mu = 0 \tag{31.16}$$

---

[5] [Eds.] See the discussion on p. 586 about the collision between canonical quantization and gauge invariance. "Choosing a gauge" is really shorthand for "choosing a gauge *condition*", but this is the language used.

[6] [Eds.] Axial gauge is also called the Arnowitt–Fickler gauge. R. L. Arnowitt and S. I. Fickler, "Quantization of the Yang–Mills Field", *Phys. Rev.* **127** (1962) 1821–1829.

[7] [Eds.] Often called the "Lorentz" gauge in the literature, but this is a misnomer. See footnote 1, p. 153.

You may be a little worried about this choice, because it does not completely fix the gauge: we can always add to A_μ a quantity $\partial_\mu\chi$, where χ is a solution to the homogeneous wave equation $\Box^2\chi = 0$. But this is not a problem. Remember that we're secretly doing all functional integrals in Euclidean space (which is hidden within our perverse notation). In Euclidean space, the homogeneous wave equation[8] is the (4-dimensional) Laplace equation. With our usual assumption that everything goes to zero at infinity, the Laplace equation has no non-trivial solutions. It's worth spending a few moments on this.

Aside: Why the Lorenz gauge condition determines A_μ uniquely

We use much the same argument with the Coulomb gauge, (31.14). That choice is also supposed to pick out a unique potential, and keep it from changing under gauge transformations. We can still make a gauge transformation

$$\mathbf{A} \to \mathbf{A}' = \mathbf{A} + \nabla\chi \tag{31.17}$$

but if $\mathbf{A}'$ is to stay in the Coulomb gauge, we must have

$$\nabla^2\chi = 0 \tag{31.18}$$

The usual boundary condition that $\mathbf{A}$ should vanish at infinity implies that $\chi \equiv 0$, because the operator ∇^2 has a unique inverse with sensible boundary conditions. The corresponding equation in the Lorenz gauge, (31.16), is

$$A_\mu \to A'_\mu = A_\mu + \partial_\mu\chi \tag{31.19}$$

If A'_μ is to stay in the Lorenz gauge, we must have

$$\Box^2\chi = 0 \tag{31.20}$$

Even if we impose boundary conditions at infinity, the homogeneous wave equation has many solutions, to wit, all the free motions of a massless scalar particle. Were we working in Minkowski space, this would be a problem; A_μ is *not* unique in the Lorenz gauge. But we are working in *Euclidean* space, and (31.20) is actually

$$\Box_E^2\chi = 0 \tag{31.21}$$

The Euclidean $\Box_E^2$ (28.46) has the same properties as ∇^2; it is just the 4-dimensional analog of the Laplacian. It has a *unique* inverse with reasonable boundary conditions, which implies that $\chi \equiv 0$ is the *only* solution: the apparent freedom expressed by (31.19) and (31.20) is illusory. We're not really doing our functional integrals in Minkowski space, we're doing them in Euclidean space where we lose information. In particular, we lose the $+i\epsilon$ prescription and the pole in the propagator when we work in Euclidean space. That pole is in Minkowski space; the pole is where the free solutions lie.

These three gauges—Coulomb, Lorenz, axial—are popular choices, but one could obviously write down many more. We could have any other gauge condition, as long it *removes the freedom to make further gauge transformations.*

[8] [Eds.] Coleman calls the homogeneous wave equation "the d'Alembert equation".

I can now state the Faddeev–Popov prescription (originally an educated guess, or *ansatz*): it is the direct generalization of the prescription (31.9) to function space. There's a normalization factor, a functional integral over all the fields, there's an $e^{i\mathcal{S}}$ as always, a delta function of whatever gauge function $G(\Phi)$ you have chosen, and finally there is a Jacobian determinant:

$$\boxed{Z = N\int (d\Phi)e^{i\mathcal{S}[\Phi]}\delta\big[G(\Phi)\big]\,\Delta} \tag{31.22}$$

$$\Delta = \det\left(\frac{\delta G}{\delta\chi}\right) \tag{31.23}$$

The variable χ is the analog of the y_b variables above; changing χ moves us along the vertical lines in Figure 31.1, from one configuration to a gauge-equivalent configuration. The quantity $\delta[G]$ is a delta function in function space, a delta *functional*, the analog of the finite-dimensional product of delta functions, $\prod_{b=1}^{m}\delta(G_b)$, such that

$$\int (df)\,\delta[G-f] = 1 \tag{31.24}$$

I want to make four remarks before going on to apply this prescription to the three gauges described above, and to show that it is equivalent to canonical quantization.

Remarks

First, whether or not the prescription is right, one thing is assured: the expression (31.22) is *guaranteed* to be gauge invariant, in the sense that it does not depend on the choice of G. We get the same value of the integral for one G as for any other, from the previous argument for the finite-dimensional case. The integrand in the Coulomb gauge will *look* very different from the integrand in the axial gauge, but the two integrals must give the same answer. So, if we can prove that the prescription is right, i.e., equivalent to canonical quantization, in any one gauge—just *one*—then it has to be right in *all* gauges.

Second, I have pulled a small swindle on you. We have assumed that $\mathcal{S}$ is gauge invariant: $\mathcal{S}(\Phi) = \mathcal{S}(\Phi')$. But typically the sort of actions $\mathcal{S}$ we've been talking about involve source terms like $J_\mu A^\mu$ or $\overline{\eta}\psi$ which break the gauge invariance. So, we must say that $\mathcal{S}$ has sources in it coupled *only* to gauge invariant operators like $F^2_{\mu\nu}$, $\overline{\psi}\psi$, $\overline{\psi}\gamma_\mu\psi$, and so on. That is, we should expect only those combinations formally gauge invariant in the *classical* theory to be independent of which gauge we do our computations in. Since we also firmly believe that the only physical observables are gauge invariant quantities, that should be sufficient to characterize the theory. If we can show that the Faddeev–Popov *ansatz* gives the same results for Green's functions of strings of gauge invariant operators, no matter what the gauge—how we choose the G—then we've shown that it defines the same physics in *any* gauge. Once we settle down in a particular gauge to do our computations, then in order to evaluate the functional integral perturbatively, it might be convenient to introduce sources coupled to A^μ and ψ, and other things like that, as an intermediate stage. But we shouldn't expect either the AA Green's function or the $\psi\overline{\psi}$ Green's function to be independent of the choice of G; we *do* expect this independence from the gauge invariant operators we construct from the A's and ψ's.

Third, (a point I should have emphasized repeatedly during this entire discussion), everything here must be taken with a grain of salt. All of this is just the formal manipulation of canonical field theory, and at the end we will have to worry about ultraviolet divergences and whether they mess up our manipulations. Functional integration is more compact than the manipulation of the canonical equations of motion, but it's no more rigorous. When we're done with all this, we'll have to see if we can put in a cutoff that preserves all the formal properties that we wanted to preserve. In particular, we will have to worry about cutting off this theory in such a way that gauge invariance is maintained. We can do it and we *will* do it.

And finally, functional integration does not *prove* anything. For proofs, we have to come down in the end to canonical quantization. The power of functional integration is that it allows us to change variables with unparalleled facility. The more changes of variables we have to worry about, the more useful the functional integral is. In gauge theories, where we have an enormous family of variable changes to worry about, the functional integral is practically the essential way of doing things.

Now let's talk about the determinant. In quantum electrodynamics we have only one gauge condition G and one gauge parameter χ. In each of the three gauges we have considered, the operator $\delta G/\delta\chi$ is a constant with respect to the fields Φ, and $\det(\delta G/\delta\chi) = \Delta$ is *independent* of Φ, so we can absorb the determinant Δ into the normalization constant N. For the Coulomb gauge, the determinant involves the Laplace operator, for the Lorenz gauge the d'Alembert operator, and for the axial gauge, the derivative of a delta function. In any of these three cases, the determinant is a constant. If we were using a more perverse choice of gauge for the Abelian case, or if we were doing a theory with a more complicated group of gauge transformations like a non-Abelian gauge field theory, in which we gauge not just electric charge conservation but say isospin conservation, then in general we would *not* be able to get rid of the determinant factor. We would have to treat it in the usual way, by putting $\det(\delta G/\delta\chi)$ into the exponential via ghost fields. I shan't do that here, because I don't need to.

31.3 Applying the prescription to QED

Let's assume the Faddeev–Popov prescription is correct, and apply it to quantum electrodynamics. Later we'll prove that it *is* correct. For this application we'll work in the Lorenz gauge, and choose

$$G(\Phi) = \partial_\mu A^\mu(x) - f(x) \tag{31.25}$$

where $f(x)$ is some fixed function. This is a generalization of the Lorenz gauge, (31.16), but the determinant Δ is still a constant, because f doesn't enter into $\delta G/\delta\chi$.

We have

$$\begin{aligned} A^\mu &\to A^\mu + \partial^\mu \delta\chi \\ \delta G &= \delta(\partial^\mu A_\mu) = \Box^2 \delta\chi \end{aligned} \tag{31.26}$$

As shown in §29.3, $\Delta = \det(\delta G/\delta\chi)$ can be handled with ghost fields:

$$\det\left(\frac{\delta G}{\delta\chi}\right) = \int (d\eta)(d\bar{\eta})\, e^{i\mathcal{S}_{\text{FP}}} \tag{31.27}$$

where the Faddeev–Popov ghost action $\mathcal{S}_{\text{FP}}$ is given by

$$\mathcal{S}_{\text{FP}} = \int d^4x\, \left(\bar{\eta}\Box^2\eta\right) = -\int d^4x\, (\partial^\mu\bar{\eta})(\partial_\mu\eta) \tag{31.28}$$

This is a constant with respect to the fields in Φ, and so it can also be absorbed into the normalization constant. We say that the ghosts **decouple**.

The Faddeev–Popov prescription says that the generating functional is

$$Z = N \int (d\Phi)\, e^{iS} \delta[\partial_\mu A^\mu - f] \tag{31.29}$$

independent of f; we get the same Z no matter what f is. (The determinant $\det(\delta G/\delta\chi)$ is included in the normalization constant N.) Since Z doesn't depend on f we can also write

$$\begin{aligned} Z &= N' \int (d\Phi)(df)\, e^{iS} \delta[\partial_\mu A^\mu - f]\, F[f] \\ &= N' \int (d\Phi)\, e^{iS} F[\partial_\mu A^\mu] \end{aligned} \tag{31.30}$$

where N' is a new normalization factor and F is a functional of f; *any* functional will do. Because the integrand in (31.29) is independent of f, we can integrate it over f with any weighting functional $F[f]$ and get the same answer, *modulo* the proportionality factor. To get nice Feynman rules, we choose F to be equal to the exponential of a quadratic form:

$$F[f] = \exp\left\{-\frac{i}{2\xi}\int d^4x\, f^2\right\} \tag{31.31}$$

with ξ real, to be chosen at our convenience.[9] Following the integration over (df) the generating functional is

$$Z = N' \int (d\Phi)\, e^{iS_{\text{eff}}} \tag{31.32}$$

where the **effective action** is

$$\mathcal{S}_{\text{eff}} = \int d^4x\, \mathscr{L}_{\text{eff}} \tag{31.33}$$

with the effective Lagrangian the sum of the original Lagrangian plus a gauge-fixing term,

$$\begin{aligned} \mathscr{L}_{\text{eff}} &= \mathscr{L} + \mathscr{L}_{\text{GF}} \\ &= -\tfrac{1}{4}(\partial_\mu A_\nu - \partial_\nu A_\mu)^2 + \overline{\psi}(i\partial\!\!\!/ - m - eA\!\!\!/)\psi - \frac{1}{2\xi}(\partial_\mu A^\mu)^2 \\ &= \tfrac{1}{2}A_\mu\left[g^{\mu\nu}\Box^2 - (1 - \tfrac{1}{\xi})\partial^\mu\partial^\nu\right]A_\nu + \overline{\psi}(i\partial\!\!\!/ - m - eA\!\!\!/)\psi \end{aligned} \tag{31.34}$$

By this device I have taken care of the earlier problem that bothered us, (30.48), the lack of a contribution from the four-dimensional longitudinal component of A_μ in $\mathcal{S}$; when we tried to find the longitudinal part of the photon propagator, we got $\frac{1}{0} = \infty$. Now we can just read the propagator off the quadratic terms of the effective Lagrangian:

$$\boxed{\widetilde{D}_\xi^{\mu\nu}(k) = \frac{i}{k^2 + i\epsilon}\left[-g^{\mu\nu} + \frac{k^\mu k^\nu}{k^2}\right] - \frac{i\xi}{k^2 + i\epsilon}\left(\frac{k^\mu k^\nu}{k^2}\right)} \tag{31.35}$$

---

[9] [Eds.] The variable ξ, commonly used for this purpose in the literature, has been substituted for Coleman's original α, which is easily confused with the fine-structure constant. See "Generalized Renomalizable Gauge Formulation of Spontaneously Broken Gauge Theories", Kazuo Fujikawa, Benjamin W. Lee, and A. I. Sanda, *Phys. Rev.* **D6** (1972) 2923–2943.

That's just the universal rule for functional integrals: invert the quadratic part of the Lagrangian to get the propagator.[10] The parameter ξ can be anything we want. This looks very much like the propagator in (30.48), but without the disastrous $\frac{1}{0}$ garbage in the last term. The gauge-fixing term has removed it and rendered the propagator finite.

We still haven't shown that the Faddeev–Popov *ansatz* is right; that it agrees with canonical quantization. But whether it's right or wrong, any gauge invariant quantity computed using one value of ξ will have the same answer using any other value of ξ. We've shown that they are all equivalent to each other.

In a slightly different use of the word "gauge" in the literature, the family of propagators (31.35) represent what are called **covariant gauges** because the propagators look covariant. If we had done the same thing with the Coulomb gauge, we would have had an extra term just involving the space part of k, because we've only got space derivatives, and it wouldn't look covariant. The covariant gauges have various names for particular choices of ξ. The two most popular choices in the literature are $\xi = 1$, called the *Feynman gauge*,[11]

$$\boxed{\widetilde{D}_F^{\mu\nu}(k) = -\frac{ig^{\mu\nu}}{k^2 + i\epsilon}} \tag{31.36}$$

and the limiting case $\xi \to 0$, called the *Landau gauge*,[12]

$$\boxed{\widetilde{D}_L^{\mu\nu}(k) = \frac{i}{k^2 + i\epsilon}\left[-g^{\mu\nu} + \frac{k^\mu k^\nu}{k^2 + i\epsilon}\right]} \tag{31.37}$$

The Feynman gauge is useful for evaluating low order Feynman graphs. Additionally, it's nice to have only $g^{\mu\nu}$ instead of keeping track of all the k_μ's. We obtain the Landau gauge in the limit $\xi \to 0$. This looks like a singular limit in (31.31) but this limit really just restores the delta function in the original form of (31.29). Its utility for certain general arguments was pointed out by Lev Landau. It is four-dimensionally transverse, $k_\mu \widetilde{D}^{\mu\nu}(k) = 0$; in a formal sense this is like setting $\partial_\mu A^\mu = 0$. A third choice, $\xi = 3$, is sometimes used. This is the *Yennie–Fried gauge*[13] with propagator

$$\widetilde{D}_Y^{\mu\nu}(k) = \frac{i}{k^2 + i\epsilon}\left[-g^{\mu\nu} - 2\frac{k^\mu k^\nu}{k^2 + i\epsilon}\right] \tag{31.38}$$

We will say nothing more about it, other than to note that it has useful infrared properties.

---

[10] [Eds.] After Fourier transforming the operator between the A's in (31.33), it can be written as $-k^2 P_T^{\mu\nu} - (k^2/\xi) P_L^{\mu\nu}$, with the projection operators in (30.46). The inverse of this operator is $-(1/k^2) P_T^{\mu\nu} - (\xi/k^2) P_L^{\mu\nu}$. By convention, this inverse is multiplied by i, and $+i\epsilon$ is added to the k^2 denominator.

[11] [Eds.] R. P. Feynman, "Space-Time Approach to Quantum Electrodynamics", *Phys. Rev.* **76** (1949) 769–789. Reprinted in Schwinger *QED*.

[12] [Eds.] L. D. Landau, A. A. Abrikosov and I. M. Khalatnikov, "Асимптотическое Выражение для Грииновской Функции Электрона в Квантовой Электродинамике", (An Asymptotic Expression for the Electron Green's Function in Quantum Electrodynamics) *Dokl. Akad. Nauk SSSR* **95** (1954) 773–776; L. D. Landau and I. M. Khalatnikov, "The Gauge Transformation of the Green Function for Charged Particles", *J. Exper. Theor. Phys. USSR* **29** (1955) 89–93; English trans. *Soviet Physics JETP* **2** (1956) 69–72. Republished in *Collected Papers of L. D. Landau*, ed. Dirk ter Haar, Pergamon Press, 1965, pp. 659–664.

[13] [Eds.] H. M. Fried and D. R. Yennie, "New Techniques in the Lamb Shift Calculation", *Phys. Rev.* **112** (1958) 1391–1404.

All of these gauges give the same answer for any gauge invariant quantity. However, non-gauge invariant quantities will look different in the different gauges. The photon propagator is one such, since A_μ looks different in different gauges.[14]

31.4 Equivalence of the Faddeev–Popov prescription and canonical quantization

I've shown that all these gauges, and indeed billions of other gauges, are equivalent. I will now show that the Faddeev–Popov *ansatz* is equivalent to canonical quantization in a particular gauge, the axial gauge $A_3(x) = 0$. I choose this gauge because canonical quantization in the axial gauge is super-easy. Once I've shown that it's true in one gauge, I know it's true in all other gauges.

Our first step to canonically quantize the theory in the axial gauge is just to impose the condition $A_3(x) = 0$, to fix the gauge. We find a set of p's and q's, and we write everything in terms of them. Finally we show that the resulting expression is equivalent to the Faddeev–Popov prescription in the axial gauge. In the following, the Latin indices i, j, etc., take the values 1 and 2 only; I will explicitly write the terms with 0 and 3.

As usual, the clue is the first-order form of the Lagrangian

$$\begin{aligned}\mathscr{L}_{\text{1st}} &= \tfrac{1}{4}F_{\mu\nu}F^{\mu\nu} - \tfrac{1}{2}F_{\mu\nu}(\partial^\mu A^\nu - \partial^\nu A^\mu) + \overline{\psi}(i\partial\!\!\!/ - m - eA\!\!\!/)\psi \\ &= \tfrac{1}{4}F_{ij}F^{ij} - \tfrac{1}{2}F_{ij}(\partial^i A^j - \partial^j A^i) + \tfrac{1}{2}F_{i3}F^{i3} + F_{i3}(\partial^3 A^i) + \tfrac{1}{2}F_{i0}F^{i0} \\ &\quad - F_{i0}(\partial^i A^0 - \partial^0 A^i) + \tfrac{1}{2}F_{03}F^{03} + F_{03}(\partial^3 A^0) + \overline{\psi}(i\partial\!\!\!/ - m - eA\!\!\!/)\psi\end{aligned} \tag{31.39}$$

This is messy, but canonically quantizing the first-order Lagrangian is like shooting fish in a barrel. The independent variables are the two components A_i, their canonical momenta F_{0i}, and the Fermi field ψ and its canonical momentum $\overline{\psi}$. All the other variables—F_{ij}, F_{i3}, F_{03}, and A_0—are constrained, defined in terms of the independent variables and their space derivatives only. F_{ij} and F_{i3} are trivially constrained, given in terms of the space derivatives of A_i:

$$\begin{aligned}F_{ij} &= \partial_i A_j - \partial_j A_i \\ F_{i3} &= -\partial_3 A_i\end{aligned} \tag{31.40}$$

Next, the component F_{03} is determined from the Euler–Lagrange equation obtained from (31.39),

$$\partial_\mu F^{\mu\nu} = e\overline{\psi}\gamma^\nu\psi \tag{31.41}$$

For $\nu = 0$ we have

$$\partial_1 F^{10} + \partial_2 F^{20} + \partial_3 F^{30} = e\overline{\psi}\gamma^0\psi \tag{31.42}$$

---

[14] [Eds.] These authors did not, of course, name these gauges after themselves. Walter Heitler first introduced the name "Lorentz gauge" (not "Lorenz gauge", though it should have been) and "Coulomb gauge" in his influential text, *The Quantum Theory of Radiation*, 3rd ed., Oxford U. P., 1954, p. 3; reprinted by Dover Publications, 1984. Bruno Zumino gave the names "Feynman", "Landau" and "Yennie" to the respective gauges: B. Zumino, "Gauge Properties of Propagators in Quantum Electrodynamics", *J. Math. Phys.* **1** (1960) 1–7. For more about these gauges, their properties and relations to each other, see J. D. Jackson and L. B. Okun, "Historical Roots of Gauge Invariance", *Rev. Mod. Phys.* (2001) **73**, 663–680, and N. Nakanishi, "Indefinite-Metric Quantum Field Theory", *Suppl. Prog. Theor. Phys.* **51** (1972) 1–95. Incidentally, Okun introduced (1962) the term "hadron" (from the Greek ἁδρός, "stout, fat, strong") as the antonym to "lepton" (Greek λεπτός, "slight, thin, small").

which determines F_{03} in terms of known quantities. Finally

$$F^{30} = \partial^3 A^0 \tag{31.43}$$

determines A^0. The generating functional we get from canonical quantization is

$$Z = N\int (dA_1)(dA_2)(dF_{01})(dF_{02})(d\psi)(d\overline{\psi})\ \exp\{i\mathcal{S}_{\mathscr{H}}\} \tag{31.44}$$

On the other hand, by our general rule (29.77), the constrained variables disappear from the functional integral

$$\int \prod_{\mu\nu}(dF_{\mu\nu})(dA_0)(dA_1)(dA_2)(d\psi)(d\overline{\psi})\ \exp\{i\mathcal{S}_{\text{1st}}\}$$

since their coefficients are just constants.[15] To within a new normalization constant, this is equal to the Hamiltonian form, and therefore it is another expression for Z:

$$Z = N'\int \prod_{\mu\nu}(dF_{\mu\nu})(dA_0)(dA_1)(dA_2)(d\psi)(d\overline{\psi})\ \exp\{i\mathcal{S}_{\text{1st}}\} \tag{31.45}$$

The part of the Lagrangian that remains in the action after integrating over the constrained variables is just $p\dot{q} - H$, the action written in terms of A_i and F_{0i}; this is the same argument as before. We obtain a third expression for Z by integrating the first-order version (31.45) over all the F's. As always, that eliminates the F's and brings the Lagrangian back into the second-order form, written in terms of the A's and ψ's with second derivatives:

$$\begin{aligned}
Z &= N\int (dA_1)(dA_2)(dF_{01})(dF_{02})(d\psi)(d\overline{\psi})\ \exp\{i\mathcal{S}_{\mathscr{H}}\} \\
&= N'\int \prod_{\mu\nu}(dF_{\mu\nu})(dA_0)(dA_1)(dA_2)(d\psi)(d\overline{\psi})\ \exp\{i\mathcal{S}_{\text{1st}}\} \\
&= N''\int (dA_0)(dA_1)(dA_2)(d\psi)(d\overline{\psi})\ \exp\{i\mathcal{S}_{\text{2nd}}\} \\
&= N''\int \prod_{\mu}(dA_\mu)(d\psi)(d\overline{\psi})\ \delta(A_3)\ \exp\{i\mathcal{S}_{\text{2nd}}\}
\end{aligned} \tag{31.46}$$

where the delta function allows us to integrate over all four components of A_μ. But this is precisely the Faddeev–Popov *ansatz* (31.22) for the axial gauge. Comparing (31.22) with the last line of (31.46), you may be troubled by the apparent lack of the Δ factor. But the determinant is a constant for the axial gauge and can be absorbed into N''. And since the Faddeev–Popov prescription is independent of which gauge we choose, if it is right in the axial gauge, it is right in any other gauge, and we are done. **QED**

This is a good place to write down the Feynman rules for electrodynamics with a massless photon. We'll use Feynman gauge:

---

[15] [Eds.] See §29.5, and Section 5.4 of "Secret Symmetry", in Coleman *Aspects*.

Feynman rules for QED (Feynman gauge)

For every ...		Write ...
(a) internal photon line		$\widetilde{D}_F^{\mu\nu}(k^2) = \dfrac{-ig^{\mu\nu}}{k^2 + i\epsilon}$
(b) internal fermion line	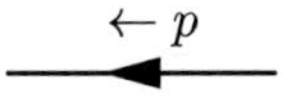	$\widetilde{S}_F(\not{p}) = \dfrac{i}{\not{p} - m + i\epsilon}$
(c) three-point vertex	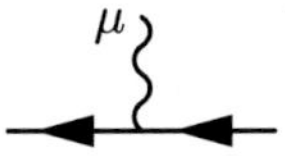	$-ie\gamma_\mu$
(d) $\left\{\begin{matrix}\text{incoming}\\ \text{outgoing}\end{matrix}\right\}$ electron	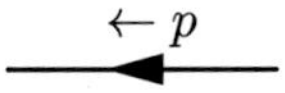	$\left\{\begin{matrix} u_p \\ \overline{u}_p \end{matrix}\right\}$
(e) $\left\{\begin{matrix}\text{incoming}\\ \text{outgoing}\end{matrix}\right\}$ positron	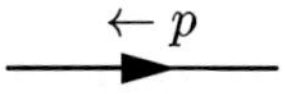	$\left\{\begin{matrix} \overline{v}_p \\ v_p \end{matrix}\right\}$
(f) $\left\{\begin{matrix}\text{incoming}\\ \text{outgoing}\end{matrix}\right\}$ photon		$\left\{\begin{matrix} \varepsilon_\mu(k) \\ \varepsilon^*_\mu(k) \end{matrix}\right\}$
(g) internal scalar line	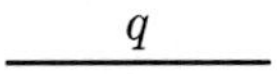	$\widetilde{\Delta}_F(q^2) = \dfrac{i}{q^2 - m^2 + i\epsilon}$
(h) three-point vertex	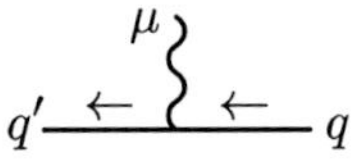	$-ie(q'_\mu + q_\mu)$
(i) seagull vertex	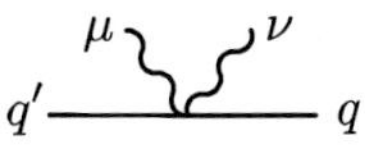	$2ie^2 g^{\mu\nu}$

As an exercise, you can try to show the equivalence between canonical quantization and the Faddeev–Popov *ansatz* in the Coulomb gauge, $\nabla \cdot \mathbf{A} = 0$. That gauge is a little harder, because we have to split $\mathbf{A}$ into transverse and longitudinal components, $\mathbf{A}_T$ and $\mathbf{A}_L$, and play the same game with $\mathbf{A}_L$ replacing A_3. But the conclusion is the same,[16] as you would expect. Because it's true for the axial gauge, it must be true for the Coulomb gauge, and for all other gauges.

The axial gauge is a terrible gauge for doing any kind of actual computation: Lorentz invariance and even rotational invariance of the theory are not manifest. But canonical quantization in the Lorenz gauge(s), with a nice propagator, is much more involved. It requires

[16] [Eds.] See P. H. Frampton, *Gauge Field Theories*, 3rd ed., Wiley-VCH Verlag BmbH, 2008, Section 2.3, pp. 59–65.

subsidiary fields and then showing they don't matter, conditions on physically allowed states, etc.[17] Canonical quantization in the axial gauge is trivial. The power of the functional integral method is that it enables us to prove that a given formula is right in a gauge where canonical quantization is simple but the Feynman rules are complicated, and then to change instantly to a different gauge, in which canonical quantization may be complicated but the Feynman rules are simple.

31.5 Revisiting the massive vector theory

All the computations in massless electrodynamics are the same ones we've done in massive electrodynamics. They are right not only for massless electrodynamics as the massless limit of massive electrodynamics, but also if we approach massless QED *ab initio* by canonical quantization. We need to address one final thing about massive electrodynamics, the $k_\mu k_\nu/\mu^2$ term in the propagator: *it doesn't make any difference.* To get rid of that term in massive electrodynamics, consider the Lagrangian interacting only with a Fermi field, augmented by a new scalar field, ϕ:

$$\mathscr{L} = -\tfrac{1}{4}(\partial_\mu A_\nu - \partial_\nu A_\mu)^2 + \tfrac{1}{2}\mu^2 A_\mu A^\mu + \overline{\psi}(i\partial\!\!\!/ - m - eA\!\!\!/)\psi + \tfrac{1}{2}a(\partial_\mu\phi)^2 - \tfrac{1}{2}b\phi^2 \tag{31.47}$$

(the interactions with a Bose field can be handled similarly).[18] The parameters a and b will be adjusted later. We just add these terms in. They don't change the physics in the slightest, because there is no coupling of ϕ to anything else; it is a free field. Because the sources to be added to the Lagrangian will couple only to ψ, $\overline{\psi}$ and A_μ, we can write the theory as a functional integral, now including $d\phi$, over the additional terms. This extra functional integration only changes the normalization factor. Since ϕ is completely unphysical, we don't have to apply any positivity constraints on a or b. If they have the wrong signs then ϕ will represent a field with negative energy or negative probability because it has the wrong sign in the propagator. But it doesn't couple to anything, so who cares?

Now make a change of variables: define ψ' by

$$\psi = \psi' e^{ie\phi} \tag{31.48}$$

$$\partial_\mu\psi = e^{ie\phi}\left(\partial_\mu\psi' + ie\psi'\partial_\mu\phi\right) \tag{31.49}$$

Trade ψ for ψ':

$$\mathscr{L} = -\tfrac{1}{4}(\partial_\mu A_\nu - \partial_\nu A_\mu)^2 + \tfrac{1}{2}\mu^2 A_\mu A^\mu + \overline{\psi'}(i\partial\!\!\!/ - m - eA\!\!\!/ - e\partial\!\!\!/\phi)\psi' + \tfrac{1}{2}a(\partial_\mu\phi)^2 - \tfrac{1}{2}b\phi^2 \tag{31.50}$$

We have introduced an *illusory* coupling between ψ' and ϕ, illusory because it doesn't affect S-matrix elements. We have cunningly arranged matters so that the only thing that comes into the fermion vertex is $A\!\!\!/ + \partial\!\!\!/\phi$. We simplify the Feynman rules by considering not the

---

[17] [Eds.] A good summary of the problems encountered in the canonical quantization of QED can be found in K. Haller and E. Lim–Lombridas, "Quantum Gauge Equivalence in QED", *Found. of Phys.* **24** (1994) 217–247.

[18] [Eds.] In Brian Hill's notes for Feb. 12, 1987, Coleman credits this trick to Stueckelberg, and following him, the field ϕ is denoted B: E. C. G. Stueckelberg, "Die Wechselwirkungskräfte in der Elektrodynamik und in der Feldtheorie der Kernkräfte (Teil II und III)" (Forces of interaction in electrodynamics and in the field theory of nuclear forces (parts II and III)), *Helv. Phys. Acta* **11** (1938) 299–328; reprinted in *E. C. G. Stueckelberg: An Unconventional Figure of Twentieth Century Physics*, J. Lacki, H. Ruegg, G. Wanders, eds., Birkhäuser, 2009; English translation by D. H. Delphenich, online at `http://neo-classical-physics.info/electromagnetism.html`.

separate propagators for A_μ and ϕ but the propagator for the *combination*, the only thing that enters the coupling to ψ'. What is the propagator for $A_\mu + \partial_\mu\phi$?

$$\overline{(A_\mu+\partial_\mu\phi)(A_\nu+\partial_\nu\phi)} = \overline{A_\mu A_\nu}+\overline{\partial_\mu\phi\,\partial_\nu\phi} = i\left[\frac{-g_{\mu\nu}+(k_\mu k_\nu/\mu^2)}{k^2-\mu^2+i\epsilon}+\frac{k_\mu k_\nu}{ak^2-b+i\epsilon}\right] \tag{31.51}$$

The parameters a and b can be anything. Different choices give different Feynman rules but they all give the same S-matrix. By choosing a and b in two different ways we can make this propagator look considerably simpler.

For example, if we choose

$$a=-\mu^2 \qquad b=-\mu^4 \tag{31.52}$$

then the last term in the propagator is

$$\frac{-k_\mu k_\nu}{\mu^2k^2-\mu^4+i\epsilon}=-\frac{(k_\mu k_\nu/\mu^2)}{k^2-\mu^2+i\epsilon} \tag{31.53}$$

exactly canceling the second term in the A_μ contraction. Then the propagator is

$$\widetilde{D}^P_{\mu\nu}=-\frac{ig_{\mu\nu}}{k^2-\mu^2+i\epsilon} \tag{31.54}$$

(the "Feynman gauge" (31.36); the quotes are because the "photon" has a mass; the superscript P is for Proca). Of course this massive theory has no actual gauge invariance. Nevertheless, it *looks* like the Feynman gauge propagator in genuine gauge invariant electrodynamics with a massless photon, except that its denominator is $k^2-\mu^2$, instead of k^2. Alternatively, we could choose

$$a=-\mu^2 \qquad b=0 \tag{31.55}$$

That gives

$$\begin{aligned}\widetilde{D}^P_{\mu\nu}&=i\left[-\frac{g_{\mu\nu}}{k^2-\mu^2+i\epsilon}+\frac{(k_\mu k_\nu/\mu^2)}{k^2-\mu^2+i\epsilon}-\frac{k_\mu k_\nu}{\mu^2k^2-i\epsilon}\right]\\&=\frac{i}{k^2-\mu^2+i\epsilon}\left[-g_{\mu\nu}+\frac{k_\mu k_\nu}{\mu^2}\left\{1-\frac{k^2-\mu^2}{k^2}\right\}\right]\\&=\frac{i}{k^2-\mu^2+i\epsilon}\left[-g_{\mu\nu}+\frac{k_\mu k_\nu}{k^2}\right]\end{aligned} \tag{31.56}$$

which is the Proca propagator in the "Landau gauge" (31.37). We could even get the Proca propagator in something like the "covariant gauges" by the substitution

$$a=-\mu^2 \quad b=-\xi\mu^4 \tag{31.57}$$

which gives

$$\widetilde{D}^P_{\mu\nu}=i\left[\frac{-g_{\mu\nu}+(k_\mu k_\nu/\mu^2)}{k^2-\mu^2+i\epsilon}-\frac{k_\mu k_\nu/\mu^2}{k^2-\xi\mu^2+i\epsilon}\right] \tag{31.58}$$

As $\xi\to 1$, we get the Proca propagator in the "Feynman gauge"; $\xi\to 0$ gives it in the "Landau gauge".

So we can perform exactly the same transformations in massive electrodynamics as in the massless theory. We can make the $k_\mu k_\nu$ part of the propagator look like whatever we want by this trick of adding a non-dynamical field, or by an appropriate generalization of it: we can always get rid of the troublesome $k_\mu k_\nu/\mu^2$ term in the case of a single massive vector.[19] This conclusion does *not* generalize to non-Abelian theories, involving *more* than one vector: massive Yang–Mills theories do *not* smoothly turn into massless Yang–Mills theories as the mass goes to zero.[20]

We've simulated a gauge transformation by introducing a new dynamical variable ϕ, but we haven't changed the physics, because ϕ is decoupled from everything else. It is *only* the mass term for A_μ that breaks gauge invariance. Otherwise we'd get lots of extra terms from the transformation (31.48). If there were not a conserved current—if the interactions with the Fermi fields broke gauge invariance—we couldn't make this trick work.

Were ϕ a dynamical variable, we'd have to have both a and b positive to get a physically sensible theory. Then we could absorb a into the normalization of the fields, and we'd be left with b. If a and b have opposite signs, the theory contains tachyons; if they are both negative, the propagator has the wrong sign, and we've destroyed the positivity of the inner product in the Hilbert space. But since the ϕ terms are completely decoupled, off in a world by themselves, these constants a and b can be whatever we want. (We know that to get the theory to look "Feynman gauge"-like, there must be some pathological degree of freedom in the theory somewhere, because of the signs of the poles in the propagator: three of them are right, but one is wrong.)

31.6 A first look at renormalization in QED

We've said practically all that needs to be said about the theory of a single vector field, on a formal level and in low orders in perturbation theory. We'll now start tackling the question of renormalization. Let's restrict ourselves to genuine electrodynamics, the theory without a photon mass term.

You might think that we had taken care of all the problems associated with renormalization because we've now put our propagators, in the Landau gauge or the Feynman gauge, into a form with the same high-k behavior as ordinary spinless boson propagators. They go like $1/k^2$, whether you're in the massless theory or the massive theory. Since Lorentz invariance was not involved in the statement of the BPHZ theorem,[21] we could apply it and treat the four components A_μ as four scalar mesons with $1/k^2$ propagators and one funny sign: $g_{\mu\nu}/k^2$ has four components with $1/k^2$. The individual components would couple to γ_μ's in non-Lorentz invariant ways, but who cares about that?

We could simply try the straight BPHZ renormalization procedure, starting out with renormalized (primed) fields:

$$A'_\mu = Z_3^{-1/2} A_\mu \qquad \psi' = Z_2^{-1/2}\psi \tag{31.59}$$

---

[19] [Eds.] See Problems 16 and their solutions, pp. 635–639; the $k^\mu k^\nu$ part of the propagator does not contribute to the amplitudes involving a massive vector.

[20] [Eds.] See note 22, p. 1044.

[21] [Eds.] See §25.2, and "Renormalization and symmetry: a review for non-specialists", Ch. 4, pp. 99–124 in Coleman *Aspects*.

and the Lagrangian written in terms of these fields,

$$\mathscr{L} = -\tfrac{1}{4}(F'_{\mu\nu})^2 - \frac{1}{2\xi}(\partial_\mu A'^\mu)^2 + \overline{\psi}'(i\partial\!\!\!/ - m - eA\!\!\!/')\psi' \tag{31.60}$$

and generate counterterms. (As usual, e is the physical charge, however we choose to define it—say à la BPHZ at zero momentum transfer. Likewise m is the physical mass, defined perhaps according to BPHZ.) The interaction is of renormalizable type, of dimension 4, exactly like a scalar meson Yukawa coupling. The only difference from scalar renormalization is that because A_μ has an index, Lorentz invariance will allow us to write down some more counterterms. We'll just write them down to see what the possible counterterms are, as given by the BPHZ procedure. And then we'll discover something disgusting.

Dimension	*Counterterm*	*Remarks*
0	—	constants only
1	—	none
2	$AA'_\mu A'^\mu$	photon mass term
3	$B\overline{\psi}'\psi'$	the only term possible of dim 3
4	$C\overline{\psi}'\partial\!\!\!/\psi'$	wave function term for fermions
4	$D\overline{\psi}'eA\!\!\!/\psi'$	coupling constant term
4	$E(F'_{\mu\nu})^2$	wave function term for photons
4	$F(\partial^\mu A'_\mu)^2$	renormalization of ξ term
4	$G(A'_\mu A'^\mu)^2$	like ϕ^4 term

Table 31.1: Possible counterterms in massless electrodynamics

Table 31.1 is the complete set of possible counterterms with dimension ≤ 4 (and consistent with the invariances of the theory) according to the BPHZ procedure; $\{A, B, \ldots, G\}$ are constants. (QED is parity invariant, so we've excluded $\overline{\psi}'\gamma_5\psi'$ and $\overline{\psi}'\gamma_\mu\gamma_5\psi' A^{\mu\prime}$.) These terms are just what conventional reasoning gives you, similar to meson-nucleon theory (§23.1; the constants are introduced in (23.25)). Let's talk about these terms. Some of them are disastrous.

If we generate $A(A'_\mu)^2$ or $G(A'_\mu A'^\mu)^2$ terms, we've destroyed the gauge invariance of the theory! But it was *only* the gauge invariance of the theory that told us both that the generating functional is independent of ξ, and that the Z obtained from the Faddeev–Popov theory was equivalent to the Z found with canonical quantization in the axial gauge. If we don't have gauge invariance, we don't have ξ-independence and we don't have equivalence to canonical quantization. So all hell can break loose if these counterterms are present. We don't know if our theory is physically sensible if we have non-gauge invariant terms proportional to A and G. It may be some theory with ghosts that are physically observable or lack of conservation of probability, or some other incredible nonsense. So it is *absolutely essential* that

$$A = G = 0 \tag{31.61}$$

because those terms break gauge invariance. If those terms are generated the whole thing goes into the trash; well, we had a nice formal structure, but if it's not preserved by renormalization, forget about it...

It would be very nice, although not absolutely essential, if we had

$$C = -D \tag{31.62}$$

because the ratio C/D determines what the bare charge is, aside from factors that depend on A'_μ, and A'_μ is the same for the electron and the proton. Of course we know empirically that

$$|e_{\text{electron}}| = |e_{\text{proton}}|$$

to many decimal places. In QED, with electrons only, it wouldn't matter. On the other hand, suppose we have a more complicated theory, describing not only electrons but also protons and π mesons and all their strong interactions. We know that the interactions of the proton and those of the electron are very different. If we didn't have a rule like (31.62), then we'd be in a peculiar situation. We would find the electric charge renormalization for the proton and that for the electron would be different, and we would have to say that the empirical equality of the electric charges is a *coincidence.* The bare charges would be completely different, by amounts that depend on the cutoff in the strong interaction coupling constant, and God knows what else, but apparently they have been so cunningly adjusted by God while creating the universe as to make the size of the physical charges exactly equal. Who would believe that? Well, it's a possibility; maybe God *is* that nice. Although at present we cannot insist that $C = -D$, let's remember that this equality would be very helpful if we are to explain the **universality of electric charge**, that the equality of the *physical* charges for two different particles should imply equality of the *bare* charges.

Finally, we also have, though it's not worth much, that

$$F = 0 \tag{31.63}$$

That is, there is no renormalization of the gauge parameter ξ; we don't need to introduce a counterterm for the ξ term.

Well, are these conditions (31.61)–(31.63) on $\{A, C, D, F, G\}$ true, or not? In fact, we will be able to prove, order by order in renormalized perturbation theory, that if our conditions

$$A = F = G = 0, \quad C = -D \tag{31.64}$$

are true to a given order, then they are true in the next order up. Thus we will show the consistency of the renormalization program with gauge invariance. The tools we will use to prove these are the **Ward identities.**[22] I'll briefly tell you what these things are.

Let's focus on (31.62). If it is true, it implies that there is some connection between the graph in Figure 31.4, which tells us the D-type counterterm, and the graph in Figure 31.5, which tells us the C-type counterterm. We will show, in the old-fashioned way, that there *is* some kind of connection between those things. Define

[22] [Eds.] What Coleman calls "the Ward identities" are conventionally known as "the Ward–Takahashi identities": J. C. Ward, "An Identity in Quantum Electrodynamics", *Phys. Rev.* **78** (1950) 182; Y. Takahashi, "On the Generalized Ward Identity", *Nuovo Cim.* **6** (1957) 371–375; Coleman *Aspects*, "Symmetry and symmetry-breaking: currents", Section 5.4, pp. 108–111; Peskin & Schroeder *QFT*, "The Ward–Takahashi Identity", Section 7.4, pp. 238–244.

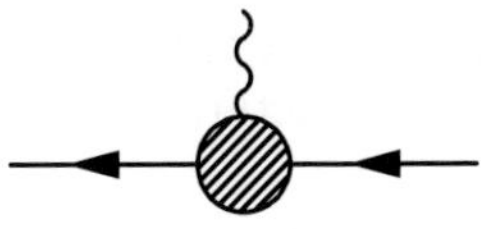

Figure 31.4: The electron-photon vertex

Figure 31.5: The electron propagator

$$j_\mu = \overline{\psi}\gamma_\mu\psi \tag{31.65}$$

At equal times

$$[j_0(\mathbf{x},t),\psi(\mathbf{y},t)] = -\delta^{(3)}(\mathbf{x}-\mathbf{y})\psi(\mathbf{y},t) \tag{31.66}$$

Consider the divergence of the time-ordered product

$$\partial_x^\mu \langle 0|T[j_\mu(x)\psi(y)\overline{\psi}(z)]|0\rangle \tag{31.67}$$

Recall the rule (27.86) for finding the time derivative of a time-ordered product: we get equal-time commutators plus a differentiated term. The differentiated term is irrelevant because of current conservation,

$$\partial^\mu j_\mu = 0 \tag{31.68}$$

so we only pick up the equal-time commutators, when $x_0 = y_0$ and $x_0 = z_0$. We get

$$\partial_x^\mu \langle 0|T[j_\mu(x)\psi(y)\overline{\psi}(z)]|0\rangle = -\delta^{(4)}(x-y)\langle 0|T[\psi(y)\overline{\psi}(z)]|0\rangle + \delta^{(4)}(x-z)\langle 0|T[\psi(y)\overline{\psi}(z)]|0\rangle \tag{31.69}$$

The space δ function comes from the commutator and the time δ function from differentiating the θ function.[23] In the next lecture we will do things in a quite different way.

But we can begin to see how we can establish a connection between some part of Figure 31.4 and some part of Figure 31.5. In some sense j_μ is the thing the photon is going to couple to when it burrows into that diagram and hits its first Fermi line. The first thing it's going to do is hit a Fermi line and then it's coupled to j_μ. So in some sense Figure 31.4 is connected with the left hand side of (31.69). On the other side of the equation we have known quantities times the fermion two-point function, which is Figure 31.5. But it's written in a terribly messy form and it's sort of awful. I haven't really got a Green's function because we need to take off the photon line to get to the j_μ. We've derived an equation for the full Green's functions but renormalization is expressed in terms of 1PI Green's functions. We would have to manipulate and manipulate and manipulate this thing to prove what we eventually want to prove, which is $C = -D$. And then we'd have to write down a whole bunch of other messy equations and manipulate them to show the other things we want to prove, $A = F = G = 0$. So we won't do it this way. It's also a mess because we've written (31.69) in terms of unrenormalized fields, and we have to figure out how to write it in terms of renormalized fields, which is ugly. Instead, we will use a method based on functional methods; we will essentially read off these equations and all the consequences of things we get by the above manipulations without going through any combinatoric work.

---

[23] [Eds.] Bjorken & Drell *Fields*, Problem 19.1, p. 376.

As a preliminary for that, we must further develop the functional integral method. In particular, we need the **generating functional for 1PI diagrams** in terms of the generating functional for full Green's functions. We will do that next time.

Problems 17

17.1 In Model 1 (§8.5), we were able to calculate (8.61), the vacuum-to-vacuum S-matrix element for a free scalar field linearly coupled to an external source $J(x)$. We now know that the reason we were able to do this so easily was that we were evaluating a Gaussian functional integral. But we also have a Gaussian functional integral if the source is coupled *quadratically* to the field.

Consider

$$\mathscr{L} = \tfrac{1}{2}\left(\partial_\mu\phi\partial^\mu\phi - \mu^2\phi^2 + J(x)\phi^2\right) \tag{P17.1}$$

With this Lagrangian, there is no linear term in the exponential, and so the functional integral for $Z[J]$ produces only the inverse square root of the determinant of the quadratic operator A,

$$A = -i(\Box^2 + \mu^2 - J) \tag{P17.2}$$

From (13.6), (28.27) and (28.50), we can derive that

$$\langle 0|\mathrm{S}|0\rangle_J = \left[\frac{\det(\Box^2 + \mu^2 - J - i\epsilon)}{\det(\Box^2 + \mu^2 - i\epsilon)}\right]^{-\frac{1}{2}} \tag{P17.3}$$

(Here I've used the fact that the matrix element is 1 when $J = 0$ to fix the normalization factor; I've also dropped some factors of i that cancel between numerator and denominator.) Show that this is the same as the answer you get by summing Feynman graphs.

If we had a complex scalar field (with coupling $J\phi^*\phi$), functional integration would give us the square of (P17.3). Can you see where the difference is in the Feynman graphs?

HINTS:

(a) I think you'll find it more convenient to write out the Feynman graphs as integrals in position space rather than in momentum space.

(b) You'll never get the right answer if you don't get the symmetry factors right.

(c) For any diagonalizable matrix A,

$$\boxed{\det A = \exp\{\mathrm{Tr}\,[\ln A]\}} \tag{P17.4}$$

This can be extended to appropriate complex matrices by analytic continuation.[1]

(1998b 4.1)

17.2 As discussed in §31.2, the Coulomb gauge (or "the radiation gauge") is defined by the gauge-fixing condition

$$\nabla\cdot\mathbf{A} = 0 \tag{31.14}$$

[1] [Eds.] See note 7, p. 608.

Just as for the covariant gauges discussed in §31.3, the Faddeev–Popov determinant Δ is a constant which we can absorb into the normalization of the functional integral. Show that in this gauge, the i–j part of the photon propagator is

$$\widetilde{D}_C^{ij}(k) = -i\left[\frac{g^{ij} + (k^i k^j/|\mathbf{k}|^2)}{k^2 + i\epsilon}\right] \tag{P17.5}$$

where, as usual, i and j are spatial indices. Compute also the i–0 and 0–0 parts of the propagator.

(1998b 5.1)

17.3 A Dirac electron is minimally coupled to a massless photon with coupling constant e. Compute to $\mathcal{O}(e^2)$ the invariant Feynman amplitude for electron–electron scattering in both the Coulomb gauge and the Feynman gauge, and show that the final answers are the same. HINT: Use the fact that the spinors obey the Dirac equation.

Remark: This is an extension of a computation that will be done in the lectures (§34.1) for the vacuum-to-vacuum amplitude in the presence of an external c-number source.

(1998b 5.2)

Solutions 17

17.1 We start with the functional integral result,

$$Z[J] = \langle 0|\mathrm{S}|0\rangle_J = \left[\frac{\det(\Box^2+\mu^2-J(x)-i\epsilon)}{\det(\Box^2+\mu^2-i\epsilon)}\right]^{-\frac{1}{2}} = \left\{\det\left[(\Box^2+\mu^2-i\epsilon)^{-1}(\Box^2+\mu^2-J(x)-i\epsilon)\right]\right\}^{-\frac{1}{2}}$$

Using the identity (P17.4), we have

$$Z[J] = \exp\left\{-\tfrac{1}{2}\mathrm{Tr}\left[\ln\left(1-\left(\Box^2+\mu^2-i\epsilon\right)^{-1}J(x)\right)\right]\right\} = \exp\left\{-\tfrac{1}{2}\mathrm{Tr}\left[\ln\left(1+\left(-\Box^2-\mu^2+i\epsilon\right)^{-1}J(x)\right)\right]\right\}$$

Since

$$\ln(1+A) = \sum_{n=1}^{\infty}\frac{(-1)^{n-1}}{n}A^n$$

we find

$$Z[J] = \exp\left\{-\tfrac{1}{2}\mathrm{Tr}\left[\sum_{n=1}^{\infty}\frac{(-1)^{n-1}}{n}\left(\left(-\Box^2-\mu^2+i\epsilon\right)^{-1}J(x)\right)^n\right]\right\}$$

Writing this out in x-space, we obtain

$$Z[J] = \exp\left\{\sum_{n=1}^{\infty}\frac{i^n}{2n}\int d^4x_1\cdots d^4x_n\Delta_F(x_n-x_1)J(x_1)\Delta_F(x_1-x_2)J(x_2)\cdots\Delta_F(x_{n-1}-x_n)J(x_n)\right\}$$

because $(-\Box_x^2-\mu^2)(-i\Delta_F(x)) = \delta^{(4)}(x)$.

Now consider the Feynman graph calculation with

$$\mathscr{L}_I = \tfrac{1}{2}J(x)\phi_I^2$$

Recall (13.11),

$$Z[J] = \exp(iW[J]),$$

where $W[J]$ is the sum of connected vacuum graphs. Any connected graph for this theory can be described as a simple loop of n alternating J interactions and ϕ propagators. The diagram for $n=5$ is shown in Figure S17.1. At order J^n, the corresponding amplitude is

$$\frac{N_D\, i^n}{2^n n!}\int d^4x_1\cdots d^4x_n\left[\Delta_F(x_n-x_1)J(x_1)\Delta_F(x_1-x_2)J(x_2)\cdots\Delta_F(x_{n-1}-x_n)J(x_n)\right]$$

where N_D is the number of distinct diagrams.

We can generate all diagrams of this pattern by permuting the n vertices and/or switching the two ϕ fields at each vertex. If each of these operations were to yield a distinct diagram, we would obtain $2^n n!$ diagrams, canceling the $2^n n!$ in the denominator. But our diagrams have an n-fold cyclic rotational symmetry and mirror symmetry (switching all pairs of ϕ's at each vertex simultaneously and arranging the vertices in reverse cyclic order). The number of distinct diagrams at order J^n is therefore

$$N_D = \frac{2^n n!}{2n}$$

We conclude that once again,

$$Z[J] = \exp\left\{\sum_{n=1}^{\infty} \frac{i^n}{2n} \int d^4x_1 \cdots d^4x_n \Delta_F(x_n - x_1)J(x_1)\Delta_F(x_1 - x_2)J(x_2)\cdots\Delta_F(x_{n-1} - x_n)J(x_n)\right\}$$

identical to the functional integral result.

If we now consider complex scalar mesons, the only difference in the Feynman graph calculation is that each propagator carries a definite direction (one vertex contributes the ϕ and the other contributes the ϕ^*). This eliminates the mirror symmetry of the real scalar theory, and the factor of $\frac{1}{2}$ disappears from the coefficient of the integral:

$$Z[J] = \exp\left\{\sum_{n=1}^{\infty} \frac{i^n}{n} \int d^4x_1 \cdots d^4x_n \Delta_F(x_n - x_1)J(x_1)\Delta_F(x_1 - x_2)J(x_2)\cdots\Delta_F(x_{n-1} - x_n)J(x_n)\right\}$$

Expressed as determinants,

$$Z[J] = \langle 0|\mathrm{S}|0\rangle_J = \left[\frac{\det(\Box^2 + \mu^2 - J(x) - i\epsilon)}{\det(\Box^2 + \mu^2 - i\epsilon)}\right]^{-1}$$

the exponent of $-\frac{1}{2}$ is replaced by -1. ■

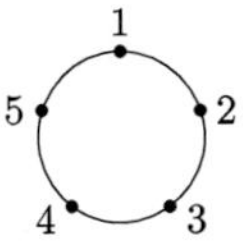

Figure S17.1: Graph for scalar field $\mathscr{L}_I = J\phi^2$, $n = 5$

17.2 We begin with the generalized Coulomb gauge constraint

$$F(x) = \nabla \cdot \mathbf{A}(x) - f(x) = 0. \tag{S17.1}$$

Under an infinitesimal gauge transformation

$$A_\mu \to A'_\mu = A_\mu + \frac{1}{e}\partial_\mu \delta\chi = A_\mu + \delta A_\mu$$

Then for the gauge-fixing function

$$\delta F(x) = \frac{1}{e}\partial^i\partial_i\delta\chi = -\frac{1}{e}\nabla^2\delta\chi(x)$$

and

$$\Delta = \det\left[\frac{\delta F(x)}{\delta\chi(y)}\right] = \det\left[-\frac{1}{e}\nabla^2\delta^{(4)}(x-y)\right]$$

This determinant is a constant (that is, it is independent of the fields over which the integration is performed) so it can be absorbed into the normalization of the functional integral.

As in §31.3, we integrate (31.31) over all possible $f(x)$ with weight

$$\exp\left\{-\frac{i}{2\xi}\int d^4x\, (\nabla \cdot \mathbf{A})^2\right\}$$

This generates an effective action of the form

$$\int d^4x \left[-\tfrac{1}{2}\partial_\mu A_\nu \partial^\mu A^\nu + \tfrac{1}{2}\partial_\mu A_\nu \partial^\nu A^\mu - \frac{1}{2\xi}(\partial_i A^i)^2\right] = \int d^4x \left[\tfrac{1}{2}A^\mu M_{\mu\nu} A^\nu\right]$$

after an integration by parts, where

$$M_{\mu\nu} = g_{\mu\nu}\,\Box^2 - \partial_\mu\partial_\nu + \frac{1}{\xi}\delta^i{}_\mu\delta^j{}_\nu\partial_i\partial_j$$

In momentum space we find

$$\widetilde{M}_{\mu\nu} = -g_{\mu\nu}k^2 + k_\mu k_\nu - \frac{1}{\xi}\delta^i{}_\mu\delta^j{}_\nu k_i k_j.$$

Using rotational invariance, we can assume that $\mathbf{k}$ points in the x-direction: $\mathbf{k} = (|\mathbf{k}|, 0, 0)$. We then have

$$\widetilde{M}_{\mu\nu} = \begin{bmatrix} |\mathbf{k}|^2 & -k^0|\mathbf{k}| & 0 & 0 \\ -k^0|\mathbf{k}| & k^2 + (1-\frac{1}{\xi})|\mathbf{k}|^2 & 0 & 0 \\ 0 & 0 & k^2 & 0 \\ 0 & 0 & 0 & k^2 \end{bmatrix} = \begin{bmatrix} \mathbf{N} & \mathbf{0} \\ \mathbf{0} & \mathbf{R} \end{bmatrix}$$

where $\mathbf{N}$ and $\mathbf{R}$ are 2×2 matrices, and $\mathbf{0}$ is a 2×2 zero matrix;

$$\mathbf{N} = \begin{pmatrix} |\mathbf{k}|^2 & -k^0|\mathbf{k}| \\ -k^0|\mathbf{k}| & k^2 + (1-\frac{1}{\xi})|\mathbf{k}|^2 \end{pmatrix} \qquad \mathbf{R} = \begin{pmatrix} k^2 & 0 \\ 0 & k^2 \end{pmatrix} \tag{S17.2}$$

The propagator is

$$\widetilde{D}_{\mu\nu} = i\widetilde{M}^{-1}_{\mu\nu} \tag{S17.3}$$

Because of the block structure of $\widetilde{M}_{\mu\nu}$, it's easy to invert:

$$\widetilde{M}^{-1}_{\mu\nu} = \begin{bmatrix} \mathbf{N}^{-1} & \mathbf{0} \\ \mathbf{0} & \mathbf{R}^{-1} \end{bmatrix} \tag{S17.4}$$

and

$$\mathbf{N}^{-1} = \frac{1}{\Delta_\mathbf{N}}\begin{pmatrix} k^2 + (1-\frac{1}{\xi})|\mathbf{k}|^2 & k^0|\mathbf{k}| \\ k^0|\mathbf{k}| & |\mathbf{k}|^2 \end{pmatrix} \qquad \mathbf{R}^{-1} = \begin{pmatrix} \frac{1}{k^2} & 0 \\ 0 & \frac{1}{k^2} \end{pmatrix}$$

where $\Delta_\mathbf{N}$ is the determinant of $\mathbf{N}$:

$$\Delta_\mathbf{N} = \det\mathbf{N} = (k^2 - k_0^2)|\mathbf{k}|^2 + \frac{\xi-1}{\xi}|\mathbf{k}|^4 = -\frac{1}{\xi}|\mathbf{k}|^4$$

Then

$$\widetilde{D}_{\mu\nu} = i\widetilde{M}^{-1}_{\mu\nu} = i\begin{bmatrix} -\frac{\xi k^2}{|\mathbf{k}|^4} + \frac{(1-\xi)}{|\mathbf{k}|^2} & -\frac{\xi k^0}{|\mathbf{k}|^3} & 0 & 0 \\ -\frac{\xi k^0}{|\mathbf{k}|^3} & -\frac{\xi}{|\mathbf{k}|^2} & 0 & 0 \\ 0 & 0 & \frac{1}{k^2} & 0 \\ 0 & 0 & 0 & \frac{1}{k^2} \end{bmatrix} \tag{S17.5}$$

In analogy with the passage to the Landau gauge (31.37), we now obtain the Coulomb gauge propagator by taking the limit $\xi \to 0$. In this limit

$$\widetilde{D}^C_{\mu\nu} = \begin{bmatrix} \frac{i}{|\mathbf{k}|^2} & 0 & 0 & 0 \\ 0 & 0 & 0 & 0 \\ 0 & 0 & \frac{i}{k^2} & 0 \\ 0 & 0 & 0 & \frac{i}{k^2} \end{bmatrix} \tag{S17.6}$$

For a general momentum $\mathbf{k}$, we have

$$\widetilde{D}^C_{00} = \frac{i}{|\mathbf{k}|^2} \tag{S17.7a}$$

$$\widetilde{D}^C_{0i} = \widetilde{D}^C_{i0} = 0 \tag{S17.7b}$$

$$\widetilde{D}^C_{ij} = -\frac{i}{k^2 + i\epsilon}\left[g_{ij} + \frac{k_i k_j}{|\mathbf{k}|^2}\right] \tag{S17.7c}$$

The 0–0 component is the Fourier transform of the static Coulomb potential (see §9.3).

Alternatively, we can use 3-space projection operators, analogous to (28.91) and (28.92), to find the propagator:

$$P^T_{ij} = g_{ij} + \frac{k_i k_j}{|\mathbf{k}|^2} \qquad P^L_{ij} = -\frac{k_i k_j}{|\mathbf{k}|^2} \tag{S17.8}$$

Expressed in terms of these,

$$\widetilde{M}_{ij} = -g_{ij}k^2 + \left[\frac{\xi - 1}{\xi}\right] k_i k_j = -P^T_{ij}k^2 - P^L_{ij}\left[\frac{\xi k^2 + (\xi - 1)|\mathbf{k}|^2}{\xi}\right] \tag{S17.9}$$

and so

$$\widetilde{D}^{-1}_{ij} = i\widetilde{M}^{-1}_{ij} = -P^T_{ij}\frac{i}{k^2 + i\epsilon} - P^L_{ij}\frac{i\xi}{\xi k^2 + (\xi - 1)|\mathbf{k}|^2} \tag{S17.10}$$

Taking the limit $\xi \to 0$, we again obtain (S17.7c).

We can express the propagators in a slightly more covariant-looking way. We replace the Coulomb gauge condition (S17.1) with

$$F(x) = n_\mu \partial^\mu (n_\nu A^\nu(x)) - f(x) = 0. \tag{S17.11}$$

where

$$n^\mu = (1, 0, 0, 0).$$

The matrix $\widetilde{M}_{\mu\nu}$ is

$$\widetilde{M}_{\mu\nu} = -k^2\left[g_{\mu\nu} - \frac{k_\mu k_\nu}{k^2}\right] - \frac{1}{\xi}\left[k_\mu k_\nu + n_\mu n_\nu (n\cdot k)^2 - n\cdot k(k_\mu n_\nu + n_\mu k_\nu)\right]$$

Inverting this and setting ξ to 0 again we find the compact form

$$\widetilde{D}^C_{\mu\nu} = -\frac{i}{k^2 + i\epsilon}\left[g_{\mu\nu} - \frac{(n\cdot k)(k_\mu n_\nu + n_\mu k_\nu) - k_\mu k_\nu}{(n\cdot k)^2 - k^2}\right] \tag{S17.12}$$

The individual matrix elements of this reproduce (S17.7).[2] ■

17.3 The diagrams for lowest nontrivial electron–electron scattering are shown below:

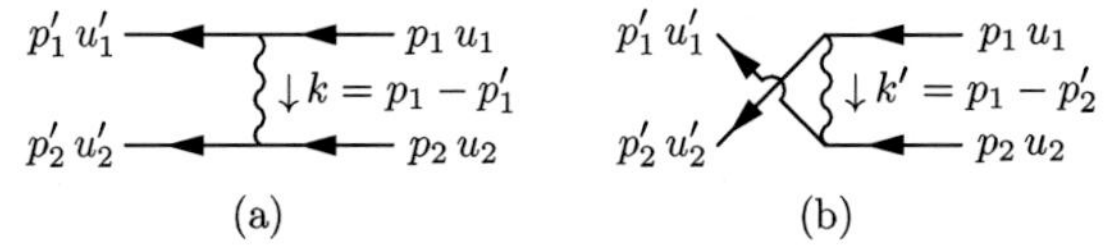

The amplitude for these together is

$$i\mathcal{A}_{fi} = -e^2\left\{J^\mu_{11}\widetilde{D}_{\mu\nu}(k)J^\nu_{22} - J^\mu_{21}\widetilde{D}_{\mu\nu}(k')J^\nu_{12}\right\} \tag{S17.13}$$

where the currents are defined as

$$J^\mu_{ab} = \overline{u}'_b \gamma^\mu u_a$$

and the momenta are

$$k = p_1 - p'_1 = p'_2 - p_2$$
$$k' = p_1 - p'_2 = p'_1 - p_2$$

The minus sign between the two contributions comes from the fermion interchange rule (see §21.4). The propagator $\widetilde{D}^C_{\mu\nu}$ in Coulomb gauge is given by (S17.7). In Feynman gauge,

$$\widetilde{D}^F_{\mu\nu} = -\frac{ig_{\mu\nu}}{k^2 + i\epsilon}$$

---

[2] [Eds.] This form is mentioned in S. Pokorski, *Gauge Field Theories*, 2nd ed., Cambridge U. P., 2000. Bjorken and Drell have a form similar to (S17.12), but with an additional $k^2 n_\mu n_\nu$ term in the numerator of the second part: *Fields*, equation (14.54), p. 79. This cancels the static Coulomb interaction, rendering $D_{00} = 0$. They continue with a discussion about why only the covariant part of the propagator matters.

The currents are conserved because the spinors obey the Dirac equation. For example, consider J^μ_{11}:

$$k_\mu J^\mu_{11} = (p_1 - p_1')_\mu \overline{u}_1' \gamma^\mu u_1 = \overline{u}_1'(\not{p}_1 u_1) - (\overline{u}_1' \not{p}_1')u_1 = m(\overline{u}_1' u_1 - \overline{u}_1' u_1) = 0 \tag{S17.14}$$

and the same is true for the others:

$$k_\mu J^\mu_{22} = k'_\mu J^\mu_{12} = k'_\mu J^\mu_{21} = 0$$

Then

$$k_0 J^0_{11} = \mathbf{k}\boldsymbol{\cdot}\mathbf{J}_{11} \qquad k_0 J^0_{22} = \mathbf{k}\boldsymbol{\cdot}\mathbf{J}_{22} \tag{S17.15}$$

$$k'_0 J^0_{12} = \mathbf{k}'\boldsymbol{\cdot}\mathbf{J}_{12} \qquad k'_0 J^0_{21} = \mathbf{k}'\boldsymbol{\cdot}\mathbf{J}_{21} \tag{S17.16}$$

Using these relations, we have

$$\begin{aligned} J^\mu_{11} g_{\mu\nu} J^\nu_{22} &= J^0_{11} J^0_{22} + J^i_{11} J^j_{22} g_{ij} = J^0_{11} J^0_{22} + J^i_{11} J^j_{22}\left[g_{ij} + \frac{k_i k_j}{|\mathbf{k}|^2} - \frac{k_i k_j}{|\mathbf{k}|^2}\right] \\ &= J^0_{11} J^0_{22}\left[1 - \frac{k_0^2}{|\mathbf{k}|^2}\right] + J^i_{11} J^j_{22}\left[g_{ij} + \frac{k_i k_j}{|\mathbf{k}|^2}\right] \\ &= J^0_{11} J^0_{22}\left[-\frac{k^2}{|\mathbf{k}|^2}\right] + J^i_{11} J^j_{22}\left[g_{ij} + \frac{k_i k_j}{|\mathbf{k}|^2}\right] \end{aligned}$$

and similarly

$$J^\mu_{12} g_{\mu\nu} J^\nu_{21} = J^0_{12} J^0_{21}\left[-\frac{k'^2}{|\mathbf{k}'|^2}\right] + J^i_{12} J^j_{21}\left[g_{ij} + \frac{k'_i k'_j}{|\mathbf{k}'|^2}\right]$$

Then the amplitude for electron–electron scattering in Feynman gauge is given by

$$\begin{aligned} i\mathcal{A}_{\text{Feynman}} &= -e^2\left\{J^\mu_{11}\left[-\frac{i g_{\mu\nu}}{k^2 + i\epsilon}\right] J^\nu_{22} - J^\mu_{12}\left[-\frac{i g_{\mu\nu}}{k'^2 + i\epsilon}\right] J^\nu_{21}\right\} \\ &= -e^2\left\{J^0_{11} J^0_{22}\left[\frac{i}{|\mathbf{k}|^2}\right] + J^i_{11} J^j_{22}\left[\frac{-i}{k^2 + i\epsilon}\right]\left[g_{ij} + \frac{k_i k_j}{|\mathbf{k}|^2}\right]\right. \\ &\qquad \left. - J^0_{12} J^0_{21}\left[\frac{i}{|\mathbf{k}'|^2}\right] - J^i_{12} J^j_{21}\left[\frac{-i}{k'^2 + i\epsilon}\right]\left[g_{ij} + \frac{k'_i k'_j}{|\mathbf{k}'|^2}\right]\right\} \\ &= i\mathcal{A}_{\text{Coulomb}} \end{aligned} \tag{S17.17}$$

While we should expect that the photon propagator is gauge-dependent (it is, after all, the Fourier transform of $\langle 0|T(A_\mu(x)A_\nu(y))|0\rangle$, and A_μ is gauge-dependent), this gauge dependence does *not* affect a scattering amplitude in a concrete calculation. ■

32

Generating functionals and Green's functions

We are going to use functional methods to analyze the structure of Green's functions. In particular we will discuss three different generating functionals.[1] These will give us

1. Full Green's functions (given by our old friend, $Z[J]$)
2. Connected Green's functions
3. One-particle irreducible (1PI) Green's functions.

The immediate reason for going through this analysis is to derive the Ward identities directly for *one-particle irreducible* Green's functions, and thus to complete the renormalization program for quantum electrodynamics. The methods are of general utility, however, and applicable in a wide variety of circumstances. There is nothing especially quantum electrodynamical about them; we could have talked about them earlier. It's simply that this is the first time that we've encountered a problem of sufficient combinatoric complexity to make it worthwhile to go through this general derivation.

32.1 The loop expansion

To simplify the notation as much as possible, we will conduct the discussion for a theory of a single scalar field ϕ. In a more general theory the only alteration is appropriately sprinkling the equations with indices and being careful of the order of terms if one is dealing with Fermi fields, and so has anti-commuting sources instead of commuting sources. The field ϕ could be the renormalized field or the unrenormalized field; as far as the combinatorics are concerned, it doesn't matter. Its dynamics are determined by some action

$$\mathcal{S}[\phi, J] = \mathcal{S}[\phi] + \int d^4x\, J(x)\phi(x) \tag{32.1}$$

where J is a c-number function of x. The generating functional for full Green's functions $G^{(n)}(x_1, \cdots, x_n)$ associated with this action is[2]

$$Z[J] = N \int (d\phi) e^{i\mathcal{S}[\phi, J]} = \sum_{n=0} \frac{i^n}{n!} \int d^4x_1 \cdots d^4x_n\, J(x_1) \cdots J(x_n)\, G^{(n)}(x_1, \cdots x_n) \tag{32.2}$$

[1] [Eds.] Peskin & Schroeder *QFT*, Section 11.5, pp. 379–383; Ryder *QFT*, Sections 6.4–6.5, pp. 196–207.

[2] [Eds.] See (13.8). Note from (13.5) that ρ in (13.8) differs from J in (32.2) by a sign.

N is a normalization factor chosen so that $Z[0] = 1$:

$$N^{-1} = \int (d\phi)\, e^{iS[\phi]} \tag{32.3}$$

Finding the generating functional for *connected* Green's functions is trivial because of our powerful theorem of general utility, (8.49), that the sum of all Feynman graphs is the exponential of the sum of all connected Feynman graphs.[3] There are no vacuum-to-vacuum graphs; they are canceled by the normalization factor N. Define the functional $W[J]$ by

$$Z[J] \equiv e^{iW[J]} \tag{32.4}$$

The i is put in by convention, so that all the i's disappear when we rotate into Euclidean space. $W[J]$ is the generating functional for *connected* Green's functions, $G_c^{(n)}(x_1, \cdots, x_n)$:

$$iW[J] = \sum_{n=0} \frac{i^n}{n!} \int d^4x_1 \cdots d^4x_n\, J(x_1) \cdots J(x_n)\, G_c^{(n)}(x_1, \cdots x_n) \tag{32.5}$$

where $G_c^{(n)}$ is the sum of all *connected* Feynman diagram with n external legs. I made this remark much earlier, when we first introduced $Z[J]$.

Before we go on to constructing the generating functional for 1PI Green's functions, let's discuss an amusing property of $W[J]$. This property, which can be described as counting loops and counting $\hbar$'s, is known as the **loop expansion**[4] or the **semi-classical expansion**. Throughout this course we have been setting $\hbar = 1$ except on occasions when it is convenient to restore it; this is one of those occasions. To see how to count loops, re-introduce $\hbar$ into the equations, writing

$$Z_\hbar[J] = N \int (d\phi)\, e^{iS[\phi,J]/\hbar} = \exp(iW_\hbar[J]) \tag{32.6}$$

Putting $\hbar$ into the action puts $\hbar$ into all our propagators and vertices, and therefore puts $\hbar$ into every Feynman graph. Let us count the powers of $\hbar$ associated with a given Feynman graph.

Every propagator or internal line will yield an $\hbar$ because the propagator is given by the inverse of the quadratic part of the action, and every quadratic part of the action has a $1/\hbar$ in it. Every vertex yields a $1/\hbar$ because the vertices are just read off from the non-quadratic part of the action. Therefore the contribution $\mathcal{G}$ of an arbitrary graph with I propagators (internal lines) and V vertices is proportional to $\hbar^{(I-V)}$:

$$\mathcal{G} \propto \hbar^{(I-V)} \tag{32.7}$$

On the other hand, for the contribution $\mathcal{G}_c$ of a *connected* graph, the number of loops L is

$$L = I - V + 1 \tag{32.8}$$

This is an old result, but I'll remind you of the derivation.[5] There is an integration for every internal line and an energy–momentum conserving delta function for every vertex. The delta

[3] [Eds.] The theorem in (8.49) is written in terms of Wick diagrams, but it holds for Feynman diagrams as well.

[4] [Eds.] Coleman *Aspects*, pp. 135–136; Ryder *QFT*, pp. 317–8.

[5] [Eds.] This derivation may have been in the lost part of the videotape of Lecture 25. It does not seem to appear elsewhere in the videos, though it is in Coleman *Aspects*, Section 3.4, pp. 135–6. See also Peskin & Schroeder *QFT*, equation (10.2), p. 316.

functions kill all the internal integrations except for one left over for overall energy–momentum conservation. So the number of free integration variables, or equivalently the number of loops L, is given by $I - V + 1$, and the power of $\hbar$ in the contribution $\mathcal{G}_c$ is

$$\mathcal{G}_c \propto \hbar^{(I-V)} = \hbar^{(L-1)} \tag{32.9}$$

Notice I emphasize "connected". If the graph were disconnected, for example if there were two components, it would have an overall energy–momentum conserving delta function for *each* of its component pieces, and the formula (32.8) would not be true. Expanding W in powers of $\hbar$ is equivalent to expanding W in the number of loops in the graph. We find

$$W_\hbar = -i \ln Z_\hbar = \underbrace{\frac{W_0}{\hbar}}_{\text{no loops}} + \underbrace{W_1}_{\text{one loop}} + \underbrace{\hbar W_2}_{\text{two loops}} + \cdots \tag{32.10}$$

This is a rather peculiar power series. The "no loop" term is called the **tree approximation**. The name is borrowed from topological network theory.[6] These graphs without loops are called **tree graphs**, and they are $\mathcal{O}(1/\hbar)$. The graphs with one loop are $\mathcal{O}(1)$, the graphs with two loops are $\mathcal{O}(\hbar)$, and so on. Actually, expansion in $\hbar$ is garbage. Although $\hbar$ is a quantity *with dimensions*, we can always choose our units so that $\hbar = 1$. So there's no particular reason to believe in an expansion in $\hbar$; the truncated series is not necessarily a good approximation. But I wanted to point out that if we *did* expand in $\hbar$, that would be equivalent to counting the number of loops.

Just to give you something definite to look at, let's look at our old friend ϕ^4 theory. In Figure 32.1 every external vertex has a J; that's indicated by the dot. Notice that there are

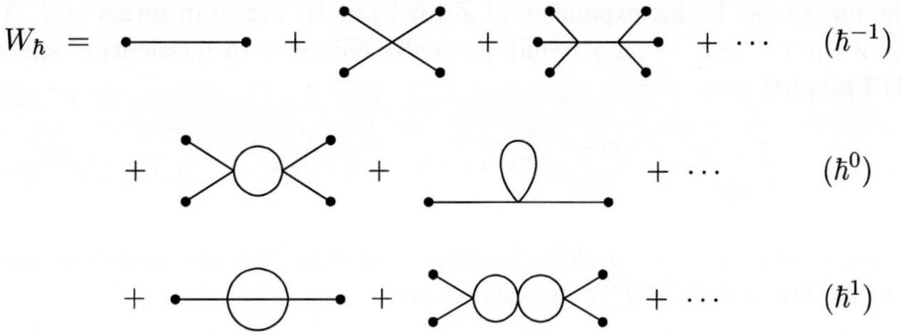

Figure 32.1: Expansion in loops is equivalent to expansion in powers of ħ

many graphs at a given order in $\hbar$. The tree approximation itself goes on forever; there are infinitely many vacuum-to-vacuum connected graphs in the presence of J with no internal loops. Of course, there are only a few 1PI graphs. At the tree level only the graphs in Figure 32.2 are 1PI. The graph 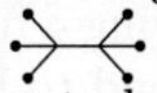 can be cut on the internal line joining the two 4-point vertices.[7] So it would be even nicer to have a generating functional for 1PI graphs.

---

[6] [Eds.] See, e.g., Figure 1.5.1, p. 12 of R. Diestel, *Graph Theory*, 3rd ed., Springer-Verlag, 2005.

[7] [Eds.] The reader may be wondering about the definition of "one-particle irreducible" (p. 321). The topology of the first tree graph, , seems to be the same as that of , with one internal line, and yet the former is 1PI, and the latter is not. Perhaps a better way to think about what makes a graph 1PI is this: if an internal line is removed, what remains? If there are separate pieces, the graph is not 1PI. In the second diagram, after

Figure 32.2: One particle irreducible tree graphs in ϕ^4 theory

32.2 The generating functional for 1PI Green's functions

Recall that 1PI graphs can't be split into two distinct pieces by cutting a single internal line. We will define the 1PI diagrams with certain cunning conventions so that everything comes out right in the end. Then we will show how to construct the generating functional for 1PI Green's functions, $\Gamma[\overline{\phi}]$, in terms of objects we already know, $Z[J]$ or (32.5) $iW[J] = \ln Z[J]$. We will write $\Gamma[\overline{\phi}]$ as a functional of a classical variable $\overline{\phi}(x)$, for reasons that will become clear shortly, rather than in terms of the c-number current J. We will call $\Gamma[\overline{\phi}]$ the **effective action**. As I will show, in the tree approximation

$$\mathcal{S}[\overline{\phi}] = \Gamma[\overline{\phi}] \tag{32.11}$$

This assertion is true, and I'll back it up with two or three diagrams. That's not enough, of course. (Let me not make invidious comparisons with my distinguished colleagues who teach this subject. One calculation done in detail is worth a hundred general arguments.)

This generating functional has a functional Taylor expansion

$$i\Gamma[\overline{\phi}] = i\sum_n \frac{1}{n!}\int d^4x_1 \cdots d^4x_n\, \Gamma^{(n)}(x_1,\cdots,x_n)\,\overline{\phi}(x_1)\cdots\overline{\phi}(x_n) \tag{32.12}$$

just like the functional Taylor expansion of Z (32.2) or W (32.5) in terms of J. The Fourier transforms of the $\widetilde{\Gamma}^{(n)}(p_1,\cdots,p_n)$ (what we really compute in momentum space when we compute 1PI graphs) are

$$\Gamma^{(n)}(x_1,\cdots,x_n) = \int \frac{d^4p_1}{(2\pi)^4}\cdots\frac{d^4p_n}{(2\pi)^4} e^{i(p_1\cdot x_1+\cdots+p_n\cdot x_n)}\widetilde{\Gamma}^{(n)}(p_1,\cdots,p_n)(2\pi)^4\delta^{(4)}(p_1+\cdots+p_n) \tag{32.13}$$

We include the energy–momentum delta function so that in later manipulations we don't have to divide it out. For $n \neq 2$ the $\widetilde{\Gamma}^{(n)}$ are defined as:

$$i\widetilde{\Gamma}^{(n)}(p_1,\cdots,p_n) = \sum(\text{all 1PI graphs with } n \text{ external lines}) = \tag{32.14}$$

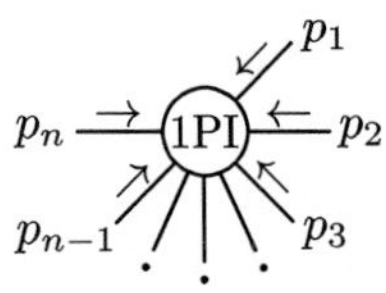

with all of our usual conventions: no energy–momentum conserving δ functions, no propagators on the external lines, just as before.[8] (The external legs are said to be **amputated**.) These 1PI Green's functions are only well-defined if the sum of the momenta is zero: $\sum p_i = 0$.

---

the internal line is removed, two pieces are left: it is not 1PI. In the case of the first diagram, *nothing* is left: it *is* 1PI. The 1PI graphs are also called **proper diagrams**. See, e.g., M. Kaku, *Quantum Field Theory: A Modern Introduction*, Oxford U. P., 1993, p. 219.

[8] [Eds.] We assume that $\widetilde{\Gamma}^{(0)} = 0$.

For $n = 2$ we will define $\widetilde{\Gamma}^{(n)}$ in a somewhat peculiar way. Recall that we had defined the full propagator (15.29) by saying

$$\widetilde{G}^{(2)}(p,p') = (2\pi)^4\delta^{(4)}(p+p')\widetilde{D}(p) = \tag{32.15}$$

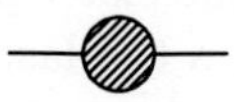

giving the renormalized or unrenormalized propagator for the renormalized or unrenormalized field, respectively. We define $\widetilde{\Gamma}^{(2)}(p,-p)$ by

$$\frac{i}{\widetilde{\Gamma}^{(2)}(p,-p)} \equiv \widetilde{D}(p) \tag{32.16}$$

(If the theory involves many fields instead of one, this formula will be expressed in terms of a matrix inverse.) This is almost our usual definition. In our old definition, $i\widetilde{\Gamma}^{(2)}$ would have been $-i\widetilde{\Pi}'(p^2)$, the self-energy operator, which was defined (15.33) as the sum of all 1PI graphs with two external lines. With the old definition, we found (15.36)

$$\widetilde{D}(p) = \frac{i}{p^2 - \mu^2 - \widetilde{\Pi}'(p^2) + i\epsilon} \tag{32.17}$$

obtained from summing the geometric series of 1PI graphs. Thus

$$i\widetilde{\Gamma}^{(2)}(p,-p) = i(p^2-\mu^2) - i\widetilde{\Pi}'(p^2) = i(p^2-\mu^2) + \text{—(1PI)—} \tag{32.18}$$

The definition of $\widetilde{\Gamma}^{(2)}(p,-p)$ differs from the definition of the other $\widetilde{\Gamma}^{(n)}(p_i)$'s by the addition of the tree-level term $(p^2-\mu^2)$, a trivial term determined by the free Lagrangian, to the old definition of the sum of the 1PI graphs with two external legs.

Now why on earth have we chosen this peculiar definition of $\Gamma^{(2)}$? The reason is simple. With these definitions, if we treat $\Gamma[\overline{\phi}]$ as an honest to goodness action, $\mathcal{S}[\overline{\phi}]$, for the field $\overline{\phi}$ —that would seem to be dumb—and compute W or equivalently, Z using *only* tree graphs, deriving Feynman rules in the tree approximation, forgetting about higher order diagrams—*double* dumb—then we get the *exact* W or Z; our errors cancel each other out! Why does this happen? That's easy: I've cooked it up so that it *should* happen.

For example, how do we get the propagator in the tree approximation? From the action, we take the coefficient of the quadratic term in $\overline{\phi}$, Fourier transform it, invert that and multiply it by i, and that's our answer. That's why we defined $\widetilde{\Gamma}^{(2)}(p,-p)$ (32.16) as we did. Using $\Gamma[\overline{\phi}]$ as the action, this procedure results in

$$\widetilde{D}(p) = \frac{i}{\widetilde{\Gamma}^{(2)}(p,-p)} \tag{32.19}$$

which is exactly right.

What about the higher order terms?[9] Let's look at, for example, the exact three-point function $\widetilde{G}^{(3)}$ in a theory with a ϕ^3 interaction, Figure 32.3. I wrote down a formula like that

[9] [Eds.] In the following we will use these graphical symbols:

Exact Green's function:  1PI Green's function: (1PI)

earlier (16.29). We follow a line through the graph. There has got to be a place where we can cut each line, as close to the central blob as we can. Everything between the dot (the J) and the cut becomes the exact propagator. Everything left over is by definition one-particle irreducible, because it's the last place we can cut the line. This is the simplest (indeed, it is the only) tree graph in such a theory: source J—propagator—vertex for all three lines. Likewise, the exact four-point function in ϕ^4 theory is shown in Figure 32.3. Once again,

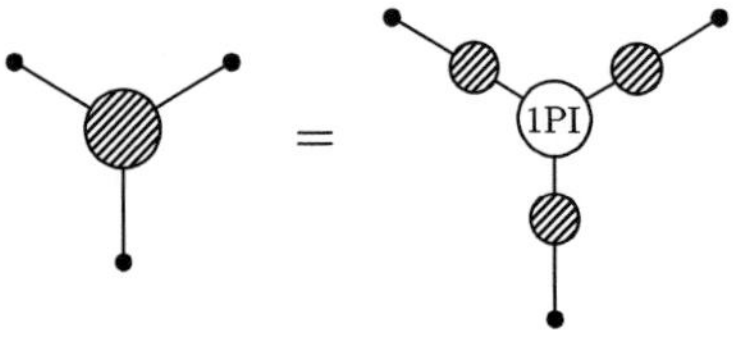

Figure 32.3: Exact three-point function $\widetilde{G}^{(3)}$ in ϕ^3 theory

these are precisely the graphs that would appear in the tree approximation, except that all

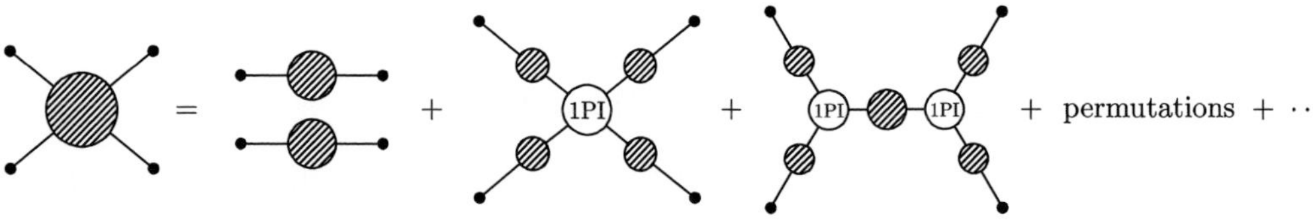

Figure 32.4: Exact four-point function $\widetilde{G}^{(4)}$ in ϕ^4 theory

the propagators are the exact propagators and all the vertices are 1PI blobs. I've got some disconnected parts, plus permutations, and those are of course what I would get in the tree approximation in a theory in which a shaded blob with only two legs is the exact propagator. Then there will be graphs where I can cut to put two lines on one side and one on the other, so there's an intermediate line. And there will be graphs where I cannot cut to produce two lines on one side and one on the other. If we treat the 1PI graphs as giving us *effective* interaction vertices, then to find the full Green's functions we only have to sum up *tree* graphs, never any loops, because all the loops have been stuffed inside the definition of the propagators and the 1PI graphs. This marvelous property of the 1PI graphs is important. Taking the 1PI graph generating functional for a quantum action enables us to turn the combinatorics of building up full Green's functions from 1PI Green's functions into an analytic statement, and we end up with the correct expressions for the full Green's functions. We're turning a topological statement of one-particle irreducibility into an analytic statement that we will find easy to handle. From this point of view it's very simple to see the rule that connects $\Gamma[\overline{\phi}]$ to $W[J]$. From the loop expansion (32.10)

$$Z[J] = \exp\left\{\frac{i}{\hbar}W[J] + \mathcal{O}(1) + \mathcal{O}(\hbar)\right\} = N\int (d\overline{\phi})\exp\left\{\frac{i}{\hbar}\left[\Gamma[\overline{\phi}] + \int d^4x\, J\overline{\phi}\right]\right\} \tag{32.20}$$

Comparing (32.6) and (32.20), we see that $\Gamma[\overline{\phi}]$ has the same general structure as $\mathcal{S}[\phi]$:

$$\frac{\Gamma[\overline{\phi}]}{\hbar} = \frac{\mathcal{S}[\overline{\phi}]}{\hbar}\left(1 + \mathcal{O}(\hbar)\right) = \underbrace{\frac{\mathcal{S}[\overline{\phi}]}{\hbar}}_{\text{tree graphs}} + \underbrace{\mathcal{S}[\overline{\phi}]\times\mathcal{O}(1)}_{\text{with loops}} \tag{32.21}$$

In the tree approximation, we drop the second term on the right, and the assertion (32.11) is proved.

Let's imagine a reader who has read this far and dutifully worked all the problems, but who is nonetheless slightly confused at this point. This is precisely how we would compute the generating functional if we treated $\Gamma[\overline{\phi}]$ as a quantum action. The term with no loops, the tree graphs, gives the correct $W[J]$. This is simply the analytic prescription for the statement about the marvelous property of 1PI Green's functions. The terms of $\mathcal{O}(1)$, $\mathcal{O}(\hbar)$ and so on, give us the graphs with loops, which don't appear in the statement; it contains only tree graphs. We've converted a topological statement, that we have the generating functional for 1PI graphs, into a analytic statement, and thence, using our previous lore about counting loops and $\hbar$'s, into an equation.

We only want the term in (32.10) which is $\mathcal{O}(1/\hbar)$. How do we evaluate the leading term in the limit of small $\hbar$? By the method of stationary phase.[10] Moreover, we don't even need to worry about the determinant in finding the stationary phase, because that is a term of $\mathcal{O}(1)$. All we have to do is look for the point of stationary phase, and solve the equation[11]

$$\frac{\delta\Gamma[\overline{\phi}]}{\delta\overline{\phi}(x)} = -J(x) \tag{32.22}$$

That is the functional equivalent of (17.37), the variational derivative of the phase in the integral with respect to $\overline{\phi}$. This determines $\overline{\phi}$ as a (nonlocal) function of J: $\overline{\phi}_J$. Once we've found the point $\overline{\phi}_J$ of stationary phase, we plug that value back into the integral and the leading term in the evaluation of the integral is its value at the point of stationary phase:

$$W[J] = \Gamma[\overline{\phi}] + \int d^4x\, J(x)\overline{\phi}_J(x) \tag{32.23}$$

This is going backwards; it's not the result we want. This is the procedure for constructing W from Γ, while we want to construct Γ from W. However, (32.23) has the form of a *Legendre transformation*, going from Γ as a functional of $\overline{\phi}$ to W as a functional of J. It is very easy to invert a Legendre transformation.[12] Starting out with W, we get J as a functional of $\overline{\phi}$:

$$\begin{aligned}\delta W &= \int d^4y \left[\frac{\delta\Gamma}{\delta\overline{\phi}(x)}\delta\overline{\phi}(y) + \delta J(y)\overline{\phi}(y) + J(y)\delta\overline{\phi}(y)\right] \\ &= \int d^4y \left[-J(y)\delta\overline{\phi}(y) + \delta J(y)\overline{\phi}(y) + J(y)\delta\overline{\phi}(y)\right] = \int d^4y\, \delta J(y)\,\overline{\phi}(y)\end{aligned} \tag{32.24}$$

---

[10] [Eds.] See (17.36)–(17.39).

[11] [Eds.] Using (28.74),

$$\frac{\delta}{\delta\overline{\phi}(x)}\int d^4y\, J(y)\overline{\phi}(y) = \int d^4y\, J(y)\delta^{(4)}(x-y) = J(x)$$

In general, though it may seem counterintuitive, for a given functional $F[\phi]$,

$$\delta F[\phi] = \int d^4x \frac{\delta F[\phi]}{\delta\phi(x)}\delta\phi(x)$$

Consider the differential of a scalar-valued function $F(\mathbf{v})$ with a vector argument:

$$dF = \nabla F \boldsymbol{\cdot} d\mathbf{v} = \frac{\partial F}{\partial x^i}dv^i$$

We have to "sum" over all the "components" of $\phi(x)$. This requires an integral.

[12] The way they invert Legendre transformations in the books would drive you crazy.

so that

$$\frac{\delta W}{\delta J(x)} = \int d^4y\, \delta^{(4)}(x-y)\overline{\phi}(y) = \overline{\phi}(x) \tag{32.25}$$

which determines $J(\overline{\phi}) \equiv J_{\overline{\phi}}$. Thus we have two Legendre transform pairs:

$$\frac{\delta \Gamma}{\delta \overline{\phi}} = -J; \qquad W[J] = \Gamma[\overline{\phi}] + \int d^4x\, J\overline{\phi}_J \tag{32.26}$$

$$\frac{\delta W}{\delta J} = \overline{\phi}; \qquad \Gamma[\overline{\phi}] = W[J] - \int d^4x\, J_{\overline{\phi}}\overline{\phi} \tag{32.27}$$

The whole procedure for finding the generating functional for 1PI graphs from the generating functional for connected graphs (I should say Green's functions, the sum of all graphs) is simply a Legendre transformation. We differentiate W with respect to J to define the new variable $\overline{\phi}$, then use the equations above. We've turned a complicated combinatoric operation involving getting into the insides of graphs, slicing them this way and slicing them that way and throwing away graphs if we can cut them in two by cutting a single internal line, into an analytic operation, simply by doing a Legendre transformation.

There's one point that will be useful later on. Because of our functional integral, it's rather trivial to differentiate W with respect to J. Differentiating (32.4) with respect to J

$$\begin{aligned}\frac{\delta Z[J]}{\delta J(x)} &= N\int (d\phi)\, e^{i[S+\int d^4y\, J\phi(y)]}\frac{\delta}{\delta J(x)}\left[i\int d^4y\, J(y)\phi(y)\right] \\ &= iN\int (d\phi)\, \phi(x)\, e^{i[S+\int d^4y\, J\phi(y)]}\end{aligned} \tag{32.28}$$

But from (32.25), it follows that

$$\frac{\delta Z[J]}{\delta J(x)} = ie^{iW[J]}\frac{\delta W[J]}{\delta J(x)} = ie^{iW[J]}\overline{\phi}(x) \tag{32.29}$$

and so

$$iN\int (d\phi)\, \phi(x)\, e^{i[S+\int d^4y\, J\phi(y)]} = i\overline{\phi}(x)N\int (d\phi)\, e^{i[S+\int d^4y\, J\phi(y)]} \tag{32.30}$$

which could be taken as a *definition* of $\overline{\phi}(x)$. Put another way,

$$\overline{\phi}(x) = \frac{\delta W[J]}{\delta J(x)} = \frac{\int (d\phi)\, \phi(x) e^{i[S+\int d^4x\, J\phi]}}{\int (d\phi)\, e^{i[S+\int d^4x\, J\phi]}} = \frac{\langle 0|\phi(x)|0\rangle_J}{\langle 0|0\rangle_J} \tag{32.31}$$

independent of the normalization N. From this equation, $\overline{\phi}(x)$ can be thought of as the mean value of ϕ averaged over function space, with J-dependent measure in function space, $d(\phi)e^{iS[\phi,J]}$. That's a good reason for calling $\overline{\phi}$ the *classical* field.[13]

---

[13] [Eds.] It will be useful to note that if $f(\phi)$ is a linear function of ϕ, $f(x) = \alpha x + \beta$ with α and β some constants, then

$$N\int (d\phi)\, f(\phi(x))e^{i[S+\int d^4y\, J\phi(y)]} = f(\overline{\phi}(x))\, N\int (d\phi)\, e^{i[S+\int d^4y\, J\phi(y)]}$$

32.3 Connecting statistical mechanics with quantum field theory

There is an amazing parallelism between what we've been doing in quantum field theory, with functional integrals, and statistical mechanics.[14] This parallelism is much exploited in mathematical physics, to steal theorems from one discipline and bring them to the other. In statistical mechanics we have a single variable, $\beta = 1/T$, in units natural for statistical mechanics (that is, the Boltzmann constant k_B is set equal to 1). We begin by defining a **partition function**, $Z(\beta)$:

$$Z(\beta) = \text{Tr}\,(e^{-\beta H}) \tag{32.32}$$

The partition function, the trace over the entire Hilbert space of $e^{-\beta H}$, is a nice thing to calculate but it's not particularly what we're interested in. We typically begin by computing the logarithm of the partition function, the **Helmholtz free energy**, F:

$$F = -\frac{1}{\beta}\ln Z \tag{32.33}$$

We differentiate $\ln Z$ with respect to β to obtain the first useful property of a statistical mechanical theory, the average value of the energy, called the **internal energy**:

$$E = -\frac{\partial(\ln Z)}{\partial\beta} = \frac{\text{Tr}\,(He^{-\beta H})}{\text{Tr}\,(e^{-\beta H})} \tag{32.34}$$

We haven't included effects of the volume or the chemical potential. Now we make a Legendre transformation, turning from a function of β to a function of E. Define the **entropy**, S:

$$S = \beta(E - F) \;\Rightarrow\; \beta = \frac{\partial S}{\partial E} \tag{32.35}$$

There are obvious parallels between these thermodynamic equations and the equations we've just derived for the generating functionals, drawn in Table 32.1.

Instead of a statistical mechanical sum, the trace, we have a sum over function space. And now you can see why people call it Z. In one case we have an infinite number of variables characterizing our system, $J(x)$ at every spacetime point x, and in the other case we have a single inverse temperature β. Apart from this, there is a nearly perfect analogy between the operations in statistical mechanics and those in quantum field theory, which is pretty much Helmholtz's statistical mechanics transformed into quantum language. Helmholtz wrote down these equations except that instead of taking a trace, he integrated over momentum and position, $\int dp\,dq$; he was summing classical systems over classical phase space. Gibbs's version includes chemical potentials characterizing the volume of the system, external magnetic fields, etc. His version has a bunch of parameters in addition to β, so it looks even more like the quantum field theory version. We can make the analogy look very close. Once we rotate our fields into Euclidean space, the analogy is perfect, because then we really are summing over all possible configurations of a classical field of a theory in four space dimensions, just as in classical statistical mechanics. Then quantum field theory is identical to the classical statistical mechanics of a classical field in four space dimensions, with $\hbar$ playing the role of T and J playing the role of an external field, such as a magnetic field $\mathbf{B}$, coupled to the system.

---

[14] [Eds.] F. Reif, *Fundamentals of Statistical and Thermal Physics*, McGraw-Hill, 1965. In the video of Lecture 32, Coleman states that the discussion of the analogy between statistical mechanics and quantum field theory was unplanned, and a few of his equations are erroneous in their factors of β. Those have been corrected here.

Statistical Mechanics	*Quantum Field Theory*
$Z(\beta) = \mathrm{Tr}\,(e^{-\beta H})$	$Z[J] = N \int (d\phi) e^{iS(\phi,J)}$
$F = -\dfrac{1}{\beta}\ln Z$	$iW = \ln Z$
$E = -\dfrac{\partial(\ln Z)}{\partial\beta}$	$\overline{\phi} = \dfrac{\delta W}{\delta J}$
$S = \beta(E - F)$	$\Gamma[\overline{\phi}] = W[J] - \int d^4x\, J\overline{\phi}$
$\dfrac{\partial S}{\partial E} = \beta$	$\dfrac{\delta \Gamma}{\delta \overline{\phi}} = -J$

Table 32.1: Parallels between statistical mechanics and quantum field theory

This is in fact not very useful. All it tells us is that classical statistical mechanics of a classical field theory in four spatial dimensions is a very complicated problem which we are not going to solve. However, it *is* a very useful analogy for people who want to prove things about field theories in fewer than four spacetime dimensions where the classical system is not so complicated, and rigorous theorems have been proved. Then they use precisely this analogy to prove rigorous theorems about the corresponding quantum field theory in 1 + 1 or 1 + 2 dimensions.[15]

32.4 Quantum electrodynamics in a covariant gauge

What are the consequences of this functional formalism for quantum electrodynamics? We can derive important relations from gauge invariance. All the fields in quantum electrodynamics have well-defined gauge transformation properties. For an infinitesimal gauge transformation

$$\begin{aligned} A_\mu &\to A_\mu + \partial_\mu \delta\chi \\ \psi &\to \psi - ie\psi\delta\chi \\ \overline{\psi} &\to \overline{\psi} + ie\overline{\psi}\delta\chi \end{aligned} \tag{32.36}$$

where $\delta\chi$ is an arbitrary (infinitesimal) function of spacetime. We'll write this in generic form by assembling all the fields, including A_μ, into a single big column vector Φ

$$\Phi = \begin{pmatrix} \psi \\ \overline{\psi} \\ A_\mu \end{pmatrix} \tag{32.37}$$

and under the action of the infinitesimal gauge transformation (32.36),

$$\Phi \to \Phi' = \Phi + A(\Phi)\delta\chi \tag{32.38}$$

[15] [Eds.] For examples of the relation between quantum field theory and statistical mechanics, see, e.g., John B. Kogut, "An introduction to lattice gauge theory and spin systems", *Rev. Mod. Phys.* **51** (1979) 659–713; and B. M. McCoy, "The Connection Between Statistical Mechanics and Quantum Field Theory", in *Statistical Mechanics and Field Theory*, V. V. Bazhanov and C. J. Burden, eds., World Scientific, (1995); pp. 26–128.

where $A(\Phi)$ is some 3×3 diagonal matrix of operators acting on $\delta\chi$, determined by (32.36). For our gauge transformation, $A(\Phi)$ is no more than *first order* in Φ, and hence first or zeroth order in the fields A_μ, ψ and $\overline{\psi}$. (If we didn't have the explicit transformation in hand, we would need to *assume* that $A(\Phi)$ was first order in Φ.) For A_μ it is the differential operator ∂_μ, and for ψ and $\overline{\psi}$, it is multiplication by $-ie\psi$ or $ie\overline{\psi}$, respectively. Since we will be using this for renormalization theory, I should tell you whether the fields and charges are renormalized or unrenormalized. In fact, *it doesn't matter.* These can be the unrenormalized fields and the bare charges. Or, they could be the renormalized fields to some finite order in perturbation theory. In that case, we are studying the effect of the gauge transformation on the action to get constraints on the divergent parts of graphs, which we will have to cancel off with counterterms in the next order.

I will assume that the action $\mathcal{S}$ consists of a gauge invariant piece $\mathcal{S}_{GI}$ and a non-invariant, gauge-fixing piece $\mathcal{S}_{GF}$. In a covariant gauge, $\mathcal{S}_{GF}$ equals the integral of the four-divergence of A_μ squared, times $-1/(2\xi)$:

$$\mathcal{S}[\Phi] = \mathcal{S}_{GI}[\Phi] + \mathcal{S}_{GF}[\Phi] = \mathcal{S}_{GI}[\Phi] - \frac{1}{2\xi}\int d^4x(\partial_\mu A^\mu)^2 \tag{32.39}$$

When we make this infinitesimal gauge transformation (32.36) on the fields, the gauge invariant part of the action doesn't change, and the gauge-fixing term changes according to

$$\begin{aligned}\mathcal{S}[\Phi] \to \mathcal{S}'[\Phi'] = \mathcal{S}[\Phi] + \delta\mathcal{S}[\Phi] = \mathcal{S}[\Phi] - \frac{1}{\xi}\int d^4x\,(\partial_\mu A^\mu)\Box^2\delta\chi \\ \equiv \mathcal{S}[\Phi] + \int d^4x\, B(\Phi)\delta\chi\end{aligned} \tag{32.40}$$

where $B(\Phi)$ is also *first order or less* in Φ. Here

$$B(\Phi) = -\frac{1}{\xi}(\partial_\mu A^\mu)\Box^2$$

The only thing we will use in what follows is that A and B obey the first order conditions as stated, and that the gauge transformation doesn't change the measure in function space:

$$(d\Phi) \to (d\Phi') = (d\Phi) \tag{32.41}$$

The gauge transformation has determinant 1, since it just adds a constant (that is, independent of Φ) to A_μ, while ψ and $\overline{\psi}$ experience a rotation proportional to $\delta\chi$. The argument will be generalizable to any case obeying these three conditions: neither A nor B is more than first order in the fields, and the integration measure is invariant.

Why do we want these three things? We want them because it means that if we make this infinitesimal transformation then *the functional integral is not going to change.* Using (32.38), (32.40) and (32.41), and expanding out to first order in $\delta\chi$,

$$\begin{aligned}e^{iW[J]} &= N\int (d\Phi)\, e^{i\{\mathcal{S}[\Phi]+\int d^4x\, J\Phi\}} \to N\int (d\Phi')\, e^{i\{\mathcal{S}'[\Phi']+\int d^4x\, J\Phi'\}} \\ &= N\int (d\Phi)\, e^{i\{\mathcal{S}[\Phi]+\int d^4x\, B(\Phi)\delta\chi+\int d^4x\, J(\Phi+A(\Phi)\delta\chi)\}} \\ &= e^{iW[J]} + N\int (d\Phi)\, e^{i\{\mathcal{S}[\Phi]+\int d^4x\, J\Phi\}}\left[i\int d^4y\,\left(B(\Phi)+JA(\Phi)\right)\delta\chi(y)\right] \\ &= e^{iW[J]}\end{aligned} \tag{32.42}$$

since the value of an integral is not changed by a change of variables. Then

$$N\int(d\Phi)\,e^{i\{S[\Phi]+\int d^4x\,J\Phi\}}\left[\int d^4y\,\left(B(\Phi)+JA(\Phi)\right)\delta\chi(y)\right]=0 \tag{32.43}$$

for arbitrary $\delta\chi$. That simply says we know how our integrand changes under a gauge transformation. Since we could interpret a gauge transformation as just a redefinition of our integration variables, it doesn't change at all.

Now we come to the key point. A and B are *linear* in Φ. When we have an integral like (32.43), its value is the mean value of a linear function. Then we can *replace* Φ *by its mean value* $\overline{\Phi}$, and write[16]

$$\begin{aligned}N\int(d\Phi)e^{i\{S[\Phi]+\int d^4x\,J\Phi\}}&\left[\int d^4y\,\left(B(\Phi)+JA(\Phi)\right)\delta\chi(y)\right]\\&=\left[\int d^4y\left(B(\overline{\Phi})+JA(\overline{\Phi})\right)\delta\chi(y)\right]N\int(d\Phi)e^{i\{S[\Phi]+\int d^4x\,J\Phi\}}\\&=e^{iW[J]}\left[\int d^4y\left(B(\overline{\Phi})+JA(\overline{\Phi})\right)\delta\chi(y)\right]\end{aligned} \tag{32.44}$$

There will be a multiplicative factor of $e^{iW[J]}$, but so what? The whole thing equals zero, so we'll just take out this factor. Then (32.43) becomes

$$\int d^4y\left(B(\overline{\Phi})+JA(\overline{\Phi})\right)\delta\chi(y)=0 \tag{32.45}$$

This follows from the linearity of A and B. (If either A or B were other than linear in Φ, e.g., quadratic, we could not substitute $\overline{\Phi}$ for Φ; $\overline{(\Phi^2)}\neq(\overline{\Phi})^2$.) We know what J is. From (32.22),

$$J(x)=-\frac{\delta\Gamma[\overline{\Phi}]}{\delta\overline{\Phi}(x)}$$

Therefore we have from (32.45)

$$\int d^4y\,\frac{\delta\Gamma[\overline{\Phi}]}{\delta\overline{\Phi}}A(\overline{\Phi})\delta\chi(y)=\int d^4y\,B(\overline{\Phi})\delta\chi(y) \tag{32.46}$$

This equation tells us how $\Gamma[\overline{\Phi}]$ transforms under gauge transformations. If we write down gauge transformations for the mean fields $\overline{\Phi}$ that are *exactly the same* as the gauge transformations (32.38) that we originally had for quantum fields,

$$\delta\overline{\Phi}=A(\overline{\Phi})\delta\chi \tag{32.47}$$

then the change in $\Gamma[\overline{\Phi}]$, which is a functional of $\overline{\Phi}$ alone, is

$$\delta\Gamma[\overline{\Phi}]=\int d^4x\,\frac{\delta\Gamma[\overline{\Phi}]}{\delta\overline{\Phi}(x)}\delta\overline{\Phi}(x)=\int d^4x\,\frac{\delta\Gamma[\overline{\Phi}]}{\delta\overline{\Phi}(x)}A(\overline{\Phi})\delta\chi(x) \tag{32.48}$$

---

[16] [Eds.] See footnote 13, p. 694.

the chain rule for differentiation.[17] That is, under the gauge transformation (32.47), the effective action $\Gamma[\overline{\Phi}]$ transforms as

$$\Gamma[\overline{\Phi}] \to \Gamma[\overline{\Phi}] + \delta\Gamma[\overline{\Phi}] = \Gamma[\overline{\Phi}] + \int d^4y\, B(\overline{\Phi})\delta\chi(y) \tag{32.49}$$

using (32.46).

Comparing (32.40) and (32.49), we see that

$$\boxed{\delta\Gamma[\overline{\Phi}] = \delta\mathcal{S}[\overline{\Phi}] = \delta\mathcal{S}_{GF}[\overline{\Phi}]} \tag{32.50}$$

(the last equality following because S_{GI} does not change under a gauge transformation). Under a gauge transformation, at most linear in the fields, *the change in the effective action equals the change in the classical action.*[18] From this general relation we will derive many identities.[19] These *Ward identities* will help establish the relations (31.64), and hence that quantum electrodynamics is renormalizable. We will call (32.50) the **generic Ward identity**.

Let's review the argument briefly. Given

$\mathcal{S}[\Phi] = \mathcal{S}_{GI}[\Phi] + \mathcal{S}_{GF}[\Phi]$	(the classical action, with S_{GF} at most quadratic in the fields)
$\Gamma[\overline{\Phi}]$	(the effective action for 1PI Green's functions)
$\overline{\Phi} \to \overline{\Phi} + \delta\overline{\Phi}$	(the gauge transformation, with $\delta\overline{\Phi}$ at most linear in the fields)

then we obtain the amazing result (32.50). We can rewrite that result as

$$\delta(\Gamma[\overline{\Phi}] - \mathcal{S}_{GF}[\overline{\Phi}]) = 0 \tag{32.51}$$

By definition, $\Gamma[\overline{\Phi}] - \mathcal{S}_{GF}[\overline{\Phi}]$ must be the gauge invariant part of $\Gamma[\overline{\Phi}]$; it doesn't change under the gauge transformation. That is,

$$\Gamma[\overline{\Phi}] = \Gamma_{GI}[\overline{\Phi}] + \mathcal{S}_{GF}[\overline{\Phi}] \tag{32.52}$$

So the generating functional is gauge invariant except for a gauge-fixing term of the same form as that in the original Lagrangian. This is *not* true for the Green's functions; for example, both the photon propagator and the electron propagator are ξ-dependent.

Let's apply (32.46) to spinor electrodynamics in the Lorenz gauge.[20] The action consists of a gauge invariant part $\mathcal{S}_{GI}$, and a gauge-fixing (non-gauge invariant) part $\mathcal{S}_{GF}$:

$$\mathcal{S} = \mathcal{S}_{GI} + \mathcal{S}_{GF} = \int d^4x \left\{ -\tfrac{1}{4}(\partial_\mu A_\nu - \partial_\nu A_\mu)^2 + \overline{\psi}(i\partial\!\!\!/ - m - eA\!\!\!/)\psi - \frac{1}{2\xi}(\partial_\mu A^\mu)^2 \right\} \tag{32.53}$$

---

[17] [Eds.] The videotape of Lecture 32 ends prematurely at this point, at 1:06:19. The lecture may have continued for another 24 minutes. Judging from the start of the next lecture, however, it appears that little more was added. The remainder of this chapter is based on notes from Coleman, Woit and the anonymous graduate student.

[18] [Eds.] John Preskill, Notes for Caltech's Physics 205 (1986–7), Ch. 4, pp. 4.55–4.57. On line at `http://www.theory.caltech.edu/~preskill/notes.html`.

[19] [Eds.] See footnote 22, p. 675.

[20] [Eds.] We omit the overbars on the fields here as they would cause confusion: ψ would be $\overline{\psi}$, $\overline{\psi}$ would be $\overline{\overline{\psi}}$, etc.

Under the infinitesimal gauge transformation

$$\delta \begin{pmatrix} \psi(x) \\ \overline{\psi}(x) \\ A_\mu(x) \end{pmatrix} = \begin{pmatrix} -ie\delta\chi(x)\psi(x) \\ ie\delta\chi(x)\overline{\psi}(x) \\ \partial_\mu \delta\chi(x) \end{pmatrix} \tag{32.54}$$

we obtain from (32.46), after an integration by parts,

$$\int d^4x\, \delta\chi(x) \left\{ -ie \frac{\delta\Gamma}{\delta\psi(x)}\psi(x) + ie\overline{\psi}(x)\frac{\delta\Gamma}{\delta\overline{\psi}(x)} - \partial^\mu \frac{\delta\Gamma}{\delta A^\mu(x)} \right\} = -\int d^4x\, \delta\chi(x) \left\{ \frac{1}{\xi}\Box^2 \partial_\mu A^\mu \right\} \tag{32.55}$$

or, since $\delta\chi$ is an arbitrary function,

$$ie\overline{\psi}(x)\frac{\delta\Gamma}{\delta\overline{\psi}(x)} - ie\frac{\delta\Gamma}{\delta\psi(x)}\psi(x) - \partial^\mu \frac{\delta\Gamma}{\delta A^\mu(x)} = -\frac{1}{\xi}\Box^2 \partial_\mu A^\mu(x) \tag{32.56}$$

This equation applies to the entire generating functional $\Gamma[\Phi]$. It encompasses a large number of equations for the 1PI Green's functions, which can be derived from the series expansion (32.12).

Next time we will apply the generic Ward identity to the renormalization of quantum electrodynamics.

33

The renormalization of QED

Last time, in a frenzy of enthusiasm I derived the generic Ward identity, (32.50). To review, suppose $\mathcal{S}$, the classical action for a gauge field theory describing a set of fields Φ, can be written in the form

$$\mathcal{S}[\Phi] = \mathcal{S}_{GI}[\Phi] + \mathcal{S}_{GF}[\Phi] \tag{33.1}$$

Here $\mathcal{S}_{GI}[\Phi]$ is invariant under the gauge transformation

$$\Phi \to \Phi + \delta\Phi \tag{33.2}$$

and $\mathcal{S}_{GF}[\Phi]$ is a (non-gauge invariant) gauge-fixing term, *at most quadratic* in Φ. Then the effective action $\Gamma[\overline{\Phi}]$, the generating functional for 1PI diagrams, has the same structure as the classical action:

$$\Gamma[\overline{\Phi}] = \Gamma_{GI}[\overline{\Phi}] + \mathcal{S}_{GF}[\overline{\Phi}] \tag{33.3}$$

with the *same* gauge-fixing term, but written in terms of the mean, or "classical", field $\overline{\Phi}$ (32.31). The fruit of the last lecture is the statement that, if $\overline{\Phi}$ is subjected to an infinitesimal gauge transformation of the form which leaves $S_{GI}[\Phi]$ invariant,

$$\overline{\Phi} \to \overline{\Phi} + \delta\overline{\Phi} \tag{33.4}$$

then we obtain the generic Ward identity,

$$\delta\Gamma[\overline{\Phi}] = \delta\mathcal{S}_{GF}[\overline{\Phi}] \tag{32.50}$$

provided that $\delta\mathcal{S}_{GF}[\overline{\Phi}]$ is at most *linear* in the fields. This allows us to replace its argument Φ with $\overline{\Phi}$. If you've seen the conventional Ward identity, you may not recognize it in this formalism. I'll explain the connection, and derive some other consequences of this result.

Equation (32.50) will enable us to complete the renormalization program for quantum electrodynamics, with or without a massive photon, by showing that all the required counterterms are gauge invariant. I will now prove this in detail for QED (or a general theory of the same form) following the BPHZ program (§25.2). Such a theory includes only gauge invariant and renormalizable interactions (with dimension ≤ 4, as described in §25.4, p. 538), apart from the gauge-fixing term which is restricted to be no more than quadratic in the dynamical variables.

33.1 Counterterms and gauge invariance

The proof proceeds inductively, or more accurately, iteratively, in a sequence of five statements.

1. We assume that we need only gauge invariant counterterms to render the theory finite to $\mathcal{O}(e^n)$, where e is some coupling constant. (If there are multiple coupling constants, $g_1, g_2, \ldots$, we assume that we have only gauge invariant counterterms of $\mathcal{O}(g_1^n)$, $\mathcal{O}(g_2^n)$, ...) We wish to prove that *only* gauge invariant counterterms are needed to make the theory finite to $\mathcal{O}(e^{n+1})$. The assertion is obviously true for $\mathcal{O}(e^0)$: *no* counterterms are needed for $n = 0$, and so these zero terms are trivially gauge invariant.

2. For a theory of this type with only renormalizable interactions, the BPHZ algorithm says that, if we've made everything finite to $\mathcal{O}(e^n)$, all divergences to $\mathcal{O}(e^{n+1})$ can be canceled by adding an additional term to the interaction

$$\mathcal{S}[\Phi'] \to \mathcal{S}[\Phi'] + e^{n+1}\mathcal{S}_{CT}^{(n+1)}[\Phi'] \tag{33.5}$$

$S_{CT}^{(n+1)}[\Phi']$ is a sum of counterterms computed to $\mathcal{O}(e^{n+1})$ with divergent coefficients depending on the cutoff (assuming that there *is* a suitable gauge invariant cutoff). The functionals are expressed in terms of the renormalized fields, Φ'. (It doesn't matter whether we're using renormalized fields or not in this argument; how the fields scale has nothing to do with this proof.) According to BPHZ, if we have only interactions of renormalizable type, then $S_{\text{CT}}^{(n+1)}[\Phi']$ is a polynomial in Φ' and $\partial_\mu\Phi'$ of dimension ≤ 4. That's simply the general BPHZ result for working in four dimensions. We add the appropriate counterterms to get rid of all divergences to the next order, and then we iterate.[1]

3. To $\mathcal{O}(e^{n+1})$, adding the counterterms gives a new term to the effective action,

$$\Gamma[\overline{\Phi}'] \to \Gamma[\overline{\Phi}'] + e^{n+1}\mathcal{S}_{CT}^{(n+1)}[\overline{\Phi}'] \tag{33.6}$$

a term of $\mathcal{O}(e^{n+1})$ coming from the terms we've added to the Lagrangian: they generate 1PI diagrams just by themselves. And in $e^{n+1}S_{CT}^{(n+1)}$ we simply replace Φ' by $\overline{\Phi}'$, just as if it were a term in the free Lagrangian. However, the counterterms will also appear as internal parts of other complicated Feynman diagrams once we've added these interactions. But since there's an explicit e^{n+1} in front, the new term produces an effect of at least $\mathcal{O}(e^{n+2})$ in those other graphs, because any complicated Feynman diagram has at least one vertex in addition to these new vertices we've added. Thus (33.6) should really be written

$$\Gamma[\overline{\Phi}'] \to \Gamma[\overline{\Phi}'] + e^{n+1}\mathcal{S}_{CT}^{(n+1)}[\overline{\Phi}'] + \mathcal{O}(e^{n+2}) \tag{33.7}$$

4. Staring at this formula (33.7), we see that, before we add the counterterms, to $\mathcal{O}(e^{n+1})$,

$$\Gamma[\overline{\Phi}'] = \Gamma_{\text{finite}} - e^{n+1}\mathcal{S}_{CT}^{(n+1)}[\overline{\Phi}'] \tag{33.8}$$

---

[1] There is a little technical sticking point here. We add the counterterms by doing a power series expansion about the point 0. If we are considering electrodynamics with a massless photon, there is the possibility of singularities at the point 0: that value sits on top of the photon mass shell. We won't worry about that here, but assume that we give the photon a small mass and afterwards consider the limit as the mass goes to zero.

where Γ_{finite} is independent of the cutoff, and so unaffected in the limit as the cutoff $\to \infty$; BPHZ says that the sum of the terms in $\Gamma[\overline{\Phi}']$ is supposed to be finite to $\mathcal{O}(e^n)$. Since we can get rid of all the divergences by adding $e^{n+1}\mathcal{S}_{CT}^{(n+1)}[\overline{\Phi}']$, which has cutoff-dependent coefficients, all of the divergences must be of the form $e^{n+1}\mathcal{S}_{CT}^{(n+1)}[\overline{\Phi}']$.

5. Now we use the Ward Identity. We have a gauge transformation

$$\overline{\Phi}' \to \overline{\Phi}' + \delta\overline{\Phi}' \tag{33.9}$$

We assume, at the nth step of the iteration, that we need only gauge invariant counterterms to $\mathcal{O}(e^n)$, so the Ward identity is valid. The Ward identity tells us that the gauge transformation leaves everything in $\Gamma[\overline{\Phi}']$ invariant except for the quadratic term $\mathcal{S}_{GF}[\overline{\Phi}']$. This term is certainly not divergent; it's something like

$$\frac{1}{2\xi}(\partial_\mu A'^\mu)^2 + \frac{1}{2}\mu^2(A'^\mu)^2, \tag{33.10}$$

It doesn't have any power series expansion in e; it's a fixed, known quantity. So there is no need to introduce counterterms for the gauge-fixing part of $\Gamma[\overline{\Phi}']$. If, apart from $\mathcal{S}_{GF}$, $\Gamma[\overline{\Phi}']$ is gauge invariant, then the cutoff-dependent part, the counterterms, must be gauge invariant:[2]

$$\mathcal{S}_{CT}^{(n+1)}[\overline{\Phi}'] \to \mathcal{S}_{CT}^{(n+1)}[\overline{\Phi}'] \tag{33.11}$$

6. Therefore, *only* gauge invariant counterterms are needed to $\mathcal{O}(e^{n+1})$, and by induction, to all orders.

We see that, to each order in perturbation theory, the generating functional of 1PI graphs is gauge invariant, aside from the non-gauge invariant terms that are exactly the same as in the classical action. This means that to each order all of the cutoff-dependent terms are gauge invariant; the need to introduce non-gauge invariant counterterms never arises. We conclude that all divergences can be removed with gauge invariant counterterms. In particular, the gauge-fixing term is unrenormalized. As we will see, gauge invariance via the Ward identity imposes relations among the counterterms.

33.2 Counterterms in QED with a massive photon

Let's do a couple of examples, spinor electrodynamics and scalar electrodynamics.

EXAMPLE. *Spinor electrodynamics*

From the BPHZ prescription (see §25.2), we have

$$\mathscr{L} = -\tfrac{1}{4}(F'_{\mu\nu})^2[1+A] + \overline{\psi}'(i\partial\!\!\!/ - eA\!\!\!/')\psi'[1+B] - \overline{\psi}'\psi'[m+C] + \tfrac{1}{2}\mu^2 A'_\lambda A'^\lambda - \frac{1}{2\xi}(\partial_\lambda A'^\lambda)^2 \tag{33.12}$$

The first three terms are each gauge invariant; for the last two terms, there is no correction, as we'll see. This Lagrangian includes all possible gauge invariant counterterms of dimension ≤ 4

[2] [Eds.] The audio of Lecture 33's videotape is unintelligible from 17:45 to 23:55. The argument has been filled in from John Preskill's "Notes for Caltech's Physics 205 (1986–7)", Ch. 5, pp. 5.60–5.61, at `http://www.theory.caltech.edu/~preskill/notes.html`, and from the anonymous graduate student's notes. The relevant sections of Coleman's own notes are missing, and Woit's are a little elliptical.

and is therefore strictly renormalizable. It gives finite answers for appropriate cutoff-dependent choices of the counterterms A, B and C to every order in a power series expansion in e.

Once we have this Lagrangian, either because it has come down to us from heaven on a golden tablet, or by the results of tedious labor to some finite order in perturbation theory, we can write it in terms of unrenormalized fields and define the values of the bare charge (and any other coupling constants, such as λ in ϕ^4 theory), the bare masses and the gauge parameter ξ. The Lagrangian in terms of unrenormalized fields is scaled so that the kinetic energies, the derivative terms, are of standard form:

$$\mathscr{L} = -\tfrac{1}{4}(F_{\mu\nu})^2 + \overline{\psi}[i\partial\!\!\!/ - e_0 A\!\!\!/ - m_0]\psi + \tfrac{1}{2}\mu_0^2 A_\lambda A^\lambda - \frac{1}{2\xi_0}(\partial_\lambda A^\lambda)^2 \tag{33.13}$$

These are the bare masses and charge. As we'll see shortly, this is also a bare ξ. We also define the quantities that give us the scale between ψ and ψ', and between A_μ and A'_μ:

$$\psi = Z_2^{1/2}\psi', \quad A_\mu = Z_3^{1/2}A'_\mu \tag{33.14}$$

There is a Z_1 that occurs in traditional treatments of electrodynamics; we won't talk about it. We can show Z_1 is equal to Z_2 because of gauge invariance.[3] These Z_i are called the **renormalization constants**.

I should say something about the gauge-dependence—or, since we're working in a theory with a massive photon, the ξ dependence—of these things. As you'll recall (§31.5), changing gauges (or equivalently, switching values of ξ, as in §31.3) by introducing an auxiliary field does not require a redefinition of A_μ, though it does require a redefinition of ψ. So we expect that Z_3 likewise would not get redefined, and hence would be ξ-independent (gauge-independent in the massless case). Z_2 on the other hand might well be ξ-dependent; we don't know.[4]

We also expect the bare masses of the particles to be ξ-independent. After all, they're masses; how can they depend on the gauge? An interpolating field may depend on the gauge; it may be a different operator in one gauge from another, but a mass is a mass, a physically observable quantity. By the same argument, the charge should be ξ-independent. To summarize:

$$\begin{aligned} &\xi\text{-dependent}: Z_2 \\ &\xi\text{-independent}: Z_3, m_0, \mu_0, e_0 \end{aligned}$$

Comment. There's a curious consequence to the possible dependence of Z_2 on ξ. We won't go through the derivation of the spectral representation for vector fields; it follows essentially the same line of reasoning as for scalar fields.[5] We can use the positivity of the weight function

[3] [Eds.] Ward's original goal was to prove the equality $Z_1 = Z_2$ conjectured by Dyson, and *this* is the identity Ward refers to in his article's title: J. C. Ward, "An Identity in Quantum Electrodynamics", *Phys. Rev.* **78** (1950) 182. Today the term "Ward identity" usually refers to a preliminary result Ward obtained, from which $Z_1 = Z_2$ follows; see note 29, p. 721. See also Peskin & Schroeder *QFT*, Section 7.4, pp. 238–244; the original identity and its consequence are on p. 243.

[4] [Eds.] In fact, Z_2 *is* gauge-dependent. See Kenneth Johnson and Bruno Zumino, "Gauge Dependence of the Wave-Function Renormalization Constant in Quantum Electrodynamics", *Phys. Rev. Lett.* **3** (1959) 351–352; Herbert M. Fried, *Modern Functional Quantum Field Theory: Summing Feynman Graphs*, World Scientific, 2014, p. 80; Greiner & Reinhardt *QED*, p. 298.

[5] [Eds.] Bjorken & Drell *Fields*, Section 16.11. But see §34.1.

to show that $Z_3 \leq 1$, our usual result. On the other hand, Z_2 is defined in terms of the ψ' field, which is the original ψ field that we started out with in a nice theory with positive definite metric in Hilbert space, and the usual good properties, times an exponential of this preposterous field with minus signs in the propagator, negative metric intermediate particles, etc.[6] That may spoil the positivity of the spectral formula for the spinor field. So don't be surprised if during a computation in one of these gauges with ξ, $Z_2 > 1$, while for some other value of ξ, $Z_2 < 1$. That's just because once we've made the change of field, we've mixed it up with this auxiliary field so there may be negative weights appearing in the spectral weight function. (I thought I should mention this, but it's just a side comment.)

Now let's work out the relations between the renormalization constants and the counterterms, **renormalization constants** by straight algebraic substitution of (33.14) into (33.13). Comparing the transformed (33.13) with the Lagrangian (33.12), we find

$1 + A = Z_3$	Uninteresting unless we know A.
$1 + B = Z_2$	Uninteresting unless we know B.
$e = Z_3^{1/2} e_0$	Very interesting; and we'll return to it.
$m_0 = Z_2^{-1}(m + C)$	Not interesting.
$\mu = Z_3^{1/2} \mu_0$	Very interesting.
$\xi_0 = Z_3 \xi$	Essentially uninteresting.

We have identified two of the equations as being interesting. First,

$$\mu = Z_3^{1/2} \mu_0 \tag{33.15}$$

This implies that as $\mu^2 \to 0$, $\mu_0^2 \to 0$ and *vice versa*, unless Z_3 develops a pole, a rather unlikely possibility. We may get some logarithms because of those intermediate photons, but a pole is rather strong. If we start with a zero bare mass for the photon we get a zero renormalized mass, or if we set the renormalized mass to zero we get a zero bare mass. So *the zero mass of the photon is preserved by renormalization.*

The second interesting equation is

$$e = Z_3^{1/2} e_0 \tag{33.16}$$

This is important because it represents the **universality of charge renormalization**. It was only laziness that kept us from writing down a theory with many more fermionic fields. However many we might have started with, we would have discovered that those with the same e would also have the same e_0. Some of the subatomic particles have, besides electromagnetism,

6 [Eds.] Crudely speaking, the function χ in the exponent in (31.11) is the line integral of a combination of the A_μ and A'_μ fields. When gauge transformations are evaluated carefully at the operator level, the ψ field picks up an exponential factor that depends on the non-transverse photon modes. See K. Haller, "Operator Gauge Transformations in Quantum Electrodynamics", *Nuc. Phys.* **B57** (1973) 589–603 and "Gauge Problems in Spinor Quantum Electrodynamics", *Acta Phys. Austr.* **42** (1975) 163–214.

strong interactions, and some of them don't. Consider a theory of electrons and protons: only the protons interact with mesons via the strong force. But the only renormalization constant that appears in the charge is Z_3, which has nothing to do with the Fermi fields and their renormalization; it concerns the *photon.* Though other fields may all have different interactions and therefore different Z_2's, they're all going to have the *same* Z_3, because there's only *one* photon. Thus *if particles have the same bare charge then they have the same renormalized charge.* This is reassuring. I raised this as a question earlier (see p. 675): Why is it that the proton and the electron seem to have exactly the same physical charge? If their renormalizations were different, if that of the proton were dependent on the strong interactions, then perhaps God had to have been incredibly kind to adjust the bare charges in such a way that the physical charges came out to be equal. He didn't have to be kind, but merely uncomplicated. If God decreed that the bare charges are equal, then automatically the renormalized charges are equal. Indeed, if the decree had been that the proton's bare charge were three times the electron's, then the proton's renormalized charge would come out three times the electron's; the ratio of the renormalized charges is the same as the ratio of the bare charges for all n of the fields:

$$e^{(a)} = Z_3^{1/2} e_0^{(a)} \qquad a = 1, \cdots, n$$
$$\frac{e^{(1)}}{e^{(2)}} = \frac{e_0^{(1)}}{e_0^{(2)}} = \cdots$$

We can imagine God deciding that there will be two charges, the proton's, $q_p = e_0$ and the up quark's, $q_u = \frac{2}{3}e_0$. He fixes the bare parameters at the Planck length. He lets the renormalization run to our scale, and *voilà*, the ratio of up quark charge to proton charge is still 2:3! More dramatically, the electron and the *antiproton* have the same charge, even though the electron does not participate in the strong interactions. This is a deeply satisfying result.

We've obtained this by an elegant but rather abstract argument, but we can understand it physically. Suppose we have a particle, say a proton, in a box and we wish to compute the expectation value of the electric charge. We know how to compute the expectation values of operators from non-relativistic quantum mechanics: we expand the actual state of the system in terms of the eigenstates of the non-interacting system and evaluate the operators in that expansion. At first glance that looks very complicated because the proton could be a bare proton with charge 1 or a bare neutron and a bare π^+ meson, also charge 1, etc. But, we really don't need to know this expansion because anything that the proton can virtually become is also a system of charge 1, a consequence of charge conservation. Electric charge differs from pseudoscalar coupling constants, for which the analogous result is not true. As a consequence of charge conservation anything the proton goes into must have charge 1. It can never be found as a bare neutron plus a bare π^-; therefore the expansion doesn't matter. The total charge in the box is 1 no matter what the proton's wave function is, because it's just a superposition of things with charge 1, or more precisely, with charge e_0. Though this argument is very comforting, on this level it makes us a bit nervous because it might indicate that there is no charge renormalization at all. And here we definitely see one, $e = Z_3^{1/2} e_0$. Why is that so? It is because we measure the physical charge (we'll give such a *gedanken* measurement shortly) by going far away from the box and looking at the long distance behavior of the electric field. We don't really construct a J_0 measuring operator and stick it inside the box; we look at the electric field at large distances.

Now in a field theory the vacuum is a dielectric.[7] A dielectric is a material from which, when we impose a constant electric field, there arises a correction to the expression for the ground state energy of the system, from the good old Maxwell expression $\frac{1}{2}|\mathbf{E}|^2$ to $\frac{1}{2}|\mathbf{E}|^2$ plus corrections. That's the defining characteristic of a dielectric. In a *quantum* field theory, if we impose a constant electric field, there are lots of complicated bubble graphs and so on contributing to the energy, and the vacuum will have dielectric properties. If we put a charge $+Q$ in a dielectric, as Faraday knew, it is shielded; the amount of the shielding depends on its dielectric constant.[8] Imagine a tiny observer *within* the dielectric, looking at the electric field some distance away from the charge. That observer does not see the charge Q that we put into the dielectric, but instead $Q' < Q$, because the charge polarizes the dielectric which in turn shields the charge, as in Figure 33.1. Of course if we are *outside* the dielectric, the

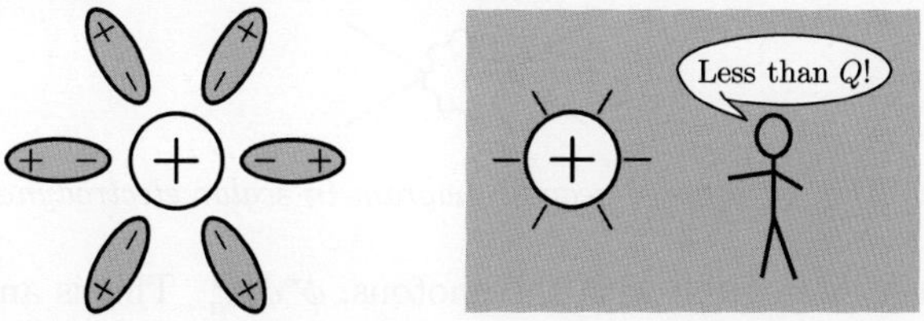

Figure 33.1: Dielectric screening of a charge, and as seen by an observer in the dielectric

missing charge appears on the surface of the dielectric. But we are *not* outside the dielectric; we are not outside the vacuum: we swim in the vacuum as fish swim in the sea. Therefore we are in the Faraday situation, inside the dielectric, and we see the charge as shielded. This does not depend on the constitution of the charge we put into the dielectric. It is a universal result that only depends on the dielectric constant of the medium. We can now interpret Z_3 as *the dielectric constant of the vacuum.* Notice that it is charge shielding: $Z_3 < 1$ and therefore $e < e_0$. The dipoles don't align themselves the other way in the dielectric: Z_3 is *not* greater than 1.

This description of course is just a metaphor. No one would accept that as a convincing argument. But it is easier to hold in our heads than the long argument we have been running through, which should be convincing. All of the answers are perfectly standard; you can find them in Bjorken and Drell or Schweber or Lurié or any other reference.[9] The methods, however, are my own.

EXAMPLE. *Scalar electrodynamics*

We've been concentrating on the electrodynamics of charged spinors. We should say a few words about charged scalars. The whole story is pretty much the same except for a technical detail. There is one additional gauge invariant counterterm of dimension 4 which might be needed in a charged scalar theory, but is not required in a spinor theory: the quartic interaction $(\phi^*\phi)^2$, just as we found in our pseudoscalar Yukawa theory, (25.71). To generate

[7] [Eds.] Peskin & Schroeder, *QFT*, Section 7.5, p. 255.

[8] [Eds.] N. W. Ashcroft and N. D. Mermin, *Solid State Physics*, Harcourt Publishers, 1976.

[9] [Eds.] Bjorken & Drell *Fields*, p. 303, equation (19.33); Schweber, *RQFT*, p. 634, equation (114) and p. 635, equation (126); Lurié, *P&F*, p. 300, equation 6(365).

such a counterterm we need to add a quartic interaction of the scalar particle with itself

$$\mathscr{L}'' = -\tfrac{1}{4!}\lambda_0(\phi^*\phi)^2 \tag{33.17}$$

to the Lagrangian (30.7) to render scalar electrodynamics renormalizable. What results is

$$\mathscr{L} = -\tfrac{1}{4}F_{\mu\nu}F^{\mu\nu} + (\partial_\mu\phi^*)(\partial^\mu\phi) - \mu_0^2\phi^*\phi \\ - ie_0\big[\phi^*\partial_\mu\phi - (\partial_\mu\phi^*)\phi\big]A^\mu + e_0^2\phi^*\phi A_\mu A^\mu - \tfrac{1}{4!}\lambda_0(\phi^*\phi)^2 \tag{33.18}$$

The electrodynamics of charged spinor particles has just one coupling constant: e_0. To be renormalizable, the electrodynamics of charged scalar particles has to include *two* coupling constants: e_0 and λ_0; we must include a scalar self-interaction. We can easily see where it comes from if we consider the scattering diagram in Figure 33.2. In the Lagrangian (33.18)

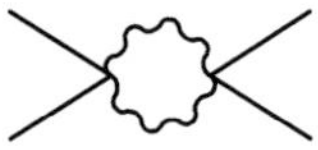

Figure 33.2: A squared seagull diagram in scalar electrodynamics

the scalars have a direct interaction with the photons, $\phi^\star\phi A_\mu^2$. This is an exceptionally simple graph because it has no derivative couplings. The integral is obviously proportional to

$$\int \frac{d^4k}{k^4}$$

at high k: $1/k^2$ comes from each photon propagator, and there are no derivatives. This graph produces a logarithmic divergence and we need the renormalization of λ_0 in order to cancel it. That's not a golden argument because there are lots of other graphs in the same order, and it requires a little checking to show that they don't cancel among themselves; they don't. For example, we have the graph shown in Figure 33.3 with derivative coupling. Each vertex

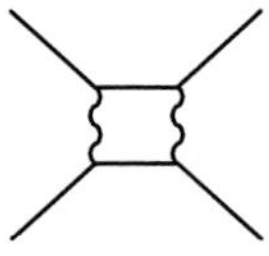

Figure 33.3: A derivative coupling diagram in scalar electrodynamics

contributes a momentum factor, and there are four propagators. At high k, this diagram is proportional to

$$\int \frac{d^4k}{k^8}k^4$$

which is also logarithmically divergent. But these two graphs do not give equal and opposite contributions, and do not cancel each other.

The need for an additional quartic interaction in scalar electrodynamics is exactly parallel to the phenomenon we ran up against in our discussion of the renormalization of the pseudoscalar Yukawa theory, where $g\bar{\psi}\gamma_5\psi\phi$ was not strictly renormalizable; there we likewise had to add a quartic interaction.[10]

---

[10] [Eds.] See the discussion following (25.69), on p. 540.

33.3 Gauge-invariant cutoffs

All of these arguments about gauge invariant counterterms make sense only if we have a way of introducing a high-energy cutoff on the Feynman integrals to regularize the counterterms in such a way that gauge invariance is preserved. We'll revisit two methods we looked at in §25.1, one of them only briefly: dimensional regularization, à la 't Hooft–Veltman, and Sirlin *et al.*,[11] and an earlier method, the regulator fields of Pauli and Villars.[12]

Dimensional regularization

As in §25.1, the basic idea of dimensional regularization is to extend the dimensionality of space from four dimensions to some unspecified number n, not necessarily an integer. The ultraviolet divergences which we encounter by integrating over all momenta are replaced by singularities related to the number of dimensions through $\Gamma(z)$, the gamma function.[13] Earlier we established (25.18) in Euclidean space, which we now rewrite as

$$\int \frac{d^n k}{(k^2+a^2)^\alpha} = \frac{\pi^{n/2}}{a^{2\alpha-n}} \frac{\Gamma(\alpha-\frac{1}{2}n)}{\Gamma(\alpha)} \tag{33.19}$$

With $n = 4$ and $\alpha = 2$, for instance, the left-hand integral is logarithmically divergent. But with n less than 4 and $\alpha \geq 2$, the integral is convergent. This suggests that we take

$$n = 4 - \epsilon \tag{33.20}$$

where ϵ is a small positive quantity. (Some authors let $n = 4-2\epsilon$.) The logarithmic divergence becomes a pole arising from the gamma function. Previously we found

$$\Gamma(\epsilon) = \frac{1}{\epsilon} - \gamma + \mathcal{O}(\epsilon) \tag{25.26}$$

($\gamma = 0.5772\ldots$ is the Euler–Mascheroni constant). From the definition (33.20), we have $\frac{1}{2}\epsilon = 2 - \frac{1}{2}n$, so from (25.26) we get

$$\Gamma(2-\tfrac{1}{2}n) = \Gamma(\tfrac{1}{2}\epsilon) = \frac{2}{\epsilon} - \gamma + \mathcal{O}(\epsilon) \tag{33.21}$$

Moreover, we have the functional equation

$$\Gamma(z+1) = z\Gamma(z) \tag{33.22}$$

and so

$$\Gamma(1-\tfrac{1}{2}n) = \frac{\Gamma(2-\frac{1}{2}n)}{1-\frac{1}{2}n} = \frac{\Gamma(2-\frac{1}{2}n)}{\frac{1}{2}\epsilon - 1} = \gamma - 1 - \frac{2}{\epsilon} + \mathcal{O}(\epsilon) \tag{33.23}$$

---

[11] [Eds.] See footnote 4 on p. 528. Also see W. J. Marciano and A. Sirlin, "Dimensional Regularization of Infrared Divergences", *Nucl. Phys.* **B88** (1975) 86–98; Peskin & Schroeder *QFT*, Section 7.5, pp. 249–251; and Ryder *QFT*, Section 9.2, pp. 313–318. For a detailed review, see G. Leibbrandt, "Introduction to the Technique of Dimensional Regularization", *Rev. Mod. Phys.* **47** (1975) 849–876.

[12] [Eds.] W. Pauli and F. Villars, "On Invariant Regularization in Relativistic Quantum Theory", *Rev. Mod. Phys.* **21** (1949) 434–444.

[13] [Eds.] Arfken & Weber *MMP*, Chapter 8, "The Gamma Function", pp. 495–533. *Warning!* Don't confuse the gamma function $\Gamma(z)$ with $\Gamma[\phi]$, the generating functional of 1PI graphs. The context should make it clear which is which.

Similarly,

$$\Gamma(3-\tfrac{1}{2}n) = (2-\tfrac{1}{2}n)\Gamma(2-\tfrac{1}{2}n) = \tfrac{1}{2}\epsilon\left[\frac{2}{\epsilon} - \gamma + \mathcal{O}(\epsilon)\right] = 1 + \mathcal{O}(\epsilon) \tag{33.24}$$

Following 't Hooft and Veltman, we adopt (33.19) for arbitrary complex n and analytically continue in n. $\Gamma(\alpha - \frac{n}{2})$ has singularities at $n = 2\alpha, 2\alpha+2, 2\alpha+4$, etc. So as long as we stay away from even integers $n > 2\alpha$, this expression is well-defined. Instead of letting an auxiliary mass go to infinity (as in the Pauli–Villars method, p. 526) for $n = 4$, we manipulate the pole in Γ, doing our renormalization in arbitrary n. Only after we have the expressions for the graphs in a convergent form (with the poles in $(n-4)$ canceling) do we let $n \to 4$. A function defined on the integers can be analytically continued (almost) uniquely in such a way that the analytic continuation is also gauge invariant.

It's fairly obvious that dimensional regularization preserves gauge invariance. Dimensional regularization starts out with formal Feynman integrals for integer dimensions and unambiguously continues them to complex dimensions, whereupon the divergences become poles. Any property which is true for integer dimensions will evidently be true for the unambiguous analytic continuation. In particular, gauge invariance does not depend on the dimension of spacetime; we could write it down in 72 dimensions. The fields may have more indices but we will still have gauge invariance. There is nothing special about four dimensions. That's more of a swindle than an argument, but nevertheless it turns out to be right. Those who want a detailed proof looking into the guts of Feynman diagrams can go to 't Hooft's lectures where he talks about matters of this kind.[14] When it comes to dimensional regularization, either we get arguments that we don't believe or we get arguments that we don't understand. That's the nature of the subject, I fear. All arguments in this area fall into two classes: those that are incredible and those that are incomprehensible. These classes are not mutually exclusive.

In general we must keep everything in n-dependent form and take the limit $n \to 4$ only after all the necessary computations have been performed. For instance, in n dimensions the metric tensor is

$$g_{\mu\nu} = \text{diagonal}\,(1,-1,-1,-1,-1,-1,\cdots); \quad g^{\mu}{}_{\mu} = n \tag{33.25}$$

From (33.19) we can easily work out other integrals. We have

$$\begin{aligned}\int d^n k\,\frac{k^2}{(k^2+a^2)^\alpha} &= \int d^n k\,\frac{1}{(k^2+a^2)^{\alpha-1}} - \int d^n k\,\frac{a^2}{(k^2+a^2)^\alpha}\\ &= (\tfrac{1}{2}n)\frac{\pi^{n/2}}{a^{2(\alpha-1)-n}}\frac{\Gamma(\alpha-1-\frac{1}{2}n)}{\Gamma(\alpha)}\end{aligned} \tag{33.26}$$

By symmetry

$$\int d^n k\,\frac{k_\mu k_\nu}{(k^2+a^2)^\alpha} = \frac{g_{\mu\nu}}{n}\int d^n k\,\frac{k^2}{(k^2+a^2)^\alpha} \tag{33.27}$$

$$\int d^n k\,\frac{k_\mu}{(k^2+a^2)^\alpha} = 0 \tag{33.28}$$

[14] [Eds.] See, for instance, G. 't Hooft and M. J. G. Veltman, *Diagrammar*, CERN publication **73-9**, 1973 (available at the CERN Document Server, `cds.cern.ch`) or M. Veltman, *Diagrammatica*, Cambridge U. P., 1994.

To craft the Dirac gamma matrices appropriate to an n-dimensional spacetime, we first create Euclidean metric Dirac matrices (later we will sprinkle in enough i's to make them obey the Minkowski metric):

$$\{\gamma_\mu, \gamma_\nu\} = 2\delta_{\mu\nu} \qquad \mu, \nu = 1, \ldots, n \qquad \gamma_\mu = \gamma_\mu^\dagger \tag{33.29}$$

We know in four dimensions, $\mathrm{Tr}\,(\not{a}\not{b}) = 4a \cdot b$. Though the dimension of spacetime is n, we don't *a priori* know the dimension of the Dirac matrices. What is the trace of the unit matrix, $\mathrm{Tr}\,(\mathbb{1})$, in this algebra? Of course,

$$\mathrm{Tr}\,(\mathbb{1})_{\text{Dirac}} = 1 + 1 + 1 + \cdots = \dim(\gamma_\mu) \tag{33.30}$$

where γ_μ is any of the Dirac matrices. We have to consider even n and odd n separately. For even n, define the set of matrices $\{a_i\}$ by

$$\begin{pmatrix} a_1 \\ a_2 \\ \vdots \\ a_{n/2} \end{pmatrix} \equiv \frac{1}{2} \begin{pmatrix} \gamma_1 + i\gamma_2 \\ \gamma_3 + i\gamma_4 \\ \vdots \\ \gamma_{n-1} + i\gamma_n \end{pmatrix} \tag{33.31}$$

with the adjoints

$$a_1^\dagger = \tfrac{1}{2}(\gamma_1 - i\gamma_2) \tag{33.32}$$

and so on. Then the algebra of these matrices is just that for fermionic simple harmonic oscillators:

$$\{a_i, a_j\} = \{a_i^\dagger, a_j^\dagger\} = 0 \tag{33.33}$$

$$\{a_i, a_j^\dagger\} = \delta_{ij} \tag{33.34}$$

The union of the sets $\{a_i\}$ and $\{a_i^\dagger\}$ forms a Clifford algebra[15] of dimension $2^{(n/2)}$, in 1-1 correspondence with the Dirac matrices. The Dirac matrices themselves form a Clifford algebra of dimension 4×4. So it seems reasonable to assign $2^{(n/2)}$ as the dimension of the Dirac identity in n dimensions:

$$\mathrm{Tr}\,(\gamma_\mu\gamma_\nu) = 2^{(n/2)}\,\delta_{\mu\nu} \tag{33.35}$$

Note that this reduces to the usual result when $n = 4$. In fact, we could just as well take[16]

$$\mathrm{Tr}\,(\mathbb{1})_{\text{Dirac}} = f(n) \tag{33.36}$$

where $f(n)$ is a smooth function of n, and $f(4) = 4$.

Nearly all the usual trace theorems generalize readily. For example,

$$\begin{aligned} &\mathrm{Tr}\,(\gamma_\mu\gamma_\kappa\gamma_\nu\gamma_\lambda) = 2^{(n/2)}[\delta_{\mu\kappa}\delta_{\nu\lambda} - \delta_{\mu\nu}\delta_{\lambda\kappa} + \delta_{\mu\lambda}\delta_{\nu\kappa}] \\ &\mathrm{Tr}\,(\text{odd number of } \gamma\text{'s}) = 0 \end{aligned} \tag{33.37}$$

There is one problem in extending the gamma matrices to n dimensions: γ_5. Recall that we found ((S11.22), p. 428)

$$\mathrm{Tr}\,(\gamma_\mu\gamma_\nu\gamma_\rho\gamma_\sigma\gamma_5) = 4i\epsilon_{\mu\nu\rho\sigma} \tag{33.38}$$

---

15 [Eds.] See note 1, p. 407.

16 [Eds.] Ryder *QFT*, p. 333.

but there is no obvious extension to higher dimensions; the Levi-Civita symbol (with four indices) is specifically 4-dimensional. While γ_{n+1} can be defined, in analogy with the definition (20.102) of γ_5 as i times the product of all the n gamma matrices, the presence of γ_5 in some currents—but fortunately, not in QED's—leads to anomalies in those field theories.

In the case of odd n, you treat $\gamma_1, \ldots, \gamma_{n-1}$ as before, with

$$\mathrm{Tr}\,(\gamma_\mu \gamma_\nu) = 2^{(n-1)/2}\,\delta_{\mu\nu} \tag{33.39}$$

and the equivalent of γ_5 is γ_n, with

$$\gamma_n \propto \pm(\gamma_1\gamma_2 \cdots \gamma_{n-1}) \tag{33.40}$$

There are two inequivalent choices, connected by parity. For parity conservation, you have to add them together. Then for theories conserving parity,

$$\mathrm{Tr}\,(\gamma_\mu \gamma_\nu) = 2^{(n+1)/2}\,\delta_{\mu\nu} \tag{33.41}$$

Regulator Fields

The method of regulator fields is somewhat more old-fashioned than dimensional regularization. But it's worth talking about because it's cute and easy to show that it works fairly well. Recall that this method was basically very simple.[17] We took a Lagrangian and added to it terms in an extra field ϕ_1:

$$\mathscr{L}_0(\phi) + \mathscr{L}'(\phi) \to \mathscr{L}_0(\phi) + \mathscr{L}_0(\phi_1) + \mathscr{L}'(\phi + i\phi_1) \tag{33.42}$$

The term $\mathscr{L}_0(\phi)$ is the free Lagrangian (including the gauge-fixing term), $\mathscr{L}'(\phi)$ is the interaction Lagrangian with counterterms, and ϕ_1 is a very heavy field with mass M; the i in $\mathscr{L}'$ is to give a relative minus sign in ϕ_1's propagator. This ϕ_1 is not a physical field but instead another kind of ghost. (If the divergences are very bad, there may be a need for more than one regulator field.) The result of these additions is to change the propagators from their usual form, by subtracting a new propagator from the original propagator:

$$\frac{1}{k^2-\mu^2} \to \frac{1}{k^2-\mu^2} - \frac{1}{k^2-M^2} \tag{33.43}$$

Every time we have a graph with a propagator we subtract an *extra* propagator with a very heavy mass, and then all the integrals become convergent. If one heavy mass is not sufficient, we subtract more of them, appropriately weighted, enough to make all of our integrals convergent:

$$\mathscr{L} \to \mathscr{L}_0(\phi,\mu) + \sum_r \mathscr{L}_0(\phi_r, M_r) + \mathscr{L}'(\phi + \sum_r c_r\phi_r) \tag{33.44}$$

$$\frac{1}{k^2-\mu^2} \to \frac{1}{k^2-\mu^2} + \sum_r \frac{c_r^2}{k^2-M_r^2} \tag{33.45}$$

The c_r need not all be real, which gives the relative minus sign in (33.43). The c_r can be chosen so that the propagators vanish as quickly as we want at large k^2.

---

[17] [Eds.] See p. 526, in particular (25.7).

We can use exactly the same trick for the photon, massive or massless, with no problem. I'll write down the spinor electrodynamic Lagrangian with the new terms for a massless photon:

$$\begin{aligned}\mathscr{L} &= \mathscr{L}_0(A_\mu, 0) + \overline{\psi}(i\partial\!\!\!/ - m - eA\!\!\!/)\psi \\ &\to \mathscr{L}_0(A_\mu, 0) + \sum_r \mathscr{L}_0(A_\mu^{(r)}, M_r) + \overline{\psi}\Big(i\partial\!\!\!/ - e(A\!\!\!/ + \sum_r c_r A\!\!\!/^{(r)}) - m\Big)\psi\end{aligned} \tag{33.46}$$

That doesn't affect the gauge invariance in any way; it just adds something to A_μ. $\mathscr{L}$ is still gauge invariant under the transformation

$$\psi \to e^{-ie\chi}\psi; \quad A_\mu \to A_\mu + \partial_\mu\chi; \quad A_\mu^{(r)} \to A_\mu^{(r)} \tag{33.47}$$

When, however, we try the same trick for the charged particles, say the fermions in the theory, we run into trouble, as Pauli and Villars pointed out. We encounter a lot of graphs that are divergent and only contain fermion propagators around the loops that are responsible for the divergence, as in Figure 33.4. This one is awful; it's quadratically divergent. No matter

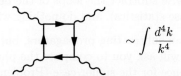

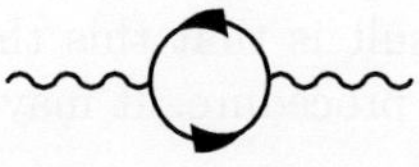

$$\sim \int \frac{d^4k}{k^2}$$

Figure 33.4: Vacuum polarization

how nice we make the photon propagator, that's not going to do anything to the divergence of that graph. Here's another example, Delbrück scattering, shown in Figure 33.5. This one is only logarithmically divergent.

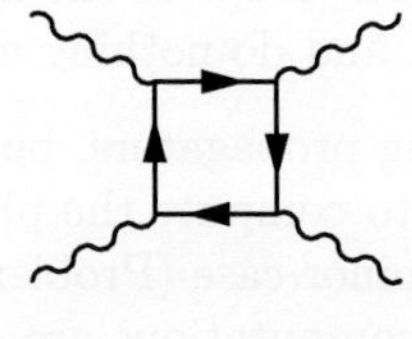

$$\sim \int \frac{d^4k}{k^4}$$

Figure 33.5: Delbrück scattering

If we try to handle these divergent fermion loops in the same way as those for the photon or the scalar fields, say by changing the term $\overline{\psi}\gamma_\mu\psi$ that appears in the interaction,

$$\overline{\psi}\gamma_\mu\psi \to (\overline{\psi} - i\overline{\psi}_1)\gamma_\mu(\psi + i\psi_1) \tag{33.48}$$

then we've made everything finite, but we have also destroyed the gauge invariance and, worse yet, broken current conservation. The divergence of the cross terms in this object is not zero, as it should be if the current is conserved; it's proportional to the difference of the masses of the two fields:

$$\partial^\mu[\overline{\psi}\gamma_\mu\psi_1 - \overline{\psi}_1\gamma_\mu\psi] = (m_1 - m)[\overline{\psi}\psi_1 + \overline{\psi}_1\psi] \tag{33.49}$$

Not only have we broken current conservation, we've done it in a disgusting way: the larger we make our cutoff mass m_1, the worse we break it. The derivative of the current is supposed to be zero, but here it is proportional to the difference between the cutoff mass and the physical mass.

Pauli and Villars thought up a clever trick to take care of this. In their method, we don't subtract the individual fields; we subtract, with appropriate coefficients to take care of all divergences, the *currents*, which certainly preserves current conservation and gauge invariance. The subtracted terms are just a bunch of fields minimally coupled. We introduce three regulator spinor fields ψ_i,

$$\overline{\psi}\gamma_\mu\psi \to \overline{\psi}\gamma_\mu\psi + \sum_{i=1}^{3} c_i\overline{\psi}_i\gamma_\mu\psi_i \tag{33.50}$$

There are no i's in the coefficients, as you might have expected from (33.48), and you may be wondering what's going on. We choose to give the ψ_i both heavy masses and strange statistics: while ψ_2 obeys Fermi statistics, ψ_1 and ψ_3 are required to obey *Bose* statistics; they are unphysical *ghost fields*. These ghosts are similar to those introduced earlier (§29.3 and §31.3), except that here we have *spinor* fields obeying *Bose* statistics. Before, we had *scalar* fields obeying *Fermi* statistics. (We're mad with power; we can do what we want!) The result of these regulator fields ψ_1 and ψ_3 obeying Bose statistics is that we don't have the usual minus sign for closed loops in which they appear. Their loops have a sign opposite to the "real" Fermi field ψ, so that by appropriately adjusting the c_i's we can make all the divergences cancel. Of course, another result is that this theory is completely unphysical, with negative energies, but it's just a cutoff procedure. It may be crazy, but it preserves both Lorentz invariance and gauge invariance.

The full regulator field prescription for putting in a gauge invariant cutoff goes like this:

(a) For photons, subtract the heavy propagators as described before.
(b) For fermions, subtract the loops of the heavy particles (don't do a thing with the propagators).
(c) For charged bosons, likewise subtract the loops of the heavy particles (these regulator fields obey traditional Bose statistics), and do nothing with the propagators.

This is a little bit less clean than subtracting propagators, but it has the great advantage of preserving gauge invariance. I will ask you to compute the photon self-energy using both of these regularization methods, once for the spinor case (Problem 18.1) and once for the scalar case (Problem 18.2), for homework. These computations are actually simple. You will learn things by doing them, and they're historically important.

Here's what happens (though the details are left to you). For the box diagram, Figure 33.5, because of the statistics of the regulator Fermi fields, the divergence is proportional to

$$\int \frac{d^4k}{k^4}\left[1|_{\text{from }\psi} - 1|_{\psi_1} + 1|_{\psi_2} - 1|_{\psi_3}\right] \tag{33.51}$$

which is zero. For the photon self-energy, Figure 33.4, the divergence is proportional to

$$\sum_{r=0}^{3}\int d^4k \frac{1}{k^2 - m_r^2 + a} \tag{33.52}$$

(note that the summation now goes from *zero* to 3) where a is a function of the external momenta and various parameters, and $m_0 = m$, the mass of the actual fermion. Expanding the denominator out in inverse powers of k^2, and taking account of the statistics of the regulator Fermi fields, we get

$$\int \frac{d^4k}{k^2}(1-1+1-1) - \int \frac{d^4k}{k^4}a(1-1+1-1) + \int \frac{d^4k}{k^4}(m^2 - m_1^2 + m_2^2 - m_3^2) + \cdots \tag{33.53}$$

The first two terms vanish, and we can choose the heavy masses m_r^2 so that the third sum also vanishes.

The history of the calculations of vacuum polarization is amusing.[18] In the late 1940's, there was a great problem with the self-mass of the photon. People didn't have any deep understanding of gauge invariance at all. Ward had not yet written down the very first Ward identity. Schwinger and Feynman were the only two people who were able to renormalize quantum electrodynamics.[19] They had the great secret. Schwinger was plugging along in Coulomb gauge because he knew what he was doing there, and Feynman was working in Feynman gauge because he didn't care if he knew what he was doing, as long as his answers were consistent. They both knew however that renormalization shouldn't require putting in a mass for the photon because that would break gauge invariance, even if they weren't quite sure how to formulate gauge invariance precisely. They both found, when they computed the vacuum polarization graph (Figure 33.4), that they needed a photon mass counterterm. That caused a great deal of irritation. Both of them made nervous remarks and swept it under the rug and said, "We have to set the photon self-energy to zero by gauge invariance," and then quickly went on to computing something observable. Well, I don't really know what Feynman said, but Schwinger actually says in one of his early papers,[20] "We just set this to zero by gauge invariance. It's divergent and therefore ambiguous and we set it equal to zero." The trouble was that neither Feynman nor Schwinger were using gauge-independent cutoffs. Pauli and Villars clarified everything by their realization that we could systematically introduce a cutoff procedure in a gauge invariant way, and then explicitly and unambiguously compute the photon self-energy, to show that it *is* zero.[21]

33.4 The Ward identity and Green's functions

As noted in (32.56) at the end of the previous chapter, the Ward Identity applies to the entire generating functional $\Gamma[\overline{\Phi}]$. It encompasses a large number of equations for the 1PI Green's functions, which can be derived from the generic series expansion (32.12). For convenience, we will (for now) stick to spinor electrodynamics, with a massive photon:

$$\mathscr{L} = \underbrace{-\tfrac{1}{4}(\partial_\mu A_\nu - \partial_\nu A_\mu)^2 + \overline{\psi}[i\partial\!\!\!/ - eA\!\!\!/ - m]\psi}_{\mathscr{L}_{GI}} - \frac{1}{2\xi}(\partial_\mu A^\mu)^2 + \tfrac{1}{2}\mu^2(A_\mu)^2 \tag{33.54}$$

I have dropped the primes and the bars on the fields. Just remember that we are talking about renormalized mean fields. We'll need the infinitesimal gauge transformation, so I'll write it

---

[18] [Eds.] Schweber *QED*, Chapter 7, pp. 335–340; Chapter 10, pp. 443–444; Crease & Mann, Chapter 6, pp. 102–108.

[19] [Eds.] Tomonaga had also figured it out, but only his colleagues in Japan knew that he had: D. Ito and K. Nishijima, "Japanese Researchers Reveal Tomonaga's Path to QED Renormalization", Letter to the Editors, *Physics Today* (**51**), 7 (1998) 15–16.

[20] [Eds.] "If the electromagnetic field is that of a light quantum, the vacuum polarization effects are equivalent to ascribing a proper mass to the photon. Previous calculations have yielded non-vanishing, divergent expressions for the light quantum proper mass. However, the latter quantity must be zero in a proper gauge invariant theory." Julian Schwinger, "Quantum Electrodynamics. I. A Covariant Formulation", *Phys. Rev.* **74** (1948) 1439–1461; see p. 1440. Wentzel found Schwinger's claim "highly objectionable": Gregor Wentzel, "New Aspects of the Photon Self-Energy Problem", *Phys. Rev.* **74** (1948) 1070–1075.

[21] [Eds.] See Problem 18.1, p. 725.

down again explicitly:

$$\begin{aligned} \delta A_\mu &= \partial_\mu \delta\chi \\ \delta\psi &= -ie\psi\delta\chi \\ \delta\overline{\psi} &= ie\overline{\psi}\delta\chi \end{aligned} \tag{33.55}$$

Γ will be gauge invariant except for the integral of the last two terms in (33.54). There are lots of Green's functions to worry about, so I will introduce a systematic notation. I will refer to

$$\Gamma^{(n,n,m)} \tag{33.56}$$

where n is the number of ψ's and the number of $\overline{\psi}$'s (these are equal unless $\Gamma \equiv 0$) and m is the number of A_μ's. These $\Gamma^{(n,n,m)}$ objects depend on a bunch of position variables, and their Fourier transforms $\widetilde{\Gamma}^{(n,n,m)}$ depend on a bunch of momentum variables. If the Green's functions involve photons, they will also have tensor indices associated with the A_μ fields.

For example, the photon propagator (to lowest order, $\mathcal{O}(e^0)$, and dropping the $i\epsilon$) is a generalization of (31.35):

$$\widetilde{D}_{\mu\nu}(k) = -\frac{i}{k^2-\mu^2}\left[g_{\mu\nu} - \frac{k_\mu k_\nu}{k^2}\right] - \frac{i}{(k^2/\xi)-\mu^2}\left[\frac{k_\mu k_\nu}{k^2}\right] + \mathcal{O}(e) \tag{33.57}$$

The quantities in the square brackets are the projection operators $P^T_{\mu\nu}$ and $P^L_{\mu\nu}$ defined earlier,

$$P^T_{\mu\nu} = g_{\mu\nu} - \frac{k_\mu k_\nu}{k^2}, \quad P^L_{\mu\nu} = \frac{k_\mu k_\nu}{k^2} \tag{30.46}$$

This makes it easy to compute $\widetilde{\Gamma}^{(0,0,2)}_{\mu\nu}$ to lowest order. Recalling the definition of $\widetilde{\Gamma}^{(2)}$

$$\frac{i}{\widetilde{\Gamma}^{(2)}(p,-p)} \equiv \widetilde{D}(p) \tag{32.16}$$

we have

$$\widetilde{\Gamma}^{(0,0,2)}_{\mu\nu}(k)\Big|_{e=0} = -(k^2-\mu^2)\left[g_{\mu\nu} - \frac{k_\mu k_\nu}{k^2}\right] - \left[(k^2/\xi)-\mu^2\right]\left[\frac{k_\mu k_\nu}{k^2}\right] \tag{33.58}$$

Both the transverse and longitudinal parts have mass μ (though the coefficient of k^2 in the longitudinal part is $1/\xi$).[22] For massive photons the most convenient gauge is the Feynman gauge, $\xi = 1$, because the pole at $k^2 = 0$, which would otherwise lead to some mild technical problems, cancels between the two terms. (The Landau gauge is also nice for certain purposes.)

We wish to study the corrections of $\mathcal{O}(e)$ and higher to $\widetilde{\Gamma}^{(0,0,2)}_{\mu\nu}(k)$. The way we study those corrections is by expanding out Γ in terms of the fields, according to (32.12) and (32.13). The term in Γ involving $\Gamma^{(0,0,2)}$ is bilinear in A_μ:

$$\Gamma = \tfrac{1}{2!}\int d^4x\, d^4y\, A^\mu(x)A^\nu(y)\Gamma^{(0,0,2)}_{\mu\nu}(x,y) + \cdots \tag{33.59}$$

---

[22] [Eds.] In the video of Lecture 33, at 1:07:33, Coleman writes ξ where (33.57) and (33.58) have $1/\xi$. As a consistency check, the propagator for the case $\mu = 0$ is consistent with $1/\xi$; see Coleman *Aspects*, Chap. 4, "Secret Symmetry", p. 164, equation (5.26) or Peskin & Schroeder *QFT*, p. 297, equation (9.58). (There was also a sign error in the second term in (33.58).)

We want to see what happens to Γ under a gauge transformation. In particular, when $A_\mu \to A_\mu + \partial_\mu \delta\chi(x)$, because δA_μ is independent of e, the associated change $\delta\Gamma$ must be as well:

$$\delta\Gamma = \delta\Gamma\big|_{e=0} \text{ when } A_\mu(x) \to A_\mu(x) + \partial_\mu \delta\chi(x) \tag{33.60}$$

Applying the infinitesimal gauge transformation to (33.59), we find, after an integration by parts,

$$\delta\Gamma = -\int d^4x\, d^4y \left(\partial_x^\mu \Gamma_{\mu\nu}^{(0,0,2)}\right) A^\nu(y)\, \delta\chi(x) + \dots \tag{33.61}$$

from which it follows

$$\partial_x^\mu \Gamma_{\mu\nu}^{(0,0,2)} = \partial_x^\mu \Gamma_{\mu\nu}^{(0,0,2)}\Big|_{e=0} \tag{33.62}$$

or, Fourier transforming both sides,

$$k^\mu \widetilde{\Gamma}_{\mu\nu}^{(0,0,2)} = k^\mu \widetilde{\Gamma}_{\mu\nu}^{(0,0,2)}\Big|_{e=0} \tag{33.63}$$

When we add a divergence to A_μ, $\delta\Gamma$ acquires only a contribution from the gauge fixing and mass terms in (33.54). In position space we add a gradient to A_μ, we integrate by parts, and pick up a term proportional to the divergence of Γ, which is like k_μ in momentum space. Since there's no change beyond the change in the zeroth order term, the kernel in (33.59), $\widetilde{\Gamma}_{\mu\nu}^{(0,0,2)}$, must obey (33.63), and

$$k^\mu \widetilde{\Gamma}_{\mu\nu}^{(0,0,2)}(k)$$

must be just the zeroth order term. Gauge invariance thus forces $k^\mu \widetilde{\Gamma}_{\mu\nu}^{(0,0,2)}(k)$ to be whatever it is at zeroth order in e, regardless of what else is going on.

Let's look at this in detail. $\widetilde{D}_{\mu\nu}(k)$ gets modified by the 1PI vacuum polarization graphs $\widetilde{\Pi}'_{\mu\nu}$ in exactly the same way as the scalar propagator $\widetilde{D}'(p^2)$ (15.36) was modified by $\widetilde{\Pi}'(p^2)$ in §15.3.[23] We expect a similar situation here, with the following modification: both $\widetilde{D}_{\mu\nu}$ and

$$\equiv i\widetilde{\Pi}'_{\mu\nu}(k^2)$$

Figure 33.6: 1PI vacuum polarization

$\widetilde{\Pi}'_{\mu\nu}$ will in general have both transverse and longitudinal parts:

$$\widetilde{D}'_{\mu\nu} = \widetilde{D}^T P^T_{\mu\nu} + \widetilde{D}^L P^L_{\mu\nu} \tag{33.64}$$

$$\widetilde{\Pi}'_{\mu\nu} = \widetilde{\Pi}^T P^T_{\mu\nu} + \widetilde{\Pi}^L P^L_{\mu\nu} \tag{33.65}$$

Because the projection operators are idempotent and orthogonal, when we string the $\widetilde{D}_{\mu\nu}$'s and the $\widetilde{\Pi}'_{\mu\nu}$'s together, the transverse parts combine with the transverse parts, the longitudinal parts combine with the longitudinal parts and there are no transverse-longitudinal cross terms:

$\widetilde{D}'_{\mu\nu}(k^2) \equiv$ 

(33.66)

[23] [Eds.] The 1PI graph in (15.33) is defined as $-i\widetilde{\Pi}'(k^2)$. The sign difference in the definitions of $\widetilde{\Pi}'$ and $\widetilde{\Pi}'^{\mu\nu}$ is to balance a corresponding sign difference between the definitions of $\widetilde{D}'$ and $\widetilde{D}'^{\mu\nu}$.

Putting these all together we get for the full propagator and its inverse

$$\widetilde{D}'_{\mu\nu}(k) = -\frac{i}{k^2-\mu^2-\widetilde{\Pi}^T}\left[g_{\mu\nu}-\frac{k_\mu k_\nu}{k^2}\right] - \frac{i}{(k^2/\xi)-\mu^2-\widetilde{\Pi}^L}\left[\frac{k_\mu k_\nu}{k^2}\right] \tag{33.67}$$

$$\widetilde{\Gamma}^{(0,0,2)}_{\mu\nu}(k) = -(k^2-\mu^2-\widetilde{\Pi}^T)\left[g_{\mu\nu}-\frac{k_\mu k_\nu}{k^2}\right] - ((k^2/\xi)-\mu^2-\widetilde{\Pi}^L)\left[\frac{k_\mu k_\nu}{k^2}\right] \tag{33.68}$$

The Ward identity applies to the full 1PI Green's function. Substituting (33.58) and (33.68) into (33.63) we find

$$-((k^2/\xi)-\mu^2-\widetilde{\Pi}^L)k_\nu = -((k^2/\xi)-\mu^2)k_\nu \tag{33.69}$$

because $k^\mu P^T_{\mu\nu} = 0$. Only the terms proportional to $P^L_{\mu\nu}$ in (33.58) and (33.68) survive contraction with k^μ. We conclude that

$$\widetilde{\Pi}^L(k^2) \equiv 0 \tag{33.70}$$

That is, *all* the possible 1PI graphs beyond zeroth order in e must contribute *only* to the part that is proportional to the *transverse* projection operator, $P^T_{\mu\nu}$. This is an important consequence of gauge invariance. Lorentz invariance by itself tells us that the propagator or its inverse $\widetilde{\Gamma}^{(0,0,2)}$ is the sum of two terms, a complicated function proportional to $P^T_{\mu\nu}$ and another such function proportional to $P^L_{\mu\nu}$. The Ward identity, on the other hand, tells us that *only* the transverse term is corrected; the zeroth order longitudinal term just sits around. Whatever the 1PI graph is, it will be proportional to $P^T_{\mu\nu}$:

$$\equiv i\widetilde{\Pi}'_{\mu\nu}(k^2) \propto \left[g_{\mu\nu}-\frac{k_\mu k_\nu}{k^2}\right] \tag{33.71}$$

We can do even better, because this expression has a pole at $k^2 = 0$, and we shouldn't expect to find a pole because we haven't got any massless particles in this theory of massive electrodynamics. Even if the photon is massless, we shouldn't expect to find a pole, because this is a 1PI graph and we've taken out the one photon pole, so we can say

$$i\widetilde{\Pi}'_{\mu\nu}(k^2) \propto \left[g_{\mu\nu}-\frac{k_\mu k_\nu}{k^2}\right]k^2 \tag{33.72}$$

to avoid the spurious pole.

The result (33.72) has the typical form of an equation that we would get if we deduced the Ward identities from current conservation. To emphasize the result, *all the corrections to the 1PI photon self-energy are purely transverse.* From this we can deduce that there is *no* photon self-mass term: a photon self-mass would require a correction[24] proportional to $g_{\mu\nu}$.

We can also obtain the right-hand side of (33.69), namely

$$k^\mu\widetilde{\Gamma}^{(0,0,2)}_{\mu\nu} = -((k^2/\xi)-\mu^2)k_\nu \tag{33.73}$$

and hence the result (33.70), from our general formula (32.56), modified to handle a massive photon:

$$ie\overline{\psi}(z)\frac{\delta\Gamma}{\delta\overline{\psi}(z)} - ie\frac{\delta\Gamma}{\delta\psi(z)}\psi(z) - \partial^\mu\frac{\delta\Gamma}{\delta A^\mu(z)} = -\left[\frac{1}{\xi}\Box^2+\mu^2\right]\partial_\mu A^\mu(z) \tag{33.74}$$

[24] [Eds.] Greiner & Reinhardt *QED*, Section 5.2, pp. 257–258.

We apply (33.74) to (33.59). The first two terms contribute nothing; we are not looking at the fermions. The rest of the equation gives

$$-\partial^{\mu}_{x}\int d^4y\,\Gamma^{(0,0,2)}_{\mu\nu}(x,y)\,A^{\nu}(y) = -\left[\frac{1}{\xi}\Box^2+\mu^2\right]\partial^{\nu}A_{\nu}(x) \tag{33.75}$$

Now we take $\delta/\delta A^{\nu}(y)$ of this expression, and get

$$-\partial^{\mu}_{x}\Gamma^{(0,0,2)}_{\mu\nu}(x,y) = -\left[\frac{1}{\xi}\Box^2+\mu^2\right]\partial_{\nu}\delta^{(4)}(x-y) \tag{33.76}$$

which in momentum space becomes (33.73):

$$k^{\mu}\widetilde{\Gamma}^{(0,0,2)}_{\mu\nu}(k) = (-(k^2/\xi)+\mu^2)k_{\nu}$$

Let's go on and study a more complicated expression. This Ward identity will describe the fundamental three-point vertex, Figure 33.7.

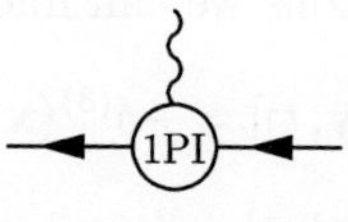

Figure 33.7: 1PI three-point function

The relevant terms in Γ are

$$\Gamma = \int d^4x\,d^4y\,\overline{\psi}(x)\psi(y)\Gamma^{(1,1,0)}(x,y) + \int d^4x\,d^4y\,d^4z\,\overline{\psi}(x)\psi(y)A^{\mu}(z)\Gamma^{(1,1,1)}_{\mu}(x,y,z) + \cdots \tag{33.77}$$

Under a gauge transformation, these two terms mix up among themselves; the $\cdots$ terms don't.

Now we apply an infinitesimal gauge transformation to these two parts together. The first part will produce a term in $\overline{\psi}\psi$, and so will the second, because A_μ just picks up a term equal to $\partial_\mu\delta\chi$. The coefficient of $\overline{\psi}\psi$ must be zero, because I know from (32.50) that the result of the gauge transformation on Γ just gives a term linear in A_μ, with no $\overline{\psi}\psi$ term in it:

$$0 = \int d^4x\,d^4y\,\overline{\psi}(x)\psi(y)\Big[\Gamma^{(1,1,0)}(x,y)[ie\delta\chi(x)-ie\delta\chi(y)] + \int d^4z\,(\partial^{\mu}\delta\chi(z))\Gamma^{(1,1,1)}_{\mu}(x,y,z)\Big] + \cdots \tag{33.78}$$

These are the only terms in $\overline{\psi}\psi$, and as Γ is gauge invariant apart from the terms quadratic in A_μ, this expression must be zero.[25] Extracting the coefficient of $\overline{\psi}(x)\psi(y)\delta\chi(z)$ (integrating by parts in the last term)[26] we obtain

$$ie\Gamma^{(1,1,0)}(x,y)[\delta^{(4)}(x-z)-\delta^{(4)}(y-z)] = \partial^{\mu}_{z}\Gamma^{(1,1,1)}_{\mu}(x,y,z) \tag{33.79}$$

We have simply applied a gauge transformation to the action, invoked the Ward identity, and said that the gauge transformation can have no effect on this term.

[25] [Eds.] In fact, the gauge transformation also produces terms proportional to $\overline{\psi}(x)\psi(y)A^{\mu}(z)$. The point is that the terms which are proportional to *only* $\overline{\psi}\psi$, with no A^{μ} factors, must vanish *separately*.

[26] [Eds.] Use the identity $\delta\chi(x) = \int d^4z\,\delta^{(4)}(x-z)\,\delta\chi(z)$.

This is, by the way, strikingly similar in structure to (31.69), an equation we found by manipulating the canonical commutation rules when we differentiated a current and two fields, despite the differences: $\Gamma_\mu^{(1,1,1)}$ is not a current; it's not a full Green's function, but a 1PI Green's function; it doesn't involve unrenormalized entities, but rather renormalized entities. Aside from these differences, it's the same equation.

Traditionally, (33.79) is derived in terms of currents;

$$\overline{\psi}\gamma_\mu\psi = j_\mu \tag{33.80}$$

(We talked about this earlier; see the discussion on p. 675.) The equations of motion (27.10) give

$$\partial^\mu F_{\mu\nu} + \mu_0^2 A_\nu = j_\nu$$

and the current is conserved:

$$\partial^\mu j_\mu = 0 \tag{33.81}$$

From the canonical commutation relations we can find

$$[j_0(\mathbf{x},t), \psi(\mathbf{y},t)] = -\delta^{(3)}(\mathbf{x}-\mathbf{y})\psi(\mathbf{y},t) \tag{33.82}$$

and a similar equation follows for $\overline{\psi}$ with the sign changed. Then

$$\partial_z^\mu \langle 0|T(j_\mu(z)\overline{\psi}(x)\psi(y))|0\rangle = \left[\delta^{(4)}(z-y) - \delta^{(4)}(z-x)\right]\langle 0|T(\overline{\psi}(x)\psi(y))|0\rangle \tag{33.83}$$

This looks a lot like equation (33.79). The traditional derivation, even when done carefully, is shorter than the modern version, starting from (32.50). But it has disadvantages. First, it is couched in terms of unrenormalized fields. Next, the Green's functions are neither full Green's functions, nor 1PI functions, and it's hard to keep straight charged scalars and charged spinors.

Again, we can get (33.79) from the general relation (33.74). Take the (left) $\overline{\psi}(x)$ derivative,[27] the (right) $\psi(y)$ derivative and set all the remaining fields to 0:

$$ie\left[\delta^{(4)}(x-z)\frac{\delta^2\Gamma}{\delta\overline{\psi}(z)\delta\psi(y)}\bigg|_0 - \delta^{(4)}(y-z)\frac{\delta^2\Gamma}{\delta\overline{\psi}(x)\delta\psi(z)}\bigg|_0\right] - \partial_z^\mu\frac{\delta^3\Gamma}{\delta\overline{\psi}(x)\delta\psi(y)\delta A^\mu(z)}\bigg|_0 = 0 \tag{33.84}$$

This equation is true in general. Applying it to (33.77), we obtain (33.79) once again:

$$ie[\delta^{(4)}(x-z) - \delta^{(4)}(y-z)]\Gamma^{(1,1,0)}(x,y) = \partial_z^\mu\Gamma_\mu^{(1,1,1)}(x,y,z)$$

In momentum space, $\widetilde{\Gamma}_\mu^{(1,1,1)} \equiv \widetilde{\Gamma}_\mu$ is the Fourier transform of the 1PI three-point function, Figure 33.7. Assign the momenta more conventionally (instead of having all the momenta going in): $p' = p + k$. Then

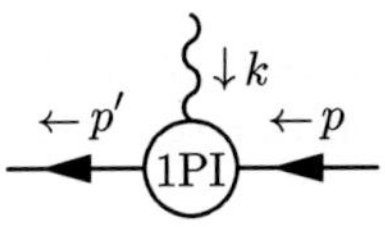

$$= \widetilde{\Gamma}_\mu(p',p,k) \equiv \widetilde{\Gamma}_\mu^{(1,1,1)}(p',p,k) \tag{33.85}$$

---

[27] [Eds.] A note for purists: To keep the signs consistent, all $\overline{\psi}$ derivatives must be taken from the left and all ψ derivatives must be taken from the right.

We already have a notation for the function $\widetilde{\Gamma}^{(1,1,0)}$, the inverse of the renormalized electron propagator:

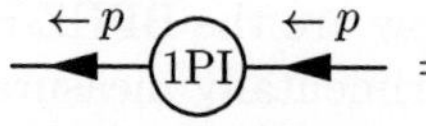

$$= i\widetilde{S}_F'^{-1}(\not{p}) \equiv \widetilde{\Gamma}^{(1,1,0)}(p) \tag{33.86}$$

We are going to Fourier transform (33.79). It's tedious to do this by hand, but it's easy to see what the general structure will be, so I'll just write down the answer. The only interesting question is what happens with the delta functions in (33.79). We know that

$$\int d^4z\, e^{ip\cdot x}e^{ik\cdot z}\,\delta^{(4)}(x-z) = e^{i(p+k)\cdot x} \tag{33.87}$$

Therefore the delta functions will give terms where the momentum carried by the propagator is either p, equal to $p'-k$; or p', equal to $p+k$, depending on which delta function we're integrating over. Instead of performing the Fourier transforms, we can just guess the result, though we might get factors of i wrong:

$$-ie[\widetilde{S}_F'^{-1}(\not{p}') - \widetilde{S}_F'^{-1}(\not{p})] = k^\mu\widetilde{\Gamma}_\mu(p',p,k) \tag{33.88}$$

We'll check our guess by demanding that the equation be right to first order in e:

$$i\widetilde{S}_F'^{-1}(\not{p}) = \not{p} - m; \qquad \widetilde{\Gamma}_\mu(p',p,k) = -e\gamma_\mu \tag{33.89}$$

If we substitute these values into (33.88), they give a correct equation linking p, p' and k:

$$-e[\not{p}' - m - (\not{p} - m)] = -e[p'^\mu - p^\mu]\gamma_\mu \overset{\checkmark}{=} -ek^\mu\gamma_\mu \tag{33.90}$$

So as it stands, (33.88) is correct; it is in fact the *original* Ward identity.[28] Diagrammatically (33.88) is

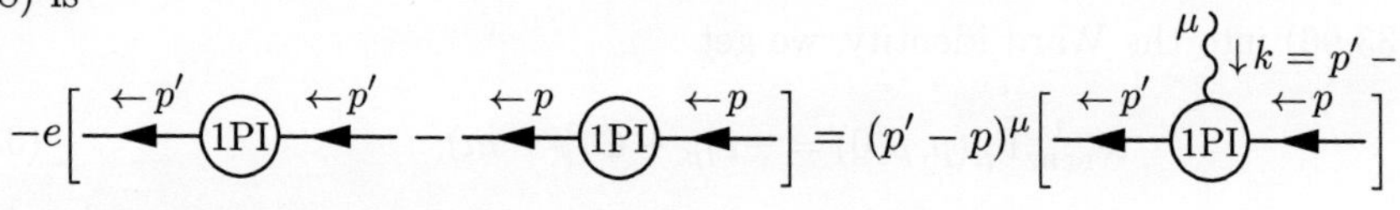

(33.91)

We can immediately write down two consequences of this relation.

We obtain the first consequence by differentiating (33.88) with respect to k_μ at $k_\mu = 0$, with p fixed. Then $\partial/\partial k^\mu = \partial/\partial p'^\mu$, and

$$-ie\frac{\partial}{\partial p'^\mu}\widetilde{S}_F'^{-1}(\not{p}')\bigg|_{p'=p} = \frac{\partial}{\partial k^\mu}\left(k^\nu\widetilde{\Gamma}_\nu(p',p,k)\right)\bigg|_{k=0} \tag{33.92}$$

so that[29]

$$-ie\frac{\partial}{\partial p^\mu}\widetilde{S}_F'^{-1}(\not{p}) = \widetilde{\Gamma}_\mu(p,p,0) \tag{33.93}$$

That is, inserting a very soft (zero-momentum) photon into an electron line is equivalent to differentiating the inverse of the electron propagator. Thus Γ_μ for a zero-momentum photon is known *completely* in terms of the electron propagator. This is an amazing result. What a surprise! It just comes out of gauge invariance, out of the Ward identity.

---

[28] [Eds.] See footnote 3, p. 704.

[29] [Eds.] This is Ward's *original* identity; Ward, *op. cit.*, note 22, p. 675.

33.5 The Ward identity and counterterms

The second consequence answers the question "How are the BPHZ-renormalized quantities e, ψ' and A'_μ, related to the renormalized and experimentally measurable physical quantities?" We can use the original Ward identity (33.88) to derive the remarkable result that

$$e_{\text{BPHZ}} = e_{\text{phys}} \tag{33.94}$$

This is only true when the photon's mass is zero, i.e., $\mu^2 = 0$. The physical charge is *defined* by the condition

$$i\overline{u}'\widetilde{\Gamma}_\mu(p', p, k)u \equiv -ie_{\text{phys}}\overline{u}'\gamma_\mu u \tag{33.95}$$

where p, p' and k are on the mass shell,

$$p^2 = p'^2 = m^2; \qquad k^2 = \mu^2; \qquad (\not{p} - m)u = 0; \qquad (\not{p}' - m)u' = 0$$

We usually need the four-momenta to be complex to satisfy these four conditions, but when $\mu^2 = 0$ we need only these conditions:

$$p' = p; \qquad p^2 = p'^2 = m^2; \qquad k = 0$$

We expand $\widetilde{S}_F'^{-1}(\not{p})$ in powers of $(\not{p} - m)$

$$\widetilde{S}_F'^{-1}(\not{p}) = -i[(\not{p} - m) + \mathcal{O}(\not{p} - m)^2] \tag{33.96}$$

The Ward identity says

$$ie_{\text{BPHZ}}^{-1}\widetilde{\Gamma}_\mu(p, p, 0) = \frac{\partial}{\partial p^\mu}\widetilde{S}_F'^{-1}(\not{p}) \tag{33.97}$$

Substituting (33.96) into the Ward identity, we get

$$ie_{\text{BPHZ}}^{-1}\widetilde{\Gamma}_\mu(p, p, 0) = -i\gamma_\mu + \mathcal{O}(\not{p} - m) \tag{33.98}$$

Sandwiching this equation between $\overline{u}'$ and u, we get, using (33.95),

$$ie_{\text{BPHZ}}^{-1}\left(-\overline{u}'\gamma_\mu u e_{\text{phys}}\right) = -i\overline{u}'\gamma_\mu u \tag{33.99}$$

because $(\not{p} - m)u = 0$. Therefore

$$e_{\text{BPHZ}} = e_{\text{phys}} = Z_3^{1/2}e_0 \tag{33.100}$$

Neat! This would *not* be true for massive vector boson theory. For $\mu^2 \neq 0$

$$e_{\text{phys}} = e_{\text{BPHZ}} + \mathcal{O}(e_{\text{BPHZ}}^3) \tag{33.101}$$

(It can be shown that terms $\mathcal{O}(e_{\text{BPHZ}}^3)$ are also $\mathcal{O}(\mu^2/m^2)$.)

We can see how (33.93) also leads to an earlier result (31.62) that we got (much more cheaply) about the conspiracy of counterterms. If we wish to compute the counterterms to some order in perturbation theory, we write $\widetilde{S}_F'^{-1}(\not{p})$ in a power series

$$i\widetilde{S}_F'^{-1}(\not{p}) = A + B\not{p} + \mathcal{O}(p^2) \tag{33.102}$$

Likewise, (33.89)

$$\widetilde{\Gamma}_{\mu}^{(1,1,1)}(p', p, k) = Ce\gamma_{\mu} + \mathcal{O}(p) + \mathcal{O}(k) \tag{33.103}$$

$B\not{p}$ fixes the $\overline{\psi}i\not{\partial}\psi$ counterterm, and $Ce\gamma_{\mu}$ fixes the $e\overline{\psi}\not{A}\psi$ counterterm.

Applying the identity (33.93) to these expansions, we have at $p = 0$

$$-e\gamma_{\mu}B = Ce\gamma_{\mu} \tag{33.104}$$

Thus B and C are connected in exactly the way our earlier argument[30] said they would be: $B = -C$. It's much more complicated to establish this through the Ward identity than with our earlier method. Nevertheless, it's reassuring to see it done another way, as here. It also tells us something else that the previous derivation did not. That is, we could obtain the exact same relationship between the two counterterms if we didn't renormalize at 0 for the electron but put the electron on the mass shell. If we put the electron on the mass shell and defined our coupling constants that way, the corresponding equation would not be (33.102), but instead

$$i\widetilde{S}_F'^{-1} = A + B(\not{p} - m) + \mathcal{O}((\not{p} - m)^2) \tag{33.105}$$

and a corresponding equation for Γ_{μ} as a power series in $\not{p} - m$. We would of course find exactly the same thing by applying the Ward identity, now not at $\not{p} = 0$ but at $\not{p} = m$. The differentiations are step-by-step the same.

So we can preserve our subtractive procedure even if we put the electron on the mass shell, rather than putting it at the BPHZ point of zero momentum transfer. We still have a perfect matching between the counterterms required of the charge renormalization type, like C, and those required of the electron wave function renormalization type, like B. On the other hand, as a very important point, even if the photon has a mass, we still have to keep the photon at zero momentum transfer, because (33.93) is true only when $k_{\mu} = 0$. We can, with no loss and perfect matching of the counterterms, keep the electron on the mass shell instead of at the BPHZ point. But the photon, whether it's massive or massless, has to be kept *at zero momentum transfer* to get that perfect matching.

Of course the divergent parts of the counterterms will still match, since the question of what counterterms we have to add to the Lagrangian to purge the answer of divergent quantities is independent of what subtraction point we use, and how we parameterize the theory after we've gotten rid of the infinities. But in general, the finite parts of the counterterms, e.g., the coupling constants, will be different if we have a subtractive renormalization scheme where all the particles are on the mass shell, unless the photon has mass zero.

This has definite physical consequences. Some people, for example J. J. Sakurai, said that the ρ meson was just like a photon except that it was heavy, and it coupled to the isospin current instead of the coupling to the electromagnetic current.[31] Sakurai's theory of strong interactions was a minimally coupled theory, with certain complications caused by the non-Abelian nature of the isospin group. In this theory, the ρ_0 meson couples to the I_3 current

---

[30] [Eds.] See Table 31.1, p. 674. The renormalization constants were labeled differently there, $C\overline{\psi}\not{\partial}\psi$ and $eD\overline{\psi}\not{A}\psi$ in place of $B\overline{\psi}\not{\partial}\psi$ and $eC\overline{\psi}\not{A}\psi$, but their roles within the Lagrangians and their relationship to each other are the same in both places.

[31] J. J. Sakurai, "Theory of Strong Interactions", *Ann. Phys.* **11** (1960) 1–48. In the end, of course, Sakurai's influential ideas did not provide the framework for a gauge theory of the strong interactions, which was instead realized in quantum chromodynamics, with massless vectors.

instead of the electromagnetic current, and it has a strong coupling constant, of order 1, rather than of order $1/\sqrt{137}$. Otherwise, he said, everything was exactly the same. Similarly he wanted the ω meson to be coupled to the hypercharge current, in effect having two photons, one which is strongly interacting and massive called the ω, and one which is weakly interacting and massless, the real photon. Therefore we get universal ω coupling, just as we get universal photon coupling. But we only get universal ω coupling when the ω is extrapolated to zero momentum transfer. As the mass of the ω is 782 MeV, that's an extrapolation of nearly 0.8 GeV, which is a long way off the mass shell, especially for the strong interactions. So the idea was hard to check even if we could compute the $\omega - NN$ coupling constant $f_{\omega NN}$, which is not easy. (The ω particles are not particularly stable, though they're more stable than most.) Even if we could have compared $f_{\omega NN}$ with, say, the $\omega - \pi\pi$ coupling constant $f_{\omega\pi\pi}$, and we found that they were 40% off, one from the other, Sakurai would still have been happy. He would just say, well, that's the error we make because we're extrapolating from a physical ω on the mass shell down to zero momentum transfer. If we have a real massive photon, the consequences on physically observable quantities are hard to check, unless the coupling constant is weak, where we can check everything by doing perturbative computations. And that's what we'll do next time, when I finally get to the anomalous magnetic moment of the electron.

Problems 18

18.1 In the theory of a charged Dirac field minimally coupled to a massless photon, compute the renormalized photon self-energy, $\widetilde{\Pi}'_{\mu\nu}(k^2)$, to lowest nontrivial order in perturbation theory, $\mathcal{O}(e^2)$. Write the answer as an integral over a single Feynman parameter. Handle divergences by the Pauli–Villars method of regulator fields as explained in §33.3, pp. 712–715. From the Fermi loop integral, subtract identical loop integrals with heavy Fermi masses times coefficients chosen to cancel both the quadratic and the logarithmic divergences in the integral. Use the BPHZ procedure to fix the counterterm: choose it to cancel the second-order term in the expansion about $k^2 = 0$. Verify that even before you send the masses to infinity, the Green's function is proportional to

$$g_{\mu\nu}k^2 - k_\mu k_\nu$$

(k_μ is the photon momentum) as the Ward identity tells us it should be; see (33.67) through (33.71).

HISTORICAL NOTE: As mentioned in Chapter 33, this problem was a famous technical pain-in-the-neck in the late 1940's. If you just blithely manipulate divergent integrals, it looks like a photon self-mass counterterm is needed. Pauli and Villars invented their gauge invariant cutoff to show that this apparent contradiction of gauge invariance is just a consequence of slovenliness, not a sign of deep sickness in the theory. See note 18, p. 715.

(1998b 7.1); historical note from (1987b 9)

18.2 Perform the same computation for a charged, spinless meson, but this time use dimensional regularization instead of the Pauli–Villars method. *Warning:* In n dimensions, $g^{\mu}{}_{\mu} = n$.

(1998b 7.2)

18.3 Even in quantum electrodynamics, it is possible (though not usual) to work in a gauge where ghost fields are needed. For example, this is a valid form of the electrodynamic Lagrangian:

$$\mathscr{L} = \mathscr{L}_{\text{em}} - \tfrac{1}{2}\lambda\left(\partial_\mu A^\mu + \sigma A_\mu A^\mu\right)^2 + \mathscr{L}_{\text{ghost}}$$

Here $\mathscr{L}_{\text{em}}$ is the standard Lagrangian, with neither gauge-fixing nor ghost terms, and λ and σ are arbitrary real numbers.

(a) What is $\mathscr{L}_{\text{ghost}}$?

(b) What is the ghost propagator?

(c) What are the vertices involving ghost fields?

(1998b 6.1)

Solutions 18

18.1 To lowest nontrivial order, and ignoring for the moment the contributions of the counterterm and the regulator fields, we have

$$-i\widetilde{\Pi}_{\mu\nu}(k^2) =$$ 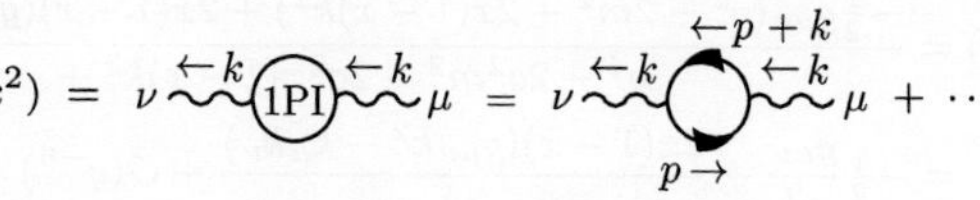 (S18.1)

and to the same order,

$$-i\widetilde{\Pi}'_{\mu\nu}(k^2) =$$ 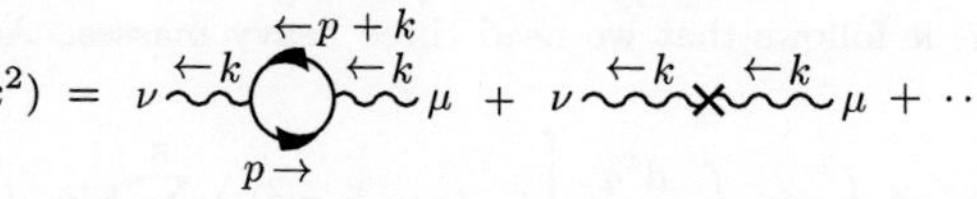 (S18.2)

The unrenormalized self-energy is, following the Feynman rules (described in the boxes on p. 443, p. 443, and p. 645),

$$-i\widetilde{\Pi}_{\mu\nu}(k^2) = -(-ie)^2 \int \frac{d^4p}{(2\pi)^4} \mathrm{Tr}\left[\gamma_\mu \frac{i(\not{p} + \not{k} + m)}{(p+k)^2 - m^2 + i\epsilon} \gamma_\nu \frac{i(\not{p} + m)}{p^2 - m^2 + i\epsilon}\right] + \cdots \tag{S18.3}$$

(the first minus sign comes from the rule for fermion loops). We can ease the evaluation of the integral by the substitution

$$q = p + \tfrac{1}{2}k$$

This substitution will make the denominator even in q, so that we will be able to discard terms odd in q in the numerator (at least when part of a convergent combination, which we assume):

$$-i\widetilde{\Pi}_{\mu\nu}(k^2) = -e^2 \int \frac{d^4q}{(2\pi)^4} \mathrm{Tr}\left[\gamma_\mu \frac{(\not{q} + \frac{1}{2}\not{k} + m)}{(q + \frac{1}{2}k)^2 - m^2 + i\epsilon} \gamma_\nu \frac{(\not{q} - \frac{1}{2}\not{k} + m)}{(q - \frac{1}{2}k)^2 - m^2 + i\epsilon}\right] + \cdots \tag{S18.4}$$

The product of the numerators gives nine terms, but four (the terms linear in m) contain an odd number of gamma matrices, and so have zero trace. Of the remaining five, two are linear in q and so are odd functions, and will vanish upon integration (in a convergent combination). That leaves in the numerator

$$\mathrm{Tr}\,[\gamma^\mu \not{q} \gamma^\nu \not{q}] - \tfrac{1}{4}\mathrm{Tr}\,[\gamma^\mu \not{k} \gamma^\nu \not{k}] + m^2 \mathrm{Tr}\,[\gamma^\mu \gamma^\nu] = 4g_{\mu\nu}(m^2 - q^2 + \tfrac{1}{4}k^2) + 8(q_\mu q_\nu - \tfrac{1}{4}k_\mu k_\nu)$$

using the identities $\mathrm{Tr}\,[\gamma^\mu\gamma^\nu] = 4g^{\mu\nu}$, $\mathrm{Tr}\,[\gamma^\mu \not{a} \gamma^\nu \not{a}] = 4(2a_\mu a_\nu - g_{\mu\nu}a^2)$. Then

$$-i\widetilde{\Pi}_{\mu\nu}(k^2) = -e^2 \int \frac{d^4q}{(2\pi)^4} \frac{4g_{\mu\nu}(m^2 - q^2 + \frac{1}{4}k^2) + 8(q_\mu q_\nu - \frac{1}{4}k_\mu k_\nu)}{\left[(q + \frac{1}{2}k)^2 - m^2 + i\epsilon\right]\left[(q - \frac{1}{2}k)^2 - m^2 + i\epsilon\right]} + \cdots \tag{S18.5}$$

Let's combine the denominators with the Feynman parametrization:

$$\frac{1}{ab} = \int_0^1 dx \, \frac{1}{[ax + b(1-x)]^2}$$

with

$$a = (q + \tfrac{1}{2}k)^2 - m^2 + i\epsilon \qquad b = (q - \tfrac{1}{2}k)^2 - m^2 + i\epsilon$$

Then

$$ax + b(1-x) = q^2 + 2(x - \tfrac{1}{2})q \cdot k + \tfrac{1}{4}k^2 - m^2 + i\epsilon$$

and

$$-i\widetilde{\Pi}_{\mu\nu}(k^2) = -e^2 \int_0^1 dx \int \frac{d^4q}{(2\pi)^4} \frac{4g_{\mu\nu}(m^2 - q^2 + \frac{1}{4}k^2) + 8(q_\mu q_\nu - \frac{1}{4}k_\mu k_\nu)}{(q^2 + 2(x - \frac{1}{2})q \cdot k + \frac{1}{4}k^2 - m^2 + i\epsilon)^2} + \cdots \tag{S18.6}$$

To get rid of the cross-terms, let

$$q' = q + (x - \tfrac{1}{2})k$$

Then, dropping the prime on the q's, as well as terms linear in q (which will integrate to zero),

$$-i\widetilde{\Pi}_{\mu\nu}(k^2) = -4e^2 \int_0^1 dx \int \frac{d^4q}{(2\pi)^4} \frac{g_{\mu\nu}(m^2 - q^2 + x(1-x)k^2) + 2(q_\mu q_\nu - x(1-x)k_\mu k_\nu)}{(q^2 + x(1-x)k^2 - m^2 + i\epsilon)^2} + \cdots \tag{S18.7}$$

As in (34.62), the quantity $q_\mu q_\nu$ can be replaced by $\frac{1}{4}q^2 g_{\mu\nu}$. We now have

$$-i\widetilde{\Pi}_{\mu\nu}(k^2) = -4e^2 \int_0^1 dx \int \frac{d^4q}{(2\pi)^4} f_{\mu\nu}(q, x, k, m^2) + \cdots \tag{S18.8}$$

where

$$f_{\mu\nu}(q, x, k, m^2) = \frac{g_{\mu\nu}(m^2 - \frac{1}{2}q^2) + x(1-x)(g_{\mu\nu}k^2 - 2k_\mu k_\nu)}{(q^2 + x(1-x)k^2 - m^2 + i\epsilon)^2} \tag{S18.9}$$

To investigate the divergences of this expression, consider its large q behavior. It's helpful to rewrite it as

$$\begin{aligned} f_{\mu\nu}(q, x, k, m^2) &= \frac{-\frac{1}{2}g_{\mu\nu}(q^2 - 2m^2 + 2x(1-x)k^2) + 2x(1-x)(g_{\mu\nu}k^2 - k_\mu k_\nu)}{q^4 - 2q^2m^2 + 2q^2x(1-x)k^2 + \cdots} \\ &= -\frac{1}{2}\frac{g_{\mu\nu}}{q^2} + \frac{2x(1-x)(g_{\mu\nu}k^2 - k_\mu k_\nu)}{q^4} + \mathcal{O}(q^{-6}) \end{aligned} \tag{S18.10}$$

Integrating $f_{\mu\nu}$ over d^4q, the first term is quadratically divergent, and the second is logarithmically divergent. From the chart on p. 527, it follows that we need three heavy masses. Adding the contribution from the regulator fields, we get

$$-i\widetilde{\Pi}_{\mu\nu}(k^2) = -e^2 \int_0^1 dx \int \frac{d^4q}{(2\pi)^4} \left[f_{\mu\nu}(q, x, k, m^2) + \sum_{i=1}^{3} b_i f_{\mu\nu}(q, x, k, M_i^2) \right] \tag{S18.11}$$

choosing the coefficients b_i and the masses M_i in accord with (33.53),

$$1 + b_1 + b_2 + b_3 = 0$$

$$m^2 + b_1 M_1^2 + b_2 M_2^2 + b_3 M_3^2 = 0$$

In the following we choose $b_1 = -b_2 = b_3 = -1$. With these choices, the integrand in (S18.11) goes as $\mathcal{O}(q^{-6})$ as $q \to \infty$, and so the integral is convergent. Applying the integral formulae (I.4) and (I.3) from the box on p. 330,

$$\int \frac{d^4q}{(2\pi)^4} \frac{1}{(q^2 + a)^2} = -\frac{i}{16\pi^2} \ln(-a) + \cdots \tag{I.4}$$

$$\int \frac{d^4q}{(2\pi)^4} \frac{1}{q^2 + a} = \frac{i}{16\pi^2} a \ln(-a) + \cdots \tag{I.3}$$

(the dots indicating divergent terms that cancel when two such terms are subtracted, provided the total integrand vanishes for high q faster than q^{-4}), we find

$$\int \frac{d^4q}{(2\pi)^4} \frac{q^2}{(q^2 + a)^2} = \int \frac{d^4q}{(2\pi)^4} \frac{q^2 + a}{(q^2 + a)^2} - \int \frac{d^4q}{(2\pi)^4} \frac{a}{(q^2 + a)^2} = \frac{i}{8\pi^2} a \ln(-a) + \cdots \tag{S18.12}$$

so that, with $a = x(1-x)k^2 - m^2 + i\epsilon$,

$$\begin{aligned} \int \frac{d^4q}{(2\pi)^4} f_{\mu\nu} &= -\tfrac{1}{2}g_{\mu\nu} \int \frac{d^4q}{(2\pi)^4} \frac{q^2}{(q^2 + a)^2} + [g_{\mu\nu}(m^2 + x(1-x)k^2) - 2x(1-x)k_\mu k_\nu] \int \frac{d^4q}{(2\pi)^4} \frac{1}{(q^2 + a)^2} \\ &= -\tfrac{1}{2}g_{\mu\nu} \frac{i}{8\pi^2} a \ln(-a) + [g_{\mu\nu}(m^2 + x(1-x)k^2) - 2x(1-x)k_\mu k_\nu] \frac{-i}{16\pi^2} \ln(-a) + \cdots \\ &= -\frac{i}{8\pi^2} x(1-x)(g_{\mu\nu}k^2 - k_\mu k_\nu) \ln(-k^2 x(1-x) + m^2 - i\epsilon) + \cdots \end{aligned} \tag{S18.13}$$

There are two notable features of this result. The quantity is transverse, even *before* we include the regulator fields, in agreement with (33.70). Also, this result has *no* mass dependence except in the argument of the logarithm; the mass in the coefficient cancels. Including the regulator terms, we have

$$-i\widetilde{\Pi}_{\mu\nu} = \frac{ie^2}{2\pi^2}(g_{\mu\nu}k^2 - k_\mu k_\nu)\int_0^1 dx\, x(1-x)\ln\left[\frac{(-k^2x(1-x)+m^2-i\epsilon)(-k^2x(1-x)+M_2^2-i\epsilon)}{(-k^2x(1-x)+M_1^2-i\epsilon)(-k^2x(1-x)+M_3^2-i\epsilon)}\right] \quad \text{(S18.14)}$$

The counterterm is determined by the BPHZ prescription. The superficial degree of divergence[1] of the original graph (S18.1) is $D = 2$, indicating that it is quadratically divergent. We must, à la BPHZ, add counterterms to $\mathscr{L}$ to cancel the terms in the Taylor expansion of $\widetilde{\Pi}_{\mu\nu}$ up to order D in k. Because the integral (S18.14) is well-defined for $k = 0$ and is multiplied by an explicitly $\mathcal{O}(k^2)$ factor, there are no zeroth or first-order terms in k. Thus the counterterm diagram makes a contribution[2]

$$-iE(g_{\mu\nu}k^2 - k_\mu k_\nu)$$

which is to be added to (S18.14). Our renormalization conditions require (33.72) that the renormalized self-energy $\widetilde{\Pi}'_{\mu\nu}$ satisfies

$$-i\widetilde{\Pi}'_{\mu\nu} = i(g_{\mu\nu}k^2 - k_\mu k_\nu)\,\widetilde{\Pi}'^T(k^2) \quad \text{(S18.15)}$$

with

$$\widetilde{\Pi}'^T(k^2)\Big|_{k^2=0} = 0 \quad \text{(S18.16)}$$

The constraint (S18.16) follows automatically from gauge invariance of the 1PI diagrams. We choose E as

$$E = \frac{e^2}{2\pi^2}\int_0^1 dx\, x(1-x)\ln\left[\frac{m^2M_2^2}{M_1^2M_3^2}\right]$$

The renormalized self-energy then becomes

$$-i\widetilde{\Pi}'_{\mu\nu} = \frac{ie^2}{2\pi^2}(g_{\mu\nu}k^2 - k_\mu k_\nu)\int_0^1 dx\, x(1-x)\ln\left[\frac{-k^2x(1-x)+m^2-i\epsilon)(-k^2x(1-x)+M_2^2-i\epsilon)M_1^2M_3^2}{(-k^2x(1-x)+M_1^2-i\epsilon)(-k^2x(1-x)+M_3^2-i\epsilon)m^2M_2^2}\right]$$

Taking the limits $M_1, M_2, M_3 \to \infty$, we find

$$-i\widetilde{\Pi}'_{\mu\nu} = \frac{ie^2}{2\pi^2}(g_{\mu\nu}k^2 - k_\mu k_\nu)\int_0^1 dx\, x(1-x)\ln\left[\frac{-k^2x(1-x)+m^2-i\epsilon}{m^2}\right]$$

which is both finite and transverse, as expected. Note that the counterterm is added into the Lagrangian via the term

$$E(F'_{\mu\nu})^2$$

and so preserves gauge invariance. There is *no* need of a gauge-breaking photon mass term to renormalize the self-energy, which remains gauge invariant:

$$k^\mu\widetilde{\Pi}'_{\mu\nu} = 0$$

Additionally, we see that to this order,

$$\widetilde{\Pi}'^T(k^2) = \frac{e^2}{2\pi^2}\int_0^1 dx\, x(1-x)\ln\left[\frac{-k^2x(1-x)+m^2-i\epsilon}{m^2}\right]$$

so that

$$\lim_{k^2\to 0}\widetilde{\Pi}'^T(k^2) = \frac{e^2}{2\pi^2}\int_0^1 dx\, x(1-x)\ln\left[1 - i\frac{\epsilon}{m^2}\right] = 0$$

as required. ■

18.2 The $\mathcal{O}(e^2)$ diagrams contributing to the photon self-energy in the scalar case are shown below:

$-i\widetilde{\Pi}_{\mu\nu}(k^2)$ = 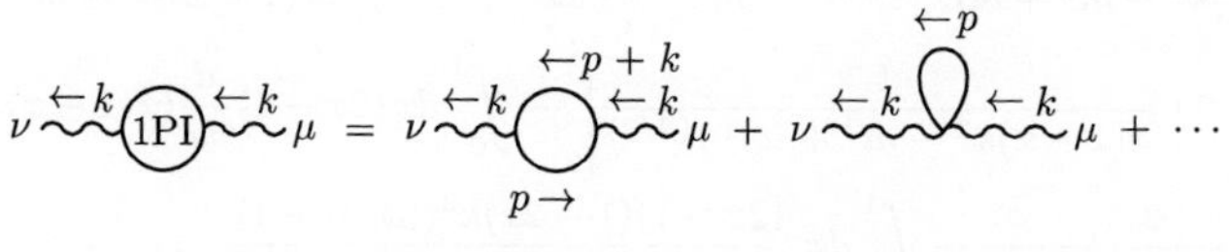 (S18.17)

---

[1] [Eds.] See §25.4.

[2] [Eds.] See the table on p. 674; E is the photon wave function renormalization constant.

and to the same order,

$$-i\widetilde{\Pi}'_{\mu\nu}(k^2) = $$ 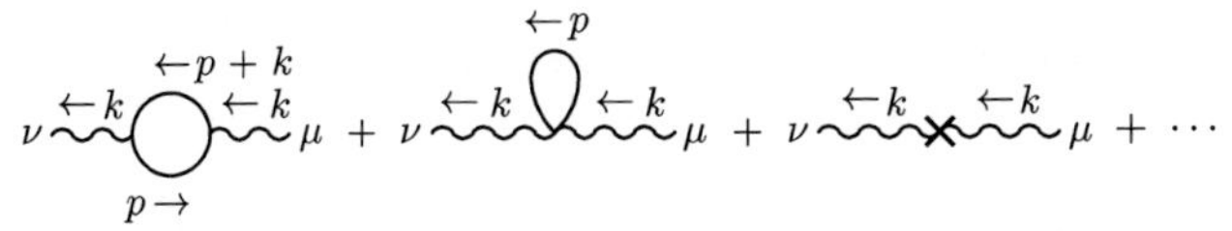 (S18.18)

Using the Feynman rules in the box on p. 644 (extended to d spacetime dimensions),

$$-i\widetilde{\Pi}_{\mu\nu}(k^2) = (-ie)^2 \int \frac{d^dp}{(2\pi)^d} \frac{i^2(p+(p+k))_\mu(p+(p+k))_\nu}{(p^2-m^2+i\epsilon)((p+k)^2-m^2+i\epsilon)} + 2ie^2 g_{\mu\nu} \int \frac{d^dp}{(2\pi)^d} \frac{i}{p^2-m^2+i\epsilon} \tag{S18.19}$$

(Note that we are *not* writing $e\nu^{4-d}$ for the coupling constant, as in (25.23). This is not necessary since we are using mass-shell renormalization conditions, and all dependence on ν drops out.)

As before, we combine the denominators in the first term with Feynman parametrization:

$$-i\widetilde{\Pi}_{\mu\nu}(k^2) = (-ie)^2 \int_0^1 dx \int \frac{d^dp}{(2\pi)^d} \frac{i^2(2p+k)_\mu(2p+k)_\nu}{[p^2+2xp\cdot k+xk^2-m^2+i\epsilon]^2} + 2ie^2 g_{\mu\nu} \int \frac{d^dp}{(2\pi)^d} \frac{i}{p^2-m^2+i\epsilon} \tag{S18.20}$$

We get rid of the cross-term by shifting the momentum variable: let $p = p' - xk$. Then

$$-i\widetilde{\Pi}_{\mu\nu}(k^2) = (-ie)^2 \int_0^1 dx \int \frac{d^dp'}{(2\pi)^d} \frac{i^2(2p'+k(1-2x))_\mu(2p'+k(1-2x))_\nu}{[p'^2+x(1-x)k^2-m^2+i\epsilon]^2} + 2ie^2 g_{\mu\nu} \int \frac{d^dp}{(2\pi)^d} \frac{i}{p^2-m^2+i\epsilon} \tag{S18.21}$$

Linear terms in p' vanish upon integration, leaving

$$-i\widetilde{\Pi}_{\mu\nu}(k^2) = e^2 \int_0^1 dx \int \frac{d^dp'}{(2\pi)^d} \frac{4p'_\mu p'_\nu + (1-2x)^2 k_\mu k_\nu}{[p'^2+x(1-x)k^2-m^2+i\epsilon]^2} - 2e^2 g_{\mu\nu} \int \frac{d^dp}{(2\pi)^d} \frac{1}{p^2-m^2+i\epsilon} \tag{S18.22}$$

As in the previous problem, we can replace $p'_\mu p'_\nu$ by a constant times $g_{\mu\nu}p'^2$; the constant is $1/d$ instead of $\frac{1}{4}$. Then

$$-i\widetilde{\Pi}_{\mu\nu}(k^2) = e^2 \int_0^1 dx \int \frac{d^dp'}{(2\pi)^d} \frac{(4/d)g_{\mu\nu}p'^2 + (1-2x)^2 k_\mu k_\nu}{[p'^2+x(1-x)k^2-m^2+i\epsilon]^2} - 2e^2 g_{\mu\nu} \int \frac{d^dp}{(2\pi)^d} \frac{1}{p^2-m^2+i\epsilon} \tag{S18.23}$$

From the Euclidean integral (33.19), we have the Minkowski space integral formula

$$\int \frac{d^dp'}{[p'^2-a+i\epsilon]^\alpha} = \frac{i\pi^{(d/2)}\Gamma(\alpha-(d/2))}{\Gamma(\alpha)(-a+i\epsilon)^{\alpha-(d/2)}} \tag{S18.24}$$

and similarly from the Euclidean integral (33.26), we get the Minkowski version

$$\int d^dp' \frac{p'^2}{[p'^2-a+i\epsilon]^\alpha} = \frac{i\pi^{(d/2)}}{(-a+i\epsilon)^{\alpha-1-(d/2)}} \frac{(d/2)\Gamma(\alpha-1-(d/2))}{\Gamma(\alpha)} \tag{S18.25}$$

Using these formulae, we get

$$-i\widetilde{\Pi}_{\mu\nu} = \frac{i\pi^{(d/2)}}{(2\pi)^d} e^2 \left[\int_0^1 dx \left\{ \frac{2g_{\mu\nu}\Gamma(1-(d/2))}{(x(1-x)k^2-m^2+i\epsilon)^{1-(d/2)}} + \frac{k_\mu k_\nu (1-2x)^2 \Gamma(2-(d/2))}{(x(1-x)k^2-m^2+i\epsilon)^{2-(d/2)}} \right\} - \frac{2g_{\mu\nu}\Gamma(1-(d/2))}{(-m^2+i\epsilon)^{1-(d/2)}} \right] \tag{S18.26}$$

The first term can be transformed with an integration by parts:

$$\begin{aligned}
\int_0^1 dx \frac{2}{(x(1-x)k^2-m^2+i\epsilon)^{1-(d/2)}} &= \int_0^1 dx \left(\frac{d}{dx}(2x-1)\right) \frac{1}{(x(1-x)k^2-m^2+i\epsilon)^{1-(d/2)}} \\
&= (2x-1)\frac{1}{(x(1-x)k^2-m^2+i\epsilon)^{1-(d/2)}}\Bigg|_0^1 - \int_0^1 dx\,(2x-1)\frac{d}{dx}(x(1-x)k^2-m^2+i\epsilon)^{((d/2)-1)} \\
&= \frac{2}{(-m^2+i\epsilon)^{1-(d/2)}} - \int_0^1 dx \frac{(2x-1)(1-2x)k^2((d/2)-1)}{(x(1-x)k^2-m^2+i\epsilon)^{2-(d/2)}} \\
&= \frac{2}{(-m^2+i\epsilon)^{1-(d/2)}} - \int_0^1 dx \frac{(1-2x)^2k^2(1-(d/2))}{(x(1-x)k^2-m^2+i\epsilon)^{2-(d/2)}}
\end{aligned}$$

Substituting this last expression into the first term of (S18.26), we obtain

$$-i\widetilde{\Pi}_{\mu\nu}(k) = -\frac{i\pi^{(d/2)}e^2}{(2\pi)^d}\left(g_{\mu\nu}k^2 - k_\mu k_\nu\right)\int_0^1 dx\,\frac{(1-2x)^2\Gamma(2-(d/2))}{\left(k^2x(1-x)-m^2+i\epsilon\right)^{2-(d/2)}} \tag{S18.27}$$

using $\Gamma(n+1) = n\Gamma(n)$. The terms independent of k cancel, and the rest once again combine to make a transverse expression.

As in the previous problem, the counterterm diagram makes a contribution

$$-iE(g_{\mu\nu}k^2 - k_\mu k_\nu)$$

Writing

$$-i\widetilde{\Pi}'_{\mu\nu}(k) = -i\left(g_{\mu\nu}k^2 - k_\mu k_\nu\right)\left[\widetilde{\Pi}'^T(k^2) + E\right] \tag{S18.28}$$

the renormalization condition

$$\widetilde{\Pi}'^T(k^2)\Big|_{k^2=0} = 0$$

fixes E:

$$\lim_{k^2\to 0}\left\{\frac{\pi^{(d/2)}e^2}{(2\pi)^d}\int_0^1 dx\,\frac{(1-2x)^2\Gamma(2-(d/2))}{\left(k^2x(1-x)-m^2+i\epsilon\right)^{2-(d/2)}} + E\right\} = 0$$

so that

$$E = -\frac{\pi^{(d/2)}e^2}{(2\pi)^d}\int_0^1 dx\,\frac{(1-2x)^2\Gamma(2-(d/2))}{\left(-m^2+i\epsilon\right)^{2-(d/2)}}$$

and

$$-i\widetilde{\Pi}'_{\mu\nu}(k) = -\frac{i\pi^{(d/2)}e^2}{(2\pi)^d}\left(g_{\mu\nu}k^2 - k_\mu k_\nu\right)\int_0^1 dx\left[\frac{(1-2x)^2\Gamma(2-(d/2))}{\left(k^2x(1-x)-m^2+i\epsilon\right)^{2-(d/2)}} - \frac{(1-2x)^2\Gamma(2-(d/2))}{\left(-m^2+i\epsilon\right)^{2-(d/2)}}\right] \tag{S18.29}$$

We now set $d = 4-\delta$, and use the expansions

$$a^\delta = (e^{\ln a})^\delta = 1 + \delta\ln a + \mathcal{O}(\delta^2)$$
$$\Gamma(\delta) = \frac{1}{\delta} - \gamma + \mathcal{O}(\delta) \tag{25.26}$$

(the value of the Euler–Mascheroni constant, γ, doesn't matter here, because it is multiplied by δ). To $O(\delta^0)$,

$$-i\widetilde{\Pi}'_{\mu\nu}(k) = \frac{ie^2}{16\pi^2}\left(g_{\mu\nu}k^2 - k_\mu k_\nu\right)\int_0^1 dx\,(1-2x)^2\ln\left[\frac{-k^2x(1-x)+m^2-i\epsilon}{m^2}\right] \tag{S18.30}$$

which again is both finite and transverse. Again, we see that $\widetilde{\Pi}'^T(k^2)\to 0$ as $k^2\to 0$, as required. ■

18.3 The effective Lagrangian is the result of averaging over gauge-fixing constraints of the form

$$F(A) = \partial_\mu A^\mu + \sigma A_\mu A^\mu - f = 0$$

The variation of the vector field under an infinitesimal gauge transformation

$$A_\mu \to A'_\mu = A_\mu + (1/e)\partial_\mu\delta\chi \tag{S18.31}$$

produces a variation in $F(A)$:

$$\delta F = \partial_\mu\delta A^\mu + 2\sigma A_\mu\delta A^\mu = (1/e)\left[\partial_\mu\partial^\mu(\delta\chi) + 2\sigma A_\mu\partial^\mu(\delta\chi)\right] \tag{S18.32}$$

We then have

$$\frac{\delta F(x)}{\delta\chi(x')} = (1/e)\left(\partial_\mu\partial^\mu + 2\sigma A_\mu\partial^\mu\right)\delta^{(4)}(x-x'). \tag{S18.33}$$

We turn the determinant, $\det[\delta F/\delta\chi]$, into a functional integral over complex ghost fields:

$$\det\left[\frac{\delta F}{\delta\chi}\right] = \int[d\eta]\,[d\overline{\eta}]\,e^{iS_{\text{ghost}}} \tag{S18.34}$$

(up to a phase that can be absorbed into the functional integral normalization), where

$$\mathcal{S}_{\text{ghost}} = \int d^4x\,d^4y\,\overline{\eta}(x)\left[\Box_x^2 + 2\sigma A^\mu\partial_\mu^x\right]\delta^{(4)}(x-y)\eta(y) = \int d^4x\,\overline{\eta}(x)\left[\Box^2 + 2\sigma A^\mu\partial_\mu\right]\eta(x) \tag{S18.35}$$

(we can absorb the coupling constant $1/e$ into the normalization of the ghost fields; *cf.* Peskin & Schroeder *QFT*, p. 514). Then

$$\det\left[\frac{\delta F}{\delta\chi}\right] \propto \int [d\eta][d\overline{\eta}] \exp\left[i\int d^4x\overline{\eta}\left(\Box^2 + 2\sigma A_\mu\partial^\mu\right)\eta\right]$$

The ghost Lagrangian is

$$\mathscr{L}_{\text{ghost}} = \overline{\eta}\left(\Box^2 + 2\sigma A_\mu\partial^\mu\right)\eta \tag{S18.36}$$

From this, we can read off the ghost propagator,

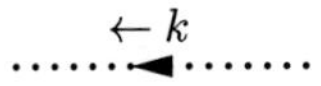

$$\frac{-i}{k^2 + i\epsilon} \tag{S18.37}$$

and the ghost-photon vertex,

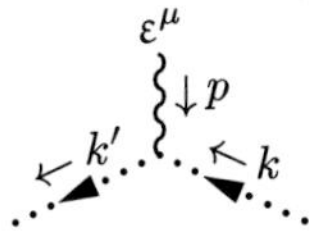

$$i(2\sigma)(ik^\mu) = -2\sigma k^\mu \tag{S18.38}$$

The first i comes from Dyson's formula; the second from the derivative on η, a field that annihilates the incoming ghost. The ghost field couples to the photon with polarization $\varepsilon^\mu(k - k')$. The asymmetric appearance of the k's is odd, but correct.[3] ■

[3] [Eds.] Cheng & Li *GT*, pp. 262–263: "One should preserve a consistent convention of entering the momentum of either the left or the right ghost line at every vertex. The ghost only enters in closed loops." Peskin & Schroeder *QFT*, p. 515 write only the outward k on the ghost lines.

34

Two famous results in QED

We now begin to discuss a sequence of problems involving interactions of quantum electrodynamics with an external conserved, c-number current distribution J^μ:

$$\mathscr{L} \to \mathscr{L} - eJ^\mu(x)A'_\mu(x), \quad \partial_\mu J^\mu(x) = 0 \tag{34.1}$$

We will restrict ourselves to the case of the real photon, with zero mass: $\mu^2 = 0$. Problems of a quantum electrodynamic system subject to an external charge distribution are quite common. They are not realistic: there are no external classical charges in the world, as far as we know, controlled by God and not by the motion of particles. But in a typical problem in which we have an electron whirling around a synchrotron, it's quite reasonable to take the distribution of currents inside the synchrotron magnets as external and given, and not worry about solving for the motions of all those electrons. We will obtain two famous results.

34.1 Coulomb's Law

The first thing we want to check is the vacuum-to-vacuum amplitude to second order in J. That's an experiment where we take two external charge or current distributions, and see what the interaction energy is between them. We want to confirm that we've calibrated everything correctly and that e really is the e measured by Monsieur Coulomb in his famous experiment.[1] The vacuum-to-vacuum amplitude to order J^2 (assuming the charges are weak, so that's the leading order) is very simple:

$$\langle 0|S|0\rangle = e^{iW} = \exp\Big($$ 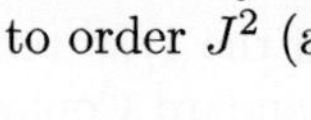 $$\Big) = 1 + \qquad + \mathcal{O}(J^2) \tag{34.2}$$

The current makes a photon, the photon goes into electron-positron pairs (or whatever), and then comes back again, reassembling into the renormalized propagator $\widetilde{D}'_{\mu\nu}$. The black dot is the interaction with the external current. Both momenta are directed inward, and

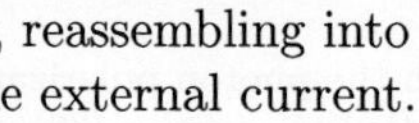

$$= (2\pi)^4\delta^{(4)}(k+k')\widetilde{D}'_{\mu\nu}(k) \tag{34.3}$$

[1] C. A. Coulomb, "Premier Mémoir sur l'Electricité & le Magnétism" (First memoir on electricity and magnetism), *Histoire de l'Académie Royale des Sciences* (1785) 569–578; the second and third memoirs follow immediately at pp. 578–612 and pp. 612–640. These are freely available online at `gallica.bnf.fr`.

The photon is massless by our renormalization conditions; the BPHZ prescription (at zero mass) and the mass shell renormalization agree. The exact photon propagator can be written as

$$\widetilde{D}'_{\mu\nu}(k) = \underbrace{-\,i\frac{[g_{\mu\nu}-(k_\mu k_\nu/k^2)]}{k^2}}_{\text{pole term with residue 1}} + \underbrace{\int_0^\infty da^2\left[\sigma(a^2)\frac{(-i)[g_{\mu\nu}-(k_\mu k_\nu/k^2)]}{k^2-a^2}\right]}_{\substack{\text{continuous superposition}\\ \text{of Yukawa potentials}}} \underbrace{-\,i\frac{\xi}{k^2}\frac{k_\mu k_\nu}{k^2}}_{\substack{\text{gauge-fixing}\\ \text{term}}} \tag{34.4}$$

We will systematically suppress the $i\epsilon$ from the one-photon intermediate states and our normalization is such that the residue at the photon pole is 1. The first term in (34.4) is the contribution from the one-photon state. The continuous contribution in the second term, *à la* the Lehmann spectral representation,[2] contains contributions from multi-particle intermediate states with $(\text{mass})^2 = a^2$. Then there is the final gauge-dependent term (§31.3), which we know suffers no **radiative corrections.**[3] This is the same derivation as for the scalar case (§15.3) with just a couple of extra indices floating around.

The Fourier transform $\widetilde{J}_\mu(k)$ of the external current distribution is defined by

$$J_\mu(x) = \int \frac{d^4k}{(2\pi)^4} e^{-ik\cdot x}\widetilde{J}_\mu(k) \tag{34.5}$$

In lowest order, the amplitude for emitting a photon of momentum k is $\widetilde{J}_\mu(k)$; that for absorbing a photon is $\widetilde{J}_\mu(-k)$. Current conservation implies

$$k_\mu \widetilde{J}^\mu(k) = 0 \tag{34.6}$$

The graph in question is easy to compute:

$$\begin{aligned} &= \frac{(-ie)^2}{2}\int \frac{d^4k}{(2\pi)^4}\widetilde{J}^\mu(k)\widetilde{D}'_{\mu\nu}(k)\widetilde{J}^\nu(-k) \\ &= \frac{ie^2}{2}\int \frac{d^4k}{(2\pi)^4}\widetilde{J}_\mu(k)\widetilde{J}^\mu(-k)\left[\frac{1}{k^2}+\int_0^\infty da^2\sigma(a^2)\frac{1}{k^2-a^2}\right] \end{aligned} \tag{34.7}$$

The $k^\mu \widetilde{J}_\mu$ terms drop out by current conservation, (34.6).

The significant feature is the coefficient of the $1/k^2$ term. We know from the study of one photon exchange[4] that this term gives the standard Coulomb force. If the $\widetilde{J}^\mu(k)$'s correspond to static charge distributions, the coefficient is 1, just as it would be in the free theory. If we take two charges that are far apart, the Yukawa potentials (coming from the integral over a^2 in (34.4)) fall off with distance, and the surviving force is the Coulomb force. This reproduces

---

[2] [Eds.] This is the spectral representation for scalars with the photon polarization sum in the numerator; see (15.30); Schweber *RQFT*, Section 17b, pp. 659–677, in particular equation (66); and Bjorken & Drell *Fields*, Section 16.11, pp. 166–170, and the (unnumbered) equation following equation (16.173).

[3] [Eds.] Given a process in lowest order of e^2, "radiative corrections" are higher order contributions to that process, typically described by diagrams with loops or the emission of extra photons in the final state (*bremsstrahlung*, or "braking radiation"). See Peskin & Schroeder *QFT*, Ch. 6, p. 175. The Ward Identity guarantees that any radiative corrections arising from the gauge-fixing term in the Lagrangian will vanish when contracted with k^μ; see (31.63). (M. Headrick, private communication).

[4] [Eds.] See the discussion following (30.22), pp. 648–650, and note 11, p. 650.

(albeit with blackboard and chalk) M. Coulomb's experiment, that the force between two charges widely separated (on the atomic scale) is e^2/r^2. And it *is* e^2; this is the proper way to couple in an external current distribution if e is to be what M. Coulomb measured.

What about the Lehmann spectral distribution? In §10.4 we studied two spinless particles exchanging a spinless particle. Then, as shown earlier[5] we found that the associated propagator

$$\frac{1}{k^2 - \mu^2}$$

corresponds to a Yukawa force. It gives a potential

$$\frac{e^{-\mu r}}{r}$$

Likewise, $1/k^2$, the Yukawa potential with $\mu = 0$, corresponds to a Coulomb potential, the first term in (34.7). Incidentally, the sign here is different from the case of scalar exchange; in (9.29) we have

$$\widetilde{\Delta}_F(p) = \frac{i}{p^2 - \mu^2 + i\epsilon} \tag{34.8}$$

whereas here we have

$$\widetilde{D}'_{\mu\nu} = -\frac{ig_{\mu\nu}}{k^2} + \cdots \tag{34.9}$$

And that's right. Scalar exchange is *attractive* between identical sources, but we know that the Coulomb force is *repulsive* between identical sources. The forces are different because the residues of the poles in the propagator have different signs. Take two large, external macroscopic charge distributions, idealized as classical. Put them at different locations and measure the force between them to order J^2, in the limit of weak charges (so we don't have to worry about nonlinear effects). For a time-independent external source, the sum of the vacuum graphs is related to the energy shift in the ground state of the theory,[6] so we can determine how the ground state energy is changed by the force between these external charge distributions. It's exactly how we determine an internuclear force in molecular theory, by measuring the ground state energy of the electron system with fixed nuclear positions. Here the analog of the electron system is the entire quantum electrodynamical vacuum state.

In the spectral distribution, physically a^2 is the squared mass of the intermediate state. Contributions from states with $a^2 > 0$ drop off faster than the Coulomb force, due to their Yukawa form. I should mention that the lowest value of a^2 is not the squared mass of the lightest *charged* particle intermediate state (that would be $(2m)^2$), because we could have an intermediate state of $\gamma \to 3\gamma$. (By charge conjugation, we can't have a two-photon state; the photon is odd under charge conjugation; $A_\mu |0\rangle$ produces an odd charge conjugation state which cannot be two photons.) This process only occurs through charged particles, as in Figure 34.1. But that doesn't matter. The fact that the transition matrix element from the vacuum to that state of A_μ involves graphs with internal electron lines is not the point. The lowest value of a^2 is the mass of the three-photon state, which is 0. I don't know how that three-photon state contributes, if it gives $1/r^5$ or $1/r^{10}$ or something else, but it goes to zero more quickly than $1/r^2$.

---

[5] [Eds.] See (9.38) through (9.41).

[6] [Eds.] See Section 3.7 in Chap. 5, "Secret Symmetry" in Coleman *Aspects*; Peskin & Schroeder *QFT*, pp. 96–98.

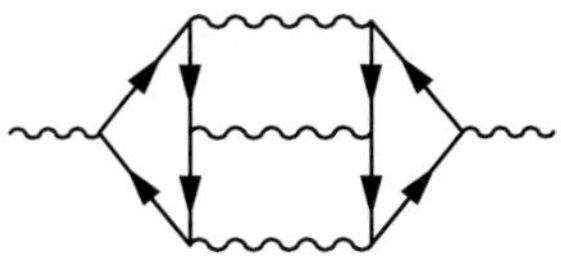

Figure 34.1: Intermediate state with three photons

34.2 The electron's anomalous magnetic moment in quantum mechanics

We've calibrated the one-photon exchange so that the resulting Coulomb potential's e^2 has the right size. Now let's turn to something more complicated, and consider a current distribution scattering an electron. This will lead to a theoretical determination of the anomalous magnetic moment of the electron. This famous calculation was carried out independently by Feynman and Schwinger,[7] and it made everyone believe in quantum field theory.[8] Feynman and Schwinger did this to one loop. We'll look at this first in the context of quantum mechanics. Analogous to the one-loop calculation in QED, we'll consider a first-order quantum mechanical

[7] [Eds.] The result was first discussed at a small Washington D.C. conference in November 1947, (Schweber *QED*, p. 317) and submitted as a letter to *Physical Review* a month later, though the explicit calculation was not carried out in the letter: Julian Schwinger, "On Quantum-Electrodynamics and the Magnetic Moment of the Electron", *Phys. Rev.* **73** (1948) 416–7. Incidentally, the result is printed erroneously, as $(\frac{1}{2}\pi)e^2/\hbar c$; the correct value is $(1/2\pi)(e^2/\hbar c) = \alpha/(2\pi)$. For Schwinger's calculation, see equation (1.122) and footnote 3 in Julian Schwinger, "Quantum Electrodynamics III. The Electromagnetic Properties of the Electron–Radiative Corrections to Scattering", *Phys. Rev.* **76** (1949) 790–817. Though Feynman does not seem to have published his calculation at the time, it follows easily from equation (24) in R. P. Feynman, "Space-Time Approach to Quantum Electrodynamics", *Phys. Rev.* **76** (1949) 769–789 (reprinted in Schwinger *QED*); see R. P. Feynman, *Quantum Electrodynamics*, W. A. Benjamin, Inc., 1962, p. 145, where Feynman derives the result from equation (28-10), identical to his article's equation (24).

[8] [Eds.] "Unlike Bethe, Weisskopf and most of the other people at the Shelter Island conference [June 2–4, 1947], Schwinger's imagination was captured not by the Lamb shift but by the discrepancy in the magnetic behavior of the electron... 'That was much more shocking,' Schwinger said. The Lamb effect, as Bethe showed, could be accounted for almost entirely without the use of relativity. 'The magnetic moment of the electron, which came from Dirac's relativistic theory, was something that *no* non-relativistic theory could describe correctly... To be told (a) that the physical answer was not what Dirac's theory gave; and (b) that there was no simpleminded way of thinking about it, that was the real challenge. That's the one I jumped on.'", Crease & Mann *SC*, p. 132. Schweber (*QED*, p. 318) writes, "The importance of Schwinger's calculation [worked through during a five hour (!) talk at the Pocono Conference, March 30, 1948] cannot be underestimated [*sic*]. In the course of theoretical developments there sometimes occur important calculations that alter the way the community thinks about particular approaches. Schwinger's calculation is one such instance. By indicating, as Feynman had noted [in a letter to his friend Herbert Corben, after the Washington conference] that the 'discrepancy in the hyperfine-structure of the hydrogen atom noted by Rabi can be explained *on the same basis* as that of the electromagnetic self-energy, as can the shift of Lamb' [emphasis added by Schweber], Schwinger had transformed the perception of quantum electrodynamics. He had made it into an effective, coherent, and consistent computational scheme to order e^2." Coleman did not discuss the Lamb shift—a tiny difference ($\sim$ 1060 MHz, with $\Delta\omega/\omega \sim 1 \times 10^{-6}$) between the $2s_{1/2}$ and $2p_{1/2}$ energy levels of hydrogen, degenerate in the Dirac theory—in the course. The measurement was carried out (1946–47) by Willis Lamb (a student of Oppenheimer) and Robert C. Retherford at Columbia University: Willis E. Lamb, Jr. and Robert C. Retherford, "Fine Structure of the Hydrogen Atom by a Microwave Method", *Phys. Rev.* **72** (1947) 241-243. Lamb shared the 1955 Nobel Prize (with his Columbia colleague, Polykarp Kusch, who'd measured the electron's magnetic moment to high precision) for this work. Bethe's *non-relativistic* derivation of this result, famously carried out while traveling by train to Schenectady from New York after the Shelter Island conference (2–4 June 1947), and based on Hendrik Kramers' idea of mass renormalization presented there, pointed the way to further progress: H. A. Bethe, "The Electromagnetic Shift of Energy Levels", *Phys. Rev.* **72** 339-341. This was the first successful application of renormalization theory to QED: Laurie M. Brown, *Renormalization: From Lorentz to Landau (and Beyond)*, Springer, 1993, p. 4. See also J. J. Sakurai, *Advanced Quantum Mechanics*, Addison-Wesley, 1967, Section 2.8, pp. 64–72 for a beautifully clear treatment of Bethe's calculation.

calculation with a weak external current distribution.

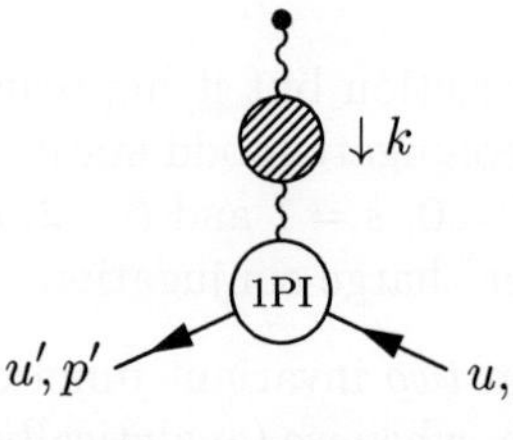

Figure 34.2: The electron-photon vertex

We have an incoming spinor u with momentum p, an outgoing spinor u' with momentum p', and a photon of momentum k, with (see Figure 34.2)

$$k = p' - p \tag{34.10}$$

To lowest order, the amplitude for this scattering process is

$$\mathcal{A} = (-ie\widetilde{J}_\mu(k)) \left[\frac{-i}{k^2 + i\epsilon}\right] \left\{-ie\overline{u}'\gamma^\mu u + \mathcal{O}(e)^3\right\} \tag{34.11}$$

which can be written as the (generic) form (suppressing the $i\epsilon$ of the photon propagator)

$$i\mathcal{A} = ie\widetilde{J}_\mu(k) \left[\frac{1}{k^2}\right] F^\mu(k) \tag{34.12}$$

$F^\mu(k)$ is a *current matrix element.* It is a function with the spinors u and $\overline{u}'$, some Dirac matrices and some functions of the momenta. To lowest order, $F^\mu(k)$ is just the factor within the curly brackets in (34.11):

$$F^\mu(k) = -ie\overline{u}'(p')\gamma^\mu u(p) + \mathcal{O}(e)^3 \tag{34.13}$$

Let's try to count the number of (Lorentz) invariant amplitudes there can be in $F^\mu(k)$, and then try to construct them.

The easiest way to count them is to go into the cross channel, reading the diagram straight down from the dot, and consider the process

$$\gamma \text{ (off-shell)} \to e^+e^- \tag{34.14}$$

We can use standard angular momentum arguments to compute how many terms there are. We have a current making an e^+e^- pair. In the center of momentum frame, $k^\mu = (k^0, \mathbf{0})$ is timelike. The current has the same properties as a $J^P = 1^-$ particle, described by a spin 1 spatial vector with parity minus.[9] So we wish to build a 1^- state out of an electron and a positron. Since e^+ and e^- are particle and antiparticle and fermions, they have opposite intrinsic parities, and therefore the 1^- state must have even ℓ.[10] There are two possibilities:

$$\ell = 0,\ s = 1$$
$$\ell = 2,\ s = 1$$

[9] [Eds.] See §6.3 and §22.1.

[10] [Eds.] See note 2, p. 460.

Those can each be put together in a unique way to make a state of total angular momentum 1 and parity minus.

Charge conjugation is also a restriction but it happens to give us no further constraint. The current of course makes charge conjugation odd states, but since the two constituents are fermion and anti-fermion, both the $\ell = 0$, $s = 1$ and $\ell = 2$, $s = 1$ states are symmetric in both space and spin, and so are odd under charge conjugation.

Therefore there are no more than *two* invariant functions of k^2 required to describe this process: the invariant functions which, when we (analytically) continue to timelike k^μ, represent the amplitudes for making the $\ell = 0$ state and the $\ell = 2$ state. With that knowledge, we will write down two functions, $F_1(k^2)$ and $F_2(k^2)$, which satisfy all the constraints of parity, charge conjugation, Lorentz invariance, etc. These two functions[11] completely characterize $F^\mu(k^2)$:

$$F^\mu(k) = e(\overline{u}'\gamma^\mu u)F_1(k^2) + \frac{ie}{2m}(\overline{u}'\sigma^{\mu\nu}u)k_\nu F_2(k^2) \tag{34.15}$$

Both the bilinear products $\overline{u}'\gamma_\mu u$ and $\overline{u}'\sigma_{\mu\nu}u$ are charge conjugation odd.[12] Both of the terms in (34.15) are vectors, and both F_1 and F_2 are real functions. They are called **form factors**. This is a subject where things bear the names of distinguished physicists, and these functions, F_1 and F_2, are known respectively as the **Dirac form factor** and the **Pauli form factor**. The factors of i and $e/2m$ in the second term are for convenience, as you'll see.

From our renormalization conventions and (34.13), we have one piece of information:[13]

$$F_1(0) = 1 \tag{34.16}$$

This is the condition on the renormalization of the electric charge: that e is the 1PI function at $k^2 = 0$. We know nothing about F_1 at any other value of k^2, except to lowest order in e, and we know nothing at all about F_2 (short of actually calculating them). This analysis is special to a spin-½ particle but not to the electron; it could be any spin-½ particle. It could be a proton, in which case there would be all sorts of strong interactions[14] inside the 1PI graph.

Aside.

This is a side remark, but it's so important physically that I am not ashamed to devote a minute to it. *Even though* the proton has these strong interactions, the fact is that we can analyze, for example, electron–proton scattering,

$$e + p \to e + p$$

at $\mathcal{O}(e^2)$, Figure 34.3, in terms of two functions like these. The blob is unknown; it sums up the effects of the strong interactions.[15] We may not know what those functions F_1 and F_2 for the proton are until we can calculate with the strong interactions, but we know that

---

[11] [Eds.] See Peskin & Schroeder *QFT* Section 6.2, pp. 185–6, or Greiner & Reinhardt *QED*, Exercise 3.5, "Rosenbluth's Formula", pp. 113–114, for the derivation of the general form in (34.15).

[12] [Eds.] See the chart on p. 469. Reminder: $\sigma_{\mu\nu} = \frac{i}{2}[\gamma_\mu, \gamma_\nu]$; (20.98), p. 419.

[13] [Eds.] S. D. Drell and F. Zachariasen, *Electromagnetic Structure of Nucleons*, Oxford U. Press, 1961, End Note 6, pp. 105–106.

[14] [Eds.] Drell and Zachariasen, *op. cit.*

[15] [Eds.] In the video of Lecture 34, Coleman adds: "Well, the effects are unknown, unless you can solve the strong interaction problem. In that case, what are you doing sitting in this class?"

there are just two of them. The electron only interacts with electromagnetism (and the weak interaction, but that's just *bubkes*[16]). Therefore we have a great simplification in studying

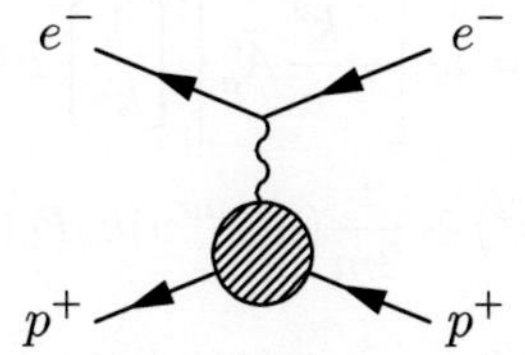

Figure 34.3: Electron–proton scattering

electron–proton scattering. In principle the scattering amplitude would be a bunch of bispinor covariant products times arbitrary functions of two variables, say the energy and the angle. We have turned it into an expression involving just two unknown functions, F_1 and F_2, of a single variable, k, and that is progress no matter how you slice it. We could also turn this argument on its head, and say that by doing electron–proton scattering we get information on F_1 and F_2 and perhaps learn something about the strong interactions, because these make F_1 and F_2 for the proton what they are. It's good both ways. We can say that we've reduced our ignorance of electron–proton scattering even in the absence of understanding the strong interactions. Or we can say that we use electrons as a probe to investigate the strong interactions of the proton in a very simple situation, instead of looking at, say, a nucleus with 42 protons interacting with each other.

I have described the utility of these form factors. Now I will discuss their physical interpretation. Here's why we've singled out k^2 in the formula to define F_μ. Let's suppose that we are going to solve Maxwell's equations for the given current distribution J_μ. In the Lorenz gauge, for example, we would have to solve the equation

$$\Box^2 A^c_\mu = eJ_\mu \tag{34.17}$$

Here A^c_μ is a classical solution. In Fourier space,

$$\widetilde{A}^c_\mu = -\frac{e}{k^2}\widetilde{J}_\mu \tag{34.18}$$

Likewise, the classical electromagnetic field $F^c_{\mu\nu}$ associated with this classical potential

$$F^c_{\mu\nu} = \partial_\mu A^c_\nu - \partial_\nu A^c_\mu \tag{34.19}$$

becomes, in Fourier space,

$$\widetilde{F}^c_{\mu\nu} = -i(k_\mu \widetilde{A}^c_\nu - k_\nu \widetilde{A}^c_\mu) = \frac{ie}{k^2}(k_\mu \widetilde{J}_\nu - k_\nu \widetilde{J}_\mu) \tag{34.20}$$

[16] [Eds.] Yiddish, *bubkes* (various pronunciations; often "BOOPkiss", to rhyme with "put this" or "BUPkiss", to rhyme with "up this" and sometimes spelled *bupkes*), "a contemptibly insignificant quantity". See Rosten *Joys*, p. 44.

This enables us to give a new meaning to our interaction amplitude. Substituting (34.15) and (34.18) into (34.12), and using (34.20) we find

$$\begin{aligned} i\mathcal{A} &= ie\widetilde{J}_\mu(k)\left[\frac{1}{k^2}\right]F^\mu(k) = ie\left[-\frac{k^2}{e}\widetilde{A}^c_\mu\right]\left[\frac{1}{k^2}\right]F^\mu(k) = -i\widetilde{A}^c_\mu F^\mu(k) \\ &= -ie\widetilde{A}^c_\mu\left[(\overline{u}'\gamma^\mu u)F_1(k^2) + \frac{i}{2m}(\overline{u}'\sigma^{\mu\nu}u)k_\nu F_2(k^2)\right] \\ &= -ie\left[(\overline{u}'\gamma^\mu u)\widetilde{A}^c_\mu F_1(k^2) + \frac{i}{4m}(\overline{u}'\sigma^{\mu\nu}u)(k_\nu\widetilde{A}^c_\mu - k_\mu\widetilde{A}^c_\nu)F_2(k^2)\right] \\ &= -ie\left[(\overline{u}'\gamma^\mu u)\widetilde{A}^c_\mu F_1(k^2) + \frac{1}{4m}(\overline{u}'\sigma^{\mu\nu}u)\widetilde{F}^c_{\mu\nu}F_2(k^2)\right] \end{aligned} \tag{34.21}$$

In the third step we can write an antisymmetric product because of the summation with the antisymmetric matrix $\sigma^{\mu\nu}$, and divide by 2 to avoid the double counting.

The first term, with the Dirac form factor, is exactly the same interaction with the classical field as would be produced if we had a fundamental coupling $e(\overline{\psi}\gamma^\mu\psi)A_\mu$. It differs in that it has a k^2 dependence through $F_1(k^2)$, rather than a simple factor of e alone. Because of its k^2 dependence, it's not a constant, as it would be for coupling to a point charge: a constant in momentum space is a delta function in position space. So we can say that the effect of the interactions of the electron with the electromagnetic field (or the effect of the strong interactions with the proton) is to "spread out" the particle. That is why this is called a *form factor*: it tells us the form of the electron, the way in which the interaction is spread out.

The second term, the Pauli form factor, is an interaction of a new type, a spin-dependent interaction, the kind that would arise if we had a Pauli term[17]

$$\mathscr{L}' = (\overline{\psi}\sigma^{\mu\nu}\psi)F_{\mu\nu} \tag{34.22}$$

in the Lagrangian. That sort of interaction is gauge invariant and consistent with charge conjugation, though it's non-minimal; it would lead to something like the second term in (34.21), with a constant F_2. Of course it can't be there as a *fundamental* interaction because it's nonrenormalizable. It's of dimension 5, in four dimensions; the Dirac bilinear has dimension 3, while the derivative and the field A^μ each have dimension 1. Nevertheless the effects that would be made by such an interaction can arise, not as a point coupling but in a spread out way as a result of the quantum electrodynamic correction that make F_2.

The only sure result so far is (34.15). We're just playing with these objects F_1 and F_2 to try to get some idea of their physical meaning. We can go farther by using a cute identity due to Walter Gordon (of Klein–Gordon fame), the **Gordon decomposition.**[18] We'll start by doing something very stupid: we'll complicate the simple expression

$$\overline{u}'\gamma^\mu u \tag{34.23}$$

Using the free particle Dirac equations

$$\not{p}u = mu \qquad \overline{u}'\not{p}' = m\overline{u}' \tag{34.24}$$

[17] [Eds.] This sort of interaction appeared previously. See (27.69), p. 585, and the discussion following.

[18] [Eds.] W. Gordon, "Der Strom der Diracschen Elektronentheorie" (The current in Dirac electron theory), *Zeit. Phys.* **50** (1928) 630–632; Peskin & Schroeder *QFT*, Problem 3.2, p. 72.

the simple expression becomes

$$\overline{u}'\gamma^\mu u = \frac{1}{2m}\overline{u}'(\not{p}'\gamma^\mu + \gamma^\mu\not{p})u \tag{34.25}$$

Now

$$\gamma^\mu\not{p} = \tfrac{1}{2}\{\gamma^\mu, \not{p}\} + \tfrac{1}{2}[\gamma^\mu, \not{p}] = p^\mu - i\sigma^{\mu\nu}p_\nu \tag{34.26}$$

Likewise

$$\not{p}'\gamma^\mu = p'^\mu + i\sigma^{\mu\nu}p'_\nu \tag{34.27}$$

Assembling everything we get *the Gordon decomposition*

$$\overline{u}'(p')\gamma^\mu u(p) = \frac{1}{2m}\overline{u}'(p')[(p+p')^\mu + i\sigma^{\mu\nu}(p'-p)_\nu]u(p) \tag{34.28}$$

This decomposition has amusing consequences and gives us further insight into the physical meaning of the form factors. Recalling $k_\nu = p'_\nu - p_\nu$, we can write (34.21) as

$$i\mathcal{A} = -ie\left[\left(\overline{u}'(p')u(p)\right)\frac{(p+p')^\mu}{2m}\widetilde{A}^c_\mu F_1(k^2) + \frac{1}{4m}\left(\overline{u}'(p')\sigma^{\mu\nu}u(p)\right)\widetilde{F}^c_{\mu\nu}\left(F_1(k^2)+F_2(k^2)\right)\right] \tag{34.29}$$

The first term, with $(p+p')^\mu\widetilde{A}^c_\mu$, looks like the coupling of a *spinless* charged particle ϕ to an external electric field $\widetilde{A}^c_\mu$ in lowest order, through its antisymmetrized current $\phi^*\partial_\mu\phi - (\partial_\mu\phi^*)\phi$. This is as close to a spin-independent coupling as a relativistic spin-½ particle can get: $\overline{u}u$ with no γ matrices inside. So the first term *looks* spin-independent (though of course there are spin-dependent factors inside the bispinor product). Spin-independent terms cannot contribute to the magnetic moment, so this first term is irrelevant to its calculation. The second term, however, *is* spin-dependent, at least in the non-relativistic limit, and *will* contribute to the magnetic moment.

Let's go immediately to the non-relativistic limit to learn about the spin-dependent coupling in the low velocity regime. The free particle Dirac spinor in the non-relativistic limit is (see (20.27) and (20.33))

$$u = \begin{pmatrix} U \\ 0 \end{pmatrix} \tag{34.30}$$

U is a two-component spinor with the non-relativistic normalization

$$U^\dagger U = 1 \tag{34.31}$$

For $k^2 \approx 0$,

$$F_1(k^2) \approx F_1(0) = 1, \qquad F_2(k^2) \approx F_2(0) \tag{34.32}$$

but we don't know the value of $F_2(0)$. In the extreme non-relativistic limit, for spinors at rest,[19]

$$\begin{aligned} u'^\dagger\sigma_{0i}u &= 0 \\ u'^\dagger\sigma_{ij}u &= \epsilon_{ijk}U'^\dagger\sigma^k U \end{aligned} \tag{34.33}$$

[19] [Eds.] Antisymmetric 4-tensors like $F^{\mu\nu}$ and $\sigma^{\mu\nu}$ can be described simply as a 3-vector and an axial 3-vector.

The Lorentz generators, σ_{0i}, are pure off-diagonal; they mix up large components with small components.[20] On the other hand, the rotation generators σ_{ij} are simply the Pauli spin matrices written as a vector, $\boldsymbol{\sigma}$; ϵ_{ijk} turns a vector into an antisymmetric tensor: $\sigma_{ij} \equiv \epsilon_{ijk}\sigma_k$. From the field strength tensor $F_{\mu\nu}$ ((34.19); see also (S2.17) on p. 103) we have

$$F_{ij} = -\epsilon_{ijk}B_k \tag{34.34}$$

Now we can very easily study the spin-dependent term in (34.29) in the non-relativistic limit. The only terms that contribute are where μ and ν are i and j. We get a factor of 2 because we sum over everything twice, once in the order ij and once in the order ji. The spin-dependent term becomes in the non-relativistic limit[21]

$$\begin{aligned}\lim_{(v/c)\to 0}\left\{-\frac{ie}{4m}\overline{u}'\sigma^{\mu\nu}u\widetilde{F}^{c}_{\mu\nu}\left(F_1(k^2)+F_2(k^2)\right)\right\} &= \frac{ie}{4m}U'^{\dagger}\sigma_k\,U\epsilon^{ijk}\epsilon_{ijl}B_l\left(1+F_2(0)\right) \\ &= \frac{ie}{2m}\Big(1+F_2(0)\Big)U'^{\dagger}\boldsymbol{\sigma}\cdot\mathbf{B}\,U\end{aligned} \tag{34.35}$$

In non-relativistic quantum mechanics, the analog of what we call $i\mathcal{A}$ is given by a standard formula in first-order perturbation theory:[22]

$$i\mathcal{A} = -i\,\langle f|H'|i\rangle \tag{34.36}$$

That is, the effective non-relativistic interaction is

$$H' = -\frac{e}{2m}\Big(1+F_2(0)\Big)\boldsymbol{\sigma}\cdot\mathbf{B} \tag{34.37}$$

Recall that the **magnetic moment operator**[23] μ is defined by a charged particle's coupling to an external magnetic field:

$$H' = -\tfrac{1}{2}\mu\boldsymbol{\sigma}\cdot\mathbf{B} \tag{34.38}$$

For example,

$$F^{\mu\nu} = \begin{pmatrix} 0 & -E_x & -E_y & -E_z \\ E_x & 0 & -B_z & B_y \\ E_y & B_z & 0 & -B_x \\ E_z & -B_y & B_x & 0 \end{pmatrix} = (F^{0i}, F^{ij}) \equiv (-\mathbf{E}, \mathbf{B})$$

and $F_{\mu\nu} = (F_{0i}, F_{ij}) = (\mathbf{E}, \mathbf{B})$. Likewise it can be shown that

$$\sigma^{\mu\nu} = (\sigma^{0i}, \sigma^{ij}) \equiv (i\boldsymbol{\alpha}, -\boldsymbol{\Sigma})$$

where $\boldsymbol{\alpha}$ is given in (20.11) and $\boldsymbol{\Sigma}$ is given by (20.7). Consequently, $\frac{1}{2}\sigma^{\mu\nu}F_{\mu\nu} = i\boldsymbol{\alpha}\cdot\mathbf{E} - \boldsymbol{\Sigma}\cdot\mathbf{B}$. See V. B. Berestetskiĭ, E. M. Lifshitz, and L. P. Pitaevskiĭ, *Relativistic Quantum Theory Part I*, Pergamon Press, 1971, Problem 1, p. 67 and p. 100.

[20] [Eds.] In the standard representation, the bottom two-component spinor is, in the non-relativistic regime, proportional to $(v/c)^2$ times the top two-component spinor. The lower spinor consists of the "small" components (going to zero as $(v/c) \to 0$) and the upper spinor contains the "large" components. See Bjorken & Drell *RQM*, p. 12.

[21] [Eds.] Recall the identity $\epsilon_{ijk}\epsilon_{ilm} = \delta_{jl}\delta_{km} - \delta_{jm}\delta_{kl}$, so $\epsilon^{ijk}\epsilon_{ijm} = 2\delta^k{}_m$. See also (37.47), p. 812.

[22] [Eds.] See Ch. 6, "Perturbation Theory", pp. 129–157, equation's (38.2), (38.5), and (40.5) in Landau & Lifshitz *QM*.

[23] [Eds.] D. J. Griffiths, *Introduction to Quantum Mechanics*, Second Edition, Prentice Hall, 2005, Section 4.4.2, pp. 181–182.

(the non-relativistic spin operator is $\frac{1}{2}\boldsymbol{\sigma}$). Comparing (34.37) and (34.38) we see that the magnetic moment is[24]

$$\mu = \frac{e}{m}\left(1 + F_2(0)\right) = 2\left(1 + F_2(0)\right)\frac{e}{2m} \equiv g\mu_B \tag{34.39}$$

where μ_B is the **Bohr magneton**,

$$\mu_B \equiv \frac{e}{2m} \tag{34.40}$$

and g is the ratio of a particle's magnetic moment μ to a Bohr magneton. It was introduced[25] by Alfred Landé in 1921.

The first term in μ, e/m, is called the **Dirac moment**. It is an excellent approximation to the magnetic moment of the electron, which asserts that for the electron, $g = 2$. It is found in all the books on non-relativistic quantum mechanics, where the explanation is postponed to relativistic quantum mechanics. Well, here it is. For the electron there is a small correction because F_2 is non-zero. To lowest order, the correction $F_2(0)$ is $\mathcal{O}(e^2)$. This is the **anomalous moment**. That's all we can say using only quantum mechanics. Now we proceed to its quantum electrodynamic calculation.

34.3 The electron's anomalous magnetic moment in QED

Our task is to compute $F_2(0)$ to $\mathcal{O}(e^2)$ in an orgy of Feynman computations. This is a Nobel Prize-class calculation. At the New York meeting of the American Physical Society in 1948 (the community of physicists was so small at that time that the sessions could be held in the classrooms of Columbia University), Julian Schwinger got a standing ovation for the computation.[26] (This is not to be considered a hint. He did it for the first time. I'm doing it for perhaps the 700th time, and indeed, the fourth time in the past twenty-four hours, until I got the signs right!)

The lowest order contribution, Figure 34.4, has amplitude

$$e\overline{u}'\gamma_\mu u \tag{34.41}$$

This tells us which factors we have to subtract out. Next, to order e^3 (the graphs are only of $\mathcal{O}(e^2)$, but there is an overall factor of e) we have four diagrams, as shown in Figure 34.5. Graph (a) is the vacuum polarization on the external photon. Graph (b) is the photon wave function counterterm; the $\times$ indicates the counterterm computed to $\mathcal{O}(e^2)$ so that the whole graph is $\mathcal{O}(e^3)$. Graph (c) is the vertex correction. Graph (d) is the charge renormalization

---

[24] [Eds.] The classical, non-relativistic treatment gives (in the units used here) the electron's magnetic moment equal to one Bohr magneton. The empirical value of g for an electron is very close to 2. It was a great success of the Dirac equation that it predicted $g = 2$ exactly. See also Jackson *CE*, Section 11.8, "Thomas Precession", pp. 548–553.

[25] [Eds.] A. Landé, "Über den anomalen Zeemaneffekt (Teil I)", (On the anomalous Zeeman effect (Part I)), *Zeits. f. Phys.* **5** (1921) 231–240.

[26] [Eds.] Freeman Dyson was in the audience. He later wrote his parents, "The great event came on Saturday morning [Jan. 31], and was an hour's talk by Schwinger, in which he gave a masterly survey of the new theory... There were tremendous cheers when he announced that the crucial experiment had supported his theory: the magnetic splitting of two of the spectral lines of gallium... were found to be in the ratio 2 times 1.00114 to 1; the old theory gave for this ratio exactly 2 to 1, while the Schwinger theory gave 2 times 1.00116 to 1." Schweber *QED*, p. 320.

Figure 34.4: Lowest order contribution to the magnetic moment

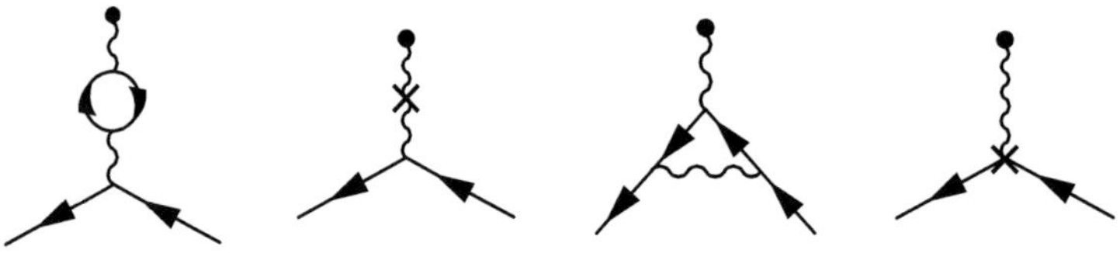

Figure 34.5: $\mathcal{O}(e^3)$ contributions to F^μ

counterterm; here, the $\times$ means the counterterm must be computed to $\mathcal{O}(e^3)$. Though we went through all that work with renormalization theory, in fact we don't have to calculate a counterterm for this process. Graphs (a), (b) and (d) are all proportional to $\overline{u}'\gamma_\mu u$. They all have only a γ_μ at the vertex regardless of what happens upstairs; they contribute *only* to F_1. So we only need to worry about graph (c); that gives troubles enough. $F_2(k^2)$ should come out to be finite without any worries about counterterms or subtractions.[27] The relevant part

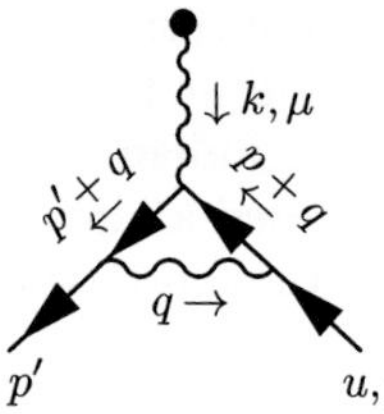

Figure 34.6: $\mathcal{O}(e^3)$ contribution to the anomalous magnetic moment

of F_μ, (34.12), is, in the Feynman gauge

$$F_\mu = \int \frac{d^4q}{(2\pi)^4}\overline{u}'(p')(-ie\gamma^\lambda)\frac{i(\not{p}' + \not{q} + m)}{(p'+q)^2 - m^2 + i\epsilon}(e\gamma_\mu)\frac{i(\not{p} + \not{q} + m)}{(p+q)^2 - m^2 + i\epsilon}(-ie\gamma^\nu)\frac{-ig_{\lambda\nu}}{q^2 + i\epsilon}u(p)$$
$$= -ie^3 \int \frac{d^4q}{(2\pi)^4}\frac{N_\mu}{D} \tag{34.42}$$

where

$$N_\mu = \overline{u}'\gamma_\lambda(\not{p}' + \not{q} + m)\gamma_\mu(\not{p} + \not{q} + m)\gamma^\lambda u \tag{34.43}$$

and

$$D = [(p'+q)^2 - m^2][(p+q)^2 - m^2][q^2] \tag{34.44}$$

(there is no $(-i)$ at the external field vertex, because it was factored out in the definition of F_μ). We'll evaluate N_μ and D separately, taking the numerator first. We won't need the $i\epsilon$'s, so we'll drop them.

[27] There are *no* corrections on the external legs. The electrons are on-shell, so any corrections to the external legs are simply canceled out by the counterterms, and therefore we don't bother to write either of them down.

Begin by moving $\not{p}'$ through γ_λ, recalling $\overline{u}'\not{p}' = m\overline{u}'$:

$$\overline{u}'\gamma_\lambda\gamma_\sigma p'^\sigma = \overline{u}'[2g_{\lambda\sigma} - \gamma_\sigma\gamma_\lambda]p'^\sigma = 2\overline{u}'p'_\lambda - \overline{u}'\not{p}'\gamma_\lambda = 2\overline{u}'p'_\lambda - m\overline{u}'\gamma_\lambda \tag{34.45}$$

Do the same for $\not{p}\gamma^\lambda u$ in the second fermion propagator:

$$p_\sigma\gamma^\sigma\gamma^\lambda u = p_\sigma[2g^{\sigma\lambda} - \gamma^\lambda\gamma^\sigma]u = 2p^\lambda u - \gamma^\lambda\not{p}u = 2p^\lambda u - m\gamma^\lambda u \tag{34.46}$$

The $-m$ terms cancel the $+m$ terms, and

$$N_\mu = \overline{u}'(2p'_\lambda + \gamma_\lambda\not{q})\gamma_\mu(2p^\lambda + \not{q}\gamma^\lambda)u \tag{34.47}$$

Multiply this out:

$$N_\mu = 4p'\cdot p\,(\overline{u}'\gamma_\mu u) + 2(\overline{u}'\gamma_\mu\not{q}\not{p}'u) + 2(\overline{u}'\not{p}\not{q}\gamma_\mu u) + (\overline{u}\gamma_\lambda\not{q}\gamma_\mu\not{q}\gamma^\lambda u) \tag{34.48}$$

The first term on the right we can simply drop; it is proportional to γ_μ and so contributes to F_1, not F_2. We're looking only for terms proportional to $\sigma_{\mu\nu}$. Leave the numerator for now, and let's turn our attention to the denominator, which is rather simple.

The electrons are on-shell:

$$p^2 = p'^2 = m^2 \tag{34.49}$$

so

$$\begin{aligned}(p+q)^2 - m^2 &= p^2 + q^2 + 2p\cdot q - m^2 = q^2 + 2p\cdot q \\ (p'+q)^2 - m^2 &= p'^2 + q^2 + 2p'\cdot q - m^2 = q^2 + 2p'\cdot q\end{aligned} \tag{34.50}$$

and the denominator simplifies to

$$D = (q^2 + 2p\cdot q)(q^2 + 2p'\cdot q)q^2 \tag{34.51}$$

We also have the marvelous denominator-combining formula using Feynman parameters (16.16),

$$\frac{1}{abc} = 2\int_\Delta dx\,dy\,\frac{1}{[ax + by + c(1-x-y)]^3} \tag{34.52}$$

The integration is over the triangular region Δ defined in Figure 34.7. Thus

$$\Delta\colon x, y \geq 0,\ x + y \leq 1$$

$\Delta =$ 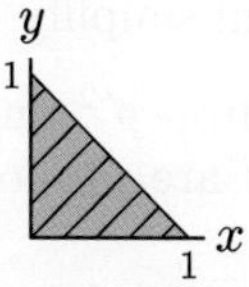

Figure 34.7: Region of integration

$$\begin{aligned}\frac{1}{D} = \frac{1}{(q^2 + 2p\cdot q)(q^2 + 2p'\cdot q)q^2} &= 2\int_\Delta dx\,dy\,\frac{1}{[x(q^2 + 2p'\cdot q) + y(q^2 + 2p\cdot q) + q^2(1-x-y)]^3} \\ &= 2\int_\Delta dx\,dy\,\frac{1}{[q^2 + 2q\cdot(xp' + yp)]^3}\end{aligned} \tag{34.53}$$

Complete the square:

$$\begin{aligned} q^2 + 2q\cdot(xp' + yp) &= (q + (xp' + yp))^2 - (xp' + yp)^2 \\ &= (q + (xp' + yp))^2 - 2xy(p\cdot p') - m^2(x^2 + y^2) \end{aligned} \tag{34.54}$$

Shift the integration variable q to q':

$$q' \equiv q + xp' + yp, \quad q = q' - xp' - yp \tag{34.55}$$

so

$$\frac{1}{D} = 2\int_\Delta dx\,dy\,\frac{1}{[q'^2 - 2xy(p\cdot p') - m^2(x^2 + y^2)]^3} \tag{34.56}$$

We can simplify this, using $k = p' - p$:

$$(p - p')^2 = 2m^2 - 2(p\cdot p') = k^2, \text{ so } (p\cdot p') = m^2 - \tfrac{1}{2}k^2 \tag{34.57}$$

We are not interested in terms of $\mathcal{O}(k^2)$; they just give corrections to F_2 as k^2 moves away from 0. Therefore,

$$2xy(p\cdot p') = 2xym^2 - xyk^2 = 2xym^2 + \mathcal{O}(k^2) \tag{34.58}$$

(and the same goes for all subsequent terms in $p\cdot p'$). We are left with

$$\frac{1}{D} = 2\int_\Delta dx\,dy\,\frac{1}{[q'^2 - m^2(x + y)^2]^3} + \mathcal{O}(k^2) \tag{34.59}$$

That completes the simplification of the denominator. It's important that the denominator is even in q'. It may seem dull to you now but that's because the standards of drama have changed; in 1948 it drew cheers.

Now back to the numerator, N_μ. We must substitute the expression for q in terms of q', (34.55), into (34.48). We will get terms with no power in q', terms linear in q' and terms quadratic in q'. The terms linear in q' are odd functions (the denominator is even) and so vanish upon integration; we can go ahead and drop them. What's left is

$$\begin{aligned} N_\mu = &-2\overline{u}'\gamma_\mu(x\not{p}' + y\not{p})\not{p}'u - 2\overline{u}'\not{p}(x\not{p}' + y\not{p})\gamma_\mu u \\ &+ \overline{u}'\gamma_\lambda(x\not{p}' + y\not{p})\gamma_\mu(x\not{p}' + y\not{p})\gamma^\lambda u + \overline{u}'\gamma_\lambda\not{q}'\gamma_\mu\not{q}'\gamma^\lambda u \end{aligned} \tag{34.60}$$

This is its most horrendous form; it will simplify drastically.

In the first and second terms we drop $\not{p}'^2$ and $\not{p}^2$, respectively,[28] since they both equal m^2 by (34.49) and hence these terms are proportional to γ_μ (and contribute only to F_1). Rewriting,

$$N_\mu = -2\overline{u}'(y\gamma_\mu\not{p}\not{p}' + x\not{p}\not{p}'\gamma_\mu)u + \overline{u}'\gamma_\lambda(x\not{p}' + y\not{p})\gamma_\mu(x\not{p}' + y\not{p})\gamma^\lambda u + \overline{u}'\gamma_\lambda\not{q}'\gamma_\mu\not{q}'\gamma^\lambda u \tag{34.61}$$

The only term quadratic in q' will involve an integrand $(\overline{u}'\gamma_\lambda\gamma_\rho\gamma_\mu\gamma_\sigma\gamma^\lambda u)q'^\rho q'^\sigma$. But

$$\int d^4q'\,f(q'^2)q'^\rho q'^\sigma = \tfrac{1}{4}g^{\rho\sigma}\int d^4q'\,f(q'^2)q'^2 \tag{34.62}$$

[28] [Eds.] $\not{p}\not{p} = p_\alpha p_\beta\gamma^\alpha\gamma^\beta = \frac{1}{2}p_\alpha p_\beta\{\gamma^\alpha, \gamma^\beta\} = p_\alpha p_\beta g^{\alpha\beta} = p^2$.

because the integration is Lorentz invariant. We can check that the $\frac{1}{4}$ is correct by taking the trace of both sides and seeing that they are equal. Therefore we can make the replacement

$$\overline{u}'\gamma_\lambda \not{q}'\gamma_\mu \not{q}'\gamma^\lambda u \rightarrow \tfrac{1}{4}q'^2\left(\overline{u}'\gamma_\lambda\gamma_\sigma\gamma_\mu\gamma^\sigma\gamma^\lambda u\right) \tag{34.63}$$

The second identity we will use is

$$\gamma_\lambda\gamma_\mu\gamma^\lambda = -2\gamma_\mu \tag{34.64}$$

This is easy to check.[29] For any given γ_μ, one of the four γ_λ's commutes with it, and three anti-commute, (while $\gamma^0\gamma_0 = \gamma^1\gamma_1 = \gamma^2\gamma_2 = \gamma^3\gamma_3 = 1$), so we are left with $1 - 3 = -2$ factors of γ_μ. Using (34.64) twice in the right-hand side of (34.63), we see that (the two 2's cancel the $\frac{1}{4}$)

$$\overline{u}'\gamma_\lambda \not{q}'\gamma_\mu \not{q}'\gamma^\lambda u \rightarrow q'^2\overline{u}'\gamma_\mu u \tag{34.65}$$

We drop this term since it's proportional to γ_μ.

We have now reached an important point. Because we have eliminated all the q'^2 terms from the numerator, the integral is manifestly convergent. It goes like

$$\int \frac{d^4q'}{q'^6} \tag{34.66}$$

The q'^2 in the numerator, had it not been proportional to γ_μ, could have given a logarithmically divergent integral and hence the wrong anomalous magnetic moment in a very drastic way, to wit, a divergent one. The answer we get may be right or it may be wrong, but it will certainly be finite.

We are still left with the remaining two God-awful terms in (34.61). In the first term, anti-commute $\not{p}$ through $\not{p}'$:

$$\begin{aligned}\gamma_\mu \not{p}\not{p}' u &= \gamma_\mu(2(p\cdot p') - \not{p}'\not{p})u \\ &= \gamma_\mu((2m^2 - k^2) - m\not{p}')u \\ &\rightarrow -m\gamma_\mu\not{p}' u\end{aligned} \tag{34.67}$$

The second equation follows from (34.57); in the third equation we drop terms $\mathcal{O}(k^2)$ and proportional to γ^μ. Likewise

$$\overline{u}'\not{p}\not{p}'\gamma_\mu = \overline{u}'(2p\cdot p' - m\not{p})\gamma_\mu \rightarrow -\overline{u}'m\not{p}\gamma_\mu \tag{34.68}$$

In both of these expressions we drop the $(p\cdot p')$ term because it is proportional to γ_μ.

The remaining term can be reduced from five γ matrices to three, with the aid of this little wonder of an identity, stated without proof:[30]

$$\gamma_\lambda \not{a}\not{b}\not{c}\gamma^\lambda = -2\not{c}\not{b}\not{a} \tag{34.69}$$

It's proved along the same lines as the previous identity. So

$$N_\mu = 2\overline{u}'\left[my\gamma_\mu\not{p}' + mx\not{p}\gamma_\mu - (x\not{p}' + y\not{p})\gamma_\mu(x\not{p}' + y\not{p})\right]u \tag{34.70}$$

29 [Eds.] We have $\gamma_\lambda\gamma_\mu\gamma^\lambda = (2g_{\lambda\mu} - \gamma_\mu\gamma_\lambda)\gamma^\lambda = 2\gamma_\mu - 4\gamma_\mu = -2\gamma_\mu$.

30 [Eds.] Bjorken & Drell *Fields*, Appendix A, p. 284.

Use (34.10) to turn the p''s on the right over to p's and the p's on the left to p''s:

$$p' = p + k, \quad p = p' - k$$

leaving us with $\not{k}$'s still floating around on the right or the left. For instance,

$$\overline{u}'\gamma_\mu \not{p}' u = \overline{u}'\gamma_\mu(\not{p} + \not{k})u = \overline{u}'\gamma_\mu(m + \not{k})u \to \overline{u}'\gamma_\mu \not{k} u \tag{34.71}$$

dropping the m term because that's proportional to γ_μ. For the last term in (34.70), we find

$$\begin{aligned} -2\overline{u}'\left(x\not{p}' + y\not{p}\right)\gamma_\mu\left(x\not{p}' + y\not{p}\right)u &= -2\overline{u}'\left((x+y)\not{p}' - y\not{k}\right)\gamma_\mu\left((x+y)\not{p} + x\not{k}\right)u \\ &= -2\overline{u}'\left(m(x+y) - y\not{k}\right)\gamma_\mu\left((x+y)m + x\not{k}\right)u \\ &\to -2mx(x+y)(\overline{u}'\gamma_\mu\not{k}u) + 2my(x+y)(\overline{u}'\not{k}\gamma_\mu u) + \mathcal{O}(k^2) \end{aligned} \tag{34.72}$$

once again dropping the term proportional to γ_μ. We're looking for $F_2(0)$, so we ignore $\mathcal{O}(k^2)$ terms. Putting all the pieces together, we have

$$N_\mu = 2\overline{u}'\left\{my\gamma_\mu\not{k} - mx\not{k}\gamma_\mu - mx(x+y)\gamma_\mu\not{k} + my(x+y)\not{k}\gamma_\mu\right\}u \tag{34.73}$$

We can simplify things by the following observation. We are integrating over a region symmetric under the exchange $x \leftrightarrow y$. If there are parts of the integrand antisymmetric under this exchange, they will vanish. Consequently, we can go ahead and make the replacements

$$x \to \tfrac{1}{2}(x+y) \qquad y \to \tfrac{1}{2}(x+y) \tag{34.74}$$

This has no effect on the denominator (34.59), but the numerator becomes (using $[\gamma_\mu, \not{k}] = -2i\sigma_{\mu\nu}k^\nu$)

$$\begin{aligned} N_\mu &= \overline{u}'\left\{m(x+y)[\gamma_\mu, \not{k}] - m(x+y)^2)[\gamma_\mu, \not{k}]\right\}u \\ &= -2im\left(\overline{u}'\sigma_{\mu\nu}u\right)k^\nu[(x+y) - (x+y)^2] \end{aligned} \tag{34.75}$$

At last we're done with spinor algebra. We have from (34.15), (34.43), (34.44), and (34.59), and dropping the prime on q',

$$\begin{aligned} F^\mu(k^2)\Big|_{k^2\approx 0} &= \frac{ie}{2m}\left(\overline{u}'\sigma^{\mu\nu}u\right)k_\nu F_2(0) = -ie^3\int\frac{d^4q}{(2\pi)^4}\frac{N_\mu}{D} \\ &= -ie^3\left[-2im\left(\overline{u}'\sigma^{\mu\nu}u\right)k_\nu\right]\int\frac{d^4q}{(2\pi)^4}\int_\Delta dx\,dy\,\frac{2[(x+y)-(x+y)^2]}{[q^2 - m^2(x+y)^2]^3} \end{aligned} \tag{34.76}$$

We can now extract $F_2(0)$, the coefficient of $(ie/2m)(\overline{u}'\sigma^{\mu\nu}u)k_\nu$:

$$F_2(0) = 8ie^2m^2\int_\Delta dx\,dy[(x+y) - (x+y)^2]\int\frac{d^4q}{(2\pi)^4}\,\frac{1}{[q^2 - m^2(x+y)^2]^3} \tag{34.77}$$

The table from Chapter 15 tells us how to do any integral of this form.[31] The relevant formula is (I.2):

$$\int\frac{d^4q}{(2\pi)^4}\frac{1}{(q^2 - a^2)^3} = -\frac{i}{32\pi^2a^2} \tag{34.78}$$

---

[31] [Eds.] See the box on p. 330.

Then

$$
\begin{aligned}
F_2(0) &= 8ie^2m^2 \int_\Delta dx\,dy\,\frac{-i[(x+y)-(x+y)^2]}{32\pi^2m^2(x+y)^2} = \frac{e^2}{4\pi^2}\int_0^1 dx\int_0^{1-x} dy\left[\frac{1}{x+y}-1\right] \\
&= \frac{e^2}{4\pi^2}\int_0^1 dx\,\left(\ln(x+y)-y\right)\Big|_0^{1-x} = \frac{e^2}{4\pi^2}\int_0^1 dx\,\left[\ln(x+1-x)-(1-x)-\ln(x)\right] \\
&= \frac{e^2}{4\pi^2}\int_0^1 dx\,\left[x-1-\ln(x)\right] = \frac{e^2}{4\pi^2}\left[\tfrac{1}{2}x^2-x-x(\ln x-1)\right]\Big|_0^1 = \frac{e^2}{8\pi^2}
\end{aligned}
\tag{34.79}
$$

Therefore the magnetic moment of the electron, to the order we are working in, is (see 34.39))

$$
\mu = \frac{e}{m}\Big(1+F_2(0)\Big) = \frac{e}{m}\left(1+\frac{1}{2\pi}\frac{e^2}{4\pi}\right) = \frac{e}{m}\left(1+\frac{\alpha}{2\pi}\right) = 2\mu_B\left(1+\frac{\alpha}{2\pi}\right) \tag{34.80}
$$

In rationalized units it's

$$
\boxed{\alpha \equiv \frac{e^2}{4\pi}} \tag{34.81}
$$

that appears in Coulomb's Law. The quantity α is called the **fine-structure constant**,

$$
\alpha \approx \frac{1}{137.036} = 0.00729735 \tag{34.82}
$$

From (34.80),

$$
g_{\text{theory}} = 2\left(1+\frac{\alpha}{2\pi}\right) \Rightarrow \frac{(g_{\text{theory}}-2)}{2} = \frac{\alpha}{2\pi} = 1161.41\times 10^{-6} \tag{34.83}
$$

The current experimental value[32] is

$$
\frac{(g_{\text{expt}}-2)}{2} = (1159.65218091 \pm 0.00000026)\times 10^{-6} \tag{34.84}
$$

The agreement between the experimental and theoretical result,[33] calculated only to first order, is to within 0.16%.

Next time we'll look at higher order corrections to the electron's magnetic moment, and also that of the muon. These comparisons with experiment will lead us to consider the electromagnetic interactions of hadrons.

---

[32] [Eds.] *PDG* 2016; `http://pdg.lbl.gov/2016/tables/rpp2016-sum-leptons.pdf`.

[33] [Eds.] Julian Schwinger is buried in Mt. Auburn Cemetery, about a mile west of Harvard Square. Above his name, his tombstone bears the inscription $\dfrac{\alpha}{2\pi}$.

35

Confronting experiment with QED

In the last lecture, to general jubilation and relief on my part, I derived the correct $\mathcal{O}(\alpha)$ formula for the magnetic moment of the electron

$$\mu = \frac{e}{m}\left[1 + \frac{\alpha}{2\pi}\right] \tag{35.1}$$

where α is the fine-structure constant

$$\alpha \equiv \frac{e^2}{4\pi} \tag{35.2}$$

Today I will discuss other experiments involving this formula and, in a qualitative but not quantitative way (except by quoting other people's results) the higher order corrections to the electron magnetic moment.[1]

35.1 Higher order contributions to the electron's magnetic moment

To get an idea of the size of these effects we have to know what α is. The current experimental value is[2]

$$\alpha^{-1} = 137.035\,999\,139\,(31) \tag{35.3}$$

That is, with a standard deviation of 31 in the last two digits displayed. The fact that α is not known exactly means that we don't know the first-order correction to μ exactly. The uncertainty in α^{-1}, 31×10^{-9}, leads to an uncertainty in μ,

$$\left|\frac{\delta\alpha}{2\pi}\right| = \left|\frac{\alpha^2\delta(\alpha^{-1})}{2\pi}\right| \approx 2.6 \times 10^{-13} \tag{35.4}$$

We can make a rough guess about the size of the higher order corrections, which we have not computed but of course are in the literature:

$$\left(\frac{\alpha}{\pi}\right)^2 \approx 5 \times 10^{-6}; \qquad \left(\frac{\alpha}{\pi}\right)^3 \approx 1 \times 10^{-8}; \qquad \left(\frac{\alpha}{\pi}\right)^4 \approx 3 \times 10^{-11} \tag{35.5}$$

[1] [Eds.] In the video of Lecture 35, Coleman quotes many experimental numbers, some as "the best available." These numbers were the best available during the years of the videos, 1975–1976. In these lectures the editors have endeavored to quote the best experimental numbers available to them, *circa* 2016.

[2] [Eds.] *PDG* 2016, `http://pdg.lbl.gov/2016/reviews/rpp2016-rev-phys-constants.pdf`.

Plugging this number (35.3) in, we find to $\mathcal{O}(e^2)$

$$1 + \frac{\alpha}{2\pi} = 1.001\ 161\ 41 \cdots \tag{35.6}$$

I've only carried it out this far, even though the error in α is known much better than that, because the anticipated size of the $(\alpha/\pi)^2$ correction would be perhaps 5 in the sixth digit. The current experimental value is[3]

$$1 + F_2(0) = 1.001\ 159\ 652\ 180\ 91\ (26) \tag{35.7}$$

The uncertainty is 26 in the last two displayed digits. As we see, to our expected level of agreement with experiment, i.e., up to the point where we expect the $(\alpha/\pi)^2$ corrections to come in, the agreement is perfect. This is impressive. This is already five decimal places of agreement. The calculation took 45 minutes; the experiment took 20 years. The theoretical calculation (including higher order corrections) gives[4]

$$1 + F_2(0) = 1.001\ 159\ 652\ 181\ 13\ (11)(37)(02)(77) \tag{35.8}$$

(the uncertainties coming from the eighth-order QED term, an estimate of the tenth-order term, the hadronic and electroweak contributions, and the uncertainty in α, respectively). To within a couple of standard deviations, the agreement between theory and experiment is perfect, a few parts in 10^{13}. The most significant source of error is in the uncertainty in α. There is also a theoretical uncertainty in the last term because there are so many graphs to compute.[5] Instead of bothering to compute some of them, we just estimate that they are less than a certain amount. But we don't know if they are going to cancel or add together, so we have a purely theoretical uncertainty, caused by lack of strength on the part of theoreticians, in the $\mathcal{O}(\alpha/\pi)^3$ terms.

35.2 The anomalous magnetic moment of the muon

There is another number for which we can use exactly the same formula, the magnetic moment of the **muon**. The muon is just a heavy electron:

$$m_e \approx 0.511\,\text{MeV}; \quad m_\mu \approx 106\,\text{MeV} \approx 200\,m_e \tag{35.9}$$

But as far as we know, in all of its interactions, the muon is exactly the same as the electron. This is one of the great unsolved mysteries. In the elegant formulation of I. I. Rabi, "Who ordered *that*?"[6] It's a good question, and still unanswered. In any event, there it is, so we can

---

[3] [Eds.] *Ibid.*, `http://pdg.lbl.gov/2016/reviews/rpp2016-list-electron.pdf`.

[4] [Eds.] T. Aoyama *et al.*, "Tenth-Order QED Lepton Anomalous Magnetic Moment: Eighth-Order Vertices Containing a Second-Order Vacuum Polarization", *Phys. Rev. D* **85** (2012) 033007. The theoretical value is cited in equation (3).

[5] [Eds.] Over *12,000* in $\mathcal{O}(e^{10})$. T. Aoyama, *op. cit.*; D. Styer, "Calculation of the anomalous magnetic moment of the electron", `www.oberlin.edu/physics/dstyer/StrangeQM/Moment.pdf`.

[6] [Eds.] Crease & Mann *SC*, p. 169, endnote 112 (text on p. 440): "Neither Rabi nor anyone else can remember where this now-famous remark was first made, but Rabi thinks it was an American Physical Society meeting in New York City." Isidor Isaac Rabi, at Columbia from 1929 until his death in 1988, was the winner of the 1944 Nobel Prize in physics for his discovery of nuclear magnetic resonance, and Julian Schwinger's thesis advisor. Though discovered in 1935, the muon's leptonic nature was not recognized until 1947.

test our theory against the experimental results for the muon. The magnetic moment of the muon is known experimentally to surprising accuracy:[7]

$$\left(\frac{m_\mu}{e}\right)\mu_\mu = 1 + F_2(0)\big|_{\text{muon}} = 1.001\,165\,920\,89\ (54)(33) \tag{35.10}$$

(the parenthetical numbers are the uncertainty in the μ^+ and μ^- moments, respectively). The theoretical QED value is[8]

$$1 + F_2(0)\big|_{\text{muon}} = 1.001\,165\,847\,18\,09\ (18) \tag{35.11}$$

This is agreement to within 1 part in 10^8. That is the situation comparing theory and experiment, and we all agree that this is heartening. At the end of this lecture we'll be doing a computation in which we'll be overjoyed to get agreement to within 20%.

There are some questions that should be asked. Why are the two magnetic moments different? Why is the muon moment larger than the electron moment? Of course, out to the fifth decimal place, the computation of the muon is identical to the computation for the electron. But obviously something is happening in higher orders that is different for the muon and for the electron. What is that something? Also, why is the theoretical error figure for the muon different from the one for the electron? Is it just that the theoretical physicists who work on computing the muon moment are less energetic than those who work on computing the electron moment? These questions turn out to have the same answer.

Many of the graphs that contribute to the anomalous magnetic moment of the electron are exactly the same as, and in one-to-one correspondence with, the graphs that contribute to the anomalous magnetic moment of the muon. As far as these graphs go, the number

$$\left(\frac{m}{e}\right)\mu$$

(μ is the lepton's magnetic moment, m is the relevant lepton's mass) is exactly the same for the electron and the muon: it's a dimensionless number, and the mass of the lepton is irrelevant. At the fourth order however, we begin to encounter one and only one graph that is different for the electron and the muon. This graph involves two leptons (and it's the first such we've seen), one on the external line and another on the internal loop. Here we can have a difference between muon and electron graphs, because one type of lepton can be inside, and the other can be outside.

These are qualitatively different graphs. For the muon moment, we have a heavy particle going through the external line and a light particle running around the internal line. In the other case, the electron moment, the roles of heavy and light are reversed. (If we have the same particle on both fermion lines, electron–electron or muon–muon, the computations would be identical by dimensional analysis.) In fact, although fourth order corrections are in general difficult to work out, it's not especially difficult to compute this particular graph.

[7] [Eds.] A. Hoecker and W. J. Marciano, "The Muon Anomalous Magnetic Moment", pp. 583–587 in J. Beringer *et al.* (Particle Data Group), "Review of Particle Physics", *Phys. Rev.* **D86** (2012) 010001; equation (3). This is the average for μ^+ and μ^-.

[8] [Eds.] *Ibid.*, equation (6). Including all standard model contributions (from electroweak and hadronic interactions), the theoretical value of the muon magnetic moment is given in equation (14) as 1.001 165 918 02 (2)(42)(26), differing by 2 parts in 10^{10} from the experimental result (35.10).

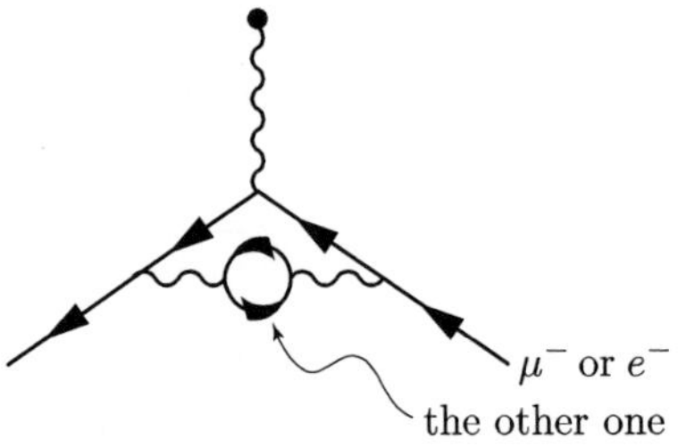

Figure 35.1: Magnetic moment diagram with two leptons

We won't do it explicitly, but it's sort of simple in its structure because it breaks up into two computations which we've already done. In the internal photon loop we have the correction to the photon propagator (last week's set of homework problems[9]); once we've put in that corrected propagator, we have the graph we did last time, for the electron moment, with one slight modification, which I will now describe.

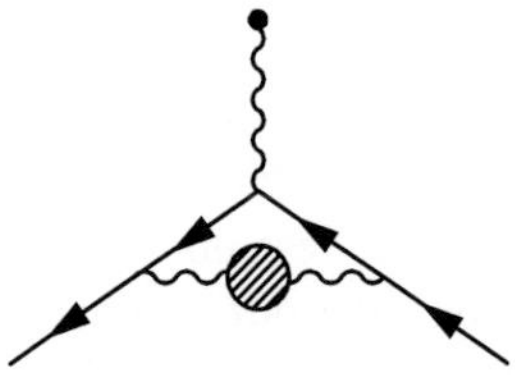

Figure 35.2: Higher order contributions to the magnetic moment

To $\mathcal{O}(e^4)$, we might as well replace the graph in Figure 35.1 (and all of its friends) by the one in Figure 35.2, with the corrected photon propagator in place of the fermion loop. The changes caused by the corrected photon propagator's including all sorts of other things besides the lepton loop are going to come in at $\mathcal{O}(e^6)$. (There will also be a host of counterterms.) The actual relationship may be written

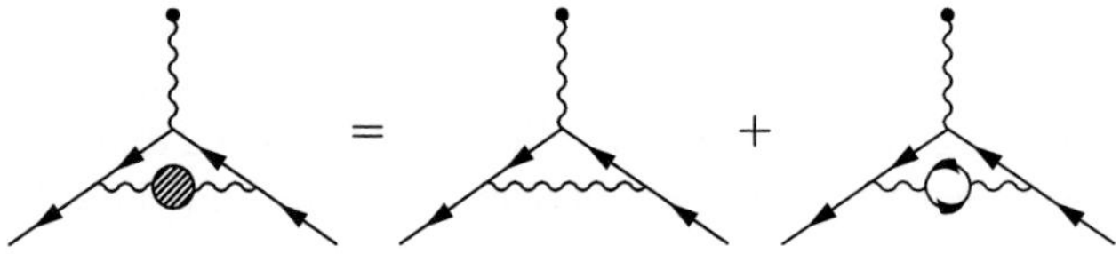

$$= \quad + \quad + \text{(counterterms)} + \mathcal{O}(e^6) \tag{35.12}$$

We know the corrected photon propagator can be written in the spectral form (34.4):

$$\widetilde{D}'_{\mu\nu}(k^2) = -i\left[g_{\mu\nu} - \frac{k_\mu k_\nu}{k^2}\right]\left[\frac{1}{k^2+i\epsilon} + \int da^2\,\frac{\sigma(a^2)}{k^2-a^2+i\epsilon}\right] + \begin{pmatrix}\text{gauge-dependent}\\ \text{terms}\end{pmatrix} \tag{35.13}$$

The first term on the right reproduces the first graph on the right hand side in (35.12), which we calculated in §34.3. Then we have a continuous superposition of the same graphs with heavy photons, whose squared mass equals a^2; $\sigma(a^2)$ is the **photon spectral function.**[10] Finally we have gauge-dependent terms, which are irrelevant if we're working in Landau gauge.

---

[9] [Eds.] Problem 18.1, p. 725.

[10] [Eds.] Bjorken & Drell *Fields*, Section 16–11, pp. 166–170. See the equation following (16.173); $\sigma(a^2)$ is denoted $\Pi(M^2)$.

If we knew how to compute the photon spectral function to the relevant order (and indeed we do, because we have the photon self-energy, or, by a trivial manipulation of the homework problem you just did, the muon intermediate state contribution to the photon spectral function), and if we knew how to compute the anomalous magnetic moment to the lowest order, with a heavy photon instead of a zero-mass photon as we did before, then we would just put these two things together. Thus we would be able to compute (perhaps in terms of an integral we'd have to do numerically) the contribution of the graphs in Figure 35.1 to the anomalous moment, without having to worry about any complications from renormalization or overlapping integrals or anything fancy.

Contribution to $F_2(0)$ from a "photon" with $\mu^2 = a^2$

This is a trivial generalization of what we did last time. Most of that work involved manipulating the numerator. We don't have to do that again, because the numerator doesn't give a damn about what's going on in the denominator, which is the only place that the photon mass appears. Recall that last time we ended up with an expression for $F_2(0)$, the Pauli form factor, as in the first line of (34.79):

$$F_2(0) = \frac{e^2 m^2}{4\pi^2} \int_\Delta dx\, dy\, \frac{(x+y)(1-x-y)}{m^2(x+y)^2} \tag{35.14}$$

That was our old denominator. It came from the fermion mass factors when we applied Feynman's formula. If the photon carries a squared mass equal to a^2, the only difference comes in the denominator. When we parametrize the integral with Feynman's trick, there will be a term proportional to the photon mass squared. Formerly we had the integral with x for one electron propagator, y on the other and $1-x-y$ for the photon propagator. So the change to massive photons leads to

$$F_2(0) = \frac{e^2 m^2}{4\pi^2} \int_\Delta dx\, dy\, \frac{(x+y)(1-x-y)}{m^2(x+y)^2 + (1-x-y)a^2} \tag{35.15}$$

That's the answer. This integral is elementary. Switching to the independent variables $x-y$ and $x+y$, the $x-y$ integral is trivial. The remainder becomes a polynomial over a quadratic form which we can look up in an integral table, so it's not particularly difficult; I won't bother.

I will however make some remarks about this expression (35.15). First, $F_2(0)$ is positive. Second, it is a monotonic decreasing function of a^2; the heavier the photon, the less the contribution it makes. Third, it becomes the standard result as $a^2 \to 0$:

$$\lim_{a^2 \to 0} F_2(0) = \frac{e^2}{8\pi^2} = \frac{\alpha}{2\pi} \tag{35.16}$$

Fourth, as $a^2 \to \infty$, the $m^2(x+y)^2$ term becomes negligible, so we drop it. Then the $(1-x-y)$ in the numerator cancels the $(1-x-y)$ in the denominator, so

$$\lim_{a^2 \to \infty} F_2(0) = \frac{e^2 m^2}{4\pi^2 a^2} \int_\Delta dx\, dy\, (x+y) = \frac{\alpha}{3\pi} \frac{m^2}{a^2} \tag{35.17}$$

To summarize, the integral for $F_2(0)$

1. is positive
2. is a monotonic, decreasing function of a^2
3. goes to the earlier result, $\alpha/2\pi$, as $a^2 \to 0$

4. goes as $\mathcal{O}(m^2/a^2)$ for $a^2 \gg m^2$

From these conclusions, we can see that the muon contribution to the electron moment is going to be very small. The contribution of the lepton intermediate states to the spectral function, that is, from the graph

will be zero if $a^2 < 4m_\ell^2$, where m_ℓ is the mass of the lepton in the loop. That is, $\sigma(a^2) = 0$ for $a^2 < 4m_\ell^2$. So

$$(\text{muon contribution to electron moment}) = (\alpha/\pi)^2\,\mathcal{O}((m_e/2m_\mu)^2) \approx \mathcal{O}((\alpha/\pi)^4) \qquad (35.18)$$

as $(m_e/2m_\mu) \approx 1/400 \approx (\alpha/\pi)$. Because of this suppression factor, the muon contribution is on the order of $(\alpha/\pi)^4$; it's negligible, and simply not worth computing.

On the other hand, the electron contribution to the muon moment is going to be very much larger. The electron is very light, so the loop makes a large contribution to the spectral function, as if the equivalent photon, with $a^2 = 4m_e^2$, were *massless*; it's effectively massless on the scale of the muon. We can get an estimate of this function very quickly, forgetting about all the numerical coefficients, simply by asking, "What if the electron were massless?" Just by dimensional analysis, the spectral weight function $\sigma(a^2)$ in (35.13) would have to go like $1/a^2$, since there is no mass in the problem; it's being integrated over da^2, and $\sigma(a^2)da^2$ has dimensions of 1. If we're integrating $1/a^2$ over da^2 we get a logarithm. That is,

$$(\text{electron contribution to muon moment}) = (\alpha/\pi)^2 \ln((m_\mu/m_e)^2) \approx 10(\alpha/\pi)^2 \qquad (35.19)$$

This expression has a logarithmic divergence as the electron mass goes to zero, though it's certainly not divergent if the electron has a non-zero mass. Since this is a number on the order of $2\ln(200) \approx 10$ times $(\alpha/\pi)^2$, it will give a rather large contribution, much larger than the muon contributes to the electron's moment. The difference between these will make the muon moment larger than the electron moment, because the contribution is positive. When we look at the experimental numbers, we find that they differ precisely at order $(\alpha/\pi)^2 \approx 5\times10^{-6}$. To this order

$$F_2(0)\big|_{e^-} : 1.001\,159 \qquad F_2(0)\big|_{\mu^-} : 1.001\,165 \qquad (35.20)$$

They start differing in the fifth decimal place, at $\mathcal{O}((\alpha/\pi)^2)$, with a rather large coefficient, 5 or 6, in line with the theoretical estimate.

The hadronic contribution to the leptons' magnetic moments

Where do the hadrons come into the game? To a first approximation our quantitative knowledge of the strong interactions is zero. Therefore when we consider higher order terms, we'll eventually have graphs with strongly interacting particles appearing, produced by the photon. The hadrons first appear at $\mathcal{O}(e^4)$, where the internal muon or electron loop in Figure 35.1 could be replaced by, for example, a pion pair, $\pi^+\pi^-$. Once a strongly interacting particle gets into our graph, we're cooked, because the strong interactions can't be analyzed by perturbation theory: the coupling constant is too large, $g \approx 15$. We can complicate the graph enormously by inserting any number of strong interactions, without adding any power of e. For example, the $\pi^+\pi^-$ loop could generate a proton-neutron loop, as shown in Figure 35.3. That doesn't give us a higher power of e, it gives us powers of g. Well, perhaps not

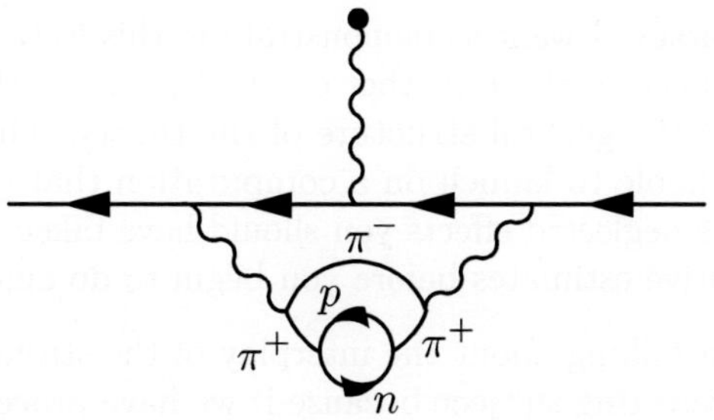

Figure 35.3: Pion loop with proton-neutron loop

$g = 15$, but we know it will not be negligible compared to what we have. Once we open the door to a hadron pair, we're going to have to introduce all the effects of the strong interactions. Fortunately, at least to $\mathcal{O}(e^4)$, the same rules apply as in Figure 35.2. The effects of all of these things up to $\mathcal{O}(e^4)$ is simply to give account of the hadrons' strong interaction corrections to the photon propagator. Now we can already see, unless something very funny is going on, that for the electrons the strong interaction corrections are going to be negligible, because the lightest hadron, the pion, is roughly as heavy as the muon.[11] And therefore we'll get the same suppression factor. The effects of the hadron intermediate states are not small, but the hadrons are heavy. So for the electron, the effects of the strong interactions are $\mathcal{O}(e^8)$, negligible.

But the situation for the muon is different. The muon belongs to the lepton family, a "lightweight" particle, but of course it's almost as heavy as the pion, which *is* a hadron. So in principle we have to take the strong interaction corrections seriously for the muons. They will be $\mathcal{O}((\alpha/\pi)^2)$ surely, but whether the factor in front is large or small is something that requires computation to determine. (There's a homework problem[12] for you to see how a specific computation is done. You won't have to do the integral; you won't need to. But you will see how one determines *experimentally* the strong interaction corrections to the spectral density function $\sigma(k^2)$.)

The result is that in fact these corrections *are* negligible. The hadronic contributions to the muon moment[13] are about 7×10^{-11}. If we look into the guts of the computation, we can see what's happening. The real reason is that those two pions essentially have no effect until they're resonant with a ρ state, and the ρ is about six times more massive than the pion: 770 MeV *vs.* 135 MeV. That brings in a suppression factor of $(2m_\pi/m_\rho)^2 \approx \frac{1}{8}$, which helps a great deal. The effects, in any event, are quite small, just on the verge of being experimentally measurable.

This concludes the discussion of experiment and the anomalous magnetic moments of the muon and the electron. I want you to be impressed by something other than what people are usually impressed by in this discussion, namely, that nine decimal place accuracy between experiment and theory. And indeed, that's very impressive. But after all, it requires a lot of

---

[11] [Eds.] In fact, it's about a third again as heavy: $m_\mu = 105.66\,\text{MeV}$; $m_\pi = 139.57$ MeV. *PDG* 2016, p. 32; p. 37.

[12] [Eds.] Problem 20.2, p. 817. Note that in this problem, $\sigma(k^2)$ is denoted $\rho(k^2)$ to avoid (some) confusion between the spectral function and the total cross-section σ_T for e^+-$e^- \to$ hadrons.

[13] [Eds.] Hoecker and Marciano, *op. cit.* See p. 584, equation (13).

hard work to get that accuracy. I want to demonstrate in this lecture how we can understand the *qualitative* nature of some of those further decimal places without doing much work at all, just by thinking about the general structure of the theory. That's just as important. If you can't do that, you're liable to launch on a computation that gives you a number that's meaningless because you've neglected effects you should have taken into account. You have to learn how to make qualitative estimates before you begin to do quantitative computations.[14]

We have already begun talking about the interplay of the strong interactions and electromagnetism. This is an interesting subject because if we have processes that we can consider as purely electromagnetic, like the anomalous moment of the electron up to order e^6, then in principle we know everything, if we are willing to work hard enough. And if we have processes that are purely strong, we know nothing. Well, we know quite a bit more than we knew once. The interesting half world is where we have strong interaction corrections to electromagnetic process or equivalently electromagnetic corrections to strong interaction processes, where we know half of what's going on. Does that enable us to tell anything about experiment? Or does the ignorance of the strong interactions corrupt everything so that we know nothing about nothing?

This is not a systematic subject, but rather a subject where people have one clever idea after another, and each of them gives us a little bit of knowledge. We will discuss two such topics: *a low-energy theorem* (due to Francis Low[15] for elastic photon–hadron scattering, principally Compton scattering off a proton or a neutron, and *selection rules* following from the quantum numbers conserved under the strong interactions: isospin $\mathbf{I}$, hypercharge Y, G-parity $G = e^{i\pi I_y}C$ for hadrons emitting one photon, two photons, etc. We will obtain selection rules close in spirit to those in atomic spectroscopy arising from the effects of spin-orbit terms in the LS coupling model. In that case it's hard to compute those terms, but we can make lots of statements about the rotational transformation properties. A similar statement will hold for photon–hadron processes.

35.3 A low-energy theorem

Let ω be the photon energy in the center of momentum frame (in which the energies of the incoming photon and outgoing photon are equal). There are two processes of interest. The first is

$$\gamma + p \to \gamma + p$$

(here p can be any charged spin-½ particle; the proton, for example). It has a differential cross-section which in the center of momentum frame can be written as (12.26)

$$\frac{d\sigma}{d\Omega} = \frac{1}{64\pi^2 E_T}|\mathcal{A}|^2 \tag{35.21}$$

where $\mathcal{A}$ is the amplitude, *not* averaged over anything. As ω, the energy of the photon, goes to zero, E_T^2 goes to $4m^2$. We'll see what $\mathcal{A}$ looks like shortly. The result, which we'll derive, is

[14] [Eds.] This admonition recalls John A. Wheeler's "First Moral Principle": *Never start a calculation before you know the answer.* Wheeler (1911–2008) was a postdoc and colleague of Bohr's, and the research supervisor of at least 46 Princeton PhD students, including Kip S. Thorne and Richard Feynman. He is credited for reviving the study of general relativity in the US after World War II, and popularizing the term "black hole."

[15] [Eds.] F. E. Low, "Scattering of Light of Very Low Frequency by Systems of Spin 1/2", *Phys. Rev.* **96** (1954) 1428–1432. Note that this is a "low-energy theorem", not a "Low energy theorem"! See also Bjorken & Drell *Fields* Sect. 19–13, pp. 357–362.

Low's theorem on soft photons:

$$\frac{d\sigma}{d\Omega} = (\text{known in terms of } e) + \omega \times (\text{known in terms of } e \text{ and } \mu) + \mathcal{O}(\omega^2) \tag{35.22}$$

The first two terms are known in terms of e and μ, the charge and static magnetic moment of the proton, including their full angular dependence and everything else. The third $\mathcal{O}(\omega^2)$ term is unknown; it involves the inner working of the strong interactions. In fact, if we don't consider the question of infrared divergences (caused by internal photons), the expression (35.22) can be shown to hold to *all* orders in e with, possibly, the second terms having some logs of ω in it. However, we will just work to lowest nontrivial order in e; that is, e^2. The second process is

$$\gamma + n \to \gamma + n$$

(n can be any spin-½, electrically neutral particle, for instance, a neutron), has a differential cross-section that can be expressed as

$$\frac{d\sigma}{d\Omega} = \omega^2 \times (\text{known in terms of } \mu) + \mathcal{O}(\omega^3) \tag{35.23}$$

The ω term is a charge-magnetic moment cross term, which vanishes because n is uncharged. There will be higher powers in ω, and we don't know what the $\mathcal{O}(\omega^3)$ term is. I don't claim that (35.23) is a convergent power series in ω; just that the terms vanish as $\omega \to 0$ at least as fast as ω^2.

We begin by considering photon–proton scattering, as shown in Figure 35.4.

$$\gamma(\varepsilon, k) + p(u, p) \to \gamma(\varepsilon', k') + p(u', p') \tag{35.24}$$

The ε and ε' are the photon polarizations, k and k' are the photon momenta, u and u' are the proton spinors and p and p' are the proton momenta. Energy–momentum conservation says

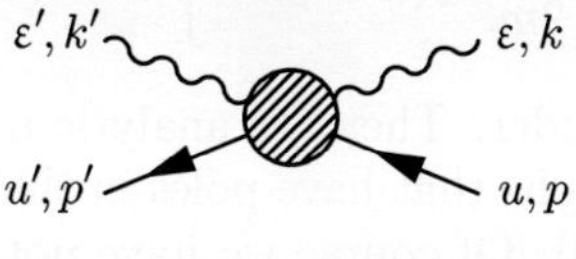

Figure 35.4: Photon-nucleon scattering

$$k + p = k' + p' \tag{35.25}$$

We put the protons on the mass shell:

$$p^2 = p'^2 = m^2 \tag{35.26}$$

We'll keep the photon masses completely free, but at a later stage in the computation we will let them go to zero:

$$k^2 \neq 0; \quad k'^2 \neq 0 \tag{35.27}$$

We might imagine a world in which there are heavy photons, or even two different kinds of heavy photons with two different masses. We could scatter a photon of the first kind and pull out a photon of the second kind. Or the incident photon could be virtual, emitted from an e^+–e^- pair.

The amplitude $\mathcal{A}$ for this process is going to be

$$\mathcal{A} = \varepsilon_\nu'^{*} \mathcal{A}^{\mu\nu} \varepsilon_\mu \tag{35.28}$$

where $\mathcal{A}^{\mu\nu}$ is constructed out of u and $\overline{u}'$ and p's and k's, etc. We break $\mathcal{A}^{\mu\nu}$ up into two terms

$$\mathcal{A}_{\mu\nu} = \mathcal{A}^B_{\mu\nu} + \mathcal{A}^A_{\mu\nu} \tag{35.29}$$

where $\mathcal{A}^B_{\mu\nu}$ is the **pole term** (or the **Born term**) and $\mathcal{A}^A_{\mu\nu}$ is analytic. I'll explain what we mean by $\mathcal{A}^B_{\mu\nu}$, and then you'll know what we mean by $\mathcal{A}^A_{\mu\nu}$. Graphically,

$i\mathcal{A}^B_{\mu\nu} =$ 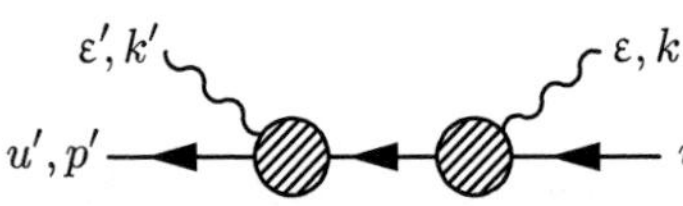

$+$ 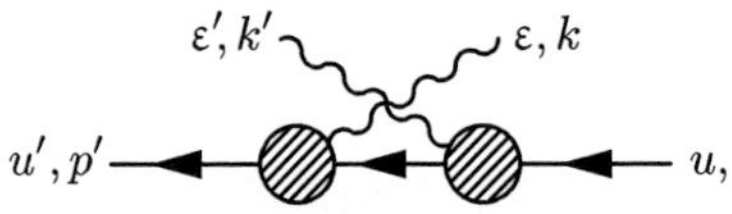

(35.30)

The blob is the term that gives us the residue at the pole that we know is present in these two graphs, at $s = m^2$ and $u = m^2$, respectively. That is, it is the value of this three-point function with everything on the mass shell[16]

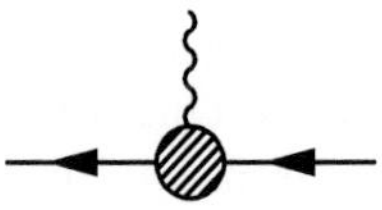 $= -ie\left[\gamma^\mu F_1(k^2) + \frac{i}{2m}F_2(k^2)\sigma^{\mu\nu}k_\nu\right]$ (35.31)

for each blob. The pole comes from the nucleon–antinucleon–photon vertex, when the nucleon is on the mass shell. So the pole term in $\mathcal{A}_{\mu\nu}$ is

$$\begin{aligned} i\mathcal{A}^B_{\mu\nu} = -ie^2\overline{u}'\left[\gamma_\nu F_1(k'^2) + \frac{i}{2m}F_2(k'^2)\sigma_{\nu\rho}k'^\rho\right]\left(\frac{1}{\not p + \not k - m + i\epsilon}\right) \\ \times\left[\gamma_\mu F_1(k^2) + \frac{i}{2m}F_2(k^2)\sigma_{\mu\lambda}k^\lambda\right]u \,+\, (k \leftrightarrow -k') \end{aligned} \tag{35.32}$$

The analytic terms are the remainder. They are analytic in the sense that we have extracted out the total residue of all the graphs that have poles in them. So $\mathcal{A}^A_{\mu\nu}$ has a Taylor expansion in k and k' near $k = 0$ and $k' = 0$. Of course we have not gone through a discussion of the analytic properties of the Feynman graphs, but it's plausible that those other graphs should be analytic at this point. After all, this was the same reasoning we used in our earlier discussion of strong interactions, when we discussed how to compute the πNN coupling constant g in Model 3 by looking for the pole in pion-nucleon scattering.[17] There is a potentially singular term, $\mathcal{A}^B_{\mu\nu}$ with poles in it; all the rest is non-singular at that point.

So far this is nothing new; we haven't introduced anything special about electrodynamics. The specifically electrodynamic part comes from the conservation of current, whatever the photon mass is: if we replace the photon polarization vector ε'^ν by k'^ν, the photon momentum, or ε^μ by k^μ, we get 0:[18]

$$k^\mu \mathcal{A}^A_{\mu\nu} = k'^\nu \mathcal{A}^A_{\mu\nu} = 0 \tag{35.33}$$

---

[16] [Eds.] See note 11 on p. 738.

[17] [Eds.] §16.3, particularly (16.37).

[18] [Eds.] This is yet another statement of the Ward identity; see Peskin & Schroeder *QFT*, equation (5.79), p. 160 and Section 7.4, pp. 238–244.

Our not having $k^2 = 0$ and $k'^2 = 0$ means that we can treat k and k' as independent variables; the only constraint is (35.25)

$$k + p = k' + p'$$

together with the requirement that the protons must be on their mass shell. Written another way,

$$k - k' = p' - p \tag{35.34}$$

As long as $k - k'$ is kept spacelike (which will still allow us to vary k and k' in four independent directions), we can keep p and p' on the mass shell with no problems.[19] Therefore, we can differentiate $\mathcal{A}^A_{\mu\nu}$ independently with respect to k and to k', treating k and k' each as four independent variables. From (35.33),

$$0 = \frac{\partial}{\partial k^\rho}(k^\nu \mathcal{A}^A_{\mu\nu})\bigg|_{k=0} = \mathcal{A}^A_{\mu\rho}\bigg|_{k=0} + \cancelto{0}{k^\nu \frac{\partial \mathcal{A}^A_{\mu\nu}}{\partial k^\rho}}\bigg|_{k=0} \tag{35.35}$$

By construction, $\mathcal{A}^A_{\mu\nu}$ is analytic at $k = 0$, and so is its derivative. Multiplying its derivative by k and sending $k \to 0$ causes the second term to vanish. Therefore,

$$\mathcal{A}^A_{\mu\nu} = \mathcal{O}(k) \tag{35.36}$$

It has no term of zeroth order in k. By exactly the same reasoning

$$\mathcal{A}^A_{\mu\nu} = \mathcal{O}(k') \tag{35.37}$$

Now if $\mathcal{A}^A_{\mu\nu}$ is $\mathcal{O}(k)$ and $\mathcal{O}(k')$, then the first term in the power series must be $\mathcal{O}(kk')$:

$$\mathcal{A}^A_{\mu\nu} = \mathcal{O}(kk') \tag{35.38}$$

There can't be a term with k but not k', because that wouldn't be zero when $k' \to 0$; there can't be a k' with no k because that wouldn't be zero when $k \to 0$; there can be a kk' term. Thus the conservation of charge, which implies (35.33), plus the analysis of singularities in terms of extracting out the poles, has given us something more powerful than either would have given us independently. If we just did the singularity analysis all we would know would be that $\mathcal{A}^A_{\mu\nu}$ is analytic (non-singular). Now we know much more. We know it vanishes as kk' as k and k' independently go to zero. This is the low-energy theorem of F. E. Low:

$$\lim_{k\to 0} \mathcal{A}^A_{\mu\nu} = \lim_{k'\to 0} \mathcal{A}^A_{\mu\nu} = 0 \tag{35.39}$$

If we did a similar singularity analysis for a process involving 72 photons we would get something vanishing like the product of all 72 photon momenta.

Armed with this knowledge, we return to the case where $k^2 = k'^2 = 0$. Now both k and k' are of order ω, the photon energy: the photon is on the mass shell, and the space parts of k and k' are the same magnitude as their time parts. Therefore, in this particular case, the low-energy theorem becomes

$$i\mathcal{A} = i\mathcal{A}^B + \mathcal{O}(\omega^2) \tag{35.40}$$

[19] [Eds.] $0 > (k - k')^2 = (p' - p)^2 = 2m^2 - 2p \cdot p' \;\Rightarrow\; p \cdot p' > m^2$.

where the Born term is given by

$$i\mathcal{A}^B = -i\varepsilon_\nu'^*\overline{u}'\left[e\gamma^\nu F_1(0) - \frac{ie}{2m}F_2(0)\sigma^{\nu\lambda}k'_\lambda\right]\left(\frac{\not{p}+\not{k}+m}{(p+k)^2-m^2}\right)$$
$$\times\left[e\gamma^\mu F_1(0) + \frac{ie}{2m}F_2(0)\sigma^{\mu\rho}k_\rho\right]u\varepsilon_\mu + \text{(the cross-term, } k \leftrightarrow -k') \tag{35.41}$$

The diagram in (35.31) is a typical Feynman graph except that at the vertex, in place of the bare coupling, we've put the effect of all the renormalization corrections, which as far the residue of the pole goes is just summed up in this expression for $i\mathcal{A}^B$. The $\mathcal{O}(\omega^2)$ comes into (35.40) because it doesn't matter which components of k and k' we have in (35.38); the product will be $\mathcal{O}(\omega^2)$.

We want now to count powers of ω, so as to establish the results in (35.22) and (35.23). Since both p and k are now on the mass shell, the denominator is

$$(p+k)^2 - m^2 = p^2 + k^2 + 2p\cdot k - m^2 = 2p\cdot k \tag{35.42}$$

In the proton's rest frame, $p = (m, \mathbf{0})$, and

$$2p\cdot k = 2m\omega \tag{35.43}$$

which goes to zero as the photon energy goes to zero. It looks like we get a factor of ω in the denominator of (35.41), coming from the cross term $F_1(0)$ with $F_1(0)$, charge with charge:

$$-ie^2\gamma^\nu\varepsilon_\nu'^*\frac{\not{p}+m}{2m\omega}\gamma^\mu\varepsilon_\mu$$

All the other terms have explicit powers of k in the numerator and are not $\mathcal{O}(\omega^{-1})$. So we have to look at possible terms of $\mathcal{O}(\omega^{-1})$ to count how many powers of ω^{-1} we get from the Born term; we will have to look at the cross terms between the Born term and the analytic term to see what we don't know about the total cross section.

Possible terms of $\mathcal{O}(\omega^{-1})$

To $\mathcal{O}(\omega)$, (35.41) is

$$\overline{u}'\not{\varepsilon}'^*\frac{\not{p}+m}{2k\cdot p}\not{\varepsilon}u \tag{35.44}$$

In general,

$$\varepsilon\cdot k = 0 \tag{35.45}$$

We can always choose ε to have only two components, perpendicular not only to the four-dimensional k but perpendicular to the space part of k. The space part of k is aligned with the space part of p in the center of momentum frame. Therefore we also have

$$\varepsilon\cdot p = 0 \tag{35.46}$$

since the time part of k is *a fortiori* aligned with the time part of p. So

$$\not{p}\not{\varepsilon} = -\not{\varepsilon}\not{p} + 2p\cdot\varepsilon = -\not{\varepsilon}\not{p} \tag{35.47}$$

and (35.44) becomes

$$\overline{u}'\not{\varepsilon}'^*\frac{\not{p}+m}{2k\cdot p}\not{\varepsilon}u = \overline{u}'\not{\varepsilon}'^*\not{\varepsilon}\frac{(-\not{p}+m)}{2k\cdot p}u = 0 \tag{35.48}$$

Therefore in the actual scattering amplitude the term of $\mathcal{O}(\omega^{-1})$ vanishes; the scattering amplitude begins at $\not{k}/\omega \sim \mathcal{O}(1)$.

We can now extract the results stated in (35.22) and (35.23). Staring at (35.41) and the corresponding expression for the cross-term, we have a term of $\mathcal{O}(1)$ proportional to e^2 coming from the $\not{k}$ term in the middle; we're not going to get rid of that. We'll also have terms of $\mathcal{O}(\omega)$ and $\mathcal{O}(\omega^2)$ coming from the F_1F_2 and F_2F_2 terms, respectively, which we could compute (but it's tedious). Hence we have established (35.22):

$$\frac{d\sigma}{d\Omega} = \mathcal{O}(1) + \mathcal{O}(\omega) + \mathcal{O}(\omega^2) \quad \textbf{QED}$$

What if, instead of a proton, we were considering a neutron (or indeed any electrically neutral spin-½ particle)? For a neutron,

$$e = 0; \quad \frac{e}{m}F_2(0) \equiv \mu \quad \text{(the magnetic moment of the neutron)} \tag{35.49}$$

In that case the $e\not{\varepsilon}'^*$ and the $e\not{\varepsilon}$ terms in (35.41) are completely missing from the amplitude. We see that the amplitude is $\mathcal{O}(k)$ because there are two k's in the two $F_2(0)$ terms and a k in the denominator. The denominator pole survives in this case, but of course it's killed by the powers of k in the numerator:

$$\frac{d\sigma}{d\Omega} = \mathcal{O}(\omega) + \cdots$$

Therefore the amplitude itself is a known term of $\mathcal{O}(\omega)$ plus an unknown term of $\mathcal{O}(\omega^2)$. We square that to get the cross section, and obtain the result stated in (35.23). The leading term is the Born approximation, and gives the famous Thomson formula for low-energy scattering.[20]

I left something out of the argument. I said that $k^\nu \mathcal{A}^A_{\mu\nu} = 0$, but actually we only know $k^\nu \mathcal{A}_{\mu\nu} = 0$ for the full $\mathcal{A}_{\mu\nu}$. Well, it's easy to check that the Born (pole) amplitudes by themselves satisfy this equation, $k^\nu \mathcal{A}^B_{\mu\nu} = 0$, because they are exactly the Born amplitudes that would arise in a completely gauge invariant theory with anomalous magnetic moment couplings; and by conservation of charge, $k^\nu \mathcal{A}^B_{\mu\nu} = 0$. Since $k^\nu \mathcal{A}_{\mu\nu} = 0$ and $k^\nu \mathcal{A}^B_{\mu\nu} = 0$, we must have $k^\nu \mathcal{A}^A_{\mu\nu} = 0$ also.

35.4 Photon-induced corrections to strong interaction processes (via symmetries)

We now turn to our second topic. The first class of processes we will consider are of $\mathcal{O}(e)$ in amplitude: they involve one interaction of the photon with strongly interacting particles. For instance

$$i \to f + \gamma \tag{35.50}$$

or (see Figure 35.5)

$$i \to f + e^+ + e^- \tag{35.51}$$

where i is the initial hadronic state, f is the final hadronic state, γ is a photon and e^+ and e^- form an electron–positron pair. (The second process is of $\mathcal{O}(e^2)$ but as far as the strong interaction end of the graph goes, there is only one photon involved.) Equivalently, we look at

---

20 [Eds.] Bjorken & Drell *Fields*, pp. 361–362: their equation (19.137) is $\left.\dfrac{d\sigma}{d\Omega}\right|_{\text{Thomson}} = \dfrac{\alpha^2}{m^2}(\varepsilon \cdot \varepsilon')^2$ for $k \to 0$.

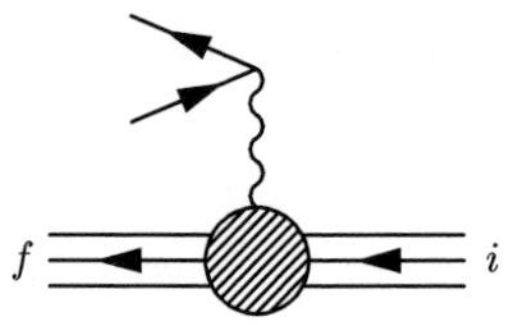

Figure 35.5: Second order electromagnetic correction to strong interaction process

hadron electromagnetic form factors, all to lowest nontrivial order in e.

All of these processes are governed by the matrix element of the electromagnetic current between the initial and final hadronic states:

$$\langle f|j_\mu^{\text{em}}|i\rangle$$

We will assume that the electromagnetic current is constructed by the minimal coupling prescription, whatever the theory of hadrons may be. We know the *Gell-Mann–Nishijima relation*[21]

$$Q = I_z + \tfrac{1}{2}Y \tag{35.52}$$

so that

$$j_\mu^{\text{em}} = j_\mu^{I_z} + \tfrac{1}{2}j_\mu^{Y} \tag{35.53}$$

I_z and Y are commuting quantities. The currents $j_\mu^{I_z}$ and j_μ^Y are the currents we would get if the photon were coupled exclusively to I_z or Y, respectively. The strong interactions[22] strictly conserve both isospin and hypercharge, and the electromagnetic interactions conserve I_z and Y (but *not* I), so we have

$$\Delta I_z = \Delta Y = 0 \;\Rightarrow\; \Delta Q = 0 \tag{35.54}$$

Now I_z, the integrated time component of the isotopic current $j_\mu^{I_z}$, is part of an isotriplet. If we count the quantum numbers of the initial and final hadron states, from the $j_\mu^{I_z}$ term in the electromagnetic current we have $\Delta I = 1$. But Y, the integrated time component of the hypercharge current j_μ^Y, is an isosinglet, so that gives us $\Delta I = 0$. That is, in electromagnetic processes,

$$\Delta I_z = \Delta Y = 0\colon \begin{cases} \Delta I = 1 \\ \Delta I = 0 \end{cases} \tag{35.55}$$

The G-parities are different for these two cases.[23] G is the product of charge conjugation and a 180° rotation about the y (or 2) axis in isospin space:

$$G = e^{i\pi I_y}C \tag{35.56}$$

---

[21] [Eds.] See note 10, p. 520. For notational convenience, we're using I_z in place of I_3.

[22] [Eds.] Here, at 1:12:50 in the video of Lecture 35, there is hissing for about 15 seconds so Coleman cannot be heard. The end of this sentence comes from Coleman's own notes.

[23] [Eds.] T. D. Lee and C. N. Yang, "Charge Conjugation, a New Quantum Number G, and Selection Rules Concerning a Nucleon-Antinucleon System", *Nuovo Cim.* **3** (4) (1956) 749–753; T. D. Lee, *Particle Physics and Introduction to Field Theory*, Harwood Academic Publisher, New York, 1981, Section 11.2, "G-Parity", pp. 225–230; Section 11.3, "Applications to Mesons and Baryons", pp. 230–240. See also §24.4.

Thus, for example, using the usual conventions,[24] with

$$C\colon \begin{pmatrix}\pi^+\\ \pi^-\\ \pi^0\end{pmatrix} \to \begin{pmatrix}\pi^-\\ \pi^+\\ \pi^0\end{pmatrix} \tag{35.57}$$

we have

$$C\colon \begin{pmatrix}\pi^1\\ \pi^2\\ \pi^3\end{pmatrix} \to \begin{pmatrix}\pi^1\\ -\pi^2\\ \pi^3\end{pmatrix} \quad \text{and so} \quad G\colon \begin{pmatrix}\pi^1\\ \pi^2\\ \pi^3\end{pmatrix} \to \begin{pmatrix}-\pi^1\\ -\pi^2\\ -\pi^3\end{pmatrix} \tag{35.58}$$

That is, under G-parity the isovector $\boldsymbol{\pi} = (\pi^1, \pi^2, \pi^3)$ transforms very simply:

$$G\colon \boldsymbol{\pi} \to -\boldsymbol{\pi} \tag{35.59}$$

The Lagrangian describing hadrons' electromagnetic interactions is invariant under charge conjugation, and the electromagnetic current shows up in it as the product $j_\mu^{\text{em}} A^\mu$. Since the photon is charge-conjugation odd (the electromagnetic field changes sign when positive and negative charges are interchanged), so is j_μ^{em}.[25] From (35.53), *both* j_μ^Y and $j_\mu^{I_z}$ must be odd under charge conjugation:

$$C\colon \begin{pmatrix} j_\mu^Y \\ j_\mu^{I_z}\end{pmatrix} \to \begin{pmatrix} -j_\mu^Y \\ -j_\mu^{I_z}\end{pmatrix} \tag{35.60}$$

Charge conjugation changes the hypercharge of the hadrons. But these currents have opposite G-parities. Since j_μ^Y is an isosinglet, I_y (or indeed any rotation in isospin space) has no effect on it, so its G-parity is the same as its charge conjugation: odd. On the other hand, while the current $j_\mu^{I_z}$ has the same charge conjugation properties as the hypercharge current, when we rotate $j_\mu^{I_z}$ by 180° about the 2 axis in isospin space, it changes sign once more. So $j_\mu^{I_z}$ is G even, and we have

$$G\colon \begin{pmatrix} j_\mu^Y \\ j_\mu^{I_z}\end{pmatrix} \to \begin{pmatrix} -j_\mu^Y \\ j_\mu^{I_z}\end{pmatrix} \tag{35.61}$$

Thus we get the selection rules for two types of hadronic reactions that can be induced by the emission of a single photon

$$\Delta I_z = \Delta Y = 0\colon \begin{cases} \Delta I = 1 \;\text{ and }\; \Delta G = 0 & \text{due to } j_\mu^{I_z} \\ \Delta I = 0 \;\text{ and }\; \Delta G \neq 0 & \text{due to } j_\mu^Y \end{cases} \tag{35.62}$$

Of course the ΔG rule is only useful if the initial and final states are G eigenstates.[26] For example, it doesn't tell us anything about $p \to$ (something) except to connect it to $\overline{p} \to$ (something), where $\overline{p}$ is an antiproton. Let's look at some examples.

EXAMPLE 1: γ *decays*

The famous decay[27]

$$\Sigma^0 \to \Lambda + \gamma \tag{35.63}$$

---

[24] [Eds.] See (S14.20), p. 551.

[25] [Eds.] M. Gell-Mann and A. Pais, "Behavior of Neutral Particles under Charge Conjugation", *Phys. Rev.* **97** (1955) 1387–1389.

[26] [Eds.] See note 12 on p. 523.

[27] [Eds.] *PDG* 2016, p. 94.

is allowed. The Σ^0 is the $I_z = 0$ component of an isotriplet, and the neutral Λ is an isosinglet. Both have $Y = 0$. So

$$\Delta I_z = \Delta Y = 0, \ \Delta I = 1 \tag{35.64}$$

The $\Delta I = 1$ is permitted. (G-parity is irrelevant here, because it can only be checked if we know the relative phase of the process $\overline{\Sigma^0} \to \overline{\Lambda} + \gamma$.) The decay (35.63) is indeed an allowed process, coming from the $\Delta I = 1$, $\Delta G = 0$ part of the electromagnetic current, $j_\mu^{I_z}$.

For the emission of two photons, apply the same rule twice; it is a *second-order* process. Consider the decay

$$\pi^0 \to \gamma + \gamma \tag{35.65}$$

Again, π^0 is the $I_z = 0$, $Y = 0$ component of an isotriplet:

$$\pi^0 \colon I^G(J^{PC}) = 1^-(0^{-+})$$

so again we have

$$\Delta I_z = \Delta Y = 0, \ \Delta I = 1 \tag{35.66}$$

The total change in G is 1, because the final hadronic state is the vacuum, which has $G = 0$, and the initial state is a single pion, which has $G = -1$. The (absolute value of the) total change in isospin is *also* 1. Therefore one photon must come from *each* of the currents in (35.62). Notice that if these rules are correct, a neutral G-odd isosinglet (such as the ω) would *not* be allowed to decay into two γ's.

EXAMPLE 2: *Magnetic moments within an isomultiplet*

Say we have an isomultiplet, and all of its members have magnetic moments. Well, are they all independent, or are they connected in some way? The magnetic moment is connected by kinematic factors (which we have worked out for one electron) to the matrix element of j_μ^{em}:

$$\langle a, \cdots | j_\mu^{\text{em}}(0) | b, \cdots \rangle$$

Suppose the states $|a, \cdots\rangle$ and $|b, \cdots\rangle$ are members of an isomultiplet. In fact the current only has diagonal matrix elements, but it's useful to consider anything on the right-hand side and anything on the left-hand side. The electromagnetic current is the sum (35.53) of two parts, one of which transforms like an isoscalar and whose matrix elements must therefore be proportional to δ_{ab}, and one of which transforms like the z-component (or the 3 component) of an isovector and therefore, by the **Wigner–Eckart theorem**[28], must be proportional to I_z. For example,

$$\langle \Sigma^0 | j_\mu^{em} | \Sigma^0 \rangle = \alpha \langle \Sigma^0 | I_z | \Sigma^0 \rangle + \beta \tag{35.67}$$

Therefore we have the following rule for the magnetic moments:

$$\mu = \alpha I_z + \beta \tag{35.68}$$

α and β are constants, α coming from the isovector part of the current, β from the isoscalar part. We must be able to solve the strong interaction problem to actually compute them.

Unfortunately, this is a useless formula for practical purposes, because the only isomultiplet for which we have measured all the magnetic moments is the neutron and proton, which

[28] [Eds.] J. J. Sakurai, *Modern Quantum Mechanics*, Addison-Wesley, 1994, pp. 238–242.

has two magnetic moments. It's no great feat to fit two experimental numbers with two adjustable parameters! The Σ moments would be a test, but unfortunately the Σ^0 moment is hard to measure because it is so damned unstable, decaying rapidly into $\gamma + \Lambda$ in a time of $\approx 7.4 \times 10^{-20}$ seconds.[29] (The $\Sigma^{\pm}$ lifetimes are much longer, $\approx 1.48 \times 10^{-10}$ seconds.) So at the moment this is a beautiful formula which everyone believes, but which is totally untested. However, in a later lecture[30] we will obtain a similar formula by identical reasoning based on SU(3), which relates the magnetic moments of a larger group of particles. That formula has more constraints and *has* been tested; it's pretty good.[31]

EXAMPLE 3: *Second-order processes and $\mathcal{O}(e^2)$ corrections in $i \to f$.*

These are processes where some initial hadronic state goes into some final hadronic state via electromagnetic interactions. They are processes of $\mathcal{O}(e^2)$, although they may not involve any explicit photons or currents. Typically, the only cases in which these can be measured experimentally are those in which the process is forbidden by the selection rules for the strong interactions, so they arise only because of the electromagnetic corrections. Examples of such processes include these two decay modes of the η into three pions:

$$\eta \to \begin{cases} \pi^0 + \pi^0 + \pi^0 \\ \pi^0 + \pi^+ + \pi^- \end{cases} \tag{35.69}$$

Because the η is even under G-parity it should not decay (via the strong interactions) into three pions, which have odd G-parity. However, the decay *does* occur, presumably due to electromagnetic corrections. The η is electrically neutral and the decay

$$\eta \to \gamma + \gamma$$

is allowed. But if there were a G-odd version of the η, a scalar particle with $I^G = 0^-$, it could not decay into two γ's.

Another example of such a process is the electromagnetic mass-splitting within an isomultiplet. In that case i and f are just a single particle each. In these processes, if we understand quantum electrodynamics but not necessarily the strong interactions, we have a blob with strong interaction mysteries going on within. From the inside of this blob we pluck, as if pulling the string on a violin, two pairs of charged fermions (quarks, perhaps) connected by a photon, as in Figure 35.6. If there are fundamental charged *bosons* lurking inside the blob, then we can also have graphs like Figure 35.7, where the charged boson comes out and the photon forms a loop, because a charged boson couples directly to two photons. This second kind of graph can be eliminated by choosing a photon propagator which is a bit different from the ones we've considered before:

$$\widetilde{D}_{\mu\nu}(k) = \frac{-i}{k^2 + i\epsilon}\left[g_{\mu\nu} - 4\frac{k_\mu k_\nu}{k^2}\right] \tag{35.70}$$

---

[29] [Eds.] *PDG* (2016), p. 94. The quark model predicts that the Σ^0 moment is the average of the Σ^+ and Σ^- moments: P. Pal, *An Introductory Course of Particle Physics*, Taylor and Francis, 2015, Section 10.8.2, pp. 283–288.

[30] [Eds.] §38.3, pp. 835–839.

[31] [Eds.] Coleman adds: "Good enough to get me my first job at Harvard, and tenure. That, and my charm." He is referring to his first publication, written with Sheldon L. Glashow; see note 40, p. 841.

Figure 35.6: Second order electromagnetic correction to strong interaction process: fermions

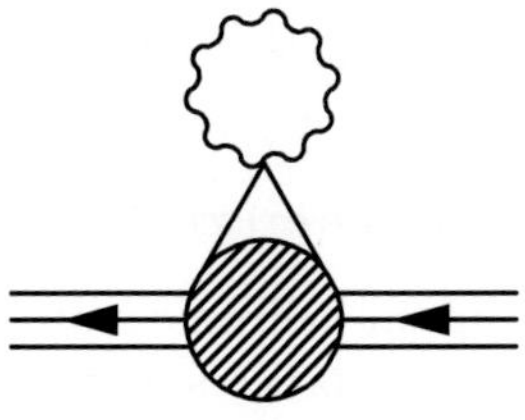

Figure 35.7: Second order electromagnetic correction to strong interaction process: bosons

It's just another choice of α; it's as good a gauge as any other. It has the advantage that

$$\widetilde{D}^{\lambda}{}_{\lambda} = 0 \tag{35.71}$$

and therefore Figure 35.7, proportional to $\widetilde{D}^{\lambda}{}_{\lambda}$, vanishes. The remaining graph, Figure 35.6, can be thought of (aside from kinematic factors) as proportional to

$$\int d^4x\, d^4y\, \langle f|\, T\Big[j^{\rm em}_{\mu}(x) j^{\rm em}_{\nu}(y)\Big]\,|i\rangle\, D^{\mu\nu}(x-y) \tag{35.72}$$

Therefore, the transformation properties of the first graph are the transformation properties of the products of two currents. Remember that the individual currents are the sum (35.53) of two parts; for the product of two currents we simply apply the reasoning leading to (35.62) twice. From the product of the two j^{Y}_{μ} currents, we combine $\Delta I = 0$ and $\Delta G = 1$ with $\Delta I = 0$ and $\Delta G = 1$ to get overall $\Delta I = 0$ and $\Delta G = 0$. Continuing in this way, and mindful of the way isospin $\mathbf{I}$ adds, we obtain the following:

$$\begin{aligned} (\Delta I = 0)\times(\Delta I = 0)&\colon\ \Delta I = 0,\ \Delta G = 0 \\ (\Delta I = 0)\times(\Delta I = 1)&\colon\ \Delta I = 1,\ \Delta G = 1 \\ (\Delta I = 1)\times(\Delta I = 1)&\colon\ \begin{cases} \Delta I = 0,\ \Delta G = 0 \\ \Delta I = 2,\ \Delta G = 0 \\ \Delta I = 1,\ \Delta G = 0 \end{cases} \end{aligned} \tag{35.73}$$

The last of these, $\Delta I = 1$ and $\Delta G = 0$, is out. If we combine two objects that transform as I_z's, we can't make a product that behaves as I_z in an isospin 1 state: the isospin 1 state is antisymmetric, so that Clebsch–Gordan coefficient vanishes. It's just like the cross-product of two identical vectors; you get zero. That's the complete list of transformation properties. The isospin can change by 0, 1 or 2. If it changes by an even amount, it must be $\Delta G = 0$; if it changes by an odd amount, it must be $\Delta G = 1$.

Let's return to η decay. The η is G-even:

$$\eta\colon\ I^G(J^{PC}) = 0^+(0^{-+})$$

Its mass is 547.86 MeV. The masses of the π^0 and $\pi^{\pm}$ are 134.98 MeV and 139.57 MeV, respectively, each less than ⅓ of the η's mass. So the decay of the η into three or fewer pions is energetically possible, though some processes may be otherwise forbidden:

$$\eta \to \pi^0 \qquad \text{(forbidden by 4-momentum conservation)} \tag{35.74}$$

$$\eta \to \begin{cases} \pi^0 + \pi^0 \\ \pi^+ + \pi^- \end{cases} \qquad \text{(forbidden by parity; two } \pi\text{'s in an } s \text{ state have even parity)} \tag{35.75}$$

$$\eta \to \begin{cases} \pi^0 + \pi^0 + \pi^0 \\ \pi^+ + \pi^- + \pi^0 \end{cases} \qquad \text{(allowed in } \textit{second} \text{ order; these modes account for } \sim 55\% \text{ of } \eta \text{ decays)} \tag{35.76}$$

Note the non-conservation of isospin in these last reactions.

In the three-pion decays, G changes by 1 so this decay must have

$$\Delta I = 1$$

Since the initial hadronic state, the η, has isospin 0, the final state of three π's, must be an $I = 1$ state with $I_z = 0$. Thus, although at the moment we know nothing about the momentum distribution of the three final π's, we know quite a bit about the isospin dependence of the three-π wave function: it must be a state of total isospin 1. This should enable us, with a little work, to calculate such things as the ratio of η decay rates:

$$\frac{\Gamma(\eta \to 3\pi^0)}{\Gamma(\eta \to \pi^+\pi^-\pi^0)} \tag{35.77}$$

Next time, I will begin to establish such a connection and test it with experiment. I will then begin a new topic (and pursue that for several more lectures), one that is qualitatively different from what we have been doing in the last few weeks. Instead of fancy field theory, we'll start fancy group theory: I will talk about SU(3).

Problems 19

19.1 We've worked out some general properties of a charged Dirac field minimally coupled to a massless photon. We

(a) derived the Ward identity for the photon–spinor–antispinor 1PI vertex (33.79);

(b) verified the identity in the tree approximation (33.90);

(c) used the identity to prove that the physical charge, defined as the quantity that appears in the gauge transformation of the physically renormalized photon field, is the same as the physical charge defined as the quantity that appears in the vertex with everything on the mass shell (33.100).

We then went on to analyze the kinematic structure of the vertex with the two spinors on the mass shell, but with the photon carrying arbitrary momentum q. We

(d) showed that there were only two independent form factors F_1 and F_2;

(e) constructed the explicit expressions (34.15) defining $F_1(q^2)$ and $F_2(q^2)$;

(f) observed that our result (c) implied (34.16) that $F_1(0) = 1$.

Do the parallel constructions for the case where the charged particle is a *scalar*.

Comment: This is an *easy* problem. Step (a) is trivial, since I never used the spin of the charged field in the derivation. The only change is a notational one, replacing $\widetilde{S}'$ by $\widetilde{D}'$. All the other steps, though non-trivial, are much easier when you don't have to worry about spin and γ matrices.

(1998b 8.1)

19.2 Two electrically neutral Dirac fields, ψ_1 and ψ_2, of masses m_1 and m_2, respectively, interact with a massless photon through the coupling

$$\mathscr{L}' = g\overline{\psi}_2\sigma_{\mu\nu}\psi_1 F^{\mu\nu} + \text{h. c.} \tag{P19.1}$$

where g is a real number and "h. c." stands for Hermitian conjugate. These fields have no other interactions. If m_1 is greater than m_2, the decay

$$\psi_1 \to \psi_2 + \gamma \tag{P19.2}$$

is kinematically allowed. Compute the decay width Γ for this process (summed over final spins and averaged over initial spins) to lowest nontrivial order in perturbation theory.

Comments:

(a) This theory is nonrenormalizable, but for this problem, it doesn't matter, since we're only working in tree approximation.

(b) You are actually computing the decay

$$\Sigma^0 \to \Lambda^0 + \gamma \tag{P19.3}$$

to lowest order in electromagnetism, but to *all* orders in the strong interactions. Just as in the class discussion of electron–proton scattering (§34.2, in particular the aside starting on p. 738) all the effects of the strong interactions can be summed up in terms of two form factors, F_1 and F_2. (Because the incoming and outgoing hadrons have different masses, the detailed definition of F_1 is a little different than in class, but this is not important here.) At $q^2 = 0$ one can show that $F_1(q^2) = 0$; otherwise one would have a very strange inverse-square force between neutral particles. So all the difficult-to-compute effects of the strong interactions are summed up in a single number, $F_2(0)$.

(1991b 6.2)

Solutions 19

19.1 (a) Following the definition (33.56), let $\widetilde{\Gamma}'^{(n_1,n_2,m)}$ stand for the full Green's functions, with n_1 initial scalar particles, n_2 final scalar particles, and m photons. As in (33.88), the Ward identity (actually the Ward–Takahashi identity) is

$$e^{-1}q^\mu\widetilde{\Gamma}'^{(1,1,1)}_\mu(p',p,q) = \left[\widetilde{\Gamma}'^{(1,1,0)}(p,p) - \widetilde{\Gamma}'^{(1,1,0)}(p',p')\right] \tag{S19.1}$$

where $p' = p+q$. For scalar particles we have (32.16)

$$\widetilde{\Gamma}'^{(1,1,0)}(p,p) = \frac{i}{\widetilde{D}'(p^2)} \tag{S19.2}$$

and so

$$ie^{-1}q^\mu\widetilde{\Gamma}'^{(1,1,1)}_\mu(p',p,q) = \frac{1}{\widetilde{D}'(p'^2)} - \frac{1}{\widetilde{D}'(p^2)} \tag{S19.3}$$

(b) In the tree approximation, we have

$$ie^{-1}q^\mu\widetilde{\Gamma}'^{(1,1,1)}_\mu(p',p,q) = -iq^\mu(p_\mu + p'_\mu) = -i(p'^2 - p^2) \tag{S19.4}$$

as well as

$$\begin{aligned}\frac{1}{\widetilde{D}'(p'^2)} - \frac{1}{\widetilde{D}'(p^2)} &= -i((p'^2 - m^2) - (p^2 - m^2)) = -i(p'^2 - p^2) \\ &\stackrel{\checkmark}{=} ie^{-1}q^\mu\widetilde{\Gamma}'^{(1,1,1)}_\mu(p',p,q)\end{aligned} \tag{S19.5}$$

and the identity is verified at the tree level; *cf.* (33.90).

(c) The physical charge e_{phys} is defined by the condition

$$i\widetilde{\Gamma}'^{(1,1,1)}_\mu(p',p,q) = -ie_{\text{phys}}(p'_\mu + p_\mu) \tag{S19.6}$$

with everything on the mass shell: $p^2 = p'^2 = m^2$, $q^2 = 0$. Differentiating (S19.3) with respect to q^μ, we get

$$\frac{\partial}{\partial q^\mu}\left[ie^{-1}q^\nu\widetilde{\Gamma}'^{(1,1,1)}_\nu(p',p,q)\right] = \frac{\partial}{\partial q^\mu}\left[\frac{1}{\widetilde{D}'(p'^2)} - \frac{1}{\widetilde{D}'(p^2)}\right] \tag{S19.7}$$

In the limits $p^2 \to m^2$, $p'^2 \to m^2$, $q^2 \to 0$, we find

$$\begin{aligned}ie^{-1}\widetilde{\Gamma}'^{(1,1,1)}_\mu(p,p,0) &= \frac{\partial}{\partial q^\mu}\left[i(p^2 - m^2) - i(p'^2 - m^2)\right]\Bigg|_{q^\mu=0} = \frac{\partial}{\partial q^\mu}\left[-iq^\nu(p_\nu + p'_\nu)\right]\Bigg|_{q^\mu=0} \\ &= -2ip_\mu\end{aligned} \tag{S19.8}$$

In these same limits, (S19.6) becomes

$$i\widetilde{\Gamma}'^{(1,1,1)}_\mu(p,p,0) = -2ie_{\text{phys}}p_\mu \tag{S19.9}$$

Comparing (S19.8) with (S19.6), we obtain as in (33.100)

$$e = e_{\text{phys}} \tag{S19.10}$$

(d) In §34.1, we used crossing symmetry and angular momentum arguments to consider the photon as decaying into a charged particle–antiparticle pair. Since the photon carries quantum numbers $J^{PC} = 1^{--}$, and the charged particles are scalars, the final state must have $j = \ell = 1$. Thus there is only *one* invariant amplitude, and hence only one vertex function, unlike in spinor electrodynamics.

(e) Analogous to rule (c) in the box of Feynman rules for scalar electrodynamics on p. 644, define the sole invariant vertex function by

$$i\widetilde{\Gamma}'^{(1,1,1)}_{\mu}(p', p, q) = -ie(p'_\mu + p_\mu)F(q^2) \tag{S19.11}$$

for $p^2 = p'^2 = m^2$. From (S19.6) and (S19.10), we know

$$i\widetilde{\Gamma}'^{(1,1,1)}_{\mu}(p, p, 0) = -2ie(p_\mu) \tag{S19.12}$$

Taking the limit of (S19.11) as $q \to 0$ gives

$$i\widetilde{\Gamma}'^{(1,1,1)}_{\mu}(p, p, 0) = -2iep_\mu F(0) = -2iep_\mu \;\Rightarrow\; F(0) = 1 \tag{S19.13}$$

which was to be shown. ■

19.2 The relevant Feynman diagram is:

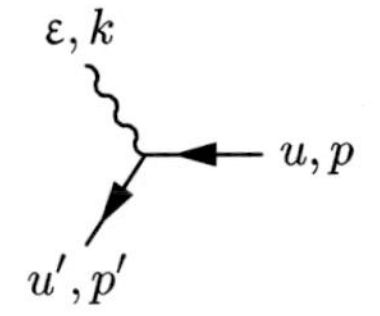

We can always choose ε^μ such that

$$\varepsilon \cdot p = \varepsilon \cdot p' = \varepsilon \cdot k = 0 \tag{S19.14}$$

With this choice,

$$\{\not{\varepsilon}, \not{p}\} = \{\not{\varepsilon}, \not{p}'\} = \{\not{\varepsilon}, \not{k}\} = 0 \tag{S19.15}$$

Also,

$$\not{\varepsilon}\not{\varepsilon} = -1; \qquad \not{k}\not{k} = 0 \tag{S19.16}$$

Then

$$i\mathcal{A} = g\overline{u}'[\not{k}\not{\varepsilon} - \not{\varepsilon}\not{k}]u = 2g\overline{u}'\not{k}\not{\varepsilon}u \tag{S19.17}$$

Averaging over initial spins and summing over final spins, as well as the two polarization states gives

$$\sum_{\substack{\text{final spins,}\\ \text{polarizations}}} |\mathcal{A}|^2 = \tfrac{1}{2}\sum_{r=1}^{2}(4g^2)\,\text{Tr}\left[(\not{p}' + m)\not{k}\not{\varepsilon}^{(r)}(\not{p} + m)\not{\varepsilon}^{(r)}\not{k}\right] = (4g)^2\,\text{Tr}\left[(\not{p}' + m)\not{k}(\not{p} - m)\not{k}\right] \tag{S19.18}$$

The trace of an odd number of γ's is zero, and $\not{k}\not{k} = 0$:

$$\begin{aligned}
\text{Tr}\left[(\not{p}' + m)\not{k}(\not{p} - m)\not{k}\right] &= \text{Tr}\left[\not{p}'\not{k}\not{p}\not{k}\right] \\
&= 4[(p' \cdot k)(p \cdot k) - (p' \cdot p)(k \cdot k) + (p' \cdot k)(p \cdot k)] \\
&= 8(p' \cdot k)(p \cdot k)
\end{aligned} \tag{S19.19}$$

using (S11.16), p. 428 for the trace of four γ's. We can reduce the dot products because $p = p' + k$, and so

$$\begin{aligned}
(p' + k)^2 &= p^2 = m_1^2 \;\Rightarrow\; 2p' \cdot k = m_1^2 - m_2^2 \\
(p - k)^2 &= p'^2 = m_2^2 \;\Rightarrow\; 2p \cdot k = m_1^2 - m_2^2
\end{aligned} \tag{S19.20}$$

Then

$$\sum_{\text{final spins}} |\mathcal{A}|^2 = 8g^2(m_1^2 - m_2^2)^2 \tag{S19.21}$$

The decay width is given by (12.33),

$$\Gamma = \frac{1}{2m_1}\int \sum_{\text{final spins}} |\mathcal{A}|^2 \frac{1}{16\pi^2}\frac{|\mathbf{p}_f|\,d\Omega_f}{E_T} \tag{S19.22}$$

The integration over the solid angle gives 4π. In the center of momentum frame, $E_T = m_1$, $p \cdot k = m_1 k^0$, and

$$|\mathbf{p}_f| = |\mathbf{p} - \mathbf{p}'| = |\mathbf{k}| = k^0 = \frac{p \cdot k}{m_1} = \frac{(m_1^2 - m_2^2)}{2m_1}$$

so that finally

$$\Gamma = \frac{g^2}{2\pi} \frac{(m_1^2 - m_2^2)^3}{m_1^3} \tag{S19.23}$$

is the decay width. ■

36

Introducing SU(3)

At the end of the last lecture, we were discussing the phenomenology of η decays. I have a little more to say about that topic, and then we'll begin a discussion of the approximate symmetry group SU(3).

36.1 Decays of the η

Recall that the η is a spin-0 meson with negative parity and positive G-parity:

$$\eta\colon\ J^{PG} = 0^{-+} \tag{36.1}$$

The pion has $I^G = 1^-$ (35.59). The η decay

$$\eta \to 3\pi \tag{36.2}$$

must be into a state with $I = 1, I_z = 0$, because the only part of the second-order electromagnetic interaction that can change G carries $\Delta I = 1$; see (35.73). (Remember, I_z is conserved by electromagnetic interactions, but $\mathbf{I}$ itself is not.) The final state must be the $I_z = 0$ member of an isotriplet. This enables us to connect the decays $\eta \to 3\pi^0$ and $\eta \to \pi^+\pi^-\pi^0$ by Clebsch–Gordan considerations of the isospin. We'll look at that process in some detail to get an idea of what sort of restrictions the Clebsch–Gordan arguments impose. We'll have to introduce a set of variables for the three-pion system. We talked about the kinematics of three-particle decays in §12.5. We'll also make use of Dalitz plots (see the discussions following Figure 11.7 on p. 235, and Figure 12.5 on p. 256).[1]

Convenient variables are the energies of the three pions. Those are not all independent, of course, because

$$E_1 + E_2 + E_3 = m_\eta = 548\,\text{MeV} \tag{36.3}$$

Thus the allowed points in the diagram can either be in the E_1, E_2 or E_3, etc., plane, which form the Dalitz plot (here, $\{1, 2, 3\}$ label the three pions). For our purposes it will be useful

[1] [Eds.] The detailed analysis of Dalitz plots is a highly technical subject. See J. D. Jackson and D. R. Tovey, Section 46, "Kinematics", in *PDG* 2016, `http://pdg.lbl.gov/2016/reviews/rpp2016-rev-kinematics.pdf`.

to introduce new variables

$$\epsilon_i = E_i - \tfrac{1}{3}m_\eta, \quad \sum_{i=1}^{3} \epsilon_i = 0 \tag{36.4}$$

The center of the Dalitz plot, assuming we ignore the electromagnetic mass differences between the pions, is the point where all the ϵ's vanish. Treating the three pions for the moment as distinguishable particles (which we certainly can, except for a set of measure zero in the Dalitz plot) we will introduce three isospin unit vectors

$$\{\hat{\mathbf{e}}_1,\, \hat{\mathbf{e}}_2,\, \hat{\mathbf{e}}_3\} \tag{36.5}$$

just like the polarization vectors we had for photons, except that in this case they measure the directions of the three one-particle states in isospin space. For the different pion states

$$\pi^0:\ \hat{\mathbf{e}} = \hat{\mathbf{z}} \tag{36.6}$$

$$\pi^\pm:\ \hat{\mathbf{e}} = \tfrac{1}{\sqrt{2}}(\hat{\mathbf{x}} \pm i\hat{\mathbf{y}}) \quad \text{(never mind which is which)} \tag{36.7}$$

How many amplitudes can we construct that have $I = 1$? This is pretty easy. Label a representation of the rotation group by its spin. If we put together two pions, without worrying about statistics or anything, we can construct a state of isospin 0, 1, or 2:

$$1 \otimes 1 = 0 \oplus 1 \oplus 2 \tag{36.8}$$

(This is the Clebsch–Gordan series (18.73) applied to isospin. That is, the combined states have $I_1 + I_2$, $I_1 + I_2 - 1$, $\cdots$ $|I_1 - I_2|$). If we put in a third pion, we get isospin 1 three times, one from each of the factors:

$$[1 \otimes 1] \otimes 1 = [0 \oplus 1 \oplus 2] \otimes 1 = (2 \otimes 1) \oplus (1 \otimes 1) \oplus (0 \otimes 1) = 1 \oplus 1 \oplus 1 \oplus \cdots \text{(non-1 part)} \tag{36.9}$$

We don't care about the non-1 part. Thus we should be able to construct three functions of the ϵ_i's, linear in the three $\mathbf{e}$'s, that transform like an isovector under isotopic rotations. It is easy to see what those functions are:

$$\begin{aligned}\boldsymbol{\mathcal{A}}(\hat{\mathbf{e}}_1, \hat{\mathbf{e}}_2, \hat{\mathbf{e}}_3) = \hat{\mathbf{e}}_1(\hat{\mathbf{e}}_2 \bullet \hat{\mathbf{e}}_3)\, F(\epsilon_1, \epsilon_2, \epsilon_3)& \\ + \hat{\mathbf{e}}_2(\hat{\mathbf{e}}_3 \bullet \hat{\mathbf{e}}_1)\, G(\epsilon_1, \epsilon_2, \epsilon_3) + \hat{\mathbf{e}}_3(\hat{\mathbf{e}}_1 \bullet \hat{\mathbf{e}}_2)\, H(\epsilon_1, \epsilon_2, \epsilon_3)&\end{aligned} \tag{36.10}$$

where F, G and H are functions of the ϵ_i's. The amplitude $\boldsymbol{\mathcal{A}}$ is the sum of three linearly independent amplitudes that transform like isovectors. The generalized exclusion principle tells us that the total amplitude must be fully symmetric when we interchange space, spin (not relevant here), and isospin variables. So the first thing we note is

$$F(\epsilon_1, \epsilon_2, \epsilon_3) = F(\epsilon_1, \epsilon_3, \epsilon_2) \tag{36.11}$$

because $\hat{\mathbf{e}}_1(\hat{\mathbf{e}}_2 \cdot \hat{\mathbf{e}}_3)$ is already symmetric under the interchange of $\hat{\mathbf{e}}_2$ and $\hat{\mathbf{e}}_3$. Likewise G and H must be the *same* function as F: the amplitude must have the form

$$\begin{aligned}\boldsymbol{\mathcal{A}}(\hat{\mathbf{e}}_1, \hat{\mathbf{e}}_2, \hat{\mathbf{e}}_3) = \hat{\mathbf{e}}_1(\hat{\mathbf{e}}_2 \bullet \hat{\mathbf{e}}_3)\, F(\epsilon_1, \epsilon_2, \epsilon_3)& \\ + \hat{\mathbf{e}}_2(\hat{\mathbf{e}}_3 \bullet \hat{\mathbf{e}}_1)\, F(\epsilon_2, \epsilon_3, \epsilon_1) + \hat{\mathbf{e}}_3(\hat{\mathbf{e}}_1 \bullet \hat{\mathbf{e}}_2)\, F(\epsilon_3, \epsilon_1, \epsilon_2)&\end{aligned} \tag{36.12}$$

The first entry in each F is the preferred position for that case. We've now constructed the most general decay amplitude that has the proper isospin transformation properties consistent with Bose statistics. Of course, the actual decay final state is the $I_z = 0$ member of this triplet, because electromagnetism preserves I_z. We project it out by dotting the amplitude $\boldsymbol{\mathcal{A}}$ into a unit vector pointing in the z-direction:

$$\mathcal{A} = \hat{\mathbf{z}} \cdot \boldsymbol{\mathcal{A}} \tag{36.13}$$

The next stage in the analysis is to make an approximation, which will introduce a small error. The ϵ's are all fairly small. Even in the case of the $3\pi^0$ decay, the mass of the neutral pion is 135 MeV, so the rest energy of three neutral pions is 405 MeV. Thus we only have 143 MeV available for the decay, which has to be split somehow among the three pions. In the case of charged pion decay, $\pi^+\pi^-\pi^0$, the situation is even worse, with even less energy: the charged pions are each 4.6 MeV heavier than the neutral pion, so we have 9.2 MeV less than the $3\pi^0$ decay. This means that the effects of electromagnetic mass differences, usually negligible, are in fact significant for this process, because they cut down the available amount of phase space by 10 to 15%. They are large effects if we have very little energy available, and even a small mass difference can be a large effect. But we ignore that and live with this possible error.

What we will not ignore is this: because there is a small amount of energy available for the decay, the ϵ's tend to be small. Therefore we will make a *linear* approximation for the function F. That is, in phase space Γ, we will ignore

$$\int_D d\epsilon_1\, d\epsilon_2\, |\epsilon_1|^2 \tag{36.14}$$

where D is the region of the Dalitz plot. This could have been $|\epsilon_i|^2$; it doesn't matter because it's all symmetric. We will expand F only to first order in ϵ and then ignore, in computing the total cross-section, quadratic effects, because we expect ϵ to be small compared to a typical strong interaction mass over the entirety of the Dalitz plot; therefore effects of order ϵ^2 should be negligible.

Thus we will approximate F as

$$F(\epsilon_1, \epsilon_2, \epsilon_3) = A + B\epsilon_1 + C(\epsilon_2 + \epsilon_3) + \mathcal{O}(\epsilon^2) \approx A + B'\epsilon_1 \tag{36.15}$$

where $B' = B - C$ since $\epsilon_1 + \epsilon_2 + \epsilon_3 = 0$. A is a constant, the value at the center of the Dalitz plot. The energies ϵ_2 and ϵ_3 must have the same coefficient, because $F(\epsilon_1, \epsilon_2, \epsilon_3)$ must be symmetric under interchange of ϵ_2 and ϵ_3. We've certainly made some error by this approximation, and we expect that error to be about 10 to 15%.

Now we're in business. We'll first consider

$$\eta \to 3\pi^0 \tag{36.16}$$

In this case

$$\hat{\mathbf{e}}_1 = \hat{\mathbf{e}}_2 = \hat{\mathbf{e}}_3 = \hat{\mathbf{z}} \tag{36.17}$$

The $\hat{\mathbf{z}} \cdot \hat{\mathbf{e}}_i$ and $\hat{\mathbf{e}}_i \cdot \hat{\mathbf{e}}_j$ factors in all three terms contribute 1, and we have a decay amplitude

$$\mathcal{A} = 3A + B'(\epsilon_1 + \epsilon_2 + \epsilon_3) = 3A \tag{36.18}$$

In the linear approximation, the distribution of points in the Dalitz plot is completely flat for $3\pi^0$ decays, independent of position in the Dalitz plot.[2]

---

[2] [Eds.] G. Barton and S. P. Rosen, "Dalitz Plot for the Decay $\eta \to \pi^+\,\pi^-\,\pi^0$", *Phys. Rev. Lett.* **8** (1962) 414–416.

For the other decay,

$$\eta \to \pi^0\pi^+\pi^- \tag{36.19}$$

it doesn't matter how we choose the $\hat{\mathbf{e}}$'s. Let

$$\hat{\mathbf{e}}_1 = \hat{\mathbf{z}}, \quad \hat{\mathbf{e}}_2 = \tfrac{1}{\sqrt{2}}(\hat{\mathbf{x}} + i\hat{\mathbf{y}}), \quad \hat{\mathbf{e}}_3 = \tfrac{1}{\sqrt{2}}(\hat{\mathbf{x}} - i\hat{\mathbf{y}}) \tag{36.20}$$

From (36.10) only the first term contributes since $\hat{\mathbf{e}}_1$ is orthogonal to $\hat{\mathbf{e}}_2$ and $\hat{\mathbf{e}}_3$, so we obtain for the decay amplitude just the first term

$$\mathcal{A} = A + B'\epsilon_1 \tag{36.21}$$

Thus the distribution in the Dalitz plot in the linear approximation is symmetric, independent of the π^+ and π^- energies, and linearly dependent on how much energy is given to the π^0.

Now let's compute the total decay rates. At first glance it looks like we can't compare the total decay rates, because one of them depends only on A and the other one depends on both A and B':

$$\begin{aligned}\Gamma(\eta \to \pi^0\pi^+\pi^-) &\propto \int_D d\epsilon_1\, d\epsilon_2\, |A + B'\epsilon_1|^2 \\ &= \int_D d\epsilon_1\, d\epsilon_2 \left[|A|^2 + (A^*B' + AB'^*)\epsilon_1 + \mathcal{O}(\epsilon_1^2)\right] \\ &\approx \int_D d\epsilon_1\, d\epsilon_2 \left[|A|^2 + 2\,\mathrm{Re}(A^*B')\epsilon_1\right]\end{aligned} \tag{36.22}$$

Ignoring the mass differences between the pions, the Dalitz plot is completely symmetric under the interchange $\epsilon_1 \leftrightarrow \epsilon_2$. Therefore the integral over the Dalitz plot of ϵ_1 is the same as the integral over the Dalitz plot of ϵ_2 or ϵ_3 or equivalently ⅓ the integral of the sum. But the sum is zero and therefore its integral is zero, so the ϵ_1 term in (36.22) integrates to 0. Therefore, although in this approximation we see a linear term in the distribution of points in the Dalitz plot, the actual total number of points in the Dalitz plot is unaffected except by quadratic terms. Points are shifted to one side or the other of the line where $\epsilon_1 = 0$, but the same number go to one side as to the other.

The $3\pi^0$ decay is much more straightforward. From (36.18),

$$\Gamma(\eta \to 3\pi^0) = \frac{1}{3!}\int_D d\epsilon_1\, d\epsilon_2\, |3A|^2 = \tfrac{3}{2}\Gamma(\eta \to \pi^0\pi^+\pi^-) \tag{36.23}$$

The $(1/3!)$ is to ensure that we don't count the same experimental event 6 times. The total decay rate is difficult to measure but the branching ratio is well-known. The experimental numbers are[3]

$$\frac{\Gamma(\eta \to 3\pi^0)}{\Gamma(\eta \to \pi^0\pi^+\pi^-)} = \frac{32.68 \pm 0.23}{22.92 \pm 0.28} = 1.426 \pm 0.012 \tag{36.24}$$

which is in agreement within the expected theoretical error caused by neglecting the three pion mass differences. It's about 5% off from the theoretical value of $\frac{3}{2} = 1.5$.

It could have been that the linear approximation was poor, but it's not. If we actually look at the density of points in the Dalitz plot, the linear approximation fits them pretty well. To

[3] [Eds.] *Ibid.* See also *PDG* 2106, p. 37.

improve the approximation, we would have to look at how the size of the Dalitz plot changes with available energy, and make a correction for that. We'd just have less phase space to decay into. That would enhance the $3\pi^0$ decay because there's more phase space around; the π^0's are lighter.

The big problem in η decay is not phase space but determining the absolute rate. This is very tricky, involving something called the Primakoff effect.[4] It involves η production from a photon interacting with the Coulomb field of a nucleus. This is a difficult experiment and the results keep fluctuating. The big theoretical problem is that unless the experiments are very badly wrong, the amplitude is embarrassingly larger than it should be on the basis of crude estimates of the size of an electromagnetic effect. With that, we leave η decays, and begin a new topic.

36.2 An informal historical introduction to SU(3)

We've seen that we can get a lot of results about properties that have the strong interactions entering into them in a complicated way, even in almost total ignorance of the strong interactions, just by exploiting the known symmetries of the strong interactions and the known transformation properties of the other interactions under the strong interaction symmetry group (which, however, was not known for a long time). Here is a short historical survey of how we came to SU(3).[5]

In the mid-to-late 1950's some very smart people, including Murray Gell-Mann and Julian Schwinger, began thinking that maybe one could play this game even more daringly. What principally made them think this was, first, the introduction by Gell-Mann and Nishijima a few years earlier, of strangeness or hypercharge,[6] which indicated that isospin seemed to be a good quantum number for all the new particles that had been coming out of the new generation of high-energy machines (low-energy machines by today's standards); exotic particles like Λ's and Σ's, K mesons, ρ's and ω's, all fit into isospin multiplets. And they were beginning to be assembled into even larger families. In particular there seemed to be eight so-called baryons, strongly interacting particles with spin ½, that were pretty much like the nucleons. All eight were assigned baryon number +1, and parity plus,

$$J^P = \tfrac{1}{2}^+,\ B = 1 \tag{36.25}$$

though they had different isospins and hypercharges. They were rather close together in mass by the scale of these things, as shown in Table 36.1. These particles seemed vaguely similar in their properties. As far as the experimental evidence went, they seemed to have the same

---

4 [Eds.] H. Primakoff, "Photo-Production of Neutral Mesons in Nuclear Electric Fields and the Mean Life of the Neutral Meson", *Phys. Rev.* **81** (1951) 899; A. Halprin, C. M. Andersen and H. Primakoff, "Photonic Decay Rates and Nuclear-Coulomb-Field Coherent Production Processes", *Phys. Rev.* **152** (1966) 1295–1303.

5 [Eds.] "An introduction to unitary symmetry", Chapter 1 in Coleman *Aspects*; S. Coleman, "Fun with SU(3)" in *High-Energy Physics and Elementary Particles*, ed. C. Fronsdal, IAEA, Vienna, 1965; M. Gell-Mann and Y. Ne'eman, *The Eightfold Way*, W. A. Benjamin Publishers, 1964; reprinted by Westview Press, 2000. Harry Lipkin suggests that SU(3) was called the "Eightfold Way" because it took people eight years (1953–1961) to figure things out; H. Lipkin, "Quark Models and Quark Phenomenology", invited talk at the *Third Symposium on the History of Particle Physics*, Stanford Linear Accelerator Center, June 24–27, 1992.

6 [Eds.] M. Gell-Mann, "The Interpretation of the New Particles as Displaced Charged Multiplets," *Nuovo Cim.* **4** Suppl., (1956) 848–866; K. Nishijima, "Charge Independence Theory of V Particles," *Prog. Theor. Phys.* **13** (1955) 285–304. See also §24.4.

Particle	I	Y	M (MeV)	*Count*
N	½	1	939	2
Λ	0	0	1116	1
Σ	1	0	1193	3
Ξ	½	-1	1318	2

Table 36.1: The eight baryons: $J^P = \frac{1}{2}^+$, $B = 1$

parity, they unquestionably had spin ½ and baryon number 1 and they were all relatively close together in mass, the mass splittings between the heaviest and the lightest being of the order of 15 to 20% of the mean mass of this collection of particles.

This led them to an idea. Back in the 1930's, when the only strongly interacting particles known were the proton and neutron, Heisenberg and others had suggested that if we neglected electromagnetism, then because of the strong interactions that remained ("nuclear forces", as they were then called), the world would be much more symmetric than it was in reality.[7] In particular it would possess isospin symmetry. Therefore, Schwinger and Gell-Mann, at around the same time, said maybe the same thing could be done with the strong interactions. Maybe they split into two families, very strong and medium strong, which we will ignore:

$$\text{Strong} \begin{cases} \text{very strong} & \\ \text{\sout{medium strong}} & \text{(ignore)} \end{cases} \tag{36.26}$$

Guided by the principle that ignoring electromagnetism leads to the neutron and proton having the same mass, they hypothesized that if the medium strong interactions were ignored—a much bolder step than ignoring electromagnetism—then all eight of these particles would have the same mass and would be part of a degenerate multiplet of a larger symmetry group than isospin. This hypothetical larger symmetry wouldn't be as good as isospin; a 10 to 20% error is much worse than a 1% error or a 0.1% error. But it would still better than nothing. It's a lot easier to try this group theory idea than to attempt to solve the dynamics of the strong interactions.

Criteria on G

It was clear what the problem was. In mathematical language, we want some internal symmetry group, G, that first of all contains a product of the SU(2) of isospin and the U(1) of hypercharge:

$$\mathrm{SU}(2) \otimes \mathrm{U}(1) \subset \mathrm{G} \tag{36.27}$$

We want the new symmetry group to include the old symmetries when we don't ignore the medium strong interactions. Next, G must have an 8-dimensional, unitary, irreducible representation,[8] to accommodate the eight observed baryons in a single representation of G. And because that representation is irreducible they will all have the same mass. Otherwise SU(2) ⊗ U(1) (which lacks such a representation) would solve the problem. Finally, when

[7] [Eds.] See note 2, p. 507.

[8] [Eds.] H. M. Georgi, *Lie Algebras in Particle Physics*, Addison-Wesley Publishing Co., 1982, pp. 5–6; R. N. Cahn, *Semi-Simple Lie Algebras and Their Representations*, Benjamin/Cummings Publishing Co., 1984; Zee *GTN*, pp. 122–123.

we reduce G to this subgroup, $G \to SU(2) \otimes U(1)$, we don't want just any 8-dimensional irreducible representation, but one that decomposes into *these* observed particles in Table 36.1 *exactly*:

$$\text{8-dimensional irreducible representation} \in G \to \{\Xi, \Sigma, \Lambda, N\} \tag{36.28}$$

(that is, $(I = \frac{1}{2}, Y = -1)$, $(I = 1, Y = 0)$, $(I = 0, Y = 0)$, and $(I = \frac{1}{2}, Y = 1)$, respectively).

At that time nobody knew anything about group theory except Wigner, and he wasn't talking.[9] They were sort of desperate, so they played around, they guessed at it; Schwinger[10] and Gell-Mann[11] made the same guess, called *global symmetry*. There wasn't much data at the time so it wasn't immediately obvious, but after a year or two it became clear that this idea was dead wrong. People tried other things. There was a whole school of thought. Saul Barshay said the Λ had opposite parity from the Σ; they were just bad experiments and we should look for a *seven*-dimensional representation; the Λ was coincidentally sitting there in the middle.[12] There were people who fiddled around with the idea that maybe there were *nine* particles and we just hadn't found the ninth one yet because it was little bit heavier than the Ξ; perhaps we should have been looking for a group with a *nine*-dimensional representation. People played this game for a while. But it became clear, and it's certainly known now in retrospect, that there *aren't* any other particles around in the same mass range. Insofar as the weak interactions allow us to define the relative parities, the parities of these eight baryons *are* all the same. Therefore, the problem as originally framed, with the three criteria above, is correctly posed.

A few years later, around 1960, Gell-Mann guessed the right group. It's a very interesting anecdote to me, because I was present at the time as a graduate student at Caltech. At that time Gell-Mann and Shelly Glashow, who was a post-doctoral fellow at Caltech, were working on a Yang–Mills theory, not knowing what it was good for.[13] (Steve Weinberg discovered that a few years later.) They wanted to learn something about Lie groups (which are involved in Yang–Mills theories). At that time Lie group theory was considered *recherché*, like fiber bundles, not something a respectable physicist knew about. I then had a totally undeserved reputation for mathematical sophistication. They asked me, "Do you know anything about Lie group theory?" I replied, "Who, me? I can tell an ϵ from a δ but that doesn't mean I'm André Weil; of *course* I don't."[14] Well, fortunately at that time the Caltech mathematics department was in the same building as the physics department. Murray went upstairs and found a mathematician, Richard Block,[15] who was willing to talk to him. Block told him

---

[9] [Eds.] E. P. Wigner, *Group Theory and its Application to the Quantum Mechanics of Atomic Spectra*, Academic Press, 1959.

[10] [Eds.] J. Schwinger, "A Theory of the Fundamental Interactions", *Ann. Phys.* **2** (1957) 407–434.

[11] [Eds.] M. Gell-Mann, "Model of the Strong Couplings", *Phys. Rev.* **106** (1957) 1296–1300. See equation (10) and the discussion following equation (12); "global symmetry" was the name given to the hypothesis that the pion–baryon coupling constant was the same for all the baryons. "Global" symmetry today means something very different. It is used in the context of gauge theories. If the group parameters depend on x^μ, the symmetry is called *local*; if not, the symmetry is called *global.*

[12] [Eds.] S. Barshay, "Hyperon–Antihyperon Production in Nucleon–Antinucleon Collisions and the Relative $\Sigma - \Lambda$ Parity", *Phys. Rev.* **113** (1959) 349–351.

[13] [Eds.] S. L. Glashow and M. Gell-Mann, "Gauge Theories of Vector Particles", *Ann. Phys.* **15** (1961) 437–460.

[14] [Eds.] André Weil (1906–1998), influential French mathematician and brother of the philosopher and mystic Simone Weil (1909–1943). He was one of the founders of the team writing mathematics under the group pseudonym "Nicolas Bourbaki".

[15] [Eds.] Crease & Mann *SC*, pp. 266–268.

to consult a book[16] in French which was a précis of the main results on Lie group theory.[17] Murray's fluent in French so he did just that. And two days later, he returned in a state of great excitement and said, "There *is* a group with an 8-dimensional representation called SU(3)." And we said, "SU(3)? 8-dimensional representation? Ah, go away." It turned out to work, as we all discovered very shortly. That was how it in fact happened historically.

A second physicist, Yuval Ne'eman[18] made the discovery independently.[19] Yuval had been trained as an engineer and he had wanted to become a physicist since 1947, but the military situation kept him from going to graduate school. And finally, after having served as acting head of Israeli military intelligence during the Sinai campaign (besides other wartime experience), he was able to talk the Israeli general staff into getting him a half-time job as a military attaché in London. The other half of his time he spent at graduate school at Imperial College under Abdus Salam. At first, the British Foreign Office was reluctant about giving him permission to do this, because they confused high energy physics with building bombs, as if he wanted to be a spy. But he finally got the head of the Israeli general staff, Moshe Dayan,[20] to write a letter explaining that Ne'eman's objective was education, not espionage. And Yuval subsequently described himself as the only graduate student accepted by Salam on the strength of a letter of recommendation from Moshe Dayan.[21] Salam put him on this problem, and he also came up with SU(3) at around the same time as Gell-Mann.[22] I remember Murray rushing into the department with a preprint, exclaiming "Some Israeli colonel has made the same discovery!" Enough of stories from my youth. (The golden age of a physicist is 23...)

Later, people began to wonder if something had slipped through the net, and started to look at the problem in a systematic way, too late as always, and to investigate the mathematics to reach the solution. That's what we'll do in lieu of the historical order; we'll turn to the mathematical answer. I won't *prove* this is the answer, because it's not a sophisticated proof, it's just tedious. Then we'll systematically explore the possibilities given by this answer. We'll see that nothing works except SU(3), and then we'll spend a lot of time on SU(3).

The answer is as follows.[23] Every group G satisfying these three criteria

- It has an 8 dimensional irreducible representation, $D(\mathrm{G})$
- It contains $\mathrm{SU}(2) \otimes \mathrm{U}(1)$ as a subgroup

---

[16] [Eds.] *Séminaire "Sophus Lie"*, École Normale Supérieure, Paris. Volume 1: 1954–1955; Volume 2: 1955–1956.

[17] [Eds.] Sophus Lie, Norwegian mathematician 1842–1899. For background on Lie's life and work, see D. J. Struik, *A Concise History of Mathematics*, G. Bell and Sons, London, 1954; reprinted by Dover Publications, New York, 1987; B. Fritzsche, "Sophus Lie: A Sketch of his Life and Work", *J. Lie Theory*, **9** (1) (1999) 1–38. Many physicists of the time learned Lie group theory from a little book by Harry J. Lipkin, *Lie Groups for Pedestrians*, North-Holland, 1965, reprinted by Dover Publications, 2002.

[18] [Eds.] Yuval Ne'eman (1925–2006), Israeli physicist, soldier and politician. Ne'eman and Gell-Mann shared the 1969 Nobel Prize in Physics for their work on SU(3).

[19] [Eds.] Crease & Mann *SC*, pp. 269–272.

[20] [Eds.] Moshe Dayan (1915–1981), Israeli military leader and politician.

[21] [Eds.] "Salam laughed at the recommendation from Dayan, and told Ne'eman to bring a recommendation from a physicist. Ne'eman never did, but Salam accepted him anyway—partly, he has said, to repay a debt incurred by Islamic science, which in its medieval heyday owed much to Jewish scholars." Crease & Mann *SC*, p. 270.

[22] [Eds.] Y. Ne'eman, "Derivation of Strong Interactions from a Gauge Invariance", *Nucl. Phys.* **26** (1961) 222–229.

[23] [Eds.] This theorem is due to Coleman, proved by him in his PhD thesis: S. Coleman, *The Structure of Strong Interaction Symmetries*, Caltech, 1962, `http://thesis.library.caltech.edu/2386/1/Coleman_sr_1962.pdf`.

- $D(\mathrm{G}) \to \{\Xi, N, \Sigma, \Lambda\}$ *exactly* as $\mathrm{G} \to \mathrm{SU}(2) \otimes \mathrm{U}(1)$

contains as a subgroup either

A. a group G_0 satisfying *minimal global symmetry*, equating some meson–baryon coupling constants. (Recall that the original erroneous suggestion of Schwinger and Gell-Mann, "global symmetry", suggested *all* pion–baryon coupling constants were equal.) Later Lee and Yang[24] pointed out that the Gell-Mann–Schwinger group contained a subgroup (to be described below) that was a better fit with experiments: it gave almost the same results of the original group, but with fewer wrong predictions.
B. SU(3), the group of all 3×3 unitary unimodular (determinant $= 1$) matrices.

Of course there are many groups containing SU(3). One answer to the problem is obviously SU(8), the group of all 8×8 unitary unimodular matrices. That contains everything that satisfies the problem. But these groups either contain minimal global symmetry, or SU(3). We will systematically investigate these two possibilities.

First let's consider option A, a group G_0 that satisfies minimal global symmetry.[25] G_0 is the product of three SU(2) factors:

$$\mathrm{G}_0 = \mathrm{SU}(2) \otimes \mathrm{SU}(2) \otimes \mathrm{SU}(2) \tag{36.29}$$

People thought to try this, because at the time the only Lie group they knew about was isospin's SU(2) Thus its generators can be written as the three commuting triplets of isospin generators:

$$\mathbf{I}^{(1)}, \quad \mathbf{I}^{(2)}, \quad \mathbf{I}^{(3)} \tag{36.30}$$

How are the isospin and hypercharge embedded as subgroups? $\mathbf{I}$ is the simultaneous rotation in the 1–2 isospin space

$$\mathbf{I} = \mathbf{I}^{(1)} + \mathbf{I}^{(2)} \tag{36.31}$$

and Y is twice the z-component of $I^{(3)}$:

$$Y = 2I^{(3)}_z \tag{36.32}$$

These commute and obviously obey the $\mathrm{SU}(2) \otimes \mathrm{U}(1)$ algebra.

The representation $D(\mathrm{G}_0)$ to which the eight baryons are assigned is in fact a *reducible* representation of this group. A general representation is labeled by three spins, s_1, s_2 and s_3, for the three isospins, $D^{(s_1,s_2,s_3)}$, just as two spins came into our analysis of the Lorentz group.[26] The representation in question is the direct sum

$$D^{(\frac{1}{2},\frac{1}{2},0)} \oplus D^{(\frac{1}{2},0,\frac{1}{2})} \quad (\mathbf{I} \oplus \mathrm{Y}) \tag{36.33}$$

G_0 is therefore *not* a solution to the problem because this is not irreducible. But the theorem says G_0 is *contained* in many solutions to the problem, not that it is itself the solution. If we reduce to isospin and hypercharge alone

$$\mathrm{G} \to \mathrm{SU}(2) \otimes \mathrm{U}(1) \tag{36.34}$$

---

[24] [Eds.] T. D. Lee and C. N. Yang, "Some Considerations on Global Symmetry", *Phys. Rev.* **122** (1961) 1954–1961. Lee and Yang cite A. Pais, "Note on Relations between Baryon–Meson Coupling Constants", *Phys. Rev.* **110** (1958) 1480–1481, which suggested that the Gell-Mann–Schwinger group corresponded to the direct product of three unitary unimodular groups; see footnote 2 in Pais. In fact the Lee–Yang group included also a discrete symmetry R (their equation (10)) to make the representations irreducible.

[25] [Eds.] "An introduction to unitary symmetry", Ch. 1 in Coleman *Aspects*, pp. 3–5.

[26] [Eds.] §18.5.

then the first factor in (36.33), $D^{(\frac{1}{2},\frac{1}{2},0)}$, has the sum of two isospin-½'s added together and we obtain isospin 1 and isospin 0:

$$I = \begin{Bmatrix} 0 \\ 1 \end{Bmatrix} \tag{36.35}$$

both with $Y = 0$, since $D^{(\frac{1}{2},\frac{1}{2},0)}$ is a singlet under the third isospin group and therefore carries zero hypercharge. This is the Σ and the Λ in Table 36.1. In the second factor, $D^{(\frac{1}{2},0,\frac{1}{2})}$ we only have isospin $\frac{1}{2}+0=\frac{1}{2}$, so both factors have $I=\frac{1}{2}$, but because I_z can have the two values $\frac{1}{2}$ and $-\frac{1}{2}$, we have $Y = 2I_z^{(3)} = \pm 1$; that's the N and the Ξ.

Indeed, the first thing people said when they looked at this pattern was, "Well, obviously there is some sort of reflection or something at work here, some symmetry operation that exchanges the N and the Ξ, and changes the sign of the hypercharge. They look the same, they're at opposite ends of the chart, flip it around." In fact, that guess turned out to be dead wrong, as we will demonstrate immediately. It was however very plausible, and it led to the group (36.29).

Well, this group is in direct contradiction to many experimental results, although of course none of them was known at the time it was proposed. The easiest way[27] to establish a contradiction is to define a group element called R:

$$R = e^{i\pi I_y^{(3)}} \tag{36.36}$$

It is a rotation by 180° about the y-axis in the third isospin group. R, belonging to $\mathbf{I}^{(3)}$, commutes with isospin, which is constructed from the first and second isospin groups, (36.31). But it changes the sign of hypercharge, because that's proportional to $I_z^{(3)}$:

$$R\mathbf{I}R^\dagger = \mathbf{I}; \quad RYR^\dagger = -Y \tag{36.37}$$

From the existence of the group element R, one can derive an almost endless string of contradictions with experiment. For example, for every hadron (stable or unstable) with a given non-zero hypercharge, there must be another hadron of the same mass (within 10 to 20%) with the opposite hypercharge, obtained by applying R, which changes the sign of the hypercharge while leaving isospin alone. And of course R commutes with baryon number, so this second hadron is not the antiparticle of the first; it's something else.

Now there's the well-known $\Delta(1232)$, a big, fat[28] resonance in pion–nucleon scattering with $I=\frac{3}{2}$ and $Y=1$, so there should be another big, fat resonance around the same mass with $I=\frac{3}{2}$ and $Y=-1$. You can search the Rosenfeld tables[29] to your heart's content, but you will not find such an object; it is not there. Of course at the time they didn't know it wasn't there.

Another argument that (36.29) is the wrong group comes from the analysis of magnetic moments that I described last time (Example 2, p. 766). The Λ is a singlet under the third

[27] [Eds.] Coleman *op. cit.*, p. 4.

[28] [Eds.] The Δ has a width $\approx$ 117 MeV. *PDG* 2016, p. 91.

[29] [Eds.] Now the *Particle Data Group* tables. Named for Arthur H. Rosenfeld (1926–2017), American physicist. The tables started as unpublished data tables to support a long review article with Gell-Mann: M. Gell-Mann and A. H. Rosenfeld, "Hyperons and Heavy Mesons (Systematics and Decay)", *Ann. Rev. Nucl. Sci.*, **7** (1957) 407–478.

isospin, so R doesn't do anything to the Λ. If we have a one-Λ intermediate state, we get the same state after operating on it with R:

$$R\,|\Lambda\rangle = |\Lambda\rangle \tag{36.38}$$

On the other hand, if we look at the Y-component of the electromagnetic current, the part that comes from j^μ_Y, R changes its sign, because it changes the sign of Y:

$$Rj^\mu_Y R^\dagger = -j^\mu_Y \tag{36.39}$$

If we look at the electromagnetic current between two Λ states, from isospin considerations, only j^μ_Y contributes: $j^\mu_{I_z}$ is a component of an isovector, and cannot have a non-vanishing matrix element between two isospin-zero states. But by R this matrix element is zero. Apply $R^\dagger R$ on the two sides of j^μ_Y; the Λ doesn't change sign, but the current does, so the matrix element vanishes:

$$\begin{aligned}\langle\Lambda|j^\mu_{em}|\Lambda\rangle &= \langle\Lambda|j^\mu_Y|\Lambda\rangle = \langle\Lambda|R^\dagger R j^\mu_Y R^\dagger R|\Lambda\rangle \\ &= \langle\Lambda|R j^\mu_Y R^\dagger|\Lambda\rangle = -\,\langle\Lambda|j^\mu_Y|\Lambda\rangle = 0\end{aligned} \tag{36.40}$$

by (36.38) and (36.39). Thus we are led to the conclusion that the magnetic moment of the Λ must be zero:

$$\mu_\Lambda = 0 \tag{36.41}$$

The magnetic moment of the Λ wasn't measured until 1963, so the prediction (36.41) didn't bother anyone at the time. But we know now the Λ moment is about a third of the neutron moment,[30] a typical hadron magnetic moment, not particularly small in any reasonable sense.

I could go on. We could run through the Particle Data Group tables, demonstrating that any symmetry including a hypercharge reflection operator of this kind is guaranteed to be wrong. The universe is *not* symmetric under a change of sign of hypercharge, while keeping the sign of isospin and the baryon number unchanged. Thus, G_0 is out.[31]

Therefore, if we accept this theorem (stated without proof), the last best guess is SU(3): either that works or the game is up. Well, of course it *does* work, otherwise I wouldn't be giving this lecture. Rather than doing things directly for SU(3), I'd like to devote some time to mathematical preliminaries in which we construct the representations of SU(3). After we have our mathematical machinery set up, we'll apply it to a variety of physics problems. It's worth the effort, because we know in advance that it's going to be good for physics.

36.3 Tensor methods for SU(n)

We start with the hadrons in Table 36.1 as an eight-dimensional representation of some group and an embedding of SU(2)$\otimes$U(1) in that group. It's extremely tedious. You look in books on Lie group theory and count up all the Lie groups with 8-dimensional irreducible representations, (or 4-dimensional representations, because it could be two 4's connected by a discrete element).

---

30 [Eds.] In the videotape of Lecture 36, Coleman says "half" instead of "a third". The current values are $\mu_\Lambda = (-0.613 \pm 0.004)\,\mu_N$, $\mu_n = (-1.913)\,\mu_N$; *PDG* 2016, p. 92.

31 [Eds.] Another candidate for G_0 was the exceptional group G_2: R. E. Behrends, J. Dreitlein, C. Fronsdal, and W. Lee, "Simple Groups and Strong Interaction Symmetries", *Rev. Mod. Phys.* **34** (1962) 1–38, and references therein. (J. L. Rosner, private communication.) Incidentally, the author "W. Lee" seems to be Benjamin W. Lee.

You write them down and figure out all possible ways of fitting in isospin and hypercharge, and you are left with, I believe, 13 possible groups. And then you figure out which contains which one, you make a big diagram with boxes and trees, and you find that they all end up either in G_0 or in SU(3). I did it for my thesis;[32] believe me, it was dull. At the end of it, though, I knew Lie group theory. If I had known Lie group theory a year before... Ah, well.

So we will begin by making some preliminary remarks about the representations of SU(n), for arbitrary n. Then we will specialize, first all the way down to SU(2), just so we can check that the methods work in a case where we already know the answer, and then to SU(3). This is in fact the method introduced by Hermann Weyl in his book on the classical groups.[33]

SU(n) is the group of all $n \times n$ unitary matrices U with determinant 1:

$$U^{\dagger}U = 1, \ \det U = 1 \tag{36.42}$$

We already know one representation of SU(n), to wit, an n-dimensional representation, where the group is represented by the matrices themselves. A complex n-vector x transforms under the action of the group according to the rule

$$x \to Ux \tag{36.43}$$

It is convenient to write these transformations out in index form as if they were Lorentz tensors, except they're not, they're SU(n) vectors:

$$x^i \to U^i{}_j x^j \tag{36.44}$$

For the moment the reason that I put one of those indices downstairs is just perverse, but I am adopting the summation convention.

Another representation that we know off-hand is the *complex conjugate representation.* The complex conjugate vectors, $(y^i)^*$, form the basis for the conjugate representation. We indicate the components of a conjugate vector by a subscript:

$$(y^i)^* \equiv y_i \tag{36.45}$$

Given the representation (36.44), then the complex conjugate is also a representation:

$$y_i \to (U_i{}^j)^* y_j \equiv (U^{\dagger})^j{}_i \, y_j \tag{36.46}$$

using the complex conjugate matrix. It may or may not be equivalent to (36.44). These two representations are equivalent in SU(2) but not in SU(3) or higher SU(n). We use a notation that mimics that of ordinary four-dimensional tensor analysis. The mimicry is introduced for a reason: it is to remind us that if we take one vector that transforms as the first representation and the second according to the conjugate representation and sum them up, then the summation of upper and lower indices is an invariant operation; that's just the definition of a unitary matrix. It's precisely the object that preserves the quadratic form:

$$x^i y_i \to U^i{}_k x^k (U^{\dagger})^j{}_i \, y_j = (U^{\dagger})^j{}_i U^i{}_k \, x^k y_j = \delta^j{}_k x^k y_j = x^j y_j \tag{36.47}$$

---

[32] [Eds.] See note 23, p. 784.

[33] [Eds.] H. Weyl, *The Classical Groups*, Princeton U. P., 1953. See also S. Coleman, "Fun with SU(3)", *op. cit.*, and J. Mathews and R. Walker, *Mathematical Methods of Physics*, 2nd ed., Addison-Wesley, 1969, Chapter 16, pp. 424–470. In the preface the authors state, "Much of Chapter 16 grew out of fruitful conversations with Dr. Sidney Coleman."

Thus we have *two* kinds of vectors, upper-index vectors and lower-index vectors, just as in ordinary tensor analysis. What we don't have is a metric tensor that allows us to raise and lower indices. At this level they're just two different kinds of objects that transform in two different ways. We may form tensors with arbitrary numbers of upper and lower indices by taking direct products of vectors and conjugate vectors:

$$x^{i_1\cdots i_m}{}_{j_1\cdots j_n} \tag{36.48}$$

Upper indices transform as if they're upstairs vectors, lower indices as downstairs, conjugate vectors. I won't take the time to write it out, because you can see how it goes: there's a U for every upstairs index and a U^* for every downstairs index. These tensors form a bunch of representations of our group SU(n). Of course they're not guaranteed to be irreducible, but they *are* guaranteed to be representations.

There are all sorts of manipulations we can do on these tensors that are invariant operations. For one thing, symmetrizing or antisymmetrizing on a pair of upper or lower indices is an invariant operation. That's because any two upper indices transform in the same way. So if we break a tensor up into a part that's symmetric on its first two upper indices and a part that's antisymmetric on the first two upper indices, then if we make the transformation the symmetric part goes into something symmetric and the antisymmetric part goes into something antisymmetric. Likewise for lower indices. On the other hand, it's pointless to symmetrize a tensor between an upper index and a lower index; it's allowed, but if we make a transformation on that tensor, the symmetry won't be preserved: the upper index transforms differently than the lower index. Another invariant operation is *contraction*: summing an upper and a lower index (this is the *trace* of the matrix over the two summed indices):

$$\delta^i{}_j x^{j\;k\ldots}_{\;i\;l\ldots} = x^{i\;k\ldots}_{\;i\;l\ldots} \tag{36.49}$$

We cannot, however, sum two lower indices together, nor two upper indices, because there's no metric tensor.

There are also some invariant tensors around; tensors, which when transformed according to the rules, don't transform at all. One is the Kronecker delta, $\delta^i{}_j$:

$$\delta^i{}_j \to U^i{}_k\,\delta^k{}_m (U^\dagger)^m{}_j = U^i{}_k\,(U^\dagger)^k{}_j = \delta^i{}_j \tag{36.50}$$

That of course is just the statement that our matrices are unitary or equivalently, that summing an upper and a lower index is an invariant operation. Another invariant tensor is the Levi–Civita ϵ. Using the identity

$$\epsilon_{ijk\cdots p} A^i{}_1 A^j{}_2 A^k{}_3 \cdots A^p{}_n = \epsilon_{123\cdots n} \det A \tag{36.51}$$

for an $n \times n$ matrix $A^i{}_j$, we see that under the action of SU(n),

$$\epsilon_{ijk\cdots p} \to \epsilon_{ijk\cdots p} \det U^\dagger = \epsilon_{ijk\cdots p} \tag{36.52}$$

because $\det U^\dagger = 1$. Likewise, $\epsilon^{ijk\cdots p}$ is invariant:

$$\epsilon^{ijk\cdots p} \to \epsilon^{ijk\cdots p} \det U = \epsilon^{ijk\cdots p} \tag{36.53}$$

I will now explain Hermann Weyl's program for SU(n), which for $n > 3$ requires exquisite knowledge of the representations of the permutation group.[34] Thankfully, no such expertise is needed for either SU(2) or SU(3).

Weyl's program attempts to find all the representations in the following way. Take all these tensors; symmetrize and antisymmetrize and multiply by epsilon tensors like crazy, until we've gone as far as we can, constructing invariant subspaces from the set of n-index tensors that are irreducible (we hope). We then prove that these are in fact irreducible representations, crossing our fingers that some clever graduate student won't come along and say "Ha! You forgot about contracting this index with that index," so we can get an even smaller subspace. If we can *prove* that you can't reduce the subspace, and hence that these representations are irreducible, then we will have constructed a complete set of irreducible representations of SU(n). That's the program. The amazing thing is that the program works (if you're Hermann Weyl). Even if you're not, it's pretty easy to make the program work for SU(2) and SU(3).

Weyl's program for SU(2)

We will work through the process for SU(2). In this case each of the epsilon tensors has only two indices. That means we can use them to raise or lower indices, just like the metric tensor:

$$\epsilon_{ij} y^j \equiv y_i; \quad y^j \equiv \epsilon^{ji} y_i \tag{36.54}$$

Of course this is *not* like the (symmetric) metric tensor, $g^{\mu\nu}$; the ϵ's are antisymmetric, but they are still invariant objects with two indices. Given any tensor with a bunch of upper and lower indices, we can always convert it into a tensor with only upper indices by raising indices with the aid of the epsilon tensor, or only lower indices by lowering with the epsilon. Thus we need only look at tensors with all indices either upper or lower.[35]

Next, we can always write tensors with more than one index as a sum of a symmetric and an antisymmetric part. Let's take a tensor x^{ij} of rank 2. We can break that up into its symmetric and antisymmetric parts:

$$\begin{aligned} x^{ij} &= x^{\{i,j\}} + x^{[i,j]} \\ x^{\{i,j\}} &= \tfrac{1}{2}[x^{ij} + x^{ji}] \quad \text{(symmetric)} \\ x^{[i,j]} &= \tfrac{1}{2}[x^{ij} - x^{ji}] \quad \text{(antisymmetric)} \end{aligned} \tag{36.55}$$

Since it only has two indices, the antisymmetric part must be proportional to the epsilon tensor (two indices, antisymmetric):

$$x^{[i,j]} = x\epsilon^{ij} \tag{36.56}$$

Here x is a scalar. (In the analysis of the Lorentz group, we similarly (18.91) split up a symmetric tensor into a part equal to its trace times $g^{\mu\nu}$, and a traceless, symmetric part.)

Thus if we have a tensor of mixed symmetry, we can always symmetrize it. The antisymmetric part can be written in terms of tensors of lower rank (with fewer indices), with the ϵ's

[34] [Eds.] Chapter III, §7, pp. 136–140; Chapter V, §12–14, pp. 347–369 in H. Weyl, *The Theory of Groups and Quantum Mechanics*, trans. H. P. Robertson (of the Robertson-Walker metric in general relativity), E. P. Dutton, New York, 1931; reprinted by Dover Publications, 1950. Originally published in German as *Gruppentheorie und Quantenmechanik*, Verlag von S. Hirzel, Leipzig, 1928.

[35] [Eds.] In the videos, and in *Aspects*, Chapter 1, "An introduction to unitary symmetry", Coleman uses lower indices.

taking up the missing indices. As we systematically examine bigger and bigger tensors looking for new representations, the only tensors we have to consider are those which have only upper indices and are fully symmetric.

What can we do to simplify those tensors? Let's guess. Our guess, which we will try to verify, is that the irreducible representations are generated by transformations on the space of these tensors. We will, with the benefit of hindsight, describe the number of their indices by the integer $2s$:

$$x^{i_1 i_2 \cdots i_{2s}} \quad (s = 0, \tfrac{1}{2}, 1, \tfrac{3}{2}, \cdots) \tag{36.57}$$

(adopted so we can see the connection with standard notation for SU(2)). Call the representation $D^{(s)}(U)$, a unitary unimodular matrix or, for short, (s):

$$D^{(s)}(U) \equiv (s) \tag{36.58}$$

What is the dimension of (s)? How many independent tensors there are with the desired symmetry properties? Since the tensor is completely symmetric, the only significant feature about a component is how many 1's it has and how many 2's it has. How they're distributed is completely irrelevant; that's what complete symmetry means. So the question is: if we have $2s$ objects and we want to put them into two boxes, how many ways are there of doing it? One box will hold the 1 indices, the other holds the 2's.

This is easy.[36] We imagine the $2s$ objects are written out on a line. Imagine we have a wall which we put down somewhere in between the dots. Everything to the left will be a 1, everything to the right will be a 2; that's the two boxes. For a tensor with $s = 3$ and r, the number of 1's, equal to 2, the diagram looks like this:

$$s = 3,\ r = 2\colon\ \bullet\ \bullet\ |\ \bullet\ \bullet\ \bullet\ \bullet \tag{36.59}$$

There are only $(2s+1)$ places we can put the wall, starting to the left of all the dots and ending to the right of all the dots. So for SU(2)

$$\dim(s) = 2s + 1, \quad s \text{ a non-negative integer} \tag{36.60}$$

We've seen that factor $(2s+1)$ before, in the context of angular momentum, and we know it's right for SU(2).[37] (For SU(3) there will be *three* boxes in the corresponding computation.) Writing

$$x^i = \begin{pmatrix} x^1 \\ x^2 \end{pmatrix} \tag{36.61}$$

---

[36] [Eds.] This is an illustration of the "sticks and stones" or the "balls and urns" or the "stars and bars" method. W. Feller, *An Introduction to Probability Theory and Its Applications*, 3rd ed., v. 1, John Wiley and Sons, 1950, Chap. II, Section 5, "Application to Occupancy Problems", pp. 38–40.

[37] [Eds.] Perhaps a reminder of the relationship between SU(2) and SO(3) would be helpful. Let $X = \mathbf{x}\cdot\boldsymbol{\sigma} = x_i\sigma_i$ (sum on i), and let $U(R) = \exp\{-i(\theta/2)\hat{\mathbf{n}}\cdot\boldsymbol{\sigma}\}$ where $\hat{\mathbf{n}}$ is a unit vector; $U(R) \in$ SU(2) because $\hat{\mathbf{n}}\cdot\boldsymbol{\sigma}$ is traceless and Hermitian. Then under the transformation $X \to X' = UXU^\dagger$, it is easy to show $X' = \mathbf{x}'\cdot\boldsymbol{\sigma}$ with $\mathbf{x}'$ given by Rodrigues' formula (see note 9, p. 374)

$$\mathbf{x}' = \mathbf{x}\cos\theta + (\hat{\mathbf{n}}\times\mathbf{x})\sin\theta + \hat{\mathbf{n}}(\hat{\mathbf{n}}\cdot\mathbf{x})(1-\cos\theta) = R(\hat{\mathbf{n}}\theta)\mathbf{x}$$

using the identities $\sigma_i\sigma_j = \delta_{ij} + i\epsilon_{ijk}\sigma_k$, and $U(R) = \cos(\theta/2)\mathbb{1} - i\sin(\theta/2)\hat{\mathbf{n}}\cdot\boldsymbol{\sigma}$ (here, $\mathbb{1}$ is a 2×2 identity matrix). That is, there is a double-valued $(\pm U)$ homomorphism between SU(2) and SO(3), and a rotation of a 3-vector $\mathbf{x}$ about a unit axis $\hat{\mathbf{n}}$ through θ is a rotation through $(\theta/2)$ of the associated operator X in spinor space. In more careful language, SU(2) is the *covering group* of SO(3): the two groups share the same Lie *algebra* $[L_i, L_j] = i\epsilon_{ijk}L_k$, and SU(2) is *locally* isomorphic to SO(3).

and with $I_z = \frac{1}{2}\sigma_z$ (the usual Pauli matrix), the eigenvalues are the familiar

$$\begin{aligned} I_z x^1 &= \tfrac{1}{2} x^1 \\ I_z x^2 &= -\tfrac{1}{2} x^2 \end{aligned} \tag{36.62}$$

For the subgroup of pure I_z rotations,[38] with $I_z = \frac{1}{2}\sigma_z$, for any angle θ,

$$U = \begin{pmatrix} e^{i\theta/2} & 0 \\ 0 & e^{-i\theta/2} \end{pmatrix} \tag{36.63}$$

That's how we embed I_z rotations in SU(2). It's a unitary matrix of determinant 1. (Note that in spinor space, the identity corresponds to an I_z rotation through an angle of 4π.) Then

$$U\colon \begin{pmatrix} x^1 \\ x^2 \end{pmatrix} \to \begin{pmatrix} e^{i\theta/2}\, x^1 \\ e^{-i\theta/2}\, x^2 \end{pmatrix} \tag{36.64}$$

If we make an isospin rotation for this particular U on a tensor component with r 1's and $(2s-r)$ 2's,

$$U\colon x^{11\cdots 122\cdots 2} \to e^{i\theta r/2} e^{-i\theta(2s-r)/2} x^{1\cdots 2} = e^{i\theta(r-s)} x^{1\cdots 2} \tag{36.65}$$

where $r = 0, 1, 2, \ldots, 2s$. Then

$$I_z x^{11\cdots 122\cdots 2} = \left(\tfrac{1}{2} r - \tfrac{1}{2}(2s-r)\right) x^{11\cdots 122\cdots 2} = (r-s) x^{11\cdots 122\cdots 2} \tag{36.66}$$

so that the eigenvalues of I_z are $-s, -s+1, \ldots, s-1, s$, which is of course the correct I_z content of a representation of SU(2). Getting them was no great triumph, but we wanted to check that this method works.

36.4 Applying tensor methods in SU(2)

Let's work out how the field theory of pions and nucleons would look in this notation. The pions have isospin 1 and the nucleons have isospin ½. We begin with the representation $D^{(\frac{1}{2})} = (\frac{1}{2})$, which is supposed to be the nucleon field. Instead of calling it x, we'll call it N: it's the nucleon field with space and spin dependence suppressed:

$$x^i \to N^i \equiv \begin{pmatrix} p \\ n \end{pmatrix} \quad \begin{aligned} i &= 1 \text{ (proton)} \\ i &= 2 \text{ (neutron)} \end{aligned} \tag{36.67}$$

Under a transformation $U \in \mathrm{SU}(2)$ we expect $N \to UN$, i.e.,

$$N^i \to U^i_{\ j} N^j \tag{36.68}$$

The conjugate fields transform according to the conjugate representation:

$$\overline{N} = (\overline{p}, \overline{n}) \to \overline{N} U^\dagger \tag{36.69}$$

---

[38] [Eds.] Greiner & Müller, *Quantum Mechanics—Symmetries*, Chapter 5, "The Isospin Group (Isobaric Spin)", pp. 95–98. See also note 37, p. 791; in particular, $U(R_z(\theta)) = \cos(\theta/2)\mathbb{1} - i\sin(\theta/2)\sigma_z$.

The definition of the conjugate representation gives us the complex conjugate. We write the vector on the left rather than on the right; this switches the indices and gives us the transpose. Taken together, we get the Hermitian adjoint, $U^\dagger$. For example, the bispinor product that we use to make an invariant Lagrangian is

$$\mathscr{L}_0 = \overline{N}(i\partial\!\!\!/ - m)N \tag{36.70}$$

which is manifestly SU(2)-invariant; no surprise.

Aside. You can make a representation that transforms like a row vector, from one that transforms as a column vector, by using the ϵ tensor:

$$N_i = \epsilon_{ij}N^j \tag{36.71}$$

Although that doesn't involve conjugate fields, it would transform like a row vector, just like the conjugate fields. And because

$$\epsilon_{12} = -\epsilon_{21} = 1$$

we would have

$$N_i = \epsilon_{ij}N^j = (n,\, -p) \tag{36.72}$$

That's why when we build an isospin singlet two-nucleon state, it is

$$N_iN^i = pn - np \tag{36.73}$$

The minus sign from the ϵ tensor is doing the job. That's the rule for putting two spin-½ objects together to make a spin-0 object (the singlet configuration, if we were talking about ordinary spin, rather than isospin). The spin-0 combination is *antisymmetric*; the spin-1 combination is *symmetric.* We either complex conjugate to lower the index (which makes *anti*nucleons) or multiply by the ϵ tensor to lower the index.

Now let's turn to the pions (or the Σ's, or any other system with isospin 1) with the aim of finding invariant quantities suitable for a Lagrangian. The pion corresponds to a *symmetric* tensor with two indices:

$$\phi^{ij} = \phi^{ji} \tag{36.74}$$

An equivalent expression (and as it turns out, a more convenient choice) is obtained by lowering one of the indices with the ϵ tensor:

$$\phi^i{}_j = \epsilon_{jk}\phi^{ik} \tag{36.75}$$

Of course, the mixed object is no longer symmetric but it still has a constraint in it; it is traceless:

$$\phi^i{}_i = \epsilon_{ik}\phi^{ik} = 0 \tag{36.76}$$

because ϕ^{ik} is symmetric. Because we've lowered an index this transforms as the outer product $x^i \otimes y_j$ of one row vector y_j and one column vector x^i. Under the action of U we can write things in matrix form

$$U\colon \phi \to U\phi U^\dagger \tag{36.77}$$

This transforms not like the inner product of row times column, which is a scalar, but as the outer product of column times row, which is an object with two indices. If we do things this way, it is obvious that it's consistent with the group to impose the condition that

$$\phi = \phi^\dagger \tag{36.78}$$

We'll do that for the pions, because we want three real fields. That corresponds to a traceless 2×2 Hermitian matrix. If we were dealing with the Σ's, where we'd want three complex fields, we wouldn't impose (36.78). (Physically, the antiparticle of the π^+ is the π^-, but the antiparticle of the Σ^+ is *not* the Σ^-, but an entirely different particle with baryon number -1.)

We can see how the 2×2 matrix ϕ transforms by multiplying

$$N\overline{N} = \begin{pmatrix} p \\ n \end{pmatrix} (\overline{p}, \overline{n}) \tag{36.79}$$

and working out the properties of the individual components. This will help us identify the components of the ϕ tensor. $N\overline{N}$ is a typical two-index object, one upper and one lower, that transforms like ${\phi^i}_j$. In the 1-1 spot we have $p\overline{p}$, an object which has zero charge and carries $I_z = 0$. Therefore ${\phi^1}_1$ must be some number α, times the neutral pion field ϕ_0. In the 1-2 spot, $p\overline{n}$ carries charge 1 and has $I_z = 1$, so ${\phi^1}_2$ must be some multiple of the positive pion field. We'll just call it ϕ_+ since how we scale our fields is a matter of taste. By conjugation we must have ϕ_- in the ${\phi^2}_1$ or $n\overline{p}$ spot, with $I_z = -1$, and by the requirement that ${\phi^i}_j$ is traceless, we have $-\alpha\phi_0$ in the ${\phi^2}_2$ or $n\overline{n}$ spot, also with zero charge and $I_z = 0$:

$$\phi = \begin{pmatrix} \alpha\phi_0 & \phi_+ \\ \phi_- & -\alpha\phi_0 \end{pmatrix} \tag{36.80}$$

How we fix α is again a matter of taste. We can normalize any one of our independent dynamical variables any way we please. But it's convenient to take (as a possible term in a Lagrangian) the expression

$$\tfrac{1}{2}\mathrm{Tr}(\phi^2)$$

which is invariant under U:

$$\mathrm{Tr}(\phi^2) \to \mathrm{Tr}(U\phi U^\dagger U\phi U^\dagger) = \mathrm{Tr}(U\phi^2 U^\dagger) = \mathrm{Tr}(U^\dagger U\phi^2) = \mathrm{Tr}(\phi^2) \tag{36.81}$$

Now

$$\tfrac{1}{2}\mathrm{Tr}(\phi^2) = \phi_+\phi_- + \alpha^2\phi_0^2 \tag{36.82}$$

Earlier, with

$$\phi_\pm = \frac{1}{\sqrt{2}}(\phi_1 \mp i\phi_2) \tag{24.21}$$

and $\phi_0 = \phi_3$, we found (24.27) the invariant mass term contained the expression

$$\tfrac{1}{2}\left(\phi_1^2 + \phi_2^2 + \phi_3^2\right) = \phi_+\phi_- + \tfrac{1}{2}\phi_0^2 \tag{36.83}$$

So we'll choose

$$\alpha = \frac{1}{\sqrt{2}} \tag{36.84}$$

to ensure that the mass term will come out right. Then[39]

$$\phi = \begin{pmatrix} \frac{1}{\sqrt{2}}\phi_0 & \phi_+ \\ \phi_- & -\frac{1}{\sqrt{2}}\phi_0 \end{pmatrix} = \frac{1}{\sqrt{2}}(\tau_1\phi_1 + \tau_2\phi_2 + \tau_3\phi_3) = \frac{1}{\sqrt{2}}(\boldsymbol{\tau}\boldsymbol{\cdot}\boldsymbol{\phi}) \tag{36.85}$$

[39] [Eds.] It's traditional to describe isospin in terms of the matrices τ_i, even though in the case of $I = \frac{1}{2}$, these are exactly the Pauli matrices σ_i. Here the τ_i imbed the pion isovector into the isospinor space of the nucleons.

Therefore, the free Lagrangian for the pion can be written as

$$\mathscr{L}_0 = \tfrac{1}{2}\text{Tr}(\partial_\mu\phi\,\partial^\mu\phi) + \tfrac{1}{2}\mu^2\text{Tr}(\phi^2) \tag{36.86}$$

After taking the traces, this is the standard expression (24.27), but here written in a form that is manifestly SU(2)-invariant.

The nice thing about doing things this way is that it's easy to write down couplings that are manifestly SU(2)-invariant. Well, it was easy to write them before, but don't forget that last time, in §24.1, we spent some time, from (24.11) to (24.14), working out the Clebsch–Gordan coefficients for the rotation group. But now we can do it in one line. We have a matrix ϕ. We have a row vector and a column vector, the nucleon fields and the antinucleon fields. We want to form an object that is an SU(2) scalar. If we're talking about Yukawa coupling, it's very easy to see that it must be

$$\mathscr{L}' = ig\overline{N}\gamma_5\phi N = i\frac{g}{\sqrt{2}}\overline{N}\gamma_5(\boldsymbol{\phi}\cdot\boldsymbol{\tau})N \tag{36.87}$$

our old friend. This product—row vector, matrix, column vector—is a scalar: the U's cancel the $U^\dagger$'s and the whole thing is obviously invariant.

We are using our prior knowledge of SU(2), in particular that the representations are irreducible, which we showed in §18.2. We don't have any prior knowledge of SU(3), so when we begin studying it, we will have to work to show that our representations are irreducible. The form on the left-hand side of (36.87) is equivalent to dotting the generators τ_i with the vector ϕ_i. It's exactly the same form. All those $1/\sqrt{2}$'s that we got last time in the isospin-invariant interaction come out automatically here. It's just (24.20), to within a factor of $\sqrt{2}$:

$$\mathscr{L}' = \frac{g}{\sqrt{2}}\left(\sqrt{2}\phi_+\overline{p}i\gamma_5 n + \sqrt{2}\phi_-\overline{n}i\gamma_5 p + \phi_0\overline{p}i\gamma_5 p - \phi_0\overline{n}i\gamma_5 n\right) \tag{36.88}$$

There is only one SU(2)-invariant Yukawa interaction. Since both methods are right, both methods must give the same result. Instead of fiddling with raising and lowering operators, it now comes from the condition that we want the trace of the square of the matrix to be properly normalized. As before (25.77), we'll need a ϕ^4 interaction for renormalizability:

$$\mathscr{L}' = ig\overline{N}\gamma_5\phi N - \tfrac{1}{4!}\lambda(\text{Tr}(\phi^2))^2 \tag{36.89}$$

That's all I want to say about SU(2), in this matrix and vector notation. It's inferior to the other way of doing things. It's nice if we're working with isospin-½ and isospin-1, but when we go to an object with higher isospin, say to isospin-3/2, we can certainly write it as a 3-index tensor, but there's no nice way of writing a 3-index object as a matrix. Still, the tensor notation is useful. First, it is completely general, with all those indices. It's like van der Waerden's method of treating the Lorentz group as a product of SU(2) ⊗ SU(2), where he has two kinds of indices, called "dotted" and "undotted", much used in the literature.[40] Second, the matrix trick is nice if we're only worried about certain selected representations of low dimensionality, which will be our situation in SU(3).

---

[40] [Eds.] This notation is not so common today, though it sometimes appears in supersymmetric theories; it was more frequently used forty years ago. See B. L. van der Waerden, "Spinoranalyse" ("Spinor Analysis"), *Nachrichten-Akad. der Wiss. Göttingen, Math.-Phys. Kl.* (1929) 100–109; B. L. van der Waerden, *Group Theory and Quantum Mechanics*, Springer, 1974, §23, "The Representations of the Lorentz-Group", pp. 114–117; Ryder *QFT*, pp. 433–439.

36.5 Tensor representations of SU(3)

In SU(3) everything is the same as in SU(2), except that the epsilon tensor ϵ^{ijk} has *three* indices. This makes all the difference in the world. For one thing, we can't make the epsilon tensor act as an ersatz metric tensor to raise and lower indices. We can lower an index with it but at the same time we raise the rank of the tensor by 1, which is bad. On the other hand, we can still write a given antisymmetric tensor in terms of a vector:

$$x^{[i,j]} = \epsilon^{ijk} x_k \tag{36.90}$$

a familiar trick from the three-dimensional rotation group, where we write an antisymmetric 3×3 matrix as an axial vector. We can thus get rid of antisymmetric parts, blithely reducing them as we move to tensors of higher rank to objects of lower rank which we have presumably already investigated in our iterative procedure. But we cannot get rid of either the upper or the lower indices; we're going to have to live with both of them. Therefore the sort of object we arrive at is a tensor with n upper indices, m lower indices, completely symmetric in both sets because we can always get rid of the antisymmetric parts using ϵ:

$$x^{i_1 \cdots i_n}_{j_1 \cdots j_m} \tag{36.91}$$

And, since summing on indices is an invariant operation, we can arrange that our tensors are fully traceless: if we sum any upper index with any lower index we get 0:

$$\underbrace{x^{i_1 \cdots i_l \cdots i_n}_{j_1 \cdots i_l \cdots j_m}}_{\text{sum on } i_l} = 0 \tag{36.92}$$

by subtracting out a bunch of terms proportional to $\delta^i{}_j$.

That's all we can do, generalizing what we did for SU(2). This set of tensors defines a representation. It may or may not be irreducible, but there they are, these symmetric, traceless tensors. We'll assume for now that they are irreducible. We know how the group acts on all tensors with n upstairs indices, m downstairs indices and no trace anywhere in between. The technique of decomposing representations into symmetric and antisymmetric parts gets really messy for SU(4), SU(5), etc., and is not the method of choice. For groups of higher dimension than SU(3), we have to use the permutation group and Young tableaux,[41] which were introduced into this subject by Hermann Weyl, cursed be his name.[42]

Thus we have defined representations which we will call

$$D^{(n,m)}(U) \equiv (n, m) \tag{36.93}$$

In the next lecture, we will call these representations IR's for short, an acronym for *irreducible representation.* Next time we will deduce their properties, their dimensions, their isospin content, and what happens when we multiply them together in analogy to the matrix tricks we saw here. And finally we will prove that they are in fact *a complete and inequivalent set of irreducible representations*, thus putting our knowledge of SU(3) on the same solid footing as our knowledge of SU(2). In the lecture after that, I'll show you four applications.

---

[41] [Eds.] Arfken & Weber *MMP*, Section 4.4, pp. 274–276.

[42] [Eds.] Weyl, *op. cit.*, §13, pp. 358–362.

37

Irreducible multiplets in SU(3)

Last time, we made a guess about the irreducible representations of SU(3).[1] In SU(3), as in SU(2), we used invariant tensors (the Kronecker delta $\delta^i{}_j$ and the antisymmetric epsilons, ϵ_{ijk} and ϵ^{ijk}) to reduce the rank of tensors. With the former, we could reduce a tensor's rank by two, by summing over an upper and a lower index to form the trace; and with the latter, we could reduce its rank by one, trading two upper indices for a lower, or *vice versa*, by summing over two upper or two lower indices. The guess was that the irreducible representations should have as their basis the set of all tensors with n upper indices and m lower indices

$$x^{i_1\cdots i_n}_{j_1\cdots j_m} \tag{37.1}$$

which are totally symmetric in both the upper and lower indices, and traceless in every pair of an upper and a lower index. Otherwise, contraction with either an epsilon or a delta could lower the tensor's rank. We call these representations

$$D^{(n,m)}(g) \equiv (n,m) \tag{37.2}$$

where $g \in \mathrm{SU}(3)$.

I've certainly defined a representation: a tensor of this kind *does* go into a linear combination of other tensors of this kind under the action of the group. In the course of this lecture we will answer the following questions:

- Are these representations irreducible?
- Are they inequivalent for different n and m?
- Are these *all* of the irreducible representations?

I'm going to answer these in counter-mathematical order, but in perfect rigor, insofar as I can achieve it. First I will deduce all sorts of useful properties of these representations. Then I will apply these properties to prove that these representations are in fact inequivalent, irreducible and complete. We will call these representations "IR"'s (for irreducible representations).

[1] [Eds.] This chapter is largely a reworking of two prior talks: Trieste, 1965 (S. Coleman, "Fun with SU(3)" in *High-Energy Physics and Elementary Particles*, C. Fronsdal, ed., IAEA, Vienna, 1965) and Erice, 1966 (reprinted in Coleman, *Aspects*, Chapter 1, "An introduction to unitary symmetry").

37.1 The irreducible representations q and $\overline{q}$

The properties of the IR's that I want to examine are the usual things to consider when one encounters a new group and its representations, the same properties we investigated for the Lorentz group in §18.3, or for the rotation group in quantum mechanics.

Conjugate representations

From (36.44), (36.45) and (36.46), it's easy to see that

$$(n,m)^* \sim (m,n) \tag{37.3}$$

The conjugates of upper indices transform like lower indices, and *vice versa*; that's the way we've defined things. So the representations and their complex conjugates (with $m \leftrightarrow n$) are equivalent. The matrices may not be the same but they can be turned into each other by a change of basis.

Dimensions

How many independent components do these tensors (37.1) have? That requires a little more work. The key formula is

$$(n,0)\otimes(0,m) \sim (n,m)\oplus[(n-1,0)\otimes(0,m-1)] \tag{37.4}$$

so that

$$\dim(n,0)\times\dim(0,m) = \dim(n,m) + \big[\dim(n-1,0)\times\dim(0,m-1)\big] \tag{37.5}$$

If you know the meaning of the symbols, it's easy to see why (37.4) is true. Recall that $(n,0)$ describes a completely symmetric tensor with only upper indices, and $(0,m)$ a completely symmetric tensor with only lower indices. We cannot impose the traceless condition on either of these varieties of tensors; they have only one type of index. When we take the product of one of each sort, we obtain completely symmetric tensors with a bunch of upper and lower indices, but there's no guarantee of tracelessness. Equation (37.4) is simply the mathematical statement that a general tensor, fully symmetric in both upper and lower indices, can be written as the sum of a fully symmetric, traceless tensor plus a general, fully symmetric tensor with one less upper index and one less lower index (times a string of invariant Kronecker delta's). For example,

$$x^{ij}_k = x^{ij}_k\Big|_{\text{traceless}} + \tfrac{1}{4}\left(\delta^i{}_k\delta^j{}_n + \delta^j{}_k\delta^i{}_n\right)x^{nm}_m \tag{37.6}$$

If we take the trace of both sides (by setting $j=k$ and summing), we find

$$x^{ik}_k = x^{ik}_k\Big|_{\text{traceless}} + x^{ik}_k, \quad \text{so } x^{ik}_k\Big|_{\text{traceless}} \overset{\checkmark}{=} 0$$

(the same is true if we set $i=k$). We can determine the dimensions of (n,m) if we can find the dimensions of $(n,0)$ and $(0,m)$.

Let's begin by computing the dimension of $(n,0)$. This is just combinatorics. As $(n,0)$ has only one kind of index and it's completely symmetric, the problem may be restated: given n objects, how many ways can we put them into three boxes labeled 1, 2 and 3? That will tell us how many 1's, how many 2's, and how many 3's there are. Put the n objects in a line and draw two barriers, creating three boxes. I've chosen $n=6$ for convenience. I'll put them in

$\bullet\bullet \mid \bullet \mid \bullet\bullet\bullet$

1's 2's 3's

Figure 37.1: One arrangement of six objects in three boxes

boxes by drawing two lines somewhere between them (Figure 37.1). Everything to the left of the leftmost line will be 1's. Everything to the right of the rightmost line will 3's. The 2's will be those between the lines. There are $(n+1)$ places for the first barrier, just as for SU(2). There are $(n+2)$ places for the second barrier, because we have the choice of putting the second barrier in front of or behind the first barrier (but which barrier we call the first and which the second is irrelevant). Therefore[2] the number of completely symmetric tensors with n upper indices or n lower indices (the combinatorics are indifferent to type)—the number of ways of putting n objects into three boxes—is

$$\dim(n,0) = \dim(0,n) = \tfrac{1}{2}(n+1)(n+2) \tag{37.7}$$

The $\frac{1}{2}$ comes because the two barriers are indistinguishable.

$\bullet \mid \;\mid \qquad \mid \bullet \mid \qquad \mid \;\mid \bullet$

Figure 37.2: One object in three boxes

As a check, how many ways can we put one object into three boxes? Three, of course; and that's what this formula gives (see Figure 37.2):

$$\dim(1,0) = \dim(0,1) = \tfrac{1}{2}(1+1)(1+2) = 3$$

For another example, consider two objects into three boxes, as in Figure 37.3. That also works out:

$$\dim(2,0) = \tfrac{1}{2}(2+1)(2+2) = 6$$

$\bullet\bullet \mid \;\mid \qquad \mid \bullet\bullet \mid \qquad \mid \;\mid \bullet\bullet \qquad \bullet \mid \bullet \mid \qquad \bullet \mid \;\mid \bullet \qquad \mid \bullet \mid \bullet$

Figure 37.3: Two objects in three boxes

Solving (37.5) for $\dim(n,m)$ we find

$$\begin{aligned} \dim(n,m) &= [\dim(n,0)\times\dim(0,m)] - [\dim(n-1,0)\times\dim(0,m-1)] \\ &= \tfrac{1}{4}[(n+1)(n+2)(m+1)(m+2) - n(n+1)m(m+1)] \\ &= \tfrac{1}{2}(n+1)(m+1)(n+m+2) \end{aligned} \tag{37.8}$$

This is a more complicated formula than $2s+1$, the corresponding formula for SU(2), because SU(3) is a more complicated group. But it is still relatively straightforward.

For example, let's work out the dimensions of some low-lying representations. See Table 37.1:

---

[2] [Eds.] See note 36, p. 791.

IR	*Dim*	*Name*
(1, 0)	3	$\mathbf{3}$
(0, 1)	3	$\overline{\mathbf{3}}$
(2, 0)	6	$\mathbf{6}$
(0, 2)	6	$\overline{\mathbf{6}}$
(1, 1)	8	$\mathbf{8}$
(3, 0)	10	$\mathbf{10}$
(0, 3)	10	$\overline{\mathbf{10}}$
(2, 2)	27	$\mathbf{27}$

Table 37.1: SU(3) *representations and their dimensions*

Aha! The **8**'s include the baryons; the **10** will describe the representation in which appear the Δ and its friends, among them the Ω^-, famed in song and story. There is no particle as far as I know that has been assigned to a 27-plet, but the **27** will come into our theory for certain operators.

The convention for the vulgar[3] name is to label the IR by its dimensions (I'll use bold type to distinguish representations from ordinary numbers), adding a bar if the second index is greater than the first, to distinguish between complex conjugate pairs of representations. This labeling is unfortunately not unique. For instance, as one can check,

$$\dim(1, 2) = \dim(0, 4) = 15$$

but neither of these representations occurs frequently in the literature, and most people call a representation **3** or $\overline{\mathbf{3}}$ etc., rather than (1, 0) or (0, 1).

Isospin and hypercharge

How should we embed the SU(2) $\otimes$ U(1) subgroup inside SU(3)? We can decide that once we've made a decision about the isospin and hypercharge of the three-dimensional representations, **3** and $\overline{\mathbf{3}}$. That will be determined by the interplay of mathematics and physics. The mathematics will tell us the *possible* ways to do it; the physics will tell us the *right* way to do it so that the baryon octet comes out as it should.

For the moment I will worry about embedding SU(2), the isospin subgroup inside SU(3), and take care of hypercharge shortly. Let's begin with the fundamental triplet representation, $(1, 0) = \mathbf{3}$. That's a three-dimensional representation and therefore when we restrict SU(3) to isospin there are three possibilities:

$$(1,0) \rightarrow \begin{cases} (1) & \text{(a) Three objects in an isotriplet} \\ \left(\frac{1}{2}\right) \oplus (0) & \text{(b) Isodoublet plus isosinglet} \\ (0) \oplus (0) \oplus (0) & \text{(c) Three isosinglets} \end{cases} \tag{37.9}$$

Those are the only distinct ways of partitioning 3 into a sum of positive integers: 3, 2 + 1, 1 + 1 + 1.

[3] [Eds.] Coleman adds: "In the sense of botany, with no pejorative connotation." See D. Gledhill, *The Names of Plants*, 4th ed., Cambridge U. P., 2008.

Possibility (c) is no good for both mathematical and physical reasons. If everything in the triplet transforms like an isosinglet, everything in the triplet bar also transforms like an isosinglet and everything in all the IR's transforms like an isosinglet. That's a trivial embedding of SU(2) inside SU(3). We want a non-trivial embedding.

Possibility (a) is also no good. It says that the triplet **3** has isospin 1, the conjugate $\overline{\mathbf{3}}$ has isospin 1, and therefore everything made by taking direct products would contain only integer isospins. This leaves no room for such friendly particles as the nucleons, with half-integer isospin. So possibility (a) is also no good. Mathematically it's fine, but physically it fails to accomodate the baryon octet.

The only remaining possibility is (b), an isodoublet plus an isosinglet. We will represent this graphically by writing the fundamental three-dimensional representation as a column vector. I have the freedom to make similarity transformations, so I will arrange things such that the first two entries are the isodoublet and the third entry is the isosinglet. Thus, SU(2) would consist of that subgroup of SU(3) which leaves the third unit vector unchanged. These are conventionally labeled

$$\begin{pmatrix} u \\ d \\ s \end{pmatrix} \qquad I_z = \begin{cases} +\frac{1}{2} \\ -\frac{1}{2} \\ 0 \end{cases} \tag{37.10}$$

The u, conventionally called "up", and d, called "down", form the isodoublet with its isospin up and down; s is the isosinglet; s originally stood for "singlet", or according to some wags, "sideways". For historical reasons it is called "strange".[4] We can consider this triplet as hypothetical states of some unknown particles called **quarks** (we love to give names to the unknowns).[5] All other hadrons can be built out of **3**'s and $\overline{\mathbf{3}}$'s, i.e., out of quarks and antiquarks. Alternatively, we can consider the **3** representation as triplet fields for some particles that have never been observed.[6] In any case, we'll denote the **3** representation by q (in honor of the quark model, though I won't discuss that theory until later), and $\overline{\mathbf{3}}$ by $\overline{q}$.

How are we going to assign hypercharge to the q representation? Hypercharge commutes with isotopic spin and therefore the two elements of the doublet must have the *same* hypercharge, which I will call α. The matrix Y whose eigenvalues are the hypercharges is also the generator of the U(1) symmetry. That is, the matrices $g \in$ U(1) are exponentials of Y:

$$g = \exp(i\chi Y) \tag{37.11}$$

where χ is some real parameter; if g is to be unitary, Y must be Hermitian. We want these matrices g to be elements of SU(3) as well, and so we need the exponential of the hypercharge matrix to have determinant 1. Using the formula[7]

$$\det A = \exp(\mathrm{Tr}\,[\ln A]) \tag{37.12}$$

---

[4] [Eds.] Crease & Mann *SC* pp. 171-177.

[5] [Eds.] F. Halzen and A. D. Martin, *Quarks and Leptons: An Introductory Course in Modern Particle Physics*, John Wiley and Sons, 1984; Crease & Mann *SC*, Chapter 15, pp. 280-285. This paragraph is taken nearly verbatim from the 1976 video of Coleman's lecture 37 (starting at 0:26:50), for its historical interest.

[6] [Eds.] Physical quarks were introduced independently by Gell-Mann and George Zweig a couple of years after people started playing with SU(3). They were widely disbelieved until around 1968, when deep inelastic scattering of electrons off protons at the Stanford Linear Accelerator (SLAC) revealed structure inside the proton. See note 28, p. 859 and note 13, p. 1096.

[7] [Eds.] See note 7, p. 608, and Problem 17.1, (P17.4), p. 679.

it follows that

$$\det g = \exp\{i\chi \,\mathrm{Tr}\, Y\} \tag{37.13}$$

which means the hypercharge matrix Y itself must have trace 0. The trace of the matrix is the sum of its eigenvalues. As the doublet elements u and d each have hypercharge α, the singlet must have hypercharge -2α. We will fix α by physical considerations. Once we have assigned isospin and hypercharge values to the representation $(1,0) = q$, then we also know the isospin and hypercharge content of the representation $(0,1) = \overline{q} = (\overline{u}, \overline{d}, \overline{s})$. These values are summarized in Table 37.2.

Quark	I_z	Y
u	$+\frac{1}{2}$	α
d	$-\frac{1}{2}$	α
s	0	-2α
$\overline{u}$	$-\frac{1}{2}$	$-\alpha$
$\overline{d}$	$+\frac{1}{2}$	$-\alpha$
$\overline{s}$	0	2α

Table 37.2: Isospin and hypercharge for the q and $\overline{q}$ representations

Our table of the dimensions of the representations led us to suspect that the baryons must be put into the $\mathbf{8} = (1,1)$ representation, the product of a row vector and a column vector with the trace subtracted out. We will label representations of the subgroup by their isospin and hypercharge like this (with the hypercharge as a superscript):

$$(I)^Y \tag{37.14}$$

In this notation, we write (37.9)

$$\begin{aligned} (1,0) &\to \left(\tfrac{1}{2}\right)^{\alpha} \oplus (0)^{-2\alpha} \\ (0,1) &\to \left(\tfrac{1}{2}\right)^{-\alpha} \oplus (0)^{2\alpha} \end{aligned} \tag{37.15}$$

From (37.4)

$$(1,0) \otimes (0,1) \sim (1,1) \oplus \big[(0,0) \otimes (0,0)\big] = (1,1) \oplus (0,0) \tag{37.16}$$

because $(0,0)$ is a singlet. In the vulgar notation,

$$\mathbf{3} \otimes \overline{\mathbf{3}} \sim \mathbf{8} \oplus \mathbf{1} \tag{37.17}$$

which makes sense: nine states on either side. On the other hand, we have

$$\begin{aligned} (1,0) \otimes (0,1) &= \left[\left(\tfrac{1}{2}\right)^{\alpha} \oplus (0)^{-2\alpha}\right] \otimes \left[\left(\tfrac{1}{2}\right)^{-\alpha} \oplus (0)^{2\alpha}\right] \\ &\sim \left[\left(\tfrac{1}{2}\right) \otimes \left(\tfrac{1}{2}\right)\right]^{0} \oplus \left[\left(\tfrac{1}{2}\right) \otimes (0)\right]^{3\alpha} \oplus \left[\left(\tfrac{1}{2}\right) \otimes (0)\right]^{-3\alpha} \oplus \left[(0) \otimes (0)\right]^{0} \\ &\sim (0)^0 \oplus (1)^0 \oplus \left(\tfrac{1}{2}\right)^{3\alpha} \oplus \left(\tfrac{1}{2}\right)^{-3\alpha} \oplus (0)^0 \end{aligned} \tag{37.18}$$

(using the Clebsch–Gordan series (18.73 for SU(2)) so that, from (37.16)

$$(1,1) \oplus (0,0) \sim (0)^0 \oplus (1)^0 \oplus \left(\tfrac{1}{2}\right)^{3\alpha} \oplus \left(\tfrac{1}{2}\right)^{-3\alpha} \oplus (0)^0 \tag{37.19}$$

The trivial representation of the group, $(0)^0$, is an isosinglet with hypercharge 0. It is the same as the representation $(0,0)$ and can be dropped on both sides. That is,

$$(1,1) \sim (0)^0 \oplus (1)^0 \oplus \left(\tfrac{1}{2}\right)^{3\alpha} \oplus \left(\tfrac{1}{2}\right)^{-3\alpha} \tag{37.20}$$

We observe that this octet matches up nicely with the eight $J^P = \frac{1}{2}^+$ baryons[8] of Table 36.1: $(0)^0$ is in a jolly position to be the Λ, the $(1)^0$ is well-suited to be the Σ, while the $(\frac{1}{2})^{3\alpha}$ and $(\frac{1}{2})^{-3\alpha}$ can be identified with the nucleon and the cascade (Ξ), respectively, if we choose

$$\alpha = \tfrac{1}{3} \tag{37.21}$$

With this identification,

$$(1,1) \sim \begin{cases} (0)^0 & \oplus \quad (1)^0 \quad \oplus \quad \left(\tfrac{1}{2}\right)^1 \quad \oplus \quad \left(\tfrac{1}{2}\right)^{-1} \\ \Lambda & \oplus \quad \Sigma \quad \oplus \quad N \quad \oplus \quad \Xi \end{cases} \tag{37.22}$$

There is of course the option to choose $\alpha = -1/3$ and change the identification of the nucleon and the cascade, but this is just a matter of convention. If we choose $\alpha = -1/3$ we would be switching the left and right entries in the representation:

$$(1,0) \rightleftarrows (0,1)$$

We would have the same hypercharge assignments for the s that we originally had for the quarks. All that would change would be the convention about what we call a quark and an antiquark. Our equations would look different but the physics would be the same. For the benefit of mathematical purists, this corresponds to the existence of an outer automorphism of the group SU(3) induced by complex conjugation.[9] However, if the physics is clear then which sign we choose is just a matter of convention.

We have finally for the quarks:

Quark	I_z	Y	$Q = I_z + \frac{1}{2}Y$
u	$+\frac{1}{2}$	$+\frac{1}{3}$	$+\frac{2}{3}$
d	$-\frac{1}{2}$	$+\frac{1}{3}$	$-\frac{1}{3}$
s	0	$-\frac{2}{3}$	$-\frac{1}{3}$

Table 37.3: The quarks and their properties

The quarks have *fractional charges.*[10]

---

[8] [Eds.] Table 36.1 is on p. 782.

[9] [Eds.] See Art. 64, p. 48 in Allan Clark, *Elements of Abstract Algebra*, Wadsworth, 1971; reprinted by Dover Publications, 1984. *Inner* automorphisms are the similarity transformations familiar to every physicist.

[10] [Eds.] Evidently it was the fractional charges that discouraged Gell-Mann from initially suggesting that quarks were physical entities; Crease & Mann *SC*, p. 281.

37.2 Matrix tricks with SU(3)

Baryons

Following the SU(2) tricks of the last lecture, we will work out a matrix representation for the octet representation $(1,1)$ which we had identified with the $J^P = \frac{1}{2}^+$ baryons. This is a traceless tensor $\psi^i{}_j$ with one upper index and one lower index:

$$\psi^i{}_i = 0$$

Recall that under a transformation $g \in \mathrm{SU}(3)$

$$g\colon \psi \to g\psi g^\dagger \tag{37.23}$$

We can consider $\psi^i{}_j$ as a matrix which transforms as if it were the outer product of a q and a $\overline{q}$, minus $\frac{1}{3}$ the trace times the identity matrix:

$$\psi^i{}_j = q^i \otimes \overline{q}_j - \tfrac{1}{3}\delta^i{}_j \mathrm{Tr}\,(q\otimes\overline{q}) \tag{37.24}$$

This is similar to what we did earlier, when we considered (36.79) the pions transforming as an outer product of an N and an $\overline{N}$. That's the way that outer products of two vectors typically transform, like a rank 2 tensor. Remember that our q consists of a 2×1 block, an isodoublet with $Y = +\frac{1}{3}$, and a 1×1 isosinglet block with $Y = -\frac{2}{3}$; likewise $\overline{q}$ has a 1×2 isodoublet block with $Y = -\frac{1}{3}$ and a 1×1 isosinglet block with $Y = \frac{2}{3}$:

$$q = \begin{pmatrix} \begin{bmatrix} 2\times 1 \\ I=\frac{1}{2}\end{bmatrix}^{\frac{1}{3}} \\ \begin{bmatrix} 1\times 1 \\ I=0\end{bmatrix}^{-\frac{2}{3}} \end{pmatrix} \qquad \overline{q} = \begin{pmatrix} \begin{bmatrix} 1\times 2 \\ I=\frac{1}{2}\end{bmatrix}^{-\frac{1}{3}} & \begin{bmatrix} 1\times 1 \\ I=0\end{bmatrix}^{\frac{2}{3}} \end{pmatrix} \tag{37.25}$$

The outer product $q\otimes\overline{q}$ gives

$$\psi = \begin{pmatrix} \begin{bmatrix} 2\times 2 \\ (I=1)\oplus(I=0)\end{bmatrix}^{0} & \begin{bmatrix} 2\times 1 \\ I=\frac{1}{2}\end{bmatrix}^{1} \\ \begin{bmatrix} 1\times 2 \\ I=\frac{1}{2}\end{bmatrix}^{-1} & \begin{bmatrix} 1\times 1 \\ I=0\end{bmatrix}^{0} \end{pmatrix} \tag{37.26}$$

The diagonal elements are connected by the traceless condition:

$$\mathrm{Tr}\,(\psi) = 0 \tag{37.27}$$

When we restrict ourselves to the SU(2) subgroup, those transformations leave the third component of the vector unchanged. We also see that the entries have the given hypercharge assignments, (37.20). Thus we can write down the matrix

$$\psi = \begin{pmatrix} \frac{1}{\sqrt{2}}\Sigma^0 + \beta\Lambda & \Sigma^+ & p \\ \Sigma^- & -\frac{1}{\sqrt{2}}\Sigma^0 + \beta\Lambda & n \\ \Xi^- & -\Xi^0 & -2\beta\Lambda \end{pmatrix} \tag{37.28}$$

The Σ terms in the upper 2×2 block follow the pattern of (36.85). The scale factor β is to be determined; the -2β in the lower right is required by the trace condition. The third row has $-\Xi^0$, not $+\Xi^0$, in order to have the conventional (Condon and Shortley)[11] phase relations between the members of an isodoublet: we use ϵ_{ij} to lay the 2×1 block (upper right) on its side.

The expression (37.28) for ψ is the representation of the eight baryons as a 3×3 matrix. Choosing a value for β is a matter of taste, but we normally arrange things to avoid SU(3)-violating wave function renormalizations. Taking a hint from the trick (36.86) we used with SU(2), we want to look at $\text{Tr}\,(\overline{\psi}\psi)$, which appears in the free Lagrangian's mass term; $\overline{\psi}$ is the (Dirac) adjoint matrix of ψ, transforming under SU(3) as

$$\overline{\psi} \to (g\overline{\psi}g^\dagger)^\dagger = g^{\dagger\dagger}\overline{\psi}g^\dagger = g\overline{\psi}g^\dagger$$

Then $\text{Tr}\,(\overline{\psi}\psi)$ is invariant:

$$\text{Tr}(\overline{\psi}\psi) \to \text{Tr}(g\overline{\psi}g^\dagger g\psi g^\dagger) = \text{Tr}(g\overline{\psi}\psi g^\dagger) = \text{Tr}(g^\dagger g\overline{\psi}\psi) = \text{Tr}(\overline{\psi}\psi) \tag{37.29}$$

The fields in ψ are Dirac fields. When we form $\overline{\psi}$, the fields appear in it as their Dirac adjoints, e.g., $\overline{p}$ instead of p. The trace is a sum of squares:

$$\text{Tr}\,(\overline{\psi}^i{}_m\psi^m{}_j) = \overline{p}p + \overline{n}n + \cdots + \beta^2\overline{\Lambda}\Lambda(1+1+4) \tag{37.30}$$

from which we determine

$$\beta = \frac{1}{\sqrt{6}} \tag{37.31}$$

We could choose $\beta = -\frac{1}{\sqrt{6}}$ but that's just a matter of convention on the phase of the Λ, which thus far is undetermined. The final result for ψ is

$$\psi = \begin{pmatrix} \frac{1}{\sqrt{2}}\Sigma^0 + \frac{1}{\sqrt{6}}\Lambda & \Sigma^+ & p \\ \Sigma^- & -\frac{1}{\sqrt{2}}\Sigma^0 + \frac{1}{\sqrt{6}}\Lambda & n \\ \Xi^- & -\Xi^0 & -\frac{2}{\sqrt{6}}\Lambda \end{pmatrix} \tag{37.32}$$

The free Lagrangian involving this mass degenerate octet[12] can now be written

$$\mathscr{L}_0 = \text{Tr}\,[\overline{\psi}(i\partial\!\!\!/ - m)\psi] \tag{37.33}$$

Mesons

As it happens, there are also eight pseudoscalar mesons, $J^P = 0^-$, of low mass; see Table 37.4.

In fact there's a ninth pseudoscalar meson, the η', with $(I)^Y = (0)^-$ and mass 960 MeV, that most people think is a singlet. It doesn't mix up much with these eight even with SU(3)-violating interactions. It is possible that medium strong interactions mix members of different multiplets. Note that the absolute value of the pseudoscalar meson mass splitting (410 MeV between the π's and the η) is comparable to that of the baryon octet (380 MeV between the nucleons and the Ξ's; see Table 36.1 on p.782). That suggests that medium strong splittings may all be of the same order, but at different mass levels (ignoring the η').

---

[11] [Eds.] E. U. Condon and G. H. Shortley, *The Theory of Atomic Spectra*, Cambridge U. P. 1935; reprinted with corrections 1951; reprinted 1991. The phases are discussed in Chapter 3, section 3, pp. 48–49.

[12] [Eds.] If SU(3) invariance were to hold exactly, the masses of a multiplet like the baryon octet would all be the same. The different masses indicate that nature breaks SU(3) invariance.

Meson	$(I)^Y$	M (MeV)
$\pi^\pm, \pi^0$	$(1)^0$	140
$K^0, \overline{K}^0$	$(\frac{1}{2})^\pm$	498
$K^\pm$	$(\frac{1}{2})^\pm$	494
η	$(0)^0$	550

Table 37.4: The eight pseudoscalar mesons: $J^P = 0^-$

We can represent the eight pseudoscalar mesons by a matrix ϕ of exactly the same form as that for the baryon matrix ψ. Since the $(1,1)$ representation is self-conjugate, $(1,1)^* \sim (1,1)$, so is the meson octet representation. We can, if we wish, impose the condition

$$\phi = \phi^\dagger \tag{37.34}$$

for the 0^- mesons. The choice that ϕ be Hermitian is invariant under SU(3) transformations. That will give us eight real fields instead of eight complex fields. We don't *need* to do that; we certainly don't want to do it for the baryons unless we want to write things in terms of Majorana fields[13], which is a bad move: the baryons are *not* their own antiparticles. But we can do it for the mesons and we choose to do so, because there are only eight pseudoscalars, not 16. The ϕ matrix looks, with one small difference, exactly the same as ψ:

$$\phi = \begin{pmatrix} \frac{1}{\sqrt{2}}\pi^0 + \frac{1}{\sqrt{6}}\eta & \pi^+ & K^+ \\ \pi^- & -\frac{1}{\sqrt{2}}\pi^0 + \frac{1}{\sqrt{6}}\eta & K^0 \\ K^- & \overline{K}^0 & -\frac{2}{\sqrt{6}}\eta \end{pmatrix} \tag{37.35}$$

There is no minus sign in front of the $\overline{K}^0$ (as there is before the Ξ^0 in the baryon matrix) because of a clash of conventions. We have two conventions we want to follow. One is that the phase relations between an isodoublet should be as found in Condon and Shortley[14]; the other is that $\overline{K}^0$ should be the conjugate field to K^0. For the $\overline{K}^0$ we adopt the second convention, and disobey the Condon–Shortley phase convention. When the first papers were written, this was the choice everyone made, and it's now standard.

We can now write down an SU(3)-invariant (indeed, it's SO(8)-invariant) meson Lagrangian. Adding it to (37.33),

$$\mathscr{L}_0 = \mathrm{Tr}\left[\overline{\psi}(i\partial\!\!\!/ - m)\psi\right] + \tfrac{1}{2}\mathrm{Tr}\left[\partial_\mu\phi\partial^\mu\phi - \mu^2\phi^2\right] \tag{37.36}$$

That's just the sum of the squares of the fields, all normalized the same way. We can also make a guess about invariant Yukawa interactions. Both ψ and ϕ transform according to (37.23), so the trace of any product of ψ's and ϕ's will be SU(3)-invariant. We have one ψ, one $\overline{\psi}$ and one ϕ. By the cyclic invariance of the trace we can always put the $\overline{\psi}$ in the first position, but then there are two possibilities, whether the ϕ precedes the ψ or follows the ψ. In fact in the

[13] [Eds.] See §22.2, and note 3, p. 464. Majorana fermions are their own antiparticles, $\psi^C = \psi$, as opposed to Dirac fermions, which are not; Ryder, *QFT*, p. 429; Palash B. Pal, "Dirac, Majorana, and Weyl Fermions", *Am. J. Phys.* **79** (2011) 485–498. Majorana fermions are used in supersymmetric theories.

[14] [Eds.] See note 11, p. 805.

literature the symmetric and antisymmetric combinations of these two possibilities are used. We write[15]

$$\mathscr{L}' = g_D \mathrm{Tr}\left[\overline{\psi} i\gamma_5\{\phi,\psi\}\right] + g_F \mathrm{Tr}\left[i\overline{\psi} i\gamma_5[\phi,\psi]\right] \tag{37.37}$$

Of course nothing we have said so far demonstrates that these are the *only* couplings, but we will shortly find that they are. For the moment they are two Lorentz-invariant, SU(3)-invariant couplings which one can write down.

37.3 Isospin and hypercharge decomposition

Earlier, we worked out the isospin and hypercharge values of the octet $(1,1)$. To understand the isospin and hypercharge decomposition of a *general* representation (n,m), we need a digression on some other matrix tricks. Return to (37.4):

$$(n,0)\otimes(0,m) \sim (n,m)\oplus[(n-1,0)\otimes(0,m-1)]$$

If we can find out the isospin and hypercharge decompositions of $(n,0)$, we can get those of $(0,m)$ by complex conjugation. Then we'll know what isospins and hypercharges are in $(n,0)\otimes(0,m)$ and in $(n-1,0)\otimes(0,m-1)$, and by subtraction we can find the isospins and hypercharges in (n,m). So our immediate goal is to work out the isospin and hypercharge decomposition of $(n,0)$.

This is fairly easy: $(n,0)$ is the completely symmetric product of n $(1,0)$ (or q, or **3**) representations. The quarks contain a doublet with hypercharge $\frac{1}{3}$ and a singlet with hypercharge $-\frac{2}{3}$, Table 37.3. When we take the direct product, we obtain only a certain number of representations, according to the Clebsch–Gordan series for SU(2):

$$(n,0) = \prod_{\otimes}^{n}\left[\left(\tfrac{1}{2}\right)^{1/3}\oplus(0)^{-2/3}\right] \tag{37.38}$$

This will be a sum of terms. For the first term we can take the completely symmetric product of n isospin $\frac{1}{2}$'s, with all the isospins aligned and their hypercharges added, so that gives us

$$(I)^Y = \left(\tfrac{1}{2}n\right)^{n/3}$$

---

[15] [Eds.] Coleman adds, "The subscripts D and F on the coupling constants have no particular meaning. It's one of those things Murray Gell-Mann found in the *Séminaire "Sophus Lie"*, so perhaps they stand for famous French politicians." Another possibility comes from the algebra of SU(3), whose eight generators are typically written λ_a (called the **Gell-Mann matrices**):

$$\lambda_i = \begin{pmatrix}\sigma_{i11} & \sigma_{i12} & 0\\ \sigma_{i21} & \sigma_{i22} & 0\\ 0 & 0 & 0\end{pmatrix};\ \lambda_{j+3} = \begin{pmatrix}\sigma_{j11} & 0 & \sigma_{j12}\\ 0 & 0 & 0\\ \sigma_{j21} & 0 & \sigma_{j22}\end{pmatrix};\ \lambda_{j+5} = \begin{pmatrix}0 & 0 & 0\\ 0 & \sigma_{j11} & \sigma_{j12}\\ 0 & \sigma_{j21} & \sigma_{j22}\end{pmatrix};\ \lambda_8 = \frac{1}{\sqrt{3}}\begin{pmatrix}1 & 0 & 0\\ 0 & 1 & 0\\ 0 & 0 & -2\end{pmatrix}$$

where σ_i are the Pauli matrices, $i=\{1,2,3\}$, and $j=\{1,2\}$. These λ_a are to SU(3) what the Pauli matrices are to SU(2); like the σ_i, the λ_a matrices are traceless and Hermitian. Analogous to the Pauli matrix algebra, the $\{\lambda_a\}$ satisfy

$$[\lambda_a,\lambda_b] = 2if_{abc}\lambda_c\,; \qquad \{\lambda_a,\lambda_b\} = \tfrac{4}{3}\delta_{ab} + 2id_{abc}\lambda_c$$

One may then define the matrices $(F^a)_{bc}\equiv(F_a)_{bc}\equiv -if_{abc}$, in which case $[F_a,F_b] = if_{abc}F_c$. Similarly one defines the matrices $(D^a)_{bc}\equiv(D_a)_{bc}\equiv d_{abc}$. As in the coupling, the matrices F are antisymmetric, and the D's are symmetric; see P. Carruthers, *Introduction to Unitary Symmetry*, Interscience Publishers, 1966, Sections 2.2, p. 30, and 2.6, pp. 50–52; or Table I, in M. Gell-Mann, "The Eightfold Way", Caltech Synchrotron Radiation Laboratory report CTSL-20, 1961 (unpublished); reprinted in *The Eightfold Way*, M. Gell-Mann and Y. Ne'eman, Benjamin, 1964. Coleman never writes down the λ_a explicitly.

The next term we can have has $(n-1)$ isospin $\frac{1}{2}$'s and one isospin 0:

$$(I)^Y = \left(\tfrac{1}{2}(n-1)\right)^{(n/3)-1}$$

The hypercharge is down by 1 since we've traded a doublet with $Y = +\frac{1}{3}$ for a singlet with $Y = -\frac{2}{3}$. Continuing along, at the end we have simply the n isospin 0 piece with hypercharge $-\frac{2}{3}n$:

$$(I)^Y = (0)^{-(2/3)n}$$

Thus the isospin-hypercharge content we get when reducing SU(3) down to SU(2) ⊗ U(1) is

$$\begin{aligned}(n,0) &\to \left(\tfrac{1}{2}n\right)^{n/3} \oplus \left(\tfrac{1}{2}(n-1)\right)^{(n/3)-1} \oplus \cdots \oplus (0)^{-(2/3)n} \\ (0,m) &\to \left(\tfrac{1}{2}m\right)^{-m/3} \oplus \left(\tfrac{1}{2}(m-1)\right)^{-(m/3)+1} \oplus \cdots \oplus (0)^{(2/3)m}\end{aligned} \tag{37.39}$$

For instance,

$$\begin{aligned}(2,0) &\to (1)^{2/3} \oplus \left(\tfrac{1}{2}\right)^{-1/3} \oplus (0)^{-4/3} \\ (3,0) &\to \left(\tfrac{3}{2}\right)^{1} \oplus (1)^{0} \oplus \left(\tfrac{1}{2}\right)^{-1} \oplus (0)^{-2}\end{aligned} \tag{37.40}$$

If we want to find out what's in $(n,0)\otimes(0,m)$ we make a large rectangular array, n by m blocks, as in Figure 37.4. In each block of the table we multiply together the two representations, using the conventional isospin combining rules. Then we fill up the entire array, add all the blocks together, and that's the product $(n,0)\otimes(0,m)$. However, to get the representation (n,m), we have to subtract out what's in $(n-1,0)\otimes(0,m-1)$. That's in fact exactly the same series except that it begins one stage further on. The hypercharge is a little bit off in $(n-1,0)$ but is a little bit off in the opposite direction in $(0,m-1)$ so it doesn't matter in the product. The content of $(n-1,0)\otimes(0,m-1)$ is the entire array aside from the border—the shaded area. Therefore the content of (n,m) is simply the entries on the top and left border of the rectangular array. That is our algorithm.

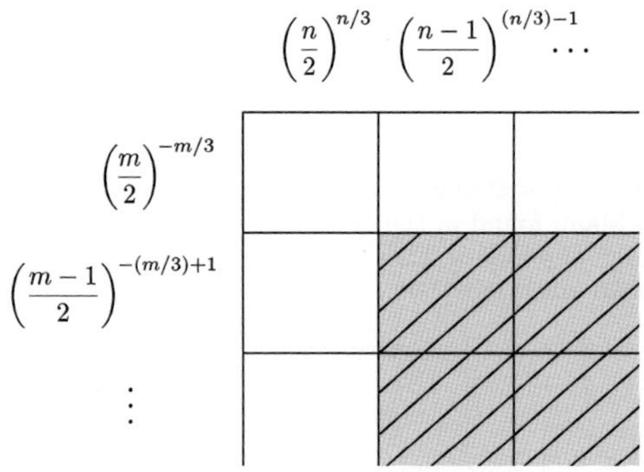

Figure 37.4: Graphical decomposition of (n,m)

Let's check this by working out the answer to something we already know, $(1,1)$. This is a 2×2 table, shown in Figure 37.5.

This is just the Σ and the Λ, the nucleon N, and the Ξ, the same answer (37.20) we obtained before.

	$\left(\frac{1}{2}\right)^{1/3}$	$(0)^{-2/3}$
$\left(\frac{1}{2}\right)^{-1/3}$	$(1)^0 \oplus (0)^0$	$\left(\frac{1}{2}\right)^{-1}$
$(0)^{2/3}$	$\left(\frac{1}{2}\right)^{1}$	

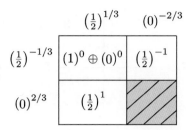

Figure 37.5: Graphical decomposition of $(1,1)$

	$(1)^{2/3}$	$\left(\frac{1}{2}\right)^{-1/3}$	$(0)^{-4/3}$
$(1)^{-2/3}$	$(2)^0 \oplus (1)^0 \oplus (0)^0$	$\left(\frac{3}{2}\right)^{-1} \oplus \left(\frac{1}{2}\right)^{-1}$	$(1)^{-2}$
$\left(\frac{1}{2}\right)^{1/3}$	$\left(\frac{3}{2}\right)^{1} \oplus \left(\frac{1}{2}\right)^{1}$		
$(0)^{4/3}$	$(1)^2$		

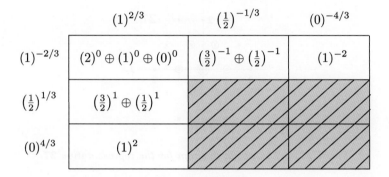

Figure 37.6: Graphical decomposition of $(2,2)$

Now let's go after bigger game. I will work out $(2,2)$, the 27-plet. That's a little bit more ambitious, a 3×3 box, Figure 37.6. Check the dimensions, summing along the row and the column:

$$\dim(2,2) = 5+3+1+4+2+3+4+2+3 = 27 \tag{37.41}$$

Aha! It's exactly what we should have found.

The only way to learn these algorithms is to pick a representation and work things out for it. I've not assigned a homework problem on this topic, but you should work out the isospin content of $(3,3)$ or $(3,4)$.

In the literature the isospin-hypercharge contents of these representations are frequently depicted on **weight diagrams**. A dot is placed on the weight diagram for every particle with a given I_z and Y. The weight diagrams for the representations **3**, $\overline{\mathbf{3}}$ and **8** are shown in Figure 37.7. A dot in a circle indicates two particles with the same values of I_z and Y; a dot in two circles indicates three. The term "weight diagram" comes from Cartan's general theory of the representations of semi-simple Lie groups.[16]

There is a deep reason why these things come out to be beautiful geometric figures rather than random arrays of dots, but that has to do with the structure theory of Lie groups and we won't go into it here. As another drill, work out the weight diagrams for some other representations, such as $(3,0)$. For those of you of Pythagorean inclinations, I leave it as an

[16] [Eds.] H. Georgi, *Lie Algebras in Particle Physics*, Addison-Wesley, 1982; 2nd ed., Perseus Books, 1999; for Élie Cartan, see note 11, p. 1017.

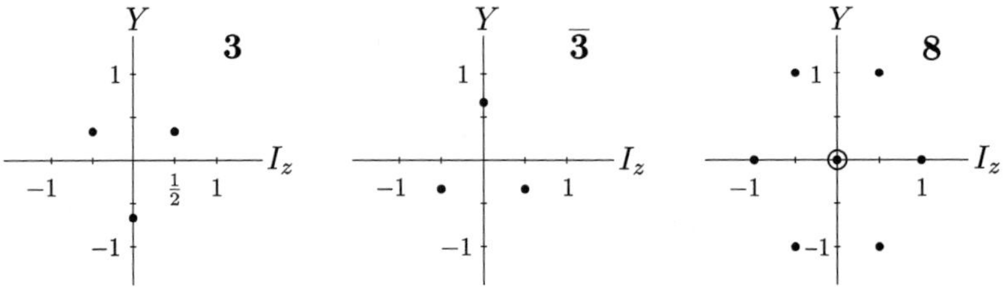

Figure 37.7: The weight diagrams for the representations **3**, $\overline{\mathbf{3}}$ *and* **8**

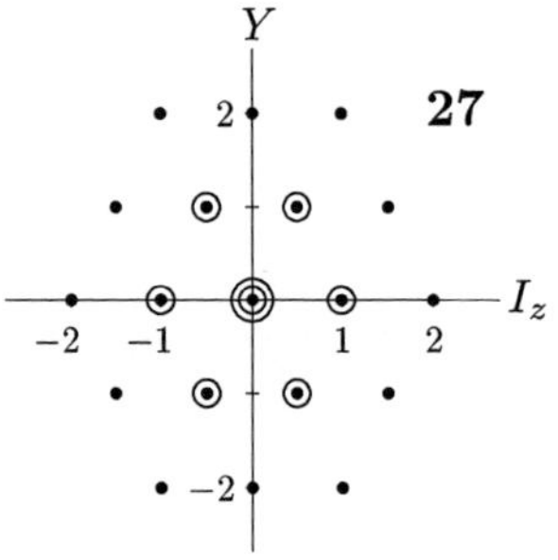

Figure 37.8: The weight diagram for the representation **27**

exercise to demonstrate that (n, n) always makes a hexagon and $(n, 0)$ or $(0, n)$ always makes a triangle. As far as I'm concerned, these pretty diagrams are not useful.[17]

37.4 Direct products in SU(3)

How do we decompose the direct product $(n, m) \otimes (n', m')$? This is itself certainly a representation. It should be expressible as a direct sum of IR's. Which ones? What are the SU(3) analogs of the familiar vector addition algorithm and Clebsch–Gordan series of SU(2)? To write it as an equation,

$$(n, m) \otimes (n', m') \sim ? \tag{37.42}$$

We know the answer for a product of $(1, 0)$'s and $(0, 1)$'s, but not for the general expression.

I will show you a non-standard algorithm for computing these direct products. It's one I invented[18] around 1964. It seems wonderfully simple and elegant to me; everyone else just looks things up in tables.[19] These tables are typically produced using a much more complicated algorithm (invented by Hermann Weyl) that is good for a general Lie group. The algorithm I'm going to show you is a special trick that works for SU(3) *only*. It will enable us to compute the direct product of anything with anything in SU(3), using only the back of an envelope and elementary arithmetic.

[17] [Eds.] This opinion is not widely shared. The editors have inserted weight diagrams where they were thought to be helpful.

[18] [Eds.] Coleman, "Fun with SU(3)", *op. cit.*; S. Coleman, "The Clebsch–Gordan Series for SU(3)", *J. Math. Phys.* **5** (1964) 1343–1344.

[19] [Eds.] J.J. de Swart, "The Octet Model and its Clebsch–Gordan Coefficients", *Rev. Mod. Phys.* **35** (1963) 916–939; reprinted in M. Gell-Mann and Y. Ne'eman, *The Eightfold Way*, Benjamin, 1964.

The algorithm has two stages. First, we reduce the product $(n,m)\otimes(n',m')$ to the sum of certain special reducible representations (which I will define) by removing all remaining traces. Second, we reduce these special reducible representations to a sum of IR's by getting rid of all antisymmetric parts.

The first stage in the algorithm involves turning a direct product $(n,m)\otimes(n',m')$ into a sum of tensors of the form

$$D^{(n,n';m,m')} \equiv (n,n';m,m') = x^{i_1\cdots i_n\, i_{n+1}\cdots i_{n+n'}}_{j_1\cdots j_m\, j_{m+1}\cdots j_{m+m'}} \tag{37.43}$$

This tensor is completely symmetric in *four sets* of indices, but not otherwise: in the first upper n indices, in the last upper n' indices, in the first lower m indices, and in the last lower m' indices. It is *completely* traceless. Roughly speaking it is what we would get if we took the direct product $(n,m)\otimes(n',m')$ with all traces removed, but without any symmetrization among either the n and n' indices or the m and m' indices. We go from the direct product to the form (37.43) by removing traces.

We start out with two traceless tensors, (n,m) and (n',m'). In the direct product $(n,m)\otimes(n',m')$ we needn't bother with traces formed from the "first" indices, the n's with the n''s, or the "last" indices, the m's with the m''s; that's already done. We need separate out only those tensors that can be obtained by contracting, in all possible ways, indices from the "outside" sets, the n's with the m''s, and from the "inside" sets, the n''s with the m's. So we get a double direct sum

$$\begin{aligned}(n,m)\otimes(n',m') \sim\ & (n,n';m,m') && \text{(no contractions; traceless)}\\ & \oplus(n-1,n';m,m'-1) && \text{(one "outside" contraction)}\\ & \oplus(n,n'-1;m-1,m') && \text{(one "inside" contraction)}\\ & \oplus(n-1,n'-1;m-1,m'-1) && \text{(two contractions)}\\ & \oplus\cdots\end{aligned} \tag{37.44}$$

The process terminates whenever we run out of indices to contract; i.e., whenever a zero appears in the series on the right. In a more compact form we have

$$(n,m)\otimes(n',m') \sim \oplus \sum_{r=0}^{\min(n,m')} \sum_{s=0}^{\min(n',m)} (n-r,n'-s;m-s,m'-r) \tag{37.45}$$

We can peel indices off of either the two "outside" indices n, m' or the two "inside" indices n', m, but we have to strip off the same number in either case, because we're contracting indices. That takes care of all the traces.

In the second stage of the algorithm, we remove the antisymmetric parts of the terms $(n,n';m,m')$ in (37.44). How do we do that? We can express any tensor as the sum of a symmetric and an antisymmetric tensor, the latter in terms of the three index ϵ's. Consider the pair of indices i_1 and i_{n+1} in the tensor

$$x^{\underline{i_1}\cdots \underline{i_{n+1}}\cdots i_{n+n'}}_{j_1\cdots j_{m+m'}}$$

We can write the antisymmetric part in these indices as a tensor $s^{\cdots}_{\cdots}$ of lower rank:

$$s^{i_2\cdots i_n\, i_{n+2}\cdots i_{n+n'}}_{kj_1\cdots j_{m+m'}} = \epsilon_{ki_1 i_{n+1}} x^{i_1\cdots i_{n+1}\cdots i_{n+n'}}_{j_1\cdots j_{m+m'}} \tag{37.46}$$

Which i_l's we pick doesn't matter because those pairs are completely symmetric. We've picked two upper indices just for simplicity; we could just as well have picked two lower indices.

Now an amazingly helpful fact keeps us from getting involved in a lengthy calculation: the tensor $s^{\cdots}_{\cdots}$ in (37.46) is *already* symmetric in all of its lower indices $k, j_1, \cdots j_{m+m'}$. This is *not* obvious, but I will prove it. That means in our systematic splitting into symmetric and antisymmetric parts, we can only turn two upper indices into a lower index, or four upper indices into two lower indices, etc. And once we start turning pairs of upper indices into lower indices, we can't turn around and start contracting two lower indices into an upper index, because any such contraction will vanish.

The proof is in fact quite straightforward. We prove it by contracting $s^{\cdots}_{\cdots}$ with an ϵ tensor in any chosen pair of lower indices, say j_1 and j_{m+1}, and showing that the result is zero. The product of two ϵ tensors can be written in terms of the products of three Kronecker deltas:

$$\epsilon^{abc}\epsilon_{def} = \det\begin{vmatrix} \delta^a_d & \delta^a_e & \delta^a_f \\ \delta^b_d & \delta^b_e & \delta^b_f \\ \delta^c_d & \delta^c_e & \delta^c_f \end{vmatrix} = \delta^a_d\,\delta^b_e\,\delta^c_f - \delta^a_d\,\delta^c_e\,\delta^b_f + \cdots \tag{37.47}$$

There are six terms altogether, positive or negative depending on the permutation of the indices. Then

$$\begin{aligned} \epsilon^{rj_1j_{m+1}}s^{i_2\cdots i_n\, i_{n+2}\cdots i_{n+n'}}_{kj_1\cdots j_{m+m'}} &= \epsilon^{rj_1j_{m+1}}\epsilon_{ki_1i_{n+1}}x^{i_1\cdots i_{n+1}\cdots i_{n+n'}}_{j_1\cdots j_{m+1}\cdots j_{m+m'}} \\ &= \left(\delta^r_k\,\delta^{j_1}_{i_1}\delta^{j_{m+1}}_{i_{n+1}} - \cdots\right)x^{i_1\cdots i_{n+1}\cdots i_{n+n'}}_{j_1\cdots j_{m+1}\cdots j_{m+m'}} \end{aligned} \tag{37.48}$$

Notice that the first term involves a Kronecker delta on i_{n+1} and j_{m+1}. But we started with a tensor that's completely traceless, and so

$$\delta^r_k\,\delta^{j_1}_{i_1}\delta^{j_{m+1}}_{i_{n+1}}x^{i_1\cdots i_{n+1}\cdots}_{j_1\cdots j_{m+1}\cdots} = \delta^r_k\,\delta^{j_1}_{i_1}\,x^{i_1\cdots i_{n+1}\cdots}_{j_1\cdots i_{n+1}\cdots} = 0 \tag{37.49}$$

Indeed, every term in the permutation (37.47) involves some Kronecker delta that sums over two of our original indices, because there are only two free indices, k and r; k could be on one of the Kronecker δ's, r could be on another, but the third has to involve a pair of our original indices. So every one of the six terms will take a trace of an object that is traceless, and thus vanish. Therefore the s tensor is already symmetric in the lower indices once we begin contraction of the uppers with ϵ's. **QED**

So, here is the second stage of our algorithm (amusin' and confusin', but all the same very practical),

$$\begin{aligned} (n,n';m,m') \sim\ & (n+n', m+m') && \text{(already symmetric)} \\ & \oplus \sum_{r=1}^{\min(n,n')} (n+n'-2r, m+m'+r) && \begin{pmatrix}\text{contract pairs of upper}\\ \text{indices to one lower}\end{pmatrix} \\ & \oplus \sum_{s=1}^{\min(m,m')} (n+n'+s, m+m'-2s) && \begin{pmatrix}\text{contract pairs of lower}\\ \text{indices to one upper}\end{pmatrix} \end{aligned} \tag{37.50}$$

That's the end of the story. Start with a tensor and choose which indices, upper or lower, to contract. Once you've finished contracting ϵ's with pairs of those indices, you're done with that tensor, because it will already be symmetric in its other indices.

The algorithm as stated here is incredibly simple. We just multiply two representations (n, m) and (n', m') together. Then we do these operations as many times as we can. When any term gets to 0 we stop:

1. Take indices from the inside, m with n', and from the outside, n with m';
2. Take two indices off the left (the uppers) and make one on the right (the lowers), or *vice versa*.

Let me give some examples of this algorithm in action. First let's look at one we've already done, to check that it works.

EXAMPLE 1: $(1,0) \otimes (0,1)$ (or $\mathbf{3} \otimes \overline{\mathbf{3}}$)

We already know (37.4) that $(1,0) \otimes (0,1) \sim (1,1) \oplus (0,0)$. What does the algorithm say?

Stage 1: Following (37.45), we have

$$\begin{aligned}(1,0) \otimes (0,1) &\sim \oplus \sum_{r=0}^{\min(1,1)} \sum_{s=0}^{\min(0,0)} (1-r, 0-s; 0-s, 1-r) \\ &\sim (1,0;0,1) \oplus (0,0;0,0)\end{aligned} \tag{37.51}$$

Stage 2: According to (37.50),

$$\begin{aligned}(1,0;0,1) &\sim (1+0, 0+1) \oplus \sum_{r=1}^{\min(1,0)} (1+0-2r, 0+1+r) \oplus \sum_{s=1}^{\min(0,1)} (1+0+s, 0+1-2s) \\ &\sim (1,1)\end{aligned} \tag{37.52}$$

The ϵ contractions require two upper or lower indices, and these terms have only one: the sums contribute nothing. Likewise

$$\begin{aligned}(0,0;0,0) &\sim (0,0) \quad \text{so} \\ (1,0) \otimes (0,1) &\sim (1,0;0,1) \oplus (0,0;0,0) \sim (1,1) \oplus (0,0)\end{aligned} \tag{37.53}$$

as required. In terms of dimensions the result is

$$\mathbf{3} \otimes \overline{\mathbf{3}} \sim \mathbf{8} \oplus \mathbf{1} \tag{37.54}$$

as we already knew. Of course the algorithm is trivial in this case; nevertheless we got the right answer. Armed with this confidence, let's go on to a more interesting case.

EXAMPLE 2: *Scattering amplitudes for the octet,* $\mathbf{8} \otimes \mathbf{8}$ *(or* $(1,1) \otimes (1,1)$*)*

We know the critical importance of isospin invariance in pion-nucleon scattering. Isospin tells us that our amplitudes are either $I = \frac{3}{2}$ or $I = \frac{1}{2}$ and that connects a lot of data.[20] SU(3) is a much more powerful group than SU(2). We've got an octet of mesons and an octet of

[20] [Eds.] See §24.3, and Problem 14.3, p. 546.

baryons. What is the corresponding statement for meson–baryon scattering in SU(3)? Do we have two amplitudes, as with isospin? 17? 121? And what are their transformation properties?

Well, we know how to compute the answer: apply the algorithm. The first stage is:

$$(1,1)\otimes(1,1)\sim(1,1;1,1)\oplus\underbrace{(0,1;1,0)}_{\text{contract outside}}\oplus\underbrace{(1,0;0,1)}_{\text{contract inside}}\oplus\underbrace{(0,0;0,0)}_{\text{contract both}} \tag{37.55}$$

This is (37.45): r runs from 0 to 1, s runs from 0 to 1, giving four terms. For the second stage:

$$\begin{gathered}(1,1;1,1)\sim(2,2)\oplus\sum_{r=1}^{\min(1,1)}(2-2r,2+r)\oplus\sum_{s=1}^{\min(1,1)}(2+s,2-2s)\sim(2,2)\oplus(0,3)\oplus(3,0)\\ (0,1;1,0)\sim(1,1)\oplus\sum_{r=1}^{\min(0,1)}(2-2r,2+r)\oplus\sum_{s=1}^{\min(1,0)}(2+s,2-2s)\sim(1,1)\\ (1,0;0,1)\sim(1,1)\quad\text{(by the same reasoning)}\\ (0,0;0,0)\sim(0,0)\end{gathered} \tag{37.56}$$

Therefore

$$(1,1)\otimes(1,1)=(2,2)\oplus(3,0)\oplus(0,3)\oplus(1,1)\oplus(1,1)\oplus(0,0) \tag{37.57}$$

To check that we haven't double counted or under counted in developing our algorithms, let's write this out in the vulgar notation, where we label the representations by their dimensions and see if the dimensions come out right. Recall $\dim(n,m)=\frac{1}{2}(n+1)(m+1)(n+m+2)$, so $\dim(2,2)=27$, $\dim(3,0)=\dim(0,3)=10$, and

$$\underbrace{\mathbf{8}\otimes\mathbf{8}}_{64}=\underbrace{\mathbf{27}\oplus\mathbf{10}\oplus\overline{\mathbf{10}}\oplus\mathbf{8}\oplus\mathbf{8}\oplus\mathbf{1}}_{64} \tag{37.58}$$

Unlike in SU(2), the same representation can occur twice in a direct product. In the SU(2)-invariant theory of pion-nucleon scattering we have two amplitudes, but in the SU(3) theory of meson–baryon scattering—and that includes *anything*: K's off of Ξ^-'s, not that anyone has done that experiment, η's off of Λ's, you name it—there are more.

There is a tricky point in finding how many amplitudes there are. Because if the initial meson–baryon state is in the $8\otimes 8$ representation, then the final meson–baryon state is represented by the same formula (37.58):

$$\begin{array}{ccccccccc}\mathbf{27}&\oplus&\mathbf{10}&\oplus&\overline{\mathbf{10}}&\oplus&\mathbf{8}\oplus\mathbf{8}&\oplus&\mathbf{1}\\ \downarrow&&\downarrow&&\downarrow&&\downarrow\!\!\times\!\!\downarrow&&\downarrow\\ \mathbf{27}&\oplus&\mathbf{10}&\oplus&\overline{\mathbf{10}}&\oplus&\mathbf{8}\oplus\mathbf{8}&\oplus&\mathbf{1}\end{array} \tag{37.59}$$

The arrows mean non-zero S-matrix elements. So a **27** can scatter only into a **27**, a $\overline{\mathbf{10}}$ can scatter only into a $\overline{\mathbf{10}}$, etc. There are six vertical arrows, six possible amplitudes. There are two ways for each of the **8**'s to scatter. The two crossed arrows for $\mathbf{8}\to\mathbf{8}$ are trivially related to each other by time-reversal. So, assuming time-reversal invariance, the two crossed arrows count as only one, giving a total of *seven* independent amplitudes for meson–baryon scattering. When we can build two octet states for the final states of the baryon and the meson, we could have in principle (not counting time reversal) *four* amplitudes connecting octet with octet (two vertical arrows, two crossed). Just as in the theory of scattering of particles with spin by a spin-dependent force, if we can build several $J = \frac{3}{2}$ states, say $\ell = 1$ and $\ell = 2$, we can have $\ell = 1$ and $\ell = 2$ crossed matrix elements.[21]

37.5 Symmetry and antisymmetry in the Clebsch–Gordan coefficients

In our discussion of both isospin and the rotation group (group theoretically the same thing), and of the Lorentz group, we had to deal with the question of the symmetry and antisymmetry of elements of the direct product under exchange of the two objects, if the two representations are identical. Put another way, we had to worry about the symmetry and antisymmetry of the Clebsch–Gordan coefficients.

For example, if we are putting together two pions in an s-wave state, in general the two pions can have $I = 0, 1$, or 2. $I = 1$ is excluded because that is antisymmetric in isospin, and it won't match with the s-wave state, which is symmetric in space and (trivially) in spin (the complete state would be antisymmetric—illegal for bosons). Given two identical representations, which terms in the series generated by this algorithm are symmetric, and which are antisymmetric? We can think we're putting together two particles in an s-wave or a p-wave or a $J = \frac{3}{2}$ or a triplet 1 state or whatever. Or, we're multiplying two fields at the same spacetime point, and we're only going to get either the symmetric or the antisymmetric combinations, depending on whether the fields satisfy Bose or Fermi statistics, respectively.

The direct product $(n, m) \otimes (n, m)$ generates a set of terms. The whole algorithm is set up in such a way that the symmetry is manifest. The algorithm begins with stage 1 (37.45):

$$(n, m) \otimes (n, m) \sim \oplus \sum_{r=0}^{\min(n,m)} \sum_{s=0}^{\min(n,m)} (n - r, n - s; m - s, m - r) \tag{37.60}$$

If we exchange the two n's and the two m's on the left-hand side, that exchanges r and s in the sum:

$$(n - r, n - s; m - s, m - r) \to (n - s, n - r; m - r, m - s) \tag{37.61}$$

That is, if $r = s$ the terms are symmetric. If $r \neq s$ the two terms change places, and by forming sums and differences we can form both a symmetric and an antisymmetric combination:

$$\begin{array}{ll} r = s & \text{symmetric under the exchange} \\ r \neq s & \text{terms exchange places} \end{array} \tag{37.62}$$

In stage 2, every time we peel off a pair of indices from one side to put one on the other, we use an ϵ tensor—an antisymmetric object. Therefore whenever we use an ϵ tensor we change symmetric to antisymmetric and *vice versa*; if we use it twice we restore the status quo, and so on. So, in stage 2, successive terms change symmetry. Of course we don't have to worry about these terms in stage 1. If we're just counting the number of symmetric and antisymmetric objects, the fact that the signs keep flipping (so we have to take the sums once and the differences the next time to get a symmetric object) is irrelevant; if $r \neq s$, the *numbers* of symmetric and antisymmetric objects stay the same. It's only the terms where $r = s$ that we have to keep track of the sign changes. Thus we see in (37.55) that under exchange of two

---

[21] [Eds.] J. R. Taylor, *Scattering Theory: The Quantum Theory of Nonrelativistic Collisions*, J. Wiley & Sons, 1972, Section 6-g.

objects, $(1,1;1,1)$ and $(0,0;0,0)$ are symmetric, and that $(0,1;1,0)$ and $(1,0;0,1)$ exchange places.

$$(1,1;1,1) \to (1,1;1,1) \qquad (0,0;0,0) \to (0,0;0,0) \qquad (0,1;1,0) \leftrightarrow (1,0;0,1)$$

EXAMPLE 3: *Coupling two identical octets*

To see what sort of representations we'd get in this case, apply the algorithm to (37.57):

$$\begin{aligned} &(2,2): \text{completely symmetrized, symmetric under exchange} \\ &(0,3): \text{used } \epsilon \text{ once, antisymmetric} \\ &(3,0): \text{used } \epsilon \text{ once, antisymmetric} \\ &(1,1): \text{the two octets exchange places; write as two linear combinations,} \\ &\qquad\;\; \text{one symmetric, the other antisymmetric} \\ &(0,0): \text{no } \epsilon \text{ used, symmetric} \end{aligned} \tag{37.63}$$

Thus, in coupling two **8**'s to make a sequence of representations of SU(3), the **27**, one of the octets and the singlet are symmetric; the **10**, the $\overline{\mathbf{10}}$ and the other octet are antisymmetric.

If we were considering meson-meson scattering in the *s*-wave, then Bose statistics would tell us that the overall wave function has to be symmetric. In the *s*-wave, the space part is already symmetric. The only states that the two mesons can occupy are the **27**, the *symmetric* octet (not two octets as in the meson–baryon case) and the singlet. Therefore, for meson-meson scattering in the *s*-wave, although there are eight kinds of initial mesons and eight kinds of final mesons, there are only *three* SU(3)-invariant amplitudes:

$$\begin{aligned} \mathbf{27} &\to \mathbf{27} \\ \mathbf{8} &\to \mathbf{8} \qquad \text{(one and only one; antisymmetric octet excluded)} \\ \mathbf{1} &\to \mathbf{1} \end{aligned} \tag{37.64}$$

Next time, we will take care of the last part of the program announced at the beginning of this lecture and prove the irreducibility, inequivalency and completeness of the IR's. Instead of losing you all at the end, I will lose you all at the beginning. (Those of you who are not group theory *mavens*[22] may come to the lecture fifteen minutes late; you will miss nothing.) The remainder of that lecture will be the beginning of applications: the Gell-Mann–Okubo mass formula, electromagnetic moments, electromagnetic mass differences, and some other related things.

[22] [Eds.] A *maven* (rhymes with "raven") is a connoisseur or expert; Rosten *Joys*, p. 221.

Problems 20

20.1 One day someone suggests to you that, in addition to the ordinary photon, there is a second, heavy photon, with exactly the same interactions but with a mass M, very much larger than the muon mass and, *a fortiori*, the electron mass. You decide to investigate the possible existence of this particle by seeing whether its effects on the anomalous magnetic moments of the electron and the muon are detectable. What lower bound on M do you deduce from the fact that conventional theory fits $1 + F_2(0)$ for the electron with an error of no more than 3×10^{-11}? For the muon with an error of no more than 8×10^{-9}?

Comment: If you carry the answer out to more than 10% accuracy, you don't understand the meaning of experimental error.

(1998b 8.2)

20.2 In class I said (see the paragraph following (35.13), and also note 12, p. 757) that we could calculate the first $\mathcal{O}(e^4)$ effects of strong interactions on lepton magnetic moments if we knew $\sigma(q^2)$, (here written as $\rho(q^2)$ to avoid confusion with the cross-section σ, which plays a leading role in this problem) in the spectral representation (34.4) of the renormalized photon propagator:

$$D'_{\mu\nu}(k^2) = -i\left[g_{\mu\nu} - \frac{k_\mu k_\nu}{k^2}\right]\left\{\frac{1}{k^2} + \int da^2\,\frac{\rho(a^2)}{k^2 - a^2}\right\} + \text{gauge-dependent term} \tag{P20.1}$$

By arguments identical to those given for a scalar field theory (§15.2) leading to (15.12), $\rho(k^2)$ is given by

$$-2\pi\left[g_{\mu\nu} - \frac{k_\mu k_\nu}{q^2}\right]\rho(k^2) = \sum_n \langle 0|A'_\mu(0)|n\rangle\,\langle n|A'_\nu(0)|0\rangle\,(2\pi)^4\delta^{(4)}(k - p_n) \tag{P20.2}$$

where the sum runs over all states except the one-photon state, and p_n is the total momentum of the n-particle state. From this formula it is clear that $\rho(k^2)$ is $\mathcal{O}(e^2)$.

Let $\sigma_T(a^2)$ be the total cross-section for electron-positron annihilation into hadrons (see Figure P20.1), averaged over initial spins, with total center of momentum energy a. Show that

$$\rho_H(a^2) = Ke^{-2}\sigma_T(a^2) + \mathcal{O}(e^4) + \mathcal{O}\Big(\frac{e^2 m_e^2}{a^2}\Big) \tag{P20.3}$$

and find the constant K. The subscript H on the ρ indicates the contribution of hadronic intermediate states *only*.

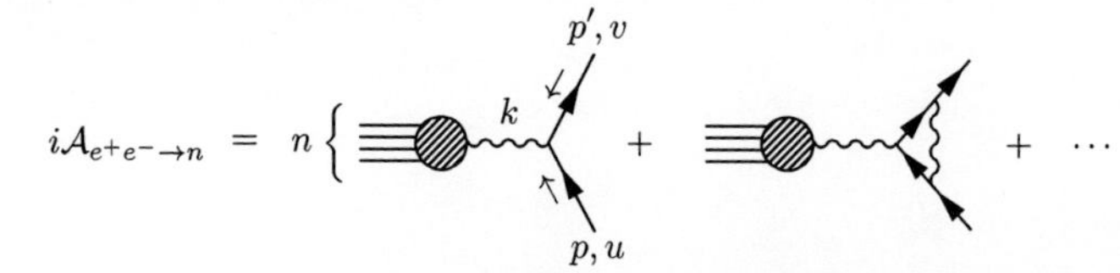

Figure P20.1: Amplitude for $e^+ + e^- \to$ hadrons (figure from B. Grossman's solution (1979b 12))

(1979b 12; 1998b 8.3)

Solutions 20

20.1 The heavy photon would make an additional contribution $\delta F_2(0)$, much like (34.42), except that the heavy photon propagator has a denominator of $(q^2 - M^2)$. The penultimate result is given in (35.15); all we have to do is substitute M for a. But let's go through the steps. There is no change to the numerator, but the new propagator results in a denominator D':

$$D' = [(p' + q)^2 - m^2][(p+q)^2 - m^2][q^2 - M^2] = (q^2 + 2p' \cdot q)(q^2 + 2p \cdot q)(q^2 - M^2) \tag{S20.1}$$

because the electrons are on their mass shells; $p^2 = p'^2 = m^2$. As in (34.53) we combine the denominators with Feynman's trick, and the denominator becomes

$$\frac{1}{D'} = 2\int_\Delta dx\, dy \frac{1}{[(q^2 + 2p' \cdot q)x + (q^2 + 2p \cdot q)y + (q^2 - M^2)(1 - x - y)]^3} \tag{S20.2}$$

Following (34.54) and (34.55), we effect the same completion of the square and exactly the same shift $q = q' - xp' - yp$, and obtain

$$\frac{1}{D'} = 2\int_\Delta dx\, dy \frac{1}{[q'^2 - (xp' + yp)^2 - M^2(1 - x - y)]^3} \tag{S20.3}$$

As in the original calculation, we use the relation $k = p' - p$ (k is the momentum of the external photon), so $k^2 = 2m^2 - 2p \cdot p'$, and

$$(xp' + yp)^2 = x^2m^2 + y^2m^2 + (2m^2 - k^2)xy = m^2(x + y)^2 + \mathcal{O}(k^2)$$

We drop the second-order term in k. Following the same steps as the original calculation, we extract the contribution $\delta F_2(0)$ as

$$\delta F_2(0) = 8ie^2m^2 \int_\Delta dx\, dy\, (x + y)(1 - x - y) \int \frac{d^4q'}{(2\pi)^4} \frac{1}{[q'^2 - m^2(x + y)^2 - M^2(1 - x - y)]^3} \tag{S20.4}$$

We evaluate the q' integral using the table in the box on p. 330 and get

$$\begin{aligned} \delta F_2(0) &= 8ie^2m^2 \int_\Delta dx\, dy\, (x + y)(1 - x - y) \frac{i}{32\pi^2} \frac{1}{(-m^2(x + y)^2 - M^2(1 - x - y))} \\ &= \frac{e^2m^2}{4\pi^2} \int_\Delta dx\, dy \frac{(x + y)(1 - x - y)}{(m^2(x + y)^2 + M^2(1 - x - y))} \end{aligned} \tag{S20.5}$$

which is exactly (35.15), with $a \to M$. We are told $M \gg m$, so we can approximate

$$\begin{aligned} \delta F_2(0) &\approx \frac{e^2m^2}{4\pi^2} \int_\Delta dx\, dy \frac{(x + y)(1 - x - y)}{M^2(1 - x - y)} \\ &= \frac{e^2m^2}{4\pi^2M^2} \int_0^1 dx \int_0^{1-x} dy\, (x + y) = \frac{e^2m^2}{4\pi^2M^2}\left(\frac{1}{3}\right) = \frac{\alpha m^2}{3\pi M^2} \end{aligned} \tag{S20.6}$$

Let ϵ be the disagreement between experiment and the usual theory in the measurement of $1 + F_2(0)$. Then

$$\epsilon \geq \frac{\alpha m^2}{3\pi M^2} \quad \text{so} \quad M \geq m\sqrt{\frac{\alpha}{3\pi\epsilon}} \tag{S20.7}$$

For the electron ($m \approx 0.5\,\text{MeV}$), $M \geq 3\,\text{GeV}$; for the muon ($m \approx 100\,\text{MeV}$), $M \geq 30\,\text{GeV}$. ■

20.2 The expression cited, (P20.2), is the vector version of the Kallén–Lehmann spectral representation (15.12). The hadronic contribution $\rho_H(k^2)$ to $\rho(k^2)$ is

$$-2\pi\left[g_{\mu\nu} - \frac{k_\mu k_\nu}{k^2}\right]\rho_H(k^2) = \sum_n \langle 0|A'_\mu(0)|n\rangle\,\langle n|A'_\nu(0)|0\rangle\,(2\pi)^4\delta^{(4)}(k-p_n) \tag{S20.8}$$

where the sum is over all intermediate states with at least one hadron. We could use the LSZ reduction formula to relate the matrix elements $\langle n|A'_\mu(0)|0\rangle$ to the Green's function $\langle 0|T[\phi'_a(x_1)\cdots\phi'_n(x_n)A'_\mu(0)]|0\rangle$, but that's not necessary for this problem.

We want to relate $\rho_H(k^2)$ to the total cross-section $\sigma_T(k^2)$ for e^+-e^- annihilation into hadrons, averaged over initial fermion spins. First, we calculate the amplitude for a given hadronic final state (see Figure P20.1):

$$i\mathcal{A}_{e^+e^-\to n} = -ie\overline{v}^s\gamma^\mu u^r\,\langle n|A'_\mu(0)|0\rangle + \mathcal{O}(e^4) \tag{S20.9}$$

The spin-averaged hadronic cross-section in the center of momentum frame of the e^+-e^- pair is, from (12.13) and (21.114),

$$\sigma_T(k^2) = \frac{e^2}{4E_T|\mathbf{p}|}\sum_n (2\pi)^4\delta^{(4)}(p_n-k)\,\langle 0|A'_\mu(0)|n\rangle\,\langle n|A'_\nu(0)|0\rangle\,\tfrac{1}{4}\sum_{r,s}(\overline{u}^r\gamma^\mu v^s)(\overline{v}^s\gamma^\nu u^r) + \mathcal{O}(e^6) \tag{S20.10}$$

Let $k^\mu = p^\mu + p'^\mu$ be the 4-momentum of the photon in the center of momentum frame:

$$\begin{aligned}
p^\mu &= (p^0, \mathbf{p}) \qquad p'^\mu = (p^0, -\mathbf{p})\\
k^\mu &= (2p^0, 0) = (E_T, 0) \equiv (a, 0)\\
p^2 &= p'^2 = m_e^2 = \tfrac{1}{4}a^2 - |\mathbf{p}|^2\\
k^2 &= a^2 = (p+p')^2 = 2m_e^2 + 2p\cdot p'\\
p\cdot k &= p'\cdot k = \tfrac{1}{2}a^2\\
4E_T|\mathbf{p}| &= 2a^2\sqrt{1-\frac{4m_e^2}{a^2}}
\end{aligned}$$

The spin sum is worked out using Casimir's trick, (21.115)

$$\begin{aligned}
\tfrac{1}{4}\sum_{r,s}(\overline{u}^r\gamma^\mu v^s)(\overline{v}^s\gamma^\nu u^r) &= \tfrac{1}{4}\sum_{r,s}\text{Tr}\left[\gamma^\mu v^s\overline{v}^s\gamma^\nu u^r\overline{u}^r\right] = \tfrac{1}{4}\text{Tr}\left[\gamma^\mu(\not{p}' - m_e)\gamma^\nu(\not{p}+m_e)\right]\\
&= p'^\mu p^\nu + p'^\nu p^\mu - g^{\mu\nu}(p'\cdot p + m_e^2) = p'^\mu p^\nu + p'^\nu p^\mu - \tfrac{1}{2}g^{\mu\nu}a^2
\end{aligned}$$

Putting the pieces together, we have

$$\sigma_T = \frac{e^2}{2a^2\sqrt{1-\dfrac{4m_e^2}{a^2}}}\left[p'^\mu p^\nu + p'^\nu p^\mu - \tfrac{1}{2}g^{\mu\nu}a^2\right]M_{\mu\nu} + \mathcal{O}(e^6) \tag{S20.11}$$

where

$$M_{\mu\nu} = \sum_n (2\pi)^4\delta^{(4)}(p_n-k)\,\langle 0|A'_\mu(0)|n\rangle\,\langle n|A'_\nu(0)|0\rangle \tag{S20.12}$$

We are told to drop terms of $\mathcal{O}\left(\dfrac{e^2m_e^2}{a^2}\right)$, and

$$\frac{e^2}{2a^2\sqrt{1-\dfrac{4m_e^2}{a^2}}} = \frac{e^2}{2a^2}\left[1+\mathcal{O}\left(\frac{m_e^2}{a^2}\right)\right] = \frac{e^2}{2a^2} + \mathcal{O}\left(\frac{e^2m_e^2}{a^4}\right)$$

so we can replace the square root by 1, and (S20.12) becomes (recalling that $M_{\mu\nu}$ is $\mathcal{O}(e^2)$)

$$\sigma_T = \frac{e^2}{2a^2}\left[p'^\mu p^\nu + p'^\nu p^\mu - \tfrac{1}{2}g^{\mu\nu}a^2\right]M_{\mu\nu} + \mathcal{O}\left(\frac{e^4m_e^2}{a^2}\right) + \mathcal{O}(e^6) \tag{S20.13}$$

We turn now to the hadronic spectral density, ρ_H. It satisfies the constraint (P20.2)

$$-2\pi\left[g_{\mu\nu} - \frac{k_\mu k_\nu}{k^2}\right]\rho_H(k^2) = M_{\mu\nu} \tag{S20.14}$$

Contracting this constraint with $p'^{\mu}p^{\nu} + p'^{\nu}p^{\mu} - \frac{1}{2}g^{\mu\nu}a^2$ gives

$$\begin{aligned}\left[p'^{\mu}p^{\nu} + p'^{\nu}p^{\mu} - \tfrac{1}{2}g^{\mu\nu}a^2\right] M_{\mu\nu} &= -2\pi\left[p'^{\mu}p^{\nu} + p'^{\nu}p^{\mu} - \tfrac{1}{2}g^{\mu\nu}a^2\right]\left[g_{\mu\nu} - \frac{k_\mu k_\nu}{k^2}\right]\rho_H \\ &= 2\pi\left(a^2 + 2m_e^2\right)\rho_H\end{aligned} \tag{S20.15}$$

Recalling ρ_H is $\mathcal{O}(e^2)$, we have

$$\sigma_T = \frac{e^2}{2a^2}\left[2\pi\left(a^2 + 2m_e^2\right)\rho_H\right] + \mathcal{O}\Big(\frac{e^4 m_e^2}{a^2}\Big) + \mathcal{O}(e^6) = \pi e^2 \rho_H + \mathcal{O}\Big(\frac{e^4 m_e^2}{a^2}\Big) + \mathcal{O}(e^6) \tag{S20.16}$$

and finally we obtain what was to be shown,

$$\rho_H = \frac{1}{\pi e^2}\sigma_T + \mathcal{O}(e^4) + \mathcal{O}\Big(\frac{e^2 m_e^2}{a^2}\Big) \tag{S20.17}$$

with $K = 1/\pi$. ■

38

SU(3): Proofs and applications

Last time I said I would prove that the representations we constructed are irreducible, inequivalent and complete. I will redeem that pledge now.[1] After the proofs, I will move on to applications.

38.1 Irreducibility, inequivalence, and completeness of the IR's

In order to prove these properties of the IR's, I will have to steal two general theorems from group theory; the proofs are in Tinkham's book or Wigner's book.[2] At least the first of them should be obvious.

Let G be some compact[3] Lie group and let $g \in$ G be an element of G. Representations of compact groups are always equivalent to *unitary* representations,[4] which are always completely reducible[5] into direct sums of finite-dimensional, inequivalent irreducible representations; we never need worry about infinite-dimensional ones.[6] Thus given a representation $D(g)$ of a compact group G, we can write

$$D(g) \sim \oplus \sum_r n_r D^{(r)}(g) \tag{38.1}$$

[1] [Eds.] §38.1, from the video of Lecture 38, is again largely a reworking of the references cited in note 1, p. 797.

[2] [Eds.] E. P. Wigner, *Group Theory and its Application to the Quantum Mechanics of Atomic Spectra*, Academic Press, 1959. M. Tinkham, *Group Theory and Quantum Mechanics*, McGraw-Hill, 1964. Two more recent books on group theory for physicists are: Howard Georgi, *Lie Algebras in Particle Physics: From Isospin to Unified Theories*, 2nd ed., Westview Press, 1999; and A. Zee, *Group Theory in a Nutshell for Physicists*, Princeton U. P., 2016, hereafter Zee *GTN*.

[3] [Eds.] "Compact" means a finite volume or parameter space; mathematically, a compact set is one which is closed and bounded. The rotation group is compact; the Lorentz group is not. See note 15, p. 379.

[4] [Eds.] Wigner, *op. cit.*, Theorem 1, p. 74.

[5] [Eds.] See Chapter III, Section 4, p. 123 in H. Weyl, *Group Theory and Quantum Mechanics*, trans. H. P. Robertson, reprinted by Dover Publications, 1953; Tinkham, *op. cit.*, Section 3–5, pp. 29–30.

[6] [Eds.] This is a consequence of the Peter–Weyl theorem: A. Barut and R. Raczka, *Theory of Group Representations and Applications*, World Scientific, 1986; A. Wawrzyńczyk, *Group Representations and Special Functions*, D. Reidel, 1984; reprinted by Springer, 1986.

where $\{D^{(r)}(g)\}$ is a complete set of inequivalent irreducible representations, r is some index (or perhaps a multiplet of indices), and the integers n_r are the number of times the $D^{(r)}(g)$'s appear in the decomposition. For any group we have the one-dimensional trivial representation $D^{(0)}(g) = 1$ for every element $g \in G$. If we consider the direct product of any two representations

$$\overline{D}^{(r)} \otimes D^{(s)}$$

this is itself a (unitary) representation and so it is equivalent to a sum of irreducible representations:

$$\overline{D}^{(r)} \otimes D^{(s)} \sim \oplus \sum_t n_{rs}^t D^{(t)} \tag{38.2}$$

n_{rs}^t is the number of times the representation $D^{(t)}$ occurs.[7]

Theorem 38.1.

$$n_{rs}^0 = \delta_{rs} \tag{38.3}$$

That is, the number of times the trivial representation occurs is once if $D^{(r)} = D^{(s)}$, and not at all if $D^{(r)} \neq D^{(s)}$.[8] This is in a sense a fact we all know, sometimes called **Schur's lemma.**[9] In field theoretic language it is the statement that if we have a set of fields that transforms according to an irreducible representation of the group, we can make *one and only one mass term* from the field and its conjugate. If you foolishly tried to make an invariant mass term from a field that transforms one way and the conjugate of a field that transforms the other way, say from an isovector and an isotensor, you couldn't make it at all: there is no such invariant. Equivalently we could consider $D^{(s)}$ as labeling a set of states on the left of the S-matrix and $D^{(r)}$ as labeling a set of states on the right of the S-matrix. Then the statement is that if r and s are different, there is no invariant S-matrix element: they cannot scatter into each other; and if the states do transform the same way, $r = s$, there is only *one* invariant S-matrix element. You can take Theorem 38.1 on trust or look it up in the books; we are going to exploit it. This theorem has a corollary that gives us a trivial test for irreducibility.

Corollary 1. *$D(g)$ is irreducible if and only if $\overline{D} \otimes D$ contains $D^{(0)}$ once and only once.*

Proof:

$$\overline{D} \otimes D \sim \left(\oplus \sum_r n_r \overline{D}^{(r)} \right) \otimes \left(\oplus \sum_s n_s D^{(s)} \right) \sim \oplus \sum_t \sum_r \sum_s n_r n_s n_{rs}^t D^{(t)} \tag{38.4}$$

If D is reducible, then when I multiply it by its conjugate, I'll get a sum of terms as in (38.2). Every irreducible component $D^{(i)}$ in D will be multiplied by its conjugate $\overline{D}^{(i)}$ in $\overline{D}$, and I'll

---

[7] [Eds.] See Section 16-3, equation (16-22), p. 436 in J. Mathews and R. Walker, *Mathematical Methods of Physics*, Addison-Wesley, 1969.

[8] [Eds.] Statements about groups are often easily grasped if you consider them in terms of the quantum theory of angular momentum. Recall that a direct product of two different angular momentum states (the irreducible representations of SU(2)) with ℓ_1 and ℓ_2 will give new states with ℓ bounded by $\ell_1 + \ell_2 \geq \ell \geq |\ell_1 - \ell_2|$. Note that ℓ will not equal 0 unless $\ell_1 = \ell_2$. (In SU(2) there is no distinction between $\overline{D}(g)$ and $D(g)$; all the tensors can be written with either upper or lower indices only.)

[9] [Eds.] Wigner, *op. cit.*, Theorem 2, pp. 75–76. Coleman states in "Fun with SU(3)" (*op. cit.*, footnote 3, p. 342): "Actually, this is not in [Wigner] in precisely this form; however it is a trivial corollary of Schur's lemma, and the fact that every representation of a compact group is equivalent to a unitary representation. (It can also be derived simply from the orthogonality relations.)" See note 12, p. 827 for a proof.

obtain as many $D^{(0)}$'s as there are irreducible components. The only way that $D^{(0)}$ could appear once and only once is if D contains only one component, i.e., it is irreducible. **QED**

Thus all I have to do to confirm that the (n, m)'s are irreducible is to check how many times a direct product of an IR with its conjugate contains a trivial representation of the group. Since I don't know yet that the putative IR's (n, m) *are* irreducible, I first have to find out how many times each representation (n, m) contains $D^{(0)}$. If I'm lucky, the answer will be that *only* $(0, 0)$ contains $D^{(0)}$. But I haven't proved that.

So the first step is to determine which (n, m)'s of SU(3) contain $D^{(0)}$, i.e., which ones contain an object that is invariant under all group representations. If an (n, m) contains such an invariant object, it must have zero isospin and zero hypercharge:

$$I = Y = 0 \tag{38.5}$$

We happen to have a handy algorithm (§37.3) for determining the isospin-hypercharge content of any (n, m). From that algorithm it's clear that only (n, n) is a possibility. Let's look at this in more detail. From the block diagrams discussed in §37.3, if we have different n's and different m's, when the isospin adds up to zero the hypercharge will not. And when the hypercharge adds up to zero the isospins will be different; see Figure 37.4. The only time[10] both $I = 0$ and $Y = 0$ is when $n = m$. So we've only got to look at (n, n), which contains only one state with $I = 0$, $Y = 0$. For example, if we look at $(1, 1)$, the thing with $I = 0$, $Y = 0$ is $a^3{}_3$, the 3-3 component of the tensor $a^i{}_j$. That's obvious: isospin acts only on indices with value 1 or 2, and a tensor with an equal number of upper and lower indices, all of which have

[10] [Eds.] In the lecture, this statement is not proved. Here is a proof. In the decomposition of (n, m) into $(I)^Y$ IR's (Figure 37.4) the general term along the top edge has

$$I_t = \tfrac{1}{2}(n - j), \quad Y_t = \tfrac{1}{3}n - j, \quad j = 0, 1, \ldots, n$$

while the general term along the left edge has

$$I_l = \tfrac{1}{2}(m - k), \quad Y_l = -\tfrac{1}{3}m + k, \quad k = 0, 1, \ldots, m$$

The $(I)^Y$ IR's come out of the direct product $(I_t)^{Y_t} \otimes (I_l)^{Y_l}$. Those have $Y = \frac{1}{3}(n - m) + (k - j)$, and values of I given by the Clebsch–Gordan series,

$$I = \tfrac{1}{2}[(n + m) - (j + k)] - \ell \qquad \ell = 0, 1, \ldots, \ell_{max}$$

where ℓ_{max} is determined by the requirement that I_{min} be non-negative:

$$\ell_{max} = \begin{cases} \frac{1}{2}[(n + m) - (j + k)] & (n + m) \text{ is even} \\ \frac{1}{2}[(n + m) - (j + k) - 1] & (n + m) \text{ is odd} \end{cases}$$

Set both Y and I equal to zero, and solve for n and m,

$$n = \ell + 2j - k \qquad m = \ell + 2k - j$$

Both n and m have to be non-negative, and so

$$\ell \geq k - 2j$$
$$\ell \geq j - 2k$$

Multiply the top equation by j, the bottom by k, and subtract:

$$\ell(j - k) \geq 2(k^2 - j^2) \ \Rightarrow\ \ell \geq -2(j + k)$$

Subtract this from the equation $\ell \geq j - 2k$ to obtain $0 \geq 3j$. But j is a non-negative integer, so $j = 0$. Similarly $k = 0$, and consequently $n = \ell = m$. **QED**

only the value 3, has $Y = 0$. Likewise, if we look at $(2,2)$, the $I = 0$, $Y = 0$ piece would be the component $a^{33}{}_{33}$ of the tensor $a^{ij}{}_{mn}$; that part is unchanged under isospin and hypercharge transformations. Can this component be invariant under *all* group operations? No, because SU(3) contains, in particular, a transformation which switches the third basis vector with the second. So there is a group element g such that

$$g \in \mathrm{SU}(3)\colon\ a^{33}{}_{33} \to a^{22}{}_{22} \tag{38.6}$$

The component $a^{22}{}_{22}$ is *not* invariant under the isospin-hypercharge subgroup. The only possibility, then, is $(0,0)$, which contains only a single element, and nothing changes under isospin and hypercharge transformations. To our question "Which IR's contain $D^{(0)}$?", we now have an answer: *only* $(0,0)$. That comes simply from the isospin-hypercharge block algorithm. So the algorithm is not only useful for computing things, it's useful for proving general theorems.

Therefore, to check for irreducibility, we have only to compute how many times the direct product of a given representation and its conjugate contains $(0,0)$. If we know how many times it contains $(0,0)$, we know how many times it contains $D^{(0)}$ and then we'll know whether or not it's irreducible. We are trying to show that the representations (n,m) are irreducible, so we consider the direct product $\overline{(n,m)} \otimes (n,m)$.

Theorem 38.2. *In* $\overline{(n,m)} \otimes (n,m)$, *the representation* $(0,0)$ *appears exactly once.*

Proof: We will use our algorithm to count representations in the decomposition of the direct product. Let's begin at the end: we want $(0,0)$ to come out when we're done. From (37.50) and (37.4), the only four-index symbol that leads to $(0,0)$ in stage 2 of the algorithm is $(0,0;0,0)$:

$$(0,0;0,0) \overset{\text{unique}}{\to} (0,0) \tag{38.7}$$

In our algorithm for reducing the four-index symbols we always take two off of one set of indices (upper or lower) and add one to the other set as in (37.50). Well, we're never going to get zeros by adding ones to some positive number. The only way we're going to get zeros is from $(0,0;0,0)$, produced in stage 1 of the algorithm. How many times is $(0,0;0,0)$ produced from $(m,n) \otimes (n,m)$? Recall from (37.45) that in stage 1, we take a single index from the outside pair (in this case, reducing the m's to $m-1$'s) and from the inside pair (turning the n's to $n-1$'s). We can do the former operation m times and the latter n times. And that's the *one and only* time that the four-symbol $(0,0;0,0)$ will be produced, when we take m indices off the m's and n indices off the n's:

$$\underbrace{(m, \overbrace{n) \otimes (n}^{\text{take n off}}, m)}_{\text{take m off}} \tag{38.8}$$

Consequently, the representation $D^{(0)}$ appears once and once only in the direct product $\overline{(n,m)} \otimes (n,m)$. **QED**

Theorem 38.3. *The IR's (n, m) are irreducible.*

Proof: By the corollary, (n,m) is irreducible: $D^{(0)}$ appears but once in $\overline{(n,m)} \otimes (n,m)$. **QED**

Up to now I hadn't proven that the representations (n, m), which I have cavalierly referred to as IR's, are in fact irreducible. One of them might have been the direct sum of 17 irreducible representations, including $D^{(0)}$. We had to prove that *only* $(0,0)$ contains $D^{(0)}$ before we could establish that the (n, m)'s really are irreducible representations, and thus deserving of the label "IR".

Next, I will demonstrate that the IR's (n, m) are *inequivalent* (for different n's and m's). That comes easily from the Theorem 38.3. Say that the IR (n', m') is equivalent to (n, m). Consider

$$\overline{(n', m')} \otimes (n, m) = (m', n') \otimes (n, m) \tag{38.9}$$

then how may times does this contain $(0,0)$? It will contain $(0,0)$ after the second stage of our algorithm only if it contains a term $(0,0;0,0)$ with four zeros after the first stage. Because we subtract equal numbers of indices from the outer and inner indices and stop when we reach a zero, the only way to reach four zeros is if $n = n'$ and $m = m'$. Thus we have:

Theorem 38.4. *The representations* (n, m) *and* (n', m') *are equivalent only if* $n = n'$ *and* $m = m'$.

Earlier we were concerned that as $(4,0)$ and $(2,1)$ were both 15-dimensional, they might secretly be equivalent. But it's not so: multiplying $(0,4) \otimes (2,1)$ as in (38.2), we'd have to get $n^0_{rs} = 1$ for them to be equivalent. But $(0,0)$ *does not appear* in the direct product; n^0_{rs} ain't one, it's zero. So $(4,0)$ and $(2,1)$ are inequivalent, despite having the same dimension. So far we have a set of representations that are guaranteed to be both irreducible and inequivalent. Have we found *all* of the irreducible representations? That is, are they *complete*? We know that when we used the tensor trick for SU(2) in §36.3 we got all of them. On the other hand if we had tried the same trick for SO(3), we would have missed the spin-½ representations.[11] So have we found them all, or are we missing some?

We'll now steal another theorem from group theory, the so-called **orthogonality theorem**[12]

---

[11] [Eds.] See Zee *GTN*, Section IV.1, pp. 185–195, for the application of these tensor methods to SO(3): only those IR's with dimension equal to $2j+1$ (with j a non-negative integer) are found.

[12] [Eds.] See Wigner, *op. cit.*, Theorem 4, equation (9.31), p. 79 for discrete groups; for continuous groups, see equation (10.12), p. 101. Incidentally this theorem affords a quick proof of Theorem 38.1. The orthogonality relations (Wigner's equation (9.31)) say

$$\sum_{g\in G} \left(\overline{D}^{(i)} \otimes D^{(j)}\right)_{ab} = \delta^{ij}\delta_{ab}\frac{h}{\ell}$$

where h is the order of the group and ℓ is the dimension of the matrices $D^{(i)}$ (and also of the Kronecker delta). But from (38.2) we have

$$\sum_{g\in G} \left(\overline{D}^{(i)} \otimes D^{(j)}\right)_{ab} = \sum_t n^{ij}_k \sum_{g\in G} D^{(k)}_{ab} \sim \sum_t n^{ij}_k \sum_{g\in G} \left(\overline{D}^{(0)} \otimes D^{(k)}\right)_{ab}$$

because $\overline{D}^{(0)} = D^{(0)} = 1$. Using the orthogonality relations on both sides gives

$$\delta^{ij}\delta_{ab}\frac{h}{\ell} = \sum_k n^{ij}_k \delta^{0k}\delta_{ab}\frac{h}{\ell} = n^{ij}_0 \delta_{ab}\frac{h}{\ell}$$

Canceling the common factors, we obtain $n^{ij}_0 = \delta^{ij}$. **QED** SU(3) is continuous, and the sum over $g \in G$ should be an integral, but the theorem goes through with integrals just the same.

Theorem 38.5. *Let G be a compact group and as before, let $D^{(r)}(g)$ be a complete set of inequivalent irreducible representations. Then*

$$\int_G dg\,\overline{D}_{ij}^{(r)}(g)\,D_{kl}^{(s)}(g) = 0 \quad \textit{if } r \neq s \tag{38.10}$$

The subscripts on the D's indicate matrix elements. We put coordinates on the group and we have a little Jacobian determinant there; we integrate over the whole group. (We also know what the integral is when $r = s$ but we don't need that for the theorem.[13]) It's the statement that, for U(1) or SO(2), for example, where the irreducible representations are all one-dimensional, $e^{in\theta}$, that

$$\int_0^{2\pi} d\theta\, e^{ir\theta} e^{-is\theta} = 0 \quad \text{if } r \neq s \tag{38.11}$$

It happens to be true in general.[14]

Consider the representation $(1,0)$, which has[15] eight independent matrix elements $D_{\alpha ij}^{(1,0)}$, $\alpha = 1, 2, \ldots, 8$. For that representation we'll consider all the matrix elements together and write them (and their conjugates, $D_\alpha^{(0,1)}(g)$) as

$$\begin{aligned} D_\alpha^{(1,0)}(g) &= y_\alpha(g) \\ D_\alpha^{(0,1)}(g) &= \overline{y}_\alpha(g) \end{aligned} \tag{38.12}$$

The set $\{y_\alpha\}$ are coordinates in group space. If we know the y_α's and the $\overline{y}_\alpha$'s, we know what the group element is. When I take the direct product of two representations I get matrix elements which are simply the ordinary numerical products of the matrix elements of the original representations. So, direct products have matrix elements that are monomials in the y_α's and the $\overline{y}_\alpha$'s.

Let us now prove by contradiction[16] that we have all the representations of SU(3). Assume that there is some irreducible representation, $D^{(?)}(g)$, which we have missed. By the orthogonality theorem, (38.10), its matrix elements are orthogonal to those of all the representations that *are* in our list. In particular, we have

$$\int_G dg\, y_\alpha\, D_{ij}^{(?)}(g) = \int_G dg\, \overline{y}_\alpha\, D_{ij}^{(?)}(g) = 0 \tag{38.13}$$

---

[13] [Eds.] For completeness,

$$\int_G dg\,\overline{D}_{ij}^{(r)}(g)\,D_{kl}^{(s)}(g) = \frac{\delta_{ik}\delta_{jl}\delta^{rs}}{\dim r}\int_G dg$$

where $\dim r$ is the dimension of the representations. Wigner, *op. cit.*, p. 101, equation (10.12).

[14] [Eds.] Compare also $\int d\Omega\, Y_\ell^{m*}\, Y_{\ell'}^{m'} = \delta_{\ell\ell'}\,\delta^{mm'}$; the spherical harmonics $Y_\ell^m(\theta,\phi)$ form an irreducible representation of SO(3) on the unit sphere.

[15] [Eds.] This is the representation generated by the eight traceless, Hermitian 3×3 Gell-Mann matrices $\{\lambda_\alpha\}$; see note 15, p. 807. The representation $D^{(1,0)}(g) = \exp\{(i/2)\theta^\alpha\lambda_\alpha\}$ where θ^α are eight parameters. Similarly $D^{(0,1)}(g) = \exp\{-(i/2)\theta^\alpha\lambda_\alpha\}$. The determination of $\lambda_{\alpha ij}$ as the matrix elements $2\langle q_i|F_\alpha|q_j\rangle$ where $\langle q_i|q_j\rangle = \delta_{ij}$ and $\{F_\alpha\}$ are the elements of the Lie algebra of SU(3) is worked out in Greiner & Müller *QMS*, Exercise 8.1, pp. 221–224.

[16] [Eds.] Coleman *Aspects*, "An introduction to unitary symmetry", pp. 16–17; Coleman, "Fun with SU(3)", *op. cit.*, pp. 343–344.

Then $D^{(?)}_{ij}(g)$ must be orthogonal to all linear combinations of the y_α and $\overline{y}_\alpha$. But it must be also orthogonal to the matrix elements of $(1,0)\otimes(1,0)$, $(1,0)\otimes(0,1)$, and $(0,1)\otimes(0,1)$, i.e., to $y_\alpha y_\beta$, $y_\alpha\overline{y}_\beta$, and $\overline{y}_\alpha\overline{y}_\beta$. All the representations in our list contain everything that can be made out of direct products, and so $D^{(?)}_{ij}(g)$ is orthogonal to every polynomial $P(y_\alpha,\overline{y}_\alpha)$ in the y_α's and the $\overline{y}_\alpha$'s:

$$\int_G dg\,\overline{P}(y_\alpha,\overline{y}_\alpha)\,D^{(?)}_{ij}(g) = 0 \tag{38.14}$$

This is because $D^{(?)}_{ij}(g)$ is not in our list, and so it has to be orthogonal to the others. Now by the approximation theorem of Weierstrass, given a complete set of coordinates on any space, anything that is orthogonal to all the polynomials in the coordinates has to be *zero*.[17] Therefore,

$$D^{(?)}_{ij}(g) = 0 \tag{38.15}$$

That is the unique function orthogonal to all the polynomials. Then $D^{(?)}_{ij}(g)$ can't be a representation because every representation must equal 1 when g equals the identity element. **QED** This orthogonality proof is a very nice method. It's practically the only trick in the past few lectures which has not been stolen from Hermann Weyl or Claude Chevalley.[18]

That is the end of the mathematics part of this lecture. If you didn't follow the math, don't worry. It's pretty, it's fun, and you'll understand things more deeply if you understand these arguments, but this discussion was not particularly about quantum field theory. Suffice it to say that in order to make myself an honest man I have explicitly proved to you that the representations (n,m) are indeed what I have been acting as if they were: a *complete set* of *inequivalent*, *irreducible* representations of SU(3).

38.2 The operators I, Y and Q in SU(3)

You'll recall that in recent lectures (§24.3, §§35.3–35.4, §36.1) we derived all sorts of electromagnetic relations between form factors and magnetic moments, assuming that the SU(2) of isospin was perfect, and treating electromagnetism only to lowest order.

Now we will do the same thing with SU(3): I will neglect the effects of the medium-strong interactions and assume that SU(3) is perfect. It's a bigger group so we should get more relations, perhaps something a bit more useful than the one (35.68) that connected the magnetic moment of the Σ^+ to that of the Σ^- (sadly, far beyond the reach of current experiment). We hope to connect theory with something that we can actually measure. The first thing I'll look at are electromagnetic formulas in the limit of perfect SU(3). Or, going to all orders of the medium-strong interactions, to first order in electromagnetism and zeroth order in the cross terms between electromagnetism and the medium-strong interactions.

---

[17] [Eds.] $D^{(?)}$ can be approximated by a sequence of polynomials $P_n(y_\alpha,\overline{y}_\alpha)$ which converge uniformly to $D^{(?)}$, and so uniform convergence gives $\int dg\,\overline{P}_n(y_\alpha,\overline{y}_\alpha)\,D^{(?)}(g) \to \int dg\,|D^{(?)}(g)|^2 = 0$, and thus $D^{(?)}(g) = 0$. See Harold and Bertha S. Jeffreys, *Methods of Mathematical Physics*, Cambridge U. P., 1946, Sections 14.08–14.081, pp. 417–418. In the context of integration over group spaces, an identical argument applied to the spherical harmonics (for SO(3)) is given by Charles Loewner, *Theory of Continuous Groups*, MIT Press, 1971; see Lecture VIII, p. 62. Republished by Dover Publications, 2008.

[18] [Eds.] H. Weyl, *The Classical Groups*, Princeton U. P., 1939, 1973 (paperback ed., 1997); C. Chevalley, *Theory of Lie Groups*, Princeton U. P., 1946, 1999.

We expect to make errors on the order of 10–20% by assuming that SU(3) is perfect. After all, in the particular case of the baryon octet, the individual baryons lie within 15 or 20% of the mean mass of the octet. That's what we've got to live with, until we have a complete dynamical theory of the SU(3) symmetry breakdown. Then we could take care of the medium-strong interactions with Feynman diagrams.

You will recall that in our isospin analysis, the key point was that the electromagnetic charge was a generator of the group, one of the symmetries of the strong interactions. And therefore the electromagnetic current, by the minimal coupling prescription, transformed like this generator, the charge, which by the Gell-Mann–Nishijima relationship was (35.53) a linear combination of the z-component of the isospin, an isovector, and hypercharge, an isosinglet. We don't know yet how the generators of SU(3) transform under the action of SU(3). Before we can start applying the same techniques, we'll have to figure out how they transform.Are they a $(3,0)$, a $(1,1)$ or something else? We need some preliminary work to determine the SU(3) generators.

Let me just remind you how we deduce the angular momentum transformation properties under the action of the rotation group.[19] In classical mechanics we have a three-dimensional vector $\mathbf{x}$ which undergoes an infinitesimal rotation defined[20] by a three-dimensional rotation matrix R:

$$\mathbf{x} \to R\mathbf{x} = \mathbf{x} + \delta\theta(\hat{\mathbf{n}} \times \mathbf{x}) + \mathcal{O}(\delta\theta^2) \tag{38.16}$$

The rotation R is specified by an axis $\hat{\mathbf{n}}$ and an angle θ. It is an element of the group SO(3):

$$R \in \mathrm{SO}(3)\colon R = R(\hat{\mathbf{n}}\theta) \quad 0 \leq \theta \leq \pi \tag{18.18}$$

An infinitesimal rotation can be represented as an operator D specified by R:

$$D(R) = 1 - i\delta\theta\,\hat{\mathbf{n}}\cdot\mathbf{J} + \mathcal{O}(\delta\theta^2) \tag{38.17}$$

$\mathbf{J}$ is a vector whose components are operators. Under the action of $D(R)$

$$D(R)^{-1}\mathbf{J}D(R) = R\mathbf{J} \tag{38.18}$$

Since $D(R)$ is unitary, $\mathbf{J}$ is Hermitian:

$$D(R)^{-1} = D(R)^\dagger \;\Rightarrow\; \mathbf{J} = \mathbf{J}^\dagger \tag{38.19}$$

We can apply the same analysis to SU(3). First we have a Hilbert space in which we have our quantum mechanical theory. We have isospin $\mathbf{I}$, a three-vector composed of operators; we have a unitary operator $U(R)$ in the Hilbert space associated with isospin transformations; R is an isospin rotation matrix acting on the operators $\mathbf{I}$. Here the operator is an element of SU(3):

$$U(R) \in \mathrm{SU}(3) \tag{38.20}$$

The analog of (38.18) is

$$U(R)\mathbf{I}U^\dagger(R) = R\mathbf{I} \tag{38.21}$$

That's the statement that the three generators of isospin transform like an isovector. It comes out as a vector because isospin is an SU(2) subgroup of SU(3), and SU(2) is the covering group[21] of SO(3). Again we have an infinitesimal rotation, now in isospin space, labeled by an

---

[19] [Eds.] See Goldstein *et al. CM*, Section 4.8, pp. 163–171; Mathews and Walker, *op. cit.*, Section 16-7, pp. 461–466.

[20] [Eds.] See §18.2. Again, Coleman uses $\hat{\mathbf{e}}$ to denote the axis of rotation, and again we have changed his notation to $\hat{\mathbf{n}}$ to avoid confusion with the base of the natural logarithms.

[21] [Eds.] See note 37, p. 791.

axis $\hat{\mathbf{n}}$ and an infinitesimal angle $\delta\theta$. It's close to the identity and it goes about some axis by an infinitesimal angle. The corresponding U for the infinitesimal rotation is

$$U = 1 - i\hat{\mathbf{n}}\bullet\mathbf{I}\,\delta\theta \tag{38.22}$$

linear in the components of $\mathbf{I}$ since we're only going to first order in $\delta\theta$. That's the primary definition of $\mathbf{I}$: $\mathbf{I}$ is an isovector because it's dotted into $\hat{\mathbf{n}}$, which is an isovector, and $\hat{\mathbf{n}}\delta\theta$ is what labels the rotation. Very shortly I will go through the same analysis for SU(3). We'll see how to label an infinitesimal SU(3) transformation and then we'll know how the generators of SU(3) must transform by the same reasoning.

Before we do that we need to know how to represent isospin generators as matrices. The three-dimensional rotation group is the same, at least locally, as the two-dimensional unitary group and we know that the isospin generators transform like the three pions. We went through considerable labor to find out how to write the three pions as a 2×2 traceless matrix. Recall (36.85) that we found

$$\phi = \begin{pmatrix} \frac{1}{\sqrt{2}}\phi_0 & \phi_+ \\ \phi_- & -\frac{1}{\sqrt{2}}\phi_0 \end{pmatrix} \tag{38.23}$$

For convenience, I'll multiply ϕ by $\sqrt{2}$; everything will still transform the right way.

$$\phi \to \sqrt{2}\phi = \begin{pmatrix} \phi_0 & \sqrt{2}\phi_+ \\ \sqrt{2}\phi_- & -\phi_0 \end{pmatrix} \tag{38.24}$$

From this we can read off the 2×2 matrix that corresponds to the three isospin generators. I'll write it down and it will look like there are a pair of algebraic errors but in fact it's the right answer:

$$I = \begin{pmatrix} I_z & I_- \\ I_+ & -I_z \end{pmatrix} \tag{38.25}$$

Looking at (38.24), apparently I've made two algebraic errors: I've transposed the plus and minus components and I've left out the $\sqrt{2}$.

In fact, I've done neither. The raising and lowering operators for isospin are

$$I_\pm = I_x \pm iI_y \tag{38.26}$$

whereas the charged pion fields are defined with $\sqrt{2}$ in the denominators. Therefore, $I_\pm$ is what corresponds to $\sqrt{2}\phi_\mp$, not what corresponds to $\phi_\pm$. That's where the $\sqrt{2}$ went.

Why have the plus and minus components of I switched places from ϕ? Because ϕ_+ is the field that *annihilates* positively charged pions, that is, *lowers* the isospin, and I_- is the operator that lowers the isospin, just as p annihilates the $|p\rangle$ state. I'm sorry, but that's the way life is! Minus and plus are used in different ways when defining isospin raising and lowering operators and charged pion fields. We have to live with that convention clash. That's why the minus and plus components have changed places (as in (24.21)) and why the $\sqrt{2}$ has disappeared: I_- is the isospin lowering *operator* while ϕ_+ is the isospin lowering *field*, because it annihilates a π^+ which has positive I_z.

Now we'll work out the generators of SU(3). If g is an SU(3) matrix it obeys two equations:

$$\begin{aligned} gg^\dagger &= 1 \quad \text{(to satisfy the "U" in SU(3))} \\ \det g &= 1 \quad \text{(to satisfy the "S")} \end{aligned} \tag{38.27}$$

We want to consider an infinitesimal SU(3) transformation, which has the form

$$g = 1 - i\epsilon\delta\theta \tag{38.28}$$

with ϵ a 3×3 matrix and $\delta\theta$ an infinitesimal angle. All the "direction" part of the transformation is in ϵ, just as all the direction in SU(2) lies in the choice of $\hat{\mathbf{n}}$.

$$gg^\dagger = 1 - i\epsilon\delta\theta + i\epsilon^\dagger\delta\theta + \mathcal{O}((\delta\theta)^2) = 1 + \mathcal{O}((\delta\theta)^2) \tag{38.29}$$

We deduce that ϵ is a 3×3 *Hermitian* matrix:

$$\epsilon = \epsilon^\dagger \tag{38.30}$$

To find $\det g$, look in a frame where ϵ is diagonal:

$$\epsilon = \begin{pmatrix} \epsilon_1 & 0 & 0 \\ 0 & \epsilon_2 & 0 \\ 0 & 0 & \epsilon_3 \end{pmatrix} \tag{38.31}$$

Then (38.28) becomes

$$g = \begin{pmatrix} 1 - i\epsilon_1\delta\theta & 0 & 0 \\ 0 & 1 - i\epsilon_2\delta\theta & 0 \\ 0 & 0 & 1 - i\epsilon_3\delta\theta \end{pmatrix} \tag{38.32}$$

The determinant of this matrix is the product of the diagonal terms. We ignore the terms of order $(\delta\theta)^2$ and higher. So

$$\det g = 1 - i\delta\theta(\epsilon_1 + \epsilon_2 + \epsilon_3) = 1 - i\delta\theta\,\mathrm{Tr}(\epsilon) = 1 \tag{38.33}$$

Though we have computed the determinant in a coordinate frame in which ϵ is diagonal, the trace is independent of what coordinate basis we use. Thus

$$\mathrm{Tr}(\epsilon) = 0 \tag{38.34}$$

The group SU(2) has $2^2 - 1 = 3$ generators, and so needs three parameters, characterized by a vector, to describe a group element; SU(3) has $3^2 - 1 = 8$ generators, so its parameters are conveniently characterized by a traceless 3×3 matrix.

Parallel to (38.22), we write

$$U(g) = 1 - i\delta\theta\,\mathrm{Tr}[\epsilon G] \tag{38.35}$$

and $\mathrm{Tr}[\epsilon G]$ is a linear function of ϵ; ϵ is a matrix, G is a matrix of operators just as $\mathbf{I}$ is a vector with operator components. In order that $U(g)$ be unitary, G must be traceless and Hermitian:

$$G = G^\dagger, \quad \mathrm{Tr}(G) = 0 \tag{38.36}$$

Just as we deduced from (38.21) that $\mathbf{I}$ transforms as a vector, so we deduce here that G is a 3×3 matrix of operators that transforms as an octet, to wit,

$$U(g)GU^\dagger(g) = gGg^\dagger \tag{38.37}$$

where $g \in$ SU(3). This equation has exactly the same form as (38.21). In the center of the left-hand side we have a 3×3 matrix of operators; $U(g)$ is a unitary operator that implements SU(3) on some Hilbert space, likewise $U^\dagger(g)$; and g and $g^\dagger$ are 3×3 matrices of numbers. We have the correspondence.

Let's try to figure out the explicit form of G. Here's what we know so far:

SU(2)	SU(3)
R	g
θ	θ
$\hat{\mathbf{n}}$	ϵ
$\mathbf{I}$	G

Table 38.1: Correspondence between SU(2) *and* SU(3)

1. In the upper left 2×2 block where the pions sat in the pseudoscalar octet, (37.35), the isospin generators must sit, as in (38.25). Those are the things that transform as an isovector, just like the pions.
2. On the diagonal, where we would have the η in the pseudoscalar mesons, we must have the isosinglet symmetry generator, the hypercharge Y (with a multiplicative constant α that we'll have to determine).
3. In the other spots we will have some generators that transform like the kaons: strangeness-changing, hypercharge-changing generators. We won't study them here. They are however very important in weak interaction theory where they have names like λ_5 and λ_6, again because of historical conventions.[22] We'll fill parts of the matrix we aren't considering with an asterisk, $*$.

Therefore the matrix G looks like this:

$$G = \begin{pmatrix} I_z + \alpha Y & I_- & * \\ I_+ & -I_z + \alpha Y & * \\ * & * & -2\alpha Y \end{pmatrix} \tag{38.38}$$

To check the normalization, consider the 3×3 defining ("fundamental") representation of the group, $(1,0)$ or $\mathbf{3}$; the quarks. After all, this is a matrix of generators for any representation; in particular, it should be true for the quarks. (For convenience, Table 37.3 of quark properties is reprinted below.) For I_z, the ϵ that corresponds to an infinitesimal I_z rotation is determined by the condition

$$\epsilon_{I_z} q = I_z q \Rightarrow \epsilon_{I_z} \begin{pmatrix} u \\ d \\ s \end{pmatrix} = \begin{pmatrix} \frac{1}{2}u \\ -\frac{1}{2}d \\ 0 \end{pmatrix} \Rightarrow \epsilon_{I_z} = \begin{pmatrix} \frac{1}{2} & 0 & 0 \\ 0 & -\frac{1}{2} & 0 \\ 0 & 0 & 0 \end{pmatrix} \tag{38.39}$$

Quark	I_z	Y	$Q = I_z + \frac{1}{2}Y$
u	$+\frac{1}{2}$	$+\frac{1}{3}$	$+\frac{2}{3}$
d	$-\frac{1}{2}$	$+\frac{1}{3}$	$-\frac{1}{3}$
s	0	$-\frac{2}{3}$	$-\frac{1}{3}$

Table 37.3: The quarks and their properties

[22] [Eds.] See note 15, p. 807. Note that the operator for Y is proportional to λ_8, and I_i to λ_i, $i = \{1, 2, 3\}$.

That is what an infinitesimal I_z rotation does to the defining (1,0) representation. Multiplying (38.39) by (38.38) we find

$$\epsilon_{I_z} G = \begin{pmatrix} \frac{1}{2}(I_z + \alpha Y) & \frac{1}{2}I_- & * \\ -\frac{1}{2}I_+ & -\frac{1}{2}(-I_z + \alpha Y) & * \\ 0 & 0 & 0 \end{pmatrix} \Rightarrow \mathrm{Tr}[\epsilon_{I_z} G] = I_z \tag{38.40}$$

That's jolly good; we didn't make some dumb mistake with the normalization.

Now let's determine α. An infinitesimal rotation in the Y direction is determined as in (38.39), and we find

$$\epsilon_Y = \begin{pmatrix} \frac{1}{3} & 0 & 0 \\ 0 & \frac{1}{3} & 0 \\ 0 & 0 & -\frac{2}{3} \end{pmatrix} \tag{38.41}$$

Multiplying (38.41) by (38.38) we obtain

$$\epsilon_Y G = \begin{pmatrix} \frac{1}{3}(I_z + \alpha Y) & \frac{1}{3}I_- & * \\ \frac{1}{3}I_+ & \frac{1}{3}(-I_z + \alpha Y) & * \\ 0 & 0 & \frac{4}{3}\alpha Y \end{pmatrix} \Rightarrow \mathrm{Tr}[\epsilon_Y G] = 2\alpha Y \tag{38.42}$$

But we want the trace to equal Y, the generator of infinitesimal hypercharge rotations, not 17 times the hypercharge. So we set

$$\alpha = \tfrac{1}{2} \tag{38.43}$$

Therefore

$$G = \begin{pmatrix} I_z + \frac{1}{2}Y & I_- & * \\ I_+ & -I_z + \frac{1}{2}Y & * \\ * & * & -Y \end{pmatrix} \tag{38.44}$$

We'll be studying electromagnetism shortly so we need to find the ϵ corresponding to a Q rotation:

$$Q = I_z + \tfrac{1}{2}Y \tag{38.45}$$

This ϵ_Q is a linear combination of the corresponding ϵ's that we already have, (38.39) and (38.41):

$$\epsilon_Q = \epsilon_{I_z} + \tfrac{1}{2}\epsilon_Y = \begin{pmatrix} \frac{2}{3} & 0 & 0 \\ 0 & -\frac{1}{3} & 0 \\ 0 & 0 & -\frac{1}{3} \end{pmatrix} \tag{38.46}$$

Naturally, the charges of the three quarks are the eigenvalues of ϵ_Q. Finally, since we will be doing three problems involving the baryon octet, we'll again write down that matrix, (37.32), with B taking the place of ψ:

$$B = \begin{pmatrix} \frac{1}{\sqrt{2}}\Sigma^0 + \frac{1}{\sqrt{6}}\Lambda & \Sigma^+ & p \\ \Sigma^- & -\frac{1}{\sqrt{2}}\Sigma^0 + \frac{1}{\sqrt{6}}\Lambda & n \\ \Xi^- & -\Xi^0 & -\frac{2}{\sqrt{6}}\Lambda \end{pmatrix} \tag{38.47}$$

We now have all the machinery we will need. We have the ϵ matrix that corresponds to the electric charge, we have the baryon octet matrix and we have the matrix of generators, which we won't need for a while.

38.3 Electromagnetic form factors of the baryon octet

We are first going to study the electromagnetic form factors of the baryon octet.[23] Aside from the neutron and proton, the only form factors that have been measured (for five of the other six baryons)[24] are the magnetic moments, and the electric form factors $F_1(0)$.

Consider the matrix element of a general current in the octet j_μ (a matrix made out of currents just like the generator matrix G) between some final baryon state described by B' and an initial baryon state in B,

$$\langle B'|j_\mu|B\rangle$$

(all space and spin indices have been suppressed). These currents will be involved in a variety of reactions. We might want to look at the hypercharge form factor or the isospin form factor; they might be different. We could look at strangeness-changing currents; those turn out to play an important role in weak interaction theory, although they're not the only currents. For the moment, we are concerned with the electromagnetic current, and therefore we will be interested in

$$\mathrm{Tr}\left[\epsilon_Q \langle B'|j_\mu^{\mathrm{em}}|B\rangle\right] \tag{38.48}$$

where ϵ_Q is the charge matrix, (38.46); that choice of ϵ picks out the electromagnetic current: $\epsilon_Q j_\mu = \epsilon_Q j_\mu^{\mathrm{em}}$.

For a general matrix element made from an octet current between two octet baryons, how many independent matrix elements are there, apart from the functions[25] F_1 and F_2? That is, out of an **8** ($|B\rangle$) and an **8** (j_μ), how many **8**'s ($\langle B'|$) can we make? We know the answer. From (37.58), we see that we can make *two* 8 multiplets:

$$\mathbf{8}\otimes\mathbf{8} = \mathbf{1}\oplus\underbrace{\mathbf{8}\oplus\mathbf{8}}_{\substack{\text{two}\\ \text{possible}\\ \text{couplings}}}\oplus\mathbf{10}\oplus\overline{\mathbf{10}}\oplus\mathbf{27} \tag{38.49}$$

Thus, the general matrix element, for any baryon on the right, any baryon on the left and any ϵ, is given in terms of just *two* quantities, neglecting space and spin dependence: two F_1's and two F_2's. In particular, this means that if we know, in the idealized limit of perfect SU(3) symmetry, the electromagnetic form factors F_1 and F_2 of the proton and the neutron, then we know F_1 and F_2 for every baryon in the octet. And furthermore we know the matrix elements of the strangeness-changing currents and the hypercharge currents and any other linear combination we want.

It's exactly like meson–nucleon coupling, (37.37): an octet coupled to two octets. We can write the most general form

$$3\alpha\,\mathrm{Tr}\left[\overline{B}'\epsilon B\right] + 3\beta\,\mathrm{Tr}\left[\overline{B}'B\epsilon\right] \tag{38.50}$$

where α and β are scalar coefficients, B is the 3×3 matrix of incoming baryons and $\overline{B}'$ is the 3×3 matrix of outgoing baryons; the 3 is to tidy up the denominators in the charge matrix

---

[23] [Eds.] Sidney Coleman and Sheldon Lee Glashow, "Electrodynamic Properties of Baryons in the Unitary Symmetry Scheme", *Phys. Rev. Lett.* **6** (1961) 423–425. See also note 40, p. 841.

[24] [Eds.] *PDG* 2016, pp. 88–95. Conspicuous by its absence is the Σ^0. The theoretical value is $\frac{1}{2}(\alpha+\beta) = -\frac{1}{2}\mu_n$ (see Table 38.2 on p. 837). The Σ^0's lifetime, $\sim 10^{-20}$ s, has thus far precluded the measurement of its magnetic moment.

[25] [Eds.] See (34.15).

(38.46). The coefficients α and β are of course actually functions of space and spin, they're $F_1(q^2)$ and $F_2(q^2)$ with all sorts of spinor factors; if we're just looking at the magnetic moment they're simply numbers. The B and B' describe which baryon we're looking at. For example, if the initial and final states are both protons,

$$B = \begin{pmatrix} 0 & 0 & 1 \\ 0 & 0 & 0 \\ 0 & 0 & 0 \end{pmatrix} \qquad \overline{B}' = \begin{pmatrix} 0 & 0 & 0 \\ 0 & 0 & 0 \\ 1 & 0 & 0 \end{pmatrix} \tag{38.51}$$

The particular ϵ that we pick, (38.46), (38.41) or (38.39), determines which form factor we get. These forms occur because we have employed minimal coupling, which goes through for SU(3) just as it does for isospin; an 8 operator acting on an 8 state going into an 8 state. With the choice of $\epsilon = \epsilon_Q$, (38.50) will tell you all the magnetic moments. I will evaluate this formula shortly for the specific case of interest, the electromagnetic current matrix element.

In principle, how many observable form factors are there? There are eight baryons and they can all have different magnetic moments, even in the limit of perfect SU(3) symmetry. From (38.50) we get eight objects, arising from matrix elements of the form $\langle b|j_\mu^{\text{em}}|b\rangle$, without cross terms $\langle b'|j_\mu^{\text{em}}|b\rangle$ with $b \neq b'$. But there *is* one possible cross term, between the Σ^0 and the Λ: since both lie on the diagonal in the baryon octet, (38.47), the electromagnetic current can have a non-zero matrix element between them. Moreover, the selection rules (35.62) for isospin and charge conjugation allow it (35.63). That's a good thing, because the principal decay mode of the Σ^0 is just this:

$$\Sigma^0 \to \Lambda + \gamma \tag{38.52}$$

and the reaction is extremely fast, $\sim 10^{-20}$ s. (Even after studying this decay for 20 years all we have is an upper bound for the lifetime[26].) That means it's as low order in electromagnetism as it can be, to wit, first. And therefore there had better be a non-vanishing electromagnetic current matrix element between the Σ^0 and the Λ.

So there are nine observable quantities here: the eight magnetic moments of the baryons and the $\Sigma^0 \to \Lambda + \gamma$ transition matrix element. (In the limit of pure SU(3) symmetry, the latter is F_2-dominated, since F_1 between Σ^0 and Λ vanishes: F_1 is the matrix element of the charge operator and both have zero charge.) Our formula enables us to deduce these things in terms of two parameters, α and β. We can then solve for α and β in terms of the proton and neutron moments and predict the other moments. In principle, we will find in the literature all nine of these things computed in terms of the magnetic moments of the proton and the neutron. But I will be somewhat less ambitious and just compute the ones that I can find in the Particle Data Group tables.[27] If you want to compute others and make prophecies about future tables you are encouraged to do so. The current table has the measurements of seven baryon magnetic moments in it.[28] The proton and the neutron are known precisely; the others have relatively large uncertainties.

We'll start with the proton. The important thing to remember is that when we multiply a matrix by a diagonal matrix, life is very simple. If the diagonal matrix is on the left, every

[26] [Eds.] In the four decades since this lecture was given, our knowledge of the Σ^0 has gotten better. It's been established that the Σ^0 has a mean life of $7.4 \pm 0.7 \times 10^{-20}$ s, and this decay mode is responsible for $\sim 100\%$ of Σ^0 decays: *PDG* 2016, p. 94.

[27] [Eds.] *PDG* 2016.

[28] [Eds.] Coleman said "five" in 1976, and listed only p, n, Σ^+, Λ and Ξ^-.

row of the other matrix gets multiplied by the diagonal entry; if we multiply it on the right every column of the other matrix gets multiplied by the diagonal entry. For the proton, from the 3α term we get

$$3\alpha\,\mathrm{Tr}\left[\overline{B}'\epsilon B\right] = 3\alpha\,\mathrm{Tr}\left[\begin{pmatrix}0&0&0\\0&0&0\\1&0&0\end{pmatrix}\begin{pmatrix}\frac{2}{3}&0&0\\0&-\frac{1}{3}&0\\0&0&-\frac{1}{3}\end{pmatrix}\begin{pmatrix}0&0&1\\0&0&0\\0&0&0\end{pmatrix}\right] = 2\alpha$$

And from the 3β term we get $-\beta$. Continuing with the other baryons, we obtain[29] Table 38.2. The theoretical values for μ_p and μ_n are not listed, since they were used[30] to fix α and β. The predictions are in an ideal world where there are no mass differences between the baryons. In reality we expect them to be off by 20-30%.

Baryon	*Coefficients*	*Magnetic Moment*	*Theory*	*Experiment*
p	$2\alpha-\beta$	μ_p	—	$2.792847351 \pm 9 \times 10^{-9}$
n	$-\alpha-\beta$	μ_n	—	$-1.9130427 \pm 5 \times 10^{-7}$
Σ^+	$2\alpha-\beta$	$\mu_{\Sigma^+} = \mu_p$	2.8	2.458 ± 0.010
Σ^-	$-\alpha+2\beta$	$\mu_{\Sigma^-} = -(\mu_p+\mu_n)$	-0.9	-1.160 ± 0.025
Ξ^0	$-\alpha-\beta$	$\mu_{\Xi^0} = \mu_n$	-1.9	-1.250 ± 0.014
Ξ^-	$-\alpha+2\beta$	$\mu_{\Xi^-} = -(\mu_p+\mu_n)$	-0.9	-0.6507 ± 0.025
Λ	$-\frac{1}{2}\alpha-\frac{1}{2}\beta$	$\mu_\Lambda = \frac{1}{2}\mu_n$	-0.95	-0.613 ± 0.004

Table 38.2: The baryon magnetic moments expressed in nuclear magnetons, $\mu_N = \dfrac{e\hbar}{2m_pc}$

These coefficients, α and β, could be chosen to fit the magnetic moments, the charges, or indeed any linear combination of the two form factors. Just to confirm that we haven't made any algebraic errors let's check that things are right if we use the charge $F_1(0)$ instead of the magnetic moment. If we set

$$2\alpha-\beta = Q_p = 1 \qquad -\alpha-\beta = Q_n = 0 \tag{38.53}$$

then we replace the μ's in the above chart by Q's. We find

$$\begin{aligned}
Q_{\Sigma^+} &= Q_p: & 1 &\overset{\checkmark}{=} 1\\
Q_{\Sigma^-} &= -(Q_p+Q_n): & -1 &\overset{\checkmark}{=} -(1+0)\\
Q_{\Xi^0} &= Q_n: & 0 &\overset{\checkmark}{=} 0\\
Q_{\Xi^-} &= -(Q_p+Q_n): & -1 &\overset{\checkmark}{=} -(1+0)\\
Q_\Lambda &= \tfrac{1}{2}Q_n: & 0 &\overset{\checkmark}{=} 0
\end{aligned} \tag{38.54}$$

[29] [Eds.] The experimental values in this table, taken from *PDG* 2016 and rounded to four places, differ from the 1976 values.

[30] [Eds.] In the more modern approach, the magnetic moments of the baryons are given in terms of the magnetic moments of the up, down and strange quarks. The agreement between theory and experiment is much better. See Griffiths *EP*, Table 5.5, p. 190; D. Perkins, *Introduction to High Energy Physics*, 4th ed., Cambridge U. P., 2000; and *PDG* 2016.

Everything checks out.

The proton and neutron moments, ever since they were first measured by Rabi, have been determined to a fare-thee-well. Compared to the hyperon moments[31] they have practically no experimental uncertainties at all.[32] In Table 38.2 I wrote down the *theoretical* answers with no experimental errors; p and n errors—and hence those in α and β—are negligible.

The units are nuclear magnetons, using the proton mass for the magnetic moment:

$$\mu_N = \frac{e\hbar}{2m_p c} = 3.152 \times 10^{-18}\,\text{MeV/G} = 5.05 \times 10^{-27}\,\text{J/T} \tag{38.55}$$

Shouldn't we use the hyperon's mass instead of the proton's? Who knows? There's a 20% difference. This is after all a computation for perfect SU(3); it would be cheating to make a decision on that. We'll use the proton mass because that's how they're expressed in the literature. Over time the numbers in Table 38.2 have gone up and down like the Dow Jones average. We have to be genuinely sophisticated to know the meaning of the experimental uncertainty; that's the secret wisdom of the theorist. The standard deviation in modern high energy experiments is a unit like the 'league' in medieval Europe: a German league was three and a half times as long as an English league.[33] There is just as vast a difference in standard deviations. Even if we take the standard deviations dead seriously, which I would not advise doing, the error bars stay narrow but the number leaps up and down from year to year. This is good.[34] The only thing you can decide, if you make a ten-year average, is that it's something with a 1% error, but that 'something' we know only to within 50%. These are very hard experiments, because these particles don't live long. Rabi got the Nobel Prize (1944) for measuring the proton's magnetic moment. The Σ^+ is a lot trickier. Though we haven't succeeded in measuring the Σ^0's magnetic moment, we do have a measurement[35] of the size (but not the sign) of the transition moment in $\Sigma^0 \to \Lambda + \gamma$:

$$|\mu_{\Sigma\Lambda}| = 1.61 \pm 0.08\,\mu_N$$

We can compute the transition moment, with B' given by the Λ's matrix and B by the Σ^0's, and we find

$$\mu_{\Sigma\Lambda} = \tfrac{\sqrt{3}}{2}(\alpha + \beta) = -\tfrac{\sqrt{3}}{2}\,\mu_n = 1.66$$

The magnitude is quite close, even if the sign is not yet established.

The Λ moment was measured in a precession experiment.[36] You don't have to polarize it;

---

[31] [Eds.] The term "hyperon" is defined in note 11, p. 521.

[32] [Eds.] G. Breit and I. I. Rabi, "On the Interpretation of Present Values of Nuclear Moments", *Phys. Rev.* **46** (1934) 230–231; I. I. Rabi, J. M. B. Kellogg and J. R. Zacharias, "The Magnetic Moment of the Proton", *Phys. Rev.* **46** (1934) 157–163; "The Magnetic Moment of the Deuton", [*sic*; nowadays "deuteron"], *Phys. Rev.* **46** (1934) 163–165.

[33] [Eds.] J. B. Friedman, K. M. Figg, S. D. Westrem, and G. G. Guzman, *Trade, Travel and Exploration in the Middle Ages*, Garland, 2000.

[34] [Eds.] Perhaps in the sense that the fluctuations keep us honest? Your guess is as good as ours.

[35] [Eds.] P. C. Petersen *et al.*, "Measurement of the $\Sigma - \Lambda$ Transition Magnetic Moment", *Phys. Rev. Lett.*, **57** (1986) 949–952.

[36] [Eds.] R. L. Cool, E. W. Jenkins, T. F. Kycia, D. A. Hill, L. Marshall and R. A. Schluter, "Measurement of the Magnetic Moment of the Λ^0 Hyperon", *Phys. Rev.* **127** (1962) 2223–2230; L. Schachinger *et al.*, "Precise Measurement of the Λ^0 Magnetic Moment", *Phys. Rev. Lett.* **41** (1978) 1348–1351.

you can tell its spin by a fluke. It decays very asymmetrically into a nucleon and a pion

$$\Lambda \to \begin{cases} p\pi^- & (63.9 \pm 0.5)\% \\ n\pi^0 & (35.8 \pm 0.5)\% \end{cases} \tag{38.56}$$

There's a large correlation between the spin of the Λ and the direction of the decay products, so you can tell how it's spinning from their trajectories; it's also produced preferentially in a certain spin state. You make a beam of Λ's, send them through a magnetic field, watch them decay and measure the precession of the magnetic moment. This is not easy. Even so, the agreement is within 20%, rather good even if you take the experimental errors seriously. Improving these results is difficult. But getting them from symmetry arguments is easy. This comparison looks very promising.

38.4 Electromagnetic mass splittings of the baryon octet

We can also use SU(3) to study second-order electromagnetic effects, just as we did[37] for η decay in SU(2). As an example of that, we will study electromagnetic mass differences between members of the same octet; for instance, between Σ^+, Σ^0 and Σ^-. That is a second-order electromagnetic effect, we believe, although nobody can compute it because it diverges. However, since all we're going to get are linear relations between things, we don't care what makes it finite. And in fact this will offer us not only a cute way of testing SU(3) but a cute way of testing the idea that the mass difference is purely electromagnetic, since we can't test it by computing the proton/neutron mass difference.[38]

The first thing we've got to do is count the number of invariants we have, in order to see if we can make any prediction. The second-order electromagnetic effect transforms like the product of two currents, as far as its internal symmetry properties go:[39]

$$\langle B' | j_\mu^{\text{em}}(x) j_\nu^{\text{em}}(y) | B \rangle \tag{38.57}$$

The product of the two currents thus can be regarded as the known direct product (37.58) of an octet with an octet:

$$\mathbf{8} \otimes \mathbf{8} = \mathbf{27} \oplus \mathbf{10} \oplus \overline{\mathbf{10}} \oplus \mathbf{8} \oplus \mathbf{8} \oplus \mathbf{1} \tag{38.58}$$

This is the product of *any* two currents: any current from the octet with any other current from the octet. Of course we're interested in the case where both of them are *electromagnetic* currents or, more to the point, both currents are the same. That means that the antisymmetric (under parity) combinations *cannot* appear. So one of the $\mathbf{8}$'s, the $\mathbf{10}$, and the $\overline{\mathbf{10}}$ are out by antisymmetry (37.63). We have the initial and final baryons, B and B', respectively, that have to be hooked together with this product in an SU(3)-invariant way. That's also $\mathbf{8} \otimes \mathbf{8}$, but with no particular symmetry or antisymmetry. Here are the states and the possible coupling to the current product:

$$\begin{array}{rccccccccccc} j_\mu^{\text{em}} \otimes j_\nu^{\text{em}}: & \mathbf{1} & \oplus & \mathbf{8} \oplus \cancel{\mathbf{8}} & \oplus & \cancel{\mathbf{10}} & \oplus & \cancel{\overline{\mathbf{10}}} & \oplus & \mathbf{27} \\ & \downarrow & & \downarrow\searrow & & & & & & \downarrow \\ B' \otimes B: & \mathbf{1} & \oplus & \mathbf{8} \oplus \mathbf{8} & \oplus & \mathbf{10} & \oplus & \overline{\mathbf{10}} & \oplus & \mathbf{27} \end{array} \tag{38.59}$$

---

[37] [Eds.] See §35.4 and §36.1.

[38] [Eds.] But see note 5, p. 508.

[39] [Eds.] See Example 3 in §35.4, p. 767.

(I've crossed out the antisymmetric representations). We can make an invariant by hooking a **27** to a **27**, the **8** in $j_\mu^{\rm em} \otimes j_\nu^{\rm em}$ to either of the **8**'s in $B' \otimes B$, and the **1** to the **1**. However, the singlet to the single is irrelevant: it just shifts *all* the masses by the same amount. It's an electromagnetic mass *shift*, but it doesn't produce an electromagnetic mass *difference*. Therefore we have three unknown constants and there are four observed electromagnetic differences: one within the neutron–proton, one within the cascade and two within the Σ's ($\Sigma^+ - \Sigma^0$ and $\Sigma^- - \Sigma^0$) so the computation is worth doing. With four experimental quantities and three free parameters we can make one prediction.

Now we have to write down the three invariants. They will involve ϵ_Q twice because they involve the current twice. We will just write down three linearly independent SU(3)-invariant terms for the electromagnetic contribution Δm_{em} to the mass splitting:

$$\Delta m_{em} = \alpha\,\mathrm{Tr}(\overline{B}'\epsilon_Q^2 B) + \beta\,\mathrm{Tr}(\overline{B}' B\epsilon_Q^2) + \gamma\,\mathrm{Tr}(\overline{B}'\epsilon_Q B\epsilon_Q) \tag{38.60}$$

This expression has the right properties: it's linear in B, antilinear in B', involves two ϵ's for the two currents, and is an SU(3) invariant.

I could now begin to calculate but it's useful to simplify the matrix algebra by observing that

$$\epsilon_Q = P - \tfrac{1}{3}\mathbb{I} \tag{38.61}$$

where $\mathbb{I}$ is a 3×3 identity matrix, and P is the projection operator

$$P = \begin{pmatrix} 1 & 0 & 0 \\ 0 & 0 & 0 \\ 0 & 0 & 0 \end{pmatrix}; \qquad P^2 = P \tag{38.62}$$

Then

$$\epsilon_Q^2 = P^2 - \tfrac{2}{3}P + \tfrac{1}{9}\mathbb{I} = \tfrac{1}{3}P + \tfrac{1}{9}\mathbb{I} \tag{38.63}$$

Therefore we could write Δm_{em} as

$$\Delta m_{em} = a\,\mathrm{Tr}(\overline{B}'PB) + b\,\mathrm{Tr}(\overline{B}'BP) + c\,\mathrm{Tr}(\overline{B}'PBP) + \cancel{d\,\mathrm{Tr}(\overline{B}'B)} \tag{38.64}$$

The last term, $d\,\mathrm{Tr}(\overline{B}'B)$, is a mass shift that affects all the baryons equally, and we can drop it. It comes from all the $\mathbb{I}$'s in the product. We went through this little trick because it's easier to compute the terms for a matrix that's mainly zeros than for a matrix that's full of $\frac{1}{3}$'s and $\frac{2}{3}$'s. We would get the same result using (38.60).

Let's write down all the things we're going to worry about. Multiplying by P on the left just multiplies the first row by 1 and annihilates all the other rows; multiplying by P on the right multiplies the first column by 1 and annihilates all the other columns. So we get an a for the p, an a for the Σ^+ and a $\frac{1}{2}a$ for the Σ^0. In this way we get Table 38.3.

Now we can form linear combinations of the observable differences that are independent of a, b and c and are therefore zero. The difference that is least well measured is $m_{\Xi^-} - m_{\Xi^0}$:

$$m_{\Xi^-} - m_{\Xi^0} = b \tag{38.65}$$

We don't want to introduce the Σ^0; we have no other information on c. Instead, write $b = (b-a) + a$:

$$m_{\Xi^-} - m_{\Xi^0} = (m_{\Sigma^-} - m_{\Sigma^+}) + \big(m_p - m_n\big) \tag{38.66}$$

Baryon	Δm_{em}
p	a
n	0
Σ^+	a
Σ^0	$\frac{1}{2}(a+b+c)$
Σ^-	b
Ξ^0	0
Ξ^-	b

Table 38.3: Electromagnetic contributions to the mass shifts, Δm_{em}

This is the desired formula.[40] How does it compare with the experimental data? The observed mass splittings[41] are (in MeV):

$$\begin{aligned} m_p - m_n &= -1.29 \\ m_{\Sigma^-} - m_{\Sigma^+} &= 8.08 \pm 0.08 \end{aligned} \tag{38.67}$$

From this we compute the mass difference between Ξ^- and Ξ^0:

$$m_{\Xi^-} - m_{\Xi^0} = \begin{cases} 6.85 \pm 0.21 & \text{(exp't)} \\ 6.79 \pm 0.08 & \text{(theory)} \end{cases} \tag{38.68}$$

The prediction is pretty good. Actually the agreement is *surprisingly* good, when we recall the differences between the predicted and experimental values of the magnetic moments.

Aside.

Some day there will be one other result deduced from (38.64). The electromagnetic corrections to the Hamiltonian have an allowed off-diagonal term that can connect Σ^0 to Λ. Therefore, although I didn't compute it here, there is a small amount of mixing between the Σ^0 and the Λ induced by this allowed off-diagonal term. Equivalently, there is a tiny transition vertex, which is computable from the off-diagonal term and is of the same order as all these other things, 4 or 5 MeV. A Σ^0 comes in, something electromagnetic goes on, and a Λ comes out.[42]

This will yield a correction to the $\mathbf{8} \otimes \mathbf{8}$ baryon propagator. It is a second-order electromagnetic interaction, just like the mass shift. It is *not* analogous to the magnetic moment: the photon is only a virtual photon. One may ask: How can a neutral particle have an

[40] [Eds.] The relation (38.66) is known in the literature as the **Coleman–Glashow mass formula**: Sidney Coleman and Sheldon Lee Glashow, "Electrodynamic Properties of Baryons in the Unitary Symmetry Scheme", *Phys. Rev. Lett.* **6** (1961) 423–425. The paper covers both the magnetic moments and the mass splittings of the baryon octet. In the 1990 lectures, Coleman said at this point, "Modesty forbids me, but honesty compels me to tell you that this was my first published paper."

[41] [Eds.] *PDG* 2016.

[42] [Eds.] This effect is *second* order, and not to be confused with the first order, *diagonal* mixing between Σ^0 and Λ which was discussed in the paragraph preceding (38.52).

Figure 38.1: Electromagnetic correction to $\Sigma^0 \to \Lambda$ decay

electromagnetic mass shift? The answer is that the neutral particle decomposes virtually into into charged particles which then recombine. There are all these charged baryons and mesons in the theory: Σ^0 can become (for instance) a Σ^+ and a π^-. That's why the neutron has a magnetic moment. The blob has all sorts of things inside; we don't know what. We don't understand the details of how electromagnetism combines with the strong interactions. If we did, we could compute the proton–neutron mass difference, which we can't.[43] But we can explore this hypothesis. This marvelous agreement with experiment (38.68) not only tests SU(3), it checks the idea that the mass differences are electromagnetic.

Now, how do we measure this? It's not easy. It introduces a small amount of mixing of Σ^0 and Λ or, equivalently, by time reversal, of Λ and Σ^0. So the Σ^0's we see coming out are not 100% Σ^0 (the neutral member of the isotriplet); they have a tiny admixture of the isosinglet Λ and *vice versa*. But that's a hard thing to look for.

Some years ago Richard Dalitz made a suggestion for measuring this quantity. But it doesn't give a good check, so I didn't bother to look up the numbers in the literature; the uncertainties are still too large. It has to do with things called *hypernuclei.*[44] Every once in a while, when a Λ goes into a detector, it gets captured by a proton and forms something like a deuteron, but made out of a proton and a Λ. And then, because the Λ is unstable, it decays and we see this hypernucleus exploding. It could also happen with heavy nuclei. So if we know something about nuclear forces, we obtain some idea of the nature of the force between the Λ and a nucleon.

Now because the Λ is an isosinglet, pion exchange (the usual mechanism for the proton–neutron interaction), cannot occur: there is no $\pi\Lambda\Lambda$ vertex—it wouldn't conserve isospin. The only thing that can happen, in fact, if we don't take account of electromagnetic effects, is this process:

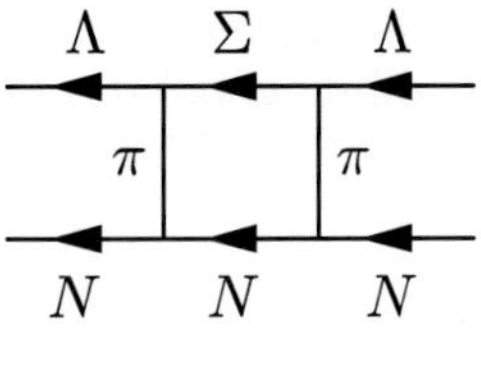

Figure 38.2: Λ capture by nucleons

That's allowed, and that process leads to the principal force between nucleons and Λ's. But it's a force of somewhat shorter range than the normal nuclear force, because instead of

[43] [Eds.] See note 5, p. 508.

[44] [Eds.] R. H. Dalitz, "The ΛΛ-Hypernucleus and the Λ–Λ Interaction", *Physics Letters* **5** (1963) 53–56; R. H. Dalitz and G. Rajasekaran, "The Binding of ΛΛ-Hypernuclei", *Nucl. Phys.* **50** (1964) 450–464; A. Gal, "The Hypernuclear Physics Heritage of Dick Dalitz" in J. Pochodzalla and T. Walcher, *Proceedings of the IXth International Conference on Hypernuclear and Strange Particle Physics*, Springer, 2007.

exchanging one pion we're exchanging two. The range is bounded by the mass of the exchanged particles:

$$(\text{range})^2 < \frac{1}{4m_\pi^2} \tag{38.69}$$

The lightest thing we can possibly exchange has a mass of $2m_\pi$. However, with the electromagnetic vertex (the blob), we can get another process, shown in Figure 38.3. The Λ comes along and becomes a Σ^0. Equivalently the Λ coming out of the beam is not pure Λ; it's a mixture of Λ and Σ^0. From then on it's the same story: Σ^0 emits a π and turns into a Λ and the π interacts with the nucleon (the vertex can occur in either location). We would normally think this would be a very small correction to the Λ-nucleon interaction (it's electromagnetic, not strong). But this is in fact not so, for two reasons. First, it is a longer range force than the first one,

$$(\text{range})^2 < \frac{1}{m_\pi^2} \tag{38.70}$$

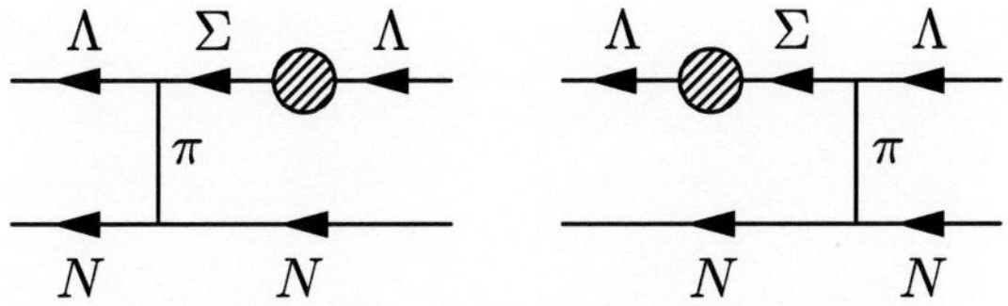

Figure 38.3: Λ–Σ *electromagnetic vertex*

So it can catch Λ's that make glancing collisions. Second, it's not as small as you might suppose. By a fluke, the Λ and the Σ are very close together in mass, about 75 MeV apart. Thus the denominator of the Σ propagator is rather small, for small momentum transfer. So it's not typically electromagnetic *in size*; that is, not down by $\frac{1}{137}$, but just down by something like $5/75 \sim \frac{1}{15}$, because we have this small denominator amplifying a small vertex.

Thus, as Dalitz suggested in the mid-1960's, by a close study of hypernuclei we should be able to detect this force and estimate its coefficient. Since we know everything in the diagram, we should be able to define the correction due to this force, deduce the Σ^0–Λ mixing matrix element, and thereby get another check of SU(3).[45]

There is another process that runs by mixing matrix elements where the mixing matrix element is not electromagnetic but medium-strong. But before we discuss that we will have to discuss the medium-strong interactions and the famous **Gell-Mann–Okubo formula**. We'll begin with that next time.

[45] [Eds.] H. Mueller and J. Shepard, "Λ–Σ^0 Mixing in Finite Nuclei", *J. Phys. G* **26** (2000) 1049–1064.

39

Broken SU(3) and the naive quark model

Thus far we've talked about a world with perfect SU(3) symmetry, broken only by the effects of electromagnetism, which we treated perturbatively. The mysterious medium-strong interactions (stronger than electromagnetism) which break SU(3) symmetry were ignored. Now we'll turn the tables, treating those medium-strong interactions as a perturbative effect on top of the strong interactions while ignoring electromagnetism.[1]

39.1 The Gell-Mann–Okubo mass formula derived

We have to start farther back. By hypothesis we assume that the Hamiltonian for the strong interactions is the sum of a very strong part, invariant under SU(3), and a medium-strong part which breaks SU(3):

$$H = \underbrace{H_{VS}}_{\substack{\text{SU(3)} \\ \text{symmetric}}} + \underbrace{H_{MS}}_{\substack{\text{SU(3)} \\ \text{breaking}}} \tag{39.1}$$

We don't know nearly as much about the medium-strong interactions as we do about electromagnetism: the latter is mediated by photons, and the requirement of renormalizability selected out minimal coupling. That enabled us to show that the current transformed the same way as the charges, and much else. We have no such handle here. If we're to make any progress along similar lines, we have to make guesses, either about the dynamical theory of the medium-strong interactions or about the predictions of such a dynamical theory insofar as pure symmetry arguments go. We'd have to guess how the medium-strong interactions transform under SU(3). Such a venture is not *a priori* guaranteed to be successful. It's possible that SU(3) is a good symmetry of nature, but the medium-strong interactions are very complicated. Perhaps these medium-strong interactions transform under SU(3) as sums with more or less equal weights of pieces that behave like every conceivable SU(3) representation. In that case we would not get a sum rule analogous to our electromagnetic ones, e.g., (38.66).

---

[1] [Eds.] At the beginning of his lectures, Coleman always asked if there were any questions. In the video of Lecture 39, a student asks about the field theory involved in these SU(3) predictions. Coleman gives a lengthy answer, in the end admitting that there isn't much field theory involved. With a smile, he asks the student, "Why did you ask that question? Was it an implicit criticism, 'What are these lectures doing in a course on quantum field theory?' This is a course on relativistic quantum mechanics," echoing his first sentence in the first lecture.

However, it turns out *a postiori* that the simplest guess anyone (and in particular, Gell-Mann) would have been tempted to make, that the medium-strong interactions have simple transformation properties under SU(3), fits experiment very well. Gell-Mann guessed[2] that

H_{MS} transforms under SU(3) like a member of $(1,1)$, the octet representation.

But *which* member? That is uniquely determined by the fact that the medium-strong interactions preserve isospin and hypercharge. So it must be that

$$H_{MS} \sim \text{the } (I=0, Y=0) \text{ member of an octet } (1,1) \text{ of operators} \tag{39.2}$$

(the Λ-like member, if you will).[3] I stress once again that this is pure hypothesis. If this guess did not work, it wouldn't necessarily mean that SU(3) is wrong. We'd try something else; maybe it transforms like a member of $(2,2)$ or the sum of members from $(1,1)$ and a $(2,2)$. However, this is certainly the simplest guess one could make that would give baryon mass differences to lowest order in perturbation theory, and therefore it is the thing to try. If it works, it is evidence for both Gell-Mann's guess and the general idea of SU(3) symmetry. Historically, Gell-Mann made this guess (39.2) before the η was discovered (by a couple of weeks).[4] So it wasn't a guess made after the fact, but before.

We're going to use this hypothesis and first order perturbation theory to compute the medium-strong mass differences within SU(3) multiplets. Well, there's a little finagle that's traditionally used: We compute δm *for fermions* but δm^2 *for bosons*. It's not really that important; it doesn't matter whether we use δm or δm^2 in first order, because $\delta m^2 = 2m\,\delta m$. If the splitting for m is small it wouldn't matter if we obtain the relationships for the shifts in m^2 or the shifts in m. The splittings are not really that small. There seems to be some small improvement obtained by using the rule of δm^2 for bosons, so that's what we'll do. This finagle was inspired by field theoretic ideas in which we think of corrections to the self-energy operator of a fermion making a shift in the mass, while those to a boson's self-energy shift the *square* of the mass.[5]

An amusing thing is that once you've made this hypothesis, you can write down the formulas for the mass shifts within a general SU(3) multiplet (an octet, a decuplet, and so on) in closed form. We don't have to multiply matrices and tensors tediously for a given multiplet. I will count how many unknown constants there are for the masses within any given multiplet. and construct that many operators.

Our task is to find out how many times $(n,m)\otimes(m,n)$ contains $(1,1)$. (Remember, (m,n) is the conjugate of (n,m).) That will tell us how many ways there are to couple an octet to a $\overline{\psi}$ and a ψ, and therefore how many terms will be in our mass formula. I will prove to you that

$$(1,1) \in (n,m)\otimes(m,n) \begin{cases} \text{twice} & \text{if } nm \neq 0 \\ \text{once} & \text{if } \textit{only one} \text{ of } m \text{ or } n = 0 \\ \text{not at all} & \text{if } m = n = 0 \end{cases} \tag{39.3}$$

---

[2] [Eds.] M. Gell-Mann, "Model of the Strong Couplings", *Phys. Rev.* **106** (1957) 1296–1299; M. Gell-Mann, "The Eightfold Way", Caltech Synchrotron Radiation Laboratory report CTSL-20, 1961 (unpublished); reprinted in *The Eightfold Way*, M. Gell-Mann and Y. Ne'eman, Benjamin, 1964.

[3] [Eds.] See (37.22).

[4] [Eds.] A. Pevsner *et al.*, "Evidence for a Three-Pion Resonance Near 550 MeV", *Phys. Rev. Lett.* **7** (1961) 421–423.

[5] [Eds.] In 1990, Coleman gave Feynman credit for the idea.

The proof can exploit our algorithm for the reduction of a direct product, but it can also be shown directly, by manipulating tensors. I'll use the latter method.

Suppose I have some traceless symmetric tensor $\psi^{i_1\cdots i_n}_{j_1\cdots j_m}$ and its conjugate $\overline{\psi}^{k_1\cdots k_m}_{l_1\cdots l_n}$, the former representing an object that transforms according to the representation (n, m), and the latter according to (m, n). We want to sum up the indices on $\overline{\psi}\psi$ in such a way that we are left with an octet; i.e., a traceless symmetric tensor with one upper index and one lower index, $\phi^{i_1}{}_{j_1}$. There are only two ways we can do it in general without using the ϵ tensor. (We will come back later to use the ϵ tensor.)

$$\overline{\psi}^{\overbrace{k_1\cdots k_m}^{A}}_{\underbrace{l_1\cdots l_n}_{a}}\psi^{\overbrace{i_1\cdots i_n}^{B}}_{\underbrace{j_1\cdots j_m}_{b}} \tag{39.4}$$

We sum all of set A with all of set b, and all but one of set B with all but one of set a, leaving alone one upper index and one lower index. Then I can take out the trace to make the octet. Alternatively, I could sum all of set B with all of set a, and all but one of set A with all but one of set b, and take out the trace. Those are the two ways of making an octet.

Using the ϵ tensor does me no good in this case. If I use only one ϵ tensor I'll get two more lower indices than upper indices, or *vice versa*, and there's no way of summing that will leave us with one upper index and one lower index. If we use two ϵ tensors that's the same as the string of three δ's in various permutations, and therefore the same as the original thing we explored. Thus there are, at most, *two* ways of constructing an octet. If I have only upper indices on one side and only lower indices on the other (i. e., if n or m is zero), one of the possibilities is no possibility at all; there would only be *one*.

So, if I can construct two operators such that in any representation they are octets, or equivalently, octets plus singlets (an additional singlet term is not going to bother us; that can be absorbed into the overall mass before I turn on the medium-strong interactions), then I say that the mass splitting is proportional to the matrix elements of those operators. I will now construct two such operators.

One of them is trivial. Remember our generator matrix, (38.44):

$$G = \begin{pmatrix} I_z + \frac{1}{2}Y & I_- & * \\ I_+ & -I_z + \frac{1}{2}Y & * \\ * & * & -Y \end{pmatrix} \tag{39.5}$$

The $*$'s indicate strangeness-changing objects, which we don't care about. This is a traceless symmetric matrix that transforms like an octet. Therefore one operator whose matrix element would follow the octet rule would be the $I = Y = 0$ component:

$$G^3{}_3 = -Y \tag{39.6}$$

This is similar to the reasoning that establishes the Wigner–Eckart theorem.[6] For example, the matrix element of every *vector* operator is proportional to the matrix elements of the

[6] [Eds.] E. Merzbacher, *Quantum Mechanics*, 3rd ed., John Wiley, 1998, Chap. 17, pp. 432–437; Arfken & Weber *MMP*, p. 273.

angular momentum: there's only one way to couple angular momentum 1 to a representation times its conjugate; therefore the *same* Clebsch–Gordan coefficients must occur in the matrix elements of the total angular momentum as occur in the matrix elements of the operator we are studying, and thus the two are proportional.

Likewise here there are two possible independent Clebsch–Gordan coefficients, so I need to find the two possible octet operators. I have found one of them; I will now find the other. Given any matrix A, its *cofactor matrix* $\mathrm{cof}[A]$ is also a matrix.[7] The cofactor matrix is the matrix made up of the determinants of the minors that enter into the expression for the inverse:

$$\mathrm{cof}[A]^i{}_j = \tfrac{1}{2}\epsilon^{ikl}\epsilon_{jrs}\{A^k{}_r, A^l{}_s\} \tag{39.7}$$

Since the A's are operators we had better take the symmetric combination; the curly brackets indicate the anticommutator. The object $\mathrm{cof}[A]$ transforms like a matrix if A does; in this case, like a combination of an octet and a singlet as it's not necessarily traceless. Therefore we will construct $\mathrm{cof}[G]^3{}_3$, the determinant on the minor of the 33 element:

$$\mathrm{cof}[G]^3{}_3 = -\{I_-, I_+\} + \{I_z + \tfrac{1}{2}Y, -I_z + \tfrac{1}{2}Y\} = 2\Big(\tfrac{1}{4}Y^2 - \mathbf{I}^2\Big) \tag{39.8}$$

That is the second operator that transforms like the 33 element of an octet. All the other cofactors carry nonzero isospin and/or hypercharge, and are ruled out by Gell-Mann's guess.

We have arrived at the **Gell-Mann–Okubo (GMO) mass formula**[8], a linear combination of these two operators and a constant term:

$$\left.\begin{array}{r} m(\text{fermion}) \\ m^2(\text{boson}) \end{array}\right\} = a + bY + c\Big[I(I+1) - \tfrac{1}{4}Y^2\Big] \tag{39.9}$$

The additive constant a is the mass present before the medium-strong interactions are turned on; a, b and c have to be fitted to experiment within each supermultiplet. For representations where either n or m is zero, the b and c terms are proportional; in that case, the GMO formula reduces to the first two terms alone:[9]

$$\left.\begin{array}{r} m(\text{fermion}) \\ m^2(\text{boson}) \end{array}\right\} = A + BY \quad \text{for IR's of the form } (n,0) \text{ or } (0,m) \tag{39.10}$$

The original proposal was by Gell-Mann. He worked out the Clebsch–Gordanry only for the octet. Later Okubo showed, by a different argument than the one here, how to write it for any representation. The cute argument I've given is due to a Russian named Smorodinskiĭ,[10] who showed it to me at the Dubna conference in 1964.

---

[7] [Eds.] Arfken & Weber *MMP*, p. 168.

[8] [Eds.] M. Gell-Mann, *op. cit.* (the formula appears as equation (4.8) in the Caltech report); S. Okubo, "Note on Unitary Symmetry in Strong Interactions", *Prog. Theo. Phys.* **27** (1962) 949–966; (reprinted in Gell-Mann and Ne'eman, *op. cit.*); S. Okubo, "Note on Unitary Symmetry in Strong Interaction II: Excited States of Baryons", *Prog. Theo. Phys.* **28** (1962) 24–32.

[9] [Eds.] For representations for which $n = 0$ or $m = 0$, $I(I+1) - \frac{1}{4}Y^2 = AY + B$. For $m = 0$, $A = \frac{1}{2} + \frac{1}{3}n$, and $B = \frac{1}{3}n(\frac{1}{3}n + 1)$. For the $\mathbf{10} = (3,0)$, we have $(I + \frac{1}{2})^2 = \frac{1}{4}(Y+3)^2$, or $I = \frac{1}{2}Y + 1$. For $n = 0$, $A \to -A$ and $n \to m$. See J. McL. Emmerson, *Symmetry Principles in Particle Physics*, Oxford U. P., 1970, pp. 121–123. Note that there is an overall sign error in equation (5.12).

[10] [Eds.] Yakov A. Smorodinskiĭ (1917–1992), Russian mathematical physicist coauthor with Lev Landau of *Lectures on Nuclear Theory*, Dover Publications, 2011. See M. Shifman, ed., *Under the Spell of Landau*, World Scientific, 2013, Chapter 4.

39.2 The Gell-Mann–Okubo mass formula applied

There are a lot of complete SU(3) representations around where all the particles have been discovered. Let's begin with the famous one, the **baryon octet**: the **8** with $J^P = \frac{1}{2}^+$. The results are shown in Table 39.1 (masses in MeV).

Baryon	I	Y	m_{theory}	$m_{\text{exp't}}$
N	$\frac{1}{2}$	1	$a+b+\frac{1}{2}c$	939
Λ	0	0	a	1116
Σ	1	0	$a+2c$	1193
Ξ	$\frac{1}{2}$	-1	$a-b+\frac{1}{2}c$	1318

Table 39.1: Gell-Mann–Okubo mass splitting in the baryon octet

Thus we obtain the formula originally written down by Gell-Mann:

$$\begin{aligned} 2a + c &= \tfrac{1}{2}(a + 2c + 3a) \\ m_N + m_\Xi &= \tfrac{1}{2}(m_\Sigma + 3m_\Lambda) \end{aligned} \tag{39.11}$$

It gives one relationship among these baryon masses.

Now, what about experiment? Given the things we are neglecting, we expect the accuracy of this formula to be second order in the medium-strong interactions. And although the medium-strong interactions are *medium* strong, even their second order contributions are larger than electromagnetic effects, so we won't bother to take account of the electromagnetic mass shifts within a multiplet. We take the average over all the electromagnetic masses. The uncertainties in these masses are negligible compared to the errors expected in this formula.[11] With the masses in MeV, we have

$$\begin{aligned} m_N + m_\Xi &= \tfrac{1}{2}(m_\Sigma + 3m_\Lambda) \quad \text{(theory)} \\ 939 + 1318 &= \tfrac{1}{2}(1193 + 3347) \quad \text{(exp't)} \\ 2257 &= 2270 \end{aligned} \tag{39.12}$$

This is pretty good; the agreement is better than we would expect. Even the most enthusiastic SU(3) fan has to admit that this agreement is fortuitous because, for heaven's sake, the error is less than or on the order of a typical *electromagnetic* mass splitting, and we cannot expect it to be that good.

The next multiplet we'll look at among the fermions is the well-known $J^P = \frac{3}{2}^+$ **resonance decuplet** in pion–nucleon scattering. In a perfectly SU(3)-symmetric world, that must occur as part of some representation of the direct product of an octet times an octet, $\mathbf{8} \otimes \mathbf{8}$:

$$\mathbf{8} \otimes \mathbf{8} = \mathbf{1} \oplus \mathbf{8} \oplus \mathbf{8} \oplus \mathbf{10} \oplus \overline{\mathbf{10}} \oplus \mathbf{27} \tag{37.58}$$

Now if all we knew about was the Δ, with $I = \frac{3}{2}$, $Y = 1$, we could still say that there are only a small number of possibilities here. It certainly can't be in a singlet because that doesn't

[11] [Eds.] *PDG* 2016.

contain any object with $I = \frac{3}{2}$, nor do the two octets; they contain (37.20) only $I = \{1, \frac{1}{2}, 0\}$. It can't be in the $\overline{\mathbf{10}}$ because that contains an object with $I = \frac{3}{2}$ but it has $Y = -1$, not $+1$. So the only possibilities are that the Δ is part of the **10** or part of the **27**.

In the early days of SU(3), some people thought the Δ was in the **27**. Glashow and Sakurai wrote a long paper called "The 27-Fold Way"[12] in which they gave a variety of convincing arguments about why it should be part of a **27**. But it turns out in fact to be part of a **10**. Its relevant energy range in all hypercharge channels has been extensively explored. None of the other objects that would be there in a **27** fits the things that would be there in a **10**.

There was a famous incident at the 1962 Rochester Conference. [13] The Δ had been known, the Σ^* (or $\Sigma(1385)$, in modern usage) had been discovered, and the discovery of the Ξ^* ($\Xi(1530)$) was announced at the conference. (Today we name the baryon resonances according to their mass, isospin, and hypercharge, instead of giving them individual names like Σ^* and K^*, as we used to.)[14] Gell-Mann looked at it and said "That's a decuplet." He then predicted the Ω^-, and gave its mass. We will see how well he predicted it.

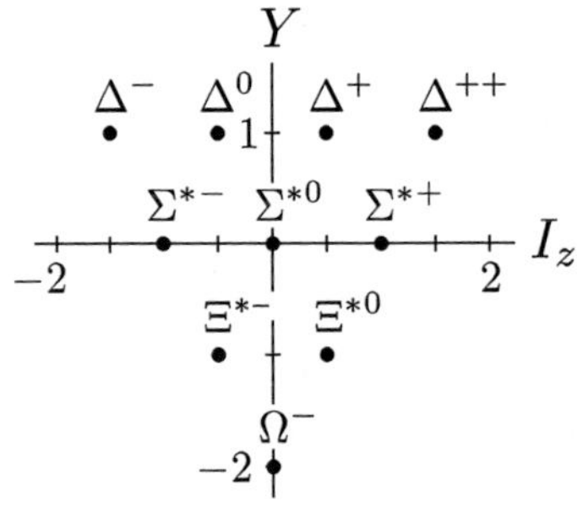

Figure 39.1: The weight diagram for the baryon decuplet, with the predicted Ω^-

The decuplet is given in Table 39.2; the masses are in MeV. The Δ ($\Delta(1232)$) is a broad bump; where we put the mass is a matter of taste. The $\Sigma(1385)$ (formerly Σ^*) was named after the familiar particle with the same I and Y. The $\Xi(1530)$ (or the Ξ^*) is an excited state of the Ξ. As this multiplet is a decuplet, $(3, 0)$, the GMO formula simplifies to (39.10),

$$m = A + BY$$

That means that the differences in mass should be proportional to Y, and a graph of M *vs.* Y should give a straight line; see Figure 39.2.

According to (39.10), the mass splitting in the decuplet is proportional to the hypercharge difference. The first three baryons are separated by $\Delta Y = 1$, and so we expect equal spacing in the mass splitting:

$$\begin{aligned} m_{\Sigma^*} - m_\Delta &= 153\,\text{MeV} \\ m_{\Xi^*} - m_{\Sigma^*} &= 145\,\text{MeV} \end{aligned} \tag{39.13}$$

---

[12] [Eds.] S. Glashow and J. Sakurai, "The 27-Fold Way and Other Ways: Symmetries of Meson-Baryon Resonances", *Nuovo Cim.* **25** (1962) 337-354.

[13] [Eds.] The 11th International Conference on High Energy Physics (Rochester Conference), July 1962, was held at CERN. According to Crease & Mann *SC*, pp. 273–274, Gell-Mann predicted the Ω^- would have a mass of 1685 MeV. It was discovered in early 1964, at Brookhaven, within 1% of Gell-Mann's estimate: V. E. Barnes *et al.*, "Observation of a Hyperon with Strangeness Minus Three", *Phys. Rev. Lett.* **12** (1964) 204–206. After this discovery, "there was no doubt, SU(3) was in." A. Pais, *Inward Bound*, Oxford U. P. 1986, p. 557.

[14] [Eds.] *PDG* 2016.

Baryon	*I*	*Y*	*Mass*
Δ	$\frac{3}{2}$	1	1232
Σ^*	1	0	1385
Ξ^*	$\frac{1}{2}$	-1	1530
Ω^-	0	-2	1672

Table 39.2: The $\frac{3}{2}^+$ baryon decuplet

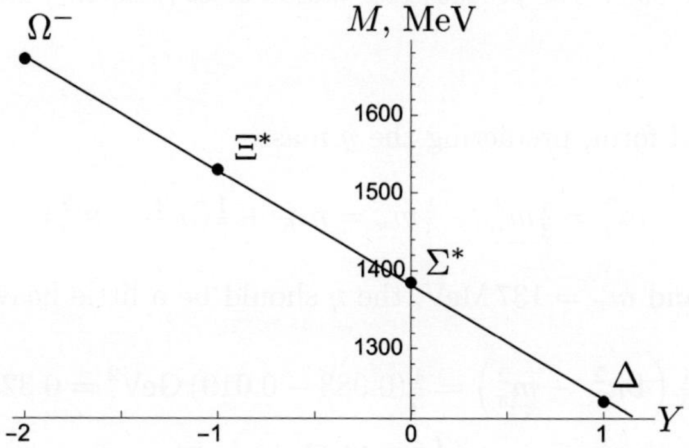

Figure 39.2: Gell-Mann–Okubo mass splitting for the $\frac{3}{2}^+$ decuplet

Thus Gell-Mann was able to predict the mass of the then-unknown tenth baryon as

$$m_{\Omega^-} \stackrel{?}{=} m_{\Xi^*} + \Delta m_{\text{avg}} = (1530 + 149)\,\text{MeV} \approx 1680\,\text{MeV} \tag{39.14}$$

The measured mass[15] is 1672.45 $\pm$0.29 MeV.

Actually we have *two* predictions, unlike the single prediction we had for the baryon octet: once we have *one* difference, say between the masses of the Σ^* and the Δ, we can predict the other two. You see that the agreement is very good, considering that we are only going to the lowest order in the medium-strong interactions. Gell-Mann's guess seems to be holding up remarkably well, but it's always dangerous policy to attempt to deduce deep physics from something like that. We'll look at another application which also works well. Then we'll come to one that doesn't work, and there's some interesting history attached to that.

Next, the **pseudoscalar meson octet,** $J^P = 0^-$ (see Table 39.3). In this case we can just copy down the baryon formula, except now the *squared* masses appear in the GMO formula because they're bosons.[16] The analog of both the nucleon and the cascade (the Ξ) is the kaon. So we obtain, analogous to (39.11),

$$2m_K^2 = \tfrac{1}{2}(3m_\eta^2 + m_\pi^2) \tag{39.15}$$

[15] [Eds.] *PDG* 2016, p. 96.

[16] [Eds.] The coefficient b in the general form (39.9) is 0, because the K has both $Y = 1$ and $Y = -1$. The same is true for the vector meson octet in Table 39.4; see (39.18).

Meson	I_z	Y	$m_{\text{exp't}}$
π	$\pm 1, 0$	0	140
K^0, $\overline{K}^0$	$\mp\frac{1}{2}$	± 1	498
K^+, K^-	$\pm\frac{1}{2}$	± 1	494
η	0	0	548
η'	0	0	958

Table 39.3: The pseudoscalar meson octet (plus one, the η')

Writing it in its original form, predicting the η mass,

$$m_\eta^2 = \tfrac{4}{3}m_K^2 - \tfrac{1}{3}m_\pi = m_K^2 + \tfrac{1}{3}(m_K^2 - m_\pi^2) \tag{39.16}$$

With $m_K = 496\,\text{MeV}$ and $m_\pi = 137\,\text{MeV}$, the η should be a little heavier than the kaon, and it is:[17]

$$\begin{aligned} m_\eta^2 &= \tfrac{1}{3}\left(4m_K^2 - m_\pi^2\right) = \tfrac{1}{3}(0.983 - 0.019)\,\text{GeV}^2 = 0.321\,\text{GeV}^2 \\ m_\eta &= \begin{cases} 548\,\text{MeV} & (\text{exp't}) \\ 567\,\text{MeV} & (\text{theory}) \end{cases} \end{aligned} \tag{39.17}$$

Again this is better than we would expect: not the typical 20% error but about 5% error or less. (The η' mixes only a little bit with the η, because of their very different masses.) This is really surprising because the pion mass is so far out of line with the other masses that it's amazing the formula works at all. Indeed, one of the reasons people spent four years exploring the dead end of global symmetry (§36.2) was because, as the product of three SU(2)'s, it enabled us to put the pions in a representation of the product of the three SU(2)'s all by themselves, $D^{(0,1,0)}$ in the notation of (36.33). Everyone said, "Well, obviously the eight baryons must be part of a multiplet. But the pions and the kaons are *so* different in mass that we should be able to put the pions in a multiplet by themselves. It's just too preposterous to imagine there is a symmetry which puts the pions and the kaons together. They look too different."

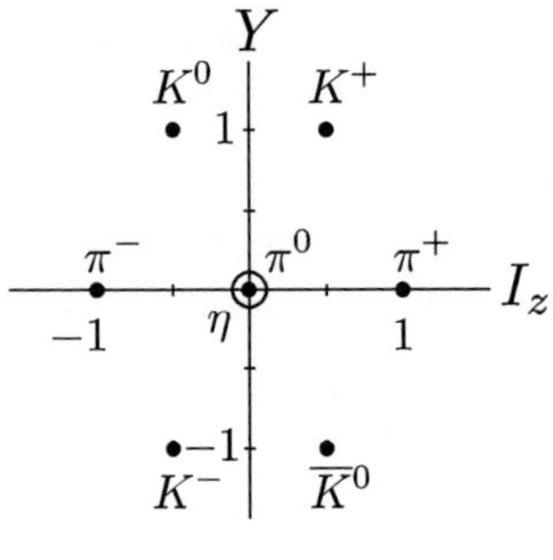

Figure 39.3: The weight diagram for the pseudoscalar meson octet

[17] [Eds.] *PDG* 2016.

I also remember, with pain at the time, and delight in retrospect, not long after Glashow and I published our electromagnetic mass formula,[18] there was a paper[19] by Sakurai proposing a different theory of electromagnetic masses. It contained the scathing footnote, "Formulas in the SU(3) symmetry scheme have recently been derived by Coleman and Glashow. The reader should understand that these formulas are valid only in the approximation $(m_K/m_\pi)^2 = 10 = 1$." That was a well-taken criticism. Nevertheless, the formula works.

39.3 The Gell-Mann–Okubo mass formula challenged

Let's go on to the next octet discovered in those golden days of 1961, **the vector bosons**, $J^P = 1^-$. The ρ had just been discovered. Those of you who think of the ρ as a permanent part of our universe should remember that the original Rosenfeld tables contained a subtable labeled "Resonances" that had only *one* entry in it. There was a time when the ρ was a great discovery, and it was followed soon after by the ω and the K^*, in January of 1961 or so. They were obviously an octet, although that turned out to be a little bit wrong, for reasons I'll explain shortly. We have the corresponding GMO formula for these mesons:

$$m^2 = a + c\left[I(I+1) - \tfrac{1}{4}Y^2\right] \tag{39.18}$$

and we get the results shown in Table 39.4. The K^* has the same quantum numbers as the kaon, except that it is a vector particle.

Meson	I_z	Y	m^2_{theory}	$m_{\text{exp't}}$
ρ	$0, \pm 1$	0	$a + 2c$	775
$K^{*0}, \overline{K}^{*0}$	$\mp\frac{1}{2}$	± 1	$a + \frac{1}{2}c$	896
K^{*+}, K^{*-}	$\pm\frac{1}{2}$	± 1	$a + \frac{1}{2}c$	892
ω	0	0	a	783

Table 39.4: The vector meson octet

We can combine the results in the table and obtain the same formula (39.16) as for the η mass, (using the average K^* mass of 894 MeV) and we find (in GeV^2)

$$\begin{aligned} m_\omega^2 &= m_{K^*}^2 + \tfrac{1}{3}(m_{K^*}^2 - m_\rho^2) \\ 0.613 &= 0.799 + \tfrac{1}{3}(0.199) = 0.865\,(?!) \end{aligned} \tag{39.19}$$

This is not good. It's a *disaster*, especially when compared with the results in the earlier cases. Well, Murray Gell-Mann is not a man to give up an idea lightly. He said, at the time: Suppose that, just by chance, in the absence of SU(3) breaking there happens to be a vector SU(3) singlet very close in mass to the ω. This is what we would compute: 0.865 GeV^2 would be the squared mass m_8^2 of the octet if the singlet weren't around. But because the singlet *is* around, things are going to be different.

---

[18] [Eds.] See note 40, p. 841.

[19] [Eds.] J. J. Sakurai, "New Resonances and Strong Interaction Symmetry", *Phys. Rev. Lett.* **7** (1961) 428–428. Note 15 reads in part, "It is easy to see, however, that most statements made in [Coleman and Glashow's] paper are expected to be accurate only up to a factor $(m_K/m_\pi)^2 \approx 13$."

Let's focus attention on the subspace of the big Hilbert space of the world which is a two-dimensional space spanned by a state of the octet vector meson ω_8, and the hypothetical singlet vector meson ω_1. I'll write down the 2×2 mass squared matrix for that. Gell-Mann said the ω_1 will acquire some mass, both because it has some mass in the absence of the medium-strong interactions, and because the medium-strong interactions may give us some correction. The ω_8 will acquire a mass, which we've just computed. But an octet operator can not only connect octet with octet, it can connect octet with a *singlet.* And exactly the same interaction that puts in the diagonal elements m_1^2 and m_8^2 could put in a cross-term, x:

$$M^2 = \begin{pmatrix} m_1^2 & x \\ x & m_8^2 \end{pmatrix} \tag{39.20}$$

(The phase between ω_1 and ω_8 is at our disposal so we can make x real; ordinarily we'd have x in one corner and x^* in the other.) The off-diagonal matrix element should be on the same order of magnitude as the diagonal matrix elements. Therefore, if m_1 and m_8 were very widely separated, by say 0.5 or 0.6 GeV^2, and x was on the order of the things we've been computing, around 0.1 GeV^2, then the effects of the off-diagonal term would be negligible. On the other hand, if by some fluke m_1 and m_8 happen to be fairly close together, then this off-diagonal term can have a very large effect. If we have two nearly degenerate levels in atomic physics, and we introduce a mixing matrix element, a symmetry-breaking Hamiltonian that connects the two, then one of them is pushed up and the other is pushed down, and the amount of the pushing can be comparable to the size of the mixing matrix element, i.e., around 0.2 GeV or so on the scale we're working with; see Figure 39.4. Gell-Mann hypothesized this was the situation here, an instance of nearly degenerate perturbation theory. In (39.19) we see the ω_8 mass lower than the predicted result. The other one, ω_1, presumably is higher. We can't predict the exact amount of pushing up and down unless we know both m_1^2 and m_8^2. But if Gell-Mann's idea is right, we can say there must be an isosinglet meson, $J^P = 1^-$, with a squared mass somewhere between 1.1 and 1.2 times greater than 0.865 GeV^2.

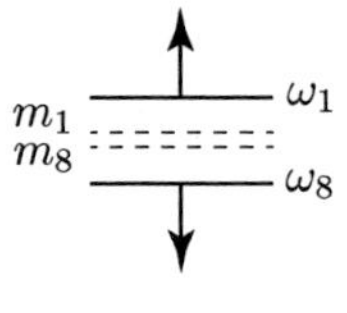

Figure 39.4: Level shift of the ω_1 and ω_8

That was his prediction; not a precise number, but that there should *be* such a particle in nature. Among the graduate students then at Caltech, this particle was known as the *fudge-on.*[20] And we were uniformly surprised when, not long afterwards, the ϕ was discovered, with exactly the predicted properties, $J^P = 1^-$, and a squared mass in the right ballpark:

$$m_\phi^2 = 1.04\,\text{GeV}^2 = 0.865\,\text{GeV}^2 \times 1.20 \tag{39.21}$$

Gell-Mann's prediction was not quantitative; it could have differed by 10% one way or the other and it would still have been all right. And in any case, there was the particle.

Now you might say there is no predictive power (other than qualitative) in this scheme, because we don't obtain a precise number from it. We have for this system three unknown

[20] [Eds.] For readers unfamiliar with the term, a *fudge factor* is an *ad hoc* quantity introduced into a calculation or measurement, ostensibly to account for error, to bring the number obtained closer to a desired value.

quantities, m_1^2, m_8^2 and x. One of them we know from the GMO formula, m_8^2, but we don't know *a priori* either m_1^2 or x. We have two experimental numbers, m_ω and m_ϕ, and therefore if we wanted to, we could deduce m_1^2 and x. But that's hardly a prediction; we can't go to God and ask, "What values did You assign to m_1^2 and x?" to check it.

However, we *can* predict the so called **mixing angle**, which is, in a sense, experimentally measurable. The mixing angle is the angle that tells us how much of the physical ω is ω_1 and ω_8 and how much of the physical ϕ is ω_1 and ω_8. I will first go through the computation of the mixing angle, then I'll explain how to measure it experimentally.

If we stick to the two-dimensional Hilbert space (39.20), consisting of ω and ϕ at rest (with the spin degrees of freedom suppressed), we would diagonalize the matrix and the eigenstates of M^2 would represent the actual physical particles, according to conventional, nearly degenerate perturbation theory. Since the M^2 matrix is real, its eigenvectors and eigenvalues must be real combinations of ω_1 and ω_8. Therefore we can define the angle θ by

$$\begin{aligned} |\phi\rangle &= \cos\theta\,|\omega_8\rangle + \sin\theta\,|\omega_1\rangle \\ |\omega\rangle &= -\sin\theta\,|\omega_8\rangle + \cos\theta\,|\omega_1\rangle \end{aligned} \tag{39.22}$$

(It turns out that ϕ is mostly ω_8, so this choice minimizes the mixing angle.) That's conventional perturbation theory. We could determine the angle θ in terms of physically observable quantities, the masses. We could do it by first finding m_1^2 and x, but we can do it much more directly. Any matrix can be written in terms of its eigenvectors and its eigenvalues:

$$M^2 = m_\phi^2\,|\phi\rangle\,\langle\phi| + m_\omega^2\,|\omega\rangle\,\langle\omega| \tag{39.23}$$

Computing from the GMO as if there were no ω_1,

$$\begin{aligned} \langle\omega_8|M^2|\omega_8\rangle &= m_8^2 = m_\phi^2\cos^2\theta + m_\omega^2\sin^2\theta \\ &= (m_\omega^2 - m_\phi^2)\sin^2\theta + m_\phi^2 \\ \Rightarrow\ \sin^2\theta &= \frac{m_\phi^2 - m_8^2}{m_\phi^2 - m_\omega^2} = \frac{1.04 - 0.865}{1.04 - 0.613} = 0.41,\ \ \theta \approx 40^\circ \end{aligned} \tag{39.24}$$

We see that θ is closer to 0° than to 90°, so our choice was suitable: according to this the ϕ is mainly ω_8 and the ω is mainly ω_1. In a moment I will explain why this number is useful.

Let's apply this mixing theory to leptonic decays of the vector bosons.

$$\left.\begin{matrix} \omega \\ \rho \\ \phi \end{matrix}\right\} \to e^+e^- \tag{39.25}$$

The vectors do not often decay into e^+e^-, but such decays *do* occur. We understand electromagnetism very well, so we know the kind of diagram that is responsible for this process:

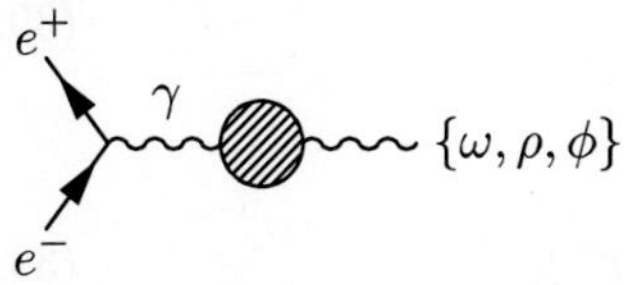

$$\mathcal{A}(V \to e^+e^-) \propto \langle 0|j_\mu^{em}|V\rangle\,, \qquad V \in \{\omega,\rho,\phi\} \tag{39.26}$$

A vector meson comes into a mysterious strong interaction blob (about which we can say nothing), a virtual photon comes out and decays into an e^+-e^- pair. The amplitude is proportional to the matrix element of the electromagnetic current between the vacuum and the one vector meson state, V. Once we peel off the electron–positron pair the matrix element is what's left.

This is a very simple example of a one (virtual) photon process, in which we have to take the matrix element of an octet operator between an octet state (if this is an octet vector meson), or a singlet state (if it is a singlet vector meson). An octet operator cannot, in the limit of strict SU(3) symmetry, connect a singlet state to a singlet state: both this singlet vector meson and the vacuum are SU(3) singlets. Therefore, in the SU(3)-symmetric world, in which the eigenstates are ω_1 and ω_8, the ω_1 decay amplitude vanishes:

$$\mathcal{A}(\omega_1 \to e^+e^-) = 0 \tag{39.27}$$

To investigate the octet decays, we'll introduce a matrix (as in (38.47) for the baryon octet) which we'll call V for vector meson. The amplitude should be proportional to

$$\mathcal{A} \propto \mathrm{Tr}(V\epsilon_Q) \tag{39.28}$$

where ϵ_Q is the 3×3 matrix (38.46) for the electromagnetic current,

$$\epsilon_Q = \frac{1}{3}\begin{pmatrix} 2 & 0 & 0 \\ 0 & -1 & 0 \\ 0 & 0 & -1 \end{pmatrix} \tag{39.29}$$

Of course the amplitude $\mathcal{A}$ is trivially zero for other than the two neutral members of the octet,[21] ω_8 and ρ_0. Let's compute it for these two neutral mesons.

The ω_8, assumed to be the $I = 0$ member of an octet, should behave like a Λ, so (38.47) its matrix V_ω is

$$V_\omega = \frac{1}{\sqrt{6}}\begin{pmatrix} -1 & 0 & 0 \\ 0 & -1 & 0 \\ 0 & 0 & 2 \end{pmatrix} \tag{39.30}$$

Multiplying the matrices and taking the trace,

$$\mathrm{Tr}(V_\omega\epsilon_Q) = \frac{1}{3\sqrt{6}}\left[-2+1-2\right] = -\frac{1}{\sqrt{6}} \tag{39.31}$$

The ρ_0, the $I = 1, I_z = 0$ member of the vector meson octet, acts like its opposite number, π_0, of the pseudoscalar octet, so (37.35) its matrix is given by

$$V_\rho = \frac{1}{\sqrt{2}}\begin{pmatrix} 1 & 0 & 0 \\ 0 & -1 & 0 \\ 0 & 0 & 0 \end{pmatrix} \tag{39.32}$$

and

$$\mathrm{Tr}(V_\rho\epsilon_Q) = \frac{1}{3\sqrt{2}}\left[2+1\right] = \frac{1}{\sqrt{2}} \tag{39.33}$$

Thus in a world with perfect SU(3) symmetry, we'd have

$$|\mathcal{A}(\rho_0 \to e^+e^-)| = \sqrt{3}\,|\mathcal{A}(\omega_8 \to e^+e^-)| \tag{39.34}$$

[21] [Eds.] The octet meson decays into an e^+-e^- pair, so it must be neutral.

There would be a ρ_0, an ω_1 and an ω_8. The ω_1 would not decay into e^+e^- at all. The ω_8 would decay into e^+e^- and the ρ_0 would decay into e^+e^- three times faster, because the decay rate is the square of the amplitude.

In the real world the SU(3) symmetry is not perfect. In this case because of the small energy denominators, the mixing effects are much larger than other effects of symmetry breaking. The mass eigenstates are not ω_1 and ω_8, but ϕ and ω. Taking account only of the mixing effect (not of the other supposedly small effects of SU(3) symmetry breaking), which is anomalously large because ω_1 and ω_8 are close in mass, we find the prediction for the branching ratios, compared here with the experimental numbers:[22]

$$\begin{aligned}\frac{\Gamma(\phi \to e^+e^-)}{\Gamma(\rho_0 \to e^+e^-)} &= \tfrac{1}{3}\cos^2\theta = \begin{cases}0.186 & \text{(exp't)}\\ 0.193 & \text{(theory)}\end{cases}\\ \frac{\Gamma(\omega \to e^+e^-)}{\Gamma(\rho_0 \to e^+e^-)} &= \tfrac{1}{3}\sin^2\theta = \begin{cases}0.085 & \text{(exp't)}\\ 0.140 & \text{(theory)}\end{cases}\end{aligned} \tag{39.35}$$

I have not taken any account of phase space. We are not close to a threshold; the e^+e^- threshold is very low compared to the masses of the ϕ, ω, and ρ, so phase space correction factors will be on the order of other effects of the medium-strong interactions which we are systematically neglecting. And if we *do* include phase space, we apparently make things worse. For example, we'd make the first prediction a little less than 0.3.

The important point is that that these ratios are in the right ballpark, mostly due to the 3's in the denominators. That 3 is a pure SU(3) Clebsch–Gordan coefficient. If we didn't have SU(3) it would be *a priori* as plausible to put the 3 in the numerator as in the denominator, which would change our result by a factor of 9. Then the predictions would be very different. So this is a non-trivial result. It's not simply that they come out right *qualitatively.*

I've gone through many more predictions in the past two lectures than throughout the rest of the course. Just because these arguments haven't involved deep thoughts and field theoretic concepts that make your head feel funny, don't think this isn't the real stuff. We've seen how all these were predicted:

- two magnetic moments in the baryon octet (μ_Σ and μ_Ξ, Table 38.2, p. 837)
- one electromagnetic mass splitting in the baryon octet (Coleman–Glashow, (38.66), p. 840)
- one baryon octuplet medium-strong mass splitting (GMO, (39.12), p. 849)
- two baryon decuplet medium-strong mass splittings (GMO, (39.13), p. 850)
- one previously unknown member of the baryon decuplet (the Ω^- from GMO, (39.14), p. 851)
- one pseudoscalar meson medium-strong mass splitting (GMO, (39.17), p. 852)
- one previously unknown vector meson singlet (the ϕ, to explain a deviation from GMO, (39.21), p. 854)
- two ratios of electromagnetic decay rates (for the vector bosons, (39.35), p. 857)

and they've all been borne out by experiment with good agreement. This concludes the numerical application and the actual experimental checks that we are going to make of SU(3), although there are many others. The literature is chock-a-block with them.

---

22 [Eds.] F. Nichitiu, "Introduction to the Vector Meson", Laboratori Nazionali de Frascati publication, *LNF-95/056* (1995); available from `www.iaea.org`; *PDG* 2016.

39.4 The naive quark model (and how it grew)

I'm now going to talk about some SU(3)-related ideas. I'd like to say a few words, although I'm not going to make any predictions or do any numerical work, about the famous **quark model.**[23] It's a real rags-to-riches story. The quark model started out as universally scorned, but it was gradually accepted by even the most snooty of us. All rags-to-riches tales establish a trajectory, sometimes with the hero ascending from poverty smoothly to success, as in Horatio Alger's novels, and sometimes his path is Dickensian, with reversals of fortune. And the quark model may fall from grace once more. I will describe the basic ideas of the naive quark model from the viewpoint of SU(3) symmetry, without talking about quark dynamics, which is a much more disputed subject. This is going to be more informal and semi-popular in structure than what has come before. I'll give you a sort of historical outline and make comments occasionally about how SU(3) ideas come in. After discussing the naive quark model I will make a few remarks about the gauge field theory of the strong interactions.

From the very first days of SU(3), and indeed even before the quark papers were written, people realized that there might be some physics in the group theoretical statement that all SU(3) representations could be built out of $\mathbf{3}$'s and $\overline{\mathbf{3}}$'s. The bosons, after all, came in octets and singlets and the fermions came in octets and decuplets, with some singlets at higher energy. This suggested a composite model. In particular, there was something extremely attractive in the formula

$$\begin{gathered}(1,0)\otimes(0,1)=(1,1)\oplus(0,0)\\ \mathbf{3}\otimes\overline{\mathbf{3}}=\mathbf{8}\oplus\mathbf{1}\end{gathered} \tag{39.36}$$

If there is a fundamental triplet and an anti-triplet, and if particles are bound states of these triplets, we get bosons that come in octets and singlets *only*. Some people suggested that maybe those triplets were *not* just mathematical figments of the imagination. Perhaps there really *are* particles that transform like SU(3) triplets: funny particles with charges like $\frac{2}{3}$ and $-\frac{1}{3}$ and fractional strangeness and so on.[24] And then the real mesons and baryons, the octets and decuplets that we see, are bound states of these triplets. There was a wave when people would investigate alternative triplet models. They said, "Well, if you give the triplet some sort of baryon number, then of course this octet and singlet are going to have baryon number 0, a particle and an anti-particle. It's fine for the mesons but it's no good for the baryons." Other people said, "Well, maybe you need two kinds of triplets, one that carries baryon number and one that doesn't. Then you'd make a baryon by binding together a baryon number-carrying triplet with a non-baryon number-carrying triplet." And some people said, "Maybe you need a particle around that's a fundamental singlet that carries baryon number and you'd make a baryon by putting together three of these things."

The big discovery, which is rather trivial but important at the time, was made simultaneously by Gell-Mann[25] and Zweig,[26] who said, "Suppose we put together three $(1,0)$'s. What do you

---

[23] [Eds.] F. Halzen and A. D. Martin, *Quarks and Leptons: An Introductory Course in Modern Particle Physics*, John Wiley, 1984; Griffiths *EP*; Harry J. Lipkin, "Quarks for Pedestrians", *Phys. Lett.* **C8** (*Physics Reports*) (1973) 173–268.

[24] [Eds.] Crease & Mann *SC*, Chapter 15, "The King and his Quarks", pp. 280–308.

[25] [Eds.] M. Gell-Mann, "A Schematic Model of Baryons and Mesons", *Phys. Lett.* **8**, 214-215 (1964). This is the first use of the term "quark" in the literature (see note 6 in the article).

[26] [Eds.] G. Zweig, "An SU(3) Model for Strong Interaction Symmetry and its Breaking", CERN Report 8182/Th 401 (1964), unpublished. Version 2 reprinted in *Developments in the Quark Theory of Hadrons*, eds. D. Lichtenberg and S. Rosen, Hadronic Press, Nonantum, MA, 1981, Vol. 1, pp. 22–101; also available as

get?" In an equation, the question is

$$(1,0) \otimes (1,0) \otimes (1,0) = \mathbf{3} \otimes \mathbf{3} \otimes \mathbf{3} = ? \tag{39.37}$$

Let's work out what this is with our algorithm (§37.4). The first step (37.45) gives

$$(1,0) \otimes (1,0) = (1,1;0,0) \tag{39.38}$$

and that is the only term we obtain; there is no possibility of contracting an inner or outer pair of indices. In the second step (37.50), we get

$$(1,1;0,0) = (2,0) \oplus (0,1) \tag{39.39}$$

or in terms of their dimensions,

$$\mathbf{3} \otimes \mathbf{3} = \overline{\mathbf{3}} \oplus \mathbf{6} \tag{39.40}$$

We have to carry out the last product:

$$\big[(1,0) \otimes (1,0)\big] \otimes (1,0) = [(2,0) \oplus (0,1)] \otimes (1,0) = \big[(2,0) \otimes (1,0)\big] \oplus \big[(1,0) \otimes (0,1)\big] \tag{39.41}$$

We already know (37.4) that $(0,1) \otimes (1,0)$ is $(1,1) \oplus (0,0)$. We have to compute the first product:

$$(2,0) \otimes (1,0) = (2,1;0,0) \tag{39.42}$$

Again there is no possibility of peeling indices off the inner or outer pairs. The second step gives

$$(2,1;0,0) = (3,0) \oplus (1,1) \tag{39.43}$$

so that finally

$$(1,0) \otimes (1,0) \otimes (1,0) \sim (3,0) \oplus (1,1) \oplus (1,1) \oplus (0,0) \tag{39.44}$$

In common language,

$$\mathbf{3} \otimes \mathbf{3} \otimes \mathbf{3} = \mathbf{10} \oplus \mathbf{8} \oplus \mathbf{8} \oplus \mathbf{1} \tag{39.45}$$

Now things get very intriguing if we take this idea seriously—if we say that there *are* fundamental objects called *aces* by Zweig and *quarks* by Gell-Mann.[27] Gell-Mann's name caught on and Zweig left the field.[28] Hadrons are made up of bound states of quarks, and if we say quarks have baryon number $\frac{1}{3}$ (why not, if they have charge $\frac{1}{3}$ and hypercharge $\frac{1}{3}$), then we're in pretty good shape. Things begin to look a little less artificial: mesons are supposed to be quark–antiquark bound states, $\mathbf{3} \otimes \overline{\mathbf{3}}$, octets and singlets. And sure enough, all the mesons that have ever been discovered are octets and singlets. We've also got three quark bound states; they're singlets and octets and decuplets. Sure enough, all those objects have baryon number 1; their antiparticles are three bound states with baryon number -1; those also go into singlets and octets and decuplets. No 27-plets or 35-plets or other exotic objects around,

---

CERN-TH-412 from the CERN document server, `cds.cern.ch`.

[27] [Eds.] The term "quark" comes from James Joyce's infamously difficult *Finnegans Wake*, a novel unlikely to have been read by many other physicists besides Gell-Mann when he appropriated the term. See the reference in note 25, p.858.

[28] [Eds.] According to Crease & Mann *SC*, p. 285, both men submitted their epochal articles to the same journal (CERN's *Physics Letters*), but Zweig's was rejected: "It was all right for someone of Gell-Mann's stature to advocate the lunatic notion that most of matter was made up of ineffable entities that were invisible to experiment; having no reputation to protect him, Zweig was denied an appointment at a major university because the head of the department thought he was a 'charlatan'."

and no $\overline{\mathbf{10}}$'s with baryon number 1, although there are $\overline{\mathbf{10}}$'s with baryon number -1, their antiparticles.

So it looks as if there's a glimmer of truth, maybe, in this idea. It seems to agree with the phenomenology of the observed hadrons. On the other hand it also looks sort of silly. We say a nucleus in made up of neutrons and protons. The way we establish that is not by studying nuclear structure, but by bashing a nucleus and watching neutrons and protons fly out. At that time, and indeed ever since, no one has ever observed a free quark. No matter how hard we hit a hadron, more hadrons fly out (mostly pions), but not any quarks. The reason for that is perhaps the quarks are very heavy. Since the hadrons are not very heavy, the quarks must be very tightly bound; then it's an extreme relativistic system. The binding energy is comparable to the $E = mc^2$ energy of the quarks. Therefore we would expect that there would be a large probability for virtual quark–antiquark pairs. Why should it be three quarks any more than five quarks, or seven, or 10? If we have all that energy around, why should it look like a simple non-relativistic system? We'd expect the wave function to have a large amplitude for having all sorts of pairs in it.

Also, what kind of force is it that can bind together a quark and an antiquark, or three quarks, but doesn't bind together *two* quarks? Why can't we have a two-quark bound state? That would be a sextuplet plus an anti-triplet, $\mathbf{6} \oplus \overline{\mathbf{3}}$, particles with fractional charge. Those multiplets would be every bit as easy to see as a quark itself. Why aren't they around?

So there was all this argument back and forth; nothing ever got anywhere. There were those who believed in the quark model and those who didn't. Those who believed in it used essentially non-relativistic reasoning to work things out because they didn't care what the kinematics were, it looked non-relativistic. And the results were sort of good and sort of bad, and it was a big mess.[29]

The people pursuing the naive quark model went even further, saying, "If we can treat this as a non-relativistic system, we can use good old non-relativistic quantum mechanics, including the ideas of spin-independent forces, even though we know that's ludicrous from the viewpoint of relativistic quantum mechanics." "We'll just be very bold," they said, "and we'll use very naïve reasoning, including the picture of things in an attractive potential with spin-independent forces, to figure out not just the SU(3) assignments for these things, but their spins."

We'll describe this work, beginning with the $q\bar{q}$ system, the mesons. That's sort of easy because there's no Pauli antisymmetrization to worry about. We'll assume the quarks are SU(3) triplets with baryon number $\frac{1}{3}$ and spin $\frac{1}{2}$. That's the simplest assumption if we're going to wind up with baryon states with half-integral angular momentum. The most tightly bound state will be an s-wave, as always for a central potential. Of course, as previously established

---

[29] [Eds.] Coleman adds, "It still is a big mess in many respects. We're starting to get an inkling of how to incorporate these ideas into a relativistic quantum field theory. That we must save for your post-Physics 253 studies. I'm just giving you an overview at this moment." He was speaking in 1976, already three years after the pioneering work of David Politzer, David Gross, and Frank Wilczek established that quantum chromodynamics is asymptotically free: H. D. Politzer, "Reliable Perturbative Results for Strong Interactions?", *Phys. Rev. Lett.* **30** (1973) 1346–1349; D. J. Gross, and F. Wilczek, "Ultraviolet Behavior of Non-Abelian Gauge Theories", *Phys. Rev. Lett.* **30** (1973) 1343–1346. The three men shared the 2004 Nobel Prize for this work; Politzer cited "Sidney Coleman, my beloved teacher from graduate school", in his acceptance speech: H. David Politzer, "The Dilemma of Attribution", on-line at `https://www.nobelprize.org/nobel_prizes/physics/laureates/2004/politzer-lecture.pdf`; the citation is the first sentence of the second paragraph.

by SU(3) analysis, the product of a quark and an antiquark has to produce $8 \oplus 1$. These mesons have to have parity -1, because they are made from fermions and antifermions.[30] The spins can be aligned to make spin 1 or spin 0: $J^P = 1^-$ or 0^-. That is, we should obtain vector bosons and pseudoscalar mesons:

$$q\overline{q} = \mathbf{3} \otimes \overline{\mathbf{3}} \sim \mathbf{8} \oplus \mathbf{1}\colon J^P = 0^-,\, 1^- \qquad (39.46)$$

Of course the 0^- and 1^- multiplets don't both have the same mass, but that's presumably due to the spin-orbit interaction. On a qualitative level it looks good; it looks like we've explained not only why the lightest bosons are octets and singlets, but also why they're pseudoscalars and vectors, not scalars and axial vectors, for example.[31] Of course you would expect scalars and axial vectors to come up eventually; they would be p-states or d-states in this imagined potential. But the lightest ones should be pseudoscalar and vector, and indeed they are. It looks good. It also looks crazy, but by God it's organizing the data correctly! So we go on. We don't ask critical questions: how can this happen or that happen? Because I don't know the answer; I'm just trying to explain things.

39.5 What can you build out of three quarks?

Let's go on to the system of three bound quarks, qqq. Here again we expect some sort of interactions, so the lightest bound state we would expect with all three quarks as close to each other as can be with no centrifugal barrier, any pair in an s-wave, the space part of the wave function being totally symmetric. This means the $\text{SU}(2)_{\text{spin}} \otimes \text{SU}(3)$ part must be *anti*symmetric, because the quarks are supposed to be fermions. But this doesn't work. Therefore the quark modelers, every last one of them utterly mad, said: "Well, let's try the alternative hypothesis: treat the quarks as if they were *bosons*." In fact they aren't bosons, if they exist at all. A little later on we'll see how that difficulty is actually resolved, with a much better solution than bosonic quarks.

Let us work out what happens if one assumes that the $\text{SU}(2)_{\text{spin}} \otimes \text{SU}(3)$ part of the qqq wave function is totally *symmetric*. (That's a nice little group theory exercise, whether or not you believe this nonsense about bosonic quarks.) Given a three-particle system of fermions, we want to work out how we put spin and SU(3) wave functions together to make something that is symmetric under permutations of the q's. (For the moment, we are *ignoring* the Pauli principle. We'll come back to it.) This requires a small group theoretical digression on the permutation group on three objects, called S_3 by its friends.[32]

A brief digression on the group S_3

S_3 is a finite group with $3! = 6$ elements, because there are $3!$ permutations on three

[30] [Eds.] The quark q and the antiquark $\overline{q}$ have opposite parities, and the parity of the meson is the product of these, which has to be -1: $P_{\text{meson}} = P_q P_{\overline{q}} = -1$. See note 2, p. 460.

[31] [Eds.] If the quarks had spin 0, the lightest particles would have $J^P = 0^+$ and the next lightest would be $J^P = 1^-$, i.e., scalars and axial vectors.

[32] [Eds.] J. Mathews and R. Walker, *Mathematical Methods of Physics*, 2nd ed., Addison-Wesley, 1969, pp. 425–433.

objects:

$$S_3 = \left\{ \begin{pmatrix} 1 & 2 & 3 \\ 1 & 2 & 3 \end{pmatrix}, \begin{pmatrix} 1 & 2 & 3 \\ 1 & 3 & 2 \end{pmatrix}, \begin{pmatrix} 1 & 2 & 3 \\ 2 & 1 & 3 \end{pmatrix}, \begin{pmatrix} 1 & 2 & 3 \\ 3 & 2 & 1 \end{pmatrix}, \begin{pmatrix} 1 & 2 & 3 \\ 2 & 3 & 1 \end{pmatrix}, \begin{pmatrix} 1 & 2 & 3 \\ 3 & 1 & 2 \end{pmatrix} \right\} \tag{39.47}$$

The top row is the initial arrangement of the three objects, and the bottom is the final arrangement. The first group element is the identity (that is, $1 \to 1$, $2 \to 2$, and $3 \to 3$), which we can write as I. The second element leaves 1 alone and swaps 2 with 3: it is more conveniently written as (2 3). Similarly the other elements are written, in order, as (1 2), (1 3), (1 2 3) (that is, $1 \to 2$, $2 \to 3$, and $3 \to 1$) and finally (1 3 2):

$$S_3 = \{I, (2\ 3), (1\ 2), (1\ 3), (1\ 2\ 3), (1\ 3\ 2)\} \tag{39.48}$$

The permutation (2 3) is called *odd*, because the number of steps (moving a bottom number to the left or right by one position) needed to bring 132 back to the arrangement 123 is *one*, an odd number. The second, third and fourth elements are all odd; the first, fifth and last, requiring zero and two steps, respectively, are *even*. If we only have two objects and we were flipping them around, then we would know there would be two irreducible representations of the group, either *symmetric* (the trivial representation), or *antisymmetric*. With three objects we can get only one other irreducible representation that is called the *mixed representation*,[33] and that has dimension 2. I summarize these in Table 39.5. Note that the number of the elements of the IR is given by the square of its dimension. The action of the representation (s) is to multiply a group element by 1. The representation (a) multiplies the group element by $+1$ if the permutation is even, and by -1 if odd. And then there is a two dimensional "mixed" representation, (m). An elementary result in the theory of finite groups states[34]

$$\sum_r d_r^2 = N(G) \tag{39.49}$$

where d_r is the number of rows or columns in the rth irreducible representation's (square) matrix, $N(G)$ is the order of the group G (how many elements it has), and the sum is over all the irreducible representations of G. It's easy to check this equation for S_3:

$$1^2 + 1^2 + 2^2 = 6 = 3!$$

I can't take the time to work out the theory of finite groups, but I will try to show at least that there is a two-dimensional irreducible representation of S_3.

Let us consider a three-dimensional vector space with axes labeled x, y and z, Figure 39.5. This space forms a representation in particular of the permutation group on three objects, just permuting x, y and z. There is obviously an invariant one-dimensional subspace, not under

[33] [Eds.] For finite groups, there are only as many irreducible representations as conjugacy classes (two elements a, b are conjugate under a similarity transformation; $a \to gag^{-1} = b$). There are three such for S_3: under conjugation the identity turns into the identity; a two-cycle turns into a two-cycle, and a three-cycle turns into a three-cycle. So there are only three irreducible representations: (s) (every element represented by 1), (a) (even elements represented by 1, and odd elements by -1), and (m), (each element is represented by a 2×2 matrix).

[34] [Eds.] K. Riley and M. Hobson, *Mathematical Methods for Physics and Engineering*, 2nd ed., Cambridge U. P., Section 25.7.1; Zee *GTN*, pp. 104–108.

Name	*Type*	*Dimension*
(s)	symmetric	1
(a)	antisymmetric	1
(m)	mixed	2

Table 39.5: The type and dimensions of the IR's of S_3

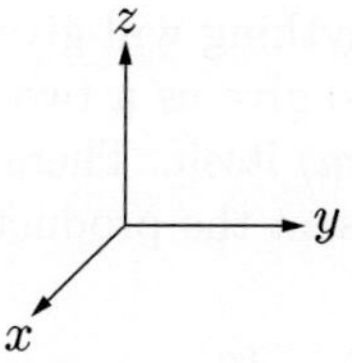

Figure 39.5: Axes for representations of S_3

the full group of rotations but under the group of permutations, which is

$$\hat{\mathbf{e}} = \frac{1}{\sqrt{3}}(\hat{\mathbf{x}} + \hat{\mathbf{y}} + \hat{\mathbf{z}}) \tag{39.50}$$

That's a one-dimensional invariant subspace of the three-dimensional space. It forms a basis for the trivial symmetric representation of the group.

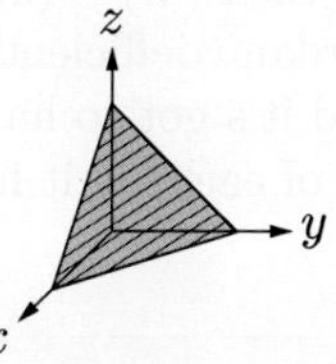

Figure 39.6: Invariant two-dimensional subspace for the mixed representations of S_3

Now let us consider all vectors orthogonal to this. That's a two-dimensional subspace, the plane passing through the origin. It's hard to draw, so I'll sketch Figure 39.6 instead; this is a plane parallel to the one we mean and that intersects the x, y and z axes at 1; it is displaced from the one we want. This two-dimensional subspace cannot be split into two invariant one-dimensional subspaces; we can't reduce things any further. How could we have a vector in this plane that was invariant under permutations? To be invariant under permutations it has to point just as much in the x-direction and the y-direction as in the z-direction. There's no way to do that here: it's like a Mandelstam plot[35]—in what direction will such a vector point? The two-dimensional vectors on this subspace form an irreducible representation of the group: we can't break it down further into two one-dimensional subspaces. That is the mixed representation.

To build states out of three quarks, we have to consider products of the representations of SU(3) and the representations of $\text{SU}(2)_{\text{spin}}$. Because we are proceeding under the foolish assumption that the wave functions are symmetric, we are really only interested in the

[35] [Eds.] See Figure 11.5, p. 233.

symmetric products. We can work out what these representations give when multiplied together by considering the products of representations of S_3. For example,

$$\begin{aligned}(s)\otimes(s) &\sim (s)\\ (s)\otimes(a) &\sim (a)\\ (a)\otimes(a) &\sim (s)\\ (a)\otimes(m) &\sim (m)\end{aligned} \tag{39.51}$$

and so on. We know that (s) times anything will give us that same anything, and (a) times (a) is (s). We get that $(a)\otimes(m)$ has to give us a two-dimensional irreducible representation, and there's only one around, namely (m) itself. There are six independent products (because the $\otimes$ is symmetric), and we can represent the products neatly in a product table, Table 39.6.

$\otimes$	(s)	(a)	(m)
(s)	(s)	(a)	(m)
(a)	(a)	(s)	(m)
(m)	(m)	(m)	$(s)\oplus(a)\oplus(m)$

Table 39.6: The $\otimes$ table of the IR's of S_3

The last product, $(m)\otimes(m)$, you have to take on trust. But I can make it plausible. You can see that the dimensions check out: $2\times 2 = 1+1+2$. We can work out $(m)\otimes(m)$ from the fact that if $(a)\otimes(m)$ contains (m) then $(m)\otimes(m)$ must contain (a). (This is the famous permutation symmetry of Clebsch–Gordan coefficients.[36]) Similarly, $(m)\otimes(m)$ must contain (s) because $(s)\otimes(m)$ contains (m); and it's got to have something left over; it can't be an (s) and it can't be an (a), there's only one of each, so it has to be (m). That's all the background we need on S_3.

If we are going to make baryons out of three quarks, we've got to put together their three spins and their three SU(3) quantities. (I remind you that we are proceeding under the ridiculous assumption that the $\text{SU}(2)_{\text{spin}}\otimes\text{SU}(3)$ part of the qqq wave function is *symmetric.*) I'll start with the spins.[37]

$$\left(\tfrac{1}{2}\right)\otimes\left(\tfrac{1}{2}\right)\otimes\left(\tfrac{1}{2}\right) = [(0)\oplus(1)]\otimes\left(\tfrac{1}{2}\right) = \underbrace{\left(\tfrac{1}{2}\right)\oplus\left(\tfrac{1}{2}\right)}_{(m)}\oplus\underbrace{\left(\tfrac{3}{2}\right)}_{(s)} \tag{39.52}$$

The state $\left(\frac{3}{2}\right)$ where all the spins are aligned is symmetric under the permutation group. On the other hand, the states $\left(\frac{1}{2}\right)$ and $\left(\frac{1}{2}\right)$ get mixed up with each other when we permute the three spins, so together they form a mixed representation of the permutation group, the sum of two two-dimensional representations of the rotation group. This direct sum forms a single irreducible representation of the direct product of the rotation group and the permutation group, $\text{SU}(2)_{\text{spin}}\otimes S_3$.

[36] [Eds.] A. Messiah, *Quantum Mechanics*, Volume 2, North Holland Publishing, 1962, Appendix C; reprinted (in a combined edition of both volumes together) by Dover Publications, 2017.

[37] [Eds.] L. Schiff, *Quantum Mechanics*, McGraw-Hill, New York, 1968, pp. 218–225.

Now for the SU(3) contents of the products of the three quarks:

$$\mathbf{3} \otimes \mathbf{3} \otimes \mathbf{3} = \mathbf{1} \oplus \mathbf{8} \oplus \mathbf{8} \oplus \mathbf{10} \tag{39.45}$$

What are their symmetries under S_3? If we think of each of the **3**'s as a vector with one index ($\mathbf{3} = (1,0)$) then the three-index tensor that transforms like a **10** is the one that is totally symmetric under permutations of the three indices. That's our irreducible representation $(3,0)$, totally symmetric under permutation of the three upper indices.[38] So the **10** is guaranteed to be symmetric, by construction. What is the singlet? There is an object that is SU(3)-invariant that has three indices, ϵ_{ijk}. Is there an object totally antisymmetric under permutation of the three indices? Yes, the same ϵ_{ijk}. So the singlet $(\epsilon_{ijk}q^i q^j q^k)$ is antisymmetric. The 8's get flipped among themselves. They're mixed; you can take it on trust or construct the wave functions.

$$\mathbf{3} \otimes \mathbf{3} \otimes \mathbf{3} = \underbrace{\mathbf{1}}_{(a)} \oplus \underbrace{\mathbf{8} \oplus \mathbf{8}}_{(m)} \oplus \underbrace{\mathbf{10}}_{(s)} \tag{39.53}$$

Let's see what we can put together from (39.52) and (39.53) to make a totally symmetric wave function. There's nothing in (39.52) that we can combine with the 1 (the (a) representation in (39.53)) to make a symmetric wave function: $(a)_{\text{SU(3)}}$ with either $(s)_{\text{spin}}$ or $(m)_{\text{spin}}$ makes an antisymmetric wave function according to our times table, and we're imagining that only the symmetric ones work. The antisymmetric singlet in (39.53) is ruled out. There are no singlets. And that's right! There are no low-lying baryon singlets.

We can put together a mixed object with a mixed object (i.e., two elements from the (m) representation) with appropriate Clebsch–Gordan coefficients to make a symmetric object. Therefore we can have an octet with $J^P = \frac{1}{2}^+$:

$$\mathbf{8} = (1,1)\colon\ J^P = \left(\tfrac{1}{2}\right)^+ \tag{39.54}$$

It's assumed to be + because it's three identical particles in a totally symmetric spatial wave function. That's right! We have such an octet, our old friends $\{N, \Sigma, \Xi, \Lambda\}$ in Table 36.1, p. 782. And we can put together the symmetric spin states (39.52), $J = \frac{3}{2}$, with the symmetric SU(3) states (39.52) in 10, to make a decuplet with $J^P = \frac{3}{2}^+$:

$$\mathbf{10} = (3,0) : J^P = \left(\tfrac{3}{2}\right)^+ \tag{39.55}$$

That fits experiment also! It's Gell-Mann's decuplet (Table 39.2, p. 851). You can't make a decuplet with $J^P = \left(\frac{1}{2}\right)^+$ nor an octet with $J^P = \left(\frac{3}{2}\right)^+$, if the wave functions are to be symmetric. All you can make is a decuplet with $J^P = \left(\frac{3}{2}\right)^+$ and an octet with $J^P = \left(\frac{1}{2}\right)^+$. Isn't that peculiar? This is why, as I said a little earlier, the $\text{SU(2)}_{\text{spin}} \otimes \text{SU(3)}$ part can't be antisymmetric: it simply doesn't fit the observed spectrum. (Of course we could make them in this model in excited states where the spatial wave function is not totally symmetric; one of the two quark pairs is in a p-wave or something. We'd expect that to have higher energy.) But good heavens, this is wonderful and nutty at the same time. It's wonderful because when we put together wave functions for states describing three fermionic quarks, we get just the right spectrum. And it's nutty because these three-quark wave functions have to be totally

---

[38] [Eds.] Recall that our tensors are totally symmetric within both the upper indices and the lower indices. See (36.57).

symmetric, in violation of the Pauli principle, if (as we believe) the quarks are fermions. That's impossible with *only* the quark characteristics we've considered thus far, at least for the lowest energy, s-wave states. (And if these are not the lowest energy states, then where are they? Thus far none lower have been observed.) Somehow a way has to be found around the Pauli principle and the notion of fermionic quarks to allow us these symmetric wave functions.[39]

Now this is an exciting moment so I will tell you the answer we think we have. Here "we" means *everyone*; it's a universally accepted idea. It began as a mad speculation and became established dogma without ever passing through the test of experiment. The idea is a new quantum number called **color**. We use it to explain two curious facts—not the fact that a non-relativistic model works well, but two other facts: Why should the three quark wave function be totally symmetric in $\mathrm{SU}(3) \otimes \mathrm{SU}(2)_{\mathrm{spin}}$; and why should there be quark–antiquark and three quark bound states, but nothing else, in particular, no quark–quark bound states? And maybe it could explain a third curious fact: if non-relativistic reasoning is good, then why can't we knock quarks out of the hadrons? Why do we never see a free quark? The idea is this. We've going to increase the number of quarks by giving them this new quantum number, *color*. Formerly there were three quarks, now there will be *nine*: each of $\{u, d, s\}$ will come in three colors. We'll see that more quarks can solve a lot of problems. I will sketch out what color is, and some of its consequences.[40]

39.6 A sketch of quantum chromodynamics

The actual symmetry group G_S of the strong interactions is going to be $\mathrm{SU}(3) \otimes \mathrm{SU}(3)$. The quarks are going to be triplets under both of these groups: $q_{i\alpha}$. The $i = \{1, 2, 3\}$ is for the first SU(3) that we know and love, and $\alpha = \{1, 2, 3\}$ is for the mysterious second SU(3). The first SU(3), the up, down and strange, is called **flavor** (a joke of Nambu's)[41]; this is our old friend, the quark *type*: $\{u, d, s\}$. The second SU(3) goes with the new quantum number, color: *red*, *green*, and *blue*.[42]

$$G_S = \mathrm{SU}(3)_{\mathrm{flavor}} \otimes \mathrm{SU}(3)_{\mathrm{color}} \tag{39.56}$$

Now, particles with color have never been seen. That's led to a hypothesis that there are very, very strong interactions between colored particles so that *the only physically observable states are color singlets.* That is, the observable baryons and mesons are "colorless", and transform trivially, as scalars under $\mathrm{SU}(3)_{\mathrm{color}}$. This is sometimes called **color confinement**. We've learned how to do this, though there is some dispute.

---

[39] [Eds.] The construction of the properly antisymmetrized quark wave functions, including color, is intricate and nontrivial for baryons: Griffiths *EP*, Section 5.6.1, pp. 181–188; see also Problem 21.3, p. 871.

[40] [Eds.] The idea of a new quantum number for quarks is due independently to Greenberg (who called it "parastatistics"), and Han and Nambu: O. W. Greenberg, "Spin and Unitary-Spin Independence in a Paraquark Model of Baryons and Mesons", *Phys. Rev. Lett.* **13** (1964) 598–602; M. Y. Han and Y. Nambu, "Three-Triplet Model with Double SU(3) Symmetry", *Phys. Rev.* **139** (1965) B1006–1010. The new quantum number was given the name "color" by Gell-Mann; M. Gell-Mann, "Quarks", *Acta Phys. Austriaca Suppl.* **9** (1972) 733–761. The term is introduced on p. 736.

[41] [Eds.] Yōichirō Nambu (1921-2015) was a Japanese-American physicist at the University of Chicago for sixty years. He had been a student of Tomonaga's in Tokyo during and immediately after World War II. He made many groundbreaking contributions to condensed matter and particle theory, work recognized by the Physics Nobel Prize in 2008 for his research into spontaneous symmetry breaking. Bruno Zumino famously joked that he once had the idea to talk to Nambu to get ten years ahead of the crowd, "but by the time I figured out what he said, ten years had passed." M. Mukerjee, "Profile: Yoichiro Nambu", *Sci. Amer.* Feb. 1995, pp. 37–39.

[42] [Eds.] Coleman used red, white and blue in this lecture. Perhaps to avoid nationalist controversies, the community switched to the primary colors of light not long after QCD became well-known.

We come to a second idea, based on color, bearing the elegant name of **quantum chromodynamics**, or QCD for short.[43] It's another wonderful name from Caltech. There is always something new out of Pasadena, to paraphrase Pliny,[44] and always a joke. The concept of a quantum field theory arising from color came from Nambu and Gell-Mann originally, but many others, notably Ken Wilson, Lenny Susskind, David Politzer, Frank Wilczek, and David Gross, made significant contributions to the theory early on.[45]

The general idea is to treat the force associated with color SU(3) just as we treated electromagnetism. In QED, electromagnetic force is mediated between charged particles by the exchange of photons coupled to charge. In much the same way, in QCD a force will be mediated between colored particles—in particular, between quarks—by the exchange of "color photons" coupled to color. As we will see (Chapter 47), there is a procedure for obtaining the field theory corresponding to a given symmetry group by making that group a **local symmetry**. The local symmetry group is called a **gauge group**; $SU(3)_{\text{color}}$ is the unbroken gauge group of QCD.[46] Each generator of the gauge group is associated with one massless

---

43 [Eds.] Peskin & Schroeder *QFT*; P. H. Frampton, *Gauge Field Theories*, 3rd enlarged and improved ed., Wiley-VCH, 2008; Greiner *et. al QCD*; Chris Quigg, "Gauge Theories of the Strong, Weak and Electromagnetic Interactions", Benjamin-Cummings, 1983, Chapter 8, pp. 193–268. The Greek word for "color" is $\chi\rho\tilde{\omega}\mu\alpha$, "chrōma". According to David Gross, the first appearance of the term "quantum chromodynamics" in the literature was in the review article by W. Marciano and H. Pagels, "Quantum Chromodynamics", *Phys. Reps.* **36C** (1978) 137–276; the authors ascribe the name to Murray Gell-Mann: David Gross, "Asymptotic Freedom and the Emergence of QCD", in *The Rise of the Standard Model: Particle Physics in the 1960s and 1970s*, eds. Lillian Hoddeson, Laurie M. Brown, Michael Riordan, and Max Dresden, Cambridge U. P., 1997, pp. 199–233. Note however that Susskind's 1976 Les Houches lectures (see note 45, p. 867) antedate the Marciano and Pagels review article by two years. The Oxford English Dictionary credits Dietrick E. Thomsen, "Chromodynamics", *Science News* **109**, June 26, 1976, 408–409 with the first print citation.

44 [Eds.] Pliny's text reads *unde etiam vulgare Graciae dictum semper aliquid novi Africam adferre* (whence the proverb, even common in Greece, that "Africa is always bringing forth something new"), Gaius Plinius Secundus (CE 23–79), *Historia Naturalis*, book 8, section 42; *Natural History*, v. III: Books 8–11, ed. H. Rackham, (dual language edition) Loeb Classical Library, Harvard U. P., 1940. Often quoted from the *Adagia* (1500) of Erasmus of Rotterdam, *Ex Africa semper aliquid novi* (Out of Africa, always something new). Naturalist, scholar and statesman, Pliny the Elder died while successfully rescuing a friend and her family during the Vesuvius eruption that destroyed Pompeii and Herculaneum, according to his nephew Pliny the Younger, who witnessed the events. For a history of the phrase, see Harvey M. Feinberg and Joseph B. Sodolow, "Out of Africa", *Jour. Afric. Hist.* **43** (2002) 255–261.

45 [Eds.] Two useful references are Crease & Mann *SC*, pp. 296–299 and pp. 327–336, and Close *IP*, pp. 257–279. Historically, the development of $SU(3)_{\text{color}}$ seems to have had three separate phases. If quarks were physical entities and fermionic, some way had to be found to deal with the Pauli principle, which seemed at odds with parts of the baryon spectrum: Greenberg, *op. cit.*; Han and Nambu, *op. cit.* Then, once color's value had been established (e.g., multiplying some theoretical expressions by factors of three, vastly improving agreement with experiment), it was quickly realized that these color charges could serve as a source of a new interaction, which might in turn bind the quarks together: Y. Nambu, "A Systematics of Hadrons in Subnuclear Physics", in *Preludes in Theoretical Physics in Honor of V. F. Weisskopf*, eds. A. de-Shalit, H. Feshbach, and L. Van Hove, North-Holland, 1966, p. 133–142; M. Gell-Mann, "Quarks", *Acta Phys. Austriaca Suppl.* **9** (1972) 733–761; H. Fritzsch and M. Gell-Mann, *Proc. XVI Intern. Conf. on High Energy Physics*, Chicago, 1972, v. 2, p. 135–165, available as `hep-ph/0208010v1` at `https://arXiv.org`; W. A. Bardeen, H. Fritzsch, and M. Gell-Mann, in *Scale and Conformal Symmetry in Hadron Physics*, ed. R. Gatto, Wiley, 1973, p. 139–151; H. Fritzsch, M. Gell-Mann and H. Leutwyler, "Advantages of the Color Octet Gluon Picture", *Phys. Lett.* **47B** (1973) 365–368. Finally, the consequences of the Yang–Mills nature of QCD began to be investigated: the work of Politzer and Gross & Wilczek on asymptotic freedom cited earlier (note 29, p. 860) and H. D. Politzer , "Asymptotic Freedom: An Approach to Strong Interactions", *Phys. Lett.* **14C** (1974) 129–180; K. G. Wilson, "Confinement of Quarks", *Phys. Rev.* **D10** (1974) 2445–2459; L. Susskind, "Coarse Grained Quantum Chromodynamics", pp. 208–308 in *Weak and Electromagnetic Interactions at High Energies: Les Houches 1976*, R. Balian and C. H. Llewellyn Smith, eds., North-Holland Press, 1977. Quantum chromodynamics is now a firmly established theory, supported by an immense body of experimental and theoretical results.

46 [Eds.] See note 5, p. 646, and note 54, p. 869.

vector boson, called a **gauge boson**. That means we have eight massless vector bosons for the eight generators of $\mathrm{SU}(3)_{\mathrm{color}}$. But it's not like electromagnetism in that the force is *very* strong. A necessary consequence of this is that color non-singlet states will be much heavier than color singlet states, because color non-singlet states, having a non-zero color, will have non-zero values of these gauge fields at infinity. Therefore when we add up the energy that's stored in the gauge fields, outside the particle, we get a positive answer. On the other hand, if we have a color singlet, the gauge fields vanish at infinity; then at least we don't have that positive contribution. Whether you can make color non-singlet states *infinitely* heavier than color singlet states is an open question. It's much easier to answer the question in two dimensions, and there the answer is "yes".[47]

In QCD, the eight gauge bosons—the "color photons"—are *also* colored, because $\mathrm{SU}(3)_{\mathrm{color}}$ is a *non-Abelian* group. These gauge bosons have zero bare mass but they might acquire a mass through the color force; they are coupled to each other. Everything has a funny name here.[48] The gauge bosons are called **gluons** because they glue things together. The gluons transform like the octet representation of the group, like its generators. And just as we can't observe free quarks because they are not color singlets, we cannot observe gluons. We could of course observe bound states made up of gluons just like we could observer bound states make up of quarks; these are called **glueballs**. But at the moment there is no definite meson that has been identified as a glueball; quarks seem to be sufficient.[49]

Let's see how $\mathrm{SU}(3)_{\mathrm{color}}$ solves our problems. First, although we can make a color singlet out of a quark and an antiquark, $\mathbf{3}_{\mathrm{color}} \otimes \mathbf{3}_{\mathrm{color}}$ (37.17), we can't make a color singlet out of two quarks (39.40):

$$\mathbf{3}_{\mathrm{color}} \otimes \mathbf{3}_{\mathrm{color}} = \overline{\mathbf{3}}_{\mathrm{color}} \oplus \mathbf{6}_{\mathrm{color}}$$

so it explains why there are no two quark bound states. Indeed, no bound states of two quarks and an antiquark, or two antiquarks and a quark, by similar computations. We simply can't make a color singlet this way. But (39.45)

$$\mathbf{3}_{\mathrm{color}} \otimes \mathbf{3}_{\mathrm{color}} \otimes \mathbf{3}_{\mathrm{color}} = \mathbf{1}_{\mathrm{color}} \oplus \mathbf{8}_{\mathrm{color}} \oplus \mathbf{8}_{\mathrm{color}} \oplus \mathbf{10}_{\mathrm{color}}$$

So we *can* put together *three* quarks to make a singlet, and it is antisymmetric in color (39.53). The generalized Pauli principle is that the wave function has to be antisymmetric in the product of *all* the degrees of freedom:[50]

$$(\text{space variables}) \otimes \mathrm{SU}(2)_{\mathrm{spin}} \otimes \mathrm{SU}(3)_{\mathrm{flavor}} \otimes \mathrm{SU}(3)_{\mathrm{color}}$$

By our rule that it has to be a color singlet, it is *forced* to be antisymmetric in color: the **1** is antisymmetric. By the ideas of the naive quark model, the space part of the wave function is symmetric, and therefore $\mathrm{SU}(2)_{\mathrm{spin}} \otimes \mathrm{SU}(3)_{\mathrm{flavor}}$ is symmetric also. So the whole thing will be antisymmetric, and there's *no* violation of the Pauli principle, as there would be without the

[47] [Eds.] "Charge Shielding and Quark Confinement in the Massive Schwinger Model", Sidney Coleman, R. Jackiw, and Leonard Susskind, *Ann. Phys.* **93** (1975) 267–275.

[48] [Eds.] Coleman adds, "I love everything about this subject except its nomenclature, which makes me feel like Bozo the Clown whenever I deliver a lecture on it."

[49] [Eds.] Coleman adds, "This is a guided tour of the land of mists and fogs. The castle in the distance may turn out to be a mere *fata morgana*. We should have at least one lecture like that in this course, off the high road and into the swamps..." That was a fair description in 1976. Forty years later, the castle of QCD is no longer distant. It is no mirage, but a stately fortress, beautifully built from the finest granite and marble.

[50] [Eds.] See note 39, p. 866.

new quantum number, color. This new quantum number answers two questions, but as yet it has not really answered the third: why can't we knock quarks out of hadrons? People believe that quark confinement can be shown from QCD (it has been shown[51] in lattice QCD), but it's an open question.[52]

Here are some facts about quantum chromodynamics, stated without proof.[53]

- Color is a gauge symmetry. There is an octet of massless vector bosons called gluons, each a Yang–Mills field associated with color.
- Color is an unbroken symmetry, because if a gauge symmetry is broken all hell breaks loose.[54]
- Color singlets are favored, for the same reason that electromagnetic charge neutrality is favored: we get electric fields if positive and negative electric charges are separated.
- q and $\overline{q}$ are connected by a flux tube. The quark–antiquark potential grows linearly with distance. (Any non-Abelian gauge theory of vector fields has this property.)

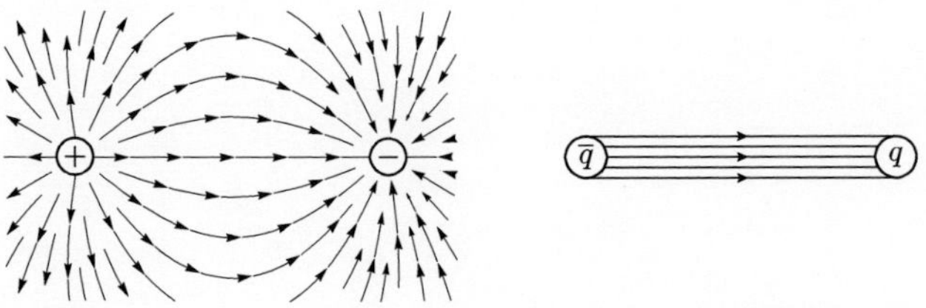

Figure 39.7: Electric dipole and quark–antiquark flux tube

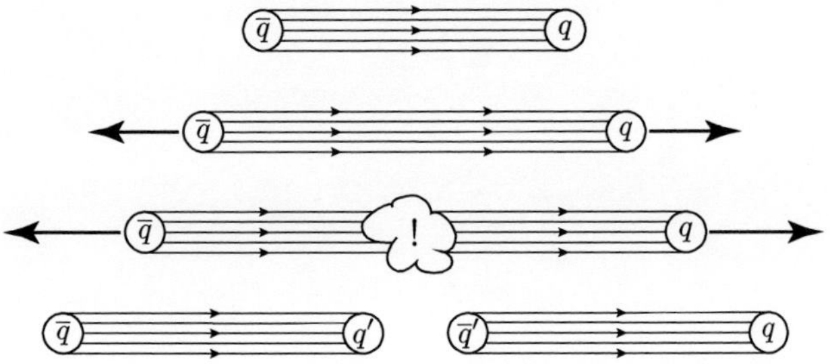

Figure 39.8: Pulling a quark and antiquark apart makes another meson

Let's contrast an electric dipole with a quark–antiquark system, as in Figure 39.7. Look at a π meson, made of a $\overline{q}q$ pair. We go to the hardware store and buy a couple of quark hooks to pull the quark and the antiquark apart. This puts in energy just like separating charges puts energy into the electric field. Eventually there is enough energy stored between the quark and

51 [Eds.] Wilson, *op. cit.*

52 [Eds.] The Clay Mathematics Institute has offered a substantial cash prize for the demonstration of a closely related result: `http://www.claymath.org/millennium-problems/yangвЃYmills-and-mass-gap`.

53 [Eds.] In 1990, Coleman told the class: "Believe them because Sidney is glamorous, not because his arguments are convincing."

54 [Eds.] Gauge symmetry *does* break, and then provides a renormalizable mechanism (due to Higgs, Englert and Brout, and others) for the generation of gauge boson masses, as in the electro-weak theory of Glashow, Salam and Weinberg. But thus far the gluons appear not to have a mass generated by this symmetry breaking. The mechanism will be described in Chapter 46.

the antiquark to make a new quark–antiquark pair $\overline{q}'q'$ between the original pair; we wind up with *two* mesons, $\overline{q}q'$ and $\overline{q}'q$. It's like breaking a string: if we want a one-ended piece of rope, we could tie one end of a two-ended piece of rope to something that won't move, and pull the other end until it snaps, as in Figure 39.8. But we do not get a one-ended piece of rope; we get two ropes each with two ends. It doesn't work.

Next time we will begin the investigation of *current algebra.*

Problems 21

21.1 (The first part of this problem is solved in old lecture notes[1] handed out in class, but you might have fun working it out yourself.)

(a) At first glance, it looks as if there are *two* possible SU(3)-invariant quartic self-couplings of the pseudoscalar octet, $\text{Tr}(\phi^4)$ and $(\text{Tr}(\phi^2))^2$. Show that these are in fact proportional to each other, and find the constant of proportionality. (The most straightforward way to do this is to write ϕ in diagonal form. Don't forget to use the fact that $\text{Tr}(\phi) = 0$.)

(b) A true story: Some years ago the theory group at Saclay was investigating bound-state approximations in quantum field theory, and decided to use as a theoretical laboratory the theory of a pseudoscalar meson octet with the most general renormalizable SU(3)-invariant parity-invariant quartic self-couplings. (There were no baryons or other fields in the theory.) Of course, they found that the lightest two-meson bound states were s-waves and, since their theory was SU(3)-invariant and since an s-wave is symmetric is space, these were singlets, octets and 27-plets. However, to their surprise, the octets and 27-plets were degenerate in mass. Can you explain why?

Possibly useful information: (1) In the lectures I explained how our method of finding representations of SU(n) became much more complicated for $n > 3$. However, the complications don't emerge until we start considering tensors with *rank* greater than 2. (2) Our methods work as well for SO(n) as for SU(n); the only difference is that for SO(n) there's no distinction between upper and lower indices. (3) For SO(n),

$$\text{(vector)} \otimes \text{(vector)} = \text{(scalar)} \oplus \text{(antisymmetric tensor)} \oplus \text{(traceless symmetric tensor)} \qquad \text{(P21.1)}$$

For $n > 4$, the three representations on the right-hand side are irreducible and inequivalent.

(1998b 9.1)

21.2 In class I worked out the magnetic moments (Table 38.2, p. 837) and electromagnetic mass splittings (Table 38.3, p. 841) of the baryon octet. Things turn out to be even simpler for the decuplet. As in class, assume perfect SU(3) symmetry, broken only by electromagnetism. Show that all magnetic moments within the decuplet are proportional to the charge, and in particular, that neutral elements have vanishing magnetic moments. Find a correspondingly simple statement for the electromagnetic mass splittings.

(1998b 9.2)

21.3 SU(3) allows only *one* possible coupling of the electromagnetic current to a quark and an antiquark. Thus (by the same reasoning used for the decuplet in the previous problem), in the limit of perfect SU(3) symmetry, if quarks are observable, their magnetic moments would be proportional to their charges. In the non-relativistic limit,

$$\boldsymbol{\mu} = \kappa q \boldsymbol{\sigma}, \qquad \text{(P21.2)}$$

where κ is an unknown constant, q is the electric charge of the quark in question, and $\boldsymbol{\sigma}$ is the vector of Pauli spin matrices.

[1] [Eds.] "An Introduction to Unitary Symmetry", the Erice notes from the summer of 1966, originally published in *Strong and Weak Interactions – Present Problems*, Academic Press, 1966, and reprinted in Coleman *Aspects*.

In the naive quark model discussed in class, the baryons are considered as non-relativistic three-quark bound states with no spin-dependent interactions. Thus, as in atomic physics, we can compute the baryon moments in terms of the quark moments, that is, in terms of the single unknown constant κ, if we know the baryon wave function. For the lightest baryon octet, the one that contains the proton and the neutron, there is no orbital contribution to the magnetic moments because each quark has zero orbital angular momentum. Thus all we need is the spin-flavor-color part of the wave function. Of course, since the assumption of perfect SU(3) already gives all the baryon moments in terms of the proton and neutron moments, the only new information we get from this analysis is the ratio of these moments. Compute the ratio and compare it to experiment.

Remark. It's clear from the way the calculation is set up that it's the *total* moment you will be computing, not the anomalous moment. Be careful that you don't use the anomalous moments when you make the computation.

HINT: You will need the spin-flavor part of the wave functions for both the proton and the neutron to do this problem. Here is an easy way to construct them without resorting to tables of $3j$ symbols. It is trivial to construct the wave function for the $I_z = J_z = \frac{3}{2}$ state of the Δ; it is $|u\uparrow, u\uparrow, u\uparrow\rangle$, with all three quarks being up quarks, and all three spinning up. If we apply both an isospin lowering operator and a spin lowering operator to this, we obtain the $I_z = J_z = \frac{1}{2}$ state of the Δ. The $J_z = \frac{1}{2}$ state of the proton must be orthogonal to this. The $J_z = \frac{1}{2}$ state of the neutron (up to an irrelevant phase) is obtained from the proton state by exchanging u and d.

(1998b 9.3)

Solutions 21

21.1 (a) There are at least two ways to solve this problem. First, the straightforward method. The 3×3 matrix
λ_2, and λ_3. Because the matrix is traceless and can be diagonalized, it follows

$$\lambda_3 = -\lambda_1 - \lambda_2 \tag{S21.1}$$

The trace of a matrix is invariant under similarity transformations, so $\mathrm{Tr}[\phi^4]$ and $(\mathrm{Tr}[\phi^2])^2$ are unchanged, and we can take ϕ to be diagonal. We then have

$$\begin{aligned}\mathrm{Tr}(\phi^4) &= \lambda_1^4 + \lambda_2^4 + (-\lambda_1 - \lambda_2)^4 \\ &= 2\lambda_1^4 + 4\lambda_1^3\lambda_2 + 6\lambda_1^2\lambda_2^2 + 4\lambda_1\lambda_2^3 + 2\lambda_2^4\end{aligned} \tag{S21.2}$$

and

$$\begin{aligned}(\mathrm{Tr}(\phi^2))^2 &= (\lambda_1^2 + \lambda_2^2 + (-\lambda_1 - \lambda_2)^2)^2 = 4(\lambda_1^2 + \lambda_2^2 + \lambda_1\lambda_2)^2 \\ &= 4(\lambda_1^4 + 2\lambda_1^3\lambda_2 + 3\lambda_1^2\lambda_2^2 + 2\lambda_1\lambda_2^3 + \lambda_2^4) \\ &= 2\,\mathrm{Tr}(\phi^4)\end{aligned} \tag{S21.3}$$

The two expressions are proportional, and the constant of proportionality is 2.

A slicker solution is found in Section 2.13 in Coleman *Aspects*. One way to write the characteristic equation of the matrix ϕ is

$$(\phi - \lambda_1)(\phi - \lambda_2)(\phi - \lambda_3) = 0 \tag{S21.4}$$

Expanding this out, we find

$$\phi^3 - \phi^2(\lambda_1 + \lambda_2 + \lambda_3) + \phi(\lambda_1\lambda_2 + \lambda_1\lambda_3 + \lambda_2\lambda_3) - \lambda_1\lambda_2\lambda_3 = 0 \tag{S21.5}$$

which can be written as

$$\phi^3 - \phi^2\,\mathrm{Tr}(\phi) + \tfrac{1}{2}\phi\left[(\mathrm{Tr}(\phi))^2 - \mathrm{Tr}(\phi^2)\right] - \det\phi = 0 \tag{S21.6}$$

Because ϕ is traceless, this becomes

$$\phi^3 - \tfrac{1}{2}\phi\mathrm{Tr}(\phi^2) - \det\phi = 0 \tag{S21.7}$$

Multiply by ϕ and take the trace to obtain

$$\mathrm{Tr}(\phi^4) - \tfrac{1}{2}(\mathrm{Tr}(\phi^2))^2 = 0 \tag{S21.8}$$

which is what we found earlier. ■

(b) The degeneracy of the **8** and the **27** of SU(3) is presumably due to a larger hidden symmetry G of the Lagrangian for which the **8** and the **27** are parts of *one* multiplet. But what is the larger group? Since there are 8 pseudoscalar mesons, an obvious guess (reinforced by the "possibly useful information") is SO(8). Let ϕ denote the matrix of fields

$$\phi = \sum_{a=1}^{8} T_a\phi_a, \tag{S21.9}$$

where T_a are the 8 matrix generators[2] of SU(3), normalized to satisfy

$$\text{Tr}(T_a T_b) = \tfrac{1}{2}\,\delta_{ab} \tag{S21.10}$$

From part (a) there is only one invariant quartic coupling,

$$\left[\text{Tr}(\phi^2)\right]^2 = \left[\text{Tr}\left[\sum_a T_a\phi_a\right]^2\right]^2 = \left(\sum_{a,b}\phi_a\phi_b\text{Tr}(T_aT_b)\right)^2 = \left(\tfrac{1}{2}\sum_a \phi_a^2\right)^2 \tag{S21.11}$$

For a pseudoscalar meson theory, the most general renormalizable Lagrange density which is Lorentz invariant, parity invariant, and SU(3) invariant is thus

$$\mathscr{L} = \sum_a \left(\tfrac{1}{2}\partial^\mu\phi_a\partial_\mu\phi_a - \tfrac{1}{2}\mu^2\phi_a^2\right) - \lambda\left(\sum_a \phi_a^2\right)^2 \tag{S21.12}$$

This $\mathscr{L}$ *does* have a larger symmetry group, SO(8), under which ϕ transforms as a real 8-dimensional vector. Physical states of this theory must therefore form representations of SO(8), and the s-wave bound states lie in a representation contained in the symmetric product of two 8-dimensional vectors.

We've already considered ((37.64),(38.58), and (38.59)) the symmetry of the direct product of two 8-dimensional vectors. We found that the symmetric parts of this product were

$$\mathbf{8}\otimes\mathbf{8}\big|_{\text{symmetric}} = \mathbf{1}\oplus\mathbf{8}\oplus\mathbf{27} \tag{S21.13}$$

In the decomposition of $\mathbf{8}\otimes\mathbf{8}$, we had another $\mathbf{8}$, a $\mathbf{10}$ and an $\overline{\mathbf{10}}$, but those were all antisymmetric.

Let's now come at this product from SO(8). From the "possibly useful information" we have that the product of two vectors $A^a \otimes B^b$ in SO(8) gives a scalar, an antisymmetric tensor and a traceless symmetric tensor. There are $\frac{1}{2}8(8+1) = 36$ symmetric products, and $\frac{1}{2}8(8-1) = 28$ antisymmetric ones. Taking the trace from the symmetric tensor with 36 elements,$A^iB^j + A^jB^i$, we decompose it into a $\mathbf{1}$ and a $\mathbf{35}$, so

$$\mathbf{8}\otimes\mathbf{8}\big|_{\text{SO}(8)} = \mathbf{1}\big|_{\text{SO}(8)} \oplus \mathbf{28}\big|_{\text{SO}(8)} \oplus \mathbf{35}\big|_{\text{SO}(8)} \tag{S21.14}$$

We know the $\mathbf{35}$ is a symmetric, irreducible multiplet of SO(8), into which can fit snugly the $\mathbf{8}$ and the $\mathbf{27}$ of SU(3). The octet and 27-plet are degenerate, because they belong to the *same* multiplet, the $\mathbf{35}$, in SO(8). ■

21.2 (a). *Electromagnetic form factors* are associated (§38.3) with the matrix element

$$\langle D|j_\mu^{\text{em}}|D\rangle$$

where D is a member of the decuplet and j_μ^{em} is the electromagnetic current. We are looking for the SU(3)-invariant couplings between $\overline{D}$, D and j_μ^{em}. $\overline{D}$ and D transform as $\overline{\mathbf{10}}$ and $\mathbf{10}$, respectively; the electromagnetic current is a member of an octet. The number of singlets in $(3,0)\otimes(0,3)\otimes(1,1)$ is the number of unknowns needed to determine the magnetic moments of the decuplet. That in turn is determined by the number of $\mathbf{8}$'s in $\overline{\mathbf{10}}\otimes\mathbf{10}$, or the number of $\mathbf{10}$'s in $\overline{\mathbf{10}}\otimes\mathbf{8}$, etc.; a tensor product of two representations includes an invariant (a singlet) if the two representations are identical, and then the product includes exactly *one* singlet. Using the algorithm in §37.4 we have

$$\begin{aligned}\overline{\mathbf{10}}\otimes\mathbf{10} &= (0,3)\otimes(3,0) = (3,0;0,3)\oplus(2,0;0,2)\oplus(1,0;0,1)\oplus(0,0;0,0)\\ &= (3,3)\oplus(2,2)\oplus(1,1)\oplus(0,0)\\ &= \mathbf{64}\oplus\mathbf{27}\oplus\mathbf{8}\oplus\mathbf{1}\end{aligned} \tag{S21.15}$$

Since the tensor product $\overline{\mathbf{10}}\otimes\mathbf{10}$ contains exactly *one* $\mathbf{8}$, there is exactly *one* invariant term in $\langle D|j_\mu^{\text{em}}|D\rangle$. By the Wigner–Eckart theorem, matrix elements of tensor operators in the same representation are proportional to each other. As both j_μ^{em} and the charge operator, Q transform as members of an octet, the current is proportional to Q, so we can write

$$\langle D|j_\mu^{\text{em}}|D\rangle = a\,\langle D|Q|D\rangle = aq(D) \tag{S21.16}$$

Then $F_1 = a_1 q$, $F_2 = a_2 q$.

(b) *Electromagnetic mass splitting* is associated (§38.4) with the matrix element

$$\langle D|(j_\mu^{\text{em}}j_\nu^{\text{em}})^S|D\rangle$$

[2] [Eds.] These $\{T_a\}$ are half the Gell-Mann λ matrices; $T_a = \frac{1}{2}\lambda_a$; see note 15, p. 807. Unfortunately, if we used $\frac{1}{2}\lambda_a$ for the generators, it would be very easy to confuse them with the eigenvalues λ_i.

where the superscript S denotes the symmetric product. We already know

$$\mathbf{8}\otimes\mathbf{8}|_{\text{symmetric}} = \mathbf{1}\oplus\mathbf{8}\oplus\mathbf{27} \tag{S21.13}$$

$$\overline{\mathbf{10}}\otimes\mathbf{10} = \mathbf{1}\oplus\mathbf{8}\oplus\mathbf{27}\oplus\mathbf{64} \tag{S21.15}$$

Among the terms of the product $(\mathbf{8}\otimes\mathbf{8}|_{\text{symmetric}})\otimes(\overline{\mathbf{10}}\otimes\mathbf{10})$ are three invariant couplings arising from the products $\mathbf{1}\otimes\mathbf{1}$, $\mathbf{8}\otimes\mathbf{8}$ and $\mathbf{27}\otimes\mathbf{27}$.

For the product $\mathbf{1}\otimes\mathbf{1}$, the operator $\mathbb{1}$ is a tensor operator, and we get a contribution

$$\langle D|(j_\mu^{\text{em}}j_\nu^{\text{em}})_{\mathbf{1}}^S|D\rangle = a\,\langle D|\mathbb{1}|D\rangle = a \tag{S21.17}$$

This term gives equal contributions to the masses of all the decuplet; it's a *shift*, not a splitting.

Now we need, for the other two invariants, an operator which transforms like the symmetric product of two currents. One candidate is Q^2. It has the right component of $(2,2)$ in it. It probably also has some $(1,1)$ and $(0,0)$ components. The latter doesn't matter, but the $(1,1)$ components could screw things up if it were not the right member of the octet. Fortunately, it *is* the right member, as an easy argument shows. A full SU(2) $\otimes$ U(1) subgroup of SU(3) commutes with Q^2, and hence with any $(1,1)$ piece of Q^2. This determines, up to a factor, which member of the octet this piece is. Since the same argument holds for the current–current Hamiltonian, it and Q^2 both contain the same member of the octet. In fact, this argument applies to Q itself: it must be the same member of an octet as the octet piece of the current–current electromagnetic Hamiltonian. So the matrix element of the current–current Hamiltonian must be a linear combination of Q and Q^2:

$$\langle D|(j_\mu^{\text{em}}j_\nu^{\text{em}})_{\mathbf{8}}^S|D\rangle + \langle D|(j_\mu^{\text{em}}j_\nu^{\text{em}})_{\mathbf{27}}^S|D\rangle = b\,\langle D|Q|D\rangle + c\,\langle D|Q^2|D\rangle = bq(D) + c(q(D))^2 \tag{S21.18}$$

The EM contributions to the masses of the baryon decuplet should be fit by the formula $m_D = a + bq + cq^2$. ∎

20.3 We are asked to compute the ratio of the magnetic moments of the proton and the neutron. The magnetic moment of a baryon in state $|B\rangle$ is computed from the magnetic moments of the three constituent quarks:

$$\mu_B = \langle B|\mu_{1z} + \mu_{2z} + \mu_{3z}|B\rangle \tag{S21.19}$$

where, according to the problem,

$$\boldsymbol{\mu} = \kappa q\boldsymbol{\sigma} \tag{P21.2}$$

Following the advice given in the problem, we start with the wave function for the Δ^{++}:

$$|\Delta^{++}\rangle = |uuu\rangle\otimes|\uparrow\uparrow\uparrow\rangle \tag{S21.20}$$

Applying I_- and J_- operators to this, we obtain the wave function for the Δ^+:

$$|\Delta^{+}\rangle = \tfrac{1}{3}\left[|uud\rangle + |udu\rangle + |duu\rangle\right]\otimes\left[|\uparrow\uparrow\downarrow\rangle + |\uparrow\downarrow\uparrow\rangle + |\uparrow\uparrow\downarrow\rangle\right] \tag{S21.21}$$

The proton wave function must be orthogonal to $|\Delta^+\rangle$. Start with the piece proportional to $|uud\rangle$. It is symmetric in *flavor* for the first two quarks and so must also be symmetric in *spin* for them:

$$|p\rangle \propto |uud\rangle\otimes\left[2\,|\uparrow\uparrow\downarrow\rangle - |\uparrow\downarrow\uparrow\rangle - |\downarrow\uparrow\uparrow\rangle\right] + \cdots \tag{S21.22}$$

The pieces proportional to $|udu\rangle$ and $|duu\rangle$ can be found in the same way. The normalized wave function is then

$$|p\rangle = \frac{1}{3\sqrt{2}}\left\{\begin{array}{l} |uud\rangle\otimes\left[2\,|\uparrow\uparrow\downarrow\rangle - |\uparrow\downarrow\uparrow\rangle - |\downarrow\uparrow\uparrow\rangle\right] \\ +\,|udu\rangle\otimes\left[2\,|\uparrow\downarrow\uparrow\rangle - |\downarrow\uparrow\uparrow\rangle - |\uparrow\uparrow\downarrow\rangle\right] \\ +\,|duu\rangle\otimes\left[2\,|\downarrow\uparrow\uparrow\rangle - |\uparrow\uparrow\downarrow\rangle - |\uparrow\downarrow\uparrow\rangle\right] \end{array}\right\} \tag{S21.23}$$

The neutron wave function is just the same, but with $u \leftrightarrow d$:

$$|n\rangle = \frac{1}{3\sqrt{2}}\left\{\begin{array}{l} |ddu\rangle\otimes\left[2\,|\uparrow\uparrow\downarrow\rangle - |\uparrow\downarrow\uparrow\rangle - |\downarrow\uparrow\uparrow\rangle\right] \\ +\,|dud\rangle\otimes\left[2\,|\uparrow\downarrow\uparrow\rangle - |\downarrow\uparrow\uparrow\rangle - |\uparrow\uparrow\downarrow\rangle\right] \\ +\,|udd\rangle\otimes\left[2\,|\downarrow\uparrow\uparrow\rangle - |\uparrow\uparrow\downarrow\rangle - |\uparrow\downarrow\uparrow\rangle\right] \end{array}\right\} \tag{S21.24}$$

The proton's magnetic moment is given by

$$\mu_p = \langle p|\kappa q\sigma_z|p\rangle \tag{S21.25}$$

and the neutron's similarly. Looking only at the first part of the proton wave function,

$$|p\rangle = \frac{1}{3\sqrt{2}}\left(2\,|u(\uparrow)u(\uparrow)d(\downarrow)\rangle - |u(\uparrow)u(\downarrow)d(\uparrow)\rangle - |u(\downarrow)u(\uparrow)d(\uparrow)\rangle + \cdots\right) \tag{S21.26}$$

Applying the operator $\kappa q\sigma_z$ to this wave function gives

$$\kappa q\sigma_z\,|p\rangle = \frac{\kappa}{3\sqrt{2}}\Big(2\left[\tfrac{2}{3}\cdot\tfrac{1}{2}+\tfrac{2}{3}\cdot\tfrac{1}{2}+(-\tfrac{1}{3})\cdot(-\tfrac{1}{2})\right]|u(\uparrow)u(\uparrow)d(\downarrow)\rangle$$
$$-\left[\tfrac{2}{3}\cdot\tfrac{1}{2}+\tfrac{2}{3}\cdot(-\tfrac{1}{2})+(-\tfrac{1}{3})\cdot\tfrac{1}{2}\right]|u(\uparrow)u(\downarrow)d(\uparrow)\rangle \tag{S21.27}$$
$$-\left[\tfrac{2}{3}\cdot(-\tfrac{1}{2})+\tfrac{2}{3}\cdot\tfrac{1}{2}+(-\tfrac{1}{3})\cdot\tfrac{1}{2}\right]|u(\downarrow)u(\uparrow)d(\uparrow)\rangle+\cdots\Big)$$

which, when cleaned up, is

$$\kappa q\sigma_z\,|p\rangle = \frac{\kappa}{3\sqrt{2}}\left(2\cdot\tfrac{5}{6}\,|u(\uparrow)u(\uparrow)d(\downarrow)\rangle+\tfrac{1}{6}\,|u(\uparrow)u(\downarrow)d(\uparrow)\rangle+\tfrac{1}{6}\,|u(\downarrow)u(\uparrow)d(\uparrow)\rangle+\cdots\right) \tag{S21.28}$$

so that

$$\mu_p = \langle p|\kappa q\sigma_z|p\rangle = \frac{\kappa}{9\cdot 2}\left(4(\tfrac{5}{6})-\tfrac{1}{6}-\tfrac{1}{6}+\cdots\right) = \frac{\kappa}{9\cdot 2}(3+\cdots) = 3\cdot\frac{\kappa}{18}(3) = \tfrac{1}{2}\kappa \tag{S21.29}$$

the extra factor of 3 arising because the other two kets are just permutations of the first one, and all the inner products give the same contribution (we are taking the basic kets $|q_1s_1q_2s_2q_3s_3\rangle$ to be orthonormal). In exactly the same way,

$$\kappa q\sigma_z\,|n\rangle = \frac{\kappa}{3\sqrt{2}}\left(2\cdot(-\tfrac{2}{3})\,|d(\uparrow)d(\uparrow)u(\downarrow)\rangle-\tfrac{1}{3}\,|d(\uparrow)d(\downarrow)u(\uparrow)\rangle-\tfrac{1}{3}\,|d(\downarrow)d(\uparrow)u(\uparrow)\rangle+\cdots\right) \tag{S21.30}$$

and

$$\mu_n = \langle n|\kappa q\sigma_z|n\rangle = \frac{\kappa}{9\cdot 2}\left(4(-\tfrac{2}{3})+\tfrac{1}{3}+\tfrac{1}{3}+\cdots\right) = \frac{\kappa}{9\cdot 2}(-2+\cdots) = 3\cdot\frac{\kappa}{18}(-2) = -\tfrac{1}{3}\kappa \tag{S21.31}$$

Then we predict

$$\frac{\mu_p}{\mu_n} = -1.5 \quad \text{(theory)} \tag{S21.32}$$

From the 2016 Particle Data Group tables, in units of the nuclear magneton μ_N,

$$\mu_p = 2.79\mu_N, \quad \mu_n = -1.91\mu_N \tag{S21.33}$$

so that

$$\frac{\mu_p}{\mu_n} = -\frac{2.79}{1.91} = -1.46 \quad \text{(exp't)} \tag{S21.34}$$

in excellent agreement with the prediction. ■

(Incidentally, this problem is worked out as Example 5.3 in Griffiths *EP*, pp. 189–190.)

40

Weak interactions and their currents

We're going to begin a completely new topic.[1] The subject goes under the general name of *current algebra*,[2] although the actual algebra will not emerge until the next lecture. I will try to keep the conventions consistent within the lectures on this topic, although not necessarily in agreement with the literature (or even with previous lectures). In order to explain the subject I will begin by giving a lightning summary of the weak interactions.[3] Not the weak interactions as we know them today, with CP violation, neutral currents, renormalizable spontaneously broken gauge theories of the weak and electromagnetic interactions, and all that, but the weak interactions as they were in my childhood: the low-energy phenomenology of the **Fermi theory**[4] as improved by Feynman and Gell-Mann, *circa* 1965, when nobody had yet seen an intermediate vector boson. That's still a pretty good theory for practically all weak interaction processes below a few GeV.

40.1 The weak interactions circa 1965

In the mid 1960's, the interaction Lagrangian responsible for the weak interactions took the form of a universal **Fermi constant** G_F, governing the strength of all weak interactions, divided by the square root of 2 by convention, times the product of some currents J_λ:

$$\mathscr{L} = \frac{G_F}{\sqrt{2}} J_\lambda J^{\lambda\dagger} \tag{40.1}$$

[1] [Eds.] Lectures 40–45 in the videotapes concern *dispersion relations*. Owing to length and time constraints, the editors have decided not to include these six lectures. (The most serious casualty of this omission is the Adler-Weisberger relation.) This (short) chapter represents the last 48 minutes of (a very long) Lecture 45, starting at 1:00:15. This book's last nine chapters coincide with the last nine videotaped lectures.

[2] [Eds.] "Soft Pions", Chapter 2, pp. 36–66 in Coleman *Aspects*; Stephen L. Adler and Roger F. Dashen, *Current Algebras and Applications to Particle Physics*, W. A. Benjamin, 1968; Jeremy Bernstein, *Elementary Particles and Their Currents*, W. H. Freeman, 1968.

[3] [Eds.] E. D. Commins and P. H. Bucksbaum, *Weak Interactions of Leptons and Quarks*, Cambridge University Press, Cambridge, 1983; Greiner & Müller *GTWI*; J. C. Taylor, *Gauge Theories of Weak Interactions*, rev. ed., Cambridge U. P., 1979; Abers & Lee *GT*.

[4] [Eds.] E. Fermi, "Tentativo di una teoria dei raggi β" (A provisional theory of beta radiation), *Nuovo Cim.* **11** (1934) 1–19; reprinted in *The Collected Papers of Enrico Fermi, v.1: Italy, 1921–1938*, ed. Emilio Segrè, U. Chicago P., 1962. English translation in C. Strachan, *The Theory of Beta Decay*, Pergamon, 1969.

The current J_λ will be described in detail below. Briefly, it is made up of all the fields that describe the weakly interacting particles under consideration. G_F is small in comparison to the scale of $\alpha = e^2/4\pi \approx 1/137$:

$$G_F = 1.16637 \pm 0.00001 \times 10^{-5}\,\text{GeV}^{-2} \approx \frac{10^{-5}}{m_p^2} \tag{40.2}$$

It is determined from measurement of the μ lifetime.[5] G_F has units; it has to. The space integral of a conventionally normalized current, a "charge", is dimensionless, so a current has dimensions L^{-3} or M^3, and a product of two currents as in (40.1) has dimensions of M^6. In order to make a Lagrange density of dimension M^4 the Fermi constant must have dimensions M^{-2}. The interaction is therefore *non-renormalizable*,[6] and that was one of the great problems with weak interaction theories in the mid-1960's. The theory enabled people to compute everything at low energy with dazzling accuracy. But whenever they tried to compute higher-order corrections they got divergences and infinities that could not be absorbed into a renormalization. It was frustrating that the weak interactions were so weak. If only the weak interactions had been a little bit stronger, we could have seen the second-order effects easily in feasible experiments. And then we might have gotten some idea about what's going on. But we couldn't, and so we had to rely upon genius to figure out the weak interactions. We think genius came through and delivered the answers. But we still aren't sure, because the second-order effects remain hard to measure.

The weak current J_λ is the sum of a hadronic part and a leptonic part:

$$J^\lambda = J_h^\lambda + J_\ell^\lambda \tag{40.3}$$

While the *theory* is charge conserving, both of the *currents* carry charge; J^λ is a positively charged (+1) current that creates positively charged particles when acting on the vacuum, and its Hermitian adjoint is negatively charged. Thus the current's matrix elements $\langle f|J^\lambda|i\rangle$ produce a change of charge

$$\Delta Q = 1 \tag{40.4}$$

and their adjoints also change charge:

$$\Delta Q = -1 \tag{40.5}$$

The interaction is set up to be *CP*-conserving. (It does not include the small *CP*-violating effects observed in the neutral kaon decays.[7])

This form (40.1) of the Lagrangian can describe three kinds of interactions. *Purely leptonic* weak interactions, such as muon decay or high-energy neutrino scattering off electrons to make muons, come from the product of lepton currents. From the study of muon decay,

$$\mu^- \to \nu_\mu + e^- + \overline{\nu}_e \tag{40.6}$$

---

[5] [Eds.] D. B. Chitwood *et al.*, "Improved Measurement of the Positive-Muon Lifetime and Determination of the Fermi Constant", *Phys. Rev. Lett.* **99** (2007) 032001; *PDG* 2016, p. 627, "Gauge & Higgs Boson Particle Listings".

[6] [Eds.] The dimensionality of the coupling constant is a quick indicator of whether the interaction is renormalizable or not. If the coupling constant has negative mass dimension or positive length dimension, as G does, the interaction is non-renormalizable; see §16.4.

[7] [Eds.] Greiner & Müller *GTWI*, Section 8.2; also see note 9, p. 240.

we know that the proper form of the leptonic weak current is given by

$$J_\ell^\lambda = \overline{\nu}_e \gamma^\lambda (1-\gamma_5) e + \overline{\nu}_\mu \gamma^\lambda (1-\gamma_5)\mu \tag{40.7}$$

(We use the notation in which the labels of the fermions stand for the four-component Dirac spinors associated with them.) The current J_ℓ^λ annihilates an electron or a muon, negatively charged particles. This current fits muon decay, a beautiful process, and it sets the scale of J^λ as well as the size (40.2) of the Fermi constant, by providing the scale of the leptonic part.[8] Muon decay proceeds through weak interactions so the lowest-order perturbation theory should be absolutely reliable. The particles involved in muon decay have no interactions worth worrying about aside from the weak interactions. Electromagnetism is present, but typically it does not make important corrections to muon decay, and when it does, we know how to compute them. From experiments we had learned enough essentially to read off the Lagrangian, and (40.1), with (40.3) and (40.7) express what we had determined up to 1967, when things changed dramatically. Note that the current (40.7) is neither pure vector, $\psi\gamma^\mu\psi$, nor pure axial vector, $\psi\gamma^\mu\gamma_5\psi$, but the *difference* of these, a "$V - A$" form.

We can also have the product of hadronic currents which give us *purely hadronic* weak interactions, such as

$$\Lambda \to p + \pi^- \tag{40.8}$$

And for that, the evidence for the current-current form of the interaction was *zero*, since we knew nothing about the strong interactions. They contaminate the process in complicated ways, making it essentially impossible to read off the interaction. The hadronic part J_h^λ of the current was conjectured (based on symmetry) in the 1960s to have a form like (40.7). About the hadronic part we will say no more now, except that it is made up only of hadronic fields. We will shortly describe it more fully.

The more interesting things are the so-called *semi-leptonic decays*, where a hadron h goes into another hadron (or hadrons), h', or possibly the vacuum, plus a pair of leptons:

$$h \to h' + \ell\colon \quad \begin{cases} n \to p + e^- + \overline{\nu}_e \\ \pi^- \to \mu^- + \overline{\nu}_\mu \\ \pi^- \to e^- + \overline{\nu}_e \end{cases} \tag{40.9}$$

The matrix element governing this process is

$$\langle h'\ell | J_{\lambda h} J_\ell^{\lambda\dagger} | h\rangle \tag{40.10}$$

if the lepton combination is negatively charged. (If they're positively charged we study the complex conjugate of this matrix element.) This factors into hadronic and leptonic parts because leptons and hadrons interact only weakly (ignoring electromagnetism), so in lowest order we simply get

$$\langle h'\ell | J_{\lambda h} J_\ell^{\lambda\dagger} | h\rangle = \langle h' | J_{\lambda h} | h\rangle \, \langle \ell | J_\ell^{\lambda\dagger} | 0\rangle \tag{40.11}$$

Thus the situation is rather like that of an electron scattering off a proton, where the matrix element factors into a known part and a mysterious part.[9]

[8] [Eds.] Bjorken & Drell *RQM*, Section 10.11, pp. 247–257; Greiner & Müller *GTWI*, Section 6.2, pp. 208–211.

[9] [Eds.] Commins and Bucksbaum, *op. cit.*, Section 4.7, pp. 156–159.

We know all the dependence on the leptonic part, so we can write the entire matrix element in terms of a one-hadron matrix element of J_h^λ, if h and h' are both single hadrons. We can parameterize this process in terms of **weak interaction form factors**, analogs of electromagnetic form factors. This is as big an improvement as what we get by writing electron scattering off a proton in terms of the proton form factors.[10] Instead of a function of many kinematic variables we have a function of only *one* kinematic variable, the momentum transfer to the current. In an atypical case like pion decay where the π^+ goes into leptons (with no final hadrons), things are even simpler. We have a matrix element of the current between a one pion state and the vacuum. Instead of form factors we just have a number, since there are no free kinematic variables. I will discuss these in more detail shortly.

40.2 The conserved vector current hypothesis

By studying these semi-leptonic decay processes we have learned a lot about J_h^λ. The leptonic current (40.7) is *parity-violating*.[11] It's the difference of a vector current V_ℓ^λ and an axial vector current A_ℓ^λ; following Feynman and Gell-Mann[12] we assume the same is true of the complete current J^λ:

$$J^\lambda = V^\lambda - A^\lambda \tag{40.12}$$

Under parity, V^λ transforms as a true vector and A^λ as an axial vector, so the sum or difference of these cannot conserve parity. On the other hand, both parts taken together (in the combination (40.11)) are charge-conserving; J_h^λ has $\Delta Q = +1$ (40.4), but $J_\ell^{\lambda\dagger}$ has $\Delta Q = -1$.

Both the vector and axial vector parts of the hadronic current act alike, though there are two separate cases. In the first case, neither the vector nor the axial vector part changes the hypercharge, and both obey the selection rule $\Delta I = 1$. The currents in the first case transform under isospin like the positively-charged component of an isovector; they have the isospin transformation properties of the π^+ state. In the second case, the two parts of the hadronic current change the hypercharge by $+1$ and the total isospin by $\frac{1}{2}$. This is the famous $\Delta I = \frac{1}{2}$ rule for semi-leptonic decays.[13] That is, this current has the same transformation properties as the K^+ state:

$$\Delta Q = 1\colon \begin{cases} \Delta Y = 0,\ \Delta I = 1\colon\ J_h^\lambda \sim \pi^+ \\ \Delta Y = 1,\ \Delta I = \frac{1}{2}\colon\ J_h^\lambda \sim K^+ \end{cases} \tag{40.13}$$

We knew something else, even in the late 1950s, about the vector part of this current with $\Delta Y = 0$. This is the famous **conserved vector current** hypothesis of Feynman and Gell-Mann, or **CVC** for short.[14]

---

[10] [Eds.] Commins and Bucksbaum, *op. cit.*, Bjorken & Drell *RQM*, pp. 242–246; §34.2, p. 738.

[11] [Eds.] T. D. Lee and C. N. Yang, "Question of Parity Conservation in Weak Interactions", *Phys. Rev.* **104** (1956) 254–258; C. S. Wu, E. Ambler, R. W. Hayward, D. D. Hoppes, and R. P. Hudson, "Experimental Test of Parity Conservation in Beta Decay", *Phys. Rev.* **105** (1957) 1413–1415.

[12] [Eds.] R. P. Feynman and M. Gell-Mann, "Theory of the Fermi Interaction", *Phys. Rev.* **109** (1958) 193–198.

[13] [Eds.] For instance, $\Lambda \to p + e^- + \overline{\nu}_e$. Quang Ho-Kim and Pham Xuan Uem, *Elementary Particles and Their Interactions*, Springer, 1998; §6.6.2 and §16.1.4.

[14] [Eds.] Feynman and Gell-Mann, *op. cit.*; S. S. Gershtein and Y. B. Zel'dovich, "Meson Corrections in the Theory of Beta Decay", *Zh. Eksp. Teor. Fiz.* **29** (1955) 698–699. [*Sov. Phys. JETP* **2** (1956) 576–578.] E. C. G. Sudarshan and R. E. Marshak, "Chirality Invariance and the Universal Fermi Interaction", *Phys. Rev.* **109** (1958) 1860-1861; Greiner & Müller *GTWI*, p. 209.

Consider $J_{h\,\lambda}^{\Delta Y=0}$, the part of the hadronic current that doesn't change the strangeness; it contributes for example to neutron β decay. It had been known for a long time that the vector part of that, $V_{h\,\lambda}^{\Delta Y=0}$, obeyed *universality*: the β decay constant seemed to be about the same as the coupling constant in muon decay. The matrix element of the vector current at small momentum transfers (the proton is so close to the neutron that only small momentum transfers are relevant) seemed to be pretty close to 1, in the scale at which the vector current, $V_{\ell\,\lambda}$, had matrix element 1 between electron and neutrino. Feynman and Gell-Mann argued that this couldn't be a coincidence. Even if $V_{h\,\lambda}^{\Delta Y=0}$ started out initially with matrix element 1, the strong interactions were going to get into our computation and change things from 1 to $\frac{3}{2}$ or $\frac{1}{2}$ or something. How can it stay 1? Well, they continued, there is only one case we know in which the matrix element of a current at small momentum transfer is not affected by the strong interactions: the current must be *conserved*. The premier example of this is electromagnetism, where $F_1(0)$ stays firmly fixed at 1. It has no strong interaction corrections. The proton has an anomalous magnetic moment but it doesn't have an anomalous *charge*. That argument was true for the lowest order in electromagnetism and to all orders in the strong interactions. Perhaps there would be a parallel between the electromagnetic current and any other conserved current. So they guessed that this current $V_{h\,\lambda}^{\Delta Y=0}$ has got to be a *conserved* current.

Gell-Mann and Feynman knew of only one conserved current, the positively-charged *isospin* current. So they guessed, in accordance with (40.13), that $V_{h\,\lambda}^{\Delta Y=0}$ was proportional to the charge-raising (and hence isospin raising) isospin current, whose integral is I_+:

$$V_{h\,\lambda}^{\Delta Y=0} = \alpha_w I_\lambda^+ \tag{40.14}$$

with α_w a constant to be determined. In this way they explained the so-called **universality of the weak interactions**: the vector part of the matrix element for neutron β decay, for small momentum transfers, was precisely[15] the same as that for muon decay, 1. (It was actually 1 within 1 or 2% but they said the difference could be due to an electromagnetic vertex correction or something.) The vector part of the $\Delta Y = 0$ current was to be proportional to just the $\Delta Q = 1$ conserved isospin current, not with any funny Pauli-type terms, e.g., $\partial_\mu \sigma^{\mu\lambda}$ times some other factor, but *exactly* that. This was a bold guess. They could have included many other conserved currents, by adding divergences of antisymmetric tensors, but they thought that made the interaction too ugly.

The physicists of the time could check this guess. The strong interactions are isospin invariant, so the form factors for these decays, F_1 and F_2, should be related just by an isospin rotation to the form factors for the I_z current. Those we know from the Gell-Mann–Nishijima relation (35.52): I_z is part of the electromagnetic current, which we get by taking the difference between the proton and the neutron form factors. Measuring these weak interaction form factors is difficult. The least difficult to measure is the analog of F_2 at zero, and that is not easy. You have to look at a cunningly chosen nuclear β decay so that the F_1 form factor obeys the wrong selection rules and can't play a role. Then only the F_2 form factor is involved. After you've divided out the nuclear physics matrix elements, you end up with, in principle, a measurement of $F_2(0)$ for this weak interaction current. That should be related by an isospin rotation to the electromagnetic $F_2(0)$, the difference between the proton and neutron. This is called **weak magnetism**[16]. Because this is not a course in the weak interactions I'm not

---

[15] [Eds.] Greiner & Müller *GTWI*, Section 6.2, pp. 208–209.

[16] [Eds.] Commins and Bucksbaum, *op. cit.*, pp. 166–167; pp. 189–200.

going to give all the details. This is a testable hypothesis which has been tested, and it works. (We're not however going to use it in our current algebra discussions.)

40.3 The Cabibbo angle

Since we have devoted a lot of time to talking about SU(3), and since this is a lightning summary of the weak interactions, I should tell you about Cabibbo's work in 1963; he fit the weak interactions together with SU(3).[17] He said, "Ha! Feynman and Gell-Mann have told us that $V_{h\,\lambda}^{\Delta Y=0}$, the $\Delta I = 1$, strangeness-conserving part of the vector current is the isospin current J_{λ}^{I+}. What about the *other* part, with $\Delta I = \frac{1}{2}$, the part that *changes* strangeness?" We know the isospin current is part of an SU(3) octet; I_z is one of the eight generators of SU(3). We know there is *another* positively-charged current in the same octet, the pseudoscalar mesons.[18] If the top entry in (40.13) is π^+-like, there's also the bottom entry, K^+-like. That's the only other positively-charged object in the octet. We also know that the bottom entry in (40.13) has the same isospin and strangeness properties as the K^+. Therefore if you label SU(3) octet currents by the transformation properties of the appropriate meson, so we have a vector current that transforms like the π^+, the positively-charged isospin current, and a current that transforms like the K^+, the positively-charged strangeness-changing current, it seems very natural to imagine that the total vector weak interaction current is simply the sum of these two things:[19]

$$V_h^{\lambda} \sim V_{\pi^+}^{\lambda} + V_{K^+}^{\lambda} \tag{40.15}$$

After all, in the world of perfect SU(3), who can tell the difference between a π^+ and a K^+? Indeed, Cabibbo suggested that the combinations are weighted together in such a way that the sum of the squares of the coefficients is 1:

$$V_h^{\lambda} = V_{\pi^+}^{\lambda} \cos\theta_C + V_{K^+}^{\lambda} \sin\theta_C \tag{40.16}$$

The angle θ_C is called the **Cabibbo angle**. The Cabibbo angle θ_C must be close to zero, so that $\cos\theta_C$ will be 1 within a few percent, to agree with our earlier discussion.

---

[17] [Eds.] Nicola Cabibbo, "Unitary Symmetry and Leptonic Decays", *Phys. Rev. Lett.* **10** (1963) 531–533; Greiner & Müller *GTWI*, §6.4.

[18] [Eds.] Table 39.3, p. 852.

[19] [Eds.] The currents are easily described in terms of the eight 3×3 Gell-Mann λ matrices (note 15, p. 807), and (37.35). The first three λ's are

$$\lambda_i = \begin{pmatrix} \sigma_{i\,11} & \sigma_{i\,12} & 0 \\ \sigma_{i\,21} & \sigma_{i\,22} & 0 \\ 0 & 0 & 0 \end{pmatrix}$$

where σ_i are the Pauli matrices, generate the isospin subgroup of SU(3). In particular, $I_+ = \lambda_1 + i\lambda_2$ has exactly the *one* non-zero matrix element occupied by π^+ in (37.35). Similarly,

$$\lambda_{j+3} = \begin{pmatrix} \sigma_{j\,11} & 0 & \sigma_{j\,12} \\ 0 & 0 & 0 \\ \sigma_{j\,21} & 0 & \sigma_{j\,22} \end{pmatrix}$$

for $j = \{1, 2\}$. And sure enough, $\lambda_4 + i\lambda_5$ has exactly *one* non-zero matrix element, the same as that occupied by K^+ in (37.35). If the hypercharge matrix Y is given by (38.41) $\equiv \frac{1}{\sqrt{3}}\lambda_8$, it's easy to show that $\lambda_4 + i\lambda_5 = Y^+$, i.e., $[Y, Y^+] = Y^+$. Given an octet of vector currents, $\{V_\lambda^a\}$, $a = \{1, \dots 8\}$ transforming like a (1,1) or **8** representation of SU(3), we can make the assignments

$$V_{h\,\lambda}^{\Delta Y=0} = V_\lambda^{\pi^+} = V_\lambda^1 + iV_\lambda^2\,; \quad V_{h\,\lambda}^{\Delta Y=1} = V_\lambda^{K^+} = V_\lambda^4 + iV_\lambda^5$$

See D. H. Lyth, *An introduction to current algebra*, Oxford U. P., (1970), p. 6.

The things that Feynman and Gell-Mann thought were electromagnetic corrections were really an electromagnetic correction plus terms of order $(\theta_C)^2$.

Let that be for a moment. Cabibbo's general idea was that God said, "Let there be weak interactions and let there be medium-strong interactions that break SU(3)", apparently without looking to make sure they were in the same direction. If there were *no* medium-strong interactions you could, with an SU(3) rotation, turn $V^\lambda_{K^+}$ into $V^\lambda_{\pi^+}$ without affecting electromagnetism. And then there would be no strangeness-changing *at all*, by definition, since you can define strangeness as you wish if there are no SU(3)-violating interactions. The angle θ_C is a measure of mismatch of the directions in SU(3) space chosen by the medium-strong interactions and chosen by the weak interactions. It just happened that the directions didn't quite match.

That was Cabibbo's idea. And being a bold man, he said exactly the same thing should be true for the *axial* vector currents, with the *same* angle:

$$A^\lambda_h = A^\lambda_{\pi^+} \cos\theta_C + A^\lambda_{K^+} \sin\theta_C \tag{40.17}$$

(Some people tried a different angle θ for the axial vectors, but the best fit to experiment is $\theta = \theta_C$.) Cabibbo postulated another octet of axial vector operators that formed the axial, nonconserved currents. The positively charged members were to be put together with the same angle. People looked at (40.16) (Shelly Glashow and I among others) and said, "What a random guess. What's the experimental evidence for that?" At the time, nobody appreciated how attractive a guess this was.

Cabibbo's guess gives us a lot of information about semi-leptonic decays of the baryon octet. There are a lot of these,[20] nine or ten. How many unknown constants do we have in this matrix element? All of these decays proceed at relatively small momentum transfer so we really have to know only the various form factors in the vector and axial vector currents at zero momentum transfer. We know the Fermi constant G_F from muon decay. That is not a free parameter; I'll put it in parentheses. We need to determine the Cabibbo angle θ_C. We have the matrix elements of the vector and axial vector currents but we know those at zero momentum transfer in the SU(3) limit because they are SU(3) conserved currents. And we have the axial vector currents, assumed to be an octet. (We call them "currents", even though they aren't associated with any conservation laws; they're just vector operators.) They can couple, octet to octet, to the baryon octet with some unknown constants d and f.[21] Thus we have four parameters with which we can fit all semi-leptonic baryon decays, three of them free:

$$(G_F), \theta_C, d, f$$

There are a lot of decays that we can fit with these. We know the vector matrix elements in terms of the parameter θ_C, and the axial vector matrix elements in terms of the parameters d and f. And it fits; it's the right theory.

[20] [Eds.] See Table 10 in A. Faessler, T. Gutsche, Barry R. Holstein, Mikhail A. Ivanov, Jürgen G. Körner, and Valery E. Lyubovitskij, "Semileptonic decays of the light $J^P = 1/2^+$ ground state baryon octet", *Phys. Rev.* **D78** (2008) 094005.

[21] [Eds.] See note 15, p. 807.

40.4 The Goldberger–Treiman relation

I want to say a little more about the various semi-leptonic matrix elements and the matrix elements of the vector and axial vector hadronic currents, in particular for the processes (40.9) of nuclear β decay

$$n \to p + e^- + \overline{\nu}_e \tag{40.18}$$

and pion decay. For neutron decay we need the matrix element of the hadronic current (at point x) between a proton and a neutron:

$$\langle p|J^h_\lambda(x)|n\rangle \tag{40.19}$$

Define the momentum k as the difference of the neutron and proton momenta:

$$k = p_n - p_p \tag{40.20}$$

For $n \to p$ this should be quite small. The only term that survives from the vector current at really small momentum transfers is the analog of an F_1 form factor, called $g_V(k^2)$. There also is a $\sigma_{\mu\nu}$ form factor and other stuff like that, but that's got powers of k in it.[22] From the axial vector current, because of the $V - A$ definition, there is also a $g_A(k^2)$ term. And then there is some other junk, which I will write down in more detail shortly:

$$\langle p|J^h_\lambda(x)|n\rangle = e^{-ik\cdot x}\overline{u}_p[\gamma_\lambda g_V(k^2) - \gamma_\lambda\gamma_5 g_A(k^2) + \cdots]u_n \tag{40.21}$$

These are the dominant elements in low energy neutron decay, where the momentum transfer is very small (a few MeV). The other terms are all proportional to powers of k, and are killed off in the limit $k \to 0$.

At zero momentum transfer, the value of $g_V(k^2)$, which we'll call g_V, should be $\cos\theta_C$, if we accept Feynman and Gell-Mann as modified by Cabibbo. The Cabibbo angle is rather small,[23] about 13°. So to the order in which we're working, we'll just ignore strangeness-changing weak interactions and set $\cos\theta_C$ to 1:

$$g_V(0) \equiv g_V = \cos\theta_C \approx 1 \tag{40.22}$$

That introduces an error of a few percent, but we're not going to get any formulas accurate even to a few percent in the remainder of this lecture. The other term, $g_A(0)$, has been measured in neutron decay,[24] and this value will become significant to us. In the literature typically one finds the ratio g_A/g_V but as g_V is essentially 1, the reported value[25] can be taken for g_A:

$$g_A(0) \equiv g_A = -1.2723 \pm 0.0023 \tag{40.23}$$

---

[22] [Eds.] For the general form of the axial current matrix element, see Exercise 3.3, p. 88–91 in Greiner *et. al* *QCD*.

[23] [Eds.] The Cabibbo angle is expressed in terms of the Wolfenstein parameter λ; $\lambda = V_{us} = \sin\theta_C$. Various experiments have established $V_{us} \approx 0.225$, giving $\theta_C \approx 13°$; *PDG* 2016, "V_{ud}, V_{us}, the Cabibbo Angle and CKM Unitarity", pp. 1011–1013. In 1975 Coleman quoted a value of about 15°. The Cabibbo angle now finds a home in the CKM matrix, from the work of Makoto Kobayashi and Toshihide Maskawa, who extended Cabibbo's ideas to include a third generation of quarks (t, top and b, bottom). M. Kobayashi and T. Maskawa, "CP-Violation in the Renormalizable Theory of Weak Interaction", *Prog. Theo. Phys.* **49** (1973) 652–657. Kobayashi and Maskawa shared the 2008 Physics Nobel Prize for this work (Nambu was also honored in 2008, but for spontaneous symmetry breaking; see Chapter 43).

[24] [Eds.] J. Liu *et al.*, "Determination of the Axial-Vector Weak Coupling Constant with Ultracold Neutrons", *Phys. Rev. Lett.* **105** (2010) 181803.

[25] [Eds.] *PDG* 2016, "Baryon Particle Listings", p. 1516.

The other process to consider is π^- decay:[26]

$$\pi^- \to \begin{cases} \mu^- + \overline{\nu}_\mu & > 99.98\% \\ \mu^- + \overline{\nu}_\mu + \gamma & 2.00 \pm 0.25 \times 10^{-4} \\ e^- + \overline{\nu}_e & 1.230 \pm 0.004 \times 10^{-4} \\ e^- + \overline{\nu}_e + \gamma & 7.39 \pm 0.05 \times 10^{-7} \end{cases} \tag{40.24}$$

The π^- is a pseudoscalar particle and it's easy to see using parity that the vector current matrix element must vanish between a pseudoscalar particle and the vacuum:

$$\langle 0|V_\mu|\pi^-\rangle = 0 \tag{40.25}$$

Only the axial vector current has a nonzero matrix element:

$$\langle 0|A_\mu(x)|\pi^-\rangle = i\frac{F_\pi}{\sqrt{2}}p_\mu e^{-ip\cdot x} \tag{40.26}$$

The form of the right-hand side is not hard to explain. The only vector around is the pion momentum. There must be a factor of $e^{-ip\cdot x}$. The remainder is a well-known number (divided by $\sqrt{2}$) called the **pion decay constant** and denoted F_π; it is measured from the pion decay rate. It doesn't depend on k^2 (by analogy with (40.20)) because there is only one momentum here. The $\sqrt{2}$ is there to make subsequent equations look simple.[27] We put in an i by convention.

We can connect F_π to $g_A(0)$ (i.e., pion decay to neutron decay). The error in the pion constant is considerably less than the error in $g_A(0)$. Without any error bars[28] it is:

$$F_\pi = 0.19656\, m_p \tag{40.27}$$

This is straight phenomenology. The relation (40.26) is just the statement that charged pion decay occurs.[29] The form of the matrix element is completely determined by parity and other constraints. You measure the rate of pion decay in the process $\pi^+ \to \mu^+ + \nu_\mu$ to establish the value of F_π.

Now comes a deep insight. Take the divergence of (40.26):

$$\langle 0|\partial^\mu A_\mu(x)|\pi^-\rangle = \frac{F_\pi}{\sqrt{2}}p^2 e^{-ip\cdot x} = \frac{F_\pi m_\pi^2}{\sqrt{2}}\, e^{-ip\cdot x} \tag{40.28}$$

It isn't zero. The divergence of the axial vector current, a pseudoscalar field, has nonzero matrix elements between the one-pion state and the vacuum. The only new piece of information is that pions decay. Back in §14.2, when we went around in circles with the LSZ reduction formula, I said that one of its consequences was that *any* local field was as good as any other for computing S-matrix elements. It didn't matter what we used for a local field, so long as it had nonzero amplitude for connecting the particle in question to the vacuum. Therefore we have the freedom to *define* the π^- field as

$$\phi_{\pi^-}(x) \equiv \frac{\sqrt{2}}{F_\pi m_\pi^2}\partial^\mu A_\mu(x) \tag{40.29}$$

---

[26] [Eds.] *PDG* 2016, p. 37.

[27] [Eds.] Many authors define this matrix element without the $\sqrt{2}$.

[28] [Eds.] *PDG* 2016, p. 1112.

[29] [Eds.] Neutral pions decay primarily into two γ's. There is no lighter hadron to decay into.

That is a *definition.* I do not say this $\phi_{\pi^-}(x)$ is a canonical field; God forbid! But it is a perfectly legitimate local operator that will serve as a pion field. It may or may not be equal to the canonical pion field that appears in the Lagrangian of a theory with fundamental pions in it.

Whether the pion is a fundamental object or not, the matrix element can connect a neutron and a proton: there *is* a strong interaction. I'll factor out the pion pole; there's an i from the pion propagator and an i from the pion–nucleon vertex that gives a -1:

$$\langle p|\phi_{\pi^-}(x)|n\rangle = -\frac{\sqrt{2}}{k^2 - m_\pi^2}\overline{u}_p i\gamma_5 u_n g(k^2)e^{-ik\cdot x} \tag{40.30}$$

The $\sqrt{2}$ is to take care of the $\sqrt{2}$ convention in the pion–nucleon coupling constant. The factor $g(k^2)$ is a form factor. If we chose a different candidate for the pion field we would get a different form factor. Whatever we choose, the residue at the pion pole is always going to be the same: $g(m_\pi^2)$ will be the conventionally defined strong interaction constant g, and the experimental value of g is known, again with a negligible error on the scale in which we are working:[30]

$$g(m_\pi^2) = g = 13.3 \tag{40.31}$$

The best way of determining g is from the forward pion–nucleon scattering.

Now let's look a bit more closely at the axial vector matrix elements between proton and neutron. There are three possible invariants.[31]

$$\langle p|A_\mu(x)|n\rangle = e^{-ik\cdot x}\overline{u}_p[-\gamma_\mu\gamma_5 g_A(k^2) + i\sigma_{\mu\nu}\gamma_5 k^\nu g_M(k^2) - k_\mu\gamma_5 g_p(k^2)]u_n \tag{40.32}$$

The first term we wrote down before (40.21); in the second, $g_M(k^2)$ is the analog of the magnetic form factor; it will disappear in the computation we are going to do. The $g_p(k^2)$ form factor is in the axial vector current but not in the vector current; the axial vector current isn't conserved (as is obvious from (40.28)). It's easy to see that all these form factors have to be there. The first two terms are just the analog of the computation we did for the vector current with γ_5 inserted. The last term contributes to the longitudinal part of A_μ. It's like the divergence of a scalar which produces a factor k_μ.

Making the sensible kinematic approximation that

$$m_p = m_n \tag{40.33}$$

one can trivially compute the divergence of (40.32). The $\not{k}$'s and the γ_5's go together and as usual they hit the spinors on the right and the left. One gets (recall $k = p_n - p_p$)

$$\langle p|\partial^\mu A_\mu(x)|n\rangle = e^{-ik\cdot x}\overline{u}_p(i\gamma_5)u_n[-2m_p g_A(k^2) + k^2 g_p(k^2)] \tag{40.34}$$

The first and third terms in (40.32) are trivial; the middle term drops out because

$$k^\mu k^\nu \sigma_{\mu\nu} = 0$$

[30] [Eds.] T. Ericson *et al.*, "Determination of the Pion–Nucleon Coupling Constant and Scattering Lengths", *Phys. Rev.* **C66** (2002) 014005; V. A. Babenko and N. M. Petrov, "Study of the Charge Dependence of the Pion–Nucleon Coupling Constant on the Basis of Data on Low-Energy Nucleon-Nucleon Interactions", *Phys. Atom. Nucl.* **79** (2016) 67–71. The latter give $g_{\pi^0}^2 = 13.55(13)$, and $g_{\pi^\pm}^2 = 14.55(13)$.

[31] [Eds.] Coleman *Aspects*, "Soft Pions", §3, p. 41. The subscript M is for "magnetic", p for "pseudoscalar".

Comparing (40.34) with (40.30) and the definition of ϕ_{π^-} we obtain

$$\frac{\sqrt{2}}{F_\pi m_\pi^2}[-2m_p g_A(k^2) + k^2 g_p(k^2)] = -\frac{\sqrt{2}}{k^2 - m_\pi^2} g(k^2) \tag{40.35}$$

This is nothing but a *definition.* It is completely empty of any physical content. It simply connects two ways of parameterizing the same matrix element, the matrix element for the divergence of the axial vector current.

Now we introduce *physics* into this equation with the following hypothesis: *The function $g(k^2)$ is free of singularities up to the three-pion threshold.* We have extracted out the one-pion pole. Furthermore, if our experience with the electromagnetic form factors, with which this object is closely analogous, is any guide, we don't expect a lot of variation even at the three-pion threshold. Experimentally, the electromagnetic form factors don't exhibit big changes at twice the mass squared of the pion; it's only when you get up to the ρ mass that they have gigantic bumps in them. So even at the three-pion threshold, if you say something like the ρ mass or the Ω mass or the mass of some axial vector meson is the thing to look at, you don't expect $g(k^2)$ to vary enormously over the region between $k^2 = 0$ and $k^2 = m_\pi^2$. That's a nice analytic region. The threshold at $9m_\pi^2$ probably has a small discontinuity, if the electromagnetic form factors are a guide. If so, m_π^2 is a small fraction of the way, about 10%, to the nearest singularity. Therefore we make the hypothesis that in a region of analyticity, small compared to the distance to the nearest singularity and small compared to the characteristic length involved in the problem, we would expect

$$g = g(m_\pi^2) \approx g(0) \tag{40.36}$$

This is a *physical* hypothesis: that this matrix element varies *in the same way* as every other matrix element we can measure; i.e., not much over a distance of m_π^2 once we've extracted out the pion singularity. The first part of the next lecture we will explain this hypothesis in four different ways, because it is so critical.

Evaluating both sides of (40.35) at $k^2 = 0$, and using the hypothesis of (40.36), we see that the m_π^2's and factors of $\sqrt{2}$ cancel. Multiplying both sides by F_π, we find

$$F_\pi g = -2m_p g_A(0) \tag{40.37}$$

This is the famous (and at one time, notorious) **Goldberger–Treiman relation.**[32] Please notice that the only thing that has gone into this is kinematics and the *single* hypothesis about the rate of variation of $g(0)$; there was no added physics besides that.

The result (40.37) is remarkable. It connects the pion decay constant F_π, the strong interaction pion–nucleon coupling constant g and the nucleon axial vector decay constant, g_A. It was very strange, because in those days people thought nucleons were fundamental, so maybe pions are bound states (nucleon plus antinucleon). They had a notion that pion decay was caused by nucleon decay:[33] a pion comes along, becomes a nucleon-antinucleon pair and these β decay, as shown in Figure 40.1.

[32] [Eds.] M. Goldberger and S. Treiman, "Decay of the Pi Meson", *Phys. Rev.* **110** (1958) 1178–1184. The original relation is their equation (24).

[33] [Eds.] Bernstein, *op. cit.*, p. 171 has a close match to Figure 40.1, for the decay of a π^+. He draws the axial lepton current A^μ as the lepton vertex with antimuon and muon neutrino, as shown:

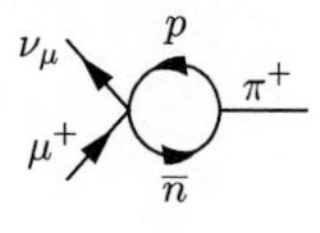

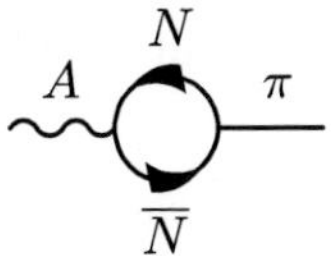

Figure 40.1: An old and erroneous view of pion decay

Notice that this picture leads to a relationship which is the wrong way around: you're connecting the decay constant F_π to the product of g at the $\pi N\overline{N}$ vertex and g_A at the $AN\overline{N}$ vertex; the g would wind up on the *right* side of (40.37), and give the erroneous

$$F_\pi \sim g g_A \quad (\textit{wrong!}) \tag{40.38}$$

I also emphasize that the factors are of completely different magnitude. On the left-hand side, we have this enormous number g, 13.3, and nothing else is so large. If this works out, we have a right to be proud.

Now the experimental situation. Well, what is the answer? Putting numbers into (40.37) we get

$$\begin{aligned} F_\pi g &= -2m_p g_A(0) \\ (0.19656\, m_p)(13.3) &\stackrel{?}{=} -2m_p(-1.2723) \\ 2.61 &\stackrel{?}{=} 2.54 \end{aligned} \tag{40.39}$$

The left-hand side is 2.61, and the right-hand side, is 2.54, in units of m_p. The agreement is, by any standard, excellent. We will have much more to say about this next time.

41

Current algebra and PCAC

Last time we derived the *Goldberger–Treiman relation*:[1]

$$F_\pi g = -2m_p g_A(0) \tag{40.37}$$

a nontrivial equality between two things of quite different orders of magnitude, the large dimensionless constant g (~ 13.3) from pion–nucleon strong interactions, and the comparatively small coupling constant g_A (~ -1.25) from nuclear beta decay. It has been confirmed within experimental error.

It will be helpful for what is to come[2] to summarize how we came to this result. We started with the matrix element for neutron beta decay (40.21), and looked specifically at the axial vector part:

$$\langle p|A_\mu(x)|n\rangle = e^{-ik\cdot x}\overline{u}_p[-\gamma_\mu\gamma_5 g_A(k^2) + i\sigma_{\mu\nu}\gamma_5 k^\nu g_M(k^2) - k_\mu\gamma_5 g_p(k^2)]u_n \tag{40.32}$$

Then we considered the weak decay of the pion, and wrote down (40.26) the matrix element of the axial current between a π^- state and the vacuum, in terms of the pion momentum and F_π ($\sim 0.196m_p$), the pion decay constant. We took the divergence of that equation to obtain

$$\langle 0|\partial^\mu A_\mu(x)|\pi^-\rangle = \frac{F_\pi m_\pi^2}{\sqrt{2}}\, e^{-ip\cdot x} \tag{40.28}$$

from which we *defined*

$$\phi_{\pi^-}(x) \equiv \frac{\sqrt{2}}{F_\pi m_\pi^2}\partial^\mu A_\mu(x) \tag{40.29}$$

On the other hand, we know that the pion field can connect a proton and a neutron. Factoring out the pion pole, we wrote the matrix element for this process in terms of a form factor, $g(k^2)$:

$$\langle p|\phi_{\pi^-}(x)|n\rangle = -\frac{\sqrt{2}}{k^2 - m_\pi^2}\overline{u}_p i\gamma_5 u_n g(k^2)e^{-ik\cdot x} \tag{40.30}$$

---

[1] [Eds.] M. Goldberger and S. Treiman, "Decay of the Pi Meson", *Phys. Rev.* **110** (1958) 1178–1184.

[2] [Eds.] Much of this lecture duplicates material in Coleman's 1967 Erice lecture, "Soft Pions", republished in Coleman *Aspects*, Chapter 2, pp. 36–66.

Taking the divergence of (40.32) and comparing it with (40.30), we found

$$\frac{\sqrt{2}}{F_\pi m_\pi^2}[-2m_p g_A(k^2) + k^2 g_p(k^2)] = -\frac{\sqrt{2}}{k^2 - m_\pi^2} g(k^2) \tag{40.35}$$

This equation is really nothing but a *definition.* We put physics into it by making the hypothesis that the value of the form factor $g(k^2)$ at $k^2 = m_\pi^2$ was the strong coupling constant g, and that $g(m_\pi^2) \approx g(0)$:

$$g = g(m_\pi^2) \approx g(0) \tag{40.36}$$

The assumption $g(m_\pi^2) \approx g(0)$ was plausible because we have extracted out the only singularity that we come near when extrapolating from zero momentum to m_π^2, the pion pole, in the definition of $g(k^2)$. Using this hypothesis and taking the limit as $k^2 \to 0$ of both sides of (40.35), we arrived at the Goldberger–Treiman relation, (40.37).

When it was first derived, it was said to be good only to within 10%, but no one was particularly disturbed by that. You'd expect errors on the order of $(m_\pi/m_\rho)^2$ or something, taking the ρ as a typical hadron, and we're making an extrapolation over a distance of one m_π^2, perhaps 5%, within experimental error. The 10% error was due to a bad measurement of neutron beta decay. It's rather tricky to extract the axial vector contribution to neutron beta decay from angular correlations, but with modern measurements the relation (40.37) is right on the nose (40.39). It is so good that it's mysterious. It's an astonishing result.

The equation (40.29) is a key step in the derivation of the Goldberger–Treiman relation. It's important to understand what this equation is saying. There are many interpretations floating around in the literature. To these we now turn.

41.1 The PCAC hypothesis and its interpretation

Equation (40.29) is known by a peculiar acronym, **PCAC**. (I will explain what PCAC stands for in a moment.) There are four interpretations that one comes across, in the literature or in conversations. Two are perfectly acceptable, one is silly, and one is wrong.

One way of looking at the equation is to say that certain matrix elements $\langle f|\phi_{\pi^-}(0)|i\rangle$ involving off-mass shell pions are slowly varying, and so can be successfully approximated by constants as the momentum transferred goes from 0 to the pion mass shell. This is sometimes phrased as "the slow variation of the matrix elements as a function of momentum transfer". I would prefer to say *normal* rather than *slow* variation, because the momentum dependence we need to justify the Goldberger–Treiman relation is pretty much the same as we find in the F_1 and F_2 form factors, and other off-shell matrix elements of local operators which we can measure. This meaning is acceptable. (I say it's acceptable because it's the one *I* choose to adopt!)

There was far more confusion about the equation in its early days than today. Another statement which you will find in some otherwise quite profound papers goes like this: "We assume that the derivative of the axial current is proportional to the pion field," or

$$\partial^\mu A_\mu \propto \phi_{\pi^-} \tag{41.1}$$

Adler's classic paper[3] begins this way. This is one of the reasons the equation is called PCAC;

[3] [Eds.] Stephen L. Adler, "Consistency Conditions on the Strong Interactions Implied by a Partially Conserved Axial-Vector Current", *Phys. Rev.* **137** (1965) B1022–B1033. See Adler's equation (3).

the acronym stands for **partially conserved axial current**. The idea is *almost* like the conserved vector current hypothesis, except that we assume that the divergence of the axial current is not zero, but is instead the canonical pion field, and call this "partial conservation". This interpretation is silly, because from the viewpoint of the LSZ reduction formula, *any* field that has a non-zero matrix element between the pion state and the vacuum is as good a pion field as any other. There is absolutely no reason to assume that the canonical pion field in a strongly interacting field theory has matrix elements any more slowly or any more rapidly varying than any other randomly selected local operator. So this is an assumption without any apparent content. Nevertheless, we will have occasion very shortly to worry a bit about inventing models that obey this relation, despite the fact that I have identified it as silly.[4]

Sometimes in the mid-1960's, when I was going around lecturing on these topics, people in the audience, typically otherwise intelligent experimenters, would suggest another way to think about PCAC ((though this point of view never found its way into the literature). They'd say, "Aren't you making a mystery out of something very simple? Isn't it just **pion pole dominance**?" (The process is shown in Figure 41.1.) "After all," they said, "here's a neutron coming in, a proton going out, here are two leptons coming out. You could say the total energy of the leptons is very small, so if we imagine calculating the right dispersion relation in some cross channel in terms of the lepton energy or something like that, it would seem very reasonable to dominate this by the pole diagram for the pion." It might seem very reasonable to just throw in the one-pion pole and say that's what dominates at low energies. But whether it's reasonable or not, it's dead wrong! It's not merely a silly explanation; it's worse than that. The pion couples derivatively to the leptons; it's the only way it can couple so its matrix

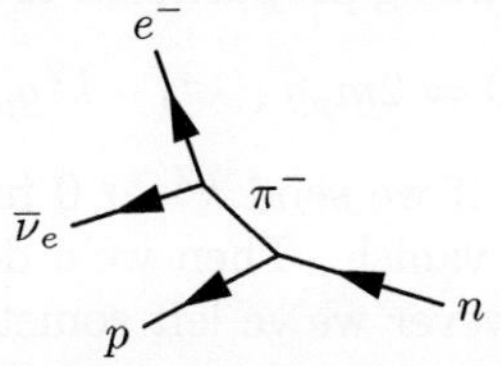

Figure 41.1: Pion pole dominance

element has a total momentum. The process is proportional to k_μ because the pion matrix element in the diagram is proportional to k_μ. The g_p term in (40.32) has a big fat factor of momentum transfer k_μ sticking out in front, but the g_A term does not. That is, if we were to write down the contribution of the pion pole diagram we would have to say

$$g_A = 0 \tag{41.2}$$

We would also predict, if you work it out,

$$g_p = -\frac{gF_\pi}{k^2 - m_\pi^2} \tag{41.3}$$

[4] [Eds.] In response to a student's question about what this interpretation means, Coleman responds: "I wouldn't have brought it up but ... It's embedded in the literature. It *is* silly. There's nothing to understand ... Ten years ago I remember, in this very seminar room, Francis Low, Steve Adler and I screaming at each other about whether this was meaningless or not. You've been brainwashed by me, and so you see it's meaningless. But if you look up all those papers from the mid-1960s, the golden age of current algebra, you will find people saying the key is that the divergence of the axial vector current is the canonical pion field. And that's just a silly statement. You can't derive *anything* from that."

(The definitions come together with our conventions to give a minus sign.[5]) Now this is neither the Goldberger–Treiman relation nor experimentally verified. So this interpretation is worse than silly; it's simply wrong.

The fourth interpretation is based on an idea that was the great discovery of Nambu.[6] At the time the idea seemed rather peculiar, but its growth led to the whole theory of spontaneous symmetry breakdown in field theory. If we imagine a world almost the same as ours, except the pion mass in that world is zero, $m_\pi^2 \to 0$, then the axial vector current (in that world) would be conserved,

$$\partial_\mu A^\mu = 0 \tag{41.4}$$

in the sense that from (40.29) we have

$$\partial_\mu A^\mu \propto F_\pi m_\pi^2 \phi_{\pi^-} \tag{41.5}$$

If you assume this holds, then as $m_\pi^2 \to 0$, $\partial_\mu A^\mu \to 0$. This is another meaning of *partially conserved axial current*: it's partially conserved in that its conservation is broken only by a *very small* parameter, one of the smallest parameters in hadron physics, perhaps the *only* small parameter: the ratio of the pion mass to a typical hadron mass. This looks like a very different assumption than what we have been talking about; it certainly doesn't look as if there's any conceivable way, in a world with massless pions, of connecting pion decay matrix elements to nucleon decay matrix elements. Nevertheless, using nothing but the equations above, one can show that this leads to exactly the same conclusion.

Since the axial vector current is supposed to be strictly conserved in a world with zero pion mass, we get, instead of something being proportional to a pion matrix element,

$$\partial^\mu A_\mu = 0 \Leftrightarrow 2m_p g_A(k^2) - k^2 g_p(k^2) = 0 \tag{41.6}$$

It looks like we'll run into trouble if we send $k^2 \to 0$ in this expression, as we did before. The second term would appear to vanish. Then we'd deduce $g_A = 0$, which is hardly the Goldberger–Treiman relation. However we've left something out. Now we're working in a hypothetical world in which there *is* a massless pion. Thus (41.3) g_p has a pion pole in it. Setting the pion mass to 0 in that expression gives

$$g_p = -\frac{gF_\pi}{k^2} + \text{non-pole terms} \tag{41.7}$$

The residue at the pion pole is unambiguous. If the pion is assumed to be the only massless particle in this hypothetical world, there will also be the non-pole terms. (There may be a three-pion term that produces a cut near $k^2 = 0$.) Thus, when we take the limit as $k^2 \to 0$, we get, from the first term in (41.6), $2mg_A(0)$ and from the second term we get, not zero, but the residue of the pole at $k^2 = 0$, which is $-gF_\pi$, by the preceding calculation. This is nothing but the Goldberger–Treiman relation again:

$$2mg_A(0) + gF_\pi = 0 \tag{41.8}$$

So Nambu's interpretation looks good, although how it connects with our other ways of reasoning is at the moment a trifle mysterious. The connection will not be revealed until we discuss other topics, many lectures from now.

---

[5] [Eds.] Drawing neutron β decay as in Figure 41.1 amounts to replacing $\langle p|J_\mu^h|n\rangle$ with $\langle p|\phi_-|n\rangle\,\langle 0|A_\mu|\pi_-\rangle$.

[6] [Eds.] Y. Nambu, "Axial Vector Current Conservation in Weak Interactions", *Phys. Rev. Lett.* **4** (1960) 380–383.

41.2 Two isotriplet currents

Before I go on I would like to make a tiny notational change, since the weak interactions are going to recede into the background for a while. I'll write things in an isospin-symmetric form. If we restrict our attention to the hypercharge-conserving current, the plus and minus components of an isotriplet are the only ones that contribute to the weak interactions.[7] But there is nothing to prevent us from considering a complete isotriplet of axial vector currents, A^a_μ, $a = \{1, 2, 3\}$, and likewise an isotriplet of vectors, V^a_μ, to go with the isotriplet of pions, ϕ^a. If we scale A^a_μ properly, the $\sqrt{2}$ in the original definition (40.29) of ϕ^- disappears. As usual (24.21),

$$\phi_{\pi^\pm} = \tfrac{1}{\sqrt{2}}\left(\phi_1 \mp i\phi_2\right) \tag{41.9}$$

so if we now set

$$A^\mu = \tfrac{1}{2}\left(A^\mu_1 + iA^\mu_2\right) \tag{41.10}$$

then

$$\partial_\mu\left(A^\mu_1 + iA^\mu_2\right) = F_\pi m^2_\pi\left(\phi_1 + i\phi_2\right) \tag{41.11}$$

We can thus write an isospin-covariant version of the PCAC equation, and define the isotriplet pion field simply by

$$\partial^\mu A^a_\mu = m^2_\pi F_\pi \phi^a \tag{41.12}$$

To maintain the "$V - A$" form of the currents

$$J_\mu = V_\mu - A_\mu \tag{41.13}$$

we write, parallel to (41.10)[8]

$$V^\mu = \tfrac{1}{2}\left(V^\mu_1 + iV^\mu_2\right) \tag{41.14}$$

We extend the CVC hypothesis of Feynman and Gell-Mann (40.14) to the whole isotriplet of vector fields:

$$V^a_\mu = \alpha_w I^a_\mu \tag{41.15}$$

Letting $\overline{\psi}_N$ be the adjoint isospinor $(\overline{p}, \overline{n})$, and ψ_N the corresponding column isospinor, we have for beta decay

$$I^a_\mu = \overline{\psi}_N \gamma_\mu \tau^a \psi_N \tag{41.16}$$

and in particular,

$$I^+_\mu = \tfrac{1}{2}\overline{p}\gamma_\mu n \tag{41.17}$$

Finally we can determine α_w (40.14). For small momentum transfers, (40.21) says

$$\langle p|V_\mu|n\rangle = e^{-ik\cdot x}\overline{u}_p\gamma_\mu g_V u_n = \tfrac{1}{2}\alpha_w e^{-ik\cdot x}\overline{u}_p\gamma_\mu u_n \tag{41.18}$$

and so

$$\alpha_w = 2g_V \tag{41.19}$$

---

[7] [Eds.] Coleman is discounting neutral weak interactions, which had been seen at CERN two years earlier: F. J. Hasert *et al.*, "Search for Elastic Muon-Neutrino Electron Scattering", *Phys. Lett.* **46B** (1973) 121–124; F. J. Hasert *et al.*, "Observation of Neutrino-like Interactions without Muon or Electron in the Gargamelle Neutrino Experiment", *Phys. Lett.* **46B** (1973) 138–140.

[8] [Eds.] Remember, all this is for the $\Delta Y = 0$ currents.

Then V_μ^3 equals $2g_V$ times the current associated with the third component of isospin:

$$V_\mu^3 = 2g_V I_\mu^3 \tag{41.20}$$

Since g_V, according to Cabibbo, is

$$g_V = \cos\theta \approx 1 \tag{41.21}$$

the normalization (41.20) is very convenient. We have scaled our current so that the third component of the vector partner of the axial vector triplet has matrix elements $+1$ in a proton state and -1 in a neutron state, as I_z has eigenvalues $\pm\frac{1}{2}$ for the proton and neutron. That's suitable for our purposes. I put the $\sqrt{2}$ into the definition (40.29) originally to get this normalization at the end.

Let's return for a moment to the second definition of PCAC. Before people really understood pion β decay, way back in the early Paleolithic era, they had an idea: neutrons and protons were fundamental and everything else was somehow made up of neutrons and protons, the so-called **Fermi–Yang model.**[9] Pions were bound states of a nucleon and an antinucleon; the π^- was really a neutron and an antiproton. It was the right idea except they used neutrons and protons instead of quarks. They had a picture of decay, of say the pion, in which the pion comes in, makes a proton–neutron pair somehow, and then the proton–neutron pair β-decays into leptons: only the axial current contributes. That picture leads to the conclusion that F_π

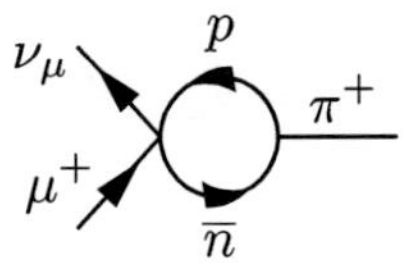

Figure 41.2: Pion decay in the Fermi–Yang model

equals, aside from some proportionality constant, the strong interaction coupling constant times the axial vector β decay coupling constant

$$F_\pi \propto g g_A \tag{41.22}$$

exactly the opposite of the Goldberger–Treiman relation.

It was originally thought that the Goldberger–Treiman relation depended on the *strength* of the strong interactions. Goldberger and Treiman first derived their relation in 1958, by a method not nearly as simple as the one given here,[10] but by an incredibly complicated method that involved deriving 42 dispersion relations in a row and making all sorts of unreliable approximations about what π-π scattering was like at low energy in order to estimate certain terms. I remember Fred Zachariasen came to Caltech in 1960 to give a sequence of lectures on dispersion relations. The great triumph of dispersion theory was then considered to be

[9] [Eds.] E. Fermi and C. N. Yang, "Are Mesons Elementary Particles?", *Phys. Rev.* **76** (1949) 1739–1743. See also note 33, p. 887, and the discussion following Figure 40.1, p. 888.

[10] [Eds.] The simplification is due to Gell-Mann and collaborators: M. Gell-Mann and M. Lévy, "The Axial Vector Current in Beta Decay", *Nuovo Cim.* **16** (1960) 705–726; J. Bernstein, S. Fubini, M. Gell-Mann, and W. Thirring, "On the Decay Rate of the Charged Pion", *Nuovo Cim.* **17** (1960) 757–766. Gell-Mann and Lévy describe the Goldberger–Treiman approximations as "violent . . . [and] not really justified." See the discussion following their equation (12).

the Goldberger–Treiman relation. It took him two full hours of lectures in order to derive it, assuming the audience already knew all the stuff about weak interaction phenomenology. People looked at that and said "Boy, this is really an impressive result, because we've somehow cracked a strong interaction problem. Look, instead of the g you get from lowest order perturbation theory [i.e., the mistaken (41.22)], there's a $1/g$ [correct]. That result must be telling us something important about the strength of the strong interactions, to turn this g into $1/g$." In fact, the assumptions we have made—slow (or normal) variation, PCAC—and the Goldberger–Treiman relation itself, have *nothing whatsoever* to do with the strong interactions being strong. One way of demonstrating that would be to create a model that obeys all of our assumptions *except* that the strong interaction coupling constant is not strong but weak, and seeing that we draw the same conclusions. Let's do just that.

41.3 The gradient-coupling model

If we can find a model that satisfies all of our assumptions then it should obey all of our conclusions, unless we've made an error in the argument. We can then see what happens in lowest order perturbation theory. The easiest way to construct such a model is to arrange matters so that (41.12) is true for canonical pion fields in our theory. In lowest order perturbation theory a canonical field is sharply distinguished from any other operator, and the form factors for canonical fields are typically trivial; they're constants or perhaps powers of the momenta. If we arrange our theory in a sensible way we can make them constants. In that case we will not only have *slow* variation, or *normal* variation, we will have *no* variation, in lowest order perturbation theory. It will be true manifestly that all of our assumptions hold.

To check that this result does not depend on mysterious facts about the strong interactions being strong, I propose to construct a model field theory, not meant to be realistic in any way, save that it obeys our key assumptions. First, the theory has an axial current. This could become the axial vector current for β decay. If the argument is sound, the theory is guaranteed to obey the Goldberger–Treiman relation, to lowest order in the strong interaction coupling constant.

We think that the correct model of hadrons is probably the quark model. Of course, if the coupling constants are weak, the quarks will have no bound states; the theory will describe only free quarks. But then we can hardly define quantities like g_A, g_p, F_π and other quantities of that sort, because we won't have any nucleons or pions. If I want to build a model to test the logic of this argument, I need it to have pions and nucleons in it at lowest order in perturbation theory. That means a model with fundamental pion and nucleon fields. The model will provide a theoretical laboratory where we can see if all of our assumptions work out, for things we can explicitly compute. In particular, we want to see if both PCAC and the Goldberger–Treiman relation emerge in our model, notwithstanding the weakness of our "strong" interactions.

We know how to construct conserved currents in a Lagrangian field theory. Let me review the procedure.[11] Suppose we have a set of fields $\phi^a(x)$ that transform (under some operation) into $\phi^a(x, \lambda)$, specified by a single parameter λ

$$\lambda\colon \phi^a(x) \to \phi^a(x, \lambda) \tag{41.23}$$

[11] [Eds.] See §5.3, p. 82.

with the condition that for $\lambda = 0$,

$$\phi^a(x, 0) = \phi^a(x) \tag{41.24}$$

Define the first-order change in the field ϕ^a as usual (5.21):

$$D\phi^a = \left.\frac{\partial \phi^a(x,\lambda)}{\partial \lambda}\right|_{\lambda=0} \tag{41.25}$$

We assume this is an internal symmetry, so we don't have to worry about adding total divergences to the Lagrangian.[12] Define a canonical momentum vector as

$$\Pi^\mu_a = \frac{\partial \mathscr{L}}{\partial(\partial_\mu \phi^a)} \tag{41.26}$$

The change in the Lagrange density is

$$D\mathscr{L} = \frac{\partial \mathscr{L}}{\partial \phi^a} D\phi^a + \Pi_{\mu a} D(\partial^\mu \phi^a) = \frac{\partial \mathscr{L}}{\partial \phi^a} D\phi^a + \Pi_{\mu a}\partial^\mu(D\phi^a) \tag{41.27}$$

because differentiation with respect to x and differentiation with respect to λ commute. Using the Euler–Lagrange equations of motion, we can rewrite the first term, and

$$D\mathscr{L} = (\partial^\mu \Pi_{\mu a})D\phi^a + \Pi_{\mu a}\,\partial^\mu(D\phi^a) = \partial^\mu\big(\Pi_{\mu a}\, D\phi^a\big) = \partial^\mu J_\mu \tag{41.28}$$

where the current J_μ associated with this symmetry is

$$J_\mu \equiv \Pi_{\mu a} D\phi^a \tag{41.29}$$

Then

$$D\mathscr{L} = \partial^\mu J_\mu \tag{41.30}$$

and the Lagrangian is invariant if the current is conserved:

$$D\mathscr{L} = 0 \;\Rightarrow\; \partial^\mu J_\mu = 0 \tag{41.31}$$

On the other hand, if the Lagrangian is *not* invariant (perhaps because we've added to it some small term that breaks the invariance), then we can deduce the divergence of the current from (41.30). This is the formula we will use to create a model that displays naïve PCAC, and therefore yields the Goldberger–Treiman relation to lowest order in perturbation theory, once we've chosen an appropriate symmetry group, G.

The theory we will consider is the **gradient-coupling model** of pions and nucleons:[13]

$$\mathscr{L} = \overline{N}(i\partial\!\!\!/ - m_N)N + \tfrac{1}{2}(\partial_\mu \boldsymbol{\phi}\boldsymbol{\cdot}\partial^\mu \boldsymbol{\phi} - m_\pi^2 \boldsymbol{\phi}^2) + \frac{g}{2m_N}\overline{N}\gamma_\mu\gamma_5 \boldsymbol{\tau} N\boldsymbol{\cdot}\partial^\mu \boldsymbol{\phi} \tag{41.32}$$

---

[12] [Eds.] Divergences of antisymmetric tensors are occasionally added to the quantity F^μ to obtain conserved currents with particularly desirable features, as in §5.4. In the introduction to Chapter 6, p. 105, Coleman defines an *internal symmetry* as one which does not relate fields at *different* space-time points, but only transforms fields at the *same* point. Thus derivatives do not arise in the change $D\mathscr{L}$ of the Lagrangian for an *internal* symmetry: $D\mathscr{L}$ is *zero*, and so is F^μ.

[13] [Eds.] In the video of Lecture 46, Coleman reminds the class that they've seen the one meson, one nucleon version of this model as a final examination question in the fall of 1975. It appears in this book as Problem 14.4, p. 546; see (P14.7).

N is the nucleon field, an isospinor composed of two Dirac 4-spinors, and $\boldsymbol{\phi} = \{\phi^a\}$ is an isotriplet of pseudoscalar pions. Since we're only working to lowest order in perturbation theory, we won't bother with bare masses and unrenormalized fields; in fact, the interaction term has dimension 5 (§25.4) and is thus nonrenormalizable. I've written the coefficient of the interaction term as the coupling constant g divided by twice the mass m_N of the isospinor. If I were to compute the matrix element of a pion field *off* the mass shell between two nucleons *on* their mass shell, the γ_μ and the derivative operators would act on the nucleon spinors and give us a factor of $2m_N$, one from the spinor on the right and one from the spinor on the left, leaving us with just a $g(k^2)$ as defined before,

$$g(k^2) = g + \mathcal{O}(k^2) \tag{41.33}$$

There will be higher order corrections. But we're going to ignore them, since we are assuming the strong interactions are *weak*. It's clear that this is a theory that has in lowest order perturbation theory a slowly-varying—indeed, constant—coupling $g(k^2)$.

The question is, does it contain naïve PCAC? The trick is to find the right group G and the right transformation $T \in \mathrm{G}$. Since we want an isotriplet of conserved currents, the transformation in this case is going to have three components; an isotriplet of transformations:

$$T\colon \begin{cases} \phi^a \to \phi^a + \lambda^a \\ N \to N \end{cases} \tag{41.34}$$

where a is now the isospin index; it runs over the three pions, $a = 1, 2, 3$. The transformations just add a constant isovector $\boldsymbol{\lambda}$ to the pion field $\boldsymbol{\phi}$ and do nothing to the isospinor N, so that part of the Lagrangian is invariant, while both the kinetic term of the $\boldsymbol{\phi}$ field and the interaction term (which depend on $\boldsymbol{\phi}$ only through its derivative) are unchanged, unlike a pseudoscalar Yukawa interaction. Only one term in the Lagrangian is not invariant under this transformation: the pion mass term.

It is easy to determine the change in the Lagrangian, which is just the infinitesimal change in the pion mass term. D is written as a vector D^a, because it is one of three objects:

$$D^a \mathscr{L} = -m_\pi^2 \phi^a \tag{41.35}$$

The change in the Lagrangian is just the pion field, times minus the square of its mass. We've deduced the divergence of the current before we've found the current itself, but the current is easy enough to compute from (41.29). The canonical momentum of the nucleon does not contribute because the nucleon field does not change. The canonical momentum of the pion is the current, because $D^a\phi^b$ is just the Kronecker delta:

$$D^a \phi^b = \frac{\partial \phi^a(x,\lambda)}{\partial \lambda^b}\bigg|_{\lambda^b=0} = \delta^{ab} \tag{41.36}$$

and

$$\Pi_\mu^a = \partial_\mu \phi^a + \frac{g}{2m_N} \overline{N} \gamma_\mu \gamma_5 \tau^a N \tag{41.37}$$

so the current is

$$\mathscr{A}_\mu^a = \Pi_\mu^b D^a \phi^b = \Pi_\mu^a \tag{41.38}$$

I write the current as $\mathscr{A}_\mu^a$ rather than as the usual axial current A_μ^a for a reason that will become clear in a moment. And just to write it down again,

$$\partial^\mu \mathscr{A}_\mu^a = \partial^\mu \Pi_\mu^a = -m_\pi^2 \phi^a \tag{41.39}$$

which certainly satisfies Nambu's formulation of PCAC: if $m_\pi = 0$, the axial current is conserved.

Why did I call it $\mathscr{A}_\mu^a$ rather than A_μ^a? Because we have no normalization condition on this current. We are considering λ^a here, but we could just as well have written $7\lambda^a$, in which case we would have obtained 7 times the current. If I'm going to make this current mimic, in the limit that $k^\mu \to 0$, the ordinary axial vector current (40.32), I will have to scale $\mathscr{A}_\mu^a$ so that A_μ^a's one-nucleon matrix element $\langle p|A_\mu^a|n\rangle$ agrees with (40.32). With this in mind, define

$$A_\mu^a \equiv -\frac{2m_N}{g} g_A \mathscr{A}_\mu^a = -g_A \overline{N}\gamma_\mu\gamma_5\tau^a N - \frac{2m_N g_A}{g}\partial_\mu\phi^a \tag{41.40}$$

so that

$$A_\mu = -\frac{2m_N}{g} g_A \mathscr{A}_\mu = -g_A \overline{p}\gamma_\mu\gamma_5 n - \frac{\sqrt{2}\,m_N g_A}{g}\partial_\mu\phi_{\pi^-} \tag{41.41}$$

With this definition, the nucleon term $-g_A\overline{N}\gamma_\mu\gamma_5\tau^a N$ has the right one-nucleon matrix elements (40.32) in lowest order perturbation theory; the second term $-(2m_N g_A/g)\partial_\mu\phi^a$ gives us the pion contribution (40.28).

Now it is easy to see that the Goldberger–Treiman relation works. If we take the matrix element of this axial vector current (normalized to have the right *nucleon* matrix element) between the vacuum and a one-pion state, in lowest order perturbation theory, the only thing that contributes is the pion field. The nucleon term in lowest order makes a nucleon–antinucleon pair, but to turn that into a pion requires higher powers of the strong interaction coupling constant g.[14] The matrix element of the axial vector, that is, the matrix element of its pion term, $-(\sqrt{2}m_N g_A/g)\partial_\mu\phi_{\pi^-}$, between the vacuum and π^- gives you ik_μ times $F_\pi e^{-ik\cdot x}$ divided by $\sqrt{2}$:

$$\langle 0|\frac{-\sqrt{2}\,m_N g_A}{g}\partial_\mu\phi_{\pi^-}|\pi^-\rangle = i\frac{k_\mu F_\pi}{\sqrt{2}}e^{-ik\cdot x} \tag{41.42}$$

Taking the divergence of each side and using the lowest order equations of motion for the ϕ_{π^-} field gives, in the limit as $k \to 0$,

$$-\frac{2m_N g_A}{g} = F_\pi \tag{41.43}$$

which *is* the Goldberger–Treiman relation again. So the Goldberger–Treiman relation has nothing to do with the strong interactions being strong. We can construct a model in which the strong interactions are *weak*, and we get exactly the same result.

Let's compare the gradient-coupling model to the Fermi–Yang model. The Feynman diagram coming out of the latter in Figure 41.2 implies $F_\pi \propto g_A g$, but that's wrong. The physics ain't like that: the pion doesn't really go into a proton and a neutron for pion beta decay.[15] In the gradient-coupling model, the pion *does* contribute to nucleon beta decay; the axial vector current comes out having a nucleon part and a pion part. They're linked together by the PCAC condition. If we changed the ratio of the two terms in (41.40) we would no longer have an equation that guarantees (in lowest order perturbation theory) that the matrix

---

[14] [Eds.] See note 15, p. 898.

[15] [Eds.] Fermi and Yang's 1949 article (*op. cit.*) neither discusses beta decay nor includes any diagrams, but the process $\pi^+ \to \overline{n} + p$ in Figure 41.2 is implicit in their Lagrangian. The same diagram also arises in the gradient-coupling model, but its contribution is only part of other processes second order in g, and so has nothing to do with the Goldberger–Treiman relation.

element is a constant. And then the process wouldn't run. That ratio is linked together in this model by having, in Nambu's way of looking at things, an almost conserved (or, to use the jargon, a *partially* conserved) axial vector current that *would* be conserved if the pion mass were zero.[16] That's how we've set up this model, so that the current would be conserved if the pion mass vanished. That tells us there must be a certain relation between the pion and the nucleon parts of the current. The process embodied in the diagram Figure 41.2 from the 1949 theory is just so much garbage.

41.4 Adler's Rule for the emission of a soft pion

Actually, this model has another use, beyond being an instructive example of a theory that embodies both good definitions of PCAC (the one following from the hypothesis of slowly varying pion–nucleon matrix elements, or the Nambu definition that says the current must be conserved when the pion mass goes to zero). Since it satisfies all of our assumptions in lowest order perturbation theory, it must yield all of our conclusions. It plays a role in a famous rule due to Adler.[17]

We consider some hadronic scattering process in which an initial state goes into a final state plus a pion carrying momentum k:

$$i \to f + \pi^a(k) \tag{41.44}$$

(where $k = p_f - p_i$). We wish to consider this process in the limit when all four components of k go to zero, that is, to relate it to the process $i \to f$. This is of course a totally *unphysical* limit; the pion is off the mass shell if $k^\mu \to 0$. Nevertheless we want to develop a rule analogous to the Goldberger–Treiman relation, for studying such a process in this limit. Depending on the case at hand and what kinematic regime we are in, we may or may not be able to extrapolate back to a physically observable region and obtain an interesting result. I won't go into the details here; this is just supposed to be a survey of current algebra methods. I will use it later on in this lecture, but for now I only want to show Adler's method.

The essential idea is that the amplitude for the one-pion emission matrix element, by the LSZ reduction formula (§14.2), is

$$\mathcal{A}_{fi} \propto \langle f|\pi^a(0)|i\rangle\,(k^2 - m_\pi^2) \tag{41.45}$$

Since these are momentum eigenstates we might as well consider the amplitude at $x = 0$, and forget about the $e^{-ik\cdot x}$; the factor of $(k^2 - m_\pi^2)$ comes from the reduction formula. On the other hand, PCAC says (41.5)

$$\langle f|\pi^a(0)|i\rangle\,(k^2 - m_\pi^2) = \frac{(k^2 - m_\pi^2)}{F_\pi m_\pi^2}\,\partial^\mu\,\langle f|A_\mu^a(0)|i\rangle \tag{41.46}$$

Since ∂^μ is $-ik^\mu$, with k the pion momentum, we can write

$$\mathcal{A}_{fi} \propto \frac{(k^2 - m_\pi^2)}{F_\pi m_\pi^2}(-ik^\mu)\,\langle f|A_\mu^a(0)|i\rangle \tag{41.47}$$

[16] [Eds.] Nambu, *op. cit.*

[17] [Eds.] S. L. Adler, "Consistency Conditions on the Strong Interactions Implied by a Partially Conserved Axial-Vector Current. II", *Phys. Rev.* **139B** (1965) 1638–1642; Y. Nambu and D. Lurié, "Chirality Conservation and Soft Pion Production", *Phys. Rev.* **125** (1962) 1429–1436; Coleman *Aspects*, p. 50; Cheng& Li *GT*, p. 155.

It looks at first glance as if this matrix element goes to 0 as $k \to 0$. This is in general not the case; sometimes it is, and sometimes it isn't. The reason why it does not *necessarily* go to zero can be best shown by a specific example.

Suppose we are considering nucleon–nucleon scattering. Some graphs that arise in perturbation theory illustrate the argument. Here is one, shown in Figure 41.3, of sufficient complexity to make the point. We have the axial vector current on one nucleon vertex although there's

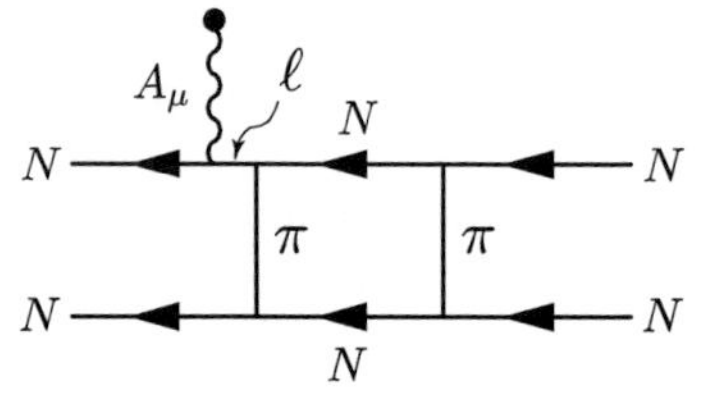

Figure 41.3: Pole graph, showing line ℓ

no propagator associated with it; I'll draw a little wiggly line with a terminal dot for the axial vector current, A_μ. Consider that graph, where the axial vector current attaches to an *external* line. Let ℓ be the line from the lower vertex of the axial current to the upper vertex of the leftmost pion. As the momentum transferred by the axial vector current goes to 0, the line ℓ goes onto the mass shell because it has the same momentum as the external line. Thus this graph will produce a pole from the ℓ propagator as $k \to 0$ in the matrix element of the axial vector current. I will call these graphs, where the axial vector current attaches to one of the external lines, **pole graphs**.

Pole graphs are not the complete set of graphs one can obtain by decorating this process. We could also have the axial vector current connecting somewhere in the middle of the diagram. I'll just call these **guts graphs**.

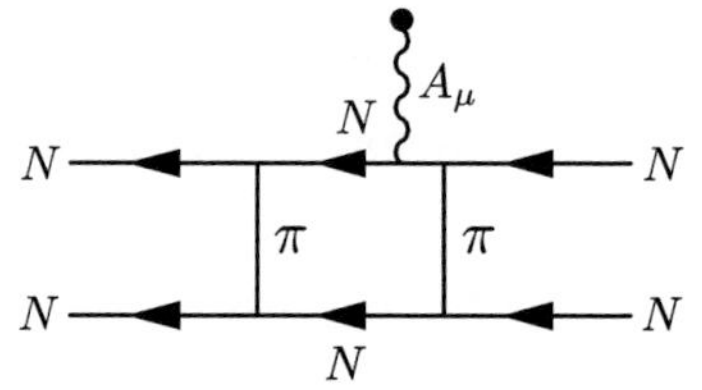

Figure 41.4: Guts graph

The important point about the guts graphs is that, except at special kinematic configurations, they will in general *not* develop singularities as $k \to 0$. The presence of singularities is governed by the **Landau rules**.[18] The particles on external lines are real, physical particles, but the others, on internal lines, aren't. The Landau rules are connected to how you assign internal momenta to a physical process in a diagram. If you have a vertex where a current carrying zero momentum meets an internal line, that doesn't change anything; it has absolutely no effect on the process. After all, the virtual particles are not on-shell, and so cannot produce

[18] [Eds.] Bjorken & Drell, *Fields*, Section 18.6, pp. 231–242; L. D. Landau, "On Analytic Properties of Vertex Parts in Quantum Field Theory", *Nuc. Phys.* **13** (1959) 181–192; James D. Bjorken, "Experimental tests of quantum electrodynamics and spectral representations of Green's functions in perturbation theory", thesis, Stanford University, 1959. Coleman discussed the Landau rules in the lectures on dispersion relations, regrettably not included in this book.

a pole as $k \to 0$. A particle goes along and at some point absorbs momentum zero. *Nu*, it keeps on going. Even at the places where in fact there are singularities, such as thresholds, one would expect from this analysis that the singularities that develop would be perhaps square root or logarithmic singularities, the sort associated with the beginnings of a cut. They wouldn't be of large enough power to kill this linear factor of k_μ out in front. So certainly at every point except thresholds, and possibly even there, the guts graphs contribute nothing to the emission of the soft pion (the pion with zero momentum). On the other hand, the pole graphs may or may not contribute; that depends on the particular process.

The pole graphs are exactly computable in terms of strong interaction processes *not* involving the emission of a pion because at this vertex we simply have the matrix element of the axial vector current at zero momentum transfer and everything to the right of that point is simply nucleon–nucleon scattering with no pion emitted. So the guts graphs contribute nothing and the pole graphs contribute something that's computable in terms of the strong interaction process *without* the soft pion.

There is a simple rule that summarizes this result. From Figure 41.3 we can find the residue of the pole, the axial vector current at zero momentum transfer: $\gamma_\mu\gamma_5$. We multiply that by k_μ and use the pion reduction, (41.47). Explicitly,

$$i\mathcal{A}_{fi} = \text{pole diagrams} + \mathcal{O}(k) = -\frac{g_A}{F_\pi}k^\mu \overline{u}'_f\gamma_\mu\gamma_5\tau^a u_i + \cdots = \frac{g}{2m_N}k^\mu \overline{u}'_f\gamma_\mu\gamma_5\tau^a u_i + \cdots \tag{41.48}$$

This is *precisely* the Feynman diagram contribution we would get in that preposterous nonrenormalizable gradient-coupling theory (41.32). We don't have to keep track of all the factors; they've got to come out the same as in the gradient-coupling theory, because that theory, though no one takes it seriously, obeys all of our assumptions. The upshot of this reasoning is **Adler's rule**: to lowest order on external lines,

Gradient-coupling theory is *exact* for the emission of a soft pion.

That's a compact way of stating it. In greater detail, Adler's rule says: to calculate a strong interaction matrix element for the emission of a soft pion, find the matrix element *without* the pion emission, and using gradient-coupling, sum all the terms obtained by attaching a pion to each external line.[19] You need only apply the gradient-coupling rule to the external lines; the contributions from the guts graphs vanish automatically. You still have to worry about the extrapolation. There may be singularities other than these that you have to consider. But as the pion momentum goes to zero, if you define the pion field to be the divergence of the axial vector current, then this is an exact statement. The typical applications are therefore near threshold, so that when you get to a physical pion, all of the invariants involving the pion momentum will be small; not just k^2 but also $k \cdot p$ where p is any other momentum in the problem.[20]

The application of this was first derived in another context by Nambu and Lurié. They applied it to pion production in nucleon–nucleon scattering near the pion production threshold,

---

[19] [Eds.] Coleman *Aspects*, p. 50.

[20] [Eds.] In Woit's notes, Coleman remarks that the Goldberger–Treiman relation follows as a special case of Adler's rule for a one nucleon initial and final state. This is shown explicitly in Weinberg *QTF2*, p. 190.

and related that to nucleon–nucleon scattering.[21] We're going to give an application of this shortly in which we won't have to worry about threshold singularities. Threshold singularities present special problems and there's a little song and dance of dubious plausibility to take care of them.

Now, if you can do it with *one* soft pion, why can't you do it with two? That would be better, and give us even more information. Or perhaps three or four? That would enable us, maybe, to discuss pion–nucleon scattering. We know what happens with one pion when the four-momentum goes to zero. If we also know what happens when two pion four-momenta simultaneously go to zero then we know an awful lot about pion–nucleon scattering. However, if we are studying two soft pions this way, it is the *time-ordered product*

$$\langle f|T\Big[\Big(\partial^\mu A^a_\mu(x)\Big)\Big(\partial^\nu A^b_\nu(y)\Big)\Big]|i\rangle \tag{41.49}$$

that goes into the reduction formula. These 4-divergences inside the time-ordered product are no help at all; they've got to be outside the time-ordered product where they can act on the Fourier transform factors through integration by parts and turn into momenta. Therefore we've got to pull the derivatives out of the time-ordered product. But as you may remember from our discussion of gradient-coupling theories (when we were talking about Feynman rules for derivative interactions, or indeed from an early homework problem[22]), when you pull a gradient operator out of a time-ordered product, you wind up with an extra equal-time commutator, due to the gradient acting on the θ function needed for time-ordering. Written symbolically,

$$\langle f|T\Big[\Big(\partial^\mu A^a_\mu(x)\Big)\Big(\partial^\nu A^b_\nu(y)\Big)\Big]|i\rangle = \partial^\mu\partial^\nu\,\langle f|T[A^a_\mu(x)A^b_\nu(y)]|i\rangle + \begin{pmatrix}\text{equal-time}\\ \text{commutator}\end{pmatrix} \tag{41.50}$$

Therefore, if we hope to discuss processes involving two soft pions, we have to know something about the commutation relations of the vector and axial vector currents. The vector currents don't look like they'll come in here, but in fact they *must*; they're required to close the algebraic structure. This is why we now begin to study current commutators and why the whole set of methods that I'm describing now is called **current algebra.**[23]

41.5 Equal-time current commutators

I will focus on the commutators of V^a_0, the vector current, and A^a_0, the axial vector current. It's actually only the temporal components that we will need because it's only the zero components that have time derivatives hooked on them, in (41.50).

First, the vector currents. As you'll recall, these are proportional (41.15) to the isospin currents (that's just the CVC hypothesis of Feynman and Gell-Mann), with a constant of proportionality equal to $2g_V$ (41.19). At equal times

$$[V^a_0(\mathbf{x},t), V^b_0(\mathbf{y},t)] = i\epsilon^{abc}2g_V V^c_0(\mathbf{x},t)\delta^{(3)}(\mathbf{x}-\mathbf{y}) \tag{41.51}$$

---

[21] [Eds.] Nambu and Lurié, *op. cit.*

[22] [Eds.] Problem 1.2, p. 49.

[23] [Eds.] Coleman *Aspects*, pp. 50–52; S. Treiman, R. Jackiw, and D. J. Gross, *Lectures on Current Algebra and its Applications*, Princeton U. P., 1972; S. L. Adler and R. F. Dashen, *Current Algebras and Applications to Particle Physics*, W. A. Benjamin, 1968; S. Treiman, R. Jackiw, B. Zumino, and E. Witten, *Current Algebra and Anomalies*, Princeton U. P., 1985.

This is the isospin algebra scaled up by $2g_V$. Integration over all space and division by $2g_V$ gives the isospin charges,[24] which are the isospin generators:

$$I^a = \frac{1}{2g_V}\int d^3\mathbf{x}\, V_0^a(\mathbf{x},t) \tag{41.52}$$

So the isospin charges obey the right commutators (24.38):

$$[I^a, I^b] = i\epsilon^{abc}I^c \tag{41.53}$$

The δ function in (41.51) ensures that the currents $\{V_0^a\}$ commute for spacelike separations. Actually the δ function is a bit of a swindle. We could have made the same statement if we put a lot of ugly terms on the right-hand side involving derivatives of δ functions. They would all go away when we integrate the commutators to obtain the isospin algebra. But (41.51) is certainly the structure one gets in the simplest models, where you just get δ functions from commuting the canonical fields and canonical momenta—the currents are proportional to ϕ_π^a, so we're going to get δ functions at equal times. In more complicated models there might be terms proportional to the gradient of a δ function; we will ignore them here. In fact when we consider a specific application, we will see the possible presence of such terms is irrelevant. Anyway, these are the commutators that would hold in a simple model such as any isospin-symmetric theory of pions and nucleons with renormalizable interactions, or the quark model. So I'll just write down these simplest forms:

$$[V_0^a(\mathbf{x},t), A_0^b(\mathbf{y},t)] = i\epsilon^{abc}2g_V A_0^c(\mathbf{x},t)\delta^{(3)}(\mathbf{x}-\mathbf{y}) \tag{41.54}$$

This is just the statement that the axial currents transform like an isovector, and therefore the isospin generators applied to them rotate them as an isovector should be rotated.

Fortunately neither of these two is what we really want. The one we *want* is

$$[A_0^a(\mathbf{x},t), A_0^b(\mathbf{y},t)] = \,? \tag{41.55}$$

This commutator has nothing to do with the isospin group, and it's *dependent on the model.*. For example, in the gradient-coupling model the axial currents commute with each other, and the commutator is zero. They commute with each other because they're canonical momentum densities for the three pion fields, up to a proportionality factor, and the canonical momenta commute at equal times. On the other hand, in the quark model, this commutator is *not* zero. So the question of what we should put for the commutator depends on what model we pick.

This is very nice, for the following reason. Up to now we've been doing things that were almost totally model-independent. In the strong interactions we always worry about model-dependent predictions because we can't compute anything in zeroth approximation in the strong interactions. We don't know whether one theory makes a given prediction or another theory makes a different prediction. We can't do perturbation theory. But (41.55) is something which we can compute in any given model without having to solve all the equations of motion, if we know what the axial currents are. Coupling constants are irrelevant: strong, weak, small; it doesn't matter. So this offers us something we can abstract from a model and write down. And then, if we are able to carry through the same sorts of computations for two

[24] [Eds.] Despite appearances, the integral is independent of time. See §6.2, and in particular (6.57) and the discussion following.

soft pions that we've just done for one soft pion, we can check that prediction without having to solve a problem of strong interaction dynamics. This is one of the rare instances in the theory of the strong interactions where you can extract something from the particular form of a Lagrangian that is true for some theories and not true for others, and use it to make experimentally verifiable predictions. Well, we had one other instance of that: the symmetries of the Lagrangian could be used that way. But this is something that goes beyond simple symmetry. It is not an *experimentum crucis* in the sense of scientific method as described by 19th century philosophers of science. There may be 40 billion models that give the same answer for the question posed by the commutator (41.55). And there may be 40 billion that give different answers. But still, if *one* answer gives us the right predictions, we can reject all the others.

It's fairly straightforward to compute this commutator, (41.55), in any given model. The currents are expressed as functions of the canonical fields. You know the commutators of the canonical fields and use the Jacobi identity.[25] There may be technical difficulties, because even in a renormalizable theory the product of two fields at exactly the same space-time point, which is what goes into a current, is a divergent and ill-defined object, and therefore one may require some special care. Normally the way to do that is to *split the points*, give them slightly different values but all at the same time, compute the commutators, which is perfectly kosher, and then let the splitting go to zero.[26] That's a very clean way of doing it, and sometimes it will reveal terms, called **Schwinger terms**, proportional to derivatives of δ functions, which you would have missed if you had been naïve.[27] If you are worried about ultraviolet divergences screwing you up, you might try to prove these commutators order-by-order in perturbation theory. That's a job for a very nervous person.

The terms discovered by Schwinger are proportional to a derivative of a δ function in the V_0-V_i commutator. He found them originally in electrodynamics, where it looks, naïvely, as if there should not be such terms, but if you're a little bit more careful, in particular, if you split points in the way described, they appear. The phrase *Schwinger term* is sometimes used in the literature to describe "anomalous" terms and commutators generically, terms that are not there if you're sloppy but are there if you make a more careful investigation. In fact the Schwinger term in the V_0-V_i commutator is a bit mysterious in that it has a divergent coefficient. When you get to the point of deriving Ward identities, it washes out in the end, so that term is irrelevant. They're not really mysterious, they're very well understood; they're just troublesome. It's just one of those things like remembering to zip up your fly if you're a man; you've got to remember to check for possible Schwinger terms. They're not going to be relevant to what I'm going to do, so I won't talk about them here. If you were to take a course on current algebra, you'd hear a lecture on how to deal with Schwinger terms.

To do calculations with axial currents, we have to make a guess for the question mark in (41.55); we have to have some model of the strong interactions from which we can abstract that commutator. I'll take a simple model, the quark model I described earlier.[28] Nobody knows how to compute anything with the quark model because in lowest order perturbation theory

---

[25] [Eds.] $[A,[B,C]] + [B,[C,A]] + [C,[A,B]] = 0$.

[26] [Eds.] For a brief discussion of the point-splitting technique, see Peskin & Schroeder *QFT*, Section 19.1.

[27] [Eds.] Julian Schwinger, "Field Theory Commutators", *Phys. Rev. Lett.* **3** (1959) 296–297; Itzykson & Zuber, *QFT*, p. 224 and p. 530.

[28] [Eds.] Cheng & Li *GT*, Section 4.4, pp. 113–124; Griffiths *EP*, Section 1.8, pp. 37–44.

none of the particles we know are present, there are just the damned quarks.[29] Nevertheless it's a well-defined Lagrangian field theory and we can compute the equal-time commutators. And if they don't involve any brand new objects that we need special dynamics to compute, so much the better.

In the quark model, hadrons are made up out of quarks, which are Fermi–Dirac fields. The quarks are held together by some forces, mediated by gluons. But gluons won't contribute to the isospin or hypercharge currents, because they couple to *color*, which commutes with ordinary (flavor) SU(3). And therefore the gluons carry no charge, no hypercharge, no isospin, no *nothing*, except color. In particular, they aren't going to contribute to the weak interaction currents. The currents in the quark model will just be quark bilinears, $\overline{q}M\gamma_\mu(\gamma_5)q$: q is the quark field, and M is some isospin matrix depending on what current you're looking at. A quark current J^μ has the form

$$V_\mu \pm A_\mu = \overline{q}M\gamma_\mu(1\pm\gamma_5)q \tag{41.56}$$

Which current you're looking at will tell you what matrix M acts on the (here suppressed) isospin or SU(3) indices of the quark. It's very easy to compute the commutators of these currents just by using the equal-time anticommutators of the quark fields, but it's even easier to do the computation using a little trick which I will now describe.

Suppose the quarks were massless and non-interacting. That's a preposterous assumption, but since we're only computing an equal-time commutator and those statements don't affect the equal-time commutators, it doesn't matter. The $1\pm\gamma_5$ are helicity projection operators,[30] akin to (20.124a). In a theory of free massless fermions, a Dirac spinor splits into two uncoupled Weyl spinors (§19.1) which are eigenstates of γ_5 with eigenvalues ±1, because γ_5 anti-commutes with $\not{p}$. So $1\pm\gamma_5$ are the projection operators on two uncoupled Weyl spinors; + for one and − for the other. The same thing works on the other side. If we drag $1\pm\gamma_5$ through the γ_μ the γ_5 changes sign. But since γ_5 is anti-self bar (20.103), when it acts to the left, on $\overline{q}$, its sign changes again. If the quarks were massless and non-interacting, one of these currents with the + sign would couple left-handed quarks only to left-handed quarks, or positive helicity to positive helicity. The other would couple negative helicity only to negative helicity. Thus the currents with the + and − signs would deal with two *completely decoupled* dynamical systems having nothing to do with each other. In particular, at equal times they would have to commute. Therefore we have for *any* quark currents

$$[V_0^a + A_0^a, V_0^b - A_0^b]\Big|_{\text{equal times}} = 0 \tag{41.57}$$

This would be a trivial statement if the quarks were massless and non-interacting. If that were the case, the commutators would be zero not only at equal times but at *all* times: you would have two separate worlds of right-handed quarks and left-handed quarks, and they would never see each other. On the other hand we also know these commutators can be computed just from what we already have, from the equal-time Dirac algebra of the quark fields, and so they have nothing to do with whether or not the quarks are massless and non-interacting. The point is that this must be true regardless of the existence of the quark interactions. In the *gradient-coupling* model, $V+A$ and $V-A$ *don't* commute, so the commutator $[A_0, A_i]$

[29] [Eds.] Happily, this is no longer true. It was not really true in 1976, but we have learned a great deal more since then.

[30] [Eds.] Peskin & Schroeder *QFT*, p. 142.

you derive from that assumption is not correct. The reason is that the axial vector current has a term in it linear in the gradient of the pion field. The vector current, the conventional isospin current, does *not* have a gradient term. Work it out and see for yourself.

We can compute the axial vector commutator in five minutes by using the equal-time commutators of q with $\overline{q}$, which aren't affected by either the quarks' masses or their interactions. We will call this principle (abstracted from the quark model, but in fact true in a much larger class of models) the **chirality principle**. The word *chirality* is often associated with current algebras. ("Kheir" is the Greek word for "hand", as in *chiromancy* which is what Madam Selena practices down in Harvard Square, reading palms.[31]) This is because chiral symmetry is associated with left-handed and right-handed spin-½ particles. An algebraic statement like (41.57) is called a statement about *chiral algebra.*[32]

We can now fill in the question mark, for the quark model, by elementary algebra:

$$[(V_0^a + A_0^a)(\mathbf{x},t), (V_0^b - A_0^b)(\mathbf{y},t)] = 0 = [V_0^a(\mathbf{x},t), V_0^b(\mathbf{y},t)] - [A_0^a(\mathbf{x},t), A_0^b(\mathbf{y},t)] \tag{41.58}$$

because the cross-terms cancel (the commutators are symmetric in $\mathbf{x}-\mathbf{y}$, but antisymmetric in $\{a,b\}$.) This completes our current algebra:

$$\begin{aligned}
[V_0^a(\mathbf{x},t), V_0^b(\mathbf{y},t)] &= i\epsilon^{abc} 2g_V V_0^c(\mathbf{x},t)\delta^{(3)}(\mathbf{x}-\mathbf{y}) \\
[V_0^a(\mathbf{x},t), A_0^b(\mathbf{y},t)] &= i\epsilon^{abc} 2g_V A_0^c(\mathbf{x},t)\delta^{(3)}(\mathbf{x}-\mathbf{y}) \\
[A_0^a(\mathbf{x},t), A_0^b(\mathbf{y},t)] &= i\epsilon^{abc} 2g_V V_0^c(\mathbf{x},t)\delta^{(3)}(\mathbf{x}-\mathbf{y})
\end{aligned} \tag{41.59}$$

We have an algebraically closed system. We can now compute arbitrary numbers of commutators of these objects, and therefore the sorts of things that would involve an arbitrary numbers of soft pion emissions.

The associated charges (the space integrals of the currents) obey a similar algebra without the δ functions. In fact it's easy to see what that algebra is. Going back to (41.57), if $V+A$ and $V-A$ commute with each other, then $V+A$ obeys an isospin algebra all by itself and $V-A$ obeys an isospin algebra all by itself. It's rather like the decomposition we made (§18.3, p. 376) for the Lorentz group into two rotation groups, except there's no need to insert an i into half of the operators. This is the algebra of SU(2) ⊗ charges, so we don't have to worry about the δ function. The charges obey an SU(2) ⊗ SU(2) algebra. This is sometimes called the SU(2) ⊗ SU(2) *chiral algebra* and, along with PCAC, is one of the two pillars of all current algebra computations.

A nice way of getting the same thing was pointed out by Gell-Mann and Ne'eman.[33] There is a vague idea that the weak interactions are pretty much the same for leptons as they are for hadrons. This vague idea has a dignified name, called **lepton-hadron universality**. It was

[31] [Eds.] Greek χείρ, "kheir", (hand). The adjective "chiral" was evidently introduced into science by Lord Kelvin in his Robert Boyle Lecture, Oxford University, May 16, 1893: "I call any geometrical figure, or group of points, 'chiral', and say that it has 'chirality' if its image in a plane mirror, ideally realized, cannot be brought to coincide with itself. Two equal and similar right hands are homochirally similar. Equal and similar right and left hands are heterochirally similar or 'allochirally' similar (but heterochirally is better). These are also called 'enantiomorphs,' after a usage introduced, I believe, by German writers. Any chiral object and its image in a plane mirror are heterochirally similar." Lord Kelvin (Sir William Thomson), *The molecular tactics of a crystal,* Oxford U. P., 1894, §22, note [8]. Available at `http://www.gutenberg.org`, book 54976.

[32] [Eds.] Cheng & Li *GT*, pp. 132–136.

[33] [Eds.] Murray Gell-Mann and Yuval Ne'eman, "Current-Generated Algebras", *Ann. Phys.* **30** (1964) 360–369.

an idea that was floating around for 20 years, that all weak interactions have the same strength. The question is, how can you compare leptons and hadrons? Leptons are electrons, muons, and taus, and their neutrinos. Hadrons are nucleons and pions and all the others in those big fat Particle Data Group tables. Or, if you take the other attitude, hadrons are quarks and gluons. Those still don't look much like leptons.[34] They've got strong interactions and the leptons don't. If you consider color, there are a lot more quarks than there are leptons. What does it mean that the weak interactions are pretty much the same for the quarks as for leptons? (I'll ignore for the moment the strangeness-changing currents, but I'll make a remark about them later.) Here we have obtained an algebraic structure that we could generate completely from the weak interaction currents. We take the positively charged weak interaction current and its adjoint. These obey the same algebra as the isospin-raising and lowering operators (as is to be expected from the CVC hypothesis). We commute the charged weak currents to get the I_z-like operator, giving us the three components of the isotriplet. We can break the whole thing up into parity-conserving and parity-violating parts and generate this entire algebraic structure by successive commutations just from the weak interaction current. Or if we wanted to, we wouldn't have to do the parity breakup; we could just look at the $V - A$ part of the current and say that's the whole thing. That will give us an SU(2) algebra.

Now if we did the same thing for leptons, it would be the same computation. And it would give the same answer as is shown from this quark model example. Because instead of building the currents from quarks in (41.56), we could use lepton pairs: the electron and its neutrino, the muon and its neutrino, or the tau and its neutrino. If we consider the electron and electron neutrino to be some sort of isodoublet, then the weak interaction current is the matrix element of an isospin-raising current. It's the same sort of structure:

$$\overline{\nu}_e\gamma_\mu(1 \pm \gamma_5)e \tag{41.60}$$

is just like

$$\overline{p}\gamma_\mu(1 \pm \gamma_5)n \tag{41.61}$$

Therefore Gell-Mann and Ne'eman suggested that the right way to make the comparison was to state: you have a lepton current and a hadron current. From the lepton current you generate an algebra. You take the current and its adjoint and commute and commute until the whole thing closes. From the hadron current you generate an algebra; you take the whole thing and commute and commute until the whole thing closes. The precise statement of lepton-hadron universality is that *these two algebras are the same;* they're *isomorphic algebraic structures.*[35] As we have demonstrated, this is a statement that is consistent with leptons being different from hadrons. There doesn't have to be a precise parallelism, as many quarks as there are leptons or anything like that. You could have fundamental scalar mesons among the hadrons, and so on. It wouldn't matter. The right statement of universality is that the algebraic structures are the same. I state without proof that this is also true if you take account of strangeness-changing currents, as in the Cabibbo theory. As far as the commutators go, the direction chosen by the medium-strong interactions is irrelevant. You can make an SU(3) rotation that will turn the Cabibbo currents into pure isospin-raising and lowering currents, whence the algebraic structure is the same. So the Cabibbo theory is also consistent with this form of universality; the algebra is exactly the same.

---

[34] [Eds.] This is a matter of taste. Many others see a striking resemblance between quarks and leptons.

[35] [Eds.] D. Burton, *An Introduction to Abstract Mathematical Structures*, Addison–Wesley, 1965, p. 57; Michael Artin, *Algebra*, Prentice-Hall, 1991, Section 2.3, pp. 48–51.

Now, we will apply these current commutators to study a pion with momentum k and isospin a scattering off of some initial hadronic state h with momentum p, going into a final pion state with momentum q and isospin b plus a final state h' with momentum p', as in Figure 41.5. In this particular example I will assume h is a nucleon. That's the case for which we have most experimental information. Sometimes I will use facts that depend on a specific feature, such as its spin. But in fact the arguments will be general and h could be any hadronic target.

$$\pi^a(k) + h(p) \to \pi^b(q) + h'(p') \tag{41.62}$$

The general technique is as follows. I will obtain constraints on the form of the amplitude

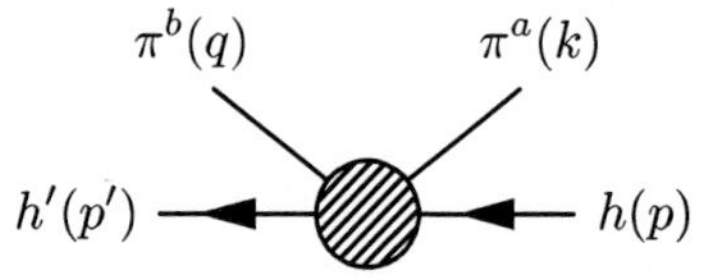

Figure 41.5: Pion–hadron scattering

$\mathcal{A}^{ba}$ for the process (41.62) at (or near) the point $q = k = 0$. (We will actually get a derivative at that point.) I will then make a power series expansion, keeping track of only the terms I know, and extrapolate it to the physical point where the pion has the smallest possible energy, that is, to threshold. One might be a little nervous about that because a threshold is the beginning of a cut, but I'm sorry, it's the best I can do, and the best anyone can do. We will just have to cross our fingers and hope it works out. So the game is to make a power series expansion about the point $q = k = 0$, extrapolate to threshold and get the threshold amplitudes, which are proportional to the scattering lengths,[36] and compare them with the experimentally observed scattering lengths. I will systematically neglect terms of $\mathcal{O}(m_\pi^2)$, because it will turn out that I can't go beyond first order in this expansion, and m_π^2 is second-order in certain invariants.

In order to do the extrapolation we had better count how many independent invariants we have, so we will know how many terms to write down in the power series. There's p^2 and p'^2, which are both equal to the square of the target mass, m_T:

$$p^2 = p'^2 = m_T^2 \tag{41.63}$$

We will not take the target off the mass shell; there is no need to do that. If the pions had a fixed mass there would be two invariants we would have to deal with. One would be for example $k \cdot p$, related to the pion mass in the center of energy frame.[37] The other would be $k \cdot q$, related to momentum transfer from the pions to the target. Finally, there are the masses of the two pions, k^2 and q^2. That's a complete list of independent invariants. In our entire range of extrapolation, k^2 and q^2 are of $\mathcal{O}(m_\pi^2)$, so we don't even have to worry about first-order terms in the power series expansion in them. So is $k \cdot q = \mathcal{O}(m_\pi^2)$; when the pion is at threshold, $k = q$ and then the product is k^2. Therefore we have in fact only *one* coefficient to expand our power series in, $k \cdot p$.

[36] [Eds.] Landau & Lifshitz, *QM*, p. 502; A. Messiah, *Quantum Mechanics*, North Holland Publishing, 1962, p. 408 and p. 861; M. L. Goldberger and K. M. Watson, *Collision Theory*, John Wiley, 1964, pp. 287–298. The scattering length a is closely related to the s-wave phase shift; see Problem 22.1. Note that different authors define the scattering length with different signs; see note 1, p. 920.

[37] [Eds.] See the paragraph following (5.91), p. 96.

The next step to construct the power series for our amplitude. There will be a constant term, the value when k and q are 0. There will be a first-order term times $k \cdot p$. We'll have to keep that because when we extrapolate to threshold, it's of order m_π. There will be all sorts of terms of order m_π^2, which we are going to neglect because we don't know what else to do with them (but at least the neglect is systematic; benign neglect, if you will). And there will be pole terms which, just as in the discussion of Adler's rule, will have to be taken account of separately. These terms can vary rapidly since they have poles in their denominators. Thus

$$\mathcal{A}^{ba} = \mathcal{A}_0^{ba} + \mathcal{A}_1^{ba}(k \cdot p) + \mathcal{O}(m_\pi^2) + \text{pole terms} \tag{41.64}$$

Here is an advanced glance at what is going to occupy us for the first half of the next lecture. We're going to look systematically at these terms, one after the other. We will show that the pole terms are also of order m_π^2, so they are in fact negligible. We will show that $\mathcal{A}_0^{ba}$ is likewise negligible; it is of order m_π^2. These steps will follow without any specific assumption about the form of the axial-axial commutators. They will be in that sense model-independent. We will just use the assumptions that we had before we filled in the question mark. Finally we will get the $\mathcal{A}_1^{ba}$ term from the current commutators. It, thank God, will *not* be of order m_π^2 (although it would be interesting if the pion–nucleon scattering lengths were zero, up to order m_π^2). We'll actually get a precise expression for that term from the current commutator. I will then assemble the whole thing and compare it with experiment.

Problems 22

22.1 Consider the scattering (below inelastic threshold) of two distinct spinless particles. A two-particle state with definite total momentum and vanishing center-of-momentum angular momentum is necessarily an eigenstate of the S-matrix; the associated eigenvalue is defined to be

$$e^{2i\delta}$$

where δ is the s-wave phase shift. The s-wave scattering length, a, is defined to be the leading term in the expansion of δ near threshold:

$$\delta = ak + \cdots \tag{P22.1}$$

where k is the center-of-momentum momentum of either particle. Find the relation between a and the invariant Feynman amplitude, $\mathcal{A}$, evaluated at threshold. How (if at all) does this relation change if the two particles are identical? (There is no loss of generality in considering spinless particles; near threshold, s-wave amplitudes dominate all others, and thus spin angular momentum and (vanishing) orbital angular momentum are *independently* conserved.)

Comment: You could have done this exercise at any time within the past few months. It appears now because we shall shortly be computing invariant Feynman amplitudes at threshold and comparing them to experimental measurements of s-wave scattering lengths.

(1982b 14)

22.2 Consider the following theory of the interactions of a massless Dirac field, ν (the neutrino), a charged massive Dirac field, e (the electron), and a massive, charged (i.e., complex) vector field, W_μ:

$$\mathscr{L} = \overline{\nu}(i\partial\!\!\!/)\nu + \overline{e}(i\partial\!\!\!/ - m)e - \tfrac{1}{2}F^{\mu\nu}F^*_{\mu\nu} + \mu^2 W^\mu W^*_\mu + g(W_\mu \overline{e}\gamma^\mu(1+\gamma_5)\nu + W^*_\mu \overline{\nu}(1-\gamma_5)\gamma^\mu e) \tag{P22.2}$$

where $F_{\mu\nu} = \partial_\mu W_\nu - \partial_\nu W_\mu$, and g, μ and m are positive numbers.

For the process

$$\nu + \overline{\nu} \to W + \overline{W}$$

there are several independent amplitudes at fixed energy and angle. We don't have to worry about the helicity of the neutrino, because the factor of $(1+\gamma_5)$ guarantees that only one helicity state participates in dynamics; however, each W can have helicity (spin along the direction of motion) of $\{1, 0, -1\}$, and thus there are nine amplitudes. An interesting limit in which to consider these amplitudes is that of high center-of-momentum energy, with center-of-momentum scattering angle θ fixed, but with $\theta \neq 0$ and $\theta \neq \pi$. (This last restriction guarantees that all three Mandelstam invariants—s, t, and u—grow with energy.)

(a) To lowest non-trivial order of perturbation theory, $\mathcal{O}(g^2)$, some of the nine helicity amplitudes approach (angle-dependent) constants in the high-energy fixed-angle limit described above; we will call these amplitudes "nice". Others, however, grow as a power of the energy; these amplitudes are "nasty". Which are the nasty amplitudes? Find the explicit high-energy forms of the nasty amplitudes, retaining the leading power of the energy only. (Don't worry about getting the phase or the sign right; in any case I haven't defined the phase of helicity eigenstates.)

(b) Now consider adding another term to the Lagrangian, involving a second massive Dirac field, e', of *opposite* charge (possibly a positron, but its mass M need not be the same as the electron's):

$$\mathscr{L} \to \mathscr{L} + \overline{e}'(i\partial\!\!\!/ - M)e' + f(W_\mu^* \overline{e}'\gamma^\mu(1+\gamma_5)\nu + W_\mu \overline{\nu}(1-\gamma_5)\gamma^\mu e') \tag{P22.3}$$

where M and f are positive numbers. A traveller once told me that if f were chosen proportional to g, *some* of the nasty amplitudes in this process would become nice. But I've forgotten which ones, and what the constant of proportionality is. Find out for me.

Possibly useful information: At some stage in this computation, you may want the Dirac matrices in the standard representation ($\mathbb{1}$ is a 2×2 identity matrix, $\mathbb{O}$ is a 2×2 zero matrix, and σ^i are the three Pauli matrices):

$$\gamma^0 = \begin{pmatrix} \mathbb{1} & \mathbb{O} \\ \mathbb{O} & -\mathbb{1} \end{pmatrix} \qquad \gamma^i = \begin{pmatrix} \mathbb{O} & \sigma^i \\ -\sigma^i & \mathbb{O} \end{pmatrix} \qquad \gamma_5 = \begin{pmatrix} \mathbb{O} & \mathbb{1} \\ \mathbb{1} & \mathbb{O} \end{pmatrix}$$

(1981 253b Final, Problem 3)

Solutions 22

22.1 (a) We work in the center of momentum (CM) frame. Let $|\mathbf{k}\rangle$ be a two particle state with momentum $\mathbf{k}$; asymptotically this ket describes two plane waves. Near the threshold for inelastic scattering, the scattering of spinless particles is isotropic; above threshold a bound state knows no distinguished direction and decays isotropically. Thus

$$\langle \mathbf{k}'|\mathrm{S}-1|\mathbf{k}\rangle = \mathcal{A}\,\delta(E'-E) \tag{S22.1}$$

where $\mathcal{A}$ is the invariant Feynman amplitude. Now let $|k\rangle$ be a s-wave state, and thus rotationally invariant, with definite linear momentum $k = |\mathbf{k}|$. Then

$$|k\rangle = \frac{1}{\sqrt{4\pi}k}\int d^3\mathbf{k}'\,\delta(k-|\mathbf{k}'|)\,|\mathbf{k}'\rangle \tag{S22.2}$$

and $\langle k'|k\rangle = \delta(k-k')$. Also,

$$\langle k'|\mathrm{S}-1|k\rangle = 4\pi k^2\mathcal{A}\,\delta(E'-E) \tag{S22.3}$$

The ket $|k\rangle$ is obviously an eigenstate of S: momentum is conserved and the final state must be in an s-state as well. The eigenvalue is $e^{2i\delta}$ since S is unitary. That is,

$$(\mathrm{S}-1)\,|k\rangle = (e^{2i\delta}-1)\,|k\rangle = 2i\delta\,|k\rangle = 2iak\,|k\rangle \tag{S22.4}$$

for small k. Now below threshold,

$$E = \frac{|\mathbf{k}|^2}{2m_1} + \frac{|\mathbf{k}|^2}{2m_2} + m_1 + m_2 = \frac{|\mathbf{k}|^2}{2\mu} + m_1 + m_2 \tag{S22.5}$$

where μ is the reduced mass. But (note 8, p. 9)

$$\delta(E'-E) = \delta\left(\frac{|\mathbf{k}'|^2}{2\mu} - \frac{|\mathbf{k}|^2}{2\mu}\right) = \frac{1}{||\mathbf{k}|/\mu|}\delta(|\mathbf{k}|'-|\mathbf{k}|) = \frac{\mu}{k}\delta(k'-k) \tag{S22.6}$$

Substituting this expression into (S22.3),

$$\langle k'|\mathrm{S}-1|k\rangle = 4\pi k\mu\,\mathcal{A}\,\delta(k'-k) \tag{S22.7}$$

Taking the inner product of both sides of (S22.3) with $|k'\rangle$,

$$\langle k'|\mathrm{S}-1|k\rangle = 2iak\,\langle k'|k\rangle = 2iak\,\delta(k'-k) \tag{S22.8}$$

Comparing this equation with the previous equation we obtain $a = -2\pi i\mu\mathcal{A}$. ■

22.2 The process to be considered is

$$\nu(p,u) + \overline{\nu}(p',v) \to \overline{W}(q,\varepsilon) + W(q',\varepsilon') \tag{S22.9}$$

described by this Feynman diagram:

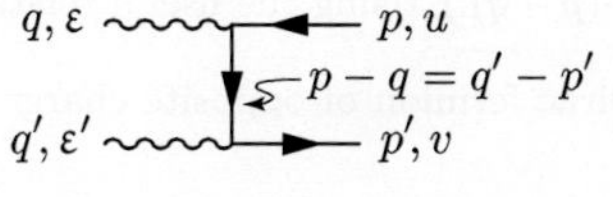

Note that there is *no* crossed graph, because there is *no* $W^*\overline{e}\nu$ vertex. The invariant amplitude is given by

$$i\mathcal{A} = (-ig)^2\overline{v}(1-\gamma_5)\not\varepsilon'^*\frac{i}{\not p - \not q - m}\not\varepsilon^*(1+\gamma_5)u \tag{S22.10}$$

We write down everything in the CM frame:

$$\begin{aligned} p &= E(1,0,0,1); & p' &= E(1,0,0,-1) \\ q &= (E, |\mathbf{q}|\sin\theta, 0, |\mathbf{q}|\cos\theta); & q' &= (E, -|\mathbf{q}|\sin\theta, 0, |\mathbf{q}|\cos\theta) \end{aligned} \tag{S22.11}$$

where $|\mathbf{q}| = \sqrt{E^2-\mu^2} \to E$ as $E\to\infty$. Similarly

$$(p-q)^2 - m^2 = -2q\cdot p + q^2 - m^2 \to -2E^2(1-\cos\theta) \text{ as } E\to\infty \tag{S22.12}$$

So the mass term in the Feynman denominator is negligible in this limit if $\theta \neq 0$.

Now for the spinors. The neutrino spinor u obeys these conditions:

$$\not p u = 0 = E(\gamma^0 - \gamma^3)u; \qquad \tfrac{1}{2}(1+\gamma_5)u = u; \qquad u^\dagger u = 2E \tag{S22.13}$$

The solution to these is

$$u = \sqrt{E}\begin{pmatrix}1\\0\\1\\0\end{pmatrix}; \qquad \text{likewise, because } \not p' v = 0, \quad v = \sqrt{E}\begin{pmatrix}0\\1\\0\\1\end{pmatrix} \tag{S22.14}$$

These spinors obey a useful relation (which you can check easily):

$$\overline{v}\gamma^\mu u = 2E(0,1,i,0) \tag{S22.15}$$

The polarization vectors, when $\theta = 0$, are given by (see (26.77) and (26.78), p. 567)

$$\begin{aligned} \varepsilon^{(\pm)} &= \frac{1}{\sqrt{2}}(0,1,\pm i,0) \\ \varepsilon^{(0)} &= \frac{1}{\mu}(|\mathbf{q}|,0,0,E) = \frac{q}{\mu} + \mathcal{O}(\frac{\mu}{E}) \end{aligned} \tag{S22.16}$$

For $\theta \neq 0$, we rotate by θ for $\overline{W}$,

$$\begin{aligned} \varepsilon^{(\pm)} &= \frac{1}{\sqrt{2}}(0,\cos\theta,\pm i,-\sin\theta) \\ \varepsilon^{(0)} &= \frac{q}{\mu} + \mathcal{O}(\frac{\mu}{E}) \end{aligned} \tag{S22.17}$$

the superscripts denoting the $\overline{W}$'s helicity, $\overline{h}$. For W's polarization vectors when $\theta \neq 0$, we rotate by $\theta + \pi$:

$$\begin{aligned} \varepsilon'^{(\pm)} &= \frac{1}{\sqrt{2}}(0,-\cos\theta,\pm i,\sin\theta) \\ \varepsilon'^{(0)} &= \frac{q'}{\mu} + \mathcal{O}(\frac{\mu}{E}) \end{aligned} \tag{S22.18}$$

the superscripts denoting the W's helicity, h.

Just by counting powers, we see that the amplitudes are nice unless $h\overline{h} = 0$, i.e., at least one of the vectors has zero helicity. Even in that case, we get nasty amplitudes only from the leading term in the high-E expression from $\varepsilon^{(0)}$ or $\varepsilon'^{(0)}$. From (S22.10) the computation gives, for $\overline{h} = 0$ and $h \neq 0$,

$$\mathcal{A} = -g^2\overline{v}(1-\gamma_5)\not\varepsilon'^*\frac{\not q/\mu}{\not p - \not q}(1+\gamma_5)u = +\frac{g^2}{\mu}\overline{v}(1-\gamma_5)\not\varepsilon'^*\frac{\not p - \not q}{\not p - \not q}(1+\gamma_5)u = \frac{4g^2}{\mu}\overline{v}\not\varepsilon'^* u \tag{S22.19}$$

where we've used $\not p u = 0$ in the second step, and $(1+\gamma_5)u = 2u$ in the third. For $h = 0$,

$$\mathcal{A} = -\frac{4g^2}{\mu}\overline{v}\not\varepsilon^* u \tag{S22.20}$$

(The sign changes, because $p' - q' = -(p-q)$.) Using the useful relation (S22.15), we find that there are five nasty amplitudes:

(b) Now we introduce e', a massive Dirac fermion of opposite charge. This interaction leads to the *crossed* graph

$\overline{h}$	h	$\mathcal{A}$
0	± 1	$\dfrac{4\sqrt{2}g^2E}{\mu}[\cos\theta \mp 1]$
± 1	0	$\dfrac{4\sqrt{2}g^2E}{\mu}[\cos\theta \pm 1]$
0	0	$-\dfrac{8g^2E^2}{\mu^2}\sin\theta$

Table S22.1: The five nasty amplitudes in $\overline{\nu}\nu \to \overline{W}W$

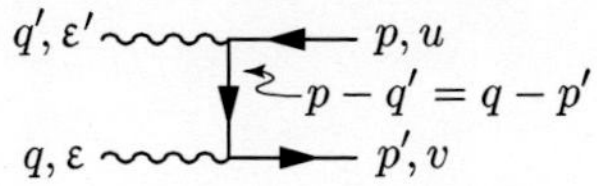

Everything goes as before, except $g \to f$, $m \to M$, $e \leftrightarrow e'$, $q \leftrightarrow q'$. For $\overline{h} = 0$, in addition to the amplitude (S22.19), we get a new term:

$$\mathcal{A}' = -f^2\overline{v}(1-\gamma_5)\frac{\not{q}/\mu}{\not{q}-\not{p}'}\not{\varepsilon}'^*(1+\gamma_5)u = -\frac{4f^2}{\mu}\overline{v}\not{\varepsilon}'^*u \tag{S22.21}$$

Adding this amplitude $\mathcal{A}'$ to the result for $\mathcal{A}$, the nasty terms cancel if $f^2 = g^2$. A similar result holds for $h = 0$. That is, *all* the nasty amplitudes become nice if $f^2 = g^2$. ■

42

Current algebra and pion scattering

We continue our investigation of pion scattering. In principle the target could be anything, although at a later stage we will make it a proton.

42.1 Pion–hadron scattering without current algebra

Consider the process

$$\pi + T \rightarrow \pi' + T' \tag{42.1}$$

in which a pion plus some target hadron T goes into another pion plus T', drawn from the same isospin multiplet as T; for example, these reactions:

$$\begin{aligned} \pi^0 + p &\rightarrow \pi^+ + n \\ \pi^+ + \pi^- &\rightarrow \pi^0 + \pi^0 \\ \pi^0 + \Sigma^+ &\rightarrow \pi^+ + \Sigma^0 \end{aligned} \tag{42.2}$$

A pion comes in with momentum k and isospin a and a pion goes out with momentum q and isospin b. The target T comes in with momentum p and a product hadron T' goes out with momentum p'. (I'll explicitly display the isospin indices only for the pions.) We are

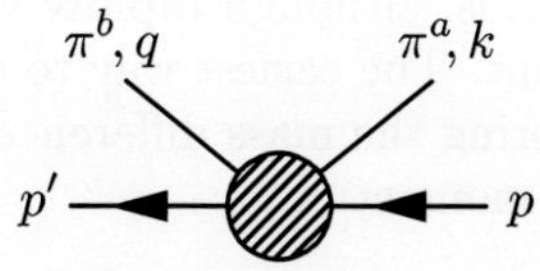

Figure 42.1: Pion–hadron scattering

interested in expanding the amplitude about the point where both pions are soft, $k = q = 0$, and extrapolating to the closest physically accessible point we can get to: threshold. We are systematically going to neglect the $\mathcal{O}(m_\pi^2)$ terms. In this region, the invariant products of pion momenta are all $\mathcal{O}(m_\pi^2)$,

$$k^2 = q^2 = k \cdot q = \mathcal{O}(m_\pi^2) \tag{42.3}$$

We will neglect them because we're only going to keep terms of $\mathcal{O}(1)$ and $\mathcal{O}(m_\pi)$. Thus the only invariant we have to play with, other than p^2 and p'^2 (which are kept on the mass shell) is $k \cdot p$. You may ask, "Why are you keeping $k \cdot p$, but ignoring $q \cdot p$?" Because they're the same, to $\mathcal{O}(m_\pi^2)$. From the conservation equation

$$p' = p + k - q \tag{42.4}$$

Squaring this we find

$$m_{T'}^2 = m_T^2 + 2(k-q)\cdot p + \mathcal{O}(m_\pi^2) \tag{42.5}$$

in the region of interest. Neglecting the (electromagnetic) mass differences between $m_{T'}^2$ and m_T^2 (presumed small in comparison with m_π^2),

$$q \cdot p = k \cdot p + \mathcal{O}(m_\pi^2) \tag{42.6}$$

This has several consequences for the expansion of the amplitude. Recall that we took the invariant amplitude to be a constant plus something times $k \cdot p$, plus $\mathcal{O}(m_\pi^2)$, plus pole terms:

$$\mathcal{A}^{ba} = \mathcal{A}_0^{ba} + \mathcal{A}_1^{ba}(k\cdot p) + \mathcal{O}(m_\pi^2) + \text{pole terms} \tag{41.64}$$

The process is invariant under crossing symmetry, which exchanges pion isospins and momenta; the incoming pion of isospin a and momentum k becomes an outgoing pion of isospin b and momentum $-q$:

$$\left.\begin{array}{c} a \leftrightarrow b \\ k \leftrightarrow -q \end{array}\right\} : \ \mathcal{A}^{ab} = \mathcal{A}^{ba} \tag{42.7}$$

At the level of this expansion, $k \cdot p$ is odd under crossing of k and q. This means $\mathcal{A}_1^{ba}$ must be isospin odd:

$$\mathcal{A}_0^{ba} = \mathcal{A}_0^{ab}; \quad \mathcal{A}_1^{ba} = -\mathcal{A}_1^{ab} \tag{42.8}$$

These are the essential kinematic facts, so the whole amplitude will be invariant under crossing. Now the pole term, which I'll evaluate by exploiting Adler's rule (p. 901). Applied to this process, Adler's rule says that in the limit of one pion's momentum going to zero, the complete amplitude should be given *exactly* by the gradient-coupling theory. I'm going to do this backwards from the usual way it's done in the literature, and start out at the end. We'll see what we can discover about $\mathcal{A}^{ba}$ from the information we already have, *before* making explicit use of the information gained from current algebra, i.e., the commutators of the axial vector currents. For this part of the discussion, I will assume we have a $J^P = {}^1\!/\!_2{}^+$ target, the nucleon or maybe the Σ or Λ. The pole terms, although rapidly varying, will turn out to be $\mathcal{O}(m_\pi^2)$ at threshold, and therefore irrelevant. The easiest way to see that is to inspect the Feynman diagram. At threshold (again, ignoring the mass difference between T and T'), the incoming and outgoing pions have the same momentum, q:

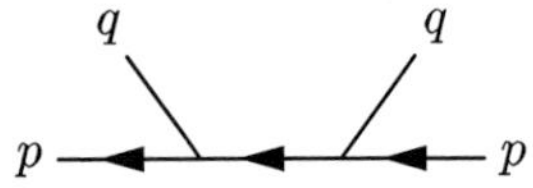

$$\propto \ \overline{u}'\not{q}\gamma_5 \frac{1}{\not{p}+\not{q}-m_T}\not{q}\gamma_5 u \ = \ \overline{u}'\not{q}\frac{1}{\not{p}+\not{q}+m_T}\not{q}u \tag{42.9}$$

The pions are derivatively coupled. These are the pole terms where all the derivatives have been pulled outside of the time-ordering symbol. At threshold

$$\not{q} = \frac{m_\pi}{m_T}\not{p} \tag{42.10}$$

because, from the point of view of the target, the kinetic energy of the pion is zero. That means there is a Lorentz frame in which

$$p^\mu = (m_T, \mathbf{0}) \qquad q^\mu = (m_\pi, \mathbf{0}) \tag{42.11}$$

and hence

$$q^\mu = \frac{m_\pi}{m_T} p^\mu \tag{42.12}$$

Since this is a covariant equation, it must be true in *every* frame. Because the target is on the mass shell,

$$\not{p} u = m_T u \tag{42.13}$$

so $\not{p}$ commutes with $\not{q}$. We don't even have to bother rationalizing the denominator to evaluate (42.9). It's just a bunch of commuting matrices acting on the eigenvectors u and $\overline{u}$, and we get

$$\overline{u}' \not{q} \frac{1}{\not{p} + \not{q} + m_T} \not{q} u = \overline{u}' u \frac{m_\pi^2}{2m_T + m_\pi} \tag{42.14}$$

which is indeed $\mathcal{O}(m_\pi^2)$. So the first lesson we learn from Adler's rule is that the pole terms in this particular process are totally irrelevant at threshold. (They may be large somewhere else in the region of extrapolation.)

I will now take care of the $\mathcal{A}_0^{ba}$ by showing that it is also $\mathcal{O}(m_\pi^2)$, leaving us with just the $\mathcal{A}_1^{ba}$ term to compute. Consider the special case where one pion has zero four-momentum and the other is on the mass shell:

$$k = 0, \quad q^2 = m_\pi^2 \tag{42.15}$$

In this case

$$\mathcal{A}^{ba} = 0 \tag{42.16}$$

(aside from pole terms, which we've already taken account of). That's Adler's rule: only the pole terms are important when one pion is soft. Because $k = 0$ means $k \cdot p = 0$, (41.64) becomes

$$\mathcal{A}^{ba} = \mathcal{A}_0^{ba} + \mathcal{A}_1^{ba}(k \cdot p) + \mathcal{O}(m_\pi^2) = 0 \;\Rightarrow\; \mathcal{A}_0^{ba} = \mathcal{O}(m_\pi^2) \tag{42.17}$$

and we conclude

$$\mathcal{A}^{ba} = \mathcal{A}_1^{ba}(k \cdot p) + \mathcal{O}(m_\pi^2) \tag{42.18}$$

If $\mathcal{A}_1^{ba}$ were also $\mathcal{O}(m_\pi^2)$, we would be uncomfortable. It will turn out to be just $\mathcal{O}(m_\pi)$ because of the explicit $k \cdot p$. We have simplified our analysis enormously.

What about scattering off of other targets? Pions are the *one* hadronic target to which this kinematic analysis does not apply. In that case everything is of $\mathcal{O}(m_\pi^2)$, and we have to keep track of these terms. (We will look at π–π scattering shortly.) For π-$e^\pm$ scattering we just compute one-photon exchange; there's no need to consider the strong interactions.

From (42.18) we can evaluate the amplitude at threshold, neglecting the terms of $\mathcal{O}(m_\pi^2)$:

$$\mathcal{A}_{\text{thr}}^{ba} = \mathcal{A}_1^{ba} m_\pi m_T \tag{42.19}$$

The amplitude at threshold is equal[1] to the scattering length a^{ba}, times a constant[2] proportional to the sum of the masses of the two particles:

$$\mathcal{A}^{ba}_{\text{thr}} = a^{ba}[8\pi(m_\pi + m_T)] \tag{42.20}$$

Everything else is suppressed; consider it a matrix on the initial and final states.

We know from (42.8) that $\mathcal{A}^{ba}_1$ is antisymmetric in b and a. Therefore in particular it must be proportional to ϵ^{bac} times something that depends on the index c. If I look at its matrix element between some initial target and some final target state, which I'll indicate just by brackets, it must be an isovector operator, something to take up the c index, evaluated between the initial target state and the final target state. Since the initial and final target states are states of an irreducible isospin multiplet, by the Wigner–Eckart theorem,[3]

$$\mathcal{A}^{ba}_1 = 2i\epsilon^{bac}\langle I^c_T\rangle B \tag{42.21}$$

That's forced on us by antisymmetry and isospin invariance: it must be some vector operator by isospin invariance, and it must be proportional to the matrix element of the isospin operator by the Wigner–Eckart theorem. So we are left with just one unknown numerical constant to evaluate. I will call[4] this constant $2iB$.

The expression (42.21) can be made even simpler by observing that the matrix element of the pion's isospin is proportional to exactly the same ϵ symbol that appears above; by the right-hand rule for isotriplet states, it's

$$\langle b|I^c_\pi|a\rangle = -i\epsilon^{bac} \tag{42.22}$$

Let

$$a^{ba} = \langle f|\,a\,|i\rangle \tag{42.23}$$

where the initial state and the final state, each being two-particle pion-target states, are given by

$$|i\rangle = |p,k,a\rangle \qquad |f\rangle = |p',q,b\rangle \tag{42.24}$$

The matrix element $\langle f|\,a\,|i\rangle$ can be found by some straightforward algebra. Combining (42.19), (42.20) and (42.21), we have

$$\langle f|\,a\,|i\rangle = \frac{m_\pi m_T}{4\pi(m_\pi + m_T)} i\epsilon^{bac}\langle I^c_T\rangle B \tag{42.25}$$

---

[1] [Eds.] The scattering length a is closely related to the s-wave phase shift δ_0. Two definitions of the scattering length, differing in sign, occur in the literature. The first, $\delta_0 = -ka$, appears in many quantum mechanics texts: K. Gottfried, *Quantum Mechanics, Volume 1: Fundamentals*, W. Benjamin, 1966, equation (40), p. 393; Landau & Lifshitz, *QM*, p. 501, equation (130.9); and note (‡), p. 502; A. Messiah, *Quantum Mechanics*, v. 1, North Holland Publishing, 1962; reprinted by Dover Publications, 2014, equation (X.47), p. 392. The second, $\delta_0 = +ka$, is widely used in the phenomenological analysis of high energy π-N scattering data: M. L. Goldberger and K. M. Watson, *Collision Theory*, John Wiley, 1964; reprinted by Dover Publications, 2004, equation (296), p. 287. This latter definition is used in (42.20).

[2] [Eds.] The derivation of this constant is given in Appendix 3 of "Soft Pions", Chapter 2 in Coleman *Aspects*, pp. 64–65.

[3] [Eds.] See note 28, p. 766 and note 6, p. 847.

[4] [Eds.] In the video of Lecture 47 and in *Aspects*, this constant B is defined by $\mathcal{A}^{ba}_1 = \epsilon^{bac}\langle I^c_T\rangle B$. However, later on a *second* constant B is introduced. With the definition (42.21), the two B's are one and the same.

Substituting (42.22) for the Levi–Civita symbol, we obtain

$$\langle f|a|i\rangle = -\frac{m_\pi m_T}{4\pi(m_\pi + m_T)} B \langle f|\mathbf{I}_\pi \cdot \mathbf{I}_T|i\rangle \tag{42.26}$$

The total isospin of the two-particle state is $I(I+1)$:

$$\mathbf{I}^2 = I(I+1) = \mathbf{I}_\pi^2 + \mathbf{I}_T^2 + 2\mathbf{I}_\pi \cdot \mathbf{I}_T \tag{42.27}$$

(A similar argument is used in calculating spin-orbit coupling.) Assuming we're looking at the expectation value in a state of specific total isospin I, I can write $\mathbf{I}_\pi \cdot \mathbf{I}_T$ in terms of the total isospin:

$$\langle f|\mathbf{I}_\pi \cdot \mathbf{I}_T|i\rangle = \tfrac{1}{2}[I(I+1) - I_T(I_T+1) - 2] \tag{42.28}$$

(the isospin $I_\pi(I_\pi + 1)$ contributes the 2, independent of the pion identity). Assembling all of this together we find the scattering length in a state of specific I as

$$a_I = -\frac{m_\pi m_T}{8\pi(m_\pi + m_T)} B[I(I+1) - I_T(I_T+1) - 2] \tag{42.29}$$

We still need to determine the constant B, defined in (42.21).

Kinematics and isospin analysis, though pedestrian, have provided us with a great deal of information. Independent of the current commutators, we have found that the scattering length associated with a pion hitting a general baryonic target is given in terms of a single coefficient, B. (There may be several scattering lengths because there may be several possible initial isospins of the two-particle state.)

As a check on our work, let's compute (42.29) for pion-nucleon scattering, and see if it's right before we go any further. After all, this equation makes a definite prediction. If that turns out to be wrong, there's no point in trying to compute the coefficient B. There are two possible isospins, $I = \frac{1}{2}$ and $I = \frac{3}{2}$; the measured scattering lengths are[5]

$$\begin{aligned} a_{1/2} &= 0.17\, m_\pi^{-1} \\ a_{3/2} &= -0.09\, m_\pi^{-1} \end{aligned} \tag{42.30}$$

Their ratio is

$$a_{3/2} : a_{1/2} = -0.53 \quad \text{(exp't)} \tag{42.31}$$

The isospin factor in our formula evaluates in these two cases to

$$I(I+1) - \tfrac{3}{4} - 2 = \begin{cases} -2, & I = \frac{1}{2} \\ 1, & I = \frac{3}{2} \end{cases} \tag{42.32}$$

Therefore the theory predicts

$$a_{3/2} : a_{1/2} = -\tfrac{1}{2} \quad \text{(theory)} \tag{42.33}$$

which is in good agreement with experiment. We hope to do better, but this is encouraging.

[5] [Eds.] J. Hamilton and W. S. Woolcock, "Determination of Pion–Nucleon Parameters and Phase Shifts by Dispersion Relations", *Rev. Mod. Phys.*, **35** (1963) 737–787; D. V. Bugg, A. A. Carter, and J. R. Carter, "New Values of Pion–Nucleon Scattering Lengths and f_2", *Phys. Lett.* **B44** (1973) 278–280.

42.2 Pion–hadron scattering and current algebra

To predict not merely the ratio of the scattering lengths but their magnitudes, we've got to determine the value of that coefficient B. It could still be a disaster. If we compute that $B = 0$, we would absolutely have no reason to trust (42.29), because then the observed scattering lengths would be entirely due to terms we have neglected. That the ratio of the two terms we have retained is $-1 : 2$ is irrelevant if they're both zero. Fortunately, we'll find that B is *not* zero.

The computation of B is tedious, so let's organize it carefully. The amplitude $i\mathcal{A}$ is given by the reduction formula (14.36),

$$i(2\pi)^4\delta^{(4)}(q+p'-k-p)\mathcal{A}^{ba} = \\ = (-i)^2(k^2-m_\pi^2)(q^2-m_\pi^2)\int d^4x\, d^4y\, \langle p'|T\left\{\phi_\pi^b(x)\phi_\pi^a(y)\right\}|p\rangle\, e^{iq\cdot x}e^{-ik\cdot y} \qquad (42.34)$$

We are interested only in $\mathcal{A}_1^{ba}$, the coefficient of the $p\cdot k$ term; we can extract that coefficient within our approximation by studying only forward scattering.[6] That will simplify the kinematics a bit. Near the end of the calculation I will say

$$q = k, \quad p = p' \qquad (42.35)$$

I'll keep the relativistic notation, but I want to point out that we can do this calculation in a frame in which the target is at rest, with the pion also at rest, although possibly not on the mass-shell:

$$p = (m_T, \mathbf{0}), \quad k = (k_0, \mathbf{0}) \qquad (42.36)$$

Although we're going to use those commutators (41.59) from the last lecture, with the δ function in them, in fact all we need are the commutators of the integrated axial charges. As the pion is carrying off zero three-momentum, the reduction formula will involve only a trivial space integral.

Now we'll use PCAC (41.12) to write the pion fields in terms of the divergence of the axial current. Then

$$i(2\pi)^4\delta^{(4)}(q+p'-k-p)\,\mathcal{A}^{ba} = \left[\frac{(-i)^2(k^2-m_\pi^2)(q^2-m_\pi^2)}{(F_\pi m_\pi^2)^2}\right]I^{ba} \qquad (42.37)$$

where

$$I^{ba} = \int d^4x\, d^4y\, e^{i(q\cdot x-k\cdot y)}\,\langle p'|\,T\left\{(\partial^\mu A_\mu^b(x))(\partial^\nu A_\nu^a(y))\right\}|p\rangle \qquad (42.38)$$

and a, b are the incoming and outgoing pions, respectively. We now have to pull out the derivatives. Our general rule for extracting a derivative from a time-ordered product of two fields A, B is always the same:

$$\partial_0^x T\left\{A(x)B(y)\right\} = T\{(\partial_0^x A(x))B(y)\} + \delta(x^0-y^0)\left[A(x), B(y)\right] \qquad (27.86)$$

We've got to apply this identity twice. We first take out the y derivative. The order within the product is irrelevant because of the time ordering.

$$T\left\{\partial_x^\mu A_\mu^b(x)\,\partial_y^\nu A_\nu^a(y)\right\} = \partial_y^\nu T\left\{A_\nu^a(y)\,\partial_x^\mu A_\mu^b(x)\right\} - \delta(y^0-x^0)\left[A_0^a(y), \partial_x^\mu A_\mu^b(x)\right] \qquad (42.39)$$

---

[6] [Eds.] Forward scattering occurs when the initial and final states are the same; Itzykson & Zuber *QFT*, §5.3.1.

The $\delta(y^0 - x^0)$ term is not in our current algebra; while we know $[A_0^a(\mathbf{y},0), A_0^b(\mathbf{x},0)]$, we don't know what the commutator of A_0^a with $\partial^\mu A_\mu^b$ is. So we're going to have to worry about that. However, we don't really have to worry very much, because I'll show you that we can throw away the second term.

In the second term, the spatial part $\partial_x^i A_i^b(x)$ of $\partial_x^\mu A_\mu^b(x)$ can be converted through integration by parts to $\mathbf{k}$, and dropped, because the pion is ultimately at rest; $k^\mu = (k^0, \mathbf{0})$. The relevant part inside the integral (42.38) is the spatial integral of the commutator with the time derivative at equal times:

$$\int d^3\mathbf{x}\, d^3\mathbf{y} \left[A_0^a(\mathbf{y},t), \partial^0 A_0^b(\mathbf{x},t)\right] \tag{42.40}$$

because of the $\delta(y^0 - x^0)$. We don't know this commutator, but we do know

$$[A_0^a(\mathbf{x},t), A_0^b(\mathbf{y},t)] = i\epsilon^{abc} 2g_V V_0^c(\mathbf{x},t)\delta^{(3)}(\mathbf{x}-\mathbf{y}) \tag{41.59}$$

Thus

$$\begin{aligned}\partial^0 \int d^3\mathbf{x}\, d^3\mathbf{y}\, [A_0^a(\mathbf{y},t), A_0^b(\mathbf{x},t)] &\propto i\epsilon^{abc}\partial^0 \int d^3\mathbf{x}\, d^3\mathbf{y}\, V_0^c(\mathbf{x},t)\delta^{(3)}(\mathbf{x}-\mathbf{y}) \\ &\propto i\epsilon^{abc}\partial^0 I^c = 0\end{aligned} \tag{42.41}$$

The CVC hypothesis (40.2) is that the vector V_μ^c is the isospin current. Taking that as true, ∂_0 of this integral is proportional to ∂_0 of the total isospin, which is zero. Even without CVC, the time derivative of the commutator is zero:

$$\begin{aligned}\partial^0 \int d^3\mathbf{x}\, d^3\mathbf{y}\, [A_0^a(\mathbf{y},t), A_0^b(\mathbf{x},t)] &= \int d^3\mathbf{x}\, d^3\mathbf{y}\, \partial^0 \Big([A_0^a(\mathbf{x},t), A_0^b(\mathbf{y},t)]\Big) \\ &= \int d^3\mathbf{x}\, d^3\mathbf{y}\, [\partial^0 A_0^a(\mathbf{x},t), A_0^b(\mathbf{y},t)] + \int d^3\mathbf{x}\, d^3\mathbf{y}\, [A_0^a(\mathbf{x},t), \partial^0 A_0^b(\mathbf{y},t)] \\ &= \int d^3\mathbf{x}\, d^3\mathbf{y}\, [\partial^0 A_0^a(\mathbf{x},t), A_0^b(\mathbf{y},t)] - \int d^3\mathbf{y}\, d^3\mathbf{x}\, [\partial^0 A_0^b(\mathbf{x},t), A_0^a(\mathbf{y},t)] \\ &= \int d^3\mathbf{x}\, d^3\mathbf{y}\, \Big\{[\partial^0 A_0^a(\mathbf{x},t), A_0^b(\mathbf{y},t)] - [\partial^0 A_0^b(\mathbf{x},t), A_0^a(\mathbf{y},t)]\Big\} = 0\end{aligned} \tag{42.42}$$

Thus the antisymmetric part of the second term of (42.39) vanishes. Therefore if the second term is not zero, it must be *symmetric*:

$$\int d^3\mathbf{x}\, d^3\mathbf{y}\, [A_0^a(\mathbf{y},t), \partial^0 A_0^b(\mathbf{x},t)] \text{ is either zero or symmetric under } (a \leftrightarrow b) \tag{42.43}$$

If it is zero, there's nothing more to say. If it is symmetric, it contributes only to $\mathcal{A}_0^{ba}$, which we have already demonstrated is $\mathcal{O}(m_\pi^2)$, and therefore irrelevant to the antisymmetric $\mathcal{A}_1^{ba}$. In either case we can ignore it.

Onwards to the first term in (42.39). Once more we bring a derivative outside, now with respect to x:

$$\partial_y^\nu T\left\{\partial^\mu A_\mu^b(x) A_\nu^a(y)\right\} = \partial_y^\nu \partial_x^\mu T\left\{A_\mu^b(x) A_\nu^a(y)\right\} - \partial_y^\nu\Big(\delta(x_0 - y_0)[A_0^b(x), A_\nu^a(y)]\Big) \tag{42.44}$$

We can neglect the first term: it will have no contribution except from pole terms. All the k's are on the outside, so we just get a term of $\mathcal{O}(m_\pi^2)$. The second term in (42.44) is going to be the whole package. That's of course the commutator we know. More precisely, we know it for $\nu = 0$, but that's all we really need; the ∂_y^ν derivative is going to be turned into a momentum and, as before, the spatial parts of all the pion momenta are zero. Just to keep things looking covariant, though, we'll write it as

$$\partial_y^\nu T\left\{\partial^\mu A_\mu^b(x)A_\nu^a(y)\right\} = -\partial_y^\nu\left(\delta^{(4)}(x-y)2ig_V\epsilon^{bac}V_\nu^c(y)\right) \tag{42.45}$$

simply for notational convenience; the space components of this will not contribute to the expression we're going to feed it into. The ∂^ν acting on the $V_\nu(y)$ gives us 0 because the vector current is conserved. Only its action on the δ function is relevant:

$$\partial_y^\nu T\left\{\partial^\mu A_\mu^b(x)A_\nu^a(y)\right\} = -\left[\partial_y^\nu\delta^{(4)}(x-y)\right]2ig_V\epsilon^{bac}V_\nu^c(y) \tag{42.46}$$

Now we're in business. We substitute (42.46), the only term that survives in the double divergence (42.39), into the integral (42.38) and compute it:

$$\begin{aligned}\int d^4x\,d^4y\,e^{i(q\cdot x-k\cdot y)}\,\langle p'|T\left\{(\partial_x^\mu A_\mu^b(x))(\partial_y^\nu A_\nu^a(y))\right\}|p\rangle &=\\ = -2ig_V\epsilon^{bac}\int d^4x\,d^4y\,e^{i(q\cdot x-k\cdot y)}\left[\partial_x^\nu\delta^{(4)}(x-y)\right]\langle p'|V_\nu^c(y)|p\rangle\end{aligned} \tag{42.47}$$

Integrating the right-hand side by parts,

$$\begin{aligned}\int d^4x\,d^4y\,e^{i(q\cdot x-k\cdot y)}\left[\partial_y^\nu\delta^{(4)}(x-y)\right]\langle p'|V_\nu^c(y)|p\rangle &=\\ = ik^\nu\int d^4x\,d^4y\,e^{i(q\cdot x-k\cdot y)}\,\delta^{(4)}(x-y)\,\langle p'|V_\nu^c(y)|p\rangle\end{aligned} \tag{42.48}$$

If we now set $\mathbf{k} = \mathbf{q}$, the integral in the reduction formula, normally a mess, becomes trivial:[7]

$$\begin{aligned}\int d^4x\,d^4y\,e^{i(q\cdot x-k\cdot y)}&\left[\partial_y^\nu\delta^{(4)}(x-y)\right]\langle p'|V_\nu^c(y)|p\rangle\\ &= ik^\nu\int dx^0\,e^{-i(k_0-q_0)x^0}\,\langle p'|\int d^3\mathbf{x}\,V_\nu^c(\mathbf{x})\,|p\rangle\\ &= ik^\nu\,(2\pi\delta(k_0-q_0))\,(2\pi)^3 2p_\nu\delta^{(3)}(\mathbf{p}'-\mathbf{p})(2g_V\,\langle I_T^c\rangle)\end{aligned} \tag{42.49}$$

because the kets $|p\rangle$ are relativistically normalized (1.57), and the integral of the currents gives $2g_V$ times the isospin operators (41.52). By overall momentum conservation, $k_0 - q_0 = p_0' - p_0$, so

$$ik^\nu\,(2\pi\delta(k_0-q_0))\,(2\pi)^3 2p_\nu\delta^{(3)}(\mathbf{p}'-\mathbf{p})(2g_V\,\langle I_T^c\rangle) = i(2\pi)^4\delta^{(4)}(p'-p)(k\cdot p)(4g_V\,\langle I_T^c\rangle) \tag{42.50}$$

---

[7] [Eds.] There are a couple of non-obvious steps in (42.49) going from the first expression on the right to the last. First, $V_\mu^c(x)$ can be replaced by $V_\mu^c(\mathbf{x})$ because it is a conserved current (see note 24, p. 903). This allows the dx^0 integration to be done only over the remaining exponential, giving the delta function $2\pi\delta(x_0 - y_0)$. Next, the relativistic normalization says that $\langle p|p'\rangle = (2\pi)^3 2p_0\delta^{(3)}(\mathbf{p}-\mathbf{p}')$. That means

$$\langle p|\int d^3\mathbf{x}\,V_0^c(\mathbf{x},t)|p'\rangle = (2\pi)^3 2p_0\delta^{(3)}(\mathbf{p}-\mathbf{p}')(2g_V)\,\langle I_T^c\rangle$$

One then argues by Lorentz invariance that if the index 0 is replaced by ν on both sides, the equation remains true.

We now substitute (42.50) into (42.37) via (42.38), and set $k = q$:

$$i(2\pi)^4\delta^{(4)}(p'-p)\,\mathcal{A}^{ba} = \left[\frac{-i(k^2-m_\pi^2)}{F_\pi m_\pi^2}\right]^2 \left(8g_V^2\epsilon^{bac}\right)(2\pi)^4\delta^{(4)}(p'-p)(k\cdot p)\,\langle I_T^c\rangle \tag{42.51}$$

If we're only interested in expanding in the range near $k = 0$, keeping linear terms, we can replace k^2 by 0, whence the m_π^2 in the numerator cancels the m_π^2 in the denominator. The i^2 gives me a minus sign. We're going near $k = 0$ to get the coefficient of the $k\cdot p$ term. Canceling the common factors, we arrive at the amplitude:

$$i\mathcal{A}^{ba} = -\frac{8g_V^2}{F_\pi^2}\epsilon^{bac}(k\cdot p)\,\langle I_T^c\rangle + \mathcal{O}(m_\pi^2) \tag{42.52}$$

From (42.18) and (42.21) we have

$$\mathcal{A}^{ba} = 2i\epsilon^{bac}(k\cdot p)\,\langle I_T^c\rangle\, B \tag{42.53}$$

And so[8]

$$B = \frac{4g_V^2}{F_\pi^2} \tag{42.54}$$

A constant useful in these calculations is

$$L = \frac{g_V^2 m_\pi}{2\pi F_\pi^2} = \frac{g^2 m_\pi}{8\pi m_p^2}\left(\frac{g_V}{g_A}\right)^2 \approx 0.1\, m_\pi^{-1} \tag{42.55}$$

(the second equality coming from the Goldberger–Treiman relation (40.37).) In terms of L,

$$B = 8\pi L m_\pi^{-1} \approx 2.5\, m_\pi^{-2} \tag{42.56}$$

We can now plug this expression for B into (42.29) and get an expression for the scattering of a pion off of any hadronic target with the exception of anther pion; that one we can't do. Thus we obtain the universal expression for the scattering length, called the **Weinberg–Tomozawa** formula,[9]

$$a_I = -\frac{g_V^2}{2\pi F_\pi^2}\left[\frac{m_\pi m_T}{m_\pi + m_T}\right][I(I+1) - I_T(I_T+1) - 2] \tag{42.57}$$

Note that the coefficient of the isospin factor is a universal number depending only on the mass of the target.[10]

The actual predictions are useful only for the pion–nucleon case, where we have good measurements of the scattering lengths. Plugging in the numbers, we find for pion–nucleon scattering

$$a_{1/2} = \begin{cases} 0.17\, m_\pi^{-1} & \text{(exp't)} \\ 0.17\, m_\pi^{-1} & \text{(theory)} \end{cases} \qquad a_{3/2} = \begin{cases} -0.09\, m_\pi^{-1} & \text{(exp't)} \\ -0.09\, m_\pi^{-1} & \text{(theory)} \end{cases} \tag{42.58}$$

---

8 [Eds.] In the video of Lecture 47, Coleman now spends several minutes trying to repair what he erroneously believes to be a sign error. The confusion, due to different sign conventions in the scattering lengths a, was resolved in the next lecture.

9 [Eds.] S. Weinberg, "Pion Scattering Lengths", *Phys. Rev. Lett.* **17** (1966) 616–621; Y. Tomozawa, "Axial-Vector Coupling Constant Renormalization and the Meson–Baryon Scattering Lengths", *Nuovo Cim.* **46A** (1966) 707–717.

10 [Eds.] The value for a proton target is about $0.085\, m_\pi^{-1}$, using $F_\pi = 0.197\, m_p$, $m_\pi = 140$ MeV, $m_p = 939$ MeV, and $g_V \sim 1$.

This is excellent agreement,[11] and a triumph of current algebra. We have verified the A_0 with A_0 commutators; indeed we have used them to explain experiment.

The physics is everywhere: in crossing symmetry, in the assumption of Goldberger–Treiman-like smoothness (which enables us to assume the amplitude has a smooth extrapolation, a polynomial fit, up to threshold, once the two-pion poles are extracted), and finally in the current algebra commutators. The only new information above and beyond the sort of reasoning we used in the Goldberger–Treiman relation is the current commutators. As we tried to emphasize by the way we've organized this lecture, or disorganized it, even without the commutators we get (42.33) the $-2:1$ ratio between the two scattering lengths. The current commutators just serve to fix the scale of the scattering length.

42.3 Pion–pion scattering

Let's go on and consider the one case which has slipped through our net, π–π scattering, following a famous analysis of Weinberg.[12] The analysis is beautiful because it involves no new physics and no new integrals to be computed. But it requires practically everything we know about the π–π system. It uses our extrapolation techniques where we assume there is a polynomial fit. It uses current algebra every inch of the way. It uses some new information about a commutator that we're going to have to extract from the quark model. And it uses crossing and analyticity and everything else. It's a beautiful calculation.

We will end up predicting two scattering lengths. The two pions can be in an $I = 0$, an $I = 1$ or an $I = 2$ state. By Bose statistics the $I = 1$ state is p-wave, so it has no defined scattering length.[13] We will end up deducing a_0 and a_2. The trick will be to consider the process in which a pion comes in with momentum k_1 and isospin a and goes out with momentum k_1' and isospin b, and a pion with momentum k_2 and isospin c comes in and goes out with k_2' and isospin d. We will investigate this process near the point where all the momenta

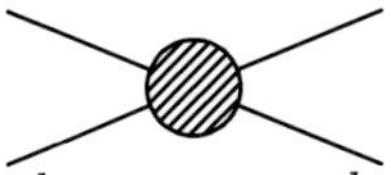

Figure 42.2: Pion–pion scattering

vanish and retain all the terms of $\mathcal{O}(m_\pi^2)$. We will then extrapolate to threshold as before. The nice thing about this problem is that there are no pole terms. A pole term—a three-pion pole, or more properly an axial vector current π–π pole, as in Figure 42.3—is forbidden by parity.[14] Since everything is going to go off mass shell in this computation, we have a straight expansion in terms of six invariants. It's convenient to begin with the over-complete set of *seven* invariants, consisting of the individual momenta squared, k_i^2 and $k_i'^2$, $i = \{1, 2\}$, and our

[11] [Eds.] Bugg *et al.*, *op. cit.*

[12] [Eds.] Weinberg, *op. cit.* (note 9, p. 925); Coleman *Aspects*, "Soft Pions", pp. 57–59. Much of this section follows Weinberg's argument very closely. See also Weinberg *QTF2*, pp. 197–202.

[13] [Eds.] The scattering length is defined in terms of the s-wave phase shift. See note 1, p. 920.

[14] [Eds.] Figure 42.3 does not appear in the video of Lecture 47 (upon which this chapter is based) nor in any lecture notes. This is only a guess at what Coleman meant by the term "three-pion pole".

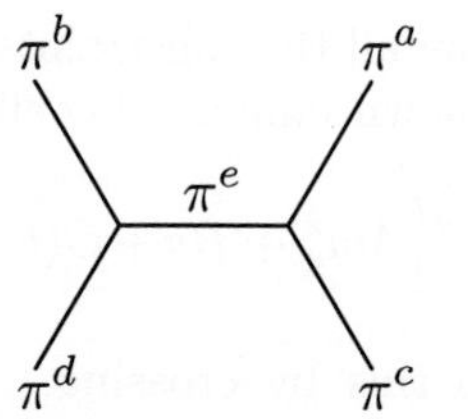

Figure 42.3: Pion pole in pion–pion scattering, forbidden by parity conservation

usual Mandelstam variables s, t and u, which, just to remind you, are

$$\begin{aligned} s &= (k_1 + k_2)^2 \\ t &= (k_1 - k_1')^2 \\ u &= (k_1 - k_2')^2 \end{aligned} \tag{42.59}$$

This set is linearly related by the kinematic identity

$$\sum_i (k_i^2 + k_i'^2) = s + t + u \tag{42.60}$$

Now let's write down the isospin-invariant amplitude. It has to carry the isospin indices of the four pions, and it has to be invariant under the interchanges $a \leftrightarrow c$, $b \leftrightarrow d$, $a \leftrightarrow b$, and $c \leftrightarrow d$. Thus it must take the form

$$\mathcal{A} = \delta^{ac}\delta^{bd}[\,\cdots\,] + \delta^{ab}\delta^{cd}[\,\cdots\,] + \delta^{ad}\delta^{bc}[\,\cdots\,] \tag{42.61}$$

The brackets are to be filled in. The things in them are all connected to each other by crossing. When we know the coefficient that appears in the first one we'll know the coefficient that appears in the other two. We'll go to first order in our seven invariants.

Let's first worry about whether we can have possible coefficients of k_1^2 in the first term. By Bose statistics, since this process is symmetric under interchange of a and c, if we have a k_1^2 then we must also have a k_2^2. By time reversal (or equivalently, by crossing ab into cd), if we have k_1^2 and k_2^2, we've got to have a $k_1'^2$ and a $k_2'^2$ with the same coefficient:

$$\mathcal{A} = \delta^{ac}\delta^{bd}[k_1^2 + k_2^2 + k_1'^2 + k_2'^2] + \delta^{ab}\delta^{cd}[\,\cdots\,] + \delta^{ad}\delta^{bc}[\,\cdots\,] \tag{42.62}$$

That can be written in terms of $s + t + u$.

$$\mathcal{A} = \delta^{ac}\delta^{bd}[s + t + u] + \delta^{ab}\delta^{cd}[\,\cdots\,] + \delta^{ad}\delta^{bc}[\,\cdots\,] \tag{42.63}$$

So rather than write the bracket contents in terms of the k's, I'll start over and use all possible linear combination of s, t and u.

What can we have? We can certainly have a constant term which, to make it have the same dimensions as everything else, we'll call Am_π^2. Then we can have a term proportional to s in the term where a and c are symmetric, interchanging particle a and particle c, and keeping everything else fixed; this exchanges t and u, so we must have a coefficient times $t + u$. All the other terms will involve st or momentum transferred squared or momentum squared times s or t or u, or s^2 or t^2 or u^2. But those terms are $\mathcal{O}(m_\pi^4)$, so we can neglect them. In the amplitude (42.61), when we apply all the constraints imposed by Bose statistics, crossing, and isospin we have only three unknown numerical coefficients to determine:

$$\mathcal{A} = \delta^{ac}\delta^{bd}\Big[Am_\pi^2 + Bs + C(t+u)\Big] + \cdots \tag{42.64}$$

The other terms are connected to this by crossing. In the δ^{ab} term we permute[15] the Mandelstam variables as $(s\ t\ u)$, and in the δ^{ad} term we permute them as $(s\ u\ t)$. Thus we have

$$\begin{aligned}\mathcal{A} &= \delta^{ac}\delta^{bd}[Am_\pi^2 + Bs + C(t+u)] \\ &+ \delta^{ab}\delta^{cd}[Am_\pi^2 + Bt + C(u+s)] \\ &+ \delta^{ad}\delta^{bc}[Am_\pi^2 + Bu + C(s+t)]\end{aligned} \tag{42.65}$$

Thus the entire amplitude, excluding terms of $\mathcal{O}(m_\pi^4)$, is known in terms of three coefficients. The task now is to compute them. We're going to need three equations to solve for these coefficients.

The first comes from Adler's rule. Imagine that one incident pion, say with momentum k_1, is soft, so

$$k_1 = 0\colon\ k_1'^2 = k_2^2 = k_2'^2 = m_\pi^2 \tag{42.66}$$

According to Adler's rule, the amplitude for soft pions vanishes except for pole terms. But there are no poles in this process, so the amplitude at this point must vanish:

$$\mathcal{A} = 0 \tag{42.67}$$

as in the previous analysis. This is also the point where

$$s = t = u = m_\pi^2 \tag{42.68}$$

from the definitions of the variables (42.59). All of the quantities in square brackets are equal, so we obtain our first equation:

$$A + B + 2C = 0 \tag{42.69}$$

Next, consider what happens when we look in the forward direction, with pion 1 (with momentum k_1) soft, and pion 2 (with momentum k_2) on mass shell:

$$k_1, k_1' \to 0; \quad k_2 = k_2'; \quad k_2^2, k_2'^2 \to m_\pi^2 \tag{42.70}$$

We know in that case we get terms of $\mathcal{O}(k^2)$ which come from that symmetric commutator, (42.42). Let's write it down. There will be a constant term about which we can say nothing. There is a term which we explicitly computed from the current commutator,

$$\mathcal{A} = (\text{const.}) + (k_1\cdot k_2)\epsilon^{baf}\,\langle I_T^f\rangle\left[i\frac{8g_V^2}{F_\pi^2}\right] \tag{42.71}$$

where the target in I_T^f is now just the second pion. That's simply our earlier computation (42.52). If we are allowed to make k and k' as small as we want, that part of the computation

[15] [Eds.] Recall that the notation $(a\ b\ c)$ means the cyclic permutation $a \to b$, $b \to c$, $c \to a$. See the paragraph before (39.48), p. 862.

works whatever the target particle is, even a pion. A little isospin algebra is necessary. By the previous analysis (42.22)

$$\langle I_T^f \rangle = -i\epsilon^{dcf} \tag{42.72}$$

where c and d are the initial and final pions, respectively. The product of two ϵ's is easy to compute (37.47), and $\mathcal{A}$ takes the following form:

$$\mathcal{A} = (\text{const.}) + (k_1 \cdot k_2)\left[\frac{8g_V^2}{F_\pi^2}\right](\delta^{ac}\delta^{bd} - \delta^{ad}\delta^{bc}) \tag{42.73}$$

We now compare this to (42.65). In forward scattering the amplitude is easy to compute. We have for the Mandelstam variables with $k_1 = k_1'$

$$t = 0 \tag{42.74}$$

and retaining terms of first order in k_1 and k_1' (and using $k_2 = k_2'$),

$$s = m_\pi^2 + 2k_1 \cdot k_2; \quad u = m_\pi^2 - 2k_1 \cdot k_2 \tag{42.75}$$

In this régime the amplitude (42.65) is

$$\begin{aligned}\mathcal{A} = \delta^{ac}\delta^{bd}&[(A+B+C)m_\pi^2 + 2(B-C)(k_1\cdot k_2)] \\ &+ \delta^{ab}\delta^{cd}[(A+2C)m_\pi^2] \\ &+ \delta^{ad}\delta^{bc}[(A+B+C)m_\pi^2 - 2(B-C)(k_1\cdot k_2)]\end{aligned} \tag{42.76}$$

Is this consistent with (42.73)? Sure enough, it is. There's a constant part, and there is a term proportional to $(k_1 \cdot k_2)$ multiplying $(\delta^{ac}\delta^{bd} - \delta^{ad}\delta^{bc})$, exactly as predicted. Comparing the coefficients in (42.73) and (42.76), we obtain our second equation:

$$B - C = \frac{4g_V^2}{F_\pi^2} \tag{42.77}$$

Our third equation comes from a constraint on the constant term in (42.71). In our earlier analysis, we found three terms in the amplitude $\mathcal{A}_1$. One had two derivatives explicitly on the outside, which, in the absence of pole terms, vanishes when two of the moments vanish. Another was the commutator term, which we've just taken care of. Finally, we had that funny term (42.40), which we argued was symmetric in a and b (42.43),

$$\int d^3\mathbf{x}\, d^3\mathbf{y}\, [A_0^a(\mathbf{x},t), \partial^\mu A_\mu^b(\mathbf{y},t)] \tag{42.78}$$

We're going to have to study this term in the context of a particular model and see what happens. We have one obvious model at hand. Let's see what we can determine about this thing from the quark model, (§39.4 and §41.5).

The currents are *bilinear* in the quarks (41.56), which carry isospin and color. Color interacts with the gluons, about which I've said next to nothing. Gluons are supposed to carry color *only*; they are presumed to be isospin neutral and flavor neutral. The divergence of A_μ is not going to be zero, but it surely is going to be a bilinear form in the quark fields; one derivative on a (Dirac) quark field equals the quark field multiplied by something. So $\partial^\mu A_\mu^b(y)$ is going to be bilinear in the quark fields. $A_0^a(x)$ is likewise bilinear in the quark fields; it's just the quark current. When you commute two bilinear fields you obtain another bilinear. That's the way it has to be, because the commutator produces one field out of the difference of two products of two fields. In any version of the quark model we can think of where the gluons have only color, (42.78) must be bilinear in the quark fields.[16] The up and down quarks carry isospin ½, while the strange quark is isospin 0. So the bilinear resulting from (42.78), built from u and d quarks only, must be $I = 0$ or $I = 1$. But $I = 1$ is ruled out; that would give an ϵ symbol[17] and we proved earlier that (42.78) is symmetric under the interchange of a and b. Therefore within the context of any sensible quark model the constant in the power series expansion (42.73) must be pure $I = 0$ in the indices a and b:

$$(\text{const.}) \propto \delta^{ab} \tag{42.79}$$

That's the only piece of information we will extract from the quark model: the constant term in the power series expansion (42.76) for the forward scattering amplitude $\mathcal{A}$, with one pion soft, another pion on the mass shell, must be proportional to δ^{ab}. That means that the $\delta^{ac}\delta^{bd}$ and the $\delta^{ad}\delta^{bc}$ terms which *aren't* proportional to δ^{ab} have got to vanish. Thus we find our third equation,

$$A + B + C = 0 \tag{42.80}$$

Putting together all three equations (42.69), (42.77), and (42.80), we quickly solve for the constants A, B, and C:

$$\begin{aligned} A &= -\frac{4g_V^2}{F_\pi^2} \\ B &= \frac{4g_V^2}{F_\pi^2} \\ C &= 0 \end{aligned} \tag{42.81}$$

We have determined everything.[18]

Our final expression for the scattering amplitude (42.65), which we know is valid near zero and hope is valid all the way up to threshold, is

$$\mathcal{A} = B\left[\delta^{ac}\delta^{bd}(s - m_\pi^2) + \delta^{ab}\delta^{cd}(t - m_\pi^2) + \delta^{ad}\delta^{bc}(u - m_\pi^2)\right] \tag{42.82}$$

---

16 [Eds.] In response to a student's question, Coleman replies, "This hypothesis was originally made in the context of the sigma model, which we have not discussed yet. Nobody knew about quarks with flavor and color then. They knew about quarks but they didn't understand why they had funny statistics. And they didn't take the quark model seriously." (The sigma model is discussed in §§45.3–45.4, pp. 993–1002.)

17 [Eds.] The argument can perhaps be fleshed out a little. Weinberg (see note 9, p. 925) writes his equation (5) in the context of both the σ and free-quark models,

$$\delta(x^0 - y^0)[A_a^0(x), \partial_\nu A_b^\nu(y)] = ig_V\sigma_{ab}(x)\delta^{(4)}(x - y) + \text{S.T.}$$

where $\sigma_{ab}(x)$ is "some scalar field which may or may not have something to do with a real 0^+ π–π resonance or enhancement, and 'S.T.' means possible Schwinger terms." The only two two-index isospin tensors are δ_{ab} and ϵ_{ab}, so $\sigma_{ab}(x) \sim \delta_{ab}f(x) + \epsilon_{ab}g(x)$. Under parity, $f(x)$ has to have $I = 0$ and $g(x)$ has to have $I = 1$. As the commutator was earlier shown to be symmetric, we take $g(x) = 0$. (In *Aspects*, p. 59, Coleman states that without the assumption that the commutator is proportional to δ_{ab}, it "could be any combination of $I = 0$ and $I = 2$".)

18 [Eds.] The signs of A and B here are reversed from those in the video of Lecture 47, apparently due to the differences in the definitions of the scattering length. They agree however with the signs in Coleman *Aspects*, "Soft Pions", with the exchange $B \leftrightarrow C$. Note that this value (42.81) of B agrees with that in (42.54).

with the coefficients given by (42.81). Notice we needed the extra assumption (42.79) which we have extracted as an unambiguous consequence from the quark model.[19] Equal time commutators are among the few things we can compute without solving strong interaction dynamics.[20]

Let's find the presumed actual π–π scattering lengths, to compare with experiment. We need to evaluate the amplitude at threshold. Which term we choose to evaluate at threshold is irrelevant because the amplitude has crossing symmetry, but we'll choose the usual one,

$$s = 4m_\pi^2; \quad t = u = 0 \tag{42.83}$$

This gives the threshold amplitude

$$\mathcal{A}_{\text{thr}} = \frac{4g_V^2 m_\pi^2}{F_\pi^2}[3\delta^{ac}\delta^{bd} - \delta^{ab}\delta^{cd} - \delta^{ad}\delta^{bc}] \tag{42.84}$$

That's the final answer for the amplitude! To write this in terms of scattering lengths, all that remains is some easy kinematics and a little isospin analysis to find out what this obviously isospin-invariant 9×9 operator, turning initial states into final states at threshold, is in terms of its eigenvalues; that is, in terms of its $I = 2$ eigenvalue and its $I = 0$ eigenvalue. First I'll do the isospin analysis.

42.4 Some operators and their eigenvalues

We already know one of the operators. Recall that

$$\delta^{ac}\delta^{bd} - \delta^{ad}\delta^{bc} = -[-i\epsilon^{baf}][-i\epsilon^{dcf}] = -\langle f|\mathbf{I}_\pi \bullet \mathbf{I}_T|i\rangle \tag{42.85}$$

Other than a minus sign, this is the product of two isospin matrices. We may call the top line (with isospin indices a and b) in Figure 42.2 "the π" and the bottom line (indices c and d) "the target", as is our privilege. Hence we have from (42.28)

$$-\langle f|\mathbf{I}_\pi \bullet \mathbf{I}_T|i\rangle = -\tfrac{1}{2}[I(I+1) - 4] \tag{42.86}$$

Note that the last term inside the brackets is -4, because now both the pion and the target have isospin 1; $I(I+1)$ is 2 for each of them. Had the target been a proton, that -4 would have been $-\frac{1}{2}(\frac{1}{2}+1) - 2 = -2\frac{3}{4}$. We can write down the eigenvalues of the operator in (42.85) from (42.86), where I is now the total s-channel isospin. This operator will have eigenvalues depending on I. The possible s-channel values for I (from the sum of the two pions' isospin) are $I = \{0, 2\}$; $I = 1$ is ruled out by parity invariance. Then

$$\delta^{ac}\delta^{bd} - \delta^{ad}\delta^{bc} = -\tfrac{1}{2}[I(I+1) - 4] = \begin{cases} 2, & I = 0 \\ -1, & I = 2 \end{cases} \tag{42.87}$$

---

[19] [Eds.] In the original derivation, Weinberg *op. cit.* says (before his equation (4)) that the commutator is "suggested by the σ model, and the free-quark model". See note 17, p. 930.

[20] [Eds.] The video of Lecture 47 ends here. The rest of this chapter is taken from the first 21 minutes of Lecture 48's video, in order to keep this material on soft pions in one chapter, and begin Chapter 43 with the Lecture 48 material on spontaneously broken symmetry.

That's one operator that occurs in this decomposition (42.84). It's the difference of one of the three $\delta^{ac}\delta^{bd}$ and $\delta^{ad}\delta^{bc}$. It happens our states are s-wave states at threshold, so they are symmetric under the exchange $a \leftrightarrow c$. If we swap a and c we get the operator

$$\delta^{ac}\delta^{bd} - \delta^{ad}\delta^{bc}\Big|_{a\leftrightarrow c} = \delta^{ac}\delta^{bd} - \delta^{ab}\delta^{cd} \tag{42.88}$$

which, acting on states symmetric in a and c, is of course the same thing as (42.87), and therefore it has exactly the same eigenvalues; it is the *same* operator on these states. That takes care of two of the three operators in (42.84).

The last operator we have to deal with is the remaining $\delta^{ac}\delta^{bd}$. This is obviously a pure $I = 0$ operator: if you rotate just the isospins of the original a and c and do nothing to the final states, it doesn't change: it is isospin-invariant. Therefore its $I = 2$ eigenvalue is 0. And its $I = 0$ eigenvalue is easily obtained. Take an isospin 0 initial state. It has some isospin wave function

$$\langle ac|I = 0\rangle \propto \delta^{ac} \tag{42.89}$$

If we apply this last operator $\delta^{ac}\delta^{bd}$ to the matrix element (42.89), and sum over the initial indices

$$(\delta^{ac}\delta^{bd})\delta^{ac} = 3\delta^{bd} \tag{42.90}$$

because $\delta^{ac}\delta^{ac}$ summed over a and c is 3. So it reproduces that isospin 0 state in the final state as it should, being an isospin-invariant operator, with a coefficient that is 3 times the original coefficient of δ^{ac} in (42.89); the eigenvalue here is 3.

We can now write the isospin-dependent part of (42.84) as

$$[3\delta^{ac}\delta^{bd} - \delta^{ab}\delta^{cd} - \delta^{ad}\delta^{bc}] = \{\delta^{ac}\delta^{bd} - \delta^{ad}\delta^{bc}\} + \{\delta^{ac}\delta^{bd} - \delta^{ab}\delta^{cd}\} + \{\delta^{ac}\delta^{bd}\} \tag{42.91}$$

I summarize the eigenvalues of these operators, and the amplitude, in Table 42.1. The

Operator	$I = 0$	$I = 2$
$\delta^{ac}\delta^{bd} - \delta^{ab}\delta^{cd}$	2	−1
$\delta^{ac}\delta^{bd} - \delta^{ad}\delta^{bc}$	2	−1
$\delta^{ac}\delta^{bd}$	3	0
$f(I)$	7	−2

Table 42.1: Eigenvalues of all isospin operators

amplitude at threshold is just the sum $f(I)$ of these three operators, times the constant Bm_π^2:

$$\mathcal{A}_{\text{thr},\,I} = \frac{4g_V^2 m_\pi^2}{F_\pi^2} f(I) \tag{42.92}$$

These are not numbers[21] of which we can say, "Well, obviously they have to be −2 and 7!"

The function $f(I)$ does not have the

$$I(I+1) - I_T(I_T+1) - 2$$

[21] [Eds.] Weinberg, *op. cit.*; Lee, *op. cit.*, Section 10b, pp. 76–77.

structure (42.29) that we found for the scattering lengths for any hadronic target *other* than a pion. In (42.86), the $(-\mathbf{I}_\pi \bullet \mathbf{I}_T)$ term, equal to $-\frac{1}{2}[I(I+1)-4]$, *does* follow this pattern (with $I=1$ for a pion target), but the $\delta^{ac}\delta^{bd}$ term in Table 42.1, which makes a large contribution to the $I=0$ amplitude, does *not*. That's because their origins are very different. The first two entries in Table 42.1 came from the current algebra commutators. The third entry in the table comes from the symmetric term (42.43) we neglected in our previous analysis, because it was of $\mathcal{O}(m_\pi^2)$.

There is a small correction[22] in the relation (42.20) connecting the amplitudes $\mathcal{A}$ with the scattering lengths a in the situation where the target and the incoming particle are the same kind. Now the constant $8\pi(m_\pi + m_T)$ is increased by a factor of 2, so that

$$\mathcal{A} = 16\pi(m_\pi + m_\pi)a = 32\pi m_\pi a \tag{42.93}$$

With this correction,

$$a_I = \frac{1}{32\pi m_\pi}\left(\frac{4g_V^2 m_\pi^2}{F_\pi^2}\right) f(I) = \frac{L}{4} f(I) \tag{42.94}$$

where $L \approx 0.1\, m_\pi^{-1}$ is given by (42.55). So the theoretical predictions for the $I=0$ and $I=2$ scattering lengths are

$$a_0 = \begin{cases} 0.175\, m_\pi^{-1} & \text{(theory)} \\ 0.235 \pm 0.03\, m_\pi^{-1} & \text{(exp't)} \end{cases} \qquad a_2 = \begin{cases} -0.050\, m_\pi^{-1} & \text{(theory)} \\ -0.031 \pm 0.007\, m_\pi^{-1} & \text{(exp't)} \end{cases} \tag{42.95}$$

Are the experimental numbers in the ratio of $-7:2$? The old values of $0.20\, m_\pi^{-1}$ and $-0.06\, m_\pi^{-1}$ certainly were, to within roundoff errors; the current values, though consistent with the theoretical predictions, are not so nice.[23]

Unfortunately, π–π scattering is not an easy process to study experimentally. The best handle on it is obtained by considering pion production in pion–nucleon scattering,[24] as shown in Figure 42.4. A π comes in, a nucleon comes in and you look at the pion pole. You try

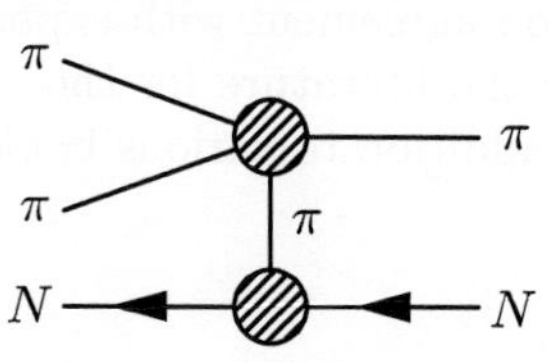

Figure 42.4: Pion production

[22] [Eds.] Coleman *Aspects*, "Soft Pions", Appendix 3.

[23] [Eds.] These values, especially a_2, have changed markedly over the years. For more recent determinations of the scattering lengths, see M. V. Olsson, "Rigorous Pion–Pion Scattering Lengths from Threshold $\pi N \to \pi\pi N$ Data", *Phys. Lett.* **B410** (1997) 311–314; S. Gevorkyan, "Pion–Pion Scattering Lengths Determination from Kaon Decays" in 7th *International Workshop on Chiral Dynamics*, Newport News, Virginia, 2012; I. Caprini, "Theoretical Aspects of Pion–Pion Interaction" in *International Conference on QCD and Hadronic Physics*, Beijing, 2005. For a history to pion scattering lengths, from Yukawa's original hypothesis up through 2008, see J. Gasser, "On the History of Pion–Pion Scattering" in *International Workshop on Effective Field Theories: from the pion to the upsilon*, Valencia, Spain, 2009. Weinberg *QTF2*, p. 202 gives $a_0 = (0.26 \pm 0.05)m_\pi$, $a_2 = (-0.028 \pm 0.012)m_\pi$.

[24] [Eds.] Olsson, *op. cit.*

to extrapolate to that pion pole, which is not in the physical region, and from that deduce something about low-energy π–π scattering. The fairest thing to say is that what we obtain is consistent with what we deduce on theoretical grounds, although there is lots of room for error and even more room for arguments about the validity of the method.

This terminates our discussion of current algebra. There are many more things that could be said, and many more good things you can do with current algebra, many more soft processes you can analyze involving soft pions. Some of the nicest are analyses of β decay processes involving the emission of one pion, where you can obtain again soft pion rules. They are a bit different in structure from what we have done here because you explicitly want to use the structure of the weak interactions. So you study a matrix element

$$\int d^4x \; \langle a, \pi | H_w | b \rangle \propto \int d^4x \; \langle a | T[A_\mu H_w] | b \rangle$$

Here H_w is the non-leptonic Hamiltonian if it's a non-leptonic process, or the weak current if it's a leptonic process. And you could relate it to scattering lengths by exactly the same tricks we have been using here. So in a typical analysis of this kind, instead of one pion and a commutator of two axial vector currents, you have to worry about the commutator of an axial vector current and a weak interaction Hamiltonian. But otherwise the tricks are much the same and the analysis is much the same. In this manner it is possible to learn quite a lot about leptonic decays of kaons[25] which go into leptons plus pions, either one π or two; both are observed. And it is possible to learn a great deal about s-wave nonleptonic hyperon decays since the process is parity-violating. In a process like

$$\Lambda \to N + \pi$$

the pion can appear in either the s-wave or the p-wave. About the p-wave process we learn nothing by these methods because that automatically vanishes when the pion momentum goes to zero. Despite statements in the early literature to the contrary, we know as little from current algebra about p-wave nonleptonic hyperon decays as we do about p-wave scattering lengths in π–π scattering, to wit: nothing. But the s-wave ones can be computed by these methods, and the results are in good agreement with experiment. But we won't have time to go into that in these lectures. See the literature for these results; there are numerous good review articles widely available in addition to various books.[26]

---

[25] [Eds.] Coleman *Aspects*, "Soft Pions", Section 10, pp. 60–62.

[26] [Eds.] For a brief introduction to current algebra, see D. H. Lyth, *An Introduction to Current Algebra*, Oxford U. P., 1970. See also note 23, p. 902.

43

A first look at spontaneous symmetry breaking

I would now like to begin a brand new subject, one that will come back to current algebra, as well as leading us in new directions towards renormalizable theories of the weak interactions. The subject (which will occupy us for several lectures) is *spontaneous symmetry breaking.*[1] I'll begin with a few general remarks, and a parable.

43.1 The man in a ferromagnet

It is a truism in non-relativistic quantum mechanics that the ground state of a system may not be invariant under the symmetry group of the Hamiltonian of the system. In the case of the hydrogen atom, the potential is rotationally invariant and the ground state is also; it's an s-state. But for more complicated systems involving many particles, this need not be so. For example, nuclear forces are rotationally invariant, but it is not true that all nuclear ground states are rotationally invariant.[2]

For a nucleus this is no problem. But for a system of infinite spatial extent it can lead to strange things. The standard example is the **Heisenberg ferromagnet**. In an idealized sense, a ferromagnet is an infinite crystalline array of little dipoles.

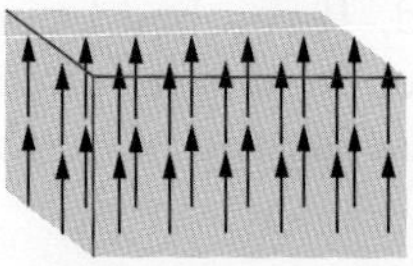

Figure 43.1: Ground state in a Heisenberg ferromagnet

[1] [Eds.] "Secret Symmetry", (Erice, 1973), Chapter 5 in Coleman *Aspects*, pp. 113–121; Peskin & Schroeder *QFT*, pp. 347–352; Ryder *QFT*, pp. 282–293; Zee *QFTN*, pp. 223–230; Abers & Lee *GT*; Jeremy Bernstein, "Spontaneous Symmetry Breaking, Gauge Theories, the Higgs Mechanism and All That", *Rev. Mod. Phys.* **46** (1974) 7–259.

[2] [Eds.] The ground state of the deuteron, for instance, has an non-zero quadrupole moment due to higher angular momentum states. J. M. B. Kellogg, I. I. Rabi, N. F. Ramsey, Jr., and J. R. Zacharias, "An Electric Quadrupole Moment of the Deuteron", *Phys. Rev.* **57** (1940) 677–695.

The law of interaction between the dipoles (the **Heisenberg exchange force**) says that neighboring dipoles like to line up; the Hamiltonian is given by[3]

$$H = -\sum_{i,j} J_{ij}\boldsymbol{\sigma}_i \cdot \boldsymbol{\sigma}_j \tag{43.1}$$

(the sum is only between nearest neighbors and the coefficients J_{ij} are exchange coupling constants); the energy is minimized when the spins all align. This interaction is rotationally invariant: it depends only on the relative angle between the spins and so is unchanged if we rotate all the spins together. At zero temperature the ground state of the ferromagnet looks like Figure 43.1—going on forever, if we imagine the ferromagnet to be infinite in extent—with the **net magnetization**[4] pointing in some direction. The ground state is a state of maximum spin. If there are N spin-$\frac{1}{2}$ dipoles in the ferromagnet (with N typically on the order of Avogadro's number), the spin of the ground state is $\ell = N \cdot \frac{1}{2}$, and so the state is $2\ell + 1 = 2(N \cdot \frac{1}{2}) + 1 = (N+1)$-fold degenerate. In the limit of an infinite ferromagnet we can think of the ground state as infinitely degenerate, labeled by a continuous vector that can point in any spatial direction, not a quantum vector with only a finite number of directions.[5] Although the interaction (43.1) *is* rotationally invariant the ground state is *not*; the total spin points in some preferred direction. The symmetry of the interaction is SO(3); once the direction of the magnetization is set, the symmetry of the ground state is reduced to SO(2), for rotations about the direction of the ground state:

$$\text{SO}(3) \rightarrow \text{SO}(2) \tag{43.2}$$

The parable[6] involves a man who lives inside the ferromagnet. He's considerably larger than an individual dipole so he can't see the granular structure; it looks like a continuum to him. But he's considerably smaller than the ferromagnet, which may be a million light years on a side. We are in telephone communication with this man. We tell him, "Physicists in the outside world have made a sensational discovery: the laws of nature are rotationally invariant." And he says: "You're crazy! There's this enormous magnetic field pointing north; it tries to pull the fillings[7] out of my teeth! The laws of nature are definitely *not* rotationally invariant; there's a preferred direction—north—and nature could not be *more* asymmetric." And we say: "No, no, no! You think that because you're living in a big ferromagnet. The *laws* of nature are really rotationally invariant, even the laws of nature for a ferromagnet. It's just that neighboring dipoles like to point in the same direction. Your experiences are influenced by the resulting strong magnetic field. If you had been living in a different ferromagnet, all the dipoles might have chosen to line up in another direction, east for example. Then the force acting on your fillings would point east instead of north."

---

[3] [Eds.] Neil W. Ashcroft and N. David Mermin, *Solid State Physics*, Harcourt College Publishers, 1976, equation (32.20), p. 681.

[4] [Eds.] E. P. O'Reilly, *Quantum Theory of Solids*, Taylor and Francis, 2003, p. 128.

[5] [Eds.] By the way, the Ising model also displays spontaneous symmetry breaking. See Robert Brout, *Phase Transitions*, W. A. Benjamin, 1965, Chapter 2, pp. 7–29. Had he not died in 2011, Brout would surely have shared the 2013 Physics Nobel with his colleague François Englert.

[6] [Eds.] Coleman included this story in a brilliant article describing the work behind the 1979 Physics Nobel Prize, won by Steven Weinberg, Abdus Salam and Sheldon L. Glashow: Sidney Coleman, "The 1979 Nobel Prize in Physics", *Science* **20** (1979) 1290–1292.

[7] [Eds.] Coleman adds: "They make dental fillings out of iron in the ferromagnet world; it's an easily available material."

The man thinks for a moment, and replies. "All right, I'll try to verify that experimentally." So he decides to change the orientations of the neighboring dipoles. If there is no preferred direction, then having all the dipoles in a new direction should be just as good a ground state with just the same energy, and the total cost of moving to that ground state should be zero. Figure 43.2 shows a few rotated dipoles.

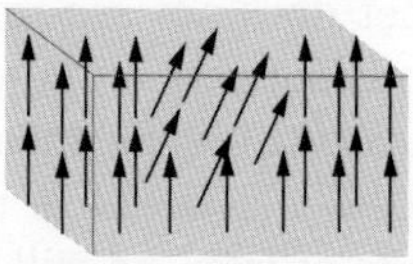

Figure 43.2: The ground state perturbed by rotating some dipoles

Of course it takes some energy to rotate each dipole. The magnet is in three dimensions so there are many boundaries in which some of the dipoles are changed in direction, and therefore the state gets a new energy. He keeps on working. His altered domain keeps growing in surface area, which is where he's losing energy, until he reaches the boundaries of the ferromagnet. Things don't look too good. If the ferromagnet is infinite, he can never reverse all of the dipoles and get his invested energy back. If the ferromagnet is finite, say with periodic boundary conditions so we don't have to worry about sharp boundary effects, he begins gaining energy again when he's reversed half the dipoles in the ferromagnet but that's still quite a lot of dipoles to reverse.

To the man in the magnet, the universe doesn't look at all as though it is rotationally invariant. There's no easy experiment he can do that will reveal this rotational invariance to him. If he understands a lot of deep physics and the laws of ferromagnets, then he can say "Oh yes, the universe *is* rotationally invariant. It just happens that I'm living in a ferromagnet which has settled down in a particular direction." But if he doesn't understand the physics he'll never believe it.[8] The rotational invariance of the Hamiltonian that governs the magnet in which he lives is completely hidden from its occupant. As far as he is concerned, it's just as if there were a large rotation-violating term in his Hamiltonian. That is how symmetry is normally broken. But that's not what happens here: the *Hamiltonian* is perfectly symmetric; it's the *dynamics* that causes the ground state to be asymmetric. The standard terminology is unfortunate. We *don't* describe the symmetry as *hidden*; instead we call this situation **spontaneous symmetry breaking**. In these circumstances we say, "The symmetry is spontaneously broken".

As you may remember, in the beginning of this course I said we always need an assumption when quantizing a theory: that the ground state, the vacuum, is *symmetric* under whatever group the Hamiltonian is.[9] The assumption is easy enough to confirm in the case of a free field, but rather difficult to check otherwise. Now we see why the man in the ferromagnet is a parable. The little man is *us* and the ferromagnet is the *universe.* It could be, if the dynamics turned out right, that there is some symmetry that is possessed by the Lagrangian but *not* by the ground state. It would be no easier for us to determine that such a symmetry held for the laws of nature than for the man in the ferromagnet to discover his physical laws are rotationally invariant. That symmetry would be completely hidden from us. Of course

---

[8] [Eds.] Other examples of spontaneous symmetry breakdown are a thin metal rod bending under pressure, in Ryder *QFT*, pp. 282–283; and Salam's banquet table, in M. Kaku, *Hyperspace*, Oxford U. P., 1994, p. 211.

[9] [Eds.] See the discussion of (2.63), p. 29.

the symmetry in question is not one we're familiar with. The symmetries we know about are all *manifest*; they're not hidden. The hidden ones are not rotational invariance or Lorentz invariance or isospin invariance. Maybe there's something else that nobody has ever thought of, at least until now, because, despite not being symmetries of the ground state, they are in fact symmetries of nature, though hidden.

Whether the man in the ferromagnet does not see rotational invariance at all, or perceives it as a weakly broken symmetry, depends entirely on how much he interacts with the magnetic field. If he and his apparatus and everything he measures are made of polyethylene, for example, then for all practical purposes his world is rotationally invariant. On the other hand, if he and everything else are made of iron, he will probably never suspect the rotational invariance of his world. So spontaneous symmetry breaking can simulate a Hamiltonian with everything from total asymmetry to weakly broken symmetry.

How might we discover these hidden symmetries? Three avenues come to mind. We could consider *high energy*, ω. At high momentum transfer, we might see the symmetries we don't see now. Or we could appeal to *high temperature*, T. For example, the magnetization of the ferromagnet is lost above its Curie temperature (when the spins are randomized) and its rotational symmetry is restored. We might not be able to do it in the laboratory, but we could look at the early universe. Finally, we could apply *high IQ*. This is the route chosen historically by Weinberg, Salam, and Glashow, among many others.

43.2 Spontaneous symmetry breaking in field theory: Examples

We'll begin investigating the possibility of a symmetric Lagrangian with a non-symmetric ground state, in the most primitive way. We will look at some classical field theories involving only scalar fields, and discuss the classical analogs to the phenomena we've been alluding to here. The examples we will discuss will be *classical* field theories involving a set of real scalar fields, which we will assemble into a big n-vector

$$\mathbf{\Phi} = \begin{pmatrix} \phi_1 \\ \phi_2 \\ \vdots \\ \phi_n \end{pmatrix} \tag{43.3}$$

with a non-derivative interaction $U(\mathbf{\Phi})$:

$$\mathscr{L} = \tfrac{1}{2}(\partial_\mu \mathbf{\Phi})\cdot(\partial^\mu \mathbf{\Phi}) - U(\mathbf{\Phi}) \tag{43.4}$$

Although these will be classical theories I will use quantum language: I will call the state of lowest energy "the vacuum" and the parameters that govern the spectrum of small oscillations about the vacuum as "particle masses", etc. I will later redeem this abuse of language by showing that for this class of theories, these classical descriptions can be thought of as the first term in a systematic quantum perturbative expansion.

The energy density $\mathscr{E}$ of this theory is

$$\mathscr{E} = \tfrac{1}{2}(\partial_0 \mathbf{\Phi})^2 + \tfrac{1}{2}(\nabla \mathbf{\Phi})^2 + U(\mathbf{\Phi}) \tag{43.5}$$

where the $(\partial_0 \mathbf{\Phi})^2$ and the $(\nabla \mathbf{\Phi})^2$ terms are summed over all the fields in $\mathbf{\Phi}$, of which there can be many. If the theory has a state of lowest energy, U must be bounded below. I will add a constant to U so that it's always greater than or equal to zero, and attains zero for some value of $\mathbf{\Phi}$:

$$U(\mathbf{\Phi}) \geq 0 \tag{43.6}$$

From (43.5) we see that the minimum energy state will occur when the non-negative terms $(\partial_0 \mathbf{\Phi})^2$ and $(\nabla \mathbf{\Phi})^2$ are as small as they can be, i.e., the ground state $\mathbf{\Phi}$ is a *constant*, independent of both time and space. We will denote this state by

$$\mathbf{\Phi} = \langle \mathbf{\Phi} \rangle = \text{constant} \tag{43.7}$$

and refer to it as the **vacuum expectation value**, or **VEV** for short, of the scalar field $\mathbf{\Phi}$, even though we are dealing with a *classical* theory. The ground state $\langle \mathbf{\Phi} \rangle$ must be a minimum of U, and hence a zero of U because of the way we've defined the potential:

$$U(\langle \mathbf{\Phi} \rangle) = 0 \tag{43.8}$$

If U has a unique minimum or zero then the ground state is unique. If it has several minima there are several possible ground states.

EXAMPLE 1: *Discrete symmetry*

Let's consider an extremely simple example, where $\mathbf{\Phi}$ is a single field ϕ.

$$U(\phi) = \tfrac{1}{4}\lambda\phi^4 + \tfrac{1}{2}\mu^2\phi^2 + \text{constant} \tag{43.9}$$

(λ is a coupling constant; the unspecified constant will be determined so that $U(\phi) \geq 0$). The symmetry group[10] is just the cyclic group Z_2; the symmetry is reflection, $\phi \to -\phi$. The ground state is invariant under this symmetry and the symmetry is manifest; it is not spontaneously broken. In order that the potential be bounded below, λ must be positive:

$$U \geq 0 \Rightarrow \lambda > 0 \tag{43.10}$$

Despite the choice of symbol, μ, which suggests a real, positive mass, we will put no restriction on μ^2 and consider either positive or negative μ^2:

$$\mu^2 > 0 \quad \text{or} \quad \mu^2 < 0 \tag{43.11}$$

Case 1: $\mu^2 > 0$. The potential is strictly concave up as shown in Figure 43.3 and the ground state is unique;

$$\langle \phi \rangle = 0 \tag{43.12}$$

If we do small oscillations about the ground state we find the energy-momentum relationship, or rather (since we are doing the classical theory) the frequency-wave vector relationship, characteristic of a particle of mass μ. Here, μ actually *is* the mass of the "meson":

$$m^2 = \mu^2 \tag{43.13}$$

Case 2: $\mu^2 < 0$. The situation is dramatically different. As shown in Figure 43.4, in this case the potential points down at the origin, because μ^2 is negative, but eventually it turns up.

---

[10] [Eds.] The cyclic group $Z_2 = \{1, -1\}$ is introduced on p. 41 of Zee *GTN*, and used frequently in the rest of Chapter I.

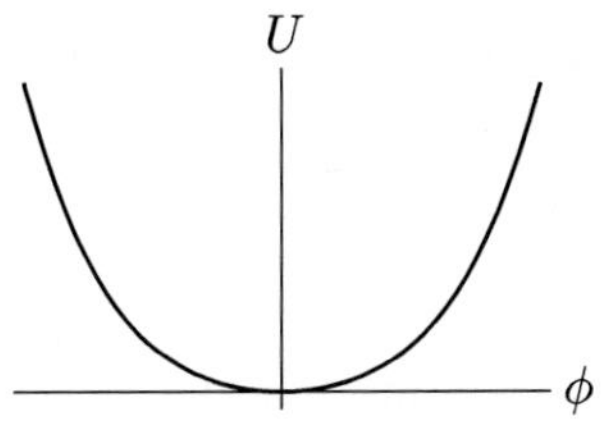

Figure 43.3: Single well

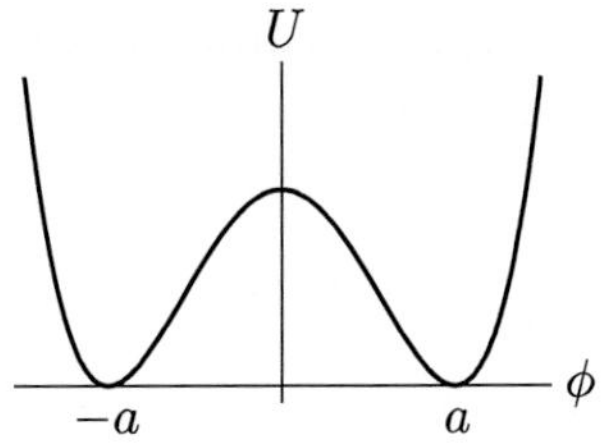

Figure 43.4: Double well

There are two points, $\pm a$ (which we will compute shortly), where the minima occur. These points occur symmetrically because of the invariance of U under $\phi \to -\phi$. We can thus write

$$U(\phi) = \tfrac{1}{4}\lambda(\phi^2 - a^2)^2 \tag{43.14}$$

The constant in (43.9) has been used to shift U up so that $U_{\text{min}} = 0$;

$$\text{const} = U(0) = \tfrac{1}{4}\lambda a^4$$

The value of a can easily be determined by comparing the quadratic terms of (43.14) and (43.9):

$$U(\phi) = \tfrac{1}{4}\lambda(\phi^2 - a^2)^2 = \tfrac{1}{4}\lambda\phi^4 - \tfrac{1}{2}\lambda a^2\phi^2 + \tfrac{1}{4}\lambda a^4 = \tfrac{1}{4}\lambda\phi^4 + \tfrac{1}{2}\mu^2\phi^2 + \tfrac{1}{4}\lambda a^4 \tag{43.15}$$

so that

$$\lambda a^2 = -\mu^2 \;\Rightarrow\; \langle\phi\rangle = a = \sqrt{\frac{-\mu^2}{\lambda}} \tag{43.16}$$

Remember, μ^2 is a negative number so this is perfectly reasonable. We have two degenerate minima. Which of these we choose to be the ground state (about which we do perturbations) is arbitrary. Any statement that we can make about the physics as seen by a man living at $+a$ or as seen by a man living at $-a$ are easily transposed, one into the other, by symmetry. But whichever one we choose, the symmetry is *spontaneously broken*. The ground state of the theory is not invariant under the symmetry group of the Hamiltonian.[11]

[11] [Eds.] The double-humped potential was used to illustrate spontaneous symmetry breaking in Goldstone's landmark paper: J. Goldstone, "Field Theories with 'Superconductor' Solutions", *Nuovo Cim.* **19** (1961) 154–164; see Figure 7, p. 162. Coleman remarks that Russian physicists know the graph in the context of I. M. Lifshitz's work on disordered semiconductors, and sometimes refer to it rudely as "Lifshitz's buttocks". Ilya M. Lifshitz (1917–1982), the brother of Lev Landau's co-author Evgeniĭ M. Lifshitz, was an outstanding condensed matter theorist. For the Lifshitz diagram, see, e.g., I. Z. Kostadinov and B. Alexandrov, "Lifshitz correlation in the hopping conductivity of high-temperature superconductors in the localized state", *Physica C* **201** (1992) 126–130; *cf.* their Figure 1, p. 127.

We can explore the consequences of spontaneous symmetry breaking by shifting the field. Then we can read off what happens for small oscillations about the ground state. Which minimum we choose is irrelevant. I'll shift to $\langle\phi\rangle = +a$. Define a new field, ϕ' (not to be confused with the renormalized field in the sense of quantum field theory):

$$\phi' = \phi - a \tag{43.17}$$

Then

$$U = \tfrac{1}{4}\lambda(\phi'^2 + 2a\phi')^2 \tag{43.18}$$

When we expand this we have no terms linear in ϕ', so we are indeed at a minimum of U. The actual mass squared of the meson is the coefficient of $\frac{1}{2}\phi'^2$ (the term that governs the energy-momentum spectrum of small oscillations). It is obtained by expanding (43.18):

$$U = \lambda a^2\phi'^2 + \lambda a\phi'^3 + \tfrac{1}{4}\lambda\phi'^4 \;\Rightarrow\; m^2 = 2\lambda a^2 = -2\mu^2 \tag{43.19}$$

So μ^2 is *not* the square of a mass, although the mass that eventually emerges is connected to it. Also we see that someone living in the minimum of this potential would see nothing like a $\phi \to -\phi$ symmetry; this symmetry is hidden from him. He would say, "I live in a very simple world in which there is only one meson, one kind of particle. It doesn't appear to obey any particular symmetry; $\phi \to -\phi$ is certainly not a symmetry here, because there's a cubic coupling in (43.19) of the meson with itself."

EXAMPLE 2: *Continuous symmetry and Goldstone bosons*

Let's consider a theory based on exactly the same principles but with a continuous internal SO(2) symmetry. We begin with two scalar fields, ϕ_1 and ϕ_2. We'll look at something we have already cooked up so that spontaneous breakdown is guaranteed to occur (again, $\mu^2 < 0$)

$$U = \tfrac{1}{4}\lambda(\phi_1^2 + \phi_2^2)^2 + \tfrac{1}{2}\mu^2(\phi_1^2 + \phi_2^2) + \text{constant} = \tfrac{1}{4}\lambda(\phi_1^2 + \phi_2^2 - a^2)^2 \tag{43.20}$$

The surface of revolution of this potential is a "Mexican hat" potential, sometimes called the "wine bottle" (or "champagne bottle") potential.

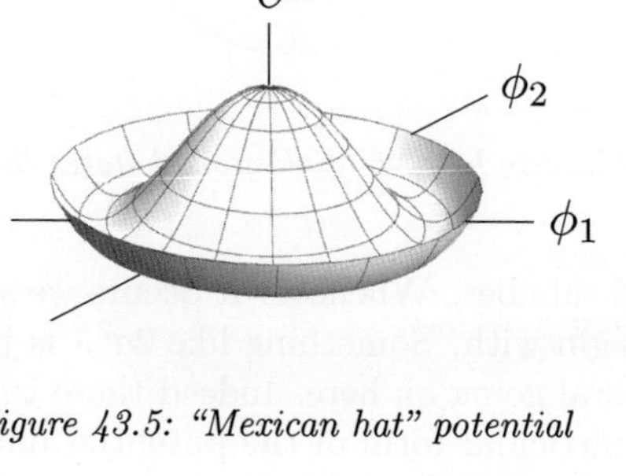

Figure 43.5: "Mexican hat" potential

This is invariant not just under $\phi_i \to -\phi_i$ but under *any* rotation in the ϕ_1-ϕ_2 space:

$$\begin{aligned}\phi_1 &\to \phi_1\cos\theta + \phi_2\sin\theta\\ \phi_2 &\to \phi_2\cos\theta - \phi_1\sin\theta\end{aligned} \tag{43.21}$$

The minimum occurs for any ϕ_1 and ϕ_2 that lie on a circle of radius a:

$$\langle\phi_1\rangle^2 + \langle\phi_2\rangle^2 = a^2 \tag{43.22}$$

See Figure 43.6, where the possible ground states, the points in the trough, are indicated on the dotted circle. *Which* point on the circle we choose to be *the* ground state of the theory is arbitrary. The ground states are *degenerate*; they all have the same energy. But once we make our choice, the ground state *loses* its rotational invariance: we have *spontaneous breakdown* of the symmetry.

For convenience, we will choose the ground state to be

$$\langle \phi_1 \rangle = a; \quad \langle \phi_2 \rangle = 0 \tag{43.23}$$

and define shifted fields

$$\phi_1' = \phi_1 - a; \quad \phi_2' = \phi_2 \tag{43.24}$$

The algebra is the same as before:

$$U = \tfrac{1}{4}\lambda(\phi_1'^2 + 2a\phi_1' + \phi_2'^2)^2 \tag{43.25}$$

The symmetry has been completely obscured. Someone arriving in this world would have no idea that it has a hidden internal SO(2) invariance. If we compute the masses, m_1^2 is the same as before; m_2^2 however is a surprise: there is *no* term quadratic in ϕ_2' in (43.25), because there's no constant for ϕ_2' to be multiplied by when we take the square:

$$m_1^2 = 2a^2\lambda; \qquad m_2^2 = 0 \tag{43.26}$$

That is, ϕ_2 describes a *massless* particle!

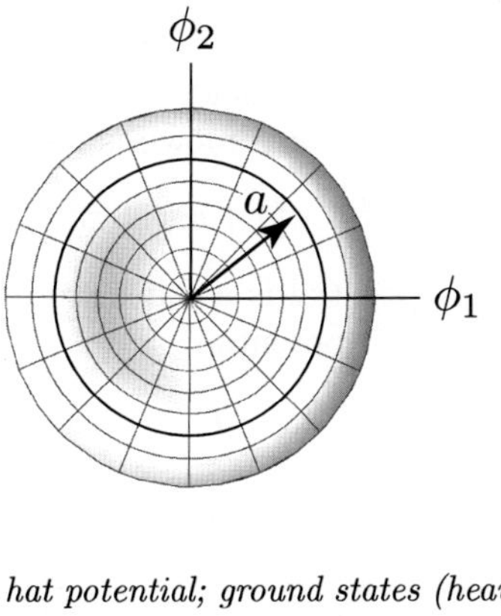

Figure 43.6: Top view of the Mexican hat potential; ground states (heavy circle) at $a = \sqrt{\langle\phi_1\rangle^2 + \langle\phi_2\rangle^2}$

Zero smells like a sacred number. Whenever it occurs we should sit up and pay attention, unless we've put it in to begin with. Something like $2a^2\lambda$ is just a number. But zero carries the scent of something general going on here. Indeed there is. We will demonstrate that this happens not just for this particular form of the potential but for any rotationally invariant potential involving ϕ_1 and ϕ_2. And then we will generalize it even further.

If ϕ did not point along one of the axes, we would obtain $\phi_1\phi_2$ cross terms in the Hamiltonian's quadratic form. We'd need to diagonalize that for the small oscillations in ϕ_1' and ϕ_2', but we'd obtain the same results: one would be massive and one would be massless. That's guaranteed, since the fields are connected by the symmetry. All the convention-independent physics must be the same. It's just the labels, what we call the massive particle and what the massless, that may differ.

The easiest way to see that there *is* something general going on is to make a change to angular variables. Define new fields, ρ and θ:

$$\phi_1 = \rho\cos\theta; \quad \phi_2 = \rho\sin\theta \tag{43.27}$$

This is a lousy choice of coordinates were we interested in expanding things about $\rho = 0$: it's singular at $\rho = 0$. But we're *not*; and for studying small vibrations about any other point, it's as good as any *other* choice.

The SO(2) symmetry is characterized by some angle α. In these variables, the symmetry is realized as

$$\theta \to \theta + \alpha; \quad \rho \to \rho \tag{43.28}$$

The Lagrangian looks rather non-canonical:

$$\mathscr{L} = \tfrac{1}{2}(\partial_\mu\rho)^2 + \tfrac{1}{2}\rho^2(\partial_\mu\theta)^2 - U(\rho) \tag{43.29}$$

The potential U is now some general rotationally invariant function; it depends only on ρ and not on θ. We'll assume

$$U(\rho) \geq 0; \quad U(a) = 0 \tag{43.30}$$

where a is some number determined by the shape of the potential. The ground state is

$$\langle\rho\rangle = a; \quad \langle\theta\rangle = \text{anything} \tag{43.31}$$

It lies anywhere on the circle $\rho^2 = a^2$ just as before.

Define new variables

$$\rho = \rho' + a; \quad \theta' = \theta \tag{43.32}$$

Then the Lagrangian is written in terms of the ρ' and θ' fields, each describing a scalar meson:

$$\mathscr{L} = \tfrac{1}{2}(\partial_\mu\rho')^2 + \tfrac{1}{2}(\rho' + a)^2(\partial_\mu\theta')^2 - U(\rho' + a) \tag{43.33}$$

We have cunningly chosen our coordinates so that the quadratic parts of $\mathscr{L}$ turn out to be diagonal. We see that the ρ' meson has some mass that depends on the second derivative of U at the minimum of the potential, about which we can't say anything until we know what the potential is. The θ' meson is *guaranteed* to be massless. If we only keep the quadratic term $\frac{1}{2}a^2(\partial_\mu\theta')^2$, it's a perfectly normal meson. It has a funny a^2 factor, but that's a constant and we can always absorb it into θ' by the classical analog of wave function renormalization. But it has no mass for the excellent reason that it never appeared in the potential in the first place, so it doesn't reappear there after we've done the shift:

$$m_{\theta'}^2 = 0; \quad m_{\rho'}^2 = \left.\frac{\partial^2 U}{\partial\rho'^2}\right|_{\rho'=0} \tag{43.34}$$

We do not need to know that the symmetry was spontaneously broken to write that down. But if there were no shift, if a were zero, we wouldn't be able to interpret the physics of the theory easily because there would be no kinetic term at all in the θ' variable. It's only because of the shift that we get an $a^2(\partial_\mu\theta')^2$. Without the shift we would say "Whoops! We're working in a terrible coordinate system that's singular at the point we're investigating. We'd better change to some other coordinate system," and then we'd wind up back at ϕ_1 and ϕ_2. This coordinate system is good for investigating small perturbations around spontaneous symmetry breakdown but bad in general, because the shift from Cartesian to angular coordinates is *singular* at the origin. But we're investigating physics somewhere on the circle, which is far from the origin.

We can in fact arrive at this result—a massless meson when a continuous symmetry is spontaneously broken—without using any specific model, just by waving our hands, now that we see what's going on. The masses of the mesons in this classical language (I keep saying masses but I really mean the things that govern the spectrum of small oscillations) are determined by the second derivatives of the potential U at its minimum. There's alway going to be a massless meson if the second derivative matrix has a vanishing eigenvalue. The appearance of massless particles is a *generic* property for rotationally symmetric potentials where the potential's minimum is not at the center. Because things are rotationally invariant there's one direction along which the potential is guaranteed to be constant, to wit the direction *along* the circle. It takes *no* energy to create tangential excitations in the trough.[12] All derivatives along *that* direction and in particular the second derivatives evaluated at the point we have chosen to be a ground state, will be zero, so we are guaranteed to have a massless meson. It is a consequence simply of the spontaneous breakdown of a continuous symmetry.

This phenomenon was discovered in a particular model by Nambu and Jona-Lasinio,[13] but they did not recognize that massless mesons were a universal feature of spontaneously broken symmetry. That realization is due to Jeffrey Goldstone.[14] The massless mesons that inevitably appear as a result of spontaneous symmetry breaking are therefore called **Goldstone bosons**. They are the signature of the spontaneous breakdown of a *continuous* symmetry.

The geometric argument we gave says there's a direction in which the potential is flat so the second derivative in that direction is zero. Using that as a starting point we can investigate what happens in the general case, where there's a general continuous internal symmetry that is spontaneously broken.

EXAMPLE 3: *Fermions and Yukawa coupling with spontaneous symmetry breakdown*

Write a Fermi field ψ as (19.81)

$$\psi = \begin{pmatrix} u_+ \\ u_- \end{pmatrix} \tag{43.35}$$

where u_+ and u_- are the two-component Weyl spinors that transform according to the $D^{(0,\frac{1}{2})}$ and $D^{(\frac{1}{2},0)}$ representations of the Lorentz group (§19.1). They have helicity ½ and −½, respectively;[15] they are also known as "right-handed" and "left-handed" spinors.[16] Recall also (19.4) that u_+ and u_- transform in the same way under rotations but with opposite signs under boosts. In the Weyl representation,

$$\gamma_5 = \begin{pmatrix} 1 & 0 \\ 0 & -1 \end{pmatrix} \tag{43.36}$$

---

[12] [Eds.] Zee *QFTN*, p. 226, contrasts the mode of oscillation in the θ direction, "rolling along the gutter", with that in the ρ direction, "climbing the wall"; the latter requires more energy.

[13] [Eds.] Y. Nambu and G. Jona-Lasinio, "Dynamical Model of Elementary Particles Based on an Analogy with Superconductivity I", *Phys. Rev.* **122** (1961) 345–358; "Dynamical Model of Elementary Particles Based on an Analogy with Superconductivity II", *Phys. Rev.* **124** (1961) 246–264.

[14] [Eds.] Goldstone, *op. cit.*; J. Goldstone, A. Salam, and S. Weinberg, "Broken Symmetries", *Phys. Rev.* **127** (1962) 965–970.

[15] [Eds.] See the discussion on p. 400; Peskin & Schroeder *QFT*, p. 47.

[16] [Eds.] See the paragraph following (19.62), p. 401.

so that the projections operators

$$P_\pm \equiv \tfrac{1}{2}(1 \pm \gamma_5) \tag{43.37}$$

project ψ onto $u_\pm$.

The ordinary Dirac Lagrangian

$$\mathscr{L}_D = \overline{\psi}(i\partial\!\!\!/ - m)\psi \tag{43.38}$$

is invariant under a constant phase transformation T_α:

$$T_\alpha\colon \psi \to e^{-i\alpha}\psi; \qquad T_\alpha\colon u_\pm \to e^{-i\alpha}u_\pm \tag{43.39}$$

However, because γ_5 anti-commutes with γ^μ (20.103), the Dirac Lagrangian is *not* invariant under the chiral transformation,[17] T_χ:

$$T_\chi\colon \psi \to e^{-i\gamma_5\theta/2}\psi; \qquad T_\chi\colon u_\pm \to e^{\mp i\theta/2}u_\pm \tag{43.40}$$

Specifically, while the kinetic term $\overline{\psi}i\partial\!\!\!/\psi$ is invariant under T_χ,

$$T_\chi\colon \overline{\psi}i\partial\!\!\!/\psi \to \overline{\psi}i\partial\!\!\!/\psi \tag{43.41}$$

the mass term is not:

$$T_\chi\colon m\overline{\psi}\psi \to m\overline{\psi}e^{-i\gamma_5\theta}\psi \tag{43.42}$$

The mass term breaks chiral invariance. Of course that's to be expected, because the bilinear $\overline{\psi}\psi$ mixes helicity states. Only the *massless* Dirac Lagrangian is invariant under (43.40). We have

$$T_\chi\colon \begin{cases} u_-^\dagger u_+ \\ u_+^\dagger u_- \end{cases} \to \begin{cases} e^{-i\theta}u_-^\dagger u_+ \\ e^{i\theta}u_+^\dagger u_- \end{cases} \tag{43.43}$$

and

$$\begin{aligned} \overline{\psi}\left[\tfrac{1}{2}(1+\gamma_5)\right]\psi &= u_-^\dagger u_+ \\ \overline{\psi}\left[\tfrac{1}{2}(1-\gamma_5)\right]\psi &= u_+^\dagger u_- \end{aligned} \tag{43.44}$$

so that, under (43.40)

$$T_\chi\colon\ \overline{\psi}\left[\tfrac{1}{2}(1\pm\gamma_5)\right]\psi \ \to\ e^{\mp i\theta}\overline{\psi}\left[\tfrac{1}{2}(1\pm\gamma_5)\right]\psi \tag{43.45}$$

This is an example of *chiral symmetry*; it acts differently on right- and left-handed fermions.

To construct an invariant Lagrangian with massless spinors coupled to two charged scalars, the scalars must transform chirally as

$$T_\chi\colon\ \phi_1 \pm i\phi_2 \ \to\ e^{\pm i\theta}(\phi_1 \pm i\phi_2) \tag{43.46}$$

The Lagrangian is

$$\begin{aligned} \mathscr{L} = {} & \tfrac{1}{2}(\partial_\mu\phi_1)^2 + \tfrac{1}{2}(\partial_\mu\phi_2)^2 - \tfrac{1}{2}\mu^2(\phi_1^2+\phi_2^2) - \tfrac{1}{4}\lambda(\phi_1^2+\phi_2^2)^2 + \overline{\psi}i\partial\!\!\!/\psi \\ & - \tfrac{1}{2}g\overline{\psi}(1+\gamma_5)\psi(\phi_1+i\phi_2) - \tfrac{1}{2}g\overline{\psi}(1-\gamma_5)\psi(\phi_1-i\phi_2) \end{aligned} \tag{43.47}$$

---

[17] [Eds.] See note 31, p. 906.

The $\frac{1}{4}\lambda$ term is the only possible chirally invariant quartic coupling. The Yukawa coupling can be written as

$$U_{\text{Yukawa}} = -g[\overline{\psi}\psi\phi_1 + \overline{\psi}(i\gamma_5)\psi\phi_2] \tag{43.48}$$

This looks like a very restrictive symmetry.

But if $\mu^2 < 0$ the scalar potential becomes as before

$$U(\phi) = \tfrac{1}{4}\lambda[(\phi_1^2 + \phi_2^2) - a^2]^2 + \text{const} \tag{43.49}$$

Again, the minima lie on a circle of radius a and exhibit $U(1)$ symmetry; see Figure 43.6. Choose as the ground state

$$\langle\phi_1\rangle = a; \quad \langle\phi_2\rangle = 0 \tag{43.50}$$

and write

$$\phi_i = \phi_i' + \langle\phi_i\rangle \tag{43.51}$$

The Yukawa coupling written in terms of ϕ_i' is

$$U_{\text{Yukawa}} = -g[\overline{\psi}\psi\phi_1' + \overline{\psi}(i\gamma_5)\psi\phi_2' + a\overline{\psi}\psi] \tag{43.52}$$

The mesons wind up with the same masses as in Example 2. However, the fermion *also* acquires a mass!

$$m_1^2 = 2a^2\lambda; \quad m_2^2 = 0; \quad m_f = ag \tag{43.53}$$

The vacuum state is *not* chirally symmetric, but the rest of the theory *is*.

43.3 Spontaneous symmetry breaking in field theory: The general case

Assume that we have a set of n real scalar fields $\{\phi_i\}$, for the moment the only fields in the world, assembled into a big vector $\mathbf{\Phi}$. We have an N-parameter group G with elements $g \in G$, characterized by real parameters λ_k, $k = 1, \ldots, N$ (and not to be confused with the Gell-Mann matrices λ_a, the generators of SU(3)). Under the action of the group[18] ($\mathbf{\Phi}^*$ is the adjoint of $\mathbf{\Phi}$)

$$G \ni g\colon \begin{cases} \mathbf{\Phi} \\ \mathbf{\Phi}^* \end{cases} \rightarrow \begin{cases} \mathscr{D}(g)\mathbf{\Phi} = e^{-i\lambda_k T_k}\mathbf{\Phi} \\ \mathbf{\Phi}^*\mathscr{D}(g)^* = \mathbf{\Phi}^* e^{i\lambda_k T_k} \end{cases} \tag{43.54}$$

where T_k are Hermitian matrices that generate the group, perhaps the isospin matrices or some generalization of them. With g near the identity (with λ_k small)

$$\mathscr{D}(g)\mathbf{\Phi} = \mathbf{\Phi} - i\lambda_k T_k \mathbf{\Phi} + \mathcal{O}(\lambda^2) \tag{43.55}$$

and the kinetic term $\partial_\mu \mathbf{\Phi}^* \cdot \partial^\mu \mathbf{\Phi}$ is unchanged under (43.54), because the matrices T_k are Hermitian.

---

[18] [Eds.] In the video of Lecture 48, Coleman uses real matrices T_k, since $\mathbf{\Phi}$ is real. To keep a single notation throughout the book, we use Hermitian matrices for the T_k.

The value of N is equal to the rank of the group, i.e., the number of generators: 3 if it's isospin, 6 if it's $SU(2) \otimes SU(2)$, etc. The group generators form a closed Lie algebra: the commutator of two generators must be another generator:[19]

$$[T_k, T_j] = ic_{kj\ell} T_\ell \tag{43.56}$$

where $c_{kj\ell}$ are the **structure constants** of the group G.

The infinitesimal change $D_k\mathbf{\Phi}$ in the field $\mathbf{\Phi}$ is, from (5.21), obtained by differentiating with respect to the kth parameter:

$$D_k\mathbf{\Phi} = \frac{\partial}{\partial\lambda_k}\left(e^{-i\lambda_j T_j}\mathbf{\Phi}\right)\bigg|_{\lambda_k=0} = -iT_k\mathbf{\Phi} \tag{43.57}$$

I'll assume the Lagrangian is of the general form

$$\mathscr{L} = \tfrac{1}{2}(\partial_\mu\mathbf{\Phi}^*)\bullet(\partial^\mu\mathbf{\Phi}) - U(\mathbf{\Phi}) \tag{43.58}$$

The potential function $U(\mathbf{\Phi})$ is assumed to be invariant under the group. U has minima at some points. We pick one of them to be our vacuum, which we will call $\langle\mathbf{\Phi}\rangle$. In general, as the above examples show, there is no reason to believe that $\langle\mathbf{\Phi}\rangle$ is invariant under the group; maybe it is, maybe not. Let's consider the case where it is invariant only under a *subgroup* of the group. Define the subgroup H, the **unbroken group**

$$H \subset G \tag{43.59}$$

as the set of all transformations that leave $\langle\mathbf{\Phi}\rangle$ unchanged: $h \in H$ if and only if

$$\mathscr{D}(h)\langle\mathbf{\Phi}\rangle = \langle\mathbf{\Phi}\rangle \tag{43.60}$$

The remaining generators are the **spontaneously broken generators.**

For instance if we had a theory with SO(3) invariance, the SO(3) generalization of the SO(2) theory we developed above, then $\langle\mathbf{\Phi}\rangle$ would be a vector that has a fixed length but points in an arbitrary direction. If we chose it to point in the 3-direction, then the group H would consist of rotations about the 3-axis.

We can arrange the generators of G so that the first ones we come across are the generators of H:

$$\begin{aligned} H &= \{T_1, T_2, \ldots, T_m\} \\ G &= \{T_1, T_2, \cdots, T_m, T_{m+1}, \cdots T_N\} \end{aligned} \tag{43.61}$$

G is the largest symmetry of the theory; H is the largest subgroup that leaves $\langle\mathbf{\Phi}\rangle$ unchanged:

$$T_k\langle\mathbf{\Phi}\rangle = 0 \quad (k = 1, \cdots, m) \tag{43.62}$$

All the other generators of G, by definition, do not leave $\langle\mathbf{\Phi}\rangle$ unchanged. That is to say, if a sum of the spontaneously broken generators acting on the minimum $\langle\mathbf{\Phi}\rangle$ equals zero,

$$\sum_{k=m+1}^{N} \alpha_k T_k \langle\mathbf{\Phi}\rangle = 0 \tag{43.63}$$

[19] [Eds.] Zee *GTN*, Chapter VI.3, "Lie Algebras in General", pp. 364–375. Incidentally, the structure constants $c_{kj\ell}$ themselves form a representation of the Lie algebra of the group G called the *adjoint representation*, defined by $(T_k)_{j\ell} = -ic_{kj\ell}$; see Zee *GTN*, p. 365.

then we must have

$$\alpha_k = 0 \tag{43.64}$$

No linear combination of the broken generators can leave $\langle\mathbf{\Phi}\rangle$ invariant. If there were such a linear combination, it would be in H and we've already got all of H. Thus, the generators in (43.63) generate an $(N-m)$-dimensional manifold, $(N-m)$ independent directions, starting from a point $\langle\mathbf{\Phi}\rangle$ which is somewhere in our big space of field strength values. There is a hyperplane of dimension $(N-m)$ tangent to that point obtained by applying the generators $(m+1), \cdots, N$, in which the potential is a constant, because U is supposed to be invariant. We can count the Goldstone bosons on this multi-dimensional Mexican hat.[20] By exactly the same arguments as before, there are

$$\dim G - \dim H = N - m \tag{43.65}$$

massless scalar Goldstone bosons, for which the second derivative matrix of U has to be zero. That is, *there is one Goldstone boson for every linearly independent spontaneously broken symmetry generator.* The number of Goldstone bosons is equal to the dimension of the group minus the dimension of the unbroken subgroup.[21]

There are $N-m$ Goldstone bosons because (43.63) and (43.64) define a manifold going through the point of the minimum, which is $N-m$ dimensional. We have a tangent direction for every α_k on which the potential is a constant, because the potential is supposed to be invariant under the group. Therefore when we take the second derivative in those directions we will get zero; the second derivative matrix projected onto that manifold will just be bunch of zeros, for the same reason that it was zero along the circle in Figure 43.6. The symmetries that are unbroken don't give us any new directions; we just apply them to $\langle\mathbf{\Phi}\rangle$ and it sits there. It's the other ones, the broken symmetries, that sweep out a part of the space. It's in those directions that we're guaranteed that the second derivative matrix about $\langle\mathbf{\Phi}\rangle$ will be zero.

There could of course be other massless bosons in the theory. Even in an ordinary theory with no spontaneous symmetry breaking we can pick the parameters so that the mass happens to be zero. But these would not be Goldstone bosons. In our simple case with a single field, with $U(\phi) = \lambda\phi^4 + \mu^2\phi^2$, we could have chosen μ^2 to be zero. Then $\langle\phi\rangle = 0$ would be the minimum of the potential because of the ϕ^4, and we would have a massless boson, but not as a consequence of the Goldstone phenomenon. The theorem doesn't say this is the *only* way to get mass zero particles. That would be obviously false, as the example I've just given shows. But spontaneous symmetry breakdown in these models is the way we *inevitably* get zero mass particles.

EXAMPLE 4: *A multiplet of scalar fields*

Let $\mathbf{\Phi}$ be an n-vector of scalar fields, in a potential which is a multi-dimensional Mexican hat

$$U = \tfrac{1}{4}\lambda(\mathbf{\Phi}^* \bullet \mathbf{\Phi} - a^2)^2 \tag{43.66}$$

and the group G is

$$G = \mathrm{SO}(n) \tag{43.67}$$

[20] [Eds.] Coleman quips that such a potential would be "fashionable attire" for a multi-dimensional *caballero*.

[21] [Eds.] The demonstration of (43.65), as well as the patterns of symmetry breaking for general rotation and unitary groups, is given in Ling-Fong Li, "Group Theory of the Spontaneously Broken Gauge Symmetries", *Phys. Rev.* **D9** (1974) 1723–1738.

The dimension of G, i.e., the number of independent planes in N-space, is

$$\dim G = \tfrac{1}{2}n(n-1) \tag{43.68}$$

As before, the ground state satisfies

$$\langle \mathbf{\Phi}^* \rangle \cdot \langle \mathbf{\Phi} \rangle = a^2 \tag{43.69}$$

I can pick a generalized North Pole,

$$\begin{aligned} \langle \phi_n \rangle &= a \\ \langle \phi_k \rangle &= 0, \ (k < n) \end{aligned} \tag{43.70}$$

I'll denote by $\mathbf{\Phi}_\perp$ the $n-1$ dimensional vector $(\phi_1, \phi_2, \ldots, \phi_{n-1})$, and then define $\mathbf{\Phi}'$ by

$$\mathbf{\Phi} = \mathbf{\Phi}' + \langle \mathbf{\Phi} \rangle \tag{43.71}$$

Using (43.70) the potential becomes

$$U = \tfrac{1}{4}\lambda(\phi_n^{*\prime}\phi_n' + 2a\phi_n' + \mathbf{\Phi}_\perp^{*\prime} \cdot \mathbf{\Phi}'_\perp)^2 \tag{43.72}$$

The masses of the mesons are

$$m_N^2 = 2a^2\lambda, \qquad m_\perp^2 = 0 \tag{43.73}$$

The subgroup H is SO$(n-1)$ with dimension

$$\dim H = \tfrac{1}{2}(n-1)(n-2) \tag{43.74}$$

There are $n-1$ Goldstone bosons:

$$\begin{aligned} \dim G - \dim H &= \tfrac{1}{2}n(n-1) - \tfrac{1}{2}(n-1)(n-2) \\ &= \tfrac{1}{2}(n-1)[n-(n-2)] = n-1 \end{aligned} \tag{43.75}$$

43.4 Goldstone's Theorem

I will now engage in a bit of hopscotch. I'm going to skip over some other models for the moment to approach the subject from another viewpoint, using the sort of general field theoretic arguments which arise in *axiomatic field theory*.[22] We'll find that consequences similar to the classical phenomena emerge. I'll then return to the classical fields to look at two other examples of spontaneous symmetry breaking, the *sigma model*,[23] which is connected with current algebra, and the famous *Abelian Higgs model*.[24] Once we know what to expect from the general arguments, we will bridge this enormous gap between classical fields and axiomatic quantum fields by looking at perturbation theory to verify that the conclusions hold to all

[22] [Eds.] R. Streater and A. Wightman, *PCT, Spin and Statistics and All That*, W. A. Benjamin Publishers, 1964; republished by Princeton U. P., 2000.

[23] [Eds.] The sigma model is discussed in §§45.3–45.4, pp. 993–1002.

[24] [Eds.] The Abelian Higgs model is discussed in §46.1.

orders. The classical field will also serve as the zeroth order of a systematic approximation scheme.

Let's dive into the general arguments.[25] I'll assume I know practically nothing except the most general things about quantum fields, and see what we can prove in the way of rigorous theorems. Suppose I have a local scalar field $\phi(x)$ and some local, conserved current $j^\mu(x)$:

$$\partial_\mu j^\mu(x) = 0 \tag{43.76}$$

If I have a theory with a conserved current, I would normally say I have a symmetry: if the integral of $j^0(x)$ over all space exists and is non-zero, that integral is a charge Q, which generates a symmetry.

For reasons that will become clear shortly, I want to be particularly cautious. I hesitate to integrate $j^0(x)$ over all space because I'm not sure that the integral will converge. So I will define a rotationally-invariant function $f(\mathbf{x})$ of compact support:

$$f(\mathbf{x}) = \begin{cases} 1, & \text{for } |\mathbf{x}| \le 1 \\ 0, & \text{for } |\mathbf{x}| > 1+\epsilon \end{cases} \tag{43.77}$$

The graph of this function is shown in Figure 43.7. Define the quantity

$$Q_R(t) = \int d^3\mathbf{x}\, f(\mathbf{x}/R)\, j_0(\mathbf{x},t) \tag{43.78}$$

(If I wanted to be a real purist, I could also smear the integral of $j_0(\mathbf{x},t)$ out in time, but I won't do that.) Because the integrand goes to zero outside a bounded region, there's no question of the integral blowing up at infinity. As R gets bigger and bigger I integrate over a larger and larger region. Formally the charge Q is defined as the limit as $R \to \infty$ of Q_R. You will see shortly why I'm being careful here.

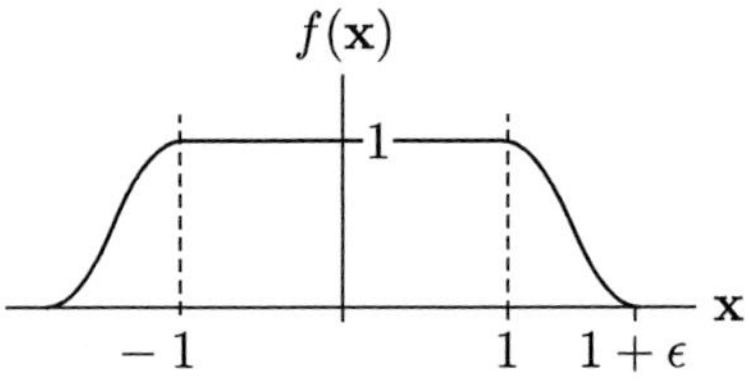

Figure 43.7: The function $f(\mathbf{x})$

Consider a generalization of the commutator (6.26),

$$D\phi(y) \equiv \lim_{R\to\infty} i\Big[Q_R(t), \phi(y)\Big] \tag{43.79}$$

I claim that the limit of the commutator with Q_R *always* exists, whether or not $\lim_{R\to\infty} Q_R$ exists. The reason is very simple. The fields are supposed to be *local*, and must commute

[25] [Eds.] Coleman's treatment follows very closely the presentation in G. S. Guralnik, C. R. Hagen and T. W. B. Kibble, "Broken Symmetries and the Goldstone Theorem", pp. 567–708 in *Advances in Particle Physics* v. 2, R. Cool and R. Marshak, eds., John Wiley, 1968. The function $f_R(\mathbf{x})$, corresponding to Coleman's $f(\mathbf{x}/R)$, is defined in their footnote 4, p. 706, and graphed in their Figure 1, p. 580.

for spacelike separations. As R increases into the region where Q_R might grow without limit, more and more of the Q_R integral will be spacelike separated from the point y, and in the limit, the commutator is finite. By the standard arguments you use to prove that the integral of a conserved current is independent of the time, you can show that this object is independent of the time. It doesn't matter what time you integrate over. Just do the usual integration by parts; you don't have to worry about the boundary terms because they're all spacelike separated with respect to y, and therefore they all vanish. The limit of the commutator exists even if the limit of $Q_R(t)$ did not exist; that's guaranteed by the fact that things commute for spacelike separations. For example, if this were the ϕ_1-ϕ_2 rotation current in the two-field model, (43.20), then $\phi(y)$ would be the field ϕ_1 and $D\phi(y)$ would be the field ϕ_2.

The unmistakable hallmark of spontaneous symmetry breaking is that, for some vacuum state $|0\rangle$,

$$\langle 0|D\phi(y)|0\rangle \neq 0 \tag{43.80}$$

(there may be several vacua in the theory). If the symmetry were manifest, the field $D\phi(y)$ would have vanishing vacuum expectation value. We sandwich (43.79) between vacua on the right and the left, the charge annihilates the vacuum ($Q\,|0\rangle = 0$) and the commutator is zero. This is a rigorous definition of what we mean (in the particular case of scalar fields) by spontaneous symmetry breaking, using only objects we are sure exist, assuming we have a local field and a local current. That is, the characteristic sign of spontaneous symmetry breaking (43.80) can be restated as

$$Q\,|0\rangle \neq 0 \iff \langle 0|\phi|0\rangle \neq 0 \tag{43.81}$$

If the ground state were symmetric, $Q\,|0\rangle = 0$, the symmetry would be manifest and the field's vacuum expectation value would be zero.

Of course, spontaneous symmetry breaking *could* occur without this particular hallmark, which has emerged in our simple models with only scalar fields. We could imagine a more complicated theory without scalar fields, but with some other (perhaps non-local) object in the theory with a non-vanishing VEV (which would vanish if the symmetry were manifest), but $Q\,|0\rangle = 0$. We will however look only at the case in which the charge fails to annihilate the vacuum, as in (43.80). Certainly *if* this happens then there is spontaneous symmetry breaking.

I will now prove the following

Theorem 43.1 (Goldstone[26]). *If for a given continuous symmetry*

$$\langle 0|D\phi|0\rangle \neq 0 \tag{43.82}$$

then there is a zero mass particle in the theory.

I'll prove the contrapositive proposition,[27] which is logically equivalent: if, except for vacua, every state has $P^\mu P_\mu \geq \epsilon > 0$, that is, the theory's particle spectrum has no massless particles, then $\langle 0|D\phi|0\rangle = 0$. (The quantity ϵ is called a *mass gap.*) I assume all the usual things: the

[26] [Eds.] Goldstone, *op. cit.*; Goldstone, Salam, and Weinberg *op. cit.*; Guralnik, Hagen and Kibble, *op. cit.*
[27] [Eds.] The contrapositive of "if A then B" is "if (not B) then (not A)".

theory doesn't have tachyons, all P^2 and all energies are positive, Lorentz invariance, and so on.

The proof is extremely simple. Consider

$$\langle 0|j_\mu(x)\phi(y)|0\rangle \tag{43.83}$$

We can write a spectral representation for this matrix element by the usual tricks, which we have used several times:[28]

$$\langle 0|j_\mu(x)\phi(y)|0\rangle = \int d^4k\, \sigma(k^2)\theta(k_0)k_\mu e^{ik\cdot(x-y)} \tag{43.84}$$

The $\theta(k_0)$ means we include only positive energy intermediate states in the integral. The vacua do not contribute to the sum of states because by Lorentz invariance,

$$\langle 0|j_\mu(x)|0\rangle = 0 \tag{43.85}$$

(there is no Lorentz-covariant vector in the theory).[29] So there are only non-vacuum states in the complete sum of intermediate states within the spectral density σ:

$$k_\mu\sigma(k^2)\theta(k_0) \sim \sum_n \langle 0|j_\mu(0)|n\rangle\, \langle n|\phi(0)|0\rangle\, \delta^{(4)}(k-P_n) \tag{43.86}$$

modulo factors of 2π, etc. By assumption we have

$$\sigma(k^2) = 0 \quad \text{for } k^2 < \epsilon \tag{43.87}$$

because the vacuum does not contribute, and all non-vacuum states have energy larger than ϵ. Taking the derivative of (43.84), by current conservation we find

$$\partial_\mu j^\mu = 0 \Rightarrow k^2\sigma(k^2) = 0 \tag{43.88}$$

because the Fourier transform of zero is zero. Within σ, $k^2 > \epsilon$. So we can divide by k^2 with confidence:

$$\sigma(k^2) = 0 \tag{43.89}$$

Then

$$\langle 0|j_\mu(x)\phi(y)|0\rangle = 0 \tag{43.90}$$

Notice that this argument would *not* work if there were mass zero particles in the theory; i.e., if we could not be sure that $P^\mu P_\mu > 0$. Then $\sigma(k^2)$ could have a delta function $\delta(k^2)$ in it, and we'd have $k^2\delta(k^2) = 0$ even though $\delta(k^2) \neq 0$.

By exactly the same reasoning, putting the two fields in the other order

$$\langle 0|\phi(y)j_\mu(x)|0\rangle = 0 \tag{43.91}$$

---

[28] [Eds]. As in the Källen–Lehmann representation in §15.2; while discussing the full propagators $\widetilde{S}'$ and $\widetilde{D}'_{\mu\nu}$ for the spin-½ field (23.42) and the photon (34.4), respectively; and Problem 19.3, p. 817)

[29] [Eds.] It's here that gauge theories provided a loophole for Higgs *et al.* to evade the Goldstone theorem. See note 6, p. 1014.

The order has nothing to do with the proof. We find $\sigma^*(k^2) = 0$ instead of $\sigma(k^2) = 0$, but that doesn't matter. If two things are zero, their difference is zero, so

$$\langle 0|\left[j_\mu(x), \phi(y)\right]|0\rangle = 0 \tag{43.92}$$

$D\phi$ is defined from this commutator through linear operations (6.26) and (6.57):

$$D\phi = i[Q,\phi] = i\int d^3\mathbf{x}\,[j_0(\mathbf{x},t),\phi] \tag{43.93}$$

so we have

$$\langle 0|D\phi|0\rangle = 0 \quad \textbf{QED} \tag{43.94}$$

Heuristically, we say

$$\text{If}\begin{cases} Q\,|0\rangle = 0, \text{ the symmetry is manifest (Wigner–Weyl realization)} \\ Q\,|0\rangle \neq 0, \text{ the symmetry is spontaneously broken (Nambu–Goldstone)}\end{cases} \tag{43.95}$$

Given a continuous symmetry,

$$\phi(x) \to \phi(x,\lambda)$$

we can derive many equations when the symmetry is manifest. Will they still be true if the symmetry breaks spontaneously? I summarize the results with a table.

Equation	*Origin*	*True with SSB?*
$\phi(x,\lambda) = U^\dagger(\lambda)\phi(x)U(\lambda)$	canonical quantization, Wigner's theorem[a]	Yes, for big enough Hilbert space
$U(\lambda) = e^{-i\lambda Q}$ $i[Q,\phi] = \left.\dfrac{\partial\phi}{\partial\lambda}\right\vert_{\lambda=0} \equiv D\phi$	Hilbert space theory, Stone's theorem[b]	Yes
$j^\mu = \pi^\mu D\phi$ $\partial^\mu j_\mu = 0$	Hamilton's Principle, Noether's theorem	Yes, *modulo* operator ordering and renormalization
$i\int d^3\mathbf{x}\,[j_0(\mathbf{x},t),\phi] = D\phi$	canonical quantization	Yes
$Q = \int d^3\mathbf{x}\; j_0(\mathbf{x},t)$	"Nothing goes wrong at ∞", no surface terms	**No!**

[a] [Eds.] See note 10, p. 129.
[b] [Eds.] Michael Reed and Barry Simon, *Functional Analysis*, Academic Press, 1972, pp. 264–267.

I've made a number of assumptions in this proof, and I should say a little bit about them. I have assumed that the VEVs of the local fields are what the mathematicians call *tempered distributions* because their Fourier transforms are tempered distributions. For purists I should say they are Schwartz distributions with test functions defined only over compact sets; from that and the positivity of the energy we can prove that they are tempered distributions.[30] That involves fancy mathematics. I assumed only that these fields have the property that if we integrate them with an infinitely differentiable function that vanishes outside some finite region, then the VEV we get is a finite quantity. Otherwise we have no business talking about the VEV's of $j_\mu(x)$, $\phi(y)$, or their product; they might not exist. With that assumption, we can prove that (43.84) is a tempered distribution, its Fourier transform is a tempered distribution, and the proof goes through.

I've also assumed that Lorentz invariance is not spontaneously broken. It *is* possible to build perfectly reasonable models in which Lorentz invariance is spontaneously broken, but that has nothing to do with the real world. We already assumed that when we wrote (43.84); that is a Lorentz-invariant expression. One of the hardest tasks for people like Arthur Jaffe and Konrad Osterwalder and their friends (who want to prove things rigorously) is to show that their ground state is Lorentz invariant (after they've gone through a sequence of limiting operations to construct the ground state in the first place). If we don't have Lorentz invariance to begin with, in general the theorem is not true. I know of no general theorem that says that if the Lagrangian is Lorentz invariant then the *ground state* has to be Lorentz invariant. It seems to be true in all the models we've looked at; it's certainly true to all orders in perturbation theory. It's also true in all the models that have been studied rigorously, like ϕ^4 in 2 and in 3 dimensions and the Yukawa model in 3 dimensions. Nothing is known rigorously about 4-dimensional theories. But it is known in the real world that Lorentz invariance is not spontaneously broken.

I should also say that the theorem doesn't show that there is a massless *field* here, only that there's a massless *particle*. What makes the particle from the vacuum is presumably the field ϕ, because it has to come into the set of intermediate states in $\langle 0|j_\mu(x)\phi(y)|0\rangle$, and ϕ by assumption is a scalar field that makes that particle from the vacuum. By refinement of the analysis we can show it has to be a spin-0 particle, but in this case it's obvious: the particle is made from the vacuum by ϕ. It has to be, otherwise it wouldn't come into the sum over intermediate states. It's connected with the vector nature of the current, which is spontaneously broken. If we were dealing with something where we had a spontaneously broken current with *two* indices, the massless Goldstone particle would end up being a *vector*.

EXAMPLE 5: *A simple model displaying Goldstone's theorem*

Here is a very simple example that will help you understand the general theorem much better. It's a rigorously solvable field theory in which the massless Goldstone particle and the non-zero vacuum expectation value emerge naturally. Consider the Lagrangian

$$\mathscr{L} = \tfrac{1}{2}(\partial_\mu\phi)^2 \tag{43.96}$$

which possesses neither a potential nor even a mass term. The theory possess a symmetry

$$\phi \to \phi + \alpha \tag{43.97}$$

with infinitesimal transformation

$$D\phi = \left.\frac{\partial\phi}{\partial\alpha}\right|_{\alpha=0} = 1 \tag{43.98}$$

[30] [Eds.] W. Appel, *Mathematics for Physics and Physicists*, Princeton U. P., 2007, p. 300.

The associated conserved current is

$$j^\mu = \pi^\mu D\phi = \partial^\mu \phi \tag{43.99}$$

The current conservation equation is obvious; it happens to be the equation of motion for a free massless scalar field:

$$\partial_\mu j^\mu = \Box^2 \phi = 0 \tag{43.100}$$

The j^0 component of the current is simply the canonical momentum,

$$j^0(x) = \frac{\partial \mathscr{L}}{\partial(\partial_0 \phi)} = \pi(x) \tag{43.101}$$

As expected, at equal times

$$i[j^0(\mathbf{x},t), \phi(\mathbf{y},t)] = \delta^{(3)}(\mathbf{x}-\mathbf{y})D\phi = \delta^{(3)}(\mathbf{x}-\mathbf{y}) \tag{43.102}$$

as our general theorems assert; integration of (43.102) would, in the absence of spontaneous symmetry breaking, reproduce (6.26). Following (43.79),

$$\lim_{R\to\infty} i[Q_R(t), \phi(\mathbf{y},t)] = D\phi = 1 \tag{43.103}$$

$$\langle 0|D\phi|0\rangle = \langle 0|1|0\rangle = 1 \neq 0 \tag{43.104}$$

Therefore we have both a massless particle and a non-vanishing VEV in the theory: the Goldstone phenomenon. Yes, it's a silly model, but with it we can compute what $Q_R(t)$ does to the vacuum, because it's a linear function of a free field. We can get a good idea of what's going on with these fancy general arguments by explicit calculation, and we can see whether or not the charge really exists.[31]

Acting on the vacuum, the charge Q_R makes only one-particle states, $|\mathbf{k}\rangle$, describing Goldstone bosons. Consider $\langle \mathbf{k}|Q_R|0\rangle$:

$$\begin{aligned}\langle \mathbf{k}|Q_R(t=0)|0\rangle &= \langle \mathbf{k}| \int d^3\mathbf{x}\, f(\mathbf{x}/R)\, j_0(\mathbf{x},0)|0\rangle \\ &= \frac{1}{(2\pi)^{3/2}}\frac{1}{\sqrt{2|\mathbf{k}|}} \int d^3\mathbf{x}\, i|\mathbf{k}| f(\mathbf{x}/R) e^{-i\mathbf{k}\cdot\mathbf{x}}\end{aligned} \tag{43.105}$$

The ∂_0 in (43.99) gives us the $i|\mathbf{k}|$. If we now scale $\mathbf{x} \to R\mathbf{x}$, this expression can be written as

$$\langle \mathbf{k}|Q_R|0\rangle \propto \sqrt{|\mathbf{k}|}R^3 \widetilde{f}(\mathbf{k}R) \tag{43.106}$$

As R gets bigger and bigger, the function $\widetilde{f}(\mathbf{k}R)$ (the Fourier transform of $f(\mathbf{x}/R)$) gets more and more sharply peaked around $|\mathbf{k}| = 0$; it gets most of its support from smaller and smaller $|\mathbf{k}|$. For huge R it behaves much like a delta function;[32] see Figure 43.8.

---

[31] [Eds.] On general grounds, it can be shown that the state $Q\,|0\rangle$ is not normalizable if $Q\,|0\rangle \neq 0$. Following Guralnik *et al.*, *op. cit.*, p. 573 and Bernstein *op. cit.*, p. 11, the argument goes like this. A vacuum state $|0\rangle$ is translation invariant, and so is $Q\,|0\rangle$. But then

$$\langle 0|QQ|0\rangle = \int d^3\mathbf{x}\, \langle 0|Qj(\mathbf{x})|0\rangle = \int d^3\mathbf{x}\, \langle 0|Qe^{-i\mathbf{P}\cdot\mathbf{x}}j(\mathbf{0})e^{i\mathbf{P}\cdot\mathbf{x}}|0\rangle = \int d^3\mathbf{x}\, \langle 0|Qj(\mathbf{0})|0\rangle = \langle 0|Qj(\mathbf{0})|0\rangle \int d^3\mathbf{x}$$

If $Q\,|0\rangle \neq 0$, this diverges. Consequently, for spontaneous symmetry breaking, $Q\,|0\rangle$ must have a divergent

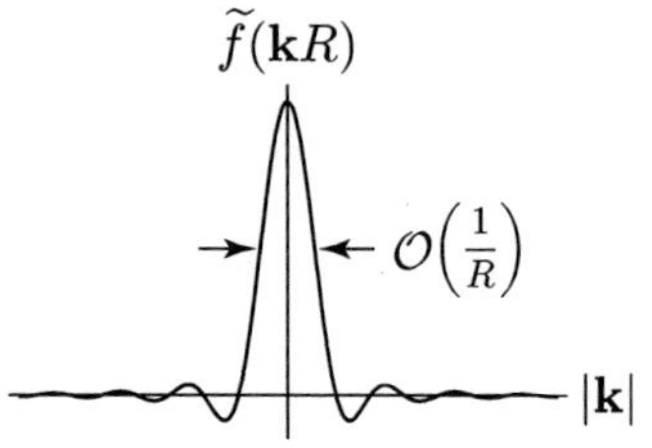

Figure 43.8: The function $\widetilde{f}(\mathbf{k}R)$

The norm of $Q_R\,|0\rangle$ can be written

$$\|\, Q_R\,|0\rangle\|^2 = \int d^3\mathbf{k}\ \langle 0|Q_R|\mathbf{k}\rangle\,\langle\mathbf{k}|Q_R|0\rangle \propto R^6 \int d^3\mathbf{k}\,|\mathbf{k}||\widetilde{f}(\mathbf{k}R)|^2 \tag{43.107}$$

Rescale $\mathbf{k}R \to \mathbf{k}$ to get rid of four powers of R:

$$\|\, Q_R\,|0\rangle\|^2 \propto R^2 \int d^3\mathbf{k}\,|\mathbf{k}||\widetilde{f}(\mathbf{k})|^2 \tag{43.108}$$

The integral no longer has any dependence[33] on R. That is,

$$\|\, Q_R\,|0\rangle\| \propto R \tag{43.109}$$

---

norm. In what follows, Coleman demonstrates $Q\,|0\rangle$'s infinite norm in this simple model.

[32] [Eds.] It's easy to see that the Fourier transform $\widetilde{f}(\mathbf{k}R) \to (2\pi)^3\,\delta^{(3)}(\mathbf{k})$. Approximating the function by

$$f(\mathbf{x}) = \begin{cases} 1, & |\mathbf{x}| \le 1 \\ 0, & \text{otherwise} \end{cases}$$

it follows that $f(\mathbf{x}/R) \to 1$ as $R \to \infty$; and the (three-space) Fourier transform of 1 is $(2\pi)^3\delta^{(3)}(\mathbf{k})$. More directly, an elementary integration (with $\mu \equiv \cos\theta$) gives

$$\widetilde{f}(\mathbf{k}R) = \int_{-\infty}^{\infty} d^3\mathbf{x}\, f(\mathbf{x}/R)e^{-i\mathbf{k}\cdot\mathbf{x}} = 2\pi\int_0^R r^2 dr\int_{-1}^{1} d\mu\, e^{-i|\mathbf{k}|r\mu} = \frac{4\pi R}{|\mathbf{k}|^2}\left[-\cos(|\mathbf{k}|R) + \frac{\sin(|\mathbf{k}|R)}{|\mathbf{k}|R}\right] \tag{$*$}$$

Figure 43.8 shows the graph of $\widetilde{f}(\mathbf{k}R)$ for $R = 2$. The height of the peak is given by the limit of the function as $|\mathbf{k}| \to 0$:

$$\lim_{|\mathbf{k}|\to 0} \widetilde{f}(\mathbf{k}R) = \tfrac{4}{3}\pi R^3 \sim R^3$$

as expected; we know $\delta^{(3)}(\mathbf{k}) \sim 1/|\mathbf{k}|^3$, and $R \sim 1/|\mathbf{k}|$. (The figure's peak is $32\pi/3$ units tall.) Like a delta function, its area (divided by $(2\pi)^3$) equals 1:

$$\int_{-\infty}^{\infty} \frac{d^3\mathbf{k}}{(2\pi)^3}\widetilde{f}(\mathbf{k}R) = 4\pi\int_0^{\infty} \frac{Rk^2 dk\, d\Omega}{(2\pi)^3 k^2}\left(-\cos kR + \frac{\sin kR}{kR}\right) = \frac{2}{\pi}\int_0^{\infty} dx\,\frac{\sin x}{x} = 1$$

independent of R. (The cosine integral doesn't really converge, but the Riemann–Lebesgue lemma suggests we can ignore it.) Alternatively,

$$\int_{-\infty}^{\infty} \frac{d^3\mathbf{k}}{(2\pi)^3}\widetilde{f}(\mathbf{k}R) = \int_{-\infty}^{\infty} d^3\mathbf{x}\, f(\mathbf{x}/R)\left[\int_{-\infty}^{\infty}\frac{d^3\mathbf{k}}{(2\pi)^3}e^{-i\mathbf{k}\cdot\mathbf{x}}\right] = \int_{-\infty}^{\infty} d^3\mathbf{x}\, f(\mathbf{x}/R)\,\delta^{(3)}(\mathbf{x}) = 1$$

[33] [Eds.] If one makes the replacement $\mathbf{k}R \to \mathbf{k}$ in the explicit form ($*$) (note 32, p. 956) of the approximate $\widetilde{f}(\mathbf{k}R)$, one finds that the R dependence of $\widetilde{f}(\mathbf{k})$ does not go away; instead, it depends on R^3. Take instead $\widetilde{f}(\mathbf{k})$, after the change of variables in the integral, as the special case of $\widetilde{f}(\mathbf{k}R)$ with $R = 1$.

the norm of the state $Q_R|0\rangle$ is proportional to R. The dependence (43.109) can also be shown easily by dimensional analysis. We have

$$[\phi] = L^{-1},\ [j^0] = L^{-2},\ [Q_R] = L,\ \text{and}\ [\,|0\rangle\,] = L^0$$

so $[Q_R|0\rangle] = L$, and the only length available is R itself.

Therefore

$$\lim_{R\to\infty} \|\, Q_R|0\rangle\|^2 = \|\, Q|0\rangle\|^2 = \infty \tag{43.110}$$

In the limit of large R the norm of the state $Q_R|0\rangle$ blows up; the state $Q|0\rangle$ is *non-normalizable.* The care above was justified, even in this simple example: the fourth component of the current integrated over all space *does* diverge (quadratically, like R^2) when applied to the vacuum state, so it's a good thing I was a purist.

Naive arguments were once made that the charge Q associated with a continuous symmetry must annihilate the vacuum, and as the charge in spontaneous symmetry breaking fails to do that, it must be that the space integral of $j^0(\mathbf{x},0)$ doesn't even exist; computations with a nonexistent charge are meaningless.[34] We get around this argument because of the Goldstone boson; the massless particle gives rise (43.106) to what amounts to a delta function at the origin, a gigantic peak at $|\mathbf{k}| = 0$. As we have seen, the space integral of the density $j^0(\mathbf{x},0)$ *applied to the vacuum* indeed blows up. We can nevertheless compute with the charge Q itself if we are careful. For example, we write the commutator of the charge and a field as the well-defined space integral of the commutator of $j^0(\mathbf{x},0)$ with the field, as in (43.103) and (43.104), where we found the commutator $[Q,\phi]$ painlessly. If you don't understand the Goldstone theorem, look at this example and see how it works.

Next time we will go on with this general analysis to deal with one more question: Is there really only *one* vacuum when there is spontaneous symmetry breakdown, or are there *many* vacua? After all, part of the time we're saying there's a unique vacuum, part of the time we're saying there are lots of vacuum states connected by the symmetry group. Which is the right way of thinking about things? We will demonstrate that *either* way is the right way. We will then return to the classical analysis and discuss how we can look at it as the zeroth order in a systematic quantum expansion.

---

[34] [Eds.] See Bernstein, *op. cit.*, Section II., pp. 10–11: "[T]he state $Q|0\rangle$ is not normalizable. This is difficult to live with but not impossible, since in all applications we will consider commutators involving $J_0(\mathbf{x},t)$ and then integrate safely later... Clearly this is a subject in which common sense will have to guide the passage between the Scylla of mathematical Talmudism and the Charybdis of mathematical nonsense." (Scylla and Charybdis were two sea monsters, the first multiheaded (deadly to some of the crew) and the other generating a whirlpool (fatal for the ship and all aboard). Odysseus and his companions had to navigate between them. For the Talmud, see Rosten *Joys*, pp. 565–576.)

Problems 23

23.1 A scalar meson ψ of mass m and charge e is minimally coupled to electromagnetism. In addition there is a massless neutral pseudoscalar meson, ϕ, with a nonminimal electromagnetic coupling:

$$\mathscr{L} = -\tfrac{1}{4}F^{\mu\nu}F_{\mu\nu} + \left|(\partial_\mu + ieA_\mu)\psi\right|^2 - m^2|\psi|^2 + \tfrac{1}{2}(\partial_\mu\phi)^2 + g\phi\epsilon^{\mu\nu\lambda\sigma}F_{\mu\nu}F_{\lambda\sigma} \tag{P23.1}$$

Here g is a positive number, $F_{\mu\nu} = \partial_\mu A_\nu - \partial_\nu A_\mu$, and $\epsilon^{0123} = +1$. Compute, to lowest nonvanishing order in perturbation theory (this is $\mathcal{O}(e^2g^2)$) the differential cross-section $d\sigma/d\Omega$, averaged over initial photon polarizations, for the process

$$\gamma + \psi \rightarrow \phi + \psi \tag{P23.2}$$

Work in the center of momentum frame, and express your result in terms of the center-of-mass total energy and the center-of-mass scattering angle.

Comments: (1) This theory isn't renormalizable, but that doesn't matter here, since you're only working in tree approximation. (2) Massless pseudoscalar mesons with this peculiar coupling appear in some of the extensions of the standard model. In these models, an experimental upper bound on g comes from studying the conversion of photons to ϕ's deep inside a star. After they are produced, the (weakly-coupled) ϕ's escape the star, stealing away energy with potentially drastic effects on stellar dynamics. This problem is a simplified version of this calculation, with the spin-½ charged particles inside the star (electrons and protons) replaced by scalar mesons. (3) I didn't ask you to compute the total cross-section because the integral that defines σ diverges at $\theta = 0$. (In case you're curious, inside the star the divergence is cut off by the shielding of the Coulomb field by the electron–proton plasma; this has an effect roughly similar to giving the exchanged photon a small rest mass.)

(1998 253b Final, Problem 1)

23.2 Example 1 of Chapter 43,

$$\mathscr{L} = \tfrac{1}{2}\partial^\mu\phi\partial_\mu\phi - \tfrac{1}{4}\lambda(\phi^2 - a^2)^2 \tag{P23.3}$$

was a theory with two ground states, $\phi = \pm a$, connected by a discrete symmetry. Such theories are in bad repute, for reasons linked to cosmology. Very early on, when the temperature of the universe is very high, the discrete symmetry in such a theory is unbroken, just as in the ferromagnet discussed (briefly) in class. As the temperature falls, the symmetry suffers spontaneous breakdown, and ϕ goes to one of its two allowed values. However, there can be no correlations between regions of space that are causally disconnected, that is to say, that are so far apart that a light signal could not have gone from one region to another in the time since the Big Bang. (Cosmological sophisticates may substitute "the end of the inflationary epoch" for "the Big Bang" in the preceding sentence.) Therefore, if ϕ is a in one of these regions, it is equally likely to be a or $-a$ in another. We thus have a picture of alternating regions of positive and negative ϕ, separated by transition zones, "domain walls". As you shall see when I finally get around to stating the problem, these domain walls typically have microphysical thicknesses and energy densities; if they're around, stretching across the universe, they mess up all sorts of things in cosmology. (Since by pointing our telescopes in different directions, we can see causally disconnected regions even now, there should be at least one domain wall currently stretching across the visible universe, causing problems not just for cosmology but for observational astronomy.)

In this problem you are asked to work out the explicit form of the simplest domain wall, one that is time-independent and flat, in Example 1. Find a solution of the field equations, $\phi(z)$, depending on the z coordinate, such that $\phi(\pm\infty) = \pm a$. (You may have to resort to an integral table.) Find the energy per unit area of this domain wall, as a function of a and λ. (*Note:* This is a problem in classical physics.)

HINT: The differential equation you'll encounter will closely resemble the Newtonian equation of motion for a point particle, with ϕ replacing the particle position and z the time. You should be able to go a long way towards solving the equation by writing down the analog of the conservation of energy.

Something to think about but not to hand in: Why isn't Example 2 in similar bad repute?

(1998b 10.2)

Solutions 23

23.1 The diagram for $\psi+\gamma \to \psi+\phi$ is (arrows denote the scalar ψ fields; the plain line denotes the pseudoscalar ϕ):

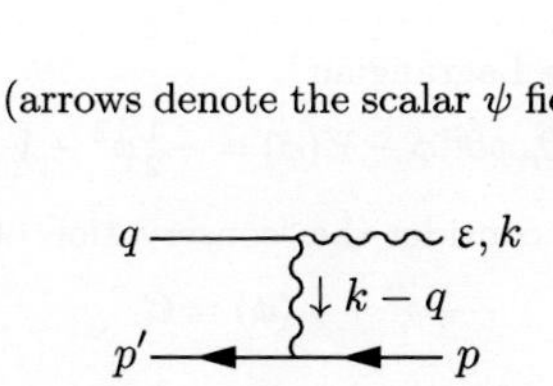

The Feynman rules for QED are given in the box on p. 670, but we have to work out the vertex corresponding to the term in the Lagrangian coupling the photon to the pseudoscalar:

$$g\phi\epsilon^{\mu\nu\lambda\sigma}F_{\mu\nu}F_{\lambda\sigma} = 4g\phi\epsilon^{\mu\nu\lambda\sigma}\partial_\mu A_\nu \partial_\lambda A_\sigma \tag{S23.1}$$

Take both photons as incoming, and Fourier transform the term. Every derivative becomes $-ip_\mu$ for an incoming momentum, and $+ip_\mu$ for an outgoing momentum (see comment (3), Problem 8, p. 309). This term becomes

$$-4g\epsilon^{\mu\nu\lambda\sigma}\widetilde{\phi}(q)k_\mu \widetilde{A}_\nu(k)k'_\lambda \widetilde{A}_\sigma(k') \tag{S23.2}$$

By functional differentiation $\delta^3/\delta\widetilde{\phi}\,\delta\widetilde{A}_\alpha\delta\widetilde{A}_\beta$, and including the usual factor $+i$ from Dyson's formula, this expression leads to a vertex

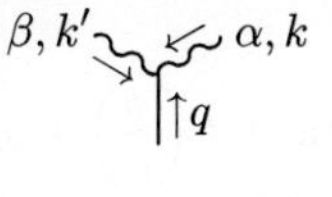

$$= -8ig\epsilon^{\mu\alpha\lambda\beta}k_\mu k'_\lambda = -8ig\epsilon^{\alpha\beta\mu\nu}k_\mu q_\nu \tag{S23.3}$$

using momentum conservation and the antisymmetry of the ϵ. The squared amplitude for $\psi+\gamma \to \psi+\phi$ is given by

$$|\mathcal{A}|^2 = \left|\varepsilon_\alpha\left(8ig\epsilon^{\alpha\beta\mu\nu}k_\mu q_\nu\right)\frac{-ig_{\beta\lambda}}{(k-q)^2+i\epsilon}\left(-ie(p+p')^\lambda\right)\right|^2 = e^2g^2\left|8\epsilon^{\alpha\beta\mu\nu}\varepsilon_\alpha k_\mu q_\nu (p+p')_\beta \frac{1}{2k\cdot q}\right|^2 \tag{S23.4}$$

using $k^2 = q^2 = 0$. It's convenient to define

$$p_t \equiv p + k = p' + q \tag{S23.5}$$

Then $p + p' = 2p_t - k - q$, and

$$\epsilon^{\alpha\beta\mu\nu}\varepsilon_\alpha k_\mu q_\nu (p+p')_\beta = 2\epsilon^{\alpha\beta\mu\nu}\varepsilon_\alpha k_\mu q_\nu (p_t)_\beta \tag{S23.6}$$

The advantage of writing things in terms of p_t is that these vectors have inner products that are simply expressed in terms of center of momentum variables:

$$p_t^2 = E_t^2 \equiv E^2, \qquad p_t\cdot k = p_t\cdot q = Ek^0 \equiv E\omega, \qquad k\cdot q = \omega^2(1-\cos\theta) \tag{S23.7}$$

To obtain the differential cross-section, we need to sum over the final spins. Writing

$$|\mathcal{A}|^2 = |M_\alpha \varepsilon^\alpha|^2 \tag{S23.8}$$

we have from (30.44)

$$\sum_{r=1}^{2}\left|M_\alpha \varepsilon^{\alpha(r)}\right|^2 = -M_\alpha^* M^\alpha \tag{S23.9}$$

In this sum we have to calculate the square of the four-dimensional Levi–Civita tensor. In analogy with (37.47) we have

$$\epsilon^{\alpha\beta\mu\nu}\epsilon_{\alpha\lambda\sigma\rho} = \delta^\beta_\lambda\delta^\mu_\rho\delta^\nu_\sigma + \delta^\beta_\sigma\delta^\mu_\lambda\delta^\nu_\rho + \delta^\beta_\rho\delta^\mu_\sigma\delta^\nu_\lambda - \left(\delta^\beta_\lambda\delta^\mu_\sigma\delta^\nu_\rho + \delta^\beta_\sigma\delta^\mu_\rho\delta^\nu_\lambda + \delta^\beta_\rho\delta^\mu_\lambda\delta^\nu_\sigma\right) \tag{S23.10}$$

(recalling $\epsilon_{\mu\alpha\beta\gamma} = -\epsilon^{\mu\alpha\beta\gamma}$). This sum would give us six terms, but three vanish because $k^2 = q^2 = 0$, and two of the others are identical. Then averaging over the initial spins and summing over the final spins,

$$\tfrac{1}{2}\sum_{\text{spins}}|\mathcal{A}|^2 = \frac{32e^2g^2}{(k\cdot q)^2}\left[2(k\cdot q)(p_t\cdot k)^2 - p_t^2(k\cdot q)^2\right] = \frac{32e^2g^2E^2(1+\cos\theta)}{(1-\cos\theta)} \tag{S23.11}$$

Finally, from (12.26),

$$\frac{d\sigma}{d\Omega} = \frac{e^2g^2}{2\pi^2}\left[\frac{1+\cos\theta}{1-\cos\theta}\right] \tag{S23.12}$$

As claimed, the integral that defines σ_T diverges at $\theta = 0$. ■

23.2 We start with the general form of the Lagrangian

$$\mathscr{L} = \tfrac{1}{2}\partial_\mu\phi\partial^\mu\phi - V(\phi) = -\tfrac{1}{2}\dot\phi^2 - V(\phi) \tag{S23.13}$$

where $\dot\phi = d\phi/dz$. Using the suggestion to consider the "conservation of energy",

$$-\tfrac{1}{2}\dot\phi^2 + V(\phi) = C \tag{S23.14}$$

Solving for $\dot\phi$ gives

$$\frac{d\phi}{dz} = \sqrt{2}\sqrt{V(\phi) - C} = \sqrt{\tfrac{1}{2}\lambda(\phi^2 - a^2)^2 - 2C} \tag{S23.15}$$

Solve for $dz/d\phi$:

$$\frac{dz}{d\phi} = \frac{1}{\sqrt{\tfrac{1}{2}\lambda(\phi^2 - a^2)^2 - 2C}} \tag{S23.16}$$

Integrating with respect to ϕ,

$$z(\phi) = \int_0^\phi \frac{d\varphi}{\sqrt{\tfrac{1}{2}\lambda(\varphi^2 - a^2)^2 - 2C}} + z_0 \tag{S23.17}$$

The problem states that $|\phi(z)| \to |a|$ as $|z| \to \infty$. Thus the integral in (S23.17) must diverge as $|\phi| \to |a|$. We conclude that $C = 0$, and

$$z(\phi) = \sqrt{\frac{2}{\lambda}}\int_0^\phi \frac{d\varphi}{|\varphi^2 - a^2|} + z_0 = \frac{\sqrt{2}}{a\sqrt{\lambda}}\tanh^{-1}\left(\frac{\phi}{a}\right) + z_0 \tag{S23.18}$$

(Gradshteyn & Ryzhik *TISP*, integrals 2.143.2 and 2.143.3.) Inverting,

$$\phi(z) = a\tanh\left(a\sqrt{\frac{\lambda}{2}}(z - z_0)\right) \tag{S23.19}$$

The energy density is

$$\mathscr{H} = \tfrac{1}{2}\left(\frac{d\phi}{dt}\right)^2 + \tfrac{1}{2}(\boldsymbol{\nabla}\phi)^2 + V(\phi) = \tfrac{1}{2}\left(\frac{d\phi}{dz}\right)^2 + V(\phi) \tag{S23.20}$$

From (S23.14) we have $-\dot\phi^2 + V(\phi) = C = 0$, so

$$\mathscr{H} = 2V(\phi) = \tfrac{1}{2}\lambda a^4\left(\tanh^2\left(a\sqrt{\frac{\lambda}{2}}(z - z_0)\right) - 1\right)^2 = \tfrac{1}{2}\lambda a^4\,\text{sech}^4\left(a\sqrt{\frac{\lambda}{2}}(z - z_0)\right) \tag{S23.21}$$

The energy per unit area is

$$\int_{-\infty}^{\infty} dz\,\mathscr{H}(z) = \tfrac{1}{2}\lambda a^4\int_{-\infty}^{\infty} dz\,\text{sech}^4\left(a\sqrt{\frac{\lambda}{2}}(z - z_0)\right) = \tfrac{1}{2}\lambda a^4\int_{-\infty}^{\infty} dz\,\text{sech}^4\left(az\sqrt{\frac{\lambda}{2}}\right) = \frac{2a^3\sqrt{2\lambda}}{3} \tag{S23.22}$$

(Gradshteyn & Ryzhik *TISP*, integral 2.423.12.) ■

44

Perturbative spontaneous symmetry breaking

In this chapter we are going to look at two questions concerning spontaneous symmetry breaking. First, as we have seen, the vacuum is not unique in theories with spontaneous symmetry breaking. Well, is this a problem or not? Second, how does perturbation theory affect spontaneous symmetry breaking? Thus far I have considered spontaneous symmetry breaking only in the context of classical fields. But (by an abuse of language) I have described the results with quantum terminology and notation, e.g., writing the value of the classical field ϕ that minimizes the potential $U(\phi)$ as a vacuum expectation value, $\langle\phi\rangle$. I've done this to smooth the connection between the classical results and what happens in quantum field theory. As I'll show, the classical results are the lowest order in a perturbation theory expansion. Do the conclusions we found in the lowest order of perturbation theory survive in higher order? Might there be corrections to a Mexican hat potential? Or maybe there's some other potential, corrections to which cause the symmetry to break spontaneously? I'll address these two questions in turn.

44.1 One vacuum or many?

We have seen several examples of field theories where a symmetry is spontaneously broken and a single vacuum state develops into several equivalent, equally valid degenerate vacuum states. The **Ising model**[1] is another such theory. It is a model of ferromagnetism similar to the Heisenberg model I talked about last time.

In the Heisenberg model the Hamiltonian (43.1) is rotationally invariant. The net magnetization in the ground state can point in *any* direction, breaking the symmetry from SO(3) of the Hamiltonian down to SO(2) of the ground state:

$$\mathrm{SO}(3) \to \mathrm{SO}(2) \tag{44.1}$$

In the simplest form of the Ising model the spins can only point up or down along one direction,

[1] [Eds.] See note 5, p. 936: the Hamiltonian in (44.2) appears as Brout's equation (33.52), p. 713; and John Preskill, Notes for Caltech's Physics 205 (1986–7), Ch. 6, pp. 6.9–6.12; online at http://www.theory.caltech.edu/~preskill/notes.html.

say the z-direction. The interaction Hamiltonian

$$H = -\tfrac{1}{2}\sum_{i,j} v_{ij}\sigma_{iz}\sigma_{jz} - \sum_i \sigma_{iz}B \tag{44.2}$$

(B is the magnetic field, v_{ij} is the exchange interaction) is *not* rotationally invariant. There are only two possible vacua: all the spins pointing up or all the spins pointing down:

$$|up\rangle\,; \quad |down\rangle \tag{44.3}$$

The cousin of the man in the Heisenberg ferromagnet lives in an Ising ferromagnet. He cannot change all of the spins in an infinite system by any finite set of local operations. These vacua are not invariant under the symmetry operation

$$|up\rangle \leftrightarrow |down\rangle \tag{44.4}$$

but they are orthogonal:

$$\begin{aligned} \langle up|\,\text{string of } m\,\sigma\text{'s}\,|up\rangle &= m \\ \langle down|\,\text{string of } m\,\sigma\text{'s}\,|down\rangle &= -m \\ \langle down|\,\text{string of } m\,\sigma\text{'s}\,|up\rangle &= 0 \end{aligned} \tag{44.5}$$

What about linear combinations of these two vacua? They are also vacua:

$$|even\rangle = \frac{1}{\sqrt{2}}\left(|up\rangle + |down\rangle\right); \quad |odd\rangle = \frac{1}{\sqrt{2}}\left(|up\rangle - |down\rangle\right) \tag{44.6}$$

These linear combinations *are* symmetric under the interchange (44.4) (*modulo* the overall sign in $|odd\rangle$), but they are *not* orthogonal:

$$\langle odd|\sigma\text{'s}|even\rangle = \tfrac{1}{2}\left(\langle up|\sigma\text{'s}|up\rangle - \langle down|\sigma\text{'s}|down\rangle\right) = \tfrac{1}{2}\left(m - (-m)\right) = m \neq 0 \tag{44.7}$$

We will focus on the orthogonal vacua and the different Hilbert spaces built upon them, and we will elevate the above discussion to the status of a theorem.

Assume that there is a finite number N of vacuum states $|0,\alpha\rangle$ with zero momenta:

$$\mathbf{P}\,|0,\alpha\rangle = \mathbf{0} \tag{44.8}$$

These states are normalized so that

$$\langle 0,\alpha|0,\beta\rangle = \delta_{\alpha\beta} \tag{44.9}$$

where $\alpha, \beta = 1, 2, \cdots, N$. There are no other normalizable momentum eigenstates. That distinguishes them from other $\mathbf{P} = \mathbf{0}$ states, such as two-particle states in the center-of-momentum frame; the other states are in the continuum and so are not normalizable. From (44.8), for a translation by some displacement $\mathbf{a}$,

$$e^{\pm i\mathbf{P}\cdot\mathbf{a}}\,|0,\alpha\rangle = |0,\alpha\rangle \tag{44.10}$$

Consider the algebra $\mathscr{A}$ of quasilocal Hermitian operators

$$A \in \mathscr{A}(R) \Rightarrow A = \int d^4x_1 \cdots d^4x_n\, f(x_1, \cdots x_n)\phi_1(x_1)\cdots\phi_n(x_n) \tag{44.11}$$

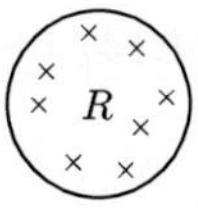

Figure 44.1: Bounded region R

with $f = 0$ if *any* of the x_i is outside some bounded region R. There is a basis where all the vacua are independent and cannot be connected by only quasilocal operators. These are called **good vacua**. That is, *all the* $A \in \mathscr{A}$ *are diagonal.* And the converse is true: If the A's are diagonal, then there will always be good vacua; these are the states that diagonalize the A's. Formally we have the following theorem:

Theorem 44.1. *There exists a basis for the vacuum states* $|0,\alpha\rangle$ *such that if* $A \in \mathscr{A}(R)$ *then*

$$\langle 0,\alpha|A|0,\beta\rangle = 0 \quad \textit{if } \alpha \neq \beta \tag{44.12}$$

The theorem says that it doesn't matter how many vacua there are. We can consider *all* of them or only *one* particular vacuum, the one we happen to live in, and just worry about that one. We never have to worry about all the other vacua because nothing we can ever do with *local* operators can ever get us to any of the other vacua. In our Ising example, we can't change all the spins in all of space. The good vacua are *globally* distinct.

After this big song and dance, the proof of the theorem turns out to be fairly simple. It depends only on translation invariance and causality, and follows easily from two lemmas:

Lemma 1: Let A, B be any two elements in $\mathscr{A}(R)$. Then

$$\lim_{\mathbf{a}\to\infty} \left[A, e^{i\mathbf{P}\cdot\mathbf{a}}Be^{-i\mathbf{P}\cdot\mathbf{a}}\right] = 0 \tag{44.13}$$

The reason is that $e^{i\mathbf{P}\cdot\mathbf{a}}f(\mathbf{x})e^{-i\mathbf{P}\cdot\mathbf{a}} = f(\mathbf{x}+\mathbf{a})$, since $e^{-i\mathbf{P}\cdot\mathbf{a}}$ is a spatial translation operator. A is associated with some region $R_A \subset R$, B is associated with some region $R_B \subset R$. When we apply the spatial translation operator we translate B by some finite value of $\mathbf{a}$, eventually to a position where it's separated from A by a spacelike interval. At that separation, they can no longer influence each other: the commutator must be zero.

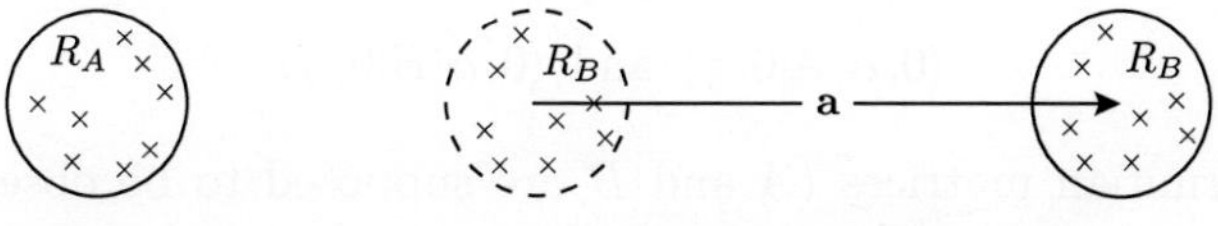

Figure 44.2: Two bounded regions, spacelike separated

Lemma 2:

$$\lim_{\mathbf{a}\to\infty} \langle 0,\alpha|Ae^{i\mathbf{P}\cdot\mathbf{a}}Be^{-i\mathbf{P}\cdot\mathbf{a}}|0,\beta\rangle \to \langle 0,\alpha|AP_0B|0,\beta\rangle \tag{44.14}$$

where P_0 is the projection operator onto the vacua,

$$P_0 = \sum_\gamma |0,\gamma\rangle\,\langle 0,\gamma| \tag{44.15}$$

Proof: Evaluate the right-hand side of (44.14) by inserting a complete set of intermediate momentum eigenstates:

$$\begin{aligned}\langle 0,\alpha|Ae^{i\mathbf{P}\cdot\mathbf{a}}Be^{-i\mathbf{P}\cdot\mathbf{a}}|0,\beta\rangle &= \sum_n \langle 0,\alpha|A|n\rangle\,\langle n|e^{i\mathbf{P}\cdot\mathbf{a}}Be^{-i\mathbf{P}\cdot\mathbf{a}}|0,\beta\rangle \\ &= \sum_n \langle 0,\alpha|A|n\rangle\,\langle n|e^{i\mathbf{P}_n\cdot\mathbf{a}}B|0,\beta\rangle\end{aligned} \tag{44.16}$$

using (44.10). The only normalizable momentum eigenstates are the vacua, with zero momentum eigenvalues. All the other states have continuous momentum eigenvalues. In the limit $\mathbf{a}\to\infty$, the continuum contributes nothing; the phases cancel out by the Riemann–Lebesgue lemma. (That's the same argument we used when discussing the reduction formula.[2]) The only contribution comes from the vacuum states:

$$\lim_{\mathbf{a}\to\infty}\langle 0,\alpha|Ae^{i\mathbf{P}\cdot\mathbf{a}}Be^{-i\mathbf{P}\cdot\mathbf{a}}|0,\beta\rangle = \sum_\gamma \langle 0,\alpha|A|0,\gamma\rangle\,\langle 0,\gamma|B|0,\beta\rangle \tag{44.17}$$

That proves the second lemma.

Obviously the order doesn't matter: with B in front of A it's the same argument. Putting these two things together we get, using the first lemma,

$$\begin{aligned}0 &= \lim_{\mathbf{a}\to\infty}\langle 0,\alpha|[A,e^{i\mathbf{P}\cdot\mathbf{a}}Be^{-i\mathbf{P}\cdot\mathbf{a}}]|0,\beta\rangle \\ &= \sum_\gamma[\langle 0,\alpha|A|0,\gamma\rangle\,\langle 0,\gamma|B|0,\beta\rangle - \langle 0,\alpha|B|0,\gamma\rangle\,\langle 0,\gamma|A|0,\beta\rangle]\end{aligned} \tag{44.18}$$

one term with A and B in one order and another with the order reversed. But

$$\sum_\gamma \langle 0,\alpha|A|0,\gamma\rangle\,\langle 0,\gamma|B|0,\beta\rangle \equiv A_{\alpha\gamma}B_{\gamma\beta}$$

The summation is nothing but the product of two matrices (summation over γ). Consequently the right-hand side of (44.18) says

$$A_{\alpha\gamma}B_{\gamma\beta} - B_{\alpha\gamma}A_{\gamma\beta} = 0 \tag{44.19}$$

That is, for any A and any B

$$\langle 0,\alpha|A|0,\gamma\rangle \text{ and } \langle 0,\alpha|B|0,\gamma\rangle \tag{44.20}$$

are commuting Hermitian matrices (A and B are supposed to be observables, and hence Hermitian).

Thus, with every observable A within $\mathscr{A}$ we associate a matrix consisting of its matrix elements between the different vacuum states. These matrices all commute with each other. If we have a family of commuting Hermitian matrices, they can all be simultaneously diagonalized by one and the same unitary transformation.[3] In this basis none of them have off-diagonal matrix elements. **QED**

[2] [Eds.] See §13.4, in particular note 4, p. 278.
[3] [Eds.] See Arfken & Weber, *MMP*, Section 3.5, pp. 215–231, and Problem 3.5.8, p. 227.

I've proved the theorem for a finite number of vacua so that these are finite-dimensional matrices, but it's also true if the number of vacua is infinite. If α is a continuous index, then the generalization of (44.9) is

$$\langle 0,\alpha|0,\beta\rangle = \delta(\alpha-\beta)$$

and the theorem generalizes to

$$\langle 0,\alpha|A|0,\beta\rangle = A_\alpha\delta(\alpha-\beta) \tag{44.21}$$

To speak a little in sophisticated mathematical talk, it could be that the big Hilbert space is not a direct sum of little Hilbert spaces but a direct integral. But that hardly matters. It's similar to what we do in going to the center of momentum frame: the big Hilbert space spanned by eigenfunctions of all total momenta is a direct integral, not a direct sum, of spaces with fixed momenta, but who cares?

It's a cunning theorem. I don't know who first proved it, nor where to find it in the literature. Arthur Wightman[4] showed it to me in 1973. The significance of the theorem is this: it doesn't matter if you say there's one vacuum or many; there are *always* good vacua. It shows that, even if we don't know anything about spontaneous symmetry breaking, and we've chosen a bad set of vacua, by a systematic constructive procedure we can always find a good choice of bases for the vacuum subspace, such that *no local operator can connect one vacuum to another.*

I don't know whether this theorem was motivated by spontaneous symmetry breaking or not; it may predate the Goldstone–Nambu ideas. Its origin may lie in statistical mechanics, where similar things occur. In my experience, when this sort of argument appears in statistical mechanics, it's usually the product of a similar argument in field theory. But there has also been a flow in the other direction, from statistical mechanics into field theory, involving people like David Ruelle.[5] I would guess this argument originated in axiomatic field theory. Maybe somebody asked, "What happens if we assume there are a lot of vacua?" And that person, or someone else, worked hard and showed that it didn't make any difference, without necessarily thinking about the application to spontaneous symmetry breaking.

44.2 Perturbative spontaneous symmetry breaking in the general case

I want to return to making a bridge between simple classical arguments and rigorous quantum arguments. (The bridge doesn't go all the way; the constructive field theorists are trying to finish the job.) Now the classical analysis of Chapter 43 will be redeemed as the leading term in a systematic perturbation theory expansion. Does the spontaneous breakdown survive to all orders in perturbation theory, for appropriate choices of the parameters (e.g. a negative

---

[4] [Eds.] Arthur S. Wightman (1922–2013) was an American mathematical physicist, a founder of axiomatic quantum field theory and originator of the Wightman axioms. A student of John A. Wheeler's, Wightman spent most of his career at Princeton. He is perhaps best known for his book*PCT, Spin and Statistics, and All That*, written with R. F. Streater, Addison-Wesley, 1964, republished by Princeton U. P., 2000.

[5] [Eds.] David Ruelle is a Belgian-French mathematical physicist, well known for his work on statistical mechanics and dynamical systems. Many others, notably Yōichirō Nambu, Philip Anderson, and Kenneth Wilson—Nobel winners all—have made major contributions to field theory using ideas from statistical mechanics and condensed matter theory.

mass squared term in our simple ϕ^4 theory)?[6] Unfortunately, as always when we're making general perturbation theory arguments, we have to use the fearsome generating functionals[7] (which I love but many students hate). Bear with them; you'll see how helpful they are. Later I'll do some specific calculations to put tangible flesh on bare and abstract bones. But first you'll have to suffer through some unavoidable (and unrelieved) formalism.

Let's recall a few facts about generating functionals. (This will be just an *aide-mémoire*; a recapitulation of earlier statements, without proofs.) To keep the notation simple I assume that I have a Lagrangian describing a single scalar field, ϕ, with some mass term and self-interaction, and, if these are really renormalized fields, also a counterterm:

$$\mathscr{L} = \tfrac{1}{2}(\partial_\mu\phi)^2 - U(\phi) + \mathscr{L}_{CT}(\phi) \tag{44.22}$$

(The argument is trivially generalizable.) If there is a mass term, it appears in $U(\phi)$. Define the action in the presence of an arbitrary external c-number function of space and time, $J(x)$:

$$\mathcal{S}[\phi, J] = \int d^4x \; (\mathscr{L} + J\phi) \tag{44.23}$$

and define $Z[J]$ and $W[J]$, the generating functionals for full and connected Green's functions, respectively, by

$$Z[J] = e^{iW[J]} = N\int [d\phi]\, e^{i\mathcal{S}[\phi,J]} = N'\, \langle 0|U(\infty,-\infty)|0\rangle_J \tag{44.24}$$

(*cf.* (13.14), (28.27), and (32.4)), where N and N' are normalization factors. Define $\overline{\phi}$ as the vacuum expectation value of ϕ in the presence of J:

$$\overline{\phi}(x) \equiv \frac{\delta W[J]}{\delta J(x)} = \frac{{}^{\text{out}}\langle 0|\phi(x)|0\rangle^{\text{in}}_J}{{}^{\text{out}}\langle 0|0\rangle^{\text{in}}_J} \tag{44.25}$$

if J is time independent.[8] The state $|0\rangle^{\text{in}}$ is the vacuum in the far past, and $|0\rangle^{\text{out}}$ is the vacuum in the far future; the vacuum is *not* the free field vacuum when there's a J around, so we have to do it this way. I now make a Legendre transformation (32.27) and define $\Gamma[\overline{\phi}]$ by

$$\Gamma[\overline{\phi}] = W[J] - \int d^4x\, \overline{\phi}(x)J(x) \tag{44.26}$$

with the equation

$$\frac{\delta\Gamma[\overline{\phi}]}{\delta\overline{\phi}(x)} = -J(x) \tag{44.27}$$

Recall (§32.2) that $i\Gamma[\overline{\phi}]$ is the generating functional for one-particle irreducible (1PI) graphs. We exploited this fact repeatedly in our investigations (§33.4) of the Ward identities in quantum electrodynamics.

---

[6] [Eds.] Much of the rest of this chapter comes from Sidney Coleman and Erick Weinberg, "Radiative Corrections as the Origin of Spontaneous Symmetry Breaking", *Phys. Rev.* **D7** (1973) 1888–1910, and "Secret Symmetry" (Erice 1973) in Coleman *Aspects*, pp. 113–184.

[7] [Eds.] See Chapter 28.

[8] [Eds.] Earlier, in §32.2, $\overline{\phi}$ was simply a classical field, the argument of the effective action $\Gamma[\overline{\phi}]$, (32.11), and not necessarily the value that minimized a potential. In Coleman *Aspects*, this value is written as ϕ_c: "Secret Symmetry", p. 312, equation (3.12), likewise in Coleman and Weinberg, *op. cit.*, equation (2.4), p. 1890.

There is a systematic way of expanding $\Gamma[\overline{\phi}]$ that corresponds to an expansion in powers of $\hbar$ if we stick an $\hbar$ back into $\mathcal{S}$. This is the *semi-classical* or *loop expansion*: expanding in no-loop graphs, one-loop graphs, etc., a natural kind of perturbation theory (32.21) for $\Gamma[\overline{\phi}]$. In the tree (no-loop) approximation $\Gamma[\overline{\phi}]$ is just the classical action, $\mathcal{S}[\overline{\phi}]$:

$$\Gamma[\overline{\phi}] = \underbrace{\int d^4x\, \mathscr{L}(\overline{\phi})}_{\mathcal{S}[\overline{\phi}]} + \hbar\,(\text{one loop}) + \hbar^2\,(\text{two loops}) + \cdots \tag{44.28}$$

(Remember that the tree approximation is what we get when we sum up the tree graphs, with no-loop corrections.[9] If it's a ϕ^4 theory, the 1PI graph with four external lines gives the ϕ^4 term and the inverse propagator gives the $(\partial_\mu\phi)^2 - \mu^2\phi^2$ term, etc.)

We are now in a position to use this formalism. You thought it was set up to facilitate the study of the Ward identities and renormalization theory in quantum electrodynamics. That's true, but it was also designed to be used in spontaneous symmetry breaking.

To take a definite example, let's consider our ϕ^4 model. The condition that the symmetry breaks spontaneously is that $\overline{\phi}$ (44.25) is non-zero even though J is zero. This equation,

$$\frac{\delta\Gamma}{\delta\overline{\phi}} = 0 \tag{44.29}$$

if it has solutions, will tell us whether or not spontaneous symmetry breaking occurs. If the theory wants to have $\overline{\phi} = 0$, this equation will tell us that the solution will be $\overline{\phi} = 0$; likewise, if the theory wants $\overline{\phi} = \pm a$. In that case (44.29) is the statement that a nonzero expectation value of ϕ is tolerable with $J = 0$. As before I'll denote $\overline{\phi}$ by $\langle\phi\rangle$, now with much more justification than in the classical theory. I can shift the field, just as in the classical analysis, and define a new quantum field ϕ':

$$\phi' = \phi - \langle\phi\rangle\,; \quad \overline{\phi}' = \overline{\phi} - \langle\phi\rangle = 0 \tag{44.30}$$

I can re-express Γ as

$$\Gamma = \Gamma[\overline{\phi}' + \langle\phi\rangle] \tag{44.31}$$

Now $\overline{\phi}' = 0$ (in the ground state of the theory) because of the way we've constructed it. I expand $\Gamma[\overline{\phi}' + \langle\phi\rangle]$ about $\overline{\phi}' = 0$ to obtain the 1PI Green's functions when the symmetry is spontaneously broken.

Everything I did in the classical theory of spontaneous symmetry breaking earlier goes through without alteration in the quantum theory (i.e., using perturbation theory), but with the *effective* action $\Gamma[\overline{\phi}]$ substituted for the *classical* action $\mathcal{S}[\phi]$. Instead of trying to find minima by finding the stationary points of the classical action, I find ground states by looking at the stationary points of the effective action; instead of finding effective coupling constants and masses by expanding about the minima of the classical action, I find 1PI Green's functions by expanding about the minima of the effective action. It's exactly the same game in the quantum and classical theories (see Table 44.1; the penultimate pair of equations in the table will be explained presently).

---

[9] [Eds] See note 6, p. 689.

Classical SSB	*Quantum SSB*
$\dfrac{\delta S}{\delta \phi} = 0$	$\dfrac{\delta \Gamma}{\delta \overline{\phi}} = 0$
$\phi = \langle \phi \rangle$	$\overline{\phi} = \langle \overline{\phi} \rangle$
$\partial_\mu \phi = 0$	$\partial_\mu \overline{\phi} = 0$
$\phi \to \text{const}$	$\overline{\phi} \to \text{const}$
$U(\langle \phi \rangle)$	$V(\overline{\phi}) = U(\langle \phi \rangle) + \text{loops}$
$\lim_{\phi \to \text{const}} S[\phi] = -U(\langle \phi \rangle) L^3 T$	$\lim_{\overline{\phi} \to \text{const}} \Gamma[\overline{\phi}] = -V(\overline{\phi}) L^3 T$
$\dfrac{dU}{d\phi} = 0 \Rightarrow \langle \phi \rangle$	$\dfrac{dV}{d\overline{\phi}} = 0 \Rightarrow \overline{\phi}$

Table 44.1: Classical and Quantum (Perturbative) SSB

Note that I have set

$$\begin{aligned} \partial_\mu \phi = 0 \text{ for } \phi = \langle \phi \rangle \quad &\text{(classical)} \\ \partial_\mu \overline{\phi} = 0 \text{ for } \overline{\phi} = \langle \phi \rangle \quad &\text{(quantum)} \end{aligned} \tag{44.32}$$

because I'm not interested in the spontaneous breakdown of translational symmetry. There are certain kinematic simplifications coming from the fact that in theories we're interested in, the ground state is spatially homogeneous: $\overline{\phi}$ is translationally invariant, a constant. I'll come back to this later.

There's no reason why translation invariance should *not* be spontaneously broken in a theory that describes the real world. It occurs in statistical mechanics, for example, where the phenomenon is called *crystallization*. There, instead of changing the square of the mass to cause the manifest symmetry to break spontaneously, one changes the temperature. Let's take a typical material such as iron, and imagine an iron universe, spatially infinite. If the temperature is above a certain point, the ground state (in the sense of statistical mechanics) is spatially homogeneous; it's iron vapor. We lower the temperature below the freezing point of iron, and the ground state becomes an infinite iron crystal, which does not have spatial homogeneity. If we now consider the rotation of a crystal somewhere in the frozen iron, how it rotates depends on its position relative to a central lattice point. That's an example of spontaneous symmetry breakdown of translational invariance.

I want to make three points that are absolutely critical. First, *the shift* $\overline{\phi}' = \overline{\phi} - \langle \phi \rangle$ *commutes with the loop expansion.* That's because the loop expansion is an expansion in powers of a parameter that multiplies the *total* action. It is therefore completely indifferent as to how we break up the action into a free and an interacting part. One way may be natural before we make the shift, and another way may be natural after we make the shift, but that's irrelevant. One-loop diagrams are not shifted into two-loop diagrams. Second, *the analysis of the quantum theory in the tree approximation recreates that of the classical theory.* That is because in the tree approximation Γ *is* the classical action. What we did with classical fields was not simply pedagogically useful, but in fact is the zeroth stage of a systematic quantum expansion. All the words get changed, but the equations are exactly the same. And we know how to compute the quantum corrections to this zeroth order term. We just compute the one-loop corrections to Γ and then go through the same algebra as before. Finally, and most critically, *spontaneous symmetry breaking does not affect renormalization.* The renormalization counterterms in a theory with spontaneous symmetry breaking are of *exactly the same type* as if there were no spontaneous symmetry breakdown. For example, if we consider a single scalar field with quartic self-interactions, the only counterterms we need to compute in Γ are a ϕ^2, a ϕ^4 and a $(\partial_\mu\phi)^2$ counterterm. We could make all of our renormalizations in the computation of Γ before we do any shift (though sometimes it's not the most expedient approach). Then we certainly won't need a ϕ^3 counterterm because, before we do the shift, there are no ϕ-ϕ-ϕ 1PI diagrams in ϕ^4 theory to be canceled out. After we do the shift, of course, a ϕ'^3 interaction *will* appear in the effective action, but we still don't need a ϕ'^3 counterterm, because we've already gotten rid of all infinities in computing Γ *before* we've made the shift. The shift is a purely algebraic operation without a single integration over internal momenta, and therefore cannot possibly introduce new ultraviolet infinities.

The value of $\overline{\phi}$ is a function of the masses and the coupling constants (or whatever renormalized parameters we choose) in the original process. If we choose to make our renormalizations before we've made our shift, we probably won't choose to renormalize on the mass shell, because that's the *wrong mass*; the mass squared is a negative number. After we make the shift and get to the physical theory, the one we really see, we might choose to make a further finite renormalization to turn things into physical parameters for the shifted theory.

The renormalization of ϕ itself is basically the wave function renormalization. In ϕ^4 theory, you fix three finite parameters (in any manner you choose: BPHZ, or maybe some fancy renormalization convention of your own, or ...) and you've fixed the theory. Those three parameters—conditions on the renormalization of the field (wave function), on the two-point function (mass), and on the four-point function (coupling constant)—are enough to absorb the infinities. The vacuum expectation value of the field depends on the choice of the three renormalized parameters. If what *you* call the mass and the coupling constant are not what *I* call the mass and the coupling constant, we'll get different analytic expressions, *but we'll be describing the same physics.* Any way we renormalize is as good as any other. It's just a matter of convention, something like the medieval disputes over the length of a standard foot: was it to be based on the foot of England's king or Belgium's? It doesn't matter, so long as we stick with our conventions.

The program we have set out is beautiful in its conceptual simplicity. But it's rather complicated to carry out, because we've got to compute the effective action. That's a messy thing to compute to all powers of ϕ, even in one loop, because of the arbitrary external momenta on all the lines.[10] Considerable simplification is made if we use the fact that, in most of the cases we are interested in, $\overline{\phi}(x)$ is independent of x, just a constant:

$$\overline{\phi}(x) = \overline{\phi} = \text{constant} \tag{44.33}$$

Investigating the effective action for *constant* $\overline{\phi}$'s is much simpler than for general $\overline{\phi}$'s. A constant field in position space has a zero derivative, and so in momentum space we need consider only graphs with zero external momenta, because that corresponds to a constant field

[10] [Eds.] "Secret Symmetry", Sections 3.4 and 3.5, pp. 135–138, in Coleman *Aspects*.

in position space. In the classical case the Lagrangian $\mathscr{L}$ for a constant field has a vanishing kinetic term and a constant potential:

$$\lim_{\phi \to \text{const}} S[\phi] = -U(\langle\phi\rangle) \int d^4x \tag{44.34}$$

$U(\langle\phi\rangle)$ is the energy density of the ground state. Usually it's set to 0, but if there are two local minima (as in Figure 44.4), $U(\langle\phi\rangle)$ will tell you which is which. The factor $\int d^4x$, formally infinite and equal to the volume L^3T of the space-time box, takes care of translational invariance.[11]

In the same way from the effective action $\Gamma[\overline{\phi}]$ we define a quantity $V(\overline{\phi})$ called the **effective potential**:

$$\lim_{\overline{\phi} \to \text{const}} \Gamma[\overline{\phi}] = -V(\overline{\phi}) \int d^4x \tag{44.35}$$

In tree approximation

$$\Gamma[\overline{\phi}] = \int d^4x \, \mathscr{L}(\overline{\phi}) + \text{loops} \tag{44.36}$$

and

$$\mathscr{L}(\phi) = \tfrac{1}{2}(\partial_\mu \phi)^2 - U(\phi) + \mathscr{L}_{CT} \tag{44.37}$$

$\mathscr{L}_{CT}$ includes the loop corrections. Then

$$V(\overline{\phi}) = U(\langle\phi\rangle) + \hbar\,(\text{one loop}) + \hbar^2\,(\text{two loops}) + \cdots \tag{44.38}$$

We defined $V(\overline{\phi})$ so that it corresponds to the ordinary potential in the tree approximation and then has corrections. $V(\overline{\phi})$ is called the *effective potential* for the same reason that $\Gamma[\overline{\phi}]$ is called the effective action. It is a generating *function*, not a generating *functional*; it doesn't depend on a variable field $\overline{\phi}(x)$ but on a single number, $\overline{\phi}$. Since $i\Gamma$ is the generating functional of 1PI graphs, $-iV(\overline{\phi})$ is the generating *function* of 1PI graphs with all the external momenta equal to zero and with the $(2\pi)^4\delta^{(4)}(0)$ from overall energy-momentum conservation divided out. That's just the Fourier space equivalent[12] of the integral $\int d^4x$.

The rule for computing $V(\overline{\phi})$ is very simple. You don't have to worry about any external momentum. You just have external lines each carrying zero momentum. Sum up all those graphs to one loop or two loops or however many loops you're going to do. I will do that summation in front of your very eyes for a general $U(\phi)$. We will get the effective potential $V(\overline{\phi})$. The condition that determines whether the symmetry is spontaneously broken or not is then

$$\frac{dV(\overline{\phi})}{d\overline{\phi}(x)} = 0 \tag{44.39}$$

—an *ordinary* derivative, not a functional derivative, because it's just a function of a number; we treat Γ as if it were S, the action, and $V(\overline{\phi})$ as if it were $U(\langle\phi\rangle)$, the potential.

If $\overline{\phi}$ is not a constant and you imagine expanding $\overline{\phi}$ in a Fourier series, any terms with non-zero momenta have to enter at least quadratically for the momenta to cancel out: Γ is translationally invariant. Therefore if we're interested in derivatives near a constant field, we

---

[11] [Eds.] The integral $\int_{-\infty}^{\infty} d^4x$ is invariant under the translation $x \to x + a$.

[12] [Eds.] $\int d^4x = \lim_{p\to 0} \int d^4x\, e^{-ip\cdot x} = \lim_{p\to 0} (2\pi)^4\delta^{(4)}(p) = (2\pi)^4\delta^{(4)}(0)$.

only need to know the value of the function for the constant field. The variational derivatives with respect to the non-zero Fourier components of $\overline{\phi}$ will automatically be zero if evaluated at a constant $\overline{\phi}$. Further expanding Γ we get something like[13]

$$\Gamma[\overline{\phi}] = \int d^4x[-V(\overline{\phi}) + W(\overline{\phi})(\partial_\mu\overline{\phi})(\partial^\mu\overline{\phi}) + \cdots] \tag{44.40}$$

where $W(\overline{\phi})$ would take all graphs and evaluate them to second order in the external momenta, picking up terms of order k^2, either second order in one momentum or first order in one and first order in another, as well as terms of order 0. Think of $V(\overline{\phi})$ not as the first term in an expansion but as a general functional evaluated for a constant field. It's defined for arbitrary fields, so in particular it's defined for a constant field. Never mind whether there's an expansion about that point or not.

44.3 Calculating the effective potential

Let's now work out $V(\overline{\phi})$ in a particular case. Actually, if we're only interested in qualitative information there's hardly any point in computing it, because the loop expansion is the expansion in powers of a coupling constant if there's only one coupling constant in the theory—loop graphs have more powers of the coupling constant than tree graphs. Therefore if we're interested only in the qualitative behavior of the theory, and if the coupling constant is small (the only case in which we have any right to use a diagrammatic expansion), there's hardly any point to the calculation. The moral has already come through: nothing qualitative will change. There will just be a small correction to the picture we've already developed. Nor will there be any problem with renormalization. So we're not going to learn anything qualitatively new in this sample calculation. (There are special cases where the tree approximation does not give an unambiguous answer. In such cases we *do* learn something new, and I'll talk about those later.) But we'll get some feeling for the structure of the argument by doing this calculation.

We begin with (44.22):

$$\mathscr{L} = \tfrac{1}{2}(\partial_\mu\phi)^2 - U(\phi) + \mathscr{L}_{CT}(\phi) \tag{44.41}$$

where $U(\phi)$ is or dimension 4 or less; otherwise the theory is not renormalizable. $\mathscr{L}_{CT}$ already contains whatever counterterms we need. (Here, these will be quadratic and quartic.) To one-loop order the full expression for $-iV(\overline{\phi})$ is

$$\underbrace{-iV(\overline{\phi})}_{\text{Sum of graphs}} = -\underbrace{iU(\overline{\phi})}_{\text{No loops}} - \underbrace{iU^{(1)}_{CT}(\overline{\phi})}_{\text{Computed to one loop}} + \cdots \tag{44.42}$$

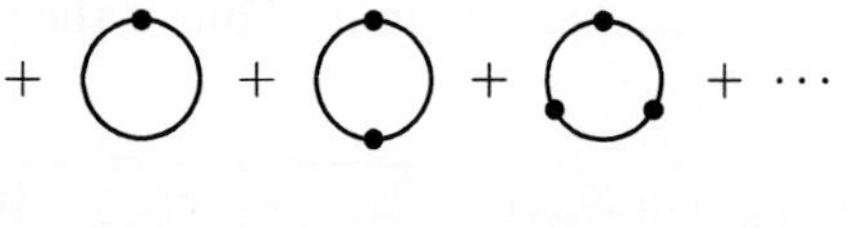

There will be an infinite sum of graphs, for which we adopt a special notation. The heavy black dot means the following: I'm going to expand in powers of U (or equivalently, in powers of the coupling constant); I'm not even going to put a mass term into the propagator. In the loop expansion it doesn't matter how we split things up, so the propagators are all going to to be[14]

$$\frac{i}{k^2 + i\epsilon} \tag{44.43}$$

[13] [Eds.] See equation (2.8), p. 1890 in Coleman and Weinberg, *op. cit.*

[14] [Eds.] In Coleman and Weinberg, *op. cit.*, the scalar field was taken to be massless; see equation (3.1), p. 1892.

The heavy dot is going to consist of everything that can go on a vertex:

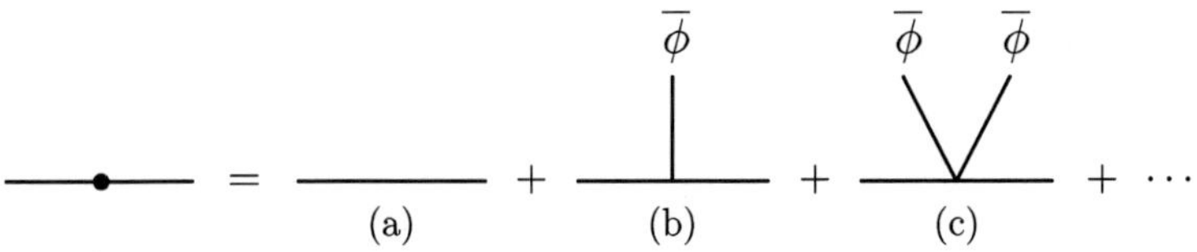

(44.44)

diagram (a) for the mass term; diagram (b) for a ϕ^3 term if present, with one external line and a number, $\overline{\phi}$, multiplying it (this is supposed to be the generating function that, when differentiated with respect to $\overline{\phi}$, gives us 1PI graphs); and diagram (c), corresponding to a ϕ^4 term, with two external lines carrying zero momentum and multiplied by $\overline{\phi}$. For example, given

$$U(\phi) = \sum_{n\geq 2}^{4} a_n \phi^n$$

then the diagrams (a), (b) and (c) are equal to $-2ia_2$, $-3\cdot 2ia_3\overline{\phi}$, and $-4\cdot 3ia_4\overline{\phi}^2$, respectively. There could be other terms if I were foolish enough to consider a non-renormalizable theory. These are all the possible interactions that can go on the dot, either with no external lines, one external line, two external lines, or what have you. All the external lines (multiplied by $\overline{\phi}$) carry momentum zero; the other two lines are going around the loop. We don't need to do any fancy summation for $U_{CT}^{(1)}$; we just look at the divergent graphs. There will be counterterms at the one-loop level, but not at the tree level; terms linear in $\overline{\phi}$ do not contribute to one-loop 1PI diagrams.[15]

It's very easy to get a rule for the heavy dot. Let's compute for example the value of the vertex for the potential

$$U = \frac{1}{n!}\lambda\phi^n \tag{44.45}$$

Only one term will appear in (44.44), the term with $(n-2)$ external lines coming off an internal line. The internal line is made of contracting two fields; there are n possible choices about the first field, and $(n-1)$ choices for the second. All the other fields are supposed to carry zero momentum. They are indistinguishable from each other so we don't have to worry about which is which. Each dot, however, carries a factor of $\lambda\overline{\phi}$ to the power $(n-2)$, because there are $(n-2)$ of the fields left. Finally there's a $-i$ because the Feynman rules are derived from $\exp(i\mathscr{L})$ and $\mathscr{L}$ has $-U$ in it. Thus if the potential has the form (44.45), the vertex is

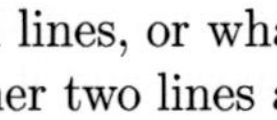

$$= -\frac{i\lambda}{(n-2)!}\overline{\phi}^{\,n-2} \tag{44.46}$$

If this is the value of the heavy dot for ϕ^n then the value for a general U is the second derivative with respect to the argument:

$$= -i\frac{d^2U}{d\phi^2}\bigg|_{\phi=\overline{\phi}} = -iU''(\overline{\phi}) \tag{44.47}$$

This reproduces (44.46). That's the heavy dot vertex with all of those lines summed up, no matter how much "hair" is sticking out of the heavy dot.

[15] [Eds.] Coleman *Aspects*, "Secret Symmetry", Section 3.5, p. 136.

It is trivial to sum up the loops: it's an infinite power series. To one loop order

$$-iV(\overline{\phi}) = -iU(\overline{\phi}) - iU_{CT}^{(1)}(\overline{\phi}) + \sum_{n=1}^{\infty} \int \frac{d^4k}{(2\pi)^4} \frac{1}{2n} \left(\frac{i}{k^2 + i\epsilon}\right)^n \left(-iU''(\overline{\phi})\right)^n \tag{44.48}$$

where n is the number of heavy dots. Here's where the factors come from: Each of the n propagators carries the same momentum k because all of the external $\overline{\phi}$ lines carry zero momentum. Each vertex contributes the same amount, $-iU''(\overline{\phi})$. The combinatoric factor, $1/(2n)$, arises because where we start, and the order in which we go around, are unimportant—if we take an n-legged polygon, we get exactly the same graph if we rotate it by $(2\pi/n)$, and also if we reflect it; neither operation leads to a new term in the Wick expansion. So the factor $(1/n!)$ from Dyson's formula is not completely canceled.[16] (The ultraviolet divergences will be soaked up in the $-iU_{CT}^{(1)}(\overline{\phi})$ term.)

Please note that we are *not* normal ordering our interactions. In general when discussing questions of symmetry, and in particular complicated invariances like gauge invariance in quantum electrodynamics, it's a very bad idea to normal order things, despite what it says in elementary books. That leads to confusion, because normal ordering does *not* commute with gauge transformations or with shifts. There are some places where it won't hurt you; you just generate new counterterms which you soak up in the old counterterms. But there are many situations where the un-normal ordered expression *is* symmetric and the normal ordered form is *not* symmetric. In those cases you certainly don't want to normal order carelessly. If you are worried about the infrared problem here because of a lack of m^2 in the propagator, stop worrying. It will disappear when we sum the series. I know it will disappear, and you do, too: I told you that it doesn't matter how I split things up, so I could always add an m^2 to k^2 in the propagators, and subtract it from $U''(\overline{\phi})$.

Let's evaluate (44.48). The i^n and the $(-i)^n$ cancel, and we multiply both sides by i. The sum in (44.48) is just the logarithmic series for $-\ln(1-x)$ with a ½ in front. We rotate to Euclidean space, which gives us another i on the right-hand side:

$$\begin{aligned} V(\overline{\phi}) &= U(\overline{\phi}) + U_{CT}^{(1)}(\overline{\phi}) + \frac{i}{2}\sum_{n=1}^{\infty} \int \frac{d^4k}{(2\pi)^4} \frac{1}{n} \left(\frac{U''(\overline{\phi})}{k^2 + i\epsilon}\right)^n \\ &= U(\overline{\phi}) + U_{CT}^{(1)}(\overline{\phi}) - \frac{i}{2} \int \frac{d^4k}{(2\pi)^4} \ln\left[1 - \frac{U''(\overline{\phi})}{k^2 + i\epsilon}\right] \\ &= U(\overline{\phi}) + U_{CT}^{(1)}(\overline{\phi}) + \frac{1}{2} \int \frac{d^4k_E}{(2\pi)^4} \ln\left[1 + \frac{U''(\overline{\phi})}{k_E^2 - i\epsilon}\right] \end{aligned} \tag{44.49}$$

You'll notice that I'm keeping the $i\epsilon$ even in Euclidean space. That's just for safety's sake. We'll see later on that it's a good thing to do. If I write the integrand as

$$\ln[k_E^2 + U''(\overline{\phi}) - i\epsilon] - \ln[k_E^2 - i\epsilon] \tag{44.50}$$

the second term integrates to a constant (i.e., independent of $U''(\overline{\phi})$). It's quadratically divergent, but to hell with that; it can be absorbed into the renormalization. What remains is an elementary integral (you can find it in the standard tables).[17] I'll put in a brutal cut-off

[16] [Eds.] Coleman *Aspects*, p. 137.

[17] [Eds.] Gradshteyn & Ryzhik *TISP*. The relevant integral is number 2.729.2.

and integrate from $k_E = 0$ to $k_E = \Lambda$. The factor $U''(\overline{\phi})$ corresponds to all those lines carrying zero momenta. Though a function of $\overline{\phi}$, it's a constant, because $\overline{\phi}$ is a constant field. Making the substitution (44.50), the integral becomes [18]

$$\begin{aligned} V(\overline{\phi}) &= U(\overline{\phi}) + U_{CT}^{(1)}(\overline{\phi}) + \frac{1}{2}\int \frac{d^4k_E}{(2\pi)^4} \ln\Big[k_E^2 + U''(\overline{\phi}) - i\epsilon\Big] + \mathcal{O}(\Lambda^2) \\ &= U(\overline{\phi}) + U_{CT}^{(1)}(\overline{\phi}) + \frac{1}{32\pi^2}\Big[\Lambda^2 U''(\overline{\phi}) + \tfrac{1}{2}\Big[U''(\overline{\phi})\Big]^2\Big(\ln\frac{U''(\overline{\phi}) - i\epsilon}{\Lambda^2} - \tfrac{1}{2}\Big)\Big] + \mathcal{O}(\Lambda^2) \end{aligned} \tag{44.51}$$

(the term $\mathcal{O}(\Lambda^2)$ is a constant).

We get the expected divergent terms, with the counterterm $U_{CT}^{(1)}(\overline{\phi})$ evaluated to one loop order, but if U is of quartic order or less, these are already accounted for: the divergent terms are of the same form as terms in the original Lagrangian—the U'' term is at most a quadratic function, which tells us we need a quadratically divergent counterterm, proportional to $\overline{\phi}^2$; and $(U'')^2$ is a quartic function which tells us we need a logarithmically divergent counterterm proportional to $\overline{\phi}^4$. We already have precisely those counterterms in our original $\mathscr{L}_{CT}$ and therefore we can absorb all the Λ-dependent terms into $U_{CT}^{(1)}$. If I had been so foolish as to investigate a non-renormalizable theory, say one with a ϕ^5 term in U, then $(U'')^2 \ln \Lambda^2$ would give a term proportional to ϕ^6, and I would be stuck: I have no counterterm to absorb it.[19] Non-renormalizable theories are sick no matter how you look at them; they're no healthier from this vantage point.

We can absorb the divergent constants into the counterterms, leaving perhaps a residual finite part of the counterterm (depending on what the renormalization conditions are). Thus we are left with

$$V(\overline{\phi}) = U(\overline{\phi}) + U_{CT}^{(1,f)}(\overline{\phi}) + \frac{1}{64\pi^2}\Big[U''(\overline{\phi})\Big]^2 \ln\Big[U''(\overline{\phi}) - i\epsilon\Big] \tag{44.52}$$

where $U_{CT}^{(1,f)}$ is the finite part of the counterterm (to first order). I can't specify it without knowing what the renormalization conditions are.

When you look at this formula (44.52) you'll say, "That's $\ln U''(\overline{\phi})$ over *what*?" After all, $U''(\overline{\phi})$ is something with the dimensions of a mass squared. One term in it *is* the mass squared, for example. Well, it doesn't matter what we choose as a denominator for $U''(\overline{\phi})$. If we change the denominator in the argument of the logarithm, we merely pick up a finite term proportional to $[U''(\overline{\phi})]^2$ and that's absorbed in the finite counterterms $U_{CT}^{(1,f)}$. You tell me the renormalization conditions and I'll tell you the denominator in $\ln[U''(\overline{\phi})/(\text{what})]$. Putting in an unspecified M^2 for the denominator, we can write this as

$$V(\overline{\phi}) = U(\overline{\phi}) + U_{CT}^{(1,f)}(\overline{\phi}) + \frac{1}{64\pi^2}\Big[U''(\overline{\phi})\Big]^2 \ln\left[\frac{U''(\overline{\phi}) - i\epsilon}{M^2}\right] \tag{44.53}$$

[18] [Eds.] Coleman's value in the video of Lecture 49 (at 0:58:42) is incorrect. The value (44.51) agrees with Coleman *Aspects*, "Secret Symmetry", equation (3.33), p. 138 and with the anonymous graduate student's notes, as well as with equation (3.4) in Coleman and Weinberg, *op. cit.* The evaluation of (44.51) is a bit tricky; see Problem 24.1, p. 1003.

[19] [Eds.] §16.4; "Renormalization and Symmetry: A Review for Non-Specialists" in Coleman *Aspects*, Section 4, pp. 104–106.

This concludes our sample computation. I hope it has put some flesh on the idea of the effective potential. I wanted to show you how to compute it, and to emphasize the point that the only renormalization constants needed are just those that would be present if the symmetry were *not* spontaneously broken (the third of the three points made earlier).

This formula can be immediately generalized to the case of many scalar fields, ϕ_a, $a = 1, 2, \cdots n$. In this case, the heavy black dot in (44.44) is labeled by the two indices on the two ϕ fields going in and out of the dot. It can also be extended to the case of spinor and vector fields, but we will postpone that for a future lecture.

Suppose there were many fields ϕ^a, as in the case of our model with Goldstone bosons. Then we define a matrix

$$U''_{ab}(\overline{\phi}) = \frac{\partial^2 U(\overline{\phi})}{\partial \overline{\phi}_a \partial \overline{\phi}_b} = a \longrightarrow\!\!\!\!\!\!\bullet\!\!\!\longrightarrow b \tag{44.54}$$

I would have to consider the same loop diagrams as in (44.42), except now each internal line could be of a different kind. For example, we would get for the loop shown in Figure 44.3 the

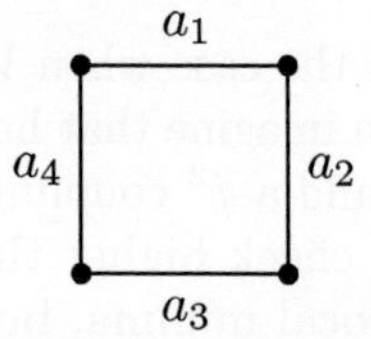

Figure 44.3: Multi-scalar loop

amplitude for a_1 going into a_2 in the presence of the external field, followed by the amplitude for a_2 going into a_3, etc., summed over all the fields, summed over repeated indices:

$$U''_{a_1 a_2} U''_{a_2 a_3} U''_{a_3 a_4} U''_{a_4 a_1} = \text{Tr}\left[(U'')^4\right] \tag{44.55}$$

I did it for four lines, just as an example. I could have done it for n lines; I'd get exactly the same result. With the U'' matrix defined by (44.54), the formula (44.53) generalizes to

$$V(\overline{\phi}) = U(\overline{\phi}) + U_{CT}^{(1,f)}(\overline{\phi}) + \frac{1}{64\pi^2}\text{Tr}\Big[\Big(U''(\overline{\phi})\Big)^2 \ln\Big(U''(\overline{\phi}) - i\epsilon\Big)\Big] \tag{44.56}$$

As before, we can't specify $U_{CT}^{(1,f)}(\overline{\phi})$ until we know the renormalization conventions. The last term here is the trace of the product of the indicated matrices. As $U''_{ab}(\overline{\phi})$ is a symmetric matrix of real quantities, it is Hermitian, and so is $\ln U''$: the logarithm of a matrix is a matrix. There's no problem defining it. Every loop integral is exactly the same as before.

This formula was first derived by Coleman and Weinberg: this Coleman and the other Weinberg, Erick Weinberg.[20] Steve Weinberg refers to this work as "that paper with pseudo-Goldstone bosons and a pseudo-Weinberg". We'll learn what a pseudo-Goldstone boson is in the next lecture.[21] The generalization to fermions will turn out to be almost exactly the same,

[20] [Eds.] Coleman and Weinberg, *op. cit.*, equation (6.3), p. 1900. Erick Weinberg was Coleman's student.
[21] [Eds.] Coleman adds that after hearing about this description, Jeffrey Goldstone asked Steven Weinberg, "Who is pseudo-Goldstone?"

as will the inclusion of gauge fields—even non-Abelian gauge fields, which I have not yet talked about.[22]

44.4 The physical meaning of the effective potential

I've introduced the effective potential $V(\overline{\phi})$ as, in some sense, the quantum generalization of the classical field theory potential $U(\phi)$. It *is* $U(\phi)$ to lowest order, and then it gets quantum corrections. The potential $U(\phi)$ has the *mathematical* meaning that its stationary points determine the ground states of the theory.[23] But it has a *physical* meaning as well: its *value* at the stationary points $\{\langle\phi\rangle\}$ is the energy per unit volume, the energy density, of the ground state (or states) for which the field takes the value $\langle\phi\rangle$. I want to show that $V(\overline{\phi})$ has *exactly* the same meaning:[24] that $V(\overline{\phi})$ is the energy density (43.5), $\mathscr{E}_0 = E/L^3$, for a state of lowest energy with $\langle\phi\rangle$ restricted to be $\overline{\phi}$. We normally consider the true ground state of the theory as the state of lowest energy without any restriction. Suppose we put a restriction on it, that the expectation value of ϕ is to be fixed at some number $\overline{\phi}$. I will demonstrate that the answer to the question "What is the lowest energy the system can have with the restriction that $\langle\phi\rangle$ must equal $\overline{\phi}$?" is $V(\overline{\phi})$.

The question becomes important in the case when $V(\overline{\phi})$ has two local minima, only one of which is an absolute minimum. One can imagine that happening even in tree approximation. If I wrote down a theory with both a ϕ^4 and a ϕ^3 coupling, then instead of those nice symmetric Russian buttocks,[25] I would find one cheek higher than the other. As before (when there was no ϕ^3 term) there would be two local minima, but now only *one* would be an absolute minimum. From the viewpoint of perturbation theory it looks like I could expand about

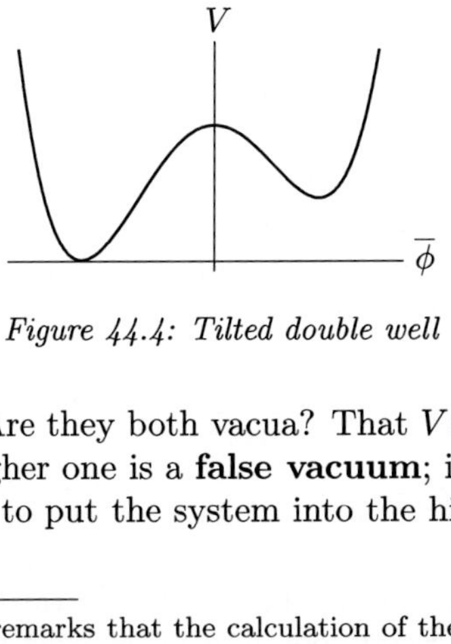

Figure 44.4: Tilted double well

either minimum equally well. Are they both vacua? That $V(\overline{\phi})$ is the energy density of the ground state says "No": the higher one is a **false vacuum**; it has a higher energy than the lower state.[26] If we attempted to put the system into the higher state, we would expect it

---

[22] [Eds.] In Woit's notes, Coleman remarks that the calculation of the effective potential can be done via functional integrals and the method of steepest descent, as in his Erice 1977 lectures, reprinted as "The Uses of Instantons", pp. 265–350 in Coleman *Aspects.* For an explicit calculation with functional integrals, see R. Jackiw, "Functional evaluation of the effective potential", *Phys. Rev.* **D9** (1974) 1686–1701; Jackiw's equation (3.5a) coincides with (44.51) for the case $n = 1$.

[23] [Eds.] See Section 3.7 in "Secret Symmetry", Coleman *Aspects.*

[24] [Eds.] In "Secret Symmetry", *Aspects*, note 16, p. 139, Coleman states that this result is due to Symanzik: K. Symanzik, "Renormalizable Models with Simple Symmetry Breaking", *Comm. Math. Phys.* **16** (1970) 48–80. As in "Secret Symmetry", p. 140, Coleman uses L^3 instead of V for a volume, to avoid confusion with $V(\overline{\phi})$.

[25] [Eds.] See note 11, p. 940.

[26] [Eds.] S. Coleman, "Fate of the False Vacuum: Semiclassical Theory", *Phys. Rev.* **D15** (1977) 2929–2936; "The Uses of Instantons", Section 6, pp. 327–340 in Coleman *Aspects.*

to eventually decay to the lower state. It's not something we'd see in any finite order in perturbation theory because it's a barrier penetration problem. Such problems involve the exponentials of terms proportional to $(-1/\hbar)$ and are therefore not seen in any order in a perturbation expansion in powers of $\hbar$, to wit, the loop expansion. Nevertheless, on simple energetic grounds, the higher minimum is an imposter: a *false* vacuum.

Another way of talking about the false vacuum is to consider the stationary points. With $U(\phi)$ we had to look for minima. For $V(\overline{\phi})$ we just have to look for *stationary points*, not necessarily minima. Well, what's wrong with the maximum between the two minima? Its derivative certainly vanishes there. What's wrong with it is that it's unstable, and not just through barrier penetration.

I've claimed that $V(\overline{\phi})$ is an energy density; in particular,

$$V(\overline{\phi})\Big|_{\overline{\phi}=\langle\phi\rangle} = \mathscr{E}_0 \tag{44.57}$$

where $\mathscr{E}_0$ is the energy density of the ground state. Greater insight is to be gained by demonstrating why this is so. It follows from minimizing the effective action, $\Gamma[\overline{\phi}]$, which will amount to minimizing the effective potential, $V(\overline{\phi})$. The argument is simple. I will look at the corresponding problem in ordinary quantum mechanics—determining the minimum of a perturbed Hamiltonian—find the answer, and then generalize it to field theory, by inserting integrals at appropriate places and replacing energies by energy densities.

Let's consider the related problem in quantum mechanics, to find the state ψ such that $\langle\psi|H|\psi\rangle$ is a minimum, subject to the constraint $\langle\psi|\psi\rangle = 1$. This problem is often solved using the *Rayleigh–Ritz* method.[27]

How do we solve a variational problem with a constraint? We can either deal with it directly, in this case by using only normalized trial states; or we can introduce a Lagrange multiplier. That is what I shall do here. Instead of minimizing $\langle\psi|H|\psi\rangle$ I will introduce a Lagrange multiplier, E, and minimize the quantity

$$\langle\psi|H|\psi\rangle - E\,\langle\psi|\psi\rangle = \langle\psi|H - E|\psi\rangle \tag{44.58}$$

I call the Lagrange multiplier E for the obvious reason: vary this quantity and you find that E is nothing but the energy eigenvalue for H:

$$\delta\,\langle\psi|H - E|\psi\rangle = 0 \;\Rightarrow\; (H - E)\,|\psi\rangle = 0 \tag{44.59}$$

Now our field theory problem has a *different* constraint, namely that $\langle 0|\phi|0\rangle$ is to equal a fixed value, $\overline{\phi}$. To find a quantum mechanical problem corresponding to the minimization of the effective potential (44.35), it's necessary to impose a second condition in addition to $\langle\psi|\psi\rangle = 1$. Let A be some operator (it doesn't matter what it is). Then impose

$$\langle\psi|A|\psi\rangle = \overline{A} \tag{44.60}$$

where $\overline{A}$ is a fixed value. The quantity to be minimized now becomes

$$\langle\psi|H|\psi\rangle - J\,\langle\psi|A|\psi\rangle - E\,\langle\psi|\psi\rangle = \langle\psi|H - JA - E|\psi\rangle \tag{44.61}$$

---

[27] [Eds.] Arfken & Weber *MMP*, Section 17.8, pp. 1072–1074; E. Butkov, *Mathematical Physics*, Addison-Wesley, 1968, Section 13.5, pp. 565–567; F. Mandl, *Quantum Mechanics*, J. Wiley and Sons, 1982, Chapter 8, pp. 186–193. In quantum mechanics, the Rayleigh–Ritz method is also known as "the variational method". The method is sometimes posed as the variation of the ratio $\langle\psi|H|\psi\rangle\,/\,\langle\psi|\psi\rangle$ for a trial function $\psi(x)$.

With malice aforethought I call the second Lagrange multiplier J. We solve this variational problem with arbitrary J, and then eliminate J from the problem to satisfy the constraint condition.

Define

$$-\overline{\mathcal{W}}[J] = \langle\psi|(H - JA)|\psi\rangle \tag{44.62}$$

at the minimum. The notation is beginning to make this quantum mechanical expression $-\overline{\mathcal{W}}[J]$ look a lot like the corresponding field theoretic expression for $W[J]$ in (44.26):

$$-W[J] = -\Gamma[\overline{\phi}] - \int d^4x\, J(x)\overline{\phi}(x) \tag{44.63}$$

(The sign difference has to do with $U(x)$ appearing with a positive sign in the Hamiltonian H, and $V(\overline{\phi})$ with a negative sign in $\Gamma(\overline{\phi})$, derived from the Lagrangian.) Notice that $-\overline{\mathcal{W}}[J]$ is the ground state energy for the altered Hamiltonian $H - JA$, because (44.61) is the Rayleigh–Ritz variational problem for $H - JA$. By a standard theorem,

$$\frac{d\overline{\mathcal{W}}}{dJ} = \langle\psi|A|\psi\rangle = \overline{A} \tag{44.64}$$

We know from non-relativistic quantum mechanics that if you vary the expectation value $\langle\psi|H|\psi\rangle$ of a Hamiltonian H with a parameter in it with respect to that parameter, you get the expectation value of the parameter's coefficient. (The term that comes from varying ψ in $\langle\psi|H - JA|\psi\rangle$ is zero, because $\overline{\mathcal{W}}[J]$ is a minimum.) We have to solve (44.64) to eliminate the Lagrange multiplier J. The energy is obtained from

$$E(\overline{A}) = \langle\psi|H|\psi\rangle = -\overline{\mathcal{W}}[J] + J\overline{A} \tag{44.65}$$

You compute the function $\overline{\mathcal{W}}[J]$, you differentiate it to obtain $\overline{A}$, you solve the resulting equation to obtain J in terms of $\overline{A}$, and finally you compute the quantity $E(\overline{A})$, the desired result. This is an elementary exercise in non-relativistic quantum mechanics.

You will notice a certain similarity, stressed by the notation, between $E(\overline{A})$ in (44.65) and $-\Gamma[\overline{\phi}]$ defined by a Legendre transformation,

$$\Gamma[\overline{\phi}] = W[J] - \int d^4x\, J(x)\overline{\phi}(x) \tag{44.26}$$

Here, $W[J]$ corresponds to the generating functional for a constant external J, since we're only dealing with constant fields; the sum of all vacuum-to-vacuum diagrams where we changed the Hamiltonian (44.23) by adding to it a term

$$-\int d^4x\, J\phi = -J\int d^4x\, \phi$$

much as we changed H by adding the term $-JA$. This $W[J]$ is the generating functional for connected vacuum-to-vacuum graphs (13.11). That is, $W[J]$ evaluated for a particular J is proportional to the sum of all the *connected* vacuum-to-vacuum graphs for this altered Hamiltonian, whose expectation value is the ground state energy density[28] in the presence

[28] [Eds.] §32.2.

of the source term J. So this quantum mechanical $-\overline{\mathcal{W}}[J]$ is exactly analogous to our field theoretic $W[J]$ (44.24): it *is* the ground state energy in the perturbed Hamiltonian, just as the field theoretic $W[J]$ is the ground state energy density. We differentiate this $\overline{\mathcal{W}}[J]$ with respect to J to define $\overline{A}$, just as we differentiated the earlier $W[J]$ with respect to J to define $\overline{\phi}$. Then we make a Legendre transformation, which will give us in this case $-\Gamma[\overline{\phi}]$ evaluated for a constant field, or (44.35) $+V(\overline{\phi})$.

Working out the quantum mechanical problem of determining the ground state energy with a restriction reproduces every step, including an equivalent Legendre transformation, used to define $\Gamma[\overline{\phi}]$, and hence $V(\overline{\phi})$. It is the same argument, aside from the substitutions of A for ϕ and energy for energy density, (because the connected vacuum-to-vacuum graphs give an energy *density*). Therefore we have proved that, if ϕ is the field which minimizes V and for which $\langle\phi\rangle = \overline{\phi}$, then

$$V(\overline{\phi}) = \mathscr{E}_0 \tag{44.66}$$

The quantity $\mathscr{E}_0$ is the lowest energy density subject to the constraint. In principle, $\overline{\phi}$ would be the state of lowest energy density if we obtained from $W[J]$ the state of lowest energy density in the presence of the external source J. We may not, since we're computing $W[J]$ perturbatively; we may run into trouble if level crossing takes place. When the coupling constants are weak, another state that is not the ground state may come up and cross that energy level, and we may find ourselves following the wrong state as we sum up our Feynman graphs. If perturbation theory cannot tell us the true ground state energy, then we won't get the true ground state energy for the constrained problem, either. On the other hand if perturbation theory serves to give the true ground state energy without constraint, it will also give us the true ground state energy *with* constraints.

There is a more direct way to establish (44.66). From (44.24), in the presence of a time-independent J,

$$\begin{aligned}\exp\left\{iW[J]\right\} &= N'\left\langle 0|U(\infty,-\infty)|0\right\rangle_J = N\int[d\phi]\,\exp\left\{i\mathcal{S}[\phi,J]\right\}\\ &= N''\exp\left\{-i\mathscr{E}_0L^3T\right\}\end{aligned} \tag{44.67}$$

because for the ground state, (44.34)

$$\lim_{\phi\to\langle\phi\rangle}\mathcal{S} = -\int d^4x\,U(\langle\phi\rangle) = -\mathscr{E}_0\int d^4x = -\mathscr{E}_0L^3T \tag{44.68}$$

On the other hand,

$$\lim_{\phi\to\overline{\phi}}\exp\left\{iW[J]\right\} = \lim_{\phi\to\overline{\phi}}\exp\left\{i\Gamma[\phi] - \int d^4x\,J\phi(x)\right\} = \exp\left\{i\Gamma[\overline{\phi}]\right\} \tag{44.69}$$

because for $\phi = \overline{\phi}$, $J = 0$, and from (44.40) and (44.35),

$$\Gamma[\overline{\phi}] = -V(\overline{\phi})L^3T \tag{44.70}$$

That is,

$$\exp\left\{iW[J]\right\} = N''\exp\left\{-i\mathcal{E}_0L^3T\right\} = \exp\left\{i\Gamma[\overline{\phi}]\right\} = \exp\left\{-iV(\overline{\phi})L^3T\right\} \tag{44.71}$$

so that, with $N'' = 1$ for the ground state, (44.66) follows.

Next time I will use this result to interpret V in another way, to explain why V sometimes develops an imaginary part and therefore why it was a good idea to keep the $-i\epsilon$ in (44.52). I'll show that the $-i\epsilon$ gives that imaginary part the right sign. I will also discuss V in terms of something we threw away around the third lecture, the zero-point energy of the ground state. I'll show that V is just another way of writing down the zero-point energy in an external field. Then I will discuss, on a much more lowbrow level, a particular model in tree approximation. We won't be missing anything, because we have learned the one-loop approximation won't make any changes. This is the famous *sigma model.*[29] It will serve as a laboratory for some of the current algebra ideas we were discussing earlier, in Chapter 41.

[29] [Eds.] See note 23, p. 994; the sigma model is covered in §§45.3–45.4.

45

Topics in spontaneous symmetry breaking

This chapter is a miscellany of three topics. First I'll discuss the role of the negative imaginary part of the energy (which came from the Feynman prescription for the propagators in (44.49)) in the effective potential. Next, I'll extend the effective potential to theories containing fermion fields. Finally, I'll construct the famous sigma model of four scalar fields: an isospin singlet, the sigma, and the pion triplet. It incorporates two kinds of symmetry breaking, both spontaneous and explicit "soft" symmetry breaking. The model is constructed so that PCAC is satisfied and gives the Goldberger–Treiman relation. More importantly, it provides a mechanism for the observed small mass of the pions.

45.1 Three heuristic aspects of the effective potential

Before proceeding let's review some things.[1] To construct the effective potential, we add a constant source term to the Lagrangian. This changes the Hamiltonian:

$$H \to H - J\int d^3x\,\phi \tag{45.1}$$

That Hamiltonian is well-defined. It's time-independent and it has a ground state $|0\rangle$, and $\langle 0|\phi(x)|0\rangle = \overline{\phi}$. We don't have to put an "in" or an "out" on the vacua, because if J is independent of time, $|0\rangle^{\text{out}}$ is the same as $|0\rangle^{\text{in}}$. The ground state just lies there; it doesn't scatter. The energy without the source term is the volume of space (if we put everything in a box) times $V(\overline{\phi})$. The prescription for the source term is: add $J\phi$ to the Lagrangian such that we obtain a ground state in which $\phi(x)$ has the desired expectation value $\overline{\phi}$. In that state there will be a certain energy, $V(\overline{\phi})$:

$$\langle 0|H|0\rangle = L^3 V(\overline{\phi})$$

At the minimum of V we have

$$\frac{dV}{d\overline{\phi}} = J$$

[1] [Eds.] "Secret Symmetry", Section 3.7, pp. 139–142 in Coleman *Aspects*.

So at the *actual minimum* of $V(\overline{\phi})$ we don't need a source to produce that ground state. That's the general picture.

I'm going to put a little more flesh on this general picture of the effective potential by telling you how the interpretation of the effective potential as an energy density explains something peculiar that could happen. Recall the master formula we got last time for the case of a single field:

$$V(\overline{\phi}) = U(\overline{\phi}) + U_{CT}^{(1,f)}(\overline{\phi}) + \frac{1}{64\pi^2}\Big[U''(\overline{\phi})\Big]^2 \ln\Big[U''(\overline{\phi}) - i\epsilon\Big] \tag{44.52}$$

(I'm proud of that $1/(64\pi^2)$.) But there's something peculiar about this formula:

$$\text{If } U''(\overline{\phi}) < 0 \text{ then } \mathrm{Im}\{V(\overline{\phi})\} < 0 \tag{45.2}$$

When the real part of the argument of the logarithm is negative, the $-i\epsilon$ (which I've carefully retained for just this purpose) gives you a negative imaginary part.[2] (By the way, $U''(\overline{\phi})$ in our standard model, pictured in Figure 43.4, becomes negative for an interval near the origin.) But how can you have an energy with an imaginary part, whatever its sign? Well, to make ϕ have the desired vacuum expectation value $\overline{\phi}$, we apply an external perturbation, $J(x)$. But it may not be possible for $\langle\phi\rangle$ to equal $\overline{\phi}$.

Consider classical electrodynamics (this involves vector fields rather than scalars, but the principle is the same). We can apply an external charge distribution such that the electric field in some region has a given desired value. (This is analogous to adding a J in (45.1) to make $\langle\phi\rangle$ a given value.) In particular, we can arrange that the electric field has absolutely any value we want within that region, independent of space and time, by bracketing the region between the charged plates of a large condenser.[3] That gives us a constant electric field. In quantum electrodynamics, however, we cannot fill a region with a field this way: the vacuum suffers *dielectric breakdown.*[4] If we have erected these giant condenser plates, even at opposite ends of the galaxy, and applied external charges on them such that a constant electric field arises over the whole extent of the galaxy, it will be energetically favorable for an electron–positron pair to materialize from the vacuum. Though that costs $2mc^2$, the system gains energy when the electron files to the positively charged plate and the positron flies to the negatively charged plate. That's the product of the electron's charge, the size of the electric field and the distance between the condenser plates. The kinetic energy gained can be much greater than the $2mc^2$ lost in creating the pair. In that case the vacuum boils off pairs until the charge on the condenser plates is neutralized, just as an ordinary dielectric in a real condenser breaks down because of the atoms in it ionizing. You end up with zero electric field, no matter what charge you try to put on the condenser plates. After all, the vacuum is a dielectric and can be polarized; that's the statement that the photon self-energy operator is not zero. If the region of space in which an electric field exists is large enough, as long as the field magnitude is non-zero, this will happen. This is an example of the famous *totalitarian*

---

[2] [Eds.] For $x > 0$, $\ln(-x) = \ln(x) + i\pi$ (choosing the principle branch), but if both x and ϵ are positive, $\lim_{\epsilon\to 0^+} \ln(-x - i\epsilon) = \ln(x) - i\pi$.

[3] [Eds.] In British English, "condenser"; more typically in American English, "capacitor".

[4] [Eds.] For reference, the field strength required for the dielectric breakdown of air is 3×10^6 V/m. J. S. Rigden, *Macmillan Encyclopedia of Physics*, Simon and Schuster, 1996.

selection principle: "Everything that is not forbidden is compulsory."[5] If it's energetically allowed, it's going to happen.

Something similar might happen with the scalar field. If we try to maintain a given value of the scalar field, it may well be that some phenomenon akin to the boiling off of electron–positron pairs could occur to neutralize the scalar field. If so, a configuration with a fixed expectation value of the scalar field will not be stable; it will decay. For example, consider a theory of protons and neutrons and π mesons, all with arbitrary masses. We define the energy of the neutron and expect its mass to be the pole in the neutron propagator. Suppose we alter the parameters of the theory, and say that it's not required to be isospin invariant. The neutron's mass could become larger than the sum of the proton's mass and the π^- meson's mass. Should we reach that point, we'd find that our nice energy formula had developed an imaginary part because the pole would move onto the second sheet.[6] This argument strongly suggests that an imaginary part in the energy density is a sign of instability, just as when we follow real energy in the neutron example up to a certain point; it develops an imaginary part which is connected to the neutron lifetime.[7] I'll now demonstrate that this negative imaginary part is equal to half the probability Γ of the neutron's decay per unit time.[8]

Recall that we found (44.67) for a time-independent source J within a box of volume L^3 over a time T

$$\exp\left\{iW[J]\right\} = \exp\left\{-i\mathscr{E}_0 L^3 T\right\} \tag{45.3}$$

where $\mathscr{E}_0$ is the energy density of the ground state of the perturbed system. If ϕ is the field which minimizes V and for which $\langle\phi\rangle = \overline{\phi}$, then

$$V(\overline{\phi}) = \mathscr{E}_0 \tag{44.66}$$

If V has a negative imaginary part (the $-i\epsilon$ as in (45.2)), the amplitude develops a certain probability of disappearing by boiling off pairs in that box. Thus it is important that the imaginary part is negative. That's a consistency check on this picture, because just as the neutron energy moves onto the second sheet, so the energy density here should move onto the second sheet:

$$\mathscr{E}_0 \to \mathscr{E}_0 - i\frac{\Gamma}{2L^3} \tag{45.4}$$

The probability $\Gamma/2$ of decay of this state per unit volume, the imaginary part of the energy, is the imaginary part of $V(\overline{\phi})$ when $U''(\overline{\phi}) < 0$:

$$\begin{aligned} \text{Im}\{V(\overline{\phi})\} &= \frac{1}{64\pi^2}\Big[U''(\overline{\phi})\Big]^2 \text{Im}\Big\{\ln\Big[U''(\overline{\phi}) - i\epsilon\Big]\Big\} = \frac{1}{64\pi^2}\Big[U''(\overline{\phi})\Big]^2 \{-\pi\} = -\frac{\Gamma}{2L^3} \\ \tfrac{1}{2}\Gamma &= \frac{1}{64\pi}\Big[U''(\overline{\phi})\Big]^2 L^3 \quad (\text{for } U''(\overline{\phi}) < 0) \end{aligned} \tag{45.5}$$

---

[5] [Eds.] Technically, the "Principle of Compulsory Strong Interactions": M. Gell-Mann, "The Interpretation of the New Particles as Displaced Charge Multiplets", *Nuovo Cim.* **4**, *Supp.* 2 (1956) 848–866, footnote on p. 859; and T. H. White, *The Once and Future King*, Ace Books, 1966, p. 121.

[6] [Eds.] R. J. Eden, P. V. Landshoff, D. I. Olive, and J. C. Polkinghorne, *The Analytic S-Matrix*, Cambridge U. P., 1966, Section 4.4, pp. 205–211; §§17.1–17.2, pp. 356-361.

[7] [Eds.] The video for Lecture 50 has a minute-long gap at 13:50, and the lecture notes describing this topic are either incomplete or refer to "Secret Symmetry" in Coleman *Aspects*. What follows from here to the words "a certain probability of disappearing" is an interpolation based on "Secret Symmetry", p. 142.

[8] [Eds.] Adding a negative imaginary term to the potential is a standard device for representing unstable particles. See, e.g., David J. Griffiths, *Introduction to Quantum Mechanics*, 2nd ed., Cambridge U. P., 2016, Problem 1.15, p. 22. Don't confuse the decay probability Γ with the effective action Γ.

plus higher loop corrections. The bigger the box the larger the chance it can decay, because there are more places for it to decay into. This decay can occur, but it is exponentially damped.[9] It's not like barrier penetration, because it occurs at the one loop level; it's like a particle balanced on top of the peak in Figure 43.4. On the other hand, if the potential were asymmetric, as in Figure 44.4, the higher minimum is unstable because of tunneling, described by the factor $\exp\{-(\Gamma/2)t\}$. We couldn't tell at the one loop order that the higher potential was the wrong choice. In such an instance we expect the maximum to reveal itself at $\mathcal{O}(\hbar)$, two loops (32.10). If this were a potential in classical mechanics I could balance a particle on the peak. In quantum mechanics, as Heisenberg pointed out, I can't do that, because quantum fluctuations will cause it to move laterally away from the peak. And as soon as a quantum fluctuation brings it to one side, off it falls. The behavior of $V(\overline{\phi})$ when $U''(\overline{\phi})$ is negative provides a fuller picture of the effective potential as an energy density and explains what happens when it acquires an imaginary part.

It's also instructive to go from a field theory in four dimensions to a particle system to one dimension. Instead of ϕ we'll call the dynamical variable x, a function of a single variable which we'll call t. The Lagrangian is:

$$L = \frac{1}{2}\left(\frac{dx}{dt}\right)^2 - U(x) \tag{45.6}$$

(Note that we are using a unit mass.) There's no need for renormalizations here; nevertheless we could define a $V(\overline{x})$ (with $\overline{x}$ the average value of x), and the formula would be exactly the same:

$$V(\overline{x}) = U(\overline{x}) + \frac{1}{2}\int \frac{dk}{2\pi} \ln\Big[k^2 + U''(\overline{x}) - i\epsilon\Big] + \text{constant} \tag{45.7}$$

We threw away constants rather blithely last time (44.51) because they were absorbed in the renormalizations in our four-dimensional theory. In a one-dimensional theory there is no need for any infinite renormalizations, but unfortunately we get a constant here.[10] The integral is easy to do using the boundary condition that $V = 0$ if $U = 0$, a reasonable assumption: a free particle should not have any potential energy. Let

$$F(U'') = \int_{-\infty}^{\infty} \frac{dk}{2\pi} \ln\Big[k^2 + U''\Big] \tag{45.8}$$

Then

$$\frac{dF(U'')}{dU''} = 2\int_0^\infty \frac{dk}{2\pi}\frac{1}{k^2+U''} = \frac{1}{\pi\sqrt{U''}}\tan^{-1}\left(\frac{k}{\sqrt{U''}}\right)\Bigg|_0^\infty = \frac{1}{2\sqrt{U''}} = \frac{d}{dU''}\left(\sqrt{U''}\right) \tag{45.9}$$

and so

$$F(U''(\overline{x}) - i\epsilon) = \sqrt{U''(\overline{x}) - i\epsilon} + \text{const} \tag{45.10}$$

---

[9] [Eds.] J. Schwinger, "On Gauge Invariance and Vacuum Polarization", *Phys. Rev.* **82** (1951) 664–679; see the discussion following equation (6.30).

[10] [Eds.] Gradshteyn & Ryzhik *TISP*. The relevant integral is number 2.733.1:

$$\int dx \ln(x^2 + a^2) = x\ln(x^2+a^2) - 2x + 2a\tan^{-1}(x/a)$$

The integral's limits are $-\infty$ to ∞. Discarding the infinite constants, the result is $2\pi a$, in agreement with (45.10). In Woit's notes, Coleman carries out the integral in (45.9) with contour integration; in the video of Lecture 50 he describes this calculation.

which gives, choosing the constant appropriately,

$$V(\overline{x}) = U(\overline{x}) + \tfrac{1}{2}\sqrt{U''(\overline{x}) - i\epsilon} \tag{45.11}$$

This result is very satisfying. Remember that the calculation is to one loop order. The loop expansion can be thought of as a systematic expansion in powers of $\hbar$. You have a particle of unit mass sitting in a potential $U(x)$, assumed symmetric so we know the ground state is at the origin; see Figure 43.3. In classical mechanics, its energy would be $U(0)$ at $x = 0$, the origin of the potential. What is the first quantum correction to the energy? The particle moves as if it were a harmonic oscillator, in a quadratic potential whose frequency is given by the square root of U''; we approximate the potential at its minimum:

$$\omega = \sqrt{U''} \tag{45.12}$$

You all know this game from the study of molecules: you add the zero point energy of the harmonic oscillator with an $\hbar$ which has been suppressed here (we're using units where $\hbar = 1$):

$$E = U(0) + \tfrac{1}{2}\hbar\omega = U(0) + \tfrac{1}{2}\hbar\sqrt{U''} \tag{45.13}$$

This is the classical energy plus the zero point energy of a harmonic oscillator about the classical minimum. That's exactly right, just what you would expect for the first quantum correction to the energy. We also see from this formalism that by combining these two analyses, we could solve Heisenberg's problem: What is the probability per unit time that a particle sitting at the origin of Figure 43.4 falls off the peak? That is left as an exercise.

So our expression for the one loop effective potential looks good. We understand the imaginary part when we go to a system that we know well, particle quantum mechanics. It gives an intuitively right answer. In fact we can get an idea of where (45.11) comes from by going back to four dimensions and doing the integral in a different way. Take the four dimensional expression

$$\frac{1}{2}\int \frac{d^4k_E}{(2\pi)^4}\ln\Big[k_E^2 + U''(\overline{\phi}) - i\epsilon\Big] + \text{constant} \tag{45.14}$$

and break it up into a time part and a space part.

$$\frac{d^4k_E}{(2\pi)^4} = \frac{dk_0}{2\pi}\frac{d^3\mathbf{k}}{(2\pi)^3} \tag{45.15}$$

Let's do the time part using the result (45.10):

$$\frac{1}{2}\int \frac{d^4k_E}{(2\pi)^4}\ln\Big[k_E^2 + U''(\overline{\phi}) - i\epsilon\Big] = \frac{1}{2}\int \frac{d^3\mathbf{k}}{(2\pi)^3}\sqrt{\mathbf{k}^2 + U''(\overline{\phi}) - i\epsilon} \tag{45.16}$$

What is this equation saying? Remember way back: I said a free field theory was like a system of harmonic oscillators. We blithely threw away a contribution to the energy, namely the zero point energy of the oscillators summed over all the oscillators.[11] What we've got here is the analog of that expression. How did we get (45.11)? We said

$$U(x) = U(\overline{x}) + \tfrac{1}{2}U''(\overline{x})(x - \overline{x})^2 + \cdots \tag{45.17}$$

[11] [Eds.] See the discussion following (2.13), p. 21.

We neglected the higher order terms, solved the harmonic oscillator problem and got a zero-point energy. Here we are doing something analogous for ϕ:

$$U(\phi) = \underbrace{U(\overline{\phi})}_{\substack{0^{\text{th}} \text{ order,} \\ \text{with no loops}}} + \tfrac{1}{2}U''(\overline{\phi})(\phi - \overline{\phi})^2 + \cdots \tag{45.18}$$

In (45.16) we're computing the zero-point energy of the system, the energy of each oscillator in a free field theory with squared mass $U''(\overline{\phi})$, summed over all the oscillators. Our mysterious one-loop computation (44.51), which involved all those fancy summations and Feynman diagrams, is revealed to be *exactly the same* in every factor, including the $\frac{1}{2}$ for the oscillator, as simply computing the zero-point energy of a free scalar field.

These three points are only heuristic: the $-i\epsilon$ helps explain what happens when U'' is negative; in one dimension, the formula recapitulates the first quantum correction to the energy; and the result (44.51) is equivalent to the calculation of the zero-point energy of a free field. They're not essential to doing any computations—the essential formula for calculations is (44.56)—but they help to provide physical meaning to the effective potential.

45.2 Fermions and the effective potential

I turn now to a different technical problem that will lead to some new physics. I'd like to compute the effects of fermions on the effective potential, because we will eventually have to deal with field theories that contain both bosons and fermions. I don't intend to generalize the effective potential to be a function of Fermi fields. That's silly: a Fermi field never gets a vacuum expectation value, by Lorentz invariance. Nevertheless there are 1PI graphs with external lines restricted to bosons that have fermions running around the loop, for instance Figure 45.1. I want to compute these fermion loops to get the one-loop corrections to the

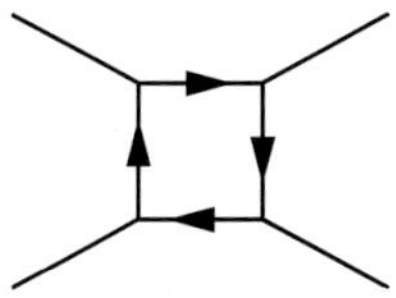

Figure 45.1: Fermion loop with external boson lines

effective potential from any fermions in the theory. I know how to take care of all the spinless bosons; we have that master formula. I now want to include the fermions.

We'll have a Lagrangian that's as before, plus a bunch of Fermi fields which we'll indicate by an index a:

$$\mathscr{L} = \cdots + \overline{\psi}^a i\partial\!\!\!/ \psi^a - \overline{\psi}^a m_{ab}\psi^b - g_{abc}\overline{\psi}^a\psi^b\phi_S^c - f_{abc}\overline{\psi}^a i\gamma_5\psi^b\phi_P^c + \mathscr{L}_{CT} \tag{45.19}$$

The ϕ_S^a in the g_{abc} term are scalar fields, the ϕ_P^c in the f_{abc} term are pseudoscalar fields; collectively we'll refer to the fields $\{\phi_S^a, \phi_P^b\}$ as ϕ. Typically we choose the **mass matrix** m_{ab} to be diagonal but I'll work in an arbitrary frame; it doesn't matter. The relevant counterterms for these kinds of graphs will be those which are functions of Bose fields only. The Fermi counterterms will not come in since those correspond to graphs with external Fermi line, and do not appear in this calculation. This is the most general possible renormalizable Lagrangian involving spinless bosons and fermions.

There are some constraints on the matrix m_{ab}. A Lagrangian should be real, so m_{ab} must be Hermitian, and since the ϕ's are real fields, the coefficients g and f must also be Hermitian with respect to a and b:

$$m_{ab} = m^*_{ba} \qquad g_{abc} = g^*_{bac} \qquad f_{abc} = f^*_{bac} \tag{45.20}$$

This makes it convenient to write $\mathscr{L}$ by writing the fermions in matrix form, assembling them into a big vector:

$$\mathscr{L} = \cdots + \overline{\psi} i\partial\!\!\!/ \psi - \overline{\psi} m(\phi)\psi \tag{45.21}$$

where $m(\phi)$ is the matrix

$$m_{ab}(\phi) = m_{ab} + g_{abc}\phi^c + f_{abc} i\gamma_5 \phi^c_P \tag{45.22}$$

This matrix $m_{ab}(\phi)$ is *not* Hermitian because of the i in the γ_5 term: γ_5 is Hermitian, $i\gamma_5$ is not (20.103). However, it does obey a very nice relation, which we'll exploit shortly:

$$m(\phi)\gamma_\mu = \gamma_\mu m^\dagger(\phi) \tag{45.23}$$

Commuting γ_μ with γ_5 gives us a minus sign that takes care of the minus sign introduced by the i. The matrix $m(\phi)$ is well-named because it is the mass the fermions would have if you replaced the quantum fields ϕ by constant c-numbers. That would be the fermion mass in tree level, if the particular values of ϕ happen to give the tree value minimum of the potential.

We can sum up the contribution to the effective potential from the fermion loops. It's exactly the same computation as before (44.42). I draw the fermion loops with heavy black dots, each representing a factor of $-im(\overline{\phi})$, $-m(\overline{\phi})$ from $\mathscr{L}$ and i from the vertex:

$$V(\overline{\phi}) = \cdots +$$

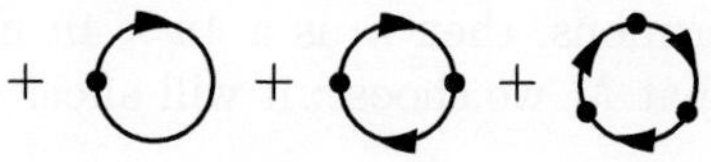

$$+ \cdots \tag{45.24}$$

That tells you how the fermions couple with the external field ϕ. There will be the usual terms in the sum of the graphs coming from the boson loops and the counterterms. From the fermion loops, we'll have an integral over all momenta,

$$\int \frac{d^4k}{(2\pi)^4}$$

and a factor of the fermion propagator (once again splitting off the mass, as with the previous calculation)[12]

$$\frac{i k\!\!\!/}{k^2 - i\epsilon}$$

times $-im(\overline{\phi})$, both raised to the nth power. Then there will be combinatoric factors, in this case $1/n$ rather than $1/(2n)$, because if we reflect the diagram we get a different graph in which the line runs around the other way. We sum over n and take the trace over everything, the Dirac indices as well as the things that label the fermions, and add a minus sign for the fermion loop:

$$V(\overline{\phi}) = \cdots - \text{Tr}\left\{\sum_n \frac{1}{n}\int \frac{d^4k}{(2\pi)^4}\left[\frac{i k\!\!\!/}{k^2 - i\epsilon}\Big(-im(\overline{\phi})\Big)\right]^n\right\} \tag{45.25}$$

---

12 [Eds.] See note 14, p. 973.

Only the even terms in the series contribute, because the trace of the product of an odd number of gamma matrices is always zero, whether or not there are γ_5's infiltrating.[13] We take account of that by replacing n by $2n$, indicating we're only going to sum over the *even* terms of the series.

$$V(\overline{\phi}) = \cdots - \text{Tr}\left\{\sum_n \frac{1}{2n}\int \frac{d^4k}{(2\pi)^4}\left[\frac{i\not{k}}{k^2 - i\epsilon}\left(-im(\overline{\phi})\right)\right]^{2n}\right\} \tag{45.26}$$

Let's examine one of those terms by itself. Looking at the quantity in the square brackets, the i cancels the $-i$ so we don't have to worry about them. We will have to raise to the nth power the quantity

$$\frac{\not{k}}{k^2 - i\epsilon}m(\overline{\phi})\frac{\not{k}}{k^2 - i\epsilon}m(\overline{\phi}) = \frac{k^2}{(k^2 - i\epsilon)^2}m(\overline{\phi})^\dagger m(\overline{\phi}) = \frac{1}{(k^2 - i\epsilon)}m(\overline{\phi})^\dagger m(\overline{\phi}) \tag{45.27}$$

where we have used (45.23) and canceled the k^2 in the numerator against one in the denominator. Aside from the minus sign in front, the series is the same as the boson series (44.49), with $m^\dagger m$ in place of U''. So we just write down the answer. (It will turn out to be just as easy for a very complicated theory with non-Abelian gauge fields when we get to that.) This fermion contribution has a factor of (-1) for the loops, the same factor of $1/(64\pi^2)$, and the trace over $\{(m^\dagger(\overline{\phi})m(\overline{\phi}))^2 \ln[m^\dagger(\overline{\phi})m(\overline{\phi}) - i\epsilon]\}$, but otherwise it's identical to the boson contribution:

$$V(\overline{\phi}) = \cdots - \frac{1}{64\pi^2}\text{Tr}\left\{\left[m^\dagger(\overline{\phi})m(\overline{\phi})\right]^2 \ln\left[\frac{m^\dagger(\overline{\phi})m(\overline{\phi}) - i\epsilon}{M^2}\right]\right\} \tag{45.28}$$

(If there are n species of fermions, then m is a $4n \times 4n$ matrix.) As with the boson series (44.53), it doesn't matter what M we choose; it will affect the size of the counterterm, but it won't affect the sum.

We can understand this form in exactly the same way that we understood the scalar case—as a shift in the zero-point energy. I haven't talked about it, but you've probably all read about Dirac's old electron theory, in which he had a bunch of negative energy levels, the **Dirac sea**[14], which he filled up. He said the real vacuum is the state in which the negative energy levels are all occupied, and we tacitly agreed. We threw away a constant when we normal ordered the Hamiltonian (21.26)—that constant was the sum of the energies over all the *negative* energy levels.

Let's suppose we have only a single type of Fermi field and m is a constant; then $m^\dagger m$ is m^2. We'll get a minus sign and *four times* the corresponding results for bosons. In the Dirac theory you don't have zero-point energies, you have all the negative energy levels; you get a minus sign from filling up those levels. When you increase the mass, the negative energy levels for a given k get *more* negative. The 4 comes from the trace, but there's a more physical

[13] [Eds.] This is easy to show using the "gamma-5 trick". See the solution to Problem 11.2, (S11.8), p. 428. Note (20.102) that $\gamma_5 \equiv i\gamma^0\gamma^1\gamma^2\gamma^3$ *itself* counts as an even product.

[14] [Eds.] Ryder *QFT*, Section 2.4, pp. 44–45; M. Kaku, *Quantum Field Theory*, Oxford U. P., 1993, p. 90; M. Srednicki, *Quantum Field Theory*, Cambridge U. P., 2007, p. 9. Weinberg describes the shortcomings of this idea, and quotes Schwinger: "The picture of an infinite sea of negative energy electrons is now best regarded as a historical curiosity, and forgotten," Weinberg *QTF1*, pp. 11–14. For a recent, more positive view of the sea and its sound mathematical foundation, see J. Dimock "The Dirac Sea", *Lett. Math. Phys.* **98** (2011) 157–166, `https://arxiv.org/pdf/1011.5865.pdf`.

way of thinking about it. Before you had $\frac{1}{2}\omega$ from the zero-point energy. But when you fill a negative energy level, you need ω, not $\frac{1}{2}\omega$. That's one factor of 2. The extra factor of 2 arises because the Dirac particle has two directions of spin; you've got twice as many energy levels for Dirac particles of a given mass as you do for Bose particles. That's why you get (-4) times the earlier result.

This is not only pretty physics, it has a practical application. I told you that there was really no point in computing the effective potential because, after all, we're doing a one-loop correction, which is just a small coupling constant approximation. It is not quite perturbation theory, but like perturbation theory, if it's good for anything, it can be useful only for small coupling constants. Unless we're computing something like the anomalous moment of the electron, where we have both high precision experiments and a theory in which we are confident, why should we bother with anything beyond tree approximation? The one-loop correction just shifts things around a little bit; it doesn't change the qualitative picture which is all we're interested in. There are in fact two cases in which the one-loop corrections are important.

The first case is a theory with both bosons and fermions, with two coupling constants: λ, a quartic coupling constant, and g, a Yukawa coupling constant. The small loop expansion is good only if both of these dimensionless quantities are small compared to 1:

$$\lambda \ll 1, \quad g \ll 1 \tag{45.29}$$

These are constraints when we do perturbation theory, including loop expansions. On the other hand, g never appears on the level of the tree approximation for the effective potential; we are not considering external fermion lines. There's a perfectly reasonable possibility that somewhere in the range of coupling constants that we're interested in, both of these conditions are true but g is much greater than λ:

$$\lambda \ll g \ll 1 \tag{45.30}$$

If that's so, the dominant terms in the effective potential, even though all the coupling constants are small, would be the term we just computed,

$$\mathrm{Tr}\Big\{\Big[m^\dagger(\bar{\phi})m(\bar{\phi})\Big]^2\cdots\Big\}$$

despite the fact that it occurs on the one-loop level. So what? It's the first term that has a g in it. The qualitative features of spontaneous symmetry breakdown will be dominated not by the tree approximation but by this monster (45.28). Though that possibility is *recherché*, it's nevertheless within our abilities to investigate it. One thing we surely know how to do is perturbation theory, so we can investigate the domain of coupling constants in (45.30). To do that we have to look at (45.28), because this is the first time we see g.

The second case is something proposed by Steve Weinberg, called **accidental symmetry**. It is best explained by an example.[15] Let's go back to our old friend, the SU(3)-invariant meson–nucleon theory. SU(3) is not spontaneously broken but it will make a good example. We have ϕ and ψ, boson and fermion SU(3) octets, respectively. The Lagrangian is a free Lagrangian plus a quartic term. It may appear at first glance that you could write down two

[15] [Eds.] S. Weinberg, "Approximate Symmetries and Pseudo-Goldstone Bosons", *Phys. Rev. Lett.* **29** (1972) 1698–1701; Coleman *Aspects*, "Secret Symmetry", Section 3.8, pp. 142–144.

quartic couplings, $\mathrm{Tr}(\phi^4)$ and $(\mathrm{Tr}(\phi^2))^2$. In fact there is only one; the first equals half the second.[16] There's some coupling constant λ, a D-type coupling constant g_D and an F-type coupling constant[17] g_F:

$$\mathscr{L} = \mathscr{L}_0 - \tfrac{1}{2}\lambda(\mathrm{Tr}(\phi^2))^2 - g_D \mathrm{Tr}\Big[\{\overline{\psi}, \phi\} i\gamma_5 \psi\Big] - g_F \mathrm{Tr}\Big[[\overline{\psi}, \phi] i\gamma_5 \psi\Big] \tag{45.31}$$

Now if we do tree approximation this theory is not just SU(3) symmetric, it's SO(8) symmetric: $(\mathrm{Tr}(\phi^2))^2$ involves the sum of the squares of the eight real meson fields and is invariant under orthogonal transformations of the fields. The real theory is by no means SO(8) invariant, but only SU(3) invariant, as real theories tend to be. There's no way of defining an SO(8) transformation on the ψ's. Even so, in tree approximation we'll find an effective potential which *is* SO(8) invariant. If we introduce a negative mass squared term to induce spontaneous symmetry breaking, we'll be completely stuck: we'll have an SO(8)-invariant family of minima rather than an SU(3)-invariant family of minima. In that situation there's no way of telling which among the minima is the *true* vacuum. If it weren't for the Fermi terms, this wouldn't matter; the theory really would be SO(8)-invariant, and whichever vacuum you take would be as good as any other. But the Fermi terms are going to make a difference. They're *not* SO(8)-invariant, and in this vast smooth field of possible vacua they're going to introduce little hills and valleys no matter how small the Fermi coupling constants are. Those little hills and valleys are going to determine the true vacuum. So even if the Fermi coupling constants g_F and g_D are much, much smaller than the quartic coupling constant λ, you have to include the term in (45.28) in order to find out qualitatively what the true vacuum is.

This is called *accidental symmetry* because by an accident the effective potential in tree approximation admits a large symmetry group that has nothing to do with the real symmetry of the theory, or even the symmetry of the effective potential *beyond* the tree approximation. You have to go beyond the tree approximation to gain even qualitative information about the vacuum state. Once you've determined the true vacuum, you'll find that small oscillations about it are peculiar. There will be some directions that break the SO(8) symmetry, in which the potential is curved up rather sharply by a λ-dependent amount. There will be other directions where the fermion terms $\mathrm{Tr}(\{\overline{\psi}, \phi\} i\gamma_5 \psi)$ and $\mathrm{Tr}([\overline{\psi}, \phi] i\gamma_5 \psi)$ put a small dimple in the potential, curving it up only a little if the Yukawa couplings are small. Therefore when you explore the second derivatives of the effective potential to get the masses of the particles, or more precisely the inverse propagators evaluated at zero momentum transfer, you'll find that some particles, those that correspond to directions where only the fermion terms keep the potential from being flat, will be much, much less massive than others. These small mass particles are called **pseudo-Goldstone bosons** because, in comparison with every other particle in the theory, their masses are not so far from zero (*genuine* Goldstone bosons have zero mass).[18] Pseudo-Goldstone bosons would appear as massless Goldstone bosons in tree approximation, except that—*only* because of the Fermi loop corrections—they acquire a small mass. I will not present any actual examples where fermions do either of these things, but I will shortly give an example where *vector mesons* do *both* of these things. I will discuss

---

[16] [Eds.] See Problem 20.1 and its solution, pp. 871–873, in particular (S21.3); Ling-Fong Li, "Group Theory of the Spontaneously Broken Gauge Symmetries", *Phys. Rev.* **D9** (1974) 1723–1738.

[17] [Eds.] See note 15, p. 807.

[18] [Eds.] Coleman adds: "As Jeffrey Goldstone asked Weinberg, 'Who is Pseudo-Goldstone?' He sounds like someone who turns up in epigraphy, like Pseudo-Dionysus, who wrote that nice book on angels." (Pseudo-Dionysus the Areopagite, Christian Neoplatonist theologian, late fifth – early sixth century CE.) See also note 21, p. 977.

the famous lower bound on the mass of the Higgs meson in Weinberg's theory of the weak interactions which comes from precisely this phenomenon.

45.3 Spontaneous symmetry breaking and soft pions: the sigma model

So far I've discussed spontaneous symmetry breaking in a general way. I'm now going to turn to specific examples. I will first investigate a particular model where I'll apply the ideas of spontaneous symmetry breaking to our old current algebra soft pion computation. Recall that I described four different meanings that one assigns to PCAC.[19] One was the statement that ultimately derives from Nambu.[20] He said that you could derive the Goldberger–Treiman relation by asserting that in the limit that the pion mass goes to zero, the axial vector current is conserved:

$$\lim_{m_\pi^2 \to 0} \partial^\mu A_\mu^a = 0 \tag{45.32}$$

Nobody knew how to make sense out of this; at the time it seemed like an orphic statement. The computations based on this limit agreed with experiment, but the meaning of the limit was mysterious. Later on, Jeffrey Goldstone and Nambu himself unraveled it. There *is* in fact an instance where the vanishing of a particle's mass is associated with conservation of the current: spontaneous symmetry breaking, with the particle a Goldstone boson. This suggests interpreting Nambu's statement as follows: *chirality*, the symmetry generated by the axial vector current, is an *approximate* symmetry. But this approximate symmetry is *not* associated with explicit symmetry-breaking terms in the Lagrangian; instead, the approximate symmetry would be realized in the Nambu–Goldstone mode (43.95) associated with spontaneous symmetry breaking.[21] If that mode of symmetry breaking occurs, the limit runs the other way:

$$\lim_{\partial^\mu A_\mu^a \to 0} m_\pi^2 = 0 \tag{45.33}$$

In this case, the three pions, which have precisely the right quantum numbers to be the Goldstone bosons associated with the axial vector mesons, would become massless. That is, the pions are *almost* Goldstone bosons.

Consider a Lagrangian which has a chiral-invariant part and a chiral-breaking part. When the chiral-breaking part goes to zero you don't have to wind up with perfect manifest symmetry; you could get unbroken symmetry for the isospin subgroup of SU(2) ⊗ SU(2), while the symmetry of the chiral generators (from the axial vector currents) could be spontaneously broken. That is, for the six-parameter group SU(2) ⊗ SU(2) (or equivalently, SO(4)), we will have a three parameter subgroup manifest and unbroken, and another three parameter subgroup's symmetry spontaneously broken, leading to three Goldstone bosons. We'll construct a model with a small, explicit chiral symmetry-breaking term in the Lagrangian, small because the pion mass *is* small compared to the other hadron masses. The idea is that if that term were to vanish then the symmetry would be exact, but it wouldn't be manifest; it would be hidden—spontaneously broken.

---

[19] [Eds.] §41.1, pp. 890–892.

[20] [Eds.] Y. Nambu, "Axial Vector Current Conservation in Weak Interactions", *Phys. Rev. Lett.* **4** (1960) 380–383.

[21] [Eds.] See notes 13 and 14, p. 944.

Another clue in that direction is our old gradient-coupling model (41.32), with the axial vector current (41.40) given by[22]

$$A^a_\mu = \partial_\mu \phi^a + \text{(nucleon terms)} \tag{45.34}$$

though we're not interested in the nucleon terms. Remember that this current would be conserved (its 4-divergence would equal zero) if the pion mass were zero. It is however an instance of spontaneous symmetry breaking. We have

$$[A^a_0(\mathbf{x},t), \phi^b(\mathbf{y},t)] = -i\delta^{ab}\delta^{(3)}(\mathbf{x}-\mathbf{y}) \tag{45.35}$$

because that's $\partial_0\phi$ with ϕ. It is in fact very close to our simplest example (43.102) of spontaneous symmetry breaking and the Goldstone theorem, the free scalar field where the current is $\partial_\mu\phi$. In hindsight, we were investigating spontaneous symmetry breaking in the gradient-coupling model, although we didn't know it. Its conserved current (41.40) is like $\partial_\mu\phi$ in the free scalar theory. The gradient-coupling current has a correction coming from the nucleon, but otherwise it provided an example of spontaneous symmetry breaking: as the pion mass went to zero we had a conserved current, whose commutator with the pion field had a non-vanishing vacuum expectation value. The gradient-coupling model is unsatisfactory in many ways. Although it's perfectly fine in tree approximation, it is non-renormalizable so we can't use it to investigate higher-order corrections. In addition, its current algebra is not the one we believe is true in the real world: the axial vector currents commute with themselves.

I will now use the ideas of spontaneous symmetry breaking to create a renormalizable model of the interaction of nucleons and pions which has both the right algebra and a symmetry in tree approximation (which is all we can really investigate) realized in the Nambu–Goldstone mode, if the pion mass were zero. (This won't involve strange particles, just pions and nucleons.) We'll presume we know how the axial vector currents act on the nucleons. I'll construct the simplest renormalizable model with these characteristics, including some scalar fields to break the symmetry spontaneously. This will end up being the **sigma model**. The model will involve nucleons, isodoublets of Dirac spinors; N, in the standard notation of Gell-Mann and Lévy:[23]

$$N = \begin{pmatrix} p \\ n \end{pmatrix} \tag{45.36}$$

and an isotriplet of pion fields π^a; and some other scalar or pseudoscalar fields. With the latter we can make the symmetry break down spontaneously; we know how to do that by choosing the quadratic term in the Lagrangian appropriately. And we want the axial vector currents A^a_μ to obey the correct current algebra.

---

[22] [Eds.] See "Soft Pions", pp. 36–56 in Coleman *Aspects*, and note 13, p. 896.

[23] [Eds.] M. Gell-Mann and M. Lévy, "The Axial Vector Current in Beta Decay", *Nuovo Cim.* **16** (1960) 705–726; J. Schwinger, "A Theory of the Fundamental Interactions", *Ann. Phys.* **2** (1957) 407–434; Peskin & Schroeder *QFT*, pp. 347–363; Cheng & Li *GT*, pp. 149–151; B. W. Lee, "Chiral Dynamics", in *Cargèse Lectures in Physics, Volume 5*, D. Bessis (editor), Gordon and Breach, 1972, pp. 119–178; printed in a separate volume as *Chiral Dynamics*, Gordon and Breach, 1972. Ben Lee, a distinguished Korean-American physicist, was professor at SUNY Stony Brook 1966–1973, and thereafter until his death both head of the theory group at Fermilab and a professor at the University of Chicago. He was killed in June 1977 while driving from Chicago to Colorado with his family. A driver in the opposite lane lost control of his truck as a result of a blown tire, crossed the divider, and smashed into Lee's car. He was 42. His wife and two children survived. Lee made many contributions to particle theory and taught the community about the GSW model, non-Abelian gauge theories, the Faddeev-Popov method, and all that with his review article with Ernest Abers (Abers & Lee *GT*).

Let's begin by examining the transformation properties of the nucleon. We know the contribution to the vector current, normalized to be the isospin current:

$$V^a_\mu = \tfrac{1}{2}\overline{N}\gamma_\mu T^a N + \cdots \tag{45.37}$$

the $\frac{1}{2}$ appearing because the fermions have isospin $\frac{1}{2}$ (the T^a are the three isospin analogs of the Pauli matrices). That's the CVC hypothesis. There may be other terms involving the other fields, but since we don't know what they are at the moment, we will keep them loose. Under an infinitesimal vector transformation

$$D^a_V N = -\tfrac{1}{2}iT^a N, \qquad D^a_V \overline{N} = \tfrac{1}{2}i\overline{N}T^a \tag{45.38}$$

This group of transformations may be denoted $\text{SU}(2)_V$. To keep from filling up the equations with $2g_V$'s, I will assume

$$2g_V = 1 \tag{45.39}$$

(You can rescale the currents (45.37) by $2g_V$ if you like; then it will obey the isospin relations and fulfill our current algebra prescriptions.)

We know how to make an axial vector current that has the right commutators (we get to these not with nucleons, but with quarks):

$$A^a_\mu = \tfrac{1}{2}\overline{N}\gamma_\mu\gamma_5 T^a N + \cdots \tag{45.40}$$

which corresponds to an infinitesimal transformation

$$D^a_A N = -\tfrac{1}{2}iT^a\gamma_5 N, \qquad D^a_A \overline{N} = -\tfrac{1}{2}i\overline{N}T^a\gamma_5 \tag{45.41}$$

(Both of these have minus signs because the T^a's are Hermitian and (20.103) $\overline{\gamma_5} = -\gamma_5$.) This group of transformations acts like SU(2), but unlike $\text{SU}(2)_V$, the signs of these transformations on $\overline{N}$ and N are *the same*; because of the γ_5's, this is an axial version of SU(2), which may be denoted $\text{SU}(2)_A$. The two currents, vector and axial vector, can be obtained from the Lagrangian

$$\mathscr{L} = \overline{N}i\partial\!\!\!/ N \tag{45.42}$$

if the transformations are symmetries of this Lagrangian.

Let's check that these *are* symmetries in the case where the nucleon is massless. The vector symmetry is trivial, because $\mathscr{L}$ is isospin invariant:

$$\begin{aligned} D^a_V(\overline{N}\gamma_\mu N) &= (D^a_V\overline{N})\gamma_\mu N + \overline{N}\gamma_\mu(D^a_V N) \\ &= (-\tfrac{1}{2}i\overline{N}T^a)\gamma_\mu N + \overline{N}\gamma_\mu(\tfrac{1}{2}iT^a N) = 0 \end{aligned} \tag{45.43}$$

so all we really have to check is the axial symmetry:

$$\begin{aligned} D^a_A(\overline{N}\gamma_\mu N) &= (D^a_A\overline{N})\gamma_\mu N + \overline{N}\gamma_\mu(D^a_A N) \\ &= (-\tfrac{1}{2}i\overline{N}T^a)\gamma_5\gamma_\mu N + \overline{N}\gamma_\mu(-\tfrac{1}{2}iT^a\gamma_5 N) = 0 \end{aligned} \tag{45.44}$$

The anticommutation of γ_5 and γ_μ supplies the relative minus sign. Likewise the derivatives are invariant:

$$D^a_V(\overline{N}i\partial\!\!\!/ N) = 0; \quad D^a_A(\overline{N}i\partial\!\!\!/ N) = 0 \tag{45.45}$$

So the (rather trivial) theory of free massless nucleons does indeed possess these symmetries. We haven't put any scalar fields in the picture so we don't yet have the possibility of spontaneous symmetry breakdown.

But we run into trouble if we try to add a nucleon mass term. There's no problem with the isospin, (45.38):

$$D_V^a(\overline{N}N) = 0 \tag{45.46}$$

The axial vector transformation (45.41) is another story:

$$D_A^a(\overline{N}N) = -i\overline{N}\gamma_5 T^a N \tag{45.47}$$

We don't get the cancellation in the axial transformation, we simply get the two terms adding together. That's to be expected; we've seen this before.[24] Once again, a fermion mass term breaks the chiral invariance: we can't have an invariance with γ_5 in it if we've got a fermion mass in the theory. Because I want to keep this chiral symmetry, the model Lagrangian cannot include an *explicit* nucleon mass term.

Perhaps you've recognized this set of transformation laws. Recall that there's a local 2-to-1 isomorphism between the SU(2)⊗SU(2) current algebra group and SO(4), the four-dimensional rotation group.[25]

$$G = \text{SU}(2) \otimes \text{SU}(2) \cong \text{SO}(4) \tag{45.48}$$

It's exactly the same analysis we went through for the Lorentz group, SO(3, 1), except there are no minus signs. So we get SO(4) instead. The infinitesimal vector (isospin) transformations D_V^a correspond to rotations in the i-j plane, where $i, j = 1, 2, 3$. The axial vector transformations D_A^a are the analogs of the Lorentz boosts; they're rotations in the i-4 planes. These transformation laws are those of an infinitesimal four-vector (in $\mathbb{R}^4$, not Minkowski space) with $\overline{N}N$ being the fourth component and $\overline{N}\gamma_5 T^a N$ being the three standard space components (in $\mathbb{R}^3$). By a trivial computation

$$D_A^a(i\overline{N}\gamma_5 T^b N) = \tfrac{1}{2}\overline{N}\{T^a, T^b\}N = \delta^{ab}\overline{N}N \tag{45.49}$$

exactly what we would expect[26] for an analog of Lorentz boosts with 4-vectors: the fourth component goes into one of the three space components (45.47); the three space components go into the fourth component (45.49) with a relative minus sign to keep the sum of the squares fixed:

$$(\overline{N}N, i\overline{N}\gamma_5 T^a N) \quad \text{transforms as a 4-vector in } \mathbb{R}^4 \text{ under SO(4).} \tag{45.50}$$

This suggests that we can very simply add in Yukawa-type couplings by introducing a quartet of mesons $(\sigma, \boldsymbol{\pi})$ that have the same transformation properties as the nucleon bilinear products; that is, as a Euclidean four-vector under the group. We'll have a singlet σ that will transform as the fourth component of a vector, and three vector components π^a that go into the singlet under the axial transformations:

$$D_A^a \sigma = -\pi^a; \quad D_A^a \pi^b = \delta^{ab}\sigma \tag{45.51}$$

The three pion fields π^a we will regard as having the properties of real pions, and so they will end up being pseudoscalar; the σ is an unobserved particle. These transformations are summarized in Table 45.1.

---

[24] [Eds.] See Example 3 in Chapter 43, p. 944.

[25] [Eds.] See §18.3, particularly the paragraph following (18.49).

[26] [Eds.] See §5.6, in particular (5.70), p. 93.

Quantity	*Transformation* D_A^a
σ	$-\pi^a$
π^b	$\delta^{ab}\sigma$
$\overline{N}N$	$-i\overline{N}\gamma_5 T^a N$
$i\overline{N}\gamma_5 T^b N$	$\delta^{ab}\overline{N}N$

Table 45.1: Axial transformations of the sigma model's fields

We can now write down a much more complicated Lagrangian. The nucleons are still massless but we can add an invariant Yukawa interaction, a coupling constant times the fourth component of one vector times the fourth component of another plus (*not* minus!) the dot product of the respective vectors' three-vector components:

$$\mathscr{L} = \overline{N}i\partial\!\!\!/ N - g[\overline{N}N\sigma + \overline{N}i\gamma_5 \mathbf{T}N\boldsymbol{\cdot}\boldsymbol{\pi}] \tag{45.52}$$

We'll write down the most general renormalizable field theory consistent with all this. We have the sum of the four components of the four-vectors in SO(4), a quartic coupling and a mass term:

$$\begin{aligned}\mathscr{L} = \overline{N}i\partial\!\!\!/ N &+ \tfrac{1}{2}(\partial_\mu\sigma)^2 + \tfrac{1}{2}(\partial_\mu\boldsymbol{\pi})^2 - \tfrac{1}{2}\mu^2(\sigma^2 + \boldsymbol{\pi}\boldsymbol{\cdot}\boldsymbol{\pi}) \\ &- g\Big[\overline{N}N\sigma + \overline{N}i\gamma_5\mathbf{T}N\boldsymbol{\cdot}\boldsymbol{\pi}\Big] - \tfrac{1}{4}\lambda(\sigma^2 + \boldsymbol{\pi}\boldsymbol{\cdot}\boldsymbol{\pi})^2\end{aligned} \tag{45.53}$$

That's the most general renormalizable, SO(4)-invariant Lagrangian of π^a and σ: its interactions are the sum of the most general quartic coupling, the most general quadratic coupling and the most general Yukawa coupling. The vector current is the ordinary isospin current,

$$V_\mu^a = \tfrac{1}{2}\overline{N}T^a\gamma_\mu N + \epsilon^{abc}\pi^b\partial_\mu\pi^c \tag{45.54}$$

The axial vector current is

$$A_\mu^a = \tfrac{1}{2}\overline{N}T^a\gamma_\mu\gamma_5 N - (\partial_\mu\sigma)\pi^a + (\partial_\mu\pi^a)\sigma \tag{45.55}$$

45.4 The physics of the sigma model

Let's make the value of μ^2 negative, and add a constant so that the SO(4) symmetry breaks spontaneously. Making this choice, the Lagrangian can be written as

$$\mathscr{L} = \overline{N}i\partial\!\!\!/ N + \tfrac{1}{2}(\partial_\mu\sigma)^2 + \tfrac{1}{2}(\partial_\mu\boldsymbol{\pi})^2 - g\Big[\overline{N}N\sigma + \overline{N}i\gamma_5\mathbf{T}N\boldsymbol{\cdot}\boldsymbol{\pi}\Big] - \tfrac{1}{4}\lambda(\sigma^2 + \boldsymbol{\pi}\boldsymbol{\cdot}\boldsymbol{\pi} - a^2)^2 \tag{45.56}$$

The possible vacua (the minima of the potential) lie on the sphere

$$\sigma^2 + \boldsymbol{\pi}\boldsymbol{\cdot}\boldsymbol{\pi} = a^2 \tag{45.57}$$

Because the theory is completely SO(4) invariant, it doesn't matter which vacuum we choose. We pick

$$\langle 0|\sigma|0\rangle = a; \quad \langle 0|\pi^a|0\rangle = 0 \tag{45.58}$$

(With any other choice for $|0\rangle$, we'd have to redefine parity.) Thus the axial symmetries are broken. The vector symmetry (isospin) remains; (45.58) are isospin-invariant statements. We define as usual

$$\sigma' = \sigma - a = \sigma - \langle\sigma\rangle\,; \quad \pi'^a = \pi^a \tag{45.59}$$

Expressed in terms of the shifted fields the Lagrangian becomes

$$\begin{aligned}\mathscr{L} = \overline{N}(i\partial\!\!\!/)N + \tfrac{1}{2}(\partial_\mu\sigma')^2 + \tfrac{1}{2}(\partial_\mu\boldsymbol{\pi}')^2 - ga\overline{N}N - \lambda a^2\sigma'^2 - g[\overline{N}N\sigma' \\ + \overline{N}i\gamma_5\mathbf{T}N\boldsymbol{\cdot}\boldsymbol{\pi}'] - \tfrac{1}{4}\lambda[(\sigma'^2 + \boldsymbol{\pi}'\boldsymbol{\cdot}\boldsymbol{\pi}')^2 + 4a\sigma'(\sigma'^2 + \boldsymbol{\pi}'\boldsymbol{\cdot}\boldsymbol{\pi}')]\end{aligned} \tag{45.60}$$

Note that *the nucleon acquires a mass*, $m_N = ga$, as a consequence of spontaneous symmetry breaking, despite the fact that the Lagrangian is chirally *invariant*; it comes from the $\overline{N}N\sigma$ term. The σ *also* gets a mass, while the three components π^a remain massless; they are Goldstone bosons:

$$m_N = ga; \quad m^2_{\sigma'} = 2\lambda a^2; \quad m^2_{\pi'} = 0 \tag{45.61}$$

This is just a model; the σ does not correspond to an observed particle. (Some had suggested that the σ could be identified with a broad resonance in the 0^+ channel of π–π scattering.[27])

Though the theory has all these different particles with different interactions, it is renormalizable. It has only three parameters after spontaneous symmetry breaking, because it had only those to begin with: g, λ and a. Among the particles is a massive nucleon. But the nucleon mass is not a free parameter, and that is preserved in the renormalization. The nucleon mass will get corrections under renormalizations of higher order in g and λ, of course, but they will be *finite* corrections that will not require any new counterterms: this is still a three-parameter theory.

Let's check the Goldberger–Treiman relation (40.37). We know how to do that in a world which has massless pions.[28] We'll just take A^a_μ (45.55) which has contributions from both σ and π^a, and make the shift $\sigma \to \sigma' = \sigma - a$. We'll get quadratic terms as before, and a $\partial_\mu\pi$ term as a consequence of the shift, from the term linear in σ (the term linear in $\partial_\mu\sigma$ is unaffected by the shift):

$$A^a_\mu = a\,\partial_\mu\pi^a + \tfrac{1}{2}\overline{N}T^a\gamma_\mu\gamma_5 N + \ldots \tag{45.62}$$

The PCAC statement in the form involving the divergence of the axial vector current is not applicable in a massless theory. But in the form (40.26)

$$\langle 0|A^a_\mu(0)|\pi^b\rangle = -ip_\mu F_\pi \delta^{ab} \tag{45.63}$$

it can be studied in the massless theory. That's the primary definition of F_π. In tree approximation there is only one term that contributes:

$$\langle 0|A^a_\mu(0)|\pi^b\rangle = a\,\langle 0|\partial_\mu\pi^a|\pi^b\rangle = -iap_\mu\delta^{ab} \;\Rightarrow\; a = F_\pi \tag{45.64}$$

We also see by comparison of the currents with

$$\langle p|J^h_\lambda(x)|n\rangle = e^{-ik\cdot x}\overline{u}_p[\gamma_\lambda g_V(k^2) - \gamma_\lambda\gamma_5 g_A(k^2) + \cdots]u_n \tag{40.21}$$

[27] [Eds.] V. E. Markushin and M. P. Locher, "Structure of the Light Scalar Mesons from a Coupled Channel Analysis of the s-wave $\pi\pi \to K\overline{K}$ Scattering", in *Workshop on Hadron Spectroscopy*, T. Bressani *et al.*, eds., Frascati, Italy, 1999; N. N. Achasov and G. N. Shestakov, "Phenomenological σ Models", *Phys. Rev.* **D49** (1994) 5779–5784.

[28] [Eds.] See the discussion following (41.5), pp. 892–892.

that

$$g_V = \tfrac{1}{2}; \quad g_A = -\tfrac{1}{2} \tag{45.65}$$

The assumption $2g_V = 1$ works out. But we cannot get around the equality between g_V and $-g_A$ (empirically they are in the ratio $\sim -4:5$); that comes from terms unaffected by the shift. There are some things we can't do no matter how cunning our model is.

The Goldberger–Treiman relation says that

$$F_\pi g = -2m_N g_A \tag{45.66}$$

Well, does it work? Of course it works! It *has* to work, because of (45.62); that's not PCAC, but it's close enough. From (45.61) we have $m_N = ga$, from (45.64) $F_\pi = a$, and from (45.65), $g_A = -\frac{1}{2}$. Then

$$\begin{aligned} F_\pi g &\stackrel{?}{=} -2m_N g_A \\ ag &\stackrel{?}{=} -2(ga)(-\tfrac{1}{2}) \stackrel{\checkmark}{=} ag \end{aligned} \tag{45.67}$$

This model gives us conserved vector currents, but it's *not* a good model for the real world. The dynamics of this model are not trustworthy: it has a σ and it doesn't have strange particles in it. There *are* no fundamental fields for σ, π or N; there are only fundamental fields for the quarks. The real world is nearly chiral SU(2) $\otimes$ SU(2) invariant:[29] the up and down quark masses are very small. Goldstone bosons are an inevitable consequence of the breakdown of a non-gauge symmetry. The physical pions are so much less massive than the other hadrons because they are *almost* Goldstone bosons.

Still, the sigma model provides a nice world to explore. Let's add a term to $\mathscr{L}$ that will give us PCAC:

$$\mathscr{L} \to \mathscr{L} + \mathscr{L}' \tag{45.68}$$

where $\mathscr{L}'$ is going to break the symmetry. To obtain PCAC we must choose

$$D_A^a \mathscr{L}' \propto \pi^a \tag{45.69}$$

The divergence of the axial current is this change;

$$\partial^\mu A_\mu^a = D_A^a \mathscr{L}' \tag{45.70}$$

Is there an object around in our theory which has a change $\propto \pi^a$? Yes, the σ field, with transformation (45.51). Therefore we choose

$$\mathscr{L}' = c\sigma \tag{45.71}$$

where c is a constant. The potential is then

$$U = \tfrac{1}{4}\lambda(\sigma^2 + \boldsymbol{\pi}\boldsymbol{\cdot}\boldsymbol{\pi} - a^2)^2 - c\sigma \tag{45.72}$$

This is guaranteed to put PCAC into the model. A term linear in σ has dimension 1; it is the most general such term that breaks SO(4) symmetry and preserves isospin symmetry. Does the new term spoil the renormalizability of the original model? No, because of Symanzik's

[29] [Eds] See the paragraph following (41.59), p. 906. Note also that only in the limit of massless quarks would chiral invariance hold.

rule for "soft" symmetry breaking: terms of raw dimension ≤ 3 added to a renormalizable theory do not alter its renormalizability.[30] Therefore the original sigma model with this term added is still renormalizable. But now we no longer have so many vacua. Instead of the sphere described by (45.57), we have an asymmetric surface of solutions; there is now a unique ground state because the axial symmetry has been broken. The minimum on the right of Figure 45.2 is at $\langle\sigma\rangle$, and the one on the left is now a *false vacuum*, because of the linear term. (To make it easier to draw, I'll sketch the potential with $\boldsymbol{\pi}\cdot\boldsymbol{\pi} = 0$.) The ground state is unique because the axial symmetry has been broken.

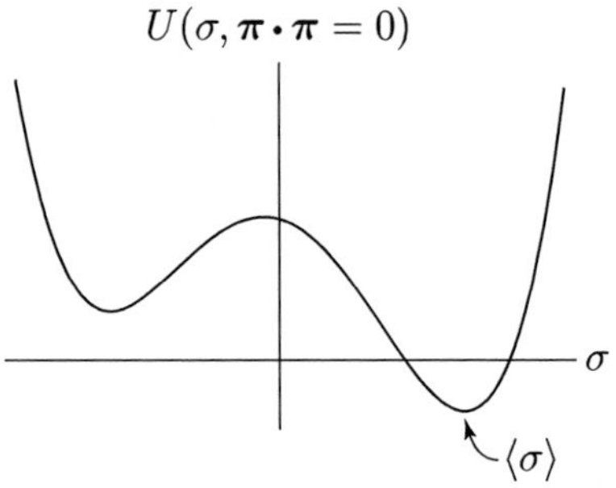

Figure 45.2: Tilted double well potential in the sigma model

There's no point in writing the Lagrangian with c as an independent parameter; I may adjust it as I want. Formerly we set $\langle\sigma\rangle = a$. I wish to replace (45.64) by

$$F_\pi = \langle\sigma\rangle \tag{45.73}$$

I want the shift to result in this equality. We still have

$$2g_V = -2g_A = 1 \tag{45.74}$$

in tree approximation. The nucleon mass becomes

$$m_N = g\,\langle\sigma\rangle \tag{45.75}$$

All that will be much the same as before, with $\langle\sigma\rangle$ now replacing a. The amount we have to shift σ will be different, which we determine by an appropriate choice of c. The only interesting nontrivial parts of the Lagrangian are the quartic and linear terms.

We're going to use c to fix $\langle\sigma\rangle$, which will no longer equal a. Shifting the fields

$$\sigma = \sigma' + \langle\sigma\rangle\,; \qquad \pi^{a\,\prime} = \pi^a \tag{45.76}$$

the Lagrangian becomes

$$\begin{aligned}\mathscr{L} = \overline{N}(i\partial\!\!\!/ - m_N)N + \tfrac{1}{2}(\partial_\mu\sigma')^2 + \tfrac{1}{2}(\partial_\mu\boldsymbol{\pi}')^2 - g\Big[\overline{N}N(\sigma') + i\overline{N}\gamma_5\mathbf{T}N\boldsymbol{\cdot}\boldsymbol{\pi}'N\Big]\\ - \tfrac{1}{4}\lambda((\sigma' + \langle\sigma\rangle)^2 + \boldsymbol{\pi}'\boldsymbol{\cdot}\boldsymbol{\pi}' - a^2)^2 + c(\sigma' + \langle\sigma\rangle)\end{aligned} \tag{45.77}$$

[30] [Eds.] See Theorem 3 in §25.4, p. 542, and "Renormalization and Symmetry: A Review for Non-Specialists" in Coleman *Aspects*, p. 107.

Let's expand the λ term:

$$\begin{aligned}-\tfrac{1}{4}\lambda(\cdots)^2 = &-\tfrac{1}{4}\lambda(\sigma'^2+\boldsymbol{\pi}'\boldsymbol{\cdot}\boldsymbol{\pi}')^2 - \lambda\langle\sigma\rangle\sigma'(\sigma'^2+\boldsymbol{\pi}'\boldsymbol{\cdot}\boldsymbol{\pi}')\\ &-\tfrac{1}{2}\lambda(\langle\sigma\rangle^2-a^2)(\sigma'^2+\boldsymbol{\pi}'\boldsymbol{\cdot}\boldsymbol{\pi}') - \lambda\langle\sigma\rangle^2\sigma'^2\\ &-\Big[\lambda\langle\sigma\rangle(\langle\sigma\rangle^2-a^2)-c\Big]\sigma' - \tfrac{1}{4}\lambda(\langle\sigma\rangle^2-a^2)^2 + c\langle\sigma\rangle\end{aligned} \tag{45.78}$$

The quartic term is still symmetric in tree approximation, and has no reference to the shift parameters c and a. The cubic term comes from the cross term and is completely determined in terms of λ and $\langle\sigma\rangle$; no new parameters are involved. The quadratic term comes from two places, cross terms and the square of σ'. We'll leave the constant terms alone, but we'll eliminate the linear terms by a particular choice for c:

$$c = \lambda\langle\sigma\rangle(\langle\sigma\rangle^2 - a^2) \tag{45.79}$$

Let's eliminate the parameter a in terms of something more physical (as we did before, when we set $a = \langle\sigma\rangle$). The pion mass comes from the coefficient of $\boldsymbol{\pi}'\boldsymbol{\cdot}\boldsymbol{\pi}'$:

$$m_{\pi'}^2 = \lambda(\langle\sigma\rangle^2 - a^2) \tag{45.80}$$

Likewise, the coefficient of σ'^2 gives the mass of the σ':

$$m_{\sigma'}^2 = \lambda(3\langle\sigma\rangle^2 - a^2) = m_{\pi'}^2 + 2\lambda\langle\sigma\rangle^2 \tag{45.81}$$

We already have the mass of the nucleon, (45.75). Rewriting the potential,

$$U(\sigma',\boldsymbol{\pi}'\boldsymbol{\cdot}\boldsymbol{\pi}') = \tfrac{1}{2}m_{\pi'}^2\boldsymbol{\pi}'\boldsymbol{\cdot}\boldsymbol{\pi}' + \tfrac{1}{2}m_{\sigma'}^2\sigma'^2 - \lambda\langle\sigma\rangle\sigma'(\sigma'^2+\boldsymbol{\pi}'\boldsymbol{\cdot}\boldsymbol{\pi}') + \tfrac{1}{4}\lambda(\sigma'^2+\boldsymbol{\pi}'\boldsymbol{\cdot}\boldsymbol{\pi}')^2 \tag{45.82}$$

plus a constant, which we can drop. Rewriting the Lagrangian one last time,

$$\begin{aligned}\mathscr{L} = &\overline{N}(i\partial\!\!\!/ - m_N)N + \tfrac{1}{2}(\partial_\mu\sigma')^2 + \tfrac{1}{2}(\partial_\mu\boldsymbol{\pi}')^2 - \tfrac{1}{2}m_{\pi'}^2(\sigma'^2+\boldsymbol{\pi}'\boldsymbol{\cdot}\boldsymbol{\pi}')\\ &- g\overline{N}(\sigma' + i\gamma_5\mathbf{T}\boldsymbol{\cdot}\boldsymbol{\pi})N - \tfrac{1}{4}\lambda(\sigma'^2+\boldsymbol{\pi}'\boldsymbol{\cdot}\boldsymbol{\pi}' + 2\langle\sigma\rangle\sigma')^2\end{aligned} \tag{45.83}$$

That's it. Even after the shift, this is *still* a four-parameter theory. The original version of this theory had parameters a, c, λ and g. We'll keep λ and g, but we have traded a and c for $\langle\sigma\rangle$ and $m_{\pi'}^2$:

$$c = m_{\pi'}^2\langle\sigma\rangle; \qquad a = \sqrt{\langle\sigma\rangle^2 - (m_{\pi'}^2/\lambda)} \tag{45.84}$$

The assignments to c and $\langle\sigma\rangle$ assure that PCAC in the form (41.5) is satisfied.

To look at this in a slightly different way, the potential U can be written as

$$U = U_0 - c\sigma \tag{45.85}$$

where U_0 is the σ model potential before the addition of the c term. The previous condition that determined the expectation value of σ

$$\left.\frac{dU}{d\sigma}\right|_{\langle\sigma\rangle} = 0 \tag{45.86}$$

is replaced by

$$\left.\frac{dU_0}{d\sigma}\right|_{\langle\sigma\rangle} = c \tag{45.87}$$

This equation says "You give me a $\langle\sigma\rangle$, and I'll give you a c that will satisfy that equation." It agrees with (45.79). It is true *to all orders* in perturbation theory because we're adding only a linear term to the Lagrangian. That just adds a corresponding linear term to the effective action and therefore to the effective potential:

$$V = V_0 - c\sigma \tag{45.88}$$

That's the only c-dependent term in the effective action. V_0 is the effective potential for the σ-model without an *explicit* symmetry-breaking term, evaluated to whatever order in perturbation theory you wish. We can always forget about c and say that $\langle\sigma\rangle$ is a free parameter.

This theory looks absurdly complicated. It's got a quartic interaction, a cubic interaction, mass terms, the σ mass is connected in a fancy way to F_π and to λ, the quartic coupling constant, and the nucleon mass is connected to F_π and to g, the Yukawa coupling constant. What a funny Lagrangian! If I had just written it down to start with and said "Here is a model that obeys all the current algebra constraints", you would have laughed me out of the room. Not only does it obey all the current algebra constraints, it's renormalizable, needing only four renormalization constants plus wave function renormalization. To check things out, we would want to take this grotesque model and compute (in tree approximation) some of the things we computed with our general assumptions to make sure that it agrees with the earlier results. Then we could be confident that we know what is going on.

Next time we come to the *Higgs mechanism*, and you'll see how to make Goldstone bosons *disappear*.

Problems 24

24.1 Verify (44.51)

$$\frac{1}{2}\int \frac{d^4k_E}{(2\pi)^4}\ln\Big[k_E^2 + U''(\overline{\phi}) - i\epsilon\Big] = \frac{1}{32\pi^2}\Big[\Lambda^2 U''(\overline{\phi}) + \frac{1}{2}[U''(\overline{\phi})]^2\Big(\ln\frac{U''(\overline{\phi}) - i\epsilon}{\Lambda^2} - \frac{1}{2}\Big)\Big] + \mathcal{O}(\Lambda^4) \qquad \text{(P24.1)}$$

by carrying out the integration and making the appropriate approximations.

(Eds.)

24.2 In Chapter 24 we discussed the isospin-invariant Yukawa theory of mesons and nucleons,

$$\mathscr{L} = \overline{N}(i\partial\!\!\!/ - m)N + \tfrac{1}{2}\partial^\mu\mathbf{\Phi}\bullet\partial_\mu\mathbf{\Phi} - \tfrac{1}{2}\mu^2\mathbf{\Phi}\bullet\mathbf{\Phi} - ig\overline{N}\gamma_5\boldsymbol{\tau}\bullet\mathbf{\Phi}N - \tfrac{1}{4}\lambda(\mathbf{\Phi}\bullet\mathbf{\Phi})^2 + \mathscr{L}_{CT} \qquad \text{(P24.2)}$$

(This Lagrangian is the sum of terms (24.27), (24.28), (24.29), and (25.77).) Here N is the nucleon isodoublet, $\mathbf{\Phi}$ is the pion isotriplet, m, μ, g, and λ are positive numbers, with $\mu < 2m$, $\boldsymbol{\tau}$ is the vector of Pauli matrices, and $\mathscr{L}_{CT}$ is the usual counterterm Lagrange density.

Now consider the same theory minimally coupled to electromagnetism (with massless photons). It is easy to see that there is a contribution to $F_2(0)$ proportional to g^2. Compute this contribution, for both the proton and the neutron. It suffices to present the answers in terms of integrals over Feynman parameters.

A note on the definition of F_2: In §34.2 we considered the scattering of an electron off a weak external current, J_μ; $\mathscr{L} = -eJ^\mu A'_\mu$, where e is the electron charge. The incoming electron has spinor u and four-momentum p, the outgoing electron, u' and $p' = p + k$. Then

$$\langle f|\text{S} - 1|i\rangle = \frac{ie}{k^2}\widetilde{J}_\mu(k)F^\mu(k) \qquad \text{(P24.3)}$$

where

$$F^\mu = e\overline{u}'\left[\gamma^\mu F_1(k^2) + \frac{i}{2m}\sigma^{\mu\nu}k_\nu F_2(k^2)\right]u \qquad \text{(P24.4)}$$

and (20.98) $\sigma^{\mu\nu} = \frac{i}{2}[\gamma^\mu, \gamma^\nu]$. (The coefficients have been chosen such that $F_1(0) = 1$.) Use the same definitions for the proton, with e the proton charge and m the nucleon mass. For the neutron, it clearly won't do to use the neutron charge, so let e be the proton charge here also. (Of course, for the neutron $F_1(0) = 0$.)

Comment: Electromagnetism breaks isospin invariance, so you might be worried about the presence of isospin-violating renormalization counterterms, like different wave-function renormalization constants for the neutron and proton. This is not a problem here. The asymmetric counterterms are all at least of order e^2 (times powers of g), and finding $F_2(0)$ to order g^2 only involves computing the nucleon–nucleon-photon vertex to order eg^2.

(1998 253b Final, Problem 2)

24.3 Example 2 of Chapter 43 (p. 941) was a theory with spontaneous breakdown of U(1) internal symmetry. The particle spectrum of the theory consisted of a massless Goldstone boson and a massive neutral scalar. Furthermore, although I did not discuss it in class, this term in the Lagrangian

$$\tfrac{1}{2}\rho^2(\partial_\mu\theta)^2 = \tfrac{1}{2}a^2(\partial_\mu\theta)^2 + a\rho'(\partial_\mu\theta)^2 + \tfrac{1}{2}\rho'^2(\partial_\mu\theta)^2 \qquad \text{(P24.5)}$$

gives rise to the decay of the massive meson into two Goldstone bosons, with an invariant Feynman amplitude proportional to a^{-1}. (This is not a misprint: before reading the decay amplitude from the Lagrangian, we must first rescale θ to put the free Lagrangian in standard form.) Now consider the theory minimally coupled instead to a massive photon with mass μ_0 (before symmetry breaking). What is the photon mass after the symmetry breaks? Does the Goldstone boson survive? If it does, what is its mass? What about the decay amplitude discussed above?

Comment: The Abelian Higgs model is the same theory minimally coupled to a *massless* photon. As we will see in Chapter 46, the Goldstone boson disappears, and the photon acquires a mass, which we will compute.

(1998b 11.1)

Solutions 24

24.1 We need to show

$$\tfrac{1}{2}\int_0^\Lambda \frac{d^4q}{(2\pi)^4}\ln\left(q^2+\alpha\right) = \frac{1}{32\pi^2}\left[\Lambda^2\alpha + \tfrac{1}{2}\alpha^2\left(\ln\frac{\alpha}{\Lambda^2}-\tfrac{1}{2}\right)\right] + \mathcal{O}(\Lambda^4) \tag{S24.1}$$

where $\alpha = U''(\overline{\phi}) - i\epsilon$, and the Euclidean subscript E has been suppressed. Using the general rule

$$\int d^4q\, f(q^2) = \pi^2\int_0^\infty zdzf(z) \tag{15.72}$$

the integral becomes

$$\tfrac{1}{2}\int_0^\Lambda \frac{d^4q}{(2\pi)^4}\ln\left(q^2+\alpha\right) = \frac{1}{32\pi^2}\int_0^{\Lambda^2} z\ln(z+\alpha)\,dz \tag{S24.2}$$

This integral can be found in Gradshteyn & Ryzhik *TISP* (2.729.2), but it's not hard to do. By parts

$$\int z\ln(z+\alpha)\,dz = \tfrac{1}{2}z^2\ln(z+\alpha) - \tfrac{1}{2}\int dz\,\frac{z^2}{z+\alpha}$$

The last integral is elementary:

$$\int dz\,\frac{z^2}{z+\alpha} = \int du\,\frac{(u-\alpha)^2}{u} = \tfrac{1}{2}u^2 - 2\alpha u + \alpha^2\ln u = \tfrac{1}{2}z^2 - \alpha z + \alpha^2\ln(z+\alpha) - \tfrac{3}{2}\alpha^2$$

We drop the constant last term, and find

$$\int z\ln(z+\alpha)\,dz = \tfrac{1}{2}(z^2-\alpha^2)\ln(z+\alpha) - \tfrac{1}{4}(z^2-2\alpha z) \tag{S24.3}$$

which agrees with Gradshteyn & Ryzhik *TISP* 2.729.2. Then

$$\frac{1}{32\pi^2}\int_0^{\Lambda^2} z\ln(z+\alpha)\,dz = \frac{1}{32\pi^2}\left\{\tfrac{1}{2}(\Lambda^4-\alpha^2)\ln(\Lambda^2+\alpha) - \tfrac{1}{4}\Lambda^4 + \tfrac{1}{2}\alpha\Lambda^2 + \tfrac{1}{2}\alpha^2\ln(\alpha)\right\} \tag{S24.4}$$

The first term can be approximated:

$$\begin{aligned}\ln(\Lambda^2+\alpha) &= \ln\left(\Lambda^2\left[1+(\alpha/\Lambda^2)\right]\right) = \ln(\Lambda^2) + \ln\left[1+(\alpha/\Lambda^2)\right]\\ &= \ln(\Lambda^2) + \frac{\alpha}{\Lambda^2} - \frac{\alpha^2}{2\Lambda^4} + \mathcal{O}(\Lambda^{-6})\end{aligned} \tag{S24.5}$$

Then

$$\tfrac{1}{2}(\Lambda^4-\alpha^2)\ln(\Lambda^2+\alpha) = \tfrac{1}{2}\Lambda^4\ln(\Lambda^2) + \tfrac{1}{2}\alpha\Lambda^2 - \tfrac{1}{4}\alpha^2 - \tfrac{1}{2}\alpha^2\ln(\Lambda^2) + \mathcal{O}(\Lambda^{-2}) \tag{S24.6}$$

and the stuff in the curly brackets becomes

$$\{\cdots\} = \alpha\Lambda^2 + \tfrac{1}{2}\alpha^2\ln\left[\frac{\alpha}{\Lambda^2}\right] - \tfrac{1}{4}\alpha^2 + \mathcal{O}(\Lambda^4) \tag{S24.7}$$

which was to be shown. ■

24.2 The interaction Lagrangian is

$$\mathscr{L}' = -ig\overline{N}\gamma_5\boldsymbol{\tau}\bullet\boldsymbol{\Phi}N = -ig(\phi_0\overline{p}\gamma_5 p - \phi_0\overline{n}\gamma_5 n + \sqrt{2}\phi_+\overline{p}\gamma_5 n + \sqrt{2}\phi_-\overline{n}\gamma_5 p) \tag{S24.8}$$

with

$$\phi_0 = \phi_3, \quad \phi_\pm = \frac{1}{\sqrt{2}}(\phi_1 \mp i\phi_2)$$

Let's do p and n in order; each will closely resemble the class computation for the electron (§34.3), from which we can steal some formulae.

For the proton:

Figure S24.1: $\mathcal{O}(eg^2)$ contributions to the proton form factor F_μ

From graph A:

$$F_\mu = \int \frac{d^4q}{(2\pi)^4}\overline{u}'g\gamma_5\frac{i(\not{p}'+\not{q}+m)}{(p'+q)^2-m^2}e\gamma_\mu\frac{i(\not{p}+\not{q}+m)}{(p+q)^2-m^2}g\gamma_5 u\frac{i}{q^2-\mu^2} = ieg^2\int\frac{d^4q}{(2\pi)^4}\frac{N_\mu}{D} \tag{S24.9}$$

where

$$N_\mu = \overline{u}'(\not{p}'+\not{q}-m)\gamma_\mu(\not{p}+\not{q}-m)u \tag{S24.10}$$

$$D = ((p'+q)^2-m^2)((p+q)^2-m^2)(q^2-\mu^2) \tag{S24.11}$$

The denominator is simplified as in class (34.57). Rewriting in terms of the photon momentum $k = p' - p$, dropping terms of $\mathcal{O}(k^2)$, and shifting the momentum using $q = q' - px - p'y$, we obtain for the denominator

$$D = [q'^2 - m^2(x+y)^2 - \mu^2(1-x-y)]^3 \tag{S24.12}$$

If we substitute this expression for q into N_μ, the terms linear in q' are odd, and vanish upon integration. Those terms in the numerator quadratic in q' contribute only to F_1. Thus for the determination of F_2, we can take

$$N_\mu = \overline{u}'(\not{p}'(1-y) - \not{p}x - m)\gamma_\mu(\not{p}(1-x) - \not{p}'y - m)u \tag{S24.13}$$

so that

$$F_\mu = ieg^2\int\frac{d^4q}{(2\pi)^4}2\int_\Delta dx\,dy\,\frac{N_\mu}{D} \tag{S24.14}$$

where $\Delta = \{x \geq 0, y \geq 0, x+y \leq 1\}$ (see Figure 34.7). As in (34.71), we write p on the left, and p' on the right, in terms of k, and use $\overline{u}'\not{p}' = \overline{u}'m$, $\not{p}u = mu$:

$$N_\mu = \overline{u}'(m(-x-y) + \not{k}x))\gamma_\mu(m(-x-y) - \not{k}y)u \tag{S24.15}$$

Keeping only terms linear in k, and symmetrizing (x, $y \to \frac{1}{2}(x+y)$, due to the region of integration), we obtain

$$N_\mu = \tfrac{1}{2}(x+y)^2 m\overline{u}'[\gamma_\mu, \not{k}]u \tag{S24.16}$$

Finally, using (I.2) in the box on p. 330, (note the sign of a^2)

$$\int\frac{d^4q}{(2\pi)^4}\frac{1}{(q^2-a^2+i\epsilon)^3} = \frac{-i}{32\pi^2 a^2} \tag{S24.17}$$

and

$$F_\mu = \cdots - \frac{e}{4m}\overline{u}'[\gamma_\mu, \not{k}]u\,F_2(0) \tag{S24.18}$$

we find for graph A,

$$F_2(0) = -\frac{g^2m^2}{8\pi^2}\int_\Delta\frac{dx\,dy\,(x+y)^2}{m^2(x+y)^2+\mu^2(1-x-y)} \tag{S24.19}$$

From graph B:

$$F_\mu = \int \frac{d^4q}{(2\pi)^4} \overline{u}' \sqrt{2} g\gamma_5 \frac{i(-\not{q}+m)}{q^2-m^2} \sqrt{2} g\gamma_5 u \frac{i}{(p+q)^2-\mu^2} e(p+p'+2q)_\mu \frac{i}{(p'+q)^2-\mu^2}$$
$$= -2ieg^2 \int \frac{d^4q}{(2\pi)^4} 2 \int_\Delta dx\, dy\, \frac{N_\mu}{D} \tag{S24.20}$$

and now (recalling (20.103) $\gamma_5^2 = 1$)

$$N_\mu = \overline{u}'(\not{q}+m)u(p+p'+2q)_\mu \tag{S24.21}$$

The same shift as before eliminates D's first-order term in q', but the zeroth order term is different:

$$D = [q'^2 - m^2(x+y)^2 + (m^2-\mu^2)(x+y) - m^2(1-x-y)]^3 = [q'^2 - m^2(1-x-y)^2 - \mu^2(x+y)]^3 \tag{S24.22}$$

Treating the numerator as before, $(x, y \to \frac{1}{2}(x+y))$

$$N_\mu = \overline{u}'(-\not{p}x - \not{p}'y + m)u(p+p'-2px-2p'y)_\mu = (1-x-y)m\overline{u}'u(p+p')_\mu(1-x-y) \tag{S24.23}$$

Using the Gordon decomposition (34.28),

$$2m\overline{u}'\gamma_\mu u = (p+p')_\mu \overline{u}'u - \tfrac{1}{2}\overline{u}'[\gamma_\mu, \not{k}]u \tag{S24.24}$$

N_μ becomes, throwing away terms linear in γ_μ which contribute only to F_1,

$$N_\mu = \tfrac{1}{2}(1-x-y)^2 m\overline{u}'[\gamma_\mu, \not{k}]u \tag{S24.25}$$

Carrying out the q' integration, and extracting the expression for $F_2(0)$ as before, we have from graph B

$$F_2(0) = \frac{g^2m^2}{4\pi^2} \int_\Delta dx\, dy\, \frac{(1-x-y)^2}{m^2(1-x-y)^2 + \mu^2(x+y)} \tag{S24.26}$$

For the neutron, things are almost the same:

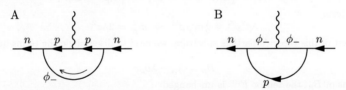

Figure S24.2: $\mathcal{O}(eg^2)$ contributions to the neutron form factor F_μ

The contribution of graph A to the neutron is *twice* that of the contribution to the proton, due to the $\sqrt{2}$ in the coupling; and graph B's contribution to the neutron is (-1) times that of the proton's graph B, because the interaction term has the opposite sign for e. We can summarize the results as follows:

$$\begin{Bmatrix} F_2(0)^P \\ F_2(0)^N \end{Bmatrix} = \begin{Bmatrix} -\frac{1}{2}g^2m^2 \\ -g^2m^2 \end{Bmatrix} \frac{1}{4\pi^2} \int_\Delta \frac{dx\, dy\, (x+y)^2}{m^2(x+y)^2 + \mu^2(1-x-y)}$$
$$+ \begin{Bmatrix} g^2m^2 \\ -g^2m^2 \end{Bmatrix} \frac{1}{4\pi^2} \int_\Delta \frac{dx\, dy\, (1-x-y)^2}{m^2(1-x-y)^2 + \mu^2(x+y)} \tag{S24.27}$$

Note: It's easy to show that

$$\int_\Delta dx\, dy\, f(x+y) = \int_0^1 dz\, z f(z)$$

This can be used to reduce the double integrals to single integrals. ■

24.3 The Lagrangian of Example 2 is given in its first form by

$$\mathscr{L} = \tfrac{1}{2}(\partial_\mu \boldsymbol{\Phi}) \bullet (\partial^\mu \boldsymbol{\Phi}) - U(\boldsymbol{\Phi}) \tag{43.4}$$

Perhaps the easiest way to couple this theory minimally to a vector field is to rewrite it as

$$\mathscr{L} = \tfrac{1}{2}(\partial_\mu \Phi)^*(\partial^\mu \Phi) - U(\Phi^*\Phi) \tag{S24.28}$$

where

$$\Phi = \phi_1 + i\phi_2 \tag{S24.29}$$

and

$$U(\Phi^*\Phi) = \tfrac{1}{4}\lambda((\phi_1^2 + \phi_2^2) - a^2)^2 \tag{S24.30}$$

Then the minimal coupling procedure gives, separating real and complex parts,

$$D_\mu\Phi = \partial_\mu\Phi + ieA_\mu\Phi \Rightarrow \begin{cases} D_\mu\phi_1 = \partial_\mu\phi_1 - eA_\mu\phi_2 \\ D_\mu\phi_2 = \partial_\mu\phi_2 + eA_\mu\phi_1 \end{cases} \tag{S24.31}$$

To study the spontaneous symmetry breaking, we parametrize the fields as in (43.27):

$$\phi_1(x) = \rho(x)\cos\theta(x), \qquad \phi_2(x) = \rho(x)\sin\theta(x) \tag{S24.32}$$

Adding the vector's kinetic and mass terms, we obtain

$$\mathscr{L}' = -\tfrac{1}{4}F_{\mu\nu}F^{\mu\nu} + \tfrac{1}{2}\mu_0^2 A_\mu A^\mu + \tfrac{1}{2}(\partial_\mu\rho)^2 + \tfrac{1}{2}(\rho)^2(\partial_\mu\theta + eA_\mu)^2 - U(\rho) \tag{S24.33}$$

(At first, it may seem strange that the ρ field is not coupled to the vector as the θ field is. Consider a U(1) gauge theory for Φ, or equivalently, a SO(2) symmetry for the real fields $\{\phi_1, \phi_2\}$. If the symmetry is a rotation, i.e., $\theta \to \theta + \alpha$, then ρ does not change, and so there's no need for it to have any gauge compensation through A_μ. Though we are considering here a massive vector theory which is *not* gauge invariant, the argument still applies.)

The symmetry breaks, and we replace

$$\rho = \rho' + a; \quad \theta' = \theta \tag{43.32}$$

The Lagrangian becomes

$$\mathscr{L}' = -\tfrac{1}{4}F_{\mu\nu}F^{\mu\nu} + \tfrac{1}{2}\mu_0^2 A_\mu A^\mu + \tfrac{1}{2}(\partial_\mu\rho')^2 + \tfrac{1}{2}(\rho' + a)^2(\partial_\mu\theta' + eA_\mu)^2 - U(\rho' + a) \tag{S24.34}$$

where

$$U(\rho' + a) = \tfrac{1}{4}\lambda((\rho' + a)^2 - a^2)^2 = \tfrac{1}{4}\lambda(\rho'^2 + 2a\rho')^2 \tag{S24.35}$$

The ρ' has a mass term

$$\lambda a^2\rho'^2 = \tfrac{1}{2}m_{\rho'}^2\rho'^2 \Rightarrow m_{\rho'}^2 = 2\lambda a^2 \tag{S24.36}$$

but to obtain the rest of this theory's particle spectrum, we need to disentangle the terms quadratic in A^μ and in θ'. Define

$$B_\mu = A_\mu - \lambda\partial_\mu\theta \tag{S24.37}$$

Writing $\mathscr{L}'$ in terms of B_μ, the tensor $F^{\mu\nu}$ is unchanged:

$$F^{\mu\nu} = \partial^\mu A^\nu - \partial^\nu A^\mu = \partial^\mu(B^\nu - \lambda\partial^\nu\theta) - \partial^\nu(B^\mu - \lambda\partial^\mu\theta) = \partial^\mu B^\nu - \partial^\nu B^\mu = F'^{\mu\nu} \tag{S24.38}$$

A judicious choice of λ will eliminate the $B^\mu\partial_\mu\theta$ cross-terms,

$$B^\mu\partial_\mu\theta'(\mu_0^2\lambda + a^2 e + a^2e^2\lambda) \tag{S24.39}$$

namely

$$\lambda = -\frac{ea^2}{\mu_0^2 + e^2a^2} \tag{S24.40}$$

The Lagrangian becomes

$$\mathscr{L}' = -\tfrac{1}{4}F'^{\mu\nu}F'_{\mu\nu} + \tfrac{1}{2}(\mu_0^2 + a^2e^2)B^\mu B_\mu + \tfrac{1}{2}\frac{a^2\mu_0^2}{\mu_0^2 + a^2e^2}(\partial^\mu\theta')^2 + \cdots \tag{S24.41}$$

so that

$$m_B^2 = \mu_0^2 + a^2e^2 \tag{S24.42}$$

$$m_{\theta'}^2 = 0 \tag{S24.43}$$

The Goldstone boson θ' *does* survive. Its kinetic term is a little strange, but we can redefine the field:

$$\vartheta = \frac{a\mu_0}{\sqrt{\mu_0^2 + a^2e^2}}\theta \tag{S24.44}$$

Then

$$\mathscr{L}' = -\tfrac{1}{4}F'^{\mu\nu}F'_{\mu\nu} + \tfrac{1}{2}m_B^2 B^\mu B_\mu + \tfrac{1}{2}(\partial^\mu\vartheta)^2 + \tfrac{1}{2}(\partial^\mu\rho')^2 - \tfrac{1}{2}m_{\rho'}^2\rho'^2 + \mathscr{L}_{\text{int}} \tag{S24.45}$$

where

$$\mathscr{L}_{\text{int}} = \tfrac{1}{2}(\rho'^2 + 2a\rho')(eB_\mu + (\mu_0/am_B)\partial_\mu\vartheta)^2 + \cdots \tag{S24.46}$$

(the dots indicate terms cubic or quadratic in ρ'). The term that governs the decay of a ρ' into two Goldstone bosons ϑ is

$$\frac{\mu_0^2}{a(\mu_0^2 + a^2e^2)}\rho'(\partial_\mu\vartheta)^2 \tag{S24.47}$$

The amplitude for the decay $\rho' \to 2\vartheta$ is proportional to

$$\frac{\mu_0^2}{a(\mu_0^2 + a^2e^2)} \tag{S24.48}$$

As $\mu_0 \to 0$, the decay amplitude goes to zero; as $\mu_0 \to \infty$, the decay amplitude becomes proportional to a^{-1}, as claimed, and there is no dependence on the gauge coupling constant e. ■

46

The Higgs mechanism and non-Abelian gauge fields

Nambu and Jona-Lasinio's investigations into spontaneous symmetry breaking were motivated by a desire to understand the nucleon's mass.[1] While the value of the nucleon mass could be obtained by this mechanism, there was an unfortunate byproduct: massless hadrons, which are not realized in nature. We've learned quite a lot about Goldstone bosons. They seem to be inevitably associated with spontaneous breakdown of a *continuous* symmetry: if we have spontaneously broken continuous symmetries we'll have Goldstone bosons. We have a very powerful and general theorem of this result due to Goldstone (§43.4), the proof of which does not depend on perturbation theory. That would seem to suggest that there is no escape; no spontaneous symmetry breaking, at least of continuous symmetries, without Goldstone bosons. But what we've learned about them so far makes it seem that their only physical application, at least in high-energy physics, is to chiral dynamics. (They have applications elsewhere in physics; things like spin waves in the ferromagnet[2] and superconductivity[3] are closely related to Goldstone bosons.) If a Lagrangian has a small symmetry-breaking term, there will be light mass spinless particles. The only light mass spinless particles in the Particle Data Group tables are the pions, and only they are possible candidates for Goldstone bosons or near-Goldstone bosons; there are *no* massless mesons. This is sad, because the idea that the universe is more symmetric than it seems to be is very attractive, as is the corollary that the apparent asymmetries in the universe are all the consequence of spontaneous symmetry breaking. It would be unfortunate if there were no loophole, no way to get a spontaneous breakdown of a continuous symmetry without a Goldstone boson.

However, there *is* a loophole. We haven't thought about the possibilities of *gauge invariant* field theories. So far we have discussed only one: quantum electrodynamics. The proof of the Goldstone theorem had two assumptions: one was *relativity*, that the theory was Lorentz invariant; the other was *positivity of the norm*, that poles in Green's functions corresponded to physically observable states. These assumptions are false for at least quantum electrodynamics,

---

[1] [Eds.] See note 13, p. 944.

[2] [Eds.] Neil W. Ashcroft and N. David Mermin, *Solid State Physics*, Saunders College Publishing, 1976, pp. 705–709.

[3] [Eds.] Ryder *QFT*, Section 8.4, pp. 296–298; I. J. R. Aitchison and A. J. G. Hey, *Gauge Theories in Particle Physics vol. II: QCD and the Electroweak Theory*, Institute of Physics Publishing, 2004, Section 17.7, pp. 218–224.

and maybe for other, similar theories. In QED, if we go to a covariant gauge like the Feynman gauge or the Landau gauge, we have poles in Green's functions corresponding to longitudinally polarized photons which are definitely *not* physical states; they are *gauge phantoms*. We can instead go to a gauge in which these gauge phantoms do not appear, for example Coulomb gauge, but the Coulomb gauge is *not* Lorentz invariant. So the assumptions underlying Goldstone's theorem do *not* hold for at least one gauge invariant theory, QED. Here could be a way out.[4]

46.1 The Abelian Higgs model

There is at least a possibility that Goldstone bosons in a theory with gauge invariance might turn out to be gauge phantoms, objects that disappear if we choose a gauge in which only physically observable states occur. One simple computation is worth two hours of argument. Let's build a simple model that has both the Goldstone phenomenon and gauge invariance, and see what happens in tree approximation (or equivalently, by doing a classical analysis). Since the only symmetry we know how to turn into a gauge symmetry is the one-parameter Lie group U(1) of QED, we'll consider a theory invariant under U(1) (or equivalently SO(2)), the model[5] we discussed in §43.2:

$$\mathscr{L} = \tfrac{1}{2}(\partial_\mu\phi_1)^2 + \tfrac{1}{2}(\partial_\mu\phi_2)^2 - \tfrac{1}{4}\lambda(\phi_1^2 + \phi_2^2 - a^2)^2 \tag{46.1}$$

The infinitesimal transformations are

$$D\phi_1 = \phi_2; \qquad D\phi_2 = -\phi_1 \tag{46.2}$$

i.e., a rotation in the ϕ_1-ϕ_2 plane. The easiest way to see what is going on is to assemble ϕ_1 and ϕ_2 into a single complex field with angular coordinates ρ and θ:

$$\phi_1 = \rho\cos\theta; \qquad \phi_2 = \rho\sin\theta \tag{46.3}$$

In this form the infinitesimal transformations become

$$D\rho = 0; \qquad D\theta = 1 \tag{46.4}$$

If we define ρ' by

$$\rho = \rho' + a \tag{46.5}$$

the Lagrangian becomes

$$\mathscr{L} = \tfrac{1}{2}(\partial_\mu\rho')^2 + \tfrac{1}{2}(\rho' + a)^2(\partial_\mu\theta)^2 - \tfrac{1}{4}\lambda(\rho'^2 + 2a\rho')^2 \tag{46.6}$$

Whether or not this theory displays spontaneous symmetry breakdown, it has a U(1) invariance, and therefore we can turn it into a gauge theory: we can couple electromagnetism to the scalars if we identify the conserved current with the electromagnetic current. The minimal coupling prescription is:

$$\mathscr{L}(\boldsymbol{\Phi}, \partial_\mu\boldsymbol{\Phi}) \rightarrow \mathscr{L}(\boldsymbol{\Phi}, D_\mu\boldsymbol{\Phi}) - \tfrac{1}{4}(F_{\mu\nu})^2 \tag{46.7}$$

[4] [Eds.] Coleman *Aspects*, "Secret Symmetry", Section 2.4, pp. 121–124; Ryder *QFT*, Section 8.3, pp. 293–296; Cheng & Li *GT*, Section 8.3, pp. 240–247; Peskin & Schroeder *QFT*, Section 20.1, pp. 690–700.

[5] [Eds.] Example 2, p. 941; see also Coleman *Aspects*, "Secret Symmetry", Section 2.2, pp. 118–119.

where

$$D_\mu \mathbf{\Phi} = \partial_\mu \mathbf{\Phi} + eA_\mu D\mathbf{\Phi}; \qquad F_{\mu\nu} = \partial_\mu A_\nu - \partial_\nu A_\mu \tag{46.8}$$

This definition of the covariant derivative is equivalent, to within a relative sign, to the earlier (27.48). We normally apply this formula only when $D\mathbf{\Phi}$ is a linear homogeneous function of $\mathbf{\Phi}$, but as it stands it's invariant under redefinitions of the fields, so long as we appropriately change $D\mathbf{\Phi}$. We could first obtain the covariant derivatives in the ϕ_1, ϕ_2 language and then transform to ρ and θ. To short-circuit a little algebra, we apply the prescription directly to the transformed fields ρ and θ (it doesn't matter in which order we do it). In the case at hand

$$D\rho' = 0; \quad D\theta = 1 \tag{46.9}$$

The field ρ' doesn't change at all because it's equal to ρ plus a constant, and ρ doesn't change. The only field that involves electromagnetic coupling is θ: $D\theta = 1$. We have from the minimal coupling prescription

$$\begin{aligned} D_\mu\theta &= \partial_\mu\theta + eA_\mu \cdot 1 = \partial_\mu\theta + eA_\mu \\ D_\mu\rho' &= \partial_\mu\rho' + eA_\mu \cdot 0 = \partial_\mu\rho' \end{aligned} \tag{46.10}$$

The λ term has no derivative and is not altered by the minimal coupling prescription. The Lagrangian becomes

$$\mathscr{L} = -\tfrac{1}{4}(F_{\mu\nu})^2 + \tfrac{1}{2}(\partial_\mu\rho')^2 + \tfrac{1}{2}(\rho' + a)^2(\partial_\mu\theta + eA_\mu)^2 - \tfrac{1}{4}\lambda(\rho'^2 + 2a\rho')^2 \tag{46.11}$$

This is just ordinary quantum electrodynamics of charged scalar particles with a quartic self-interaction, written in angular coordinates.

It's not obvious how to obtain the particle content of the theory from this Lagrangian. Usually one can read off the spectrum of small oscillations from the classical theory. But that's difficult here because there's a cross term between A_μ and $\partial_\mu\theta$ coming from the third term in $\mathscr{L}$. This difficulty can be circumvented by defining a *new* field, B_μ:

$$B_\mu \equiv A_\mu + e^{-1}\partial_\mu\theta \tag{46.12}$$

This definition looks like—and *is*—a gauge transformation of A_μ. Since $F_{\mu\nu}$ is invariant under a gauge transformation,

$$F_{\mu\nu} = \partial_\mu A_\nu - \partial_\nu A_\mu = \partial_\mu B_\nu - \partial_\nu B_\mu \tag{46.13}$$

(the two terms in $\partial_\mu\partial_\nu\theta$ cancel). We now have

$$\mathscr{L} = -\tfrac{1}{4}(F_{\mu\nu})^2 + \tfrac{1}{2}(\partial_\mu\rho')^2 + \tfrac{1}{2}e^2(\rho' + a)^2 B_\mu B^\mu - \tfrac{1}{4}\lambda(\rho'^2 + 2a\rho')^2 \tag{46.14}$$

The algebra is impeccable but the result is surprising. We now have in tree approximation, from the quadratic terms in the Lagrangian, a massive scalar meson ρ', which has the same squared mass (43.26) we computed before for the non-Goldstone boson, ϕ_1,

$$m_{\rho'}^2 = 2\lambda a^2 \tag{46.15}$$

and a vector B_μ, with the standard Proca Lagrangian for a vector boson of squared mass

$$m_B^2 = e^2 a^2 \tag{46.16}$$

The field θ has disappeared; there is *no* massless meson! The squared masses for the other mesons are positive numbers, thank God. This looks rather surprising but at least the degrees of freedom have been conserved. Let's count them.

When the symmetry does not break spontaneously, when the a^2 term and hence the bare mass are positive, we obtain two scalar mesons ϕ_1 and ϕ_2, each with one degree of freedom, and one massless vector A_μ, with two degrees of freedom (two polarization states); four in all. With spontaneous symmetry breaking we have a massive scalar with one degree of freedom, and we have a massive vector which has three degrees of freedom corresponding to the three polarization states: again four degrees of freedom. That's the difference between a massless and a massive vector boson. The photon only has the two transverse excitations but no longitudinal excitation. No mysterious particles have appeared or disappeared. Although the whole thing happens in one "swell foop", one way of thinking about what is going on is to say that the Goldstone boson appears and it is promptly *eaten* by the gauge field, which becomes heavy; the two degrees of freedom of the photon and the one degree of freedom of the Goldstone boson combine together to give the three degrees of freedom of a massive vector boson.

This fact was discovered independently by many people around the same time, but the one who understood and explained it most clearly was Peter Higgs, and therefore it is usually called the **Higgs mechanism** (or Higgs phenomenon). If I were really fair I would call it the Anderson-Brout-Englert-Guralnik-Hagen-Higgs-Kibble mechanism, but I'm not going to do that.[6] The Feynman rules for the Abelian Higgs model are summarized in the box on p. 1015.

We can get a deeper insight into what is going on by remembering the reason for introducing the minimal coupling prescription in the first place: gauge invariance. The rotation in the ϕ_1-ϕ_2 plane is a gauge transformation:

$$\theta(x) \to \theta(x) + \chi(x) \tag{46.17}$$

Only θ, the phase of the field, transforms: we change only that, in an arbitrary way *at every spacetime point.* This is an absolute invariance of the classical theory that has no effect on any physically observable quantities. The photon field A_μ also transforms, but we're not going to worry about that for the moment. There is nothing to stop us from choosing our gauge, given any field configuration, so that, following (46.17), we can make the field $\theta(x)$ disappear completely:

$$\chi = -\theta \;\Rightarrow\; \theta \to 0 \tag{46.18}$$

We can always gauge transform the phase of the field at any given point in spacetime so that the field is real and positive.[7] Once we have chosen such a gauge we see that the reason the Goldstone bosons don't appear in the final theory is that the degrees of freedom which they

---

[6] [Eds.] P. W. Anderson, "Plasmons, Gauge Invariance, and Mass", *Phys. Rev.* **130** (1963) 439–442; F. Englert and R. Brout, "Broken Symmetry and the Mass of Gauge Vector Mesons", *Phys. Rev. Lett.* **13** (1964) 321–323; P. W. Higgs, "Broken Symmetries, Massless Particles and Gauge Fields", *Phys. Lett.* **12** (1964) 132–133; "Broken Symmetries and the Masses of Gauge Bosons", *Phys. Rev. Lett.* **13** (1964) 508–509; "Spontaneous Symmetry Breakdown without Massless Bosons", *Phys. Rev.* **145** (1966) 1156-1163; G. S. Guralnik, C. R. Hagen, and T. W. B. Kibble, "Global Conservation Laws and Massless Particles", *Phys. Rev. Lett.* **13** (1964) 585–587; Peskin & Schroeder, *QFT*, pp. 690–692, 731–739; Cheng & Li, *GT*, pp. 241–243; Ryder *QFT*, 301–303. Englert and Brout, Higgs, and Guralnik, Hagen, and Kibble shared the 2010 Sakurai Prize for contributions to the electroweak theory; Peter Higgs and François Englert shared the 2013 Physics Nobel Prize for the work leading to the 2012 discovery of the Higgs boson. Sadly, Englert's colleague Robert Brout died in 2011; the Nobel Prize is not awarded posthumously.

[7] I should add: This operation would be singular if we were working near the point $\rho = 0$, because the phase θ of the field is not well-defined when the magnitude ρ of the field vanishes. As we are doing a perturbation theory expansion about $\rho = a$ we needn't worry; but in the absence of spontaneous symmetry breaking it could be a problem.

represent, to wit θ oscillations, can simply be *gauged away*, as in (46.18). The Goldstone bosons are θ oscillations and we can choose the gauge *so there ain't no* θ. In this way we can get the second form of the Lagrangian (46.14), the one involving the B field, from the first form, simply by applying the rather unconventional but perfectly legitimate gauge condition $\theta = 0$.

Feynman rules for the Abelian Higgs model

1. For every … Write …

   (a) internal vector line 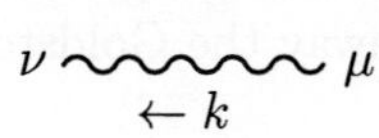

   $$\widetilde{D}_F^{\mu\nu}(k) = \left(g^{\mu\nu} - \frac{k^\mu k^\nu}{m_B^2}\right)\frac{-i}{k^2 - m_B^2 + i\epsilon}$$

   (b) internal scalar line 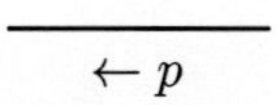

   $$\widetilde{\Delta}_F(p) = \frac{i}{p^2 - m_{\rho'}^2 + i\epsilon}$$

   (c) Three point scalar vertex 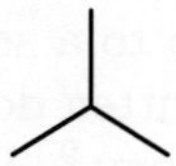 $-6i\lambda a$

   (d) Four point scalar vertex  $-6i\lambda$

   (e) Scalar-bivector vertex 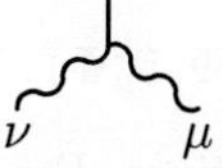

   $2ie^2 a g^{\mu\nu}$

   (f) Seagull vertex 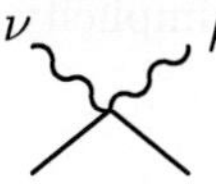

   $2ie^2 g^{\mu\nu}$

2. Ensure momentum conservation at each vertex: $(2\pi)^4\,\delta^{(4)}(\sum p_{\text{out}} - \sum p_{\text{in}})$

3. Multiply by $\displaystyle\int \frac{d^4q}{(2\pi)^4}$ and integrate over all internal momenta q.

4. Polarization factors:

   For every $\left\{\begin{matrix}\text{incoming}\\ \text{outgoing}\end{matrix}\right\}$ vector boson, a factor $\left\{\begin{matrix}\varepsilon_\mu\\ \varepsilon_\mu^{*\prime}\end{matrix}\right\}$, with $\varepsilon \cdot k = 0, \varepsilon' \cdot k' = 0$.

We normally like to apply the gauge conditions on the divergence of A_μ or something like that, but that's just prejudice and habit. We could just as well fix our gauge by declaring $\theta = 0$, whereupon the $\partial_\mu\theta$ term in (46.11) drops out and the original form of the Lagrangian becomes the form (46.14) with A replaced by B. The reason the Goldstone bosons do not appear when spontaneous symmetry breaking occurs is that *they were never there in the first place.* The degrees of freedom of the system which they represent are pure gauge degrees of freedom and can always be gauged away. That's why the world is not full of massless Goldstone bosons and massless vector bosons; the photon is the only one.

This gives us a hint on how to generalize this phenomenon to more complicated theories involving non-Abelian, noncommutative groups of symmetries, suffering spontaneous symmetry breakdown. All we have to do is figure out how to promote them to gauge symmetries, just as we promoted the U(1) symmetry here from a *global* symmetry with constant χ to a *local* symmetry with spacetime dependent $\chi(x)$. Remember from our earlier general analysis the Goldstone bosons always corresponded to degrees of freedom associated with applying infinitesimal symmetry transformations to the vacuum state. They just move us around the minimum of the Mexican hat potential.[8] If we have gauge invariance for those infinitesimal symmetry degrees of freedom, we can gauge away the Goldstone bosons and prevent them from appearing at the end.

46.2 Non-Abelian gauge field theories

We are going to generalize the gauge invariance of electrodynamics from a single gauge field, the photon, with a simple U(1) or SO(2) group to a set of gauge fields and a general non-Abelian group. This sort of theory was first written down by Yang and Mills in 1954, and is consequently called a **Yang–Mills field theory**.[9] This problem has nothing to do with spontaneous symmetry breaking. We will in due course employ these non-Abelian gauge fields in spontaneous symmetry breaking, but first we'll need to discover how to make a continuous internal symmetry into a gauge symmetry.

Let's begin with the general situation. We have a Lagrangian

$$\mathscr{L} = \mathscr{L}(\mathbf{\Phi}, \partial_\mu \mathbf{\Phi}) \tag{46.19}$$

depending on a set of fields $\mathbf{\Phi}$ and their derivatives. For the moment we will assume the $\mathbf{\Phi}$'s are real scalar fields; that's just for notational simplicity. It's trivial to generalize to the case

---

[8] [Eds.] Figure 43.5, p. 941.

[9] [Eds.] Coleman *Aspects*, Section 2.3. See note 5, p. 646. The modern development of gauge theories began with the epochal paper of Yang and Mills, which generalized the U(1) group to isospin and SU(2): C. N. Yang and R. L. Mills, "Conservation of isotopic spin and isotopic gauge invariance", *Phys. Rev.* **96** (1954) 191–195. Ronald Shaw, a doctoral student of Salam's at Cambridge, independently found an essentially identical theory in January 1954 (six months after Yang and Mills) and presented it in the last part of his PhD dissertation (1955): "Invariance under general isotopic gauge transformations", Part II, Chapter III. In retrospect, the first extension of the gauge invariance in electromagnetism was Einstein's general relativity (1915), accomplished before Weyl elucidated the modern view of gauge invariance: H. Weyl, "Elektron und Gravitation", *Zeits. f. Phys.* **56** (1929) 330–352. This point of view was promoted by Ryoyu Utiyama, who extended the gauge prescription to an arbitrary group and explicitly drew attention to the close relation between general relativity and Yang–Mills theory: R. Utiyama, "Invariant theoretical interpretation of interaction", *Phys. Rev.* **101** (1956) 1597–1607. For a collection of the fundamental articles on gauge theories, including the last part of Shaw's dissertation, Utiyama's paper, and an English translation of Weyl's article, together with a valuable historical survey, see L. O'Raifeartaigh, *The Dawning of Gauge Theory*, Princeton U. P., 1997. Gauge theories provide the framework for theories of all the fundamental forces, and the literature on them is enormous. Every modern quantum field theory book discusses the topic, e.g., Peskin & Schroeder *QFT* Chapters 14–22; Itzykson & Zuber *QFT*, Chapter 12; Ryder *QFT*, Sections 3.5, 3.6, and Chapter 7. Entire books are devoted to gauge theories, e.g., Cheng & Li *GT*, and Chris Quigg, *Gauge Theories of the Strong, Weak, and Electromagnetic Interactions*, 2nd ed., Princeton U. P., 2013. Two of the earliest surveys remain very useful: J. C. Taylor, *Gauge theory of weak interactions*, Cambridge U. P., 1976, 1979, and Abers & Lee *GT*.

where there are spinors or complex fields. We have a group of infinitesimal transformations

$$\delta\mathbf{\Phi} = -iT^a\delta\omega^a\mathbf{\Phi} \tag{46.20}$$

The $\delta\omega^a$ are infinitesimal parameters independent of position and time. The T^a's are Hermitian matrices,[10] one for each a:

$$T^{a\dagger} = T^a \tag{46.21}$$

These could, for example, be the matrices that generate SU(2) or SU(3) or SU(2) ⊗ SU(3) or whatever group you have in mind. This transformation is supposed to be an invariance of $\mathscr{L}$:

$$\delta\mathscr{L} = 0 \tag{46.22}$$

Because the T^a generate a symmetry group, they are closed under commutation

$$[T^a, T^b] = ic^{abc}T^c \tag{46.23}$$

Later on we will have to exploit certain symmetry properties of the **structure constants** c^{abc} and therefore it's useful, before we even start talking about Lagrangian field theory, to make a few remarks about a nice way to normalize these things.

I define a positive-definite quadratic form on the a-b space through the trace of a product of two generators, normalized according to[11]

$$\text{Tr}(T^aT^b) = c\,\delta^{ab}, \quad c > 0 \tag{46.24}$$

I will adopt different constants c for different groups. That is I will form linear combinations of the T^a's so they're orthogonal in terms of this so-called **trace norm**. Then c^{abc} is revealed to be

$$c^{abc} = -\frac{i}{c}\text{Tr}\Big(T^c[T^a, T^b]\Big) \tag{46.25}$$

The orthogonality property of the trace extracts out the coefficient of T^c. This is very useful because it tells us that the c^{abc}'s have essentially the same symmetry properties as the ϵ^{abc}'s do in the particular case of SU(2), being even under cyclic permutations and odd under anti-cyclic permutations:

$$c^{abc} = c^{bca} = c^{cab} = -c^{bac} = -c^{acb} = -c^{cba} \tag{46.26}$$

That's very convenient because it means we don't have to really worry about whether we get abc or bca in an expression; we may need to make a change of sign, but that's all.

I now turn to making the Lagrangian (46.19) *gauge invariant*. That is, I wish to alter the theory in such a way that the infinitesimal transformation of the fields $\mathbf{\Phi}$ involves *local*—spacetime-dependent—parameters $\delta\omega^a(x)$:

$$\delta\mathbf{\Phi}(x) = -iT^a\delta\omega^a(x)\mathbf{\Phi}(x) \tag{46.27}$$

---

[10] [Eds.] In the videotape of Lecture 51 (on which this chapter is based), Coleman uses a real representation since these are real fields. However, in order to keep a single notation throughout these lectures, we use the Hermitian representation.

[11] [Eds.] This construction, standard in Lie algebras, is called the **Cartan–Killing metric**: Élie Cartan (1869–1951), French, widely regarded one of the greatest mathematicians of the twentieth century; Wilhelm Killing (1847–1923), German geometer and algebraist. See Zee *GTN*, Section VI.3, pp. 365-366.

This is to be an invariance of the Lagrangian. The transformation (46.27) is spacetime-dependent so that we can impose gauge invariance to kill the Goldstone bosons. The problem comes up in $\delta(\partial_\mu \mathbf{\Phi})$, which consists of a term that's jolly plus a term that's just disgusting:

$$\delta(\partial_\mu \mathbf{\Phi}) \equiv \partial_\mu(\delta\mathbf{\Phi}) = -i\Big(\underbrace{T^a\delta\omega^a\partial_\mu\mathbf{\Phi}}_{\text{jolly}} + \underbrace{T^a(\partial_\mu\delta\omega^a)\mathbf{\Phi}}_{\text{disgusting}}\Big) \tag{46.28}$$

We want to arrange matters so that

$$\delta\mathscr{L} = 0 \tag{46.29}$$

The Lagrangian as it stands is not invariant under (46.27) because of the disgusting term. We're going to have to change $\mathscr{L}$ just as we did in the Abelian case by introducing new fields, one photon-like field for each $\delta\omega^a$. We're going to change the physics so that this extra term cancels out.

Use electrodynamics (27.48) as a model. From the minimal coupling prescription of electrodynamics there's a strong suggestion that we introduce a vector boson gauge field, a photon-like field, for each a, to sop things up. Let's investigate that possibility, and define a *new* covariant derivative appropriate to a *non*-Abelian gauge group:

$$D_\mu\mathbf{\Phi} \equiv \partial_\mu\mathbf{\Phi} - iA^a_\mu T^a\mathbf{\Phi} \tag{46.30}$$

where A^a_μ are some vector fields. I will concoct infinitesimal transformation laws for the fields A^a_μ so that

$$\delta(D_\mu\mathbf{\Phi}) = -iT^a\delta\omega^a D_\mu\mathbf{\Phi} \tag{46.31}$$

i.e., so that $(D_\mu\mathbf{\Phi})$ transforms in the same way that $\mathbf{\Phi}$ does. That's our goal. It's not obvious that we can arrange that. If we can, then $\mathscr{L}(\mathbf{\Phi}, D_\mu\mathbf{\Phi})$ will be gauge invariant because this Lagrangian will transform under local, spacetime-dependent transformations in the same way as it did under global, spacetime-*independent* transformations:

$$\delta\mathscr{L}(\mathbf{\Phi}, D_\mu\mathbf{\Phi}) = 0 \tag{46.32}$$

We'll still have to worry about the free Lagrangian for the A^a_μ field, but we'll get to that later. At least we'll have taken care of the minimal coupling terms. I have not put a free parameter or coupling constant in (46.30); any such are included in the A^a_μ. Later I will make these explicit, but at this stage I'll just absorb them into the A^a_μ to keep from cluttering the equations. I don't yet have a free Lagrangian for the A^a_μ fields, so I have no natural scale for them.

It's elementary (though tedious) to see what needs to be done to make (46.31) true. First we've got the term we already know, (46.28). Then we'll have the term coming from the change in A^a_μ; that's what we want to compute. This is supposed to satisfy our criterion, (46.31):

$$\delta(D_\mu\mathbf{\Phi}) = \delta(\partial_\mu\mathbf{\Phi}) - i(\delta A^a_\mu)T^a\mathbf{\Phi} - iA^a_\mu T^a\delta\mathbf{\Phi} \tag{46.33}$$

Expanding the various terms, we have

$$\begin{aligned} -iT^a\delta\omega^a[\partial_\mu\mathbf{\Phi} - iA^b_\mu T^b\mathbf{\Phi}] = \; & -i\Big(T^a\delta\omega^a\partial_\mu\mathbf{\Phi} + T^a(\partial_\mu\delta\omega^a)\mathbf{\Phi}\Big) \\ & - i(\delta A^a_\mu)T^a\mathbf{\Phi} - iA^a_\mu T^a(-iT^b\delta\omega^b\mathbf{\Phi}) \end{aligned} \tag{46.34}$$

It only takes a little algebra to see what this implies about δA^a_μ. The first term on the left-hand side of (46.34) cancels the first term on the right-hand side. We move the remaining term on the left-hand side of (46.34) to the right and bring the δA^a_μ term to the left-hand side. Finally, we swap the dummy indices a and b in the last term in (46.34), and obtain

$$i(\delta A^a_\mu)T^a\mathbf{\Phi} = -iT^a(\partial_\mu \delta\omega^a)\mathbf{\Phi} + T^aT^b\delta\omega^a A^b_\mu \mathbf{\Phi} - T^bT^a\delta\omega^a A^b_\mu\mathbf{\Phi} \tag{46.35}$$

The last two terms are just the commutator $[T^a, T^b]$ times $\delta\omega^a A^b_\mu \mathbf{\Phi}$. The right-hand side becomes

$$i(\delta A^a_\mu)T^a\mathbf{\Phi} = -iT^a(\partial_\mu \delta\omega^a)\mathbf{\Phi} + ic^{abc}\delta\omega^a A^b_\mu T^c\mathbf{\Phi} \tag{46.36}$$

The last term can be rewritten. We have, swapping the dummy indices,

$$c^{abc}\delta\omega^a A^b_\mu T^c = c^{bca}\delta\omega^b A^c_\mu T^a = c^{abc}\delta\omega^b A^c_\mu T^a \tag{46.37}$$

the last step following from the invariance of the structure constants under cyclic permutation (46.26). Thus all sums involve T^a on an arbitrary $\mathbf{\Phi}$:

$$i(\delta A^a_\mu)T^a\mathbf{\Phi} = -iT^a(\partial_\mu \delta\omega^a)\mathbf{\Phi} + ic^{abc}T^a\delta\omega^b A^c_\mu\mathbf{\Phi} \tag{46.38}$$

We can obtain the desired result if we choose

$$\boxed{\delta A^a_\mu = -\partial_\mu(\delta\omega^a) + c^{abc}\delta\omega^b A^c_\mu} \tag{46.39}$$

This is the key equation.[12]

The transformation law (46.39) for the gauge fields A^a_μ is rather complicated. Before going on to finding the possible forms of the pure gauge field Lagrangian, it might be nice to understand their physical meaning, by looking at these transformation laws in a particular special case, say SU(2), isospin, in which $\{a, b, c\}$ run over 1, 2 and 3, and the structure constants c^{abc} are ϵ^{abc}. Choose a gauge transformation associated with rotations about the third axis in isospin space:

$$\delta\omega^1 = \delta\omega^2 = 0 \tag{46.40}$$

so the only thing left is $\delta\omega^3$. What does (46.39) give for δA^1_μ and δA^2_μ? For $a = 1$,

$$\delta A^1_\mu = \epsilon^{1bc}\delta\omega^b A^c_\mu - \partial_\mu(\delta\omega^1) = \epsilon^{13c}\delta\omega^3 A^c_\mu = -\delta\omega^3 A^2_\mu \tag{46.41}$$

The derivative term is irrelevant because that's non-zero only for $a = 3$. In ϵ^{13c}, the only non-zero term has $c = 2$, and $\epsilon^{132} = -1$. By the same reasoning for $a = 2$ and $a = 3$,

$$\begin{aligned} \delta A^2_\mu &= \delta\omega^3 A^1_\mu \\ \delta A^3_\mu &= -\partial_\mu \delta\omega^3 \end{aligned} \tag{46.42}$$

In the last line the ϵ carries two 3's and is therefore zero. These transformations look a great deal simpler. These three gauge bosons in a sense transform like an isotriplet, like the generators of the group. If I restrict myself to gauge transformations corresponding to rotations about the three axis, A^1_μ and A^2_μ transform like an ordinary pair of charged particles, with $I_3 = \pm 1$; they just rotate. As far as gauge transformations along the I_3 axis goes, you can't tell that A^1_μ and A^2_μ are gauge particles, and not simply some particles that transform

---

[12] [Eds.] The sign of $\delta\omega^a$ (46.20) and the sign of the term in A^a_μ in (46.30), can be chosen in four different ways, or, if you prefer, two different classes: both the same or both different. Each of these choices yields a unique set of relative signs in (46.39). Note however that these choices *differ* from the Abelian (27.13) and (27.48).

linearly and homogeneously along the group. On the other hand, you can't tell that A^3_μ carries any internal quantum numbers. It acts just like an uncharged photon, transforming exactly as a photon would, with a gradient.

In the general case, the gauge boson corresponding to an infinitesimal transformation associated with only one symmetry generator transforms like a photon: it gets only an added gradient. The other gauge bosons transform like ordinary non-gauge fields; they just rotate among themselves according to the transformation of the particular symmetry group we're looking at (because of the c^{abc} term, they transform like the group's generators), without any gradients. That's why you need both terms. To put it another way, the first term is here so that if $\delta\omega^a$ is constant, (a possibility, after all) the second term would drop out, and the gauge bosons would mix among themselves just as the group generators do. That's obviously necessary if the expression $A^a_\mu T^a$ is not to break the symmetry. The second term is there so that if I consider gauge transformations along a particular direction in internal symmetry space, the gauge boson associated with *that* direction transforms like a photon. That's why we have this funny transformation law with two terms in it.

The next stage is to find the analog of $(F_{\mu\nu})^2$. That is what gave us a perfectly satisfactory Lagrangian in the case of electromagnetism. If we just had $\mathscr{L}(\mathbf{\Phi}, D_\mu\mathbf{\Phi})$ we'd have a pretty dumb theory. We'd have the A_μ's in the theory but not their derivatives and we wouldn't expect much dynamics to come out of that; you'd have introduced new vector fields with no associated *free* Lagrangian. I'll construct the analog to $F_{\mu\nu}$ by the following chain of reasoning. Consider the commutator of two covariant derivatives

$$D_\mu D_\nu - D_\nu D_\mu \tag{46.43}$$

I will demonstrate that one can find a function of the gauge fields and their derivatives, $F^a_{\mu\nu}$, such that

$$D_\mu D_\nu\mathbf{\Phi} - D_\nu D_\mu\mathbf{\Phi} = -iF^a_{\mu\nu}T^a\mathbf{\Phi} \tag{46.44}$$

Both sides of this equation must transform under gauge transformations the same way as $\mathbf{\Phi}$ does, because that's the property of covariant differentiation. The $F^a_{\mu\nu}$'s will have to transform in such a way that this equation doesn't break the invariance: $\delta F^a_{\mu\nu}$ must transform homogeneously and indeed like the group generators, like the A^a_μ transforms except without the derivative term:

$$\delta F^a_{\mu\nu} = c^{abc}\delta\omega^b F^c_{\mu\nu} \tag{46.45}$$

An extra term in it would screw things up. The tensors $F^a_{\mu\nu}$ are nice objects that transform linearly and homogeneously, like the group generators. They will be the non-Abelian analogs of $F_{\mu\nu}$. By playing around with them we will be able to find invariant quadratic Lagrangians. The transformation of the tensor (46.45) follows trivially from the transformation of A^a_μ (46.39); it's just the statement that the tensor should be covariant.

Before I demonstrate (46.44), again a tedious but straightforward computation, let me write the finite versions of these expressions. The transformation for finite ω^a is

$$U(\omega) = \exp(-i\omega^a T^a) \tag{46.46}$$

The various quantities transform according to:

$$\mathbf{\Phi} \to \mathbf{\Phi}' = U(\omega)\mathbf{\Phi} \tag{46.47}$$

$$D_\mu\mathbf{\Phi} \to D'_\mu\mathbf{\Phi}' = UD_\mu\mathbf{\Phi} \;\Rightarrow\; D'_\mu = UD_\mu U^{-1} \tag{46.48}$$

$$F^a_{\mu\nu}T^a\mathbf{\Phi} \to F'^a_{\mu\nu}T^a\mathbf{\Phi}' = UF^a_{\mu\nu}T^a\mathbf{\Phi} \;\Rightarrow\; F'^a_{\mu\nu}T^a = UF^a_{\mu\nu}T^aU^{-1} \tag{46.49}$$

The thing we want to compute is the commutator of two covariant derivatives:

$$\begin{aligned} D_\mu D_\nu \mathbf{\Phi} - D_\nu D_\mu \mathbf{\Phi} &= (\partial_\mu - iA^a_\mu T^a)(\partial_\nu - iA^b_\nu T^b)\mathbf{\Phi} - (\mu \leftrightarrow \nu) \\ &= \partial_\mu \partial_\nu \mathbf{\Phi} - iA^a_\mu T^a \partial_\nu \mathbf{\Phi} - iA^b_\nu T^b \partial_\mu \mathbf{\Phi} \\ &\quad - i(\partial_\mu A^b_\nu) T^b \mathbf{\Phi} - A^a_\mu T^a A^b_\nu T^b \mathbf{\Phi} - (\mu \leftrightarrow \nu) \end{aligned} \tag{46.50}$$

Most of the terms will disappear when we antisymmetrize. For example the first term, $\partial_\mu \partial_\nu \mathbf{\Phi}$, vanishes because ordinary derivatives commute. There are two terms involving one derivative of $\mathbf{\Phi}$:

$$-iA^a_\mu T^a \partial_\nu \mathbf{\Phi}, \quad -iA^b_\nu T^b \partial_\mu \mathbf{\Phi} \tag{46.51}$$

They differ only by the exchange of μ and ν; whether I sum over a or sum on b is a matter of notation. So the two $\partial \mathbf{\Phi}$ terms together are symmetric in μ and ν and so will cancel when we antisymmetrize. There are two terms

$$-A^a_\mu T^a A^b_\nu T^b \mathbf{\Phi} - (\mu \leftrightarrow \nu) \tag{46.52}$$

Exchanging μ and ν is just commuting the order of those two matrices since I can relabel the summation indices. So this equals

$$-A^a_\mu T^a A^b_\nu T^b + A^b_\nu T^b A^a_\mu T^a = -ic^{abc} A^a_\mu A^b_\nu T^c = -ic^{abc} A^b_\mu A^c_\nu T^a \tag{46.53}$$

by using the algebra of the generators and the symmetry of the structure constants (46.26). Thus I have a function of the gauge fields and their derivatives times T^a acting on $\mathbf{\Phi}$; referring to (46.44),

$$F^a_{\mu\nu} = \partial_\mu A^a_\nu - \partial_\nu A^a_\mu + c^{abc} A^b_\mu A^c_\nu \tag{46.54}$$

This field tensor is a sum of three terms. The first two, $\partial_\mu A^a_\nu - \partial_\nu A^a_\mu$, will look familiar to anyone who has studied electromagnetism. The third nonlinear term, absent from electrodynamics, is the glory, the joy, and the nightmare of non-Abelian gauge field theory, $c^{abc} A^b_\mu A^c_\nu$.

Having obtained the **field strength tensors** $F^a_{\mu\nu}$ we have to find ways to make gauge invariant objects out of them. We can square them and then sum on a; that's guaranteed to be invariant. Therefore we can write, following electrodynamics

$$\mathscr{L} = -\frac{1}{4g^2} F^a_{\mu\nu} F^{\mu\nu\, a} + \mathscr{L}(\mathbf{\Phi}, D_\mu \mathbf{\Phi}) \tag{46.55}$$

The reason for the constant g will become clear shortly; we don't know *a priori* what that coefficient should be. This structure is obviously invariant: it's the sum of squares of all these objects that transform according to some representation of the group, plus the original Lagrangian. That's certainly the simplest possible generalization of electrodynamics. The free constant is there because I have no way of knowing the relative scale of these two terms.

We'll now get rid of that constant; otherwise it leads to a quadratic term that is rather dumb, with a constant $1/g^2$ in it instead of the 1 that we want for the free theory of vector bosons. We will do it by rescaling: define

$$A^a_\mu = gA'^a_\mu \;\Rightarrow\; D_\mu \mathbf{\Phi} = \partial_\mu \mathbf{\Phi} - igA'^a_\mu T^a \mathbf{\Phi} \tag{46.56}$$

Then $F^a_{\mu\nu}$ becomes

$$F^a_{\mu\nu} = g\big(\partial_\mu A'^a_\nu - \partial_\nu A'^a_\mu + gc^{abc} A'^b_\mu A'^c_\nu\big) \equiv gF'^a_{\mu\nu} \tag{46.57}$$

The g^2 disappears from the term quadratic in the derivatives of the A fields and the Lagrangian becomes

$$\mathscr{L} = -\tfrac{1}{4}F'^{a}_{\mu\nu}F'^{\mu\nu\,a} + \mathscr{L}(\mathbf{\Phi}, D_\mu\mathbf{\Phi}) \tag{46.58}$$

The scale factor g is now sensibly located to act as a coupling constant. The quadratic terms in the Lagrangian are totally free of g's and the non-quadratic terms all have g's in them. By convention we will drop the primes from now on and always use these fields we've defined as primed fields. In particular, the covariant derivative and the Yang–Mills field tensor become

$$\boxed{D_\mu\mathbf{\Phi} = \partial_\mu\mathbf{\Phi} - igA^a_\mu T^a\mathbf{\Phi}} \tag{46.59}$$

$$\boxed{F^a_{\mu\nu} \equiv \partial_\mu A^a_\nu - \partial_\nu A^a_\mu + gc^{abc}A^b_\mu A^c_\nu} \tag{46.60}$$

This looks like a reasonable theory of a bunch of fields, *if* we can handle the problem of quantization of the gauge fields A^a_μ, which after all caused a lot of trouble for electrodynamics and may cause us troubles here. Some of the fields—the A^a_μ—are massless; the others, the $\mathbf{\Phi}$, are massive or massless depending on what $\mathscr{L}$ is like, the fields all interacting together in a complicated way governed by this coupling constant g. Even if there were no $\mathbf{\Phi}$ fields, the non-Abelian gauge fields would have inherent self-interactions. This is a striking contrast between non-Abelian gauge field theory and Abelian gauge field theory (electromagnetism). In the absence of charged particles pure electromagnetism is a free field theory; nothing could be more trivial. The pure gauge theory of Yang–Mills fields[13] has complicated interactions even if there are no electrons or π mesons or anything else in the world, except the gauge bosons.

There's a reason for this. Remember the old argument, which you were no doubt all told by your elders, about why gravity is necessarily a nonlinear field theory. The argument goes as follows. The source of gravity is energy. The graviton carries energy. Therefore the graviton must be a source of gravity. Therefore there would be a nonlinear coupling even if gravity were all that the universe contained. The same thing is true here. For simplicity let us think of the isospin example. We have three gauge fields, A^1_μ, A^2_μ and A^3_μ; A^3_μ is the gauge field for the T^3 rotations. It couples to everything that carries the third component of isospin. Among things that carry the third component of isospin are A^1_μ and A^2_μ. We have just seen that they are rotated under an T^3 rotation. Therefore A^3_μ must couple to A^1_μ and A^2_μ even if there were no other particles in the world. The amazing thing is *not* that a nonlinear coupling is necessary but that we can get by with such a simple nonlinear coupling that has only cubic and quartic terms. In gravity we have to go on *forever.* Here we can stop after the fourth order. That's amazing but that's the way things work. I have done this in a way that makes the Yang–Mills theory look as close to gravity as I can without lying to you. If you've taken Steve Weinberg's course in general relativity, or another's, you'll recognize that (46.44) is very close to the definition of the *Riemann–Christoffel tensor* as the commutator of covariant derivatives;[14] and that writing the Lagrangian in the form (46.55) with the coupling constant in front of the free Lagrangian (rather than inside the interaction) is precisely the way it's

[13] [Eds.] See note 9, p. 1016.

[14] [Eds.] S. Weinberg, *Gravitation and Cosmology: Principles and applications of the general theory of relativity*, John Wiley and Sons, 1972, equation (6.5.1), p. 140; Charles W. Misner, Kip S. Thorne, and John Archibald Wheeler, *Gravitation*, W. H. Freeman, 1970, 1971; reissued by Princeton U. P., 2017, Exercise 16.3, p. 389; A. Zee, *Einstein's Gravity in a Nutshell*, Princeton U. P., 2013, equation (4)–equation (9), pp. 341–342.

written in gravity. There you have $1/(8\pi G)$ times R, the Ricci scalar, plus a matter-energy term that's free of the gravitational coupling constant.[15]

One coupling constant, or many?

Now we have to address a tiny technical point: do we always have only one gauge coupling constant or can there be several? This is really the question of whether the only invariant we can form is $F^a_{\mu\nu}F^{\mu\nu\,a}$. Maybe we could form different ones. For example, if our group was SU(2) then that would be the only invariant we could form; the only way we could take two isovectors and put them together as a scalar is a dot product. On the other hand if our group is chiral SU(2) $\otimes$ SU(2) the generators fall into two sets, the generators of the left-handed isospin and the generators of the right-handed isospin. Now we can form two scalars: left dot left and right dot right and they can have *independent coefficients.* This is obviously the general situation if you make appropriate definitions, which we will now do.

If the generators T^a of the gauge group G transform according to an *irreducible* representation of G, then there is only one invariant that can be constructed, $F^a_{\mu\nu}F^{\mu\nu\,a}$, and therefore there is only one gauge coupling constant g. In this case, for those of you who have been reading group theory books, we say that G is **simple.**[16] On the other hand, if the T^a's transform according to a *reducible* representation of the group, as in for example chiral SU(2) $\otimes$ SU(2), then it's easy to see from the antisymmetry of the structure constants that the algebra falls into a sum of a bunch of subalgebras, none of which talk to each other; c^{abc} vanishes unless a, b and c are associated with the *same* factor (i.e., the same subgroup). G is a product, at least locally, of simple groups generated by the various irreducible representations. And then we can have one gauge coupling constant for each factor, one for the right-handed SU(2) and a completely different one for the left-handed SU(2). For example, if we were to consider the product SU(2) $\otimes$ U(1), and call the quantities coupled to the associated gauge fields "isospin" and "hypercharge", there would be a hypercharge gauge boson and there would be an isospin triplet of isospin gauge bosons and they could have *different* gauge coupling constants. The hypercharge is a SU(2) invariant and so is the square of isospin. That's in fact what happens in the *Glashow–Salam–Weinberg model of weak interactions.*[17] Our formulas for non-simple groups would be somewhat generalized. If we use the prime fields for example, $D_\mu\mathbf{\Phi}$ in (46.56) would be $\partial_\mu\mathbf{\Phi}$ plus a g that may depend on a, still summing on repeated indices even though the index appears *three* times:

$$D_\mu\mathbf{\Phi} = \partial_\mu\mathbf{\Phi} - ig_a A^a_\mu T^a\mathbf{\Phi} \tag{46.61}$$

Each g_a would be a constant, for a given factor in a product of groups, but they might be different for different group factors (in the example above, one constant for U(1) and another for SU(2)). Likewise

$$F^a_{\mu\nu} = \partial_\mu A^a_\nu - \partial_\nu A^a_\mu + g_a c^{abc} A^b_\mu A^c_\nu \tag{46.62}$$

---

[15] [Eds.] See Weinberg *op. cit.*, Section 12.4, pp. 364–365, in particular equation (12.4.2); Misner *et al.*, *op. cit.*, §21.2, pp. 491–492, specifically equation (21.18); Zee *op. cit.*, equation (9), p. 390. For a discussion of gravity as a field theory, see Zee *QFTN*, Section VIII.1, pp. 433–447. Both Feynman and DeWitt investigated Yang–Mills theories as a warm-up to a quantum theory of gravity; see the references in note 10, p. 625, and once again, note 9, p. 1016.

[16] [Eds.] F. W. Byron and R. W. Fuller, *Mathematics of Classical and Quantum Physics*, Dover Publications, New York, 1970, p. 596; Cheng and Li, *GT*, p. 87; Zee *GTN*, pp. 63–64. A subgroup S of a group G is called *normal* if it is turned into itself under the action of the elements of G: $gSg^{-1} = S$. (This doesn't mean that the elements are *individually* invariant; the subgroup Z of G whose elements *each* commute with all the elements of G is called the *center* of the group.) A normal subgroup is thus invariant under the group. A *simple* group is a group with no normal subgroups (excluding the identity and the full group itself).

[17] [Eds.] The GSW model is the subject of Chapters 48 and 49.

Now we have the complete theory, though we don't yet know if we can quantize it. And if we *can* quantize it, we don't know if it's renormalizable. But at least at a classical level we have the complete theory of non-Abelian gauge fields. We can certainly write more complicated Lagrangians just as we could in electrodynamics. We could have analogs of the Pauli coupling $F_{\mu\nu}\sigma^{\mu\nu}$; we could have terms involving the fourth power of $F_{\mu\nu}$ but that would add in more derivatives. Since we eventually hope to get a renormalizable field theory we should try to get away with as few derivatives and as few powers of the fields as possible. This is the minimum number.

When Yang and Mills first proposed this theory for the case of isospin in the mid-fifties, nobody knew anything about spontaneous symmetry breaking. It was thought that maybe these things *did* exist; perhaps there were three massless vector bosons in addition to the photon. But they had to be very weakly coupled; otherwise they would have been observed. People went through a long series of arguments involving cosmological experiments, protons, neutrons, virtual pions, long-range forces that would compete with gravity and got all sorts of bounds. It gradually became clear that this theory was hopeless; there was no chance that these massless vector gauge bosons existed in the real world. The only massless gauge boson in the real world was the photon. Around 1960 and 1961, everyone was very excited about spontaneous symmetry breaking: Nambu, Jona-Lasinio, Goldstone and the Goldstone phenomenon.[18] But it was soon discovered that Goldstone bosons were inevitable, and Goldstone's theorem was proved.[19] Then everyone said, well, that's the end of that, because there aren't any massless scalar mesons in the world except maybe the pions, with approximately zero mass. We had two things floating around that were more or less theorist's toys: Yang–Mills theories of zero-mass vectors, and spontaneous symmetry breaking with zero-mass Goldstone scalars.

The Higgs phenomenon reconciles these two problems; the two diseases turn out to be each other's cure.[20] If the group that is spontaneously broken is a *gauge group* then the Goldstone bosons are *eaten by the gauge bosons*, which become massive vector bosons. You wind up with neither unobserved massless gauge bosons nor unobserved massless Goldstone bosons. Let's see how this works out in the general case of spontaneous symmetry breaking, again only in the tree approximation. In this case we can't go farther because we can't quantize the theory yet. We will quantize it as soon as we are done with this discussion. It turns out to be just what we did for electromagnetism over again. (We went through electromagnetism that way to serve as a warm-up for the non-Abelian case.)

46.3 Yang–Mills fields and spontaneous symmetry breaking

We want to consider the same kind of Lagrangian we had before, now promoted to a gauge theory.

$$\mathscr{L} = -\tfrac{1}{4}F^{a}_{\mu\nu}F^{\mu\nu\,a} + \tfrac{1}{2}D_{\mu}\mathbf{\Phi}\cdot D^{\mu}\mathbf{\Phi} - U(\mathbf{\Phi}) \tag{46.63}$$

---

[18] [Eds.] See note 13, p. 944.

[19] [Eds.] See note 14, p. 944.

[20] [Eds.] As Higgs himself has emphasized, this idea was first suggested by Philip Anderson, who conjectured that "the Goldstone zero-mass difficulty is not a serious one, because one can probably cancel it off against an equal Yang–Mills zero-mass problem," though Anderson did not give any example of how this might happen. Peter Higgs, "My Life as a Boson: The Story of 'The Higgs'", *Int. J. Mod. Phys. A* **17** (supplement 01), (2002) 86–88; P. W. Anderson, "Plasmons, Gauge Invariance, and Mass", *Phys. Rev.* **130** (1963) 439–441. Anderson's remark is in his penultimate paragraph, p. 441.

The field $\mathbf{\Phi}$ is an N-component vector; N is the total number of generators of the group. We have the covariant derivative of $\mathbf{\Phi}$:

$$D_\mu \mathbf{\Phi} = \partial_\mu \mathbf{\Phi} - i g_a A^a_\mu T^a \mathbf{\Phi} \tag{46.64}$$

(remembering that we sum over repeated indices even when there are three of them). Just as before the state of lowest energy is found by taking $\mathbf{\Phi}$ to be a constant and setting the minimum of $U(\mathbf{\Phi})$ to 0. Again we call that state the vacuum:

$$\mathbf{\Phi} = \langle \mathbf{\Phi} \rangle \tag{46.65}$$

The whole group is now promoted to a gauge group.

We divide our group generators into two classes: those that annihilate the vacuum

$$T^a \langle \mathbf{\Phi} \rangle = 0, \qquad a = 1, 2, \cdots n \tag{46.66}$$

where $n < N$, and the remaining orthogonal set of linearly independent generators that don't:

$$T^b \langle \mathbf{\Phi} \rangle \neq 0, \quad b = n+1, \cdots N \tag{46.67}$$

Equivalently,

$$c^b T^b \langle \mathbf{\Phi} \rangle = 0 \Rightarrow c^b = 0, \quad b = n+1, \cdots N \tag{46.68}$$

The generators (46.66) that annihilate the vacuum are the **unbroken generators**, while the others (46.68) are the **spontaneously broken generators**.[21] The unbroken generators define a subgroup H of G,

$$H \subset G, \qquad (N = \dim G, \; n = \dim H) \tag{46.69}$$

The group H remains a manifest symmetry.

We now have a gauge theory and therefore we must pick a gauge. I will impose a set of gauge conditions that will *eliminate* the Goldstone bosons immediately. The Goldstone bosons correspond to oscillations determined by $T^b \langle \mathbf{\Phi} \rangle$; modes generated from $\langle \mathbf{\Phi} \rangle$ by infinitesimal transformations about the minimum:

$$\langle \mathbf{\Phi} \rangle \to \langle \mathbf{\Phi} \rangle - i T^b \delta\omega^b \langle \mathbf{\Phi} \rangle \tag{46.70}$$

They are what the spontaneously broken generators generate (in two senses of the word "generate"). By applying a gauge transformation that counteracts just that group transformation (46.70) I can arrange that

$$T^b \delta\omega^b \langle \mathbf{\Phi} \rangle = 0 \tag{46.71}$$

I will choose as the gauge condition[22]

$$\mathbf{\Phi}' \cdot T^b \langle \mathbf{\Phi} \rangle = 0 \tag{46.72}$$

where $\mathbf{\Phi}'$ is the *gauge-transformed* $\mathbf{\Phi}$ (46.47), $\langle \mathbf{\Phi} \rangle$ is a constant N vector in $\mathbf{\Phi}$ space which minimizes $U(\mathbf{\Phi})$, T^b are $N \times N$ matrices for the *broken* generators, and (46.72) is their dot

[21] [Eds.] Equations (46.66) and (46.67) are sometimes written $T^a |0\rangle = 0$ and $T^b |0\rangle \neq 0$, respectively.

[22] [Eds.] See Abers and Lee, *op. cit.* (note 9, p. 1016), p. 28, equation (3.20).

product. This condition (which can be imposed on any compact group) was first published in a paper by Weinberg.[23] It is a set of linear conditions on the fields, equivalent to the earlier condition (46.18) $\theta = 0$, which immediately eliminates the Goldstone bosons. Unphysical motions away from the ground state of the theory (corresponding to oscillations along the bottom of the trough of the potential in Figure 43.5) are canceled by this equation. The relation (46.72) expresses only $(N - n)$ gauge conditions; there are still n conditions left. For the remaining n gauge conditions we choose whatever gauge pleases us: axial gauge, Coulomb gauge, covariant gauge.

Weinberg's "proof" tells us we can always fix the gauge according to this condition (46.72). But the gauge is not uniquely determined when $\mathbf{\Phi}$ has a zero, and this general argument breaks down. If you imagine classical field configurations very far from the configuration of minimum energy, then

$$\mathbf{\Phi} = 0 \tag{46.73}$$

That doesn't bother us when we're doing perturbation theory in tree approximation, because then we're always expanding about the configuration of minimum energy; the deviations from that configuration are small. It is evident however from the way we have found the Goldstone modes that it is always possible to choose a gauge in which the Goldstone bosons are gauged away. They are precisely those modes (spanning a hypersurface in $\mathbf{\Phi}$ space) that are swept out by the action of the group generators on the ground state. And since they are made by group generators, they can be *unmade* by group generators. That's just what a gauge transformation does for you.

This particular choice of gauge is called the ***U* gauge**;[24] "U" stands for "unitary" and expresses the fact that we have eliminated the unphysical degrees of freedom, the Goldstone bosons, which we know won't be there because they will be eaten by the gauge bosons. It has the additional property that certain cross terms in $D_\mu\mathbf{\Phi} \bullet D^\mu\mathbf{\Phi}$ disappear after the shift is made. (We will see this when we discuss the Glashow–Salam–Weinberg model.) It is in general a terrible gauge for anything *except* exploring the tree approximation. You can quantize in it, but you get awful Feynman rules that lead to ostensibly non-renormalizable theories. When you sum up all the graphs, the horrible divergences cancel. But only a madman would work in a formalism where every individual piece of the computation is hideously divergent, and it's only when you sum them all up that you get convergence.[25] (There *are* people who have worked in this gauge.) Let's look at a couple of SO(n) examples in detail to see how this gauge condition is applied.[26]

EXAMPLE. SO(2): *Rotations in a plane*

---

[23] [Eds.] S. Weinberg, "General Theory of Broken Local Symmetries", *Phys. Rev.* **D7** (1973) 1068–1082.

[24] [Eds.] S. Weinberg, "Physical Processes in a Convergent Theory of the Weak and Electromagnetic Interactions", *Phys. Rev. Lett.* **27** (1971) 1688–1691; Abers and Lee, *op. cit.* (note 9, p. 1016), Section 3, pp. 20–25. A different class of gauges, the $\boldsymbol{R_\xi}$ **gauges** of G. 't Hooft and B. Lee , makes the renormalizability of the theory manifest, though its unitarity is obscured: G. 't Hooft, "Renormalizable Lagrangians for Massive Yang–Mills Fields", *Nucl. Phys.* **B35** (1971) 167–188; Benjamin W. Lee, "Renormalizable Massive Vector-Meson Theory—Perturbation Theory of the Higgs Phenomenon", *Phys. Rev.* **D5** (1972) 823–834.

[25] [Eds.] Weinberg calls such theories "cryptorenormalizable"; Weinberg, "General Theory of Broken Local Symmetries", *op. cit.*

[26] [Eds.] For reference, the number of generators of SO(n) is $\frac{1}{2}n(n-1)$, and the number of generators of SU(n) is $n^2 - 1$: Zee *GTN*, p. 80 and p. 237, respectively.

Write the fields ϕ_1-ϕ_2 in our SO(2) example, p. 941, as

$$\mathbf{\Phi} = \begin{pmatrix} \phi_1 \\ \phi_2 \end{pmatrix} \tag{46.74}$$

Earlier we chose (43.23)

$$\left.\begin{matrix} \langle\phi_1\rangle \\ \langle\phi_2\rangle \end{matrix}\right\} = \left\{\begin{matrix} a \\ 0 \end{matrix}\right. \quad \text{and so} \quad \langle\mathbf{\Phi}\rangle = \begin{pmatrix} a \\ 0 \end{pmatrix} \tag{46.75}$$

There is one generator, T, associated with rotations, and one Goldstone boson. An element of SO(2) has the form

$$R(\theta) = \begin{pmatrix} \cos\theta & \sin\theta \\ -\sin\theta & \cos\theta \end{pmatrix} \tag{46.76}$$

and from it we obtain the generator T,

$$iT = \left.\frac{dR}{d\theta}\right|_{\theta=0} = \begin{pmatrix} 0 & 1 \\ -1 & 0 \end{pmatrix} \tag{46.77}$$

In agreement with (46.67), T does not annihilate the vacuum:

$$iT\,\langle\mathbf{\Phi}\rangle = \begin{pmatrix} 0 \\ -a \end{pmatrix} \neq 0 \tag{46.78}$$

and the gauge condition (46.72) is

$$\mathbf{\Phi}' \bullet T^a\,\langle\mathbf{\Phi}\rangle = (\phi'_1, \phi'_2) \bullet \begin{pmatrix} 0 \\ -a \end{pmatrix} = 0 \Rightarrow \phi'_2 = 0 \tag{46.79}$$

Because of the definition (46.3), the gauge condition $\phi'_2 = 0$ is the same as the $\theta = 0$ condition (46.18) we applied before.

EXAMPLE. SO(3): *Rotations in three-space*

There are three generators associated with the rotations in three-space (about each axis in turn):

$$iT^1 = \begin{pmatrix} 0 & 0 & 0 \\ 0 & 0 & 1 \\ 0 & -1 & 0 \end{pmatrix}; \quad iT^2 = \begin{pmatrix} 0 & 0 & -1 \\ 0 & 0 & 0 \\ 1 & 0 & 0 \end{pmatrix}; \quad iT^3 = \begin{pmatrix} 0 & 1 & 0 \\ -1 & 0 & 0 \\ 0 & 0 & 0 \end{pmatrix} \tag{46.80}$$

Taking the vacuum to be

$$\langle\mathbf{\Phi}\rangle = \begin{pmatrix} a \\ 0 \\ 0 \end{pmatrix} \tag{46.81}$$

we see that

$$T_1\,\langle\mathbf{\Phi}\rangle = 0; \quad T_2\,\langle\mathbf{\Phi}\rangle \neq 0; \quad T_3\,\langle\mathbf{\Phi}\rangle \neq 0 \tag{46.82}$$

and hence there are two broken generators and two Goldstone bosons. There is one unbroken generator, so the remaining symmetry is SO(2), rotations in the plane about the 1-axis.

(The vacuum was chosen to have its non-zero expectation value along the 1-axis rather than the 3-axis to simplify notation in what is to come.)

With the gauge condition (46.72) we don't have to worry about the Goldstone bosons. We *do* have to worry about the gauge bosons. We'd expect that some of them, those associated with the spontaneously broken generators, are going to get a mass, and others, corresponding to the subgroup H of the unbroken generators, will remain massless; for them the theory is much like ordinary electrodynamics. In tree approximation we only have to worry about the quadratic terms that have an effect of making the shift

$$\mathbf{\Phi} = \mathbf{\Phi}' + \langle\mathbf{\Phi}\rangle \tag{46.83}$$

The only terms that do that in the Lagrangian are

$$\begin{aligned}\mathscr{L} &= \tfrac{1}{2}D_\mu\mathbf{\Phi}\cdot D^\mu\mathbf{\Phi} + \cdots \\ &= \tfrac{1}{2}(\partial_\mu\mathbf{\Phi} - ig_a A^a_\mu T^a\mathbf{\Phi})\cdot(\partial^\mu\mathbf{\Phi} - ig_b A^{\mu\,b}T^b\mathbf{\Phi}) + \cdots\end{aligned} \tag{46.84}$$

The other terms are totally irrelevant; they're not going to involve the vector bosons and won't be affected by making a shift of the scalar fields. After we make this shift, the only terms that will be both quadratic in the vector bosons, and decoupled from the scalar bosons, are those involving the $\langle\mathbf{\Phi}\rangle$ part of $\mathbf{\Phi}$, an N-vector for N boson fields. This equals, keeping only terms quadratic in the A^a_μ's,

$$\begin{aligned}\mathscr{L} &= \cdots - \tfrac{1}{4}(\partial_\mu A^a_\nu - \partial_\nu A^a_\mu)(\partial^\mu A^{\nu\,a} - \partial^\nu A^{\mu\,a}) + \tfrac{1}{2}ig_a T^a A^a_\mu\langle\mathbf{\Phi}\rangle\cdot ig_b T^b A^{\mu\,b}\langle\mathbf{\Phi}\rangle \\ &= \cdots - \tfrac{1}{4}(\partial_\mu A^a_\nu - \partial_\nu A^a_\mu)(\partial^\mu A^{\nu\,a} - \partial^\nu A^{\mu\,a}) + \tfrac{1}{2}M^2_{ab}A^a_\mu A^{\mu\,b} + \cdots\end{aligned} \tag{46.85}$$

M^2 is a vector boson **mass matrix**. The tree approximation is

$$M^2_{ab} = ig_a T^a\langle\mathbf{\Phi}\rangle\cdot ig_b T^b\langle\mathbf{\Phi}\rangle \tag{46.86}$$

(there is no summation on a and b). That's an obviously symmetric matrix. It's positive definite within the space of spontaneously broken generators (for $a, b > n$) because of (46.68), which asserts that if I put any vector c^a on the left and any vector c^b on the right I get a non-zero matrix element. So it is a positive definite symmetric matrix. Its eigenvalues, which you have to compute for any particular given $\mathbf{\Phi}$, given the generators, determine what the masses of the vector bosons are. If we choose a and b to correspond to one of the generators of H, then this matrix is zero, so indeed the vector bosons corresponding to the unbroken symmetries remain massless;

$$iT^a\langle\mathbf{\Phi}\rangle = 0 \;\Rightarrow\; M^2_{ab} = 0 \tag{46.87}$$

Writing M^2_{ab} as a block diagonal matrix we get

$$M^2_{ab} = \begin{array}{c} \begin{matrix} a<n & \quad a\geq n \end{matrix} \\ \begin{pmatrix} 0 & 0 \\ 0 & \begin{bmatrix}\text{positive}\\ \text{definite}\end{bmatrix}\end{pmatrix}\end{array} \tag{46.88}$$

The 0 blocks are from the unbroken generators, (46.66). For our SO(3) example above, $n = 1$, $N - n = 2$, and there is only one unbroken generator, T^1. The mass matrix is

$$M^2_{ab} = \begin{pmatrix} 0 & 0 & 0 \\ 0 & g^2a^2 & 0 \\ 0 & 0 & g^2a^2 \end{pmatrix} \tag{46.89}$$

The gauge bosons associated with the broken symmetry generators, T_2 and T_3, become massive by eating the corresponding Goldstone bosons. The gauge boson associated with the unbroken generator, T_1, remains massless.

We have seen how in the tree approximation the Higgs mechanism generalizes to the non-Abelian case. If we can successfully quantize the theory and prove it is renormalizable in any gauge, though perhaps not this one (which is convenient for seeing what's happening in this simple case), then all of this apparatus will, *mutatis mutandis*, carry through to the quantum theory, for exactly the same reasons that were given for a pure scalar theory. (Regrettably, I won't be showing you the proof of renormalizability, because it's a bit too complicated for the time remaining in the course.) We have a Γ, a generating functional, a function of classical $\mathbf{\Phi}$ fields and classical A fields. We go to the minimum, we expand around the minimum, we work things out. It all goes through smooth as silk. If it goes through for scalars it goes through for vectors also. Therefore our only remaining problem of principle is to quantize this theory and find out precisely what *are* the Feynman rules for non-Abelian gauge theories. That I will attack next time.

47

Quantizing non-Abelian gauge fields

We've investigated gauge field theories as classical field theories, including the effects of spontaneous symmetry breaking. From our previous experience with field theories of scalars and spinors we know that if we can construct a quantum theory, the classical theory can be reinterpreted as the first term—the tree approximation—in a systematic perturbative expansion. If we ever want to obtain higher order corrections, however, we *have* to construct the corresponding quantum theory. So we will now turn to the problem of quantizing gauge field theories.[1]

47.1 Quantization of gauge fields by the Faddeev–Popov method

We have already quantized the simplest gauge theory, electromagnetism.[2] Let me provide a brief review of that earlier treatment of electromagnetism, an *aide mémoire* for what is to come.

In the quantization of electromagnetism we used a magic method due to Faddeev and Popov. It may have seemed complicated when we were struggling through it, but it is simple compared to what we now have to confront. Recall how Faddeev–Popov quantization worked in electromagnetism. We started with a gauge invariant action $\mathcal{S}_{GI}$. We grouped all the dynamical variables—scalar, spinor, vector, whatever—together under a single symbol, Φ (31.12). The first step was to choose an appropriate equation

$$F(\Phi) = 0 \tag{47.1}$$

to fix the gauge. We had several choices. One that proved especially useful was *axial gauge*

$$A_3(x) = 0 \tag{47.2}$$

In it we were able prove *directly* that the functional in the Faddeev–Popov prescription was equivalent to canonical quantization. Other gauges, specified by the condition

$$\partial_\mu A^\mu(x) - f(x) = 0 \tag{47.3}$$

[1] [Eds.] Ryder *QFT*, Section 7.2, pp. 250–260; E. S. Abers and B. W. Lee, *Phys. Lett.* **9C** (1973) 1–141, Sections 12 and 13; Cheng & Li *GT*, Chap. 9, pp. 248–278.

[2] [Eds.] See §31.3, pp. 665–668.

(where $f(x)$ is some specified function of x) were also useful, because they gave us simple Feynman rules. By subsequent functional integration over f we were able to get $\partial_\mu A^\mu$ up into an exponential; that led to *Landau gauge* (31.37), *Feynman gauge* (31.36), etc., the so-called *covariant gauges* (31.35).[3] Once we've picked a gauge, the Faddeev–Popov prescription says that the generating functional Z is given by (31.22),

$$Z = N \int (d\Phi) e^{i\mathcal{S}_{GI}[\Phi]} \delta[F(\Phi)] \Delta \tag{47.4}$$

where $\mathcal{S}_{GI}$ is the original gauge invariant action, $\delta[F(\Phi)]$ is a *functional* delta function of $F(\Phi)$, and the gauge invariant determinant Δ is

$$\Delta = \det\left(\frac{\delta F(\Phi'(x))}{\delta\omega(x')}\right)\bigg|_{F=0} \tag{47.5}$$

Because of the delta function, we only needed to evaluate the determinant at the point $F = 0$ (ω is the parameter in the original gauge transformation). Although it does not tell us the right thing to do for a given theory, the Faddeev–Popov *ansatz* has the great advantage that it gives an expression that is manifestly gauge invariant and independent of the choice of the function F, so long as F is a well-posed gauge fixing term. Once we proved the Faddeev–Popov method was valid in one gauge (this task was easiest in axial gauge), we knew it was valid in any other gauge.

It is essentially trivial to generalize this way of doing field theory to the case in which there is a larger gauge group than in QED. Instead of gauging the U(1) group of electrodynamics, we gauge the SU(2) group or SU(3) or whatever it is, with its several gauge parameters. We have many gauge fields, so we need to impose one gauge condition for each of these fields, or, what is the same thing, one for each group generator:

$$F^a(\Phi) = 0 \tag{47.6}$$

A typical gauge condition could be the obvious generalization of (47.2)

$$F^a(\Phi) = A_3^a(x) = 0 \tag{47.7}$$

or of (47.3),

$$F^a(\Phi) = \partial^\mu A_\mu^a(x) - f^a(x) = 0 \tag{47.8}$$

We have to integrate over the surface where all the gauge conditions are held fixed, so we have a delta function for *each* gauge condition. Finally, we have the functional determinant Δ that we need to cancel out the changes in the delta function from gauge transformations; it's the Jacobian factor arising from integrating the delta function. This will be a determinant not only in function space but in *a*-*b* space as well. By trivial extension of the argument given before, this generalized Faddeev–Popov *ansatz* with n gauge conditions will be independent of the choice of gauge; we don't have to prove that all over again:

$$Z = N \int (d\Phi) e^{i\mathcal{S}_{GI}} \prod_a \delta[F^a(\Phi)] \Delta \tag{47.9}$$

[3] [Eds.] Ryder *QFT*, Section 7.1, pp. 245–250.

with

$$\Delta = \det\left(\frac{\delta F^a(\Phi'(x))}{\delta\omega^b(x')}\right)\Big|_{F^a=0} \tag{47.10}$$

The question is, can we prove that quantization according to (47.9) is equivalent to canonical quantization, along the same lines as before? What does quantization look like in the covariant gauge (47.8), which we know gives us nice Feynman rules in electrodynamics? The first step will be to show that Faddeev–Popov quantization *is* equivalent to canonical quantization. I will not go through the whole argument again, but simply point out the places where there might be differences, and show that everything goes through all right, unaltered from the Abelian case.

Here we go, in the axial gauge (47.7). Recall for the unscaled vector field,

$$\delta A^a_\mu = c^{abc}\delta\omega^b A^c_\mu - \partial_\mu(\delta\omega^a) \tag{46.39}$$

which becomes, after $A^a_\mu \to A'^a_\mu = gA^a_\mu$, and dropping the primes on the vector,

$$\delta A^a_\mu = c^{abc}\delta\omega^b A^c_\mu - g^{-1}\partial_\mu(\delta\omega^a) \tag{47.11}$$

Using (47.11), the change in A^a_3 is

$$\delta A^a_3(x) = c^{abc}\delta\omega^b(x)A^c_3(x) - g^{-1}\partial_3(\delta\omega^a(x)) = \delta F^a(\Phi'(x)) \tag{47.12}$$

This is a mess. Fortunately, since we evaluate the determinant *at the zero* of the delta function $\delta[F^a(\Phi)] = \delta[A^a_3]$, the first term can be dropped:

$$\delta A^a_3 = -g^{-1}\partial_3(\delta\omega^a) \tag{47.13}$$

It's just the same as in the Abelian case;[4] the determinant is a constant:

$$\Delta = \det\left[-\delta^{ab}g^{-1}\partial_3\delta^{(4)}(x-x')\right] = \text{constant} \tag{47.14}$$

Δ is independent of the fields over which the functional integration will be performed, and so it's a constant; we don't have to worry about it. Thus Δ is irrelevant and can be absorbed into the normalization N.

The next step in showing the equivalence between the axial gauge generating functional and canonical quantization was to write the theory in so-called *first-order form*, where we treat the F's and the A's as independent variables. The gauge fields are the only part of the Lagrangian that's relevant.

$$\mathscr{L}_{\text{1st}} = \tfrac{1}{4}F^a_{\mu\nu}F^{\mu\nu\,a} - \tfrac{1}{2}F^{\mu\nu\,a}\left(\partial_\mu A^a_\nu - \partial_\nu A^a_\mu + gc^{abc}A^b_\mu A^c_\nu\right) + \cdots \tag{47.15}$$

The other fields in the theory are not going to change much; it's going to be a bunch of normal electrodynamic-type couplings, which I won't worry about. It's only the possible nonlinear terms in (47.15) that may give us trouble. If you treat $F^a_{\mu\nu}$ as an independent variable and vary it, you obtain the defining equation for $F^a_{\mu\nu}$, the monstrous expression in parentheses:

$$F^a_{\mu\nu} = \partial_\mu A^a_\nu - \partial_\nu A^a_\mu + gc^{abc}A^b_\mu A^c_\nu \tag{47.16}$$

[4] [Eds.] See the paragraph before §31.3, p. 665, and the solution to Problem 17.2, p. 682.

This is exactly the field strength tensor that we had computed before, (46.54). Plug that back into the Lagrangian and obtain the *second-order form*,

$$\mathscr{L}_{\text{2nd}} = -\tfrac{1}{4}F^a_{\mu\nu}F^{\mu\nu\,a} + \cdots \tag{47.17}$$

So the first- and second-order forms give the same theory.

In the Abelian case we showed (§31.4) that functional integration in the axial gauge is equivalent to canonical quantization in the first-order form, by dividing the variables into three sets ($i = \{1,2\}$):

$$\begin{aligned} &\text{independent variables: } \begin{Bmatrix} A_i, \text{"coordinates"} \\ F_{0i}, \text{"momenta"} \end{Bmatrix} \\ &\text{constrained variables: } \{A_0, F_{03}, F_{i3}, F_{ij}\} \end{aligned} \tag{47.18}$$

The "coordinates" A_i and "momenta" F_{0i} were the only independent variables; all the other components of A_μ and $F_{\mu\nu}$ were superfluous variables given in terms of the coordinates and momenta at a fixed time by solving the Euler–Lagrange equations. We then calculated with the functional integral. The constrained variables entered at most quadratically and the coefficients of quadratic terms were constants independent of the fields. Thus we just integrated over the constrained variables and eliminated them. We were left with the *Hamiltonian* form ((30.1) and (31.45)) of the functional integral, which we know is equivalent to canonical quantization. Here we divide the variables up in exactly the same way, putting in an extra index a on everything, and then checking that the constrained variables enter the Lagrangian at most quadratically with constant coefficients ($i = \{1,2\}$):

$$\begin{aligned} &\text{independent variables: } \begin{Bmatrix} A^a_i, \text{"coordinates"} \\ F^a_{0i}, \text{"momenta"} \end{Bmatrix} \\ &\text{constrained variables: } \{A^a_0, F^a_{03}, F^a_{i3}, F^a_{ij}\} \end{aligned} \tag{47.19}$$

Looking at (47.15), we only have to worry about the extra trilinear terms. Let's go through the constrained variables one at a time and see what we get.

A^a_0: This term appears in the first-order Lagrangian in the combination $gF^{\mu\nu\,a}c^{abc}A^b_\mu A^c_\nu$, i.e.,

$$gF^{0\nu\,a}c^{abc}A^b_0 A^c_\nu$$

Because $gc^{abc}A^b_\mu A^c_\nu$ is an antisymmetric tensor (from the symmetry properties of the structure constants), if $A^b_\mu = A^b_0$, then A^c_ν will have to be a space index, i. So $F^{\mu\nu\,a}$ will be $F^{0i\,a}$, a momentum. That is, we have

$$gF^{0\nu\,a}c^{abc}A^b_0 A^c_\nu = gF^{0i\,a}c^{abc}A^b_0 A^c_i$$

This term is a "momentum" times a "coordinate" times the dependent variable A^b_0; that's linear in the constrained variables, so it's not a problem.

F^a_{ij}: That can appear as the term

$$gF^{ij\,a}c^{abc}A^b_i A^c_j$$

Both A^b_i and A^c_j are "coordinates", so this term is also linear in the constrained variable F^a_{ij}.

F^a_{i3}: The trilinear term is

$$gF^{i3\,a}c^{abc}A^b_iA^c_3$$

The term vanishes by the gauge condition $A^c_3 = 0$; there's no trilinear term to worry about, so that's trivially all right.

F^a_{03}: The comments about F^a_{i3} apply here as well; the trilinear term proportional to A^c_3 vanishes.

It's the same machine as before, and it runs just the same. I won't go through the proof step-by-step. I just wanted to point out that the extra terms that distinguish the non-Abelian case from the Abelian case have *absolutely no effect* on the argument. Thus the Faddeev–Popov *ansatz* works just as well in the non-Abelian theory. It is equivalent to canonical quantization by exactly the same proof we gave in the Abelian case.

47.2 Feynman rules for a non-Abelian gauge theory

The next task is to find the Feynman rules, which we'll derive from the functional (47.9). The axial gauge is wonderful for proving the theory can be canonically quantized, but it's terrible for deriving the Feynman rules; it's not even covariant. We know a better gauge for the Feynman rules:

$$F^a = \partial^\mu A^a_\mu(x) - f^a(x) \tag{47.20}$$

where f^a is an arbitrary function of x. We'll explore the consequences of the *ansatz* in one of these Lorenz-like gauges. We have to compute the change in this object F^a; f^a is a c-number function that doesn't change under a gauge transformation. Using (47.11),

$$\begin{aligned}\delta\Big(\partial^\mu A^a_\mu(x)\Big) &= \partial^\mu\Big[c^{abc}\delta\omega^b(x)A^c_\mu(x) - g^{-1}\partial_\mu(\delta\omega^a(x))\Big]\\ \frac{\delta(\partial^\mu A^a_\mu(x))}{\delta\omega^b(y)} &= g^{-1}\partial^\mu\left(\Big[\delta^{ab}\partial_\mu - gc^{abc}A^c_\mu\Big]\delta^{(4)}(x-y)\right)\end{aligned} \tag{47.21}$$

In the **adjoint representation** the structure constants themselves form a representation of the group generators:[5]

$$(T^a)^{bc} = -ic^{abc} \tag{47.22}$$

Then (47.21) becomes

$$\frac{\delta(\partial^\mu A^a_\mu(x))}{\delta\omega^b(y)} = g^{-1}\partial^\mu\left(\Big[\delta^{ab}\partial_\mu - ig(T^c)^{ab}A^c_\mu\Big]\delta^{(4)}(x-y)\right) \tag{47.23}$$

This is *exactly* the covariant derivative operator (46.30) acting on a field that transforms according to the adjoint representation of the group, i.e., like the gauge fields themselves or like the field strength tensor. Then

$$\Delta = \text{constant} \times \det\Big[\partial^\mu(D_\mu)\Big] \tag{47.24}$$

---

[5] [Eds.] See note 19, p. 947; Howard Georgi, *Lie Algebras in Particle Physics: From Isospin to Unified Theories*, 2nd ed., Perseus Books, 1999, Section 2.4, pp. 48–50.

We know how to write a determinant as an integral over ghost fields,[6] and here we have to do that, because Δ is not a constant with respect to the fields. Introducing a set of ghost fields η^a, the determinant can be written

$$\Delta = \text{constant} \times \det\left[\partial^\mu (D_\mu)\right] = \text{constant} \times \prod_a \int (d\eta^a)(d\overline{\eta}^a) e^{iS_g} \tag{47.25}$$

where the *ghost action* $\mathcal{S}_g$ is

$$\mathcal{S}_g = \int d^4x (\partial^\mu \overline{\eta}^a)(D_\mu \eta)^a = \int d^4x (\partial^\mu \overline{\eta}^a)[\delta^{ab}\partial_\mu - g c^{abc} A^c_\mu]\eta^b \tag{47.26}$$

(We first exploited this famous trick while studying derivative couplings.[7]) We don't care about the overall constant. This is the only difference between the derivation of the Feynman rules for the Abelian and non-Abelian gauge field theories. In the Abelian case the structure constants c^{abc} are zero (there is only one generator, which trivially commutes with itself) and therefore we had to find the determinant of a constant; we didn't have to introduce any ghost fields. Here the determinant *does* depend on the dynamical variables A^a_μ so we *have* to introduce ghost fields. The rest of the argument proceeds as before.

We exponentiate the argument of the delta function (31.29) by integrating with an appropriate function of f to put a term proportional to

$$(\partial^\mu A^a_\mu)^2$$

into the exponent as in the steps from (31.29) to (31.31). We arrive at the following effective Lagrangian—the thing that has to be put into the functional integral (maybe we should call it the "Feynmanian"): the original gauge invariant Lagrangian, together with the $(\partial^\mu A^a_\mu)^2$ to adjust the transverse parts of the propagator as we please, and the ghost part:

$$\mathscr{L}_{\text{eff}} = \mathscr{L}_{\text{GI}} - \frac{1}{2\xi}(\partial^\mu A^a_\mu)^2 + (\partial^\mu \overline{\eta}^a)(D_\mu \eta)^a \tag{47.27}$$

(The free parameter ξ which determines the gauge could be different for different fields.) The ghosts now have real dynamics. In the tree approximation they are massless charged ghost fields (massless because there is no mass term), interacting trilinearly with the vector boson. They have a very funny looking interaction; it doesn't look gauge invariant. Well *of course* it doesn't look gauge invariant! The whole point of the Faddeev–Popov *ansatz* is to pick a gauge which *destroys* gauge invariance. The gauge-fixing term $(\partial^\mu A^a_\mu)^2$ doesn't look gauge invariant either, and with good reason; it's *not* invariant. The ghosts are just to be treated like normal particles except that, strangely enough, they have a Fermi minus sign for every closed loop. That's their peculiar feature; it's what makes them ghosts. The whole thing's got to work out; you'll never get a negative residue in a Green's function from going around a ghost loop, because every gauge invariant quantity could just as well be computed in axial gauge where there are no ghosts. So the ghosts always have to cancel out against longitudinal photons or what have you in any specific calculation of a gauge invariant quantity. But they're there. You can't get away from them. It was a great discovery that they are necessary.

---

[6] [Eds.] See §29.3 and §31.3.

[7] [Eds.] See §29.3, pp. 625–628.

The history of the ghost fields is interesting. Feynman and Bryce DeWitt, independently and around the same time, started trying to quantize gravity by guess work; a messy problem.[8] They realized that Yang–Mills fields have many of the same features as gravity. (Recall my earlier discussion of gravity as to why nonlinear self-coupling of the fields is necessary.[9]) They saw that Yang–Mills theory is simpler than gravity, and must have said to themselves, "We only have to go up to quartic terms instead of an infinite series, so let's try to quantize Yang–Mills fields." Independently they tried to guess the right Feynman rules and then computed away. Both men discovered that their first guesses didn't work; they found a breakdown of unitarity and gauge invariance and everything else, e.g., at the one-loop level the imaginary part of the forward scattering amplitude was not given by the Optical Theorem. The reason it wasn't is that they had left out a term, which they eventually realized was a box with a ghost going around the box. Then they gave up, because they didn't know how to go beyond the one-loop level. There the matter sat for eight or nine years, until Faddeev and Popov came along and invented this method. They showed that the ghosts not only cured the problems at the one-loop level but at all levels.[10]

The effective Lagrangian (47.27) is horrendous, and we're not going to do any non-tree level computations in this theory—except for one fairly trivial calculation. One can in principle read off the Feynman rules just by standard methods: every derivative becomes a momentum, etc. But they are very cumbersome because everything is carrying so many indices, especially if you look at vertices connecting multiple vector bosons. Every one is carrying a momentum, a Lorentz index and an internal symmetry index. Since things get God-awful looking, I'll write down a few simple things and work out one complicated one. I won't even try to write down the complete set of Feynman rules; no one would remember them.[11]

---

[8] [Eds.] See note 10, p. 625.

[9] [Eds.] See p. 1022.

[10] [Eds.] See note 10, p. 625. Feynman presented ghosts at the one loop level in a talk at the 1962 Warsaw (Jabłonna) conference on gravity (known as "GR3" in the relativity community). Responding to persistent questioning by DeWitt, Feynman went into detail about the one-loop result; the transcribed talk (and the question period, following) were published (see note 8, p. 1037): *The Feynman Lectures on Gravitation*, Richard Feynman, Fernando B. Morinigo and William G. Wagner, ed. Brian Hatfield, Addison-Wesley, 1995; pp. xxviii–xxix. DeWitt extended the idea to two loops in 1964, and via a functional integral, to all orders, in the last weeks of 1965. For a variety of reasons (page charges, a dispute with a reviewer, and other work) DeWitt didn't publish this last result until 1967, about two weeks prior to the first appearance of Faddeev and Popov's results (but not their method): B. S. DeWitt, "Theory of Radiative Corrections for Non-Abelian Gauge Fields", *Phys. Rev. Lett.* **12** (1964) 742–746; B. S. DeWitt, "Quantum Theory of Gravity II. The Manifestly Covariant Theory", *Phys. Rev.* **162** (1967) 1195–1239; L. D. Faddeev and V. N. Popov, "Feynman Diagrams for the Yang–Mills Field", *Phys. Lett.* **B25** (1967) 29–30; C. DeWitt-Morette, *The Pursuit of Quantum Gravity: Memoirs of Bryce DeWitt*, Springer, 2011, pp. 20–22; p. 52; pp. 126–127. At the end of 1966, Faddeev was visiting the IHES near Paris at the same time as Stanley Deser, who introduced him to DeWitt's work. Spurred by this, Faddeev and Popov wrote up their *Physics Letters* article, and shortly thereafter produced a much longer preprint ("Теория возмущений для калибровочно-инвариантных полей", Kiev 1967, ITP-67-36). But it was never published—quantum field theory was *doctrina non grata* in the former Soviet Union: L. D. Faddeev, "Quantizing the Yang–Mills Fields", in *At the Frontier of Particle Physics: Handbook of QCD (Boris Ioffe Festscrift)*, M. Shifman, ed., World Scientific, 2001–2002. An English translation, "Perturbation Theory for Gauge-Invariant Fields", appeared only in 1972 as a Fermilab preprint (NAL-THY-57), nine years after Feynman's *Acta Physica Polonica* article. Though frequently xeroxed and passed hand to hand, this prized translation likewise was never published before it appeared in anthologies decades later: C. H. Lai, ed., *Gauge Theory of Weak and Electromagnetic Interactions*, World Scientific, 1981, pp. 213–233; L. D. Faddeev, *Forty Years in Mathematical Physics*, World Scientific, 1995, pp. 31–51; G. 't Hooft, ed., *Fifty Years of Yang–Mills Theory*, World Scientific, 2005, pp. 40–50. The Fermilab preprint is available online: `http://lss.fnal.gov/archive/1972/pub/Pub-72-057-T.pdf`.

[11] [Eds.] Though not stated in the lectures, the complete set of Feynman rules for a Yang–Mills theory is given

For pure gauge theories there is the gauge boson which carries indices μ and ν and indices a and b. It has a propagator

b,ν  a,μ
$$\widetilde{D}^{ab}_{\mu\nu}(k) = -i\delta^{ab}\left[g_{\mu\nu} - \frac{k_\mu k_\nu}{k^2}(1-\xi)\right]\frac{1}{k^2+i\epsilon} \tag{47.28}$$

That is the conventional propagator: the bosons are independent (δ^{ab}) and they always have zero mass.

We have the ghosts which are charged particles ($\eta \neq \overline{\eta}$), so we indicate them by directed, dotted lines with indices a and b, and their propagator is simply the conventional propagator for a massless scalar field:

b  a
$$\frac{i\delta^{ab}}{p^2+i\epsilon} \tag{47.29}$$

The only non-conventional thing comes in when one considers *ghost loops* and then one has to *add an extra minus sign*; these ghost fields are scalars, but they obey *Fermi* statistics.

Then there are all sorts of interactions coming from $\mathscr{L}_{\text{GI}}$. There is in particular a tri-vector interaction from the $(F^a_{\mu\nu})^2$ term. $F^a_{\mu\nu}$ has (47.16) a term linear in the derivatives of the fields plus a term quadratic in the fields. So from the cross term there will be an interaction like Figure 47.1. We will shortly compute it because that's the messiest one: it's got derivatives and internal indices and space-time indices.

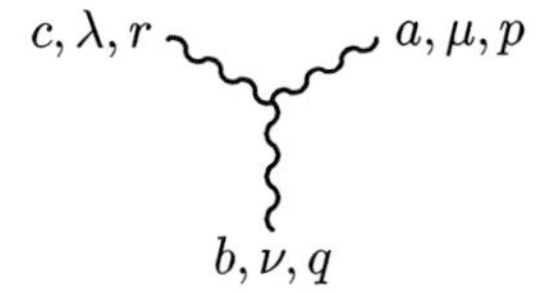

Figure 47.1: Tri-vector vertex

There is a quad-vector interaction, Figure 47.2, which comes from the square of the non-derivative term in $(F^a_{\mu\nu})^2$. That one is actually not quite so horrendous because it doesn't have any factors of momentum to keep straight.

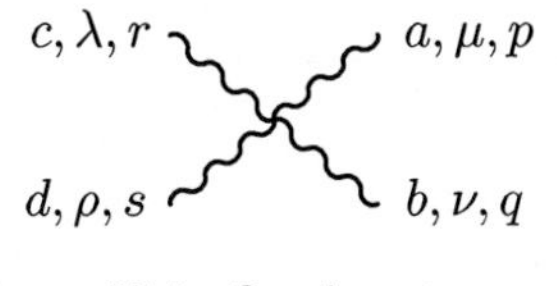

Figure 47.2: Quad-vector vertex

There is an interaction of the gauge fields with the ghosts, coming from the A^a_μ factor hiding in the covariant derivative (47.26) of the ghost field, as shown in Figure 47.3.

All three of these things have Feynman rules that are nasty to work out; I'll do just the first of them, the most dreadful of the lot, shown in Figure 47.1. Then I will make some remarks about its physical meaning. People normally scream when they see this rule in a paper; they say "Ugh, where does *that* come from?" I'll try to convince you that it's really very sensible physically.

---

in the box on p. 1042.

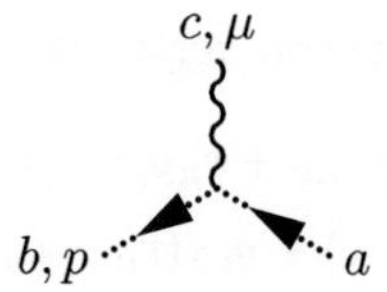

Figure 47.3: Ghost-vector vertex

The term in the Lagrangian that's going to do the dirty work for the trilinear interaction is

$$\mathscr{L} = -\tfrac{1}{2}g\left(\partial_\mu A^a_\nu - \partial_\nu A^a_\mu\right)c^{abc}A^{\mu\,b}A^{\nu\,c} \tag{47.30}$$

The original coefficient was $-\frac{1}{4}$ but the cross term doubles it. We can simplify this a little bit by first observing that $c^{abc}A^{\mu\,b}A^{\nu\,c}$ is automatically an antisymmetric tensor in μ and ν so there's no need to keep both terms in $(\partial_\mu A^a_\nu - \partial_\nu A^a_\mu)$; one of them gives the same thing in the summation as the other so the coefficient is just -1:

$$\mathscr{L} = -g\left(\partial_\alpha A^a_\beta\right)c^{abc}A^{\alpha\,b}A^{\beta\,c} \tag{47.31}$$

I used α and β in $\mathscr{L}$ because I want to save μ and ν for a different purpose.

Let's look at Figure 47.1. On one leg we have a vector boson carrying indices a, μ with momentum p; another has b, ν, momentum q; the third, c, λ, momentum r. The momenta are not independent, of course:

$$p + q + r = 0 \tag{47.32}$$

I'll choose them all directed inward. Figuring out this vertex is difficult because there are so many possibilities: which field absorbs which of the three bosons. We've got to worry about all of them. There are 3! possibilities coming up. Some factors are common: there's a $-ig$ in front; since we're always differentiating an incoming field to get a momentum, we pick up $-i$ from that; it's like an annihilation operator. There'll always be a c^{abc} in some permutation or other, and aside from minus signs one permutation is the same as another because of our cunning symmetry condition, (46.26). So we can just write it as c^{abc} and worry about whether we get plus or minus signs in various combinations.

Now comes the mess. Let's take the case where one boson, (μ, p), is absorbed by the first factor—the one carrying the a; the other bosons are absorbed by the b and c, no permutations, just as stated. The first boson is being differentiated so we get a p because the first boson carries momentum p. But p with what index? The *same* index as the *derivative* which is the index of the *second* boson, so it's p_ν. The other two indices are being summed together so I get $g_{\mu\lambda}$. There are two other terms that are trivially obtained from these by cyclic permutations; those I can just cycle around clockwise. I get q with the next index over, λ. Then I have two remaining indices to sum, $g_{\mu\nu}$. Then r with the next index over, μ, times $g_{\nu\lambda}$. Then there are the terms where I go anti-cyclic. Instead of summing each of these with the index attached to each of the momenta in the clockwise direction, I attach the index in the anti-clockwise direction, and thus c^{abc} changes sign. So we have $p_\lambda g_{\mu\nu}$, minus q with the next index up, $q_\mu g_{\nu\lambda}$, minus $r_\nu g_{\mu\lambda}$. What a mess! If you keep your wits about you you can derive it.[12] The

[12] [Eds.] A more formal but straightforward way to obtain (47.33) is to consider the effective action $\Gamma = \int d^4x\,\mathscr{L}$ for the interaction (47.31) and take the three functional derivatives $(\delta/\delta A^a_\mu(x))(\delta/\delta A^b_\nu(y))(\delta/\delta A^c_\lambda(z))\Gamma$. This yields a series of terms that reproduce (47.33).

final expression for the trilinear vector boson vertex is:

$$\begin{aligned}(-ig)(-i)c^{abc}&\left[p_\nu g_{\mu\lambda}+q_\lambda g_{\mu\nu}+r_\mu g_{\nu\lambda}-p_\lambda g_{\mu\nu}-q_\mu g_{\nu\lambda}-r_\nu g_{\mu\lambda}\right]\\ &=-gc^{abc}\left[g_{\mu\nu}(q_\lambda-p_\lambda)+g_{\nu\lambda}(r_\mu-q_\mu)+g_{\lambda\mu}(p_\nu-r_\nu)\right]\end{aligned}\tag{47.33}$$

God have mercy on anyone who tries to do a two-loop computation with these things appearing at every vertex and having to be summed over. Tini Veltman wrote a computer program, SCHOONSCHIP , to do these complicated computations in Yang–Mills theory.[13] But there are some simple calculations. I'll show you the explicit one-loop computation of the effective potential, which is fairly easy. There, by being cunning in our choice of gauge, we can make the indices take care of themselves. Similar calisthenics, although slightly less strenuous, will give you the other two kinds of fundamental vertices, but I won't bother to derive them.

Though the tri-vector boson vertex looks complicated, in fact it has a very simple physical meaning. Suppose we consider the theory of an ordinary photon coupled to charged scalar bosons. Remember from scalar electrodynamics what that vertex looks like.[14] I'll redraw it to ease comparison, with all the momenta going inward; see Figure 47.4. The photon carries an index μ and the scalars have no indices.

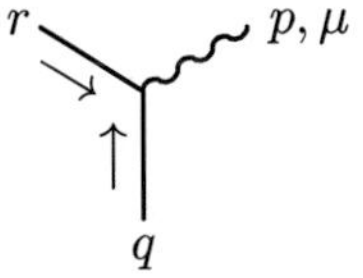

Figure 47.4: Scalar-scalar-vector vertex

As we found earlier, this diagram makes a contribution proportional to the sum of the incoming and outgoing momenta (due to the derivative coupling of the scalar and the photon). Here one is incoming and one is outgoing, so

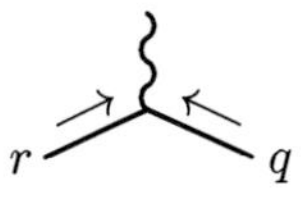

$$\propto (q_\mu - r_\mu)\tag{47.34}$$

We've got a term in (47.33) that is just like that; the coefficient of $g_{\nu\lambda}$ is $(r_\mu - q_\mu)$. So these two terms can be read as saying that the particle (a, μ, p) acts like a photon and the other two act like spinless charged particles. It has a coefficient which is given by the group structure constants, and it doesn't matter what polarization state they're in, what λ or ν is, so this is just a $g_{\lambda\nu}$. Each of the two massless gauge bosons has two polarization states. Then these four polarization states act just like four independent charged scalars. We can look at the vertex in different ways. We can say (a, μ, p) is like a photon and the others are like a charged particle, or we could say (c, λ, r) is like a photon and the others like charged particles, or we

[13] [Eds.] "Schoonschip" is Dutch for "clean ship", or loosely, "shipshape", everything neat and tidy. M. Veltman, "An IBM-7090 Program for Symbolic Evaluation of Algebraic Expressions, Especially Feynman Diagrams", CERN PRINT-65-879 (1965); M. Veltman,"SCHOONSCHIP", CERN preprint, July 1967; M. Veltman and D. Williams, "Schoonschip '91" (University of Michigan preprint UM-TH-91-18, June 9, 1991); available at `https://arxiv.org/abs/hep-ph/9306228`. Written in assembly language, Veltman's program was designed to automate the calculation of roughly 50,000 terms in radiative corrections to a process of photons interacting with a charged vector boson. It is arguably the first computer program written to perform symbolic algebra.

[14] [Eds.] See rule (h) and its diagram in the box on p. 670.

could say the third one is like a photon and the other two are like charged particles. In the case of SU(2), the 1 and 2 bosons act as charged particles when seen by the 3, which acts like the photon; they're the source of the 3. In the same way, the 1 and the 3 are the source of the 2, etc. That's the wonder of Yang–Mills theory; amusing but confusing. So the contribution of the diagram in Figure 47.1 can be thought as the sum of charged scalar-photon interactions over three permutations, depending on which one we think of as the photon.

In the same way, the four-gauge boson coupling, which also is a mess of permutations, can be thought of as just summing over permutations of the seagull diagram, Figure 47.5,

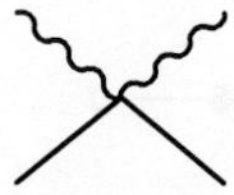

Figure 47.5: Scalar-vector seagull vertex

which is also present in scalar electrodynamics,[15] changing which pair you think of as the photons and which pair you think of as the charged particles. Although the vertex in Figure 47.2 looks more complicated than charged scalar electrodynamics, it's really not, though certainly the computations are more complicated. It's only that you have many choices as to which boson you think of as the photon. Therefore you have sums over many permutations. This point of view is something you won't find in the literature, but I think it helps you understand, at least in a semi-physical way, why these complicated structures necessarily arise.

The full Lagrangian, including coupling to fermions and scalars, is

$$\mathscr{L} = -\tfrac{1}{4}F^a_{\mu\nu}F^{\mu\nu\,a} - \frac{1}{2\xi}(\partial^\mu A^a_\mu)^2 + (\partial^\mu\overline{\eta}^a)(D_\mu\eta^a) + \overline{\psi}(i\not{D} - m)\psi + (D_\mu\phi)^\dagger D^\mu\phi - \mu^2\phi^\dagger\phi \quad (47.35)$$

where (46.59) $D_\mu = \partial_\mu - igT^aA^a_\mu$. The Feynman rules for this Lagrangian are given in the box on p. 1042.

47.3 Renormalization of pure gauge field theories

Let's consider a pure gauge vector theory, without scalar or spinor fields. All of our vector propagators, no matter what the gauge, have a denominator k^2. At every vertex, we have either four undifferentiated vector fields, or three fields, one of which is differentiated. So we immediately have the first half of a proof of renormalization. The only counterterms we will need will be monomials of the same form as the monomials appearing in the original Lagrangian, with no more fields and no more derivatives. We do not have the proof of the second part of renormalization, showing that these monomials come through with *exactly the right coefficients* to correspond to a redefinition of the parameters in the original Lagrangian. In electrodynamics, we obtain the connection between the various counterterms by lengthy arguments, systematically exploiting gauge invariance and its consequence, the *Ward identities*.[16] That argument depends on the Lagrangian's gauge-fixing terms being no more than *quadratic* in the fields, and does not hold in the non-Abelian case: the ξ term is quadratic, but the ghost term is *trilinear* in the fields. Therefore another proof is required.

---

[15] [Eds.] See Figure 27.1, p. 585, and rule (i), p. 670.

[16] [Eds.] §32.4, specifically (32.50); §33.4.

Feynman rules for a non-Abelian gauge theory

1. For every ... Write ... (all *incoming* momenta positive)

(a) internal vector line

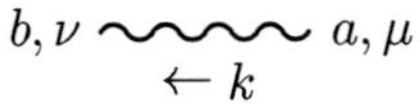

$$\tilde{D}^{ab}_{F\mu\nu}(k) = \frac{-i\delta^{ab}}{k^2+i\epsilon}\left[g_{\mu\nu} - (1-\xi)\frac{k_\mu k_\nu}{k^2}\right]$$

(b) internal ghost line

$$\tilde{\Delta}^{ab}_{G}(p) = \frac{i\delta^{ab}}{p^2+i\epsilon}$$

(c) internal fermion line

$$\tilde{S}^{ab}_{F}(p) = \frac{i(\not{p}+m)\delta^{ab}}{p^2-m^2+i\epsilon}$$

(d) internal scalar line

$$\tilde{\Delta}^{ab}_{F}(k) = \frac{i\delta^{ab}}{k^2-\mu^2+i\epsilon}$$

(e) 3-vector vertex

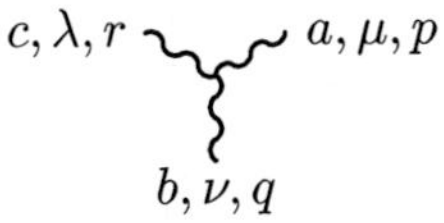

$$-gc^{abc}\left[g_{\mu\nu}(q_\lambda - p_\lambda) + g_{\nu\lambda}(r_\mu - q_\mu) + g_{\lambda\mu}(p_\nu - r_\nu)\right]$$

(f) 4-vector vertex

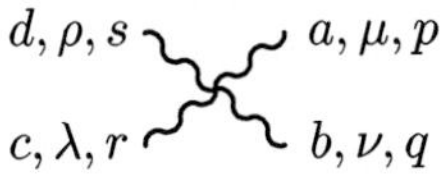

$$-ig^2\left[c^{abe}c^{cde}(g_{\mu\lambda}g_{\nu\rho} - g_{\mu\rho}g_{\nu\lambda}) + c^{ace}c^{bde}(g_{\mu\nu}g_{\lambda\rho} - g_{\mu\rho}g_{\nu\lambda}) + c^{ade}c^{cbe}(g_{\mu\lambda}g_{\nu\rho} - g_{\mu\nu}g_{\lambda\rho})\right]$$

(g) fermion-vector vertex

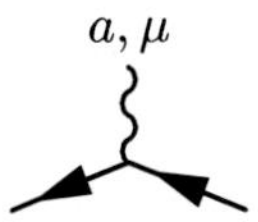

$ig\gamma^\mu T^a$

(h) ghost-vector vertex

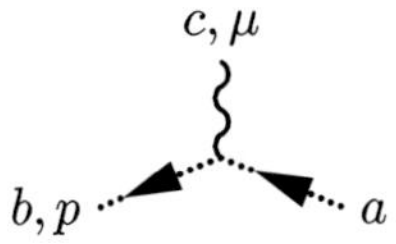

$gc^{abc}p^\mu$

(i) scalar-vector vertex

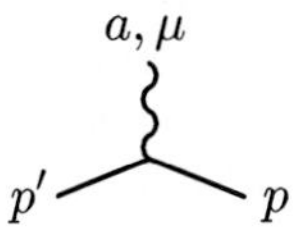

$igT^a(p^\mu + p'^\mu)$

(j) seagull vertex

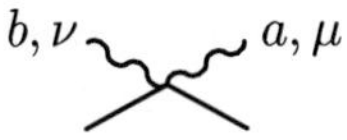

$ig^2 g_{\mu\nu}\{T^a, T^b\}$

2. Ensure momentum conservation at each vertex: $(2\pi)^4\,\delta^{(4)}(\sum p_{\text{out}} - \sum p_{\text{in}})$

3. For every fermion loop, and every ghost loop, a factor of -1.

4. Multiply by $\displaystyle\int \frac{d^4q}{(2\pi)^4}$ and integrate over all internal momenta q.

5. For every $\left\{\begin{matrix}\text{incoming}\\ \text{outgoing}\end{matrix}\right\}$ vector boson, a factor $\left\{\begin{matrix}\varepsilon^a_\mu\\ \varepsilon^{a*}_\mu\end{matrix}\right\}$

The original proofs were unbelievably horrible.[17] They were gradually improved and simplified,[18] until finally Becchi, Rouet, and Stora at Marseille found a proof that was only *believably* horrible. I refer you to those papers if you want to see the proof.[19] Take my word for it, it is possible to prove that Yang–Mills theory is strictly renormalizable.

Spontaneous symmetry breaking is irrelevant for renormalization; we've already established that. We can compute the generating functional without asking whether or not the symmetry breaks spontaneously. If it does, we go and look at the generating functional to see if we have to shift the fields. If you've proved a theory is renormalizable, that proof holds regardless of whether or not there is symmetry breaking. That conclusion follows from the same argument we gave for our scalar field theories. You may have to be concerned if you're proving things on a fine technical level. If you do your subtractions à la BPHZ, have you proved Hepp's theorem with subtraction at a Euclidean point, for example. If you don't have symmetry breaking you have massless vector bosons. These may give you bad infrared divergences that make things blow up at the BPHZ point. But on the level at which we're working it doesn't matter. In principle, in the generating functional formalism, the question of renormalizability and the occurrence of spontaneous symmetry breaking are completely separated. First we renormalize; then we look through the renormalized theory to see if the symmetry breaks.

We should also add that in certain cases, especially theories involving boson coupling to γ_5-type currents (axial vector currents), the naïve proof of the Ward identities can break down due to the occurrence of **anomalies**,[20] unless great care is taken with the cutoff. Occasionally

[17] [Eds.] G. 't Hooft, "Renormalization of Massless Yang–Mills Fields", *Nuc. Phys.* **B33** (1971) 173–199; "Renormalizable Lagrangians for Massive Yang–Mills Fields", *Nuc. Phys.* **B35** (1971) 167–188; G. 't Hooft and M. Veltman, "Regularization and Renormalization of Gauge Fields", *Nuc. Phys.* **B44** (1972) 189–213; "Combinatorics of Gauge Fields", *Nuc. Phys.* **B50** (1972) 318–353. Gerard 't Hooft and Martinus Veltman shared the 1999 Physics Nobel Prize for the proof of the renormalizability of massless and massive (via the Higgs mechanism) Yang–Mills fields.

[18] [Eds.] A. A. Slavnov, "Ward Identities in Gauge Theories", *Theo. Math. Phys.* **10** (1972) 99–104; B. W. Lee, "Renormalizable Massive Vector-Meson Theory–Perturbation Theory of the Higgs Phenomenon", *Phys. Rev.* **D5** (1972) 823–835; J. C. Taylor, "Ward Identities and Charge Renormalization of the Yang–Mills Field", *Nuc. Phys.* **B33** (1971) 436–444; J. C. Taylor, *Gauge Theories of Weak Interactions*, Cambridge U. P., 1976, 1978, Chapters 12, 13, and 14, pp. 94–127.

[19] [Eds.] "The generalized Ward–Takahashi identities for non-Abelian gauge theories were first formulated in a rather complicated way ... Fortunately the formulation of these identities has been simplified by a device due to Becchi, Rouet, and Stora." Taylor, *op. cit.*, p. 94: C. Becchi, A. Rouet and R. Stora, "The Abelian Higgs–Kibble Model. Unitarity of the S Operator", *Phys. Lett.* **B52** (1974) 344–346; "Renormalization of the Abelian Higgs–Kibble Model", *Commun. Math. Phys.* **42** (1975) 127–162; "Renormalization of Gauge Theories", *Ann. Phys.* **98** (1976) 287–321; also in *Renormalization Theory*, G. Velo and A. S. Wightman eds., Reidel, 1976; I. V. Tyutin, Lebedev Institute preprint FIAN n. 39 (1975) (unpublished); M. Z. Iofa and I. V. Tyutin, "Gauge Invariance of Spontaneously Broken Non-Abelian Theories in the Bogolyubov–Parasyuk–Hepp–Zimmerman Method", *Theo. Math. Phys.* **27** (1976) 316–322; Ryder *QFT*, Sections 7.5 and 7.6, pp. 277–282; Cheng & Li *GT*, Section 9.7, pp. 267–278; Peskin & Schroeder *QFT*, Section 16.4, pp. 517–521. The generalized Ward–Takahashi identities are often called "Slavnov–Taylor" identities, and the "device" referred to by Taylor is usually described as the "BRST transformation", after Becchi, Rouet, Stora, and Tyutin, who shared the 2009 Dannie Heineman Prize for its discovery; Itzykson & Zuber *QFT*, Section 12-4, pp. 594–606. See also Gerard 't Hooft, "Reflections on the renormalization procedure for gauge theories", *Nuc. Phys.* **B912** (2016) 4–14, a memorial issue to Raymond Stora (1930–2015).

[20] [Eds.] J. S. Bell and R. Jackiw, "A PCAC Puzzle: $\pi^0 \to \gamma\gamma$ in the σ Model", *Nuovo Cim.* **A60** (1969) 47–61; Steven L. Adler, "Axial-Vector Vertex in Spinor Electrodynamics", *Phys. Rev.* **177** (1966) 2426–2438; Barry R. Holstein, "Anomalies for Pedestrians", *Am. J. Phys.* **61** (1993) 142–147; Cheng & Li *GT*, Section 6.2, pp. 173–182. See also note 5, p. 82.

you have to worry about anomalies. But they have been cataloged; it's an exercise in *nosology*, the categorization of diseases. You only have to check that they cancel among the various axial vector currents. There is another long song and dance to deal with them.[21]

Until the work of Faddeev and Popov, the renormalizability of Yang–Mills fields was only conjectured; the Feynman rules for non-Abelian gauge theories were not known and so renormalizability could not be proven. How could you possibly tell if the divergences from all the graphs canceled before you knew what the graphs were? It gets pretty complicated. You've got to look at it the right way to make it look simple. Otherwise you will calculate to a certain level and then you'll vomit up a bunch of indices and decide to look at another problem. It's just a matter of organizing the details. Faddeev and Popov found this neat way using the functional integral to organize the theory. With their prescription, you can see everything happening at once. You don't have to drive yourself crazy computing things like (47.33).

Veltman was probably the only person concentrating on Yang–Mills theory throughout the middle 1960's. And he, being a man of great taste, must have said to himself: "If I'm going to tackle this problem I'm going to have to learn how to manipulate all those indices. It will be hard to avoid mistakes; I'm not a machine." So he wrote that computer program, SCHOONSCHIP, to do the work. But his premise was faulty. He thought, "Well, obviously the massless theory is the limit of the massive theory. I know how to quantize the massive theory. I'll just go ahead with that, so I won't have to worry about gauge invariance; there isn't any." He put in a mass term, intending to compute everything and at the end to let the mass go to zero. Unfortunately, unlike electrodynamics, the massless theory of Yang–Mills fields is *not* the smooth limit of the massive theory. That's now a well-known result,[22] but it was not known at the time that Veltman was looking at the massive Yang–Mills theory. For example, if you have an SU(2) theory you have three photons coming together at a vertex. If you work things out you can always get rid of one of the three helicity zero photons, but you can't get rid of all three of them simultaneously. There was just no way of making it work. So he was, unknown to himself and unknown to everyone else, pursuing a dead end. The way forward with the massless theory was blocked until Faddeev and Popov's paper appeared.

The Faddeev–Popov paper was not widely understood at first. It was obscurely written and it took about a year or two for its results to sink in. In fact it took 't Hooft's discovery of the theory's renormalizability for the significance of the paper to be appreciated.[23] Faddeev and Popov had obtained the Feynman rules in a very strange way. But Yang–Mills theory finally came together. Essentially independently, 't Hooft found the right Feynman rules. Even more, he discovered the crucial point: spontaneous symmetry breaking enables the construction of theories involving massive vector bosons which might provide, in a way I will describe later, a *renormalizable* weak interaction theory. He didn't know that Weinberg had proposed a similar model four years earlier. Though he had conjectured that his model would be renormalizable, Weinberg had been unable to see why it would be so. In his first papers 't Hooft presented renormalization as a formal argument, manipulating infinite quantities and imposing gauge invariance when necessary. I remember that time very well. I met Tini Veltman in Marseille[24]

---

[21] [Eds.] Weinberg *QTF2*, Chapter 22, pp. 359–420.

[22] [Eds.] A. A. Slavnov and L. D. Faddeev, "Massless and Massive Yang–Mills Fields", *Theo. Math. Phys.* **3** (1971) 312–316.

[23] [Eds.] See note 17, p. 1042.

[24] [Eds.] Presumably the two met at the Colloquium on Renormalization Theory, CNRS, Marseille, June

and he said, "A graduate student of mine has a renormalizable theory of massive charged vector bosons," and I said, "I don't believe it." He said, "It's true," and we were about to make a bet. It's a good thing we didn't because I would have lost a lot of money. That's why I'm one of the few people in the world who doesn't mispronounce 't Hooft's name, because Veltman told me his name, Gerard 't Hooft. When I got the preprint and saw how he spelled his name, my reaction was, "What a funny way to spell 'et Hoaft' instead of the reaction of everyone else, which is "What a funny way to pronounce 'tooft'".

47.4 The effective potential for a gauge theory

Let's calculate something in this theory of non-Abelian gauge fields. We'll choose a quantity that we've been computing in stages, the effective potential.[25] We're generalizing the effective potential to add gauge fields to the scalar and fermion fields we've considered previously. It's a very simple object because it only involves external scalar lines, and they all carry zero momentum. Even though we're summing up a huge number of graphs, they're very simple graphs.

The effective potential $V(\overline{\phi})$ which depends upon the classical scalar fields is going to be the sum of several terms. There's the zeroth order contribution $U(\overline{\phi})$; the contribution V_S of the scalars themselves,

$$V_S(\overline{\phi}) = \frac{1}{64\pi^2}\text{Tr}\left\{\left[U''(\overline{\phi})\right]^2 \ln\left[\frac{U''(\overline{\phi}) - i\epsilon}{M^2}\right]\right\} \tag{44.53}$$

(adding the trace over the internal group indices on the scalar fields), the contribution V_F from the fermions,

$$V_F(\overline{\phi}) = -\frac{1}{64\pi^2}\text{Tr}\left\{\left[m^\dagger(\overline{\phi})m(\overline{\phi})\right]^2 \ln\left[\frac{m^\dagger(\overline{\phi})m(\overline{\phi}) - i\epsilon}{M^2}\right]\right\} \tag{45.28}$$

where $m(\overline{\phi})$ is the fermion mass matrix, (45.22); the contribution V_G from the gauge fields which we are about to compute; and finally the contribution from the counterterms. These are finite terms of the same form as other quantities that occur in the classical potential U, and are determined by the other quantities once we fix our renormalization conditions. All together we have for the one-loop contribution

$$V(\overline{\phi}) = U(\overline{\phi}) + V_S(\overline{\phi}) + V_F(\overline{\phi}) + V_G(\overline{\phi}) + V_{CT}(\overline{\phi}) \tag{47.36}$$

Now for the contribution from the gauge fields. At first glance we get a horrible mess. We can have graphs like Figure 47.6, where there are scalar fields on the outside and gauge fields running around inside. Those we can handle by our usual techniques.

But we also have the trilinear scalar-scalar-vector interactions, and thus the possibility of a graph in which something like Figure 47.7 happens. However, by an astute choice of gauge we can make these disappear.

1971. At Veltman's invitation, 't Hooft gave a brief report of his results at the Amsterdam International Conference on Elementary Particles, 30 June–6 July 1971. See Frank Close, *The Infinity Puzzle*, Basic Books, 2011, Chapter 11, "And Now I Introduce Mr. 't Hooft".

[25] [Eds.] The effective potential is defined by (44.38), calculated for the scalar field in §44.3 and for the fermion field in §45.2; "Secret Symmetry" in Coleman *Aspects*, Section 3.5, pp. 136–138; Appendix, pp. 180–182.

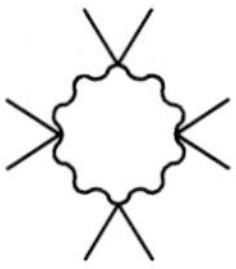

Figure 47.6: Loop with A^2-ϕ^2 couplings

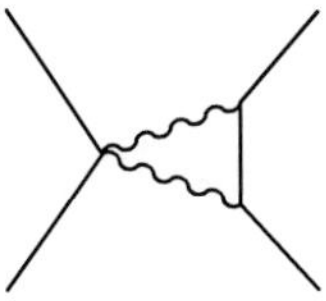

Figure 47.7: Loop with A-ϕ-$\partial\phi$ couplings

I should have emphasized this important point earlier: if a theory includes a gauge field, the effective potential is *not* a gauge invariant object because the ϕ fields are not gauge invariant. Then again, neither is the propagator. *Nothing* is gauge invariant until you put it all together and assemble physically observable quantities; *those* are gauge invariant. Any gauge should be as good as any other, so long as you don't change gauges in the middle of a computation.

We will use Landau gauge, in which the propagator is[26]

$$\widetilde{D}_{\mu\nu}(k) = -\frac{i}{k^2 + i\epsilon}\left[g_{\mu\nu} - \frac{k_\mu k_\nu}{k^2}\right] \tag{47.37}$$

Why is that a good gauge? Let's focus on what's happening at the ϕ-$\partial\phi$-A vertex.

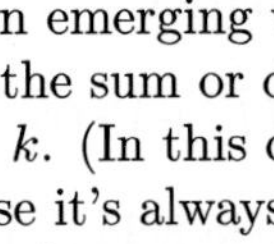

Figure 47.8: ϕ-$\partial\phi$-A vertex

Figure 47.8 shows a scalar boson coming out with momentum zero, a gauge boson coming in with momentum k and a scalar boson emerging with the same momentum k. Therefore, depending upon how you orient things, the sum or difference of the momenta carried by the internal scalar boson (in Figure 47.7) is k. (In this case the orientation doesn't matter.) At the vertex we have a factor $k_\mu A^\mu$ because it's always the sum of the momenta that occurs in item (i) in the list of vertices. In the Landau gauge k hits the propagator and kills it:

$$\left[g_{\mu\nu} - \frac{k_\mu k_\nu}{k^2}\right] k^\mu = 0 \tag{47.38}$$

So as long as a boson line has external momentum equal to zero, the vertex vanishes! We don't have to worry about the ϕ-$\partial\phi$-A; we just have to worry about graphs of the kind in Figure 47.6.

What are those graphs? They're graphs like Figure 47.9:

---

[26] [Eds.] See note 12, p. 667.

Figure 47.9: Vector boson loop

using a little black dot to indicate that four-boson interaction, just the same way as before (see (44.44) and (45.24)). That black dot comes from the term in the Lagrangian

$$\mathscr{L} = \cdots + \tfrac{1}{2}\left(i\,g_a T^a\overline{\phi}\right)\bullet\left(i\,g_b T^b\overline{\phi}\right)A^a_\mu A^{\mu\,b} \tag{47.39}$$

the $D_\mu\overline{\phi}\bullet D^\mu\overline{\phi}$ term expanded out to second order in A. This is just the mass term of the vector boson in the presence of an external ϕ field, and therefore the form of this vertex, the form of the black dot, is simply

$$b,\nu \sim\!\sim\!\bullet\!\sim\!\sim a,\mu \;=\; ig_{\mu\nu}\mu^2_{ab}(\overline{\phi}) \tag{47.40}$$

The *vector boson mass matrix* $\mu^2_{ab}(\overline{\phi})$ is defined in (47.39); it's the mass of the vector boson in a given $\overline{\phi}$ field. That acts just like the fermion case (45.22) where we had $m(\overline{\phi})$ appearing at each vertex.

We still have to deal with all those propagators running around the loops. The propagators are δ functions in ab space but they do have $\mu\nu$ indices:

$$b,\nu \sim\!\sim\!\sim\!\sim a,\mu \;=\; -\frac{i\delta^{ab}}{k^2+i\epsilon}\left[g_{\mu\nu}-\frac{k_\mu k_\nu}{k^2}\right] \tag{47.41}$$

Fortunately the $-i$ from the propagator cancels the i from the vertex, and you always have one propagator for each vertex. Also fortunately, the factor in the square brackets of the propagator (47.41) is a projection operator in $\mu\nu$ space, and thus idempotent:

$$\left[\delta_\mu{}^\lambda - \frac{k_\mu k^\lambda}{k^2}\right]\left[g_{\lambda\nu}-\frac{k_\lambda k_\nu}{k^2}\right] = \left[g_{\mu\nu}-\frac{k_\mu k_\nu}{k^2}\right] \tag{47.42}$$

Whether you multiply three lines or 17 lines or 121 lines, you just get the same thing. At the end you have to take the trace:

$$\mathrm{Tr}\left[g_{\mu\nu}-\frac{k_\mu k_\nu}{k^2}\right] = \delta^\nu{}_\nu - \frac{k^\nu k_\nu}{k^2} = 3 \tag{47.43}$$

Otherwise the computation is exactly the same as the scalar or the spinor or any other computation we've done. We've got a string of matrices running around, we have a

$$\frac{1}{k^2+i\epsilon}$$

for every internal line. All the index structure collapses and it becomes the scalar computation all over again, because of the choice of gauge. The only difference in the vector contribution to the effective potential from the scalar contribution is a factor of 3 coming from the trace of the Landau gauge propagator:

$$V_G(\overline{\phi}) = \frac{3}{64\pi^2}\mathrm{Tr}\left\{\left[\mu^2(\overline{\phi})\right]^2\ln\left[\frac{\mu^2(\overline{\phi})-i\epsilon}{M^2}\right]\right\} \tag{47.44}$$

Note that this has a definite physical meaning, just as the other contributions had. Remember that V_S was the zero point energy of a sum of independent harmonic oscillators. So is V_G. Why does it have a factor of 3? A massive vector boson has three degrees of freedom! A massive scalar has one. So we've got three times as much zero point energy. There are three virtual oscillators for every momentum state of a massive vector boson. Thus the factor of 3 is easy to see on physical grounds.

These formulas will be important to us later. At the moment we've just been accumulating them for when we finally discuss things like Weinberg's famous lower bound on the mass of the Higgs boson.[27] But they're very simple, aside from the $64\pi^2$ which one has to memorize; you figure out what they are just by counting up zero point energies. The minus sign for fermions, (45.28), is because the zero point energy goes the other way. Instead of subtracting the zero point energies of the individual oscillators you're adding the energies of the negative energy states, filling up the holes in the Dirac sea.

Aside from our discussion of the sigma model as a model of current algebra, all the stuff about gauge fields and Higgs phenomena and so on, admittedly beautiful (and also elegantly and wittily presented), are nevertheless just theoretician's toys, with no apparent connection to the real world. Next time I'll show you how all these ideas were put to work, and turn to the theory that makes all this important: the famous *Glashow–Salam–Weinberg model* of weak and electromagnetic interactions.

[27] [Eds.] See §49.3.

Problems 25

25.1 A real vector field of mass μ is coupled to a real scalar field of mass m in an unconventional way:

$$\mathscr{L} = -\tfrac{1}{4}(F_{\mu\nu})^2 + \tfrac{1}{2}\mu^2 A_\mu A^\mu + \tfrac{1}{2}(\partial_\mu\phi)^2 - \tfrac{1}{2}m^2\phi^2 + gA^\mu A_\mu\phi \tag{P25.1}$$

where g is a real number. The vector field is not coupled to a conserved current, and thus we might expect the theory to suffer from various ailments.

We will choose to study vector-scalar elastic scattering,

$$A + \phi \rightarrow A + \phi \tag{P25.2}$$

For this process there are nine independent amplitudes at fixed energy and angle, because both the incoming and outgoing vectors may have helicity (spin along the direction of motion) equal to any of $\{1, 0, -1\}$. An interesting limit in which to consider these amplitudes is that of high center-of-momentum energy, with CM scattering angle fixed, but equal to neither 0 nor π. (This restriction guarantees that all Mandelstam invariants—s, t, and u—grow with energy.)

(a) To lowest nontrivial order in perturbation theory, $\mathcal{O}(g^2)$, some of the nine helicity amplitudes approach (possibly angle-dependent, possibly vanishing) constants in the high-energy, fixed-angle limit described above. We will call those amplitudes "nice". Others, however, grow as a power of the energy; these we will call "nasty". Which are the nasty amplitudes? Find the explicit high-energy forms of the nasty amplitudes, retaining only terms that grow as positive powers of the energy. (Since I haven't defined the phase of helicity eigenstates, don't worry about getting the phase (let alone the sign) of the answer right.)

(b) Now let us add another term to the Lagrangian:

$$\mathscr{L} \rightarrow \mathscr{L} + h\phi^2 A^\mu A_\mu \tag{P25.3}$$

If we add the contribution of this term (in tree approximation) to our previous computation, then, for an appropriate choice of h, some of the nasty amplitudes become nice. Which ones? What is the appropriate choice of h? (*Cf.* Problem 22.2 and its solution.)

(1987 253b Final, Problem 3)

25.2 In class discussions of gauge field theories (§46.2), I described how the matter fields transformed under a finite gauge transformation,

$$\phi \rightarrow g\phi \tag{P25.4}$$

and also under an infinitesimal one, $g = 1 + \delta\omega$,

$$\delta\phi = \delta\omega\phi \tag{P25.5}$$

where $\delta\omega \equiv -i\omega^a T^a$ (46.27), $g \in G$, and T^a are the generators of some representation of the Lie group G. For the fields $F_{\mu\nu}$, I only described infinitesimal transformations,

$$\delta F_{\mu\nu} = [\delta\omega, F_{\mu\nu}] \tag{P25.6}$$

(the matrix form of (46.45)). It's easy to see however that this infinitesimal transformation implies that under finite transformations,

$$F_{\mu\nu} \to gF_{\mu\nu}g^{-1} \tag{P25.7}$$

(the matrix form of (46.49)). The argument runs as follows: (1) Every finite transformation can be built up as a product of infinitesimal ones. (2) The stated transformation law under finite transformations has the *group property*: the result of first applying the transformation g_1 and then applying the transformation g_2 is the same as that of applying the transformation $g_2 g_1$. (3) The stated finite transformation agrees with the known infinitesimal transformation for $g = 1 + \delta\omega$. (If you're disturbed by taking the infinite product of infinitesimal transformations to get a finite transformation, you can rephrase the whole argument in terms of integrating differential equations, but really, it's not worth the bother.)

(a) Use similar reasoning to show that the matrix form of (46.39)

$$\delta A_\mu = [\delta\omega, A_\mu] - \partial_\mu \delta\omega \tag{P25.8}$$

(where $A_\mu \equiv iA^a_\mu T^a$) implies

$$A_\mu \to A^{(g)}_\mu \equiv gA_\mu g^{-1} + g\partial_\mu g^{-1} \tag{P25.9}$$

(b) Let $x(s)$ be some path in spacetime, where the path parameter s runs from 0 to ∞. Suppose you have a unitary matrix, $U(s)$, which solves the differential equation

$$\frac{dU}{ds} = -A_\mu(x(s))\frac{dx^\mu}{ds}U(s) \tag{P25.10}$$

with the boundary condition $U(0) = 1$. (Note the similarity to interaction-picture perturbation theory.) Show that the solution to the differential equation

$$\frac{dU^{(g)}}{ds} = -A^{(g)}_\mu(x(s))\frac{dx^\mu}{ds}U^{(g)}(s) \tag{P25.11}$$

with the boundary condition $U^{(g)}(0) = 1$, is

$$U^{(g)}(s) = g(x(s))U(s)g(x(0))^{-1} \tag{P25.12}$$

(1987b 17)

25.3 In the lectures on quantum electrodynamics, we studied processes where some initial state i went into some final state f, plus a photon of momentum k' and polarization vector ε'. (Both i and f could be multiparticle states.) The invariant amplitude $\mathcal{A}$ for this process was (see (26.73) and (35.28))

$$\mathcal{A} = \varepsilon'^*_\mu M^\mu$$

for some M^μ, the matrix element of a conserved current. Thus, even when the photon was off the mass shell ($k'^2 \neq 0$), $k'_\mu M^\mu = 0$. Furthermore, we showed that this remained true even if the initial state contained an off-mass-shell photon, so long as all the other particles in the initial and final states were on the mass shell. (You may remember that this was important in our derivation of the low energy theorem for photon-nucleon scattering in §35.3.) For the purposes of this problem, "on the mass shell" means, for a Dirac particle, not only $p^2 = m^2$, but also $(\not{p} - m)u = 0$; for a gauge boson, not only $k^2 = 0$, but also $\varepsilon \cdot k = 0$.

As you have seen, non-Abelian gauge field theories are in many ways generalizations of electrodynamics. Consider a non-Abelian gauge theory with some gauge group G, with a coupling constant g and a set of N Dirac fields of mass m transforming according to some representation of G with generators T^a. The defining equations of this theory were given in §46.2, but here they are, summarized:

$$[T^a, T^b] = ic^{abc}T^c \tag{46.23}$$

$$F^a_{\mu\nu} = \partial_\mu A^a_\nu - \partial_\nu A^a_\mu + c^{abc}A^b_\mu A^c_\nu \tag{46.54}$$

$$D_\mu \psi = \partial_\mu \psi - igA^a_\mu T^a \psi \tag{46.56}$$

$$\mathscr{L} = -\tfrac{1}{4}F^a_{\mu\nu}F^{\mu\nu a} + \overline{\psi}(i\not{D} - m)\psi \tag{46.58}$$

The T's are a set of $N \times N$ matrices, acting on the internal symmetry indices of ψ only; the coefficients c^{abc} are the "structure constants" for the Lie algebra of G's generators $\{T_a\}$ with $c^{abc} = -c^{bac}$, Latin indices run from 1 to dim G, and the sum over repeated indices is implied. (Incidentally, the sign of g differs in the literature.)

Compute $k'_\mu M^\mu$ for the elastic scattering of gauge bosons off Dirac particles in the tree approximation, i.e., to order g^2. Let all the particles except the final gauge boson be on the mass shell, and investigate the circumstances when $k'_\mu M^\mu$ vanishes. Set $\varepsilon'^*_\mu = k'_\mu$ to compute $k'_\mu M^\mu$.

Comment: In this problem I found it convenient to use explicit group indices for the gauge fields, but to treat the Dirac fields as one big vector with $4N$ components. Thus the diagram shown below yields the amplitude

$$\mathcal{A} = -g^2 \overline{u}' \varepsilon'^*_\mu \gamma^\mu T^b \frac{1}{\not{p} + \not{k} - m + i\epsilon} T^a \varepsilon_\nu \gamma^\nu u$$

This is one of three diagrams you have to consider; the other two are the cross of this (as in electrodynamics) and t-channel gauge-boson exchange.

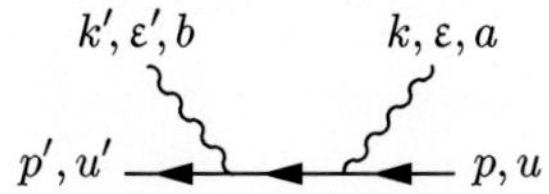

(1998b 6.2)

25.4 In the Abelian Higgs model, compute, in tree approximation, vector-scalar elastic scattering for the case in which both the initial and the final vector mesons have helicity zero, in the limit of high center-of-momentum energy, with center-of-momentum scattering angle θ fixed, but equal to neither π nor 0. (This guarantees that all three Mandelstam invariants—s, t, and u—grow with energy.) Show that in this limit, the amplitude approaches a (possibly angle-dependent) constant, even though some of the individual graphs that contribute to the amplitude grow as powers of energy. (This is the *overt* version of the Abelian Higgs model, as opposed to the *covert* version in Problem 25.1, above.)

(1998b 11.2)

Solutions 25

25.1 To lowest order in g^2, the elastic scattering of a massive vector and a scalar looks like this diagram:

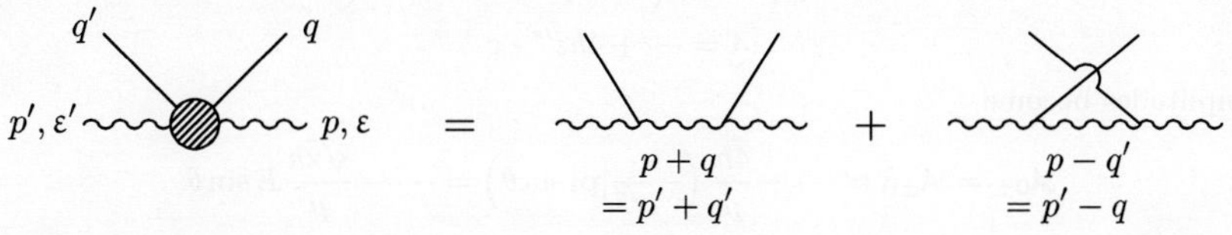

The relevant Feynman rules are for the vector-vector-scalar vertex and the vector propagator:

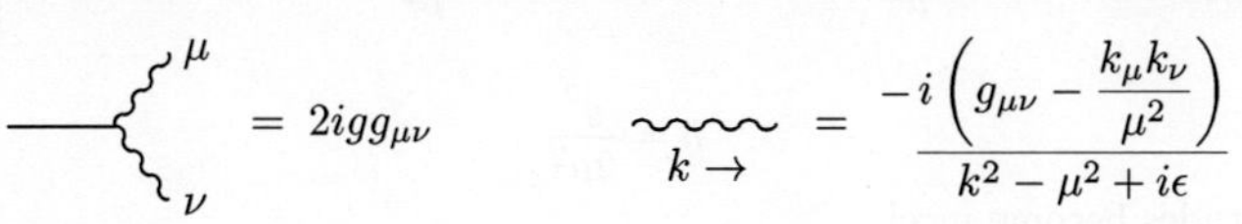

In the CM frame,

$$
\begin{gathered}
p = \left(\sqrt{|\mathbf{p}|^2 + \mu^2}, 0, 0, |\mathbf{p}|\right); \qquad q = \left(\sqrt{|\mathbf{p}|^2 + m^2}, 0, 0, -|\mathbf{p}|\right) \\
\varepsilon^{(\pm)} = \frac{1}{\sqrt{2}}(0, 1, \pm i, 0); \qquad \varepsilon^{(0)} = \frac{1}{\mu}\left(|\mathbf{p}|, 0, 0, \sqrt{|\mathbf{p}|^2 + \mu^2}\right) \\
E_{\text{CM}} \equiv E = \sqrt{|\mathbf{p}|^2 + \mu^2} + \sqrt{|\mathbf{p}|^2 + m^2}; \quad s = (p+q)^2; \quad u = (p-q')^2
\end{gathered}
\tag{S25.1}
$$

The primed quantities are obtained by rotating by θ;

$$
\begin{gathered}
p' = \left(\sqrt{|\mathbf{p}|^2 + \mu^2}, |\mathbf{p}|\sin\theta, 0, |\mathbf{p}|\cos\theta\right); \qquad q' = \left(\sqrt{|\mathbf{p}|^2 + m^2}, -|\mathbf{p}|\sin\theta, 0, -|\mathbf{p}|\cos\theta\right) \\
\varepsilon'^{(\pm)} = \frac{1}{\sqrt{2}}(0, \cos\theta, \pm i, -\sin\theta); \qquad \varepsilon'^{(0)} = \frac{1}{\mu}\left(|\mathbf{p}|, \sqrt{|\mathbf{p}|^2 + \mu^2}\sin\theta, 0, \sqrt{|\mathbf{p}|^2 + \mu^2}\cos\theta\right)
\end{gathered}
\tag{S25.2}
$$

In the graphs, the $g_{\mu\nu}$ term in the propagator makes only nice amplitudes; even in the worst case, helicity 0 to helicity 0, $\varepsilon'^* \cdot \varepsilon$ grows like E^2, but that's canceled by the denominator. Thus we need only keep track of the $k_\mu k_\nu$ term:

$$
i\mathcal{A} = (\text{nice terms}) \; + \; i(2ig)^2\left[\frac{(\varepsilon'^* \cdot q')(\varepsilon \cdot q)}{\mu^2(s - \mu^2)} + \frac{(\varepsilon'^* \cdot q)(\varepsilon \cdot q')}{\mu^2(u - \mu^2)}\right]
\tag{S25.3}
$$

where I've used the orthogonality of the vector's momenta and its polarization vectors: $\varepsilon \cdot p = \varepsilon' \cdot p' = 0$. For initial and final helicities not equal to 0, the numerator grows no faster than the denominator:

$$
\text{The amplitudes } \{\mathcal{A}_{++}, \mathcal{A}_{+-}, \mathcal{A}_{-+}, \mathcal{A}_{--}\} \text{ are all nice.}
\tag{S25.4}
$$

The other amplitudes grow with energy:

$$
\text{The amplitudes } \{\mathcal{A}_{\pm 0}, \mathcal{A}_{0\pm}, \mathcal{A}_{00}\} \text{ are all nasty.}
\tag{S25.5}
$$

Since in the worst case ($\mathcal{A}_{00}$) the amplitude grows like E^2, we can safely expand everything for high E and discard terms that are down by at least *two* powers of E compared to the leading term. Thus,

$$\begin{gathered} p \approx |\mathbf{p}|(1,0,0,1); \qquad q \approx |\mathbf{p}|(1,0,0,-1) \\ \varepsilon^{(0)} \approx \frac{|\mathbf{p}|}{\mu}(1,0,0,1) \approx \frac{p}{\mu}; \qquad \varepsilon'^{(0)} \approx \frac{p'}{\mu} \\ s-\mu^2 \approx s = E^2 \approx 4|\mathbf{p}|^2; \qquad u-\mu^2 \approx u \approx -2|\mathbf{p}|^2(1+\cos\theta) \end{gathered} \tag{S25.6}$$

The less nasty amplitudes become in the regime of large E

$$\mathcal{A}_{0\pm} = \mathcal{A}_{\pm 0} \approx -\frac{4g^2}{\mu^3}\frac{\left[|\mathbf{p}|^2(1+\cos\theta)\right]\left[\frac{1}{\sqrt{2}}|\mathbf{p}|\sin\theta\right]}{-2|\mathbf{p}|^2(1+\cos\theta)} = \frac{g^2 E\sin\theta}{\sqrt{2}\mu^3} \tag{S25.7}$$

The worst becomes

$$\mathcal{A}_{00} = -\frac{4g^2}{\mu^4}\left[\frac{(2|\mathbf{p}|^2)^2}{4|\mathbf{p}|^2} + \frac{(|\mathbf{p}|^2(1+\cos\theta))^2}{-2|\mathbf{p}|^2(1+\cos\theta)}\right] = \frac{g^2E^2}{2\mu^4}(\cos\theta - 1) \tag{S25.8}$$

(b) Now add in the new interaction. The relevant Feynman rule for the vertex is simple:

$$= 4ihg_{\mu\nu}$$

This results in a new term added to the amplitude (S25.3):

$$\mathcal{A} = \cdots + 4h\varepsilon'^* \cdot \varepsilon \tag{S25.9}$$

The nasty amplitudes become

$$\begin{aligned} \mathcal{A}_{0\pm} = \mathcal{A}_{\pm 0} &\approx \cdots + \frac{4h}{\mu}\left(-\tfrac{1}{\sqrt{2}}|\mathbf{p}|\sin\theta\right) = \cdots - \frac{\sqrt{2}h}{\mu}E\sin\theta \\ \mathcal{A}_{00} &\approx \cdots + \frac{4h}{\mu^2}|\mathbf{p}|^2(1-\cos\theta) = \cdots + \frac{h}{\mu^2}E^2(1-\cos\theta) \end{aligned} \tag{S25.10}$$

If we choose

$$h = \frac{g^2}{2\mu^2} \tag{S25.11}$$

all of the nasty amplitudes become nice!

Comment: This Lagrangian is, in disguise, just the *Abelian Higgs model*, after the symmetry breaks spontaneously:

$$\mathscr{L} = \cdots + \tfrac{1}{2}(\phi + a)^2 e^2 A^\mu A_\mu \Rightarrow \mu^2 = e^2a^2; \quad g = ae^2; \quad h = \tfrac{1}{2}e^2 = g^2/(2\mu^2)$$

This "miraculously" mild high-energy behavior is a reflection of the secret renormalizability of the theory. If we considered simple scalar field theories, this is what we would find for a renormalizable interaction like ϕ^4, but not for a nonrenormalizable one like $\phi^2(\partial_\mu\phi)^2$. We look at helicity zero states because we know from our study of vector mesons coupled to non-conserved currents (as this appears to be, if your eyes cannot pierce the veil of spontaneous symmetry breaking) that these are the states most likely to display pathological behavior. ■

25.2 (a) First, the given finite transformation can be built up as a succession of n infinitesimal transformations. Let

$$\Delta\omega = \tfrac{1}{n}\omega \tag{S25.12}$$

The infinitesimal transformation

$$\delta A_\mu = [\Delta\omega, A_\mu] - \partial_\mu\Delta\omega$$

can be written as (keeping only terms up to the first order in $\Delta\omega$)

$$A_\mu \to A_\mu^{(g)} = (1+\Delta\omega)A_\mu(1-\Delta\omega) + (1+\Delta\omega)\partial_\mu(1-\Delta\omega) \tag{S25.13}$$

Applying this twice gives

$$\begin{aligned} A_\mu^{(g^2)} &= (1+\Delta\omega)^2A_\mu(1-\Delta\omega)^2 + (1+\Delta\omega)^2\left(\partial_\mu(1-\Delta\omega)\right)(1-\Delta\omega) + (1+\Delta\omega)\partial_\mu(1-\Delta\omega) \\ &= (1+\Delta\omega)^2A_\mu(1-\Delta\omega)^2 + (1+\Delta\omega)^2\partial_\mu(1-\Delta\omega)^2 \end{aligned} \tag{S25.14}$$

using the identity $(1+\Delta\omega)(1-\Delta\omega) = 1$ to first order in $\Delta\omega$. Repeating the operation n times gives

$$\begin{aligned} A_\mu^{(g^n)} &= (1+\tfrac{1}{n}\omega)^nA_\mu(1-\tfrac{1}{n}\omega)^n + (1+\tfrac{1}{n}\omega)^n\partial_\mu(1-\tfrac{1}{n}\omega)^n \\ &\to \exp(\omega)A_\mu\exp(-\omega) + \exp(\omega)\partial_\mu\exp(-\omega) = gA_\mu g^{-1} + g\partial_\mu g^{-1} \end{aligned} \tag{S25.15}$$

Next, the stated finite transformation obeys the group property:

$$\begin{aligned}\left(A_\mu^{(g_1)}\right)^{(g_2)} &= g_2 A_\mu^{(g_1)} g_2^{-1} + g_2 \partial_\mu g_2^{-1} \\ &= g_2 (g_1 A_\mu g_1^{-1}) g_2^{-1} + g_2 (g_1 \partial_\mu g_1^{-1}) g_2^{-1} + g_2 \partial_\mu g_2^{-1} \\ &= g_2 (g_1 A_\mu g_1^{-1}) g_2^{-1} + g_2 g_1 \partial_\mu (g_1^{-1} g_2^{-1}) \\ &= (g_2 g_1) A_\mu (g_2 g_1)^{-1} + g_2 g_1 \partial_\mu (g_2 g_1)^{-1} = A_\mu^{(g_2 g_1)}\end{aligned} \tag{S25.16}$$

Finally, the finite transformation agrees with the infinitesimal transformation for $g = 1 + \delta\omega$:

$$\begin{aligned}\delta A_\mu = A_\mu^{(g)} - A_\mu &= (1 + \delta\omega) A_\mu (1 - \delta\omega) + (1 + \delta\omega)\partial_\mu (1 - \delta\omega) - A_\mu \\ &= [\delta\omega, A_\mu] - \partial_\mu \delta\omega \quad \textbf{QED}\end{aligned} \tag{S25.17}$$

(b) First, note that $U^{(g)}(s)$ satisfies the boundary condition $U^{(g)}(0)$:

$$U^{(g)}(0) = g(x(0))U(0)g(x(0))^{-1} = g(x(0))(1)g(x(0))^{-1} = g(x(0))g(x(0))^{-1} = 1 \tag{S25.18}$$

Now plug (P25.12) into (P25.11) to see if it works:

$$\begin{aligned}\frac{dU^{(g)}}{ds} &= \frac{dg(x(s))}{ds} U(s) g(x(0))^{-1} + g(x(s)) \frac{dU(s)}{ds} g(x(0))^{-1} \\ &= \partial_\mu g(x(s)) \frac{dx^\mu}{ds} U(s) g(x(0))^{-1} - g(x(s)) A_\mu(x(s)) \frac{dx^\mu}{ds} U(s) g(x(0))^{-1} \\ &= \partial_\mu g(x(s)) \left[g(x(s))^{-1} g(x(s))\right] \frac{dx^\mu}{ds} U(s) g(x(0))^{-1} \\ &\quad - g(x(s)) A_\mu(x(s)) \left[g(x(s))^{-1} g(x(s))\right] \frac{dx^\mu}{ds} U(s) g(x(0))^{-1} \\ &= \left[-g(x(s)) \partial_\mu g(x(s))^{-1} - g(x(s)) A_\mu(x(s)) g(x(s))^{-1}\right] \frac{dx^\mu}{ds} g(x(s)) U(s) g(x(0))^{-1} \\ &= -A_\mu^{(g)}(x(s)) \frac{dx^\mu}{ds} U^{(g)}(s)\end{aligned} \tag{S25.19}$$

using $(\partial_\mu g(x(s))g^{-1}(x(s)) = -g(x(s))\partial_\mu g^{-1}(x(s))$. So it *does* work. ■

25.3 The scattering can be described by this diagram:

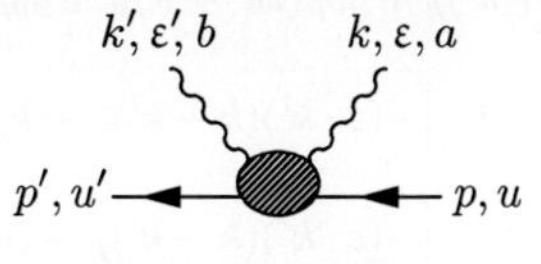

In the tree approximation, the diagram includes three graphs:

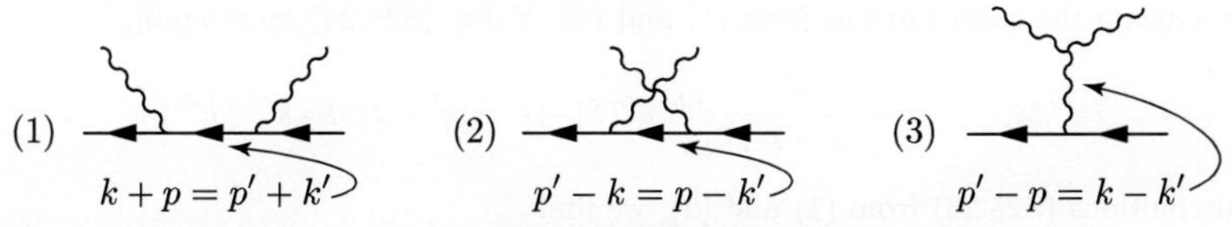

The Feynman rules can be obtained from the Lagrangian. From the term $gA_\mu^a \overline{\psi} T^a \gamma^\mu \psi$, we have the vertex (see the box on p. 1042, item (g))

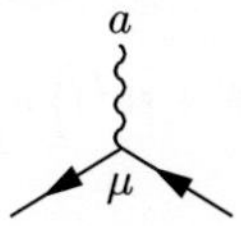

$$igT^a\gamma^\mu \tag{S25.20}$$

and from the term $gc^{abc}(\partial_\mu A_\nu^a)A^{b\mu}A^{c\nu}$ we have the vertex (box on p. 1042, item (e))

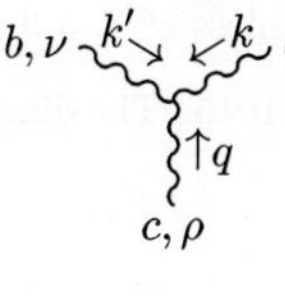

$$-gc^{abc}\left[g_{\mu\rho}(k_\nu - q_\nu) + g_{\rho\nu}(q_\mu - k'_\mu) + g_{\nu\mu}(k'_\rho - k_\rho)\right] \tag{S25.21}$$

We set $\varepsilon'^{*\nu} = k'^{\nu}$, and look at diagrams (1) and (2) together (remembering that the fermions are on their mass shell):

$$ik'^{\nu}M_\nu = -ig^2\overline{u}'\left[\frac{T^bT^a\not{k}'(\not{p}'+\not{k}'+m)\not{\varepsilon}}{k'^2+2k'\cdot p'}+\frac{T^aT^b\not{\varepsilon}(\not{p}-\not{k}'+m)\not{k}'}{k'^2-2k'\cdot p}\right]u + (3) \tag{S25.22}$$

Anticommuting $\not{p}'$ and $\not{p}$ through $\not{k}'$ and using $(\not{p}-m)u = 0 = (\not{p}'-m)u'$ yields

$$\begin{aligned}ik'^{\mu}M_\mu &= -ig^2\overline{u}'\left[T^bT^a\frac{(2p'\cdot k'+k'^2)}{k'^2+2k'\cdot p'}\not{\varepsilon}+T^aT^b\not{\varepsilon}\frac{(2p\cdot k'-k'^2)}{k'^2-2k'\cdot p}\right]u + (3)\\ &= -ig^2\,\overline{u}'\not{\varepsilon}[T^b,T^a]u + (3) = ig^2\,\overline{u}'\not{\varepsilon}[T^a,T^b]u + (3)\\ &= -g^2c^{abc}\,\overline{u}'\not{\varepsilon}T^cu + (3)\end{aligned} \tag{S25.23}$$

Now for diagram (3), which includes the vertex (S25.21) and the vector boson propagator, $D_{\mu\nu}(q)$, where $q = p'-p = k-k'$. The general covariant gauge propagator

$$\widetilde{D}^{\mu\nu}(q) = \frac{i}{q^2+i\epsilon}\left[-g^{\mu\nu}+\frac{q^\mu q^\nu}{q^2}\right]-\frac{i\xi}{q^2+i\epsilon}\left(\frac{q^\mu q^\nu}{q^2}\right) \tag{31.35}$$

is the sum of the Feynman gauge propagator plus terms in q^μ. In (3), this propagator will be contracted with the γ_μ in the fermion-meson vertex (S25.20) and sandwiched between $\overline{u}'$ and u. But these q^μ terms are irrelevant, because the fermions are on their mass shell:

$$\overline{u}'\not{q}u = \overline{u}'(\not{k}-\not{k}')u = \overline{u}'(\not{p}'-\not{p})u = 0 \tag{S25.24}$$

Thus we might as well use Feynman gauge,

$$\widetilde{D}_{\mu\nu}(q) = \frac{-ig_{\mu\nu}}{q^2+i\epsilon} = \frac{-ig_{\mu\nu}}{(k-k')^2+i\epsilon}$$

From the general form (S25.21), the upper vertex is (reversing the directions of $q = k-k'$ and k')

$$-gc^{abc}[-g_{\mu\nu}(k+k')_\rho+g_{\nu\rho}(2k'-k)_\mu+g_{\rho\mu}(2k-k')_\nu] \tag{S25.25}$$

so that (with $\varepsilon'^{*\nu} = k'^{\nu}$)

$$\begin{aligned}ik'^{\nu}M_\nu &= \cdots - g^2k'^{\nu}c^{abc}\overline{u}'[-g_{\mu\nu}(k+k')_\rho+g_{\nu\rho}(2k'-k)_\mu+g_{\rho\mu}(2k-k')_\nu]\frac{g^{\rho\sigma}}{(k-k')^2+i\epsilon}\varepsilon^\mu T^c\gamma_\sigma u\\ &= \cdots - g^2\frac{1}{(k-k')^2+i\epsilon}c^{abc}\overline{u}'T^c\left[-(\varepsilon\cdot k')(k+k')_\rho+k'_\rho(2k'-k)\cdot\varepsilon+\varepsilon_\rho(2k-k')\cdot k'\right]\gamma^\rho u\\ &= \cdots - g^2\frac{1}{(k-k')^2+i\epsilon}c^{abc}\overline{u}'T^c\left[-(\varepsilon\cdot k')(k-k')_\rho-(\varepsilon\cdot k)k'_\rho+\varepsilon_\rho(2k-k')\cdot k'\right]\gamma^\rho u\end{aligned} \tag{S25.26}$$

where the dots indicate the contributions from (1) and (2). Using (S25.24) once again,

$$ik'^{\nu}M_\nu = \cdots - \frac{g^2}{(k-k')^2}c^{abc}\overline{u}'T^c[-(\varepsilon\cdot k)\not{k}'+\not{\varepsilon}(2k\cdot k'-k'^2)]u \tag{S25.27}$$

Adding the contributions (S25.23) from (1) and (2), we find

$$\begin{aligned}ik'^{\nu}M_\nu &= -\frac{g^2}{(k-k')^2}c^{abc}\left[\overline{u}'T^c\left\{-(\varepsilon\cdot k)\not{k}'+\not{\varepsilon}(2k\cdot k'-k'^2)+\not{\varepsilon}(k-k')^2\right\}u\right]\\ &= -\frac{g^2}{(k-k')^2}c^{abc}\left[\overline{u}'T^c\left\{-(\varepsilon\cdot k)\not{k}'+\not{\varepsilon}k^2)\right\}u\right]\end{aligned} \tag{S25.28}$$

This expression does *not* vanish for off-shell incoming mesons, but it *does* for those on-shell. This is consistent with QED: in QED, $k^\mu M_\mu \neq 0$ if there are off-shell *charged* particles (whether bosons or fermions is irrelevant). With respect to the charge to which A^b_μ couples, A^a_μ is charged, unless $c^{abc} = 0$ for all c. ■

25.4 The Feynman rules for the Abelian Higgs model are given in the box on p. 1015. The diagrams responsible for scalar-vector elastic scattering at tree level are shown below:

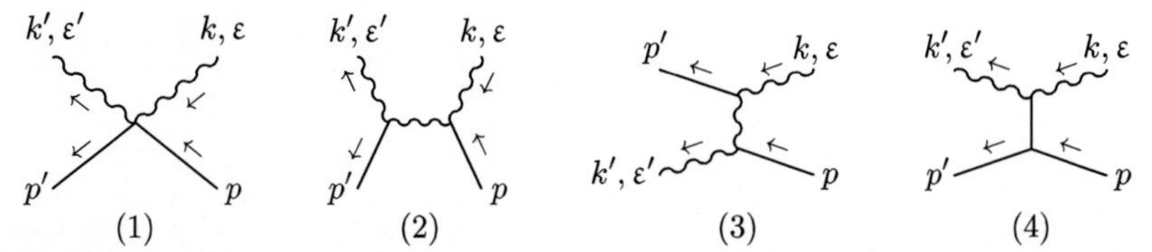

The corresponding amplitudes are (note: $M^2 = a^2e^2$ is the mass of the vector; $m^2 = 2\lambda a^2$ is the mass of the Higgs boson):

$$\begin{aligned}
i\mathcal{A}_1 &= 2ie^2\varepsilon'^*_\mu\varepsilon^\mu \\
i\mathcal{A}_2 &= (2ie^2a)^2 i\varepsilon'^*_\mu\varepsilon_\nu\left[\frac{1}{(k+p)^2-M^2}\right]\left(-g^{\mu\nu}+\frac{(k+p)^\mu(k+p)^\nu}{M^2}\right) \\
i\mathcal{A}_3 &= (2ie^2a)^2 i\varepsilon'^*_\mu\varepsilon_\nu\left[\frac{1}{(k'-p)^2-M^2}\right]\left(-g^{\mu\nu}+\frac{(k'-p)^\mu(k'-p)^\nu}{M^2}\right) \\
i\mathcal{A}_4 &= (2ie^2a)(-6i\lambda a)\varepsilon'^*_\mu\varepsilon^\mu\left[\frac{i}{(k'-k)^2-m^2}\right]
\end{aligned} \tag{S25.29}$$

Polarization vectors for helicity 0 are given in (26.78) for motion in the $\hat{\mathbf{z}}$ direction. Viewed in the center of momentum frame, let the initial vector be traveling in the $\widehat{\mathbf{k}}$ direction, and the final vector in the $\widehat{\mathbf{k}}'$ direction, with $\widehat{\mathbf{k}}\bullet\widehat{\mathbf{k}}' = \cos\theta$; θ is the center of momentum scattering angle. Then

$$\begin{aligned}
k^\mu &= (\omega, \omega\widehat{\mathbf{k}} + \mathcal{O}(\omega^{-1})), & p^\mu &= (\omega, -\omega\widehat{\mathbf{k}} + \mathcal{O}(\omega^{-1})) \\
k'^\mu &= (\omega, \omega\widehat{\mathbf{k}}' + \mathcal{O}(\omega^{-1})), & p'^\mu &= (\omega, -\omega\widehat{\mathbf{k}}' + \mathcal{O}(\omega^{-1})) \\
\varepsilon^\mu &= \frac{1}{M}(\omega, \omega\widehat{\mathbf{k}} + \mathcal{O}(\omega^{-1})), & \varepsilon'^\mu &= \frac{1}{M}(\omega, \omega\widehat{\mathbf{k}}' + \mathcal{O}(\omega^{-1}))
\end{aligned}$$

These obey the following relations:

$$k\cdot\varepsilon = k'\cdot\varepsilon' = 0, \qquad \varepsilon\cdot p = \varepsilon'\cdot p' = \frac{2\omega^2}{M}, \qquad \varepsilon\cdot\varepsilon' = \frac{2\omega^2}{M^2}\sin^2\frac{\theta}{2}, \qquad \varepsilon\cdot p' = \varepsilon'\cdot p = \frac{2\omega^2}{M}\cos^2\frac{\theta}{2}$$

The squares of the propagators' momenta are

$$(k+p)^2 = 4\omega^2, \qquad (k'-p)^2 = -4\omega^2\cos^2\frac{\theta}{2}, \qquad (k'-k)^2 = -4\omega^2\sin^2\frac{\theta}{2}$$

Using these relations, in the limit of large ω we have

$$i\mathcal{A}_1 = 4ie^2\frac{\omega^2}{M^2}\sin^2\frac{\theta}{2} \tag{S25.30}$$

$$i\mathcal{A}_2 = -4ie^2\frac{\omega^2}{M^2} + 2ie^2\sin^2\frac{\theta}{2} \tag{S25.31}$$

$$i\mathcal{A}_3 = 4ie^2\frac{\omega^2}{M^2}\cos^2\frac{\theta}{2} - 2ie^2\tan^2\frac{\theta}{2} \tag{S25.32}$$

$$i\mathcal{A}_4 = -6i\lambda \tag{S25.33}$$

The terms without any ω dependence are all $\mathcal{O}(1)$. Adding the amplitudes,

$$i\mathcal{A}_1 + i\mathcal{A}_2 + i\mathcal{A}_3 + i\mathcal{A}_4 = 4ie^2\frac{\omega^2}{M^2}\left\{\sin^2\frac{\theta}{2} - 1 + \cos^2\frac{\theta}{2}\right\} + \mathcal{O}(1) = \mathcal{O}(1) \tag{S25.34}$$

As expected, the terms that grow with energy cancel, and the total amplitude is $\mathcal{O}(1)$. ■

48

The Glashow–Salam–Weinberg Model I. A theory of leptons

Recall when I wrote down the weak interaction Lagrangian in the current-current form (40.1), I told you it was *nonrenormalizable*: we couldn't compute higher order corrections. In practice that didn't matter for most experiments, because the coupling constant is weak. Nature seemed to work in such a way that even the square of the Fermi constant times infinity was effectively a small number; it's very hard to find any conflict with experiments. For many years, it was nevertheless a *beau idéal* of theoretical physicists to concoct a renormalizable weak interaction theory. Finally, Glashow, Salam, and Weinberg did it, by constructing a gauge field theory with spontaneous symmetry breaking.[1] Because it *was* a gauge theory, with only renormalizable interactions and small coupling constants, it was *guaranteed* to be renormalizable. As the dust of spontaneous symmetry breaking settles, the interactions become very complicated, and it doesn't look like a gauge field theory at all. It's got massive vector bosons, as well as a massless one that is identified with the photon. The whole thing looks grotesque and disgustingly non-renormalizable, but that's an illusion. Just as in our discussion of the sigma model (§45.4), there are all sorts of secret relations among the coupling constants, which are preserved by renormalization because it is secretly a symmetric theory. These secret relations guarantee that when you work everything out, the theory remains renormalizable. That's the importance of our earlier comment, that renormalizability and spontaneous symmetry breaking are separable phenomena.[2]

48.1 Putting the pieces together

The **Glashow–Salam–Weinberg model** (hereafter GSW model) is supposed to describe the real world, when sufficiently generalized. There are many variants: the Georgi–Glashow model,[3] the Pati–Salam model,[4] there's this model and that model. The GSW model was

[1] [Eds.] S. L. Glashow, "Partial Symmetries of Weak Interactions", *Nucl. Phys.* **22** (1961) 579–588; S. Weinberg, "A Model of Leptons", *Phys. Rev. Lett.* **19** (1967) 1264–1266; A. Salam, "Weak and Electromagnetic Interactions", in *Elementary Particle Theory: Relativistic Groups and Analyticity.* (Eighth Nobel Symposium), N. Svartholm, ed., Almqvist and Wiksell, Stockholm, 1968. See also Cheng & Li *GT*, Chapters 11 and 12, pp. 336–400.

[2] [Eds.] See §44.2, p. 970.

[3] [Eds.] H. M. Georgi and S. L. Glashow, "Unity of All Elementary-Particle Forces", *Phys. Rev. Lett.* **32** (1974) 438–441.

[4] [Eds.] J. Pati and A. Salam, "Lepton Number as the Fourth 'Color'", *Phys. Rev.* **D10** (1974) 275–289.

the first one proposed, and it is still the simplest. These are all models that are cooked up to yield a *renormalizable theory of the weak interactions.*

What would a model describing the real world have to include? For spontaneous symmetry breaking to occur in perturbation theory, it has to have fundamental scalars. We don't want any Goldstone bosons around at the end, because they certainly aren't there in the real world. So there will have to be gauge fields present to eat the Goldstone bosons and become massive vector bosons; the only massless gauge field around is the photon. The real world also has leptons and hadrons, and possibly quarks. And although we're not going to expect perturbation theory to offer much insight into the strong interactions, we'll eventually have to extend the model to contain either fundamental baryons and mesons or colored quarks. The first version of the model we'll discuss will include only scalars, gauge fields and leptons—for simplicity, only the electron and its neutrino. Later on we'll see what happens if we put in other leptons. It's a very simple weak interaction theory, one in which there's only an effective current-current interaction between electrons and their neutrinos. (I'm leaving the muons out for the moment—we'll soon get to a theory that involves them.)

The first thing to decide on is the symmetry group of the theory. There will be a gauge group G we choose to be U(2), that is to say, SU(2) plus phase transformations.[5]

$$G = \mathrm{U}(2) = \underbrace{\mathrm{SU}(2)}_{\mathbf{I}_W} \otimes \underbrace{\mathrm{U}(1)}_{Y_W} \tag{48.1}$$

This is very much like the isospin and hypercharge of the strong interactions. We don't need to invent a new terminology; we'll just call these generators $\mathbf{I}_W$ and Y_W, the **weak isospin** and **weak hypercharge**, respectively. (These are not, of course, the familiar generators $\mathbf{I}$ and Y which occur in SU(3).) The **weak charge** is, by analogy with the Gell-Mann–Nishijima relation,[6]

$$Q_W = I_W^3 + \tfrac{1}{2}Y_W \tag{48.2}$$

Because these symmetries break spontaneously, they don't correspond to any manifest invariances of the real world.[7] There will also be an additional *global* U(1) symmetry, having nothing to do with gauge transformations, which we'll just impose on the Lagrangian as a

---

[5] [Eds.] In 1990, Coleman added: "This comes from God. If you ask why, you will be fried by a lightning bolt." In fact, Gell-Mann and Lévy in 1960 already had the weak charge-changing current as inducing transitions between members of an SU(2) doublet (private communication, Jonathan L. Rosner): M. Gell-Mann and M. Lévy, "The Axial Vector Current in Beta Decay", *Nuovo Cim.* **16** (1960) 705–726. Schwinger had earlier considered vectors mediating the weak interactions as members of a family including the photon: Julian Schwinger, "A Theory of the Fundamental Interactions", *Ann. Phys.* **2** (1957) 407–434. Schwinger writes (p. 424): "The exceptional position of the electromagnetic field in our scheme, and the formal suggestion that this field is the third component of a three-dimensional isotopic vector, encourage an affirmative answer. We are thus led to the concept of a spin one family of bosons, comprising the massless, neutral, photon and a pair of electrically charged particles that presumably carry mass..." Glashow, a student of Schwinger's, had taken up Schwinger's idea of the weak interactions mediated by massive vectors in his thesis (1959). In its appendix he states, "It is of little value to have a potentially renormalizable theory of beta processes without the possibility of a renormalizable electrodynamics. We should care to suggest that a fully acceptable theory of these interactions may *only* be achieved if they are treated together." Sheldon Lee Glashow, "The Vector Meson in Elementary Particle Decay", thesis, Harvard University, 1959. In an article published the same year, Glashow extended the ideas of his thesis and considered the group SU(2) ⊗ U(1): Sheldon L. Glashow, "The Renormalizability of Vector Meson Interactions", *Nucl. Phys.* **10** (1959) 107–117; Crease & Mann *SC*, pp. 222–223.

[6] [Eds.] See note 10, p. 520, and note 21, p.764.

[7] [Eds.] Henceforth we drop the subscript W.

phase transformation on the Fermi fields. The conserved charge associated with this symmetry will be *lepton number*.[8]

Since we have a four-parameter gauge group (three from SU(2), one from U(1)) we will have four vector bosons, one that we will call V_μ, corresponding to the weak hypercharge, and a family of three that we will call W^a_μ, $a = \{1, 2, 3\}$, corresponding to the isospin generators. As these are two independent groups, they are allowed independent gauge coupling constants.[9] Following Weinberg we will call them g' and g:

$$W^a_\mu \leftrightarrow g \qquad V_\mu \leftrightarrow \tfrac{1}{2}g' \tag{48.3}$$

(the unconventional factor of $\frac{1}{2}$ will simplify later expressions). Once we have introduced the scalar field and Fermi field content of the theory, the interactions of the vector bosons are completely determined: they follow the minimal coupling principle. What are the scalar fields and what are the Fermi fields? There is only going to be one multiplet of scalar fields ϕ. Its eigenvalues are

$$\phi\colon I = \tfrac{1}{2},\ Y = 1 \tag{48.4}$$

If this were the original $\mathbf{I}$ and Y we'd be describing the kaons. We'll write the four real scalar fields $\{\phi_i\}$, $i = 1, \dots, 4$, as a two-component, complex isospinor ($I = \frac{1}{2}$) ϕ. We will call these complex fields ϕ^+ and ϕ^0, just like the kaons (K^+ and K^0):

$$\phi = \begin{pmatrix} \phi^+ \\ \phi^0 \end{pmatrix} = \frac{1}{\sqrt{2}}\begin{pmatrix} \phi_1 + i\phi_2 \\ \phi_3 + i\phi_4 \end{pmatrix} \tag{48.5}$$

This is an abuse of language since we don't know what the electric charge is; the symmetry isn't broken yet. The scale of the generators is defined[10] so that subsequent expressions are simple, once we write down the covariant derivative of ϕ:

$$D_\mu\phi = \left(\partial_\mu - \tfrac{1}{2}ig\tau^a W^a_\mu - \tfrac{1}{2}ig' y V_\mu\right)\phi \tag{48.6}$$

The $\frac{1}{2}$ in the V_μ term has to do with how we scale the generators τ^a of the weak isospin SU(2), the ordinary Pauli matrices,

$$\left[\tfrac{1}{2}\tau^a, \tfrac{1}{2}\tau^b\right] = i\epsilon^{abc}\tfrac{1}{2}\tau^c \tag{48.7}$$

The matrix y is the generator of the Abelian weak hypercharge. As ϕ has $Y = 1$, y can be replaced here by the identity matrix:

$$D_\mu\phi = \left(\partial_\mu - \tfrac{1}{2}ig\tau^a W^a_\mu - \tfrac{1}{2}ig' V_\mu\right)\phi \tag{48.8}$$

Since ϕ is a column vector, $\phi^\dagger$ is a row vector, with covariant derivative

$$(D_\mu\phi)^\dagger = \phi^\dagger\left(\overleftarrow{\partial}_\mu + \tfrac{1}{2}ig\tau^a W^a_\mu + \tfrac{1}{2}ig' V_\mu\right) \tag{48.9}$$

---

[8] [Eds.] Griffiths *EP*, pp. 28–29.

[9] [Eds.] See p. 1023.

[10] [Eds.] The notation used here differs from that in the videotape of Lecture 52. Following Aitchison, we include the hypercharge generator y (the generator is often omitted for Abelian gauge groups): I. J. R. Aitchison, *An Informal Introduction to Gauge Theories*, Cambridge U. P., 1984, p. 108, equation (7.13). The editors have found this practice helpful in avoiding (some) sign errors. Otherwise we use Weinberg's original notation, as Coleman did in later years teaching Physics 253b. Neither Coleman nor Weinberg wrote y explicitly.

(the derivative acting to the left). These are the only scalar fields in the model.

The most general Lagrangian invariant under the group G allowing for the possibility of spontaneous symmetry breaking is

$$\mathscr{L} = \mathscr{L}_{\rm YM} + (D_\mu\phi)^\dagger \bullet D^\mu\phi - \tfrac{1}{4}\lambda[\phi^\dagger \bullet \phi - a^2]^2 \tag{48.10}$$

$\mathscr{L}_{\rm YM}$ is the pure gauge field part, just the Abelian electrodynamic part for V_μ (that is, (26.47) with $\mu^2 = 0$) and the standard form (the first term of (46.58)) for W^a_μ, the triplet. $(D_\mu\phi)^\dagger \bullet D^\mu\phi$ is the gauge invariant kinetic energy and interaction. There can't be any derivative interactions—they're not renormalizable—but we're allowed quartic and quadratic non-derivative interactions between ϕ and $\phi^\dagger$. Nor are there linear nor trilinear interactions, because you can't make a scalar with one isospinor or three isospinors. The only symmetric interaction is the one in square brackets. We've summed things together in the conventional way giving us two parameters so that the symmetry breaks spontaneously. (If the a^2 term had the opposite sign, the Lagrangian would still be invariant, but it would not lead to spontaneous symmetry breaking.) This is the most general renormalizable Lagrangian we can build from these fields. The fermions are of course very important, but let's take a preliminary look at what we have so far.

We're going to investigate this model in tree approximation, where we have the minimum value of ϕ. Because ϕ is a two-component complex vector, at the minimum that sum of the *four* squares of the (real) fields must be a^2:

$$\langle\phi^\dagger\rangle \bullet \langle\phi\rangle = a^2 \tag{48.11}$$

With the full U(2) group at our disposal we can take any two-component vector and make it one of our basis vectors. Which one we choose doesn't matter; they're all connected by the symmetries.

We will choose the symmetry breaking so that the expectation value of ϕ is

$$\langle\phi\rangle = \begin{pmatrix} 0 \\ a \end{pmatrix} \tag{48.12}$$

(with a real). The advantage of this is that $\langle\phi\rangle$ does not break electric charge conservation since it is ϕ^0 that develops an expectation value. On the other hand, the other three of the four generators of the group *are* broken. You can make a phase transformation along the "0" axis and one along the "a" axis and that's all you can do. Therefore we know already that we expect to find one massive scalar, and three Goldstone bosons which are eaten by three of the four gauge bosons to make three massive vector bosons. We also know that two of these will carry charges, plus and minus; they will be the isospin raising and lowering vector bosons, since electric charge conservation is not violated. One of the massive vectors will be neutral; it will be some linear combination which we have yet to compute, of the I_3 vector boson and the hypercharge vector boson, since there are two electrically neutral generators. The electric charge is the single remaining symmetry. This is supposed to be a realistic model, and the only massless gauge boson we know about is the photon. The other three (massive) bosons will end up being the *intermediate vector bosons*,[11] the exchange of which simulates the current-current interaction. But we haven't gotten to that yet because we haven't gotten to the leptons which source their currents.

[11] [Eds.] Cheng & Li *GT*, pp. 342–345.

We see already on this level that we have a model which at least meets the minimum criteria for a realistic model of the weak and electromagnetic interactions: the symmetry breaking is such that there is only one massless vector boson remaining. Notice that the way this model was cooked up is perfectly general. Once we have stated the symmetry transformation properties of the fields and require that the interactions have to be renormalizable, spontaneous symmetry breaking occurs in such a way that only one generator is unbroken. There's only one massless vector boson left at the end of the game.

For the fermions we do the same thing: We can have them transform under this group any way we like. Once we stipulate their transformations, we can write down the most general interaction involving them. Then we will examine the effects of spontaneous symmetry breaking on the fermions.

This is a theory which knows nothing about parity. When you first hear about spontaneous symmetry breaking you might say "Oh, that's marvelous. Parity non-conservation is going to arise as a consequence of spontaneous symmetry breaking. That's how we're going to get parity non-conservation into the weak interactions." In fact, the GSW model goes exactly the other way. It says that the original dynamics which God created *before* spontaneous symmetry breaking occurred is so ignorant of parity that it's not written in terms of Dirac four-component fields, but in terms of *Weyl fields*, two-component spinors.[12] It's not parity *non*-conservation in the *weak* interactions that's a result of dynamics; rather, it's parity *conservation* in the *electromagnetic* interactions.

Now let's introduce the Fermi fields and their transformation properties. First I have to show you a little notation to write *left-handed* and *right-handed* Weyl fields. For convenience, so we don't have to go back to that crazy σ notation and I can still use γ_μ's, we'll just take the four-component field and break it up into what we will call *left* and *right* fields; these are the γ_5 eigenstates.

$$\psi_L = \tfrac{1}{2}(1-\gamma_5)\psi, \quad \psi_R = \tfrac{1}{2}(1+\gamma_5)\psi \tag{48.13}$$

Since (20.103) γ_5 is anti-self-bar, the corresponding expression for $\overline{\psi}$ has a minus sign in it:

$$\overline{\psi}_L = \overline{\psi}\tfrac{1}{2}(1+\gamma_5), \quad \overline{\psi}_R = \overline{\psi}\tfrac{1}{2}(1-\gamma_5) \tag{48.14}$$

Of course

$$\psi = \psi_L + \psi_R \tag{48.15}$$

In a basis where γ_5 is block diagonal,

$$\gamma_5 = \begin{pmatrix} 1 & 0 \\ 0 & -1 \end{pmatrix} \tag{48.16}$$

then

$$\psi_R = \begin{pmatrix} \cdot \\ \cdot \\ 0 \\ 0 \end{pmatrix} \qquad \psi_L = \begin{pmatrix} 0 \\ 0 \\ \cdot \\ \cdot \end{pmatrix} \tag{48.17}$$

where the dots $(\cdot)$ indicate some non-zero entries. Even though they are written as four-component spinors they really have only two non-zero components; two are zero by the

[12] [Eds.] See §19.1.

equations (48.13) that define them. A trivial computation shows that

$$\overline{\psi}\gamma^{\mu}\psi = \overline{\psi}_R\gamma^{\mu}\psi_R + \overline{\psi}_L\gamma^{\mu}\psi_L \tag{48.18}$$

If we had only the kinetic energy term, the two helicity states would be dynamically independent. The mass term, however, mixes them:

$$\overline{\psi}\psi = \overline{\psi}_R\psi_L + \overline{\psi}_L\psi_R \tag{48.19}$$

Next we define the (weak) isospin and hypercharge of the Fermi fields. We will have two, L and R, each carrying lepton number. The field L is an isodoublet which is made up entirely of left-handed fields. Its eigenvalues are

$$L\colon I = \tfrac{1}{2},\ Y = -1 \tag{48.20}$$

It's like the $\{\overline{K}^0, K^-\}$ isodoublet.[13] Its covariant derivative is:

$$D_\mu L = (\partial_\mu - \tfrac{1}{2}ig\tau^a W^a_\mu - \tfrac{1}{2}ig'yV_\mu)L = (\partial_\mu - \tfrac{1}{2}ig\tau^a W^a_\mu + \tfrac{1}{2}ig'V_\mu)L \tag{48.21}$$

The last term changes sign (as compared with (48.8)) because $yL = -L$. There is also a right-handed field R which is an isosinglet. Its eigenvalues are

$$R\colon I = 0,\ Y = -2 \tag{48.22}$$

It's a little peculiar, like the Ω^-. Its covariant derivative is given by

$$D_\mu R = (\partial_\mu - \tfrac{1}{2}ig'yV_\mu)R = (\partial_\mu + ig'V_\mu)R \tag{48.23}$$

There's no $-\frac{1}{2}$ in the last term because $yR = -2R$.

The Lagrangian has the other terms as before, (48.10): the gauge invariant kinetic energy and Yukawa couplings (the only renormalizable interaction the scalar fields and other fields can have). Notice that the hypercharge of L minus the hypercharge of ϕ equals the hypercharge of R so we can have a hypercharge-conserving Yukawa interaction by coupling $\overline{L}$, ϕ, and R, with a real coupling constant, f:

$$\mathscr{L} = \cdots + \overline{L}(i\not{D})L + \overline{R}(i\not{D})R - f\overline{L}\phi R + \text{h.c.} \tag{48.24}$$

where "h. c." is the Hermitian conjugate. By a proper choice of phase we can always make f positive. If we hadn't chosen the hypercharges to allow an invariant Yukawa coupling, we would have gotten a rather trivial theory, as we would have no interaction between the fermions and the scalar bosons. This is the most general renormalizable Yukawa interaction. You might say "Couldn't I put a γ_5 in $\overline{L}\phi R$?" No, because R and L are γ_5 eigenstates, so if we put in a γ_5 that's just putting in a factor of 1 or -1; it's not an independent coupling. These terms (48.24) are all there are. There are many free parameters, but we've written down every one. The full Lagrangian is

$$\begin{aligned}\mathscr{L} = -\tfrac{1}{4}F_{\mu\nu}F^{\mu\nu} - \tfrac{1}{4}F^a_{\mu\nu}F^{\mu\nu\,a} + (D_\mu\phi)^\dagger\!\cdot D^\mu\phi - \tfrac{1}{2}\lambda[\phi^\dagger\!\cdot\phi - a^2]^2 \\ + \overline{L}(i\not{D})L + \overline{R}(i\not{D})R - f\overline{L}\phi R + \text{h.c.}\end{aligned} \tag{48.25}$$

[13] [Eds.] See Table 37.4, p. 806, and Figure 39.3, p. 852.

What are the implications of this Lagrangian? First, we're going to have electric charge left as a symmetry. We can take L and R, break them up into components, and figure out what their electric charges are. The left-handed fields L will have a negatively charged field in the bottom component, because

$$Q_{L,\,\text{bottom}} = I_3 + \tfrac{1}{2}Y = -\tfrac{1}{2} + \tfrac{1}{2}(-1) = -1 \tag{48.26}$$

With malice aforethought we will call that field the left-handed electron, e_L (more accurately, we are using only the non-zero parts of the spinor, its two lower components, for e_L). In the top component there will be a neutral field:

$$Q_{L,\,\text{top}} = I_3 + \tfrac{1}{2}Y = +\tfrac{1}{2} + \tfrac{1}{2}(-1) = 0 \tag{48.27}$$

We'll call this field (or more accurately, its non-zero components) ν_L, the left-handed electron neutrino. Then

$$L = \begin{pmatrix} \nu_L \\ e_L \end{pmatrix} \tag{48.28}$$

(This is a four-component object.) The right-handed field R is an isosinglet, and it has charge

$$Q_R = I_3 + \tfrac{1}{2}Y = 0 + \tfrac{1}{2}(-2) = -1 \tag{48.29}$$

We'll call that field the right-handed electron:

$$R = e_R \tag{48.30}$$

Let's summarize the scalar fields $\{\phi\}$ and the left and right lepton fields $\{L, R\}$ and their properties:

Field	*Charge*	*Weak I_z*	*Weak Y*
ϕ^0	0	$-\frac{1}{2}$	$+1$
ϕ^+	1	$+\frac{1}{2}$	$+1$
ν_L	0	$+\frac{1}{2}$	-1
e_L	-1	$-\frac{1}{2}$	-1
e_R	-1	0	-2

Table 48.1: The scalar and lepton fields' properties in the GSW model

The result of spontaneous symmetry breaking is to give some particles masses. It will also tell us the interactions with that remaining scalar boson. The Yukawa coupling gives the fermion masses; a gives the scale of the breakdown.

$$\mathscr{L} = \cdots - fa(\bar{e}_L e_R + \bar{e}_R e_L) \tag{48.31}$$

And there it is, a mass term for the fermions, (48.19). We started out with these three massless Weyl fields that had absolutely nothing to do with each other. One of them is a weak isodoublet, one of them is a weak isosinglet. We write down the most general renormalizable interaction Lagrangian, we make the shift, and miraculously a mass term appears! The mass of the electron is

$$m_e = fa \tag{48.32}$$

The neutrino remains massless. We've done the most general case and the neutrino mass comes out to be zero. We'll *always* be left with one massless particle in a theory of this kind. We start out with an odd number of Weyl fields. We can pair two of them together to make a mass term, but the third one is just left there. This is a consequence of there being fewer right-handed fields than left-handed fields, so somebody has to be the odd man out; we call him the neutrino. There's no way we can give the neutrino a mass with this scheme.[14] Of course, we don't know f and we don't know a, so we can't actually *calculate* the electron's mass. But we've seen how the electron gets a mass and the neutrino doesn't; other fermions get a mass and their neutrinos don't, by the same automatic mechanism, no matter what the coupling constants are.

What can we say about the vector boson masses? Three of them are massive and one of them is massless. The Lagrangian has a term

$$\mathscr{L} = \cdots + (D_\mu\phi)^\dagger \bullet D^\mu\phi + \cdots \tag{48.33}$$

Expanding the covariant derivatives and shifting the fields $\phi \to \phi' + \langle\phi\rangle$, we get a large number of terms, including cross terms of the form (T^a is a generic generator)

$$\phi^\dagger \bullet T^a \langle\phi\rangle , \quad (\partial_\mu\phi^\dagger) \bullet T^a \langle\phi\rangle , \text{ etc}$$

However, since we are working in the U gauge, (46.72), all these cross terms vanish, leaving us with terms involving only $\phi^\dagger$ and ϕ or only $\langle\phi^\dagger\rangle$ and $\langle\phi\rangle$. The masses arise as a result of the shift when the ϕ and $\phi^\dagger$ are replaced by their vacuum expectation values.

$$\mathscr{L} = \cdots + (D_\mu \langle\phi\rangle)^\dagger \bullet D^\mu \langle\phi\rangle + \cdots \tag{48.34}$$

The ordinary derivative part of D_μ will give nothing. The other part will give a term linear in the vector fields which, when squared, will give the tree approximation masses. There are two kinds of terms obviously. There's W_1 and W_2 which involve τ^1 and τ^2 and turn the lower vector in (48.12) into an upper vector, which then gets squared. And there are the two neutral ones, W_3 and V which involve τ^3 and the identity matrix, and turn the lower vector into itself. Let's write down those two terms separately.

From (48.6) and (48.9)

$$g\boldsymbol{\tau}\bullet\boldsymbol{W}^\mu + g'yV^\mu = \begin{pmatrix} gW_3^\mu + g'V^\mu & g(W_1^\mu - iW_2^\mu) \\ g(W_1^\mu + iW_2^\mu) & -gW_3^\mu + g'V^\mu \end{pmatrix} \tag{48.35}$$

$$(g\boldsymbol{\tau}\bullet\boldsymbol{W}^\mu + g'yV^\mu)\langle\phi\rangle = a\begin{pmatrix} g(W_1^\mu - iW_2^\mu) \\ -gW_3^\mu + g'V^\mu \end{pmatrix} \tag{48.36}$$

$$\langle\phi^\dagger\rangle(g\boldsymbol{\tau}\bullet\boldsymbol{W}^\mu + g'yV^\mu) = a\left(g(W_1^\mu + iW_2^\mu),\ -gW_3^\mu + g'V^\mu\right) \tag{48.37}$$

(with $y\phi$ replaced by ϕ, because ϕ has $y = 1$). Then

$$\mathscr{L} = \cdots + \tfrac{1}{4}g^2a^2[(W_\mu^1)^2 + (W_\mu^2)^2] + \tfrac{1}{4}a^2[-gW_\mu^3 + g'V_\mu]^2 \tag{48.38}$$

---

[14] [Eds.] In 1976, neutrinos were believed to be massless, but the 1988 discovery of neutrino oscillations requires the neutrinos to have a non-zero mass. The current bound is $m_\nu < 2\,\text{eV}$; *PDG* 2016, p. 758. The 2015 Nobel Prize in Physics was awarded to Takaaki Kajita of the Super-Kamiokande Collaboration and Arthur B. McDonald of the Sudbury Neutrino Observatory Collaboration for establishing that these oscillations occur. Various extensions of the standard model have been proposed to incorporate massive neutrinos. See Vernon Barger, Danny Marfatia and Kerry Whisnant, *The Physics of Neutrinos*, Princeton U. P., 2012, Chapter 9, "Model Building", pp. 99–114.

That's the vector boson mass matrix; it's pretty easy to diagonalize. I'll call the new fields $W_\mu^\pm$ and Z_μ:

$$W^{\pm\mu} = \frac{1}{\sqrt{2}}(W_1^\mu \mp iW_2^\mu), \quad Z^\mu = \frac{-gW_3^\mu + g'V^\mu}{\sqrt{g^2 + g'^2}} \tag{48.39}$$

The fields $W_\mu^\pm$ describe *charged vector bosons* $W^\pm$ made from W_μ^1 and W_μ^2. They have the same mass:

$$M_{W^\pm}^2 = \tfrac{1}{2}g^2a^2 \tag{48.40}$$

The field Z_μ describes a massive *neutral vector boson* Z^0 with mass squared greater than that of the $W_\pm$:

$$M_Z^2 = \tfrac{1}{2}a^2(g^2 + g'^2) > M_{W_\pm}^2 \tag{48.41}$$

Finally there is a remaining orthogonal *neutral vector boson*

$$A^\mu = \frac{g'W_3^\mu + gV^\mu}{\sqrt{g^2 + g'^2}} \tag{48.42}$$

That orthogonal combination has *no* mass term. That's reasonable: we have an unbroken symmetry, so we've got to have a remaining massless vector boson, the photon:

$$M_\gamma^2 = 0 \tag{48.43}$$

Inverting the expressions for the neutral vectors A_μ and Z_μ we get

$$W_3^\mu = \frac{-gZ^\mu + g'A^\mu}{\sqrt{g^2 + g'^2}}, \quad V^\mu = \frac{gA^\mu + g'Z^\mu}{\sqrt{g^2 + g'^2}} \tag{48.44}$$

Three of the four real components of ϕ have been eaten by the gauge fields to give us a charged vector doublet (of unknown mass, until we determine the parameters of the theory); a neutral massive vector boson, also of unknown mass (except that it is guaranteed to be heavier); and a massless vector boson. The last part of ϕ, corresponding to the real part of ϕ^0, remains. It is referred to in the literature as the **Higgs boson.**[15]

48.2 The electron-neutrino weak interactions

How are the weak interactions described in this theory? Let's look at the charged part of the current after we've made the shift in the scalar field. That comes just from the τ^a in the covariant derivative (48.21), in the term $\overline{L}i\not{D}L$. Those are the only charged terms. Here comes

[15] [Eds.] Many physicists independently considered the Goldstone model coupled to a massless vector, and found the mechanism whereby the vector became massive and the Goldstone boson disappeared (see note 6, p. 1014). Only Higgs predicted (1964) that there would be an *observable massive scalar* left over: "Broken Symmetries and the Masses of Gauge Bosons", *Phys. Rev. Lett.* **13** (1964) 508–509. Its properties were described in his subsequent paper (1966): "Spontaneous Symmetry Breakdown without Massless Bosons", *Phys. Rev.* **145** (1966) 1156–1163. Its discovery, confirming the mechanism, was announced at CERN on July 4, 2012. The current mass of the scalar—the Higgs boson—is 125.09 ± 0.24 GeV: *PDG* 2016, p. 30. The 2013 Nobel Prize in Physics was awarded to Peter Higgs and François Englert for their elucidation of the mechanism leading to the scalar's prediction.

the real wonder. From D_μ I have a $\frac{1}{2}$, and writing $W_1 \mp iW_2$ as $\sqrt{2}W^\pm$, I get a $\sqrt{2}$. Using (48.35),

$$\begin{aligned}\mathscr{L} &= \cdots + \overline{L}\gamma^\mu(\tfrac{1}{2}g\boldsymbol{\tau}\bullet\boldsymbol{W}_\mu)L \\ &= \cdots + \frac{g}{\sqrt{2}}\overline{\nu}_L\gamma^\mu e_L W_\mu^+ + \frac{g}{\sqrt{2}}\overline{e}_L\gamma^\mu\nu_L W_\mu^- \\ &= \cdots + \frac{g}{2\sqrt{2}}\overline{\psi}_\nu\gamma^\mu(1-\gamma_5)\psi_e W_\mu^+ + \frac{g}{2\sqrt{2}}\overline{\psi}_e\gamma^\mu(1-\gamma_5)\psi_\nu W_\mu^-\end{aligned} \tag{48.45}$$

because of the $\frac{1}{2}$ in the definition (48.13) of L. That's the unique coupling of the charged vector bosons to the fermions. Please notice: this *automatically* has the $(V-A)$ form $\gamma^\mu(1-\gamma_5)$, because $L = \frac{1}{2}(1-\gamma_5)\psi$. So the interactions are *automatically maximally parity violating.* You might say "Ha, that's nice but the weak interaction is current times current." Well, this Lagrangian leads to a four fermion interaction as a result of vector boson exchange, a plus at one end and a minus on the other, as shown in Figure 48.1. In fact, this interaction looks a

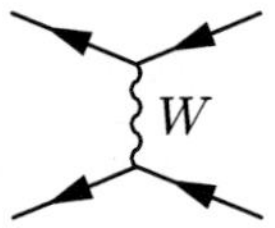

Figure 48.1: W-vector mediated four fermion interaction

great deal like Fermi's theory (40.1), particularly at low momentum transfer. Recall that the $W^\pm$ are massive vector bosons, and their propagators are

$$\widetilde{D}_{\mu\nu}(k^2) = \frac{-i}{k^2 - M^2 + i\epsilon}\left[g_{\mu\nu} - \frac{k_\mu k_\nu}{M^2}\right] \tag{48.46}$$

What happens to this if we imagine the boson is very massive compared to the mass of the electron, so k is much less than M: $k \ll M$? Then $k_\mu k_\nu/M^2$ is *bubkes*, M^2 is a constant, and we simply get

$$\widetilde{D}_{\mu\nu}(k^2) \sim i\frac{g_{\mu\nu}}{M^2} \tag{48.47}$$

That is, for small momentum transfer (small compared to the mass of the vector boson), I get effectively a point coupling just like the Fermi coupling, with the identification

$$\frac{G_F}{\sqrt{2}} = \frac{g^2}{8M_{W^\pm}^2} = \frac{1}{4a^2} \tag{48.48}$$

using the definition (48.40) for $M_{W^\pm}^2$.

Please notice that the weakness of the weak interactions is revealed to be an *illusion.* The W vector coupling constants in these theories are g and g'. We will shortly extract the coupling of the photon, and we will see that both g and g' are roughly the order of magnitude of the electromagnetic coupling constant, e. The smallness of the Fermi constant has absolutely *nothing* to do with the presence of weak dimensionless parameters; the weakness of the weak interactions is not due to weak couplings of these vector bosons. It is a consequence of the size of the parameter a, a mass that entered the original Lagrangian: a is very large compared with the electron mass, which sets the mass scale. Recall we found (48.32) the electron mass is the Yukawa coupling constant f times a. So the weak interactions are weak not because of tiny dimensionless coupling constants like 10^{-5}, but because there is a superweak *Yukawa* coupling f that makes the electron mass much smaller than the characteristic mass scale, a, the only parameter in the theory with the dimensions of mass. They're weak due to the fact that the intermediate vector bosons have large masses compared to the masses of the leptons, because the Yukawa couplings are weak. It's the weakness of the Yukawa couplings that makes the weak interactions look weaker than the electromagnetic interactions.

48.3 Electromagnetic interactions of the electron and neutrino

Instead of writing out the leptons explicitly we'll simply write

$$\mathscr{L} = \cdots + gW^3_\mu J^{3\,\mu} + \tfrac{1}{2}g' V_\mu Y^\mu \tag{48.49}$$

where $J^{3\,\mu}$ is the leptonic weak I_3 current,

$$J^3_\mu = \overline{L}\gamma_\mu \tfrac{1}{2}\tau^3 L = \tfrac{1}{2}\overline{\nu}_L\gamma_\mu\nu_L - \tfrac{1}{2}\overline{e}_L\gamma_\mu e_L \tag{48.50}$$

and Y^μ is the leptonic weak hypercharge current,

$$Y_\mu = \overline{L}\gamma_\mu y L + \overline{R}\gamma_\mu y R = -\overline{L}\gamma_\mu L - 2\overline{R}\gamma_\mu R \tag{48.51}$$

We can substitute the formulas (48.44) for W^3_μ and V_μ into the Lagrangian and find the couplings of the two mass eigenstates, A_μ and Z_μ, to the leptons, at least in tree approximation, which is all we're considering:

$$\mathscr{L} = \cdots + \frac{gg'}{\sqrt{g^2+g'^2}} A^\mu [J^3_\mu + \tfrac{1}{2}Y_\mu] - \frac{1}{\sqrt{g^2+g'^2}} Z^\mu [g^2 J^3_\mu - \tfrac{1}{2}g'^2 Y_\mu] \tag{48.52}$$

The combination that appears in the first term is nothing but the electromagnetic current, the third component of weak isospin plus half the weak hypercharge, just as in the Gell-Mann–Nishijima relation:[16]

$$J^{em}_\mu = J^3_\mu + \tfrac{1}{2}Y_\mu \tag{48.53}$$

That is unsurprising. Electromagnetism has a manifest gauge symmetry so the massless particle, the photon, must couple to the electromagnetic current. Electric charge conservation is *not* spontaneously broken. As a check, we can calculate explicitly what J^{em}_μ is:

$$\begin{aligned} J^{em}_\mu &= J^3_\mu + \tfrac{1}{2}Y_\mu \\ &= \tfrac{1}{2}\overline{\nu}_L\gamma_\mu\nu_L - \tfrac{1}{2}\overline{e}_L\gamma_\mu e_L - \tfrac{1}{2}\overline{\nu}_L\gamma_\mu\nu_L - \tfrac{1}{2}\overline{e}_L\gamma_\mu e_L - \overline{e}_R\gamma_\mu e_R \\ &= -\overline{e}_L\gamma_\mu e_L - \overline{e}_R\gamma_\mu e_R = -\overline{\psi}_e\gamma_\mu\psi_e \end{aligned} \tag{48.54}$$

which is just what it should be.

The new information from the GSW theory is that its coupling constant e, what we normally call the electron's charge, is given in terms of g and g' by

$$e = \frac{gg'}{\sqrt{g^2+g'^2}} \tag{48.55}$$

This equation can be made a bit more transparent by squaring and inverting:

$$\frac{1}{e^2} = \frac{1}{g^2} + \frac{1}{g'^2} \tag{48.56}$$

---

[16] [Eds.] See note 10, p. 520, and (35.52)–(35.53).

The solutions of this equation can be parameterized in terms of an angle assuming, as is indeed the fact, that e is a known quantity and that g and g' are unknown quantities:

$$g = \frac{e}{\sin\theta_W}, \quad g' = \frac{e}{\cos\theta_W} \tag{48.57}$$

As promised, we see that g and g' are indeed $\mathcal{O}(e)$. The parameter θ_W is called the **Weinberg angle.**[17] It was introduced by Weinberg who, with commendable modesty, called it the *weak interaction angle.*[18]

When Weinberg and Salam first proposed this model, there were no observed *weak neutral currents*. All they knew about were the Fermi constant and the electromagnetic charge. Aside from the quartic coupling constant λ (which gives (46.15) the Higgs boson mass in terms of a), the one quantity not predicted in terms of known quantities is the Weinberg angle. We can substitute this expression (48.57) for g into the formula (48.48) for the mass of the $W^\pm$ to get

$$M^2_{W^\pm} = \tfrac{1}{2}\frac{a^2e^2}{\sin^2\theta_W} \tag{48.58}$$

This formula means that it's going to be a long time[19] before anyone directly observes a $W^\pm$, because it gives a lower bound when θ_W is a multiple of $\pi/2$, and using $a^2 = \sqrt{2}/(4G_F)$, and $G_F \approx 10^{-5}/m_p^2$, this lower bound is

$$M^2_{W^\pm} = \frac{e^2}{4\sqrt{2}G_F}\frac{1}{\sin^2\theta_W} = \frac{10^5}{4\sqrt{2}}\frac{m_p^2e^2}{\sin^2\theta_W} \geq (37\,\text{GeV})^2 \tag{48.59}$$

That's a large number. Things are even worse with the neutral vector boson, the Z^0. Just plugging into the formula for the Z^0 mass and doing a little algebra

$$M_Z^2 = \frac{2a^2e^2}{\sin^2(2\theta_W)} \tag{48.60}$$

There's a 2 instead of a $\frac{1}{2}$, and we get a lower bound whose square is twice as large as the lower bound of $M^2_{W^\pm}$:

$$M_Z^2 = \frac{10^5}{\sqrt{2}}\frac{m_p^2e^2}{\sin^2(2\theta_W)} \geq (74\,\text{GeV})^2 \tag{48.61}$$

So it's even harder to see the Z^0 than the $W^\pm$. Of course these two lower bounds cannot be attained simultaneously; the first occurs when $\theta_W = \pi/2$, and the second when $\theta_W = \pi/4$.

When you go beyond tree approximation, the bounds change only by terms of order e^2 or order λ^2 and so on—small corrections if all these parameters are small. That means, by the way we defined θ_W, that θ_W cannot be close to an integer multiple of $\pi/2$; if it were,

---

[17] [Eds.] Cheng & Li *GT*, pp. 351–352. The current values are $\sin^2\theta_W \approx 0.23129(5)$ or $\theta_W \approx 28.746°$: *PDG* 2016, p. 119.

[18] [Eds.] In fact the idea of a mixing angle was introduced by Glashow: S. Glashow, "Partial Symmetries of Weak Interactions", *Nuc. Phys.* **22** (1961) 579–588 (the angle is introduced on p. 585); Crease & Mann *SC*, p. 226; Close *IP*, p. 118, pp. 292–293.

[19] [Eds.] Coleman made this statement in 1976. The $W^\pm$ and Z^0 were discovered at CERN in 1983; see note 9, p. 519. The current values of their masses are $M_{W^\pm}: 80.385 \pm 0.015$ GeV; $M_{Z^0}: 91.1876 \pm 0.0026$ GeV. See *PDG* 2016, p. 29.

then g or g' would be large. But so long as g and g' are small enough to justify perturbation theory, so long as all the dimensionless parameters of the theory—g, g', λ and f—are much less that 1, as they seem to be, this determines our scale of mass; it's irrelevant how large a is. The tree approximation should then be reliable, because it is simply the lowest order in perturbation theory. All the formulas we are writing down will obtain corrections of higher powers of the various coupling constants. This is a renormalizable theory that should be finite and computable and the corrections *will be small.* We will later worry about a case where these coupling constants are all small but differ from each other by many orders of magnitude. Then we have to worry about one-loop corrections in the large coupling constants affecting formulas that only involve the small coupling constants in zeroth order; we will see that in more detail as we go on. They have corrections which have been computed by people who wanted to check the renormalizability of these schemes. In general, corrections are small provided that you don't choose θ_W perversely, so that either g or g' is enormous.

Now let's turn to the coupling of the Z boson. Because it is necessarily very heavy, the Z has got to have a Fermi-type interaction, at least at acceptable energies. Let's write out the coupling (48.52), in a form where we can see what the interaction is. We'll split this into an electromagnetic current and a remainder. If the parameters turn out such that the Z boson is coupled only to the electromagnetic current then we get a short-range interaction obeying exactly the same selection rules as for electromagnetism and very weak to boot. At low energies, we'll never be able to distinguish the short-range interaction from higher-order electromagnetic corrections. So it's important, if the effects of the Z boson are in any way observable, that it be coupled to something other than just the electromagnetic current. This combination will be electromagnetism plus something else, and it will be the amount of the something else that will tell us the observable effects. We have the formula for the electromagnetic current, so we can eliminate Y_μ in terms of J_μ^{em} and J_μ^3. The result can be written in the following form (I will skip a few lines of trivial algebra):

$$\mathscr{L} = \cdots - \frac{\sqrt{2}}{a} M_Z Z^\mu \left[J_\mu^3 - \frac{g'^2}{g^2 + g'^2} J_\mu^{em} \right] \tag{48.62}$$

I've used the formula (48.41) for the mass of the Z to simplify the expression. Note also the identity

$$\frac{g'^2}{g^2 + g'^2} = \sin^2 \theta_W \tag{48.63}$$

We see two things. First, while the term J_μ^{em} is parity-conserving, the term J_μ^3 is maximally parity-violating; it contributes only left-handed things. Therefore the parity-violating effects which would be the signature of the presence of this object, which would be due to the cross terms between J_μ^{em} and J_μ^3 in one-Z exchange, are *proportional to* $\sin^2 \theta_W$. In this sense θ_W is very much like the Cabibbo angle.[20]

Putting the M_Z in front is a good idea because you get a $1/M_Z^2$ from the Z propagator in the exchange. You get an effective Fermi-type interaction with, aside from Clebsch–Gordan factors, a Fermi-scale strength, proportional to $1/a^2$, with the cross term between the two neutral currents. Therefore the theory inevitably predicts a *parity-violating neutral current-current* type interaction of calculable magnitude at low energies, at least in this very simplified model in which all you have in the world are electrons and their neutrinos.

---

[20] [Eds.] Peskin & Schroeder *QFT*, p. 605.

48.4 Adding in the other leptons

Let's generalize the model to include the muons, the taus and their neutrinos. The obvious thing is just to put in additional left-handed doublets and additional right-handed singlets:

$$\begin{pmatrix} \nu_\mu \\ \mu_L \end{pmatrix}, \; \mu_R; \qquad \begin{pmatrix} \nu_\tau \\ \tau_L \end{pmatrix}, \; \tau_R \tag{48.64}$$

However, saying things this way is a bit backwards. Presumably the particle structure should emerge as a consequence of spontaneous symmetry breaking. I'll start out with a model that involves three left-handed SU(2) doublets L_α, $\alpha = \{1, 2, 3\}$ with exactly the same transformation properties under $\mathrm{SU}(2) \otimes \mathrm{U}(1)$, and three singlets R_α, right-handed Weyl fields:

$$L_\alpha, \quad R_\alpha \quad \alpha = \{1, 2, 3\} \tag{48.65}$$

Let's ask the following interesting question: If we write the most general renormalizable Lagrangian consistent with these fields, what will happen as a result of spontaneous symmetry breaking? Will it inevitably turn out to be a tau and a massless neutrino, a muon and a massless neutrino and an electron and a massless neutrino, or is there a possibility that other things could happen if we choose the coupling constant properly? The only undetermined coupling constants are the Yukawa couplings (other than the coupling constants involving the ϕ fields, its self-interaction and the gauge field couplings; these are completely determined by the stated transformation properties and the minimal coupling prescription). Thus the new term in the Lagrangian is

$$\mathscr{L} = \cdots - f_{\alpha\beta} \overline{L}_\alpha \phi R_\beta + \text{h. c.} \tag{48.66}$$

where $f_{\alpha\beta}$ is in general a 3×3 matrix. If we were considering generalizations of this model with other, newly discovered, kinds of leptons, then we would have to run α and β over a larger range. Everything else in the Lagrangian is as before, completely determined.

We will demonstrate the following **diagonalization theorem**:[21] Given any $n \times n$ matrix f we can always write f in the following form.

$$f = U_1^\dagger \Delta U_2 \tag{48.67}$$

U_1 and U_2 are unitary, Δ is diagonal and positive; it's a diagonal matrix with only positive entries. This is a pure matrix theorem; it does not depend on f being Hermitian or anything like that. We will first give the application of this theorem to the GSW model and then prove it.

All of our left-handed doublets transform in exactly the same way under the gauge group, as do all of our right-handed singlets. They have the same weak hypercharge and the same weak isospin. Therefore we are perfectly free in a Lagrangian of this kind, without changing anything else, to redefine our doublets and our singlets by unitary transformation, mixing them up any way we want (to keep the kinetic energy unchanged the transformation must be unitary). In particular we can define

$$L'_\alpha = (U_1)_{\alpha\delta} L_\delta, \quad R'_\beta = (U_2)_{\beta\gamma} R_\gamma \tag{48.68}$$

[21] [Eds.] The form (48.67) is called the *singular-value decomposition.* That f can be written in this form is a well-known theorem, originating in differential geometry. See Gilbert Strang, *Introduction to Linear Algebra*, 5th ed., Wellesley-Cambridge Press, 1998, Chapter 7, pp. 364–400.

By the theorem, we can write $f = U_1^\dagger \Delta U_2$, with Δ diagonal, as follows:

$$\Delta = \begin{pmatrix} f_1 & 0 & 0 & \cdots & 0 \\ 0 & f_2 & 0 & \cdots & 0 \\ 0 & 0 & \cdot & \cdots & 0 \\ 0 & 0 & \cdot & \cdots & f_n \end{pmatrix} \tag{48.69}$$

Therefore, in terms of these transformed fields our Lagrangian involves separate Yukawa couplings summed on α:

$$\mathscr{L} = \cdots - f_\alpha \overline{L}'_\alpha \phi R'_\alpha + \text{h.c.} \tag{48.70}$$

That is to say, we can diagonalize the Yukawa couplings by *independently* shuffling around the right-handed fields and the left-handed fields. We get a sum of Yukawa systems, decoupled into mass eigenstates, each of which has exactly the same structure as the electron-neutrino system we have considered before.

The proof of the theorem goes as follows. Given any non-singular matrix f, $ff^\dagger$ is a positive definite Hermitian matrix:

$$ff^\dagger = H^2, \quad H = H^\dagger, \quad H \text{ is positive} \tag{48.71}$$

H is the unique positive square root. Because $H = H^\dagger$, it can be diagonalized by a unitary matrix U_1:

$$U_1 H U_1^\dagger = \Delta \Rightarrow H = U_1^\dagger \Delta U_1 \tag{48.72}$$

Then

$$H^2 = ff^\dagger = U_1^\dagger \Delta^2 U_1 \tag{48.73}$$

Define the matrix U_2 in the obvious way, from (48.67):

$$U_2 \equiv \Delta^{-1} U_1 f \tag{48.74}$$

Showing that U_2 is unitary will complete the proof of the theorem. But this is easy:

$$U_2 U_2^\dagger = \Delta^{-1} U_1 f f^\dagger U_1^\dagger \Delta^{-1} = \Delta^{-1} U_1 U_1^\dagger \Delta^2 U_1 U_1^\dagger \Delta^{-1} \tag{48.75}$$

On the right-hand side, there is a product $U_1 U_1^\dagger = 1$ and a Δ^{-1} on either side of the Δ^2. All the terms collapse, and[22]

$$U_2 U_2^\dagger = 1 \qquad \textbf{QED}$$

What is the significance of this result? No matter how we try to arrange the model, as long as it is consistent with the constraints of renormalizability and gauge invariance, and as long as it doesn't involve any fields other than the ones we've itemized, wc automatically get *separate* electron, muon and tau systems, as far as the Yukawa coupling is concerned, independently coupled to the ϕ field. Thus we have

$$\begin{aligned} m_e &= f_e a, \quad m_{\nu_e} = 0 \\ m_\mu &= f_\mu a, \quad m_{\nu_\mu} = 0 \\ m_\tau &= f_\tau a, \quad m_{\nu_\tau} = 0 \end{aligned} \tag{48.76}$$

[22] [Eds.] In 1976, Coleman claimed that the theorem is true even when f is singular, though care has to be taken because of possible zero eigenvalues of Δ.

The lepton masses (in MeV) are[23]

$$\begin{aligned} e&: 0.5109989461 \pm 0.0000000031 \\ \mu&: 105.6583745 \pm 0000024 \\ \tau&: 1776.86 \pm 0.12 \end{aligned} \tag{48.77}$$

The coupling constants f_e, f_μ and f_τ are diagonal matrix elements of Δ, about which we can say nothing *a priori*. Since the left-handed doublets transform in exactly the same way, all of the currents are exactly the same as before: all the things that go into electromagnetism or the weak interactions or the Z-mediated interactions, are *sums* of separate electron, muon and tau parts with identical coefficients as in (48.52). Because all of the algebra is the same this is automatic.

48.5 Summary and outlook

Let's summarize what we have found in the GSW theory. In the course of the summary we'll introduce some new language that is frequently used in discussing these things. So far we have a theory only of leptons. We have not yet put in the quarks. The theory has many good features:

• It provides a *renormalizable theory of weak interactions* (*modulo* the pesky question of anomalies,[24] which we will not discuss). That enables us to compute higher order weak corrections. Of course once we compute them, we find they're tiny. But they give the experimentalists a good excuse to ask for lots of money from their governments so they can measure them. That's a good thing.

• It *unifies electromagnetism and the weak interactions*, which is aesthetically very pleasing. These forces are two aspects of the *same* force. It is not that we have two independent field theories; we have *one*. All the coupling constants are of the same order of magnitude provided we don't choose θ_W to be exceptionally small. It's not true that there's a large vector boson coupling constant, for electromagnetism, and a small one for the weak interactions. They're all about the same size. It is spontaneous symmetry breaking that causes this completely symmetric theory to put on a false beard and appear to be two grotesquely different things, electromagnetism and the weak interactions.

• The theory exudes *naturalness*. When we have a spontaneously broken gauge theory, there are some things that are generally true no matter what values we assign to the coupling constants, provided that we choose values such that spontaneous symmetry breaking occurs. Features are said to be **natural** if they do not depend on some perverse choice of coupling constants, but are true over a wide range of values. (I use the word "natural" in a technical sense.) We write down what the fields are and how they transform under the gauge group; those are the rules of the game. We write down the most general renormalizable interaction Lagrangian satisfying those rules. Then we find certain general consequences that match experimental results, without having to fine tune the coupling constants. This is a very pleasing theory. Here are six examples of naturalness in the GSW theory, good features that emerge *automatically*, independent of the values of parameters:

---

[23] [Eds.] *PDG* 2016, p. 32.

[24] [Eds.] See note 21, p. 1044.

1. Electromagnetism *conserves parity*, and
2. *the mass of the photon is zero.* It's not possible to arrange the parameters in the theory so that the interaction is renormalizable in any other way than what we have written down. The symmetry always breaks down leaving electromagnetism with only a remaining U(1) symmetry unbroken.
3. The form of the weak interactions *takes the $(V - A)$ form* of the Fermi interaction. This comes about no matter what values the parameters have; we derived that without any assumptions about them.
4. Each of the leptons has a *separately conserved lepton number.* That is a consequence of the diagonalization theorem. The only terms that can possibly mix the different leptons and their neutrinos are off-diagonal terms in the Yukawa coupling. We did not require a diagonal Yukawa matrix *a priori*; we allowed for the possibility of an arbitrary $f_{\alpha\beta}$ and then showed that we can always choose the fields so that the off-diagonal terms disappear.
5. The leptons display *universality* in the Lagrangian: all currents are made up of electron, muon or tau parts with exactly the same coefficients. No matter how we choose the initial parameters, after spontaneous symmetry breakdown, at least to lowest order, all the leptons couple under the weak interactions with the *same* coupling constants, and they have the *same* charges.
6. The neutrinos *are massless*, no matter how we choose the parameters. (If we introduced an electrically neutral right-handed field for the leptons before spontaneous symmetry breaking, we could get a neutrino mass.)

There are two things as yet unexplained, which we would like to see addressed in the ultimate theory. They are both associated with thus far unnatural features.

• *Why is G_F small?* The reason the weak interactions look weak is because the $W^{\pm}$ and Z^0 are heavy in comparison to the other particles. As I pointed out earlier (see the paragraph following (48.48)), that is connected to the reasons, as yet unknown, for the inequalities

$$f_e,\ f_\mu,\ f_\tau \ll 1 \tag{48.78}$$

The size of the Yukawa couplings determines the masses of the leptons in terms of the sole parameter with dimensions of mass, a. Because we have the masses of the vector bosons $M_V \sim a$, and $G_F \sim 1/a^2$, the size of a has to be large, and so to obtain the observed lepton masses, the Yukawa coupling constants f must be very small. Nobody knows why the f's are so small.[25] The theory could still be analyzed perturbatively if the ratios of the lepton masses to the vector masses were on the order of 1/10; but then the weak interactions would not be so weak. That is unnatural. We can certainly *choose* these Yukawa constants to be small, but we *need* not do so. The theory doesn't explain that.

• *Why are the lepton masses so different?* We have

$$m_e \ll m_\mu \ll m_\tau, \quad f_e \ll f_\mu \ll f_\tau \tag{48.79}$$

[25] [Eds.] In 1990, Coleman added: "Steve Weinberg says that the quarks would have masses similar to the $W^{\pm}$ and Z^0 were the Yukawa couplings comparable to the gauge couplings g and g'. The real question is: why are the quarks we're made of—the u and the d—so anomalously light? The top quark for example has a mass ~ 180 GeV." The current value of m_t is 173.21 ± 0.51 GeV; *PDG* 2016, p. 36.

We can arrange matters so that these conditions are met. We simply have to choose the Yukawa coupling constant f_e to be 100 or 150 times smaller than the Yukawa coupling constant f_μ, but there is no reason why we have to choose it that way. We could choose them to be equal and then we'd get a theory with equal lepton masses. That isn't the real world, but the theory doesn't explain *why* it isn't.

These are not serious shortcomings. It would be nice to have a theory in which the electron-muon mass ratio was a computable quantity not associated with the ratio of free parameters.[26] It would be very nice to have a theory in which the weakness of the weak interactions was not merely *consistent* with the theory but *inevitable* in it. We do not yet have such a theory.

Next time we will expand the GSW model to include quarks, and thereby all strongly interacting particles made from them.

[26] [Eds.] H. Sato, "Muon-Electron Mass Ratio and CP Violation as a Quantum Effect", *Nucl. Phys.* **B148** (1979) 433–444; K. Nishijima and H. Sato, "Higgs-Kibble Mechanism and the Electron-Muon Mass Ratio", *Prog. Theor. Phys.* **59** (1978) 571–578.

49

The Glashow–Salam–Weinberg Model II. Adding quarks

Now we're going to assimilate the strongly-interacting particles into our scheme. The rules of the game will be exactly the same as before; we'll just throw in some more fields. We have perfect freedom to choose these fields as we wish, and specify how they transform under the gauge group and what the Yukawa couplings are. We will choose them cunningly (or rather, we will employ other people's cunning choices), so that after spontaneous symmetry breaking the model begins to resemble the real world. We will construct a renormalizable theory that unifies electromagnetism with the weak interactions. After we're done, we will see which properties of the semi-leptonic weak interactions emerge naturally and which do not.

49.1 A simplified quark model

We will start with a simplified quark model in which there are no *strange* particles. In this case we don't need the strange quark, s, and we can get by with just two quarks, the up quark, u, and the down quark, d. Instead of SU(3), we just have SU(2), isotopic spin. The up and down quarks have charges $Q = \frac{2}{3}$ and $Q = -\frac{1}{3}$ respectively:

Quark	*Charge*
u	$+\frac{2}{3}$
d	$-\frac{1}{3}$

Table 49.1: Quarks and their charges

We will build our hadrons out of these two quarks. Quarks also carry a color index (§39.6), having to do with *color* SU(3) that couples to the *gluons*, but we're not going to have to worry about that. Color factors out of this part of the analysis. We simply want the quark form of the currents. Those gluons are there, but they and the SU(2) ⊗ U(1) gauge group don't talk to each other, by assumption. This means that the formulas we get for quark masses, etc. will have large corrections due to the strong interactions between the quarks, even though the formulas we get for the *currents* will not be affected.[1]

By exactly the same trick we used with the leptons (see (48.28) and (48.30)), we can build a left-handed doublet and two right-handed singlets out of the quark fields:

$$L_1 = \begin{pmatrix} u_L \\ d_L \end{pmatrix} \qquad \{R_1, R_2\} = \{u_R, d_R\} \tag{49.1}$$

(The subscript on L_1 is to distinguish it from the leptonic doublet L, and also from a *second* left-handed quark doublet we are shortly going to introduce.) L_1 is a *weak isodoublet*, the R's are two *weak isosinglets*.[2] Their weak hypercharges are determined (48.2) by their charge assignments.[3] The only thing that is going to be different is the Yukawa couplings; everything else is exactly the same in its transformation properties.

Field	*Charge*	*Weak I_z*	*Weak Y*
u_L	$+\frac{2}{3}$	$+\frac{1}{2}$	$+\frac{1}{3}$
d_L	$-\frac{1}{3}$	$-\frac{1}{2}$	$+\frac{1}{3}$
u_R	$+\frac{2}{3}$	0	$+\frac{4}{3}$
d_R	$-\frac{1}{3}$	0	$-\frac{2}{3}$

Table 49.2: The up and down quark fields divided into left and right fields

We can make *two* invariant Yukawa couplings. We can have the original coupling

$$\mathscr{L} = \cdots - f\overline{L}\phi d_R \tag{49.2}$$

That's *down-right* simple. It's perfectly consistent with isospin and hypercharge conservation. (Isospin is obvious. To check hypercharge, we need only check the electric charge where the ϕ^0 couples $\overline{d}_L$ and d_R which is obviously charge conserving.) However, there is another possible Yukawa coupling, because we can put in *two* right-handed quarks. Recalling the definition of ϕ,

$$\phi = \frac{1}{\sqrt{2}}\begin{pmatrix} \phi_1 + i\phi_2 \\ \phi_3 + i\phi_4 \end{pmatrix} = \begin{pmatrix} \phi^+ \\ \phi^0 \end{pmatrix} \tag{48.5}$$

we can introduce the **charge conjugate**[4] field, ϕ^C:

$$\phi^C = \frac{1}{\sqrt{2}}\begin{pmatrix} \phi_3 - i\phi_4 \\ -\phi_1 + i\phi_2 \end{pmatrix} = \begin{pmatrix} \overline{\phi}^0 \\ -\phi^- \end{pmatrix} \tag{49.3}$$

The vacuum expectation values of ϕ and ϕ^C are

$$\langle\phi\rangle = \begin{pmatrix} 0 \\ a \end{pmatrix}, \quad \langle\phi^C\rangle = \begin{pmatrix} a \\ 0 \end{pmatrix} \tag{49.4}$$

[1] [Eds.] Cheng & Li *GT*, Chapters 11 and 12, pp. 336–400.

[2] [Eds.] To distinguish between weak isospin and the old, strong isospin, the latter will be described in this chapter as "isotopic spin".

[3] [Eds.] In parallel with the *strong* Gell-Mann–Nishijima relation linking isotopic spin, the usual hypercharge, and charge, $Q = I_z + \frac{1}{2}Y$. See note 10, p. 520, and (35.52). At this point in 1976, Coleman added, "This goes to show that God has less imagination than the high energy theorists, who have thought of many possibilities more baroque than this."

[4] [Eds.] See p. 119.

Because u_R has been assigned exactly the right transformation properties, we can put together a hypercharge invariant coupling of the following form (see (48.24)):

$$\mathscr{L} = \cdots - f_1 \overline{L}_1 \phi d_R - f_2 \overline{L}_1 \phi^C u_R + \text{h. c.} \tag{49.5}$$

When spontaneous symmetry breaking occurs the first term will give a mass to the down quark and the second term will give a mass to the up quark:

$$m_d = a f_1, \quad m_u = a f_2 \tag{49.6}$$

This extra term does not occur in the purely leptonic sector because that would require the presence of a right-handed field carrying the same charges as the upper element of L, (48.28). The upper element is a neutrino and there is no right-handed field carrying neutrino charge or with the proper assignment of weak hypercharge.[5]

Everything here goes just as before: parity-conserving electromagnetism, Fermi theory of the weak interactions, universality, because everything transforms the same way under the gauge group. So we just get the sum of a quark term plus a non-quark term. We still have three massless neutrinos, now joined by two massive quarks. We get independent conservation of individual lepton number and quark (i.e., baryon) number. Here we don't have to use the diagonalization theorem because of the way we have defined the left and right fields and arranged the coupling. There is no possible way of writing a hypercharge-invariant quark-lepton Yukawa coupling, because the quarks carry *fractional* weak hypercharge (Table 49.2), while the leptons carry *integral* weak hypercharge, as do the ϕ's (Table 48.1). There is no way of adding two integers to make a *fraction.* To say it plainly, we have *independent* conservation of quark number (or, if you prefer, baryon number), and of lepton number.

What we don't yet have are the *strange* particles.[6] There is something else in this theory that is grossly unnatural: the approximate conservation of isotopic spin (*strong*, not weak). That depends on f_1 and f_2 being approximately equal. There is no symmetry principle in this theory that would *require* f_1 and f_2 to be approximately equal; it is just a coincidence that they are. As far as I know, no one has found a model with all these other satisfactory features in which the approximate conservation of isotopic spin is a natural result. It's like the weakness of the weak interactions. We can *make* isotopic spin approximately conserved. We just have to do it "by hand", and choose f_1 approximately equal to f_2. But there is no symmetry principle forcing us to do so.

49.2 Charm and the GIM mechanism

Onward! Now we come to the bright idea[7] of Glashow, Iliopoulos and Maiani, the *GIM mechanism.* It wasn't phrased in the context of this form of the theory, but in quite another, where it was much less obvious what the right thing to do was—namely, suppressing strangeness-changing neutral currents. The bright idea is this. The reason we keep getting universality

---

[5] [Eds.] As always, Coleman is assuming the neutrinos are massless. See note 14, p. 1066.

[6] [Eds.] F. Halzen and A. D. Martin, *Quarks and Leptons*, John Wiley, 1984, Section 1.7, pp. 26–27; Section 2.9, pp. 44–46.

[7] [Eds.] S. Glashow, J. Iliopoulos and L. Maiani, "Weak Interactions with Lepton-Hadron Symmetry", *Phys. Rev.* **D2** (1970) 1285–1291; L. Maiani, "The GIM Mechanism: Origin, Predictions and Recent Uses", in *Rencontre de Moriond: Electroweak Interactions and Unified Theories*, La Thuile, Valle d'Aosta, Italy, 2013, avaliable on-line at `https://arxiv.org/abs/1303.6154`.

is because everything is a left-handed doublet. The charged fields $W_\mu^\pm$ couple to purely left-handed currents. So everything couples universally because we've simply got the same damn Pauli matrix all the time. If we're going to have some form of universality after we introduce strange quarks, then they've got to be put into a left-handed doublet also. Unfortunately if we only have three flavors of quarks, there's no way we can make two doublets. We can make a doublet and a singlet, but that's the end of the game. To carry this scheme on in any natural (in the vernacular sense) way, we're going to have to have an even number of quarks. The smallest even number greater than three is four, so we'll need (at least) four quarks. The new quark is called **charm.**[8] This new quark has a charge of $+\frac{2}{3}$, and a new quantum number, *charm*, with $C = 1$, just as the strange quark has $S = -1$.

We will arrange the two quarks $\{c, s\}$ exactly the same as $\{u, d\}$, as shown in Table 49.3 (compare Table 49.2). What results in two left-handed weak isodoublets, with the upper field having $Q = +\frac{2}{3}$ and the lower having $Q = -\frac{1}{3}$, and four right-handed weak isosinglets, two with $Q = +\frac{2}{3}$ and two with $Q = -\frac{1}{3}$. Then we have all sorts of possible Yukawa couplings. It's now not a 2×2 matrix but a 4×4 matrix, since there are four things on the right and four things on the left of the Yukawa coupling. Rather than choose the Yukawa couplings and see what mass eigenstates result, it's perhaps better to tackle the problem in reverse: we'll choose the mass eigenstates, and see what Yukawa couplings, and hence what doublets, come out of those choices. (The singlets we can mix up as we wish; they'll have to be chosen to be the mass eigenstates.)

Field	*Charge*	*Weak* I_z	*Weak* Y
c_L	$+\frac{2}{3}$	$+\frac{1}{2}$	$+\frac{1}{3}$
s_L	$-\frac{1}{3}$	$-\frac{1}{2}$	$+\frac{1}{3}$
c_R	$+\frac{2}{3}$	0	$+\frac{4}{3}$
s_R	$-\frac{1}{3}$	0	$-\frac{2}{3}$

Table 49.3: The strange and charm quark fields divided into left and right fields

We'll choose our mass eigenstates to consist of a down quark and a strange quark, which have $Q = -\frac{1}{3}$, and an up quark and a charmed quark, which have $Q = \frac{2}{3}$. They are determined only up to a phase, and we'll take advantage of that freedom.

$$\text{mass eigenstates:} \begin{cases} \{u, c\} & \text{with charge } +\frac{2}{3} \\ \{d, s\} & \text{with charge } -\frac{1}{3} \end{cases} \tag{49.7}$$

[8] [Eds.] "Aesthetic arguments led J. D. Bjorken and me to conjecture a fourth quark, more than a decade ago. Since leptons and quarks are most fundamental, and since there are four kinds of leptons, should there not also be four kinds of quarks? We called our construct *the charmed quark*, for we were fascinated and pleased by the symmetry it brought to the sub-nuclear world. The case for *charm*—or the fourth quark—became much firmer when it was realized that there was a serious flaw in the familiar three-quark theory, which predicted that strange particles would sometimes decay in ways that they did not. In an almost magical way, the existence of the charmed quark prohibits these unwanted and unseen decays, and brings the theory into agreement with experiment. Thus did my recent collaborators John Iliopoulos, Luciano Maiani, and I justify another definition of charm, as a magical device to avert evil." Sheldon L. Glashow, "The Hunting of the Quark", *The New York Times Magazine*, July 18, 1976, pp. 154, 159, 161; reprinted in *The Charm of Physics*, Sheldon L. Glashow, Copernicus Books, 1991; B. J. Bjørken and S. L. Glashow, "Elementary Particles and SU(4)", *Phys. Lett.* **11** (1964) 255–257; Glashow, Iliopoulos, and Maiani, *op. cit.* Note the way Bjorken signed the article, in "disguise", due to the whimsical character of the proposal; Crease & Mann *SC*, p. 291.

These are some linear orthogonal combinations of the original entries in our left-handed doublets. The problem is to determine what orthogonal combinations they are or, equivalently, how the doublets are made out of the mass eigenstates, the inverse of the transformation we did before for the case of the leptons.

Things are pretty much constrained. We have two identical doublets, which we can mix up as we wish by a unitary transformation. We can always choose the first to have for its upper spot the left-handed up quark, u_L. If we pick them randomly, one will have some linear combination of u_L and the left-handed charm quark, c_L, and the other will have another combination. Then we'll form a mixture so one is pure u_L. In the lower spot we must have some combination with norm 1 of the two possible things that can go in the lower spot, the left-handed down quark field d_L and the left-handed strange quark field, s_L. We'll have some phase times the cosine of some angle (not θ_W; an independent angle) times d_L plus some other phase times the sine of that angle times s_L:

$$L_{1,\,\text{lower}} = e^{i\delta_1} d_L \cos\theta + e^{i\delta_2} s_L \sin\theta \tag{49.8}$$

That's the most general thing we can build with charge $-\frac{1}{3}$ and norm 1. We can absorb the phases $e^{i\delta_1}$ and $e^{i\delta_2}$ into the d_L and the s_L. So we can always choose one of our left-handed doublets to look like this:

$$L_1 = \begin{pmatrix} u_L \\ d_L \cos\theta + s_L \sin\theta \end{pmatrix} \tag{49.9}$$

Now this is obtained by diagonalizing a Hermitian mass matrix, so the other doublet must be orthogonal to it. In the top slot we must have some phase and the only orthogonal isodoublet, charmed left, c_L. In the bottom slot we must have the vector that is orthogonal to the vector in (49.9), so

$$L_2 = \begin{pmatrix} e^{i\gamma} c_L \\ e^{i\delta}(s_L \cos\theta - d_L \sin\theta) \end{pmatrix} \tag{49.10}$$

This is forced on us by orthogonality. We can always choose the phase of c_L so that $e^{i\gamma}$ is the same as $e^{i\delta}$, upstairs and downstairs, because we haven't talked at all about the phase of c_L. Once we do that, we have a common phase factor, and we can send them both to 1 simply by changing the phase of the doublet L_2. That doesn't change its gauge transformation properties or anything else. The upshot of this is,[9] that *there are no phases that correspond*

---

[9] [Eds.] In Glashow, Iliopoulos, and Maiani *op. cit.*, p. 1287, the hadronic weak current is written as (their equation (3))

$$J^H_\mu = \overline{q} C_H \gamma_\mu (1+\gamma_5) q$$

where q is the quark column vector (c, u, d, s) (the authors use $(\mathcal{P}', \mathcal{P}, \mathcal{N}, \lambda)$, respectively). The matrix C_H must have the form (their equation (4))

$$C_H = \begin{pmatrix} \mathbb{O} & U \\ \mathbb{O} & \mathbb{O} \end{pmatrix}$$

where $\mathbb{O}$ is the 2×2 zero matrix, and U is a 2×2 matrix, if J^H_μ is to carry unit charge. After asserting that "The strong-interaction Lagrangian is supposed to be invariant under chiral SU(4), except for a symmetry-breaking term transforming, like the quark masses, according to the $(\mathbf{4}, \overline{\mathbf{4}}) \oplus (\overline{\mathbf{4}}, \mathbf{4})$ representation. This term may always be put into real diagonal form by a transformation of SU(4) $\otimes$ SU(4), so that $\mathcal{B}$ [baryon number], Q, Y, C and parity are necessarily conserved by these strong interactions," the authors state, "Nevertheless, suitable redefinitions of the relative phases of the quarks may be performed in order to make U real and orthogonal..." If, however, the two families of quarks are joined by a *third* generation, as was pointed out by Kobayashi and Maskawa, *CP*-violation *can* be incorporated into a theory of quarks: Makoto Kobayashi and Toshihide Maskawa, "*CP*-Violation in the Renormalizable Theory of Weak Interaction", *Prog. Theo. Phys.* **49** (1973) 652–657. This

to any physically observable quantities. The only unknown thing we have in the result is this angle θ. That's the result of putting in perfectly arbitrary Yukawa couplings consistent with the symmetries of the theory. We've just systematically pushed out arbitrary phases that are just matters of convention.[10] This angle θ is nothing but (40.16) the *Cabibbo angle*, θ_C.[11] Remember, the charged weak currents always involve Pauli τ matrices. They are left-handed currents which take the upper part of one of these doublets and mix it up with the lower part of one of these doublets. We haven't needed the charmed quark in our low-energy phenomenology, so it must be very heavy, not present in any of the observed particles.[12] So we don't have to worry about L_2, which will involve the so-called **charmed current**. For instance, from $\overline{L}_1\tau^1\gamma^\mu L_1$, we get both a *strangeness-conserving* current with magnitude $\cos\theta_C$,

$$\overline{d}_L\gamma^\mu u_L\cos\theta_C$$

and a *strangeness-changing* current with amplitude $\sin\theta_C$,

$$\overline{s}_L\gamma^\mu u_L\sin\theta_C$$

Their coefficients are $\cos\theta_C$ and $\sin\theta_C$, exactly as predicted by the Cabibbo theory (40.16). The extension to include four quarks should be clear. In addition to the terms in (48.25), there

---

idea is called the **CKM mechanism** (after their initials, and Cabibbo's). Two more quarks, $\{b, t\}$, "bottom" and "top"—a third generation—were observed in 1977 and 1995, respectively, at Fermilab. Kobayashi and Maskawa were awarded half of the 2008 Nobel Prize in Physics for their explanation of CP-violation via a mixing of the three generations of quarks; the mixing matrix is called the CKM matrix. (The other half went to Nambu for spontaneous symmetry breaking.) The new quarks each have a new quantum number: bottom has B, equal to -1, and the top has $T = +1$. The generalized Gell-Mann–Nishijima relation is (I_z corresponds to *strong* isotopic spin)

$$Q = I_z + \tfrac{1}{2}(\mathcal{B} + S + C + B + T)$$

where the baryon number $\mathcal{B} = \frac{1}{3}$ for all quarks ($-\frac{1}{3}$ for antiquarks); *PDG* 2016, p. 279, equation (15.1). In the video of Lecture 53, Coleman mentioned a theorem by Maiani, probably referring to the results reported in L. Maiani, "*CP* Violation in Purely Lefthanded Weak Interactions", *Phys. Lett.* **62B** (1976) 183–186, where the Kobayashi–Maskawa model of three doublets is analyzed. There it is shown that three mixed doublets reduce to two mixed doublets and an unmixed one, and hence *no CP*-violation results, when one real angle vanishes, or two quarks of the same charge are degenerate in mass (L. Maiani, private communication).

[10] [Eds.] Following Halzen and Martin, *op. cit.*, p. 283, equation (12.110), the doublets L_1 and L_2 can be written compactly as

$$L_1 = \begin{pmatrix} u \\ d' \end{pmatrix}, \quad L_2 = \begin{pmatrix} c \\ s' \end{pmatrix}$$

where the weak (primed) eigenstates are related to the mass or physical (unprimed) eigenstates by

$$\begin{pmatrix} d' \\ s' \end{pmatrix} = \begin{pmatrix} \cos\theta_C & \sin\theta_C \\ -\sin\theta_C & \cos\theta_C \end{pmatrix}\begin{pmatrix} d \\ s \end{pmatrix} = U\begin{pmatrix} d \\ s \end{pmatrix}$$

(This gives the same result as in the GIM paper, though their U appears different, due to a different ordering of the four quark fields.) The weak currents are then of the form

$$J^\mu = (\overline{u}, \overline{c})\tfrac{1}{2}\gamma^\mu(1-\gamma^5)U\begin{pmatrix} d \\ s \end{pmatrix} = (\overline{u}, \overline{c})\tfrac{1}{2}\gamma^\mu(1-\gamma^5)\begin{pmatrix} d' \\ s' \end{pmatrix}$$

[11] [Eds.] See note 17, p. 882.

[12] [Eds.] "Observed particles" as of May, 1976, when this lecture was given. The current estimate of m_c is 1.27 ± 0.03 GeV; *PDG* 2016, p. 36. The $\overline{c}c$ meson was found in November, 1974, by simultaneous discoveries at Brookhaven National Lab (headed by Sam Ting, MIT; the resonance was called "J") and SLAC (headed by Burton Richter; "ψ"); the meson is denoted J/ψ today. These two men shared the 1976 Nobel Prize for its discovery. The J/ψ's current mass is 3096.900 ± 0.006 MeV, *PDG* 2016, p. 1371. The various states of $\overline{c}c$ are collectively called "charmonium".

will be the additional kinetic terms $\overline{L}_\alpha \not{D} L_\alpha$ and $\overline{R}_\alpha \not{D} R_\alpha$, and perhaps additional Yukawa couplings of the form $\overline{L}_\alpha \phi R_\alpha$, where $\alpha = \{1, 2\}$.

What features does this model have, and are they natural? The following features in this theory are *natural*:

- *The Fermi theory* with Cabibbo-expressed *universality*: $\cos\theta_C$ times the strangeness-conserving current, $\sin\theta_C$ times the strangeness-changing current, with only left-handed $(V - A)$ currents.
- Discounting the weak interactions, *the electromagnetic interactions conserve all the independent quark numbers*, once we've defined them in terms of these mass eigenstates: four independent quantities which count the kinds of quarks, or simply the four currents $\{\overline{u}\gamma_\mu u, \overline{d}\gamma_\mu d, \overline{s}\gamma_\mu s, \overline{c}\gamma_\mu c\}$; the z-component of isospin, I_z; and hypercharge, Y.
- *There is no $\Delta Y \neq 0$ neutral current.* The Z^0 boson does not contribute to strangeness-changing decays. This is very important if the theory is to match experiment. If the Z^0 did contribute to $\Delta S \neq 0$ decays, we would instantly get the decay process

$$K^0 \to \mu^+\mu^-$$

which no one has observed. Why is this ruled out? If we look at the neutral currents, they act on fields either in the upper or the lower entry of L_2. With c_L there is no problem, you have charm with charm and nothing happens. In the bottom of L_2 we could get a cross term $\cos\theta_C \times (-\sin\theta_C)$. Remember they're universal, they're always added together from all the doublets. In the bottom part of L_1 we get the *same* term with the *opposite* sign. So these two terms *cancel* automatically, i.e., naturally. You don't have to fudge the parameters. Automatically the cross term in the bottom of L_1 is canceled by the cross term in the bottom of L_2 when we construct the currents. There is *no* strangeness-changing neutral current.[13]

Aside from continual fights with whether it agrees with detailed experiments, the model fails to explain two results naturally, and that makes this model slightly unsatisfactory:

- The mass of the up quark and the mass of the down quark are *equal to $\mathcal{O}(e^2)$*:

$$m_d = m_u + \mathcal{O}(e^2) \tag{49.11}$$

One would expect that to be so in order that we could be deluded into believing that all isotopic spin violation is electromagnetic, the standard dogma.[14] This is equivalent to saying that the appropriate Yukawa coupling constants are equal up to terms of order e^2. That is not natural. We can *choose* them to be equal up to $\mathcal{O}(e^2)$, but there is no

[13] [Eds.] Greiner & Müller *GTWI*, p. 230, footnote 15, echo Glashow's second meaning of "charm", stating that charm equates to magic, since it helps remove the unwanted currents.

[14] [Eds.] Forty years ago, masses within an isotopic multiplet were believed to be equal to within a few percent, i.e., to within $\mathcal{O}(e^2)$: all mass splitting was thought to be due to electromagnetic effects. This is no longer the case; strong interactions appear to be responsible for much of the difference, and isotopic spin invariance is now regarded as only *approximately* exact: Griffiths *EP*, pp. 135–136, and footnote on p. 135. The current values are: $m_u = 2.2^{+0.6}_{-0.4}$ MeV, $m_d = 4.7^{+0.5}_{-0.4}$ MeV; *PDG* 2016, p. 36. Lattice QCD calculations agree very well with these numbers; "Precise Charm to Strange Mass Ratio and Light Quark Masses from Full Lattice QCD", C. T. H. Davies, C. McNeile, K. Y. Wong, E. Follana, R. Horgan, K. Hornbostel, G. P. Lepage, J. Shigemitsu, and H. Trottier, *Phys. Rev. Lett.* **104** (2010) 132003 find, at an energy of 2 GeV, $m_u = 2.01(14)$ MeV and $m_d = 4.79(16)$ MeV.

reason why they *should* be approximately equal. That is the big unnatural feature of this model. The riddle is: why is isotopic spin approximately good?

- *CP violation is missing.* In this theory, *CP* symmetry is unbroken. That's the bad thing about our wonderful ability to eliminate all of these arbitrary phases by choosing our conventions properly. If we had some phases left around that we could not eliminate, we might have a chance of a *CP*-violating current. We don't have one in this model. We have written down all the renormalizable interactions there are, and none gives *CP* violation after the symmetry spontaneously breaks. You couldn't find a more *CP*-conserving model than the GSW theory.[15]

There are a thousand and one models, variations on the themes of the GSW model. The one I have described here is known as the **standard model.**[16] This is the best of a bad lot, in that most of the desirable things are natural, and the fewest desirable things are unnatural. Once you see how it is done, you too can construct a model. You just fiddle around putting in a bunch of left-handed quarks and right-handed quarks to make a larger or smaller gauge group, you throw in lots of unobserved particles, let the machine rip and deduce what happens. Most of these models either involve huge numbers of unobserved particles or involve some things that are now natural coming out as unnatural. For instance, the Georgi–Glashow model[17] has no neutral currents in it, which was thought to be an advantage at one time. Not anymore. But it had the unfortunate feature that the masslessness of the neutrino was unnatural. Certainly $e-\mu$ universality was unnatural. It required genius to invent this game but unfortunately it only requires persistence to continue playing it indefinitely. The literature is chockablock with models, for example there's the Pati–Salam model.[18] You have to be a real expert to be familiar with them all. To the extent that the experiments verify any model, however, they seem to support the original. That may change, of course, with a new generation of accelerators and physicists, or if someone playing this game comes up with an idea no one has thought of before. Perhaps someone will introduce a little extra twist, just as spontaneous symmetry breaking and the Higgs phenomenon by themselves were little extra twists, before they were incorporated into a new theory. But for the moment this model is the standard. 4f

49.3 Lower bounds on scalar boson masses

I want to show you at least one nontrivial calculation that involves a higher-order correction. This is the simplest I know. It uses the effective potential. Originally the calculation was carried out in the GSW model. But the algebra is complex, so I will discuss it in a simplified model. Trust me when I say that similar arguments, keeping track of all of the coupling constants, can be made in the full model.

---

[15] [Eds.] D. Chang, X-G. He, and B. McKellar, "Ruling out the Weinberg Model of Spontaneous CP Violation", *Phys. Rev.* **D63** (2001) 096005. But see note 9, p. 1081 for the CKM mechanism, which offers a way to explain *CP* violation with a third generation of quarks, a new pair observed only in 1977 and 1995, respectively.

[16] [Eds.] Griffiths *EP*, pp. 49–52. Today, "the standard model" usually means the GSW model of the electroweak forces, plus quantum chromodynamics with *three* generations of quarks $\{u, d; c, s; t, b\}$ and leptons $\{e, \nu_e; \mu, \nu_\mu; \tau, \nu_\tau\}$.

[17] [Eds.] Howard Georgi and S. L. Glashow, "Unity of All Elementary-Particle Forces", *Phys. Rev. Lett.* **32** (1974) 438–441.

[18] [Eds.] Jogesh C. Pati and Abdus Salam, "Lepton Number as the Fourth 'color'", *Phys. Rev.* **D10** (1974) 275–289.

In 1976 Steve Weinberg worked out the dynamics for his model and got a lower bound[19] on the mass of the single remaining scalar boson, the Higgs boson, corresponding to the real part of what we called the ϕ^0 field.[20] I will discuss it in a much simpler model where the essential physics is the same. The model is our old friend, the purely Abelian Higgs model: one gauge field, the photon, plus two real fields that form an SO(2) doublet:

$$\mathscr{L} = -\tfrac{1}{4}(F_{\mu\nu})^2 + \tfrac{1}{2}(D_\mu\phi_1)^2 + \tfrac{1}{2}(D_\mu\phi_2)^2 - \tfrac{1}{4}\lambda[\phi_1^2 + \phi_2^2 - a^2]^2 \tag{49.12}$$

We worked out the covariant derivatives some time ago:

$$D_\mu\phi = \partial_\mu\phi + ieA_\mu Q\phi \tag{27.48}$$

From (46.2)

$$D\phi_1 = -iQ\phi_1 = \phi_2; \quad D\phi_2 = -iQ\phi_2 = -\phi_1 \quad \text{so} \tag{49.13}$$

$$D_\mu\phi_1 = \partial_\mu\phi_1 - eA_\mu\phi_2; \quad D_\mu\phi_2 = \partial_\mu\phi_2 + eA_\mu\phi_1 \tag{49.14}$$

When spontaneous symmetry breaking occurs in the tree approximation we get a vector boson of mass

$$m_V^2 = e^2a^2 \tag{49.15}$$

and a scalar boson of mass

$$m_S^2 = 2\lambda a^2 \tag{49.16}$$

(If you don't believe me, differentiate the λ term twice about $\phi_1 = a, \phi_2 = 0$.) We can't say what's a big mass and what's a small mass. That depends on what our scale is. But we can talk about the *ratio*

$$\frac{m_S^2}{m_V^2} = \frac{2\lambda}{e^2} \tag{49.17}$$

It looks like m_S could be as large or as small as we want. We can keep λ and e^2 small, so perturbation theory is good, and make λ either much larger or much smaller than e^2. A similar remark can be applied in the case of practical interest, the GSW model, where perhaps we can hope to calculate this ratio. We'd be interested in knowing how large the scalar's mass is.[21]

Now I will demonstrate that in fact this is wrong, using only the physics we have: that there is a *lower bound* on the ratio

$$\frac{m_S^2}{m_V^2}$$

The critical point is this: if

$$\lambda \ll e^2 \tag{49.18}$$

then U, the scalar potential, is much less than the contribution from the gauge field loops:[22]

$$U \ll V_G \tag{49.19}$$

---

[19] [Eds.] Steven Weinberg, "Mass of the Higgs Boson", *Phys. Rev. Lett.* **36** (1976) 294–296.

[20] [Eds.] Peskin & Schroeder *QFT*, pp. 715–717.

[21] [Eds.] In 1976, Coleman said, "If its mass could be something like 10 eV, for example, it might be worth looking for." He probably meant "10 GeV". In 1990, he said that "m_H can't be too small, or we would have seen it; it can't be too large, or we'd have no right to do perturbation theory. We expect to see it at the SSC." In the anonymous graduate student's notes, Coleman has penciled in that $m_H > 58$ GeV. Alas, the American large accelerator, the Superconducting Super Collider, was canceled in 1993. As the world knows, we *did* see the Higgs, at CERN's Large Hadron Collider in 2012.

[22] [Eds.] See Figure 47.9, p. 1047.

Therefore we have no right to compute the mass just in tree approximation, without including the effects of the gauge field loops. Their effects, as we will see shortly, are $\mathcal{O}(e^4)$, but if $\lambda \ll e^4$, which is possible if λ and e are both small, we have no right to neglect them. They are the first terms involving e that appear in the spontaneous symmetry breaking problem. Therefore we will approximate the effective potential to investigate the situation, always assuming (49.18). That's the region we're interested in. From (47.36),

$$V = U + V_G + U_{CT} \tag{49.20}$$

U_{CT} is the finite part, the counterterms. I won't bother about the higher-order correction involving loops with scalar bosons running around them, because those are higher powers in λ, which is supposed to be negligible and we've already included the leading term of λ, in U.

Let's investigate this object, V. I'll first write down the gauge field part, (47.44):

$$V_G = \frac{3}{64\pi^2}\mathrm{Tr}\Big\{\mu^4(\overline{\phi})\ln\mu^2(\overline{\phi})\Big\} \tag{49.21}$$

The trace is trivial because there is only one vector boson in the game. It is convenient to define a parameter ρ, the "length" of ϕ^2, by

$$\rho^2 = \overline{\phi}_1^2 + \overline{\phi}_2^2 \tag{49.22}$$

and to simply write $\mu^2(\overline{\phi})$ as

$$\mu^2(\overline{\phi}) = e^2\rho^2, \tag{49.23}$$

the tree approximation mass. The gauge field part is

$$V_G = \frac{3}{64\pi^2}e^4\rho^4\ln(e^2\rho^2) \tag{49.24}$$

The rest of the argument is just algebra. The combination of U and the counterterms will come together and give us some coefficient which we'll figure out later. It's going to be some function of λ determined by our renormalization, but at this moment we don't care what it is. So $U + U_{CT}$ will be some coefficient α times ϕ^2, plus some other coefficient A times ϕ^4 plus a possible constant, and V will be all this, plus (49.24). We'll just subtract the constant, that's not going to make a difference, and obtain

$$V = \alpha\rho^2 + \beta\rho^4 + \frac{3}{32\pi^2}e^4\rho^4\ln(\rho) \tag{49.25}$$

We've used the identity

$$A\rho^4 + \frac{3}{64\pi^2}e^4\rho^4\ln(e^2\rho^2) = \beta\rho^4 + \frac{3}{32\pi^2}e^4\rho^4\ln(\rho) \tag{49.26}$$

Because $\rho^4\ln e^2$ is proportional to ρ^4, we've just included that term in β.

Now we have to determine these constants α and β. We could go through renormalization systematically and determine them in terms of the coupling constants and renormalization conventions. But we might as well get them directly in terms of quantities we want, and avoid a lot of algebra. I will impose two conditions. First, there is going to be spontaneous symmetry breaking, so (44.39) $V(\rho)$ has a minimum:

$$\frac{dV}{d\rho} = 0 \tag{49.27}$$

That value of ρ determines the scale of mass. At the end, we're just going to be computing a dimensionless ratio. To avoid a lot of complicated algebra, I'll simply choose $\rho = 1$; that sets the mass scale:

$$\left.\frac{dV}{d\rho}\right|_{\rho=1} = 0 \tag{49.28}$$

That will eliminate one of the two unknown quantities α and β. The second condition will be determined by the statement that at the minimum $\rho = 1$, the second derivative of V gives the mass of the scalar boson:

$$\left.\frac{d^2V}{d\rho^2}\right|_{\rho=1} = m_S^2 \tag{49.29}$$

More precisely, this gives the inverse of the scalar boson propagator at zero momentum transfer. But that's equivalent to the scalar boson mass, except for higher-order corrections, of order e^4 and so on. These are the equations that will determine α and β in terms of the quantities of interest. Since the symmetry breaking occurs at $\rho = 1$, the vector boson mass in these units is simply

$$m_V^2 = e^2 \tag{49.30}$$

(again, plus higher-order corrections that we're not interested in; we're just looking at the leading terms).

Starting from (49.25), the first step is to consider

$$\frac{d}{d\rho}[\rho^4 \ln \rho] = 4\rho^3 \ln \rho + \rho^3 \tag{49.31}$$

We can automatically eliminate one of the coupling constants, β, from (49.27) at $\rho = 1$:

$$\beta = -\tfrac{1}{2}\alpha - \frac{3e^4}{128\pi^2} \tag{49.32}$$

This determines the ρ^4 coefficient and we write

$$V = \alpha\left[\rho^2 - \tfrac{1}{2}\rho^4\right] + \frac{3e^4}{32\pi^2}\rho^4\left(\ln\rho - \tfrac{1}{4}\right) \tag{49.33}$$

We've split the ρ^4 term up into two parts so that they *individually* have vanishing derivatives at $\rho = 1$. There remains one unknown constant α, which we will determine in terms of the scalar boson mass, by differentiating twice.

Differentiating the first term twice at $\rho = 1$ is no hard job. We get 2 from the first term, -6 from the second term, so that gives -4α. In the other term the ρ^3 terms cancel out in the first derivative. The only non-zero term will come when we differentiate the logarithm:

$$\left.\frac{d^2V}{d\rho^2}\right|_{\rho=1} = m_S^2 = -4\alpha + \frac{3e^4}{8\pi^2} \tag{49.34}$$

At first glance it looks as though we could make the mass anything we want, by an appropriate choice of α. However, to make the mass go to zero we would have to choose α to be

$$\alpha = \frac{3e^4}{32\pi^2} \tag{49.35}$$

But what *is* α? Let's look back at V in (49.33). All the terms except the first have vanishing second derivatives at the origin:

$$\left.\frac{d^2V}{d\rho^2}\right|_{\rho=0} = 2\alpha \tag{49.36}$$

Therefore if we choose the mass very small, that is if α is positive, we have a potential that is concave upward at the origin and also at the minimum we are exploring, near $\rho = 1$: there is a stable point at the origin. It doesn't look like the tree approximation (Figure 43.4), but nevertheless that's what we've got. It looks like Figure 49.1. In that case we have to worry:

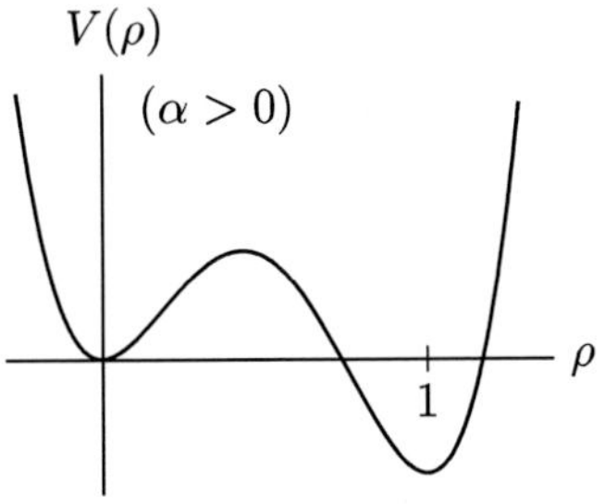

Figure 49.1: The false vacuum at $\rho = 0$

does spontaneous symmetry breaking occur? It does, if the minimum at $\rho = 1$ is *less* than the minimum at the origin, which then is a false vacuum. On the other hand if the minimum at $\rho = 1$ is *greater* than the minimum at the origin, then $\rho = 1$ corresponds to a phony vacuum. In that case we would be exploring the second derivative of the potential at a place that has absolutely nothing to do with real physics, because in fact the theory does not experience spontaneous symmetry breaking. So we have a criterion for spontaneous symmetry breaking:

$$\text{If spontaneous symmetry breaking occurs, then } V(1) \leq V(0) = 0 \tag{49.37}$$

What is $V(1)$? That's fairly easy to calculate:

$$V(1) = \tfrac{1}{2}\alpha - \frac{3e^4}{128\pi^2} \tag{49.38}$$

Since we require $V(1) \leq 0$, we now have an upper bound on α, which means the mass can't get *too* small:

$$\alpha \leq \frac{3e^4}{64\pi^2} \tag{49.39}$$

This generates (49.34) a lower bound on the scalar boson mass:

$$m_S^2 \geq \frac{3e^4}{16\pi^2} \tag{49.40}$$

Or, writing things in terms of dimensionless quantities (using (49.30)) so our peculiar scale conventions are not relevant:

$$\frac{m_S^2}{m_V^2} \geq \frac{3e^2}{16\pi^2} \tag{49.41}$$

Therefore you *cannot* make the scalar mass over the vector mass ratio as small as you please. It is obvious that similar reasoning will work in the Weinberg model. We get functions of exactly the same form. It's just that everything will be much more complicated because we've got a lot of gauge bosons to use in computing the effective potential, and therefore there's a lot more algebra involving $\sin\theta_W$ and $\cos\theta_W$. But the physics is identical.[23] In the Weinberg model, it gives an absolute lower bound on the Higgs mass on the order of 3.72 GeV.

The next lecture will be the last. In response to popular request, it will be on the *renormalization group* and its uses, and its connection with non-Abelian gauge field theory. It will be somewhat different in character from the other lectures, because I cannot cover the whole subject in ninety minutes. It will be structured more like a colloquium. I will ask you to take certain things on trust and show you other things in detail.

---

[23] [Eds.] Weinberg *op. cit.*, note 19, p. 1085. The bound Weinberg obtained is (see his equation (8))

$$m_S^2 = \frac{3\alpha^2(2+\sec^4\theta_W)}{16\sqrt{2}G_F\sin^4\theta_W} \geq \begin{cases} 4.9\,\text{GeV}, & \theta_W \approx 35^\circ \text{ (Coleman's 1976 value)} \\ 6.2\,\text{GeV}, & \theta_W \approx 27.746^\circ \text{ (}\textit{PDG}\text{ 2016, p. 119)} \end{cases}$$

Weinberg also considered a theory in which all bare masses are zero, and obtained a lower bound of about 7 GeV. The current value of the Higgs mass is 125.09 ± 0.24 GeV, well above Weinberg's lower bounds; *PDG* 2016, p. 30.

50

The Renormalization Group

We're going to take up a totally new subject and dispose of it in the course of a single lecture. The subject goes under the name of the *renormalization group*, often abbreviated as RG.[1] As you will see, this is a pretentious name for a rather simple set of ideas. It's an old idea that must have occurred to physicists many, many times: if you do experiments at sufficiently high energies, in some sense of the word energy, the masses of the elementary particles should be irrelevant. The questions are: is this true, and in what sense do we mean high energy? Obviously we don't mean it in the naïve sense. We don't mean something like total cross-section, because total cross-sections involve the imaginary parts of forward scattering amplitudes.[2] No matter how high s is, t is fixed at zero for forward scattering amplitudes,[3] and therefore we would expect at least the masses of the particles exchanged in the cross channel—the t channel—to be of critical importance, no matter how high the energy.

50.1 The renormalization group for ϕ^4 theory

Let's take a definite theory and make this idea a bit more precise. For simplicity I'll use our old friend, ϕ^4 theory with bare mass μ:

$$\mathscr{L} = \tfrac{1}{2}(\partial_\mu\phi)^2 - \tfrac{1}{2}\mu^2\phi^2 - \tfrac{1}{4!}g\phi^4 + \mathscr{L}_{CT} \tag{50.1}$$

I'll call the coupling constant g instead of λ. Let's try to find a region in which we have a better chance of masses becoming unimportant at high energies than we would, for example, for fixed-t scattering processes or total cross-sections. I'll pick a thoroughly unphysical region.

I'll consider an n-point function of n momenta, as defined in (32.14):

$$\widetilde{\Gamma}^{(n)}(p_1, \cdots, p_n) \tag{50.2}$$

---

[1] [Eds.] E. C. G. Stueckelberg and A. Petermann, "La normalisation des constantes dans la theorie des quanta" (The normalization of constants in quantum theory), *Helv. Phys. Acta* **26** (1953) 499–520; M. Gell-Mann and F. E. Low, "Quantum Electrodynamics at Small Distances", *Phys. Rev.* **95** (1954) 1300–1311; T. D. Lee, *Particle Physics and Introduction to Field Theory*, Harwood Academic Publishers, 1981, pp. 458–462; "Dilatations", pp. 79–96 and "Secret Symmetry", Sections 6.2 to 6.4, pp. 171–178 in Coleman *Aspects*; Ryder *QFT*, Section 9.4, pp. 334–339; Cheng & Li, *GT*, Chapter 3, pp. 67–85.

[2] [Eds.] The Optical Theorem, (12.49).

[3] [Eds.] See §11.3, p. 231. The Mandelstam variables s, t, and u are defined in (11.19a)–(11.19c).

The $\widetilde{\Gamma}^{(n)}$ are 1PI functions, although we could just as well worry about full Green's functions. I will consider the region where the p_i are all *Euclidean.* That is already very unphysical and rather difficult to measure, but it makes sure that I'm not going to encounter any singularities when I move around. I will also assume *no partial sum of the* p_i *is zero.* The sum of *all* the p_i's has to be zero, but I want no subset of them to add up to zero:[4]

$$\sum_i p_i = 0 \quad \text{but} \quad \sum_{\text{not all } i} p_i \neq 0 \tag{50.3}$$

Then no matter how I cut the graphs into two parts, I won't get momentum summing to zero; there will always be some non-zero momentum flowing around. I'll define an overall energy scale E (really, a pseudo-energy since we're in the Euclidean region) in terms of the sum of the p_i^2, with a minus to take care of my Euclidean metric convention:

$$E^2 = -\sum_i p_i^2 \tag{50.4}$$

I'll define some angular variables Ω_{ij}, not all independent:

$$\Omega_{ij} = \frac{p_i \cdot p_j}{E^2} \tag{50.5}$$

I will consider the **deep Euclidean region**, very far from the mass shell,[5] with the angular variables held constant:

$$E \gg \mu; \quad E \to \infty, \; \Omega_{ij} \textit{ fixed} \tag{50.6}$$

What I'm really doing is scaling up all the momenta with some overall scale E, so that any momentum passing through any part of the graph is scaled up. In that limit we would expect the mass to be unimportant. That's a guess. I won't go through complicated combinatorics of showing whether or not that guess is true order by order in perturbation theory; we haven't the time.

We can make some preliminary progress by dimensional analysis. Using only the fact that the field $\phi(x)$ has dimensions of mass, we find

$$\widetilde{\Gamma}^{(n)} = E^{(4-n)}\widehat{\Gamma}^{(n)}(\frac{\mu}{E}, g, \Omega) \tag{50.7}$$

$\widehat{\Gamma}^{(n)}$ is a dimensionless function of μ/E, the coupling constant g and all the Ω's; I'll leave out the indices. (Incidentally, the equation (50.7) is an application of *Weinberg's bound.*[6]) Let's

---

[4] [Eds.] In the literature, a set of Euclidean $\{p_i\}$ is called *unexceptional* if no *proper* subset of them sums to zero.

[5] [Eds.] Cheng & Li *GT*, pp. 73–74.

[6] [Eds.] Steven Weinberg, "High-Energy Behavior in Quantum Field Theory", *Phys. Rev.* **118** (1960) 838–849; Cheng & Li *GT*, pp. 73–74. A slightly different version is given in Bjorken & Drell *Fields*, pp. 322–324. See also Coleman *Aspects*, p. 80, equation (3.6); Weinberg's bound, that $\widetilde{\Gamma}^{(n)}$ grows no faster than E^{4-n} times a polynomial in $\ln(E/\mu)$, is given on p. 81. Briefly, the dimensional analysis goes like this: the dimension of $\phi(x)$ is $L^{-1} = E$. From the definition (13.22) of $G^{(n)}(x_1, \dots, x_n)$, it follows $G^{(n)}(x_i) \sim E^n$. Taking the Fourier transform (13.4) leads to $\widetilde{G}^{(n)}(p_i) \sim E^{-4n}E^n \sim E^{-3n}$. Finally, from the definition (32.14) we can write

$$\widetilde{G}^{(n)}(p_i)\prod_i(p_i^2 - m^2) = i^n(2\pi)^4\delta^{(4)}(p_1 + \cdots + p_n)\,\widetilde{\Gamma}^{(n)}(p_1, \dots, p_n)$$

so that $\widetilde{\Gamma}^{(n)}(p_i) \sim E^4E^{2n}E^{-3n} \sim E^{(4-n)}$.

check that: the two-point function $\widetilde{\Gamma}^{(2)}$ goes like an energy squared (32.18), and the four-point function $\widetilde{\Gamma}^{(4)}$ is dimensionless (25.29), in agreement with (50.7). Those checks suffice to fix the powers of E, times the dimensionless function $\widehat{\Gamma}^{(n)}$.

The question we want to ask about the behavior of $\widetilde{\Gamma}^{(n)}$ is: *Does it have a limit as $\mu \to 0$ for fixed values of E and the Ω's?* By dimensional analysis, that's the same as the limit $E \to \infty$ with n, Ω and g fixed. Asking whether or not the n-point functions are independent of the masses in the deep Euclidean region is equivalent to asking if a zero mass theory exists as a nice smooth limit. This is a complicated question for a general graph, and a trivial one for a tree graph. Let me take the case where it's interesting enough to have structure, but not too complicated to make this lecture infinitely long: a one-loop graph, as shown in Figure 50.1, with all momenta directed inward. It's like the graphs we considered when we were doing the

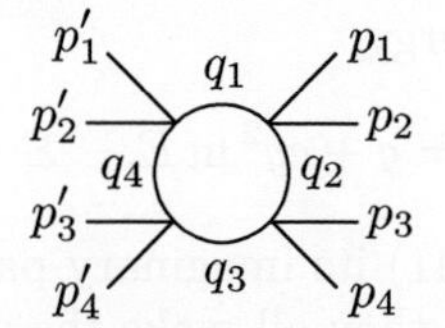

Figure 50.1: One-loop diagram

effective potential (see (44.42) and (44.44)), except now the external momenta are *non-zero*. This one-loop graph will involve an integral over the single loop momentum ℓ, times a product of four propagators over all the internal lines

$$\int d^4\ell \prod_{i=1}^{4} \frac{1}{(q_i^2 + \mu^2)} \tag{50.8}$$

(where $q_i^2 = \ell^2 +$ other stuff). I want to know if there's a limit as $\mu^2 \to 0$, which is the same as asking if there's a limit as $E \to \infty$. Naïvely I would argue as follows: there *is* a limit as long as two q's don't vanish simultaneously. After all, there are four powers of ℓ^2 in the numerator, and therefore if only *one* q_i vanishes at some particular point in the region of ℓ integration, that's not going to bother me. I can call that point the center of my ℓ integration. Then I will have

$$\int \frac{d^4\ell}{\ell^6 + \cdots}$$

which is still convergent in the ultraviolet regime. But if *two* q's vanish at the same point, I get

$$\int \frac{d^4\ell}{\ell^4 + \cdots}$$

and possible troubles from a logarithmic divergence (or worse, if more than two q's vanish).

Now, is it possible to assign the loop momentum such that two internal momenta, say q_1 and q_3 in Figure 50.1, vanish simultaneously? No, it's not. If two of the internal momenta are both equal to zero, then the sum of the four external momenta $\sum p_i'$ must be zero by energy-momentum conservation, and by hypothesis I'm in the region where no partial sum of the external momenta vanishes.[7] Provisionally it looks like, if the one-loop diagrams aren't lying to us, and more complicated things don't happen on the multi-loop level, then in the deep Euclidean region

$$\widetilde{\Gamma}^{(n)} \approx E^{(4-n)}\widehat{\Gamma}^{(n)}(0, g, \Omega) \tag{50.9}$$

This argument looks pretty good. The only problem is that it's *wrong*. We actually know what the first corrections to the four-point function $\widetilde{\Gamma}^{(4)}$ are in this theory:

$\widetilde{\Gamma}^{(4)} =$ 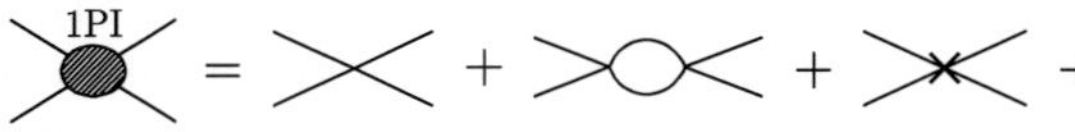 $+ \cdots +$ (crossed diagrams) (50.10)

where the dots indicate terms $\mathcal{O}(g^3)$; the $\times$ in the last graph indicates a counterterm. Consider the bubble graph in the series above. We know a lot about this graph, which we have encountered many times in the course of these lectures.[8] In particular, we know (25.29) that it grows logarithmically with the energy:

$$\widetilde{\Gamma}^{(4)} = g + cg^2 \ln E, \quad E \to \infty \tag{50.11}$$

where c is some constant, and (S15.41) its imaginary part goes to a positive, finite constant. The crossed graphs don't cancel out; they all make the same sort of contribution in the deep Euclidean region. In that region, according to (50.9), $\widetilde{\Gamma}^{(4)}$ is supposed to be a constant, but the expression in (50.11) scales like $\ln E$.

What's gone wrong? Well, I've been a little bit careless about the problem of *renormalization subtractions.* I've been treating these graphs as if they were convergent. We have two conventions at hand. Both conventions get into trouble when the mass goes to zero, because they collide into each other. One convention is to do our renormalizations *on the mass shell.* That's a disaster if the mass is zero, because the mass shell is on top of all the singularities: the one-particle pole, the two-particle cut, the three-particle cut, etc., all on top of each other. The other is to do all our subtractions with all external momenta equal to *zero*—the BPHZ prescription.[9] That's also a disaster because if all the external momenta are zero, we can get hideous *infrared* divergences when the mass goes to zero. While this sort of argument is golden before we do our renormalization subtractions, the very fact that we have to make the renormalization subtractions keeps it from working. The renormalization subtractions themselves, although they cancel the *ultraviolet* divergences, introduce new *infrared* divergences. One could argue that in the deep Euclidean region, the particle loses all knowledge of what its mass is. However, we are expressing Green's functions for renormalized fields as functions of a parameter g. What *is* g and what is the renormalized field? The renormalized field, and g, associated with a four-point function in this theory, are renormalized *on the mass shell.*

---

[7] [Eds.] To flesh out this argument, let $q_2 = \ell$. Then

$$q_1 = \ell + p_1 + p_2 = 0, \quad q_4 = \ell + p_1 + p_2 + p_1' + p_2' = p_1' + p_2',$$
$$q_3 = \ell + p_1 + p_2 + p_1' + p_2' + p_3' + p_4' = p_1' + p_2' + p_3' + p_4'$$

But if $q_3 = 0$, the partial sum $p_1' + p_2' + p_3' + p_4' = 0$, contrary to hypothesis.

[8] [Eds.] The diagram was introduced in Figure 14.3, p. 307, made a second appearance in Figure 16.9, p. 344, and was the subject of an example starting on p. 530, where the logarithmic dependence (25.29) was obtained, p. 531. It reappeared briefly on p. 536. It was also the subject of Problem 4 on the 1975 253a final examination (see Problem 15.4, p. 591, and its solution).

[9] [Eds.] See §25.2; Chapter 4, "Renormalization and Symmetry: A Review for Non-Specialists", pp. 103–104 in Coleman *Aspects*; Peskin & Schroeder, *QFT*, pp. 337–344 describe the BPHZ procedure with the four-point function in ϕ^4 theory.

Those two quantities remember the mass shell in their definition, no matter how far into the deep Euclidean region we go. This is a disease but it is very easy to cure. We won't have any problems as long as we make all of our subtractions at some fixed point in the Euclidean region that has absolutely nothing to do with the masses.

Define renormalized quantities—the renormalized wave function and charge, etc.—at some point M^2, characterized by a mass M in the Euclidean region. Then our Green's functions look more complicated because there is now another mass in the problem. They're functions of μ/E, E/M, Ω and g:

$$\widetilde{\Gamma}^{(n)}(g,(\mu/E),(E/M),\Omega) \tag{50.12}$$

This g is now a *completely different* g for the *same* physical theory, because we've defined g differently. Then the $\widetilde{\Gamma}^{(n)}$ should have a limit as $\mu \to 0$. Therefore we should be able to define a massless theory, and that should be equivalent to the massive theory in the deep Euclidean region

$$E \gg \mu, \quad M \gg \mu \tag{50.13}$$

Not only are our unrenormalized Green's functions defined far away from the mass shell, but so are the renormalization prescriptions themselves. So all of our counterterms are insensitive to the mass.

Is the trick clear? In order to avoid the renormalization conventions bringing the mass shell back in again, we've got to pick the renormalization point, where we define both the scale of the renormalized fields and the renormalized coupling constants, to be some point very far off the mass shell, so that they will not see the mass term. At this moment this is just a hope: that if we do things this way there will be a smooth zero-mass limit. I will now turn it from a hope to a flat assertion and tell you that it *can* be proven order by order in perturbation theory. I won't do that here.[10] The argument for one-loop graphs is given above; there are many complexities in analyzing multi-loop graphs.

We don't have any predictions at this stage. Before we had a beautiful prediction that everything would just go like a power of E by dimensional analysis. Even if we set μ/E to zero, we still have a dimensionless parameter E/M. So it looks like we've solved our problem in principle, but gained no practical information. In fact one gains an enormous amount of practical information by doing this. We're going to study massless theories, and in particular, develop general techniques for analysing the energy behavior of Green's functions in massless field theories, which I will write down in a more general form than this simple example.

The *renormalization group* is a technique for studying fully massless renormalizable field theories. It doesn't work for nonrenormalizable theories. (Nothing works for nonrenormalizable theories, to our knowledge; we don't even know if they exist in any real sense.) We want to study them because the behavior of such a fully massless theory will mimic the behavior of a real theory with masses when we go to the deep Euclidean region. Indeed there are other cases where we can study certain properties of a real theory with masses by studying corresponding properties of a fully massless theory. There is a long, famous, and very important analysis that shows that certain quantities associated with deep inelastic electroproduction, the so-called "moments" of the **Bjorken structure functions**, $F_i(k^2, x)$ $(i = 1, 2)$, behave as they would

---

[10] [Eds.] Coleman *Aspects*, "Dilatations", Section 4.3, "Scaling and the Operator Product Expansion", pp. 93–96; Itzykson & Zuber, *QFT*, pp. 654–656.

behave in a fully massless theory of quarks (the variable x is defined below).[11] Let me describe the physics very briefly.

Deep inelastic electroproduction

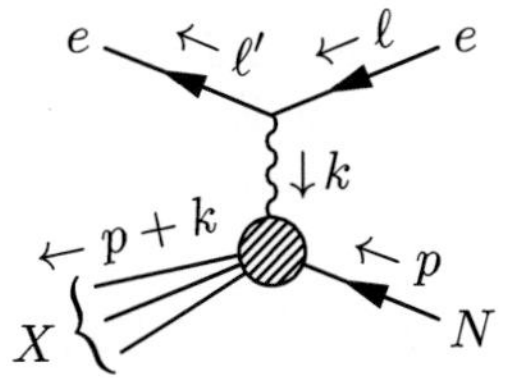

Figure 50.2: Deep inelastic electroproduction

The process in deep inelastic electroproduction, shown in Figure 50.2, is

$$e + N \to e + X \tag{50.14}$$

where N is a nucleon, typically a proton, and X is any multiparticle state. Let k equal the momentum of the virtual photon, and p the momentum of the target nucleon. If we know the momentum transfer k, we know everything, because k^2 and $k+p$ are invariants; k is spacelike, because $k = \ell' - \ell$, where ℓ and ℓ' are the electron momenta, and $k^2 = k \cdot (\ell' - \ell) < 0$ in an inelastic collision.[12] It is useful to introduce the *Bjorken scaling variable*, x;

$$x \equiv \frac{-k^2}{2k \cdot p} = -\frac{k^2}{2Em_N} \tag{50.15}$$

where E is the energy of the virtual photon in the lab frame, and m_N is the mass of the target nucleon. Because $(k+p)^2 > p^2$, elementary kinematics give

$$k^2 + 2k \cdot p \geq 0 \Rightarrow \frac{-k^2}{2k \cdot p} \in [0, 1] \tag{50.16}$$

The moments of the structure functions are defined by

$$F_i^n \equiv \int_0^1 dx\, x^n F_i(k^2, x) \to \text{constants} \tag{50.17}$$

as $-k^2 \to \infty$. In fact, as $-k^2 \to \infty$, the F_i's appear to depend *only* on x. This phenomenon is called *Bjorken scaling*. Shortly after Bjorken's discovery, it was realized that this behavior implied that the electrons scattered off *pointlike particles* inside the nucleon; moreover, these particles behaved as if they were essentially free.[13]

---

[11] [Eds.] J. D. Bjorken, "Asymptotic Sum Rules at Infinite Momentum", *Phys. Rev.* **179** (1969) 1547–1553; Coleman *Aspects*, "Secret Symmetry", Section 6.1, pp. 169–171. Be careful not to confuse the Bjorken functions F_i's with the form factors F_1 and F_2 from earlier lectures.

[12] [Eds.] Another way to see that $k = \ell' - \ell$ is spacelike: square both sides. Then $k^2 = 2m^2 - 2\ell \cdot \ell'$. Go to the center of momentum frame of the electrons, $\ell = (\ell_0, \boldsymbol{\ell})$, $\ell' = (\ell_0, -\boldsymbol{\ell})$, and $k^2 = 2m^2 - 2(\ell_0^2 + |\boldsymbol{\ell}|^2) = 4(m^2 - \ell_0^2) = -4|\boldsymbol{\ell}|^2 < 0$.

[13] [Eds.] The first direct evidence of quarks came from deep inelastic scattering experiments of electrons off protons, carried out at the Stanford Linear Accelerator (SLAC) in 1967–1970, headed by Jerome I. Friedman, Henry W. Kendall and Richard E. Taylor. These three shared the 1990 Nobel Prize in Physics for this work. See Crease & Mann *SC*, pp. 299–308.

So there are lots of things we know, beyond the fact that these simple Green's functions, in the deep Euclidean region, behave as they would behave in a fully massless theory. And then there are things that we can actually measure that are insensitive to the masses in the theory. That's an important but secondary matter. Someone else does some hard work and says "Look, the quantity zilch is insensitive to the masses as the masses go to zero," and then you say "I can use the renormalization group to study zilch." These are two separate issues.

I want to explain what I mean by a *fully massless theory.* A fully massless theory is one which has no masses in it and no parameters with the dimensions of mass. No ϕ couplings, no ϕ^3 couplings, just a set of dimensionless coupling constants. In fact we know in what sort of interactions such dimensionless coupling constants appear: quartic interactions between scalar mesons, Yukawa interactions, and gauge field interactions. That's it. Their values are not important. I'll call them

$$g^a, \quad a = 1, \dots, m \tag{50.18}$$

The theory will involve a set of fields

$$\phi^A, \quad A = 1, \dots, N \tag{50.19}$$

These fields may be the fundamental fields of the theory, whose Green's functions we want to study:

$$\widetilde{\Gamma}_{A_1,\cdots,A_S}(p_1,\dots,p_S;g^1,\dots,g^r;M) \tag{50.20}$$

Perhaps we want to look at something like the Green's functions for a string of currents in electrodynamics. Or maybe we want to investigate something peculiar, like the Green's function for seven of the ϕ's. It doesn't matter whether these are fundamental fields or not, nor what their Lorentz transformation properties are. These properties will not be relevant in our analysis.

Finally, despite the fact that it is a fully massless theory, it has *one* mass, M, which determines the *mass scale* at which we define all of our renormalization conventions. That mass *cannot* be zero. If it were, then as we make subtractions at zero, we're subtracting infrared divergent quantities. This sole mass M will define the renormalization point, the place where we subtract our propagators, and the place where we set four-point functions equal to a certain value to get the physical coupling constant.

I'll apply the method to a Green's function, but once we see how it works, we'll see that it applies to practically anything. A general Green's function in such a theory will be[14]

$$\text{F.T. } \langle 0|T\Big(\phi^{A_1}(x_1)\cdots\phi^{A_S}(x_S)\Big)|0\rangle = \sum_r \underbrace{(\text{kinematic factors})^{(r)}}_{p^2,\ \not{p},\ \text{etc.}} f^{(r)}(E/M,\Omega,g) \tag{50.21}$$

The $f^{(r)}$ are *scalar* functions, which I will choose to be dimensionless by pulling out sufficient powers of E. It's important that the $f^{(r)}$ are scalars, not the components of a 3-vector. That they are *Lorentz* scalars is going to be irrelevant. The range of r depends on what the Green's function is (how many covariants we can make). If it's a two-point function for a spinor field there'll be two of them; if it's for a scalar field there'll be one; if it involves 17 fields of very high spin there'll be all sorts of things with Dirac γ matrices and tensor indices, and then they'll break up into a bunch of scalar invariants.

---

[14] [Eds.] F.T. = Fourier transform.

I stress that it's the *physical* masses that are zero, not the *bare* masses. That is the one renormalization convention that *must* be imposed at the point 0, rather than at the point M. For vector theories, gauge invariance imposes zero physical mass if you have zero bare mass (*modulo* questions of spontaneous symmetry breaking, which aren't relevant for this kind of analysis). And for spinors, in most of the theories we're interested in, γ_5 invariance requires zero physical mass if the bare mass is zero. It's only for scalar fields that zero bare mass does *not* imply zero physical mass; in this case you have to make a subtraction. Then you may worry about whether that subtraction will give you new infrared divergences. It doesn't, but you'll have to take my word for it. (A subsidiary argument needs to be made; the wave function renormalization prevents it.)

You should really not trust me in some matters. If I tell you that something has been proved, that doesn't mean that I've actually gone through the paper and read all the details. It means that someone has sent me a preprint.[15] I look at it, and if it looks too horrible to wade through, I read the abstract which says a theorem has been proved. I tell my students it's been proved. Then two years later somebody comes by and says that this proof was no good, and I say "It's *not* been proved?" It's like that joke: "Life is *not* a fountain?"[16]

We want to study the behavior of this Green's function (50.21) as we change E. I'll suppress both Ω, since that's going to be held fixed throughout the entire argument, and also the index r since we'll just look at these things one at a time; *which* one is not particularly relevant. One feature will turn out to give us powerful information: M has an *arbitrary* value, so long as it is somewhere in the Euclidean region. That is, if I change M by a dimensionless infinitesimal amount ϵ,

$$M \to (1+\epsilon)M \tag{50.22}$$

then I can keep the *same* physical theory; I have just changed my renormalization convention. I will also have to change all the g's by an amount of order ϵ because I've changed the renormalization point. There will be some function, β^a, a function of all the g's but not of M by dimensional analysis:

$$g^a \to g^a + \epsilon\beta^a(g) \tag{50.23}$$

When I write a g inside $\beta^a(g)$ I mean the set $\{g^1, \dots, g^m\}$; the functions $\{\beta^a\}$ are functions of *all* the g's. The $\{\beta^a\}$ are called, not surprisingly, **beta functions**. Each of the coupling constants may be related to all the others in a complicated nonlinear way. I may have to rescale my fields:

$$\phi^A \to \phi^A + \epsilon\gamma^A(g)\phi^A \tag{50.24}$$

The γ^A are functions, not Dirac matrices. For reasons to be explained later they are often called **anomalous dimensions**. Under these changes (50.22)–(50.24), f will remain *fixed:*

$$f \to f \tag{50.25}$$

[15] [Eds.] This was in the days before the arXiv, when preprints arrived in an envelope. Occasionally the same result simultaneously derived by two rival groups was sent by each to the other, and crossed in the mail.

[16] [Eds.] An old joke, with many variants. A young man seeking enlightenment travels to a distant land to ask a famous wise man the meaning of life. "Life," the ancient sage tells him, "is like a fountain." The young man thanks the sage, and goes off to make his fortune. Many years later, he decides to revisit the old master in his last days. He says to him, "Master, I thank you for your wonderful advice. It has served me well through many trials. But I must confess to you that I really don't understand it." The sage reflects for a few moments, and asks the younger man, "Life is *not* like a fountain?" For another version, see Jimmy Pritchard, *The New York City Bartender's Joke Book*, Warner Books, 2002.

because it's the *same* theory.

Please note that it does not matter what the momenta are as long as the renormalization point is in the Euclidean region. (I don't want to make my renormalization subtractions on top of singularities.) Of course the massless theory is comparable to the massive theory *only* in the deep Euclidean region. But if I separate the question into two parts—how do I study the massless theory, and when can I use the massless theory to study the massive theory—the question of being in the deep Euclidean region is relevant to the second part, but not to the first.

I should say that (50.24) may be a matrix equation. We have occasionally talked about cases where we have to mix together several fields which have the same dimension and the same Lorentz transformation properties as a consequence of the renormalization, and we might get much worse things, if we're looking at objects like ϕ^4, which might get mixed up with $(\partial_\mu\phi)^2$. So really the γ in (50.24) should be thought of as a matrix. For algebraic simplicity and subsequent equations, however, I will assume it's diagonal; that the fields we are studying do not mix up with each other under renormalization. But I will make a little point here. It might be that

$$\phi^A \to \phi^A + \epsilon\gamma^{AB}(g)\phi^B \tag{50.26}$$

That is, some of our fields may mix up with each other in the course of renormalization. I just won't worry about that here. The generalization to the case where these matrices are present is fairly trivial.

I am assuming that in the deep Euclidean region, the Green's functions are continuous. If you've got the same physical theory, you may suddenly wake up and find yourself in the world of fully massless particles. You want to parametrize it. You say, "I suspect this is fully massless ϕ^4 theory (or Yang–Mills theory). I'm going to do a bunch of experiments. I'll measure some Green's functions in the deep Euclidean region," (you have terrific experimental apparatus) "and define the coupling constants."

So you find the coupling constant g for $g\phi^4$ theory and say to the outside world "I found myself in a universe of $g\phi^4$ with coupling constant equal to 0.1, for the massless theory." And they say "How did you define g?" And you say, "Oh, I defined it as a four-point function at m_e, the mass of an electron, with $s = t = u$, and with all external $p^2 = -m_e^2$." And they come back and say "That's not the standard way. We want you to define it at the mass of a Coleman," or some other arbitrary different mass. But it's the *same* theory. So you say, "Oh, all right, I'll make that measurement." And you say it's the theory with $g = \frac{3}{4}$ or $\frac{1}{10}$ or whatever. But it's still the same theory. All M tells you is how you label the Green's function, and how you scale your field. So, suppose they want the coupling constant for four fields at the mass of an electron. Well, you ask, "Which fields?" They say, "Scalar fields." You ask "How do you want it renormalized? Renormalized so that the two-point function has first derivative equal to one at $-m_e^2$?" And they reply, "No, we want it to be at $-m^2_{\text{Coleman}}$." And so on. But it's the *same theory.* The amazing thing is that by keeping our wits about us we can use this trivial fact to get nontrivial information.

50.2 The renormalization group equation

These Green's functions, or these dimensionless scalar quantities f that characterize the Green's functions, are unchanged by this trivial group of transformations, which is just the reparametrization of the theory. The set of transformations (50.22)–(50.24) is called the

renormalization group. Rarely has there been a more pretentious name in the history of physics. It's like calling classical dynamics "the study of the Hamiltonian group of time translations". Nevertheless, that's what it's called. I've written this in terms of infinitesimals, but everything I can write in terms of infinitesimals I can of course write in terms of a differential equation, and I will now do so. This differential equation says that f *does not change* under these combined things. I'll first write it down for a particular function f:

$$\boxed{\left[M\frac{\partial}{\partial M}+\beta^a(g)\frac{\partial}{\partial g^a}+\gamma\right]f(E/M,g)=0} \tag{50.27}$$

where

$$\gamma\equiv\sum_{A=1}^{N}\gamma^A \tag{50.28}$$

is the sum of the little γ's associated with whatever fields occur in the definition of the Green's function; each of the fields ϕ^A is getting rescaled by an amount $1+\gamma^A$, and that just multiplies the whole Green's function by the number $1+\gamma$. This is known variously as the **renormalization group equation** (or RGE), or as the **Callan–Symanzik** (or CS) equation.[17] This differential equation follows from the trivial statement that it doesn't matter what the mass M is: γ depends on the particular f you are studying and in this way depends on how many fields of which kind it has in it. Everything else is totally independent of what the particular function f is.

In any fully massless field theory there are, by ordinary dimensional analysis, functions of the coupling constant only (not depending on the renormalization point), the β's; one for each coupling constant. There are other functions of the coupling constants, one for each field you happen to be studying, the γ's, such that each and every Green's function will obey this differential equation (50.27). From this fact we could compute the β's and the γ's iteratively in perturbation theory, because there is no problem computing the Green's functions iteratively in perturbation theory. Thus we could fix the β's and the γ's as those coefficients that make this equation true. It looks much more complicated but it's still the same trivial statement that the value of M is irrelevant, and that the effects of infinitesimally changing M can be compensated for by effects of changing the coupling constants, with the $\{\beta^a\}$, and changing the scale of the fields, with the $\{\gamma^A\}$. There is no profound input into it, but it yields surprisingly profound output.

A nice exercise (that I leave to you) is to find the β's and the γ's by applying this equation to Green's functions at the point where the renormalization constants are defined. That makes

[17] [Eds.] K. Symanzik, "Small Distance Behavior in Field Theory and Power Counting", *Commun. Math. Phys.* **18** (1970) 227–246; Curtis G. Callan, Jr., "Broken Scale Invariance in Scalar Field Theory", *Phys. Rev.* **D2** (1970) 1541–1546; "Introduction to Renormalization Theory", pp. 42–77 in *Methods in Field Theory (Les Houches 1975)*, eds. R. Balian and J. Zinn-Justin, North-Holland, 1976; "Dilatations", p. 86 in Coleman *Aspects*; Ryder *QFT*, Section 9.4, pp. 334–339. See also the closely related article by Wilson and the review article by Huang: Kenneth G. Wilson, "Anomalous Dimensions and the Breakdown of Scale Invariance in Perturbation Theory", *Phys. Rev.* **D2** (1970) 1478–1493; K. Huang, "A Critical History of Renormalization", *Int. J. Mod. Phys. A* **28** (2013) 11330050; available online at `https://arxiv.org/abs/1310.5533`. It is perhaps worth quoting the acknowledgement in Callan's 1970 article: "It is a pleasure to acknowledge many discussions with Sidney Coleman, without which this paper could not have been written." Technically the CS equation is an inhomogeneous equation with a mass-related term on the right-hand side. It becomes the RGE in the deep Euclidean region. For a discussion of the differences between the two equations, see M. Kaku, *Quantum Field Theory: A Modern Introduction*, Oxford U. P., 1993, Section 14.7, pp. 485–488.

life particularly simple, but it doesn't matter. If you know the Green's functions for any specified value of the coupling constants, you know how you have to change the definition of the coupling constants, and rescale the fields, to keep the physics the same. Then you can compute β and γ. Thus the β's and the γ's have well-defined perturbation theory expansions. Whether they're convergent or not is of course an open question, just as it is for the Green's functions. Here are two examples to give you an idea of how these things go.

EXAMPLE 1. $\mathscr{L}' = g\phi^4$

The single constant β arises because we have to redefine the one coupling constant g when we go to a new renormalization mass. The coupling constant is defined in terms of the four-point function, and therefore we will have to redefine it only if the four-point function has a nontrivial momentum dependence. Such dependence arises at the one-loop level as shown in Figure 50.3—that's a term in the four-point function that does depend on momentum—and therefore β will first appear in order g^2 in this theory. There's only one coupling constant so

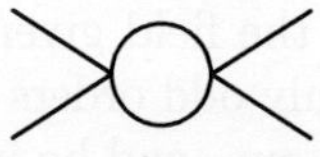

Figure 50.3: The one-loop contribution to β in ϕ^4 theory

$$\beta = cg^2 + \mathcal{O}(g^3) \tag{50.29}$$

where c is a constant. The function γ^A appears if we have to change the scale of the field when we change the renormalization point. That happens only if the *propagator* has a nontrivial momentum dependence. In this theory that also happens in order g^2, as shown in Figure 50.4. There's only one field, so there's only one γ:

$$\gamma = dg^2 + \mathcal{O}(g^3) \tag{50.30}$$

That's obvious. What's *not* obvious, and will be important to us, is the sign of c (or more

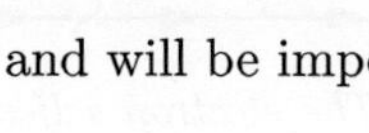

Figure 50.4: The sunset diagram contribution to the propagator in ϕ^4 theory

precisely, *the sign of* β); the sign of d will turn out to be completely irrelevant. In the interest of time, I ask you to take on trust that c is *positive*:

$$c > 0 \tag{50.31}$$

(It's trivial to verify. Just compute these graphs, which is pretty easy in a fully massless theory. You won't obtain any complicated functions at all, just $\ln(E/M)$.) That sign will be important to us later, although it's not yet clear why.

EXAMPLE 2. QED: $\mathscr{L}' = -g\overline{\psi}\gamma_\mu\psi A^\mu$

It's not always true that β and γ first appear in the same order of perturbation theory. For example, take quantum electrodynamics, with a massless electron as well as a massless

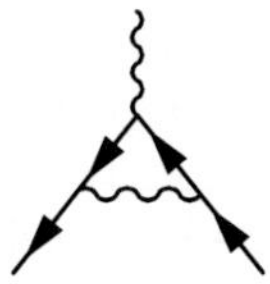

Figure 50.5: The vertex correction in QED to $\mathcal{O}(e^3)$

photon. The coupling constant is defined in terms of the three-point function and a famous diagram: The first diagram that gives the three-point function momentum dependence is shown in Figure 50.5, which is $\mathcal{O}(g^3)$ (or $\mathcal{O}(e^3)$, as we said earlier[18]):

$$\beta = c'g^3 + \mathcal{O}(g^5) \tag{50.32}$$

The amplitude for an off-shell electron to go into a physical electron and a photon has momentum dependence. It therefore will introduce momentum dependence on the renormalization point, and how you define the scale of the field given in the fixed theory. These things are defined on the mass shell, not off. It's only odd orders in this case because we have to stick on a photon in two places. It also will turn out—and be important to us later—that c' is positive,

$$c' > 0 \tag{50.33}$$

That's what we have to do a computation for. We actually have that computation in hand.[19] In QED there are two separate γ functions, γ_ψ for the electron's field and γ_A for the photon's. The first time the electron propagator starts getting momentum dependence is in Figure 50.6, and

$$\gamma_\psi = d'g^2 + \mathcal{O}(g^4) \tag{50.34}$$

We've always got to add an even number of powers. For the photon vacuum polarization, the

Figure 50.6: The electron self-energy to $\mathcal{O}(e^2)$

relevant graph is shown in Figure 50.7. We have

$$\gamma_A = d''g^2 + \mathcal{O}(g^4) \tag{50.35}$$

This is how you get the powers of g. It should be clear how we actually compute the coefficients. We just stick in the graphs at the appropriate order and then fix the coefficients so the renormalization group equations are true.[20]

Why did I bother to go through all this? I wanted to show you that M is an irrelevant parameter. It's necessary, but its value isn't important; we always get the same physics.

---

[18] [Eds.] See Figure 34.6, p. 744.

[19] [Eds.] See the discussion of the electron's anomalous magnetic moment, §34.3. The β function in QED requires input from the electron self-energy, the photon self-energy and the Ward identity in the simple form $Z_1 = Z_2$, as well as the vertex diagram. The value for c' in (50.32) is $c' = 1/(12\pi^2)$; Peskin & Schroeder *QFT*, pp. 415–416.

[20] [Eds.] Peskin & Schroeder *QFT*, Section 12.2, "The Callan–Symanzik Equation", pp. 406–418.

Figure 50.7: The vacuum polarization to $\mathcal{O}(e^2)$

Therefore no matter what Green's function I study, I know its M dependence from (50.27) if I know the β's and γ's. In fact I don't know what they are, but presumably I could calculate them in perturbation theory. There are certainly far fewer quantities than the possible number of Green's functions. If I do know them, I know the M dependence. By dimensional analysis $f^{(r)}$ depends on M only in the combination E/M. So if I know the M dependence, I know the E dependence. It's almost as good as the old case where we were being very naïve, not worrying about renormalization effects. In that case we assumed the E dependence of all these dimensionless functions was trivial: they were E-independent. Here we say: Well, we don't know them trivially. But if we know the β's and the γ's, a finite set of functions, then we'll know the E dependence of everything. I will now go through a little exercise using the *method of characteristics*.[21] I will write down the general solution of the renormalization group equation in terms of initial value data at a fixed M, show how we get the solution at a general value of M, study its properties and apply it.

50.3 The solution to the renormalization group equation

How do I solve this equation using physical intuition, assuming that I know the β's and γ's exactly (which in general I don't)? Actually I hardly have to do any work. Although it may not look like it, this has a similar structure to an equation whose solution we can almost obtain by inspection. Let me show you that second equation,[22] then I'll write down the solution. Let $\rho(x,t)$ be a scalar function for the population of bacteria in a fluid, at the position x at a given time t. The bacteria are carried with a known velocity $v(x)$ down a transparent tube, subject to a position-dependent illumination $L(x)$, also known, which determines their rate of reproduction. The bacteria move down the pipe only because the fluid is moving. They have a limitless amount of sugar to eat; all they need is light. Then they grow exponentially, depending on how much light they're exposed to. Under these conditions, $\rho(x,t)$ obeys the

[21] [Eds.] In 1990, Coleman said, "You can read about this [topic] in Courant and Hilbert, which nobody younger than me has ever held in their hands." He was referring to Richard Courant and David Hilbert, *Methods of Mathematical Physics*, v.2, John Wiley Interscience Publishers, 1962, pp. 450–463. (The editors proudly serve as counterexamples.) The work deserves to be better known by later generations. Long ago, the immense importance of Courant–Hilbert to the development of quantum mechanics was famous: "In retrospect, it seems almost uncanny how mathematics now prepared itself for its future service to quantum mechanics... [In May, 1924] Courant, utilizing Hilbert's lectures, finished in Göttingen the first volume... Published at the end of 1924, it contained precisely those parts of algebra and analysis on which the later development of quantum mechanics had to be based; its merits for the subsequent rapid growth of our theory can hardly be exaggerated. One of Courant's assistants in the preparation of this work was Pascual Jordan..." Max Jammer, *The Conceptual Development of Quantum Mechanics*, MIT Press, 1966, p. 207. The two volumes were deemed so crucial to the war effort that the U.S. government (which had seized the copyright on all German works) had Interscience publish an edition (in the original German, which nearly all American physicists of the era read) in 1943; 7000 copies were sold: Constance Reid, *Courant in Göttingen and New York: The Story of an Improbable Mathematician*, Springer-Verlag, 1976, p. 465; also published with Reid's earlier biography *Hilbert* (Springer-Verlag, 1970) in a single volume, *Hilbert–Courant*, Springer-Verlag, 1986.

[22] [Eds.] Coleman *Aspects*, Chapter 3, "Dilatations", pp. 88–90; Peskin & Schroeder *QFT*, pp. 418–420.

following differential equation:

$$\left[\frac{\partial}{\partial t} + v(x)\frac{\partial}{\partial x}\right]\rho(x,t) = L(x)\rho(x,t) \tag{50.36}$$

You can see a family resemblance to the renormalization group equation (50.27). The motion of a fluid element in a given velocity field $v(x)$ is often described by a device well known in hydrodynamics, the convective or total derivative D/Dt:

$$\frac{D}{Dt} = \frac{\partial}{\partial t} + \mathbf{v}\cdot\nabla = \frac{\partial}{\partial t} + v\frac{\partial}{\partial x} \quad \text{(in one dimension)} \tag{50.37}$$

I'll solve the equation (50.36), and then, after making a transcription between the two equations, it will be very easy to obtain the solution of the renormalization group equation. With one g they are the same equation, with v playing the role of β and $-\gamma$ playing the role of L. When you've got several g's, it's much the same story, except that instead of moving down a pipe in a given velocity field, the bacteria are moving in an n-dimensional space.

The solution is pretty simple; it requires two steps. First we find out how to describe the motion of an element of fluid, and then we work out the history of the bacteria. So step one is to solve an ordinary differential equation. One defines the function $\overline{x}$ as the solution of the equation

$$\frac{d\overline{x}}{dt} = v(\overline{x}) \tag{50.38}$$

with the boundary condition

$$\overline{x}(x,0) = x \tag{50.39}$$

That's an ordinary differential equation, not a partial differential equation. It determines x as a function of the single variable t and also of the boundary values $\overline{x}(x,0)$. Physically, $\overline{x}(x,t)$ is the position at time t of a fluid element which was at x at a time $t = 0$. The differential equation (50.38) tells us how an element of fluid moves in the given velocity field. In particular, $\overline{x}(x, t_1 - t_2)$ is the position at time t_1 of the element of fluid which reaches x at a time t_2.

In step two we study the bacteria. At time $t = 0$, the function $\rho(x,t)$ is equal to some function $P\big(\overline{x}(x,0)\big) = P(x)$ of how many bacteria were then at location x—the initial value of ρ. The bacteria multiply exponentially depending on the value of L at the point where they are now. The whole thing is time-translation invariant, so the solution to the bacteriological problem (50.36) is[23]

$$\rho(x,t) = P\big(\overline{x}(x,-t)\big)\exp\left[\int_{-t}^{0} dt'\,L\big(\overline{x}(x,t')\big)\right] \tag{50.40}$$

[23] [Eds.] Perhaps because the time was short, Coleman skipped a few steps in this derivation. At time $t = 0$, the differential equation (50.36) can be written, with the given definitions, as

$$\left.\frac{dP\big(\overline{x}(x,t)\big)}{dt}\right|_{t=0} = L\big(\overline{x}(x,t)\big)\,P\big(\overline{x}(x,t)\big)\Big|_{t=0}$$

The solution to the associated equation for *all* times is

$$P\big(\overline{x}(x,t)\big) = P\big(\overline{x}(x,0)\big)\exp\left[\int_0^t dt'\,L\big(\overline{x}(x,t')\big)\right]$$

Shift t backwards to 0, and (50.40) follows. See Peskin & Schroeder *QFT*, pp. 418–420; Coleman *Aspects*, "Dilatations", pp. 88–89.

To check this solution, note that at $t = 0$, the exponential disappears, $\overline{x}(x, 0)$ is simply x, and the solution becomes

$$\rho(x, 0) = P(x) \tag{50.41}$$

as it should.

Let's forget the charming bacteria and return to the renormalization group equation. We may have no comparable intuition for the renormalization group equation, but surely we have enough wit to make the translation from one equation to the other. The solution of the RG equation in terms of the β's and γ's is going to be the pivot point for the rest of the lecture.

Bacteria	*RGE*
t	$-\ln(E/M)$
v	β
x	g
L	$-\gamma$

Table 50.1: Translation between bacteria and renormalization group variables

Following the solution to the bacteria problem, define $\{\overline{g}^a\}$ as a set of functions that solve the simultaneous ordinary differential equations

$$\frac{d\overline{g}^a}{dt} = \beta^a(\overline{g}) \tag{50.42}$$

The function β^a depends on all the $\overline{g}$'s, with the boundary conditions

$$\overline{g}^a(g^b, 0) = g^a \tag{50.43}$$

This leads to functions $\overline{g}^a(g, t)$ depending on the initial values of all the g's and t, called **running coupling constants.**

Now to make the substitution.[24] We identify t with

$$t = -\ln(E/M) \tag{50.44}$$

Then the solution to the RG equation (transcribed from (50.40)) is

$$\boxed{f(E/M, g) = F\big(\overline{g}(g, \ln(E/M))\big) \exp\left[\int_0^{\ln(E/M)} dt' \gamma\left(\overline{g}(g, t')\right)\right]} \tag{50.45}$$

where $F\big(\overline{g}(g, 0)\big) = F(g) = f(1, g)$.

This tells us exactly what we would expect the first-order differential equation to tell us: that if we know the value of the function at any fixed E for all values of the coupling constants, and if we know the β's and γ's, then we know the function at all E's and for all coupling constants. That's a very powerful statement, but of course, it's also a trivial statement. It's the statement that M is an irrelevant parameter, exploited by straightforward calculus. We have

24 [Eds.] $M\partial/\partial M = (M/E)\partial/\partial(M/E) = \partial/\partial t$ if $t = \ln(M/E)$.

found the general solution (50.45) to the first-order *partial* differential RG equation (50.27) in terms of the solution for a *system* of first-order *ordinary* differential equations (which, by the way, is the "method of characteristics" referred to earlier).

I should make a small point. In order to keep the parallelism I had to substitute $-\ln(E/M)$ for t, but frequently in the literature, because of the way all of the minus signs come into the solution of the hydrodynamic equation, the parameter that enters is in fact $\ln(E/M)$. But it's the same prescription.

50.4 Applications of the renormalization group equation

I will show you three out of a host of applications of this equation and its general solution. The three applications will be:

- *The zeros of* β. In a one-coupling constant theory, $\beta(g)$ always has a zero at the origin (at $g = 0$). The question is: if there are zeros elsewhere, does that tell us anything about the high-energy behavior of the function? That's not a question we can answer from perturbation theory. We have to invent some non-perturbative method of analysis to see what happens when β has a zero. Nevertheless, the consequences of β having a zero are so interesting that it is worth pursuing them, even if we don't know whether or not it does.
- *Study of powers of* $\ln(E/M)$ *in perturbation theory.* This is also sometimes known as *summing the leading logarithms.*[25] When we did the four-point function in perturbation theory, we found to lowest order there was no logarithm. To $\mathcal{O}(g^2)$ there was one power of a logarithm. Does this go on? To $\mathcal{O}(g^3)$ are there two powers of logarithms? To $\mathcal{O}(g^4)$ are there three? Maybe things sometimes go bananas. When we go to $\mathcal{O}(g^{18})$ do we get 22 powers of a logarithm, or does each order introduce a single power of the logarithm? Who knows? But with the aid of this little wonder, the RG equation, we'll be able to answer that question, and without doing any work.
- **Asymptotic Freedom**. I have saved the best for last, and, in the manner of Hyman Kaplan,[26] I put asterisks around it, for reasons you will soon appreciate.

The effects of zeros of β

For simplicity I will assume we are working in a one coupling constant theory, like $g\phi^4$, so I can draw a graph of β. Figure 50.8 shows a hypothetical function for β. The only thing we know (50.11) is that this begins with a positive coefficient times g^2, so it starts out like Figure 50.8, with a quadratic root at the origin. After that we are in a state of total ignorance. Maybe some people who work with lattice quantum field theory can compute the strong coupling limit and tell us something about it, but I can't. So let's just make a guess about β and see its consequences. I have no particular reason for assuming this form; it's just a nice example for describing the zoölogy of the zeros of β. To make life interesting, I'll assume it has a zero at a point g_1 and a second zero at g_2, a third zero at g_3, and then it stays negative. Who knows?

---

[25] [Eds.] V. V. Sudakov, "Vertex Parts at Very High Energies in Quantum Electrodynamics", *Sov. Phys. JETP* **3** (1956) 65–71; Cheng & Li, *GT*, pp. 316–320.

[26] [Eds.] Leonard Ross (pseudonym of Leo Rosten), *The Education of H*Y*M*A*N K*A*P*L*A*N*, Harcourt, Brace, 1937; *The Return of H*Y*M*A*N K*A*P*L*A*N*, Harper, New York, 1959. Combined as *O K*A*P*L*A*N! My K*A*P*L*A*N!*, Harper and Row, 1976.

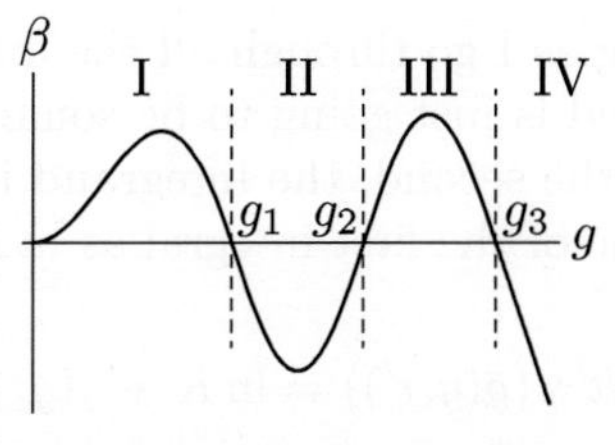

Figure 50.8: Hypothetical β function in ϕ^4 theory

That's just a guess. If you want 14 zeros or a double zero or a triple zero, you can work out the consequences. I'll work out the consequences of this one.

Now something very interesting occurs if we study what happens to $\bar{g}$ (50.42). For a theory with only one coupling constant,

$$\frac{d\bar{g}}{dt} = \beta(\bar{g}) \tag{50.46}$$

I'll assume I start out with my initial condition

$$\bar{g} = g \text{ at } t = 0, \; g \leq g_1 \tag{50.47}$$

I know the solution to this equation. If t is anywhere in region I, β is positive and therefore $\bar{g}$ increases as t increases, because the slope stays positive until g_1. But the curve $\beta(\bar{g}(g))$ can't cross at g_1 into negative β, because at g_1, $\beta = 0$ and $d\bar{g}/dt = 0$. In terms of our hydrodynamic analogy, this is a *stagnation point*; it's a *sink*. The velocities pour into it. Therefore, in region I,

$$\lim_{t\to\infty} \bar{g} = g_1 \tag{50.48}$$

No matter where we start in region I, we end up at the same place. It doesn't matter where we are on the river, we will eventually go over Niagara Falls. I've drawn it so that it's got a nice derivative at $\bar{g} = g_1$. We can also find the approach to the limit. In region I,

$$\beta(\bar{g}) = -a(\bar{g} - g_1) + \mathcal{O}(\bar{g} - g_1)^2 \tag{50.49}$$

where a is a positive constant, $a > 0$. Near the limit we can drop the $\mathcal{O}(g - g_1)^2$ terms; those will be second-order in a small quantity:

$$\frac{d\bar{g}}{dt} = -a(\bar{g} - g_1) \tag{50.50}$$

the solution to which is

$$\bar{g} = g_1 + \mathcal{O}(e^{-at}) \tag{50.51}$$

In region I, $\bar{g}$ reaches g_1 pretty quickly, like an exponential in t once it is in the neighborhood of g_1:

$$\text{Region I:} \; \lim_{t\to\infty} \bar{g}(g, t) = g_1 \tag{50.52}$$

What effect does this have on our general Green's function? Well, $\bar{g}$ is going to g_1 no matter where in region I it started from. This means that f (50.45) goes to a function $F(g_1)$ (because $\bar{g}$ is going to g_1), times an exponential which I can break up into two parts:[27]

$$\int_0^{\ln(E/M)} dt'\gamma\left(\bar{g}(g, t')\right) = \int_0^{\infty} dt'\Big[\gamma\left(\bar{g}(g, t')\right) - \gamma(g_1)\Big] + \int_0^{\ln(E/M)} dt'\gamma(g_1) \tag{50.53}$$

---

[27] [Eds.] Coleman *Aspects*, "Dilatations", Section 4.2, pp. 90–92.

In the first integral, γ is changing as I go through all the intermediate values of t in (50.45). As $\overline{g} \to g_1$ that part converges and is just going to be some multiplicative constant; I don't particularly care about that. In the second, the integrand is a constant, and the integral is trivial. Writing the constant value of the first integral as $\ln K$ for later convenience,

$$\int_0^{\ln(E/M)} dt' \gamma\left(\overline{g}(g,t')\right) = \ln K + \gamma(g_1)\ln(E/M) \tag{50.54}$$

So we have

$$f\left((E/M), g\right) \to KF(g_1)\left(E/M\right)^{\gamma(g_1)} \tag{50.55}$$

I get *simple power behavior*. No matter what Green's function I start out with, no matter which coupling constant I choose, the f goes like a simple power. It doesn't matter what the initial value of the coupling constant is; the asymptotic form is totally independent of the initial value $g < g_1$ of the coupling constant, within the range I'm studying. Only in the scaling constant K, which involves all the coupling constants, do I have any information about where I started from. And that K is trivial. I can always get rid of it by changing the normalization of my fields.

Remember that γ has the following structure:

$$\gamma = \sum_A \gamma_A \tag{50.28}$$

The A's label the various fields that go into f, which is defined for the Green's function

$$\langle 0|T(\phi^{A_1}\cdots\phi^{A_S})|0\rangle$$

in (50.21). The function f depends on what fields are there, but that's all it depends on: how many times each field comes in. So the rule for the power is indeed very simple: we have appropriate powers of (E/M), the powers determined by the value of γ at g_1 for *each* field that's in the Green's function. This is very similar to the sort of behavior we found (50.9) when we were using simple dimensional analysis, and not worrying about renormalization effects. There we also just got a simple power. In that case the simple power was 0 for a dimensionless function, in particular $\Gamma^{(4)}$. But that's not what (50.55) says. For this reason these quantities γ are sometimes called *anomalous dimensions*.[28] We obtain the same sort of scaling behavior we would get from dimensional analysis if the dimensions of the fields were something different than what we naïvely expect—if instead of the scalar field having dimension 1 it had dimension $1+\gamma$. We've shown that if $0 < g < g_1$ then as $t \to \infty$, $\overline{g} \to g_1$. Who knows if β has a zero or not? (It doesn't actually seem to have a zero in $\lambda\phi^4$ theory.) This is marvelous stuff, isn't it?

Let's go on to region II, $g_1 \leq g < g_2$. If $g = g_1$, $\beta = 0$ and $\overline{g}$ stays g_1 forever. If however $g_2 \geq g > g_1$, because β is negative, we're pushed again to g_1. That's exactly the same story in the whole of region II:

$$\text{Region II:} \lim_{t\to\infty} \overline{g}(g,t) = g_1 \tag{50.56}$$

The asymptotic behavior of the theory at high s is determined by the behavior at g_1, which is sometimes called an **ultraviolet stable fixed point**,[29] "ultraviolet" because we are going to

---

[28] [Eds.] Kenneth G. Wilson, "Renormalization Group and Strong Interactions", *Phys. Rev.* **D3** (1971) 1818–1846; Ryder *QFT*, p. 326. Incidentally, Figure 50.8 bears a strong resemblance to Wilson's Figure 1, p. 1826.
[29] [Eds.] Peskin & Schroeder *QFT*, p. 427; Ryder *QFT*, p. 327.

high energy, $\ln(E/M)$ going to ∞, and "stable" because on either side of it, g is inexorably drawn into that value, and will not budge once it gets there.

What about g_2? Well, $g = g_2$ is peculiar, an exceptional point; β vanishes at $g = g_2$, so $\overline{g} = g_2$ forever, no limit required:

$$g = g_2 \;\Rightarrow\; \overline{g} = g_2 \tag{50.57}$$

On the other hand if we're a little way to the left or right of g_2 we get drawn into either g_1 or g_3, respectively. So g_2 is called a **UV unstable**, or sometimes an **infrared stable**, fixed point. It's the same story with g_3 as we had with g_1. If g is up in region III, $g > g_2$ as I've drawn it, it increases to g_3; if it's down in region IV it decreases to g_3, so

$$\left.\begin{matrix}\text{Region III}\\ \text{Region IV}\end{matrix}\right\}: \lim_{t\to\infty} \overline{g}(g,t) = g_3 \tag{50.58}$$

We summarize these results in Table 50.2. (The infrared stable fixed points are sometimes of physical interest in statistical mechanics, but that's a long story that I don't want to go into.[30])

Region	*Range of g*	*Value of $\overline{g}$*
I, II	$0 \le g < g_2$	$\to g_1$
II ∩ III	$g = g_2$	g_2
III, IV	$g_2 < g$	$\to g_3$

Table 50.2: Possible values of $\overline{g}$ from the hypothetical β in Figure 50.8

Notice the simplicity of the asymptotic structure we get in this hypothetical model. This is a theory with an apparent free parameter g in it, a coupling constant that I can vary any way I like. But no matter what initial value I give the coupling constant, I get only three possible asymptotic values: g_1, g_2 (for the single choice $g = g_2$), or g_3. The asymptotic form is a *discontinuous* function of the initial value of the coupling constant, governed by the values of the single β in this theory: we have only three different theories here, not a continuous infinity. That's the fundamental result of the renormalization group: what the coupling constant is depends on what you choose for your scale of mass. In this model, we can get to any coupling constant between 0 and g_1 just by changing the mass; likewise between g_1 and g_2, and between g_2 and g_3. (And if we pick the mass so we start out at an IR stable fixed point, if we start out at g_2, then it will stay that way no matter where we choose the mass M.) The way we choose M doesn't depend on where we start. It's arbitrary. If I say, for example, $g_1 = 1$, there is no difference between the theory with $g = \frac{1}{10}$ and the theory with $g = 100$. We have a theory with $g = 100$ if we choose the renormalization point to be the mass of an electron. The theory with $g = \frac{1}{10}$ is obtained if we use, instead of the mass of an electron, the mass of a Coleman. It's the same theory, though with two different renormalization conventions, so of course it has the same asymptotic behavior. The two versions differ only in the mass scale, but that's trivial; we get rid of M by dimensional analysis.

We started with a naïve viewpoint. We thought that the theory depended trivially on the mass (the dimensionless functions were mass-independent), and nontrivially (in some

[30] [Eds.] Kerson Huang, *Statistical Mechanics*, 2nd ed., John Wiley & Sons, 1987, Chapter 18, pp. 441–467.

complicated way) on the coupling constant. We were deluding ourselves. This is precisely wrong! The f's depend nontrivially on the *mass* (through the γ's), and trivially on the *coupling constant*. These results are found through simple deduction, starting from the observation that we need *some* mass in a naturally massless theory, but the *value* of that mass is arbitrary; the mass and coupling constant are continuously varying quantities. We've turned our naïve viewpoint upside down.

For hypothetical purposes, we can choose the shape of the curve $\beta(\overline{g})$ any way we want; nothing is known in general. All I need to assume is that the graph is continuous and, I suppose, for this little estimate, differentiable at the point where β changes sign. If the curve stays above the g-axis as $t \to \infty$ then we will not get smooth asymptotic behavior. All those theories would be the same because we could turn one into another by changing the renormalization point, but we wouldn't get simple power behavior. The high-energy behavior would be some awful mess, depending on precisely at what rate things went to infinity, and how $\gamma(\overline{g})$ grew with the coupling constant.We wouldn't be able to simplify the integral (50.53) as we did.

Summing the leading logs

I'll put aside making guesses about the graph of β, and turn next to the study of the structure of perturbation theory and the logarithms that appear in it. A particular Green's function, some $f\big((E/M), g\big)$, will typically be a mess. If we compute things out to large order, we find some numerical coefficients, powers of g, and powers of $\ln(E/M)$:

$$f\big((E/M), g\big) = \sum_{n,m} c_{nm} g^n \left(\ln \frac{E}{M}\right)^m \tag{50.59}$$

It will in general be that complicated, but not more so. You might think, "Hmm, why isn't there a $\ln(\ln(E/M))$, or powers of (E/M)—say, $\sqrt{(E/M)}$?" I'll now show that $f\big((E/M), g\big)$ is indeed of the form (50.59), by using the renormalization group equations, again for a single coupling constant theory. It's easy enough to generalize the argument.

My starting point will be the differential equation (50.46) for the running coupling constant $\overline{g}$:

$$\frac{d\overline{g}}{dt} = \beta(\overline{g}) = \sum_{m=2}^{\infty} c_m \overline{g}^m \tag{50.60}$$

with the boundary condition $\overline{g} = g$ at $t = 0$ (50.43). That's certainly right. This is the β which we compute order by order in perturbation theory. The series starts with $m = 2$, because we know that in ϕ^4 theory, to lowest order $\beta \propto g^2$ (50.29). I will show that this equation admits a power series solution of the form

$$\overline{g}(g, t) = g + \sum_{n=2}^{\infty} \sum_{r=1}^{n} c_{nr} g^n t^{(n-r)} \tag{50.61}$$

That is, with every power n of g we get a power of t no higher than $(n-1)$. Once I have proved this, you will easily see how to organize the logarithms in an arbitrary Green's function, because t gets replaced by $\ln(E/M)$ and the γ gives us various powers from this power series (50.28). All the f's have a power series in $\overline{g}$, and γ has a power series in $\overline{g}$, and we just plug it all into the RG equation. If $\overline{g}$ has the form (50.61), then all the coefficients of the RG equation do as well, and so $f\big((E/M), g\big)$ will have the form (50.59). This will tell us in particular that we never get more than one more power of the same logarithm for each extra power of g. We'll never get log log or $\sqrt{E}$ or anything else like that.

I'll prove (50.59) in an absolutely trivial way. If I take a series of the form (50.61), with the minimum value of r equal to some integer, k, and plug it into the right-hand side of (50.60), its mth power is a series of the same form, except it has a larger minimum value of r. For example, consider $n = 3$. The least value of r in (50.61) is 1, and the term corresponding to $n = 3$ is

$$g^3(c_{31}t^2 + c_{32}t + c_{33}) \tag{50.62}$$

Now consider the term $m = 2$ in the series on the right-hand side of (50.60). When I square (50.62) I get

$$g^6(c_{31}^2 t^4 + 2c_{31}c_{32}t^3 + \dots) \equiv c'_{62}g^6 t^{(6-2)} + \dots \tag{50.63}$$

That is, the minimum r equals 2. When I cube it I'll get things like g^9t^6 which has $r = 3$, three fewer powers of t. So the nth power of the term with minimum $r = k$ winds up with $r = nk$. On the left-hand side of (50.59), the derivative $d\overline{g}/dt$ (50.60) knocks off one t but doesn't do anything to the g's, so the corresponding term in the derivative has minimum $r = k + 1$.

It's easy to see what happens next. I plug (50.61) into (50.60). On the right-hand side, there will be no terms with $r = 1$ (i.e., no term of the form $g^n t^{n-1}$) because $\beta(\overline{g})$ begins with g^2. There will be terms with $r = 2$. They'll only come from the $r = 1$ term in the expansion (50.61) plugged into the $\overline{g}^2$ term. All the other terms will have $r = 3$. On the left-hand side, the t derivative will also have no $r = 1$ terms, and $r = 2$ terms only from the $r = 1$ term in the original expansion. I'll match the terms of $\mathcal{O}(g^2)$ on either side with $r = 2$:

$$c_{21} = c_2 \tag{50.64}$$

That tells me the terms with $r = 1$ in the original expansion completely determine β to order g^2. All the terms in the original expansion with $r > 1$ have $r \geq 2$ when I plug them in, and will give terms with $r \geq 4$. I don't have to worry about them. And the g^3 terms, even from my original term, will give me terms with $r = 4$. So I know all the terms with $r = 1$ in the function if I know c_2. Likewise if I know both c_2 and c_3 I know all the terms with $r = 2$. If I know c_2, c_3, and c_4, I know all the terms with $r = 1, 2$, and 3. You see what is happening. I keep building up the power of r whenever I raise the power of the series expansion for $\overline{g}$. I keep raising the power of g relative to the power of $t = \ln(E/M)$. The two series clearly can be made equal to each other iteratively, so I've shown that $\overline{g}$ has a power series solution of the form (50.61). Second, the iterative solution demonstrates an amazing fact: if I want to know the highest power of t in any given power of g, the so-called **leading logarithms**, I need only know c_2. If I want to know next-to-leading logarithms I need only compute c_3.

So I've learned two remarkable things. First, I *do* have a power series expansion of the form (50.59) with m bounded by n for any given Green's function, bounded in a rather trivial way by how many powers of g emerge in the lowest order. Second, I can easily find the coefficients of $g^m\left(\ln(E/M)\right)^{m-1}$. If I'm studying, for example, a four-point function, and I want to look at order g^{128}, I know that it occurs as the product $g^{128}\left(\ln(E/M)\right)^{127}$. What is the coefficient of that term? I only need to know the terms of $\mathcal{O}(\overline{g}^2)$ in this expansion, and β to $\mathcal{O}(g^2)$. That is, I only need to do a one-loop computation and plug it into the renormalization group equation to get, with 100% accuracy, the coefficient of $g^{128}\left(\ln(E/M)\right)^{127}$. There are few more efficient ways of finding that coefficient. Writing down all diagrams of 128th order and studying their asymptotic form is not the right way to do it. If I want to know the coefficient of $\left(\ln(E/M)\right)^{126}$ then I have to do a two-loop calculation.

Again it's just a consequence of the earlier arguments. The only input needed for this argument is that M is an irrelevant parameter. When I change M, all the terms I generate from the power series expansion (50.59) in terms of logarithms have got to come together, and be absorbed in some way into the redefinitions of g and the overall scale. That's the secret of the magic. It means there are complicated tight relations among those coefficients, as we've seen.

Asymptotic Freedom

I will now discuss the hero of the hour (actually the hero of the *lustrum*[31]), *asymptotic freedom*. From our previous analysis, we know that at high energies perturbation theory is liable to be unreliable, even if we start out with a small coupling constant. The reason is that we not only get powers of g but powers of $\ln(E/M)$. As the validity of perturbation theory, in the most naïve sense, requires that the things that multiply your various coefficients should be small, we need not only that $|g| < 1$, but also that $g|\ln(E/M)| \ll 1$:

$$|g| \ll 1, \quad g|\ln(E/M)| \ll 1 \tag{50.65}$$

I will show how we can use the renormalization group to improve perturbation theory, to replace (50.65) with a single condition

$$|\overline{g}| \ll 1 \tag{50.66}$$

which may sometimes be valid when the two separate conditions are not met. We'll get an idea of how we can do this by summing the logarithms, as we talked about above. I'll again take as an example a simple theory with only one coupling constant, quantum electrodynamics:

$$\mathscr{L}' = -g\overline{\psi}\gamma_\mu\psi A^\mu \tag{50.67}$$

I'm going to solve the renormalization group equations approximately. Everything in (50.45) is given automatically in a power series in $\overline{g}$. If $\overline{g}$ is small, that is groovy.

Now let's see about $\overline{g}$, by solving the equation (50.60) approximately:

$$\beta(\overline{g}) = \frac{d\overline{g}}{dt} = c'\overline{g}^3 + \mathcal{O}(\overline{g}^5) \tag{50.68}$$

(we know from (50.32) that to lowest order, $d\overline{g}/dt$ in QED is $\mathcal{O}(\overline{g}^3)$). I'm going to assume that I'm working in a range where $\overline{g}$ is small, so I can neglect the higher orders. I will later check that for self-consistency. By solving the equation I'll know when $\overline{g}$ is large and when it's small. Solving this first-order differential equation equation is trivial, as easy as doing your income tax:

$$\frac{d\overline{g}}{\overline{g}^3} = c'\,dt \tag{50.69}$$

---

[31] [Eds.] In ancient Rome, the census was held every five years. At the end of the census, there was a period of penitence and public expiation ceremonies, typically involving animal sacrifice, called the *lustrum*; the word derives from the Greek verb λύω, "luo", to loosen, release, undo, or repent (it is a root of the word "analysis", and of the name of Aristophanes' heroine Lysistrata, "undoing the army"; the Greek upsilon υ is often transliterated "y"); N. G. L. Hammond and H. H. Scullard, eds., *The Oxford Classical Dictionary*, 2nd ed., Oxford U. P., 1970, "Lustration", p. 626. Coleman is using the word here in its sense of a five-year period. Asymptotic freedom was discovered in 1973, within five years of the videotaped 1976 lectures.

with the boundary condition (50.43), $\overline{g} = g$ at $t = 0$, has the solution

$$\overline{g}^2 = \frac{g^2}{1 - 2c'tg^2} \tag{50.70}$$

The whole approximation is based on the idea that $\overline{g}^2$ stays small. Certainly if g^2 is small at our starting point that's true for small t. But the question is, can we get beyond that?

Now we notice something marvelous. Recall (50.33): c' is *positive.* If $t \to -\infty$, $\overline{g}^2$ stays small. If $t \to +\infty$, we're out of luck: $\overline{g}$ becomes imaginary. Therefore, we can indeed extend perturbation theory, but only to arbitrarily large *negative* $t = \ln(E/M)$. We can't extend it to arbitrarily large *positive* $\ln(E/M)$, because the approximation becomes inconsistent. But to arbitrarily large negative $\ln(E/M)$ we can improve perturbation theory and replace, as stated, (50.65) by the single condition (50.66). This is wonderful. Unfortunately, it's also absolutely useless. The reason is that large negative $\ln(E/M)$ means very *small* E. We're not interested in the behavior of the massless theory at very small E. It's supposed to simulate the behavior of the massive theory only for very *large* E. When we go to very small E, the deep infrared region, we'll again see those masses we threw away at the very beginning of this lecture, unless we're really living in a world with fully massless electrodynamics. And in that case, we'd be able to sum up this infrared structure exactly. That's true but not particularly interesting. The problem is the positive sign of c'.

I come finally to the sensational discovery made independently by Politzer, 't Hooft, and Gross and Wilczek, the last two working collaboratively.[32] Though 't Hooft did not realize its consequences, Politzer, and Gross and Wilczek, *did*, and went crazy. They made a very simple computation (which you yourself are capable of doing with the methods I've shown you) of β for a non-Abelian Yang–Mills theory, with a multiplet of fermions, or without, it doesn't matter. It's still a theory with a single coupling constant, the gauge coupling constant; the graphs look the same. They discovered that c', and hence $\beta(g)$, is *negative* if there are not too many fermions. "Not too many" is a technical issue. For an SU(N) gauge theory with n_f species of fermions in the fundamental representation

$$\beta(g) = c'g^3 = -\frac{g^3}{(4\pi)^2}\left(\tfrac{11}{3}N - \tfrac{2}{3}n_f\right) \tag{50.71}$$

In the gauge group we associate with color, SU(3), 17 triplets of fermions in the fundamental representation are too many, but 16 are not.[33] That's the cross-over point.

[32] [Eds.] H. David Politzer, "Reliable Perturbative Results for Strong Interactions?", *Phys. Rev. Lett.* **30** (1973) 1346–1349; "Asymptotic Freedom: An Approach to Strong Interactions", *Phys. Reps.* **C14** (1974) 129–180; David J. Gross and Frank Wilczek, "Ultraviolet Behavior of Non-Abelian Gauge Theories", *Phys. Rev. Lett.* **30** (1973) 1343–1346; "Asymptotically Free Gauge Theories I", *Phys. Rev.* **D8** (1973) 3633–3652; "Asymptotically Free Gauge Theories II", *Phys. Rev.* **D9** (1974) 980–992; G. 't Hooft, unpublished remarks, *Marseille Conference on Renormalization of Yang–Mills Fields and Applications in Particle Physics*, June, 1972; Gerard 't Hooft, "When was Asymptotic Freedom Discovered? or, The Rehabilitation of Quantum Field Theory", *Nuc. Phys. B (Proc. Suppl.)* **74** (1999) 413–425; David J. Gross, "Twenty-Five Years of Asymptotic Freedom" *Nuc. Phys. B (Proc. Suppl.)* **74** (1999) 426–446; Crease & Mann *SC*, 329-335; Close*IP*, pp. 258–276. See also note 29, p. 860 and note 45, p. 867; Politzer, Gross, and Wilczek shared the 2004 Nobel Prize in Physics for this work. In his Nobel speech, Wilczek refers to Coleman, who while visiting Princeton had been very helpful to him and Gross, as "uniquely brilliant": Frank A. Wilczek, "Asymptotic Freedom: From Paradox to Paradigm", on-line at `https://www.nobelprize.org/nobel_prizes/physics/laureates/2004/wilczek-lecture.html`.

[33] [Eds.] Peskin & Schroeder *QFT*, p. 541, equation (16.135). There appear to be 6 triplets (in three **generations**): $\{d, u; s, c; b, t\}$.

Now what does this mean? Well, it means that everything I have said before is still true, except *ultraviolet replaces infrared*, because the sign of c' has changed, from positive to negative. We can now sum up the improved perturbation theory in a region that *is* of interest to us, the high-energy region where our massless theory is supposed to simulate a massive theory. The high-energy behavior of a theory of Yang–Mills particles and fermions, is *computable* at arbitrarily high Euclidean energies. We can probe these theories with electroproduction experiments. The high-energy behavior is computable, so we can predict in a Lagrangian field theory what is going on at high energies. Furthermore, it doesn't matter how large the coupling constant is initially as long as it's not *too* large. We now know β points *downward* near the origin, as shown in Figure 50.9. For all we know it may keep on going down forever.

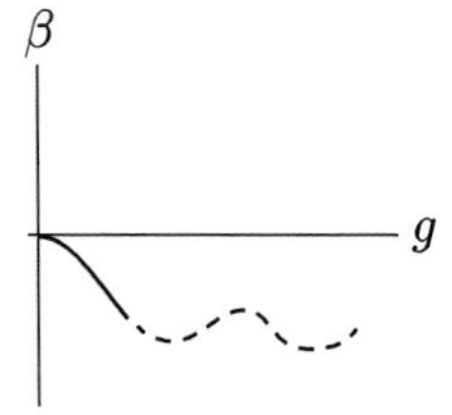

Figure 50.9: Beta function for asymptotic freedom in Yang–Mills theories

Maybe it has a zero someplace. If it does have a zero, we know that it's not a small number. If it were, we could compute it in perturbation theory. But in perturbation theory it doesn't have a zero. For any value of the coupling constant between this possible zero and the origin, by the arguments given before as $E \to \infty$, we are forced into the origin. We're not able to study what happens all the way along the g-axis, but that doesn't matter. Eventually we're coming down to the origin. We don't care how the renormalization group equation drove us there. When we're near the origin we can compute what happens using perturbation theory fixed up by the renormalization group. If this zero does exist at all, in the theory in question, it doesn't matter how large the coupling constant is. If the zero is set at $g = 17$, for any $g < 17$ the method will work.

This is called **asymptotic freedom**, because instead of being pushed towards one of those points g_1, we are pushed towards a *free* field theory, $g = 0$. Writing $b = -c'$, where $b > 0$, the running coupling constant $\overline{g}$ (50.70) becomes

$$\overline{g}^2 = \frac{g^2}{1 + 2btg^2} \to 0 \text{ as } t \to \infty \tag{50.72}$$

Asymptotically the theory is free, aside from corrections we know how to compute. They turn out to be powers of logarithms, as I'll now show. From (50.45), asymptotically f goes to $F(0)$ times the exponential of the integral of γ. Analogous to (50.35) we can say for a gluon, G,

$$\gamma_G = d''\overline{g}^2 + \mathcal{O}(g^4) = \frac{d''g^2}{1 + 2btg^2} + \mathcal{O}(g^4) \to \frac{d''}{2bt} \tag{50.73}$$

using (50.72). The integral (50.45) becomes

$$\int_0^{\ln(E/M)} dt'\, \gamma_G\left(\overline{g}(g, t')\right) = \frac{d''}{2b} \int^{\ln(E/M)} \frac{dt'}{t'} = (d''/2b) \ln\left(\ln(E/M)\right) \tag{50.74}$$

so that

$$f \sim F(0) \exp\left[\ln\left(\ln(E/M)^{d''/2b}\right)\right] = F(0) \left(\ln(E/M)\right)^{d''/2b} \tag{50.75}$$

The correction involves a power of $\ln(E/M)$, as claimed.[34] We can compute these corrections with what we have. Everything is predictable and everything looks like a free field theory, with tiny corrections that get tinier and tinier as the energy gets larger and larger, because $\overline{g} \to 0$. Of course what the coupling constant is depends on what the renormalization mass is.

Let me tell you an anecdote. Asymptotic freedom was discovered by my graduate student David Politzer.[35] I was off at Princeton on a sabbatical, and he came down from Harvard to visit me. We were working on some other (totally uninteresting) problem, trying to solve dynamical symmetry breakdown, to get the Nambu–Goldstone phenomenon without fundamental scalar fields. We thought it would be easier than it turned out to be after a year of labor. I said, "You're getting nowhere with your thesis. It would be nice to know the renormalization group functions for the Yang–Mills theory. Nobody's worked them out yet. Why don't you compute them? That's not going to be a lot of work, but it's something to do." Actually 't Hooft had computed them the summer before, but hadn't published them. He announced them at a seminar in Marseille.[36] I added, "Nobody expects them to come out negative." No one had thought in advance what the consequences would be if the beta function turned out to *be* negative.

Politzer went back to Harvard, and here's where you see the sign of genius. Not only did he follow my orders, he knew what to do with the result. He called me up one night and said, "I've computed them, and they're negative." And I said "Oh, that's interesting. This is telling us something important about the strong interactions." He was very smart; he realized what it meant. Not only did he get the right sign, he drew the right conclusion, which even someone as smart as 't Hooft didn't do. This result would explain why you apparently see free quarks inside the nucleon when you do deep inelastic scattering. Then you're probing this region of high energy, and in that region the effective coupling constant $\overline{g}$, the quantity that governs the interactions among the quarks, is small. In fact shortly thereafter, David Gross and I showed that in four dimensions, the *only* renormalizable field theories that allow for asymptotic freedom are Yang–Mills theories.[37] For everything else β is positive. The color interaction between colored quarks is due to a non-Abelian Yang–Mills theory. They look freer and freer at higher and higher energy because of this phenomenon, asymptotic freedom: $\overline{g}$ is getting smaller and smaller with higher energies.

The big test is deep inelastic electroproduction,[38] as mentioned earlier:

$$e + N \to e + X \tag{50.14}$$

It's rather complicated to fit deep inelastic electroproduction because the data changes a lot. It usually turns out that to get an accurate fit, you have to know something that the experimentalists haven't quite measured yet, the value of $\overline{g}$. It is known that $\overline{g}^2/(4\pi) \approx 0.5$

---

[34] [Eds.] See also "Secret Symmetry" in Coleman *Aspects*, pp. 174–178.

[35] [Eds.] Politzer's own account of this period is described in his Nobel lecture: H. David Politzer, "The Dilemma of Attribution", on-line at `https://www.nobelprize.org/nobel_prizes/physics/laureates/2004/politzer-lecture.html`.

[36] [Eds.] Close *IP*, pp. 261–264; 't Hooft, *op. cit.*, pp. 416–417. See note 32, p. 1113.

[37] [Eds.] Sidney Coleman and David J. Gross, "Price of Asymptotic Freedom", *Phys. Rev. Lett.* **31** (1973) 851–854. Other field theories are asymptotically free in a different number of space-time dimensions, e.g., in two dimensions, the Gross–Neveu model of Dirac fermions: David J. Gross and André Neveu, "Dynamical Symmetry Breaking in Asymptotically Free Field Theories", *Phys. Rev.* **D10** (1974) 3235–3252.

[38] [Eds.] Peskin & Schroeder *QFT*, pp. 475–479.

near 1 GeV.[39] So it's falling off fairly rapidly. It works the other way around. That's the opposite side of asymptotic freedom. As you go the other direction, towards lower energies, $\overline{g}$ gets bigger and bigger, and faster and faster. That's presumably why the quarks don't get out of the hadrons containing them: the force is getting stronger and stronger as they're getting farther and farther apart at larger distances. That's **infrared slavery**; the quarks are confined.

I would like to say one or two sentences about the content of this course. There are many topics I have not covered. I've said nothing about Regge poles, and for that I feel guilty.[40] I've said nothing about many strong interaction processes, like inclusive pion production. There's a lot of important physics which you haven't learned from this course that involve, in one way or another, field theoretical ideas. I think it is pleasant, however, that in the last few weeks, I've been able to deliberately contradict two things I previously taught as received dogma, the last time I taught the second half of this course, five years ago. One was that weak interactions are much weaker than electromagnetism. That's false. The GSW model tells us they are exactly the same strength. We were worried about non-renormalizable theories because we thought the weak interactions got stronger at higher and higher energy with the piling up of all those powers of energy. That's also false. The GSW model of the weak intereactions is a renormalizable theory. They get to electromagnetic strength and stay there. And the other thing is the marvelous reversal. Instead of believing the weak interactions get strong at high energies, we now believe the strong interactions get weak at high energies, as I've demonstrated. That is the end of the course, and I hope you've enjoyed it.[41]

[39] [Eds.] *PDG* 2016, Section 9.3.4, "Measurements of the strong coupling constant", pp. 128–131, in particular, the graph in Figure 9.4 on p. 131. Peskin & Schroeder *QFT*, p. 552, cite $\alpha_s = \overline{g}^2/(4\pi) \approx 0.4$ at 1 GeV.

[40] [Eds.] Because of space and time constraints, Coleman's six lectures on dispersion relations were not included in this book.

[41] [Eds.] At the end of the last lecture, the students honor an old academic tradition: they applaud their professor.

Concordance of videos and chapters

Nearly all of the text in the chapters comes from the editors' (RS and DD) transcriptions of the videotapes at the Harvard Physics Department's site `https://www.physics.harvard.edu/events/videos/Phys253`, with additional text from Sidney Coleman's original notes (1975–76), or from the sources named in the Preface. Occasionally the editors interpolated text from these to fill out an argument or provide an insight from later versions of the course. These interpolations are usually only a sentence or two, often relegated to the footnotes. The exceptions occur when a lecture is fragmentary, e.g., Chapter 25. Below is a concordance to aid those who might want to watch the Coleman videotapes as they read.

Chapter	*Pages*	*Video*	*Length*
1	1 – 16	1	1:35:12
2	17 – 30	2	1:19:25
3	31 – 47	2	1:19:25 – 1:35:20
		3	1:19:45
4	57 – 75	3	1:19:45 – 1:23:17
		4	1:35:16
5	77 – 97	5	1:32:14
6	105 – 130	6	1:42:54
		7	0:19:00
7	131 – 144	7	0:19:00 – 1:25:30
8	153 –173	8	1:35:36
		9	0:17:40
9	183 – 198	9	0:17:40 – 1:30:04
10	205 – 224	10	1:33:13
11	225 – 244	11	1:39:33
12	245 – 259	12	1:32:26
13	267 – 283	13	1:34:04

Chapter	*Pages*	*Video*	*Length*
14	291 – 308	14	1:28:18
15	313 – 330	15	1:33:38
16	331 – 346	16	1:38:52
17	355 – 367	17	1:33:52
18	369 – 386	18	1:29:36
		19	0:20:30
19	393 – 406	19	0:20:30 – 1:33:24
20	407 – 423	20	1:02:19 (Truncated)
21	429 – 450	21	1:24:48
22	459 – 480	22	1:41:50
23	481 – 500	23	1:54:47
24	507 – 524	24	1:30:32
25	525 – 543	25	0:19:04 (Truncated)
		26	0:35:50
26	555 – 573	26	0:35:50 – 1:46:54
27	575 – 590	27	1:09:21
28	599 – 616	27	1:09:21 – 1:39:21
		28	1:25:49
29	617 – 634	29	1:29:23
30	641 – 658	30	1:33:04
31	659 – 677	31	1:30:12
32	687 – 700	32	1:06:18 (Truncated)
33	701 – 724	33	1:38:33
34	733 – 749	34	1:38:35
35	751 – 769	35	1:30:39
36	777 – 796	36	1:32:22
37	797 – 816	37	1:29:33
38	823 – 843	38	1:35:20
39	845 – 870	39	1:40:50
40	877 – 888	45	1:00:00 – 1:48:48
41	889 – 909	46	1:34:47
42	917 – 934	47	1:33:20
		48	0:21:30
43	935 – 957	48	0:21:30 – 1:36:08
44	963 – 982	49	1:22:26
45	983 – 1002	50	1:32:11
46	1011 – 1029	51	1:21:51
47	1031 – 1048	52	1:06:00
48	1059 – 1076	52	1:06:00 – 1:44:49
		53	0:42:10
49	1077 – 1089	53	0:42:10 – 1:31:48
50	1091 – 1116	54	1:46:06

Index

Page numbers for entries occurring in a footnote are followed by an *n* and the footnote number. **Bold** page numbers indicate a term's definition or an individual's biographical sketch.

SC = Sidney Coleman.